## 1989 Recommended Dietary Allowances (RDA)

| Age (yr) | Energy (kcal) | Protein (g) | Vitamin A (μg RE) | Vitamin E (mg α-TE) | Vitamin K (μg) | Vitamin C (mg) | Iron (mg) | Zinc (mg) | Iodine (μg) | Selenium (μg) |
|---|---|---|---|---|---|---|---|---|---|---|
| **Infants** | | | | | | | | | | |
| 0.0–0.5 | 650 | 13 | 375 | 3 | 5 | 30 | 6 | 5 | 40 | 10 |
| 0.5–1.0 | 850 | 14 | 375 | 4 | 10 | 35 | 10 | 5 | 50 | 15 |
| **Children** | | | | | | | | | | |
| 1–3 | 1300 | 16 | 400 | 6 | 15 | 40 | 10 | 10 | 70 | 20 |
| 4–6 | 1800 | 24 | 500 | 7 | 20 | 45 | 10 | 10 | 90 | 20 |
| 7–10 | 2000 | 28 | 700 | 7 | 30 | 45 | 10 | 10 | 120 | 30 |
| **Males** | | | | | | | | | | |
| 11–14 | 2500 | 45 | 1000 | 10 | 45 | 50 | 12 | 15 | 150 | 40 |
| 15–18 | 3000 | 59 | 1000 | 10 | 65 | 60 | 12 | 15 | 150 | 50 |
| 19–24 | 2900 | 58 | 1000 | 10 | 70 | 60 | 10 | 15 | 150 | 70 |
| 25–50 | 2900 | 63 | 1000 | 10 | 80 | 60 | 10 | 15 | 150 | 70 |
| 51+ | 2300 | 63 | 1000 | 10 | 80 | 60 | 10 | 15 | 150 | 70 |
| | | | | | | | | | | |
| **Females** | | | | | | | | | | |
| 11–14 | 2200 | 46 | 800 | 8 | 45 | 50 | 15 | 12 | 150 | 45 |
| 15–18 | 2200 | 44 | 800 | 8 | 55 | 60 | 15 | 12 | 150 | 50 |
| 19–24 | 2200 | 46 | 800 | 8 | 60 | 60 | 15 | 12 | 150 | 55 |
| 25–50 | 2200 | 50 | 800 | 8 | 65 | 60 | 15 | 12 | 150 | 55 |
| 51+ | 1900 | 50 | 800 | 8 | 65 | 60 | 10 | 12 | 150 | 55 |
| | | | | | | | | | | |
| **Pregnancy** | +300 | 60 | 800 | 10 | 65 | 70 | 30 | 15 | 175 | 65 |
| **Lactation** | | | | | | | | | | |
| 1st 6 mo. | +500 | 65 | 1300 | 12 | 65 | 95 | 15 | 19 | 200 | 75 |
| 2nd 6 mo. | +500 | 62 | 1200 | 11 | 65 | 90 | 15 | 16 | 200 | 75 |

In addition to the values that serve as goals for nutrient intakes (presented in the adjacent tables), the Dietary Reference Intakes include a set of values called Tolerable Upper Intake Levels—the maximum amount of a nutrient that appears safe for most healthy people to consume on a regular basis. Chapter 1 and Highlight 10 provide additional details.

## Tolerable Upper Intake Levels for Selected Nutrients (per day)

| Age (yr) | Vitamin D (μg)[a] | Niacin (mg) | Vitamin $B_6$ (mg) | Folate (μg) | Choline (mg) | Calcium (mg) | Phosphorus (mg) | Magnesium (mg) | Fluoride (mg) |
|---|---|---|---|---|---|---|---|---|---|
| **Infants** | | | | | | | | | |
| 0.0–0.5 | 25 | —[b] | —[b] | —[b] | —[b] | —[b] | —[b] | —[b] | 0.7 |
| 0.5–1.0 | 25 | —[b] | —[b] | —[b] | —[b] | —[b] | —[b] | —[b] | 0.9 |
| **Children** | | | | | | | | | |
| 1–3 | 50 | 10 | 30 | 300 | 1000 | 2500 | 3000 | 65 | 1.3 |
| 4–8 | 50 | 15 | 40 | 400 | 1000 | 2500 | 3000 | 110 | 2.2 |
| 9–13 | 50 | 20 | 60 | 600 | 2000 | 2500 | 4000 | 350 | 10.0 |
| 14–18 | 50 | 30 | 80 | 800 | 3000 | 2500 | 4000 | 350 | 10.0 |
| **Adults** | | | | | | | | | |
| 19–70 | 50 | 35 | 100 | 1000 | 3500 | 2500 | 4000 | 350 | 10.0 |
| >70 | 50 | 35 | 100 | 1000 | 3500 | 2500 | 3000 | 350 | 10.0 |
| **Pregnancy** | 50 | 35 | 100 | 1000 | 3500 | 2500 | 3500 | 350 | 10.0 |
| **Lactation** | 50 | 35 | 100 | 1000 | 3500 | 2500 | 4000 | 350 | 10.0 |

NOTE: An Upper Level was not established for thiamin, riboflavin, vitamin $B_{12}$, panthothenic acid, and biotin because of a lack of data, not because these nutrients are safe to consume at any level of intake; all nutrients can have adverse effects when intakes are excessive.

[a]To convert μg to IU, multiply by 40. For example, 50 μg × 40 = 2000 IU.

[b]Upper Levels were not established for many nutrients in the infant category because of a lack of data.

Source: Adapted from the first two of the *Dietary Reference Intakes* series, National Academy Press. Copyright 1997 and 1998, by the National Academy of Sciences. Courtesy of the National Academy Press, Washington, D.C.

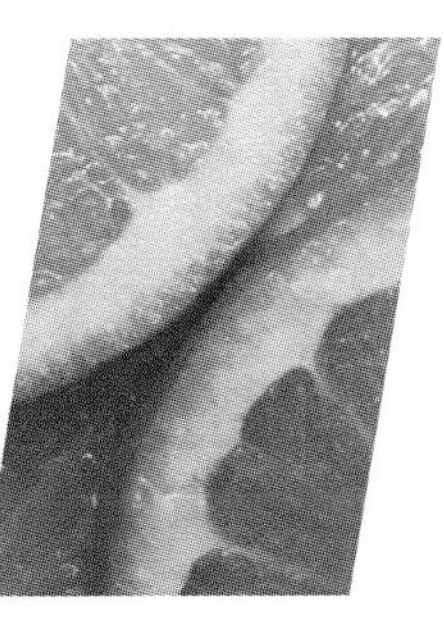

# Understanding Nutrition

Eighth Edition

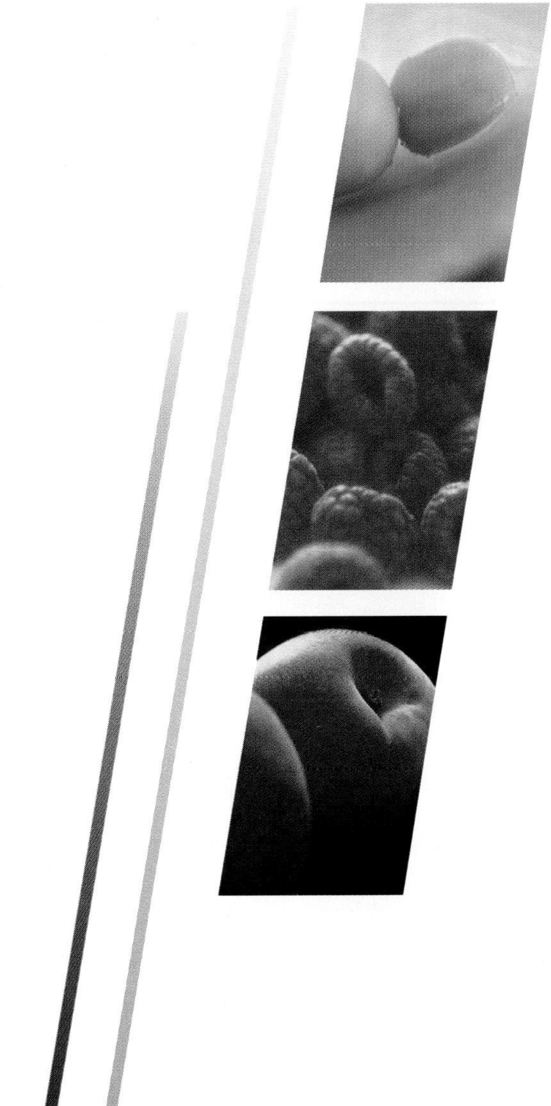

# Understanding Nutrition

Eighth Edition

Eleanor Noss Whitney

Sharon Rady Rolfes

West/Wadsworth

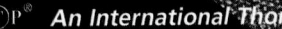

An International Thomson Publishing Company

Belmont, CA • Albany, NY • Boston • Cincinnati • Johannesburg
London • Madrid • Melbourne • Mexico City • New York
Pacific Grove, CA • Scottsdale, AZ • Singapore • Tokyo • Toronto

**Nutrition Publisher:**
Peter Marshall
**Associate Development Editor:**
Laura Graham
**Editorial Assistant:**
Tangelique Williams
**Marketing Manager:**
Becky Tollerson
**Project Editor:**
Sandra Craig
**Print Buyer:**
Barbara Britton
**Permissions Editor:**
Robert Kauser
**Production Coordinator:**
Dusty Friedman, The Book Company
**Text and Cover Design:**
Norman Baugher
**Cover Image:**
©1996 FoodPix®
**Copy Editor:**
Patricia Lewis
**Photo Research:**
Stephen Forsling
**Illustrations:**
Impact Publications,
McMahon Medical Art
**Photographer and Food Stylists:**
Michael Shay, Lucy Radys, Carol Ladd
**Index:**
Barbara Farabaugh
**Compositor:**
Parkwood Composition Service
**Printer:**
World Color Book Services/Versailles



Printed in the United States of America
4 5 6 7 8 9 10

For more information, contact Wadsworth Publishing Company, 10 Davis Drive, Belmont, CA 94002, or electronically at http://www.wadsworth.com

International Thomson Publishing Europe
Berkshire House 168-173
High Holborn
London, WC1V 7AA, England

Nelson ITP Australia
102 Dodds Street
South Melbourne
Victoria 3205 Australia

Nelson Canada
1120 Birchmount Road
Scarborough, Ontario
Canada M1K 5G4

International Thomson Publishing
Southern Africa
Building 18, Constantia Park
138 Sixteenth Road, P.O. Box 2459
Halfway House 1685 South Africa

International Thomson Editores
Seneca, 53
Colonia Polanco
11560 México D.F. México

International Thomson Publishing Asia
60 Albert Street
#15-01 Albert Complex
Singapore 189969

International Thomson Publishing Japan
Hirakawa-cho Kyowa Building, 3F
2-2-1 Hirakawa-cho Chiyoda-ku
Tokyo 102, Japan



**Library of Congress Cataloging-in-Publication Data**

Whitney, Eleanor Noss.
Understanding nutrition / Eleanor Noss Whitney, Sharon Rady Rolfes.
—8th ed.
p. cm.
Includes bibliographical references and index.
ISBN 0-534-54612-9
1. Nutrition. I. Rolfes, Sharon Rady. II. Title.
QP141.W46 1999
613.2—dc21 98-37207

This book includes photos taken from CNN videotape. See Photo and Art Credits on page 39, following the index, which is a continuation of this copyright page

*To*

*The memory*
*of my beloved*
*husband, my hero,*
*Jack Yaeger, Jr.*

*Ellie*

*To*

*My wonderful*
*husband, Tom, and*
*our delightful*
*children, Kristen,*
*Lyle, and Marni.*

*Sharon*

# About the Authors

*Eleanor Noss Whitney, Ph.D.,* received her B.A. in biology from Radcliffe College in 1960 and her Ph.D. in biology from Washington University, St. Louis, in 1970. Formerly on the faculty at Florida State University, and a dietitian registered with the American Dietetic Association, she now devotes full time to research, writing, and consulting. Her earlier publications include articles in *Science, Genetics,* and other journals. Her textbooks include *Understanding Normal and Clinical Nutrition, Nutrition Concepts and Controversies, Life Span Nutrition: Conception through Life, Nutrition and Diet Therapy,* and *Essential Life Choices* for college students and *Making Life Choices* for high school students. Her most intense interests currently include energy conservation, solar energy uses, alternatively fueled vehicles, and ecosystem restoration.

*Sharon Rady Rolfes, M.S., R.D.,* received her B.S. in psychology and criminology in 1974 and her M.S. in nutrition and food science in 1982 from Florida State University. She is a founding member of Nutrition and Health Associates, an information resource center that maintains an ongoing bibliographic database that tracks research in over 1000 nutrition-related topics. Her other publications include the textbooks *Understanding Normal and Clinical Nutrition, Understanding Clinical Nutrition, Life Span Nutrition: Conception through Life,* and *Nutrition for Health and Health Care* and a multimedia CD-ROM called *Nutrition Interactive*. In addition to writing, she also lectures at universities and at professional conferences and serves as a consultant for various educational projects. She maintains her registration as a dietician and is a member of the American Dietetic Association.

# Brief Contents

# Contents

# Preface

This eighth edition of *Understanding Nutrition* shares the same goals established over 20 years ago in writing the first edition: to provide a textbook that would both reveal the fascination of the science of nutrition and share the fun and excitement of nutrition with the reader. Readers want more than just facts—they want an understanding of how the scientific facts apply to their daily lives. While the goals for this edition remain unchanged, every chapter has been substantially revised to reflect the many changes that have occurred in the field of nutrition over the years.

To the person reading this text, it will be obvious that, like most sciences, nutrition possesses no absolute certainties. Nutrition scientists simply do not have all the answers yet; in some cases, we have not even asked all the questions yet. This is true in many areas of nutrition; it is a growing, young science dating only from around the turn of the twentieth century. One of the missions of this text, beginning in Chapter 1, is to show readers how researchers ascertain the "facts."

• ***The Chapters*** • This book presents the core information of an introductory nutrition course. Chapter 1 wastes no time in exploring why we eat the foods we do and continues with a brief overview of the nutrients, the science of nutrition, recommended nutrient intakes, assessment, and important relationships between diet and health. Chapter 2 describes the diet-planning principles and food guides used to create diets that support good health and includes instructions on how to read a food label. In Chapter 3, readers follow the journey of digestion and absorption as the body transforms foods into nutrients. Chapters 4 through 6 describe carbohydrates, fats, and proteins—their chemistry, health effects, roles in the body, and places in the diet. Then Chapter 7 shows how the body derives energy from these three nutrients. Chapters 8 and 9 continue the story with a look at energy balance, the factors associated with overweight and underweight, and the benefits and dangers of weight loss and weight gain. Chapters 10 through 13 complete the introductory lessons by describing the vitamins, the minerals, and water—their roles in the body, deficiency and toxicity symptoms, and sources.

The next seven chapters weave that basic information into practical applications, showing how nutrition influences people's lives. Chapter 14 describes how physical activity and nutrition work together to support health. Chapters 15, 16, and 17 present the special nutrient needs of people through the life cycle—pregnancy and lactation; infancy, childhood, and adolescence; and adulthood and the later years. Chapter 18 focuses on the dietary risk factors and recommendations associated with chronic diseases, and Chapter 19 addresses consumer concerns about the safety of the food supply. Chapter 20 closes the book with a look at hunger and global environmental problems and offers suggestions for establishing sustainable foodways.

• ***The Features*** • The chapters in this edition have been designed with special features to enhance learning. For example, definitions are provided whenever new terms are introduced. These definitions often include pronunciations and derivations to facilitate understanding. A glossary at the end of the text includes all defined terms.

New to this edition are notations of Web site addresses. These sites offer additional information and resources on the topic discussed in the accompanying text.

Many of the chapters include "How to" sections that guide readers through problem-solving tasks. For example, the "How to" in Chapter 1 shows readers how to calculate energy intake from the grams of carbohydrate, fat, and protein in a food; another "How to" in Chapter 13 describes how to calculate iron absorption from a meal.

Many chapters close with a "Making It Click" section. These sections reinforce the "How to" lesson and provide practice in doing nutrition-related calculations. The problems enable readers to apply their skills to hypothetical situations and then check their answers (found in Appendix K). Readers who successfully master these exercises will be well prepared for "real-life" nutrition-related problems.

Each major section within a chapter concludes with a summary paragraph that reviews the key concepts. Similarly, summary tables, figures, and margin lists cue readers to important reviews.

Also featured in this edition are the Healthy People 2000 nutrition-related priorities, which are presented whenever their subjects are discussed (Appendix G presents them in full). Healthy People 2000 is a report developed by the U.S. Department of Health and Human Services that establishes national objectives in health promotion and disease prevention for the year 2000.

Each chapter closes with study questions in essay and multiple-choice format. Study questions offer readers the opportunity to review the major concepts presented in the chapters in preparation for exams. The page numbers after each essay question refer readers to discussions that answer the question; multiple-choice answers appear in Appendix K.

• ***The Highlights*** • Every chapter is followed by a Highlight. Each Highlight provides readers with an in-depth look at a current, and often controversial, topic that relates to its companion chapter. New Highlights in this edition examine the use of the Internet in finding reliable nutrition information, the science and science fiction behind high-protein weight-loss diets, and the possible benefits and potential harms of alternative therapies.

• ***The Appendixes*** • The appendixes are valuable references for a number of purposes. Appendix A summarizes background information on the hormonal and nervous systems, complementing Appendixes B and C on basic chemistry, the chemical structures of nutrients, and major metabolic pathways. Appendix D assists readers with calculations and conversions. Appendix E provides detailed coverage on nutrition assessment, and Appendix F lists nutrition resources, including book and journal recommendations as well as addresses, phone numbers, and Websites. Appendix G presents the Recommended Dietary Allowances (1989 RDA), the nutrition-related priorities of Healthy People 2000, the United States Exchange System, and recommendations from the World Health Organization (WHO). Appendix H is a 2000-item food composition table made from the latest nutrient database assembled by ESHA Research, Inc., of Salem, Oregon. Appendix I presents information for Canadians: the Recommended Nutrient Intakes (1990 RNI), the Exchange System, and instructions on reading food labels. Appendix J describes measures of protein quality, and Appendix K presents the answers to the "Making It Click" sections and multiple-choice questions that appear at the ends of chapters.

• ***The Inside Covers*** • The inside covers put commonly used information at your fingertips. The front covers presents the current nutrient recommendations (introduced in Chapter 1); the inside back cover (left) features the nutrient values used on food labels (described in Chapter 2); and the inside back cover (right) shows the suggested weight ranges for various heights (discussed in Chapter 8).

• ***Closing Comments*** • We have tried to keep the number of notes to a minimum. Many statements that have appeared in previous editions with notes now appear without them, but every statement is backed by research, and the authors will supply references upon request. We have not provided a separate list of suggested readings, but have tried to include references that will provide readers with additional details or a good overview of the subject. Nutrition is a fascinating subject, and we hope our enthusiasm for it comes through on every page.

Eleanor Noss Whitney
Sharon Rady Rolfes
October 1998

# Acknowledgments

To produce a book requires the coordinated effort of a team of people—and, no doubt, each team member has another team of support people as well. We salute, with a big round of applause, everyone who has worked so diligently to ensure the quality of this book.

We thank Linda DeBruyne and Pam Schmidt for their valuable contributions to the fitness and consumer concerns chapters, respectively, and Margaret Hedley for her attention to the Canadian information throughout the text and in Appendixes F and I. A million thank yous to Sally Mayo for her patient attention to manuscript preparation, permissions, and a multitude of other daily tasks. To Linda Patton, a special thank you for her skilled assistance in library research. We also thank the many people who have prepared the ancillaries that accompany this text: Harry Sitren for writing and enhancing the Test Bank; Lori Turner, Mary Rhiner, and Margaret Hedley for preparing the Instructor's Manual; Charlene Hamilton for developing the electronic lecture presentations; and Lori Turner for authoring the Student Study Guide. A big thank you to Elizabeth Hands, Bob Geltz, and their staff at ESHA for their meticulous effort in creating the food composition appendix, verifying the data in figures and tables, and developing the computerized diet analysis program that accompanies this book. Our special thanks to Peter Marshall for his editorial advice and creative ideas; to Sandra Craig and Dusty Friedman for their careful attention to the many details involved in producing this book; to Becky Tollerson for her marketing talents; and to Laura Graham for her coordination of reviews and ancillaries. We also thank Norman Baugher for designing the pages to enhance our work; Sandra McMahon and David Ruppe for creating accurate and attractive artwork to complement our writing; Michael Shay for photographing foods beautifully; Pat Lewis for copyediting thousands of pages of manuscript; Barbara Farabaugh for creating a thorough and useful index. To the many others involved in typesetting, dummying, marketing, and sales, we tip our hats in appreciation.

We are especially grateful to our associates, friends, and families for their continued encouragement and support. We also thank our many reviewers for their comments and contributions.

## Reviewers of *Understanding Nutrition*

Patricia Benarducci
Miami-Dade Community College

Sharleen J. Birkimer
University of Louisville

Ellen Brennan
San Antonio College

Leah Carter
Bakersfield College

Mary Ann Cessna
Indiana University of Pennsylvania

Jo Carol Chezum
Ball State University

Jim Daugherty
Glendale Community College

Robert DiSilvestro
Ohio State University

Pam Fletcher
Albuquerque Technical Vocational Institute

Eileen Ford
University of Pennsylvania

William Forsythe
University of Southern Mississippi

Jean Fremont
Simon Fraser University

Julie Rae Friedman
State University of New York at Farmingdale

Betty J. Forbes
West Virginia University

Patricia Garrett
University of Tennessee, Chattanooga

Francine Genta
Cabrillo College

Leon Hageman
Burlington County College

Charlene Hamilton
University of Delaware

Shelley Hancock
The University of Alabama

Margaret Hedley
University of Guelph

*continued*

Tracy Horton
University of Colorado Health Sciences Center

Donna-Jean Hunt
Stephen F. Austin University

Bernadette Janas
Rutgers University

Michael Jenkins
Kent State University

Jayanthi Kandiah
Ball State University

Younghee Kim
Bowling Green State University

Kim Kline
University of Texas, Austin

Betty Larson
Concordia College

Chunhye Kim Lee
Northern Arizona University

Robert Lee
Central Michigan University

Anne Leftwich
University of Central Arkansas

Joseph Leichter
University of British Columbia

Janet Levins
Pensacola Junior College

Samantha Logan
University of Massachusetts, Amherst

Elaine M. Long
Boise State University

Bruce McDonald
University of Manitoba

Lisa McKee
New Mexico State University

Paula Netherton
Tulsa Junior College

Steven Nizielski
Texas A & M University

Marvin Parent
Oakland Community College

Linda Peck
University of Findlay, Ohio

Erwina Peterson
Yakima Valley Community College

Roseanne L. Poole
Tallahassee Community College

Robin R. Roach
University of Memphis

Janet Sass
Northern Virginia Community College

Tammy Sakanashi
Santa Rosa Junior College

Nancy Shearer
Cape Cod Community College

Wendy Stuhldreher
Slippery Rock University of Pennsylvania

Carla Taylor
University of Manitoba

Janet Thompson
University of Waterloo

Michele Trankina
St. Mary's University

Anne VanBeber
Texas Christian University

Ava Craig-Waite
Sacramento City College

Suzy Weems
Stephen F. Austin University

Lisa Young
New York University

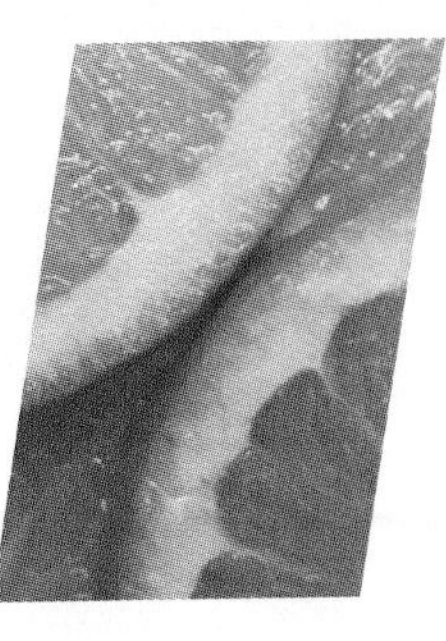

# Understanding Nutrition

Eighth Edition

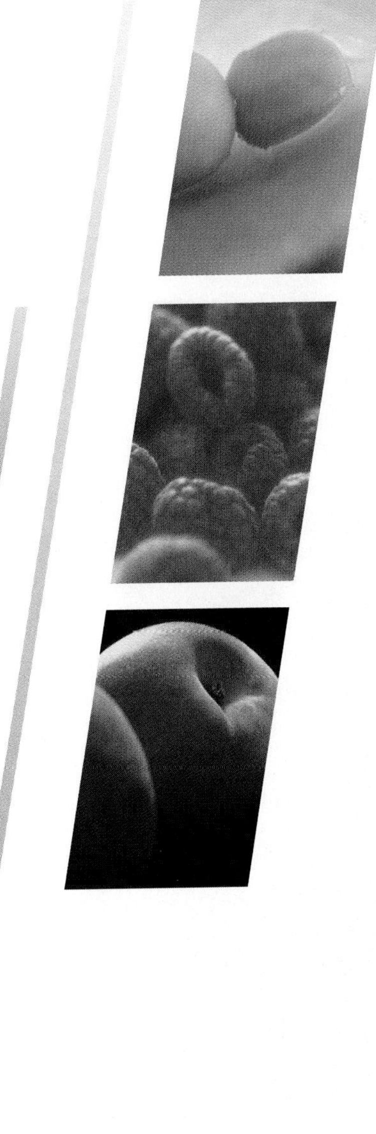

CHAPTER 1

# An Overview of Nutrition

Welcome to the world of nutrition. Nutrition has played a significant role in your life, even from before your birth, although you may not always have been aware of it. And it will continue to affect you in major ways, depending on the foods you select.

Every day, several times a day, you make food choices that influence your body's health for better or worse. Each day's choices may benefit or harm your health only a little, but when these choices are repeated over years and decades, the rewards or consequences become major. That being the case, close attention to good eating habits now can bring health benefits later. Conversely, carelessness about food choices from youth on can be a major contributor to many of today's most prevalent chronic diseases of later life, including heart disease and cancer. Of course, some people will become ill or die young no matter what choices they make, and others will live long lives despite making poor choices. For the large majority, however, the food choices they make each and every day will benefit or impair their health in proportion to the wisdom of the choices.

While most people realize that their food habits affect their health, they often choose foods for other reasons. After all, foods bring to the table a variety of pleasures, traditions, and associations as well as nourishment. The challenge, then, is to combine favorite foods and fun times with a nutritionally balanced diet.

**nutrition:** the science of foods and the nutrients and other substances they contain, and of their actions within the body (including ingestion, digestion, absorption, transport, metabolism, and excretion). A broader definition includes the social, economic, cultural, and psychological implications of food and eating.

**foods:** products derived from plants or animals that can be taken into the body to yield energy and nutrients for the maintenance of life and the growth and repair of tissues.

In general, a **chronic** disease is one of long duration that progresses slowly or with little change. By comparison, an **acute** disease develops quickly, produces sharp symptoms, and runs a short course.

- **acute** = sharp
- **chronos** = time

**diet:** the foods and beverages a person eats and drinks.

 U.S. Government
www.healthfinder.gov/searchoptions/topicsaz.htm
Search for Nutrition

# Food Choices

People decide what to eat, when to eat, and even whether to eat in highly personal ways, often based on behavioral or social motives rather than on awareness of nutrition's importance to health. Fortunately, many different food choices can be healthy ones, but nutrition awareness helps to make them so.

• ***Personal Preference*** • As you might expect, the number one reason people choose foods is that they like certain flavors.[1] Two widely shared preferences are for the sweetness of sugar and the tang of salt. Liking high-fat foods appears to be another universally common preference.[2] Other preferences might be for the hot peppers common in Mexican cooking or the curry spices of Indian cuisine. Some research suggests that genetics may influence people's food preferences.[3]

• ***Habit*** • People sometimes select foods out of habit. They eat cereal every morning, for example, simply because they have always eaten cereal for breakfast. Eating a familiar food and not having to make any decisions can be comforting.

• ***Ethnic Heritage or Tradition*** • Among the strongest influences on food choices are ethnic heritage and tradition. People eat the foods they grew up eating. Every country—and every region of a country—has its own typical foods and ways of combining foods into meals. Regional preferences reflect not only the people, but the availability of foods and the economic and political events of the time.[4] Highlight 2 shows how people can make healthful food choices within their own ethnic cuisines, and Highlight 6 examines the Mediterranean diet.

• ***Social Interactions*** • Food signifies friendliness. Meals are social events, and the sharing of food is part of hospitality. Social customs almost compel people to accept food or drink offered by a host or shared by a group. When your friends are going out for pizza or ice cream, how can you refuse to go along?

• ***Availability, Convenience, and Economy*** • People eat foods that are accessible, quick and easy to prepare, and within their financial means. Consumers today

***People enjoy companionship while eating. (Courtesy of CNN)***

*For many people, a special family dinner brings pleasant memories of the holidays.*

value convenience especially highly, as reflected in the choices of meals they can prepare and clean up quickly, recipes with few ingredients, and products they can cook in microwave ovens.[5] Many people frequently eat out, bring home ready-to-eat meals, or have food delivered, which limits their food choices to the selections on the store or restaurant's menus.

• ***Positive and Negative Associations*** • People tend to like foods with happy associations—such as hot dogs at ball games or cake and ice cream at birthday parties. By the same token, people can attach intense and unalterable dislikes to foods that they ate when they felt sick, or that were forced on them when they weren't hungry. Parents may teach their children to like and dislike certain foods by using those foods as rewards or punishments.

Sometimes foods are associated with certain uses. For example, people may believe that peanut butter is for children, or that lobster is for the rich. Then, depending on whether they permit themselves to be childlike or to indulge in luxuries, they will choose to eat or refrain from eating those foods.

• ***Emotional Comfort*** • Some people eat in response to emotional stimuli—for example, to relieve boredom or depression or to calm anxiety. A lonely person may choose to eat rather than to call a friend and risk rejection. A person who has returned home from an exciting evening out may unwind with a late-night snack. Eating in response to emotions can easily lead to overeating and obesity, but may be appropriate at times. For example, sharing food at times of bereavement serves both the giver's need to provide comfort and the receiver's need to be cared for and to interact with others, as well as to take nourishment.

• ***Values*** • Food choices may reflect people's religious beliefs, political views, or environmental concerns. For example, many Christians forgo meat during Lent, the period prior to Easter, and Jewish law includes an extensive set of dietary rules. A political activist may boycott fruit picked by migrant workers who have been exploited. People may buy vegetables from local farmers to save the fuel and environmental costs of foods shipped in from far away. Consumers may also select foods packaged in containers that can be reused or recycled.

• ***Body Image*** • Sometimes people select certain foods and supplements that they believe will improve their physical appearance and avoid those they believe might be detrimental. Such decisions can be beneficial when based on sound nutrition and fitness knowledge, but undermine good health when based on faddism or carried to extremes, as Highlights 9 and 14 point out.

• ***Nutrition*** • Finally, of course, many consumers make food choices that will benefit health. Food manufacturers and restaurant chefs have responded to scientific findings linking health with nutrition by offering an abundant selection of health-promoting foods and beverages. In some cases, the health-promoting foods are as natural and familiar as oatmeal or tomatoes. In other cases, the foods have been processed or prepared in a way that provides health benefits, perhaps by lowering the fat contents (see Highlight 5). In still other cases, manufacturers have developed functional foods, products designed specifically to promote health and prevent disease.[6] An example of a functional food is a margarine made with a certain plant sterol that lowers blood cholesterol.[7]

**functional foods:** foods or food ingredients that have been modified to provide a health benefit beyond their nutrient contributions.

International Food Information Council

ificinfo.health.org

Search for Functional foods

Consumers welcome these new foods into their diets, provided that the foods are reasonably priced, clearly labeled, easy to find in the grocery store, and convenient to prepare. These foods must also taste good—as good as the traditional choices.[8] Of course, a person need not eat any of these "special" foods to enjoy a healthy diet; ordinary foods, well chosen, serve just as well.

IN SUMMARY

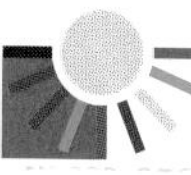

A person selects foods for a variety of reasons. Whatever those reasons may be, food choices influence health. Individual food selections neither make nor break a diet's healthfulness, but the balance of foods selected over time can make an important difference to health. For this reason, people are wise to think "nutrition" when making their food choices.

# Introducing the Nutrients

Do you ever think of yourself as a collection of carefully arranged atoms, molecules, cells, tissues, and organs? Are you aware of the activity going on within your body even as you sit still? The atoms, molecules, and cells of your body continually move and change, even though the structures of your tissues and organs and your external appearance remain relatively constant.

Your skin, which has covered you since your birth, is replaced entirely by new cells every seven years. The fat beneath your skin is not the same fat that was there a year ago. Your oldest red blood cell is only 120 days old, and the entire lining of your digestive tract is renewed every 3 days. To maintain your "self," you must continually replenish, from foods, the energy and the nutrients you deplete in maintaining your body.

## *The Six Classes of Nutrients*

Amazingly, the body can derive all the energy, structural materials, and regulating agents that it needs from the foods we eat. This section introduces the nutrients that foods bring to your body and shows how they participate in the dynamic processes that keep people alive and well.

**nutrients:** chemical substances obtained from food and used in the body to provide energy, structural materials, and regulating agents to support growth, maintenance, and repair of the body's tissues; nutrients may also reduce the risks of some diseases.

As Chapter 5 explains, most lipids are fats.

IN SUMMARY

**Six Classes of Nutrients**

- Carbohydrates.
- Lipids.
- Proteins.
- Vitamins.
- Minerals.
- Water.

• ***Composition of Foods*** • Chemical analysis of a food such as a tomato shows that it is composed primarily of water (95 percent). Most of the solid materials are carbohydrates, lipids, and proteins. If you could remove these materials, you would find a tiny residue of vitamins, minerals, and other compounds. Water, carbohydrates, lipids, proteins, vitamins, and some of the minerals found in foods are nutrients—substances the body uses for the growth, maintenance, and repair of its tissues. Other nutritionally important constituents of foods are the fibers—members of the carbohydrate family that also support good health.

• ***Composition of the Body*** • A complete chemical analysis of your body would show that it is made of materials similar to those found in foods. A healthy 150-pound body contains about 90 pounds of water and about 30 pounds of fat. The other 30 pounds are mostly compounds containing protein and carbohydrate and the major minerals of the bones. Vitamins, other minerals, and incidental extras constitute a fraction of a pound.

• ***Chemical Composition of Nutrients*** • The simplest of the nutrients are the minerals. Each mineral is a chemical element; its atoms are all alike. As a result, its identity never changes. Iron, for example, remains iron when a food is cooked, when a person eats the food, when iron becomes part of a red blood cell, when the cell is broken down, and when the iron is lost from the body by excretion. The next simplest nutrient is water, a compound made of two elements—hydrogen and oxygen. Minerals and water are inorganic nutrients—they contain no carbon.

The other four classes of nutrients (carbohydrates, lipids, proteins, and vitamins) are more complex. In addition to hydrogen and oxygen, they all contain carbon, an element found in all living things. They are therefore called organic compounds (meaning, literally, "alive"). Protein and some vitamins also contain nitrogen and may contain other elements as well (see Table 1-1).

**inorganic:** not containing carbon or pertaining to living things.
- **in** = not

**organic:** a substance or molecule containing carbon-carbon bonds or carbon-hydrogen bonds.* Some farmers call their produce "organic" if it was grown without manufactured fertilizers and pesticides (see Chapter 19), but by the definition given here, all foods are organic.

*This definition excludes coal, diamonds, and a few carbon-containing compounds that contain only a single carbon and no hydrogen, such as carbon dioxide ($CO_2$), calcium carbonate ($CaCO_3$), magnesium carbonate ($MgCO_3$), and sodium cyanide (NaCN).

• ***Essential Nutrients*** • The body can make some nutrients, but it cannot make all of them and it makes some in insufficient quantities to meet its needs. It must obtain these nutrients from foods. The nutrients that foods must supply are *essen-*

**Table 1-1**
**Elements in the Six Classes of Nutrients**

Notice that organic nutrients contain carbon.

| | Carbon | Hydrogen | Oxygen | Nitrogen | Minerals |
|---|---|---|---|---|---|
| ***Inorganic nutrients*** | | | | | |
| Minerals | | | | | ✔ |
| Water | | ✔ | ✔ | | |
| ***Organic nutrients*** | | | | | |
| Carbohydrates | ✔ | ✔ | ✔ | | |
| Lipids (fats) | ✔ | ✔ | ✔ | | |
| Proteins[a] | ✔ | ✔ | ✔ | ✔ | |
| Vitamins[b] | ✔ | ✔ | ✔ | | |

[a]Some proteins also contain the mineral sulfur.
[b]Some vitamins contain nitrogen; some contain minerals.

**essential nutrients:** nutrients a person must obtain from food because the body cannot make them for itself in sufficient quantity to meet physiological needs; also called **indispensable nutrients.** About 40 nutrients are known to be essential for human beings.

**phytochemicals** (FIE-toe-KEM-ih-cals): nonnutrient compounds found in plant-derived foods that have biological activity in the body.
- **phyto** = plant

**nonnutrients:** compounds in foods that do not fit within the six classes of nutrients.

**energy-yielding nutrients:** the nutrients that break down to yield energy the body can use:
- Carbohydrate.
- Fat.
- Protein.

**energy:** the capacity to do work. The energy in food is chemical energy. The body can convert this chemical energy to mechanical, electrical, or heat energy.

**calorie:** a unit by which energy is measured. Food energy is measured in **kilocalories** (1000 calories equal 1 kilocalorie), abbreviated **kcalories** or **kcal.** A capitalized version is also sometimes used: **Calories.** One kcalorie is the amount of heat necessary to raise the temperature of 1 kilogram (kg) of water 1°C.

1 g carbohydrate = 4 kcal.
1 g protein = 4 kcal.
1 g fat = 9 kcal.

1 g alcohol = 7 kcal.

*tial nutrients.* When used to refer to nutrients, the word *essential* means more than just "necessary"; it means "needed from outside the body"—normally, from foods.

• ***Nonnutrients*** • This book focuses mostly on the nutrients, but foods have other constituents as well—alcohols, phytochemicals, pigments, additives, and others. Some are beneficial, some are neutral, and a few are harmful. Later sections of the book touch on these nonnutrients and their significance.

## *The Energy-Yielding Nutrients*

In the body, three of the organic nutrients can be used to provide energy: carbohydrate, fat, and protein. In contrast, vitamins, minerals, and water do not yield energy in the human body.

• ***Energy Measured in kCalories*** • The energy released from carbohydrates, fats, and proteins can be measured in calories—tiny units of energy so small that a single apple provides tens of thousands of them. To ease calculations, energy is expressed in 1000-calorie units known as kilocalories (shortened to kcalories, but commonly called "calories"). When you read in popular books or magazines that an apple provides "100 calories," understand that it means 100 kcalories. This book uses the term *kcalorie* and its abbreviation *kcal* throughout, as do other scientific books and journals. The "How to" on the next page provides a few tips on how to "think metric."

A kcalorie is not a constituent of foods; it is a measure of the potential energy in foods. Thus to speak of "kcalories" in a cookie is technically incorrect, just as to speak of the inches in a person is incorrect. It is correct to speak of the *energy* available from a food (and of the *height* of a person).

• ***Energy from Foods*** • The amount of energy a food provides depends on how much carbohydrate, fat, and protein it contains. When completely broken down in the body, a gram of carbohydrate yields about 4 kcalories of energy; a gram of protein also yields 4 kcalories; and a gram of fat yields 9 kcalories.* The "How to" on p. 8 explains how to calculate the energy available from foods.

One other substance contributes energy: alcohol. Alcohol is not a nutrient because it interferes with the growth, maintenance, and repair of the body, but it does yield energy when metabolized in the body.**

Most foods contain all three energy-yielding nutrients, as well as water, vitamins, minerals, and other substances. Thus it is inaccurate to describe a food as its predominant nutrient—for example, to speak of meat as protein or of bread as carbohydrate. Meat and bread are *foods* rich in these nutrients. Meat contains

*For those using kilojoules: 1 g carbohydrate = 17 kJ; 1 g protein = 17 kJ; 1 g fat = 37 kJ.
**For those using kilojoules: 1 g alcohol = 29 kJ.

## HOW TO Think Metric

Like other scientists, nutrition scientists use metric units of measure. They measure food energy in kilocalories, people's height in centimeters, people's weight in kilograms, and the weights of foods and nutrients in grams, milligrams, or micrograms. For ease in using these measures, it helps to remember that the prefixes on the grams imply 1000. For example, a *kilo*gram is 1000 grams, a *milli*gram is 1/1000 of a gram, and a *micro*gram is 1/1000 of a milligram.

Most food labels and many recipe books provide "dual measures," listing both household measures, such as cups, quarts, and teaspoons, and metric measures, such as milliliters, liters, and grams. This practice gives people a chance to gradually learn to "think metric."

A person might begin to "think metric" by simply observing the measure—by noticing the amount of soda in a 2-liter bottle, for example. Through such experiences, a person can become familiar with a measure without having to do any conversions.

To facilitate communication, many members of the international scientific community have adopted a common system of measurement—the International System of Units (SI). In addition to using metric measures, the SI establishes common units of measurement. For example, the SI unit for measuring food energy is the joule (not the kcalorie). A joule is the amount of energy expended when 1 kilogram is moved 1 meter by a force of 1 newton. The joule is thus a measure of *work* energy, whereas the kcalorie is a measure of *heat* energy. While many scientists and journals report their findings in kilojoules (kJ), many others, particularly those in the United States, use kcalories. To convert energy measures from kcalories to kilojoules, multiply by 4.2. For example, a 50-kcalorie cookie provides 210 kilojoules:

$$50 \text{ kcal} \times 4.2 = 210 \text{ kJ}.$$

Exact conversion factors for these and other units of measure are in Appendix D.

**Volume: Liters (L)**

1 L = 1000 milliliters (mL).

0.95 L = 1 quart.

1 mL = 0.03 fluid ounces.

250 mL = 1 cup.

*A liter of liquid is approximately one U.S. quart. (Four liters are only about 5 percent more than a gallon.)*

*One cup is about 250 milliliters; a half-cup of liquid is about 125 milliliters.*

**Weight: Grams (g)**

1 g = 1000 milligrams (mg).

1 g = 0.04 ounces (oz).

1 oz = 28.35 g or ≈ 30 g.

100 g ≈ 3 ½ oz.

1 kilogram (kg) = 1000 g.

1 kg = 2.2 pounds (lb).

454 g = 1 lb.

A kilogram is slightly more than 2 lbs; conversely, a pound weighs about ½ kg.

*A half-cup of vegetables weighs about 100 grams.*

*A 5-pound bag of potatoes weighs about 2 kilograms, and a 176-pound person weighs 80 kilograms.*

## HOW TO Calculate the Energy Available from Foods

To calculate the energy available from a food, multiply the number of grams of carbohydrate, protein, and fat by 4, 4, and 9, respectively. Then add the results together. For example, 1 slice of bread with 1 tablespoon of peanut butter on it contains 16 grams carbohydrate, 7 grams protein, and 9 grams fat:

$$16 \text{ g carbohydrate} \times 4 \text{ kcal/g} = 64 \text{ kcal.}$$
$$7 \text{ g protein} \times 4 \text{ kcal/g} = 28 \text{ kcal.}$$
$$9 \text{ g fat} \times 9 \text{ kcal/g} = 81 \text{ kcal.}$$
$$\text{Total} = 173 \text{ kcal.}$$

From this information, you can calculate the percentage of kcalories each of the energy nutrients contributes to the total. To determine the percentage of kcalories from fat, for example, divide the 81 fat kcalories by the total 173 kcalories:

$$81 \text{ fat kcal} \div 173 \text{ total kcal} = 0.468 \text{ (rounded to 0.47).}$$

Then multiply by 100 to get the percentage:

$$0.47 \times 100 = 47\%.$$

Health recommendations that urge people to limit fat intake to 30 percent of kcalories refer to the day's total energy intake, not to individual foods. Still, if the proportion of fat in each food choice throughout a day exceeds 30 percent of kcalories, then the day's total surely will, too. Knowing that this snack provides 47 percent of its kcalories from fat alerts a person to the need to make lower-fat selections at other times that day.

water, fat, vitamins, and minerals as well as protein. Bread contains water, a trace of fat, a little protein, and some vitamins and minerals in addition to its carbohydrate. Only a few foods are exceptions to this rule, the common ones being sugar (pure carbohydrate) and oil (essentially pure fat).

The processes by which nutrients are broken down to yield energy or rearranged into body structures are known as **metabolism** (defined and described further in Chapter 7).

• ***Energy in the Body*** • The body uses the energy-yielding nutrients to fuel its metabolic and physical activities. When the body metabolizes the energy-yielding nutrients, the bonds between their atoms break. As the bonds break, they release energy. Some of this energy is released as heat, but some is used to send electrical impulses through the brain and nerves, to synthesize body compounds, and to move muscles. Thus the energy from food supports every activity from quiet thought to vigorous sports.

If the body does not use these nutrients to fuel metabolic and physical activities, it rearranges them into storage compounds, to be used between meals and overnight when fresh energy supplies run low. If more energy is consumed than expended, the result is weight gain. Similarly, if less energy is consumed than expended, the result is weight loss.

When consumed in excess of energy need, alcohol, too, can be converted to body fat and stored. When alcohol contributes a substantial portion of the energy in a person's diet, the harm it does extends far beyond the problems of excess body fat. (Highlight 7 describes the effects of alcohol on health and nutrition.)

The body's use (metabolism) of the energy-yielding nutrients can be summarized as follows. Carbohydrate, fat, and protein from foods are broken down, yielding energy and smaller molecules. The energy may:

- Escape as heat.
- Help build new compounds (and some energy may be stored in them).
- Help move the body (do work).

The smaller molecules may:

- Serve as building blocks for new compounds (fat, muscle, or other tissues).
- Be excreted as waste materials.

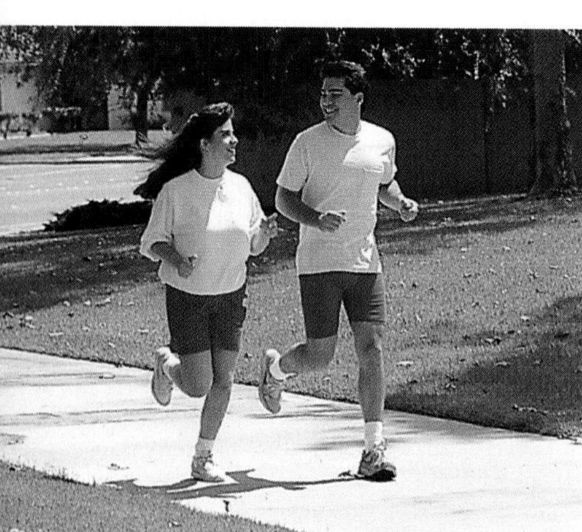

***All the energy used to keep the heart beating, the brain thinking, and the legs running comes from the carbohydrates, fats, and proteins in foods.***

• ***Other Roles of Energy-Yielding Nutrients*** • In addition to providing energy, carbohydrates, fats, and proteins provide the raw materials for building the body's tissues and regulating its many activities. In fact, protein's role as a fuel source is relatively minor compared with both the other two nutrients and its other roles. Proteins are found in structures such as the muscles and skin and help to regulate activities such as digestion and energy metabolism.

## The Vitamins

Like the first three classes of nutrients (carbohydrates, lipids, and proteins), the vitamins are vital to life, organic, and available in food.* They differ, however, in that the body does not extract usable energy from vitamins; rather, it uses vitamins as helpers in metabolic processes.

**vitamins:** organic, essential nutrients required in small amounts by the body for health.

Vitamins can function only if they are intact, but because they are complex organic molecules, they are vulnerable to destruction by heat, light, and chemical agents. This is why the body handles them carefully, and why nutrition-wise cooks do, too. The strategies of cooking foods at moderate temperatures, in or over small amounts of water, and for short times all help to preserve the vitamins.

There are 13 different vitamins, each with its own special roles to play. One vitamin enables the eyes to see in dim light, another helps protect the lungs from air pollution, and still another helps make the sex hormones—among other things. When you cut yourself, one vitamin helps stop the bleeding and another helps repair the skin. Vitamins busily help replace old red blood cells and the lining of the digestive tract. Almost every action in the body requires the assistance of vitamins.

## The Minerals

In contrast to the vitamins, which are organic compounds, the minerals are pure inorganic elements.** That means the minerals occur in the simplest of chemical forms, as atoms of a single element. Some minerals are put together in orderly arrays in such structures as bones and teeth. Some minerals are found in the fluids of the body and influence their properties. Whatever their roles, minerals are not metabolized, nor do they yield energy.

**minerals:** inorganic elements; some minerals are essential nutrients required in small amounts.

Some 16 minerals are known to be essential in human nutrition. Others are still being studied to determine whether they play significant roles in the human body. Still other minerals are *not* essential nutrients, but are important nevertheless because they are environmental contaminants that displace the nutrient minerals from their workplaces in the body, disrupting body functions. The problems caused by one of the contaminant minerals, lead, are detailed in Highlight 13.

Because they are indestructible, minerals in foods need not be handled with the special care that vitamins require. Minerals, can, however, be bound by substances that make it hard for the body to absorb them. They can also be lost during food refining processes or dissolve into water during cooking and then be discarded.

## Water

Water, indispensable and abundant, provides the environment in which nearly all the body's activities are conducted. It participates in many metabolic reactions and supplies the medium for transporting vital materials to cells and waste products away from them. Water is discussed fully in Chapter 12, but it is mentioned in

***Water itself is an essential nutrient and naturally contains many minerals, which give it flavor.***

*The water-soluble vitamins are vitamin C and the eight B vitamins: thiamin, riboflavin, niacin, vitamins $B_6$ and $B_{12}$, folate, biotin, and pantothenic acid. The fat-soluble vitamins are vitamins A, D, E, and K. The water-soluble vitamins are the subject of Chapter 10 and the fat-soluble vitamins, of Chapter 11.

**The major minerals are calcium, phosphorus, potassium, sodium, chloride, magnesium, and sulfur. The trace minerals are iron, iodine, zinc, chromium, selenium, fluoride, molybdenum, copper, and manganese. Chapters 12 and 13 are devoted to the major and trace minerals, respectively.

every chapter. If you watch for it, you cannot help but be impressed by water's participation in all life processes.

IN SUMMARY

Foods provide nutrients—substances that support the growth, maintenance, and repair of the body's tissues. Foods rich in the energy-yielding nutrients (carbohydrates, fats, and proteins) provide the major materials for building the body's tissues and yield energy for the body's use or storage. Energy is measured in kcalories. Vitamins, minerals, and water facilitate a variety of activities in the body. Without exaggeration, nutrients provide the physical and metabolic basis for nearly all that we are and all that we do.

# The Science of Nutrition

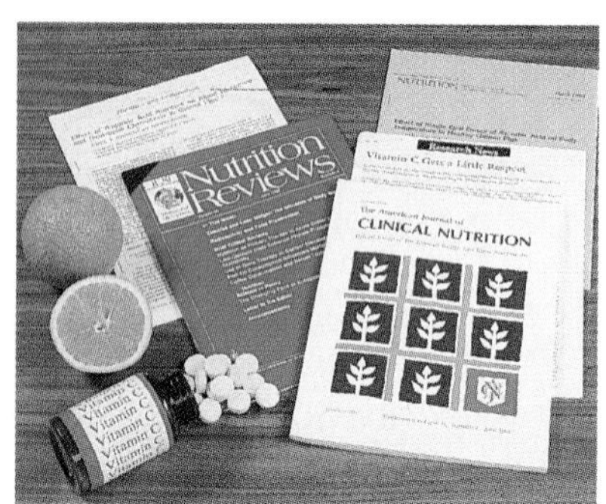


***Knowledge about the nutrients and their effects on health comes from scientific study.***

The science of nutrition is the study of the nutrients and other substances in foods and the body's handling of them. As sciences go, nutrition is a young one, but as you can see from the size of this book, much has happened in nutrition's short life. This section introduces the research methods scientists have used in uncovering the wonders of nutrition.

## *Nutrition Research*

Research always begins with a question. For example, "What foods or nutrients might protect against the common cold?" In search of an answer, scientists make educated guesses (hypotheses) and then systematically conduct research studies to test each hypothesis. Some examples of the various types of research studies follow:

- *Epidemiological studies.* Scientists observe how much and what kinds of foods a group of people eat and how healthy those people are. Their findings identify factors that might influence the incidence of a disease in various populations.
- *Case-control studies.* Researchers compare people who do and do not have a given condition such as a disease, closely matching them in age, sex, and other key variables so that differences in other factors will stand out. These differences may account for the condition in the group that has it.
- *Animal studies.* Researchers feed animals special diets that provide or omit specific nutrients and then observe any changes in health. Such studies test possible disease causes and treatments in a laboratory where all conditions can be controlled.
- *Human intervention (or clinical) trials.* Scientists ask people to adopt a new behavior (for example, eat a citrus fruit, take a vitamin C supplement, or exercise daily). These trials help determine the effects such measures have on the development or prevention of disease.

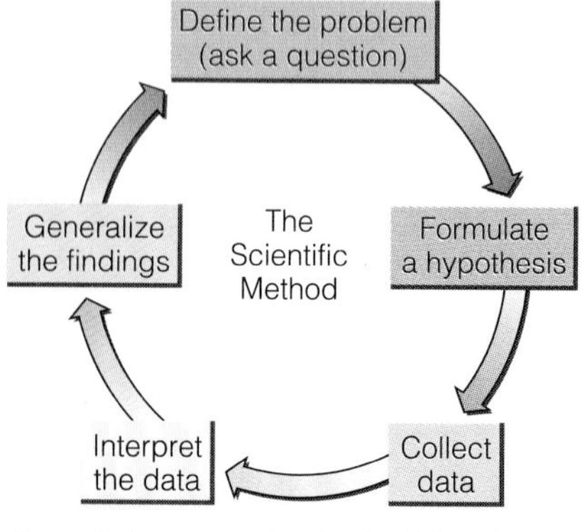


In conducting research, scientists follow these steps, which define the scientific method.

Each type of study has advantages and disadvantages. Findings must be interpreted with an awareness of the study's limitations. (See Highlight 1 for a discussion on evaluating research findings.)

In attempting to discover whether a nutrient relieves symptoms or cures a disease, all research tries to answer the same kinds of questions. Research on vitamin C and the common cold illustrates particularly well what those questions are. The accompanying glossary defines relevant terms.

• ***Controls*** • In most studies on the efficacy of vitamin C, researchers divide people into two groups. One group (the experimental group) receives a vitamin C supplement, and the other (the control group) does not. Researchers observe both groups to determine whether the vitamin C group has fewer or shorter colds than

## Glossary of Research Terms

**blind experiment:** an experiment in which the subjects do not know whether they are members of the experimental group or the control group.

**control group:** a group of individuals similar in all possible respects to the experimental group except for the treatment. Ideally, the control group receives a placebo while the experimental group receives a real treatment.

**correlation** (CORE-ee-LAY-shun): the simultaneous increase, decrease, or change in two variables. If A increases as B increases, or if A decreases as B decreases, the correlation is positive. (This does not mean that A causes B or vice versa.) If A increases as B decreases, or if A decreases as B increases, the correlation is negative. (This does not mean that A prevents B or vice versa.) Some third factor may account for both A and B.

**double-blind experiment:** an experiment in which neither the subjects nor the researchers know which subjects are members of the experimental group and which are serving as control subjects, until after the experiment is over.

**experimental group:** a group of individuals similar in all possible respects to the control group except for the treatment. The experimental group receives the real treatment.

**peer review:** a process in which a panel of scientists rigorously evaluates a research study to assure that the scientific method was followed.

**placebo** (pla-SEE-bo): an inert, harmless medication given to provide comfort and hope; a sham treatment used in controlled research studies.

**placebo effect:** the healing effect that faith in medicine, even inert medicine, often has.

**randomization** (RAN-dom-ih-ZAY-shun): a process of choosing the members of the experimental and control groups without bias.

**replication** (REP-lee-KAY-shun): repeating an experiment and getting the same results. The skeptical scientist, on hearing of a new, exciting finding, will ask, "Has it been replicated yet?" If it hasn't, the scientist will withhold judgment regarding the finding's validity.

**subjects:** the people or animals participating in a research project.

**validity** (va-LID-ih-tee): having the quality of being founded on fact or evidence.

**variable:** a factor that changes. A variable may depend on another variable (for example, a child's height depends on his age), or it may be independent (for example, a child's height does not depend on the color of her eyes). Sometimes both variables correlate with a third variable (a child's height and eye color both depend on genetics).

the control group. A number of pitfalls are inherent in an experiment of this kind and must be avoided.

First, the two groups of people must be similar in all respects (except that one group receives vitamin C). Similarity of the experimental and control groups is accomplished by randomization; that is, the members are chosen from the same population by throws of the dice or some other method involving chance.

Importantly, both groups must have the same track record with respect to colds to rule out the possibility that an observed difference might have occurred anyway. If group A would have caught twice as many colds as group B anyway and group B happens to receive the treatment, then the findings prove nothing.

In experiments involving a nutrient, the diets of both groups must also be similar, especially with respect to that nutrient. If those in group B were receiving less vitamin C from their diet, this might cancel the effects of the supplement.

• ***Sample Size*** • To ensure that chance variation between the two groups does not influence the results, the groups must be large. If one member of a group of five people catches a bad cold by chance, he will pull the whole group's average toward bad colds; but if one member of a group of 500 catches a bad cold, she will not unduly affect the group average. Statistical methods are used to determine whether differences between groups of various sizes support a hypothesis.

• ***Placebos*** • If people take vitamin C for colds and *believe* it will cure them, their chances of recovery are improved. Taking anything believed to be beneficial hastens recovery in about half of all cases. This phenomenon, the effect of faith on healing, is known as the placebo effect. In experiments designed to determine vitamin C's effect on colds, this mind-body effect must be rigorously controlled. Severity of symptoms is often a subjective measure, and people who believe they are receiving treatment may report less severe symptoms.

One way experimenters control for the placebo effect is to give pills to all participants; some pills contain vitamin C, and others of similar appearance and taste contain an inactive ingredient (placebos). This way, the effects of faith will work equally in both groups. It is not necessary to convince all subjects that they are receiving vitamin C, but the extent of belief or unbelief must be the same in both groups. A study conducted under these conditions is called a blind experiment—that is, the subjects do not know (are blind to) whether they are members of the experimental group (receiving treatment) or the control group (receiving the placebo).

• ***Double Blind*** • When both the subjects and the researchers do not know which subjects are in which group, the study is called a double-blind experiment. Being fallible human beings and having an emotional investment in a successful outcome, researchers might record and interpret results with a bias in the expected direction. To prevent such distortions, the pills are coded by a third party, who does not reveal to the experimenters which subjects were in which group until all results have been recorded.

• ***Correlations and Causes*** • Researchers often examine the relationships between two or more variables—for example, daily vitamin C intake and the number of colds. Findings sometimes suggest no correlation between the two variables (regardless of the amount of vitamin C consumed, the number of colds remains the same). Other times, studies find either a positive correlation (the more vitamin C, the more colds) or a negative correlation (the more vitamin C, the fewer colds). Correlational evidence proves only that two variables are associated, not that one is the cause of the other. People often jump to conclusions when they learn of correlations, but the conclusions are often wrong. To prove that A causes B, scientists have to find evidence of the *mechanism*—that is, to catch A in the act of causing B, so to speak. Furthermore, other scientists must confirm or disprove the findings through replication before the results are accepted into the body of nutrition knowledge. Before the findings are published, they are subjected to peer review—a process whereby a panel of scientists evaluates the study to confirm that it followed standard scientific methods.

## Research versus Rumors

In discussing these subtleties of experimental design, our intent is to show you what a far cry scientific validity is from the experience of your neighbor Mary (sample size, one; no control group), who says she takes vitamin C when she feels a cold coming on and "it works every time." She knows what she is taking, she has faith in its efficacy, and she tends not to notice when it doesn't work. Before concluding that an experiment has shown that a nutrient cures a disease or alleviates a symptom, ask these questions:

- Was there similarity between the control group and the experimental group?
- Was the sample size large enough to rule out chance variation?
- Was a placebo effectively administered (blind)?
- Was the experiment double blind?

These characteristics of well-designed research have enabled scientists to study the actions of nutrients in the body. Such research has laid the foundation for quantifying how much of each nutrient the body needs.

### IN SUMMARY

Scientists learn about nutrition by conducting experiments that follow the protocol of scientific research. Researchers take care to establish similar control and experimental groups, large sample sizes, placebos, and blind treatments. Their findings must be reviewed and replicated by other scientists before being accepted as valid.

# Dietary Reference Intakes

Defining the amounts of energy, nutrients, and other dietary components that best support health is a huge task. For more than 50 years, nutrition experts produced a set of energy and nutrient standards known as the Recommended Dietary Allowances (RDA) for the people in the United States. The Canadian equivalent of the RDA is the RNI (Recommended Nutrient Intakes). Over the years, these standards were revised periodically as new evidence became available, but each revision maintained the original goal of protecting against nutrient deficiencies. Given the abundance of research now linking diet and health, that goal has been broadened to include supporting optimal activities within the body and preventing chronic diseases as well. Previous editions also focused narrowly on nutrients known to be essential. With recent research revealing the health benefits of other dietary components such as fiber and phytochemicals, their recommended intakes need to be addressed as well.

To that end, a major revision of nutrient recommendations is currently under way. The revised recommendations are called Dietary Reference Intakes (DRI) and reflect the collaborative efforts of both the United States and Canada. The first in a series of reports was published in 1997 and presents both the framework for developing the DRI and the revised recommendations for the five nutrients that play key roles in bone health—calcium, phosphorus, magnesium, vitamin D, and fluoride.[9] A second report came out in 1998 and features revised recommendations for the eight B vitamins and a related compound choline.[10] The recommendations for these 14 nutrients appear on the inside front cover under the heading 1997–1998 Dietary Reference Intakes. For the remaining nutrients and energy, the 1989 RDA will continue to serve health professionals until DRI can be established.[11] The values for these nutrients also appear on the inside front cover under the heading 1989 Recommended Dietary Allowances.

*When people shop for foods, they are buying nutrients. (Courtesy of CNN)*

**Dietary Reference Intakes (DRI):** a set of values for the dietary nutrient intakes of healthy people in the United States and Canada. These values are used for planning and assessing diets and include:

- Estimated Average Requirements.
- Recommended Dietary Allowances.
- Adequate Intakes.
- Tolerable Upper Intake Levels.

 National Academy Press
www.nap.edu/readingroom
Search for Dietary Reference Intakes

Appendix G presents additional 1989 RDA tables, and Appendix I presents the 1991 Canadian RNI tables.

The DRI reports are produced by the Food and Nutrition Board of the Institute of Medicine in cooperation with scientists from Canada.

## *Establishing Nutrient Recommendations*

The DRI Committee consists of highly qualified scientists who base their estimates of nutrient needs on careful examination and interpretation of scientific evidence. The next paragraphs discuss specific aspects of how the committee goes about establishing the values that make up the DRI:

- Estimated Average Requirements.
- Recommended Dietary Allowances.
- Adequate Intakes.
- Tolerable Upper Intake Levels.

• ***Estimated Average Requirements*** • The committee reviews hundreds of research studies to determine how much of a nutrient is needed in the diet to maintain a specific function (for example, the amount of calcium needed to minimize bone loss in later life).[12] The committee selects a different criterion for each nutrient based on the nutrient's roles both in performing activities in the body and in reducing disease risks. From this information, the committee determines an Estimated Average Requirement for the nutrient—an amount that appears sufficient to maintain a specific body function in half of the population.

**Estimated Average Requirement:** the amount of a nutrient that will maintain a specific biochemical or physiological function in half the people of a given age and sex group.

An examination of all the available data reveals that each person's body is unique and has its own set of requirements. For example, Mr. A might need 40 units of the nutrient each day; Ms. B might need 35; Mr. C, 57. A look at enough individuals might reveal that their requirements fall into a symmetrical distribution, with most near the midpoint (shown in Figure 1-1 as 45 units) and only a few at the extremes.

**requirement:** the lowest continuing intake of a nutrient that will maintain a specified criterion of adequacy.

**Figure 1-1**
**Estimated Average Requirements and Recommended Dietary Allowances Compared**

Each of the 120 squares shown here represents a person. Some people require only a small amount of the nutrient, and some require a lot, but most fall somewhere near the middle. The text discusses three of these people: Mr. A, Ms. B, and Mr. C.

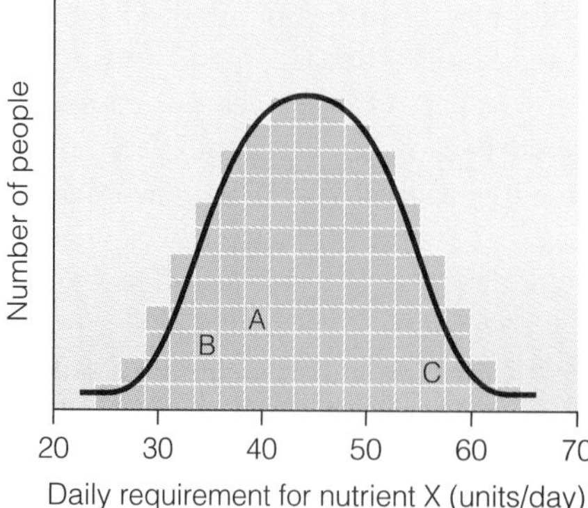


The RDA for nutrients is set well above the Estimated Average Requirement. It covers about 98% of the population.

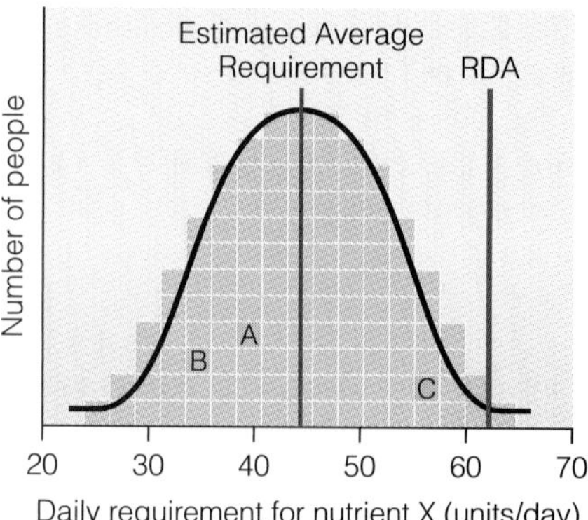


**Recommended Dietary Allowance (RDA):** the average daily amount of a nutrient considered adequate to meet the known nutrient needs of practically all healthy people; a goal for dietary intake by individuals.

• ***Recommended Dietary Allowances (RDA)*** • Then the committee must decide what intake to recommend for everybody—a Recommended Dietary Allowance (RDA). Assuming the distribution shown in Figure 1-1, the Estimated Average Requirement (shown in the figure as 45 units) for each nutrient is probably closest to everyone's need. (Actually, the data for most nutrients other than protein have a much less symmetrical distribution.) But if people consumed exactly the average requirement of a given nutrient each day, half of the population would develop deficiencies of that nutrient; in Figure 1-1, Mr. C would be among them. Recommendations should be set high enough above the Estimated Average Requirement to meet the needs of most healthy people.

In this example, a reasonable RDA might be 63 units a day (see Figure 1-1). Such a point can be calculated mathematically so that it covers about 98 percent of a population. Almost everybody—including Mr. C whose needs were higher than the average—would be covered if they met this dietary goal. Relatively few people's requirements would exceed this recommendation, and even then, they wouldn't exceed by much.

In contrast to the RDA for nutrients, the RDA for energy is not generous, but is set at the mean of the population's estimated requirements (see Figure 1-2). Although not enough energy may cause undernutrition, too much energy is as bad for health as too little because excess energy leads to obesity. In contrast, in

**Figure 1-2**
**The Nutrient RDA and the Energy RDA Compared**

The nutrient RDA are set high enough to cover nearly everyone's requirements (the boxes represent people).

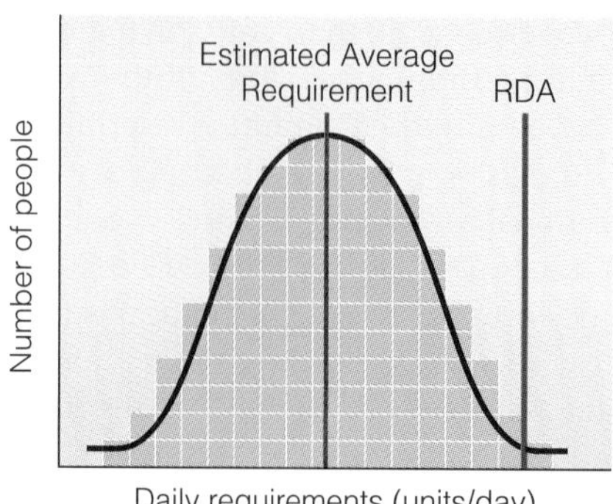


**Nutrient RDA**

The energy RDA is set at the average so that half the population's requirements fall below and half above them.

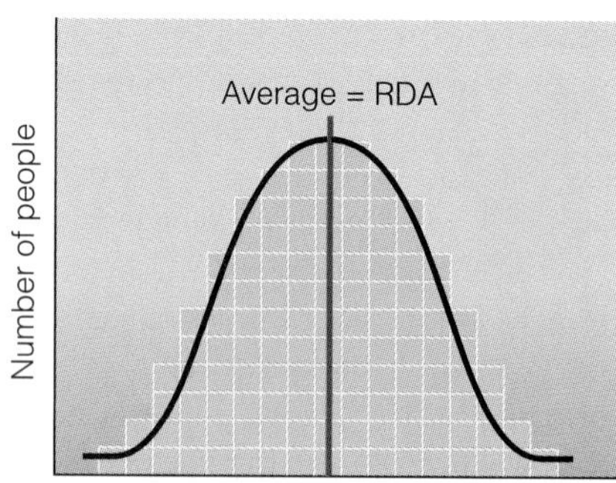


**Energy RDA**

the case of vitamins and minerals, small amounts above the daily requirement do no harm, whereas amounts below the requirement lead to health problems. When people's intakes are consistently deficient (less than the requirement), their nutrient stores decline, and over time this decline leads to poor health and deficiency symptoms. Therefore, to ensure that the vitamin and mineral RDA meet the needs of as many people as possible, these RDA are set near the top end of the range of the population's estimated requirements.

**deficient:** the amount of a nutrient below which almost all healthy people can be expected, over time, to experience deficiency symptoms.

• ***Adequate Intakes (AI)*** • For some nutrients, there is insufficient scientific evidence to determine an Estimated Average Requirement (which is needed to set an RDA). In these cases, the committee establishes an Adequate Intake (AI) instead of an RDA. An AI reflects the average amount of a nutrient that a group of healthy people consumes. Like the RDA, the AI may be used as nutrient goals for individuals.

**Adequate Intake (AI):** the average amount of a nutrient that appears sufficient to maintain a specified criterion; a value used as a guide for nutrient intake when an RDA cannot be determined.

While both RDA and AI serve as nutrient intake goals for individuals, their differences are noteworthy. An RDA for a given nutrient is based on enough scientific evidence to expect that the needs of almost all healthy people will be met. An AI, on the other hand, must rely more heavily on scientific judgments because sufficient evidence is lacking. The percentage of people covered by an AI is unknown; it is expected that an AI exceeds average requirements, but it may cover more or fewer people than an RDA would (if an RDA could be determined). For these reasons, AI values are more tentative than RDA. Later chapters will present the RDA and AI values for the vitamins and minerals.

• ***Tolerable Upper Intake Levels*** • The recommended intakes for nutrients are generous, and although they do not necessarily cover every individual for every nutrient, they probably should not be exceeded by much. People's tolerances for high doses of nutrients vary, and somewhere above the recommended intake is a Tolerable Upper Intake Level beyond which a nutrient is likely to become toxic.[13] It is naive to think of recommendations as minimum amounts. A more accurate view is to see a person's nutrient needs as falling within a range, with marginal and danger zones both below and above it (see Figure 1-3).

**Tolerable Upper Intake Level:** the maximum amount of a nutrient that appears safe for most healthy people and beyond which there is an increased risk of adverse health effects.

Upper levels are particularly useful in guarding against the overconsumption of nutrients, which is most likely to occur when people use supplements or fortified foods regularly. Later chapters discuss the dangers associated with excessively high intakes of vitamins and minerals, and Highlight 10 presents a table that includes the upper-level values for selected nutrients.

**Figure 1-3**
**Naive versus Accurate View of Nutrient Needs**

The RDA for a given nutrient represents a point within a range of appropriate and reasonable intakes that lies between toxicity and deficiency. The AI also falls within this range, but its determination is not as exact as an RDA; it may cover more or fewer people than the RDA. Both of these recommendations are high enough to provide reserves in times of short-term dietary inadequacies, but not so high as to approach toxicity. Nutrient intakes above or below this range may be equally harmful.

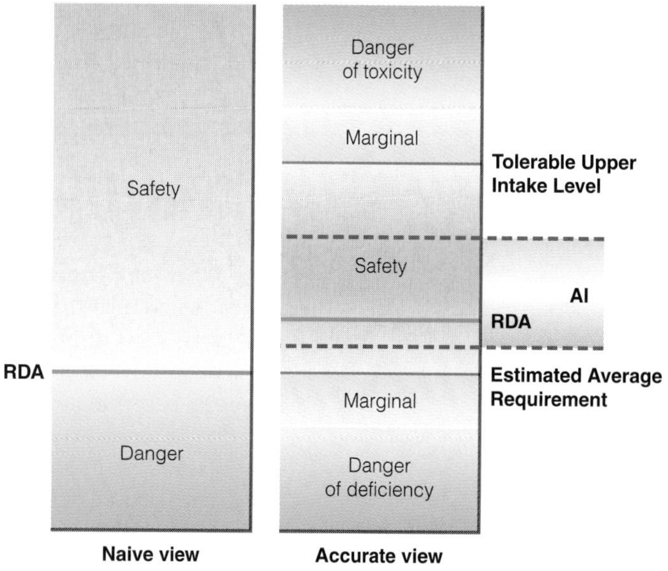


## Using Nutrient Recommendations

Although the intent of nutrient recommendations may seem simple enough, they are the subject of much misunderstanding and controversy. Perhaps the following facts will help put them in perspective. First, estimates of adequate energy and nutrient intakes apply to *healthy* people. They do not apply to malnourished people or to those with medical problems who may require supplemented or restricted intakes.

Second, these *recommendations* include a generous margin of safety. They are not minimum requirements, nor are they necessarily optimal intakes for all individuals. Registered dietitians and other qualified health professionals can help determine whether recommendations should be adjusted to meet individual needs.

Third, most nutrient goals are intended to be met through diets composed of a variety of *foods* whenever possible. Because foods contain mixtures of nutrients and nonnutrients, they deliver more than just those nutrients covered by the recommendations. Excess intakes of vitamins and minerals are unlikely when their sources are foods rather than supplements.

Fourth, recommendations apply to *average* daily intakes. To try to meet recommendations for every nutrient every day is difficult and unnecessary. The length of time over which a person's intake can deviate from the average without risk of deficiency or overdose varies for each nutrient, depending on the body's use and storage of the nutrient. For most nutrients (such as thiamin and vitamin C), deprivation would lead to rapid development of deficiency symptoms (within days or weeks); for others (such as vitamin A and vitamin $B_{12}$), deficiencies would develop much slower (over months or years).

Fifth, each of the four DRI categories serves a unique purpose. For example, the Estimated Average Requirements are most appropriately used to develop and evaluate nutrition programs for *groups* such as schoolchildren or military personnel. The RDA (or AI if an RDA is not available) can be used to set goals for *individuals*. Tolerable Upper Intake Levels help to keep nutrient intakes below the amounts that increase the risk of toxicity. With these understandings, professionals can use the DRI for a variety of purposes.

## *Comparing Nutrient Recommendations*

At least 40 different nations and international organizations have published nutrient standards similar to those used in the United States and Canada. Slight differences may be apparent, reflecting differences both in the interpretation of the data from which the standards were derived and in the food habits and physical activities of the populations they serve.

Many countries use the recommendations developed by two international groups: the Food and Agriculture Organization (of the United Nations) and the World Health Organization. The FAO/WHO recommendations are considered sufficient to support the maintenance of health in nearly all healthy people worldwide.

Nutrient recommendations from FAO/WHO are provided in Appendix G.

 Food and Agriculture Organization
www.fao.org

 World Health Organization
www.who.org

IN SUMMARY

The Dietary Reference Intakes (DRI) are a set of four nutrient intake values that can be used to plan and evaluate diets for healthy people. The Estimated Average Requirement defines the amount of a nutrient that supports a specific function in the body for *half* of the population. The Recommended Dietary Allowance (RDA) uses the Estimated Average Requirement to establish a goal for dietary intake that will meet the needs of *almost all* healthy people. An Adequate Intake (AI) serves a similar purpose when an RDA cannot be determined. The Tolerable Upper Intake Level establishes the highest amount that appears safe for regular consumption.

# Nutrition Assessment

What happens when a person doesn't get enough of a nutrient or energy or gets too much? If the deficiency or excess is significant over time, the person exhibits signs of malnutrition. With a deficiency of energy, the person may display the symptoms of undernutrition by becoming extremely thin, losing muscle tissue, and becoming prone to infection and disease. With a deficiency of a nutrient, the person may experience skin rashes, depression, hair loss, bleeding gums, muscle spasms, night blindness, or other symptoms. With an excess of energy, the person may become obese and vulnerable to diseases associated with overnutrition such as heart disease, cancer, and diabetes. With a sudden nutrient overdose, the person may experience hot flashes, yellowing skin, a rapid heart rate, low blood

**malnutrition:** any condition caused by excess or deficient food energy or nutrient intake or by an imbalance of nutrients.
- **mal** = bad

**undernutrition:** deficiency of energy or nutrients.

**overnutrition:** excess energy or nutrients.

pressure, or other symptoms. Similarly, regular intakes in excess of needs may have adverse effects.

Malnutrition symptoms are easy to miss. They resemble the symptoms of other diseases: diarrhea, skin rashes, pain, and the like. But a person who has learned how to read the signs can tell when these conditions are caused by malnutrition and can take steps to correct it. Dietitians have developed assessment techniques to detect malnutrition. This discussion presents the basics of nutrition assessment; many more details are offered in later chapters and in Appendix E.

## Nutrition Assessment of Individuals

To prepare a nutrition assessment, the assessor, usually a registered dietitian or a physician trained in clinical nutrition, uses:

- Historical information.
- Anthropometric data.
- Physical examinations.
- Laboratory tests.

Each of these methods involves collecting data in various ways and interpreting each finding in relation to the others to create a total picture.

**nutrition assessment:** a comprehensive approach, completed by a registered dietitian, to determine a person's nutrition status using health, socioeconomic, drug, and diet histories; anthropometric measurements; physical examinations; and laboratory tests.

A **registered dietitian** is a college-educated food and nutrition specialist who is qualified to evaluate people's nutritional health and needs. See Highlight 1 for more on what constitutes a nutrition expert.

• ***Historical Information*** • One step in evaluating nutrition status is to obtain information about a person's history with respect to health status, socioeconomic status, drug use, and diet. The health history may reveal a disease that interferes with the person's ability to eat or the body's use of nutrients. Socioeconomic circumstances may show a financial inability to buy foods or inadequate kitchen facilities in which to prepare them. A drug history may highlight possible drug-nutrient interactions that lead to nutrient deficiencies (as described in Highlight 17). A diet history can indicate whether the diet may be under- or oversupplying nutrients or energy.

To take a diet history, the assessor collects and analyzes data about the foods a person eats. The data may be collected by recording the foods the person has eaten over a period of 24 hours, three days, or a week or more or by asking what foods the person typically eats and how much of each. The days in a record have to be fairly typical of the person's diet, and portion sizes must be recorded accurately. To determine the amounts of nutrients consumed, the assessor usually enters the foods and their portion sizes into a computer using a diet analysis program. Alternatively, this step can be done manually by looking up each food in a table of food composition such as Appendix H in this book. Then the assessor compares the calculated nutrient intakes with recommended intakes.

Appendix E describes the tools used to obtain food intake data: the 24-hour recall, usual intake record, food frequency checklist, and food record.

An estimate of energy and nutrient intakes from a diet history, combined with other sources of information, can help confirm or rule out the *possibility* of suspected nutrition problems. A sufficient intake of a nutrient does not guarantee adequate nutrition status for an individual, and an insufficient intake does not always indicate a deficiency, but such findings warn of possible problems.

• ***Anthropometric Data*** • A second technique that may help reveal nutrition problems is the taking of anthropometric measures such as height and weight. The assessor compares measurements taken on an individual with standards specific for sex and age or with previous measures on the same individual.

**anthropometric** (AN-throw-poe-MET-rick): relating to measurement of the physical characteristics of the body, such as height and weight.
- **anthropos** = human
- **metric** = measuring

Measurements taken periodically and compared with previous measurements reveal patterns and indicate trends in a person's overall nutrition status, but they provide little information about specific nutrients. Instead, measurements out of line with expectations may reveal such problems as growth failure in children, wasting or swelling of body tissues in adults, and obesity—conditions that may reflect energy or nutrient deficiencies or excesses.

• ***Physical Examinations*** • A third nutrition assessment technique is a physical examination that looks for clues to poor nutrition status. Every part of the body

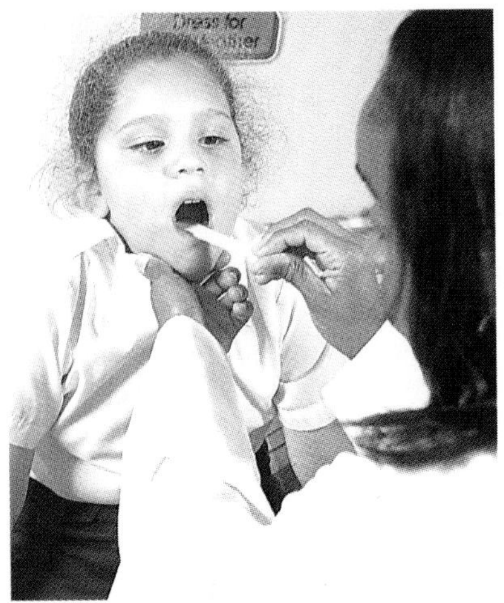

*A peek inside the mouth provides clues to a person's nutrition status.*

that can be inspected can offer such clues: the hair, eyes, skin, posture, tongue, fingernails, and others. The examination requires skill, for many physical signs can reflect more than one nutrient deficiency or toxicity or even nonnutrition conditions. Like the other assessment techniques, a physical examination does not by itself point to firm conclusions. Instead, it reveals possible nutrient imbalances for other assessment techniques to confirm, or it confirms data collected from other assessment measures.

• ***Laboratory Tests*** • A fourth way to detect a developing deficiency, imbalance, or toxicity is to take samples of blood or urine, analyze them in the laboratory, and compare the results with normal values for a similar population. A goal of nutrition assessment is to uncover early signs of malnutrition before symptoms appear, and laboratory tests are most useful for this purpose. In addition, they can confirm suspicions raised by other assessment methods.

• ***Iron, for Example*** • The mineral iron can be used to illustrate the stages in the development of a nutrient deficiency and the assessment techniques useful in detecting them. The overt, or outward, signs of an iron deficiency appear at the end of a long sequence of events. Figure 1-4 describes what happens in the body as a nutrient deficiency progresses and shows how assessment methods can reveal those changes.

First, too little iron gets into the body—either because iron is lacking in the person's diet (a primary deficiency) or because the person's body doesn't absorb or use iron normally (a secondary deficiency). A diet history provides clues to primary deficiencies; a health history provides clues to secondary deficiencies.

Then the body begins to use up its stores of iron. At this stage, the deficiency might be described as subclinical. It exists as a covert condition and might be detected by laboratory tests, but outward signs have not yet appeared.

Finally, iron stores are exhausted. Now, the body cannot make enough iron-containing red blood cells to replace those that are aging and dying. The iron in red blood cells normally carries oxygen to all the body's tissues. When iron is lacking, fewer red blood cells are made, the new ones are pale and small, and every part of the body feels the effects of an oxygen shortage. Now the overt symptoms of deficiency appear—weakness, fatigue, pallor, and headaches, reflecting the iron-deficient state of the blood. Physical examination would reveal these symptoms.

**Figure 1-4**
**Stages in the Development of a Nutrient Deficiency**

Internal changes precede outward signs of deficiencies. As a corollary, signs of sickness need not appear before a person takes corrective measures. Tests can either reveal the presence of problems in the early stages or confirm that nutrient stores are adequate.

| Here is what is happening inside the body: | How can the health care provider tell? |
|---|---|
| Primary deficiency caused by inadequate diet | Diet history |
| or Secondary deficiency caused by problem inside the body | Health history |
| ↓ Declining nutrient stores | Laboratory tests |
| ↓ Abnormal functions inside the body | Laboratory tests |
| ↓ Physical (outward) signs and symptoms | Physical examination and anthropometric measures |

The **Healthy People 2000** report sets national objectives in health promotion and disease prevention for the year 2000.[14] Healthy People 2010 is scheduled for release in January 2000. The 21 nutrition-related priorities are listed in Appendix G and appear in the text where their subjects are discussed.

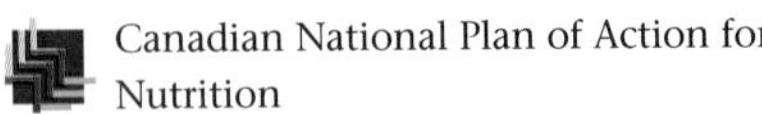

Canadian National Plan of Action for Nutrition
www.hc-sc.gc.ca/datahpsb/npu

Healthy People 2000
odphp.osophs.dhhs.gov/pubs/hp2000

Healthy People 2010
web.health.gov/healthypeople

**Healthy People 2000:** Increase to at least 75% the proportion of primary care providers who provide nutrition assessment and counseling and/or referral to qualified nutritionists or dietitians.

## *Nutrition Assessment of Populations*

To assess a population's nutrition status, researchers conduct surveys using techniques similar to those used on individuals. One kind of survey—a food consumption survey—determines the kinds and amounts of foods people eat. Then researchers calculate the energy and nutrients in the foods and compare the amounts consumed with a standard. An example of this type of survey is the Nationwide Food Consumption Survey (NFCS). Information for the third NFCS (1994–1996) was gathered from 15,000 people using food intake records for two nonconsecutive days.

Another kind of survey—a nutrition status survey—examines the people themselves, using nutrition assessment methods. The National Health and Nutrition Examination Survey (NHANES) is an example of a nutrition status survey. The third NHANES (1988–1996) gathered information from between 40,000 and 70,000 people using diet histories, anthropometric measurements, physical examinations, and laboratory tests. The data provide information on several nutrition-related conditions, such as growth retardation, heart disease, and nutrient deficiencies. Both the NFCS and the NHANES oversample high-risk groups (low-income families, infants and children, and the elderly) in order to glean an accurate estimate of their health and nutrition status.

Until 1990, findings from the nation's many nutrition surveys, including these two largest ones, were almost impossible to compare and synthesize into a single cohesive report. Then the National Nutrition Monitoring and Related Research Act was enacted to coordinate the many nutrition-related activities that had been underway within 22 different federal agencies. The law mandated that the U.S. Department of Agriculture (USDA) and the Department of Health and Human Services (DHHS) establish and implement a Ten-Year Comprehensive Plan for nutrition monitoring and related research.[15]

The resulting wealth of information is used for a variety of purposes. For example, Congress uses this information to establish public policy on nutrition education, food assistance programs, and the regulation of the food supply. Scientists use the information to establish research priorities. All major reports that examine the contribution of diet and nutrition status to the health of the people of the United States depend on information collected and coordinated by this national program.* These data provided the basis for the mid-decade report on Healthy People 2000 that showed we are not meeting many of our health goals; in fact, we are not even heading in the right direction for some goals, such as reducing the prevalence of overweight in the United States.[16]

**overt** (oh-VERT): out in the open and easy to observe.
- **ouvrir** = to open

**primary deficiency:** a nutrient deficiency caused by inadequate dietary intake of a nutrient.

**secondary deficiency:** a nutrient deficiency caused by something other than an inadequate intake such as a disease condition that reduces absorption, accelerates use, hastens excretion, or destroys the nutrient.

**subclinical deficiency:** a deficiency in the early stages, before the outward signs have appeared.

**covert** (KOH-vert): hidden, as if under covers.
- **couvrir** = to cover

**food consumption survey:** a survey that measures the amounts and kinds of foods people consume (using diet histories), estimates the nutrient intakes, and compares them with a standard.

**nutrition status survey:** a survey that evaluates people's nutrition status using diet histories, anthropometric measures, physical examinations, and laboratory tests.

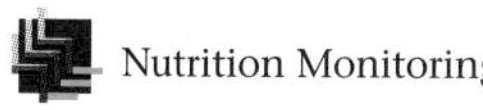


www.cdc.gov
Search for Nutrition monitoring

IN SUMMARY

People become malnourished when they get too little or too much energy or nutrients. To detect nutrition problems in individuals, health care professionals use four nutrition assessment methods. Reviewing dietary data and health information may suggest a nutrition problem in its earliest stages. Laboratory tests may detect it before it becomes overt, whereas anthropometrics and physical examinations pick up on the problem only after it is causing symptoms. Similar assessment methods are used in surveys to measure people's food consumption and to evaluate the nutrition status of populations.

# Diet and Health

Diet has always played a vital role in supporting health. Early nutrition research focused on identifying the nutrients in foods that would prevent such common diseases as rickets and scurvy, the vitamin D– and vitamin C–deficiency diseases. More recently, with nutrient deficiencies no longer a major threat, nutrition research has focused on chronic diseases associated with energy and nutrient excesses. Today, overconsumption of foods—especially foods high in fats—is a major health concern for people in the United States.

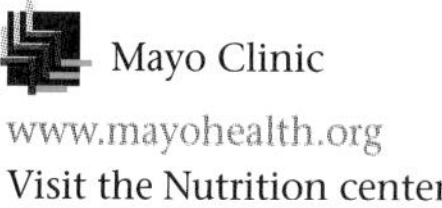


www.mayohealth.org
Visit the Nutrition center

*Such reports include:
- *Dietary Reference Intakes.*
- *Healthy People 2000: National Health Promotion and Disease Prevention Objectives.*
- *Diet and Health: Implications for Reducing Chronic Disease Risks.*
- *Surgeon General's Report on Nutrition and Health.*
- *Dietary Guidelines for Americans.*

**Table 1-2**
**Ten Leading Causes of Death in the United States**

1. **Heart disease.**
2. **Cancers.**
3. **Strokes.**
4. Chronic obstructive lung disease.
5. Unintentional injuries.
6. Pneumonia and influenza.
7. **Diabetes mellitus.**
8. HIV infection.
9. Suicide.
10. Chronic liver disease.

NOTE: The diseases in bold type have relationships with diet.

**chronic diseases:** degenerative diseases characterized by deterioration of the body organs; also called chronic, **noncommunicable diseases (NCD).** Examples include heart disease, cancer, and diabetes.

**risk factors:** factors associated with an elevated frequency of a disease but not proved to be causal.

## *Chronic Diseases*

Table 1-2 lists the ten leading causes of death in the United States. These "causes" are stated as if single conditions such as heart disease caused death, but most chronic diseases arise from multiple factors over many years. A person who died of heart failure may have had preexisting conditions such as overweight and high blood pressure, may have been a cigarette smoker, and may have spent years eating a high-fat diet and getting too little exercise.

Of course, not all people who die of heart disease fit this description, nor do all people with these characteristics die of heart disease. People who are overweight might die from the complications of diabetes instead, or those who smoke might die of cancer. They might even die from something totally unrelated to any of these factors, such as an automobile accident. Still, statistical studies have shown that certain conditions and behaviors are linked to certain diseases.

## *Risk Factors for Chronic Diseases*

Factors that increase or reduce the *risk* of developing chronic diseases are identified by analyzing statistical data. A strong association between a risk factor and a disease means that when the factor is present, the *likelihood* of developing the disease increases. It does not mean that all people with the risk factor will develop the disease. Similarly, a lack of risk factors does not guarantee freedom from a given disease. On the average, though, the more risk factors in a person's life, the greater that person's chances of developing the disease. Conversely, the fewer risk factors in a person's life, the better the chances for good health.

• ***Risk Factors Persist*** • Risk factors tend to persist over time. Without intervention, a young adult with high blood pressure will most likely continue to have high blood pressure as an older adult, for example. Thus, to minimize the damage, early intervention is most effective.

• ***Risk Factors Cluster*** • Risk factors also tend to cluster. For example, a person who is overweight is likely to be physically inactive, to have high blood pressure, and to have high blood cholesterol—all risk factors associated with heart disease. Intervention that focuses on one risk factor often benefits the others as well. For example, physical activity can help reduce weight. Then both physical activity and weight loss will help to lower blood pressure and blood cholesterol.

• ***Risk Factors in Perspective*** • The most prominent factor contributing to death in the United States is tobacco use, followed by diet and activity patterns, and alcohol use. Risk factors such as smoking, dietary habits, physical activity, and alcohol consumption are personal behaviors that can be changed. Decisions to not smoke, to eat a well-balanced diet, to engage in regular physical activity, and to drink alcohol in moderation (if at all) improve the likelihood that a person will enjoy good health. Other risk factors, such as genetics, sex, and age, also play important roles in the development of chronic diseases, but they cannot be changed. Health recommendations acknowledge the influence of such factors on the development of disease, but must focus on those that are changeable. For the two out of three Americans who do not smoke or drink alcohol excessively, the one choice that can influence long-term health prospects more than any other is diet.[17]

IN SUMMARY

The next several chapters will provide many more details about nutrients and how they support health. Whenever appropriate, they will show how diet influences each of today's major diseases. Dietary recommendations will appear again and again, as each nutrient's relationships with health are explored. Most people who follow the recommendations will benefit and can enjoy good health into their later years.

## Making It Click

Several chapters end with problems to give you practice in doing simple nutrition-related calculations. They use hypothetical situations in order to teach a lesson that can provide answers (see Appendix K). Once you have mastered these examples, you will be prepared to examine your own food choices. Be sure to show your calculations for each problem.

1. Calculate the energy provided by a food from its energy-nutrient contents. A cup of fried rice contains 5 grams protein, 30 grams carbohydrate, and 11 grams fat.

   a. How many kcalories does the rice provide from these energy nutrients?

   ________ = ___ kcal protein.
   ________ = ___ kcal carbohydrate.
   ________ = ___ kcal fat.
   Total = ___ kcal.

   b. What percentage of the energy in the fried rice comes from each of the energy-yielding nutrients?

   ________ = ___ % kcal from protein.
   ________ = ___ % kcal from carbohydrate.
   ________ = ___ % kcal from fat.
   Total = ___ %

   Note: The total should add up to 100%; 99% or 101% due to rounding is also acceptable.

   c. Calculate how many of the 146 kcalories provided by a 12-ounce can of beer come from alcohol, if the beer contains 1 gram protein and 13 grams carbohydrate. (Hint: The remaining kcalories derive from alcohol.)

   1 g protein = ___ kcal protein.
   13 g carbohydrate = ___ kcal carbohydrate.
   = ___ kcal alcohol.

   How many grams of alcohol does this represent? ___ g alcohol.

2. Even a little nutrition knowledge can help you identify some bogus claims. Consider an advertisement for a new "super supplement" that claims the product provides 15 grams protein and 10 kcalories per dose. Is this possible? ___. Why or why not? ________ = ___ kcal.

## Study Questions

These questions will help you review this chapter. You will find the answers in the discussions on the pages provided.

1. Give several reasons (and examples) why people make the food choices that they do. (pp. 3–4)
2. What is a nutrient? Name the six classes of nutrients found in foods. What is an essential nutrient? (pp. 5–6)
3. Which nutrients are inorganic, and which are organic? Discuss the significance of that distinction. (pp. 6, 9)
4. Which nutrients yield energy, and how much energy do they yield per gram? How is energy measured? (p. 6)
5. Describe how alcohol resembles nutrients. Why is alcohol not considered a nutrient? (p. 6)
6. What is the science of nutrition? Describe the types of research studies and methods used in acquiring nutrition information. (p. 10)
7. Explain how variables might be correlational but not causal. (p. 12)
8. What are the DRI? Who develops the DRI? To whom do they apply? How are they used? In your description, identify the four categories of DRI and indicate how they are related. (pp. 13–15)
9. What judgment factors are involved in setting the energy and nutrient recommendations? (p. 14)
10. What happens when people get either too little or too much energy or nutrients? Define malnutrition, undernutrition, and overnutrition. Describe the four methods used to detect energy and nutrient deficiencies and excesses. (pp. 16–18)
11. What methods are used in nutrition surveys? What kinds of information can these surveys provide? (pp. 18–19)

These questions will help you prepare for an exam. Answers can be found in Appendix K.

1. When people eat the foods typical of their families or geographical region, their choices are influenced by:
   a. habit.
   b. nutrition.
   c. personal preference.
   d. ethnic heritage or tradition.
2. Both the human body and many foods are composed mostly of:
   a. fat.
   b. water.
   c. minerals.
   d. proteins.
3. The inorganic nutrients are:
   a. proteins and fats.
   b. vitamins and minerals.
   c. minerals and water.
   d. vitamins and proteins.
4. The energy-yielding nutrients are:
   a. fats, minerals, and water.
   b. minerals, proteins, and vitamins.

c. carbohydrates, fats, and vitamins.
d. carbohydrates, fats, and proteins.

5. Studies of populations that reveal correlations between dietary habits and disease incidence are:
a. clinical trials.
b. laboratory studies.
c. case-control studies.
d. epidemiological studies.

6. An experiment in which neither the researchers nor the subjects know who is receiving the treatment is known as:
a. double blind.
b. double control.
c. blind variable.
d. placebo control.

7. RDA stands for:
a. Required Daily Average.
b. Reference Dietary Average.
c. Recommended Dietary Allowances.
d. Regulations on Deficiency and Adequacy.

8. Historical information, physical examinations, laboratory tests, and anthropometric measures are:
a. techniques used in diet planning.
b. steps used in the scientific method.
c. approaches used in disease prevention.
d. methods used in a nutrition assessment.

9. A deficiency caused by an inadequate dietary intake is a(n):
a. overt deficiency.
b. covert deficiency.
c. primary deficiency.
d. secondary deficiency.

10. Behaviors such as smoking, dietary habits, physical activity, and alcohol consumption that influence the development of disease are known as:
a. risk factors.
b. chronic causes.
c. preventive agents.
d. disease descriptors.

## Notes


1. Food Marketing Institute, *Trends in the United States: Consumer Attitudes and the Supermarket* (Chicago: Food Marketing Institute, 1996).
2. A. Drewnowski, Why do we like fat? *Journal of the American Dietetic Association* 97 (1997): S58–S62.
3. A. Drewnowski and C. L. Rock, The influence of genetic taste markers on food acceptance, *American Journal of Clinical Nutrition* 62 (1995): 506–511; G. A. Falciglia and P. A. Norton, Evidence for a genetic influence on preference for some foods, *Journal of the American Dietetic Association* 94 (1994): 154–158.
4. J. Cousminer and G. Hartman, Understanding America's regional taste preferences, *Food Technology* 50 (1996): 73–77.
5. A. E. Sloan, America's appetite '96: The top 10 trends to watch and work on, *Food Technology* 50 (1996): 55–71.
6. International Life Sciences Institute, First International Conference on East-West Perspectives on Functional Foods, *Nutrition Reviews* 54 (1996): entire issue.
7. T. A. Miettinen and coauthors, Reduction of serum cholesterol with sitostanol-ester margarine in a mildly hypercholesterolemic population, *New England Journal of Medicine* 333 (1995): 1308–1312.
8. A. Drewnowski, Taste preferences and food intake, *Annual Review of Nutrition* 17 (1997): 237–253; J. Palmer and C. Leontos, Nutrition training for chefs: Taste as an essential determinant of choice, *Journal of the American Dietetic Association* 95 (1995): 1418–1421.
9. Committee on Dietary Reference Intakes, *Dietary Reference Intakes for Calcium, Phosphorus, Magnesium, Vitamin D, and Fluoride* (Washington, D.C.: National Academy Press, 1997).
10. Committee on Dietary Reference Intakes, *Dietary Reference Intakes for Thiamin, Riboflavin, Niacin, Vitamin $B_6$, Folate, Vitamin $B_{12}$, Pantothenic Acid, Biotin, and Choline* (Washington, D.C.: National Academy Press, 1998).
11. Committee on Dietary Allowances, *Recommended Dietary Allowances,* 10th ed. (Washington, D.C.: National Academy Press, 1989).
12. Committee on Dietary Reference Intakes, 1997; J. King, The need to consider functional endpoints in defining nutrient requirements, *American Journal of Clinical Nutrition* 63 (1996): 983S.
13. W. Mertz, Risk assessment of essential trace elements: New approaches to setting Recommended Dietary Allowances and safety limits, *Nutrition Reviews* 53 (1995): 179–185.
14. *Healthy People 2000: National Health Promotion and Disease Prevention Objectives* (Washington, D.C.: U.S. Department of Health and Human Services, 1990).
15. Ten-year comprehensive plan for the national nutrition monitoring and related research program, *Federal Register,* June 11, 1993.
16. J. M. McGinnis and P. R. Lee, *Healthy People 2000* at mid decade, *Journal of the American Medical Association* 273 (1995): 1123–1129.
17. *The Surgeon General's Report on Nutrition and Health: Summary and Recommendations,* DHHS (PHS) publication no. 88-50211 (Washington, D.C.: Government Printing Office, 1988).

HIGHLIGHT 1

# Finding Nutrition Information—Sites and Sources

People learn about nutrition daily as they watch television, read newspapers, turn the pages of magazines, talk with friends, and search the Internet. They want to know how best to take care of themselves. In some cases, they are seeking miracles: tricks to help them lose weight, foods to forestall aging, and supplements to build muscles. People's heightened interest in nutrition and health translates into billions of dollars spent on services and products sold by both legitimate and fraudulent businesses. While consumers who obtain legitimate products can improve their health, those enticed into scams may lose their health, their savings, or both. Ironically, nutrition quackery prevents people from attaining the health they seek by giving them false hope and delaying the implementation of effective strategies. Furthermore, the conflicting information that results from a mixture of science and quackery confuses consumers and interferes with the development of public policy and the allocation of resources for nutrition-related legislation and programs.[1]

Science and quackery may be easy to tell apart at the extremes, but much nutrition information lies between the extremes. How can people distinguish valid nutrition information from misinformation? One excellent approach is to notice who is purveying the information. The "who" behind the information is not always evident, though, especially in the world of electronic media. Consumers need to keep in mind that *people* develop CD-ROMs and create websites on the Internet, just as *people* write books and report the news. The Internet offers endless opportunities to obtain high-quality information, but it also delivers an abundance of incomplete, misleading, or inaccurate information.[2] Simply put: anyone can publish anything.

This highlight begins by examining the unique potential and problems of using the Internet as a resource for nutrition information. It continues with a discussion of how to identify reliable nutrition information that applies to all resources, including the Internet.

National Council Against Health Fraud
www.ncahf.org

## Nutrition on the Net

Got a question? The Internet has an answer. For experienced users with nutrition knowledge, results are just a mouse click away; not only is access easy, but the information is often more current than that obtainable from other sources. For others, though, answers lie tangled in a web of information overload and questionable reliability.

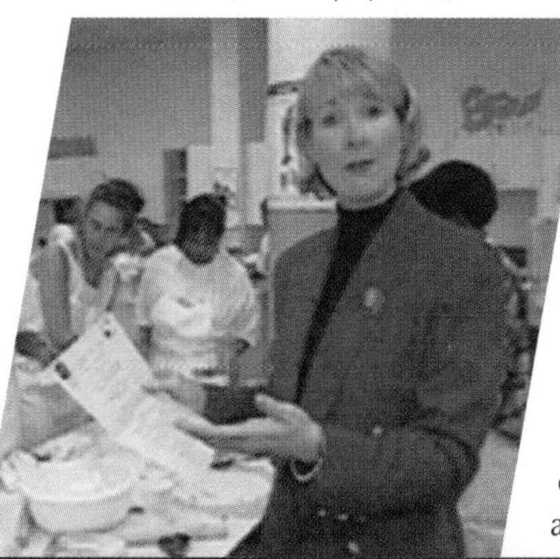

*The quality of nutrition information depends on the provider's knowledge and credentials. (Courtesy of CNN)*

With hundreds of millions of websites on the Internet, searching for nutrition information can be an overwhelming experience—much like walking into an enormous bookstore with millions of books, magazines, newspapers, and videos. And like a bookstore, the Internet offers no guarantees of the accuracy of the information found there—and much of it is pure fiction.

When using the Internet, keep in mind that the quality of information available covers a broad range. Just because you find it on the Net doesn't make it so. Websites must be evaluated for their accuracy, just like every other source.

To help users find reliable nutrition information on the Internet, Tufts University maintains an online rating and review guide called the Nutrition Navigator. The ratings reflect the opinions of a panel of nutrition experts who have scored selected websites on the basis of their accuracy, depth, and ease of use. Each website receives a score from 1 to 25, and links are provided to recommended sites. In addition to a rating, the Nutrition Navigator provides a review of the website's content and usefulness. The Nutrition Navigator is an excellent site from which to launch your ventures into nutrition cyberspace.

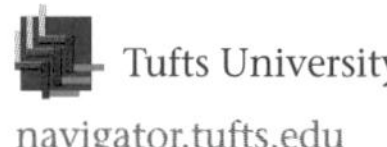

Tufts University
navigator.tufts.edu

## Identifying Nutrition Experts

Regardless of whether the medium is electronic, print, or video, consumers need to ask, is the person behind the information qualified to speak on nutrition? If the creator of a website on the Internet recommends eating three pineapples a day to lose weight, a trainer at the gym praises a high-protein diet, or a health-store clerk suggests an herbal supplement, should you believe these people? Can you distinguish between accurate news reports and sensational programs on television? Have you noticed that many televised nutrition messages are presented by celebrities, fitness experts, psychologists, food editors, and chefs—that is, almost anyone except a dietitian?[3] When you are confused or need sound dietary advice, whom should you ask?

### Physicians

Many people turn to their physicians for dietary advice, expecting them to know all about health-related matters. But are physicians the best sources of accurate and current information on *nutrition*? Only about one-fourth of all medical schools in the United States require students to take even one nutrition course.[4] Students attending these classes receive an average of 20 hours of nutrition instruction—an amount they, themselves, consider inadequate. By comparison, most students reading this text are taking a nutrition class that provides an average of 45 hours of instruction.

The American Dietetic Association (ADA) supports the inclusion of nutrition education as an essential component at all levels of medical education.[5] Furthermore, the ADA asserts that standardized nutrition education should be included in the curricula for all health care professionals: physician's assistants, dental hygienists, physical and occupational therapists, social workers, and all others who provide services directly to clients.[6] When these professionals understand the relevance of nutrition in the treatment and prevention of disease and have command of reliable nutrition information, then all the people they serve will also be better informed.

Most physicians appreciate the connections between health and nutrition and recognize potential nutrition problems. Those who have specialized in clinical nutrition are especially well qualified to speak on the subject. Few physicians, however, have the time or experience to develop diet plans and provide detailed diet instructions for clients. Often physicians wisely refer their clients to a qualified nutrition expert—a registered dietitian (RD). See the accompanying glossary for this and related terms.

## Registered Dietitians

A registered dietitian has the educational background necessary to deliver reliable nutrition advice and care. To become an RD, a person must earn an undergraduate degree requiring some 60 or so semester hours in nutrition and food science; complete a year's clinical internship or the equivalent; pass a national examination administered by the ADA; and maintain up-to-date knowledge by participating in required continuing education activities: attending seminars, taking courses, or writing professional papers.

Dietitians perform a multitude of duties in many settings in most communities.[7]* They work in the food industry, pharmaceutical companies, home health agencies, long-term care institutions, private practice, public health departments, research centers, education settings, fitness centers, and hospitals.

Depending on their work settings, dietitians can assume a number of different job responsibilities and positions. In hospitals, administrative dietitians manage the foodservice system; clinical dietitians provide client care (see Table H1-1); and nutrition support team dietitians coordinate nutrition care with other health care professionals. In the food industry, dietitians conduct research, develop products, and market services.

*To find a registered dietitian in your area, call the American Dietetic Association hotline (800-366-1655) or visit their website (www.eatright.org)

### Glossary

**accredited:** approved; in the case of medical centers or universities, certified by an agency recognized by the U.S. Department of Education.

**American Dietetic Association (ADA):** the professional organization of dietitians in the United States. The Canadian equivalent is Dietitians of Canada, which operates similarly.

**correspondence school:** a school that offers courses and degrees by mail. Some correspondence schools are accredited; others are *diploma mills.*

**dietetic technician registered (DTR):** a person with an associate's degree and training in nutrition, food science, and diet planning who works under the guidance of an RD (registered dietitian).

**dietitian:** a person trained in nutrition, food science, and diet planning. See also *registered dietitian.*

**DTR:** see *dietetic technician registered.*

**fraud** or **quackery:** the promotion, for financial gain, of devices, treatments, services, plans, or products (including diets and supplements) that alter or claim to alter a human condition without proof of safety or effectiveness. (The word *quackery* comes from the term *quacksalver,* meaning a person who quacks loudly about a miracle product—a lotion or a salve.)

**Internet (the Net):** a worldwide collection of millions of computers linked together to share information.

**license to practice:** permission under state or federal law, granted on meeting specified criteria, to use a certain title (such as dietitian) and offer certain services. Licensed dietitians may use the initials LD after their names.

**misinformation:** false or misleading information.

**nutritionist:** a person who specializes in the study of nutrition. Some nutritionists are registered dietitians, whereas others are self-described experts whose training is questionable. In states with responsible legislation, the term applies only to people who have MS or PhD degrees from properly accredited institutions.

**public health dietitian:** a dietitian who specializes in public health nutrition.

**RD:** see *registered dietitian.*

**registered dietitian (RD):** a dietitian who has graduated from a university or college after completing a program of dietetics that has been accredited by the American Dietetic Association (or Dietitians of Canada), has served in an internship or coordinated program to practice the necessary skills, has passed the association's registration examination, and maintains competency through continuing education. Many states require licensing for practicing dietitians.

**registration:** listing; with respect to health professionals, listing with a professional organization that requires specific course work, experience, and passing of an examination.

**World Wide Web (WWW, the Web):** a graphical subset of the Internet.

**Table H1-1**
**Responsibilities of a Clinical Dietitian**

- Assesses clients' nutrition status.
- Determines clients' nutrient requirements.
- Monitors clients' nutrient intakes.
- Develops, implements, and evaluates clients' nutrition care plans.
- Counsels clients to cope with unique diet plans.
- Teaches clients and their families about nutrition needs and diet plans.
- Provides training for other dietitians, nurses, interns, and dietetics students.
- Serves as liaison between clients and the foodservice department.
- Communicates with physicians, nurses, pharmacists, and other health care professionals about clients' progress, needs, and treatments.
- Participates in professional activities to enhance knowledge and skill.

Public health dietitians who work in government-funded agencies play a key role in delivering nutrition services to people in the community.[8] Among their many roles, public health dietitians help plan, coordinate, and evaluate food assistance programs; act as consultants to other agencies; manage finances; and much more. Those with advanced degrees in public health are well placed for employment in this vast field.

### *Other Dietary Employees*

In some facilities, a dietetic technician registered (DTR) assists registered dietitians in both administrative and clinical responsibilities.[9] A DTR has been educated and trained to work under the guidance of a registered dietitian.

Other dietary employees may include clerks, aides, cooks, porters, and other assistants. These dietary employees do not have extensive formal training in nutrition, and their ability to provide accurate information may be limited.

## *Identifying Fake Credentials*

In contrast to registered dietitians, thousands of people possess fake nutrition degrees and claim to be nutrition consultants or doctors of "nutrimedicine."[10] These and other such titles may sound meaningful, but most of these people lack the established credentials and training of the ADA-sanctioned dietitian. If you look closely, you can see signs of their fake expertise.

Consider educational background, for example. The minimal standards of education for a dietitian specify a bachelor of science (BS) degree in food science and human nutrition or related fields from an accredited college or university. Such a degree generally requires four to five years of study. In contrast, a fake nutrition expert may display a degree from a six-month correspondence course. Such a degree simply falls short.* In some cases, schools posing as legitimate correspondence schools offer even less—they sell certificates to anyone who pays the fees. To obtain these "degrees," a candidate need not read any books or pass any examinations.**

***Charlie displays his professional credentials.***

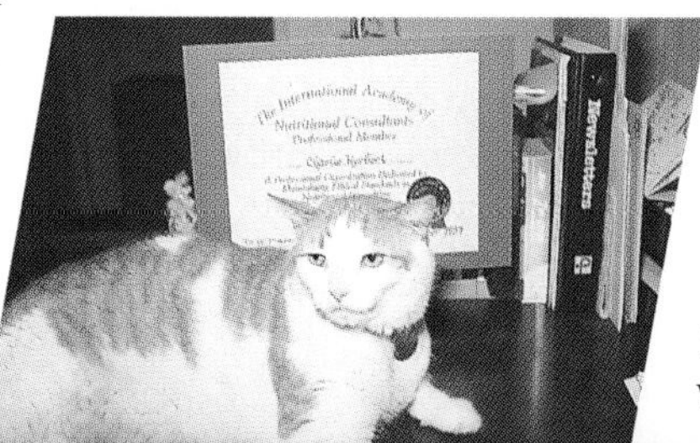

To guard educational quality, an accrediting agency recognized by the U.S. Department of Education (DOE) certifies that certain schools meet criteria established to ensure that an institution provides complete and accurate schooling. Unfortunately, fake nutrition degrees are available from schools "accredited" by more than 30 phony accrediting agencies.***

To dramatize the ease with which anyone can obtain a fake nutrition degree, one writer enrolled in a correspondence course for a fee of $82. She made every attempt to fail, intentionally answering all examination questions incorrectly. Even so, she received a "nutritionist" certificate at the end of the course. The "school" explained that it was sure she must have just misread the test.

In a similar stunt, Ms. Sassafras Herbert was named a "professional member" of a professional association. For her efforts, Sassafras has received a wallet card and is listed in a *Who's Who* publication that is distributed at health fairs and trade shows nationwide. Sassafras is a poodle; her master, Victor Herbert, MD, paid $50 to prove that she could be awarded these honors merely by sending in her name. Mr. Charlie Herbert, who is also a professional member of such an organization, is a cat.

Some states allow anyone to use the titles *dietitian* or *nutritionist,* but others allow only RDs or people with certain graduate degrees to call themselves dietitians. Many states provide a further guarantee: the license to practice.[11] Licensing identifies people who have met minimal standards of education and experience.

By knowing what qualifies someone to speak on nutrition, consumers can determine whether that person's advice might be harmful or helpful. Don't be afraid to ask for credentials. Does the personal trainer at the gym have a degree in nutrition from an accredited university? Is the creator of a nutrition website an RD or otherwise qualified to

---

*To find out whether a correspondence school is accredited, write the Distance Education and Training Council, Accrediting Commission, 1601 Eighteenth Street, N.W., Washington, D.C. 20009, call (202) 234-5100, or visit their website (www.detc.org).

**To find out whether a school is properly accredited for a dietetics degree, write the American Dietetic Association, Division of Education and Research, 216 West Jackson Boulevard, Chicago, IL 60606, call (312) 899-4870, or visit their website (www.eatright.org/caade).

***The American Council on Education publishes a directory of accredited institutions, professionally accredited programs, and candidates for accreditation in *Accredited Institutions of Postsecondary Education Programs Candidates* (available at many libraries). For additional information, write the American Council on Education, One Dupont Circle NW, Suite 800, Washington, D.C. 20036, call (202) 939-9382, or visit their website (www.acenet.edu).

write on nutrition? Have you seen the health-store clerk's license to practice as a dietitian? If not, seek a better-qualified source. After all, your health depends on it.

## Identifying Valid Information

Where do nutrition experts get their information? As Chapter 1 explained, nutrition knowledge derives from scientific research.

Researchers conduct experiments and then record and analyze their results, exercising caution in their interpretation of the findings. For example, in an epidemiological study, scientists may use a specific segment of the population—say, men 50 to 60 years old. When the scientists draw conclusions, they are careful not to generalize the findings to all people. Similarly, scientists performing research studies using animals are cautious in applying their findings to human beings. Conclusions from any one research study are always tentative and take into account findings from studies conducted by other scientists as well. As evidence accumulates, scientists gain confidence about making recommendations that affect people's health and lives. Still, their statements are worded cautiously, as in "A diet high in fruits and vegetables *may* protect against some cancers."

Quite often, as they approach an answer to one research question, scientists raise several more questions, so future research projects are never lacking. Further scientific investigation then seeks to answer questions such as "What substance or substances within fruits and vegetables provide protection?" If those substances turn out to be the vitamins A and C found so abundantly in fresh produce, then, "How much vitamin A and C is needed to offer protection?" "How do these vitamins protect against cancer?" "Is it their action as antioxidant nutrients?" "If not, might it be another action or even another substance that accounts for the protection fruits and vegetables provide against cancer?" (Highlight 11 explores the answers to these questions and reviews recent research on antioxidant nutrients, phytochemicals, and disease.)

The findings from a research study are submitted to a board of reviewers composed of other scientists who rigorously evaluate the study to assure that the scientific method was followed—a process known as peer review. The reviewers critique the study's hypothesis, methodology, statistical significance, and conclusions (Table H1-2 describes the parts of a research article). If the reviewers consider the conclusions to be well supported by the evidence, they endorse the work for publication in a scientific journal where others can read it. This raises an important point regarding information found on the Internet: much gets published without the rigorous scrutiny of peer review.[12] Consequently, readers must assume greater responsibility for examining the data and conclusions presented. Until you feel confident in critically evaluating nutrition information, you would be wise to restrict your research to one of the many online peer-reviewed journals (see the "How to" featured on p. 28 for selected website addresses).

**Table H1-2**
**Parts of a Research Article**

- *Abstract.* The abstract provides a brief overview of the article.
- *Introduction.* The introduction clearly states the purpose of the current study by proposing a hypothesis.
- *Review of literature.* A comprehensive review of the literature reveals all that science has uncovered on the subject to date.
- *Methodology.* The methodology section defines key terms and describes the instruments and procedures used in conducting the study.
- *Results.* The results report the findings and may include tables and figures that summarize the information.
- *Conclusions.* The conclusions drawn are those supported by the data and reflect the original purpose as stated in the introduction. Usually, they answer a few questions and raise several more.
- *References.* The references reflect the investigator's knowledge of the subject and should include an extensive list of relevant studies (including key studies several years old as well as current ones).

Regardless of whether an article is on the Internet or in print, readers must evaluate the study and assess the findings in light of knowledge gleaned from other studies. Figure H1-1 provides examples of reliable nutrition information.

Even when a new finding is published or released to the media, it is still only preliminary and not very meaningful by itself. Other scientists will need to confirm or disprove the findings through replication. To be accepted into the body of nutrition knowledge, a finding must stand up to rigorous, repeated testing in experiments performed by several different researchers. What we "know" in nutrition results from years of replicating study findings. Communicating the latest finding in its proper context without distorting or oversimplifying the message is a challenge for scientists and journalists alike.[13]

With each report from scientists, the field of nutrition changes a little—each finding contributes another piece to the whole body of knowledge. People who know how science works understand that single findings, like single frames in a movie, are just small parts of a larger story. Over years, the picture of what is "true" in nutrition gradually changes, and modifications in recommendations then follow.[14] Instead of eating 4 servings of fruits and vegetables as recommended by the old Four Food Group Plan, people are now encouraged to eat 2 to 4 servings of fruits and 3 to 5 servings of vegetables as suggested by the current Daily Food Guide (presented in Chapter 2).

Because science is a step-by-step, information-gathering and testing process, old research still has value. A hypothesis first advanced in 1960 that stands up to decades of validation has real strength. When it comes to scientific information, "new" does not necessarily mean "improved." In fact, any science report based on all new references is suspect, for truly strong research is based on a body of work conducted over many years. This is why, even in books published just this year, you will see references to old reports. Some studies have become classics: they were excit-

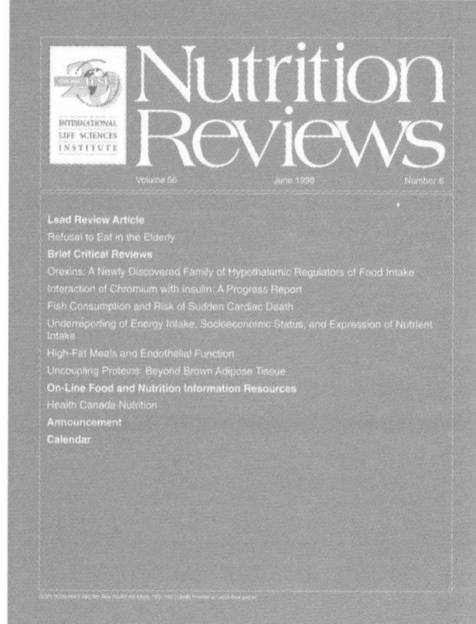


REVIEWS
Articles that examine all the major work on a subject are published in review journals like *Nutrition Reviews.* These articles provide references to all of the original work reviewed.

JOURNALS
Articles that present all the details of the methods, results, and conclusions of a particular study are published in journals like the *American Journal of Clinical Nutrition.*

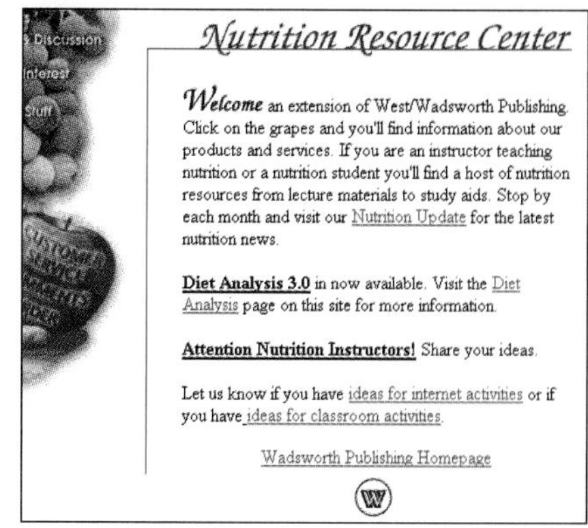


WEBSITES
Websites on the Internet developed by credible sources, such as those listed on p. 28, can provide valuable nutrition information and direct users to other resources. A quick link to many of these nutrition resources is available when you visit our website at: www.wadsworth.com/nutrition

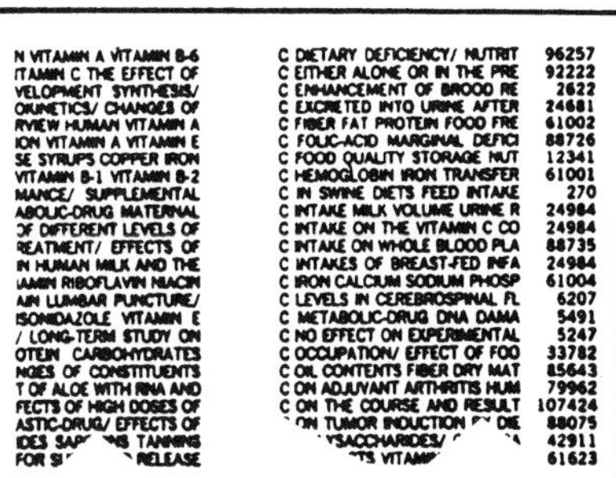


Figure H1–1

**Sources of Reliable Nutrition Information**

INDEXES
An index of abstracts directs you to many research articles on a given subject. This index from *Biological Abstracts* lists recently published works on vitamin C.

ONLINE INDEXES
Several online indexes are available, but one of the best for nutrition research is the National Library of Medicine's MEDLINE, which provides an index of more than 9 million articles from almost 4000 journals. For free access to MEDLINE visit www.nlm.nih.gov

ing when they first appeared, and they have stood up to the test of time.

## *Identifying Misinformation*

Nutrition is a hot topic, and scattered among the valid research findings are thousands of misleading and unfounded claims. The media, hungry for the latest news, often report scientific findings prematurely—without benefit of the careful interpretation, replication, and review that evaluate the findings. As a result, the public receives news quickly, but not always in perspective. Oftentimes findings from studies seem to contradict one another, and consumers feel frustrated and betrayed, when, in fact, this is simply the normal course of science at work.

People who do not understand how science operates may become distrustful as they try to learn nutrition from current news reports: "How am I supposed to know what to eat when the scientists themselves don't know?" General background knowledge about the science of nutrition is the best foundation a person can have for judging the validity of new nutrition information. (Congratulations on your decision to take this course.)

How can a person identify nutrition misinformation and quackery? Once upon a time, quacks rode into town in wooden wagons hawking snake oil for 50 cents a bottle to "cure what ails you," but those days are gone. Today's purveyors approach consumers in less obvious ways. They deliver their messages via the Internet, on glossy pages of magazines, in televised infomercials, and at social gatherings. The claims may look slick and *sound* logical, but they lack the research support found in nutrition science. The following techniques can alert consumers to misinformation:[15]

- Use of anecdotes, case histories, testimonials, and subjective evidence to support claims.
- Promises of quick, dramatic, miraculous cures.
- Use of pseudo-medical terms and jargon, which lend a false legitimacy to the claim, confuse the client, and camouflage the lack of substance.
- Fake credentials.
- Claims that most health problems are caused by poor nutrition and therefore can be corrected with proper nutrition.
- Claims that "natural" vitamins are better than synthetic ones.
- Recommendations to eat products derived from animal tissue to rejuvenate the counterpart in a human being.
- Statements that belittle medicine, science, and government regulations and offer "alternatives" that have not been proved safe or effective.
- Claims that products from health-food stores are better than those from regular grocery stores.
- Opposition to public health strategies such as fluoridating the water supply or vaccinating children against infectious diseases.
- Claims that food processing and storage destroy all nutrients.
- Claims that food additives are poisons responsible for a variety of problems—from misbehavior to murder.
- Claims that stress and other conditions raise nutrient needs higher than can be met by foods alone.
- Claims that people want to hear but that are too good to be true—such as that vitamin and mineral supplements will prevent cancer.
- Use of hair analysis and other unproven diagnostic tests to detect alleged "nutrient deficiencies."
- Claims that the only beneficial products are those whose sales happen to profit the purveyor.
- Claims that sugar is a poison.
- Diagnosis of "nutrient deficiencies" by computerized questionnaires.
- Claims that restrictive fad diets cure a variety of health conditions.
- Claims that physicians are money-grabbing, incompetent misfits who are not to be trusted.

In short, quackery respects neither science nor honesty in its pursuit of ill-gotten gains.

 Stephen Barrett's Quackwatch
www.quackwatch.com

Sales of unproven and dangerous products have always been a concern, but the Internet now provides merchants with an easy and inexpensive way to reach millions of customers around the world. Because of the difficulty in regulating the Internet, fraudulent and illegal sales of medical products have hit a bonanza.[16] As is the case with the air, no one owns the Internet, and similarly, no one has control over the pollution. Countries have different laws regarding sales of drugs, dietary supplements, and other health products, but applying these laws to the Internet marketplace is almost impossible. Even if illegal activities could be

## HOW TO Find Credible Sources of Nutrition Information

Government agencies, volunteer associations, consumer groups, and professional organizations provide consumers with reliable health and nutrition information. Credible sources of nutrition information include:

- Nutrition and food science departments at a university or community college.
- Local agencies such as the health department or County Cooperative Extension Service.
- Government health agencies such as:
  - Department of Agriculture (USDA) — www.usda.gov
  - Department of Health and Human Services (DHHS) — www.os.dhhs.gov
  - Food and Drug Administration (FDA) — www.fda.gov
  - Health Canada — www.hc-sc.gc.ca/nutrition
- Volunteer health agencies such as:
  - American Cancer Society — www.cancer.org
  - American Diabetes Association — www.diabetes.org
  - American Heart Association — www.amhrt.org
- Reputable consumer groups such as:
  - American Council on Science and Health — www.acsh.org
  - Consumer Information Center — www.pueblo.gsa.gov
  - International Food Information Council — ificinfo.health.org
- Professional health organizations such as:
  - American Dietetic Assocation — www.eatright.org
  - American Medical Association — www.ama-assn.org
  - Dietitians of Canada — www.dietitians.ca
- Journals such as:
  - *American Journal of Clinical Nutrition* — www.faseb.org/ajcn
  - *New England Journal of Medicine* — www.nejm.org
  - *Nutrition Reviews* — www.ilsi.org/pubs.html

Appendix F provides websites and addresses for these and other organizations.

defined and identified, finding the person responsible for a particular website is not always possible. Websites can open and close in a blink of a cursor. Now, more than ever, consumers must heed the caution "Buyer beware."

In summary, when you hear nutrition news, consider its source. Ask yourself these two questions: Is the person purveying the information qualified to speak on nutrition? Is the information based on valid scientific research? If not, find a better source, for your health is your most precious asset.

 U.S. Government

www.healthfinder.gov/searchoptions/topicsaz.htm
Search for Quackery

## Notes


1. Position of The American Dietetic Association: Food and nutrition misinformation, *Journal of the American Dietetic Association* 95 (1995): 705–707.
2. W. M. Silberg, G. D. Lundberg, and R. A. Musacchio, Assessing, controlling, and assuring the quality of medical information on the Internet, *Journal of the American Medical Association* 277 (1997) 1244–1245.
3. T. MacLaren, Messages for the masses: Food and nutrition issues on television, *Journal of the American Dietetic Association* 97 (1997): 733–734.
4. M. E. Shils, National Dairy Council award for excellence in medical and dental nutrition education lecture, 1994: Nutrition education in medical schools—The prospect before us, *American Journal of Clinical Nutrition* 60 (1994): 631–638.
5. Position of The American Dietetic Association: Nutrition—An essential component of medical education, *Journal of the American Dietetic Association* 94 (1994): 555–557.
6. Position of The American Dietetic Association: Nutrition education of health professionals, *Journal of the American Dietetic Association* 98 (1998): 343–346.
7. C. Gopalan, Dietetics and nutrition: Impact of scientific advances and development, *Journal of the American Dietetic Association* 97 (1997): 737–741.
8. M. C. Egan, Public health nutrition: A historical perspective, *Journal of the American Dietetic Association* 94 (1994): 298–302.
9. J. Arena and P. Walters, Do you know what a dietetic technician can do? A focus on clinical technicians and their expanded roles and responsibilities, *Journal of the American Dietetic Association* 97 (1997): S139–S141.
10. J. Raso, Nutrition-related "credentialing" organizations: The good, the bad, and the abysmal, *Priorities* 7 (1995): 31–34.
11. Update on state licensure laws and ADA regulatory remarks, *Journal of the American Dietetic Association* 97 (1997): 1251.
12. D. E. Kipp, J. D. Radel, and J. A. Hogue, The Internet and the nutritional scientist, *American Journal of Clinical Nutrition* 64 (1996): 659–662.
13. Advisory group convened by the Harvard School of Public Health and the International Food Information Council Foundation, Improving public understanding: Guidelines for communicating emerging science on nutrition, food safety, and health, *Journal of the National Cancer Institute* 90 (1998): 194–199.
14. K. McNutt, Where truth comes from, *Nutrition Today,* March/April 1994, pp. 43–48.
15. Adapted with permission from Thirty ways to spot quacks and pushers, in S. Barrett and V. Herbert, *The Vitamin Pushers: How the "Health Food" Industry Is Selling America a Bill of Goods* (Amherst, N.Y.: Prometheus Books, 1994), pp. 15–35.
16. A. A. Skolnick, WHO considers regulating ads, sale of medical products on Internet, *Journal of the American Medical Association* 278 (1997): 1723–1725.

CHAPTER 2

# Planning a Healthy Diet

Chapter 1 explained that the body's many activities are supported by the array of nutrients delivered by the foods people eat. Food choices made over years influence the body's health, and consistently poor choices increase the risks of developing chronic diseases. This chapter attempts to show how a person can select from the tens of thousands of foods available to create a diet that supports health. Fortunately, most foods provide several nutrients, so one trick for wise diet planning is to select a combination of foods that deliver a full array of nutrients. This chapter begins by introducing the diet-planning principles and dietary guidelines that assist people in selecting foods that will deliver nutrients without excess energy.

# Principles and Guidelines

How well you nourish yourself does not depend on the selection of any one food. Instead it depends on the selection of many different foods at numerous meals over days, months, and years. Diet-planning principles and dietary guidelines are key concepts to keep in mind whenever you are selecting foods—whether shopping at the grocery store, choosing from a restaurant menu, or preparing a home-cooked meal.

 U.S. Government
www.healthfinder.gov/searchoptions/topicsaz.htm
Search for Diet

## Diet-Planning Principles

Diet planners have developed several ways to select foods. Whatever plan or combination of plans they use, though, they keep in mind the six basic diet-planning principles listed in the margin.

Diet-planning principles:
- Adequacy.
- Balance.
- kCalorie (energy) control.
- Nutrient Density.
- Moderation.
- Variety.

• ***Adequacy*** • An adequate diet provides sufficient energy and enough of all the nutrients to meet the needs of healthy people. Take the essential nutrient iron, for example. Each day the body loses some iron, so people have to replace it by eating foods that contain iron. A person whose diet fails to provide enough iron-rich foods may develop the symptoms of iron-deficiency anemia: the person may feel weak, tired, and listless; have frequent headaches; and find that even the smallest amount of muscular work brings disabling fatigue. To prevent these deficiency symptoms, diet planners include foods that supply adequate iron. The same is true for all the other essential nutrients introduced in Chapter 1.

**adequacy (dietary):** providing all the essential nutrients, fiber, and energy in amounts sufficient to maintain health.

• ***Balance*** • The essential minerals calcium and iron, taken together, illustrate the importance of dietary balance. Meats, fish, and poultry are rich in iron but poor in calcium. Conversely, milk and milk products are rich in calcium but poor in iron. In fact, milk (except breast milk) and milk products are so low in iron that overuse of these foods can lead to iron-deficiency anemia by displacing iron-rich foods from the diet. Yet milk is the single most nutritious food for infants and can be an important source of calcium for people of all ages.

**balance (dietary):** providing foods of a number of types in proportion to each other, such that foods rich in some nutrients do not crowd out of the diet foods that are rich in other nutrients.

The art of balancing the diet involves using enough—but not too much—of each type of food. Use some meat or meat alternates for iron; use some milk and milk products for calcium; and save some space for other foods, too, since a diet consisting of milk and meat alone would not be adequate. For the other nutrients, people need grains, vegetables, and fruits.

Balance in the diet helps to ensure adequacy.

**kcalorie (energy) control:** management of food energy intake.

• ***kCalorie (Energy) Control*** • Clearly, designing an adequate, balanced diet without overeating requires careful planning. The discussion of weight control in Chapter 9 examines this issue in more detail, but the key to controlling energy intake is to select foods of high nutrient density.

**nutrient density:** a measure of the nutrients a food provides relative to the energy it provides. The more nutrients and the fewer kcalories, the higher the nutrient density.

Nutrient density promotes adequacy and kcalorie control.

• ***Nutrient Density*** • To eat well without overeating, select foods that deliver the most nutrients for the least food energy. Consider foods containing calcium, for example. You can get about 300 milligrams of calcium from either 1½ ounces of cheddar cheese or 8 ounces of nonfat milk, but the cheese contributes about twice as much food energy as the milk. The nonfat milk, then, is twice as calcium dense as the cheddar cheese; it offers the same amount of calcium for half the energy intake. Both foods are excellent choices for adequacy's sake alone, but to achieve adequacy while controlling kcalories, the nonfat milk is the better choice.

Just like a person who has to pay for rent, food, clothes, and tuition on a tight budget, a person whose energy allowance is limited has to obtain iron, calcium, and all the other essential nutrients on a tight energy budget. To succeed, the person has to get many nutrients for each kcalorie "dollar." In the cola and grapes example on the next page, both provide about the same number of kcalories, but the grapes deliver many more nutrients. A person who makes nutrient-dense choices such as fruit over cola can meet daily nutrient needs on a lower energy budget.

**moderation (dietary):** providing enough but not too much of a substance.

Moderation contributes to adequacy, balance, and kcalorie control.

• ***Moderation*** • Foods rich in fat and sugar provide enjoyment and energy but relatively few nutrients. In addition, they promote weight gain when eaten in excess. A person practicing moderation would eat such foods only on occasion and would regularly select foods low in fat and sugar, a practice that automatically improves nutrient density. Returning to the example of cheddar cheese and nonfat milk, the nonfat milk not only offers the same amount of calcium for less energy, but it contains far less fat than the cheese.

***To ensure an adequate and balanced diet, eat a variety of foods daily, choosing different foods from each group.***

• ***Variety*** • A diet may have all of the virtues just described and still lack variety, if a person eats the same foods day after day. People should select foods from each of the food groups daily and vary their choices within each food group from day to day for several reasons. First, different foods within the same group contain different arrays of nutrients. Among the fruits, for example, strawberries are especially rich in vitamin C while cantaloupes are rich in vitamin A. Second, no food is guaranteed entirely free of substances that, in excess, could be harmful. The strawberries might contain trace amounts of one contaminant, the cantaloupes another. By alternating fruit choices, a person will ingest very little of either contaminant. (Contamination of foods is the subject of Chapter 19.) Third, as the adage goes, variety is the spice of life. Even if a person eats beans frequently, the person can enjoy pinto beans in Mexican burritos today, garbanzo beans in Greek salad tomorrow, and baked beans with barbecued chicken on the weekend. Eating nutritious meals need never be boring.

**variety (dietary):** eating a wide selection of foods within and among the major food groups (the opposite of monotony).

Dietary Guidelines for Americans
www.nal.usda.gov/fnic/dga/dguide95.html

## *Dietary Guidelines for Americans*

Table 2-1 presents the 1995 *Dietary Guidelines for Americans*. In general, the *Dietary Guidelines* answer the question, What should an individual eat to stay healthy? The first guideline encourages people to eat a variety of foods to get the nutrients needed to support good health, while the second guideline encourages physical activity to help maintain a healthy weight. The next two guidelines urge a shift in the balance of energy nutrients: they suggest that most people should reduce their fat, saturated fat, and cholesterol intakes and increase their complex carbohydrate and fiber intakes by choosing a diet with plenty of grains, vegetables, and fruits. The last three guidelines encourage moderation in the use of sugars, salt and sodium, and alcoholic beverages for those who partake. Together, these seven

**Table 2-1**

**1995 *Dietary Guidelines for Americans***

- Eat a variety of foods.
- Balance the food you eat with physical activity; maintain or improve your weight.
- Choose a diet with plenty of grain products, vegetables, and fruits.
- Choose a diet low in fat, saturated fat, and cholesterol.
- Choose a diet moderate in sugars.
- Choose a diet moderate in salt and sodium.
- If you drink alcoholic beverages, do so in moderation.

NOTE: These guidelines are designed for healthy people over two years old.
*Source:* The *Dietary Guidelines for Americans* are developed by the U.S. Department of Agriculture and the U.S. Department of Health and Human Services.

**Table 2-2**

**Canada's Guidelines for Healthy Eating**

- Enjoy a variety of foods.
- Emphasize cereals, breads, other grain products, vegetables, and fruits.
- Choose lower-fat dairy products, leaner meats, and foods prepared with little or no fat.
- Achieve and maintain a healthy body weight by enjoying regular physical activity and healthy eating.
- Limit salt, alcohol, and caffeine.

Source: These guidelines derive from *Action Towards Healthy Eating: The Report of the Communications/Implementation Committee and Nutrition Recommendations . . . A Call for Action: Summary Report of the Scientific Review Committee and the Communications/ Implementation Committee,* which are available from Branch Publications Unit, Health Services and Promotion Branch, Department of Health and Welfare, 5th Floor, Jeanne Mance Building, Ottawa, Ontario K1A 1B4.

***This cola and bunch of grapes illustrate nutrient density. Each provides about 150 kcalories, but the grapes offer a trace of protein, some vitamins, minerals, and fiber along with the energy; the cola beverage offers only "empty" kcalories. Grapes, or any fruit for that matter, are more nutrient dense than cola beverages.***

guidelines point the way toward better health. Table 2-2 presents *Canada's Guidelines for Healthy Eating.*

Healthy People 2000: Increase to at least 90% the proportion of restaurants and institutional foodservice operations that offer identifiable low-fat, low-kcalorie food choices, consistent with the *Dietary Guidelines for Americans.*

IN SUMMARY

A well-planned diet delivers adequate nutrients, a balanced array of nutrients, and an appropriate amount of energy. It is based on nutrient-dense foods, moderate in substances that can be detrimental to health, and varied in its selections. The *Dietary Guidelines* apply these principles, offering practical advice on how to eat for good health.

# Diet-Planning Guides

To plan a diet that achieves all of the dietary ideals just outlined, a planner needs tools as well as knowledge. Two of the tools most widely used for diet planning are food group plans and exchange lists.

## *Food Group Plans*

Food group plans build a diet from clusters of foods that are similar in origin and nutrient content. Thus each group represents a set of nutrients that differs from the nutrients supplied by the other groups. Selecting foods from each of the groups eases the task of creating a balanced diet.

• ***Daily Food Guide*** • Figure 2-1 (on pp. 34–35) presents the USDA's Daily Food Guide, a food group plan that assigns foods to five major food groups. The figure lists the number of servings recommended, the most notable nutrients of each

**food group plans:** diet-planning tools that sort foods of similar origin and nutrient content into groups and then specify that people should eat certain numbers of servings from each group.

Five food groups:
- Breads, cereals, and other grain products.
- Vegetables.
- Fruits.
- Meat, poultry, fish, and alternates.
- Milk, cheese, and yogurt.

**Figure 2-1**
**The Daily Food Guide**

**Breads, Cereals, and Other Grain Products: 6 to 11 servings per day.**

These foods are notable for their contributions of complex carbohydrates, riboflavin, thiamin, niacin, iron, protein, magnesium, and fiber.
Serving = 1 slice bread; ½ c cooked cereal, rice, or pasta; 1 oz ready-to-eat cereal; ½ bun, bagel, or English muffin; 1 small roll, biscuit, or muffin; 3 to 4 small or 2 large crackers.

- ♦ Whole grains (wheat, oats, barley, millet, rye, bulgur, couscous, polenta), enriched breads, rolls, tortillas, cereals, bagels, rice, pastas (macaroni, spaghetti), air-popped corn.
- Pancakes, muffins, cornbread, crackers, cookies, biscuits, presweetened cereals, granola, taco shells, waffles, french toast.
- ♦ Croissants, fried rice, doughnuts, pastries, cakes, pies.

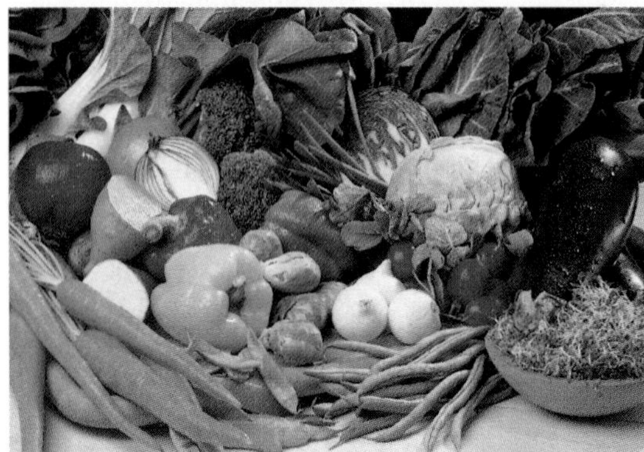

**Vegetables: 3 to 5 servings per day** (use dark green, leafy vegetables and legumes several times a week).

These foods are notable for their contributions of vitamin A, vitamin C, folate, potassium, magnesium, and fiber, and for their lack of fat and cholesterol.
Serving = ½ c cooked or raw vegetables; 1 c leafy raw vegetables; ½ c cooked legumes; ¾ c vegetable juice.

- ♦ Bamboo shoots, bok choy, bean sprouts, broccoli, brussels sprouts, cabbage, carrots, cauliflower, corn, cucumbers, eggplant, green beans, green peas, leafy greens (spinach, mustard, and collard greens), legumes, lettuce, mushrooms, okra, onions, peppers, potatoes, pumpkin, scallions, seaweed, snow peas, soybeans, tomatoes, water chestnuts, winter squash.
- Candied sweet potatoes.
- ♦ French fries, tempura vegetables, scalloped potatoes, potato salad.

**Fruits: 2 to 4 servings per day.**

These foods are notable for their contributions of vitamin A, vitamin C, potassium, and fiber, and for their lack of sodium, fat, and cholesterol.
Serving = typical portion (such as 1 medium apple, banana, or orange, ½ grapefruit, 1 melon wedge); ¾ c juice; ½ c berries; ½ c diced, cooked, or canned fruit; ¼ c dried fruit.

- ♦ Apricots, blueberries, cantaloupe, grapefruit, guava, oranges, orange juice, papaya, peaches, strawberries, plums, apples, bananas, pears, watermelon; unsweetened juices.
- Canned or frozen fruit (in syrup); sweetened juices.
- ♦ Dried fruit, coconut, avocados, olives.

**Meat, Poultry, Fish, and Alternates: 2 to 3 servings per day.**

Meat, poultry, and fish are notable for their contributions of protein, phosphorus, vitamin $B_6$, vitamin $B_{12}$, zinc, iron, niacin, and thiamin; legumes are notable for their protein, fiber, thiamin, folate, vitamin E, potassium, magnesium, iron, and zinc, and for their lack of fat and cholesterol.
Servings = 2 to 3 oz lean, cooked meat, poultry, or fish (total 5 to 7 oz per day); count 1 egg, ½ c cooked legumes, 4 oz tofu, or 2 tbs nuts, seeds, or peanut butter as 1 oz meat (or about ⅓ serving).

- ♦ Poultry (light meat, no skin), fish, shellfish, legumes, egg whites.
- Lean meat (fat-trimmed beef, lamb, pork); poultry (dark meat, no skin); ham; refried beans; whole eggs, tofu, tempeh.
- ♦ Hot dogs, luncheon meats, ground beef, peanut butter, nuts, sausage, bacon, fried fish or poultry, duck.

**Key:**

- ♦ Foods generally highest in nutrient density (good first choice).
- Foods moderate in nutrient density (reasonable second choice).
- ♦ Foods lowest in nutrient density (limit selections).

**Milk, Cheese, and Yogurt: 2 servings per day;** 3 servings per day for teenagers and young adults, pregnant/lactating women, women past menopause; 4 servings per day for pregnant/lactating teenagers.

These foods are notable for their contributions of calcium, riboflavin, protein, vitamin $B_{12}$, and, when fortified, vitamin D and vitamin A.
Serving = 1 c milk or yogurt; 2 oz process cheese food; 1½ oz cheese.

- Nonfat and 1% low-fat milk (and nonfat products such as buttermilk, cottage cheese, cheese, yogurt); fortified soy milk.
- 2% reduced-fat milk (and low-fat products such as yogurt, cheese, cottage cheese); chocolate milk; sherbet; ice milk.
- Whole milk (and whole-milk products such as cheese, yogurt); custard; milk shakes; ice cream.

*Note:* These serving recommendations were established before the 1997 DRI, which raised the recommended intake for calcium; meeting the calcium recommendation may require an additional serving from the milk, cheese, and yogurt group.

**Fats, Sweets, and Alcoholic Beverages: Use sparingly.**

These foods contribute sugar, fat, alcohol, and food energy (kcalories). They should be used sparingly because they provide food energy while contributing few nutrients. Miscellaneous foods not high in kcalories, such as spices, herbs, coffee, tea, and diet soft drinks, can be used freely.

- Foods high in fat include margarine, salad dressing, oils, lard, mayonnaise, sour cream, cream cheese, butter, gravy, sauces, potato chips, chocolate bars.
- Foods high in sugar include cakes, pies, cookies, doughnuts, sweet rolls, candy, soft drinks, fruit drinks, jelly, syrup, gelatin, desserts, sugar, and honey.
- Alcoholic beverages include wine, beer, and liquor.

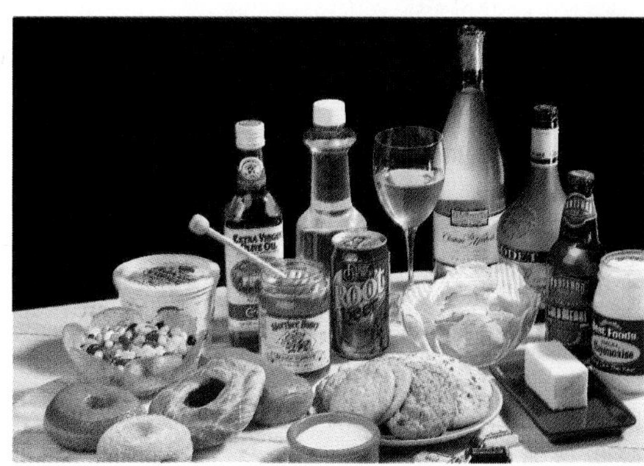

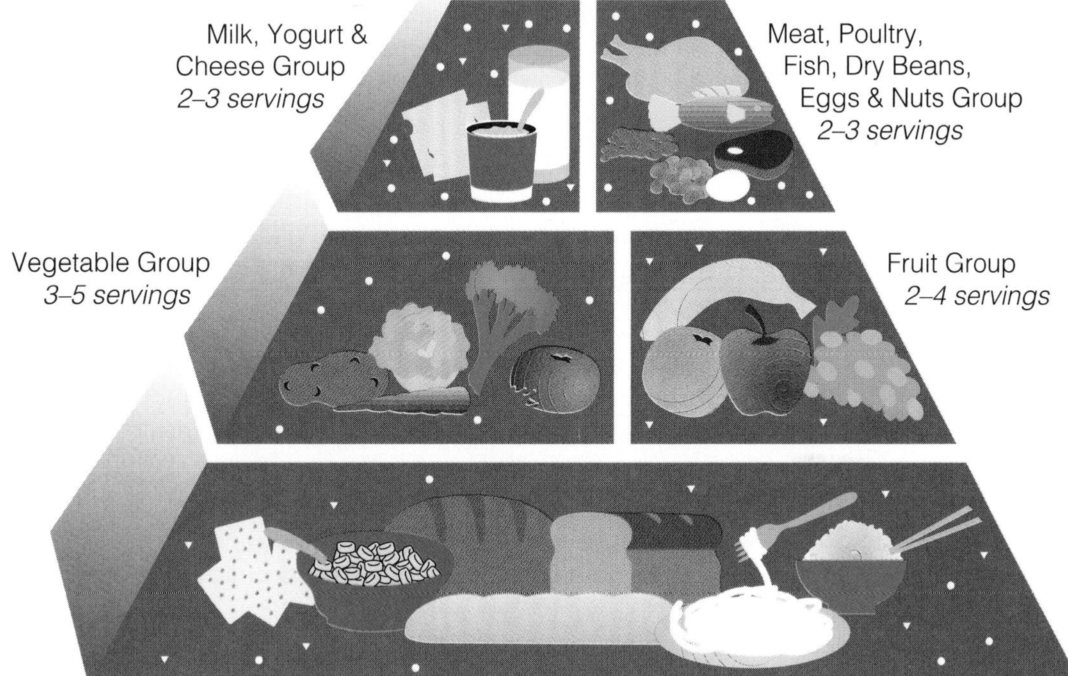


**Food Guide Pyramid**

A Guide to Daily Food Choices
The breadth of the base shows that grains (breads, cereals, rice, and pasta) deserve most emphasis in the diet. The tip is smallest: use fats, oils, and sweets sparingly.

group, the serving sizes, and the foods within each group categorized by nutrient density. It also includes an illustration of the USDA's Food Guide Pyramid, a pictorial representation of the Daily Food Guide.

*Milk "beverages" or "drinks" may taste delicious, but they lack the nutrient richness of real milk products.*

• ***Notable Nutrients*** • The beauty of the Daily Food Guide lies in its simplicity and flexibility. For example, a person can substitute cheese for milk because both supply the key nutrients for the milk group. A person following a food group plan receives not only the nutrients each group is noted for, but small amounts of other nutrients as well. For example, milk, cheese, and yogurt are notable for their calcium, protein, and riboflavin, but they also provide other nutrients. In contrast, a drink concocted from sugar, water, calcium, protein, and riboflavin lacks this nutrient richness. Milk, cheese, and yogurt are foundation foods; synthetic drinks are not.

• ***Miscellaneous Foods*** • Some foods—such as the synthetic drink just mentioned—do not fit into any of the food groups. Foods that are high in fat, sugar, or alcohol provide energy, but too few nutrients to hold a significant place in the diet. Such foods should be used sparingly and only after basic nutrient needs have been met by the foundation foods. Examples of "miscellaneous" foods include salad dressings, jams, and alcoholic beverages.

• ***Mixtures of Foods*** • Some foods—such as casseroles, soups, and sandwiches—fall into two or more food groups. With a little practice, users can begin to see the number of servings represented by each food group. From the Daily Food Guide point of view, a chicken enchilada looks like one serving from each of four different food groups if it is made with a corn tortilla from the bread group; ½ cup chopped onion, pepper, and tomatoes from the vegetable group; 3 ounces of chicken from the meat group; and 1½ ounces of shredded cheese from the milk group.

• ***Nutrient Density*** • The Daily Food Guide provides a strong foundation for a healthy diet, but it fails to specify food energy intakes. Large fat and energy differences exist within a single food group—for example, between nonfat milk and ice cream, fish and hot dogs, green beans and french fries, apples and avocados, or bread and biscuits—yet according to the Daily Food Guide, any of these substitutions would be acceptable. People who have low energy allowances are advised to select the most nutrient-dense foods within each group, whereas people with high energy needs may select some of the less nutrient-dense, higher-kcalorie foods. Notice that Figure 2-1 provides a key indicating which foods *within each group* are high, moderate, or low in nutrient density.

• ***Recommended Servings*** • As mentioned earlier, all food groups offer valuable nutrients, and people should make selections from each group daily. The recommended numbers of daily servings are:

- 6 to 11 servings of breads and cereals.
- 3 to 5 servings of vegetables.
- 2 to 4 servings of fruits.
- 2 to 3 servings of meats and meat alternates.
- 2 servings of milk and milk products. (Teenagers and young adults, women who are pregnant or breastfeeding, and women past menopause are advised to have 3 servings, and teenagers who are pregnant or breastfeeding should have 4.)

Given the 1997 DRI, which raised calcium recommendations, people may need an additional serving from the milk group.

The lower number of servings from each group provides about the right amount of food energy for sedentary women and older adults. The middle of the range is appropriate for most children, teenage girls, active women, and sedentary men. The upper end meets the needs of teenage boys, active men, and very active women. Table 2-3 provides estimated kcalorie amounts for each of these three levels. Physical activity increases a person's energy allowance and permits the person to eat more foods, or higher-kcalorie foods, to supply the needed nutrients without gaining unwanted weight.

• ***Serving Sizes*** • What counts as a serving? The answer differs for each food group and for various foods within a group. Furthermore, serving sizes may not represent the amounts people actually put on their plates. Figure 2-1 lists the serving sizes for standard foods within each group. For example, ½ cup of cooked rice is considered one serving. So, 1 cup of rice counts as 2 of the recommended 6 to 11 daily servings from the bread group. Similarly, ¼ cup counts as ½ serving.

• ***Food Guide Pyramid*** • The Food Guide Pyramid is a graphic depiction of the Daily Food Guide (see Figure 2-1 again). The illustration was designed to depict variety, moderation, and also proportions: the size of each section represents the number of daily servings recommended. The broad base at the bottom conveys the message that grains should be abundant and form the foundation of a healthy diet. Fruits and vegetables appear at the next level, showing that they have a less prominent, but still important, place in the diet. Meats and milks appear in a smaller band near the top. A few servings of each can contribute valuable nutrients, such as protein, vitamins, and minerals, without too much fat and cholesterol. Fats, oils, and sweets occupy the tiny apex, indicating that they should be used sparingly.

Alcoholic beverages do not appear in the pyramid, but they too should be limited. Items such as spices, coffee, tea, and diet soft drinks provide few, if any, nutrients, but can add flavor and pleasure to meals when used judiciously.

Icons of tiny dots and triangles are sprinkled over the food groups to represent naturally occurring and added fats and added sugars, respectively. These icons are meant to remind users that specific foods within the various groups are high in fats, sugars, or both, and so should be eaten in moderation.

The Daily Food Guide plan and Food Guide Pyramid emphasize grains, vegetables, and fruits—all plant foods. Some 75 percent of a day's servings should come from these three groups. This strategy helps all people obtain complex carbohydrates, fiber, vitamins, and minerals with little fat. It also eases diet planning for vegetarians.

• ***Vegetarian Food Guide*** • Vegetarian diets rely mainly on plant foods: grains, vegetables, legumes, fruits, seeds, and nuts. Some vegetarian diets include eggs, milk products, or both. People who do not eat meats or milk products can still use the Daily Food Guide to create an adequate diet.[1] The food groups are similar, and the number of servings remain the same. Vegetarians select *meat alternates* from the meat group—foods such as legumes, seeds, nuts, tofu, and, for those who eat them, eggs. Legumes and at least one cup of dark leafy greens help to supply the iron that meats usually provide. Vegetarians who do not drink cow's milk can use soy "milk"—a product made from soybeans that provides similar nutrients if it has been fortified with calcium, vitamin D, and vitamin $B_{12}$. Highlight 6 presents the Daily Food Guide for Vegetarians, defines vegetarian terms, and provides more information on vegetarian diets.

• ***Ethnic Food Choices*** • People can use the Food Guide Pyramid and still enjoy a diverse array of cuisines by assigning ethnic foods to their appropriate food groups.[2] For example, a person eating Mexican foods would put tortillas in the bread group, jicama in the vegetable group, and guava in the fruit group.

• ***Perceptions and Actual Intakes*** • The Daily Food Guide and Food Guide Pyramid were developed to help people choose a balanced and healthful diet. Are these plans successful? Yes, they can help people select nutrient-rich diets. In fact, one survey reports that only adults who select the recommended number of servings from each of the five food groups meet recommendations for energy, fat, fiber, vitamins, and minerals.[3] Unfortunately, only 1 percent of the more than 8000 people surveyed made such selections. More commonly, they neglected the fruit, milk, and bread groups, which raised their percentage of kcalories from fat and lowered their fiber, calcium, and zinc intakes.

Many adults *think* they are selecting foods that reflect the recommendations of the Food Guide Pyramid, when they are actually eating too many fats, sweets, and

**Table 2-3**
**Sample Diet Plans for Different Levels of Energy Intake**

| Food Group | Servings | | |
|---|---|---|---|
| Bread | 6 | 9 | 11 |
| Vegetable | 3 | 4 | 5 |
| Fruit | 2 | 3 | 4 |
| Milk[a] | 2–3[a] | 2–3[a] | 2–3[a] |
| Meat[b] | 5 | 6 | 7 |
| kCalories | 1600 | 2200 | 2800 |

NOTE: The 1600-kcalorie plan assumes a total of 53 grams of fat and allows 6 teaspoons of added sugar. The 2200-kcalorie plan assumes a total of 73 grams of fat and allows 12 teaspoons of added sugar. The 2800-kcalorie plan assumes a total of 93 grams of fat and allows 18 teaspoons of added sugar.
[a]Women who are pregnant or breastfeeding, teenagers and young adults, and women past menopause need 3 servings; teenagers who are pregnant or breastfeeding need 4 servings.
[b]Meat group amounts are in total ounces.

Food Guide Pyramid
www.nal.usda.gov/fnic/Fpyr/pyramid.html

**legumes** (lay-GYOOMS, LEG-yooms): plants of the bean and pea family, rich in high-quality protein compared with other plant-derived foods.

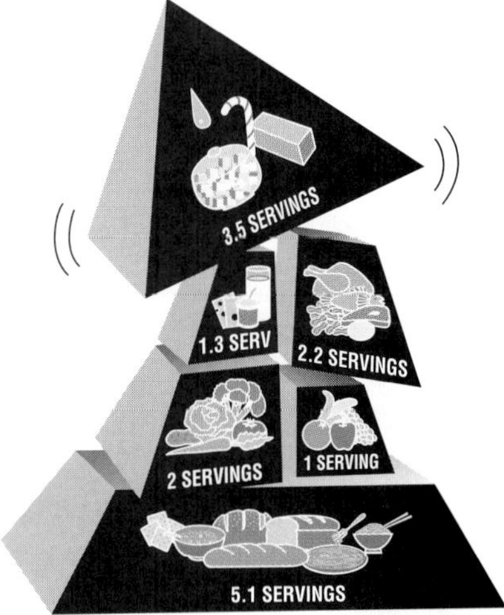


Actual Consumption Pyramid

Compared with recommendations, actual consumption resembles a precariously built, "tumbling" pyramid.

Source: National Livestock and Meat Board, courtesy of the National Cattlemen's Beef Association.

oils. In a sense, our pyramids are top-heavy and "tumbling."[4] They need more support from each of the five food groups to build a balanced diet.

## Exchange Lists

**exchange lists:** diet-planning tools that organize foods by their proportions of carbohydrate, fat, and protein. Foods on any single list can be used interchangeably.

 American Diabetes Association
www.diabetes.org
Search for Exchange lists

Appendix G gives complete details of the major exchange system used in the United States, and Appendix I provides details of the choice system used in Canada.

An exchange system:
- Names the foods on each list.
- Specifies portion sizes.
- States the amounts of carbohydrate, protein, fat, and kcalories each portion contributes.

Food group plans are particularly well suited to help the diet planner achieve dietary adequacy, balance, and variety. Exchange lists provide additional help in achieving kcalorie control and moderation. Originally developed for people with diabetes, exchange systems have proved useful for general diet planning as well.

Unlike the Daily Food Guide, which sorts foods primarily by their protein, vitamin, and mineral contents, the exchange system sorts foods into three main groups by their proportions of carbohydrate, fat, and protein. These three groups—the carbohydrate group, the fat group, and the meat and meat substitute group (protein)—organize foods into several exchange lists (see Table 2-4). The carbohydrate group covers these exchange lists:

- Starch (cereals, grains, pasta, breads, crackers, snacks, starchy vegetables, and dried beans, peas, and lentils).
- Fruit.
- Milk (nonfat, low-fat, and whole).
- Other carbohydrates (desserts and snacks with added sugars and fats).
- Vegetables.

The fat group covers this exchange list:

- Fats.

**Table 2-4**
**The Exchange Lists**

| Group/Lists | Typical Item/Portion Size | Carbohydrate (g) | Protein (g) | Fat (g) | Energy[a] (kcal) |
|---|---|---|---|---|---|
| ***Carbohydrate Group*** | | | | | |
| Starch[b] | 1 slice bread | 15 | 3 | 1 or less | 80 |
| Fruit | 1 small apple | 15 | — | — | 60 |
| Milk | | | | | |
| Nonfat | 1 c nonfat milk | 12 | 8 | 0–3 | 90 |
| Reduced-fat | 1 c reduced-fat milk | 12 | 8 | 5 | 120 |
| Whole | 1 c whole milk | 12 | 8 | 8 | 150 |
| Other carbohydrates[c] | 2 small cookies | 15 | varies | varies | varies |
| Vegetable | ½ c cooked carrots | 5 | 2 | — | 25 |
| ***Meat and Meat Substitute Group***[d] | | | | | |
| Meat | | | | | |
| Very lean | 1 oz chicken (white meat, no skin) | — | 7 | 0–1 | 35 |
| Lean | 1 oz lean beef | — | 7 | 3 | 55 |
| Medium-fat | 1 oz ground beef | — | 7 | 5 | 75 |
| High-fat | 1 oz pork sausage | — | 7 | 8 | 100 |
| ***Fat Group*** | | | | | |
| Fat | 1 tsp butter | — | — | 5 | 45 |

NOTE: The complete details of the U.S. exchange system are provided in Appendix G. Those of the Canadian system are shown in Appendix I.
[a]The energy value for each exchange list represents an approximate average for the group and does not reflect the precise number of grams of carbohydrate, protein, and fat. For example, a slice of bread contains 15 grams of carbohydrate (that's 60 kcalories), 3 grams protein (that's another 12 kcalories), and a little fat—rounded to 80 kcalories for ease in calculating. A half-cup of vegetables (not including starchy vegetables) contains 5 grams carbohydrate (20 kcalories) and 2 grams protein (8 more), which has been rounded down to 25 kcalories.
[b]The starch list includes cereals, grains, breads, crackers, snacks, starchy vegetables (such as corn, peas, and potatoes), and legumes (dried beans, peas, and lentils).
[c]The other carbohydrates list includes foods that contain added sugars and fats such as cakes, cookies, doughnuts, ice cream, potato chips, pudding, syrup, and frozen yogurt.
[d]The meat and meat substitutes list includes legumes, cheeses, and peanut butter.

The meat and meat substitute group (protein) covers these exchange lists:

- Meat and meat substitutes (very lean, lean, medium-fat, and high-fat).

• ***Portion Sizes*** • The exchange system helps people control their energy intakes by paying close attention to portion sizes. The portion sizes have been adjusted so that a portion of any food on a given list provides roughly the same amount of carbohydrate, fat, and protein and, therefore, total kcalories. Any food on a list can then be exchanged, or traded, for any other food on that same list without significantly affecting the diet's balance or total kcalories. For example, a person may select either 17 grapes or ½ grapefruit as one fruit portion, and either choice would provide roughly 60 kcalories. A whole grapefruit, however, would count as two portions.

The amount of food per serving in the exchange system is not always the same as in the Daily Food Guide, especially when it comes to meats. The exchange system lists meats and most cheeses in single ounces; that is, one *exchange* of meat is 1 ounce, whereas one *serving* in the Daily Food Guide is 2 to 3 ounces. Calculating meat by the ounce encourages the planner to keep close track of the exact amounts eaten. This in turn helps control energy and fat intakes. Be aware, too, that most people do not serve foods in carefully measured portions, nor do the amounts reflect the exchange system or Daily Food Guide serving sizes. Many restaurants, for example, offer steaks that are equivalent to four or five servings of meat. Similarly, a bakery may sell muffins or bagels that are two to three times the size of a typical bread serving.

To apply the system successfully, users must become familiar with portion sizes. A convenient way to remember the portion sizes and energy values is to keep in mind a typical item from each list. Figure 2-2 (on pp. 40–41) shows the foods on each of the exchange lists and their accurate portion sizes.

*It may look like* one, *but the large muffin counts as* two *servings.*

• ***The Foods on the Lists*** • Foods are not always on the exchange list where you might first expect to find them because they are grouped according to their energy-nutrient contents rather than by their source (such as milks), their outward appearance, or their vitamin and mineral contents. For example, cheeses are grouped with meats because, like meats, cheeses contribute energy from protein and fat but provide negligible carbohydrate. (In the food group plan presented earlier, cheeses are classed with milk because they are milk products with a similar calcium content.)

For similar reasons, starchy vegetables such as corn, green peas, and potatoes are listed on the starch list in the exchange system, rather than with the vegetables. Likewise, olives are not classed as a "fruit" as a botanist would claim; they are classified as a "fat" because their fat content makes them more similar to butter than to berries. Bacon and nuts are also on the fat list to remind users of their high fat content. These groupings highlight the characteristics of foods that are significant to energy intake.

Users of the exchange lists learn to view mixtures of foods, such as casseroles and soups, as combinations of foods from different exchange lists. They also learn to interpret food labels with the exchange system in mind (see the margin). Knowing that foods on the starch list provide 15 grams of carbohydrate and those on the vegetable list provide 5, you can count a lasagna dinner that provides 37 grams of carbohydrate as "2 starches and 1 vegetable"; knowing that foods on the meat list provide 7 grams of protein, you might count it as "3 meats"; the grams of fat suggest that the meat (and cheese) is probably medium-fat.

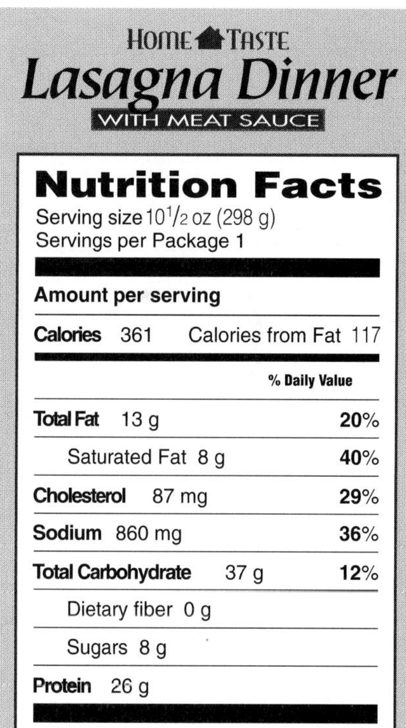


Can you "see" these exchanges in the label above?

| Exchange | Carbohydrate | Protein | Fat |
|---|---|---|---|
| 2 starches | 30 g | 6 g | — |
| 1 vegetable | 5 g | 2 g | — |
| 3 medium-fat meats | — | 21 g | 15 g |
| **Total** | **35** | **29** | **15** |

• ***Controlling Energy and Fat*** • By assigning items like bacon and avocados to the fat list, the exchange system alerts consumers to foods that are unexpectedly high in fat. Even the starch list specifies which grain products contain added fat (such as biscuits, muffins, and waffles). In addition, the exchange system encourages users to think of nonfat milk as milk and of whole milk as milk with added fat, and to think of very lean meats as meats and of lean, medium-fat, and high-

**Figure 2-2**
**The Exchange System: Example Foods, Portion Sizes, and Energy-Nutrient Contributions**

THE CARBOHYDRATE GROUP

**Starch**
1 starch exchange is like:
1 slice bread.
¾ c ready-to-eat cereal.
½ c cooked pasta, rice noodles, or bulgar.
⅓ c cooked rice.
½ c cooked beans.[a]
½ c corn, peas, or yams.
1 small (3 oz) potato.
½ bagel, English muffin, or bun.
1 tortilla, waffle, roll, taco, or matzoh.
(1 starch = 15 g carbohydrate, 3 g protein, 0–1 g fat, and 80 kcal.)

**Vegetables**
1 vegetable exchange is like:
½ c cooked carrots, greens, green beans, brussels sprouts, beets, broccoli, cauliflower, or spinach.
1 c raw carrots, radishes, or salad greens.
1 lg tomato.
(1 vegetable = 5 g carbohydrate, 2 g protein, and 25 kcal.)

**Fruits**
1 fruit exchange is like:
1 small banana, nectarine, apple, or orange.
½ large grapefruit, pear, or papaya.
½ c orange, apple, or grapefruit juice.
17 small grapes.
⅓ cantaloupe (or 1 c cubes).
2 tbs raisins.
1½ dried figs.
3 dates.
1½ carambola (star fruit).
(1 fruit = 15 g carbohydrate and 60 kcal.)

[a] ½ c cooked beans = 1 very lean meat exchange *plus* 1 starch exchange.

THE MEAT AND MEAT SUBSTITUTES GROUP (PROTEIN)

**Meat and substitutes (very lean)**
1 very lean meat exchange is like:
1 oz chicken (white meat, no skin).
1 oz cod, flounder, or trout.
1 oz tuna (canned in water).
1 oz clams, crab, lobster, scallops, shrimp, or imitation seafood.
1 oz fat-free cheese.
½ c cooked beans, peas, or lentils.
¼ c nonfat or low-fat cottage cheese.
2 egg whites (or ¼ c egg substitute).
(1 very lean meat = 7 g protein, 0–1 g fat, and 35 kcal).

**Meats and substitutes (lean)**
1 lean meat exchange is like:
1 oz beef or pork tenderloin.
1 oz chicken (dark meat, no skin).
1 oz herring or salmon.
1 oz tuna (canned in oil, drained).
1 oz low-fat cheese or luncheon meats.
(1 lean meat = 7 g protein, 3 g fat, and 55 kcal.)

**Meats and substitutes (medium-fat)**
1 medium-fat meat exchange is like:
1 oz ground beef.
1 oz pork chop.
1 egg.
¼ c ricotta.
4 oz tofu.
(1 medium-fat meat = 7 g protein, 5 g fat, and 75 kcal.)

**Other carbohydrates**
1 other carbohydrates exchange is like:
2 small cookies.
1 small brownie or cake.
5 vanilla wafers.
1 granola bar.
½ c ice cream.
(1 other carbohydrate = 15 g carbohydrate and may be exchanged for 1 starch, 1 fruit, or 1 milk. Because many items on this list contain added sugar and fat, their fat and kcalorie values vary, and their portion sizes are small.)

**Milks (nonfat and low-fat)**
1 nonfat milk exchange is like:
1 c nonfat or 1% milk.
¾ c nonfat yogurt, plain.
1 c nonfat or low-fat buttermilk.
½ c evaporated nonfat milk.
⅓ c dry nonfat milk.
(1 nonfat milk = 12 g carbohydrate, 8 g protein, 0–3 g fat, and 90 kcal.)

**Milks (reduced-fat)**
1 reduced-fat milk exchange is like:
1 c 2% milk.
¾ c low-fat yogurt, plain.
(1 reduced-fat milk = 12 g carbohydrate, 8 g protein, 5 g fat, and 120 kcal.)

**Milks (whole)**
1 whole milk exchange is like:
1 c whole milk.
½ c evaporated whole milk.
(1 whole milk = 12 g carbohydrate, 8 g protein, 8 g fat, and 150 kcal.)

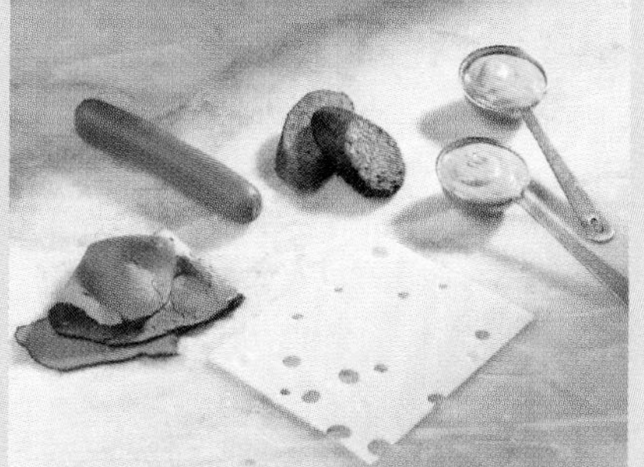

**Meats and substitutes (high-fat)**
1 high-fat meat exchange is like:
1 oz pork sausage.
1 oz luncheon meat (such as bologna).
1 oz regular cheese (such as cheddar or swiss).
1 small hot dog (turkey or chicken).[b]
2 tbs peanut butter.[c]
(1 high-fat meat = 7 g protein, 8 g fat, and 100 kcal.)

[b]A beef or pork hot dog counts as 1 high-fat meat exchange *plus* 1 fat exchange.
[c]Peanut butter counts as 1 high-fat meat exchange *plus* 1 fat exchange.

**THE FAT GROUP**

**Fats**
1 fat exchange is like:
1 tsp butter.
1 tsp margarine or mayonnaise (1 tbs reduced fat).
1 tsp any oil.
1 tbs salad dressing (2 tbs reduced fat).
8 large black olives.
10 large peanuts.
⅛ medium avocado.
1 slice bacon.
2 tbs shredded coconut.
1 tbs cream cheese (2 tbs reduced fat).
(1 fat = 5 g fat and 45 kcal.)

NOTE: Health recommendations urge people to limit their intakes of saturated fats; butter, bacon, coconut, and cream cheese contain saturated fats.

fat meats as meats with added fat. To that end, foods on the milk and meat lists are separated into categories based on their fat contents. The milk group is classed as nonfat, reduced-fat, and whole; the meat group as very lean, lean, medium-fat, and high-fat.

Control of food energy and fat intake can be highly successful with the exchange system. Exchange plans do not, however, guarantee adequate intakes of vitamins and minerals. Food group plans work better from that standpoint because the food groupings are based on similarities in vitamin-mineral content. In the exchange system, for example, meats are grouped with cheeses, yet the meats are iron-rich and calcium-poor, whereas the cheeses are iron-poor and calcium-rich. To take advantage of the strengths of both food group plans and exchange patterns, and to compensate for their weaknesses, diet planners often combine these two diet-planning tools.

## *Combining Food Group Plans and Exchange Lists*

A diet planner may find that using a food group plan together with the exchange lists eases the task of choosing foods that provide all the nutrients. The food group plan ensures that all classes of nutritious foods are included, thus promoting adequacy, balance, and variety. The exchange system classifies the food selections by their energy-yielding nutrients, thus controlling energy and fat intakes.

Table 2-5 shows how to use the Daily Food Guide plan together with the exchange lists to plan a diet. The Daily Food Guide ensures that a certain number of servings is chosen from each of the five food groups (see the first column of the table). The second column translates the number of servings (using the midpoint) into exchanges. With the addition of a small amount of fat, this sample diet plan provides about 1750 kcalories. Most people can meet their needs for all the nutrients within this reasonable energy allowance. (Table 9-4 in Chapter 9 shows patterns for other energy intakes.) The next step in diet planning is to assign the exchanges to meals and snacks. The final plan might look like the one in Table 2-6.

Next, a person could begin to fill in the plan with real foods to create a menu (use Figure 2-2 and Appendix G). For example, the breakfast plan calls for 2 starches, 1 fruit, and 1 nonfat milk. A person might select a bowl of shredded wheat with banana slices and milk (1 cup shredded wheat = 2 starches, 1 small banana = 1 fruit, and 1 cup nonfat milk = 1 milk); or a bagel and a bowl of cantaloupe pieces topped with yogurt (1 bagel = 2 starches, ⅓ cantaloupe melon = 1

**Table 2-5**

**Diet Planning with the Exchange System Using the Daily Food Guide Pattern**

| Patterns from Daily Food Guide Plan | Selections Made Using the Exchange System | Energy Cost (kcal) |
|---|---|---|
| Grains (breads and cereals)—6 to 11 servings | Starch list—select 9 exchanges | 720 |
| Vegetables—3 to 5 servings | Vegetable list—select 4 exchanges | 100 |
| Fruits—2 to 4 servings | Fruit list—select 3 exchanges | 180 |
| Meat—2 to 3 servings[a] | Meat list—select 6 lean exchanges | 330 |
| Milk—2 servings | Milk list—select 2 nonfat exchanges | 180 |
| | Fat list—select 5 exchanges | 225 |
| Total | | 1735 |

[a]In the food group plan, 1 serving is 2 to 3 ounces; in the exchange system, 1 exchange is 1 ounce. The Daily Food Guide suggests that amounts should total 5 to 7 ounces of meat daily.

**Table 2-6**
**A Sample Diet Plan**

| Exchange | Breakfast | Lunch | Snack | Dinner | Snack |
|---|---|---|---|---|---|
| 9 starch | 2 | 2 | 1 | 3 | 1 |
| 4 vegetable | | | | 4 | |
| 3 fruit | 1 | 1 | 1 | | |
| 6 lean meat | | 2 | | 4 | |
| 2 nonfat milk | 1 | | | | 1 |
| 5 fat | | 1 | | 4 | |

NOTE: This diet plan is one of many possibilities. It follows the number of servings suggested by the Daily Food Guide and meets dietary recommendations to provide 55 to 60% of its kcalories from carbohydrate, 15 to 20% from protein, and less than 30% from fat.

fruit, and ¾ cup nonfat plain yogurt = 1 milk). A person who wanted butter on the bagel could move a fat exchange or two from dinner to breakfast. If willing to use two fat exchanges at breakfast, the person could have pancakes with strawberries and milk (4 small pancakes = 2 starches plus 2 fats, 1¼ cup strawberries = 1 fruit, and a cup of nonfat milk = 1 milk). Then the person could move on to complete the menu for lunch, dinner, and snacks. As you can see, we all make countless food-related decisions daily—whether we have a plan or not. Following a plan, like the Daily Food Guide, that incorporates health recommendations and diet-planning principles helps anyone to make wise decisions.

## *From Guidelines to Groceries*

Dietary recommendations emphasize foods low in fat such as grains, fruits, vegetables, lean meats, fish, poultry, and low-fat milk products. Only you can design such a diet for yourself, but how do you begin? Start with the foods you enjoy eating. Then try to make improvements, little by little. When shopping, think of the food groups, and choose nutrient-dense foods within each group.

• ***Breads, Cereals, and Other Grain Products*** • When shopping for grain products, you will find them described as *refined, enriched,* or *whole grain* (see the accompanying glossary). These terms refer to the milling process and the making of products, and they have different nutrition implications. Refined foods may have lost many nutrients during processing; enriched products may have had some

*With its many grains (including wheat, rye, oats, corn, and rice) and types of foods (such as pastas, breads, and cereals), this group does more than its share for variety.*

### Glossary of Grain Terms

**bran:** the protective coating around the kernel similar in function to the shell of a nut; rich in nutrients and fiber.

**endosperm** (EN-doe-sperm): the bulk of the edible part of the kernel containing starch and proteins.

**enriched:** the addition of nutrients to a food to meet a specified standard. In the case of refined bread or cereal, five nutrients have been added: thiamin, niacin, folate, and riboflavin in amounts approximately equivalent to, or higher than, those originally present, and iron in amounts to alleviate the prevalence of iron-deficiency anemia.

**germ:** the nutrient-rich inner part of a grain. The germ is the seed that grows into a wheat plant, so it is especially rich in vitamins and minerals to support new life.

**gluten** (GLOO-ten): an elastic protein found in wheat and other grains that gives dough its structure and cohesiveness.

**husk:** the outer, inedible part of a grain; also called the *chaff.*

**refined:** the process by which the coarse parts of a food are removed. When wheat is refined into flour, the bran, germ, and husk are removed, leaving only the endosperm.

**unbleached flour:** a tan-colored endosperm flour with texture and nutritive qualities that approximate those of regular white flour.

**wheat flour:** any flour made from wheat, including white flour; wheat flour has been refined whereas *whole-wheat flour* has not.

**white flour:** an endosperm flour that has been refined and bleached for maximum softness and whiteness.

**whole grain:** a grain milled in its entirety (all but the husk), not refined.

**whole-wheat flour:** flour made from whole-wheat kernels; a whole-grain flour.

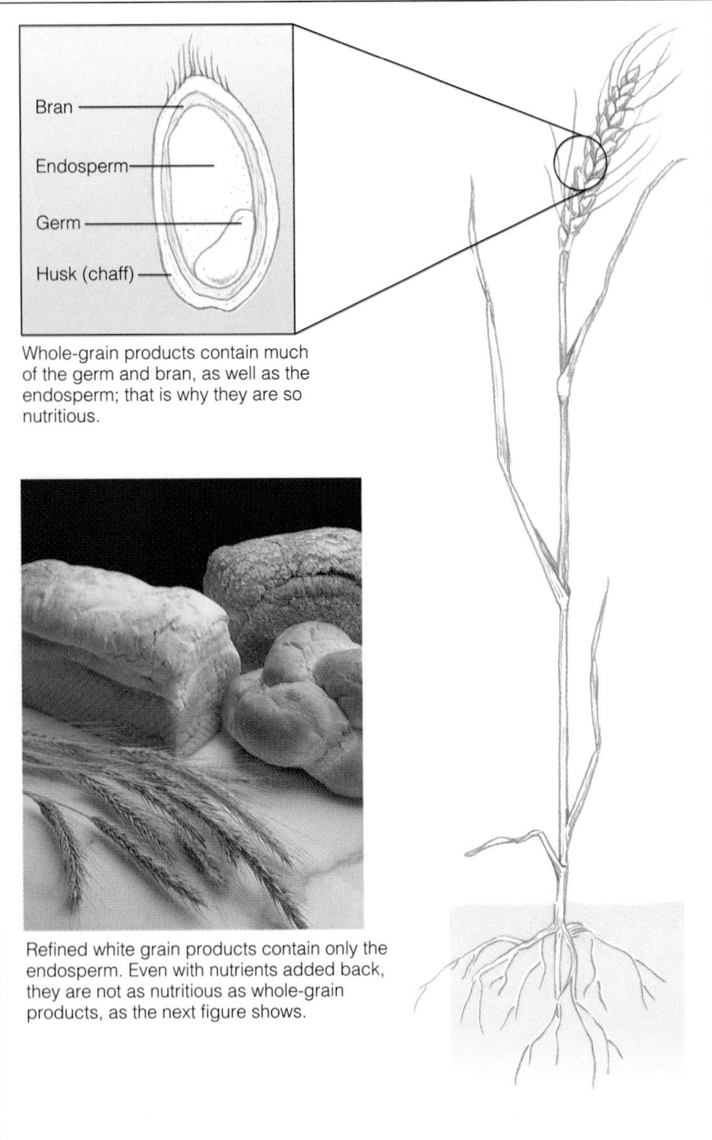


Whole-grain products contain much of the germ and bran, as well as the endosperm; that is why they are so nutritious.

Refined white grain products contain only the endosperm. Even with nutrients added back, they are not as nutritious as whole-grain products, as the next figure shows.

**Figure 2-3**
**A Wheat Plant**

The milling process breaks wheat kernels into their parts, shown here.

nutrients added back; and whole-grain products may be rich in all nutrients found in the original grain (see Figure 2-3).

When it became a common practice to refine the wheat flour used for bread by milling it and throwing away the bran and the germ, consumers suffered a tragic loss of many nutrients. As a consequence, in the early 1940s Congress passed legislation requiring that all grain products that cross state lines be enriched with iron, thiamin, riboflavin, and niacin. In 1996 this legislation was amended to include folate, a vitamin considered essential in the prevention of some birth defects. Enrichment restores these nutrients to the levels present in the original whole grain and actually raises thiamin, riboflavin, folate, and iron to higher levels. Most grain products that have been refined, such as rice, wheat pastas like macaroni and spaghetti, and cereals (both cooked and ready-to-eat types), have subsequently been enriched, and their labels say so.

Enrichment doesn't make a slice of bread rich in these added nutrients, but people who eat several slices a day obtain significantly more of these nutrients than they would from unenriched white bread. To a great extent, the enrichment of white flour helps to prevent deficiencies of these nutrients, but it fails to compensate for losses of many other nutrients and fiber. As Figure 2-4 shows, whole-grain items still outshine the enriched ones. Only *whole-grain* flour contains all of the nutritive portions of the grain. Whole-grain products, such as brown rice or oatmeal, not only provide more nutrients and fiber, but do not contain the added salt and sugar of flavored, processed rice or sweetened cereals.

Speaking of cereals, ready-to-eat breakfast cereals are the most highly fortified foods on the market. Like an enriched food, a *fortified* food has had nutrients added during processing, but in a fortified food, the added nutrients may not have been present in the original product. Some breakfast cereals made from refined flour and fortified with high doses of vitamins and minerals are actually more like supplements disguised as cereals than they are like whole grains. They may be nutritious—with respect to the nutrients added—but they still may fail to convey the full spectrum of nutrients that a whole-grain food or a mixture of such foods might provide. Furthermore, minerals (especially iron) are not as well absorbed from enriched foods as from naturally occurring sources.

**fortified:** the addition to a food of nutrients that were either not originally present or present in insignificant amounts. Fortification can be used to correct or prevent a widespread nutrient deficiency, to balance the total nutrient profile of a food, or to restore nutrients lost in processing.

*Note:* The terms "fortified" and "enriched" may be used interchangeably.

Legumes include:
- Black beans.
- Black-eyed peas.
- Garbanzo beans.
- Great northern beans.
- Kidney beans.
- Lentils.
- Navy beans.
- Peanuts.
- Pinto beans.
- Soybeans.
- Split peas.

• ***Vegetables*** • Choose fresh vegetables, especially green and yellow-orange vegetables like spinach, broccoli, and sweet potatoes. Cooked or raw, vegetables are good sources of vitamins, minerals, and fiber. Frozen and canned vegetables without added salt are acceptable alternatives to fresh. To control fat, energy, and sodium intakes, limit butter, salad dressings, and salt on vegetables.

• ***Legumes*** • Choose often from the variety of legumes available. They are an economical, low-fat, nutrient- and fiber-rich food choice.

• ***Fruit*** • Choose fresh fruits often, especially citrus fruits and yellow-orange fruits like cantaloupes and apricots. Frozen, dried, and canned fruits without added sugar are acceptable alternatives to fresh. Fruits supply valuable vitamins,

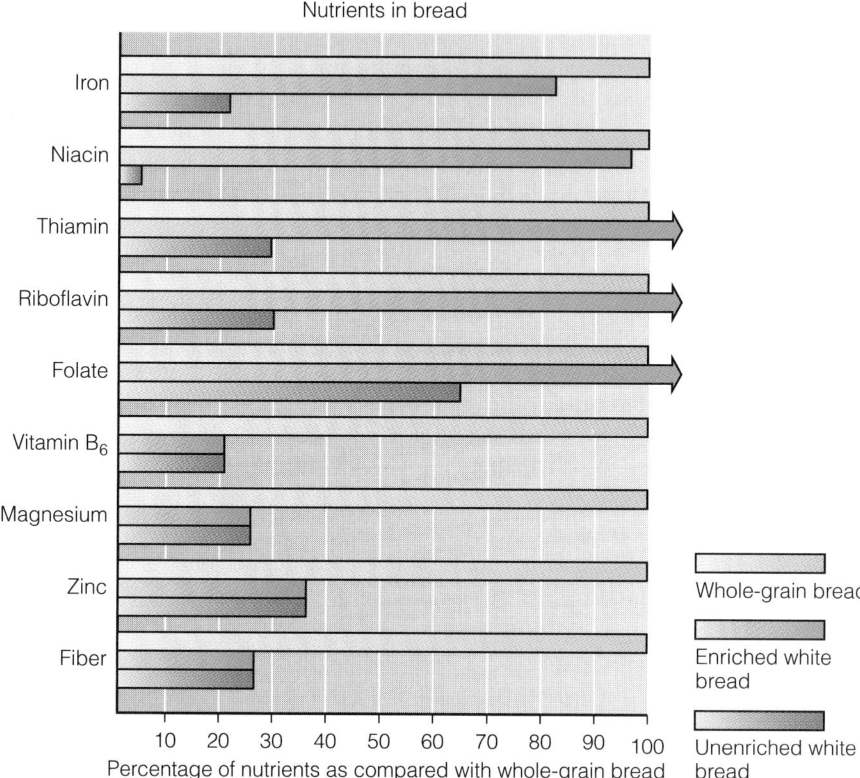


**Figure 2-4**
**Nutrients in Bread**

Whole-grain bread is more nutritious than other breads, even enriched bread. For iron, thiamin, riboflavin, niacin, and folate, enriched bread provides about the same quantities as whole-grain bread and significantly more than unenriched bread. For fiber and the other nutrients (both those shown here and those not shown), enriched bread provides less than whole-grain bread.

minerals, and fibers. They add flavors, colors, and textures to meals, and their natural sweetness makes them enjoyable as snacks or desserts.

Fruit juices are healthy beverages, but contain little dietary fiber compared with whole fruits. Whole fruits satisfy the appetite better than juices, thereby helping people to limit food energy intakes. Juices, on the other hand, are a good choice for people who need extra food energy. Be aware that sweetened fruit "drinks" or "ades" contain mostly water, sugar, and a little juice for flavor. Some may have been fortified with vitamin C, but lack any other significant nutritional value.

• ***Meat, Fish, and Poultry*** • Meat, fish, and poultry provide essential minerals, such as iron and zinc, and abundant B vitamins as well as protein. To buy and prepare these foods without excess energy, fat, and sodium takes a little knowledge and planning. When shopping in the meat department, choose fish, poultry, and lean cuts of beef and pork named "round" or "loin" (as in top round or pork tenderloin). As a guide, "prime" and "choice" cuts generally have more fat than "select" cuts. Restaurants usually serve prime cuts. Ground beef, even "lean" ground beef, derives most of its food energy from fat as the margin table shows. Have the butcher trim and grind a lean round steak instead. Alternatively, textured

**Percent kCalories Fat in Selected Meats**

| | |
|---|---|
| • Ground beef | |
| Regular | 66% |
| Lean | 57% |
| Extra lean | 54% |
| • Ground turkey | 51% |
| • Ground round (lean and trimmed) | 27% |

**textured vegetable protein:** processed soybean protein used in vegetarian products such as soy burgers.

*Combining legumes with foods from other food groups creates delicious meals.*

*Add rice to red beans for a hearty meal.*

*Enjoy a Greek salad topped with garbanzo beans for a little ethnic diversity.*

*A bit of meat and lots of spices turn kidney beans into chili con carne.*

Quick and easy estimate:
- 3 oz meat is about the size of a deck of cards.
- ¼ lb (4 oz) hamburger patty, uncooked, is about 3 oz, cooked.

Chapter 5 offers many additional strategies for lowering fat intake.

**imitation food:** a food that substitutes for and resembles another food, but is nutritionally inferior to it with respect to vitamin, mineral, or protein content. If the substitute is not inferior to the food it resembles and if its name provides an accurate description of the product, it need not be labeled "imitation."

**substitute food:** a food that is designed to replace another.

**Nonfat** milk may also be called **fat-free, skim, zero-fat,** or **no-fat.**

**Low-fat** milk refers to 1% milk.

**Reduced-fat** milk refers to 2% milk; may also be called **less-fat.**

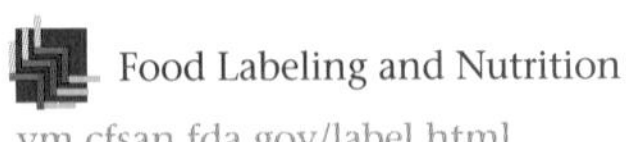
Food Labeling and Nutrition
vm.cfsan.fda.gov/label.html

***Consumers read food labels to learn about nutrition and its possible connections with health. (Courtesy of CNN)***

vegetable protein can be used instead of ground beef in a casserole, spaghetti sauce, or chili, saving fat kcalories.

Weigh meat after it is cooked and the bones and fat are removed. In general, 4 ounces of raw meat is equal to about 3 ounces of cooked meat. Some examples of 3-ounce portions of meat include 1 medium pork chop, ½ chicken breast, or 1 steak or hamburger about the size of a deck of cards. To keep fat intake down, bake, roast, broil, grill, or braise meats (but do not fry them in fat); remove the skin from poultry; trim visible fat before cooking; and drain fat after cooking.

• ***Milk*** • Shoppers will find a variety of fortified foods in the dairy case. Examples are milk, to which vitamins A and D have been added, and soy milk, to which calcium, vitamin D, and vitamin $B_{12}$ have been added. In addition, shoppers may find imitation foods (such as cheese products) and food substitutes (such as egg substitutes). As food technology advances, many such foods offer fat-free and low-fat alternatives. For example, egg substitutes help people who want to reduce their fat and cholesterol intakes. Highlight 5 gives other examples.

When shopping, choose nonfat or low-fat milk, yogurt, and cheeses. Such selections help consumers lower their fat intake to 30 percent of their daily energy intake.[5] Milk products are important sources of calcium, but can provide too much sodium and fat if not selected with care.

IN SUMMARY

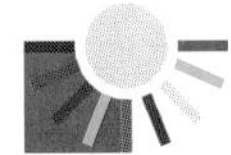

Food group plans select from different families of similar foods to provide adequacy, balance, and variety in the diet. Exchange lists define portion sizes so that foods within a given group supply similar amounts of energy nutrients, thus helping to attain kcalorie control and moderation. Together, they make it easier to plan a diet that includes abundant grains, vegetables, legumes, and fruits and moderate amounts of meats and milk products. In making any food choice, remember to view the food in the context of your total diet. It is the combination of many different foods that provides the abundance of nutrients so essential to a healthy diet.

# Food Labels

Many consumers read food labels to help them select foods with less fat, saturated fat, cholesterol, and sodium and more complex carbohydrates and dietary fiber. Food labels appear on virtually all processed foods, and posters or brochures provide similar nutrition information for fresh meats, fruits, and vegetables (see Figure 2-5). A few foods need not carry nutrition labels: those contributing few nutrients, such as plain coffee, tea, and spices; those produced by small businesses; and those prepared and sold in the same establishment. Producers of some of these items, however, voluntarily use labels. Even markets selling nonpackaged items voluntarily present nutrient information, either in brochures or on signs posted at the point of purchase. Restaurants need not supply complete nutrition information for menu items unless claims such as "low fat" or "heart healthy" have been made.[6] When ordering such items, keep in mind that restaurants tend to serve extralarge portions—two to three times standard serving sizes. A "low-fat" ice cream, for example, may have only 3 grams of fat per ½ cup, but you may be served 2 cups for a total of 12 grams of fat and all their accompanying kcalories.

**Healthy People 2000:** Achieve useful and informative nutrition labeling for virtually all processed foods and at least 40% of fresh meats, poultry, fish, fruits, vegetables, baked goods, and ready-to-eat carry-away foods.

**Figure 2-5**
**Examples of Food Labels**

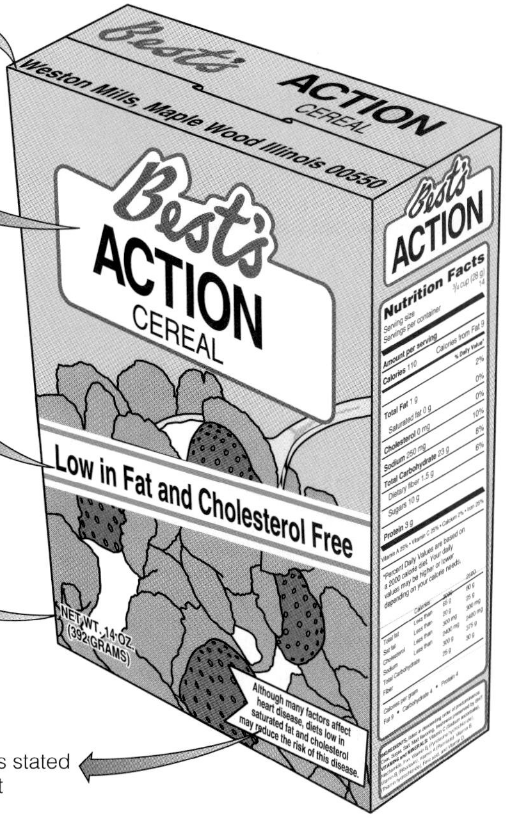

The name and address of the manufacturer, packer, or distributor
The common or usual product name
Descriptive terms if the product meets specified criteria
The net contents in weight, measure, or count
Approved health claims stated in terms of the total diet
Best's ACTION CEREAL
Weston Mills, Maple Wood Illinois 00550
Low in Fat and Cholesterol Free
NET WT. 14 OZ. (392 GRAMS)
Although many factors affect heart disease, diets low in saturated fat and cholesterol may reduce the risk of this disease.

**Nutrition Facts**

| | | |
|---|---|---|
| Serving size | | 3/4 cup (28 g) |
| Servings per container | | 14 |
| **Amount per serving** | | |
| **Calories** 110 | | Calories from fat 9 |
| | | **% Daily Value*** |
| **Total Fat** 1 g | | 2% |
| Saturated fat 0 g | | 0% |
| **Cholesterol** 0 mg | | 0% |
| **Sodium** 250 mg | | 10% |
| **Total Carbohydrate** 23 g | | 8% |
| Dietary fiber 1.5 g | | 6% |
| Sugars 10 g | | |
| **Protein** 3 g | | |

Vitamin A 25% • Vitamin C 25% • Calcium 2% • Iron 25%

*Percent Daily Values are based on a 2000 calorie diet. Your daily values may be higher or lower depending on your calorie needs.

| | Calories: | 2000 | 2500 |
|---|---|---|---|
| Total fat | Less than | 65 g | 80 g |
| Sat fat | Less than | 20 g | 25 g |
| Cholesterol | Less than | 300 mg | 300 mg |
| Sodium | Less than | 2400 mg | 2400 mg |
| Total Carbohydrate | | 300 g | 375 g |
| Fiber | | 25 g | 30 g |

Calories per gram
Fat 9 • Carbohydrate 4 • Protein 4

**INGREDIENTS**, listed in descending order of predominance: Corn, Sugar, Salt, Malt flavoring, freshness preserved by BHT. **VITAMINS and MINERALS:** Vitamin C (Sodium ascorbate), Niacinamide, Iron, Vitamin $B_6$ (Pyridoxine hydrochloride), Vitamin $B_2$ (Riboflavin), Vitamin A (Palmitate), Vitamin $B_1$ (Thiamin hydrochloride), Folic acid, and Vitamin D.

The serving size and number of servings per container

kCalorie information and quantities of nutrients per serving, in actual amounts

Quantities of nutrients as "% Daily Values" based on a 2000-kcalorie energy intake

Daily Values for selected nutrients for a 2000- and a 2500-kcalorie diet

kCalorie per gram reminder

The ingredients in descending order of predominance by weight

BONELESS
WHITE CHUNK
CHICKEN
IN WATER
6oz. DRAINED WEIGHT

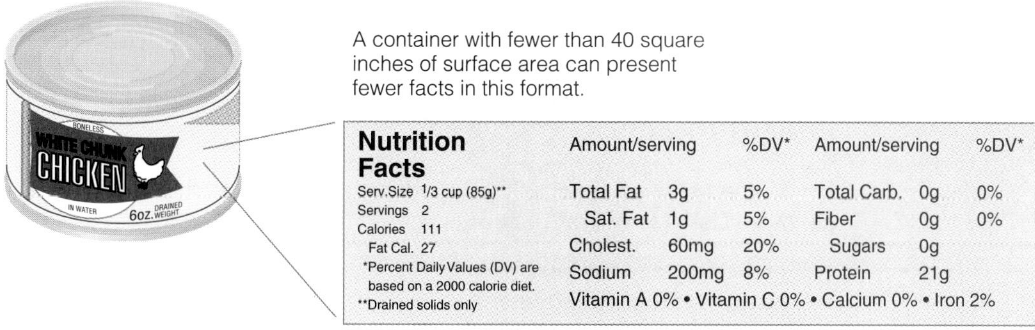

A container with fewer than 40 square inches of surface area can present fewer facts in this format.

| **Nutrition Facts** | Amount/serving | | %DV* | Amount/serving | | %DV* |
|---|---|---|---|---|---|---|
| Serv.Size 1/3 cup (85g)** | Total Fat | 3g | 5% | Total Carb. | 0g | 0% |
| Servings 2 | Sat. Fat | 1g | 5% | Fiber | 0g | 0% |
| Calories 111 | Cholest. | 60mg | 20% | Sugars | 0g | |
| Fat Cal. 27 | Sodium | 200mg | 8% | Protein | 21g | |
| *Percent Daily Values (DV) are based on a 2000 calorie diet. **Drained solids only | Vitamin A 0% • Vitamin C 0% • Calcium 0% • Iron 2% | | | | | |

Packages with fewer than 12 square inches of surface area need not carry nutrition information, but they must provide an address or telephone number for obtaining information.

CANDY RINGS

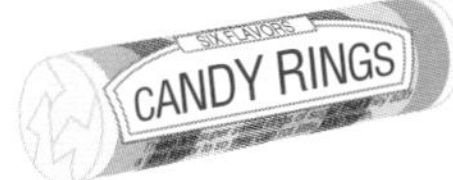

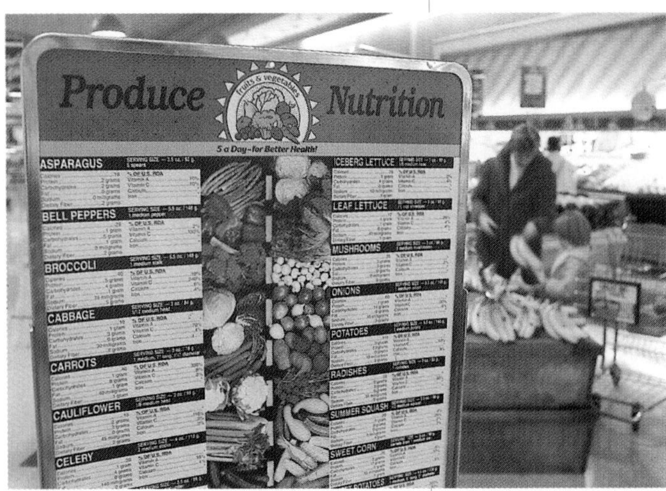


*Posters in the produce department present nutrition information for nonpackaged items such as raw fruits and vegetables.*

## The Ingredient List

All foods must list all ingredients on the label in descending order of predominance by weight. Knowing that the first ingredient predominates by weight, consumers can glean much information. Compare these products, for example:

- An orange powder that contains "sugar, citric acid, orange flavor . . ." versus a juice that contains "water, tomato concentrate, concentrated juices of carrots, celery . . . ."
- A cereal that contains "puffed milled corn, sugar, corn syrup, molasses, salt . . ." versus one that contains "100 percent rolled oats."
- A canned fruit that contains "sugar, apples, water" versus one that contains simply "apples, water."

In each comparison, consumers can tell that the second product is the more nutrient dense.

## Serving Sizes

Because labels present nutrient information per serving, they must identify the size of a serving. The Food and Drug Administration (FDA) has established specific serving sizes that reflect amounts that people customarily consume and requires that all labels for a given product use the same serving size. For example, the serving size for all ice creams is a half-cup and for all beverages, 8 fluid ounces. This facilitates comparison shopping. Consumers can see at a glance which brand has more or fewer kcalories or grams of fat. Standard serving sizes are expressed in both common household measures, such as cups, and metric measures, such as milliliters, to accommodate users of both types of measures (see Table 2-7).

**Table 2-7**
**Household and Metric Measures**

- 1 teaspoon (tsp) = 5 milliliters (ml)
- 1 tablespoon (tbs) = 15 ml
- 1 cup (c) = 240 ml
- 1 fluid ounce (fl oz) = 30 ml
- 1 ounce (oz) = 28 grams (g)

When examining the nutrition facts on a food label, consumers need to consider how the serving size compares with the actual quantity eaten. If it is not the same, they will need to adjust the quantities accordingly. For example, if the serving size is four cookies and you only eat two, then you need to cut the nutrient and kcalorie values in half; similarly, if you eat eight cookies, then you need to double the values.

## Nutrition Facts

In addition to the serving size and the servings per container, the "Nutrition Facts" panel on a label shows the quantities per serving of the following:

- Total food energy (kcalories).
- Food energy from fat (kcalories).
- Total fat (grams).
- Saturated fat (grams).
- Cholesterol (milligrams).
- Sodium (milligrams).
- Total carbohydrate, including starch, sugar, and fiber (grams).
- Dietary fiber (grams).
- Sugars (grams), including both those naturally present in and those added to the food.
- Protein (grams).

In addition, labels must present nutrient content information as compared with a standard for the following vitamins and minerals:

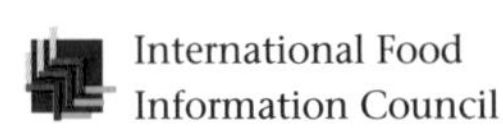
International Food Information Council

ificinfo.health.org
Search for Food labels

- Vitamin A.
- Vitamin C.
- Iron.
- Calcium.

Comparing nutrient amounts against a standard helps make them meaningful to label readers. A person might wonder, for example, whether 1 milligram of iron or calcium is a little or a lot. Well, the standard value for iron on food labels is 18 milligrams, so 1 milligram of iron is enough to take notice of: it is over 5 percent. But the standard value for calcium on food labels is 1000 milligrams, so 1 milligram of calcium is essentially nothing.

It would be nice for consumers if food labels could express a food's nutrient contents as a percentage of each individual's recommended intakes. Unfortunately, though, recommended intakes depend on age and sex. Manufacturers can't know who will be reading the label—an 8-year-old boy, a 70-year-old woman, or a pregnant teenage girl. They need one set of standard values to represent the needs of a "typical consumer." These standard values, developed by the FDA for use on food labels, are called the Daily Values.

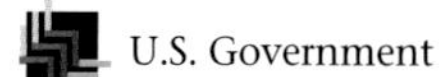
U.S. Government
www.healthfinder.gov/searchoptions/topicsaz.htm
Search for Food labels

## The Daily Values

Labels present nutrient information in two ways—in quantities (such as grams) and as percentages of Daily Values. Daily Values reflect dietary recommendations for nutrients and dietary components that have important relationships with health. The "% Daily Value" column provides a ballpark estimate of how individual foods contribute to the total diet. It compares key nutrients in a serving of food with the daily goals of a person consuming 2000 kcalories. A 2000-kcalorie diet is considered about right for moderately active women, teenage girls, and sedentary men. Older adults, children, and sedentary women may need fewer kcalories. Large labels list, at the bottom, Daily Values for both a 2000-kcalorie and a 2500-kcalorie diet, but the "% Daily Value" column on all labels applies only to a 2000-kcalorie diet. A 2500-kcalorie diet is considered about right for many men, teenage boys, and active women. People who are exceptionally active may need still higher kcalorie intakes. Labels may also provide a reminder of the kcalories in a gram of carbohydrate, fat, and protein below the Daily Value information.

**Daily Values (DV):** reference values developed by the FDA specifically for use on food labels.

A person who consumes 2000 kcalories a day can simply add up all the "% Daily Values" for a particular nutrient to see if the day's diet fits with recommendations. If the "% Daily Values" total 100 percent, then recommendations are met. People who require more or less than 2000 kcalories daily must do some calculations to see how foods compare with their personal nutrition goals. They can use the calculation column in the table shown on the inside back cover (left) or the suggestions presented in the "How to" feature on the next page.

Daily Values help consumers see easily whether a food contributes "a little" or "a lot" of a nutrient. For example, the "% Daily Value" column on a label of macaroni and cheese may say 20 percent for fat. This tells the consumer that each serving of this food contains about 20 percent of the day's allotted 65 grams of fat. A person consuming 2000 kcalories a day could simply keep track of the percentages of Daily Values from foods eaten in a day and try not to exceed 100 percent. To determine whether a particular food is a wise choice, a consumer needs to consider the other foods eaten during the day.

Daily Values also make it easy to compare foods. For example, a consumer might discover that frozen macaroni and cheese has a Daily Value for fat of 20 percent, whereas macaroni and cheese prepared from a boxed mix has a Daily Value of 15 percent. By comparing labels, consumers who are concerned about their fat intakes will be able to make informed decisions.

## HOW TO Calculate Personal Daily Values

The Daily Values on food labels are designed for a 2000-kcalorie intake, but you can calculate a personal set of Daily Values based on your energy allowance. Consider a person with a 1500-kcalorie intake, for example. To calculate a daily goal for fat, multiply energy intake by 30 percent:

$$1500 \text{ kcal} \times 0.30 \text{ kcal from fat} = 450 \text{ kcal from fat.}$$

The "kcalories from fat" are listed on food labels, so a person could then add all the "kcalories from fat" values for a day, using 450 as a goal. A person who preferred to count grams of fat could divide this 450 kcalories from fat by 9 kcalories per gram to determine the goal in grams:

$$450 \text{ kcal from fat} \div 9 \text{ kcal/g} = 50 \text{ g fat.}$$

Alternatively, a person could calculate that 1500 kcalories is 75 percent of the 2000–kcalorie intake used for Daily Values:

$$1500 \text{ kcal} \div 2000 \text{ kcal} = 0.75.$$
$$0.75 \times 100 = 75\%.$$

Then, instead of trying to achieve 100 percent of the Daily Value, a person consuming 1500 kcalories would aim for 75 percent. Similarly, a person consuming 2800 kcalories would aim for 140 percent:

$$2800 \text{ kcal} \div 2000 \text{ kcal} = 1.40 \text{ or } 140\%.$$

The inside back cover (left) includes a calculation column that can help you estimate your personal daily value for several nutrients.

## *Descriptive Terms*

The FDA defines the words a label may use to describe the nutrient content of a product (see Table 2-8). Definitions include the conditions under which each term can be used. For example, in addition to having less than 2 milligrams of cholesterol, a "cholesterol-free" product may not contain more than 2 grams of saturated fat per serving. The term "fresh" can be used only for raw food that has never been frozen; the descriptive term "freshly" (baked or prepared) can be used only if the food has been recently made and has not been frozen, heated, processed, or chemically preserved.

Some descriptions *imply* that a food contains, or does not contain, a nutrient. Implied claims are prohibited unless they meet specified criteria. For example, a claim that a product "contains no oil" *implies* that the food contains no fat. If the product is truly fat-free, then it may make the no-oil claim, but if it contains another source of fat, such as butter, it may not.

## *Health Claims*

**health claim:** any statement that characterizes the relationship between any nutrient or other substance in a food and a disease or health-related condition.

Health claims describe an association between a nutrient or food substance and a specific health problem. The FDA has approved several health claims based on specified criteria, including:

- The nutrient or food substance must be related to a disease or health condition for which most people or a specific group of people, such as the elderly, are at risk.
- The relationship between diet and health has been clearly established by scientific evidence.

Health claims on products must emphasize the importance of the total diet and not exaggerate the role of a particular food or diet in disease prevention. No one food possesses magical healing powers, and manufacturers must take care not to distort the roles of their products in promoting health.

Claims must be honest and balanced. For example, health claims can say that foods high in calcium "may" or "might" reduce the risk of osteoporosis. Claims must also explain that diseases develop in response to many factors. They may

**Table 2-8**
**Terms on Food Labels**

**General Terms**

**Free:** "nutritionally trivial" and unlikely to have a physiological consequence; synonyms include "without," "no," and "zero." A food that does not contain a nutrient naturally may make such a claim, but only as it applies to all similar foods (for example, "applesauce, a fat-free food").

**Healthy:** a food that is low in fat, saturated fat, cholesterol, and sodium and that contains at least 10% of the Daily Values for vitamin A, vitamin C, iron, calcium, protein, or fiber.

**High:** 20% or more of the Daily Value for a given nutrient per serving; synonyms include "rich in" or "excellent source."

**Less:** at least 25% less of a given nutrient or kcalories than the comparison food (see individual nutrients below); synonyms include "fewer" and "reduced."

**Light** or **lite:** any use of the term other than as defined below must specify what it is referring to (for example, "light in color" or "light in texture").

**Low:** an amount that would allow frequent consumption of a food without exceeding the Daily Value for the nutrient. A food that is naturally low in a nutrient may make such a claim, but only as it applies to all similar foods (for example, "fresh cauliflower, a low-sodium food"); synonyms include "little," "few," and "low source of."

**More:** at least 10% more of the Daily Value for a given nutrient than the comparison food; synonyms include "added" and "extra."

**Good source of:** product provides between 10 and 19% of the Daily Value for a given nutrient per serving.

**Energy**

**kCalorie-free:** fewer than 5 kcal per serving.

**Light:** one-third fewer kcalories than the comparison food.

**Low kcalorie:** 40 kcal or less per serving.

**Reduced kcalorie:** at least 25% fewer kcalories per serving than the comparison food.

**Fat and Cholesterol**

**Percent fat-free:** may be used only if the product meets the definition of *low fat* or *fat-free* and must reflect the amount of fat in 100 g (for example, a food that contains 2.5 g of fat per 50 g can claim to be "95 percent fat free").

**Fat-free:** less than 0.5 g of fat per serving (and no added fat or oil); synonyms include "zero-fat," "no-fat," and "nonfat."

**Low fat:** 3 g or less fat per serving.

**Less fat:** 25% or less fat than the comparison food.

**Saturated fat-free:** less than 0.5 g of saturated fat and 0.5 g of *trans*-fatty acids per serving.

**Low saturated fat:** 1 g or less saturated fat per serving.

**Less saturated fat:** 25% or less saturated fat than the comparison food.

**Cholesterol-free:** less than 2 mg cholesterol per serving and 2 g or less saturated fat per serving.

**Low cholesterol:** 20 mg or less cholesterol per serving and 2 g or less saturated fat per serving.

**Less cholesterol:** 25% or less cholesterol than the comparison food (reflecting a reduction of at least 20 mg per serving), and 2 g or less saturated fat per serving.

**Extra lean:** less than 5 g of fat, 2 g of saturated fat, and 95 mg of cholesterol per serving and per 100 g of meat, poultry, and seafood.

**Lean:** less than 10 g of fat, 4.5 g of saturated fat, and 95 mg of cholesterol per serving and per 100 g of meat, poultry, and seafood.

**Light:** 50% or less of the fat than in the comparison food (for example, 50% less fat than our regular cookies).

NOTE: Foods containing more than 13 g total fat per serving or per 50 g of food must indicate those contents immediately after a cholesterol claim. As you can see, all cholesterol claims are prohibited when the food contains more than 2 g saturated fat per serving.

**Carbohydrates: Fiber and Sugar**

**High fiber:** 5 g or more fiber per serving. A high-fiber claim made on a food that contains more than 3 g fat per serving and per 100 g of food must also declare total fat.

**Sugar-free:** less than 0.5 g of sugar per serving.

**Sodium**

**Sodium-free** and **salt-free:** less than 5 mg of sodium per serving.

**Low sodium:** 140 mg or less per serving.

**Light:** a low-kcalorie, low-fat food with a 50% reduction in sodium.

**Light in sodium:** no more than 50% of the sodium of the comparison food.

**Very low sodium:** 35 mg or less per serving.

even mention beneficial factors, such as exercise. For example, a health claim may state that "Development of cancer depends on many factors. A diet low in total fat may reduce the risk of some cancers." The health claim is true, it acknowledges that diet is among many factors influencing disease development, and it is phrased in terms of total diet, not in terms of the particular product. The following relationships for health claims on labels have been authorized:

- *Calcium and osteoporosis.* Foods or supplements must be high in calcium (at least 20 percent of the Daily Value) and contain no more phosphorus than calcium.

- *Sodium and hypertension (high blood pressure).* Foods must qualify as "low sodium" (see Table 2-8).
- *Dietary saturated fat and cholesterol and risk of coronary heart disease.* Foods must qualify as "low saturated fat," "low cholesterol," and "low fat" or, in the case of meat and poultry, as "extra lean" (see Table 2-8).
- *Dietary fat and cancer.* Foods must qualify as "low fat" or, in the case of meat and poultry, as "extra lean." Claims may not specify types of fat and must speak in terms of "some types of cancers" or "some cancers."
- *Fiber-containing grain products, fruits, and vegetables and cancer.* Grain products, fruits, or vegetables must qualify both as "low fat" and as "good sources" (without fortification) of dietary fiber (see Table 2-8). Claims may not specify types of fiber and must speak in terms of "some types of cancer" or "some cancers."
- *Fruits, vegetables, and grain products that contain fiber, particularly soluble fiber, and risk of coronary heart disease.* Fruits, vegetables, or grain products must qualify as "low saturated fat," "low cholesterol," and "low fat" and provide (without fortification) at least 0.6 grams of soluble fiber per serving.
- *Fruits and vegetables and cancer.* Fruits or vegetables must qualify as "low fat" and as "good sources" (without fortification) of vitamin A, vitamin C, or dietary fiber.
- *Folate and neural tube defects.* Foods must qualify as "good sources" (without fortification) of folate and contain no more than the recommended intakes of vitamin A or vitamin D.
- *Sugar alcohols and tooth decay.* Foods must be sugar-free, contain sugar alcohols, and not lower dental plaque pH below 5.7 by bacterial fermentation.
- *Soluble fiber from whole oats and heart disease; soluble fiber from psyllium seed husk and heart disease.* Products must provide at least 0.75 grams of soluble fiber from whole oats or 1.7 grams of soluble fiber from psyllium seed husk per serving and qualify as "low saturated fat," "low cholesterol," and "low fat."

Health claims on supplement labels are presented in Highlight 10.

With the exception of sugar alcohols and dental caries, all other health claims must also meet two additional criteria. First, a food making a health claim must be a *naturally* good source (containing at least 10 percent of the Daily Value) of at least one of the following nutrients: vitamin A, vitamin C, iron, calcium, protein, or fiber. Second, foods are disqualified from making health claims if a standard serving contains more than 20 percent of the Daily Value for total fat, saturated fat, cholesterol, or sodium. Thus milk, which is high in calcium, may make a calcium and osteoporosis claim if it is nonfat or low-fat milk, but not if it is whole milk because excess fat increases the risks of some cancers and heart disease.

**Table 2-9**
**Messages from the "It's All About You" Campaign**

**Make healthy choices that fit your lifestyle so you can do the things you want to do.**

- **Be realistic**
Make small changes over time in what you eat and the level of activity you do. After all, small steps work better than giant leaps. *Tip: Sprinkle shredded cheese on salads or pasta to boost your calcium intake.*
- **Be adventurous**
Expand your tastes to enjoy a variety of foods. *Tip: Try a new food or recipe once a month.*
- **Be flexible**
Go ahead and balance what you eat and the physical activity you do over several days. No need to worry about just one meal or one day. *Tip: If you eat ice cream, increase your physical activity for several days.*
- **Be sensible**
Enjoy all foods, just don't overdo it. *Tip: If your favorite food is high in fat, eat a smaller portion.*
- **Be active**
Walk the dog, don't just watch the dog walk. *Tip: Climb the stairs instead of taking the elevator or escalator.*

## Consumer Education

Labels are valuable only if people know how to use them, and so the FDA has designed several programs to educate consumers. Consumers who understand how to read labels will be best able to apply the information to achieve and maintain healthful dietary practices.

To help consumers understand and coordinate the messages from food labels, the *Dietary Guidelines,* and the Food Guide Pyramid, an alliance of health organizations, the food industry, and government agencies has developed an educational program called "It's All About You." The program is designed to deliver simple messages that will motivate consumers to think positively about making reasonable changes in their eating and physical activity habits (see Table 2-9).

**Healthy People 2000:** Increase to at least 85% the proportion of people aged 18 and older who use food labels to make nutritious food selections.

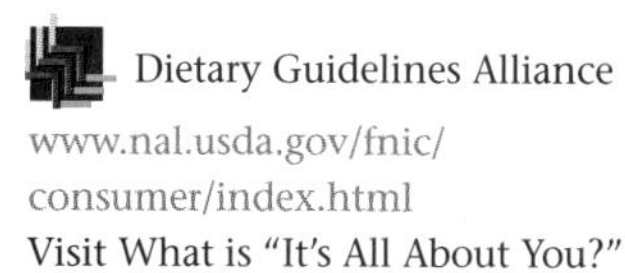


www.nal.usda.gov/fnic/consumer/index.html
Visit What is "It's All About You?"

## IN SUMMARY

Food labels provide consumers with information they need to select foods that will help them meet their nutrition and health goals.[7] Given labels with relevant information presented in a standardized, easy-to-read format, consumers are well prepared to plan and create healthful diets.

# Making It Click

These problems will give you practice in doing simple nutrition-related calculations. They use hypothetical situations in order to teach a lesson that can provide answers (see Appendix K). Be sure to show your calculations for each problem.

1. *Read a food label.* Look at the cereal label in Figure 2-5 and answer the following questions:
   a. What is the size of a serving of cereal?
   b. How many kcalories are in a serving?
   c. How much fat is in a serving?
   d. How many kcalories does this represent?
   e. What percentage of the kcalories in this product comes from fat?
   f. What does this tell you?
   g. What is the % Daily Value for fat?
   h. What does this tell you?
   i. Does this cereal meet the criteria for a low-fat product (refer to Table 2-8)?
   j. How much fiber is in a serving?
   k. Read the Daily Value chart on the lower section of the label. What is the Daily Value for fiber?
   l. What percentage of the Daily Value for fiber does a serving of the cereal contribute? Show the calculation the label-makers used to come up with the % Daily Value for fiber.
   m. What is the predominant ingredient in the cereal?
   n. Have any nutrients been added to this cereal (is it fortified)?
2. *Calculate a personal Daily Value.* The Daily Values on food labels are for people with a 2000-kcalorie intake.
   a. Suppose a person has a 1600-kcalorie energy allowance. Use the calculation factors listed in the last column of the table on the inside back cover (left) of this book to calculate a set of personal "Daily Values" based on 1600 kcalories. Show your calculations.
   b. Revise the % Daily Value chart of the cereal label in Figure 2-5 based on your "Daily Values" for a 1600-kcalorie diet.

# Study Questions

These questions will help you review the chapter. You will find the answers in the discussion on the pages provided.

1. Name the diet-planning principles and briefly describe how each principle helps in diet planning. (pp. 31–32)
2. What recommendations appear in the *Dietary Guidelines for Americans*? (p. 33)
3. Name the five food groups in the Daily Food Guide and identify several foods typical of each group. Explain how such plans group foods and what diet-planning principles the plans best accommodate. How are food group plans used, and what are some of their strengths and weaknesses? (pp. 33–38)
4. Name the exchange lists and identify a food typical of each list. Explain how the exchange system groups foods and what diet-planning principles the system best accommodates. How are exchange systems used, and what are some of their strengths and weaknesses? (pp. 38–42)
5. Review the *Dietary Guidelines*. What types of grocery selections would you make to achieve those recommendations? (pp. 33, 43–46)
6. What information can you expect to find on a food label? How can this information help you choose between two similar products? (pp. 47–49)
7. What are the Daily Values? How can they help you meet health recommendations? (p. 49)
8. What health claims have been approved by the FDA for use on labels? What criteria must all health claims meet? (pp. 50–52)

These questions will help you prepare for an exam. Answers can be found in Appendix K.

1. The diet-planning principle that provides all the essential nutrients in sufficient amounts to support health is:
   a. balance.
   b. variety.
   c. adequacy.
   d. moderation.
2. A person who chooses a chicken leg that provides 0.5 milligrams of iron and 95 kcalories instead of two tablespoons of peanut butter that also provides 0.5 milligrams of iron but 188 kcalories is using the principle of nutrient:
   a. control.
   b. density.
   c. adequacy.
   d. moderation.
3. Which of the following is consistent with the *Dietary Guidelines for Americans*?
   a. Choose a diet moderate in fat and cholesterol.
   b. Balance the food you eat with physical activity.
   c. Choose a diet with plenty of milk products and meats.
   d. Eat an abundance of foods to ensure nutrient adequacy.
4. According to the Food Guide Pyramid, which food group provides the foundation of a healthy diet?
   a. vegetables
   b. milk, yogurt, and cheese
   c. breads, cereals, rice, and pasta
   d. meat, poultry, fish, dry beans, eggs, and nuts
5. Foods within a given food group are similar in their contents of:
   a. energy.
   b. proteins and fibers.
   c. vitamins and minerals.
   d. carbohydrates and fats.
6. In the exchange system, each portion of food on any given list provides about the same amount of:
   a. energy.
   b. satiety.
   c. vitamins.
   d. minerals.
7. In the exchange system, corn and potatoes are on the:
   a. fruit list.
   b. starch list.
   c. vegetable list.
   d. meat alternate list.
8. Which of the following is *not* on the fat exchange list?
   a. bacon
   b. avocados
   c. black olives
   d. peanut butter
9. Enriched grain products are fortified with:
   a. fiber, folate, iron, niacin, and zinc.
   b. thiamin, iron, calcium, zinc, and sodium.
   c. iron, thiamin, riboflavin, niacin, and folate.
   d. folate, magnesium, vitamin $B_6$, zinc, and fiber.
10. Daily Values on food labels are based on a:
    a. 1500-kcalorie diet.
    b. 2000-kcalorie diet.
    c. 2500-kcalorie diet.
    d. 3000-kcalorie diet.

## Notes

1. Position of The American Dietetic Association: Vegetarian diets, *Journal of the American Dietetic Association* 93 (1993): 1317–1319.
2. C. Achterberg, E. McDonnell, and R. Bagby, How to put the Food Guide Pyramid into practice, *Journal of the American Dietetic Association* 94 (1994): 1030–1035.
3. S. M. Krebs-Smith and coauthors, Characterizing food intake patterns of American adults, *American Journal of Clinical Nutrition* 65 (1997): 1264S–1268S.
4. National Live Stock and Meat Board, *Eating in America Today: A Dietary Pattern and Intake Report,* 1994.
5. H. H. C. Lee, S. A. Gerrior, and J. A. Smith, Energy, macronutrient, and food intakes in relation to energy compensation in consumers who drink different types of milk, *American Journal of Clinical Nutrition* 67 (1998): 616–623.
6. Foods in menu claims must meet FDA rule, *FDA Consumer,* October 1996, p. 5.
7. Position of The American Dietetic Association: Nutrition and health information on nutrition labels, *Journal of the American Dietetic Association* 90 (1990): 583–585.

HIGHLIGHT 2

# Ethnic Cuisines and Healthy Choices

Do some foodways support health better than others? This highlight presents a few of the many ethnic foodways to show how basic diet-planning principles can apply to many different cuisines (the glossary on p. 56 defines foodways, cuisines, and related terms). A look at the traditional foods of other countries is a delightful way to learn about the world's people and is especially useful to those who advise others on nutrition. A counselor who is familiar with the cultural and religious traditions that influence a person's food choices is better able to make suggestions that fit into the person's life.

Every country, and in fact every region of a country, has its own typical foods and ways of combining them into meals. The "American diet" includes many ethnic foods from various countries, all adding variety to the diet. This is most evident when we eat out: 60 percent of our restaurants (excluding fast-food places) have an ethnic emphasis, most commonly Chinese, Italian, or Mexican.[1] While such variety helps to ensure nutrient adequacy, only moderation can control energy and fat intakes—keys to reducing the risks of chronic diseases. How can people enjoy ethnic meals and still limit their energy and fat intakes? Keep this question in mind as you read the following sections.

## Northern European Entrées

An evening meal of hearty roast beef, mashed potatoes, boiled cabbage, and bread, with fruit pie for dessert, is typical of the cuisines of Germany, England, and Ireland. Such meals are served for countless dinners across the United States, too, and variations on this plan are numerous. Even a Thanksgiving turkey dinner with all the trimmings follows this pattern.

Traditionally, people have filled their plates with meat and served starches and vegetables on the side. This eating style ensures adequate protein and other nutrients associated with meat, but delivers a lot of fat, too, and is short on fiber and the vitamins and minerals associated with fruits and vegetables. Today's health advice encourages people to load their plates with tasty grains and vegetables and to limit meat to 2- to 3-ounce portions. This way, the meal provides plenty of carbohydrate and fiber, adequate protein, and not too much fat.

*Low-fat ethnic foods, such as fajitas, have become such common restaurant fare that we often forget their origins.*

## French Foods

The French expertly combine butter, cream, eggs, herbs, and wine into classic sauces. They prepare pastries filled with seafood, cheese, or meats and covered with rich sauces for a main dish or wrapped around sweetened fruits and creams for dessert. These choices are extraordinarily high in fat. In fact, the French people eat as much fat as, and even more saturated fat than, is typical of people in the United States.[2] Yet the French have a lower-than-expected rate of heart disease. Among the dietary factors suggested as protecting the French and others of the Mediterranean region against heart disease are their increased consumption of olive oil, red wines, and a variety of foods.[3] Highlight 6 reviews some of the links between the Mediterranean diets and heart health.

French food can be low in fat when a person follows the suggestion to "keep it simple." The country cuisine of southern France along the Mediterranean Sea offers wonderful options: elegant clear soups; steamed and poached seafood; lean meats, legumes, and vegetables seasoned with lemon, herbs, or wine; fruit; and huge loaves of French bread.

## Greek Fare

Greeks are known for a robust cuisine that includes whole broiled fish and other seafoods; roasts and stews of vegetables such as eggplant with lamb, chicken, or beef; and the flavors of fresh lemons, garlic, and herbs (dill, mint, and parsley); and, always, olives and their oil, which lavishly season many dishes. Traditional Greek salads contain no lettuce but combine chunks of tomato, peppers, cucumbers, onions, feta cheese, anchovies, and cured ripe olives with a tangy olive oil and lemon dressing. Stuffed grape leaves may be topped by a famous Greek sauce of eggs and lemon juice. Gyros (pronounced YEE-roce, meaning "a circle" in Greek) is a high-fat, highly seasoned meat that is roasted over open flames as it slowly turns on a spit. When cooked, the gyros is thinly sliced, dressed with yogurt and cucumber, and served in a pita (pocket) bread. As for Greek desserts, pastries (baklava) soaked with butter and honey and layered with nuts are traditional at celebrations. Because of their intense sweetness, small servings of these desserts are usually satisfying.

The average Greek diet provides a startling 42 percent of its kcalories from fat, mostly from olive oil and olives. Yet people living in Greece have a lower incidence of cardiovascular disease than northern Europeans and North Americans and enjoy one of the longest life expectancies worldwide despite the tendency of many Greeks to carry excess body fat.[4] Some researchers suggest that a diet similar to that of Greece might be healthier and easier to follow than the limited fat diet recommended in the United States (Highlight 6 revisits this topic).

A valid question is whether Mediterranean people may be naturally resistant to heart disease, but this seems not to be the case. People who move from one place to another and adopt the dietary habits of the new locale have heart disease rates typical of people native to that area. Mediterranean populations are, however, more physically active than those in the United States and Canada. They also consume more fruits and vegetables, less meat and animal fat, more olive oil, and more of each day's kcalories early in the day. The effects of meal timing on heart disease may be important. A small but significant improvement in blood lipids has been observed in people who eat frequent small meals each day, rather than a few larger meals.[5] Many aspects of the diet affect heart health.

## *Italian Edibles*

Like Greece, Italy relies on the Mediterranean Sea for seafoods, but its cuisine is perhaps best known for its pastas. Italian cuisines—and pastas—differ from north to south. Generally, northern pastas are egg based and either ribbon shaped or stuffed with meats or cheeses (tortellini). Northern Italians eat their pasta with plenty of meat, butter, cheese, eggs, and cream. In the southern regions, the diet is more "Mediterranean" in character; wheat pastas are made without eggs, shaped more like macaroni, and served with vegetables such as artichokes, eggplants, peppers, and tomatoes. Beans appear on the table more often than meat, and all dishes are seasoned with olive oil, not butter. Unlike the Greeks, southern Italians keep total fat intakes relatively low. Compared with the United States, life expectancy is longer and heart disease is lower in Italy (and neighboring Spain), so health experts are reluctant to suggest dietary "improvements."

While the link between diet and heart disease is strong, no one knows what part of the diet might be beneficial. Some think using olive oil instead of butter or shortening is the key factor, while others point to the protective effects of both nutrients and nonnutrients found in vegetables, seafoods, and seasonings. Fat may play a role in disease development, but antioxidants from plant foods may play a more important role in defending the heart. (Antioxidants are featured in Highlight 11.)

*Ethnic meals and family gatherings nourish the spirit as well as the body.*

## *Cajun Cuisine*

The French have clearly influenced the bayou region of Louisiana, and so have the African-American ancestors of the area's present population. The settlers in southern Louisiana adapted their heritages to the local food supply, creating the Cajun style of cooking that is now popular across the nation. Many Cajun dishes are based on a roux made by browning flour and salt with oil. Cajun dishes include spicy stews (gumbo, jambalaya), sausages, seafood, hot pepper sauce, and red beans with dirty rice (rice made brown with chopped chicken livers and seasonings). Many people enjoy Cajun coffee brewed with chicory root and a sugared doughnut known as a beignet (pronounced ben-YAY).

Classic Cajun dishes are often full of flavor from expert use of spices and abundant in nutrients. A jambalaya of seafood (crawfish, shrimp, and fish), tomatoes, vegetables, and rice seasoned with a little strong-flavored sausage is packed with vitamins and minerals, adequate in protein, rich in carbohydrate, and usually low in fat. With a careful eye on the fat during food preparation and a limit on the beignets, people eating Cajun style can achieve the goals embodied in the *Dietary Guidelines*.

## Glossary

**cuisine** (kwi-ZEEN): style of cooking or preparing food.

**ethnic diets:** foodways and cuisines typical of national origins, cultural heritages, or geographic locations.

**foodways:** the sum of the food habits, customs, beliefs, and preferences of a culture.

**kosher** (KOE-sure): foods prepared according to Jewish dietary laws.

## Southern Suppers

The cuisine of the southern United States provides ample vitamins and minerals from a variety of vegetables: sweet potatoes, collards and other leafy greens, black-eyed (cow) peas, okra, tomatoes, and corn. Unfortunately, many of these traditional vegetable dishes are flavored with salt pork, smoked bacon and its fat, or lard, or they are fried in shortening. Fried green tomatoes, a southern specialty, have been dredged in spicy cornmeal and then deep-fried. Many southerners enjoy large servings of meats, such as fried chicken, fatty pork cuts or sausages, and spareribs, greatly overemphasizing protein and fat in the diet while slighting whole grains, fruits, and vegetables. Biscuits, a favored bread, are made with almost as much shortening as flour and are often served with butter or fat-rich gravy. Southern families take pride in their recipes for pecan pie, a pastry shell filled with butter, syrup, and nuts.

Such a high-fat diet is associated with many health problems, including obesity, heart disease, and stroke. The southeastern United States has the dubious distinction of being the nation's Stroke Belt because of its high incidence of cardiovascular disease. People indulging in traditional southern cuisine need to keep moderation in mind if they are to maintain their health. The trick to choosing health-promoting rural southern food is to limit the fatty, salty meats and vegetables, biscuits, and gravies. Instead select small portions of low-fat meats; prepare vegetables without added fat and salt; and eat beans, rice, and cornbread with nonfat seasonings and spreads.

## Native American Ways

Over the last 200 years, Native Americans have seen unprecedented changes in their foodways. Hunter-gatherer and agricultural lifestyles have given way to a modern culture relying on fast foods, alcohol, abundant high-fat meats, and dairy products. For Native Americans, the effects of these changes on health have been overwhelmingly negative.

The Pima Indians of central Arizona are a well-studied example of a group that has experienced the effects of a "modernized" diet. Until the 1930s, the Pima diet consisted of wild and cultivated desert legumes, cactus leaves and fruit, fish, venison (deer meat), small seeds, mesquite pods, acorns, and corn. Then the Pima largely replaced their traditional wild foods, which had become scarce, with wheat flour, lard, sugar, coffee, and ready-to-eat cereals that were easy to obtain. The result has been tragic: the Pima now suffer the highest per capita rate of diabetes known among any people in the world. Likewise, the Sioux Indians rarely suffered heart disease when consuming their traditional diets, but now have one of the highest rates of cardiovascular disease.

Both the Pima and the Sioux changed their highly active lifestyles, as well as their diets. Modern-day Pima and Sioux no longer hunt game, toil in the fields, or cook over stone fireplaces as their ancestors did. Like others in the United States, they now drive to supermarkets to purchase foods and cook in microwave ovens. The Sioux also traded their ceremonial pipes for cigarette smoking, a habit that is especially damaging to the heart.

Native diets are not perfect. They may not provide adequate amounts of some nutrients, and availability of foods depends on such unpredictable factors as weather changes and herd movements. Still, when Native Americans consumed their original high-fiber, low-fat native foods, their hearts and bodies benefited.

## Mexican Meals

Mexican restaurants in the United States typically offer tortillas filled with meats and cheeses, some fried and crisp and others baked and soft, along with flavored rice, refried beans, sour cream, guacamole (avocado sauce), and salsa (tomato sauce). These foods are of Mexican origin, but few Mexican families eat them as daily fare. A typical Mexican lunch or dinner is rather simple, consisting of a stew of beans, meat, rice, and potatoes served with tortillas or bread, tomato salsa, and lettuce salad or cooked vegetables.[6] A Mexican breakfast might include tortillas and eggs with a beverage.

If carefully chosen, the foods from both taco stands and fancier Mexican restaurants can make valuable contributions to the diet. Many traditional Mexican dishes, even the fast-food type, provide beans in abundance. Beans are rich in vitamins, minerals, and fiber while low in fat (although some restaurants prepare the refried variety with lard). Soft tortillas filled with beans, lettuce, and salsa with a side order of rice are high in nutrient density. Without careful selection, though, Mexican foods can be extraordinarily high in fat, such as a fried tortilla shell filled with high-fat ground beef and topped with cheese and sour cream.

For a healthy Mexican meal, use sour cream lightly, and skip the fried tortillas. Try a fajita—lean meats, marinated and sizzled on a grill, wrapped in soft tortillas with chili

*Tortillas filled with lean beef and fresh vegetables are a welcome alternative to sandwiches.*

salsa toppings. Salsa (made of chopped tomatoes, onions, and hot peppers) adds zest but no fat to a meal. As for guacamole, even though avocados are high in fat, they add interest and flavor to a meal, and the type of fat they contain is not implicated in heart disease. Their high fat content does mean that people who need to control their weight should eat avocados in small quantities on infrequent occasions.

## Other Hispanic Foods

Mexican foodways share similarities with those from other Hispanic countries such as Cuba and Puerto Rico, but there are also notable differences.[7] For example, for the main grain of a meal, Mexicans prefer tortillas, Cubans choose rice, and Puerto Ricans select breads. Mexican fare usually includes more protein-rich foods such as beef, legumes, and eggs and more vegetables and potatoes than do the other two cuisines. These meals contribute more energy and fat as well, which may account for the higher rates of obesity and cardiovascular disease among Mexican Americans compared with their Hispanic peers.

## Chinese Foodways

China's foodways reflect the efficiency that is essential in a country where the population density is more than 1000 people per acre, yet only 10 percent of the land can be used to grow food. China has over a billion people, and 75 percent of them are involved in agriculture. Contrast these figures with the United States: population density, 113 per acre; land in farms, about 50 percent; population, over 260 million; farmers, 1 to 2 percent.

There seems to be little malnutrition or obesity in China, even though the people consume 20 percent more food energy each day than we do. On the whole, Chinese people eating traditional foods consume three times the fiber of people eating the American way, take in about half the fat, and have blood cholesterol values about half of what they are in the United States. Only 4 out of every 100,000 men in China die of heart disease each year compared with 67 in the United States. Chinese living in China also suffer much less cancer of the colon and rectum than do Chinese Americans who have adopted a Western diet. It seems worthwhile to study the fine points of a diet so conservative of resources yet so superbly supportive of health—not to convince everyone to eat Chinese meals three times a day, but to illustrate the governing principles that can be applied to foods of all origins.

*Foodways and cuisines of all cultures can support heart health when dietary fat is limited. (Courtesy of CNN)*

Typical Chinese meals do not follow the meat-vegetable-starch pattern of northern Europe that is common in much of the United States today. Instead, meals center on a staple starch food—every diner has a dish of rice and chooses small amounts of meat (or fish or egg) and plenty of vegetables from serving dishes according to appetite. The Chinese also usually drink soup or tea throughout each meal, which slows dining to a relaxing pace.

Vegetables and fruits provide tremendous variety. Seasonings and sauces such as ginger root, scallions, rice wine, garlic, soy, hoisin, oyster, bean, and plum add tasty flavors but little fat unlike our gravies, butter, and sour cream. The Chinese mode of cooking in a wok requires just a tablespoon or two of oil for an entire dish. Chinese dishes do, however, tend to be high in sodium. Both hypertension and a high sodium intake are common in China.[8] The average sodium intake in China is 15 grams a day (about six times the recommended intake of our *Dietary Guidelines*). Diners can enjoy low-sodium meals if they are prepared without salt or monosodium glutamate (MSG) and with judicious use of soy sauce.

Cooking foods the Chinese way tends to preserve nutrients. Meats and vegetables are cut into bite-sized pieces before cooking, so cooking is quick and destroys few nutrients. No cooking water is thrown away, so nutrients are not lost that way either. The water in which the rice is cooked soaks back into the rice, so the rice retains its nutrients.

Some Chinese dishes do have nutrition drawbacks, though. In China, deep-fried foods are eaten only seldom, but Chinese restaurants in this country often feature these and other high-fat items. Another drawback is the inclusion of salted, fermented pickles, which have been linked to a high incidence of digestive tract cancers. Chinese restaurants in the United States rarely serve these fermented items because they tend not to appeal to Western tastes.

## Japanese Diet

Traditional Japanese cuisine bears similarities to the Chinese diet. Grains such as rice or millet form the foundation of most traditional Japanese meals. Vegetables and fruits are next in prominence, and seafood, eggs, poultry, and meats play a supporting role. For a quick and easy traditional meal, a person in Japan might stop by a "noodle house" for a bowl of noodles (somen) in a clear, seasoned broth—a dish of Chinese origin. A Japanese delicacy popular in the United States is sushi—vinegar-flavored rice holding bits of colorful vegetables and seafood, wrapped in seaweed and served with horseradish (wasabi) or seasoned soy sauce. In the United States, the word *sushi* has come to mean "raw fish," which may be an ingredient, but sushi actually refers to the vinegared rice, and many types of sushi are made with cooked ingredients.

***Traditional Japanese cuisine honors the body by providing foods rich in nutrients, flavors, and visual appeal.***

Since the 1950s, Japan has transformed itself from a wartorn, still largely traditional country struggling to feed its population to an industrial and economic world leader. Today, Japan's cuisine reflects the "hurry-up" lifestyle of a nation buzzing with mass communications media, high incomes, and high expenses. Time-consuming, home-cooked traditional dishes, though still favored in restaurants, have proved impractical for the two-income family of the 1990s.

In a land where rice once occupied center stage at every meal, meats, breads, and milk products now dominate. Meat consumption in Japan has jumped more than tenfold since the 1940s. Egg intakes have risen more than sixfold and margarine consumption has more than doubled in the same period. Vegetable intakes have fallen precipitously. Ice cream has replaced fruit as the preferred dessert, and instant coffee is replacing tea. High-fat snacks from hamburger places, southern fried chicken restaurants, and doughnut shops have shoved low-fat noodle houses into the background.

The health implications of such changes are two-sided. On the one side, a steady increase in heart attacks and strokes is alarming. Japanese men who grew up consuming a "Western" diet are more likely to suffer from diabetes than are men who grew up on a traditional Japanese diet, and diabetes is a risk factor for heart disease. On the other side, deaths in Japan have declined as modern sanitation and immunizations have brought infectious diseases under control, and the new diet provides much more protein than a traditional rice-based diet. Extra protein during the growing years has enabled the younger generation to grow taller and stronger than any generation before them. In general, Japan still enjoys a higher life expectancy and lower overall rate of heart disease than many other nations, but new choices present new risks.

People in the United States who wish to dine in the Japanese style can freely choose from traditional dishes, being wary only of a few battered and deep-fried meats and vegetables (tempuras). A traditional Japanese chef may toss together a mixture of mushrooms, carrots, and bamboo shoots with bits of seafood or meat; season it with fat-free (but salty) soy sauce; add sesame seeds; and serve it with a large portion of rice. Shrimp in rice-cake soup features a clear broth, mushrooms, shrimp, and spinach and is served with rice cakes (mochi). A fish-and-noodle casserole might combine a lean fish fillet and broth seasoned with sugar, soy sauce, and mirin (a syrupy rice wine) with a big bowl of thick noodles. Beware of any restaurant that claims to serve traditional Japanese meals but centers the meals around large portions of meat. Today's Japanese diners may consume such meals, but they are not traditional cuisine, and they incur the same warnings that accompany northern European foodways.

## *Religious Food Rituals*

A discussion of ethnic foodways would be incomplete without mentioning foodways practiced by religious groups. According to many religions, ritual and ceremony surrounding food can provide nourishment for the spirit as well as for the body. Like national groups, religious groups derive their distinct identity in part from special foodways.

The Jewish laws set forth an extensive set of dietary rules. Many people, on hearing the word *kosher,* think of foods such as pickles, bagels and lox, corned beef, or matzoh crackers. Kosher is not a cuisine, however, but rather a set of restrictions that Orthodox Jews place on the selection and preparation of animal-derived foods. Jews from eastern Europe, Germany, the former Soviet Union, the Middle East, or India eat different foods, but the kosher rules apply to all.[9]

Religious commitment is the sole intent of those who keep kosher. Occasionally, someone suggests that the laws of kosher originated for reasons of health—that kosher food was "clean" and therefore kept people safe from food-borne illnesses—but the rules of kosher offer no special benefits to health. These rules permit Orthodox Jews to eat beef but not pork, fish but not shellfish, and they dictate special

*Many religions include foods in their ceremonies.*

handling methods for permitted foods. Because blood is forbidden as food, kosher rules govern methods of animal slaughter, cuts that may be eaten, and preparation rituals.

Kosher law prohibits Jews from consuming milk and meat in the same meal. This law leads some kosher cooks to replace milk with nondairy creamer in meals that include meat. Nutritionally, however, creamers do not resemble milk, and they are high in saturated fat. A better choice is to use soy "milk" products formulated to resemble the nutrient and cooking qualities of milk products.

Like other cuisines, Jewish cuisines and kosher foods can be evaluated according to dietary standards. A meal might be improved by reducing the schmaltz (chicken fat) used in cooking or by frying latke (potato pancakes) in nonstick pans rather than in oil. Bagels with lox (a form of salmon) and nonfat cream cheese are an excellent breakfast choice—bagels are naturally low in fat, and lox is rich in fish oils thought to be protective against the development of heart disease.

Food symbolism abounds in most other religions as well. During certain days of Lent, the period prior to Easter, many Christians eat only vegetarian dishes, giving up meat until Easter Sunday. Eastern Orthodox Christians observe many fast days on which they consume no animal products at all. The Mormon faith allows no alcohol, coffee, or tea. Many Seventh-Day Adventists eat eggs and milk products, but consume no meat; they also shun alcohol, coffee, and tea. Their doctrine advises them to avoid strong spices such as mustard and pepper and discourages between-meal snacks. Other faiths, such as Islam, Hinduism, and Buddhism, prohibit some dietary practices while promoting others.

As you can see by now, consumers can apply the diet-planning principles introduced in Chapter 2 to any ethnic foodway. It is not the ethnic cuisine itself, but the diner's habitual selections from the many traditional choices that determine whether a diet will benefit health. Whatever the cuisine, consumers must learn to balance all foods in a way that provides adequate nourishment without excess.*

*The American Diabetes Association and the American Dietetic Association offer a series of booklets on the ethnic and regional food practices of dozens of foodways. These booklets describe food practices, customs, and holidays; present meals modified according to nutrition recommendations; provide exchange list, food composition values, and glossaries for ethnic foods; and list additional resources. See Appendix F for addresses.

## Notes


1. J. Cousminer and G. Hartman, Understanding America's regional taste preferences, *Food Technology* 50 (1996): 73–77.
2. A. Drewnowski and coauthors, Diet quality and dietary diversity in France: Implications for the French paradox, *Journal of the American Dietetic Association* 96 (1996): 663–669.
3. S. Renaud and coauthors, Cretan Mediterranean diet for the prevention of coronary heart disease, *American Journal of Clinical Nutrition* 61 (1995): 1360S–1367S; E. N. Frankel and coauthors, Inhibition of oxidation of human low-density lipoprotein by phenolic substances in red wine, *Lancet* 341 (1993): 454–457; Drewnowski and coauthors, 1996.
4. U.S. Department of Commerce, *Statistical Abstract of the United States 1994,* 114th ed. (Washington, D.C.: Government Printing Office, 1994), pp 854–855.
5. L. M. Arnold and coauthors, Effect of isoenergetic intake of three or nine meals on plasma lipoproteins and glucose metabolism, *American Journal of Clinical Nutrition* 57 (1993): 446–451.
6. S. J. Algert and T. H. Ellison, Mexican American food practices, customs, and holidays, *Ethnic and Regional Food Practices* (series) (Chicago and Alexandria, Va.: American Dietetic Association and American Diabetes Association, 1989).
7. M. F. Kuczmarski, R. J. Kuczmarski, and M. Najjar, Food usage among Mexican-American, Cuban, and Puerto Rican adults—Findings from the Hispanic HANES, *Nutrition Today* 30 (1995): 30–37.
8. E. B. Feldman, Chinese lessons to Western nutrition—And vice versa, *Nutrition Today* 30 (1995): 79–83.
9. C. Higgins and H. S. Warshaw, Jewish food practices, customs, and holidays, *Ethnic and Regional Food Practices* (series) (Chicago and Alexandria, Va.: American Dietetic Association and American Diabetes Association, 1989).

CHAPTER 3

# Digestion, Absorption, and Transport

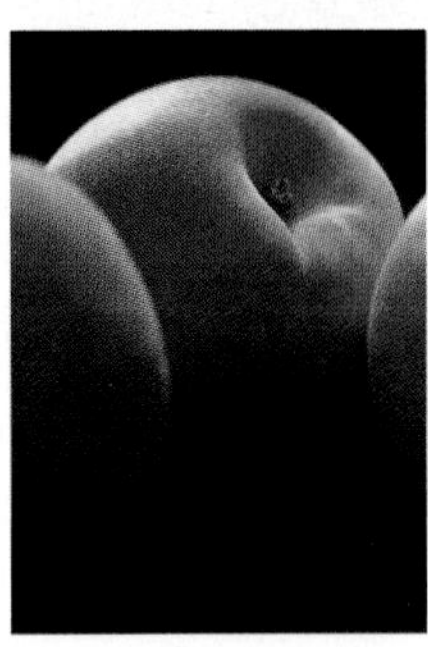

Have you ever wondered what happens to the food you eat after you swallow it? Or how your body extracts nutrients from food? Have you ever marveled how it all just seems to happen? This chapter takes you on the journey that transforms the foods you eat into the nutrients featured in the later chapters. Then it follows the nutrients as they travel through the intestinal cells and into the body to do their work. This introduction presents a general overview of the processes common to all nutrients; later chapters discuss the specifics of digesting and absorbing individual nutrients.

# Digestion

The digestive system is the body's ingenious way of breaking down foods into nutrients ready for absorption. In the process, it overcomes many obstacles for you without any conscious effort on your part. Consider these obstacles:

**digestion:** the process by which food is broken down into absorbable units.
- **digestion** = take apart

**digestive system:** all the organs and glands associated with the ingestion and digestion of food.

1. Human beings breathe, eat, and drink through their mouths. Air taken in through the mouth must go to the lungs; food and liquid must go to the stomach. The throat must be arranged so that swallowing and breathing don't interfere with each other.
2. Below the lungs lies the diaphragm, a dome of muscle that separates the upper half of the major body cavity from the lower half. Food must pass through this wall to reach the stomach.
3. To move through the system, food must be lubricated with water. Too much water would form a liquid that would flow too rapidly; too little water would form a paste too dry and compact to move at all. The amount of water must be regulated to keep the intestinal contents at the right consistency to move smoothly along.
4. When the digestive enzymes are breaking food down, they need it in finely divided form, suspended in enough water so that every particle is accessible. Once digestion is complete and the needed nutrients have been absorbed out of the tract into the body, the system must excrete the residue that remains, but excreting all the water along with the solid residue would be both wasteful and messy. Some water should be withdrawn, leaving a paste just solid enough to be smooth and easy to pass.
5. The materials within the tract should be kept moving, slowly but steadily, at a pace that permits all reactions to reach completion.
6. The enzymes of the digestive tract are designed to digest carbohydrate, fat, and protein. The walls of the tract, composed of living cells, are also made of carbohydrate, fat, and protein. These cells need protection against the action of the powerful digestive juices that they secrete.
7. Once waste matter has reached the end of the tract, it must be excreted, but it would be inconvenient and embarrassing if this function occurred continuously. Provision must be made for periodic, voluntary evacuation.

The following sections show how the body elegantly and efficiently handles these obstacles.

*The process of digestion transforms all kinds of* **foods** *into* **nutrients.**

## *Anatomy of the Digestive Tract*

The gastrointestinal (GI) tract is a flexible muscular tube from the mouth, through the esophagus, stomach, small intestine, large intestine, and rectum to the anus.

**GI tract:** the gastrointestinal tract or digestive tract; the principal organs are the stomach and intestines.
- **gastro** = stomach
- **intestinalis** = intestine

**bolus** (BOH-lus): a portion; with respect to food, the amount swallowed at one time.
- **bolos** = lump

The glossary below defines GI anatomy terms and Figure 3-1 traces the path followed by food from one end to the other. In a sense, the human body surrounds the GI tract. Only when a nutrient or other substance penetrates the GI tract's wall does it enter the body proper; many nonnutritive materials pass through the GI tract without being digested or absorbed.

• ***Mouth*** • The process of digestion begins in the mouth. As you chew, your teeth crush large pieces of food into smaller ones, and fluids blend with these pieces to ease swallowing. Fluids also help dissolve the food so that you can taste it; only particles in solution can react with taste buds. The tongue allows you not only to taste food, but also to move food around the mouth, facilitating chewing and swallowing. When you swallow a mouthful of food, it first slides across your epiglottis, bypassing the entrance to your lungs. This is the body's solution to obstacle 1: the epiglottis closes off your air passages so that you don't choke when you swallow (choking is discussed on p. 83). After a mouthful of food has been swallowed, it is called a bolus.

• ***Esophagus to the Stomach*** • Next, the bolus slides down the esophagus, which conducts it through the diaphragm (obstacle 2) to the stomach. The stomach cells produce secretions both to break down food particles and to protect themselves from being broken down (obstacle 6). The cardiac sphincter at the entrance to the stomach closes behind the bolus so that it can't slip back into the esophagus (obstacle 5). The stomach retains the bolus for a while in its upper portion. Little by little, the stomach transfers the food to its lower portion, adds juices to it, and

## Glossary of GI Anatomy Terms

These terms are listed in order from start to end of the digestive tract.

**epiglottis** (epp-ee-GLOTT-iss): cartilage in the throat that guards the entrance to the trachea and prevents fluid or food from entering it when a person swallows.
- **epi** = upon (over)
- **glottis** = back of tongue

**trachea** (TRAKE-ee-uh): the windpipe; the passageway from the mouth and nose to the lungs.

**esophagus** (ee-SOFF-uh-gus): the food pipe; the conduit from the mouth to the stomach.

**cardiac sphincter** (CARD-ee-ack SFINK-ter): the sphincter muscle at the junction between the esophagus and the stomach; also called the *lower esophageal sphincter* or the *gastroesophageal sphincter.*
**cardiac** = the heart

**sphincter** (SFINK-ter): a circular muscle surrounding, and able to close, a body opening. Sphincters are found at specific points along the GI tract and regulate the flow of food particles.
- **sphincter** = band (binder)

**stomach:** a muscular, elastic, saclike portion of the digestive tract that grinds and churns swallowed food, mixing it with acid and enzymes to form chyme.

**pyloric** (pie-LORE-ic) **sphincter:** the circular muscle that separates the stomach from the small intestine and regulates the flow of partially digested food into the small intestine; also called *pylorus* or *pyloric valve.*
- **pylorus** = gatekeeper

**liver:** the organ that manufactures bile and is the first to receive nutrients from the intestines. The liver's many other functions are described in Chapter 7.

**gallbladder:** the organ that stores and concentrates bile. When it receives the signal that fat is present in the duodenum, the gallbladder contracts and squirts bile through the bile duct into the duodenum.

**pancreas:** a gland that secretes digestive enzymes and juices into the duodenum.

**small intestine:** a 10-foot length of small-diameter intestine that is the major site of digestion of food and absorption of nutrients; its segments are the duodenum, jejunum, and ileum.

**duodenum** (doo-oh-DEEN-um, doo-ODD-num): the top portion of the small intestine (about "12 fingers' breadth" long in ancient terminology).
- **duodecim** = twelve

**jejunum** (je-JOON-um): the first two-fifths of the small intestine beyond the duodenum.

**ileum** (ILL-ee-um): the last segment of the small intestine.

**ileocecal** (ill-ee-oh-SEEK-ul) **valve:** the sphincter separating the small and large intestines.

**large intestine** or **colon** (COAL-un): the lower portion of intestine that completes the digestive process; its segments are the ascending colon, the transverse colon, the descending colon, and the sigmoid colon.
- **sigmoid** = shaped like an S (sigma in Greek)

**appendix:** a narrow blind sac extending from the beginning of the colon that stores lymph cells.

**rectum:** the muscular terminal part of the intestine, extending from the sigmoid colon to the anus.

**anus** (AY-nus): the terminal sphincter of the GI tract.

**Figure 3-1**
**The Gastrointestinal Tract**

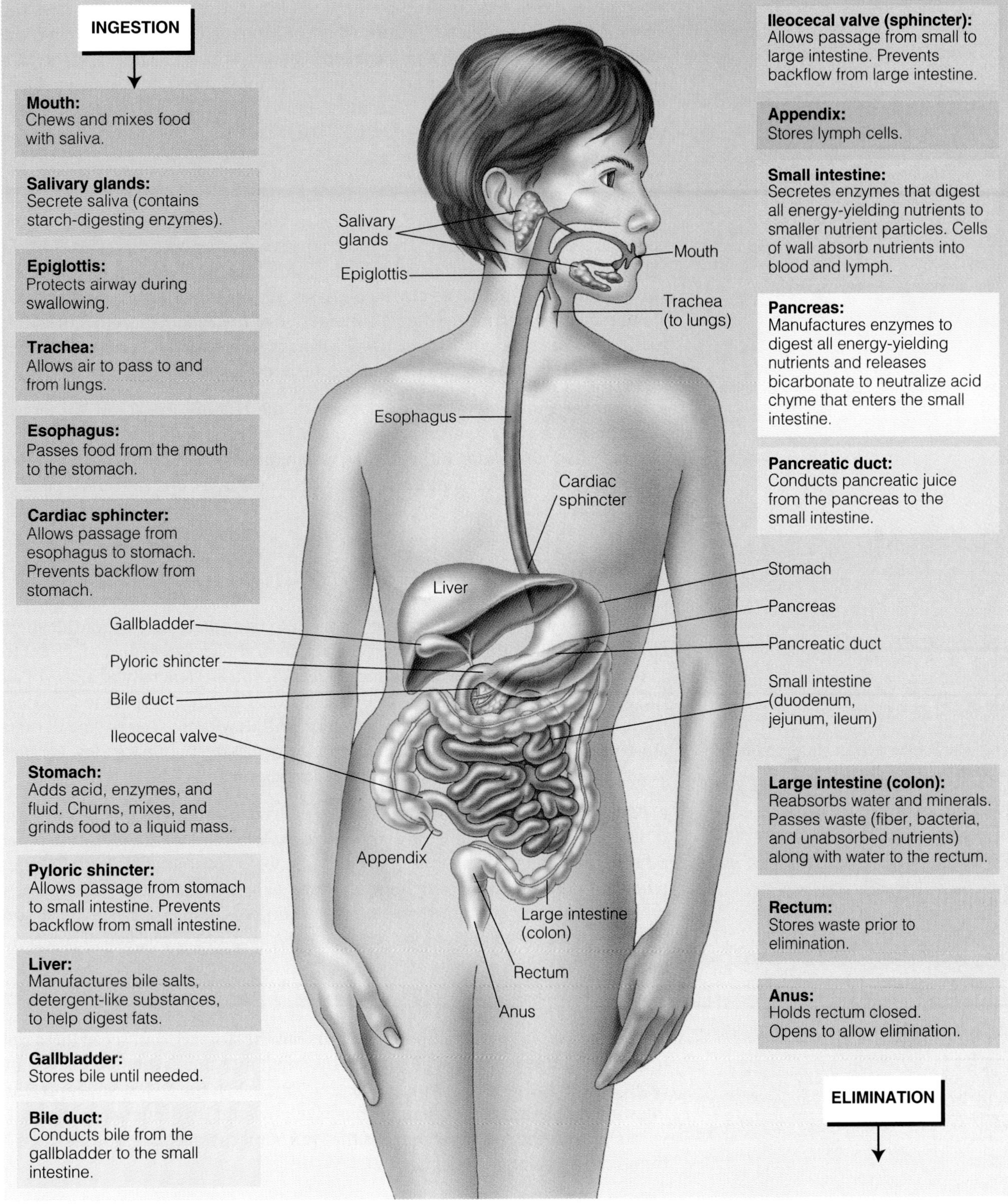
INGESTION
**Mouth:** Chews and mixes food with saliva.
**Salivary glands:** Secrete saliva (contains starch-digesting enzymes).
**Epiglottis:** Protects airway during swallowing.
**Trachea:** Allows air to pass to and from lungs.
**Esophagus:** Passes food from the mouth to the stomach.
**Cardiac sphincter:** Allows passage from esophagus to stomach. Prevents backflow from stomach.
Gallbladder
Pyloric shincter
Bile duct
Ileocecal valve
**Stomach:** Adds acid, enzymes, and fluid. Churns, mixes, and grinds food to a liquid mass.
**Pyloric shincter:** Allows passage from stomach to small intestine. Prevents backflow from small intestine.
**Liver:** Manufactures bile salts, detergent-like substances, to help digest fats.
**Gallbladder:** Stores bile until needed.
**Bile duct:** Conducts bile from the gallbladder to the small intestine.
Salivary glands
Epiglottis
Mouth
Trachea (to lungs)
Esophagus
Cardiac sphincter
Liver
Appendix
Large intestine (colon)
Rectum
Anus
**Ileocecal valve (sphincter):** Allows passage from small to large intestine. Prevents backflow from large intestine.
**Appendix:** Stores lymph cells.
**Small intestine:** Secretes enzymes that digest all energy-yielding nutrients to smaller nutrient particles. Cells of wall absorb nutrients into blood and lymph.
**Pancreas:** Manufactures enzymes to digest all energy-yielding nutrients and releases bicarbonate to neutralize acid chyme that enters the small intestine.
**Pancreatic duct:** Conducts pancreatic juice from the pancreas to the small intestine.
Stomach
Pancreas
Pancreatic duct
Small intestine (duodenum, jejunum, ileum)
**Large intestine (colon):** Reabsorbs water and minerals. Passes waste (fiber, bacteria, and unabsorbed nutrients) along with water to the rectum.
**Rectum:** Stores waste prior to elimination.
**Anus:** Holds rectum closed. Opens to allow elimination.
ELIMINATION

**chyme** (KIME): the semiliquid mass of partly digested food expelled by the stomach into the duodenum.
- **chymos** = juice

grinds it to a semiliquid mass called chyme. Then, bit by bit, the stomach releases the chyme through the pyloric sphincter, which opens into the small intestine and then closes behind the chyme.

• ***Small Intestine*** • At the top of the small intestine, the chyme bypasses the opening from the common bile duct, which is dripping fluids (obstacle 3) into the small intestine from two organs outside the GI tract—the gallbladder and the pancreas. The chyme travels on down the small intestine through its three segments—the duodenum, the jejunum, and the ileum—almost 10 feet of tubing coiled within the abdomen.[1]

• ***Large Intestine (Colon)*** • Having traveled the length of the small intestine, the chyme arrives at another sphincter (obstacle 5 again): the ileocecal valve, at the beginning of the large intestine (colon) in the lower right-hand side of the abdomen. As the chyme enters the colon, it passes another opening. Had it slipped into this opening, it would have ended up in the appendix, a blind sac about the size of your little finger. The chyme bypasses this opening, however, and travels along the large intestine up the right-hand side of the abdomen, across the front to the left-hand side, down to the lower left-hand side, and finally below the other folds of the intestines to the back side of the body, above the rectum.

• ***Rectum*** • During the chyme's passage to the rectum, the colon withdraws water from it, leaving semisolid waste (obstacle 4). The strong muscles of the rectum hold back this waste until it is time to defecate. Then the rectal muscles relax (obstacle 7), and the last sphincter in the system, the anus, opens to allow passage of the waste.

## *The Muscular Action of Digestion*

The first step in the reduction of food to a liquid takes place in the mouth, where chewing, the addition of saliva, and the action of the tongue reduce the food to a coarse mash. Then you swallow, and thereafter, you are generally unaware of all the activity that follows. As is the case with so much else that happens in the body, the muscles of the digestive tract meet internal needs without your having to exert any conscious effort. They keep things moving at just the right pace, slow enough to get the job done and fast enough to make progress.

The ability of the GI tract muscles to move is called their **motility** (moh-TIL-ah-tee).

**peristalsis** (peri-STALL-sis): wavelike muscular contractions of the GI tract that push its contents along.
- **peri** = around
- **stellein** = wrap

• ***Peristalsis*** • The entire GI tract is ringed with circular muscles that can squeeze it tightly. Surrounding these rings of muscle are longitudinal muscles. When the rings tighten and the long muscles relax, the tube is constricted. When the rings relax and the long muscles tighten, the tube bulges. These actions follow each other continuously and push the intestinal contents along (obstacle 5). (If you have ever watched a lump of food pass along the body of a snake, you have a good picture of how these muscles work.)

The waves of contraction ripple along the GI tract at varying rates and intensities depending on the part of the GI tract and on whether food is present. For example, waves occur three times per minute in the stomach, but speed up to ten times per minute when chyme reaches the small intestine. When you have just eaten a meal, the waves are slow and continuous; when the GI tract is empty, the intestine is quiet except for periodic bursts of powerful rhythmic waves. Peristalsis, along with the sphincter muscles that surround the tract at key places, keeps things moving along (see Figure 3-2).

• ***Stomach Action*** • The stomach has the thickest walls and strongest muscles of all the GI tract organs. In addition to the circular and longitudinal muscles, it has a third layer of diagonal muscles that also alternately contract and relax. These three sets of muscles work to force the chyme downward, but the pyloric sphincter usually remains tightly closed, preventing the chyme from passing into the duodenum. As a result, the chyme is churned and forced down, hits the pyloric

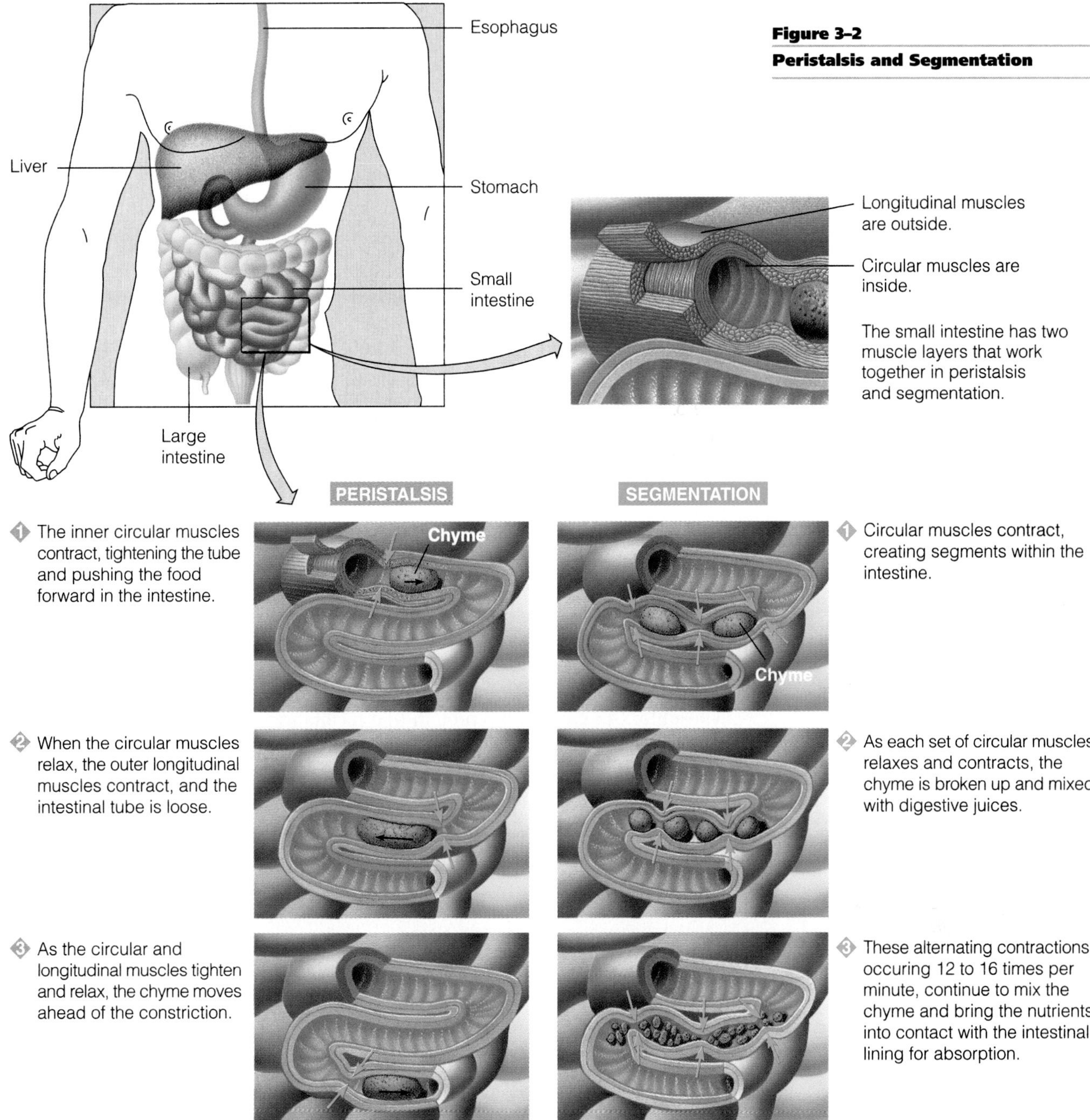

Esophagus
Liver
Stomach
Small intestine
Large intestine
Longitudinal muscles are outside.
Circular muscles are inside.
The small intestine has two muscle layers that work together in peristalsis and segmentation.
PERISTALSIS
SEGMENTATION
1 The inner circular muscles contract, tightening the tube and pushing the food forward in the intestine.
Chyme
Chyme
1 Circular muscles contract, creating segments within the intestine.
2 When the circular muscles relax, the outer longitudinal muscles contract, and the intestinal tube is loose.
2 As each set of circular muscles relaxes and contracts, the chyme is broken up and mixed with digestive juices.
3 As the circular and longitudinal muscles tighten and relax, the chyme moves ahead of the constriction.
3 These alternating contractions, occuring 12 to 16 times per minute, continue to mix the chyme and bring the nutrients into contact with the intestinal lining for absorption.


**Figure 3-2**
**Peristalsis and Segmentation**

sphincter, and remains in the stomach. Meanwhile, the stomach wall releases juices. When the chyme is completely liquefied, the pyloric sphincter opens briefly, about three times a minute, to allow small portions of chyme through. At this point, the chyme no longer resembles food in the least.

• ***Segmentation*** • The intestines periodically squeeze, as well as push, their contents. This motion, called segmentation, momentarily forces the intestinal contents back a few inches, mixing them and promoting close contact with the digestive juices and the absorbing cells of the intestinal walls before letting the contents move slowly along (see Figure 3-2).

**segmentation** (SEG-men-TAY-shun): a periodic squeezing or partitioning of the intestine at intervals along its length by its circular muscles.

• ***Sphincter Contractions*** • Four major sphincter muscles divide the GI tract into its principal divisions. At the bottom of the esophagus, the cardiac sphincter

**The Cardiac Sphincter**

When the circular muscles of a sphincter contract, the passage closes; when they relax, the passage opens.

Longitudinal muscle

Esophagus

Circular muscle

Esophagus muscles relax, opening the passageway.

Diaphragm muscles relax, opening the passageway.

Ligaments anchor the diaphragm to the esophagus.

Stomach

Esophagus muscles contract, squeezing on the inside.

Diaphragm muscles contract, squeezing on the outside.

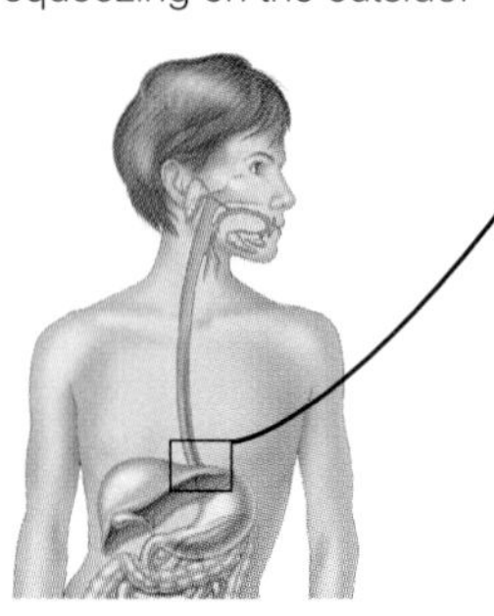

**catalyst** (CAT-uh-list): a compound that facilitates chemical reactions without itself being changed in the process.

prevents reflux of the stomach contents.[2] At the bottom of the stomach, the pyloric sphincter, which stays closed most of the time, holds the bolus in the stomach long enough so that it can be thoroughly mixed with gastric juice and liquefied. It also prevents the intestinal contents from backing up into the stomach. At the end of the small intestine, the ileocecal valve performs a similar function, emptying the contents of the small intestine into the large intestine. Finally, the tightness of the rectal muscle is a kind of safety device; together with the anus, it prevents elimination until you choose to perform it voluntarily (obstacle 7).

## *The Secretions of Digestion*

Have you ever wondered how people can eat differently yet have essentially the same body composition? It all comes down to the process of digestion, of rendering food—whatever kind of food it is to start with—into nutrients.

To break down food to small nutrients that the body can absorb, five different organs produce secretions: the salivary glands, the stomach, the pancreas, the liver (via the gallbladder), and the small intestine. (The accompanying glossary identifies some of the digestive glands and their secretions.) These secretions enter the GI tract at various points along the way, bringing an abundance of water and a variety of enzymes.

Enzymes are formally introduced in Chapter 6, but for now a simple definition will suffice. An enzyme is a protein that facilitates a chemical reaction—making a molecule from smaller parts, breaking a molecule into smaller parts, changing the arrangement of a molecule, or exchanging parts of molecules. As a catalyst, the enzyme itself remains unchanged. The enzymes involved in digestion facilitate a chemical reaction known as hydrolysis—the addition of water (hydro) to break

## Glossary of Digestive Glands and Their Secretions

These terms are listed in order from start to end of the digestive tract.

**gland:** a cell or group of cells that secretes materials for special uses in the body. Glands may be **exocrine** (EKS-oh-crin) **glands,** secreting their materials "out" (into the digestive tract or onto the surface of the skin), or **endocrine** (EN-doe-crin) **glands,** secreting their materials "in" (into the blood).

- **exo** = outside
- **endo** = inside
- **krine** = to separate

**salivary glands:** exocrine glands that secrete saliva into the mouth.

**saliva:** the secretion of the salivary glands; its principal enzyme begins carbohydrate digestion.

**gastric glands:** exocrine glands in the stomach wall that secrete gastric juice into the stomach.

- **gastro** = stomach

**gastric juice:** the digestive secretion of the gastric glands of the stomach.

**hydrochloric acid:** an acid composed of hydrogen and chloride atoms (HCl). The gastric glands normally produce this acid.

**mucus** (MYOO-kus): a slippery substance secreted by goblet cells of the GI lining (and other body linings) that protects the cells from exposure to digestive juices (and other destructive agents). The lining of the GI tract with its coat of mucus is a **mucous membrane.** (The noun is **mucus;** the adjective is **mucous.**)

**bile:** an emulsifier that prepares fats and oils for digestion; an exocrine secretion made by the liver, stored in the gallbladder, and released into the small intestine when needed.

**emulsifier** (ee-MUL-sih-fire): a substance with both water-soluble and fat-soluble portions that promotes the mixing of oils and fats in a watery solution.

**pancreatic** (pank-ree-AT-ic) **juice:** the exocrine secretion of the pancreas, containing enzymes for the digestion of carbohydrate, fat, and protein as well as bicarbonate, a neutralizing agent. The juice flows from the pancreas into the small intestine through the pancreatic duct. (The pancreas also has an endocrine function, the secretion of insulin and other hormones.)

**bicarbonate:** an alkaline secretion of the pancreas, part of the pancreatic juice. (Bicarbonate also occurs widely in all cell fluids.)

(lysis) a molecule into smaller pieces. The glossary below identifies some of the digestive enzymes and defines related terms.

• ***Saliva*** • The salivary glands squirt just enough saliva to moisten each mouthful of food so that it can pass easily down the esophagus (obstacle 3). The saliva contains water, salts, and enzymes that initiate the digestion of carbohydrates. In fact, you can taste the change if you hold a piece of starchy food like a cracker in your mouth for a few minutes without swallowing it—the cracker begins tasting sweeter as the enzyme acts on it. Saliva also protects the teeth and the linings of the mouth, esophagus, and stomach from attack by substances that might harm them.

• ***Gastric Juice*** • Cells in the stomach secrete gastric juice, a mixture of water, enzymes, and hydrochloric acid. The acid is so strong that it causes the sensation of heartburn if it happens to reflux into the esophagus. Highlight 3, following this chapter, discusses heartburn and other common digestive problems.

The strong acidity of the stomach prevents bacterial growth and kills most bacteria that enter the body with food. It would destroy the cells of the stomach as well, but for their natural defenses. To protect themselves from gastric juice, the goblet cells of the stomach wall secrete mucus, a thick, slippery, white substance that coats the cells, protecting them from the acid and enzymes that might otherwise harm them.

Figure 3-3 (on p. 70) shows how the strength of acids is measured—in pH units. Note that the acidity of gastric juice registers below "2" on the pH scale—stronger than vinegar. The stomach enzymes work most efficiently in the stomach's strong acid, but the salivary enzymes, which are swallowed with food, do not work in acid this strong. Consequently, the salivary digestion of carbohydrate gradually ceases as the stomach acid penetrates each newly swallowed bolus of food. In fact, salivary enzymes become just other proteins to be digested, and carbohydrate digestion waits to resume in the next digestive organ, the small intestine.

The major digestive event in the stomach is the partial breakdown (hydrolysis) of proteins. The acid helps to uncoil proteins, making them available for digestion. Both an enzyme and the stomach acid itself act as catalysts for this reaction. Minor events are the digestion of a very little fat by a gastric lipase, the breakdown of some carbohydrate by gastric acid, and the release of vitamin $B_{12}$ from dietary proteins and its attachment to a protein carrier.

• ***Pancreatic Juice and Intestinal Enzymes*** • By the time food leaves the stomach, digestion of all three energy nutrients has begun, and the action gains momentum in the small intestine. There the pancreas contributes digestive juices by way of ducts leading into the duodenum. The pancreatic juice contains

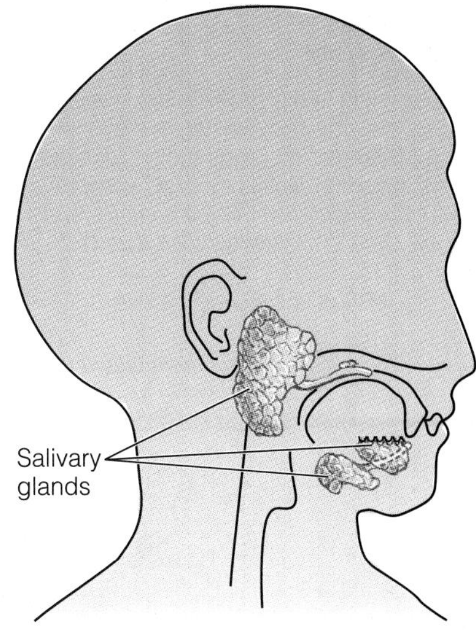


**The Salivary Glands**

The salivary glands secrete saliva into the mouth and begin the digestive process. Given the short time food is in the mouth, salivary enzymes contribute little to digestion.

**reflux:** a backward flow.
• **re** = back
• **flux** = flow

**goblet cells:** cells of the GI tract (and lungs) that secrete mucus.

**pH:** the unit of measure expressing a substance's acidity or alkalinity (Chapter 12 provides a more detailed definition).

## Glossary of Digestive Enzymes

All enzymes and some hormones are proteins, but an enzyme is not a hormone. Enzymes facilitate the making and breaking of bonds in chemical reactions; hormones act as chemical messengers, sometimes regulating enzyme action.

**digestive enzymes:** proteins found in digestive juices that act on food substances, causing them to break down into simpler compounds.

**-ase** (ACE): a word ending denoting an enzyme. Enzymes are often identified by the place they come from and the compounds they work on; *gastric lipase,* for example, is a stomach enzyme that acts on lipids, whereas *pancreatic lipase* comes from the pancreas (and also works on lipids).

**carbohydrase** (KAR-boe-HIGH-drase): an enzyme that hydrolyzes carbohydrates.

**hydrolysis** (high-DROL-ih-sis): a chemical reaction in which a major reactant is split into two products, with the addition of a hydrogen atom (H) to one and a hydroxyl group (OH) to the other (from water, $H_2O$). (The noun is **hydrolysis;** the verb is **hydrolyze.**)
• **hydro** = water
• **lysis** = breaking

**lipase** (LYE-pase): an enzyme that hydrolyzes lipids (fats).

**protease** (PRO-tee-ase): an enzyme that hydrolyzes proteins.

**Figure 3-3**

**The pH Scale**

A substance's acidity or alkalinity is measured in pH units. The pH is the negative logarithm of the hydrogen ion concentration. Each increment presents a tenfold increase in concentration of hydrogen particles. For example, a pH of 2 is 1000 times stronger than a pH of 5.

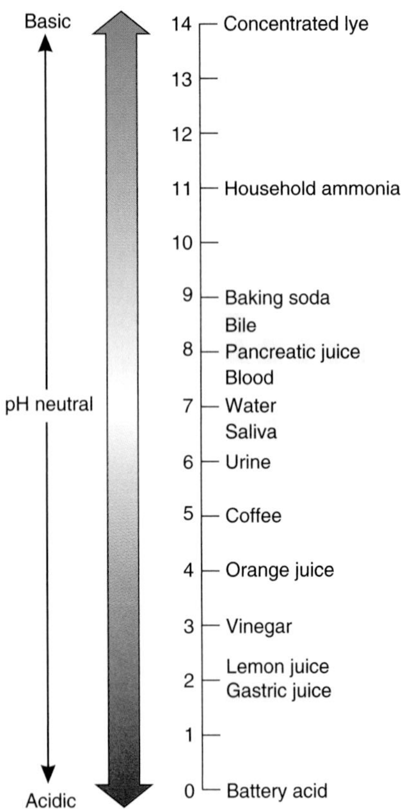


enzymes that act on all three energy nutrients, and the cells of the intestinal wall also possess digestive enzymes on their surfaces.

In addition to enzymes, the pancreatic juice contains sodium bicarbonate, which is basic or alkaline—the opposite of the stomach's acid (review Figure 3-3). The pancreatic juice thus neutralizes the acid chyme arriving in the small intestine from the stomach. From this point on, the chyme remains at a neutral or slightly alkaline pH. The enzymes of both the intestine and the pancreas work best in this environment.

• ***Bile*** • Bile also flows into the duodenum. The liver continuously produces bile, which is then concentrated and stored in the gallbladder. The gallbladder squirts the bile into the duodenum when fat arrives there. Bile is not an enzyme, but an emulsifier that brings fats into suspension in water so that enzymes can break them down into their component parts. Thanks to all these secretions, the three energy-yielding nutrients are digested in the small intestine (for a summary of digestive secretions and their actions, see the table below).

• ***Protective Factors*** • Both the small and the large intestine, being neutral in pH, permit the growth of bacteria (known as the intestinal flora). In fact, a healthy intestinal tract supports a thriving bacterial population that normally does the body no harm and may actually do some good. Bacteria in the GI tract produce a couple of vitamins, including a significant amount of vitamin K, although the amount is insufficient to meet the body's total need for that vitamin.[3]

Provided that the normal intestinal flora are thriving, infectious bacteria have a hard time getting established and launching an attack on the system. Diet is one of several factors that influence the bacterial population and its environment.[4] In addition, secretions from the GI tract—saliva, mucus, gastric acid, and digestive enzymes—not only help with digestion, but also defend against foreign invaders. The GI tract also maintains several different kinds of defending cells that confer specific immunity against intestinal diseases.[5]

## *The Final Stage*

The story of how digestion prepares food for absorption is now nearly complete. The three energy-yielding nutrients—carbohydrate, fat, and protein—have been disassembled and are ready to be absorbed. Most of the other nutrients—vitamins,

IN SUMMARY

**Summary of Digestive Secretions**

| Organ or Gland | Target Organ | Secretion | Action |
|---|---|---|---|
| Salivary glands | Mouth | Saliva | Fluid eases swallowing; salivary enzyme breaks down **carbohydrate.** |
| Gastric glands | Stomach | Gastric juice | Fluid mixes with bolus; hydrochloric acid uncoils **proteins;** enzymes break down proteins; mucus protects stomach cells. |
| Pancreas | Small intestine | Pancreatic juice | Bicarbonate neutralizes acidic gastric juices; pancreatic enzymes break down **carbohydrates, fats,** and **proteins.** |
| Liver | Gallbladder | Bile | Bile stored until needed. |
| Gallbladder | Small intestine | Bile | Bile emulsifies **fat** into small particles that enzymes can attack. |
| Intestinal glands | Small intestine | Intestinal juice | Intestinal enzymes break down **carbohydrate** and **protein** fragments; mucus protects the intestinal wall. |

minerals, and water—need no such disassembly; some vitamins and minerals are altered slightly during digestion, but most are absorbed as they are. Undigested residues, such as some fibers, are not absorbed, but continue through the digestive tract, providing a semisolid mass that helps exercise the muscles and keep them strong enough to perform peristalsis efficiently. Fiber also retains water, accounting for the stools' pasty consistency, and carries some bile acids, some minerals, and some additives and contaminants with it out of the body.

The process of absorbing the nutrients into the body presents its own problems, to be discussed in the next section. For the moment, assume that the digested nutrients simply are absorbed from the GI tract as soon as they are ready. Most are gone by the time the contents of the GI tract reach the end of the small intestine. Little remains but water, a few dissolved salts and body secretions, and undigested materials such as fiber. These enter the large intestine (colon).

In the colon, intestinal bacteria degrade some of the fiber to simpler compounds, while the colon itself retrieves all materials that the body can recycle—water and dissolved salts. The waste that is finally excreted has little or nothing of value left in it. The body has extracted all that it can use from the food. Figure 3-4 (on p. 72) summarizes digestion by following a sandwich through the GI tract and into the body.

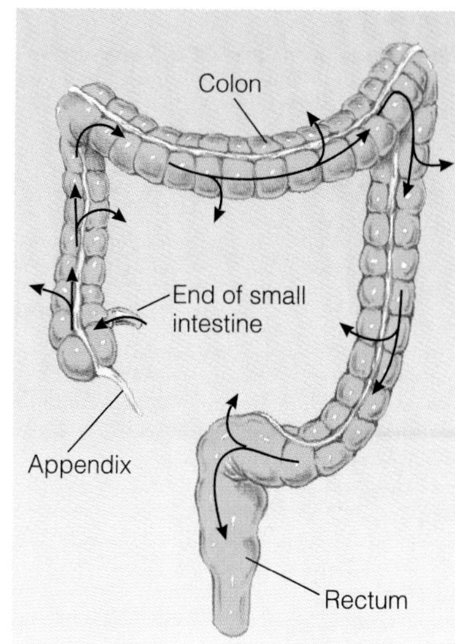


**The Colon**

The colon reabsorbs water and salts.

**stools:** waste matter discharged from the colon; also called **feces** (FEE-seez).

IN SUMMARY

As Figure 3-1 shows, food enters the mouth and travels down the esophagus and through the cardiac sphincter to the stomach, then through the pyloric sphincter to the small intestine, on through the ileocecal valve to the large intestine, past the appendix to the rectum, ending at the anus. The wavelike contractions of peristalsis and the periodic squeezing of segmentation keep things moving at a reasonable pace. Along the way, secretions from the salivary glands, stomach, pancreas, liver (via the gallbladder), and small intestine deliver fluids and digestive enzymes.

# Absorption

The problem of absorption: Given a meal that delivers millions of molecules of nutrients, provide a means by which all can enter the body simultaneously. Within three or four hours after you have eaten a dinner of beans and rice (or spinach lasagna, or steak and potatoes) with vegetable, salad, beverage, and dessert, your body must find a way to absorb—one by one—some two hundred thousand million, million, million molecules derived from carbohydrate digestion; a comparable number of molecules derived from protein and fat digestion; and many vitamin and mineral molecules as well.

**absorption:** the passage of nutrients from the GI tract into either the blood or the lymph.

The absorptive system is one of the most elegantly designed organ systems in the body. In 10 feet of small intestine, it provides a surface area equivalent to a quarter of a football field, which engulfs and absorbs the nutrient molecules. To remove the molecules rapidly and provide room for more to be absorbed, a rush of circulating blood continuously washes the underside of this surface, carrying the absorbed nutrients away to the liver and other parts of the body.

## *Anatomy of the Absorptive System*

The inner surface of the small intestine looks smooth and slippery, but viewed through a microscope, it turns out to be wrinkled into hundreds of folds. Each fold in turn is contoured into thousands of fingerlike projections, as numerous as the hairs on velvet fabric. These small intestinal projections are the villi. A single villus, magnified still more, turns out to be composed of hundreds of cells, each

**villi** (VILL-ee, VILL-eye): fingerlike projections from the folds of the small intestine; singular **villus.**

**Figure 3-4**

**The Digestive Fate of a Sandwich**

To review the digestive processes and enzymes, follow a peanut butter and banana sandwich on whole-wheat, sesame seed bread through the GI tract.

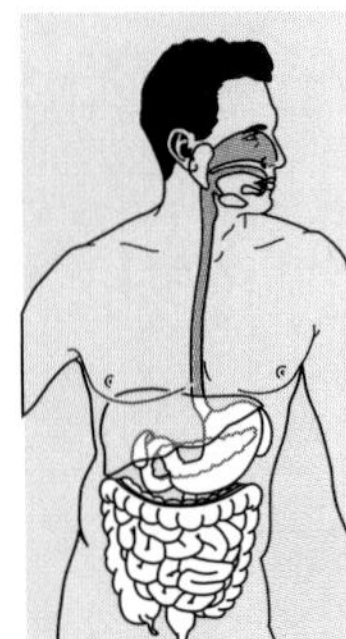

MOUTH: CHEWING AND SWALLOWING, WITH LITTLE DIGESTION

- **Carbohydrate** digestion begins as the salivary enzyme starts to break down the starch from bread and peanut butter.
- **Fiber** covering on the sesame seeds is crushed by the teeth, which exposes the nutrients inside the seeds to the upcoming digestive enzymes.

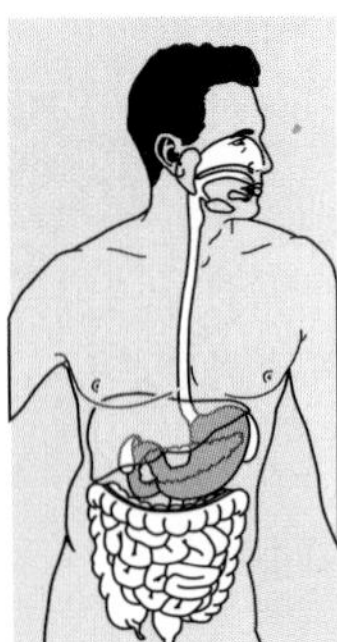

STOMACH: COLLECTING AND CHURNING, WITH SOME DIGESTION

- **Carbohydrate** digestion continues until the mashed sandwich has been mixed with the gastric juices; the stomach acid of the gastric juices inactivates the salivary enzyme.
- **Proteins** from the bread, seeds, and peanut butter begin to uncoil when they mix with the gastric acid, making them available to the gastric protease enzymes that begin to digest proteins.
- **Fat** from the peanut butter forms a separate layer on top of the watery mixture.

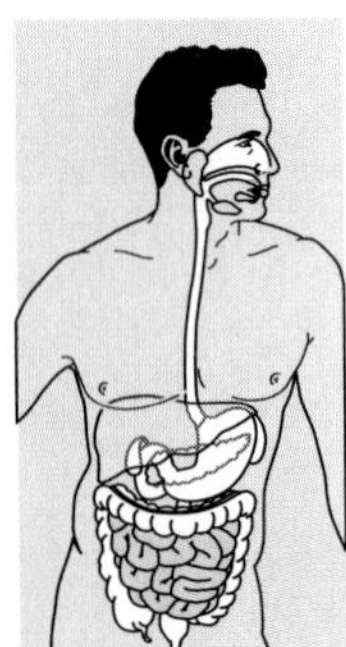

SMALL INTESTINE: DIGESTING AND ABSORBING

- **Sugars** from the banana require so little digestion that they begin to traverse the intestinal cells immediately on contact.
- **Starch** digestion picks up when the pancreas sends pancreatic enzymes to the small intestine via the pancreatic duct. Enzymes on the surfaces of the small intestinal cells complete the process of breaking down starch into small fragments that can be absorbed through the intestinal cell walls and into the blood.
- **Fat** from the peanut butter and seeds is emulsified with the watery digestive fluids by bile. Now the pancreatic and intestinal lipases can begin to break down the fat to smaller fragments that can be absorbed through the cells of the small intestinal wall and into the lymph.
- **Protein** digestion depends on the pancreatic and intestinal proteases. Small fragments of protein are liberated and absorbed through the cells of the small intestinal wall and into the blood.
- **Vitamins and minerals** are absorbed.

*Note:* Sugars and starches are members of the carbohydrate family.

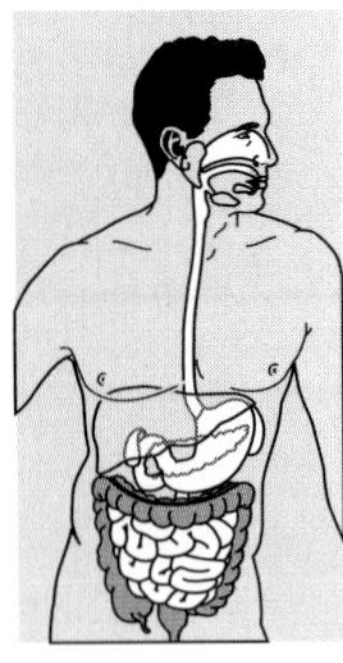

LARGE INTESTINE: REABSORBING AND ELIMINATING

- **Fluids and some minerals** are absorbed.
- **Some fibers** from the seeds, whole-wheat bread, peanut butter, and banana are partly digested by the bacteria living there, and some of these products are absorbed.
- **Most fibers** pass through the large intestine and are excreted as feces; some fat, cholesterol, and minerals bind to fiber and are also excreted.

covered with its own microscopic hairs, the microvilli (see Figure 3-5). In the crevices between the villi lie the crypts—tubular glands that secrete the intestinal juices into the small intestine.

The villi are in constant motion. Each villus is lined by a thin sheet of muscle, so it can wave, squirm, and wriggle like the tentacles of a sea anemone. Any nutrient molecule small enough to be absorbed is trapped among the microvilli that coat the cells and then drawn into the cells. Some partially digested nutrients are caught in the microvilli, digested further by enzymes there, and then absorbed

**microvilli** (MY-cro-VILL-ee, MY-cro-VILL-eye): tiny, hairlike projections on each cell of every villus that can trap nutrient particles and transport them into the cells; singular **microvillus.**

**crypts** (KRIPTS): tubular glands that lie between the intestinal villi and secrete intestinal juices into the small intestine.

**Figure 3-5**
**The Small Intestinal Villi**

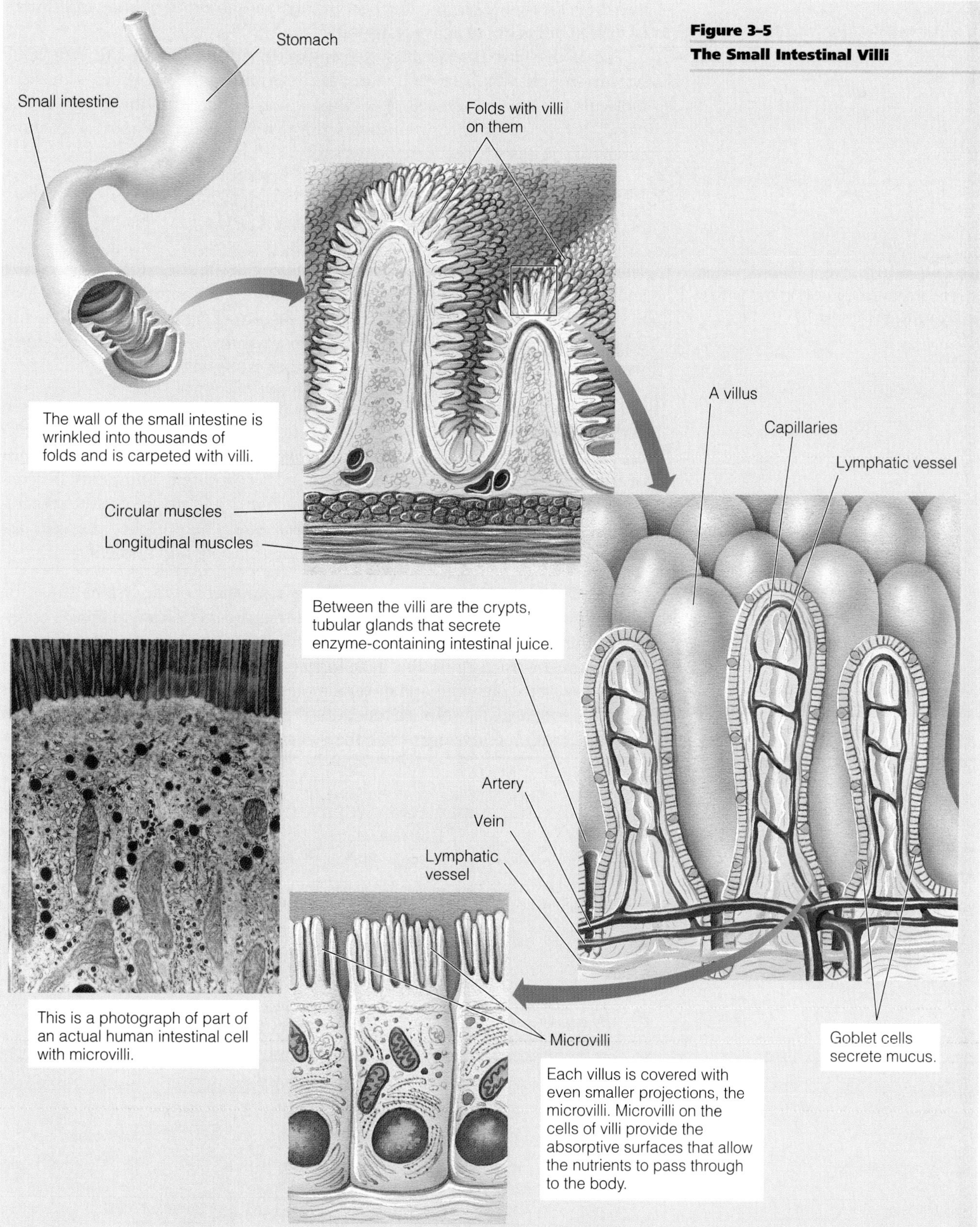
Stomach
Small intestine
Folds with villi on them
The wall of the small intestine is wrinkled into thousands of folds and is carpeted with villi.
Circular muscles
Longitudinal muscles
A villus
Capillaries
Lymphatic vessel
Between the villi are the crypts, tubular glands that secrete enzyme-containing intestinal juice.
Artery
Vein
Lymphatic vessel
This is a photograph of part of an actual human intestinal cell with microvilli.
Microvilli
Goblet cells secrete mucus.
Each villus is covered with even smaller projections, the microvilli. Microvilli on the cells of villi provide the absorptive surfaces that allow the nutrients to pass through to the body.

into the cells. Figure 3-6 describes how nutrients are absorbed by simple diffusion, facilitated diffusion, or active transport.

The body's two transport systems—the bloodstream and the lymphatic system—supply vessels to each villus, as shown in Figure 3-5. When a nutrient molecule has crossed the cell of a villus, it may enter either the lymph or the blood, but before following nutrients through the body, we must look more closely at the digestive cells themselves.

## A Closer Look at the Intestinal Cells

The problem of food contaminants, which may be absorbed defenselessly by the body, is the subject of Chapter 19.

The cells of the villi are among the most amazing in the body, for they recognize and select the nutrients the body needs and regulate their absorption. A close look at these cells is worthwhile, because it will help to explode a common misconception about nutrition: that you have to do anything to ensure that your digestive tract does its job. Nothing could be further from the truth.

• ***The Cells' Capabilities*** • As already described, each cell of a villus is coated with thousands of microvilli, which project from the cell's membrane (review Figure 3-5). In these microvilli and in the membrane lie hundreds of different kinds of enzymes and "pumps," which recognize and act on different nutrients. Descriptions of specific enzymes and "pumps" for each nutrient are presented in the following chapters where appropriate, but the point here is that the cells are equipped to handle all kinds and combinations of foods and nutrients.

• ***Specialization in the GI Tract*** • A further refinement of the system is that the cells of successive portions of the intestinal tract are specialized to absorb different nutrients. The nutrients that are ready for absorption early are absorbed near the top of the tract; those that take longer to be digested are absorbed farther down. Registered dietitians and medical professionals who treat digestive disorders learn the specialized absorptive functions of different parts of the GI tract so that if one part becomes dysfunctional, the diet can be adjusted accordingly.

**Figure 3-6**
**Absorption of Nutrients**

Absorption of nutrients into intestinal cells typically occurs by simple diffusion, facilitated diffusion, or active transport.

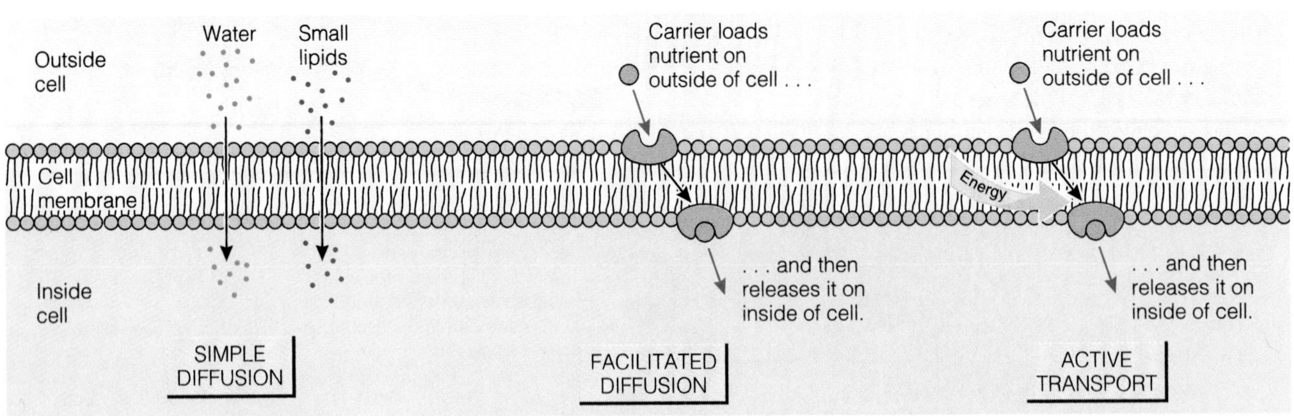


Some nutrients, such as water and small lipids, are absorbed by simple diffusion. They cross into intestinal cells freely.

Some nutrients, such as the water-soluble vitamins, are absorbed by facilitated diffusion. They need a specific carrier to transport them from one side of the cell membrane to the other. (Alternatively, facilitated diffusion may occur when the carrier changes the cell membrane in such a way that the nutrients can pass through.)

Some nutrients, such as glucose and amino acids, must be absorbed actively. These nutrients move against a concentration gradient, which requires energy.

• ***The Myth of "Food Combining"*** • The idea that people should not eat certain food combinations (for example, fruit and meat) at the same meal, because the digestive system cannot handle more than one task at a time, is a myth. The art of "food combining" (which actually emphasizes "food separating") is based on this idea, and it represents faulty logic and a gross underestimation of the body's capabilities. In fact, the contrary is often true; foods eaten together can enhance each other's use by the body. For example, vitamin C in a pineapple or other citrus fruit can enhance the absorption of iron from a meal of chicken and rice or other iron-containing foods. Many other instances of mutually beneficial interactions are presented in later chapters.

• ***Preparing Nutrients for Transport*** • Once inside the intestinal cells, the products of digestion must be released for transport to the rest of the body. The water-soluble nutrients and the smaller products of fat digestion are released directly into the bloodstream via the capillaries. The larger fats and the fat-soluble vitamins are insoluble in water, however, and blood is mostly water. The intestinal cells assemble many of the products of fat digestion into larger molecules. These larger molecules cluster together, and special proteins are inserted into their surfaces, forming chylomicrons. These chylomicrons cannot pass into the capillaries and are released into the lymphatic system instead; the chylomicrons move through the lymph and later enter the bloodstream at a point near the heart.

Chylomicrons (kye-lo-MY-cronz) are described in Chapter 5.

IN SUMMARY

The many folds and villi of the small intestine dramatically increase its surface area, facilitating nutrient absorption. Nutrients pass through the cells of the villi and enter either the blood (if they are water soluble or small fat fragments) or the lymph (if they are fat soluble).

# The Circulatory Systems

Once a nutrient has entered the bloodstream, it may be transported to any of the cells in the body, from the tips of the toes to the roots of the hair. The circulatory systems deliver nutrients wherever they are needed.

## *The Vascular System*

The vascular, or blood circulatory, system is a closed system of vessels through which blood flows continuously in a figure eight, with the heart serving as a pump at the crossover point (see Figure 3-7 on p. 76). As the blood circulates through this system, it picks up and delivers materials as needed.

All the body tissues derive oxygen and nutrients from the blood and deposit carbon dioxide and other wastes into it. The lungs exchange carbon dioxide (which leaves the blood to be exhaled) and oxygen (which enters the blood to be delivered to all cells). The digestive system supplies the nutrients to be picked up. In the kidneys, wastes other than carbon dioxide are filtered out of the blood to be excreted in the urine (see Figure 3-8 on p. 77).

Blood leaving the right side of the heart circulates by way of arteries into the lung capillaries and then back through veins to the left side of the heart. The left side of the heart then pumps the blood out through arteries to all systems of the body. The blood circulates in the capillaries, where it exchanges material with the cells, and then collects into veins, which return it again to the right side of the heart. In short, blood travels this simple route:

• Heart to arteries to capillaries to veins to heart.

The routing of the blood past the digestive system has a special feature. The blood is carried to the digestive system (as to all organs) by way of an artery, which

**arteries:** vessels that carry blood away from the heart.

**capillaries** (CAP-ill-aries): small vessels that branch from an artery. Capillaries connect arteries to veins. Exchange of oxygen, nutrients, and waste materials takes place across capillary walls.

**veins** (VANES): vessels that carry blood back to the heart.

**Figure 3-7**
**The Vascular System**

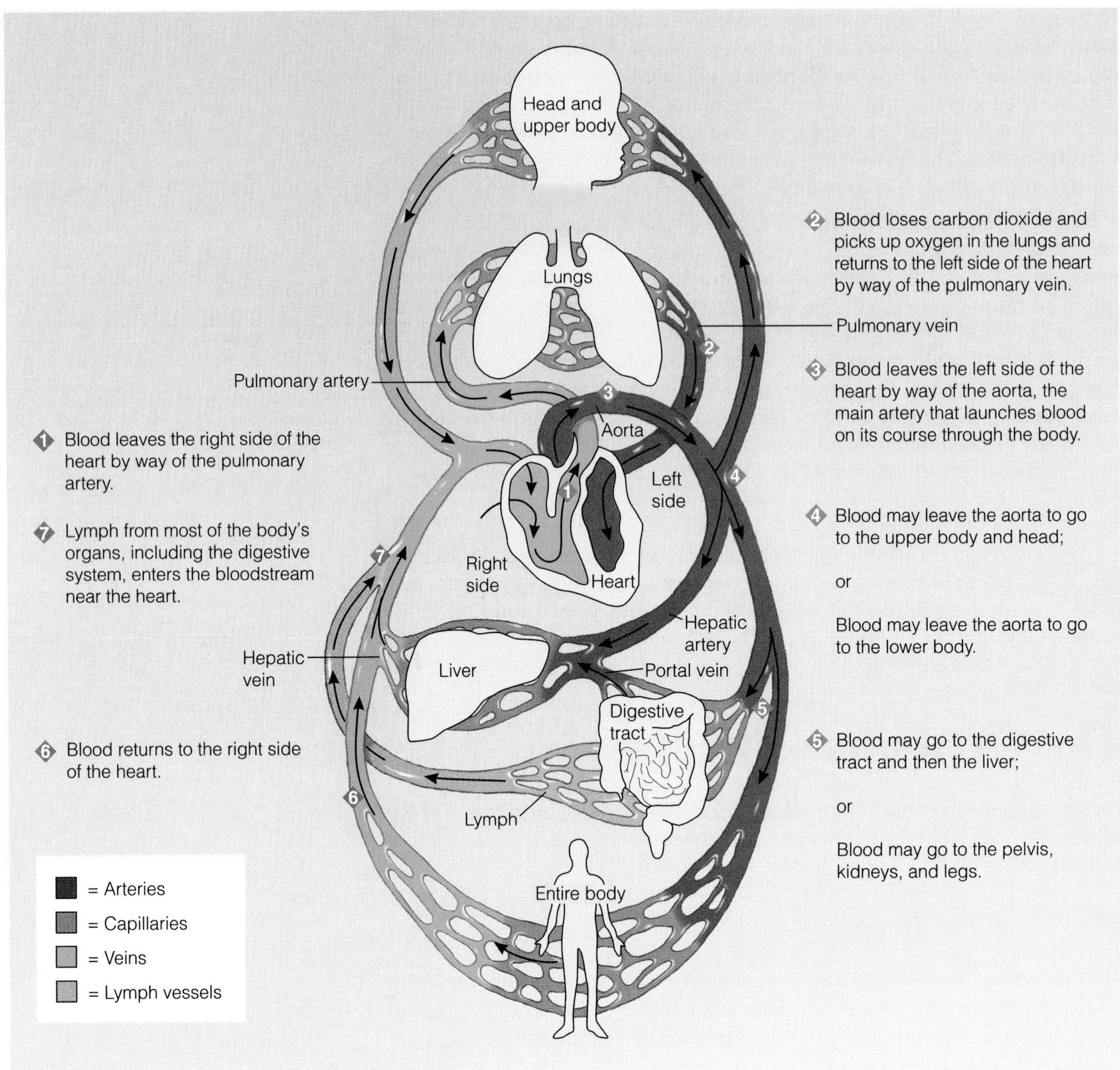
Head and upper body
Lungs
Pulmonary artery
Aorta
Left side
Right side
Heart
Hepatic artery
Portal vein
Hepatic vein
Liver
Digestive tract
Lymph
Entire body
1 Blood leaves the right side of the heart by way of the pulmonary artery.
2 Blood loses carbon dioxide and picks up oxygen in the lungs and returns to the left side of the heart by way of the pulmonary vein.
Pulmonary vein
3 Blood leaves the left side of the heart by way of the aorta, the main artery that launches blood on its course through the body.
4 Blood may leave the aorta to go to the upper body and head;
or
Blood may leave the aorta to go to the lower body.
5 Blood may go to the digestive tract and then the liver;
or
Blood may go to the pelvis, kidneys, and legs.
6 Blood returns to the right side of the heart.
7 Lymph from most of the body's organs, including the digestive system, enters the bloodstream near the heart.
= Arteries
= Capillaries
= Veins
= Lymph vessels

The vein that collects blood from the GI tract and conducts it to capillaries in the liver is the **portal vein.**
- **portal** = gateway

The vein that collects blood from the liver capillaries and returns it to the heart is the **hepatic vein.**
- **hepatic** = liver

(as in all organs) branches into capillaries to reach every cell. Blood leaving the digestive system, however, goes by way of a vein, not back to the heart, but to another organ—the liver. This vein *again* branches into *capillaries* so that every cell of the liver also has access to the blood carried by the vein. Blood leaving the liver then *again* collects into a vein, which returns to the heart.

The route is:

- Heart to arteries to capillaries (in intestines) to vein to capillaries (in liver) to vein to heart.

An anatomist studying this system knows there must be a reason for this special arrangement. The liver's placement ensures that it will be first to receive the

**Figure 3-8**

**A Nephron, One of the Kidney's Many Functioning Units**

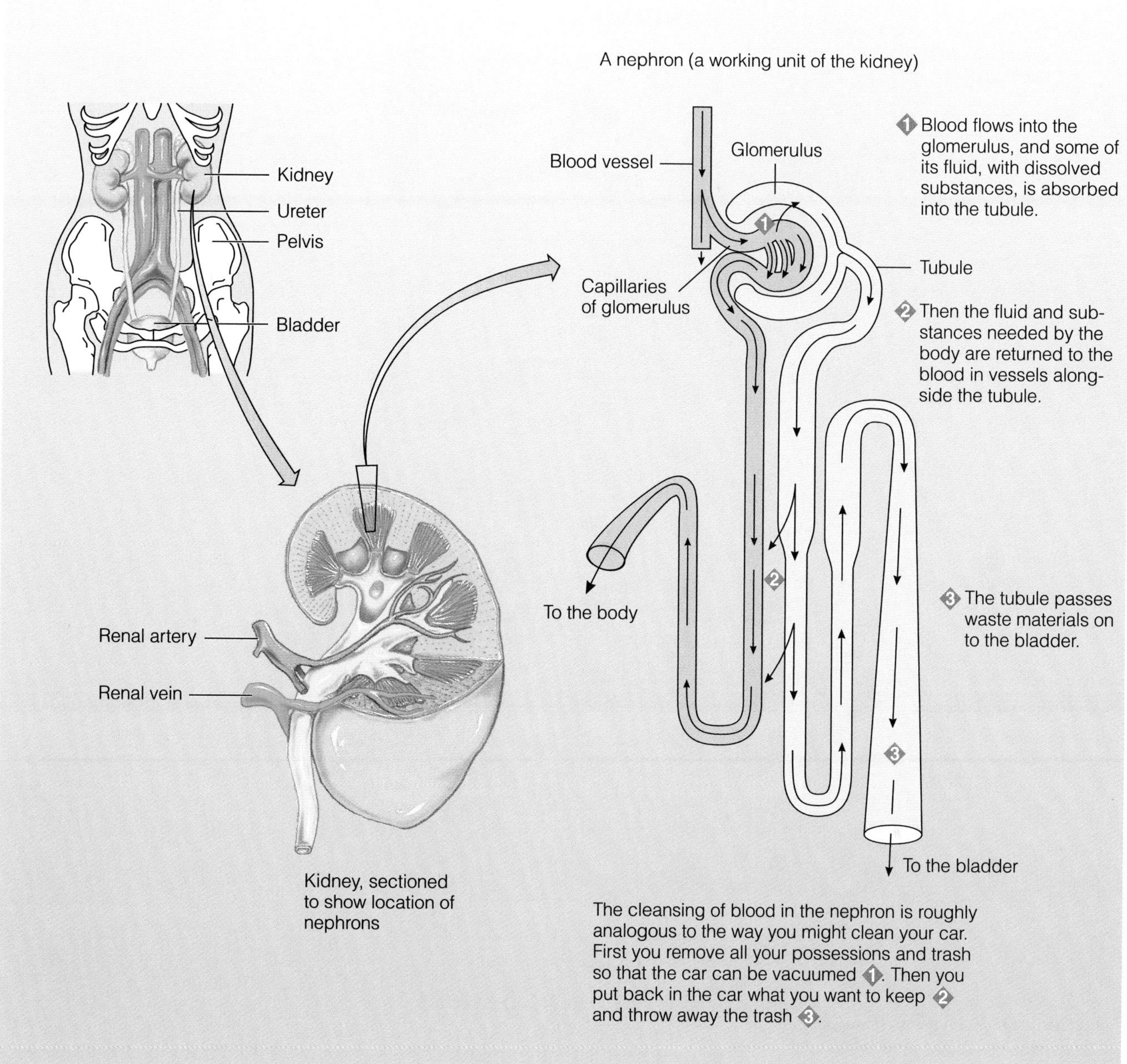
A nephron (a working unit of the kidney)
Kidney
Ureter
Pelvis
Bladder
Blood vessel
Glomerulus
Capillaries of glomerulus
Tubule
1 Blood flows into the glomerulus, and some of its fluid, with dissolved substances, is absorbed into the tubule.
2 Then the fluid and substances needed by the body are returned to the blood in vessels alongside the tubule.
3 The tubule passes waste materials on to the bladder.
To the body
To the bladder
Renal artery
Renal vein
Kidney, sectioned to show location of nephrons
The cleansing of blood in the nephron is roughly analogous to the way you might clean your car. First you remove all your possessions and trash so that the car can be vacuumed 1. Then you put back in the car what you want to keep 2 and throw away the trash 3.

materials absorbed from the GI tract. In fact, the liver has many jobs to do in preparing the absorbed nutrients for use by the body. It is the body's major metabolic organ.

You might guess that, in addition, the liver serves as a gatekeeper to defend against substances that might harm the heart or brain. This is why, when people ingest poisons that succeed in passing the first barrier (the intestinal cells), the liver quite often suffers the damage—from the hepatitis virus, from drugs such as barbiturates or alcohol, from poisons, and from contaminants such as mercury. Perhaps, in fact, you have been undervaluing your liver, not knowing what heroic tasks it quietly performs for you. Figure 3-9 shows the liver's key position in nutrient transport.

**Figure 3-9**
**The Liver**

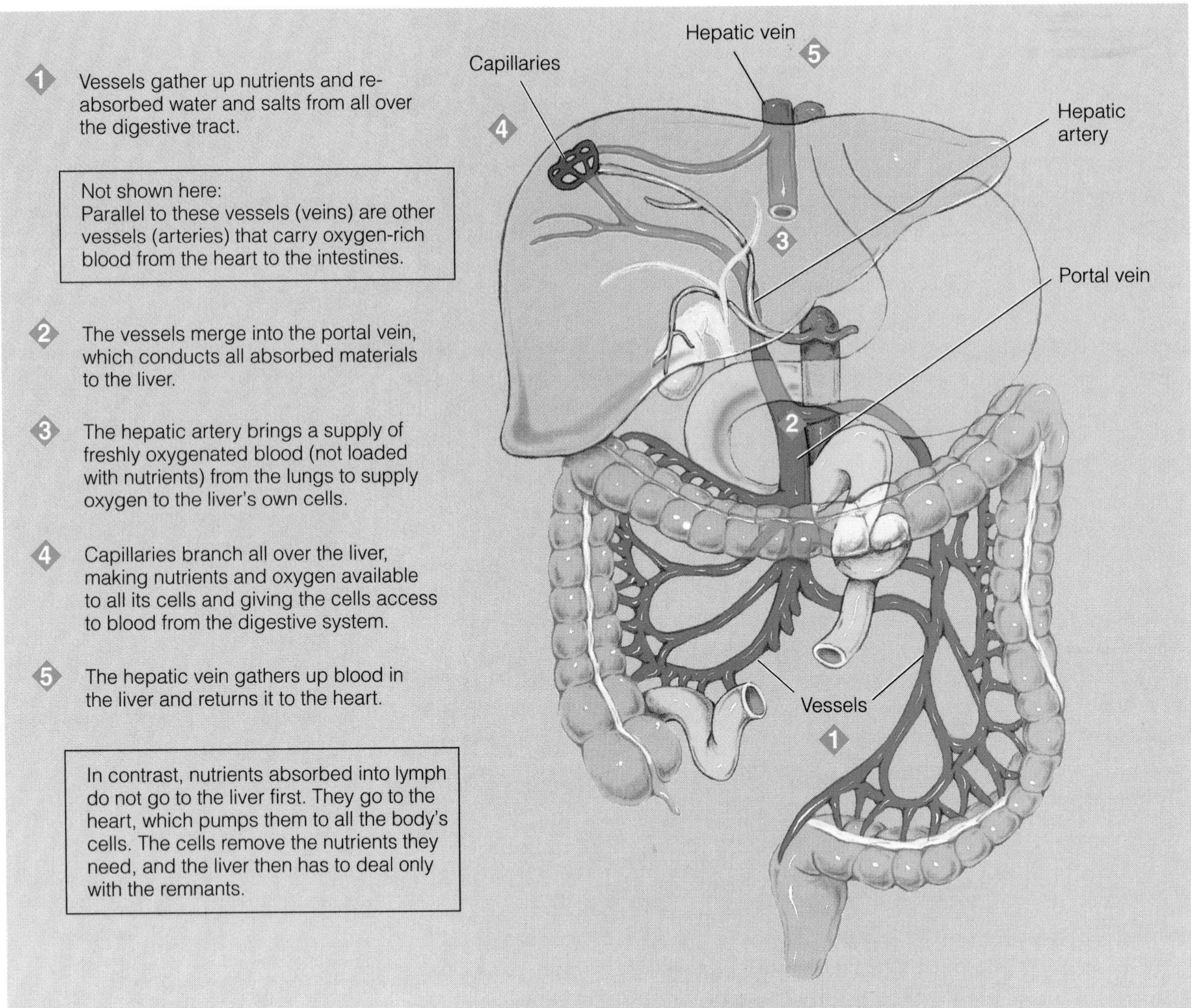
1 Vessels gather up nutrients and reabsorbed water and salts from all over the digestive tract.
Not shown here: Parallel to these vessels (veins) are other vessels (arteries) that carry oxygen-rich blood from the heart to the intestines.
2 The vessels merge into the portal vein, which conducts all absorbed materials to the liver.
3 The hepatic artery brings a supply of freshly oxygenated blood (not loaded with nutrients) from the lungs to supply oxygen to the liver's own cells.
4 Capillaries branch all over the liver, making nutrients and oxygen available to all its cells and giving the cells access to blood from the digestive system.
5 The hepatic vein gathers up blood in the liver and returns it to the heart.
In contrast, nutrients absorbed into lymph do not go to the liver first. They go to the heart, which pumps them to all the body's cells. The cells remove the nutrients they need, and the liver then has to deal only with the remnants.
Hepatic vein
5
Capillaries
4
Hepatic artery
3
Portal vein
2
Vessels
1

**lymphatic** (lim-FAT-ic) **system:** a loosely organized system of vessels and ducts that convey fluids toward the heart; the GI part of the lymphatic system carries the products of digestion into the bloodstream.

**lymph** (limf): a clear yellowish fluid that resembles blood without the red blood cells; lymph from the GI tract transports fat and fat-soluble vitamins to the bloodstream via lymphatic vessels.

The duct that conveys lymph toward the heart is the **thoracic** (thor-ASS-ic) **duct.** The **subclavian** (sub-KLAY-vee-an) **vein** connects this duct with the right upper chamber of the heart, providing a passageway by which lymph can be returned to the vascular system.

## *The Lymphatic System*

The lymphatic system provides a one-way route for fluid from the tissue spaces to enter the blood. Lymph fluid circulates between the cells of the body and collects into tiny vessels. Lymph is almost identical to blood except that it contains no red blood cells or platelets, because they cannot escape through the blood vessel walls.

The lymphatic system has no pump; instead, lymph is squeezed from one portion of the body to another like water in a sponge, as muscles contract and create pressure here and there. Ultimately, much of the lymph collects in a large duct behind the heart. This duct terminates in a vein that conducts the lymph toward the heart. Thus materials from the GI tract that enter lymphatic vessels (large fats and fat-soluble vitamins) ultimately enter the blood circulatory system, circulating through arteries, capillaries, and veins like the other nutrients, with a notable exception—they bypass the liver at first.

Once inside the vascular system, the nutrients can travel freely to any destination and can be taken into cells and used as needed. What becomes of them is described in later chapters.

IN SUMMARY

Nutrients leaving the digestive system via the blood are routed directly to the liver before being transported to the body's cells. Those leaving via the lymphatic system eventually enter the vascular system, but bypass the liver at first.

# Regulation of Digestion and Absorption

There is nothing random about digestion and absorption; they are coordinated in every detail. The ability of the digestive tract to handle its ever changing contents routinely illustrates an important physiological principle that governs the way all living things function—the principle of homeostasis. Simply stated, conditions have to stay about the same for an organism to survive; if they deviate too far from the norm, the organism must "do something" to bring them back to normal. The body's regulation of digestion is one example of homeostatic regulation. The body also regulates its temperature, its blood pressure, and all other aspects of its blood chemistry in similar ways.

The following paragraphs describe the regulation of digestion and absorption in healthy adults, but many factors can influence normal GI function. For example, peristalsis and sphincter action are poorly coordinated in newborns, so infants tend to "spit up" during the first several months of life. Older adults often experience constipation, in part because the intestinal wall loses strength and elasticity with age, which slows GI motility. Diseases can also interfere with digestion and absorption and often lead to malnutrition. Lack of nourishment, in general, and lack of certain dietary constituents such as fiber, in particular, alter the structure and function of GI cells. Quite simply, GI tract health depends on food.

**homeostasis** (HOME-ee-oh-STAY-sis): the maintenance of constant internal conditions (such as blood chemistry, temperature, and blood pressure) by the body's control systems. A homeostatic system is constantly reacting to external forces so as to maintain limits set by the body's needs.

- **homeo** = the same
- **stasis** = staying

Factors influencing GI function:

- Physical immaturity.
- Aging.
- Illness.
- Nutrition.

## *Gastrointestinal Hormones and Nerve Pathways*

Two intricate and sensitive systems coordinate all the digestive and absorptive processes: the hormonal (or endocrine) system and the nervous system. Even before the first bite of food is taken, the mere thought, sight, or smell of food can trigger a response from these systems.[6] Then, as food travels through the GI tract, it either stimulates or inhibits digestive secretions by way of messages that are carried from one section of the GI tract to another by both hormones and nerve pathways.

Notice that the kinds of regulation that will be described are all examples of *feedback* mechanisms. A certain condition demands a response. The response changes that condition, and the change then cuts off the response. Thus the system is self-corrective. Examples follow:

- *The stomach normally maintains a pH between 1.5 and 1.7. How does it stay that way?* Food entering the stomach stimulates cells in the stomach wall to release the hormone gastrin. Gastrin, in turn, stimulates the stomach glands to secrete the components of hydrochloric acid. When pH 1.5 is reached, the acid itself turns off the gastrin-producing cells. They stop releasing gastrin, and the glands stop producing hydrochloric acid. Thus the system adjusts itself.

  Nerve receptors in the stomach wall also respond to the presence of food and stimulate both the gastric glands to secrete juices and the muscles to contract. As the stomach empties, the receptors are no longer stimulated, the flow of juices slows, and the stomach quiets down.

Appendix A presents a brief summary of the body's hormonal system and nervous system.

**hormones:** chemical messengers. Hormones are secreted by a variety of glands in response to altered conditions in the body. Each hormone travels to one or more specific target tissues or organs, where it elicits a specific response to maintain homeostasis. In general, a gastrointestinal hormone is called an **enterogastrone** (EN-ter-oh-GAS-trone).*

*The term *enterogasterone* refers specifically to any hormone that inhibits gastric secretions, including secretin, cholecystokinin (CCK), and gastric-inhibitory peptide; more broadly, the term refers to any hormone released from the intestine.

**gastrin:** a hormone secreted by cells in the stomach wall. Target organ: the stomach. Response: secretion of gastric juice.

*To become part of your body, food must first be digested and absorbed. (Courtesy of CNN)*

**secretin** (see-CREET-in): a hormone produced by cells in the duodenum wall. Target organ: the pancreas. Response: secretion of bicarbonate-rich pancreatic juice.

**cholecystokinin** (coal-ee-sis-toe-KINE-in), or **CCK:** a hormone produced by cells of the intestinal wall. Target organ: the gallbladder. Response: release of bile and slowing of GI motility.

**gastric-inhibitory peptide:** a hormone produced by the intestine. Target organ: the stomach. Response: slowing of the secretion of gastric juices and of GI motility.

- *The pyloric sphincter opens to let out a little chyme, then closes again. How does it know when to open and close?* When the pyloric sphincter relaxes, acidic chyme slips through. The cells of the pyloric muscle on the intestinal side sense the acid, causing the pyloric sphincter to close tightly. Only after the chyme has been neutralized by pancreatic bicarbonate and the juices surrounding the pyloric sphincter have become alkaline can the muscle relax again. This process ensures that the chyme will be released slowly enough to be neutralized as it flows through the small intestine. This is important, because the small intestine has less of a mucous coating than the stomach does and so is not as well protected from acid.
- *As the chyme enters the intestine, the pancreas adds bicarbonate to it so that the intestinal contents always remain at a slightly alkaline pH. How does the pancreas know how much to add?* The presence of chyme stimulates the cells of the duodenum wall to release the hormone secretin into the blood. When secretin reaches the pancreas, it stimulates the pancreas to release its bicarbonate-rich juices. Thus, whenever the duodenum signals that acidic chyme is present, the pancreas responds by sending bicarbonate to neutralize it. When the need has been met, the secretin cells of the duodenal wall are no longer stimulated to release the hormone, the hormone no longer flows through the blood, the pancreas no longer receives the message, and it stops sending pancreatic juice. Nerves also regulate pancreatic secretions.
- *Pancreatic secretions contain a mixture of enzymes to digest carbohydrate, fat, and protein. How does the pancreas know how much of each type of enzyme to provide?* This is one of the most interesting questions physiologists have asked. The question awaits final answer, but clearly the pancreas does know, somehow, what its owner has been eating, and it secretes enzyme mixtures tailored to deal with the food mixtures that have been arriving lately (over the last several days). Enzyme activity changes proportionately in response to the amounts of carbohydrate, fat, and protein in the diet.[7] If a person has been eating mostly carbohydrates, the pancreas makes and secretes mostly carbohydrases; if the person's diet has been high in fat, the pancreas produces more lipases; and so forth. Presumably, hormones from the GI tract, secreted in response to meals, keep the pancreas informed as to its digestive tasks. The day or two lag between the time a person's diet changes and the time digestion of the new diet becomes efficient explains why dietary changes can "upset digestion" and should be made gradually.
- *When fat is present in the intestine, the gallbladder contracts to squirt bile into the intestine to emulsify the fat. How does the gallbladder get the message that fat is present?* Fat in the intestine stimulates cells of the intestinal wall to release the hormone cholecystokinin (CCK). This hormone, traveling by way of the blood to the gallbladder, stimulates it to contract, releasing bile into the small intestine. Once the fat in the intestine is emulsified and enzymes have begun to work on it, the fat no longer provokes release of the hormone, and the message to contract is canceled.
- *Fat takes longer to digest than carbohydrate does. When fat is present, intestinal motility slows to allow time for its digestion. How does the intestine know when to slow down?* Cholecystokinin and gastric-inhibitory peptide slow GI tract motility. By slowing the digestive process, fat helps to maintain a pace that will allow all reactions to reach completion. Gastric-inhibitory peptide also inhibits gastric acid secretion. Hormonal and nervous mechanisms like these account for much of the body's ability to adapt to changing conditions.

Once a person has started to learn the answers to questions like these, it may be hard to stop. Some people devote their whole lives to the study of physiology. For now, however, these few examples will be enough to illustrate how all the processes throughout the digestive system are precisely and automatically regulated without any conscious effort.

IN SUMMARY

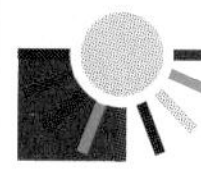

Digestion and absorption depend on the coordinated efforts of the hormonal system and the nervous system. Together, they regulate the processes of transforming foods into nutrients.

## *The System at Its Best*

This chapter has described the anatomy of the digestive tract on several levels: the sequence of digestive organs, the cells and structures of the villi, and the selective machinery of the cell membranes. The intricate architecture of the digestive system makes it sensitive and responsive to conditions in its environment. Knowing the optimal conditions will help you to promote the best functioning of the system.

One indispensable condition is good health of the digestive tract itself. This health is affected by such lifestyle factors as sleep, physical activity, and state of mind. Adequate sleep allows for repair, maintenance of tissue, and removal of wastes that might impair efficient functioning. Activity promotes healthy muscle tone. Mental state influences the activity of regulatory nerves and hormones; for healthy digestion, you should be relaxed and tranquil at mealtimes.

Another factor is the kind of meals you eat. Among the characteristics of meals that promote optimal absorption of nutrients are those mentioned in Chapter 2: balance, moderation, variety, and adequacy. Balance and moderation require having neither too much nor too little of anything. For example, too much fat is harmful, but some fat is needed to slow down intestinal motility, providing time for absorption of some of the nutrients that are slow to be absorbed.

Variety is important for many reasons, but partly because some food constituents interfere with nutrient absorption. For example, some compounds common in high-fiber foods such as whole-grain cereals, certain leafy green vegetables, and legumes bind with minerals. To some extent, then, the minerals in those foods may become "unavailable." Not that these high-fiber foods are undesirable, but people who use cereals, leafy greens, and legumes to the exclusion of other foods may be obtaining fewer minerals from their diets than they would if their choices were more varied. They might want to exercise moderation in their use of these high-fiber foods.

As for adequacy—in a sense, this entire book is about dietary adequacy. But here, at the end of this chapter, is a good place to underline the interdependence of the nutrients. It could almost be said that every nutrient depends on every other. All the nutrients work together and are all present in the cells of a healthy digestive tract. To maintain health and promote the functions of the GI tract, you should make balance, moderation, variety, and adequacy features of every day's menus.

## Study Questions

These questions will help you review the chapter. You will find the answers in the discussion on the pages provided.

1. Describe the obstacles associated with digesting food and the solutions offered by the human body. (pp. 63–64, 66, 68)
2. Describe the path food follows as it travels through the digestive system. Summarize the muscular actions that take place along the way. (pp. 65, 66–68)
3. Name five organs that secrete digestive juices. How do the juices and enzymes facilitate digestion? (pp. 68–70)
4. Describe the problems associated with absorbing nutrients and the solutions offered by the small intestine. (pp. 71–75)
5. How is blood routed through the digestive system? Which nutrients enter the bloodstream directly? Which are first absorbed into the lymph? (pp. 75–76)

6. Describe how the body coordinates and regulates the processes of digestion and absorption. (p. 79)
7. How does the composition of the diet influence the functioning of the GI tract? (pp. 79, 80)
8. What steps can you take to help your GI tract function at its best? (p. 81)

These questions will help you prepare for an exam. Answers can be found in Appendix K.

1. The semiliquid, partially digested food that travels through the intestinal tract is called:
   a. bile.
   b. lymph.
   c. chyme.
   d. secretin.
2. The muscular contractions that move food through the GI tract are called:
   a. hydrolysis.
   b. sphincters.
   c. peristalsis.
   d. bowel movements.
3. The main function of bile is to:
   a. emulsify fats.
   b. catalyze hydrolysis.
   c. slow protein digestion.
   d. neutralize stomach acidity.
4. All blood leaving the GI tract travels first to the:
   a. heart.
   b. liver.
   c. kidneys.
   d. pancreas.
5. Which nutrients leave the GI tract by way of the lymphatic system?
   a. water and minerals
   b. proteins and minerals
   c. all vitamins and minerals
   d. fats and fat-soluble vitamins
6. Digestion and absorption are coordinated by the:
   a. pancreas and kidneys.
   b. liver and gallbladder.
   c. hormonal system and the nervous system.
   d. vascular system and the lymphatic system.

## Notes


1. The length of the small intestine in living adults is almost 2½ times shorter than at death, when muscles are relaxed and elongated, in *Review of Medical Physiology*, ed. W. F. Ganong (Norwalk, Conn.: Appleton & Lange, 1993), pp. 438–465; E. A. Shaffer, Digestive system, physiology, and biochemistry, in *Encyclopedia of Human Biology* (San Diego: Academic Press, 1991), p. 76.
2. R. K. Mittal and D. H. Balaban, The esophagogastric junction, *New England Journal of Medicine* 336 (1997): 924–932.
3. Committee on Dietary Allowances, *Recommended Dietary Allowances*, 10th ed. (Washington, D.C.: National Academy Press, 1989), pp. 108–109.
4. M. B. Roberfroid and coauthors, Colonic microflora: Nutrition and health, *Nutrition Reviews* 53 (1995): 127–130; D. Kelly, R. Begbie, and T. P. King, Nutritional influences on interactions between bacteria and the small intestinal mucosa, *Nutritional Research Reviews* 7 (1994): 233–257.
5. C. Galperin and E. Gershwin, Immunopathogenesis of gastrointestinal and hepatobiliary diseases, *Journal of the American Medical Association* 278 (1997): 1946–1955.
6. R. D. Mattes, Physiological responses to sensory stimulation by food: Nutritional implications, *Journal of the American Dietetic Association* 97 (1997): 406–410, 413.
7. M. Armand and coauthors, Dietary fat modulates gastric lipase activity in healthy humans, *American Journal of Clinical Nutrition* 62 (1995): 74–80; P. M. Brannon, Adaptation of the exocrine pancreas to diet, *Annual Review of Nutrition* 10 (1990): 85–105.

HIGHLIGHT 3

# Common Digestive Problems

The facts of anatomy and physiology presented in Chapter 3 permit easy understanding of some common problems that occasionally arise in the digestive tract. Food may slip into the air passages instead of the esophagus, causing choking. Bowel movements may be loose and watery, as in diarrhea, or painful and hard, as in constipation. Some people complain about belching, while others are bothered by intestinal gas. Sometimes people develop medical problems such as an ulcer. This highlight describes some of the symptoms and strategies for preventing these common digestive problems (the glossary on p. 84 defines these terms).

 NIDDK Health Information

www.niddk.nih.gov/health/health.htm
Visit Digestive Diseases

 American College of Gastroenterology

www.acg.gi.org
Visit Digest This!

## Choking on Food

When someone chokes on food, it is because the food has slipped into the air passage and cut off breathing (see Figure H3-1 on p. 84). Food can lodge so securely in the trachea that it cuts off all air. No sound can be made, because the larynx is in the trachea and makes sounds only when air is pushed across it.

The choking scenario might read like this. A person is dining in a restaurant with friends. A chunk of food, usually meat, becomes lodged in his trachea so firmly that he cannot make a sound. Often he chooses to suffer alone rather than "make a scene in public." If he tries to communicate distress to his friends, he must depend on pantomime. The friends are bewildered by his antics and become terribly worried when the victim "faints" after a few minutes without air. They call for an ambulance. By the time the victim arrives at the hospital, however, he is dead from suffocation.

To help a person who is choking, first ask this critical question: "Can you make any sound at all?" If the victim makes a sound, relax. You have time to continue your questioning to see what you can do to help; you are not going to have to make a quick decision. But whatever you do, don't hit him on the back—the particle may become lodged more firmly in his air passage. If the victim cannot make a sound, follow the procedures described in Figure H3-2 on p. 85. You would do well to take a lifesaving course and practice these techniques, for you will have no time for hesitation once you are called on to perform this death-defying act.

Almost any food can cause choking, although some are cited more often than others: tough meats, hot dogs, nuts, grapes, carrots, hard candies, popcorn, and peanut butter. These foods are particularly difficult for young children to safely chew and swallow. Each year, more than 300 children in the United States choke to death. Always remain alert to the dangers of choking whenever young children are eating. To prevent choking, cut food into small pieces, chew thoroughly before swallowing, don't talk or laugh with food in your mouth, and don't eat when breathing hard.

## Vomiting

Another common digestive mishap is vomiting. Vomiting can be a symptom of many different diseases or may arise in situations that upsets the body's equilibrium, such as air or sea travel. For whatever reason, the waves of peristalsis reverse direction, and the contents of the stomach are propelled up through the esophagus to the mouth and expelled.

If vomiting continues long enough or is severe enough, the reverse peristalsis will extend beyond the stomach and carry the contents of the duodenum, with its green bile salts, into the stomach and then up the esophagus. Although certainly unpleasant and wearying for the nauseated person, vomiting such as this is no cause for alarm. Vomiting is one of the body's adaptive mechanisms to rid itself of something irritating. The best advice is to rest and drink small amounts of fluids as tolerated until the nausea subsides.

A physician's care may be needed, however, when large quantities of fluid are lost from the GI tract, causing dehydration. With massive fluid loss from the GI tract, all of the body's other fluids redistribute themselves so that, eventually, fluid is taken from every cell of the body. Leaving the cells with the fluid are salts that are absolutely essential to the life of the cells, and they must be replaced, which is difficult while the vomiting continues. Intravenous feedings of saline and glucose are frequently necessary while the physician is diagnosing the cause of the vomiting and instituting corrective therapy.

In an infant, vomiting is likely to become serious early in its course, and a physician should be contacted soon after onset. Infants have more fluid

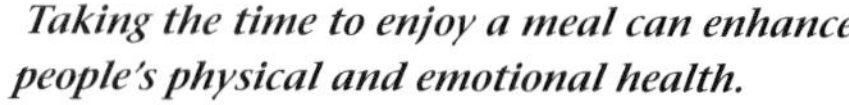
*Taking the time to enjoy a meal can enhance people's physical and emotional health.*

between their body cells than adults do, so more fluid can move readily into the digestive tract and be lost from the body. Consequently, the body water of infants becomes depleted and their body salt balance upset faster than in adults.

Self-induced vomiting, such as occurs in bulimia, also has serious consequences. In addition to fluid and salt imbalances, repeated vomiting can cause irritation and infection of the pharynx, esophagus, and salivary glands; erosion of the teeth; and dental caries. The esophagus may rupture or tear, as may the stomach. Sometimes the eyes become red from pressure during vomiting. Bulimic behavior reflects underlying problems that require intervention. (Bulimia nervosa is the subject of Highlight 9.)

**Figure H3-1**
**Normal Swallowing and Choking**

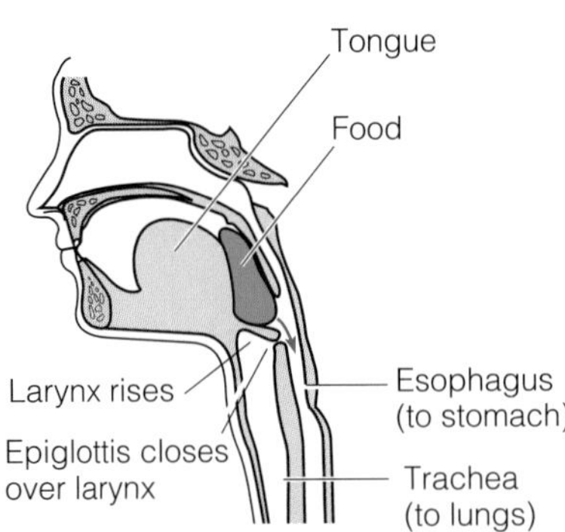


Swallowing. The epiglottis closes over the larynx, blocking entrance to the lungs via the trachea. The red arrow shows that food is heading down the esophagus normally.

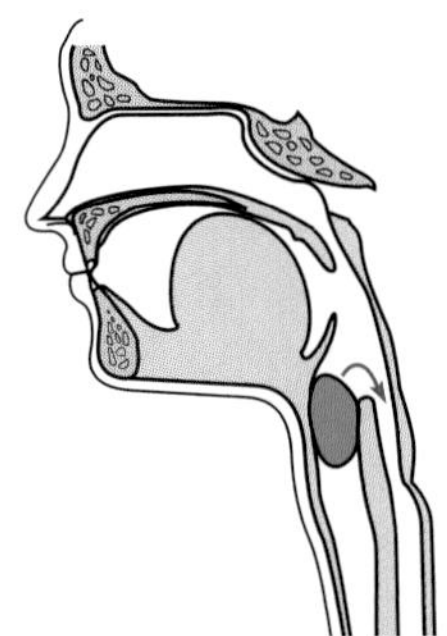

Choking. A choking person cannot speak or gasp because food lodged in the trachea blocks the passage of air. The red arrow points to where the food should have gone to prevent choking.

## Glossary

**acid controllers:** drugs used to prevent or relieve indigestion by suppressing production of acid in the stomach; also called **H2 blockers.** Common brands include Pepcid AC, Tagamet HB, Zantac 75, and Axid AR.

**antacids:** drugs used to relieve indigestion by neutralizing acid in the stomach. Common brands include Alka-Seltzer, Maalox, Rolaids, and Tums.

**belch:** the expulsion of gas from the stomach through the mouth.

**colitis** (ko-LYE-tis): inflammation of the colon.

**colonic irrigation:** the popular, but potentially harmful practice of "washing" the large intestine with a powerful enema machine.

**constipation:** the condition of having infrequent or difficult bowel movements.

**defecate** (DEF-uh-cate): to move the bowels and eliminate waste.
- **defaecare** = to remove dregs

**diarrhea:** the frequent passage of watery bowel movements.

**enema:** the insertion of solutions into the rectum and colon to stimulate a bowel movement and empty the lower large intestine.

**heartburn:** a burning sensation in the chest area caused by backflow of stomach acid into the esophagus.

**Heimlich maneuver:** a technique for removing an object from the trachea of a choking person (see Figure H3-2).

**hemorrhoids** (HEM-oh-royds): painful swelling of the veins surrounding the rectum.

**hiccups** (HICK-ups): repeated cough-like sounds and jerks that are produced when an involuntary spasm of the diaphragm muscle sucks air down the windpipe; also spelled *hiccoughs.*

**indigestion:** incomplete or uncomfortable digestion, usually accompanied by pain, nausea, vomiting, heartburn, intestinal gas, or belching.
- **in** = not

**irritable bowel syndrome:** an intestinal disorder of unknown cause; symptoms include abdominal discomfort and cramping, diarrhea, constipation, or alternating diarrhea and constipation.

**larynx:** the voice box (see Figure H3-1).

**peptic ulcer:** an erosion in the mucous membrane of either the stomach (a gastric ulcer) or the duodenum (a duodenal ulcer).

**ulcer:** an erosion in the topmost, and sometimes underlying, layers of cells in an area. See also *peptic ulcer.*

**vomiting:** expulsion of the contents of the stomach up through the esophagus to the mouth.

U.S. Government
www.healthfinder.gov/searchoptions/topicsaz.htm
Search for Choking

U.S. Government
www.healthfinder.gov/searchoptions/topicsaz.htm
Search for Vomiting

**Figure H3-2**
**First Aid for Choking**

- If the choking victim is an infant, lay her face down on your lap with her head firmly supported and held lower than her body.

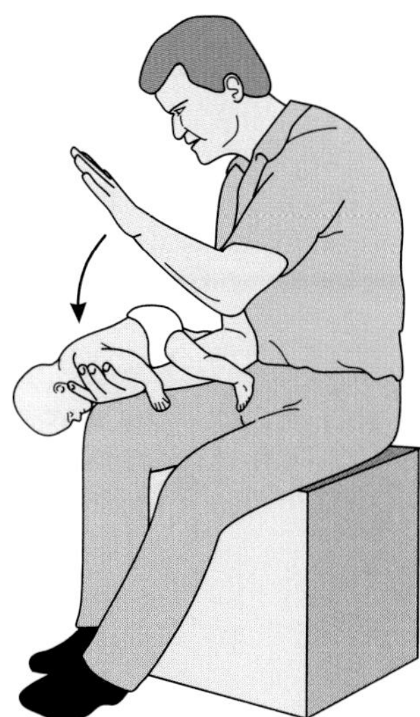

Administer five blows rapidly with the heel of your hand high between her shoulder blades.

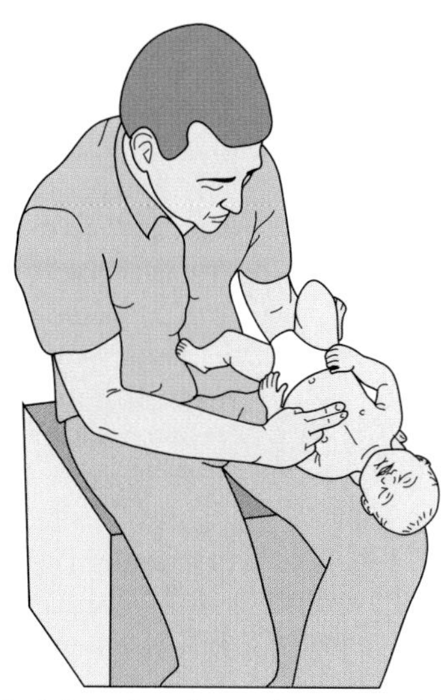

If that doesn't work, turn her over and deliver five quick downward thrusts over her sternum using two fingers.

- If the choking victim is an older child or adult, the strategy most likely to succeed is abdominal thrusts, sometimes called the Heimlich maneuver.

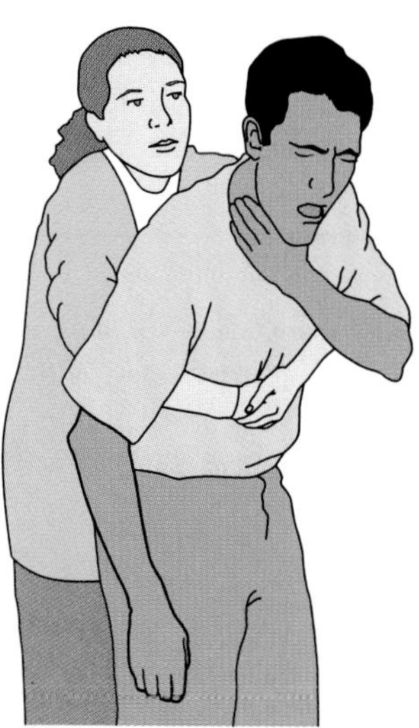

Stand behind the victim, and wrap your arms around him. Place the thumb side of one fist snugly against his body, slightly above the navel and below the rib cage. Grasp your fist with your other hand and give him a sudden strong hug inward and upward. Repeat thrusts as necessary.

To self-administer first aid, place the thumb side of one fist slightly above the navel and below the rib cage, grasp the fist with your other hand, and then press inward and upward with a quick motion. If this is unsuccessful, quickly press your upper abdomen over any firm surface such as the back of a chair, a countertop, or a railing.

- If all else fails, open the mouth by grasping both the tongue and lower jaw and lifting. Then, and only if you can see the object, use your finger to sweep it out and begin rescue breathing.

Sources: H. J. Heimlich and M. H. Uhley, The Heimlich maneuver, *Clinical Symposia* 31 (1979): 1–32; H. J. Heimlich, Self-application of the Heimlich maneuver, *New England Journal of Medicine* 318 (1988): 714–715.

Projectile vomiting is also serious. The contents of the stomach are expelled with such force that they leave the mouth in a wide arc like a bullet leaving a gun. This type of vomiting requires immediate medical attention.

## *Diarrhea*

Diarrhea is characterized by frequent, loose, watery stools. Such stools indicate that the intestinal contents have moved too quickly through the intestines for fluid absorption to take place, or that water has been drawn from the cells lining the intestinal tract and added to the food residue. Like vomiting, diarrhea can lead to considerable fluid and salt losses, but the composition of the fluids is different. Stomach fluids lost in vomiting are highly acidic, whereas intestinal fluids lost in diarrhea are nearly neutral. When fluid losses require medical attention, correct replacement is crucial.

Diarrhea is a symptom of a variety of medical conditions and treatments. It may occur abruptly in a healthy person as a result of infections (such as food poisoning) or as a side effect of medications. When used in large quantities, food ingredients such as the sugar alternative sorbitol and the fat alternative olestra may also cause diarrhea in some people. If a food is responsible, then that food must be omitted from the diet, at least temporarily. If a drug is responsible, a different drug, when possible, or a different form of the drug (injectable versus oral, for example) may alleviate the problem.

Diarrhea may also occur as a result of disorders of the GI tract, such as irritable bowel syndrome or colitis. Irritable bowel syndrome is one of the most common GI disorders and is characterized by a disturbance in the motility of the GI tract. Dietary treatment hinges on identifying and avoiding individual foods that cause intolerance. For most people, a low-fat diet provided in small meals, with a gradual increase in fiber, is helpful. People with colitis, an inflammation of the large intestine, may also suffer from severe diarrhea. They often benefit from complete bowel rest and medication. If treatment fails, surgery to remove the colon and rectum may be necessary.

As you can see, treatment for diarrhea depends on its cause and its severity.[1] Mild diarrhea may remit without treatment; simply rest and drink fluids to replace losses. If diarrhea persists, though, especially in an infant, call a physician. Severe diarrhea can lead to dehydration and electrolyte imbalances. (See Chapter 12 for information on dehydration and its therapy.)

U.S. Government

www.healthfinder.gov/searchoptions/topicsaz.htm
Search for Diarrhea

## Constipation

Unlike diarrhea, constipation is generally not a cause for immediate concern. Each person's GI tract responds to food in its own way, with its own rhythm, depending on such factors as the owner's health, the type of food eaten, and when the person's schedule allows time to defecate. Even when several days pass between movements, a person is not constipated, as long as these movements take place without discomfort and the time that has elapsed since the previous bowel movement is typical for that person. But if the time between bowel movements is much longer or if a movement is passed with difficulty, discomfort, or pain, then the person is constipated.

Often a person's lifestyle may cause constipation. Being too busy to respond to the defecation signal is a common complaint. If a person receives the signal to defecate and ignores it, the signal may not return for several hours. In the meantime, water continues to be withdrawn from the fecal matter, so when the person does defecate, the bowel movement is dry and hard. In such a case, a person's daily regimen may need to be revised to allow time to have a bowel movement when the body sends its signal. One possibility is to go to bed earlier in order to rise earlier, allowing ample time for a leisurely breakfast and a movement.

Another cause of constipation is lack of physical activity. Physical activity improves muscle tone, not just of the outer body, but also of the digestive tract.

Although constipation usually reflects lifestyle habits, in some cases it may be a side effect of medication or may reflect a medical problem such as tumors that are obstructing the passage of waste. If discomfort is associated with passing fecal matter, a physician's help should be sought to rule out disease. Once this has been done, dietary or other measures for correction can be considered.

One dietary measure that may be appropriate is to increase dietary fiber. Some fibers—those found in cereal products—help to prevent constipation by increasing fecal mass. In the GI tract, fiber attracts water, creating soft, bulky stools that stimulate bowel contractions to push the contents along. These contractions strengthen the intestinal muscles. The improved muscle tone, together with the water content of the stools, eases elimination, reducing the pressure in the rectal veins and helping to prevent hemorrhoids. Chapter 4 provides more information on fiber's role in maintaining a healthy colon and reducing the risks of colon cancer and diverticulosis.

Drinking plenty of water in conjunction with eating high-fiber foods also helps with constipation. The increased bulk physically stimulates the upper GI tract, promoting peristalsis throughout.

Eating prunes can also be helpful. Prunes are high in fiber and also contain a laxative substance.* If a morning defecation is desired, a person can drink prune juice at bedtime; if the evening is preferred, the person can drink prune juice with breakfast.

Honey can also have a laxative effect due to its incomplete absorption.[2] Although this characteristic may cause problems for people with irritable bowel syndrome, eating honey may be an easy and effective treatment for those who are constipated.

Adding fat to the diet can relieve some constipation by stimulating the hormone cholecystokinin, which summons bile into the duodenum. Bile's high salt content draws water from the intestinal wall, which stimulates peristalsis and softens the fecal matter.

These suggested changes in lifestyle or diet should correct chronic constipation without the use of laxatives, enemas, or mineral oil, although television commercials often try to persuade people otherwise. One of the fallacies often perpetrated by television commercials is that one person's successful use of a product is a good recommendation for others to use that product.

*This substance is dihydroxyphenyl isatin.

***Beans, broccoli, cabbage, and onions produce gas in many people. People troubled by gas need to determine which foods bother them and then eat those foods in moderation.***

As a matter of fact, even diet changes that relieve constipation for one person may increase the constipation of another. For instance, increasing fiber intake stimulates peristalsis and helps the person with a sluggish colon. Some people, though, have a spastic type of constipation, in which peristalsis promotes strong contractions that close off a segment of the colon and prevent passage; for these people, increasing fiber intake would be exactly the wrong thing to do.

A person who seems to need products such as laxatives should seek a physician's opinion. Advice from friends or alternative medicine practitioners may cause more harm than good. One potentially harmful but currently popular practice that is being promoted by some alternative medicine practitioners is colonic irrigation—the internal washing of the large intestine with a powerful enema machine. Such an extreme cleansing is not only unnecessary, but the force of the machine can rupture the intestine. Less extreme practices can cause problems, too. Frequent use of laxatives and enemas can lead to dependency; upset the body's fluid, salt, and mineral balances; and, in the case of mineral oil, interfere with the absorption of fat-soluble vitamins. (Mineral oil dissolves the vitamins, but is not itself absorbed; instead, it leaves the body, carrying the vitamins with it.) Table H3-1 includes strategies to prevent or alleviate constipation.

**Table H3-1**
**Strategies to Prevent or Alleviate Common GI Problems**

| GI Problem | Strategies |
|---|---|
| Choking | • Take small bites of food.<br>• Chew thoroughly before swallowing.<br>• Don't talk or laugh with food in your mouth.<br>• Don't eat when breathing hard. |
| Diarrhea | • Rest.<br>• Drink fluids to replace losses.<br>• Call for medical help if diarrhea persists. |
| Constipation | • Eat a high-fiber diet.<br>• Drink plenty of fluids.<br>• Exercise regularly.<br>• Respond promptly to the urge to defecate. |
| Belching | • Eat slowly.<br>• Chew thoroughly.<br>• Relax while eating. |
| Intestinal gas | • Eat bothersome foods in moderation. |
| Heartburn | • Eat small meals.<br>• Drink liquids between meals.<br>• Sit up while eating.<br>• Wait 1 hour after eating before lying down.<br>• Wait 2 hours after eating before exercising.<br>• Refrain from wearing tight-fitting clothing.<br>• Avoid foods, beverages, and medications that aggravate your heartburn.<br>• Refrain from smoking cigarettes.<br>• Lose weight if overweight. |
| Ulcer | • Take medicine as prescribed by your physician.<br>• Avoid coffee and caffeine- and alcohol-containing beverages.<br>• Avoid foods that aggravate your ulcer.<br>• Minimize aspirin use.<br>• Refrain from smoking cigarettes. |

 U.S. Government

www.healthfinder.gov/searchoptions/topicsaz.htm
Search for Constipation

## Belching and Gas

Many people complain of problems that they attribute to excessive gas. For some, belching is the complaint. Others blame intestinal gas for abdominal discomforts and embarrassment. Most people believe that the problems occur after they eat certain foods. This may be the case with intestinal gas, but belching results from swallowing air. The best advice for belching seems to be to eat slowly, chew thoroughly, and relax while eating.

Everyone swallows a little bit of air with each mouthful of food, but people who eat too fast may swallow too much air and then have to belch. Ill-fitting dentures, carbonated beverages, and chewing gum can also contribute to the swallowing of air with resultant belching. Occasionally, belching can be a sign of a more serious disorder, such as gallbladder disease or a peptic ulcer.

People who eat or drink too fast may also trigger hiccups, the repeated spasms that produce a cough-like sound and jerky movement. Normally, hiccups soon subside and are of no medical significance, but they can be bothersome. The most effective cure is to hold the breath for as long as possible, which helps to relieve the spasms of the diaphragm.

While expelling gas can be a humiliating experience, it is quite normal. (People experiencing painful bloating from malabsorption diseases, however, require medical treatment.) Healthy people expel several hundred milliliters of gas several times a day. Almost all (99 percent) of the gases expelled—nitrogen, oxygen, hydrogen, methane, and carbon dioxide—are odorless. The remaining "volatile" gases are the infamous ones.

Foods that produce gas usually must be determined individually. The most common offenders are foods rich in the carbohydrates—sugars, starches, and fibers. When partially digested carbohydrates reach the large intestine, bacteria digest them, giving off gas as a by-product. People can test foods suspected of forming gas by omitting them individually for a trial period and seeing if there is any improvement.

## Heartburn and "Acid Indigestion"

Almost everyone has experienced heartburn at one time or another, usually after a meal. Heartburn is the painful sensation a person feels when the cardiac sphincter fails to prevent the stomach contents from refluxing into the esophagus. This may happen if a person eats or drinks too much (or both). Tight clothing and even changes of position (lying down, bending over) can cause it, too, as can some medications and smoking. A defect of the cardiac sphincter itself is a possible, but less likely cause.

If the heartburn is not caused by an anatomical defect, treatment is fairly simple. To avoid such misery in the future, the person needs to learn to eat less at a sitting, chew food more thoroughly, and eat it more slowly. Additional tips for people suffering from heartburn are included in Table H3-1.

 U.S. Government

www.healthfinder.gov/searchoptions/topicsaz.htm
Search for Heartburn

As far as "acid indigestion" is concerned, recall from Chapter 3 that the strong acidity of the stomach is a desirable condition—television commercials for antacids and acid controllers notwithstanding. People who overeat or eat too quickly are likely to suffer from indigestion. The muscular reaction of the stomach to unchewed lumps or to being overfilled may be so violent that it causes regurgitation (reverse peristalsis). When this happens, overeaters may taste the stomach acid and feel pain. Responding to television commercials, they may reach for antacids or acid controllers. Both of these drugs were originally designed to treat GI illnesses such as ulcers. As is true of most over-the-counter medicines, antacids and acid controllers should be used only infrequently for occasional heartburn; they may mask or cause problems if used regularly, as the next section explains.

 U.S. Government

www.healthfinder.gov/searchoptions/topicsaz.htm
Search for Indigestion

## Ulcers

Ulcers of the stomach (gastric ulcers) or duodenum (duodenal ulcers) are another common digestive problem. (The term *peptic ulcer* includes both types.) An ulcer is an erosion of the top layer of cells from an area, such as the wall of the stomach or duodenum. This erosion leaves the underlying layers of cells unprotected and exposed to gastric juices. The erosion may proceed until the gastric juices reach the capillaries that feed the area, leading to bleeding, and reach the nerves, causing pain. If GI bleeding is excessive, iron deficiency may develop.[3] If the erosion penetrates all the way through the GI lining, a life-threatening infection can develop.

Many people naively believe that an ulcer is caused by stress or spicy foods, but this is not the case—at least not at first.[4] The stomach lining in a healthy person is well protected by its mucous coat. What, then, causes ulcers to form?

Three major causes of ulcers have been identified: bacterial infection with *Helicobacter pylori,* the use of certain anti-inflammatory drugs such as ibuprofen and naproxen, and disorders that cause excessive gastric acid secretion.[5] The cause of the ulcer dictates the type of drug used in treatment.[6] For example, people with ulcers caused by infection receive antibiotics, whereas those with ulcers caused by drugs discontinue their use. In addition, all treatment plans aim to relieve pain, heal the ulcer, and prevent recurrence.

Diet therapy once played a major role in ulcer treatment, but it no longer does. Current practice is simply to treat for infection, eliminate any food that routinely causes indigestion or pain, and avoid coffee and caffeine- and alcohol-containing beverages. Both regular and decaffeinated coffee stimulate acid secretion and so aggravate *existing* ulcers.

Ulcers and their treatments highlight the importance of not self-medicating when symptoms persist. People with *H. pylori* infection often take over-the-counter acid controllers to relieve the pain of their ulcers when they need physician-prescribed antibiotics instead. Suppressing gastric acidity not only fails to heal the ulcer, but actually worsens inflammation during an *H. pylori* infection.[7] Furthermore, *H. pylori* infection has been linked with stomach cancer as well, making prompt diagnosis and appropriate treatment most important.[8]

 U.S. Government

www.healthfinder.gov/searchoptions/topicsaz.htm
Search for Ulcers

Table H3-1 summarizes strategies to prevent or alleviate common GI problems. Many of these problems reflect hurried lifestyles. For this reason, many of their remedies require that people slow down and take the time to eat slowly; chew food thoroughly to prevent choking, heartburn, and acid indigestion; rest until vomiting and diarrhea subside; and heed the urge to defecate. In addition, learn how to handle life's day-to-day problems and challenges without overreacting and becoming upset; learn how to relax, to get enough sleep, and to enjoy life. Remember, "what's eating you" may cause more GI distress than what you eat.

## Notes


1. M. Donowitz, F. T. Kokke, and R. Saidi, Evaluation of patients with chronic diarrhea, *New England Journal of Medicine* 332 (1995): 725–729.
2. S. D. Ladas, D. N. Haritos, and S. A. Raptis, Honey may have a laxative effect on normal subjects because of incomplete fructose absorption, *American Journal of Clinical Nutrition* 62 (1995): 1212–1215.
3. R. Yip and coauthors, Pervasive occult gastrointestinal bleeding in an Alaska native population with prevalent iron deficiency: Role of *Helicobacter pylori* gastritis, *Journal of the American Medical Association* 277 (1997): 1135–1139.
4. Knowledge about causes of peptic ulcer disease—United States, March-April 1997, *Morbidity and Mortality Weekly Report* 46 (1997): 985–987.
5. D. Y. Graham, *Helicobacter pylori:* Its epidemiology and its role in duodenal ulcer disease, *Journal of Gastroenterology and Hepatology* 6 (1991): 105–113.

6. A. H. Soll for the Practice Parameters Committee of the American College of Gastroenterology, Medical treatment of peptic ulcer disease: Practice guidelines, *Journal of the American Medical Association* 275 (1996): 622– 629; M. S. Khuroo and coauthors, A comparison of omeprazole and placebo for bleeding peptic ulcer, *New England Journal of Medicine* 336 (1997): 1054–1058.
7. E. J. Kuipers and coauthors, Atrophic gastritis and *Helicobacter pylori* infection in patients with reflux esophagitis treated with omeprazole or fundoplication, *New England Journal of Medicine* 334 (1996): 1018–1022.
8. L. E. Hansson and coauthors, The risk of stomach cancer in patients with gastric or duodenal ulcer disease, *New England Journal of Medicine* 335 (1996): 242–249.

CHAPTER 4

# The Carbohydrates: Sugars, Starches, and Fibers

A student, quietly studying a textbook, is seldom aware that within his brain cells, billions of glucose molecules are splitting each second to provide the energy that permits him to learn. Yet glucose provides nearly all of the energy the human brain uses daily. Similarly, a marathon runner, bursting across the finish line in an explosion of sweat and triumph, seldom gives thanks to the glycogen fuel her muscles have devoured to help her finish the race. Yet, together, glucose and its storage form glycogen provide about half of all the energy muscles and other body tissues use. The other half of the body's energy comes mostly from fat.

People don't eat glucose and glycogen directly; they eat foods rich in carbohydrates. Then their bodies convert the carbohydrates mostly into glucose for immediate energy and into glycogen for reserve energy.

Carbohydrates contribute so much to the bulk of most foods that many people mistakenly think of them as "fattening" and avoid them when trying to lose weight. Actually, such a strategy may be counterproductive. People can better control body weight by selecting high-carbohydrate, high-fiber foods and limiting fat-rich foods. All unrefined plant foods—grains, vegetables, legumes, and fruits—provide ample carbohydrate and fiber with little or no fat. (Milk also contains carbohydrates. So do shellfish and organ meats such as liver, but only a little.)

**carbohydrates:** compounds composed of carbon, oxygen, and hydrogen arranged as monosaccharides or multiples of monosaccharides. Most, but not all, carbohydrates have a ratio of one carbon molecule to one water molecule: $(CH_2O)_n$.
- **carbo** = carbon (C)
- **hydrate** = with water ($H_2O$)

The dietary carbohydrate family includes the simple carbohydrates (the sugars) and the complex carbohydrates (the starches and fibers). The simple carbohydrates are those that chemists describe as:

- Monosaccharides—single sugars.
- Disaccharides—sugars composed of pairs of monosaccharides.

The complex carbohydrates are:

- Polysaccharides—large molecules composed of chains of monosaccharides.

**simple carbohydrates** (sugars)**:** monosaccharides and disaccharides.

**complex carbohydrates** (starches and fibers)**:** polysaccharides composed of straight or branched chains of monosaccharides.

# The Chemist's View of Carbohydrates

To understand the structure of carbohydrates, look at the units of which they are made. The sugars most important in nutrition are the 6-carbon monosaccharides known as hexoses. Each contains 6 carbon atoms, 12 hydrogens, and 6 oxygens (written in shorthand as $C_6H_{12}O_6$).

Each atom can form a certain number of chemical bonds with other atoms:

- Carbon atoms can form four bonds.
- Nitrogen atoms, three.
- Oxygen atoms, two.
- Hydrogen atoms, only one.

Chemists represent the bonds as lines between the chemical symbols (such as C, N, O, and H) that stand for the atoms (see Figure 4-1).

Atoms form molecules in ways that satisfy the bonding requirements of each atom. Figure 4-1 shows the structure of ethyl alcohol, the active ingredient of

**Figure 4-1**
**Atoms and Their Bonds**

The four main types of atoms found in nutrients are hydrogen (H), oxygen (O), nitrogen (N), and carbon (C). Appendix B presents basic chemistry terms and relationships.

```
H—     —O—     —N—     —C—
                |       |
1       2       3       4
```

Each atom has a characteristic number of bonds it can form with other atoms.

```
    H   H
    |   |
H—C—C—O—H
    |   |
    H   H
```

Ethyl alcohol, a simple molecule showing bonding.

**Figure 4-2**
**Chemical Structure of Glucose**

On paper, the structure of glucose has to be drawn flat, but in nature the five carbons and oxygen are roughly in a plane. The atoms attached to the ring carbons extend above and below the plane.

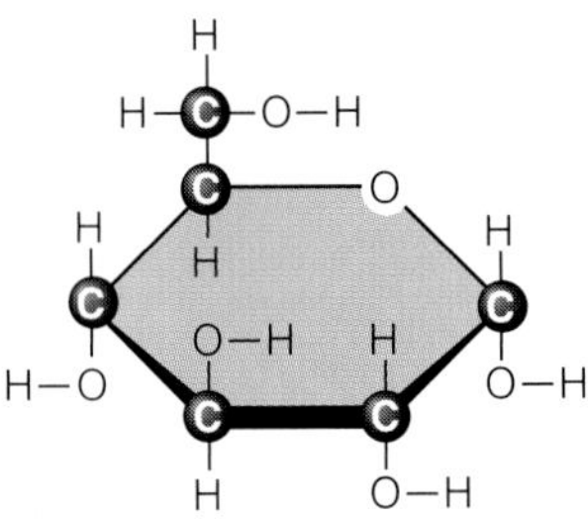

alcoholic beverages, as an example. The two carbons each have four bonds represented by lines; the oxygen has two; and each hydrogen has one bond connecting it to other atoms. An accurate drawing of a chemical structure must obey these rules because the laws of nature demand it.

IN SUMMARY

The carbohydrates are made of carbon (C), oxygen (O), and hydrogen (H). Each of these atoms can form a specified number of chemical bonds: carbon forms four, oxygen forms two, and hydrogen forms one.

# The Simple Carbohydrates

The following list of the six sugars most important in nutrition symbolizes them as hexagons and pentagons of different colors. Three are single sugars or monosaccharides:

- Glucose. • Fructose. • Galactose. 

Fructose is shown as a pentagon, but it does have 6 carbon atoms. The ring contains 4 carbons and an oxygen; 2 carbons stick out from the ring (see Figure 4-4).

Galactose occurs only as a part of lactose.

Three are double sugars or disaccharides:

- Maltose (glucose + glucose).
- Sucrose (glucose + fructose). 
- Lactose (glucose + galactose).

## *Monosaccharides*

**monosaccharide** (mon-oh-SACK-uh-ride): a carbohydrate of the general formula $C_nH_{2n}O_n$ that consists of a single ring.
- **mono** = one
- **saccharide** = sugar

See Appendix C for the complex chemical structures of the monosaccharides.

**glucose** (GLOO-kose): a monosaccharide; sometimes known as blood sugar or **dextrose.**
- **ose** = carbohydrate
- (hexagon) = glucose

The three monosaccharides important in nutrition all have the same numbers and kinds of atoms, but in different arrangements. These chemical differences account for the differing sweetness of the monosaccharides. A pinch of purified glucose on the tongue gives only a mild sweet flavor, and galactose hardly tastes sweet at all, but fructose is as intensely sweet as honey and, in fact, is the sugar primarily responsible for honey's sweetness.

• ***Glucose*** • Chemically, glucose is a larger and more complicated molecule than ethyl alcohol, but it obeys the same rules of chemistry: each carbon atom has four bonds; each oxygen, two bonds; and each hydrogen, one bond. Figure 4-2 illustrates the chemical structure of a glucose molecule.

The diagram of a glucose molecule shows all the relationships between the atoms and proves simple on examination, but chemists have adopted even simpler ways to depict chemical structures. Figure 4-3 shows that a chemical struc-

**Figure 4-3**
**Simplified Diagrams of Glucose**

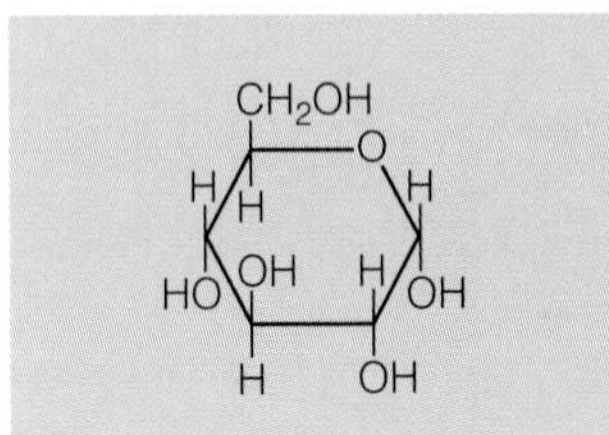

The carbons at the corners are not shown, and the formula $CH_2OH$ stands for the structure in Figure 4-2.

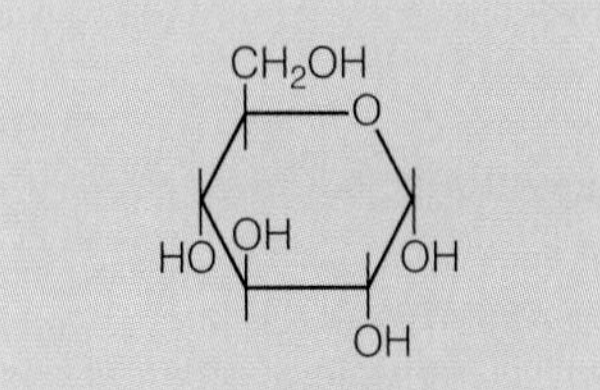

Now the single hydrogens are not shown, but lines still extend upward or downward from the ring to show where they belong.

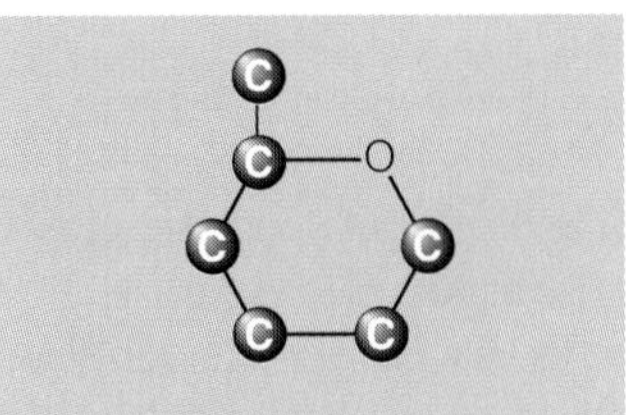

Another way to look at glucose is to notice that its six carbon atoms are all connected.

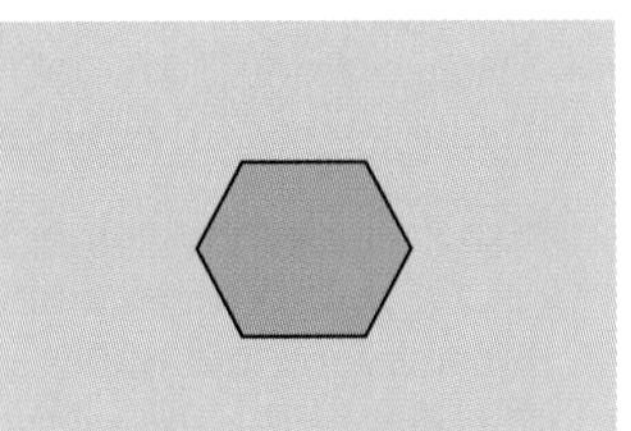

In this and other illustrations throughout this book, glucose is represented as a blue hexagon.

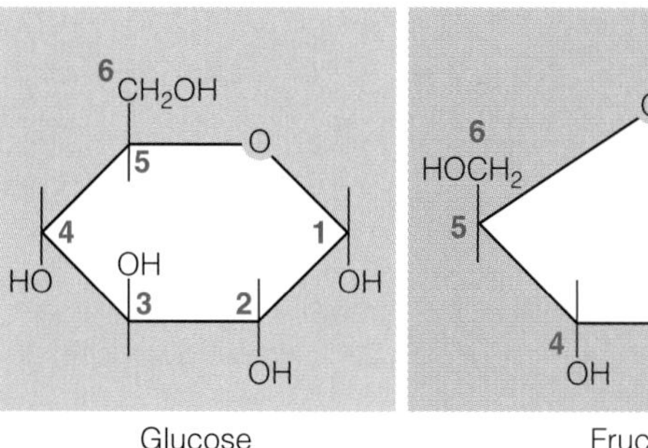


**Figure 4-4**

**Two Monosaccharides: Glucose and Fructose**

Can you see the similarities? If you learned the rules in Figure 4-3, you will be able to "see" 6 carbons (numbered), 12 hydrogens (those shown plus one at the end of each single line), and 6 oxygens in both these compounds.

ture can combine or omit a number of letters without losing the information it conveys.

The significance of glucose to nutrition is tremendous. Glucose is one of the two sugars in every disaccharide and is the unit from which the polysaccharides are made almost exclusively. One of these polysaccharides, starch, is the chief energy food of the world's people; another, glycogen, is a major storage form of energy in the body. Glucose reappears frequently throughout this chapter and all those that follow.

• ***Fructose*** • Fructose is the sweetest of the sugars. Curiously, fructose has exactly the same chemical *formula* as glucose—$C_6H_{12}O_6$—but its *structure* differs (see Figure 4-4). The arrangement of the atoms in fructose stimulates the taste buds on the tongue to produce the sweet sensation. Fructose occurs naturally in fruits and honey; food manufacturers also use it in products sweetened with high-fructose corn syrup (HFCS), a corn product used as an additive.

**fructose** (FRUK-tose or FROOK-tose): a monosaccharide; sometimes known as fruit sugar or **levulose,** fructose is found abundantly in fruits, honey, and saps.
• **fruct** = fruit
• = fructose

• ***Galactose*** • Seldom occurring free in nature, galactose binds with another monosaccharide to form the sugar in milk. Galactose has the same numbers and kinds of atoms as glucose and fructose in yet another arrangement. Figure 4-5 shows galactose beside a molecule of glucose for comparison.

**galactose** (ga-LAK-tose): a monosaccharide; part of the disaccharide lactose.
• = galactose

# Disaccharides

The disaccharides are pairs of the three sugars just discussed. Glucose occurs in all three; the second member of the pair is either fructose, galactose, or another glucose. These carbohydrates and all the other energy nutrients are put together and taken apart by similar chemical reactions.

**disaccharide** (dye-SACK-uh-ride): a pair of monosaccharides linked together. See Appendix C for the chemical structures of the disaccharides.
• **di** = two

• ***Condensation*** • To make a disaccharide, a chemical reaction known as condensation links two monosaccharides together (see Figure 4-6 on p. 94). A hydroxyl (OH) group from one monosaccharide and a hydrogen atom (H) from the other combine to create a molecule of water ($H_2O$). The two originally separate monosaccharides link together with a single oxygen (O).

**condensation:** a chemical reaction in which two reactants combine to yield a larger product.

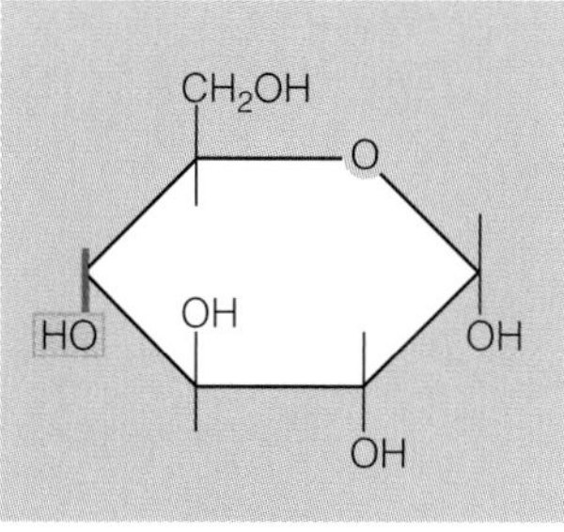


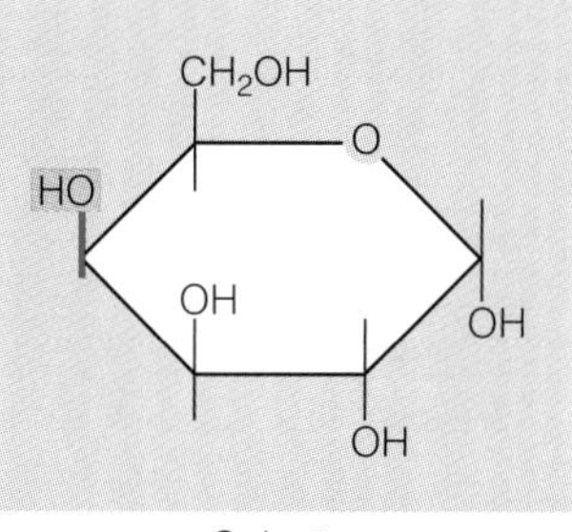


**Figure 4-5**

**Two Monosaccharides: Glucose and Galactose**

Notice the similarities and the difference.

**Figure 4-6**
**Condensation of Two Monosaccharides to Form a Disaccharide**

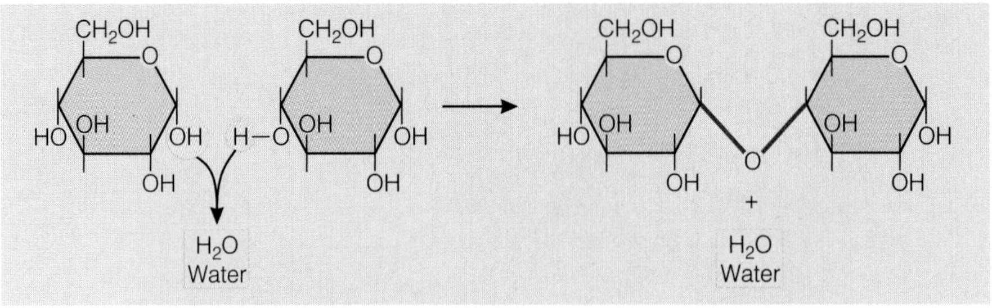

$CH_2OH$ $CH_2OH$ → $CH_2OH$ $CH_2OH$

$H_2O$ Water + $H_2O$ Water

Glucose + glucose → Maltose

An OH group from one glucose and an H atom from another glucose combine to create a molecule of $H_2O$.

The two glucose molecules bond together with a single O atom to form the disaccharide maltose.

Reminder: A *hydrolysis* reaction splits a major reactant into two products, with H added to one and OH to the other (from water).

**maltose** (MAWL-tose): a disaccharide composed of two glucose units; sometimes known as malt sugar.
- 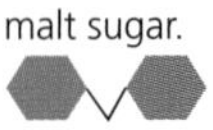 = maltose

**sucrose** (SUE-krose): a disaccharide composed of glucose and fructose; commonly known as table sugar, beet sugar, or cane sugar. Sucrose also occurs in many fruits and some vegetables and grains.
- **sucro** = sugar
- 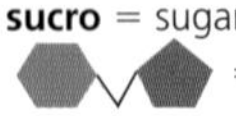 = sucrose

**lactose** (LAK-tose): a disaccharide composed of glucose and galactose; commonly known as milk sugar.
- **lact** = milk
- 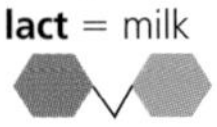 = lactose

• ***Hydrolysis*** • To break a disaccharide in two, a chemical reaction known as hydrolysis occurs (see Figure 4-7). A molecule of water splits to provide the H and OH needed to complete the resulting monosaccharides. Hydrolysis reactions commonly occur during digestion.

• ***Maltose*** • The disaccharide maltose consists of two glucose units. Maltose is produced whenever starch breaks down—as happens in plants when seeds germinate and in human beings during carbohydrate digestion. It also occurs during the fermentation process that yields alcohol. Maltose is only a minor constituent of a few foods.

• ***Sucrose*** • Fructose and glucose together form sucrose. Because the fructose is in a position accessible to the taste receptors, sucrose tastes sweet, accounting for some of the natural sweetness of fruits, vegetables, and grains. To make table sugar, sucrose is refined from the juices of sugarcane and sugar beets, then granulated. Depending on the extent to which it is refined, the product becomes the familiar brown, white, and powdered sugars available at grocery stores.

• ***Lactose*** • The combination of galactose and glucose makes the disaccharide lactose, the principal carbohydrate of milk. Known as milk sugar, lactose contributes about 5 percent of milk's weight. Depending on the milk's fat content, lactose contributes 30 to 50 percent of milk's energy.

### IN SUMMARY

Six simple carbohydrates, or sugars, are important in nutrition. The three monosaccharides (glucose, fructose, and galactose) all have the same chemical formula ($C_6H_{12}O_6$), but their structures differ. The three disaccharides (sucrose, lactose, and maltose) are pairs of monosaccharides.

**Figure 4-7**
**Hydrolysis of a Disaccharide**

Hydrolysis occurs during digestion.

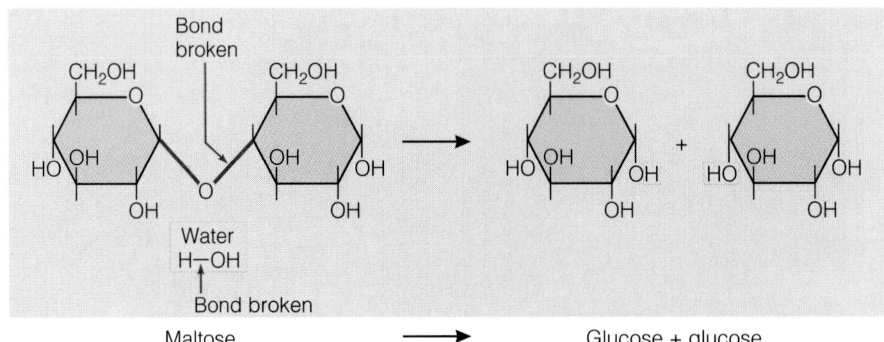


The disaccharide maltose splits into two glucose molecules with H added to one and OH to the other (from water).

The sugars derive primarily from plants, except for lactose and its component galactose, which come from milk and milk products. Two monosaccharides can be linked together by a condensation reaction to form a disaccharide and water. A disaccharide, in turn, can be broken into its two monosaccharides by a hydrolysis reaction using water.

*Fruits package their simple sugars with fibers, vitamins, and minerals, making them a sweet and healthy snack.*

# The Complex Carbohydrates

The simple carbohydrates are the sugars just mentioned: glucose, fructose, and galactose, either singly or paired with glucose. In contrast, the complex carbohydrates contain many glucose units and a few other monosaccharides strung together as polysaccharides. Three are important in nutrition: glycogen, starches, and fibers.

Glycogen is a storage form of energy in the animal body; starches play that role in plants; and the fibers of plants serve as structural elements in stems, trunks, roots, leaves, and skins. Both glycogen and starch are built of glucose units, but they are linked together differently. The fibers are composed of a variety of monosaccharides and other carbohydrate derivatives.

**polysaccharide:** many monosaccharides linked together.
- **poly** = many
- **saccharide** = sugar

## Glycogen

Glycogen is found only to a limited extent in meats and not at all in plants.* For this reason, glycogen is not a significant food source of carbohydrate, but it does perform an important role in the body. The human body stores much of its glucose as glycogen—many glucose molecules linked together in highly branched chains (see the left side of Figure 4-8). This arrangement permits rapid hydrolysis.

*Glycogen in animal muscles rapidly hydrolyzes after slaughter.

**glycogen** (GLY-co-gen): an animal polysaccharide composed of glucose; it is manufactured and stored in the liver and muscles as a storage form of glucose. Glycogen is not a significant food source of carbohydrate and is not counted as one of the complex carbohydrates in foods.
- **glyco** = glucose
- **gen** = gives rise to

**Figure 4-8**

**Glycogen and Starch Molecules Compared (Small Segments)**

Notice that the more highly branched the structure, the greater the number of ends from which glucose can be released. (These units would have to be magnified millions of times to appear at the size shown in this figure. For details of the chemical structures, see Appendix C.)

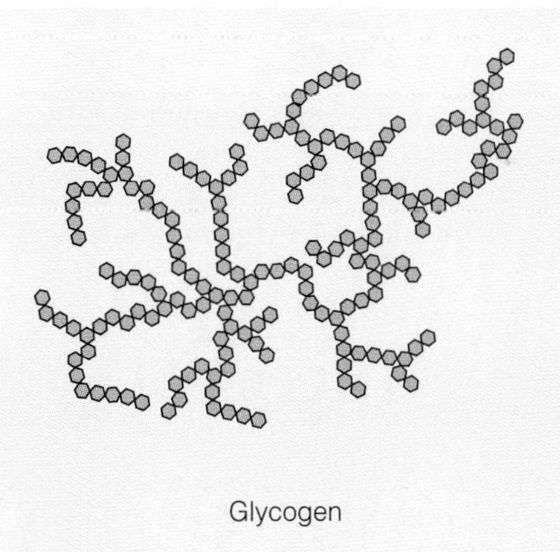


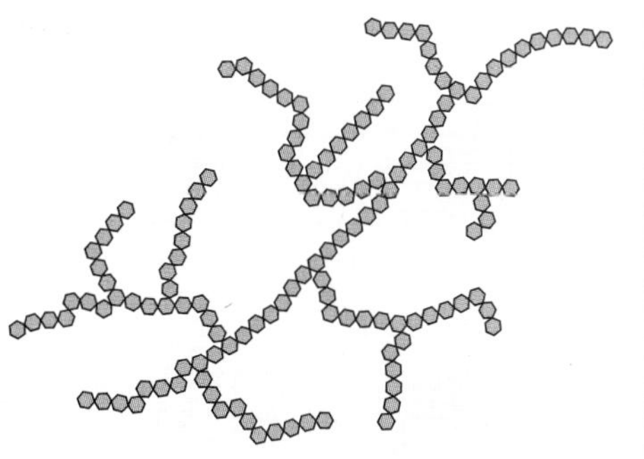


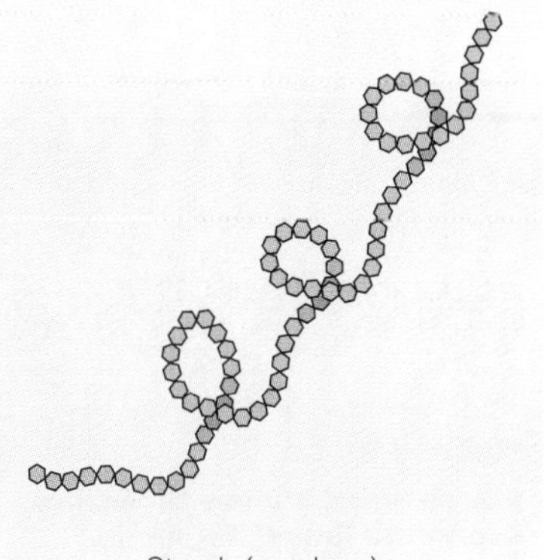


A glycogen molecule contains hundreds of glucose units in long, highly branched chains.

A starch molecule contains hundreds of glucose molecules in either occasionally branched chains (amylopectin) or unbranched chains (amylose).

*Major sources of starch include grains, legumes, and tubers (such as potatoes, yams, and cassava).*

When the hormonal message "Release energy" arrives at the storage sites in a liver or muscle cell, enzymes respond by attacking all the many branches of each glycogen simultaneously, making a surge of fuel available.*

## Starches

Just as the human body stores glucose as glycogen, plant cells store glucose as starches—long, branched or unbranched chains of hundreds or thousands of glucose molecules linked together (see the middle and right side of Figure 4-8 on p. 95). These giant molecules are packed side by side in grains such as wheat or rice, in tubers such as potatoes, and in legumes such as peas and beans. A cubic inch of food may contain as many as a million starch molecules. When you eat the plant, your body hydrolyzes the starch to glucose and uses the glucose for its own energy purposes.

All starchy foods derive from plants. Grains are the richest food source of starch, providing much of the food energy for people all over the world—rice in Asia; wheat in Canada, the United States, and Europe; corn in much of Central and South America; and millet, rye, barley, and oats elsewhere.[1] Legumes and tubers, such as potatoes, yams, and cassava, are also important sources of starch.

## Fibers

**starches:** plant polysaccharides composed of glucose.

**fibers:** in plant foods, the *nonstarch polysaccharides* that are not digested by *human* digestive enzymes, although some are digested by GI tract bacteria; fibers include cellulose, hemicelluloses, pectins, gums, and mucilages and the nonpolysaccharides lignins, cutins, and tannins.

Fibers are the structural parts of plants and thus are found in all plant-derived foods—vegetables, fruits, grains, and legumes. Most fibers are polysaccharides, but starch is not one of them; in fact, fibers are often described as nonstarch polysaccharides. Nonstarch polysaccharides include cellulose, hemicelluloses, pectins, gums, and mucilages. Fibers also include some nonpolysaccharides such as lignins, cutins, and tannins.**

Even though most are polysaccharides, fibers differ from starches in that the bonds between their monosaccharides cannot be broken down by human digestive enzymes. The bacteria of the GI tract can break some fibers down, however, and this is important to digestion and to health.

Each of the fibers has a different structure. Most contain monosaccharides, but differ in the types they contain and in the bonds that link the monosaccharides to each other. These differences produce diverse health effects.

• ***Cellulose*** • Cellulose is the primary constituent of plant cell walls and therefore occurs in all vegetables, fruits, and legumes. Like starch, cellulose is composed of glucose molecules connected in long chains. Unlike starch, however, the chains do not branch, and the bonds linking the glucose molecules together resist digestion by human enzymes (see Figure 4-9).

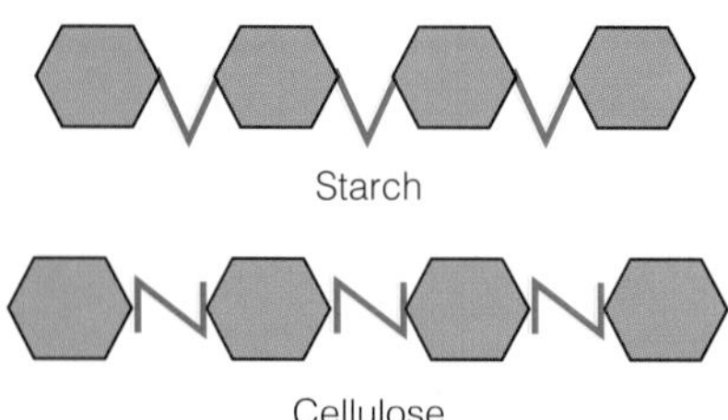


**Figure 4-9**
**Starch and Cellulose Molecules Compared (Small Segments)**

The bonds that link the glucose molecules together in cellulose are different from the bonds in starch (and glycogen). Human enzymes cannot digest cellulose. See Appendix C for chemical structures and descriptions of linkages.

*Normally, only the liver can return glucose *directly* from glycogen to the blood; muscle cells use glycogen internally to produce glucose. Muscle cells can restore the blood glucose level *indirectly*, however, as Chapter 7 explains.

**The terms *crude fiber, neutral-detergent fiber,* and *dietary fiber* reflect different methodologies used to estimate the fiber contents of foods; they do not identify different types of fiber. The structure of cellulose is shown in Appendix C; the other polysaccharide fibers are similar, but differ slightly in their bonding. Besides glucose, their component sugars may include a variety of other monosaccharides. As for the lignins, they are polymers of several dozen molecules of phenol (an organic alcohol), with strong internal bonds that make them impervious to digestive enzymes.

• ***Hemicelluloses*** • The hemicelluloses are the main constituent of cereal fibers. They are composed of various monosaccharide backbones with branching side chains of monosaccharides.* The many backbones and side chains make the hemicelluloses a diverse group; some are soluble, while others are insoluble.

• ***Pectins*** • All pectins consist of a backbone derived from carbohydrate with side chains of various monosaccharides. Commonly found in vegetables and fruits (especially citrus fruits and apples), pectins may be isolated and used by the food industry to thicken jelly, keep salad dressing from separating, and otherwise control texture and consistency. Pectins can perform these functions because they readily form gels in water.

• ***Gums and Mucilages*** • When cut, a plant secretes gums from the site of the injury. Like the other fibers, gums are composed of various monosaccharides and their derivatives. Gums such as *gum arabic* are used as additives by the food industry. Mucilages are similar to gums in structure; they include *guar* and *carrageenan,* which are added to foods as stabilizers.

• ***Lignin*** • This *nonpolysaccharide* fiber has a three-dimensional structure that gives it strength. Because of its toughness, few of the foods that people eat contain much lignin. It occurs in the woody parts of vegetables such as carrots and the small seeds of fruits such as strawberries.

• ***Other Classifications of Fibers*** • Scientists classify fibers in several ways. The previous paragraphs classified them according to their chemical properties. Fibers can also be classified according to their solubility. The effects of fibers on the body do not divide neatly along the lines of solubility, but some generalizations of significance to health can be made (see Table 4-1).

Some researchers classify fibers according to other physical properties that affect GI function and nutrient absorption. Physical properties of fibers include:

- *Water-holding capacity*—the capacity to capture water like a sponge, swelling and increasing the bulk of the intestines' contents.
- *Viscosity*—the capacity to form viscous, gel-like solutions.
- *Cation-exchange capacity*—the ability to bind minerals.
- *Bile-binding capacity*—the ability to bind to bile acids.
- *Fermentability*—the extent to which bacteria in the GI tract can break down fibers to fragments that the body can use.**

Clearly, the fibers are a diverse group of compounds. The accompanying table summarizes the carbohydrate family of compounds.

*In hemicelluloses, the most common backbone monosaccharides are xylose, mannose, and galactose; the common side chains are arabinose, glucuronic acid, and galactose (see Appendix C for structures).

**Dietary fibers are fermented by colon bacteria to short-chain fatty acids, which are absorbed and metabolized by the GI mucosa and liver.

IN SUMMARY

**Simple Carbohydrates (sugars)**

- Monosaccharides
  - Glucose
  - Fructose
  - Galactose
- Disaccharides
  - Maltose
  - Sucrose
  - Lactose

**Complex Carbohydrates**

- Polysaccharides
  - Glycogen[a]
  - Starches
  - Fibers (nonstarch polysaccharides)
    - Soluble
    - Insoluble

[a]Glycogen is a complex carbohydrate (a polysaccharide), but not a *dietary* source of carbohydrate.

**Table 4-1**
**Fibers: Their Solubilities, Sources, and Actions**

| Type of Fiber | Major Food Sources | Action in the Body |
|---|---|---|
| *Soluble fibers* | | |
| Gums, pectins, some hemicelluloses, mucilages | Fruits (apples, citrus), oats, barley, legumes | Delay GI transit.<br>Delay glucose absorption.<br>Lower blood cholesterol. |
| *Insoluble fibers* | | |
| Cellulose, many hemicelluloses, lignins | Wheat bran, corn bran, whole-grain breads and cereals, vegetables | Accelerate GI transit.<br>Increase fecal weight (promotes bowel movements).<br>Slow starch hydrolysis.<br>Delay glucose absorption. |

NOTE: These generalizations are useful, but exceptions occur. For example, insoluble rice bran also lowers blood cholesterol, and the soluble fiber of the psyllium plant effectively promotes bowel movements.

**phytic** (FYE-tick) **acid:** a nonnutrient component of plant seeds; also called **phytate** (FYE-tate). Phytic acid occurs in the husks of grains, legumes, and seeds and is capable of binding minerals such as zinc, iron, calcium, magnesium, and copper in insoluble complexes in the intestine, which the body excretes unused.

A compound not classed as a fiber but often found with it in foods is phytic acid. Because of this close association, researchers have been unable to determine whether it is the fiber, the phytic acid, or both, that binds with minerals, preventing their absorption.[2] This binding presents a risk of mineral deficiencies, but the risk is minimal when fiber intake is reasonable and mineral intake adequate. The nutrition consequences of such mineral losses are described further in Chapters 12 and 13.

IN SUMMARY

The complex carbohydrates are the polysaccharides (chains of monosaccharides): glycogen, starches, and fibers. Both glycogen and starch are storage forms of glucose—glycogen in the body, and starch in plants—and both yield energy for human use. The fibers also contain glucose (and other monosaccharides), but their bonds cannot be broken by human digestive enzymes, so they yield little, if any, energy.

*When a person eats carbohydrate-rich fruits, vegetables, and legumes, the body receives a valuable commodity—glucose.*

# Digestion and Absorption of Carbohydrates

The ultimate goal of digestion and absorption of sugars and starches is to dismantle them into small molecules that the body can absorb and use—chiefly glucose. The large starch molecules require extensive breakdown; the disaccharides need only to be hydrolyzed once. The initial splitting begins in the mouth; the final splitting and absorption occur in the small intestine; and conversion to a common energy currency (glucose) takes place in the liver. The details follow.

## *The Processes of Digestion and Absorption*

The short chains of glucose units that result from the breakdown of starch are known as **dextrins.** The word sometimes appears on food labels because dextrins can be used as thickening agents in foods.

Figure 4-10 traces the digestion of carbohydrates through the GI tract. When a person eats foods containing starch, enzymes hydrolyze the long chains to shorter chains, the short chains to disaccharides, and, finally, the disaccharides to monosaccharides. This process begins in the mouth.

**amylase** (AM-ih-lace): an enzyme that hydrolyzes amylose (a form of starch). Amylase is a carbohydrase, an enzyme that breaks down carbohydrates.

• ***In the Mouth*** • In the mouth, vigorous chewing of high-fiber foods slows eating and stimulates the flow of saliva. The salivary enzyme amylase starts to work, hydrolyzing starch to shorter polysaccharides and to maltose. Because food is in the mouth for only a short time, very little digestion takes place there.

Reminder: A *bolus* is a portion of food swallowed at one time.

• ***In the Stomach*** • The swallowed bolus mixes with the stomach's acid and protein-digesting enzymes, and these digest the salivary enzyme amylase. Thus amylase is removed from the scene before its job of starch digestion is completed. To a small extent, the stomach's acid continues breaking starch down, but its juices contain no enzymes to digest carbohydrate. Fibers linger in the stomach and delay gastric emptying, thereby providing a feeling of fullness and satiety.

**satiety** (sah-TIE-eh-tee): the feeling of fullness and satisfaction that food brings (Chapter 8 provides a more detailed definition).
• **sate** = to fill

• ***In the Small Intestine*** • The small intestine performs most of the work of carbohydrate digestion. A major carbohydrate-digesting enzyme, pancreatic amylase, enters the intestine via the pancreatic duct and continues breaking down the polysaccharides to shorter glucose chains and disaccharides. The final step takes place on the outer membranes of the intestinal cells. There specific enzymes dismantle specific disaccharides:

- Maltase breaks maltose into 2 glucose molecules.
- Sucrase breaks sucrose into 1 glucose and 1 fructose molecule.
- Lactase breaks lactose into 1 glucose and 1 galactose molecule.

**maltase:** an enzyme that hydrolyzes maltose.

**sucrase:** an enzyme that hydrolyzes sucrose.

**lactase:** an enzyme that hydrolyzes lactose.

**Figure 4-10**
**Carbohydrate Digestion in the GI Tract**

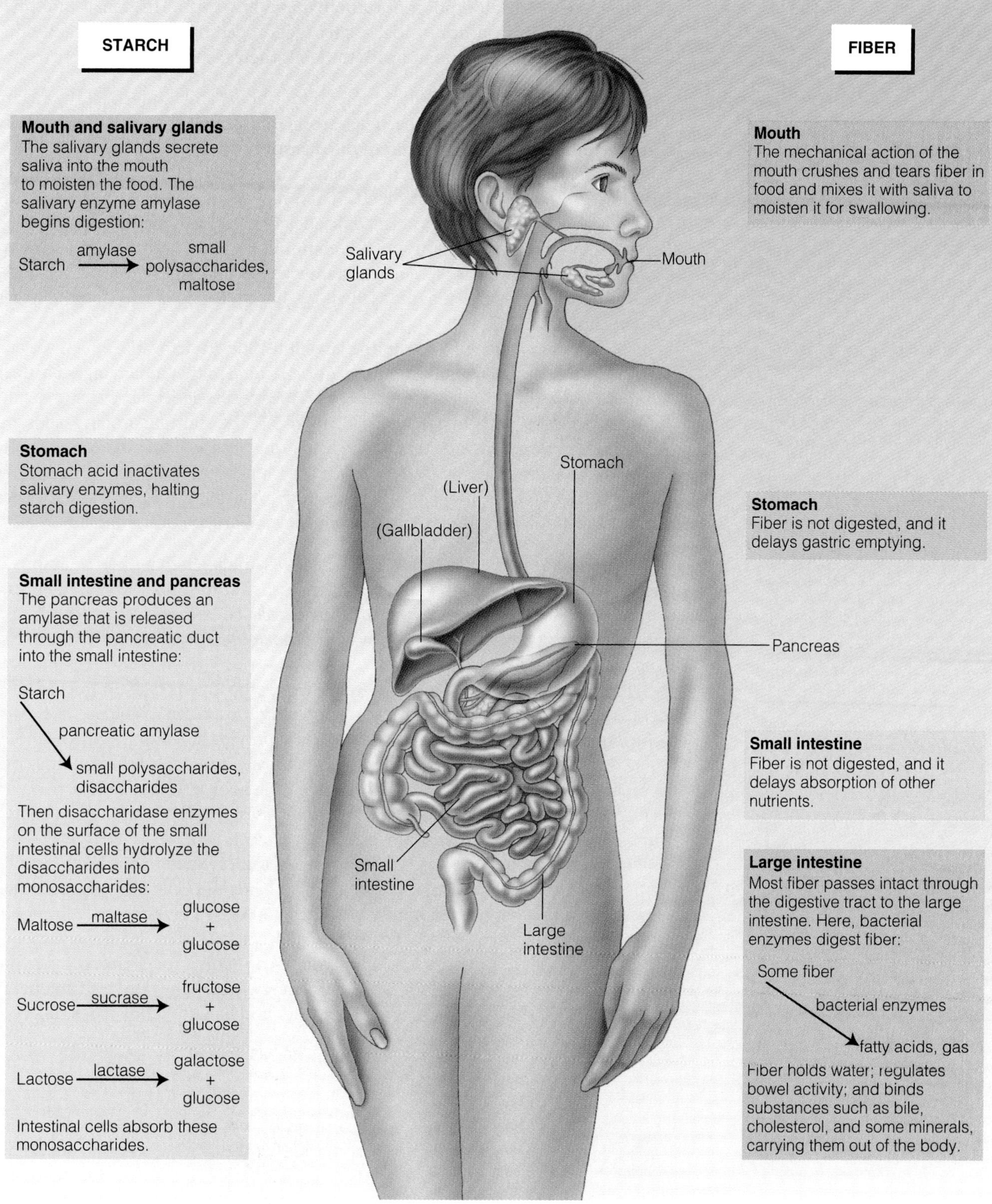

STARCH
Mouth and salivary glands
The salivary glands secrete saliva into the mouth to moisten the food. The salivary enzyme amylase begins digestion:
Starch —amylase→ small polysaccharides, maltose
Stomach
Stomach acid inactivates salivary enzymes, halting starch digestion.
Small intestine and pancreas
The pancreas produces an amylase that is released through the pancreatic duct into the small intestine:
Starch —pancreatic amylase→ small polysaccharides, disaccharides
Then disaccharidase enzymes on the surface of the small intestinal cells hydrolyze the disaccharides into monosaccharides:
Maltose —maltase→ glucose + glucose
Sucrose —sucrase→ fructose + glucose
Lactose —lactase→ galactose + glucose
Intestinal cells absorb these monosaccharides.
Salivary glands
Mouth
(Liver)
Stomach
(Gallbladder)
Pancreas
Small intestine
Large intestine
FIBER
Mouth
The mechanical action of the mouth crushes and tears fiber in food and mixes it with saliva to moisten it for swallowing.
Stomach
Fiber is not digested, and it delays gastric emptying.
Small intestine
Fiber is not digested, and it delays absorption of other nutrients.
Large intestine
Most fiber passes intact through the digestive tract to the large intestine. Here, bacterial enzymes digest fiber:
Some fiber —bacterial enzymes→ fatty acids, gas
Fiber holds water; regulates bowel activity; and binds substances such as bile, cholesterol, and some minerals, carrying them out of the body.

At this point, all disaccharides contribute at least one glucose molecule to the body. Fructose and galactose can eventually become glucose after being processed in the liver, as explained later.

Fibers delay the absorption of carbohydrates and fats in the small intestine, conferring benefits on health that a later section describes further. In addition, fibers in the intestines can bind with minerals there, as mentioned earlier.

• ***In the Large Intestine*** • Within one to four hours after a meal, all the sugars and most of the starches have been digested. Only a small fraction of the starches and the indigestible fibers remain in the digestive tract.[3]

**resistant starch:** starch that escapes digestion and absorption in the small intestine of healthy people.

The small fraction of starches that escapes digestion and absorption in the small intestine is known as resistant starch. Starch may resist digestion for several reasons, reflecting both the individual's efficiency in digesting starches and the food's physical properties.[4] Resistant starch is common in whole lentil beans, raw potatoes, and unripe bananas. Because resistant starches remain in the large intestine, they promote bowel movements as fibers do, but unlike fibers, they do not lower blood cholesterol.[5]

**ferment:** to digest in the absence of oxygen.

Like resistant starches, fibers in the large intestine attract water, which softens the stools for passage without straining. Also, bacteria in the human digestive tract ferment both fibers and resistant starches. This process generates water, gas, and short-chain fatty acids (described in Chapter 5).* The short-chain fatty acids are absorbed in the colon and yield energy when metabolized. Metabolism of short-chain fatty acids occurs in both the intestinal mucosa and the liver. Food fibers and resistant starches, therefore, do contribute some energy (about 2 kcalories per gram), depending on the extent to which they are broken down and absorbed.[6]

• ***Absorption into the Bloodstream*** • Glucose is unique in that it can be absorbed to some extent through the lining of the mouth, but for the most part, all nutrient absorption takes place in the small intestine. The monosaccharides traverse the cells lining the small intestine by active transport and are washed away in the circulating blood.†

The blood then circulates through the liver, whose cells take up fructose and galactose and convert them to other compounds, most often to glucose, as shown in Figure 4-11. Thus all disaccharides not only provide at least one glucose molecule directly, but they can also provide another one indirectly—through the conversion of fructose and galactose to glucose. (The body does not use fructose and glucose in exactly the same ways, but for purposes of this book, they are treated as being metabolically identical.)

This description of the way the body receives carbohydrate should help explode a myth perpetrated by advertisers of high-sugar foods and beverages. They describe sugar as "quick energy" and imply that when you need a pick-me-up, you should reach for a candy bar and a cola beverage. Concentrated sugars do offer energy, but clearly, the best pick-me-ups are carbohydrate-containing foods that deliver vitamins, minerals, and fiber along with their energy. Next time you need an energy boost, why not have a delicious peanut butter and banana sandwich, a tall, cool glass of milk, and a fresh, juicy orange?

## IN SUMMARY

In the digestion and absorption of carbohydrates, the body breaks down starches into disaccharides and disaccharides into monosaccharides; it then converts monosaccharides mostly to glucose to provide energy for the cells' work. The fibers help to regulate the passage of food through the GI system, but contribute little, if any, energy.

*The short-chain fatty acids produced by GI bacteria are primarily acetic acid, propionic acid, and butyric acid.

†Fructose is absorbed by facilitated diffusion.

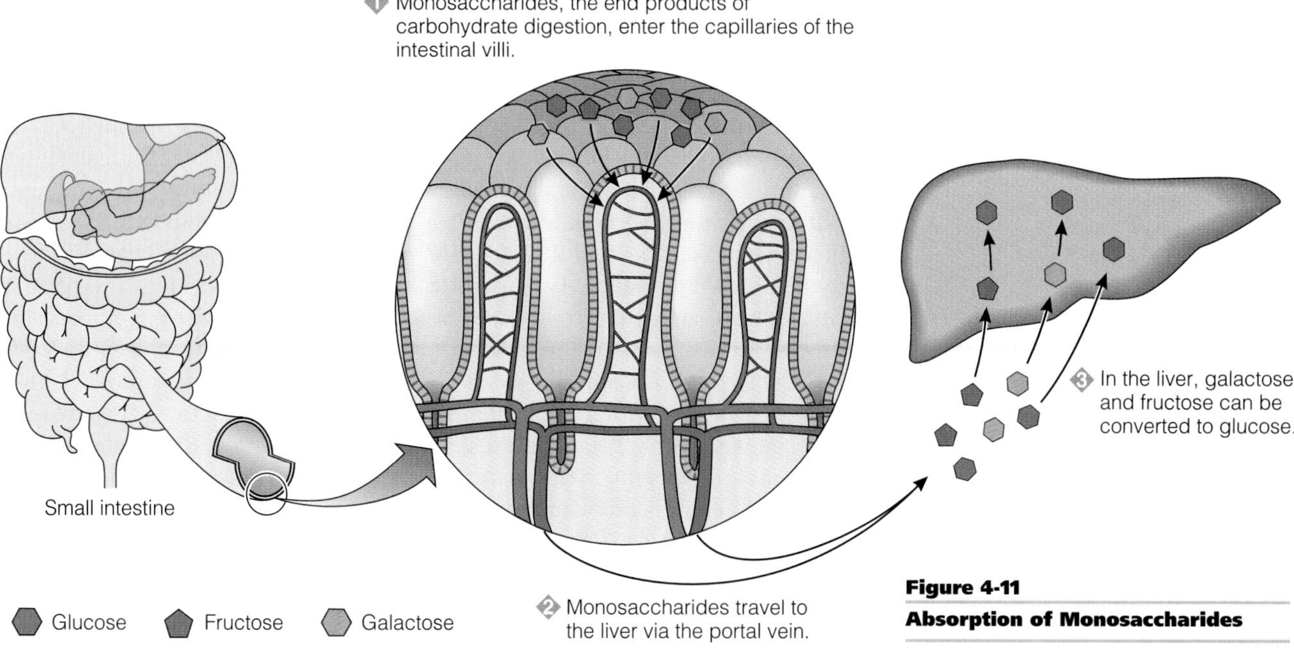
1 Monosaccharides, the end products of carbohydrate digestion, enter the capillaries of the intestinal villi.
Small intestine
2 Monosaccharides travel to the liver via the portal vein.
3 In the liver, galactose and fructose can be converted to glucose.
Glucose
Fructose
Galactose

**Figure 4-11**
**Absorption of Monosaccharides**

## *Lactose Intolerance*

Normally, the enzyme lactase ensures that the disaccharide lactose found in milk is both digested and absorbed efficiently. Lactase activity is highest immediately after birth, as befits an infant whose first and only food for a while will be breast milk or infant formula. In the great majority of the world's populations, lactase activity declines dramatically during childhood and adolescence to about 5 to 10 percent of the activity at birth.[7] Only a relatively small percentage (about 30 percent) of the people in the world retain enough lactase to digest and absorb lactose efficiently throughout adult life.

• ***Symptoms*** • When more lactose is consumed than the available lactase can handle, lactose molecules remain in the intestine undigested, attracting water and causing bloating, abdominal discomfort, and diarrhea—the symptoms of lactose intolerance. The undigested lactose becomes food for intestinal bacteria, which multiply and produce irritating acid and gas, further contributing to the discomfort and diarrhea.

• ***Causes*** • As mentioned, lactase activity commonly declines with age. Lactase deficiency may also develop when the intestinal villi are damaged by disease, certain medicines, prolonged diarrhea, or malnutrition; this can lead to temporary or permanent lactose malabsorption, depending on the extent of the intestinal damage. In extremely rare cases, an infant is simply born with a lactase deficiency.

• ***Prevalence*** • The prevalence of lactose intolerance varies widely among ethnic groups, indicating that the trait is genetically determined.[8] The prevalence of lactose intolerance is lowest among Scandinavians and other northern Europeans and highest among native North Americans and Southeast Asians.

• ***Dietary Changes*** • Managing lactose intolerance requires some dietary changes, although total elimination of milk products is usually not necessary. Excluding all milk products from the diet can lead to nutrient deficiencies, for milk

**lactose intolerance:** a condition that results from inability to digest the milk sugar lactose; characterized by bloating, gas, abdominal discomfort, and diarrhea. Lactose intolerance differs from milk allergy, which is caused by an immune reaction to the protein in milk.

**lactase deficiency:** a lack of the enzyme required to digest the disaccharide lactose into its component monosaccharides (glucose and galactose).

Estimated prevalence of lactose intolerance:
>80% Asian Americans.
80% Native Americans.
75% African Americans.
70% Mediterranean peoples.
60% Inuits.
50% Hispanics.
20% Caucasians.
<10% Northern Europeans.

Lactose in selected foods:

| | |
|---|---|
| Whole-wheat bread, 1 slice | 0.5 g |
| Dinner roll, 1 | 0.5 g |
| Cheese, 1 oz | |
| Cheddar or American | 0.5 g |
| Parmesan or cream | 0.8 g |
| Doughnut (cake type), 1 | 1.2 g |
| Chocolate candy, 1 oz | 2.3 g |
| Sherbet, 1 c | 4.0 g |
| Cottage cheese (low-fat), 1 c | 7.5 g |
| Ice cream, 1 c | 9.0 g |
| Milk, 1 c | 12.0 g |
| Yogurt (low-fat), 1 c | 15.0 g |

is a major source of several nutrients, notably the mineral calcium, the B vitamin riboflavin, and vitamin D. Fortunately, many people with lactose intolerance can consume foods containing up to 6 grams of lactose (½ cup milk) without symptoms.[9] The most successful strategies are to increase intake of milk products gradually, take them with other foods in meals, and spread their intake throughout the day.[10] A change in the GI bacteria, not the reappearance of the missing enzyme, accounts for the ability to adapt to milk products.[11]

In many cases, lactose-intolerant people can tolerate fermented milk products such as yogurt and acidophilus milk. The bacteria in these products digest lactose for their own use, leaving these foods relatively low in lactose. Hard cheeses and cottage cheese are often well tolerated because most of the lactose is removed with the whey during manufacturing. Lactose continues to diminish as the cheese ages.

Many lactose-intolerant people use commercially prepared milk products that have been treated with an enzyme that breaks down the lactose. Alternatively, they take enzyme tablets with meals or add enzyme drops to their milk. The enzyme hydrolyzes much of the lactose in milk to glucose and galactose, which lactose-intolerant people can absorb without ill effects. Most healthy adults with lactose intolerance can drink milk when the lactose content has been reduced by 50 percent.[12]

Because people's tolerance to lactose varies widely, lactose-restricted diets must be highly individualized. A completely lactose-free diet can be difficult because lactose appears not only in milk and milk products but also as an ingredient in many nondairy foods such as breads, cereals, breakfast drinks, salad dressings, and cake mixes. People on strict lactose-free diets need to read labels and avoid foods that include milk, milk solids, whey (milk liquid), and casein (milk protein, which may contain traces of lactose). They also need to check all drugs with the pharmacist because 20 percent of prescription drugs and 5 percent of over-the-counter drugs contain lactose as a filler.

People who consume few or no milk products must take care to meet riboflavin, vitamin D, and calcium needs. Later chapters on the vitamins and minerals offer help with finding good nonmilk sources of these nutrients.

U.S. Government
www.healthfinder.gov/searchoptions/topicsaz.htm
Search for Lactose intolerance

IN SUMMARY

Lactose intolerance is a common condition that occurs when there is insufficient lactase to digest the disaccharide lactose found in milk and milk products. Symptoms include GI distress. Because treatment requires limiting milk intake, other sources of riboflavin, vitamin D, and calcium must be included in the diet.

# Glucose in the Body

Starches and sugars are called **available carbohydrates** because human digestive enzymes break them down for the body's use. In contrast, fibers are called **unavailable carbohydrates** because human digestive enzymes cannot break their bonds.

The primary role of the available carbohydrates in human nutrition is to supply the body's cells with glucose to deliver the indispensable commodity, energy. Starch contributes most to the body's glucose supply, but as explained earlier, any of the sugars can also provide glucose.

Glucose plays the central role in carbohydrate metabolism. The next two sections provide an overview first of the pathways glucose can follow in the body and then of the ways the body regulates those pathways.

## *A Preview of Carbohydrate Metabolism*

This brief discussion provides just enough information about carbohydrate metabolism to illustrate that the body needs and uses glucose as a chief energy nutrient. Chapter 7 provides a full description of energy metabolism.

• ***Storing Glucose as Glycogen*** • The liver stores one-third of the body's total glycogen and releases glucose as needed. During times of plenty, blood glucose rises, and liver cells link the excess glucose molecules into long, branching chains of glycogen. When blood glucose falls, the liver cells dismantle the glycogen into single molecules of glucose and release them into the bloodstream. Thus glucose can supply energy to the central nervous system and other organs regardless of whether the person has eaten recently. Muscle cells can also store glucose as glycogen (the other two-thirds), but they hoard most of their own supply, using it just for themselves during exercise.

Glycogen holds water and therefore is rather bulky. The body can store only enough glycogen to provide energy for relatively short periods of time—during exercise, a few hours' worth at most. For its long-term energy reserves, for use over days or weeks of food deprivation, the body uses its abundant, water-free fuel, fat, as Chapter 5 describes.

***The carbohydrates of grains, vegetables, fruits, and legumes supply most of the energy in a healthful diet.***

• ***Using Glucose for Energy*** • Glucose fuels the work of most of the body's cells. Inside a cell, enzymes break glucose in half. These halves can be put back together to make glucose, or they can be further broken down into smaller fragments (never again to be reassembled to form glucose). The small fragments can yield energy when broken down completely to carbon dioxide and water, or they can be reassembled, but only into units of body fat.

As mentioned, glycogen stores last only for hours, not for days. To keep providing glucose to meet the body's energy needs, a person has to eat dietary carbohydrate frequently. Yet people who do not always attend faithfully to their bodies' carbohydrate needs still survive. How do they manage without glucose from dietary carbohydrate? Do they simply draw energy from the other two energy-yielding nutrients, fat and protein? They do draw energy, but not simply.

• ***Making Glucose from Protein*** • Body protein can be converted to glucose to some extent, but protein has jobs of its own that no other nutrient can do. Body fat cannot be converted to glucose to any significant extent, and although fat breakdown can yield energy for many of the body's cells, only glucose can provide energy for brain cells, other nerve cells, and developing red blood cells.

Thus, when a person does not replenish depleted glycogen stores by eating carbohydrate, body proteins are dismantled to make glucose to fuel these special cells. The conversion of protein to glucose is called gluconeogenesis—literally, the making of new glucose. Only adequate dietary carbohydrate can prevent this use of protein for energy, and this role of carbohydrate is known as its protein-sparing action.

**gluconeogenesis** (gloo-co-nee-oh-GEN-ih-sis): the making of glucose from a noncarbohydrate source (described in more detail in Chapter 7).
- **gluco** = glucose
- **neo** = new
- **genesis** = making

**protein-sparing action:** the action of carbohydrate (and fat) in providing energy that allows protein to be used for other purposes.

**ketone** (KEE-tone) **bodies:** the product of the incomplete breakdown of fat when glucose is not available in the cells.

**ketosis** (kee-TOE-sis): an undesirably high concentration of ketone bodies in the blood and urine.

**acid-base balance:** the equilibrium in the body between acid and base concentrations; see Chapter 12.

• ***Making Ketone Bodies from Fat Fragments*** • An inadequate supply of carbohydrate combined with an accelerated breakdown of fat can shift the body's energy metabolism in a precarious direction. With less carbohydrate available for energy, more fat may be broken down, but not all the way to energy. Instead, the fat fragments combine with each other, forming ketone bodies. Muscles and other tissues can use ketone bodies for energy, but when their production exceeds their use, they accumulate in the blood, causing ketosis, a condition that disturbs the body's normal acid-base balance, as Chapter 7 describes.

To ensure complete sparing of body protein and to prevent ketosis requires 50 to 100 grams of carbohydrate a day. Dietary recommendations urge people to select abundantly from carbohydrate-rich foods to provide for this allowance and considerably more.

• ***Converting Glucose to Fat*** • Given more carbohydrate than it needs, the body uses glucose to meet its energy needs, fills its glycogen stores to capacity, and may still have some leftover. To store the extra glucose, the liver breaks it (and energy-containing fragments from protein or fat, too) into smaller molecules and puts them together into the more permanent energy-storage compound—fat. Then the fat travels to the fatty tissues of the body for storage. Unlike the liver cells, which can store only about half a day's worth of glycogen, fat cells can store unlimited quantities of fat.

Even though excess carbohydrate can be converted to fat and stored, this is a minor pathway.[13] Storing carbohydrate as body fat is energetically expensive. Quite simply, the body uses more energy to convert dietary carbohydrate to body fat than it does to convert dietary fat to body fat. Consequently, body fat comes mainly from dietary fat.[14] A balanced diet high in complex carbohydrates actually helps control body weight. Most carbohydrate-rich foods are so bulky and naturally so low in fat that when large quantities are eaten, they tend to crowd fat out of the diet. Since carbohydrate is less energy dense than fat (with only 4 kcalories to the gram compared with fat's 9), eating a diet high in carbohydrate usually reduces energy intake and supports weight control.

## The Constancy of Blood Glucose

Every body cell depends on glucose for its fuel to some extent, and ordinarily, the cells of the brain and the rest of the nervous system depend *primarily* on glucose for their energy. The activities of these cells never cease, and they do not have the ability to store glucose. Day and night they continually draw on the supply of glucose in the fluid surrounding them. To maintain the supply, a steady stream of blood moves past these cells bringing more glucose from either the intestines (food) or the liver (glycogen).

Reminder: *Homeostasis* is the maintenance of constant internal conditions by the body's control systems.

Normal blood glucose: 80 to 120 mg/dL.

• ***Maintaining Glucose Homeostasis*** • To function optimally, the body must maintain blood glucose within limits that permit the cells to nourish themselves. If blood glucose falls below normal, the person may become dizzy and weak; if it rises above normal, the person may become confused and have difficulty breathing. Left untreated, fluctuations to the extremes—either high or low—can be fatal.

**insulin** (IN-suh-lin): a hormone secreted by special cells in the pancreas in response to (among other things) increased blood glucose concentration. The primary role of insulin is to control the transport of glucose from the bloodstream into the cells.

**glucagon** (GLOO-ka-gon): a hormone that is secreted by special cells in the pancreas in response to low blood glucose concentration and elicits release of glucose from storage.

• ***The Regulating Hormones*** • Blood glucose homeostasis is regulated primarily by two hormones: insulin, which moves glucose from the blood into the cells, and glucagon, which brings glucose out of storage when necessary. Figure 4-12 depicts these hormonal regulators at work.

After a meal, as blood glucose rises, special cells of the pancreas respond by secreting insulin into the blood.* As the circulating insulin contacts the receptors on the body's other cells, the receptors respond by ushering glucose from the blood into the cells. Most of the cells take only the glucose they can use for energy right away, but the liver and muscle cells can assemble the small glucose units into long, branching chains of glycogen for storage. The liver cells can also convert glucose to fat for export to other cells. Thus high blood glucose returns to normal as excess glucose is stored as glycogen (which can be converted back to glucose) and fat (which cannot be).

When blood glucose falls (as occurs between meals), other special cells of the pancreas respond by secreting glucagon into the blood.** Glucagon raises blood glucose by signaling the liver to dismantle its glycogen stores and release glucose into the blood for use by all the other body cells.

**epinephrine** (EP-ih-NEFF-rin): a hormone of the adrenal gland that modulates the stress response; formerly called **adrenaline.**

Another hormone that calls glucose from the liver cells is the "fight-or-flight" hormone, epinephrine. When a person experiences stress, epinephrine acts quickly, ensuring that all the body cells have energy fuel in emergencies. Like glucagon, epinephrine works to return glucose to the blood from liver glycogen.

*The *beta* (BAY-tuh) *cells,* one of several types of cells in the pancreas, secrete insulin in response to elevated blood glucose concentration.

**The *alpha cells* of the pancreas secrete glucagon in response to low blood glucose.

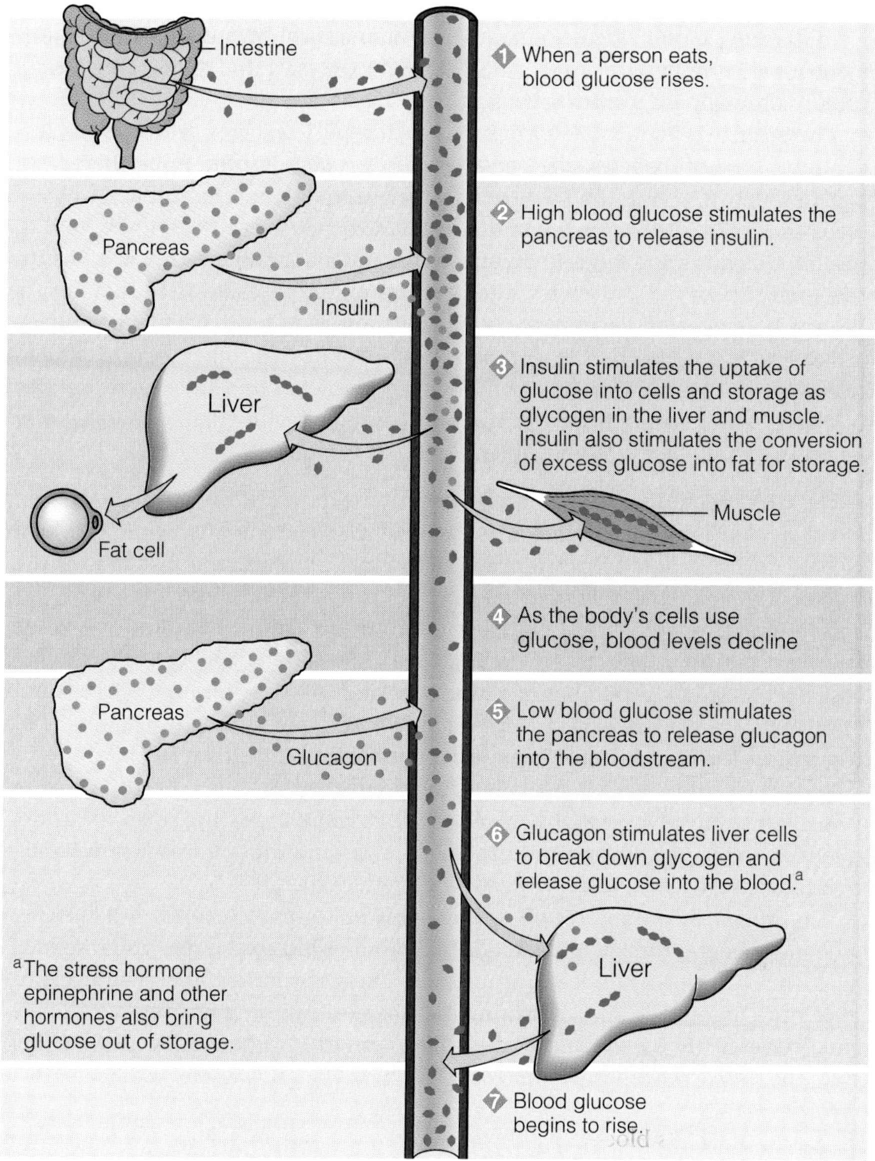

Intestine
1 When a person eats, blood glucose rises.
Pancreas
Insulin
2 High blood glucose stimulates the pancreas to release insulin.
Liver
Fat cell
3 Insulin stimulates the uptake of glucose into cells and storage as glycogen in the liver and muscle. Insulin also stimulates the conversion of excess glucose into fat for storage.
Muscle
4 As the body's cells use glucose, blood levels decline
Pancreas
Glucagon
5 Low blood glucose stimulates the pancreas to release glucagon into the bloodstream.
6 Glucagon stimulates liver cells to break down glycogen and release glucose into the blood.[a]
Liver
[a]The stress hormone epinephrine and other hormones also bring glucose out of storage.
7 Blood glucose begins to rise.


**Figure 4-12**

**Maintaining Blood Glucose Homeostasis**



• ***Balancing within the Normal Range*** • The maintenance of normal blood glucose ordinarily depends on two processes. When blood glucose falls too low, food can replenish it, or in the absence of food, glucagon can signal the liver to break down glycogen stores. When blood glucose rises too high, insulin can signal the cells to take in glucose for energy. Eating balanced meals helps the body maintain a happy medium between the extremes. Balanced meals provide abundant complex carbohydrates, including fibers, some protein, and a little fat. The fibers and fat slow down the digestion and absorption of carbohydrate, so glucose enters the blood gradually, providing a steady, ongoing supply. Dietary protein elicits the secretion of glucagon, whose effects oppose those of insulin, helping to maintain blood glucose within the normal range.

• ***Falling outside the Normal Range*** • This influence of foods on blood glucose has given rise to the oversimplification that foods *govern* blood glucose concentrations. Foods do not; the body does. In some people, however, blood glucose regulation fails. When this happens, either of two conditions can result: diabetes or hypoglycemia. People with these conditions can often use special diet patterns to help maintain their blood glucose within a normal range.

**diabetes** (DYE-ah-BEE-teez): a disorder of carbohydrate metabolism resulting from inadequate or ineffective insulin.

**hypoglycemia** (HIGH-po-gligh-SEE-me-ah): an abnormally low blood glucose concentration.

In diabetes, blood glucose remains high after a meal because insulin is either inadequate or ineffective. Thus, while *blood* glucose is central to diabetes, *dietary* carbohydrates do not cause diabetes.

There are two main types of diabetes.[15] In type 1 diabetes, which is the less common type of diabetes, the pancreas fails to make insulin; researchers hold genetics, toxins, a virus, and a disordered immune system responsible. In type 2 diabetes, which is the more common type of diabetes, the cells fail to respond to insulin; this condition tends to occur as a consequence of obesity. Because obesity can precipitate type 2 diabetes, the best preventive measure is to maintain a healthy body weight. Recommendations for those who have diabetes encourage a diet low in fat and rich in complex carbohydrates and fibers. Concentrated sweets are not strictly excluded from the diabetic diet as they once were, but can be eaten in limited amounts with meals as part of a healthy diet.[16] Diabetes and its associated problems receive full attention in Chapter 18.

**type 1 diabetes:** the less common type of diabetes in which the person produces no insulin at all; also known as **insulin-dependent diabetes mellitus (IDDM)** or **juvenile-onset diabetes** (because it frequently develops in childhood), although some cases arise in adulthood.

**type 2 diabetes:** the more common type of diabetes in which the fat cells resist insulin; also called **noninsulin-dependent diabetes mellitus (NIDDM)** or **adult-onset diabetes.** Type 2 usually progresses more slowly than type 1.

**glycemic** (gligh-SEEM-ic) **effect:** a measure of the extent to which a food, as compared with pure glucose, raises the blood glucose concentration and elicits an insulin response.

• ***The Glycemic Effect*** • The term *glycemic effect* describes the effect of food on blood glucose: how quickly glucose is absorbed after a person eats, how high blood glucose rises, and how quickly it returns to normal. Slow absorption, a modest rise in blood glucose, and a smooth return to normal are considered desirable; fast absorption, a surge in blood glucose, and an overreaction that plunges glucose below normal are undesirable. Different foods have different effects on blood glucose depending on a number of factors working together, and the effect is not always what one might expect.[17] Ice cream, for example, is a high-sugar food, but it produces less of a response than baked potatoes, a high-starch food.[18]

Most relevant to real life, a food's glycemic effect differs depending on whether it is eaten alone or as part of a mixed meal. In addition, eating small meals frequently spreads glucose absorption across the day and thus offers the same metabolic advantages as do foods with a low glycemic effect.[19]

Popular articles sometimes describe eating many small meals and snacks throughout the day as **grazing**.

The rate of glucose absorption is particularly important to people with diabetes, who may benefit from avoiding foods that produce too great a rise, or too sudden a fall, in blood glucose. Indeed, some studies have shown that taking the glycemic effect into account in meal planning is a practical way to improve glucose control.[20] Overall, though, planning should focus on total carbohydrate intake rather than the source of carbohydrate.[21]

IN SUMMARY

Dietary carbohydrates provide glucose that can be used by the cells for energy, stored by the liver and muscle as glycogen, or converted into fat if intakes exceed needs. All of the body's cells depend on glucose; those of the central nervous system are especially dependent on it. Without glucose, the body is forced to break down its protein tissues to make glucose and to alter energy metabolism to make ketone bodies from fats. Blood glucose regulation depends primarily on two pancreatic hormones: insulin to remove glucose from the blood into the cells when levels are high and glucagon to free glucose from glycogen stores and release it into the blood when levels are low.

# Health Effects and Recommended Intakes of Sugars

Ever since people first discovered honey and dates, they have enjoyed the sweetness of sugars. In the United States, the natural sugars of milk, fruits, vegetables, and grains account for about half of the sugar intake; the other half consists of sugars that have been refined and added to foods for a variety of purposes. Added sug-

## Glossary of Sugars

**brown sugar:** refined white sugar crystals to which manufacturers have added molasses syrup with natural flavor and color; 91 to 96 percent pure sucrose.

**confectioners' sugar:** finely powdered sucrose; 99.9 percent pure.

**corn sweeteners:** corn syrup and sugars derived from corn.

**corn syrup:** a syrup produced by the action of enzymes on cornstarch; contains mostly glucose. See also *high-fructose corn syrup (HFCS).*

**dextrose:** an older name for glucose.

**granulated sugar:** crystalline sucrose; 99.9 percent pure.

**high-fructose corn syrup (HFCS):** a corn-syrup sweetener made especially for use in processed foods and beverages, where it is the predominant sweetener. HFCS is mostly fructose; glucose makes up the balance.

**honey:** sugar (mostly sucrose) formed from nectar gathered by bees. An enzyme splits the sucrose into glucose and fructose. Composition and flavor vary, but honey always contains a mixture of sucrose, fructose, and glucose.

**invert sugar:** a mixture of glucose and fructose formed by the hydrolysis of sucrose in a chemical process; sold only in liquid form and sweeter than sucrose. Invert sugar is used as a food additive to help preserve freshness and prevent shrinkage.

**levulose:** an older name for fructose.

**maple sugar:** a sugar (mostly sucrose) purified from the concentrated sap of the sugar maple tree.

**molasses:** the thick brown syrup produced during sugar refining. Molasses retains residual sugar and other by-products and a few minerals; blackstrap molasses contains significant amounts of calcium and iron—the iron comes from the *machinery* used to process the sugar.

**raw sugar:** the first crop of crystals harvested during sugar processing. Raw sugar cannot be sold in the United States because it contains too much filth (dirt, insect fragments, and the like). Sugar sold as "raw sugar" domestically has actually gone through over half of the refining steps.

**turbinado** (ter-bih-NOD-oh) **sugar:** sugar produced using the same refining process as white sugar, but without the bleaching and anti-caking treatment; traces of molasses give turbinado its sandy color.

**white sugar:** pure sucrose or "table sugar," produced by dissolving, concentrating, and recrystallizing raw sugar.

ars assume various names: sucrose, invert sugar, corn sugar, corn syrups and solids, high-fructose corn syrup, and honey (see the glossary above).

The use of sweeteners in food manufacturing has risen steadily over the past two decades, reaching a record high of 139 pounds per person per year. This estimate represents all sweeteners used in the marketing system, including sugar lost or wasted, such as in the brine of sweet pickles or in jams or bakery goods that spoil before they are eaten. It also includes sugar used in pet foods and in fermentation. Estimates of *intake* indicate that on the average, each person consumes about 45 pounds of added sugar per year.[22] As a percentage of daily energy intake, this amount is roughly equivalent to current recommendations that concentrated sugars contribute no more than about 10 percent of energy intake.

As an additive, sugar:
- Enhances flavor.
- Supplies texture and color to baked goods.
- Provides fuel for fermentation, causing bread to rise or producing alcohol.
- Acts as a bulking agent in ice cream and baked goods.
- Acts as a preservative in jams.
- Balances the acidity of tomato- and vinegar-based products.

# *Health Effects of Sugars*

In moderate amounts (similar to current consumption levels), sugars add pleasure to meals without harming health. In excess, however, they can be detrimental in two ways. One, sugars can contribute to nutrient deficiencies by supplying energy (kcalories) without providing nutrients, and so dietary guidelines caution people against eating large quantities. Two, sugars contribute to tooth decay, and so dietary guidelines caution people against eating frequent snacks containing sugars and starches.

• ***Nutrient Deficiencies*** • Foods that contain lots of added sugar such as cakes, candies, and colas deliver glucose and energy with few, if any, other nutrients—they are called empty-kcalorie foods. By comparison, foods such as grains, vegetables, and fruits that contain some natural sugars and lots of starches and fibers deliver their glucose and energy along with protein, vitamins, and minerals.

**empty-kcalorie food:** a popular term used to denote foods that contribute energy but are lacking protein, vitamins, and minerals. Empty-kcalorie foods are *low–nutrient density foods.* The most notorious empty-kcalorie foods are sugar, fat, and alcohol.

A person spending 200 kcalories of a day's energy allowance on a 16-ounce cola gets little of value for those kcaloric "dollars." In contrast, a person using 200 kcalories on three slices of whole-wheat bread gets 9 grams of protein, 6 grams of fiber, plus several of the B vitamins with those kcalories. For the person who wants

*1 tsp honey = 22 kcal.*
*1 tsp sugar = 16 kcal.*

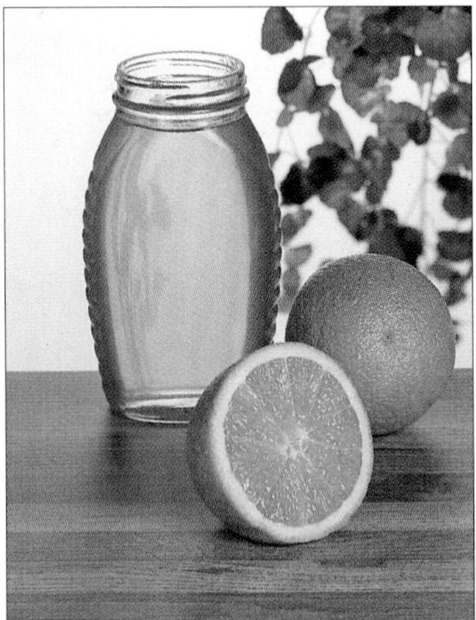

*You receive the same sugars from an orange as from honey, but the packaging makes a big nutrition difference.*

something sweet, perhaps a reasonable compromise would be to have two slices of bread with a teaspoon of jam on each. The amount of sugar a person can afford depends on how many kcalories are available beyond those needed to deliver indispensable vitamins and minerals.

With careful food selections, a person can obtain all the needed nutrients within an allowance of about 1500 kcalories. Some people have more generous energy allowances with which to "purchase" nutrients. For example, an active teenage boy may need as many as 4000 kcalories a day. If he eats mostly nutritious foods, then the "empty kcalories" of cola beverages may be an acceptable addition to his diet. On the other hand, an inactive older woman who is limited to fewer than 1500 kcalories a day cannot afford any but the most nutrient-dense foods.

Some people believe that because honey is a natural food, it is nutritious—or, at least, more nutritious than sugar. A look at their chemical structures reveals the truth. Honey, like table sugar, contains glucose and fructose. The primary difference is that in table sugar the two monosaccharides are bonded together, whereas in honey some of them are free. Whether a person eats monosaccharides individually, as in honey, or linked together, as in table sugar, they end up the same way in the body: as glucose and fructose.

Honey does contain a few vitamins and minerals, but not many, as Table 4-2 shows. Honey is denser than crystalline sugar, too, so it provides more energy per spoon.

This is not to say that all sugar sources are alike, for some are more nutritious than others. Consider a fruit, say, an orange. The fruit may give you the same amounts of fructose and glucose and the same number of kcalories as a dose of sugar or honey, but the packaging is more valuable nutritionally. The fruit's sugars arrive in the body diluted in a large volume of water, packaged in fiber, and mixed with valuable minerals and vitamins.

As these comparisons illustrate, the significant difference between sugar sources is not between "natural" honey and "purified" sugar but between concentrated sweets and the dilute, naturally occurring sugars that sweeten foods. You can suspect an exaggerated nutrition claim when someone asserts that one product is more nutritious than another because it contains honey.

Sugar can contribute to nutrient deficiencies only by displacing nutrients. For nutrition's sake, the appropriate attitude to take is not that sugar is "bad" and must be avoided, but that nutritious foods must come first. If the nutritious foods end up crowding sugar out of the diet, that is fine—but not the other way around. As always, the goals to seek are balance, variety, and moderation.

**• *Dental Caries* •** Both sugars and starches begin breaking down to sugars in the mouth and so can contribute to tooth decay. Bacteria in the mouth ferment the sugars and in the process produce an acid that dissolves tooth enamel. People can eat sugar without this happening, though, for much depends on how long acid-yielding foods stay in the mouth. Sticky foods stay on the teeth longer and keep yielding acid longer than foods that are readily cleared from the mouth. For that reason, sugar consumed quickly in a soft drink, for example, is less likely to cause dental caries than sugar in a pastry. By the same token, the sugar in sticky foods such as dried fruits is more detrimental than its quantity alone would suggest.

**dental caries:** decay of teeth.
- **caries** = rottenness

Another concern is how often people eat sugar. Bacteria produce acid for 20 to 30 minutes after each exposure. If a person eats three pieces of candy at one time, the teeth will be exposed to approximately 30 minutes of acid destruction. But, if the person eats three pieces at half-hour intervals, the time of exposure increases to 90 minutes. Likewise, slowly sipping a sugary soft drink may be more harmful than drinking quickly and clearing the mouth of sugar. Nonsugary foods can help remove sugar from tooth surfaces; hence, it is better to eat sugar with meals than between meals.

To prevent dental caries:
- Eat sugary foods with meals.
- Limit between-meal snacks containing sugars and starches.
- Brush and floss teeth regularly.
- If brushing and flossing are not possible, at least rinse with water.

**plaque, dental:** a gummy mass of bacteria that grows on teeth and can lead to dental caries and gum disease.

The development of caries depends on several factors: the bacteria that reside in the plaque, the saliva that cleanses the mouth, the minerals that form the teeth,

**Table 4-2**
**Sample Nutrients in Sugars and Other Foods**

The indicated portion of any of these foods provides approximately 100 kcalories. Notice that for a similar number of kcalories and grams of carbohydrate, milk, legumes, fruits, grains, and vegetables offer more of the other nutrients than do the sugars.

| | Size of 100 kcal Portion | Carbohydrate (g) | Protein (g) | Calcium (mg) | Iron (mg) | Vitamin A (µg RE) | Vitamin C (mg) |
|---|---|---|---|---|---|---|---|
| ***Foods*** | | | | | | | |
| Milk, 1% low-fat | 1 c | 12 | 8 | 300 | 0.1 | 144 | 2 |
| Kidney beans | ½ c | 20 | 7 | 30 | 1.6 | 0 | 2 |
| Apricots | 6 | 24 | 2 | 30 | 1.1 | 554 | 22 |
| Bread, whole wheat | 1½ slices | 20 | 4 | 30 | 1.9 | 0 | 0 |
| Broccoli, cooked | 2 c | 20 | 12 | 188 | 2.2 | 696 | 148 |
| ***Sugars*** | | | | | | | |
| Sugar, white | 2 tbs | 24 | 0 | trace | trace | 0 | 0 |
| Molasses, blackstrap | 2½ tbs | 28 | 0 | 343 | 12.6 | 0 | 0.1 |
| Cola beverage | 1 c | 26 | 0 | 6 | trace | 0 | 0 |
| Honey | 1½ tbs | 26 | trace | 2 | 0.2 | 0 | trace |

and the foods that remain after swallowing.[23] For most people, good oral hygiene will prevent dental caries.[24] In short, sugars cause dental caries, but they require a cooperating victim to do so.

U.S. Government
www.healthfinder.gov/searchoptions/topicsaz.htm
Search for Tooth decay

# *Accusations against Sugars*

Sugars have been blamed for a variety of other problems. The following paragraphs evaluate some of these accusations.

• ***Accusation: Sugar Causes Obesity*** • Foods high in added sugars are usually high in fat, too, so consumption of these foods increases total energy and fat intakes. Simultaneously, physical activity often declines. Thus sugar contributes to obesity as a companion to fat, not as the sole cause of obesity—and obesity can occur without a high-sugar consumption.[25] The notion that eating sweet foods stimulates appetite and promotes overeating has not been supported by research.[26]

• ***Accusation: Sugar Causes Heart Disease*** • Researchers agree that unusually high doses of refined sugar can alter blood lipids to favor heart disease.[27] This effect is most dramatic in "carbohydrate-sensitive" individuals—people who respond to sucrose with abnormally high insulin secretion, which promotes the making of excess fat.[28] For most people, though, moderate sugar intakes do *not* influence the risk of heart disease. To keep these findings in perspective, consider that heart disease correlates most closely with factors that have nothing to do with nutrition, such as smoking and genetics. Among dietary risk factors, several—such as total fats, saturated fats, cholesterol, and obesity—have much stronger associations with heart disease than do sugar intakes.

• ***Accusation: Sugar Causes Misbehavior in Children and Criminal Behavior in Adults*** • Sugar has been blamed for the misbehaviors of hyperactive children, delinquent adolescents, and lawbreaking adults. Such speculations have been based on personal stories and have not been confirmed by scientific research.[29] No scientific evidence supports a relationship between sugar and hyperactivity or other misbehaviors.[30] Chapter 16 provides accurate information on diet and children's behavior.

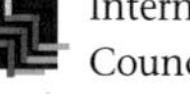
International Food Information Council
ificinfo.health.org
Search for Sugars

## Recommended Intakes of Sugars

The *Dietary Guidelines* urge people to use sugars only in moderation. Other recommendations specify that sugars should account for only 10 percent or less of the day's total energy intake. A person consuming 2000 kcalories a day, then, should receive no more then 200 kcalories (that is, 50 grams or less) from concentrated sugars.

For perspective, each of these concentrated sugars provides 200 kcal:

- 16 oz cola.
- 2 oz jelly beans.
- ¾ c gelatin dessert.

Food labels list the total grams of sugar a food provides. This total reflects both added sugars and those occurring naturally in foods. A food is likely to be high in sugars if its ingredient list starts with any of the sugars named in the glossary on p. 107 or if it includes several of them.

IN SUMMARY

As currently consumed, sugars pose no major health threat except for an increased risk of dental caries. Excessive intakes may displace needed nutrients and fiber; when accompanied by fat, sugars may contribute to obesity. If on these grounds a person decides to limit daily sugar intake, it is important to recognize that not all sugars need to be restricted, just *concentrated* sweets, which are relatively empty of other nutrients and high in kcalories. Sugars that occur naturally in fruits, vegetables, and milk are acceptable.

# Health Effects and Recommended Intakes of Starch and Fibers

Carbohydrates and fats are the two major sources of energy in the diet. When one is high, the other is usually low—and vice versa. The average fat intake in the United States is high compared with health recommendations. To lower fat intake and improve the balance between these two energy nutrients, people need to replace fatty foods with grain products, vegetables, legumes, and fruits—foods noted for their complex carbohydrates.

***Foods rich in starch and fiber offer many health benefits.***

## Health Effects of Starch and Fibers

In addition to starch and dietary fibers, grains, vegetables, legumes, and fruits supply valuable vitamins and minerals and little or no fat. The following paragraphs describe some of the health benefits of diets rich in complex carbohydrates.

• ***Weight Control*** • Foods rich in complex carbohydrates tend to be low in fat and added sugars and can therefore promote weight loss by providing less energy per bite. They also provide satiety and delay hunger. Several studies have found that people who eat a high-carbohydrate breakfast take in fewer kcalories at later meals and snacks than people who eat low-fiber or high-fat breakfasts.[31]

Many weight-loss products on the market today contain bulk-inducing fibers such as methylcellulose, but buying pure fiber compounds like this is neither necessary nor advisable. To use fiber in a weight-loss plan, select fresh fruits, vegetables, legumes, and whole-grain foods. High-fiber foods not only add bulk to the diet, but are economical and nutritious. (A note of caution, though: on baked goods, read the label—many items, including those popular large bran muffins, are high in fat and contain added sugar.)

International Food Information Council
ificinfo.health.org
Search for Fiber

• ***Heart Disease*** • High-carbohydrate diets are associated with low blood cholesterol and a low risk of heart disease.[32] Sorting out the exact reasons why can be difficult.[33] Such diets are low in animal fat and cholesterol and high in soluble fibers and vegetable proteins—all factors associated with a lower risk of heart disease.

The role of animal fat and cholesterol in heart disease is discussed in Chapter 5. The role of vegetable proteins in heart disease is discussed in Chapter 6.

Foods rich in soluble fibers (such as oat bran, barley, and legumes) lower blood cholesterol by binding with bile acids, thus increasing their excretion. With fewer bile acids available in the intestine, fewer lipids are absorbed. More importantly, the binding prevents bile acids from being returned to the liver where they can be reused to make cholesterol.[34] Consequently, the liver must use its cholesterol to make new bile acids. In addition, the bacterial by-products of fiber digestion in the colon also inhibit cholesterol synthesis in the liver. The net result is lower blood cholesterol.

Several researchers have speculated that fiber may also exert its effect by displacing fats in the diet. Even when dietary fat is low, however, high intakes of soluble fibers exert a separate and significant cholesterol-lowering effect.[35] In other words, a high-fiber diet helps to prevent heart disease independent of fat intake.[36]

Adequate fiber:
- Fosters weight control.
- Lowers blood cholesterol.
- Helps prevent colon cancer.
- Helps prevent and control diabetes.
- Helps prevent and alleviate hemorrhoids.
- Helps prevent appendicitis.
- Helps prevent diverticulosis.

• ***Cancer*** • A high-carbohydrate diet, especially one that includes plenty of green and yellow vegetables and citrus fruits, protects against some types of cancer. Again, it is unclear whether the protection derives from the fiber, the vitamins, or phytochemicals.

Populations consuming high-fiber diets generally have lower rates of colon cancer than similar populations consuming low-fiber diets. Fiber may help prevent colon cancer by diluting, binding, and rapidly removing potentially cancer-causing agents from the colon. Alternatively, the protective effect may be due to the fermentation of resistant starch and fiber in the colon, which lowers the pH.[37] A lower pH in the colon is associated with decreased colon cancer risks.[38]

The role antioxidant vitamins and phytochemicals play in cancer prevention is discussed in Highlight 11.

• ***Diabetes*** • Populations eating high-carbohydrate diets often have low rates of diabetes, most likely because such diets are low in fat. High-carbohydrate, low-fat diets help control weight, and this is the most effective way to prevent the most common type of diabetes (type 2). Furthermore, when soluble fibers trap nutrients and delay their transit through the GI tract, glucose absorption is slowed, and this helps to prevent the glucose surge and rebound that seem to be associated with diabetes onset. High-fiber foods play a key role in reducing the risk of diabetes.[39]

**diverticula** (dye-ver-TIC-you-la): sacs or pouches that develop in the weakened areas of the intestinal wall (like bulges in an inner tube where the tire wall is weak). The condition of having diverticula is **diverticulosis** (DYE-ver-tic-you-LOH-sis). About one in every six people in Western countries develops diverticulosis in middle or later life. Diverticulosis may become **diverticulitis** (DYE-ver-tic-you-LYE-tis) if the pockets become infected or inflamed.
- **divertir** = to turn aside
- **osis** = condition
- **itis** = infection or inflammation

• ***GI Health*** • Dietary fibers enhance the health of the large intestine. The healthier the intestinal walls, the better they can block absorption of unwanted constituents. Insoluble fibers such as cellulose (as in cereal brans, fruits, and vegetables) enlarge the stools, easing passage, and they speed up transit time. In this way, the undigested fibers, together with the microbial growth they stimulate, help to alleviate or prevent constipation.

Taken with ample fluids, fibers help to prevent several GI disorders. Large, soft stools ease elimination for the rectal muscles and reduce the pressure in the lower bowel, making it less likely that rectal veins will swell (hemorrhoids). Fiber prevents compaction of the intestinal contents, which could obstruct the appendix and permit bacteria to invade and infect it (appendicitis). In addition, fiber stimulates the GI tract muscles so that they retain their strength and resist bulging out into pouches known as diverticula (illustrated in Figure 4-13).

**Figure 4-13**
**Diverticula**

Diverticula may develop anywhere along the GI tract, but are most common in the colon.

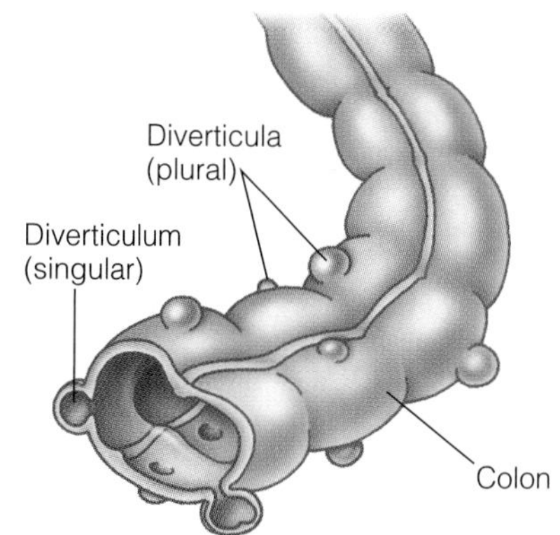


• ***Harmful Effects of Excessive Fiber Intake*** • Despite fiber's benefits to health, a diet high in fiber also has a few drawbacks. A person who has a small capacity and eats mostly high-fiber foods may not be able to take in enough food energy or nutrients. The malnourished, the elderly, and children adhering to all-plant diets are especially vulnerable to this problem.

Launching suddenly into a high-fiber diet can cause temporary bouts of abdominal discomfort, gas, and diarrhea and, more seriously, can obstruct the GI tract. To prevent such complications, a person adopting a high-fiber diet is advised to:

- Increase fiber intake gradually over several weeks to give the GI tract time to adapt.

Excess fiber:

- Displaces energy- and nutrient-dense foods.
- Causes intestinal discomfort and distention.
- Interferes with mineral absorption.

- Drink lots of fluids to soften the fiber as it moves through the GI tract.
- Select fiber-rich foods from a variety of sources—fruits, vegetables, legumes, and whole-grain breads and cereals.

Fiber can limit the absorption of some nutrients by speeding the transit of foods through the GI tract. Also, insoluble fibers can bind to minerals and interfere with their absorption. When mineral intake is adequate, however, a reasonable intake of high-fiber foods does not seem to compromise mineral balance.

Clearly, fiber is like all the nutrients in that "more" is only "better" up to a point. Too much is no better than too little. Again, the key words are balance, moderation, and variety.

## *Recommended Intakes of Starch and Fibers*

Quick and easy estimate:
To attain 55 to 60% of energy from carbohydrate, look for 14 to 15 g of carbohydrate for each 100 kcal of food.

Dietary recommendations suggest that carbohydrates provide more than half (55 to 60 percent) of the energy requirement. A person consuming 2000 kcalories a day should therefore have 1100 to 1200 kcalories of carbohydrate, or about 275 to 300 grams. The Food and Drug Administration (FDA) used this 60 percent of kcalories guideline in establishing the Daily Value on food labels for carbohydrate of 300 grams per day. For most people, this means increasing total carbohydrate intake. To this end, the *Dietary Guidelines* encourage people to choose a diet with plenty of grains, vegetables, and fruit products.

**Healthy People 2000:** Increase complex carbohydrate and fiber-containing foods in the diets of adults to five or more daily servings for vegetables (including legumes) and fruits and to six or more daily servings for grain products.

Chapter 2 described how whole-grain products retain many nutrients that are commonly lost during processing (see Figure 2-4 on p. 45).

Recommendations for fiber suggest the same foods just mentioned: whole-grains, vegetables, fruits, and legumes, which also provide minerals and vitamins. The FDA set a Daily Value on food labels for fiber at 25 grams or 11.5 grams per 1000-kcalorie energy intake. The American Dietetic Association suggests 20 to 35 grams of dietary fiber daily, which is about two times higher than the average intake in the United States.[40] An effective way to add fiber while cutting fat is to substitute plant sources of proteins (legumes) for animal sources (meats). Table 4-3 presents a list of fiber sources.

• ***Choose Wisely*** • In selecting high-fiber foods, keep in mind the principle of variety. The fibers in some foods lower cholesterol, those in other foods help promote GI tract health.

A diet following the Daily Food Guide plan, which includes 3 to 5 vegetable servings, 2 to 4 fruit servings, and 6 to 11 bread servings daily, can easily supply the recommended amount of carbohydrates and fiber. The "How to" feature on p. 114 describes an easy way to estimate the carbohydrate content of a meal. The exchange system has no sugar list, but sugars do contribute to carbohydrate and energy intake. To help estimate carbohydrate and energy intakes accurately, the list in the margin shows what concentrated sweets are equivalent to 1 teaspoon of white sugar. These sugars all provide *about* 5 grams of carbohydrate and *about* 20 kcalories per teaspoon. Some are lower (16 kcalories for table sugar), while others are higher (22 kcalories for honey), but a 20-kcalorie average is an acceptable approximation. For a person who uses catsup liberally, it may help to remember that 1 tablespoon of catsup supplies about 1 teaspoon of sugar.

1 tsp white sugar =

- 1 tsp brown sugar.
- 1 tsp candy.
- 1 tsp corn sweetener or corn syrup.
- 1 tsp honey.
- 1 tsp jam or jelly.
- 1 tsp maple sugar or maple syrup.
- 1 tsp molasses.
- 1½ oz carbonated soda.

1 tbs catsup = 1 tsp sugar.

The exchange system includes on its "other carbohydrate" list moderate servings of some foods that are high in sugar and fat, such as cakes and cookies. Table G-9 in Appendix G lists exchanges for these; notice their small portion sizes.

• ***Read Food Labels*** • Food labels list the amount, in grams, of total carbohydrate—including starch, fibers, and sugars—per serving. Fiber grams are also listed

**Table 4-3**
**Fiber in Selected Foods**

**Bread, Cereal, Rice, and Pasta Group**

Whole-grain products provide about 1 to 2 grams (or more) of fiber per serving:

- 1 slice whole-wheat, pumpernickel, rye bread.
- 1 oz ready-to-eat cereal (100% bran cereals contain 10 grams or more).
- ½ c cooked barley, bulgur, grits, oatmeal.

**Vegetable Group**

Most vegetables contain about 2 to 3 grams of fiber per serving:

- 1 c raw bean sprouts.
- ½ c cooked broccoli, brussels sprouts, cabbage, carrots, cauliflower, collards, corn, eggplant, green beans, green peas, kale, mushrooms, okra, parsnips, potatoes, pumpkin, spinach, sweet potatoes, swiss chard, winter squash.
- ½ c chopped raw carrots, peppers.

**Fruit Group**

Fresh, frozen, and dried fruits have about 2 grams of fiber per serving:

- 1 medium apple, banana, kiwi, nectarine, orange, pear.
- ½ c applesauce, blackberries, blueberries, raspberries, strawberries.
- Fruit juices contain very little fiber.

**Legumes**

Many legumes provide about 8 grams of fiber per serving:

- ½ c cooked baked beans, black beans, black-eyed peas, kidney beans, navy beans, pinto beans.

Some legumes provide about 5 grams of fiber per serving:

- ½ c cooked garbanzo beans, great northern beans, lentils, lima beans, split peas.

NOTE: Appendix H provides fiber grams for over 2000 foods.

separately, as are the grams of sugars. (With this information, you could calculate starch grams by subtracting the grams of fibers and sugars from the total carbohydrate.) Sugars reflect both added sugars and those that occur naturally in foods. Total carbohydrate and dietary fiber are also expressed as "% Daily Values" for a person consuming 2000 kcalories; there is no Daily Value for sugars.

The FDA authorizes three health claims concerning fiber on food labels: one is for "fruits, vegetables, and grain products that contain fiber, particularly soluble fiber, and risk of coronary heart disease," another is for "fiber-containing grain products, fruits, and vegetables and cancer," and the most recent is for "soluble fiber from whole oats and heart disease." Another more general health claim addresses "fruits and vegetables and cancer." Chapter 2 describes the criteria foods must meet to bear these health claims.

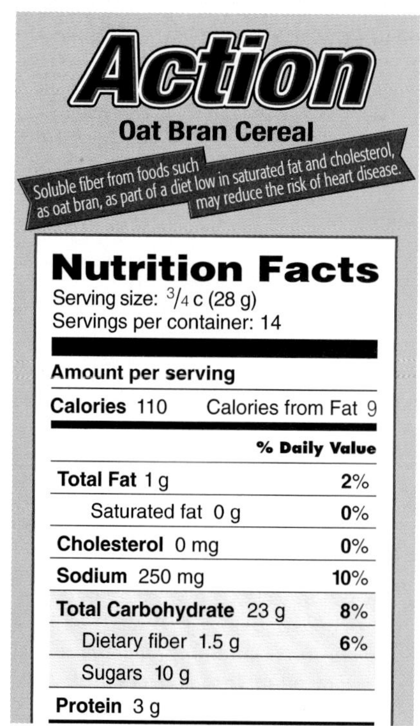


Food labels provide the quantities of total carbohydrate, dietary fiber, and sugars. Total carbohydrate and dietary fiber are also stated as "% Daily Values."

IN SUMMARY

Clearly, a diet rich in complex carbohydrates—starches and fibers—supports efforts to control body weight and prevent heart disease, cancer, diabetes, and GI disorders.[41] For these reasons, recommendations urge people to eat plenty of grains, vegetables, legumes, and fruits—enough to provide 55 to 60 percent of the daily energy intake from carbohydrate.

In today's world, there is one other reason why plant foods rich in starch and natural sugars are a better choice than animal foods or foods high in concentrated sweets: in general, the energy and resources required to grow and process plant foods are less. Chapter 20 takes a closer look at the environmental impacts of food production and use.

## HOW TO | Use the Exchange System to Estimate Carbohydrate

The exchange system described in Chapter 2 provides a convenient way to estimate carbohydrate intake because the foods within each list have a similar carbohydrate content. To use the system, you need to know the carbohydrate value for each exchange list (see the accompanying table) and the foods on that list with their portion sizes (review Figure 2-2 on pp. 40–41).

Familiarity with portion sizes makes estimations easier. For example, it helps to recognize that this bowl of cereal contains 1 cup shredded wheat with 1 cup of milk and ½ banana. Then you can translate these portions into exchanges: 2 starches, 1 milk, and ½ fruit. Finally, you can calculate 15 grams of carbohydrate for each starch, 12 grams for the milk, and 8 grams for the ½ fruit.

| One Exchange | Carbohydrate (g) |
|---|---|
| Starch | 15 |
| Fruits | 15 |
| Other carbohydrates | 15 |
| Milks | 12 |
| Vegetables | 5 |
| Meats | — |
| Fats | — |
| Sugars (1 tsp)[a] | 5 |

[a]Sugars are not officially part of the exchange system, but do have to be counted.

| Breakfast | | Exchange | Carbohydrate (g) Estimate | Carbohydrate (g) Actual |
|---|---|---|---|---|
| 1 c shredded wheat | = | 2 starches | 30 | 34 |
| 1 c milk | = | 1 milk | 12 | 12 |
| ½ banana | = | ½ fruit | 8 | 13 |
| | | | 50 | 59 |

Using the exchange system to estimate, this breakfast provides about 50 grams of carbohydrate. A computer diet analysis program came to a slightly higher conclusion (59 grams), as would a diet analysis using the values in Appendix H. Small variations between values arrived at differently may seem disconcerting, but remember that all are only estimates. Estimates save time; often only a ballpark figure is needed anyway.

Most estimates of the nutrient contents of foods are rough but serviceable approximations. A "90-kcalorie potato" actually means a "90-kcalorie plus or minus about 20 percent potato," which makes it not significantly different from a 100-kcalorie potato. For most purposes, a variation of about 20 percent is considered reasonable, which is the difference between the values in this example.

Fiber appears in only the starches, vegetables, and fruits lists. To estimate fiber, remember that most items on these lists provide at least 2 grams of fiber per serving; some provide 3 or more. Knowing this, a reasonable fiber estimate for this breakfast (with 2 starches and ½ fruit) would be 5 to 7 grams—and a diet analysis report of 5 grams would agree.

Just a few calculations of this kind will give you a feel for the carbohydrate content of a diet. Once you are aware of the major carbohydrate-contributing foods you eat, you can return to thinking simply in terms of foods, developing a sense of how much of each is enough.

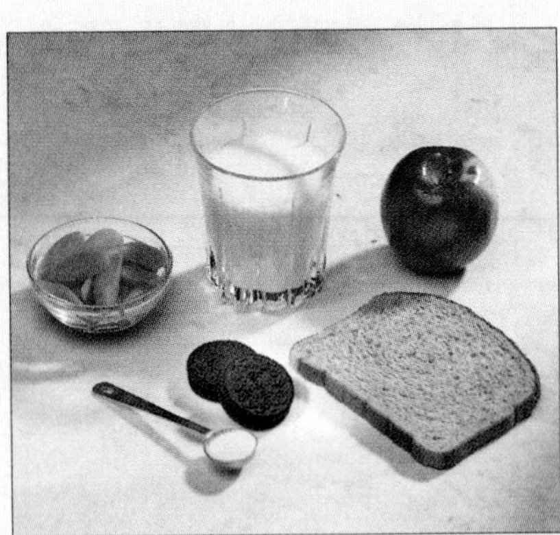

15 g in 1 slice bread (starch)

15 g in 1 fruit portion (sugars)

15 g in 1 small dessert or snack (starch/sugars)

12 g in 1 c milk (lactose)

5 g in ½ c vegetables (starch/sugars)

5 g in 1 tsp sugar (sugars)

*Carbohydrate-containing foods appear in several exchange lists: starches, fruits, milks, "others" such as desserts and snacks, and vegetables. Sugars also contribute carbohydrate.*

## Making It Click

These problems use hypothetical situations to give you practice in doing simple nutrition-related calculations (see Appendix K for answers). Be sure to show your calculations for each problem.

Health recommendations suggest that 55 to 60 percent of the daily energy intake come from carbohydrates. Stating recommendations in terms of percentage of energy intake is meaningful only if energy intake is known. The following exercises illustrate this concept.

1. Calculate the carbohydrate intake (in grams) for a student who has a high carbohydrate intake (70 percent of energy intake) and a moderate energy intake (2000 kcalories a day).

   How does this carbohydrate intake compare to the Daily Value of 300 grams? To the "over half" recommendation?
2. Now consider a professor who eats half as much carbohydrate as the student (in grams) and has the same energy intake. What percentage does carbohydrate contribute to the daily intake?

   How does carbohydrate intake compare to the Daily Value of 300 grams? To the "over half" recommendation?
3. Now consider an athlete who eats twice as much carbohydrate (in grams) as the student and has a much higher energy intake (6000 kcalories a day). What percentage does carbohydrate contribute to this person's daily intake?

   How does carbohydrate intake compare to the Daily Value of 300 grams? To the "over half" recommendation?
4. One more example. In an attempt to lose weight, a person adopts a diet that provides 150 grams of carbohydrate per day and limits energy intake to 1000 kcalories. What percentage does carbohydrate contribute to this person's daily intake?

   How does this carbohydrate intake compare to the Daily Value of 300 grams? To the "over half" recommendation?

These exercises should convince you of the importance of examining actual intake as well the percentage of energy intake.

## Study Questions

These questions will help you review the chapter. You will find the answers in the discussions on the pages provided.

1. Which carbohydrates are described as simple, and which are complex? (p. 91)
2. Describe the structure of a monosaccharide and name the three monosaccharides important in nutrition. Name the three disaccharides commonly found in foods and their component monosaccharides. In what foods are these sugars found? (pp. 92–94)
3. What happens in a condensation reaction? In a hydrolysis reaction? (pp. 93–94)
4. Describe the structure of polysaccharides and name the ones important in nutrition. How are starch and glycogen similar, and how do they differ? How do the fibers differ from the other polysaccharides? (pp. 95–97)
5. Describe carbohydrate digestion and absorption. What role does fiber play in the process? (pp. 98–100)
6. What are the possible fates of glucose in the body? What is the protein-sparing action of carbohydrate? (p. 103)
7. How does the body maintain its blood glucose concentration? What happens when the blood glucose concentration rises too high or falls too low? (pp. 104–105)
8. What are the health effects of sugars? What are the dietary recommendations regarding concentrated sugar intakes? (pp. 106–108, 110)
9. What are the health effects of starches and fibers? What are the dietary recommendations regarding these complex carbohydrates? (pp. 110–112)
10. What foods provide starches and fibers? (pp. 112–113)

These questions will help you prepare for an exam. Answers can be found in Appendix K.

1. Carbohydrates are found in virtually all foods except:
   a. milks.
   b. meats.
   c. breads.
   d. fruits.
2. Disaccharides include:
   a. starch, glycogen, and fiber.
   b. amylose, pectin, and dextrose.
   c. sucrose, maltose, and lactose.
   d. glucose, galactose, and fructose.
3. The making of a disaccharide from two monosaccharides occurs during:
   a. digestion.
   b. hydrolysis.
   c. condensation.
   d. gluconeogenesis.
4. The storage form of glucose in the body is:
   a. insulin.
   b. maltose.
   c. glucagon.
   d. glycogen.

5. The significant difference between starch and cellulose is that:
   a. starch is a polysaccharide, but cellulose is not.
   b. animals can store glucose as starch, but not as cellulose.
   c. hormones can make glucose from cellulose, but not from starch.
   d. digestive enzymes can break the bonds in starch, but not in cellulose.
6. The ultimate goal of carbohydrate digestion and absorption is to yield:
   a. fibers.
   b. glucose.
   c. enzymes.
   d. amylase.
7. The enzyme that breaks a disaccharide into glucose and galactose is:
   a. amylase.
   b. maltase.
   c. sucrase.
   d. lactase.
8. With insufficient glucose in metabolism, fat fragments combine to form:
   a. dextrins.
   b. mucilages.
   c. phytic acids.
   d. ketone bodies.
9. What does the pancreas secrete when blood glucose rises? When blood glucose falls?
   a. insulin; glucagon
   b. glucagon; insulin
   c. insulin; glycogen
   d. glycogen; epinephrine
10. What percentage of the daily energy intake should come from carbohydrates?
   a. 15 to 20
   b. 25 to 30
   c. 45 to 50
   d. 55 to 60

## Notes


1. A. M. Stephen and coauthors, Intake of carbohydrate and its components—International comparisons, trends over time, and effects of changing to low-fat diets, *American Journal of Clinical Nutrition* 62 (1995): 851S–867S.
2. W. Frølich, Bioavailability of minerals from cereals, in *Dietary Fiber in Human Nutrition,* ed. G. A. Spiller (Boca Raton, Fla.: CRC Press, 1993), pp. 209–243.
3. N. G. Asp, J. M. M. vanAmelsvoort, and J. G. A. J. Hautvast, Nutritional implications of resistant starch, *Nutrition Research Reviews* 9 (1996): 1–31; G. Annison and D. L. Topping, Nutritional role of resistant starch: Chemical structure vs physiological function, *Annual Review of Nutrition* 14 (1994): 297–320; N. Asp, Nutritional classification and analysis of food carbohydrates, *American Journal of Clinical Nutrition* 59 (1994): 679S–681S.
4. K. M. Behall and J. C. Howe, Breath-hydrogen production and amylose content of the diet, *American Journal of Clinical Nutrition* 65 (1997): 1783–1789; K. R. Silvester, H. N. Englyst, and J. H. Cummings, Ileal recovery of starch from whole diets containing resistant starch measured in vitro and fermentation of ileal effluent, *American Journal of Clinical Nutrition* 62 (1995): 403–411.
5. M. L. A. Heijnen and coauthors, Neither raw nor retrograded resistant starch lowers fasting serum cholesterol concentrations in healthy normolipidemic subjects, *American Journal of Clinical Nutrition* 64 (1996): 312–318.
6. K. M. Behall and J. C. Howe, Contribution of fiber and resistant starch to metabolize energy, *American Journal of Clinical Nutrition* 62 (1995): 1158S–1160S.
7. M. Lee and S. D. Krasinski, Human adult-onset lactase decline: An update, *Nutrition Reviews* 56 (1998): 1–8.
8. F. L. Suarez and D. A. Savaiano, Diet, genetics, and lactose intolerance, *Food Technology* 51 (1997): 74–76.
9. T. H. Vesa, R. A. Korpela, and T. Sahi, Tolerance to small amounts of lactose in lactose maldigesters, *American Journal of Clinical Nutrition* 64 (1996): 197–201; S. R. Hertzler, B. L. Huynh, and D. A. Savaiano, How much lactose is low lactose? *Journal of the American Dietetic Association* 96 (1996): 243–246.
10. L. D. McBean and G. D. Miller, Allaying fears and fallacies about lactose intolerance, *Journal of the American Dietetic Association* 98 (1998): 671–676; F. L. Suarez and coauthors, Tolerance to the daily ingestion of two cups of milk by individuals claiming lactose intolerance, *American Journal of Clinical Nutrition* 65 (1997): 1502–1506; A. O. Johnson and coauthors, Adaptation of lactose maldigesters to continued milk intakes, *American Journal of Clinical Nutrition* 58 (1993): 879–881; F. L. Suarez, D. A. Savaiano, and M. D. Levitt, A comparison of symptoms after the consumption of milk or lactose-hydrolyzed milk by people with self-reported severe lactose intolerance, *New England Journal of Medicine* 333 (1995): 1–4.
11. S. R. Hertzler and D. A. Savaiano, Colonic adaptation to daily lactose feeding in lactose maldigesters reduces lactose intolerance, *American Journal of Clinical Nutrition* 64 (1996): 232–236.
12. J. C. Brand and S. Holt, Relative effectiveness of milks with reduced amounts of lactose in alleviating milk intolerance, *American Journal of Clinical Nutrition* 54 (1991): 148–151.
13. E. Jéquier, Carbohydrates as a source of energy, *American Journal of Clinical Nutrition* 59 (1994): 682S–685S.
14. J. Hirsch, Role and benefits of carbohydrate in the diet: Key issues for future dietary guidelines, *American Journal of Clinical Nutrition* 61 (1995): 996S–1000S.
15. The Expert Committee on the Diagnosis and Classification of Diabetes Mellitus, Report on the Expert Committee on the Diagnosis and Classification of Diabetes Mellitus, *Diabetes Care* 20 (1997): 1183–1197.
16. B. Vessby, Dietary carbohydrates in diabetes, *American Journal of Clinical Nutrition* 59 (1994): 742S–746S.
17. T. M. S. Wolever and J. B. Miller, Sugars and blood glucose control, *Ameri-*

*can Journal of Clinical Nutrition* 62 (1995): 212S–227S; I. Bjorck and coauthors, Food properties affecting the digestion and absorption of carbohydrates, *American Journal of Clinical Nutrition* 59 (1994): 699S–705S.

18. K. Foster-Powell and J. B. Miller, International tables of glycemic index, *American Journal of Clinical Nutrition* 62 (1995): 871S–893S.
19. D. J. A. Jenkins and coauthors, Low glycemic index: Lente carbohydrates and physiological effects of altered food frequency, *American Journal of Clinical Nutrition* 59 (1994): 706S–709S.
20. J. C. B. Miller, Importance of glycemic index in diabetes, *American Journal of Clinical Nutrition* 59 (1994): 747S–752S.
21. Position statement: Nutrition recommendations and principles for people with diabetes mellitus, *Diabetes Care* 17 (1994): 519–522.
22. W. H. Glinsmann and Y. K. Park, Perspective on the 1986 Food and Drug Administration assessment of the safety of carbohydrate sweeteners: Uniform definitions and recommendations for future assessments, *American Journal of Clinical Nutrition* 62 (1995): 161S–169S.
23. J. M. Navia, Carbohydrates and dental health, *American Journal of Clinical Nutrition* 59 (1994): 719S–727S.
24. K. G. König and J. M. Navia, Nutritional role of sugars in oral health, *American Journal of Clinical Nutrition* 62 (1995): 275S–283S.
25. J. O. Hill and A. M. Prentice, Sugar and body weight regulation, *American Journal of Clinical Nutrition* 62 (1995): 264S–274S.
26. G. H. Anderson, Sugars, sweetness, and food intake, *American Journal of Clinical Nutrition* 62 (1995): 195S–202S.
27. L. C. Hudgins and coauthors, Human fatty acid synthesis is reduced after the substitution of dietary starch for sugar, *American Journal of Clinical Nutrition* 67 (1998): 631–639.
28. K. N. Frayn and S. M. Kingman, Dietary sugars and lipid metabolism in humans, *American Journal of Clinical Nutrition* 62 (1995): 250S–263S.
29. J. W. White and M. Wolraich, Effect of sugar on behavior and mental performance, *American Journal of Clinical Nutrition* 62 (1995): 242S–249S.
30. M. L. Wolraich, D. B. Wilson, and J. W. White, The effect of sugar on behavior or cognition in children: A meta-analysis, *Journal of the American Medical Association* 274 (1995): 1617–1621.
31. J. E. Blundell, S. Green, and V. Burley, Carbohydrates and human appetite, *American Journal of Clinical Nutrition* 59 (1994): 728S–734S; A. S. Levine and coauthors, Effect of breakfast cereals on short-term food intake, *American Journal of Clinical Nutrition* 50 (1989): 1303–1307; L. Lissner and coauthors, Dietary fat and the regulation of energy intake in human subjects, *American Journal of Clinical Nutrition* 46 (1987): 886–892.
32. A. S. Truswell, Food carbohydrates and plasma lipids—An update, *American Journal of Clinical Nutrition* 59 (1994): 710S–718S.
33. M. B. Katan, Forever fiber, *Nutrition Reviews* 54 (1996): 253–257.
34. A. Lia and coauthors, Postprandial lipemia in relation to sterol and fat excretion in ileostomy subjects given oat-bran and wheat test meals, *American Journal of Clinical Nutrition*, 66 (1997): 357–365.
35. D. J. A. Jenkins and coauthors, Effect on blood lipids of very high intakes of fiber in diets low in saturated fat and cholesterol, *New England Journal of Medicine* 329 (1993): 21–26.
36. E. B. Rimm and coauthors, Vegetable, fruit, and cereal fiber intake and risk of coronary heart disease among men, *Journal of the American Medical Association* 275 (1996): 447–451.
37. M. Noakes and coauthors, Effect of high-amylose starch and oat bran on metabolic variables and bowel function in subjects with hypertriglyceridemia, *American Journal of Clinical Nutrition* 64 (1996): 944–951.
38. I. P. Munster and coauthors, Effect of resistant starch on breath-hydrogen and methane excretion in healthy volunteers, *American Journal of Clinical Nutrition* 59 (1994): 626–630.
39. J. Salmerón and coauthors, Dietary fiber, glycemic load, and risk of non-insulin-dependent diabetes mellitus in women, *Journal of the American Medical Association* 277 (1997): 472–477.
40. Position of The American Dietetic Association: Health implications of dietary fiber, *Journal of the American Dietetic Association* 97 (1997): 1157–1159.
41. J. W. Anderson, B. M. Smith, and N. J. Gustafson, Health benefits and practical aspects of high-fiber diets, *American Journal of Clinical Nutrition* 59 (1994): 1242S–1247S.

# Alternatives to Sugar

People who want to limit their use of sugar may encounter two sets of alternative sweeteners. One set, the artificial sweeteners, provide virtually no energy and are sometimes referred to as nonnutritive sweeteners. The other set, the sugar alcohols, yield energy and are sometimes referred to as nutritive sweeteners.

## *Artificial Sweeteners*

Artificial sweeteners permit people to keep their sugar and energy intakes down, yet still enjoy the delicious sweet tastes of their favorite foods and beverages. The Food and Drug Administration (FDA) has approved the use of four artificial sweeteners—saccharin, aspartame, acesulfame potassium (acesulfame-K), and sucralose. Two others have petitioned the FDA and are awaiting approval—alitame and cyclamate. Table H4-1 and the glossary on p. 120 provide general details about each of these sweeteners.

Saccharin, acesulfame-K, and sucralose are not metabolized in the body; they pass through the kidneys unchanged. In contrast, the body digests aspartame as a protein. In fact, aspartame is *technically* classified as a nutritive sweetener because it yields energy, but for all practical purposes, that energy is negligible.

Some consumers have challenged the safety of using artificial sweeteners. Considering that all compounds are toxic at some dose, it is little surprise that large doses of artificial sweeteners (or their components or metabolic by-products) have toxic effects. The question to ask is whether their ingestion is safe for human beings in quantities people normally use (and potentially abuse). The answer is yes, except in the special case described for aspartame later.

U.S. Government
www.healthfinder.gov/searchoptions/topicsaz.htm
Search for Artificial sweeteners

### *The Safety of Saccharin*

Saccharin, used for over 100 years in the United States, is currently used by some 50 million people—primarily in soft drinks, secondarily as a tabletop sweetener. Saccharin is rapidly excreted in the urine and does not accumulate in the body.

Questions about saccharin's safety surfaced in 1977, when experiments suggested that large doses of saccharin (equivalent to hundreds of cans of diet soda daily for a lifetime) increased the risk of bladder cancer in rats. The FDA proposed banning saccharin as a result. Public outcry in favor of saccharin was so loud, however, that Congress imposed a moratorium on the ban—a moratorium that was repeatedly extended until 1991, when the FDA withdrew its proposal to ban saccharin.[1] Products containing saccharin must still carry a warning label, "Use of this product may be hazardous to your health. This product contains saccharin, which has been determined to cause cancer in laboratory animals."

Does saccharin cause cancer? The largest population study to date, involving 9000 men and women, showed overall that saccharin use did not increase the risk of cancer. Among certain small groups of the population, however, such as those who both smoked heavily and used saccharin, the risk of bladder cancer was slightly greater. Other studies involving more than 5000 people with bladder cancer showed no association between bladder cancer and saccharin use.[2] Common sense dictates that consuming large amounts of any substance is probably not wise, but at current, moderate intake levels, saccharin is assumed to be safe for most people. It has been approved for use in more than 100 countries.

### *The Safety of Aspartame*

Aspartame is one of the most studied of all food additives; extensive animal and human studies document its safety. Long-term consumption of aspartame is not associated with any adverse health effects.

The nutrients in aspartame may present a problem for certain people, however, and for this reason, aspartame also carries a warning on its label. Aspartame is a simple chemical compound made of components common to many foods: two amino acids (phenylalanine and aspartic acid) and a methyl group ($CH_3$). Figure H4-1 shows its chemical structure. The flavors of the components give no clue to the combined effect; one of them tastes bitter, and the other is tasteless, but the combination creates a product that is 180 times sweeter than sucrose.

*People wanting to limit their sugar intake can find a variety of foods and beverages made with artificial sweeteners.*

In the digestive tract, enzymes split aspartame into its three component parts. The body absorbs the two amino acids and uses them just as if they had come from food protein, which is made entirely of amino acids including these two.

Because this sweetener contributes phenylalanine, products containing aspartame must bear a warning label for people with the inherited disease phenylketonuria (PKU). People with

PKU are unable to dispose of any excess phenylalanine. The accumulation of phenylalanine and its by-products is toxic to the developing nervous system, causing irreversible brain damage. For this reason, all newborns in the United States are screened for PKU. The treatment for PKU is a special diet that must strike a balance, providing enough phenylalanine to support normal growth and health but not enough to cause harm. The question then is does aspartame raise blood phenylalanine high enough to be toxic to people with PKU? Apparently not. The little extra phenylalanine from aspartame, even in heavy users, poses only a small risk.

Still, there is a compelling reason why children with PKU need to get all their phenylalanine from foods, and not from an artificial sweetener. The PKU diet excludes such protein-rich and nutrient-rich foods as milk, meat, fish, poultry, cheese, eggs, nuts, legumes, and many bread products. Only with difficulty can these children obtain the many essential nutrients—such as calcium, iron, and the B vitamins—found along with phenylalanine in these foods. To suggest that children with PKU squander any of their limited phenylalanine allowance on the purified phenylalanine of aspartame, which contributes none of the associated vitamins or minerals essential for good health and normal growth, would open the way for poor nutrition.

Setting aside the special case of PKU, is there any reason to be concerned about the products aspartame yields in the body? During metabolism, the methyl group momentarily becomes methyl alcohol (methanol)—a potentially toxic compound (see Figure H4-2). Then enzymes convert methanol to formaldehyde, another toxic compound. Finally, formaldehyde is broken down to carbon dioxide. Before aspartame could be approved, the quantities of these products generated during metabolism had to be determined; they were found to fall below the threshold at which they would cause harm. In fact, ounce for ounce, tomato juice yields six times as much methanol as a diet soda.

Finally, aspartame breaks down to diketopiperazine, or DKP for short. Long-term studies using animals have directly tested this product and eliminated it as a source of concern.

In conclusion, except for people with PKU, aspartame is safe. Some individuals may exhibit vague, but not dangerous, symptoms due to unusual sensitivity to aspartame, but it is generally safe. Like saccharin, aspartame has been approved for use in more than 100 countries.

International Food Information Council

ificinfo.health.org

Search for Sweeteners

**Table H4-1**
**Artificial Sweeteners**

| Artificial Sweeteners | Properties | Acceptable Daily Intakes | Approved Uses |
|---|---|---|---|
| ***Approved Sweeteners*** | | | |
| Saccharin | Heat stable, long shelf life, water soluble, colorless, odorless; bitter, metallic aftertaste | 5 mg/kg body weight | Tabletop sweeteners, wide range of foods, beverages, cosmetics, and pharmaceutical products |
| Aspartame | Loses sweetness with heat, no aftertaste | 50 mg/kg body weight[a] Warning to people with PKU: Contains phenylalanine | General purpose sweetener in all foods and beverages |
| Acesulfame-K | Heat stable, long shelf life, water soluble, synergistic sweetening effect when used with other sweeteners, slight aftertaste | 15 mg/kg body weight[b] | Tabletop sweeteners, puddings, gelatins, chewing gum, candies, baked goods, desserts, alcoholic beverages |
| Sucralose | Extremely stable, water soluble | 15 mg/kg body weight | Carbonated beverages, dairy products, baked goods, coffee and tea, fruit spreads, syrups, tabletop sweeteners, chewing gum, frozen desserts, salad dressing |
| ***Sweeteners with Approval Pending*** | | | |
| Alitame | Heat stable, water soluble, no aftertaste; except in acidic foods at high temperatures, synergistic sweetening effect when used with other sweeteners | | Beverages, baked goods, tabletop sweeteners, frozen desserts |
| Cyclamate | Heat stable, water soluble, no aftertaste, synergistic sweetening effect when used with saccharin | | Tabletop sweeteners, baked goods |

[a]Recommendations from the World Health Organization and in Europe and Canada limit aspartame intake to 40 mg/kg body weight.

[b]Recommendations from the World Health Organization limit acesulfame-K intake to 9 mg/kg body weight.

**Figure H4-1**
**Structure of Aspartame**

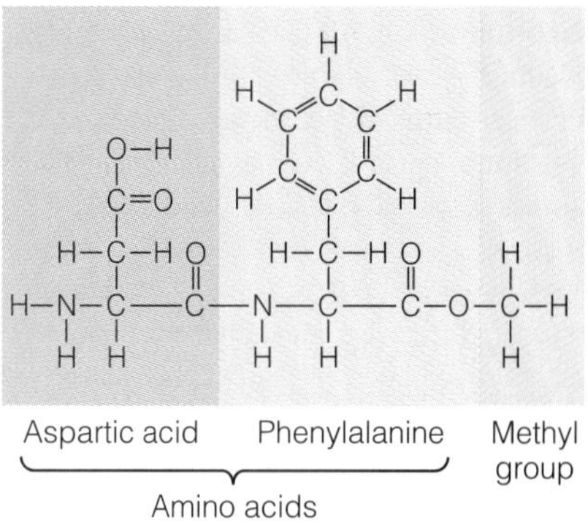

Aspartic acid — Phenylalanine — Methyl group

Amino acids

## The Safety of Acesulfame-K

The FDA approved acesulfame-K in 1988 after reviewing more than 90 safety studies conducted over 15 years. Some consumer groups believe that acesulfame-K causes tumors in rats and should not have been approved. The FDA counters that the tumors were not caused by the sweetener, but were typical of those commonly found in rat studies. Acesulfame-K has been approved for use in more than 60 countries.

## The Safety of Sucralose

Sucralose is the "new kid on the block," having received FDA approval in 1998. The FDA approval came after a review of over 110 safety studies conducted on both animals and human beings.

## The Safety of Alitame and Cyclamate

FDA approval for alitame and cyclamate is still pending. To date, no safety issues have been raised for alitame and it has been approved for use in other countries. Cyclamate, on the other hand, has been battling safety issues for 50 years. Approved by the FDA in 1949, cyclamate was banned in 1969 principally on the basis of one study indicating that it caused bladder cancer in rats.

The National Research Council has reviewed dozens of studies on cyclamate and concluded that neither cyclamate nor its metabolites cause cancer. The council did, however, recommend further research to determine the risks for heavy or long-term use. Although cyclamate does not initiate cancer, it may promote cancer development once started. The FDA currently has no policy on substances that enhance the cancer-causing activities of other substances. Consequently, the FDA is unlikely to approve cyclamate soon, if at all. Agencies in more than 50 other countries, including Canada, have approved cyclamate.

## Acceptable Daily Intake

The amount of artificial sweetener consider safe for daily use is called the Acceptable Daily Intake (ADI). The ADI represents the level of consumption that, if maintained every day throughout a person's life, would still be considered safe by a wide margin.

For example, the ADI for aspartame is 50 milligrams per kilogram of body weight. That is, the FDA approved aspartame based on the assumption that no one would consume more than 50 milligrams per kilogram of body weight in a day. This

**Figure H4-2**
**Metabolism of Aspartame**

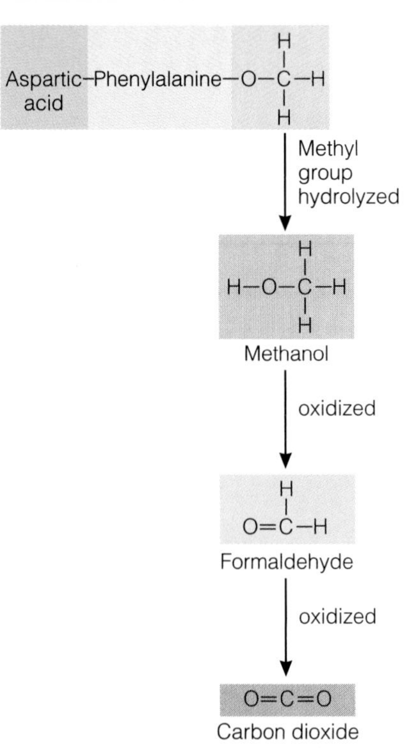


## Glossary

**acesulfame** (AY-sul-fame) **potassium:** an artificial sweetener composed of an organic salt that tastes 200 times as sweet as sucrose and provides 0 kcalories per gram; also known as acesulfame-K because K is the chemical symbol for potassium.

**ADI (Acceptable Daily Intake):** the amount of a sweetener that individuals can safely consume each day over the course of a lifetime without adverse effect. It includes a 100-fold safety factor.

**alitame** (AL-ih-tame): an artificial sweetener composed of two amino acids (alanine and aspartic acid) that tastes 2000 times as sweet as sucrose. Alitame provides 4 kcalories per gram, as does protein, but because so little is used, its energy contribution is negligible. FDA approval pending.

**artificial sweeteners:** sugar substitutes that provide negligible, if any, energy; sometimes called **nonnutritive sweeteners.**

**aspartame** (ah-SPAR-tame or ASS-par-tame): an artificial sweetener composed of two amino acids (phenylalanine and aspartic acid) that tastes 160 to 220 times as sweet as sucrose. Aspartame provides 4 kcalories per gram, as does protein, but because so little is used, its energy contribution is negligible. In powdered form it is sometimes mixed with lactose, however, so a 1-gram packet may provide 4 kcalories.

**cyclamate** (SIGH-klo-mate): an artificial sweetener that tastes 30 times as sweet as sucrose and provides 0 kcalories per gram; FDA approval pending in the United States; available in Canada as a tabletop sweetener, not as an additive.

**diketopiperazine** (dye-KEY-toe-pie-PER-a-zeen), or **DKP:** a product to which aspartame breaks down during metabolism.

**nutritive sweeteners:** sweeteners that yield energy, including both sugars and sugar alcohols.

**saccharin** (SAK-ah-ren): an artificial sweetener that tastes 200 to 700 times as sweet as sucrose and provides 0 kcalories per gram; approved in the United States. In Canada, approval in foods and beverages is pending; currently available only in pharmacies and only as a tabletop sweetener, not as an additive.

**sucralose** (SUE-kra-lose): an artificial sweetener that tastes 600 times as sweet as sucrose and provides 0 kcalories per gram.

**sugar alcohols:** sugarlike compounds that can be derived from fruits or commercially produced from dextrose; also called **polyols.** Like sugars, sugar alcohols are sweet to taste, but they provide less energy than regular sugars. Sugar alcohols are absorbed more slowly than other sugars and metabolized differently in the human body, and are not readily utilized by ordinary mouth bacteria. Examples are **maltitol, mannitol, sorbitol, xylitol, isomalt,** and **lactitol.**

maximum daily intake is indeed a lot: for a 150-pound adult, it adds up to 97 packets of Equal or 20 cans of soft drinks sweetened only with aspartame. The company that produces aspartame estimates that if all the sugar and saccharin in the U.S. diet were replaced with aspartame, 1 percent of the population would be consuming the FDA maximum. Most people who use aspartame consume less than 5 milligrams per kilogram of body weight per day. A five-year-old child who drinks four glasses of aspartame-sweetened beverages on a hot day and has five servings of other products with aspartame that day (such as pudding, chewing gum, cereal, gelatin, and frozen desserts) takes in the FDA maximum level. Although this presents no proven hazard, it seems wise to offer children other foods so as not to exceed the limit. Table H4-2 lists the average amounts of aspartame in some common foods.

For persons choosing to use artificial sweeteners, the American Dietetic Association wisely advises that they be used in moderation and only as part of a well-balanced nutritious diet.[3] The dietary principles of both moderation and variety help to reduce the possible risks associated with any food.

## Artificial Sweeteners and Weight Control

Many people eat and drink products sweetened with artificial sweeteners to help them control weight. Ironically, a few studies have reported that intense sweeteners, such as aspartame, may stimulate appetite, which could lead to weight gain.[4] Contradicting these reports, most studies find no change in feelings of hunger and no change in food intakes or body weight.[5] Adding to the confusion, some studies report lower energy intakes and greater weight losses when people eat or drink artificially sweetened products.[6]

When studying the effects of artificial sweeteners on food intake and body weight, researchers ask different questions and take different approaches. It matters, for example, whether the people used in a study are of a healthy weight and whether they are following a weight-loss diet. Motivations for using sweeteners differ, too, and this influences a person's actions. For example, a person might drink an artificially sweetened beverage now so as to be able to eat a high-kcalorie food later. This person's energy intake might stay the same or increase. On the other hand, a person trying to control food energy intake might drink an artificially sweetened beverage now and then choose a low-kcalorie food later. This plan would help reduce the person's energy intake.

In designing experiments on artificial sweeteners, researchers have to distinguish between the effects of sweetness and the effects of a particular substance. If a person is hungry shortly after eating an artificially sweetened snack, is that because the sweet taste (of all sweeteners, including sugars) stimulates appetite? Or is it because the artificial sweetener itself stimulates appetite? Research must also distinguish between the effects of food energy and the effects of the substance. If a person is hungry shortly after eating an artificially sweetened snack, is that because less food energy was available to satisfy hunger? Or is it because the artificial sweetener itself triggers hunger? Furthermore, if appetite is stimulated and a person feels hungry, does that actually lead to increased food intake?

One study tried to answer these questions by feeding normal-weight people one of four cheese samples for breakfast and then measuring their food intake at later meals throughout the day.[7] Two of the samples provided 700 kcalories: one contained sucrose, and the other aspartame with enough starch to equalize the kcalories. The other two samples provided 300 kcalories: one was plain and the other contained aspartame. Those who ate the lower-kcalorie breakfast, regardless of sweetness, were hungrier later. They ate more at lunch (100 kcalories on average), although not enough to fully compensate for the 400-kcalorie difference between the breakfasts. Because energy intake at later meals was similar, those who ate the 700-kcalorie breakfasts had higher total energy intakes.

Whether a person compensates for the energy reduction of artificial sweeteners either partially or fully depends on several factors, including the person's motivations. Using artificial sweeteners will not automatically lower energy intake; to control energy intake successfully, a person needs to make informed diet and activity decisions throughout the day (as Chapter 9 explains).

## Sugar Alcohols

Some "sugar-free" or reduced-kcalorie products such as hard candies, sugarless gums, jams, and jellies contain the sugar alcohols—mannitol, sorbitol, xylitol, maltitol, isomalt, or lactitol. They claim to be "sugar-free" on their labels, but in this case, "sugar-free" does not mean free of kcalories. Sugar alcohols occur naturally in fruits and vegetables; they are also used by manufacturers as a low-energy bulk ingredient in many products.[8] Sugar alcohols provide energy, although less than regular sugars. Table H4-3 presents a general description of several of the sugar alcohols.

Sugar alcohols evoke a low glycemic response. The body absorbs sugar alcohols slowly; consequently, they are slower to enter the bloodstream than other sugars. Side effects such as gas, abdominal discomfort, and diarrhea, however, make them less attractive than the artificial sweeteners. For this reason, regulations require food labels to state that "Excess consumption may have a laxative effect" if reasonable

**Table H4-2**
**Average Aspartame Contents of Selected Foods**

| Food | Aspartame (mg) |
|---|---|
| 12 oz diet soft drink | 170 |
| 8 oz powdered drink | 100 |
| 8 oz sugar-free fruit yogurt | 124 |
| 4 oz gelatin dessert | 80 |
| 1 packet sweetener | 35 |

consumption could result in the daily ingestion of 50 grams of a sugar alcohol.

**Table H4-3**
**Sugar Alcohols**

| Sugar Alcohols | Sweetness Compared with Sucrose | kCaloric Value (kcal/g) | Uses |
|---|---|---|---|
| Isomalt | 55% as sweet | 2.0 | Candies, chewing gum, ice cream, jams and jellies, frostings, beverages, baked goods |
| Lactitol | 40% as sweet | 2.0 | Candies, chewing gum, frozen dairy desserts, jams and jellies, frostings, baked goods |
| Maltitol | 90% as sweet | 3.0 | Particularly good for candy coating |
| Mannitol | 70% as sweet | 1.6 | Bulking agent, chewing gum |
| Sorbitol | 50–70% as sweet | 2.6 | Special dietary foods, candies, gums |
| Xylitol | 100% as sweet | 2.4 | Chewing gum, candies, pharmaceutical and oral health products |

The real benefit of using sugar alcohols is that they do not contribute to dental caries. Bacteria in the mouth cannot metabolize sugar alcohols as rapidly as sugar. They are therefore valuable in chewing gums, breath mints, and other products that people keep in their mouths for a while.

The sugar alcohols, like the artificial sweeteners, can occupy a place in the diet, and provided they are used in moderation, they will do no harm. In fact, they can help, both by providing an alternative to sugar for people with diabetes and by inhibiting caries-causing bacteria. People may find it appropriate to use all three sweeteners at times: artificial sweeteners, sugar alcohols, and sugar itself.

## Notes

1. Withdrawal of certain pre-1986 proposed rules: Final actions, *Federal Register* 56 (1991): 67422.
2. Position of The American Dietetic Association: Use of nutritive and nonnutritive sweeteners, *Journal of the American Dietetic Association* 98 (1998): 580–587.
3. Position of The American Dietetic Association, 1998.
4. J. E. Blundell and P. J. Rogers, Sweet carbohydrate substitutes (intense sweeteners) and the control of appetite: Scientific issues, in *Appetites and Body Weight Regulation: Sugar, Fat, and Macronutrient Substitutes*, eds. J. D. Fernstrom and G. D. Miller (Boca Raton, Fla.: CRC Press, 1994), pp. 113–124.
5. S. J. Gatenby and coauthors, Extended use of foods modified in fat and sugar content: Nutrition implications in a free-living female population, *American Journal of Clinical Nutrition* 65 (1997): 1867–1873; A. Drewnowski, Intense sweeteners and the control of appetite, *Nutrition Reviews* 53 (1995): 1–7.
6. G. L. Blackburn and coauthors, The effect of aspartame as part of a multidisciplinary weight-control program on short- and long-term control of body weight, *American Journal of Clinical Nutrition* 65 (1997): 409–418; M. G. Tordoff and A. M. Alleva, Effect of drinking soda sweetened with aspartame or high-fructose corn syrup on food intake and body weight, *American Journal of Clinical Nutrition* 51 (1990): 963–969.
7. A. Drewnowski and coauthors, Comparing the effects of aspartame and sucrose on motivational ratings, taste preferences, and energy intake in humans, *American Journal of Clinical Nutrition* 59 (1994): 338–345.
8. F. R. J. Bonet, Undigestible sugars in food products, *American Journal of Clinical Nutrition* 59 (1994): 763S–769S.

# The Lipids: Triglycerides, Phospholipids, and Sterols

Most people are surprised to learn that fat has some virtues. Only when people consume either too much or too little fat does ill health follow. It is true, though, that in our society of abundance, people are likely to encounter too much fat.

Fat is actually a subset of the class of nutrients known as lipids, but the term *fat* is often used to refer to all the lipids. The lipid family includes triglycerides (fats and oils), phospholipids, and sterols, all important to nutrition. The triglycerides predominate, both in foods and in the body. The triglycerides provide the body with a continuous fuel supply, keep it warm, and protect it from mechanical shock; their component fatty acids serve as starting materials for important hormonal regulators. The phospholipids and sterols contribute to the cells' structures, and the sterol cholesterol serves as the raw material for some hormones, vitamin D, and bile.

In foods, triglycerides are the solid fats and liquid oils. Triglycerides carry with them the four fat-soluble vitamins—A, D, E, and K—together with many of the compounds that give foods their flavor, texture, and palatability. Fat is responsible for the delicious aromas associated with sizzling bacon and hamburgers on the grill, onions being sautéed, or vegetables in a stir-fry. Of course, these wonderful characteristics lure people into eating too much from time to time.[1] Studies have revealed that obese people eat more fat than their normal-weight peers, but not whether the preference for fat or the obesity comes first.[2]

**lipids:** a family of compounds that includes triglycerides (fats and oils), phospholipids, and sterols.

**fats:** lipids in foods or the body, both of which are composed mostly of triglycerides.

**oils:** liquid fats (at room temperature).

Of the lipids in foods, 95% are fats and oils (triglycerides), and 5% are other lipids (phospholipids and sterols). Of the lipids stored in the body, 99% are triglycerides.

IN SUMMARY

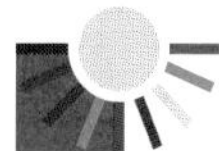

Triglycerides in foods:
- Deliver fat-soluble vitamins.
- Contribute to the sensory appeal of foods.

# The Chemist's View of Triglycerides and Fatty Acids

Like carbohydrates, triglycerides are composed of carbon (C), hydrogen (H), and oxygen (O). However, triglycerides have many more carbons and hydrogens in proportion to their oxygens, and so can supply more energy per gram (Chapter 7 provides details).

For people who think more easily in words than in chemical symbols, this *preview* of the upcoming chemistry may be helpful. The following paragraphs and diagrams demonstrate that:

1. Every triglyceride contains one molecule of glycerol (see Figure 5-1) and three fatty acids (basically chains of carbon atoms).
2. Fatty acids may be 4 to 24 (even numbers of) carbons long, the 18-carbon ones being the most common in foods and especially noteworthy in nutrition.
3. Fatty acids may also be saturated or unsaturated. The latter may have one or more points of unsaturation (may be mono- or polyunsaturated).
4. Of special importance in nutrition are the polyunsaturated fatty acids whose *first* point of unsaturation is next to the third carbon (known as omega-3 fatty acids) or next to the sixth carbon (omega-6), when counting from the methyl end ($CH_3$) of the carbon chain.
5. The 18-carbon fatty acids that fit this description are linolenic acid (omega-3) and linoleic acid (omega-6). Each is the primary member of a "family" of longer-chain fatty acids that regulate blood pressure, clotting, and other body functions important to health.

*Research scientists commonly use the term *triacylglycerols;* this book continues to use the more familiar term *triglycerides,* as do many other health and nutrition books and journals.

**triglycerides** (try-GLISS-er-rides): the chief form of fat in the diet and the major storage form of fat in the body; composed of a molecule of glycerol with three fatty acids attached; also called **triacylglycerols** (try-ay-seel-GLISS-er-ols).*
- **tri** = three
- **glyceride** = a compound of glycerol
- **acyl** = a carbon chain

**glycerol** (GLISS-er-ol): an alcohol composed of a three-carbon chain, which can serve as the backbone for a triglyceride.
- **ol** = alcohol

**Figure 5-1**
**Glycerol**

When glycerol is free, an OH group is attached to each carbon. When glycerol is part of a triglyceride, each carbon is attached to a fatty acid by a carbon-oxygen bond.

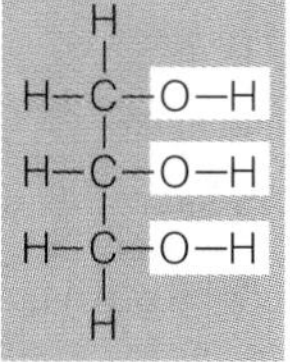

**fatty acid:** an organic compound composed of a carbon chain with hydrogens attached and an acid group (COOH) at one end. The COOH group of an organic acid can be represented this way:

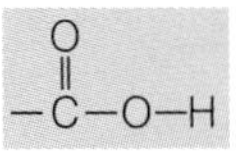

```
  O
  ‖
—C—O—H
```

Notice that this structure meets the requirement that C must have four bonds and O two. To accomplish this, O and C form a double bond.

# The Fatty Acids

A fatty acid is an organic acid—a chain of carbon atoms with hydrogens attached—that has an acid group (COOH) at one end and a methyl group ($CH_3$) at the other end. The organic acid shown in Figure 5-2 is acetic acid, the compound that gives vinegar its sour taste. Acetic acid is the simplest such acid, with a "chain" only two carbon atoms long.

• ***The Carbon Chain*** • Most naturally occurring fatty acids contain even numbers of carbons in their chains—up to 24 carbons in length. This discussion begins with the 18-carbon fatty acids, which are abundant in our food supply. Stearic acid is the simplest of the 18-carbon fatty acids; the bonds between its carbons are all alike:

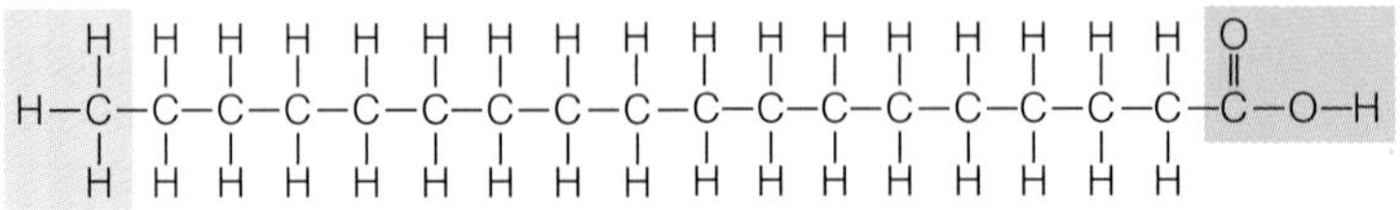

Stearic acid, an 18-carbon saturated fatty acid.

```
  H H H H H H H H H H H H H H H H H O
  | | | | | | | | | | | | | | | | | ‖
H—C—C—C—C—C—C—C—C—C—C—C—C—C—C—C—C—C—C—O—H
  | | | | | | | | | | | | | | | | |
  H H H H H H H H H H H H H H H H H
```

(As you can see, stearic acid is 18 carbons long, and each atom meets the rules of chemical bonding described in Chapter 4.) A fatty acid like stearic acid that contains only single bonds between its carbon atoms is a saturated fatty acid. The following structure also depicts stearic acid, but in a simpler way, with each "corner" on the zigzag line representing a carbon atom with two attached hydrogens:

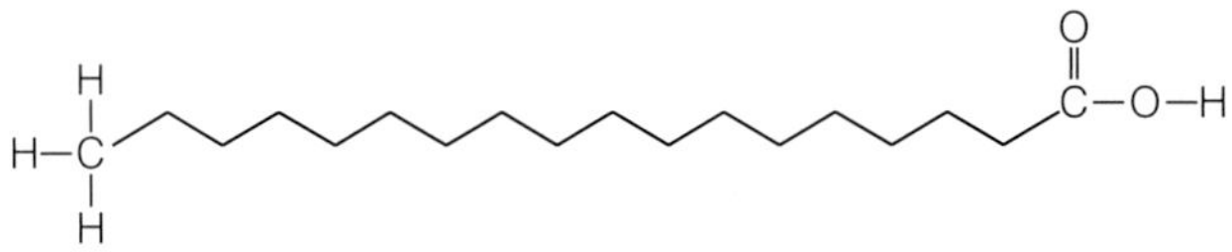

Stearic acid (simplified structure).

```
   H                                   O
   |                                   ‖
H—C/\/\/\/\/\/\/\/\/\/\/\/\/\/\/\/\/C—O—H
   |
   H
```

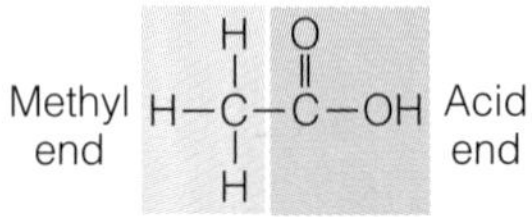

• ***Triglyceride Formation*** • Few fatty acids occur free in foods or in the body. Most often, they are incorporated into triglycerides. To make a triglyceride, three fatty acids are attached to a glycerol molecule by condensation reactions (see Figure 5-3). Each condensation reaction combines a hydrogen atom (H) from the glycerol and a hydroxyl (OH) group from a fatty acid, forming a molecule of water ($H_2O$) and leaving a bond between the other two molecules.

**Figure 5-2**
**Acetic Acid**

Acetic acid is a two-carbon organic acid.

```
         H  O
         |  ‖
Methyl H—C—C—OH  Acid
end      |       end
         H
```

• ***Degree of Saturation*** • The glossary on p. 127 defines the terms that describe fatty acids. The triglyceride shown in Figure 5-3 is a saturated fat because all three

**Figure 5-3**
**Condensation of Glycerol and Fatty Acids to Form a Triglyceride**

To make a triglyceride, three fatty acids attach to glycerol in condensation reactions:

```
  H           O                                H
  |           ‖                                |
H—C—O—H   H—O—C/\/\/\/\/\/\/\/\/\/\/\/\/\/\/\/\C—H
  |                                            |
  |                                            H
  |           O                                H
  |           ‖                                |
H—C—O—H   H—O—C/\/\/\/\/\/\/\/\/\/\/\/\/\/\/\/\C—H
  |                                            |
  |                                            H
  |           O                                H
  |           ‖                                |
H—C—O—H   H—O—C/\/\/\/\/\/\/\/\/\/\/\/\/\/\/\/\C—H
  |                                            |
  H                                            H

  H       O                                H
  |       ‖                                |
H—C—O—C/\/\/\/\/\/\/\/\/\/\/\/\/\/\/\/\C—H  + H2O
  |                                        |
  |                                        H
  |       O                                H
  |       ‖                                |
H—C—O—C/\/\/\/\/\/\/\/\/\/\/\/\/\/\/\/\C—H  + H2O
  |                                        |
  |                                        H
  |       O                                H
  |       ‖                                |
H—C—O—C/\/\/\/\/\/\/\/\/\/\/\/\/\/\/\/\C—H  + H2O
  |                                        |
  H                                        H
```

Glycerol + 3 fatty acids ⟶ Triglyceride + 3 water molecules

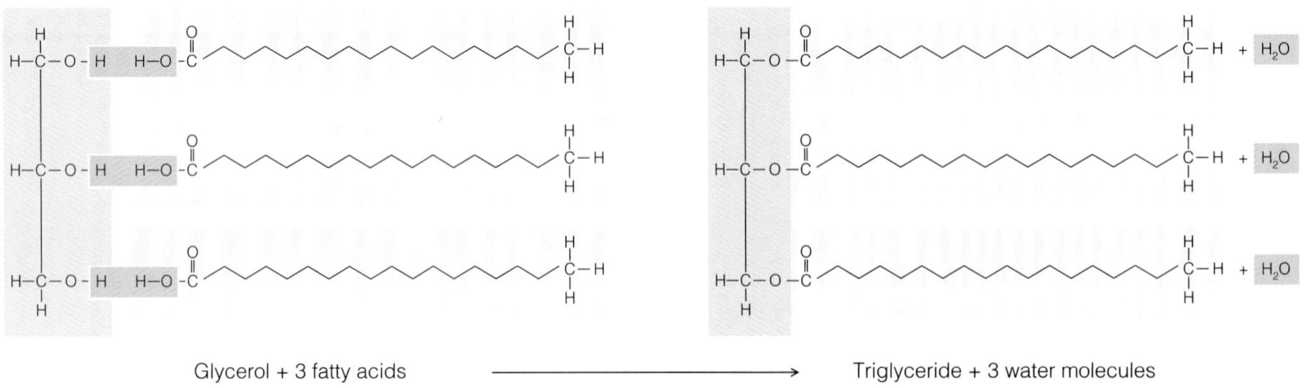

Water is removed from the glycerol and the fatty acids, forming a bond between the O on the glycerol and the C at the acid end of each fatty acid.

Three fatty acids attached to a glycerol form a triglyceride and yield water. In this example, all three fatty acids are stearic acid, but most often triglycerides contain mixtures of fatty acids.

## Glossary of Fatty Acids

These terms are listed in order from the most saturated to the most unsaturated.

**saturated fatty acid:** a fatty acid carrying the maximum possible number of hydrogen atoms—for example, stearic acid. A **saturated fat** is composed of triglycerides in which most of the fatty acids are saturated.

**unsaturated fatty acid:** a fatty acid that lacks hydrogen atoms and has at least one double bond between carbons (includes monounsaturated and polyunsaturated fatty acids). An **unsaturated fat** is composed of triglycerides in which most of the fatty acids are unsaturated.

**monounsaturated fatty acid:** a fatty acid that lacks two hydrogen atoms and has one double bond between carbons—for example, oleic acid. A **monounsaturated fat** is composed of triglycerides in which most of the fatty acids are monounsaturated.

- **mono** = one

**polyunsaturated fatty acid (PUFA):** a fatty acid that lacks four or more hydrogen atoms and has two or more double bonds between carbons—for example, linoleic acid (two double bonds) and linolenic acid (three double bonds). A **polyunsaturated fat** is composed of triglycerides in which most of the fatty acids are polyunsaturated.

- **poly** = many

fatty acids are saturated fatty acids—that is, fully loaded with hydrogen atoms. If hydrogens were missing, there would be points of unsaturation where the carbons would have to form double bonds with one another. The result would be an unsaturated, or even a polyunsaturated, fat. Consider stearic acid once more. If two hydrogens were missing from the middle of the carbon chain, the remaining structure might be:

**point of unsaturation:** the double bond of a fatty acid, where hydrogen atoms can easily be added to the structure.

```
  H H H H H H H H     H H H H H H H O
  | | | | | | | |     | | | | | | | ‖
H-C-C-C-C-C-C-C-C-C-C-C-C-C-C-C-C-C-C-O-H
  | | | | | | | | | | | | | | | | |
  H H H H H H H H H H H H H H H H H
```

An impossible chemical structure.

Such a compound cannot exist, however, because two of the carbons have only three bonds each, and nature requires that every carbon have four bonds. The two carbons therefore form a double bond:

```
  H H H H H H H H     H H H H H H H O
  | | | | | | | |     | | | | | | | ‖
H-C-C-C-C-C-C-C-C-C=C-C-C-C-C-C-C-C-C-O-H
  | | | | | | | | | | | | | | | | |
  H H H H H H H H H H H H H H H H H
```

Oleic acid, an 18-carbon monounsaturated fatty acid.

The same structure drawn more simply looks like this:*

```
   H                                   O
   |                                   ‖
H-C /\/\/\/\/\/\/\/\=/\/\/\/\/\/\/\/\ C-O-H
   |
   H
```

Oleic acid (simplified structure).

A fatty acid like this—with two hydrogens missing and a double bond—is an *un*saturated fatty acid. This one is the 18-carbon *mono*unsaturated fatty acid oleic acid, which is abundant in the triglycerides of olive oil.

A *poly*unsaturated fat contains triglycerides whose fatty acids have two or more carbon-to-carbon double bonds. The best known of these is linoleic acid, the 18-carbon fatty acid common in vegetable oils. Linoleic acid lacks four hydrogens and has two double bonds:

**linoleic** (lin-oh-LAY-ick) **acid:** an essential fatty acid with 18 carbons and two double bonds (18:2).

```
  H H H H H     H     H H H H H H H O
  | | | | |     |     | | | | | | | ‖
H-C-C-C-C-C-C=C-C-C=C-C-C-C-C-C-C-C-C-O-H
  | | | | | | | | | | | | | | | | |
  H H H H H H H H H H H H H H H H H
```

Linoleic acid, an 18-carbon polyunsaturated fatty acid.

---

*Remember that each "corner" on the zigzag line represents a carbon atom with two attached hydrogens. In addition, the actual shape bends at the double bonds and rotates around single bonds. These molecules, although drawn straight on paper, are constantly twisting and bending. At any given moment, they may be coiled, horseshoe shaped, or straight.

**Table 5-1**
**18-Carbon Fatty Acids**

| Name | Notation[a] | Number of Double Bonds | Saturation | Common Food Sources |
|---|---|---|---|---|
| Stearic acid | 18:0 | 0 | Saturated | Most animal fats |
| Oleic acid | 18:1 | 1 | Monounsaturated | Olive, canola oils |
| Linoleic acid | 18:2 | 2 | Polyunsaturated | Sunflower, safflower, corn oils |
| Linolenic acid | 18:3 | 3 | Polyunsaturated | Soybean oils |

[a]Chemists use a shorthand notation to describe fatty acids. The first number indicates the number of carbon atoms; the second, the number of double bonds.

Drawn more simply, linoleic acid looks like this:

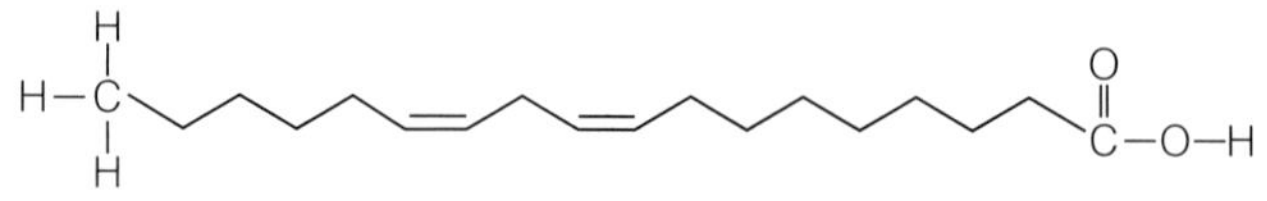

Linoleic acid (simplified structure).

A fourth 18-carbon fatty acid is linolenic acid, which has three double bonds (see Table 5-1).

**linolenic** (lin-oh-LEN-ick) **acid:** an essential fatty acid with 18 carbons and three double bonds (18:3).

Tables C-1 and C-2 in Appendix C provide the names, chain lengths, and sources of fatty acids commonly found in foods.

Having considered four of the most common fatty acids in foods, one can predict what the others will look like. They vary only in their degrees of unsaturation, the location of the double bond(s), and the lengths of their chains. The long-chain (12 to 24 carbons) fatty acids of meats and fish are most common in the diet. Smaller amounts of medium-chain (6 to 10 carbons) and short-chain (less than 6 carbons) fatty acids also occur, primarily in dairy products. Most triglycerides contain a mixture of more than one type of fatty acid (see Figure 5-4).

**Figure 5-4**
**A Mixed Triglyceride**

This mixed triglyceride includes a saturated fatty acid, a monounsaturated fatty acid, and a polyunsaturated fatty acid. Sometimes the chemical structure of a triglyceride is drawn with the second fatty acid to the left of the glycerol.

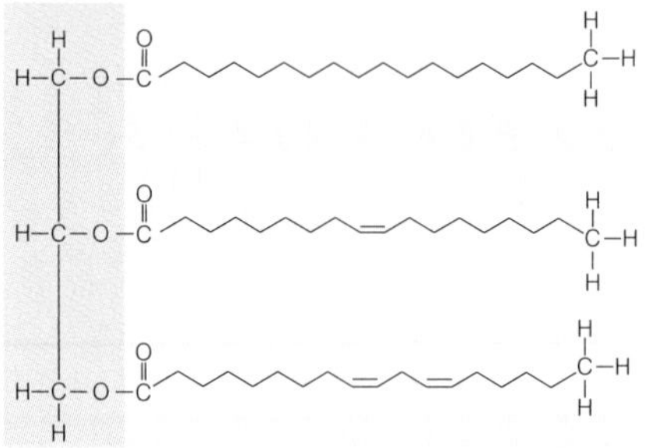

IN SUMMARY

The lipids important in nutrition are of three classes: triglycerides (commonly called fats), phospholipids, and sterols. The predominant lipids both in foods and in the body are triglycerides: glycerol backbones with three fatty acids attached. Fatty acids that are fully loaded with hydrogens are saturated; those that are missing hydrogens and therefore have double bonds are unsaturated (monounsaturated and polyunsaturated). The vast majority of triglycerides contain more than one type of fatty acid.

## Fats in Foods

Fatty acid compositions of selected fats and oils are shown in Figure 5-5, and Appendix H provides the fat and fatty acid contents of many other foods. The degree of unsaturation of fats affects health, as a later section in this chapter explains. The degree of unsaturation also influences the firmness of fats at room temperature. Generally speaking, the polyunsaturated vegetable oils are liquid at room temperature, and the more saturated animal fats are harder. Butter is harder than margarine because butter is more saturated than margarine; this is why people limiting their intakes of saturated fats use margarine. Not all vegetable oils are polyunsaturated, however. Cocoa butter and palm and coconut oils are saturated even though they are of vegetable origin; they are firmer than most vegetable oils because of their saturation, but softer than most animal fats because of their short carbon chains (only 10 and 12 carbons long, respectively). Generally, the shorter the carbon chain, the more liquid the fat is at room temperature.

The food industry often refers to palm and coconut oils as the "tropical oils."

• ***Processed Fat*** • Saturation also influences stability. All fats can become rancid when exposed to oxygen. Polyunsaturated fatty acids spoil most readily because their double bonds are unstable. The oxidation of unsaturated fats yields a variety of products that smell and taste rancid; saturated fats are more resistant to oxi-

**oxidation** (OKS-ee-day-shun): the process of a substance combining with oxygen.

**Figure 5-5**

**Comparison of Dietary Fats**

Most fats are a mixture of saturated, monounsaturated, and polyunsaturated fatty acids. See p. 131 for information on omega-6 (ω6) and omega-3 (ω3) fatty acids; the table on p. 132 lists good sources of these omega fatty acids.

| Dietary fat | Cholesterol (mg/tbs) | Polyunsaturated fats |
|---|---|---|
| Coconut oil | 0 | ω6 |
| Butter | 33 | ω6 ω3 |
| Beef tallow | 14 | ω6 ω3 |
| Palm oil | 0 | ω6 |

• Animal fats and the tropical oils of coconut and palm are mostly saturated.

| Dietary fat | Cholesterol (mg/tbs) | Polyunsaturated fats |
|---|---|---|
| Olive oil | 0 | ω6 ω3 |
| Canola oil | 0 | ω6 ω3 |
| Peanut oil | 0 | ω6 |
| Lard | 12 | ω6 ω3 |

• Some vegetable oils, such as olive and canola, are rich in monounsaturated fatty acids.

| Dietary fat | Cholesterol (mg/tbs) | Polyunsaturated fats |
|---|---|---|
| Safflower oil | 0 | ω6 ω3 |
| Sunflower oil | 0 | ω6 |
| Corn oil | 0 | ω6 ω3 |
| Soybean oil | 0 | ω6 ω3 |
| Cottonseed oil | 0 | ω6 |

• Many vegetable oils are rich in polyunsaturated fatty acids.

Saturated fats

Monounsaturated fats

Polyunsaturated fats
ω3 Linolenic acid
ω6 Linoleic acid

***At room temperature, unsaturated fats (such as those found in oil) are usually liquid, whereas saturated fats (such as those found in butter) are solid.***

dation and thus less likely to become rancid. Other types of spoilage can occur due to microbial growth.

Manufacturers can protect fat-containing products against rancidity in three ways—none of them perfect. First, products may be sealed air-tight and refrigerated—an expensive and inconvenient storage system. Second, manufacturers may add antioxidants to compete for the oxygen and thus protect the oil (examples are the additives BHA and BHT and vitamins C and E); the advantages and disadvantages of antioxidants in food processing are presented in Chapter 19. Third, manufacturers may saturate some or all of the points of unsaturation by adding hydrogen molecules—a process known as hydrogenation.

**antioxidant:** a compound that protects others from oxidation by being oxidized itself.

• ***Hydrogenation of Fats*** • The process of hydrogenation offers two advantages: it protects against oxidation (which prolongs shelf life) and alters the texture of foods. When partially hydrogenated, vegetable oils become spreadable margarine. Hydrogenated fats make pie crusts flaky and puddings creamy. A disadvantage is that hydrogenation makes polyunsaturated fats more saturated (see Figure 5-6). Consequently, any health advantages of using polyunsaturated fats instead of saturated fats are lost in hydrogenation.

**hydrogenation** (high-dro-gen-AY-shun): a chemical process by which hydrogens are added to monounsaturated or polyunsaturated fats to reduce the number of double bonds, making the fats more saturated (solid) and more resistant to oxidation (protecting against rancidity). Hydrogenation produces *trans*-fatty acids.

• ***Formation of*** **trans*-Fatty Acids*** • Another disadvantage of hydrogenation is that some of the molecules that remain unsaturated after processing change shape from *cis* to *trans*. In nature, most unsaturated fatty acids are *cis*-fatty acids—meaning that the hydrogens next to the double bonds are on the same side of the

***trans*-fatty acids:** fatty acids with an unusual configuration around the double bond.

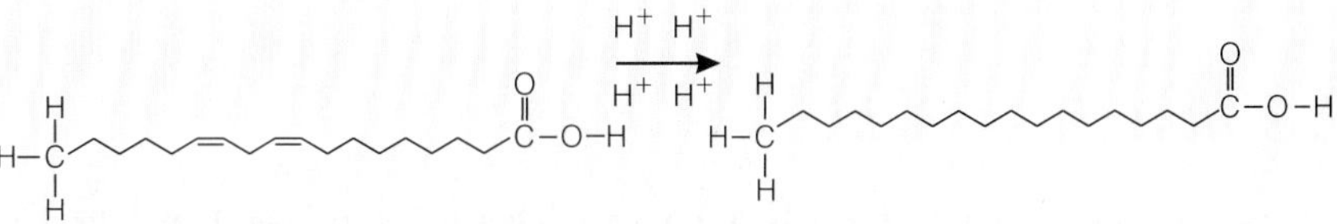


Double bonds carry a slightly negative charge and readily accept positively charged hydrogen atoms, creating a saturated fatty acid.

**Figure 5-6**

**Hydrogenation**

Hydrogenation yields a product that is more saturated, more spreadable, and more resistant to oxidation.

**Figure 5-7**

***cis-* and *trans*-Fatty Acids Compared**

Manufacturers rarely use total hydrogenation; most often a fat is partially hydrogenated, yielding a *trans*-monounsaturated fatty acid. This example shows the *cis* configuration for oleic acid and its corresponding *trans* configuration (elaidic acid).

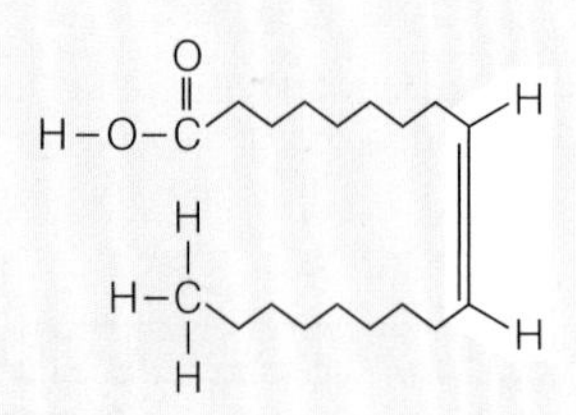

*cis*-fatty acid

A *cis*-fatty acid has its hydrogens on the same side of the double bond; *cis* molecules fold back into a U-like formation. Most naturally ocurring unsaturated fatty acids in foods are *cis*.

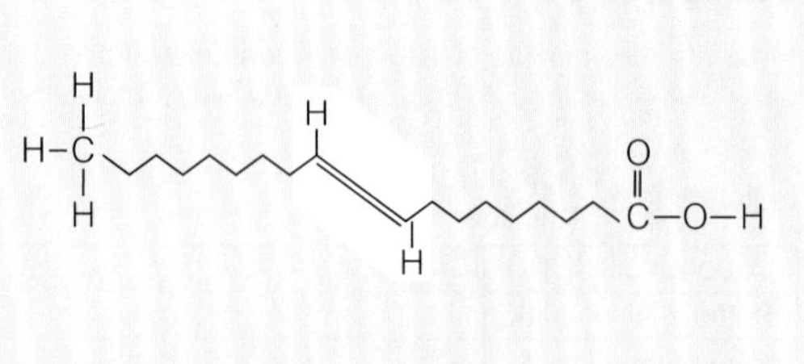

*trans*-fatty acid

A *trans*-fatty acid has its hydrogens on the opposite sides of the double bond; *trans* molecules are more linear. The *trans* form typically occurs in partially hydrogenated foods when hydrogen atoms shift around some double bonds and change the configuration from *cis* to *trans*.

Major sources of *trans*-fatty acids:
- Margarine (hard stick, soft tub).
- Cakes, cookies, doughnuts, crackers.
- Meat and dairy products.
- Snack chips.
- Peanut butter.
- Shortening (fried foods).

carbon chain. Only a few (notably those found in milk and butter) are *trans*-fatty acids—meaning that the hydrogens next to the double bonds are on opposite sides of the carbon chain (see Figure 5-7). These arrangements result in different configurations for the fatty acids, and this difference affects function.

*Trans*-fatty acids make up approximately 5 percent of the fat intake in the U.S. diet.[3] The amount of *trans*-fatty acids a person consumes depends on the amount of fat eaten and the types of foods selected. A food that lists partially hydrogenated oils among its first three ingredients usually contains substantial amounts of *trans*-fatty acids, as well as some saturated fat. The relationship between *trans*-fatty acids and heart disease has been the subject of much recent research, as a later section describes.

IN SUMMARY

Fatty acid saturation affects fats' physical characteristics, cooking qualities, storage properties, and their contributions to people's susceptibility to disease. Hydrogenation, which makes polyunsaturates more saturated, gives rise to some *trans*-fatty acids, altered fatty acids that may have health effects similar to those of saturated fatty acids.

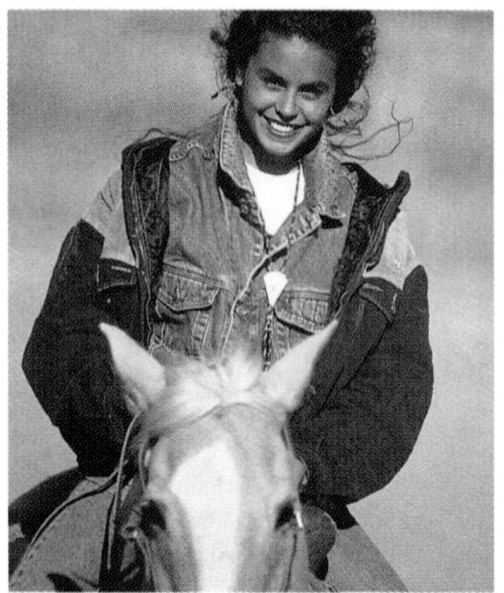

***Thanks to the body's fat pads, a horseback ride causes no serious damage to internal organs.***

IN SUMMARY

Triglycerides in the body:
- Provide energy to meet daily needs.
- Provide an energy reserve when stored in the body's fat tissue.
- Insulate against temperature extremes.
- Protect organs against shock.
- Help the body use carbohydrate and protein efficiently.

## Roles of Triglycerides and Fatty Acids

First and foremost, the triglycerides provide the body with energy. When a person dances all night, her stored triglycerides provide the fuel to keep her moving; when a person loses his appetite, his stored triglycerides fuel much of his body's work until he can eat again. Stored fat supports many of life's activities.

Stored fat also insulates the body. Fat is a poor conductor of heat, so the layer of fat beneath the skin helps keep the body warm. Fat pads also serve as shock absorbers, supporting and cushioning the vital organs.

Fat also helps the body use its two other energy nutrients—carbohydrate and protein—efficiently. Fat fragments combine with glucose fragments during energy metabolism, and fat helps spare protein, providing energy so that protein can be used for other important tasks.

## Essential Fatty Acids

**essential fatty acids:** fatty acids needed by the body, but not made by it in amounts sufficient to meet physiological needs.

The human body needs fatty acids, and it can make all but two of them—linoleic acid and linolenic acid. They must be supplied by the diet and are therefore called essential fatty acids. The body uses these essential fatty acids to maintain the structural parts of cell membranes and to make many hormonelike substances known as eicosanoids. Eicosanoids help regulate blood pressure, blood clot formation, blood lipids, and the immune response to injury and infection.

Linoleic acid and linolenic acid were introduced earlier as the 18-carbon members of the two omega families of fatty acids.* Figure 5-8 compares the structures of these two families.

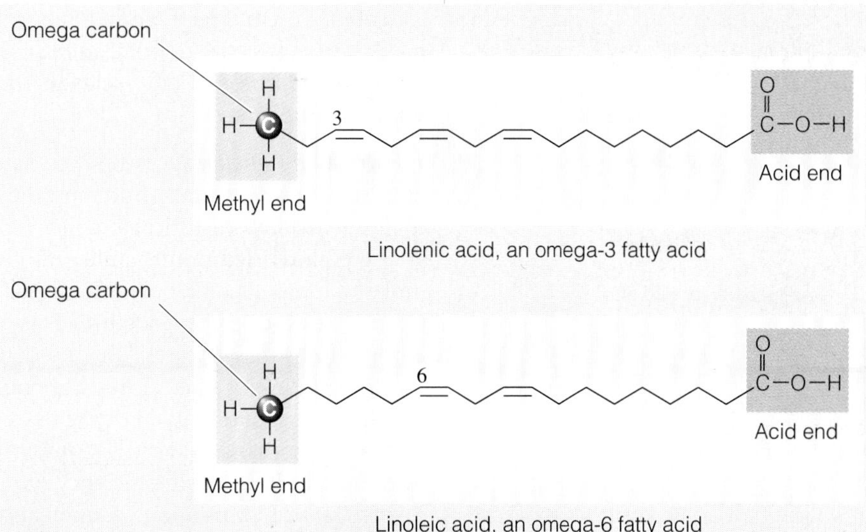


**Figure 5-8**
**Structural Formulas for Omega-3 and Omega-6 Fatty Acids**

The omega number indicates the position of the first double bond in a fatty acid, counting from the methyl ($CH_3$) end. Thus an omega-3 fatty acid's first double bond occurs three carbons from the methyl end, and an omega-6 fatty acid's first double bond occurs six carbons from the methyl end. The members of a given family may have different lengths and different numbers of double bonds, but the first double bond occurs at the same point in all of them.

• ***Linoleic Acid, an Omega-6 Fatty Acid*** • Linoleic acid is the primary member of the omega-6 family. Given linoleic acid, the body can make other members of the omega-6 family—such as the 20-carbon polyunsaturated fatty acid, arachidonic acid. Should a linoleic acid deficiency develop, arachidonic acid, and all other fatty acids that derive from linoleic acid, would also become essential and have to be obtained from the diet. Normally, vegetable oils and meats supply enough omega-6 fatty acids to meet the body's needs.

• ***Linolenic Acid, an Omega-3 Fatty Acid*** • Linolenic acid is the primary member of the omega-3 family.† Like linoleic acid, this 18-carbon acid cannot be made in the body and must be supplied by foods. Given dietary linolenic acid, the body can make the 20- and 22-carbon members of the omega-3 series, eicosapentaenoic acid (EPA) and docosahexaenoic acid (DHA). These omega-3 fatty acids are essential for normal growth and development, and they may play an important role in the prevention and treatment of heart disease, hypertension, arthritis, and cancer.

• ***A Comment on Essentiality*** • A simple definition of an essential nutrient has already been given: a nutrient that the body cannot make, or cannot make in sufficient quantities to meet its physiological needs. In the case of fatty acids, though, the body can make some fatty acids only if others are available. Also, some may be essential only for growth or for disease prevention.

The cells do not possess the enzymes to make any of the omega-6 or omega-3 fatty acids from scratch; nor can they convert an omega-6 fatty acid to an omega-3 fatty acid or vice versa. They *can* start with the 18-carbon member of a series and make the longer fatty acids of that series by forming double bonds (desaturation) and lengthening the chain two carbons at a time (elongation). This is a slow process because the two families compete for the same enzymes. Therefore, the most effective way to maintain body supplies of these polyunsaturated fatty acids

*A fatty acid has two ends, designated the methyl ($CH_3$) end and the acid (COOH) end. Chemists usually number the carbons beginning at the acid end, but make an exception for polyunsaturated fatty acids. Because the omega number refers to the location of the double bond and the body lengthens fatty acid chains by adding carbons at the acid end of the chain, the omega numbers would change as the chains grew longer. Chemists therefore number the carbons beginning at the methyl end, which eases the task of keeping track of fatty acids: when an omega-3 fatty acid is lengthened, the derivative is also an omega-3 fatty acid.

†This omega-3 linolenic acid is known as alpha-linolenic acid and is the fatty acid referred to in this discussion. Another fatty acid, also with 18 carbons and three double bonds, belongs to the omega-6 family and is known as gamma-linolenic acid.

**eicosanoids** (eye-COSS-uh-noyds): derivatives of fatty acids; hormonelike compounds that regulate blood pressure, clotting, and other body functions. They include *prostaglandins, thromboxanes,* and *leukotrienes.*

**omega:** the last letter of the Greek alphabet (ω), used by chemists to refer to the position of the first double bond from the methyl end in a fatty acid.

**omega-6 fatty acid:** a polyunsaturated fatty acid in which the first double bond is six carbons from the methyl ($CH_3$) end of the carbon chain.

**arachidonic** (a-RACK-ih-DON-ic) **acid:** an omega-6 polyunsaturated fatty acid with 20 carbons and four double bonds (20:4); synthesized from linoleic acid.

**omega-3 fatty acid:** a polyunsaturated fatty acid in which the first double bond is three carbons away from the methyl ($CH_3$) end of the carbon chain.

**eicosapentaenoic** (EYE-cossa-PENTA-ee-NO-ick) **acid (EPA):** an omega-3 polyunsaturated fatty acid with 20 carbons and five double bonds (20:5); synthesized from linolenic acid.

**docosahexaenoic** (DOE-cossa-HEXA-ee-NO-ick) **acid (DHA):** an omega-3 polyunsaturated fatty acid with 22 carbons and six double bonds (22:6); synthesized from linolenic acid.

Linoleic acid (18:2)
desaturation ↓
(18:3)
elongation ↓
(20:3)
desaturation ↓
Arachidonic acid (20:4)

NOTE: The first number indicates the number of carbons and the second, the number of double bonds. Similar reactions occur when the body makes EPA and DHA from linolenic acid.

is to obtain them directly from foods. A balanced diet that includes grains, seeds, nuts, leafy vegetables (or small amounts of vegetable oils), and fish supplies all the omega-6 and omega-3 fatty acids in abundance. Table 5-2 lists the chief dietary sources of the omega fatty acids.

• ***Fatty Acid Deficiencies*** • Essential fatty acids should make up at least 3 percent of the day's energy intake, and most diets meet this minimum requirement more than adequately. Historically, deficiencies have developed only in infants and young children fed nonfat milk and low-fat diets or in hospital clients fed formulas that provided no polyunsaturated fatty acids for long periods of time. More recently, researchers have identified essential fatty acid deficiencies in people with chronic intestinal diseases.[4] Such a deficiency creates a vicious cycle. The deficiency first develops because of the poor absorption associated with the disease, but then a lack of essential fatty acids exacerbates the disease. Essential fatty acids participate in eicosanoid activity, immunity, and GI cell growth—all important to intestinal health. Classic deficiency symptoms include growth retardation, reproductive failure, skin lesions, kidney and liver disorders, and subtle neurological and visual problems.

IN SUMMARY

Small amounts of fats are necessary to provide energy, support health, and carry fat-soluble vitamins. Linoleic acid (18 carbons, omega-6) and linolenic acid (18 carbons, omega-3) are essential nutrients; deficiencies are unlikely.

# The Chemist's View of Phospholipids and Sterols

The preceding pages have been devoted to one of the three classes of lipids, the triglycerides, and their component parts, the fatty acids. The other two classes of lipids, the phospholipids and sterols, make up only 5 percent of the lipids in the diet, but they are nevertheless interesting and important.

**Table 5-2**
**Sources of Omega Fatty Acids**

| **Omega-6** | |
|---|---|
| Linoleic acid | Leafy vegetables, seeds, nuts, grains, vegetable oils (corn, safflower, soybean, cottonseed, sesame, sunflower) |
| Arachidonic acid | Meats (or can be made from linoleic acid) |
| **Omega-3** | |
| Linolenic acid | Fats and oils (canola, soybean, walnut, wheat germ, margarine and shortening made from canola and soybean oil) |
| | Nuts and seeds (butternuts, walnuts, soybean kernels) |
| | Vegetables (soybeans) |
| EPA and DHA | Human milk |
| | Shellfish and fish[a] (mackerel, salmon, bluefish, mullet, sablefish, menhaden, anchovy, herring, lake trout, sardines, tuna) |
| | (or can be made from linolenic acid) |

[a]All of these fish except tuna provide at least 1 gram of omega-3 fatty acids in 100 grams of fish (3.5 ounces); the fish oil content of each species varies with the season and site of harvest. Tuna provides fewer omega-3 fatty acids, but because it is commonly consumed, its contribution can be significant.

## The Phospholipids

The best-known phospholipids are the lecithins. Each lecithin has a backbone of glycerol with two of its three attachment sites occupied by fatty acids like those in triglycerides. The third site is occupied by a phosphate group and a molecule of choline. The fatty acids make phospholipids soluble in fat; the phosphate-containing group enables them to dissolve in water. Such versatility enables the food industry to use phospholipids as emulsifiers, mixing fats with water in such products as mayonnaise and candy bars. A diagram of a lecithin molecule is shown in Figure 5-9.

• ***Phospholipids in Foods*** • In addition to the phospholipids used by the food industry as emulsifiers, phospholipids are also found in foods naturally. The richest food sources of lecithin are eggs, liver, soybeans, wheat germ, and peanuts.

• ***Roles of Phospholipids*** • The lecithins and other phospholipids are important constituents of cell membranes. Because phospholipids can dissolve in both water and fat, they can help lipids move back and forth across the lipid-containing cell membranes into the watery fluids on both sides. They thus allow fat-soluble substances, including vitamins and hormones, to pass easily in and out of cells. The phospholipids also act as emulsifiers in the body, helping to keep fats suspended in the blood and body fluids.

Lecithin periodically receives attention in the popular press. Its fans claim that it is a major constituent of cell membranes (true), that all cells depend on the integrity of their membranes (true), and that consumers must therefore take lecithin supplements (false). The liver makes from scratch all the lecithin a person needs. As for lecithin taken as a supplement, the digestive enzyme lecithinase in the intestine hydrolyzes most of it before it passes into the body fluids, so little lecithin reaches the body tissues intact. In other words, the lecithins are *not essential nutrients;* they are just another lipid. Like other lipids, they contribute 9 kcalories per gram to the body's energy economy—an unexpected "bonus" many people taking lecithin supplements fail to realize. Furthermore, large doses of lecithin may cause GI distress, sweating, salivation, and loss of appetite. Perhaps these symptoms are beneficial because they may warn people to stop self-dosing with lecithin.

**phospholipid:** a compound similar to a triglyceride but having a phosphate group (a phosphorus-containing salt) and choline (or another nitrogen-containing compound) in place of one of the fatty acids.

**lecithin** (LESS-uh-thin): one of the phospholipids; a compound of glycerol to which are attached two fatty acids, a phosphate group, and a choline molecule. Both nature and the food industry use lecithin as an emulsifier to combine two ingredients that do not ordinarily mix, such as water and oil.

**choline** (KOH-leen): a nitrogen-containing compound found in foods as part of lecithin and other phospholipids.

Reminder: An *emulsifier* promotes the mixing of two substances, such as oil and water, that are not mutually soluble.

Reminder: The word ending *-ase* denotes an enzyme. Hence, lecithinase is an enzyme that works on lecithin.

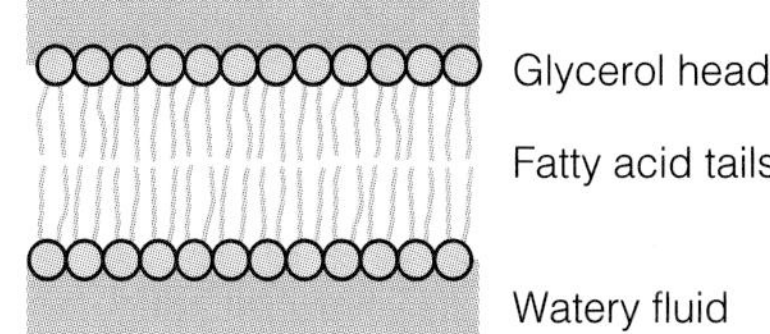


A cell membrane is made of phospholipids assembled into an orderly formation called a bilayer. The fatty acid "tails" orient themselves away from the watery fluid inside and outside of the cell. The glycerol and phosphate "heads" are attracted to the watery fluid.

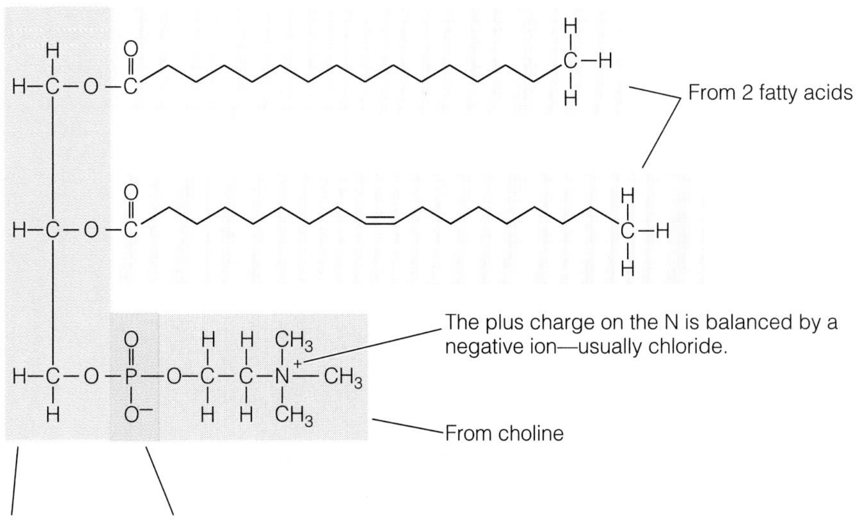


**Figure 5-9**
**A Lecithin**

This is one of the lecithins. Other lecithins have different fatty acids at the upper two positions. Notice that a molecule of lecithin is similar to a triglyceride but contains only two fatty acids. The third position is occupied by a phosphate group and a molecule of choline.

IN SUMMARY

Phospholipids, including lecithin, have a unique chemical structure that allows them to be soluble in both water and fat. In the body, phospholipids are part of cell membranes; the food industry uses phospholipids as emulsifiers.

## The Sterols

**sterols:** compounds composed of C, H, and O atoms arranged in rings, like those of cholesterol, with any of a variety of side chains attached.

**cholesterol** (koh-LESS-ter-ol): one of the sterols.

In addition to triglycerides and phospholipids, the lipids include the sterols, compounds with a multiple-ring structure. The most famous sterol is cholesterol; Figure 5-10 shows its chemical structure. The accompanying table offers a summary of the lipid family.

• ***Sterols in Foods*** • Both plant and animal foods contain sterols, but only animal foods contain cholesterol: meats, eggs, fish, poultry, and dairy products. Organ meats, such as liver and kidneys, and eggs are richest in cholesterol; cheeses and meats have less. Shellfish contain many sterols, but much less cholesterol than was thought in the past. Figure 5-11 shows the cholesterol contents of selected foods. Many more foods, with their cholesterol contents, appear in Appendix H.

IN SUMMARY

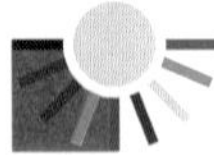

**The Lipid Family**

**Triglycerides (fats and oils)**
- Glycerol (1 per triglyceride)
- Fatty acids (3 per triglyceride)
  - Saturated
  - Monounsaturated
  - Polyunsaturated
    - Omega-3
    - Omega-6

**Phospholipids (such as lecithin)**

**Sterols (such as cholesterol)**

An egg contains just over 200 milligrams of cholesterol, all of it in the yolk. A person on a strict low-cholesterol diet must curtail the use of egg yolks, and food manufacturers have produced several nonfat, no-cholesterol egg substitutes. For most people trying to lower blood cholesterol, however, limiting saturated fat is more effective than limiting cholesterol intake. Eggs are a valuable part of the diet because they are inexpensive, useful in cooking, and a source of high-quality protein. The American Heart Association approves an intake of up to four eggs a week.

Some people, confused about the distinction between dietary and blood cholesterol, have asked which foods contain the "good" cholesterol. "Good" cholesterol is not a type of cholesterol found in foods, but refers to the way the body transports cholesterol in the blood, as explained later (p. 140).

• ***Roles of Sterols*** • Many vitally important body compounds are sterols. Among them are bile acids, the sex hormones (such as testosterone), the adrenal hormones (such as cortisol), and vitamin D, as well as cholesterol itself. Cholesterol in the body can serve as the starting material for synthesis of these compounds or as a structural component of cell membranes; more than nine-tenths of all the body's cholesterol resides in the cells. Despite popular impressions to the contrary, therefore, cholesterol is not a villain lurking in some evil foods—it is a compound the body makes and uses. Your liver is manufacturing cholesterol now, as you read. At the rate of perhaps $5 \times 10^{16}$ (50,000,000,000,000,000) molecules per second (800 to 1500 milligrams per day), the liver contributes much more cholesterol to the body's total than does the diet.

Cholesterol that is made in the body is **endogenous** (en-DOGDE-eh-nus), whereas cholesterol from outside the body (from foods) is **exogenous** (eks-ODGE-eh-nus).
- **endo** = within
- **gen** = arising
- **exo** = outside (the body)

**Figure 5-10**
**Cholesterol**

The fat-soluble vitamin D is synthesized from cholesterol; notice the many similarities. The only difference, in fact, is vitamin D's open ring, which accounts for its vitamin activity. Notice, too, how different cholesterol is from the triglycerides and phospholipids.

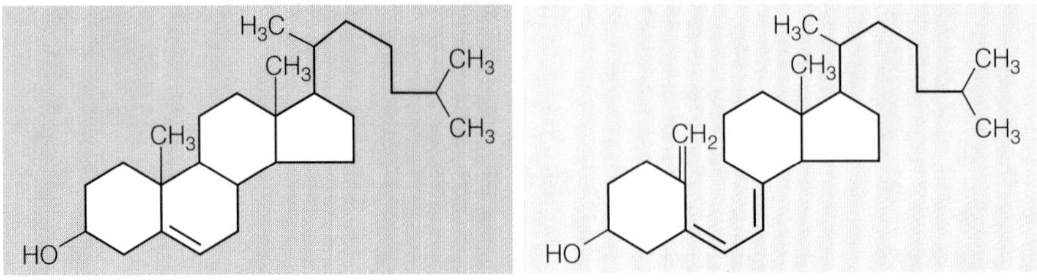

$H_3C$ $CH_3$ $CH_3$ $CH_3$ $CH_3$ HO

Cholesterol

$H_3C$ $CH_3$ $CH_3$ $CH_3$ $CH_2$ HO

Vitamin $D_3$

**Figure 5-11**
**Cholesterol in Selected Foods**

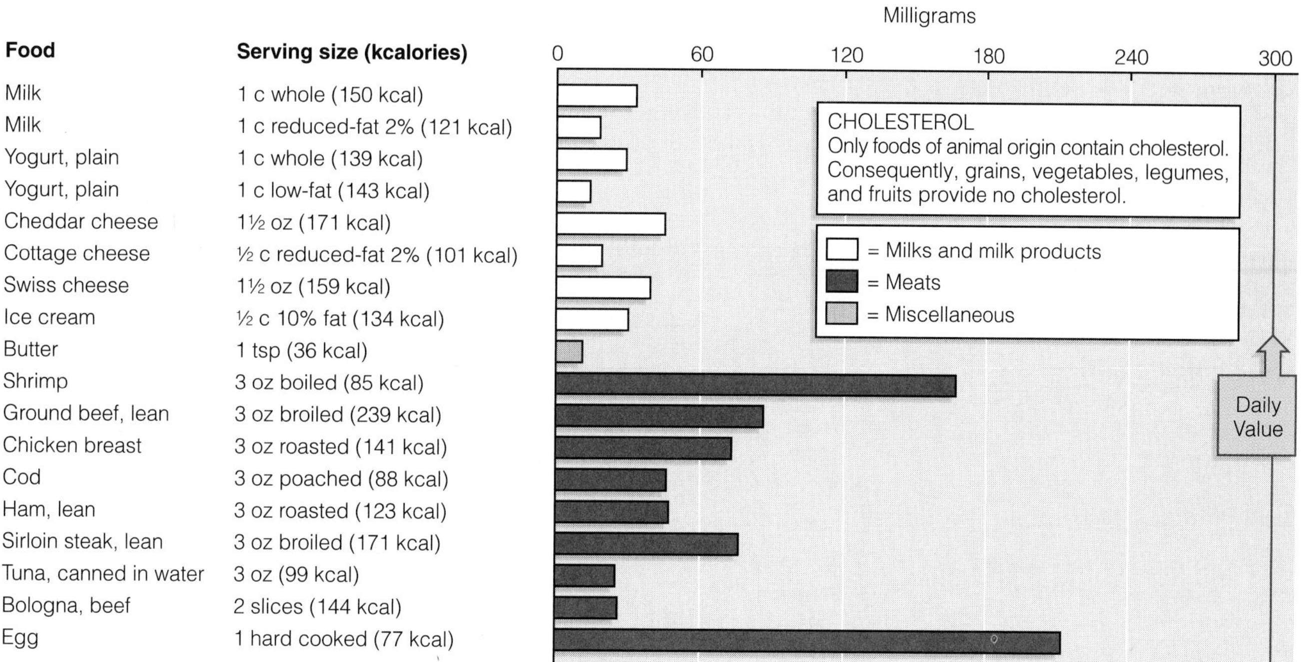


The liver can use fragments derived from carbohydrate, protein, or fat as the starting material from which to make cholesterol. Cholesterol synthesis depends not only on the availability of these raw materials, but also on the extent of bile production and the presence of regulating hormones such as insulin. When insulin concentrations remain low, as occurs when people eat many small meals, cholesterol synthesis slows. Spreading total food intake over many meals and snacks a day without an increase in energy intake is a relatively easy and effective way to lower blood cholesterol.[5] Restricting energy intake will also lower blood cholesterol by slowing cholesterol synthesis in the liver.[6]

Reminder: Eating many small meals and snacks throughout the day is sometimes called **grazing.**

Cholesterol's harmful effects in the body occur when it forms deposits in the artery walls. These deposits lead to atherosclerosis, a disease that causes heart attacks and strokes (see Chapter 18 for more details).

**atherosclerosis** (ath-er-oh-scler-OH-sis): a type of artery disease characterized by accumulations of lipid-containing material on the inner walls of the arteries (see Chapter 18).
- **athero** = porridge or soft
- **scleros** = hard
- **osis** = condition

IN SUMMARY

Sterols, including cholesterol, have a multiple-ring structure that differs from the other lipids. Sterols in the body include bile, vitamin D, and the sex hormones. Only animal-derived foods contain cholesterol.

# Digestion, Absorption, and Transport of Lipids

Each day, the GI tract receives, on average, 50 to 100 grams of triglycerides, 4 to 8 grams of phospholipids, and 300 to 450 milligrams of cholesterol. The body faces a challenge in digesting and absorbing these lipids: getting at them.

Fats are hydrophobic—that is, they tend to separate from water—whereas the enzymes for digesting fats are hydrophilic. Since the watery fluids of the GI tract settle at the bottom of the stomach while the dietary fats float on top, the fats start

**hydrophobic** (high-dro-FOE-bick): a term referring to water-fearing, or non-water-soluble, substances; also known as **lipophilic** (fat loving).
- **hydro** = water
- **phobia** = fear
- **lipo** = lipid
- **phile** = friend

**hydrophilic** (high-dro-FIL-ick): a term referring to water-loving, or water-soluble, substances.

out separated from their enzymes. The following paragraphs describe how the body mixes the fats into the watery fluids and then digests them.

## Lipid Digestion

**monoglyceride:** a molecule of glycerol with one fatty acid attached. A molecule of glycerol with two fatty acids attached is a **diglyceride.**

The goal of fat digestion is to dismantle triglycerides into small molecules that the body can absorb and use—namely, monoglycerides, fatty acids, and glycerol. Figure 5-12 traces the digestion of triglycerides through the GI tract.

**Figure 5-12**
**Triglyceride Digestion in the GI Tract**

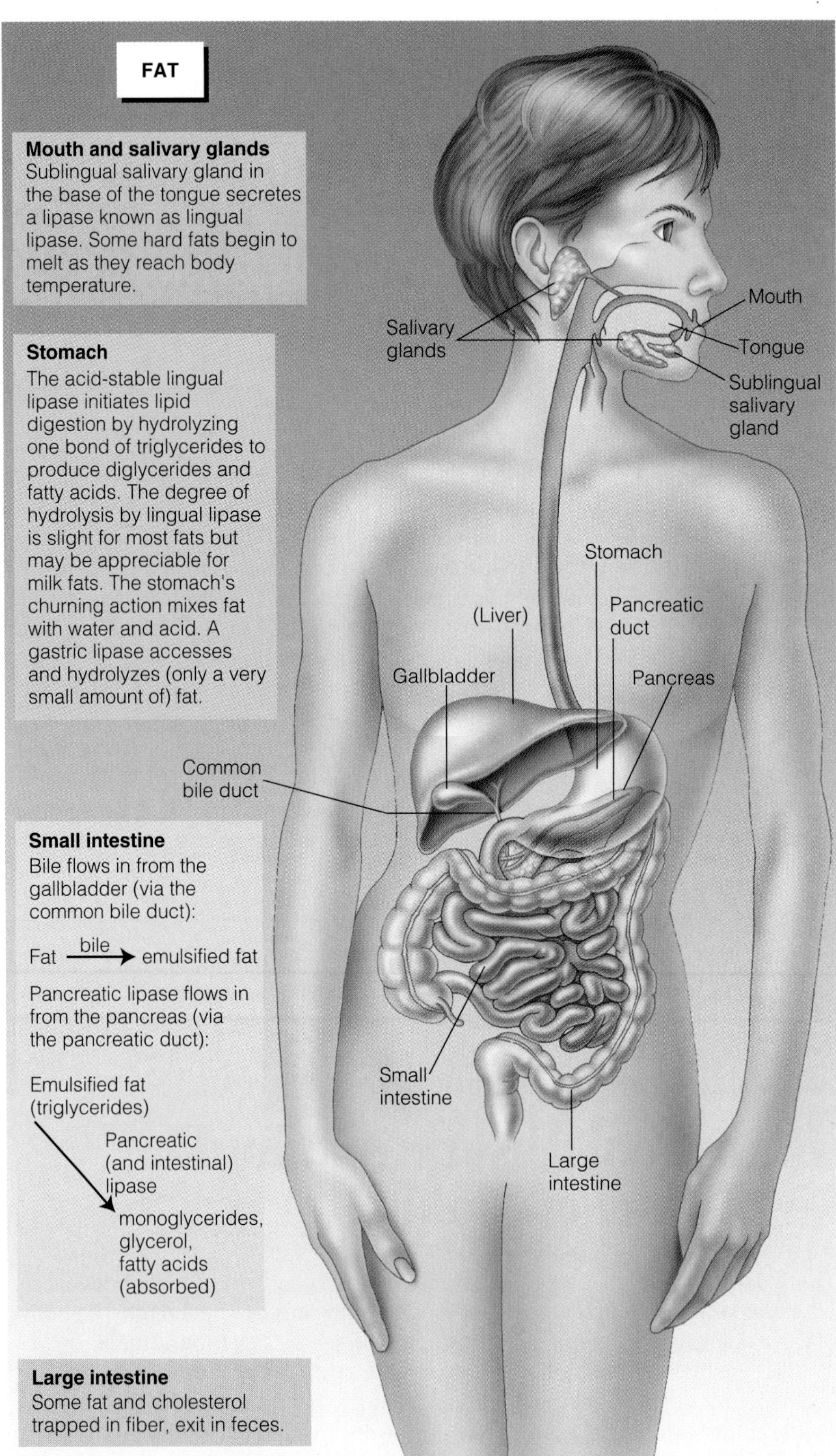

FAT
Mouth and salivary glands
Sublingual salivary gland in the base of the tongue secretes a lipase known as lingual lipase. Some hard fats begin to melt as they reach body temperature.
Stomach
The acid-stable lingual lipase initiates lipid digestion by hydrolyzing one bond of triglycerides to produce diglycerides and fatty acids. The degree of hydrolysis by lingual lipase is slight for most fats but may be appreciable for milk fats. The stomach's churning action mixes fat with water and acid. A gastric lipase accesses and hydrolyzes (only a very small amount of) fat.
Common bile duct
Small intestine
Bile flows in from the gallbladder (via the common bile duct):
Fat —bile→ emulsified fat
Pancreatic lipase flows in from the pancreas (via the pancreatic duct):
Emulsified fat (triglycerides)
Pancreatic (and intestinal) lipase
monoglycerides, glycerol, fatty acids (absorbed)
Large intestine
Some fat and cholesterol trapped in fiber, exit in feces.
Mouth
Salivary glands
Tongue
Sublingual salivary gland
Stomach
(Liver)
Pancreatic duct
Gallbladder
Pancreas
Small intestine
Large intestine

• ***In the Mouth*** • Fat digestion starts off slowly in the mouth, with some hard fats beginning to melt when they reach body temperature. The salivary glands at the base of the tongue release a lipase enzyme that plays a small role in fat digestion in adults and an active role in infants. In infants, this enzyme efficiently digests the short- and medium-chain fatty acids found in milk.

• ***In the Stomach*** • In the stomach, fat floats as a layer above the other components of swallowed food. As a result, little fat digestion takes place.

• ***In the Small Intestine*** • When fat enters the small intestine, the hormone cholecystokinin (CCK) signals the gallbladder to release its stores of bile, an emulsifier. (The liver manufactures bile acids from cholesterol, and the gallbladder stores the bile until it is needed.)

At one end of each bile acid are side chains of amino acids (units of protein) that are attracted to water, and at the other end is a sterol portion that is attracted to fat (see Figure 5-13). This structure allows bile to draw fat molecules into the surrounding watery fluids. There the fats meet lipase enzymes from the pancreas and small intestine and are fully digested. The process of emulsification is diagrammed in Figure 5-14.

Most of the hydrolysis of triglycerides occurs in the small intestine. The major fat-digesting enzymes are pancreatic lipases; some intestinal lipases are also active. These enzymes remove one, then the other, of each triglyceride's outer fatty acids, leaving a monoglyceride. Occasionally, enzymes remove all three fatty acids, leaving a free molecule of glycerol. The process of hydrolysis is shown in Figure 5-15.

Phospholipids are digested similarly—that is, their fatty acids are removed by hydrolysis. The two fatty acids and the remaining phospholipid fragment are then absorbed. Sterols can be absorbed as is; if any fatty acids are attached, they are first hydrolyzed off.

• ***Bile's Routes*** • After bile has entered the intestine and emulsified fat, it has two possible destinations, illustrated

Reminder: An enzyme that hydrolyzes lipids is called a *lipase*.

Reminder: Fat in the intestine triggers the release of the hormone *cholecystokinin (CCK)*, which signals the gallbladder to send bile.

Reminder: *Bile* is an emulsifier, as are lecithins and other phospholipids. An *emulsifier* promotes the mixing of oils and fats in a watery solution.

**Figure 5-13**
**A Bile Acid**

This is one of several bile acids the liver makes from cholesterol. It is then bound to an amino acid to improve its ability to form micelles, spherical complexes that carry fatty acids into the intestinal cells during digestion. Most bile acids occur as bile salts, usually in association with sodium, but sometimes with potassium or calcium. In addition to bile acids and bile salts, bile contains cholesterol, phospholipids (especially lecithin), antibodies, water, electrolytes, and bilirubin (a pigment resulting from the breakdown of heme).



**Figure 5-14**
**Emulsification of Fat by Bile**

Detergents are emulsifiers and work the same way, which is why they are effective in removing grease spots from clothes. Molecule by molecule, the grease is dissolved out of the spot and suspended in the water, where it can be rinsed away.

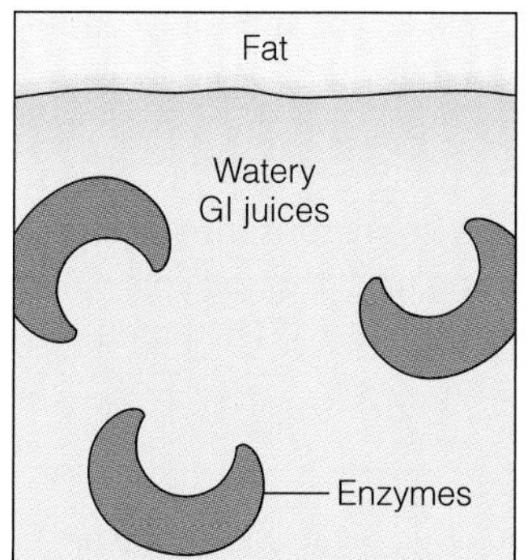


In the stomach, the fat and watery GI juices tend to separate. The enzymes are in the water and can't get at the fat.

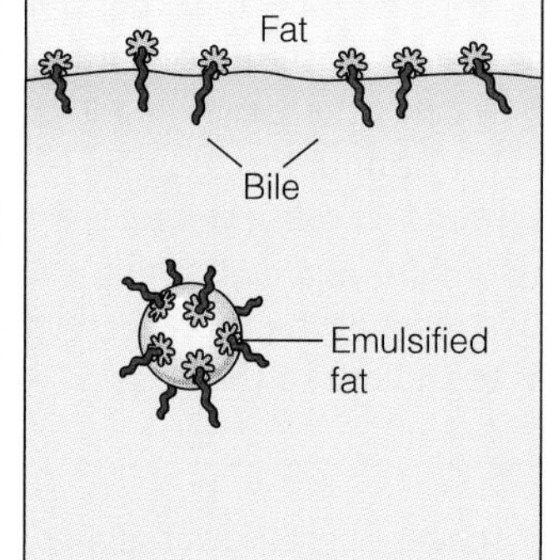


When fat enters the small intestine, the gallbladder secretes bile. Bile has an affinity for both fat and water, so it can bring the fat into the water.

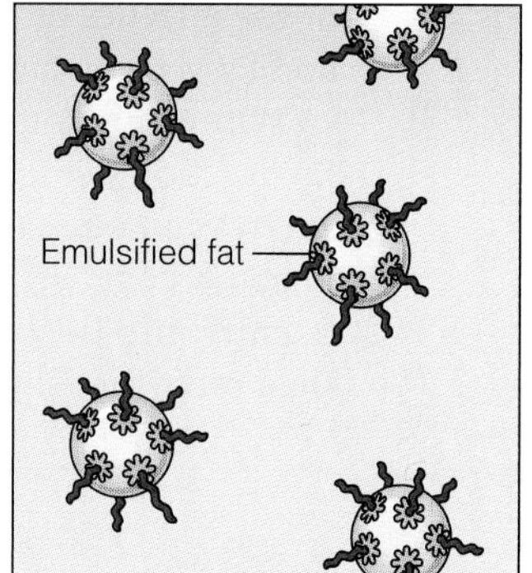


Bile's emulsifying action converts large fat globules into small droplets that repel each other.

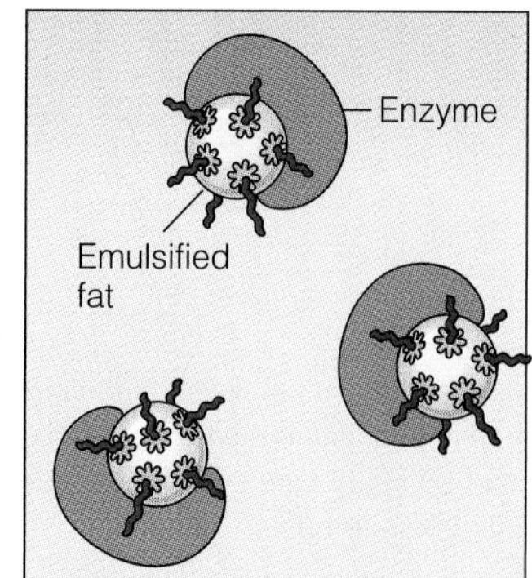


After emulsification, the enzymes have easy access to the fat droplets.

**Figure 5-15**
**Digestion (Hydrolysis) of a Triglyceride**

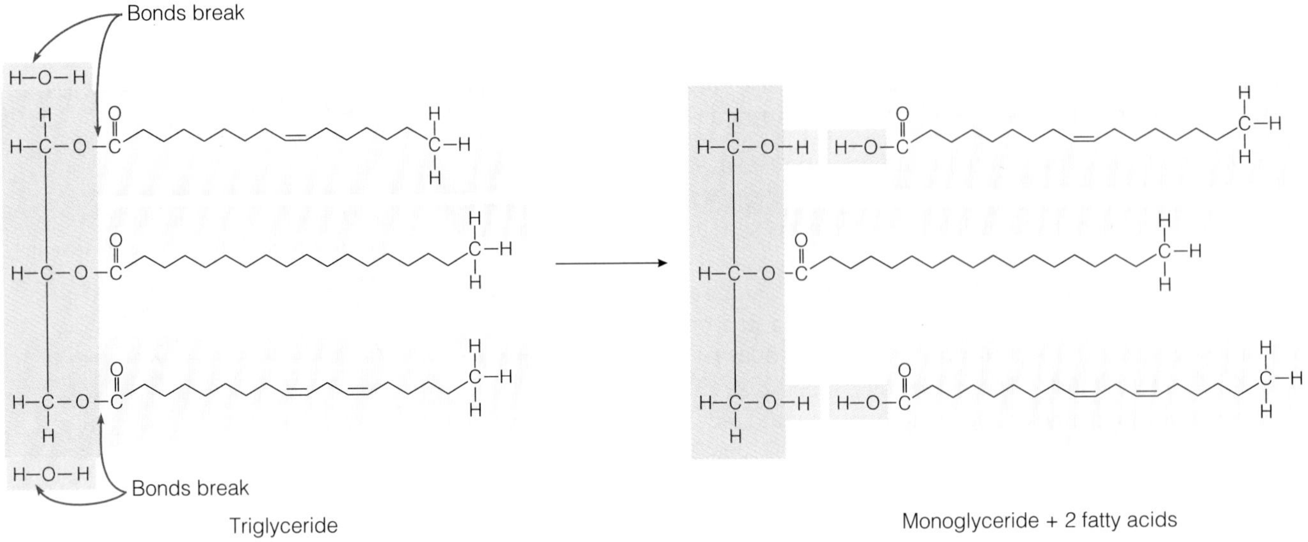

Bonds break

Bonds break

Triglyceride

The triglyceride and two molecules of water are split, and the pieces combine to give two fatty acids and a monoglyceride.

Monoglyceride + 2 fatty acids

These products may pass into the intestinal cells, but sometimes the monoglyceride is split with another molecule of water to give a third fatty acid and glycerol. Fatty acids, monoglycerides, and glycerol are absorbed into intestinal cells.

**Figure 5-16**
**Enterohepatic Circulation**

The recycling of cholesterol and bile through the intestine and liver is known as the **entero-hepatic circulation** of bile.

- **enteron** = intestine
- **hepat** = liver

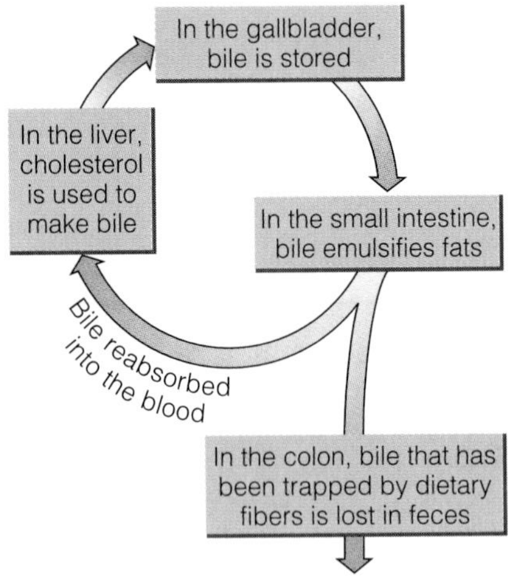
In the gallbladder, bile is stored
In the liver, cholesterol is used to make bile
In the small intestine, bile emulsifies fats
Bile reabsorbed into the blood
In the colon, bile that has been trapped by dietary fibers is lost in feces

in Figure 5-16. For one, bile can be reabsorbed from the intestine and recycled. The other possibility is that some of the bile can be trapped by dietary fibers in the large intestine and carried out of the body with the feces. Because cholesterol is needed to make bile, the excretion of bile effectively reduces blood cholesterol. The fibers most effective at lowering blood cholesterol this way are the soluble pectins and gums commonly found in fruits, oats, and legumes.

## *Lipid Absorption*

Figure 5-17 illustrates the absorption of lipids. Small molecules of digested triglycerides (glycerol and short- and medium-chain fatty acids) can diffuse easily into the intestinal cells; they are absorbed directly into the bloodstream. Larger molecules (the monoglycerides and long-chain fatty acids) merge into spherical complexes, known as micelles, which are so small that they can fit between the tiny, hairlike microvilli of a single intestinal cell. (Emulsified fat particles are 100 times larger in diameter and contain tens of thousands of molecules.) The lipid contents of the micelles diffuse into the intestinal cells. Once inside, the monoglycerides and long-chain fatty acids are reassembled into new triglycerides.

Within the intestinal cells, the newly made triglycerides and the other large lipids (cholesterol and phospholipids) are packed into transport vehicles known as chylomicrons. The intestinal cells then release the chylomicrons into the lymphatic system. The chylomicrons glide through the lymph until they reach a point of entry into the bloodstream at the thoracic duct near the heart. The blood can then carry these lipids to the rest of the body.

**micelles** (MY-cells): tiny spherical complexes that arise during fat digestion; each carries about 20 fatty acids and/or monoglycerides into intestinal cells.

Reminder: Nutrients from the GI tract that enter the lymph system bypass the liver at first.

IN SUMMARY

The body makes special arrangements to digest, absorb, transport, store, and use lipids. It provides the emulsifier bile to make them accessible to the fat-digesting lipases that dismantle triglycerides, mostly to monoglycerides and fatty acids, for absorption by the intestinal cells.

**Figure 5-17**
**Absorption of Lipids**

The end products of fat digestion are mostly monoglycerides, some fatty acids, and very little glycerol. Their absorption differs depending on their size. (In reality, molecules of fatty acid are too small to see without a powerful microscope, while villi are visible to the naked eye.)

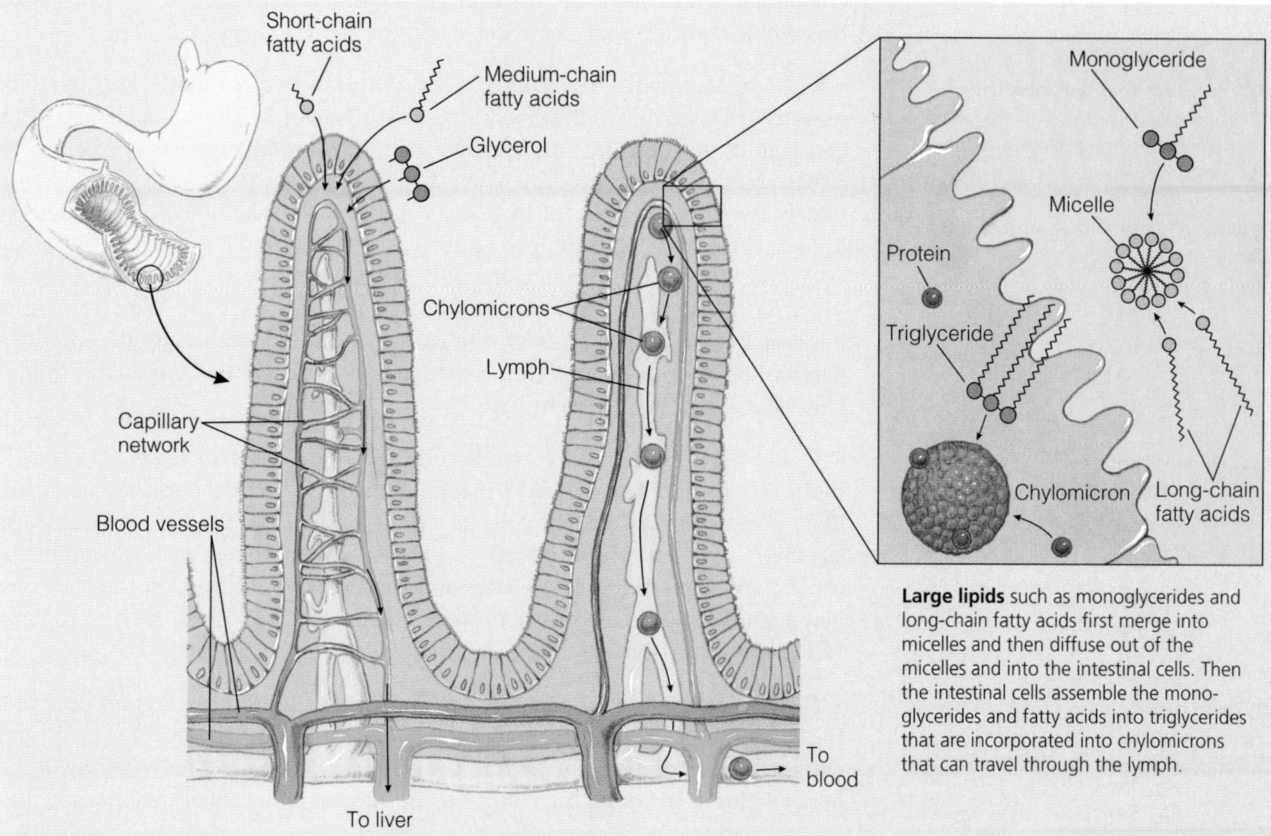

**Large lipids** such as monoglycerides and long-chain fatty acids first merge into micelles and then diffuse out of the micelles and into the intestinal cells. Then the intestinal cells assemble the monoglycerides and fatty acids into triglycerides that are incorporated into chylomicrons that can travel through the lymph.

**Glycerol and small lipids** such as short- and medium-chain fatty acids can move directly into the bloodstream.

The intestinal cells assemble freshly absorbed lipids into chylomicrons, lipid packages with protein escorts, for transport so that cells all over the body may select needed lipids from them. The margin summarizes the absorption of lipids.

## Lipid Transport

The chylomicrons are only one of several clusters of lipids and proteins that are used as transport vehicles for fats. As a group, these vehicles are known as lipoproteins, and they solve the body's problem of transporting fatty materials through the watery medium of the bloodstream. The body makes four main types of lipoproteins, distinguished by their size and density.* Each type contains different kinds of special proteins and carries different amounts of the various lipids.

• ***Chylomicrons*** • The chylomicrons are the largest and least dense of the lipoproteins. They transport *diet*-derived lipids (mostly triglycerides) from the

*The lipoproteins are distinguished by density because the chemist uses this feature to separate them in the laboratory. The chemist layers a blood sample below a thick fluid in a test tube and spins the tube in a centrifuge. The most buoyant particles (highest in lipids) rise to the top, and the densest particles (highest in proteins) remain at the bottom. Lipoproteins with a low protein-to-lipid ratio have a low density; those with a high protein-to-lipid ratio have a high density.

### IN SUMMARY

Absorbed directly into blood:
- Glycerol.
- Short-chain fatty acids.
- Medium-chain fatty acids.

Merged into micelles, moved into the intestinal cells, and made into triglycerides:
- Long-chain fatty acids.
- Monoglycerides.

Assembled into chylomicrons, absorbed into lymph, then into blood:
- Triglycerides.
- Cholesterol.
- Phospholipids.

**lipoproteins** (LIP-oh-PRO-teenz): clusters of lipids associated with proteins that serve as transport vehicles for lipids in the lymph and blood.

**chylomicrons** (kye-lo-MY-cronz): the class of lipoproteins that transport lipids from the intestinal cells into the body.

intestine to the rest of the body. Cells all over the body remove lipids from the chylomicrons as they pass by, so the chylomicrons get smaller and smaller. Within 14 hours after absorption, little is left of them but protein remnants and a few odds and ends of lipid. Special protein receptors on the membranes of the liver cells recognize and remove these remnants from the blood.[7] After collecting the chylomicron remnants, the liver cells first dismantle them and then promptly reassemble the pieces into new triglycerides.

**VLDL (very-low-density lipoprotein):** the type of lipoprotein made primarily by liver cells to transport lipids to various tissues in the body; composed primarily of triglycerides.

Chylomicrons and VLDL transport triglycerides.

• ***VLDL*** • Meanwhile, the liver cells are synthesizing other lipids. The liver cells use fatty acids arriving in the blood to make cholesterol, other fatty acids, and other compounds. At the same time, the liver cells may be making lipids from carbohydrates, proteins, or alcohol. The liver is the most active site of lipid synthesis. Ultimately, the lipids made in the liver are packaged with proteins as very-low-density lipoproteins (VLDL) and shipped to other parts of the body.

As the VLDL travel through the body, cells remove triglycerides, causing the VLDL to shrink. As they lose triglycerides, the VLDL gather cholesterol from other lipoproteins circulating through the bloodstream and eventually become low-density lipoproteins (LDL).* This exchange explains why LDL contain few triglycerides but are loaded with cholesterol.

**LDL (low-density lipoprotein):** the type of lipoprotein derived from very-low-density lipoproteins (VLDL) as cells remove triglycerides from them; composed primarily of cholesterol.

• ***LDL*** • The LDL circulate throughout the body, making their contents available to the cells of all tissues—muscle, including the heart muscle; fat stores; the mammary glands; and others. The cells take triglycerides from the LDL; they also collect cholesterol and phospholipids to build new membranes, make hormones or other compounds, or store for later use. Special LDL receptors on the liver cells play a crucial role in the control of blood cholesterol concentrations by removing LDL from circulation.

**HDL (high-density lipoprotein):** the type of lipoprotein that transports cholesterol back to the liver from the cells; composed primarily of protein.

LDL and HDL transport cholesterol.

• ***HDL*** • Fat cells may release glycerol, fatty acids, cholesterol, and phospholipids to the blood. The liver makes high-density lipoprotein (HDL) packages to carry cholesterol and phospholipids from the cells back to the liver for recycling or disposal. Figure 5-18 shows the relative compositions and sizes of the lipoproteins.

• ***Health Implications*** • The distinction between LDL and HDL has implications for the health of the heart and blood vessels. The blood cholesterol linked to heart disease is LDL cholesterol. HDL also carry cholesterol, but elevated HDL represent cholesterol returning from the arteries to the liver for breakdown and excretion. High LDL cholesterol is associated with a high risk of heart attack, whereas high HDL cholesterol seems to have a protective effect.[8] This is why some people refer to LDL as "bad," and HDL as "good," cholesterol. Keep in mind that there is only *one* kind of cholesterol, and that the differences between LDL and HDL reflect the *proportions* of lipids and proteins within them—not the type of cholesterol. The margin lists factors that influence LDL and HDL, and Chapter 18 provides many more details.

 U.S. Government
www.healthfinder.gov/searchoptions/topicsaz.htm
Search for Cholesterol

To help you remember, think of elevated HDL as **H**ealthy and elevated LDL as **L**ess healthy.

Factors that improve the LDL-to-HDL ratio:
- Weight control.
- Monounsaturated or polyunsaturated, instead of saturated, fatty acids in the diet.
- Soluble fibers (see Chapter 4).
- Antioxidants (see Highlight 11).
- Moderate alcohol consumption.
- Physical activity.

IN SUMMARY

The liver packages lipids and proteins into lipoproteins for transport around the body. All four types of lipoproteins carry all classes of lipids (triglycerides, phospholipids, and cholesterol), but the chylomicrons are the largest and the highest in triglycerides; VLDL are smaller and are about half triglycerides; LDL are smaller still and are high in cholesterol; and HDL are the smallest and are rich in protein.

*Before becoming LDL, the VLDL are first transformed into intermediate-density lipoproteins (IDL), sometimes called VLDL remnants. Some IDL may be picked up by the liver and rapidly broken down; those IDL that remain in circulation pick up cholesterol and become LDL. Researchers debate whether IDL are simply transitional particles or a separate class of lipoproteins.

**Figure 5-18**

**Sizes and Compositions of the Lipoproteins**

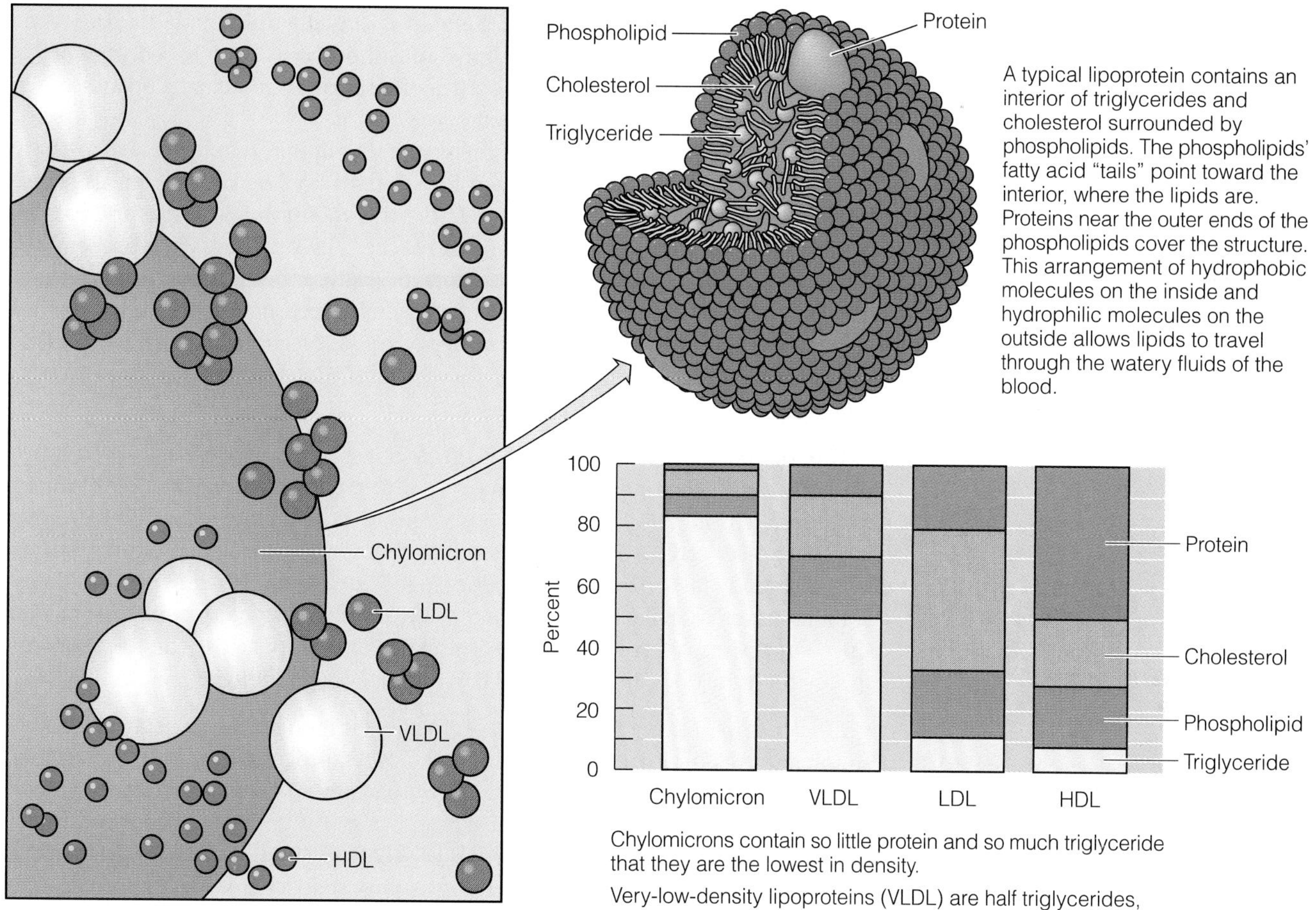


A typical lipoprotein contains an interior of triglycerides and cholesterol surrounded by phospholipids. The phospholipids' fatty acid "tails" point toward the interior, where the lipids are. Proteins near the outer ends of the phospholipids cover the structure. This arrangement of hydrophobic molecules on the inside and hydrophilic molecules on the outside allows lipids to travel through the watery fluids of the blood.

This solar system of lipoproteins shows their relative sizes. Notice how large the fat-filled chylomicron is compared with the others and how the others get progressively smaller as their proportion of fat declines and protein increases.

Chylomicrons contain so little protein and so much triglyceride that they are the lowest in density.

Very-low-density lipoproteins (VLDL) are half triglycerides, accounting for their low density.

Low-density lipoproteins (LDL) are half cholesterol, accounting for their implication in heart disease.

High-density lipoproteins (HDL) are half protein, accounting for their high density.

# Lipids in the Body

The blood carries lipids to various sites around the body. Once they arrive at their destinations, the lipids can get to work providing energy, insulating against temperature extremes, protecting against shock, and building cell structures. This section provides an overview first of the triglycerides in the blood and then of the metabolic pathways triglycerides can follow within the body's cells.

## *Triglycerides in the Blood*

The body's cells use both glucose and fat to fuel all their activities. Consequently, the blood must continuously deliver a constant supply of fat—either from the intestines' food supply or the body's fat stores. Either way, triglycerides and fatty acids are always circulating in the blood.

## *A Preview of Lipid Metabolism*

The blood delivers triglycerides to the cells for their use. This is a preview of how the cells store and release energy from fat; Chapter 7 provides details.

**Figure 5-19**
**An Adipose Cell**

An adipose, or fat, cell seems to expand almost indefinitely. The more fat it stores, the larger it grows.

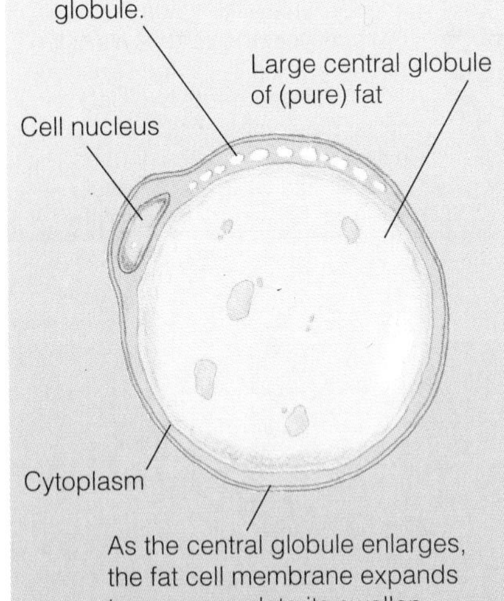
Newly imported triglycerides first form small droplets at the periphery of the cell, then merge with the large, central globule.
Large central globule of (pure) fat
Cell nucleus
Cytoplasm
As the central globule enlarges, the fat cell membrane expands to accommodate its swollen contents.

**adipose** (ADD-ih-poce) **tissue:** the body's fat tissue consists of masses of fat-storing cells.

**lipoprotein lipase (LPL):** an enzyme mounted on the surface of fat cells (and other cells) that hydrolyzes triglycerides passing by in the bloodstream and directs their parts into the cells, where they can be metabolized or reassembled for storage.

**hormone-sensitive lipase:** an enzyme inside adipose cells that responds to the body's need for fuel by hydrolyzing triglycerides so that their parts (glycerol and fatty acids) escape into the general circulation and thus become available to other cells as fuel. The signals to which this enzyme responds include epinephrine and glucagon, which oppose insulin (see Chapter 4).

1 lb body fat = 3500 kcal.

*Fat supplies most of the energy during a long-distance run.*

• ***Storing Fat as Fat*** • The triglycerides, familiar as the fat in foods and as body fat, serve the body primarily as a source of fuel. Fat provides more than twice the energy of carbohydrate and protein, making it an extremely efficient storage form of energy. The body's storage space for energy is virtually unlimited, thanks to the special cells of the adipose tissue. Unlike most body cells, which can store only limited amounts of fat, the fat cells of the adipose tissue readily take up and store fat. An adipose cell is depicted in Figure 5-19.

Adipose cells have an enzyme on their surfaces—lipoprotein lipase (LPL)—that captures circulating triglycerides from lipoproteins passing by after meals. This enzyme hydrolyzes the triglycerides to fatty acids and monoglycerides and passes these products into the cells' interiors. Inside the cells, other enzymes reassemble the pieces into triglycerides again for storage. Triglycerides pack tightly together within adipose cells, storing a lot of energy in a relatively small space. Adipose cells always store fat after meals, when a heavy traffic of chylomicrons and VLDL loaded with triglycerides passes by; they release it later when the blood needs replenishing.

• ***Making Fat from Carbohydrate or Protein*** • Earlier, Figure 5-3 showed how the body can make triglycerides from glycerol and fatty acids. Fatty acids, in turn, can be made from two-carbon fragments derived from any nutrient. (This is why most fatty acid carbon chains come in even numbers.) Thus glucose can be converted to body fat: enzymes break glucose into two-carbon fragments and then combine them to make long-chain fatty acids. Enzymes can also convert some of the components of protein (certain amino acids) to fatty acids. The food source from which the body most easily makes fat for storage, though, is fat itself.

• ***Efficiency of Making Fat from Fat*** • To convert food fats to body fat, the body simply absorbs the parts and puts them (or others) together again in storage. It requires very little energy to do this. By comparison, to convert dietary carbohydrate to body fat, the body must break starches first into disaccharides and then into monosaccharides, absorb the monosaccharides, then dismantle them, and reassemble many of their fragments into fatty acid chains. Each conversion requires energy. Thus it costs less (energetically) to store dietary fat as body fat than to convert and store dietary carbohydrate as body fat. Whether fat is stored at all depends more on total energy intake and on how much fat is eaten than on how much carbohydrate is eaten. The message is clear: to limit fat storage in the body, limit fat intake from foods. Chapter 8 discusses energy balance in more detail.

• ***Using Fat for Energy*** • Unlike the liver's glycogen stores, the body's fat stores have virtually unlimited capacity, and fat supplies 60 percent of the body's ongoing energy needs during rest.[9] During prolonged light to moderately intense exercise or extended periods of food deprivation, fat stores may make a slightly greater contribution to energy needs.

When cells demand energy, an enzyme (hormone-sensitive lipase) inside the adipose cells responds by dismantling stored triglycerides and releasing the glycerol and fatty acids directly into the blood. Energy-hungry cells anywhere in the body can then break down these components into small fragments and take them through a series of chemical reactions to yield energy, carbon dioxide, and water.

A person who fasts (drinking only water) will rapidly metabolize body fat. A pound of body fat provides 3500 kcalories, so you might think a fasting person who expends 2000 kcalories a day could lose more than half a pound of body fat each day.* Actually, the person has to obtain some energy from lean tissue because the brain, nervous system, and red blood cells need glucose, which fat cannot supply. Also, as mentioned, fat needs carbohydrate or protein to break down completely. Even on a total fast, a person cannot lose more than half a pound of pure

*The reader who knows that 1 pound = 454 grams and that 1 gram of fat = 9 kcalories may wonder why a pound of body fat does not equal $9 \times 454$ kcalories. The reason is that body fat contains some cell water and other materials; it is not quite pure fat.

fat per day. Still, in conditions of enforced starvation—say, during a siege or a famine—a fatter person can survive longer than a thinner person thanks to this energy reserve.

Although fat provides energy during a fast, it can provide very little glucose to give energy to the brain and nerves. Only the small glycerol molecule can be converted to glucose; fatty acids cannot be. After prolonged glucose deprivation, brain and nerve cells develop the ability to derive about two-thirds of their energy from the ketone bodies that the body makes from fat fragments. Ketone bodies cannot sustain life by themselves, however. As Chapter 7 explains, fasting for too long will cause death, even if the person still has ample body fat.

The small contribution that fat can make to the body's glucose supply is detailed in Chapter 7.

IN SUMMARY

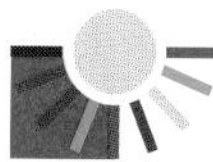

The body can easily store unlimited amounts of fat if excesses are available, and this body fat is used for energy when needed. The liver can also convert excess carbohydrate and protein into fat. Fat breakdown requires simultaneous carbohydrate breakdown for maximum efficiency; without carbohydrate, fats break down to ketone bodies, producing ketosis.

# Health Effects and Recommended Intakes of Lipids

Of all the nutrients, fat is most often linked with chronic diseases. A high-fat diet raises the risks of heart disease, some types of cancer, hypertension, diabetes, and obesity.[10] Fortunately, the same recommendation can help with all of these health problems: eat less fat.

## *Health Effects of Lipids*

Hearing a physician say, "Your blood lipid profile looks fine," is reassuring. The blood lipid profile reveals the concentrations of various lipids in the blood, notably triglycerides and cholesterol, and their lipoprotein carriers (VLDL, LDL, and HDL). This information alerts people to their disease risks and their need to change eating habits.

**blood lipid profile:** results of blood tests that reveal a person's total cholesterol, triglycerides, and various lipoproteins. Desirable blood lipid profile:

- Total cholesterol: <200 mg/dL (5.2 mmol/L).
- LDL cholesterol: <130 mg/dL (3.4 mmol/L).
- HDL cholesterol: >35 mg/dL (0.9 mmol/L).
- Triglycerides: <200 mg/dL (2.3 mmol/L).

• ***Heart Disease*** • Most people realize that elevated blood cholesterol is a major risk factor for cardiovascular disease.* Cholesterol accumulates in the arteries, restricting blood flow and raising blood pressure. The consequences are deadly; in fact, heart disease is the nation's number one killer of adults. Blood cholesterol is often used to predict the likelihood of a person's suffering a heart attack or stroke; the higher the cholesterol, the earlier and more likely the tragedy.

Commercials advertise products that are low in cholesterol, and magazine articles tell readers how to cut the cholesterol in their favorite recipes. What most people don't realize, though, is that *food* cholesterol does not raise *blood* cholesterol as dramatically as *saturated fat* does.

Other risk factors for heart disease include smoking, high blood pressure, diabetes, family history, sex, race, and obesity. Chapter 18 provides many more details about these risk factors, the development of heart disease, and dietary recommendations.

**cardiovascular disease (CVD):** a general term for all diseases of the heart and blood vessels. Atherosclerosis is the main cause of CVD. When the arteries that carry blood to the heart muscle become blocked, the heart suffers damage known as **coronary heart disease (CHD).**

- **cardio** = heart
- **vascular** = blood vessels

• ***Risks from Saturated Fats*** • Recall that LDL cholesterol raises the risk of heart disease. LDL concentrations respond to both the total amount and the type of fat in the diet.[11] Most often implicated in raising LDL cholesterol are the saturated fats, although not all saturated fats have the same cholesterol-raising effect.[12] Most notable among the saturated fatty acids that raise blood cholesterol are lauric,

*The concentration of cholesterol is similar in *blood, plasma,* and *serum;* this book uses the term *blood* cholesterol. Plasma is blood with the cells removed; serum is plasma with the clotting factors also removed.

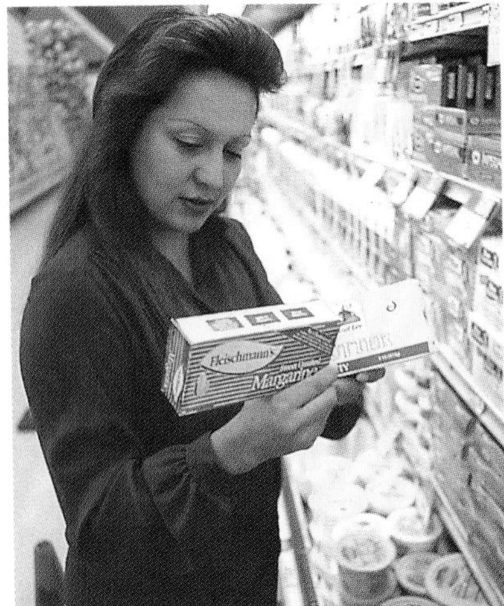

*Whether you decide to use butter or margarine, remember to use them sparingly.*

myristic, and palmitic acids (12, 14, and 16 carbons, respectively).[13] In contrast, stearic acid (18 carbons) does not seem to raise blood cholesterol.[14] Common sources of stearic acid are beef (tallow) and milk chocolate (cocoa butter). Whether a saturated fatty acid raises blood cholesterol may depend, in part, on its position on the glycerol molecule.[15]

• ***Risks from* trans*-Fatty Acids*** • In the body, *trans*-fatty acids—even the monounsaturated ones—alter blood cholesterol the same way some saturated fats do: they raise LDL and lower HDL cholesterol, although not to the same extent.[16] Epidemiological studies suggest a weak association between dietary *trans*-fatty acids and heart disease risk.[17] Evidence linking *trans*-fatty acids and cancer is generally lacking.[18]

Reports on *trans*-fatty acids have raised consumer doubts about whether margarine is, after all, a better choice than butter for heart health. The American Heart Association has stated that because butter is rich in both saturated fat and cholesterol while margarine is made from vegetable fat with no dietary cholesterol, margarine is still preferable to butter.[19] Others disagree, claiming the occasional use of butter is preferable to the use of margarine and other products containing *trans*-fatty acids.[20] In addition to strict limits on *trans*-fatty acid use, some experts are calling for food labels to state the *trans*-fatty acid amounts in foods.[21]

The health implications of *trans*-fatty acids remain an area of active research that is not yet ready for the evening news. All things considered, health risks from saturated fatty acids appear to outweigh those from *trans*-fatty acids.[22] Replacing both saturated and *trans* fats with monounsaturated and polyunsaturated fats may be more effective in preventing heart disease than reducing total fat intake.[23]

• ***Effects of Polyunsaturated and Monounsaturated Fats*** • In general, polyunsaturated fatty acids lower LDL cholesterol, and monounsaturated fatty acids have little or no independent effect.[24] Dietary cholesterol's influence on blood cholesterol is relatively minor.

Also of interest are the effects these fats have on the "good" HDL cholesterol. Some research suggests that polyunsaturated fats tend to lower both HDL and LDL, whereas monounsaturated fats raise HDL, thus improving the blood lipid profile.[25] Other research indicates that both polyunsaturated and monounsaturated fatty acids lower both LDL and HDL.[26] Whatever the particulars, most studies conclude that people should replace the saturated fats in their diets with unsaturated fats—particularly omega-3 polyunsaturated fats.[27]

Chapter 18 presents a few more details on the action of omega-3 fatty acids in preventing heart disease.

• ***Benefits from Omega-3 Fatty Acids*** • Research on the omega-3 polyunsaturated fatty acids has spotlighted the unique effects of different types of fat on blood cholesterol and heart disease. Inuit peoples of Alaska and Greenland enjoy relative freedom from heart disease despite high-energy, high-fat, and high-cholesterol diets. Why? Their foods derive primarily from fish and other marine animals and are rich in omega-3 fatty acids, particularly EPA and DHA. Research reveals that people who eat some fish each week can lower their blood cholesterol and their risk of heart attack or stroke.[28] A diet low in both fat and saturated fat, combined with regular fish consumption, produces an optimal lipid profile. In addition to improving blood lipids, fish oils prevent blood clots and may also lower blood pressure, especially in people with hypertension or atherosclerosis.[29]

**hypertension:** high blood pressure; defined further in Chapter 18.

Data from Japan seem to confirm that a diet low in fat and high in fish benefits health. The Japanese diet today has become westernized with few Japanese people eating the large quantities of rice and fish their ancestors ate. These dietary changes have been accompanied by health consequences: higher rates of cardiovascular disease and cancer.

• ***Balance Omega-3 and Omega-6 Intakes*** • The Committee on Dietary Reference Intakes has not yet established a recommended intake for omega-3 and omega-6 fatty acids. The 1990 Canadian RNI include specific amounts for both omega-3 and omega-6 fatty acids.* Many researchers believe the body's requirements depend on an optimal ratio; that is, that more omega-3 fatty acids are not necessarily better, but that an appropriate balance between the two omega families may be crucial. A ratio of between 1 to 5 and 1 to 10 (omega-3 to omega-6 fatty acids) has been suggested as appropriate.[30]

To obtain the right balance between omega-3 and omega-6 fatty acids, most people need to eat more fish and less meat. Eating fish instead of meat two or three meals a week supports heart health, especially when combined with physical activity. Even one fish meal a week may be enough to make a difference.[31] The fish may not even need to be rich in omega-3 fatty acids; one study found that farm-raised catfish (which is relatively low in omega-3 fatty acids) improves lipid profiles similarly to wild Alaskan salmon.[32] Fish provides many minerals (except iron) and vitamins and is leaner than most other animal-protein sources.

Fish oil should come from fish, not from supplements. Fish oil supplements are not recommended for a number of reasons.† Perhaps most importantly, the scientific evidence on their safety and effectiveness is not conclusive. Also, high intakes of fish oil increase bleeding time, interfere with wound healing, worsen diabetes, and impair immune function. Fish oil supplements are made from fish skins and livers, which may contain environmental contaminants. Fish oils also naturally contain large amounts of the two most potentially toxic vitamins, A and D. Lastly, supplements are expensive; money is better spent on foods that can provide a full array of nutrients. In addition to fish, other functional foods are being developed to help consumers improve their omega-3 fatty acid intake. For example, hens fed flaxseed produce eggs rich in omega-3 fatty acids.[33]

*Enjoy low-fat foods for good heart health.*

• ***Cancer*** • The evidence linking dietary fats with cancer is less conclusive than for heart disease, but it does suggest an association between total fat and some types of cancers. Dietary fat seems not to *initiate* cancer development but to *promote* cancer once it has arisen.[34] Some studies report a relationship between specific cancers and saturated fats or dietary fat from animal sources (which is mostly saturated).[35] Thus health advice to reduce cancer risks parallels that given to reduce heart disease risks: reduce total fat, especially saturated fat, intake.

Other risk factors for cancer include smoking, alcohol, and environmental contaminants. Chapter 18 provides many more details about these risk factors and the development of cancer.

The relationship between dietary fat and the risk of cancer differs for various types of cancers. In the case of breast cancer, some studies indicate little or no association between dietary fat and cancer.[36] Others find that total *energy* intake is a better predictor than percentage of kcalories from fat.[37] In the case of prostate cancer, there does appear to be a strong association with fat.[38] This association appears to be due primarily to the saturated fat from meats; fat from milk or fish has not been implicated in cancer risk.

• ***Obesity*** • As the photos in Figure 5-20 show, fat accounts for a lot of the energy in foods, and removing the fat from foods cuts energy intake dramatically. Fat contributes twice as many kcalories per gram as either carbohydrate or protein. Consequently, people who eat high-fat diets regularly may exceed their energy needs and gain weight.

Reminder: Fat is a more concentrated energy source than the other energy nutrients: 1 g carbohydrate or protein = 4 kcal, but 1 g fat = 9 kcal.

Furthermore, people who eat high-fat diets tend to store body fat efficiently and have more body fat than their energy intakes would predict. Chapter 9 revisits the issue of the fattening power of fat and concludes that low-fat, high-carbohydrate foods are most appropriate for satisfying hunger and controlling appetite.[39]

*For omega-3 fatty acids, the RNI is 0.5 percent of total energy or 0.55 grams per 1000 kcalories; for omega-6 fatty acids, the RNI is 3 percent of total energy or 3.3 grams per 1000 kcalories. These recommendations represent a 1 to 6 ratio (omega-3 to omega-6).

†In Canada, fish oil supplements require a physician's prescription.

**Figure 5-20**
**Cutting Fat Cuts kCalories**

Pork chop with a half-inch of fat (275 kcal and 19 g fat).

Potato with 1 tbs butter and 1 tbs sour cream (350 kcal and 14 g fat).

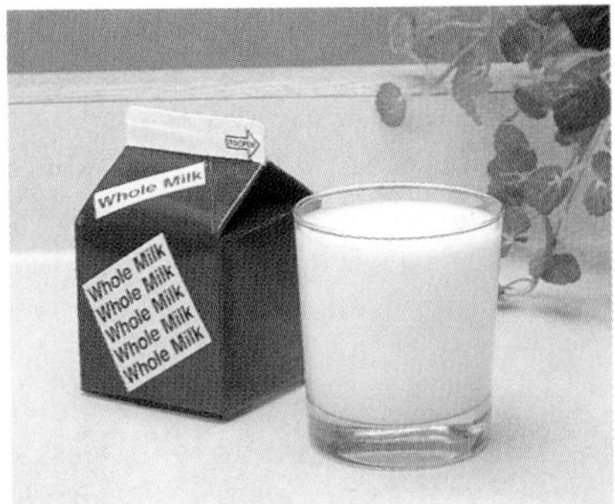


Whole milk, 1 c (150 kcal and 8 g fat).

Pork chop with fat trimmed off (165 kcal and 8 g fat).

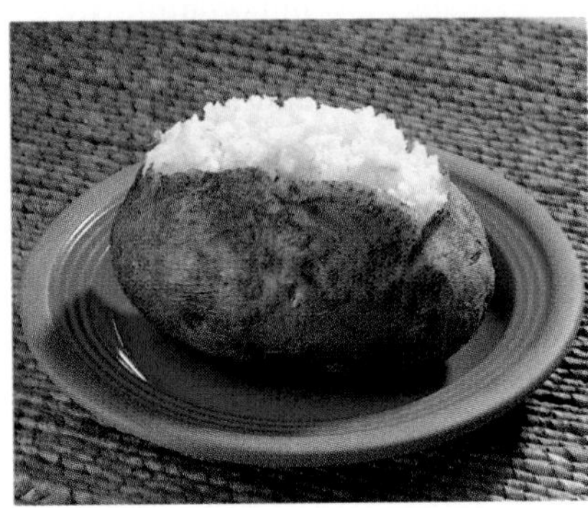

Plain potato (220 kcal and <1 g fat).

Nonfat milk, 1 c (90 kcal and 1 g fat).

The symptoms of fatty acid deficiencies are presented on p. 132.

• ***Don't Overdo Fat Restriction*** • Although it is very difficult to do, some people actually manage to eat too little fat—to their detriment. Among them are people with eating disorders, described in Highlight 9. Most adults should consume at least 15 percent of their energy intake from fat.[40] As a practical guideline, it is wise to include the equivalent of at least a teaspoon of fat in every meal—a little peanut butter on toast or mayonnaise on tuna, for example. Parents should not restrict the fat intakes of their infants and young children; dietary recommendations that limit fat were developed for healthy people over age two.

IN SUMMARY

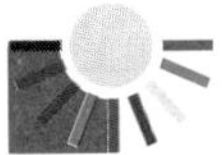

Health authorities single out high fat intakes as a major flaw in the North American diet: excess fat contributes to heart disease, obesity, and other health problems. High blood LDL cholesterol, specifically, poses a risk of heart disease, and high intakes of saturated fat contribute most to high LDL. Cholesterol itself, in foods, presents much less of a risk; *trans*-fatty acids' effects are not yet clear from research. Omega-3 fatty acids appear to be protective. High-fat diets also accelerate (but do not initiate) cancer development.

## *Recommended Intakes of Fat*

The Committee on Dietary Reference Intakes has not established a recommended intake for fat, but the Committee on Diet and Health makes the following recommendations:

- Reduce total fat intake to 30 percent or less of energy intake.
- Reduce saturated fat intake to less than 10 percent of energy intake.
- Reduce cholesterol intake to less than 300 milligrams daily.

A person consuming 2000 kcalories a day should therefore have 600 kcalories or less from fat (roughly 65 grams). Of those fat kcalories, only 200 or less should come from saturated fats (roughly 20 grams).

Daily Values for a 2000 kcal diet:
Fat: 65 g.
Saturated fat: 20 g.

**Healthy People 2000:** Reduce dietary fat intake to an average of 30% of energy or less and average saturated fat intake to less than 10% of energy among people aged two years and older.

To meet dietary fat recommendations, many people have reduced their fat intakes. Fat intake peaked at more than 40 percent of daily kcalories in the 1950s and has fallen steadily ever since.[41] According to recent surveys, adults in the United States receive about 34 percent of their total energy from fat, with saturated fat contributing about 12 percent of the total.[42] Cholesterol intakes in the United States average 250 milligrams a day for women and 350 for men.

• ***Reduce Total Fat Intake*** • Triglycerides are abundant in all fats and oils. They also accompany protein in foods derived from animals, such as meat, fish, poultry, and eggs, and carbohydrate in foods derived from plants, such as avocados and coconuts.

 U.S. Government
www.healthfinder.gov/searchoptions/topicsaz.htm
Search for Dietary fat

To reduce dietary fat, eliminate fat as a seasoning and in cooking; remove the fat from high-fat foods; replace high-fat foods with low-fat alternatives; and emphasize grains, fruits, and vegetables. The "How to" feature on p. 148 provides additional tips for reducing fat in the diet, food group by food group.

• ***Reduce Saturated Fat Intake*** • Fats from animal sources are the main sources of saturated fats in most people's diets. Some vegetable fats (coconut and palm) and hydrogenated fats provide smaller amounts of saturated fats. Selecting lean meats and nonfat milk products helps to lower saturated fat intake. Using monounsaturated margarine and cooking oil is another simple change that can make a big difference.[43]

• ***Reduce Cholesterol Intake*** • Recall that cholesterol is found only in animal products. Consequently, eating less fat from meat, eggs, and milk products will also help lower dietary cholesterol intake (as well as total and saturated fat intakes).

 ADA ABC's of Fats, Oils, and Cholesterol
www.eatright.org/nfs2.html

• ***Select Lean Meats and Nonfat Milks*** • Many foods that contain fat, saturated fat, and cholesterol—such as meats, milk, cheese, and eggs—also provide high-quality protein and valuable vitamins and minerals. They can be included in a healthy diet if a person selects lean and nonfat products and prepares them using the suggestions outlined on p. 148. Figure 2-2 on pp. 40–41 shows examples of very lean, lean, medium-fat, and high-fat meats and of nonfat, low-fat, reduced-fat, and whole-milk products.

• ***Eat Plenty of Vegetables, Fruits, and Grains*** • Choosing vegetables, fruits, cereals, and legumes also helps lower fat intake. Vegetables and fruits contain no fat, and most grains contain only trace amounts. Some grain *products* such as fried taco shells, croissants, and granola cereal are high in fat, though, so consumers need to read food labels. Similarly, many people prepare grains and vegetables with added fat. Because fruits are often eaten without added fat, a diet that includes several servings of fruit can more readily meet the recommendation for 30 percent or less of kcalories from fat.[44]

***Even well-balanced, healthy meals provide some fat. In this chicken stir-fry, only two teaspoons of oil were used in preparation, but 30 percent of the kcalories come from fat. The chicken and sesame seeds also contribute some fat.***

Because a low-fat diet is usually rich in vegetables, fruits, cereals, and legumes, it offers abundant vitamin C, folate, vitamin A, and dietary fiber—all important in supporting health. Consequently, such a diet protects against disease in two ways: reducing fat and increasing nutrients.[45]

## HOW TO Lower Fat Intake by Food Group

### Meat, Fish, and Poultry

- Fat adds up quickly, even with lean meat; limit intake to about 6 ounces (cooked weight) daily.
- Choose fish, poultry, or lean cuts of pork or beef; look for unmarbled cuts named *round* or *loin* (eye of round, top round, bottom round, round tip, tenderloin, sirloin, center loin, and top loin).
- Trim the fat from pork and beef; remove the skin from poultry.
- Grill, roast, broil, bake, stir-fry, stew, or braise meats; don't fry. When possible, place meat on a rack so that fat can drain.
- Use lean ground turkey or lean ground beef in recipes; brown ground meats without added fat, then drain off fat.
- Refrigerate meat pan drippings and broth; when it solidifies, remove the fat and use the defatted broth in recipes.
- Select tuna, sardines, and other canned meats packed in water; rinse oil-packed items with hot water to remove much of the fat.
- Fill kabob skewers with lots of vegetables and slivers of meat; create main dishes and casseroles by combining a little meat, fish, or poultry with a lot of pasta, rice, or vegetables.
- Make meatless spaghetti sauces and casseroles; use legumes often.
- Eat a meatless meal or two daily.

### Milk and Cheeses

- Switch from whole milk to reduced-fat, from reduced-fat to low-fat, and from low-fat to fat-free (nonfat).
- Use nonfat and low-fat cheeses (such as part-skim ricotta and low-fat mozzarella) instead of regular cheeses.
- Use nonfat or low-fat yogurt or sour cream instead of regular sour cream.
- Use evaporated nonfat milk instead of cream.
- Enjoy nonfat frozen yogurt, sherbet, or ice milk instead of ice cream.

### Fruits and Vegetables

- Enjoy the natural flavor of steamed vegetables for dinner and fruits for dessert.
- Use butter-flavored granules on vegetables instead of butter or margarine.
- Use nonfat yogurt or nonfat salad dressing instead of sour cream, cheese, mayonnaise, or other sauces on vegetables and in casseroles.
- Select nonfat or low-fat salad dressings, or use herbs, lemon juice, and spices instead of regular salad dressing.
- Add a little water to thick, bottled salad dressing to dilute the amount of fat each serving provides.
- Eat at least two vegetables (in addition to a salad) with dinner.
- Snack on raw vegetables or fruits instead of high-fat items like potato chips.
- Buy frozen vegetables without sauce.

### Breads and Cereals

- Use fruit butters or jellies on bread instead of butter or margarine.
- Select breads, cereals, and crackers that are low in fat (for example, bagels instead of croissants).
- Prepare pasta with a tomato sauce instead of a cheese or cream sauce.

### Other Foods and Cooking Tips

- Use a nonstick pan or coat the pan lightly with vegetable oil.
- Use egg substitutes in recipes instead of whole eggs or use 2 egg whites in place of each whole egg.
- Use half the margarine, butter, or oil called for in a recipe. (The minimum amount of fat for muffins, quick breads, and biscuits is 1 to 2 tablespoons per cup of flour; for cakes and cookies, 2 tablespoons per cup.)
- Use less butter or margarine. Select the whipped types of butter, margarine, or cream cheese for use at the table; they contain half the kcalories of the regular types.
- Use butter replacers instead of butter.
- For sandwiches and salads, use spicy mustard, nonfat salad dressing, lemon juice, flavored vinegar, salsa or the nonfat versions instead of regular mayonnaise, salad dressing, or sour cream.
- Use wine; lemon, orange, or tomato juice; herbs; spices; fruits; or broth instead of butter, margarine, or oil when cooking.
- Stir-fry in a small amount of oil; add moisture and flavor with broth, tomato juice, or wine.
- Use variety to enhance enjoyment of the meal: vary colors, textures, and temperatures—hot cooked versus cool raw foods—and use garnishes to complement food.

*Salad dressing can add more than 20 grams of fat to an otherwise low-fat salad.*

• ***Use Fats and Oils Sparingly*** • Practice moderation when using oils and fats such as butter, margarine, mayonnaise, and salad dressings. These foods offer much fat and little nourishment.

• ***Look for Invisible Fat*** • *Visible* fat, such as butter, the oil in salad dressing, and the fat trimmed from meat, is easy to see. *Invisible* fat is less apparent and can be present in foods in surprising amounts. Invisible fat "marbles" a steak or is hidden in foods like nuts, cheese, avocados, and olives. Any *fried* food contains abundant fat: potato chips, french fries, fried wontons, and fried fish. Many *baked* goods, too, are high in fat: pie crusts, pastries, crackers, biscuits, cornbread, doughnuts, sweet rolls, cookies, and cakes. Most chocolate bars contain more fat energy than sugar energy. Even cream-of-mushroom soup prepared with water derives 66 percent of its energy from fat. Abundant fat lurks on salad bars, too, not only in the dressings, but also in the potato salad, the macaroni salad, the coleslaw, and the marinated beans that are mixed with oil-based dressings. Keep invisible fats in mind when making food selections.

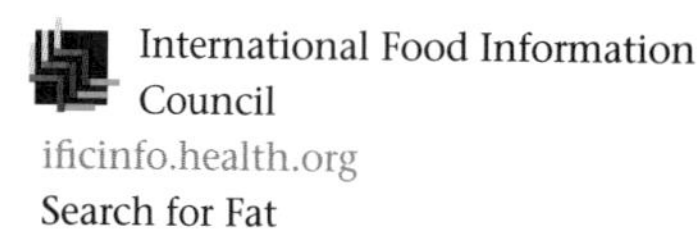
International Food Information Council
ificinfo.health.org
Search for Fat

• ***Choose Wisely*** • The *Dietary Guidelines* urge people to choose a diet low in fat, saturated fat, and cholesterol. A diet following the Daily Food Guide plan can support this goal if selections are made carefully. Only by following such a plan and making low-fat choices consistently can the goals of the *Diet and Health* recommendations be met. The "How to" feature on pp. 150–151 describes an easy way to estimate the fat content of a meal.

Consumers are finding more low-fat food choices available than ever before. In many cases, they are familiar foods presented with less fat. Animals fed special diets produce lower-fat meats, milks, and eggs, which can improve a person's blood cholesterol with little effort.[46] Cuts of meat are often trimmed of fat more closely than in the past. In the dairy case, nonfat and reduced-fat milk, yogurt, cheeses, and sour cream offer healthy alternatives to their higher-fat counterparts. Many processed foods such as salad dressings, crackers, chips, and cookies are now available with little or no fat. Such choices make low-fat eating easy. A simple switch to nonfat salad dressing can lower a person's average fat intake. Every little fat-saving step helps a person get closer to the 30 percent goal.

Many fat-free foods have been developed using familiar nutrients. Milk and egg proteins or carbohydrate derivatives, for example, are heated, acidified, or blended to simulate the properties of fat. Highlight 5 examines some of these new alternatives to fat.

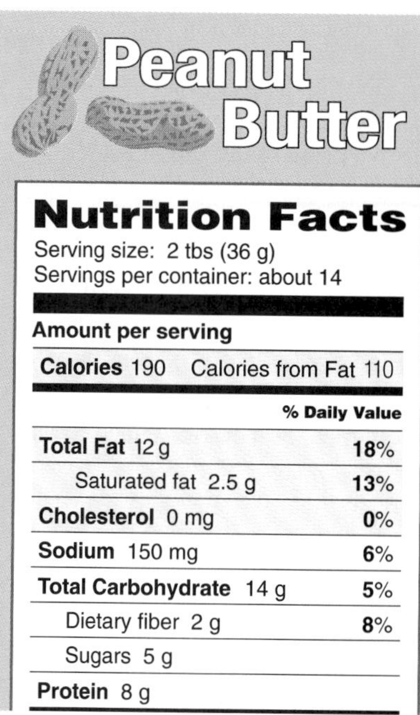


**Nutrition Facts**
Serving size: 2 tbs (36 g)
Servings per container: about 14

| Amount per serving | |
|---|---|
| **Calories** 190 | Calories from Fat 110 |
| | **% Daily Value** |
| **Total Fat** 12 g | **18%** |
| Saturated fat 2.5 g | **13%** |
| **Cholesterol** 0 mg | **0%** |
| **Sodium** 150 mg | **6%** |
| **Total Carbohydrate** 14 g | **5%** |
| Dietary fiber 2 g | **8%** |
| Sugars 5 g | |
| **Protein** 8 g | |

Food labels list the kcalories from fat and the quantities and Daily Values for fat, saturated fat, and cholesterol.

**Healthy People 2000:** Increase to at least 5000 brand items the number of processed food products that are reduced in fat and saturated fat.

• ***Read Food Labels*** • Labels list total fat, saturated fat, and cholesterol contents of foods in addition to fat kcalories per serving. (Labels do not provide information on *trans*-fatty acids; if the ingredients list includes hydrogenated oils, though, you know the food contains *trans*-fatty acids—you just don't know how much.) Because each package provides information for a single serving and serving sizes are standardized, consumers can easily compare similar products. Total fat, saturated fat, and cholesterol are also expressed as "% Daily Values" for a person consuming 2000 kcalories. People who are consuming more or less than 2000 kcalories daily can calculate their personal Daily Value for fat as described on p. 152.

Be aware that the "% Daily Value" for fat is not the same as "% kcalories from fat." Consider, for example, a piece of lemon meringue pie that provides 140 kcalories and 12 grams of fat. Because the Daily Value for fat is 65 grams for a 2000-kcalorie intake, 12 grams represent about 18 percent, or almost one-fifth, of the day's fat allowance (the pie's "% Daily Value" is 18). Uninformed consumers may mistakenly believe that this food meets the guideline to limit fat to "30 percent kcalories," but it doesn't—for two reasons. First, the pie's 12 grams of fat contribute 108 of the 140 kcalories, for a total of 77 percent kcalories from fat. Second, the "30 percent kcalories from fat" guideline applies to a day's total intake, not to an individual food. (Of course, if every selection throughout the day

*% Daily Value for fat:*
• *12 g ÷ 65 g = 0.18 × 100 = 18%.*
*% kCalories from fat:*
• *12 g × 9 kcal/g = 108 kcal.*
*108 kcal ÷ 140 kcal = 77%.*

# HOW TO Use the Exchange System to Estimate Fat

The exchange system is especially informative about the fats in foods. To use the exchange system, you need to know the fat value for each list (see the accompanying table) and the foods on that list with their portion sizes (review Figure 2-2 on pp. 40–41). Two of the lists—vegetables and fruits—contain no fat. A third list—starch—provides only a little. (Starch foods prepared with fat are counted as one starch exchange *plus* a fat exchange.) So you only need to learn the fat values for three lists: milks, meats, and fats.

The milk list offers three fat values for nonfat, reduced-fat, and whole milk. Think of nonfat milk as milk and of reduced-fat and whole milk as milk with added fat.

The meat list offers four fat values for very lean, lean, medium-fat, and high-fat products. People are often surprised to learn how much fat comes from meats and cheeses. An ounce of lean meat or low-fat cheese supplies about half of its energy from fat (28 protein kcalories and 27 fat kcalories). An ounce of high-fat meat or most cheeses supplies 72 percent of its energy from fat (28 protein kcalories and 72 fat kcalories). As for the meat alternate, peanut butter, 2 tablespoons supply 76 percent of their energy from fat (32 protein kcalories and 144 fat kcalories). Note that one meat exchange is a single ounce; to use the exchange system, learn to recognize the number of ounces in a serving.

| One Exchange | Fat (g) |
|---|---|
| Milks | |
| Nonfat (and low-fat) | 0–3 |
| Reduced-fat | 5 |
| Whole | 8 |
| Meats | |
| Very lean | 0–1 |
| Lean | 3 |
| Medium-fat | 5 |
| High-fat | 8 |
| Fats | 5 |
| Starch | 1 or less |
| Vegetables | — |
| Fruits | — |
| Other carbohydrates | varies |

Fat-containing foods appear primarily in three of the exchanges lists: milks, meats, and fats.

exceeds 30 percent kcalories from fat, you can be certain that the day's total intake will, too.)

Because recommendations apply to average daily intakes and not to individual food items, food labels do not provide the percentage of kcalories from fat. Still, you can get an idea of whether a particular food is high or low in fat. To quickly compare recommendations with the fat content of a food, use the rule of thumb that 3 grams of fat (27 kcalories of fat) represent about 30 percent of the kcalories in 100 kcalories of food. Alternatively, you can multiply the grams of fat in a serving by 30 and then compare that number to the kcalories.[47] If it is less, then the food has less than 30 percent kcalories from fat.

Quick and easy estimates:

- A food is low in fat if it has: ≤3 g fat in 100 kcal food.
- A food is low in fat if: g fat × 30 < kcal.

The fat list includes butter, margarine, and oil, of course, but it also includes bacon, olives, avocados, and many kinds of nuts. These foods are grouped together because a portion of any of them contains as much fat as a pat of butter and, like butter, offers negligible protein and carbohydrate. In Appendix G, the fat list is sorted into saturated, monounsaturated, and polyunsaturated groups, which helps people make heart-wise selections when choosing fat.

To estimate the fat in this meal, you first need to recognize that this spaghetti dinner is really 1 cup of pasta, with 1 cup of tomato sauce and 3 ounces of lean ground beef. Then you need to translate these portions into exchanges: 2 starches, 1 vegetable, and 3 lean meats, respectively. Ignore the vegetables in the salad, but count the ½ cup of garbanzo beans as 1 starch + 1 very lean meat, and the sunflower seeds and ranch dressing as 1 fat each.

| Dinner | Exchange | Fat (g) Estimate | Actual |
|---|---|---|---|
| Salad: | | | |
| 1 c raw spinach leaves, shredded carrots, and sliced mushrooms | = free | 0 | 0 |
| ½ c garbanzo beans | = 1 starch and 1 very lean meat | 2 | 2 |
| 1 tbs sunflower seeds | = 1 fat | 5 | 4 |
| 1 tbs ranch salad dressing | = 1 fat | 5 | 6 |
| Entrée: | | | |
| Spaghetti with meat sauce | | | |
| 1 c pasta (cooked) | = 2 starches | 2 | |
| 1 c tomato sauce | = 1 vegetable | — | 12 |
| 3 oz ground round | = 3 lean meats | 9 | |
| ½ c green beans | = 1 vegetable | — | 0 |
| 1 medium corn on the cob | = 1 starch | 1 | 0 |
| 2 tsp butter | = 2 fats | 10 | 8 |
| Dessert: | | | |
| 1⁄12 angel food cake | = 1 other carbohydrate | — | 0 |
| Beverage: | | | |
| 1 c 1% low-fat milk | = 1 nonfat milk | 3 | 3 |
| | | 37 | 35 |

Using the exchange system to estimate, this dinner provides about 37 grams of fat. A computer diet analysis program came to a similar conclusion (35 grams), as would a diet analysis using the values in Appendix H. To keep from underestimating fat intake, count "1 or less" and "0–1" as "1 gram of fat" and count "0–3" as 3 grams of fat.

The FDA authorizes two health claims on labels concerning fat: one for "dietary saturated fat and cholesterol and risk of coronary heart disease" and one for "dietary fat and cancer." To make these claims, foods must meet specified criteria, as described in Chapter 2.

IN SUMMARY

Health authorities recommend limiting total fat to 30 percent or less of energy intake; saturated fat to one-third of total fat, or 10 percent of energy intake; and cholesterol to less than 300 milligrams a day. They also recommend consuming relatively more polyunsaturates, particularly

## HOW TO Calculate a Personal Daily Value for Fat

The % Daily Value for fat on food labels is based on 65 grams. To know that you've met recommendations, you can either count grams until you reach 65, or add the "% Daily Values" until you reach 100 percent—if your energy intake is 2000 kcalories a day. If your energy intake is more or less, you have a couple of options.

You can calculate your personal daily fat allowance in grams. Multiply your total energy intake by 30 percent, then divide by 9. Suppose your energy intake is 1800 kcalories per day and your goal is 30 percent kcalories from fat:

$$1800 \text{ total kcal} \times 0.30 \text{ from fat} = 540 \text{ fat kcal.}$$
$$540 \text{ fat kcal} \div 9 \text{ kcal/g} = 60 \text{ g fat.}$$

Another way to calculate your personal fat allowance is to cross out the last digit of your energy intake and divide by 3.[a] For example, 1800 kcalories becomes 180; then you divide by 3:

$$180 \div 3 = 60 \text{ g fat/day.}$$

(In familiar measures, 60 grams of fat is about the same as ⅔ stick of butter or ¼ cup of oil.)

The accompanying table shows the numbers of grams of fat allowed per day for various energy intakes. With one of these numbers in mind, you can quickly evaluate the number of fat grams in foods you are considering eating.

**Recommended Grams of Fat for Different Energy Intakes**

| Energy (kcal/day) | 30% kCalories from Fat | Fat (g/day) |
|---|---|---|
| 1200 | 360 | 40 |
| 1500 | 450 | 50 |
| 1800 | 540 | 60 |
| 1900 (RDA for women 51 years and over) | 570 | 63 |
| 2000 (Daily Value for food labels) | 600 | 65 |
| 2200 (RDA for women 19 to 50 years old) | 660 | 73 |
| 2300 (RDA for men 51 years and over) | 690 | 77 |
| 2600 | 780 | 87 |
| 2900 (RDA for men 19 to 50 years old) | 870 | 97 |
| 3000 | 900 | 100 |

[a]K. McNutt, Fat traps, tips, and tricks, *Nutrition Today,* May/June 1992, pp. 47–49.

omega-3 fatty acids, than in the past from foods such as fish, not from supplements. Many purchasing and cooking strategies can help bring these goals within reach, and food labels make it easier to select foods consistent with these guidelines.

If people were to make only one change in their diets, they would be wise to limit their intakes of total fat, which would control their energy intake as well. A second change might be to specifically limit saturated fat. Chances are good that if total fat and saturated fat meet recommendations, then cholesterol intake will, too. Many guidelines suggest these changes in that order of priority: low in fat, saturated fat, and cholesterol.

Lowering fat intake can be difficult, though, because fats make foods taste delicious. To maintain good health, must a person give up all high-fat foods forever—never again to eat marbled steak, hollandaise sauce, or gooey chocolate cake? Not at all. These foods bring pleasure to a meal and can be enjoyed as part of a healthy diet when eaten in small quantities on occasion, but it is true that they are not everyday foods. The key word for fat is not deprivation, but moderation: appreciate the energy and enjoyment that fat provides, but take care not to exceed your needs.

## Making It Click

These problems will give you practice in doing simple nutrition-related calculations (see Appendix K for answers). Show your calculations for each problem.

1. Be aware of the fats in milks. Following are four categories of milk.

| | Wt (g) | Fat (g) | Prot (g) | Carb (g) |
|---|---|---|---|---|
| Milk A (1 c) | 244 | 8 | 8 | 12 |
| Milk B (1 c) | 244 | 5 | 8 | 12 |
| Milk C (1 c) | 244 | 3 | 8 | 12 |
| Milk D (1 c) | 244 | 0 | 8 | 12 |

   a. Based on *weight,* what percentage of each milk is fat (round off to a whole number)?
   b. How much energy from fat will a person receive from drinking 1 cup of each milk?
   c. How much total energy will the person receive from 1 cup of each milk?
   d. What percentage of the energy in each milk comes from fat?
   e. In the grocery store, how is each milk labeled?

2. Judge foods' fat contents by their labels.
   a. A food label says that one serving of the food contains 6.5 grams fat. What would the % Daily Value for fat be? What does the Daily Value you just calculated mean?
   b. How many kcalories from fat does a serving contain? (Round off to the nearest whole number.)
   c. If a *serving* of the food contains 200 kcalories, what percentage of the energy is from fat?

This example should show you how easy it is to evaluate foods' fat contents by reading labels and to see the difference between the % Daily Value and the percentage of kcalories from fat.

3. Now consider a piece of carrot cake. Remember that the Daily Value suggests 65 grams of fat as acceptable within a 2000-kcalorie diet. A serving of carrot cake provides 30 grams fat. What percentage of the Daily Value is that? What does this mean?

## Study Questions

These questions will help you review the chapter. You will find the answers in the discussions on the pages provided.

1. Name three classes of lipids found in the body and in foods. What features do fats bring to foods? What are some of their functions in the body? (pp. 125, 130)
2. Describe the structure of a triglyceride. What are the differences between saturated, unsaturated, monounsaturated, and polyunsaturated fats? (pp. 126–127)
3. What features distinguish fatty acids from each other? (p. 128)
4. What does hydrogenation do to fats? What are *trans*-fatty acids, and how do they influence heart disease? (pp. 129–130, 144)
5. What does the term *omega* mean with respect to fatty acids? Describe the roles of the omega fatty acids in disease prevention. (pp. 131, 144–145)
6. Which of the fatty acids are essential? Name their chief dietary sources. (pp. 128, 130–131)
7. How do phospholipids differ from triglycerides in structure? How does cholesterol differ? How do these differences in structure affect function? (pp. 133–135)
8. Trace the steps in fat digestion, absorption, and transport. (pp. 136–140)
9. What do lipoproteins do? What are the differences among the chylomicrons, VLDL, LDL, and HDL? (pp. 139–140)
10. What roles does cholesterol play in the body? (p. 134)
11. Describe the routes cholesterol takes in the body. (pp. 137–140)
12. What roles do the triglycerides and phospholipids perform in the body? (pp. 130, 132–133)
13. How does excessive fat intake influence health? What factors influence LDL, HDL, and total blood cholesterol? (pp. 133, 141–143)
14. What are the dietary recommendations regarding fat and cholesterol intake? List ways to reduce intake. (pp. 146, 148)
15. Which food lists of the exchange system supply fat in abundance? In moderation? Not at all? (pp. 150–151)
16. What is the Daily Value for fat (for a 2000-kcalorie diet)? What does this number represent? (pp. 149–150)

These questions will help you prepare for an exam. Answers can be found in Appendix K.

1. A triglyceride consists of:
   a. three glycerols attached to a lipid.
   b. three fatty acids attached to a glucose.
   c. three fatty acids attached to a glycerol.
   d. three phospholipids attached to a cholesterol.
2. Saturated fatty acids:
   a. are always 18 carbons long.
   b. have at least one double bond.
   c. are fully loaded with hydrogens.
   d. are always liquid at room temperature.

3. The difference between *cis*- and *trans*-fatty acids is:
   a. the number of double bonds.
   b. the length of their carbon chains.
   c. the location of the first double bond.
   d. the configuration around the double bond.
4. Which of the following is *not* true? Fats:
   a. supply glucose.
   b. provide energy.
   c. protect against organ shock.
   d. carry vitamins A, D, E, and K.
5. The essential fatty acids include:
   a. stearic acid and oleic acid.
   b. oleic acid and linoleic acid.
   c. palmitic acid and linolenic acid.
   d. linoleic acid and linolenic acid.
6. Which of the following is *not* true? Lecithin is:
   a. an emulsifier.
   b. a phospholipid.
   c. an essential nutrient.
   d. a constituent of cell membranes.
7. Chylomicrons are produced in the:
   a. liver.
   b. pancreas.
   c. gallbladder.
   d. small intestine.
8. Transport vehicles for lipids are called:
   a. micelles.
   b. lipoproteins.
   c. blood vessels.
   d. monoglycerides.
9. The lipoprotein most associated with a high risk of heart disease is:
   a. CHD.
   b. HDL.
   c. LDL.
   d. LPL.
10. A person consuming 2200 kcalories a day who wants to meet the *Diet and Health* recommendations should limit daily fat intake to:
    a. 25 grams or less.
    b. 73 grams or less.
    c. 98 grams or less.
    d. 123 grams or less.

## Notes


1. A. Drewnowski, Taste preferences and food intake, *Annual Review of Nutrition* 17 (1997): 237–253.
2. J. E. Blundell and J. I. Macdiarmid, Fat as a risk factor for overconsumption: Satiation, satiety, and patterns of eating, *Journal of the American Dietetic Association* 97 (1997): S63–S69.
3. E. A. Emken, *Trans* fatty acids and coronary heart disease risk: Physicochemical properties, intake, and metabolism, *American Journal of Clinical Nutrition* 62 (1995): 659S–669S.
4. E. N. Siguel and R. H. Lerman, Prevalence of essential fatty acid deficiency in patients with chronic gastrointestinal disorders, *Metabolism* 45 (1996): 12–23.
5. L. M. Arnold and coauthors, Effect of isoenergetic intake of three or nine meals on plasma lipoproteins and glucose metabolism, *American Journal of Clinical Nutrition* 57 (1993): 446–451; P. J. H. Jones, C. A. Leitch, and R. A. Pederson, Meal-frequency effects on plasma hormone concentrations and cholesterol synthesis in humans, *American Journal of Clinical Nutrition* 57 (1993): 868–874.
6. P. J. H. Jones, Regulation of cholesterol biosynthesis by diet in humans, *American Journal of Clinical Nutrition* 66 (1997): 438–446.
7. R. Havel, McCollum Award Lecture, 1993: Triglyceride-rich lipoproteins and atherosclerosis—New perspectives, *American Journal of Clinical Nutrition* 59 (1994): 795–799.
8. NIH Consensus Conference, Triglyceride, high-density lipoprotein, and coronary heart disease, *Journal of the American Medical Association* 269 (1993): 505–510.
9. J. L. Groff, S. S. Gropper, and S. M. Hunt, *Advanced Nutrition and Human Metabolism* (St. Paul, Minn.: West, 1995), pp. 466–483.
10. L. H. Kuller, Dietary fat and chronic diseases: Epidemiologic overview, *Journal of the American Dietetic Association* 97 (1997): S9–S15.
11. R. H. Knopp and coauthors, Long-term cholesterol-lowering effects of 4 fat-restricted diets in hypercholesterolemic and combined hyperlipidemic men: The Dietary Alternatives Study, *Journal of the American Medical Association* 278 (1997): 1509–1515.
12. R. P. Mensink, Effects of the individual saturated fatty acids on serum lipid and lipoprotein concentrations, *American Journal of Clinical Nutrition* (supplement) 57 (1993): 711S–714S.
13. A. M. Salter and D. A. White, Effect of dietary fat on cholesterol metabolism: Regulation of plasma LDL concentrations, *Nutrition Research Reviews* 9 (1996): 241–257.
14. S. M. Grundy, Influence of stearic acid on cholesterol metabolism relative to other long-chain fatty acids, *American Journal of Clinical Nutrition* 60 (1994): 986S–990S.
15. E. A. Decker, The role of stereospecific saturated fatty acid positions on lipid nutrition, *Nutrition Reviews* 54 (1996): 108–110.
16. M. B. Katan and P. L. Zock, *Trans* fatty acids and their effects on lipoproteins in humans, *Annual Review of Nutrition* 15 (1995): 473–493; A. H. Lichtenstein, *Trans* fatty acids and hydrogenated fat—What do we know? *Nutrition Today* 30 (1995): 102–107; J. T. Judd and coauthors, Dietary *trans* fatty acids: Effects on plasma lipids and lipoproteins of healthy men and women, *American Journal of Clinical Nutrition* 59 (1994): 861–868; R. Troisi, W. C. Willett, and S. T. Weiss, *Trans*-fatty acid intake in relation to serum lipid concentrations in adult men, *American Journal of Clinical Nutrition* 56 (1992): 1019–1024.
17. S. Shapiro, Do *trans* fatty acids increase the risk of coronary artery disease? A critique of the epidemiologic evidence, *American Journal of Clinical Nutrition* 66 (1997): 1011S–1017S; D. B. Allison, *Trans* fatty acids and coronary heart disease risk: Epidemiology, *American Journal of Clinical Nutrition* 62 (1995): 670S–678S.
18. C. Ip, Review of the effects of *trans* fatty acids, oleic acid, n-3 polyunsaturated fatty acids, and conjugated linoleic acid on mammary carcinogenesis in ani-

mals, *American Journal of Clinical Nutrition* 66 (1997): 1523S–1529S.

19. American Heart Association, Nutrition Advisory Committee, *News Release,* Trans fatty acids, May 13, 1994.
20. A. Ascherio and W. C. Willett, Health effects of *trans* fatty acids, *American Journal of Clinical Nutrition* 66 (1997): 1006S–1010S.
21. A. P. Simopoulos and coauthors, The 1st Congress of the International Society for the Study of Fatty Acids and Lipids (ISSFAL): Fatty acids and lipids from cell biology to human disease, *Nutrition Today,* July/August 1994, pp. 24–27; W. C. Willett and A. Ascherio, *Trans* fatty acids: Are the effects only marginal? *American Journal of Public Health* 84 (1994): 722–724; M. B. Katan, researcher calls for reconsideration of *trans* fatty acids, *Journal of the American Dietetic Association* 94 (1994): 1097–1098.
22. ASCN/AIN Task Force on *Trans* Fatty Acids, Position paper on *trans* fatty acids, *American Journal of Clinical Nutrition* 63 (1996): 663–670.
23. F. B. Hu and coauthors, Dietary fat intake and the risk of coronary heart disease in women, *New England Journal of Medicine* 337 (1997): 1491–1499; M. B. Katan, High-oil compared with low-fat, high-carbohydrate diets in the prevention of ischemic heart disease, *American Journal of Clinical Nutrition* 66 (1997): 974S–979S.
24. D. M. Hegsted and D. Kritchevsky, Diet and serum lipid concentrations: Where are we? *American Journal of Clinical Nutrition* 65 (1997): 1893–1896; B. V. Howard and coauthors, Polyunsaturated fatty acids result in greater cholesterol lowering and less triacylglycerol elevation than monounsaturated fatty acids in a dose-response comparison in a multiracial study group, *American Journal of Clinical Nutrition* 62 (1995): 392–402; M. B. Katan, P. L. Zock, and R. P. Mensink, Effects of fats and fatty acids on blood lipids in humans, *American Journal of Clinical Nutrition* 60 (1994): 1017S–1022S.
25. Katan, Zock, and Mensink, 1994; P. Mata and coauthors, Effects of long-term monounsaturated- vs polyunsaturated-enriched diets on lipoproteins in healthy men and women, *American Journal of Clinical Nutrition* 55 (1992): 846–850.
26. M. C. Nydahl, I. B. Gustafsson, and B. Vessby, Lipid-lowering diets enriched with monounsaturated or polyunsaturated fatty acids but low in saturated fatty acids have similar effects on serum lipid concentrations in hyperlipidemic patients, *American Journal of Clinical Nutrition* 59 (1994): 115–122.
27. M. F. Oliver, It is more important to increase the intake of unsaturated fats than to decrease the intake of saturated fats: Evidence from clinical trials relating to ischemic heart disease, *American Journal of Clinical Nutrition* 66 (1997): 980S–986S.
28. N. J. Stone, Fish consumption, fish oil, lipids, and coronary heart disease, *Circulation* 94 (1996): 2337–2340; M. L. Daviglus and coauthors, Fish consumption and the 30-year risk of fatal myocardial infarction, *New England Journal of Medicine* 336 (1997): 1046–1053; R. F. Gillum, M. E. Mussolino, and J. H. Madans, The relationship between fish consumption and stroke incidence: The NHANES I epidemiologic follow-up study, *Archives of Internal Medicine* 156 (1996): 537–542; D. S. Siscovick and coauthors, Dietary intake and cell membrane levels of long-chain n-3 polyunsaturated fatty acids and the risk of primary cardiac arrest, *Journal of the American Medical Association* 274 (1995): 1363–1367.
29. S. L. Connor and W. E. Connor, Are fish oils beneficial in the prevention and treatment of coronary artery disease? *American Journal of Clinical Nutrition* 66 (1997): 1020S–1031S.
30. WHO and FAL Joint Consultation: Fats and oils in human nutrition, *Nutrition Reviews* 53 (1995): 202–205.
31. C. M. Albert and coauthors, Fish consumption and risk of sudden cardiac death, *Journal of the American Medical Association* 279 (1998): 23–28; A. Ascherio and coauthors, Dietary intake of marine n-3 fatty acids, fish intake, and the risk of coronary disease among men, *New England Journal of Medicine* 332 (1995): 977–982.
32. D. K. Tidwell and coauthors, Comparison of the effects of adding fish high or low in n-3 fatty acids to a diet conforming to the Dietary Guidelines for Americans, *Journal of the American Dietetic Association* 93 (1993): 1124–1128.
33. L. K. Ferrier and coauthors, α-Linoleic acid– and docosahexaenoic acid–enriched eggs from hens fed flaxseed: Influence on blood lipids and platelet phospholipid fatty acids in humans, *American Journal of Clinical Nutrition* 62 (1995): 81–86.
34. D. P. Rose, Dietary fatty acids and cancer, *American Journal of Clinical Nutrition* 66 (1997): 998S–1003S; J. H. Weisburger, Dietary fat and risk of chronic disease: Mechanistic insights from experimental studies, *Journal of the American Dietetic Association* 97 (1997): S16–S23.
35. B. C.-H. Chiu and coauthors, Diet and risk of non-Hodgkin lymphoma in older women, *Journal of the American Medical Association* 275 (1996): 1315–1321.
36. D. J. Hunter and coauthors, Cohort studies of fat intake and the risk of breast cancer—A pooled analysis, *New England Journal of Medicine* 334 (1996): 356–361.
37. E. Barrett-Connor and N. J. Friedlander, Dietary fat, calories, and the risk of breast cancer in postmenopausal women: A prospective population-based study, *Journal of the American College of Nutrition* 12 (1993): 390–399.
38. K. J. Pienta and P. S. Esper, Is dietary fat a risk factor for prostate cancer? *Journal of the National Cancer Institute* 85 (1993): 1538–1540; E. Giovannuci and coauthors, A prospective study of dietary fat and risk of prostate cancer, *Journal of the National Cancer Institute* 85 (1993): 1571–1579.
39. B. J. Rolls, Carbohydrates, fats, and satiety, *American Journal of Clinical Nutrition* 61 (1995): 960S–967S.
40. WHO/FAO Joint Consultation, 1995.
41. S. M. Garn, From the Miocene to olestra: A historical perspective on fat consumption, *Journal of the American Dietetic Association* 97 (1997): S54–S57.
42. N. D. Ernst and coauthors, Consistency between US dietary fat intake and serum total cholesterol concentrations: The National Health and Nutrition Examination Surveys, *American Journal of Clinical Nutrition* 66 (1997): 965S–972S.
43. B. Matheson and coauthors, Effect on serum lipids of monounsaturated oil and margarine in the diet of an Antarctic expedition, *American Journal of Clinical Nutrition* 63 (1996): 933–938.
44. S. M. Krebs-Smith and coauthors, Characterizing food intake patterns of American adults, *American Journal of Clinical Nutrition* 65 (1997): 1264S–1268S.
45. A. F. Subar and coauthors, US dietary patterns associated with fat intake: The 1987 National Health Interview Survey, *American Journal of Public Health* 84 (1994): 359–366.
46. M. Noakes and coauthors, Modifying the fatty acid profile of dairy products through feedlot technology lowers plasma cholesterol of humans consuming the products, *American Journal of Clinical Nutrition* 63 (1996): 42–46.
47. D. Green-Burgeson, Calculating fat the easy way, *Journal of the American Dietetic Association* 94 (1994): 256.

# Alternatives to Fat

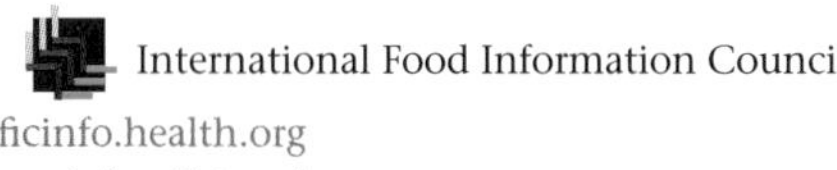

ificinfo.health.org
Search for Fat replacers

As people learn more about the health consequences of high-fat diets, they want to lower their fat intake, but they'd rather not give up their favorite foods. As the adage goes, they want to have their cake and eat it, too—both figuratively and literally. Fat replacers offer an easy way to lower fat intake and still enjoy many foods that were once high in fat. Skeptics say that people will use these fat substitutes the way they use artificial sweeteners—in *addition* to fats rather than *instead* of fats—but preliminary reports indicate that people who use foods containing fat replacers do eat less fat. They may compensate with more carbohydrate, however, so their total energy intakes remain constant.[1] Still, even if energy intakes remain steady, eating carbohydrate in place of fat improves blood lipids and helps to shift body composition toward the lean.

Food chemists have been working for decades on ways to reduce the fat in foods. Juggling the needs of the human body, the taste perceptions of consumers, and the requirements of food preparation is a complex task. For the body, products must contribute little food energy, be nontoxic and completely excreted, and not rob the body of valuable fat-soluble nutrients. To satisfy consumers, products must be attractive, feel right in the mouth, and have an acceptable flavor. Food manufacturers need a compound that remains stable while meeting a product's requirements for temperature, moisture, and texture. That's a tall order, but it looks as though food chemists are mastering the task. Today shoppers can select from thousands of fat-free, low-fat, and reduced-fat products. Many bakery goods, cheeses, frozen desserts, and other products offer less than half a gram of fat in a serving.

Some techniques for reducing food fat are quite simple. For example, manufacturers can dilute fat by adding water or whipping in air. They use nonfat milk in creamy desserts and lean meats in frozen entrées. Sometimes they simply prepare the products differently. For example, fat-free potato chips are now baked instead of fried.

Other techniques to replace some or all of the fat in foods use ingredients that derive from either carbohydrate, protein, or fat. Because these ingredients are common dietary substances, companies can ask the Food and Drug Administration (FDA) to approve them as generally recognized as safe (GRAS) substances. The body may digest and absorb some of these substances, so they may contribute some energy, although significantly less than fat's 9 kcalories per gram.[2] The glossary and Table H5-1 describe the various fat replacers in more detail.

## *Oatrim and Other Carbohydrate-Based Fat Replacers*

Manufacturers have long used carbohydrate-based compounds as thickeners and stabilizers in foods. Today carbohydrate derivatives such as dextrins, modified food starches, and gums are used as fat replacers in a variety of products. For example, maltodextrin, a carbohydrate derived from corn, melts when sprinkled on hot, moist foods such as baked potatoes, providing a flavor similar to that of butter or margarine.

The U.S. Department of Agriculture has developed two carbohydrate-based fat replacers—Oatrim, which is made from digestible oat fiber, and Z-Trim, which is made from the crushed hulls of grains. These products mimic the texture and feel of fat by forming gels when added to water. They are heat stable and can be used in baking, but not in frying.

Because the body can digest Oatrim, it provides energy—up to 4 kcalories per gram. This represents less than half of the kcalories provided by fat, but kcalories nevertheless. Z-Trim, on the other hand, is an insoluble fiber that passes through the body undigested, thus yielding 0 kcalories. Better still is the contribution Z-Trim makes to a person's fiber intake. When used to replace 100 grams of fat, Z-Trim saves 900 kcalories and provides 10 grams of fiber.

## *Simplesse and Other Protein-Based Fat Replacers*

Perhaps the best-known protein-based fat replacer, and the only one to receive FDA approval to date, is Simplesse. The FDA declared Simplesse safe for use in ice cream and frozen desserts in 1990. Simplesse is made from either egg white or milk proteins processed into mistlike particles that feel and taste like fat. Because the components of Simplesse are common in foods, safety studies have not been required.

Simplesse cannot be used for frying because it gels at high temperatures and loses its creaminess. It works fine on hot foods, though; for example, it makes a good imitation butter spread for toast or a sour cream–type topping for a baked potato. Simplesse is not available for home use.

Simplesse creates the *perception* of fat without all the kcalories. In the body, Sim-

*Low-fat and nonfat foods offer as much flavor and enjoyment as their high-fat counterparts—for less fat and fewer kcalories.*

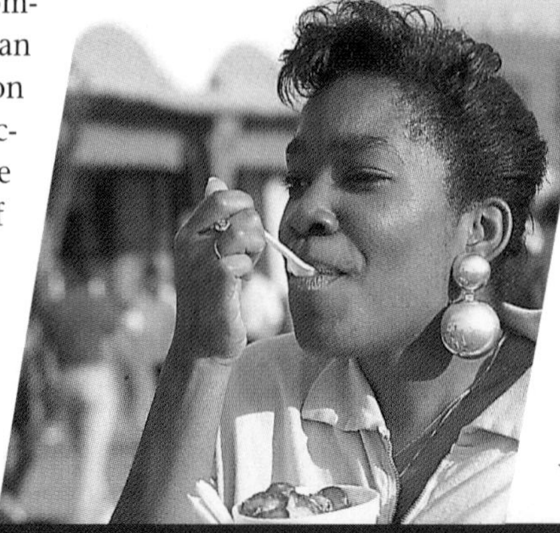

**Table H5-1**
**Fat Replacers**

| | Energy Value (kcal/g) | Properties | Uses |
|---|---|---|---|
| ***Carbohydrate-based*** | | | |
| Oatrim[a] ("hydrolyzed oat flour") | 1–4 | Creamy, retains moisture, heat stable in cooking and baking but not frying | Ice cream, frozen desserts, salad dressing, soups, cheeses, peanut butter, mayonnaise, baked goods, cereals, meats, milks |
| Z-Trim | 0 | Creamy, no flavor, adds fiber, heat stable in baking but not frying | Baked goods, cheeses, meats, milks |
| ***Protein-based*** | | | |
| Microparticulated protein[b] | 1–2 | Creamy, heat stable in cooking and baking but not frying | Frozen desserts, cheeses, yogurt, salad dressings, pie crusts, coffee creamer, cheesecake, pastries, pizza, cheese sauces, cream soups |
| ***Fat-based*** | | | |
| Salatrim[c] | 5 | Same properties as fats | Chocolate, confections, cookies, crackers, sour cream, frozen desserts, cheese |
| Caprenin | 5 | Same properties as fats | Soft candies, cocoa butter, confectionery coatings |
| ***Synthetic*** | | | |
| Olestra[d] | 0 | Same properties as fats; heat stable in frying, cooking, and baking | Potato, corn, tortilla chips; crackers; cheese puffs |

[a]Sold under the brand names of Oatrim and TrimChoice.
[b]Sold under the brand names of Simplesse and K-Blazer.
[c]Sold under the brand name of Benefat.
[d]Sold under the brand name of Olean.
Source: Adapted from N. I. Hahn, Replacing fat with food technology: A brief review of new fat replacement ingredients, *Journal of the American Dietetic Association* 97 (1997): 15–16.

plesse is digested and absorbed, contributing 1 to 2 kcalories per gram. Because 1 gram of Simplesse can replace 3 grams of fat, this represents a 25-kcalorie savings. Substituting Simplesse for fat reduces the energy values of some foods dramatically. In some cases, though, such as fat-free ice cream, so much sugar is added that the kcalorie count of a fat-free product may be as high as in the original (see Table H5-2). Replacing both the fat and the sugar in a product is difficult to do, because both contribute to flavor, texture, and stability. Substituting Simplesse for fat, however, does reduce a food's fat and cholesterol content appreciably.

Some people, such as those who are allergic or sensitive to egg or milk proteins, may have to avoid Simplesse. (FDA regulations require a product's label to identify the source of its protein in the ingredients list.) People on protein-restricted diets may have to consult with their dietitians before including Simplesse in their diets, although its use is not expected to raise protein intake by more than 2 grams daily. Companies have developed other protein-derived products similar to Simplesse but have yet to petition the FDA for approval.

## Salatrim and Other Fat-Based Fat Replacers

Fat-based fat replacers share many of the properties of regular fats, but provide fewer kcalories. Two of these products—caprenin and salatrim—are triglycerides that have been created with a specific combination of fatty acids attached to the glycerol molecule.[3] By using short-chain (8-carbon) fatty acids that provide fewer kcalories and a long-chain (22-carbon) fatty acid that is poorly absorbed, manufacturers have created a fat that provides 5 kcalories per gram instead of 9.

## Synthetic Fat Replacers

In January 1996, the FDA approved a fake fat known as olestra for use as an additive in snack foods.[4] Olestra is the first artificial fat to reach the market that can withstand the heat needed to fry foods such as potato chips or cheese puffs. Manufactured under the trade name Olean (pronounced oh-LEEN), olestra has been

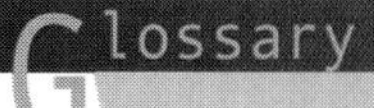


**caprenin:** a fat-based fat replacer made from a triglyceride with poorly absorbed fatty acids; provides 5 kcalories per gram.

**fat replacer:** an ingredient that replaces some or all of the functions of fat and may or may not provide energy. In this text, the term *fat replacer* is used interchangeably with **fat substitute,** which technically applies only to an ingredient that replaces all of the functions of fat and provides no energy.

**microparticulated protein:** a protein-based fat replacer made from milk, egg, or whey proteins; provides 1 to 2 kcalories per gram.

**oatrim:** a carbohydrate-based fat replacer made from oat flour; provides 1 to 4 kcalorie per gram.

**olestra:** a synthetic fat made from sucrose and fatty acids that provides 0 kcalories per gram; also known as **sucrose polyester.**

**salatrim:** a fat-based fat replacer made from **s**hort-**a**nd **l**ong-chain **a**cid **tri**glyceride **m**olecules; provides an estimated 5 kcalories per gram.

**Z-Trim:** a carbohydrate-based fat replacement made from the seed hulls of oats, soybeans, peas, and rice or from the bran of corn or wheat; provides 0 kcalories per gram.

**Table H5-2**
**Fat Content of Regular Foods and Foods Prepared with Fat Replacers**

| Food | Fat (g) | Cholesterol (mg) | Energy (kcal) |
|---|---|---|---|
| Ice cream | | | |
| Super premium (½ c) | 19 | 97 | 274 |
| Regular (½ c) | 7 | 30 | 135 |
| Ice milk (½ c) | 3 | 9 | 92 |
| Frozen dessert | | | |
| Made with Simplesse (½ c) | <1 | 14 | 120 |
| Made with Oatrim (½ c) | 1 | 4 | 135 |
| Butter (1 tsp) | 4 | 11 | 36 |
| Margarine (1 tsp) | 4 | 0 | 34 |
| Maltodextrin sprinkles (½ tsp) | <1 | 0 | 3 |
| French fries | | | |
| Fried in vegetable oil | 12.3 | 0 | 227 |
| Fried in 75% olestra blend | 3.1 | 0 | 144 |
| Chicken | | | |
| Fried in vegetable oil | 14.6 | 0 | 252 |
| Fried in 75% olestra blend | 8.6 | 0 | 198 |
| Onion rings | | | |
| Fried in vegetable oil | 19.5 | 0 | 315 |
| Fried in 75% olestra blend | 4.9 | 0 | 184 |

NOTE: Equivalent serving sizes were compared; in the case of butter and margarine, ½ teaspoon of sprinkles was compared with 1 teaspoon butter or margarine as per label directions.

approved for use in snack foods such as potato chips, crackers, and tortilla chips.

Olestra's chemical structure is similar to that of a regular fat (a triglyceride) but with important differences. A triglyceride is composed of a glycerol molecule with three fatty acids attached, whereas olestra is made of a sucrose molecule with six to eight fatty acids attached. Enzymes in the digestive tract cannot break the bonds of olestra, so unlike sucrose or fatty acids, olestra passes through the system unabsorbed. As you will see, this characteristic is responsible for both the triumphs and the troubles of this nonfat fat.

Because olestra contains several of the fatty acids common to shortening and cooking oils, it shares many of their physical properties such as appearance, color, taste, heat stability, and shelf life. Consequently, olestra looks, feels, and tastes like dietary fat and can be used in frying, cooking, and baking. Best of all, olestra performs like a fat without losing flavor, adding kcalories, or raising blood lipids. Potato chips made with olestra deliver half the kcalories of regular chips and none of the fat. Does this sound too good to be true? It may be.

The FDA's evaluation of olestra's safety answered two questions. First, is olestra toxic? Second, does it affect either nutrient absorption or the health of the digestive tract?

Regarding possible toxicity, there is little controversy. Research on both animals and human beings supports the safety of olestra as a partial replacement for dietary fats and oils. Studies on animals have reported no evidence of either cancer or birth defects caused by olestra.

Olestra's effects on nutrient absorption and digestive tract health, however, raise serious concerns.[5] One of the positive attributes of fats is that they carry the fat-soluble vitamins A, D, E, and K with them into the body. When olestra passes through the digestive tract unabsorbed, it binds with some of these vitamins and carries them out of the body, robbing the person of these valuable nutrients. To compensate for these losses, the FDA has required the manufacturer to fortify olestra with vitamins A, D, E, and K. Saturating olestra with these vitamins blocks its ability to bind with the vitamins from other foods.

Olestra also sweeps other fat-soluble substances through the digestive system. Among those lost are the carotenoids, the colorful pigments found in fruits and vegetables.[6] Carotenoids act as antioxidants and may be important in protecting against a variety of diseases, including heart disease, some cancers, and macular degeneration (an eye disorder that causes blurry vision and blindness). Because the relationships between carotenoids and disease prevention have not yet been proved, the FDA has required the manufacturer to conduct long-term studies on whether olestra use and carotenoid losses will impair health. Should this prove to be a problem, fortifying products is not a likely solution. If you consider that there are hundreds of carotenoids commonly found in foods and then think of the hundreds of *non*carotenoid substances that may also be beneficial to health, then you will quickly realize that fortification is not feasible. The FDA will review olestra's status when new research findings become available.

Consumers may not perceive any immediate ill effects of a diet depleted of its fat-soluble vitamins and carotenoids, but they will surely notice the digestive distress that sometimes accompanies olestra consumption: cramps, gas, bloating, and diarrhea. For some people, eating as little as 2 ounces of olestra-containing potato chips produces "fecal urgency" (the immediate need for a bathroom) and "anal leakage" (noted by stained underwear); for others, the incidence and severity of GI symptoms are no different than when eating regular potato chips.[7] The FDA considers these symptoms "unpleasant" but not "medically significant." Whether consumers agree and accept such "annoyances" in trade for a bag of fat-free chips with lunch remains an unanswered multimillion dollar question.

In approving olestra, the FDA has required foods made with it to carry a label warning that "olestra may cause

Nutrition Facts
Ingredients: Potatoes, Olestra (OLEAN® brand), Salt, alpha-Tocopheryl Acetate (Vitamin E), Vitamin A Palmitate, Tocopherols (to protect flavor), Vitamin K, and Vitamin D.
FRITO-LAY, INC.
This Product Contains Olestra. Olestra may cause abdominal cramping and loose stools. Olestra inhibits the absorption of some vitamins and other nutrients. Vitamins A,D,E and K have been added.
OLEAN is a registered trademark of The Procter & Gamble Company.

***Whether consumers are willing to trade a little GI distress for potato chips that deliver half of the kcalories of regular chips and none of the fat remains to be seen.***

abdominal cramping and loose stools" and that it "inhibits the absorption of some vitamins and other nutrients." Those who read food labels may hesitate to purchase products with these warnings; others may not even notice the warnings.

Will people become slimmer by munching on chips and cookies made with artificial fats? Or will they simply feel free to eat more chips and cookies? Interestingly, one study found that women did eat more lunch after having eaten a low-fat yogurt than after a high-fat yogurt when the yogurt was labeled as such.[8] If they did not know which yogurt they had eaten, the women eating the low-fat yogurt consumed less lunch than those eating the high-fat yogurt. Simply knowing the fat content of the yogurt influenced the energy intake of their next meal. Consumers need to keep in mind that low-fat and fat-free foods still deliver kcalories. Decades ago, consumers hailed the arrival of artificial sweeteners as a weight-loss wonder, but in reality, kcalories saved by using artificial sweeteners were readily replaced by kcalories from other foods. The alternatives to fat discussed in this highlight can help to lower fat intake and support weight loss only when they actually *replace* fat and energy in the diet.[9]

 FDA

www.fda.gov

Visit foods
Search for Fat substitutes

## Notes

1. S. N. Gershoff, Nutrition evaluation of dietary fat substitutes, *Nutrition Reviews* 53 (1995): 305–313.
2. A. M. Miraglio, Nutrient substitutes and their energy values in fat substitutes and replacers, *American Journal of Clinical Nutrition* 62 (1995): 1175S–1179S.
3. S. J. Bell and coauthors, The new dietary fats in health and disease, *Journal of the American Dietetic Association* 97 (1997): 280–286.
4. Olestra: Approved with special labeling, *FDA Consumer,* April 1996, p. 11.
5. H. Blackburn, Olestra and the FDA, *New England Journal of Medicine* 334 (1996): 984–986.
6. J. A. Westrate and K. H. van het Hof, Sucrose polyester and plasma carotenoid concentrations in healthy subjects, *American Journal of Clinical Nutrition* 62 (1995): 591–597.
7. L. J. Cheskin and coauthors, Gastrointestinal symptoms following consumption of olestra or regular triglyceride potato chips: A controlled comparison, *Journal of the American Medical Association* 279 (1998): 150–152.
8. D. J. Shide and B. J. Rolls, Information about the fat content of preloads influences energy intake in healthy women, *Journal of the American Dietetic Association* 95 (1995): 993–998.
9. Position of The American Dietetic Association, *Journal of the American Dietetic Association* 98 (1998): 463–468.

CHAPTER 6

# Protein: Amino Acids

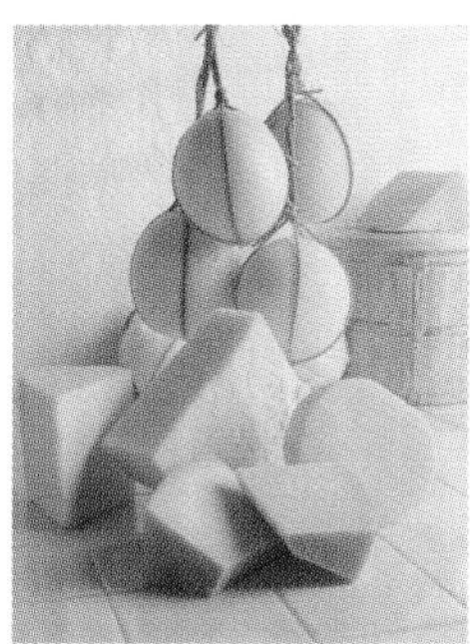

People commonly associate protein with strength and meat with protein. Consequently, they eat steak to build their muscles, but their thinking is only partly correct. Protein is a vital structural and working substance in all cells, not just muscle cells. Meat is a good source of protein, but so are milk, eggs, legumes, and many grains and vegetables. People who overvalue protein may overemphasize meat in their diets, sometimes at the expense of other, equally important nutrients and foods. Protein is important, but it is only one of the nutrients needed to maintain the body's health.

# The Chemist's View of Proteins

Chemically, proteins contain the same atoms as carbohydrates and lipids—carbon (C), hydrogen (H), and oxygen (O)—but proteins also contain nitrogen (N) atoms. These nitrogen atoms give the name *amino* (nitrogen containing) to the amino acids—the links in the chains of proteins. Also, proteins assume extraordinary and unique shapes, which enable them to play their vital roles in the body.

**proteins:** compounds composed of carbon, hydrogen, oxygen, and nitrogen atoms, arranged into amino acids linked in a chain. Some amino acids also contain sulfur atoms.

## Amino Acids

All amino acids have the same basic structure—a central carbon (C) atom with a hydrogen (H), an amino group ($NH_2$), and an acid group (COOH) attached to it. Carbon atoms need to form four bonds, though, so a fourth attachment is necessary, and it is this fourth site that distinguishes each amino acid from the others. Attached to the carbon atom at the fourth bond is a distinct atom, or group of atoms, known as the *side group* or *side chain* (see Figure 6-1).

**amino** (a-MEEN-oh) **acids:** building blocks of proteins; each contains an amino group, an acid group, a hydrogen atom, and a distinctive side group, all attached to a central carbon atom.
- **amino** = containing nitrogen

Reminder:
- H forms 1 bond.
- O forms 2 bonds.
- N forms 3 bonds.
- C forms 4 bonds.

• ***Unique Side Groups*** • The side groups on amino acids vary from one amino acid to the next, making proteins more complex than either carbohydrates or lipids. A polysaccharide (starch, for example) may be several thousand units long, but every unit is a glucose molecule just like all the others. A protein, on the other hand, is made up of about 20 different amino acids, each with a different side group. Table 6-1 lists the amino acids most common in proteins.*

The simplest amino acid, glycine, has a hydrogen atom as its side group. A slightly more complex amino acid, alanine, has an extra carbon with three hydrogen atoms. Other amino acids have more complex side groups (see Figure 6-2 on p. 162 for examples). Thus, although all amino acids share a common structure, they differ in size, shape, electrical charge, and other characteristics because of differences in these side groups.

**Figure 6-1**
**Amino Acid Structure**

All amino acids have a carbon (known as the alpha-carbon), with an amino group ($NH_2$), an acid group (COOH), a hydrogen (H), and a side group attached. The side group is a unique chemical structure that differentiates one amino acid from another.

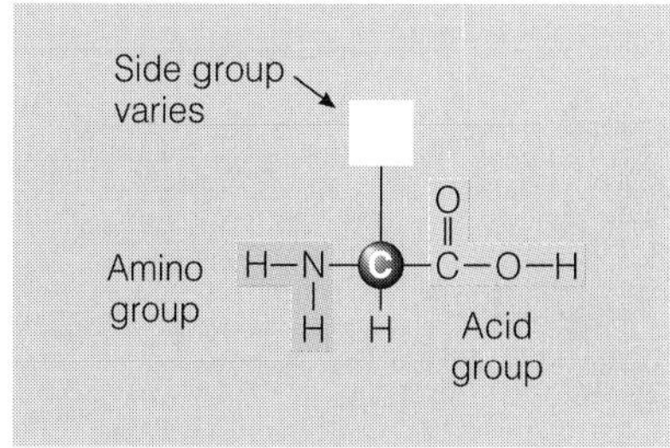


• ***Nonessential Amino Acids*** • The body can synthesize more than half of the amino acids for itself, if it is given nitrogen to form the amino group and fragments from carbohydrate and fat to form the rest of the structure. Proteins in foods usually deliver these amino acids, but it is not essential that they do so.

• ***Essential Amino Acids*** • There are nine amino acids that the human body either cannot make at all or cannot make in sufficient quantity to meet its needs. These nine amino acids must be supplied by the diet; they are essential.

**essential amino acids:** amino acids that the body cannot synthesize in amounts sufficient to meet physiological needs (see Table 6-1). Some researchers refer to essential amino acids as **indispensable** and to nonessential amino acids as **dispensable.**

*Some amino acids occur in related forms (for example, proline can acquire an OH group to become hydroxyproline). Besides the 20 common amino acids, which can all be components of proteins, others occur individually (for example, taurine and ornithine), and still others can be made by chemists.

**Table 6-1**
**Amino Acids**

Proteins are made up of about 20 common amino acids. The first column lists the *essential* amino acids for human beings (those the body cannot make—that must be provided in the diet).

| Essential Amino Acids | | Nonessential Amino Acids | |
|---|---|---|---|
| Histidine | (HISS-tuh-deen) | Alanine | (AL-ah-neen) |
| Isoleucine | (eye-so-LOO-seen) | Arginine | (ARJ-ih-neen) |
| Leucine | (LOO-seen) | Asparagine | (ah-SPAR-ah-geen) |
| Lysine | (LYE-seen) | Aspartic acid | (ah-SPAR-tic acid) |
| Methionine | (meh-THIGH-oh-neen) | Cysteine | (SIS-teh-een) |
| Phenylalanine | (fen-il-AL-ah-neen) | Glutamic acid | (GLU-tam-ic acid) |
| Threonine | (THREE-oh-neen) | Glutamine | (GLU-tah-meen) |
| Tryptophan | (TRIP-toe-fan, TRIP-toe-fane) | Glycine | (GLY-seen) |
| | | Proline | (PRO-leen) |
| Valine | (VAY-leen) | Serine | (SEER-een) |
| | | Tyrosine | (TIE-roe-seen) |

NOTE: In special cases, some nonessential amino acids may become conditionally essential (see the text).

**conditionally essential amino acid:** an amino acid that is normally nonessential, but must be supplied by the diet in special circumstances when the need for it exceeds the body's ability to produce it.

• ***Conditionally Essential Amino Acids*** • Sometimes a nonessential amino acid becomes essential under special circumstances. For example, the body normally makes tyrosine (a nonessential amino acid) from the essential amino acid phenylalanine. But if the diet fails to supply enough phenylalanine, or if the body cannot make the conversion for some reason (as happens in the inherited disease phenylketonuria), then tyrosine becomes *conditionally* essential. Similarly, glutamine, the most abundant amino acid in the body, becomes conditionally essential in advanced liver disease.[1]

## Proteins

**peptide bond:** a bond that connects the acid end of one amino acid with the amino end of another, forming a link in a protein chain.

**dipeptide** (dye-PEP-tide): two amino acids bonded together.
- **di** = two
- **peptide** = amino acid

**tripeptide:** three amino acids bonded together.
- **tri** = three

**polypeptide:** many (ten or more) amino acids bonded together. An intermediate string of four to nine amino acids is an **oligopeptide** (OL-ee-go-PEP-tide).
- **poly** = many
- **oligo** = few

Cells link amino acids end-to-end in a virtually infinite variety of sequences to form thousands of different proteins. Each amino acid is connected to the next by a peptide bond.

• ***Amino Acid Chains*** • Condensation reactions create the bonds between amino acids, just as they combine monosaccharides to form disaccharides, and fatty acids with glycerol to form triglycerides.* Two amino acids bonded together form a dipeptide (see Figure 6-3). By another such reaction, a third amino acid can be added to the chain to form a tripeptide. As additional amino acids join the chain, a polypeptide is formed. Most proteins are a few dozen to several hundred amino acids long. Figure 6-4 provides an example—insulin.

*Later in the chapter, Figure 6-6 shows how each protein's sequence is dictated by the genetic code in DNA, and the text describes how the sequence shapes the protein.

**Figure 6-2**
**Examples of Amino Acids**

Note that all amino acids have a common chemical structure but that each has a different side group. Appendix C presents the chemical structures of the 20 amino acids most common in proteins.

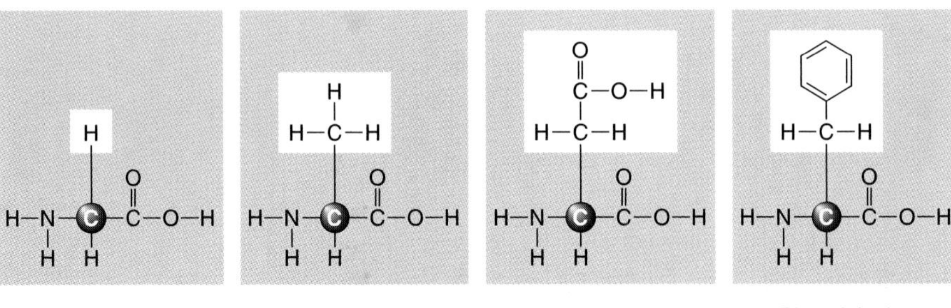

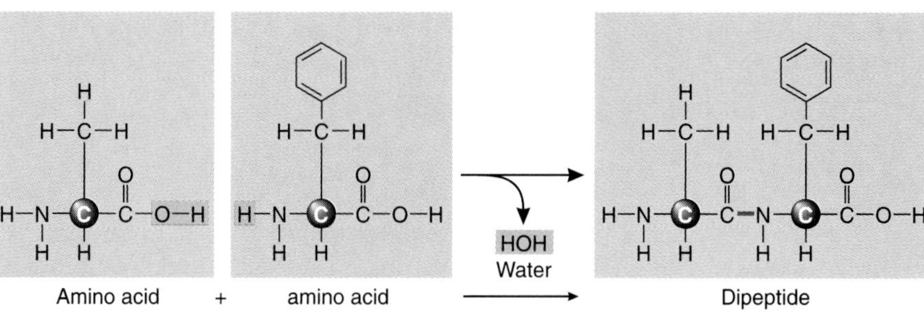


An OH group from the acid end of one amino acid and an H atom from the amino group of another join to form a molecule of water.

A peptide bond (highlighted in red) forms between the two amino acids, creating a dipeptide.

**Figure 6-3**

**Condensation of Two Amino Acids to Form a Dipeptide**

• ***Amino Acid Sequences*** • If a person could walk along a carbohydrate molecule like starch, the first stepping stone would be a glucose. The next stepping stone would also be a glucose, and it would be followed by a glucose, and yet another glucose. But if a person were to walk along a polypeptide chain, each stepping stone would be one of 20-odd different amino acids. The first stepping stone might be the amino acid methionine. The second might be an alanine. The third might be a glycine, and the fourth a tryptophan, and so on. Walking along another polypeptide path, a person might step on a phenylalanine, then a valine, and a glutamine. In other words, amino acid sequences within proteins vary.

The amino acids can act somewhat like the letters in an alphabet. If you had only the letter *G,* all you could write would be a string of Gs: G–G–G–G–G–G–G. But with 20 different letters available, you could create poems, songs, or novels. The 20 amino acids can be linked together in an even greater variety of sequences than are possible for letters in a word or words in a sentence. Thus the variety of possible sequences for polypeptide chains is tremendous.

The shape of a protein depends on the sequence of its amino acids, the bonds linking the amino acids, and the interactions of the side chains with each other and with the surrounding molecules.

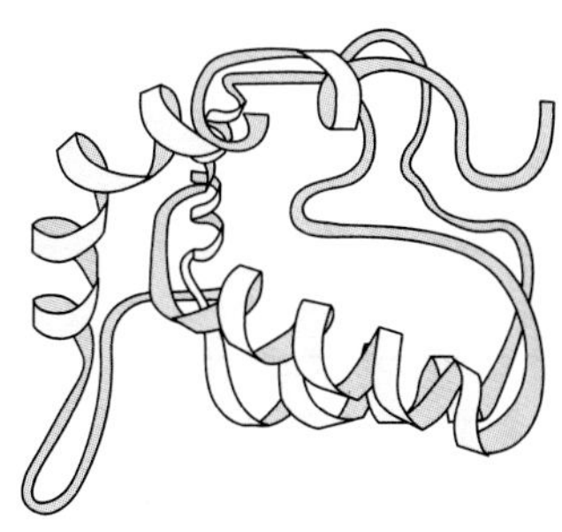

• ***Protein Shapes*** • Polypeptide chains twist into a variety of complex, tangled shapes, depending on their amino acid sequences. Each amino acid has a unique chemical character that attracts it to, or repels it from, the surrounding fluids and other amino acids. Some amino acid side chains carry electrical charges that are attracted to water molecules (they are hydrophilic). Other side chains are neutral and are repelled by water (they are hydrophobic). As amino acids are strung together to make a polypeptide, the chain folds so that its charged hydrophilic side chains are on the outer surface near water; the neutral hydrophobic groups tuck themselves inside, away from water. The intricate, coiled shape the polypeptide finally assumes gives it maximum stability in the body's watery fluids.

**hemoglobin** (HE-moh-GLOW-bin): the globular protein of the red blood cells that carries oxygen from the lungs to the cells throughout the body.
- **hemo** = blood
- **globin** = globular protein

• ***Protein Functions*** • The different shapes of proteins enable them to perform various tasks in the body. Some form hollow balls that can carry and store materials within them, and some, such as those of tendons, are more than ten times as long as they are wide, forming strong, rodlike structures. Some polypeptides are functioning proteins as they are; others need to associate with other polypeptides to form larger working complexes. Some proteins require minerals to activate them. One molecule of hemoglobin—the large, globular protein molecule that, by the billions, packs the red blood cells and carries oxygen—is made of four associated polypeptide chains, each holding the mineral iron. Figure 6-4 shows the two chains of insulin.

**Figure 6-4**

**Amino Acid Sequence of Human Insulin**

Human insulin is a relatively small protein that consists of 51 amino acids in two short polypeptide chains. (For amino acid abbreviations, see Appendix C.) Two bridges link the two chains. A third bridge spans a section within the short chain.

Known as disulfide bridges, these links always involve the amino acid cysteine (Cys), whose side group contains sulfur (S). Cysteines connect to each other when bonds form between these side groups.

• ***Protein Denaturation*** • When proteins are subjected to heat, acid, or other conditions that disturb their stability, they undergo denaturation—that is, they uncoil and lose their shapes and, consequently, their functions. Past a certain point, denaturation is

**denaturation** (dee-NAY-chur-AY-shun): the change in a protein's shape brought about by heat, acid, base, alcohol, heavy metals, or other agents.

irreversible. Familiar examples of denaturation include the hardening of an egg when it is cooked, the curdling of milk when acid is added, and the stiffening of egg whites when they are whipped.

IN SUMMARY

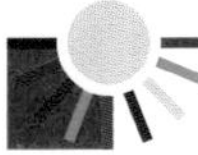

Chemically speaking, proteins are more complex than carbohydrates or lipids, being made of some 20 different amino acids, 9 of which the body cannot make (they are essential). Each amino acid contains an amino group, an acid group, a hydrogen atom, and a distinctive side group. Cells link amino acids together in a series of condensation reactions to create proteins. The distinctive sequence of amino acids in each protein determines its unique shape and function.

# Digestion and Absorption of Protein

Proteins in foods do not become body proteins, but supply the amino acids from which the body makes its own proteins. When a person eats foods containing protein, enzymes break the long polypeptide strands into shorter strands, the short strands into tripeptides and dipeptides, and, finally, the tripeptides and dipeptides into amino acids.

## The Process of Digestion

Figure 6-5 illustrates the digestion of protein through the GI tract. Proteins are crushed and moistened in the mouth, but the real action begins in the stomach.

The inactive form of an enzyme is called a **proenzyme.**
- **pro** = before

**pepsin:** a gastric protease. Pepsin is secreted in an inactive form, **pepsinogen,** which is activated by hydrochloric acid in the stomach.

Reminder: An enzyme that hydrolyzes protein is a *protease* (PRO-tee-ace).

• ***In the Stomach*** • In the stomach, hydrochloric acid uncoils (denatures) each protein's tangled strands so that digestive enzymes can attack the peptide bonds. The hydrochloric acid also converts the inactive form of the enzyme pepsinogen to its active form, pepsin. Pepsin cleaves proteins—large polypeptides—into smaller polypeptides and some amino acids.

• ***In the Small Intestine*** • When polypeptides enter the small intestine, pancreatic and intestinal proteases hydrolyze them further into short peptide chains (oligopeptides), tripeptides, dipeptides, and amino acids. Figure 6-5 includes the details of digestive enzyme action for dietary protein. A number of distinct carriers transport these protein pieces into the intestinal cells.

## The Process of Absorption

**peptidase:** a digestive enzyme that hydrolyzes peptide bonds. *Tripeptidases* cleave tripeptides; *dipeptidases* cleave dipeptides. *Endopeptidases* cleave peptide bonds *within* the chain to create smaller fragments, whereas *exopeptidases* cleave bonds at the *ends* to release free amino acids.
- **tri** = three
- **di** = two
- **endo** = within
- **exo** = outside

The cells of the small intestine absorb amino acids and have peptidase enzymes on their surfaces that split most of the dipeptides and tripeptides into single amino acids. A few dipeptides, tripeptides, and even larger molecules sometimes escape digestion and cross the digestive tract wall to enter the bloodstream.

Some nutrition faddists fail to realize that most proteins are broken down to amino acids before absorption. They urge consumers to "Eat enzyme A. It will help you digest your food." Or "Don't eat food B. It contains enzyme C, which will digest cells in your body." In reality, though, enzymes in foods are digested, just as all proteins are. Only the digestive enzymes, whose design prevents them from being denatured or digested, can work in such an environment.

Another misconception is that eating predigested proteins (amino acid supplements) saves the body from having to digest proteins and keeps the digestive system from "overworking." Such a belief grossly underestimates the body's abil-

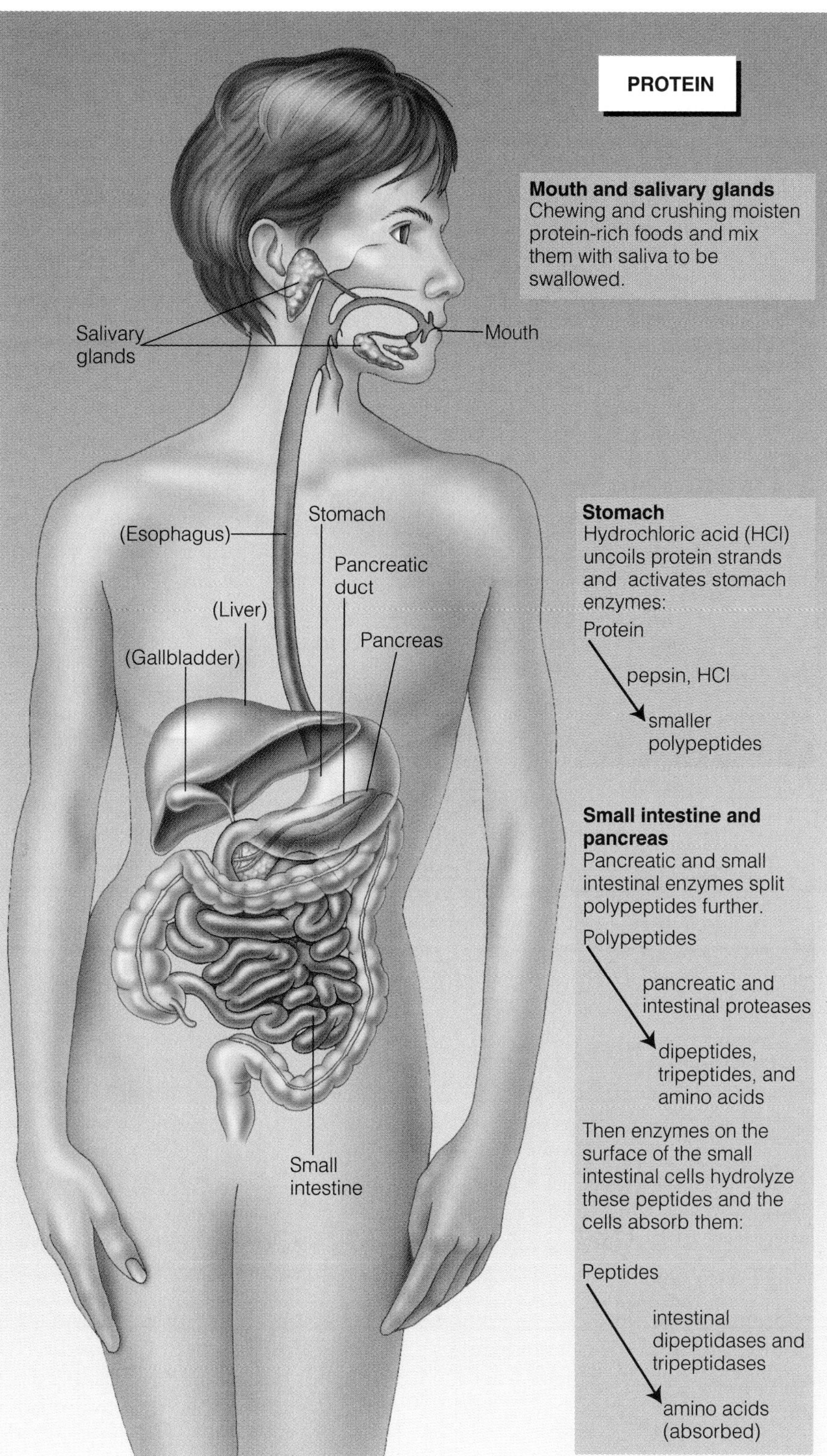

PROTEIN
Mouth and salivary glands
Chewing and crushing moisten protein-rich foods and mix them with saliva to be swallowed.
Salivary glands
Mouth
(Esophagus)
Stomach
Pancreatic duct
(Liver)
(Gallbladder)
Pancreas
Small intestine
Stomach
Hydrochloric acid (HCl) uncoils protein strands and activates stomach enzymes:
Protein
pepsin, HCl
smaller polypeptides
Small intestine and pancreas
Pancreatic and small intestinal enzymes split polypeptides further.
Polypeptides
pancreatic and intestinal proteases
dipeptides, tripeptides, and amino acids
Then enzymes on the surface of the small intestinal cells hydrolyze these peptides and the cells absorb them:
Peptides
intestinal dipeptidases and tripeptidases
amino acids (absorbed)


**Figure 6-5**

**Protein Digestion in the GI Tract**

**In the Stomach:**

Hydrochloric acid (HCl)

- Denatures protein structure.
- Activates pepsinogen to pepsin.

Pepsin

- Cleaves proteins to smaller polypeptides and some free amino acids.
- Inhibits pepsinogen synthesis.

**In the Small Intestine:**

Enteropeptidase[a]

- Converts pancreatic trypsinogen to trypsin.

Trypsin

- Inhibits trypsinogen synthesis.
- Cleaves peptide bonds next to the amino acids lysine and arginine.
- Converts pancreatic procarboxypeptidases to carboxypeptidases.
- Converts pancreatic chymotrypsinogen to chymotrypsin.

Chymotrypsin

- Cleaves peptide bonds next to the amino acids phenylalanine, tyrosine, tryptophan, methionine, asparagine, and histidine.

Carboxypeptidases

- Cleave amino acids from the acid (carboxyl) ends of polypeptides.

Elastase and collagenase

- Cleave polypeptides into smaller polypeptides and tripeptides.

Aminopeptidases

- Cleave amino acids from the amino ends of small polypeptides (oligopeptides).

Tripeptidases

- Cleave tripeptides to dipeptides and amino acids.

[a]Enteropeptidase was formerly known as *enterokinase.*

ities. As a matter of fact, the digestive system handles whole proteins *better* than predigested ones because it dismantles and absorbs the amino acids at rates that are optimal for the body's use. (The last section of this chapter discusses amino acid supplements further.)

IN SUMMARY

Via digestion facilitated mostly by the stomach's acid and enzymes, the body first denatures dietary proteins, then cleaves them into polypeptides, then oligo-, tri-, and dipeptides, and some amino acids. Intestinal enzymes split these further, mostly to single amino acids. Then carriers in the membranes of intestinal cells transport the amino acids into the cells, where they are released into the bloodstream.

# Proteins in the Body

The human body contains an estimated 10,000 to 50,000 different kinds of proteins. Of these, about 1000 have been studied. Only about 10 are described in this chapter—but these should be enough to illustrate proteins' versatility, uniqueness, and importance. As you will see, each protein has a specific function and that function is determined during protein synthesis.

## *Protein Synthesis*

***Growing children end each day with more bone, blood, muscle, and skin cells than they had at the beginning of the day.***

Each human being is unique because of minute differences in the body's proteins. These differences are determined by the amino acid sequences of proteins, which, in turn, are determined by genetics. The following paragraphs describe in words the ways cells synthesize proteins; Figure 6-6 provides a pictorial description.

The instructions for making every protein in a person's body are transmitted by way of the genetic information received at conception. This body of knowledge, which is filed in the DNA within the nucleus of every cell, never leaves the nucleus.

• ***Delivering the Instructions*** • To inform a cell of the sequence of amino acids for a needed protein, a stretch of DNA serves as a template for making a strand of RNA that carries a code, listing in order the amino acids that will be needed to make a given protein. Known as messenger RNA, this molecule escapes through the nuclear membrane. Messenger RNA seeks out and attaches itself to one of the ribosomes (a protein-making machine, which is itself composed of RNA and protein). Thus situated, messenger RNA presents its list, specifying the sequence in which the amino acids are to line up to make a strand of protein.

• ***Lining Up the Amino Acids*** • Other forms of RNA, called transfer RNA, collect amino acids from the cell fluid and bring them to the messenger. Each of the 20 amino acids has a specific transfer RNA. Thousands of transfer RNA, each carrying its amino acid, cluster around the ribosomes, awaiting their turn to unload. When the messenger's list calls for a specific amino acid, the transfer RNA carrying that amino acid moves into position. Then the next loaded transfer RNA moves into place and then the next and the next. Thus the amino acids line up in the sequence that is called for, and enzymes bind them together. Finally, the completed protein strand is released, the messenger is degraded, and the transfer RNA are freed to return for another load of amino acids.

• ***Sequencing Errors*** • The sequence of amino acids in each protein determines its configuration, which supports a specific function. If a genetic error alters the amino acid sequence of a protein, or if a mistake is made in copying the sequence,

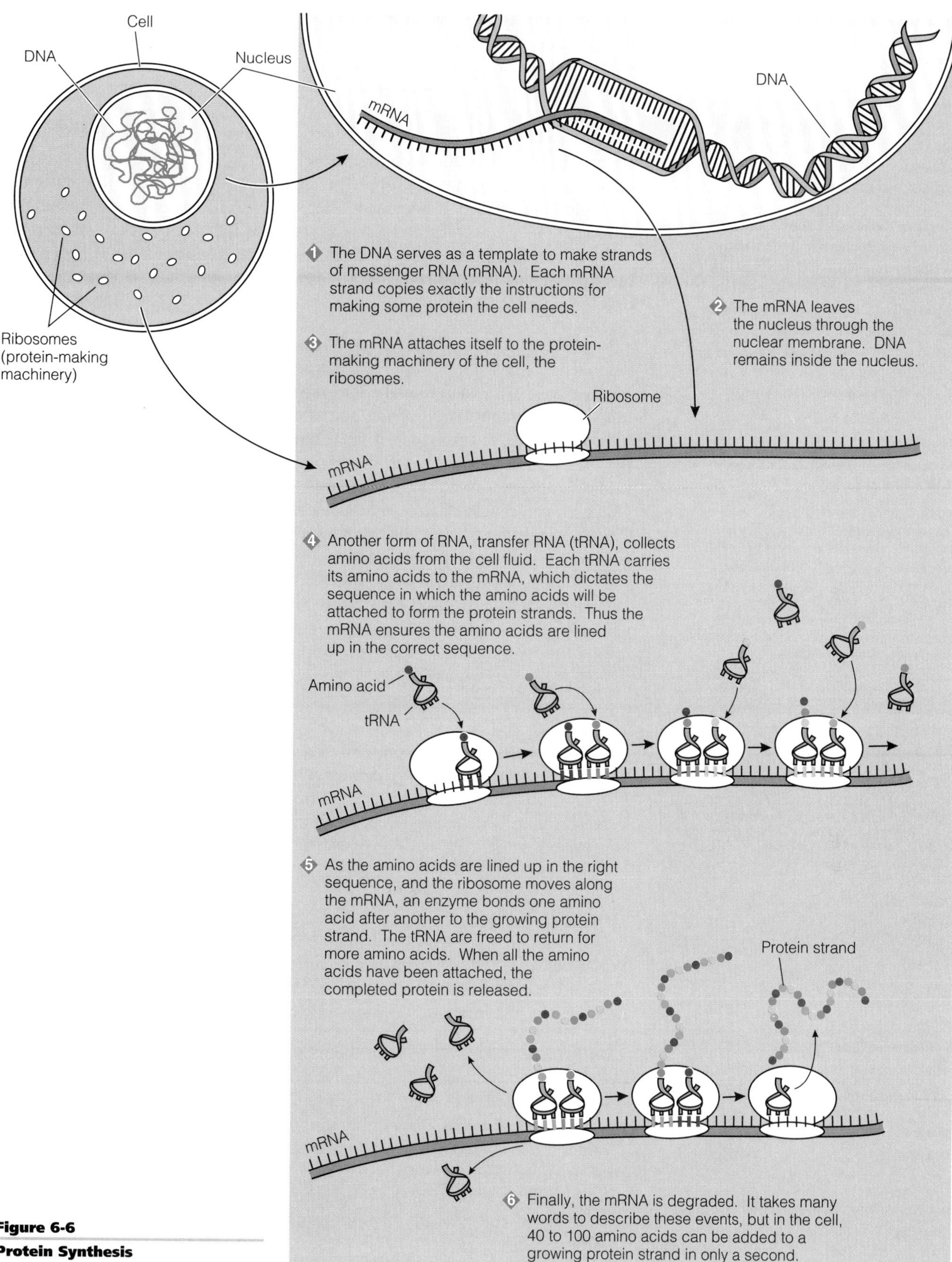

Cell
DNA
Nucleus
Ribosomes
(protein-making
machinery)
mRNA
DNA
1 The DNA serves as a template to make strands of messenger RNA (mRNA). Each mRNA strand copies exactly the instructions for making some protein the cell needs.
2 The mRNA leaves the nucleus through the nuclear membrane. DNA remains inside the nucleus.
3 The mRNA attaches itself to the protein-making machinery of the cell, the ribosomes.
Ribosome
mRNA
4 Another form of RNA, transfer RNA (tRNA), collects amino acids from the cell fluid. Each tRNA carries its amino acids to the mRNA, which dictates the sequence in which the amino acids will be attached to form the protein strands. Thus the mRNA ensures the amino acids are lined up in the correct sequence.
Amino acid
tRNA
mRNA
5 As the amino acids are lined up in the right sequence, and the ribosome moves along the mRNA, an enzyme bonds one amino acid after another to the growing protein strand. The tRNA are freed to return for more amino acids. When all the amino acids have been attached, the completed protein is released.
Protein strand
mRNA
6 Finally, the mRNA is degraded. It takes many words to describe these events, but in the cell, 40 to 100 amino acids can be added to a growing protein strand in only a second.


**Figure 6-6**
**Protein Synthesis**

**sickle-cell anemia:** a hereditary form of anemia characterized by abnormal sickle- or crescent-shaped red blood cells. Sickled cells interfere with oxygen transport and blood flow. Symptoms include hemolytic anemia (red blood cells burst), fever, and severe pain in the joints and abdomen; they are precipitated by dehydration and insufficient oxygen (as may occur at high altitudes).

*Note:* Anemia is not a disease, but a symptom of various diseases. In the case of sickle-cell anemia, a defect in the hemoglobin molecule changes the shape of the red blood cells. Later chapters describe how vitamin and mineral deficiencies change the size and color of the red blood cells. In all cases, the abnormal blood cells are unable to meet the body's oxygen demands.

**Figure 6-7**

**Normal Red Blood Cells Compared with Sickle Cells**

Normally, red blood cells are disc-shaped; in the inherited disorder sickle-cell anemia, red blood cells are sickle- or crescent-shaped. This alteration in shape occurs because valine replaces glutamic acid in the amino acid sequence of hemoglobin's polypeptide chain. As a result of this one amino acid's being in the wrong place, the hemoglobin has a diminished capacity to carry oxygen.

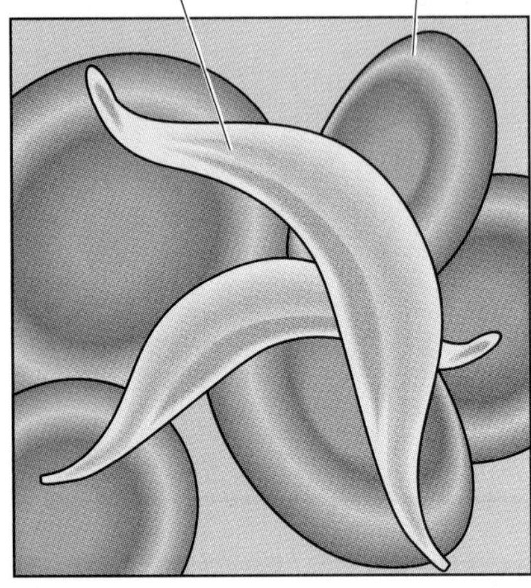


Amino acid sequence of normal hemoglobin:
Val–His–Leu–Thr–Pro–Glu–Glu

Amino acid sequence of sickle-cell hemoglobin:
Val–His–Leu–Thr–Pro–Val–Glu

an altered protein will result, sometimes with dramatic consequences. The protein hemoglobin offers one example of such a genetic variation. In a person with sickle-cell anemia, two of hemoglobin's four polypeptide chains (described earlier on p. 163) have the normal sequence of amino acids, but the other two chains do not—they have the amino acid valine in a position that is normally occupied by glutamic acid (see Figure 6-7). This single alteration in the amino acid sequence changes the character and shape of the protein so much that hemoglobin loses its ability to carry oxygen effectively.[2] The red blood cells filled with this abnormal hemoglobin stiffen into elongated sickle, or crescent, shapes instead of maintaining their normal pliable disc shape—hence the name, sickle-cell anemia. Sickle-cell anemia causes many medical problems and can be fatal. Caring for children with sickle-cell anemia includes diligent attention to their water needs; dehydration can trigger a crisis.[3]

• ***Nutrients and Gene Expression*** • When a cell makes a protein as described earlier, scientists say that the gene for that protein has been "expressed." Cells can regulate gene expression to make the type of protein, in the amounts and at the rate, they need. Nearly all of the body's cells possess the genes for making all human proteins, but each type of cell makes only the proteins it needs. For example, only cells of the pancreas express the gene for insulin; in other cells, that gene is idle. Similarly, the cells of the pancreas do not make the protein hemoglobin, which is needed only by the red blood cells.

Recent research has unveiled some of the fascinating ways nutrients regulate gene expression and protein synthesis. These discoveries have begun to explain some of the relationships between nutrients, genes, and disease development. The benefits of polyunsaturated fatty acids in defending against heart disease, for example, are partially explained by their role in influencing gene expression for lipid enzymes. Later chapters provide additional examples of how nutrients influence gene expression.

IN SUMMARY

Cells synthesize proteins according to the genetic information provided by the DNA in the nucleus of each cell. This information dictates the order in which amino acids must be linked together to form a given protein. Sequencing errors occasionally occur, sometimes with significant consequences.

## Roles of Proteins

Whenever the body is growing, repairing, or replacing tissue, proteins are involved. Sometimes their role is to facilitate or to regulate; other times it is to become part of a structure. Yes, versatility is a key feature of proteins.

• ***As Building Materials*** • From the moment of conception, proteins form the building blocks of most body structures. For example, to build a bone or a tooth, cells first lay down a matrix of the protein collagen and then fill it with crystals of calcium, phosphorus, magnesium, fluoride, and other minerals.

The protein collagen is also the material of ligaments and tendons and the strengthening glue between the cells of the artery walls that enables the arteries to withstand the pressure of the blood surging through them with each heartbeat. Also made of collagen are scars that knit the separated parts of torn tissues together.

As old skin cells fall off, new cells made largely of protein grow from underneath to compensate. Cells in the deeper skin layers synthesize new proteins to go into hair and fingernails. GI tract cells are replaced every three days. Both inside and outside, then, the body constantly deposits protein into new cells that replace those that have been lost.

**matrix** (MAY-tricks): the basic substance that gives form to a developing structure; in the body, the formative *cells* from which teeth and bones grow.

**collagen** (KOL-ah-jen): the protein material from which connective tissues such as scars, tendons, ligaments, and the foundations of bones and teeth are made.

• ***As Enzymes*** • Digestive enzymes have appeared in every chapter since Chapter 3, but digestion is only one of the many processes enzymes facilitate. Enzymes not only break down substances, they also build substances and transform one substance into another. Figure 6-8 diagrams a synthesis reaction.

An analogy may help to clarify the role of enzymes. Enzymes are comparable to the clergy and judges who make and dissolve marriages. When a minister marries two people, they become a couple, with a new bond between them. They are joined together—but the minister remains unchanged. The minister represents synthetase enzymes that make large compounds from smaller ones. One minister can perform thousands of marriage ceremonies, just as one enzyme can perform billions of synthetic reactions.

Similarly, a judge who lets married couples separate may decree many divorces before retiring or dying. The judge represents enzymes that hydrolyze larger compounds to smaller ones; for example, the digestive enzymes. The point is that, like the minister and the judge, enzymes themselves are not altered by the reactions they facilitate. They are catalysts, permitting reactions to occur more quickly and efficiently than if substances depended on chance encounters alone.

**enzymes:** proteins that facilitate chemical reactions without being changed in the process; protein catalysts.

**synthetase** (SIN-the-tase): an enzyme that enables two or more substances to form a more complex structure.

Reminder: A *catalyst* facilitates chemical reactions without itself being changed in the process.

• ***An Example of Enzyme Action*** • The chemical structures in the margin and the paragraphs that follow provide an example of enzyme action. This single biochemical pathway illustrates how one compound encounters an enzyme, is converted to another compound that encounters another enzyme, and so forth until the final product is entirely different from the starting material. The details are offered only to give you insight into the kinds of processes that take place in the daily lives of the body's cells.

In the breakdown of glucose (a 6-carbon compound), enzymes add two phosphate groups, alter the arrangement of the atoms, and then split the molecule in half, leaving two 3-carbon compounds. One of these is compound A and the other is converted to compound A, so the two halves derived from glucose follow the same path from that point on.

Compound A floats around until it encounters an enzyme that recognizes it. This enzyme removes hydrogens from molecules of compound A. Without hydrogens, carbon and oxygen must form a double bond; thus compound B is created. Compound B is released from this enzyme and encounters another enzyme that removes an oxygen and substitutes an amino group in its place; the result is compound C. The next enzyme removes the phosphate group and replaces it with a hydrogen, leaving compound D.

The characteristics of compound D become apparent upon close examination and may not surprise some readers, but this example takes the process one step further before revealing the identity of compound D. Another enzyme, whose function is to remove $CH_2OH$ groups from molecules, forms compound E.

Compound E appeared earlier in this chapter. It has an amino group at one end, an acid group at the other, and a central carbon carrying two hydrogen atoms. It is the amino acid glycine (introduced in Figure 6-2).

Compound A

Compound B

Compound C

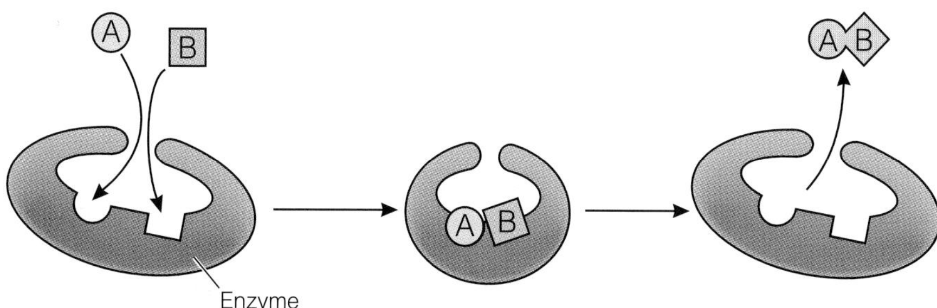


Two separate compounds, A and B, are attracted to the enzyme's active site, making a reaction likely.

The enzyme forms a complex with A and B.

The enzyme is unchanged, but A and B have formed a new compound, AB.

**Figure 6-8**
**Enzyme Action**

Each enzyme facilitates a specific chemical reaction. In this diagram, an enzyme enables two compounds to make a more complex structure, but the enzyme itself remains unchanged.

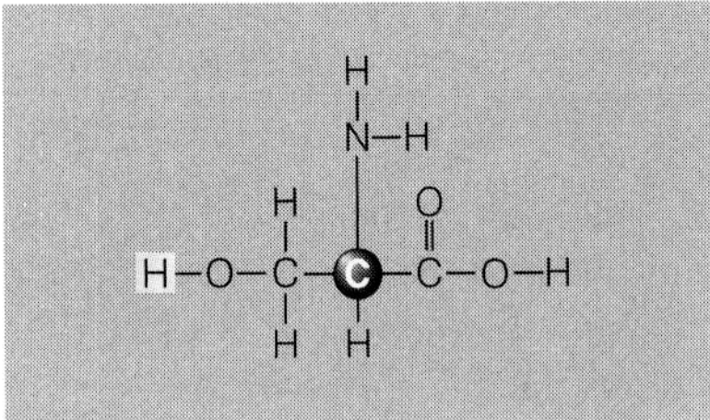

Compound D

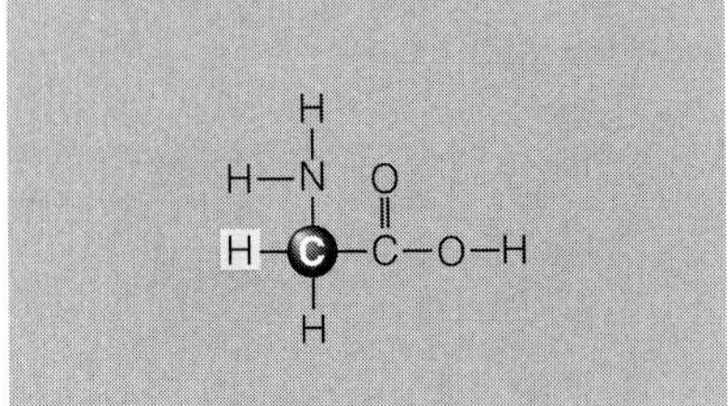

Compound E

Amazing! The cellular machinery started with a molecule of glucose (a derivative of dietary carbohydrate), made one small change after another, and transformed it into an amino acid (a member of the protein family). The lesson of this sequence of events is that the body can use glucose and nitrogen-containing compounds to make many of the amino acids needed to build body proteins. The nonessential amino acid glycine is just one example. Compound D, which precedes glycine on the pathway, is another example: the nonessential amino acid serine. Thus, among the thousands of tasks that enzymes perform, they even manufacture many of the nonessential amino acids they themselves are made of.

Perhaps you have realized by now that the protein story is circular. To follow the circle in nutrition, start with a person eating food proteins. The proteins are broken down by proteins (digestive enzymes) into amino acids. The amino acids enter the body cells, where proteins (synthetases) link them into long chains whose sequences are specified by DNA. The chains twist and fold forming proteins, some of which are enzymes. Some enzymes break apart compounds; others put compounds together. Day by day, in billions of reactions, these processes repeat themselves, and life goes on. Only living systems are capable of such self-renewal. A car cannot make another car; a toaster cannot fix another toaster. Only living creatures and the parts they are composed of—the cells—can duplicate and repair themselves.

• ***As Hormones*** • Cells can switch their protein machinery on or off in response to the body's needs. Often hormones do the switching, with marvelous precision. The body's many hormones are messenger molecules, and *some* hormones are proteins. Various glands in the body release hormones in response to changes in the internal environment. The blood carries the hormones to their target tissues, where they elicit the appropriate responses to restore normal conditions.

The hormone insulin provides a familiar example. When blood glucose rises, the pancreas releases its insulin. Insulin stimulates the cells' transport proteins to pump glucose into the cells faster than it can leak out. (After acting on the message, the cells destroy the insulin.) Then, as blood glucose falls, the pancreas reduces its insulin output. Many other proteins act as hormones, maintaining the distribution of hundreds of substances in the body (see Table 6-2).

• ***As Regulators of Fluid and Electrolyte Balance*** • Proteins help to maintain the body's fluid and electrolyte balance. As Figure 6-9 shows, the body's fluids are contained inside the blood vessels (intravascular), within the cells (intracellular), and between the cells (intercellular). Fluids can flow freely between these compartments, but the cells can't move fluids directly. They can manufacture proteins, though. Being large, proteins cannot pass freely across membranes; they are trapped on one side where they attract water. By making and keeping proteins, cells can retain fluids. Similarly, the cells can ship proteins out into the blood and intercellular spaces to maintain the fluid volume there. Should this system fail, too much fluid would collect outside the cells, causing edema.

Proteins help to regulate the composition of body fluids, as well as their quantity. Special transport proteins maintain equilibrium in the surrounding fluids by

**fluid and electrolyte balance:** maintenance of the proper types and amounts of fluid and minerals in each compartment of the body fluids (see also Chapter 12).

Minerals also help regulate fluid distribution; Chapter 12 provides more details.

**edema** (eh-DEEM-uh): the swelling of body tissue caused by excessive amounts of fluid in the interstitial spaces; seen in protein deficiency (among other conditions).

**Table 6-2**
**Examples of Hormones and Their Actions**

| Hormones | Actions |
|---|---|
| Growth hormone | Promotes growth. |
| Insulin and glucagon | Regulate blood glucose (see Chapter 4). |
| Thyroxin | Regulates the body's metabolic rate (see Chapter 8). |
| Calcitonin and parathormone | Regulate blood calcium (see Chapter 12). |
| Antidiuretic hormone | Regulates fluid and electrolyte balance (see Chapter 12). |

NOTE: *Hormones* are chemical messengers that are secreted by endocrine glands in response to altered conditions in the body. Each travels to one or more specific target tissues or organs, where it elicits a specific response. For descriptions of many hormones important in nutrition, see Appendix A.

moving molecules into and out of cells. Most of these proteins reside in cell membranes and act as "pumps," picking up compounds on one side of the membrane and depositing them on the other. In doing so, transport proteins enable cells to take up and release substances as needed. Each transport protein is specific for a certain compound or group of related compounds. Figure 6-10 illustrates how a membrane-bound transport protein helps to maintain the sodium and potassium concentrations in the fluids inside and outside of cells. The balance of these two electrolytes is critical to neural transmissions and muscle contractions; any disturbance triggers a major medical emergency. Such imbalances can cause irregular heartbeats, muscular weakness, kidney failure, and even death.

• *As Acid-Base Regulators* • Proteins also help to maintain the balance between acids and bases within the body fluids. Normal body processes continually produce acids and bases, which the blood carries to the kidneys and lungs for excretion. The challenge is to do this without upsetting the blood's acid-base balance.

In an acid solution, hydrogen ions abound; the more hydrogen ions, the more concentrated the acid. Proteins, which have negative charges on their surfaces, attract hydrogen ions, which have positive charges. By accepting and releasing hydrogen ions, proteins act as buffers, maintaining the acid-base balance of the blood and body fluids.

The blood's acid-base balance is tightly controlled. The extremes of acidosis and alkalosis lead to coma and death, largely because they denature working proteins. Disturbing a protein's shape renders it useless. To give just one example, hemoglobin, when denatured, loses its capacity to carry oxygen.

• *As Transporters* • Some transport proteins are not attached to membranes, but move about in the body fluids, carrying nutrients and other molecules. The protein hemoglobin carries oxygen from the lungs to the cells. The lipoproteins transport lipids around the body. Special proteins carry vitamins and minerals.

The transport of the mineral iron provides an especially good illustration of these proteins' specificity and precision. When iron enters an intestinal cell, it is captured by a protein that will not let go unless the body needs iron. Before leaving the cell to enter the bloodstream, iron is attached to a carrier protein. The carrier, in turn, can pass iron on to a storage protein in the bone marrow or other tissues, which will hold the iron until it is needed. When it is needed, iron is incorporated into proteins in the red blood cells and muscles that assist in oxygen transport and use.

• *As Antibodies* • Proteins also defend the body against disease. A virus—whether it is one that causes flu, smallpox, measles, or the common cold—enters the cells and multiplies there. One virus may produce 100 replicas of itself within an hour or so. Each replica can then burst out and invade 100 different cells, soon

**Figure 6-9**
**One Cell and Its Associated Fluids**

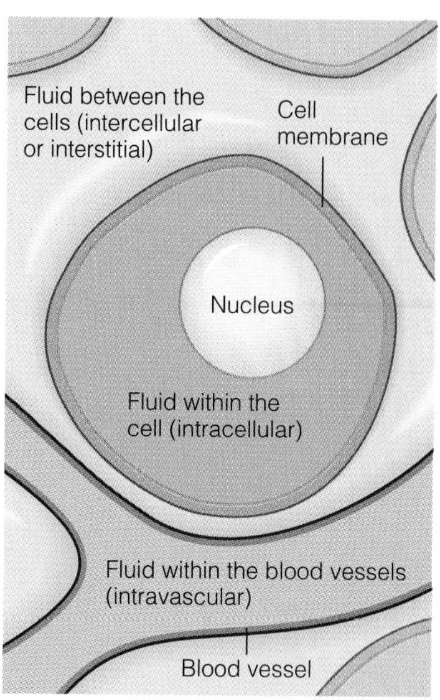


**acids:** compounds that release hydrogen ions in a solution.

**bases:** compounds that accept hydrogen ions in a solution.

**buffers:** compounds that help keep a solution's acidity or alkalinity constant.

**acidosis** (assi-DOE-sis): above-normal acidity in the blood and body fluids.

**alkalosis** (alka-LOE-sis): above-normal alkalinity (base) in the blood and body fluids.

The protein in the cells of the intestinal wall is **ferritin;** the carrier protein, **transferrin;** the storage protein, **ferritin** again; the red blood cell protein, **hemoglobin;** and the muscle cell protein, **myoglobin.**

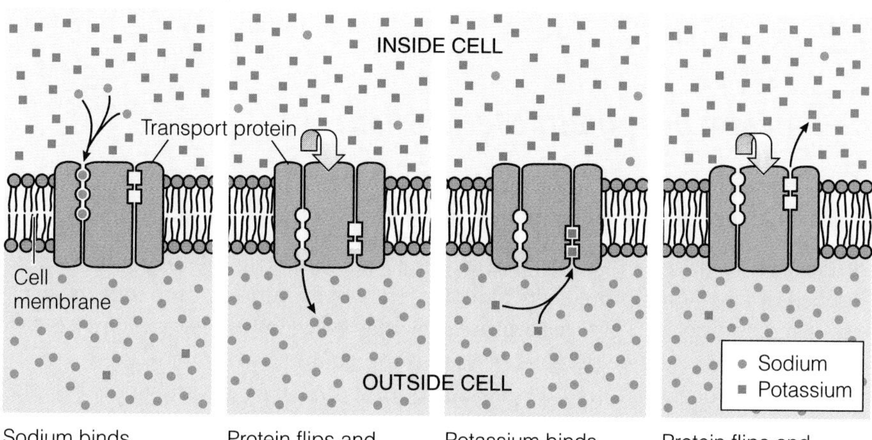


**Figure 6-10**
**Transport Proteins**

A transport protein within a cell membrane picks up substances on one side of the membrane and carries them to the other side without leaving the membrane. The substances being transported here are sodium and potassium. Maintaining a high concentration of potassium and a low concentration of sodium within the cells requires energy.

**antibodies:** large proteins of the blood and body fluids, produced by the immune system in response to the invasion of the body by foreign molecules (usually proteins called antigens); antibodies combine with and inactivate the foreign invaders, thus protecting the body.

**antigen:** a substance that elicits the formation of antibodies or an inflammation reaction from the immune system. A bacterium, a virus, a toxin, and a protein in food that causes allergy are all examples of antigens.

**immunity:** the body's ability to recognize and eliminate foreign invaders; see Chapter 18.

yielding 10,000 virus particles, which invade 10,000 cells. Left free to do their worst, they will soon overwhelm the body with disease.

Fortunately, when the body detects invaders, it manufactures antibodies, giant protein molecules designed specifically to combat them. The antibodies work so swiftly and efficiently that in a normal, healthy individual, most diseases never have a chance to get started. Without sufficient protein, though, the body cannot maintain its resistance to disease.

Each antibody is designed to destroy just one invader. Once the body has manufactured antibodies against a particular antigen (such as the measles virus), it remembers how to make them. Consequently, the next time the body encounters that same invader, it will produce antibodies even more quickly. In other words, the body develops a molecular memory, known as immunity.

• ***Other Roles*** • As mentioned earlier, proteins form integral parts of most body structures such as skin, muscles, and bones. They also participate in some of the body's most amazing activities such as blood clotting and vision. When a tissue is injured, a rapid chain of events leads to the production of fibrin, a stringy, insoluble mass of protein fibers that forms a clot from liquid blood. Later, more slowly, the protein collagen forms a scar to replace the clot and permanently heal the cut. The light-sensitive pigments in the cells of the retina are molecules of the protein opsin. Opsin responds to light by changing its shape, thus initiating the nerve impulses that convey the sense of sight to higher brain centers.

IN SUMMARY

The protein functions discussed here are summarized in the accompanying table. They are only a few of the many roles proteins play, but they convey some sense of the immense variety of proteins and their importance in the body.

**Summary of Proteins' Functions**

| | |
|---|---|
| **Growth and maintenance** | Proteins form integral parts of most body structures such as skin, tendons, membranes, muscles, organs, and bones. As such, they support the growth and repair of body tissues. |
| **Enzymes** | Proteins facilitate chemical reactions. |
| **Hormones** | Proteins regulate body processes. (Some, but not all, hormones are made of protein.) |
| **Antibodies** | Proteins inactivate foreign invaders, thus protecting the body against diseases. |
| **Fluid and electrolyte balance** | Proteins help to maintain the volume and composition of body fluids. |
| **Acid-base balance** | Proteins help maintain the acid-base balance of body fluids by acting as buffers. |
| **Transportation** | Proteins transport substances, such as lipids, vitamins, minerals, and oxygen, around the body. |
| **Energy** | Proteins provide some fuel for the body's energy needs. |

## *A Preview of Protein Metabolism*

This section previews protein metabolism; Chapter 7 provides a full description. Cells have several metabolic options, depending on their amino acid needs.

**protein turnover:** the degradation and synthesis of endogenous protein.

**endogenous** (en-DODGE-eh-nus) **protein:** the protein in the body. In contrast, protein in foods is **exogenous** (eks-ODGE-eh-nus) **protein.**
- **endo** = within
- **gen** = arising
- **exo** = outside (the body)

• ***Protein Turnover*** • Within each cell, proteins are constantly being made and broken down. When proteins break down, they free amino acids to join the general circulation. Some of these amino acids may be promptly recycled into other proteins; others may be stripped of their nitrogen and used for energy. Together the constant degradation and synthesis of body proteins are known as protein turnover,[4] and the protein that participates in this flux is called endogenous protein.

• ***Nitrogen Balance*** • If the body maintains the same *amount* of protein in its tissues from day to day, it is in nitrogen balance. If the body adds protein, nitrogen status becomes positive; if it loses protein, nitrogen status becomes negative.

Normally, healthy adults receive enough protein to meet their needs, and they dispose of any excess. Their nitrogen intake equals their nitrogen output, and they are said to be in zero nitrogen balance, or nitrogen equilibrium. Nitrogen status is positive in growing infants and children, pregnant women, and people recovering from protein deficiency or illness; their nitrogen intake exceeds their nitrogen output. They are building protein tissues—adding new blood, bone, skin, and muscle cells to their bodies. In contrast, nitrogen status is negative in people who are starving or suffering other severe stresses such as burns, injuries, infections, and fever; their nitrogen output exceeds their nitrogen intake. During these times, the body loses protein as it breaks down body proteins for energy.

**nitrogen balance:** the amount of nitrogen consumed (N in) as compared with the amount of nitrogen excreted (N out) in a given period of time.*

Nitrogen equilibrium (zero nitrogen balance): N in = N out.
Positive nitrogen: N in > N out.
Negative nitrogen: N in < N out.

• ***Using Amino Acids to Make Proteins or Nonessential Amino Acids*** • Cells can assemble amino acids into the proteins they need to do their work. If a particular nonessential amino acid is not readily available, cells can dismantle another amino acid and combine the amino group with carbon fragments from glucose to make the needed one. If an essential amino acid is missing, the body may break down some of its own proteins to obtain it.

• ***Using Amino Acids to Make Other Compounds*** • Cells can also use amino acids to make other compounds. For example, the amino acid tyrosine is used to make the neurotransmitters norepinephrine and epinephrine, which relay nervous system messages throughout the body. Tyrosine can also be made into the pigment melanin, which is responsible for brown hair, eye, and skin color, or into the hormone thyroxin, which helps to regulate the metabolic rate. For another example, the amino acid tryptophan serves as a precursor for the neurotransmitter serotonin and the vitamin niacin.

**neurotransmitters:** chemicals that are released at the end of a nerve cell when a nerve impulse arrives there; they diffuse across the gap to the next cell and alter the membrane of that second cell to either inhibit or excite it.

• ***Using Amino Acids for Energy*** • Even though amino acids are needed to do the work that only they can perform—build vital proteins—they will be sacrificed to provide energy and glucose if need be. Without energy, cells die; without glucose, the brain and nervous system falter. When glucose or fatty acids are limited, cells are forced to use amino acids for energy and glucose. The body does not make a specialized storage form of protein as it does for carbohydrate and fat. Glucose is stored as glycogen in the liver and fat as triglycerides in adipose tissue, but protein in the body is available only as the working and structural components of the tissues. When the need arises, the body dismantles its tissue proteins and uses them for energy. Thus, over time, energy deprivation (starvation) always incurs wasting of lean body tissue as well as fat loss. An adequate intake of carbohydrates and fats spares amino acids from being used for energy and allows them to perform their unique roles.

Reminder: The making of glucose from noncarbohydrate sources such as amino acids is *gluconeogenesis.* The action of carbohydrate and fat in providing enough energy to allow amino acids to be used to build body proteins is known as the *protein-sparing action* of carbohydrate and fat.

• ***Deaminating Amino Acids*** • When amino acids are broken down (as occurs when they are used for energy), they are first deaminated—stripped of their nitrogen-containing amino groups. Deamination produces ammonia, which the cells release into the bloodstream. The liver picks up the ammonia, converts it into urea (a less toxic compound), and returns the urea to the blood. The kidneys filter urea out of the blood; thus the amino nitrogen ends up in the urine. Urea is produced from both exogenous and endogenous amino acids. The remaining carbon fragments may enter a number of metabolic pathways—for example, they may be used to make fat.

**deamination** (dee-AM-eh-NAY-shun): removal of the amino ($NH_2$) group from a compound such as an amino acid.

Urea metabolism is described in Chapter 7.

• ***Using Amino Acids to Make Fat*** • If a person eats more protein than the body needs, the amino acids are deaminated, the nitrogen is excreted, and the remaining

*The genetic materials DNA and RNA contain nitrogen, but the quantity is insignificant compared with the amount in protein. The average amino acid weighs about 6.25 times as much as the nitrogen it contains, so scientists can estimate the amount of protein in a sample of food, body tissue, or other material by multiplying the weight of the nitrogen in it by 6.25.

carbon fragments are converted to fat and stored for later use.* In this way, valuable, expensive, protein-rich foods can contribute to obesity.

IN SUMMARY

Proteins are constantly being synthesized and broken down as needed. The body's assimilation of amino acids into proteins and release of amino acids via protein degradation and excretion can be tracked by measuring nitrogen balance, which should be positive during growth and steady in adulthood. An energy deficit or an inadequate protein intake may force the body to use amino acids as fuel, creating a negative nitrogen balance. Protein eaten in excess of need is degraded and stored as body fat.

# Protein in Foods

In the United States, where nutritious foods are abundant, people eat protein in such large quantities that even if its amino acid balance is not perfect, they receive all the amino acids they need. Where people eat only marginal amounts of protein-rich foods, however, the *quality* of the protein becomes crucial to their health. Hence, the protein quality of the diet is of great concern when making nutrition recommendations in countries where malnutrition is widespread.

## *Protein Quality*

Food proteins that provide an unbalanced assortment of amino acids are poor-quality proteins. In countries where food is scarce, or where the people receive marginal or inadequate amounts of protein, the quality of the dietary protein determines, in large part, how well the children grow and how well the adults maintain their health.

***Black beans and rice, a favorite Hispanic combination, together provide a full array of amino acids.***

• ***Limiting Amino Acids*** • To make proteins, a cell must have all the needed amino acids available simultaneously. The liver can produce any nonessential amino acid that may be in short supply so that the cells can continue linking amino acids into protein strands. If an essential amino acid is missing, though, a cell must dismantle its own proteins to obtain it. Therefore, to prevent protein breakdown, dietary protein must supply at least the nine essential amino acids plus enough nitrogen-containing amino groups and energy for the synthesis of the others. If the diet supplies too little of any essential amino acid, protein synthesis will be limited. The body makes whole proteins only; if one amino acid is missing, the others cannot form a "partial" protein. The body has no storage site for extra amino acids and is forced to either waste them or use them for another purpose. An essential amino acid supplied in less than the amount needed to support protein synthesis is called a *limiting* amino acid.

**limiting amino acid:** the essential amino acid found in the shortest supply relative to the amounts needed for protein synthesis in the body. Four amino acids are most likely to be limiting:
- Lysine.
- Methionine.
- Threonine.
- Tryptophan.

**complete protein:** a dietary protein containing all the essential amino acids in relatively the same amounts that human beings require; it may also contain nonessential amino acids.

• ***Complete Protein*** • A complete dietary protein contains all the essential amino acids in relatively the same amounts as human beings require; it may or may not contain all the nonessential amino acids. Generally, proteins derived from animals (meat, fish, poultry, cheese, eggs, and milk) are complete, although gelatin is an exception (it lacks tryptophan and cannot support growth and health as a diet's sole protein). Proteins from plants (vegetables, grains, and legumes) have more

*Chemists sometimes classify amino acids according to the destinations of their carbon fragments after deamination. If the fragment leads to the production of glucose, the amino acid is called "glucogenic"; if it leads to the formation of ketone bodies, fats, and sterols, the amino acid is called "ketogenic." There is no sharp distinction between glucogenic and ketogenic amino acids, however. A few are both; most are considered glucogenic; only one (leucine) is clearly ketogenic.

diverse amino acid patterns, and some tend to be limiting in one or more essential amino acids. Some plant proteins (for example, corn protein) are notoriously incomplete. Others (for example, soy protein) are complete.

• ***Complementary Proteins*** • In general, plant proteins are of lower quality than animal proteins, and plants also offer less protein per unit (either weight or measure) of food. For this reason, many vegetarians combine plant-protein foods with different but complementary amino acid patterns to improve the quality of proteins in their diets. This strategy is called mutual supplementation, and it yields complementary proteins that contain all the essential amino acids in quantities sufficient to support health. The protein quality of the combination is greater than for either food alone (see Figure 6-11).

Many people have long believed that mutual supplementation at every meal is critical to protein nutrition. For most healthy vegetarians, though, it is not necessary to balance amino acids at each meal when protein intake is varied and energy intake is sufficient.[5] Vegetarians can receive all the amino acids they need over the course of a day, if they eat a variety of grains, legumes, seeds, nuts, and vegetables. Protein deficiency will develop, however, when fruits and certain vegetables make up the core of the diet, severely limiting the *quantity* and *quality* of protein. Highlight 6 shows how to plan a nutritious vegetarian diet.

• ***Digestibility*** • Ideally, a protein is both complete and easily digestible, so that enough amino acids are available for protein synthesis. Such a protein is a high-quality protein. Digestibility depends on a protein's configuration, other foods eaten with it, and reactions that influence the release of amino acids.

• ***Reference Protein*** • One of the most complete and digestible proteins is egg protein. Until the early 1990s, egg protein was used as the standard for measuring protein quality; it was assigned a value of 100, and the quality of other food proteins was determined based on how they compared with egg. Such a standard is called a reference protein. Now the Food and Agriculture Organization (FAO) of the United Nations and the World Health Organization (WHO) have established a new standard for the reference protein: the essential amino acid requirements of preschool-age children.

**Figure 6-11**
**An Example of Mutual Supplementation**

In general, legumes provide plenty of isoleucine (Ile) and lysine (Lys), but fall short in methionine (Met) and tryptophan (Trp). Grains have the opposite strengths and weaknesses, making them a perfect match for legumes.

| | Ile | Lys | Met | Trp |
|---|---|---|---|---|
| Legumes | | | | |
| Grains | | | | |
| Together | | | | |

**mutual supplementation:** the strategy of combining two protein foods in a meal so that each food provides the essential amino acid(s) lacking in the other. Mutual supplementation is the dietary strategy that brings complementary proteins together in a meal.

**complementary proteins:** two or more proteins whose amino acid assortments complement each other in such a way that the essential amino acids missing from one are supplied by the other.

**protein digestibility:** a measure of the amount of amino acids absorbed from a given protein intake.

**high-quality protein:** an easily digestible, complete protein.

**reference protein:** a standard against which to measure the quality of other proteins.

IN SUMMARY

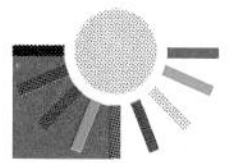

A diet short in any of the essential amino acids limits protein synthesis. The best guarantee of amino acid adequacy is to eat foods containing complete proteins or mixtures of foods containing incomplete proteins so that each can supply the amino acids missing in the other. Vegetarians can meet their protein needs by eating a variety of whole grains, legumes, seeds, nuts, and vegetables.

*Vegetarians obtain their protein from whole grains, legumes, nuts, vegetables, and, in some cases, eggs and milk products.*

# *Measures of Protein Quality*

Researchers have developed several methods for evaluating the quality of food proteins. The object of all of these methods is to identify high-quality proteins—that is, proteins that are easily digestible and contain all of the essential amino acids in relatively the same proportion as human beings require. Proteins that are low in an essential amino acid cannot, by themselves, support protein synthesis. The following paragraphs briefly describe these measures; Appendix J provides more detail.

• ***Amino Acid Scoring*** • The simplest way to evaluate a food protein's quality is to determine its amino acid composition and compare it with a reference protein. Scientists can easily identify the limiting amino acid—it is the one that falls shortest compared with the reference. If the test protein's limiting amino acid is 70 percent of the amount found in the reference protein, it receives a score of 70. Such calculations fail to estimate digestibility, however.

**amino acid scoring:** a method of evaluating protein quality by comparing a test protein's amino acid pattern with that of a reference protein; sometimes called **chemical scoring.**

**biological value (BV):** the amount of protein nitrogen that is retained for growth and maintenance, expressed as a percentage of the protein nitrogen that has been digested and absorbed; a measure of protein quality.

**net protein utilization (NPU):** the amount of protein nitrogen that is retained from a given amount of protein nitrogen eaten; a measure of protein quality.

**protein efficiency ratio (PER):** a measure of protein quality assessed by determining how well a given protein supports weight gain in growing rats; used to establish the protein quality for infant formulas and baby foods.

**protein digestibility–corrected amino acid score (PDCAAS):** a measure of protein quality assessed by comparing the amino acid score of a food protein with the amino acid requirements of preschool-age children and then correcting for the true digestibility of the protein; recommended by the FAO/WHO and used to establish protein quality of foods for Daily Value percentages on food labels.

• ***Biological Value*** • The biological value (BV) of a protein measures its efficiency in supporting the body's needs. Scientists feed a given food protein to experimental animals as the sole protein in their diet and measure the animals' retention and loss of nitrogen. The more nitrogen retained, the higher the protein quality. (Recall that when an essential amino acid is missing, protein synthesis stops, and the remaining amino acids are deaminated and the nitrogen excreted.)

• ***Net Protein Utilization*** • Like BV, net protein utilization (NPU) measures nitrogen retention. Instead of measuring retention of absorbed nitrogen (as in BV), NPU measures retention of food nitrogen.

• ***Protein Efficiency Ratio*** • The protein efficiency ratio (PER) measures the weight gain of a growing animal and compares it to the animal's protein intake. Until recently, the PER was generally accepted in the United States and Canada as the official method for assessing protein quality.

• ***PDCAAS*** • The protein digestibility–corrected amino acid score, or PDCAAS, method compares the amino acid contents of a protein with human amino acid requirements and corrects for digestibility. The protein's amino acid score is determined as described earlier, and then it is compared against the amino acid requirements of preschool-age children. This comparison reveals the most limiting amino acid. The rationale behind using the requirements of this age group is that if a protein will effectively support a young child's growth and development, then it will meet or exceed the requirements of older children and adults. Thus the PDCAAS method evaluates dietary protein quality for all age groups except infants. (The PER method described earlier is used to evaluate proteins for infants.)

To arrive at the PDCAAS, the amino acid score is multiplied by the food's protein digestibility percentage. Because the digestibility of many foods is similar in human beings and in rats, values for protein digestibility in rats are commonly used. Appendix J provides an example of how to calculate the PDCAAS and a table that lists the PDCAAS values of selected foods.

## *Protein Regulations for Food Labels*

The Food and Drug Administration's (FDA) labeling regulations use the PDCAAS method to assess protein quality in foods intended for people over age one. For infant formulas and baby foods, the PER method using casein as a standard is used to measure protein quality.

All food labels must state the *quantity* of protein in grams. The "% Daily Value" for protein is not mandatory on all labels, but is required whenever a food makes a protein claim or is intended for consumption by children under four years old.* Whenever the Daily Value percentage is declared, researchers must calculate in the *quality* of the protein by using the PDCAAS method. Thus when a % Daily Value is stated for protein, it reflects both quantity and quality.

IN SUMMARY

The quality of protein is measured by its amino acid content, its digestibility, and its ability to support growth. Such measures are of great importance in dealing with malnutrition worldwide, but in the United States and Canada, where protein deficiency is not common, protein quality scores of individual foods deserve little emphasis.

*For labeling purposes, the Daily Values for protein are as follows: for infants, 14 grams; for children under age four, 16 grams; for older children and adults, 50 grams; for pregnant women, 60 grams; and for lactating women, 65 grams.

# Health Effects and Recommended Intakes of Protein

As you know by now, protein is indispensable to life. It should come as no surprise that protein deficiency can have devastating effects on people's health. But like the other nutrients, protein in excess can also be harmful. This section examines the health effects and recommended intakes of protein.

*Donated food saves some people from starvation, but it is usually insufficient to meet nutrient needs or even to provide a full belly for every person who is hungry.*

## Protein-Energy Malnutrition

When people are deprived of protein, energy, or both, the result is protein-energy malnutrition (PEM). Although PEM touches many adult lives, it most often strikes early in childhood. It is the most widespread form of malnutrition in the world, afflicting over 500 million children. Most of the 33,000 children who die each day are malnourished.[6]

Inadequate food intake leads to poor growth in children and to weight loss and wasting in adults. Children who are thin for their height may be suffering from acute PEM (recent severe food deprivation), whereas children who are short for their age have experienced chronic PEM (long-term food deprivation). Poor growth due to PEM is easy to overlook because a small child may look quite normal, but it is the most common sign of malnutrition.

PEM is most prevalent in Africa, Central America, South America, the Middle East, and East and Southeast Asia. In the United States, homeless people and those living in substandard housing in inner cities and rural areas have been diagnosed with PEM. In addition to those living in poverty, elderly people who live alone and adults who are addicted to drugs and alcohol are frequently victims of PEM. Adult PEM is also seen in people hospitalized with infections such as AIDS or tuberculosis; infections deplete body proteins, demand extra energy, induce nutrient losses, and alter metabolic pathways. PEM is also common in those suffering from the eating disorder anorexia nervosa. Prevention emphasizes frequent, nutrient-dense, energy-dense meals and, equally important, resolution of the underlying causes of PEM—poverty, infections, and illness.

**protein-energy malnutrition (PEM),** also called **protein-kcalorie malnutrition (PCM):** a deficiency of both protein and energy; the world's most widespread malnutrition problem, including kwashiorkor, marasmus, and instances in which they overlap.

**acute PEM:** protein-energy malnutrition caused by recent severe food restriction; characterized in children by thinness for height (wasting).

**chronic PEM:** protein-energy malnutrition caused by long-term food deprivation; characterized in children by short height for age (stunting).

 WHO
www.who.org
Search for Protein energy malnutrition

• ***Classifying PEM*** • PEM occurs in two forms: marasmus and kwashiorkor, which differ in their clinical features (see Table 6-3). Appendix E describes the assessment and classification of PEM, and the following paragraphs present the three clinical syndromes—marasmus, kwashiorkor, and the combination of the two.

**marasmus** (ma-RAZ-mus): a form of PEM that results from a severe deprivation, or impaired absorption, of energy, protein, vitamins, and minerals.

• ***Marasmus*** • Marasmus reflects a severe deprivation of food over a long time (chronic PEM) and therefore is caused by an inadequate energy *and* protein intake (and by inadequate essential fatty acids, vitamins, and minerals as well). Marasmus occurs most commonly in children from 6 to 18 months of age in all the overpopulated urban slums of the world. Children in impoverished nations simply do not have enough to eat and subsist on diluted cereal drinks that supply scant energy and protein of low quality; such food can barely sustain life, much less support growth. Consequently, marasmic children look like little old people—just skin and bones.

Without adequate nutrition, muscles, including the heart, waste and weaken. Because the brain normally grows to almost its full adult size within the first two years of life, marasmus impairs brain development and learning ability. Reduced synthesis of key hormones slows metabolism and lowers body temperature. There is little or no fat under the skin to insulate against cold. Hospital workers find that children with marasmus need to be wrapped up and kept warm. They also need love because they have often been deprived of parental attention as well as food.

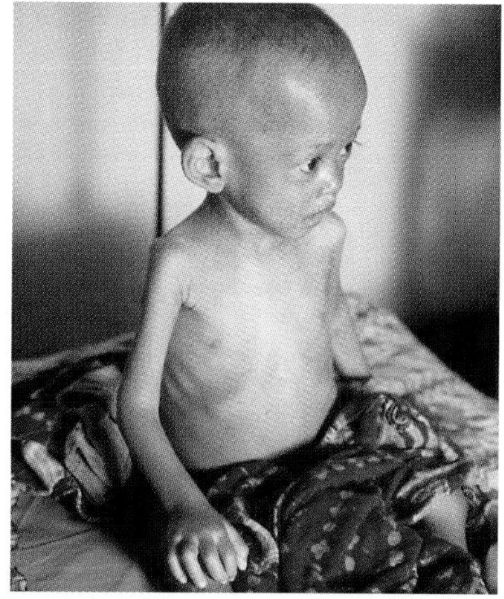

*The extreme loss of muscle and fat characteristic of marasmus is apparent in this child's "matchstick" arms and legs.*

**Table 6-3**
**Features of Marasmus and Kwashiorkor in Children**

Separating PEM into two classifications oversimplifies the condition, but at the extremes, marasmus and kwashiorkor exhibit marked differences. Marasmus-kwashiorkor mix presents symptoms common to both marasmus and kwashiorkor. In all cases, children are likely to develop diarrhea, infections, and multiple nutrient deficiencies.

| Marasmus | Kwashiorkor |
|---|---|
| Infancy (less than 2 yr) | Older infants and young children (1 to 3 yr) |
| Severe deprivation, or impaired absorption, of protein, energy, vitamins, and minerals | Inadequate protein intake or, more commonly, infections |
| Develops slowly; chronic PEM | Rapid onset; acute PEM |
| Severe weight loss | Some weight loss |
| Severe muscle wasting, with fat | Some muscle wasting, with retention of some body fat |
| Growth: <60% weight-for-age | Growth: 60 to 80% weight-for-age |
| No detectable edema | Edema |
| No fatty liver | Enlarged fatty liver |
| Anxiety, apathy | Apathy, misery, irritability, sadness |
| Good appetite possible | Loss of appetite |
| Hair is sparse, thin, and dry; easily pulled out | Hair is dry and brittle; easily pulled out; changes color; becomes straight |
| Skin is dry, thin, and easily wrinkles | Skin develops lesions |

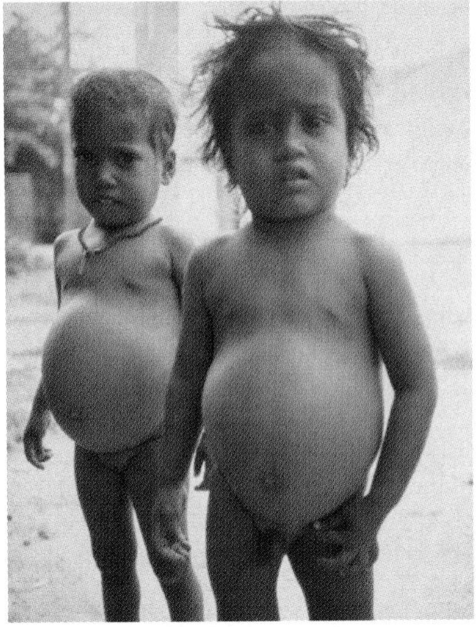

***The edema and enlarged liver characteristic of kwashiorkor is apparent in these children's swollen bellies. Malnourished children commonly have an enlarged abdomen from parasites as well.***

**kwashiorkor** (kwash-ee-OR-core, kwash-ee-or-CORE): a form of PEM that results either from inadequate protein intake or, more commonly, from infections.

**aflatoxin:** a potent cancer-causing toxin produced by the mold *Aspergillus flavus* that infects grains and peanuts. The USDA tests grains and peanuts grown in the United States for aflatoxin contamination.

Reminder: *Edema* is the swelling of body tissue caused by excessive fluid in the interstitial spaces, seen in protein deficiency (among other conditions).

The starving child faces this threat to life by engaging in as little activity as possible—not even crying for food. The body musters all its forces to meet the crisis, so it cuts down on any expenditure of protein not needed for the functioning of the heart, lungs, and brain. Growth ceases; the child is no larger at age four than at age two. Digestive enzymes are in short supply, the GI tract lining deteriorates, and absorption fails. The child can't assimilate what little food is eaten.

• ***Kwashiorkor*** • Kwashiorkor typically reflects a sudden and recent deprivation of food (acute PEM). Kwashiorkor was originally a Ghanaian word meaning "the evil spirit that infects the first child when the second child is born." When a mother who has been nursing her first child bears a second child, she weans the first child and puts the second one on the breast. The first child, suddenly switched from nutrient-dense, protein-rich breast milk to a starchy, protein-poor cereal, soon begins to sicken and die. Kwashiorkor typically sets in between 18 months and two years.

Kwashiorkor usually develops rapidly as a result of protein deficiency or, more commonly, is precipitated by an illness such as measles or other infection.[7] Possibly, kwashiorkor is a form of food poisoning superimposed on malnutrition. One supporting piece of evidence is that kwashiorkor seems to appear only in rainy, tropical communities. Many temperate regions have experienced widespread famine, yet have not had kwashiorkor. Another clue is that under hot, humid conditions, a common mold, *Aspergillus flavus,* produces aflatoxin, a toxin that inhibits protein synthesis. When malnourished children are forced to eat moldy grain for lack of other food, their weakened bodies cannot defend against the toxin.

The loss of weight and body fat is usually not as severe in kwashiorkor as in marasmus, but there may be some muscle wasting. Proteins and hormones that previously maintained fluid balance diminish, and fluid leaks into the interstitial spaces. The child's limbs and face become swollen with edema, a distinguishing feature of kwashiorkor. Also contributing to the edema is the inflammatory response that accompanies kwashiorkor.[8] The lack of the protein carriers that transport fat out of the liver causes the belly to bulge with a fatty liver. The fatty liver lacks enzymes to clear poisons from the body, so their toxic effects are prolonged. Without sufficient tyrosine to make melanin, the child's hair loses its color; inadequate protein synthesis leaves the skin patchy and scaly, often with sores that fail to heal. The lack of proteins to carry or store iron leaves iron free. Unbound iron is common in children with kwashiorkor and may contribute to their illnesses and deaths by promoting bacterial growth and free-radical damage.[9]

Free-radical damage is discussed fully in Highlight 11.

• ***Marasmus-Kwashiorkor Mix*** • The combination of marasmus and kwashiorkor is characterized by the edema of kwashiorkor with the wasting of marasmus. Most often, the child is suffering the effects of both malnutrition and infections. Some researchers believe that kwashiorkor and marasmus are two stages of the same disease. They point out that kwashiorkor and marasmus often exist side by side in the same community where children consume the same diet. They note that a child who has marasmus can later develop kwashiorkor. Some research indicates that marasmus represents the body's adaptation to starvation and that kwashiorkor develops when adaptation fails.

• ***Infections*** • In PEM, antibodies to fight off invading bacteria are degraded to provide amino acids for other uses, leaving the malnourished child vulnerable to infections. Blood proteins, including hemoglobin, are no longer synthesized, so the child becomes anemic and weak. Dysentery, an infection of the digestive tract, causes diarrhea, further depleting the body of nutrients. In the marasmic child, once infection sets in, kwashiorkor often follows.[10]

**dysentery** (DISS-en-terry): an infection of the digestive tract that causes diarrhea.

The combination of infections, fever, electrolyte imbalances, and anemia often leads to heart failure and occasionally sudden death. Infections combined with malnutrition are responsible for two-thirds of the deaths of young children in developing countries. Measles, which might make a healthy child sick for a week or two, kills a child with PEM within two or three days.

• ***Rehabilitation*** • If caught in time, the life of a starving child may be saved by careful nutrition therapy. Diarrhea will have depleted the body's potassium and disturbed other electrolyte balances. Careful correction of fluid and electrolyte imbalances usually raises the blood pressure and strengthens the heartbeat. After the first 24 to 48 hours, protein and food energy may be given in small quantities, with intakes gradually increased as tolerated.

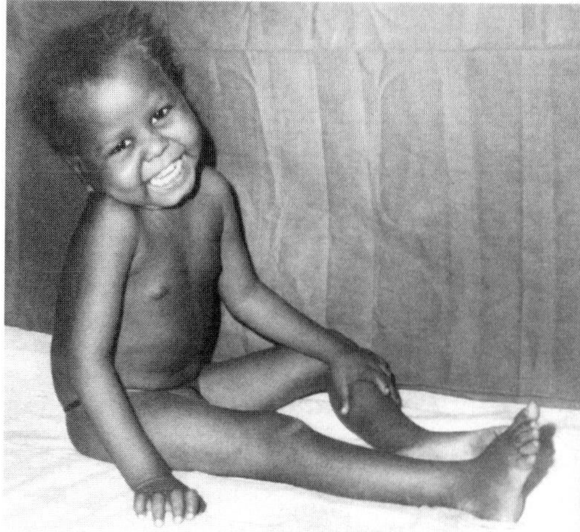

***Given appropriate nutrition care, this child has successfully recovered from kwashiorkor.***

Experts assure us that we possess the knowledge, technology, and resources to end hunger. Programs that have involved the local people in the process of identifying problems and devising solutions have met with some success. But until those who have the food, technology, and resources make fighting hunger a priority, the war on hunger will not be won (see Chapter 20 for more on hunger).

## *Health Effects of Protein*

While many of the world's people struggle to obtain enough food energy and protein, in developed countries both are so abundant that problems of excess are seen. Overconsumption of protein offers no benefits and may pose health risks.

The relationships between excess protein and chronic diseases are not clearly evident. Population studies have difficulty determining whether diseases correlate with animal proteins or with their accompanying saturated fats, for example. Studies that rely on data from vegetarians must sort out the many lifestyle factors, other than a "no-meat diet," that might explain relationships between protein and health.

• ***Heart Disease*** • As mentioned, foods rich in animal protein tend to be rich in saturated fats. Consequently, it is not surprising to find a correlation between animal-protein intake and heart disease, although no independent effect has been demonstrated. On the other hand, substituting soy protein for animal protein lowers blood cholesterol, especially in those with high blood cholesterol.[11]

Recent research suggests that the amino acid homocysteine may be an independent risk factor for heart disease.[12] When compared with others, men with elevated homocysteine were three times as likely to have heart attacks.[13] Researchers do not yet know the exact role homocysteine plays in heart disease, nor do they understand what raises homocysteine in the blood, but they are following several leads in pursuit of the answers.[14] Interestingly, blacks seem to metabolize homocysteine more efficiently than whites, which may partially explain the lower incidence of heart disease among blacks despite their high blood pressure and smoking habits.[15] Coffee's role in heart disease has been controversial, but recent research suggests it is among the most influential factors in raising homocysteine, which may explain some of the adverse health effects of heavy consumption.[16] Homocysteine is also elevated with suboptimal intakes of B vitamins and can usually be corrected with vitamin $B_{12}$, vitamin $B_6$, and folate supplements.[17] Whether a high intake of these vitamins reduces the risk of heart attacks remains unclear, but preliminary research seems to suggest that it does.[18]

• ***Cancer*** • As in heart disease, the effects of protein and fats cannot be easily separated. Population studies suggest a correlation between high intakes of animal proteins and some types of cancer (notably, cancer of the colon, breast, kidneys, pancreas, and prostate).

Other risk factors for adult bone loss (osteoporosis) are sex, age, and race, as Highlight 12 explains.

• ***Adult Bone Loss (Osteoporosis)*** • Do high protein intakes accelerate bone loss? Calcium excretion rises as protein intake increases. Whether excess protein depletes the bones of their chief minerals is controversial and may depend upon the ratio of calcium to protein intakes.[19] An ideal ratio has not been established, but a young woman whose intake meets recommendations for both nutrients has a calcium-to-protein ratio of more than 20 to 1 (milligrams to grams). For most women in the United States, however, average calcium intakes are lower and protein intakes are higher, yielding a 9-to-1 ratio, which may produce calcium losses that compromise bone health.[20] In contrast, moderate increases in physical activity and calcium intake may protect against such losses. In establishing calcium recommendations, the Committee on Dietary Reference Intakes considered protein's effect on calcium retention, but did not find sufficient evidence to warrant an adjustment.[21]

• ***Weight Control*** • Protein-rich foods are often fat-rich foods that contribute to obesity with its accompanying health risks. As Highlight 8 explains, weight-loss gimmicks that encourage a high-protein diet are rarely effective; overweight people have better success with diets that provide adequate protein, minimal fat, and ample energy from carbohydrates. The higher a person's intake of protein-rich foods such as meat and milk, the more likely that fruits, vegetables, and grains will be crowded out, making the diet inadequate in other nutrients.

IN SUMMARY

Protein deficiencies arise from both energy-poor and protein-poor diets and lead to the devastating diseases of marasmus and kwashiorkor. Together, these diseases are known as PEM (protein-energy malnutrition), the major form of malnutrition causing death in children worldwide. Excesses of food energy and protein offer no advantage; in fact, overconsumption of protein-rich foods may incur health risks as well.

## *Recommended Intakes of Protein*

As mentioned earlier, the body continuously breaks down and loses its proteins and cannot store amino acids. To replace protein, the body needs dietary protein for two reasons: first, food protein is the only source of the *essential* amino acids; and second, it is the only practical source of *nitrogen* with which to build the nonessential amino acids and other nitrogen-containing compounds.

The *Diet and Health* report recommends that people's fat intakes should contribute 30 percent or less of total food energy, and carbohydrate, 55 percent or more—which leaves about 15 percent for protein. Current intakes in the United States and Canada, though higher than recommendations, do not seem to be high enough to cause harm. The *Diet and Health* report advises people to maintain moderate protein intakes—between the RDA and twice the RDA.

• ***Protein RDA*** • The protein RDA for healthy adults is 0.8 grams per kilogram of appropriate body weight per day. For infants and children, the RDA is higher. When compared to total energy intake, however, the protein RDA for infants and children is similar to that for adults as Table 6-4 shows. The RDA generously covers the needs for replacing worn-out tissue, so it increases for larger people; it also covers the needs for building new tissue during growth, so it increases for infants, children, and pregnant women. The accompanying "How to" shows how to calculate your RDA for protein.

In setting the RDA, the committee assumes that people are healthy and do not have unusual metabolic needs for protein; that the protein eaten will be of mixed quality; and that the body will use the protein about as efficiently as it uses reference proteins. In addition, the committee assumes that the protein is consumed along with sufficient carbohydrate and fat to provide adequate energy and that other nutrients in the diet are adequate.

• ***Adequate Energy*** • Note the qualification "adequate energy" in the preceding statement, and consider what happens if energy intake falls short of needs. An intake of 50 grams of protein, which is equal to 200 kcalories, provides about 10 percent of the total energy from protein, if the person receives 2000 kcalories a day. But if the person cuts energy intake drastically—to, say, 800 kcalories a day—then an intake of 200 kcalories from protein is suddenly 25 percent of the total; yet it's still the same number of grams. The protein intake is reasonable, but the energy intake is not; the low energy intake will force the body to use the protein to meet energy needs rather than to replace lost body protein. Similarly, if the person's energy intake is high—say, 4000 kcalories—the 50-gram protein intake will represent only 5 percent of the total; yet it *still* is a reasonable protein intake. Again, the energy intake is unreasonable for most people, but in this case, it will permit the protein to be used to meet the body's needs.

Be careful when judging protein intake as a percentage of energy. Always ascertain the number of grams as well, and compare it with the RDA or another standard stated in grams. A recommendation stated as a percentage of energy intake is useful only if the energy intake is within reason.

• ***Protein in Abundance*** • Many people tend to overvalue protein, perhaps because they have been so impressed with its many critical roles in the body. They think they need *lots* of protein, when, in fact, they are already receiving plenty. A high protein intake increases the work of the kidneys.[22] Excretion of the end products of protein metabolism depends, in part, on an adequate fluid intake and healthy kidneys.

**Table 6-4**

**Protein RDA as a Percentage of Energy RDA**

When expressed as a percentage of energy intake, the protein requirement represents about 10 percent of the energy RDA.

| Age (yr) | Protein RDA (g/kg) | Protein RDA (in kCalories) as a Percentage of Energy RDA (%) |
|---|---|---|
| 0 to ½ | 2.2 | 8.0 |
| ½ to 1 | 1.6 | 6.5 |
| 1 to 3 | 1.2 | 4.9 |
| 4 to 6 | 1.1 | 5.3 |
| 7 to 10 | 1.0 | 5.6 |
| Males | | |
| 11 to 14 | 1.0 | 7.2 |
| 15 to 18 | 0.9 | 7.9 |
| 19 to 24 | 0.8 | 8.0 |
| 25 to 50 | 0.8 | 8.7 |
| 51 + | 0.8 | 11.0 |
| Females | | |
| 11 to 14 | 1.0 | 8.4 |
| 15 to 18 | 0.8 | 8.0 |
| 19 to 24 | 0.8 | 8.4 |
| 25 to 50 | 0.8 | 9.1 |
| 51 + | 0.8 | 10.5 |

## HOW TO / Calculate Recommended Protein Intakes

To figure your protein RDA:

- Look up the appropriate weight for a person of your height (inside back cover). If your present weight falls within that range, use it for the following calculations. If your present weight falls outside the range, use the midpoint of the acceptable weight range as your reference weight.
- Convert pounds to kilograms, if necessary (pounds divided by 2.2 equals kilograms).
- Multiply kilograms by 0.8 to get your RDA in grams per day. (Males 18 years old and younger, multiply by 0.9.)

Example:

Weight = 150 lb.

150 lb ÷ 2.2 lb/kg = 68 kg (rounded off).

68 kg × 0.8 g/kg = 54 g protein (rounded off).

Chapter 14 discusses athletes' protein needs further.

Most people in developed countries such as the United States and Canada receive much more protein than they need. Even athletes typically don't need to increase their protein intakes. This is not surprising considering the abundance of food eaten and the central role meats hold in the North American diet. A single ounce of meat delivers about 7 grams of protein, so one 8-ounce serving of meat alone supplies more than the RDA for an average-sized person. Besides meat, well-fed people eat many other nutritious foods, many of which also contain protein.

To illustrate how easy it is to overconsume protein, consider the *minimum* recommended servings for the Daily Food Guide. Six servings from the bread, cereal, rice, and pasta group provide about 18 grams of protein; 3 servings of vegetables deliver about 6 grams; 2 servings of milk offer 16 grams; and 2 servings (about 5 ounces) of meat contain about 35 grams. This totals 75 grams of protein—higher than recommendations for most people and slightly lower than the average intake of people in the United States. (The accompanying "How to" feature describes how to estimate protein in foods.)

Just think how much more protein people receive when they eat additional servings. No wonder most people in the United States and Canada get more protein than they need. If they have an adequate *food* intake, they have a more-than-adequate protein intake. The key diet-planning principle to emphasize for protein is moderation. Even though most people receive plenty of protein, some feel compelled to take supplements as well, as the next section describes.

IN SUMMARY

Optimally, the diet will be adequate in energy from carbohydrate and fat and will deliver 0.8 grams of protein per kilogram of normal body weight each day. U.S. and Canadian diets are typically more than adequate in this respect, and protein or amino acid supplements are superfluous.

## Protein and Amino Acid Supplements

Health food stores and popular magazine articles advertise a wide variety of protein supplements, and people take these supplements for many different reasons, all of them unfounded. Athletes take them to build muscle. Dieters take them to spare their bodies' protein while losing weight. Women take them to strengthen their fingernails. People take individual amino acids, too—to cure herpes, to make themselves sleep better, to lose weight, and to relieve pain and depression.* Like many other magic solutions to health problems, protein and amino acid supplements don't work these miracles, and they can be harmful.

Muscle work builds muscle; protein supplements do not, and athletes do not need them. Instead, athletes need a well-balanced diet that provides sufficient dietary protein and adequate food energy. Food energy spares body protein; carbohydrate and fat serve this purpose equally well, and carbohydrate is safer. Fingernails are not affected by protein supplements, provided the diet is adequate.

Furthermore, protein supplements are expensive, less completely digested than protein-rich foods, and, when used as replacements for such foods, often downright dangerous. The "liquid protein" diet, advocated some years ago for weight loss, caused deaths in many users; even some physician-supervised protein-sparing fasts based on liquid protein have caused abnormal heart rhythms. The FDA warns that their use as a total diet without medical supervision "may cause serious illness or death."

Single amino acids do not occur naturally in foods and offer no benefit to the body; in fact, they can be harmful. The body was not designed to handle the high concentrations and unusual combinations of amino acids found in supplements. An excess of one amino acid can create such a demand for a carrier that it prevents

Use of amino acids as dietary supplements is *inappropriate,* especially for:

- All women of childbearing age.
- Pregnant or lactating women.
- Infants, children, and adolescents.
- Elderly people.
- People with inborn errors of metabolism that affect their bodies' handling of amino acids.
- Smokers.
- People on low-protein diets.
- People with chronic or acute mental or physical illnesses who take amino acids without medical supervision.

*Canada allows single amino acid supplements to be sold only as drugs or as food additives.

## HOW TO Use the Exchange System to Estimate Protein

| Exchange | Protein (g) |
|---|---|
| Milks | 8 |
| Meats | 7 |
| Starch | 3 |
| Vegetables | 2 |
| Fruits | — |
| Fats | — |

The exchange system provides an easy way to estimate dietary protein. The foods on the milk and meat lists supply protein in abundance: a cup of milk provides 8 grams of protein; an ounce of meat, 7 grams. The starch and vegetable lists contribute small amounts of protein, but they can add up to significant quantities; fruits and fats provide no protein.

To estimate the protein in this meal, you first need to recognize that this burrito contains about ½ cup pinto beans and ½ ounce shredded cheese wrapped in a tortilla, 1 cup of milk and an apple. Then you need to translate these portions into exchanges: 1½ meats, 1 starch, 1 milk, and 1 fruit, respectively.

| Lunch | Exchange | Protein (g) Estimate | Protein (g) Actual |
|---|---|---|---|
| ½ c pinto beans | = 1 meat | 7 | |
| ½ oz cheese | = ½ meat | 4 | 14 |
| 1 tortilla | = 1 starch | 3 | |
| 1 c milk | = 1 milk | 8 | 8 |
| 1 apple | = 1 fruit | — | — |
| | | 22 | 22 |

***Milks and meats provide lots of protein; starch and vegetables contain a little; fruits and fats have none.***

Using the exchange system to estimate, this lunch provides about 22 grams of protein. A computer diet analysis program calculated the same. The exchange system sometimes over- or underestimates the protein contents of individual foods, but for most, its estimates of daily intakes are close. In any case, for nutrients eaten in such large quantities as protein, a difference of a few grams in a day's total is insignificant.

the absorption of another amino acid, leading to a deficiency. Those amino acids winning the competition enter in excess, creating the possibility of a toxicity. Toxicity of single amino acids in animal studies raises concerns about their use in human beings. Anyone considering taking amino acid supplements should check with a physician first.

In two cases, recommendations for single amino acid supplements have led to widespread public use—lysine to prevent or relieve the infections that cause herpes cold sores on the mouth or genital organs, and tryptophan to relieve pain, depression, and insomnia. In both cases, enthusiastic popular reports preceded careful scientific experiments. A review of the research indicates that lysine may suppress herpes infections in some individuals and appears safe (up to 3 grams per day) when taken in divided doses with meals.[23]

Tryptophan is also effective with respect to pain and sleep, but its use for these purposes is still experimental. More than 1500 people who elected to take tryptophan supplements developed a rare blood disorder known as eosinophilia-myalgia syndrome (EMS). EMS is characterized by severe muscle pain, extremely high fever, and, in over three dozen cases, death. Treatment usually involves physical therapy and low doses of corticosteroids to relieve symptoms temporarily. Early evidence suggested that a major tryptophan processing plant may have introduced contaminants that caused the disease, but later research indicated that multiple factors were involved; the exact causes of EMS remain unknown. The FDA issued a recall of all products containing tryptophan.

### IN SUMMARY

Normal, healthy people never need protein or amino acid supplements. It is safest to obtain lysine, tryptophan, and all other amino acids in protein-rich foods, eaten with carbohydrate to facilitate their use in the body. With all that we know about science, it is hard to improve on nature.

## Making It Click

These problems will give you practice in doing simple nutrition-related calculations using hypothetical situations (see Appendix K for answers). Once you have mastered these examples, you will be prepared to examine your own protein needs. Be sure to show your calculations for each problem.

1. Compute recommended protein intakes for people of different sizes. Refer to the "How to" on p. 181 and compute the protein recommendation for the following people. The intake for a woman 5 feet 8 inches tall is computed for you as an example.

   A woman 5 feet 8 inches tall is 68 inches tall. From the table on the inside back cover, the midpoint in the green area for this woman is 144 pounds.

   $$144 \text{ lb} \div 2.2 \text{ lb/kg} = 65 \text{ kg.}$$
   $$0.8 \text{ g/kg} \times 65 \text{ kg} = 52 \text{ g protein per day.}$$

   a. A woman 5 feet 1 inch tall.
   b. A man (18 years) 6 feet 4 inches tall.

2. The chapter warns that recommendations based on percentage of energy intake are not always appropriate. Consider a man 26 years old who is 5 feet 10 inches tall, weighs 163 pounds, is moderately active, and eats 3500 kcalories/day with 10 percent of the kcalories from protein.
   a. What is this man's protein intake? Show your calculations.
   b. Is his protein intake appropriate? Too high? Too low? Justify your answer.

This exercise should help you develop a perspective on protein recommendations.

## Study Questions

These questions will help you review the chapter. You will find the answers in the discussions on the pages provided.

1. How does the chemical structure of proteins differ from the structures of carbohydrates and fats? (pp. 161–163)
2. Describe the structure of amino acids, and explain how their sequence in proteins affects the proteins' shapes. What are essential amino acids? (pp. 161, 163)
3. Describe protein digestion and absorption. (pp. 164–166)
4. Describe protein synthesis. (pp. 166–168)
5. Describe some of the roles proteins play in the human body. (pp. 168–172)
6. What are enzymes? What roles do they play in chemical reactions? Describe the differences between enzymes and hormones. (pp. 169–170)
7. How does the body use amino acids? What is deamina-

tion? Define nitrogen balance. What conditions are associated with zero, positive, and negative balance? (p. 173)
8. What factors affect the quality of dietary protein? What is a complete protein? (pp. 174–175)
9. How can vegetarians meet their protein needs without eating meat? (p. 175)
10. What are the health consequences of ingesting inadequate protein and energy? Describe marasmus and kwashiorkor. How can the two conditions be distinguished, and in what ways do they overlap? (pp. 177–179)
11. How might protein excess, or the type of protein eaten, influence health? (pp. 179–180)
12. What factors are considered in establishing recommended protein intakes? (p. 181)
13. Which food lists of the exchange system supply protein in abundance? In moderation? Not at all? (p. 183)
14. What are the benefits and risks of taking protein and amino acid supplements? (pp. 182–184)

These questions will help you prepare for an exam. Answers can be found in Appendix K.

1. Which part of its chemical structure differentiates one amino acid from another?
   a. its side group
   b. its acid group
   c. its amino group
   d. its double bonds
2. Isoleucine, leucine, and lysine are:
   a. proteases.
   b. polypeptides.
   c. essential amino acids.
   d. complementary proteins.
3. In the stomach, hydrochloric acid:
   a. denatures proteins and activates pepsin.
   b. hydrolyzes proteins and denatures pepsin.
   c. emulsifies proteins and releases peptidase.
   d. condenses proteins and facilitates digestion.
4. Proteins that facilitate chemical reactions are:
   a. buffers.
   b. enzymes.
   c. hormones.
   d. antigens.
5. If an essential amino acid that is needed to make a protein is unavailable, the cells must:
   a. deaminate another amino acid.
   b. substitute a similar amino acid.
   c. break down proteins to obtain it.
   d. synthesize the amino acid from glucose and nitrogen.
6. Eating two foods together so that each provides an amino acid that the other lacks is known as:
   a. dual deamination.
   b. random limitation.
   c. mutual supplementation.
   d. double complementation.
7. The protein efficiency ratio and PDCAAS are two methods used to:
   a. determine protein quality.
   b. assess protein-energy malnutrition.
   c. estimate the weight of nitrogen in a food.
   d. calculate the percentage kcalories from protein.
8. Marasmus develops from:
   a. too much fat clogging the liver.
   b. megadoses of amino acid supplements.
   c. inadequate protein and energy intake.
   d. excessive fluid intake causing edema.
9. The protein RDA for a healthy adult who weighs 180 pounds is:
   a. 50 milligrams/day.
   b. 65 grams/day.
   c. 180 grams/day.
   d. 2000 milligrams/day.
10. Which of these foods contains the least protein per serving?
   a. rice
   b. broccoli
   c. pinto beans
   d. orange juice

## Notes


1. J. C. Teran, K. D. Mullen, and A. J. McCullough, Glutamine—A conditionally essential amino acid in cirrhosis? *American Journal of Clinical Nutrition* 62 (1995): 897–900.
2. H. F. Bunn, Pathogenesis and treatment of sickle cell disease, *New England Journal of Medicine* 337 (1997): 762–769.
3. Committee on Genetics, Health supervision for children with sickle cell diseases and their families, *Pediatrics* 98 (1996): 467–472; E. M. Chiocca, Sickle cell crisis, *American Journal of Nursing* 96 (1996): 49.
4. J. C. Waterlow, Whole-body protein turnover in humans—Past, present, and future, *Annual Review of Nutrition* 15 (1995): 57–92.
5. V. R. Young and P. L. Pellett, Plant proteins in relation to human protein and amino acid nutrition, *American Journal of Clinical Nutrition* 59 (1994): 1203S–1212S; Position of The American Dietetic Association: Vegetarian diets, *Journal of the American Dietetic Association* 97 (1997): 1317–1321.
6. D. G. Schroeder and R. Martorell, Enhancing child survival by preventing malnutrition, *American Journal of Clinical Nutrition* 65 (1997): 1080–1081.
7. J. C. Waterlow, Childhood malnutrition in developing nations: Looking back and looking forward, *Annual Review of Nutrition* 14 (1994): 1–19.
8. R. W. Sauerwein and coauthors, Inflammatory mediators in children with protein-energy malnutrition, *American Journal of Clinical Nutrition* 65 (1997): 1534–1539.
9. W. S. Dempster and coauthors, Misplaced iron in kwashiorkor, *European Journal of Clinical Nutrition* 49 (1995): 208–210.

10. L. Lewinter-Suskind and coauthors, The malnourished child, in *Textbook of Pediatric Nutrition,* 2nd ed., eds. R. M. Suskind and L. Lewinter-Suskind (New York: Raven Press, 1993), pp. 127–140.
11. J. W. Anderson, B. M. Johnstone, and M. E. Cook-Newell, Meta-analysis of the effects of soy protein intake on serum lipids, *New England Journal of Medicine* 333 (1995): 276–282; D. Kritchevsky, Dietary protein, cholesterol, and atherosclerosis: A review of the early history, *Journal of Nutrition* 125 (1995): 589S–593S.
12. J. Selhub and coauthors, Association between plasma homocysteine concentrations and extracranial carotid-artery stenosis, *New England Journal of Medicine* 332 (1995): 286–291; M. J. Stampfer and M. R. Malinow, Can lowering homocysteine levels reduce cardiovascular risk? *New England Journal of Medicine* 332 (1995): 328–329; J. B. Ubbink, Homocysteine—An atherogenic and a thrombogenic factor? *Nutrition Reviews* 53 (1995): 323–332.
13. M. J. Stampfer and coauthors, A prospective study of plasma homocyst(e)ine and risk of myocardial infarction in U.S. physicians, *Journal of the American Medical Association* 268 (1992): 877–881.
14. J. S. Stamler and A. Slivka, Biological chemistry of thiols in the vasculature and in vascular-related disease, *Nutrition Reviews* 54 (1996): 1–30.
15. J. B. Ubbink and coauthors, Effective homocysteine metabolism may protect South African blacks against coronary heart disease, *American Journal of Clinical Nutrition* 62 (1995): 802–808.
16. O. Nygård and coauthors, Coffee consumption and plasma total homocysteine: The Hordeland Homocysteine Study, *American Journal of Clinical Nutrition* 65 (1997): 136–143.
17. N. Pancharuniti and coauthors, Plasma homocyst(e)ine, folate, and vitamin B-12 concentrations and risk for early-onset coronary artery disease, *American Journal of Clinical Nutrition* 59 (1994): 940–948; J. B. Ubbink and coauthors, Vitamin B-12, vitamin B-6, and folate nutritional status in men with hyperhomocysteinemia, *American Journal of Clinical Nutrition* 57 (1993): 47–53; J. Selhub and coauthors, Vitamin status and intake as primary determinants of homocysteinemia in an elderly population, *Journal of the American Medical Association* 270 (1993): 2693–2698.
18. E. B. Rimm and coauthors, Folate and vitamin $B_6$ from diet and supplements in relation to risk of coronary heart disease among women, *Journal of the American Medical Association* 279 (1998): 359–364.
19. Committee on Dietary Allowances, *Recommended Dietary Allowances,* 10th ed. (Washington, D.C.: National Academy Press, 1989), pp. 72–73; C. D. Arnaud and S. D. Sanchez, The role of calcium in osteoporosis, *Annual Review of Nutrition* 10 (1990): 397–414; R. P. Heaney, Protein intake and the calcium economy, *Journal of the American Dietetic Association* 93 (1993): 1259–1260.
20. J. A. Metz, J. J. B. Anderson, and P. N. Gallagher, Intakes of calcium, phosphorus, and protein, and physical activity level are related to radial bone mass in young adult women, *American Journal of Clinical Nutrition* 58 (1993): 537–542; Heaney, 1993.
21. Committee on Dietary Reference Intakes, *Dietary Reference Intakes for Calcium, Phosphorus, Magnesium, Vitamin D, and Fluoride* (Washington, D.C.: National Academy Press, 1997), p. 4-4.
22. E. Brändle, H. G. Sieberth, and R. E. Hautmann, Effect of chronic dietary protein intake on the renal function in healthy subjects, *European Journal of Clinical Nutrition* 50 (1996): 734–740.
23. N. W. Flodin, The metabolic roles, pharmacology, and toxicology of lysine, *Journal of the American College of Nutrition* 16 (1997): 7–21.

HIGHLIGHT 6

# Vegetarian, Mediterranean, and Other Meat-Restricted Foodways

The waiter presents this evening's specials: a fresh spinach salad topped with mandarin oranges, raisins, and sunflower seeds, served with a bowl of pasta smothered in a mushroom and tomato sauce and topped with grated parmesan cheese. Then this one: a salad made of chopped parsley, scallions, celery, and tomatoes mixed with bulgur wheat and dressed with olive oil and lemon juice, served with a spinach and feta cheese pie. Do these meals sound good to you? Or is something missing . . . a pork chop or ribeye, perhaps?

Would vegetarian fare be acceptable to you some of the time? Most of the time? Ever? Perhaps it is helpful to recognize that dietary choices fall along a continuum—from one end, where people eat no meat or foods of animal origin, to the other end, where they eat generous quantities daily. Meat's place in the diet has been the subject of much research and controversy, as this highlight will reveal. One of the missions of this highlight, in fact, is to identify the *range* of meat intakes most compatible with health.

People who choose to exclude meat and other animal-derived foods from their diets today do so for many of the same reasons the Greek philosopher Pythagoras cited in the sixth century B.C.: physical health, ecological responsibility, and philosophical concerns. They might also cite world hunger issues, economic reasons, ethical concerns, or religious beliefs as motivating factors.

Vegetarians generally are categorized, not by their motivations, but by the foods they choose not to eat (see the glossary on p. 188). Some exclude red meat only; some also exclude chicken or fish; others also exclude eggs; and still others exclude milk and milk products as well. As you will see, though, the foods a person *excludes* are not nearly as important as the foods a person *includes* in the diet. Most vegetarian diets include a variety of grains, vegetables, legumes, and fruits, which offer abundant complex carbohydrates and fibers, an assortment of vitamins and minerals, and little fat—characteristics that reflect current dietary recommendations aimed at reducing obesity and the risks of several chronic diseases such as hypertension, heart disease, and cancer. Vegetarian diets that are well planned can offer sound nutrition and health benefits to adults.[1]

This highlight first looks at the health benefits and potential problems of vegetarian diets and then shows how to plan a well-balanced vegetarian diet. It closes with a description of the Mediterranean diet—an ethnic pattern of eating that includes very little meat and exceeds current dietary recommendations for fat and alcohol, yet still seems to support good health.

***A balanced meal need not include meat to be nutritious.***

## Health Benefits of Vegetarian Diets

Research on the health impacts of vegetarianism would be relatively easy if vegetarians differed from other people only in not eating meat. Many vegetarians, however, have adopted lifestyles that differentiate them from others: they typically maintain a healthy weight, use no tobacco or illicit drugs, use little (if any) alcohol, and are physically active.[2] Researchers must account for these lifestyle differences before they can determine which aspects of health correlate just with diet. Even then, *correlations* merely reveal what health factors *go with* the vegetarian diet, not what health effects may be *caused by* the diet. Without more evidence, conclusions remain tentative. Still, with all these qualifications, research findings seem to suggest that vegetarian diets offer some health benefits.

### Weight Control

In general, vegetarians maintain a healthier body weight than nonvegetarians. Since obesity impairs health in a number of ways, this gives vegetarians a health advantage.

### Blood Pressure

Appropriate body weight helps to maintain a healthy blood pressure, as does a diet low in total fat and saturated fat and high in fiber, fruits, and vegetables.[3] Lifestyle factors also seem to influence blood pressure: smoking and alcohol intake raise blood pressure, and physical activity lowers it.

### Coronary Artery Disease

Fewer vegetarians than meat eaters suffer from diseases of the heart and arteries. The dietary factor most directly related to coronary artery disease is saturated fat, and in general, vegetarian diets are lower in total fat, saturated fat, and cholesterol than typical meat-based diets. Vegetarian diets are also higher in dietary fiber, another factor that helps control blood lipids.

When vegetarians are fed meat, which contains saturated fat, their blood lipid profiles change for the worse; when meat eaters are fed a low-fat vegetarian diet, their blood lipid profiles and blood pressure improve.[4]

## Glossary

**lactovegetarians:** people who include milk and milk products, but exclude meat, poultry, fish, seafood, and eggs from their diets.
- **lacto** = milk

**lacto-ovo-vegetarians:** people who include milk, milk products, and eggs, but exclude meat, poultry, fish, and seafood from their diets.
- **ovo** = egg

**macrobiotic diets:** extremely restrictive diets limited to a few grains and vegetables; based on metaphysical beliefs and not on nutrition.

**meat replacements:** products formulated to look and taste like meat, fish, or poultry; usually made of textured vegetable protein.

**omnivores:** people who have no formal restriction on the eating of any foods.
- **omni** = all
- **vores** = to eat

**semivegetarians:** people who include some, but not all, groups of animal-derived foods in their diets; they usually exclude red meat, but may occasionally include poultry, fish, and seafood; sometimes called **partial vegetarians.**

**tempeh** (TEM-pay): a fermented soybean food, rich in protein and fiber.

**textured vegetable protein:** processed soybean protein used in vegetarian products such as soy burgers; see *meat replacements.*

**tofu** (TOE-foo): a curd made from soybeans, rich in protein and often fortified with calcium; used in many Asian and vegetarian dishes in place of meat.

**vegans** (VAY-guns or VEJ-ans): people who exclude all animal-derived foods (including meat, poultry, fish, eggs, and dairy products) from their diets; also called **pure vegetarians, strict vegetarians,** or **total vegetarians.**

**vegetarians:** a general term used to describe people who exclude meat, poultry, fish, or other animal-derived foods from their diets.

Similarly, semivegetarians who eat one to three servings of meat per week have blood lipids between the low blood lipids of vegetarians and the higher lipids of nonvegetarians.[5]

### Cancer

Seventh-Day Adventists, a religious group whose foodways center on a lacto-ovo-vegetarian diet, have a significantly lower mortality rate from cancer than the rest of the population, even after all the cancers attributed to smoking and alcohol are discounted.[6] Their low cancer rates may be due to their vegetarian diets; evidence is overwhelming that high intakes of fruits and vegetables reduce the risks of cancer.[7]

Some scientific findings indicate that vegetarian diets are associated not only with lower cancer mortality in general, but with lower incidence of cancer at specific sites as well, most notably, colon cancer.[8] People with colon cancer seem to eat more meat, more saturated fat, and less fiber than others without cancer. High-protein, high-fat, low-fiber diets create an environment in the colon that promotes the development of cancer in some people. A high-meat diet has been associated with other cancers as well.[9]

In general, then, adults who eat vegetarian diets can reduce their risks of several chronic diseases, including obesity, high blood pressure, heart disease, and cancer. But there is nothing mysterious about the vegetarian diet; it simply includes ample fruits, vegetables, whole grains, and legumes—foods that are higher in fiber, richer in antioxidant vitamins, and lower in fats than meat-based diets.[10] Some people find it easier to meet today's dietary recommendations for health by following a vegetarian diet than by eating meals with meat. A meat eater can gain some of the same advantages by limiting meat intake to the recommended 5 to 7 ounces daily and selecting lean cuts, as well as including abundant grains, fruits, and vegetables.

Conversely, both vegetarian and meat-based diets can be detrimental to health when overloaded with fat. A vegetarian who dines on cheddar cheese, butter sauces, sour cream, and deep-fried vegetables invites the same health hazards as the person who overeats high-fat meats. And both diets, if not properly balanced, can lack nutrients. Poorly planned vegetarian diets typically lack iron, zinc, calcium, vitamin $B_{12}$, and vitamin D; without planning, the meat eater's diet may lack vitamin A, vitamin C, folate, and fiber, among others.

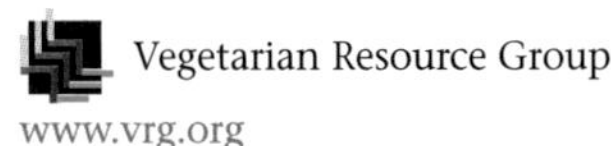

www.vrg.org

## Problems Associated with Vegetarian Diets

The negative health aspects of any diet, including vegetarian diets, reflect poor diet planning. Careful attention to energy intake and specific problem nutrients can ensure adequacy. Diet planning during pregnancy, lactation, infancy, childhood, and illness, in particular, must provide for the increases in energy and nutrients needed during those times—when the consequences of poor nutrition can be great.

### Adequacy of Most Vegetarian Diets

Vegetarians who include milk products and eggs can meet recommendations for most nutrients about as easily as nonvegetarians.[11] Such diets provide enough energy, protein, and other nutrients to support the health of adults and the growth of children and adolescents.

### Inadequacy of Strict Vegetarian Diets

Achieving adequate energy and nutrient intakes may be difficult for the vegan who excludes all animal products, and particularly for growing children and pregnant and lactating women. Foods of plant origin generally offer much less energy per bite than foods of animal origin. While a diet that delivers a lot of food with relatively little energy may be advantageous for many adults, it can be detrimental for children who need energy-dense foods for growth. Vegan diets can fail to provide sufficient energy to support the growth of a child within a quantity of food small enough for the child to eat. A child's small stomach can hold only so much food, and a vegan child may feel full before eating enough to meet nutrient and energy needs. A vegan child's diet should emphasize cereals, legumes, and nuts to meet protein and energy needs in a small volume. Meat, which contains abundant protein, iron, and food energy in less bulk, supports the growth of children more efficiently. Compared with meat-eating children, vegan children tend to be smaller in height and lighter in weight; their low energy intakes can impair growth.[12]

When vegan children get their protein only from plant foods, they may need protein intakes higher than the RDA for normal growth and health. The standard protein recommendations may be inadequate to support the growth of vegan children, but specific recommendations have not been established.

Approximately 2 out of every 15 households include one or more vegetarians. These people number some 12 million nationwide, representing an eightfold increase over the past two decades.[13] Those who plan their diets carefully easily obtain all the nutrients they need to support good health.

## Vegetarian Diet Planning

The vegetarian has the same meal-planning task as any other person—using a variety of foods that will deliver all the needed nutrients within an energy allowance that maintains a healthy body weight. An added challenge is to do so with fewer foods.

Well-planned vegetarian meals can provide adequate amounts of all the nutrients a person needs for good health. Vegetarians can follow the Daily Food Guide presented in Chapter 2 with a few modifications (see Table H6-1). Those who include milk products and eggs can follow the regular plan, using legumes and products made from them, such as peanut butter, tempeh, and tofu, in place of meat. Those who do not use milk can use soy milk fortified with calcium, vitamin D, and vitamin $B_{12}$. Vegetarian adults should include at least one cup of dark green vegetables daily to help meet iron needs and legumes to help meet zinc needs. In general, these tactics ensure adequate intakes of the main nutrients vegetarian diets might otherwise lack: iron, zinc, calcium, vitamin $B_{12}$, and vitamin D. In contrast, most vegetarians easily obtain large quantities of the nutrients that are abundant in plant foods: thiamin, riboflavin, folate, and vitamins $B_6$, C, A, and E.

 FDA
www.fda.gov
Visit foods
Search for Vegetarian

### Protein

Protein is not the problem it was once thought to be for vegetarian diets. People who use animal-derived foods such as milk and eggs receive high-quality proteins and are unlikely to develop protein deficiencies. Even those who eat only plant-derived foods are unlikely to develop protein deficiencies provided that energy intakes are adequate and the protein sources varied.[14] The proteins of whole grains, legumes, seeds, nuts, and vegetables can provide adequate amounts of all the amino acids. An advantage of many vegetarian protein foods is that they are generally lower in saturated fat than meats and are often higher in fiber and richer in some vitamins and minerals.

To ease meal preparation, vegetarians sometimes use meat replacements made of textured vegetable protein (soy protein). These foods are formulated to look and taste like meat, fish, or poultry. Many of these products are designed to match the known nutrient contents of animal-protein foods, but sometimes they fall short. A wise vegetarian does

**Table H6–1**
**Daily Food Guide for Vegetarian Meal Planning**

| Food Group | Suggested Daily Servings | Serving Sizes |
|---|---|---|
| Breads, cereals, rice, pasta, and other grain products | 6 to 11 | 1 slice bread<br>½ bun, bagel, or English muffin<br>½ c cooked cereal, rice, or pasta<br>1 oz ready-to-eat cereal |
| Vegetables | 3 to 5[a] | ½ c cooked or chopped raw vegetables<br>1 c raw leafy vegetables |
| Fruits | 2 to 4 | 1 medium-sized piece fresh fruit<br>¾ c fruit juice<br>½ c canned, cooked, or chopped raw fruit<br>¼ c dried fruit |
| Legumes, nuts, seeds, eggs, and other meat substitutes | 2 to 3 | ½ c cooked legumes<br>¼ c tofu or tempeh<br>1 c soy milk<br>2 tbs peanut butter, nuts, or seeds (these tend to be high in fat, so use sparingly)<br>1 egg or 2 egg whites |
| Milk, yogurt, cheese, and other milk products | 2 to 3[b] | 1 c milk<br>1 c yogurt<br>1½ oz cheese |

[a]Include 1 cup of dark green vegetables daily to help meet iron requirements.
[b]People who do not use milk or milk products: use soy milk fortified with calcium, vitamin D, and vitamin $B_{12}$. Other nonmilk calcium-rich food sources are provided in Chapter 12.
Source: Adapted from Position of The American Dietetic Association: Vegetarian diets, *Journal of the American Dietetic Association* 97 (1997): 1320.

not rely on these products too heavily, but learns to use a variety of whole foods instead. Vegetarians may also use soybeans in the form of bean curds, or tofu, to bolster protein intake.

## Iron

Getting enough iron can be a problem even for meat eaters, and those who eat no meat must pay special attention to their iron intake. The iron in plant foods such as legumes, dark green leafy vegetables, iron-fortified cereals, and whole-grain breads and cereals is not readily absorbed. Iron absorption is enhanced by vitamin C, though, and vegetarians typically eat many vitamin C–rich fruits and vegetables, so they suffer no more iron-deficiency anemia than other people do.[15]

## Zinc

Zinc is similar to iron in that meat is its richest food source and zinc from plant sources is not well absorbed. In addition, soy, which is commonly used as a meat alternate, interferes with zinc absorption. Nevertheless, most vegetarian adults are not zinc deficient.[16] Perhaps the best advice to vegetarians regarding zinc is to eat a variety of nutrient-dense foods; include grains, nuts, and legumes such as black-eyed peas, pinto beans, and kidney beans; and maintain an adequate energy intake. For vegetarians who include seafood, oysters, crabmeat, and shrimp are rich in zinc.

## Calcium

The calcium intakes of lactovegetarians are similar to those of the general population, but people who use no milk risk deficiency. Careful planners select calcium-rich foods, such as calcium-fortified juices or soy milk, in ample quantities regularly. This is especially important for children. Soy formulas for infants are fortified with calcium and can be used in cooking, even for adults. Other good calcium sources include calcium-set tofu, some legumes, some green vegetables such as broccoli and turnip greens, some nuts such as almonds, and certain seeds such as sesame seeds.[17*] The choices should be varied because binders in some plant foods may limit absorption.

## Vitamin $B_{12}$

The requirement for vitamin $B_{12}$ is small, but this vitamin is found only in animal-derived foods. Fermented soy products such as tempeh may contain some vitamin $B_{12}$ from the bacteria that did the fermenting, but unfortunately, much of the vitamin $B_{12}$ found in these products may be an inactive form. Seaweeds such as nori and chlorella supply some vitamin $B_{12}$, but not much, and excessive intakes can lead to iodine toxicity.[18] To defend against vitamin $B_{12}$ deficiency, vegans must rely on vitamin $B_{12}$-fortified sources (such as soy milk or breakfast cereals) or supplements.

## Vitamin D

For people who do not use vitamin D–fortified milk and do not receive enough exposure to sunlight to synthesize adequate vitamin D, supplements may be warranted. This is particularly important for children and older adults. In northern climates during winter months, young children on vegan diets can readily develop rickets, the vitamin D–deficiency disease.

As you can see, vegetarianism is not a religion like Buddhism or Hinduism, but merely an eating plan that selects plant foods to deliver needed nutrients. The quality of the diet depends not on whether it includes meat, but on whether the food choices are nutritionally sound. Health experts would quickly add that one should also limit intakes of substances such as fat and alcohol that are harmful in excess—and vegetarians in North America typically do. Interestingly, the Mediterranean diet—a predominantly vegetarian diet with an ethnic flair—breaks these rules on moderation, yet still seems to have health advantages.

# An Almost-Vegetarian Diet: The Mediterranean Diet

Coastal populations that share the bounty of the Mediterranean waters may also share some important health advantages: The incidence of chronic diseases is low and life expectancy is high.[19] While popular sources report the marvels of the "Mediterranean diet," scientists who have attempted to define that diet or its health benefits have run into problems. One problem is that many countries border the Mediterranean Sea: Italy, Spain, Portugal, France, Greece, Syria, Lebanon, Israel, Turkey, Egypt, Algeria, and more. Consequently, there is no single "Mediterranean diet." Also, some of the data backing claims about causes of death in Mediterranean countries were collected in the 1960s, when the majority of people still consumed traditional diets. Even with these limitations, the links between Mediterranean diets and health are worth pondering.

Although each of the many countries that border the Mediterranean Sea has its own culture, traditions, and dietary habits, similarities are also evident.[20] The people dine on crusty breads, grains, potatoes, and pastas; a variety of vegetables and legumes; feta and mozzarella cheeses and yogurt; and fruit (especially grapes and figs).[21] They eat some fish, other seafood, and poultry, a few eggs, and very little meat. Their principal source of fat is olive oil, and they typically drink wine with meals. Consequently, traditional Mediterranean diets are:[22]

- Low in saturated fat.
- Rich in monounsaturated fat.

*Calcium salts are often added during processing to coagulate the tofu.

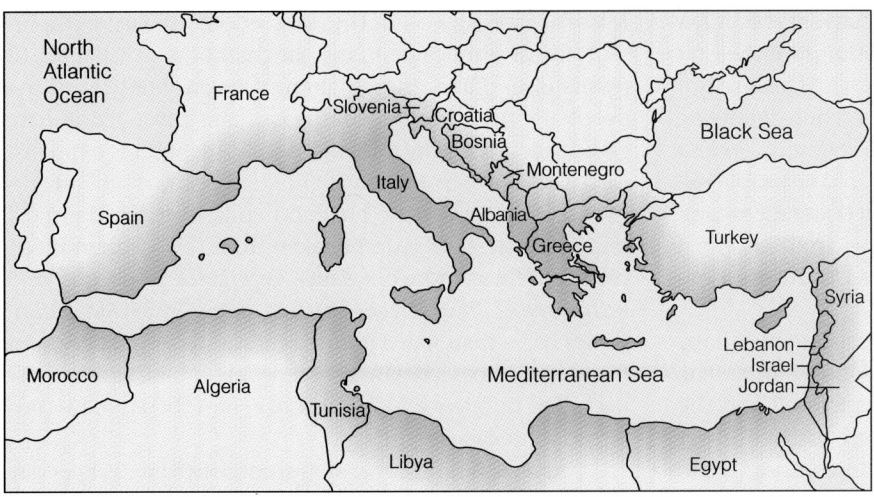

The people of the Mediterranean area eat plenty of fruits, vegetables, legumes, and grains; some dairy products, fish, and poultry; and very little red meat. Olive oil is their principal source of dietary fat.

- Rich in carbohydrate and fiber.
- Rich in nutrients and nonnutrients that support good health.

Furthermore, because processed foods are used modestly, intakes of salt, refined sugars, and *trans*-fatty acids are low.[23] All in all, the Mediterranean diet has gained a reputation for its health benefits as well as its delicious flavors.[24]

## The Mediterranean Diet Pyramid

A few nutrition experts were so impressed with the Mediterranean diet and its health benefits that they created a renegade food pyramid.* Like the official USDA Food Guide Pyramid introduced in Chapter 2, their pyramid is based on breads, cereals, rice, pasta, and other grains, and it places vegetables and fruits on the next level up. The Mediterranean pyramid introduces a small difference at this level in that it includes legumes with the vegetable group instead of with the meats, but greater differences become apparent farther up the pyramid. Olive oil sits just above the fruits, vegetables, and legumes, with cheese and yogurt above that; these foods are to be included daily. Fish, poultry, eggs, and sweets come next and are to be eaten a few times per week. Lean red meats sit at the tip of the pyramid, to be eaten only a few times per month. Figure H6-1 compares the two pyramids.

This Mediterranean pyramid contradicts many diet and health recommendations and is worth examining because it raises interesting issues. For example, although current dietary recommendations restrict fat to no more than 30 percent of daily kcalories, the traditional Mediterranean diet can deliver as much as 40 percent of a day's kcalories from fat. The Mediterranean plan does not restrict total fat, but it

*The Mediterranean diet pyramid was developed by the Harvard School of Public Health, the European office of the World Health Organization, and the Oldways Preservation & Exchange Trust in Boston.

**Figure H6-1**
**Food Pyramids Compared**

**Mediterranean Diet Pyramid**

This pyramid is based on the dietary traditions of Crete around 1960, structured in light of current nutrition research.

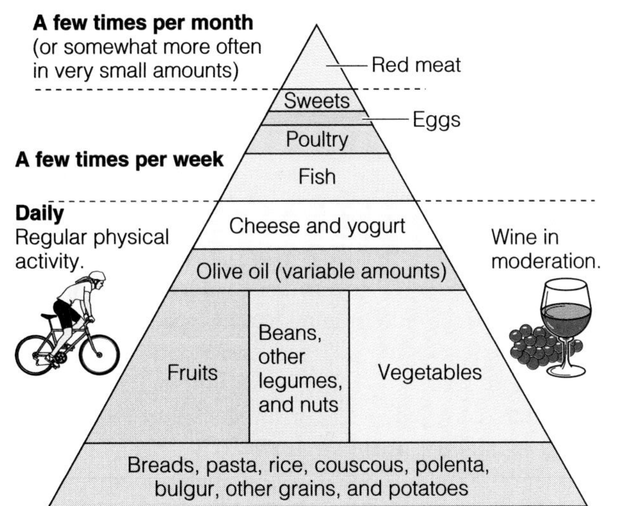


**USDA Pyramid**

This pyramid is based on the dietary guidelines established in 1992 by the U.S. Department of Agriculture.

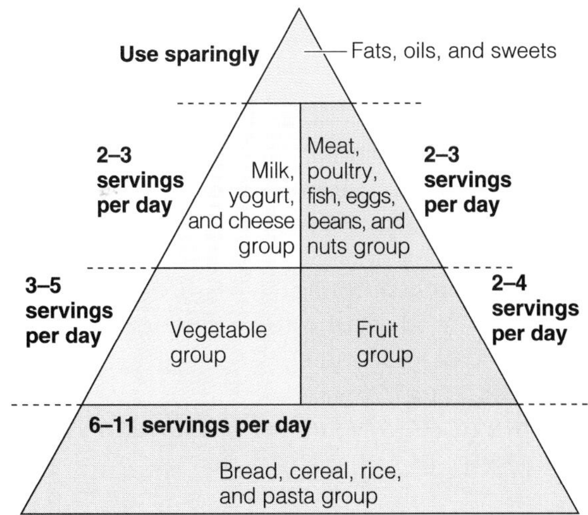


Sources: 1994 Oldways Preservation & Exchange Trust, U.S. Department of Agriculture.

does limit *animal* fat—a distinction not made in the USDA plan. People following the USDA plan get most of their fat from red meat, poultry, eggs, milk, yogurt, and cheese; consequently, much of their fat is saturated fat. Those following the Mediterranean plan use olive oil abundantly, and so receive most of their fat as monounsaturated vegetable oil. Their limited consumption of dairy products and meats provides less than 10 percent of their kcalories from saturated fats—a goal both plans agree on, but the USDA plan typically fails to meet.

The distinctions between types of fat have implications for chronic diseases, as Chapter 5 pointed out. The monounsaturated fats of olive oil and canola oil and the omega-3 polyunsaturated fats of fish may actually lower blood lipids and benefit heart health;[25] in contrast, most, but not all, saturated fats are detrimental. Substituting unsaturated fats such as olive oil for saturated fats or *trans*-fatty acids improves blood lipids and reduces the risks of cardiovascular disease.[26] In addition to its beneficial fatty acid composition (high in monounsaturated and low in saturated fatty acids), olive oil contains vitamin E and other antioxidant compounds that protect against heart disease.[27]

These distinctions in types of fat are not evident in the USDA pyramid and dietary recommendations. People are simply advised to cut back on all fat so that they will cut back on saturated fat—the primary suspect.

Many Mediterranean people drink wine with each meal, and the Mediterranean pyramid includes wine in moderation. Moderate alcohol consumption reduces the risk of cardiovascular disease and seems to be compatible with a healthy lifestyle.[28] The USDA pyramid does not address alcoholic beverages directly, but most diet and health recommendations advise people to drink alcoholic beverages in moderation, if at all.

Perhaps the hallmark of the Mediterranean diet is its abundance of vegetables, fruits, legumes, and whole grains—foods associated with lower risks of cardiovascular disease and cancer.[29] The protective effects of these plant-derived foods are attributed not only to their lack of fat, but also to their abundance of nutrients and nonnutrients, many of which act as antioxidants (see Highlight 11).

## Some Concerns about the Mediterranean Plan

Critics of the Mediterranean plan have expressed concerns that it may be inadequate in calcium and iron—two problem nutrients for many people, especially women. Because these nutrients are typically lacking in many people's diets, it seems unwise to restrict calcium selections to cheeses and yogurt and iron-rich meat consumption to a few times a month.

## An Implication of the Mediterranean Plan: Limit Meat

Is it appropriate to suggest that people in the United States should adopt Mediterranean eating habits and begin indulging in olive oil and wine? Not really, for at least two reasons. First, diet is not the only, or even the most important, factor implicated in disease causation, as Chapter 18 points out. Many other differences in the lifestyles of the people living in the Mediterranean and those living here could account for the differences in life expectancy and disease risks. Furthermore, as Highlight 2 pointed out, all ethnic food patterns have pros and cons. Perhaps the most important suggestion to be taken from the Mediterranean plan is to focus more on grains, vegetables, and fruits, and less on meats. On average, a person in the United States consumes more than half a pound of meat per *day;* a person in the Mediterranean region consumes about half a pound per *week.* The difference in meat intake, and therefore in saturated fat intake, is significant.

In general, at most, two 3-ounce servings of meat per day are needed. This amount of meat alone provides most of a person's daily recommended protein intake—and other foods together can provide a similar amount. Some researchers argue that this much meat eaten daily is not compatible with good health; if any meat is eaten, they suggest that it be eaten infrequently and in small portions.[30] With the evidence pointing to the health advantages of a meat-restricted diet, perhaps between 0 and 6 ounces of meat daily would best serve the needs of most people; the USDA pyramid suggests 5 to 7 ounces of meat, poultry, or fish a day.

## Other Ethnic Pyramids

The Mediterranean pyramid was the first ethnic pyramid introduced by this group of nutrition educators, but it wasn't the last. Asian and Latin American pyramids have followed and an African-American pyramid is being developed.[31] Because the various sections of each pyramid reflect the culinary staples of its ethnic cuisine, the pyramids differ, but they share a common theme. Each emphasizes grains, vegetables, legumes, fruits, and oils daily; recommends eggs, milk products, fish, and poultry less frequently; and suggests beef and pork even less often.

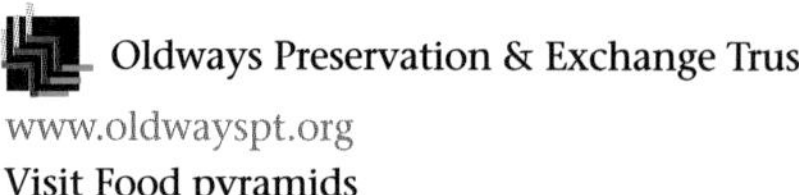
Oldways Preservation & Exchange Trust
www.oldwayspt.org
Visit Food pyramids

# Pyramids—Official and Otherwise

What about these renegade pyramids? It seems almost sacrilegious to oppose the government's official word on nutrition, but it can be enlightening to take a peek at the politics involved in developing such recommendations.

For more than a century, government agencies have issued statements advising consumers about food choices. Dietary guidelines may originally have been developed purely for the public good, but now they underlie national policy in many areas. They are used to define curricula for nutrition education, establish regulations for food labels, develop new food products, regulate institutional foodservices, provide commodity foods, and create school menus. And because the guidelines encourage people to eat more of some foods and less of others, they exert a profound effect on food purchases. Inevitably, therefore, politics has become involved in the dietary guidelines.

Food producers did not complain when early dietary recommendations urged people to "Eat more" of their products to help prevent nutrient deficiencies. However, when recommendations began to urge people to "Eat less" of some products to help prevent chronic diseases, food producers became aroused. Now lobbyists representing the food industry scurry about Capitol Hill trying to protect their interests and influence national policies that affect dietary intakes. Their efforts have been successful.

The influence of the meat industry on government policy makers provides a notable example.[32] In 1977 a dietary goal was to "Decrease consumption of meat." The "Eat less" guideline was changed in 1980 to "Choose lean meat." By 1990 the recommendation was stated even more favorably, "Have two or three servings, with a daily total of about 6 ounces." By 1992, when the Food Guide Pyramid was created, the daily total had been revised to "5 to 7 ounces." The design of the pyramid itself was delayed by a year and cost an additional million dollars, in large part because of protests from the meat industry.

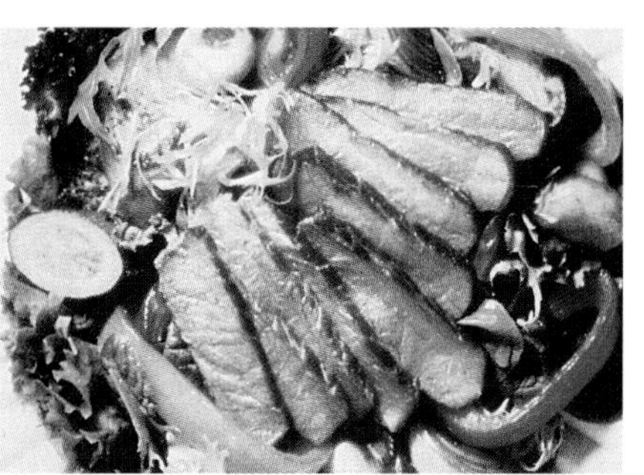

***Two meat servings of the size depicted here represent the maximum daily meat intake suggested by the Daily Food Guide as health promoting.***

The preceding paragraph was not written to pick on the meat industry. Lobbyists representing the dairy industry, the egg industry, the sugar industry, and every other food manufacturer try to influence dietary recommendations. The point is that people outside the world of nutrition science profoundly influence our nation's diet. Shifting our diet toward a healthier plan would require major changes in our agricultural and food manufacturing policies and practices.[33] One wonders what the government's nutrition advice would be if it were untainted by politics.

Having learned some of the relationships between diet and health, many people may discover that their strategies for planning meals need to change. In the past, they decided what cut of beef, ham, pork, lamb, poultry, or fish to prepare and then filled in the menu with an accompanying "starch" (potato, rice, or noodles), salad or other vegetable, and bread. Now they fill their dinner plates with legumes, grains, vegetables, and fruits. Then they add small quantities of milk products, eggs, lean meat, fish, or poultry.

For the most part, it seems that nonmeat and low-meat diets can both support good health. Keep in mind, too, that diet is only one factor influencing health. Whatever a diet consists of, its context is also important: no smoking; alcohol consumption in moderation, if at all; regular physical activity; adequate rest; and medical attention when needed all contribute to a healthy life. Establishing these healthy habits early in life seems to be the most important step one can take to reduce the risks of later diseases (as Highlight 16 explains).[34]

## Notes


1. Position of The American Dietetic Association: Vegetarian diets, *Journal of the American Dietetic Association* 97 (1997): 1317–1321.
2. P. Walter, Effects of vegetarian diets on aging and longevity, *Nutrition Reviews* 55 (1997): S61–S68.
3. L. J. Beilin and V. Burke, Vegetarian diet components, protein and blood pressure: Which nutrients are important? *Clinical and Experimental Pharmacology and Physiology* 22 (1995): 195–198.
4. J. McDougall and coauthors, Rapid reduction of serum cholesterol and blood pressure by a twelve-day, very low fat, strictly vegetarian diet, *Journal of the American College of Nutrition* 14 (1995): 491–496.
5. C. L. Melby, M. L. Toohey, and J. Cebrick, Blood pressure and blood lipids among vegetarian, semivegetarian, and nonvegetarian African Americans, *American Journal of Clinical Nutrition* 59 (1994): 103–109.
6. P. K. Mills and coauthors, Cancer incidence among California Seventh-Day Adventists, 1976–1982, *American Journal of Clinical Nutrition* 59 (1994): 1136S–1142S.
7. W. C. Willett, Micronutrients and cancer risk, *American Journal of Clinical Nutrition* 59 (1994): 1162S–1165S.
8. R. Frentzel-Beyme and J. Chang-Claude, Vegetarian diets and colon cancer: The German experience, *American Journal of Clinical Nutrition* 59 (1994): 1143S–1152S.
9. B. C.-H. Chiu and coauthors, Diet and risk of non-Hodgkin lymphoma in older women, *Journal of the American Medical Association* 275 (1996): 1315–1321.
10. M. Thorogood, The epidemiology of vegetarianism and health, *Nutrition Research Reviews* 8 (1995): 179–192; A.-L. Rauma and coauthors, Antioxidant status in long-term adherents to a strict uncooked vegan diet, *American Journal of Clinical Nutrition* 62 (1995): 1221–1227.
11. K. C. Janelle and S. I. Barr, Nutrient intakes and eating behavior score of vegetarian and nonvegetarian women, *Journal of the American Dietetic Association* 95 (1995): 180–189.
12. T. A. B. Sanders and S. Reddy, Vegetarian diets and children, *American Journal of Clinical Nutrition* 59 (1994): 1176S–1181S.
13. P. K. Johnson, Preface to the Second International Congress on Vegetarian Nutrition, *American Journal of Clinical Nutrition* (supplement) 59 (1994): vii.
14. V. R. Young and P. L. Pellett, Plant proteins in relation to human protein and amino acid nutrition, *American Journal of Clinical Nutrition* 59 (1994): 1203S–1212S; Position of The American Dietetic Association, 1997.
15. W. J. Craig, Iron status of vegetarians, *American Journal of Clinical Nutrition* 59 (1994): 1233S–1237S.

16. R. J. Gibson, Content and bioavailability of trace elements in vegetarian diets, *American Journal of Clinical Nutrition* 59 (1994): 1223S–1232S.
17. C. M. Weaver and K. L. Plawecki, Dietary calcium: Adequacy of a vegetarian diet, *American Journal of Clinical Nutrition* 59 (1994): 1238S–1241S.
18. A.-L. Rauma and coauthors, Vitamin B-12 status of long-term adherents of a strict uncooked vegan diet ("Living Food Diet") is compromised, *Journal of Nutrition* 125 (1995): 2511–2515.
19. A. Keys, Mediterranean diet and public health: Personal reflections, *American Journal of Clinical Nutrition* 61 (1995): 1321S–1323S.
20. E. Helsing, Traditional diets and disease patterns of Mediterranean, circa 1960, *American Journal of Clinical Nutrition* 61 (1995): 1329S–1337S.
21. W. C. Willett and coauthors, Mediterranean diet pyramid: A cultural model for healthy eating, *American Journal of Clinical Nutrition* 61 (1995): 1402S–1406S.
22. A. P. Simopoulos, The Mediterranean Food Guide—Greek column rather than an Egyptian pyramid, *Nutrition Today* 30 (1995): 54–61.
23. W. P. T. James, Nutrition science and policy research: Implications for Mediterranean diets, *American Journal of Clinical Nutrition* 61 (1995): 1324S–1328S.
24. B. Haber, The Mediterranean diet: A view from history, *American Journal of Clinical Nutrition* 66 (1997): 1053S–1057S; A. Trichopoulou and P. Lagiou, Healthy traditional Mediterranean diet: An expression of culture, history, and lifestyle, *Nutrition Reviews* 55 (1997): 383–389.
25. A. Trichopoulou and P. Lagiou, Worldwide patterns of dietary lipids intake and health implications, *American Journal of Clinical Nutrition* 66 (1997): 961S–964S; N. Mekki and coauthors, Effects of lowering fat and increasing dietary fiber on fasting and postprandial plasma lipids in hypercholesterolemic subjects consuming a mixed Mediterranean-Western diet, *American Journal of Clinical Nutrition* 66 (1997): 1443–1451; M. de Lorgeril and coauthors, Effect of a Mediterranean type of diet on the rate of cardiovascular complications in patients with coronary artery disease: Insights into the cardioprotective effect of certain nutriments, *Journal of the American College of Cardiology* 28 (1996): 1103–1108.
26. M. B. Katan, P. L. Zock, and R. P. Mensink, Dietary oils, serum lipoproteins, and coronary heart disease, *American Journal of Clinical Nutrition* 61 (1995): 1368S–1373S.
27. F. Visioli and C. Galli, The effect of minor constituents of olive oil on cardiovascular disease: New findings, *Nutrition Reviews* 56 (1998): 142–147.
28. E. B. Rimm and R. C. Ellison, Alcohol in the Mediterranean diet, *American Journal of Clinical Nutrition* 61 (1995): 1378S–1382S.
29. A. Tavani and C. LaVecchia, Fruit and vegetable consumption and cancer risk in a Mediterranean population, *American Journal of Clinical Nutrition* (1995): 1374S–1377S; L. H. Kushi, E. B. Lenart, and W. C. Willett, Health implications of Mediterranean diets in light of contemporary knowledge. 1. Plant foods and dairy products, *American Journal of Clinical Nutrition* 61 (1995): 1407S–1415S.
30. L. H. Kushi, E. B. Lenart, and W. C. Willett, Health implications of Mediterranean diets in light of contemporary knowledge. 2. Meat, wine, fats, and oils, *American Journal of Clinical Nutrition* 61 (1995): 1416S–1427S.
31. Oldways Preservation & Exchange Trust, www.oldwayspt.org, visited July 23, 1998.
32. M. Nestle, Editorial: The politics of dietary guidance—A new opportunity, *American Journal of Public Health* 84 (1994): 713–715.
33. P. O'Brien, Dietary shifts and implications for US agriculture, *American Journal of Clinical Nutrition* 61 (1995): 1390S–1396S.
34. V. Fønnebø, The healthy Seventh-Day Adventist lifestyle: What is the Norwegian experience? *American Journal of Clinical Nutrition* 59 (1994): 1124S–1129S.

CHAPTER 7

# Metabolism: Transformations and Interactions

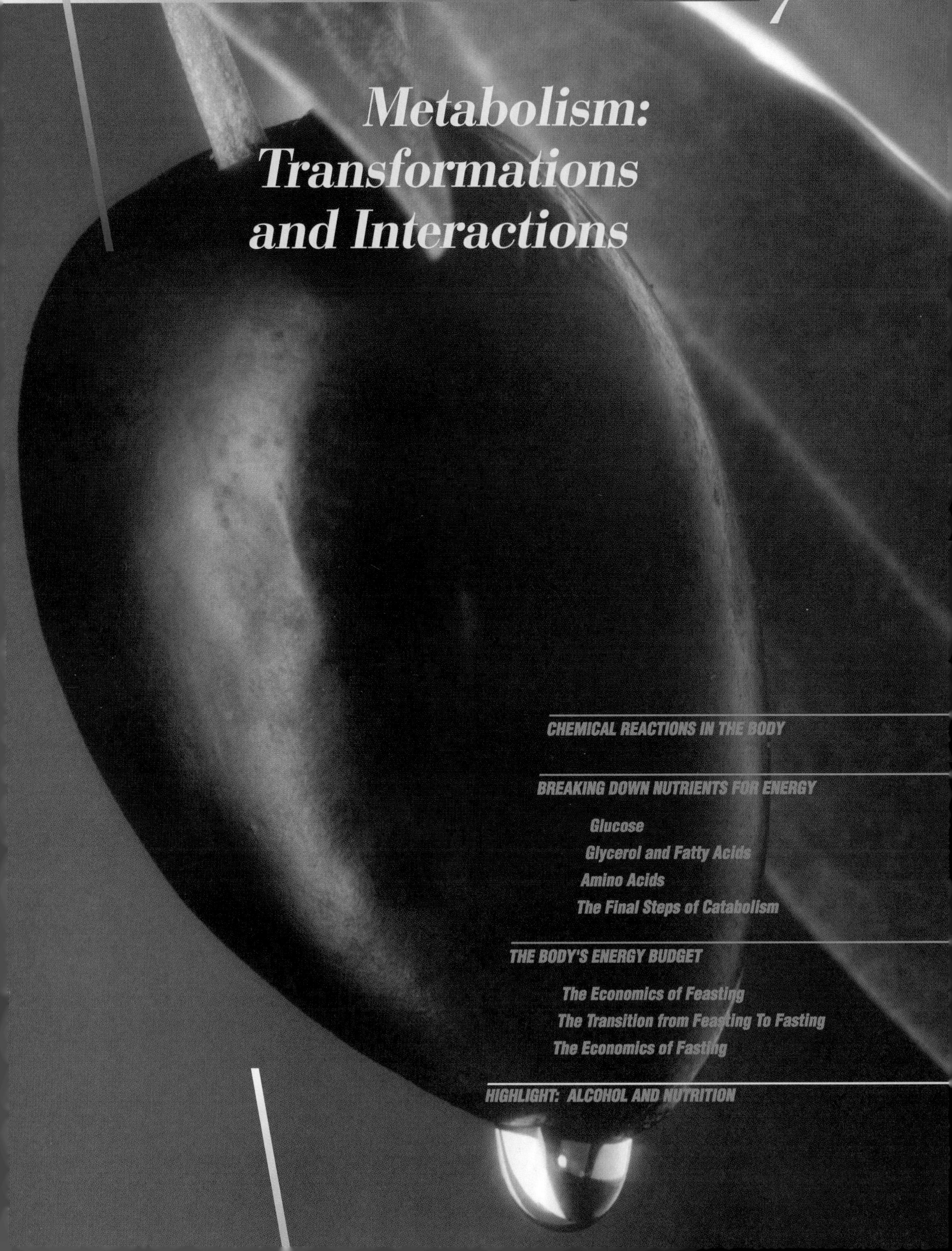

CHEMICAL REACTIONS IN THE BODY

BREAKING DOWN NUTRIENTS FOR ENERGY

- Glucose
- Glycerol and Fatty Acids
- Amino Acids
- The Final Steps of Catabolism

THE BODY'S ENERGY BUDGET

- The Economics of Feasting
- The Transition from Feasting To Fasting
- The Economics of Fasting

HIGHLIGHT: ALCOHOL AND NUTRITION

Almost all living things depend on the sun's energy. Plants rely directly on the sun to provide the light energy that drives the reactions of photosynthesis—the process by which plants make carbohydrate from carbon dioxide and water using the energy from sunlight. That energy from the sun is captured in the energy that holds atoms together—the energy of chemical bonds. We humans, and all other animals, use the sun indirectly, for we cannot photosynthesize. We depend on plants, or on animals that eat plants, for the food that gives us our energy-yielding nutrients.

This chapter answers the question, How do we obtain energy from foods? It describes the processes in the human body that *release* energy from the chemical bonds in nutrients. As the bonds break, they release energy in a controlled version of the process by which wood burns in a fire. Both wood and food have the potential to provide energy. When wood burns in the presence of oxygen, it generates heat and light (energy), steam (water), and some carbon dioxide and ash (waste). Similarly, during the body's metabolism of nutrients, energy, water, and carbon dioxide are released.

Energy derived from the metabolism of nutrients enables people to ride bicycles, compose music, and do everything else they do. An excess of food energy makes people fat, though most people do not understand exactly how it does this. Nor do most people understand exactly how physical activity speeds up energy use and fat loss. By studying metabolism, readers who are interested in losing weight will discover which foods contribute most to body fat and which to select when trying to lose weight safely. Physically active readers will discover which foods best support endurance activities and which to select when trying to build lean body mass.

***We receive energy from the sun by way of the foods we eat.***

**photosynthesis:** the process by which green plants make carbohydrates from carbon dioxide and water using the green pigment chlorophyll to trap the sun's energy.
- **photo** = light
- **synthesis** = put together (making)

# Chemical Reactions in the Body

Earlier chapters introduced some of the body's chemical reactions: the making and breaking of the bonds in carbohydrates, lipids, and proteins. The sum of these and all the other chemical reactions that go on in living cells is known as metabolism; and *energy* metabolism includes all the ways the body obtains and spends energy from food.

Chapters 4, 5, and 6 laid the groundwork for the study of metabolism; a brief review may be helpful. During digestion, the body breaks down the three energy-yielding nutrients—carbohydrates, lipids, and proteins—into four basic units that can be absorbed into the blood:

- From carbohydrates—glucose.*
- From lipids (triglycerides)—glycerol and fatty acids.
- From proteins—amino acids.

Amino acids are primarily building blocks for proteins, but they can flow into energy pathways if needed or if eaten in excess, so they are included.

Look for these four basic units to appear again and again in the metabolic reactions described in this chapter. Alcohol also enters many of the metabolic pathways; Highlight 7 focuses on how alcohol disrupts metabolism and how the body handles it.

Appendix B provides an overview of basic chemistry concepts.

**metabolism:** the sum total of all the chemical reactions that go on in living cells; **energy metabolism** includes all the reactions by which the body obtains and spends the energy from food.
- **meta** = among
- **bole** = change

*This chapter features glucose because of its central role in carbohydrate metabolism, but the monosaccharides fructose and galactose also enter the metabolic pathways.

**anabolism** (an-ABB-o-lism): reactions in which small molecules are put together to build larger ones. Anabolic reactions require energy.
- **ana** = up

**catabolism** (ca-TAB-o-lism): reactions in which large molecules are broken down to smaller ones. Catabolic reactions usually release energy.
- **kata** = down

**coupled reactions:** pairs of chemical reactions in which energy released from the breakdown of one compound is used to create a bond in the formation of another compound.

**ATP** or **adenosine** (ah-DEN-oh-seen) **triphosphate** (try-FOS-fate): a common high-energy compound composed of a purine (adenine), a sugar (ribose), and three phosphate groups.

• ***Building Reactions—Anabolism*** • The cells can use the basic units of energy-yielding nutrients to build body compounds. Glucose molecules may be joined together to make glycogen chains. Glycerol and fatty acids may be assembled into triglycerides. Amino acids may be linked together to make proteins. Each of these reactions starts with small, simple compounds and uses them as building blocks to form larger, more complex structures. Such reactions involve doing work and so require energy. The building up of body compounds is known as anabolism; this book represents anabolic reactions, wherever possible, with "up" arrows in chemical diagrams (such as those shown in Figure 7-1).

• ***Breakdown Reactions—Catabolism*** • The breaking down of body compounds is known as catabolism; catabolic reactions usually release energy and are represented, wherever possible, by "down" arrows in chemical diagrams (as in Figure 7-1). Catabolic reactions include the breakdown of glycogen to glucose, of triglycerides to fatty acids and glycerol, and of protein to amino acids. When the body needs energy, it breaks down any or all of these four basic units into even smaller units, as described later.

• ***The Transfer of Energy in Reactions*** • When a chemical bond breaks, energy can be released as heat, captured in another chemical bond, or both. Often, as one compound is broken apart, some of the energy is released as heat, and some is used to put together another compound. Such reactions, in which the breakdown of one compound provides energy for the building of another, are known as coupled reactions.

The energy released during catabolism is often captured by go-between molecules that can easily transfer that energy to other compounds. These molecules are sometimes called the body's "common energy currency," or "high-energy compounds." One such compound is ATP (adenosine triphosphate). The breakdown of energy-nutrient molecules is coupled to the making of many ATP molecules, which capture much of the released energy in their bonds.

**Figure 7-1**
**Anabolic and Catabolic Reactions Compared**

Note: You need not memorize a color code to understand the figures in this chapter but you may find it helpful to know that blue is used for carbohydrates, yellow for fats, and red for proteins.

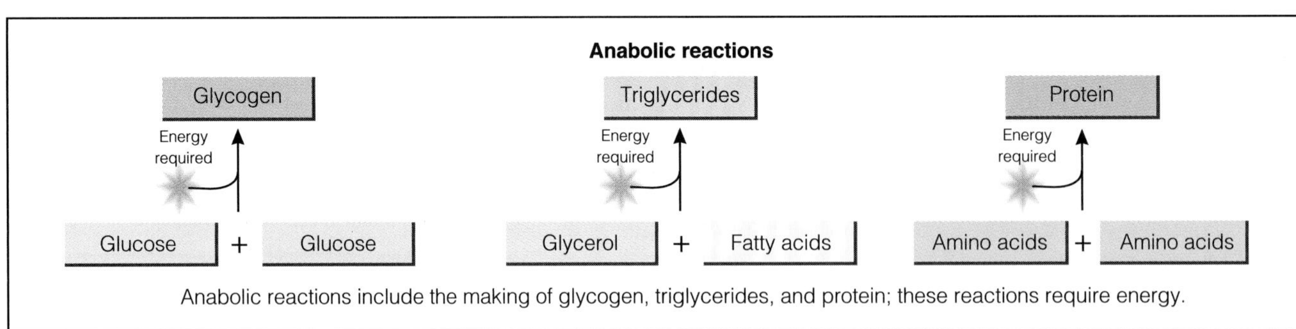


Anabolic reactions include the making of glycogen, triglycerides, and protein; these reactions require energy.

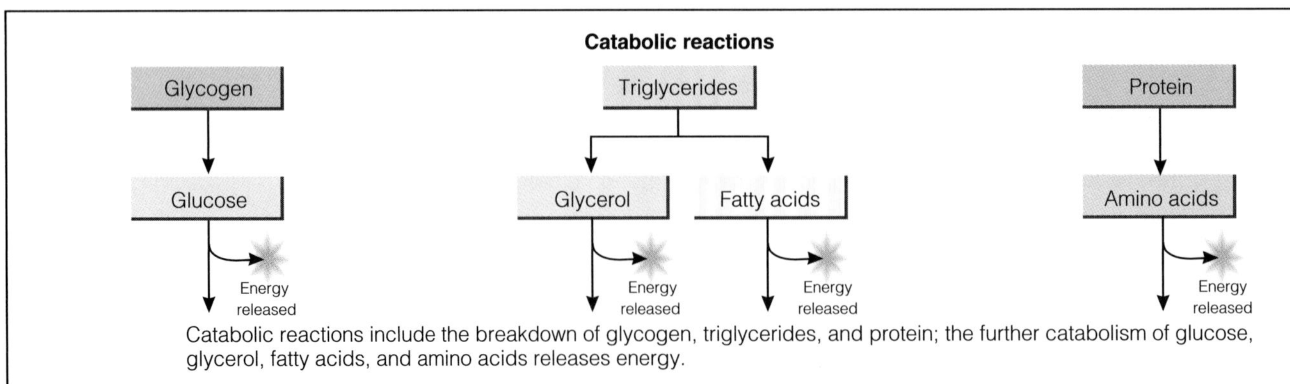


Catabolic reactions include the breakdown of glycogen, triglycerides, and protein; the further catabolism of glucose, glycerol, fatty acids, and amino acids releases energy.

Adenosine + 3 phosphate groups

**Figure 7-2**
**ATP (Adenosine Triphosphate)**

ATP is one of the body's quick-energy molecules. Notice that the bonds connecting the three phosphate groups have been drawn as wavy lines, indicating a high-energy bond. When these bonds are broken, a large amount of energy is released.

ATP, as its name indicates, contains three phosphate groups (see Figure 7-2). The energy in the bonds between each phosphate group is greater than the energy in most other chemical bonds. When energy is needed, hydrolysis reactions readily break these high-energy bonds, splitting off one or two phosphate groups and releasing their energy. These reactions, in turn, are coupled to other reactions that use that energy. Thus the body uses ATP to transfer the energy produced during catabolic reactions to power its anabolic reactions. Figure 7-3 illustrates how the

ATP = A–P~P~P.
(Each ~ denotes a "high-energy" bond.)

Reminder: *Hydrolysis* reactions break a molecule with the addition of water.

**Figure 7-3**
**Transfer of Energy by ATP**

Before the transfer of energy:

Glucose and fat have broken down, and some of their energy has been used to attach phosphate groups to molecules of adenosine diphosphate (ADP), building ATP.[a]

Enzymes are present that can hydrolyze ATP.

Building blocks are available to build compounds.[b]

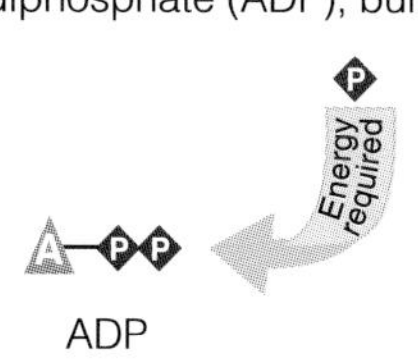


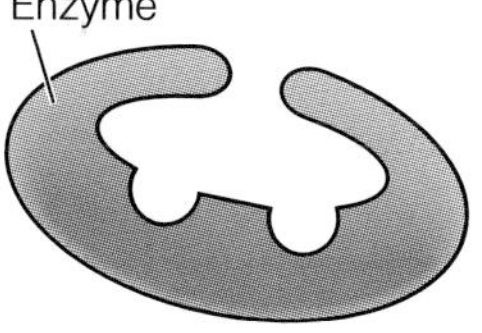


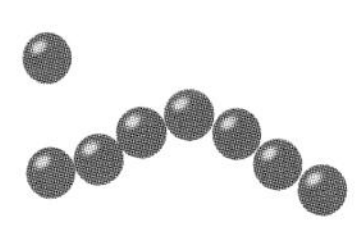

During the transfer of energy:

The enzyme hydrolyzes ATP, splitting off a phosphate group. Energy is released.

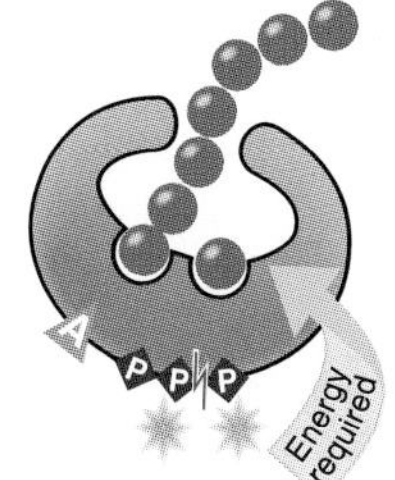


The enzyme uses that energy to attach a building block to a growing molecule.[c]

After the transfer of energy:

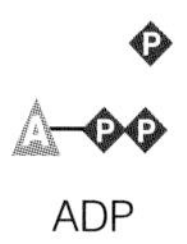


ADP and a phosphate group remain. More energy from nutrients will be required to regenerate ATP.

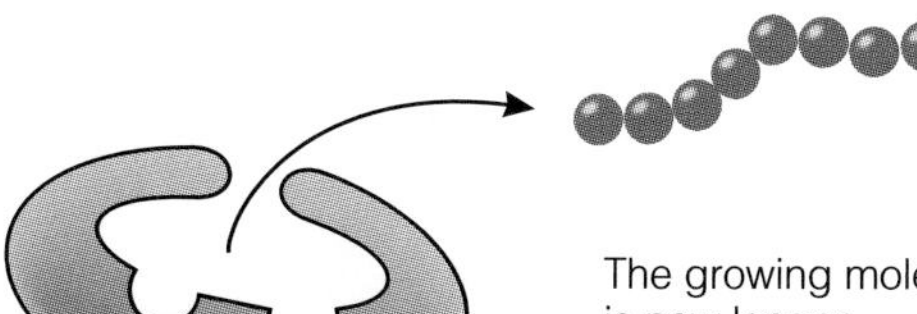

The enzyme complex is now ready to work again.

The growing molecule is now longer.

[a]ADP (adenosine diphosphate) is lower in energy than ATP; AMP (adenosine monophosphate) is even lower.
[b]Compounds that ATP energy might be used to build include glycogen, fat, proteins, and hormones, among others.
[c]In all such reactions, half or more of the total original energy is lost as heat, accounting for the temperature-raising effect of metabolism. ATP can also break apart without doing work and release all of its energy as heat if needed. The breakdown of ATP to release heat only, with no energy captured for use in another reaction, is an *uncoupled reaction.*

body uses ATP to carry its energy currency, build body structures, do other work, or generate heat, as needed.

Appendix A presents a brief summary of the structure and function of the cell.

• ***The Site of Reactions—Cells*** • Metabolic work is going on all the time within all the body's trillions of cells. Figure 7-4 depicts a typical cell and shows where the major reactions of energy production take place. The type and extent of metabolic activity vary depending on the type of cell, but of all the body's cells, the liver cells are the most versatile and metabolically active. Table 7-1 offers insights into the liver's work.

Reminder: An *enzyme* is a special protein that serves as a catalyst for a chemical reaction and is not altered in the process.

**coenzymes:** small organic molecules that work with enzymes to facilitate the enzymes' activity. Many coenzymes have B vitamins as part of their structures (Figure 10-1 in Chapter 10 illustrates coenzyme action).
• **co** = with

• ***The Helpers in Reactions—Enzymes and Coenzymes*** • Metabolic reactions almost always require enzymes to facilitate their action. In some cases, the enzymes need assistants to help them. Enzyme helpers are called coenzymes.

Coenzymes are small organic molecules that associate closely with most enzymes, but are not proteins themselves. The relationships between coenzymes and enzymes differ in detail, but one thing is true of all: without its coenzyme, an enzyme cannot function. Some of the B vitamins serve as coenzymes to the enzymes that release energy from glucose, glycerol, fatty acids, and amino acids. These B vitamin coenzymes stand alongside the metabolic pathways, so to speak, and help to keep the disassembly lines moving. Chapter 10 provides more details on the coenzyme actions of the B vitamins.

IN SUMMARY

During digestion the energy-yielding nutrients—carbohydrates, lipids, and proteins—are broken down to glucose, glycerol, fatty acids, and amino acids. Aided by enzymes and coenzymes, the cells use these products of digestion to build more complex compounds (anabolism) or break them down further to release energy (catabolism). The energy released during catabolism is often captured by high-energy compounds such as ATP.

**Figure 7-4**
**A Typical Cell (Simplified Diagram)**

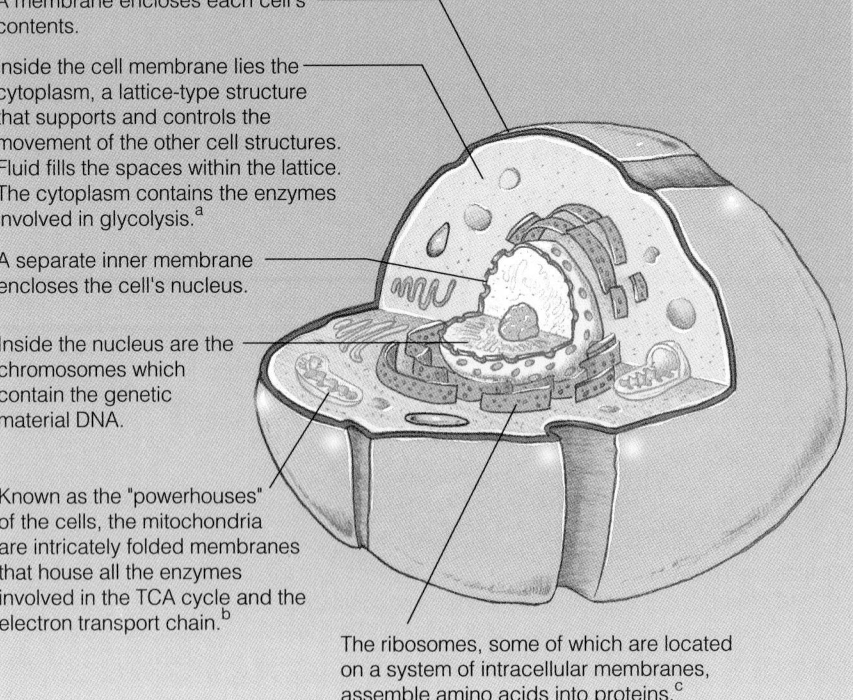
A membrane encloses each cell's contents.
Inside the cell membrane lies the cytoplasm, a lattice-type structure that supports and controls the movement of the other cell structures. Fluid fills the spaces within the lattice. The cytoplasm contains the enzymes involved in glycolysis.[a]
A separate inner membrane encloses the cell's nucleus.
Inside the nucleus are the chromosomes which contain the genetic material DNA.
Known as the "powerhouses" of the cells, the mitochondria are intricately folded membranes that house all the enzymes involved in the TCA cycle and the electron transport chain.[b]
The ribosomes, some of which are located on a system of intracellular membranes, assemble amino acids into proteins.[c]

[a]Glycolysis is described on pp. 202–203.
[b]The TCA cycle and electron transport chain are described on pp. 211–212.
[c]Figure 6-6 on p. 167 describes protein synthesis.

**Table 7-1**
**Metabolic Work of the Liver**

The liver is the most active processing center in the body. When nutrients enter the body, the liver receives them first; then it metabolizes, packages, stores, or ships them out for use by other organs. When alcohol, drugs, or poisons enter the body, they are also sent directly to the liver; here they are detoxified and their by-products shipped out for excretion. An enthusiastic anatomy and physiology professor once remarked that given the many vital activities of the liver, we should express our feelings for others by saying, "I love you with all my liver," instead of with all my heart. Granted, this declaration lacks romance, but it makes a valid point. Here are just *some* of the many jobs performed by the liver.

**Carbohydrates:**

- Converts fructose and galactose to glucose.
- Makes and stores glycogen.
- Breaks down glycogen and releases glucose.
- Breaks down glucose for energy when needed.
- Makes glucose from some amino acids and glycerol when needed.

**Lipids:**

- Builds and breaks down triglycerides, phospholipids, and cholesterol as needed.
- Breaks down fatty acids for energy when needed.
- Packages extra lipids in lipoproteins for transport to other body organs.
- Manufactures bile to send to the gallbladder for use in fat digestion.
- Makes ketone bodies when necessary.

**Proteins:**

- Manufactures nonessential amino acids that are in short supply.
- Removes from circulation amino acids that are present in excess of need and deaminates them or converts them to other amino acids.
- Removes ammonia from the blood and converts it to urea to be sent to the kidneys for excretion.
- Makes other nitrogen-containing compounds the body needs (such as bases used in DNA and RNA).
- Makes plasma proteins such as clotting factors.

**Other:**

- Detoxifies alcohol, other drugs, and poisons; prepares waste products for excretion.
- Helps dismantle old red blood cells and captures the iron for recycling.
- Stores most vitamins and many minerals.
- Forms lymph.

To renew your appreciation for this remarkable organ, you might want to review Figure 3-9 on p. 78.

# Breaking Down Nutrients for Energy

With these introductory remarks in mind, it is time to enter a cell and follow the various paths that glucose, glycerol, fatty acids, and amino acids take to yield energy. As you will see, each starts down a different path, but they all reach a common destination. At a certain point, they lose their individuality and most of their options—during catabolism all roads lead to energy.

Glucose, glycerol, fatty acids, and amino acids are the basic units derived from food, but a molecule of each of these compounds is made of still smaller units, the atoms—carbons, nitrogens, oxygens, and hydrogens. During catabolism, the body separates these atoms from one another. To follow this action, recall how many

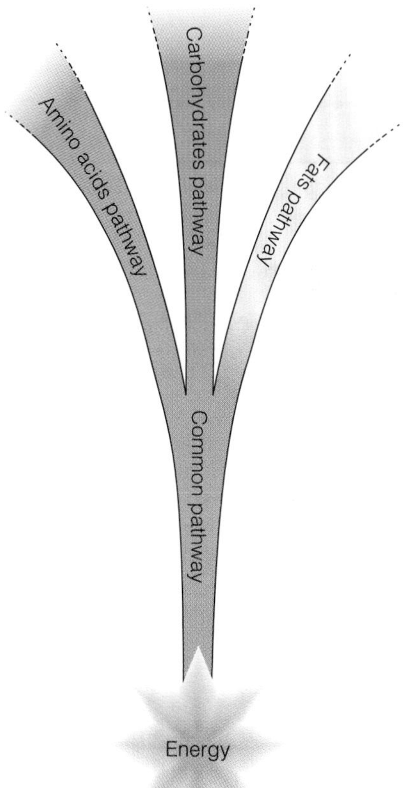


This simple overview introduces the metabolism that is presented in the upcoming text and detailed in the summary figure on p. 214.

carbons are in the "backbones" of these compounds:

- Glucose has 6 carbons:

- Glycerol has 3 carbons:

- A fatty acid usually has an even number of carbons, commonly 18 carbons:

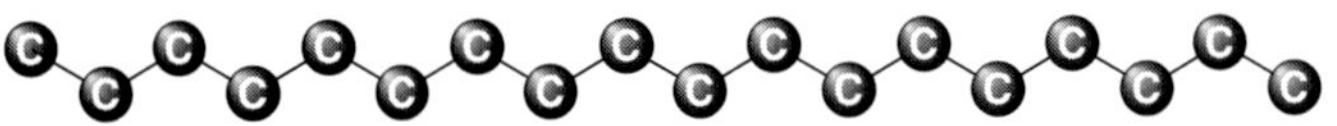

- An amino acid has 2, 3, or more carbons with a nitrogen attached:*

Full chemical structures and reactions appear both in the earlier chapters and in Appendix C; this chapter diagrams the reactions using just the compounds' carbon and nitrogen backbones.

What happens to these compounds inside cells can best be understood by starting with glucose. Two new names appear—pyruvate (a 3-carbon structure) and acetyl CoA (a 2-carbon structure with a coenzyme attached)—and the rest of the story falls into place around them.† A major point to notice in the following discussion is that all compounds that can be converted to pyruvate can be used to make glucose. Compounds that are converted directly to acetyl CoA cannot make glucose, however.

**pyruvate** (PIE-roo-vate): pyruvic acid, a 3-carbon compound that, in metabolism, can be derived from glucose, certain amino acids, or glycerol.

$CH_3$
|
C=O
|
COOH

**acetyl CoA** (ASS-eh-teel, or ah-SEET-il, coh-AY): a 2-carbon compound (**acetate,** or **acetic acid,** shown in Figure 5-2 on p. 126) to which a molecule of CoA is attached.

**CoA** (coh-AY): coenzyme A; the coenzyme derived from the B vitamin pantothenic acid and central to the energy metabolism of nutrients.

**glycolysis** (gligh-COLL-ih-sis): the metabolic breakdown of glucose to pyruvate. Glycolysis does not require oxygen (anaerobic).

- **glyco** = glucose
- **lysis** = breakdown

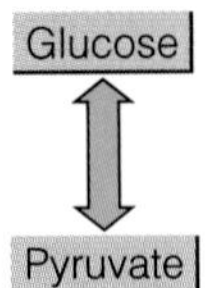


## Glucose

The first pathway glucose takes on its way to yield energy is called glycolysis (glucose splitting).‡ Figure 7-5 shows a simplified drawing of glycolysis, which actually involves several steps and several enzymes (see Appendix C for details). Along the way, the 6-carbon glucose is split in half, forming two 3-carbon compounds. These 3-carbon compounds continue along the pathway until they are converted to pyruvate. Thus the net yield of one glucose molecule is two pyruvate molecules. If they continue breaking down, both pyruvate molecules will release much of their energy to form ATP molecules and some of their energy as heat.

• ***Glucose-to-Pyruvate, and Back Again*** • After splitting glucose to pyruvate, a cell can make glucose again from pyruvate in a process similar to the reversal of glycolysis. Making glucose requires energy, however, and a few different enzymes. Still, glucose is retrievable from pyruvate, so the arrows between glucose and pyruvate could point up as well as down.

• ***Glucose-to-Pyruvate, an Anaerobic Pathway*** • To start the process of splitting glucose to pyruvate, the cell must use a little energy, but then it produces more energy than it had to invest initially.§ No oxygen has been required thus far—that

*The figures in this chapter usually show amino acids as compounds of 2, 3, or 5 carbons arranged in a straight line, but in reality amino acids may contain other numbers of carbons and assume other structural shapes (see Appendix C).
†The term *pyruvate* means a salt of *pyruvic acid.* (Throughout this book, the ending *-ate* is used interchangeably with *-ic acid;* for our purposes they mean the same thing.)
‡Glycolysis takes place in the cytoplasm of the cell (see Figure 7-4).
§The cell uses 2 ATP to begin the breakdown of glucose to pyruvate, but then gains 4 ATP for a net gain of 2 ATP.

The 6-carbon compound glucose is split into two interchangeable 3-carbon compounds that are converted to pyruvate in a series of reactions.

All of the other monosaccharides can enter the pathway at various points.

Glycolysis ends with the production of pyruvate (unless there is a shortage of oxygen, in which case pyruvate is converted to lactic acid, as a later section of the text describes).

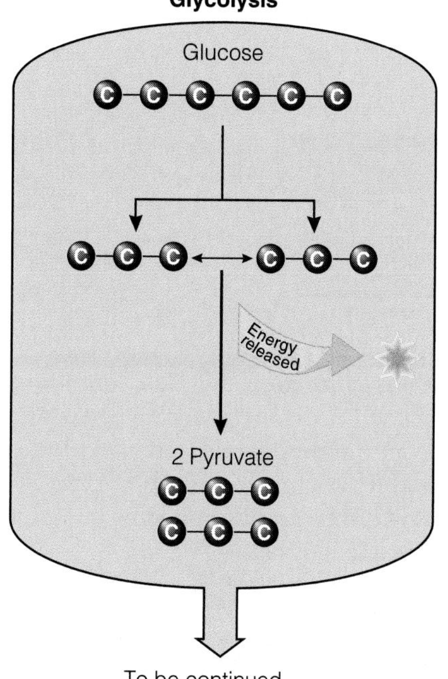


**Figure 7-5**

**Glycolysis: Glucose-to-Pyruvate Pathway (Anaerobic)**

Glucose splits to two 3-carbon compounds that become pyruvate. The pathway is called glycolysis (glucose splitting) and occurs in anaerobic conditions (does not require oxygen).

NOTE: These arrows point down indicating the breakdown of glucose to pyruvate during energy metabolism. Alternatively, the arrows could point up indicating the making of glucose from pyruvate, but that is not the focus of this discussion.

is, glycolysis is an anaerobic pathway. More energy can be released by taking pyruvate through additional metabolic reactions, but oxygen is needed for these reactions (they are aerobic).*

• ***Pyruvate-to-Acetyl CoA*** • If the cell needs energy and oxygen is available, it removes a carbon group (COOH) from the 3-carbon pyruvate to produce the 2-carbon acetyl CoA. The carbon group from pyruvate becomes carbon dioxide, which is released into the blood, circulated to the lungs, and breathed out. The remaining 2-carbon compound bonds with a molecule of CoA, becoming acetyl CoA. Figure 7-6 diagrams the pyruvate-to-acetyl CoA reaction.

• ***Glucose Retrieval via the Cori Cycle*** • Alternatively, when less oxygen is available, pyruvate is converted to lactic acid. This anaerobic reaction occurs to a limited extent even at rest, but increases dramatically during high-intensity exercise—that is, whenever exertion exceeds the capacity of the heart and lungs to clear carbon dioxide from the muscles. With limited oxygen available and limited carbon dioxide clearance, lactic acid accumulates in muscles, causing burning pain and fatigue. (To relieve this pain, relax the muscles frequently so that the circulating blood can carry the lactic acid away to the liver.) The liver can convert lactic acid to glucose in a recycling process called the Cori cycle. (Muscle cells cannot recycle lactic acid to glucose because they lack a necessary enzyme.)

• ***Muscles' Needs for Oxygen*** • Whenever carbohydrates, fats, or proteins are broken down to provide energy, oxygen is always ultimately involved in the process. The role of oxygen in metabolism is worth noticing, for it helps our understanding of physiology and metabolic reactions. Chapter 14 will describe the body's use of the energy nutrients to fuel physical activity, but the facts just presented offer a sneak preview. The first pathway in glucose metabolism (glycolysis) yields some energy without oxygen (it is anaerobic), but anaerobic metabolism cannot be sustained for long. Conversely, the later pathways require oxygen (they are aerobic) and can be sustained for a long time. Aerobic metabolism yields by far the *most energy* and so is crucial for endurance activities.

*With sufficient oxygen, pyruvate molecules enter the mitochondria of the cell (see Figure 7-4) where they will be converted to acetyl CoA.

**anaerobic** (AN-air-ROE-bic): not requiring oxygen.
• **an** = not

**aerobic** (air-ROE-bic): requiring oxygen.

**lactic acid:** an acid produced from pyruvate during anaerobic metabolism.

**Cori cycle:** the path from muscle glycogen to glucose to pyruvate to lactic acid (which travels to the liver) to glucose (which can travel back to the muscle) to glycogen; named after the scientist who elucidated this pathway.

**Figure 7-6**

**Pyruvate-to-Acetyl CoA (Aerobic)**

Each pyruvate loses a carbon as carbon dioxide and picks up a molecule of CoA, becoming acetyl CoA. The arrow goes only one way (down), because the step is not reversible. Result (from 1 glucose): 2 carbon dioxide and 2 acetyl CoA.

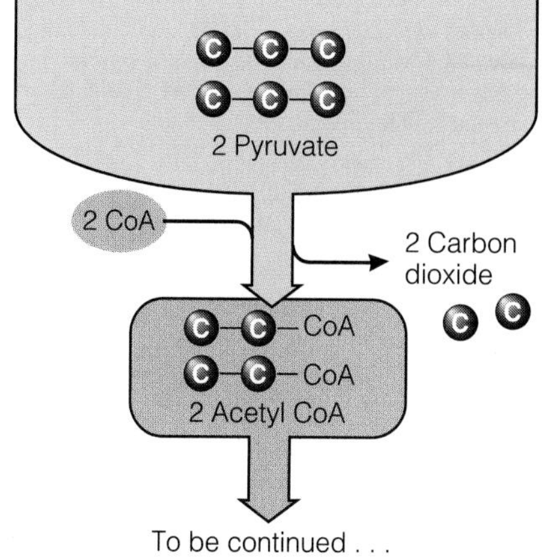

**Figure 7-7**
**The Paths of Pyruvate and Acetyl CoA**

Pyruvate and acetyl CoA may follow several reversible paths, but the path from pyruvate to acetyl CoA is irreversible.

Amino acids that can be used to make glucose are called *glucogenic;* amino acids that are converted to acetyl CoA are called *ketogenic.*

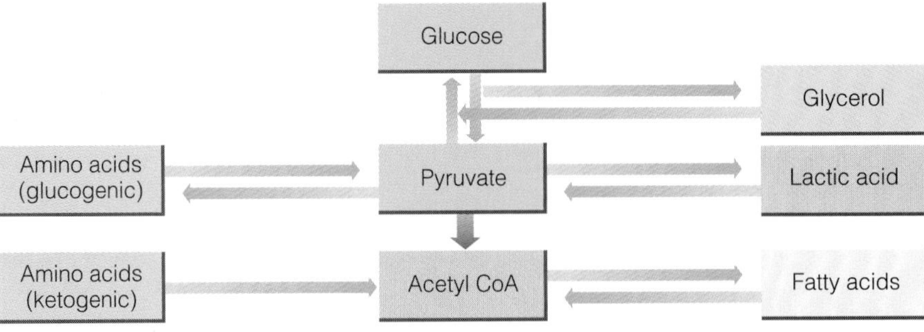


• ***Pyruvate-to-Acetyl CoA, an Irreversible Step*** • The step from pyruvate to acetyl CoA is metabolically irreversible: a cell cannot retrieve the shed carbons from carbon dioxide to remake pyruvate, and then glucose. It is a one-way step and is therefore shown with only a "down" arrow in Figure 7-7. Notice that acetyl CoA can be used as a building block for fatty acids, but it cannot be used to remake glucose.

**TCA cycle:** a series of metabolic reactions that break down molecules of acetyl CoA to carbon dioxide and hydrogen atoms; more details are provided later in the text.

• ***Acetyl CoA-to-Carbon Dioxide: The TCA Cycle*** • Once made, acetyl CoA has the option of taking different metabolic paths, depending on the cell's needs. If the cell needs energy, acetyl CoA may proceed through a series of reactions known as the TCA cycle. The TCA cycle converts the 2-carbon acetyl CoA to two carbon dioxide molecules and frees its coenzyme (CoA) to be reused (see Figure 7-8). In the process, much more energy is made available than during glycolysis (more details are given later).

• ***Acetyl CoA-to-Fat*** • If energy is not needed, acetyl CoA will not enter the TCA cycle, but will be used to make fatty acids instead. This explains how carbohydrate, eaten in excess of the body's needs, can lead to fat deposition. As you will see, fat or protein eaten in excess of immediate energy needs can take the same pathway to body fat.

**Figure 7-8**
**The Breakdown of Acetyl CoA**

The complete oxidation of acetyl CoA is accomplished through the reactions of the **TCA** (tricarboxylic acid) **cycle,** or **Krebs cycle** (named for the biochemist who elucidated them), and the **electron transport chain.** In the TCA cycle, the acetyl CoA carbons are converted to carbon dioxide. Each CoA returns to pick up another acetate (coming from glucose, glycerol, fatty acids, and amino acids).

The net result is that acetyl CoA splits, the carbons combine with oxygen, and the energy originally in the acetyl CoA is stored in ATP and similar compounds, thus becoming available for the body's use. Chapter 10 describes how the B vitamin coenzymes participate in these metabolic pathways. For more details, see the text and Appendix C.

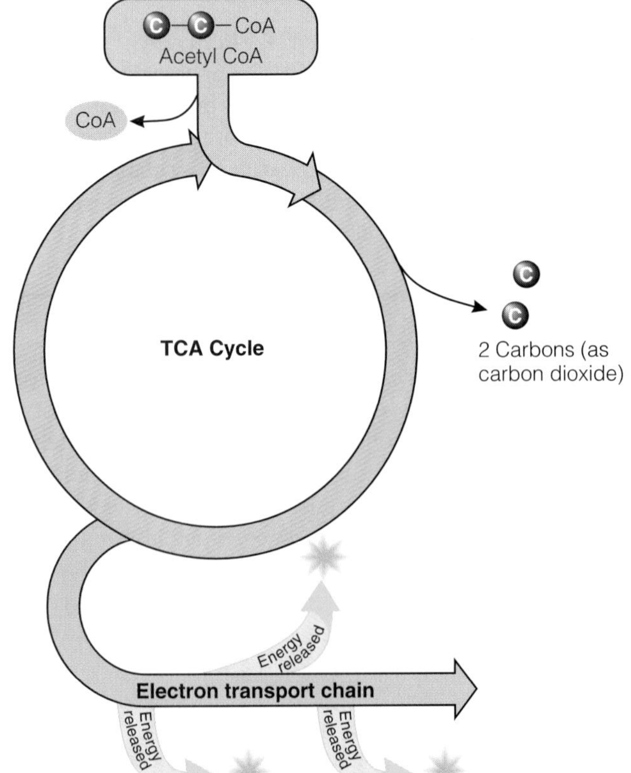

IN SUMMARY

Figure 7-9 combines Figures 7-5, 7-6, and 7-8 and shows the whole sequence of steps in glucose breakdown. In summary, the main steps in the catabolism of glucose are:

Glucose to
pyruvate to
acetyl CoA to
carbon dioxide.

The first step—glucose-to-pyruvate—is anaerobic, but the later steps are aerobic and require oxygen. Keep in mind that glucose can be retrieved only from pyru-

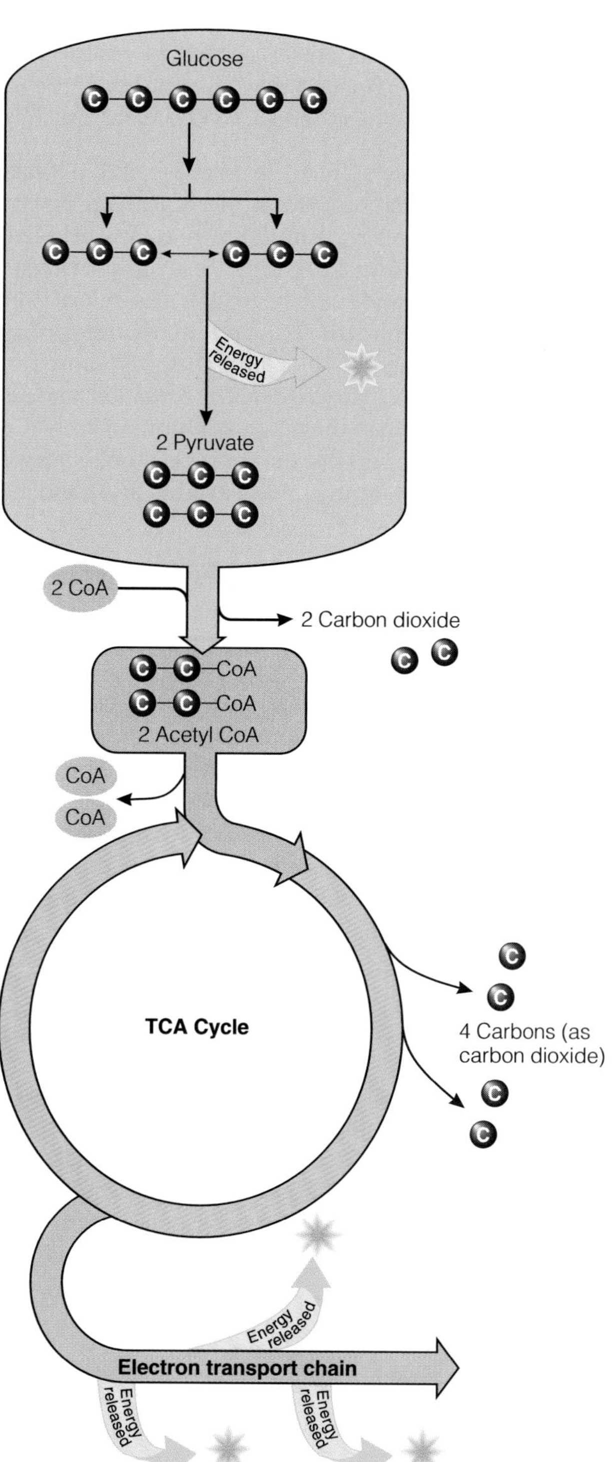


**Figure 7-9**
**Glucose-to-Energy Pathway**

Through these processes, energy from glucose is made available to do the cells' work. Ultimately, glucose is completely disassembled to single-carbon fragments, and the fragments are combined with oxygen to form carbon dioxide. Much of the energy released is trapped and stored in ATP. Details of the TCA cycle and the electron transport chain are given later and in Appendix C.

vate or compounds earlier in the pathway. Once the commitment to acetyl CoA is made, glucose is not retrievable; acetyl CoA can go on to carbon dioxide, fat, or other compounds but not back to glucose.

## Glycerol and Fatty Acids

Once glucose breakdown is understood, fat and protein breakdown are easily learned, for all three share a common metabolic pathway. Recall that triglycerides can break down to glycerol and fatty acids. Figure 7-10 repeats the pathway that glucose follows and shows how glycerol and fatty acids enter into it.

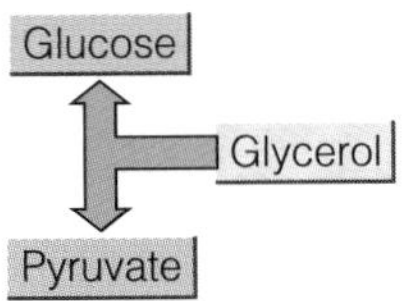


• ***Glycerol-to-Pyruvate*** • Glycerol (a 3-carbon compound like pyruvate, but with a different arrangement of H and OH on the C) is easily converted to another 3-carbon compound. This compound may go either "up" the pathway to form glucose or "down" to form pyruvate and acetyl CoA and, finally, carbon dioxide.

**fatty acid oxidation:** the metabolic breakdown of fatty acids to acetyl CoA; also called *beta oxidation.*

• ***Fatty Acids-to-Acetyl CoA*** • Unlike glycerol, which is a 3-carbon compound that can be converted to 3-carbon pyruvate, fatty acids are taken apart 2 carbons at a time in a series of aerobic reactions known as fatty acid oxidation.* Figure 7-11 (on p. 208) illustrates fatty acid oxidation and shows that in the process, each 2-carbon fragment splits off and combines with a molecule of CoA to make acetyl CoA. Each acetyl CoA then enters the TCA cycle in the same manner as acetyl CoA from glucose does (review Figure 7-10). A little energy is released each time a 2-carbon fragment breaks off from a fatty acid during oxidation, and nearly three times as much energy is released when these 2-carbon units of acetyl CoA are fully oxidized. If the cell does not need energy, the acetyl CoA molecules will combine with each other to make body fat, in the same way acetyl CoA produced from excess carbohydrate does.

• ***Glucose Not Retrievable from Fatty Acids*** • Cells can make glucose from pyruvate and other 3-carbon compounds, as mentioned earlier, but they cannot make glucose from the 2-carbon fragments of fatty acids. In chemical diagrams, the arrow between pyruvate and acetyl CoA always points only one way—down—and fatty acid fragments enter the metabolic path below this arrow (review Figure 7-7). Thus fatty acids cannot be used to make glucose.

The significance of this is that fat, for the most part, normally cannot provide energy for red blood cells or the brain and nervous system, which require glucose as fuel. Remember that almost all dietary fats are triglycerides, and that triglycerides contain only one small molecule of glycerol (3 carbons) with three fatty acids. The glycerol can yield glucose, but that represents only 3 of the 50 or so carbon atoms in the molecule—about 5 percent of its weight (see Figure 7-12 on p. 208). Thus fat is an insignificant source of glucose; about 95 percent of fat cannot be converted to glucose.

Reminder: The making of glucose from the glycerol of triglycerides (or from amino acids) is *gluconeogenesis*. About 5% of fat (the glycerol portion of a triglyceride) and most amino acids can be converted to glucose (review Figure 7-7).

IN SUMMARY

The body can convert the small glycerol portion of a triglyceride to either pyruvate (and then glucose) or acetyl CoA. The fatty acids of a triglyceride, on the other hand, cannot make glucose, but they can provide acetyl CoA. Acetyl CoA from either source may then enter the TCA cycle to produce energy or combine with other molecules of acetyl CoA to make body fat.

*Oxidation of fatty acids occurs in the mitochondria of the cells (see Figure 7-4).

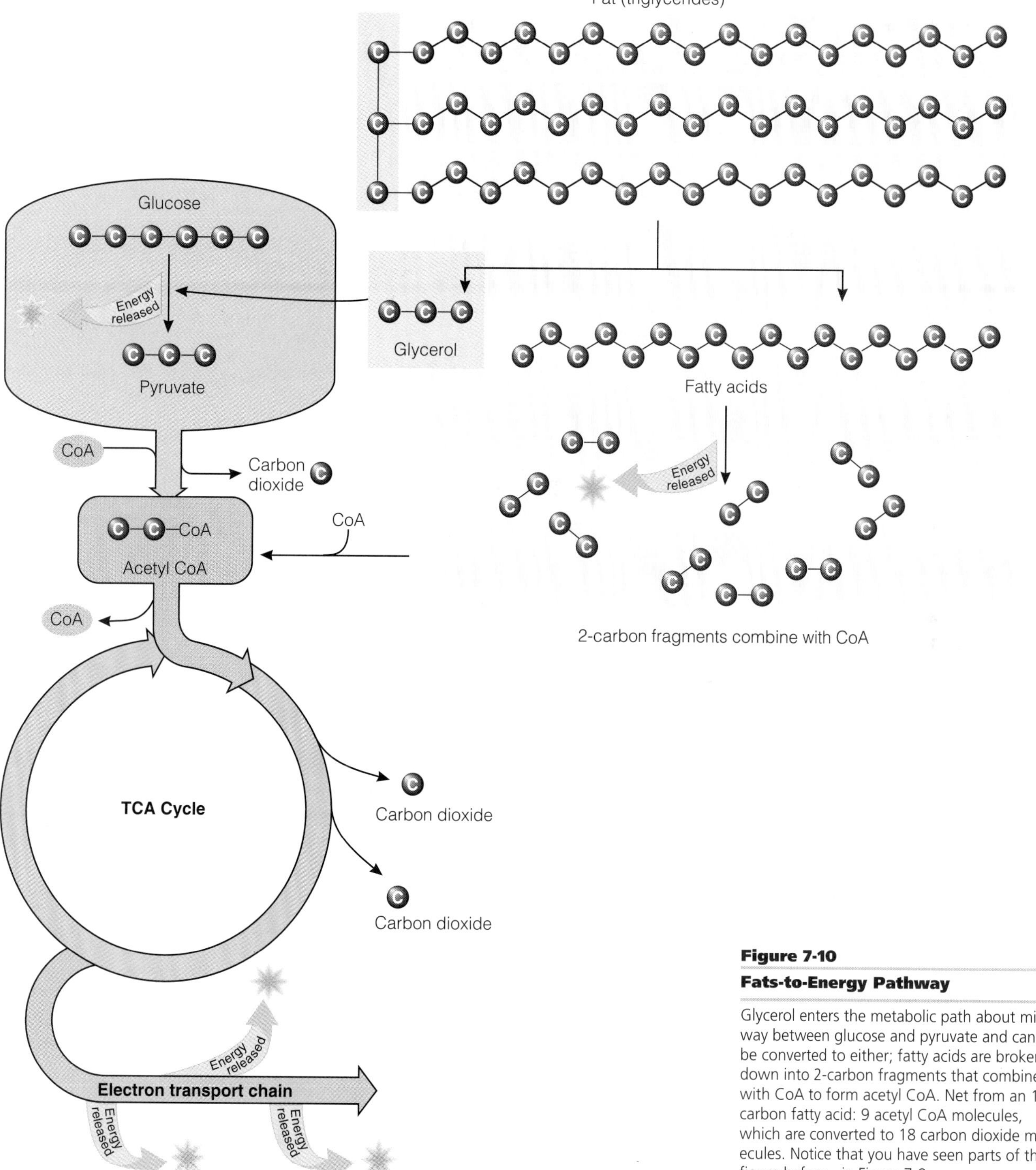


**Figure 7-10**

**Fats-to-Energy Pathway**

Glycerol enters the metabolic path about midway between glucose and pyruvate and can be converted to either; fatty acids are broken down into 2-carbon fragments that combine with CoA to form acetyl CoA. Net from an 18-carbon fatty acid: 9 acetyl CoA molecules, which are converted to 18 carbon dioxide molecules. Notice that you have seen parts of this figure before—in Figure 7-9.

## Amino Acids

The preceding two sections have shown how the breakdown of carbohydrate and fat provides energy for the body's use. One energy-yielding nutrient remains: protein or, rather, the amino acids of protein.

• ***Amino Acid Catabolism*** • If amino acids are needed for energy, or if they are consumed in excess of the need to synthesize protein, they enter the metabolic pathway as shown in Figure 7-13 (on p. 209). First, amino acids are deaminated

Reminder: *Deamination* is the reaction that removes the nitrogen-containing amino group from an amino acid.

**Figure 7-11**
**Fatty Acid Oxidation**

During oxidation, fatty acids are taken apart to 2-carbon fragments that combine with CoA to make acetyl CoA. Fatty acid oxidation is a series of aerobic reactions.

The fatty acid is first activated by coenzyme A.

A little energy is released each time a carbon-carbon bond is cleaved.

Another CoA joins the chain, and the bond at the second carbon (the beta-carbon) weakens. Acetyl CoA splits off, leaving a fatty acid that is two carbons shorter.

The shorter fatty acid enters the pathway and the cycle repeats. The molecules of acetyl CoA enter the TCA cycle, yielding abundant energy.

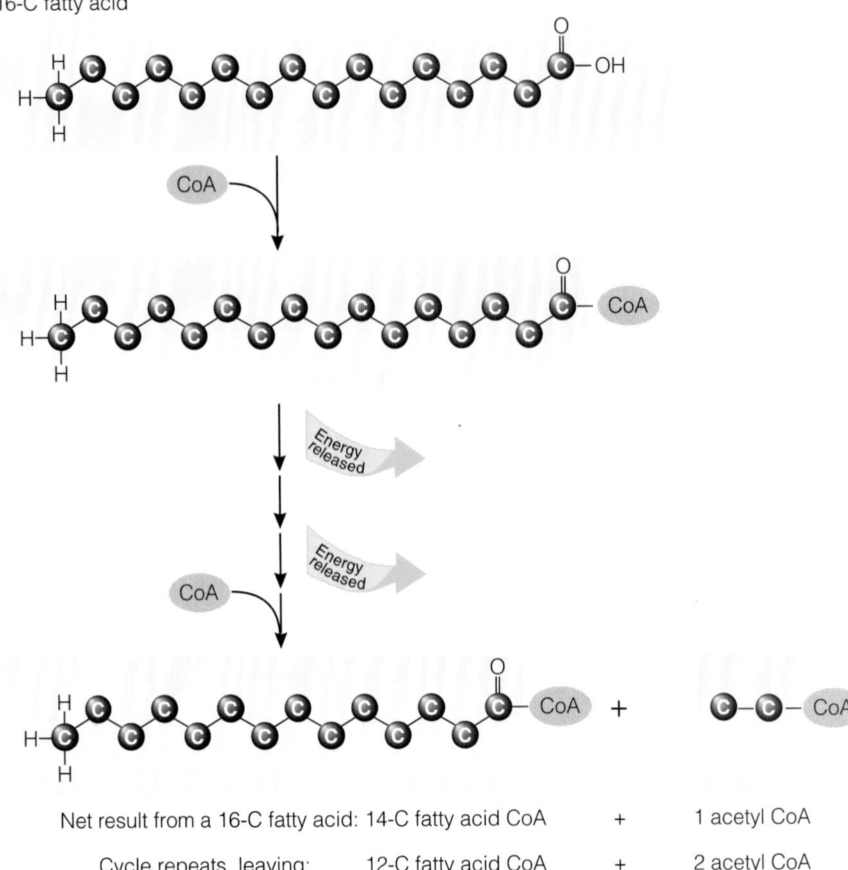


| | | | |
|---|---|---|---|
| Net result from a 16-C fatty acid: | 14-C fatty acid CoA | + | 1 acetyl CoA |
| Cycle repeats, leaving: | 12-C fatty acid CoA | + | 2 acetyl CoA |
| Cycle repeats, leaving: | 10-C fatty acid CoA | + | 3 acetyl CoA |
| Cycle repeats, leaving: | 8-C fatty acid CoA | + | 4 acetyl CoA |
| Cycle repeats, leaving: | 6-C fatty acid CoA | + | 5 acetyl CoA |
| Cycle repeats, leaving: | 4-C fatty acid CoA | + | 6 acetyl CoA |
| Cycle repeats, leaving: | 2-C fatty acid CoA* | + | 7 acetyl CoA |

*Notice that 2-C fatty acid CoA = acetyl CoA, so that the final yield from a 16-C fatty acid is 8 acetyl CoA.

(that is, they lose their nitrogen as described in the next section), and then they are catabolized in a variety of ways. Some amino acids can be converted to pyruvate; others are converted to acetyl CoA; and still others enter the TCA cycle directly as compounds other than acetyl CoA.

• ***Glucose Retrievable from Amino Acids*** • As you might expect, amino acids that are used to make pyruvate can provide glucose for the body, whereas those used to make acetyl CoA can provide additional energy or make body fat but cannot make glucose. Amino acids entering the TCA cycle directly can continue in

**Figure 7-12**
**The Carbons of a Typical Triglyceride**

A typical triglyceride contains only one small molecule of glycerol (3 C), but has three fatty acids (each about 18 C on the average, or about 54 C). Only the glycerol portion of a triglyceride can yield glucose.

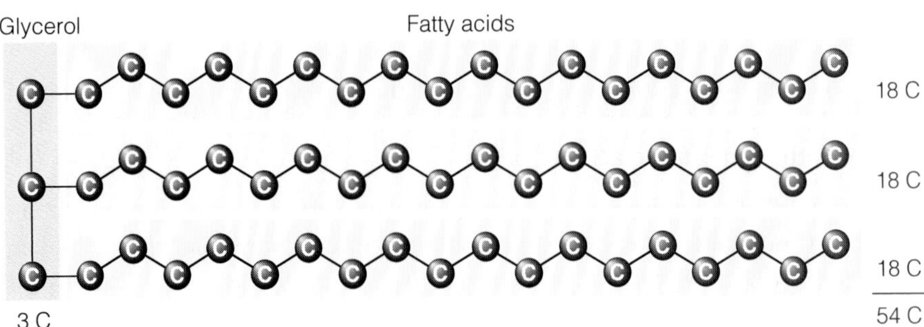

**Figure 7-13**

**Amino Acids-to-Energy Pathway**

Notice that you have seen parts of this figure before—in Figures 7-9 and 7-10. Note: The arrows from pyruvate and the TCA cycle to amino acids are possible only for *nonessential* amino acids; remember, the body cannot make essential amino acids.

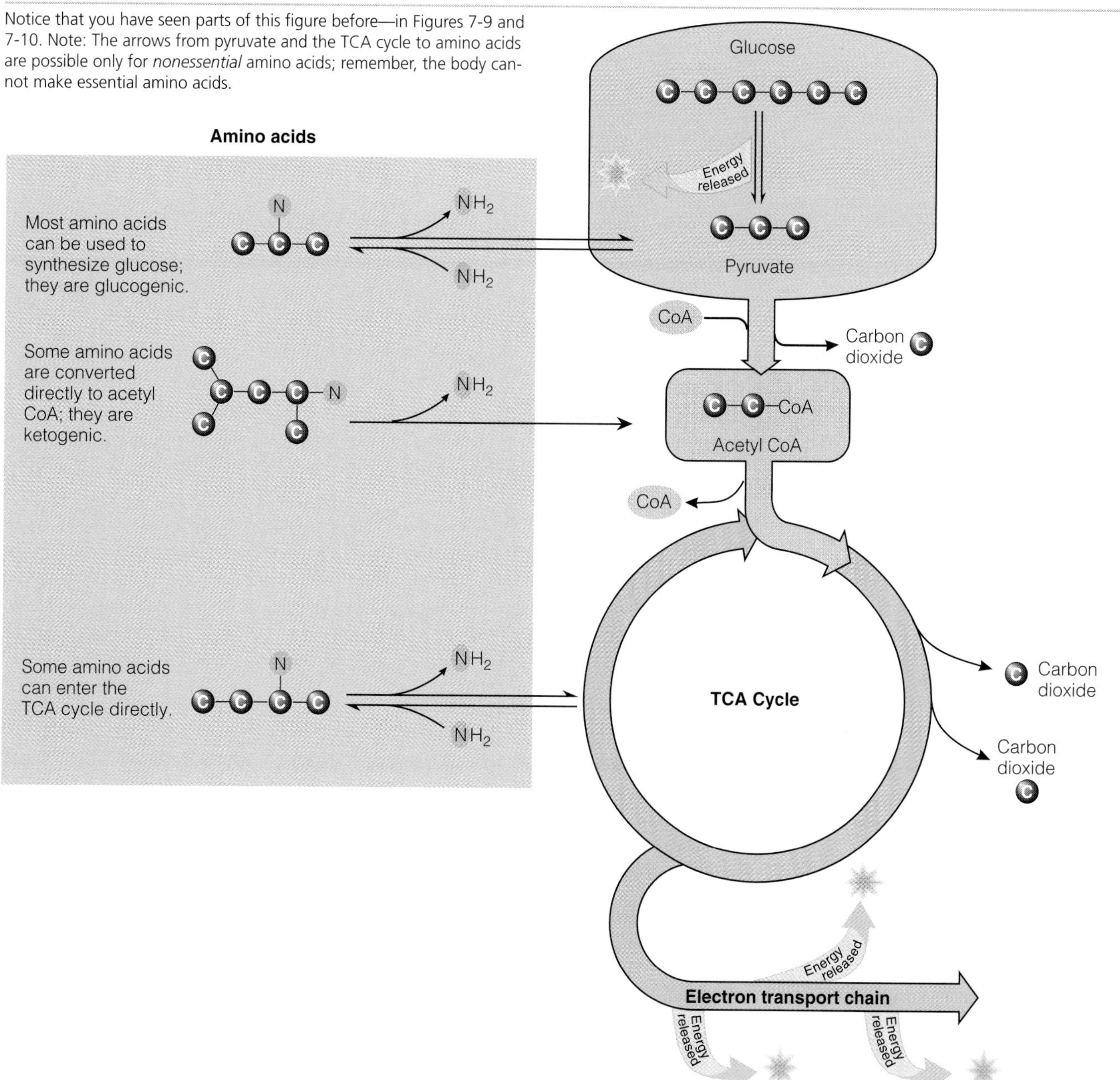


the cycle and generate energy; alternatively, they can generate glucose.[1] Thus protein, unlike fat, is a fairly good source of glucose when carbohydrate is not available; and like fat and carbohydrate, it is converted to body fat when consumed beyond the body's needs.

A key to understanding these metabolic pathways is learning which fuels can be converted to glucose and which cannot. The parts of protein and fat that can be converted to pyruvate *can* provide glucose for the body, whereas the parts that are converted to acetyl CoA *cannot* provide glucose, but can readily provide fat. You must have glucose to fuel the activities of the central nervous system and red blood cells. If you don't obtain glucose from food, your body will devour its own lean tissue to provide it. Therefore, to keep this from happening, you need to supply fuels that can provide glucose—primarily carbohydrate. If you feed your body only fat, which delivers mostly acetyl CoA, you put your body in the position of having to break down protein tissue for glucose. If you feed your body only protein, you put your body in the position of having to convert protein to glucose.

**Figure 7-14**
**Keto Acids**

The deamination of an amino acid produces ammonia ($NH_3$) and a keto acid.

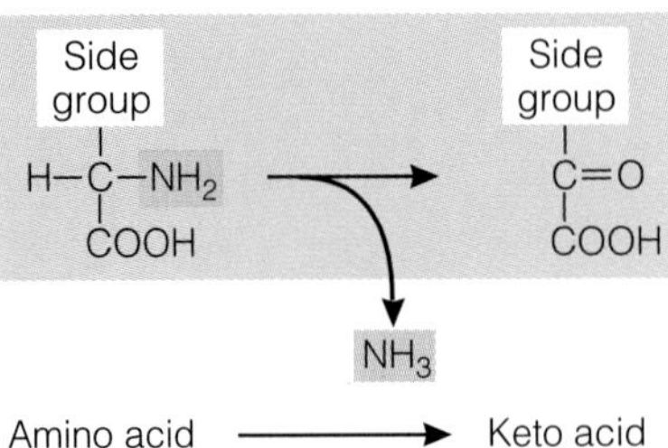


Given a source of $NH_3$, the body can make nonessential amino acids from keto acids.

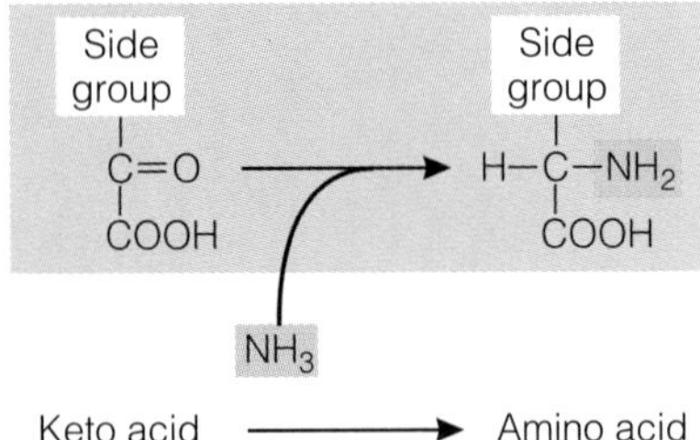


Reminder: Diet and health recommendations advise that daily energy intake provide:
- 55–60% carbohydrate.
- ≤ 30% fat.
- 10–15% protein.

Clearly, the best diet supplies some protein, some fat, and abundant carbohydrate.

• ***Amino Acids-to-Fat*** • Once amino acids have been converted to acetyl CoA, if energy is not needed, fatty acids are made and stored as triglycerides in adipose tissue. (Recall from Chapter 6 that the body cannot store surplus amino acids as such; it has to convert them to other compounds.) Thus protein can also add to fat stores if eaten in excess.

People who eat huge portions of meat and other protein-rich foods may wonder why they have weight problems. Not only does the fat in those foods lead to fat storage, but the protein can, too, when energy intake exceeds energy needs. Many fad weight-loss diets encourage high protein intakes based on the false assumption that protein builds only muscle, not fat (see Highlight 8 for more details).

Products of deamination:
- Keto acid.
- Ammonia.

**keto** (KEY-toe) **acid:** an organic acid that contains a carbonyl group (C=O).

**ammonia:** a compound with the chemical formula $NH_3$; produced during the deamination of amino acids.

**transamination** (TRANS-am-ih-NAY-shun): the transfer of an amino group from one amino acid to a keto acid, producing a new nonessential amino acid and a new keto acid.

**urea** (you-REE-uh): the principal nitrogen-excretion product of metabolism. Two ammonia fragments are combined with carbon dioxide to form urea.

• ***Deamination*** • When amino acids are metabolized for energy or used to make fat, they must be deaminated first. Two products result from deamination. One is, of course, the structure without its amino group—often a keto acid (see Figure 7-14). The other product is ammonia, a toxic compound chemically identical to the strong-smelling ammonia in bottled cleaning solutions. Ammonia is a base, and if the body produces larger quantities than it can handle, the blood's critical acid-base balance becomes upset.

• ***Transamination*** • As the discussion of protein in Chapter 6 pointed out, only some amino acids are essential; others can be made in the body, given a source of nitrogen. The body does this by transferring an amino group from one amino acid to its corresponding keto acid, producing a new amino acid and a new keto acid, as shown in Figure 7-15. Through many such reactions, involving many different keto acids, the liver cells can synthesize the nonessential amino acids.

• ***Ammonia-to-Urea in the Liver*** • The liver continuously produces small amounts of ammonia in deamination reactions. Some of this ammonia provides the nitrogen needed for the synthesis of nonessential amino acids. The liver quickly combines any unused ammonia with carbon dioxide to make urea, a much less toxic compound (see Figure 7-16). The diagram greatly oversimplifies the reactions; details are shown in Appendix C.

**Figure 7-15**
**Transamination to Make a Nonessential Amino Acid**

The body can transfer amino groups ($NH_2$) from an amino acid to a keto acid, forming a new nonessential amino acid and a new keto acid.

Transamination reactions require the vitamin $B_6$ coenzyme.

Side group / C=O / COOH + Side group / H–C–$NH_2$ / COOH → Side group / H–C–$NH_2$ / COOH + Side group / C=O / COOH

Keto acid A + Amino acid B → Amino acid A + Keto acid B

• ***Urea Excreted via the Kidneys*** • Liver cells release urea into the blood, where it circulates until it passes through the kidneys (see Figure 7-17 on p. 212). The kidneys then remove urea from the blood for excretion in the urine. Normally, the liver efficiently scoops up all the ammonia, makes urea from it, and releases the urea into the blood; then the kidneys clear all the urea from the blood. This division of labor allows easy diagnosis of diseases of both organs. If the liver is sick, blood ammonia will be high; if the kidneys are sick, blood urea will be high.

• ***Water Needed to Excrete Urea*** • Urea is the body's principal vehicle for excreting unused nitrogen, and the amount produced increases with protein intake. To keep urea in solution, the body needs water. For this reason, a person who regularly consumes a high-protein diet (say, 100 grams a day or more) must drink more water than usual; without extra water, the person risks an accumulation of urea in the blood. In fact, the weight loss from water loss makes high-protein diets *appear* to be effective, but water loss, of course, is of no value to the person who wants to lose body fat (as Highlight 8 explains).

**Figure 7-16**

**Urea Synthesis**

When amino nitrogen is stripped from amino acids, ammonia is produced. The liver detoxifies ammonia before releasing it into the bloodstream by combining it with another waste product, carbon dioxide, to produce urea. See Appendix C for details.

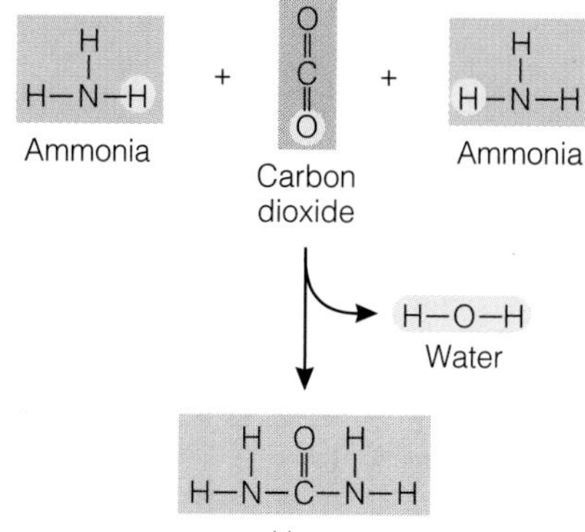


IN SUMMARY

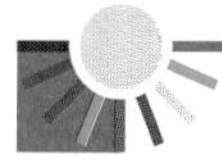

The body can use some amino acids to produce glucose, while others can be used either to generate energy or to make fat. Before an amino acid enters one of these metabolic pathways, its nitrogen-containing amino group is removed through deamination. Some of the nitrogen may be used to make nonessential amino acids and other nitrogen-containing compounds; the rest is cleared from the body via urea synthesis in the liver and excretion in the kidneys.

# *The Final Steps of Catabolism*

To review the ways the body can use the energy-yielding nutrients, see the summary table below. To obtain energy, the body uses glucose and fatty acids as its primary fuels, although it can use amino acids to provide energy if need be. To make glucose, the body can use all carbohydrates and most amino acids, but can convert only 5 percent of fat (the glycerol portion) to glucose. To make body proteins, the body needs amino acids. It can use glucose to make some nonessential amino acids when nitrogen is available; it cannot use fats to make body proteins. Finally, when energy is consumed beyond the body's needs, the body can convert all three energy-yielding nutrients to fat for storage.

• ***The TCA Cycle*** • Thus far the discussion has followed each of the energy-yielding nutrients to the point where acetyl CoA enters the TCA cycle.* The TCA

*The TCA cycle reactions take place in the mitochondria of the cell (see Figure 7-4).

IN SUMMARY

**The Body's Use of Energy-Yielding Nutrients**

| Nutrient | Yields Energy | Yields Glucose | Yields Amino Acids and Body Proteins | Yields Fat Stores |
|---|---|---|---|---|
| Carbohydrates (glucose) | Yes | Yes | Yes—when nitrogen is available, can yield *nonessential* amino acids | Yes |
| Lipids (triglycerides) | Yes | No—glycerol provides minimal amount | No | Yes |
| Proteins (amino acids) | Yes—if needed | Yes—when carbohydrate is unavailable | Yes | Yes |

cycle serves as a busy traffic center through which these 2-carbon acetyl CoA molecules pass on their way to carbon dioxide, releasing their energy to other compounds as they go.

The TCA cycle is called a cycle, but that doesn't mean it regenerates acetyl CoA. Acetyl CoA goes one way only—to carbon dioxide, releasing energy as it goes. The TCA cycle is a circular path, though, in the sense that a 4-carbon carbohydrate-like compound does cycle around and around.* This compound picks up acetyl CoA (a 2-carbon compound), drops off one carbon (as carbon dioxide), then another carbon (as carbon dioxide), and returns to pick up another acetyl CoA. As for the acetyl CoA, its carbons go only one way—to carbon dioxide (see Appendix C for additional details).

As acetyl CoA molecules break down to carbon dioxide, hydrogen atoms with their electrons are removed from the compounds in the cycle. Coenzymes of the B vitamins niacin and riboflavin receive the hydrogens and their electrons and transfer them to the electron transport chain.

**electron transport chain (ETC):** the final pathway in energy metabolism where the electrons from hydrogen are passed to oxygen and the energy released is trapped in the bonds of ATP.

• ***The Electron Transport Chain*** • The electron transport chain (ETC) consists of a series of proteins that serve as electron "carriers." These carriers are mounted in sequence on a membrane inside the energy-generating structures within the cell known as mitochondria (review Figure 7-4). As each carrier receives electrons, it releases a little energy and passes the electrons on to the next carrier. While some of the energy is released as heat, much of it is captured in the bonds of ATP molecules. These electron-transferring molecules continue passing electrons and giving up energy until, at the end of the chain, any usable energy has been captured in the body's ATP molecules. In the last step, the low-energy electrons with their hydrogen atoms (H) combine with oxygen (O), forming water ($H_2O$), from which the body cannot extract any more energy. Everyone knows that oxygen is essential to life—now you understand why. The TCA cycle and the ETC represent the body's most efficient means of capturing the energy from nutrients and transferring it into the bonds of ATP. Figure 7-18 provides a simple diagram of the ETC; see Appendix C for details.

**Figure 7-17**
**Urea Excretion**

The liver and kidneys both play a role in disposing of excess nitrogen. Can you see why the person with liver disease has high blood ammonia, while the person with kidney disease has high blood urea? (Figure 3-8 provides details of how the kidneys work.)

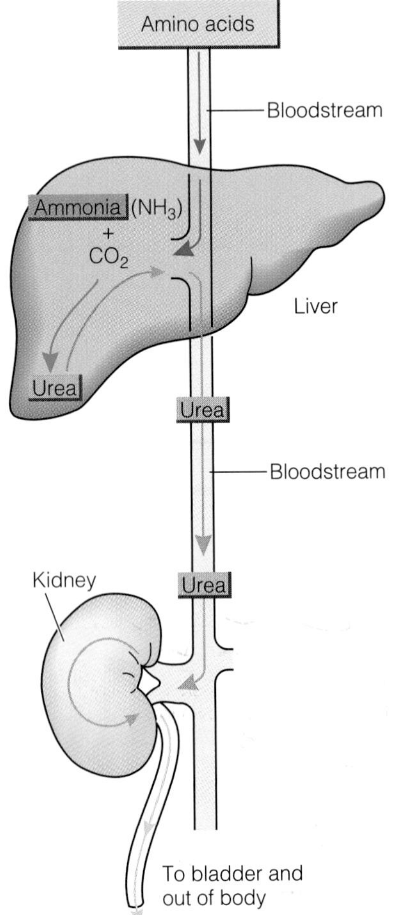


• ***The kCalories per Gram Secret Revealed*** • Of the three energy-yielding nutrients, fat provides the most energy for its weight. The reason this is may be apparent from Figure 7-19, which compares a fatty acid molecule with a glucose molecule. Notice that nearly all the bonds in a fatty acid molecule are between carbons and hydrogens. Oxygen can be added to all of them (forming carbon dioxide with the carbons, and water with the hydrogens). As this happens, the energy in the fat is released. In glucose, on the other hand, an oxygen is already bonded to each carbon; thus there is less potential for oxidation, and less energy will become available when the remaining bonds are broken.

Because fat contains many carbon-hydrogen bonds that can be readily oxidized, it generates abundant ATP during oxidation. This explains why fat yields more kcalories per gram than carbohydrate or protein. (Remember that each ATP

---

*Actually, the 4-carbon compound does not cycle around as the same structure throughout; instead it travels through a series of reactions. On picking up acetyl CoA, it becomes a 6-carbon compound. On dropping off carbon dioxide, it becomes a 5- and then a 4-carbon compound. Each reaction changes the structure slightly until finally the original 4-carbon compound forms again and picks up another acetyl CoA, starting the series of reactions over again.

The carbons that enter the cycle in acetyl CoA may not be the ones that are given off as carbon dioxide. In one of the steps of the cycle, a 6-carbon compound of the cycle becomes symmetrical, both ends being identical. Thereafter it loses carbons to carbon dioxide at one end or the other. Thus only half of the carbons from acetyl CoA are given off as carbon dioxide in any one turn of the cycle; the other half become part of the compound that returns to pick up another acetyl CoA. It is true to say, though, that for each acetyl CoA that enters the TCA cycle, 2 carbons are given off as carbon dioxide. It is also true that with each turn of the cycle the energy equivalent of one acetyl CoA is released.

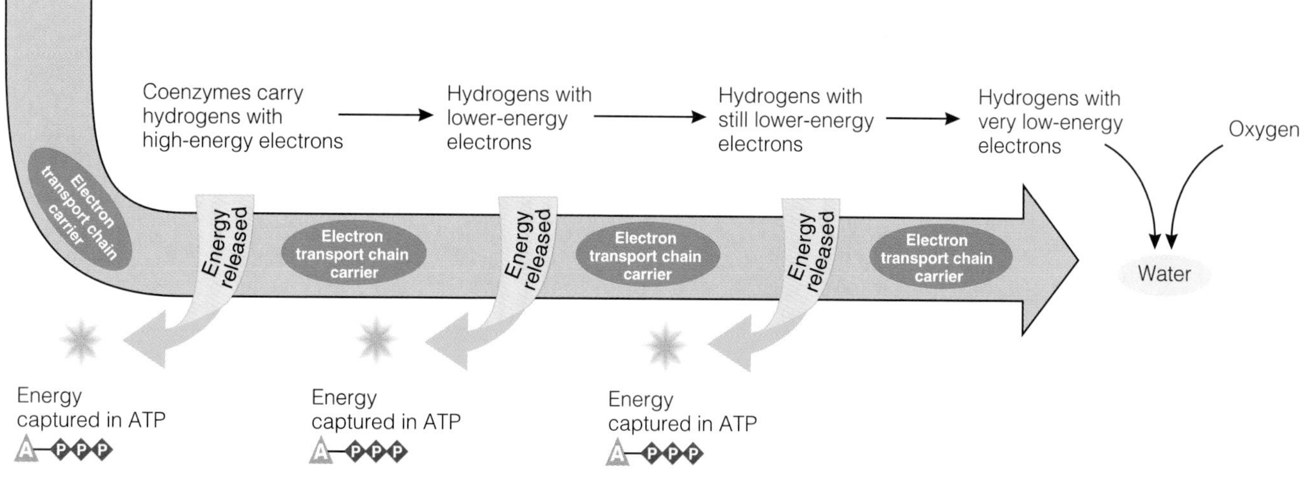


**Figure 7-18**

**Electron Transport Chain**

An important concept to remember is that an electron is not a fixed amount of energy. The electrons that bond the hydrogens to the B vitamin coenzymes have a relatively large amount of energy. In the series of reactions that follow, they lose this energy in small amounts, until at the end the electrons with their hydrogens (H) are attached to oxygen (O) to make water ($H_2O$). In some of the steps, the energy the electrons lose is captured into ATP in coupled reactions. Appendix C provides a more detailed explanation.

holds energy and that kcalories measure energy; thus the more ATP generated, the more kcalories have been collected.) For example, one glucose molecule will yield 38 ATP when completely oxidized. In comparison, one 16-carbon fatty acid molecule will yield 129 ATP when completely oxidized. Gram for gram, fat can pack much more energy than either of the other two energy-yielding nutrients, making it the body's preferred form of energy storage.

## IN SUMMARY

All the details this chapter has presented so far are combined in the summary figure on p. 214. After a balanced meal, the body handles the nutrients as shown. The digestion of *carbohydrate* yields glucose; some is stored as glycogen, and some is broken down to pyruvate and acetyl CoA to provide energy. The acetyl CoA can then enter the TCA cycle and ETC to provide more energy. The digestion of *fat* yields glycerol and fatty acids; some are reassembled and stored as fat, and others are broken down to acetyl CoA, enter the TCA cycle and ETC, and provide energy. The digestion of *protein* yields amino acids, some of which are used to build body protein. If there is a surplus, however, or if not enough carbohydrate and fat are available to meet energy needs, some amino acids are broken down through the same pathways as glucose to provide energy. Other amino acids enter directly into the TCA cycle, and these, too, can be broken down to yield energy. In summary, although carbohydrate, fat, and protein enter the TCA cycle by different routes, the energy-generating pathways that follow are common to all energy-yielding nutrients.

**Figure 7-19**

**Chemical Structures of a Fatty Acid and Glucose Compared**

The structure shown here for glucose is not the ring structure shown in Chapter 4, but an alternative way of drawing its chemical structure.

Fatty acid

Glucose

IN SUMMARY

**The Central Pathways of Energy Metabolism**

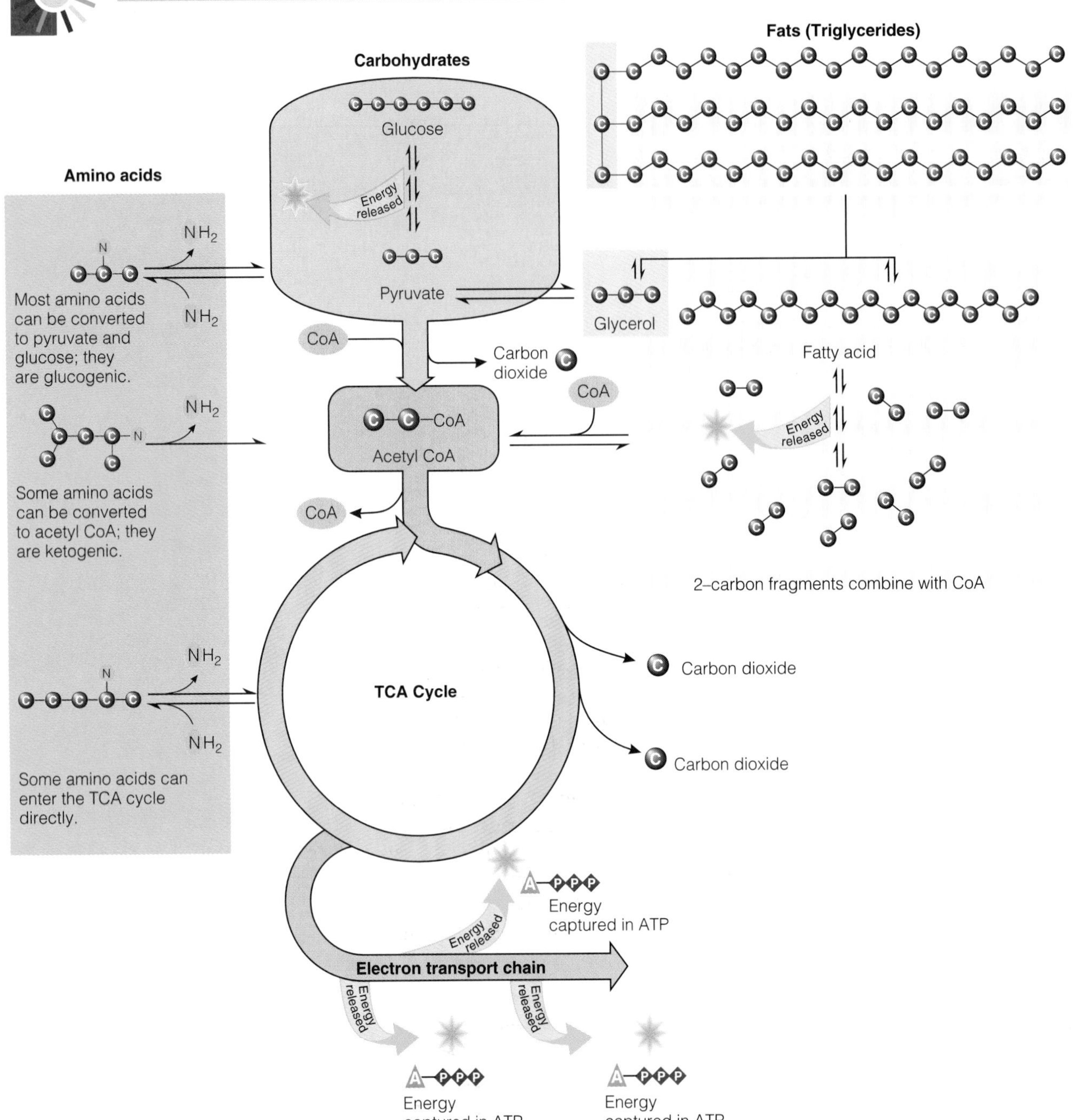


# The Body's Energy Budget

The average person takes in close to a million kcalories a year and expends more than 99 percent of them, maintaining a stable weight for years on end. This remarkable achievement, which many people manage without even thinking about it, could be called the economy of maintenance. The body's energy budget is balanced. Some people, however, eat too much and get fat; others eat too little and get thin. The metabolic details have already been described; the next sections will review them from the perspective of the body fat gained or lost. The possible reasons why people gain or lose weight are explored in Chapter 8.

## The Economics of Feasting

When a person eats too much, metabolism favors fat formation. Fat cells enlarge and multiply regardless of whether the excess derives from protein, carbohydrate, or fat. The pathway from dietary fat to body fat, however, is the most direct (requiring only a few metabolic steps) and the most efficient (costing only a few kcalories). To convert a dietary triglyceride to a triglyceride in adipose tissue, the body removes two of the fatty acids from the glycerol backbone, absorbs the parts, and puts them (and others) together again. By comparison, to convert a molecule of sucrose, the body has to split glucose from fructose, absorb them, dismantle them to pyruvate and acetyl CoA, assemble many acetyl CoA molecules into fatty acid chains, and finally attach fatty acids to a glycerol backbone to make a triglyceride for storage in adipose tissue. Quite simply, the body uses much less energy to convert dietary fat to body fat than it does to convert dietary carbohydrate to body fat. On average, storing excess energy from dietary fat in body fat uses only 5 percent of the ingested energy intake, but storing excess energy from dietary carbohydrate in body fat requires an expenditure of 25 percent of ingested energy intake.

The pathways from excess protein and excess carbohydrate to body fat are not only indirect and inefficient, but also less preferred (having other priorities). Before entering fat storage, protein must first tend to its many roles in the body's lean tissue, and carbohydrate must fill its glycogen stores. Simply put, making fat is a low priority for these two nutrients.

This chapter has described each of the energy-yielding nutrients individually, but cells use a mixture of these fuels. How much of which nutrient is in the fuel mix depends, in part, on its availability from the diet. (The proportion of each fuel also depends on activity, as Chapter 14 explains.) Dietary protein and dietary carbohydrate influence the mixture of fuel used during energy metabolism. Usually, protein's contribution to the fuel mix is relatively minor and fairly constant; protein oxidation increases only slightly, if at all, when protein is eaten in excess. Dietary carbohydrate, however, when eaten in excess, significantly enhances carbohydrate oxidation. In contrast, fat oxidation does not respond very well to dietary fat intake, especially when dietary changes occur abruptly. The more protein or carbohydrate in the fuel mix, the less fat contributes to the fuel mix. Instead of being oxidized, fat accumulates in storage. Details follow.

***People can enjoy bountiful meals such as this without storing body fat, provided that they spend as much energy as they take in. (Courtesy of CNN)***

*Note:* The *oxidation* of energy nutrients refers to the metabolic reactions that lead to the production of energy.

• ***Surplus Protein*** • Recall from Chapter 6 that extra protein is not stored as protein in the body. Contrary to popular opinion, a person cannot grow muscle simply by overeating protein. Lean tissue such as muscle develops in response to a stimulus such as hormones or physical activity. When a person overeats protein, the body uses the surplus by first replacing normal daily losses and then increasing protein oxidation slightly.[2] The body achieves protein balance this way, but any increase in protein oxidation displaces fat in the fuel mix. Any additional protein is then deaminated and the remaining carbons used to make fatty acids. Thus a person can grow fat by eating too much protein.

• ***Surplus Carbohydrate*** • Compared with protein, the proportions of carbohydrate and fat in the fuel mix change more dramatically when a person overeats. The body handles abundant carbohydrate by first storing it as glycogen, but glycogen storage areas are limited and fill quickly. Because maintaining glucose balance is critical, the body uses glucose frugally when the diet provides only small amounts and freely when stores are abundant. In other words, glucose oxidation rapidly adjusts to the dietary intake of carbohydrate.[3]

Excess glucose can be converted to fat directly, but this is a minor pathway.[4] As mentioned earlier, converting glucose to fat is energetically expensive and does not occur until after glycogen stores have been filled.[5] Even then, little or no new fat is made from carbohydrate.[6]

Nevertheless, excess dietary carbohydrate may lead to weight gain when extra carbohydrate displaces fat in the fuel mix.[7] When this occurs, carbohydrate spares both dietary fat and body fat from oxidation. Excess carbohydrate suppresses fat oxidation, leaving the fat to accumulate in adipose tissue. The net result: excess carbohydrate contributes to obesity.

• ***Surplus Fat*** • Unlike excess protein and carbohydrate, which both enhance their own oxidation, eating too much fat does not promote fat oxidation.[8] Instead, excess dietary fat moves efficiently into the body's fat stores; almost all of the excess is stored.

IN SUMMARY

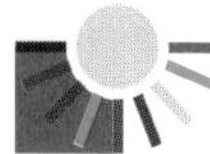

If energy intake exceeds the body's energy needs, the result will be weight gain—regardless of whether the excess intake is from protein, carbohydrate, or fat. The difference is that the body is much more efficient at storing energy when the excess derives from dietary fat. Given the same kcaloric excess, fat contributes more body fat than does carbohydrate or protein.

## *The Transition from Feasting to Fasting*

After a meal, glucose, glycerol, and fatty acids from foods are either used or stored. Later, as the body shifts from a fed state to a fasting one, it begins drawing on these stores. Glycogen and fat are released from storage to provide more glucose, glycerol, and fatty acids to produce energy.

Energy is needed all the time. Even when a person is asleep and totally relaxed, the cells of many organs are hard at work. In fact, this work—the cells' work that maintains all life processes without any conscious effort—represents about two-thirds to three-fourths of the total energy a person spends in a day. The small remainder is the work that a person's muscles perform voluntarily during waking hours.

The body's top priority is to meet the cells' needs for energy, and it normally does this by periodic refueling—that is, by eating several times a day. When food is not available, the body turns to its own tissues for other fuel sources. If people choose not to eat, we say they are fasting; if they have no choice, we say they are starving. The body makes no such distinction. In either case, the body is forced to switch to a wasting metabolism, drawing on its reserves of carbohydrate and fat and, within a day or so, on its vital protein tissues as well. Figure 7-20 shows the metabolic pathways operating in the body as it shifts from feasting (part A) to fasting (parts B and C).

## *The Economics of Fasting*

As Figure 7-20 shows, during fasting, all paths lead to energy—fuel must be delivered to every cell. As the fast begins, glucose from the liver's stored glycogen and fatty acids from the body's stored fat are both flowing into cells, then breaking down to yield acetyl CoA, and delivering energy to power the cells' work. Several hours later, however, most of the glucose is used up—liver glycogen is exhausted and blood glucose begins to fall. Low blood glucose serves as a signal that promotes further fat breakdown.

• ***Glucose Needed for the Brain*** • At this point, most of the cells are depending on fatty acids to continue providing their fuel. But red blood cells and the cells of the nervous system need glucose. Glucose is their major energy fuel, and even when other energy fuels are available, glucose must be present to permit the energy-metabolizing machinery of the nervous system to work. Normally, the

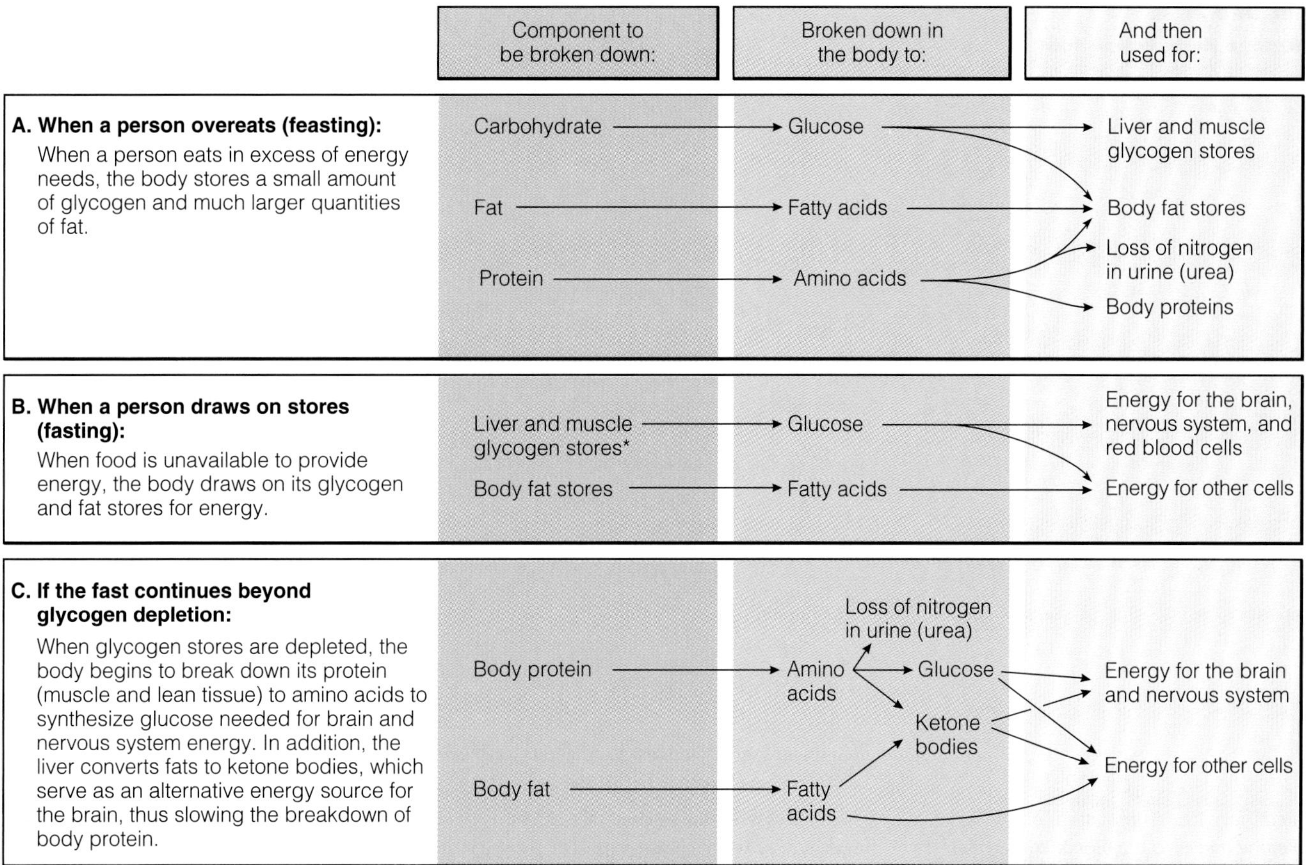

Component to be broken down:
Broken down in the body to:
And then used for:
A. When a person overeats (feasting):
When a person eats in excess of energy needs, the body stores a small amount of glycogen and much larger quantities of fat.
Carbohydrate
Glucose
Liver and muscle glycogen stores
Fat
Fatty acids
Body fat stores
Protein
Amino acids
Loss of nitrogen in urine (urea)
Body proteins
B. When a person draws on stores (fasting):
When food is unavailable to provide energy, the body draws on its glycogen and fat stores for energy.
Liver and muscle glycogen stores*
Glucose
Energy for the brain, nervous system, and red blood cells
Body fat stores
Fatty acids
Energy for other cells
C. If the fast continues beyond glycogen depletion:
When glycogen stores are depleted, the body begins to break down its protein (muscle and lean tissue) to amino acids to synthesize glucose needed for brain and nervous system energy. In addition, the liver converts fats to ketone bodies, which serve as an alternative energy source for the brain, thus slowing the breakdown of body protein.
Loss of nitrogen in urine (urea)
Body protein
Amino acids
Glucose
Energy for the brain and nervous system
Ketone bodies
Energy for other cells
Body fat
Fatty acids
*The muscles' stored glycogen provides glucose only for the muscle in which the glycogen is stored.

**Figure 7-20**
**Feasting and Fasting**

brain and nerve cells consume about two-thirds of the total *glucose* used each day—about 400 to 600 kcalories' worth. About one-fifth of the *energy* the body uses when it is at rest is used for the brain.

• ***Protein Called on to Meet Glucose Needs*** • The red blood cells' and brain's special requirements for glucose pose a problem for the fasting body. The body can use its stores of fat, which may be quite generous, to furnish most of its cells with energy, but the red blood cells are completely dependent on glucose, and the brain and nerves prefer energy in the form of glucose. For this reason, body protein tissues such as muscle and liver always break down to some extent during fasting. Amino acids that yield pyruvate can be used to make glucose; and to obtain them, body proteins must be broken down. The amino acids that can't be used to make glucose are used as an energy source for other body cells.

• ***Fat's Small Glucose Contribution from Glycerol*** • The breakdown of body protein is an expensive way to obtain glucose. In the first few days of a fast, body protein provides about 90 percent of the needed glucose; glycerol, about 10 percent. If body protein losses were to continue at this rate, death would ensue within three weeks, regardless of the quantity of fat a person had stored. Fortunately, fat breakdown also increases with fasting—in fact, fat breakdown almost doubles, providing energy for other body cells and glycerol for glucose production.[9]

• ***The Shift to Ketosis*** • As the fast continues, the body finds a way to use its fat to fuel the brain. It adapts by condensing together acetyl CoA fragments derived from fatty acids to produce an alternate energy source, ketone bodies (see Figure 7-21). Normally produced and used only in small quantities, ketone bodies can provide fuel for some brain cells. Ketone body production rises until, after about 10 days of fasting, it is meeting much of the nervous system's energy needs. Still,

Reminder: *Condensation* is the process by which two molecules are joined together with the removal of water.

Reminder: The group of ketones that are formed during the incomplete oxidation of fatty acids are *ketone bodies*. A *ketone* is a compound that contains a carbonyl group (C=O) between other carbons.

**Figure 7-21**
**Ketone Body Formation**

1. The first step in the formation of ketone bodies is the condensation of two molecules of acetyl CoA and the removal of the CoA to form a compound that is converted to the first ketone body.

$$\mathrm{CH_3{-}CO{-}CoA}\ (\text{Acetyl CoA}) + \mathrm{CH_3{-}CO{-}CoA}\ (\text{Acetyl CoA}) + H_2O \xrightarrow{-\,2\ \text{CoA}} \mathrm{CH_3{-}CO{-}CH_2{-}COOH}$$

A ketone, acetoacetate

2. This ketone body may lose a molecule of carbon dioxide to become another ketone.

$$\xrightarrow{-\,CO_2} \mathrm{CH_3{-}CO{-}CH_3}$$

A ketone, acetone

3. Or, the acetoacetate may add two hydrogens, becoming another ketone body (beta-hydroxybutyrate). See Appendix C for more details.

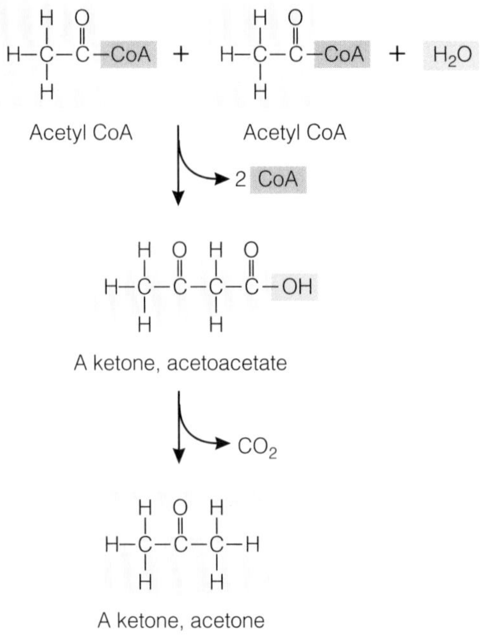

Reminder: The combination of elevated ketone bodies in the blood (ketonemia) and in the urine (ketonuria) is *ketosis.*

many areas of the brain rely exclusively on glucose, and body protein continues to be sacrificed to produce it.

When ketone bodies contain a COOH (acid) group, they are called keto acids. Small amounts of keto acids are a normal part of the blood chemistry; but when their concentration rises, the pH of the blood declines and ketone bodies spill into the urine. This is ketosis, and it is a sign that the body's chemistry is going awry.

• ***Suppression of Appetite*** • The starvation that produces ketosis also causes loss of appetite. Researchers have theorized that having a reduced appetite is an advantage to a person without access to food, because the search for food would be a waste of energy. When the person finds food and eats again, the body shifts out of ketosis, the hunger center gets the message that food is again available, and the appetite returns. This chain of events has served as justification for weight-loss routines that induce ketosis, such as fasting and low-carbohydrate diets. However, any kind of food restriction, with or without ketosis, leads a person to adapt by losing appetite. A well-balanced low-kcalorie diet can induce the same effect. Therefore ketosis-producing diets offer no special advantage in terms of appetite suppression, and because ketosis can disrupt the body's acid-base balance, other weight-loss regimens are preferred to ketogenic diets. Highlight 8 includes a discussion of the risks of ketogenic diets in its review of popular weight-loss diets.

• ***Slowing of Metabolism*** • In any case, while the body is shifting to the use of ketone bodies, it simultaneously reduces its energy output and conserves both its fat and its lean tissue. As the lean (protein-containing) organ tissues shrink in mass, they perform less metabolic work, reducing energy expenditures. As the muscles waste, they can do less work and so demand less energy, reducing expenditures further. The hormones of fasting slow metabolism even further in the effort to conserve lean body mass for as long as possible. Because of the slowed metabolism, the loss of fat falls to a bare minimum—less, in fact, than the fat that would be lost on a low-kcalorie diet. Thus, although *weight* loss during fasting may be quite dramatic, *fat* loss may be less than when at least some food is eaten.

• ***Symptoms of Starvation*** • The adaptations just described—slowing of energy output and reduction in fat loss—occur in the starving child, the hungry homeless adult, the fasting religious person, the adolescent with anorexia nervosa, and the malnourished hospital client. Such adaptations help to prolong their lives and explain the physical symptoms of energy deprivation: wasting, slowed metabolism, lowered body temperature, and reduced resistance to disease.

The body's adaptations to fasting are sufficient to maintain life for a long time. Mental alertness need not be diminished, and even physical energy may remain unimpaired for a surprisingly long time. Still, fasting presents hazards. The same alterations in metabolism occur on a low-carbohydrate diet, as Highlight 8 explains.

### IN SUMMARY

When fasting, the body makes a number of adaptations: increasing the breakdown of fat to provide energy for most of the cells, using glycerol and amino acids to make glucose for the red blood cells and central nervous system, producing ketones to fuel the brain, suppressing the appetite, and slowing metabolism. All of these activities conserve energy and minimize losses. In fact, metabolism slows to such an extent that the loss of fat eventually slows to less than would be achieved with a low-kcalorie diet.

This chapter has probed the intricate details of metabolism at the level of the cells, exploring the transformations of nutrients to energy and to storage compounds. Several chapters and highlights to come build on this information. The highlight that follows this chapter shows how alcohol disrupts normal metabolism. Chapter 8 describes how a person's intake and expenditure of energy are reflected in body weight and body composition. Chapter 9 examines the consequences of unbalanced energy budgets—overweight and underweight—and what to do about them. Chapter 10 shows the vital roles the B vitamins play as coenzymes assisting all the metabolic pathways described here. And Chapter 14 revisits metabolism to show how it supports the work of physically active people and how athletes can best apply that information in their choices of foods to eat.

Indeed, the sun's energy sparks every move that we make. And our beautifully designed bodies make use of it in astonishing ways.

## Study Questions

These questions will help you review the chapter. You will find the answers in the discussions on the pages provided.

1. Define metabolism, anabolism, and catabolism; give an example of each. (pp. 197–198)
2. Name one of the body's quick-energy molecules, and describe how is it used. (pp. 198–199)
3. What are coenzymes, and what service do they provide in metabolism? (p. 200)
4. Name the four basic units, derived from foods, used by the body in metabolic transformations. How many carbons are in the "backbones" of each? (pp. 201–202)
5. Define aerobic and anaerobic metabolism. How does insufficient oxygen influence metabolism? (p. 203)
6. How does the body dispose of excess nitrogen? (p. 210)
7. Summarize the main steps in the metabolism of glucose, glycerol, fatty acids, and amino acids. (p. 214)
8. Describe how a surplus of the three energy nutrients contributes to body fat stores. (pp. 215–216)
9. What adaptations does the body make during a fast? What are ketone bodies? Define ketosis. (pp. 216–219)
10. Distinguish between a loss of *fat* and a loss of *weight,* and describe how each might happen. (pp. 218–219)

These questions will help you prepare for an exam. Answers can be found in Appendix K.

1. Hydrolysis is an example of a(n):
   a. coupled reaction.
   b. anabolic reaction.
   c. catabolic reaction.
   d. synthesis reaction.
2. During metabolism, released energy is captured and transferred by:
   a. enzymes.
   b. pyruvate.
   c. acetyl CoA.
   d. adenosine triphosphate.
3. Glycolysis:
   a. requires oxygen.
   b. generates abundant energy.
   c. converts glucose to pyruvate.
   d. produces ammonia as a by-product.
4. The pathway from pyruvate to acetyl CoA:
   a. produces lactic acid.
   b. is known as gluconeogenesis.
   c. is metabolically irreversible.
   d. requires more energy than it produces.

5. For complete oxidation, acetyl CoA enters:
   a. glycolysis and the TCA cycle.
   b. glycolysis and the Cori cycle.
   c. the TCA cycle and electron transport chain.
   d. the Cori cycle and electron transport chain.
6. Deamination of an amino acid produces:
   a. vitamin $B_6$ and energy.
   b. pyruvate and acetyl CoA.
   c. ammonia and a keto acid.
   d. carbon dioxide and water.
7. Before entering the TCA cycle, each of the energy-yielding nutrients is broken down to:
   a. ammonia.
   b. pyruvate.
   c. electrons.
   d. acetyl CoA.
8. The body stores energy for future use in:
   a. proteins.
   b. acetyl CoA.
   c. triglycerides.
   d. ketone bodies.
9. During a fast, when glycogen stores have been depleted, the body begins to synthesize glucose from:
   a. acetyl CoA.
   b. amino acids.
   c. fatty acids.
   d. ketone bodies.
10. During a fast, the body produces ketone bodies by:
   a. hydrolyzing glycogen.
   b. condensing acetyl CoA.
   c. transaminating keto acids.
   d. converting ammonia to urea.

## Notes


1. J. L. Groff, S. S. Gropper, and S. M. Hunt, *Advanced Nutrition and Human Metabolism* (St. Paul, Minn.: West Publishing, 1995), pp. 174–175.
2. S. A. Jebb and coauthors, Changes in macronutrient balance during over- and underfeeding assessed by 12-d continuous whole-body calorimetry, *American Journal of Clinical Nutrition* 64 (1996): 259–266.
3. T. J. Horton and coauthors, Fat and carbohydrate overfeeding in humans: Different effects on energy storage, *American Journal of Clinical Nutrition* 62 (1995): 19–29; P. S. Shetty and coauthors, Alterations in fuel selection and voluntary food intake in response to isoenergetic manipulation of glycogen stores in humans, *American Journal of Clinical Nutrition* 60 (1994): 534–543.
4. M. K. Hellerstein, Regulation of hepatic de novo lipogenesis in humans, *Annual Review of Nutrition* 16 (1996): 523–557.
5. J. P. Flatt, McCollum Award Lecture, 1995: Diet, lifestyle, and weight maintenance, *American Journal of Clinical Nutrition* 62 (1995): 820–836.
6. C. Prosperpi and coauthors, Ad libitum intake of a high-carbohydrate or high-fat diet in young men: Effects on nutrient balances, *American Journal of Clinical Nutrition* 66 (1997): 539–545; Jebb and coauthors, 1996; Horton and coauthors, 1995; B. Swinburn and E. Ravussin, Energy balance or fat balance? *American Journal of Clinical Nutrition* 57 (1993): 766S–771S.
7. Hellerstein, 1996.
8. E. Ravussin and A. Tataranni, Dietary fat and human obesity, *Journal of the American Dietetic Association* 97 (1997): S42–S46; J. P. Flatt, Use and storage of carbohydrate and fat, *American Journal of Clinical Nutrition* 61 (1995): 952S–959S; Horton and coauthors, 1995.
9. J. E. Ati, C. Beji, and J. Danguir, Increased fat oxidation during Ramadan fasting in healthy women: An adaptive mechanism for body weight maintenance, *American Journal of Clinical Nutrition* 62 (1995): 302–307; M. G. Carlson, W. L. Snead, and P. J. Campbell, Fuel and energy metabolism in fasting humans, *American Journal of Clinical Nutrition* 60 (1994): 29–36.

HIGHLIGHT 7

# Alcohol and Nutrition

From backyard barbecues to formal weddings, many social gatherings are occasions for offering beverages that contain alcohol, and people must choose whether to drink them. Most people who drink manage their relationships with alcohol relatively safely.[1] Unfortunately, some 14 million people in the United States abuse alcohol to the point that their personal relationships, work, and health become impaired. With the understanding of metabolism gained from Chapter 7, you are in a position to understand how the body handles alcohol, how alcohol interferes with metabolism, and how alcohol impairs health and nutrition. The possible benefits of alcohol consumption are presented in Chapter 18.

## Alcohol in Beverages

To the chemist, *alcohol* refers to a class of organic compounds containing hydroxyl (OH) groups. The glycerol to which fatty acids are attached in triglycerides is an example of an alcohol to a chemist. To most people, though, *alcohol* refers to the intoxicating ingredient in beer, wine, and hard liquor (distilled spirits). The chemist's name for this particular alcohol is *ethyl alcohol,* or *ethanol.* Glycerol has 3 carbons with 3 hydroxyl groups attached; ethanol has only 2 carbons and 1 hydroxyl group (see Figure H7-1). The remainder of this highlight talks about the particular alcohol, ethanol, but refers to it simply as *alcohol.* (The glossary on p. 222 defines related terms.)

Alcohols affect living things profoundly, partly because they act as lipid solvents. Their ability to dissolve lipids out of cell membranes allows alcohols to penetrate rapidly into cells, destroying cell structures and thereby killing the cells. For this reason, most alcohols are toxic in relatively small amounts; by the same token, because they kill microbial cells, they are useful as disinfectants.

Ethanol is less toxic than the other alcohols. Sufficiently diluted and taken in small enough doses, its action in the brain produces euphoria—a pleasing effect that people seek—not with zero risk, but with a low enough risk (if the doses are low enough) to be tolerable. Used to achieve this effect, alcohol is a drug—that is, a substance that modifies body functions. Like all drugs, alcohol offers both benefits and hazards. It must be used with caution, if used at all.

Beer, wine, and liquor deliver different amounts of alcohol. The amount of alcohol in distilled liquor is stated as *proof:* 100 proof liquor is 50 percent alcohol, 80 proof is 40 percent alcohol, and so forth. Wine (at 8 to 14 percent) and beer (at 4 to 6 percent) have less alcohol than distilled liquor, although some fortified wines and beers have more alcohol than the regular varieties.

*Shared conversations and meals sometimes include alcoholic beverages.*

Taken in moderation, alcohol can be compatible with good health. The term *moderation* is important in describing alcohol use. How many drinks constitute moderate use, and how much is "a drink"? First, a drink is any alcoholic beverage that delivers ½ ounce of *pure ethanol:*

- 4 to 5 ounces of wine.
- 10 ounces of wine cooler.
- 12 ounces of beer.
- 1½ ounce of distilled liquor (80 proof whiskey, scotch, rum, or vodka).

Second, because people have different tolerances to alcohol, it is impossible to name an exact amount of alcohol per day that is appropriate for everyone. Authorities have attempted to set limits that are acceptable for most healthy people. An accepted definition of moderation is not more than two drinks a day for the average-sized man and not more than one drink a day for the average-sized woman. Notice that this advice is stated as a maximum, not as an average; seven drinks one night a week would not be considered moderate, even though one a day would be. Doubtless some people could consume slightly more; others could not handle nearly so much without risk. The amount a person can drink safely is highly individual, depending on genetics, health condition, sex, body composition, age, and family history.

**Figure H7-1**

**Two Alcohols: Glycerol and Ethanol**

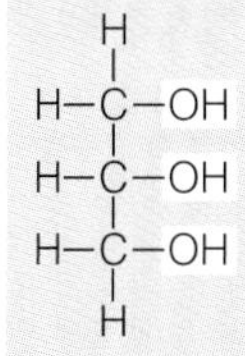



Glycerol is the alcohol used to make triglycerides.

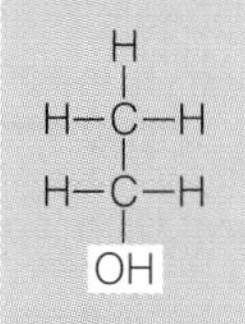



Ethanol is the alcohol in beer, wine, and distilled spirits.

## Alcohol in the Body

From the moment an alcoholic beverage enters the body, it is treated as if it has special privileges. Unlike foods, which require time for digestion, alcohol needs no digestion and is quickly absorbed. About 20 percent is absorbed directly across the walls of an empty stomach and can reach the brain within a minute. Consequently, a person can immediately feel euphoric when drinking, especially on an empty stomach.

When the stomach is full of food, alcohol has less chance of touching the walls and diffusing through, so its influence on the brain is slightly delayed. This information leads to a practical tip: eat snacks when drinking alcoholic beverages. Carbohydrate snacks slow alcohol absorption and high-fat snacks slow peristalsis, keeping the alcohol in the stomach longer. Salty snacks make a person thirsty; to quench thirst, drink water instead of more alcohol.

The stomach begins to break down alcohol with its alcohol dehydrogenase enzyme.[2] This action can reduce the amount of alcohol entering the blood by about 20 percent. Women produce less of this stomach enzyme than men, which partially explains why women become more intoxicated on less alcohol than men. Women absorb about one-third more alcohol than men of the same size who drink the same amount of alcohol.

Alcohol is rapidly absorbed in the small intestine. From this point on, alcohol receives VIP (Very Important Person) treatment: it gets absorbed and metabolized before most nutrients. Alcohol's priority status helps to ensure a speedy disposal and reflects two facts: alcohol cannot be stored in the body, and it is potentially toxic.

## Glossary

**acetaldehyde** (ass-et-AL-duh-hide): an intermediate in alcohol metabolism.

**alcohol:** a class of organic compounds containing hydroxyl (OH) groups.

**alcohol dehydrogenase** (dee-high-DROJ-eh-nayz): an enzyme active in the stomach and the liver that converts ethanol to acetaldehyde.

**alcoholism:** disease characterized by loss of control over drinking and dependence on alcohol that harms health, family and social relations, and ability to work.

**antidiuretic hormone (ADH):** a hormone produced by the pituitary gland in response to dehydration (or a high sodium concentration in the blood); it stimulates the kidneys to reabsorb more water and therefore to excrete less. This ADH should not be confused with the enzyme alcohol dehydrogenase, which is sometimes also abbreviated ADH.

**beer:** an alcoholic beverage brewed by fermenting malt and hops.

**cirrhosis** (seer-OH-sis): advanced liver disease in which liver cells turn orange, die, and harden, permanently losing their function; often associated with alcoholism.

- **cirrhos** = an orange

**distilled liquor:** an alcoholic beverage made by fermenting and distilling grains; sometimes called *distilled spirits* or *hard liquor.*

**drink:** a dose of any alcoholic beverage that delivers ½ oz of pure ethanol:

- 4 to 5 oz of wine.
- 10 oz of wine cooler.
- 12 oz of beer.
- 1½ oz of hard liquor (80 proof whiskey, scotch, rum, or vodka).

**drug:** a substance that can modify one or more of the body's functions.

**ethanol:** a particular type of alcohol found in beer, wine, and distilled spirits; also called *ethyl alcohol* (see Figure H7-1). Ethanol is the most widely used—and abused—drug in our society. It is also the only legal, nonprescription drug that produces euphoria.

**euphoria** (you-FORE-eh-uh): a feeling of great well-being, which people often seek through the use of drugs such as alcohol.

- **eu** = good
- **phoria** = bearing

**fatty liver:** an early stage of liver deterioration seen in several diseases, including kwashiorkor and alcoholic liver disease. Fatty liver is characterized by an accumulation of fat in the liver cells.

**fibrosis** (fye-BROH-sis): an intermediate stage of liver deterioration seen in several diseases, including viral hepatitis and alcoholic liver disease. In fibrosis, the liver cells lose their function and assume the characteristics of connective tissue cells (fibers).

**gout** (GOWT): a painful condition in which uric acid crystals form in the joints.

**MEOS** or **microsomal** (my-krow-SO-mal) **ethanol-oxidizing system:** a system of enzymes in the liver that oxidize not only alcohol, but also several classes of drugs. (The *microsomes* are tiny particles of membranes with associated enzymes that can be collected from broken-up cells.)

- **micro** = tiny
- **soma** = body

**moderation:** in relation to alcohol consumption, not more than two drinks a day for the average-sized man and not more than one drink a day for the average-sized woman.

**NAD (nicotinamide adenine dinucleotide):** the main coenzyme form of the vitamin niacin; its reduced form is NADH.

**narcotic** (nar-KOT-ic): any drug that dulls the senses, induces sleep, and becomes addictive with prolonged use.

**proof:** a way of stating the percentage of alcohol in distilled liquor. Liquor that is 100 proof is 50% alcohol; 90 proof is 45%, and so forth.

**wine:** an alcoholic beverage made by fermenting grape juice.

## *Alcohol Arrives in the Liver*

The capillaries of the digestive tract merge into veins that carry the alcohol-laden blood to the liver. These veins branch and rebranch into capillaries that touch every liver cell. Liver cells are the only other cells in the body that can make enough of the enzyme alcohol dehydrogenase to oxidize alcohol at an appreciable rate. The routing of blood through the liver cells gives them the chance to dispose of some alcohol before it moves on.

Alcohol affects every organ of the body, but the most dramatic evidence of its disruptive behavior appears in the liver. If liver cells could talk, they would describe alcohol as demanding, egocentric, and disruptive of the liver's efficient way of running its business. For example, liver cells normally prefer fatty acids as their fuel, and they like to package excess fatty acids into triglycerides and ship them out to other tissues. When alcohol is present, however, the liver cells are forced to metabolize alcohol and let the fatty acids accumulate, sometimes in huge stockpiles. Alcohol metabolism also permanently changes liver cell structure, which impairs the liver's ability to metabolize fats. This explains why heavy drinkers develop fatty livers.

**Figure H7-2**
**Alcohol Metabolism**

The conversion of alcohol to acetyl CoA requires the B vitamin niacin in its role as NAD. When the enzymes oxidize alcohol, they remove H atoms and attach them to NAD. Thus NAD is used up, and NADH accumulates. (Note: More accurately, $NAD^+$ is converted to NADH + $H^+$. For simplicity's sake, the process has been described here as if one hydrogen were added to HAD, but, in reality, two are added.)

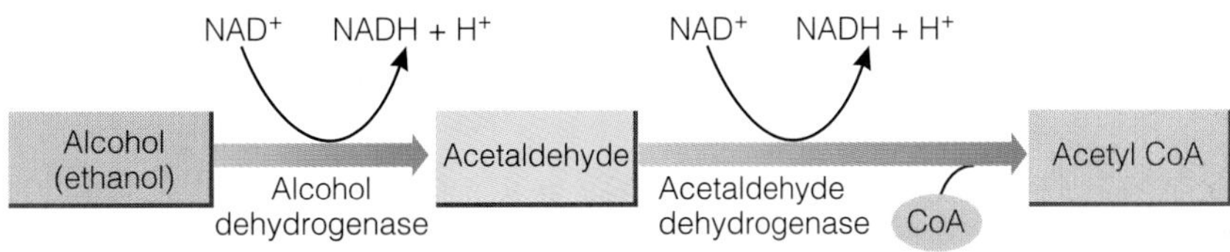


The liver can process about ½ ounce *ethanol* per hour (the amount in a typical drink), depending on the person's body size, previous drinking experience, food intake, and general health. This maximum rate of alcohol breakdown is set by the amount of alcohol dehydrogenase available. If more alcohol arrives at the liver than the enzymes can handle, the extra alcohol travels to all parts of the body, circulating again and again until liver enzymes are finally available to process it. Another practical tip derives from this information: drink slowly enough to allow the liver to keep up—no more than 1 drink per hour.

The amount of alcohol dehydrogenase enzyme present in the liver varies with individuals, depending on the genes they have inherited and on how recently they have eaten. Fasting for as little as a day forces the body to degrade its proteins, including the alcohol-processing enzyme, and this can slow the rate of alcohol metabolism by half. Drinking on an empty stomach thus causes the drinker to feel the effects more promptly for two reasons: rapid absorption and slowed breakdown. By maintaining higher blood alcohol concentrations for longer times, alcohol can anesthetize the brain more completely.

The alcohol dehydrogenase enzyme breaks down alcohol by removing hydrogens in two steps. (Figure H7-2 provides a simplified diagram of alcohol metabolism; Appendix C provides the chemical details.) In the first step, alcohol dehydrogenase oxidizes alcohol to acetaldehyde. High concentrations of acetaldehyde in the brain and other tissues are responsible for many of the punishing effects of alcohol abuse.

In the second step, a related enzyme, acetaldehyde dehydrogenase, oxidizes the acetaldehyde to acetyl CoA, the "crossroads" compound that can enter the TCA cycle to generate energy. These reactions produce hydrogen ions ($H^+$). The B vitamin niacin (in its role as the coenzyme NAD) helpfully picks up these hydrogen ions (becoming NADH). Thus, whenever the body breaks down alcohol, NAD diminishes and NADH accumulates.

## *Alcohol Disrupts the Liver*

During alcohol metabolism, NAD becomes unavailable for the multitude of other metabolic processes for which it is required, including glycolysis, the TCA cycle, and the electron transport chain. Its presence is sorely missed in these energy pathways because it is the chief carrier of the hydrogens that travel with their electrons along the electron transport chain. Without NAD, the energy pathway is blocked. Traffic either backs up, or an alternate route is taken. Such changes in the normal flow from glucose to energy have striking physical consequences.

For one, the accumulation of hydrogen ions during alcohol metabolism shifts the body's acid-base balance toward acid. For another, the accumulation of NADH slows the TCA cycle, so pyruvate and acetyl CoA build up. Excess acetyl CoA then takes the route to fatty acid synthesis (as Figure H7-3 illustrates), and fat clogs the liver.

As you might expect, a liver clogged with fat cannot function properly. Liver cells become less efficient at performing a number of tasks. Much of this inefficiency impairs a person's nutritional health in ways that cannot be corrected by diet alone. For example, the liver has difficulty activating vitamin D, as well as producing and releasing bile. To overcome such problems, a person needs to stop drinking alcohol.

The synthesis of fatty acids accelerates with exposure to alcohol. Fat accumulation can be seen in the liver after a single night of heavy drinking. Fatty liver, the first stage of liver deterioration seen in heavy drinkers, interferes with the distribution of nutrients and oxygen to the liver cells. If the

**Figure H7-3**
**Alternate Route for Acetyl CoA: To Fat**

Acetyl CoA molecules are blocked from getting into the TCA cycle by the high level of NADH. Instead of being used for energy, the acetyl CoA molecules become builing blocks for fatty acids.

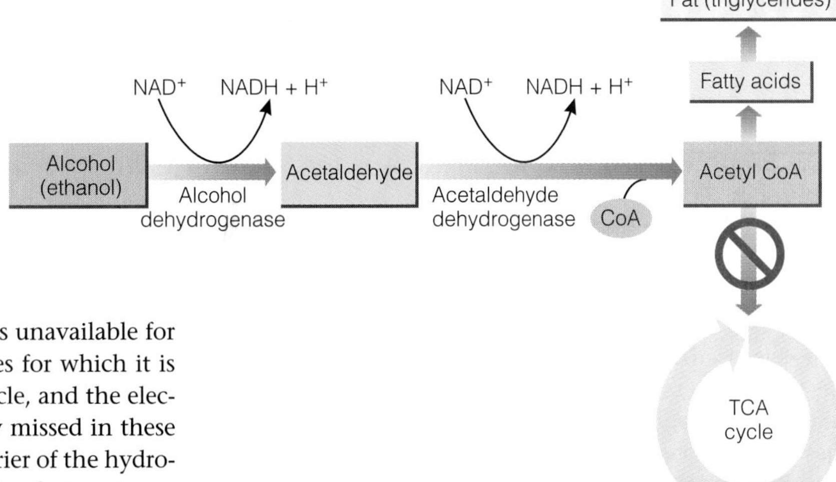

condition lasts long enough, the liver cells will die and form fibrous scar tissue—the second stage of liver deterioration, called fibrosis. Some liver cells can regenerate with good nutrition and abstinence from alcohol, but in the most advanced stage, cirrhosis, damage is the least reversible.

The fatty liver has difficulty generating glucose from protein. The lack of glucose together with the overabundance of acetyl CoA sets the stage for ketosis. The body uses the acetyl CoA to make ketone bodies, which push the acid-base balance further toward acid.

Excess NADH also promotes the making of lactic acid from pyruvate. The conversion of pyruvate to lactic acid uses the hydrogens from NADH and restores some NAD, but a lactic acid buildup has serious consequences of its own—it adds still further to the body's acid burden and interferes with the excretion of another acid, uric acid, causing inflammation of the joints.

Alcohol alters both amino acid and protein metabolism. Synthesis of proteins important in the immune system slows down, weakening the body's defenses against infection. Protein deficiency can develop, both from a diminished synthesis of protein and from a poor diet. Normally, the cells would at least use the amino acids from the protein foods a person eats, but the drinker's liver deaminates the amino acids and uses the carbon fragments to make fat or ketones. Eating well does not protect the drinker from protein depletion; a person has to stop drinking alcohol.

The liver's VIP treatment of alcohol affects its handling of drugs as well as nutrients. In addition to the dehydrogenase enzyme already described, the liver possesses an enzyme system that metabolizes *both* alcohol and several types of other drugs. Called the MEOS (microsomal ethanol-oxidizing system), this system handles about one-fifth of the total alcohol a person consumes. At high blood alcohol concentrations, however, or if repeatedly exposed to alcohol, the MEOS grows larger.

As a person's blood alcohol rises, alcohol competes with—and wins out over—other drugs whose metabolism relies on the MEOS. If a person drinks and uses another drug at the same time, the drug will be metabolized more slowly and will therefore exert greater effects. The MEOS is busy disposing of alcohol, so the drug cannot be handled until later; the dose may build up so that its effects are greatly amplified—sometimes to the point of being fatal.

In contrast, once a heavy drinker stops drinking and alcohol is no longer competing with other drugs, the enlarged MEOS metabolizes drugs much faster than before. As a result, determining the correct dosages of medications can be confusing and tricky. A skilled anesthesiologist always asks clients about their drinking patterns before sedating them.

This discussion has emphasized the major way that the blood is cleared of alcohol—metabolism by the liver—but there is another way. About 10 percent of the alcohol leaves the body through the breath and in the urine. This is the basis for the breath and urine tests for drunkenness. The amounts of alcohol in the breath and in the urine are in proportion to the amount still in the bloodstream and brain. In nearly all states, legal drunkenness is set at 0.10 percent or less, reflecting the relationship between alcohol use and industrial and traffic accidents.

## Alcohol Arrives in the Brain

Alcohol is a narcotic. People used it for centuries as an anesthetic because it can deaden pain. But alcohol was a poor anesthetic because one could never be sure how much a person would need and how much would be a fatal dose. Consequently, new, more predictable anesthetics have replaced alcohol. However, alcohol continues to be used today as a kind of social anesthetic to help people relax or to relieve anxiety. People think that alcohol is a stimulant because it seems to relieve inhibitions. Actually, though, it accomplishes this by sedating *inhibitory* nerves, which are more numerous than excitatory nerves. Ultimately, alcohol acts as a depressant and affects all the nerve cells. Figure H7-4 describes alcohol's effects on the brain.

It is lucky that the brain centers respond to a rising blood alcohol concentration in the order described in Figure H7-4 because a person usually passes out before managing to drink a lethal dose. It is possible, though, to drink so fast that the effects of alcohol continue to accelerate after the person has gone to sleep. Occasionally, a person dies from drinking enough to stop the heart before passing out. Table H7-1 shows the blood alcohol levels that correspond to progressively greater intoxication, and Table H7-2 on p. 226 shows the brain responses that occur at these blood levels.

Like liver cells, brain cells die with excessive exposure to alcohol. Liver cells may be replaced, but not all brain cells can regenerate. Thus some heavy drinkers suffer permanent brain damage.

People who drink alcoholic beverages may notice that they urinate more, but they may be unaware of the vicious cycle that results. Alcohol depresses production of antidiuretic hormone (ADH), a hormone produced by the pituitary gland that retains water. Loss of body water leads to thirst, and thirst leads to more drinking. The only fluid that will relieve dehydration is water, but the thirsty drinker may drink alcohol instead. This only worsens the problem. Such information provides another practical tip: drink water when thirsty and before each alcoholic drink.

Water loss is accompanied by the loss of important minerals. As Chapters 12 and 13 will explain, these minerals are vital to the body's fluid balance and to many chemical reactions in the cells, including muscle action. Detoxification treatment includes restoration of mineral balance as quickly as possible.

## Alcohol and Malnutrition

For many moderate drinkers, alcohol does not suppress food intake, and in some cases, it may actually stimulate appetite. Moderate drinkers usually consume alcohol as *added* energy—on top of their normal food intake. In addition,

**Figure H7-4**
**Alcohol's Effects on the Brain**

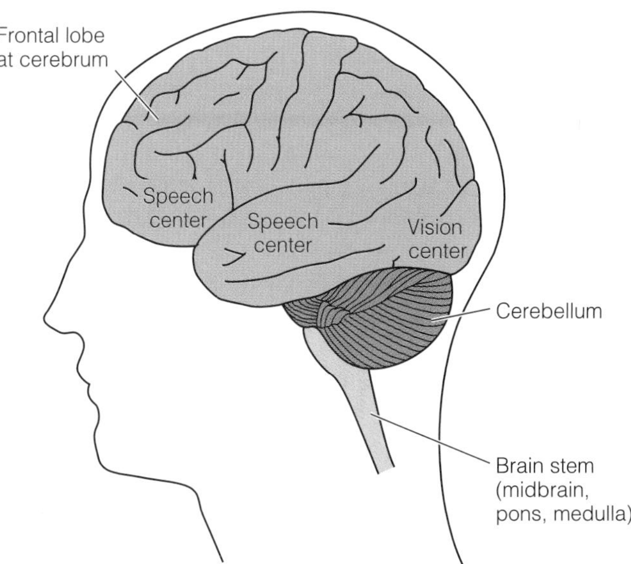


1. Judgment and reasoning centers are most sensitive to alcohol. When alcohol flows to the brain, it first sedates the frontal lobe, the reasoning part. As the alcohol molecules diffuse into the cells of these lobes, they interfere with reasoning and judgment.

2. Speech and vision centers are affected next. If the drinker drinks faster than the rate at which the liver can oxidize the alcohol, blood alcohol concentrations rise: the speech and vision centers of the brain become sedated.

3. Voluntary muscular control is then affected. At still higher concentrations, the cells in the cerebellum responsible for coordination of voluntary muscles are affected including those used in speech, eye, and limb movements. At this point people under the influence stagger or weave when they try to walk, or they may slur their speech.

4. Respiration and heart action are the last to be affected. Finally, the conscious brain is completely subdued, and the person passes out. Now the person can drink no more; this is fortunate because higher doses have an anesthetic effect that could reach the deepest brain centers, which control breathing and heartbeat, and the person could die.

alcohol in moderate doses is efficiently metabolized. Consequently, alcohol can contribute to body fat.[3] Metabolically, alcohol behaves like fat in promoting obesity; each ounce of alcohol represents about a half ounce of fat.[4] Alcohol's contribution to body fat is most evident in the central obesity that commonly accompanies alcohol consumption, popularly—and appropriately—known as the "beer belly."[5] Heavy drinkers, on the other hand, usually consume alcohol as *substituted* energy—instead of their normal food intake. Alcohol produces euphoria, which depresses appetite, so heavy drinkers tend to eat poorly and suffer malnutrition.

**Table H7-1**
**Alcohol Doses and Blood Levels**

| Number of Drinks[a] | Percentage of Blood Alcohol by Body Weight | | | | |
|---|---|---|---|---|---|
| | *100 lb* | *120 lb* | *150 lb* | *180 lb* | *200 lb* |
| 2 | 0.08 | 0.06 | 0.05 | 0.04 | 0.04 |
| 4 | 0.15 | 0.13 | 0.10 | 0.08 | 0.08 |
| 6 | 0.23 | 0.19 | 0.15 | 0.13 | 0.11 |
| 8 | 0.30 | 0.25 | 0.20 | 0.17 | 0.15 |
| 12 | 0.45 | 0.36 | 0.30 | 0.25 | 0.23 |
| 14 | 0.52 | 0.42 | 0.35 | 0.34 | 0.27 |

NOTE: In some states driving under the influence is proven when an adult's blood contains 0.08 percent alcohol, and in others, 0.10. Many states have adopted a "zero-tolerance" policy for drivers under age 21, using 0.02 percent as the limit.
[a]Taken within an hour or so; each drink equivalent to ½ ounce pure ethanol.

Alcohol in heavy doses is not efficiently metabolized, generating more heat than fat.

Alcohol is rich in energy (7 kcalories per gram), but like pure sugar or fat, the kcalories are empty of nutrients. The more alcohol people drink, the less likely that they will eat enough food to obtain adequate nutrients. The more kcalories spent on alcohol, the fewer kcalories available to spend on nutritious foods. Table H7-3 shows the kcalorie amounts of typical alcoholic beverages.

Chronic alcohol abuse not only displaces nutrients from the diet but also interferes with the body's metabolism of nutrients. Most dramatic is alcohol's effect on the B vitamin folate. The liver loses its ability to retain folate, and the kidneys increase their excretion of it.[6] Alcohol abuse creates a folate deficiency that devastates digestive system function. The intestine normally releases and retrieves folate continuously, but it becomes damaged by folate deficiency and alcohol toxicity, so it fails to retrieve its own folate and misses any that may trickle in from food as well. Alcohol also interferes with the action of folate in converting homocysteine to the nonessential amino acid methionine. The result is an excess of homocysteine, which has been linked to heart disease, and an inadequate supply of methionine, which slows the production of new cells, especially the rapidly dividing cells of the intestine and the blood.[7] The combination of poor folate status and alcohol consumption has also been implicated in promoting colorectal cancer.[8]

The inadequate food intake and impaired nutrient absorption that accompany chronic alcohol abuse frequently lead

**Table H7-2**
**Alcohol Blood Levels and Brain Responses**

| Blood Alcohol Concentration | Effect on Brain |
|---|---|
| 0.05 | Impaired judgment, relaxed inhibitions, altered mood, increased heart rate |
| 0.10 | Impaired coordination, delayed reaction time, exaggerated emotions, impaired peripheral vision, impaired ability to operate a vehicle |
| 0.15 | Slurred speech, blurred vision, staggered walk, seriously impaired coordination and judgment |
| 0.20 | Double vision, inability to walk |
| 0.30 | Uninhibited behavior, stupor, confusion, inability to comprehend |
| 0.40 to 0.60 | Unconsciousness, shock, coma, death (cardiac or respiratory failure) |

NOTE: Blood alcohol concentration depends on a number of factors, including alcohol in the beverage, the rate of consumption, the person's gender, and body weight. For example, a 100-pound female can become legally drunk (0.10 concentration) by drinking three beers in an hour, whereas a 220-pound male consuming that amount at the same rate would have a 0.05 blood alcohol concentration.

to a deficiency of another B vitamin—thiamin. In fact, the cluster of thiamin-deficiency symptoms commonly seen in chronic alcoholism has its own name—the Wernicke-Korsakoff syndrome. This syndrome is characterized by paralysis of the eye muscles, poor muscle coordination, impaired memory, and damaged nerves, and is treated with thiamin supplements.

Acetaldehyde, an intermediate in alcohol metabolism, interferes with nutrient use, too. For example, acetaldehyde dislodges vitamin $B_6$ from its protective binding protein so that it is destroyed, causing a vitamin $B_6$ deficiency and, thereby, lowered production of red blood cells.

Malnutrition occurs not only because of lack of intake and altered metabolism, but because of direct toxic effects as well.[9] Alcohol causes stomach cells to oversecrete both gastric acid and histamine, an agent of the immune system that produces inflammation. Beer in particular stimulates gastric acid secretion, irritating the stomach and esophagus linings and making them vulnerable to ulcer formation.

Nutrient deficiencies are virtually inevitable in alcohol abuse, not only because alcohol displaces food but also because alcohol directly interferes with the body's use of nutrients, making them ineffective even if they are present. Intestinal cells fail to absorb B vitamins, notably thiamin, folate, and vitamin $B_{12}$. Liver cells lose efficiency in activating vitamin D. Cells in the retina of the eye, which normally process the alcohol form of vitamin A (retinol) to its aldehyde form needed in vision, find themselves processing ethanol to acetaldehyde instead.

Regardless of dietary intake, excessive drinking over a lifetime creates deficits of all the nutrients mentioned in this discussion and more. No diet can compensate for the damage caused by heavy alcohol consumption.

## *Alcohol's Short-Term Effects*

Heavy or binge drinking (defined as at least 4 to 5 drinks in a row) is widespread on college campuses and poses serious health and social consequences to drinkers and nondrinkers alike.[10]* In fact, binge drinking can kill: the respiratory center of the brain becomes anesthetized, and breathing stops. Binge drinking is most common among college students who live in a fraternity or sorority house, attend parties frequently, engage in other risky behaviors, and have a history of binge drinking in high school.[11] Alcohol is responsible for most accidental deaths, including automobile fatalities. Compared with nondrinkers or moderate drinkers, people who frequently binge drink (at least three times within two weeks) are more likely to engage in unprotected sex, have multiple sex partners, damage property, and assault others.[12]

*This definition of binge drinking, without specification of time elapsed, is consistent with standard practice in alcohol research among college students.

**Table H7-3**
**kCalories in Alcoholic Beverages and Mixers**

| Beverage | Amount (oz) | Energy (kcal) |
|---|---|---|
| Beer | | |
| Regular | 12 | 150 |
| Light | 12 | 78–131 |
| Nonalcoholic | 12 | 32–82 |
| Distilled liquor (gin, rum, vodka, whiskey) | | |
| 80 proof | 1½ | 100 |
| 86 proof | 1½ | 105 |
| 90 proof | 1½ | 110 |
| Liqueurs | | |
| Coffee liqueur, 53 proof | 1½ | 175 |
| Coffee and cream liqueur, 34 proof | 1½ | 155 |
| Crème de menthe, 72 proof | 1½ | 185 |
| Mixers | | |
| Club soda | 12 | 0 |
| Cola | 12 | 150 |
| Cranberry juice cocktail | 8 | 145 |
| Diet drinks | 12 | 2 |
| Ginger ale or tonic | 12 | 125 |
| Grapefruit juice | 8 | 95 |
| Orange juice | 8 | 110 |
| Tomato or vegetable juice | 8 | 45 |
| Wine | | |
| Dessert | 3½ | 110–135 |
| Nonalcoholic | 8 | 14 |
| Red or rosé | 3½ | 75 |
| White | 3½ | 70 |
| Wine cooler | 12 | 170 |

Binge drinking is not limited to college campuses, of course, but that environment seems most accepting of such behavior despite its problems. Social acceptance may make it difficult for binge drinkers to recognize themselves as problem drinkers. For this reason, interventions must focus both on individuals and on the whole population. The damage alcohol causes only becomes worse if the pattern is not broken. Alcohol addiction sets in much more quickly in young people than in adults.[13] Those who start drinking at an early age more often suffer from alcoholism than people who start later on. Table H7-4 lists the key signs of alcoholism.

National Clearinghouse for Alcohol and Drug Information (NCADI)

www.health.org

**Table H7-4**
**Signs of Alcoholism**

- Tolerance—the person needs higher and higher intakes of alcohol to achieve intoxication.
- Withdrawal—the person who stops drinking experiences anxiety, agitation, increased blood pressure, or seizures, or seeks alcohol to relieve these symptoms.
- Impaired control—the person intends to have 1 or 2 drinks, but has 9 or 10 instead, or the person tries to control or quit drinking, but fails.
- Disinterest—the person neglects important social, family, job, or school activities because of drinking.
- Time—the person spends a great deal of time obtaining and drinking alcohol or recovering from excessive drinking.
- Impaired ability—the person's intoxication or withdrawal symptoms interfere with work, school, or home.
- Problems—the person continues drinking despite physical hazards or medical, legal, psychological, family, employment, or school problems.

The presence of three or more of these conditions is required to make a diagnosis.

Source: Adapted from *Diagnostic and Statistical Manual of Mental Disorders,* 4th ed. (Washington, D.C.: American Psychiatric Association, 1994).

## *Alcohol's Long-Term Effects*

By far the longest-term effect of alcohol is the damage done to a child whose mother abused alcohol during pregnancy. The devastating effects of alcohol on the unborn, and the message that pregnant women should not drink alcohol, are presented in Highlight 15.

For nonpregnant adults, a drink or two sets in motion many destructive processes in the body, but the next day's abstinence reverses them. As long as the doses are moderate, time between them is ample, and nutrition is adequate, recovery is probably complete.

If the doses of alcohol are heavy and the time between them short, complete recovery cannot take place. Repeated onslaughts of alcohol gradually take a toll on all parts of the body (see Table H7-5). Compared with nondrinkers, heavy drinkers have significantly greater risks of dying from all causes.[14]

National Council on Alcoholism and Drug Dependence (NCADD)

www.ncadd.org

## *Personal Strategies*

One obvious option available to people attending social gatherings is to enjoy the conversation, eat the food, and drink nonalcoholic beverages. Several nonalcoholic beverages are available that mimic the look and taste of their alcoholic counterparts. For those who enjoy champagne or beer, sparkling ciders and beers without alcohol are available. Instead of drinking a cocktail, a person can sip tomato juice with a slice of lime and a stalk of celery or just a plain cola beverage. Any of these drinks can ease conversation.

The person who chooses to drink alcohol should sip each drink slowly with food. The alcohol should arrive at the liver cells slowly enough that the enzymes can handle the load. It is best to space drinks, too, allowing about an hour or so to metabolize each drink.

**Table H7-5**
**Health Effects of Alcohol Consumption**

| Health Problem | Effects of Alcohol |
|---|---|
| Arthritis | Increases the risk of inflamed joints. |
| Cancer | Increases the risk of cancer of the liver, pancreas, rectum, and breast; increases the risk of cancer of the mouth, pharynx, larynx, and esophagus, where alcohol interacts synergistically with tobacco. |
| Fetal alcohol syndrome | Causes physical and behavioral abnormalities in the fetus (see Highlight 15). |
| Heart disease | In heavy drinkers, raises blood pressure, blood lipids, and the risk of stroke and heart disease; when compared with those who abstain, heart disease risk is generally lower in light-to-moderate drinkers (see Chapter 18). |
| Hyperglycemia | Raises blood glucose. |
| Hypoglycemia | Lowers blood glucose, especially in people with diabetes. |
| Kidney disease | Enlarges the kidneys, alters hormone functions, and increases the risk of kidney failure. |
| Liver disease | Causes fatty liver, alcoholic hepatitis, and cirrhosis. |
| Malnutrition | Increases the risk of protein-energy malnutrition; low intakes of protein, calcium, iron, vitamin A, vitamin C, thiamin, vitamin $B_6$, and riboflavin; and impaired absorption of calcium, phosphorus, vitamin D, and zinc. |
| Nervous disorders | Causes neuropathy and dementia; impairs balance and memory. |
| Obesity | Increases energy intake, but is not a primary cause of obesity. |
| Psychological disturbances | Causes depression, anxiety, and insomnia. |

NOTE: This list is by no means all-inclusive. Alcohol has direct toxic effects on all body systems.

If you want to help sober up a friend who has had too much to drink, don't bother walking arm in arm around the block. Walking muscles have to work harder, but muscle cells can't metabolize alcohol; only liver cells can. Remember that each person has a limited amount of alcohol dehydrogenase that clears the blood at a steady rate. In short, time alone will do the job.

Nor will it help to give your friend a cup of coffee. Caffeine is a stimulant, but it won't speed up alcohol metabolism. The police say ruefully, "If you give a drunk a cup of coffee, you'll just have a wide-awake drunk on your hands." Table H7-6 presents other alcohol myths.

Don't drive too soon after drinking. The lack of glucose for the brain's function and the length of time needed to clear the blood of alcohol make alcohol's adverse effects linger long after its blood concentration has fallen to zero. Driving coordination is still impaired the morning *after* a night of drinking, even if the drinking was moderate. Responsible aircraft pilots know that they must allow 24 hours for their bodies to clear alcohol completely, and they refuse to fly any sooner. The Federal Aviation Administration and major airlines enforce this rule.

Society also pays a high price when a drinker gets behind the wheel of an automobile—and that occurs an estimated 123 million times a year.[15] Traffic accidents are the number one cause of death among young people (ages 5 to 32), and almost half of all traffic fatalities involve alcohol. On the average, a person is killed in an alcohol-related traffic accident every 30 minutes.[16] This rate is actually lower than in past decades thanks to the educational efforts of MADD (Mothers Against Drunk Driving), the implementation of designated driver programs, a higher minimum drinking age in many states, and the severe legal consequences of driving while under the influence (DUI). In addition to traffic fatalities, alcohol use has been implicated in most of the other deaths among young people, including drownings, falls, suicides, and homicides.[17]

Look again at the drawing of the brain in Figure H7-4 and note that when someone drinks, judgment fails first. Judgment might tell a person to limit alcohol consumption to two drinks at a party, but if the first drink takes judgment away, many more drinks may follow. The failure to stop drinking as planned, on repeated occasions, is a danger sign warning that the person should not drink at all. Appendix F provides addresses for organizations that offer information about alcohol and alcohol abuse.

 Alcoholic's Anonymous (AA)

www.aa.org

Ethanol interferes with a multitude of chemical and hormonal reactions in the body—many more than have been enumerated here. With heavy alcohol consumption, the potential for harm is great. The best way to escape the harmful effects of alcohol is, of course, to refuse alcohol altogether. If you do drink, do so with care, and in moderation.

 U.S. Government

www.healthfinder.gov/searchoptions/topicsaz.htm
Search for Alcohol

**Table H7-6**
**Myths and Truths concerning Alcohol**

| | |
|---|---|
| Myth: | Hard liquors such as rum, vodka, and tequila are more harmful than wine and beer. |
| Truth: | The damage caused by alcohol depends largely on the *amount* consumed. Compared with hard liquor, beer and wine have relatively low percentages of alcohol, but they are often consumed in larger quantities. |
| Myth: | Consuming alcohol with raw seafood diminishes the likelihood of getting hepatitis. |
| Truth: | Alcohol is a disinfectant and can destroy some organisms. In fact, people have eaten contaminated oysters while drinking alcoholic beverages and not gotten as sick as those who were not drinking. But do not be misled: hepatitis is too serious an illness for anyone to depend on alcohol for protection. |
| Myth: | Alcohol stimulates the appetite. |
| Truth: | For some people, alcohol may stimulate appetite, but it seems to have the opposite effect in heavy drinkers. Heavy drinkers tend to eat poorly and suffer malnutrition. |
| Myth: | Drinking alcohol reduces the risk of heart disease. |
| Truth: | Moderate alcohol consumption is associated with a lower risk for heart disease in some people (see Chapter 18 for more details). Higher intakes, however, raise the risks for high blood pressure, stroke, heart disease, some cancers, accidents, violence, suicide, birth defects, and deaths in general. Furthermore, excessive alcohol consumption damages the liver, pancreas, brain, and heart. No authority recommends that nondrinkers begin drinking alcoholic beverages to obtain health benefits. |
| Myth: | Wine increases the body's absorption of minerals. |
| Truth: | Wine may increase the body's absorption of potassium, calcium, phosphorus, magnesium, and zinc, but the alcohol in wine also promotes the body's excretion of these minerals, so no benefit is gained. |
| Myth: | Alcohol is legal and, therefore, not a drug. |
| Truth: | Alcohol is legal for adults 21 years old and older, but it is also a drug—a substance that alters one or more of the body's functions. |
| Myth: | A shot of alcohol warms you up. |
| Truth: | Alcohol diverts blood flow to the skin making you *feel* warmer, but it actually cools the body. |
| Myth: | Wine and beer are mild; they do not lead to addiction. |
| Truth: | Wine and beer drinkers worldwide have high rates of death from alcohol-related illnesses. It's not what you drink, but how much that makes the difference. |
| Myth: | Mixing different types of drinks gives you a hangover. |
| Truth: | Too much alcohol in any form produces a hangover. |
| Myth: | Alcohol is a stimulant. |
| Truth: | People think alcohol is a stimulant because it seems to relieve inhibitions, but it does so by depressing the activity of the brain. Alcohol is medically defined as a depressant drug. |
| Myth: | Beer is a great source of carbohydrate, vitamins, minerals, and fluids. |
| Truth: | Beer does provide some carbohydrate, but most of its kcalories come from alcohol. The few vitamins and minerals in beer cannot compete with rich food sources. And the diuretic effect of alcohol causes the body to lose more fluid in urine than is provided by the beer. |

# Notes


1. Secretary of Health and Human Services, Eighth Special Report to the U.S. Congress on Alcohol and Health (Rockville, Md.: U.S. Department of Health and Human Services, 1993).
2. H. K. Seitz and C. M. Oneta, Gastrointestinal alcohol dehydrogenase, *Nutrition Reviews* 56 (1998): 52–60.
3. P. M. Suter, E. Häsler, and W. Vetter, Effects of alcohol on energy metabolism and body weight regulation: Is alcohol a risk factor for obesity? *Nutrition Reviews* 55 (1997): 157–171.
4. J. P. Flatt, Body weight, fat storage, and alcohol metabolism, *Nutrition Reviews* 50 (1992): 267–270.
5. B. B. Duncan and coauthors, Association of the waist-to-hip ratio is different with wine than with beer or hard liquor consumption, *American Journal of Epidemiology* 142 (1995): 1034–1038.
6. L. Feinman and C. S. Lieber, Nutrition and diet in alcoholism, in *Modern Nutrition in Health and Disease,* eds. M. E. Shils, J. A. Olson, and M. Shike (Philadelphia: Lea & Febiger, 1994), pp. 1081–1101.
7. M. L. Cravo and coauthors, Hyperhomocysteinemia in chronic alcoholism: Correlation with folate, vitamin B-12, and vitamin B-6 status, *American Journal of Clinical Nutrition* 63 (1996): 220-224.
8. Folate, alcohol, methionine, and colon cancer risk: Is there a unifying theme? *Nutrition Reviews* 52 (1994): 18–20.
9. C. S. Lieber, Herman Award Lecture, 1993: A personal perspective on alcohol, nutrition, and the liver, *American Journal of Clinical Nutrition* 58 (1993): 430–442.
10. H. Wechsler and coauthors, Health and behavioral consequences of binge drinking in college: A national survey of students at 140 campuses, *Journal of the American Medical Association* 272 (1994): 1672–1677.
11. H. Wechsler and coauthors, Binge drinking, tobacco, and illicit drug use and involvement in college athletes: A survey of students at 140 American colleges, *Journal of American College Health* 45 (1995): 195–200.
12. Wechsler and coauthors, 1994; K. L. Graves, Risky sexual behavior and alcohol use among young adults: Results from a national survey, *American Journal of Health Promotion* 10 (1995): 27–36; D. M. Ferguson and M. T. Lynskey, Alcohol misuse and adolescent sexual behaviors and risk taking, *Pediatrics* 98 (1996): 91–96.
13. Committee on Substance Abuse, Alcohol use and abuse: A pediatric concern, *Pediatrics* 95 (1995): 439–442.
14. A. L. Klatsky, M. A. Armstrong, and G. D. Friedman, Alcohol and mortality, *Annals of Internal Medicine* 117 (1992): 646–654.
15. S. Liu and coauthors, Prevalence of alcohol-impaired driving: Results from a national self-reported survey of health behaviors, *Journal of the American Medical Association* 277 (1997): 122–125.
16. Alcohol-related traffic fatalities, *FDA Consumer,* March 1993, p. 26.
17. Committee on Substance Abuse, 1995.

CHAPTER 8

# Energy Balance and Body Composition

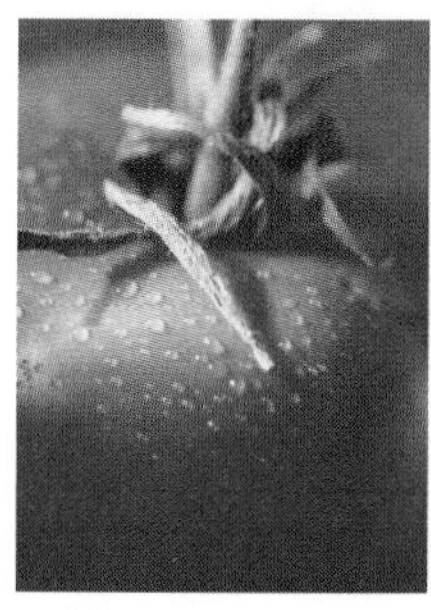

The body's remarkable machinery can cope with many extremes of diet. As you have seen, it can convert both carbohydrate (glucose) and protein (amino acids) to fat. To some extent, it can convert amino acids to glucose. To a very limited extent, it can even convert fat (the glycerol portion) to glucose. But a grossly unbalanced diet imposes hardships on the body. If energy intake is too low or if too little carbohydrate or protein is supplied, the body must degrade its own lean tissue to meet its glucose and protein needs. If energy intake is too high or if fat is abundant, the body stores fat.

Overfatness and underweight both result from unbalanced energy budgets. The simple picture is as follows. Overfat people have consumed more food energy than they have spent and have banked the surplus in their body fat. To reduce body fat, they need to spend more energy than they take in from food. In contrast, underweight people have consumed too little food energy to support their bodies' activities and so have depleted their bodies' fat stores and possibly some of their lean tissues as well. To gain weight, they need to take in more food energy than they expend. As you will see, though, the details of the body's weight regulation are quite complex. This chapter describes energy balance and body composition and examines the problems associated with having too much or too little body fat; the next chapter presents strategies toward resolving these problems.

The term *overfat* refers to an excess of body fat, which is not necessarily the same as *overweight,* as a later section of the chapter explains.

# Energy Balance

People spend energy continuously and eat periodically to refuel. Ideally, their energy intakes cover their energy expenditures without too much excess. Excess energy is stored as fat, and stored fat is used for energy between meals. The amount of body fat a person deposits in, or withdraws from, "storage" on any given day depends on the energy balance for that day—the amount consumed (energy in) versus the amount expended (energy out). When a person is maintaining weight, energy in equals energy out.

When energy in balances with energy out, a person's body weight is stable.

Most people maintain a steady energy balance over time. On any given day, they may eat a little more or a little less than usual, and their weight may go up or down a pound or two, but for the most part, they stay in balance. When the balance shifts, their weight changes.

A pound of body fat stores about 3500 kcalories. It stands to reason that a person who eats 3500 extra kcalories should gain a pound, and that a person who cuts 3500 kcalories should lose a pound, but this does not always happen. When a person overeats, much of the excess energy is stored, but some energy is spent to maintain the heavier body. Furthermore, people seem to gain more body fat when they eat extra fat kcalories than when they eat extra carbohydrate kcalories, and they seem to lose body fat most efficiently when they limit kcalories specifically from fat. Whether a person chooses extra potatoes or extra butter may make a great difference to body weight and body composition.

1 lb body fat = 3500 kcal.
Body fat, or adipose tissue, is composed of a mixture of mostly fat, some protein, and water. A pound of body fat (454 g) is approximately 87% fat, or (454 × 0.87) 395 g, and 395 g × 9 kcal/g = 3555 kcal.

The *Dietary Guidelines* suggest a reasonable rate of weight loss for overweight people of ½ to 1 pound a week; many obesity experts agree that weight loss should not exceed an average of 2 pounds per week. Even for obese people, a reasonable weight-loss rate is only 1 percent of body weight per week. Such gradual weight losses are more likely to be maintained than rapid losses and can be achieved with a reasonable energy intake of about 10 kcalories per pound of body weight. If food energy is restricted too severely, dieters lose lean tissue and may not receive

Energy intake for weight loss:
10 kcal/lb body weight.

When food is burned, the chemical bonds between the carbons and hydrogens are broken, and energy is released in the form of heat. The amount of heat generated provides a direct measure of the amount of energy stored in the food's chemical bonds.

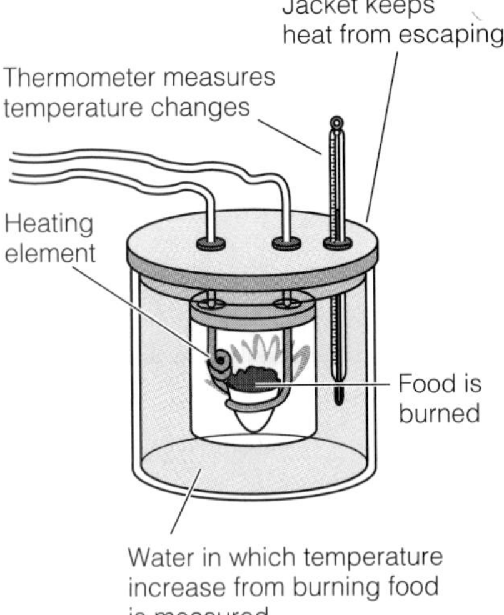


**Figure 8-1**
**Bomb Calorimeter**

**bomb calorimeter** (KAL-oh-RIM-eh-ter): an instrument that measures the *heat* energy released when foods are burned, thus providing an estimate of the potential energy of foods.
- **calor** = heat
- **metron** = measure

Food energy values can be determined by:
- **Direct calorimetry**, which measures the amount of heat released.
- **Indirect calorimetry**, which measures the amount of oxygen consumed.

Reminder: A *kcalorie* is a unit of *heat* energy. One kcalorie is the amount of heat necessary to raise the temperature of 1 kilogram of water 1°C.

 USDA Database
www.nal.usda.gov/fnic/foodcomp

The number of kcalories that the body derives from a food, as contrasted with the number of kcalories determined by calorimetry, is the **physiological fuel value.**

Reminder:
- 1 g carbohydrate = 4 kcal.
- 1 g fat = 9 kcal.
- 1 g protein = 4 kcal.
- 1 g alcohol = 7 kcal.

enough nutrients. In addition, restrictive eating may set in motion the unhealthy cycle of restrictive dieting and binge eating.

Besides, quick changes in weight are not just changes in fat. Weight gained or lost rapidly includes some fat, large amounts of fluid, and some lean tissues such as muscles and bone minerals. (Because water constitutes about 60% of an adult's body weight, retention or loss of water influences body weight.) Even over the long term, the composition of weight gained or lost is normally about 75 percent fat and 25 percent lean. During starvation, losses of fat and lean are about equal. Invariably, though, *fat* gains and losses are gradual.

IN SUMMARY

When the energy consumed equals the energy expended, the person is in energy balance and body weight is stable. If more energy is taken in than is expended, the person gains weight. If more energy is spent than is taken in, the person loses weight. The next two sections examine the two sides of the energy-balance equation: energy in and energy out.

# Energy In: The kCalories Foods Provide

Foods and beverages are the "energy in" part of the energy-balance equation. How much energy a person receives depends on the composition of the foods and beverages and on how much the person eats and drinks.

## Food Composition

To find out how many kcalories a food provides, a scientist can burn food in a bomb calorimeter (see Figure 8-1). When the food burns, the chemical bonds between the carbon and hydrogen atoms break, releasing energy in the form of heat. The amount of heat given off provides a *direct* measure of the food's energy value (remember that kcalories are units of heat energy). In addition to releasing heat, these reactions generate carbon dioxide and water—just as the body's cells do when they metabolize the energy-yielding nutrients. When the food burns and the chemical bonds break, the carbons (C) and hydrogens (H) combine with oxygen (O) to form carbon dioxide ($CO_2$) and water ($H_2O$). The amount of oxygen consumed gives an *indirect* measure of the amount of energy released.

A bomb calorimeter measures the available energy in foods but overstates the amount of energy that the human body derives from foods. The body is less efficient than a calorimeter and cannot metabolize all of a food's energy-yielding nutrients all the way to carbon dioxide and water. Researchers can correct for this discrepancy mathematically to make useful tables of the energy values of foods (such as Appendix H). These values provide reasonable estimates, but do not reflect the *precise* amount of energy a person will derive from the foods consumed.

The energy values of foods can also be computed from the amounts of carbohydrate, fat, and protein (and alcohol, if present) in the foods.* For example, a food containing 8 grams protein, 12 grams carbohydrate, and 5 grams fat would provide 32 protein kcalories, 48 carbohydrate kcalories, and 45 fat kcalories, for a total of 125 kcalories.

*Some of the food energy values in the table of food composition in Appendix H were derived by bomb calorimetry, and many were calculated from their energy-yielding nutrient contents.

# Food Intake

To achieve energy balance, the body must meet its needs without taking in too much or too little energy. Somehow the body decides how much and how often to eat—when to start eating and when to stop. As you will see, human appetite reflects a combination of many signals that initiate or delay eating.

**appetite:** the integrated response to the sight, smell, thought, or taste of food that initiates or delays eating.

**hunger:** the physiological drive for food that initiates food-seeking behavior.

• ***Hunger*** • People eat for a variety of reasons, most obviously (although not necessarily most commonly), because they are hungry. Most people recognize hunger as an irritating feeling that prompts thoughts of food and motivates them to start eating. The stomach is ideally designed to handle periodic batches of food, and people typically eat meals at roughly four-hour intervals. Four hours after a meal, most, if not all, of the food has left the stomach and been absorbed by the intestine. Most people do not feel like eating again until the stomach is either empty or almost so. Even then, a person may not feel hungry for quite a while.

The body seems to adapt its hunger response to accommodate changes in food intake. People who restrict their food intakes may feel pangs of hunger for the first few days, but these sensations diminish with time. After the body has adapted to less food, eating a large, energy-rich meal makes the person feel uncomfortable, in part because the stomach's capacity has diminished.[1] People can adapt to eating excessive amounts of food as well; repeated binge eating enlarges the stomach's capacity. Consequently, the person may feel less satisfied after a normal-size meal, setting the stage for another binge.

Receptors in the GI tract also adapt and change their responses depending on whether nutrient intake has been high or low. Consider that after two weeks on a high-fat diet, digestion and absorption of fat are accelerated.[2] The GI tract adapts to handle the increase in fat intake efficiently, thus conserving energy. Such findings have interesting implications for the development of obesity.

Hunger is only one of the signals determining whether a person will eat. A person may eat without hunger, for example, when presented with a hot piece of homemade apple pie after having eaten a large dinner. In contrast, a person may feel hungry but have no desire for food when faced with unfamiliar foods, a stressful situation, or illness; in such circumstances, eating becomes a chore.

• ***Satiation*** • During the course of a meal, as hunger diminishes, satiation develops. Satiation depends on the presence of food in the GI tract. Receptors in the stomach stretch, and the person begins to feel too uncomfortable to continue eating. Nutrients in the small intestine trigger the release of GI hormones (such as cholecystokinin) and the stimulation of nerves. Together, gastric distension, nutrients in the small intestine, and GI hormones send messages about the amount of food eaten and the kinds of nutrients received to the hypothalamus. The response: satiation occurs and people stop eating.

**satiation** (say-she-AY-shun): the feeling of satisfaction and fullness that occurs during a meal and halts eating. Satiation determines how much food is consumed during a meal.

• ***Satiety*** • After a meal, the feeling of satiety continues to suppress hunger and allows a person to not eat again for a while. Whereas *satiation* informs us of when to "stop eating," *satiety* maintains the signal to "not start eating again." Figure 8-2 summarizes the relationships among hunger, satiation, and satiety. Of course, people can override these signals, especially when presented with stressful situations or favorite foods.

**satiety** (sah-TIE-eh-tee): the feeling of satisfaction that occurs after a meal and inhibits eating until the next meal. Satiety determines how much time passes between meals.

• ***Overriding Hunger and Satiety Signals*** • Not surprisingly, eating can be triggered by signals other than hunger, even when food is not needed. Some people experience food cravings when they are bored or anxious. In fact, they may eat in response to any kind of stress, negative or positive. ("What do I do when I'm grieving? Eat. What do I do when I'm celebrating? Eat!") Some people respond to external stimuli such as the time of day ("It's time to eat") or the availability, sight, and taste of food ("I'd love a piece of chocolate even though

Eating in response to arousal is called **stress eating.**

Some people eat in response to such **external cues** as the presence of food or the time of day rather than to such internal signals as hunger.

**Figure 8-2**
**Hunger, Satiation, and Satiety**

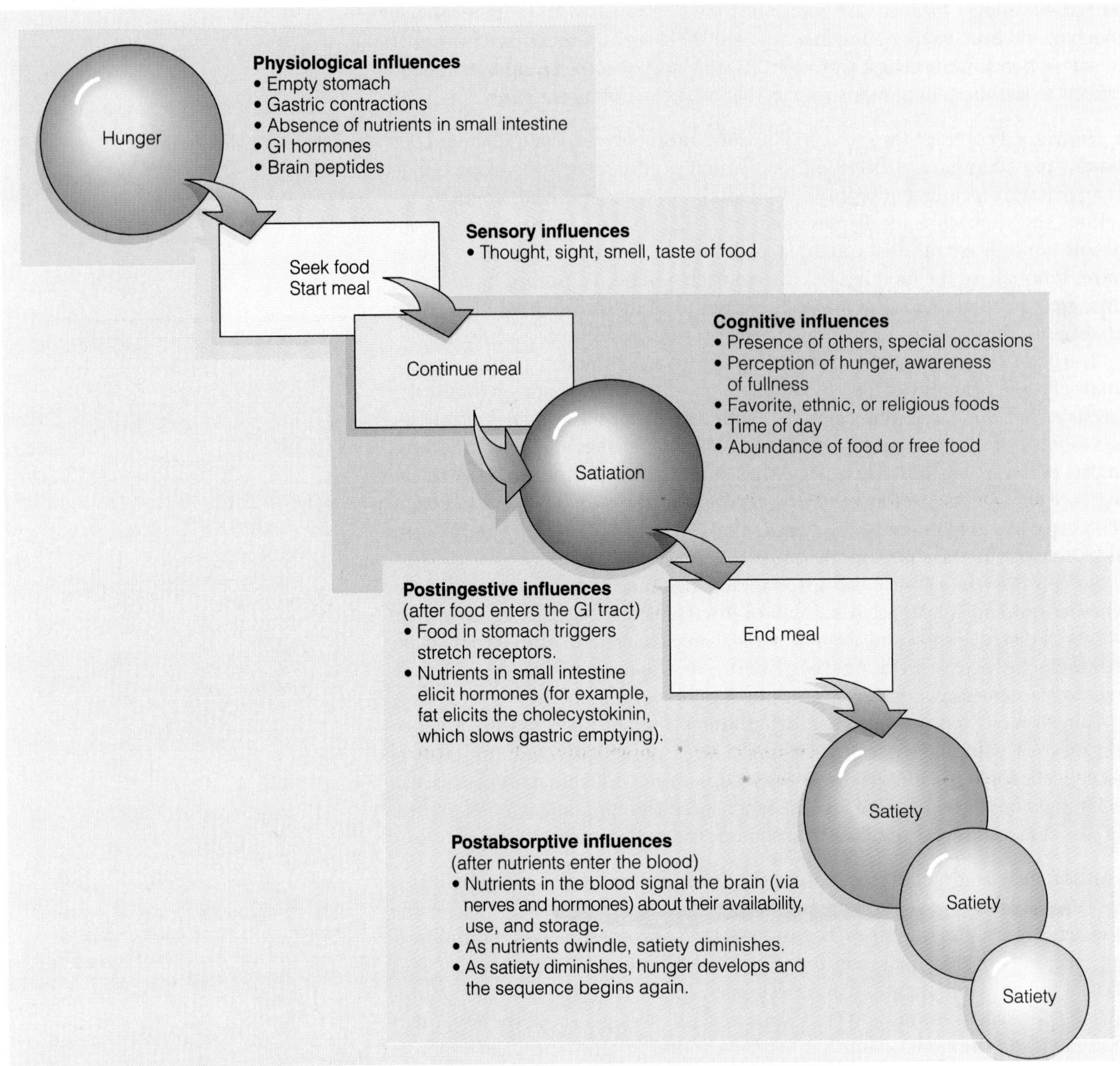


I'm stuffed"). Being presented with a variety of foods also stimulates eating; people eating one food until satisfied may begin eating enthusiastically again when offered a fresh selection of different foods. Such behavior can easily lead to weight gain.

Highlight 9 features eating disorders.

Eating can also be suppressed by signals other than satiety, even when a person is hungry. People with the eating disorder anorexia nervosa, for example, use tremendous discipline to ignore the pangs of hunger. Some people simply cannot eat during times of stress, negative or positive. ("I'm too sad to eat. I'm too excited to eat!") Why some people overeat in response to stress and others cannot eat at all remains a bit of a mystery. Factors that appear to be involved include how the person perceives the stress and whether normal eating behaviors are restrained.

**satiating:** the power to suppress hunger and inhibit eating.

• ***Nutrients, Satiation, and Satiety*** • The extent to which foods produce satiation and sustain satiety depends in part on the nutrient composition of a meal. Of the three energy-yielding nutrients, protein is the most satiating.[3] Foods rich in carbohydrates and fibers also effectively extend the duration of satiety by filling the stomach and delaying the absorption of nutrients.[4] In contrast, fat has a weak satiating effect; consequently, eating high-fat foods leads to passive overconsumption.[5] High-fat foods offer flavor, which entices people to eat more, and energy density, which delivers more kcalories per bite, but little satiation during a meal and much less satiety than protein or carbohydrate after a meal. Eating high-fat foods while trying to limit energy intake can leave a person feeling hungry.

People attempting to control their weight may find it easier to limit energy intake by selecting foods based on satiety. Researchers have developed a "satiety index" of common foods.[6] In doing so, they confirmed that the satiating effect of fat is weak and that of protein and fiber strong. High-fat foods such as croissants, doughnuts, peanuts, and potato chips please the palate and stimulate the appetite, but score low on the satiety index; by comparison, simple, whole foods such as potatoes, apples, oranges, whole-grain pastas, fish, and steak are highly satiating—and they provide a rich array of nutrients. Instead of feeling deprived by having to eat small servings of high-fat foods, a person can feel satisfied eating large servings of high-protein and high-fiber foods. Serving sizes correlate directly with a food's satiety index, and Figure 8-3 illustrates how fat influences serving size.

• ***Message Central—The Hypothalamus*** • Eating is a complex behavior controlled by a variety of psychological, social, metabolic, and physiological factors.[7] The hypothalamus appears to be the control center, integrating messages about energy intake, expenditure, and storage from other parts of the brain and from the mouth, GI tract, and liver. Some of these messages influence satiation, controlling the size of a meal; others influence satiety, determining the frequency of meals. Metabolic messages reflecting the balance between energy intake and energy needs over extended periods, such as in a pregnant woman or an athlete, may influence food intake for several months.

Researchers have identified dozens of chemicals in the brain that may participate in appetite control. Some are specific to particular nutrient cravings; for example, one of these chemicals, neuropeptide Y, causes carbohydrate cravings, and another, galanin, underlies a yen for fat. By understanding the action of these

**hypothalmus** (high-po-THAL-ah-mus): a brain center that controls activities such as maintenance of water balance, regulation of body temperature, and control of appetite.

**Figure 8-3**
**How Fat Influences Serving Size**

837 kcal, 71 g fat — 55 kcal, 3 g fat

For the same size serving, peanuts deliver more than 15 times the kcalories and 20 times the fat of popcorn.

100 kcal, 9 g fat — 100 kcal, 5 g fat

Popcorn offers twice the satiety of peanuts. For the same number of kcalories, a person can have a few high-fat peanuts or almost 2 cups of high-fiber popcorn. (This comparison used oil-based popcorn; using air-popped popcorn would double the amount of popcorn in this example.)

**Figure 8-4**
**Components of Energy Expenditure**

The amount of energy spent in a day differs for each individual, but in general, basal metabolism is the largest component of energy expenditure (60 to 65 percent), and the thermic effect of food is the smallest (only 10 percent). The amount spent in voluntary physical activities has the greatest variability, depending on a person's activity patterns.

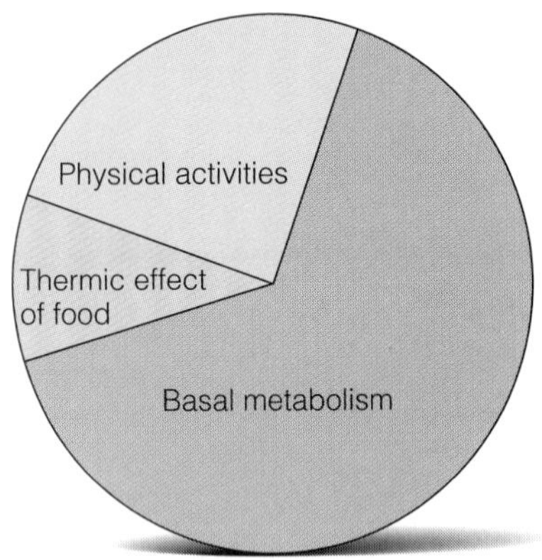


**thermogenesis:** the generation of heat; used in physiology and nutrition studies as an index of how much energy the body is spending. The total energy a body spends reflects three main categories of thermogenesis:
- Basal thermogenesis (metabolism).
- Exercise-induced thermogenesis (physical activity).
- Diet-induced thermogenesis (thermic effect of food).

A fourth category is sometimes involved:
- Adaptive thermogenesis (energy of adaptation).

**direct calorimetry** (kal-oh-RIM-eh-tree): the measurement of energy output as heat energy.

**indirect calorimetry:** the estimation of energy output from measures of the amount of oxygen used and carbon dioxide eliminated.

**basal metabolism:** the energy needed to maintain life when a body is at complete rest after a 12-hour fast (to exclude the thermic effect of the previous meal).

**basal metabolic rate (BMR):** the rate of energy use for metabolism under basal conditions, usually expressed as kcalories per kilogram body weight per hour. (Table 8-3 on p. 239 provides equations for estimating BMR.)

A similar measure of energy output is the **resting energy expenditure (REE).** The REE measure is usually less precise than the BMR because the criteria for rest and fasting are less strict, but the difference is usually less than 10% and can be discounted for most purposes.

**lean body mass:** the weight of the body minus the fat.

brain chemicals, researchers may one day be able to control appetite. Their greatest challenge is in sorting out the many actions of these brain chemicals. For example, in addition to being responsible for carbohydrate cravings, neuropeptide Y also initiates eating, decreases energy expenditure, and increases fat storage—all factors favoring a positive energy balance and weight gain.

IN SUMMARY

A mixture of signals governs people's eating behaviors. Hunger, satiation, and satiety each result from several stimuli generated by the nervous and hormonal systems. Superimposed on these are complex factors involving emotions, habits, and other aspects of human behavior.

# Energy Out: The kCalories the Body Spends

The body converts the energy of food to the energy currency of ATP molecules with about 50 percent efficiency, radiating the rest as heat. Then, when ATP energy is used to do work, again about 50 percent is lost as heat. Thus the overall efficiency of the human body in converting food energy to work is 25 percent; the other 75 percent is released as heat. The work itself, as it is done, generates heat as well, so a body's total heat production reflects the amount of energy it is spending.

The body's generation of heat is known as thermogenesis, and it can be measured to determine the amount of energy expended. This measurement of heat output is known as *direct calorimetry.* Alternatively, a person's energy expenditure can be calculated by *indirect calorimetry,* which involves measuring the amount of oxygen consumed and carbon dioxide expelled.

## *Components of Energy Expenditure*

People spend energy when they are physically active, of course, but they also spend energy when they are resting quietly. In fact, quiet metabolic activities account for the lion's share of most people's energy expenditures, as Figure 8-4 shows.

• ***Basal Metabolism*** • At least two-thirds of the energy the average person spends in a day supports the body's metabolic activities. Metabolic activities maintain the body temperature and keep the lungs inhaling and exhaling air, the bone marrow making new red blood cells, the heart beating 100,000 times a day, the kidneys filtering wastes—in short, they support all the basic processes of life.

The basal metabolic rate (BMR) is the rate at which the body spends energy for these maintenance activities. The rate may vary dramatically from person to person and may vary for the same individual with a change in circumstance or physical condition.[8] The rate is slowest when a person is sleeping undisturbed, but it is usually measured in a room with a comfortable temperature when the person is lying still after a restful sleep and is not digesting any food.

In general, the more a person weighs, the more *total* energy is required, but the amount of energy *per pound* of body weight may be lower. For example, an adult's BMR might be 1500 kcalories and an infant's only 500, but compared to body weight, the infant's BMR is more than twice as fast. Similarly, a normal-weight adult may have a metabolic rate one and a half times that of an obese adult when compared to body weight.

Table 8-1 summarizes the factors that raise and lower the BMR. For the most part, the BMR is highest in people with considerable lean body mass (growing

**Table 8-1**
**Factors That Affect the BMR**

| Factor | Effect on BMR |
|---|---|
| Age | Lean body mass diminishes with age, slowing the BMR.[a] |
| Height | In tall, thin people, the BMR is higher.[b] |
| Growth | In children and pregnant women, the BMR is higher. |
| Body composition | The more lean tissue, the higher the BMR (which is why males usually have a higher BMR than females). The more fat tissue, the lower the BMR. |
| Fever | Fever raises the BMR.[c] |
| Stresses | Stresses (including many diseases and certain drugs) raise the BMR. |
| Environmental temperature | Both heat and cold raise the BMR. |
| Fasting/starvation | Fasting/starvation lowers the BMR.[d] |
| Malnutrition | Malnutrition lowers the BMR. |
| Hormones | The thyroid hormone thyroxin, for example, can speed up or slow down the BMR.[e] |
| Smoking | Nicotine increases energy expenditure. |
| Caffeine | Caffeine increases energy expenditure. |
| Sleep | BMR is lowest when sleeping. |

[a]The BMR begins to decrease in early adulthood (after growth and development cease) at a rate of about 2 percent/decade. A reduction in voluntary activity as well brings the total decline in energy expenditure to 5 percent/decade.
[b]If two people weigh the same, the taller, thinner person will have the faster metabolic rate, reflecting the greater skin surface, through which heat is lost by radiation, in proportion to the body's volume (see margin drawing).
[c]Fever raises the BMR by 7 percent for each degree Fahrenheit.
[d]Prolonged starvation reduces the total amount of metabolically active lean tissue in the body, although the decline occurs sooner and to a greater extent than body losses alone can explain. More likely, the neural and hormonal changes that accompany fasting are responsible for changes in the BMR.
[e]The thyroid gland releases hormones that travel to the cells and influence cellular metabolism. Thyroid hormone activity can speed up or slow down the rate of metabolism by as much as 50 percent.

Notice that each of these structures is made of 8 blocks. They weigh the same, but they are arranged differently. If you were to count the sides of these structures, you would find that the short, wide one has 24 sides and the tall, thin one has 34. Because the tall, thin structure has a greater surface area, it will lose more heat (expend more energy) than the short, wide one. Similarly, two people of different heights might weigh the same, but the taller, thin one will have a higher BMR (expending more energy) because of the greater skin surface.

children, physically active people, pregnant women, and males). One way to increase the BMR then is to participate in endurance and strength-building activities regularly to maximize lean body tissue.[9] The BMR is also high in people who are tall and so have a large surface area for their weight, in people with fever or under stress, and in people with highly active thyroid glands.

The BMR declines during adulthood as lean body mass diminishes. This change in body composition occurs, in part, because some hormones that influence metabolism become more, or less, active as a person ages. Voluntary activity tends to be reduced as well, bringing the average decline in energy expenditure to about 5 percent per decade. The decline in the BMR that occurs when a person reduces voluntary activity reflects the loss of lean body mass and may be prevented with ongoing physical activity. The BMR also slows down during fasting and malnutrition.

• ***Physical Activity*** • The second component of a person's energy output is physical activity: voluntary movement of the skeletal muscles and support systems. Physical activity is the most variable component of energy expenditure. Consequently, its influence on both weight gain and weight loss can be significant.

During physical activity, the muscles need extra energy to move, and the heart and lungs need extra energy to deliver nutrients and oxygen and dispose of wastes. The amount of energy needed for any activity, whether playing tennis or studying for an exam, depends on three factors: muscle mass, body weight, and activity. The larger the muscle mass required and the heavier the weight of the body part being moved, the more energy is spent. Table 8-2 gives average energy expenditures for people of different body weights engaged in various activities and

**voluntary activities:** conscious and deliberate muscular work—walking, lifting, climbing, or other physical activity. In contrast, **involuntary activities** occur independently, without conscious will or knowledge—heart beating, lungs breathing, and other activities critical to maintaining life.

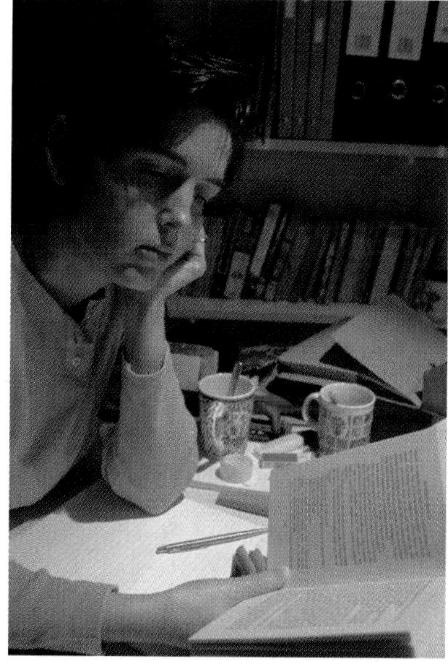

***It feels like work and it may make you tired, but studying requires only a kcalorie or two per minute.***

Chapter 14 describes how the activity's duration, frequency, and intensity also influence the body's fuel mix of carbohydrate and fat.

shows that a heavy person usually uses more energy per minute to perform a task than a light person does. The activity's duration, frequency, and intensity also influence energy cost: the longer, the more frequent, and the more intense the activity, the more kcalories spent.

• ***Thermic Effect of Food*** • The body uses some energy to process food. When a person eats, the GI tract muscles speed up their rhythmic contractions, and the cells that manufacture and secrete digestive juices begin their tasks. This acceleration of activity produces heat and is known as the thermic effect of food (TEF).

**Table 8-2**
**Energy Spent on Various Activities**

| Activity | kCal/lb/min[a] | kCalories per Minute at Different Body Weights | | | | |
|---|---|---|---|---|---|---|
| | | *110 lb* | *125 lb* | *150 lb* | *175 lb* | *200 lb* |
| Aerobic dance (vigorous) | .062 | 6.8 | 7.8 | 9.3 | 10.9 | 12.4 |
| Basketball (vigorous, full court) | .097 | 10.7 | 12.1 | 14.6 | 17.0 | 19.4 |
| Bicycling | | | | | | |
| 13 mph | .045 | 5.0 | 5.6 | 6.8 | 7.9 | 9.0 |
| 15 mph | .049 | 5.4 | 6.1 | 7.4 | 8.6 | 9.8 |
| 17 mph | .057 | 6.3 | 7.1 | 8.6 | 10.0 | 11.4 |
| 19 mph | .076 | 8.4 | 9.5 | 11.4 | 13.3 | 15.2 |
| 21 mph | .090 | 9.9 | 11.3 | 13.5 | 15.8 | 18.0 |
| 23 mph | .109 | 12.0 | 13.6 | 16.4 | 19.0 | 21.8 |
| 25 mph | .139 | 15.3 | 17.4 | 20.9 | 24.3 | 27.8 |
| Cross-country skiing | | | | | | |
| 8 mph | .104 | 11.4 | 13.0 | 15.6 | 18.2 | 20.8 |
| Golf (carrying clubs) | .045 | 5.0 | 5.6 | 6.8 | 7.9 | 9.0 |
| Handball | .078 | 8.6 | 9.8 | 11.7 | 13.7 | 15.6 |
| Horseback riding (trot) | .052 | 5.7 | 6.5 | 7.8 | 9.1 | 10.4 |
| Rowing (vigorous) | .097 | 10.7 | 12.1 | 14.6 | 17.0 | 19.4 |
| Running | | | | | | |
| 5 mph | .061 | 6.7 | 7.6 | 9.2 | 10.7 | 12.2 |
| 6 mph | .074 | 8.1 | 9.2 | 11.1 | 13.0 | 14.8 |
| 7.5 mph | .094 | 10.3 | 11.8 | 14.1 | 16.4 | 18.8 |
| 9 mph | .103 | 11.3 | 12.9 | 15.5 | 18.0 | 20.6 |
| 10 mph | .114 | 12.5 | 14.3 | 17.1 | 20.0 | 22.9 |
| 11 mph | .131 | 14.4 | 16.4 | 19.7 | 22.9 | 26.2 |
| Soccer (vigorous) | .097 | 10.7 | 12.1 | 14.6 | 17.0 | 19.4 |
| Studying | .011 | 1.2 | 1.4 | 1.7 | 1.9 | 2.2 |
| Swimming | | | | | | |
| 20 yd/min | .032 | 3.5 | 4.0 | 4.8 | 5.6 | 6.4 |
| 45 yd/min | .058 | 6.4 | 7.3 | 8.7 | 10.2 | 11.6 |
| 50 yd/min | .070 | 7.7 | 8.8 | 10.5 | 12.3 | 14.0 |
| Table tennis (skilled) | .045 | 5.0 | 5.6 | 6.8 | 7.9 | 9.0 |
| Tennis (beginner) | .032 | 3.5 | 4.0 | 4.8 | 5.6 | 6.4 |
| Walking (brisk pace) | | | | | | |
| 3.5 mph | .035 | 3.9 | 4.4 | 5.2 | 6.1 | 7.0 |
| 4.5 mph | .048 | 5.3 | 6.0 | 7.2 | 8.4 | 9.6 |

[a]To calculate kcalories spent per minute of activity for your own body weight, multiply kcal/lb/min by your exact weight and then multiply that number by the number of minutes spent in the activity. For example, if you weigh 142 pounds, and you want to know how many kcalories you spent doing 30 minutes of vigorous aerobic dance: 0.062 × 142 = 8.8 kcalories per minute; 8.8 × 30 (minutes) = 264 total kcalories spent.

The thermic effect of food is proportional to the food energy taken in and is usually estimated at 10 percent of energy intake. Thus a person who ingests 2000 kcalories in a day probably spends about 200 kcalories on the thermic effect of food. Because the thermic effect of food reflects the body's digestion and absorption activities, it is influenced by factors such as meal size, frequency, and composition; in general, the thermic effect of food is greater for high-carbohydrate foods than for high-fat foods and for a meal eaten all at once rather than spread out over a couple of hours. For most purposes, however, the thermic effect of food can be ignored because its contribution to total energy output is smaller than the probable errors involved in estimating overall energy intake and output.

**thermic effect of food (TEF):** an estimation of the energy required to process food (digest, absorb, transport, metabolize, and store ingested nutrients); also called the **specific dynamic effect (SDE)** of food or the **specific dynamic activity (SDA)** of food. The sum of the TEF and any increase in the metabolic rate due to overeating is known as **diet-induced thermogenesis (DIT).**

• ***Adaptive Thermogenesis*** • Some additional energy is spent when a person must adapt to dramatically changed circumstances (adaptive thermogenesis). When the body has to adapt to physical conditioning, cold, overfeeding, starvation, trauma, or other types of stress, it has extra work to do, building the tissues and producing the enzymes and hormones necessary to cope with the demand. In some circumstances this energy makes a considerable difference in the total energy spent. Because this component of energy expenditure is so variable and specific to individuals, it is not included when calculating the energy RDA.

**adaptive thermogenesis:** adjustments in energy expenditure related to changes in environment such as cold and to physiological events such as overfeeding, trauma, and changes in hormone status.

## *Estimating Energy Requirements*

In calculating the energy RDA, the following components of energy expenditure are considered:

- Energy spent on basal metabolism.
- Energy spent on physical activities.
- Energy spent on digesting and metabolizing food.

These three components vary, depending on a person's age, sex, body size, heredity, state of health, and other factors. First energy spent on basal metabolism for each age-sex group is estimated. Then increments for physical activity are added, assuming the average person would be lightly to moderately active. Finally, increments for the influence of food are added, assuming that each person would meet energy needs by eating a mixed diet of ordinary foods.

To estimate energy spent on basal metabolism, equations that consider age, sex, and weight are used, as shown in Table 8-3. (The "How to" on p. 241 shows a sample calculation.)

To estimate the energy spent on physical activity, individual values such as those presented in Table 8-2 cannot be used. This process is too time-consuming and impractical to be useful for estimating the energy needs of a population. Instead various activities are clustered according to the typical intensity of a day's efforts (under the headings of light, moderate, and heavy activity). Then an "activity factor" for each level of intensity for each sex is determined (see Table 8-4). Again, the "How to" on p. 241 shows a sample calculation.

**Table 8-3**
**Equations for Estimating BMR from Body Weight**

| Sex and Age (yr) | Equation to Derive BMR in kCal/day |
|---|---|
| **Males** | |
| 0–3 | (60.9 × wt[a]) − 54 |
| 3–10 | (22.7 × wt) + 495 |
| 10–18 | (17.5 × wt) + 651 |
| 18–30 | (15.3 × wt) + 679 |
| 30–60 | (11.6 × wt) + 879 |
| >60 | (13.5 × wt) + 487 |
| **Females** | |
| 0–3 | (61.0 × wt) − 51 |
| 3–10 | (22.5 × wt) + 499 |
| 10–18 | (12.2 × wt) + 746 |
| 18–30 | (14.7 × wt) + 496 |
| 30–60 | (8.7 × wt) + 829 |
| >60 | (10.5 × wt) + 596 |

[a]Weight expressed in kilograms.
Source: Reprinted with permission from *Recommended Dietary Allowances,* 10th edition. Copyright 1989 by National Academy of Sciences. Published by the National Academy Press, Washington, D.C.

IN SUMMARY

A person takes in energy from food and, on average, spends most of it on basal metabolic activities, some of it on physical activities, and a little on the thermic effect of food. Because energy requirements vary from person to person, estimates of requirements must consider such factors as age, sex, and weight when calculating energy spent on basal metabolism and the intensity and duration of the activity when calculating expenditures on physical activities.

# Body Weight, Body Composition, and Health

A person 5 feet 10 inches tall who weighs 150 pounds may carry only about 30 of those pounds as fat. The rest is mostly water and lean tissues—muscles, organs such as the heart and liver, and the bones of the skeleton. Direct measures of body composition are impossible in living human beings; instead, researchers assess body composition indirectly based on the following assumption:

**body composition:** the proportions of muscle, bone, fat, and other tissue that make up a person's total body weight.

$$\text{Body weight} = \text{fat} + \text{lean tissue (including water)}.$$

Weight gains and losses tell us nothing about how the body's composition may have changed, yet weight is the measure most people use to judge their "fatness." For many people, overweight means overfat. This is not always the case, though. Athletes with dense bones and well-developed muscles may be overweight by an arbitrary standard such as weight-for-height tables, but have little body fat. Conversely, inactive people may seem to have acceptable weights, when, in fact, they may have too much body fat.

*At 6 feet 3 inches tall and 245 pounds, Mike O'Hearn would be considered overweight by most height-weight standards. Yet he is clearly not overfat.*

## *Defining Healthy Body Weight*

How much should a person weigh? How can a person know if her weight is appropriate for her height and age? How can a person know if his weight is jeopardizing his health? Such questions seem so simple, yet even the experts can't agree on the answers.[10] Most often, they try to identify the weights associated with lowest mortality. With this in mind, healthy body weight is defined by three criteria:

- A weight within the suggested range for height and age, as shown in Table 8-5.
- A fat distribution pattern that is associated with a low risk of illness or death.
- Freedom from all medical conditions that would suggest a need for weight loss.

People who meet all of these criteria may not gain any health advantage by changing their weights. Those who mistakenly think of themselves as overweight even though they meet these criteria for healthy weight may need to revise their self-

**Table 8-4**
**Estimating Daily Energy Expenditure at Various Levels of Physical Activity**

| Level of Intensity | Type of Activity | Activity Factor (× BMR) | Energy Expenditure (kcal/kg/day) |
|---|---|---|---|
| Very light | Seated and standing activities, painting trades, driving, laboratory work, typing, sewing, ironing, cooking, playing cards, playing a musical instrument | 1.3 (men)<br>1.3 (women) | 31<br>30 |
| Light | Walking on a level surface at 2.5 to 3 mph, garage work, electrical trades, carpentry, restaurant trades, housecleaning, child care, golf, sailing, table tennis | 1.6 (men)<br>1.5 (women) | 38<br>35 |
| Moderate | Walking 3.5 to 4 mph, weeding and hoeing, carrying a load, cycling, skiing, tennis, dancing | 1.7 (men)<br>1.6 (women) | 41<br>37 |
| Heavy | Walking with a load uphill, tree felling, heavy manual digging, basketball, climbing, football, soccer | 2.1 (men)<br>1.9 (women) | 50<br>44 |
| Exceptional | Training in professional or world-class athletic events | 2.4 (men)<br>2.2 (women) | 58<br>51 |

Source: Adapted with permission from *Recommended Dietary Allowances,* 10th edition. Copyright 1989 by the National Academy of Sciences. Published by the National Academy Press, Washington, D.C.

## HOW TO Estimate Daily Energy Output

### Basal Metabolism

One way to estimate your energy output for basal metabolism is to use Table 8-3 on p. 239.[a] For example, a 20-year-old male who weighed 160 pounds would select the equation appropriate for his sex and age range:

$$(15.3 \times \text{wt}) + 679.$$

First, he would convert his weight from pounds to kilograms:

$$160 \text{ lb} \div 2.2 \text{ lb/kg} = 72.7 \text{ kg}.$$

Then, he would insert his weight into the equation:

$$(15.3 \times 72.7 \text{ kg}) + 679 = 1791 \text{ kcal/day}.$$

The estimated energy expenditure to cover basal metabolism for a 20-year-old male who weighs 160 pounds is 1791 kcalories/day.

A shortcut method uses an easy-to-remember formula for estimating basal energy needs. Round off 72.7 kg to 73 and use the factor 1 kcal/kg/hr for men (or 0.9 for women). For example:

$$1 \text{ kcal} \times 73 \text{ kg} \times 24 \text{ hr} = 1752 \text{ kcal/day}.$$

The difference between 1791 and 1752 is insignificant and acceptable in estimations such as these.

### Basal Metabolism and Voluntary Physical Activity

To account for the energy used in physical activities as well, review the activities listed in Table 8-4 (p. 240) and determine which level of intensity typifies your average daily activity. Then multiply the selected activity factor by your value for basal metabolism. For example, if the man introduced above engages in mostly light activity, his activity factor would be 1.6. Multiply this factor by his basal metabolism kcalories:

$$1.6 \times 1791 = 2866 \text{ kcal/day}.$$

The result, 2866 kcalories/day, expresses his *total* daily energy needs.

Alternatively, total energy expenditure can be estimated in one step based on body weight as shown in the last column of Table 8-4. As an example, for a 160-pound man engaged in mostly light activity:

$$38 \text{ kcal} \times 73 \text{ kg} = 2774 \text{ kcal/day}.$$

Keep in mind that these estimates of daily energy output are just that—*estimates*. The difference between 2866 and 2774 is insignificant and acceptable. Either way, the man's total energy needs are about 2800 kcalories/day.

[a]In the United States, many researchers use another set of equations known as the Harris-Benedict method to estimate energy output for BMR. The values calculated from the two sets of equations do not differ significantly. Harris-Benedict equations:

For men: BMR = 66 + (13.7 × wt in kg) + (5 × ht in cm) − (6.8 × age in yr).
For women: BMR = 655 + (9.6 × wt in kg) + (1.8 × ht in cm) − (4.7 × age in yr).

See Appendix D for equations to convert kg and cm.

Quick and easy estimate for energy needs:
- Men: kg × 24 = kcal/day.
- Women: kg × 23 = kcal/day.

image. Such people may still want to improve their eating and exercise habits, but they should do so to reap the rewards of being physically fit, not for the sake of weight loss. Anyone who does not meet all of the above criteria may want to consult with a health care professional, who should carefully consider each criterion in relation to the others. The rest of the chapter examines these three criteria in more detail.

• ***Body Weight and Its Standards*** • Health care professionals often compare people's weights with standard weight-for-height tables, such as Table 8-5. Normally, the assessor uses the midpoint of the weight range for a person of a given height and assumes a medium build. If the person's actual weight is 10 to 20 percent above that, then the person is considered overweight; if 20 percent or more above the standard, the person is obese; and if 10 percent below the standard, the person is underweight.

Standards for desirable weights have changed over the years. Authorities debate which weight standards are most appropriate. Some approve, and some disapprove, of the current weight table for not specifying recommendations by sex; the table simply states that higher weights in the ranges generally apply to men and lower weights more often apply to women. Similarly, some criticize, and some

Quick and easy estimate for midpoint weight:[11]
- Women: 119 lb for 5' + 3 lb per additional inch.
- Men: 135 lb for 5'3" + 3 lb per additional inch.

**overweight:** body weight above some standard of acceptable weight that is usually defined in relation to height (such as the weight-for-height tables).

**underweight:** body weight below some standard of acceptable weight that is usually defined in relation to height (such as the weight-for-height tables).

praise, the current table for not adjusting standards according to age; the guidelines state that "health risks due to excess weight appear to be the same for older as for younger adults," but some research suggests that health risks are greater among overweight young adults.[12] Current weight standards for all adults are the same as those issued in 1990 for young adults, but such standards may be unrealistic without a substantial increase in physical activity over the years.[13]

As long as there have been tables of recommended weights, debates have raged over their validity and usefulness. Recommended weights are often based on insurance data, which underrepresent the lower socioeconomic class, minorities, and the elderly. Insurance-based weight-for-height tables are especially inadequate for identifying the weights most closely associated with minimal health risks.

**body mass index (BMI):** an index of a person's weight in relation to height, determined by dividing the weight (in kilograms) by the square of the height (in meters).

- BMI <18.5 = underweight.
- BMI 18.5 to 24.9 = normal.
- BMI 25 to 29.9 = overweight.
- BMI ≥30 = obese.

• ***Body Mass Index*** • Many health professionals and obesity researchers prefer to use a standard derived by manipulating the height and weight measures mathematically—the body mass index (BMI):

$$\text{BMI} = \frac{\text{weight (kg)}}{\text{height (m)}^2}$$

A person who takes measurements in pounds and inches can convert them to metric units or can use this modified equation:*

$$\text{BMI} = \frac{\text{weight (lb)}}{\text{height (in)}^2} \times 705.$$

The lower end of the suggested weight range in Table 8-5 was calculated at a BMI of 19, and the upper end at a BMI of 25 for adults; the midpoint reflects a BMI of 22. The upper end of the range represents a healthy target either for overweight people to achieve or for others to not exceed.[14] Obesity-related diseases become evident beyond this upper limit. The inside back cover presents weights associated with various BMI values, and the accompanying "How to" describes how to determine an appropriate body weight based on BMI. The average BMI of adults in the United States is 26.3.[15]

Keep in mind that BMI reflects height and weight measures and not body composition. Consequently, a bodybuilder may be classified as overweight by BMI standards and not be overfat; the model on p. 240 has a BMI greater than 30. Similarly, a petite Chinese gymnast may be considered underweight, but not be unhealthy. Striking differences in body composition are apparent among people of various ethnic groups.[16]

**Table 8-5**
**Healthy Weights for Adults**

| Height[a] | Weight (lb)[a] | |
|---|---|---|
| | Midpoint | Range |
| 4'10" | 105 | 91–119 |
| 4'11" | 109 | 94–124 |
| 5'0" | 112 | 97–128 |
| 5'1" | 116 | 101–132 |
| 5'2" | 120 | 104–137 |
| 5'3" | 124 | 107–141 |
| 5'4" | 128 | 111–146 |
| 5'5" | 132 | 114–150 |
| 5'6" | 136 | 118–155 |
| 5'7" | 140 | 121–160 |
| 5'8" | 144 | 125–164 |
| 5'9" | 149 | 129–169 |
| 5'10 | 153 | 132–174 |
| 5'11" | 157 | 136–179 |
| 6'0" | 162 | 140–184 |
| 6'1" | 166 | 144–189 |
| 6'2" | 171 | 148–195 |
| 6'3" | 176 | 152–200 |
| 6'4" | 180 | 156–205 |
| 6'5" | 185 | 160–211 |
| 6'6" | 190 | 164–216 |

NOTE: The higher weights in the ranges generally apply to men, who tend to have more muscle and bone; the lower weights more often apply to women, who have less muscle and bone.
[a]Without shoes or clothes.
Source: *Report of the Dietary Guidelines Advisory Committee on the Dietary Guidelines for Americans* (Washington, D.C.: Government Printing Office, 1995).

IN SUMMARY

Controversy surrounds the setting of standards for body weight. Current standards are based on a person's weight in relation to height, called the body mass index (BMI), and reflect disease risks. Although weight measures are inexpensive, easy to take, and highly accurate, they fail to reveal two valuable pieces of information in assessing disease risk: how much of the weight is fat and where the fat is located.

## *Body Fat and Its Distribution*

The ideal amount of body fat depends partly on the person. A normal-weight man may have from 12 to 20 percent body fat; a woman, because of her greater quantity of indispensable fat, 20 to 30 percent.[17]

• ***Some People Need Less*** • For many athletes, a lower percentage of body fat may be ideal—just enough fat to provide fuel, insulate and protect the body, assist in

*The conversion factor 705 was selected because it is a whole number that is relatively easy to remember; it does overestimate BMI by 0.06 percent, however, and some experts have suggested that 704.5 might be a better value.

## HOW TO / Determine Body Weight Based on BMI

A person whose BMI reflects an unacceptable health risk can choose a desired BMI and then calculate an appropriate body weight. For example, a woman who is 5 feet 5 inches (1.65 meters) tall and weighs 180 pounds (82 kilograms) has a BMI of 30:

$$\text{BMI} = \frac{82 \text{ kg}}{1.65 \text{ m}^2} \text{ or}$$

$$\frac{180 \text{ lb} \times 705}{65 \text{ in}^2} = 30.$$

| Height | Factor | Height | Factor | Height | Factor |
|---|---|---|---|---|---|
| 4'7" | 0.232 | 5'3" | 0.177 | 5'11" | 0.139 |
| 4'8" | 0.224 | 5'4" | 0.172 | 6'0" | 0.136 |
| 4'9" | 0.216 | 5'5" | 0.166 | 6'1" | 0.132 |
| 4'10" | 0.209 | 5'6" | 0.161 | 6'2" | 0.128 |
| 4'11" | 0.202 | 5'7" | 0.157 | 6'3" | 0.125 |
| 5'0" | 0.195 | 5'8" | 0.152 | 6'4" | 0.122 |
| 5'1" | 0.189 | 5'9" | 0.148 | 6'5" | 0.119 |
| 5'2" | 0.183 | 5'10" | 0.143 | 6'6" | 0.116 |

Source: R. P. Abernathy, Body mass index: Determination and use. Copyright the American Dietetic Association. Reprinted by permission from *Journal of the American Dietetic Association* 91 (1991): 843.

A reasonable target for most overweight people is a BMI 2 units below their current one. To determine a desired goal weight based on a BMI of 28, for example, the woman could divide the desired BMI by the factor appropriate for her height from the accompanying table:

Desired BMI ÷ factor = goal weight.

$$28 \div 0.166 = 169 \text{ lb.}$$

To reach a BMI of 28, this woman would need to lose 11 pounds. Such a calculation can help a person to determine realistic weight goals using health risk as a guide. Alternatively, a person could search the table on the inside back cover for the weight that corresponds to his or her height and the desired BMI.

nerve impulse transmissions, and support normal hormone activity, but not so much as to burden the muscles with excess weight to carry. For athletes, then, ideal body fat might be 5 to 10 percent for men and 15 to 20 percent for women. (You may want to review the photo on p. 240 to appreciate what 8 percent body fat looks like.)

• ***Some People Need More*** • For an Alaskan fisherman, a higher percentage of body fat is probably beneficial because fat provides an insulating blanket to prevent excessive loss of body heat in cold climates. A woman starting a pregnancy needs sufficient body fat to support conception and fetal growth. Below a certain threshold for body fat, hormone synthesis falters, and individuals may become infertile, develop depression, experience abnormal hunger regulation, or become unable to keep warm. These thresholds differ for each function and for each individual; much remains to be learned about them.

*A healthy body contains enough lean tissue to support health and the right amount of fat to meet body needs.*

• ***The Criterion of Health*** • In asking what is ideal, people often mistakenly turn to fashion for the answer. Keep in mind that fashion is fickle; body shapes that society values change with time and have little in common with health. Fashion models whose careers depend on body shape often develop eating disorders.

Clearly, the most important criterion for determining how much a person should weigh and how much body fat a person needs is health. Ideally, a person has enough fat to meet basic needs but not so much as to incur health risks. Researchers find health problems develop when body fat exceeds 22 percent in younger men, 25 percent in older men, 32 percent in younger women, and 35 percent in older women; these are the values used to define obesity, and age 40 is the dividing line.

• ***Fat Distribution*** • The distribution of fat on the body may be more critical than fatness alone. Intra-abdominal fat that is stored around the organs of the

**intra-abdominal fat:** fat stored within the abdominal cavity in association with the internal abdominal organs, as opposed to the fat stored directly under the skin (subcutaneous fat).

**central obesity:** excess fat around the trunk of the body; also called **abdominal fat** or **upper-body fat.**

abdomen presents a greater risk to health than fat elsewhere on the body and increases the risk of premature death. This distribution of fat is referred to as central obesity or upper-body fat and, independently of total body fat, is associated with increased risks of heart disease, stroke, diabetes, hypertension, and some types of cancer.

Abdominal fat is common in women past menopause and even more common in men. Even when total body fat is similar, men have more abdominal fat than either premenopausal or postmenopausal women.[18] Regardless of menopausal status, the risks of cardiovascular disease and mortality are increased for women with abdominal fat, just as they are for men.[19]

Interestingly, people with central obesity smoke more and drink alcohol more than the average. A smoker may weigh less than the average nonsmoker, but the smoker's central obesity may be greater, suggesting that smoking may directly affect body fat distribution.[20] Exercise, in contrast, correlates negatively with central obesity.

Popular articles sometimes call bodies with upper-body fat "apples" and those with lower-body fat, "pears." Researchers sometimes refer to upper-body fat as "android" (manlike) obesity and to lower-body fat as "gynoid" (womanlike) obesity.

Fat around the hips and thighs, sometimes referred to as lower-body fat, is most common in women in their reproductive years and seems relatively harmless. In fact, people who are overweight, but who do not have excessive fat around the abdomen "seem robust" and less susceptible to health problems than overweight people with central obesity; theirs is a benign obesity.

Exactly how abdominal fat influences disease development remains unknown. Researchers are studying the links between abdominal fat stores, blood lipids, blood pressure, and glucose metabolism. Abdominal fat seems to be more active than lower-body fat. When mobilized, abdominal fat goes directly to the liver rather than emptying into the general circulation, as other fat does. The liver then packages this fat into VLDL, and these become LDL, the lipoprotein most implicated in heart disease. As blood lipids rise, the nervous system responds by releasing hormones and neurotransmitters that accelerate the heart rate and raise the blood pressure, aggravating heart problems and hypertension. Fat metabolism interferes with the liver's ability to clear insulin from the bloodstream. As a consequence, blood glucose and insulin levels remain elevated, setting the stage for diabetes.

**fatfold measure:** a clinical estimate of total body fatness in which the thickness of a fold of skin on the back of the arm (over the triceps muscle), below the shoulder blade (subscapular), or in other places is measured with a caliper. (The older, less preferred, term is **skinfold test.**)

• ***Fatfold Measures*** • Health care professionals use several techniques to estimate body fat and its distribution. Fatfold measures provide a good estimate of total body fat and a fair assessment of the fat's location.[21] About half of the fat in the body lies directly beneath the skin, so the thickness of this subcutaneous fat reflects total body fat. On some parts of the body, such as the back and the back of the arm over the triceps muscle, this fat is loosely attached; a skilled assessor can measure its thickness and then compare the measurement with standards (see Appendix E).

If a person gains body fat, the fatfold increases proportionately; if the person loses fat, it decreases. Measures taken from central-body sites (around the abdomen) better reflect changes in fatness than those taken from upper sites (arm and back).

A major limitation of the fatfold test is that fat may be thicker under the skin in one area than in another. This limitation can be overcome by taking fatfold measurements at three or more different places on the body (including both central- and lower-body sites) and comparing each measurement with standards for that site. Most often, however, the triceps fatfold measurement alone is used because it is easily accessible.

Fatfold measurements correlate directly with the risk of heart disease. They assess central obesity and its associated risks better than do weight measures such as the BMI.

**waist circumference:** an anthropometric measurement used to assess a person's abdominal fat.

• ***Waist Circumference*** • Another valuable indicator of fat distribution and abdominal fat is the waist circumference.[22] In general, women with a waist circumference of greater than 35 inches and men with a waist circumference of greater than 40 inches have a high risk of central obesity-related health problems.[23]

Alternatively, some clinicians measure both the waist and the hips. The waist-to-hip ratio also assesses abdominal obesity, but provides no more information than using the waist circumference alone. In general women with a waist-to-hip ratio of 0.80 or greater and men with a waist-to-hip ratio of 0.90 or greater have a high risk of health problems.

To calculate the waist-to-hip ratio, divide the waistline measurement by the hip measurement. For example, a woman with a 28-inch waist and 38-inch hips would have a ratio of:

$28 \div 38 = 0.75.$

• ***Other Measures of Body Composition*** • Other techniques for estimating body fat include hydrodensitometry and bioelectrical impedance analysis. To estimate body density using hydrodensitometry, the person is weighed twice—first on land and then again when submerged under water. The difference between the person's actual weight and underwater weight provides a measure of the body's volume. A mathematical equation using two measurements (volume and actual weight) allows the assessor to calculate body density, from which the percentage of body fat can be estimated. Underwater weighing usually generates a good estimate of body fat and is useful in research, although the technique has drawbacks: it requires bulky, expensive, and nonportable equipment. Furthermore, submerging some people (especially those who are very young, very old, ill, or fearful) underwater is not always practical.

**hydrodensitometry** (HI-dro-DEN-see-TOM-eh-tree): a method of measuring body density in which the person is weighed and then weighed again while submerged in water.

To measure body fat using the bioelectrical impedance technique, a very-low-intensity electrical current is briefly sent through the body by way of electrodes placed on the wrist and ankle. Since electrolyte-containing fluids, which readily conduct an electrical current, are found primarily in lean body tissues, the leaner the person, the less resistance to the current. The measurement of electrical resistance is then used in a mathematical equation to estimate the percentage of body fat. As is true of other anthropometric techniques, bioelectrical impedance requires standardized procedures and calibrated instruments to provide reliable results.[24] Appendix E provides more details and includes many of the tables and charts routinely used in assessment procedures.*

**bioelectrical impedance** (im-PEE-dans): a method for estimating body fat using low-intensity electrical current.

IN SUMMARY

The ideal amount of body fat varies from person to person, but researchers have found that body fat in excess of 22 percent for younger men and 32 percent for younger women (the levels rise

*Researchers sometimes estimate body composition using these methods: total body water, radioactive potassium count, dual-energy X-ray absorptiometry, near-infrared spectrophotometry, ultrasound, computed tomography, and magnetic resonance imaging. Each has advantages and disadvantages with respect to cost, technical difficulty, and precision of estimating body fat (see Appendix E for a comparison).

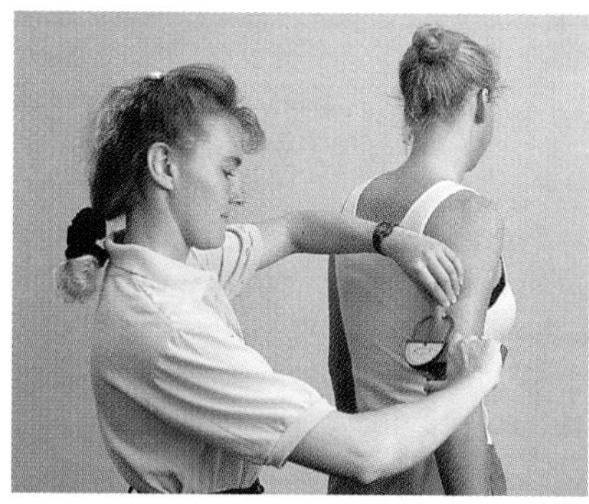

*Obtaining an accurate fatfold measure requires training in the use of a caliper that has been calibrated.*

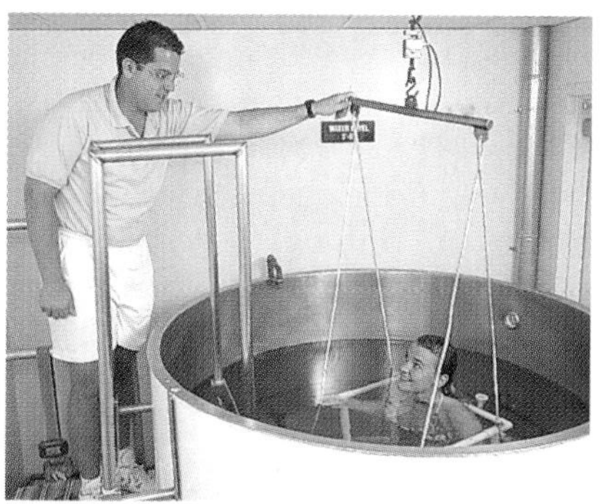

*Researchers may use hydrodensitometry to estimate the percentage of body fat.*

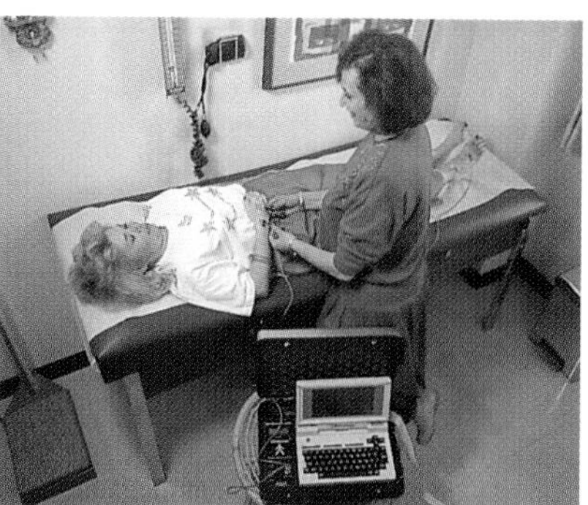

*Bioelectrical impedance provides a simple and painless way to estimate body fat.*

**Figure 8-5**
**Body Mass Index and Mortality**

Both underweight and overweight present risks of a premature death. This J-shaped curve describes the relationship between body mass index (BMI) and mortality and shows that optimal BMI is between 21 and 25 (some researchers extend this range from 19 to 27).

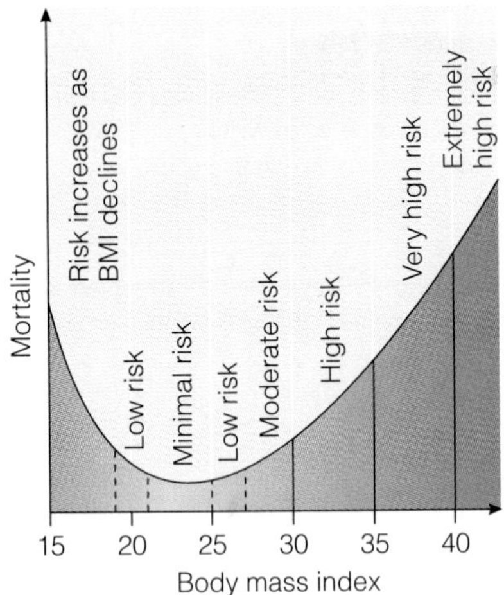

www.shapeup.org

slightly with age) poses health risks. Central obesity in which excess fat is distributed around the trunk of the body presents greater health risks than excess fat distributed on the lower body. Researchers have devised a number of techniques to assess body composition including fatfold measures, wait-to-hip ratio, hydrodensitometry, and bioelectrical impedance.

## *Health Risks Associated with Body Weight and Body Fat*

BMI values correlate with disease risks. Most people with a BMI between 19 and 25 have few health risks; risks increase as BMI falls below 19 or rises above 25, indicating that both too little and too much body fat impair health. [25]Factors such as blood pressure or smoking habits raise risks independently of BMI.

Similarly, epidemiological data show a J-shaped relationship between body weights and mortality (see Figure 8-5).[26] People who are underweight or extremely overweight carry high risks of early deaths; people whose weights fall within the acceptable to slightly overweight range may live longest.[27] Equally important, both central obesity and weight gains of more than 20 pounds between early and middle adulthood correlate with increased mortality.[28] For health's sake, adults should limit their weight gains to 10 or so pounds.[29]

• ***Health Risks of Underweight*** • Some underweight people enjoy an active, healthy life, but others are underweight because of smoking habits or poor health.[30] An underweight person, especially an older adult, may be unable to preserve lean tissue during the fight against a wasting disease such as cancer or a digestive disorder, especially when accompanied by malnutrition. Without adequate nutrient and energy reserves, an underweight person will have a particularly tough battle against such medical stresses. In fact, many people with cancer die, not from the cancer itself, but from malnutrition. Underweight women become infertile. Exactly how infertility develops is unclear, but contributing factors include not only body weight, but also restricted energy and fat intake and depleted body fat stores. Those who do conceive may give birth to unhealthy infants. An underweight woman can improve her chances of having a healthy infant by gaining weight prior to conception, during pregnancy, or both. For all these reasons, underweight people may benefit from enough of a weight gain to provide an energy reserve and protective amounts of all the nutrients that can be stored.

• ***Health Risks of Overweight*** • As for excessive body fat, the health risks are so many that it has been declared a disease: obesity. Among the health risks of obesity are diabetes, hypertension, cardiovascular disease, sleep apnea (abnormal ceasing of breathing during sleep), osteoarthritis, abdominal hernias, some cancers, varicose veins, gout, gallbladder disease, respiratory problems (including Pickwickian syndrome, a breathing blockage linked with sudden death), liver malfunction, complications in pregnancy and surgery, flat feet, and even a high accident rate. Each year, these obesity-related illnesses cost our nation billions of dollars.[31] The cost in terms of lives is also great: an estimated 300,000 people die each year from obesity-related diseases.[32] In fact, obesity is second only to tobacco in causing preventable illnesses and premature deaths. Mortality increases as excess weight increases; people with a BMI greater than 35 are twice as likely to die prematurely as others. Independently of obesity itself, fluctuations in body weight, as typically occur with "yo-yo" dieting, also increase the risks of chronic diseases and premature death (Chapter 9 provides more details).[33]

• ***Cardiovascular Disease*** • The relationship between obesity and cardiovascular disease risk is strong, with links to both blood cholesterol and blood pressure.[34] Central obesity may raise the risk of heart attack and stroke as much as the leading three risk factors (high blood cholesterol, hypertension, and smoking) do. In addition to body fat and its distribution, weight gain also increases the risk of cardiovascular disease.[35] Weight loss, on the other hand, effectively lowers both blood cholesterol and blood pressure in obese people.[36] Of course, lean and normal-weight people may also have high blood lipids and blood pressure, and these factors are just as dangerous in lean people as in obese people.

Cardiovascular disease risk factors associated with obesity:
- High LDL cholesterol.
- Low HDL cholesterol.
- High blood pressure (hypertension).
- Diabetes.

Chapter 18 provides many more details.

• ***Diabetes*** • Diabetes (type 2) is three times more likely to develop in an obese person than in a nonobese person. Furthermore, the person with type 2 diabetes often has central obesity.[37] Central-body fat cells appear to be larger and more insulin-resistant than lower-body fat cells, and insulin resistance is a major risk factor for the development of type 2 diabetes.[38]

Reminder: *Type 2 diabetes* refers to noninsulin-dependent diabetes, the most common form of diabetes in which the body's cells fail to respond to insulin (insulin resistance).

Diabetes appears to be influenced not only by body weight, but by weight gains as well. A weight gain of 11 to 24 pounds since age 18 doubles the risk of developing diabetes, even in women of average weight.[39] In contrast, weight loss is effective in improving glucose tolerance and insulin resistance.[40]

• ***Cancer*** • The risk of cancer increases with both body weight and weight gain, but researchers do not fully understand the relationship.[41] One possible explanation may be that obese people have elevated levels of hormones that could influence cancer development. For example, adipose tissue is the major site of estrogen synthesis in women, obese women have elevated levels of estrogen, and estrogen has been implicated in the development of cancers of the female reproductive system. These cancers account for half of all cancers in women.

IN SUMMARY

The weight appropriate for an individual depends largely on factors specific to that individual, including body fat distribution, family health history, occupation, and current health status. At the extremes, both overweight and underweight carry clear risks to health. The next chapter explores weight control and the benefits of achieving and maintaining a healthy weight.

## Making It Click

These problems give you practice in estimating energy needs. Once you have mastered these examples, you will be prepared to examine your own energy intakes and energy expenditures. Be sure to show your calculations for each problem and check Appendix K for answers.

1. Estimate various people's basal metabolic energy needs per day using Table 8-3 on p. 239. For example, calculate the energy needs of a 10-year-old boy who weighs 75 pounds (34 kilograms):

   Using the range for ages 10–18: $(17.5 \times 34) + 651 = 1246$ kcal/day.

   Using the range for ages 3–10: $(22.7 \times 34) + 495 = 1267$ kcal/day.

   These two answers are similar. The boy needs about 1250 kcal/day for basal metabolism.

   a. A 10-year-old girl of the same weight. (Use either range. If you try them both, you'll find about a 100-kcalorie difference, but 10-year-old girls' energy needs vary by even more than this.)
   b. An 18-year-old man who weighs 150 pounds (68 kilograms). Again, use either range, and again, if you use both, you'll find about a 100-kcalorie difference.
   c. A 35-year-old man who weighs 200 pounds (91 kilograms).
   d. A 50-year-old woman who weighs 115 pounds (52 kilograms).

2. Compare the energy a person might spend on various physical activities. Refer to Table 8-2 on p. 238, and compute how much energy a person who weighs 142 pounds would spend doing each of the following. Show your calculations. The first example is done for you. You may want to compare various activities based on your weight.

   30 min vigorous aerobic dance:

   $0.062 \text{ kcal/lb/min} \times 142 \text{ lb} = 8.8 \text{ kcal/min}$.

   $8.8 \text{ kcal/min} \times 30 \text{ min} = 264 \text{ kcal}$.

   a. 2 hours golf, carrying clubs.
   b. 20 minutes running at 9 mph.
   c. 45 minutes swimming at 20 yd/min.
   d. 1 hour walking at 3.5 mph.

## Making It Click (continued)

3. Consider the effect of age on BMR. An infant who weighs 20 pounds has a BMR of 500 kcalories/day; an adult who weighs 170 pounds has a BMR of about 1500. Based on body weight, who has the faster BMR (show your calculations)?
4. Compute daily energy needs for a woman, age 20, who is 5 feet 6 inches tall, weighs 130 pounds, and is lightly active.
   a. From Table 8-3 on p. 239, what is her energy need for basal metabolism (show your calculations)?
   b. From Table 8-4 on p. 240, estimate her daily energy expenditure, using her activity factor and using her weight.
   c. What percentage of her daily energy expenditure is used for basal metabolism?
5. Discover what weight is needed to achieve a desired BMI. Refer to the table on p. 243 and consider a person who is 5 feet 4 inches tall. Suppose this person wants to have a BMI of 21. What should this person weigh? Show your calculations. Does this agree with the table on the inside back cover?
6. Calculate safe weight-loss rates for people of different sizes using the recommended rate of 1 percent of body weight per week.
   a. What would the safe rate be for a person who weighs 120 pounds? Show your calculations.
   b. What would the safe rate be for a person who weighs 250 pounds? Show your calculations.
7. Calculate the daily energy intakes appropriate for weight loss using the suggested minimum energy intake of 10 kcalories per pound of body weight per day.
   a. How many kcalories for a person who weighs 130 pounds?
   b. How many kcalories for a person who weighs 250 pounds?
8. Convert body fat into kcalorie values. Suppose a man is 5 feet 3 inches tall and weighs 150 pounds, and 30 percent of that weight is fat.
   a. How many pounds of fat does he have?
   b. Assuming that the energy value of body fat is 3500 kcalories per pound, how many kcalories does this person have stored in his body fat?
   c. Suppose he loses 15 pounds. For purposes of this question, assume that he exercised enough to retain virtually all of his lean tissue. Now he weighs 135 pounds. How many pounds of this is fat? What percentage of his body weight is now fat (show your calculations)?
   d. How many kcalories does a 15-pound loss of fat represent?
   e. If a diet and exercise plan provides a deficit of 500 kcalories/day (that is, 500 kcalories less each day is eaten than is spent), how many days (or weeks or months) will it take to lose that much fat?

## Study Questions

These questions will help you review the chapter. You will find the answers in the discussions on the pages provided.

1. What are the consequences of an unbalanced energy budget? (p. 231)
2. Define hunger, appetite, and satiety and describe how each influences food intake. (pp. 233–236)
3. Describe each component of energy expenditure. What factors influence each? How can energy expenditure be estimated? (pp. 236–239)
4. Distinguish between body weight and body composition. What assessment techniques are used to measure each? (pp. 240–242, 244–245)
5. What problems are involved in defining "ideal" body weight? (pp. 240–241, 242–243)
6. What is central obesity, and what is its relationship to disease? (pp. 243–244)
7. What risks are associated with excess body weight and excess body fat? (pp. 246–247)

These questions will help you prepare for an exam. Answers can be found in Appendix K.

1. A person who consistently consumes 1700 kcalories a day and spends 2200 kcalories a day for a month would be expected to:
   a. lose ½ to 1 pound.
   b. gain ½ to 1 pound.
   c. lose 4 to 5 pounds.
   d. gain 4 to 5 pounds.
2. A bomb calorimeter measures:
   a. physiological fuel.
   b. energy available from foods.
   c. kcalories a person derives from foods.
   d. heat a person releases in basal metabolism.
3. The psychological desire to eat that accompanies the sight, smell, or thought of food is known as:
   a. hunger.
   b. satiety.
   c. appetite.
   d. palatability.
4. A person watching television after dinner reaches for a snack during a commercial in response to:
   a. external cues.

b. hunger signals.
c. stress arousal.
d. satiety factors.

5. The largest component of energy expenditure is:
   a. basal metabolism.
   b. physical activity.
   c. indirect calorimetry.
   d. thermic effect of food.
6. The major factor influencing BMR is:
   a. sex.
   b. food intake.
   c. body composition.
   d. physical activity.
7. The thermic effect of an 800-kcalorie meal is about:
   a. 8 kcalories.
   b. 80 kcalories.
   c. 160 kcalories.
   d. 200 kcalories.
8. For health's sake, a person with a BMI of 21 might want to:
   a. lose weight.
   b. maintain weight.
   c. gain weight.
9. Which of the following reflects height and weight?
   a. body mass index
   b. fatfold measures
   c. waist-to-hip ratio
   d. bioelectrical impedance
10. Which of the following increases disease risks?
   a. BMI 19–21
   b. BMI 22–25
   c. lower-body fat
   d. central obesity

## Notes


1. A. Geliebter and coauthors, Reduced stomach capacity in obese subjects after dieting, *American Journal of Clinical Nutrition* 63 (1996): 170–173.
2. K. M. Cunningham and coauthors, Gastrointestinal adaptation to diets of differing fat composition in human volunteers, *Gut* 32 (1991): 483–486; S. J. French and coauthors, Adaptation to high-fat diets: Effects on eating behaviour and plasma cholecystokinin, *British Journal of Nutrition* 73 (1995): 179–189.
3. J. E. Blundell and coauthors, Control of human appetite: Implications for the intake of dietary fat, *Annual Review of Nutrition* 16 (1996): 285–319; R. J. Stubbs and coauthors, Breakfasts high in protein, fat or carbohydrate: Effect on within-day appetite and energy balance, *European Journal of Clinical Nutrition* 50 (1996): 409–417.
4. H. J. DeLargy and coauthors, Effects of different soluble:insoluble fibre ratios at breakfast on 24-h pattern of dietary intake and satiety, *European Journal of Clinical Nutrition* 49 (1995): 754–766; A. Raben and coauthors, Decreased postprandial thermogenesis and fat oxidation but increased fullness after a high-fiber meal compared with a low-fiber meal, *American Journal of Clinical Nutrition* 59 (1994): 1386–1394; J. E. Blundell, S. Green, and V. Burley, Carbohydrates and human appetite, *American Journal of Clinical Nutrition* 59 (1994): 728S–734S; W. H. Turnbull, J. Walton, and A. R. Leeds, Acute effects of mycoprotein on subsequent energy intake and appetite variables, *American Journal of Clinical Nutrition* 58 (1993): 507–512; S. J. French and N. W. Read, Effect of guar gum on hunger and satiety after meals of differing fat content: Relationship with gastric emptying, *American Journal of Clinical Nutrition* 59 (1994): 87–91.
5. Stubbs and coauthors, 1996; J. E. Blundell and J. I. Macdiarmid, Fat as a risk factor for overconsumption: Satiation, satiety, and patterns of overeating, *Journal of the American Dietetic Association* 97 (1997): S63–S69; Blundell and coauthors, 1996.
6. S. H. A. Holt and coauthors, A satiety index of common foods, *European Journal of Clinical Nutrition* 49 (1995): 675–690.
7. A. S. Levine and C. J. Billington, Why do we eat? A neural systems approach, *Annual Review of Nutrition* 17 (1997): 597–619; N. Read, S. French, and K. Cunningham, The role of the gut in regulating food intake in man, *Nutrition Reviews* 52 (1994): 1–10; P. Norton, G. Falciglia, and D. Gist, Physiologic control of food intake by neural and chemical mechanisms, *Journal of the American Dietetic Association* 93 (1993): 450–454.
8. L. O. Schulz and D. A. Scheller, A compilation of total daily energy expenditures and body weights in healthy adults, *American Journal of Clinical Nutrition* 60 (1994): 676–681.
9. T. J. Horton and C. A. Geissler, Effect of habitual exercise on daily energy expenditure and metabolic rate during standardized activity, *American Journal of Clinical Nutrition* 59 (1994): 13–19.
10. S. M. Garn, Fractionating healthy weight, *American Journal of Clinical Nutrition* 63 (1996): 412S–414S.
11. M. A. Miller, A calculated method for determination of ideal body weight, *Nutritional Support Services* 5 (1985): 31–33.
12. J. Stevens and coauthors, The effect of age on the association between body-mass index and mortality, *New England Journal of Medicine* 338 (1998): 1–7.
13. P. T. Williams, Evidence for the incompatibility of age-neutral overweight and age-neutral physical activity standards from runners, *American Journal of Clinical Nutrition* 65 (1997): 1391–1396.
14. J. G. Meisler and S. St. Jeor, Summary and recommendations from the American Health Foundation's Expert Panel on Healthy Weight, *American Journal of Clinical Nutrition* 63 (1996): 474S–477S.
15. R. J. Kuczmarski and coauthors, Increasing prevalence of overweight among US adults, *Journal of the American Medical Association* 272 (1994): 205–211.
16. K. J. Ellis, Body composition of a young, multiethnic, male population,

*American Journal of Clinical Nutrition* 66 (1997): 1323–1331.

17. R. P. Abernathy and D. R. Black, Healthy body weights: An alternative approach, *American Journal of Clinical Nutrition* 63 (1996): 448S–451S.
18. S. Lemieux and coauthors, Sex differences in the relation of visceral adipose tissue accumulation to total body fatness, *American Journal of Clinical Nutrition* 58 (1993): 463–467; C. J. Ley, B. Lees, and J. C. Stevenson, Sex- and menopause-associated changes in body-fat distribution, *American Journal of Clinical Nutrition* 55 (1992): 950–954.
19. A. C. Perry and coauthors, Relation between anthropometric measures of fat distribution and cardiovascular risk factors in overweight pre- and postmenopausal women, *American Journal of Clinical Nutrition* 66 (1997): 829–836; M. J. Williams and coauthors, Regional fat distribution in women and risk of cardiovascular disease, *American Journal of Clinical Nutrition* 65 (1997): 855–860; J. E. Manson and coauthors, Body weight and mortality among women, *New England Journal of Medicine* 333 (1995): 677–685.
20. R. J. Troisi, Cigarette smoking, dietary intake, and physical activity: Effects on body fat distribution—The Normative Aging Study, *American Journal of Clinical Nutrition* 53 (1991): 1104–1111.
21. C. Orphanidou and coauthors, Accuracy of subcutaneous fat measurement: Comparison of skinfold calipers, ultrasound, and computed tomography, *Journal of the American Dietetic Association* 94 (1994): 855–858.
22. M. E. J. Lean, T. S. Han, and C. E. Morrison, Waist circumference as a measure for indicating need for weight management, *British Medical Journal* 311 (1995): 158–161; S. Lemieux and coauthors, A single threshold value of waist girth identifies normal-weight and overweight subjects with excess visceral adipose tissue, *American Journal of Clinical Nutrition* 64 (1996): 685–693.
23. National Institutes of Health Obesity Education Initiative, *Clinical Guidelines on the Identification, Evaluation, and Treatment of Overweight and Obesity in Adults* (Washington, D.C.: U.S. Department of Health and Human Services, 1998).
24. Bioelectrical impedance analysis in body composition measurement: National Institutes of Health Technology Assessment Conference Statement, *American Journal of Clinical Nutrition* 64 (1996): 524S–532S.
25. Committee on Diet and Health, Food and Nutrition Board, *Diet and Health: Implications for Reducing Chronic Disease Risk* (Washington, D.C.: National Academy Press, 1989), pp. 563–592; Meisler and St. Jeor, 1996.
26. T. B. VanItallie, Body weight, morbidity, and longevity, in *Obesity*, eds. P. Björntorp and B. N. Brodoff (Philadelphia: J. B. Lippincott, 1992), pp. 361–369; M. Stern, Epidemiology of obesity and its link to heart disease, *Metabolism: Clinical and Experimental* 44 (1995): 1–3.
27. R. P. Troiano and coauthors, The relationship between body weight and mortality: A quantitative analysis of combined information from existing studies, *International Journal of Obesity* 20 (1996): 63–75.
28. C. G. Solomon and J. E. Manson, Obesity and mortality: A review of the epidemiologic data, *American Journal of Clinical Nutrition* 66 (1997): 1044S–1050S; Manson and coauthors, 1995.
29. A. M. Wolf and G. A. Colditz, Social and economic effects of body weight in the United States, *American Journal of Clinical Nutrition* 63 (1996): 466S–469S.
30. W. C. Willett, Weight loss in the elderly: Cause or effect of poor health? *American Journal of Clinical Nutrition* 66 (1997): 737–738.
31. Wolf and Colditz, 1996.
32. J. M. McGinnis and W. H. Foege, Actual causes of death in the United States, *Journal of the American Medical Association* 270 (1993): 2207–2212.
33. R. W. Jeffrey, Does weight cycling present a health risk? *American Journal of Clinical Nutrition* 63 (1996): 452S–455S.
34. W. B. Kannel, R. B. D'Agostino, and J. L. Cobb, Effect of weight on cardiovascular disease, *American Journal of Clinical Nutrition* 63 (1996): 419S–422S; D. A. McCarron and M. E. Reusser, Body weight and blood pressure regulation, *American Journal of Clinical Nutrition* 63 (1996): 423S–425S.
35. T. B. Harris and coauthors, Carrying the burden of cardiovascular risk in old age: Associations of weight and weight change with prevalent cardiovascular disease, risk factors, and health status in the Cardiovascular Health Study, *American Journal of Clinical Nutrition* 66 (1997): 837–844; K. M. Roxrode and coauthors, A prospective study of body mass index, weight change, and risk of stroke in women, *Journal of the American Medical Association* 277 (1997): 1539–1545.
36. I. Katzel and coauthors, Effects of weight loss vs aerobic exercise training on risk factors for coronary disease in healthy, obese, middle-aged and older men, *Journal of the American Medical Association* 274 (1995): 1915–1921; J. Wylie-Rosett and coauthors, Trial of Antihypertensive Intervention and Management: Greater efficacy with weight reduction than with a sodium-potassium intervention, *Journal of the American Dietetic Association* 93 (1993): 408–415; S. A. Corrigan and coauthors, Weight reduction in the prevention and treatment of hypertension: A review of representative clinical trials, *American Journal of Health Promotion* 5 (1991): 208–214.
37. V. J. Carey and coauthors, Body fat distribution and risk of non-insulin-dependent diabetes mellitus in women, *American Journal of Epidemiology* 145 (1997): 614–619.
38. S. Lillioja and coauthors, Insulin resistance and insulin secretory dysfunction as precursors of non-insulin-dependent diabetes mellitus: Prospective studies of Pima Indians, *New England Journal of Medicine* 329 (1993): 1988–1992.
39. G. A. Colditz and coauthors, Weight gain as a risk factor for clinical diabetes mellitus in women, *Annals of Internal Medicine* 122 (1995): 481–486.
40. F. X. Pi-Sunyer, Weight and non-insulin-dependent diabetes mellitus, *American Journal of Clinical Nutrition* 63 (1996): 426S–429S.
41. Z. Huang and coauthors, Dual effects of weight and weight gain on breast cancer risk, *Journal of the American Medical Association* 278 (1997): 1407–1411; R. Ballard-Barbash and C. A. Swanson, Body weight: Estimation of risk for breast and endometrial cancers, *American Journal of Clinical Nutrition* 63 (1996): 437S–441S; M. Shike, Body weight and colon cancer, *American Journal of Clinical Nutrition* 63 (1996): 442S–444S.

HIGHLIGHT 8

# The Latest and Greatest Weight-Loss Diet—Again

To paraphrase William Shakespeare, "a fad diet by any other name would still be a fad diet." And the names are legion: the Atkins New Diet Revolution, the Calories Don't Count Diet, the Protein Power Diet, the Lo-Carbo Diet, the Healthy for Life Diet, the Zone Diet. Year after year, "new and improved" diets appear on bookstore shelves and circulate among friends. People of all sizes eagerly try the best diet on the market ever, hoping that this one will really work. And sometimes it seems to work for a while, but more often than not, its success is short-lived. And then another fad diet takes the spotlight. Here's how Dr. K. Brownell, an obesity expert at Yale University, describes this phenomenon: "When I get calls about the latest diet fad, I imagine a trick birthday cake candle that keeps lighting up and we have to keep blowing it out."

Realizing that fad diets do not offer a safe and effective plan for weight loss, health professionals speak out, but they never get the candle blown out permanently. New fad diets can keep making the same claims because no one requires their advocates to prove what they say. They do not have to conduct credible research on the benefits or dangers of their diets. They can simply make unsubstantiated statements that fall far short of the truth, but sound impressive to the uninformed. They often offer distorted bits of legitimate research. They may start with one or more actual facts, but then leap from one erroneous conclusion to the next. Anyone who wants to believe them is forced to wonder how the thousands of scientists working on obesity research over the past century could possibly have missed such obvious connections.

No matter what their names are, most fad diets espouse essentially the same high-protein, low-carbohydrate diet. After all, diets may come in all flavors, but only in three proportions: high fat, high carbohydrate, or high protein. Few consumers would believe that high-fat diets could lead to weight loss; contrary to such an outrageous claim, dietary fat does not promote fat oxidation. Consumers already hear from many free sources that high-carbohydrate diets support good health, so peddling that idea would not be a profitable venture. That leaves high-protein diets, and they surface regularly in various guises as the best way to lose weight. High-protein diets are by design relatively low in carbohydrate. This highlight examines some of the science and the science fiction behind high-protein, low-carbohydrate fad diets.

*The cautious consumer distinguishes between loss of fat and loss of weight.*

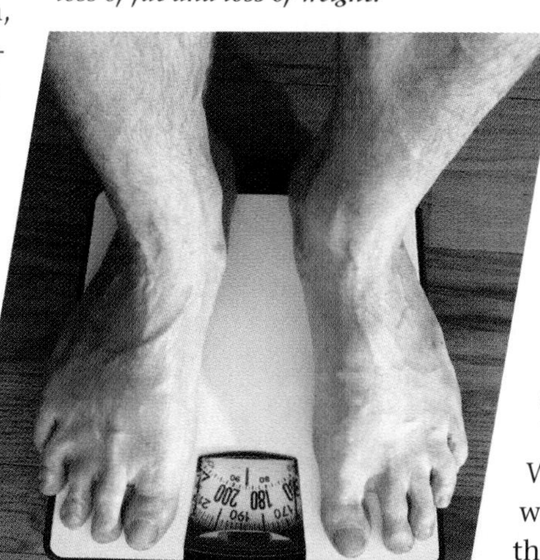

## The Diet's Appeal

Perhaps the greatest appeal of a high-protein, low-carbohydrate diet is that it turns current diet recommendations upside down. Foods such as meats and milk products that need to be selected and measured carefully to limit fat can now be eaten with abandon. Grains, vegetables, and fruits that we are told to eat in abundance can now be ignored. For some people, this is a dream come true: steaks without the potatoes, ribs without the coleslaw, and meatballs without the pasta. Who can resist the promise of weight loss when allowed to eat freely from a list of favorite foods?

To lure dieters in, proponents of high-protein diets often blame the currently recommended high-carbohydrate, low-fat diet for our obesity troubles. They claim that the incidence of obesity is rising because we are eating less fat. Such a claim may impress the naive, but it sends skeptical people running for the facts. True, the incidence of obesity has risen dramatically from 25 to 35 percent over the past two decades.[1] True, our intake of fat has dropped from 36 to 34 percent of daily energy intake. Such facts might seem to imply that lowering fat intake leads to obesity, but this is an erroneous conclusion. The *percentage* declined only because average energy intakes increased by 200 kcalories a day (from about 2000 kcalories a day to 2200).[2] Actual fat intake *increased* by 4 grams a day (from 82 grams to 86). Furthermore, fewer than half of us engage in regular physical activity.[3] Obesity experts blame our high energy intakes and low energy outputs for the increase in obesity. Weight loss, after all, depends on a negative energy balance.

If high-protein diets were as successful as some people claim, then consumers who tried them would lose lots of weight, and our obesity problems would be solved. Obviously, this is not the case. Similarly, if high-protein diets were as worthless as others claim, then consumers would eventually stop pursuing them. Clearly, this is not happening either. These diets have enough going for them that they work for some people at least for a short time, but they fail to produce long-lasting results for most people. The following sections examine some of the apparent achievements and shortcomings of high-protein diets.

## The Diet's Achievements

With one-third of our nation's adults overweight and many more concerned about their weight, the market for a weight-loss book, product, or program is huge (no pun intended). Even a plan that offers only minimal weight-loss success easily attracts a following. High-protein diet plans offer a little success to some people for a short time. Here's why.

### *Don't Count kCalories*

Who wants to count kcalories? Even experienced dieters find counting kcalories burdensome, not to mention timeworn. They want a new, easy way to lose weight, and high-protein diet plans seem to offer this boon. But while these diets often claim to disregard kcalories, their design typically ensures a low energy intake. They advise dieters to stop counting kcalories, but then recommend three meals "not to exceed 500 kcalories each and two snacks of less than 100 kcalories each." Even when it is truly not necessary to count kcalories, the total tends to be low simply because food intake is so limited. Without its refried beans, tortilla wrapping, and chopped vegetables, a burrito is reduced to a pile of ground beef. Weight loss occurs because of the low energy intake—not the proportion of energy nutrients.[4] Success, then, depends on a restricted intake, not on protein's magical powers or carbohydrate's evil forces.

### *Satisfy Hunger*

As Chapter 8 mentioned, of the three energy-yielding nutrients, protein produces the strongest feelings of satiety.[5] People feel full after eating even small quantities of high-protein foods. When a diet leaves a person feeling satisfied instead of deprived, the overall quality of life improves.[6]

### *Follow a Plan*

People need specific instructions and examples to make dietary changes.[7] Fad diets offer dieters a plan. The user doesn't have to decide what foods to eat, how to prepare them, or how much to eat. Unfortunately, these instructions serve short-term weight-loss needs only. They do not provide for long-term weight maintenance or health goals.

## *The Diet's Shortcomings*

People who have followed high-protein diet plans for several months have lost weight. But can these diets also be harmful?

### *Too Much Fat*

Some fad diets focus so intently on promoting protein and curbing carbohydrate that they fail to account for the fat that accompanies many high-protein foods. A breakfast of bacon and eggs, lunch of ham and cheese, and dinner of barbecued short ribs would provide 100 grams of protein—and 121 grams of fat! Yet this day's meals, even with a snack of peanuts, provide only 1600 kcalories. Without careful selection, protein-rich diets can be extraordinarily high in fat, saturated fat, and cholesterol—all dietary risk factors for heart disease.

### *Not Enough Fiber, Vitamins, and Minerals*

Without fruits, vegetables, and grains, these diets lack not only carbohydrate, but fiber, vitamins, and minerals as well—all dietary factors protective against disease. To help shore up some of these inadequacies, fad diets often recommend a daily multivitamin-mineral supplement. Conveniently, many of the companies selling fad diets also peddle supplements. But as Highlights 10 and 11 explain, foods offer many more health benefits than any supplement can provide.

### *Too Little Variety*

Diets that omit hundreds of foods and several food groups lack variety. Some people lose interest in eating, which further reduces energy intake. Others "cheat" to experience a broader array of flavors. Even if the allowed foods are favorites, eating the same foods week after week can become monotonous.

## *The Body's Perspective*

When a person consumes a low-carbohydrate diet, a metabolism similar to that of fasting prevails (see Chapter 7 for a review of fasting). With little dietary carbohydrate coming in, the body uses its glycogen to provide glucose for the cells of the brain, central nervous system, and blood. Once the body depletes its glycogen reserves, it turns to its only significant remaining source of glucose—protein. A low-carbohydrate diet may provide abundant protein from food, but some protein still derives from body tissue.

Dieters can know this wasting process has begun by monitoring their urine. Whenever glycogen or protein is broken down, water is released and urine production increases. Signs of ketosis can also be detected in the urine. Ketones form whenever glucose is lacking and fat breakdown is incomplete.

Low-carbohydrate diets induce ketosis, and many fad diets regard ketosis as the key to losing weight. People in ketosis may experience a loss of appetite and a dramatic weight loss within the first few days. They would be disillusioned if they were aware that much of this weight loss reflects the loss of glycogen and protein together with large quantities of body fluids and important minerals. They need to learn to appreciate the difference between loss of *fat* and loss of *weight.* Fat losses on ketogenic diets are no greater than on other diets providing the same number of kcalories. Once the dieter returns to well-balanced meals that provide adequate energy, carbohydrate, fat, protein, vitamins, and minerals, the body avidly retains these needed nutrients. The weight will zoom back, quite often to

**Table H8-1**

**Adverse Side Effects of Low-Carbohydrate, Ketogenic Diets**

- Nausea
- Fatigue
- Constipation
- Low blood pressure
- Elevated uric acid (which may cause inflammation of the joints in those predisposed to gout)
- Stale, foul taste in the mouth

higher than the starting point. Table H8-1 lists other consequences of a ketogenic diet.

Table H8-2 offers guidelines for identifying fad diets and other weight-loss scams. Diets that fall short, if used only for a little while, may not harm healthy people, but they cannot support optimal health for long. The ideal weight-control diet is one you can live with for the rest of your life. Keep that criterion in mind when you evaluate the next "latest and greatest weight-loss diet" that comes along.

**Table H8-2**

**Guidelines for Identifying Fad Diets and Other Weight-Loss Scams**

1. They promise dramatic, rapid weight loss. Weight loss should be gradual and not exceed 2 pounds per week.
2. They promote diets that are nutritionally unbalanced or extremely low in kcalories. Diets should provide:
   - A reasonable number of kcalories (not fewer than 1200 kcalories per day).
   - Enough, but not too much, protein (between the RDA and twice the RDA).
   - Enough, but not too much fat (between 20 and 30 percent of daily energy intake from fat).
   - Enough carbohydrate to spare protein and prevent ketosis (at least 100 grams per day) and 20 to 30 grams of fiber from food sources.
   - A balanced assortment of vitamins and minerals from a variety of foods from each of the food groups.
   - At least 1 liter (about 1 quart) water daily or 1 milliliter per kcalorie daily—whichever is more.
3. They use liquid formulas rather than foods. Foods should accommodate a person's ethnic background, taste preferences, and financial means.
4. They attempt to make clients dependent upon special foods or devices. Programs should teach clients how to make good choices from the conventional food supply.
5. They fail to encourage permanent, realistic lifestyle changes. Programs should provide physical activity plans that involve spending at least 300 kcalories a day and behavior-modification strategies that help to correct poor eating habits.
6. They misrepresent salespeople as "counselors" supposedly qualified to give guidance in nutrition and/or general health. Even if adequately trained, such "counselors" would still be objectionable because of the obvious conflict of interest that exists when providers profit directly from products they recommend and sell.
7. They collect large sums of money at the start or require that clients sign contracts for expensive, long-term programs. Programs should be reasonably priced and run on a pay-as-you-go basis.
8. They fail to inform clients of the risks associated with weight loss in general or the specific program being promoted. They should provide information about dropout rates, the long-term success of their clients, and possible side effects.
9. They promote unproven or spurious weight-loss aids such as human chorionic gonadotropin hormone (HCG), starch blockers, diuretics, sauna belts, body wraps, passive exercise, ear stapling, acupuncture, electric muscle-stimulating (EMS) devices, spirulina, amino acid supplements (e.g., arginine, ornithine), glucomannan, methylcellulose (a "bulking agent"), "unique" ingredients, and so forth.
10. They fail to provide for weight maintenance after the program ends.

Sources: Adapted from American College of Sports Medicine, *ACSM's Guidelines for Exercise Testing and Prescription* (Baltimore: Williams & Wilkins, 1995), pp. 218–219; J. T. Dwyer, Treatment of obesity: Conventional programs and fad diets, in *Obesity,* eds. P. Björntorp and B. N. Brodoff (Philadelphia: J. B. Lippincott, 1992), p. 668; *National Council Against Health Fraud Newsletter,* March/April 1987, National Council Against Health Fraud, Inc.

## Notes


1. Update: Prevalence of overweight among children, adolescents, and adults—United States, 1988–1994, *Morbidity and Mortality Weekly* 46 (1997): 199–202.
2. N. D. Ernst and coauthors, Consistency between US dietary fat intake and serum total cholesterol concentrations: The National Health and Nutrition Examination Surveys, *American Journal of Clinical Nutrition* 66 (1997): 965S–972S.
3. *Physical Activity and Health—A Report of the Surgeon General Executive Summary* (Washington, D.C.: Government Printing Office, 1996).
4. A. Golay and coauthors, Similar weight loss with low- or high-carbohydrate diets, *American Journal of Clinical Nutrition* 63 (1996): 174–178; B. B. Alford, A. C. Blankenship, and R. D. Hagen, The effects of variations in carbohydrate, protein and fat content of the diet upon weight loss, blood values and nutrient intake of adult obese women, *Journal of the American Dietetic Association* 90 (1990): 534–540.
5. J. E. Blundell and coauthors, Control of human appetite: Implications for the intake of dietary fat, *Annual Review of Nutrition* 16 (1996): 285–319.
6. M. Shah, Comparison of a low-fat, ad libitum complex-carbohydrate diet with a low-energy diet in moderately obese women, *American Journal of Clinical Nutrition* 59 (1994): 980–984.
7. R. R. Wing, Food provision in dietary intervention studies, *American Journal of Clinical Nutrition* 66 (1997): 421–422.

CHAPTER 9

# Weight Control: Overweight and Underweight

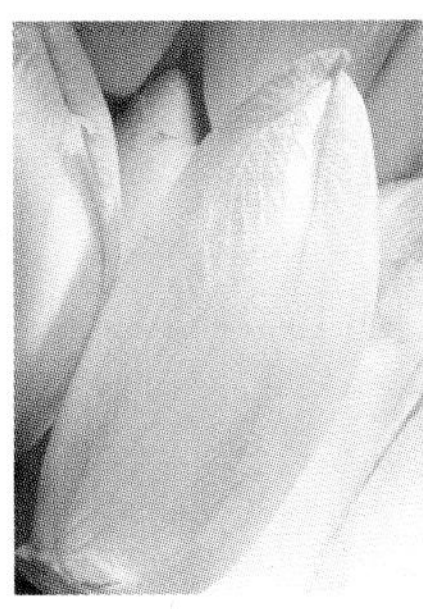

Are you pleased with your body weight? If so, you are a rare individual. Nearly all people in our society think they should weigh more or less (mostly less) than they do. Usually, their primary reason is appearance, but they often perceive, correctly, that physical health is also somehow related to body weight. At the extremes, both overweight and underweight present health risks.

This chapter emphasizes overweight, partly because it has been more intensively studied and partly because it is a widespread health problem in developed countries and a growing concern in developing countries. Information on underweight is presented wherever appropriate. The highlight that follows this chapter delves into the eating disorders anorexia nervosa and bulimia nervosa.

U.S. Government
www.healthfinder.gov/searchoptions/topicsaz.htm
Search for Obesity

# Overweight

Despite our preoccupation with body image and weight loss, the prevalence of overweight in the United States continues to rise dramatically (see Figure 9-1). Approximately one out of three adults, one out of nine teenagers, and one out of seven children in the United States are now overweight.[1] The prevalence of overweight has been increasing and is especially high among women, the poor, and some ethnic groups. Half of all African American women are overweight as are half of all Mexican American women. If the prevalence of obesity continues to rise at the present rate, some experts predict that by the year 2230, every adult in the United States will be overweight.[2] Such a dramatic statement may sound preposterous, but it speaks to the reality: obesity is a major public health problem without a viable solution. Before examining the suspected causes of obesity and the myriad treatments used to overcome it, it may be helpful to understand the development and metabolism of body fat.

**Figure 9-1**
**Prevalence of Overweight among Adults in the United States**

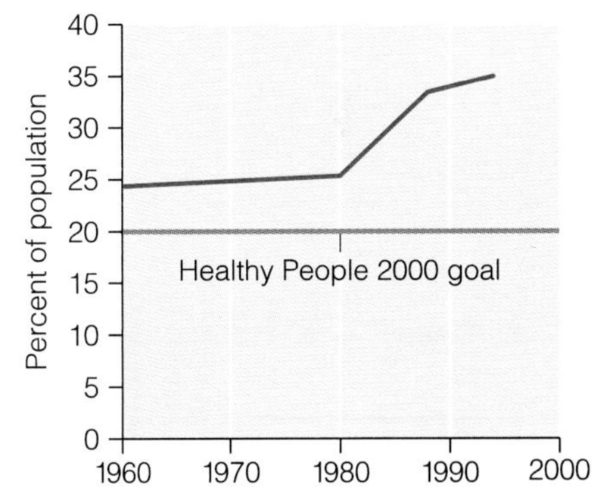

Healthy People 2000: Reduce overweight to a prevalence of no more than 20% among people aged 20 years and older and maintain prevalence at no more than 15% among adolescents aged 12 through 19 years.

## *Fat Cell Development*

When more energy is consumed than is spent, much of the excess energy is stored in fat cells. The amount of fat in a person's body reflects both the *number* and the *size* of the fat cells. The number of fat cells increases most rapidly during the growing years of late childhood and early puberty. Fat cell number increases more rapidly in obese children than in lean children, and obese children entering their teen years may already have as many fat cells as do adults of normal weight.

The fat cells expand in size as they fill with fat droplets (review Figure 5-19 on p. 142). When the cells reach their maximum size, they may also divide. Thus obesity develops when a person's fat cells increase in number, in size, or quite often both. Figure 9-2 illustrates fat cell development.

With fat loss, the size of fat cells dwindles, but not their number. People with extra fat cells tend to regain lost weight rapidly; they can shrink their fat cells, but

Obesity due to an increase in the *number* of fat cells is **hyperplastic obesity.** Obesity due to an increase in the *size* of fat cells is **hypertrophic obesity.**

Overweight is defined as BMI ≥27.8 for men and ≥27.3 for women.

**Figure 9-2**
**Fat Cell Development**

Fat cells are capable of increasing their size by 20-fold and their number by several thousandfold.

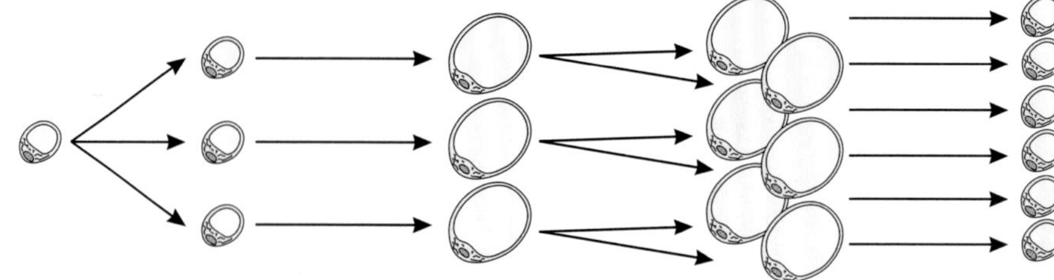

During growth, fat cells increase in number.

When energy intake exceeds expenditure, fat cells increase in size.

When fat cells have reached their maximum size and energy intake continues to exceed energy expenditure, fat cells increase in number again.

With fat loss, the size of the fat cells shrinks, but not the number.

they cannot make the cells disappear. With weight gain, their many fat cells readily fill. In contrast, people with an average number of enlarged fat cells may be more successful in maintaining weight losses; when their cells shrink, both cell size and number are normal. Prevention of obesity is most critical, then, during the growing years when fat cells increase in number.

## Fat Cell Metabolism

Reminder: The enzyme *lipoprotein lipase (LPL)* promotes fat storage.

The enzyme lipoprotein lipase (LPL) promotes fat storage in both fat and muscle cells. People with high LPL activity store fat especially efficiently. Since LPL is mounted on fat cell membranes, obese people generally have much more LPL activity in their fat cells than lean people (their muscle cell LPL activity is similar though).[3] Consequently, even modest excesses in energy intake have a more dramatic impact on obese people than on lean people.

The activity of LPL is partially regulated by sex-specific hormones—estrogen in women and testosterone in men. In women, fat cells in the breasts, hips, and thighs produce abundant LPL, putting fat away in those body sites; in men, fat cells in the abdomen produce abundant LPL. This activity explains why men tend to develop central obesity whereas women more readily develop lower-body fat.

Differences are also apparent in the activity of the enzymes controlling fat breakdown in various parts of the body. The lower body is less active than the upper body in releasing fat from storage.[4] Consequently, people tend to have a more difficult time losing fat from the hips and thighs than from around the chest and stomach.

Enzyme activity may also explain why people who lose weight so easily regain it. After weight loss, LPL activity rises, and it rises most dramatically in people who had been fattest prior to weight loss. Apparently, weight loss serves as a signal to the gene that produces the LPL enzyme, saying "Make more enzyme to store fat." People easily regain weight after having lost it because they are battling against enzymes that want to store fat. The activity of these enzymes provides an explanation for the observation that some inner mechanism seems to set a person's weight or body composition at a fixed point, and that if the person tries to change it, the body will adjust to restore the set point.

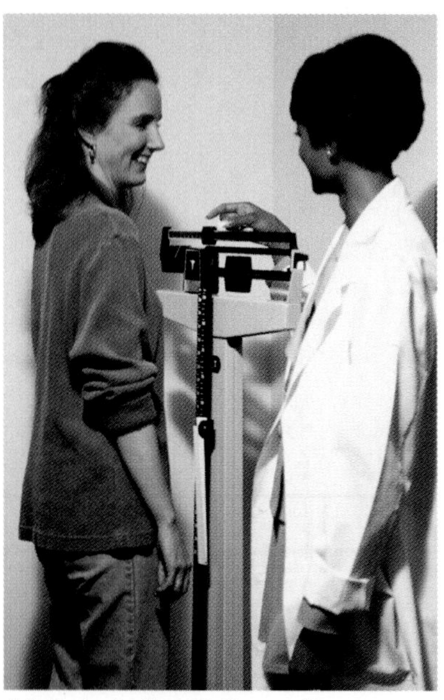

***Adults typically gain ½ pound per year between the ages of 25 and 55.***

## Set-Point Theory

**set point:** the point at which controls are set (for example, on a thermostat). The set-point theory that relates to body weight proposes that the body tends to maintain a certain weight by means of its own internal controls.

Many internal physiological variables, such as blood glucose, blood pH, and body temperature, remain fairly stable under a variety of conditions. The hypothalamus and other regulatory centers constantly monitor and delicately adjust conditions so as to maintain homeostasis. The stability of such complex systems may depend on set-point regulators that maintain variables within specified limits.

Researchers have confirmed that after weight gains or losses, the body adjusts its metabolism so as to restore the original weight. Energy expenditure increases after weight gain and decreases after weight loss.[5] These changes in energy expenditure exceed those predicted based on body composition and help to explain why it is so difficult for an obese person to maintain weight losses.

IN SUMMARY

Fat cells develop by increasing in number and size. Prevention of excess weight gain depends on maintaining a reasonable number of fat cells. With gains or losses, the body attempts to return to its previous status.

# Causes of Obesity

Why do people accumulate excess body fat? The obvious answer is that they take in more food energy than they spend. But that answer falls short of explaining why they do this. Is it genetic? Environmental? Cultural? Behavioral? Socioeconomic? Psychological? Metabolic? All of these? Most likely, obesity has many interrelated causes; experts in the field speak of many different *obesities*. Why an imbalance between energy intake and energy expenditure occurs remains a bit of a mystery; the next sections summarize possible explanations.

## Genetics

Prior to the twentieth century, people were far more likely to lack food than to have it in abundance, and only about 10 percent of adults were obese—most likely, for genetic reasons.[6] Today, with food plentiful and labor-saving devices common, the incidence of obesity has increased dramatically, but the population's gene pool has changed little. For most of these people, genes do not *cause* their obesity, but they may influence food intake and activity patterns that lead to it.[7]

When both parents are obese, the chances that their children will be obese are high (80 percent), whereas when neither parent is obese, the chances are small (less than 10 percent). Researchers have also found that adopted children tend to be similar in weight to their biological parents, not to their adoptive parents. Studies of twins yield similar findings: identical twins are twice as likely to weigh the same as fraternal twins—even when reared apart. These findings suggest an important role for genetics in determining a person's *susceptibility* to obesity.[8]

Clearly, something makes a person more or less likely to gain or lose weight when overeating or undereating.[9] Some people gain more weight than others on comparable energy intakes. Given an extra 1000 kcalories a day for 100 days, some pairs of identical twins gain less than 10 pounds while others gain up to 30 pounds. Within each pair, the amounts of weight gained, percentages of body fat, and locations of fat deposits are similar. Thus, some people lose more weight than others on comparable exercise regimens.

Researchers have been examining several genes in search of answers to obesity questions.[10] As Chapter 6's section on protein synthesis described, each cell expresses only the genes for the proteins it needs, and each protein performs a unique function. The following paragraphs describe some recent research involving proteins that might influence obesity development.

• ***Leptin*** • Researchers have identified an obesity gene, called *ob,* that is expressed in the fat cells and codes for the protein leptin.[11] Leptin acts as a hormone, primarily in the hypothalamus. Preliminary research suggests that leptin promotes a negative energy balance by suppressing appetite and increasing energy expenditure (see Figures 9-3a and 9-3b ); changes in energy expenditure primarily reflect changes in basal metabolism, but may also include changes in physical activity patterns. Mice with a defective *ob* gene do not produce leptin and can weigh up

**Figure 9-3a**

**Leptin's Action in the Body**

Leptin maintains energy homeostasis. When the body gains fat, the increase in leptin shifts energy balance toward the negative: eat less and spend more energy. Such a scenario would ensure that all fat gains were followed by losses, but this is not the case in reality, of course. Most obese people have high levels of leptin, but their energy balance does not automatically shift to the negative, suggesting a resistance to leptin's action in obesity. Figure 9-3b on page 258 shows what happens when leptin balance decline.

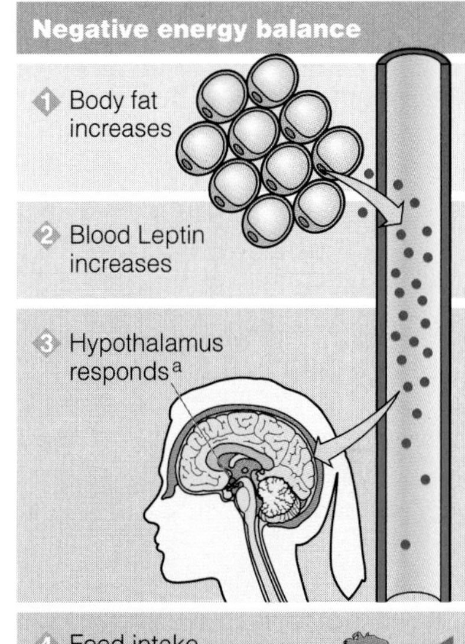
Negative energy balance
1 Body fat increases
2 Blood Leptin increases
3 Hypothalamus responds[a]

4 Food intake decreases and energy expenditure increases

[a] Among the compounds produced by the hypothalamus is neuropeptide y, introduced in Chapter 8. Neuropeptide y initiates eating and slows energy expenditure, which can lead to obesity. One of the leptin's actions is to inhibit neuropeptide y production.

**leptin:** a protein produced by fat cells under direction of the *ob* gene that decreases appetite and increases energy expenditure; sometimes called the *ob protein.*
• **leptos** = thin

**Figure 9-3b**
**Leptin's Action in the Body**

When the body loses fat, the decrease in leptin shifts energy balance toward the positive: eat more and spend less energy.

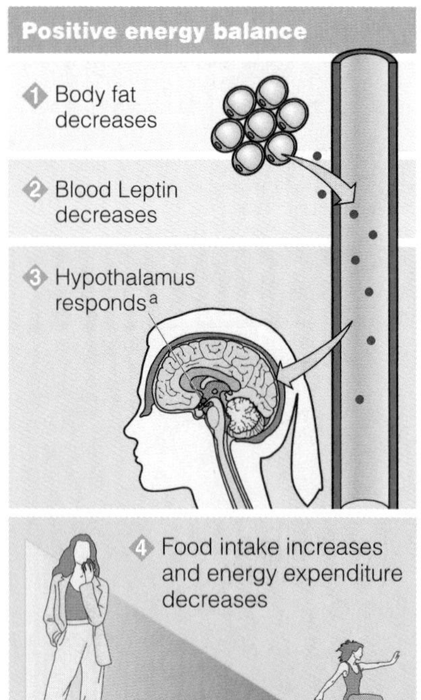
Positive energy balance
1 Body fat decreases
2 Blood Leptin decreases
3 Hypothalamus responds[a]
4 Food intake increases and energy expenditure decreases

[a] Among the compounds produced by the hypothalamus is neuropeptide y, introduced in Chapter 8.

**brown adipose tissue:** masses of specialized fat cells packed with pigmented mitochondria that produce heat instead of ATP.

to three times as much as normal mice and have five times as much body fat (see Figure 9.4). When injected with a synthetic form of leptin, the mice rapidly lose body fat. (Because leptin is a protein, it would be destroyed during digestion if given orally; consequently, it must be given by injection.) The fat cells not only lose fat, but they self-destruct, which may explain why weight gains are delayed when the mice are fed again.[12]

Although extremely rare, a genetic deficiency of leptin has been identified in human beings as well.[13] An error in the gene that codes for leptin was discovered in two extremely obese children with barely detectable blood levels of leptin. Without leptin, their appetite control is impaired; the children are constantly hungry and eat considerably more than their siblings or peers.

Very few obese people have a leptin deficiency, however. In fact, blood levels of leptin usually correlate directly with body fat: the more body fat, the more leptin.[14] Obese people generally have high leptin levels, and researchers have found that when people with low leptin levels gain weight, their leptin concentrations increase.[15] The researchers speculate that leptin rises in an effort to suppress appetite and inhibit fat storage, but its action is ineffective in obesity; obesity appears to be associated with an insensitivity or resistance to leptin.[16] Perhaps leptin or its receptors are defective or other signals override its action; all these possibilities are being considered in research.[17] The picture that seems to be developing is that leptin may appear in two scenarios in obesity as insulin does in diabetes. Relatively few people have an insulin deficiency (type 1 diabetes); many others have elevated insulin but are resistant to its action (type 2 diabetes).

• ***Uncoupling Proteins*** • Other genes code for proteins involved in energy metabolism. These proteins may influence obesity by storing or spending energy with different efficiencies or in different forms of fat. The body has both white and brown adipose tissue. White adipose tissue stores fat for other cells to use for energy; brown adipose tissue releases stored energy as heat. Recall from Chapter 7 that when fat is oxidized, some of the energy is released in heat and some is captured in ATP. In brown adipose tissue, oxidation may be uncoupled from ATP formation; it produces heat only. Radiating energy away as heat enables the body to spend, rather than store, energy. Heat production is particularly important in newborns, in adults who live in extremely cold climates, and in animals who hibernate; they have plenty of brown adipose tissue. In contrast, most human adults have small amounts of brown fat in strategic locations, and its role in body weight regulation is just beginning to be understood.

Recently, researchers working on the protein that uncouples reactions in brown adipose tissue discovered a gene that codes for a second uncoupling protein.[18] This protein is active not only in brown fat, but in white fat and many other tissues as well.[19] Its actions seem to influence the basal metabolic rate (BMR) and oppose the development of obesity.[20] Animals with abundant amounts of this uncoupling protein resist weight gain, whereas those with minimal amounts gain weight easily. Whether the body dissipates the energy from an ice cream sundae as heat or stashes it in body fat has major consequences, of course, for the person's body weight.

**Figure 9-4**
**Mice with and without Leptin Compared**

Both of these mice have a defective *ob* gene. Consequently, they do not produce leptin. They both became obese, but the one on the right received daily injections of leptin, which suppressed food intake and increased energy expenditure, resulting in weight loss.

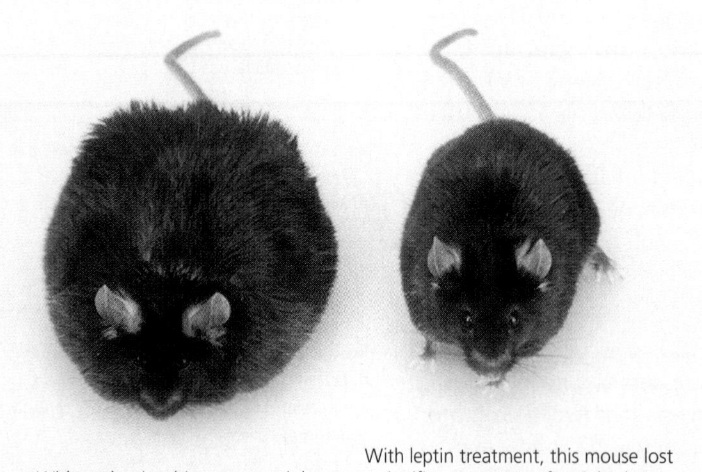

Without leptin, this mouse weighs almost three times as much as a normal mouse.

With leptin treatment, this mouse lost a significant amount of weight, but still weighs almost one and a half times as much as a normal mouse.

## *Overeating*

One obvious, although not necessarily satisfactory, explanation for obesity is that overweight people overeat. Whether this is true has not been determined, however.

• ***Energy Intake*** • Diet histories from obese people often report energy intakes that are similar to, or even less than, those of others. Diet histories may not be accurate records of actual intakes, however. Nonobese as well as obese people commonly underestimate their dietary intakes.[21]

Studies on rats do not depend on diet histories, of course, and reveal interesting findings. Genetically obese rats eat much more than their nonobese littermates during their early months of growth and rapidly gain weight. When their weights stabilize, however, all rats' intakes become similar, even though the obese rats continue to weigh twice as much as the others. When researchers control the rats' energy intakes, the obese ones gain less than when allowed to eat freely, but they still gain much more weight than the lean rats. Apparently, overeating contributes to obesity, but does not fully explain it.

• ***Fat Intake*** • A high-fat diet promotes obesity, especially in people with a genetic predisposition.[22] In short, high-fat foods taste delicious, so people eat more, but fats have little satiating power—so people eat even more.[23] Fat's 9 kcalories per gram quickly add up, amplifying people's energy intakes and enlarging their body fat stores.[24] Furthermore, excess dietary fat prompts the body's metabolism to shift into a fat-storing mode, as Chapter 7 described.

Using food frequency questionnaires, researchers studied healthy men consuming the "typical American diet"—about 16 percent of total kcalories from protein, 45 percent from carbohydrate, 36 percent from fat, and 3 percent from alcohol.[25] Total food energy intakes (2600 kcalories per day) fell short of current recommendations (2900 kcalories per day), which might suggest that the men were not exceeding their energy needs. Yet, when the researchers compared the men's diet records to their body weights and body composition measures, they made two interesting observations. They found no correlation between total food energy intakes and the men's body fat measurements. They did, however, find a positive correlation between the men's *dietary* fat and their *body* fat: the more fat a man ate, the greater his body fat. The researchers also found a negative correlation between body fat and complex carbohydrates and fiber: the less legumes, grains, fruits, and vegetables a man ate, the greater his body fat. These findings suggest that when people eat diets high in fat and low in carbohydrate, they tend to store body fat efficiently, even with moderate food energy intakes.

Like adults, children who eat higher-fat foods have higher body fat.[26] Again and again, experimental data suggest that dietary fat influences body fatness, and fat people tend to eat more fat. The implication is that reducing fat in the diet should play a prominent part in weight loss, as a later section of the chapter explains.

Dietary fat promotes obesity because:
- It is palatable (increases food intake).
- It produces less satiety (increases food intake).
- It provides more kcalories per gram (increases energy intake).
- It induces more efficient metabolism (increases body fat stores).

## Physical Inactivity

People may be obese, not because they eat too much, but because they spend too little energy.[27] More than one-third of the overweight population report no physical activity during their leisure time.[28] Some obese people are so extraordinarily inactive that even when they eat less than lean people, they still have an energy surplus. Reducing their food intake further would jeopardize health and incur nutrient deficiencies. Physical activity, then, is a necessary component of nutritional health. People must be physically active if they are to eat enough food to deliver all the nutrients they need without unhealthy weight gain.

One hundred years ago, 30 percent of the energy used in farm and factory work came from muscle power; today only 1 percent does. Modern technology has replaced physical activity at home, at work, and in transportation. Underactivity is probably the single most important contributor to obesity. In turn, television watching may contribute most to physical inactivity.[29]

Watching television contributes to obesity in several ways. First, television viewing requires little energy beyond the resting metabolic rate. Second, it replaces time spent in more vigorous activities. Third, watching television influences family food purchases and correlates with between-meal snacking on the high-

***Lack of physical activity fosters obesity.***

kcalorie, high-fat foods most heavily advertised on programs. Nonnutritious foods and beverages appear not only in commercials, but also within the television programs themselves. People, especially children, who see television stars eating and drinking these foods and remaining thin may miss the message that such behavior will bring about weight gain.

IN SUMMARY

In recent years, the view has been gaining ground that obesity is much more complicated than mere undisciplined gluttony. Most likely, obesity has many causes and different combinations of causes in different people. Some causes, such as overeating and physical inactivity, may be within a person's control, and some, such as genetics, may be beyond it.

# Controversies in Obesity Treatment

Clinical Guidelines on the Identification, Evaluation, and Treatment of Overweight and Obesity in Adults

www.nhlbi.nih.gov/nhlbi/cardio/obes/prof/guidelns/ob_home.htm

An estimated 30 to 40 percent of all U.S. women (and 20 to 25 percent of U.S. men) are trying to lose weight at any given time, spending up to $40 billion each year to do so. Some of these people do not even need to lose weight. Others need to lose weight, but are not successful; few succeed, and even fewer succeed permanently.

## *Elusive Goals*

Many people assume that every overweight person can achieve slenderness and should pursue that goal. First consider that most overweight people cannot—for whatever reason—become slender: only 5 percent of all people who successfully lose weight maintain their losses. Then consider the prejudice involved in that assumption. People come with varying weight tendencies just as they come with varying potentials for height and degrees of health, yet we do not expect tall people to shrink or healthy people to get sick in an effort to become "normal."

Large segments of our society place such enormous value on thinness that obese people face prejudice and discrimination: they are judged on their appearance more than on their character. Socially, obese people are stereotyped as lazy, stupid, and lacking in self-control. Psychologically, they may suffer embarrassment when others treat them with hostility and contempt, and some have even learned to view their own bodies as grotesque and loathsome. Parents and friends may scold them for lacking the discipline to resolve their weight problems. Health care professionals, including dietitians, are among the chief offenders.[30] All of this hurts self-esteem. Such a critical view of overweight is not prevalent in many other cultures, including segments of our society. Instead, overweight is embraced as a sign of robust health and beauty. Many overweight people today are tired of our obsession with weight control and simply want to be accepted as they are. To free our society of its obsession with body weight and prejudice against obesity, we must first learn to judge others for who they are and not for what they weigh.[31]

Still, based on the associated health risks, medical experts continue to urge all obese people to lose weight. Obese people die younger from a host of causes, including heart attacks, strokes, some cancers, and complications of diabetes (type 2). Even after these diagnosed diseases are discounted, the risk of death remains twice as high for obese people, especially those with lifelong obesity, than would otherwise be expected.

Encouraging weight loss may be justified when health benefits are clear. For example, a 30-year-old man with a body mass index (BMI) of 40 might be able to prevent or control the diabetes that runs in his family by losing 75 pounds. The

For reference, a man with a BMI of 40 might be:
- 5 ft 8 in, 265 lb.
- 5 ft 10 in, 280 lb.
- 6 ft, 295 lb.

effort required to do so may be great, but it is far less than the effort and consequences of living with diabetes. Sometimes health benefits appear with even less weight loss. A 60-year-old man with a BMI of 40 might gain relief from the arthritis in his knees by losing just 25 pounds.[32] In this case, losing more weight might not bring added health benefits.

Often a person's motivations for weight loss have nothing to do with health. A young woman with a BMI of 23 might want to lose a few pounds for spring break, but might be healthier *not* losing weight. In the case of a person with anorexia nervosa with a BMI of 18, losing weight might incur devastating medical consequences (see Highlight 9).

For reference, a woman with a BMI of 23 might be:

- 5 ft 3 in, 130 lb.
- 5 ft 5 in, 139 lb.
- 5 ft 7 in, 146 lb.

And a woman with a BMI of 18 might be:

- 5 ft 4 in, 105 lb.
- 5 ft 6 in, 112 lb.
- 5 ft 8 in, 118 lb.

See the inside back cover for additional examples of heights and weights for various BMI.

### IN SUMMARY

The question whether a person should lose weight depends on many factors: the extent of overweight, age, health, and genetic makeup among them. Not all obesity will cause disease or shorten life expectancy. Just as there are unhealthy, normal-weight people, there are healthy, obese people. Weight-loss advice, then, does not apply equally to all overweight people. Some people may risk more in the process of losing weight than in remaining overweight.

## *Dangers of Weight Loss*

People attach so many dreams of happiness to weight loss that they willingly risk huge sums of money for the slightest chance of success. As a result, weight-loss schemes flourish. Of tens of thousands of claims, treatments, and theories for losing weight, few are effective—and many are downright dangerous. The negative effects must be carefully considered before embarking on any weight-loss program. Physical problems may arise from fad diets and "yo-yo" dieting, and psychological problems may emerge from repeated "failures."

Many states have developed consumer bills of rights to help protect potential weight-loss clients. Such documents explain the risks associated with weight-loss programs and provide honest predictions of success (see Table 9-1).

• ***Fad Diets*** • Fad diets often sound good, but typically fall short of delivering on their promises. They espouse exaggerated or false theories of weight loss and advise consumers to follow inadequate diets. Some fad diets are more hazardous to health than obesity itself. Adverse reactions can be as minor as headaches, nausea, and dizziness or as serious as death. Highlight 8 offers guidelines for identifying unsound weight-loss schemes and diets.

**Table 9-1**

**Weight-Loss Consumer Bill of Rights (An Example)**

1. *WARNING:* Rapid weight loss may cause serious health problems. Rapid weight loss is weight loss of more than 1½ to 2 pounds per week or weight loss of more than 1 percent of body weight per week after the second week of participation in a weight-loss program.
2. Consult your personal physician before starting any weight-loss program.
3. Only permanent lifestyle changes, such as making healthful food choices and increasing physical activity, promote long-term weight loss and successful maintenance.
4. Qualifications of this provider are available upon request.
5. *YOU HAVE A RIGHT TO:*
   - Ask questions about the potential health risks of this program and its nutritional content, psychological support, and educational components.
   - Receive an itemized statement of the actual or estimated price of the weight-loss program, including extra products, services, supplements, examinations, and laboratory tests.
   - Know the actual or estimated duration of the program.
   - Know the name, address, and qualifications of the dietitian or nutritionist who has reviewed and approved the weight-loss program.

**weight cycling:** repeated cycles of weight loss and gain. The weight-cycling pattern is popularly called the **ratchet effect** or **yo-yo effect** of dieting.

**Figure 9-5**
**The Weight Cycling Effect of Repeated Dieting**

Each round of dieting is followed by a rebound of weight to a higher level than before.

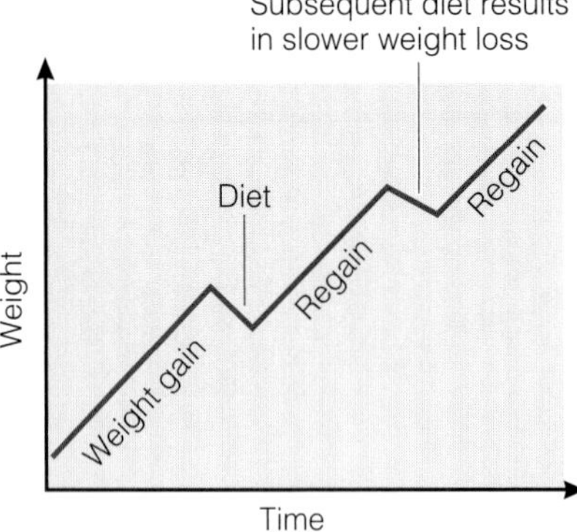


Losing 10 lb and keeping them off may be beneficial, but losing 100 lb and regaining them may be quite harmful.

• ***Weight Cycling*** • Many people who try to lose weight become trapped in weight cycling, the endless repeating rounds of weight loss and regain from "yo-yo" dieting (see Figure 9-5). When people repeatedly lose weight only to regain it, their bodies become very efficient at making and storing fat. This increased efficiency shows itself in a way familiar to dieters who have lost and gained—and lost and gained again. With each attempt, it becomes harder and takes longer to lose weight and easier and quicker to gain it back. In fact, previous weight-cycling history can predict a person's success (or lack thereof) in maintaining weight loss.[33]

Such fluctuations in body weight appear to increase the risks of chronic diseases and even premature death, independently of obesity itself.[34] Maintaining a stable weight, even if it is overweight, may be less harmful to health than repeated bouts of weight gain and loss. Such concerns should not deter obese people who want to lose weight from trying, but rather should encourage them to commit to lifelong changes that will maintain weight losses.[35]

• ***Psychological Problems*** • Some of the nation's most popular diet books and weight-loss programs have misled consumers with unsubstantiated claims and deceptive testimonials. Furthermore, they fail to provide an assessment of the short- and long-term results of their treatment plans, even though such evaluations are possible and would permit consumers to make informed decisions.[36] Of course, some weight-loss programs are better than others in terms of cost, approach, and customer satisfaction, but few are particularly successful in helping people keep lost weight off.

Most programs assume that the problem can be solved simply by applying willpower and hard work. If determination were the only factor involved, though, the success rate would be far greater than 5 percent. Overweight people may readily assume the blame for their failures to lose weight and maintain the losses when, in fact, the programs have failed. Ineffective treatment and its associated sense of failure add to a person's psychological burden. Figure 9-6 illustrates how the devastating psychological effects of obesity and dieting perpetuate themselves.

IN SUMMARY

Weight-loss efforts are often misguided. When pursued via unwise weight-loss techniques, they can be physically and psychologically damaging.

**Figure 9-6**
**The Psychology of Weight Cycling**

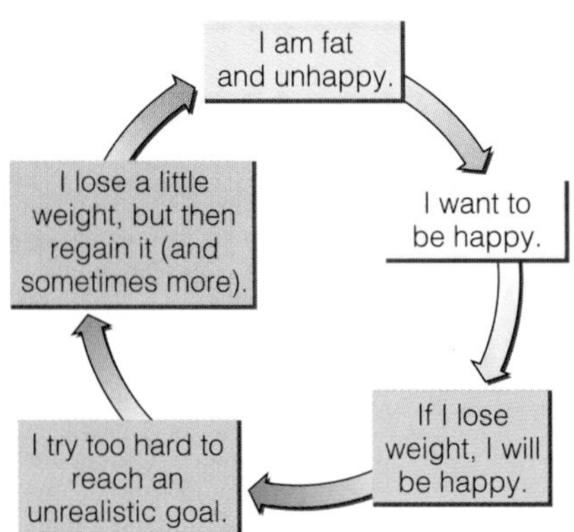


Source: Adapted with permission from J. P. Foreyt and G. K. Goodrick, *Living without Dieting* (Houston: Harrison Publishing, 1992).

# Aggressive Treatments of Obesity

Sometimes the problems associated with weight loss reflect inappropriate efforts. A person with a BMI of 25 may need to improve eating habits and increase physical activity, but someone with a BMI greater than 30 may need more aggressive treatment options.

## *Drugs*

Several drugs for weight loss have been tried over the years. When used as part of a long-term, comprehensive, weight-loss program, drugs can help obese people to lose approximately 10 percent of their weight and maintain that loss for at least a year.[37] Because weight regain commonly occurs with the discontinuation of drug therapy, treatment must be long term. Yet the long-term use of drugs poses risks. We don't yet know whether a person would benefit more from maintaining a 20-pound excess or from taking a drug for a decade to keep the 20 pounds off. Physicians must prescribe drugs appropriately, inform clients of the potential risks, and monitor side effects carefully.

• ***Diuretics*** • Temporary water retention may add several pounds to a person's weight on the scale, but water does not cause obesity. Nor does obesity cause water retention.* In fact, obese people have a *smaller* percentage of body water than people of normal weight. When people take diuretics, they lose water, not fat. The weight loss lasts only half a day or so, and the price is dehydration and mineral imbalances.

**diuretics** (dye-you-RET-ics): drugs that promote water excretion; popularly, a "water pill."
- **dia** = through
- **ure** = urine

• ***Amphetamines*** • Decades ago, physicians routinely prescribed dexamphetamine and methamphetamine to speed up metabolism, but today amphetamines are no longer approved by the Food and Drug Administration (FDA) for weight loss. They were ineffective, and worse, they were also highly addictive. Withdrawal causes extreme fatigue, mental depression, and sleep abnormalities. Common side effects associated with amphetamine use include dizziness, irritability, blurred vision, nausea, vomiting, and diarrhea.

**amphetamine** (am-FET-ah-mean): a central nervous system stimulant used to treat narcolepsy and some types of depression. Use of amphetamines to control appetite in the treatment of obesity has not proved effective, and prolonged use causes dependency. Amphetamines are marketed under the trade names *Benzedrine* and *Amphedrine*.

• ***Progress in Drug Research*** • For several decades after the disappointments with amphetamines, drug treatment for obesity floundered amidst negative publicity, government regulations, and ineffective and dangerous outcomes. Then, with new understandings of obesity's genetic basis and an acceptance of its classification as a chronic disease, drug treatment research rapidly gained ground. Experts reasoned that if obesity is a chronic disease, it should be treated as such—and the treatment of most chronic diseases includes drugs.[38] The challenge, then, is to develop an effective drug that can be used over time without adverse side effects or the potential for abuse. No such drug currently exists.

• ***Dexfenfluramine and the Fen-Phen Combination*** • In 1996, the FDA approved a new drug for use in obesity treatment: dexfenfluramine. The public emphatically demanded it. Drug companies eagerly produced it. Physicians indiscriminately prescribed it. And the media enthusiastically lauded it. Finally, there was a "miracle" drug that would suppress appetite and allow people to eat less and lose weight. Its success, however, was short-lived.

Dexfenfluramine and its sister drug fenfluramine enhance blood levels of the neurotransmitter serotonin in the brain. This action suppresses appetite—not without risk, but at a risk deemed acceptable given the consequences of obesity itself. The FDA intended the treatment to be used only for truly obese people—those with a BMI of 30 or greater (or 27 or greater if other risk factors were also present).

**dexfenfluramine** (DEKS-fen-FLOOR-ah-mean): a drug used in the treatment of obesity that triggers the release of serotonin in the brain, thus suppressing appetite; marketed under the trade name *Redux*. Its sister drug **fenfluramine** (fen-FLOOR-ah-mean) acts similarly and is marketed under the trade name *Pondimin*.

**serotonin** (SER-oh-tone-in): a neurotransmitter important in sleep regulation, appetite control, and sensory perception among other roles. Serotonin is synthesized in the body from the amino acid tryptophan with the help of vitamin $B_6$.

For reference, consider a person 5 ft 6 in:
- BMI of 30 = ≥186 lb.
- BMI of 27 = ≥167 lb.

The balance between the benefits and the risks shifted when physicians began prescribing either dexfenfluramine or the combination of fenfluramine and phentermine (commonly known as fen-phen) to almost 5 million people, many of whom were only slightly overweight. Like dexfenfluramine and fenfluramine, phentermine suppresses appetite, but not quite as effectively. The fen-phen combination was more effective than either fenfluramine or phentermine alone, and it could be prescribed in lower doses, which reduced the side effects and lowered the costs. Fenfluramine and phentermine were each approved by the FDA years ago for short-term use, but not in combination or for an extended length of time.

**phentermine** (FEN-ter-mean): a drug used in the treatment of obesity that suppresses appetite; marketed under the trade names *Adipex*, *Fastin*, and *Ionamin*.

With prolonged use, dexfenfluramine and fenfluramine raise blood pressure in the lungs, causing pulmonary hypertension.[39] In addition, they damage the valves of the heart, causing shortness of breath, fatigue, and heart murmurs.[40] Both pulmonary hypertension and valvular heart disease are usually rare but life-threatening conditions that can lead to heart failure. Within a year, the FDA had received almost 150 reports of people who had taken these drugs and

**pulmonary hypertension:** abnormally high blood pressure in the lungs; a rare, but life-threatening condition associated with dexfenfluramine, fenfluramine, phentermine, or combinations of these drugs.

**valvular heart disease:** abnormal changes in the valves of the heart; a rare, but life-threatening condition associated with dexfenfluramine, fenfluramine, phentermine, or combinations of these drugs.

*Many women experience temporary water retention around the time of the menstrual period. Oral contraceptives may also cause water retention and may even promote fat gain in some women. A woman who has this problem should consult her physician about switching brands.

Center for Drug Evaluation and Research

www.fda.gov/cder
Search for Fen-phen
Search for Sibutramine

**sibutramine** (sigh-BYOO-tra-mean): a drug used in the treatment of obesity that slows the reabsorption of serotonin in the brain, thus suppressing appetite and creating a feeling of fullness; marketed under the trade name *Meridia*.

Phenylpropanolamine is marketed under the trade names:

- Acutrim.
- Dex-A-Diet.
- Dexatrim.
- Permathene.
- Phenoxine.
- Phenyldrine.
- Prolamine.
- Super Ordinex.
- Thin Z.
- Unitrol.

Benzocaine is marketed under the trade names:

- Diet Ayds (candy).
- Slim Mint (gum).

*So many promises, so little success.*

experienced symptoms of heart disease.[41] Furthermore, when the FDA requested heart examinations of people *without* symptoms, one-third had abnormal results.[42] This prevalence of heart valve damage among users is substantially higher than that seen in the general population. By 1997, the FDA had determined that dexfenfluramine and fenfluramine presented an unacceptable risk and recalled them from the market. Consumers who had used the drugs were urged to see their physicians for lung and heart exams.

• ***Other Drugs*** • Researchers have not given up on finding a safe and effective weight-loss drug. In fact, within three months of the dexfenfluramine and fenfluramine recall, the FDA approved another new drug: sibutramine. Like dexfenfluramine and fenfluramine, sibutramine suppresses appetite by enhancing serotonin, but by a different mechanism. Because sibutramine raises blood pressure, the FDA advises those with high blood pressure not to use it and others to monitor their blood pressure. As Table 9-2 shows, several other drugs are currently under study.

• ***Over-the-Counter Drugs*** • The FDA has approved only two over-the-counter medications to help with weight loss. One contains phenylpropanolamine, which suppresses appetite and enhances weight loss when used with a low-kcalorie diet.[43] Reported side effects include dry mouth, rapid pulse, nervousness, sleeplessness, hypertension, irregular heartbeats, kidney failure, seizures, and strokes. The other over-the-counter product contains benzocaine (in a candy or gum form), which anesthetizes the tongue, reducing taste sensations.

• ***Herbal Products*** • In their search for weight-loss magic, some consumers turn to "natural" herbal products. St. John's Wort, for example, contains substances that enhance serotonin and thus suppress appetite as do the drugs mentioned earlier. In addition to the many cautions that accompany the use of any herbal remedies, consumers should be aware that St. John's Wort is often prepared in combination with the herbal stimulant ephedrine. Ephedrine, in combination with caffeine and/or aspirin, has shown some promise as a treatment for obesity in a limited number of research studies. Without further ado, manufacturers began to market "natural" combinations of ephedrine extracted from the Chinese plant ma huang, caffeine from coffee beans, and acetosalicyclic acid (aspirin) from willow bark. These ephedrine-containing supplements have been implicated in several cases of heart attacks and seizures and have been linked to more than a dozen deaths.*

*Ma huang (ephedrine) is illegal in Canada.

**Table 9-2**
**Experimental Drugs for the Treatment of Obesity**

| Drugs | Actions | Side Effects |
|---|---|---|
| Acarbose | Reduces carbohydrate digestion | Intestinal gas, cramping, and diarrhea |
| Cholecystokinin | Reduces food intake | Effectiveness diminishes |
| Enterostatin | Reduces energy intake, specifically preferences for fat | None reported |
| Ephedrine (with caffeine and/or aspirin) | Enhances fat breakdown and decreases fat storage | Tremors, cardiac events |
| Leptin | Decreases appetite and increases energy expenditure | None reported |
| Orlistat | Reduces fat absorption | Intestinal gas, cramping, and diarrhea |

NOTE: None of these drugs has received FDA approval for obesity treatment; acarbose has been approved for diabetes treatment only.
Source: Adapted from R. L. Atkinson, Use of drugs in the treatment of obesity, *Annual Review of Nutrition* 17 (1997): 383–403.

Herbal laxatives containing senna, aloe, rhubarb root, cascara, castor oil, and buckthorn (or various combinations) are commonly sold as "dieter's tea." Such concoctions commonly cause nausea, vomiting, diarrhea, cramping, and fainting and may have contributed to the deaths of four women who had drastically reduced their food intakes.[44] Consumers mistakenly believe that laxatives will diminish nutrient absorption and save kcalories, but remember that absorption occurs primarily in the upper small intestine and these laxatives act on the lower large intestine.

Highlight 18 explores the possible benefits and potential dangers of herbal products and other alternative therapies.

• ***Other Gimmicks*** • Other gimmicks don't help with weight loss either. Hot baths do not speed up metabolism so that pounds can be lost in hours. Steam and sauna baths do not melt the fat off the body, although they may dehydrate people so that they lose water weight. Brushes, sponges, wraps, creams, and massages intended to move, burn, or break up "cellulite" do nothing of the kind, because there is no such thing as cellulite.

**cellulite** (SELL-you-light or SELL-you-leet): supposedly, a lumpy form of fat; actually, a fraud. Fatty areas of the body may appear lumpy when the strands of connective tissue that attach the skin to underlying muscles pull tight where the fat is thick. The fat itself is the same as fat anywhere else in the body. So, if the fat in these areas is lost, the lumpy appearance disappears.

## *Very-Low-kCalorie Diets*

During the 1980s, obesity treatments aggressively strived to reduce body weight to the ideals set in standard weight-for-height tables. To that end, medical centers offered obese people very-low-kcalorie diets (VLCD).

VLCD plans provide 800 kcalories, at least 1 gram of high-quality protein, per kilogram of body weight, little or no fat, and a minimum of 50 grams of carbohydrate (not enough to spare protein). Clients receive an assortment of vitamins and minerals from supplements. Meals consist of a limited number of foods (primarily lean meats, fish, and poultry) each day, a powdered formula available by prescription, or a combination of the two.

VLCD formulas are designed to be nutritionally adequate, but the body responds to this severe energy restriction as if the person were starving—conserving energy and preparing to regain weight at the first opportunity. As Chapter 7 described, several changes occur in hormone concentrations, metabolic activities, fluid and electrolyte balances, and organ functions in the effort to meet the challenge of living on a much-less-than-adequate energy intake. For these reasons, a VLCD is appropriate only for short-term use (four months) and under close medical supervision. Table 9-3 on p. 266 lists common side effects of VLCD.

Without doubt, weight losses on VLCD are dramatic. Unfortunately, weight regains are almost certain. With weight loss comes a slower BMR and slower fat oxidation—conditions that favor weight gain. Such rapid losses followed by steady gains are detrimental to both physical and psychological health.

1980s definition of successful weight loss: reduction of body weight to the ideal established in weight-for-height tables.

Recommended criteria for people on VLCD:

- Motivated.
- BMI >30.
- No heart, kidney, gallbladder, or liver disease; cancer; or alcoholism.
- No psychiatric disorders (including eating disorders).

## *Surgery*

Surgery as an approach to weight loss is justified in some specific cases of clinically severe obesity. Two gastric partitioning procedures have gained wide acceptance and are illustrated in Figure 9-7 on p. 267. Both procedures effectively limit food intake by reducing the size of the stomach. They reduce the size of the outlet as well, so they delay the passage of food from the stomach into the intestine for digestion and absorption.

The long-term safety and effectiveness of gastric surgery depend, in large part, on compliance with dietary instructions. Common immediate postsurgical complications include infections, nausea, vomiting, and dehydration; in the long term, vitamin and mineral deficiencies and psychological problems are common. Lifelong medical supervision is necessary for those who choose the surgical route, but in suitable candidates the benefits of weight loss prove worth the risks.

Another surgical procedure is used, not to treat obesity, but to remove the evidence. Plastic surgeons can extract some fat deposits by suction lipectomy, or "liposuction." This cosmetic procedure has little effect on body weight, but can alter body shape slightly in specific areas.

**clinically severe obesity:** a BMI of 40 or greater or 100 lb or more overweight for an average adult. A less preferred term used to describe the same condition is *morbid obesity.*

**gastric partitioning:** a surgical procedure used to treat clinically severe obesity. The operation limits food intake by reducing the size of the stomach and delays gastric emptying by restricting the outlet.

**Table 9-3**
**Possible Physical Consequences of Very-Low-kCalorie Diets**

| Blood | Immunity |
|---|---|
| • Blood carotene concentrations increase.<br>• Blood cholesterol concentrations increase.<br>• Blood urea concentrations increase. | • Immune response diminishes.<br>• White blood cells decrease in number. |
| **Cardiovascular/Respiratory** | **Metabolic** |
| • Blood pressure declines.<br>• Carbon dioxide production declines.<br>• Cardiac output declines.<br>• Heart muscle atrophies.<br>• Heartbeat becomes irregular.<br>• Oxygen consumption declines.<br>• Pulse rate declines.<br>• Respiratory rate declines. | • Basal metabolism declines.<br>• Bone mineral content shifts.<br>• Cold intolerance occurs.<br>• Dehydration may occur.<br>• Gout may occur.<br>• Ketosis develops.<br>• Lean body tissues are lost.<br>• Mineral and electrolyte imbalances occur.<br>• Nitrogen balance becomes negative. |
| **Digestive** | **Other** |
| • Gallstones and kidney stones form.<br>• GI tract motility declines.<br>• Liver inflammation and fibrosis develop.<br>• Nausea, vomiting, diarrhea, abdominal discomfort, and constipation occur. | • Body and breath odor (from ketone excretion) may become apparent.<br>• Hair falls out.<br>• Headaches occur.<br>• Lethargy, fatigue, and loss of stamina set in.<br>• Skin dries out.<br>• Sleeplessness may occur.<br>• Sudden death becomes possible. |
| **Hormonal** | |
| • Menstrual irregularity develops.<br>• Sex drive is lost. | |

Sources: Evidence on metabolic rate, lean body tissue, liver, gallstones, heartbeat, bones, and nitrogen balance, from various authors in *American Journal of Clinical Nutrition* (supplement) 56 (1992); immune failure reported in C. J. Field, R. Gougeon, and E. B. Marliss, Changes in circulating leukocytes and mitogen responses during very-low-energy all-protein reducing diets, *American Journal of Clinical Nutrition* 54 (1991): 123–129; reduced oxygen consumption reported in K. N. Pavlou and coauthors, Exercise as an adjunct to weight loss and maintenance in moderately obese subjects, *American Journal of Clinical Nutrition* 49 (1989): 1115–1123; low blood pressure, headache, atrophy of heart muscle, and hormonal effects from R. L. Atkinson, Low calorie diets and obesity, in *Biotechnology and Nutrition*, eds. D. D. Bills and S. D. Kung (Boston: Butterworth-Heinemann, 1992), pp. 29–45.

IN SUMMARY

Obese people with high risks of medical problems may need aggressive treatment, including drugs, VLCD, or surgery. Others may benefit most from improving eating and exercise habits.

# Reasonable Treatments of Obesity

1990s definition of successful weight loss: reduction in body weight to achieve health goals.

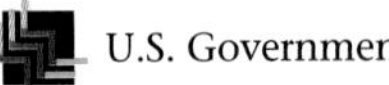
U.S. Government
www.healthfinder.gov/searchoptions/topicsaz.htm
Search for Weight control

In the 1990s, experts on obesity treatments revised their thinking. They now embrace small changes, moderate losses, and reasonable goals. They realize that a 200-pound woman who loses 10 to 20 pounds in a year is much more likely to maintain losses and reap health benefits than if she were to drop to 130 pounds as suggested in weight-for-height tables. This philosophy prompted the *Dietary Guidelines* to suggest that for good health, a person should "maintain or improve your weight." The focus is not on weight loss per se, but on health gains. In fact, the *Guidelines* go on to say, "if you are overweight and cannot lose weight, try not to gain weight."

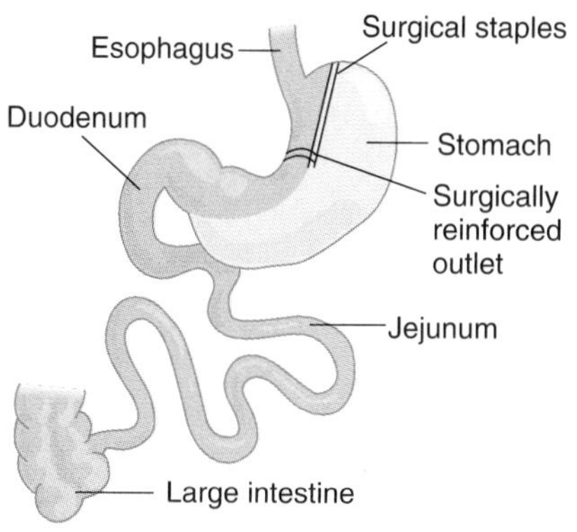


In vertical banded gastroplasty, the surgeon constructs a small gastric pouch and restricts the outlet from the stomach to the intestine.

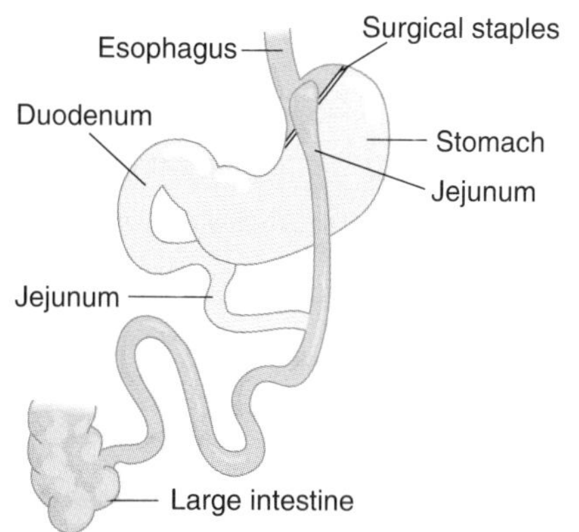


In gastric bypass, the surgeon constructs a small gastric pouch and creates an outlet directly to the jejunum.

**Figure 9-7**

**Surgical Procedures Used in the Treatment of Severe Obesity**

The dark pink areas highlight the flow of food through the GI tract. Notice that the first procedure maintains a relatively normal flow whereas the second one bypasses most of the stomach, all of the duodenum, and some of the jejunum. The pale pink areas indicate the sections that have been bypassed.

Modest weight loss, even when a person is still overweight, can improve control of diabetes and reduce the risks of heart disease by lowering blood pressure and blood cholesterol, especially for those with upper-body fat. For these reasons, parameters such as blood pressure, blood cholesterol, or even self-esteem are more useful than body weight in marking success.

Of course, the same eating and exercise habits that improve health often lead to a healthier body weight and composition as well.[45] A loss of 10 to 15 pounds can improve a person's BMI by 2 units, which can significantly improve health—even if the person is still overweight. Successful weight loss, then, is not defined by weight-for-height tables, but by reductions in disease risks.[46] People less concerned with disease risks may prefer to set goals for personal fitness, such as being able to play with children or climb stairs without becoming short of breath.

Whether the goal is health or fitness, weight-loss expectations need to be reasonable. Unreachable targets ensure frustration and failure. If goals are achieved or exceeded, there will be rewards instead of disappointments.

Findings from a recent study highlight the great disparity between lofty expectations and reasonable success.[47] Before beginning a weight-loss program, obese women identified the weights they would describe as "dream," "happy," "acceptable," and "disappointing" (see Figure 9-8). All of these weights were below their starting weight. Their goal weights reflected a 32 percent loss of initial weight, far exceeding the 5 to 10 percent recommended by experts, or even the 15 percent reported by the most successful weight-loss studies. Even the "disappointing" weights exceeded recommended goals. Close to a year later, and after an average loss of 35 pounds, almost half of the women did not achieve even their "disappointing" weights. They did, however, experience more physical, social, and psychological benefits than they had predicted for that weight. Still, in a culture that overvalues thinness, these women were not satisfied with a 16 percent reduction in weight—not because their efforts were unsuccessful, but because their expectations were unrealistic.

Realistic goals include reasonable time frames—at least six months for a 10 percent loss of initial weight. Keep in mind that pursuing good health is a lifelong journey. Most adults are keenly aware of their body weights and shapes and realize that what they eat and what they do can make a difference to some extent. Those who are

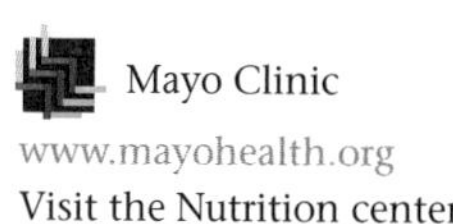


www.mayohealth.org
Visit the Nutrition center

• Weight-loss tip: Adopt reasonable expectations about health and weight goals and about how long it will take to achieve them.

**Figure 9-8**

**Reasonable Weight Goals and Expectations Compared**

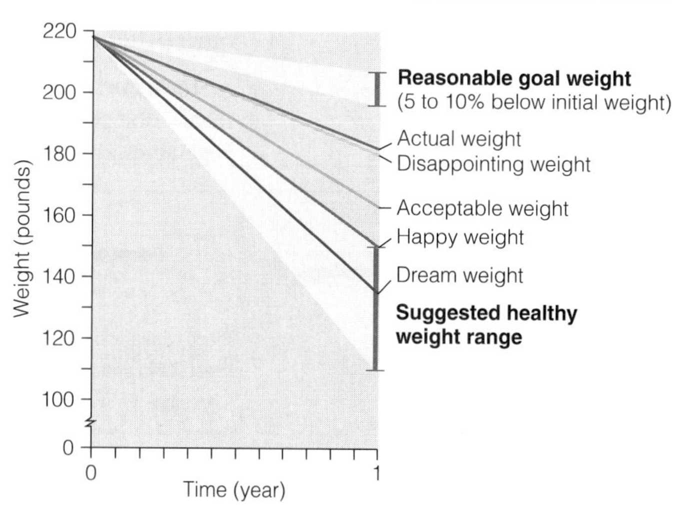


Source: Adapted from G. D. Foster and coauthors, What is a reasonable weight loss? Patients' expectations and evaluations of obesity treatment outcomes, *Journal of Consulting and Clinical Psychology* 65 (1997): 79–85.

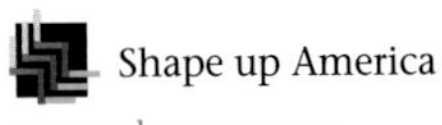
Shape up America
www.shapeup.org

most successful at preventing weight gain or maintaining weight loss seem to have fully incorporated healthful eating and physical activity into their daily lives.

**Healthy People 2000:** Increase to at least 50% the proportion of overweight people aged 12 years and older who have adopted sound dietary practices combined with regular physical activity to attain an appropriate body weight.

## Eating Plans

• Weight-loss tip: Be involved in planning.

No one food plan is magical, and no specific food must be included or avoided in a weight-control plan. In designing a plan, people need only consider foods that they like or can learn to like, that are available, and that are within their means.

• ***Realistic Energy Intake*** • The main characteristic of a weight-loss diet is that it provides less energy than the person needs to maintain present body weight. Restricting energy intake too severely, however, may be counterproductive. Rapid weight loss usually means excessive loss of lean tissue and a rapid weight gain to follow.

• Weight-loss tip: Restrict energy to 10 kcal/lb body weight.

Energy intake should provide nutritional adequacy without excess—that is, somewhere between deprivation and complete freedom to eat everything in sight. A reasonable suggestion is that a person needs at least 10 kcalories per pound of current weight each day to lose fat efficiently while retaining lean tissue. For example, a 180-pound woman would start by setting her energy intake at 1800 kcalories a day. Then, as she lost weight, she could adjust her energy intake downward.

• Weight-loss tip: Make nutritional adequacy a high priority.

• ***Nutritional Adequacy*** • Nutritional adequacy is difficult to achieve on fewer than 1200 kcalories a day, and most healthy adults need never consume any less than that. An intake of 1200 kcalories a day would allow even a 120-pound adult to lose weight, yet what 120-pound adult would need to lose weight? People weighing more than 120 pounds can lose weight on higher energy intakes. A plan that provides an adequate intake supports a healthier and more successful weight loss than a restrictive plan that creates feelings of starvation and deprivation, which can lead to an irresistible urge to binge.[48]

NIDDK Health Information
www.niddk.nih.gov/health/nutrit/nutrit.htm

Take a look at Table 9-4 and notice that the 1200-kcalorie food plan represents the minimum servings suggested in the Daily Food Guide (introduced in Chapter 2) and allows a teaspoon of fat at each of three meals. Such an intake would allow most people to lose weight and still meet their nutrient needs with careful, nutrient-dense food selections. (Women might need an iron supplement.) The other patterns provide for higher energy intakes.

• Weight-loss tip: Emphasize nutrient-dense foods.

• Weight-loss tip: Eat small portions of foods at each meal.

• ***Small Portions*** • Overweight people may need to learn to eat less food at each meal—one piece of chicken for dinner instead of two, a teaspoon of butter on the vegetables instead of a tablespoon, and one cookie for dessert instead of six. The

**Table 9-4**
**Recommended Number of Servings for Different Energy Intakes**

| Food Group | Energy Level (kcal) | | | | | | |
|---|---|---|---|---|---|---|---|
| | *1200* | *1500* | *1800* | *2000* | *2200* | *2600* | *3000* |
| Bread, cereal, rice, and pasta | 6 | 7 | 8 | 9 | 11 | 13 | 15 |
| Meat (lean) and meat alternates[a] | 4 | 5 | 6 | 6 | 6 | 7 | 8 |
| Vegetable | 3 | 4 | 5 | 5 | 5 | 6 | 6 |
| Fruit | 2 | 3 | 4 | 4 | 4 | 5 | 6 |
| Milk and milk products (nonfat) | 2 | 2 | 2 | 3 | 3 | 3 | 3 |
| Fat (tsp) | 3 | 5 | 6 | 7 | 8 | 10 | 12 |

NOTE: These patterns follow the Daily Food Guide plan and supply less than 30 percent of kcalories as fat.
[a]Meat servings are given in ounces.

goal is to eat enough food for energy, nutrients, and pleasure, but not more. This amount should leave a person feeling satisfied—not necessarily full. Keep in mind that even low-fat foods can deliver a lot of kcalories when a person eats large quantities. A low-fat cookie or two can be a sweet treat even on a weight-loss diet, but a whole box is clearly excessive.

• Weight-loss tip: Share a restaurant meal with a friend or take home half for lunch tomorrow.

• Weight-loss tip: Limit low-fat treats to the serving size on the label.

• ***Eat Carbohydrates*** • Healthy meals and snacks center on complex carbohydrate foods. Fresh fruits, vegetables, legumes, and whole grains offer abundant vitamins, minerals, and fiber but little fat. High-fiber foods also require effort to eat—an added bonus. People who eat these foods in abundance spontaneously eat for longer times and take in fewer kcalories than when eating foods of high energy density. The satiety signal indicating fullness is sent after a 20-minute lag, so a person who slows down and savors each bite eats less before the signal reaches the brain.

• Weight-loss tip: Make legumes, grains, vegetables, and fruits central to your diet plan.

• Weight-loss tip: Eat slowly.

• ***Restrict Fats*** • Satiety plays a key role in weight-loss diets. As Chapter 8 mentioned, fat has a weak satiating effect. Consequently, a person eating a high-fat diet tends to overeat, which raises energy intake by adding both more food and more fat kcalories. A person who eats a low-fat diet (high in carbohydrate and adequate in protein)—even without intentionally limiting energy intake—eats less food, satisfies hunger, fills the stomach, and diminishes the desire to eat.[49] Whereas most people on weight-loss diets feel deprived of food, those following a low-fat, high carbohydrate diet complain of having to eat much more food than they are accustomed to eating.[50] How much weight is lost on an unrestricted, low-fat diet depends on the extent of fat reduction, the degree of obesity, and the intensity of physical activity.[51] A person focusing only on limiting energy intake will also lose weight, but maintaining that weight loss will be easier with a low-fat diet.[52] While total fat losses may be similar, losses from lower body fat appear greater with the low-fat diet.[53]

• Weight-loss tip: Limit high-fat foods.

Clearly, then, to lose weight and improve body composition, measure fat with extra caution. A slip of the butter knife adds more kcalories than a slip of the sugar spoon. Less fat in the diet means less fat in the body (review p. 148 for strategies to lower fat in the diet). Be careful not to take this advice to the extreme, however; too little fat in the diet or in the body carries health risks as well.

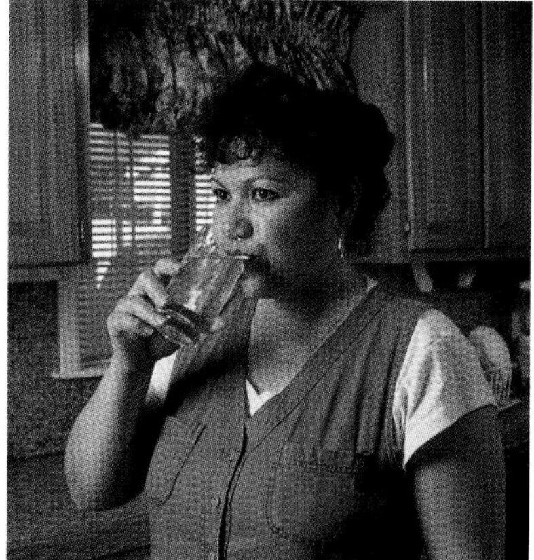

***Drinking water is a healthy habit.***

• ***Other Empty kCalories*** • Speaking of empty kcalories, a person trying to achieve or maintain a healthy weight needs to pay attention not only to fat, but to sugar and alcohol, too. Using them for pleasure on occasion is compatible with health as long as most daily choices are of nutrient-dense foods.

• Weight-loss tip: Limit concentrated sweets and alcoholic beverages.

• ***Adequate Water*** • Learn to satisfy thirst with water. Water fills the stomach between meals and dilutes the metabolic wastes generated from the breakdown of fat. It meets the water need that was formerly met by eating extra food (remember that foods provide water). Water also helps the GI tract adapt to a high-fiber diet.

• Weight-loss tip: Drink plenty of water (8 glasses or more a day).

IN SUMMARY

Adopt a lifelong "eating plan for good health" rather than a "diet for weight loss." That way, you will be more likely to keep the lost weight off.

• Weight-loss tip: Learn, practice, and follow a healthful eating plan for the rest of your life.

## Physical Activity

The best approach to weight control combines diet and physical activity.[54] People who combine diet and exercise are more likely to lose more fat, retain more muscle, and regain less weight than those who only diet.[55] People who include physical activity in their weight-control program do not necessarily lose more weight, but they seem to follow their diet plans more closely and maintain their losses better than those who do not exercise.[56] Consequently, they benefit both from taking in a little less energy and from expending a little more energy in physical activity.

• Fitness tip: Participate in some form of physical activity regularly.

***Regular physical activity, such as bicycling or walking briskly, will help to burn fat. (Courtesy of CNN)***

• ***Activity and Energy Expenditure*** • Table 8-2 (on p. 238) shows how much energy each of several activities uses. The table also shows that the number of kcalories spent in an activity depends more upon body weight than on how fast the exercise is done. For example, a person who weighs 150 pounds and runs a mile in 6 minutes spends about 103 kcalories. That same person walking a mile in 15 minutes uses almost the same amount—about 92 kcalories. Similarly, a 220-pound person spends about 150 kcalories on the 6-minute mile, and only a little less—about 135 kcalories—on the 15-minute walk. Whether a person chooses to run or walk the distance, the same energy will be spent; walking will just take longer. To lose fat, exercise as intensely, and expend as much energy, as your time allows.[57]

Remember to adjust estimates of energy costs of activities after losing weight. After a significant weight loss, formerly obese people spend less energy on the same activity than they spent previously. The new expenditure is less than would be expected given their lower body weight.[58]

• ***Activity and Basal Metabolism*** • Activity also contributes to energy expenditure in an indirect way—by speeding up basal metabolism.[59] It does this both immediately and over the long term. On any given day, basal metabolism remains elevated for several hours after intense and prolonged exercise. Over the long term, daily vigorous activity for many weeks gradually changes body composition toward more lean tissue. Metabolic rate rises accordingly, and this supports continued weight loss or maintenance.

The metabolic rate remains elevated for as long as the person keeps exercising regularly. The more energy expended in metabolic activities, the greater the energy requirement. Consequently, a person can eat more food without gaining weight, which in turn brings both pleasure and nutrients.

• ***Activity and Appetite Control*** • Physical activity also helps to control appetite. Many people think that exercising will make them eat more, but this is not entirely true.[60] Active people do have healthy appetites, but immediately after an intense workout, most people do not feel like eating. They may be thirsty and want to shower, but they are not hungry. The reason is that the body has released fuels from storage to support the exercise, so glucose and fatty acids are abundant in the blood. At the same time, the body has suppressed its digestive functions. Hard physical work and eating are not compatible. A person must calm down, put energy fuels back in storage, and relax before eating. Thus exercise may actually help curb appetite, especially the inappropriate appetite that accompanies boredom, anxiety, or depression. Weight-control programs encourage people who feel the urge to eat when not hungry to go out and exercise instead. The activity passes time, relieves anxiety, and prevents inappropriate eating.

Benefits of physical activity in a weight-control program:

- Short-term increase in energy expenditure (from exercise and from a slight rise in BMR).
- Long-term increase (slight) in BMR.
- Appetite control.
- Stress reduction and control of stress eating.
- Physical, and therefore psychological, well-being.
- High self-esteem.

• ***Activity and Psychological Benefits*** • Activity also helps reduce stress. Since stress itself cues inappropriate eating for many people, activity can help here, too. Activity offers still more psychological advantages. The fit person looks and feels healthy and, as a result, gains self-esteem. High self-esteem motivates a person to persist in seeking good health and fitness, which keeps the beneficial cycle going.

• ***Choosing Activities*** • Clearly, physical activity is a plus in a weight-control program. What kind of physical activity is best? People seeking to lose weight should choose activities that they enjoy and are willing to do regularly. Sustained physical activities of moderate intensity (aerobic exercises) are more effective in weight control than short bursts of vigorous exercise.

In addition to exercise, a person can incorporate hundreds of energy-spending activities into daily routines: take the stairs instead of the elevator, walk to the neighbor's apartment instead of making a phone call, and rake the leaves instead of using a blower. Remember that sitting uses more kcalories than lying down, standing uses more kcalories than sitting, and moving uses more kcalories than standing. A 175-pound person who replaces a 30-minute television program with a 2-mile walk a day can spend enough energy to lose (or not gain) 18 pounds in a year. Walk a mile. Run a race. Swim a lap. Dance a jig. Ride a bike. Climb a mountain. Do whatever you enjoy doing—and do it often.

• ***Spot Reducing*** • People sometimes ask about "spot reducing." Unfortunately, muscles do not "own" the fat that surrounds them. Fat cells all over the body release fat in response to the demand of physical activity for use by whatever muscles are active. No exercise can remove the fat from any one particular area.

Exercise can help with trouble spots in another way, though. The "trouble spot" for most men is the abdomen, their primary site of fat storage. During aerobic exercise, abdominal fat readily releases its stores, providing fuel to the physically active body. With regular exercise and weight loss, men will deplete these abdominal fat stores before those in the lower body.[61] Women may also deplete abdominal fat with exercise, but their "trouble spots" are more likely to be their hips and thighs.

In addition to aerobic activity, strength training can help to improve the tone of muscles in a trouble area; and stretching to gain flexibility can help with associated posture problems. A combination of aerobic, strength, and flexibility workouts best improves fitness and physical appearance.

IN SUMMARY

Physical activity should be an integral part of a program for weight loss or maintenance. Physical activity can increase energy expenditure, help control appetite, reduce stress and stress eating, and enhance physical and psychological well-being.

## Behavior and Attitude

Behavior modification once held a key position in weight-loss programs, but its status has diminished in recent years.[62] Still, behavior and attitude play an important role in supporting efforts to achieve and maintain appropriate body weight and composition. Changing the hundreds of small behaviors of overeating and underexercising that lead to, and perpetuate, obesity requires time and effort. A person must commit to take action.

**behavior modification:** the changing of behavior by the manipulation of *antecedents* (cues or environmental factors that trigger behavior), the behavior itself, and *consequences* (the penalties or rewards attached to behavior).

It also helps to adopt a positive, matter-of-fact attitude. Healthy eating and activity choices are an essential part of healthy living and should simply be incorporated into the day—much like brushing one's teeth or wearing a safety belt.

• ***Become Aware of Behaviors*** • To solve a problem, a person must first identify all the behaviors that created the problem. Keeping a record will help to identify eating and exercise behaviors that may need changing (see Figure 9-9 on p. 272). It will also establish a baseline against which to measure future progress.

• Behavior change tip: Keep a record of diet and exercise habits; it reveals problem areas, the first step toward improving behaviors.

• ***Change Behaviors*** • The "How to" on p. 273 describes strategies that support weight control. These strategies encourage desired eating and exercise behaviors and eliminate unwanted behaviors.

With so many possible behavior changes available, a person can choose where to begin. Start simply and don't try to master them all at once. Attempting too many changes at one time invites failure; a person must set priorities. Pick one trouble area that is manageable and start there. Practice a desired behavior until it becomes routine. Then select another trouble area to work on, and so on. Another bit of advice along the same lines: don't try to tackle weight loss during a particularly stressful time of life.

• ***Personal Attitude*** • For many people, overeating and being overweight have become an integral part of their identity. Those who fully understand their personal relationships with food are best prepared to make healthful changes in eating and exercise behaviors.

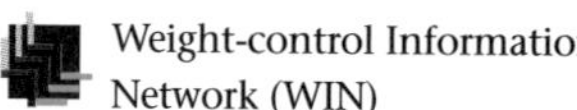

Weight-control Information Network (WIN)

www.niddk.nih.gov/health/nutrit/win.htm

Sometimes habitual behaviors that are hazardous to health, such as smoking or drinking alcohol, contribute positively by helping people adapt to stressful

**Figure 9-9**
**Food Record**

The entries in a food record should include the times and places of meals and snacks, the types and amounts of foods eaten, the persons present when food is eaten, and a description of the individual's feelings when eating. The diary should also record physical activities: the kind, the intensity level, the duration, and the person's feelings about them.

| Time | Place | Activity or food eaten | People present | Mood |
|---|---|---|---|---|
| 10:30– | School | 6 peanut | by myself | Starved |
| 10:40 | vending | butter crackers | | |
| | machine | and 12 oz. cola | | |
| | | | | |
| 12:15– | Restaurant | Sub sandwich | friends | relaxed |
| 12:30 | | and 12 oz. cola | | & friendly |
| | | | | |
| 3:00– | Gym | Weight | work out | tired |
| 3:45 | | training | partner | |
| | | | | |
| 4:00– | Snack bar | Small | by myself | OK |
| 4:10 | | frozen | | |
| | | yogurt | | |

situations. Similarly, many people overeat to cope with the stresses of life. To break out of that pattern, they must first identify the particular stressors that trigger the urge to overeat. Then, when faced with these situations, they must learn and practice problem-solving skills. These skills will help them to respond appropriately to difficult situations.

All this is not to imply that psychological therapy holds the magic answer to a weight problem. Still, efforts to improve one's general well-being may result in healthy eating habits even when weight loss is not the primary goal. When the problems that trigger the urge to overeat are resolved in alternative ways, people may find they eat less. They may begin to respond appropriately to internal cues of hunger rather than inappropriately to external signals of stress. Sound emotional health supports a person's ability to take care of physical health in all ways—including nutrition, weight control, and fitness.

• Behavior change tip: Learn alternative ways to deal with emotions and stresses.

• ***Support Groups*** • Group support is important when making life changes. Some people find it helpful to join a group that provides support in efforts to lose weight, such as Take Off Pounds Sensibly (TOPS), Weight Watchers (WW), Overeaters Anonymous (OA), or others. A modest expenditure for health is well worthwhile, but people need to avoid rip-offs, of course. Many dieters find it helpful to form their own self-help groups.

• Behavior change tip: Attend support groups regularly or develop supportive relationships with others.

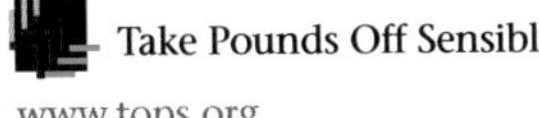
Take Pounds Off Sensibly
www.tops.org

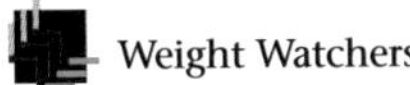
Weight Watchers
www.weightwatchers.com

**Healthy People 2000:** Increase to at least 50% the proportion of worksites with 50 or more employees that offer nutrition education and/or weight management programs for employees.

## IN SUMMARY

A surefire remedy for obesity has yet to be found, although many people find a combination of the approaches just described to be most effective. Diet and exercise shift energy balance so that more energy is being spent than is taken in. Physical activity maintains or even builds the lean body so that fat is preferentially lost and metabolic energy needs remain high. Behavior modification retrains habits so that once weight is lost, it will not return; and an improvement in inner self helps a person to manage life without a dependency on food. This treatment package requires time, individualization, and sometimes the assistance of skilled health care professionals.

## HOW TO Change Behaviors to Support Weight Loss

1. To eliminate inappropriate eating cues:
   - Buy foods that are low in fat.
   - Shop when you are not hungry.
   - Serve low-fat meals.
   - Let other family members buy, store, and serve their own sweets (monitor children's intakes).
   - Change channels or look away when food commercials appear on television.
   - Shop only from a list and stay away from convenience stores.
   - Carry appropriate snacks from home and avoid vending machines.
2. To suppress the cues you cannot eliminate:
   - Eat only in one place (at a table), and in one room; use plates, bowls, and eating utensils.
   - Clear plates directly into the garbage.
   - Create obstacles to the eating of problem foods (for example, make it necessary to unwrap, cook, and serve each one separately).
   - Minimize contact with excessive food (serve individual plates, don't put serving dishes on the table, and leave or clean the table when you have finished eating).
   - Make small portions of food look large by spreading food out and serving on small plates.
   - Control deprivation (eat regular meals, don't skip meals, avoid getting tired, avoid boredom by keeping cues to fun activities in sight).
3. To strengthen the cues to appropriate eating and exercise:
   - Encourage others to eat appropriate foods with you.
   - Keep your favorite appropriate foods in the front of the refrigerator.
   - Learn appropriate portion sizes and prepare one portion at a time.
   - Establish specific times for meals and snacks.
   - Prepare foods attractively.
   - Keep your walking shoes (ski poles, tennis racket) by the door.
4. To engage in desired eating or exercise behaviors:
   - Eat only at planned times; plan not to eat after a specified time (say, 7:00 or 8:00 P.M.).
   - Slow down (pause several times during a meal, put down utensils between mouthfuls, chew thoroughly before swallowing, swallow before reloading the fork, always use utensils).
   - Leave some food on the plate.
   - Engage in no other activities while eating (such as reading or watching television).
   - Move more (shake a leg, pace, fidget, flex your muscles).
   - Join in and exercise with a group of active people.
5. To arrange or emphasize negative consequences of inappropriate eating:
   - Eat your meals with other people.
   - Ask that others respond neutrally when you deviate from your plan (make no comment). This is a negative consequence because it withholds attention.
   - If you slip, don't punish yourself.
6. To arrange or emphasize positive consequences of appropriate behaviors:
   - Update records of food intake, exercise, and weight change regularly.
   - Arrange for rewards for each unit of behavior change or weight loss.
   - Ask family and friends for reinforcement (praise and encouragement).

## *Weight Maintenance*

People who are successful often lose much of their weight within half a year and then reach a plateau.[63] This slowdown can be disappointing, but should be recognized as an opportunity for the body to adjust to its new weight. It also gives the person relief from the distraction of weight-loss dieting.[64] An appropriate goal at this point is to continue eating and activity behaviors that will maintain the initial loss. Attempting to lose additional weight at this point would require heroic efforts and would almost certainly meet with failure.

Be aware that obesity is not "cured" by simply attaining a reasonable body weight; eating wisely and staying active must continue to be part of life's daily routines. If, on arriving at goal weight after months of self-discipline, the victorious weight loser resumes old habits, then maintenance will not be successful.

Those who are successful in maintaining their weight loss have established vigorous exercise regimens and careful eating patterns, consuming less energy and a lower percentage of kcalories from fat than the national average.[65] They realize that they do not have the same flexibility in their food and activity habits as their friends who have never been overweight. With weight loss, metabolism shifts downward so that formerly overweight people require about 15 percent less energy than might be expected given their current body weight and body composition. Consequently, to keep weight off, they must either eat less or exercise more than people the same size who have never been obese.

• Behavior change tip: Adopt permanent lifestyle changes to achieve and maintain a healthy weight.

Physical activity plays a key role in maintaining weight loss. Those who continue exercising vigorously are far more successful than those who remain inactive.[66] On average, weight maintenance requires a person to exercise either moderately (such as brisk walking) for 80 minutes a day or vigorously (such as fast bicycling) for 35 minutes a day.

## Prevention

To prevent excessive weight gain:[67]
- Eat regular meals and limit snacking.
- Drink water instead of high-kcalorie beverages.
- Select low-fat foods regularly and limit dietary fat to 30% of daily kcalorie intake.
- Become physically active and limit television viewing time.

Given the information presented up to this point in the chapter, the adage "An ounce of prevention is worth a pound of cure" seems particularly apropos. Being obese is unhealthy, and losing weight is challenging and often temporary.[68] Strategies for preventing weight gain are very similar to those for losing weight, with one exception: they begin early and continue throughout life. Over the years, they become an integral part of a person's life. It is much easier for a person to resist doughnuts for breakfast if he rarely eats them. Similarly, a person will have little trouble walking 5 miles each morning if she has always been active.

IN SUMMARY

Preventing weight gains and maintaining weight losses requires vigilant attention to diet and physical activity. Taking care of oneself is a lifelong responsibility.

# Underweight

Reminder: *Underweight* is a body weight so low as to have adverse health effects; it is generally defined as 10% or more below the standard or BMI <20.

Underweight is a far less prevalent problem than overweight, affecting no more than 10 percent of U.S. adults. Whether the underweight person needs to gain weight is a question of health and, like weight loss, a highly individual matter. People who are healthy at their present weights may stay there; those who are malnourished or ill should try to gain. Medical advice can help make the distinction.

Some people are unalterably thin by reason of heredity or early physical influences. They may find gaining weight difficult. People who wish to gain weight for appearance's sake or to improve their athletic performance need to be aware that healthful weight gains can be achieved only by physical conditioning combined with high energy intakes. On a high-kcalorie diet alone, a person will gain weight, but it will be mostly fat. Even if the gain improves appearance, it can be as detrimental to health as being slightly underweight. For an athlete, such a weight gain might impair performance. Therefore, in weight gain, as in weight loss, physical activity and energy intake are essential components of a sound plan.

## Problems of Underweight

The causes of underweight may be as diverse as those of overweight—hunger, appetite, and satiety irregularities; psychological traits; metabolic factors; and hereditary tendencies. Habits learned early in childhood, especially food aversions, may perpetuate themselves.

The demand for energy to support physical activity and growth often contributes to underweight. An active, growing boy may need more than 4000 kcalories a day to maintain his weight and may be too busy to take time to eat. Underweight people find it hard to gain weight—due, in part, to their expenditure of energy in adaptive thermogenesis. So much energy may be spent adapting to a higher food intake that at first as many as 750 to 800 extra kcalories a day may be needed to gain a pound a week. Like those who want to lose weight, people who want to gain must learn new habits and learn to like new foods. They are also similarly vulnerable to potentially harmful schemes and would be wise to review the consumer bill of rights on p. 261, using the term "weight gain" instead of "weight loss" where appropriate.

• Weight-gain tip: Expect weight gain to take time (1 lb per month would be reasonable).

An extreme underweight condition known as anorexia nervosa sometimes develops in people who employ heroic self-denial in order to control their weight. They go to such extremes that they become severely undernourished, achieving final body weights of 70 pounds or even less. The distinguishing feature of a person with anorexia nervosa, as opposed to other underweight people, is that the starvation is intentional. Anorexia nervosa is a major eating disorder seen in our society today. Another is bulimia nervosa—compulsive overeating and purging. Eating disorders are the subject of the highlight that follows this chapter.

## *Weight-Gain Strategies*

Weight-gain strategies center on eating foods that provide many kcalories in a small volume and exercising to build muscle. Conventional advice to the bodybuilder is to eat enough foods to provide about 700 to 1000 kcalories a day above normal energy needs and to exercise to build lean tissue. Table 9-4 (on p. 268) provides diet patterns for energy intakes up to 3000 kcalories a day. These patterns follow the Daily Food Guide plan and incorporate many of the principles for planning a healthy diet.

• ***Energy-Dense Foods*** • Energy-dense foods (the very ones eliminated from a successful weight-loss diet) hold the key to weight gain. Pick the highest-kcalorie items from each food group—that is, milk shakes instead of nonfat milk, salmon instead of snapper, avocados instead of cucumbers, a cup of grape juice instead of a small apple, and whole-wheat muffins instead of whole-wheat bread. Because fat contains more than twice as many kcalories per teaspoon as sugar does, fat adds kcalories without adding much bulk.

• Weight-gain tip: Eat energy-dense foods regularly.

Be aware that health experts routinely recommend a low-fat diet because the biggest health problems in the United States involve overweight and heart disease. Eating high-kcalorie, high-fat foods is not healthy for most people, but may be essential for an underweight individual who needs to gain weight. An underweight person who is physically active and eating a nutritionally adequate diet can afford a few extra kcalories from fat. Well-trained endurance athletes can maintain a favorable lipid profile on a diet providing 42 percent of kcalories from fat.[69] For health's sake, it would be wise to select foods with monounsaturated and polyunsaturated fats instead of those with saturated fats: for example, sautéeing vegetables in olive oil instead of butter.

• ***Regular Meals Daily*** • People who are underweight need to make meals a priority and take the time to plan, prepare, and eat each meal. They should eat at least three healthy meals every day and learn to eat more food within the first 20 minutes of a meal. Another suggestion is to eat meaty appetizers or the main course first and leave the soup or salad until later.

• Weight-gain tip: Eat at least three meals a day.

• ***Large Portions*** • Underweight people need to learn to eat more food at each meal. Put extra slices of ham and cheese on the sandwich for lunch, drink milk from a larger glass, and eat cereal from a larger bowl.

• Weight-gain tip: Eat large portions of foods and expect to feel full.

The person should expect to feel full. Most underweight individuals are accustomed to small quantities of food. When they begin eating significantly more, they feel uncomfortable. This is normal and passes over time.

• Weight-gain tip: Eat snacks between meals.

• ***Extra Snacks*** • Since a substantially higher energy intake is needed each day, in addition to eating more food at each meal, it is necessary to eat more frequently. Between-meal snacking offers a solution. For example, a student might make three sandwiches in the morning and eat them between classes in addition to the day's three regular meals. Snacking on dried fruit, nuts, and seeds will also add kcalories easily.

• Weight-gain tip: Drink plenty of juice and milk.

• ***Juice and Milk*** • Beverages provide an easy way to increase energy intake. Consider that 6 cups of cranberry juice add almost 1000 kcalories to the day's intake. kCalories can be added to milk by mixing in powdered milk or packets of instant breakfast.

For people who are underweight due to illness, concentrated liquid formulas are often recommended because a weak person can swallow them easily. A physician or registered dietitian can recommend high-protein, high-kcalorie formulas to help the underweight person maintain or gain. Used in addition to regular meals, these can help considerably.

• Weight-gain tip: Exercise and eat to build muscles.

• ***Exercising to Build Muscles*** • To gain weight, use strength training primarily and increase energy intake to support that exercise. Eating extra food will then support a gain of both muscle and fat. About 700 to 1000 kcalories a day above normal energy needs is enough to support both the exercise and the building of muscle.

### IN SUMMARY

Both the incidence of underweight and health problems associated with it are less prevalent than overweight. To gain weight, a person must train physically and increase energy intake by selecting energy-dense foods, eating regular meals, taking larger portions, and consuming extra snacks and beverages.

## Making It Click

These problems give you practice in doing simple energy balance calculations (see Appendix K for answers). Once you have mastered these examples, you will be prepared to examine your own food choices. Be sure to show your calculations for each problem.

1. Critique a commercial weight-loss plan. Consumers spend billions of dollars a year on weight-loss programs such as Slim-Fast, Sweet Success, Weight Watchers, Nutri/System, Jenny Craig, Optifast, Medifast, and Formula One. One such plan calls for a milk shake in the morning, noon, and afternoon snack, and "a sensible, balanced, low-fat dinner" in the evening. One shake mixed in 8 ounces of vitamin A– and D–fortified nonfat milk offers 190 kcalories; 32 grams of carbohydrate, 13 grams of protein, and 1 gram of fat; at least one-third of the Daily Value for all vitamins and minerals; plus 2 grams of fiber.
   a. Calculate the kcalories and grams of carbohydrate, protein, and fat that three shakes provide.
   b. How do these values compare with the criteria listed in item 2 in Table H8-2 on p. 253?
   c. Plan "a sensible, balanced, low-fat dinner" that will help make this weight-loss plan adequate and balanced. Now, how do the day's totals compare with the criteria in item 2 in Table H8-2 on p. 253?
   d. Critique this plan using the other criteria described in Table H8-2 on p. 253 as a guide.
2. Evaluate a weight-gain attempt. People attempting to gain weight sometimes have a hard time because they choose low-kcalorie, high-bulk foods that make it hard to consume enough energy. Consider the following lunch: a chef's salad consisting of 2 cups iceberg lettuce, 1 whole tomato, 1 ounce swiss cheese, 1 ounce roasted ham (lean and fat), 1

## Making It Click (continued)

hard-boiled egg, ½ cucumber, and ¼ cup mayonnaise-type salad dressing. If you weighed these foods, you'd find that they totaled 551 grams. This is a pretty filling meal.

a. How much does this meal weigh in pounds?
b. The meal provides 541 kcalories. What is the energy density of this meal, expressed in kcalories per gram?
c. To gain weight, this person is advised to eat an additional 500 kcalories at this meal. Using foods with this same energy density, how much more chef's salad will this person have to eat?
d. Suppose a person simply can't do this. Try to reduce the bulk of this meal by replacing some of the lettuce with more energy-dense foods. Delete 1 cup lettuce from the salad and add 1 ounce roast beef and 1 ounce cheddar cheese. Show how these changes influence the weight and kcalories of this meal.

| Item No./Food | Weight (g) | Energy (kcal) |
|---|---|---|
| Original totals: | 551 | 541 |
| Minus: | | |
| # 867 Lettuce, 1 c | – | – |
| Plus: | | |
| # 603 Roast beef, 1 oz | + | + |
| #37 Cheddar cheese, 1 oz | + | + |
| Totals: | | |

e. How many kcalories did the changes add?
f. How much more *weight* of food did these changes add?

This exercise should reveal why people attempting to gain weight are advised to add high-fat items, within reason, to their daily meals.

## Study Questions

These questions will help you review the chapter. You will find the answers in the discussions on the pages provided.

1. Describe how body fat develops and suggest some reasons why it is difficult for an obese person to maintain weight loss. (pp. 255–257)
2. What factors contribute to obesity? (pp. 257–260)
3. List several aggressive ways to treat obesity, and explain why such methods are not recommended for every overweight person. (pp. 262–266)
4. Discuss reasonable dietary strategies for achieving and maintaining a healthy body weight. (pp. 266–272)
5. What are the benefits of increased physical activity in a weight-loss program? (pp. 269–270)
6. Describe the behavioral strategies for changing an individual's dietary habits. What role does personal attitude play? (pp. 271–272)
7. Describe strategies for successful weight gain. (pp. 275–276)

These questions will help you prepare for an exam. Answers can be found in Appendix K.

1. With weight loss, fat cells:
   a. decrease in size only.
   b. decrease in number only.
   c. decrease in both number and size.
   d. decrease in number, but increase in size.
2. Obesity is caused by:
   a. overeating.
   b. inactivity.
   c. defective genes.
   d. multiple factors.
3. The protein produced by the fat cells under the direction of the *ob* gene is called:
   a. leptin.
   b. serotonin.
   c. sibutramine.
   d. phentermine.
4. The biggest problem associated with the drugs dexfenfluramine and fenfluramine is their:
   a. cost.
   b. chronic dosage.
   c. ineffectiveness.
   d. adverse side effects.
5. Weight loss is successful when an obese person reduces body weight:
   a. down to the weight he or she was at age 25.
   b. down to the ideal weight in the weight-for-height tables.
   c. by 5 percent and maintains that loss for at least 1 year.
   d. by 15 percent and maintains that loss for at least 3 months.
6. A nutritionally sound weight-loss diet might restrict daily energy intake to about:
   a. 1 kcalorie per pound of current body weight.
   b. 1 kcalorie per kilogram of current body weight.
   c. 10 kcalories per pound of current body weight.
   d. 10 kcalories per kilogram of current body weight.

7. Successful weight loss depends on:
   a. avoiding fats and limiting water.
   b. taking supplements and drinking water.
   c. increasing proteins and restricting carbohydrates.
   d. reducing energy intake and increasing physical activity.
8. Physical activity does not help a person to:
   a. lose weight.
   b. retain muscle.
   c. maintain weight loss.
   d. lose fat in trouble spots.
9. Which strategy would *not* help an overweight person to lose weight?
   a. Exercise.
   b. Eat slowly.
   c. Limit high-fat foods.
   d. Eat energy-dense foods regularly.
10. Which strategy would *not* help an underweight person to gain weight?
    a. Exercise.
    b. Drink plenty of water.
    c. Eat snacks between meals.
    d. Eat large potions of foods.

# Notes


1. Update: Prevalence of overweight among children, adolescents, and adults—United States, 1988–1994, *Morbidity and Mortality Weekly Report* 46 (1997): 199–202.
2. J. Foreyt and K. Goodrick, The ultimate triumph of obesity, *The Lancet* 346 (1995): 134–135.
3. P. A. Kern, Potential role of TNFα and lipoprotein lipase as candidate genes for obesity, *Journal of Nutrition* 127 (1997): 1917S–1922S.
4. M. D. Jensen, Lipolysis: Contribution from regional fat, *Annual Review of Nutrition* 17 (1997): 127–139.
5. R. L. Leibel, M. Rosenbaum, and J. Hirsch, Changes in energy expenditure resulting from altered body weight, *New England Journal of Medicine* 332 (1995): 621–628; J. M. Kinney, Influence of altered body weight on energy expenditure, *Nutrition Reviews* 53 (1995): 265–268; P. Pasquet and M. Apfelbaum, Recovery of initial body weight and composition after long-term massive overfeeding in men, *American Journal of Clinical Nutrition* 60 (1994): 861–863.
6. J. Albu and coauthors, Obesity solutions: Report of a meeting, *Nutrition Reviews* 55 (1997): 150–156.
7. B. L. Heitmann and coauthors, Are genetic determinants of weight gain modified by leisure-time physical activity? A prospective study of Finnish twins, *American Journal of Clinical Nutrition* 66 (1997): 672–678.
8. J. P. Foreyt and W. S. C. Poston II, Diet, genetics, and obesity, *Food Technology* 51 (1997): 70–73; C. Bouchard, Human variation in body mass: Evidence for a role of the genes, *Nutrition Reviews* 55 (1997): S21–S30.
9. C. Bouchard and A. Tremblay, Genetic influences on the response of body fat and fat distribution to positive and negative energy balances in human identical twins, *Journal of Nutrition* 127 (1997): 943S–947S.
10. C. D. Berdanier, The candidate gene approach in obesity research, in C. D. Berdanier, *Nutrients and Gene Expression: Clinical Aspects* (Boca Raton, Fla.: CRC Press, 1996), pp. 39–49.
11. Y. Zhang and coauthors, Positional cloning of the mouse *obese* gene and its human homologue, *Nature* 372 (1994): 425–431.
12. H. Qian and coauthors, Brain administration of leptin causes deletion of adipocytes by apoptosis, *Endocrinology* 139 (1998): 791–794.
13. C. T. Montague and coauthors, Congenital leptin deficiency is associated with severe early-onset obesity in humans, *Nature* 387 (1997): 903–908.
14. J. M. Friedman, Leptin, leptin receptors, and the control of body weight, *Nutrition Reviews* 56 (1998): S38–S46; R. V. Considine, Serum immunoreactive-leptin concentrations in normal-weight and obese humans, *New England Journal of Medicine* 334 (1996): 292–295; S. G. Hassink and coauthors, Serum leptin in children with obesity: Relationship to gender and development, *Pediatrics* 98 (1996): 201–203.
15. E. Ravussin and coauthors, Relatively low plasma leptin concentrations precede weight gain in Pima Indians, *Nature Medicine* 3 (1997): 238–240.
16. Albu and coauthors, 1997.
17. M. Rosenbaum, R. L. Leibel, and J. Hirsch, Obesity, *New England Journal of Medicine* 337 (1997): 396–407; G. Wolf, Neuropeptides responding to leptin, *Nutrition Reviews* 55 (1997): 85–88.
18. C. Fleury and coauthors, Uncoupling protein-2: A novel gene linked to obesity and hyperinsulinemia, *Nature Genetics* 15 (1997): 269–272.
19. J. S. Flier and B. B. Lowell, Obesity research springs a proton leak, *Nature Genetics* 15 (1997): 223–224.
20. G. Wolf, A new uncoupling protein: A potential component of the human body weight regulation system, *Nutrition Reviews* 55 (1997): 178–179.
21. S. B. Heymsfield and coauthors, The calorie: Myth, measurement, and reality, *American Journal of Clinical Nutrition* 62 (1995): 1034S–1041S.
22. B. L. Heitmann and coauthors, Dietary fat intake and weight gain in women genetically predisposed for obesity, *American Journal of Clinical Nutrition* 61 (1995): 1213–1217.
23. J. E. Blundell and J. I. Macdiarmid, Fat as a risk factor for overconsumption: Satiation, sateity, and patterns of eating, *Journal of the American Dietetic Association* 97 (1997): S63–S69.
24. A. M. Prentice and S. D. Poppitt, Importance of energy density and macronutrients in the regulation of energy intake, *International Journal of Obesity and Related Metabolic Disorders* 20 (1996): S18–S23; C. Proserpi and coauthors, Ad libitum intake of a high-carbohydrate or high-fat diet in young men: Effects on nutrient balances, *American Journal of Clinical Nutrition* 66 (1997): 539–545; D. E. Larson and coauthors, Ad libitum food intake on a "cafeteria diet" in Native American women: Relations with body composition and 24-h energy expenditure, *American Journal*

*of Clinical Nutrition* 62 (1995): 911–917.

25. L. H. Nelson and L. A. Tucker, Diet composition related to body fat in a multivariate study of 203 men, *Journal of the American Dietetic Association* 96 (1996): 771–777.
26. L. A. Tucker, G. T. Seljaas, and R. L. Hager, Body fat percentage of children varies according to their diet composition, *Journal of the American Dietetic Association* 97 (1997): 981–986; V. T. Nguyen and coauthors, Fat intake and adiposity in children of lean and obese parents, *American Journal of Clinical Nutrition* 63 (1996): 507–513.
27. R. Rising and coauthors, Determinants of total daily energy expenditure: Variability in physical activity, *American Journal of Clinical Nutrition* 59 (1994): 800–804.
28. Prevalence of physical inactivity during leisure time among overweight persons—1994, *Morbidity and Mortality Weekly Report* 45 (1996): 183–188.
29. S. L. Gortmaker, W. H. Dietz, Jr., and L. W. Y. Cheung, Inactivity, diet, and the fattening of America, *Journal of the American Dietetic Association* 90 (1990): 1247–1255; E. Obarzanek and coauthors, Energy intake and physical activity in relation to indexes of body fat: The National Heart, Lung, and Blood Institute Growth and Health Study, *American Journal of Clinical Nutrition* 60 (1994): 15–22.
30. H. Oberrieder and coauthors, Attitude of dietetics students and registered dietitians toward obesity, *Journal of the American Dietetic Association* 95 (1995): 914–915.
31. J. S. Cassell, Social anthropology and nutrition: A different look at obesity in America, *Journal of the American Dietetic Association* 95 (1995): 424–427.
32. D. T. Felson, Weight and osteoarthritis, *American Journal of Clinical Nutrition* 63 (1996): 430S–432S.
33. G. Haus and coauthors, Key modifiable factors in weight maintenance: Fat intake, exercise, and weight cycling, *Journal of the American Dietetic Association* 94 (1994): 409–413.
34. R. W. Jeffrey, Does weight cycling present a health risk? *American Journal of Clinical Nutrition* 63 (1996): 452S–455S.
35. National Task Force on the Prevention and Treatment of Obesity, Weight cycling, *Journal of the American Medical Association* 272 (1994): 1196–1202.
36. Committee to Develop Criteria for Evaluating the Outcomes of Approaches to Prevent and Treat Obesity, Food and Nutrition Board, Institute of Medicine, National Academy of Sciences, *Journal of the American Dietetic Association* 95 (1995): 96–105.
37. National Task Force on the Prevention and Treatment of Obesity, Long-term pharmacotherapy in the management of obesity, *Journal of the American Medical Association* 276 (1996): 1907–1915.
38. R. L. Atkinson, Use of drugs in the treatment of obesity, *Annual Review of Nutrition* 17 (1997): 383–403.
39. L. Abenhaim and coauthors, Appetite-suppressant drugs and the risk of primary pulmonary hypertension, *New England Journal of Medicine* 335 (1996): 609–616.
40. H. M. Connolly and coauthors, Valvular heart disease associated with fenfluramine-phentermine, *New England Journal of Medicine* 337 (1997): 581–588.
41. Cardiac valvulopathy associated with exposure to fenfluramine or dexfenfluramine: U.S. Department of Health and Human Services Interim Public Health Recommendations, November 1997, *Journal of the American Medical Association* 278 (1997): 1729–1731.
42. K. Fackelmann, Diet drug debacle: How two federally approved weight-loss drugs crashed, *Science News* 152 (1997): 252–253.
43. D. E. Schteingart, Phenylpropanolamine in the management of moderate obesity, in *Obesity: New Directions in Assessment and Management,* eds. T. B. VanItallie and A. P. Simopoulos (Philadelphia: Charles Press, 1995), pp. 220–226.
44. P. Kurtzweil, Dieter's brews make tea time a dangerous affair, *FDA Consumer,* July/August 1997, pp. 6–11.
45. R. P. Abernathy and D. R. Black, Healthy body weights: An alternative perspective, *American Journal of Clinical Nutrition* 63 (1996): 448S–451S.
46. J. G. Meisler and S. St. Jeor, Summary and recommendations from the American Health Foundation's Expert Panel on Healthy Weight, *American Journal of Clinical Nutrition* 63 (1996): 474S–477S.
47. G. D. Foster and coauthors, What is a reasonable weight loss? Patients' expectations and evaluations of obesity treatment outcomes, *Journal of Consulting and Clinical Psychology* 65 (1997): 79–85.
48. J. Polivy, Psychological consequences of food restriction, *Journal of the American Dietetic Association* 96 (1996): 589–592.
49. A. Astrup and coauthors, The role of low-fat diets and fat substitutes in body weight management: What have we learned from clinical studies? *Journal of the American Dietetic Association* 97 (1997): S82–S87; M. Shah and coauthors, Comparison of a low-fat ad libitum complex-carbohydrate diet with a low-energy diet in moderately obese women, *American Journal of Clinical Nutrition* 59 (1994): 980–984.
50. L. R. Roust, K. D. Hammel, and M. D. Jensen, Effects of isoenergetic, low-fat diets on energy metabolism in lean and obese women, *American Journal of Clinical Nutrition* 60 (1994): 470–475.
51. H. E. Carmichael and coauthors, Lower fat intake as a predictor of initial and sustained weight loss in obese subjects consuming an otherwise ad libitum diet, *Journal of the American Dietetic Association* 98 (1998): 35–39.
52. S. Toubro and A. Astrup, *Ad libitum* low-fat, high-carbohydrate diet *versus* calorie-counting for weight maintenance after major weight loss in obese patients, *British Medical Journal* 314 (1997): 29–34.
53. K. Z. Walker and coauthors, Body fat distribution and non-insulin-dependent diabetes: Comparison of a fiber-rich, high-carbohydrate, low-fat (23%) diet and a 35% fat diet high in monounsaturated fat, *American Journal of Clinical Nutrition* 63 (1996): 254–260.
54. S. N. Blair, Diet and activity: The synergistic merger, *Nutrition Today* 30 (1996): 108–112; J. H. Wilmore, Increasing physical activity: Alterations in body mass and composition, *American Journal of Clinical Nutrition* 63 (1996): 456S–460S.
55. A. Geliebter and coauthors, Effects of strength or aerobic training on body composition, resting metabolic rate, and peak oxygen consumption in obese dieting subjects, *American Journal of Clinical Nutrition* 66 (1997): 557–563; B. L. Marks and coauthors, Fat-free mass is maintained in women following a moderate diet and exercise program, *Medicine and Science in Sports and Exercise* 27 (1995): 1243–1251; K. P. G. Kempen, W. H. M. Saris, and K. R. Westerterp, Energy balance during an 8-wk energy-restricted diet with and without exercise in obese women, *American Journal of Clinical Nutrition* 62 (1995): 722–729; R. Ross,

H. Pedwell, and J. Rissanen, Effects of energy restriction and exercise on skeletal muscle and adipose tissue in women as measured by magnetic resonance imaging, *American Journal of Clinical Nutrition* 61 (1995): 1179–1185; S. B. Racette and coauthors, Effects of aerobic exercise and dietary carbohydrate on energy expenditure and body composition during weight reduction in obese women, *American Journal of Clinical Nutrition* 61 (1995): 486–494; D. D. Hensrud and coauthors, A prospective study of weight maintenance in obese subjects reduced to normal body weight without weight-loss training, *American Journal of Clinical Nutrition* 60 (1994): 688–694.

56. S. B. Racette and coauthors, Exercise enhances dietary compliance during moderate energy restriction in obese women, *American Journal of Clinical Nutrition* 62 (1995): 345–349; M. L. Skender and coauthors, Comparison of 2-year weight loss trends in behavioral treatments of obesity: Diet, exercise, and combination interventions, *Journal of the American Dietetic Association* 96 (1996): 342–346.
57. P. D. Wood, Clinical applications of diet and physical activity in weight loss, *Nutrition Reviews* 54 (1996): S131–S135; M. Grediagin and coauthors, Exercise intensity does not effect body composition change in untrained, moderately overfat women, *Journal of the American Dietetic Association* 95 (1995): 661–665.
58. G. D. Foster and coauthors, The energy cost of walking before and after significant weight loss, *Medicine and Science in Sports and Exercise* 27 (1995): 888–894.
59. A. M. Sjödin and coauthors, The influence of physical activity on BMR, *Medicine and Science in Sports and Exercise* 28 (1996): 85–91.
60. N. A. King, A. Tremblay, and J. E. Blundell, Effects of exercise on appetite control: Implications for energy balance, *Medicine and Science in Sports and Exercise* 29 (1997): 1076–1089.
61. B. C. Nindl and coauthors, Regional fat placement in physically fit males and changes with weight loss, *Medicine and Science in Sports and Exercise* 28 (1996): 786–793.
62. Albu and coauthors, 1997.
63. Astrup and coauthors, 1997.
64. M. W. Green and P. J. Rogers, Impaired cognitive functioning during spontaneous dieting, *Psychological Medicine* 25 (1995): 1003–1010.
65. S. M. Shick and coauthors, Persons successful at long-term weight loss and maintenance continue to consume a low-energy, low-fat diet, *Journal of the American Dietetic Association* 98 (1998): 408–413; M. L. Klem and coauthors, A descriptive study of individuals successful at long-term maintenance of substantial weight loss, *American Journal of Clinical Nutrition* 66 (1997): 239–246; Albu and coauthors, 1997.
66. D. A. Schoeller, K. Shay, and R. F. Kushner, How much physical activity is needed to minimize weight gain in previously obese women? *American Journal of Clinical Nutrition* 66 (1997): 551–556.
67. C. Bouchard, Can obesity be prevented? *Nutrition Reviews* 54 (1996): S125–S130.
68. National Institutes of Health Obesity Education Initiative, *Clinical Guidelines on the Identification, Evaluation, and Treatment of Overweight and Obesity in Adults* (Washington, D.C.: U.S. Department of Health and Human Services, 1998).
69. J. Leddy and coauthors, Effect of a high or a low fat diet on cardiovascular risk factors in male and female runners, *Medicine and Science in Sports and Exercise* 29 (1997): 17–25.

HIGHLIGHT 9

# Eating Disorders

For some people, dieting to lose weight progresses to a dangerous and obsessive point. An estimated 2 million people in the United States, primarily girls and young women, suffer from the eating disorders anorexia nervosa and bulimia nervosa (the glossary on p. 282 defines these and related terms). Many more suffer from unspecified eating disorders—conditions that do not meet the strict criteria for anorexia nervosa or bulimia nervosa, but still imperil a person's well-being.

Why do so many people in our society suffer from eating disorders? Excessive pressure to be thin is at least partly to blame. By making thinness the ideal, society pushes people to view the healthy body of normal weight as too fat. Healthy people then adopt unhealthy eating behaviors to battle this imaginary problem. Societal pressure to be thin is no doubt a factor in the development of eating disorders, but most experts agree that the causes are multifactorial: sociocultural, psychological, and perhaps neurochemical. Athletes are particularly likely to develop eating disorders.

## *The Female Athlete Triad*

At age 14, Suzanne was a top contender for a spot on the state gymnastics team. Each day her coach reminded team members that they must weigh no more than their assigned weights in order to qualify for competition. The coach chastised gymnasts who gained weight, and Suzanne was terrified of being singled out. She was convinced that the less she weighed, the better she would perform. She weighed herself several times a day to confirm that she had not exceeded her 80-pound limit. Driven to excel in her sport, Suzanne kept her weight down by eating very little and training very hard. Unlike many of her friends at school, Suzanne never began menstruating. A few months before her fifteenth birthday, Suzanne's coach dropped her back to the second-level team. Suzanne blamed her poor performance on a slow-healing stress fracture. Mentally stressed and physically exhausted, she quit gymnastics and began overeating between her periods of self-starvation. Suzanne had developed the dangerous combination of problems—disordered eating, amenorrhea, and osteoporosis—collectively known as the female athlete triad (see Figure H9-1).[1]

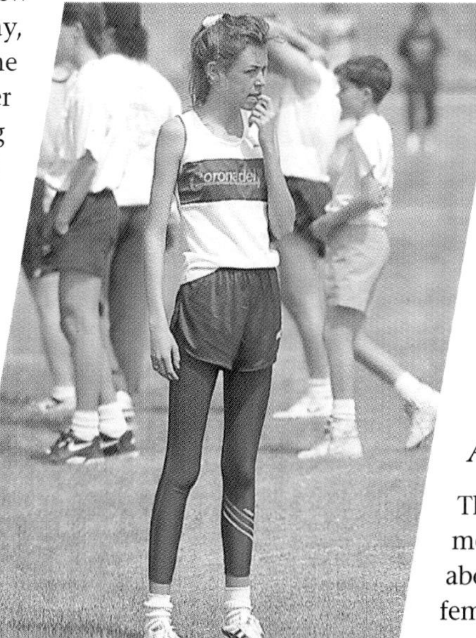

*People with anorexia nervosa see themselves as fat, even when they are dangerously underweight.*

### *Disordered Eating*

At least part of the reason many athletes engage in self-destructive eating behaviors is that they and their coaches have adopted unsuitable weight standards. An athlete's body must be heavier for height than a nonathlete's body because the athlete's bones and muscles are denser. When athletes consult standard weight-for-height tables and see that they are on the heavy side, they may mistakenly believe that they are too fat. Weight standards that may be appropriate for others are inappropriate for athletes. Measures such as fatfold measures yield more useful information about body composition.

Many young athletes severely restrict energy intakes to improve performance, enhance the aesthetic appeal of their performance, or meet the weight guidelines of their specific sports.[2] They fail to realize that the loss of lean tissue that accompanies energy restriction actually impairs their physical performance. The increasing incidence of abnormal eating habits among athletes, especially young women, is causing concern. Male athletes, especially wrestlers and gymnasts, are affected by these disorders as well, but research shows that females are most vulnerable.[3] Risk factors for eating disorders among athletes include the following:

**Figure H9-1**
**The Female Athlete Triad**

Eating Disorder
- Restrictive dieting (inadequate energy and nutrient intake)
- Overexercising
- Weight loss
- Lack of body fat

Osteoporosis
- Loss of calcium from bones

Amenorrhea
- Diminished hormones

- Young age (adolescence).
- Pressure to excel at a chosen sport.
- Focus on achieving or maintaining an "ideal" body weight or body fat percentage.
- Participation in endurance sports or competitions that judge performance on aesthetic appeal such as gymnastics, figure skating, or dance.
- Dieting at an early age.
- Unsupervised dieting.

### *Amenorrhea*

The prevalence of amenorrhea among premenopausal women in the United States is about 2 to 5 percent overall, but among female athletes, it may be as high as 66 percent.[4] Contrary to previous notions, amenorrhea is *not* a normal adaptation to strenuous physical training: it is a symptom of something going wrong.[5] Amenorrhea is characterized by low blood estrogen, infertility, and often bone mineral losses. Some research seems to indicate that depleted body fat contributes to amenorrhea; other studies indicate that the percentage of body fat is not associated with normal menstruation in athletes. However amenorrhea develops, amenorrheic athletes are more likely to suffer bone loss than other women. These losses remain significant even after menstruation resumes.[6]

## Osteoporosis

In general, weight-bearing physical activity, dietary calcium, and the hormone estrogen protect against the bone loss of osteoporosis, but in women with disordered eating and amenorrhea, strenuous activity may impair bone health. Vigorous training combined with low food energy intakes and other life stresses seems to trigger amenorrhea and promote bone loss. Low estrogen leads to diminished bone mass and increased bone fragility. Although athletes should have dense, strong bones, many amenorrheic athletes have decreased bone density, similar to that of older women with osteoporosis.[7] Amenorrheic athletes should be encouraged to consume at least 1500 milligrams of calcium each day, to eat nutrient-dense foods, and to obtain enough energy to support weight gain and cover the energy expended in physical activity. Future research will focus on the benefits of hormone replacement therapy for these women.

## Other Dangerous Practices of Athletes

Only female athletes experience the triad of disordered eating, amenorrhea, and osteoporosis, of course, but many male athletes practice dubious dietary and exercise habits as well. Each week throughout the season, 20 to 30 percent of competitive wrestlers drastically restrict their food and fluid intake before a match in an effort to "make weight." In at least three cases in 1997, the consequences were deadly.[8] Wrestlers and their coaches believe that competing in a lower weight class will give them a competitive advantage over smaller opponents. To that end, the wrestlers practice in rubber suits, sit in steam rooms, and take diuretics to lose 4 to 6 pounds. Many of them hope to replenish the lost fluids, glycogen, and lean tissue during the hours between their weigh-in and competition, but the body needs more time to correct this metabolic mayhem. Reestablishing fluid and electrolyte balances may take a day or two, replenishing glycogen stores may take two to three days, and replacing lean tissue may take even longer.

Ironically, the combination of food deprivation and dehydration impairs physical performance by reducing muscle strength, decreasing anaerobic power, reducing endurance capacity, and lowering oxygen consumption.[9] For optimal performance, wrestlers need to first achieve their competitive weight during the off-season and then eat well-balanced meals and drink plenty of fluids during the competitive season.[10]

Some athletes go to extreme measures to *gain* weight. People afflicted with muscle dysmorphia eat high-protein diets, take dietary supplements, weight train for hours at a time, and often abuse steroids in an attempt to bulk up. Their bodies are large and muscular, yet they see themselves as 90-pound weaklings. Like others with distorted body images, people with muscle dysmorphia weigh themselves frequently and center their lives on diet and exercise.

## Preventing Eating Disorders in Athletes

To prevent eating disorders in athletes and dancers, both the performers and their coaches must be educated about inappropriate body weight ideals, improper weight-loss techniques, eating disorder development, proper nutrition, and safe weight-control methods. Coaches and dance instructors should never encourage unhealthy weight loss to qualify for competition or to conform with distorted artistic ideals. Athletes who

## Glossary

**amenorrhea** (ay-MEN-oh-REE-ah): the absence of or cessation of menstruation. **Primary amenorrhea** is menarche delayed beyond 16 years of age. **Secondary amenorrhea** is the absence of three to six consecutive menstrual cycles.

**anorexia** (an-oh-RECK-see-ah) **nervosa:** an eating disorder characterized by a refusal to maintain a minimally normal body weight and a distortion in perception of body shape and weight; most commonly seen in teenage girls and young women.

- **an** = without
- **orex** = mouth
- **nervos** = of nervous origin

**bulimia** (byoo-LEEM-ee-ah) **nervosa:** an eating disorder characterized by repeated episodes of binge eating usually followed by self-induced vomiting, misuse of laxatives or diuretics, fasting, or excessive exercise.

- **buli** = ox

**cathartic** (ka-THAR-tik): a strong laxative.

**eating disorder:** a disturbance in eating behavior that jeopardizes a person's physical or psychological health.

**emetic** (em-ETT-ic): an agent that causes vomiting.

**female athlete triad:** a potentially fatal triad of medical problems: disordered eating, amenorrhea, and osteoporosis.

**muscle dysmorphia** (dis-MORE-fee-ah): a newly coined psychiatric disorder characterized by a preoccupation with building body mass.

**stress fractures:** bone damage or breaks caused by stress on bone surfaces during exercise.

**unspecified eating disorder:** eating disorders that do not meet the defined criteria for specific eating disorders.

truly need to lose weight should try to do so during the off-season and under the supervision of a health care professional.[11] Frequent weighings can push young people who are striving to lose weight into a cycle of starving to confront the scale, then bingeing uncontrollably afterward. The erosion of self-esteem that accompanies these events can interfere with the normal psychological development of the teen years and set the stage for serious problems later on.

Table H9-1 includes some suggestions to help athletes and dancers protect themselves against developing eating disorders. The remaining sections describe eating disorders that anyone, athlete or nonathlete, may experience.

## Anorexia Nervosa

Julie is 18 years old and a superachiever in school. She watches her diet with great care, and she exercises daily, maintaining a heroic schedule of self-discipline. She is thin, but she is determined to lose more weight. She is 5 feet 6 inches tall and weighs 85 pounds. She has anorexia nervosa.

Julie is unaware that she is undernourished, and she sees no need to obtain treatment. She developed amenorrhea several months ago and has become moody and chronically depressed. She insists that she is too fat, although her eyes are sunk in deep hollows in her face. Julie denies that she is ever tired, although she is close to physical exhaustion and no longer sleeps easily. Her family is concerned, and though reluctant to push her, they have finally insisted that she see a psychiatrist. Julie's psychiatrist has diagnosed anorexia nervosa (see Table H9-2) and prescribed group therapy as a start, but warns that if she does not begin to gain weight soon, she will need to be hospitalized.

U.S. Government

www.healthfinder.gov/searchoptions/topicsaz.htm
Search for Anorexia

### Characteristics of Anorexia Nervosa

As mentioned in the introduction, most anorexia nervosa victims are females; males account for only about 1 in 20 cases. Underlying family conflicts often play an important role in the development of anorexia nervosa. Parents may oppose one another's authority and vacillate between defending and condemning the anorexic child's behavior. Parental control is often rigid and overprotective. In the extreme, parents may even be abusive. Julie is a perfectionist, and her parents expect perfection. She cannot get in touch with her own identity and rejects food as a means of gaining control.

How can a person as thin as Julie continue to starve herself? Julie uses tremendous discipline against her hunger to strictly limit her portions of low-kcalorie foods. She will deny her hunger, saying she is full after eating only a half-dozen carrot sticks. She can recite the kcalorie contents of dozens of foods and the kcalorie costs of as many exercises. If she feels that she has gained an ounce of weight, she runs or jumps rope until she is sure she has exercised it off. If she fears that the food she has eaten outweighs the exercise, she takes laxatives to hasten the passage of food from her system. She is desperately hungry. In fact, she is starving, but she doesn't eat because her need for self-control dominates.

**Table H9-1**
**Tips for Combating Eating Disorders**

**General Guidelines**

- Never restrict food servings to below the numbers suggested for adequacy by the Food Guide Pyramid.
- Eat frequently. People often do not eat frequent meals because of time constraints, but eating can be incorporated into other activities, such as snacking while studying or commuting. The person who eats frequently never gets so hungry as to allow hunger to dictate food choices.
- If not at a healthy weight, establish a reasonable weight goal based on a healthy body composition.
- Allow a reasonable time to achieve the goal. A reasonable loss of excess fat can be achieved at the rate of about 10 percent of body weight in six months.
- Establish a weight-maintenance support group with people who share interests.

**Specific Guidelines for Athletes and Dancers**

- Remember that eating disorders impair physical performance. Seek confidential help in obtaining treatment if needed.
- Restrict weight-loss activities to the off-season.
- Focus on proper nutrition as an important facet of your training, as important as proper technique.

**Table H9-2**
**Criteria for Diagnosis of Anorexia Nervosa**

A person with anorexia nervosa demonstrates the following:

A. Refusal to maintain body weight at or above a minimal normal weight for age and height (e.g., weight loss leading to maintenance of body weight less than 85 percent of that expected; or failure to make expected weight gain during period of growth, leading to body weight less than 85 percent of that expected).

B. Intense fear of gaining weight or becoming fat, even though underweight.

C. Disturbance in the way in which one's body weight or shape is experienced, undue influence of body weight or shape on self-evaluation, or denial of the seriousness of the current low body weight.

D. In females past puberty, amenorrhea, i.e., the absence of at least three consecutive menstrual cycles. (A woman is considered to have amenorrhea if her periods occur only following hormone, e.g., estrogen, administration.)

Two types:

*Restricting type:* During the episode of anorexia nervosa, the person does not regularly engage in binge eating or purging behavior (i.e., self-induced vomiting or the misuse of laxatives, diuretics, or enemas).

*Binge eating/purging type:* During the episode of anorexia nervosa, the person regularly engages in binge eating or purging behavior (i.e., self-induced vomiting or the misuse of laxatives, diuretics, or enemas).

Source: Reprinted with permission from the *Diagnostic and Statistical Manual of Mental Disorders,* 4th ed. (Washington, D.C.: American Psychiatric Association, 1994).

Many people, on learning of this disorder, say they wish they had "a touch" of it to get thin. They mistakenly think that people with anorexia nervosa feel no hunger. They also fail to recognize the pain of the associated psychological and physical trauma.

Central to the diagnosis of anorexia nervosa is a distorted body image that overestimates personal body fatness. When Julie looks at herself in the mirror, she sees a "fat" 85-pound body. The more Julie overestimates her body size, the more resistant she is to treatment, and the more unwilling to examine her faulty values and misconceptions. Vitamin deficiencies are known to affect brain functioning and judgment in this way, causing lethargy, confusion, and delirium.

*People with anorexia nervosa may reluctantly eat small amounts of low-kcalorie foods.*

Anorexia nervosa cannot be self-diagnosed. Nearly everyone in our society is engaged in the pursuit of thinness, and denial runs high among people with anorexia nervosa. Some women have all the attitudes and behaviors associated with the condition, but without the dramatic weight loss.

Anorexia nervosa damages the body much as starvation does. In fact, after a few months, most people with anorexia nervosa have protein-energy malnutrition (PEM) that is similar to marasmus.[12] Victims are dying to be thin—quite literally. In young people, growth ceases and normal development falters. They lose so much lean tissue that basal metabolic rate slows, an effect that may remain even after treatment and regain of weight.[13] In addition, the heart pumps inefficiently and irregularly, the heart muscle becomes weak and thin, the chambers diminish in size, and the blood pressure falls. Electrolytes that help to regulate heartbeat become unbalanced. Many deaths occur due to multiple organ system failure.

Starvation brings other physical consequences as well: impaired immune response, anemia, and a loss of digestive functions that worsens malnutrition. Peristalsis becomes sluggish, the stomach empties slowly, and the lining of the intestinal tract atrophies. The deteriorated GI tract fails to provide sufficient absorptive surfaces and digestive enzymes for handling any food the victim may eat. The pancreas slows its production of digestive enzymes. The person may suffer from diarrhea, further worsening malnutrition.

Other effects of starvation include altered blood lipids, high blood vitamin A and vitamin E, low blood proteins, dry thin skin, abnormal nerve functioning, reduced bone density, low body temperature, low blood pressure, and the development of fine body hair (the body's attempt to keep warm). The electrical activity of the brain becomes abnormal, and insomnia is common. Both women and men lose their sex drives.

Women with anorexia nervosa develop amenorrhea (it is one of the diagnostic criteria). Sometimes that symptom precedes the weight loss. Anorexia nervosa delays the onset of menstruation in young girls. Menstrual periods typically resume with recovery, although some women never restart even after they have gained weight. Should an underweight woman with anorexia nervosa become pregnant, she is likely to give birth to an underweight baby—and low-birthweight babies face many health problems (as Chapter 15 explains).

## *Treatment in Anorexia Nervosa*

Table H9-3 lists principles of nutrition intervention in anorexia nervosa. Treatment of anorexia nervosa requires a multidisciplinary approach. Teams of physicians, nurses, psychiatrists, family therapists, and dietitians work together to resolve two sets of issues and behaviors: those relating to food and weight, and those involving relationships with oneself and others.[14] The first dietary objective is to stop weight loss while establishing regular eating patterns. Appropriate diet is crucial to recovery and must be tailored individually to each client's needs. Because body weight is low and fear of weight gain is high, initial food intake may be small. As eating becomes more comfortable, energy intake should increase gradually. Physicians and clients need to realize that weight gain may be difficult, especially during the first week of treatment, perhaps because both the

**Table H9-3**
**Principles of Nutrition Intervention in Anorexia Nervosa**

- Include foods from each of the food groups, with portion sizes increasing as energy intake increases.
- Increase food energy intake gradually (adding 200 kcalories/week).
- Prescribe well-balanced diets, with *some* individual variations according to client preferences (e.g., vegetarian).
- Give multiple vitamin-mineral supplements to restore nutrient losses.
- Enhance elimination with dietary fiber from grain sources.
- Reduce sensations of bloating with small, frequent feedings.
- In behavioral programs, link rewards to food energy intake, not to weight gain.
- Use liquid supplements when the client cannot achieve desired intake with solid food.
- Reduce satiety sensations by offering cold or room-temperature foods and finger foods (e.g., snacks).
- Provide interactive nutrition counseling as an ongoing process.
- Reduce excessive caffeine intake.
- Provide IV nutritional support only in severe states of ill health, malnutrition, and wasting.

Source: Adapted from C. L. Rock and J. Yager, Nutrition and eating disorders: A primer for clinicians, *International Journal of Eating Disorders* 6 (1987): 276, as cited in *Nutrition and the M.D.*, July 1988, with permission. Reprinted with permission of John Wiley & Sons, Inc., copyright 1987.

resting metabolic rate and the thermic effect of food in people with anorexia nervosa are so high.[15] Seldom are clients willing to eat for themselves, but if they are, chances are they can recover without other interventions.

Because anorexia nervosa is like starvation physically, health care professionals classify clients based on indicators of PEM.* Low-risk clients need nutrition counseling. Intermediate-risk clients may need supplements such as high-kcalorie, high-protein formulas in addition to regular meals, but they may not have to be hospitalized. High-risk clients may require hospitalization and may need to be force-fed by tube at first to prevent death. This step causes psychological trauma. Drugs are commonly prescribed, but play a limited role in treatment.

Denial runs high among those with anorexia nervosa. Few seek treatment on their own. Almost half of the women who are treated can maintain their body weight within 15 percent of a healthy weight; at that weight, many of them begin menstruating again. The other half have poor or fair treatment outcomes; two-thirds of those treated continue a mental battle with recurring morbid thoughts about food and body weight.[16] Many relapse into abnormal eating behaviors to some extent.[17] Over 10 percent of those treated for anorexia nervosa die—most commonly from starvation, electrolyte imbalance, or suicide.

Before drawing conclusions about someone who is extremely thin or who eats very little, remember that diagnosis requires professional assessment. Several national organizations offer information for people who are seeking help with anorexia nervosa, either for themselves or for others.†

 Anorexia Nervosa and Related Eating Disorders

www.anred.com

## Bulimia Nervosa

Kelly is a charming, intelligent, 20-year-old flight attendant of normal weight who thinks constantly about food. She alternatively starves herself and secretly binges; when she has eaten too much, she makes herself vomit. Most readers recognize these symptoms as those of bulimia nervosa.

Bulimia nervosa is distinct from anorexia nervosa and is more prevalent, although the true incidence is difficult to establish because bulimia nervosa is not as physically apparent. More men suffer from bulimia nervosa than from anorexia nervosa, but bulimia nervosa is still more common in women than in men.[18] The secretive nature of bulimic behaviors makes recognition of the problem difficult, but once it is recognized, diagnosis is based on the criteria listed in Table H9-4.

*Indicators of protein-energy malnutrition: a low percentage of body fat, low serum albumin, low serum transferrin, and impaired immune reactions.

†Internet sites, phone numbers, and addresses are in Appendix F.

Parents of bulimic people have often suffered from depression and alcohol abuse. While their expectations for their child's achievements are high, parental involvement is minimal.[19] Any changes in the family structure often meet with resistance, even when the changes would greatly benefit the person with bulimia nervosa. In addition, the family may have secrets that are hidden from outsiders; many people with bulimia nervosa report having been abused sexually or physically by family members or family friends.

 U.S. Government

www.healthfinder.gov/searchoptions/topicsaz.htm

Search for Bulimia

### Characteristics of Bulimia Nervosa

Like the typical person with bulimia nervosa, Kelly is single, female, and white. She is well educated and close to her ideal body weight, although her weight fluctuates over a range of 10 pounds or so every few weeks. As a flight attendant, she prefers to weigh less than the weight that her body maintains naturally.

Kelly seldom lets her eating disorder interfere with work or other activities, although a third of all victims do. From early childhood she has been a high achiever and emotionally dependent on her parents. As a young teen, Kelly cycled on and off crash diets but could never maintain an appropriate weight. Kelly feels anxious at social events and

**Table H9-4**

**Criteria for Diagnosis of Bulimia Nervosa**

A person with bulimia nervosa demonstrates the following:

A. Recurrent episodes of binge eating. An episode of binge eating is characterized by both of the following:
   1. Eating, in a discrete period of time (e.g., within any two-hour period), an amount of food that is definitely larger than most people would eat during a similar period of time and under similar circumstances.
   2. A sense of lack of control over eating during the episode (e.g., a feeling that one cannot stop eating or control what or how much one is eating).

B. Recurrent inappropriate compensatory behavior in order to prevent weight gain, such as self-induced vomiting; misuse of laxatives, diuretics, enemas, or other medications; fasting; or excessive exercise.

C. Binge eating and inappropriate compensatory behaviors both occur, on average, at least twice a week for three months.

D. Self-evaluation unduly influenced by body shape and weight.

E. The disturbance does not occur exclusively during episodes of anorexia nervosa.

Two types:

*Purging type:* The person regularly engages in self-induced vomiting or the misuse of laxatives, diuretics, or enemas.

*Nonpurging type:* The person uses other inappropriate compensatory behaviors, such as fasting or excessive exercise, but does not regularly engage in self-induced vomiting or the misuse of laxatives, diuretics, or enemas.

Source: Reprinted with permission from *Diagnostic and Statistical Manual of Mental Disorders,* 4th ed. (Washington, D.C.: American Psychiatric Association, 1994).

cannot easily establish close personal relationships. She is usually depressed, is often impulsive, and has low self-esteem. Some people with bulimia nervosa abuse drugs, steal compulsively (kleptomania), or are sexually promiscuous.

Like the person with anorexia nervosa, the person with bulimia nervosa spends much time thinking about her body weight and food. Her preoccupation with food manifests itself in secretive binge-eating episodes, which usually progress through several emotional stages: anticipation and planning, anxiety, urgency to begin, rapid and uncontrollable consumption of food, relief and relaxation, disappointment, and finally shame or disgust.

*Bulimic binges are often followed by self-induced vomiting and feelings of shame or disgust.*

A bulimic binge is unlike normal eating. Food is consumed for its emotional comfort, not its nutritional value. Eating is not primarily a response to hunger; it is a compulsion. A typical binge occurs periodically, in secret, usually at night, and lasts an hour or more. Because a binge frequently follows a period of rigid dieting, eating is accelerated by intense hunger. Energy restriction followed by bingeing can set in motion a pattern of weight cycling, which may make weight loss and maintenance more difficult over time.[20]

During a binge, Kelly consumes thousands of kcalories of easy-to-eat, low-fiber, high-fat, and, especially, high-carbohydrate foods. Typically, she chooses cookies, cakes, and ice cream—and she eats the entire bag of cookies, the whole cake, and every last spoonful in a carton of ice cream. After the binge, Kelly pays the price with swollen hands and feet, bloating, fatigue, headache, nausea, and pain.

To purge the food from her body, Kelly may use a cathartic—a strong laxative that can injure the lower intestinal tract. Or she may induce vomiting, using an emetic—a drug intended as first aid for poisoning. These purging behaviors are often accompanied by feelings of shame or guilt. Hence a vicious cycle develops: negative self-perceptions followed by dieting, bingeing, and purging, which in turn lead to negative self-perceptions (see Figure H9-2).[21]

**Figure H9-2**

**The Vicious Cycle of Restrictive Dieting and Binge Eating**

Negative self-perceptions → Dieting → Binge eating → Purging → Negative self-perceptions

Bulimic behaviors often begin in late adolescence or early adulthood after a long series of various unsuccessful weight-reduction diets. People with bulimia nervosa commonly follow a pattern of restrictive dieting interspersed with binge-eating and purging behaviors and experience weight fluctuations of more than 10 pounds up and down over short periods of time.

On first glance, purging seems to offer a quick and easy solution to the problems of unwanted kcalories and body weight. Many people perceive such behavior as neutral or even positive, when, in fact, binge eating and purging have serious physical consequences. Signs of subclinical malnutrition are evident in a compromised immune system.[22] Fluid and electrolyte imbalances caused by vomiting or diarrhea can lead to abnormal heart rhythms and injury to the kidneys, which have to cope with the altered balance. Urinary tract infections can lead to kidney failure. Vomiting causes irritation and infection of the pharynx, esophagus, and salivary glands; erosion of the teeth; and dental caries. The esophagus may rupture or tear, as may the stomach. Sometimes the eyes become red from pressure during vomiting. The hands may be calloused or cut by the teeth while inducing vomiting.[23] Overuse of emetics depletes potassium concentrations and can lead to death by heart failure.[24]

Unlike Julie, Kelly is aware that her behavior is abnormal, and she is deeply ashamed of it. She wants to recover, and this makes recovery more likely for her than for Julie, who clings to denial. Feeling inadequate ("I can't even control my eating"), Kelly tends to be passive and to look to others, primarily men, for confirmation of her sense of worth. When she experiences rejection, either in reality or in her imagination, her bulimia nervosa becomes worse. If Kelly's depression deepens, she may seek solace in drug or alcohol abuse or other addictive behaviors. Clinical depression is common in people with bulimia nervosa, and the rates of alcohol, marijuana, and cigarette abuse are high.[25]

## *Treatment in Bulimia Nervosa*

To gain control over food and establish regular eating patterns, Kelly needs to adhere to a structured eating plan. Restrictive weight-loss dieting almost always precedes and may even trigger a binge. Weight maintenance, rather than cyclic weight gains and losses, is the treatment goal. Many a victim has taken a major step toward recovery by learning to eat enough food to satisfy hunger needs (at least 1600 kcalories a day). Table H9-5 offers diet strategies to correct the eating problems of bulimia nervosa.

A mental health professional should be on the treatment team to help clients with their depression and addictive behaviors. Some physicians prescribe the antidepressant drug fluoxetine in the treatment of bulimia nervosa.* Another drug that may be useful in the management of bulimia nervosa is naloxone, an opiate antagonist that

*Fluoxetine is marketed under the trade name Prozac.

**Table H9-5**
**Diet Strategies for Combatting Bulimia Nervosa**

- Avoid finger foods; eat foods that require the use of utensils.
- Enhance satiety by eating warm foods.
- Include vegetables, salad, and/or fruit at meals to prolong eating time.
- Choose whole-grain and high-fiber breads and cereals to maximize bulk.
- Eat a well-balanced diet and meals consisting of a variety of foods.
- Use foods that are naturally divided into portions, such as potatoes (rather than rice or pasta); 4- and 8-ounce containers of yogurt or cottage cheese; precut steak or chicken parts; and frozen entreés.
- Include foods containing ample complex carbohydrates (for satiety) and some fat (to slow gastric emptying).
- Eat meals and snacks sitting down.
- Plan meals and snacks, and record plans in a food diary prior to eating.

Source: Adapted from C. L. Rock and J. Yager, Nutrition and eating disorders: A primer for clinicians, *International Journal of Eating Disorders* 6 (1987): 276, as cited in *Nutrition and the M.D.,* July 1988, with permission. Reprinted with permission of John Wiley & Sons, Inc., copyright 1987.

suppresses the consumption of sweet and high-fat foods in binge-eaters.[26]

Anorexia nervosa and bulimia nervosa are distinct eating disorders, yet they sometimes overlap in important ways.[27] Anorexia victims may purge, and victims of both disorders share an overconcern with body weight and the tendency to drastically undereat. Many perceive foods as "forbidden" and "give in" to an eating binge. The two disorders can also appear in the same person, or one can lead to the other.

 American Anorexia/Bulimia Association
member.aol.com/amanbu

## *Binge-Eating Disorder*

Cheryl is a 40-year-old schoolteacher who has been overweight all her life. Her friends and family are forever encouraging her to lose weight, and she has come to believe that if she only had more willpower, dieting would work. She periodically gives dieting her best shot—restricting energy intake for a day or two only to succumb to uncontrollable cravings, especially for high-fat foods. Like Cheryl, up to half of the obese people who try to lose weight periodically binge; unlike people with bulimia nervosa, however, they typically do not purge. Such an eating disorder does not meet the criteria for either anorexia nervosa or bulimia nervosa—yet such compulsive overeating is a problem and occurs in people of normal weight as well as those who are severely overweight.[28] Table H9-6 lists criteria for unspecified eating disorders, including binge eating. Obesity alone is not an eating disorder.

Clinicians note differences between people with bulimia nervosa and those with binge-eating disorder. People with binge-eating disorder consume less during a binge, rarely purge, and exert less restraint during times of dieting. Similarities also exist, including feeling out of control, disgusted, depressed, embarrassed, guilty, or distressed because of their self-perceived gluttony.[29]

There are also differences between obese binge-eaters and obese people who do not binge. Those with the binge-eating disorder report higher rates of self-loathing, disgust about body size, depression, and anxiety.[30] Their eating habits differ as well. Obese binge-eaters tend to consume more kcalories and more dessert and snack-type foods

**Table H9-6**
**Unspecified Eating Disorders, including Binge-Eating Disorder**

**Criteria for Diagnosis of Unspecified Eating Disorders, in General**

Many people have eating disorders but do not meet all the criteria to be classified as having anorexia nervosa or bulimia nervosa. Some examples include those who:

A. Meet all of the criteria for anorexia nervosa, except irregular menses.
B. Meet all of the criteria for anorexia nervosa, except that their current weights fall within the normal ranges.
C. Meet all of the criteria for bulimia nervosa, except that binges occur less frequently than stated in the criteria.
D. Are of normal body weight and who compensate inappropriately for eating small amounts of food (example: self-induced vomiting after eating two cookies).
E. Repeatedly chew food, but spit it out without swallowing.
F. Have recurrent episodes of binge eating but who do not compensate as do those with bulimia nervosa.

**Criteria for Diagnosis of Binge-Eating Disorder, Specifically**

A person with a binge-eating disorder demonstrates the following:

A. Recurrent episodes of binge eating. An episode of binge eating is characterized by both of the following:
   1. Eating, in a discrete period of time (e.g., within any two-hour period) an amount of food that is definitely larger than most people would eat in a similar period of time under similar circumstances.
   2. A sense of lack of control over eating during the episode (e.g., a feeling that one cannot stop eating or control what or how much one is eating).
B. Binge-eating episodes are associated with at least three of the following:
   1. Eating much more rapidly than normal.
   2. Eating until feeling uncomfortably full.
   3. Eating large amounts of food when not feeling physically hungry.
   4. Eating alone because of being embarrassed by how much one is eating.
   5. Feeling disgusted with oneself, depressed, or very guilty after overeating.
C. The binge eating causes marked distress.
D. The binge eating occurs, on average, at least twice a week for six months.
E. The binge eating is not associated with the regular use of inappropriate compensatory behaviors (e.g., purging, fasting, excessive exercise) and does not occur exclusively during the course of anorexia nervosa or bulimia nervosa.

Source: Reprinted with permission from American Psychiatric Association, *Diagnostic and Statistical Manual of Mental Disorders,* 4th ed. (Washington, D.C.: American Psychiatric Association, 1994).

during regular meals and binges than obese people who do not binge.

Binge eating is a behavioral disorder that can be resolved with treatment—even a placebo.[31] Resolving such behavior may not bring weight loss, but it may make participation in weight-control programs easier. It also improves physical health, mental health, and the chances of success in breaking the cycle of rapid weight losses and gains.

## Eating Disorders in Society

Proof that society plays a role in eating disorders is found in their demographic distribution—they are known only in developed nations, and they become more prevalent as wealth increases and food becomes plentiful. Some people point to the vomitoriums of ancient times and claim that bulimia nervosa is not new, but the two are actually distinct. Ancient people were eating for pleasure, without guilt, and in the company of others; they vomited so that they could rejoin the feast. Bulimia nervosa is a disorder of isolation and is often accompanied by low self-esteem.

A food-centered society that favors thinness puts people in a bind. Families may encourage hearty eating and socializing around the dinner table. Party hosts take pride in the delicacies they serve, and guests are obliged to indulge. A child raised in such a setting and also encouraged to aspire to a thin ideal may see little alternative but to celebrate by indulging in food and then to vomit, crash diet, or starve to "undo" possible weight gain. Then, starving and guilty, but still reluctant to appear to be a glutton, the child may begin eating uncontrollably to relieve a desperate hunger.

There is no doubt that our society sets unrealistic ideals for body weight, especially in women, and devalues those who do not conform to them. Even professionals, including physicians and dietitians, are prone to praise people for losing weight and to suggest weight loss to people who do not need it for their health. As a result, even beautiful, normal-weight preteen girls are already worried that they are too fat. Two-thirds of adolescent girls and one-third of adolescent boys are dissatisfied with their body weight and shape.[32] Characteristics of disordered eating such as restrained eating, fasting, binge eating, purging, fear of fatness, and distortion of body image are extraordinarily common among young, white, middle- and upper-class girls. Most are "on diets," and many are poorly nourished. Some eat too little food to support normal growth; thus they miss out on their adolescent growth spurts and may never catch up. Many eat so little that hunger propels them into binge-purge cycles. Magazines, newspapers, and television all convey the message that to be thin is to be beautiful and happy. Anorexia nervosa and bulimia nervosa are not a form of rebellion against these unreasonable expectations, but rather an exaggerated acceptance of them.

Perhaps a person's best defense against these disorders is to learn to appreciate his or her own uniqueness. When people discover and honor the body's real needs, they become unwilling to sacrifice health for conformity. To respect and value oneself may be lifesaving.

 U.S. Government

www.healthfinder.gov/searchoptions/topicsaz.htm
Search for Eating disorders

## Notes


1. American College of Sports Medicine, Position stand: The Female Athlete Triad, *Medicine and Science in Sports and Exercise* 29 (1997): i–ix.
2. J. H. Wilson, Nutrition, physical activity and bone health in women, *Nutrition Research Reviews* 7 (1994): 67–91.
3. K. K. Yeager and coauthors, The female athlete triad: Disordered eating, amenorrhea, osteoporosis, *Medicine and Science in Sports and Exercise* 25 (1993): 775–777.
4. Yeager and coauthors, 1993.
5. C. L. Otis, American College of Sports Medicine's Ad Hoc Task Force on Women's Issues in Sports Medicine, as quoted in A. A. Skolnick, "Female athlete triad" risk for women, *Journal of the American Medical Association* 270 (1993): 921–923.
6. L. K. Micklesfield and coauthors, Bone mineral density in mature, premenopausal ultramarathon runners, *Medicine and Science in Sports and Exercise* 27 (1995): 688–696.
7. M. L. Rencken, C. H. Chesnut, and B. L. Drinkwater, Bone density at multiple skeletal sites in amenorrheic athletes, *Journal of the American Medical Association* 276 (1996): 238–240.
8. Hyperthermia and dehydration-related deaths associated with intentional rapid weight loss in three collegiate wrestlers—North Carolina, Wisconsin, and Michigan, November–December 1997, *Morbidity and Mortality Weekly Report* 47 (1998): 105–108.
9. American College of Sports Medicine, Position stand: Weight loss in wrestlers, *Medicine and Science in Sports and Exercise* 28 (1996): ix–xii.
10. E. Coleman, "Making weight" for wrestling: Six recommendations, *Sports Medicine Digest* 18 (1996): 13, 22–23.
11. Committee on Sports Medicine and Fitness, Promotion of healthy weight-control practices in young athletes, *Pediatrics* 97 (1996): 752–753.
12. P. Barbe and coauthors, Sex-hormone-binding globulin and protein-energy malnutrition indexes as indicators of nutritional status in women with anorexia nervosa, *American Journal of Clinical Nutrition* 57 (1993): 319–322.
13. L. Scalfi and coauthors, Bioimpedance analysis and resting energy expenditure in undernourished and refed anorectic patients, *European Journal of Clinical Nutrition* 47 (1993): 61–67.
14. Position of The American Dietetic Association: Nutrition intervention in the treatment of anorexia nervosa, bulimia nervosa, and binge eating, *Journal of the American Dietetic Association* 94 (1994): 902–907.
15. M. Moukaddem and coauthors, Increase in diet-induced thermogenesis at the start of refeeding in severely malnourished anorexia nervosa patients, *American Journal of Clinical Nutrition* 66 (1997): 133–140; M. V. Solanto and coauthors, Rate of weight gain of inpatients with anorexia nervosa

under two behavioral contracts, *Pediatrics* 93 (1994): 989–991; E. Obarzanek, M. D. Lesem, and D. C. Jimerson, Resting metabolic rate of anorexia nervosa patients during weight gain, *American Journal of Clinical Nutrition* 60 (1994): 666–675.

16. American Psychiatric Association Workgroup on Eating Disorders, Practice guidelines for eating disorders, I. Disease definition, epidemiology, and natural history, *American Journal of Psychiatry* 150 (1993): 212–228.
17. American Psychiatric Association Workgroup on Eating Disorders, 1993.
18. American Psychiatric Association, *DSM-IV* (Washington, D.C.: American Psychiatric Association, 1994); M. J. Devlin and B. T. Walsh, Anorexia nervosa and bulimia, in *Obesity,* eds. P. Björntorp and B. N. Brodoff (Philadelphia: J. B. Lippincott, 1992), pp. 436–444.
19. C. G. Fairburn and coauthors, Risk factors for bulimia nervosa: A community-based case-control study, *Archives of General Psychiatry* 54 (1997): 509–517.
20. D. Neumark-Sztainer, Excessive weight preoccupation, *Nutrition Today,* March/April 1995, pp. 68–74.
21. D. Newmark-Sztainer, R. Butler, and H. Palti, Dieting and binge eating: Which dieters are at risk? *Journal of the American Dietetic Association* 95 (1995): 586–589.
22. A. Marcos, Evaluation of immunocompetence and nutritional status in patients with bulimia nervosa, *American Journal of Clinical Nutrition* 57 (1993): 65–69.
23. A. Daluiski, B. Rahbar, and R. A. Meals, Russell's sign: Subtle hand changes in patients with bulimia nervosa, *Clinical Orthopaedics and Related Research* 343 (1997): 107–109.
24. J. E. Mitchell, Medical complications of bulimia nervosa, in *Eating Disorders and Obesity: A Comprehensive Handbook,* eds. K. D. Brownell and C. G. Fairburn (New York: Guilford Press, 1995), pp. 271–275.
25. J. A. Bushnell and coauthors, Bulimia comorbidity in the general population and in the clinic, *Psychological Medicine* 24 (1994): 605–611.
26. A. Drewnowski and coauthors, Naloxone, an opiate blocker, reduces the consumption of sweet high-fat foods in obese and lean female binge eaters, *American Journal of Clinical Nutrition* 61 (1995): 1206–1212.
27. C. G. Fairburn and G. T. Wilson, Binge eating: Definition and classification, in *Binge Eating: Nature, Assessment, and Treatment,* eds. C. G. Fairburn and G. T. Wilson (New York: Guilford Press, 1993), pp. 3–14.
28. B. Bruce and D. Wilfley, Binge eating among the overweight population: A serious and prevalent problem, *Journal of the American Dietetic Association* 96 (1996): 58–61.
29. J. P. Foreyt and G. K. Goodrick, Weight management without dieting, *Nutrition Today,* March/April 1993, pp. 4–9; R. L. Spitzer and coauthors, Binge eating disorder: A multisite field trial of the diagnostic criteria, *International Journal of Eating Disorders* 11 (1992): 191–203.
30. American Psychiatric Association, 1994.
31. A. Stunkard and coauthors, Binge eating disorder and the night-eating syndrome, *International Journal of Obesity* 20 (1996): 1–6.
32. D. C. Moore, Body image and eating behavior in adolescents, *Journal of the American College of Nutrition* 12 (1993): 505–510.

CHAPTER 10

# The Water-Soluble Vitamins: B Vitamins and Vitamin C

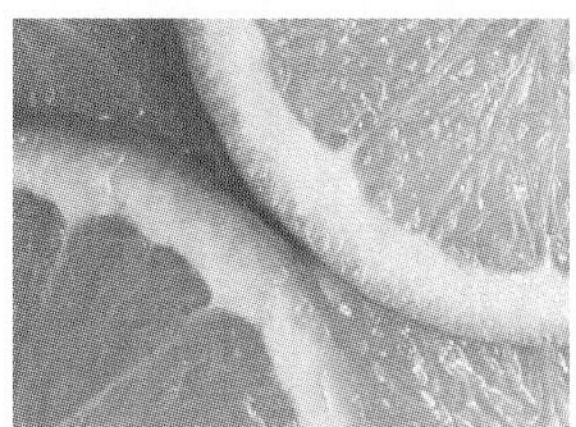

Earlier chapters focused on the energy-yielding nutrients, which play leading roles in the body. The vitamins and minerals are their supporting cast. This chapter begins with an overview of the vitamins and then examines each of the water-soluble vitamins; the next chapter features the fat-soluble vitamins.

# The Vitamins—An Overview

The vitamins are powerful substances, as their absence attests. Vitamin A deficiency can cause blindness; a lack of niacin can cause symptoms of mental confusion; and a lack of vitamin D can retard bone growth. The consequences of deficiencies are so dire, and the effects of restoring the needed vitamins so dramatic, that people spend billions of dollars every year in the belief that vitamin pills will cure a host of ailments (see Highlight 10). Vitamins certainly contribute to sound nutritional health, but they do not cure all ills. Furthermore, vitamin supplements do not offer the many benefits that come from vitamin-rich foods.

Reminder: The *vitamins* are organic, essential nutrients required in minute amounts to perform specific functions that promote growth, reproduction, or the maintenance of health and life.
- **vita** = life
- **amine** = containing nitrogen (the first vitamins discovered contained nitrogen)

The *presence* of the vitamins also attests to their power. Vitamin C not only prevents the deficiency disease scurvy, but also seems to protect against certain types of cancer. Similarly, vitamin E seems to help protect against some facets of cardiovascular disease. The B vitamin folate helps to prevent birth defects. As you will see, the vitamins' roles in supporting optimal health extend far beyond preventing deficiency diseases. In fact, some of the credit given to low-fat diets in preventing disease actually belongs to the vitamins that such diets deliver (see Highlight 11 for more on vitamins in disease prevention). A diet that includes plenty of vegetables, fruits, and grain products is low in fat *and* provides vitamins in abundance. Both attributes help to maintain health and slow the progression of disease. Chapter 18 highlights the roles of vitamins in supporting a strong immune system.

The vitamins differ from the carbohydrates, fats, and proteins in the following ways:

- *Structure.* Vitamins are individual units; they are not linked in long chains.
- *Function.* Vitamins do not yield usable energy when broken down; they assist the enzymes that release energy from carbohydrates, fats, and proteins.
- *Food contents.* The amounts of vitamins people ingest daily and the amounts they require are measured in *micrograms* (μg) or *milligrams* (mg), rather than grams (g).

1 g = 1000 mg.
1 mg = 1000 μg.
For perspective, a dollar bill weighs about 1 g; 28 g equal approximately 1 ounce.

The vitamins are similar to the energy-yielding nutrients, though, in that they are vital to life, organic, and available in foods.

• ***Bioavailability*** • The availability of vitamins from foods depends on two factors: the quantity provided by a food and the amount absorbed and used by the body (the vitamins' *bio*availability). Researchers analyze foods to determine their vitamin contents and publish the results in tables of food composition such as Appendix H. Determining the bioavailability of a vitamin is a more complex task because it depends on many factors, including:[1]

- Efficiency of digestion and time of transit through the GI tract.
- Previous nutrient intake and nutrition status.
- Other foods consumed at the same time.
- Method of food preparation (raw, cooked, processed).
- Source of the nutrient (synthetic, fortified, or naturally occurring).

Experts consider these factors when estimating recommended intakes.

**bioavailability:** the rate and extent to which a nutrient is absorbed and used.

 American Dietetic Association
www.eatright.org
Search for Vitamins

Be aware that many websites on the Internet are peddling vitamin supplements, not accurate information.

**precursors:** substances that precede others; with regard to vitamins, compounds that can be converted into active vitamins; also known as **provitamins.**

• ***Precursors*** • Some of the vitamins are available from foods in inactive forms known as precursors, or provitamins. Once inside the body, precursors are changed chemically to an active form of the vitamin. Thus, in measuring a person's vitamin intake, it is important to count both the amount of the active vitamin and the potential amount available from its precursors. The summary tables throughout this chapter and the next indicate which vitamins have precursors.

• ***Organic Nature*** • Being organic, vitamins can be destroyed and rendered unable to perform their duties. Therefore, they must be handled with care during storage and in cooking. Prolonged heating may destroy as much as 40 percent of the thiamin in food. Riboflavin can be destroyed by the ultraviolet rays of the sun or by fluorescent light; foods stored in transparent glass containers are most likely to lose riboflavin. Oxygen destroys vitamin C, so losses are closely related to the extent to which foods are cut or broken and thereby exposed to air.

The body makes special provisions to absorb vitamins, supplying most of them with special protein carriers. It also provides special enzymes to activate vitamins so that they can perform different roles.

IN SUMMARY

**The Vitamins**

**Water-Soluble Vitamins:**
- B vitamins.
- Vitamin C.

**Fat-Soluble Vitamins:**
- Vitamin A.
- Vitamin D.
- Vitamin E.
- Vitamin K.

• ***Solubility*** • As you may recall, carbohydrates and proteins are hydrophilic and lipids are hydrophobic. The vitamins divide along the same lines—the hydrophilic, water-soluble ones are the B vitamins and vitamin C; the hydrophobic, fat-soluble ones are vitamins A, D, E, and K.

Solubility is apparent in the food sources of the different vitamins, and it affects their absorption, transport, storage, and excretion by the body. The water-soluble vitamins are found in the watery compartments of foods; the fat-soluble vitamins usually occur together in the fats and oils of foods. On being absorbed, the water-soluble vitamins move directly into the blood; like fats, the fat-soluble vitamins must first enter the lymph, then the blood. Once in the blood, many of the water-soluble vitamins travel freely; many of the fat-soluble vitamins require protein carriers for transport. Upon reaching the cells, water-soluble vitamins freely circulate in the water-filled compartments of the body; fat-soluble vitamins are held in fatty tissues and the liver until needed. The kidneys, monitoring the blood that flows through them, detect and remove excess water-soluble vitamins; fat-soluble vitamins tend to remain in fat-storage sites in the body rather than being excreted, and so are more likely to reach toxic levels when consumed in excess.

Because the body stores fat-soluble vitamins, they can be eaten in large amounts once in a while and still meet the body's needs over time. Water-soluble vitamins are retained for varying periods in the body; a single day's omission from the diet does not bring on a deficiency, but still, the water-soluble vitamins must be eaten more regularly than the fat-soluble vitamins.

• ***Toxicity*** • Knowledge about some of the amazing roles of vitamins has prompted many people to begin taking supplements, assuming that more is better. But just as an inadequate intake can cause harm, so can an excessive intake. Even some of the water-soluble vitamins have adverse effects when taken in large doses.

That a vitamin can be both essential and harmful may seem surprising, but the same is true of most nutrients. The effects of every substance depend on its dose, and this is one reason consumers should not self-prescribe vitamins for their ailments. See the accompanying "How to" on p. 293 for a perspective on doses.

The Dietary Reference Intakes (DRI) report addresses the possibility of adverse health effects from large amounts of nutrients with its Tolerable Upper Intake Levels. An Upper Level defines the highest amount of a nutrient that is likely not to cause harm for most healthy people when consumed daily. The risk of harm increases as intakes rise above the Upper Level. Of the nutrients discussed in this chapter, niacin, vitamin $B_6$, folate, and choline have had Upper Levels set, and these values are presented in their respective summary tables. Data were lacking to establish Upper Levels for the remaining B vitamins, but this does not mean that excessively high intakes would be without risk.

## HOW TO Understand Dose Levels and Effects

A substance may have a beneficial or harmful effect, but a critical thinker would not conclude that the substance itself was beneficial or harmful without first asking what dose was used. Figure 10-1 shows three possible relationships between dose levels and effects. The third diagram represents the situation with nutrients—more is better up to a point, but beyond that point, still more is harmful.

**Figure 10-1**
**Dose Levels and Effects**

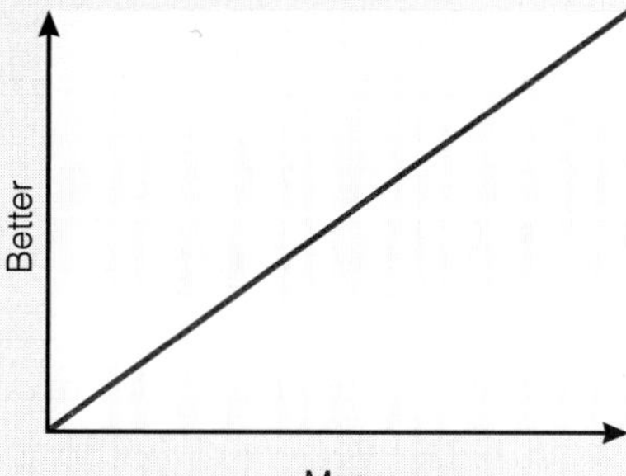


As you progress in the direction of more, the effect gets better and better, with no end in sight (real life is seldom, if ever, like this).

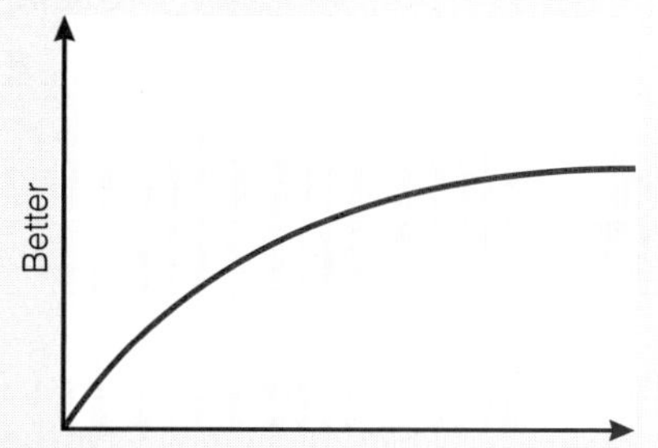


As you progress in the direction of more, the effect reaches a maximum and then a plateau, becoming no better with higher doses.

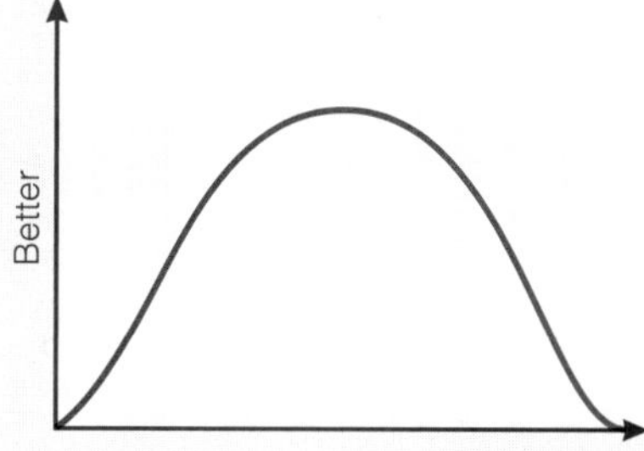


As you progress in the direction of more, the effect reaches an optimum at some intermediate dose and then declines, showing that more is better up to a point and then harmful. That too much is as harmful as too little represents the situation with nutrients.

### IN SUMMARY

The vitamins are essential nutrients that are needed in tiny amounts in the diet both to prevent deficiency diseases and to support optimal health. The water-soluble vitamins are the B vitamins and vitamin C; the fat-soluble vitamins are vitamins A, D, E, and K. The accompanying table summarizes the differences between the water-soluble and fat-soluble vitamins.

**Water-Soluble and Fat-Soluble Vitamins Compared**

| | Water-Soluble Vitamins: B Vitamins and Vitamin C | Fat-Soluble Vitamins: Vitamins A, D, E, and K |
|---|---|---|
| **Absorption** | Directly into the blood. | First into the lymph, then the blood. |
| **Transport** | Travel freely. | Many require protein carriers. |
| **Storage** | Freely circulate in water-filled parts of the body. | Trapped in the cells associated with fat. |
| **Excretion** | Kidneys detect and remove excess in urine. | Less readily excreted; tend to remain in fat-storage sites. |
| **Toxicity** | Possible to reach toxic levels when consumed from supplements. | Likely to reach toxic levels when consumed from supplements. |
| **Requirements** | Needed in frequent doses (perhaps 1 to 3 days). | Needed in periodic doses (perhaps weeks or even months). |

NOTE: Exceptions occur, but these differences between the water-soluble and fat-soluble vitamins are valid generalizations.

As each vitamin was discovered, it was given a name and sometimes a letter and number as well. Many of the water-soluble vitamins have multiple names, which has led to some confusion. The table below lists the standard names; summary tables throughout this chapter provide the common alternative names.

The discussion of B vitamins that follows begins with a brief description of each of them, then offers a look at the ways they work together. Thus a preview of the "trees" is followed by a survey of the "forest."

IN SUMMARY

**The Water-Soluble Vitamins**

- B vitamins
  Thiamin, Riboflavin, Niacin, Biotin, Pantothenic acid, Vitamin $B_6$, Folate, Vitamin $B_{12}$
- Vitamin C

# The B Vitamins—As Individuals

Despite supplement advertisements that claim otherwise, the vitamins do not provide the body with fuel for energy. The energy-yielding nutrients—carbohydrate, fat, and protein—are used for fuel; the B vitamins help the body to use that fuel. It is true, though, that without B vitamins the body would lack energy. Several of the B vitamins—thiamin, riboflavin, niacin, pantothenic acid, and biotin—form part of the coenzymes that enable enzymes to release energy from carbohydrate, fat, and protein. Other B vitamins play other indispensable roles in metabolism. Vitamin $B_6$ assists enzymes that metabolize amino acids; folate and vitamin $B_{12}$ help cells to multiply. Among these cells are the red blood cells and the cells lining the GI tract—cells that deliver energy to all the others.

Reminder: A *coenzyme* is a small organic molecule that associates closely with enzymes; many B vitamins form an integral part of coenzymes.

The vitamin portion of a coenzyme allows a chemical reaction to take place; the remaining portion of the coenzyme binds to the enzyme. Without its coenzyme, an enzyme cannot function. Thus symptoms of B vitamin deficiencies directly reflect the disturbances of metabolism incurred by a lack of coenzymes. Figure 10-2 illustrates coenzyme action.

**Figure 10-2**
**Coenzyme Action**

Some vitamins form part of the coenzymes that enable enzymes either to synthesize compounds (as illustrated by the lower enzyme in this figure) or to dismantle compounds (as illustrated by the upper enzyme).

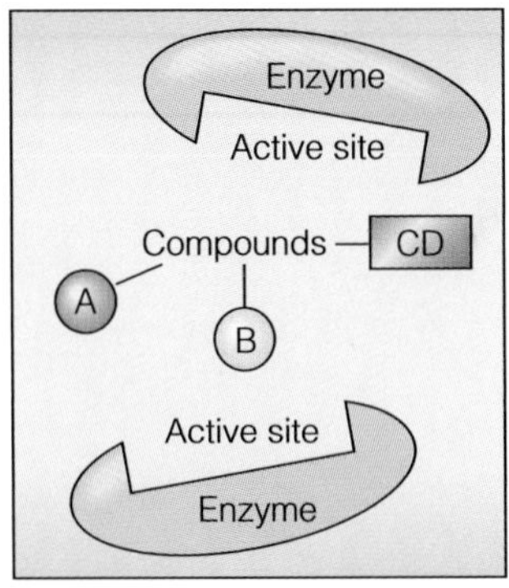


Without coenzymes, compounds A, B, and CD don't respond to their enzymes.

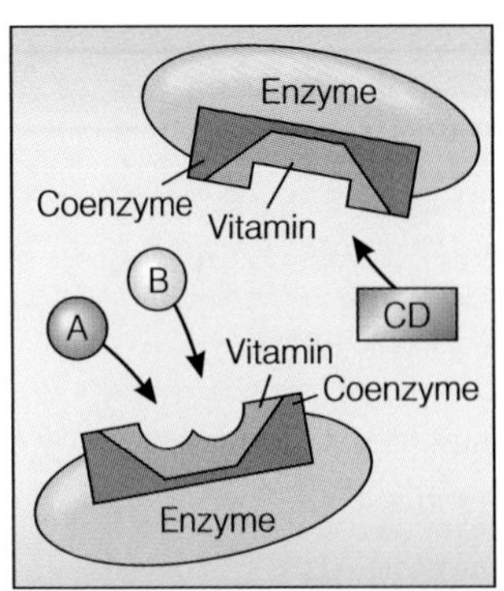


With the coenzymes in place, compounds are attracted to their sites on the enzymes . . .

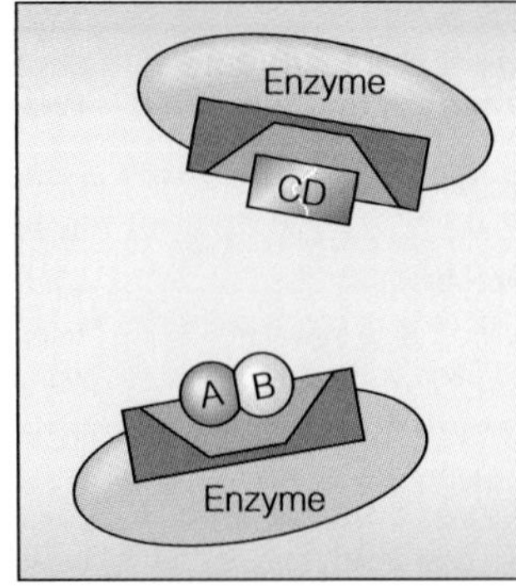


. . . and the reactions proceed instantaneously. The coenzymes often donate or accept electrons, atoms, or groups of atoms.

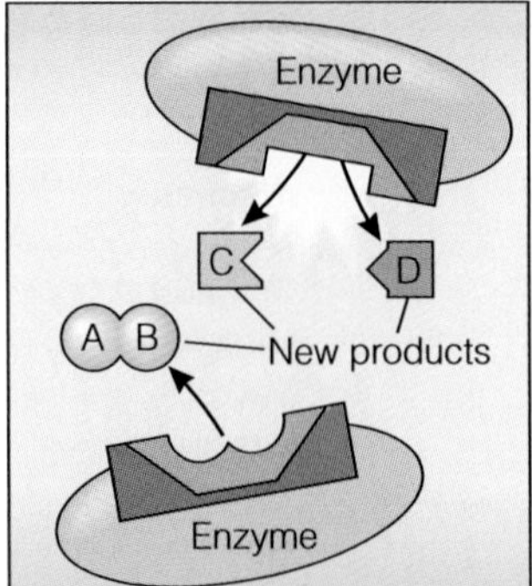


The reactions are completed with either the formation of a new product, AB, or the breaking apart of a compound into two new products, C and D, and the release of energy.

The following sections describe individual B vitamins and note many coenzymes and metabolic pathways. Keep in mind that a later section will assemble these pieces of information into a whole picture.

The following sections also present the recommendations, deficiency and toxicity symptoms, and food sources for each vitamin. The recommendations for the B vitamins reflect the 1998 DRI.[2] For thiamin, riboflavin, niacin, vitamin $B_6$, folate, and vitamin $B_{12}$, sufficient data were available to establish an RDA; for pantothenic acid, biotin, and choline, an Adequate Intake (AI) was set. Their values appear in the summary tables and figures that follow.

**Figure 10-3**

**Thiamin-Deficiency Symptom—The Edema of Beriberi**

Beriberi may be characterized as "wet" (referring to edema) or "dry" (with muscle wasting, but no edema). Physical examination confirms that this woman has wet beriberi. Notice how the impression of the physician's thumb remains on her leg.

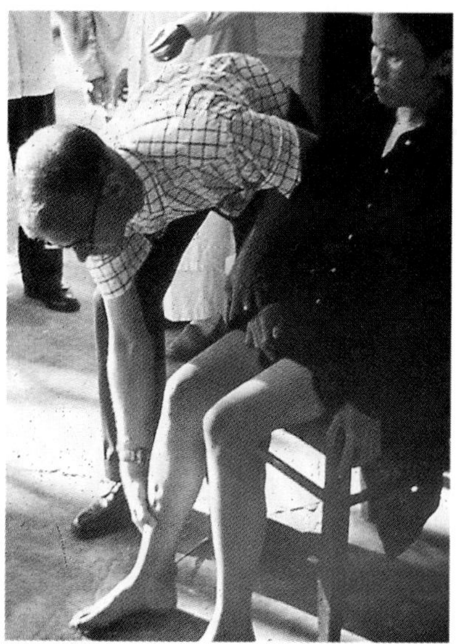

# Thiamin

Researchers first isolated the coenzyme form of thiamin in 1926. Since then, the world of vitamins and enzyme action has opened up dramatically.

Thiamin is the vitamin part of the coenzyme TPP, which assists in energy metabolism. The TPP coenzyme promotes the conversion of pyruvate to acetyl CoA. The reaction removes one carbon from the 3-carbon pyruvate to make the 2-carbon acetyl CoA and carbon dioxide. Later TPP promotes a similar step in the TCA cycle where it helps convert a 5-carbon compound to a 4-carbon compound. Besides playing these pivotal roles in the energy metabolism of all cells, thiamin occupies a special site on the membranes of nerve cells. Processes in nerves and in their responding tissues, the muscles, depend heavily on thiamin.

**thiamin** (THIGH-ah-min): a B vitamin; the coenzyme form is TPP (thiamin pyrophosphate).

• ***Thiamin Recommendations*** • Because thiamin participates in energy metabolism, previous RDA were stated in milligrams per 1000 kcalories of food energy. Finding no evidence to support this relationship between energy intake and thiamin requirements, the 1998 DRI committee focused primarily on thiamin's role in enzyme activity in establishing the new RDA. Generally, if a person eats enough food to meet energy needs and obtains that energy from nutritious foods, thiamin needs will be met. The average thiamin intake in the United States and Canada meets or exceeds this new recommendation.

• ***Thiamin Deficiency*** • People who fail to eat enough food to meet energy needs risk nutrient deficiencies, including thiamin deficiency. Inadequate thiamin intakes have been reported among the nation's malnourished and homeless people.[3] Similarly, people who derive most of their energy from empty-kcalorie items, like alcohol, risk thiamin deficiency. Alcohol contributes energy, but provides few, if any, nutrients and often displaces food. In addition, alcohol enhances thiamin excretion in the urine, doubling the risk of deficiency.

The most serious form of thiamin deficiency in alcohol abusers is the **Wernicke-Korsakoff** (VER-nee-key KORE-sah-kof) **syndrome.** Symptoms include disorientation, loss of short-term memory, jerky eye movements, and staggering gait.

Prolonged thiamin deficiency can result in the disease beriberi, which was first observed in East Asia when the custom of polishing rice became widespread. Rice provided 80 percent of the energy intake of the people of that area, and rice hulls were their principal source of thiamin. When the hulls were removed, beriberi spread like wildfire.

**beriberi:** the thiamin-deficiency disease.
- **beri** = weakness
- **beriberi** = "I can't, I can't"

Beriberi was first believed to be caused by an infectious agent. Medical researchers wasted much time and energy seeking a microbial cause before they realized that the cause of beriberi was not something that was in the food, but something that was absent from it.

Because thiamin participates in nerve processes, paralysis sets in when it is lacking. The symptoms of beriberi include damage to the nervous system as well as to the heart and other muscles. Figure 10-3 presents one of the symptoms of beriberi.

• ***Thiamin Food Sources*** • Before examining Figure 10-4, you may want to read the "How to" on p. 296, which describes the many features found in this and similar figures in this chapter and the next three chapters. When you look at Figure 10-4, notice that thiamin occurs in small quantities in many nutritious foods; highly refined foods contain almost no thiamin. The long red bars near the

## HOW TO Evaluate Foods for Their Nutrient Contributions

Figure 10-4 is the first of a series of figures in this and the next three chapters that present the vitamins and minerals in foods. Each figure presents the same 45 foods, which were selected to ensure a variety of choices representative of each of the food groups as suggested by the Daily Food Guide Pyramid. From its base, for example, a bread, a cereal, a rice, and a pasta were chosen. The suggestion to include a variety of vegetables was also considered: dark green, leafy vegetables (spinach, broccoli); deep yellow vegetables (carrots, sweet potatoes); starchy vegetables (potatoes, corn, green peas); legumes (navy, pinto, kidney, and garbanzo beans); and other vegetables (green beans). The selection of fruits followed the Daily Food Guide Pyramid's suggestions to use whole fruits (apples, bananas); citrus fruits (oranges, grapefruit juice); melons (watermelon); and berries (strawberries). Items were selected from the milk and meat groups in a similar way. In addition to the 45 foods that appear in all of the figures, five different foods were selected for each of the nutrients. These five foods were chosen to add variety and often reflect excellent, and sometimes unusual, sources of the specific nutrient.

Notice that the figures list the food, the serving size, and the food energy (kcalories) on the left and graph the amount of the nutrient per serving on the right along with the RDA (or AI) for adults, so you can see how many servings would be needed to meet recommendations. Serving sizes reflect those used by the Daily Food Guide plan. The colored bars show at a glance which food groups best provide a nutrient: yellow for breads and cereals; green for vegetables; purple for fruits; white for milk and milk products; brown for legumes; and red for meat, fish, and poultry. (Because the Daily Food Guide Pyramid mentions legumes with both the meat group and the vegetable group and because legumes are especially rich in many vitamins and minerals, they have been given their own color to highlight their nutrient contributions.) Notice how the bar graphs shift in the various figures. Careful study of all of the figures taken together will confirm that variety is the key to nutrient adequacy.

Another way to evaluate foods for their nutrient contributions is to consider their nutrient density (their calcium *per 100 kcalories,* for example). Quite often, vegetables rank higher on a nutrient-per-100 kcalories list than they do on a nutrient-per-serving list. Both listings offer valuable information, though, especially when combined with a realistic appraisal. For example, turnip greens provide more calcium per kcalorie than milk, but milk offers more calcium per serving. What matters most is which are you more likely to consume—1½ cups of turnip greens or 1 cup of milk? Both provide about 300 milligrams of calcium, but the greens save you about 50 kcalories. The left column in the figure highlights in yellow the foods that offer the best deal for your energy "dollar" (the kcalorie). Notice how many of them are vegetables.

Realistically, people cannot eat for single nutrients. Fortunately, most foods deliver more than one nutrient, allowing diet planners to combine foods into nourishing meals.

***Pork is the richest source of thiamin, but enriched or whole-grain products typically make the greatest contribution to a day's intake because of the quantities eaten. Sunflower seeds and legumes such as split peas are also valuable sources of thiamin.***

bottom of the graph represent meats that are exceptionally rich in thiamin—those in the pork family.

As mentioned earlier, prolonged cooking can destroy thiamin. Also, like other water-soluble vitamins, thiamin leaches into water when foods are boiled or blanched. Cooking methods that require little or no water such as steaming and microwave heating conserve thiamin and other water-soluble vitamins. The table on p. 298 summarizes thiamin's main functions, food sources, and deficiency symptoms.

## *Riboflavin*

Like thiamin, riboflavin facilitates the release of energy from nutrients in all body cells. The coenzyme forms of riboflavin are FMN and FAD; both can accept and then donate two hydrogens (see Figure 10-5 on p. 298). During energy metabolism, FAD picks up two hydrogens (with their electrons) from the TCA cycle and delivers them to the electron transport chain (see Chapter 7).

**Figure 10-4**
**Thiamin in Selected Foods**

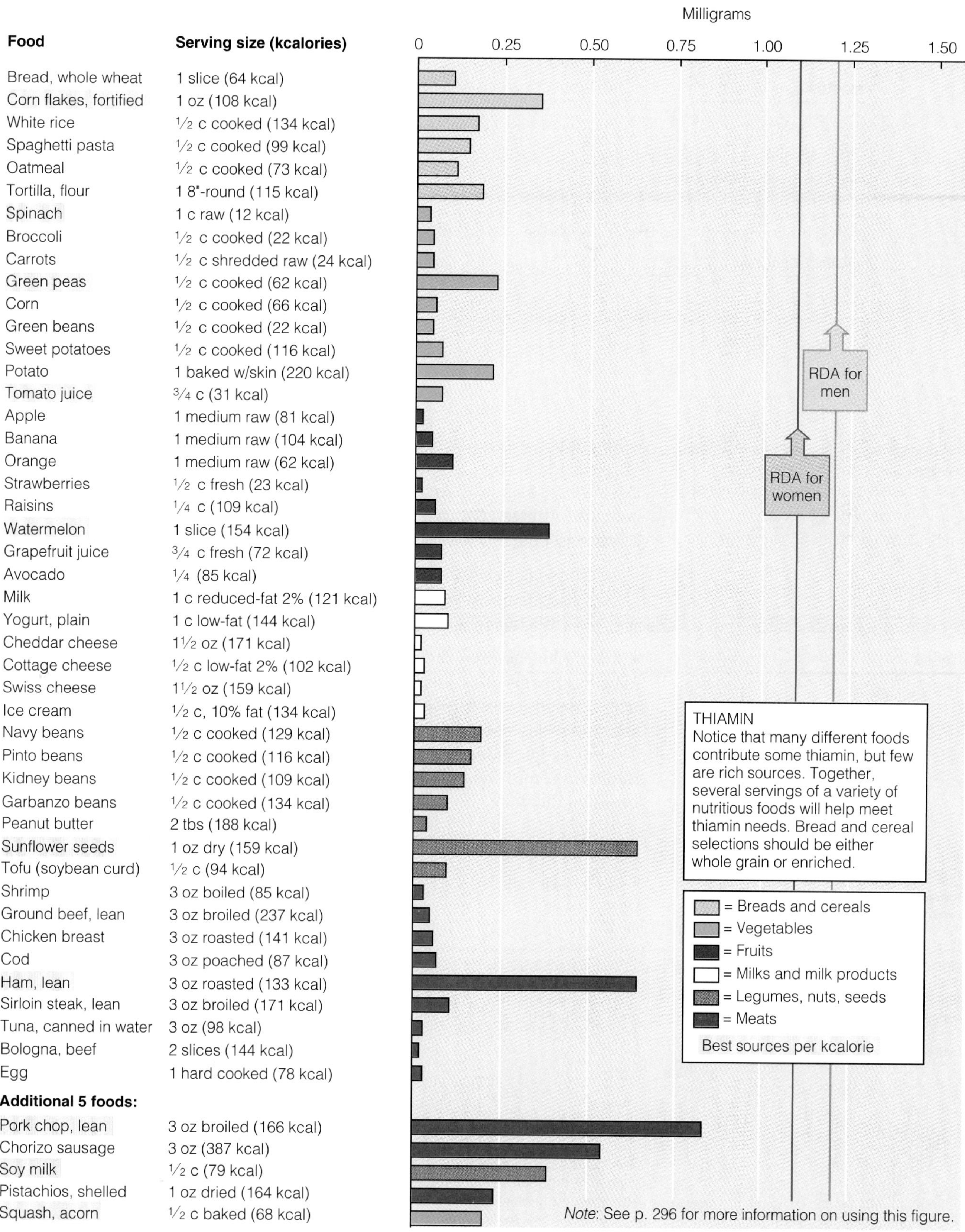
Milligrams
0
0.25
0.50
0.75
1.00
1.25
1.50
Food
Serving size (kcalories)
Bread, whole wheat 1 slice (64 kcal)
Corn flakes, fortified 1 oz (108 kcal)
White rice ½ c cooked (134 kcal)
Spaghetti pasta ½ c cooked (99 kcal)
Oatmeal ½ c cooked (73 kcal)
Tortilla, flour 1 8"-round (115 kcal)
Spinach 1 c raw (12 kcal)
Broccoli ½ c cooked (22 kcal)
Carrots ½ c shredded raw (24 kcal)
Green peas ½ c cooked (62 kcal)
Corn ½ c cooked (66 kcal)
Green beans ½ c cooked (22 kcal)
Sweet potatoes ½ c cooked (116 kcal)
Potato 1 baked w/skin (220 kcal)
Tomato juice ¾ c (31 kcal)
Apple 1 medium raw (81 kcal)
Banana 1 medium raw (104 kcal)
Orange 1 medium raw (62 kcal)
Strawberries ½ c fresh (23 kcal)
Raisins ¼ c (109 kcal)
Watermelon 1 slice (154 kcal)
Grapefruit juice ¾ c fresh (72 kcal)
Avocado ¼ (85 kcal)
Milk 1 c reduced-fat 2% (121 kcal)
Yogurt, plain 1 c low-fat (144 kcal)
Cheddar cheese 1½ oz (171 kcal)
Cottage cheese ½ c low-fat 2% (102 kcal)
Swiss cheese 1½ oz (159 kcal)
Ice cream ½ c, 10% fat (134 kcal)
Navy beans ½ c cooked (129 kcal)
Pinto beans ½ c cooked (116 kcal)
Kidney beans ½ c cooked (109 kcal)
Garbanzo beans ½ c cooked (134 kcal)
Peanut butter 2 tbs (188 kcal)
Sunflower seeds 1 oz dry (159 kcal)
Tofu (soybean curd) ½ c (94 kcal)
Shrimp 3 oz boiled (85 kcal)
Ground beef, lean 3 oz broiled (237 kcal)
Chicken breast 3 oz roasted (141 kcal)
Cod 3 oz poached (87 kcal)
Ham, lean 3 oz roasted (133 kcal)
Sirloin steak, lean 3 oz broiled (171 kcal)
Tuna, canned in water 3 oz (98 kcal)
Bologna, beef 2 slices (144 kcal)
Egg 1 hard cooked (78 kcal)
Additional 5 foods:
Pork chop, lean 3 oz broiled (166 kcal)
Chorizo sausage 3 oz (387 kcal)
Soy milk ½ c (79 kcal)
Pistachios, shelled 1 oz dried (164 kcal)
Squash, acorn ½ c baked (68 kcal)
RDA for men
RDA for women
THIAMIN
Notice that many different foods contribute some thiamin, but few are rich sources. Together, several servings of a variety of nutritious foods will help meet thiamin needs. Bread and cereal selections should be either whole grain or enriched.
= Breads and cereals
= Vegetables
= Fruits
= Milks and milk products
= Legumes, nuts, seeds
= Meats
Best sources per kcalorie
Note: See p. 296 for more information on using this figure.

IN SUMMARY

**Thiamin**

**Other Names**

Vitamin $B_1$

**1998 RDA**

Men: 1.2 mg/day
Women: 1.1 mg/day

**Chief Functions in the Body**

Part of the coenzyme TPP (thiamin pyrophosphate) used in energy metabolism; supports normal appetite and nerve function

**Significant Sources**

Occurs in all nutritious foods in moderate amounts; pork products, liver, whole-grain or enriched breads and cereals, legumes, nuts

Easily destroyed by heat

**Deficiency Disease Name**

Beriberi

**Deficiency Symptoms**

***Blood/Circulatory System***

Edema, enlarged heart, abnormal heart rhythms, heart failure

***Nervous/Muscular Systems***

Degeneration, wasting, muscle weakness, painful calf muscles, low morale, loss of appetite, difficulty walking, loss of ankle and knee-jerk reflexes, mental confusion, paralysis

**riboflavin** (RYE-boh-flay-vin): a B vitamin; the coenzyme forms are FMN (flavin mononucleotide) and FAD (flavin adenine dinucleotide).

• ***Riboflavin Recommendations*** • Like thiamin's RDA, riboflavin's RDA is no longer stated in terms of energy intake. Instead, other criteria were used to establish the 1998 RDA, with small adjustments to reflect differences in energy use and body size. Athletes may have an increased requirement for riboflavin, but the RDA is generous enough to cover the needs of most physically active people.

• ***Riboflavin Deficiency*** • No one disease is associated with riboflavin deficiency. Lack of the vitamin affects the facial skin, eyes, and GI tract. The summary table on p. 299 lists riboflavin's chief roles, food sources, and deficiency symptoms.

• ***Riboflavin Food Sources*** • Milk, milk products such as cheese, and liver dominate the riboflavin list (see Figure 10-6 on p. 300). The need for riboflavin is a major reason for including milk and milk products daily; no other commonly eaten food can make such a substantial contribution in a single serving.

Most people easily meet their riboflavin RDA. The greatest contributions of riboflavin come from milk and milk products; whole-grain or enriched bread and cereal products are also valuable contributors because of the quantities typically

**Figure 10-5**
**Riboflavin Coenzyme, Accepting and Donating Hydrogens**

This figure shows the chemical structure of the riboflavin portion of the coenzyme only; the remainder of the coenzyme structure is represented by dotted lines (see Appendix C for the complete chemical structure of FAD and FMN). The reactive sites that accept and donate hydrogens are highlighted in white.

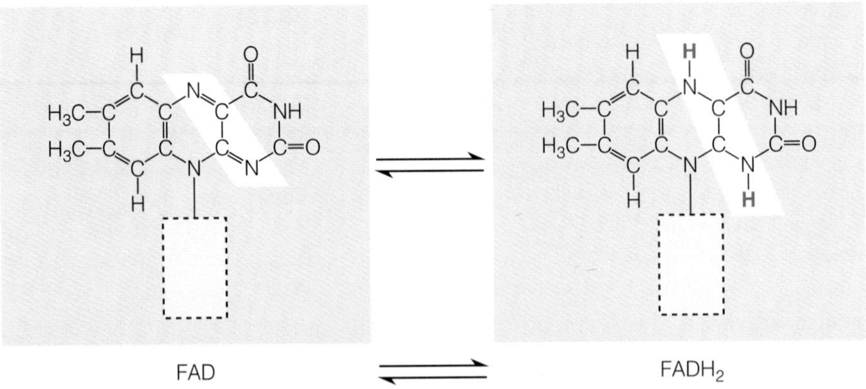

FAD

During the TCA cycle, compounds release hydrogens, and the riboflavin coenzyme FAD picks up two of them. As it accepts two hydrogens, FAD becomes $FADH_2$.

$FADH_2$

$FADH_2$ carries the hydrogens to the electron transport chain. At the end of the electron transport chain, the hydrogens are accepted by oxygen, creating water, and $FADH_2$ becomes FAD again. For every $FADH_2$ that passes through the electron transport chain, 2 ATP are generated.

IN SUMMARY

**Riboflavin**

**Other Names**

Vitamin $B_2$

**1998 RDA**

Men: 1.3 mg/day

Women: 1.1 mg/day

**Chief Functions in the Body**

Part of coenzymes FMN (flavin mononucleotide) and FAD (flavin adenine dinucleotide) used in energy metabolism; supports normal vision and skin health

**Significant Sources**

Milk, yogurt, cottage cheese, meat, leafy green vegetables, whole-grain or enriched breads and cereals

Easily destroyed by ultraviolet light and irradiation

**Deficiency Disease Name**

Ariboflavinosis (ay-RYE-boh-FLAY-vin-oh-sis)

**Deficiency Symptoms**

***Mouth, Gums, Tongue***

Cracks and redness at corners of mouth;[a] painful, smooth, purplish red tongue[b]

***Nervous System and Eyes***

Inflamed eyelids and sensitivity to light,[c] reddening of cornea

***Skin***

Inflammation characterized by lesions covered with greasy scales

[a]Cracks at the corners of the mouth are termed *cheilosis* (kee-LOH-sis).
[b]Smoothness of the tongue is caused by loss of its surface structures and is termed *glossitis* (gloss-EYE-tis).
[c]Hypersensitivity to light is *photophobia* (FOE-toe-FOE-bee-ah).

consumed. When riboflavin sources are ranked per 100 kcalories, many dark green, leafy vegetables (such as broccoli, turnip greens, asparagus, and spinach) appear high on the list (notice those foods highlighted in yellow in the left column of Figure 10-6). Vegetarians and others who don't use milk must rely on ample servings of dark greens and enriched grains for riboflavin. Nutritional yeast is another rich source.

Ultraviolet light and irradiation destroy riboflavin. For these reasons, milk is sold in cardboard or opaque plastic containers, and precautions are taken when vitamin D is added to milk by irradiation.* In contrast, riboflavin is stable to heat, so cooking does not destroy it; heat stability diminishes in alkaline solutions.

## Niacin

The name niacin describes two chemical structures: nicotinic acid and nicotinamide (also known as niacinamide). The body can easily convert nicotinic acid to nicotinamide, which is the major form of niacin in the blood.

The two coenzyme forms of niacin, NAD and NADP, are central in energy-transfer reactions, especially the metabolism of glucose, fat, and alcohol. NAD is similar to the riboflavin coenzymes in that it carries hydrogens (and their electrons) during metabolic reactions, including the pathway from the TCA cycle to the electron transport chain.

• ***Niacin Recommendations*** • Niacin is unique among the B vitamins in that the body can make it from the amino acid tryptophan. To make 1 milligram of niacin requires approximately 60 milligrams of dietary tryptophan. For this reason, recommended intakes are stated in "niacin equivalents." A food containing 1 milligram of niacin and 60 milligrams of tryptophan provides the equivalent of 2 milligrams of niacin, or 2 niacin equivalents (NE). The RDA for niacin allows for

*Vitamin D can be added to milk by feeding cows irradiated yeast or by irradiating the milk itself.

***All of these foods are rich in riboflavin, but milk and milk products provide much of the riboflavin in the diets of most people.***

**niacin** (NIGH-a-sin): a B vitamin; the coenzyme forms are NAD (nicotinamide adenine dinucleotide) and NADP (the phosphate form of NAD). Niacin can be eaten preformed or made in the body from its precursor, tryptophan, one of the amino acids.

**niacin equivalents (NE):** the amount of niacin present in food, including the niacin that can theoretically be made from its precursor, tryptophan, present in the food.

1 NE = 1 mg niacin.

**Figure 10-6**
**Riboflavin in Selected Foods**

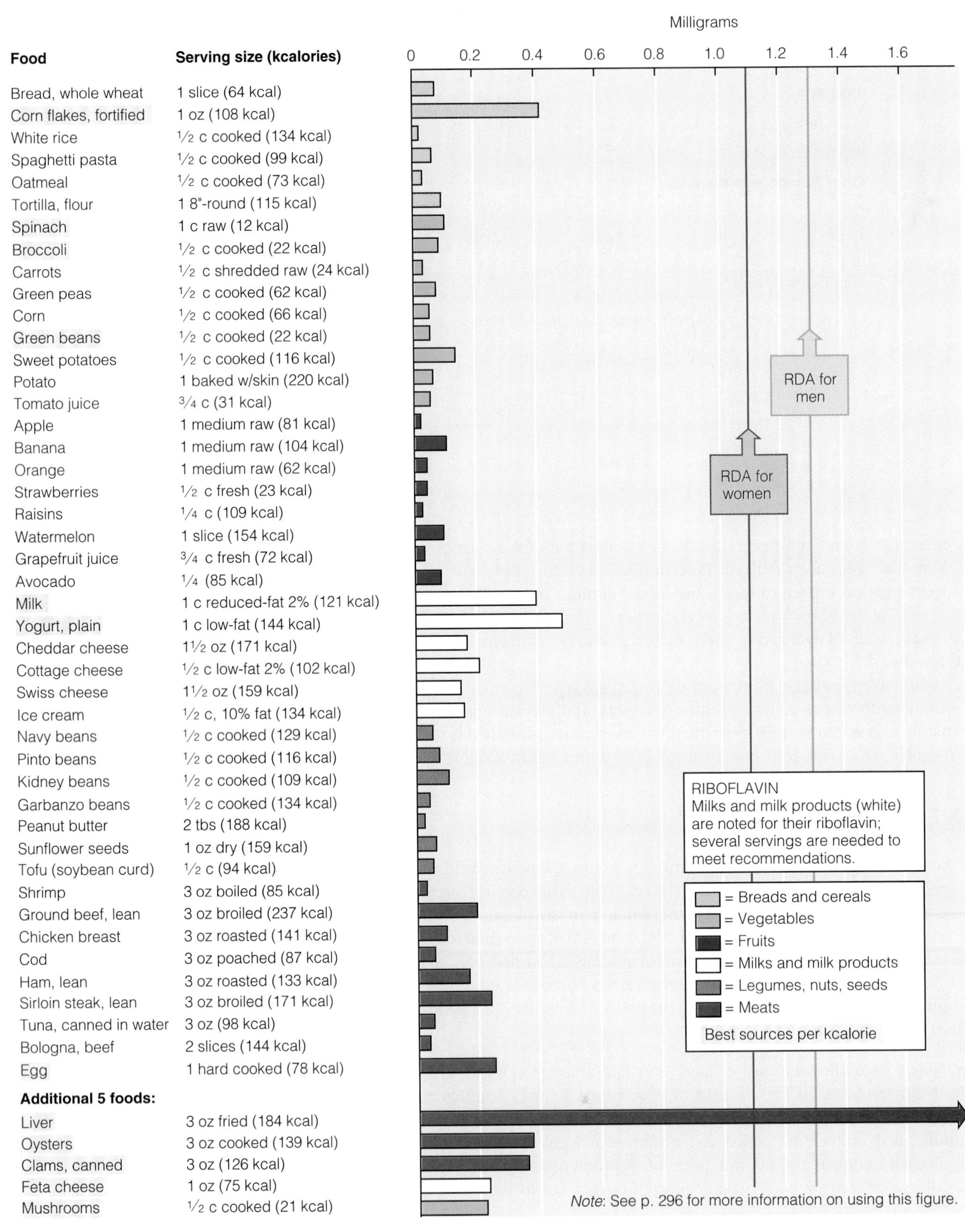
Milligrams
0
0.2
0.4
0.6
0.8
1.0
1.2
1.4
1.6
Food
Serving size (kcalories)
Bread, whole wheat — 1 slice (64 kcal)
Corn flakes, fortified — 1 oz (108 kcal)
White rice — 1/2 c cooked (134 kcal)
Spaghetti pasta — 1/2 c cooked (99 kcal)
Oatmeal — 1/2 c cooked (73 kcal)
Tortilla, flour — 1 8"-round (115 kcal)
Spinach — 1 c raw (12 kcal)
Broccoli — 1/2 c cooked (22 kcal)
Carrots — 1/2 c shredded raw (24 kcal)
Green peas — 1/2 c cooked (62 kcal)
Corn — 1/2 c cooked (66 kcal)
Green beans — 1/2 c cooked (22 kcal)
Sweet potatoes — 1/2 c cooked (116 kcal)
Potato — 1 baked w/skin (220 kcal)
Tomato juice — 3/4 c (31 kcal)
Apple — 1 medium raw (81 kcal)
Banana — 1 medium raw (104 kcal)
Orange — 1 medium raw (62 kcal)
Strawberries — 1/2 c fresh (23 kcal)
Raisins — 1/4 c (109 kcal)
Watermelon — 1 slice (154 kcal)
Grapefruit juice — 3/4 c fresh (72 kcal)
Avocado — 1/4 (85 kcal)
Milk — 1 c reduced-fat 2% (121 kcal)
Yogurt, plain — 1 c low-fat (144 kcal)
Cheddar cheese — 11/2 oz (171 kcal)
Cottage cheese — 1/2 c low-fat 2% (102 kcal)
Swiss cheese — 11/2 oz (159 kcal)
Ice cream — 1/2 c, 10% fat (134 kcal)
Navy beans — 1/2 c cooked (129 kcal)
Pinto beans — 1/2 c cooked (116 kcal)
Kidney beans — 1/2 c cooked (109 kcal)
Garbanzo beans — 1/2 c cooked (134 kcal)
Peanut butter — 2 tbs (188 kcal)
Sunflower seeds — 1 oz dry (159 kcal)
Tofu (soybean curd) — 1/2 c (94 kcal)
Shrimp — 3 oz boiled (85 kcal)
Ground beef, lean — 3 oz broiled (237 kcal)
Chicken breast — 3 oz roasted (141 kcal)
Cod — 3 oz poached (87 kcal)
Ham, lean — 3 oz roasted (133 kcal)
Sirloin steak, lean — 3 oz broiled (171 kcal)
Tuna, canned in water — 3 oz (98 kcal)
Bologna, beef — 2 slices (144 kcal)
Egg — 1 hard cooked (78 kcal)
Additional 5 foods:
Liver — 3 oz fried (184 kcal)
Oysters — 3 oz cooked (139 kcal)
Clams, canned — 3 oz (126 kcal)
Feta cheese — 1 oz (75 kcal)
Mushrooms — 1/2 c cooked (21 kcal)
RDA for men
RDA for women
RIBOFLAVIN
Milks and milk products (white) are noted for their riboflavin; several servings are needed to meet recommendations.
= Breads and cereals
= Vegetables
= Fruits
= Milks and milk products
= Legumes, nuts, seeds
= Meats
Best sources per kcalorie

*Note*: See p. 296 for more information on using this figure.

this conversion and is stated in niacin equivalents. (The "How to" on p. 302 describes how to calculate niacin equivalents in the diet.)

• ***Niacin Deficiency*** • The niacin-deficiency disease, pellagra, produces the symptoms of diarrhea, dermatitis, dementia, and eventually death. In the early 1900s, pellagra caused widespread misery and some 87,000 deaths in the U.S. South, where many people subsisted on a low-protein diet centered on corn. This diet supplied neither enough niacin nor enough tryptophan. At least 70 percent of the niacin in corn is bound to complex carbohydrates and small peptides, making it unavailable for absorption. Furthermore, corn is high in the amino acid leucine, which interferes with the tryptophan-to-niacin conversion, thus further contributing to the development of pellagra. Figure 10-7 illustrates the dermatitus of pellagra, and the summary table includes both deficiency and toxicity symptoms as well as niacin's various names, functions, and food sources.

• ***Niacin Toxicity*** • Large doses of niacin exert a druglike effect on the nervous system and on blood lipids and blood glucose. When niacin in the form of nicotinic acid is taken in doses only three to four times the RDA, it dilates the capillaries and causes a tingling sensation that can be painful, an effect known as the "niacin flush." The nicotinamide form does not produce this effect.

Physicians can effectively lower blood cholesterol with large doses of niacin, but such therapy must be closely monitored because of its adverse side effects (liver damage and peptic ulcers, among others). Pharmacological doses of nicotinamide to prevent diabetes (type 1) are currently being tested in a large international study.[4] Nicotanamide inhibits the destruction of the insulin-producing cells of the pancreas, thus early intervention is critical.

**pellagra** (pell-AY-gra): the niacin-deficiency disease.
- **pellis** = skin
- **agra** = rough

**niacin flush:** a burning, tingling, and itching sensation that occurs when a person takes a large dose of nicotinic acid; often accompanied by a headache and reddened face, arms, and chest.

When a normal dose of a nutrient (levels commonly found in foods and not exceeding 150% of the RDA) provides a normal blood concentration, the nutrient is having a **physiological effect.** When a large dose (two to ten times greater than the RDA) overwhelms some body system and acts like a drug, the nutrient is having a **pharmacological effect.**
- **physio** = natural
- **pharma** = drug

Reminder: Type 1 diabetes refers to insulin-dependent diabetes mellitus (IDDM).

## IN SUMMARY

### Niacin

**Other Names**

Nicotinic acid, nicotinamide, niacinamide, vitamin $B_3$; precursor is dietary tryptophan (an amino acid)

**1998 RDA**

Men: 16 mg NE/day

Women: 14 mg NE/day

**Upper Level**

Adults: 35 mg/day

**Chief Functions in the Body**

Part of coenzymes NAD (nicotinamide adenine dinucleotide) and NADP (its phosphate form) used in energy metabolism; supports health of skin, nervous system, and digestive system

**Significant Sources**

Milk, eggs, meat, poultry, fish, whole-grain and enriched breads and cereals, nuts, and all protein-containing foods

**Deficiency Disease Name**

Pellagra

| Deficiency Symptoms | Toxicity Symptoms |
|---|---|
| ***Digestive System*** | |
| Diarrhea, abdominal pain, vomiting | Diarrhea, heartburn, nausea, ulcer, irritation, vomiting |
| ***Mouth, Gums, Tongue*** | |
| Inflamed, swollen, smooth, bright red tongue[a] | |
| ***Nervous System*** | |
| Depression, apathy, fatigue, loss of memory | Fainting, dizziness, fatigue, headache |
| ***Skin*** | |
| Bilateral symmetrical dermatitis, especially on areas exposed to sun | Painful flush, hives, and rash ("niacin flush"); excessive sweating |
| ***Other*** | |
| | Liver damage, low blood pressure |

[a]Smoothness of the tongue is caused by loss of its surface structures and is termed *glossitis* (gloss-EYE-tis).

**Figure 10-7**

**Niacin-Deficiency Symptom—The Dermatitis of Pellagra**

In the dermatitis of pellagra, the skin darkens and flakes away as if it were sunburned. Kwashiorkor also produces a "flaky paint" dermatitis, but the two are easily distinguished. The dermatitis of pellagra is bilateral and symmetrical and occurs only on those parts of the body exposed to the sun.

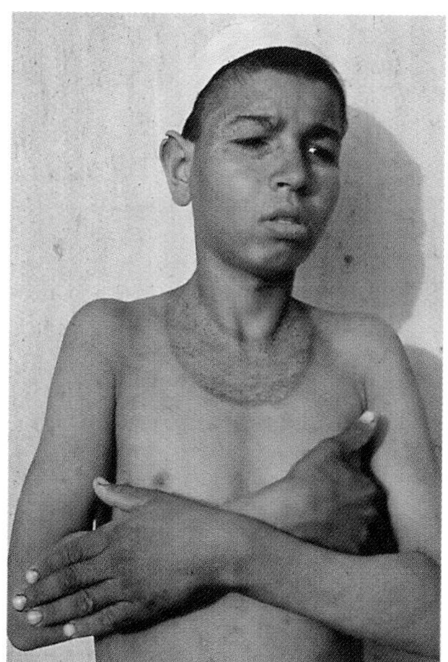

## HOW TO Determine Niacin Equivalents

To obtain a rough approximation of niacin equivalents:

1. Calculate total protein consumed (grams).
2. Assuming that the RDA amount of protein will be used first to make body protein, subtract the RDA to obtain "leftover" protein available to make niacin (grams). (Actually, the RDA provides a generous protein allowance, so "leftover" protein may be even greater than this.)
3. About 1 gram of every 100 grams of protein is tryptophan, so divide by 100 to obtain the tryptophan in this leftover protein (grams).
4. Multiply by 1000 to express this amount of tryptophan in milligrams.
5. Divide by 60 to get niacin equivalents (milligrams).
6. Finally, add the amount of preformed niacin obtained in the diet (milligrams).

***Protein-rich foods such as meat, fish, poultry, and peanut butter contribute much of the niacin in people's diets. Enriched breads and cereals and a few vegetables are also rich in niacin.***

• ***Niacin Food Sources*** • Tables of food composition typically list preformed niacin only, but people also obtain the vitamin from protein, which almost invariably contains the niacin precursor, tryptophan. Hence diets that are high in protein are never deficient in niacin. The accompanying "How to" shows how to calculate the total amount of niacin available from the diet. Dietary tryptophan could meet about half the daily need for most people, but the average diet easily supplies enough preformed niacin.

The predominance of red bars in Figure 10-8 explains why meat, poultry, and fish contribute about half the niacin equivalents most people need. About a fourth of most people's niacin comes from enriched breads and cereals. Mushrooms, asparagus, and leafy green vegetables are among the richest vegetable sources (per kcalorie) and can provide abundant niacin when eaten in generous amounts.

Niacin is less vulnerable to losses during food preparation and storage than other water-soluble vitamins. Being fairly heat-resistant, niacin can withstand reasonable cooking times, but like other water-soluble vitamins, it will leach into cooking water.

## Biotin

Biotin plays an important role in metabolism as a coenzyme that carries carbon dioxide. This role is critical in the TCA cycle: biotin delivers a carbon to 3-carbon pyruvate, thus replenishing the 4-carbon compound needed to combine with acetyl CoA to keep the TCA cycle turning.* The biotin coenzyme also serves crucial roles in gluconeogenesis, fatty acid synthesis, and the breakdown of certain fatty acids and amino acids.

**biotin** (BY-oh-tin): a B vitamin that functions as a coenzyme in the metabolism of carbohydrates and fats.

The protein **avidin** (AV-eh-din) in egg whites binds biotin.
• **avid** = greedy

• ***Biotin Recommendations*** • Biotin is needed in very small amounts. Instead of an RDA, an Adequate Intake (AI) has been set.

• ***Biotin Deficiency*** • Biotin deficiencies rarely occur. Researchers can induce a biotin deficiency in animals or human beings by feeding them raw egg whites, which contain a protein that binds biotin and thus prevents its absorption. Biotin-deficiency symptoms include scaly dermatitis, hair loss, loss of appetite, nausea, hallucinations, and depression. More than two dozen egg whites must be consumed daily for several months to produce these effects, however, and the eggs have to be raw; cooking denatures the binding protein.

*This 4-carbon intermediate of the TCA cycle is called oxaloacetate.

**Figure 10-8**
**Niacin in Selected Foods**

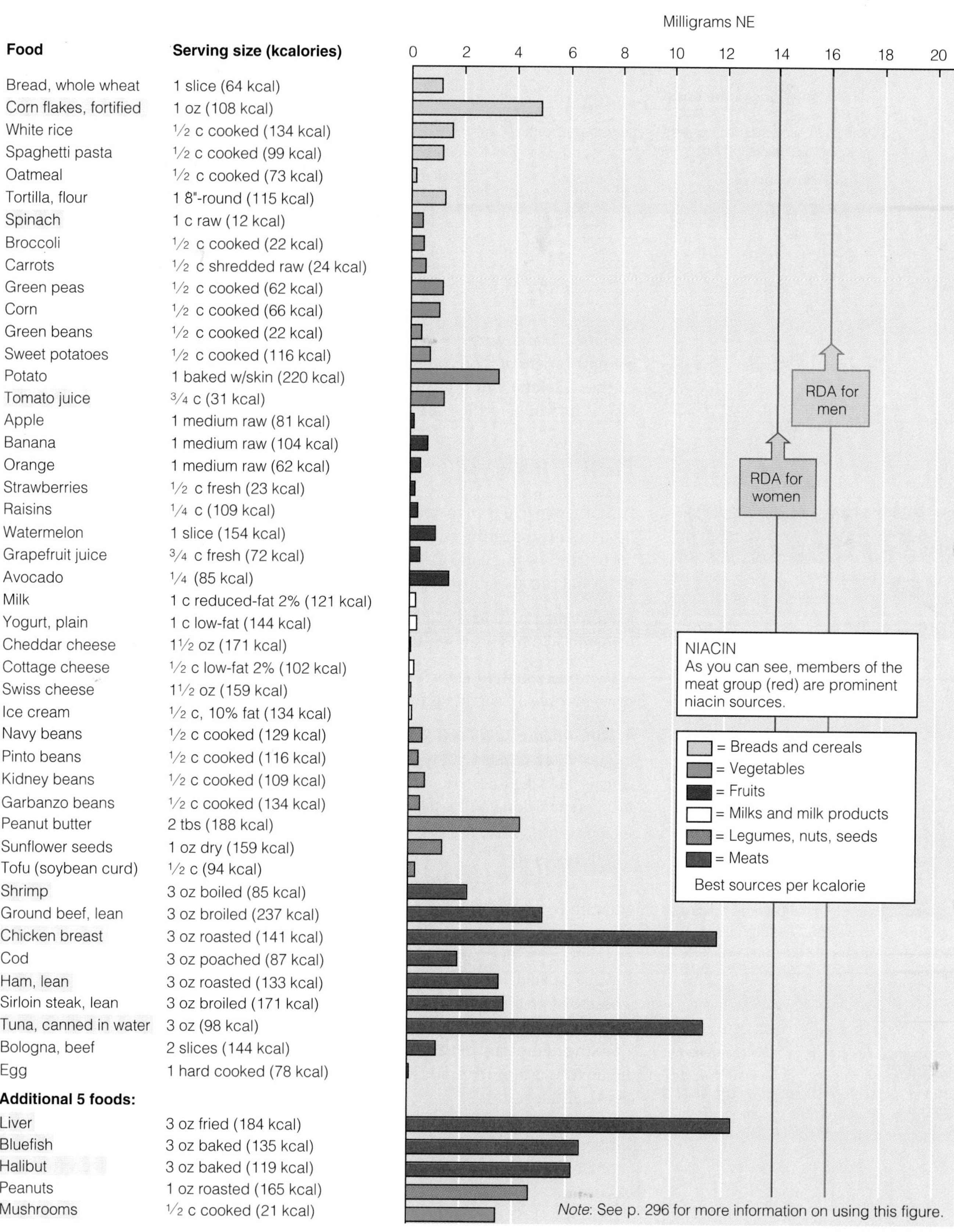

IN SUMMARY

**Biotin**

**1998 Adequate Intake (AI)**

Adults: 30 µg/day

**Chief Functions in the Body**

Part of a coenzyme used in energy metabolism, fat synthesis, amino acid metabolism, and glycogen synthesis

**Significant Sources**

Widespread in foods; organ meats, egg yolks, soybeans, fish, whole grains

**Deficiency Symptoms**

***Digestive System***

Loss of appetite, nausea

***Nervous/Muscular Systems***

Depression, lethargy, hallucinations, muscle pain, weakness, fatigue, seizures

***Skin***

Drying, scaly dermatitis, hair loss

***Biotin Food Sources*** • Biotin is widespread in foods (including egg yolks), so eating a variety of foods protects against deficiencies. Biotin is also synthesized by GI tract bacteria, but how much of it is absorbed is unknown. A review of biotin facts is provided in the summary table.

## Pantothenic Acid

**pantothenic** (PAN-toe-THEN-ick) **acid:** a B vitamin; the principal active form is part of coenzyme A, called "CoA" throughout Chapter 7.
• **pantos** = everywhere

Pantothenic acid is involved in more than 100 different steps in the synthesis of lipids, neurotransmitters, steroid hormones, and hemoglobin. It serves as part of coenzyme A—the same CoA that forms acetyl CoA, the "crossroads" compound in several metabolic pathways, including the TCA cycle.

• ***Pantothenic Acid Recommendations*** • An Adequate Intake (AI) for pantothenic acid has been set. It reflects the amount needed to replace daily losses.

• ***Pantothenic Acid Deficiency*** • Pantothenic acid deficiency is rare. Its symptoms involve a general failure of all the body's systems (see the summary table).

• ***Pantothenic Acid Food Sources*** • Pantothenic acid is widespread in foods, and typical diets seem to provide adequate intakes. Meat, fish, poultry, whole-grain cereals, and legumes are particularly good sources. Pantothenic acid loss during food preparation can be substantial because it is readily destroyed by heat.

## Vitamin $B_6$

**vitamin $B_6$:** a family of compounds—pyridoxal, pyridoxine, and pyridoxamine. The primary active coenzyme form is PLP (pyridoxal phosphate).

**serotonin** (SER-oh-tone-in): is a neurotransmitter important in appetite control, sleep regulation, and sensory perception, among other roles; it is synthesized from the amino acid tryptophan with the help of vitamin $B_6$.

Vitamin $B_6$ occurs in three forms—pyridoxal, pyridoxine, and pyridoxamine. All three can be converted to the coenzyme PLP, which is active in amino acid metabolism. Because PLP can transfer amino groups, the body can synthesize nonessential amino acids when amino groups are available (review Figure 7-15 on p. 210). The ability to add and remove amino groups makes PLP valuable in protein and urea metabolism as well. The conversion of the amino acid tryptophan to niacin or to the neurotransmitter serotonin also depends on PLP as does the synthesis of heme (the nonprotein portion of hemoglobin), nucleic acids (such as DNA and RNA), and lecithin.

A surge of vitamin $B_6$ research in the last decade has revealed that vitamin $B_6$ influences cognitive performance, immune function, and steroid hormone activity.[5] Unlike other water-soluble vitamins, vitamin $B_6$ is stored extensively in muscle tissue.

Among the many drugs that interact with vitamin $B_6$, alcohol stands out. As Highlight 7 described, when the body breaks down alcohol, it first produces acetaldehyde. If allowed to accumulate, acetaldehyde has toxic effects, so it must quickly be broken down further. Acetaldehyde impairs vitamin $B_6$ status by dis-

IN SUMMARY

**Pantothenic Acid**

| 1998 Adequate Intake (AI) | Deficiency Symptoms | Toxicity Symptoms |
|---|---|---|
| Adults: 5 mg/day | ***Digestive System*** | |
| **Chief Functions in the Body** | Vomiting, nausea, stomach cramps | Occasional diarrhea |
| Part of coenzyme A, used in energy metabolism | ***Nervous System*** | |
| **Significant Sources** | Insomnia, fatigue, depression, irritability, restlessness | |
| Widespread in foods; organ meats, mushrooms, avocados, broccoli, whole grains | ***Other*** | |
| Easily destroyed by heat | | Water retention (rare) |

lodging the PLP coenzyme from its enzymes; once loose, PLP breaks down and is excreted. Thus alcohol contributes to the destruction and loss of vitamin $B_6$ from the body.

Another drug that acts as a vitamin $B_6$ antagonist is INH, a drug that inhibits the growth of the tuberculosis bacterium.* INH has saved countless lives, but as a vitamin $B_6$ antagonist, it binds and inactivates the vitamin, inducing a deficiency. Whenever INH is used to treat tuberculosis, vitamin $B_6$ supplements must be given to protect against deficiency.

**antagonist:** a competing factor that counteracts the action of another factor. When a drug displaces a vitamin from its site of action, the drug renders the vitamin ineffective and thus acts as a vitamin antagonist.

• ***Vitamin $B_6$ Recommendations*** • Because the vitamin $B_6$ coenzymes play many roles in amino acid metabolism, previous RDA were expressed in terms of protein intakes; the current RDA for vitamin $B_6$, however, is not. Research does not support claims that large doses of vitamin $B_6$ enhance muscle strength or physical endurance. Pills cannot compete with a nutritious diet.

• ***Vitamin $B_6$ Deficiency*** • People given a vitamin $B_6$–deficient diet first experience weakness, irritability, and insomnia. Advanced symptoms include growth failure, impaired motor function, and convulsions. Immune function is also impaired in vitamin $B_6$ deficiency.

Women using oral contraceptives may have concerns about their vitamin $B_6$ status, but this may be unwarranted. Early studies reported signs of deficiency in oral contraceptive users, but that was when the pills contained estrogen at three to five times the quantities used today. Estrogen stimulates the breakdown of tryptophan, which requires vitamin $B_6$, thus creating a shortage of the vitamin.

• ***Vitamin $B_6$ Toxicity*** • The first major report of vitamin $B_6$ toxicity appeared in 1983. Until that time, everyone (including researchers and dietitians) believed that, like the other water-soluble vitamins, vitamin $B_6$ could not reach toxic concentrations in the body. The affected people had been taking more than 2 grams of vitamin $B_6$ daily (20 times the current Upper Level) for two months or more.

Some women use vitamin $B_6$ supplements in an attempt to treat the symptoms of premenstrual syndrome (PMS). PMS is a cluster of physical, emotional, and psychological symptoms that some women experience prior to menstruation. In about 5 percent of women with PMS, at least one physical or psychological symptom can reach such severity as to be temporarily disabling.

Specific PMS symptoms vary from woman to woman, but their timing is predictable: they begin seven to ten days prior to menstruation and wane after menstruation begins. The cause of PMS remains undefined, although researchers generally agree that the hormonal changes of the menstrual cycle must be responsible. Without a full understanding of PMS causes, medical treatments flounder,

*INH stands for *isonicotinic acid hydrazide.*

Common PMS symptoms:
- Headaches.
- Breast swelling and tenderness.
- Water retention.
- Weight gain.
- Irritability.
- Anxiety.
- Fatigue.
- Depression.
- Appetite changes and food cravings.
- Backaches.
- Acne.
- Constipation.

 U.S. Government

www.healthfinder.gov/searchoptions/topicsaz.htm

Search for Premenstrual syndrome

*Most protein-rich foods such as meat, fish, and poultry provide ample vitamin $B_6$; some vegetables and fruits are good sources, too.*

**carpal tunnel syndrome:** a pinched nerve at the wrist, causing pain or numbness in the hand.

**folate** (FOLE-ate): a B vitamin; also known as folic acid, folacin, or pteroylglutamic (tare-o-EEL-glue-TAM-ick) acid (PGA). The coenzyme forms are DHF (dihydrofolate) and THF (tetrahydrofolate).

and quack treatments abound. Among nutritional approaches, the taking of vitamin $B_6$ has received much attention, but seems to have done more harm than good.

Some people have taken vitamin $B_6$ supplements in an attempt to cure carpal tunnel syndrome and sleep disorders. Findings from one study suggest no correlation between blood levels of vitamin $B_6$ and symptoms of carpal tunnel syndrome, but those from another indicate that vitamin $B_6$ may be effective in treating carpal tunnel syndrome.[6] Self-prescribing is ill-advised, however, because large doses of vitamin $B_6$ taken for months or years may cause irreversible nerve degeneration. The summary table lists common symptoms of both deficiency and toxicity as well as the chief functions and food sources of vitamin $B_6$.

• ***Vitamin $B_6$ Food Sources*** • As you can see from the colored bars in Figure 10-9, meats, fish, and poultry (red), potatoes and a few other vegetables (green), and fruits (purple) offer vitamin $B_6$. As is true of most of the other vitamins, vegetables would rank considerably higher if foods were ranked by nutrient density (vitamin $B_6$ per 100 kcalories). Several servings of vitamin $B_6$–rich foods are needed to meet recommended intakes.

Foods lose vitamin $B_6$ when heated. Information is limited, but vitamin $B_6$ bioavailability from plant-derived foods seems to be lower than from animal-derived foods; fiber does not appear to hamper absorption.

## Folate

Folate, also known as folacin or folic acid, has a chemical name that would fit a flying dinosaur: pteroylglutamic acid (PGA for short). Its primary coenzyme form, THF, serves as part of an enzyme complex that transfers one-carbon compounds that arise during metabolism. Folate's coenzyme action helps convert vitamin $B_{12}$ to one of its coenzyme forms and helps synthesize the DNA required for all rapidly growing cells.

IN SUMMARY

### Vitamin $B_6$

**Other Names**

Pyridoxine, pyridoxal, pyridoxamine

**1998 RDA**

Adults (19–50 yr): 1.3 mg/day

**Upper Level**

Adults: 100 mg/day

**Chief Functions in the Body**

Part of coenzymes PLP (pyridoxal phosphate) and PMP (pyridoxamine phosphate) used in amino acid and fatty acid metabolism; helps to convert tryptophan to niacin and to serotonin; helps to make red blood cells

**Significant Sources**

Green and leafy vegetables, meats, fish, poultry, shellfish, legumes, fruits, whole grains

Easily destroyed by heat

| Deficiency Symptoms | Toxicity Symptoms |
|---|---|
| ***Blood/Circulatory System*** | |
| Anemia (small-cell type)[a] | Bloating |
| ***Mouth, Gums, Tongue*** | |
| Smooth tongue,[b] cracked corners of the mouth[c] | |
| ***Nervous/Muscular Systems*** | |
| Abnormal brain wave pattern, irritability, insomnia, muscle twitching, convulsions | Depression, fatigue, irritability, headaches, nerve damage causing numbness and muscle weakness leading to an inability to walk |
| ***Skin*** | |
| Irritation of sweat glands, scaly dermatitis | |
| ***Other*** | |
| Kidney stones | Bone pain |

[a]Small-cell–type anemia is *microcytic anemia*.
[b]Smoothness of the tongue is caused by loss of its surface structures and is termed *glossitis* (gloss-EYE-tis).
[c]Cracks at the corners of the mouth are termed *cheilosis* (kee-LOH-sis).

**Figure 10-9**
**Vitamin $B_6$ in Selected Foods**

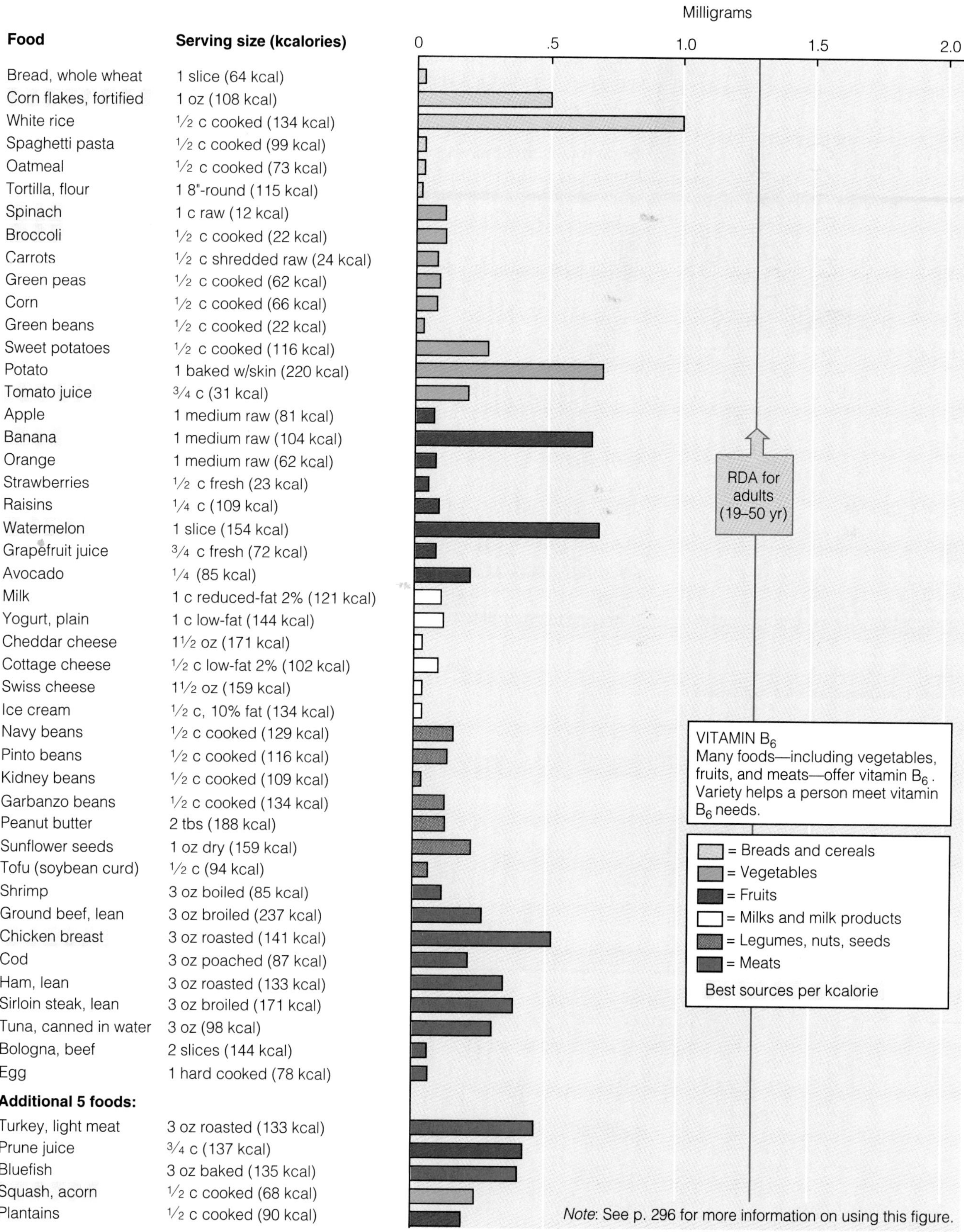


*Note*: See p. 296 for more information on using this figure.

Foods deliver folate mostly in the "bound" form—that is, combined with a string of amino acids (glutamate), known as polyglutamate (see Appendix C for the chemical structure). The intestine prefers to absorb the "free" folate form—folate with only one glutamate attached (the monoglutamate form). Enzymes on the intestinal cell surfaces hydrolyze the polyglutamate to monoglutamate and then attach a methyl group ($CH_3$). Special transport systems deliver the monoglutamate form to the liver and other body cells.

To store folate, cells add glutamate, converting it back to the polyglutamate form. To release it, they hydrolyze it back to monoglutamate again. To dispose of excess folate, the liver secretes most of it into bile and ships it to the gallbladder. Thus folate returns to the intestine in an enterohepatic circulation route like that of bile itself (review Figure 5-16 on p. 138).

In order for the folate coenzyme to function, the methyl group needs to be removed from methyl-THF. The enzyme that removes the methyl group requires the help of vitamin $B_{12}$. Without that help, folate becomes trapped inside cells in its methyl form, unavailable to support DNA synthesis and cell growth (see Figure 10-10).

This complicated system for handling folate is vulnerable to GI tract injuries. Since folate is actively secreted back into the intestinal tract with bile, it has to be reabsorbed repeatedly. If the GI tract cells are damaged, then folate is rapidly lost from the body. Such is the case in alcohol abuse; folate deficiency rapidly develops and, ironically, damages the GI tract further. The folate coenzymes, remember, are active in cell multiplication—and the cells lining the GI tract are among the most rapidly renewed cells in the body. Unable to make new cells, the GI tract deteriorates and not only loses folate, but also fails to absorb other nutrients.

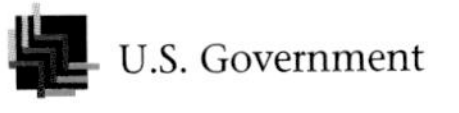
U.S. Government
www.healthfinder.gov/searchoptions/topicsaz.htm
Search for Folic acid

• ***Folate Recommendations*** • The bioavailability of folate ranges from 50 percent for foods to 100 percent for supplements taken on an empty stomach. These differences in bioavailability were considered in establishing the folate RDA. Naturally occurring folate from foods is given full credit. Synthetic folate from for-

**Figure 10-10**
**Folate's Absorption and Activation**

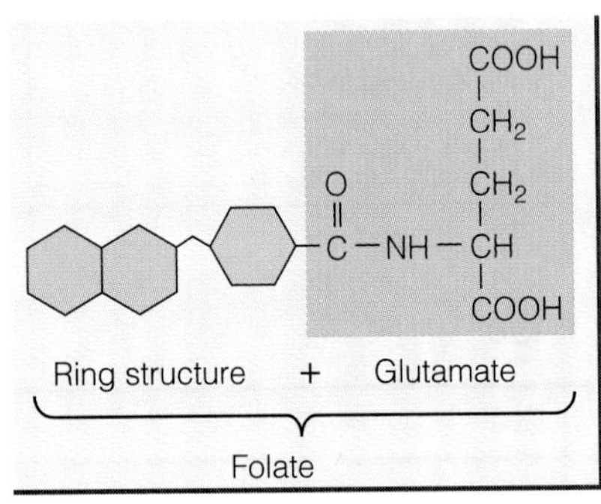
COOH
CH2
CH2
O
C—NH—CH
COOH
Ring structure + Glutamate
Folate

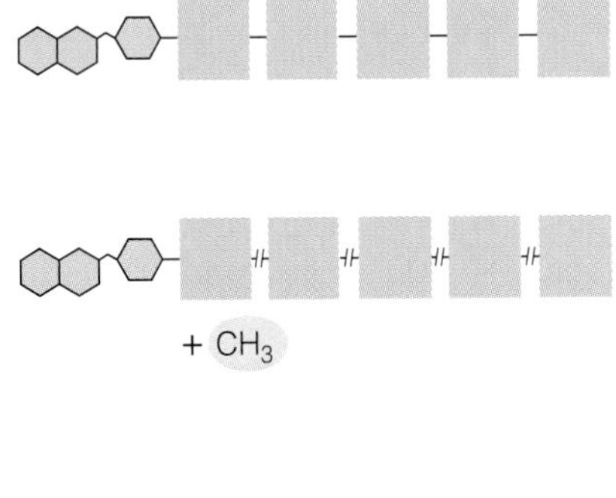
+ CH3

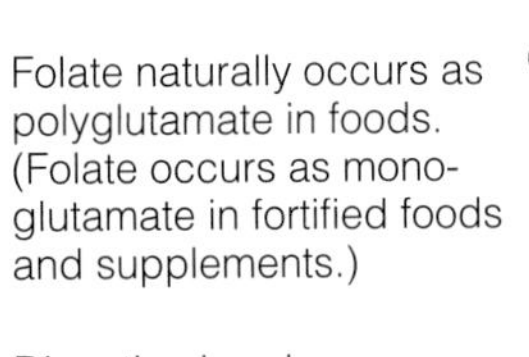
Folate naturally occurs as polyglutamate in foods. (Folate occurs as monoglutamate in fortified foods and supplements.)
Digestion breaks glutamates off . . .

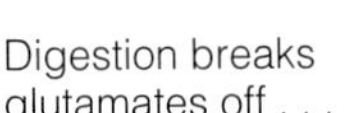
and adds a methyl group.

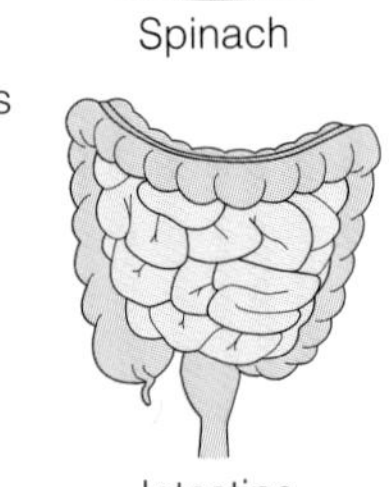
Spinach
Intestine

–CH3
Folate is absorbed and delivered to cells where it is trapped in its inactive form.

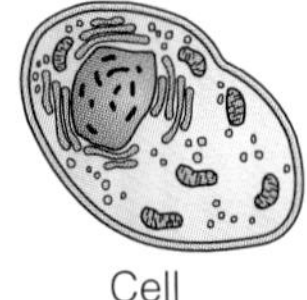
Cell

B12
CH3
Vitamin B12 removes and keeps the methyl group.

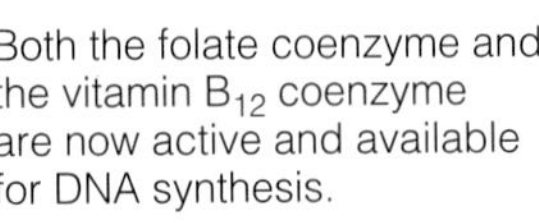
B12 – CH3
Both the folate coenzyme and the vitamin B12 coenzyme are now active and available for DNA synthesis.

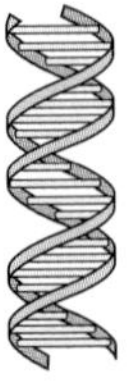
DNA

tified foods and supplements is given extra credit because, on average, it is 1.7 times more available than naturally occurring food folate. Thus a person consuming 100 micrograms of folate from foods and 100 micrograms from a supplement receives 270 dietary folate equivalents (DFE). (Chapter 15 presents a "How to" feature on how to estimate dietary folate equivalents on p. 479.) The need for folate rises considerably during pregnancy and whenever cells are multiplying, so the recommendations for pregnant women are considerably higher than for other adults.

**dietary folate equivalents (DFE):** the amount of folate available to the body from naturally occurring sources, fortified foods, and supplements, accounting for differences in the bioavailability from each source. Use the following equation to calculate DFE:

μg food folate + (1.7 × μg synthetic folate).

Using the example in the text:

100 μg/food
+ 170 μg/supplement (1.7 × 100 μg)
270 μg DFE

• ***Folate and Neural Tube Defects*** • Several research studies have confirmed the importance of folate in reducing the risks of neural tube defects.[7] The brain and spinal cord develop from the neural tube, and defects in its orderly formation during the early weeks of gestation may result in various central nervous system disorders and death.

**neural tube defects:** malformations of the brain, spinal cord, or both during embryonic development. The two main types of neural tube defects are spina bifida (literally, "split spine") and anencephaly ("no brain").

Folate supplements taken one month before conception and continued throughout the first trimester of pregnancy can help prevent neural tube defects. For this reason, the Public Health Service has recommended that all women of childbearing age who are capable of becoming pregnant should take 0.4 milligrams (400 micrograms) of folate daily.[8] This amount of folate can be met through a diet that includes at least five servings of fruits and vegetables daily, but many women typically fail to do so and receive only 0.2 milligrams of folate from foods daily. Furthermore, because the bioavailability of folate is enhanced, supplementation or fortification improves folate status more significantly than does a dietary intake that meets recommendations.[9]

Chapter 15 provides photos of neural tube development.

Spina Bifida Association of America
www.sbaa.org/Folic.htm

Women of childbearing age (15 to 45 yr) should:
- Eat folate-rich foods or
- Eat folate-fortified foods or
- Take a multivitamin daily.

Because many pregnancies are unplanned and because neural tube defects occur early in development before most women realize they are pregnant, the Food and Drug Administration (FDA) has mandated that grain products be fortified to deliver folate to the U.S. population.* Labels on fortified products may claim that "adequate intake of folate has been shown to reduce the risk of neural tube defects." Fortification is expected to prevent half of the 4000 neural tube defects that have occurred each year, but folate fortification raises safety concerns as well. Because high intakes of folate complicate the diagnosis of a vitamin $B_{12}$ deficiency, folate consumption should not exceed 1 milligram daily.[10] Whether it is wise to fortify our food supply with folate, but not vitamin $B_{12}$, is the subject of much debate.[11]

Recently, researchers have found a genetic flaw that induces a folate deficiency and predisposes women to bear infants with neural tube defects.[12] But folate's exact role in preventing neural tube defects remains unclear. Some women whose infants develop neural tube defects are not deficient in folate, and others with severe folate deficiencies do *not* give birth to infants with neural tube defects. Researchers continue to look for other factors that must also be involved.

• ***Folate and Heart Disease*** • The FDA's decision to fortify grain products with folate was strengthened by research indicating an important role for folate in preventing heart disease.[13] As Chapter 6 mentioned, research indicates that high levels of the amino acid homocysteine and low levels of folate increase the risk of fatal heart disease.[14] One of folate's key roles in the body is to break down homocysteine. Without folate, homocysteine accumulates, which seems to enhance blood clot formation and arterial wall deterioration. Cereal fortified with folate raises blood folate and reduces blood homocysteine levels, but whether fortification helps to prevent heart disease remains to be seen.[15]

• ***Folate Deficiency*** • Folate deficiency impairs cell division and protein synthesis—processes critical to growing tissues. In a folate deficiency, the replacement of red blood cells and GI tract cells falters. Not surprisingly, then, two of the first symptoms of a folate deficiency are anemia and GI tract deterioration.

**anemia:** literally, "too little blood." Anemia is any condition in which too few red blood cells are present, or the red blood cells are immature (and therefore large) or too small or contain too little hemoglobin to carry the normal amount of oxygen to the tissues. It is not a disease itself but can be a symptom of many different disease conditions, including many nutrient deficiencies, bleeding, excessive red blood cell destruction, and defective red blood cell formation.
- **an** = without
- **emia** = blood

*Bread products, flour, corn grits, cornmeal, farina, rice, macaroni, and noodles must be fortified with 1.4 milligrams of folate per 100 grams of food.

Large-cell anemia is known as **macrocytic** or **megaloblastic** anemia.
- **macro** = large
- **cyte** = cell
- **mega** = large

The anemia of folate deficiency is characterized by large, immature blood cells. Without folate, DNA synthesis slows and the cells lose their ability to divide. The nucleus of the cell is not released as normally occurs during development. As a result, the immature blood cells are enlarged and oval-shaped. They cannot carry oxygen or travel through the capillaries as efficiently as normal red blood cells. The table below provides a summary of information about folate.

Folate deficiencies may develop from inadequate intake and have been reported in infants fed goat's milk, which is notoriously low in folate. Folate deficiency may also result from impaired absorption or an unusual metabolic need for the vitamin. Metabolic needs increase wherever cell multiplication must speed up: in pregnancies involving twins and triplets; in cancer; in skin-destroying diseases such as chicken pox and measles; and in burns, blood loss, GI tract damage, and the like.

Highlight 17 discusses nutrient-drug interactions and includes a figure illustrating the similarities between the vitamin folate and the drug methotrexate (on p. 558).

Of all the vitamins, folate appears to be most vulnerable to interactions with drugs, which can lead to a secondary deficiency. Some drugs, notably anticancer drugs, have a chemical structure similar to folate and can displace the vitamin from enzymes and interfere with normal metabolism. Cancer cells, like all cells, need the real vitamin to multiply; without it, they die. Unfortunately, other cells in the body also need folate, and vitamin deficiency develops.

Aspirin and antacids also interfere with the body's handling of folate. Healthy adults who use these drugs to relieve an occasional headache or upset stomach need not be concerned, but people who rely heavily on aspirin or antacids should be aware of the nutrition consequences. Oral contraceptives may also impair folate status, as may smoking. Abnormalities in the cervical cells of oral contraceptive users and in the lung cells of smokers seem to indicate a "localized" folate deficiency. Early studies reported that folate supplements appeared to cause a regression in the development of the cervical cancer cells in these oral contraceptive users, but follow-up studies have not confirmed such a benefit.[16]

• ***Folate Food Sources*** • Figure 10-11 shows that folate is especially abundant in legumes and vegetables. The vitamin's name suggests the word *foliage,* and indeed,

## IN SUMMARY

### Folate

**Other Names**

Folic acid, folacin, pteroylglutamic acid (PGA)

**1998 RDA**

Adults: 400 μg/day

**Upper Level**

Adults: 1000 μg/day

**Chief Functions in the Body**

Part of coenzymes THF (tetrahydrofolate) and DHF (dihydrofolate) used in DNA synthesis and therefore important in new cell formation

**Significant Sources**

Leafy green vegetables, legumes, seeds, liver

Easily destroyed by heat and oxygen

| Deficiency Symptoms | Toxicity Symptoms |
|---|---|
| ***Blood/Circulatory System*** | |
| Anemia (large-cell type)[a] | |
| ***Digestive System*** | |
| Heartburn, diarrhea (loss of villi and their enzymes), constipation | |
| ***Immune System*** | |
| Suppression, frequent infections | |
| ***Mouth, Gums, Tongue*** | |
| Smooth, red tongue[b] | |
| ***Nervous System*** | |
| Depression, mental confusion, weakness, fainting, fatigue, insomnia, irritability | |
| ***Other*** | |
| | Masks vitamin $B_{12}$–deficiency symptoms |

[a]Large-cell–type anemia is known as either *macrocytic* or *megaloblastic anemia.*

[b]Smoothness of the tongue is caused by loss of its surface structures and is termed *glossitis* (gloss-EYE-tis).

**Figure 10-11**
**Folate in Selected Foods**

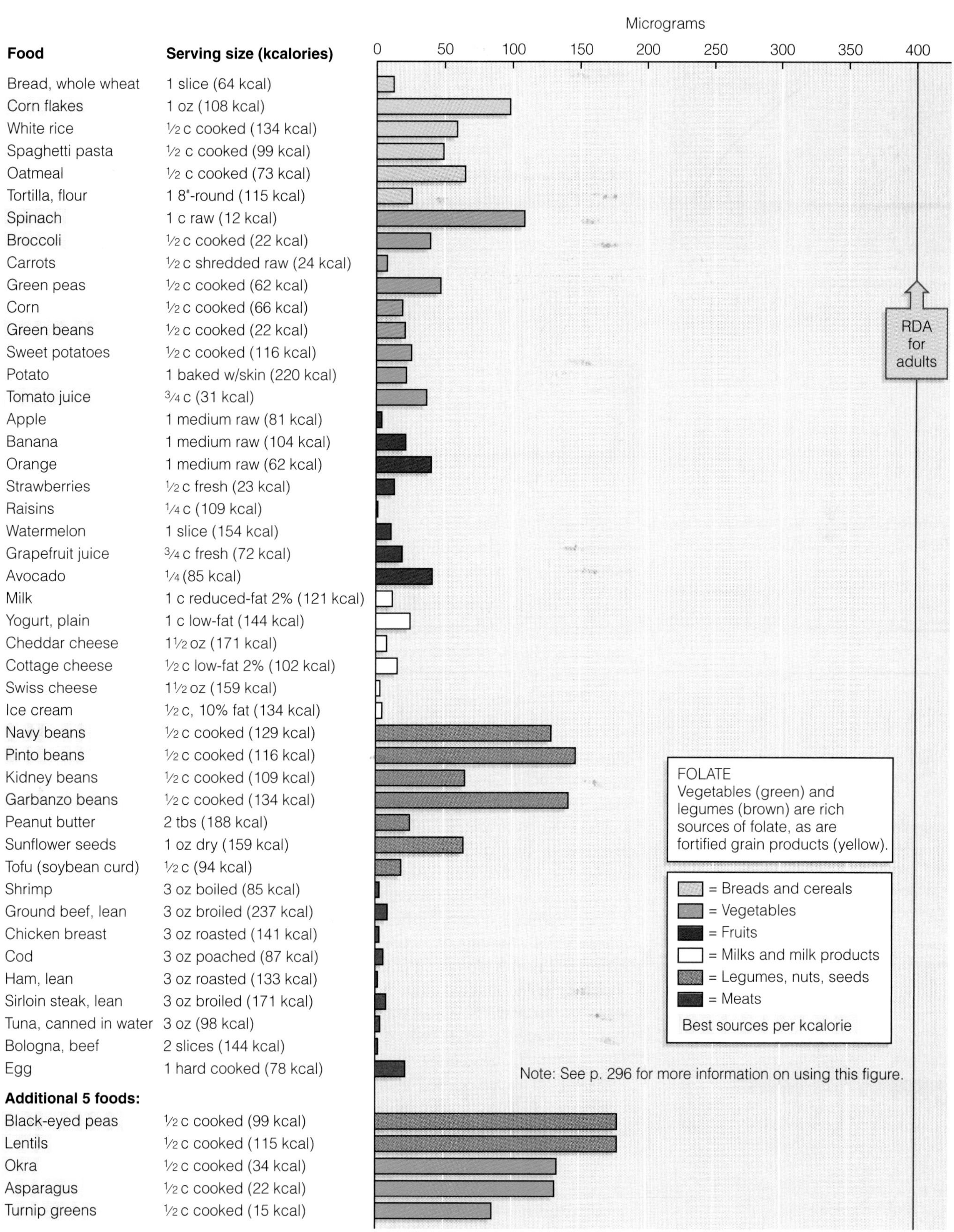

*Leafy green vegetables (such as spinach and broccoli), legumes (such as black beans, kidney beans, and black-eyed peas), liver, and some fruits (notably oranges) are naturally rich in folate.*

leafy green vegetables are outstanding sources. With fortification, grain products also contribute folate; the bioavailability of added folate is good, too, suggesting that consumers will benefit from the enrichment program.[17] The lack of red and white bars in Figure 10-11 illustrates that meats, milk, and milk products are poor folate sources. Heat and oxidation during cooking and storage can destroy as much as half of the folate in foods.

## Vitamin $B_{12}$

Vitamin $B_{12}$ and folate are closely related: each depends on the other for activation. Recall that vitamin $B_{12}$ removes a methyl group to activate the folate coenzyme; when folate donates its methyl group, the vitamin $B_{12}$ coenzyme becomes activated (review Figure 10-10). The regeneration of the amino acid methionine and the synthesis of DNA and RNA depend on both folate and vitamin $B_{12}$.* In addition, without any help from folate, vitamin $B_{12}$ maintains the sheath that surrounds and protects nerve fibers and promotes their normal growth. Bone cell activity and metabolism also seem to depend on vitamin $B_{12}$.

In the stomach, hydrochloric acid and the enzyme pepsin release vitamin $B_{12}$ from the proteins to which it was attached in foods. Then the vitamin binds with an "intrinsic factor" that is synthesized in the stomach. After the intrinsic factor attaches to vitamin $B_{12}$, the complex passes to the small intestine, where the vitamin is gradually absorbed into the bloodstream. Transport of vitamin $B_{12}$ in the blood depends on specific binding proteins.

Like folate, vitamin $B_{12}$ follows the enterohepatic circulation route. It is continually secreted into bile and delivered to the intestine, where it is reabsorbed. Because most of this vitamin $B_{12}$ is reabsorbed, healthy people rarely develop a deficiency even when their intake is minimal.

**vitamin $B_{12}$:** a B vitamin characterized by the presence of cobalt (see Figure 13-10 in Chapter 13); the active forms of coenzyme $B_{12}$ are methylcobalamin and deoxyadenosylcobalamin.

**intrinsic:** inside the system. The **intrinsic factor** is a glycoprotein (a protein with short polysaccharide chains attached) manufactured in the stomach that aids in the absorption of vitamin $B_{12}$.

• ***Vitamin $B_{12}$ Recommendations*** • The RDA for adults is only 2.4 micrograms of vitamin $B_{12}$ a day—just over two-millionths of a gram. The ink in the period at the end of this sentence may weigh about 2.4 micrograms. But tiny though this amount appears to the human eye, it contains billions of molecules of vitamin $B_{12}$, enough to provide coenzymes for all the enzymes that need its help.

• ***Vitamin $B_{12}$ Deficiency*** • Most vitamin $B_{12}$ deficiencies reflect inadequate absorption, not poor intake. Inadequate absorption typically occurs for one of two reasons: a lack of hydrochloric acid or a lack of intrinsic factor. Many people, especially those over 60, develop atrophic gastritis, a common condition in older people that damages the cells of the stomach. Atrophic gastritis may also develop in response to iron deficiency or infection with *Helicobacter pylori,* the bacterium implicated in ulcer formation. Without healthy stomach cells, production of hydrochloric acid and intrinsic factor diminishes. Even with adequate intake from foods, vitamin $B_{12}$ status suffers. Without hydrochloric acid, the vitamin is not released from the dietary proteins and so is not available for binding with the intrinsic factor. Without the intrinsic factor, the vitamin cannot be absorbed.

**atrophic gastritis:** chronic inflammation of the stomach accompanied by a diminished size and functioning of the mucosa and glands.
- **atrophy** = wasting
- **gastro** = stomach
- **itis** = inflammation

Some people inherit a defective gene for the intrinsic factor. In such cases, or when the stomach has been injured and cannot produce enough of the intrinsic factor, vitamin $B_{12}$ must be injected to bypass the need for intestinal absorption. The vitamin $B_{12}$ deficiency caused by atrophic gastritis and a lack of intrinsic factor is known as pernicious anemia.[18]

Because vitamin $B_{12}$ is required to convert folate to its active form, one of the most obvious vitamin $B_{12}$–deficiency symptoms is the anemia of folate deficiency.

**pernicious** (per-NISH-us) **anemia:** a blood disorder that reflects a vitamin $B_{12}$ deficiency caused by lack of intrinsic factor and characterized by abnormally large and immature red blood cells. Other symptoms include muscle weakness and irreversible neurological damage.
- **pernicious** = destructive

*In the body, methionine serves as a methyl ($CH_3$) donor. In doing so, methionine can be converted to other amino acids. Although some of these amino acids can regenerate methionine, methionine is still an essential amino acid that is needed in the diet.

This anemia is characterized by large, immature red blood cells, which are indicative of slow DNA synthesis and an inability to divide (see Figure 10-12). When folate is trapped in its inactive (methyl folate) form due to vitamin $B_{12}$ deficiency, or is unavailable due to folate deficiency itself, DNA synthesis slows.

First to be affected in vitamin $B_{12}$ or folate deficiency are the rapidly growing blood cells. Either vitamin $B_{12}$ or folate will clear up the anemia, but if folate is given when vitamin $B_{12}$ is needed, the result is disastrous: devastating neurological symptoms. Remember that vitamin $B_{12}$, but not folate, maintains the sheath that surrounds and protects nerve fibers and promotes their normal growth. Folate "cures" the *blood* symptoms of a vitamin $B_{12}$ deficiency, but allows the *nerve* symptoms to progress. By doing so, folate "masks" a vitamin $B_{12}$ deficiency. A deficiency of vitamin $B_{12}$ causes a creeping paralysis of the nerves and muscles, which begins at the extremities and works inward and up the spine. Early detection and correction are necessary to prevent permanent nerve damage and paralysis. With sufficient folate in the diet, the neurological symptoms of vitamin $B_{12}$ deficiency can develop without evidence of anemia. Such interactions between folate and vitamin $B_{12}$ highlight some of the safety issues surrounding the use of supplements and fortification of the food supply. The table on p. 314 provides a summary of information about vitamin $B_{12}$.

• ***Vitamin $B_{12}$ Food Sources*** • Vitamin $B_{12}$ is unique among the vitamins in being found almost exclusively in foods derived from animals. Anyone who eats reasonable amounts of meat is guaranteed an adequate intake, and vegetarians who use milk products or eggs are also protected from deficiency. Fermented soy products such as miso (a soybean paste) and sea algae such as spirulina do *not* provide vitamin $B_{12}$ in its active form. Extensive research shows that the amounts listed on the labels of these plant products are inaccurate and misleading, because the vitamin $B_{12}$ is in an inactive, unavailable form. Vegans need a reliable source, such as vitamin $B_{12}$–fortified soy "milk," meat replacements, or vitamin $B_{12}$ supplements. Yeast that is grown on a vitamin $B_{12}$–enriched medium and mixed with that medium provides some vitamin $B_{12}$, but yeast itself does not contain active vitamin $B_{12}$.

Reminder: *Meat replacements* are textured vegetable-protein products formulated to look and taste like meat, fish, or poultry and often fortified with nutrients commonly found in meats.

People who stop eating foods containing vitamin $B_{12}$ may take up to 20 years to develop deficiencies because the body recycles much of its vitamin $B_{12}$, reabsorbing it over and over again. Even when the body fails to absorb vitamin $B_{12}$, deficiency may take up to three years to develop because the body conserves its supply.

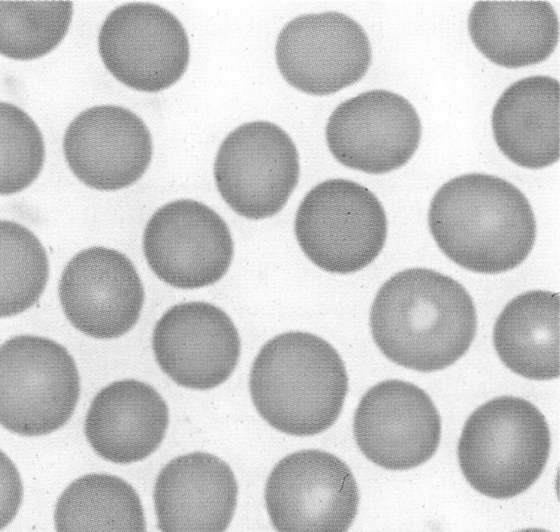

Normal blood cells. The size, shape, and color of the red blood cells show that they are normal. Mature red blood cells have lost their nuclei.

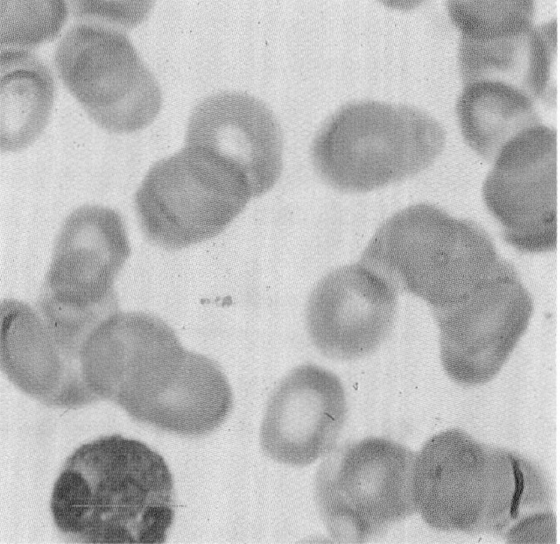

Blood cells in pernicious anemia (megaloblastic). Megaloblastic blood cells are arrested at an immature stage of development, so they still have their nuclei. For this reason, they are slightly larger than normal red blood cells, and their shapes are irregular.

**Figure 10-12**
**Normal and Anemic Blood Cells**

The anemia of folate deficiency is indistinguishable from that of vitamin $B_{12}$ deficiency. Appendix E describes the biochemical tests used to differentiate the two conditions.

IN SUMMARY

**Vitamin $B_{12}$**

| Other Names | Deficiency Disease Name |
|---|---|
| Cobalamin (and related forms) | Pernicious anemia[a] |
| **1998 RDA** | **Deficiency Symptoms** |
| Adults: 2.4 µg/day | ***Blood/Circulatory System*** <br> Anemia (large-cell type)[b] |
| **Chief Functions in the Body** | ***Mouth, Gums, Tongue*** <br> Smooth tongue[c] |
| Part of coenzymes methylcobalamin and deoxyadenosylcobalamin used in new cell synthesis; helps to maintain nerve cells; reforms folate coenzyme; helps to break down some fatty acids and amino acids | ***Nervous System*** <br> Fatigue, degeneration of peripheral nerves progressing to paralysis |
| **Significant Sources** | ***Skin*** |
| Animal products (meat, fish, poultry, shellfish, milk, cheese, eggs) <br> Easily destroyed by microwave cooking | Hypersensitivity |

[a]The name *pernicious anemia* refers to the vitamin $B_{12}$ deficiency caused by atrophic gastritis and a lack of intrinsic factor, but not to that caused by inadequate dietary intake.
[b]Large-cell–type anemia is known as either *macrocytic* or *megaloblastic anemia.*
[c]Smoothness of the tongue is caused by loss of its surface structures and is termed *glossitis* (gloss-EYE-tis).

As mentioned earlier, the water-soluble vitamins are particularly vulnerable to losses in cooking. For most of these nutrients, microwave heating minimizes losses as well as, or better than, traditional cooking methods. Such is not the case for vitamin $B_{12}$, however. Microwave heating inactivates vitamin $B_{12}$.[19] To preserve this vitamin, use the oven or stovetop instead of a microwave to cook meats and milk products (major sources of vitamin $B_{12}$).

## Non-B Vitamins

Nutrition scientists debate whether other dietary compounds might also be considered vitamins. In some cases, the compounds may be conditionally essential—that is, needed by the body from foods when synthesis becomes insufficient to support normal growth and metabolism. In other cases, the compounds may be vitamin impostors—not needed under any circumstances.

**choline** (KOH-leen): a nitrogen-containing compound found in foods and made in the body from the amino acid methionine. Choline is used to make the phospholipid lecithin and the neurotransmitter acetylcholine.

• ***Choline*** • The essentiality of choline has been blurry for decades, in part because the body can make choline from the amino acid methionine.[20] Furthermore, choline is commonly found in many foods, as part of the lecithin molecule (review Figure 5-9 on p. 133). Consequently, choline deficiencies are rare. Without any dietary choline, however, synthesis alone appears to be insufficient to meet the body's needs, making it a conditionally essential nutrient. For this reason, the 1998 DRI report established an Adequate Intake (AI) for choline. The body uses choline to make the neurotransmitter acetylcholine and the phospholipid lecithin. The accompanying table summarizes key choline facts.

**inositol** (in-OSS-ih-tall): a nonessential nutrient that can be made in the body from glucose. Inositol is used in cell membranes.

**carnitine** (CAR-neh-teen): a nonessential nutrient made in the body from the amino acid lysine.

• ***Inositol and Carnitine*** • Like choline, inositol and carnitine can be made by the body, but unlike choline, no recommendations have been established. Researchers continue to explore the possibility that these substances may be essential. Even if they are essential, though, supplements are unnecessary because these compounds are widespread in foods.

Some vitamin companies include choline, inositol, and carnitine in their formulations to make their vitamin pills look more "complete" than others, but these

IN SUMMARY

**Choline**

**1998 Adequate Intake (AI)**

Men: 550 mg/day
Women: 425 mg/day

**Upper Level**

Adults: 3500 mg/day

**Chief Functions in the Body**

Needed for the synthesis of the neurotransmitter acetylcholine and the phospholipid lecithin

**Deficiency Symptoms**

Liver damage

**Toxicity Symptoms**

Body odor, sweating, salivation, reduced growth rate, low blood pressure, liver damage

**Significant Sources**

Milk, liver, eggs, peanuts

compounds confer no advantage. For a rational way to compare different vitamin-mineral supplements, read Highlight 10.

• ***Vitamin Impostors*** • Other substances have been mistaken for essential nutrients for human beings because they are needed for growth by bacteria or other forms of life. Among them are PABA (para-aminobenzoic acid), the bioflavonoids (vitamin P or hesperidin), pyrroloquinoline quinone (methoxatin), orotic acid, lipoic acid, and ubiquinone (coenzyme $Q_{10}$). Other names associated wrongly with vitamins are "vitamin $B_5$" (another name for pantothenic acid), "vitamin $B_{15}$" (also called "pangamic acid," a hoax), and "vitamin $B_{17}$" (laetrile, an alleged "cancer cure" and not a vitamin or a cure by any stretch of the imagination—in fact, laetrile is a potentially dangerous substance).

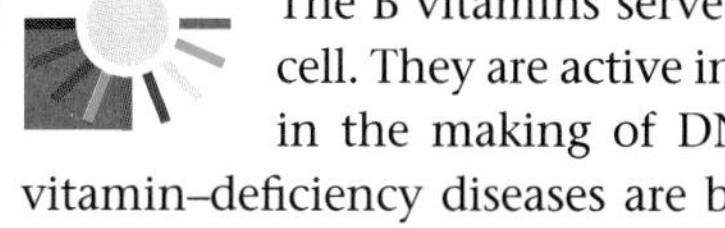

IN SUMMARY

The B vitamins serve as coenzymes that facilitate the work of every cell. They are active in carbohydrate, fat, and protein metabolism and in the making of DNA and thus new cells. Historically famous B vitamin–deficiency diseases are beriberi (thiamin), pellagra (niacin), and pernicious anemia (vitamin $B_{12}$). Pellagra can be prevented by adequate protein because the amino acid tryptophan can be converted to niacin in the body. A high intake of folate can mask the blood symptom of a vitamin $B_{12}$ deficiency, but it will not prevent the associated nerve damage. Vitamin $B_6$ participates in amino acid metabolism and can be toxic in excess. Biotin and pantothenic acid serve important roles in energy metabolism and are abundant in food. Many substances that people claim as B vitamins are not.

# The B Vitamins—In Concert

This chapter has described some of the impressive ways that vitamins work individually, as if their many actions in the body could easily be disentangled. In fact, oftentimes it is difficult to tell which vitamin is truly responsible for a given effect because the nutrients are interdependent; the presence or absence of one affects another's absorption, metabolism, and excretion. You have already seen this interdependence with folate and vitamin $B_{12}$.

Riboflavin and vitamin $B_6$ are another example of a B vitamin relationship. One of the riboflavin coenzymes, FMN, assists the enzyme that converts vitamin $B_6$ to its coenzyme form PLP. Consequently, a severe riboflavin deficiency can impair vitamin $B_6$ activity. Thus a deficiency of one nutrient may alter the action of another. Furthermore, a deficiency of one nutrient may create a deficiency of another. For example, a vitamin $B_6$ deficiency hinders calcium absorption and enhances magnesium excretion. These interdependent relationships are evident in many of the roles of B vitamins in the body.

## B Vitamin Roles

Figure 10-13 is intended to convey an *impression* of the ways B vitamins busily work in metabolic pathways all over the body. Metabolism is the body's work, and the B vitamin coenzymes are indispensable to every step. In scanning the pathways of metabolism depicted in the figure, note the abbreviations for the coenzymes that keep the processes going.

Look at the first step in the now-familiar pathway of glucose breakdown. To break down glucose to pyruvate, the cells must have certain enzymes. For the enzymes to work, they must have the niacin coenzyme NAD. To make NAD, the cells must be supplied with niacin (or enough of the amino acid tryptophan to make niacin). They can make the rest of the coenzyme without dietary help.

The next step in glucose catabolism is the breakdown of pyruvate to acetyl CoA. The enzymes involved in this step require NAD plus the thiamin coenzyme TPP. The cells can manufacture the TPP they need from thiamin, if thiamin is in the diet.

Another coenzyme needed for this step is CoA. Predictably, the cells can make CoA except for an essential part that must be obtained in the diet—pantothenic acid. Another coenzyme requiring biotin serves the enzyme complex involved in converting pyruvate to a compound that can combine with acetyl CoA in the TCA cycle.

These and other coenzymes are involved throughout all the metabolic pathways. When the diet provides riboflavin, the body synthesizes FAD—a needed coenzyme in the TCA cycle. Vitamin $B_6$ is an indispensable part of PLP—a coenzyme required for many amino acid conversions, for a crucial step in the making of the iron-containing portion of hemoglobin for red blood cells, and for many other reactions. Folate becomes THF—the coenzyme required for the synthesis of new genetic material and therefore new cells. The vitamin $B_{12}$ coenzyme, in turn, regenerates THF to its active form; thus vitamin $B_{12}$ is also necessary for the formation of new cells.

Thus each of the B vitamin coenzymes is involved, directly or indirectly, in energy metabolism. Some are facilitators of the energy-releasing reactions themselves; others help build new cells to deliver the oxygen and nutrients that permit the energy pathways to run.

## B Vitamin Deficiencies

Now suppose the body's cells lack one of these B vitamins—niacin, for example. Without niacin, the cells cannot make NAD. Without NAD, the enzymes involved in every step of the glucose-to-energy pathway cannot function. Then, because all the body's activities require energy, literally everything begins to grind to a halt. This is no exaggeration. The deadly disease pellagra, caused by niacin deficiency, produces the "devastating *Ds*": dermatitis, which reflects a failure of the skin; dementia, a failure of the nervous system; diarrhea, a failure of digestion and absorption; and eventually, as would be the case for any severe nutrient deficiency, death. These symptoms are the obvious ones, but a niacin deficiency affects all other organs, too, because all are dependent on the energy pathways. In short, niacin is like the horseshoe nail for want of which a war was lost.

For want of a nail, a horseshoe was lost.
For want of a horseshoe, a horse was lost.
For want of a horse, a soldier was lost.
For want of a soldier, a battle was lost.
For want of a battle, the war was lost,
And all for the want of a horseshoe nail!
—Mother Goose

All the vitamins are like horseshoe nails. With any B vitamin deficiency, many body systems become deranged, and similar symptoms may appear. Removing "horseshoe nails" can have disastrous and far-reaching effects.

Deficiencies of single B vitamins seldom show up in isolation, however. After all, people do not eat nutrients singly; they eat foods, which contain mixtures of

**Figure 10-13**
**Metabolic Pathways Involving B Vitamins**

These metabolic pathways were introduced in Chapter 7 and are presented here to highlight the many coenzymes that facilitate the reactions. These coenzymes depend on the following vitamins:

- NAD and NADP: niacin.
- TPP: thiamin.
- CoA: pantothenic acid.
- $B_{12}$: vitamin $B_{12}$.
- FMN and FAD: riboflavin.
- THF: folate.
- PLP: vitamin $B_6$.
- Biotin.

Pathways leading toward acetyl CoA and the TCA cycle are catabolic, and those leading toward amino acids, glycogen, and fat are anabolic. For further details, see Appendix C.

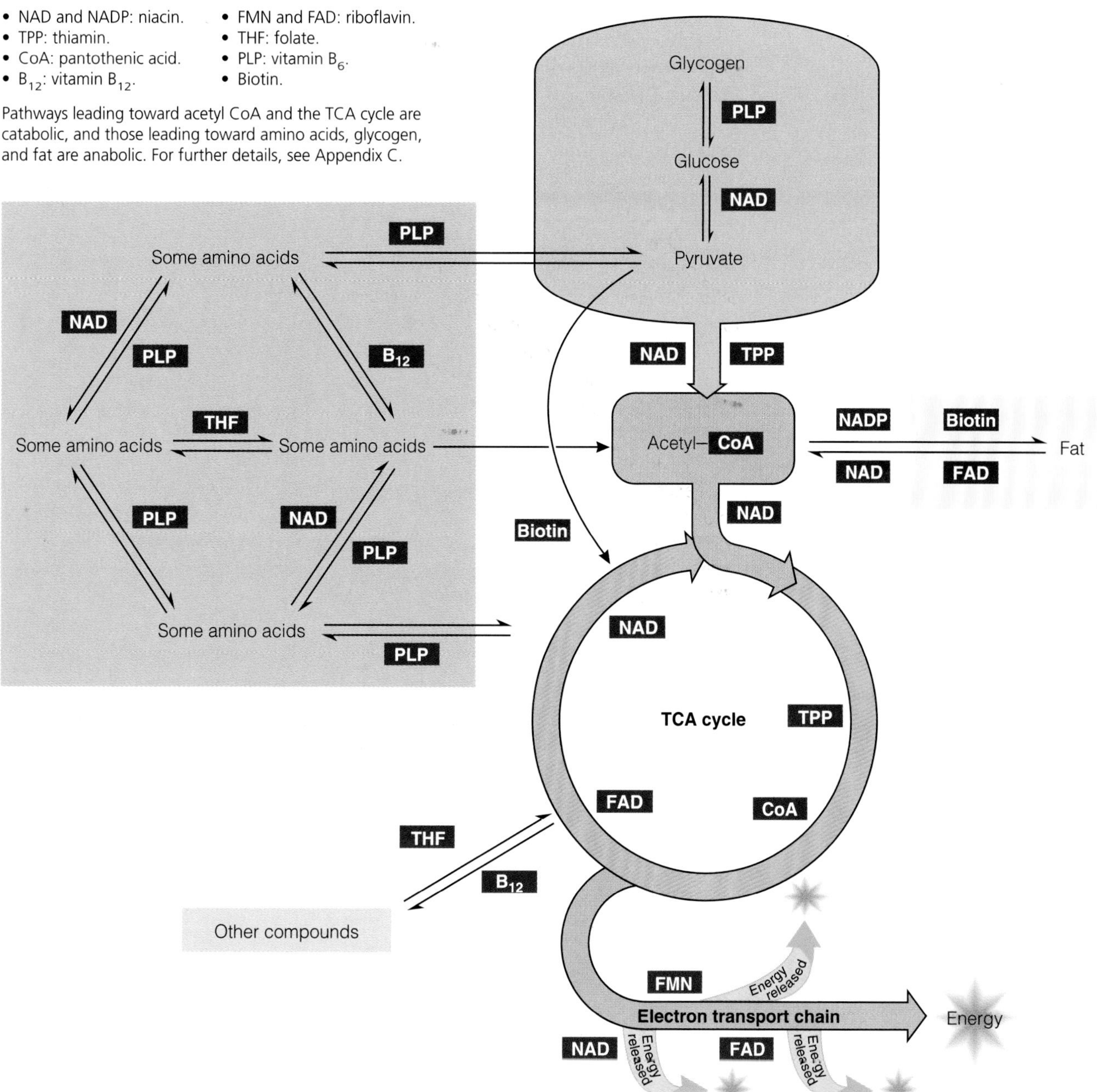


nutrients. Only in two cases described earlier—beriberi and pellagra—have dietary deficiencies associated with single B vitamins been observed on a large scale in human populations. Even in these cases, the deficiencies were not pure. Both diseases were attributed to deficiencies of single vitamins, but both likely were deficiencies of several vitamins in which one vitamin stood out above the rest. When foods containing the vitamin known to be needed were provided, the vitamins that may have been in short supply came as part of the package.

Major deficiency diseases of epidemic proportions such as pellagra and beriberi are no longer seen in the United States and Canada, but lesser deficiencies of nutrients, including the B vitamins, sometimes occur in people whose food choices are poor because of poverty, ignorance, illness, or poor health habits like

**Figure 10-14**

**B Vitamin–Deficiency Symptom—The Smooth Tongue of Glossitis**

In a B vitamin deficiency, the tongue becomes smooth due to atrophy of the tissue (glossitis).

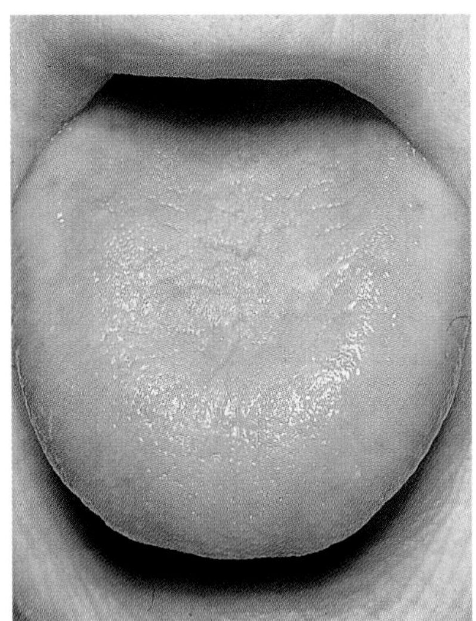

alcohol abuse. (Review Highlight 7 to fully appreciate how alcohol induces vitamin deficiencies and interferes with energy metabolism.) Remember from Chapter 1 that deficiencies can arise not only from deficient intakes (primary causes), but also for other (secondary) reasons.

In identifying deficiencies, it is important not to assume that a particular symptom always has the same cause. The skin and the tongue (shown in Figure 10-14) appear to be especially sensitive to B vitamin deficiencies, but isolating these body parts in the summary tables earlier in this chapter gives them undue emphasis. Both the skin and the tongue are readily visible in a physical examination. If the skin is degenerating, other tissues beneath it may be, too. Similarly, the mouth and tongue are the visible part of the digestive system; if they are abnormal, most likely the rest of the GI tract is, too. The physician sees and reports the deficiency's outward manifestations, but the impact of a vitamin deficiency occurs inside the cells of the body. The "How to" feature below offers other insights into symptoms and their causes.

## B Vitamin Toxicities

Toxicities of the B vitamins from foods alone are unknown, but they can occur when people overuse supplements. With supplements, the quantities can quickly overwhelm the cells. Consider that one small capsule can easily deliver 2 milligrams of vitamin $B_6$, but it would take more than 3000 bananas, 6600 cups of rice, or 3600 chicken breasts to supply an equivalent amount. When the cells become oversaturated with a vitamin, they must work to eliminate the excess. The cells dispatch water-soluble vitamins to the urine for excretion, but sometimes they fail to regain homeostasis.

## B Vitamin Food Sources

Chapter 2 describes the enrichment of grain products and compares the nutrient content of refined, enriched, and whole-grain breads (see Figure 2-4 on p. 45).

Significantly, the deficiency diseases of beriberi and pellagra were eliminated by supplying foods—not pills. Vitamin pill advertisements make much of the fact that vitamins are indispensable to life, but human beings obtained their nour-

### HOW TO / Distinguish Symptoms and Causes

It is more and more apparent that no one can observe a symptom and automatically jump to a conclusion regarding its cause. The summary tables in this chapter show that deficiencies of riboflavin, niacin, and vitamin $B_6$ can all cause skin rashes. But so can a deficiency of protein, linoleic acid, and vitamin A. Because skin is on the outside and easy to see, it is a useful indicator of things-going-wrong-inside-cells. But by itself, a skin symptom says nothing about its possible cause.

The same is true of anemia. Anemia is often caused by iron deficiency, but it can also be caused by a folate or vitamin $B_{12}$ deficiency; by digestive tract failure to absorb any of these nutrients; or by such nonnutritional causes as infections, parasites, cancer, or loss of blood. Again, no specific nutrient will always cure a given symptom.

A person who feels chronically tired may be tempted to self-diagnose iron-deficiency anemia and self-prescribe an iron supplement. But this will relieve tiredness only if the cause is indeed iron-deficiency anemia. If the cause is a folate deficiency, taking iron will only prolong the fatigue. A person who is better informed may decide to take a vitamin supplement with iron, covering the possibility of a vitamin deficiency. But the symptom may have a nonnutritional cause. If the cause of the tiredness is actually hidden blood loss due to cancer, the postponement of a diagnosis may be fatal. When fatigue is caused by a lack of sleep, of course, no nutrient or combination of nutrients can replace a good night's rest. A person who is chronically tired should see a physician rather than self-prescribe. If the condition is nutrition related, a registered dietitian should be consulted as well.

ishment from foods for centuries before vitamin pills existed. If the diet lacks a vitamin, the first solution is to adjust food intake to obtain that vitamin.

Manufacturers of so-called *natural* vitamins boast that their pills are purified from real foods rather than synthesized in a laboratory. Think back on the course of human evolution; it is not *natural* to take any kind of pill. In reality, the finest, most complete vitamin "supplements" available are grains, vegetables, fruits, meat, fish, poultry, eggs, legumes, nuts, and milk and milk products.

The food figures presented in this chapter, taken together, sing the praises of the balanced diet. The cereal and bread group delivers thiamin, riboflavin, niacin, and folate. The fruit and vegetable groups excel in folate. The meat group serves thiamin, niacin, vitamin $B_6$, and vitamin $B_{12}$ well. The milk and milk products group stands out for riboflavin and vitamin $B_{12}$. A diet that offers a variety of foods from each group, prepared with reasonable care, serves up ample B vitamins.

IN SUMMARY

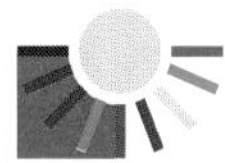

The B vitamin coenzymes work together in energy metabolism. Some are facilitators of the energy-releasing reactions themselves; others help build cells to deliver the oxygen and nutrients that permit the energy pathways to run. These vitamins depend on each other to function optimally; a deficiency of any of them creates multiple problems. Fortunately, a variety of foods from each of the five food groups will provide an adequate supply of all of the B vitamins.

# Vitamin C

Two hundred and fifty years ago, any man who joined the crew of a seagoing ship knew he had only half a chance of returning alive—not because he might be slain by pirates or die in a storm, but because he might contract the dread disease scurvy. As many as two-thirds of a ship's crew might die of scurvy on a long voyage. Only men on short voyages, especially around the Mediterranean Sea, were free of scurvy. No one knew the reason: that on long ocean voyages, the ship's cook used up the fresh fruits and vegetables early and then served cereals and meats until the return to port.

**scurvy:** the vitamin C–deficiency disease.

The first nutrition experiment ever performed on human beings was devised in 1747 to find a cure for scurvy. James Lind, a British physician, divided 12 sailors with scurvy into six pairs. Each pair received a different supplemental ration: cider, vinegar, sulfuric acid, seawater, oranges and lemons, or a purgative mixed with spices. Those receiving the citrus fruits quickly recovered, but sadly, it was 50 years before the British navy required all vessels to provide every sailor with lime juice daily. This tradition gave British sailors the nickname "limeys."

**purgative:** a strong laxative.

The antiscurvy "something" in limes and other foods was dubbed the antiscorbutic factor. Nearly 200 years later, the factor was isolated from lemon juice and found to be a six-carbon compound similar to glucose; it was named ascorbic acid. Shortly thereafter, it was synthesized, and today hundreds of millions of vitamin C pills are produced in pharmaceutical laboratories each year and sold for a few dollars a bottle.

**antiscorbutic** (AN-tee-skor-BUE-tik) **factor:** the original name for vitamin C.
- **anti** = against
- **scorbutic** = causing scurvy

**ascorbic acid:** one of the two active forms of vitamin C (see Figure 10-15). Many people refer to vitamin C by this name.
- **a** = without
- **scorbic** = having scurvy

## *Vitamin C Roles*

Vitamin C parts company with the B vitamins in its mode of action. In some settings, vitamin C helps a specific enzyme perform its job, but in others, it acts in a more general way as an antioxidant.

• ***As an Antioxidant*** • An antioxidant is any substance that prevents or inhibits the oxidation of another substance. In doing so, the antioxidant becomes

**antioxidant:** a compound that protects others from oxidation by being oxidized itself. An antioxidant donates electrons to another substance; that substance becomes reduced as the antioxidant simultaneously becomes oxidized. Chemists describe the antioxidant action of vitamin C as maintaining the "oxidation-reduction equilibrium," or "redox state."

**Figure 10-15**

**Active Forms of Vitamin C**

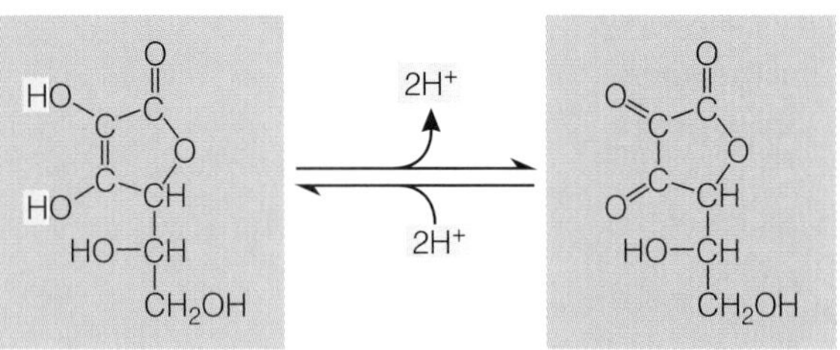


Ascorbic acid is the reduced form of vitamin C. Ascorbic acid can easily give up two hydrogens with their electrons, thereby becoming dehydroascorbic acid. Molecules with unpaired electrons (free radicals) combine with antioxidants such as vitamin C instead of causing oxidative damage to the cells.

Dehydroascorbic acid is the oxidized form of vitamin C. The reversibility of this reaction is key to vitamin C's role as an antioxidant.

oxidized itself, but this is useful because it protects the other substance from being altered or even destroyed by oxidation. Vitamin C is like a bodyguard for water-soluble substances; it stands ready to sacrifice its own life to save theirs. Figure 10-15 illustrates how vitamin C's structure can change, so that it can serve as an antioxidant.

Because of vitamin C's antioxidant property, manufacturers sometimes add it to foods as a preservative. In the cells and body fluids, vitamin C helps to prevent damage to tissues, which may be important in preventing disease. In the intestines, vitamin C protects iron from oxidation and so enhances iron absorption.

**• *In Collagen Formation* •** Vitamin C helps to form the fibrous structural protein of connective tissues known as collagen. Collagen serves as the matrix on which bones and teeth are formed. When a person is wounded, collagen glues the separated tissues together, forming scars. Cells are held together largely by collagen; this is especially important in the artery walls, which must expand and contract with each beat of the heart, and in the thin capillary walls, which must withstand a pulse of blood every second or so without giving way.

The body makes all proteins by stringing together chains of amino acids. In collagen, the amino acids proline and lysine appear in abundance. During the synthesis of collagen, each time a proline or lysine is added to the growing protein chain, an enzyme hydroxylates it (adds an OH group to it), making the amino acid hydroxyproline or hydroxylysine, respectively. Figure 10-16 illustrates the conversion of proline to hydroxyproline. This hydroxylase enzyme requires both vitamin C and iron. Iron works as a cofactor in the reaction, and vitamin C maintains iron in the form that allows it to do so. Without them, the hydroxylation step does not occur. Hydroxyproline and hydroxylysine facilitate the binding together of collagen fibers to make strong, ropelike structures.

Highlight 11 discusses the role of antioxidant nutrients in disease prevention in more detail. Chapter 13 provides more details on the relationship between vitamin C and iron.

Reminder: *Collagen* is the protein material from which connective tissues such as scars, tendons, ligaments, and the foundations of bones and teeth are made.

**cofactor:** a mineral element that, like a coenzyme, works with an enzyme to facilitate a chemical reaction. The cofactor maintains the structural integrity of the enzyme and may also facilitate the enzyme's catalytic activity.

**• *In Stress* •** The adrenal glands contain more vitamin C than any other organ in the body, and during stress, these glands release the vitamin, together with hormones, into the blood. The vitamin's exact role in the stress reaction remains unclear. Psychological stress alone does not appear to raise needs above the RDA, but some physical stresses such as infections, wound healing, and exposure to cold raise vitamin C needs. When immune system cells are called into action, they use a lot of oxygen and produce oxidants that can damage the cells themselves.[21] Thus vitamin C is used as an antioxidant whenever the immune system becomes active. The hormone thyroxin (made with vitamin C's help) regulates the metabolic rate, which speeds up under extreme stress and also when the body needs to produce extra heat—for example, in fever or cold weather.

**Figure 10-16**

**Vitamin C's Role in Hydroxyproline Synthesis**

Collagen is unique among body proteins because it contains large amounts of the amino acid hydroxyproline. The hydroxylase enzyme, which forms hydroxyproline by adding a hydroxyl group to the amino acid proline, requires vitamin C and iron.

Proline

Hydroxylase + vitamin C and iron

Hydroxyproline

**• *As a Cure for the Common Cold* •** Newspaper headlines touting vitamin C as a cure for colds have appeared frequently over the years, but research supporting such claims has been conflicting and controversial. A major review of the research on vitamin C in the treatment and prevention of the common cold revealed a significant difference in duration of less than a day per cold in favor of those taking a daily dose of at least 1 gram vitamin C.[22] The term *significant* means that *statistical* analysis suggests that the findings probably didn't arise from a chance event, but from the experimental treatment being tested. Is a day enough savings to warrant routine daily supplementation? Supplement users seem to think so.

Interestingly, researchers in one study found that those who received the placebo *but thought they were receiving vitamin C* had fewer colds than the group who received vitamin C *but thought they were receiving the placebo.* (Never underestimate the healing power of faith!)

Discoveries of the ways vitamin C works in the body provide possible links between the vitamin and the common cold. Anyone who has ever had a cold knows the discomfort of a runny or stuffed-up nose. Nasal congestion develops in response to elevated blood histamine, and people commonly take antihistamines for relief. Like an antihistamine, vitamin C comes to the rescue and deactivates histamine.

**histamine** (HISS-tah-mean, or HISS-tah-men): a substance produced by cells of the immune system as part of a local immune reaction to an antigen; participates in causing inflammation.

• ***In Disease Prevention*** • The role of vitamin C in the prevention of, or therapy for, cancer and other diseases is still being studied, and findings are presented in Highlight 11. Researchers conducting an epidemiological study of more than 11,000 U.S. adults reported an inverse relationship between all causes of death and vitamin C intake up to a few hundred milligrams.[23] The relationship was stronger for men than for women and remained apparent after controlling for variables such as age, sex, cigarette smoking, disease history, race, and education.

## Vitamin C Recommendations

How much vitamin C does a person need? Allowances set by different nations are based on similar research findings, but range from 30 milligrams per day in Great Britain and in Canada (for nonsmoking women) to 60 in the United States and 100 in Japan.[24] As Figure 10-17 illustrates, all the different recommendations fall within a fairly narrow range of safe intakes.

The requirement—the amount needed to prevent the overt symptoms of scurvy—is only 10 milligrams daily. However, 10 milligrams a day does not saturate all the body tissues; larger intakes increase the body's total vitamin C. At about 60 milligrams per day, the tissues in the average person stop responding to further increases in intake, and at 100 milligrams per day, 95 percent of the population probably reaches tissue saturation. After the tissues are saturated, excess vitamin C is readily excreted.

The RDA for vitamin C, like all the RDA, is intended to maintain health in healthy people, not to restore health in sick people. A variety of physical stresses deplete the body's vitamin C supply and may make higher intakes desirable. Among the stresses known to increase vitamin C needs are infections; burns; extremely high or low temperatures; intakes of toxic heavy metals such as lead, mercury, and cadmium; the chronic use of certain medications, including aspirin, barbiturates, and oral contraceptives; and cigarette smoking. Cigarette smoke contains oxidants, which greedily deplete this potent antioxidant.[25] Exposure to cigarette smoke, especially when accompanied by low intakes of vitamin C, depletes the body's pool in both active and passive smokers.[26] Similarly, people who chew tobacco have low levels of vitamin C as well.[27] Whereas the RDA for nonsmokers is 60 milligrams a day, the RDA for people who smoke cigarettes regularly is 100 milligrams; the Canadian RNI provides a 50 percent increase for those who smoke. Some researchers suggest a higher recommendation (perhaps 200 milligrams daily) for the next revision of the RDA.[28]

After oral surgery, dentists may prescribe supplemental vitamin C to hasten healing. After major operations or extensive burns, when scar tissue is forming, the amount needed may be as high as 1000 milligrams (1 gram) a day or even more. In individual cases, a physician may prescribe vitamin C supplements for certain conditions; self-medication is not recommended.

**Figure 10-17**
**Vitamin C Intake (mg)**

Recommendations differ, but all are generously above the minimum requirement and below the toxicity level. In contrast, megadoses of 2 grams (2000 milligrams) a day are clearly way up in the clouds.

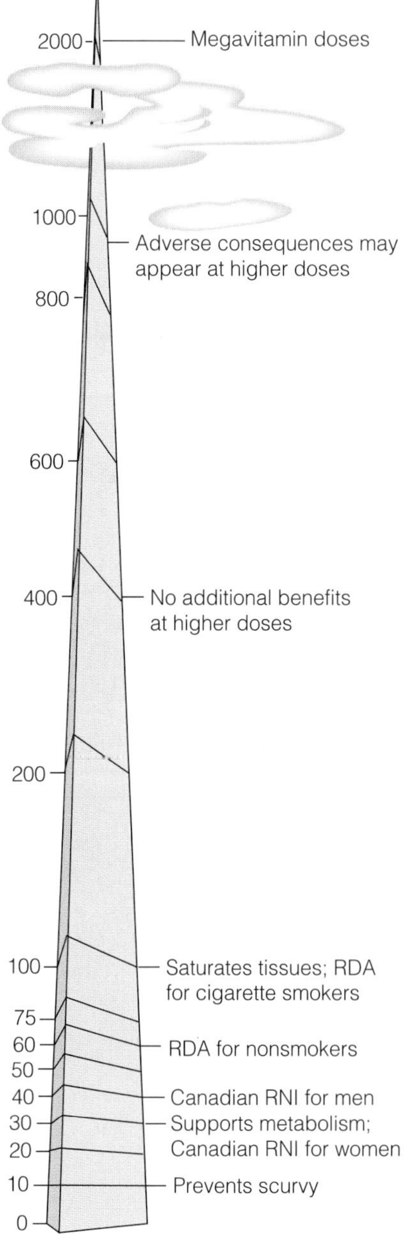


## Vitamin C Deficiency

Two of the most notable signs of a vitamin C deficiency reflect its role in maintaining the integrity of blood vessels. The gums bleed easily around the teeth, and capillaries under the skin break spontaneously, producing pinpoint hemorrhages (see Figure 10-18). Atherosclerotic plaques grow rapidly in the arteries. The summary table on p. 323 reviews deficiency symptoms and other vitamin C information.

When the vitamin C pool falls to about a fifth of its optimal size (this may take several weeks on a diet lacking vitamin C), scurvy symptoms begin to appear. Failure to promote normal collagen synthesis causes further hemorrhaging. Muscles, including the heart muscle, degenerate. The skin becomes rough, brown, scaly, and dry. Wounds fail to heal because scar tissue will not form. Bone rebuilding falters; the ends of the long bones become softened, malformed, and painful, and fractures occur. The teeth become loose as the cartilage around them weakens. Anemia and infections are common. There are also characteristic psychological signs, including hysteria and depression. Sudden death is likely, occasioned by severe atherosclerosis or by massive bleeding into the joints and body cavities.

Once diagnosed, scurvy is readily reversed by vitamin C. Moderate doses in the neighborhood of 100 milligrams per day are sufficient, curing the scurvy within about five days. Such an intake is easily achieved by including vitamin C–rich foods in the diet.

## Vitamin C Toxicity

The easy availability of vitamin C supplements and the publication of books recommending vitamin C to prevent colds and cancer have led thousands of people to take large doses of vitamin C. Not surprisingly, instances of vitamin C's causing harm have surfaced.

Toxic effects such as nausea, abdominal cramps, and diarrhea are often reported. Several instances of interference with medical regimens are also known. Large amounts of vitamin C excreted in the urine obscure the results of tests used to detect diabetes, giving a false positive result in some instances and a false negative in others. People taking anticlotting medications may unwittingly counteract the effect if they also take massive doses of vitamin C.* Those who have a tendency toward gout and those who have a genetic abnormality that alters vitamin C's breakdown to its excretion products are prone to forming kidney stones

**false positive:** a test result indicating that a condition is present (positive) when in fact it is not (therefore false).

**false negative:** a test result indicating that a condition is *not* present (negative) when in fact it is present (therefore false).

Reminder: *Gout* is a metabolic disease in which uric acid crystals precipitate in the joints.

*Vitamin C interferes with such anticoagulants as warfarin, dicumarol, heparin, and coumadin. It is unclear whether vitamin C inhibits the absorption or the action of these drugs.

**Figure 10-18**

**Vitamin C–Deficiency Symptoms—Scorbutic Gums, Scorbutic Pose, and Pinpoint Hemorrhages**

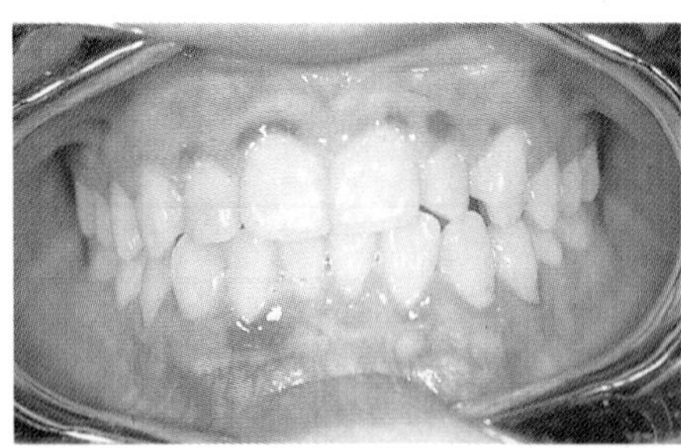

Scorbutic gums. Unlike other lesions of the mouth, scurvy presents a symmetrical appearance without infection.

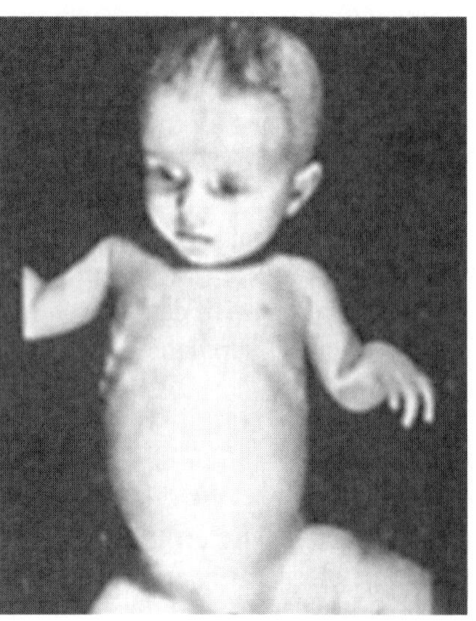

Infant scurvy. This is the characteristic "scorbutic pose," with legs bent and thighs rotated open. The infant's joints are painful, and she will cry if made to move.

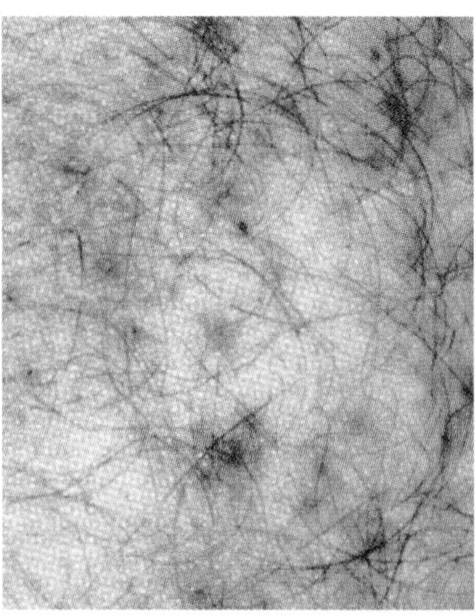

Pinpoint hemorrhages. Small red spots appear in the skin, indicating spontaneous bleeding internally.

IN SUMMARY

**Vitamin C**

**Other Names**

Ascorbic acid

**1989 RDA**

Adults: 60 mg/day

Smokers: 100 mg/day

**Chief Functions in the Body**

Collagen synthesis (strengthens blood vessel walls, forms scar tissue, provides matrix for bone growth), antioxidant, thyroxin synthesis, amino acid metabolism, strengthens resistance to infection, helps in absorption of iron

**Significant Sources**

Citrus fruits, cabbage-type vegetables, dark green vegetables, cantaloupe, strawberries, peppers, lettuce, tomatoes, potatoes, papayas, mangoes

Easily destroyed by heat and oxygen

**Deficiency Disease Name**

Scurvy

| Deficiency Symptoms | Toxicity Symptoms |
|---|---|
| ***Blood/Circulatory System*** | |
| Anemia (small-cell type),[a] atherosclerotic plaques, pinpoint hemorrhages | |
| ***Digestive System*** | |
| | Nausea, abdominal cramps, diarrhea |
| ***Immune System*** | |
| Suppression, frequent infections | |
| ***Mouth, Gums, Tongue*** | |
| Bleeding gums, loosened teeth | |
| ***Nervous/Muscular Systems*** | |
| Muscle degeneration and pain, hysteria, depression | Headache, fatigue, insomnia |
| ***Skeletal System*** | |
| Bone fragility, joint pain | |
| ***Skin*** | |
| Rough skin, blotchy bruises | Hot flashes, rashes |
| ***Other*** | |
| Failure of wounds to heal | Interference with medical tests, aggravation of gout symptoms, urinary tract problems, kidney stones[b] |

[a]Small-cell–type anemia is *microcytic anemia*.

[b]People who have a tendency toward gout and those who have a genetic abnormality that alters the breakdown of vitamin C are prone to forming kidney stones. Vitamin C is inactivated and degraded by several routes, sometimes producing oxalate, which can form stones in the kidneys.

if they take large doses of vitamin C.* Vitamin C supplements are dangerous for people with iron overload; vitamin C enhances iron absorption and releases iron from body stores.[29]

A person who has taken large doses of vitamin C for a long time (say, ten times the RDA daily for several weeks) may adapt by limiting absorption and destroying and excreting more of the vitamin than usual. If the person then suddenly reduces intake to normal, the accelerated disposal system may not be able to put on its brakes fast enough to avoid destroying too much of the vitamin. It has been suggested that adults who stop taking large doses may develop scurvy on intakes that would protect most adults, but evidence is scanty on this point. If scurvy does develop, the situation is similar to the withdrawal reaction seen in drug and alcohol abusers when they discontinue drug use. When people cease taking vitamin C in excessive doses, they might be wise to do so gradually.

**withdrawal reaction:** a reaction to removal of a substance (usually, a drug) that reveals that the user has become dependent. An infant whose mother had taken massive doses of vitamin C developed **rebound scurvy** on an intake that would have been adequate for most infants.

Few instances warrant consuming more than 100 to 300 milligrams of vitamin C a day. For adults who dose themselves with 1 to 2 grams a day, the risks may not be great; those taking more than 2 grams, and especially those taking amounts above 8 grams per day, should be aware of the distinct possibility of harm.

*Vitamin C is inactivated and degraded by several routes, and sometimes oxalate, which can form kidney stones, is produced along the way. People may also develop oxalate crystals in their kidneys regardless of vitamin C status.

*When dietitians say "vitamin C," people think "oranges."*

To protect the vitamin C in foods:

- Store cut produce in airtight wrappers and juices in closed containers (it's easily oxidized).
- Refrigerate produce, and avoid high temperatures and long cooking times (it's vulnerable to heat).
- Use a microwave oven, or steam vegetables in a small amount of water (it's lost in the liquid).
- Add vegetables *after* water has come to a boil (warm water activates vitamin C–destroying enzymes before the water gets hot enough to inactivate them).

*But these foods are also rich in vitamin C.*

In conclusion, the range of safe vitamin C intakes seems to be broad, as is typical for water-soluble vitamins. Between the absolute minimum of 10 milligrams a day and a reasonable maximum of perhaps 300 milligrams, nearly everyone can find a suitable intake. People who venture outside these limits may be taking health risks.

## Vitamin C Food Sources

Fruits and vegetables can easily provide a generous amount of vitamin C. A cup of orange juice at breakfast, a salad for lunch, and a stalk of broccoli and a potato for dinner alone provide more than 300 milligrams. Clearly, a person making such food choices needs no vitamin C pills.

Figure 10-19 shows the amounts of vitamin C in various common foods. The overwhelming abundance of purple and green bars reveals not only that the citrus fruits are justly famous for being rich in vitamin C, but that other fruits and vegetables are in the same league. A single serving of broccoli, green pepper, cauliflower, cantaloupe, or strawberries provides more than 50 milligrams of the vitamin (and an array of other nutrients) for less than 60 kcalories. Because vitamin C is vulnerable to heat, raw fruits and vegetables usually have a higher nutrient density than their cooked counterparts.

The potato is an important source of vitamin C, not because one potato by itself meets the daily need, but because potatoes are such a common staple that they make significant contributions.[30] In fact, scurvy was unknown in Ireland until the potato blight of the mid-1840s when some two million people died of malnutrition and infection.[31] Potatoes provide about 20 percent of all the vitamin C in the U.S. diet.

The lack of yellow, white, brown, and red bars in Figure 10-19 confirms that grains, milk (except breast milk), legumes, and meats are notoriously poor sources of vitamin C. Organ meats (liver, kidneys, and others) and raw meats contain some vitamin C, but most people don't eat large quantities of these. Raw meats and fish contribute enough vitamin C to be significant in parts of Alaska, Canada, and Japan, but elsewhere fruits and vegetables are necessary to supply sufficient vitamin C.

As mentioned earlier, food manufacturers sometimes add vitamin C or one of its close relatives to foods to prevent oxidation and spoilage. Some beverages and most cured meats, such as luncheon meats, use a variation of vitamin C. This compound safely preserves these food products, but it does not provide vitamin C activity in the body.[32] Simply put, "ham and bacon cannot replace fruits and vegetables."[33]

### IN SUMMARY

*Vita* means life. After this discourse on the vitamins, who could dispute that they deserve their name? Their regulation of metabolic processes makes them vital to normal growth, development, and maintenance of the body. The remarkable roles of the vitamins continue in the next chapter.

**Figure 10-19**
**Vitamin C in Selected Foods**

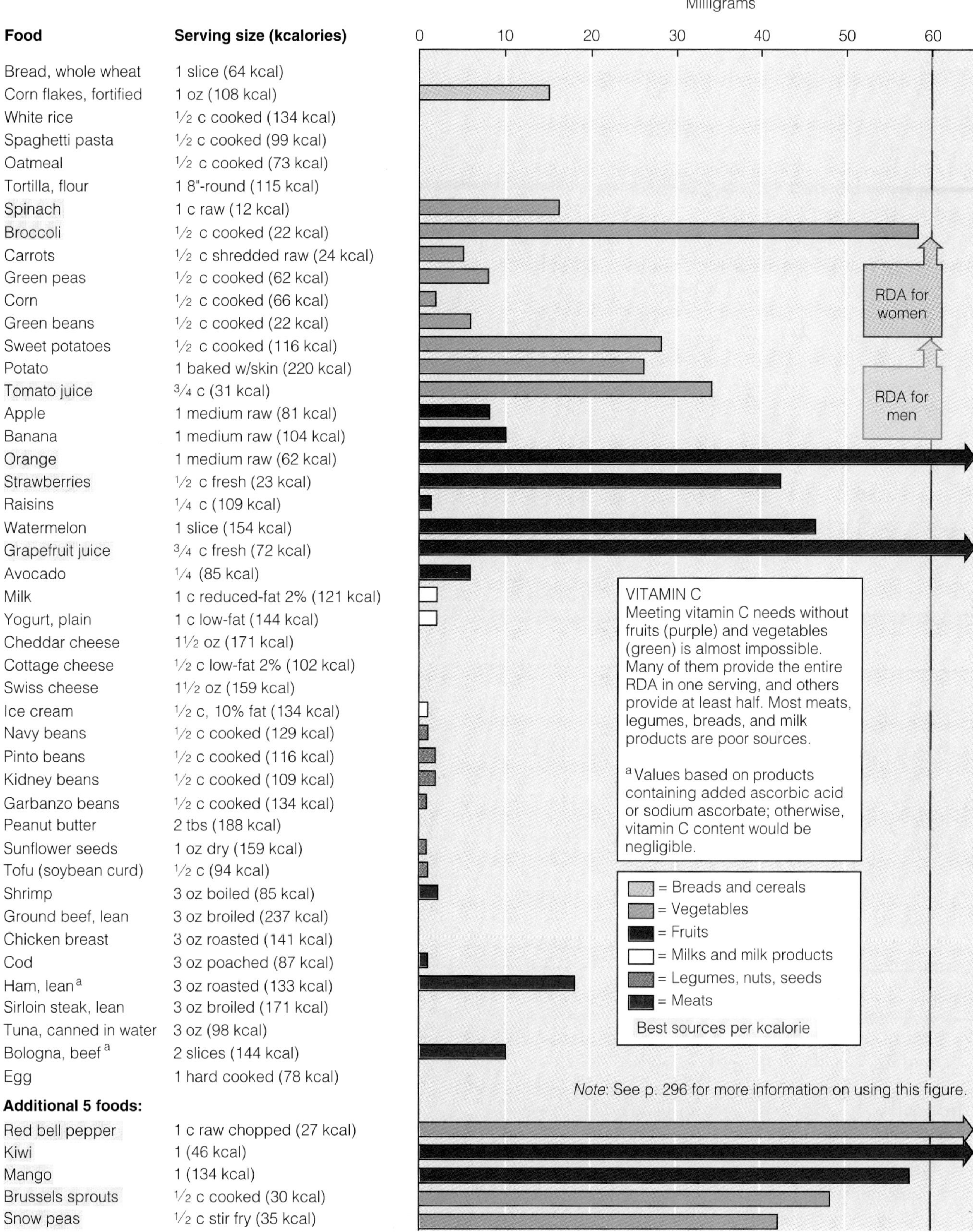

## Making It Click

These problems give you practice in doing simple vitamin-related calculations (answers are provided in Appendix K). Be sure to show your calculations for each problem.

1. Review the units in which vitamins are measured (a spot check).
   a. For each of these vitamins, note the unit of measure:

      Thiamin — Folate
      Riboflavin — Vitamin $B_{12}$
      Niacin — Vitamin C
      Vitamin $B_6$

   b. Recall from the chapter's description of people's self-dosing with vitamin $B_6$ that people who suffer toxicity symptoms may be taking more than 2 grams a day, whereas the RDA is less than 2 *milli*grams. How much higher than 2 milligrams is 2 grams?
   c. Vitamin $B_{12}$ is measured in micrograms. How many micrograms are in a gram? How many grams are in a teaspoon of a granular powder? How many micrograms does that represent? What is your RDA for vitamin $B_{12}$?

This exercise should convince you that the amount of vitamins a person needs is indeed quite small—yet still essential.

2. Be aware of how niacin intakes are affected by dietary protein availability.
   a. Refer to the "How to" on p. 302, and calculate how much niacin a woman receives from a diet that delivers 90 grams protein and 9 milligrams niacin. (Assume her RDA for protein is 46 grams/day.) Show your calculations.
   b. Is this woman getting her RDA of niacin (15 milligrams NE)?

This exercise should demonstrate that protein helps meet niacin needs.

## Study Questions

These questions will help you review the chapter. You will find the answers in the discussions on the pages provided.

1. How do the vitamins differ from the energy nutrients? (p. 291)
2. Describe some general differences between fat-soluble and water-soluble vitamins. (p. 293)
3. Which B vitamins are involved in energy metabolism? Protein metabolism? Cell division? (p. 294)
4. For thiamin, riboflavin, niacin, biotin, pantothenic acid, vitamin $B_6$, folate, vitamin $B_{12}$, and vitamin C, state:
   - Its chief function in the body.
   - Its characteristic deficiency symptoms.
   - Its significant food sources. (see respective summary tables)
5. What is the relationship of tryptophan to niacin? (pp. 299, 302)
6. Describe the relationship between folate and vitamin $B_{12}$. (pp. 306, 308, 312)
7. What risks are associated with high doses of niacin? Vitamin $B_6$? Vitamin C? (pp. 301, 305–306, 322–323)

These questions will help you prepare for an exam. Answers can be found in Appendix K.

1. Vitamins:
   a. are inorganic compounds.
   b. yield energy when broken down.
   c. are soluble in either water or fat.
   d. perform best when linked in long chains.
2. The rate and extent to which a vitamin is absorbed and used in the body is known as its:
   a. bioavailability.
   b. intrinsic factor.
   c. physiological effect.
   d. pharmacological effect.
3. Many of the B vitamins serve as:
   a. coenzymes.
   b. antagonists.
   c. antioxidants.
   d. serotonin precursors.
4. With respect to thiamin, which of the following is the most nutrient dense?
   a. 1 slice whole-wheat bread (69 kcalories and 0.1 milligram thiamin)
   b. 1 cup yogurt (144 kcalories and 0.1 milligram thiamin)
   c. 1 cup snow peas (69 kcalories and 0.22 milligram thiamin)
   d. 1 chicken breast (141 kcalories and 0.06 milligram thiamin)
5. The body can make niacin from:
   a. tyrosine.
   b. serotonin.
   c. carnitine.
   d. tryptophan.
6. The vitamin that protects against neural tube defects is:
   a. niacin.
   b. folate.

c. riboflavin.
d. vitamin $B_{12}$.

7. A lack of intrinsic factor may lead to:
   a. beriberi.
   b. pellagra.
   c. pernicious anemia.
   d. atrophic gastritis.
8. Which of the following is a B vitamin?
   a. inositol
   b. carnitine
   c. vitamin $B_{15}$
   d. pantothenic acid
9. Vitamin C serves as a(n):
   a. coenzyme.
   b. antagonist.
   c. antioxidant.
   d. intrinsic factor.
10. The RDA for vitamin C is highest for:
   a. smokers.
   b. athletes.
   c. alcoholics.
   d. the elderly.

## Notes


1. M. J. Jackson, The assessment of bioavailability of micronutrients: Introduction, *European Journal of Clinical Nutrition* 51 (1997): S1–S2.
2. Committee on the Dietary Reference Intakes, *Dietary Reference Intakes for Thiamin, Riboflavin, Niacin, Vitamin $B_6$, Folate, Vitamin $B_{12}$, Pantothenic Acid, Biotin, and Choline* (Washington, D.C.: National Academy Press, 1998).
3. C. K. Austin, C. E. Goodman, and L. L. Van Halderen, Absence of malnutrition in a population of homeless veterans, *Journal of the American Dietetic Association* 96 (1996): 1283–1285.
4. J. J. Cunningham, Micronutrients as nutriceutical interventions in diabetes mellitus, *Journal of the American College of Nutrition* 17 (1998): 7–10; M. T. Behme, Nicotinamide and diabetes prevention, *Nutrition Reviews* 53 (1995): 137–139.
5. J. E. Leklem, Vitamin B-6, in *Present Knowledge,* eds., E. E. Ziegler and L. J. Filer (Washington, D.C.: International Life Sciences Institute Press, 1996), pp. 174–183; K. M. Riggs and coauthors, Relations of vitamin B-12, vitamin B-6, folate, and homocysteine to cognitive performance in the Normative Aging Study, *American Journal of Clinical Nutrition* 63 (1996): 306–314.
6. A. Franzblau and coauthors, The relationship of vitamin $B_6$ status to median nerve function and carpal tunnel syndrome among active industrial workers, *Journal of Occupational and Environmental Medicine* 38 (1996): 485–491; A. L. Bernstein and J. S. Dineson, Effect of pharmacologic doses of vitamin $B_6$ on carpal tunnel syndrome, electroencephalographic results, and pain, *Journal of the American College of Nutrition* 12 (1993): 73–76.
7. C. E. Butterworth, Jr., and A. Bendich, Folic acid and the prevention of birth defects, *Annual Review of Nutrition* 16 (1996): 73–97.
8. From the Centers for Disease Control and Prevention, Recommendations for use of folic acid to reduce number of spina bifida cases and other neural tube defects, *Journal of the American Medical Association* 269 (1993): 1233–1238.
9. G. J. Cuskelly, H. McNulty, and J. M. Scott, Effect of increasing dietary folate on red-cell folate: Implications for prevention of neural tube defects, *Lancet* 347 (1996): 657–659.
10. Folate and neural tube defects: US policy evolves, *Nutrition Reviews* 51 (1993): 358–361.
11. V. Herbert and J. Bigaouette, Call for endorsement of a petition to the Food and Drug Administration to always add vitamin B-12 to any folate fortification or supplement, *American Journal of Clinical Nutrition* 65 (1997): 572–573.
12. A. M. Molloy and coauthors, Thermolabile variant of 5,10methylenetetrahydrofolate reductase associated with low red-cell folates: Implications for folate intake recommendations, *Lancet* 349 (1997): 1591–1593.
13. J. B. Ubbink, P. J. Becker, and W. J. H. Vermaak, Will an increased dietary folate intake reduce the incidence of cardiovascular disease? *Nutrition Reviews* 54 (1996): 213–216.
14. H. I. Morrison and coauthors, Serum folate and risk of fatal coronary heart disease, *Journal of the American Medical Association* 275 (1996): 1893–1896; O. Nygard and coauthors, Plasma homocysteine levels and mortality in patients with coronary artery disease, *New England Journal of Medicine* 337 (1997): 230–236; J. Selhub and coauthors, Association between plasma homocysteine concentrations and extracranial carotid-artery stenosis, *New England Journal of Medicine* 332 (1995): 286–291; M. J. Stampfer and M. R. Malinow, Can lowering homocysteine levels reduce cardiovascular risk? *New England Journal of Medicine* 332 (1995): 328–329; J. B. Ubbink, Homocysteine—An atherogenic and a thrombogenic factor? *Nutrition Reviews* 53 (1995): 323–332; J. S. Stamler and A. Slivka, Biological chemistry of thiols in the vasculature and in vascular-related disease, *Nutrition Reviews* 54 (1996): 1–30.
15. M. R. Malinow and coauthors, Reduction of plasma homocyst(e)ine levels by breakfast cereal fortified with folic acid in patients with coronary heart disease, *New England Journal of Medicine* 338 (1998): 1009–1015.
16. A. R. Giuliano and S. Gapstur, Can cervical dysplasia and cancer be prevented with nutrients? *Nutrition Reviews* 56 (1998): 9–16.
17. C. M. Pfeiffer and coauthors, Absorption of folate from fortified cereal-grain products and of supplemental folate consumed with or without food determined by using a dual-label stable-isotope protocol, *American Journal of Clinical Nutrition* 66 (1997): 1388–1397.
18. B. H. Toh, I. R. van Driel, and P. A. Gleeson, Pernicious anemia, *New England Journal of Medicine* 337 (1997): 1441–1448.

19. F. Watanabe and coauthors, Effects of microwave heating on the loss of vitamin $B_{12}$ in foods, *Journal of Agricultural and Food Chemistry* 46 (1998): 206–210.
20. D. J. Canty and S. H. Zeisel, Lecithin and choline in human health and disease, *Nutrition Reviews* 52 (1994): 327–339; E. P. Shronts, Essential nature of choline with implications for total parenteral nutrition, *Journal of the American Dietetic Association* 97 (1997): 646–649.
21. G. Wolf, Uptake of ascorbic acid by human neutrophils, *Nutrition Reviews* 51 (1993): 337–338.
22. H. Hemilä and Z. S. Herman, Vitamin C and the common cold: A retrospective analysis of Chalmer's review, *Journal of the American College of Nutrition* 14 (1995): 116–123.
23. J. E. Enstrom, L. E. Kanim, and M. A. Klein, Vitamin C intake and mortality among a sample of the United States population, *Epidemiology* 3 (1992): 194–202.
24. S. N. Gershoff, Vitamin C (ascorbic acid): New roles, new requirements, *Nutrition Reviews* 51 (1993): 313–326.
25. J. Lykkesfeldt and coauthors, Ascorbic acid and dehydroascorbic acid as biomarkers of oxidative stress caused by smoking, *American Journal of Clinical Nutrition* 65 (1997): 959–963.
26. D. L. Tribble, L. J. Giuliano, and S. P. Fortmann, Reduced plasma ascorbic acid concentrations in nonsmokers regularly exposed to environmental tobacco smoke, *American Journal of Clinical Nutrition* 58 (1993): 886–890.
27. D. W. Giraud, H. D. Martin, and J. A. Driskell, Plasma and dietary vitamin C and E levels of tobacco chewers, smokers, and nonusers, *Journal of the American Dietetic Association* 95 (1995): 798–800.
28. M. Levine and coauthors, Vitamin C pharmacokinetics in healthy volunteers: Evidence for a recommended dietary allowance, *Proceedings from the National Academy of Sciences* 93 (1996): 3704–3709.
29. V. Herbert, Vitamin C supplements are dangerous for iron-overloaded persons, *Journal of the American Dietetic Association* 93 (1993): 526–527.
30. G. B. Forbes, Potatoes: A reliable source of vitamin C (Scorbutus nauticus cured without citrus), *Nutrition Today*, January/February 1993, pp. 33–35.
31. A. Nikiforuk, The Irish famine: A blighted fable, in *The Fourth Horseman: A Short History of Epidemics, Plagues, Famine and Other Scourges* (New York: M. Evans & Company, 1993): pp 110–125.
32. H. E. Sauberlich and coauthors, Effects of erythorbic acid on vitamin C metabolism in young women, *American Journal of Clinical Nutrition* 64 (1996): 336–346.
33. M. Levine, Fruits and vegetables: There is no substitute, *American Journal of Clinical Nutrition* 64 (1996): 381–382.

HIGHLIGHT 10

# Vitamin and Mineral Supplements

More than half of the U.S. population takes vitamin and mineral supplements regularly, spending billions of dollars on them each year.[1] Many people take supplements as dietary insurance—in case they are not meeting their nutrient needs from foods alone. Others take supplements as health insurance—to protect against certain diseases.

One out of every five people takes multinutrient pills daily.[2] Others take large doses of single nutrients, most commonly, vitamin C, vitamin E, beta-carotene, iron, and calcium. In many cases, taking supplements is a costly but harmless practice; sometimes, it is both costly and harmful to health.

For the most part, people self-prescribe supplements, taking them on the advice of friends, television, or books that may or may not be reliable. Sometimes, they take supplements on the recommendation of a physician. When such advice follows a valid nutrition assessment, supplementation may be warranted, but even then the preferred course of action is to improve food choices and eating habits. Without an assessment, the advice to take supplements may be inappropriate. A registered dietitian can help with the decision.

When people think of supplements, they often think only of vitamins, but minerals are important, too, of course. People whose diets lack vitamins, for whatever reason, probably lack several minerals as well. Vitamin-mineral supplements may be appropriate in some circumstances.

This highlight asks several questions related to supplement taking (the glossary on p. 330 defines supplements and related terms). What are the arguments for taking supplements? What are the arguments against taking them? Finally, if people do take supplements, how can they choose the appropriate ones?

*Over 3000 different vitamin and mineral supplements are available in a number of formulations, shapes, and flavors, but none offers the full array of nutrients that a variety of foods can provide. (Courtesy of CNN)*

## Arguments for Supplements

Supplements do have appropriate uses. In some cases, they can correct deficiencies; in others, they can reduce the risk of diseases.

### Correct Overt Deficiencies

In the United States and Canada, adults rarely suffer nutrient deficiency diseases such as scurvy, pellagra, and beriberi, but they do still occur. Correcting an overt deficiency disease may require therapeutic doses two to ten times the RDA (or AI) of a nutrient. When doses exceed the amounts of nutrients commonly found in foods, the supplements are being used more as drugs than as foods.

### Improve Nutrition Status

In contrast to the classical deficiencies, which present a multitude of symptoms and are easy to recognize, subclinical deficiencies are subtle and easy to overlook—and they are also more likely to occur. People who do not eat enough food to deliver the needed amounts of nutrients, such as habitual dieters and the elderly, risk developing subclinical deficiencies. Similarly, vegetarians who restrict their use of entire food groups without appropriate substitutions may fail to fully meet their nutrient needs. If there is no way for these people to eat enough nutritious foods to meet their needs, then vitamin-mineral supplements may be appropriate to help prevent nutrient deficiencies.

### Reduce Disease Risks

Highlight 11 reviews the relationships between supplement use and disease prevention. It describes some of the accumulating evidence suggesting that intakes of certain nutrients at levels much higher than can be attained from foods alone may be beneficial in reducing disease risks. It also presents research confirming the risks of taking supplements. Clearly, consumers must be cautious in taking supplements to reduce disease risks.

Many people, especially postmenopausal women and those who are intolerant to lactose or allergic to milk, may not receive enough calcium to forestall the bone degeneration of old age, osteoporosis. For them, nonmilk calcium-rich foods are especially valuable, but calcium supplements may also be appropriate (Highlight 12 provides more details).

## Support Increased Nutrient Needs

As Chapters 15–17 explain, nutrient needs increase during certain stages of the life cycle, making it difficult to meet some of those needs without supplementation. For example, women who lose a lot of blood and therefore a lot of iron during menstruation each month may need an iron supplement. Women of childbearing age may need folate supplements to reduce the risks of neural tube defects. Similarly, pregnant women and women who are breastfeeding their infants have exceptionally high nutrient needs and so usually need special supplements. Newborns routinely receive a single dose of vitamin K at birth to prevent abnormal bleeding. Infants may need other supplements as well, depending on whether they are breastfed or receiving formula, and on whether their water contains fluoride.

## Improve Body's Defenses

Health care professionals may provide special supplementation to people being treated for addictions to alcohol or other drugs and to people with prolonged illnesses, extensive injuries, or other severe stresses such as surgery. Illnesses that interfere with appetite, eating, or nutrient absorption limit nutrient intakes, yet nutrient needs are often heightened by diseases or drugs. In all these cases, supplements are appropriate.

## Who Needs Supplements?

In summary, the following list acknowledges that in these specific conditions, these people may need to take supplements:

- People with nutrient deficiencies.
- People with low food energy intakes (less than 1200 kcalories per day).
- People who eat all-plant diets (vegans).
- Women who bleed excessively during menstruation.
- People whose calcium intakes are too low to forestall extensive bone loss.
- People in certain stages of the life cycle who have increased nutrient needs (for example, infants, women of childbearing age, and pregnant women).
- People who have diseases, infections, or injuries or who have undergone surgery that interferes with the intake, absorption, metabolism, or excretion of nutrients.
- People taking medications that interfere with the body's use of specific nutrients.

Except for people in these circumstances, most adults can normally get all the nutrients they need by eating a varied diet of nutrient-dense foods. Even athletes can meet their nutrient needs without the help of supplements, as Chapter 14 explains.

# Arguments against Supplements

Foods rarely cause nutrient imbalances or toxicities, but supplement taking is risky.[3] The higher the dose, the greater the risk of harm. People's tolerances for high doses of nutrients vary, just as their risks of deficiencies do. Amounts that some can tolerate may be harmful for others, and no one knows who falls where along the spectrum. It is difficult to determine just how much of a nutrient is enough—or too much. The Tolerable Upper Intake Levels of the DRI answer the question how much is too much by defining the highest amount that appears safe for most healthy people (see the table on the inside front cover, right). Table H10-1 presents suggested limits for selected vitamins and minerals.

## Toxicity

The extent and severity of supplement toxicity remain unclear. Only a few alert health care professionals can recognize toxicity, even when it is acute. When it is chronic, with the effects developing subtly and progressing slowly, it often goes unrecognized. In one case, a woman took just 1000 RE (5000 IU) of vitamin A, an amount typically found in vitamin-mineral supplements, daily for ten years. She was diagnosed with liver disease. Only when she discontinued the supplement did the condition clear up.[4] In view of the potential hazards, some authorities believe supplements should bear warning labels, advising consumers that large doses may be toxic.

Toxic overdoses of vitamins and minerals in children are more readily recognized and, unfortunately, fairly common. In 1996, poison control centers received more than 50,000 reports of children under the age of six swallowing excessively large doses of supplements.[5] Fruit-flavored, chewable vitamins shaped like cartoon characters entice young children to eat them like candy in amounts that can cause poisoning. High-potency iron supplements (30 milligrams of iron or more per tablet) are especially toxic and are the leading cause of accidental ingestion fatalities among children. Even mild

### Glossary

**high potency:** 100% or more of the Daily Value for the nutrient in a single supplement and for at least two-thirds of the nutrients in a multinutrient supplement.

**nutraceuticals:** substances that supposedly provide medical and health benefits; a term not recognized by the FDA.

**supplements:** pills, capsules, tablets, liquids, or powders that contain vitamins, minerals, herbs, or amino acids; intended to increase dietary intake of these substances.

**Table H10-1**
**Vitamin and Mineral Intakes for Adults**

| Nutrient | Safe Intake/Day | Average Multivitamin-Mineral Supplement | Single-Nutrient Supplement |
|---|---|---|---|
| **Vitamins** | | | |
| Vitamin A | 3000 μg RE (10,000 IU) | 5000 IU | 8000 to 10,000 IU |
| Vitamin D | 50 μg (2000 IU) | 400 IU | 400 IU |
| Vitamin E | 200 to 800 mg α-TE (130 to 530 IU) | 30 IU | 100 to 1000 IU |
| Thiamin | —[a] | 1.5 mg | 50 mg |
| Riboflavin | —[a] | 1.7 mg | 25 mg |
| Niacin (as niacinamide) | 35 mg | 20 mg | 100 to 500 mg |
| Vitamin $B_6$ | 100 mg | 2 mg | 100 to 200 mg |
| Folate | 1000 μg | 400 μg | 400 μg |
| Vitamin $B_{12}$ | —[a] | 6 μg | 100 to 1000 μg |
| Pantothenic acid | —[a] | 10 mg | 100 to 500 mg |
| Biotin | —[a] | 30 μg | 300 to 600 μg |
| Vitamin C | 1000 mg | 10 mg | 500 to 2000 mg |
| Choline | 3500 mg | 10 mg | 250 mg |
| **Minerals** | | | |
| Calcium | 2500 mg | 160 mg | 250 to 600 mg |
| Phosphorus | 4000 mg | 110 mg | —[c] |
| Magnesium | 350 mg | 100 mg | 250 mg |
| Iron | 10 to 40 mg | 18 mg | 18 to 30 mg |
| Zinc | 10 to 25 mg | 15 mg | 10 to 100 mg |
| Iodine | —[b] | 150 μg | —[c] |
| Selenium | 200 μg | 10 μg | 50 to 200 μg |
| Fluoride | 10 mg | — | —[c] |

[a]Thiamin, riboflavin, vitamin $B_{12}$, pantothenic acid, and biotin have been evaluated by the DRI committee for Upper Tolerable Intake Levels, but none were established for these nutrients because of insufficient data. No adverse effects have been reported with intakes of these nutrients at levels typical of supplements, but caution is still advised, given the potential for harm that accompanies excessive intakes.
[b]Not recommended in supplemental form.
[c]Available as a single supplement by prescription.
Sources: Values for niacin, vitamin $B_6$, folate, and choline reflect Tolerable Upper Intake Levels as established in Committee on Dietary Reference Intakes, *Dietary Reference Intakes for Thiamin, Riboflavin, Niacin, Vitamin $B_6$, Folate, Vitamin $B_{12}$, Pantothenic Acid, Biotin, and Choline* (Washington, D.C.: National Academy Press, 1998); values for vitamin D, calcium, phosphorus, magnesium, and fluoride reflect Tolerable Upper Intake Levels as established in Committee on Dietary Reference Intakes, *Dietary Reference Intakes for Calcium, Phosphorus, Magnesium, Vitamin D, and Fluoride* (Washington, D.C.: National Academy Press, 1997); values for vitamin A, vitamin C, and selenium have been adapted from J. N. Hathcock, Vitamins and minerals: Efficacy and safety, *American Journal of Clinical Nutrition* 66 (1997): 427–437; values for vitamin E were adapted from R. J. Sokol, Vitamin E, in *Present Knowledge in Nutrition,* eds. E. E. Ziegler and L. J. Filer (Washington, D.C.: International Life Sciences Institute Press, 1996), pp. 130–136.

overdoses cause GI distress, nausea, and black diarrhea that reflects gastric bleeding. Severe overdoses result in bloody diarrhea, shock, liver damage, coma, and death.

## *Life-Threatening Misinformation*

Another problem arises when people who are ill come to believe that high doses of vitamins or minerals can be therapeutic. Not only can high doses be toxic, but such action may prevent the person from seeking medical help. Furthermore, there are no guarantees that the supplements will be effective. Marketing materials for supplements often make health statements that are required to be "truthful and not misleading," but often fall far short of both. Instead of demanding scientific support of unapproved health statements, federal regulations only require supplements to state on their labels: "This statement has not been evaluated by the Food and Drug Administration. This product is not intended to diagnose, treat, cure or prevent any disease." Highlight 18 revisits this topic and includes a discussion of herbal preparations and other alternative therapies.

## *Unknown Needs*

Another argument against the use of supplements is that no one knows exactly how to formulate the "ideal" supplement. What nutrients should be included? How much of each? On whose needs should the choices be based? Surveys have repeatedly shown little relationship between the supplements people take and the nutrients they actually need. Often people take supplements containing the nutrients they need least and miss out on the nutrients their diets are failing to deliver.

## *False Sense of Security*

Another argument against supplement use is that it may lull people into a false sense of security. A person might eat irresponsibly, thinking, "My supplement will cover my needs." Or, experiencing a warning symptom of a disease, a person might postpone seeking a diagnosis, thinking, "I probably need a nutrient supplement to make this go away." Such self-diagnosis is potentially dangerous.

## *Other Invalid Reasons*

Other invalid reasons why people might take supplements include:

- The belief that the food supply or soil contains inadequate nutrients.

- The belief that supplements can provide energy.
- The belief that supplements can enhance athletic performance or build lean body tissues without physical work or faster than work alone.
- The belief that supplements will help a person cope with stress.
- The belief that supplements can prevent, treat, or cure conditions ranging from the common cold to cancer.

Ironically, people with health problems are more likely to take supplements than other people, yet today's health problems are more likely to be due to overnutrition and poor lifestyle choices than to nutrient deficiencies. The truth—that most people need to change their eating and exercise habits—is harder to swallow than a supplement.

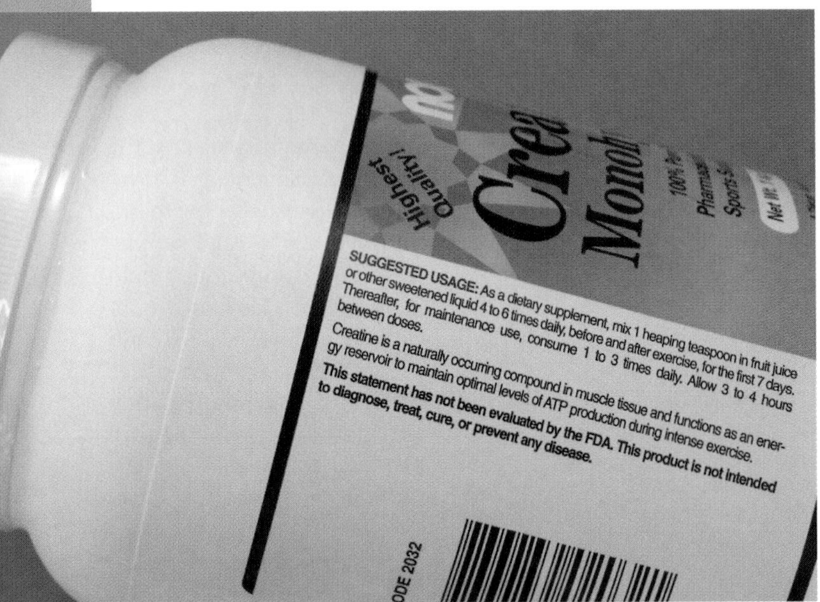


***This FDA disclaimer on supplement labels should raise suspicion about the accuracy of the health information provided.***

## Bioavailability and Antagonistic Actions

In general, the body absorbs nutrients best from foods in which the nutrients are diluted and dispersed among other ingredients that may facilitate their absorption. Taken in pure, concentrated form, nutrients are likely to interfere with one another's absorption or with the absorption of nutrients in foods eaten at the same time. Documentation of these effects is particularly extensive for minerals: zinc hinders copper and calcium absorption, iron hinders zinc absorption, calcium hinders magnesium and iron absorption, and magnesium hinders the absorption of calcium and iron. The same interference takes place when people use foods that are fortified with added minerals; thus the consumer who wants the benefits of optimal absorption of nutrients should use ordinary foods, selected for nutrient density and variety.

Although minerals provide the most familiar and best-documented examples, interference among vitamins is now being seen as supplement use increases. The vitamin A precursor beta-carotene, long thought to be completely nontoxic, interferes with vitamin E metabolism when taken over the long term as a dietary supplement. Vitamin E, on the other hand, antagonizes vitamin K activity and so should not be used by people being treated for blood-clotting disorders.

In view of all the negatives associated with supplement taking, several professional nutrition societies have indicated that people ordinarily should *not* use supplements. Whenever a person's diet is inadequate, the person should first attempt to improve it so as to obtain the needed nutrients from foods. If that is truly impossible, then the person needs a multivitamin-mineral supplement that supplies between 50 and 150 percent of the RDA (or AI) amount for each of the nutrients. These amounts reflect the ranges commonly found in foods and therefore are compatible with the body's normal handling of nutrients (its physiologic tolerance). Some pointers can assist in the selection of an appropriate supplement.

## *Selection of Supplements*

Whenever a physician or dietitian recommends a supplement, follow the directions carefully. When selecting a supplement yourself, look for a single, balanced vitamin-mineral supplement.

If you decide to take a vitamin-mineral supplement, ignore the eye-catching art and meaningless claims. Pay attention to the form the supplements are in, the list of ingredients, and the price. Here's where the truth lies, and from it you can make a rational decision based on facts. You have two basic questions to answer.

### Form

The first question: What form do you want—chewable, liquid, or pills? If you'd rather drink your supplements than chew them, fine. (If you choose a chewable form, though, be aware that chewable vitamin C can dissolve tooth enamel.)

If you choose pills, look for statements about the disintegration time. The U.S. Pharmacopoeia (USP) suggests that supplements should completely disintegrate within 30 to 45 minutes.* Obviously, supplements that don't dissolve have little chance of entering the bloodstream, so look for a brand that claims to meet USP disintegration standards.

 U.S. Pharmacopeia
www.usp.org/pubs/just_ask
Visit Vitamin and Mineral Supplement Products

*The USP establishes standards for quality, strength, and purity of supplements.

### Contents

The second question: What vitamins and minerals do *you* need? Generally, an appropriate supplement provides vitamins and minerals in amounts smaller than, equal to, or very close to recommended intakes. Avoid supplements that, in a daily dose, provide more than the recommended amounts of vitamin A, vitamin D, or any mineral or more than ten times the recommended amount for *any* nutrient. Avoid preparations with more than 10 milligrams of iron per dose, except as prescribed by a physician. Iron is hard to get rid of once it's in the body, so an excess of iron can cause problems, just as a deficiency can (see Chapter 13).

### Misleading Claims

Ignore "organic" or "natural" claims. Such supplements are no better than standard types and often cost more. The word *synthetic* may sound like "fake," but to synthesize just means to put together. Whether vitamins are synthesized in a laboratory or synthesized by plants and animals, your body uses them similarly. Only your wallet can tell the difference.

Avoid products that make "high potency" claims. More is not better (review the "How to" on p. 293). Remember that foods are also providing these nutrients. Nutrients can build up and cause unexpected problems. For example, a man who takes vitamins and begins to lose his hair may think his hair loss means he needs *more* vitamins, when in fact it may be the early sign of a vitamin A overdose. (Of course, it may be completely unrelated to nutrition as well.)

Be wise to fake vitamins and preparations that contain items not needed in human nutrition, such as carnitine and inositol. Such ingredients reveal a marketing strategy aimed at your pocket, not at your health. The manufacturer wants you to believe that its pills contain the latest "new" nutrient that other brands omit, but in reality, these substances are not known to be needed by human beings.

Realize that the claim that supplements "relieve stress" is another marketing ploy. If you give even passing thought to what people mean by "stress," you'll realize manufacturers could never design a supplement to meet everyone's needs. Is it stressful to take an exam? Well, yes. Is it stressful to survive a major car wreck with third-degree burns and multiple bone fractures? Definitely yes. The body's responses to these stresses are different. The body does use vitamins and minerals in mounting a stress response, but a body fed a well-balanced diet can meet the needs of most minor stresses. As for the major ones, medical intervention is needed. In any case, taking a vitamin supplement won't make life any less stressful.

Other marketing tricks to sidestep are "green" pills that contain dehydrated, crushed parsley, alfalfa, and other fruit and vegetable extracts. The nutrients and phytochemicals advertised can be obtained from a serving of vegetables more easily and for less money. Such pills may also provide enzymes, but these are inactivated in the stomach during digestion.

Be aware that most geriatric "tonics" are poor in vitamins and minerals, yet so high in alcohol as to threaten inebriation. The liquids designed for infants are more complete.

Recognize the latest nutrition buzzwords. Manufacturers were marketing "antioxidant" supplements before the print had time to dry on the first scientific reports of antioxidant vitamins' action in preventing cancer and cardiovascular disease. Remember, too, that high doses can alter a nutrient's action in the body. An antioxidant in physiological quantities may be beneficial, but in pharmacological quantities, it may produce harmful by-products.[6] Highlight 11 explores antioxidants and supplement use in more detail.

### Cost

When shopping for supplements, remember that local or store brands may be just as good as nationally advertised brands. If they are less expensive, it may be because the price does not have to cover the cost of national advertising.

## Regulation of Supplements

The Dietary Supplement Health and Education Act of 1994 was intended to enable consumers to make informed choices about nutrient supplements. The act subjects supplements to the same general labeling requirements that apply to foods.[7] Specifically:

- Nutrition labeling for dietary supplements is required.
- Labels may describe nutrient contents (as "high" or "low") according to specific criteria.
- The FDA authorizes health claims that are supported by significant scientific agreement and are not brand specific. To date, the following health claims have been approved for supplements: folate and neural tube defects, calcium and osteoporosis, soluble fiber from whole oats or psyllium husks and cardiovascular disease, and sugar alcohols and dental caries.
- Health claims have *not* been authorized for the following nutrient-disease relationships: dietary fiber and cancer, dietary fiber and cardiovascular disease, antioxidant vitamins and cancer, omega-3 fatty acids and coronary heart disease, and zinc and immune deficiency in the elderly.
- Products may not claim to diagnose, treat, cure, or relieve a specific disease.
- Labels may describe the role a nutrient plays in the body, explain how the nutrient performs its function, and indicate that consuming the nutrient is associated with general well-being.

Figure H10-1 on p. 334 provides an example of a supplement label that complies with the mandatory requirements.

In effect, the Dietary Supplement Health and Education Act resulted in a deregulation of the supplement industry.[8] Unlike food additives or drugs, supplements do not need the FDA's approval before being marketed. Manufacturers alone decide whether their products are safe and effective. Should a problem arise, the burden falls to the FDA to prove

**Figure H10-1**

**An Example of a Supplement Label**

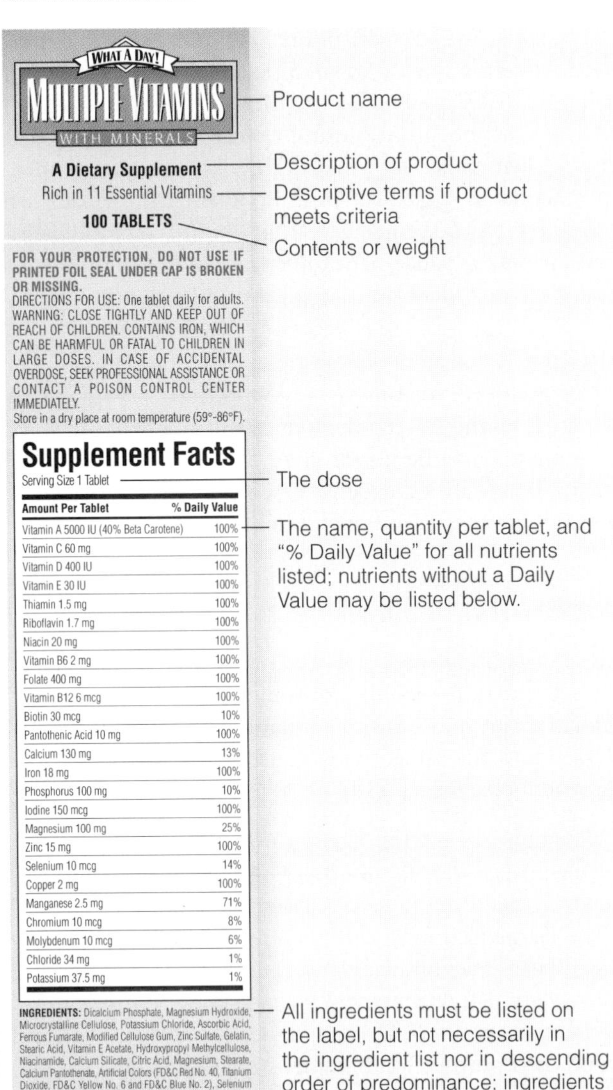


that the supplement poses an unreasonable risk and should be removed from the market.

The $4-billion-a-year supplement industry continues to fight against regulations. Part of the battle involves a proposal to create a new regulatory category for "nutraceuticals"—a term not legally or scientifically defined.[9] To those in the food industry, "nutraceuticals" refers to foods, beverages, and supplements with added or engineered health-promoting properties.[10] Supplements (in pill, powder, or other medicinal form) make up only a small fraction of the $77 billion nutraceutical market. The lion's share belongs to meals, snacks, and beverages designed to satisfy consumers' quest for health through foods. Liquid meals, once used only by people with illnesses to boost their nutrient intakes or by overweight people to replace a meal, are now marketed as "complete on-the-run meals for today's active lifestyle." They provide all the vitamins and minerals of a supplement in an easy-to-swallow form. Similarly, energy bars can supposedly deliver the "ideal" proportion of nutrients in a tasty bite or two; herbal concoctions promise phytochemicals and antioxidants in a few sweet sips. Consumers can buy orange juice with calcium or peanut butter with all the essential vitamins and minerals. The options are endless, and those in the food industry sit ready and willing to market whatever ingredient is currently being studied in nutrition research.

Proponents of nutraceuticals are urging the FDA to loosen its criteria for health claims and to allow manufacturers to make exclusive claims based on their own research—without being required to make the research public. Such exclusivity of health claims on products based on private research would be contrary to the open sharing of information that is fundamental to nutrition science. It would also be confusing and misleading for consumers. If a "nutraceutical" company could claim that its Brand X vitamin E prevents lung cancer, then consumers would need to be aware that other brands have a similar effect and that many foods provide vitamin E as well. Furthermore, consumers would need to know that such a claim is backed by sound scientific research—and that's what FDA approval provides.

If all the nutrients we need can come from food, why not just eat food? Foods have so much more to recommend them than supplements do. Nutrients in foods come in an infinite variety of combinations with a multitude of different carriers and absorption facilitators. They come with water, fiber, and a host of beneficial and interesting nonnutrients. Foods stimulate the GI tract to keep it healthy. They provide energy, and since you need energy each day, why not ask nutritious foods to deliver it? They offer pleasure, satiety, and opportunities for socializing while eating. In no way can nutrient supplements hold a candle to foods as a means of meeting human health needs. For further proof, read Highlight 11.

 American Dietetic Association

www.eatright.org
Search for Supplements

 FDA Center for Food Safety and Applied Nutrition

vm.cfsan.fda.gov/~dms/supplmnt.html

## Notes


1. Report of the Commission on Dietary Supplement Labels (Washington, D.C.: Department of Health and Human Services, 1997).
2. M. J. Slesinski, A. F. Subar, and L. L. Kahle, Trends in use of vitamin and mineral supplements in the United States: The 1987 and 1992 National Health Interview Surveys, *Journal of the American Dietetic Association* 95 (1995): 921–923.
3. Position of The American Dietetic Association: Vitamin and mineral supplementation, *Journal of the American Dietetic Association* 96 (1996): 73–77.
4. R. Oren and Y. Ilan, Reversible hepatic injury induced by long-term vitamin A ingestion, *American Journal of Medicine* 93 (1992): 703–704.
5. T. L. Litovitz and coauthors, 1996 Annual Report of the American Association of Poison Control Center Toxic Exposure Surveillance System, *American Journal of Emergency Medicine* 15 (1997): 447–500.
6. V. Herbert, The antioxidant supplement myth, *American Journal of Clinical Nutrition* 60 (1994): 157–158.
7. Dietary supplements: Recent chronology and legislation, *Nutrition Reviews* 53 (1995): 31–36.
8. M. C. Nesheim, Regulation of dietary supplements, *Nutrition Today* 33 (1998): 62–68.
9. J. R. Hunt, Nutritional products for specific health benefits—Foods, pharmaceuticals, or something in between? *Journal of the American Dietetic Association* 94 (1994): 151–153.
10. A. E. Sloan, The top 10 trends to watch and work on, *Food Technology* 50 (1996): 55–71.

CHAPTER 11

# The Fat-Soluble Vitamins: A, D, E, and K

The fat-soluble vitamins A, D, E, and K differ from the water-soluble vitamins in several significant ways (review the table on p. 293). The fat-soluble vitamins are found in the fats and oils of foods. They are insoluble in water, so they require bile for absorption. Upon absorption, fat-soluble vitamins travel through the lymphatic system within chylomicrons before entering the bloodstream, where many of them require protein carriers for transport. The fat-soluble vitamins are stored in the liver and adipose tissue until they are needed; they are not readily excreted, as most of the water-soluble vitamins are. Having stored the fat-soluble vitamins, people can eat less than their daily need for days, weeks, or even months or years without ill effects. The body maintains blood concentrations by retrieving the vitamins from storage as needed; thus a person need only ensure that over time *average* daily intakes approximate recommendations. By the same token, because fat-soluble vitamins are stored, the risk of toxicity is greater than it is for the water-soluble vitamins.

IN SUMMARY

**The Fat-Soluble Vitamins:**

- Vitamin A.
- Vitamin D.
- Vitamin E.
- Vitamin K.

# Vitamin A and Beta-Carotene

Vitamin A was the first fat-soluble vitamin to be recognized. More than 75 years later, vitamin A and its precursor, beta-carotene, continue to intrigue researchers with their diverse roles and profound effects on health.

Three different forms of vitamin A are active in the body: retinol, retinal, and retinoic acid. Collectively, these compounds are known as retinoids. Foods derived from animals provide compounds (retinyl esters) that are easily hydrolyzed to retinol in the intestine. Foods derived from plants provide carotenoids, some of which have vitamin A activity. The most important of the carotenoids is beta-carotene, which can be split to form retinol in the intestine and liver. Beta-carotene's absorption and conversion are less efficient than those of the retinoids. Figure 11-1 illustrates the structural similarities and differences of these vitamin A compounds.

The cells can convert retinol and retinal to the other active forms as needed. The oxidation of retinol to retinal is reversible, but the further oxidation of retinal

**vitamin A:** all naturally occurring compounds with the biological activity of retinol (RET-ih-nol), the alcohol form of vitamin A.

**beta-carotene** (BAY-tah KARE-oh-teen): an orange pigment and vitamin A precursor found in plants. A *precursor* is a compound that can be converted into an active vitamin.

**retinoids** (RET-ih-noyds): chemically related compounds with biological activity similar to retinol; metabolites of retinol.

**carotenoids** (kah-ROT-eh-noyds): pigments commonly found in plants and animals, some of which have provitamin A activity. Carotenoids are among the best-known **phytochemicals** (see Highlight 11).

**vitamin A activity:** a term referring to both the active forms of vitamin A and the precursor forms (carotenes) in foods without distinguishing between them.

**Figure 11-1**
**Forms of Vitamin A**

In this diagram, corners represent carbon atoms, as in all previous diagrams in this book. A further simplification here is that methyl groups ($CH_3$) are understood to be at the ends of the lines extending from corners. (See Appendix C for complete structures.)

Retinol, the alcohol form

Retinal, the aldehyde form

Retinoic acid, the acid form

Cleavage at this point can yield two molecules of vitamin A*

Beta-carotene, a precursor

*Sometimes cleavage occurs at other points as well, so that one molecule of beta-carotene may yield only one molecule of vitamin A. Furthermore, not all beta-carotene is converted to vitamin A, and absorption of beta-carotene is not as efficient as vitamin A. For these reasons, 6 μg of beta-carotene are equivalent to 1 μg of vitamin A. Conversion of other carotenoids to vitamin A is even less efficient.

**Figure 11-2**
**Conversion of Vitamin A Compounds**

Notice that the conversion from retinol to retinal is reversible, whereas the pathway from retinal to retinoic acid is not.

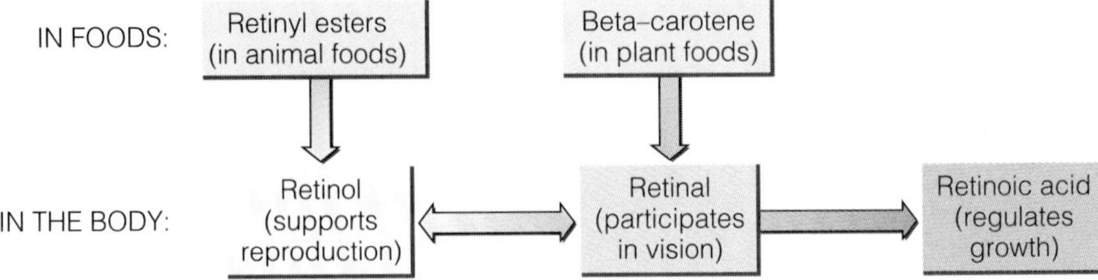


**retinol-binding protein (RBP):** the specific protein responsible for transporting retinol.

to retinoic acid is irreversible. This irreversibility is significant because each form of vitamin A performs a function that the others cannot (see Figure 11-2).

A special transport protein, retinol-binding protein (RBP), picks up vitamin A from the liver, where it is stored, and carries it in the blood. Cells that will use vitamin A have special protein receptors for it, as if the vitamin were fragile and had to be passed carefully from hand to hand without being dropped. Each form of vitamin A has its own receptor protein (retinol has several) within the cells.[1]

## *Roles in the Body*

Vitamin A is a versatile vitamin. Its major roles include:

- Promoting vision.
- Participating in protein synthesis and cell differentiation (and thereby maintaining the health of epithelial tissues and skin).
- Supporting reproduction and growth.

Each form of vitamin A performs specific tasks. Retinol supports reproduction and is the major transport and storage form of the vitamin. Retinal is active in vision and is also an intermediate in the conversion of retinol to retinoic acid. Retinoic acid acts like a hormone, regulating cell differentiation, growth, and embryonic development. Animals raised on retinoic acid as their sole source of vitamin A can grow normally, but they become blind, because retinoic acid cannot be converted to retinal.

• ***Vitamin A in Vision*** • Vitamin A plays two indispensable roles in the eye: it helps maintain a crystal-clear outer window, the cornea, and it participates in the conversion of light energy into nerve impulses at the retina (see Figure 11-3). Light passes through the cornea of the eye and strikes the cells of the retina. Inside the cells, pigment molecules called rhodopsin absorb light (the glossary on p. 339 defines rhodopsin and related terms). Each rhodopsin molecule is composed of a protein called opsin bonded to a molecule of retinal. When light energy enters the eye, rhodopsin responds by changing shape and becoming bleached. Simultaneously, the retinal shifts from a *cis* to a *trans* configuration as fatty acids do during hydrogenation (see p. 129). Retinal in the *trans* form cannot remain bonded to opsin. When retinal is released, opsin changes shape, thereby disturbing the membrane of the cell and generating an electrical impulse that travels along the cell's length. At the other end of the cell, the impulse is transmitted to a nerve cell, which conveys the message to the brain. Much of the retinal is then converted back to its active *cis* form and combined with the opsin protein to regenerate the pigment rhodopsin. Some retinal, however, may be oxidized to retinoic acid, a biochemical dead end for the visual process.

A genius could not have designed a better system. Light itself cannot be conducted through the solid material of the brain, but nerve impulses can be. To preserve the information in the different colors, or wavelengths, that light comes in, the eye uses the color-sensitive cone cells to receive them. Blue light is absorbed by one set of cells, green by another, and yellow-red by a third. By day, cones receive these colors and convey the full range of color vision to the optic center

**Figure 11-3**
**Vitamin A's Role in Vision**

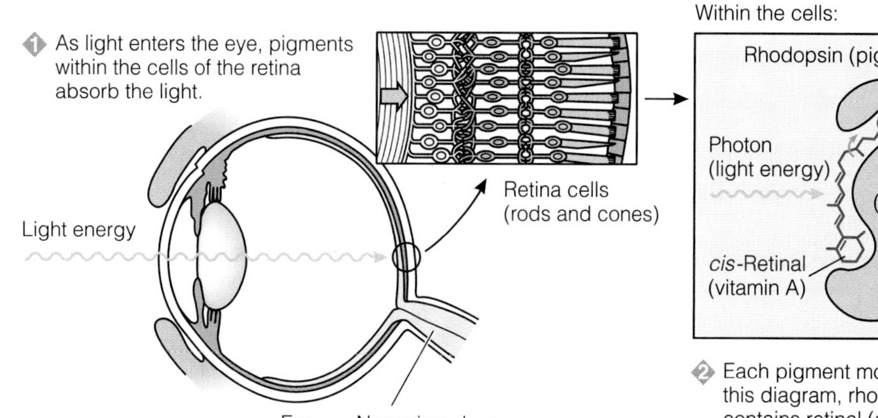


1. As light enters the eye, pigments within the cells of the retina absorb the light.

Within the cells:

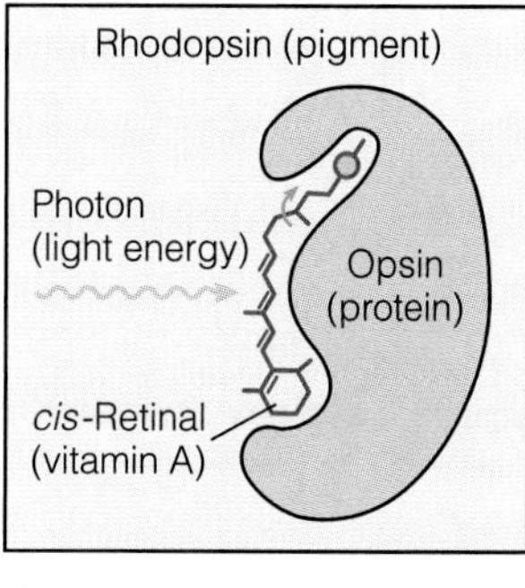


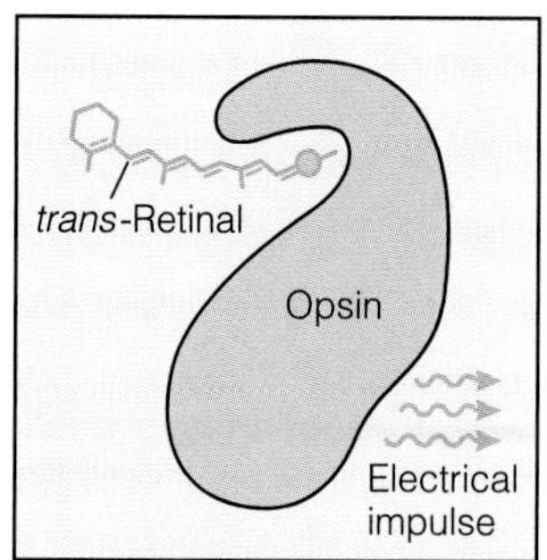


2. Each pigment molecule (in this diagram, rhodopsin) contains retinal (an active form of vitamin A) and a protein (in this diagram, opsin).

3. When rhodopsin absorbs light, its shape and color change, generating nerve impulses that travel into the brain.

in the brain. By night, the light entering the eye is of low intensity and can be received only by the rod cells; so by night a person can normally discern only the presence of light, not its color.

About 6 to 7 million cone cells and 100 million rod cells reside in the retina, and each contains about 30 million molecules of retinal-containing visual pigments. Visual activity leads to repeated small losses of retinal and necessitates its constant replenishment from retinol in the blood, which brings a new supply from the body stores. Ultimately, foods supply all the retinal in the pigments of the eye.

A lot of retinal can be destroyed at night. If the body's vitamin A stores are marginal, using the eyes at night can lead to a vitamin A deficiency and the early, telltale symptom of night blindness. The eye is especially vulnerable to retinal destruction at night for three reasons. First, the pupil opens wide at night to allow as much light as possible to enter the eye. Second, a shadowing pigment that protects the rods by day withdraws at night, leaving them exposed. Third, there are many more rods than cones. Hence, if a bright light suddenly shines at night through the wide-open pupil onto the unprotected rods, much of the pigment in them is bleached and momentarily inactivated. More retinal than usual is released, and more is lost. A moment passes before the pigments regenerate and sight returns. You no doubt have been temporarily "blinded" on occasion by a light

**night blindness:** slow recovery of vision after flashes of bright light at night or an inability to see in dim light; an early symptom of vitamin A deficiency.

## Glossary of Vision Terms

**cones:** the cells of the retina that respond to bright light and are responsible for color vision.

**cornea** (KOR-nee-uh): the transparent membrane covering the outside of the eye.

**opsin** (OP-sin): the protein portion of the visual pigment molecule.

**pigment:** a molecule capable of absorbing certain wavelengths of light, so that it reflects only those that we perceive as a certain color.

**retina** (RET-in-uh): the layer of light-sensitive nerve cells lining the back of the inside of the eye; consists of rods and cones.

**rhodopsin** (ro-DOP-sin): the light-sensitive pigment of the rods in the retina; it contains the retinal form of vitamin A and the protein opsin. Like the rods, the cones of the retina contain a light-sensitive pigment—**iodopsin** (eye-oh-DOP-sin)—that contains retinal, but its protein differs from rhodopsin's.
- **rhod** = rod-shaped cells
- **opsin** = visual protein

**rods:** the cells of the retina that respond to dim light and convey black-and-white vision.

shining directly into your eyes. Normally, of course, you quickly recover your ability to see.

• ***Vitamin A in Protein Synthesis and Cell Differentiation*** • Despite its important role in vision, only one-thousandth of the body's vitamin A is in the retina. Much more is in the cells lining the body's surfaces, where the vitamin participates in protein synthesis and cell differentiation.[2]

All body surfaces, both inside and out, are covered by layers of cells known as epithelial cells. The epithelial tissue on the outside of the body is, of course, the skin. The epithelial tissues that line the inside of the body are the mucous membranes: the linings of the mouth, stomach, and intestines; the linings of the lungs and the passages leading to them; the linings of the urinary bladder and urethra; the linings of the uterus and vagina; and the linings of the eyelids and sinus passageways. Within the body, the mucous membranes of the GI tract alone line an area larger than a quarter of a football field, and vitamin A helps to maintain their integrity (see Figure 11-4).

Vitamin A promotes differentiation of both epithelial cells and goblet cells, one-celled glands that synthesize and secrete mucus. Mucus coats and protects the epithelial cells from invasive microorganisms and other harmful substances, such as gastric juices.

• ***Vitamin A in Reproduction and Growth*** • As mentioned, vitamin A also supports reproduction and growth. In men, retinol participates in sperm development, and in women, vitamin A supports normal fetal development during pregnancy.[3] Growth failure is common in children with vitamin A deficiency. When they are given vitamin A supplements, they gain weight and grow taller.

The growth of bones illustrates that growth is a complex phenomenon of remodeling. To convert a small bone into a large bone, the bone-remodeling cells must "undo" some parts of the bone as they go, and vitamin A participates in the undoing.[4] The cells that break down bone contain sacs of degradative enzymes. With the help of vitamin A, these enzymes eat away at selected sites in the bone, removing the parts that are not needed.

• ***Beta-Carotene as an Antioxidant*** • In the body, beta-carotene serves as a vitamin A precursor, but it also has roles of its own. Not all dietary beta-carotene is converted to active vitamin A. Some beta-carotene acts as an antioxidant capable of protecting the body against disease. Highlight 11 discusses some recent findings related to this protection.

## *Vitamin A Deficiency*

Vitamin A status depends mostly on the adequacy of vitamin A stores, 90 percent of which are in the liver. Vitamin A status also depends on a person's protein status, because retinol-binding proteins serve as the vitamin's carriers for inside-the-

**differentiation:** development of specific functions different from those of the original.

**epithelial** (ep-i-THEE-lee-ul) **cells:** cells on the surface of the skin and mucous membranes.

**epithelial tissues:** the layers of the body that serve as selective barriers between the body's interior and the environment (examples are the cornea, the skin, the respiratory lining, and the lining of the digestive tract).

**mucous** (MYOO-kus) **membranes:** the membranes, composed of mucus-secreting cells, that line the surfaces of body tissues.

**urethra** (you-REE-thruh): the tube through which urine from the bladder passes out of the body.

Reminder: *Goblet cells* in the epithelium of the GI tract and lungs secrete mucus.

Reminder: *Mucus* is a slippery substance secreted by the goblet cells of mucous membranes.

**remodeling:** the dismantling and re-formation of a structure, in this case, bone.

The cells that destroy bone during growth are **osteoclasts;** those that build bone are **osteoblasts.**
- **osteo** = bone
- **clast** = break
- **blast** = build

The sacs of degradative enzymes are **lysosomes** (LYE-so-zomes).

**Figure 11-4**
**Mucous Membrane Integrity**

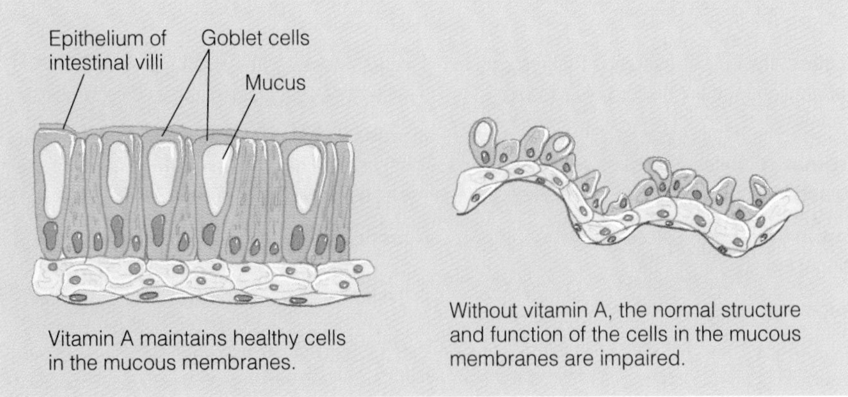
Epithelium of intestinal villi
Goblet cells
Mucus
Vitamin A maintains healthy cells in the mucous membranes.
Without vitamin A, the normal structure and function of the cells in the mucous membranes are impaired.

body transport. If a healthy adult were to stop eating vitamin A–rich foods, deficiency symptoms would not begin to appear until after stores were depleted—one to two years for an adult but less time for a growing child. Then, however, the consequences would be profound and severe.

Vitamin A deficiency is one of the developing world's major nutrition problems. More than 100 million children worldwide have some degree of vitamin A deficiency, and so are vulnerable to infectious diseases and blindness.

• ***Infectious Diseases*** • The World Health Organization (WHO) and UNICEF (the United Nations International Children's Emergency Fund) have made the control of vitamin A deficiency a major goal in their quest to improve child survival throughout the developing world. In developing countries around the world, measles is a devastating infectious disease, killing as many as 2 million children each year. The severity of the illness often correlates with the degree of vitamin A deficiency; deaths are usually due to related infections such as pneumonia and severe diarrhea.[5] Providing large doses of vitamin A to children with measles helps them recover from pneumonia and other infections and reduces their risk of dying.[6]

Of historical interest is a trial carried out in London in 1932 that showed remarkably consistent results: children hospitalized with measles who were given daily doses of cod liver oil (a rich source of vitamin A) died at a rate only half that of similar children not given the oil.[7] WHO recommends routine vitamin A supplementation for all children with measles in areas where vitamin A deficiency is a problem or where the measles death rate is high.[8] In the United States, the American Academy of Pediatrics recommends vitamin A supplementation for certain groups of measles-infected infants and children.[9]

WHO recommendations for children with measles:

- <1 yr: 100,000 IU vitamin A.
- >1 yr: 200,000 IU vitamin A.

Vitamin A recommendations are expressed in retinol equivalents (RE) or international units (IU)—see p. 343.

• ***Night Blindness*** • Night blindness is one of the first detectable signs of vitamin A deficiency and permits early diagnosis of the condition. In night blindness, the blood bathing the cells of the retina does not supply sufficient retinal to rapidly regenerate visual pigments bleached by light. The person loses the ability to recover promptly from the temporary blinding that follows a flash of bright light at night or simply is unable to see after the lights go out. In many parts of the world, after the sun goes down, vitamin A–deficient children become night-blind: they cannot find their shoes or toys and often cling to others or sit still, afraid that they may trip and fall or lose their way if they try to walk alone. In many developing countries, night blindness due to vitamin A deficiency is so common that the people have special words to describe it. In Indonesia, the term is *buta ayam,* which means "chicken eyes" or "chicken blindness." (Chickens do not have rods in their eyes and therefore cannot see at night.) Figure 11-5 shows the eyes' slow recovery in response to a flash of bright light in night blindness.

• ***Blindness (Xerophthalmia)*** • Beyond night blindness lies total blindness—failure to see at all. Night blindness is caused by a lack of vitamin A at the back of the eye, the retina; total blindness is caused by a lack at the front of the eye, the cornea. Vitamin A deficiency is the major cause of childhood blindness in the world, causing more than half a million preschool children to lose their sight each year. Blindness due to vitamin A deficiency, known as xerophthalmia, develops in stages. First, the cornea becomes dry and hard, a condition known as xerosis. Corneal xerosis can quickly progress to keratomalacia, the softening of the cornea that leads to irreversible blindness. Dietary vitamin A reduces the risk of xerophthalmia significantly.[10]

**xerophthalmia** (zer-off-THAL-mee-uh): progressive blindness caused by vitamin A deficiency.

- **xero** = dry
- **ophthalm** = eye

**xerosis** (zee-ROW-sis): abnormal drying of the skin and mucous membranes; a sign of vitamin A deficiency.

**keratomalacia** (KARE-ah-toe-ma-LAY-shuh): softening of the cornea seen in severe vitamin A deficiency that leads to irreversible blindness.

• ***Keratinization*** • Elsewhere in the body, vitamin A deficiency affects other surfaces. Without vitamin A, the goblet cells in the stomach and intestines diminish in number and activity, limiting the secretion of mucus. With less mucus, normal digestion and absorption of nutrients falter, and this, in turn, worsens the deficiency by limiting the absorption of whatever vitamin A the diet may deliver. Similar changes in the cells of other epithelial tissues weaken defenses, making

**Figure 11-5**
**Vitamin A-Deficiency Symptom—Night Blindness**

These photographs illustrate the eyes' slow recovery in response to a flash of bright light at night. In animal research studies, the response rate is measured with electrodes.

In dim light, you can make out the details in this room. You are using your rods for vision.

A flash of bright light momentarily blinds you as the pigment in the rods is bleached.

You quickly recover and can see the details again in a few seconds.

With inadequate vitamin A, you do not recover but remain blinded for many seconds.

**keratin** (KERR-uh-tin): a water-insoluble protein; the normal protein of hair and nails. Keratin-producing cells may replace mucus-producing cells in vitamin A deficiency.

**keratinization:** accumulation of keratin in a tissue; a sign of vitamin A deficiency.

**teratogenic** (ter-AT-oh-jen-ik): causing abnormal fetal development and birth defects.
- **terato** = monster
- **genic** = to produce

For perspective, 10,000 IU ≈ 3000 μg RE, almost four times the RDA for women.

**acne:** a chronic inflammation of the skin's follicles and oil-producing glands, which leads to an accumulation of oils inside the ducts that surround hairs; usually associated with the maturation of young adults.

**Figure 11-6**
**Vitamin A-Deficiency Symptom—The Rough Skin of Keratinization**

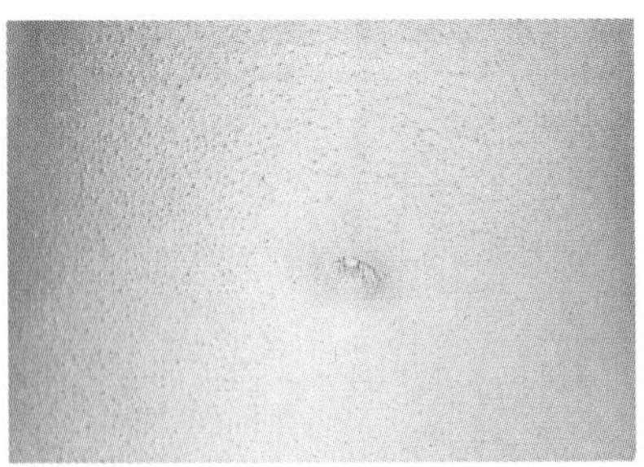

In vitamin A deficiency, the epithelial cells secrete the protein keratin in a process known as **keratinization.** (Keratinization doesn't occur in the GI tract, but mucus-producing cells dwindle, and mucus production declines.) The progression of this condition to the extreme is **hyperkeratinization** or **hyperkeratosis.** When keratin accumulates around each hair follicle, the condition is known as **follicular hyperkeratosis.**

infections of the respiratory tract, the GI tract, the urinary tract, the vagina, and possibly the inner ear likely. On the body's outer surface, the epithelial cells change shape and begin to secrete the protein keratin—the hard, inflexible protein of hair and nails. As Figure 11-6 shows, the skin becomes dry, rough, and scaly as lumps of keratin accumulate around each hair follicle (keratinization).

## Vitamin A Toxicity

A deficiency of vitamin A affects all body systems, as does a toxicity. Symptoms begin to develop when all the binding proteins are swamped, and free vitamin A damages the cells. Such effects are unlikely when a person depends on a balanced diet for nutrients, but if the person takes large amounts of the preformed vitamin from animal foods or supplements, toxicity is a real possibility. Children are most vulnerable to toxicity because they need less and are more sensitive to overdoses.

Beta-carotene, which is found in a wide variety of plant foods, is not converted efficiently enough in the body to cause vitamin A toxicity; instead, it is stored in fat depots under the skin. Overconsumption of beta-carotene may turn the skin yellow, but this is not harmful.

• ***Birth Defects*** • Excessive vitamin A poses a *teratogenic* risk.[11] Researchers in one study found that among women who took more than 10,000 IU of supplemental vitamin A daily, approximately 1 out of every 57 infants was born with a malformation attributable to high vitamin A intake.[12] Intake before the seventh week appeared to be the most damaging. For this reason, vitamin A is not given as a supplement in the first trimester of pregnancy unless there is specific evidence of deficiency, which is rare.

• ***Not for Acne*** • Adolescents need to know that massive doses of vitamin A have no corrective effect on acne, but may cause toxicity. The prescription medicine Accutane is made from vitamin A but is chemically altered. Taken orally, Accutane is effective against the deep lesions of cystic acne. It is highly toxic, however, especially during growth, and has caused birth defects in infants when women have taken it during their pregnancies. For this reason, women taking Accutane must begin using an effective form of contraception at least a month before starting to take the drug and continue using contraception a month after discontinuing its use.

Another vitamin A relative, Retin-A, fights acne, the wrinkles of aging, and other skin disorders. Applied topically, this ointment smooths and softens skin; it also lightens skin that has become darkly pigmented after inflammation.[13] During treatment, the skin becomes red and tender and peels.

## Vitamin A Recommendations

Vitamin A intakes can range widely before deficiency or toxicity symptoms appear (see Figure 11-7). Recommended intakes in both the United States and Canada are set at about double the minimum necessary to prevent deficiency. Doubtless, many people need not consume amounts this high; other countries and international agencies recommend lower values.[14] The exact upper limit of safety has not been determined, in part because people's tolerances to overdoses vary. Several authorities agree that the RDA offers the best guideline for safety and that higher intakes provide no benefits.

Because the body can derive vitamin A from various retinoids and carotenoids, its contents in foods and its recommendations are expressed as retinol equivalents or RE. Some products report their vitamin A contents using international units (IU), an older system used to measure fat-soluble vitamins before direct chemical analysis was possible.

**RE (retinol equivalent):** a measure of vitamin A activity; the amount of retinol that the body will derive from a food containing preformed retinol or its precursor beta-carotene.

1 RE = 1 μg retinol.
= 6 μg beta-carotene.
= 12 μg of other vitamin A precursor carotenes.

1 RE = 3.33 IU.
1 IU = 0.3 μg RE.

**preformed vitamin A:** dietary vitamin A in its active form.

## Vitamin A in Foods

The richest sources of preformed vitamin A are foods of animal origin—liver, fish liver oils, milk and milk products, butter, and eggs. Plants contain no preformed vitamin A, but many vegetables and some fruits contain vitamin A precursors—the carotenoids, red and yellow pigments of plants. Only a few carotenoids have vitamin A activity. The carotenoid with the greatest vitamin A activity is beta-carotene.

• ***The Colors of Vitamin A Foods*** • Recommendations to eat dark green and deep orange vegetables and fruits help people meet their vitamin A needs (see Figure 11-8). A 1-cup serving of carrots, sweet potatoes, or dark greens such as spinach provides such liberal amounts of carotenoids that even allowing for inefficient absorption and conversion, the intake is sufficient for many days. A diet including more or larger servings of medium sources also ensures an ample intake.

Most foods with vitamin A activity are brightly colored—green, yellow, orange, and red. Any plant food with significant vitamin A activity must have some color, since carotene is a rich, deep yellow, almost orange compound. The dark green, leafy vegetables contain abundant amounts of the green pigment chlorophyll, which masks the carotene in them. An attractive meal includes foods of different colors and most likely supplies vitamin A as well.

Bright color is not always a sign of vitamin A activity, however. Beets and corn, for example, derive their colors from the red and yellow xanthophylls, which have

**chlorophyll** (KLO-row-fil): the green pigment of plants, which absorbs light and transfers the energy to other molecules, thereby initiating photosynthesis.

**xanthophylls** (ZAN-tho-fills): pigments found in plants; responsible for the color changes seen in autumn leaves.

**Figure 11-7**
**Vitamin A Deficiency and Toxicity**

As the dose increases from zero, normalcy is reached. Intakes are safe over a wide range, and then toxicity is reached.

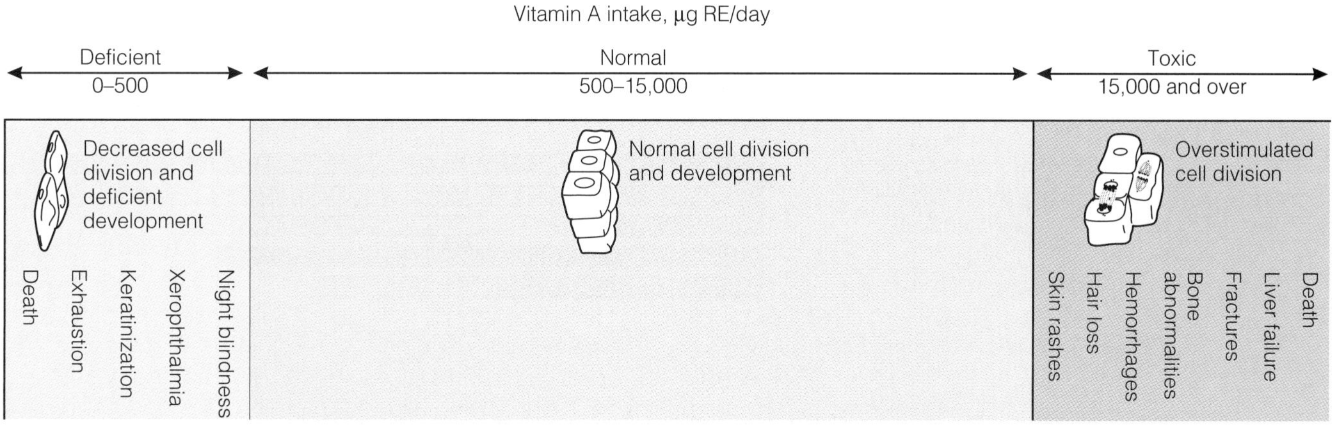

**Figure 11-8**
**Vitamin A in Selected Foods**

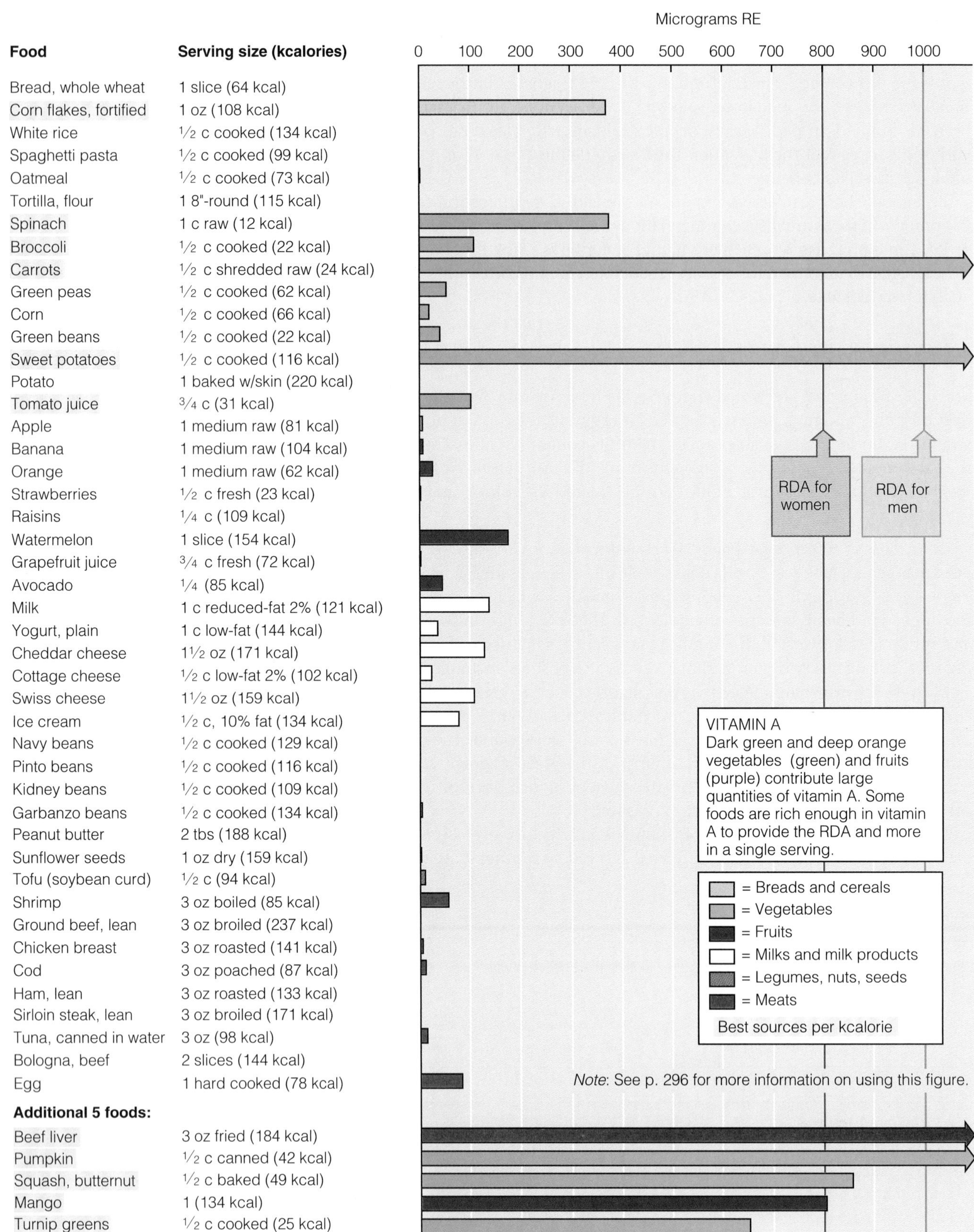

no vitamin A activity. As for white plant foods such as potatoes, cauliflower, pasta, and rice, they also possess little or no vitamin A activity.

• ***Typical Intakes*** • In the typical U.S. diet, about half of the vitamin A activity comes from vegetables and fruits, and half of this comes from the dark leafy greens (like spinach—not celery or cabbage) and the rich yellow or deep orange vegetables and fruits (such as winter squash, cantaloupe, carrots, and sweet potatoes—not corn or bananas). The other half of the vitamin A activity in diets is from preformed vitamin A in milk, cheese, butter, and other dairy products; eggs; and liver. Since vitamin A is fat soluble, it is lost when milk is skimmed. To compensate, nonfat milk is often fortified so as to supply about 40 percent of the RDA per quart.* Margarine is usually fortified so as to provide the same amount of vitamin A as butter.

• ***Vitamin A–Poor Fast Foods*** • Fast foods often lack vitamin A. Anyone who dines frequently on hamburgers, french fries, and colas is advised to emphasize colorful vegetables and fruits at other meals.

• ***Vitamin A–Rich Liver*** • People sometimes wonder if eating liver too frequently can cause vitamin A toxicity. Liver is a rich source of preformed vitamin A, because vitamin A is stored there in animals, just as in humans.** Arctic explorers who have eaten large quantities of polar bear liver have become ill with symptoms suggesting vitamin A toxicity. Closer to home, vitamin A toxicity symptoms were reported in young children who regularly ate a chicken liver spread that provided a daily average of up to three times their recommended intake.[15] Liver offers many nutrients, and eating it periodically may improve a person's nutrition status, but caution is warranted not to eat too much too often, especially for pregnant women. With one ounce of beef liver providing more than three times the RDA for vitamin A, intakes can rise quickly.

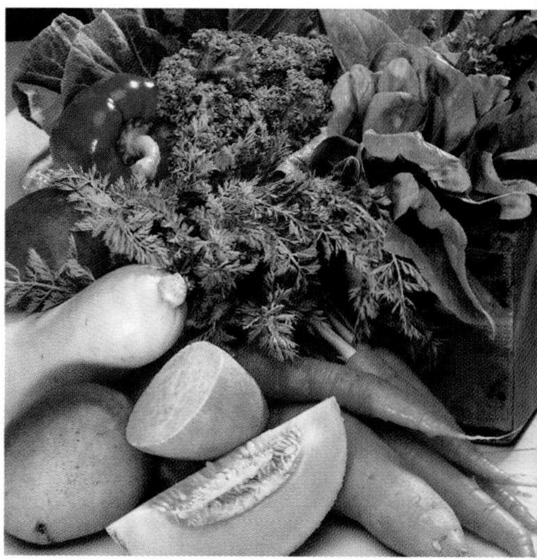

***The carotenoids bring colors to meals; the retinoids allow us to see them.***

IN SUMMARY

Vitamin A is found in the body in three forms: retinol, retinal, and retinoic acid. Together, they are essential to vision, healthy epithelial tissues, and growth. Vitamin A deficiency is a major health problem worldwide, leading to infections, blindness, and keratinization of epithelial tissues. Toxicity can also cause problems and is most often associated with supplement abuse. Preformed vitamin A is found primarily in animal-derived foods such as liver and milk, whereas the precursor beta-carotene is found in brightly colored plant foods such as spinach, carrots, and pumpkins. In addition to providing vitamin A, beta-carotene acts as an antioxidant in the body. The table on p. 346 summarizes deficiency symptoms, toxicity symptoms, functions in the body, and food sources.

# Vitamin D

Vitamin D is different from all the other nutrients in that the body can synthesize it, with the help of sunlight, from a precursor that the body makes from cholesterol. Therefore, vitamin D is not an essential nutrient: given enough time in the sun, people need no vitamin D from foods.

Figure 11-9 diagrams the pathway for making and activating vitamin D. Ultraviolet rays from the sun hit the precursor in the skin and convert it to previtamin $D_3$. This compound works its way into the body and slowly, over the next 36 hours, is converted to vitamin $D_3$ with the help of the body's heat. The biological activity of the active vitamin is 500- to 1000-fold higher than that of its precursor.

**Figure 11-9**
**Vitamin D Synthesis and Activation**

The precursor of vitamin D is made in the liver from cholesterol (see Figure 5-10 on p. 134 and Appendix C). The hydroxylation of vitamin D to its active form is a closely regulated process. The final product, active vitamin D, is 1,25-dihydroxycholecalciferol (or calcitriol).

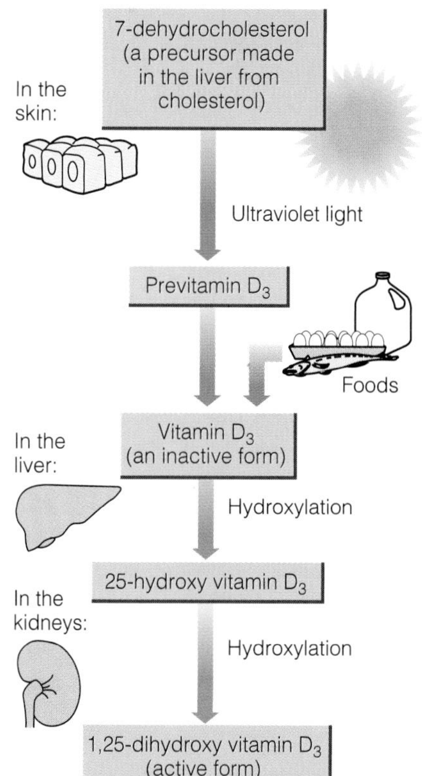


*Similarly, in Canada all milk that has had fat removed must be fortified with vitamin A.
**The liver is not the only organ that stores vitamin A. The kidneys, adrenals, and other organs do, too, but the liver stores the most and is the one most commonly eaten.

## IN SUMMARY

### Vitamin A

**Other Names**

Retinol, retinal, retinoic acid; precursor is provitamin A carotenoids such as beta-carotene

**1989 RDA**

Men: 1000 μg RE/day

Women: 800 μg RE/day

**Chief Functions in the Body**

Vision; maintenance of cornea, epithelial cells, mucous membranes, skin; bone and tooth growth; reproduction; immunity

**Significant Sources**

Retinol: fortified milk, cheese, cream, butter, fortified margarine, eggs, liver

Beta-carotene: spinach and other dark leafy greens; broccoli, deep orange fruits (apricots, cantaloupe) and vegetables (squash, carrots, sweet potatoes, pumpkin)

| Deficiency Disease Name | Toxicity Disease Name |
|---|---|
| Hypovitaminosis A | Hypervitaminosis A[a] |
| **Deficiency Symptoms** | **Toxicity Symptoms** |
| ***Bones/Teeth*** | |
| Cessation of bone growth, painful joints, impaired enamel formation, cracks in teeth, tendency to decay, atrophy of dentin-forming cells | Increased activity of osteoclasts[b] causing decalcification, joint pain, fragility, stunted growth, and thickening of long bones; increase of pressure inside skull, mimicking brain tumor; headaches |
| ***Blood*** | |
| Anemia, often masked by dehydration | Loss of hemoglobin and potassium by red blood cells, cessation of menstruation, slowed clotting time, easily induced bleeding |
| ***Eyes[c]*** | |
| Night blindness, changes in epithelial tissue (hyperkeratinization), drying (xerosis), triangular gray spots on eye (Bitot's spots), softening of the cornea (keratomalacia), and corneal degeneration and blindness (xerophthalmia) | |
| ***Skin*** | |
| Plugging of hair follicles with keratin, forming white lumps (hyperkeratosis) | Dryness; itching; peeling; rashes; dry, scaling lips; cracking and bleeding of lips; nosebleeds; loss of hair; brittle nails |
| ***Digestive System*** | |
| Changes in lining, diarrhea | Nausea, vomiting, abdominal pain, diarrhea, weight loss |
| ***Immune System*** | |
| Suppression of immune reactions; frequent respiratory, digestive, bladder, vaginal, and kidney infections | Overstimulation of immune reactions |
| ***Nervous/Muscular Systems*** | |
| Brain and spinal cord growth too fast for stunted skull and spine | Loss of appetite, irritability, fatigue, insomnia, restlessness, headaches, blurred vision, nausea, vomiting, muscle weakness |
| ***Other*** | |
| Kidney stones | Amenorrhea,[d] jaundice,[e] enlargement of liver[f] and spleen, massive accumulation of fat and vitamin A in liver, birth defects |

[a] A related condition, *hypercarotenemia,* is caused by the accumulation of too much of the vitamin A precursor beta-carotene in the blood, which turns the skin noticeably yellow. Hypercarotenemia is not, strictly speaking, a toxicity symptom.

[b] *Osteoclasts* are the cells that destroy bone during its growth. Those that build bone are *osteoblasts.*

[c] The eyes' symptoms of vitamin A deficiency are collectively known as *xerophthalmia.*

[d] Elevated serum carotene concentrations are associated with amenorrhea.

[e] *Jaundice* (JAWN-dice) is a symptom of liver disease, in which bile and related pigments spill into the bloodstream and the skin yellows.

[f] If liver impairment is severe, the "classic" signs seen in skin and hair may be masked.

Regardless of whether the body manufactures vitamin $D_3$ or obtains it from food, two hydroxylation steps must occur before the vitamin becomes fully active. First, the liver adds an OH group, and then the kidneys add another OH group to produce the active vitamin. A review of Figure 11-9 shows how diseases affecting either the liver or the kidneys can interfere with the activation of vitamin D and produce symptoms of deficiency.

## Roles in the Body

Though called a vitamin, vitamin D is actually a hormone—a compound manufactured by one organ of the body that affects another part. Like vitamin A, vitamin D has a binding protein that carries the vitamin to its target organs—most notably, the intestines, the kidneys, and the bones. All respond to vitamin D by making calcium available for bone growth.

• ***Vitamin D in Bone Growth*** • Vitamin D is a member of a large and cooperative bone-making and maintenance team composed of nutrients and other compounds, including vitamins A, C, and K; the hormones parathormone and calcitonin; the protein collagen; and the minerals calcium, phosphorus, magnesium, and fluoride. Vitamin D's special role in bone growth is to make calcium and phosphorus available in the blood that bathes the bones. The bones grow denser and stronger as they absorb and deposit these minerals.

Vitamin D raises blood concentrations of these minerals in three ways. It stimulates their absorption from the GI tract, their retention by the kidneys, and their withdrawal from the bones into the blood. The vitamin may work alone, as it does in the GI tract, or in combination with parathormone, as it does in the bones and kidneys. Vitamin D is the director, but the star of the show is calcium itself. Details of calcium balance appear in Chapter 12.

• ***Vitamin D in Other Roles*** • Scientists have discovered many other vitamin D target tissues, including the brain and nervous system, pancreas, skin, muscles and cartilage, reproductive organs, and many cancer cells.[16] These discoveries suggest that vitamin D has numerous functions and may be valuable in treating a number of disorders, including cancer.

**Figure 11-10**
**Vitamin D–Deficiency Symptom—The Bowed Legs of Rickets**

This child has the bowed legs commonly seen in rickets.

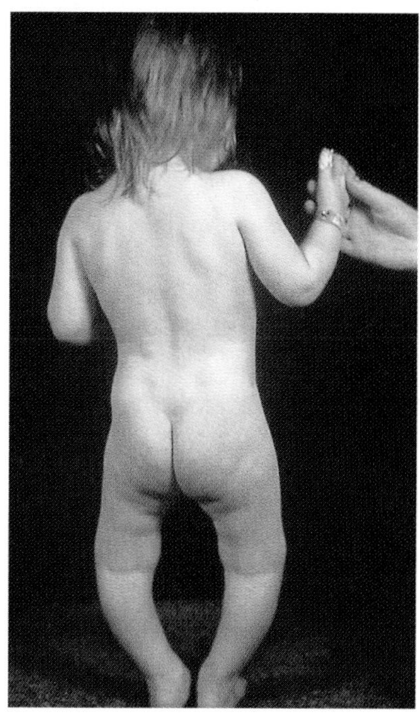

## Vitamin D Deficiency

In vitamin D deficiency, production of the protein that binds calcium in the intestinal cells slows. Thus, even when calcium in the diet is adequate, it passes through the GI tract unabsorbed, leaving the bones undersupplied. The symptoms of a vitamin D deficiency are those of calcium deficiency, shown in the summary table on p. 348.

• ***Rickets*** • Worldwide, the vitamin D–deficiency disease rickets still afflicts large numbers of children. The bones fail to calcify normally, causing growth retardation and skeletal abnormalities. The bones become so weak that they bend when they have to support the body's weight (see Figure 11-10). A child with rickets who is old enough to walk characteristically develops bowed legs, often the most obvious sign of the disease. Another sign is the protruding belly that results from lax abdominal muscles.

**rickets:** the vitamin D–deficiency disease in children characterized by inadequate mineralization of bone (manifested in bowed legs or knock-knees, outward-bowed chest, and knobs on ribs). A rare type of rickets, not caused by vitamin D deficiency, is known as **vitamin D–refractory rickets.**

• ***Osteomalacia*** • The adult form of rickets, osteomalacia, occurs most often in women who have low calcium intakes and little exposure to sun and who go through repeated pregnancies and periods of lactation. Given this combination of risk factors, the bones of the legs may soften to such an extent that a young woman who is tall and straight at 20 may become bent, bowlegged, and stooped before she is 30.

**osteomalacia** (OS-tee-oh-ma-LAY-shuh): a bone disease characterized by softening of the bones; symptoms include bending of the spine and bowing of the legs. The disease occurs most often in adult women.
- **osteo** = bone
- **malacia** = softening

• ***Osteoporosis*** • Any failure to synthesize adequate vitamin D sets the stage for a loss of calcium from the bones, which can result in fractures. Highlight 12

IN SUMMARY

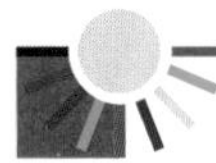

## Vitamin D

**Other Names**

Calciferol (kal-SIF-er-ol), 1,25-dihyroxy vitamin D (calcitriol); the animal version is **vitamin $D_3$** or **cholecalciferol** (KO-lee-kal-SIF-er-ol); the plant version is **vitamin $D_2$** or **ergocalciferol** (er-go-kal-SIF-er-ol); precursor is the body's own cholesterol

**1997 Adequate Intake (AI)**

Adults: 5 μg/day (19–50 yr)
10 μg/day (51 –70 yr)
15 μg/day (>70 yr)

**Upper Level**

50 μg/day

**Chief Functions in the Body**

Mineralization of bones (raises blood calcium and phosphorus by increasing absorption from digestive tract, withdrawing calcium from bones, stimulating retention by kidneys)

**Significant Sources**

Synthesized in the body with the help of sunlight; fortified milk, margarine, butter, cereals, and chocolate mixes; veal, beef, egg yolks, liver, fatty fish (herring, salmon, sardines) and their oils

**Deficiency Disease Names**

Rickets, osteomalacia

**Toxicity Disease Name**

Hypervitaminosis D

| | Deficiency Symptoms: Rickets in Children | Deficiency Symptoms: Osteomalacia in Adults | Toxicity Symptoms |
|---|---|---|---|
| ***Bones/Teeth*** | Faulty calcification, resulting in misshapen bones (bowing of legs) and retarded growth; enlargement of ends of long bones (knees, wrists); deformities of ribs (bowed, with beads or knobs);[a] delayed closing of fontanel, resulting in rapid enlargement of head (see figure); slow eruption of teeth; malformed, decay-prone teeth | Softening effect: deformities of limbs, spine, thorax, and pelvis; demineralization; pain in pelvis, lower back, and legs; bone fractures | Increased calcium withdrawal |
| ***Digestive System*** | | | Nausea, vomiting |
| ***Blood*** | | Decreased calcium and/or phosphorus concentration, increased alkaline phosphatase[b] | Increased calcium and phosphorus concentration |
| ***Nervous/Muscular Systems*** | Lax muscles resulting in protrusion of abdomen; muscle spasms | Involuntary twitching, muscle spasms | Loss of appetite, headache, muscle weakness, joint pain, fatigue, excessive thirst, irritability, apathy |
| ***Excretory System*** | Increased excretion of calcium in stools, decreased calcium in urine | | Increased excretion of calcium in urine, kidney stones, irreversible kidney damage |
| ***Other*** | Abnormally high secretion of parathormone | | Calcification of soft tissues (blood vessels, kidneys, heart, lungs, tissues around joints), frequent urination, death |

**Fontanel**
A fontanel is an open space in the top of a baby's skull before the bones have grown together. In rickets, closing of the fontanel is delayed.

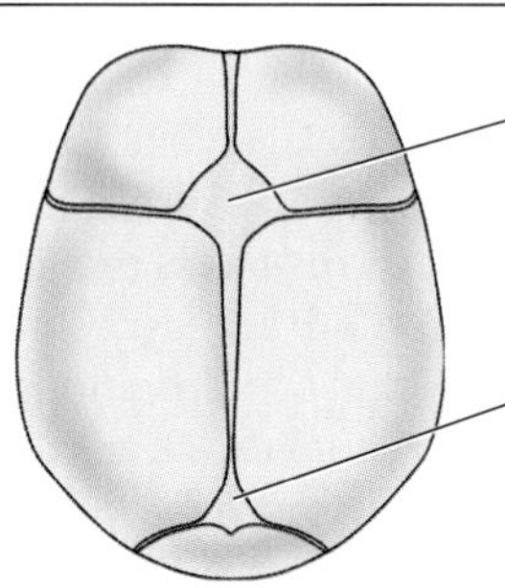
Anterior fontanel normally closes by the end of the second year
Posterior fontanel normally closes by the end of the first year

[a]Bowing of the ribs causes the symptoms known as *pigeon breast*. The beads that form on the ribs resemble rosary beads; thus this symptom is known as *rachitic* (ra-KIT-ik) rosary ("the rosary of rickets").
[b]Alkaline phosphatase is an enzyme in the blood that rises during bone resorption.

describes the many factors that lead to osteoporosis, a condition of reduced bone density.

## Vitamin D Toxicity

Whereas vitamin D deficiency depresses calcium absorption, blood calcium, and bone mineralization, an excess of the vitamin does the opposite, as shown in the summary table on p. 348. It enhances calcium absorption, produces high blood calcium, and promotes the return of bone calcium into the blood. Excess blood calcium tends to precipitate in the soft tissue, forming stones, especially in the kidneys where calcium is concentrated in the effort to excrete it. Calcification may also harden the blood vessels and is especially dangerous in the major arteries of the heart and lungs, where it can cause death.

**mineralization:** the process in which calcium, phosphorus, and other minerals crystallize on the collagen matrix of a growing bone, hardening the bone.

High blood calcium is known as **hypercalcemia** and may develop from a variety of disorders, including vitamin D toxicity. It does *not* develop from a high calcium intake.

The liver's storage capacity is smaller for vitamin D than for vitamin A. Consequently, the range of safe intakes of vitamin D is narrower than that of vitamin A. In fact, vitamin D is the most likely of the vitamins to have toxic effects when consumed in amounts above recommendations on a continuous basis. The amounts of vitamin D in foods available in the United States and Canada are well within safe limits, but supplements containing the vitamin in concentrated form should be kept out of the reach of children and used cautiously by adults. During the warm months of the year, when sunlight exposure may be frequent, vitamin D supplements can harm healthy children and adults who drink at least 2 glasses of vitamin D–fortified milk per day. Vitamin D toxicity has also been reported in people who drank milk that was fortified with too much vitamin D by mistake.[17]

## Vitamin D Recommendations and Sources

Only a few foods contain vitamin D naturally. Fortunately, the body can make all the vitamin D it needs with the help of a little sunshine. In setting the Adequate Intake (AI), however, the committee assumed that no vitamin D was available from this synthesis.

• ***Vitamin D in Foods*** • Only a few foods supply significant amounts of vitamin D, notably those derived from animals: egg yolks, liver, fatty fish, butter, and fortified milk. The fortification of milk with vitamin D is the best guarantee that people will meet their needs and underscores the importance of milk in a well-balanced diet.* For those who use margarine in place of butter, fortified margarine is a significant source. A plant version of vitamin D may yield an active compound on irradiation, but its contribution is minor. Without adequate sunshine, fortification, or supplementation, a strict vegetarian diet cannot meet vitamin D needs.[18]

Vitamin D activity was previously expressed in international units (IU), but is now expressed in micrograms of cholecalciferol. To convert, use the following factor: 1 IU = 0.025 μg cholecalciferol. For example:
- 100 IU = 2.5 μg (100 IU × 0.025 μg).
- 400 IU = 10 μg (400 IU × 0.025 μg).

Most adults, especially in sunny regions, need not make special efforts to obtain vitamin D from food. People who are not outdoors much or who live in northern or predominantly cloudy or smoggy areas are advised to drink at least 2 cups of vitamin D–fortified milk a day.

• ***Vitamin D from the Sun*** • Most of the world's population relies on natural exposure to sunlight to maintain adequate vitamin D nutrition. The sun imposes no risk of vitamin D toxicity; prolonged exposure to sunlight degrades the vitamin D precursor in the skin, preventing its conversion to the active vitamin. Even lifeguards on southern beaches are safe from vitamin D toxicity from the sun.

Prolonged exposure to sunlight does, however, prematurely wrinkle the skin and present the risk of skin cancer. Sunscreens help reduce these risks, but unfortunately, sunscreens with sun protection factors (SPF) of 8 and above also

***A cool glass of milk refreshes as it replenishes the bone-building nutrients.***

*Vitamin D fortification of milk in the United States is 10 micrograms cholecalciferol (400 IU) per quart; in Canada, 9.6 micrograms (385 IU) per liter.

prevent vitamin D synthesis. A strategy to avoid this dilemma is to apply sunscreen after enough time has elapsed to provide sufficient vitamin D synthesis. For most people, exposing hands, face, and arms on a clear summer day for 10 to 15 minutes a few times a week should be sufficient to maintain vitamin D nutrition.

Dark-skinned people require longer sunlight exposure than light-skinned people: heavily pigmented skin arrives at the same plateau of vitamin D synthesis in three hours as fair skin in 30 minutes. The ultraviolet (UV) rays of the sun that promote vitamin D synthesis are blocked by heavy clouds, smoke, or smog. Differences in skin pigmentation and smog may account for the finding that dark-skinned people in northern, smoggy cities are more prone to rickets. For these people, and for those who are unable to go outdoors frequently, dietary vitamin D is most important. Deficiency is especially likely in older adults because they typically drink little or no milk, their exposure to sunlight is limited, and the skin, liver, and kidneys lose their ability to make and activate vitamin D with advancing age.

Factors that may limit sun exposure and, therefore, vitamin D synthesis:

- Geographical location.
- Season of the year.
- Air pollution.
- Clothing.
- Tall buildings.
- Indoor living.
- Sunscreens.

Depending on the UV radiation used, the UV rays from tanning lamps and tanning booths may also stimulate vitamin D synthesis, but the hazards outweigh any possible benefits.* The Food and Drug Administration (FDA) warns that if the lamps are not properly filtered, people using tanning booths risk burns, damage to the eyes and blood vessels, and skin cancer.

*Smog filters out ultraviolet rays of the sun.*

IN SUMMARY

Vitamin D can be synthesized in the body with the help of sunlight or obtained from animal foods. It sends signals to three primary target sites: the GI tract to absorb more calcium and phosphorus, the bones to release more, and the kidneys to retain more. These actions support bone formation and maintenance. A deficiency causes rickets in childhood and osteomalacia in later life. Fortified milk is an important food source.

# Vitamin E

In 1922, researchers discovered a component of vegetable oils necessary for reproduction in rats. This antisterility factor was named tocopherol, which means "to bring forth offspring." A few years later, the compound was named vitamin E. When chemists isolated four different tocopherol compounds, they designated them by the first four letters of the Greek alphabet: alpha, beta, gamma, and delta. The tocopherols consist of a complex ring structure and a long saturated side chain (see Appendix C).† The numbers and positions of methyl groups on the ring distinguish one tocopherol from another. The most abundant and biologically active tocopherol in nature is alpha-tocopherol.

**tocopherol** (tuh-KOFF-er-ol): a general term for several chemically related compounds, most of which have vitamin E activity (see Appendix C for chemical structures).

**alpha-tocopherol:** the most biologically active vitamin E compound.

## *Vitamin E as an Antioxidant*

Reminder: An *antioxidant* is a compound that protects other compounds from oxidation by being oxidized itself.

Vitamin E is a fat-soluble antioxidant and one of the body's primary defenders against oxidation, protecting the lipids and other vulnerable components of the

*The best wavelengths for vitamin D synthesis are UV-B rays between 290 and 310 nanometers. Some tanning parlors advertise "UV-A rays only, for a tan without the burn," but in fact, UV-A rays can damage the skin. D. A. Bender, *Nutritional Biochemistry of the Vitamins* (Cambridge: Cambridge University Press, 1992), p. 56 and A. Greeley, Dodging the rays, *FDA Consumer,* July/August 1993, pp. 30–33.

†Another group of chemically related compounds, the tocotrienols (TOE-koe-try-EEN-ols), have unsaturated side chains. The tocotrienols are less abundant, less active, and less important in nutrition than the tocopherols are.

cells and their membranes from destruction. Most notably, vitamin E is effective in preventing the oxidation of the polyunsaturated fatty acids (PUFA), but it protects other lipids and related compounds (for example, vitamin A) as well. Vitamin E acts synergistically with the mineral selenium, another participant in the body's antioxidant activities. Together they are especially effective in protecting cell membranes against oxidative damage.

Accumulating evidence suggests that vitamin E may reduce the risk of heart disease by protecting LDL against oxidation.[19] The oxidation of LDL has been implicated as a key factor in the development of heart disease. People who eat foods rich in vitamin E or take supplements of vitamin E have a reduced risk of heart disease.[20] Highlight 11 provides many more details on how vitamin E and other antioxidants protect against chronic diseases, such as heart disease and cancer.[21]

Vitamin E exerts an especially important antioxidant effect in the lungs, where the exposure of cells to oxygen is maximal. Several kinds of cells benefit from the vitamin's protection: the red and white blood cells that pass through the lungs and the cells of the lung tissue itself. Vitamin E also protects the lungs from air pollutants such as nitrogen dioxide and ozone that can initiate damaging reactions.

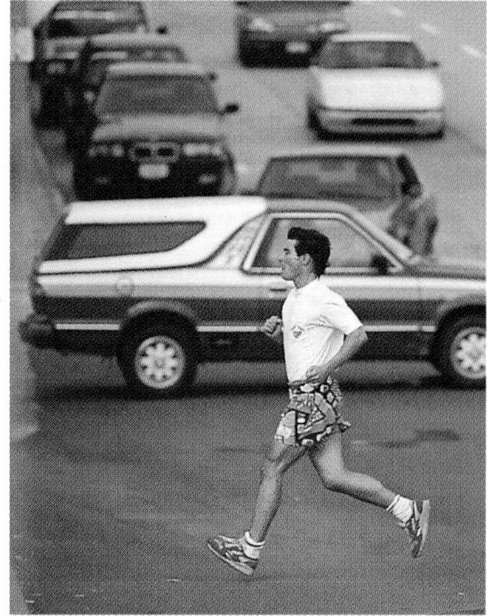

***Vitamin E helps to protect the lungs against air pollutants, especially when a person is breathing hard during exercise.***

While research continues to reveal possible roles for vitamin E, it also has clearly discredited claims that vitamin E improves physical performance, enhances sexual performance, or cures sexual dysfunction in males. Vitamin E does not slow or prevent the processes of aging such as hair turning gray or skin wrinkling. Nor does it slow the progression of Parkinson's disease.[22]

## *Vitamin E Deficiency*

In human beings, dietary vitamin E deficiency is rare; deficiency is usually associated with diseases of fat malabsorption such as cystic fibrosis. Without vitamin E, the red blood cells break open and spill their contents, probably due to oxidation of the PUFA in their membranes. This classic sign of vitamin E deficiency, known as erythrocyte hemolysis, is seen in premature infants, born before the transfer of vitamin E from the mother to the infant that takes place in the last weeks of pregnancy. Vitamin E treatment corrects hemolytic anemia.

Prolonged vitamin E deficiency also causes neuromuscular dysfunction involving the spinal cord and retina of the eye. Common symptoms include loss of muscle coordination and reflexes and impaired vision and speech. Vitamin E treatment corrects these neurological symptoms of vitamin E deficiency, but it does *not* prevent or cure the hereditary muscular dystrophy that afflicts children. Children with this condition do not benefit from vitamin E treatment and usually die at an early age when their respiratory muscles deteriorate.

Two other conditions seem to respond to vitamin E therapy, although results are inconsistent. One is a nonmalignant breast disease (fibrocystic breast disease), and the other is an abnormality of blood flow that causes cramping in the legs (intermittent claudication).

**erythrocyte** (eh-RITH-ro-cite) **hemolysis** (he-MOLL-uh-sis): the breaking open of red blood cells (erythrocytes); a symptom of vitamin E–deficiency disease in human beings.
- **erythro** = red
- **cyte** = cell
- **hemo** = blood
- **lysis** = breaking

**hemolytic** (HE-moh-LIT-ick) **anemia:** the condition of having too few red blood cells as a result of erythrocyte hemolysis.

**muscular dystrophy** (DIS-tro-fee): a hereditary disease in which the muscles gradually weaken; its most debilitating effects arise in the lungs.

**fibrocystic breast disease:** a harmless condition in which the breasts develop lumps, sometimes associated with caffeine consumption. In some, it responds to treatment by abstinence from caffeine; in others, it can be treated with vitamin E.
- **fibro** = fibrous tissue
- **cyst** = closed sac

**intermittent claudication:** severe calf pain caused by inadequate blood supply; it occurs when walking and subsides during rest.
- **intermittent** = at intervals
- **claudicare** = to limp

**D, L:** *D* stands for *dextro,* or "right-handed," and *L,* for *levo,* or "left-handed," referring to molecules that are identical to each other physically and chemically, but rotate light oppositely.

## *Vitamin E Toxicity*

Vitamin E supplement use has risen in recent years as its protective actions against diseases have been recognized. Still, toxicity is not as common, and its effects are not as detrimental, as with vitamins A and D. Extremely high doses of vitamin E may interfere with the blood-clotting action of vitamin K and enhance the effects of drugs used to oppose blood clotting, causing hemorrhage.

## *Vitamin E Recommendations*

Tocopherols occur in two forms, D and L; the D form is more active. The most active vitamin E compound is D-alpha-tocopherol. Different tocopherols vary in

**tocopherol equivalents (TE):** the units in which vitamin E activity is measured. One TE equals the amount of vitamin E activity in 1 milligram of D-alpha-tocopherol.

their vitamin E activity; to reconcile them, recommended intakes are expressed in tocopherol equivalents (TE). One TE equals the amount of vitamin E activity in 1 milligram of D-alpha-tocopherol.*

A person who consumes a large amount of PUFA needs more vitamin E. Fortunately, vitamin E and PUFA tend to occur together in the same foods.

## Vitamin E in Foods

Vitamin E is widespread in foods (see Figure 11-11). Much of the vitamin E in the diet comes from vegetable oils and products made from them, such as margarine, salad dressings, and shortenings, and from fruits and vegetables. Wheat germ oil is especially rich in vitamin E. Corn and soybean oils rank second; a tablespoon of either supplies more than the recommended intake of the vitamin and three times as much as some other oils, such as peanut oil. Animal fats such as meat and milk fat contain little or no vitamin E.

Vitamin E is readily destroyed by heat processing (such as deep-fat frying) and oxidation, so fresh or lightly processed foods are preferable sources. Most processed and convenience foods do not contribute enough vitamin E to ensure an adequate intake.

### IN SUMMARY

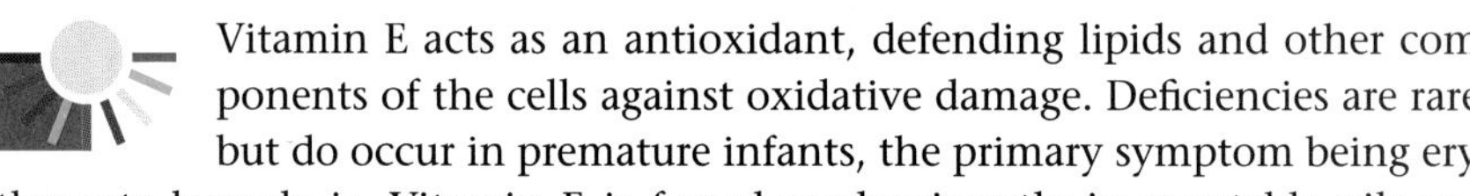

Vitamin E acts as an antioxidant, defending lipids and other components of the cells against oxidative damage. Deficiencies are rare, but do occur in premature infants, the primary symptom being erythrocyte hemolysis. Vitamin E is found predominantly in vegetable oils and appears to be one of the least toxic of the fat-soluble vitamins. The summary table reviews vitamin E's functions, deficiency symptoms, toxicity symptoms, and food sources.

*The activities of beta- and gamma-tocopherol and alpha-tocotrienol are one-half, one-tenth, and one-third the activity of D-alpha-tocopherol, respectively.

### IN SUMMARY

**Vitamin E**

**Other Names**

Alpha-tocopherol, tocopherol, tocotrienol

**1989 RDA**

Men: 10 mg α-TE/day

Women: 8 mg α-TE/day

**Chief Functions in the Body**

Antioxidant [stabilization of cell membranes, regulation of oxidation reactions, protection of polyunsaturated fatty acids (PUFA) and vitamin A]

**Significant Sources**

Polyunsaturated plant oils (margarine, salad dressings, shortenings), leafy green vegetables, wheat germ, whole-grain products, liver, egg yolks, nuts, seeds

Easily destroyed by heat and oxygen

| Deficiency Symptoms | Toxicity Symptoms |
|---|---|
| ***Blood/Circulatory System*** | |
| Red blood cell breakage,[a] anemia | Augments the effects of anticlotting medication |
| ***Nervous/Muscular Systems*** | |
| Degeneration, weakness, difficulty walking, leg cramps | |

[a]The breaking of red blood cells is called *erythrocyte hemolysis.*

**Figure 11-11**
**Vitamin E in Selected Foods**

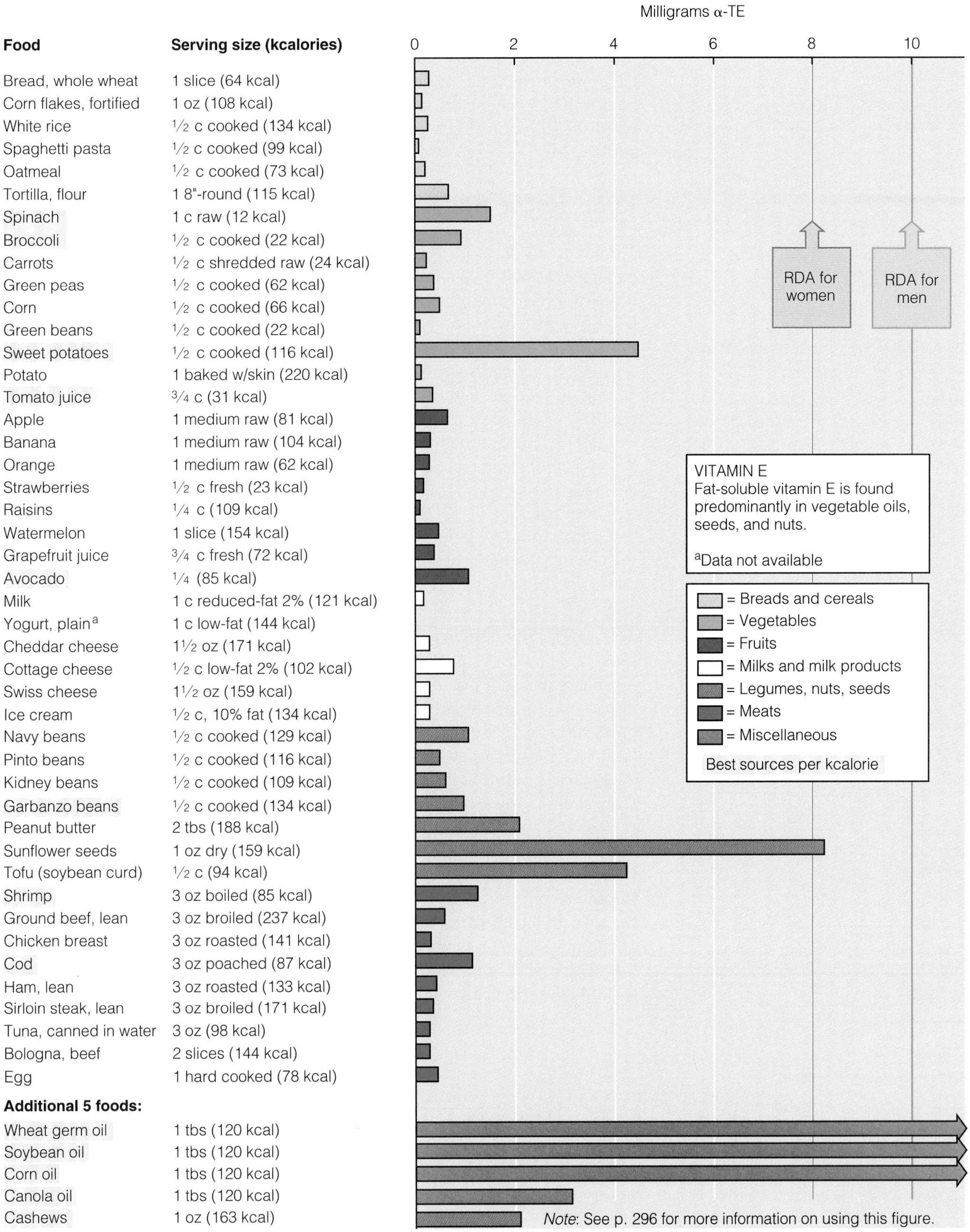
Milligrams α-TE
0 2 4 6 8 10
Food
Serving size (kcalories)
Bread, whole wheat 1 slice (64 kcal)
Corn flakes, fortified 1 oz (108 kcal)
White rice 1/2 c cooked (134 kcal)
Spaghetti pasta 1/2 c cooked (99 kcal)
Oatmeal 1/2 c cooked (73 kcal)
Tortilla, flour 1 8"-round (115 kcal)
Spinach 1 c raw (12 kcal)
Broccoli 1/2 c cooked (22 kcal)
Carrots 1/2 c shredded raw (24 kcal)
Green peas 1/2 c cooked (62 kcal)
Corn 1/2 c cooked (66 kcal)
Green beans 1/2 c cooked (22 kcal)
Sweet potatoes 1/2 c cooked (116 kcal)
Potato 1 baked w/skin (220 kcal)
Tomato juice 3/4 c (31 kcal)
Apple 1 medium raw (81 kcal)
Banana 1 medium raw (104 kcal)
Orange 1 medium raw (62 kcal)
Strawberries 1/2 c fresh (23 kcal)
Raisins 1/4 c (109 kcal)
Watermelon 1 slice (154 kcal)
Grapefruit juice 3/4 c fresh (72 kcal)
Avocado 1/4 (85 kcal)
Milk 1 c reduced-fat 2% (121 kcal)
Yogurt, plain[a] 1 c low-fat (144 kcal)
Cheddar cheese 1 1/2 oz (171 kcal)
Cottage cheese 1/2 c low-fat 2% (102 kcal)
Swiss cheese 1 1/2 oz (159 kcal)
Ice cream 1/2 c, 10% fat (134 kcal)
Navy beans 1/2 c cooked (129 kcal)
Pinto beans 1/2 c cooked (116 kcal)
Kidney beans 1/2 c cooked (109 kcal)
Garbanzo beans 1/2 c cooked (134 kcal)
Peanut butter 2 tbs (188 kcal)
Sunflower seeds 1 oz dry (159 kcal)
Tofu (soybean curd) 1/2 c (94 kcal)
Shrimp 3 oz boiled (85 kcal)
Ground beef, lean 3 oz broiled (237 kcal)
Chicken breast 3 oz roasted (141 kcal)
Cod 3 oz poached (87 kcal)
Ham, lean 3 oz roasted (133 kcal)
Sirloin steak, lean 3 oz broiled (171 kcal)
Tuna, canned in water 3 oz (98 kcal)
Bologna, beef 2 slices (144 kcal)
Egg 1 hard cooked (78 kcal)
Additional 5 foods:
Wheat germ oil 1 tbs (120 kcal)
Soybean oil 1 tbs (120 kcal)
Corn oil 1 tbs (120 kcal)
Canola oil 1 tbs (120 kcal)
Cashews 1 oz (163 kcal)
RDA for women
RDA for men
VITAMIN E
Fat-soluble vitamin E is found predominantly in vegetable oils, seeds, and nuts.
[a]Data not available
= Breads and cereals
= Vegetables
= Fruits
= Milks and milk products
= Legumes, nuts, seeds
= Meats
= Miscellaneous
Best sources per kcalorie
Note: See p. 296 for more information on using this figure.

# Vitamin K

Blood has a remarkable ability to remain a liquid, but can turn solid within seconds when the integrity of that system is disturbed. (If blood did not clot, a single pinprick could drain the entire body of all its blood, just as a tiny hole in a bucket makes the bucket forever useless for holding water.) Vitamin K acts primarily in blood clotting, where its presence can make the difference between life and death.

*K* stands for the Danish word *koagulation* ("coagulation" or "clotting").

## Roles in the Body

More than a dozen different proteins and the mineral calcium are involved in making a blood clot. Vitamin K is essential for the activation of several of these proteins, among them prothrombin, made by the liver as a precursor of the protein thrombin (see Figure 11-12).[23] When any of the blood-clotting factors is lacking, hemorrhagic disease results. If an artery or vein is cut or broken, bleeding goes unchecked. (Of course, this is not to say that hemorrhaging is always caused by vitamin K deficiency. Another cause is hemophilia, which is not curable with vitamin K.)

**hemorrhagic** (hem-oh-RAJ-ik) **disease:** a disease characterized by excessive bleeding.

**hemophilia** (HE-moh-FEEL-ee-ah): a hereditary disease that is caused by a genetic defect and has no relation to vitamin K; the blood is unable to clot because it lacks the ability to synthesize certain clotting factors.

Vitamin K also participates in the synthesis of bone proteins. Without vitamin K, the bones produce an abnormal protein that cannot bind to the minerals that normally form bones. The role of vitamin K in protecting against osteoporosis remains unclear.[24]

Like vitamin D, vitamin K can be obtained from a nonfood source. Bacteria in the GI tract synthesize vitamin K that the body can absorb, but this source alone meets only about half of a person's needs.

## Vitamin K Deficiency

A primary deficiency of vitamin K is rare, but a secondary deficiency may occur in two circumstances. First, whenever fat absorption falters, as occurs when bile production fails, vitamin K absorption diminishes. Second, some drugs disrupt vitamin K's synthesis and action in the body: antibiotics kill the vitamin K–producing bacteria in the intestine, and anticoagulant drugs interfere with vitamin K metabolism and activity. When vitamin K deficiency does occur, it can be fatal.

Reminder: A *primary deficiency* develops in response to an inadequate intake whereas a *secondary deficiency* occurs for other reasons.

Newborn infants present a unique case of vitamin K nutrition because they are born with a sterile digestive tract, and the vitamin K–producing bacteria take weeks to establish themselves in the intestines. At the same time, plasma prothrombin concentrations are low (this makes fatal blood clotting unlikely during the stress of birth). To prevent hemorrhagic disease in the newborn, a single dose of vitamin K (usually as the naturally occurring form, phylloquinone) is given at birth either orally or by intramuscular injection. Concerns that vitamin K given at birth raises the risks of childhood cancer are unproved and unlikely.[25]

**sterile:** free of microorganisms, such as bacteria.

A synthetic form of vitamin K is **menadione** (men-uh-DYE-own); see Appendix C.

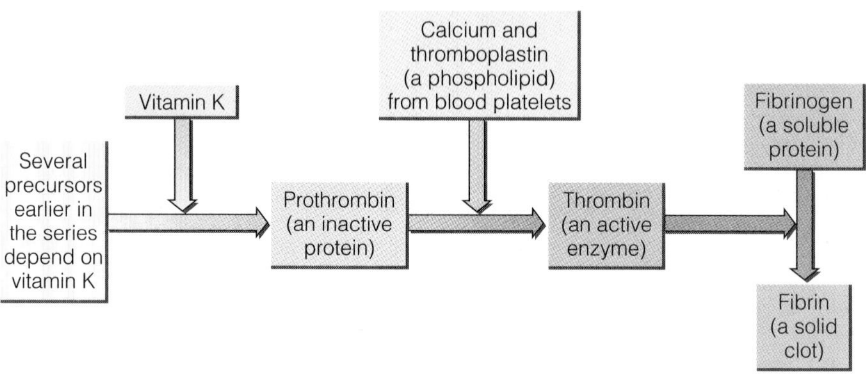


**Figure 11-12**
**Blood-Clotting Process**

When blood is exposed to air, foreign substances, or secretions from injured tissues, platelets (small, cell-like structures in the blood) release a phospholipid known as thromboplastin. Thromboplastin catalyzes the conversion of the inactive protein prothrombin to the active enzyme thrombin. Thrombin then catalyzes the conversion of the precursor protein fibrinogen to the active protein fibrin that forms the clot.

IN SUMMARY

**Vitamin K**

| Other Names | Deficiency Symptoms | Toxicity Symptoms |
|---|---|---|
| Phylloquinone, menaquinone, menadione, naphthoquinone | ***Blood/Circulatory System*** | |
| | Hemorrhaging | Interference with anticlotting medication; vitamin K analogues may cause jaundice, red blood cell hemolysis, and brain damage |

**1989 RDA**

1 µg/kg body weight/day

Men: 70 µg/day (19–24 yr)
80 µg/day (25 and over)

Women: 60 µg/day (19–24 yr)
65 µg/day (25 and over)

**Chief Functions in the Body**

Synthesis of blood-clotting proteins and a bone protein that regulates blood calcium

**Significant Sources**

Bacterial synthesis in the digestive tract;[a] liver; leafy green vegetables, cabbage-type vegetables; milk

[a]Vitamin K needs cannot be met from bacterial synthesis alone; however, it is a potentially important source in the small intestine, where absorption efficiency ranges from 40 to 70 percent.

## Vitamin K Toxicity

Toxicity is not common but can result when vitamin K supplements are prescribed, especially to infants or pregnant women. High doses of vitamin K can reduce the effectiveness of anticoagulant drugs used to prevent blood clotting. People taking these drugs should eat vitamin K–rich foods in moderation and keep their intakes consistent from day to day. Toxicity symptoms include red blood cell hemolysis, jaundice, and brain damage.

**jaundice** (JAWN-dis): yellowing of the skin due to spillover of the bile pigment **bilirubin** (bill-ee-ROO-bin) from the liver into the general circulation; also known as **hyperbilirubinemia** (HIGH-per-BILL-eh-roo-bin-EE-me-ah). When these pigments invade the brain, the condition is **kernicterus** (ker-NICK-ter-us). Jaundice may be caused by obstruction of bile passageways, hemolysis, or dysfunctional liver cells.

## Vitamin K Recommendations and Sources

As mentioned earlier, vitamin K is made in the GI tract by the billions of bacteria that normally reside there. Once synthesized, vitamin K is absorbed and stored in the liver. The total need for vitamin K cannot be met by bacterial synthesis alone.[26] People eating vitamin K–rich foods such as liver, leafy green vegetables, and members of the cabbage family can easily meet their needs. Milk, meats, eggs, cereals, fruits, and vegetables provide smaller, but still significant, amounts.

*Notable food sources of vitamin K include milk, eggs, brussels sprouts, liver, cabbage, spinach, and broccoli.*

IN SUMMARY

Vitamin K helps with blood clotting, and its deficiency causes uncontrolled bleeding. Bacteria in the GI tract can make the vitamin; people typically receive about half of their requirements from bacterial synthesis and half from foods such as liver, leafy green vegetables, and members of the cabbage family. Because people depend on bacterial synthesis for vitamin K, deficiency is most likely in newborn infants and in people taking antibiotics.

# The Fat-Soluble Vitamins—In Summary

The four fat-soluble vitamins play many specific roles in the growth and maintenance of the body. Their presence affects the health and function of the eyes, skin,

GI tract, lungs, bones, teeth, nervous system, and blood; their deficiencies become apparent in these same areas. Toxicities of the fat-soluble vitamins are possible, especially when people use supplements, because the body stores excesses.

As with the water-soluble vitamins, the function of one fat-soluble vitamin often depends on the presence of another. Recall that vitamin E protects vitamin A from oxidation. In vitamin E deficiency, vitamin A absorption and storage are impaired. Three of the four fat-soluble vitamins—A, D, and K—play important roles in bone growth and remodeling. As mentioned, vitamin K helps synthesize a specific bone protein, and vitamin D regulates that synthesis. Vitamin A, in turn, may control which bone-building genes respond to vitamin D.[27]

Fat-soluble vitamins also interact with minerals: vitamin D and calcium cooperate in bone formation; and zinc is required for the synthesis of vitamin A's transport protein, retinol-binding protein. Zinc also assists the enzyme that regenerates retinal from retinol in the eye.

The roles the fat-soluble vitamins play differ from those of the water-soluble vitamins, and they appear in different foods, yet they are just as essential to life. The need for them underlines the importance of eating a wide variety of nourishing foods daily.

## Making It Click

These exercises will help you learn the best food sources for the vitamins and prepare you to examine your own food choices. Appendix K provides answers.

1. Review the units in which vitamins are measured (a spot check). For each of these vitamins, note the unit of measure:

   | | |
   |---|---|
   | Vitamin A | Vitamin D |
   | Vitamin E | Vitamin K |

2. Analyze the vitamin contents of foods. Review the figures, photos, and food sources sections in Chapters 10 and 11 and list the food group(s) that contributed the most of each vitamin. Which food groups offer the most thiamin? The most riboflavin? The most niacin? The most vitamin $B_6$? The most folate? The most vitamin $B_{12}$? The most vitamin C? The most vitamin A? The most vitamin D? The most vitamin E?

   List the groups that provided "the most" and compare them with the Food Guide Pyramid in Chapter 2.

This exercise should convince you that each of the food groups provides some, but not all, of the vitamins needed daily. For a full array, a person needs to eat a variety of foods from each of the food groups regularly.

## Study Questions

These questions will help you review the chapter. You will find the answers in the discussions on the pages provided.

1. List the fat-soluble vitamins. What characteristics do they have in common? How do they differ from the water-soluble vitamins? (p. 337)
2. Summarize the roles of vitamin A and the symptoms of its deficiency. (pp. 337–342)
3. What is meant by vitamin precursors? Name the precursors of vitamin A, and tell in what classes of foods they are located. Give examples of foods with high vitamin A activity. (pp. 337, 343–344)
4. How is vitamin D unique among the vitamins? What is its chief function? What are the richest sources of this vitamin? (pp. 345, 347, 349)
5. Describe vitamin E's role as an antioxidant. What are the chief symptoms of vitamin E deficiency? (pp. 350–351)
6. What is vitamin K's primary role in the body? What conditions may lead to vitamin K deficiency? (p. 354)

These questions will help you prepare for an exam. Answers can be found in Appendix K.

1. Fat-soluble vitamins:
   a. are easily excreted.
   b. seldom reach toxic levels.
   c. require bile for absorption.
   d. are not stored in the body's tissues.

2. The form of vitamin A active in vision is:
   a. retinal.
   b. retinol.
   c. rhodopsin.
   d. retinoic acid.
3. Vitamin A deficiency symptoms include:
   a. rickets and osteomalacia.
   b. hemorrhaging and jaundice.
   c. night blindness and keratomalacia.
   d. fibrocystic breast disease and erythrocyte hemolysis.
4. Good sources of vitamin A include:
   a. oatmeal, pinto beans, and ham.
   b. apricots, turnip greens, and liver.
   c. whole-wheat bread, green peas, and tuna.
   d. corn, grapefruit juice, and sunflower seeds.
5. To keep minerals available in the blood, vitamin D targets:
   a. the skin, the muscles, and the bones.
   b. the kidneys, the liver, and the bones.
   c. the intestines, the kidneys, and the bones.
   d. the intestines, the pancreas, and the liver.
6. Vitamin D can be synthesized from a precursor that the body makes from:
   a. bilirubin.
   b. tocopherol.
   c. cholesterol.
   d. beta-carotene.
7. Vitamin E's most notable role is to:
   a. protect lipids against oxidation.
   b. activate blood-clotting proteins.
   c. support protein and DNA synthesis.
   d. enhance calcium deposits in the bones.
8. The classic sign of vitamin E deficiency is:
   a. rickets.
   b. xeropthalmia.
   c. muscular dystrophy.
   d. erythrocyte hemolysis.
9. Without vitamin K:
   a. muscles atrophy.
   b. bones become soft.
   c. skin rashes develop.
   d. blood fails to clot.
10. A significant amount of vitamin K comes from:
   a. vegetable oils.
   b. sunlight exposure.
   c. bacterial synthesis.
   d. fortified grain products.

## Notes

1. E. Li and A. W. Norris, Structure/function of cytoplasmic vitamin A–binding proteins, *Annual Review of Nutrition* 16 (1996): 205–234.
2. L. M. DeLuca, Vitamin A in epithelial differentiation and skin carcinogenesis, *Nutrition Reviews* (supplement) 52 (1994): S45–S52.
3. D. R. Soprano and K. J. Soprano, Retinoids as teratogens, *Annual Review of Nutrition* 15 (1995): 111–132.
4. S. L. Teitelbaum and coauthors, Cellular and molecular mechanisms of bone resorption, *Mineral and Electrolyte Metabolism* 21 (1995): 193–196.
5. J. C. Butler and coauthors, Measles severity and serum retinol (vitamin A) concentration among children in the United States, *Pediatrics* 91 (1993): 1176–1181.
6. W. W. Fawzi and coauthors, Vitamin A supplementation and child mortality: A meta-analysis, *Journal of the American Medical Association* 269 (1993): 898–903; L. J. Machlin and H. E. Sauberlich, New views on the function and health effects of vitamins, *Nutrition Today,* January/February 1994, pp. 25–29.
7. J. B. Ellison, Intensive vitamin therapy in measles, *British Medical Journal* 2 (1932): 708–711.
8. The joint WHO/UNICEF statement on vitamin A for measles, *Lancet* 1 (1987): 1067–1068.
9. American Academy of Pediatrics, Committee on Infectious Diseases, Vitamin A treatment of measles, *Pediatrics* 91 (1993): 1014–1015.
10. W. W. Fawzi and coauthors, Vitamin A supplementation and dietary vitamin A in relation to the risk of xerophthalmia, *American Journal of Clinical Nutrition* 58 (1993): 385–391.
11. Soprano and Soprano, 1995.
12. K. J. Rothman and coauthors, Teratogenicity of high vitamin A intake, *New England Journal of Medicine* 333 (1995): 1369–1373.
13. S. M. Bulengo-Ransby and coauthors, Topical tretinoin (retinoic acid) therapy for hyperpigmented lesions caused by inflammation of the skin in black patients, *New England Journal of Medicine* 328 (1993): 1438–1443.
14. J. A. Olson, 1992 Atwater Lecture: The irresistible fascination of carotenoids and vitamin A, *American Journal of Clinical Nutrition* 57 (1993): 833–839.
15. T. O. Carpenter and coauthors, Severe hypervitaminosis A in siblings: Evidence of variable tolerance to retinol intake, *Journal of Pediatrics* 111 (1987): 507–512.
16. A. W. Norman and coauthors, Differing shapes of 1α,25-dihydroxyvitamin $D_3$ function as ligands for the D-binding protein, nuclear receptor and membrane receptor: A status report, *Journal of Steroid Biochemistry and Molecular Biology* 56 (1996): 13–22.
17. C. H. Jacobus and coauthors, Hypervitaminosis D associated with drinking milk, *New England Journal of Medicine* 326 (1992): 1173–1177; S. Blank and coauthors, An outbreak of hypervitaminosis D associated with the overfortification of milk from a home-delivery dairy, *American Journal of Health Promotion* 85 (1995): 656–659.
18. C. Lamberg-Allardt and coauthors, Low serum 25-hydroxyvitamin D concentrations and secondary hyperparathyroidism in middle-aged white strict vegetarians, *American Journal of Clinical Nutrition* 58 (1993): 684–689.
19. J. Regnström and coauthors, Inverse relation between the concentration of low-density lipoprotein vitamin E and severity of coronary artery disease, *American Journal of Clinical Nutrition* 63 (1996): 377–385; K. G. Losonczy, T. B. Harris, and R. J. Havlik, Vitamin E and vitamin C supplement use and risk of all-cause and coronary heart

disease mortality in older persons: The Established Populations for Epidemiologic Studies of the elderly, *American Journal of Clinical Nutrition* 64 (1996): 190–196; L. H. Kushi and coauthors, Dietary antioxidant vitamins and death from coronary heart disease in postmenopausal women, *New England Journal of Medicine* 334 (1996): 1156–1162; J. P. Flaather and coauthors, The antioxidant vitamins and cardiovascular disease: A critical review of epidemiologic and clinical trial data, *Annals of Internal Medicine* 123 (1995): 860–872.

20. Kushi and coauthors, 1996; M. J. Stampfer and coauthors, Vitamin E consumption and the risk of coronary disease in women, *New England Journal of Medicine* 328 (1993): 1444–1449; E. B. Rimm and coauthors, Vitamin E consumption and the risk of coronary disease in men, *New England Journal of Medicine* 328 (1993): 1450–1456.
21. M. G. Traber, Vitamin E in humans: Demand and delivery, *Annual Review of Nutrition* 16 (1996): 321–347.
22. W. Koller (and other members of the Parkinson Study Group), Effects of tocopherol and deprenyl on the progression of disability in early Parkinson's disease, *New England Journal of Medicine* 328 (1993): 176–183.
23. P. Dowd and coauthors, The mechanism of action of vitamin K, *Annual Review of Nutrition* 15 (1995): 419–440.
24. N. C. Binkley and J. W. Suttie, Vitamin K nutrition and osteoporosis, *Journal of Nutrition* 125 (1995): 1812–1821.
25. American Academy of Pediatrics, Vitamin K Ad Hoc Task Force, Controversies concerning vitamin K and the newborn, *Pediatrics* 91 (1993): 1001–1003; M. A. Klebanoff and coauthors, The risk of childhood cancer after neonatal exposure to vitamin K, *New England Journal of Medicine* 329 (1993): 905–908.
26. J. W. Suttie, The importance of menaquinones in human nutrition, *Annual Review of Nutrition* 15 (1995): 399–417.
27. R. T. Franceschi, Nuclear signaling pathways for 1,25-dihydroxyvitamin $D_3$ are controlled by the vitamin A metabolite, 9-*cis*-retinoic acid, *Nutrition Reviews* 51 (1993): 303–305.

# Antioxidant Nutrients and Nonnutrients in Disease Prevention

Count on supplement manufacturers to proclaim the day's hot topics in nutrition. The moment bits of research news surface, new supplements appear—and terms like "antioxidants" become household words. Friendly faces in TV commercials try to persuade us that these antioxidants hold the magic in the fight against aging, disease, and death. Then new antioxidant supplements hit the market and cash registers ring. Vitamin C, long the leading single nutrient supplement, gains new popularity, and sales of beta-carotene and vitamin E supplements soar as well.

In the meantime, scientists and medical experts around the world continue their work to clarify and confirm the roles of these antioxidant nutrients in preventing chronic diseases.[1] This highlight summarizes some of the accumulating evidence on nutrients, *non*nutrients, and their antioxidant actions. It also revisits the advantages of foods over supplements.

## Free Radicals and Disease

The body's cells use oxygen in metabolic reactions. In the process, oxygen sometimes reacts with body compounds and produces highly unstable molecules known as free radicals (the glossary on p. 361 defines free radicals and related terms). In addition to normal body processes, environmental factors such as radiation, pollution, tobacco smoke, and a high-fat diet generate free radicals.[2]

A free radical is a molecule with one or more unpaired electrons.* An electron without a partner is unstable and highly reactive. To regain its stability, the free radical quickly finds a stable but vulnerable compound from which to steal an electron (see Figure H11-1 on p. 360).

With the loss of an electron, the formerly stable molecule becomes a free radical itself and steals an electron from another nearby molecule. Thus, an electron-snatching chain reaction is under way. Antioxidants neutralize free radicals by donating one of their own electrons, thus ending the chain reaction. When they lose electrons, antioxidants do not become free radicals because they are stable in either form. (Review Figure 10-15 on p. 320 to see how ascorbic acid can give up two hydrogens with their electrons and become dehydroascorbic acid.)

Sometimes the action of free radicals is helpful. For example, cells of the immune system use free radicals as ammunition in an "oxidative burst" that demolishes disease-causing viruses and bacteria. Other times free radicals lead to widespread damage. They commonly attack the polyunsaturated fatty acids in lipoproteins and in cell membranes, disrupting the transport of substances into and out of cells. Free radicals also damage cell proteins, altering their functions, and DNA, creating mutations.

The body's natural defense and repair systems try to control the damage caused by free radicals, but these systems are not 100 percent effective. In fact, they become less effective with age, and the unrepaired damage accumulates. To some extent, dietary antioxidants defend the body against oxidative stress, but if antioxidants are unavailable, or if free-radical production becomes excessive, health problems may develop.[3] Oxygen-derived free radicals may cause diseases not only by indiscriminately destroying the valuable components of cells, but also by serving as signals for specific activities within the cells.[4] Scientists have implicated oxidative stress in the aging process and in the development of diseases such as cancer, arthritis, cataracts, and heart disease.

## Defending against Free Radicals

The body maintains a couple of lines of defense against free-radical damage. A system of enzymes disarms the most harmful oxidants.† The action of these enzymes depends on the minerals selenium, copper, manganese, and zinc. If the diet fails to provide adequate supplies of these minerals, this line of defense weakens. The body also provides a small team of antioxidant compounds, but its primary source of antioxidants is the diet.

*People who eat generous amounts of fruits and vegetables daily are helping their bodies to fight disease.*

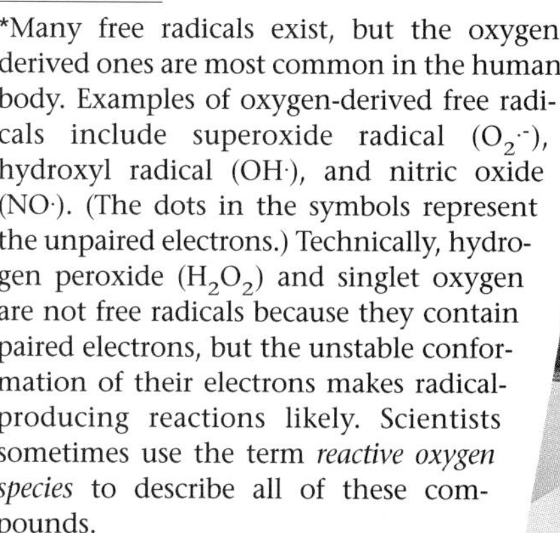

*Many free radicals exist, but the oxygen-derived ones are most common in the human body. Examples of oxygen-derived free radicals include superoxide radical ($O_2^{\cdot -}$), hydroxyl radical ($OH\cdot$), and nitric oxide ($NO\cdot$). (The dots in the symbols represent the unpaired electrons.) Technically, hydrogen peroxide ($H_2O_2$) and singlet oxygen are not free radicals because they contain paired electrons, but the unstable conformation of their electrons makes radical-producing reactions likely. Scientists sometimes use the term *reactive oxygen species* to describe all of these compounds.

†These enzymes include glutathione peroxidases, superoxide dismutases, and catalase.

**Figure H11-1**

**The Actions of Free Radicals and Antioxidants**

1 FREE-RADICAL FORMATION

During normal energy metabolism, hydrogens and electrons are added to oxygen in a series of reactions known as the electron transport chain (introduced in Chapter 7). This sequence eventually produces water, but some of the intermediate compounds inevitably created during the process are free radicals. Reminder: the dot in the symbols represents the unpaired electrons.

$O_2$ (Oxygen) —$e^-$→ $O_2^{\bullet -}$ (Superoxide radical) —$e^-$→ —$H^+$→ —$H^+$→ $H_2O_2$ (Hydrogen peroxide) —$e^-$→ —$H^+$→ $OH^\bullet$ (Hydroxyl radical) + $H_2O$ (Water)

Occasionally, oxygen gains an extra electron from the electron transport chain . . .

. . . which generates the free radical called superoxide radical (a molecule of oxygen with an extra, unpaired electron).

The superoxide radical can gain another electron (again, from the electron transport chain) and react with two hydrogen ions . . .

. . . to form hydrogen peroxide. Hydrogen peroxide can react with an electron and hydrogen . . .

. . . to form another free radical called a hydroxyl radical . . .

. . . and water.

2 FREE-RADICAL CHAIN REACTION AND DAMAGE

Hydroxyl radicals are highly reactive, wanting to match their unpaired electrons. For example, they might take electrons from the lipids in a cell membrane, which causes damage that gives rise to degenerative diseases.

Lipid + $OH^\bullet$ (Hydroxyl radical) → Lipid$^\bullet$ (Lipid radical) + $H_2O$ (water)

When a hydroxyl radical takes a hydrogen atom from a lipid (such as a polyunsaturated fatty acid) . . .

. . . it generates a lipid radical . . .

. . . and water.

The lipid radical can, in turn, react with oxygen to form another lipid radical, which can, in turn, remove hydrogen atoms from other lipids, producing new radicals, thereby initiating a chain reaction.

3 ANTIOXIDANT PROTECTION

Antioxidants interact with free radicals and break the destructive chain reaction that damages tissues.

Active vitamin E + Lipid$^\bullet$ → Inactive vitamin E$^\bullet$ + Lipid

Inactive vitamin E$^\bullet$ + Vitamin C (ascorbic acid with its H atoms) = Active vitamin E (with its H atom) + Vitamin C (dehydroascorbic acid without its H atoms)

Vitamin E gives up one of its hydrogens to a lipid radical.*

The result is that vitamin E is no longer active, but it has successfully stopped the radicals from causing more damage and generating more radicals.

Vitamin E can be reactivated by accepting a hydrogen atom from fellow antioxidant vitamin C. Vitamin C's two structures are presented in Figure 10–15 on p. 320.

*The compound is actually a lipid peroxyl radical.

Among the antioxidant vitamins, vitamin E and beta-carotene defend the body's lipids. Vitamin E efficiently breaks the free radical chain reaction at a rate 200 times faster than BHT,* a commercial antioxidant added to baked goods to prevent rancidity from fat oxidation. Vitamin C protects the body's watery components, such as the fluid of the blood, against free-radical attacks. Vitamin C seems especially adept at neutralizing free radicals from polluted air and cigarette smoke; it also has the ability to restore oxidized vitamin E to its active state.

## *Defending against Cancer*

Cancers arise when cellular DNA is damaged—sometimes by free-radical attacks.[5] Antioxidant nutrients may reduce

*BHT is butylated hydroxytoluene.

cancer risks by protecting DNA from this damage. Many researchers conducting epidemiological studies have reported that people with high intakes of vegetables and fruits rich in antioxidant nutrients have low rates of cancer. Preliminary reports suggest an inverse relationship between DNA damage and vegetable intake and a direct relationship with beef and pork intake.[6] Laboratory studies with animals and with cells in tissue culture seem to support such findings.

### The Carotenoids and the Retinoids

Studies report a consistent relationship between high intakes of vegetables and fruits rich in beta-carotene and the other carotenoids and low rates of lung cancer. Blood samples also reflect that low concentrations of beta-carotene and other carotenes correlate with the development of lung, mouth, cervical, and breast cancers. The retinoids have also proved effective in the treatment and prevention of some cancers.[7]

### Vitamin C

High intakes of foods rich in vitamin C also seem to protect against certain types of cancers, especially those of the mouth, larynx, and esophagus. Such a correlation may reflect the benefits of a diet rich in fruits and vegetables and low in fat and does not necessarily support the taking of vitamin C supplements to treat or prevent cancer.

### Vitamin E

Evidence that vitamin E helps guard against cancer is less consistent than for beta-carotene and vitamin C. Still, people with low blood levels of vitamin E have high rates of some cancers.

 National Cancer Institute
rex.nci.nih.gov/INFO_CANCER/FACTS_INDX.htm
Search for Vitamin

## Defending against Heart Disease

High blood cholesterol carried in low-density lipoproteins (LDL) is a major risk factor for cardiovascular disease, but how do LDL exert their damage? One scenario is that free radicals within the arterial walls oxidize LDL, changing their structure and function. The oxidized LDL then accelerate the formation of artery-clogging plaques.[8] These free radicals also oxidize the polyunsaturated fatty acids of the cell membranes, sparking additional changes in the arterial walls, which impede the flow of blood. Susceptibility to such oxidative damage within the arterial walls is heightened by a high-fat diet or cigarette smoke.[9] Supplementation with vitamins E and C eliminates the free-radical action within the arterial wall that typically follows a high-fat meal.[10] In fact, blood flow through the arteries is similar to that seen after a low-fat meal.

### Vitamin E

Research suggests that antioxidants, especially vitamin E, may protect against cardiovascular disease.[11] Researchers conducting epidemiological studies report that people who eat foods rich in vitamin E have low rates of death from heart disease.[12] Other researchers report a similar association between large doses of vitamin E supplements and reduced risks of heart disease.[13]

### Vitamin C

Some studies suggest that vitamin C protects against LDL oxidation, raises HDL, lowers total cholesterol, and improves blood pressure.[14] Vitamin C may also protect the arteries against oxidative damage.[15] Research suggests a synergism between vitamin C and vitamin E in defending against LDL oxidation: vitamin C regenerates vitamin E from its oxidized form, making it available to act as an antioxidant once again.

## Foods versus Supplements

In the process of scavenging and quenching free radicals, antioxidants themselves become oxidized. To some extent, they can be regenerated, but still, losses occur and free radicals attack continuously. To maintain defenses, a person must replenish these antioxidant nutrients regularly. The question arises, replenish antioxidants from foods or from supplements?

Some research suggests a protective effect from as little as a daily glass of orange juice and carrot juice (rich

### Glossary

**free radical:** an unstable and highly reactive atom or molecule that has one or more unpaired electron(s) in the outer orbital (see Appendix B for a review of basic chemistry concepts).

**oxidant** (OK-see-dant): a compound (such as oxygen itself) that oxidizes other compounds. Compounds that prevent oxidation are called *anti*oxidants, whereas those that promote it are called *pro*oxidants.
- **anti** = against
- **pro** = for

**oxidative stress:** a condition in which the production of oxidants and free radicals exceeds the body's ability to defend itself.

**peroxidation:** the production of unstable molecules containing more than the usual amount of oxygen. Hydrogen peroxide, $H_2O_2$, for example, may be produced from water, $H_2O$.

Reminder: *Nonnutrients* are compounds in foods that do not fit into the six classes of nutrients.

*Phytochemicals* are nonnutrients found in plants that have biological activity in the body.

sources of vitamin C and beta-carotene, respectively).[16] Other intervention studies, however, have used levels of nutrients that far exceed current recommendations and can only be achieved by taking supplements. In making their new recommendations, members of the Dietary Reference Intakes committees are considering whether these studies support substantially higher intakes to help protect against chronic diseases. But research has not yet proved that taking vitamin pills is truly more beneficial than eating a healthy diet alone.

While awaiting final answers, should people anticipate the go-ahead and start taking antioxidant supplements now?[17] Most scientists agree that it is too early to make such a recommendation. While fruits and vegetables that contain many antioxidant nutrients have been associated with a diminished risk of many cancers, supplements of beta-carotene and vitamins C and E have not always proved beneficial.[18] In fact, sometimes the benefits are more apparent when the vitamins come from foods than from supplements.[19] Without data to confirm the benefits of supplements, we cannot accept the potential risks. And the risks are real.

***Cruciferous vegetables, such as cauliflower, broccoli, and brussels sprouts, contain nutrients and nonnutrients that may inhibit cancer development.***

Consider the findings from a study to determine whether daily supplements of vitamin E, beta-carotene, or both would reduce the incidence of lung cancer among smokers.[20] After five to eight years of supplementation, there was no reduction in the incidence of lung cancer; in fact, the researchers found a higher incidence of lung cancer among those receiving the beta-carotene. Another group of researchers reported similar findings: smokers and asbestos workers receiving beta-carotene and vitamin A supplements for four years had a higher incidence of lung cancer and risk of death than those taking a placebo.[21] These findings brought the study to an end much earlier than planned. Given the association between high intakes of dietary beta-carotene and low rates of lung cancer reported in earlier epidemiological studies, findings of increased risk were surprising, to say the least. Clearly, remedies to life-threatening diseases such as lung cancer are not as simple as taking daily pills. Smokers are much wiser to stop smoking than to rely on supplements to protect them from lung cancer.

Even if research clearly proves that a particular nutrient is the ultimate protective ingredient in foods, supplements would not be the answer because their contents are limited. Vitamin E supplements, for example, usually contain alpha-tocopherol, but foods provide an assortment of tocopherols, many of which provide valuable protection against free-radical damage.[22] Supplements shortchange users.

Furthermore, much more research is needed to define optimal and dangerous levels of intake. This much we know: antioxidants behave differently under various conditions.[23] At physiological levels typical of a healthy diet, they act as antioxidants, but at pharmacological doses typical of supplements, they may act as *prooxidants,* stimulating the production of free radicals.[24] This is especially likely when metal ions such as iron are present. Until the optimum intake of these nutrients can be determined, the risks of supplement use remain unclear. The best way to supplement antioxidant nutrients is to eat generous servings of fruits and vegetables, especially citrus fruits and green and yellow vegetables.

## Foods Making Health Claims

Results of clinical studies may one day support the use of selected supplements to prevent disease, but until then people will want to select foods rich in all of the vitamins and minerals—particularly the antioxidants. Which foods to select?

The Food and Drug Administration (FDA) examined the available scientific evidence concerning antioxidant vitamins and cancer to determine whether a health claim on food labels was appropriate. The agency concluded that *diets* high in fruits and vegetables, which are good sources of two antioxidant vitamins (vitamin A as beta-carotene and vitamin C), are strongly associated with reduced risks of several types of cancer. Still, the FDA rejected the antioxidant health claim, stating that the reduction in risk could not be attributed directly and solely to the antioxidant effect of the vitamins. Therefore labels may not claim an association between antioxidant vitamins and cancer; the health claim must be stated in terms of "fruits and vegetables and cancer." National campaigns to "Eat 5 a Day" in the United States and "Reach for It" in Canada encourage consumers to select several servings of fruits and vegetables daily.

National 5 A Day Program
www.dcpc.nci.nih.gov/5aday

## Nonnutrients in Disease Prevention

Clearly, the protective effect of fruits and vegetables must depend on more than the antioxidant nutrients alone.[25] Other, nonnutrient compounds must also be involved.

The nonnutrient compounds found in plants are called phytochemicals, and they have been the focus of much recent research. In foods, these compounds may impart flavors and colors, but in the body, they can have profound physiological effects, including suppression of the development of cancer.[26] Table H11-1 summarizes the common food sources and actions of selected phytochemicals.

This book has focused primarily on the nutrients, but foods deliver thousands of other chemicals. For this reason, researchers must be careful in giving credit for particular health benefits to any one nutrient. Diets rich in whole grains, legumes, vegetables, and fruits seem to be protective against cancer, but identifying *the* specific foods or components of foods that are responsible is difficult. Green leafy vegetables such as spinach and kale, for example, contain lutein, a carotenoid with more antioxidant activity than beta-carotene. The anticancer benefits of green leafy vegetables may be due to beta-carotene, but they may be due to lutein—or to another as yet unidentified component. Perhaps credit even belongs to the unique *combination* of chemicals found in leafy greens. We simply do not have all the answers. Similarly, compounds in tomatoes and soybeans appear to have anticancer activity, and those in tea and onions protect against heart disease.[27] Naturally occurring salicylates may provide the same protective effects as low doses of aspirin.[28] Spices such as curry, paprika, and thyme are especially good sources of salicylates; many fruits, vegetables, teas, and candies flavored with wintergreen (methyl salicylate) also contribute to a day's intake.

Other foods with antioxidant activity include beverages such as wine, spices such as oregano, and oils such as olive oil. These foods contain phytochemicals that may offer important health benefits and explain, in part, why people who drink wine and eat the Mediterranean diet have reduced risks of heart disease.[29]

Everyone eats a variety of phytochemicals in small quantities every day. This approach may be more beneficial than taking large doses of any one phytochemical.[30] In large doses, some phytochemicals can be toxic. The regulation of phytochemicals depends on how they are used.[31] Consider garlic, for example. A clove of garlic is a food. The FDA

**Table H11-1**
**Phytochemicals—Their Food Sources and Actions**

| Food Source | Name | Action in the Body |
|---|---|---|
| Deeply pigmented fruits and vegetables (carrots, sweet potatoes, tomatoes, spinach, broccoli, cantaloupe, pumpkin, apricots) | Carotenoids[a] (including beta-carotene and lycopene) | Act as antioxidants, reducing the risk of cancer. |
| Citrus fruits | Limonene | Triggers enzyme production to facilitate carcinogen excretion. |
| | Phenols | Inhibit lipid oxidation; block formation of carcinogenic nitrosamines in the body. |
| Garlic, onions, leeks, chives | Allyl sulfides | Trigger enzyme production to facilitate carcinogen excretion. |
| Broccoli and other cruciferous vegetables (cauliflower, cabbage, kale, brussels sprouts) | Sulforaphane<br>Dithiolthiones | Protects against cancer.<br>Trigger enzyme production to block carcinogen damage to cells' DNA. |
| | Indoles | Trigger enzymes to inhibit estrogen action, reducing the risk of breast cancer. |
| | Isothiocyanates | Trigger enzyme production to block carcinogen damage of cells' DNA. |
| Grapes | Ellagic acid | Scavenges carcinogens. |
| Soy/legumes | Protease inhibitors | Suppress enzyme production in cancer cells, slowing tumor growth. |
| | Phytosterols | Inhibit cell reproduction in GI tract, preventing colon cancer. |
| | Isoflavones[b] | Block estrogen activity in cells, reducing the risk of breast and ovarian cancer. |
| | Saponins | Interfere with DNA reproduction, preventing cancer cell multiplication. |
| Flaxseed | Lignans[b] | Block estrogen activity in cells, reducing the risk of breast and ovarian cancer. |
| Fruits (blueberries, prunes, grapes), oats, soybeans | Caffeic acid | Triggers enzyme production to make carcinogens water soluble, facilitating excretion. |
| | Ferulic acid | Binds to nitrates in stomach, preventing the conversion to nitrosamines. |
| Grains | Phytic acid | Binds to minerals, preventing cancer-causing free-radical formation. |
| Fruits, vegetables, tea, wine, oregano | Flavonoids (including quercetin) | Act as antioxidants, reducing the risk of cancer and heart disease. |

[a]In addition to beta-carotene, other carotenoids include alpha-carotene, beta-cryptoxanthin, lutein, zeaxanthin, and lycopene.
[b]Isoflavones and lignans are types of phytoestrogens—compounds that bind to estrogen receptors and reduce estrogen activity.

***Many cancer-fighting products are available now at your local produce counter. (Courtesy of CNN)***

classifies dehydrated garlic and garlic extracts as generally recognized as safe (GRAS) substances. A product derived from garlic that makes a special health claim, on the other hand, is classified as a drug.

Of course, manufacturers have already begun to market phytochemicals as supplements. (Highlight 10 explained how some manufacturers tried to market supplements as "nutraceuticals" that fight against disease.) It should be clear by now, though, that we cannot know the identity and action of every chemical in every food. Even if we did, why create a supplement to replicate a food? Why not eat foods and enjoy the pleasure, nourishment, and health benefits they provide? The beneficial constituents in foods are widespread among plants.[32] Don't try to single out one particular food for its magic phytochemical. Instead, eat a wide variety of fruits and vegetables in generous quantities every day—and get *all* the magic compounds these foods have to offer.

## Notes


1. J. E. Buring and C. H. Hennekens, Antioxidant vitamins and cardiovascular disease, *Nutrition Reviews* 55 (1997): S53–S60; B. Halliwell, Antioxidants and human disease: A general introduction, *Nutrition Reviews* 55 (1997): S44–S52; C. L. Rock, R. A. Jacob, and P. E. Bowen, Update on the biological characteristics of the antioxidant micronutrients: Vitamin C, Vitamin E, and the carotenoids, *Journal of the American Dietetic Association* 96 (1996): 693–702; Health promotion and disease prevention: The role of antioxidant vitamins, *American Journal of Medicine* 97 (supplement 3A) (1994): 1S–28S.
2. B. Halliwell, Antioxidants in human health and disease, *Annual Review of Nutrition* 16 (1996): 35–50; J. D. Morrow and coauthors, Increase in circulating products of lipid peroxidation ($F_2$-isoprostanes) in smokers—Smoking as a cause of oxidative damage, *New England Journal of Medicine* 332 (1995): 1198–1203; L. Langseth, *Oxidants, Antioxidants, and Disease Prevention* (Brussels: International Life Sciences Institute, 1995).
3. Halliwell, 1996; R. A. Jacob and B. J. Burri, Oxidative damage and defense, *American Journal of Clinical Nutrition* 63 (1996): 985S–990S.
4. H. J. Palmer and K. E. Paulson, Reactive oxygen species and antioxidants in signal transduction and gene expression, *Nutrition Reviews* 55 (1997): 353–361.
5. I. T. Johnson, G. Williamson, and S. R. R. Musk, Anticarcinogenic factors in plant foods: A new class of nutrients? *Nutrition Research Reviews* 7 (1994): 175–204; B. N. Ames, M. K. Shigenaga, and T. M. Hagen, Oxidants, antioxidants, and the degenerative diseases of aging, *Proceedings of the National Academy of Sciences* 90 (1993): 7915–7922.
6. Z. Djuric and coauthors, Oxidative DNA damage levels in blood from women at high risk for breast cancer are associated with dietary intakes of meats, vegetables, and fruits, *Journal of the American Dietetic Association* 98 (1998): 524–528.
7. M. S. Tallman and coauthors, All-*trans*-retinoic acid in acute promyelocytic leukemia, *New England Journal of Medicine* 337 (1997): 1021–1028; S. Zhang and coauthors, Measurement of retinoids and carotenoids in breast adipose tissue and a comparison of concentrations in breast cancer cases and control subjects, *American Journal of Clinical Nutrition* 66 (1997): 626–632.
8. B. Halliwell, Oxidation of low-density lipoproteins: Questions of initiation, propagation, and the effect of antioxidants, *American Journal of Clinical Nutrition* 61 (1995): 670S–677S; P. D. Reaven and J. L. Witztum, Oxidized low density lipoproteins in atherogenesis: Role of dietary modification, *Annual Review of Nutrition* 16 (1996): 51–71.
9. Reaven and Witztum, 1996; C. E. Cross and M. G. Traber, Cigarette smoking and antioxidant vitamins: The smoke screen continues to clear but has a way to go, *American Journal of Clinical Nutrition* 65 (1997): 562–563.
10. G. D. Plotnick, M. C. Corretti, and R. A. Vogel, Effect of antioxidant vitamins on the transient impairment of endothelium-dependent brachial artery vasoactivity following a single high-fat meal, *Journal of the American Medical Association* 278 (1997): 1682–1686.
11. C. Bonithon-Kopp and coauthors, Combined effects of lipid peroxidation and antioxidant status on carotid atherosclerosis in a population aged 59–71 y: The EVA study, *American Journal of Clinical Nutrition* 65 (1997): 121–127; L. H. Kushi and coauthors, Dietary antioxidant vitamins and death from coronary heart disease in postmenopausal women, *New England Journal of Medicine* 334 (1996): 1156–1162; C. C. Tangney, Vitamin E and cardiovascular disease, *Nutrition Today* 32 (1997): 13–22; M. Abbey, The importance of vitamin E in reducing cardiovascular risk, *Nutrition Reviews* 53 (1995): S28–S32; T. Byers, Vitamin E supplements and coronary heart disease, *Nutrition Reviews* 51 (1993): 333–336.
12. P. Knekt and coauthors, Antioxidant vitamin intake and coronary mortality in a longitudinal population study, *American Journal of Epidemiology* 139 (1994): 1180–1189; K. F. Gey and coauthors, Increased risk of cardiovascular disease at suboptimal plasma concentrations of essential antioxidants: An epidemiological update with special attention to carotene and vitamin C, *American Journal of Clinical Nutrition* 57 (1993): 787S–797S.
13. M. J. Stampfer and coauthors, Vitamin E consumption and the risk of coronary disease in women, *New England Journal of Medicine* 328 (1993): 1444–1449; E. B. Rimm and coau-

thors, Vitamin E consumption and the risk of coronary disease in men, *New England Journal of Medicine* 328 (1993): 1450–1456.

14. D. Harats and coauthors, Citrus fruit supplementation reduces lipoprotein oxidation in young men ingesting a diet high in saturated fat: Presumptive evidence for an interaction between vitamins C and E in vivo, *American Journal of Clinical Nutrition* 67 (1998): 240–245; J. P. Moran and coauthors, Plasma ascorbic acid concentrations relate inversely to blood pressure in human subjects, *American Journal of Clinical Nutrition* 57 (1993): 213–217.
15. G. N. Levine and coauthors, Ascorbic acid reverses endothelial vasomotor dysfunction in patients with coronary artery disease, *Circulation* 93 (1996): 1107–1113.
16. M. Abbey, M. Noakes, and P. J. Nestel, Dietary supplementation with orange and carrot juice in cigarette smokers lowers oxidation products in copper-oxidized low-density lipoproteins, *Journal of the American Dietetic Association* 95 (1995): 671–675.
17. D. Steinberg, Antioxidant vitamins and coronary heart disease, *New England Journal of Medicine* 328 (1993): 1487–1489.
18. J. M. Rapola and coauthors, Effect of vitamin E and beta carotene on the incidence of angina pectoris: A randomized, double-blind, controlled trial, *Journal of the American Medical Association* 275 (1996): 693–698; E. R. Greenberg and coauthors, Mortality associated with low plasma concentration of beta carotene and the effect of oral supplementation, *Journal of the American Medical Association* 275 (1996): 699–703; C. H. Hennekens and coauthors, Lack of effect of long-term supplementation with beta carotene on the incidence of malignant neoplasms and cardiovascular disease, *New England Journal of Medicine* 334 (1996): 1145–1149; E. R. Greenberg and coauthors, A clinical trial of antioxidant vitamins to prevent colorectal adenoma, *New England Journal of Medicine* 331 (1994): 141–147.
19. Kushi and coauthors, 1996.
20. O. P. Heinonen, J. K. Huttunen, and D. Albanes (and other participants in the alpha-tocopherol, beta carotene cancer prevention study group), The effect of vitamin E and beta carotene on the incidence of lung cancer and other cancers in male smokers, *New England Journal of Medicine* 330 (1994): 1029–1035.
21. G. S. Omenn and coauthors, Effects of a combination of beta carotene and vitamin A on lung cancer and cardiovascular disease, *New England Journal of Medicine* 334 (1996): 1150–1155.
22. S. Christen and coauthors, $\tau$-Tocopherol traps mutagenic electrophiles such as $NO_x$ and complements $\alpha$-tocopherol: Physiological implications, *Proceedings of the National Academy of Sciences* 94 (1997): 3217–3222.
23. E. A. Decker, Phenolics: Prooxidants or antioxidants? *Nutrition Reviews* 55 (1997): 396–407.
24. K. M. Brown, P. C. Morrice, and G. G. Duthie, Erythrocyte vitamin E and plasma ascorbate concentrations in relation to erythrocyte peroxidation in smokers and nonsmokers: Dose response to vitamin E supplementation, *American Journal of Clinical Nutrition* 65 (1997): 496–502.
25. H. Priemé and coauthors, No effect of supplementation with vitamin E, ascorbic acid, or coenzyme Q10 on oxidative DNA damage estimated by 8-oxo-7,8-dihydro-2'-deoxyguanosine excretion in smokers, *American Journal of Clinical Nutrition* 65 (1997): 503–507.
26. L. W. Wattenberg, Inhibition of carcinogenesis by minor dietary constituents, *Cancer Research* 52 (1992): 2085s–2091s.
27. S. K. Clinton, Lycopene: Chemistry, biology, and implications for human health and disease, *Nutrition Reviews* 56 (1998): 35–51; E. Giovannucci and coauthors, Intake of carotenoids and retinol in relation to risk of prostate cancer, *Journal of the National Cancer Institute* 87 (1995): 1767–1776; M. Messina and S. Barnes, The role of soy products in reducing risk of cancer, *Journal of the National Cancer Institute* 83 (1991): 541–546; P. C. H. Hollman and coauthors, Absorption of dietary quercetin glycosides and quercetin in healthy ileostomy volunteers, *American Journal of Clinical Nutrition* 62 (1995): 1276–1282.
28. C. A. Perry and coauthors, Health effects of salicylates in foods and drugs, *Nutrition Reviews* 54 (1996): 225–240.
29. F. Visioli and C. Galli, The effect of minor constituents of olive oil on cardiovascular disease: New findings, *Nutrition Reviews* 56 (1998): 142–147; Dietary flavonoids and risk of coronary heart disease, *Nutrition Reviews* 52 (1994): 59–61.
30. L. U. Thompson, Antioxidants and hormone-mediated health benefits of whole grains, *Critical Reviews in Food Science and Nutrition* 34 (1994): 473–497.
31. J. N. Hathcock, Safety and regulatory issues for phytochemical sources: "Designer foods," *Nutrition Today,* November/December 1993, pp. 23–25.
32. Position of The American Dietetic Association: Phytochemicals and functional foods, *Journal of the American Dietetic Association* 95 (1995): 493–496; E. A. Decker, The role of phenolics, conjugated linoleic acid, carnosine, and pyrroloquinoline quinone as nonessential dietary antioxidants, *Nutrition Reviews* 53 (1995): 49–58.

CHAPTER 12

# Water and the Major Minerals

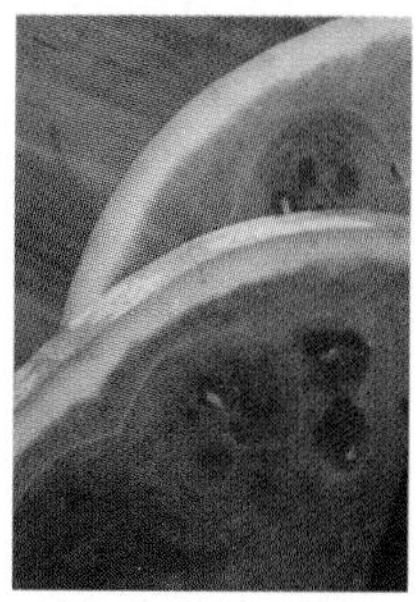

Water is an essential nutrient, as important to life as any of the others. In fact, you can survive only a few days without water, whereas a deficiency of the other nutrients may take weeks, months, or even years to develop.

This chapter begins with a look at water and the body's fluids. The body maintains an appropriate balance and distribution of water with the help of another class of nutrients—the minerals. In addition to introducing the minerals that help regulate body fluids, the chapter describes many of the other important functions minerals perform in the body. Highlight 19 revisits water as a beverage and addresses consumer concerns about its safety.

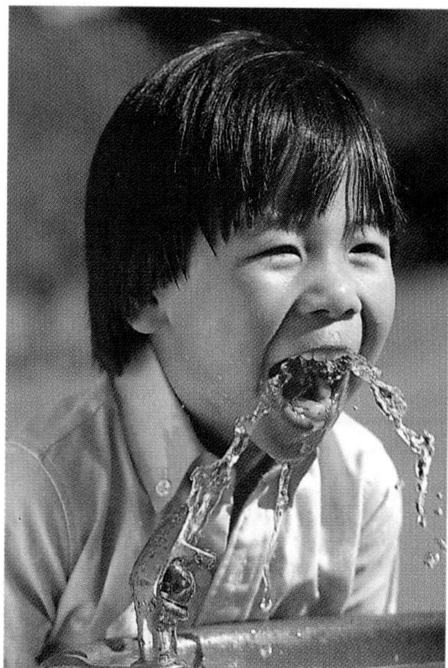

*Water is the most indispensable nutrient.*

# Water and the Body Fluids

In the body, water becomes the fluid in which all life processes occur. Every cell contains fluid of the exact composition that is best for that cell (intracellular fluid) and is bathed externally in another such fluid (interstitial fluid). (Figure 6-9 on p. 171 illustrates a cell and its associated fluids.) These fluids continually lose and replace their components, yet the composition in each compartment remains remarkably constant at all times. The entire system of cells and fluids remains in a delicate but controlled state of homeostasis. The water in the body fluids:

- Carries nutrients and waste products throughout the body.
- Maintains the structure of large molecules such as proteins and glycogen.
- Participates in chemical reactions.
- Serves as the solvent for minerals, vitamins, amino acids, glucose, and many other small molecules.
- Acts as a lubricant and cushion around joints and inside the eyes, the spinal cord, and, in pregnancy, the amniotic sac surrounding the fetus in the womb.
- Aids in the regulation of body temperature.
- Maintains blood volume.

To support these and other vital functions, the body actively maintains an appropriate water balance.

**intracellular fluid:** fluid within the cells, usually high in potassium and phosphate. Intracellular fluid accounts for approximately two-thirds of the body's water.
- **intra** = within

**interstitial** (IN-ter-STISH-al) **fluid:** fluid between the cells, usually high in sodium and chloride. Interstitial fluid is a large component of **extracellular fluid** (fluid outside the cells), which also includes plasma and the water of structures such as the skin and bones. Extracellular fluid accounts for approximately one-third of the body's water.
- **inter** = in the midst, between
- **extra** = outside

Reminder: *Homeostasis* is the maintenance of relatively constant conditions within the body's systems.

**water balance:** the balance between water intake and output (losses).
Water balance = intake − output.

## *Water Balance and Recommended Intakes*

Water constitutes about 60 percent of an adult's body weight and a higher percentage of a child's. Because water makes up about three-fourths of the weight of lean tissue and less than one-fourth of the weight of fat, a person's body composition influences how much of the body's weight is water. The proportion of water is generally smaller in females, obese people, and the elderly because of their smaller proportion of lean tissue. Because imbalances can be devastating, the body attempts to restore homeostasis as promptly as possible, adjusting both water intake and excretion as needed.

• ***Water Intake*** • Thirst and satiety influence water intake, apparently in response to changes sensed by the mouth, hypothalamus, and nerves. When the blood becomes concentrated (having lost water, but not the dissolved substances within it), the mouth becomes dry and the hypothalamus initiates drinking behavior. Stretch receptors in the stomach send signals to stop drinking as do receptors in the heart that monitor blood volume.

**thirst:** a conscious desire to drink.

Reminder: The *hypothalamus* is a brain center that controls activities such as maintenance of water balance, regulation of body temperature, and control of appetite.

Appendix A presents a diagram of the human body illustrating the hypothalamus and other glands that regulate body processes.

**Table 12-1**
**Adverse Effects of Dehydration**

| Body Weight Lost (%) | Symptoms |
|---|---|
| 1–2 | Thirst, fatigue, weakness, vague discomfort, loss of appetite |
| 3–4 | Impaired physical performance, dry mouth, reduction in urine, flushed skin, impatience, apathy |
| 5–6 | Difficulty in concentrating, headache, irritability, sleepiness, impaired temperature regulation, increased respiratory rate |
| 7–10 | Dizziness, spastic muscles, loss of balance, delirium, exhaustion, collapse |

NOTE: The onset and severity of symptoms at various percentages of body weight lost depend on the activity, fitness level, degree of acclimation, temperature, and humidity.

Thirst drives a person to seek water, but it lags behind the body's need. A water deficiency that develops slowly can switch on drinking behavior in time to prevent serious dehydration, but a deficiency that develops quickly may not. Also, thirst itself does not remedy a water deficiency; a person must pay attention to the thirst signal and take the time to get a drink. The long-distance runner, the gardener in hot weather, the child busy playing, and the elderly person whose thirst sensation may be blunted can experience serious dehydration if they fail to drink promptly in response to their need for water.

Dehydration may easily develop with either water deprivation or excessive water losses. The symptoms progress rapidly from thirst, to weakness, to exhaustion and delirium and end in death if not corrected (see Table 12-1). Water intoxication, on the other hand, is rare but can occur with excessive water ingestion and kidney disorders that reduce urine output. The symptoms may include confusion, convulsion, and even death in extreme cases.

**dehydration:** the condition in which body water output exceeds water input. Symptoms include thirst, dry skin and mucous membranes, rapid heartbeat, low blood pressure, and weakness.

Chapter 14 revisits dehydration and the fluid needs of athletes.

**water intoxication:** the rare condition in which body water contents are too high.

The amount of water the body has to excrete each day to dispose of its wastes is the **obligatory** (ah-BLIG-ah-TORE-ee) **water excretion**—about 500 ml (about 2 c, or a pint).

• ***Water Sources*** • The obvious dietary sources of water are water itself and other beverages, but nearly all foods also contain water. Most fruits and vegetables contain up to 90 percent water; many meats and cheeses contain at least 50 percent (see Table 12-2 for selected foods and Appendix H for many more). Water is also generated during metabolism. Recall that when the energy-yielding nutrients break down, their carbons and hydrogens combine with oxygen to yield carbon dioxide ($CO_2$)—and water ($H_2O$). As Table 12-3 shows, the water derived daily from these three sources averages about 2½ liters (roughly 2½ quarts).

**Table 12-2**
**Percentage of Water in Selected Foods**

| | |
|---|---|
| 100% | Water, diet sodas |
| 90–99% | Nonfat milk, strawberries, watermelon, lettuce, cabbage, celery, spinach, broccoli |
| 80–89% | Fruit juice, yogurt, apples, grapes, oranges, carrots |
| 70–79% | Shrimp, bananas, corn, potatoes, avocados, cottage cheese, ricotta cheese |
| 60–69% | Pasta, legumes, salmon, ice cream, chicken breast |
| 50–59% | Ground beef, hot dogs, feta cheese |
| 40–49% | Pizza |
| 30–39% | Cheddar cheese, bagels, bread |
| 20–29% | Pepperoni sausage, cake, biscuits |
| 10–19% | Butter, margarine, raisins |
| 1–9% | Crackers, cereals, pretzels, taco shells, peanut butter, nuts |
| 0% | Oils |

• ***Water Losses*** • The body must excrete a minimum of about 500 milliliters of water each day as urine—enough to carry away the waste products generated by a day's metabolic activities. Above this amount, excretion adjusts to balance intake. If a person drinks more water, the urine becomes more dilute. In addition to urine, water is lost from the lungs as vapor and from the skin as sweat; some is also lost in feces.* The losses from all of these sources total about 2½ liters a day on the average. Table 12-3 shows how water excretion balances intake.

• ***Water Recommendations*** • Water needs vary, depending primarily on diet, activity, environmental temperature, and humidity. Accordingly, a general water requirement is difficult to establish. Recommendations for adults are expressed in proportion to the amount of energy expended under average environmental conditions.[1] A person who expends 2000 kcalories a day needs about 2 to 3 liters of water (about 7 to 11 cups).

Fluid needs are best met by water, but milk and juices can account for part of the day's recommended intake.[2] In addition to their high water content, these beverages deliver valuable nutrients. Alcoholic beverages and those containing caf-

*Water lost from the lungs and skin accounts for almost one-half of the daily losses even when a person is not visibly perspiring; these losses are commonly referred to as *insensible water losses*.

**Table 12-3**
**Water Balance**

| Water Sources | Amount (ml) | Water Losses | Amount (ml) |
|---|---|---|---|
| Liquids | 550 to 1500 | Kidneys | 500 to 1400 |
| Foods | 700 to 1000 | Skin | 450 to 900 |
| Metabolic water | 200 to 300 | Lungs | 350 |
| | | Feces | 150 |
| | 1450 to 2800 | | 1450 to 2800 |

NOTE: These values reflect data from several sources and are compatible with those cited in many other references. For further information, see L. Sherwood, *Fundamentals of Physiology: A Human Perspective* (St. Paul, Minn.: West Publishing, 1995), pp. 396–417; Committee on Dietary Allowances, *Recommended Dietary Allowances*, 10th ed. (Washington, D.C.: National Academy Press, 1989), pp. 247–261; J. L. Groff, S. S. Gropper, and S. M. Hunt, *Advanced Nutrition and Human Metabolism* (St. Paul, Minn.: West Publishing, 1995), pp. 423–438.

feine, such as coffee, tea, and sodas, however, are not good substitutes for water. Both alcohol and caffeine act as diuretics, causing the body to lose fluids.

Water recommendation for adults:

- 1.0 to 1.5 ml/kcal expended.
- 4.2 to 6.3 ml/kJ expended.

Water recommendation for infants:

- 1.5 ml/kcal expended.

Note: 1 ml = 0.03 fluid ounce.
125 ml ≈ ½ c.
Easy estimation: ½ c per 100 kcal expended.

IN SUMMARY

Water makes up about 60 percent of the body's weight. It assists with the transportation of nutrients and waste products throughout the body, participates in chemical reactions, acts as a solvent, serves as a shock absorber, and regulates body temperature. To maintain water balance, intake from liquids, foods, and metabolism must equal losses from kidneys, skin, lungs, and feces.

# Blood Volume and Blood Pressure

Water balance maintains the blood volume, which in turn influences blood pressure. If the body loses too much water, blood volume and blood pressure fall.

• ***ADH and Water Retention*** • Whenever the blood becomes too concentrated, or whenever blood volume or blood pressure falls too low, the hypothalamus signals the pituitary gland to release the antidiuretic hormone (ADH). ADH stimulates the kidneys to reabsorb water, rather than excrete it. Consequently, the more water you need, the less your kidneys excrete.

**ADH (antidiuretic hormone):** a hormone released by the pituitary gland in response to highly concentrated blood. The kidneys respond by reabsorbing water, thus preventing water loss. In addition to its antidiuretic effect, ADH also elevates blood pressure and is called **vasopressin.**

- **anti** = against
- **dia** = through
- **ure** = urine
- **vaso** = vessel
- **press** = pressure

Recall from Highlight 7 how alcohol depresses ADH activity, thus promoting fluid losses and dehydration.

• ***Angiotensin and Blood Vessel Constriction*** • Cells in the kidneys respond to low blood pressure by releasing an enzyme called renin. Through a complex series of events, renin causes the kidneys to reabsorb sodium. Sodium reabsorption, in turn, is always accompanied by water retention, which helps to restore blood volume and blood pressure. Renin also activates the blood protein angiotensinogen to angiotensin. Angiotensin is a powerful vasoconstrictor: it narrows the diameters of blood vessels, thereby raising the blood pressure.

**renin** (REN-in): an enzyme from the kidneys that activates angiotensin.

**angiotensin:** a hormone involved in blood pressure regulation. Its precursor protein is called **angiotensinogen.**

**vasoconstrictor:** a substance that constricts or narrows the blood vessels.

• ***Aldosterone and Sodium Retention*** • Angiotensin also mediates the release of the hormone aldosterone from the adrenal glands. Aldosterone causes the kidneys to retain more sodium (and thus more water). Again, the effect is that when more water is needed, less is excreted.

**aldosterone** (al-DOS-ter-own): a hormone secreted by the adrenal glands that stimulates the reabsorption of sodium by the kidneys; aldosterone also regulates chloride and potassium concentrations.

**adrenal glands:** glands adjacent to, and just above, each kidney.

IN SUMMARY

In response to low blood volume or highly concentrated blood, these three actions combine to effectively restore homeostasis (see Figure 12-1):

- ADH causes water retention.
- Angiotensin constricts blood vessels.
- Aldosterone causes sodium retention.

**Figure 12-1**
**How the Body Regulates Water Excretion**

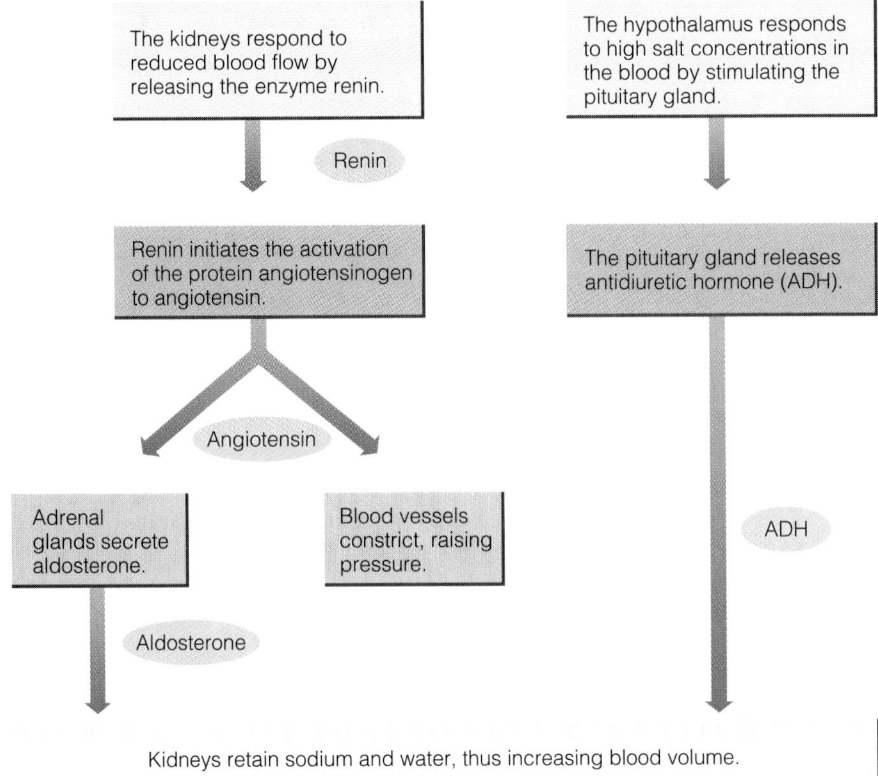
The kidneys respond to reduced blood flow by releasing the enzyme renin.
The hypothalamus responds to high salt concentrations in the blood by stimulating the pituitary gland.
Renin
Renin initiates the activation of the protein angiotensinogen to angiotensin.
The pituitary gland releases antidiuretic hormone (ADH).
Angiotensin
Adrenal glands secrete aldosterone.
Blood vessels constrict, raising pressure.
ADH
Aldosterone
Kidneys retain sodium and water, thus increasing blood volume.

These three mechanisms can maintain water balance only if a person drinks enough water.

## Fluid and Electrolyte Balance

Reminder: *Fluid and electrolyte balance* is the maintenance of the proper amounts and kinds of fluid and minerals in each compartment of the body fluids.

**salt:** a compound composed of a positive ion other than $H^+$ and a negative ion other than $OH^-$. An example is sodium chloride ($Na^+Cl^-$).
Na = sodium.
Cl = chloride.

**dissociation** (dis-SO-see-AY-shun): the physical separation of a compound into ions.

**ions** (EYE-uns): atoms or molecules that have gained or lost electrons and therefore have electrical charges. Examples include the positively charged sodium ion ($Na^+$) and the negatively charged chloride ion ($Cl^-$). For a closer look at ions, see Appendix B.

**cations** (CAT-eye-uns): positively charged ions.

**anions** (AN-eye-uns): negatively charged ions.

**electrolytes:** salts that dissolve in water and dissociate into charged particles called ions.

**electrolyte solutions:** solutions that can conduct electricity.

Maintaining a balance of about two-thirds of the body fluids inside the cells and one-third outside is vital to the life of the cells. If too much water were to enter the cells, it might rupture them; if too much water were to leave, they would collapse. To control the movement of water, the cells direct the movement of the major minerals.

• ***Dissociation of Salt in Water*** • When a mineral salt such as sodium chloride (NaCl) dissolves in water, it separates (dissociates) into ions—positively and negatively charged particles ($Na^+$ and $Cl^-$). The positive ions are cations; the negative ones are anions. Unlike pure water, which conducts electricity poorly, ions dissolved in water carry electrical current. For this reason, ions are called electrolytes, and the fluids of the body, which contain water and dissociated salts, are electrolyte solutions.

In all electrolyte solutions, anion and cation concentrations balance. If a fluid contains 1000 negative charges, it must contain 1000 positive charges, too. If an anion enters the fluid, a cation must accompany it or another anion must leave so that electroneutrality will be maintained.

Regardless of whether the body fluids are inside or outside the cells, the number of positive and negative charges will be equal. Table 12-4 shows that, indeed, the positive and negative charges inside and outside cells are perfectly balanced even though the numbers of each kind of ion differ over a wide range. For example, it is not necessary to have the same number of $Na^+$ and $Cl^-$ ions in a body fluid. $Na^+$ ions can leave a cell, provided that some other positive ions enter: potassium ($K^+$) ions, for example.

**Table 12-4**
**Important Body Electrolytes**

| Electrolytes | Intracellular (inside cells) Concentration (mEq/L) | Extracellular (outside cells) Concentration (mEq/L) |
|---|---|---|
| **Cations (positively charged ions)** | | |
| Sodium ($Na^+$) | 10 | 142 |
| Potassium ($K^+$) | 150 | 5 |
| Calcium ($Ca^{++}$) | 2 | 5 |
| Magnesium ($Mg^{++}$) | 40 | 3 |
| | 202 | 155 |
| **Anions (negatively charged ions)** | | |
| Chloride ($Cl^-$) | 2 | 103 |
| Bicarbonate ($HCO_3^-$) | 10 | 27 |
| Phosphate ($HPO_4^=$) | 103 | 2 |
| Sulfate ($SO_4^=$) | 20 | 1 |
| Organic acids (lactate, pyruvate) | 10 | 6 |
| Proteins | 57 | 16 |
| | 202 | 155 |

NOTE: The number of positive and negative charges in a given fluid is the same. For example, in extracellular fluid, the cations and anions both equal 155 milliequivalents per liter (mEq/L). Of the cations, sodium ions make up 142 mEq/L; and potassium, calcium, and magnesium ions make up the remainder. Of the anions, chloride ions number 103 mEq/L; bicarbonate ions number 27; and the rest are provided by phosphate ions, sulfate ions, organic acids, and protein.

Inside the cells, the positive charges total 202 and the negative charges balance these perfectly. Outside the cells, the amounts and proportions of the ions differ from those inside, but again the positive and negative charges balance. (Chemists count these charges in milliequivalents, mEq.)

• ***Electrolytes Attract Water*** • Electrolytes attract water molecules. Each water molecule has a net charge of zero, but the oxygen side of the molecule is slightly negatively charged, and the hydrogens are slightly positively charged. Figure 12-2 shows the result in an electrolyte solution: both positive and negative ions attract clusters of water molecules around them. This attraction dissolves salts in water and enables the body to move fluids into appropriate compartments.

• ***Water Follows Electrolytes*** • Some electrolytes reside primarily outside the cells (notably, sodium and chloride), while others are predominantly inside the cells (notably, potassium, magnesium, phosphate, and sulfate). Cells can move the electrolytes in and out, and water will follow them.

**milliequivalents (mEq):** the concentration of electrolytes in a volume of solution. Milliequivalents are a useful measure when considering ions, because the number of charges reveals characteristics about the solution that are not evident when the concentration is expressed in terms of weight.

A neutral molecule that has opposite charges spatially separated within the molecule is **polar**; see Appendix B for more details.

The word ending **-ate** denotes a salt of the mineral.

+
H
O H +
–
–

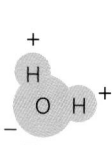

The negatively charged electrons that bond the hydrogens to the oxygen spend most of their time near the oxygen atom. As a result, the oxygen is slightly negative, and the hydrogen is slightly positive (see Appendix B).

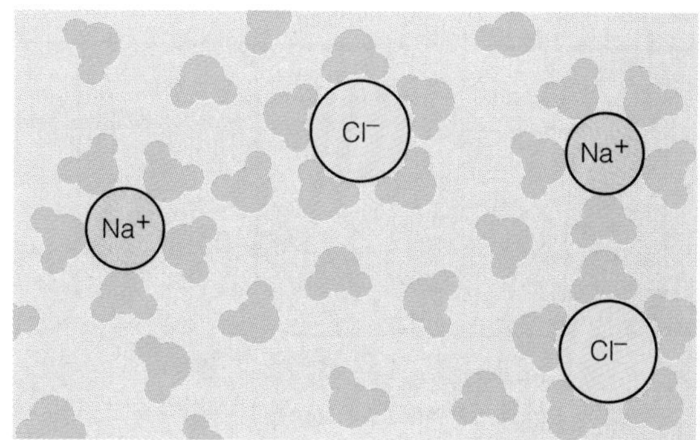


In an electrolyte solution, water molecules are attracted to both anions and cations. Notice that the negative oxygen atoms of the water molecules are drawn to the sodium cation ($Na^+$), while the positive hydrogen atoms of the water molecules are drawn to the chloride ions ($Cl^-$).

**Figure 12-2**
**Water Dissolves Salts and Follows Electrolytes**

The structural arrangement of the two hydrogen atoms and one oxygen atom enables water to dissolve salts. Water's role as a solvent is one of its most valuable characteristics.

**solutes** (SOLL-yutes): the substances that are dissolved in a solution.

**osmotic pressure:** the force that moves water, but not the solutes, across a membrane when the two solutions differ in concentration. Water flows *toward* the side in which the solutes are more concentrated.

The pump actively exchanges sodium for potassium across the cell membrane, maintaining a strong *concentration gradient* of each. Known as the **sodium-potassium pump,** it uses ATP as an energy source and the enzyme **sodium-potassium ATPase** (A-T-P-ace) to release that energy from ATP.

When water follows electrolytes, a force moves water toward concentrated solutes. This force, known as osmotic pressure, moves water across a membrane whenever the solute concentrations on the two sides are not equal and the solutes themselves cannot cross the membrane. Figure 12-3 shows this principle in operation.

**• *Proteins Regulate Flow of Fluids and Ions* •** Chapter 6 described how proteins help to regulate fluid movement. Transport proteins in the cell membranes also regulate the passage of positive ions and other substances from one side of the membrane to the other. Negative ions follow positive ions, and water flows toward the more concentrated solution.

A well-understood protein that regulates the flow of fluids and ions is the sodium-potassium pump, an enzyme that pumps sodium out of cells faster than it can diffuse back in. Simultaneously, the enzyme pumps potassium ions the other way, into the cell. Figure 6-10 on p. 171 illustrates this action.

**• *Regulation of Fluid and Electrolyte Balance* •** The amounts of various minerals in the body must remain nearly constant. Regulation occurs chiefly at two sites: the GI tract and the kidneys.

The GI tract continuously pours minerals out into its upper portions (stomach and small intestine) in the digestive juices and bile it secretes. It then reabsorbs these minerals and those from foods in its lower segment (the colon) as needed. In a day, 8 liters of fluids and associated minerals are recycled this way, providing ample opportunity for the regulation of electrolyte balance.

The kidneys' control of the body's *water* content by way of the hormone ADH has already been described. To regulate the *electrolyte* contents, the kidneys depend on the adrenal glands, which send out messages by way of the hormone aldosterone. If the body's sodium is low, aldosterone stimulates sodium reabsorption from the kidneys. As sodium is reabsorbed, potassium (another positive ion) is excreted in accordance with the rule that total positive charges must remain in balance with total negative charges. (See Figure 3-8 on p. 77 for a review of kidney function.)

## Fluid and Electrolyte Imbalance

Normally, the body defends itself successfully against fluid and electrolyte imbalances. Certain situations, however, may overwhelm the body's ability to compensate. Vomiting, diarrhea, heavy sweating, burns, wounds, and the like may incur such great fluid and electrolyte losses as to precipitate a medical emergency.

**Figure 12-3**

**Osmotic Pressure**

Water flows in the direction of the higher concentration of solute.

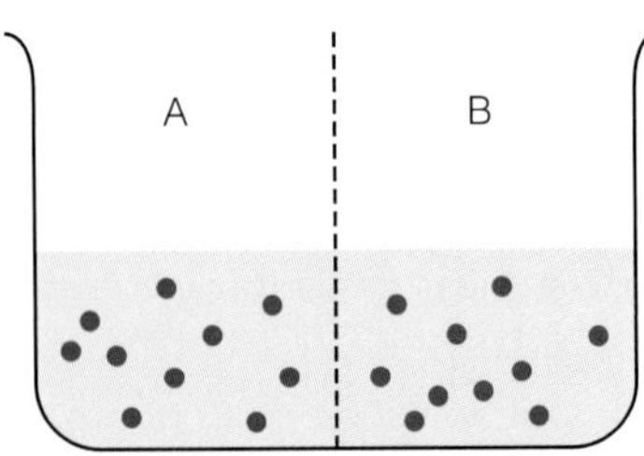


1. With equal numbers of solute particles on both sides, the concentrations are equal, and the tendency of water to move in either direction is about the same.

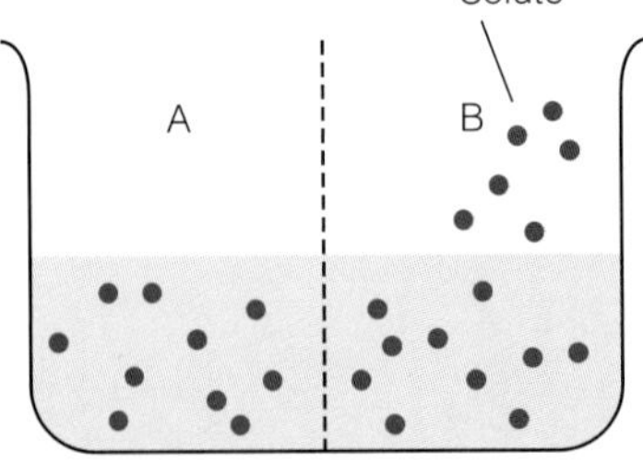


2. Now additional solute is added to side B. Solute cannot flow across the divider (in the case of a cell, its membrane).

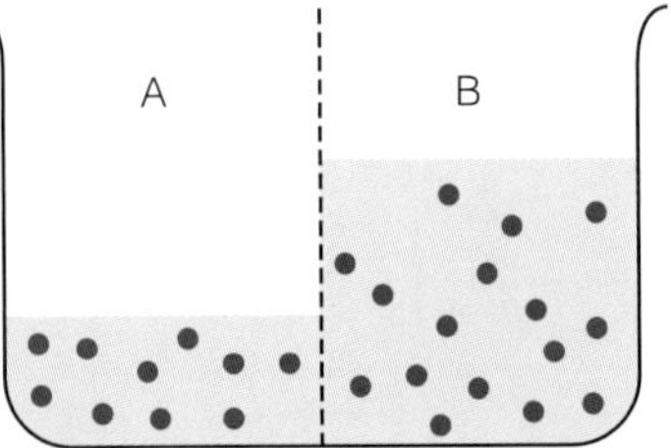


3. Water can flow both ways across the divider, but has a greater tendency to move from side A to side B, where there is a greater concentration of solute. The volume of water becomes greater on side B, and the concentrations on sides A and B become equal.

• ***Sodium and Chloride Most Easily Lost*** • Because sodium and chloride are the body's principal extracellular cation and anion, they are first to be lost when fluid is lost by sweating, bleeding, or excretion. It is no coincidence that after sweating excessively or losing fluid in other ways, a person craves salty foods and refreshing drinks.

• ***Different Solutes Lost by Different Routes*** • If fluid is lost by vomiting or diarrhea, sodium is lost indiscriminately. If the adrenal glands oversecrete aldosterone, as occurs when a tumor develops, the kidneys may excrete too much potassium. And the person with uncontrolled diabetes may lose a solute not normally excreted: glucose, and with it, large amounts of water. All three situations bring on dehydration, but drinking water alone cannot restore balance. In each case, medical intervention is required.

• ***Replacing Lost Fluids and Electrolytes*** • In many cases, people can replace the fluids and minerals lost in sweat or in a temporary bout of diarrhea by drinking plain cool water and eating regular foods. Some cases, however, demand rapid replacement of fluids and electrolytes—for example, when diarrhea threatens the life of a malnourished child.

Caretakers around the world have learned to use simple formulas to treat mild-to-moderate cases of diarrhea. These lifesaving formulas do not require hospitalization and can be prepared from ingredients available locally. Caretakers need only learn to measure ingredients carefully and use sanitary water. Once rehydrated, children can begin eating foods.

***Physically active people must remember to replace their body fluids.***

Chapter 14 presents the advantages and disadvantages of sport drinks, and Highlight 3 discusses diarrhea.

Health care workers use **oral rehydration therapy (ORT)**—a simple solution of sugar, salt, and water, taken by mouth—to treat dehydration caused by diarrhea. A simple ORT recipe:

1 c boiling water.
2 tsp sugar.
A pinch of salt.

## Acid-Base Balance

The body uses its ions not only to help maintain water balance, but also to regulate the acidity (pH) of its fluids. The pH scale of Chapter 3 is repeated here, in Figure 12-4, with the normal and abnormal pH ranges of the blood added.

**pH:** a measure of the concentration of $H^+$ ions (see Appendix B). The lower the pH, the higher the $H^+$ ion concentration and the stronger the acid. A pH above 7 is alkaline, or base (a solution in which $OH^-$ ions predominate).

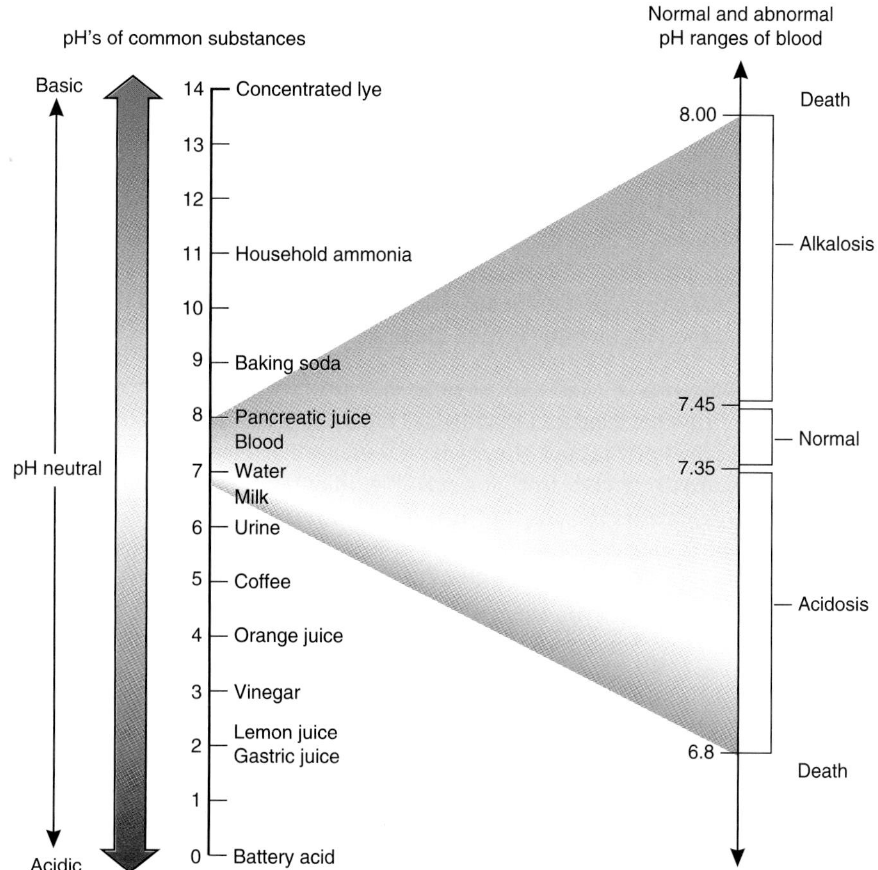


**Figure 12-4**
**The pH Scale**

Note: Each step is ten times as concentrated in base (1/10 as much acid, or $H^+$) as the one below it.

Reminder: *Buffers* are compounds that help keep a solution's acidity or alkalinity constant. The buffering action of proteins is described in Chapter 6.

• ***Regulation by Buffers*** • Some of the electrolyte mixtures in the body fluids, as well as some of the proteins, protect the body against changes in acidity by acting as buffers—substances that can neutralize acids or bases. The body's buffer systems serve as a first line of defense against changes in the fluids' acid-base balance.

**carbonic acid:** a compound with the formula $H_2CO_3$ that results from the combination of carbon dioxide ($CO_2$) and water ($H_2O$); of particular importance in the body's buffer system.

• ***Regulation by Excretion*** • The lungs, skin, GI tract, and kidneys provide other defenses. Carbon dioxide, which is formed all the time by cellular respiration, forms carbonic acid in the blood, pushing the balance toward acid. If too much acid builds up, the respiration rate speeds up, and more carbon dioxide is exhaled. If base builds up, the respiration rate slows; carbon dioxide is retained and forms more carbonic acid. The skin can excrete acid in sweat, and the specialized tear ducts can alter the composition of tears. These are of minor, although not negligible, importance; the kidneys play the primary role in maintaining acid-base balance.

• ***Regulation by the Kidneys*** • The kidneys adjust the acid-base balance by selecting which ions to retain and which to excrete. Their work is complex, but its net effect is easy to sum up. The *body's* total acid burden remains nearly constant; to a great extent, what a person ingests affects the acidity, not of the body, but of the *urine.*

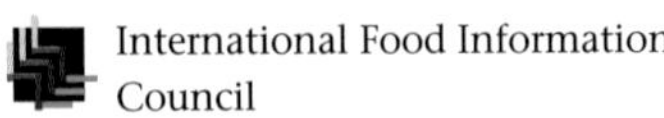

International Food Information Council

ificinfo.health.org/insight/waterref.htm

IN SUMMARY

Electrolytes (charged minerals) in the fluids help distribute the fluids inside and outside of cells, thus ensuring the appropriate water balance and acid-base balance to support all life processes. Excessive losses of fluids and electrolytes upset these balances; the kidneys play a key role in restoring homeostasis.

# The Minerals—An Overview

Reminder: *Minerals* are inorganic elements; some minerals are required in small amounts and therefore are essential nutrients.

**major minerals:** essential mineral nutrients found in the human body in amounts larger than 5 g; sometimes called **macrominerals.**

**trace minerals:** essential mineral nutrients found in the human body in amounts less than 5 g; sometimes called **microminerals.**

Figure 12-5 shows the amounts of the major minerals and, for comparison, some of the trace minerals found in the body. The distinction between the major and trace minerals does not reflect the importance of one group over the other—all minerals are vital. The major minerals are so named because they are present, and needed, in the largest amounts in the body. They are shown at the top of the figure and are discussed in this chapter. The trace minerals (shown at the bottom) are discussed in Chapter 13.

A few generalizations pertain to all of the minerals and distinguish them from the vitamins. Especially notable is their chemical nature.

• ***Inorganic Elements*** • Unlike the organic vitamins, which are easily destroyed, minerals are inorganic elements that always retain their chemical identity. Once minerals enter the body proper, they remain there until excreted; they cannot be changed into anything else. Iron, for example, may temporarily combine with other charged elements in salts, but it is always iron. Neither can minerals be destroyed by heat, air, acid, or mixing; only a little care is needed to preserve minerals during food preparation. In fact, the ash that remains when a food is burned contains all the minerals that were in the food originally. Minerals can be lost from food only when they leach into water that is then thrown away.

• ***The Body's Handling of Minerals*** • The minerals also differ from the vitamins in the amounts the body can absorb and in the extent to which they must be specially handled. Some minerals, such as potassium, are easily absorbed into the blood, transported freely, and readily excreted by the kidneys, much like the water-soluble vitamins. Some minerals, such as calcium, are more like fat-soluble

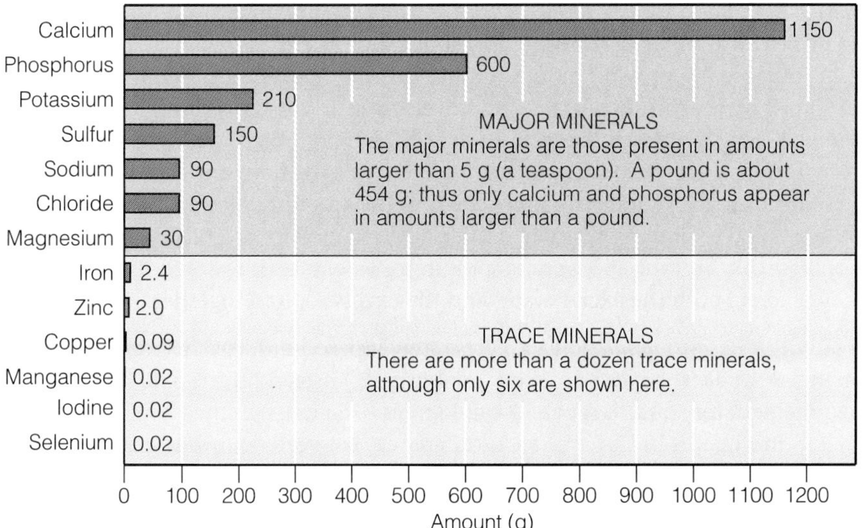


**Figure 12-5**
**Minerals in a 60-kilogram (132-pound) Human Body**

vitamins in that they must have carriers to be absorbed and transported. And, like the fat-soluble vitamins, minerals taken in excess can be toxic.

• ***Variable Bioavailability*** • The bioavailability of minerals varies. Some foods contain binders that combine chemically with minerals, preventing their absorption and carrying them out of the body with other wastes. Examples of binders include phytic acid, which is found primarily in legumes and grains, and oxalic acid, which is present in rhubarb and spinach, among other foods. These foods contain more minerals than the body actually receives for use.

Reminder: *Bioavailability* refers to the rate and extent to which a nutrient is absorbed and used. Some nutrients are not readily released from foods during digestion or are not efficiently absorbed, which reduces their bioavailability.

**binders:** chemical compounds occurring in foods that can combine with nutrients (especially minerals) to form complexes the body cannot absorb. Examples of such binders include **phytic** (FIGHT-ic) **acid** and **oxalic** (ox-AL-ic) **acid.**

• ***Varied Roles*** • While all the major minerals help to maintain the body's fluid balance described earlier, sodium, chloride, and potassium are most noted for that role. For this reason, these three minerals are discussed first here. Later sections describe the minerals most noted for their roles in bone growth and maintenance—calcium, phosphorus, and magnesium.

IN SUMMARY

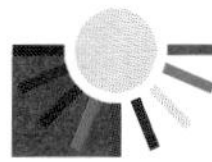

The major minerals are found in larger quantities in the body, whereas the trace minerals occur in smaller amounts. Minerals are inorganic elements that retain their chemical identities; they usually receive special handling and regulation in the body; and they may bind with other substances, thus limiting their absorption.

# Sodium

People have held salt (sodium chloride) in high regard throughout recorded history. We say "you are the salt of the earth" to someone we admire and "you are not worth your salt" to someone we consider worthless. Even the word *salary* comes from the Latin word for salt.

Cultures vary in their use of salt, but most people find its taste innately appealing.[3] Salt brings its own tangy taste and enhances other flavors, most likely by suppressing the bitter flavors.[4] You can taste this effect for yourself: tonic water with its bitter quinine tastes sweeter with a little salt added.

**sodium:** the principal cation in the extracellular fluids of the body; critical to the maintenance of fluid balance, nerve transmissions, and muscle contractions.

• ***Sodium Roles in the Body*** • Sodium is the principal cation of the extracellular fluid and the primary regulator of its volume. Sodium also helps maintain acid-base balance and is essential to nerve transmission and muscle contraction.* The table below summarizes information about sodium.

Foods usually provide more sodium than the body needs. Sodium is readily absorbed by the intestinal tract and travels freely in the blood until it reaches the kidneys, which filter all the sodium out of the blood; then, with great precision, they return to the bloodstream the exact amount the body needs. Normally, the amount excreted is approximately equal to the amount ingested on a given day. When blood sodium rises, as when a person eats salted foods, thirst signals the person to drink until the appropriate sodium-to-water ratio is restored. Then the kidneys excrete both the excess water and the excess sodium together.

Salt (sodium chloride) is about 40% sodium.
1 g salt contributes 400 mg sodium.
5 g salt = 1 tsp.
1 tsp salt contributes 2000 mg sodium.

• ***Sodium Recommendations*** • Diets rarely lack sodium. For this reason, recommended intakes have not been established; instead, the *minimum* sodium requirement for adults is set at 500 milligrams (0.5 grams) in the United States and at 115 milligrams in Canada. Such differences between countries' recommendations typically reflect differences in judgment more than differences in research data. The lower recommendation reflects the minimum average requirement for adults without active sweating, whereas the higher recommendation accommodates a wider variety of physical activities and climates.

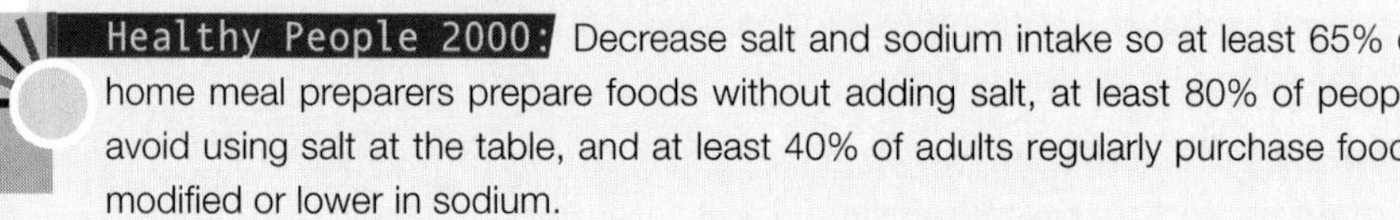

**Healthy People 2000:** Decrease salt and sodium intake so at least 65% of home meal preparers prepare foods without adding salt, at least 80% of people avoid using salt at the table, and at least 40% of adults regularly purchase foods modified or lower in sodium.

• ***Sodium and Hypertension*** • For years, a high *sodium* intake was considered *the* primary factor responsible for high blood pressure. Then research pointed to *salt* (sodium chloride) as the dietary culprit. Salt has a greater effect on blood pressure than either sodium or chloride alone or in combination with other ions.[5] A low salt intake diminishes the risk of hypertension.[6] Consequently, health recommendations advise limiting daily *salt* intake to less than 6 grams (the equivalent of 2.4 grams or 2400 milligrams of *sodium*).[7]

Some individuals are genetically sensitive and experience high blood pressure from excesses in salt intake.[8] People with chronic renal disease, diabetes, hypertension, African Americans, and people over 50 years of age are most likely to be salt sensitive.[9†] Salt restriction may help lower blood pressure, especially in salt-sensitive individuals, but weight loss, either alone or in combination with salt restriction, may be the most effective dietary treatment for hypertension.[10]

*One of the ways the kidneys regulate acid-base balance is by excreting hydrogen ions in exchange for sodium ions.

†Salt-sensitive individuals have elevated concentrations of renin in their blood, compared with others.

## IN SUMMARY

**Sodium**

| 1989 Estimated Minimum Requirement | Chief Functions in the Body | Deficiency Symptoms | Toxicity Symptoms | Significant Sources |
|---|---|---|---|---|
| Adults: 500 mg/day | Maintains normal fluid and electrolyte balance; assists in nerve impulse transmission and muscle contraction | Muscle cramps, mental apathy, loss of appetite | Edema, acute hypertension | Table salt, soy sauce; moderate amounts in meats, milks, breads, and vegetables; large amounts in processed foods |

Whether people with normal blood pressure benefit from a salt-restricted diet remains controversial.[11]

• ***Sodium and Osteoporosis*** • A high sodium intake has also been associated with calcium excretion and bone losses.[12] Dietary advice to prevent osteoporosis might suggest eating more calcium-rich foods while eating fewer foods high in sodium.

• ***Sodium Intakes*** • Men in the United States consume an average of 3300 milligrams of sodium (equivalent to 8 grams of salt) a day, which is slightly more than the average intake for women and health recommendations. Asian people, whose staple sauces and flavorings are based on soy sauce and monosodium glutamate (MSG), consume the equivalent of about 30 to 40 grams of salt per day. Interestingly, the prevalence of high blood pressure in China, Japan, and Korea is equal to or greater than that in the United States.[13]

• ***Sodium in Foods*** • In general, processed foods have the most sodium, while unprocessed foods such as fresh fruits and vegetables have the least. In fact, as much as 75 percent of the sodium in people's diets comes from salt added to foods by manufacturers; about 15 percent comes from salt added during cooking and at the table; and only 10 percent comes from the natural content in foods.

Because processed foods may contain sodium without chloride, as in additives such as sodium bicarbonate or sodium saccharin, they do not always taste salty. Most people are surprised to learn that 1 ounce of cornflakes contains more sodium than 1 ounce of salted peanuts—and that ½ cup of instant chocolate pudding contains still more. (A reason the peanuts taste saltier is that the salt is all on the surface, where the tongue's sensors immediately pick it up.)

Figure 12-6 shows that processed foods contain not only more sodium but also less potassium than their less-processed counterparts. Low potassium may be as

FDA

www.fda.gov/fdac/foodlabel/sodium.html

**Figure 12-6**

**What Processing Does to the Sodium and Potassium Contents of Foods**

People who eat foods high in salt often happen to be eating fewer potassium-containing foods at the same time. Note how the *same* food loses potassium and gains sodium as it goes through processing, so that its potassium-to-sodium ratio falls dramatically. Even when potassium isn't lost, the addition of sodium still lowers the potassium-to-sodium ratio. Limiting sodium intake may help in two ways, then—by lowering blood pressure in salt-sensitive individuals and by indirectly raising potassium intakes in all individuals.

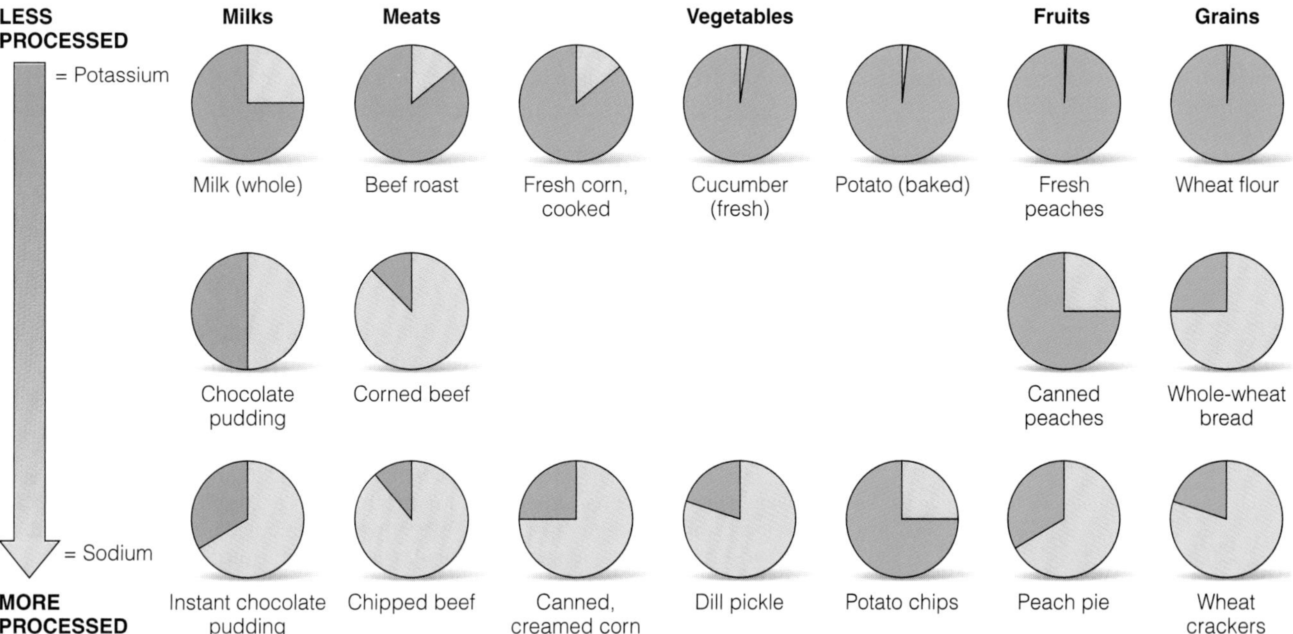

*Fresh herbs, such as basil, add flavor to a recipe without adding salt. (Courtesy of CNN.)*

significant as high sodium when it comes to blood pressure regulation, so these foods have two strikes against them. The "How to" featured on the next page offers strategies for cutting sodium/salt intake. Chapter 18 reviews the research on sodium, potassium, and hypertension in relation to heart disease.

• ***Sodium Deficiency*** • If blood sodium drops, as may occur in vomiting, diarrhea, or heavy sweating, both sodium and water must be replenished. Under normal conditions of sweating due to exercise, salt losses can easily be replaced later in the day with ordinary foods. Salt tablets are not recommended because too much salt, especially if taken with too little water, can induce dehydration.

During intense activities, such as ultra-endurance events, athletes can lose so much sodium and drink so much water that they develop hyponatremia—too little sodium in the blood. Beverages with sodium and glucose and salty foods will help restore sodium balance.

• ***Sodium Toxicity and Excessive Intakes*** • The immediate symptoms of acute sodium toxicity are edema and hypertension, but such toxicity poses no problem as long as water needs are met. Prolonged excessive sodium intake may contribute to hypertension in some people, as explained earlier.

IN SUMMARY

Sodium is the main cation outside cells and one of the electrolytes primarily responsible for maintaining fluid balance. Dietary deficiency is rare, and excesses may aggravate hypertension in some people. For this reason, health professionals advise a diet moderate in salt and sodium.

# Chloride

**chloride** (KLO-ride): the major anion in the extracellular fluids of the body. Chloride is the ionic form of chlorine, $Cl^-$; see Appendix B for a description of the chlorine-to-chloride conversion.

The element *chlorine* ($Cl_2$) is a poisonous gas. When chlorine reacts with sodium or hydrogen, however, it forms the negative chloride ion ($Cl^-$). *Chloride* is required in the diet.

• ***Chloride Roles in the Body*** • Chloride is the major anion of the extracellular fluids, where it occurs mostly in association with sodium (the table on p. 379 summarizes information on chloride). Chloride can move freely across membranes and so also associates with potassium inside cells. Like sodium, chloride maintains fluid and electrolyte balance.

In the stomach, the chloride ion is part of hydrochloric acid, which maintains the strong acidity of the gastric juice. One of the most serious consequences of vomiting is the loss of this acid from the stomach, which upsets the acid-base balance.* Such imbalances are commonly seen in bulimia nervosa, as Highlight 9 describes.

Reminder: The loss of acid can lead to *alkalosis,* an above-normal alkalinity in the blood and body fluids.

• ***Chloride Recommendations and Intakes*** • Chloride is abundant in foods (especially processed foods) as part of sodium chloride and other salts. A recommended intake has not been established for chloride; instead a minimum requirement for adults has been estimated.

• ***Chloride Deficiency and Toxicity*** • Diets rarely lack chloride. Chloride losses may occur in conditions such as heavy sweating, chronic diarrhea, and vomiting. The only known cause of high blood chloride concentrations is dehydration due

*Hydrochloric acid secretion into the stomach involves the addition of bicarbonate ions to the plasma. These bicarbonate ions are neutralized by hydrogen ions from the gastric secretions that are reabsorbed into the plasma. When hydrochloric acid is lost during vomiting, these hydrogen ions are no longer available for reabsorption, which in effect increases the concentrations of bicarbonate ions in the plasma. In this way, excessive vomiting of acidic gastric juices leads to *metabolic alkalosis.*

## HOW TO Cut Salt Intake

Most people eat more salt and sodium than they need, and some people can lower their blood pressure by avoiding highly salted foods and removing the saltshaker from the table. Foods eaten without salt may seem less tasty at first, but with repetition, people can learn to enjoy the natural flavors of many unsalted foods. Strategies to cut salt intake include:

- Cook with only small amounts of added salt.
- Prepare foods with sodium-free spices such as basil, bay leaves, curry, garlic, ginger, lemon, mint, oregano, pepper, rosemary, and thyme.
- Add little or no salt at the table; taste foods before adding salt.
- Read labels with an eye open for salt. (See Table 2–8 on p. 51 for terms used to describe the sodium contents of foods on labels.)
- Select low-salt or salt-free products when available.
- Eat high-salt foods in moderation.

Use these foods sparingly:

- Foods prepared in brine, such as pickles, olives, and sauerkraut.
- Salty or smoked meats, such as bologna, corned or chipped beef, frankfurters, ham, lunch meats, salt pork, sausage, and smoked tongue.
- Salty or smoked fish, such as anchovies, caviar, salted and dried cod, herring, sardines, and smoked salmon.
- Snack items such as potato chips, pretzels, salted popcorn, salted nuts, and crackers.
- Bouillon cubes; seasoned salts; soy, Worcestershire, and barbeque sauces.
- Cheeses, especially processed types.
- Canned and instant soups.
- Prepared horseradish, catsup, and mustard.

to water deficiency. In both cases, consuming ordinary foods and beverages can restore chloride balance.

IN SUMMARY

Chloride is the major anion outside cells, and it associates closely with sodium. In addition to its role in fluid balance, chloride is part of the stomach's hydrochloric acid, which facilitates protein digestion and iron absorption.

# Potassium

Like sodium, potassium is a positively charged ion. In contrast to sodium, potassium is the body's principal cation *inside* the body cells.

**potassium:** the principal cation within the body's cells; critical to the maintenance of fluid balance, nerve transmissions, and muscle contractions.

• ***Potassium Roles in the Body*** • Potassium plays a major role in maintaining fluid and electrolyte balance and cell integrity. During nerve transmission and muscle contraction, potassium and sodium briefly trade places across the cell

IN SUMMARY

**Chloride**

| 1989 Estimated Minimum Requirement | Chief Functions in the Body | Deficiency Symptoms | Toxicity Symptoms | Significant Sources |
|---|---|---|---|---|
| Adults: 750 mg/day | Maintains normal fluid and electrolyte balance; part of hydrochloric acid found in the stomach, necessary for proper digestion | Do not occur under normal circumstances | Vomiting | Table salt, soy sauce; moderate amounts in meats, milks, eggs; large amounts in processed foods |

*Fresh fruits and vegetables provide potassium in abundance.*

membrane. The cell then quickly pumps them back into place. Controlling potassium distribution is a high priority for the body because it affects many aspects of homeostasis, including a steady heartbeat.

• ***Potassium Recommendations and Intakes*** • As for sodium and chloride, a minimum potassium requirement for adults has been estimated. Potassium is abundant in all living cells, both plant and animal. Because cells remain intact unless foods are processed, the richest sources of potassium are *fresh* foods of all kinds—as Figure 12-7 shows. People who emphasize fresh fruits and vegetables in their diets have high intakes of potassium, but toxicity is normally not a concern when the source is foods.

Fresh foods, especially fruits, contain much more potassium than sodium. In contrast, most processed foods such as canned vegetables, ready-to-eat cereals, and luncheon meats contain more sodium and less potassium (recall Figure 12-6).

• ***Potassium and Hypertension*** • Low-potassium diets seem to play an important role in the development of high blood pressure. Diets low in potassium raise blood pressure, whereas high potassium intakes appear to both prevent and correct hypertension.[14]

• ***Potassium Deficiency*** • Potassium deficiency occurs more often due to excessive losses than to deficient intakes. Conditions such as diabetic acidosis, dehydration, or prolonged vomiting or diarrhea can create a potassium deficiency, as can the regular use of certain drugs, including diuretics, steroids, and strong laxatives.* For this reason, many physicians prescribe potassium supplements along with these potassium-wasting drugs. One of the earliest symptoms of deficiency is muscle weakness. The table below summarizes facts about potassium.

• ***Potassium Toxicity*** • Potassium toxicity can result from the overuse of potassium salt, especially in an infant or a person with heart disease; it does not result from overeating foods high in potassium. Given more potassium than the body needs, the kidneys accelerate their excretion. If the GI tract is bypassed, however, and potassium is injected rapidly into a vein, it can stop the heart.

IN SUMMARY

Potassium, like sodium and chloride, is an electrolyte that plays an important role in maintaining fluid balance. Potassium is the primary cation inside cells; fresh fruits and vegetables are its best sources.

*People using diuretics to control hypertension should know that some cause potassium excretion and can induce a deficiency. Those using these drugs must be particularly careful to include rich sources of potassium in their daily diets. (Some diuretics are designed to spare potassium.)

IN SUMMARY

**Potassium**

| 1989 Estimated Minimum Requirement | Chief Functions in the Body | Deficiency Symptoms[a] | Toxicity Symptoms | Significant Sources |
|---|---|---|---|---|
| Adults: 2000 mg/day | Maintains normal fluid and electrolyte balance; facilitates many reactions; supports cell integrity; assists in nerve impulse transmission and muscle contractions | Muscular weakness, paralysis, confusion | Muscular weakness; vomiting; if given into a vein, can stop the heart | All whole foods: meats, milks, fruits, vegetables, grains, legumes |

[a]Deficiency accompanies dehydration.

**Figure 12-7**
**Potassium in Selected Foods**

| Food | Serving size (kcalories) |
|---|---|
| Bread, whole wheat | 1 slice (64 kcal) |
| Corn flakes, fortified | 1 oz (108 kcal) |
| White rice | 1/2 c cooked (134 kcal) |
| Spaghetti pasta | 1/2 c cooked (99 kcal) |
| Oatmeal | 1/2 c cooked (73 kcal) |
| Tortilla, flour | 1 8"-round (115 kcal) |
| Spinach | 1 c raw (12 kcal) |
| Broccoli | 1/2 c cooked (22 kcal) |
| Carrots | 1/2 c shredded raw (24 kcal) |
| Green peas | 1/2 c cooked (62 kcal) |
| Corn | 1/2 c cooked (66 kcal) |
| Green beans | 1/2 c cooked (22 kcal) |
| Sweet potatoes | 1/2 c cooked (116 kcal) |
| Potato | 1 baked w/skin (220 kcal) |
| Tomato juice | 3/4 c (31 kcal) |
| Apple | 1 medium raw (81 kcal) |
| Banana | 1 medium raw (104 kcal) |
| Orange | 1 medium raw (62 kcal) |
| Strawberries | 1/2 c fresh (23 kcal) |
| Raisins | 1/4 c (109 kcal) |
| Watermelon | 1 slice (154 kcal) |
| Grapefruit juice | 3/4 c fresh (72 kcal) |
| Avocado | 1/4 (85 kcal) |
| Milk | 1 c reduced-fat 2% (121 kcal) |
| Yogurt, plain | 1 c low-fat (144 kcal) |
| Cheddar cheese | 1 1/2 oz (171 kcal) |
| Cottage cheese | 1/2 c low-fat 2% (102 kcal) |
| Swiss cheese | 1 1/2 oz (159 kcal) |
| Ice cream | 1/2 c, 10% fat (134 kcal) |
| Navy beans | 1/2 c cooked (129 kcal) |
| Pinto beans | 1/2 c cooked (116 kcal) |
| Kidney beans | 1/2 c cooked (109 kcal) |
| Garbanzo beans | 1/2 c cooked (134 kcal) |
| Peanut butter | 2 tbs (188 kcal) |
| Sunflower seeds | 1 oz dry (159 kcal) |
| Tofu (soybean curd) | 1/2 c (94 kcal) |
| Shrimp | 3 oz boiled (85 kcal) |
| Ground beef, lean | 3 oz broiled (237 kcal) |
| Chicken breast | 3 oz roasted (141 kcal) |
| Cod | 3 oz poached (87 kcal) |
| Ham, lean | 3 oz roasted (133 kcal) |
| Sirloin steak, lean | 3 oz broiled (171 kcal) |
| Tuna, canned in water | 3 oz (98 kcal) |
| Bologna, beef | 2 slices (144 kcal) |
| Egg | 1 hard cooked (78 kcal) |
| **Additional 5 foods:** | |
| Acorn squash | 1/2 c baked (68 kcal) |
| Soybeans | 1/2 c cooked (149 kcal) |
| Artichoke | 1 (60 kcal) |
| Pomegranate | 1 (105 kcal) |
| Buttermilk, nonfat | 1 c (99 kcal) |

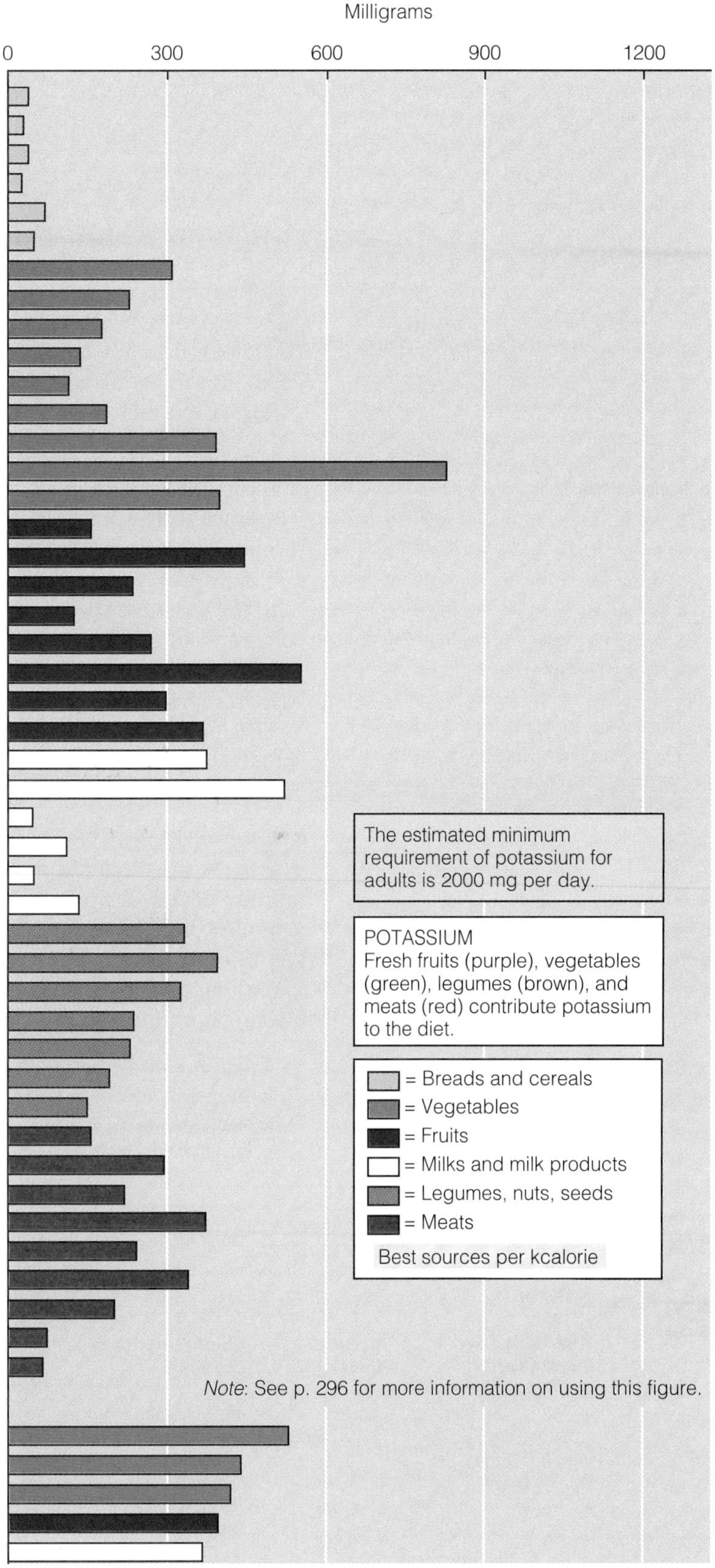

*Note*: See p. 296 for more information on using this figure.

# Calcium

**calcium:** the most abundant mineral in the body; found primarily in the body's bones and teeth.

Calcium is the most abundant mineral in the body. It receives much emphasis in this chapter and in the highlight that follows because an adequate intake helps grow a healthy skeleton in early life and helps minimize bone loss in later life. Calcium's roles, deficiency symptoms, and food sources appear in the summary table below.

## *Calcium Roles in the Body*

Ninety-nine percent of the body's calcium is in the bones (and teeth), where it plays two roles. First, it is an integral part of bone structure, providing a rigid frame that holds the body upright and serves as attachment points for muscles, making motion possible. Second, it serves as a calcium bank, offering a readily available source of the mineral to the body fluids should a drop in blood calcium occur.

**hydroxyapatite** (high-drox-ee-APP-ah-tite): crystals made of calcium and phosphorus.

• ***Calcium in Bones*** • As bones begin to form, calcium salts form crystals, called hydroxyapatite, on a matrix of the protein collagen. As the crystals become denser, they give strength and rigidity to the maturing bones. As a result, the long leg bones of children can support their weight by the time they have learned to walk.

Many people have the idea that once a bone is built, it is inert like a rock. Actually, the bones are gaining and losing minerals continuously in an ongoing process of remodeling. Growing children gain more bone than they lose, and healthy adults maintain a reasonable balance. When withdrawals substantially exceed deposits, problems such as osteoporosis develop (see Highlight 12).

The formation of teeth follows a pattern similar to that of bones. The turnover of minerals in teeth is not as rapid as in bone, however; fluoride hardens and stabilizes the crystals of teeth, opposing the withdrawal of minerals from them.

• ***Calcium in Body Fluids*** • The 1 percent of the body's calcium that circulates in the fluids as ionized calcium is vital to life. The calcium ion participates in the regulation of muscle contraction, the clotting of blood, the transmission of nerve impulses, the secretion of hormones, and the activation of some enzyme reactions.

**calmodulin** (cal-MOD-you-lin): an inactive protein that becomes active when bound to calcium; then it becomes a messenger that tells other proteins what to do. The system serves as an interpreter for hormone- and nerve-mediated messages arriving at cells.

Calcium also activates a protein called calmodulin. This protein relays messages from the cell surface to the inside of the cell. Several of these messages help to maintain normal blood pressure.

• ***Calcium and Disease Prevention*** • Calcium may be useful in protecting against hypertension, even in salt-sensitive people.[15] An adequate calcium intake can lower blood pressure, superseding the effects of a high sodium intake.[16] For this rea-

IN SUMMARY

**Calcium**

| 1997 Adequate Intake (AI) | Upper Level | Chief Functions in the Body | Deficiency Symptoms | Toxicity Symptoms | Significant Sources |
|---|---|---|---|---|---|
| Adults (19–50 yr): 1000 mg/day<br>Adults (51 and older): 1200 mg/day | Adults: 2500 mg/day | Mineralization of bones and teeth; also involved in muscle contraction and relaxation, nerve functioning, blood clotting, blood pressure, and immune defenses | Stunted growth in children; bone loss (osteoporosis) in adults | Constipation; increased risk of urinary stone formation and kidney dysfunction; interference with absorption of other minerals | Milk and milk products, small fish (with bones), tofu (bean curd), greens (broccoli, chard), legumes |

son, restricting sodium to treat hypertension is narrow advice, especially considering that low-sodium diets are often low in calcium as well. Instead, recommendations should also focus on raising calcium intakes.[17] Some evidence also suggests relationships between dietary calcium and blood cholesterol, diabetes, and some cancers. Highlight 12 explores calcium's role in preventing osteoporosis.

• ***Calcium Balance*** • Calcium homeostasis is one of the body's highest priorities and involves a system of hormones and vitamin D. Whenever blood calcium falls too low or rises too high, three organ systems respond: the intestines, bones, and kidneys. Figure 12-8 (on p. 384) illustrates how vitamin D and the hormones parathormone and calcitonin return blood calcium to normal.

The hormones **parathormone** (PAIR-ah-THOR-moan) from the parathyroid glands and **calcitonin** (KAL-see-TOE-nin) from the thyroid gland, as well as vitamin D, regulate calcium balance. Parathormone and vitamin D raise blood calcium, while calcitonin lowers it.

Note: Parathormone is also known as **parathyroid hormone.**

The calcium in bone provides a nearly inexhaustible bank of calcium for the blood. The blood borrows and returns calcium as needed, so that even with a dietary deficiency, blood calcium remains normal—even as bone calcium diminishes. Blood calcium changes only in response to abnormal regulatory control, not to diet. This makes a developing calcium deficiency completely silent. A person can have an inadequate calcium intake for years and suffer no noticeable symptoms. Only late in life does it become apparent that the integrity of the bones has been compromised.

Blood calcium above normal results in calcium rigor: the muscles contract and cannot relax. Similarly, blood calcium below normal causes calcium tetany—also characterized by uncontrolled muscle contraction. These conditions do *not* reflect a *dietary* excess or lack of calcium; they are caused by a lack of vitamin D or by abnormal secretion of the regulatory hormones. A chronic *dietary* deficiency of calcium, or a chronic deficiency due to poor absorption over the years, depletes the savings account in the bones. Again: it is the *bones,* not the blood, that are robbed by calcium deficiency.

**calcium rigor:** hardness or stiffness of the muscles caused by high blood calcium concentrations.

**calcium tetany** (TET-ah-nee): intermittent spasm of the extremities due to nervous and muscular excitability caused by low blood calcium concentrations.

• ***Calcium Absorption*** • Many factors affect calcium absorption, but on the average, adults absorb about 30 percent of the calcium they ingest. The stomach's acidity helps to keep calcium soluble, and absorption is enhanced when calcium supplements are consumed with a meal. Vitamin D helps the absorptive cells of the GI tract to make the necessary calcium-binding protein. This explains why calcium-rich milk is the best food for fortification with vitamin D. The lactose in milk also enhances calcium absorption. Also, calcium seems to be better absorbed when accompanied by an approximately equal amount of phosphorus, as occurs in milk.

Factors that enhance calcium absorption:

- Stomach acid.
- Vitamin D.
- Lactose.
- Phosphorus in an optimal ratio.
- Growth hormones.

The body regulates calcium absorption by increasing its production of the calcium-binding protein when calcium is needed. The result is obvious in the case of a pregnant woman, who absorbs 50 percent of the calcium from the milk she drinks. Similarly, growing children absorb 50 to 60 percent of ingested calcium. Then, when bone growth slows or stops, absorption falls to the adult level of about 30 percent. In addition, absorption becomes more efficient during times of inadequate intakes.[18]

**calcium-binding protein:** a protein in the intestinal cells, made with the help of vitamin D, that facilitates calcium absorption.

Many of the conditions that enhance calcium absorption inhibit its absorption when they are absent. For example, sufficient vitamin D supports absorption, while a deficiency impairs it. In addition, fiber, in general, and the binders phytate and oxalate, in particular, interfere with calcium absorption, but their effects are relatively minor at intakes typical of U.S. diets. Vegetables with oxalates and whole grains with phytates are nutritious foods, of course, but they are not useful calcium sources. The margin summarizes factors that influence calcium balance.

Factors that inhibit calcium absorption:

- Lack of stomach acid.
- Vitamin D deficiency.
- High phosphorus intake.
- High-fiber diet.
- Phytates (in seeds, nuts, grains).
- Oxalates (in beets, rhubarbs, spinach).

## Calcium Recommendations and Sources

Calcium is unlike most other nutrients, in that its blood concentration does not reflect the body's calcium status. Calcium recommendations are therefore based

**Figure 12-8**
**Calcium Balance in Bone**

Blood calcium is regulated in part by vitamin D and two hormones—calcitonin and parathormone. Bone serves as a reservoir when blood calcium is high and as a source of calcium when blood calcium is low. Osteoclasts break down bone and release calcium into the blood; osteoblasts build new bone using calcium from the blood.

| | IF BLOOD CALCIUM IS TOO HIGH | IF BLOOD CALCIUM IS TOO LOW |
|---|---|---|
| 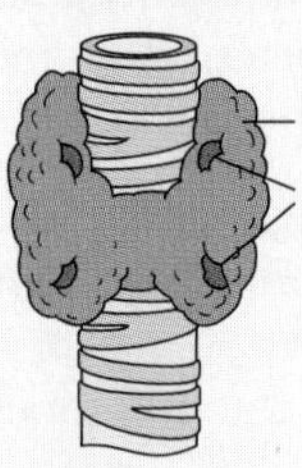 Thyroid; Parathyroid (imbedded in the thyroid) | Rising blood calcium signals the thyroid gland to secrete calcitonin. | Falling blood calcium signals the parathyroid glands to secrete parathormone. |
| 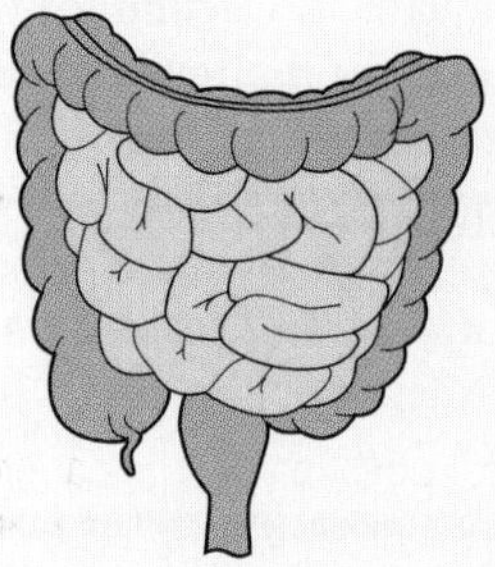 | Calcitonin limits calcium absorption in the intestines. | Vitamin D enhances calcium absorption in the intestines. |
| 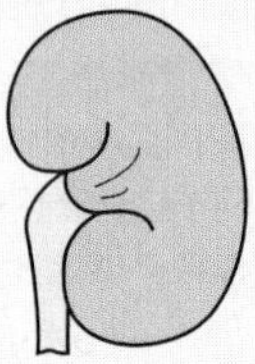 | Calcitonin inhibits the activation of vitamin D. | Parathormone stimulates the activation of vitamin D. |
| | Calcitonin stimulates calcium excretion in the kidneys. | Vitamin D and parathormone stimulate calcium reabsorption in the kidneys. |
| 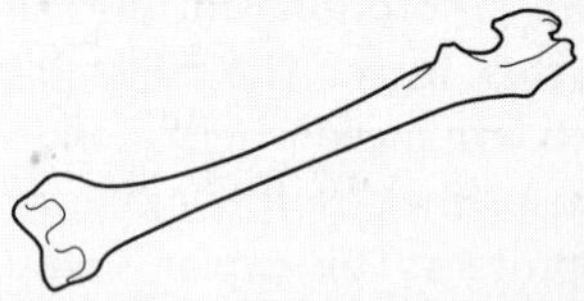 | Calcitonin inhibits osteoclast cells from breaking down bone, preventing a rise in blood calcium. | Vitamin D and parathormone stimulate osteoclast cells to break down bone, releasing calcium into the blood. |
| | All these actions lower blood calcium, which inhibits calcitonin secretion. | All these actions raise blood calcium, which inhibits parathormone secretion. |

on balance studies, which measure daily intake and excretion. An optimal intake reflects the amount needed to retain the most calcium. By retaining the most calcium possible, the bones can develop to their fullest potential in size and density, within genetic limits.

Without sufficient evidence to establish an RDA, the Committee on Dietary Reference Intakes estimated calcium recommendations as an Adequate Intake (AI) value (see Chapter 1).

• ***Calcium Recommendations*** • Because obtaining enough calcium during early life helps to ensure that the skeleton will be strong and dense, recommendations have been set high at 1300 milligrams daily for adolescents up to the age of 18 years. Between the ages of 19 and 50, recommendations are lowered to 1000 milligrams a day; for later life, recommendations are raised again to 1200 milligrams a day to minimize the bone loss that tends to occur late in life. Some authorities advocate as much as 1500 milligrams a day for women over 50. Many people in the United States and Canada, particularly women, have calcium intakes far below current recommendations.

**Healthy People 2000:** Increase calcium intake so that at least 50% of people aged 25 years and older consume two or more servings of calcium-rich foods daily.

High intakes of both dietary protein and sodium increase calcium losses, but whether this impairs bone development remains unclear. In establishing an Adequate Intake for calcium, the committee considered these nutrient interactions, but did not adjust dietary recommendations based on this information.

• ***Calcium Sources*** • Figure 12-9 (on p. 386) shows that calcium is found most abundantly in a single class of foods—milk and milk products. Unfortunately, many people, especially women, perceive milk as fattening and omit it from their diets in their attempts to lose weight. Whole milk and many cheeses are high in fat, but nonfat, low-fat, and reduced-fat options are available. Such choices help a person meet calcium needs within a reasonable energy and fat allowance.

Suggested minimum daily milk servings:
- Children: 2 c.
- Teenagers: 3 c.
- Adults: 2 c.
- Pregnant or lactating women: 3 c.
- Pregnant or lactating teens: 4 c.

The person who doesn't like to drink milk may prefer to eat cheese or yogurt. Alternatively, milk and milk products can be concealed in foods. Powdered nonfat milk can be added to casseroles, meat loaf, and other mixed dishes in preparation; 5 heaping tablespoons offer the equivalent of a cup of milk. This simple step is an excellent way for older women to obtain not only extra calcium, but more protein, vitamins, and minerals as well.[19]

• ***Nonmilk Sources*** • Some cultures do not use milk in their cuisines; some vegetarians exclude milk as well as meat; and some people are allergic to milk protein or are lactose intolerant. These people need to find nonmilk sources of calcium to help meet their calcium needs. Some brands of tofu, corn tortillas, some nuts (such as almonds), and some seeds (such as sesame seeds) can supply calcium for the person who doesn't use milk products. Wheat bread contains only about one-tenth of the calcium found in milk, but can be a major source for people who eat a lot of it because the calcium is well absorbed. Among the vegetables, mustard and turnip greens, bok choy, kale, parsley, watercress, and broccoli are good sources of available calcium. So are some seaweeds such as the nori popular in Japanese cooking. Some dark green, leafy vegetables—notably spinach, rhubarb, and Swiss chard—appear to be calcium-rich but actually provide little, if any, calcium to the body because of the binders they contain. The margin drawing ranks selected foods according to the bioavailability of their calcium.

People with lactose intolerance may be able to consume small quantities of milk, as Chapter 4 explains.

**Bioavailability of Calcium from Selected Foods**

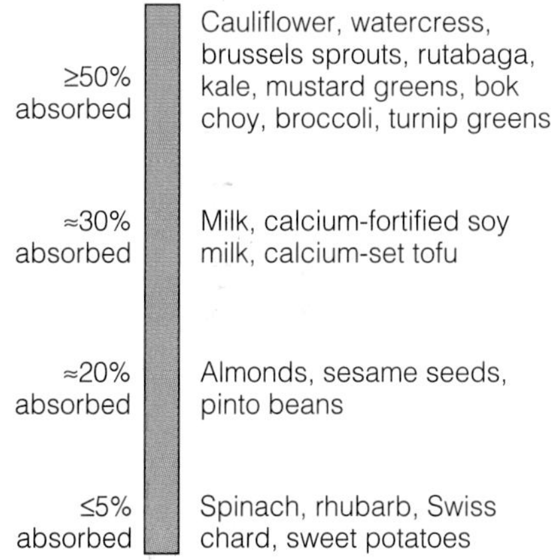


Oysters are also a rich source of calcium, as are small fish eaten with their bones, such as canned sardines. Many Asians prepare a stock from bones that helps account for their adequate calcium intake without the use of milk. They soak the cracked bones from chicken, turkey, pork, or fish in vinegar and then slowly boil the bones until they become soft. The bones release calcium into the acid medium, and most of the vinegar taste boils off. Cooks then use the stock, which contains more than 100 milligrams of calcium per tablespoon, in place of water to prepare soup, vegetables, rice, or stew. Similarly, cooks in the Navajo tribe use an ash prepared from the branches and needles of the juniper tree in their recipes.[20] One teaspoon of juniper ash provides about as much calcium as a cup of milk.

Some mineral waters provide as much as 500 milligrams of calcium per liter, offering a convenient way to meet both calcium and water needs.[21] Similarly, calcium-fortified orange juice allows a person to meet both calcium and vitamin C needs easily. Other examples of calcium-fortified foods include high-calcium milk (milk with extra calcium added) and calcium-fortified cereals.

Calcium supplements are discussed in Highlight 12.

A generalization that has been gaining strength throughout this book is supported by the information given here about calcium. A balanced diet that supplies a variety of foods is the best assurance of adequacy for all essential nutrients. All food groups should be included, and none should be overused. In our culture, calcium is usually lacking wherever milk is underemphasized in the diet—whether through ignorance, poverty, simple dislike, fad dieting, lactose intolerance, or allergy. By contrast, iron is usually lacking whenever milk is overemphasized, as Chapter 13 will show.

**Figure 12-9**
**Calcium in Selected Foods**

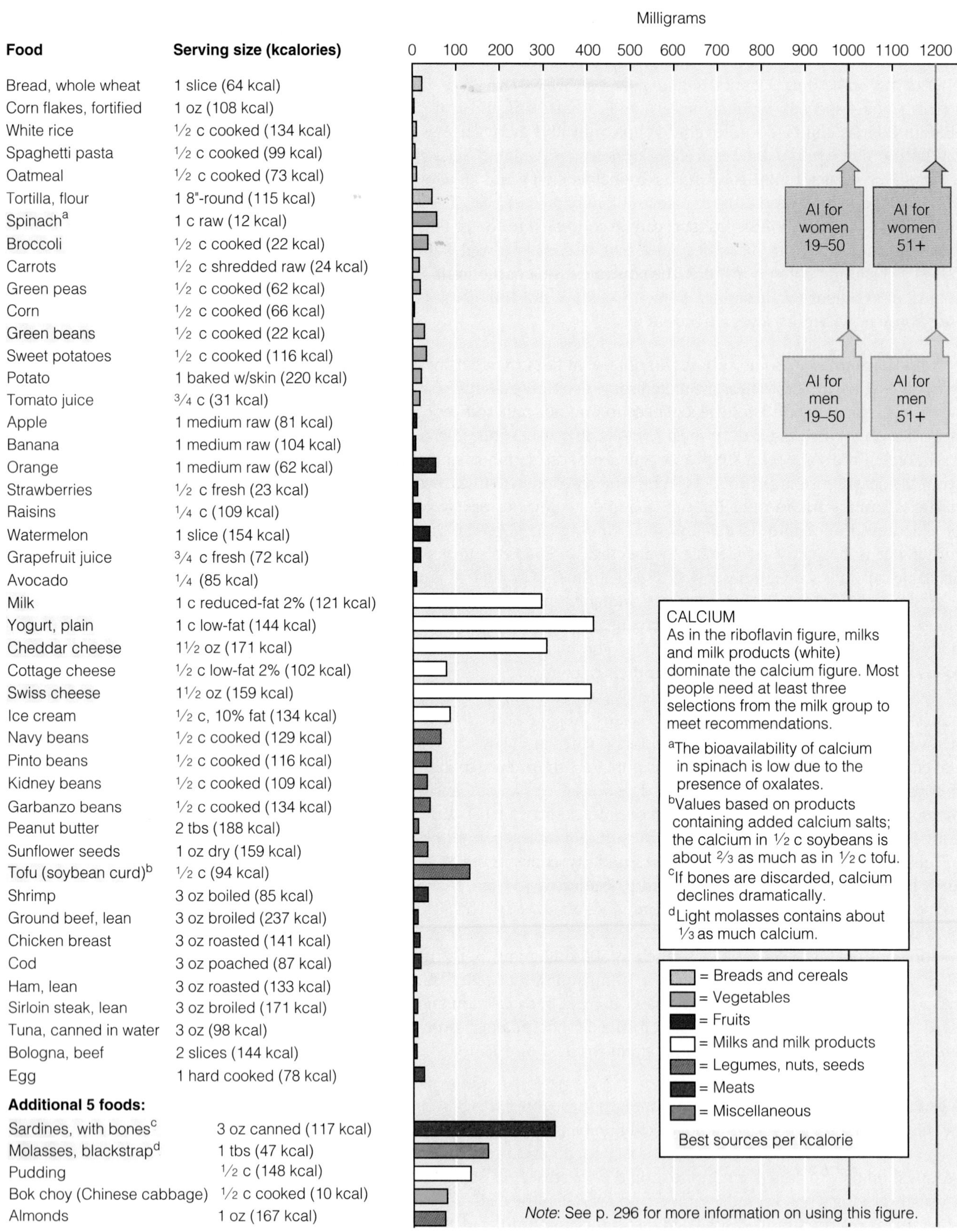

## Calcium Deficiency

A low calcium intake during the growing years limits the bones' ability to achieve an optimal mass and density. Most people achieve a peak bone mass by about age 30, and dense bones best protect against age-related bone loss and fracture. All adults lose bone as they grow older, beginning before they are 40. When bone loss reaches the point at which bones fracture under common, everyday stresses, the condition is known as osteoporosis. Osteoporosis afflicts more than 25 million people in the United States, mostly older women.

Unlike many diseases that make themselves known through symptoms such as pain, shortness of breath, skin lesions, tiredness, and the like, osteoporosis is silent. The body sends no signals saying bone loss is occurring. Blood samples offer no clues because blood calcium remains normal regardless of bone content, and measures of bone density are rarely taken. Highlight 12 suggests strategies to protect against bone loss, of which eating calcium-rich foods is only one.

**peak bone mass:** the highest attainable bone density for an individual, developed during the first three decades of life.

**osteoporosis** (OS-tee-oh-pore-OH-sis): a condition of older persons in which the bones become porous and fragile due to a loss of minerals; also called **adult bone loss.**
- **osteo** = bone
- **porosis** = porous

IN SUMMARY

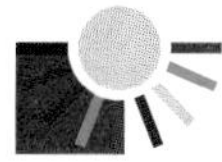

Most of the body's calcium is in the bones where it provides a rigid structure and a reservoir of calcium for the blood. Blood calcium participates in muscle contraction, blood clotting, and nerve impulses and is closely regulated by a system of hormones and vitamin D. Calcium is found predominantly in milk and milk products, but some other foods including certain vegetables and tofu also provide calcium. Even when a calcium deficiency exists, blood calcium remains normal, but at the expense of bone loss, which can lead to osteoporosis.

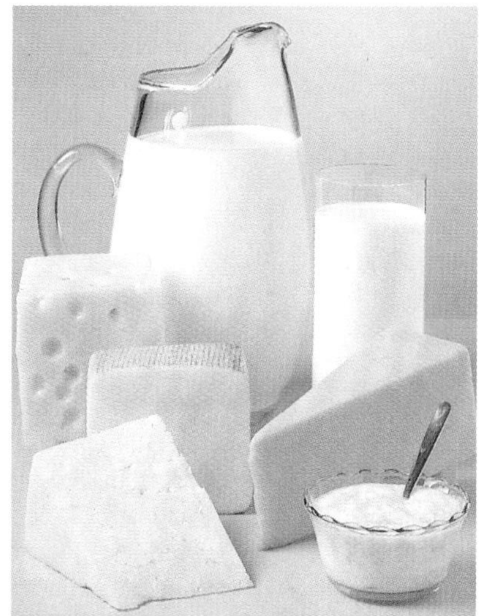

*Milk and milk products are rightly famous for their calcium contents.*

# Phosphorus

Phosphorus is the second most abundant mineral in the body. About 85 percent of it is found combined with calcium in the hydroxyapatite crystals of bones and teeth.

**phosphorus:** a major mineral found mostly in the body's bones and teeth.

• ***Phosphorus Roles in the Body*** • Phosphorus salts (phosphates) are found not only in bones and teeth, but in all body cells as part of a major buffer system (phosphoric acid and its salts). Phosphorus is also part of DNA and RNA and is therefore necessary for all growth.

Phosphorus assists in energy metabolism. Many enzymes and the B vitamins become active only when a phosphate group is attached. ATP itself, the energy carrier of the cells, uses three phosphate groups to do its work.

Lipids containing phosphorus as part of their structures (phospholipids) help to transport other lipids in the blood. Phospholipids are also the major structural components of cell membranes, where they affect the transport of nutrients into and out of the cells. Some proteins, such as the casein in milk, contain phosphorus as part of their structures (phosphoproteins). The summary table on p. 388 lists functions and other information about phosphorus.

• ***Phosphorus Recommendations*** • Recommended intakes of phosphorus were once the same as for calcium, but that changed when they were recently revised. Diets that provide adequate energy and protein also supply adequate phosphorus.

• ***Phosphorus Intakes*** • Dietary deficiencies of phosphorus are unknown. As Figure 12-10 (on p. 388) shows, foods rich in proteins are the best sources of phosphorus. In addition to legumes and foods from the milk and meat groups, processed foods (including soft drinks) are usually high in phosphorus. Phosphorus from additives in processed foods can add significantly to people's intakes.

**Figure 12-10**
**Phosphorus in Selected Foods**

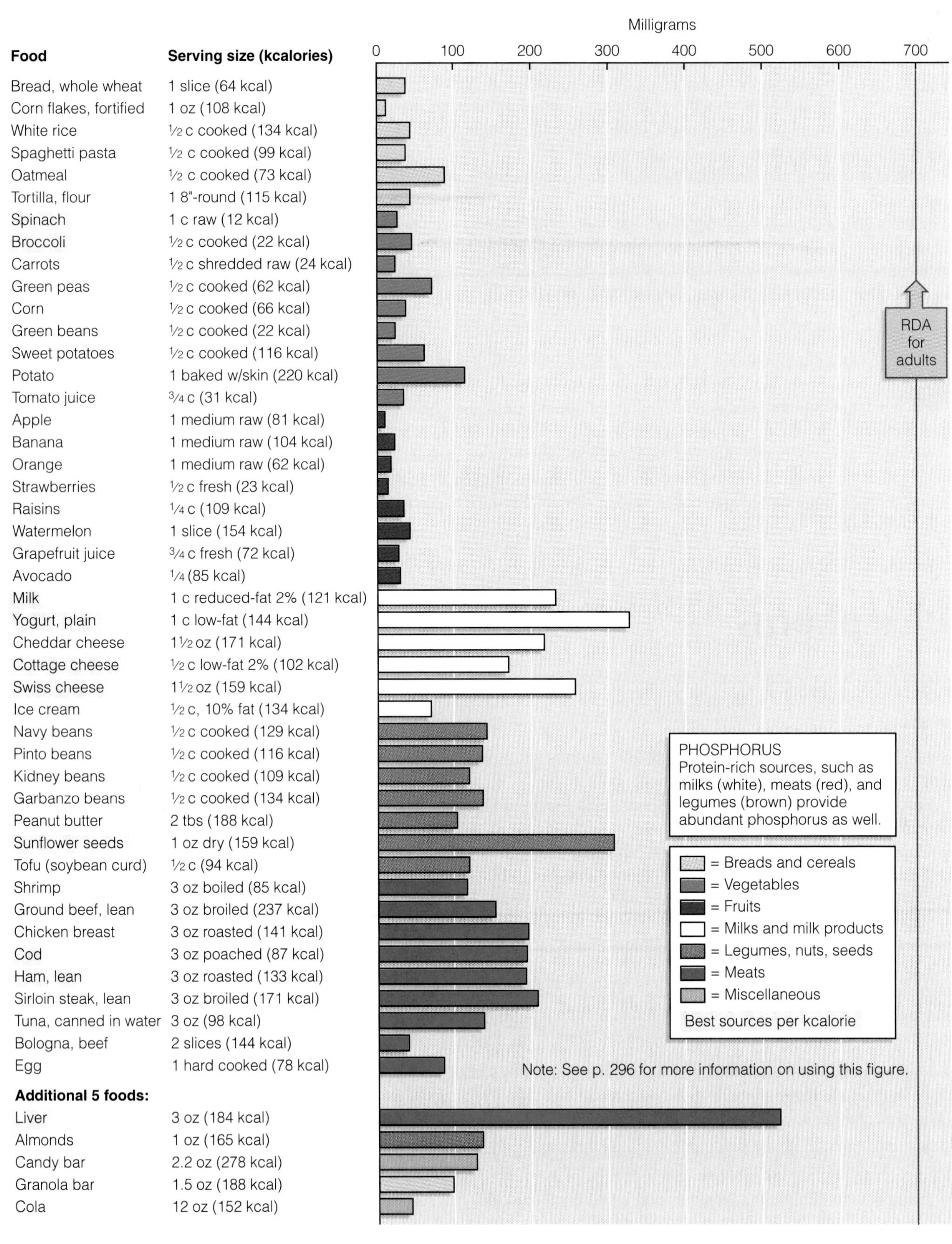

IN SUMMARY

**Phosphorus**

| 1997 RDA | Upper Level | Chief Functions in the Body | Deficiency Symptoms | Toxicity Symptoms | Significant Sources |
|---|---|---|---|---|---|
| Adults: 700 mg/day | Adults (19–70 yr): 4000 mg/day | Mineralization of bones and teeth; part of every cell; important in genetic material, part of phospholipids, used in energy transfer and in buffer systems that maintain acid-base balance | Weakness, bone pain[a] | Low blood calcium levels | All animal tissues (meat, fish, poultry, eggs, milk) |

[a]Dietary deficiency rarely occurs, but some drugs can bind with phosphorus making it unavailable and resulting in bone loss that is characterized by weakness and pain.

In the past, researchers emphasized the importance of an ideal calcium-to-phosphorus ratio from the diet to support calcium metabolism, but there is little or no evidence to support this concept.[22] A wide range of ratios has proved to have no effect on calcium balance. The quantities of calcium and phosphorus in the diet are far more important than their ratio to each other.

IN SUMMARY

Phosphorus accompanies calcium both in the crystals of bone and in many foods such as milk. Phosphorus is also important in energy metabolism, as part of phospholipids, and as part of the genetic materials DNA and RNA.

# Magnesium

Magnesium barely qualifies as a major mineral: only about 1 ounce of magnesium is present in the body of a 130-pound person. Over half of the body's magnesium is in the bones. Most of the rest is in the muscles and soft tissues, with only 1 percent in the extracellular fluid. Bone magnesium seems to be a reservoir to ensure that some will be on hand for vital reactions, regardless of recent dietary intake.

**magnesium:** a cation within the body's cells, active in many enzyme systems.

• ***Magnesium Roles in the Body*** • Magnesium acts in all the cells of the soft tissues, where it forms part of the protein-making machinery and is necessary for energy metabolism. It participates in hundreds of enzyme systems. A major role is as a catalyst in the reaction that adds the last phosphate to the high-energy compound ATP. As a required component for ATP metabolism, magnesium is essential to the body's use of glucose; the synthesis of protein, fat, and nucleic acids; and the cells' membrane transport systems. Together with calcium, magnesium is involved in muscle contraction and blood clotting: calcium promotes the processes, whereas magnesium inhibits them. This dynamic interaction between the two minerals helps regulate blood pressure and the functioning of the lungs. Magnesium also helps prevent dental caries by holding calcium in tooth enamel. Like many other nutrients, magnesium supports the normal functioning of the immune system. The table on p. 390 offers a summary.

• ***Magnesium Intakes*** • Dietary magnesium intakes average about three-quarters of the recommended intake for U.S. adults. Dietary intake data, however, do not

IN SUMMARY

**Magnesium**

| 1997 RDA | Upper Level | Chief Functions in the Body | Deficiency Symptoms | Toxicity Symptoms | Significant Sources |
|---|---|---|---|---|---|
| Men (19–30 yr): 400 mg/day<br>Women (19–30 yr): 310 mg/day | Adults: 350 mg nonfood magnesium/day | Bone mineralization, building of protein, enzyme action, normal muscle contraction, nerve impulse transmission, maintenance of teeth, and functioning of immune system | Weakness; confusion; if extreme, convulsions, bizarre muscle movements (especially of eye and face muscles), hallucinations, and difficulty in swallowing; in children, growth failure[a] | Not known; large doses have been taken in the form of the laxative Epsom salts without ill effects except diarrhea | Nuts, legumes, whole grains, dark green vegetables, seafood, chocolate, cocoa |

[a]A still more severe deficiency causes tetany, an extreme, prolonged contraction of the muscles similar to that caused by low blood calcium.

include the contribution made by water. In some areas, the water contains calcium and magnesium ("hard" water) and contributes significantly to intakes.

The brown bars in Figure 12-11 indicate that legumes, seeds, and nuts make significant magnesium contributions. Magnesium is part of the chlorophyll molecule, so leafy green vegetables are also good sources.

• ***Magnesium Deficiency*** • Even when average magnesium intakes are below recommendations, deficiency symptoms rarely appear except with diseases. Magnesium deficiency may develop in cases of alcohol abuse, protein malnutrition, kidney or endocrine disorders, and prolonged vomiting or diarrhea. People using diuretics may also show symptoms. A severe magnesium deficiency causes a tetany similar to the calcium tetany described earlier. Magnesium deficiencies also impair central nervous system activity and may be responsible for the hallucinations experienced by people withdrawing from alcohol intoxication.

• ***Magnesium and Hypertension*** • Magnesium is critical to heart function and seems to protect against hypertension and heart disease. Interestingly, people living in areas of the country with "hard" water, which contains high concentrations of calcium and magnesium, tend to have low rates of heart disease. With magnesium deficiency, the walls of the arteries and capillaries tend to constrict, a possible mechanism for the hypertensive effect.

IN SUMMARY

Like calcium and phosphorus, magnesium supports bone mineralization. Magnesium is also involved in numerous enzyme systems and in heart function. It is found abundantly in legumes and leafy green vegetables and, in some areas, in water.

# Sulfur

**sulfur:** a mineral present in the body as part of some amino acids.

The body does not use sulfur by itself as a nutrient (see the summary table on p. 392). Sulfur is mentioned here because it occurs in essential nutrients that the body does use, such as the B vitamin thiamin and the amino acids methionine and cysteine. Sulfur plays a well-known role in determining the contour of protein molecules. The sulfur-containing side chains in cysteine molecules can link to each other, forming disulfide bridges, which stabilize the protein structure (see the drawing of insulin with its disulfide bridges on p. 163). Skin, hair, and nails contain some of the body's more rigid proteins, which have a high sulfur content.

**Figure 12-11**
**Magnesium in Selected Foods**

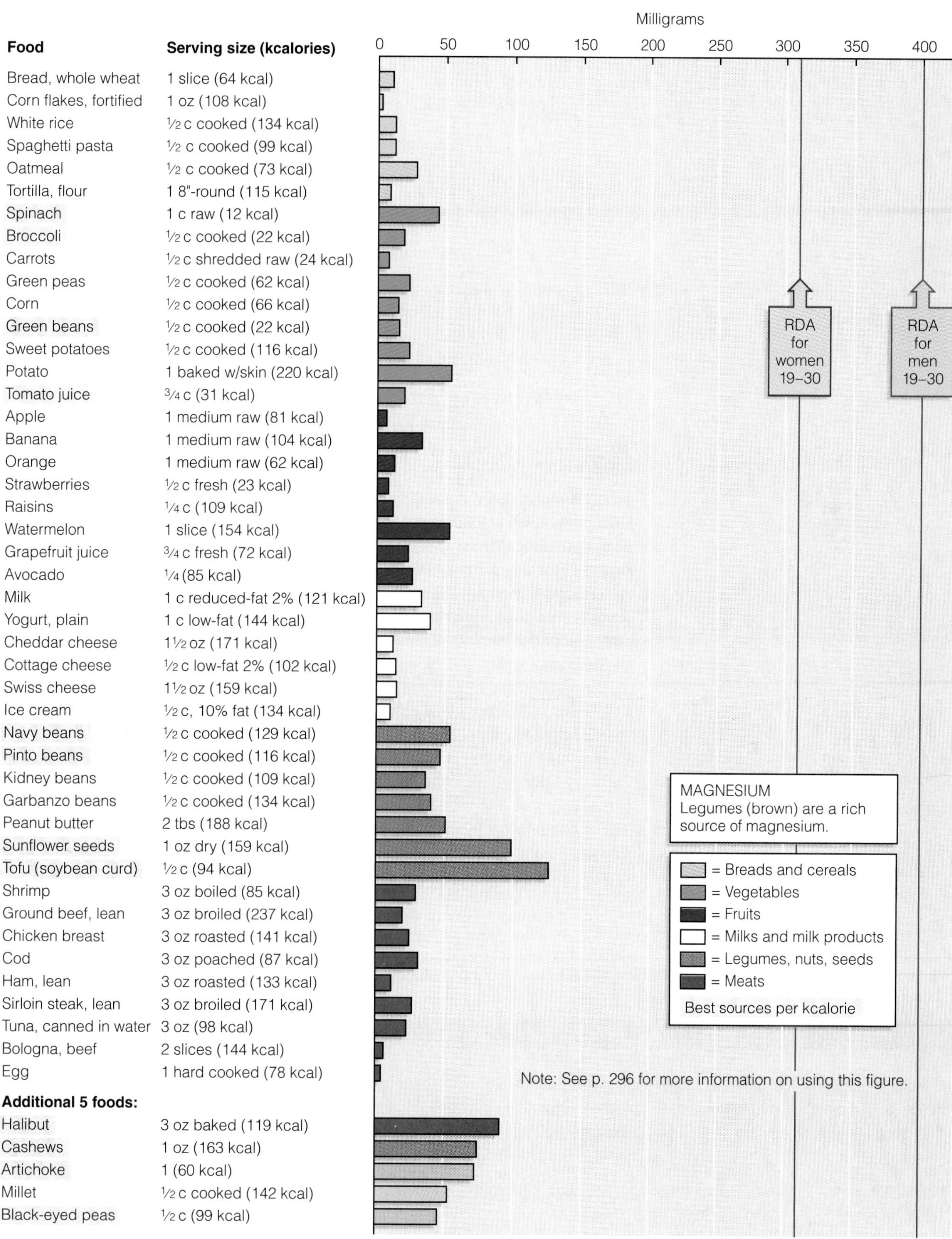

IN SUMMARY

**Sulfur**

| Chief Functions in the Body | Deficiency Symptoms | Toxicity Symptoms | Significant Sources |
|---|---|---|---|
| As part of proteins, stabilizes their shape by forming disulfide bridges; part of the vitamins biotin and thiamin and the hormone insulin | None known; protein deficiency would occur first | Toxicity would occur only if sulfur-containing amino acids were eaten in excess; this (in animals) depresses growth | All protein-containing foods (meats, fish, poultry, eggs, milk, legumes, nuts) |

There is no recommended intake for sulfur, and no deficiencies are known. Only when people lack protein to the point of severe deficiency will they lack the sulfur-containing amino acids.

IN SUMMARY

Like the other nutrients, the minerals' actions are coordinated to get the body's work done. The major minerals, especially sodium, chloride, and potassium, influence the body's fluid balance; whenever an anion moves, a cation moves—always maintaining homeostasis. Sodium, chloride, potassium, calcium, and magnesium are key members of the team of nutrients that direct nerve transmission and muscle contraction; they are also the primary nutrients involved in regulating blood pressure.[23] Phosphorus and magnesium participate in many reactions involving glucose, fatty acids, amino acids, and the vitamins. Calcium, phosphorus, and magnesium combine to form the structure of the bones and teeth. Each major mineral also plays other specific roles in the body.

With all of the tasks these minerals perform, they are of great importance to life. Consuming enough of each of them every day is easy, given a variety of foods from each of the food groups. Whole-grain breads supply magnesium; fruits, vegetables, and legumes also provide magnesium and potassium, too; milks offer calcium and phosphorus; meats also offer phosphorus and sulfur as well; all foods provide sodium and chloride, excesses being more problematic than inadequacies. The message is quite simple and has been repeated throughout this text: for an adequate intake of all the nutrients, including the major minerals, choose different foods from each of the five food groups. And drink plenty of water.

## Making It Click

These problems give you practice in nutrition-related math and an appreciation for the minerals in foods. Be sure to show your calculations for each problem (see Appendix K for answers).

1. For each of these minerals, note the unit of measure:
   Calcium Magnesium Phosphorus
   Potassium Sodium

2. Learn to appreciate calcium-dense foods. Following is a list of foods ranked in order of their calcium contents per serving.
   a. Which foods offer the most calcium per kcalorie? To calculate calcium density, divide calcium (mg) by energy (kcal). Record your answer in the table (round your answers); the first one is done for you.

## Making It Click (continued)

| Food | Calcium (mg) | Energy (kcal) | Calcium Density (mg/kcal) |
|---|---|---|---|
| Sardines, 3 oz canned | 325 | 176 | 1.85 |
| Milk, nonfat, 1 c | 301 | 85 | |
| Cheddar cheese, 1 oz | 204 | 114 | |
| Salmon, 3 oz canned | 182 | 118 | |
| Broccoli, cooked from fresh, chopped, ½ c | 36 | 22 | |
| Sweet potato, baked in skin, 1 ea | 32 | 140 | |
| Cantaloupe melon, ½ | 29 | 93 | |
| Whole-wheat bread, 1 slice | 21 | 64 | |
| Apple, 1 medium | 15 | 125 | |
| Sirloin steak, lean, 3 oz | 9 | 171 | |

b. The top five items ranked in order of calcium contents per serving are sardines > milk > cheese > salmon > broccoli. What are the top five items in order of calcium contents per kcalorie?

This information should convince you that milk, milk products, fish eaten with their bones, and dark green vegetables are the best choices for calcium.

3. a. Consider how the rate of absorption influences the amount of calcium available for the body's use. Use the drawing on p. 385 to determine how much calcium the body actually receives from the following foods by multiplying the milligrams of calcium in the food by the percentage absorbed. The first one is done for you.

| Food | Calcium in the Food (mg) | Absorption Rate (%) | Calcium in the Body (mg) |
|---|---|---|---|
| Cauliflower, ½ c cooked, fresh | 10 | ≥ 50 | ≥ 5 |
| Broccoli, ½ c cooked, fresh | 36 | | |
| Milk, 1 c 1% low-fat | 300 | | |
| Almonds, 1 oz | 75 | | |
| Spinach, 1 c raw | 55 | | |

b. To appreciate how the absorption rate influences the amount of calcium available to the body, compare broccoli with almonds. Which provides more calcium in foods and to the body?

c. To appreciate how the calcium content of foods influences the amount of calcium available to the body, compare cauliflower with milk. How much cauliflower would a person have to eat to receive an equivalent amount of calcium as from one cup of milk? How does your answer change when you account for differences in their absorption rates?

## Study Questions

These questions will help you review the chapter. You will find the answers in the discussions on the pages provided.

1. List the roles of water in the body. (p. 367)
2. List the sources of water intake and routes of water excretion. (pp. 367–368)
3. What is ADH? Where does it exert its action? What is aldosterone? How does it work? (p. 369)
4. How does the body use electrolytes to regulate fluid balance? (pp. 370–372)
5. What do the terms *major* and *trace* mean when describing the minerals in the body? (p. 374)
6. Describe some characteristics of minerals that distinguish them from vitamins. (pp. 374–375)
7. What is the major function of sodium in the body? Describe how the kidneys regulate blood sodium. Is a dietary deficiency of sodium likely? Why? (pp. 375–378)
8. List calcium's roles in the body. How does the body keep blood calcium constant regardless of intake? (pp. 382–383)
9. Name significant food sources of calcium. What are the consequences of inadequate intakes? (pp. 385–387)
10. List the roles of phosphorus in the body. Discuss the relationships between calcium and phosphorus. Is a dietary deficiency of phosphorus likely? Why? (pp. 387, 389)
11. State the major functions of chloride, potassium, magnesium, and sulfur in the body. Are deficiencies of these nutrients likely to occur in your own diet? Why? (pp. 378, 379–380, 388, 390, 392)

These questions will help you prepare for an exam. Answers can be found in Appendix K.

1. The body generates water during the:
   a. buffering of acids.
   b. dismantling of bone.
   c. metabolism of minerals.
   d. breakdown of energy nutrients.
2. Regulation of fluid and electrolyte balance and acid-base balance depends primarily on the:
   a. kidneys.
   b. intestines.
   c. sweat glands.
   d. specialized tear ducts.

3. The distinction between the major and trace minerals reflects the:
   a. ability of their ions to form salts.
   b. amounts of their contents in the body.
   c. importance of their functions in the body.
   d. capacity to retain their identity after absorption.
4. The principal cation in extracellular fluids is:
   a. sodium.
   b. chloride.
   c. potassium.
   d. phosphorus.
5. The role of chloride in the stomach is to help:
   a. support nerve impulses.
   b. convey hormonal messages.
   c. maintain a strong acidity.
   d. assist in muscular contractions.
6. Which would provide the most potassium?
   a. bologna
   b. potatoes
   c. pickles
   d. whole-wheat bread
7. Calcium homeostasis depends on:
   a. vitamin K, aldosterone, and renin.
   b. vitamin K, parathormone, and renin.
   c. vitamin D, aldosterone, and calcitonin.
   d. vitamin D, calcitonin, and parathormone.
8. Calcium absorption is hindered by:
   a. lactose.
   b. oxalates.
   c. vitamin D.
   d. stomach acid.
9. Phosphorus assists in many activities in the body, but *not:*
   a. energy metabolism.
   b. the clotting of blood.
   c. the transport of lipids.
   d. bone and teeth formation.
10. Most of the body's magnesium can be found in the:
   a. bones.
   b. nerves.
   c. muscles.
   d. extracellular fluids.

## Notes


1. Committee on Dietary Allowances, *Recommended Dietary Allowances,* 10th ed. (Washington, D.C.: National Academy Press, 1989), pp. 247–261.
2. Pamphlet from The American Dietetic Association, Water: The beverage of life, 1994.
3. R. D. Mattes, The taste of salt in humans, *American Journal of Clinical Nutrition* 65 (1997): 692S–697S.
4. P. A. S. Breslin and G. K. Beauchamp, Salt enhances flavour by suppressing bitterness, *Nature* 387 (1997): 563.
5. T. A. Kotchen and J. M. Kotchen, Dietary sodium and blood pressure: Interactions with other nutrients, *American Journal of Clinical Nutrition* 65 (1997): 708S–711S.
6. T. C. Beard and coauthors, Association between blood pressure and dietary factors in the Dietary and National Survey of British Adults, *Archives of Internal Medicine* 157 (1997): 234–238.
7. The Sixth Report of the Joint National Committee on Prevention, Detection, Evaluation, and Treatment of High Blood Pressure (NIH publication, November 1997), p. 20.
8. A. W. Cowley, Jr., Genetic and nongenetic determinants of salt sensitivity and blood pressure, *American Journal of Clinical Nutrition* 65 (1997): 587S–593S.
9. F. C. Luft and M. H. Weinberger, Heterogeneous responses to changes in dietary salt intake: The salt-sensitivity paradigm, *American Journal of Clinical Nutrition* 65 (1997): 612S–617S.
10. P. K. Whelton and coauthors, Sodium restriction and weight loss in the treatment of hypertension in older persons: A randomized controlled Trial of Nonpharmacologic Interventions in the Elderly (TONE), *Journal of the American Medical Association* 279 (1998): 839–846; P. K. Whelton and coauthors, Efficacy of nonpharmacologic interventions in adults with high-normal blood pressure: Results from phase 1 of the Trials of Hypertension Prevention, *American Journal of Clinical Nutrition* 65 (1997): 652S–660S.
11. J. P. Midgley and coauthors, Effect of reduced dietary sodium on blood pressure: A meta-analysis of randomized controlled trials, *Journal of the American Medical Association* 275 (1996): 1590–1597; Letters: Dietary sodium and blood pressure, *Journal of the American Medical Association* 276 (1996): 1467–1470.
12. A. Devine and coauthors, A longitudinal study of the effect of sodium and calcium intakes on regional bone density in postmenopausal women, *American Journal of Clinical Nutrition* 62 (1995): 740–745; R. Itoh and Y. Suyama, Sodium excretion in relation to calcium and hydroxyproline excretion in a healthy Japanese population, *American Journal of Clinical Nutrition* 63 (1996): 735–740.
13. Committee on Diet and Health, Food and Nutrition Board, *Diet and Health: Implications for Reducing Chronic Disease Risk* (Washington, D.C.: National Academy Press, 1989), pp. 99–135.
14. P. K. Whelton and coauthors, Effects of oral potassium on blood pressure: Meta-analysis of randomized controlled clinical trial, *Journal of the American Medical Association* 277 (1997): 1624–1632.
15. H. C. Bucher and coauthors, Effects of dietary calcium supplementation on blood pressure: A meta-analysis of randomized controlled trials, *New England Journal of Medicine* 275 (1996): 1016–1022; D. A. McCarron and D. Hatton, Dietary calcium and lower blood pressure—We can all benefit, *New England Journal of Medicine* 275 (1996): 1128–1129; C. G. Osborne and coauthors, Evidence for the relationship of calcium to blood pressure, *Nutrition Reviews* 54 (1996): 365–381.
16. D. A. McCarron, Role of adequate dietary calcium intake in the prevention and management of salt-sensitive hypertension, *American Journal of Clinical Nutrition* 65 (1997): 712S–716S.

17. W. A. Levey and coauthors, Blood pressure responses of white men with hypertension to two low-sodium metabolic diets with different levels of dietary calcium, *Journal of the American Dietetic Association* 95 (1995): 1280–1287.

18. K. O. O'Brien and coauthors, Increased efficiency of calcium absorption during short periods of inadequate calcium intake in girls, *American Journal of Clinical Nutrition* 63 (1996): 579–583.

19. A. Devine, R. L. Prince, and R. Bell, Nutritional effect of calcium supplementation by skim milk powder or calcium tablets on total nutrient intake in postmenopausal women, *American Journal of Clinical Nutrition* 64 (1996): 731–737.

20. N. K. Christensen and coauthors, Juniper ash as a source of calcium in the Navajo diet, *Journal of the American Dietetic Association* 98 (1998): 333–334.

21. F. Couzy and coauthors, Calcium bioavailability from calcium- and sulfate-rich mineral water, compared with milk, in young adult women, *American Journal of Clinical Nutrition* 62 (1995): 1239–1244.

22. Committee on Dietary Reference Intakes, *Dietary Reference Intakes for Calcium, Phosphorus, Magnesium, Vitamin D, and Fluoride* (Washington, D.C.: National Academy Press, 1997), pp. 5-6–5-7.

23. M. E. Reusser and D. A. McCarron, Micronutrient effects on blood pressure regulation, *Nutrition Reviews* 52 (1994): 367–375.

# Osteoporosis and Calcium

Osteoporosis develops without warning. It often first becomes apparent when someone's hip suddenly gives way. People say, "She fell and broke her hip," but in fact the hip may have been so fragile that it broke *before* she fell. Even bumping into a table may be enough to shatter a porous bone into fragments so numerous and scattered that they cannot be reassembled. Removing them and replacing them with an artificial joint requires major surgery. About a fifth of the patients die of complications within a year. Half of those who survive will never walk independently again. Their quality of life slips downward.[1]

This highlight examines osteoporosis, one of the most prevalent diseases of aging, affecting more than 25 million people in the United States—most of them women. It reviews the many factors that contribute to the 1 million breaks in the bones of the hips, wrists, arms, and ankles each year.[2] And it presents strategies to reduce the risks, paying special attention to the role of dietary calcium.

## Bone Development and Disintegration

Bone has two compartments: the outer, hard shell of cortical bone, and the inner, lacy structural matrix of trabecular bone. Both can lose minerals, but in different ways and at different rates. (The accompanying glossary defines relevant terms.) The opening photograph shows a human leg bone sliced lengthwise, exposing the lacy, calcium-containing crystals of trabecular bone. These crystals give up calcium to the blood when the day's supply from the diet runs short, and they take up calcium again when the dietary supply is plentiful. For people who have eaten calcium-rich foods throughout the bone-forming years of their childhood and young adulthood, these deposits provide an abundant source of calcium.

Surrounding and protecting the trabecular bone is a dense, ivorylike exterior shell—the cortical bone. Cortical bone composes the shafts of the long bones, and a thin cortical shell caps the end of the bone, too. Both compartments confer strength on bone: cortical bone provides the sturdy outer wall, while trabecular bone provides support along the lines of stress.

The two types of bone play different roles in calcium balance and osteoporosis. Supplied with blood vessels and metabolically active, trabecular bone is sensitive to hormones that govern day-to-day deposits and withdrawals of calcium. It readily gives up minerals whenever blood calcium needs replenishing. Losses of trabecular bone start becoming significant for men and women in their 30s, although losses can occur whenever calcium withdrawals exceed deposits.

*Trabecular bone is the lacy network of calcium-containing crystals that fills the interior. Cortical bone is the dense, ivorylike bone that forms the exterior shell.*

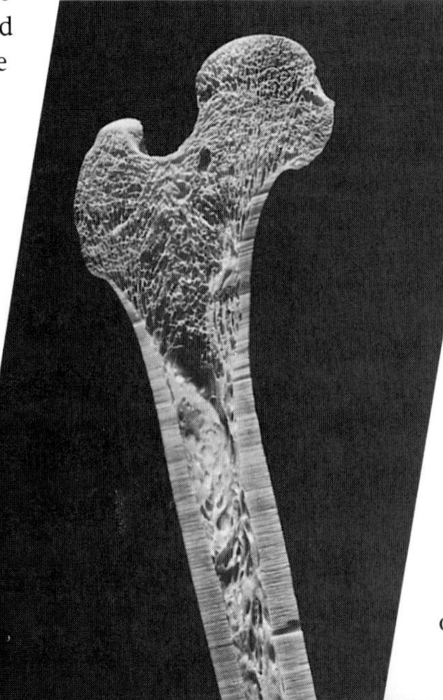

Cortical bone also gives up calcium, but slowly and at a steady pace. Cortical bone losses typically begin at about 40 years of age and continue slowly but surely thereafter.

Losses of trabecular and cortical bone reflect two types of osteoporosis, which cause two types of bone breaks. Type I osteoporosis involves losses of trabecular bone (see Figure H12-1). These losses sometimes exceed three times the expected rate, and bone breaks may occur suddenly. Trabecular bones become so fragile that even the body's own weight can overburden the spine—vertebrae may suddenly disintegrate and crush down, painfully pinching major nerves. Wrists may break as bone ends weaken, and teeth may loosen or fall out as the trabecular bone of the jaw recedes. Women are most often the victims of this type of osteoporosis, outnumbering men six to one. Taking estrogen for at least seven years after menopause is the most effective preventive measure against this type of osteoporosis.[3]

In type II osteoporosis, the calcium of both cortical and trabecular bone is drawn out of storage, but slowly over the years. As old age approaches, the vertebrae may compress into wedge shapes, forming what is often called "dowager's hump," the posture many older people assume as they "grow shorter." Figure H12-2 (on p. 397) shows the effect of compressed spinal bone on a woman's height and posture. Because both the cortical shell and the trabecular interior weaken, breaks most often occur in the hip, as in the opening example. A woman is twice as likely as a man to suffer type II osteoporosis. Table H12-1 (on p. 398) summarizes the differences between the two types of osteoporosis.

Whether a person develops osteoporosis seems to depend partly on heredity and partly on other factors, including nutrition. The strongest predictor of bone density is age. Of the quarter million hip fractures in the United States each year, 90 percent occur in people over the age of 50.[4]

**Figure H12-1**

**Healthy and Osteoporotic Trabecular Bones**

Electron micrograph of healthy trabecular bone.

Electron micrograph of trabecular bone affected by osteoporosis.

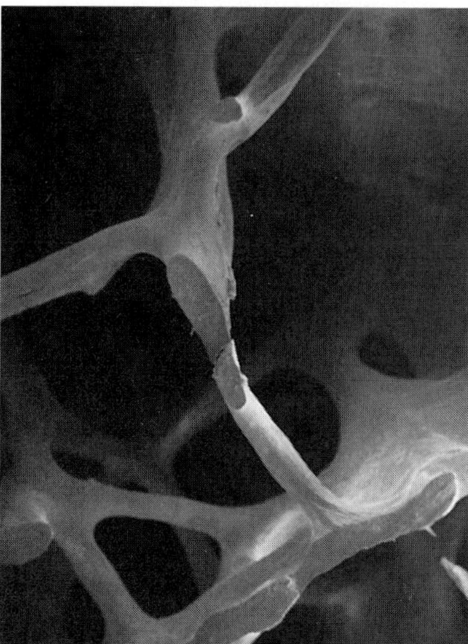

**Figure H12-2**

**Loss of Height in a Woman Caused by Osteoporosis**

The woman on the left is about 50 years old. On the right, she is 80 years old. Her legs have not grown shorter: only her back has lost length, due to collapse of her spinal bones (vertebrae). Collapsed vertebrae cannot protect the spinal nerves from pressure that causes excruciating pain.

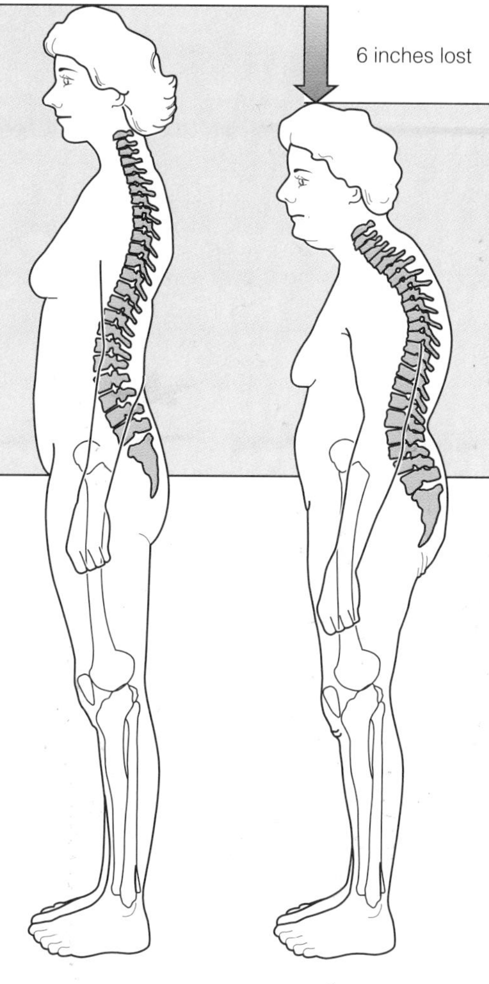


## *Age and Bone Calcium*

During the first two decades of life, the skeleton grows stronger and denser as it accumulates minerals. Then, in the late 20s to early 30s, the bones stop growing. As the years pass, the cells that build bone gradually become less active, while those that dismantle bone continue working. Thus, with advancing age, bones lose strength and density (see Figure H12-3 on p. 398).

*Young adults wisely drink milk now to defend against bone loss later.*

Calcium intakes of older adults are typically low, and calcium absorption declines after about the age of 65 years. The kidneys do not activate vitamin D as well as they did earlier (recall that active vitamin D enhances calcium absorption). Also, sunlight is needed to form vitamin D, and many older people spend little or no time outdoors in the sunshine. For these reasons, and because intakes of vitamin D are typically low anyway, blood vitamin D decreases. Supplementation with calcium and vitamin D reduces bone loss and the risk of fractures.[5] One leading researcher estimates that as much as half of our osteoporosis problem could be eliminated with adequate calcium and vitamin D intakes.[6]

Some of the hormones that regulate bone and calcium metabolism also change with age and accelerate

### Glossary

**bone density:** a measure of bone strength. When minerals fill the bone matrix (making it dense), they give it strength.

**cortical bone:** the very dense bone tissue that forms the outer shell surrounding trabecular bone and comprises the shaft of a long bone.

**trabecular** (tra-BECK-you-lar) **bone:** the lacy inner structure of calcium crystals that supports the bone's structure and provides a calcium storage bank.

**type I osteoporosis:** osteoporosis characterized by rapid bone losses, primarily of trabecular bone.

**type II osteoporosis:** osteoporosis characterized by gradual losses of both trabecular and cortical bone.

**Table H12-1**
**Types of Osteoporosis Compared**

| | Type I | Type II |
|---|---|---|
| Other name | Postmenopausal osteoporosis | Senile osteoporosis |
| Age of onset | 50 to 70 years old | 70 years and older |
| Bone loss | Trabecular bone | Both trabecular and cortical bone |
| Fracture sites | Wrist and spine | Hip |
| Gender incidence | 6 women to 1 man | 2 women to 1 man |
| Primary causes | Rapid loss of estrogen in women following menopause; loss of testosterone in men with advancing age | Reduced calcium absorption, increased bone mineral loss, increased propensity to fall |

Source: Adapted from C. Niewoehner, Calcium and osteoporosis, *Cereal Foods World* 33 (1988): 784–787.

bone mineral withdrawal.* Together, these age-related factors probably contribute to bone loss: inefficient bone remodeling, reduced calcium intakes, impaired calcium absorption, poor vitamin D status, physical inactivity, and hormonal changes that favor bone mineral withdrawal.

## Sex and Hormones

After age, sex is the next strongest predictor of loss of bone density with aging: men have greater bone density than women at maturity, and women have greater losses than men in later life. Consequently, women account for four out of five cases of osteoporosis. Menopause imposes special perils on women's bones. Bone dwindles rapidly when the hormone estrogen diminishes and menstruation ceases. Accelerated losses continue for 6 to 8 years following menopause, then taper off, so that women again lose bone at the same rate as men their age (see Figure H12-4 on p. 399). Losses of bone minerals continue throughout the remainder of a woman's lifetime, but not at the free-fall pace of the menopause years.

When *young* women experience reduced estrogen secretion and cease menstruating, they, too, lose bone rapidly. Highlight 9 described how women who overexercise and unreasonably restrict their body weights develop amenorrhea and become susceptible to bone fractures. The combination of irregular or absent menstrual periods and low body weights explains much of the bone loss seen in young athletes. Estrogen therapy can help nonmenstruating women prevent further bone loss and reduce the incidence of fractures.[7]

If estrogen deficiency is a major cause of osteoporosis in women, what is the cause of bone loss in men? Men produce only a little estrogen, yet they resist osteoporosis better than women. Male hormones must also play a role because men suffer more fractures after removal of the testes (in cases of disease) or when their testes lose functional ability with aging. Men who have delayed puberty also appear to have more fractures, suggesting that timing of puberty influences peak bone density. Thus both male and female sex hormones appear to play roles in the development of osteoporosis.

One more complication of the estrogen theory is that some women lose bone tissue in middle age before they reach menopause. Clearly, hormones are only one of several factors affecting bones. Still, being female and experiencing menopause remain prime risk factors for the development of osteoporosis.

## Genetic Inheritance and Race

Studies of mothers and daughters confirm that heredity plays an influential role in bone density.[8] Most likely, inheritance influences both the peak bone mass achieved during growth and the extent of bone loss during menopause. The extent to which a given genetic potential is realized, however, depends on many outside factors.[9] Nutrition and physical activity, for example, can maximize peak bone density during growth, whereas alcohol and tobacco abuse can accelerate bone losses later in life.

Risks of osteoporosis appear to run along racial lines and reflect genetic differences in bone metabolism between African Americans and Caucasians.[10] African Americans use and conserve calcium more efficiently than Caucasians. Consequently, even though their calcium intakes are typically lower, black people have denser bones than white people do. Greater bone density expresses itself in a lower rate of osteoporosis among blacks. Hip fractures, for example, are

**Figure H12-3**
**Phases of Bone Development throughout Life**

The active growth phase occurs from birth to approximately age 20. The next phase of peak bone mass development occurs between the ages of 12 and 40. The final phase, when bone resorption exceeds formation, begins between age 30 and 40 and continues throughout the remainder of life.

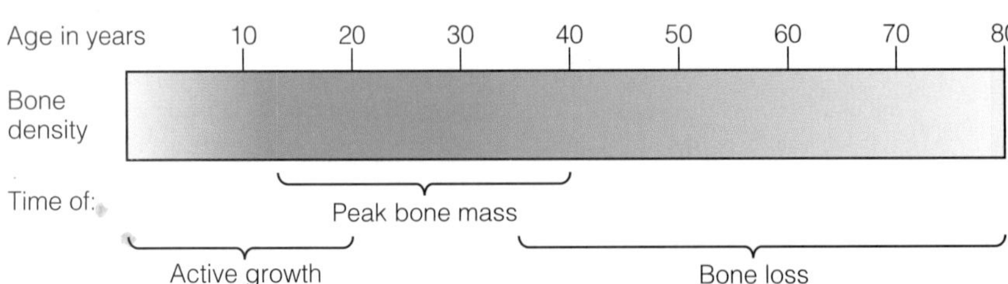


*Among the hormones suggested as influential are parathormone, calcitonin, and estrogen.

**Figure H12-4**

**High Peak Bone Mass Early in Life Postpones Osteoporosis**

Over a lifetime, women lose 30 to 40 percent of their bone mass and men, 20 to 30 percent. Can you see why it is so important for young people to consume enough calcium to attain a maximum bone mass?

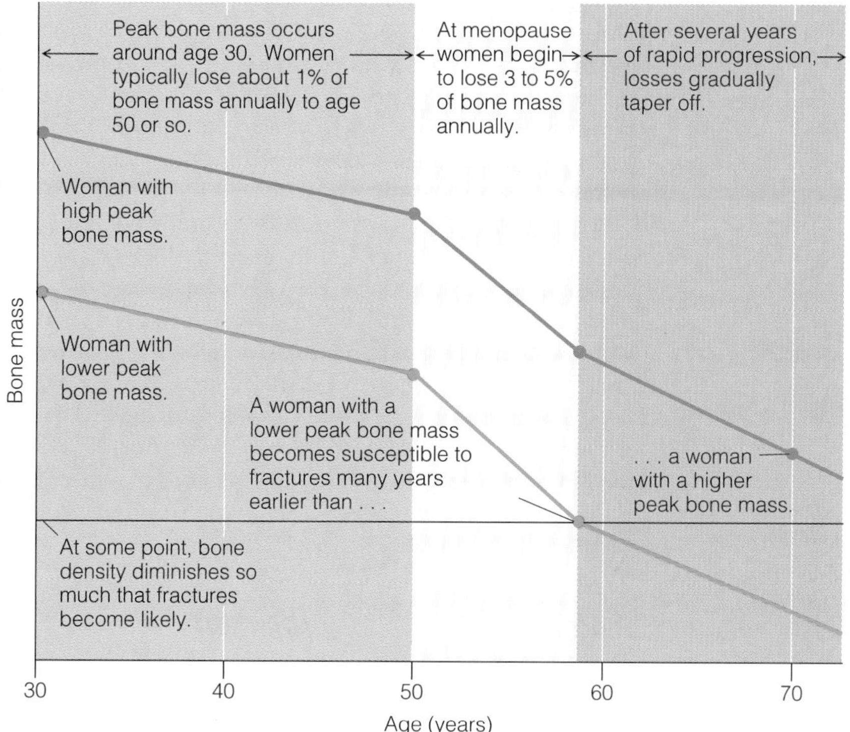
Peak bone mass occurs around age 30. Women typically lose about 1% of bone mass annually to age 50 or so.
At menopause women begin to lose 3 to 5% of bone mass annually.
After several years of rapid progression, losses gradually taper off.
Woman with high peak bone mass.
Woman with lower peak bone mass.
A woman with a lower peak bone mass becomes susceptible to fractures many years earlier than . . .
. . . a woman with a higher peak bone mass.
At some point, bone density diminishes so much that fractures become likely.
Bone mass
30
40
50
60
70
Age (years)

about three times more likely at 80 years of age in white women than in African American women.

Other ethnic groups have *less* dense bones than do many people of European heritage. Asians from China and Japan, Mexican Americans, Hispanic people from Central and South America, and Inuit people from St. Lawrence Island all have lower bone density than do people with a northern European background. One might predict that these groups would suffer more bone fractures, but this is not always the case. Again, genetic differences may explain why. Asians, for example, generally have small, compact hips, which makes them less susceptible to fractures.[11]

Genetics alone do not explain how certain ethnic groups can use no dairy products and have low calcium intakes, yet still maintain calcium balance. Part of the answer is the body's adaptation to low calcium intakes. Another part of the answer is that bone loss is not always apparent. Women in regions of China where calcium intakes are high have greater bone densities than women in other areas.[12]

Findings from around the world demonstrate that although a person's genes may lay the groundwork, environmental factors influence the genes' ultimate expression—and calcium nutrition is one of those environmental factors. Others include physical activity, body weight, smoking, and alcohol—all factors within a person's control.

## Physical Activity and Body Weight

When people are inactive—for example, when they are confined to bed—their bones lose strength just as their muscles do. Astronauts who live with reduced gravity for weeks or months at a time also lose bone and muscle strength.[13]

Muscle strength and bone strength tend to go together. When muscles work, they pull on the bones, stimulating them to develop more trabeculae and grow denser. Also, when muscles work, the hormones that promote new muscle growth also favor the building of bone. As a result, active bones are denser than sedentary bones.[14] Weight-bearing physical activity, such as walking, dancing, or jogging, is especially effective. Even modest increases in physical activity and calcium intake help to maximize bone gains in young adulthood and minimize further losses that occur with inactivity.[15]

Heavier body weights and weight gains place a similar stress on the bones and promote their maintenance. In fact, body weight, weight changes, and, to some extent, body fatness are significant and consistent predictors of bone density and risk of fractures.[16] As mentioned earlier, the combination of underweight, severely restricted energy intake, extreme daily exercise, and amenorrhea reliably predict bone loss.[17]

Cells respond, with the help of the necessary regulators, to the demands put upon them. Then they select the nutrients they need from what is offered. To increase bone density, put a demand on the bones, make them work, and then provide the raw materials from which they can grow strong: calcium, other minerals, all the nutrients in the right balance. The combination of regular physical activity and adequate calcium intake best supports bone density.[18]

## Smoking and Alcohol

Smokers experience more fractures from slight injury than do nonsmokers. Women who smoke a pack of cigarettes a day throughout adulthood lose an extra 5 to 10 percent of their bone density by menopause.[19] Although the mechanism of action remains undefined, both the lower body weights of smokers and the earlier menopause of female smokers may be factors.[20]

People who abuse alcohol often suffer from osteoporosis and experience more bone breaks than others. Several factors appear to be involved: alcohol enhances fluid excretion

**Table H12-2**
**Risk and Protective Factors for Osteoporosis**

| Risk Factors | Protective Factors |
|---|---|
| ***High Correlation*** | |
| Advanced age; postmenopausal | African American |
| Alcohol abuse | Estrogens, long-term use |
| Anorexia nervosa | |
| Caucasian or East Asian | |
| Chronic steroid use | |
| Female sex | |
| Rheumatoid arthritis | |
| Surgical removal of ovaries | |
| Thinness (<120 lb) | |
| ***Moderate Correlation*** | |
| Chronic thyroid hormone use | Having given birth |
| Cigarette smoking | High body weight |
| Diabetes (type 1) | |
| Early menopause | |
| Excessive antacid use | |
| Low-calcium diet | |
| Sedentary lifestyle or immobility | |
| Vitamin D deficiency | |
| ***Probably Important But Not Yet Proved*** | |
| Caffeine use | Alcohol taken in moderation |
| Family history of osteoporosis | High-calcium diet |
| High-fiber diet | Regular physical activity |
| High-protein diet | |
| High-sodium diet | |

that leads to excessive calcium losses in the urine; upsets the hormonal balance required for healthy bones; slows bone formation, leading to lower bone density; and increases the risk of falling. Alcohol in lesser amounts, however, does not appear to be as damaging to the bones. Bone density is actually greater in adults who consume alcohol in moderation than in those who abstain.[21]

Table H12-2 summarizes the risk factors and protective factors covered so far and includes some others, among them nutrition factors, discussed next. The more risk factors that apply to a person, the greater the chances of bone loss. Notice that several factors are more influential than diet in the development of osteoporosis.

## Calcium Nutrition

As important as calcium may be to bone health, osteoporosis is not a calcium-deficiency disease comparable to iron-deficiency anemia. In iron-deficiency anemia, high iron intakes reliably reverse the condition; in osteoporosis, high calcium intakes alone during adulthood may prevent further deterioration, but they do little or nothing to reverse bone loss.

Bone strength later in life depends most on how well the bones are developed and maintained during youth, and adequate calcium nutrition during the growing years is essential to achieving optimal peak bone mass.[22] To that end, the Committee on Dietary Reference Intakes recommends 1300 milligrams of calcium per day for everyone 9 through 18 years of age. Once the bone-losing years of middle age are reached, those who formed dense bones during youth have the advantage: they can afford to lose more bone tissue before beginning to suffer ill effects. After bone loss has begun, a person can still do a few things to maintain the bones, as discussed earlier.

Unfortunately, few girls meet the recommendations for calcium during their bone-forming years. (Boys generally obtain intakes close to those recommended because they eat more food.) Consequently, most girls start their adult lives with less than optimal bone density. As for adults, women rarely meet their recommended intakes of 1000 to 1200 milligrams from food within their energy allowances. Some authorities suggest 1500 milligrams of calcium for postmenopausal women who are not receiving estrogen, but warn that intakes exceeding 2000 milligrams a day could cause health problems.[23]

## A Perspective on Supplements

People who do not consume milk products or other calcium-rich foods in amounts that provide even half the recommended calcium may benefit from calcium supplements. During the menopausal years, calcium supplements of 1 gram may slow, but cannot fully prevent, the inevitable bone loss.[24] Supplements are commonly used as a part of therapy for osteoporosis, along with gentle exercise and, for women, estrogen replacement. Supplements should not, however, be used as a substitute for estrogen.

Anyone contemplating the use of calcium supplements should do so only on a physician's advice. Taking calcium supplements may present risks, as described in Table H12-3. If these risks are deemed acceptable, the consumer still has several decisions to make when selecting a calcium supplement.[25] Incidentally, multivitamin-mineral pills contain little or no calcium and cannot be used as calcium supplements. The label may list a few milligrams of calcium, but remember that the recommended intake is a gram or more for adults.

Supplements are available in three forms. Simplest are the purified calcium compounds, such as calcium carbonate, citrate, gluconate, lactate, malate, or phosphate, and compounds of calcium with amino acids (called amino acid chelates). Also available are mixtures of calcium with other compounds, such as calcium carbonate with magnesium carbonate, with aluminum salts (as in some antacids), or with vitamin D. Then there are powdered, calcium-rich materials such as bone meal, powdered bone, oyster shell, or dolomite (limestone). (See Table H12-4 for a description of calcium supplement terms.)

**Table H12-3**
**Problems Arising from Calcium Supplementation**

People who take calcium supplements risk:

- Impaired iron status. (Calcium inhibits iron absorption.)
- Accelerated calcium loss. (Calcium-containing antacids that also contain aluminum and magnesium hydroxide cause a net calcium loss.)
- Urinary tract stones or kidney damage in susceptible individuals. (People who have a history of kidney stones need to be monitored by a physician and to use calcium citrate supplements, which are most soluble.)
- Exposure to contaminants. (Some preparations of bone meal and dolomites are contaminated with hazardous amounts of arsenic, cadmium, mercury, and lead.)
- Vitamin D toxicity. (Vitamin D is needed to enhance calcium absorption, but continued high intakes of vitamin D, which is present in many calcium supplements, can be toxic. Users must eliminate other concentrated vitamin D sources and take enough, but not too much, vitamin D to normalize calcium absorption.)
- Excess blood calcium. (This complication is seen only with doses of calcium fourfold or more greater than customarily prescribed.)
- Milk alkali syndrome. (This condition is characterized by high blood calcium, metabolic alkalosis, and renal failure. Early symptoms include irritability, headaches, and apathy.)
- Other nutrient interactions. (Calcium inhibits absorption of magnesium, phosphorus, and zinc.)
- Drug interactions. (Calcium and tetracycline form an insoluble complex that impairs both mineral and drug absorption.)
- GI distress. (Constipation, intestinal bloating, and excess gas are especially common in older people.)

The first question to ask is how well the body absorbs and uses the calcium from various supplements. Most healthy people absorb calcium equally well—and as well as from milk—from any of these supplements: amino acids chelated with calcium; calcium phosphate dibasic; or calcium acetate, carbonate, citrate, gluconate, or lactate. People absorb calcium less well from a mixture of calcium and magnesium carbonate, oyster shell calcium fortified with inorganic magnesium, a chelated calcium-magnesium combination, or calcium carbonate fortified with vitamins and iron.

The next question to ask is how much calcium the supplement provides. To be safe, calcium intakes should not exceed 2000 milligrams a day. And remember foods also provide calcium. Read the label to find out how much a dose supplies. Calcium carbonate is 40 percent calcium, whereas calcium gluconate is only 9 percent. The user should select a low-dose supplement and take it several times a day rather than taking a large-dose supplement all at once. Taking supplements in doses of 500 milligrams or less improves absorption.

Then consider that when manufacturers compress large quantities of calcium into small pills, the stomach acid has difficulty penetrating the pill. To test a supplement's absorbability, drop it into a 6-ounce cup of vinegar, and stir occasionally. A high-quality formulation will dissolve within half an hour.

Finally, having chosen a supplement, a person must take it regularly, but when should you take it? To circumvent adverse nutrient interactions, take calcium supplements between, not with, meals. To enhance calcium absorption, take supplements with meals. (If such contradictory advice drives you crazy, reconsider the benefits of food sources of calcium.)

Consider another benefit of food. Some people absorb calcium better from milk and milk products than from even the most absorbable supplements. Indeed, experts agree that it remains highly desirable to adjust food and beverage intakes to provide calcium. The Consensus Conference on Osteoporosis recommends milk. The American Society for Bone and Mineral Research recommends foods as a source of calcium in preference to supplements. The *Diet and Health* report urges people to eat low- or nonfat dairy products and dark green vegetables to meet their calcium needs and concludes that current research does not justify the use of calcium supplements.[26] Seldom is such a consensus seen among researchers and dietitians.

## *Some Closing Thoughts*

Unfortunately, many of the strongest risk factors for osteoporosis are beyond people's control: age, sex, genetics, and race. But several factors within people's control can help to reduce the risk: a calcium-rich diet, an adequate vitamin D status, daily physical activity, abstinence from cigarette smoking, and moderation in, or abstinence from, alcohol use. Women should be evaluated for possible estrogen

**Table H12-4**
**Calcium Supplements**

- **Amino acid chelates** (KEY-lates) are compounds of minerals (such as calcium) combined with amino acids in a form that favors their absorption. A *chelating agent* is a molecule that surrounds another molecule and can then either promote or prevent its movement from place to place; *chele* means claw.
- **Antacids** are acid-buffering agents used to counter excess acidity in the stomach. Calcium-containing preparations (such as Tums) contain available calcium. Antacids with aluminum or magnesium hydroxides (such as Rolaids) can accelerate calcium losses.
- **Bone meal** or **powdered bones** are crushed or ground bone preparations intended to supply calcium to the diet. Calcium from bone is not well absorbed and is often contaminated with toxic materials such as arsenic, mercury, lead, and cadmium.
- **Calcium compounds** such as calcium carbonate, citrate, gluconate, lactate, malate, or phosphate are the simplest forms of purified calcium. These supplements vary in the amount of calcium they contain, so read the labels carefully. A 500-milligram tablet of calcium gluconate may provide only 45 milligrams of calcium, for example.
- **Dolomite** is a compound of minerals (calcium magnesium carbonate) found in limestone and marble. Dolomite is powdered and is sold as a calcium-magnesium supplement, but may be contaminated with toxic minerals, is not well absorbed, and interacts adversely with absorption of other essential minerals.
- **Oyster shell** is a product made from the powdered shells of oysters that is sold as a calcium supplement, but is not well absorbed.

replacement therapy at menopause. Other drug therapies may also be effective both in preventing bone loss and restoring lost bone.[27] The reward is the best possible chance of preserving bone health throughout life.

 National Osteoporosis Foundation

www.nof.org

 Osteoporosis and Related Bone Diseases—National Resource Center

www.osteo.org

## Notes


1. T. D. Galsworthy and P. L. Wilson, Osteoporosis: It steals more than bone, *American Journal of Nursing* 96 (1996): 27–32.
2. Incidence and costs to Medicare of fractures among Medicare beneficiaries aged ≥65 years—United States, July 1991–June 1992, *Morbidity and Mortality Weekly Report* 45 (1996): 877–883.
3. D. T. Felson and coauthors, The effect of postmenopausal estrogen therapy on bone density in elderly women, *New England Journal of Medicine* 329 (1993): 1141–1146; L. G. Tolstoi and R. M. Levin, Osteoporosis—The treatment controversy, *Nutrition Today,* July/August 1992, pp. 6–12.
4. J. D. Zuckerman, Hip fracture, *New England Journal of Medicine* 334 (1996): 1519–1525.
5. B. Dawson-Hughes and coauthors, Effect of calcium and vitamin D supplementation on bone density in men and women 65 years of age or older, *New England Journal of Medicine* 337 (1997): 670–676.
6. R. P. Heaney, Bone mass, nutrition, and other lifestyle factors, *Nutrition Reviews* 54 (1996): S3–S10.
7. D. L. Schneider, E. L. Barrett-Connor, and D. J. Morton, Timing of postmenopausal estrogen for optimal bone mineral density, *Journal of the American Medical Association* 277 (1997): 543–547; L. Speroff and coauthors, The comparative effect on bone density, endometrium, and lipids of continuous hormones as replacement therapy (CHART Study): A randomized controlled trial, *Journal of the American Medical Association* 276 (1996): 1397–1403; The Writing Group for the PEPI Trial, Effects of hormone therapy on bone mineral density: Results from the Postmenopausal Estrogen/Progestin Interventions (PEPI) Trial, *Journal of the American Medical Association* 276 (1996): 1389–1396.
8. S. R. Cummings and coauthors, Risk factors for hip fractures in white women, *New England Journal of Medicine* 332 (1995): 767–773; J. Lutz and R. Tesar, Mother-daughter pairs: Spinal and femoral bone densities and dietary intakes, *American Journal of Clinical Nutrition* 52 (1990): 878–888.
9. C. M. Ulrich and coauthors, Bone mineral density in mother-daughter pairs: Relations to lifetime exercise, lifetime milk consumption, and calcium supplements, *American Journal of Clinical Nutrition* 63 (1996): 72–79.
10. H. M. Perry and coauthors, A preliminary report of vitamin D and calcium metabolism in older African Americans, *Journal of the American Geriatric Society* 41 (1993): 612–616.
11. R. P. Heaney, Osteoporosis, in *Nutrition in Women's Health,* eds. D. A. Kummel and P. M. Kris-Etherton (Gaithersburg, Md.: Aspen Publishers, 1996), pp. 418–439.
12. J. F. Hu and coauthors, Dietary calcium and bone density among middle-aged and elderly women in China, *American Journal of Clinical Nutrition* 58 (1993): 219–227.
13. S. M. Smith and coauthors, Nutrition in space, *Nutrition Today* 32 (1997): 6–12.
14. L. Alekel and coauthors, Contributions of exercise, body composition, and age to bone mineral density in premenopausal women, *Medicine and Science in Sports and Exercise* 27 (1995): 1477–1485; I. Vuori, Peak bone mass and physical activity: A short review, *Nutrition Reviews* 54 (1996): S11–S14.
15. ACSM Position stand on osteoporosis and exercise, *Medicine and Science in Sports and Exercise* 27 (1995): i–vii; R. R. Recker and coauthors, Bone gain in young adult women, *Journal of the American Medical Association* 268 (1992): 2403–2408.
16. J. A. Langlois and coauthors, Weight change between age 50 years and old age is associated with risk of hip fracture in white women aged 67 years and older, *Archives of Internal Medicine* 156 (1996): 989–994; M. M. Hla and coauthors, A multicenter study of the influence of fat and lean mass on bone mineral content: Evidence for differences in their relative influence at major fracture sites, *American Journal of Clinical Nutrition* 64 (1996): 354–360; J. F. Aloia and coauthors, To what extent is bone mass determined by fat-free or fat mass? *American Journal of Clinical Nutrition* 61 (1995): 1110–1114; Cummings and coauthors, 1995; S. L. Edelstein and E. Barrett-Connor, Relation between body size and bone mineral density in elderly men and women, *American Journal of Epidemiology* 138 (1993): 160–169.
17. B. L. Drinkwater, B. Bruemner, and C. H. Chestnut III, Menstrual history as a determinant of current bone density in young athletes, *Journal of the American Medical Association* 263 (1990): 545–548; J. H. Wilson, Nutrition, physical activity and bone health in women, *Nutrition Research Reviews* 7 (1994): 76–91.
18. S. Suleiman and coauthors, Effect of calcium intake and physical activity level on bone mass and turnover in healthy, white, postmenopausal women, *American Journal of Clinical Nutrition* 66 (1997): 937–943.
19. J. L. Hopper and E. Seeman, The bone density of female twins discordant for tobacco use, *New England Journal of Medicine* 330 (1994): 387–392.
20. C. W. Slemenda, Cigarettes and the skeleton, *New England Journal of Medicine* 330 (1994): 430–431.
21. D. T. Felson and coauthors, Alcohol intake and bone mineral density in elderly men and women: The Framingham Study, *American Journal of Epidemiology* 142 (1995): 485–492.
22. R. P. Heaney, Nutritional factors in osteoporosis, *Annual Review of Nutrition* 13 (1993): 287–316; F. Bronner, Calcium and osteoporosis, *American Journal of Clinical Nutrition* 60 (1994): 831–836.
23. S. J. Whiting and R. J. Wood, Adverse effects of high-calcium diets in humans, *Nutrition Reviews* 55 (1997): 1–9.
24. I. R. Reid and coauthors, Effect of calcium supplementation on bone loss in postmenopausal women, *New England Journal of Medicine* 328 (1993): 460–464.
25. D. I. Levenson and R. S. Bockman, A review of calcium preparations, *Nutrition Reviews* 52 (1994): 221–232.

26. Committee on Diet and Health, Food and Nutrition Board, *Diet and Health: Implications for Reducing Chronic Disease Risk* (Washington, D.C.: National Academy Press, 1989), p. 17.
27. D. Hosking and coauthors, Prevention of bone loss with alendronate in postmenopausal women under 60 years of age, *New England Journal of Medicine* 338 (1998): 485–492; P. D. Delmas and coauthors, Effects of raloxifene on bone mineral density, serum cholesterol concentrations, and uterine endometrium in postmenopausal women, *New England Journal of Medicine* 337 (1997): 1641–1647; U. A. Liberman and coauthors, Effect of oral alendronate on bone mineral density and the incidence of fractures in postmenopausal osteoporosis, *New England Journal of Medicine* 333 (1995): 1437–1443.

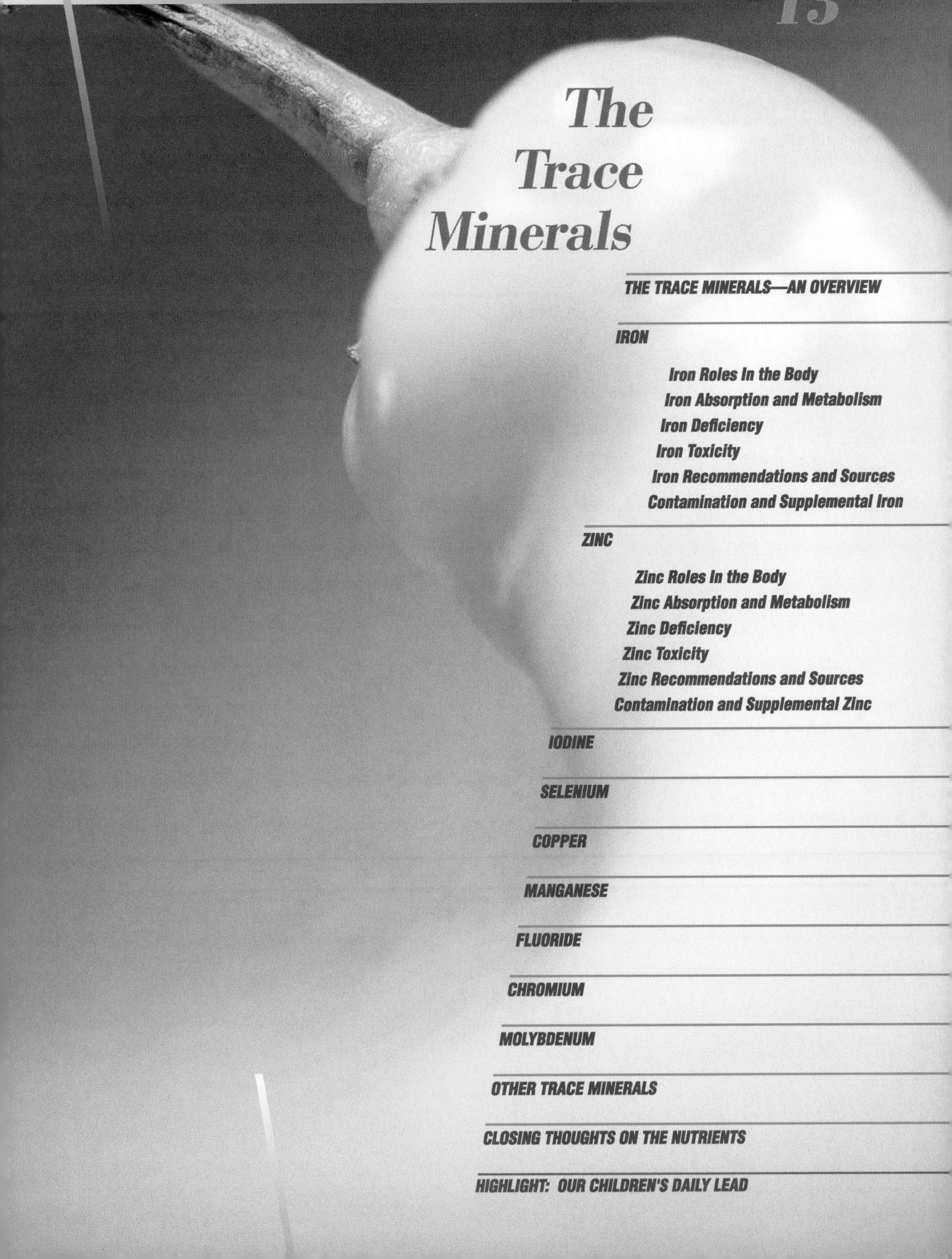

CHAPTER 13

# The Trace Minerals

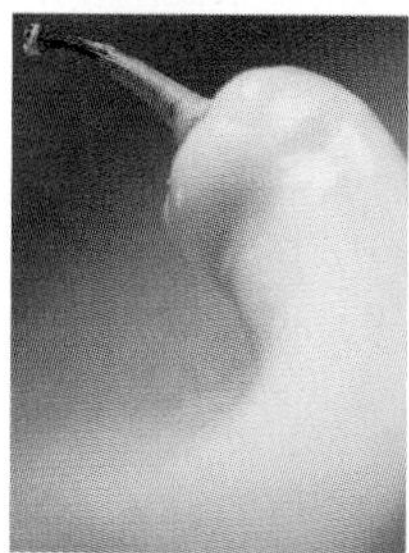

Figure 12-5 in the last chapter (p. 375) showed the tiny quantities of trace minerals in the human body. All together, they would produce only a bit of dust, hardly enough to fill a teaspoon. Yet each of the trace minerals performs a vital role. A deficiency of any of them may be fatal, and an excess of many is equally deadly. Remarkably, people's diets normally supply just enough of these minerals to maintain health.

# The Trace Minerals—An Overview

The body requires the trace minerals in minuscule quantities. They all assist enzymes in diverse tasks all over the body. In addition, each one has special duties that only it can perform.

• ***Food Sources*** • The trace mineral contents of foods depend on soil and water composition and on how foods are processed. Furthermore, many factors in the diet and within the body affect the minerals' bioavailability. Still, outstanding food sources for each of the trace minerals, just like those for the other nutrients, would include a wide variety of foods, especially unprocessed, whole foods.

• ***Deficiencies*** • Severe deficiencies of the better-known minerals are easy to recognize. Deficiencies of the others may be harder to diagnose, and for all minerals, mild deficiencies are easy to overlook. In general, the most common result of a deficiency is failure of children to grow and thrive, for the minerals are active in all the body systems—the GI tract, cardiovascular system, blood, muscles, bones, and central nervous system.

• ***Toxicities*** • The trace elements are toxic at an intake not far above the estimated requirements. Thus it is important not to habitually exceed the upper end of the range of recommended intakes.

• ***Supplements*** • Many vitamin-mineral pills contain trace minerals, making it easy for users to exceed their needs. The Food and Drug Administration (FDA) is not permitted to limit the amounts of trace minerals in supplements; consumers have demanded the freedom to choose their own doses of nutrients.* Individuals who take supplements must therefore be aware of the possible dangers and select supplements that contain no more than 100 percent of the Daily Value. It would be easier and safer to select a variety of foods than to combine an assortment of pills to meet nutrient needs without causing toxicity.

• ***Interactions*** • Interactions among the trace minerals are common and often lead to nutrient imbalances. An excess of one may cause a deficiency of another. (A slight manganese overload, for example, may aggravate an iron deficiency.) A deficiency of one may open the way for another to cause a toxic reaction. (Iron deficiency, for example, makes the body vulnerable to lead poisoning.) A deficiency of one may exacerbate the problems associated with the deficiency of another. (A combined iodine and selenium deficiency, for example, reduces thyroid hormone production more than an iodine deficiency alone.)[1] These examples highlight the need to balance intakes and to steer clear of supplement use. A good food source of one nutrient may be a poor food source of another; and factors that enhance the action of some trace minerals may hinder others.

*Canada limits the amounts of trace minerals in supplements.

(Vitamin C, for example, enhances the absorption of iron but hinders that of copper.) Research on the trace minerals is active, suggesting that we have much more to learn about them.

IN SUMMARY

To recap, although the body uses only tiny amounts of the trace minerals, they are vital to health. Because so little is required, the trace minerals can be toxic at levels not far above estimated requirements—a consideration for supplement users. Like the other nutrients, the trace minerals are best obtained by eating a variety of whole foods.

# Iron

Iron is an essential nutrient, vital to many of the cells' activities, but it poses a problem for millions of people: some people simply don't eat enough iron-containing foods to support their health optimally, while others have so much iron that it threatens their well-being. The principle that too little or too much of a nutrient is harmful seems particularly apropos for iron.

## Iron Roles in the Body

Iron's two ionic states:
- **Ferrous iron** (reduced): $Fe^{++}$.
- **Ferric iron** (oxidized): $Fe^{+++}$.

For details about ions, oxidation, and reduction, see Appendix B.

Reminder: A *cofactor* is a mineral element that works with an enzyme to facilitate a chemical reaction.

Reminder: *Hemoglobin* is the oxygen-carrying protein of the red blood cells that transports oxygen from the lungs to tissues throughout the body; hemoglobin accounts for 80% of the body's iron.

**myoglobin:** the oxygen-holding protein of the muscle cells.
- **myo** = muscle

Iron has the knack of switching back and forth between two ionic states. In the reduced state, iron has lost two electrons and therefore has a net positive charge of two; it is known as ferrous iron. In the oxidized state, iron has lost a third electron, has a net positive charge of three, and is known as ferric iron. Because it can exist in these different ionic states, iron can serve as a cofactor to enzymes involved in oxidation-reduction reactions. In every cell, iron works with several of the electron-transport-chain proteins that perform the final steps of the energy-yielding metabolic pathways.* These proteins transfer hydrogens and electrons from energy-yielding nutrients to oxygen, forming water, and in the process make ATP for the cell's use.

Most of the body's iron is found in two proteins: hemoglobin in the red blood cells and myoglobin in the muscle cells. In both, iron helps accept, carry, and then release oxygen. Iron is also found in many enzymes that oxidize compounds—reactions so widespread in metabolism that they occur in all cells. Iron is also required by enzymes involved in the making of amino acids, hormones, and neurotransmitters.

## Iron Absorption and Metabolism

The body conserves iron zealously and handles it carefully (see Figure 13-1). Because it is difficult to excrete iron once it is in the body, balance is maintained primarily through absorption: more iron is absorbed when stores are empty and less when they are full.[2]

**Mucosal ferritin** (FERR-ih-tin) holds iron in the cell, and **mucosal transferrin** (trans-FERR-in) passes the iron on to **blood transferrin.**

• ***Iron Absorption*** • Special proteins help the body absorb iron from food. One protein, called *mucosal ferritin,* receives iron from the GI tract and stores it in the mucosal cells of the small intestine. When the body needs iron, mucosal ferritin releases some iron to another protein, called *mucosal transferrin.* Mucosal trans-

*The iron-containing proteins at the end of the metabolic pathway include several TCA cycle enzymes and the electron carriers of the electron transport chain—known as *cytochromes.* See Appendix C for these pathways.

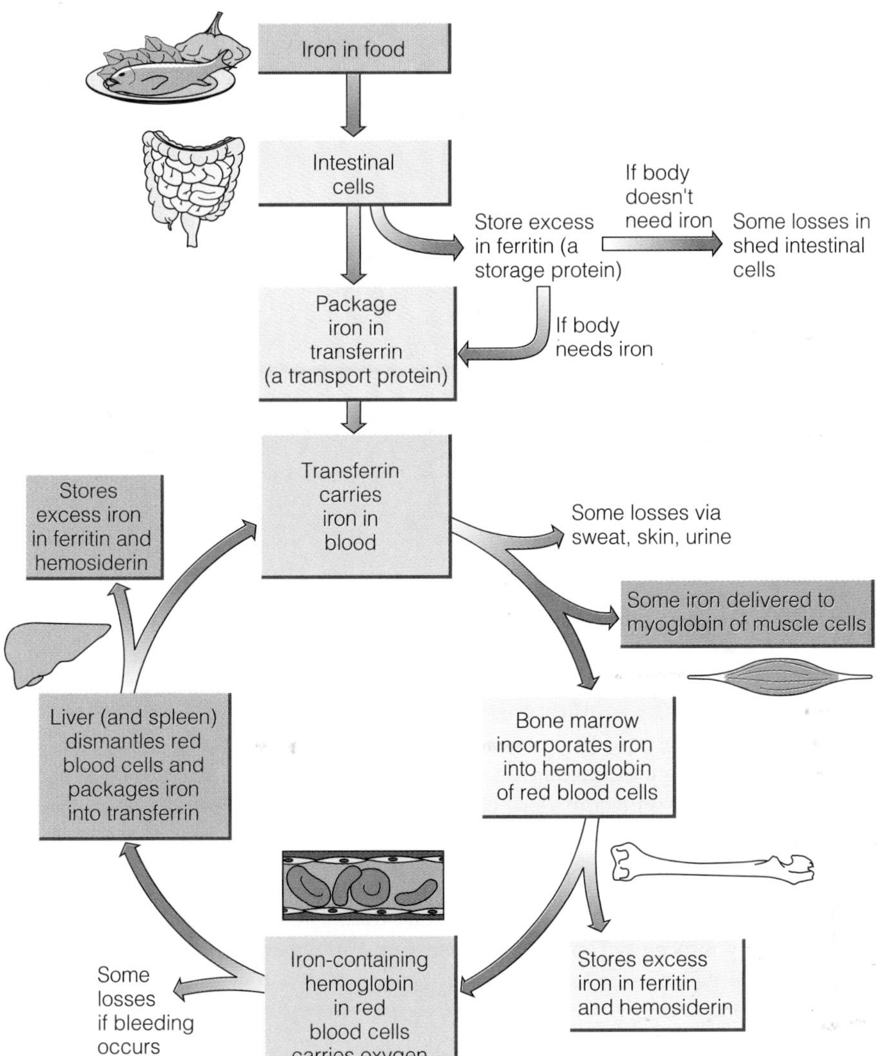


**Figure 13-1**
**Iron Routes in the Body**

Most iron is recycled. Some is lost with body tissues and must be replaced by eating iron-containing food.

ferrin transfers the iron to another protein, *blood transferrin,* which transports the iron to the rest of the body. Intestinal mucosal cells are replaced about every three days; when the cells are shed and excreted in the feces, they carry some iron out with them. By holding iron temporarily, these cells can either deliver iron when the day's intake falls short or dispose of it when intakes exceed needs.[3]

• ***Heme and Nonheme Iron*** • Iron absorption depends in part on its source. Iron occurs in two forms in foods: as heme iron, which is found only in foods derived from the flesh of animals, such as meats, poultry, and fish; and as nonheme iron, which is found in both plant-derived and animal-derived foods (see Figure 13-2 on p. 408). On the average, about 10 percent of the iron a person consumes in a day comes from heme iron. Even though heme iron accounts for only a small proportion of the intake, it is so well absorbed that it contributes significant iron: about 23 percent of heme iron is absorbed. By comparison, only 2 to 20 percent of nonheme iron is absorbed, depending on other dietary factors and the body's iron stores. People with severe iron deficiencies absorb both heme and nonheme iron more efficiently and are more sensitive to absorption enhancing factors than people with better iron status.

**heme** (HEEM): the iron-holding part of the hemoglobin and myoglobin proteins. About 40% of the iron in meat, fish, and poultry is bound into heme; the other 60% is nonheme iron.

• ***Absorption-Enhancing Factors: MFP and Vitamin C*** • Meat, fish, and poultry contain not only the well-absorbed heme iron, but also a factor (MFP factor) that promotes the absorption of nonheme iron from other foods eaten at the

**MFP factor:** a factor associated with the digestion of meat, fish, and poultry that enhances iron absorption.

**Figure 13-2**
**Heme and Nonheme Iron in Foods**

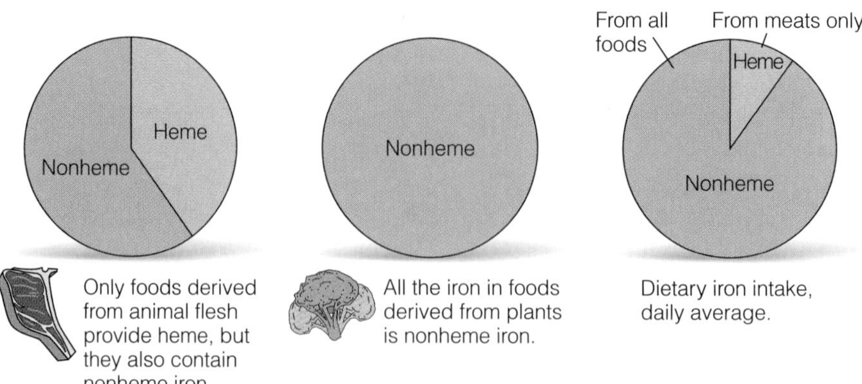
Heme
Nonheme
Nonheme
From all foods
From meats only
Heme
Nonheme
Only foods derived from animal flesh provide heme, but they also contain nonheme iron.
All the iron in foods derived from plants is nonheme iron.
Dietary iron intake, daily average.

Factors that enhance nonheme iron absorption:
- MFP factor.
- Vitamin C (ascorbic acid).
- Citric acid and lactic acid from foods and HCl acid from the stomach.
- Sugars (including the sugars in wine).

Factors that inhibit iron absorption:
- Phytates and fibers.
- Calcium and phosphorus (milk).
- EDTA (in food additives).
- Tannic acid (and other polyphenols).

The iron storage proteins are **ferritin** (FERR-ih-tin) and **hemosiderin** (heem-oh-SID-er-in).

same meal. Vitamin C also enhances nonheme iron absorption from foods eaten in the same meal by capturing the iron and keeping it in the reduced ferrous form, ready for absorption. Some acids and sugars also enhance nonheme iron absorption. A system of calculating the iron absorbed from a meal has been developed and reveals some considerations worthy of attention (see the accompanying "How to" feature).[4]

• ***Absorption Inhibitors*** • Some dietary factors bind with nonheme iron, inhibiting absorption. These factors include the phytates and fibers in soy products, whole-grain cereals, and nuts; the calcium and phosphorus in milk; the EDTA in food additives;* and the tannic acid in tea, coffee, nuts, and some fruits and vegetables.

• ***Adaptability of Absorption*** • Overall, only about 10 to 15 percent of dietary iron is absorbed, although this amount may vary widely. Absorption can be as low as 2 percent in a person with GI disease or as high as 35 percent in a rapidly growing, healthy child. Absorption increases when a person's iron intake falls short, or when the need increases for any reason (such as pregnancy). The body makes more mucosal transferrin to absorb more iron from the intestines and more blood transferrin to carry more iron around the body.

• ***Iron Transport and Storage*** • Blood transferrin delivers iron to the bone marrow and other tissues. The bone marrow uses large quantities to make new red blood cells, whereas other tissues use less. Surplus iron is stored in the protein ferritin, primarily in the liver, but also in the bone marrow and spleen. When dietary iron has been plentiful, ferritin is constantly and rapidly made and broken down, providing an ever-ready supply of iron. When iron concentrations become abnormally high, the liver converts some ferritin into another storage protein called hemosiderin. Hemosiderin releases iron more slowly than ferritin does. By storing excess iron, the body protects itself: free iron acts as a free radical, attacking cell lipids, DNA, and protein.[5] (See Highlight 11 for more information on free radicals.)

• ***Iron Recycling*** • The average red blood cell lives about four months; then the spleen and liver cells remove it from the blood, take it apart, and prepare the degradation products for excretion or recycling. The iron is salvaged: the liver attaches it to blood transferrin, which transports it back to the bone marrow to be

*EDTA is ethylenediamine tetra acetate, a chelating agent that is used in food processing to retard crystal formation and promote color retention.

## HOW TO Calculate Iron Absorbed from Meals

Three factors go into the calculation of the amount of iron absorbed from a meal: first, how much of the iron in the meal was heme and how much was nonheme iron; second, how much vitamin C was in the meal; and third, how much total meat, fish, and poultry (MFP) was consumed. Iron stores are assumed to be moderate and are not taken into consideration. Record the foods you eat at a typical meal, use a diet analysis computer program or look up their iron content in Appendix H, and then answer these questions:

1. How much iron was from meat, fish, and poultry? ____ mg.
2. On average, 40% of the amount in step 1 is heme iron:

   ____ mg (step 1) × 0.40 =
   ____ mg heme iron.
3. How much iron was from other sources? ____ mg.
4. The amount in step 3 plus 60% of step 1 is nonheme iron:

   ____ mg (step 3) +
   ____ mg (step 1 × 0.60) =
   ____ mg nonheme iron.
5. How much vitamin C was in the meal? ____ mg. Less than 25 mg is low; 25 to 75 mg is medium; more than 75 mg is high.
6. How much MFP was in the meal? ____ oz. Less than 1 oz is low; 1 to 3 oz is medium; more than 3 oz is high.*

Now calculate: You absorbed 23% of the heme iron, or:

____ mg (step 2) × 0.23 =
____ mg heme iron absorbed.

*This chili dinner provides heme and nonheme iron and MFP from meat, nonheme iron from legumes, and vitamin C from tomatoes. The combination of heme iron, nonheme iron, MFP, and vitamin C helps to achieve maximum iron absorption.*

Now, take your best score from steps 5 and 6. If either vitamin C or MFP was high or if both were medium, the availability of your nonheme iron was high. If neither was high, but one was medium, the availability of your nonheme iron was medium. If both were low, your nonheme iron had poor availability. You absorbed:

- High availability: 8% of the nonheme iron.
- Medium availability: 5% of the nonheme iron.
- Poor availability: 3% of the nonheme iron.

Now calculate: You absorbed ____ % of the nonheme iron, or:

____ mg (step 4) × ____ % =
____ mg total nonheme iron absorbed.

Add the two together:

____ mg heme iron absorbed +
____ mg nonheme iron absorbed =
____ mg total iron absorbed.

The RDA assumes you will absorb 10 percent of the iron you ingest. Thus, if you are a man of any age or a woman over 50 years old (RDA 10 mg), you need to absorb 1 mg per day; if you are a woman 11 to 50 years old (RDA 15 mg), you need to absorb 1.5 mg per day. If you have higher menstrual losses than the average woman, you may need still more.

*In question 6, we adapted the calculation of Monsen and coauthors, stating it in ounces. Actual numbers are less than 23 grams cooked meat, low; 23 to 46 grams, medium; and 69 grams or more, high.

Source: E. R. Monsen and coauthors, Estimation of available dietary iron, *American Journal of Clinical Nutrition* 31 (1978): 134–141.

reused in making new red blood cells. Thus, although red blood cells live for only about four months, the iron recycles through each new generation of cells. The body loses some iron daily via the GI tract and, if bleeding occurs, in blood; only tiny amounts of iron are lost in urine, sweat, and shed skin.*

*The adult male loses about 0.9 to 1.0 milligram of iron per day. Because of women's smaller surface area, their basal losses are 0.7 to 0.8 milligram per day, but women lose additional iron in menses. Menstrual losses vary considerably, but over a month, they average about 0.5 milligram per day.

# Iron Deficiency

**iron deficiency:** the state of having depleted iron stores.

Worldwide, iron deficiency is the most common nutrient deficiency, affecting more than one billion people. In developing countries, one-third of the children and women of childbearing age suffer from iron-deficiency anemia.[6] In the United States, iron-deficiency anemia is less prevalent, but still affects 10 percent of toddlers, adolescent girls, and women of childbearing age; preventing and correcting iron deficiency are high priorities.[7]

High risk for iron deficiency:
- Women in their reproductive years.
- Pregnant women.
- Infants and young children.
- Teenagers.

• ***Vulnerable Stages of Life*** • Some stages of life both demand more iron and provide less, making deficiency likely. Women are especially prone to iron deficiency during their reproductive years because of repeated blood losses during menstruation. Pregnancy demands additional iron to support the added blood volume, growth of the fetus, and blood loss during childbirth. Infants and young children receive little iron from their high-milk diets, yet need extra iron to support their rapid growth. The rapid growth of adolescence, especially for males, and the menstrual losses of females also demand extra iron that a typical teen diet may not provide. An adequate iron intake is especially important during these stages of life.

**Healthy People 2000:**
Reduce iron deficiency to less than 3% among children aged 1 to 4 and women of childbearing age.

The iron content of blood is about 0.5 mg/100 mL blood. A person donating a pint of blood (approximately 500 mL) loses about 2.5 mg of iron.

• ***Blood Losses*** • Bleeding from any site incurs iron losses. In some cases, as in an active ulcer, the bleeding may not be obvious, but even small chronic blood losses significantly deplete iron reserves.[8] In developing countries, blood loss is often brought on by parasitic infections of the GI tract. People who donate blood regularly also incur losses and may benefit from iron supplements.[9] As mentioned, regular menstrual losses make women's iron needs much greater than men's.

Stages of iron deficiency:
- Iron stores diminish.
- Transport iron decreases.
- Hemoglobin production falls.

• ***Assessment of Iron Deficiency*** • Iron deficiency develops in stages. This section provides a brief overview of how to detect these stages, and Appendix E provides more details. In the first stage of iron deficiency, iron stores diminish. Measures of serum ferritin reflect iron stores and are most valuable in assessing iron status.

The second stage of iron deficiency is characterized by a decrease in iron being transported within the body: serum iron falls, and the iron-carrying protein transferrin *increases* (an adaptation that enhances iron absorption). Together, these two measures can determine the severity of the deficiency—the more transferrin and the less iron in the blood, the more advanced the deficiency is.

The third stage of iron deficiency occurs when the lack of iron limits hemoglobin production. Now the hemoglobin precursor, erythrocyte protoporphyrin, begins to accumulate as hemoglobin and hematocrit values decline.

**erythrocyte protoporphyrin** (PRO-toe-PORE-fe-rin): a precursor to hemoglobin.

**hematocrit** (hee-MAT-oh-krit): measurement of the volume of the red blood cells packed by centrifuge in a given volume of blood.

Hemoglobin and hematocrit tests are easy, quick, and inexpensive, so they are the tests most commonly used in evaluating iron status; their usefulness is limited, however, because they are late indicators of iron deficiency. Furthermore, other nutrient deficiencies and medical conditions can influence their values.

**iron-deficiency anemia:** severe depletion of iron stores that results in low hemoglobin and small, pale, red blood cells.

Iron-deficiency anemia is a **microcytic** (my-cro-SIT-ic) **hypochromic** (high-po-KROME-ic) **anemia.**
- **micro** = small
- **cytic** = cells
- **hypo** = too little
- **chrom** = color

• ***Iron Deficiency and Anemia*** • Iron deficiency and iron-deficiency anemia are not the same: people may be iron deficient without being anemic. The term *iron deficiency* refers to depleted body iron stores without regard to the degree of depletion or to the presence of anemia. The term *iron-deficiency anemia* refers to the severe depletion of iron stores that results in a low hemoglobin concentration. In iron-deficiency anemia, red blood cells are pale and small (see Figure 13-3). They can't carry enough oxygen from the lungs to the tissues, so energy metabolism in the cells falters. The result is fatigue, weakness, headaches, apathy, pallor, and poor resistance to cold temperatures. Since hemoglobin is the bright red pigment of the

**Figure 13-3**
**Normal and Anemic Blood Cells**

Both size and color are normal in these blood cells.

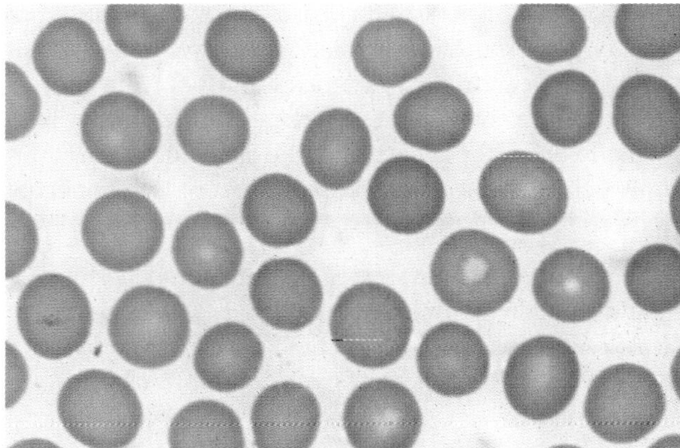

Blood cells in iron-deficiency anemia are small (microcytic) and pale (hypochromic) because they contain less hemoglobin.

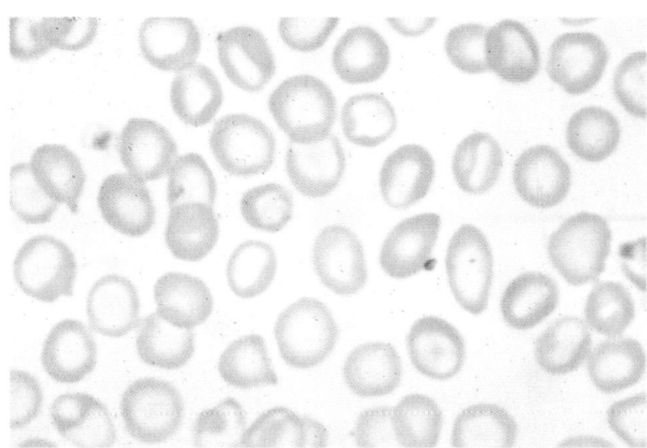

blood, the skin of a fair person who is anemic may become noticeably pale. In a dark-skinned person, the eye lining, normally pink, will be very pale. The summary table includes a list of the symptoms of iron deficiency.

• ***Iron Deficiency and Behavior*** • Long before the red blood cells are affected and anemia is diagnosed, a developing iron deficiency affects behavior. Even at slightly lowered iron levels, the complete oxidation of pyruvate is impaired, reducing physical work capacity and productivity. With reduced energy available to work,

The iron needs of physically active people and the special iron deficiency known as runner's anemia are discussed in Chapter 14.

## IN SUMMARY

**Iron**

**1989 RDA**

Men: 10 mg/day
Women: 15 mg/day (19–50 yr)
10 mg/day (51+)

**Chief Functions in the Body**

Part of the protein hemoglobin, which carries oxygen from place to place in the body; part of the protein myoglobin in muscles, which makes oxygen available for muscle contraction; necessary for the utilization of energy as part of the cells' metabolic machinery

**Significant Sources**

Red meats, fish, poultry, shellfish, eggs, legumes, dried fruits

| Deficiency Symptoms | Toxicity Symptoms |
|---|---|
| ***Eyes*** | |
| Blue sclera[a] | |
| ***Immune System*** | |
| Reduced resistance to infection (lowered immunity) | Infections |
| ***Nervous/Muscular Systems*** | |
| Reduced work productivity, tolerance to work, and voluntary work; reduced physical fitness; weakness; fatigue; impaired cognitive function (children); reduced learning ability; increased distractibility (inability to pay attention); impaired visual discrimination; impaired reactivity and coordination (infants) | Lethargy, joint disease |
| ***Skin*** | |
| Itching; pale nailbeds, eye membranes, and palm creases; concave nails; impaired wound healing | Pigmentation, loss of hair |
| ***General*** | |
| Reduced resistance to cold, inability to regulate body temperature, pica (clay eating, ice eating) | Death by accidental poisoning in children; organ damage, enlarged liver, amenorrhea, impotence |

[a]*Sclera* is a tough fibrous tissue that covers the "white" of the eye; *blue sclera* has an abnormal degree of blueness.

plan, think, play, sing, or learn, people simply do these things less. They have no obvious deficiency symptoms; they just appear unmotivated, apathetic, and less physically fit.

The effects of iron deficiency on children's behavior are discussed in Chapter 16.

Many of the symptoms associated with iron deficiency are easily mistaken for behavioral or motivational problems. A restless child who fails to pay attention in class might be thought contrary. An apathetic homemaker who has let housework pile up might be thought lazy. No responsible dietitian would ever claim that all behavioral problems are caused by nutrient deficiencies, but poor nutrition is always a possible contributor to problems like these. When investigating a behavioral problem, it makes sense to check the adequacy of the diet and to seek a routine physical examination before undertaking more expensive, and possibly harmful, treatment options.

**pica** (PIE-ka): a craving for nonfood substances. Also known as **geophagia** (gee-oh-FAY-gee-uh) when referring to clay eating and **pagophagia** (pag-oh-FAY-gee-uh) when referring to ice craving.
- **picus** = woodpecker or magpie
- **geo** = earth
- **phagein** = to eat
- **pago** = frost

• ***Iron Deficiency and Pica*** • A curious behavior seen in some iron-deficient people, especially in women and children of low-income groups, is pica—an appetite for ice, clay, paste, and other nonfood substances. These substances contain no iron and cannot remedy a deficiency; in fact, clay actually inhibits iron absorption, which may explain the iron deficiency that accompanies such behavior.

## Iron Toxicity

In general, even a diet that includes fortified foods poses no special risk for iron toxicity.[10] The body normally absorbs less iron when its stores are full, but some individuals are poorly defended against iron toxicity. Once considered rare, iron overload has emerged as an important disorder of iron metabolism and regulation.[11]

**iron overload:** toxicity from excess iron.

**hemochromatosis** (HE-moh-KRO-ma-toe-sis): a hereditary defect in iron metabolism characterized by deposits of iron-containing pigment in many tissues, with tissue damage.

**hemosiderosis** (HE-mo-sid-er-OH-sis): a condition characterized by the deposition of hemosiderin in the liver and other tissues.

Iron Overload Diseases Association
www.ironoverload.org

• ***Iron Overload*** • Iron overload is known as hemochromatosis and is usually caused by a genetic disorder that enhances iron absorption. Hereditary hemochromatosis is the most common genetic disorder in the United States, affecting some 1.5 million people.[12] Other causes of iron overload include repeated blood transfusions, massive doses of supplementary iron, and other rare metabolic disorders. Long-term overconsumption of iron may cause hemosiderosis, a condition characterized by large deposits of the iron storage protein hemosiderin in the liver and other tissues.

Some of the signs and symptoms of iron overload are similar to those of iron deficiency: apathy, lethargy, and fatigue. Therefore, taking iron supplements before assessing iron status is clearly unwise; hemoglobin tests alone would fail to make the distinction.

Iron overload is characterized by tissue damage, especially in iron-storing organs such as the liver. Infections are likely because bacteria thrive on iron-rich blood. Symptoms are most severe in alcohol abusers because alcohol damages the intestine, further impairing its defenses against absorbing excess iron. Untreated hemochromatosis aggravates the risks of diabetes, liver cancer, heart disease, and arthritis.

Iron overload is more common in men than in women, and is twice as prevalent among men as is iron deficiency. The widespread fortification of foods with iron makes it difficult for people with hemochromatosis to follow a low-iron diet, but greater dangers lie in the indiscriminate use of iron and vitamin C supplements. Vitamin C not only enhances iron absorption, but releases iron from ferritin, allowing free iron to wreak the damage typical of free radicals.[13] This is an example of how vitamin C acts as a *pro*oxidant when taken in high doses (see Highlight 11).

• ***Iron and Heart Disease*** • Research suggesting a link between heart disease and elevated iron stores is inconclusive.[14] In one study, researchers found that only smoking predicted heart attacks better than iron status.[15] In another study, researchers reported a reduced risk of heart attacks in blood donors, perhaps because of their repeated iron losses.[16] Too much iron poses no problem for healthy people, but in certain diseases, free radicals attack ferritin, causing it to release its iron.[17] Left free, iron acts as an oxidant that can trigger free-radical reactions. Iron's oxidation of LDL may explain its role in the development of heart disease.

Prior to these findings, scientists had speculated that premenopausal women were protected against heart disease by their estrogen (women's rate of heart disease is lower and begins to approach that of men only after menopause). Now scientists are asking whether low iron stores from repeated menstrual losses might be the protective factor. (Women's iron stores tend to catch up with men's after menopause.)

See Highlight 11 for a discussion of free radicals and their effects on disease development.

• ***Iron and Cancer*** • There also appears to be a positive association between iron and cancer.[18] Explanations for how iron might be involved in causing cancer focus on its free-radical activity, which can damage DNA (see Highlight 11). One of the benefits of a high-fiber diet may be that its phytates bind iron, making it less available for such reactions.

• ***Iron Poisoning*** • Ingestion of iron-containing supplements remains the leading cause of accidental poisoning in small children. Symptoms of intoxication include nausea, vomiting, diarrhea, a rapid heartbeat, a weak pulse, dizziness, shock, and confusion. As few as 5 iron tablets containing as little as 200 milligrams of iron have caused the deaths of dozens of young children. If you suspect iron poisoning, call the nearest poison control center or a physician immediately.

Keep iron-containing tablets out of the reach of children.

U.S. Government
Iron Poisoning in Children
www.healthfinder.gov/searchoptions/topicsaz.htm
Search for Iron

## Iron Recommendations and Sources

To obtain enough iron, people must first select iron-rich foods and then eat so as to maximize iron absorption. This discussion begins by identifying iron-rich foods, then reviews factors affecting absorption.

• ***Recommended Iron Intakes*** • The usual diet in the United States provides only about 6 to 7 milligrams of iron for every 1000 kcalories. The recommended daily intake for men is 10 milligrams, and most men eat more than 2000 kcalories a day, so men can meet their iron needs with little effort. Women during their childbearing years, however, need 15 milligrams a day. (The "How to" on p. 414 explains how to calculate the recommended intake.) Because women have higher iron needs and may need fewer than 2000 kcalories per day, they sometimes have trouble obtaining sufficient iron. On the average, women receive only 10 to 11 milligrams iron per day, not enough until after menopause. To meet their iron needs from foods, premenopausal women must emphasize the most iron-rich foods in every food group at every meal.

• ***Iron in Foods*** • Figure 13-4 (on p. 415) shows the amounts of iron in selected foods. Meats, fish, and poultry contribute the most iron; other protein-rich foods such as legumes and eggs are also good sources. Foods in the milk group are as poor in iron as they are rich in calcium. Although an indispensable part of the diet, milk products should not be overemphasized. Grain foods vary, with whole-grain and enriched breads and cereals providing the most iron. Finally, dark greens and some fruits (especially when dried) contribute some iron.

• ***Iron-Enriched Foods*** • Iron is one of the enrichment nutrients for grain products. One serving of enriched bread or cereal provides only a little iron, but because people eat many servings of these foods, the contribution can be significant. Iron added to foods is not absorbed as well as naturally occurring iron, but when eaten with absorption-enhancing foods, enrichment iron can make a difference. In some cases, enrichment may even contribute to iron overload, at least in men.

## HOW TO Estimate the Recommended Daily Intake for Iron

To calculate the recommended daily iron intake, a number of factors need to be considered. For example, for an adult woman:

- Losses from shed skin: 1.0 milligram.
- Losses through menstruation (about 15 milligrams total averaged over 30 days): 0.5 milligram.
- Average daily need (total): 1.5 milligrams.
- Average iron intake: 10 to 11 milligrams per day (provides adequate stores for most women).
- Added margin of safety to cover the needs of essentially all adult women.

Assuming that an average of 10 to 15 percent of ingested iron is absorbed, the RDA is set at 15 milligrams.

*An old-fashioned iron skillet adds iron to foods.*

**contamination iron:** iron found in foods as the result of contamination by inorganic iron salts from iron cookware, iron-containing soils, and the like.

A **chelate** (KEY-late) is a substance that can grasp the positive ions of a metal.
- **chele** = claw.

• ***Maximizing Iron Absorption*** • In general, the bioavailability of iron in meats, fish, and poultry is high; in grains and legumes, intermediate; and in most vegetables, especially those high in oxalate such as spinach, low. The amount of iron ultimately absorbed from a meal depends on the interplay between enhancing and inhibiting factors. For maximum absorption of nonheme iron, eat meat for MFP and fruits or vegetables for vitamin C. The iron of baked beans, for example, will be enhanced by the MFP in a piece of ham served with them; the iron of bread will be enhanced by vitamin C in a slice of tomato on a sandwich.

## *Contamination and Supplemental Iron*

In addition to the iron from foods, contamination iron from nonfood sources of inorganic iron salts can contribute to the day's intakes. People can also ingest iron in supplement form.

• ***Contamination Iron*** • Foods cooked in iron cookware take up iron salts. The more acidic the food, and the longer it is cooked in iron cookware, the higher the iron content. The iron content of eggs can triple in the short time it takes to scramble them in an iron pan. Similarly, dried peaches or raisins contain more iron than the fresh fruits do, because they have been dried in iron pans. For example, a dried fig provides more than twice as much iron as a fresh fig. (Ounce for ounce, a dried fig provides more than six times as much iron as a fresh fig, reflecting the concentration of iron with the removal of water.) Admittedly, the absorption of this iron may be poor (perhaps only 1 to 2 percent), but every little bit counts.

• ***Iron Supplements*** • People who are iron deficient may need supplements as well as an iron-rich, absorption-enhancing diet. Many physicians routinely recommend iron supplements to pregnant women, infants, and young children. Iron from supplements is less well absorbed than that from food, so the doses have to be high. The absorption of iron taken as ferrous sulfate or as an iron chelate is better than that from other iron supplements. Absorption also improves when supplements are taken between meals or at bedtime on an empty stomach, and with liquids other than milk, tea, or coffee, which inhibit absorption. There is no benefit to taking iron supplements with orange juice because vitamin C does not enhance absorption from supplements as it does for dietary iron. (Vitamin C helps iron absorption by converting insoluble ferric iron in foods to the more soluble ferrous iron, and supplemental iron is already in the ferrous form.) Constipation is a common side effect of iron supplementation; a plentiful fluid intake may help to relieve this problem.

**Figure 13-4**
**Iron in Selected Foods**

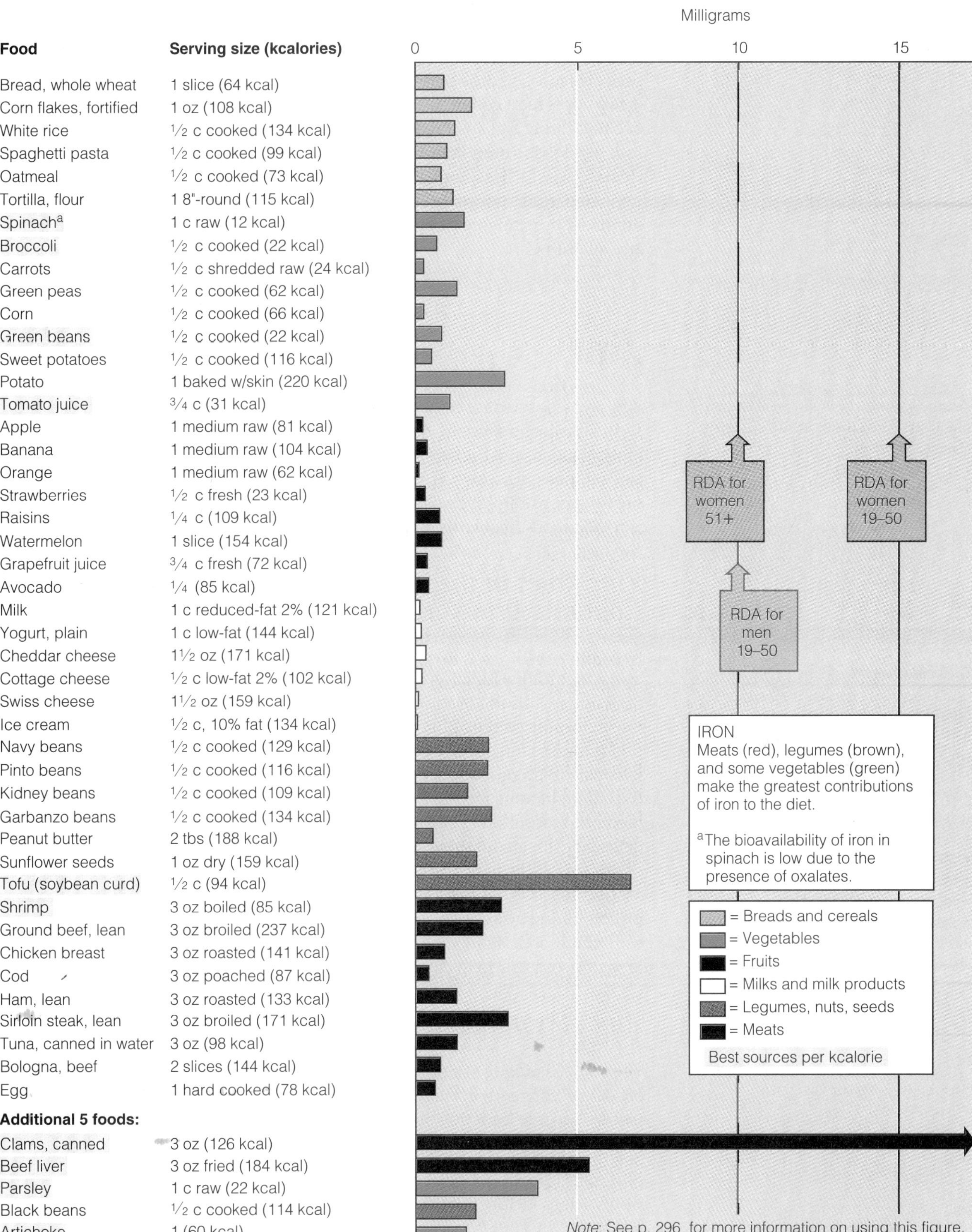

Milligrams
0
5
10
15
Food
Serving size (kcalories)
Bread, whole wheat 1 slice (64 kcal)
Corn flakes, fortified 1 oz (108 kcal)
White rice ½ c cooked (134 kcal)
Spaghetti pasta ½ c cooked (99 kcal)
Oatmeal ½ c cooked (73 kcal)
Tortilla, flour 1 8"-round (115 kcal)
Spinach[a] 1 c raw (12 kcal)
Broccoli ½ c cooked (22 kcal)
Carrots ½ c shredded raw (24 kcal)
Green peas ½ c cooked (62 kcal)
Corn ½ c cooked (66 kcal)
Green beans ½ c cooked (22 kcal)
Sweet potatoes ½ c cooked (116 kcal)
Potato 1 baked w/skin (220 kcal)
Tomato juice ¾ c (31 kcal)
Apple 1 medium raw (81 kcal)
Banana 1 medium raw (104 kcal)
Orange 1 medium raw (62 kcal)
Strawberries ½ c fresh (23 kcal)
Raisins ¼ c (109 kcal)
Watermelon 1 slice (154 kcal)
Grapefruit juice ¾ c fresh (72 kcal)
Avocado ¼ (85 kcal)
Milk 1 c reduced-fat 2% (121 kcal)
Yogurt, plain 1 c low-fat (144 kcal)
Cheddar cheese 1½ oz (171 kcal)
Cottage cheese ½ c low-fat 2% (102 kcal)
Swiss cheese 1½ oz (159 kcal)
Ice cream ½ c, 10% fat (134 kcal)
Navy beans ½ c cooked (129 kcal)
Pinto beans ½ c cooked (116 kcal)
Kidney beans ½ c cooked (109 kcal)
Garbanzo beans ½ c cooked (134 kcal)
Peanut butter 2 tbs (188 kcal)
Sunflower seeds 1 oz dry (159 kcal)
Tofu (soybean curd) ½ c (94 kcal)
Shrimp 3 oz boiled (85 kcal)
Ground beef, lean 3 oz broiled (237 kcal)
Chicken breast 3 oz roasted (141 kcal)
Cod 3 oz poached (87 kcal)
Ham, lean 3 oz roasted (133 kcal)
Sirloin steak, lean 3 oz broiled (171 kcal)
Tuna, canned in water 3 oz (98 kcal)
Bologna, beef 2 slices (144 kcal)
Egg 1 hard cooked (78 kcal)
Additional 5 foods:
Clams, canned 3 oz (126 kcal)
Beef liver 3 oz fried (184 kcal)
Parsley 1 c raw (22 kcal)
Black beans ½ c cooked (114 kcal)
Artichoke 1 (60 kcal)
RDA for women 51+
RDA for women 19–50
RDA for men 19–50
IRON
Meats (red), legumes (brown), and some vegetables (green) make the greatest contributions of iron to the diet.
[a]The bioavailability of iron in spinach is low due to the presence of oxalates.
= Breads and cereals
= Vegetables
= Fruits
= Milks and milk products
= Legumes, nuts, seeds
= Meats
Best sources per kcalorie
Note: See p. 296 for more information on using this figure.

IN SUMMARY

Most of the body's iron is in hemoglobin and myoglobin where it carries oxygen for use in energy metabolism; some iron is also a cofactor for enzymes involved in a variety of reactions. Special proteins assist with iron absorption, transport, and storage—all helping to maintain an appropriate balance, because both too little and too much iron can be damaging. Iron deficiency is most common among infants and young children, teenagers, women of childbearing age, and pregnant women; symptoms include fatigue and anemia. Iron overload is most common in men and has been linked with heart disease. Heme iron, which is found only in meat, fish, and poultry, is better absorbed than nonheme iron, which occurs in most foods. Nonheme iron absorption is improved by eating iron-containing foods with foods containing the MFP factor and vitamin C.

# Zinc

Reminder: A *cofactor* is a mineral element that works with an enzyme to facilitate a chemical reaction.

Zinc is a versatile trace element required as a cofactor by more than 100 enzymes. Virtually all cells contain zinc, but the highest concentrations are in bone, the prostate gland, and the eyes. Muscle contains the highest proportion of total body zinc (60 percent), however, because it accounts for most of the body's mass. Tissues do not readily give up their zinc when blood levels fall, so a person must eat zinc-rich foods frequently.

## Zinc Roles in the Body

**metalloenzyme** (meh-TAL-oh-EN-zime): an enzyme that contains one or more minerals as part of its structure.

Enzymes that zinc assists:
- Help make parts of the genetic materials DNA and RNA.
- Manufacture heme for hemoglobin.
- Participate in essential fatty acid metabolism.
- Release vitamin A from liver stores.
- Metabolize carbohydrates.
- Synthesize proteins.
- Metabolize alcohol in the liver.
- Dispose of damaging free radicals.

Zinc supports the work of numerous proteins in the body, including the metalloenzymes, which are involved in a variety of metabolic processes.* Zinc also assists in immune function and in growth and development. Zinc associates with the hormone insulin in the pancreas, although it does not appear to play a direct role in insulin's action. Zinc interacts with platelets in blood clotting, affects thyroid hormone function, and influences behavior and learning performance. It is necessary to produce the active form of vitamin A (retinal) in visual pigments and the retinol-binding protein that transports vitamin A. It is essential to normal taste perception, wound healing, the making of sperm, and fetal development. A zinc deficiency impairs all these and other functions, underlining the vast importance of proteins as the body's working machines.

Zinc, like iron, helps protect the body from heavy metal poisoning—for example, poisoning by lead. This is especially important during fetal development and early childhood. Highlight 13 reveals the damage heavy metals can do and shows how iron and zinc help defend against it.

## Zinc Absorption and Metabolism

The body's handling of zinc resembles that of iron in some ways and differs in others. A key difference is that the mucosal cells in the intestine provide a two-way passage for zinc from the intestine to the blood and back again.

• ***Zinc Absorption*** • The rate of zinc absorption varies from about 15 to 40 percent, depending on a person's zinc status: if more is needed, more is absorbed. Also, dietary factors influence zinc absorption. For example, fiber and phytates bind zinc, thus limiting its bioavailability.

*Among the metalloenzymes requiring zinc are carbonic anhydrase, deoxythymidine kinase, DNA and RNA polymerase, and alkaline phosphatase.

Upon absorption into an intestinal cell, zinc has several options. It may become involved in the metabolic functions of the cell itself. Alternatively, it may be retained within the cell by metallothionein, a special binding protein similar to the iron storage protein, mucosal ferritin.

The synthesis of metallothionein in the intestinal cells helps to regulate zinc absorption. When zinc intakes are high, more metallothionein is made; it holds zinc in reserve, thus inhibiting absorption. (Similarly, metallothionein in the liver binds zinc until other body tissues signal a need for it.) When the body needs zinc, intestinal metallothionein releases it into the blood where it can be transported around the body. Some zinc eventually reaches the pancreas.

**metallothionein** (meh-TAL-oh-THIGH-oh-neen): a sulfur-rich protein that avidly binds with metals such as zinc.
- **metallo** = containing a metal
- **thio** = containing sulfur
- **ein** = a protein

• ***Enteropancreatic Circulation*** • Many of the digestive enzymes released from the pancreas into the intestine at mealtimes contain zinc. The intestine thus receives two doses of zinc with each meal—one from ingested foods and the other from the zinc-rich pancreatic secretions. The circulation of zinc in the body from the pancreas to the intestine and back to the pancreas is referred to as the enteropancreatic circulation of zinc. As this zinc circulates through the intestine, it may be refused entry by the intestinal cells or retained in them on any of its times around (see Figure 13-5).

**enteropancreatic** (EN-ter-oh-PAN-kree-AT-ik) **circulation:** the circulatory route from the pancreas to the intestine and back to the pancreas.

• ***Zinc Transport by Albumin*** • Zinc's main transport vehicle in the blood is the protein albumin, which is a major determinant of zinc absorption. This may

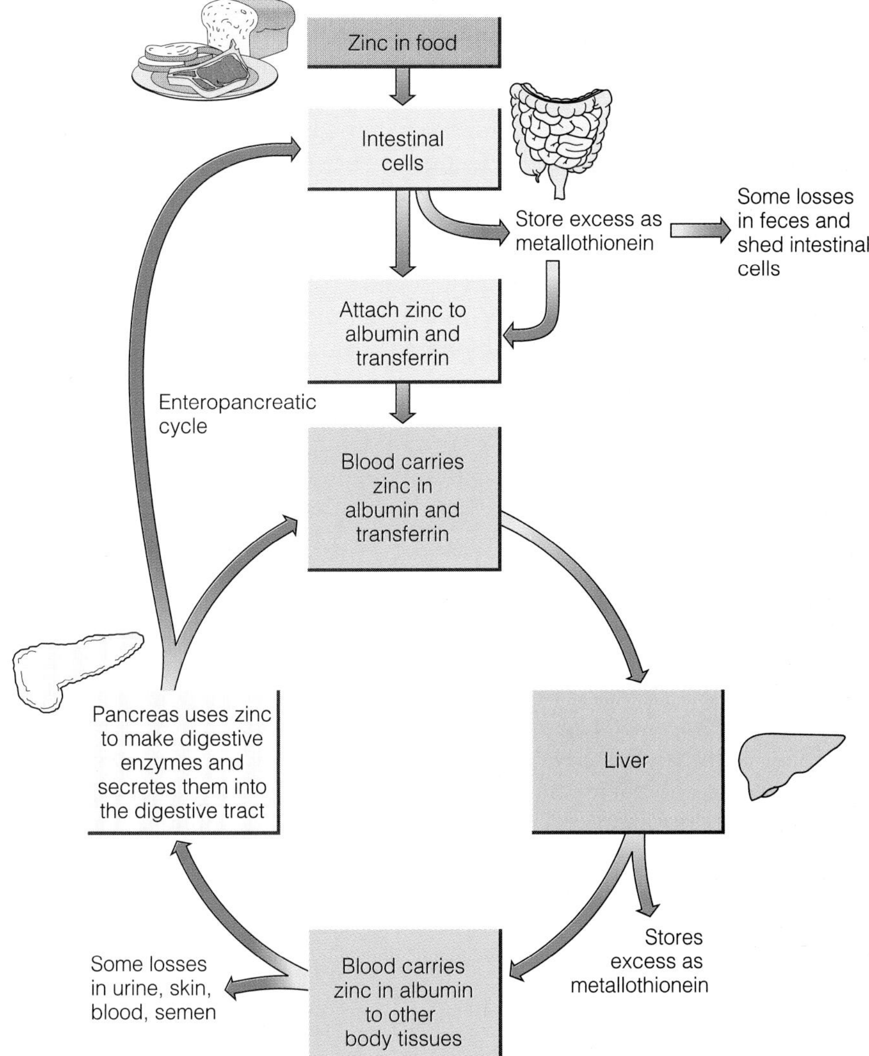


**Figure 13-5**
**Zinc's Routes in the Body**

Notice the enteropancreatic circulation of zinc from the intestines through the blood to the pancreas and back to the intestines.

account for observations that zinc absorption declines in conditions that lower albumin concentrations—for example, pregnancy and protein-energy malnutrition.

• ***Zinc Interactions with Iron and Copper*** • Some plasma zinc also binds to transferrin—the same transferrin that carries iron in the blood. In healthy individuals, transferrin is usually less than 50 percent saturated with iron, but in iron overload, it is more saturated. Dietary iron-to-zinc ratios of greater than 2 to 1 leave too few transferrin-binding sites available for zinc and thereby interfere with zinc absorption. The converse is also true: large doses of zinc inhibit iron absorption.

Large doses of zinc create a similar problem with another essential mineral, copper. Recall that when zinc intakes are high, the intestinal cells synthesize large amounts of the binding protein metallothionein. This protein binds copper more strongly than zinc and captures copper in a nonabsorbable form. This binding impairs copper absorption and may result in a copper deficiency. These nutrient interactions highlight one of the many reasons why people should use supplements conservatively: supplementation can easily create imbalances.

• ***Zinc Losses*** • Zinc exits the body primarily in feces. Smaller losses occur in urine, shed skin, hair, sweat, menstrual fluids, and semen.

## *Zinc Deficiency*

Human zinc deficiency was first reported in the 1960s in children and adolescent boys in Egypt, Iran, and Turkey. Children have especially high zinc needs because they are growing rapidly and synthesizing many zinc-containing proteins; the native diets among those populations were not meeting these needs. Middle Eastern diets are typically low in the richest zinc source, meats, and the staple foods are legumes, unleavened breads, and other whole-grain foods—all high in fiber and phytates, which inhibit zinc absorption.*

**Figure 13-6**

**Zinc-Deficiency Symptom—The Stunted Growth of Dwarfism**

The Egyptian man on the right is an adult of average height. The Egyptian boy on the left is 17 years old but is only 4 feet tall, like a 7-year-old in the United States. His genitalia are like those of a 6-year-old. The retardation, known as dwarfism, is rightly ascribed to zinc deficiency because it is partially reversible when zinc is restored to the diet.

• ***Zinc-Deficiency Symptoms*** • Figure 13-6 shows the severe growth retardation and arrested sexual maturation that characterize zinc deficiency. In addition, zinc deficiency hinders digestion and absorption, causing diarrhea, which worsens malnutrition not only for zinc, but for all nutrients. It impairs the immune response, making infections likely—among them, infections of the GI tract, which worsen malnutrition, including zinc malnutrition (a classic downward spiral of events). Chronic zinc deficiency damages the central nervous system and brain functioning. Because zinc deficiency directly impairs vitamin A metabolism, vitamin A–deficiency symptoms often appear. Zinc deficiency also disturbs thyroid function and the metabolic rate. It alters taste, causes anorexia, and slows wound healing—in fact, its symptoms are so all-pervasive that generalized malnutrition and sickness are more likely to be the diagnosis than simple zinc deficiency. The accompanying summary table includes a list of zinc-deficiency symptoms.

• ***Vulnerable Stages of Life*** • Severe zinc deficiencies are not widespread in developed countries, but they do occur in vulnerable groups—pregnant women, young children, the elderly, and the poor. Even a mild zinc deficiency can result in poor growth, poor appetite, impaired immune response, abnormal taste, and abnormal vision in darkness. Supplementation does not appear to improve growth in children with zinc deficiencies, but it can reduce their incidence of infectious diseases.[19]

*Unleavened bread contains no yeast, which normally breaks down phytates during fermentation.

IN SUMMARY

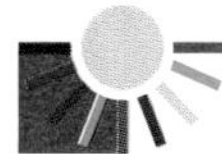

**Zinc**

**1989 RDA**

Men: 15 mg/day

Women: 12 mg/day

**Chief Functions in the Body**

Part of many enzymes; associated with the hormone insulin; involved in making genetic material and proteins, immune reactions, transport of vitamin A, taste perception, wound healing, the making of sperm, and the normal development of the fetus

**Significant Sources**

Protein-containing foods: meats, fish, poultry, whole grains, vegetables

| Deficiency Symptoms[a] | Toxicity Symptoms |
|---|---|
| ***Blood*** | |
| High ammonia, low alkaline phosphatase, low insulin | Anemia: reduced hemoglobin production |
| ***Bones*** | |
| Growth retardation, abnormal collagen synthesis | Growth in length, but without normal zinc content |
| ***Cells/Metabolism*** | |
| Slow DNA synthesis, impaired cell division and protein synthesis | Raised LDL, lowered HDL |
| ***Digestive System*** | |
| Weak sense of smell, poor sensitivity to the taste of salt, weight loss, delayed glucose absorption, diarrhea, nausea, impaired folate absorption | Diarrhea, vomiting, decreased calcium and copper absorption |
| ***Eyes*** | |
| Night blindness | |
| ***Glandular System*** | |
| Delayed onset of puberty, small gonads in males, decreased synthesis and release of testosterone, abnormal glucose tolerance, reduced synthesis of adrenocortical hormones, altered thyroid function | |
| ***Immune System*** | |
| Altered skin test responses, low white blood cell count, few antibody-forming cells, thymus atrophy, susceptibility to infection | Fever, elevated white blood cell count |
| ***Kidney*** | |
| | Renal failure |
| ***Liver/Spleen*** | |
| Enlargement | |
| ***Nervous/Muscular Systems*** | |
| Anorexia (poor appetite), mental lethargy, irritability | Muscular pain and incoordination, heart muscle degeneration, exhaustion, dizziness, drowsiness |
| ***Reproductive System*** | |
| Impaired reproductive function (rats), low sperm counts | Reproductive failure |
| ***Skin*** | |
| Generalized hair loss; lesions; rough, dry appearance; slow healing of wounds and burns | |

[a]A rare inherited disease of zinc malabsorption, *acrodermatitis enteropathica,* causes additional and more severe symptoms.

## *Zinc Toxicity*

Accidental high doses (2 grams or more) of zinc may cause vomiting, diarrhea, fever, exhaustion, and other symptoms (see the summary table). A dose of just a few milligrams above the recommended intake, especially when taken regularly over time, lowers the body's copper content—an effect that, in animals, leads to degeneration of the heart muscle. High doses also appear to accelerate the development of atherosclerosis.

## Zinc Recommendations and Sources

Zinc recommendations assume that 20 percent of dietary zinc is absorbed. Average intakes in the United States are about 10 milligrams per day, somewhat below recommendations. Requirements for infants and children are relatively higher than for adults due to zinc's role in normal growth and development.

Figure 13-7 shows zinc amounts in foods per serving. Zinc is highest in protein-rich foods such as shellfish (especially oysters), meats, poultry, and liver. Legumes and whole-grain products are good sources of zinc if eaten in large quantities; in typical U.S. diets, phytate intake from grains is not high enough to impair zinc absorption. Vegetables vary in zinc content depending on the soil in which they are grown.

## Contamination and Supplemental Zinc

**galvanized:** a term referring to metals that have been treated with a zinc-containing coating to prevent rust.

In earlier times, galvanized cooking pots and pipes used in plumbing contributed zinc to people's intakes. With today's use of stainless steel and plastic, these sources of zinc have been largely eliminated.

Zinc supplements are known to be useful in two instances: to remedy an accurately diagnosed zinc deficiency and to displace other ions in unusual medical circumstances. Otherwise, it should be possible to obtain enough zinc from the diet without resorting to supplements.

The use of zinc lozenges to treat the common cold has been controversial and inconclusive, with half the studies finding effectiveness and the other half not.[20] The differences in study results may reflect the formulations of the lozenges more than the effectiveness of zinc itself. Studies using zinc gluconate report positive findings, whereas those using other combinations of zinc or zinc gluconate bound to a flavor-enhancing chelator report negative results. Zinc gluconate shortens the duration of cold symptoms such as coughing, nasal congestion, headache, and sore throat.[21] Common side effects of zinc lozenges include nausea and bad taste reactions.

*Zinc is highest in protein-rich foods such as oysters, beef, and poultry; and legumes and nuts.*

IN SUMMARY

Zinc assists enzymes in a multitude of reactions affecting growth, vitamin A activity, and pancreatic digestive enzyme synthesis, among others. Both dietary zinc and zinc-rich pancreatic secretions (via enteropancreatic circulation) are available for absorption. Absorption is monitored by a special binding protein (metallothionein) in the intestine. Protein-rich foods derived from animals are the best sources of bioavailable zinc. Fiber and phytates in cereals bind zinc, limiting absorption. Growth retardation and sexual immaturity are hallmark symptoms of zinc deficiency.

# Iodine

Traces of the iodine ion are indispensable to life. In the GI tract, iodine from foods is converted to iodide; this chapter uses *iodine* when referring to the nutrient in foods and *iodide* when referring to it in the body. Iodide occurs in the body in a tiny quantity, but its principal role in human nutrition is well known, and the amount needed is well established.

The two hormones from the thyroid gland are **triiodothyronine** ($T_3$), which is the active form, and **tetraiodothyronine** ($T_4$), which is more commonly known as *thyroxin*.

• ***Iodide Roles in the Body*** • Iodide is an integral part of two hormones released by the thyroid gland that regulate body temperature, metabolic rate, reproduction, growth, blood cell production, nerve and muscle function, and more. These hormones control the rate at which the cells use oxygen, which controls the rate at which energy is released during metabolism.

**Figure 13-7**
**Zinc in Selected foods**

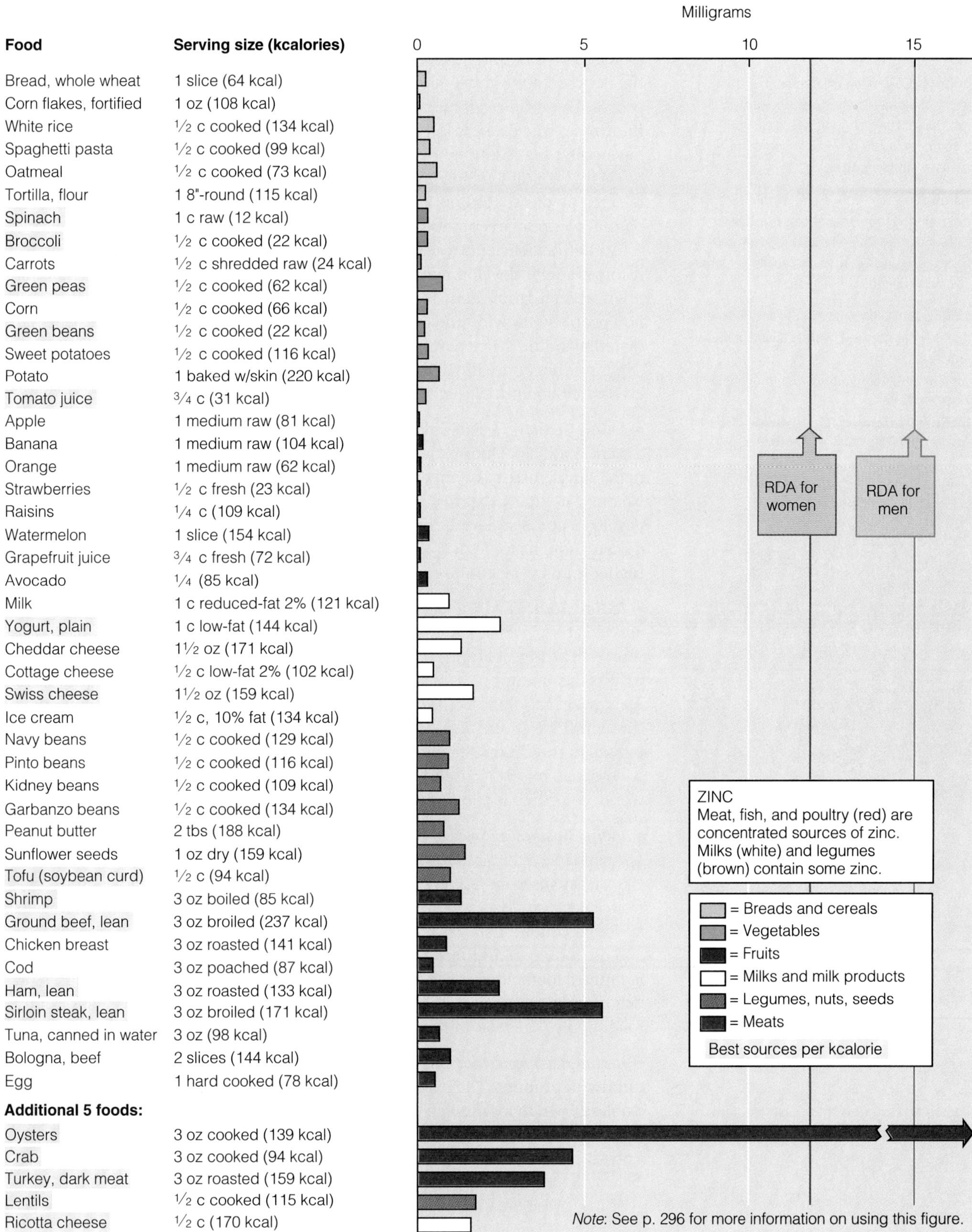

**goiter** (GOY-ter): an enlargement of the thyroid gland due to an iodine deficiency, malfunction of the gland, or overconsumption of a goitrogen. Goiter caused by iodine deficiency is **simple goiter.**

**goitrogen** (GOY-troh-jen): a thyroid antagonist found in food; causes **toxic goiter.** Goitrogens are found in such foods as cabbage, kale, brussels sprouts, cauliflower, broccoli, and kohlrabi.

**cretinism** (CREE-tin-ism): an iodine-deficiency disease characterized by mental and physical retardation.

**Figure 13-8**

**Iodine-Deficiency Symptom—The Enlarged Thyroid of Goiter**

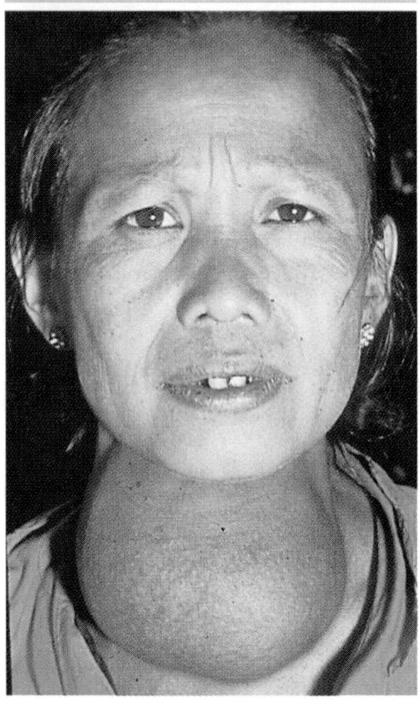

In iodine deficiency, the thyroid gland enlarges—a condition known as simple goiter.

2 g iodized salt (less than ½ tsp) contains the RDA for iodine.

• ***Iodine Deficiency*** • The hypothalamus regulates the thyroid hormones by controlling the release of the pituitary's thyroid-stimulating hormone (TSH). With iodine deficiency, thyroid hormone declines, and the body responds by secreting more TSH in a futile attempt to accelerate iodide uptake by the thyroid gland. If a deficiency persists, the cells of the thyroid gland enlarge, so as to trap as much iodide as possible. Sometimes the gland enlarges until it makes a visible lump in the neck, a simple goiter (shown in Figure 13-8). Goiter afflicts about 200 million people the world over, many of them in Asia and Africa. In all but 4 percent of these cases, the cause is iodine deficiency. As for the 4 percent (8 million), most have goiter because they overconsume plants of the cabbage family and other foods that contain an antithyroid substance (goitrogen) whose effect is not counteracted by dietary iodine. The goitrogens present in plants remind us that even natural components of foods can cause harm when eaten in excess.

An iodine deficiency causes sluggishness and weight gain. During pregnancy, it impairs fetal development, causing the extreme and irreversible mental and physical retardation known as cretinism. Cretinism affects approximately 6 million people worldwide and can be averted by the early diagnosis and treatment of maternal iodine deficiency. If treatment comes too late or not at all, the child may live his or her entire life with an IQ as low as 20 (100 is average).[22] Children with even a mild iodine deficiency typically have goiters and perform poorly in school.[23]

• ***Iodine Toxicity*** • Excessive intakes of iodine can enlarge the thyroid gland, just as deficiency can. During pregnancy, exposure to excessive iodine from foods, prenatal supplements, or medications is especially damaging to the developing infant. An infant exposed to toxic amounts of iodine during gestation may develop a goiter so severe as to block the airways and cause suffocation. The toxic dose is thought to be over 2000 micrograms per day for an adult—several times higher than average intakes.

• ***Iodine Sources*** • The ocean is the world's major source of iodine. In coastal areas, seafood, water, and even iodine-containing sea mist are dependable iodine sources. Further inland, the amount of iodine in foods is variable and generally reflects the amount present in the soil in which plants are grown or on which animals graze. Landmasses that were once under the ocean have soils rich in iodine; those in flood-prone areas where water leaches iodine from the soil are poor in iodine. In the United States and Canada, the iodization of salt has eliminated the widespread misery caused by iodine deficiency during the 1930s. Around the world, countries add iodine to bread, fish paste, or drinking water.

• ***Iodine Intakes*** • Average consumption of iodine in the United States exceeds recommendations, but falls below toxic levels as well. Some of the excess iodine in the U.S. diet stems from fast foods, which use iodized salt liberally. Some iodine comes from bakery products and from milk. The baking industry uses iodates (iodine salts) as dough conditioners, and most dairies feed cows iodine-containing medications and use iodine to disinfect milking equipment. Now that these sources have been identified, food industries have reduced their use of these compounds, but the sudden emergence of this problem points to a need for continued surveillance of the food supply. The table at the top of p. 423 provides a summary of iodine.

• ***Iodine Recommendations*** • The recommended intake of iodine for adults is a minuscule amount. The need for iodine is easily met by consuming seafood, vegetables grown in iodine-rich soil, and iodized salt. In the United States, labels indicate whether salt is iodized; in Canada, all table salt is iodized.

IN SUMMARY

Iodide, the ion of the mineral iodine, is an essential component of two thyroid hormones. An iodine deficiency can lead to simple goiter—enlargement of the thyroid gland—and can impair fetal development, causing cretinism. Iodization of salt has largely eliminated iodine deficiency in the United States and Canada.

IN SUMMARY

**Iodine**

| 1989 RDA | Deficiency Disease Name | |
|---|---|---|
| Adults: 150 µg/day | Simple goiter, cretinism | |
| **Chief Functions in the Body** | **Deficiency Symptoms** | **Toxicity Symptoms** |
| A component of two thyroid hormones that help to regulate growth, development, and metabolic rate | Enlargement of the thyroid gland, weight gain, mental and physical retardation of an infant | Enlargement of the thyroid gland, depressed thyroid activity |
| **Significant Sources** | | |
| Iodized salt, seafood, bread, dairy products, plants grown in iodine-rich soil and animals fed those plants | | |

# Selenium

The essential mineral selenium is one of the body's antioxidants, working closely with the enzyme glutathione peroxidase. Glutathione peroxidase prevents free-radical formation, thus blocking the chain reaction before it begins. (Highlight 11 describes free-radical formation, chain reactions, and antioxidant action in detail.) Glutathione peroxidase and vitamin E work in concert: if free radicals do form and a chain reaction starts, vitamin E halts it. Selenium also works closely with the enzyme that converts thyroid hormone to its active form.

**selenium** (se-LEEN-ee-um): a trace element.

• ***Selenium Deficiency*** • Selenium deficiency is associated with a heart disease that is prevalent in regions of China where the soil and foods are lacking selenium. The primary cause of this heart disease is probably a virus, but selenium deficiency appears to predispose people to it, and adequate selenium seems to prevent it.

The heart disease associated with selenium deficiency is named **Keshan** (KESH-an) **disease** for one of the provinces of China where it was studied. Keshan disease is characterized by heart enlargement and insufficiency; fibrous tissue replaces the muscle tissue that normally composes the middle layer of the walls of the heart.

• ***Selenium and Cancer*** • In other parts of the world, selenium-poor soil correlates with a high incidence of certain kinds of cancer. Some research suggests that selenium supplements may reduce the incidence of some types of cancers, but given the potential for harm and the lack of additional evidence, recommendations to take selenium supplements would be premature.[24]

U.S. Government

www.healthfinder.gov/searchoptions/topicsaz.htm

Search for Selenium

• ***Selenium Intakes*** • Some regions in the United States and Canada produce crops on selenium-poor soil, but the people are protected from deficiency, because they eat supermarket foods transported from other regions and selenium-rich meats. Because selenium is associated with the protein in foods, meats and other animal products are reliable sources.

• ***Selenium Toxicity*** • High doses (a milligram or more daily) of selenium are toxic. Selenium toxicity causes vomiting, diarrhea, loss of hair and nails, and lesions of the skin and nervous system. See the table below for a summary of selenium.

IN SUMMARY

**Selenium**

| 1989 RDA | Chief Functions in the Body | Deficiency Symptoms | Toxicity Symptoms | Significant Sources |
|---|---|---|---|---|
| Men: 70 µg/day<br>Women: 55 µg/day | Part of an enzyme system that works with vitamin E to protect body compounds from oxidation | Predisposition to heart disease characterized by cardiac tissue becoming fibrous | Digestive system disorders, loss of hair and nails, skin lesions, nervous system disorders, tooth damage | Seafood, meat, grains |

IN SUMMARY

Selenium is an antioxidant that works closely with the glutathione peroxidase enzyme and vitamin E. Selenium is found in association with protein in foods. Deficiencies are associated with a predisposition to a type of heart disease and possibly with some kinds of cancer.

# Copper

The body contains about 100 milligrams of copper. About one-fourth is in the muscles, one-fourth is in the liver, brain, and blood, and the rest is in the bones, kidneys, blood, and other tissues.

• ***Copper Roles in the Body*** • Copper serves as a constituent of several enzymes.[25] The copper-containing enzymes have diverse metabolic roles with one common characteristic: all involve reactions that consume oxygen or oxygen radicals. For example, copper-containing enzymes catalyze the oxidation of ferrous iron to ferric iron.* Copper's role in iron metabolism makes it a key factor in hemoglobin synthesis. Another copper- and zinc-containing enzyme functions as an antioxidant.† Still another copper enzyme helps to manufacture collagen and heal wounds.‡ Copper, like iron, is needed in many of the metabolic reactions related to the release of energy.§

• ***Copper Deficiency and Toxicity*** • Copper deficiency is rare, but has been seen in malnourished children.[26] Dietary factors such as phytates and high intakes of vitamin C, zinc, and iron interfere with copper absorption and can lead to deficiency.[27] Copper deficiency in animals raises blood cholesterol and damages blood vessels, leading researchers to explore whether low dietary copper might contribute to cardiovascular disease.[28] Some genetic disorders create a copper toxicity, but excessive intakes from foods are unlikely.[29]

Two rare genetic disorders affect copper status in opposite directions. In Menkes disease, the intestinal cells absorb copper, but cannot release it into circulation, causing a life-threatening deficiency. In Wilson's disease, copper accumulates in the liver and brain, creating a life-threatening toxicity. Wilson's disease can be controlled with chelating agents such as penicillamine or zinc, which interfere with copper absorption.

The use of chelation in health care is mentioned in Highlight 18's discussion of alternative therapies.

• ***Copper Recommendations and Intakes*** • The richest food sources of copper are legumes, whole grains, nuts, shellfish, organ meats, and seeds. Over half of the copper taken in food is absorbed, and the major route of elimination appears to be bile.[30] Water may also provide copper, depending on the type of plumbing pipe and the hardness of the water.[31] See the table at the top of p. 425 for a summary of copper facts.

IN SUMMARY

Copper is a component of several enzymes, all of which are involved in some way with oxygen or oxidation. Some act as antioxidants; others are essential to iron metabolism. Legumes, whole grains, and shellfish are good sources of copper.

*The copper-requiring enzymes ceruloplasmin and ferroxidase II participate in the oxidation of ferrous iron to ferric iron.
†The copper-requiring enzyme superoxide dismutase protects cell membranes against free-radical damage.
‡The copper-requiring enzyme lysyl oxidase helps synthesize connective tissues.
§The copper-requiring enzyme cytochrome C oxidase is part of the electron transport chain.

IN SUMMARY

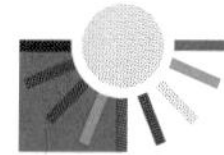

**Copper**

| 1989 Estimated Safe and Adequate Intake | Chief Functions in the Body | Deficiency Symptoms | Toxicity Symptoms | Significant Sources |
|---|---|---|---|---|
| Adults: 1.5–3.0 mg/day | Necessary for the absorption and use of iron in the formation of hemoglobin; part of several enzymes | Anemia, bone abnormalities (rare in human beings) | Vomiting, diarrhea | Seafood, nuts, grains, seeds, legumes |

# Manganese

The human body contains a tiny 20 milligrams of manganese, mostly in the bones and metabolically active organs such as the liver, kidneys, and pancreas. Manganese acts as a cofactor for many enzymes that facilitate dozens of different metabolic processes. For example, manganese metalloenzymes assist in urea synthesis, the conversion of pyruvate to a TCA cycle compound, and the prevention of lipid peroxidation by free radicals.

• ***Manganese Deficiency*** • Manganese requirements are low, and many plant foods contain significant amounts of this trace mineral, so deficiencies are rare.[32] As is true of other trace minerals, however, dietary factors such as phytates inhibit its absorption.[33] In addition, high intakes of iron and calcium limit manganese absorption, so people who use supplements of these minerals regularly may experience depressed manganese status.

• ***Manganese Toxicity*** • Toxicity is more likely to occur from an environment contaminated with manganese than from dietary intake. Miners who inhale large quantities of manganese dust on the job over prolonged periods show symptoms of a brain disease, along with abnormalities of appearance and behavior. A summary of manganese appears in the table below.

IN SUMMARY

Manganese serves as a cofactor for many enzymes involved in various metabolic processes. It is widespread in plant foods, so deficiencies are rare, although regular use of calcium and iron supplements may limit manganese absorption.

IN SUMMARY

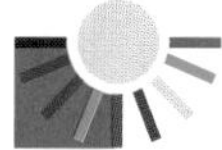

**Manganese**

| 1989 Estimated Safe and Adequate Intake | Chief Functions in the Body | Deficiency Symptoms | Toxicity Symptoms | Significant Sources |
|---|---|---|---|---|
| Adults: 2–5 mg/day | Facilitator, with enzymes, of many cell processes | In experimental animals: poor growth, nervous system disorders, reproductive abnormalities | Nervous system disorders | Widely distributed in foods |

# Fluoride

Fluoride is present in virtually all soils, water supplies, plants, and animals. Only a trace of fluoride occurs in the human body, but with this amount, the crystalline deposits in bones and teeth are larger and more perfectly formed. The table below summarizes fluoride information.

• ***Fluoride Roles in the Body*** • During the mineralization of bones and teeth, calcium and phosphorus form crystals called hydroxyapatite. Then fluoride replaces the hydroxyl (OH) portions of the hydroxyapatite crystal, forming fluorapatite, which makes the bones stronger and the teeth more resistant to decay.

**fluorapatite** (floor-APP-uh-tite): the stabilized form of bone and tooth crystal, in which fluoride has replaced the hydroxyl groups of hydroxyapatite.

• ***Fluoridation and Dental Caries*** • Dental caries ranks as the nation's most widespread health problem: an estimated 95 percent of the population have decayed, missing, or filled teeth. By interfering with a person's ability to chew and eat a wide variety of foods, dental problems can quickly lead to a multitude of nutrition problems. Where fluoride is lacking, dental decay is common. By fluoridating the drinking water, a community offers its residents, particularly the children, a safe, economical, practical, and effective way to defend against dental caries.[34]

 U.S. Government
www.healthfinder.gov/searchoptions/topicsaz.htm
Search for Fluoride

Drinking water is usually the best source of fluoride. The National Research Council of the National Academy of Sciences recommends fluoridation of drinking water to raise the concentration to about 1 part fluoride per 1 million parts water. Water with 1 part per million (1 ppm) fluoride offers the greatest protection against dental caries at virtually no risk of toxicity.

1 ppm = 1 mg per liter.

• ***Fluoride Toxicity*** • Fluoride poisoning has been reported in communities where the public water system failed and allowed fluoride concentrations to reach 150 parts per million.[35] Symptoms of fluoride poisoning include nausea, vomiting, diarrhea, abdominal pain, and numbness or tingling of the face and extremities.

• ***Fluorosis*** • Too much fluoride can damage the teeth, causing fluorosis. In mild cases, the teeth develop small white specks; in severe cases, the enamel becomes pitted and permanently stained (as shown in Figure 13-9). Fluorosis occurs only during tooth development and cannot be reversed, making its prevention a high priority.

**fluorosis** (floor-OH-sis): discoloration and pitting of tooth enamel caused by excess fluoride during tooth development.

To prevent fluorosis:
• Monitor the fluoride content of the local water supply.
• Supervise toddlers when they brush their teeth and use only a little toothpaste (pea-size amount).
• Use fluoride supplements only as prescribed by a physician.

• ***Fluoride Intakes*** • About half of the U.S. population has access to water with an optimal fluoride concentration, which typically delivers about 1 milligram per person per day. Fish and most teas contain appreciable amounts of natural fluoride.

IN SUMMARY

Fluoride makes bones stronger and teeth more resistant to decay. Fluoridation of public water supplies can significantly reduce the incidence of dental caries, but an excess of fluoride during tooth development can cause fluorosis—discolored and pitted tooth enamel.

IN SUMMARY

**Fluoride**

| 1997 Adequate Intake (AI) | Upper Level | Chief Functions in the Body | Deficiency Symptoms | Toxicity Symptoms | Significant Sources |
|---|---|---|---|---|---|
| Men: 3.8 mg/day<br>Women: 3.1 mg/day | Adults: 10 mg/day | Involved in the formation of bones and teeth; helps to make teeth resistant to decay | Susceptibility to tooth decay | Fluorosis (discoloration of teeth), nausea, diarrhea, chest pain, itching, vomiting | Drinking water (if fluoride containing or fluoridated), tea, seafood |

# Chromium

Chromium is an essential mineral that participates in carbohydrate and lipid metabolism. Like iron, chromium can have different charges. In the case of chromium, the $Cr^{+++}$ ion seems to be the most effective.

• ***Chromium Roles in the Body*** • Chromium helps maintain glucose homeostasis by enhancing the activity of the hormone insulin.[36] Consequently, less insulin is needed to control blood glucose. When chromium is lacking, a diabeteslike condition may develop with elevated blood glucose and impaired glucose tolerance, insulin response, and glucagon response.

• ***Chromium Recommendations and Intake*** • Chromium is present in a variety of foods. Unrefined foods are the best sources, particularly liver, brewer's yeast, whole grains, nuts, and cheeses. The more refined foods people eat, the less chromium they ingest. The table below provides a summary of chromium.

• ***Chromium Picolinate Supplements*** • Supplement advertisements have succeeded in convincing consumers that they can lose fat and build muscle by taking chromium picolinate. The manufacturers claim their formulations are better absorbed and more effective than plain chromium even though no scientific research supports such claims. Neither does sound research support the claims that chromium—picolinate or plain—supplements reduce body fat or improve muscle strength. Highlight 14 revisits chromium picolinate and other supplements athletes use in the hopes of improving their performance.

IN SUMMARY

Chromium enhances insulin's action. A deficiency can result in a diabeteslike condition. Chromium is widely available in unrefined foods including brewer's yeast, whole grains, and liver.

**Figure 13-9**

**Fluoride-Toxicity Symptom—The Mottled Teeth of Fluorosis**

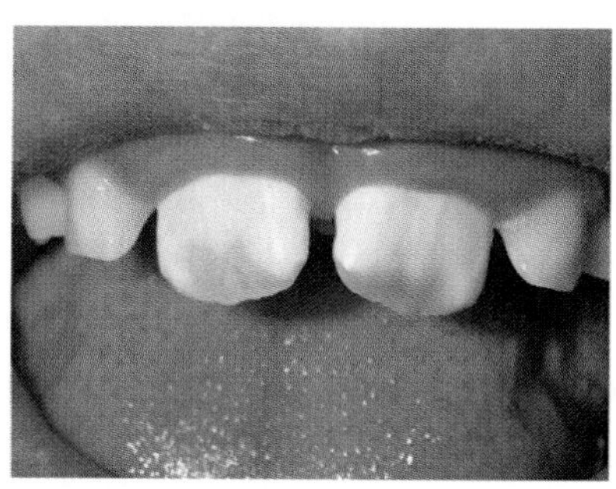

Small organic compounds that enhance insulin's action are called **glucose tolerance factors (GTF).** Some glucose tolerance factors contain chromium.

# Molybdenum

Molybdenum acts as a working part of several metalloenzymes. Deficiencies of molybdenum are unknown because the amounts needed are minuscule—as little as 0.1 part per million parts of body tissue. Legumes, breads and other grains, leafy green vegetables, milk, and liver are molybdenum-rich foods. Average daily intakes fall within the suggested range of intakes.

Molybdenum toxicity is rare, but has been reported in workers exposed to its dust. Characteristics include goutlike symptoms in human beings. For a summary of molybdenum facts, see the table on p. 428.

**molybdenum** (mo-LIB-duh-num): a trace element.

IN SUMMARY

**Chromium**

| 1989 Estimated Safe and Adequate Intake | Chief Functions in the Body | Deficiency Symptoms | Toxicity Symptoms | Significant Sources |
|---|---|---|---|---|
| Adults: 50–200 µg/day | Associated with insulin and required for the release of energy from glucose | Diabeteslike condition marked by an inability to use glucose normally | Unknown as a nutrition disorder; occupational exposures damage skin and kidneys | Meat, unrefined foods, fats, vegetable oils |

IN SUMMARY

**Molybdenum**

| 1989 Estimated Safe and Adequate Intake | Chief Functions in the Body | Deficiency Symptoms | Toxicity Symptoms | Significant Sources |
|---|---|---|---|---|
| Adults: 75–250 µg/day | Facilitator, with enzymes, of many cell processes | Unknown | Enzyme inhibition, goutlike symptoms | Legumes, cereals, organ meats |

# Other Trace Minerals

Research to determine whether other trace minerals are essential is difficult, both because their quantities in the body are so small and because human deficiencies are unknown. Much of the available knowledge comes from research using animals.

Nickel is recognized as important for the health of many body tissues; deficiencies harm the liver and other organs. Silicon is needed for healthy bones, brains, and blood vessels in animals; some researchers believe it may be essential for human beings as well.[37] Tin is necessary for growth in animals. Vanadium, too, is necessary for growth and bone development and also for normal reproduction. Cobalt is a key mineral in the large vitamin $B_{12}$ molecule (see Figure 13-10), but it is not an essential nutrient and no recommendation has been established. Boron may play a key role in bone development and the prevention of osteoporosis.[38] In the future many other trace minerals may turn out to play key nutritional roles. Even arsenic—famous as a poison used by murderers and known to be a carcinogen—may turn out to be essential for human beings in tiny quantities.

**Figure 13-10**
**Cobalt with Vitamin $B_{12}$**

The intricate vitamin $B_{12}$ molecule contains one atom of the mineral cobalt. The alternative name for vitamin $B_{12}$, cobalamin, reflects the presence of cobalt in its structure.

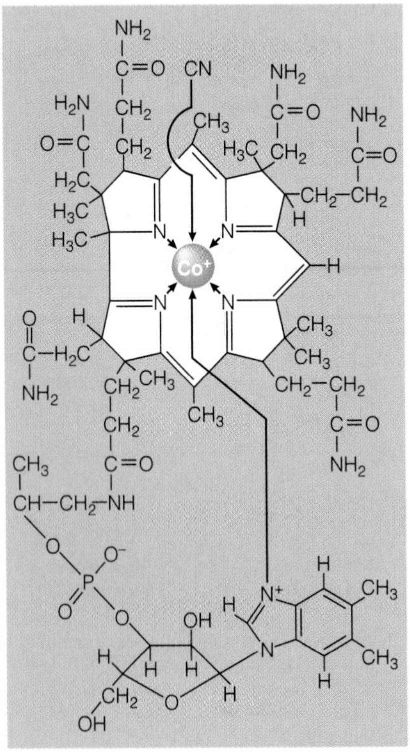

# Closing Thoughts on the Nutrients

This chapter completes the introductory lessons on the nutrients. Each nutrient from the amino acids to zinc has been described rather thoroughly—its chemistry, roles in the body, sources in the diet, symptoms of deficiency and toxicity, and influences on health and disease. Such a detailed examination is informative, but it can also be misleading. It is important to step back from the myopic study of the individual nutrients to look at them as a whole. After all, people eat foods, not nutrients, and most foods deliver dozens of nutrients. Furthermore, nutrients work cooperatively with each other in the body; their actions are most often *interactions*. This chapter alone mentioned how iron depends on vitamin C to keep it in its active form and copper to incorporate it into hemoglobin; how zinc is needed to activate and transport vitamin A; and how both iodine and selenium are used in the synthesis of thyroid hormones.

Estimates of how much of each particular nutrient the body needs fall between intakes that are inadequate and cause illness and intakes that are excessive and cause illness. Between deficiency and toxicity lies a wide range of intakes that support health—to varying degrees. In the past, nutrient needs were determined by how much was needed to cure deficiency symptoms. If lack of a nutrient caused illness, it was defined as essential.

Today, nutrient needs are based on how much is needed to support optimal health. The amount of vitamin C needed to prevent scurvy is much less than the amount correlated with reducing the risk of cancer, for example. Furthermore, nutrients are being examined within the context of the whole diet. Health benefits are not credited to vitamin C alone, but to the vitamin C–rich fruits and vegetables that also provide many other nutrients—and nonnutrients (phytochemicals)—important to health.

People can also improve their health with physical activity. Energy expenditure is the opposite of money expenditure: it is desirable to *spend* energy, not to save it (within reason, of course). The more energy people spend, the more food they can afford to eat—food delivering both nutrients and pleasure. The next chapter describes nutrition and physical activity.

## Making It Click

Once you have mastered these examples, you will understand minerals a little better and be prepared to examine your own food choices. Be sure to show your calculations for each problem. (see Appendix K for answers.)

1. Appreciate foods for their iron density. Following is a list of foods with the energy amount and the iron content per serving.

| Food | Iron (mg) | Energy (kcal) | Iron Density (mg/kcal) |
|---|---|---|---|
| Milk, nonfat, 1 c | 0.10 | 85 | ____ |
| Cheddar cheese, 1 oz | 0.19 | 114 | ____ |
| Broccoli, cooked from fresh, chopped, 1 c | 1.31 | 44 | ____ |
| Sweet potato, baked in skin, 1 ea | 0.51 | 117 | ____ |
| Cantaloupe melon, ½ | 0.56 | 93 | ____ |
| Carrots, from fresh, ½ c | 0.48 | 35 | ____ |
| Whole-wheat bread, 1 slice | 0.87 | 64 | ____ |
| Green peas, cooked from frozen, ½ c | 1.26 | 62 | ____ |
| Apple, medium | 0.38 | 125 | ____ |
| Sirloin steak, lean, 4 oz | 3.81 | 228 | ____ |
| Pork chop, lean, broiled, 1 ea | 0.66 | 166 | ____ |

   a. Rank these foods by iron per serving.
   b. Calculate the iron density (divide milligrams by kcalories) for these foods and rank them by their iron per kcalorie.
   c. Name three foods that are higher on the second list than they were on the first list.
   d. What do these foods have in common?

2. Some people find it difficult to eat enough iron-rich foods and absorb enough iron to meet their needs.
   a. Calculate the iron that a woman of childbearing age might absorb from the following meal:

| Food | Iron (mg) | Vitamin C (mg) |
|---|---|---|
| Sirloin steak, lean, 4 oz | 3.81 | 0 |
| Green peas, cooked from frozen, ½c | 1.26 | 8 |
| Brown rice, cooked, 1 c | 0.82 | 0 |
| Iced tea, instant, sweetened, 1 c | 0.05 | 0 |

   b. Use the "How to" on p. 409 to calculate total iron absorbed. Show your calculations.
   c. The RDA assumes that a person absorbs 10 percent of the iron eaten. Was that true in this case? What percentage of iron did this person absorb?
   d. Does the meal include any factor that might reduce the absorption of iron as calculated here?
   e. Setting this factor aside for the moment, and assuming the person eats three meals similar to this one, is a woman of childbearing age likely to meet her iron RDA for the day? Show your calculations.

## Study Questions

These questions will help you review the chapter. You will find the answers in the discussions on the pages provided.

1. Distinguish between heme and nonheme iron. Discuss the factors that enhance iron absorption. (pp. 407–408)
2. Distinguish between iron deficiency and iron-deficiency anemia. What are the symptoms of iron-deficiency anemia? (pp. 410–411)
3. What causes iron overload? What are its symptoms? (p. 412)
4. Describe the similarities and differences in the absorption and regulation of iron and zinc. (pp. 406–407, 416–417)

5. Discuss possible reasons for a low intake of zinc. What factors affect the bioavailability of zinc? (pp. 418, 420)
6. Describe the principal functions of iodide, selenium, copper, manganese, fluoride, chromium and molybdenum in the body. (pp. 420, 423, 424, 425, 426, 427)
7. What public health measure has been used in preventing simple goiter? What measure has been recommended for protection against tooth decay? (pp. 422, 426)
8. Discuss the importance of balanced and varied diets in obtaining the essential minerals and avoiding toxicities. (pp. 405, 426)
9. Describe some of the ways trace minerals interact with each other and with other nutrients. (pp. 405, 426)

These questions will help you prepare for an exam. Answers can be found in Appendix K.

1. Iron absorption is impaired by:
   a. heme.
   b. phytates.
   c. vitamin C.
   d. MFP factor.
2. Which of these people is *least* likely to develop an iron deficiency?
   a. 3-year-old boy
   b. 52-year-old man
   c. 17-year-old girl
   d. 24-year-old woman
3. Which of the following would *not* describe the blood cells of a severe iron deficiency?
   a. anemic
   b. microcytic
   c. pernicious
   d. hypochromic
4. Which provides the most absorbable iron?
   a. 1 apple
   b. 1 c milk
   c. 3 oz steak
   d. ½ c spinach
5. The intestinal protein that helps to regulate zinc absorption is:
   a. albumin.
   b. ferritin.
   c. hemosiderin.
   d. metallothionein.
6. A classic sign of zinc deficiency is:
   a. anemia.
   b. goiter.
   c. mottled teeth.
   d. growth retardation.
7. Cretinism is caused by a deficiency of:
   a. iron.
   b. zinc.
   c. iodine.
   d. selenium.
8. The mineral best known for its role as an antioxidant is:
   a. copper.
   b. selenium.
   c. manganese.
   d. molybdenum.
9. Fluorosis occurs when fluoride:
   a. is excessive.
   b. is inadequate.
   c. binds with phosphorus.
   d. interacts with calcium.
10. Which mineral enhances insulin activity?
   a. zinc
   b. iodine
   c. chromium
   d. manganese

## Notes


1. G. J. Beckett and coauthors, Effects of combined iodine and selenium deficiency on thyroid hormone metabolism in rats, *American Journal of Clinical Nuirition* 57 (1993): 240S–243S.
2. E. Beutler, How little we know about the absorption of iron, *American Journal of Clinical Nutrition* 66 (1997): 419–420.
3. J. L. Beard, H. Dawson, and D. J. Piñero, Iron metabolism: A comprehensive review, *Nutrition Reviews* 54 (1996): 295–317.
4. E. R. Monsen and coauthors, Estimation of available dietary iron, *American Journal of Clinical Nutrition* 31 (1978): 134–141.
5. H. Munro, The ferritin genes: Their response to iron status, *Nutrition Reviews* 51 (1993): 65–73.
6. C. E. West, Strategies to control nutritional anemia, *American Journal of Clinical Nutrition* 64 (1996): 789–790.
7. A. C. Looker and coauthors, Prevalence of iron deficiency in the United States, *Journal of the American Medical Association* 277 (1997): 973–976; Recommendations to prevent and control iron deficiency in the United States, *Morbidity and Mortality Weekly Report* 47 (1998): supplement.
8. R. Yip and coauthors, Pervasive occult gastrointestinal bleeding in an Alaska native population with prevalent iron deficiency: Role of *helicobacter pylori* gastritis, *Journal of the American Medical Association* 277 (1997): 1135–1139; D. C. Rockey and J. P. Cello, Evaluation of the gastrointestinal tract in patients with iron-deficiency anemia, *New England Journal of Medicine* 329 (1993): 1691–1695.
9. P. J. Garry, K. M. Koehler, and T. L. Simon, Iron stores and iron absorption: Effects of repeated blood donations, *American Journal of Clinical Nutrition* 62 (1995): 611–620.
10. L. Hallberg, L. Hultén, and E. Gramatkovski, Iron absorption from the whole diet in men: How effective is the regulation of iron absorption? *American Journal of Clinical Nutrition* 66 (1997): 347–356.
11. J. C. Fleet, Discovery of the hemochromatosis gene will require rethinking the regulation of iron metabolism, *Nutrition Reviews* 54 (1996): 285–292.
12. Iron overload disorders among Hispanics—San Diego, California, 1995,

*Morbidity and Mortality Weekly Report* 45 (1996): 991–993; D. H. G. Crawford and coauthors, Factors influencing disease expression in hemochromatosis, *Annual Review of Nutrition* 16 (1996): 139–160; C. E. McLaren and coauthors, Prevalence of heterozygotes for hemochromatosis in the white population of the United States, *Blood* 84 (1995): 2121–2127.

13. V. Herbert, S. Shaw, and E. Jayatilleke, Vitamin C supplements are harmful to lethal for over 10% of Americans with high iron stores, *FASEB Journal* 8 (1994): A678.
14. C. T. Sempos, A. C. Looker, and R. F. Gillum, Iron and heart disease: The epidemiologic data, *Nutrition Reviews* 54 (1996): 73–84.
15. J. T. Salonen and coauthors, High stored iron levels are associated with excess risk of myocardial infarction in eastern Finnish men, *Circulation* 86 (1992): 803–811.
16. T. P. Tuomainen and coauthors, Cohort study of relation between donating blood and risk of myocardial infarction in 2682 men in eastern Finland, *British Medical Journal* 314 (1997): 793–794.
17. J. M. McCord, Effects of positive iron status at a cellular level, *Nutrition Reviews* 54 (1996): 85–88.
18. R. L. Nelson and coauthors, Body iron stores and risk of colonic neoplasia, *Journal of the National Cancer Institute* 86 (1994): 455–460.
19. J. L. Rosado and coauthors, Zinc supplementation reduced morbidity, but neither zinc nor iron supplementation affected growth or body composition of Mexican preschoolers, *American Journal of Clinical Nutrition* 65 (1997): 13–19.
20. Zinc lozenges reduce the duration of common cold symptoms, *Nutrition Reviews* 55 (1997): 82–88.
21. S. B. Mossad and coauthors, Zinc gluconate lozenges for treating the common cold: A randomized, double-blind, placebo-controlled study, *Annals of Internal Medicine* 125 (1996): 81–88.
22. N. Bleichrodt and coauthors, The benefits of adequate iodine intake, *Nutrition Reviews* 54 (1996): S72–S78; C. Xue-Yi and coauthors, Timing of vulnerability of the brain to iodine deficiency in endemic cretinism, *New England Journal of Medicine* 331 (1994): 1739–1744.
23. B. D. Tiwari and coauthors, Learning disabilities and poor motivation to achieve due to prolonged iodine deficiency, *American Journal of Clinical Nutrition* 63 (1996): 782–786.
24. L. C. Clark and coauthors, Effects of selenium supplementation for cancer prevention in patients with carcinoma of the skin—A randomized controlled trial, *Journal of the American Medical Association* 276 (1996): 1957–1963; G. A. Colditz, Selenium and cancer prevention—Promising results indicate further trials required, *Journal of the American Medical Association* 276 (1996): 1984–1985; Letters from V. Herbert, L. H. Kuller, J. S. Parker, and L. C. Clark, Selenium supplementation and cancer rates, *Journal of the American Medical Association* 277 (1997): 880–881.
25. R. Uauy, M. Olivares, and M. Gonzalez, Essentiality of copper in humans, *American Journal of Clinical Nutrition* 67 (1998): 952S–959S.
26. A. Cordano, Clinical manifestations of nutritional copper deficiency in infants and children, *American Journal of Clinical Nutrition* 67 (1998): 1012S–1016S.
27. R. A. Wapnir, Copper absorption and bioavailability, *American Journal of Clinical Nutrition* 67 (1998): 1054S–1060S; B. Lönnerdal, Bioavailability of copper, *American Journal of Clinical Nutrition* 63 (1996): 821S–829S.
28. G. E. Bunce, Hypercholesterolemia of copper deficiency is linked to glutathione metabolism and regulation of hepatic MHG-CoA reductase, *Nutrition Reviews* 51 (1993): 305–307; Decreased dietary copper impairs vascular function, *Nutrition Reviews* 51 (1993): 188–189; Low-copper diets increase aortic lipid peroxides in rats, *Nutrition Reviews* 51 (1993): 88–89.
29. Z. L. Harris and J. D. Gitlin, Genetic and molecular basis for copper toxicity, *American Journal of Clinical Nutrition* 63 (1996): 836S–841S.
30. M. C. Linder and M. Hazegh-Azam, Copper biochemistry and molecular biology, *American Journal of Clinical Nutrition* 63 (1996): 797S–811S.
31. D. J. Fitzgerald, Safety guidelines for copper in water, *American Journal of Clinical Nutrition* 67 (1998): 1098S–1102S.
32. S. Fairweather-Tait and R. F. Hurrell, Bioavailability of minerals and trace elements, *Nutrition Research Reviews* 9 (1996): 295–324.
33. L. Davidson and coauthors, Manganese absorption in humans: The effect of phytic acid and ascorbic acid in soy formula, *American Journal of Clinical Nutrition* 62 (1995): 984–987.
34. Position of The American Dietetic Association: The impact of fluoride on dental health, *Journal of the American Dietetic Association* 94 (1994): 1428–1431.
35. B. D. Gessner and coauthors, Acute fluoride poisoning from a public water system, *New England Journal of Medicine* 330 (1994): 95–99; D. E. Leland, K. E. Powell, and R. S. Anderson, Jr., A fluoride overfeed incident at Harbor Springs, Mich., *Journal of the American Water Works Association* 72 (1980): 238–243; L. R. Petersen and coauthors, Community health effects of a municipal water supply hyperfluoridation accident, *American Journal of Public Health* 78 (1988): 711–713; Acute fluoride poisoning—North Carolina, *Morbidity and Mortality Weekly Report* 23 (1974): 199.
36. Fairweather-Taite and Hurrell, 1996.
37. C. D. Seaborn and F. H. Nielsen, Silicon: A nutritional beneficence for bones, brains, and blood vessels? *Nutrition Today,* July/August 1993, pp. 13–18.
38. H. McCoy and coauthors, Relation of boron to the composition and mechanical properties of bone, *Environmental Health Perspectives* (supplement) 102 (1994): 49–53.

# Our Children's Daily Lead*

At nine months, Joey crawled about exploring the world around him—touching and tasting everything, as all infants do. He chewed on table legs, toys, the spindles of railings with flaky paint—whatever was in his reach. He eagerly drank his morning bottle of formula, which his mother prepared with the first water drawn from the tap. Not until he was two did he toddle about while his parents watched proudly. At four, he amused his parents when he chased after balls tossed his way, but he couldn't catch them. By age five, he was a cautious, quiet preschooler who clung tightly to stair railings with both hands as he slowly climbed up or down.

Joey was late in walking, small for his age, seldom played as vigorously as other children, and was prone to health disturbances such as diarrhea, irritability, and lethargy. His kindergarten teacher reported that Joey had difficulty hearing and that his progress was slower than expected. While his health quietly deteriorated, his parents thought these subtle symptoms were within the range of normal variations seen in children. Finally, a pediatrician detected lead toxicity and started treating Joey with lead-scavenging drugs.[1†] Joey is now growing normally and playing vigorously, although he still has minor learning disabilities. His physician expects the deficits in brain function to persist into adulthood.

Even one year of lead exposure can permanently injure the brain, nervous system, and psychological functioning. Furthermore, the effects occur with even low exposure.[2] The amount of lead in the blood recognized to cause harm is estimated to be 10 micrograms per 100 milliliters of blood; earlier it was thought to be 25. Health agencies point to lead poisoning as the most serious environmental threat to young children. The FDA has proposed reducing the acceptable level of lead in foods tenfold—from its 1958 limit of 10 parts per million to 0.5 to 1.0 part per million.[3]

This highlight describes how lead disrupts body processes and impairs nutrition status and then points out sources of lead in the environment. With awareness, we can make changes that will safeguard the health of our children.

## Lead in the Body

Chapters 12 and 13 told of the many ways minerals serve the body—maintaining fluid and electrolyte balance, providing structural support to the bones, transporting oxygen, and assisting enzymes. In contrast to those minerals that the body requires, lead impairs the body's growth, work capacity, and general health.

Like other minerals, lead is indestructible; the body cannot change its chemistry. Chemically similar to nutrient minerals like iron, calcium, and zinc (cations with two positive charges), lead displaces them from some of the metabolic sites they normally occupy, but is then unable to perform their roles. For example, lead interferes with the enzymes that facilitate heme formation (see Figure H13-1). Excess lead in the blood also deranges the structure of red blood cell membranes, making them leaky and fragile. Lead interacts with white blood cells, too, impairing their ability to fight infection, and it binds to antibodies, thereby impairing the body's resistance to disease.

In addition to its effects on the blood, lead damages many body systems, particularly the vulnerable nervous system, kidneys, and bone marrow. It impairs such normal activities as growth by interfering with hormone activity. In short, lead's interactions with a variety of substances in the body have profound adverse effects.[4] The greater the exposure, the more damaging the effects. Table H13-1 lists symptoms of lead toxicity.

## Lead and Growing Children

The body readily absorbs lead, especially during times of rapid growth. Thereafter, it hoards lead possessively. Lead is not easily excreted and accumulates mainly in the bones, but also in the brain, teeth, and kidneys.

During pregnancy, lead readily moves across the placenta, inflicting severe damage on the developing fetal nervous system. In addition, infants exposed to low levels of lead during gestation weigh less at birth and consequently struggle to survive.[5]

Infants and young children absorb five to ten times as much lead as adults do. One out of every six children from six months to five years old and one out of every nine fetuses are exposed to harmful doses of lead. Lead toxicity is most prevalent among children under six—as many as 1.7 million children (10 to 15 percent of all preschoolers) may have blood lead concentrations high enough to cause mental, behavioral, and other health problems.[6]

Lead poisoning in infants most often comes from infant formula made with con-

*Old, lead-based paint threatens the health of an exploring child.*

*Title borrowed from M. A. Wessel and A. Dominski, Our children's daily lead, *American Scientist* 65 (1977): 294–298.

†The majority of children with high blood lead levels (45 µg/dL) are treated using a process called chelation, which involves using drugs (most often succimer or calcium-disodium EDTA) that bind to lead in the blood and carry it out in the urine.

**Figure H13-1**
**Lead Displaces Iron**

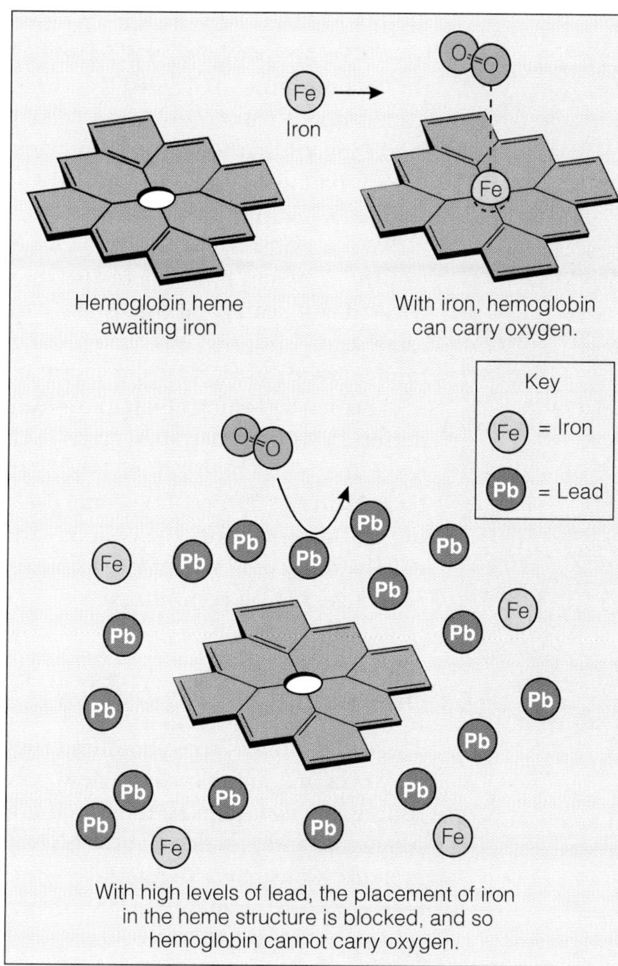
Fe
Iron
O=O
Fe
Hemoglobin heme awaiting iron
With iron, hemoglobin can carry oxygen.
Key
Fe = Iron
Pb = Lead
O=O
Pb
Fe
With high levels of lead, the placement of iron in the heme structure is blocked, and so hemoglobin cannot carry oxygen.

taminated water. The water, in turn, receives its lead burden from lead-soldered plumbing. The first water drawn from the tap each day is highest in lead—therefore, a person living in a house with old, lead-soldered plumbing should let the water run a few minutes before drinking or using it to prepare formula or food.

Lead intoxication in young children comes from their own behaviors and activities—putting their hands in their mouths, playing in dirt, and eating nonfood items (see Figure H13-2 on p. 434). The toddler years see a marked rise in blood lead. Tragically, a child's neuromuscular system is maturing at precisely the same time. No wonder children with high blood lead experience impairment of balance, motor development, and the relaying of nerve messages to and from the brain. Among children two and three years old, those with the highest blood lead suffer the greatest developmental delays at age four. Parents and teachers of older children with high bone lead levels report antisocial and delinquent behavior among these children.[7] Researchers studying children's development and behavior must consider the possibility that lead poisoning may affect their results.

**Table H13-1**
**Symptoms of Lead Toxicity**

**In children:**
- Learning disabilities (reduced short-term memory; impaired concentration)
- Low IQ
- Behavior problems
- Slow growth
- Iron-deficiency anemia
- Dental caries
- Sleep disturbances (night waking, restlessness, head banging)
- Nervous system disorders; seizures
- Slow reaction time; poor coordination
- Impaired hearing

**In adults:**
- Hypertension
- Reproductive complications
- Kidney failure

## *The Malnutrition-Lead Connection*

Nutrient deficiencies enhance lead toxicity.[8] Children who are malnourished are most vulnerable to lead poisoning. Children absorb more lead if their stomachs are empty, if they have low calcium or zinc intakes, and, of greatest concern because it is so common, if they have iron deficiencies.[9]

As Chapter 13 mentioned, lead poisoning can cause iron deficiency, and iron deficiency weakens the body's defenses against lead absorption. In fact, the interactions between lead poisoning and iron deficiency are so strong that lead poisoning is considered an adverse consequence of iron deficiency. A child with adequate iron stores is not immune, but a child with iron-deficiency anemia is three times as likely to have high blood lead. Common to both iron deficiency and lead poisoning are a low socioeconomic background and a lack of immunizations against infectious diseases.[10] Another common factor is pica—a craving for nonfood items. Many children with lead poisoning eat dirt or newspapers, two common sources of lead.

The anemia brought on by lead poisoning may be mistaken for a simple iron deficiency and therefore may be incorrectly treated. Like iron deficiency, mild lead toxicity has nonspecific effects, including diarrhea, irritability, and fatigue. The symptoms are not reversible by adding iron to the diet; exposure to lead must stop. With further exposure, the signs become more pronounced: children lose cognitive, verbal, and perceptual abilities and develop learning disabilities and behavior problems. Still more severe lead toxicity can cause irreversible nerve damage, paralysis, mental retardation, and death.

Just as iron deficiency enhances lead uptake, an inadequate calcium intake enhances lead absorption and retention. Zinc deficiency also enhances both tissue accumulation of lead and sensitivity to its effects; serum zinc is frequently low in children with high blood lead. Prevention of lead toxicity rests primarily on reduced exposure, but parents can protect their children to some degree, by making sure that they eat foods rich in iron, calcium, zinc, and other nutrients.

## *Lead in the Environment*

All foods contain some lead. Most of it derives from industrial pollution. People are exposed to lead in some types of gasoline, paint, newspaper ink, batteries, shotgun

**Figure H13-2**

**Sources of Lead Exposure**

Lead finds its way into the bodies of children when they ingest lead-containing foods, water, dust, or paint chips, or when they breathe lead-laden air.

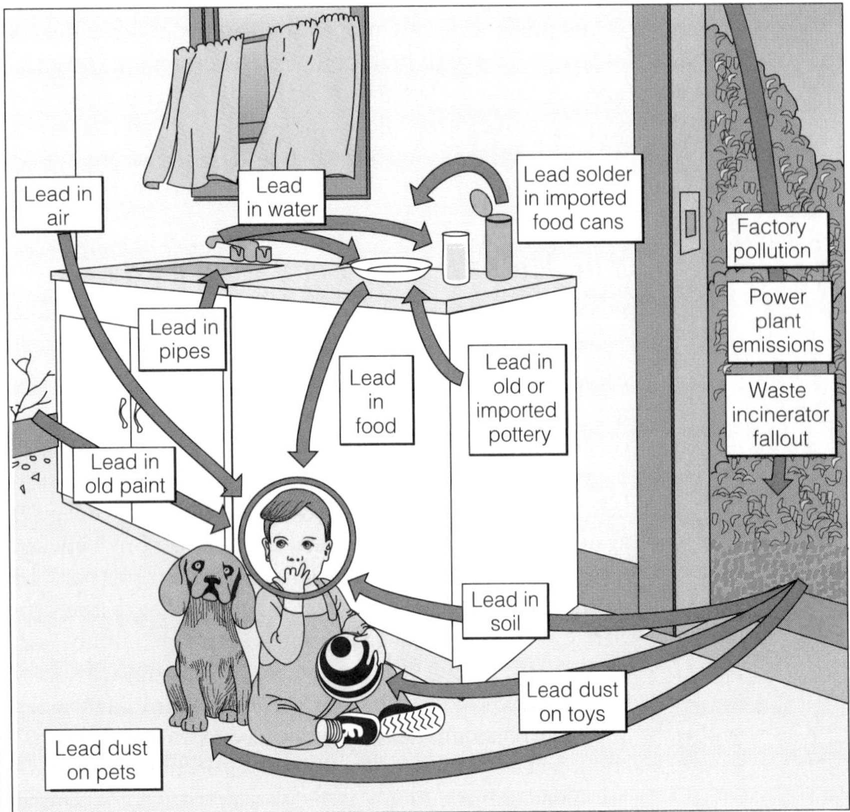


ammunition, hair dyes, and pesticides as well as in the air and water that carry lead from industrial processes and landfills. Lead works its way through rainfall and soil into plants and animals that people use for food. Lead also enters food from containers such as tin cans sealed with lead solder. Manufacturers in the United States no longer use lead solder in canning, but many in foreign countries still do. Even though the FDA has banned imported canned foods sealed with lead solder, they still appear on the shelves of small ethnic grocery stores.[11] Similarly, herbal remedies imported from other countries may contain lead.[12] Old, handmade, or imported pottery decorated with lead glazes can also leach lead into foods.

Faucets and pipes soldered with lead or coupled with brass fixtures also release lead into drinking water, in which it is the nation's most significant contaminant. Unsafe levels of lead are most likely in older communities along the nation's east coast; in urban and industrial areas; near highways; and in slums where old leaded paint peels from the buildings. People suffering the effects of lead exposure are often black, male, and from low-income families, but it is also seen in upper-middle-class families as they move into inner-city areas and renovate older homes.

Federal law has mandated reductions in leaded gasolines, lead-based solder, and other products over the past three decades. These efforts have helped to reduce the amounts of lead in the environment—and in children's blood. The decline in blood lead in children since the 1970s has paralleled the decline in the nation's use of leaded gasoline, leaded house paint, and lead-soldered food cans. In that time, the percentage of our nation's children poisoned by lead has fallen from 25 to 1.

Paint remains the primary source; 70 percent of homes built before 1960 are covered with lead paint that is likely to cause lead poisoning, especially during times of renovation. A routine question asked when screening children for lead poisoning is whether they have ever lived in a house built before 1960.[13] If leaded surfaces in these homes are peeling and deteriorating, the children are either already poisoned or face immediate danger of lead poisoning.

## *Strategies for Protection*

Three major discoveries about lead toxicity occurred simultaneously: lead poisoning has *subtle* effects; the effects are *permanent;* and they occur at *low levels of exposure.* Consumers should take ultraconservative measures to protect themselves, and especially their infants and young children, from lead poisoning. Defensive strategies include:

- The American Academy of Pediatrics recommends testing children for lead poisoning; effective screening with an appropriate questionnaire helps to identify high-risk children and thus treat the devastating effects.[14] About half the pediatricians surveyed report universal screening.[15]
- In contaminated environments, keep small children from putting dirty or old painted objects in their mouths, and make sure children wash their hands before eating. Similarly, keep small children from eating any nonfood items. Lead poisoning has been reported in young children who have eaten pool cue chalk.[16]
- Be aware that other countries do not have the same regulations protecting consumers against lead. Children have been poisoned by eating crayons made in China and drinking fruit juice canned in Mexico.
- Make infant formula from lead-free ingredients. Do not use lead-contaminated water.
- Once you have opened canned food, immediately move it to a lead-free storage container to prevent lead migration into the food.
- Do not store acidic foods or beverages (such as vinegar or

orange juice) in ceramic dishware or alcoholic beverages in pewter or crystal decanters.

- Many manufacturers are now making lead-safe products.* Old, handmade, or imported ceramic cups and bowls may contain lead and should not be used to heat coffee or tea or acidic foods such as tomato soup.
- U.S. wineries have stopped using lead in their foil seals, but older bottles may still be around and other countries may still use lead; to be safe, wipe the foil-sealed rim of a wine bottle with a clean wet cloth before removing the cork.
- Feed children nutritious meals regularly (see Chapter 16 for more details).
- Before using your newspaper to wrap food, mulch garden plants, or add to your compost, confirm with the publisher that the paper uses no lead in its ink.

The Environmental Protection Agency (EPA) also publishes a booklet, *Lead and Your Drinking Water,* in which the following cautions appear:

- Have the water in your home tested by a competent laboratory.
- Use only cold water for drinking, cooking, and making formula (cold water absorbs less lead).
- When water has been standing in pipes for more than two hours, flush the cold-water pipes by running water through them for 30 seconds before using it for drinking, cooking, or mixing formulas.
- If lead contamination of your water supply seems probable, obtain additional information and advice from the EPA and your local public health agency.

By taking these steps, parents can protect themselves and their children from this preventable danger.†

This highlight may appear to have been just about lead, but its actions typify the ways all heavy metals behave in the body: they interfere with nutrients that are trying to do their jobs. The "good guy" nutrients are shoved aside by the "bad guy" contaminants. Then the contaminants—whether lead, mercury, cadmium, or some other—cannot perform the roles of the nutrients, and health declines. To safeguard our health, we must defend ourselves against contamination by eating nutrient-rich foods and preserving a clean environment.

U.S. Government

www.healthfinder.gov/searchoptions/topicsaz.htm

Search for Lead

National Safety Council

www.nsc.org

Visit the Lead program

*A Shopper's Guide to Low-Lead China is available from the Environmental Defense Fund, 257 Park Avenue South, New York, NY 10010; telephone (800) 284-3322.

†The National Lead Information Center provides two hotlines; call (800) LEAD-FYI (532-3394) for general information or (800) 424-LEAD (424-5323) with specific questions.

Centers for Disease Control

www.cdc.gov/nceh/programs/lead.htm

Environmental Protection Agency
National Lead Information Center

www.epa.gov/opptintr/lead

## Notes


1. American Academy of Pediatrics, Committee on Drugs, Treatment guidelines for lead exposure in children, *Pediatrics* 96 (1995): 155–160.
2. K. N. Dietrich, O. G. Berger, and P. A. Succop, Lead exposure and the motor developmental status of urban six-year-old children in the Cincinnati Prospective Study, *Pediatrics* 97 (1993): 301–307.
3. FDA seeks lower lead levels in food additives and GRAS ingredients, *Journal of the American Dietetic Association* 94 (1994): 495.
4. P. Mushak, A. Crocetti, Lead and nutrition: Biologic interactions of lead with nutrients, *Nutrition Today* 31 (1996): 12–17.
5. T. González-Cossío and coauthors, Decrease in birth weight in relation to maternal bone-lead burden, *Pediatrics* 100 (1997): 856–862.
6. Update: Blood lead levels—United States, 1991–1994, *Morbidity and Mortality Weekly Report* 46 (1997): 141–146.
7. H. L. Needleman and coauthors, Bone lead levels and delinquent behavior, *Journal of the American Medical Association* 275 (1996): 363–369.
8. R. A. Goyer, Nutrition and metal toxicity, *American Journal of Clinical Nutrition* 61 (1995): 646S–650S.
9. R. A. Goyer, Toxic and essential metal interactions, *Annual Review of Nutrition* 17 (1997): 37–50; Mushak, Crocetti, 1996.
10. W. G. Adams and coauthors, Anemia and elevated lead levels in underimmunized inner-city children, *Pediatrics* 101 (1998): 462 [available in full online at http://www.pediatrics.org/cgi/content/full/101/3/e6].
11. Sixth-grader opens lid for FDA investigation, *FDA Consumer,* September/October 1997, pp. 34–35.
12. S. B. Markowitz and coauthors, Lead poisoning due to *Hai Ge Fen:* The porphyrin content of individual erythrocytes, *Journal of the American Medical Association* 271 (1994): 932–934.
13. H. J. Binns, Is there lead in the suburbs? Risk assessment in Chicago pediatric practices, *Pediatrics* 93 (1994): 164–171.
14. American Academy of Pediatrics, Committee on Environmental Health, Lead poisoning: From screening to primary prevention, *Pediatrics* 92 (1993): 176–183; S. J. Schaffer and coauthors, Lead poisoning risk determination in a rural setting, *Pediatrics* 97 (1996): 84–90; M. N. Haan, M. Gerson, and B. A. Zishka, Identification of children at risk for lead poisoning: An evaluation of routine pediatric blood lead screening in an HMO-insured population, *Pediatrics* 97 (1996): 79–83; D. C. Snyder and coauthors, Development of a population-specific risk assessment to predict elevated blood lead levels in Santa Clara County, CA, *Pediatrics* 96 (1995): 643–648.
15. J. R. Campbell and coauthors, Blood lead screening practices among U.S. pediatricians, *Pediatrics* 97 (1996): 372–377.
16. Pool cue chalk: A source of environmental lead, *Pediatrics* 97 (1996): 916–917.

CHAPTER 14

# Fitness: Physical Activity, Nutrients, and Body Adaptations

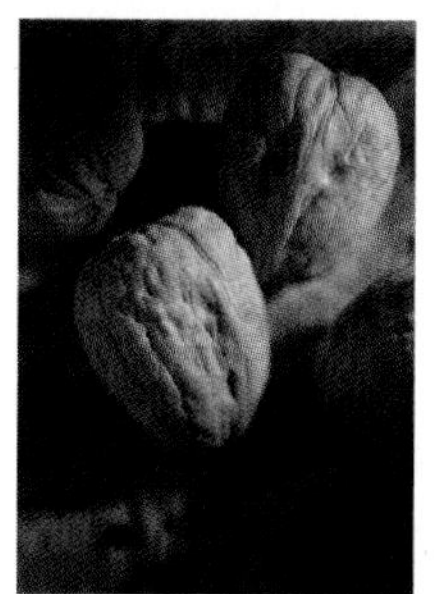

Are you physically fit? If so, the following description applies to you. You are graceful and move with ease. You are strong and meet physical challenges without strain. You have endurance, and your energy lasts for hours. You meet daily physical challenges and have plenty of energy in reserve to handle emergencies. What's more, you are prepared to meet mental and emotional challenges, too—for physical fitness supports mental and emotional energy and resilience as well.

Or perhaps you are not yet physically fit. Regardless, this chapter is written for "you," whoever you are and whatever your goals—whether you want to hone your athletic skills, improve your health, prepare for a career in health care, ensure your position on a sports team, lose weight, or become physically active.

The chapter begins by defining fitness and presenting its benefits. It goes on to explain how the body uses energy nutrients to fuel physical activity and finally describes how nutrition supports fitness.

*Physical activity, or its lack, exerts a significant and pervasive influence on everyone's nutrition and overall health. (Courtesy of CNN)*

# Fitness

Fitness depends on a certain minimum amount of physical activity. In 1990, the American College of Sports Medicine (ACSM) updated its position statement on the quantity and quality of physical activity recommended for developing and maintaining fitness in healthy adults (see Table 14-1).[1] The main objective of these guidelines is to outline the types and amounts of physical activity needed to improve *physical fitness*. These familiar guidelines help adults develop programs to improve their cardiorespiratory endurance and body composition. The types and amounts of physical activity needed to promote *fitness,* however, may differ from those needed to obtain *health* benefits of reduced disease risk. For health's sake, the ACSM specifies that people need only spend an accumulated minimum of 30 minutes in some sort of physical activity on most days of each week.[2] Eight minutes spent climbing up stairs, another 10 spent pulling weeds, and 12 more spent walking the dog all contribute to the day's total. The guidelines for developing fitness are still optimal, though, because improving fitness provides additional health

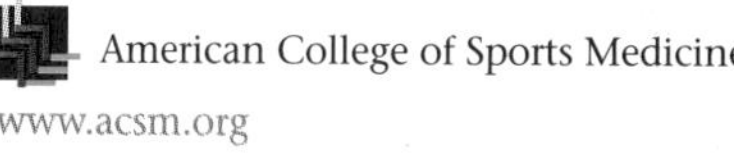
American College of Sports Medicine
www.acsm.org

**cardiorespiratory endurance:** the ability to perform large-muscle, dynamic exercise of moderate-to-high intensity for prolonged periods.

Reminder: *Body composition* refers to the proportions of muscle, bone, fat, and other tissue that make up a person's total body weight.

**Table 14-1**
**Physical Activity Guidelines**

**Guidelines for developing and maintaining *physical fitness:***

- **Frequency of activity:** three to five days per week.
- **Intensity of activity:** 50 to 90% of maximum heart rate.
- **Duration of activity:** 20 to 60 minutes of continuous activity.
- **Mode of activity:** any activity that uses large muscle groups.
- **Resistance activity:** strength training of moderate intensity at least two times per week.

**Guidelines for obtaining *health* benefits:**

- **Frequency of activity:** every day.
- **Intensity of activity:** moderate.
- **Duration of activity:** at least 30 minutes total of activity (can be intermittent).
- **Mode of activity:** any activity.

NOTE: Duration and intensity are inversely related. To obtain similar fitness benefits, a person may exercise either at a low intensity for a long duration or at a high intensity for a short duration. For example, a person may choose to walk briskly for 40 to 50 minutes (lower intensity and longer duration) or to jog for 20 to 30 minutes (higher intensity and shorter duration).

benefits (further reduction of cardiovascular disease risk and improved body composition, for example).[3]

Physical activity leads to fitness, and fitness, in turn, makes activity easy, a beneficial cycle. Activity and fitness are so closely connected that this chapter makes no distinction between them.

## Definitions of Fitness

**fitness:** the characteristics that enable the body to perform physical activity; more broadly, the ability to meet routine physical demands with enough reserve energy to rise to a physical challenge; or the body's ability to withstand stress of all kinds.

Narrowly defined, fitness refers to *the characteristics that enable the body to perform physical activity.* These characteristics include flexibility of the joints; strength and endurance of the muscles, including the heart muscle; and a healthy body composition. A broader definition of fitness is *the ability to meet routine physical demands with enough reserve energy to rise to a sudden challenge.* This definition shows how fitness relates to everyday life. Ordinary tasks such as carrying heavy suitcases, opening a stuck window, or climbing four flights of stairs, which might strain an unfit person, are easy for a fit person. Still another definition is *the body's ability to withstand stress,* meaning both physical and psychological stresses. These definitions do not contradict each other; all three describe the same wonderful condition of the body.

**sedentary:** physically inactive (literally, "sitting down a lot").

The opposite of a physically active life is a sedentary life, which means literally "sitting down a lot." Today's world fosters inactivity by providing people with escalators, cars, golf carts, and other labor-saving devices. As people go through life exerting minimal physical effort, they become weak and unfit and begin to feel unwell. In fact, a sedentary lifestyle fosters the development of several chronic diseases.

## Benefits of Fitness

Surgeon General's Report on Physical Activity and Health

www.cdc.gov/nccdphp/sgr/sgr.htm

The following comparisons reflect similar differences in the risks associated with chronic disease and death:

- Vigorous exercise vs. minimal exercise.
- Ideal weight vs. 20% overweight.
- Nonsmoking vs. smoking (one pack a day).

Extensive evidence confirms that regular physical activity promotes health and prevents disease.[4] Still, despite an increasing awareness of the health benefits that physical activity confers, more than 60 percent of adults in the United States are either irregularly active or completely inactive.[5] Physical inactivity is linked to the major degenerative diseases—heart disease, cancer, stroke, diabetes, and hypertension—that are the primary killers of adults in developed countries.[6]

People don't have to run marathons to reap the health rewards of physical activity. Most experts agree that any physical activity, even moderate activity, provides health benefits. In fact, people who are extremely inactive stand to gain the greatest health benefits by engaging in regular, moderate-intensity, endurance-type activity.[7] The authors of an extensive study on fitness and mortality concluded that "moderate levels of physical fitness that are attainable by most adults appear to be protective against early mortality."[8] It makes sense, then, to encourage the least active people to participate in whatever activities they can readily perform since they may benefit most. The ACSM advises that public health efforts focus on "getting more people more active more of the time" rather than dictating a specific activity level for people to attain.[9]

Activities that promote fitness are themselves enjoyable, and they quickly lead to rewards in terms of physical improvements. In general, physically fit people enjoy:

- *Restful sleep.* Rest and sleep occur naturally after periods of physical activity. During rest, the body repairs injuries, disposes of wastes generated during activity, and builds new physical structures.
- *Nutritional health.* Physical activity spends energy and thus allows people to eat more food. If they choose wisely, active people will consume more nutrients and be less likely to develop nutrient deficiencies.
- *Optimal body composition.* A balanced program of physical activity limits body fat and maintains lean tissue. Physically active people have relatively less body fat than sedentary people at the same body weight.[10]

- *Optimal bone density.* Weight-bearing physical activity builds bone strength and protects against osteoporosis.[11]
- *Resistance to colds and other infectious diseases.* Fitness enhances immunity.[12]*
- *Low risks of some types of cancers.* Lifelong physical activity may help to protect against colon cancer, breast cancer, and others.[13]
- *Strong circulation and lung function.* Physical activity that challenges the heart and lungs slows the aging of the circulatory system.
- *Low risk of cardiovascular disease.* Physical activity lowers blood pressure, slows resting pulse rate, and lowers blood cholesterol, thus reducing the risks of heart attack and strokes.[14] Some research suggests that physical activity may reduce the risk of cardiovascular disease in another way as well—by reducing intra-abdominal fat stores.[15]
- *Low risk of diabetes.* Physical activity normalizes glucose tolerance, especially via the secretion of insulin.[16]
- *Low incidence and severity of anxiety and depression.* Compared with sedentary people, physically active people deal better with psychological stress.
- *Strong self-image.* The sense of achievement that comes from meeting physical challenges promotes self-confidence.
- *Long life and high quality of life in the later years.* Active people have a lower mortality rate than sedentary people.[17] Even a two-mile walk daily can add years to a person's life.[18] In addition to extending longevity, physical activity supports independence and mobility in later life by reducing the risk of falls and minimizing the risk of injury should a fall occur.[19]

As a person becomes physically fit, the health of the entire body improves.

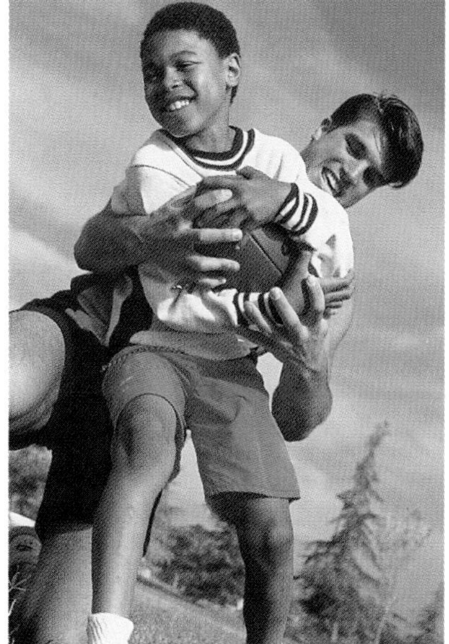

***Physical activity helps you look good, feel good, and have fun, and it brings many long-term health benefits as well.***

## Components of Fitness

To be physically fit, a person needs to develop enough flexibility, muscle strength and endurance, and cardiorespiratory endurance to meet the everyday demands of life with some to spare and to achieve a reasonable body weight and body composition. Flexibility allows the joints to move with less chance of injury. Muscle strength and endurance enable muscles to work harder and longer without fatigue. Cardiorespiratory endurance supports the ongoing action of the heart and lungs. Physical activity supports desirable lean body tissue and eliminates excess body fat. A person who practices a physical activity *adapts* by becoming better able to perform it after each session—more flexible, stronger, and more enduring.

**flexibility:** the capacity of the joints to move through a full range of motion; the ability to bend and recover without injury.

**muscle strength:** the ability of muscles to work against resistance.

**muscle endurance:** the ability of a muscle to contract repeatedly without becoming exhausted.

**conditioning:** the physical effect of *training;* improved flexibility, strength, and endurance.

**training:** practicing an activity regularly, which leads to conditioning. (Training is what you do; conditioning is what you get.)

**progressive overload principle:** the training principle that a body system, in order to improve, must be worked at frequencies, durations, or intensities that gradually increase physical demands.

**frequency:** the number of occurrences per unit of time (for example, the number of activity sessions per week).

**intensity:** the degree of exertion while exercising (for example, the amount of weight lifted or the speed of running).

**duration:** length of time (for example, the time spent in each activity session).

## Conditioning by Training

Whatever component of fitness a person seeks to develop—flexibility, strength, or endurance—the principles of conditioning apply. During conditioning, the body adapts microscopically to perform the work asked of it. The way to achieve conditioning is by training, primarily by applying the progressive overload principle—that is, by asking a little more of the body in each training session.

• ***The Overload Principle*** • You can apply the progressive overload principle in several different ways. You can perform the activity more often—that is, increase its frequency. You can perform it more strenuously—that is, increase its intensity. Or you can do it for longer times—that is, increase its duration. All three strategies, individually or in combination, work well. The rate of progression depends on individual characteristics such as fitness level, health status, age, and

*Moderate physical activity can stimulate immune function. Intense, vigorous, prolonged activity such as marathon running, however, may compromise immune function. J. A. Smith, Guidelines, standards, and perspectives in exercise immunology, *Medicine and Science in Sports and Exercise* 27 (1995): 497–506.

preference. If you enjoy your workout, do it more often. If you do not have much time, increase intensity. If you hate hard work, take it easy and go longer. If you want continuous improvements, remember to overload progressively as you reach higher levels of fitness.

• ***Applying Overload*** • When increasing the frequency, intensity, or duration of a workout, exercise to a point that only *slightly* exceeds the comfortable capacity to work. It is better to progress too slowly than to risk serious injury by overexertion. Other tips include:

- Be active all week, not just on the weekends.
- Use proper equipment and attire.
- Perform approved exercises using proper form.
- Within each activity session, include warm-up and cool-down activities. Warming up helps to prepare muscles, ligaments, and tendons for the upcoming activity and mobilizes fuels to support strength and endurance activities; cooling down reduces muscle cramping and allows the heart rate to slow gradually.
- Train hard enough to challenge your strength or endurance once or twice a week, not every time you work out. Between challenges, do moderate workouts and include at least one day of rest each week.
- Pay attention to body signals. Symptoms such as abnormal heartbeats, dizziness, lightheadedness, cold sweat, confusion, or pain or pressure in the middle of the chest, teeth, jaw, neck, or arm demand immediate medical attention.

**warm-up:** 5 to 10 minutes of light activity, such as easy jogging or cycling, prior to a workout to prepare the body for more vigorous activity.

**cool-down:** 5 to 10 minutes of light activity, such as walking or stretching, following a vigorous workout to return the body's core gradually to near-normal temperature.

Work out wisely. Do not start with activities so demanding that pain stops you within two days. Learn to enjoy small steps toward improvement. Fitness builds slowly.

• ***Cautions on Starting*** • Before beginning a fitness program, make sure it is safe for you to do so. The ACSM classifies individuals into three groups based on major coronary risk factors: "apparently healthy" individuals have no more than one of the risk factors listed in the margin, "individuals at higher risk" have two or more of the risk factors and/or symptoms suggestive of disease, and "individuals with disease" are known to have cardiac, pulmonary, or metabolic disease.[20] Most apparently healthy people can begin moderate exercise programs such as walking or increasing daily activities without medical examination, but people in either of the other two classifications need medical advice.

Major coronary risk factors:

- Age (men >45 years; women >55 years).
- Family history of heart disease.
- Cigarette smoking.
- Hypertension.
- Serum cholesterol >200 mg/dL.
- Diabetes.
- Sedentary lifestyle.

**moderate exercise:** activity that can be sustained comfortably for 60 minutes or so.

• ***The Body's Response to Physical Activity*** • Fitness develops in response to demand and wanes when demand ceases. Muscles gain size and strength after being made to work repeatedly, a response called hypertrophy. Conversely, without activity, muscles diminish in size and lose strength, a response called atrophy.

Hypertrophy and atrophy are adaptive responses to the muscles' greater and lesser work demands, respectively. Thus cyclists often have strong, well-developed legs but less arm or chest strength; a tennis player may have one superbly strong arm, while the other is just average. For balanced muscular development, people should work different muscle groups from day to day. This strategy provides a day or two of rest for different muscle groups, giving them time to replenish nutrients and to repair any minor damage incurred by the activity.

**hypertrophy** (high-PER-tro-fee): of muscles, growing larger; an increase in size in response to use.

**atrophy** (AT-ro-fee): of muscles, becoming smaller; a decrease in size because of disuse, undernutrition, or wasting diseases.

• ***Weight Training*** • Weight training builds lean body mass, develops strength and endurance of muscles, and benefits health and overall fitness. Strong muscles in the back and abdomen improve posture and reduce the risk of back injury. Weight training can also help prevent the decline in physical mobility that often accompanies aging.[21] Older adults who participate in weight training programs not only gain muscle strength, but also improve their muscle endurance, which enables them to walk significantly longer before exhaustion. Leg strength and walking endurance are powerful indicators of an older adult's physical abilities.[22]

**weight training** (also called **resistance training**): the use of free weights or weight machines to provide resistance for developing muscle strength and endurance. A person's own body weight may also be used to provide resistance as when a person does push-ups, pull-ups, or abdominal crunches.

Weight training enhances performance in other sports, too. Swimmers can develop a more efficient stroke and tennis players, a more powerful serve, when they train with weights.

The ACSM advises that weight training to improve muscle strength and endurance may also help maximize and maintain bone mass.[23] Research supports this advice. Young women participating in a weight training program were able to increase the bone density of their spines.[24]

Depending on the technique, weight training can emphasize either muscle strength or muscle endurance. To emphasize muscle strength, combine high resistance (heavy weight) with a low number of repetitions. To emphasize muscle endurance, combine less resistance (lighter weight) with more repetitions.

***People's bodies are shaped by the activities they perform.***

## *Cardiorespiratory Endurance*

The length of time a person can remain active with an elevated heart rate—that is, the ability of the heart, lungs, and blood to sustain a given demand—defines a person's cardiorespiratory endurance. Cardiorespiratory endurance training can improve a person's ability to sustain a vigorous activity such as running, brisk walking, or swimming. Such training enhances the ability of the heart, lungs, and blood to deliver oxygen to, and remove waste from, the body's cells. The benefits of this training are not just physical, though, because all of the body's cells, including the brain cells, require oxygen to function. When the cells receive more oxygen more readily, both the body and the mind benefit.

Working muscles need a lot of oxygen to produce energy. Cardiorespiratory endurance training requires the heart and lungs to work hard for a sustained period to deliver oxygen to the muscle cells. Cardiorespiratory endurance training, therefore, is *aerobic*. As the cardiorespiratory system gradually adapts to the demands of aerobic activity, the body delivers oxygen more efficiently.

• ***Cardiorespiratory Conditioning*** • The changes brought about by aerobic workouts are called cardiorespiratory conditioning. Among its benefits, cardiac output increases, thus enhancing oxygen delivery. The heart becomes larger and stronger, and each beat pumps more blood. Because the heart pumps more blood with each beat, fewer beats are necessary, and the resting heart rate slows down. The average resting pulse rate for adults is around 70 beats per minute, but people who achieve cardiorespiratory conditioning may have resting pulse rates of 50 or even lower. The muscles that work the lungs become stronger, too, so breathing becomes more efficient. Circulation through the arteries and veins improves. Blood moves easily, and blood pressure falls.

**cardiorespiratory conditioning:** improvements in heart and lung function and increased blood volume, brought about by aerobic training.

**cardiac output:** the volume of blood discharged by the heart each minute; it is determined by multiplying the stroke volume by the heart rate. The **stroke volume** is the amount of oxygenated blood the heart ejects toward the tissues at each beat.

Cardiac output (volume/minute) =
stroke volume (volume/beat)
× heart rate (beats/minute).

Cardiorespiratory endurance is the physical achievement that many people appropriately prize the most highly because it reflects the health of the heart and circulatory system, on which all other body systems depend. Figure 14-1 shows the major relationships among the heart, circulatory system, and lungs.

To improve your cardiorespiratory endurance, the activity you choose must be sustained for 20 minutes or longer and use most of the large muscle groups of the body (legs, buttocks, and abdomen). You must also train at an intensity that elevates your heart rate.

Cardiorespiratory conditioning:

- Increases cardiac output and oxygen delivery.
- Increases stroke volume.
- Slows resting pulse.
- Increases breathing efficiency.
- Improves circulation.
- Reduces blood pressure.

An activity's intensity can be determined by the maximum rate of either oxygen consumption ($VO_2$ max) or heart beats per minute, but a person's own perceived effort is usually a reliable indicator. In general, when you're working out, do so at an intensity that raises your heart rate, but still leaves you able to talk comfortably. If you are more competitive and want to work to your limit on some days, a treadmill test can reveal your maximum heart rate. You can work out safely at up to 90 percent of that rate. The ACSM guidelines for developing and maintaining cardiorespiratory fitness were given in Table 14-1 on p. 437.

**$VO_2$ max:** the maximum rate of oxygen consumption by an individual at sea level.

• ***Muscle Conditioning*** • A fringe benefit of cardiorespiratory training is that fit muscles use oxygen more efficiently than less fit muscles, reducing the heart's

**Figure 14-1**

**Delivery of Oxygen by the Heart and Lungs to the Muscles**

The cardiorespiratory system responds to the muscles' demand for oxygen by building up its capacity to deliver oxygen. Researchers can measure cardiorespiratory fitness by measuring the maximum amount of oxygen a person consumes per minute while working out, a measure called $VO_2$ max.

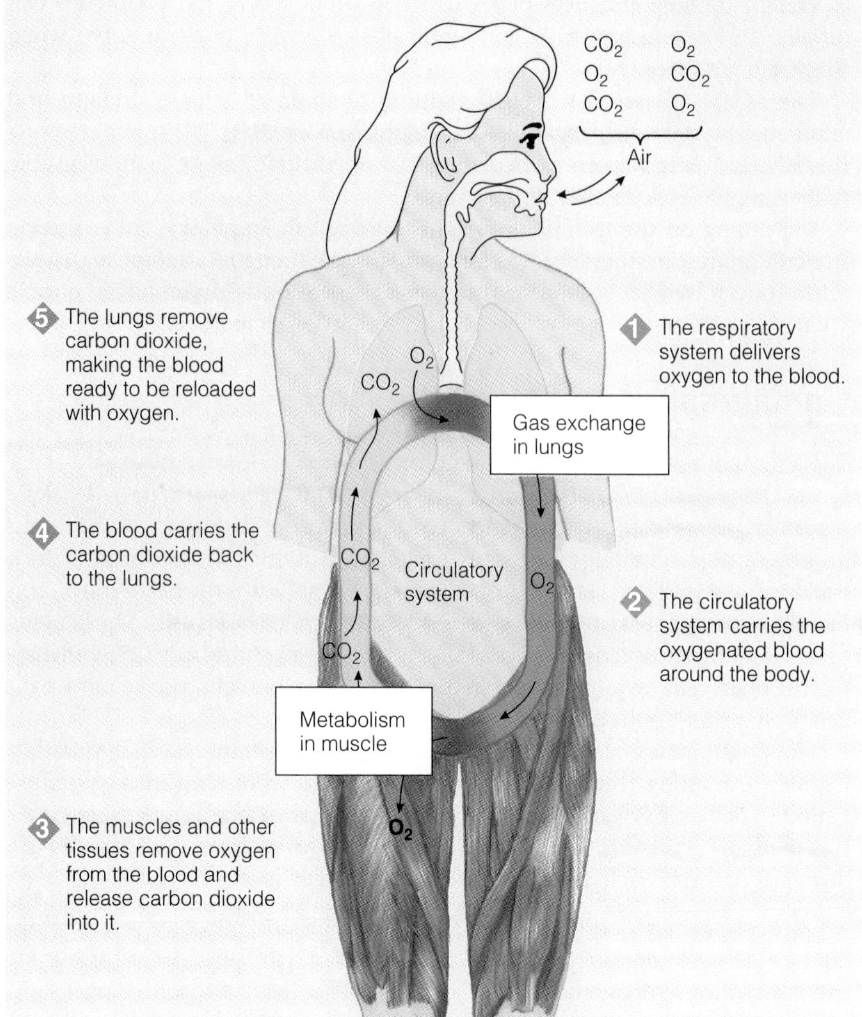

CO2 O2
O2 CO2
CO2 O2
Air
1 The respiratory system delivers oxygen to the blood.
Gas exchange in lungs
2 The circulatory system carries the oxygenated blood around the body.
Circulatory system
Metabolism in muscle
3 The muscles and other tissues remove oxygen from the blood and release carbon dioxide into it.
4 The blood carries the carbon dioxide back to the lungs.
5 The lungs remove carbon dioxide, making the blood ready to be reloaded with oxygen.
O2
CO2


**Table 14-2**

**A Sample Balanced Fitness Program (45 Minutes a Day)**

Monday, Wednesday, Friday:

- 5 minutes of warm-up activity.
- 30 minutes of aerobic activity.
- 10 minutes of cool-down activity and stretching.

Tuesday, Thursday:

- 5 minutes of warm-up activity.
- 30 minutes of weight training.
- 10 minutes of cool-down activity and stretching.

Saturday or Sunday:

- Softball, walking, hiking, biking, or swimming.

Highlight 9 describes some of the harmful dietary practices of athletes and the combination of problems—disordered eating, amenorrhea, and osteoporosis—collectively known as the female athlete triad.

workload. An added bonus is that muscles that use oxygen efficiently can burn fat longer—a plus for body composition and weight control.

• ***A Balanced Fitness Program*** • In a balanced fitness program, aerobic activity improves cardiorespiratory fitness, stretching enhances flexibility, and weight training or calisthenics develops muscle strength and endurance. Table 14-2 provides an example of a balanced fitness program.

IN SUMMARY

Physical activity brings positive rewards: good health, long life, and freedom from disease. To develop fitness—whose components are flexibility, muscle strength and endurance, and cardiorespiratory endurance—a person must condition the body, through training, to adapt to the activity performed. (Pursued in excess, however, intense physical activity combined with poor eating habits can undermine health, as Highlight 9 explains.)

# Energy Systems, Fuels, and Nutrients to Support Activity

Nutrition and physical activity go hand in hand. Activity demands carbohydrate and fat as fuel; protein to build and maintain lean tissues; vitamins and minerals to support both energy metabolism and tissue building; and water to help distribute the fuels and to dissipate the resulting heat and wastes. This section describes how nutrition supports a person who decides to get up and go.

## *The Energy Systems of Physical Activity—ATP and CP*

Muscles contract fast. When called upon, they respond quickly without taking time to metabolize fat or carbohydrate for energy. In the first fractions of a second, muscles starting to move depend on their supplies of quick-energy compounds to power their movements. Exercise physiologists know these compounds by their abbreviations, ATP and CP.

• ***ATP*** • As Chapter 7 described, all of the energy-yielding nutrients—carbohydrate, fat, and protein—can transfer energy to make the high-energy compound ATP (adenosine triphosphate). ATP is present in small amounts in all body tissues all the time, and it can deliver energy instantaneously. In the muscles, ATP provides the chemical driving force for contraction. When ATP is split, its energy is released, and the muscle cells channel some of that energy into mechanical movement and most of it into heat—heat the exerciser can feel building up. A tiny but essential pool of ATP is always ready to meet the cells' sudden demands for movement.

• ***CP*** • Unlike a single reflexive muscle jerk, prolonged activity involves sustained or repeated muscle contractions that require ongoing use of ATP. This creates a demand: more must be made.

Immediately after the onset of the demand, before muscle ATP pools dwindle, a muscle enzyme begins to break down another high-energy compound that is stored in the muscle, CP, or creatine phosphate. CP is made from creatine, a compound commonly found in muscles, with a phosphate group attached, and it can split (anaerobically) to release phosphate and replenish ATP supplies. Supplies of CP in a muscle last for only about 20 seconds, but can produce enough quick energy without oxygen for a 100-meter dash.

When activity ceases and the muscles are resting, ATP feeds energy back to CP by giving up one of its phosphate groups to creatine. Thus CP is produced during rest by reversing the process that occurs during muscular activity.

**CP, creatine phosphate** (also called **phosphocreatine**): a high-energy compound in muscle cells that acts as a reservoir of energy that can maintain a steady supply of ATP; CP provides the energy for short bursts of activity.

Highlight 14 includes creatine supplements in its discussion of substances commonly used by athletes.

During activity: CP → ATP + creatine.
During rest: ATP + creatine → CP.

• ***The Energy-Yielding Nutrients*** • To meet more prolonged demands, the muscles keep generating ATP from the more abundant fuels: carbohydrate, fat, and protein. The breakdown of these nutrients generates ATP all day every day, and so maintains the supply indefinitely.

During rest, the body derives slightly more than half of its ATP from fatty acids and most of the rest from carbohydrate, along with a small percentage from amino acids. During physical activity, the body adjusts its mixture of fuels. Muscles never use just one fuel. How much of which fuels they use during physical activity depends on an interplay among the fuels available from the diet, the intensity and duration of the activity, and the degree to which the body is conditioned to perform that activity. The next sections explain these relationships by examining each of the energy-yielding nutrients individually, but keep in mind that fuel use is not an all-or-none process. One fuel may predominate at a given time, but the other two will still be active.

As you read about each of the energy-yielding nutrients, notice how their contributions to the fuel mixture shift depending on whether the activities are anaerobic or aerobic. Anaerobic activities are associated with strength, agility, and split-second surges of power. The jump of the basketball player, the slam of the tennis serve, the heave of the weight lifter at the barbells, and the blast of the fullback through the opposing line involve anaerobic work. Such high-intensity, short-duration activities depend mostly on glucose breakdown without oxygen as the chief energy fuel.

Fuel mixture during activity depends on:
- Diet.
- Intensity and duration of activity.
- Training.

***Split-second surges of power as in the heave of a barbell or jump of a basketball player involve* anaerobic work.**

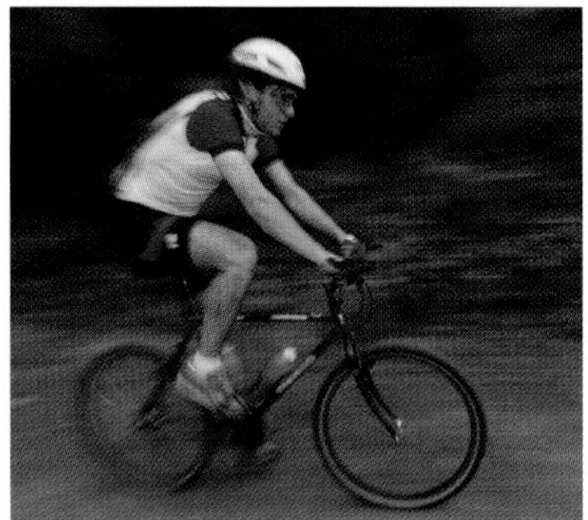

*Sustained muscular efforts as in a long-distance bike ride or cross-country run involve* aerobic *work.*

Endurance activities of low-to-moderate intensity and long duration depend more on fat to provide energy aerobically. The ability to continue swimming to the shore, to keep on hiking to the top of the mountain, or to continue pedaling all the way home reflects aerobic capacity. As mentioned earlier, aerobic capacity is also crucial to maintaining the health of the heart and circulatory system. The relationships among fuels and physical activity bear heavily on what foods best support your chosen activities.

## Glucose Use during Physical Activity

Glucose, stored in the liver and muscles as glycogen, is vital to physical activity. During exertion, the liver releases its glucose into the bloodstream. The muscles use both this glucose and their own private glycogen stores to fuel their work. Glycogen supplies can easily support everyday activities, but are limited. The more glycogen the muscles store, the longer the stores will last during physical activity.

• Fitness tip: To fill glycogen stores, eat plenty of carbohydrate-rich foods.

• ***Diet Affects Glycogen Storage and Use*** • The body constantly uses and replenishes its glycogen. How much carbohydrate a person eats influences how much glycogen is stored, which in turn influences performance. When glycogen is depleted, the muscles become fatigued.

A classic study compared fuel use during activity among three groups of runners on different diets. For several days before testing, one group consumed a normal mixed diet, a second group consumed a high-carbohydrate diet, and the third group consumed a no-carbohydrate diet. As Figure 14-2 shows, the high-carbohydrate diet allowed the runners to keep going longer before exhaustion. This study and many others that followed have confirmed that high-carbohydrate diets enhance endurance by enlarging glycogen stores.

• ***Intensity of Activity Affects Glycogen Use*** • How long an exercising person's glycogen will last depends not only on diet, but also on the intensity of the activity. The most intense activities—the kind that make it difficult "to catch your breath," such as a quarter-mile run—use glycogen quickly. Glycogen depletion usually occurs within two hours from the onset of intense activity. Other, less intense activities, such as jogging, during which breathing is steady and easy, use glycogen more slowly.

During *moderate* physical activity, the lungs and circulatory system have no trouble keeping up with the muscles' need for oxygen. The individual breathes easily, and the heart beats steadily—the activity is aerobic. The muscles derive their energy from both glucose and fatty acids. By depending partly on fatty acids, mod-

**Figure 14-2**
**The Effect of Diet on Physical Endurance**

A high-carbohydrate diet can increase an athlete's endurance. In this study, the fat and protein diet provided 94 percent of kcalories from fat and 6 percent from protein; the normal mixed diet provided 55 percent of kcalories from carbohydrate; and the high-carbohydrate diet provided 83 percent of kcalories from carbohydrate.

**Table 14-3**
**Fuels Used for Activities of Different Intensities and Durations**

| Activity Intensity | Activity Duration | Preferred Fuel Source | Oxygen Needed? | Activity Example |
|---|---|---|---|---|
| Extreme[a] | Less than 30 sec | ATP-CP (immediate availability) | No | 100-yard dash, shot put |
| Very high | 30 sec to 3 min | ATP from carbohydrate (lactic acid) | No (anaerobic) | ¼-mile run at maximal speed |
| High | 3 min to 20 min | ATP from carbohydrate | Yes (aerobic) | Cycling, swimming, or running |
| Moderate | More than 20 min | ATP from fat | Yes (aerobic) | Hiking |

[a]All levels of activity intensity use the ATP-CP system initially; extremely intense short-term activities rely solely on the ATP-CP system.

erate aerobic activity conserves glycogen. Table 14-3 shows how fuel use changes according to the intensity of the activity.

*Intense* activity presents a different metabolic situation. Whenever a person exercises at a rate that exceeds the capacity of the heart and lungs to supply oxygen to the muscles, aerobic metabolism cannot meet energy needs. Instead, the muscles must draw more heavily on glucose, which they can use anaerobically, breaking it down to pyruvate and producing some energy (ATP) without requiring oxygen.

Reminder: Oxygen is needed for pyruvate to take the metabolic pathway to acetyl CoA and on through the TCA cycle (see Chapter 7).

• ***Lactic Acid*** • When the rate of activity exceeds the ability to provide enough oxygen, the accumulating pyruvate molecules are converted to lactic acid in an anaerobic process (see the first part of Figure 14-3). Lactic acid can produce ATP during intense activity for only 1 to 3 minutes (as in a 400- to 800-meter race or a boxing match). At low intensities, lactic acid is readily cleared from the blood, but at higher intensities, lactic acid accumulates.

Reminder: *Lactic acid* is the anaerobic breakdown product of pyruvate. When lactic acid accumulates, exhaustion sets in and the exerciser cannot continue.

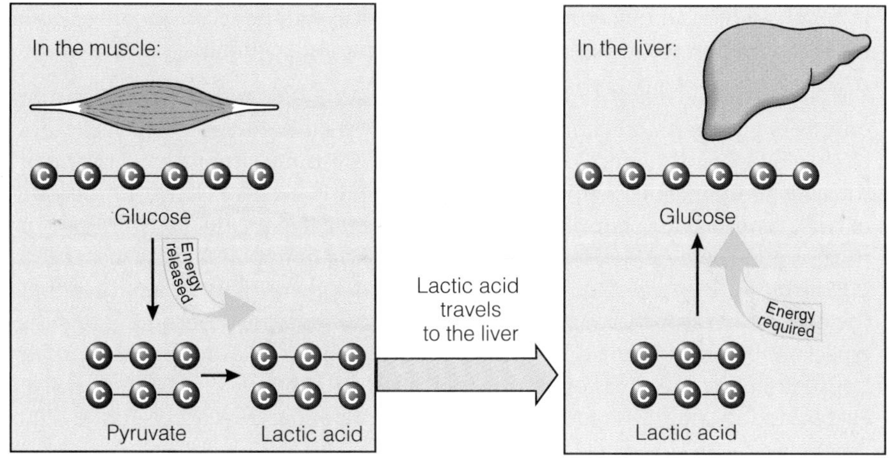
In the muscle:
Glucose
Energy released
Pyruvate
Lactic acid
Lactic acid travels to the liver
In the liver:
Glucose
Energy required
Lactic acid
Without ample oxygen, working muscles break down most of their glucose molecules anaerobically to pyruvate, releasing a little energy. Pyruvate is then converted to lactic acid.
Liver enzymes can convert lactic acid to glucose, but this reaction requires a lot of energy.

**Figure 14-3**
**Recycling of Glucose during Anaerobic Metabolism**

Note: Lactic acid production serves the body's metabolic needs well during anaerobic times. When pyruvate is converted to lactic acid, an enzyme uses the niacin coenzyme NADH and produces $NAD^+$. $NAD^+$ can then be used to break down glucose to pyruvate, thus permitting glucose to continue providing energy anaerobically for a while.

Reminder: The recycling process that regenerates glucose from lactic acid is known as the *Cori cycle*.

Muscle cells cannot accommodate much lactic acid; they release it, and it travels in the blood to the liver. There, liver enzymes convert it back into glucose (as shown in the second part of Figure 14-3). Glucose can then return to the muscles to fuel additional activity. (This cycle, known as the Cori cycle, was first introduced in Chapter 7.)

• ***Duration of Activity Affects Glycogen Use*** • Glycogen use depends not only on the intensity of an activity, but also on its duration. Within the first 20 minutes or so of moderate activity, a person uses mostly glycogen for fuel—about one-fifth of the available glycogen. As the muscles devour their own glycogen, they become ravenous for more glucose. The liver responds by emptying out its glycogen stores for use by the exercising muscles, which increase their uptake of blood glucose 20-fold or more. The muscles' accelerated uptake keeps blood glucose from rising too high and it soon begins to decline.

After 20 minutes, a person who continues exercising moderately (mostly aerobically) begins to use less and less glycogen and more and more fat for fuel (review Table 14-3 on p. 445). Still, glycogen use continues, and if the activity lasts long enough and is intense enough, muscle and liver glycogen stores will be depleted. Physical activity can continue for a short time thereafter only because the liver scrambles to produce, from lactic acid and certain amino acids, the minimum amount of glucose needed to briefly forestall total depletion.

• ***Glucose Depletion*** • After a couple of hours of strenuous activity, glucose stores are depleted. When depletion hits, it brings nervous system function to a near halt, making continued exertion almost impossible. Marathon runners refer to this point of glucose exhaustion as "hitting the wall."

To avoid such debilitation, endurance athletes try to maintain their blood glucose for as long as they can. To maximize glucose supply, endurance athletes:

- Eat a high-carbohydrate diet (approximately 8 grams of carbohydrate per kilogram of body weight or about 70 percent of energy intake) regularly.*
- Take glucose (usually in diluted fruit juice or other sweet beverages) periodically during activity that lasts for an hour or more.
- Eat carbohydrate-rich foods (approximately 60 grams of carbohydrate) following activity.
- Train the muscles to store as much glycogen as possible.

For perspective, snack ideas providing 60 g carbohydrate:
- 16 oz sports drink and a bagel.
- 16 oz milk and 4 oatmeal cookies.
- 8 oz cranberry juice and a granola bar.

The last section of this chapter, "Diets for Physically Active People," discusses how to design a high-carbohydrate diet for performance, and the accompanying "How to" describes how to maximize glycogen stores for long endurance competitions.

• ***Glucose during Activity*** • Muscles can obtain the carbohydrate they need, not only from glycogen stores but also from sugar taken during activity, which elevates blood glucose and enhances endurance. Normally, insulin stimulates all tissues of the body to drain glucose from the blood and stow it away—exactly the opposite of what is needed for performance. During physical activity, the body's release of the hormone epinephrine keeps insulin from rising in response to glucose entering the blood. Physical activity also enhances muscle sensitivity to insulin so that the muscles become the primary recipient of blood glucose. Consuming sugar is especially useful during exhausting endurance activities (lasting more than an hour). Endurance athletes often run short of glucose by the end of competitive events, and they are wise to take light carbohydrate snacks or drinks (under 200

Reminder: *Epinephrine* is one of the stress hormones that is secreted whenever emergency action is called for; it readies body systems for fast action and mobilizes fuel to support that action.

*Percentage of energy intake is meaningful only when total energy intake is known. Consider that at high energy intakes (say, 5000 kcalories/day), even a moderate carbohydrate diet (40 percent of energy intake) supplies 500 grams of carbohydrate—enough for a 137-pound athlete in heavy training. By comparison, at a moderate energy intake (2000 kcalories/day), a high carbohydrate intake (70 percent of energy intake) supplies 350 grams—plenty of carbohydrate for most people, but not enough for athletes in heavy training.

## HOW TO Maximize Glycogen Stores: Carbohydrate Loading

Some athletes use a technique called carbohydrate loading to trick their muscles into storing extra glycogen before a competition. Ideally, the athlete in training eats a high-carbohydrate diet. During the first four days of the week before competition, the athlete trains moderately hard (one to two hours per day) and eats a diet that is moderate in carbohydrate. During the three days before competition, the athlete gradually cuts back on activity and eats a very-high-carbohydrate diet.

Extra glycogen gained this way can benefit an athlete who must keep going for 90 minutes or longer. Those who exercise for shorter times simply need a regular high-carbohydrate diet. In a hot climate, extra glycogen confers an additional advantage: as glycogen breaks down, it releases water, which helps to meet the athlete's fluid needs.

**carbohydrate loading:** a regimen of moderate exercise followed by the consumption of a high-carbohydrate diet that enables muscles to store glycogen beyond their normal capacities; also called **glycogen loading** or **glycogen super compensation.**

kcalories) periodically during activity.[25] During the last stages of an endurance competition, when glycogen is running low, glucose consumed during the event can make its way slowly from the digestive tract to the muscles and augment the body's supply of glucose enough to forestall exhaustion.

• Fitness tip: To help delay exhaustion during long, competitive events, eat or drink a light carbohydrate-rich snack during the event.

• ***Glucose after Activity*** • Research indicates that eating high-carbohydrate foods *after* physical activity also enlarges glycogen stores. A high-carbohydrate meal eaten within 15 minutes after physical activity accelerates the rate of glycogen storage by 300 percent. After two hours, the rate of glycogen storage declines by almost half. Despite this slower rate of glycogen restoration, muscles will still attain high glycogen concentrations as long as athletes eat carbohydrate-rich foods within two hours following activity.[26] This is particularly important to athletes who train hard more than once a day. A practical tip: after your next workout, enjoy a bagel or a glass of juice for your glycogen's sake.

• ***Training Affects Glycogen Use*** • Training, too, affects how much glycogen muscles will store. Muscle cells that repeatedly deplete their glycogen through hard work adapt to store greater amounts of glycogen to support that work.

• Fitness tip: To make glycogen, muscles need carbohydrate, but they also need rest, so vary daily exercise routines to work different muscles on different days.

Conditioned muscles also rely less on glycogen and more on fat for energy, so glycogen breakdown and glucose use occur more slowly in trained than in untrained individuals at a given work intensity.[27] A person attempting an activity for the first time uses much more glucose than an athlete who is trained to perform it. Oxygen delivery to the muscles by the heart and lungs plays a role, but equally importantly, trained muscles are better equipped to use the oxygen because their cells contain more mitochondria. Untrained muscles depend more heavily on anaerobic glucose breakdown, even when physical activity is just moderate.

Reminder: The *mitochondria* are the structures within a cell responsible for producing ATP aerobically (see Figure 7-4 on p. 200).

## *Fat Use during Physical Activity*

As Figure 14-2 showed, researchers have long recognized the importance of a high-carbohydrate diet for endurance performance. Recently, though, a few researchers have begun to question this long-held premise. Some research suggests that high-fat diets may benefit endurance performance and that severe dietary fat restriction may be detrimental for some athletes.[28] Additional research indicates that high-fat diets have the same effect on endurance as a high-carbohydrate diet—at least over the short term (one month).[29]

Of course, high-fat diets carry risks of heart disease and cannot be recommended without careful consideration. During endurance activity, the higher oxygen requirement for fat metabolism produces more cardiovascular stress than does the oxidation of carbohydrate.[30] Furthermore, when researchers examined heart disease risk factors in athletes eating a high-fat diet, they found that blood cholesterol concentrations rose.[31] A later study, however, found no adverse effects.[32] Clearly, more studies are needed before any conclusions about the benefits of high-fat diets and endurance performance can be made.

Physical activity offers some protection against cardiovascular disease, but even athletes can suffer heart attacks and strokes. Eating a high-fat diet for a prolonged time warrants medical supervision of blood lipids and other risk factors. Most nutrition experts agree that the potential for adverse health effects of prolonged high-fat diets continues to outweigh any possible benefit to performance.

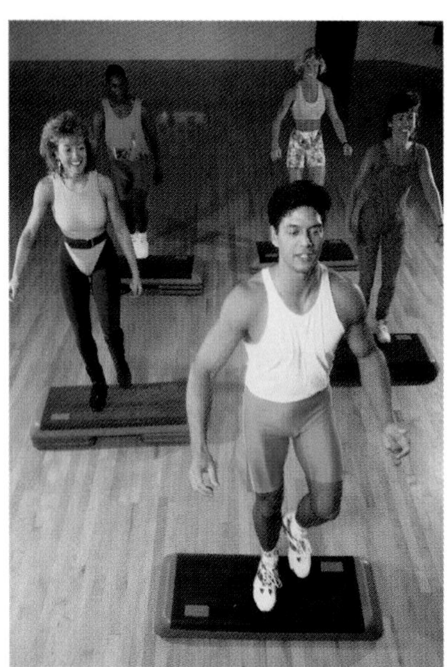

***Abundant energy from the breakdown of fat can come only from aerobic metabolism.***

In contrast to *dietary* fat, *body* fat stores are of tremendous importance during physical activity, as long as the activity is not too intense. Unlike glycogen stores, fat stores can fuel hours of activity without running out.

The fat used in physical activity is liberated as fatty acids from the internal fat stores and from the fat under the skin. Areas that have the most fat to spare donate the greatest amounts of fatty acids to the blood (although they may not be the areas that appear fattiest). This is why "spot reducing" doesn't work—muscles do not own the fat that surrounds them. Fat cells release fatty acids into the blood, not into the underlying muscles. Then the blood gives to each muscle the amount of fat that it needs. Proof of this is found in a tennis player's arms—the fatfolds measure the same in both arms, even though the muscles of one arm work much harder and may be larger than those of the other. A balanced fitness program that includes strength training, however, will tighten muscles underneath the fat, improving the overall appearance. Keep in mind that some body fat is essential to good health.

• Fitness tip: To burn fat, exercise for 20 minutes or more.

• ***Duration of Activity Affects Fat Use*** • Early in an activity, as the muscles draw on fatty acids, blood levels fall. If the activity continues for more than a few minutes, the hormone epinephrine is released, and in response, the fat cells begin rapidly breaking down their stored triglycerides and liberating fatty acids into the blood. After about 20 minutes of physical activity, the blood fatty acid concentration surpasses the normal resting concentration. Thereafter, sustained, moderate activity uses body fat stores as its major fuel.

• Fitness tip: To burn fat, exercise at a low-to-moderate intensity.

• ***Intensity of Activity Affects Fat Use*** • The intensity of physical activity also affects fat use. As the intensity of activity increases, fat makes less and less of a contribution to the mix of fuels used. Remember that fat can be broken down for energy only by aerobic metabolism. In fact, the use of fatty acids for energy requires more oxygen, even on a per-kcalorie basis, than does the use of glucose. This is because fatty acid oxidation generates so many acetyl CoA molecules that enter the TCA cycle and because each turn of the cycle generates so many hydrogens that travel through the electron transport chain to oxygen. For fat to fuel activity, then, oxygen must be abundantly available. If a person is breathing easily during activity, the muscles are getting all the oxygen they need and are able to use more fat in the fuel mixture.

• Fitness tip: To burn more fat, train aerobically.

• ***Training Affects Fat Use*** • Training—repeated aerobic activity—produces the adaptations that permit the body to draw heavily on fat for fuel. Training stimulates the muscle cells to manufacture more and larger mitochondria, the cellular structures that conduct aerobic metabolism. Another adaptation: the heart and lungs become stronger and better able to deliver oxygen to muscles at high activity intensities. Still another: hormones in the body of a trained person slow glucose release from the liver and speed up fat use instead. These adaptations reward not only trained athletes but all active people; a person who trains aerobically becomes well suited to the task.

• ***Recommended Intensities and Durations*** • Health care professionals frequently advise people who want to control their body weight and lose fat to engage in activities of low-to-moderate intensity for a long duration, such as an hour-long fast-paced walk. The reasoning behind such advice is that people exercising at low-to-moderate intensity are likely to stick with their activity for longer times and are less likely to injure themselves. In addition, some research suggests that the longer the duration of activity, the greater the contribution fat will make to the fuel mixture, and consequently, the more body fat will be lost—but this conclusion is controversial.[33] Some research refutes the notion that a person "burns more fat during low-intensity activity" and suggests that weight-control benefits are the same from either low- or high-intensity activity.[34] Regardless of the contribution fat makes to the fuel mix, people who engage in regular, vigorous physical activities have less body fat than those who engage in moderately intense activities.[35] Fat use may continue at an accelerated rate for some time after vigorous physical activity has ceased. The conditioned body that is adapted to strenuous and prolonged aerobic activity uses more fat all day long, not just during activity.[36] The bottom line on physical activity and weight and/or fat loss seems to be that total energy expenditure is the main factor, regardless of how you do it.[37]

President's Council on Physical Fitness and Sports

www.whitehouse.gov/WH/PCPFS/html/fitnet.html

Shape Up America

www.shapeup.org/sua

• Fitness tip: To lose weight, spend more energy in physical activity than you consume from foods.

• ***Choosing an Activity*** • The intensity and type of physical activities that are best for one person may not be good for another. The intensity to choose depends on your present fitness: work so as to breathe fast, but not so fast as to incur an oxygen debt. A rule of thumb is that you should be breathing easily enough to talk but not sing. If you can sing, pick up the pace; if you have to huff and puff to talk, slow down. If you have been sedentary for the past few years, the activity intensity that will initially make you breathe slightly fast will differ dramatically from the intensity at which a fit person will breathe slightly fast.

• Fitness tip: Select an activity and an intensity level that are challenging, but not overwhelming.

The type of physical activity that is best for you depends, too, on what you want to achieve and what you enjoy doing. If you are looking for health benefits, such as reducing your disease risks and lowering your blood cholesterol, then you might want to spend at least 30 minutes each day doing some kind of physical activity. If you are looking to lose weight and improve body composition, then choose an activity that you can sustain for 45 minutes or more at least three days a week. Choose an activity you enjoy: some people love walking, while others prefer to dance or ride a bike. If you want to be stronger and firmer, lift weights or do calisthenics. And remember, muscle is more metabolically active than body fat, so the more muscle you have, the more energy you'll burn.

Fitness tip: Select an activity that will help you meet your goals.

***The key to regular physical activity is finding an activity that you enjoy. (Courtesy of CNN)***

## *Protein Use during Physical Activity—and between Times*

Table 14-3 on p. 445 summarized the fuel uses discussed so far, but did not include the third energy-yielding nutrient, protein, because protein is not a major fuel for physical activity. Nevertheless, physically active people use protein just as other people do—to build muscle and other lean tissue structures and, to some extent, to fuel activity. The body does, however, handle protein differently during activity than during rest.

• ***Protein Used in Muscle Building*** • Synthesis of body proteins is suppressed during activity and for several hours afterward. In the hours following this period, though, protein synthesis accelerates beyond normal resting levels. Remember that the body adapts and builds the molecules, cells, and tissues it needs for the next period of activity. Whenever the body remodels a part of itself, it also tears down old structures to make way for new ones. Repeated activity, with just a slight overload, triggers the protein-dismantling and protein-synthesizing equipment of each muscle cell to make needed changes—that is, to adapt.

• Fitness tip: To conserve protein, eat a diet adequate in energy and rich in carbohydrate.

The physical work of each muscle cell acts as a signal to its DNA and RNA to begin producing the kinds of proteins that will best support that work. Take jogging, for example. In the first difficult sessions, the body is not yet equipped to perform aerobic work easily, but with each session, the cells' genetic material gets the message that an overhaul is needed. In the hours that follow the session, the genes send molecular messages to the protein-building equipment that tell it what old structures to break down and what new structures to build, and within the limits of its genetic potential, it responds. Among the new structures are more mitochondria to facilitate efficient aerobic metabolism. Over a few weeks' time, remodeling occurs and jogging becomes easier.

Such remodeling requires protein. During active muscle-building phases of training, an athlete may add between ¼ ounce and 1 ounce (between 7 and 28 grams) of body protein to existing muscle mass each day. This increase occurs only during periods of *building*—not times of maintenance—when the athlete exercises at high intensities.

• ***Protein Used as Fuel*** • Not only do athletes retain more protein in their muscles, they also use more protein as fuel: muscles speed up their use of amino acids for energy during physical activity, just as they speed up their use of fat and carbohydrate.[38] Still, protein contributes at most about 10 percent of the total fuel used, both during activity and during rest. The most active people of all, endurance athletes, use up enormous amounts of all energy fuels, including protein, during performance, but such athletes also eat more food and therefore usually consume enough protein.

• ***Diet Affects Protein Use during Activity*** • The factors that affect how much protein is used during activity seem to be the same three that influence the use of fat and carbohydrate—for one, diet. People who consume diets adequate in energy and rich in *carbohydrate* use less protein than those who eat protein- and fat-rich diets. Recall that carbohydrates spare proteins from being broken down to make glucose when needed. Since physical activity requires glucose, a diet lacking in carbohydrate necessitates the conversion of amino acids to glucose. So does a diet high in fat, because fatty acids can never provide glucose.

• ***Intensity and Duration of Activity Affect Protein Use during Activity*** • A second factor, the intensity and duration of activity, also modifies protein use.[39] Endurance athletes who train for over an hour a day, engaging in aerobic activity of moderate intensity and long duration, may deplete their glycogen stores by the end of their workouts and become somewhat more dependent on body protein for energy. The protein needs of bodybuilders and weight lifters are higher than those of sedentary people, but certainly not as high as the protein intakes many bodybuilders consume.

• ***Training Affects Protein Use*** • A third factor that influences a person's use of protein during physical activity is the extent of training. Predictably, the higher the degree of training, the less protein a person uses during an activity.

• ***Protein Recommendations for Active People*** • As mentioned earlier, all active people, and especially those who work like athletes, probably need more protein than do sedentary people. Endurance athletes, such as long-distance runners and cyclists, use more protein for fuel than power athletes do, and they retain some, especially in the muscles used for their sport. Power athletes, such as bodybuilders and football players, use less protein for fuel, but they still use some and retain much more. Therefore, *all* athletes in training should attend to protein needs, but should back up the protein with ample carbohydrate. Otherwise, they will burn off as fuel the very protein that they wish to retain in muscle.

How much protein, then, should an active person consume? A joint position paper from the American Dietetic Association (ADA) and the Canadian Dietetic Association (CDA) recommends 1.0 to 1.5 grams of protein per kilogram of body weight each day, an amount somewhat higher than the RDA for the general population.[40] Another authority suggests different protein intakes for athletes pursuing different activities.[41] Table 14-4 lists these recommendations and translates them into daily intakes for active people. Athletes who want to build muscle mass should first meet their energy needs with adequate carbohydrate intakes and then check to ensure that they have met protein needs as well. A later section translates protein recommendations into a diet plan and shows that no one needs protein supplements, or even large servings of meat, to obtain the highest recommended protein intakes.

Chapter 6 concluded that most people receive more than enough protein without supplements and reviewed the potential dangers of using protein and amino acid supplements.

## Vitamins and Minerals to Support Activity

Many of the vitamins and minerals assist in releasing energy from fuels and in transporting oxygen. This knowledge has led many people to believe, mistakenly, that vitamin and mineral *supplements* offer physically active people both health benefits and athletic advantages. (Highlight 10 focuses on vitamin and mineral supplements, and Highlight 14 explores the many other tricks athletes use in the hope of enhancing performance.)

• ***Supplements*** • Nutrient supplements do not enhance the performance of well-nourished people. Deficiencies of vitamins and minerals, however, do impede performance. In general, active people who eat enough nutrient-dense foods to meet energy needs also meet their vitamin and mineral needs. After all, active people eat more food; it stands to reason that with the right choices, they'll get more nutrients.

Some athletes mistakenly believe that taking vitamin or mineral supplements directly before competition will enhance performance. These beliefs are contrary to scientific reality. Most vitamins and minerals function as small parts of larger working units. After entering the blood, they have to wait for the cells to combine them with their appropriate other parts so that they can do their work. This takes time—hours or days. Vitamins or minerals taken right before an event are useless for improving performance, even if the person is actually suffering deficiencies of them.

In general, then, active people who eat well-balanced meals need no vitamins or minerals in supplement form. Iron may be an exception to this rule, however, as the next section explains.

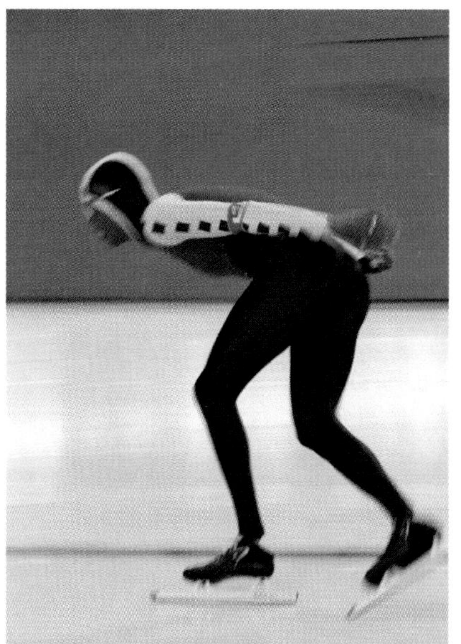

***For perfect functioning, every nutrient is needed.***

**Table 14-4**

**Recommended Protein Intakes for Athletes**

| | Recommendations (g/kg/day) | Protein Intakes (g/day) | |
|---|---|---|---|
| | | ***Males*** | ***Females*** |
| RDA for adults | 0.8 | 56 | 44 |
| ADA/CDA recommended intake | 1.0–1.5 | 70–105 | 55–83 |
| Recommended intake for power (strength or speed) athletes | 1.2–1.7 | 84–119 | 66–94 |
| Recommended intake for endurance athletes | 1.2–1.4 | 84–98 | 66–77 |
| U.S. average intake | | 95 | 65 |

NOTE: Daily protein intakes are based on a 70-kilogram (154-pound) man and 55-kilogram (121-pound) woman.
Sources: Committee on Dietary Allowances, *Recommended Dietary Allowances,* 10th ed. (Washington, D.C.: National Academy Press, 1989); Position of The American Dietetic Association and The Canadian Dietetic Association: Nutrition for physical fitness and athletic performance for adults, *Journal of the American Dietetic Association* 93 (1993): 691–695; P. W. R. Lemon, Effect of exercise on protein requirements, in *Foods, Nutrition and Sports Performance: An International Scientific Consensus,* eds. C. Williams and J. T. Devlin (London: E & FN Spon, 1992), pp. 65–86.

• ***Iron*** • Physically active young women, especially those who engage in endurance activities such as distance running, are prone to iron deficiency. Physical activity can affect iron status in several ways. For one thing, iron losses in sweat can contribute to iron deficiency.[42] For another, red blood cell destruction can lead to iron loss; blood cells are squashed when body tissues (such as the soles of the feet) make high-impact contact with an unyielding surface (such as the ground). Perhaps more significant than these losses are deficits caused by poor iron absorption in some athletes and the high demands of muscles for the iron-containing molecules of the mitochondria and the muscle protein myoglobin. In addition, physical activity may cause small blood losses through the digestive tract, at least in some athletes.

• ***Iron Deficiency*** • Iron deficiency affects more young women than men, and physically active people are no exception. Habitually low intakes of iron-rich foods and high iron losses through menstruation, as well as through the other routes mentioned, can cause iron deficiency in physically active young women. Even short-term moderate aerobic activity may compromise women's iron status.[43]

• ***Iron-Deficiency Anemia*** • Evidence is equivocal as to whether marginal iron deficiency impairs physical performance.[44] Iron-deficiency anemia, however, clearly impairs physical performance because the hemoglobin in the red blood cells is indispensable for delivering oxygen to the cells for energy metabolism. Without adequate oxygen, an active person cannot use fat for fuel, cannot perform aerobic activities, and tires easily.

**sports anemia:** a transient condition of low hemoglobin in the blood, associated with the early stages of sports training or other strenuous activity.

• ***Sports Anemia*** • Early in training, athletes may develop low blood hemoglobin for a while. This condition, sometimes called "sports anemia," is not a true iron-deficiency condition. Strenuous aerobic activity promotes destruction of the more fragile, older red blood cells, and the resulting cleanup work reduces the blood's iron content temporarily. Strenuous activity also expands the blood's plasma volume, thereby reducing the red blood cell count per unit of blood, but the red blood cells do not diminish in size or number as in anemia, so oxygen-carrying capacity is not hindered. Most researchers view sports anemia as an *adaptive,* temporary response to endurance training. Iron-deficiency anemia requires iron supplementation, but sports anemia does not respond to it.

• ***Iron Recommendations for Athletes*** • The best strategy concerning iron depends on the individual. Many menstruating women probably border on iron deficiency even without the iron losses incurred by physical activity. Active teens of both sexes also have high iron needs because they are growing. Especially for women and teens, then, prescribed supplements may be needed to correct deficiencies of iron, but medical testing should guide decisions on supplementation. Chapter 13 provides many more details about iron and the tests used in assessing its status.

***To prevent dehydration and the fatigue that accompanies it, drink plenty of liquids before, during, and after physical activity.***

## *Fluids and Electrolytes to Support Activity*

The body's need for water far surpasses its need for any other nutrient. The body relies on watery fluids as the medium for all of its life-supporting activities, and if it loses too much water, its well-being becomes compromised.

Obviously, the body loses water via sweat. Breathing uses water, too, exhaled as vapor. During physical activity, water losses from both routes are significant, and dehydration becomes a threat. Dehydration's first symptom is fatigue: a water loss of even 1 to 2 percent of body weight can reduce a person's capacity to do muscular work.[45] With a water loss of about 7 percent, a person is likely to collapse.[46]

• ***Fluid Losses via Sweat*** • Recall that working muscles produce heat as a by-product of energy metabolism. During intense activity, muscle heat production can be 15 to 20 times greater than at rest.[47] The body cools itself by sweating.

Each liter of sweat dissipates almost 600 kcalories of heat, preventing a rise in body temperature of almost 10 degrees on the Celsius scale. The body routes its blood supply through the capillaries just under the skin, and the skin secretes sweat to evaporate and cool the skin and the underlying blood. The blood then flows back to cool the deeper body chambers.

Note: 10 degrees on the Celsius scale is about 18 degrees on the Fahrenheit scale.

• ***Hyperthermia*** • In hot, humid weather, sweat doesn't evaporate well because the surrounding air is already laden with water. Body heat builds up and triggers maximum sweating, but without sweat evaporation, little cooling takes place. In such conditions, active people must take precautions to prevent heat stroke. To reduce the risk of heat stroke, drink enough fluid before and during the activity, rest in the shade when tired, and wear lightweight clothing that allows evaporation.[48] (Hence the danger of rubber or heavy suits that supposedly promote weight loss during physical activity—they promote profuse sweating, prevent sweat evaporation, and invite heat stroke.) If you ever experience any of the symptoms of heat stroke listed in the margin, stop your activity, sip fluids, seek shade, and ask for help. Heat stroke can be fatal, young people often die of it, and these symptoms demand attention.

**hyperthermia:** an above-normal body temperature.

**heat stroke:** the dangerous accumulation of body heat with accompanying loss of body fluid.

• Fitness tip: To prevent heat stroke, drink fluids, rest when tired, and wear appropriate clothing.

Symptoms of heat stroke:
- Headache.
- Nausea.
- Dizziness.
- Clumsiness.
- Stumbling.
- Excessive or insufficient sweating.
- Confusion or other mental changes.

• ***Hypothermia*** • In cold weather, *hypo*thermia, or low body temperature, can pose as serious a threat as heat stroke. Inexperienced, slow runners participating in long races on cold or wet, chilly days are especially vulnerable to hypothermia. Slow runners who produce little heat can become too cold if clothing is inadequate. Early symptoms of hypothermia include shivering and euphoria. As body temperature continues to fall, shivering may stop, and weakness, disorientation, and apathy may occur. Each of these symptoms can impair a person's ability to act against a further drop in body temperature. Even in cold weather, however, the active body still sweats and still needs fluids. The fluids should be warm or at room temperature to help protect against hypothermia.

**hypothermia:** a below-normal body temperature.

• Fitness tip: To prevent hypothermia, drink fluids and wear appropriate clothing.

• ***Fluid Replacement via Hydration*** • Endurance athletes can easily lose 1.5 liters or more of fluid during *each hour* of activity. To prepare for fluid losses, a person must hydrate before activity. To replace fluid losses, the person must rehydrate during and after activity. (Table 14-5 presents one schedule of hydration for physical activity.) Even then, in hot weather, the GI tract may not be able to absorb enough water fast enough to keep up with sweat losses, and some degree of dehydration may be inevitable.

Athletes who are preparing for competition are often advised to drink extra fluids in the *days* immediately before the event, especially if they are still training. The extra water is not stored in the body, but drinking extra water ensures maximum hydration at the start of the event. Full hydration is imperative for every athlete both in training and in competition. The athlete who arrives at an event even slightly dehydrated arrives with a disadvantage. Drinking extra water does no harm and may be protective.

Water recommendation for adults:
- 1.0 to 1.5 ml/kcal expended.

Note: 1 ml = 0.03 fluid oz.
125 ml ≈ ½ c.
Easy estimation: ≈ ½ c per 100 kcal expended.

What is the best fluid for an exercising body? For noncompetitive, everyday active people, plain, cool water is recommended, especially in warm weather, for

**Table 14-5**
**Hydration Schedule for Physical Activity**

| When to Drink | Approximate Amount of Fluid |
|---|---|
| 2 hr before activity | 3 c |
| 10 to 15 min before activity | 2 c |
| Every 15 min during activity | 1 c |
| After activity | 2 c |

Source: D. C. Nieman, *Fitness and Sports Medicine: An Introduction* (Palo Alto, Calif.: Bull Publishing, 1990), p. 234.

two reasons: it rapidly leaves the digestive tract to enter the tissues where it is needed, and it cools the body from the inside out. For endurance athletes, other beverages may be appropriate. Fluid ingestion during the event has the dual purposes of replenishing water lost through sweating and providing a source of carbohydrate to supplement the body's limited glycogen stores. Carbohydrate depletion brings on fatigue in the athlete, but as already mentioned, fluid loss and the accompanying buildup of body heat can be life-threatening. Thus the first priority for endurance athletes should be to replace fluids. Many good-tasting drinks are marketed for active people; the accompanying "How to" compares them with water.

Gatorade Sports Science Institute
www.gssiweb.com

• ***Electrolyte Losses and Replacement*** • When a person sweats, small amounts of electrolytes—the electrically charged minerals sodium, potassium, chloride, and magnesium—are lost from the body along with water. Losses are greatest in beginners; training improves electrolyte retention.

To replenish lost electrolytes, a person ordinarily needs only to eat a regular diet that meets energy and nutrient needs. In extremely demanding endurance events lasting more than three hours, sports drinks may be needed to replace fluids and electrolytes. Salt tablets can worsen dehydration and impair performance. They increase potassium losses, irritate the stomach, and cause vomiting.

## *Poor Beverage Choices: Caffeine and Alcohol*

Athletes, like others, sometimes drink beverages that contain caffeine or alcohol. Each of these substances can influence physical performance.

• ***Caffeine*** • Caffeine is a stimulant, and athletes sometimes use it to enhance performance as Highlight 14 explains. Caffeine is also a diuretic that induces fluid losses. It can be particularly hazardous when people competing in hot environments drink caffeine-containing beverages in place of water or other beverages. Carbonated soft drinks, whether they contain caffeine or not, may not be a wise choice for athletes: bubbles make a person feel full quickly and so limit fluid intake.

• ***Alcohol*** • Some athletes mistakenly believe that they can replace fluids and load up on carbohydrates by drinking beer. A 12-ounce beer provides 13 grams of carbohydrate—one-third the amount of carbohydrate in a glass of orange juice the same size. In addition to carbohydrate, beer also contains alcohol, of course. Energy from alcohol breakdown generates heat, but does not fuel muscle work because alcohol is metabolized in the liver.

Beer facts:
- *Beer is not carbohydrate-rich.* Beer is *kcalorie*-rich, but only ⅓ of its kcalories are from carbohydrates. The other ⅔ are from alcohol.
- *Beer is mineral-poor.* Beer contains a few minerals, but to replace those lost in sweat, athletes need good sources such as fruit juices.
- *Beer is vitamin-poor.* Beer contains tiny traces of some B vitamins, but it cannot compete with food sources.
- *Beer causes fluid losses.* Beer is a fluid, but alcohol is a diuretic and causes the body to lose more fluid in urine than is provided by the beer.

It is hard to overstate alcohol's detrimental effects on physical activity. Alcohol's diuretic effect impairs the body's fluid balance, making dehydration likely; after physical activity, a person needs to replace fluids, not lose them by drinking beer. Alcohol impairs the body's ability to regulate its temperature, making hypothermia or heat stroke much more likely.

Alcohol also alters perceptions; slows reaction time; reduces strength, power, and endurance; and hinders accuracy, balance, eye-hand coordination, and coordination in general—all opposing optimal athletic performance. In addition, it deprives people of their judgment, thereby compromising their safety in sports; many sports-related fatalities and injuries involve alcohol or other drugs.

Clearly, alcohol impairs performance, but physically active people do drink on occasion. A word of caution: do not drink alcohol before exercising, and drink plenty of water after exercising before drinking alcohol.

IN SUMMARY

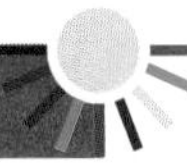

The mixture of fuels the body uses during physical activity depends on diet, the intensity and duration of the activity, and training. During intense activity, the muscles use glucose primarily; during less intense, moderate activity, fat makes a greater energy contribution, and glycogen

## HOW TO Evaluate Sports Drinks

Hydration is critical to optimal performance. Water best meets the fluid needs of most people, yet manufacturers market many good-tasting sports drinks for active people. More than 20 "power beverages" compete for their share of the $1 billion market. What do sports drinks have to offer?

- *Fluid.* Sports drinks offer fluids to help offset the loss of fluids during physical activity, but plain water can do this, too. Alternatively, fruit juices can be diluted (by one-half to one-third), if preferred to plain water.
- *Glucose.* Sports drinks offer simple sugars or glucose polymers that help maintain hydration and blood glucose and enhance performance as effectively as, or maybe even better than, water. Such measures are especially beneficial for strenuous endurance activities lasting longer than an hour. Most sports drinks contain about 7 percent glucose (about half the sugar of ordinary soft drinks, or about 5 teaspoons in each 12 ounces). Less than 6 percent may not enhance performance, and more than 10 percent may cause abdominal cramps, nausea, and diarrhea. Fluid transport to the tissues from beverages containing up to 10 percent glucose is rapid.[a]

While glucose does enhance endurance performance in strenuous competitive events, for the moderate exerciser, it can be counterproductive if weight loss is the goal. Glucose is sugar, and like candy, it provides only empty kcalories—no vitamins or minerals. Most sports drinks provide between 50 and 100 kcalories per cup.

- *Sodium and other electrolytes.* Sports drinks offer sodium and other electrolytes to help replace those lost during physical activity. Sodium in sports drinks also helps to increase the rate of fluid absorption from the GI tract and maintain plasma volume during activity and recovery.

Most physically active people do not need to replace the minerals lost in sweat immediately; a meal eaten within hours of competition replaces these minerals soon enough. Most sports drinks are relatively low in sodium, however, so those who choose to use these beverages run little risk of excessive intake.

In strenuous, world-class competitions lasting 6 hours or more, heavy sweating coupled with consumption of large amounts of plain water dangerously dilutes blood sodium. In these few cases, intravenous fluid and electrolyte repletion is needed.[b]

- *Good taste.* Manufacturers reason that if a drink tastes good, people will drink more, thereby ensuring adequate hydration. For athletes who prefer the flavors of sports drinks over water, it may be worth paying for good taste to replace lost fluids.
- *Psychological edge.* Sports drinks provide a psychological edge for some people who associate the drinks with athletes and sports. The need to belong is valid. If the drinks boost morale and are used with care, they may do no harm.

For trained endurance athletes who exercise for an hour or more, sports drinks may provide a slight advantage over water. For most physically active people, though, water is the best fluid to replenish lost fluids. The most important thing to do is drink—even if you don't feel thirsty.

**glucose polymers:** compounds that supply glucose, not as single molecules, but linked in chains somewhat like starch. The objective is to attract less water from the body into the digestive tract (osmotic attraction depends on the number, not the size of particles).

[a]C. V. Gisolfi, Fluid balance for optimal performance, *Nutrition Reviews* 54 (1996): S159–S168.
[b]N. Clark, J. Tobin, and C. Ellis, Feeding the ultraendurance athlete: Practical tips and a case study, *Journal of the American Dietetic Association* 92 (1992): 1258–1262.

use is slower. With the possible exception of iron, well-nourished active people do not need nutrient supplements. Active people need to drink plenty of water, especially during training or competition.

# Diets for Physically Active People

No one diet best supports physical performance. Active people who choose foods within the framework of the diet-planning principles presented in Chapter 2 can design many excellent diets.

Sport Medicine and Science Council of Canada

www.smscc.ca

## Choosing a Diet to Support Fitness

First, remember that water is depleted more rapidly than any other nutrient. A diet to support fitness must provide water, energy, and all the other nutrients.

• ***Water*** • Even casual exercisers must attend conscientiously to their fluid needs. Physical activity blunts the thirst mechanism, especially in cold weather. During activity, thirst signals come too late, so don't wait to feel thirsty before drinking. To find out how much water is needed to replenish activity losses, weigh yourself before and after the activity—the difference is almost all water. One pound equals roughly 2 cups (500 milliliters) fluid.

• ***Nutrient Density*** • A healthful diet is based on nutrient-dense foods—foods that supply adequate vitamins and minerals for the energy they provide. Active people need to eat both for nutrient adequacy and for energy. They are not immune to heart disease and cancer and so must limit fats. A diet that is high in carbohydrate (60 percent of total kcalories or more), low in fat (25 percent or less), and adequate in protein (12 to 15 percent) ensures full glycogen and other nutrient stores.

Carbohydrate recommendation for athletes in heavy training: 8 g/kg body weight.

• ***Carbohydrate*** • On two occasions, the active person's regular high-carbohydrate, fiber-rich diet may require temporary adjustment. Both of these exceptions involve training for competition rather than fitness. During intensive training, energy needs may be so high as to outstrip the person's capacity to eat enough food to meet them. At that point, added sugar and fat may be needed. The person can add concentrated carbohydrate foods such as dried fruits, sweet potatoes, nectars, and even high-fat foods such as avocados, nuts, cookies, and ice cream. Still, a nutrient-rich diet remains central for adequacy's sake. Though vital, energy alone is not enough to support performance. The other special occasion is the pregame meal, when fiber-rich, bulky foods are best avoided. The pregame meal is discussed in a later section.

Highlight 8 examines some of the pros and cons of the high-protein, low-carbohydrate diets popular among athletes today.

• ***Protein*** • In addition to carbohydrate and some fat (and the energy they provide), physically active people need protein. How much of what kinds of foods supply enough protein to meet their needs? Meats and milk products are rich protein sources, but to recommend that active people emphasize these foods would be narrow advice for many reasons. For one thing, all people must protect themselves from heart disease, and even lean meats and reduced- or low-fat milk products contain fat, much of it saturated fat. For another, as emphasized repeatedly, active people need diets high in carbohydrate, and of course, meats have none to offer. Legumes, grains, and vegetables provide protein with abundant carbohydrate and little fat. Table 14-4 (on p. 451) showed some possible protein intakes for active people.

***A variety of foods is the best source of nutrients for athletes.***

• ***A Performance Diet Example*** • A person weighing 70 kilograms who engages in vigorous physical activity on a daily basis could likely require more than 3000 kcalories per day. To meet this need, the person can choose a variety of nutrient-dense foods. Figure 14-4 shows one example of meals that provide 3300 kcalories. These meals supply over 130 grams of protein, more than even the highest recommended intake for such a person. Obviously, the more energy a person requires, the more protein that person will receive, assuming the foods chosen are nutrient dense. This relationship between energy and protein intakes breaks down only when people meet their energy needs with high-fat, high-sugar confections. The meals shown in Figure 14-4 provide almost 550 grams of carbohydrate, or over 60 percent of total kcalories. Athletes who train exhaustively for endurance events may want to aim for somewhat higher carbohydrate intakes. Beyond these specific concerns of total energy, protein, and carbohydrate, the diet most beneficial to athletic performance is remarkably similar to the diet recommended for most people.

Breakfast:

1 c shredded wheat.
1 c 1% low-fat milk.
1 small banana.
2 slices whole-wheat toast.
4 tsp jelly.
1½ c orange juice.

Snack:

3 c plain popcorn.
A smoothie made from:
1½ c apple juice.
1½ frozen banana.

Lunch:

2 turkey sandwiches.
1½ c 1% low-fat milk.
Large bunch of grapes.

Dinner:

Salad: 1 c spinach, carrots, and mushrooms.
½ c garbanzo beans.
1 tbs sunflower seeds.
1 tbs ranch salad dressing.
1 c spaghetti with meat sauce.
1 c green beans.
1 corn on the cob.
2 slices Italian bread.
4 tsp butter.
1 piece angel food cake.
1¼ c fresh strawberries.
1 tbs whipping cream.
1 c 1% low-fat milk.

**Total kcal: 3300**

63% kcal from carbohydrate
22% kcal from fat
15% kcal from protein

All vitamin and mineral intakes exceed the RDA for both men and women.

**Figure 14-4**
**An Athlete's Meal Selections**

## *Meals before and after Competition*

No single food improves speed, strength, or skill in competitive events, although some *kinds* of foods do support performance better than others as already explained. Still, a competitor may eat a particular food before or after an event for psychological reasons. One eats a steak the night before wrestling, another takes some honey five minutes after diving. As long as these practices remain harmless, they should be respected.

• ***Pregame Meals*** • Science indicates that the pregame meal or snack should include plenty of fluids and be light and easy to digest. It should provide between 300 and 800 kcalories, primarily from carbohydrate-rich foods that are familiar and well tolerated by the athlete. The meal should end two to five hours before competition to allow time for the stomach to empty before exertion.

Breads, potatoes, pasta, and fruit juices—that is, carbohydrate-rich foods low in fat and fiber—form the basis of the best pregame meal. Bulky, fiber-rich foods such as raw vegetables or high-bran cereals, although usually desirable, are best avoided just before competition. Fiber in the digestive tract attracts water out of the blood and can cause stomach discomfort during performance. Liquid meals are easy to digest, and many such meals are commercially available. Alternatively, athletes can mix nonfat milk or juice, frozen fruits, and flavorings in a blender.

• ***Postgame Meals*** • As mentioned earlier, eating high-carbohydrate foods *after* physical activity enhances glycogen storage. Since people are usually not hungry immediately following physical activity, carbohydrate-containing beverages such

High-carbohydrate, liquid pregame meal ideas:

- Apple juice, frozen banana, and cinnamon.
- Papaya juice, frozen strawberries, and mint.
- Nonfat milk, frozen banana, and vanilla.

as the sports drinks discussed earlier may be preferred. If an active person does feel hungry after an event, then foods high in carbohydrate and low in protein, fat, and fiber are the ones to choose—the same ones recommended prior to competition. Foods high in protein and fat should be avoided during the first few hours after activity as these foods may suppress hunger and thus limit carbohydrate intake.[49]

## IN SUMMARY

The person who wants to excel physically will apply accurate nutrition knowledge along with dedication to rigorous training. A diet that provides ample fluid and includes a variety of nutrient-dense foods in quantities to meet energy needs will enhance not only athletic performance, but overall health as well. Carbohydrate-rich foods that are light and easy-to-digest are recommended for both the pregame and the postgame meal. Training and genetics being equal, who will win a competition—the athlete who habitually consumes inadequate amounts of needed nutrients or the competitor who arrives at the event with a long history of full nutrient stores and well-met metabolic needs?

Some athletes learn that nutrition can support physical performance and turn to pills and powders instead of foods. In case you need further convincing that a healthful diet surpasses such potions, the following highlight addresses this issue.

U.S. Government
www.healthfinder.gov/searchoptions/topicsaz.htm
Search for Physical fitness

## Study Questions

These questions will help you review the chapter. You will find the answers in the discussions on the pages provided.

1. Define fitness, and list its benefits. (pp. 438–439)
2. Explain the overload principle. (pp. 439–440)
3. What types of activity are anaerobic? Which are aerobic? (pp. 443–444)
4. Define cardiorespiratory conditioning and list some of its benefits. (p. 441)
5. Describe the relationships among energy expenditure, type of activity, and oxygen use. (p. 443)
6. What factors influence the body's use of glucose during physical activity? How? (pp. 444–447)
7. What factors influence the body's use of fat during physical activity? How? (pp. 447–449)
8. What factors influence the body's use of protein during physical activity? How? (pp. 449–451)
9. Why are some athletes likely to develop iron-deficiency anemia? Compare iron-deficiency anemia and sports anemia, explaining the differences. (p. 452)
10. Discuss the importance of hydration during training, and list recommendations to maintain fluid balance. (pp. 452–454)
11. Describe the components of a healthy diet for athletic performance. (p. 456)

These questions will help you prepare for an exam. Answers can be found in Appendix K.

1. Physical inactivity is linked to all of the following diseases except:
   a. cancer.
   b. diabetes.
   c. emphysema.
   d. hypertension.
2. The progressive overload principle can be applied by performing:
   a. an activity less often.
   b. an activity with more intensity.
   c. an activity in a different setting.
   d. a different activity each day of the week.
3. The process that regenerates glucose from lactic acid is known as the:
   a. Cori cycle.
   b. ATP-CP cycle.
   c. adaptation cycle.
   d. cardiac output cycle.
4. "Hitting the wall" is a term runners sometimes use to describe:
   a. dehydration.
   b. competition.
   c. indigestion.
   d. glucose depletion.
5. The technique endurance athletes use to maximize glycogen stores is called:
   a. aerobic training.
   b. muscle conditioning.
   c. carbohydrate loading.
   d. progressive overloading.
6. Conditioned muscles rely less on ______ and more on ______ for energy.
   a. protein; fat
   b. fat; protein
   c. glycogen; fat
   d. fat; glycogen

7. Vitamin or mineral supplements taken right before an event are useless for improving performance because the:
   a. athlete sweats the nutrients out during the event.
   b. stomach can't digest supplements during physical activity.
   c. nutrients are diluted by all the fluids the athlete drinks.
   d. body needs hours or days for the nutrients to do their work.
8. Physically active young women, especially those who are endurance athletes, are prone to:
   a. energy excess.
   b. iron deficiency.
   c. protein overload.
   d. vitamin A toxicity.
9. The body's need for ______ far surpasses its need for any other nutrient.
   a. water
   b. protein
   c. vitamins
   d. carbohydrate
10. A recommended pregame meal includes plenty of fluids and provides between:
   a. 300 and 800 kcalories, mostly from fat-rich foods.
   b. 50 and 100 kcalories, mostly from fiber-rich foods.
   c. 1000 and 2000 kcalories, mostly from protein-rich foods.
   d. 300 and 800 kcalories, mostly from carbohydrate-rich foods.

## Notes


1. American College of Sports Medicine, The recommended quality and quantity of exercise for developing and maintaining fitness in healthy adults, *Medicine and Science in Sports and Exercise* 22 (1990): 265–274.
2. U.S. Centers for Disease Control and Prevention and American College of Sports Medicine, Summary statement: Workshop on physical activity and public health, *Sports Medicine Bulletin* 28 (1993): 7.
3. D. A. Leaf, D. L. Parker, and D. Schaad, Changes in $VO_{2max}$, physical activity, and body fat with chronic exercise: Effects on plasma lipids, *Medicine and Science in Sports and Exercise* 29 (1997): 1152–1159; P. T. Williams, Relationship of distance run per week to coronary heart disease risk factors in 8283 male runners, *Archives of Internal Medicine* 157 (1997): 191–198.
4. U.S. Department of Health and Human Services, *Physical Activity and Health: A Report of the Surgeon General Executive Summary* (Washington, D.C.: Government Printing Office, 1996); R. R. Pate and coauthors, Physical activity and public health: A recommendation from the Centers for Disease Control and Prevention and the American College of Sports Medicine, *Journal of the American Medical Association* 273 (1995): 402–407; R. S. Paffenbarger and coauthors, The association of changes in physical-activity level and other lifestyle characteristics with mortality among men, *New England Journal of Medicine* 328 (1993): 538–545; L. Sandvik and coauthors, Physical fitness as a predictor of mortality among healthy, middle-aged Norwegian men, *New England Journal of Medicine* 328 (1993): 533–537.
5. U.S. Department of Health and Human Services, 1996.
6. S. N. Blair, Physical inactivity and cardiovascular disease risk in women, *Medicine and Science in Sports and Exercise* 28 (1996): 9–10; I. Thune and coauthors, Physical activity and the risk of breast cancer, *New England Journal of Medicine* 336 (1997): 1269–1275; NIH Consensus Development Panel on Physical Activity and Cardiovascular Health, Physical activity and cardiovascular health, *Journal of the American Medical Association* 276 (1996): 241–246; Pate and coauthors, 1995; American Heart Association Position Statement on Exercise: Benefits and recommendations for physical activity programs for all Americans, *Circulation* 86 (1992): 340–344.
7. NIH Consensus Development Panel on Physical Activity and Cardiovascular Health, 1996; W. L. Haskell, Health consequences of physical activity: Understanding and challenges regarding dose-response, *Medicine and Science in Sports and Exercise* 26 (1994): 649–660.
8. S. N. Blair and coauthors, Changes in physical fitness and all-cause mortality, *Journal of the American Medical Association* 273 (1995): 1093–1098.
9. American College of Sports Medicine, *ACSM's Guidelines for Exercise Training and Prescription*, 5th ed. (Philadelphia: Williams & Wilkins, 1995), pp. 3–11.
10. J. H. Wilmore, Increasing physical activity: Alterations in body mass and composition, *American Journal of Clinical Nutrition* 63 (1996): 456S–460S.
11. D. Teegarden and coauthors, Previous physical activity relates to bone mineral measures in young women, *Medicine and Science in Sports and Exercise* 28 (1996): 105–113; ACSM Position Stand on Osteoporosis and Exercise, *Medicine and Science in Sports and Exercise* 27 (1995): i–iv; B. P. Conroy and coauthors, Bone mineral density in elite junior Olympic weightlifters, *Medicine and Science in Sports and Exercise* 25 (1993): 1103–1109.
12. D. C. Nieman, Exercise, upper respiratory tract infection, and the immune system, *Medicine and Science in Sports and Exercise* 26 (1994): 128–139.
13. Thune and coauthors, 1997; M. M. Kramer and C. L. Wells, Does physical activity reduce risk of estrogen-dependent cancer in women? *Medicine and Science in Sports and Exercise* 28 (1996): 322–334; J. A. Woods and J. M. Davis, Exercise, monocyte/macrophage function, and cancer, *Medicine and Science in Sports and Exercise* 26 (1994): 147–157.
14. A. R. Folsom and coauthors, Physical activity and incidence of coronary heart disease in middle-aged women and men, *Medicine and Science in Sports and Exercise* 29 (1997): 901–909; Blair, 1996; G. B. M. Mensink and coauthors, Intensity, duration, and frequency of physical activity and coronary risk factors, *Medicine and Science in Sports and Exercise* 29 (1997):

1192–1198; NIH Consensus Development Panel on Physical Activity and Cardiovascular Health, 1996; P. T. Williams, High-density lipoprotein cholesterol and other risk factors for coronary heart disease in female runners, *New England Journal of Medicine* 334 (1996): 1298–1303; A. L. Macnair, Physical activity, not diet, should be the focus of measures for the primary prevention of cardiovascular disease, *Nutrition Research Reviews* 7 (1994): 43–65; Paffenbarger and coauthors, 1993.

15. G. R. Hunter and coauthors, Fat distribution, physical activity, and cardiovascular risk factors, *Medicine and Science in Sports and Exercise* 29 (1997): 362–369; A. Goulding and coauthors, More exercise, less central fat distribution in women, *Journal of the American Medical Association* 276 (1996): 193–194.
16. G. Perseghin and coauthors, Increased glucose transport-phosphorylation and muscle glycogen synthesis after exercise training in insulin-resistant subjects, *New England Journal of Medicine* 335 (1996): 1357–1362; S. N. Blair and coauthors, Physical activity, nutrition, and chronic disease, *Medicine and Science in Sports and Exercise* 28 (1996): 335–349.
17. U. M. Kujala and coauthors, Relationship of leisure-time physical activity and mortality: The Finnish Twin Cohort, *Journal of the American Medical Association* 279 (1998): 440–444; L. H. Kushi and coauthors, Physical activity and mortality in postmenopausal women, *Journal of the American Medical Association* 277 (1997): 1287–1292.
18. A. A. Hakim and coauthors, Effects of walking on mortality among nonsmoking retired men, *New England Journal of Medicine* 338 (1998): 94–99.
19. L. DiPietro, The epidemiology of physical activity and physical function in older people, *Medicine and Science in Sports and Exercise* 28 (1996): 596–600; L. E. Voorrips and coauthors, The physical conditions of elderly women differing in habitual physical activity, *Medicine and Science in Sports and Exercise* 25 (1993): 1152–1157.
20. American College of Sports Medicine, 1995, pp. 12–26.
21. P. A. Ades and coauthors, Weight training improves walking endurance in healthy elderly persons, *Annals of Internal Medicine* 124 (1996): 568–572.
22. J. M. Guralnik and coauthors, Lower-extremity function in persons over the age of 70 years as a predictor of subsequent disability, *New England Journal of Medicine* 332 (1995): 556–561.
23. American College of Sports Medicine, Position stand: Osteoporosis and exercise, *Medicine and Science in Sports and Exercise* 27 (1995): i–vii.
24. C. M. L. Snow-Harter, Effects of resistance and endurance exercise on bone mineral status of young women: A randomized exercise intervention trial, *Journal of Bone Mineral Research* 7 (1992): 761–769.
25. American College of Sports Medicine, Position stand, Exercise and fluid replacement, *Medicine and Science in Sports and Exercise* 28 (1996): i–vii; G. McConell, K. Kloot, and M. Hargreaves, Effect of timing of carbohydrate ingestion on endurance exercise performance, *Medicine and Science in Sports and Exercise* 28 (1996): 1300–1304; P. R. Below and coauthors, Fluid and carbohydrate ingestion independently improve performance during 1 h of intense exercise, *Medicine and Science in Sports and Exercise* 27 (1995): 200–210.
26. J. A. M. Parkin and coauthors, Muscle glycogen storage following prolonged exercise: Effect of timing of ingestion of high glycemic index food, *Medicine and Science in Sports and Exercise* 29 (1997): 220–224.
27. A. R. Coggan, Plasma glucose metabolism during exercise: Effect of endurance training in humans, *Medicine and Science in Sports and Exercise* 29 (1997): 620–627.
28. D. M. Muoio and coauthors, Effect of dietary fat on metabolic adjustments to maximal $VO_2$ and endurance in runners, *Medicine and Science in Sports and Exercise* 26 (1994): 81–88.
29. J. W. Helge, B. Wulff, and B. Kiens, Impact of a fat-rich diet on endurance in man: Role of the dietary period, *Medicine and Science in Sports and Exercise* 30 (1998): 456–461.
30. W. M. Sherman and N. Leenders, Fat loading: The next magic bullet? *International Journal of Sports Nutrition* 5 (1995): S1–S12.
31. S. D. Phinney and coauthors, The human metabolic response to chronic ketosis without caloric restriction: Preservation of submaximal exercise capacity with reduced carbohydrate oxidation, *Metabolism* 32 (1983): 769–776.
32. J. Leddy and coauthors, Effect of a high or a low fat diet on cardiovascular risk factors in male and female runners, *Medicine and Science in Sports and Exercise* 29 (1997): 17–25.
33. P. Arnos, F. Andres, and K. Drowatzky, Fat oxidation and RPE at varied exercise intensities, *Medicine and Science in Sports and Exercise* (supplement) 25 (1993): S9; F. A. Kulling and coauthors, Identification and evaluation of the exercise intensity which maximizes fat oxidation in young women, *Medicine and Science in Sports and Exercise* (supplement) 25 (1993): S179.
34. M. A. Grediagin and coauthors, Exercise intensity does not effect body composition change in untrained, moderately overfat women, *Journal of the American Dietetic Association* 95 (1995): 661–665.
35. A. Tremblay and coauthors, Effect of intensity of physical activity on body fatness and fat distribution, *American Journal of Clinical Nutrition* 51 (1990): 153–157.
36. T. J. Horton and C. A. Geissler, Effect of habitual exercise on daily energy expenditure and metabolic rate during standardized activity, *American Journal of Clinical Nutrition* 59 (1994): 13–19.
37. Grediagin and coauthors, 1995.
38. F. Carraro and coauthors, Alanine kinetics in humans during low-intensity exercise, *Medicine and Science in Sports and Exercise* 26 (1994): 348–353.
39. P. R. Lemon, Is increased dietary protein necessary or beneficial for individuals with a physically active lifestyle? *Nutrition Reviews* 54 (1996): S169–S175.
40. Position of The American Dietetic Association and The Canadian Dietetic Association: Nutrition for physical fitness and athletic performance for adults, *Journal of the American Dietetic Association* 93 (1993): 691–695.
41. Lemon, 1996.
42. M. F. Waller and E. M. Haymes, The effects of heat and exercise on sweat iron loss, *Medicine and Science in Sports and Exercise* 28 (1996): 197–203.
43. R. M. Lyle and coauthors, Iron status in exercising women: The effect of oral iron therapy vs. increased consumption of muscle foods, *American Journal of Clinical Nutrition* 56 (1992): 1049–1055.
44. Y. I. Zhu and J. D. Haas, Iron depletion without anemia and physical performance in young women, *American*

*Journal of Clinical Nutrition* 66 (1997): 334–341; J. J. LaManca and E. M. Haymes, Effects of iron repletion on $VO_2$ max, endurance, and blood lactate in women, *Medicine and Science in Sports and Exercise* 25 (1993): 1386–1392.

45. C. V. Gisolfi, Fluid balance for optimal performance, *Nutrition Reviews* 54 (1996): S159–S168.
46. J. E. Greenleaf, Problem: Thirst, drinking behavior, and involuntary dehydration, *Medicine and Science in Sports and Exercise* 24 (1992): 645–656.
47. Gisolfi, 1996.
48. American College of Sports Medicine, Position stand: Heat and cold illnesses during distance running, *Medicine and Science in Sports and Exercise* 28 (1996): i–x.
49. E. F. Coyle, Timing and method of increased carbohydrate intake to cope with heavy training, competition, and recovery, in *Foods, Nutrition, and Sports Performance: An International Scientific Consensus,* eds. C. Williams and J. T. Devlin (London: E & FN Spon, 1992), pp. 37–61.

# Supplements and Ergogenic Aids Athletes Use

Athletes gravitate to promises that they can enhance their performance by taking pills, powders, or potions. Unfortunately, they often hear such promises from their coaches and peers, who advise them to use nutrient supplements, take drugs, or follow procedures that claim to deliver results without effort. When such aids are harmless, they are only a waste of money; when they impair performance or harm health, they waste athletic potential and cost lives. This highlight looks at some promises of magic to improve physical performance.

## Ergogenic Aids

Many substances or treatments claim to be *ergogenic,* meaning work enhancing. The glossary defines this term and several of the commonly used ergogenic aids discussed in this highlight. Table H14-1 (on pp. 464–465) lists many more substances promoted as ergogenic aids. For the large majority of these substances, research findings do not support those claims. Athletes who hear that a product is ergogenic should ask who is making the claim and who will profit from the sale.

Sometimes it is difficult to distinguish valid claims from bogus ones. Fitness magazines are particularly troublesome because many of them present both valid and invalid nutrition articles alongside slick advertisements for nutrition products. Advertisements often feature colorful anatomical figures, graphs, and tables that appear scientific. Some ads even include "reviews of literature" citing such credible sources as the *American Journal of Clinical Nutrition* and the *Journal of the American Medical Association.* Such ads create the illusion of credibility to gain readers' trust. Keep in mind, however, that the ads are created not to teach, but to sell. A careful reading of the cited research might reveal that the ads have presented the research findings out of context. In one such case, ad writers cited an article to support the invalid conclusion that their supplement was "critical to maximum anabolic utilization." Researchers reporting in the cited article had reached another conclusion: regular exercise has a definite anabolic effect, and the use of supplements does not appear to enhance that effect. Scientific facts had been twisted and created to promote sales. Highlight 1 describes ways to recognize quackery.

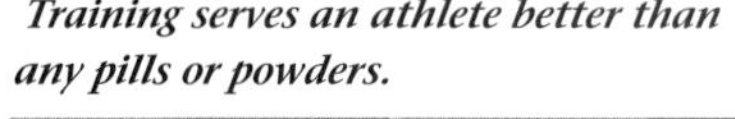

***Training serves an athlete better than any pills or powders.***

## Dietary Supplements

A variety of supplements make claims based on misunderstood nutrition principles. The claims may sound good, but for the most part, they have little or no factual basis.

### Protein Powders

Protein powders are among the most common supplements athletes use.[1] Protein powders can supply amino acids to the body, but nature's protein sources—lean meat, milk, and legumes—supply all these amino acids and more. Because the body builds muscle protein from amino acids, many athletes take protein powders with the false hope of stimulating muscle growth. Purified protein preparations, however, contain none of the other nutrients needed to support the building of muscle, and the protein they supply is not needed by athletes who eat food. It is excess protein, and the body dismantles it and uses it for energy or stores it as body fat. The deamination of excess amino acids places an extra burden on the kidneys to excrete unused nitrogen.

### Amino Acid Supplements

Chapter 6 (pp. 182–184) describes how the body cannot handle the high concentrations and unusual combinations of amino acids from supplements. Amino acids compete for carriers, which can limit the absorption of some amino acids, creating a deficiency, and can enhance the absorption of others, creating the possibility of toxicity. Most healthy athletes eating well-balanced diets do not need amino acid supplements. Advertisers point to research that identifies the branched-chain amino acids as the main ones used as fuel by exercising muscles. What the ads leave out is that compared to glucose and fatty acids, branched-chain amino acids provide almost no fuel and that ordinary foods provide them in abundance anyway.

Large doses of branched-chain amino acids can raise plasma ammonia concentrations, which can be toxic to the brain.[2] Branched-chain amino acid supplements are neither effective nor safe and are not recommended.

### Carnitine

Carnitine, a nonprotein amino acid, is a popular supplement among endurance athletes. The athletes believe carnitine will help them burn more fat, thereby sparing glycogen during endurance events. Carnitine is also promoted to bodybuilders as a "fat burner."

In the body, carnitine facilitates the transfer of fatty acids across the mitochondrial membrane. Supplement manufacturers suggest that with more carnitine available, fat oxidation will be enhanced, but this does not seem to be the case. Carnitine supplementation for 7 to 14 days neither raised muscle carnitine concentrations nor influenced fat or carbohydrate oxidation.[3] It did, however, produce diarrhea in half of the men tested. Milk and meat products are good sources of carnitine, and supplements are not needed.

## Vitamin E

Some research shows a relationship between physical activity and oxidative stress.[4] Many athletes and active people are taking megadoses of vitamin E in hopes of preventing oxidative damage to muscles. Vitamin E and other antioxidant nutrients may be especially effective for athletes exercising in extreme environments, such as heat, cold, and high altitudes. Studies examining the long-term effects of vitamin E supplementation are lacking, and researchers examining oxidative stress and physical activity report inconsistent findings. Some research suggests that prolonged, high-intensity activity enhances production of damaging free radicals in the body.[5] Another study shows that prolonged, intense physical activity reduces susceptibility to oxidative stress.[6] Clearly, more research is needed, but in the meantime, active people can benefit by eating generous servings of antioxidant-rich fruits and vegetables.

## Chromium Picolinate

Chapter 13 introduced chromium as an essential trace mineral involved in carbohydrate and lipid metabolism. Advertisements in bodybuilding magazines claim that chromium picolinate, which is supposed to be more easily absorbed than chromium alone, builds muscle, enhances energy, and burns fat. Such claims derive from one or two initial studies reporting that men who weight trained while taking chromium picolinate supplements increased lean body mass and reduced body fat.[7] Most studies of chromium picolinate and strength training that have followed, however, show no effects of chromium picolinate supplementation on strength, lean body mass, or body fat.[8] Whether high doses of chromium picolinate affect iron status remains unclear.[9]

## Complete Nutrition Supplements

Several drinks and candy bars appeal to athletes by claiming to provide "complete" nutrition. These products usually taste good and provide extra food energy, but fall short of providing "complete" nutrition. They can be useful as a pregame meal or a between-meal snack but they should not replace regular meals.

A nutritionally "complete" drink may help a nervous athlete who cannot tolerate solid food on the day of an event. A liquid meal two or three hours before competition can supply some of the fluid and carbohydrate needed in a pregame meal, but a shake of nonfat milk or juice (such as apple or papaya) and ice milk or frozen fruit (such as strawberries or bananas) can do the same thing less expensively.

## Creatine

Interest in—and use of—creatine monohydrate supplements to enhance performance during intense activity has grown dramatically in the last few years.[10] Power athletes such as weight lifters use creatine supplements to enhance stores of the high-energy

## Glossary

**anabolic steroids:** drugs related to the male sex hormone, testosterone, that stimulate the development of lean body mass.
- **anabolic** = promoting growth
- **sterols** = compounds chemically related to cholesterol

**branched-chain amino acids:** the amino acids leucine, isoleucine, and valine, which are present in large amounts in skeletal muscle tissue; falsely promoted as fuel for exercising muscles.

**caffeine:** a natural stimulant found in many common foods and beverages, including coffee, tea, and chocolate; may enhance endurance by stimulating fatty acid release but also causes fluid losses. Overdoses cause headaches, trembling, rapid heart rate, and other undesirable side effects.

**carnitine** (CAR-ne-teen): a nonprotein amino acid made in the body from lysine that helps transport fatty acids across the mitochondrial membrane. Carnitine supposedly "burns" fat and spares glycogen during endurance events, but in reality it does neither.

**chromium** (CROW-mee-um) **picolinate:** a trace mineral supplement; falsely promoted as building muscle, enhancing energy, and burning fat. **Picolinate** (pick-oh-LYN-ate) is a derivative of the amino acid tryptophan that seems to enhance chromium absorption.

**creatine** (KREE-ah-tin): a nitrogen-containing compound that combines with phosphate to form the high-energy compound creatine phosphate (or phosphocreatine) in muscles. Claims that creatine enhances energy use and muscle strength need further confirmation.

**DHEA** (dehydroepiandrosterone): a hormone made in the adrenal glands that serves as a precursor to the male hormone testosterone; falsely promoted as burning fat, building muscle, and slowing aging. Side effects include acne, aggressiveness, and liver enlargement.

**ergogenic** (ER-go-JEN-ick) **aids:** substances or techniques used in an attempt to enhance physical performance.
- **ergo** = work
- **genic** = gives rise to

**hGH (human growth hormone):** a hormone produced by the brain's pituitary gland that regulates normal growth and development; also called *somatotropin.* Some athletes misuse this hormone to increase their height and strength.

compound creatine phosphate (or phosphocreatine) in muscles. Theoretically, the more creatine phosphate in muscles, the higher the intensity at which an athlete can train. High-intensity training stimulates the muscles to adapt, which, in turn, improves performance.

The results of some studies suggest that creatine supplementation enhances performance of high-intensity strength activity such as weight lifting.[11] Other research findings, however, conflict with reports that creatine supplements improve strength performance.[12]

Some medical and fitness experts voice concern that, like many performance enhancement supplements before it, creatine is being taken in huge doses (5 to 30 grams per day) before evidence of its value has been ascertained.[13] Even people who eat red meat, which is a creatine-rich food, do not consume near the amount athletes are taking. Athletes who take megadoses of creatine risk possible long-term side effects such as organ and muscle damage. Despite the uncertainties, creatine supplements are not illegal in international competition.

## *Caffeine*

Although some research findings support the use of caffeine to enhance endurance, other studies suggest that caffeine has no effect on athletic performance. If caffeine does enhance endurance, it probably does so by stimulating fatty acid release, thereby slowing glycogen use. Of course, physically active people can make their work seem easier without consuming caffeine by warming up with light activity. Light activity before a workout stimulates fat release, as does caffeine, but the activity also warms the muscles and connective tissues, making them flexible and resistant to injury. Caffeine does not offer these added benefits.

Caffeine is a stimulant that elicits a number of physiological and psychological effects in the body. (The table at the start of Appendix H provides a list of common caffeine-containing items and the doses they deliver.) The possible benefits of caffeine use must be weighed against its adverse effects—stomach upset, nervousness, irritability, headaches, and diarrhea. Caffeine-containing beverages should be used in moderation, if at all, and *in addition* to other fluids, not as a substitute for them. College, national, and international athletic competitions prohibit the use of caffeine in amounts greater than the equivalent of 5 or 6 cups of coffee consumed in a two-hour period prior to competition. Urine tests that detect more caffeine than this disqualify athletes from competition.

## ***Hormonal Supplements***

The dietary supplements discussed this far may or may not help athletic performance, but in the doses commonly taken, they seem to cause little harm. The remaining discussion features supplements that are clearly damaging.

## *Anabolic Steroids*

Among the most dangerous and illegal ergogenic practices is the taking of anabolic steroids. These drugs are derived from the male sex hormone testosterone, which promotes the development of male characteristics and lean body mass. Athletes take steroids to stimulate muscle bulking.

To athletes struggling to excel, the promise of bigger, stronger muscles than training alone can produce has been

**Table H14-1**
**Substances Promoted as Ergogenic Aids**

- **Arginine:** a nonessential amino acid falsely promoted as enhancing the secretion of human growth hormone, the breakdown of fat, and the development of muscle.
- **Bee pollen:** a product consisting of bee saliva, plant nectar, and pollen that supposedly aids in weight loss and boosts athletic performance; it does neither and may cause an allergic reaction in individuals sensitive to it.
- **Boron:** a nonessential mineral that is promoted as a "natural" steroid replacement.
- **Brewer's yeast:** a preparation of yeast cells, containing a concentrated amount of B vitamins and some minerals; falsely promoted as an energy booster.
- **Cell salts:** a preparation of minerals supposedly harvested from living cells, sold as a health-promoting supplement.
- **Coenzyme Q10:** a lipid found in cells (mitochondria) shown to improve exercise performance in heart disease patients, but not effective in improving performance of healthy athletes.
- **Desiccated liver:** dehydrated liver powder that supposedly contains all the nutrients found in liver in concentrated form; possibly not dangerous, but it has no particular nutritional merit and is considerably more expensive than fresh liver.
- **DNA (deoxyribonucleic acid):** the genetic material of cells necessary in protein synthesis; falsely promoted as an energy booster.
- **Epoetin** (eh-poy-EE-tin): a drug derived from the human hormone erythropoietin and marketed under the trade name Epogen; illegally used to increase oxygen capacity.
- **Gelatin:** a soluble form of the protein collagen, used to thicken foods; sometimes falsely promoted as a strength enhancer.
- **Ginseng:** a plant whose extract supposedly boosts energy; side effects of chronic use include nervousness, confusion, and depression.
- **Glycine:** a nonessential amino acid, promoted as an ergogenic aid because it is a precursor of the high-energy compound creatine phosphate. Other amino acids commonly packaged for athletes that are equally useless include tryptophan, ornithine, arginine, lysine, and the branched-chain amino acids.
- **Growth hormone releasers:** herbs or pills that supposedly regulate hormones; falsely promoted for enhancing athletic performance.
- **Guarana:** a reddish berry found in Brazil's Amazon valley that is used as an ingredient in carbonated sodas and taken in powder or tablet form. Guarana is marketed as an ergogenic aid to enhance speed and endurance, an aphrodisiac, a "cardiac tonic," an "intestinal disinfectant," and a smart drug that supposedly improves memory and concentration and wards off senility. Because guarana contains seven times as much caffeine as its relative the coffee bean, there are concerns that high doses can stress the heart and cause panic attacks.
- **Herbal steroids** or **plant sterols:** curious mixtures of herbs, "adaptogens," and "aphrodisiacs" that supposedly enhance hormone activity. Products marketed as herbal steroids include astragalus, damiana, dong quai, fo ti teng, ginseng root, licorice root, palmetto berries, sarsaparilla, schizardra, unicorn root, yohimbe bark, and yucca.

tempting. Athletes who lack superstar genetic material and who normally would not be able to break into the elite ranks can, with the help of steroids, suddenly compete with true champions. Especially in professional circles, where monetary rewards for excellence are sky-high, steroid use is common despite its illegality and side effects.

The American Academy of Pediatrics and the American College of Sports Medicine condemn athletes' use of anabolic steroids, and the International Olympic Committee bans their use.[14] These authorities cite the known toxic side effects and maintain that taking these drugs is a form of cheating. Other athletes are put in the difficult position of either conceding an unfair advantage to competitors who use steroids or taking them and accepting the risk of harmful side effects. Young athletes should not be forced to make such a choice.

The list of adverse reactions to steroids is long and continues to grow amid only a slight decline in use of the drugs. Table H14-2 (on p. 466) lists the side effects of steroids.

The price for the potential competitive edge that steroids confer is high—sometimes it is life itself. Steroids are not simple pills that build bigger muscles, but complex chemicals to which the body reacts in many ways, particularly when bodybuilders and other athletes take large amounts. The safest, most effective way to build muscle has always been through hard training and a sound diet, and—despite popular misconceptions—it still is.

Some manufacturers push specific herbs as legal substitutes for steroid drugs. They falsely claim that these herbs contain hormones, enhance the body's hormonal activity, or both. In some cases, an herb may contain plant sterols, such as oryzanol, but these compounds are poorly absorbed. Even if absorption occurs, the body cannot convert herbal compounds to anabolic steroids. None of these products has any proven anabolic steroid activity, none enhances muscle strength, and some contain natural toxins. In short, "natural" does not mean "harmless."

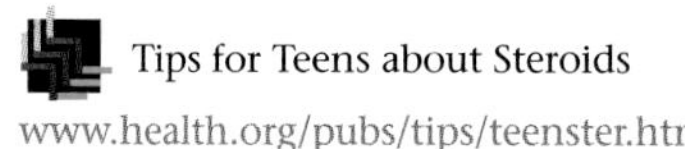
Tips for Teens about Steroids
www.health.org/pubs/tips/teenster.htm

- **Inosine:** an organic chemical that is falsely said to "activate cells, produce energy, and facilitate exercise," but actually has been shown to reduce the endurance of runners.
- **Ma huang:** an evergreen plant derivative that supposedly boosts energy and helps with weight control. Ma huang contains ephedrine, a cardiac stimulant, and has been associated with high blood pressure, rapid heart rate, nerve damage, muscle injury, psychosis, stroke, and memory loss.
- **Niacin:** a B vitamin that when taken in excess rushes blood to the skin, producing vascularity and a red tint—physical attributes bodybuilders strive to attain prior to performance. These attributes do not enhance performance, and excess niacin can cause headaches and nausea.
- **Octacosanol:** an alcohol isolated from wheat germ; often falsely promoted to enhance athletic performance.
- **Ornithine:** a nonessential amino acid falsely promoted as enhancing the secretion of human growth hormone, the breakdown of fat, and the development of muscle.
- **Oryzanol:** a plant sterol that supposedly provides the same physical responses as anabolic steroids without the adverse side effects; also known as *ferulic acid, ferulate,* or *FRAC.*
- **Pangamic acid:** also called vitamin $B_{15}$ (but not a vitamin, nor even a specific compound—it can be anything with that label); falsely claimed to speed oxygen delivery.
- **Phosphate pills:** a product demonstrated to increase the levels of a metabolically important phosphate compound (diphosphoglycerate) in red blood cells and the potential of the cells to deliver oxygen to the body's muscle cells; however, it does not extend endurance nor increase efficiency of aerobic metabolism and may cause calcium losses from the bones if taken in excess.
- **Pyruvate:** a 3-carbon compound derived during the metabolism of glucose, certain amino acids, and glycerol; falsely promoted as burning fat and enhancing endurance. Common side effects include intestinal gas and diarrhea.
- **RNA (ribonucleic acid):** the genetic material of cells necessary for protein synthesis; falsely promoted to enhance athletic performance.
- **Royal jelly:** the substance produced by worker bees and fed to the queen bee; falsely promoted to increase strength and enhance performance.
- **Sodium bicarbonate:** baking soda; an alkaline salt believed to neutralize blood lactic acid and thereby to reduce pain and enhance possible workload. "Soda loading" may cause intestinal bloating and diarrhea.
- **Spirulina:** a kind of alga ("blue-green manna") that supposedly contains large amounts of protein and vitamin $B_{12}$, suppresses appetite, and improves athletic performance; it does none of these things and is potentially toxic.
- **Succinate:** a compound synthesized in the body and involved in the TCA cycle; falsely promoted as a metabolic enhancer.
- **Superoxide dismutase (SOD):** an enzyme that protects cells from oxidation. When it is taken orally, the body digests and inactivates this protein; it is useless to athletes.
- **Wheat germ oil:** the oil from the wheat kernel; often falsely promoted as an energy aid.

## *Insulin*

Some athletes have been using the hormone insulin in conjunction with their weight training routines to develop muscles. They may know that insulin's normal action is to promote muscle growth, but they fail to realize that insulin in excess of the body's needs can be extremely harmful, causing blood glucose to fall dangerously low. The consequences of using insulin inappropriately include weakness, confusion, rapid pulse, convulsions, coma, and death.

## *DHEA*

Some athletes use DHEA as an alternative to anabolic steroids. DHEA is a hormone (dehydroepiandrosterone), made in the adrenal glands, that serves as a precursor to the male hormone testosterone. Advertisements claim it "burns fat," "builds muscle," and "slows aging," but evidence to support such claims is lacking.

DHEA's short-term side effects include oily skin, acne, body hair growth, liver enlargement, and aggressive behavior.[15] Long-term effects of DHEA use remain to be seen and may take years to become evident. The potential for harm from DHEA supplements is great, and athletes, as well as others, should avoid it.

DHEA is banned by the International Olympic Committee and the National Collegiate Athletic Association.

## Human Growth Hormone

Some short or average-sized athletes seek to use human growth hormone (hGH) to build lean tissue and increase their height if they are still in their growing years. Athletes in power sports such as weight lifting and judo are most likely to experiment with hGH, believing the injectable hormone will provide the benefits of anabolic steroids without the dangerous side effects.

Abuse of hGH is not as extensive as abuse of steroids or other such drugs, in part because of the cost. A dose of hGH that will produce the effect sought might cost $2000 a week on the black market. As with other drugs sold on the black market, athletes often do not get what they think they are buying.

Taken in large quantities, hGH causes the disease acromegaly, in which the body becomes huge and the organs and bones overenlarge. Other effects include diabetes, thyroid disorder, heart disease, menstrual irregularities, diminished sexual desire, and shortened life span. The U.S. Olympic Committee bans hGH use, but tests cannot distinguish between naturally occurring hGH and hGH used as a drug. The committee maintains that use of hGH is a form of cheating that undermines the quest for physical excellence and that its use is coercive to other athletes.

**Table H14-2**
**Anabolic Steroids: Side Effects and Adverse Reactions**

**Mind**

- Extreme aggression with hostility ("steroid rage"); mood swings; anxiety; dizziness; drowsiness; unpredictability; insomnia; psychotic depression; personality changes, suicidal thoughts

**Face and Hair**

- Swollen appearance; greasy skin; severe, scarring acne; mouth and tongue soreness; yellowing of whites of eyes (jaundice)
- In females, male-pattern hair loss and increased growth of face and body hair

**Voice**

- In females, irreversible deepening of voice

**Chest**

- In males, breathing difficulty; breathing stoppage; breast development
- In females, breast atrophy

**Heart**

- Heart disease; elevated or reduced heart rate; heart attack; stroke; hypertension; increased LDL; drastic reduction in HDL

**Abdominal Organs**

- Nausea; vomiting; bloody diarrhea; pain; edema; liver tumors (possibly cancerous); liver damage, disease, or rupture leading to fatal liver failure (peliosis hepatitis)[a]; kidney stones and damage; gallstones; frequent urination; possible rupture of aneurysm or hemorrhage

**Blood**

- Blood clots; high risk of blood poisoning; those who share needles risk contracting HIV (the AIDS virus) or other disease-causing organisms; septic shock (from injections)

**Reproductive System**

- In males, permanent shrinkage of testes; prostate enlargement with increased risk of cancer; sexual dysfunction; loss of fertility; excessive and painful erections
- In females, loss of menstruation and fertility; permanent enlargement of external genitalia; fetal damage, if pregnant

**Muscles, Bones, and Connective Tissues**

- Increased susceptibility to injury with delayed recovery times; cramps; tremors; seizurelike movements; injury at injection site
- In adolescents, failure to grow to normal height

**Other**

- Fatigue; increased risk of cancer

[a]In peliosis hepatitis, excess buildup of bile causes destruction of liver cells. Blood pools form, and liver failure causes death.

Sources: K. L. Ropp, No-win situation for athletes, *FDA Consumer,* December 1992, pp. 8–12; National Academy of Sports Medicine policy statement and position paper: Anabolic androgenic steroids, growth hormones, stimulants, ergogenics, and drug use in sports, in B. Goldman and R. Klatz, *Death in the Locker Room II: Drugs and Sports* (Chicago: Elite Sports Medicine Publications, 1992), pp. 328–373.

The search for a single food, nutrient, drug, or technique that will safely and effectively enhance athletic performance will no doubt continue as long as people strive to achieve excellence in sports. So far, when athletic performance does improve after use of an ergogenic aid, the improvement can usually be attributed to the placebo effect, which is strongly at work in athletes. Even if a reliable source reports a performance boost from a newly tried product, give the effect time to fade away. Chances are excellent that it simply reflects the power of the mind over the body.

The overwhelming majority of potions sold for athletes are frauds. Wishful thinking will not substitute for talent, hard training, adequate diet, and mental preparedness in competition. But don't discount the power of mind over body for a minute—it is formidable, and sports psychologists dedicate their work to harnessing it. You can use it by imagining yourself a winner and visualizing yourself excelling in your sport. You don't have to buy magic to obtain a winning edge; you already possess it—your mind.

 S. Barrett Quackwatch
www.quackwatch.com
Search for Topic of interest

## Notes


1. E. A. Applegate and L. E. Grivetti, Search for the competitive edge: A history of dietary fads and supplements, *Journal of Nutrition* 127 (1997): 869S–873S.
2. E. Coleman, Branched-chain amino acids and fatigue, *Sports Medicine Digest* 18 (1996): 44.
3. M. Vukovich, D. L. Costill, and W. J. Fink, Carnitine supplementation: Effect on muscle carnitine and glycogen content during exercise, *Medicine and Science in Sports and Exercise* 26 (1994): 1122–1129.
4. E. W. Askew, Environmental and physical stress and nutrient requirements, *American Journal of Clinical Nutrition* 61 (1995): 631S–637S; H. M. Alessio, Exercise induced oxidative stress, *Medicine and Science in Sports and Exercise* 25 (1993): 218–224; R. R. Jenkins and A. Goldfarb, Introduction: Oxidative stress, aging, and exercise, *Medicine and Science in Sports and Exercise* 25 (1993): 210–212.
5. D. A. Leaf and coauthors, The effect of exercise intensity on lipid peroxidation, *Medicine and Science in Sports and Exercise* 29 (1997): 1036–1039; M. Kanter, Free radicals and exercise: Effects of nutritional antioxidant supplementation, *Exercise and Sports Science Review* 23 (1995): 375–397.
6. G. S. Ginsburg and coauthors, Effects of a single bout of ultra endurance exercise on lipid levels and susceptibility of lipids to peroxidation in triathletes, *Journal of the American Medical Association* 276 (1996): 221–225.
7. G. W. Evans, The effect of chromium picolinate on insulin-controlled parameters in humans, *Journal of Biosocial Medicine Research* 11 (1989): 163–180.
8. H. C. Lukaski and coauthors, Chromium supplementation and resistance training: Effects on body composition, strength, and trace element status of men, *American Journal of Clinical Nutrition* 63 (1996): 954–965; M. A. Hallmark and coauthors, Effects of chromium and resistance training on muscle strength and body composition, *Medicine and Science in Sports and Exercise* 28 (1996): 139–144.
9. W. W. Campbell and coauthors, Chromium picolinate supplementation and resistive training by older men: Effects on iron-status and hematologic indexes, *American Journal of Clinical Nutrition* 66 (1997): 944–949; Lukaski and coauthors, 1996.
10. Applegate and Grivetti, 1997.
11. J. S. Volek and coauthors, Creatine supplementation enhances muscular performance during high-intensity resistance exercise, *Journal of the American Dietetic Association* 97 (1997): 765–770; S. M. Tolar, Creatine is an ergogen for anaerobic exercise, *Nutrition Reviews* 55 (1997): 21–23; C. P. Earnest and coauthors, The effect of creatine monohydrate ingestion on anaerobic power indices, muscular strength, and body composition, *Acta Physiologica Scandinavica* 153 (1995): 207–209.
12. L. M. Odland and coauthors, Effect of oral creatine supplementation on muscle [PCr] and short-term maximum power output, *Medicine and Science in Sports and Exercise* 29 (1997): 216–219.
13. T. Noakes, as quoted in M. Gaie, Olympic athletes face heat, other health hurdles, *Journal of the American Medical Association* 276 (1996): 178–180.
14. D. H. Catlin and T. H. Murray, Performance-enhancing drugs, fair competition and Olympic sport, *Journal of the American Medical Association* 276 (1996): 231–237.
15. E. Coleman, DHEA—An anabolic aid? *Sports Medicine Digest* 18 (1996): 140–141.

***"I lied to a lot of people for a lot of years when I said I didn't use steroids. . . . If you're on steroids or human growth hormone, stop. I should have." Lyle Alzado, former NFL football player who died of cancer in May of 1992; as quoted by S. Smith, I'm sick and I'm scared,*** **Sports Illustrated,** ***July 8, 1991, pp. 21–25.***

CHAPTER 15

# Life Cycle Nutrition: Pregnancy and Lactation

All people—pregnant and lactating women, infants, children, adolescents, and adults—need the same nutrients, but the amounts they need vary depending on their stage of life. This chapter focuses on nutrition in preparation for, and support of, pregnancy and lactation.

# Growth and Development during Pregnancy

A whole new life begins at conception. Organ systems develop rapidly, and nutrition plays many supportive roles. This section describes placental development and fetal growth, paying close attention to times of intense developmental activity.

**conception:** the union of the male sperm and the female ovum; fertilization.

 Mayo Health Clinic
www.mayohealth.org
Visit Pregnancy and Child health center

## *Placental Development*

In the early days of pregnancy, a spongy structure known as the placenta develops in the uterus. Two associated structures also form (see Figure 15-1). One is the amniotic sac, a fluid-filled balloonlike structure that houses the developing fetus. The other is the umbilical cord, a ropelike structure containing fetal blood vessels that extends through the fetus's "belly button" (the umbilicus) to the placenta.

**placenta** (plah-SEN-tuh): the organ that develops inside the uterus early in pregnancy, through which the fetus receives nutrients and oxygen across the placenta and returns carbon dioxide and other waste products to be excreted.

**Figure 15-1**
**The Placenta and Associated Structures**

To understand how placental villi absorb nutrients without maternal and fetal blood interacting directly, think of how the intestinal villi work. The GI side of the intestinal villi is bathed in a nutrient-rich fluid (chyme). The intestinal villi absorb the nutrient molecules and release them into the body via capillaries. Similarly, the maternal side of the placental villi is bathed in nutrient-rich maternal blood. The placental villi absorb the nutrient molecules and release them to the fetus via fetal capillaries.

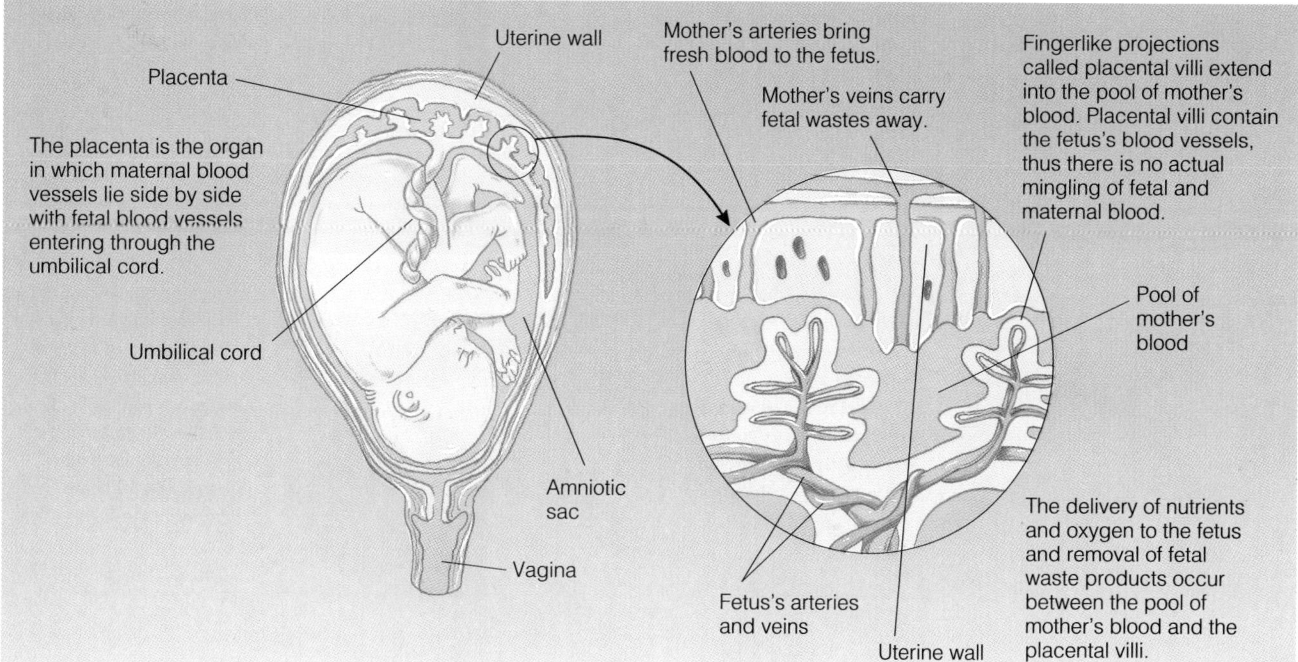
Uterine wall
Placenta
The placenta is the organ in which maternal blood vessels lie side by side with fetal blood vessels entering through the umbilical cord.
Umbilical cord
Amniotic sac
Vagina
Mother's arteries bring fresh blood to the fetus.
Mother's veins carry fetal wastes away.
Fingerlike projections called placental villi extend into the pool of mother's blood. Placental villi contain the fetus's blood vessels, thus there is no actual mingling of fetal and maternal blood.
Pool of mother's blood
The delivery of nutrients and oxygen to the fetus and removal of fetal waste products occur between the pool of mother's blood and the placental villi.
Fetus's arteries and veins
Uterine wall

**uterus** (YOU-ter-us): the muscular organ within which the infant develops before birth.

**amniotic** (am-nee-OTT-ic) **sac:** the "bag of waters" in the uterus, in which the fetus floats.

**umbilical** (um-BILL-ih-cul) **cord:** the ropelike structure through which the fetus's veins and arteries reach the placenta; the route of nourishment and oxygen into the fetus and the route of waste disposal from the fetus. The scar in the middle of the abdomen that marks the former attachment of the umbilical cord is the **umbilicus** (um-BILL-ih-cus), commonly known as the "belly button."

**ovum** (OH-vum): the female reproductive cell, capable of developing into a new organism upon fertilization; commonly referred to as an egg.

**sperm:** the male reproductive cell, capable of fertilizing an ovum.

**zygote** (ZY-goat): the product of the union of ovum and sperm; so-called for the first two weeks after fertilization.

**implantation:** the stage of development in which the zygote embeds itself in the wall of the uterus and begins to develop; occurs during the first two weeks after conception.

**embryo** (EM-bree-oh): the developing infant from two to eight weeks after conception.

These three structures serve crucial roles during pregnancy and then are expelled from the uterus during childbirth.

The placenta develops as an interweaving of fetal and maternal blood vessels embedded in the uterine wall. The maternal blood transfers oxygen and nutrients to the fetus's blood and picks up fetal waste products. By exchanging oxygen, nutrients, and waste products, the placenta performs the respiratory, absorptive, and excretory functions that the fetus's lungs, digestive system, and kidneys will provide after birth.

The placenta is a versatile, metabolically active organ. Like all body tissues, the placenta uses energy and nutrients to support its work. Like a gland, it produces an array of hormones that maintain pregnancy and prepare the mother's breasts for lactation (making milk). A healthy placenta is essential for the developing fetus to attain its full potential.[1]

## Fetal Growth and Development

Fetal development begins with the fertilization of an ovum by a sperm. Three stages follow: the zygote, the embryo, and the fetus (see Figure 15-2).

• ***The Zygote*** • The newly fertilized ovum, or zygote, begins as a single cell and divides to become many cells during the days after fertilization. Within two weeks, the zygote embeds itself in the uterine wall—a process known as implantation. Cell division continues—each set of cells divides into many other cells. As development proceeds, the zygote becomes an embryo.

• ***The Embryo*** • The embryo develops at an amazing rate. At first, the number of cells in the embryo doubles approximately every 24 hours; later the rate slows, and only one doubling occurs during the final ten weeks of pregnancy. The embryo's size changes very little, but at eight weeks, the 1¼-inch embryo has a

**Figure 15-2**
**Stages of Embryonic and Fetal Development**

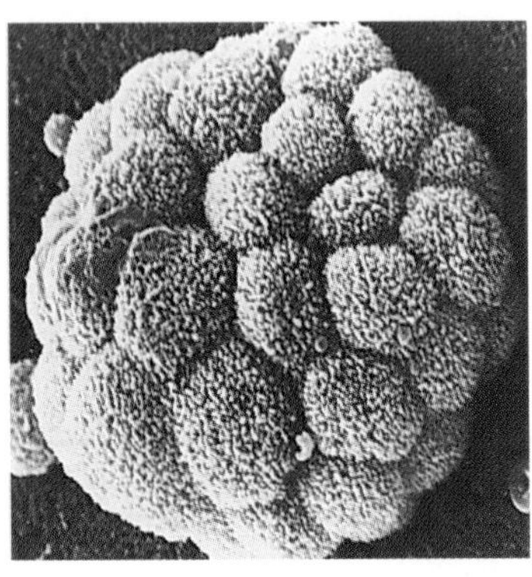

1. *A newly fertilized ovum is about the size of the period at the end of this sentence. This zygote at less than one week after fertilization is not much bigger and is ready for implantation.*

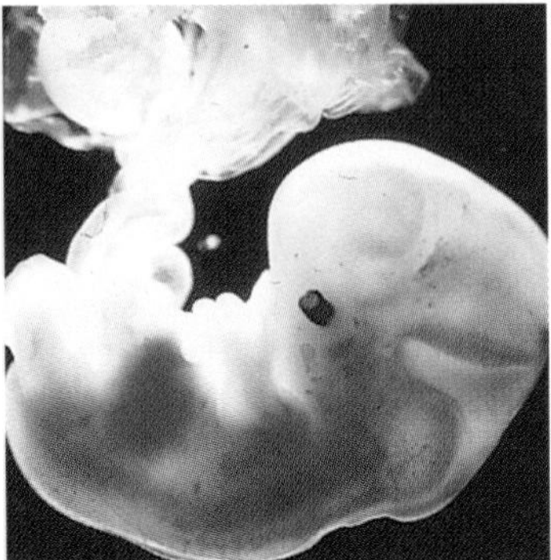

2. *After implantation, the placenta develops and begins to provide nourishment to the developing embryo. An embryo five weeks after fertilization is about ½ inch long.*

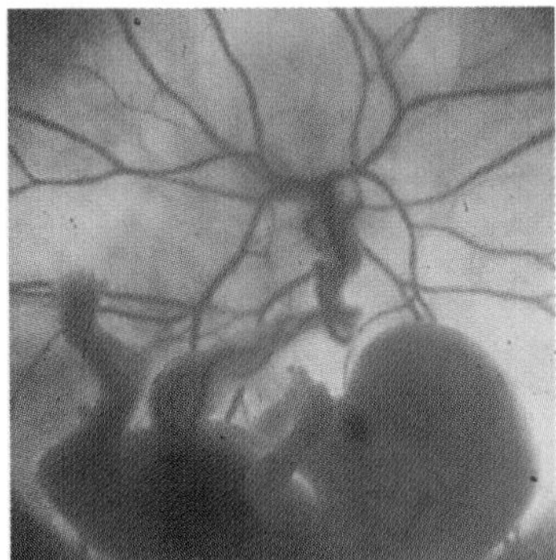

3. *A fetus after 11 weeks of development is just over an inch long. Notice the umbilical cord and blood vessels connecting the fetus with the placenta.*

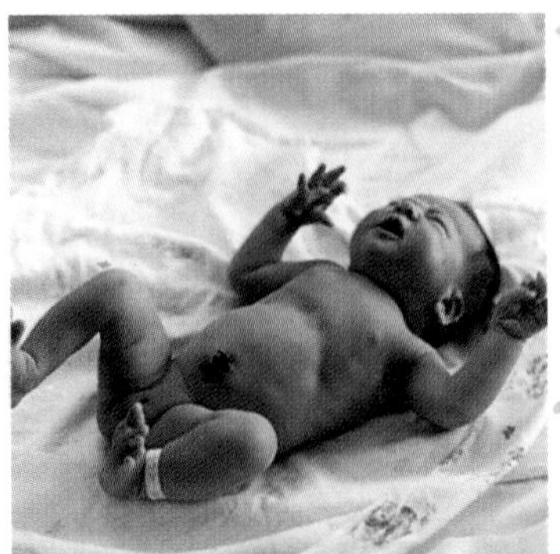

4. *A newborn infant after nine months of development measures close to 20 inches in length. From eight weeks to term, this infant grew 20 times longer and 50 times heavier.*

complete central nervous system, a beating heart, a digestive system, well-defined fingers and toes, and the beginnings of facial features.

• *The Fetus* • During the next seven months, each organ grows to maturity according to its own schedule, with greater intensity at some times than at others. As Figure 15-2 shows, fetal growth is phenomenal: weight increases from less than an ounce to about 7½ pounds (3500 grams).

**fetus** (FEET-us): the developing infant from eight weeks after conception until term.

## Critical Periods

Times of intense development and rapid cell division are called critical periods—critical in the sense that the events scheduled for those times can occur only at those times. If cell division and the final cell number achieved in an organ are limited during a critical period, full recovery will not occur (see Figure 15-3).

Each organ and tissue is most vulnerable to nutrient deficiencies or toxins during its own critical period. The critical period for neural tube development, for example, is from 17 to 30 days gestation.[2] Consequently, neural tube development is most vulnerable to nutrient deficiencies, nutrient excesses, or toxins during this critical time—when most women do not even realize that they are pregnant. Any abnormal development of the neural tube or its failure to close completely can cause a major defect in the central nervous system. Figure 15-4 (on p. 472) shows photos of neural tube development in the early weeks of gestation.

**critical periods:** finite periods during development in which certain events may occur that will have irreversible effects on later developmental stages. In a body organ, a critical period is usually a period of rapid cell division.

**gestation** (jes-TAY-shun): the period from conception to birth; for human beings gestation lasts from 38 to 42 weeks. Pregnancy is often divided into thirds, called **trimesters.**

 March of Dimes
www.modimes.org
Visit Having a Healthy Baby
Visit Birth Defects Information

 U.S. Government
www.healthfinder.gov/searchoptions/topicsaz.htm
Search for Birth Defects

• *Neural Tube Defects* • In the United States, approximately 1 of every 1000 newborns has a neural tube defect; some 2500 to 3000 infants are affected each year.* Many other pregnancies with neural tube defects end in abortion or stillbirths.

One of the most common types of neural tube defects is spina bifida, a disorder characterized by incomplete closure of the spinal cord and its bony encasement. The membranes covering the spinal cord often protrude as a sac, which may rupture and lead to meningitis, a life-threatening inflammation of the membranes. Spina bifida is accompanied by varying degrees of paralysis, depending on the extent of spinal cord damage. Mild cases may not even be noticed, but severe cases lead to death. Common problems include clubfoot, dislocated hip, kidney disorders, curvature of the spine, muscle weakness, mental handicaps, and motor and sensory losses.

**Figure 15-3**
**The Concept of Critical Periods**

Critical periods occur early in development. An adverse influence felt early can have a much more severe and prolonged impact than one felt later on.

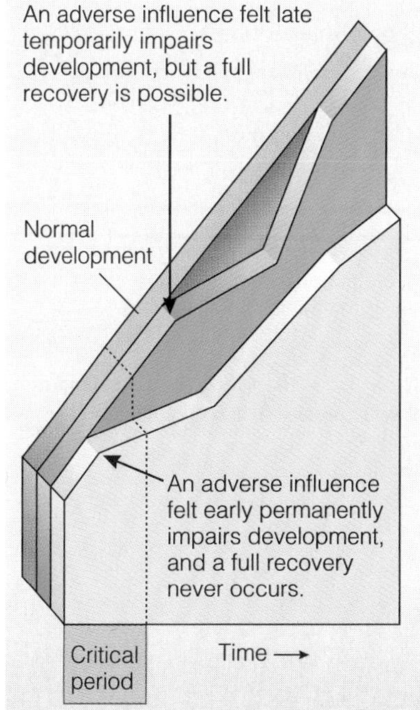
An adverse influence felt late temporarily impairs development, but a full recovery is possible.
Normal development
An adverse influence felt early permanently impairs development, and a full recovery never occurs.
Critical period
Time →

• *Folate Supplementation* • Chapter 10 described how folate supplements taken one month before conception and continued throughout the first trimester can reduce the risks of neural tube defects.[3] For this reason, the Public Health Service recommends that all women of childbearing age who are capable of becoming pregnant take 0.4 milligrams of folate daily. Supplements offer women a convenient way to ingest sufficient folate regularly and continuously enough to benefit pregnancy.[4] Most over-the-counter multivitamin supplements contain 0.4 milligram of folate; prenatal supplements usually contain at least 0.8 milligram. A woman who has previously had an infant with a neural tube defect may be advised by her physician to take folate supplements in doses ten times larger—4 milligrams daily. The risks associated with high doses of folate are not all known, but they can mask the pernicious anemia of a vitamin $B_{12}$ deficiency. For this reason, quantities of 1 milligram or more require a prescription.

To deliver folate to the U.S. population, the Food and Drug Administration (FDA) has mandated fortification of grain products. This decision carefully weighed the benefits of fortification against the risks of overconsumption. On the one hand, an adequate folate intake is expected to reduce the incidence of neural tube defects by 50 percent. On the other hand, if vitamin $B_{12}$ deficiency is masked

*Worldwide, some 300,000 to 400,000 infants are born with neural tube defects each year.

**Figure 15-4**

**Neural Tube Development**

The neural tube is the beginning structure of the brain and spinal cord. Any failure of the neural tube to close or to develop normally results in central nervous system disorders such as spina bifida and anacephaly. Successful development of the neural tube depends, in part, on the vitamin folate.

**spina** (SPY-nah) **bifida** (BIFF-ih-dah): one of the most common types of neural tube defects characterized by the incomplete closure of the spinal cord and its bony encasement.
- **spina** = spine
- **bifida** = split

**anencephaly** (AN-en-SEF-a-lee): an uncommon and always fatal type of neural tube defect characterized by the absence of a brain.
- **an** = not (without)
- **encephalus** = brain

Source: L. Nilsson, *A Child Is Born* (New York: Delacorte Press/Seymour Lawrence, 1977), pp. 44, 64.

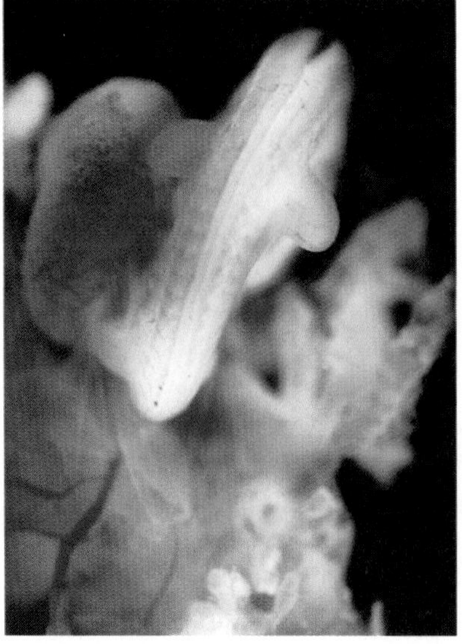

At four weeks, the neural tube has yet to close (notice the gap at the top).

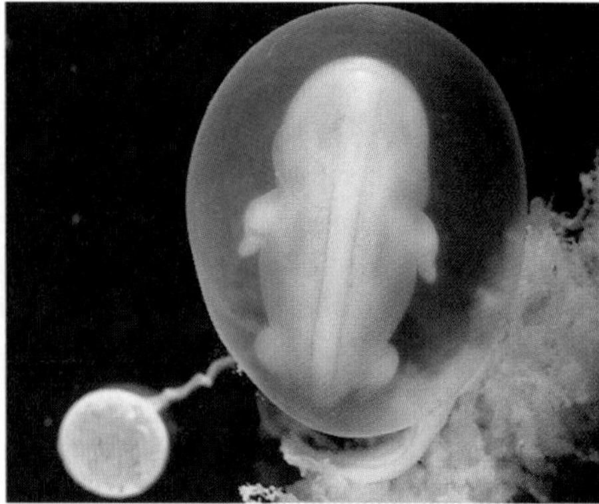

At six weeks, the neural tube (outlined by the delicate red vertebral arteries) has successfully closed.

Folate RDA:
- For women: 400 μg (0.4 mg)/day.
- During pregnancy: 600 μg (0.6 mg)/day.

by folate and left untreated, irreversible nerve damage may occur. The FDA regulation requires manufacturers to add folate to cereal, pasta, flour, rolls, buns, farina, grits, cornmeal, and rice.*

IN SUMMARY

Maternal nutrition before and during pregnancy affects both the mother's health and the infant's growth. As the infant develops through its three stages—the zygote, embryo, and fetus—its organs and tissues grow, each on its own schedule. Times of intense development are critical periods that depend on nutrients to proceed smoothly. Without folate, for example, the neural tube fails to develop completely during the first month of pregnancy, prompting recommendations that all women of childbearing age take folate daily.

Because critical periods occur throughout pregnancy, a woman should continuously take good care of her health. That care should include achieving and maintaining a healthy body weight prior to pregnancy and gaining sufficient weight during pregnancy to support a healthy infant.

U.S. Government
www.healthfinder.gov/searchoptions/topicsaz.htm
Search for Pregnancy

# Maternal Weight

Reminder: Weight-for-height measures are often expressed in terms of BMI (body mass index).

Birthweight is the most reliable indicator of an infant's health. As a later section of this chapter explains, an underweight infant is more likely to have physical and mental defects, become ill, and die than a normal-weight infant. In general, higher birthweights present lower risks for infants. Two characteristics of the mother's weight influence an infant's birthweight: her weight prior to conception and her weight gain during pregnancy.

*These products must be fortified with 1.4 milligrams of folate per 100 grams of food.

## Weight prior to Conception

A woman's weight prior to conception influences fetal growth. Even with the same weight gain during pregnancy, underweight women tend to have smaller babies than heavier women.

• ***Underweight*** • An underweight woman has a high risk of having a low-birthweight infant, especially if she is unable to gain sufficient weight during pregnancy. In addition, the rates of preterm births and infant deaths are higher for underweight women.[5] An underweight woman improves her chances of having a healthy infant by gaining sufficient weight prior to conception or by gaining extra pounds during pregnancy. To increase food energy intake, an underweight woman can follow the dietary recommendations for pregnant women (described in Table 15-2 on p. 478).

Underweight is defined as BMI <19.8.

**preterm** (infant): an infant born prior to the 38th week of pregnancy; also called a **premature** infant. A **term** infant is born between the 38th and 42nd week of pregnancy.

• ***Overweight*** • Overweight also creates problems related to pregnancy and childbirth. Overweight women have an especially high risk of medical complications such as hypertension, gestational diabetes, and postpartum infections.[6] Compared with other women, overweight women are also more likely to have stillbirths and other complications of labor and delivery.[7]

Overweight is defined as BMI 26 to 29. Obese is defined as BMI >29.

Infants of overweight women are likely to be born post term and to weigh more than 9 pounds. Abnormally large newborns increase the likelihood of a difficult labor and delivery, birth trauma, and cesarean section. Consequently, these infants have a greater risk of poor health and death than infants of normal weight.[8] Overweight women are less likely to have premature infants, but if they do, the infants may be large for their gestational age.

**post term** (infant): an infant born after the 42nd week of pregnancy.

The term **macrosomia** (mak-row-SO-me-ah) is sometimes used to describe high-birthweight infants (roughly 9 lb, or 4000 g, or more); macrosomia results from prepregnancy obesity, excess weight gain during pregnancy, or uncontrolled diabetes.
- **macro** = large
- **soma** = body

Of greater concern than infant birthweight is the poor development of infants born to obese mothers. Some research suggests that obesity may double the risk for neural tube defects in the infant.[9]

Weight-loss dieting during pregnancy is never advisable. Overweight women should try to achieve a healthy body weight before becoming pregnant, avoid excessive weight gain during pregnancy, and postpone weight loss until after childbirth. Weight loss is best achieved by eating moderate amounts of nutrient-dense foods and exercising to lose body fat.

## Weight Gain during Pregnancy

All pregnant women must gain weight—fetal growth and maternal health depend on it. Maternal weight gain during pregnancy correlates closely with infant birthweight, which is a strong predictor of the health and subsequent development of the infant.

• ***Recommended Weight Gains*** • Table 15-1 presents recommended weight gains for various prepregnancy weights. The recommended gain for a woman who begins pregnancy at a healthy weight and is carrying a single fetus is 25 to 35 pounds.[10] An underweight woman needs to gain between 28 and 40 pounds; and an overweight woman, between 15 and 25 pounds. Some women should strive for gains at the upper end of the target range, notably, black women and adolescents who are still growing themselves. Short women (5 feet 2 inches and under) should strive for gains at the lower end of the target range. Women who are carrying twins should aim for a weight gain of 35 to 45 pounds. If a woman gains more than is recommended early in pregnancy, she should not restrict her energy intake later in order to lose weight. A large weight gain over a short time, however, indicates

**Table 15-1**
**Recommended Weight Gains Based on Prepregnancy Weight Status**

| Prepregnancy Weight Status | Recommended Weight Gain |
|---|---|
| Underweight[a] | 12.5 to 18.0 kg (28 to 40 lb) |
| Normal weight[b] | 11.5 to 16.0 kg (25 to 35 lb) |
| Overweight[c] | 7.0 to 11.5 kg (15 to 25 lb) |
| Obese[d] | 6.8 kg minimum (15 lb minimum) |

[a]Underweight defined as BMI <19.8.
[b]Normal weight defined as BMI 19.8 to 26.0.
[c]Overweight defined as BMI 26.0 to 29.0.
[d]Obese defined as BMI >29.0.
Source: Committee on Nutritional Status during Pregnancy and Lactation, Food and Nutrition Board, *Nutrition during Pregnancy* (Washington, D.C.: National Academy Press, 1990), pp. 10, 12.

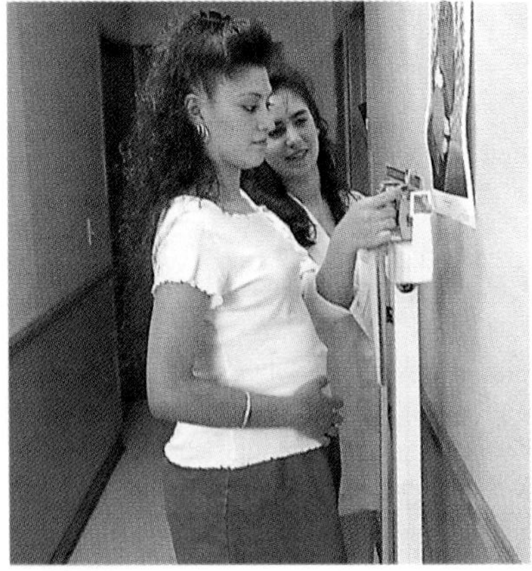

*Fetal growth and maternal health depend on a sufficient weight gain during pregnancy.*

excessive fluid retention and may be the first sign of the serious medical complication preeclampsia discussed later.

• ***Weight-Gain Patterns*** • For the normal-weight woman, weight gain ideally follows a pattern of 3½ pounds during the first trimester and 1 pound per week thereafter. Health care professionals monitor weight gain using a prenatal weight-gain grid (see Figure 15-5).

• ***Components of Weight Gain*** • Women often express concern about the weight gain that accompanies a healthy pregnancy. They may find comfort in a reminder that most of the gain supports the growth and development of the placenta, uterus, blood, and breasts, as well as an optimally healthy 7½-pound infant. A small amount goes into maternal fat stores, and even that fat is there for a special purpose: to provide energy for labor and lactation. Figure 15-6 shows the components of a typical 30-pound weight gain.

• ***Weight Loss after Pregnancy*** • The pregnant woman loses some weight at delivery. In the following weeks, she loses more as her blood volume returns to normal and she sheds accumulated fluids. The typical woman does not, however, return to her prepregnancy weight. In general, the more weight a woman gains beyond what she needs for pregnancy, the more she will retain. Even with an average weight gain, though, most women tend to retain a couple of pounds with each pregnancy.[11]

**Figure 15-5**

**Recommended Prenatal Weight Gain Based on Prepregnancy Weight**

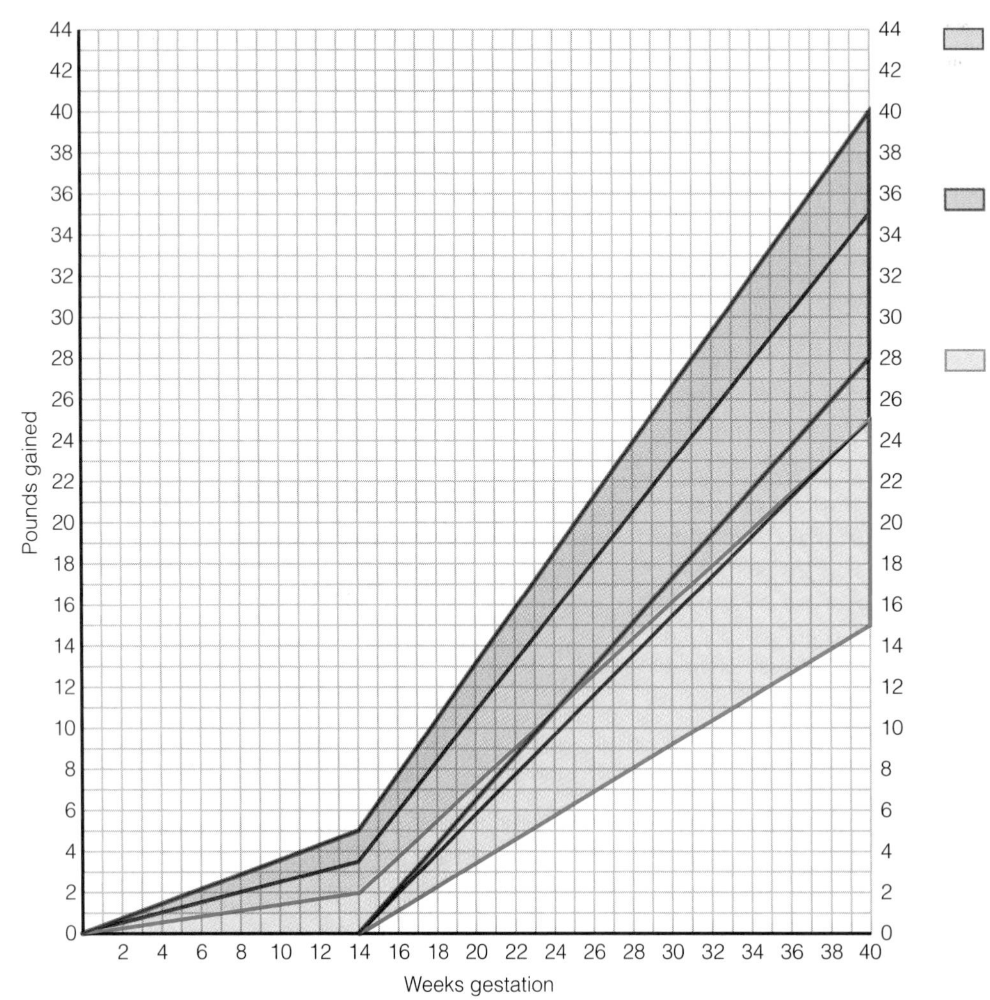
Pounds gained
0
2
4
6
8
10
12
14
16
18
20
22
24
26
28
30
32
34
36
38
40
42
44
2 4 6 8 10 12 14 16 18 20 22 24 26 28 30 32 34 36 38 40
Weeks gestation
Normal-weight women should gain about 3½ pounds in the first trimester and just under 1 pound/week thereafter, achieving a total gain of 25 to 35 pounds by term.
Underweight women should gain about 5 pounds in the first trimester and just over 1 pound/week thereafter, achieving a total gain of 28 to 40 pounds by term.
Overweight women should gain about 2 pounds in the first trimester and ⅔ pound/week thereafter, achieving a total gain of 15 to 25 pounds.

## Exercise during Pregnancy

An active, physically fit woman experiencing a normal pregnancy can continue to exercise throughout pregnancy, adjusting the duration and intensity as the pregnancy progresses. Staying active can improve fitness, prevent gestational diabetes, facilitate labor, and reduce stress.[12] Women who exercise during pregnancy report fewer discomforts throughout their pregnancies.[13] Regular exercise develops the strength and endurance a woman needs to carry the extra weight through pregnancy and to labor through an intense delivery. It also maintains the habits that help a woman lose excess weight and get back into shape after the birth.

A pregnant woman should participate in "low-impact" activities and avoid sports in which she might fall or be hit by other people or objects. For example, playing tennis with one person on each side of the net is safer than a fast-moving game of racquetball in which the two competitors can collide. Swimming is ideal because it allows the body to remain cool and move freely with the water's support. Figure 15-7 (on p. 476) provides some guidelines for exercise during pregnancy.[14] Several of the guidelines are aimed at preventing excessively high internal body temperature and dehydration, both of which can harm fetal development. To this end, pregnant women should also stay out of saunas, steam rooms, and hot whirlpools.

**Figure 15-6**
**Components of Weight Gain during Pregnancy**

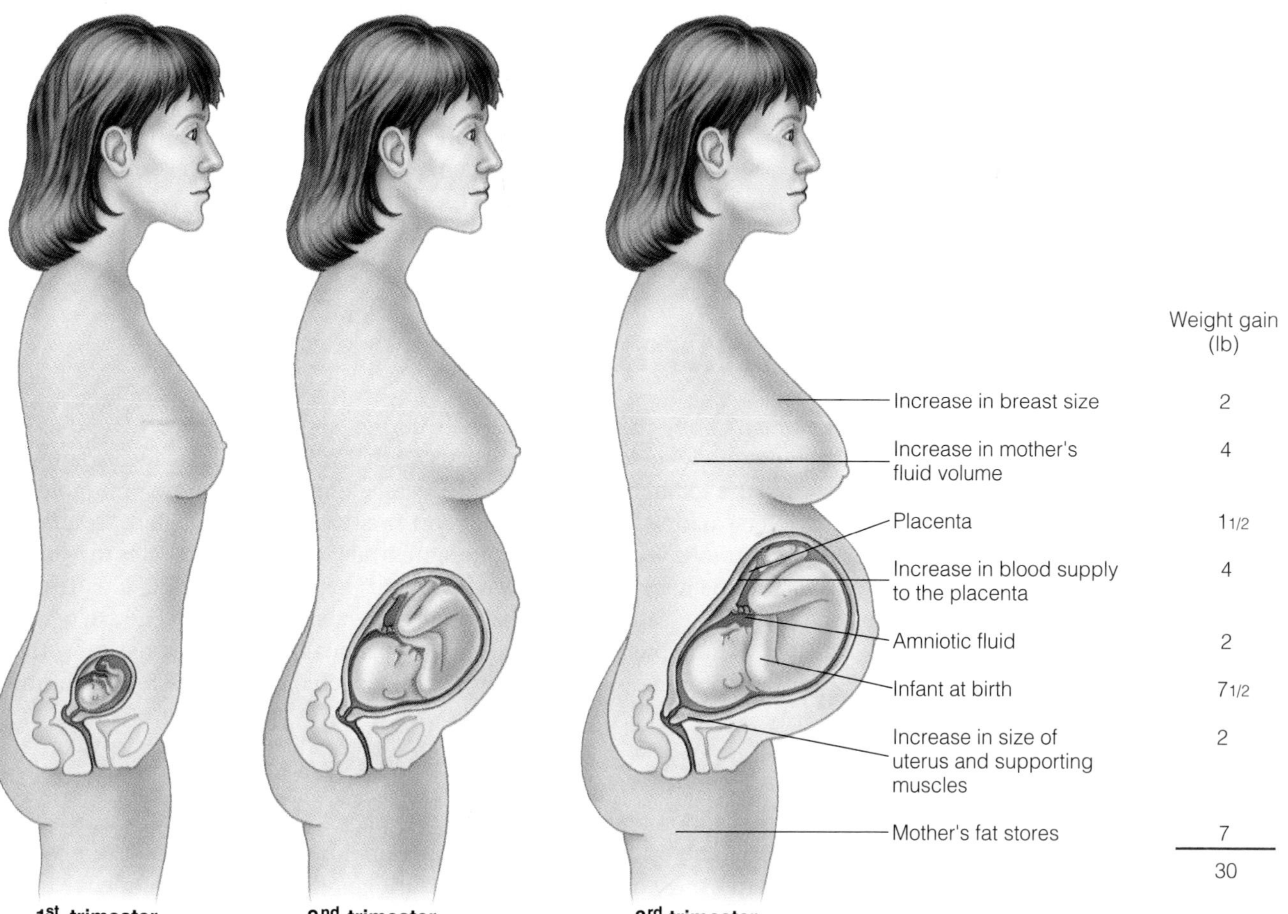

**Figure 15-7**
**Exercise Guidelines during Pregnancy**

| DO | | DON'T |
|---|---|---|
| Do exercise regularly (at least three times a week).<br>Do warm up with 5 to 10 minutes of light activity.<br>Do exercise for 20 to 30 minutes at your target heart rate.<br>Do cool down with 5 to 10 minutes of slow activity and gentle stretching.<br>Do drink water before, after, and during exercise.<br>Do eat enough to support the additional needs of pregnancy plus exercise. | <br>Pregnant women can enjoy the benefits of exercise. | Don't exercise vigorously after long periods of inactivity.<br>Don't exercise in hot, humid weather.<br>Don't exercise when sick with fever.<br>Don't exercise while lying on your back after the first trimester of pregnancy or stand motionless for prolonged periods.<br>Don't exercise if you experience any pain or discomfort.<br>Don't participate in activities that may harm the abdomen or involve jerky, bouncy movements. |

IN SUMMARY

A healthy pregnancy depends on a sufficient weight gain. Women who begin their pregnancies at a healthy weight need to gain about 30 pounds, which covers the growth and development of the placenta, uterus, blood, breasts, and infant. By remaining active throughout pregnancy, a woman can develop the strength she needs to carry the extra weight and maintain habits that will help her lose it after the birth.

# Nutrition during Pregnancy

The table on the inside front cover provides separate listings for women during pregnancy and lactation, reflecting their heightened nutrient needs.

A woman's body changes dramatically during pregnancy. Her uterus and its supporting muscles increase in size and strength; her blood volume increases by half to carry the additional nutrients and other materials; her joints become more flexible in preparation for childbirth; her feet swell in response to high concentrations of the hormone estrogen, which promotes water retention and helps to ready the uterus for delivery; and her breasts grow in preparation for lactation. The hormones that mediate all these changes may influence her mood. She can best prepare to handle these changes given a nutritious diet, regular physical activity, plenty of rest, and caring companions. This section highlights the role of nutrition.

## *Energy and Nutrient Needs during Pregnancy*

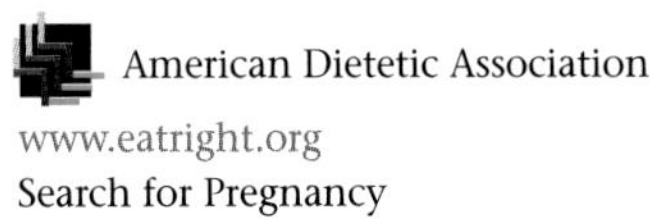

www.eatright.org
Search for Pregnancy

From conception to birth, all parts of the infant—bones, muscles, organs, blood cells, skin, and other tissues—are made from nutrients in the foods the mother eats. For most women, nutrient needs during pregnancy and lactation are higher than at any other time (see Figure 15-8).

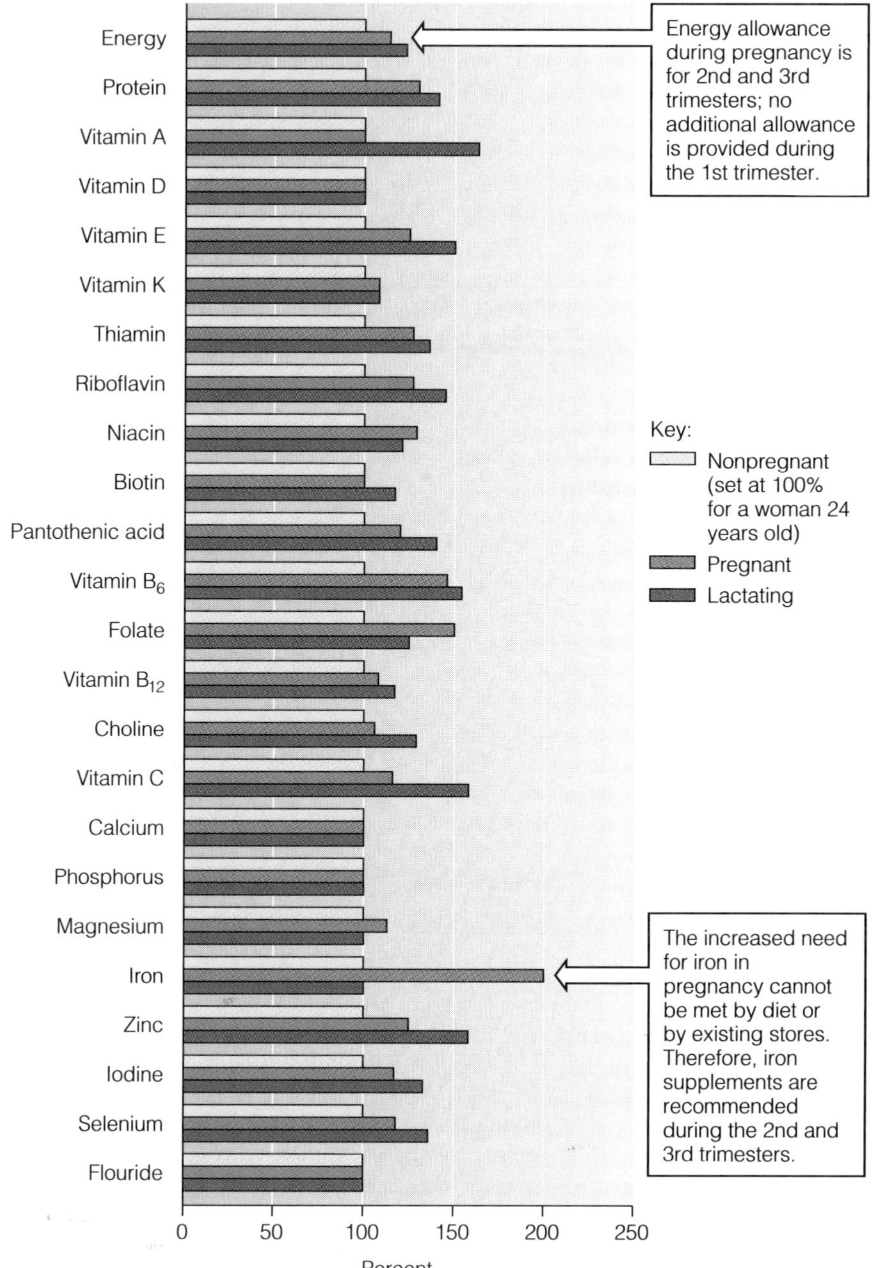


**Figure 15-8**

**Comparison of Nutrient Recommendations for Nonpregnant, Pregnant, and Lactating Women**

For actual values, turn to the table on the inside front cover.

• ***Energy*** • A pregnant woman needs extra food energy, but only a little extra—300 kcalories above the allowance for nonpregnant women—and only during the second and third trimesters. A woman can easily get 300 kcalories with just one extra serving from each of the five food groups—a slice of bread, a serving of vegetables, an ounce of lean meat, a piece of fruit, and a cup of nonfat milk (see Table 15-2 on p. 478). Pregnant teenagers, underweight women, and physically active women may require more.

For a 2000-kcalorie daily intake, 300 kcalories represent about 15 percent more food energy than before pregnancy. The increase in nutrient needs is greater than this, so nutrient-dense foods should supply the 300 kcalories: foods such as whole-grain breads and cereals, legumes, dark green vegetables, citrus fruits, nonfat milk and milk products, and lean meats, fish, poultry, and eggs. Ample carbohydrate (ideally, 250 grams or more per day and certainly no less than 100 grams) is necessary to spare the protein needed for growth.

Energy RDA during pregnancy (2nd and 3rd trimesters):
+300 kcal/day.
Canadian RNI during pregnancy:
+100 to 300 kcal/day.*

*For all Canadian RNI values during pregnancy, the lower value indicates recommendations for the first trimester, and the higher value indicates those for the second and third trimesters.

**Table 15-2**
**Daily Food Choices for Pregnant and Lactating Women**

| Food Group | Number of Servings | |
|---|---|---|
| | *Adult* | *Pregnant or Lactating Women* |
| Breads/cereals | 6 to 11 | 7 to 11 |
| Vegetables | 3 to 5 | 4 to 5 |
| Fruits | 2 to 4 | 3 to 4 |
| Meat/meat alternatives | 2 to 3 | 3 |
| Milk/milk products | 2 | 3 to 4 |

This sample meal plan follows the Daily Food Guide for pregnant and lactating women and provides about 2500 kcalories (55% from carbohydrate, 20% from protein, and 25% from fat).

**Breakfast**
1 English muffin
2 tbs peanut butter
1 c low-fat vanilla yogurt
½ c fresh strawberries
1 c orange juice

**Midmorning snack**
1 c cranberry juice
1 oz pretzels

**Lunch**
Sandwich (tuna salad on whole-wheat bread)
½ carrot (sticks)
1 c low-fat milk

**Dinner**
Chicken cacciatore
4 oz chicken
¾ c stewed tomatoes
1 c rice
¾ c summer squash
1½ c salad (spinach, mushrooms, onions)
1 tbs salad dressing
2 slices Italian bread
2 tsp butter or margarine
1 c low-fat milk

**Evening snack**
1 c low-fat milk
3 oatmeal cookies

NOTE: Figure 2-1 in Chapter 2 provides a detailed summary of the Daily Food Guide.

Protein RDA during pregnancy:
+10 g/day.
Canadian RNI during pregnancy:
+5 to 24 g/day.*

• ***Protein*** • The protein RDA for pregnancy is 10 grams per day higher than for nonpregnant women. Because people in the United States typically exceed the RDA, most women need not add 10 grams of protein to their diets. In fact, pregnant women in the United States—even those with low incomes who are not participating in food assistance programs—generally receive between 75 and 110 grams of protein a day.[15] Pregnant vegetarian women who meet their energy needs by eating ample servings of protein-containing plant foods such as legumes, whole grains, nuts, and seeds meet their protein needs as well. Use of high-protein supplements during pregnancy can be harmful and is discouraged. Among the problems associated with high-protein supplement use during pregnancy are high rates of low birthweights, preterm births, and deaths.[16]

Thiamin RDA during pregnancy:
1.4 mg/day.

Riboflavin RDA during pregnancy:
1.4 mg/day.

Niacin RDA during pregnancy:
18 mg NE/day.

Vitamin $B_6$ RDA during pregnancy:
1.9 mg/day.

Folate RDA during pregnancy:
600 µg DFE/day.

*For the first trimester, the RNI is an additional 5 grams/day; for the second trimester, it is an additional 20 grams/day; and for the third trimester, it is 24 grams/day.

• ***Nutrients Associated with Energy and Protein Metabolism*** • The RDA during pregnancy is set slightly above the nonpregnant woman's RDA for thiamin, riboflavin, niacin, and vitamin $B_6$. The usual intake of these nutrients is adequate for most pregnant women in the United States.

• ***Nutrients for Blood Production and Cell Growth*** • New cells are laid down at a tremendous pace as the fetus grows and develops. At the same time, the mother's red blood cell mass expands. All nutrients are important in these processes, but the needs for folate, vitamin $B_{12}$, iron, and zinc are especially great due to their key roles in the synthesis of DNA and new cells.

The requirement for folate increases dramatically during pregnancy. It is best to obtain sufficient folate from a combination of supplements, fortified foods, and a diet that includes fruits, juices, green vegetables, and whole-grain products.[17] The "How to" featured on p. 479 describes how folate from each of these sources contributes to a day's intake.

## HOW TO Estimate Dietary Folate Equivalents

The folate RDA during pregnancy is 600 μg DFE (dietary folate equivalents) per day and is best met from a combination of supplements, fortified foods, and diet. As Chapter 10 explains, folate is expressed in terms of DFE because synthetic folate from supplements and fortified foods is absorbed at almost twice (1.7 times) the rate of naturally occurring folate from other foods.

Consider, for example, a pregnant woman who takes a supplement and eats a bowl of fortified cornflakes, 2 slices of fortified bread, and a cup of fortified pasta:

| | |
|---|---|
| Supplement | 100 μg folate |
| Fortified cornflakes | 100 μg folate |
| Fortified bread | 40 μg folate |
| Fortified pasta | 60 μg folate |
| | 300 μg folate |

To calculate the DFE, multiply the sum by 1.7:

$$300\ \mu g \times 1.7 = 510\ \mu g\ DFE.$$

Now add the folate from the other foods in her diet—in this example, another 90 μg of folate.

$$510\ \mu g\ DFE + 90\ \mu g = 600\ \mu g\ DFE.$$

Notice that if we had not converted synthetic folate from supplements and fortified foods to DFE, then this woman's intake would appear to fall short of the 600 μg recommendation (300 μg + 90 μg = 390 μg). But as our example shows, her intake does meet the recommendation. At this time, supplement and fortified food labels list folate in μg only, not μg DFE, making such calculations necessary.

The pregnant woman also has a slightly greater need for the B vitamin that activates the folate enzyme—vitamin $B_{12}$. Generally, even modest amounts of meat, fish, eggs, or milk products together with body stores easily meet the need for vitamin $B_{12}$. Strict vegetarians who exclude all foods of animal origin, however, may need daily supplements of vitamin $B_{12}$ to prevent deficiency.

Vitamin $B_{12}$ RDA during pregnancy: 2.6 μg/day.

Pregnant women need iron to support their enlarged blood volume and to provide for placental and fetal needs. The developing fetus draws on maternal iron stores to create stores of its own to last through the first four to six months after birth when milk, which is poor in iron, will be its sole food. Also, the blood losses inevitable at birth, especially during a cesarean delivery, can further drain the mother's supply.*

**cesarean delivery:** a surgically assisted birth involving removal of the fetus by an incision into the uterus, usually by way of the abdominal wall.

During pregnancy, the body makes several adaptations to help meet the exceptionally high need for iron. Menstruation, the major route of iron loss in women, ceases, and iron absorption nearly triples due to an increase in blood transferrin, the body's iron-absorbing and iron-carrying protein. Without sufficient replacement, though, iron stores would quickly dwindle.

Iron RDA during pregnancy: 30 mg/day.
Canadian RNI during pregnancy: +0 to 10 mg/day.

Few women enter pregnancy with adequate iron stores, so a daily iron supplement is recommended during the second and third trimesters for all pregnant women.[18] To enhance absorption, the supplement should be taken between meals or at bedtime on an empty stomach and with liquids other than milk, coffee, or tea, which inhibit iron absorption.[19] Vitamin C does not enhance iron absorption from supplements as it does from foods; vitamin C enhances iron absorption by converting iron from ferric to ferrous, but supplemental iron is already in the ferrous form.[20]

Zinc is required for DNA and RNA synthesis and thus for protein synthesis and cell development. Typical zinc intakes for pregnant women are lower than

Zinc RDA during pregnancy: 15 mg/day.
Canadian RNI during pregnancy: +6 mg/day.

*The average blood loss during a cesarean delivery is almost twice that occurring during the average vaginal delivery of a single fetus.

*A pregnant woman's food choices support both her health and her infant's growth and development. (Courtesy of CNN)*

recommendations, but routine supplementation is not advised.[21] Women taking iron supplements (more than 30 milligrams per day), however, may need zinc supplementation because large doses of iron can interfere with the body's absorption and use of zinc.

• ***Nutrients for Bone Development*** • Vitamin D and the bone-building minerals calcium, phosphorus, magnesium, and fluoride are in great demand during pregnancy. Insufficient intakes may produce abnormal fetal bones and teeth.

Vitamin D plays a vital role in calcium absorption and utilization. Consequently, severe maternal vitamin D deficiency interferes with normal calcium metabolism, resulting in rickets in the fetus and osteomalacia in the mother. Exposure to sunlight and vitamin D–fortified milk is usually sufficient to provide the recommended amount of vitamin D during pregnancy. Routine supplementation is not recommended because of the toxicity risk. Vegetarians who avoid milk, eggs, and fish may receive enough vitamin D from daily exposure to sunlight or from fortified soy milk.

Calcium absorption more than doubles early in pregnancy, and the mother's bones store the mineral. During the last trimester, as the fetal bones begin to calcify, a dramatic shift of calcium across the placenta occurs. Whether calcium added to the mother's bones early in pregnancy is withdrawn to provide sufficient calcium to the fetus later in gestation is unclear.[22] In the final weeks of pregnancy, over 300 milligrams a day are transferred to the fetus. Recommendations to ensure an adequate calcium intake during pregnancy are aimed at conserving maternal bone while supplying fetal needs.

Most pregnant women tend to drink more milk than other women, but still their calcium intakes typically fall below recommendations. Because bones are still actively depositing minerals until about age 25, adequate calcium is especially important for young women. Pregnant women under age 25 who receive less than 600 milligrams of dietary calcium daily need to increase their intake of milk, cheese, yogurt, and other calcium-rich foods. Alternatively, and less preferably, they may need a daily supplement of 600 milligrams of calcium.[23]

**Healthy People 2000:**
Increase calcium intake so at least 50% of pregnant and lactating women consume three or more servings daily of foods rich in calcium.

• ***Other Nutrients*** • The nutrients mentioned here are those most intensely involved in blood production, cell growth, and bone growth. Of course, other nutrients are also needed during pregnancy to support the growth and health of both fetus and mother. Even with adequate nutrition, repeated pregnancies less than a year apart deplete nutrient reserves: fetal growth may be protected, but maternal health may decline.

• ***Nutrient Supplements*** • Women who make wise food choices during pregnancy can meet most of their nutrient needs, except for iron. As mentioned, iron supplements (30 milligrams per day) are recommended during the second and third trimesters of pregnancy. Daily multivitamin-mineral supplements are recommended for women who do not eat adequately and for those in high-risk groups: women carrying multiple fetuses, cigarette smokers, and alcohol and drug abusers. The use of prenatal supplements can help reduce the risks of preterm delivery and low infant birthweights.[24] Table 15-3 lists recommended amounts for supplements.

**Table 15-3**
**Nutrient Supplements during Pregnancy[a]**

| Nutrient | Amount |
|---|---|
| Folate | 300 µg |
| Vitamin $B_6$ | 2 mg |
| Vitamin C | 50 mg |
| Vitamin D | 5 µg |
| Calcium | 250 mg |
| Copper | 2 mg |
| Iron | 30 mg |
| Zinc | 15 mg |

[a]For pregnant women at nutritional risk (see Table 15-5).
Source: Reprinted with permission from *Nutrition during Pregnancy* © by the National Academy of Sciences. Published by the National Academy Press, Washington, D.C., 1990.

## Common Nutrition-Related Concerns of Pregnancy

Nausea, constipation, heartburn, and food sensitivities are common nutrition-related concerns during pregnancy. A few simple strategies can help alleviate the discomfort (see Table 15-4).

• ***Nausea*** • Not all women have uneasy stomachs in the early months of pregnancy, but many do. The nausea of "morning" (actually, anytime) sickness ranges from mild queasiness to debilitating nausea and vomiting. Severe and continued vomiting may require hospitalization if it results in acidosis, dehydration, or excessive weight loss. The hormonal changes of early pregnancy seem to be responsible for a woman's sensitivities to the appearance, texture, or smell of foods. Traditional strategies for quelling nausea are listed in Table 15-4, but some women benefit most from simply eating the foods they want when they feel like eating.[25] They may also find comfort in a cleaner, quieter, and more temperate environment.[26]

• ***Constipation and Hemorrhoids*** • As the hormones of pregnancy alter muscle tone and the growing fetus crowds intestinal organs, an expectant mother may experience constipation. She may also develop hemorrhoids (swollen veins of the rectum). These can be painful, and straining during bowel movements may cause bleeding. She can gain relief by following the strategies listed in Table 15-4.

• ***Heartburn*** • Heartburn is another common complaint during pregnancy. As the growing fetus puts increasing pressure on a woman's stomach, acid may back up in the lower esophagus and create a burning sensation near the heart. Tips to help relieve heartburn are listed in Table 15-4.

• ***Food Cravings and Aversions*** • Some women develop cravings for, or aversions to, particular foods and beverages during pregnancy. These likes and dislikes are fairly common, but do not seem to reflect real physiological needs. In other words, a woman who craves pickles does not necessarily need salt, nor does a woman who craves chocolate need caffeine or fat. Similarly, cravings for ice cream are common in pregnancy, but do not signify a calcium deficiency. Food cravings and aversions that arise during pregnancy are most likely due to hormone-induced changes in sensitivity to taste and smell.

**food craving:** a deep longing for a particular food.

**food aversion:** a strong desire to avoid a particular food.

• ***Nonfood Cravings*** • Some pregnant women develop cravings for nonfood items such as laundry starch, clay, dirt, or ice—a practice known as pica. Pica is especially common among African American women and is often associated with iron-deficiency anemia.[27] Pica is a cultural phenomenon that reflects a society's folklore, not a response to the physiological need for a nutrient such as iron; neither clay nor ice provides iron. Pica may lead to anemia by interfering with iron absorption and displacing iron-rich foods from the diet.

Reminder: The craving for nonfood items such as clay or ice is known as *pica*.

IN SUMMARY

Energy and nutrient needs are high during pregnancy. A balanced diet that includes an extra serving from each of the five food groups can usually meet these needs, with the exception of iron (supplements are

**Table 15-4**
**Strategies to Alleviate Maternal Discomforts**

| To Alleviate the Nausea of Pregnancy | To Prevent or Alleviate Constipation | To Prevent or Relieve Heartburn |
|---|---|---|
| • On waking, arise slowly.<br>• Eat dry toast or crackers.<br>• Chew gum or suck hard candies.<br>• Eat small, frequent meals.<br>• Avoid foods with offensive odors.<br>• When nauseated, do not drink citrus juice, water, milk, coffee, or tea. | • Eat foods high in fiber (fruits, vegetables, and whole-grain cereals).<br>• Exercise regularly.<br>• Drink at least eight glasses of liquids a day.<br>• Respond promptly to the urge to defecate.<br>• Use laxatives only as prescribed by a physician; do not use mineral oil, because it interferes with absorption of fat-soluble vitamins. | • Relax and eat slowly.<br>• Chew food thoroughly.<br>• Eat small, frequent meals.<br>• Drink liquids between meals.<br>• Avoid spicy or greasy foods.<br>• Sit up while eating; elevate the head while sleeping.<br>• Wait an hour after eating before lying down.<br>• Wait two hours after eating before exercising. |

recommended). The nausea, constipation, and heartburn that sometimes accompany pregnancy can usually be alleviated with a few simple strategies. Food cravings do not typically reflect physiological needs.

# High-Risk Pregnancies

**high-risk pregnancy:** a pregnancy characterized by indicators that make it likely the birth will be surrounded by problems such as premature delivery, difficult birth, retarded growth, birth defects, and early infant death.

**low-risk pregnancy:** a pregnancy characterized by indicators that make a normal outcome likely.

Some pregnancies jeopardize the life and health of the mother and infant. Table 15-5 identifies several characteristics of a high-risk pregnancy. A woman with none of these risk factors is said to have a low-risk pregnancy. The more factors that apply, the higher the risk. All pregnant women, especially those in high-risk categories, need prenatal care, including dietary advice.[28] The section at the top of the next page describes government efforts to provide food assistance to pregnant women in the United States.

Nutrition advice in prenatal care:

- Eat well-balanced meals.
- Take prenatal supplements as prescribed.
- Stop drinking alcohol.
- Gain enough weight to support fetal growth.

## Malnutrition and Pregnancy

Good nutrition clearly supports a pregnancy. In contrast, malnutrition interferes with the ability to conceive, the likelihood of implantation, and the subsequent development of a fetus should conception and implantation occur.

**fertility:** the capacity of a woman to produce a normal ovum periodically and of a man to produce normal sperm; the ability to reproduce.

Reminder: *Amenorrhea* is the temporary or permanent absence of menstrual periods. Amenorrhea is normal before puberty, after menopause, during pregnancy, and during lactation; otherwise it is abnormal.

• ***Malnutrition and Fertility*** • The nutrition habits and lifestyle choices people make can influence the course of a pregnancy they are not even planning at the time. Severe malnutrition and food deprivation can reduce fertility: women may develop amenorrhea, and men may lose their ability to produce viable sperm. Furthermore, both men and women lose sexual interest during times of starvation. Starvation arises predictably during famines, wars, and droughts, but can also occur amidst peace and plenty. Many young women who diet excessively and exercise intensely are starving and suffering from malnutrition (see Highlight 9).

**Table 15-5**
**High-Risk Pregnancy Factors**

| Factor | Condition That Raises Risk |
|---|---|
| Maternal weight | |
| Prior to pregnancy | Prepregnancy BMI either <19.8 or >26.0 |
| During pregnancy | Insufficient or excessive pregnancy weight gain |
| Maternal nutrition | Nutrient deficiencies or toxicities; eating disorders |
| Socioeconomic status | Poverty, lack of family support, low level of education, limited food available |
| Lifestyle habits | Smoking, alcohol or other drug use |
| Age | Teens, especially 15 years or younger; women 35 years or older |
| Previous pregnancies | |
| Number | Many previous pregnancies (3 or more to mothers under age 20; 4 or more to mothers age 20 or older) |
| Interval | Short intervals between pregnancies (<1 year) |
| Outcomes | Previous history of problems |
| Multiple births | Twins or triplets |
| Birthweight | Low- or high-birthweight infants |
| Maternal health | |
| High blood pressure | Development of pregnancy-related hypertension |
| Diabetes | Development of gestational diabetes |
| Chronic diseases | Diabetes; heart, respiratory, and kidney disease; certain genetic disorders; special diets and drugs |

## Food Assistance Programs for Pregnant Women, Infants, and Children

WIC (the Special Supplemental Food Program for Women, Infants, and Children) provides nutrition education and nutritious foods to low-income pregnant women and their young children. WIC provides eggs, milk, cereal, juice, cheese, legumes, peanut butter, and infant formula to infants, children up to age five, and pregnant and breastfeeding women who qualify financially and have a high risk of medical or nutritional problems. The program is both remedial and preventive: services include health care referrals, nutrition education, and food packages or vouchers for specific foods to supply nutrients known to be lacking in the diets of the target population. Prenatal WIC participation can effectively reduce infant mortality, low birthweight, and newborn medical costs.[a] For every dollar spent on WIC, an estimated three dollars in medical costs are saved. In 1992, participation in WIC reduced first-year medical expenses for infants by $1.19 billion.[b]

www.usda.gov/fcs
Visit the WIC Program

[a]P. A. Buescher and coauthors, Prenatal WIC participation can reduce low birth weight and newborn medical costs: A cost-benefit analysis of WIC participation in North Carolina, *Journal of the American Dietetic Association* 93 (1993): 163–166.
[b]S. Avruch and A. P. Cackley, Savings achieved by giving WIC benefits to women prenatally, *Public Health Reports* 110 (1995): 27–34.

• ***Malnutrition and Early Pregnancy*** • If a malnourished woman does become pregnant, she faces the challenge of supporting both the growth of a baby and her own health with inadequate nutrient stores. Malnutrition prior to and around conception prevents the placenta from developing fully. A poorly developed placenta cannot deliver optimum nourishment to the fetus, and the infant will be born small and possibly with physical and cognitive abnormalities.[29] If this small infant is a female, she may develop poorly and have an elevated risk of developing a chronic condition that could impair her ability to give birth to a healthy infant. Thus a woman's malnutrition can adversely affect not only her children but her *grandchildren.*

American College of Obstetricians and Gynecologists

www.acog.org

• ***Malnutrition and Fetal Development*** • Without adequate nutrition during pregnancy, fetal growth and infant health are compromised. In general, consequences of malnutrition during pregnancy include:

- Fetal growth retardation.
- Congenital malformations (birth defects).
- Spontaneous abortion and stillbirth.
- Premature birth.
- Low infant birthweight.

Birthweight is most frequently used as a predictor of an infant's survival and health. Malnutrition, coupled with low birthweight, is the underlying or associated cause of more than half of all deaths of children under four years of age worldwide.[30]

## The Infant's Birthweight

A high-risk pregnancy may produce a low-birthweight infant. Low-birthweight infants, defined as infants who weigh 5½ pounds or less, are classified according to gestational age. Preterm, or premature, infants are born before they are fully developed; they are often underweight and have trouble breathing because their lungs are immature. Preterm infants may be small, but if their size and weight are appropriate for their age, they can catch up in growth given adequate nutrition

**low birthweight (LBW):** a birthweight of 5½ lb (2500 g) or less; indicates probable poor health in the newborn and poor nutrition status in the mother during pregnancy, before pregnancy, or both. Normal birthweight for a full-term baby is 6½ to 8¾ lb (about 3000 to 4000 g).

Some preterm infants are of a weight **appropriate for gestational age (AGA)**; others are **small for gestational age (SGA)**, often reflecting malnutrition. The latter type are also called **small-for-date** babies.

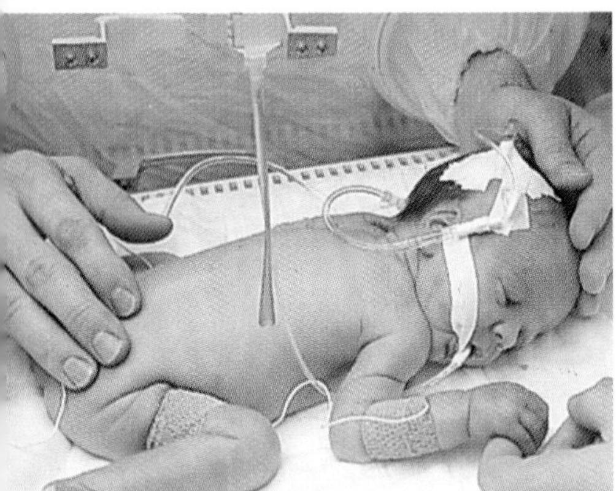

*Low-birthweight babies need special care and nourishment.*

support. In contrast, small-for-gestational-age infants have suffered growth failure in the uterus and do not catch up as well. For the most part, survival improves with increased gestational age and birthweight.

Low-birthweight infants are more likely to experience complications during delivery than normal-weight babies. They also have a statistically greater chance of having physical and mental birth defects, contracting diseases, and dying early in life. Of infants who die before their first birthdays, about two-thirds are low-birthweight babies.

A strong relationship has been established between socioeconomic disadvantage and low birthweight. Low socioeconomic status impairs fetal development by causing stress and by limiting access to medical care and to nutritious foods. Low socioeconomic status often accompanies teen pregnancies, smoking, and alcohol and drug abuse—all predictors of low birthweight.

## The Mother's Health Status

Medical disorders can threaten the life and health of both mother and fetus. If diagnosed and treated early, many diseases can be managed to ensure a healthy outcome—another strong argument for early prenatal care.

• ***Preexisting Diabetes*** • Whether diabetes presents risks depends on how well it is controlled before and during pregnancy. Without proper management of maternal diabetes, women face high infertility rates, and those who do conceive may experience episodes of severe hypoglycemia or hyperglycemia, spontaneous abortions, and pregnancy-related hypertension. Infants may be larger than normal and suffer physical and mental abnormalities and other complications such as severe hypoglycemia or respiratory distress, both of which can be fatal. Ideally, a woman with diabetes will receive the prenatal care needed to achieve glucose control before conception and continued glucose control throughout pregnancy.

**gestational diabetes:** the appearance of abnormal glucose tolerance during pregnancy, with subsequent return to normal postpartum. Gestational diabetes is associated with cesarean delivery, birth trauma, and high birthweight.

Risk factors for gestational diabetes:
- Age 35 or older.
- Overweight (BMI >25) or excessive weight gain.
- Complications in previous pregnancies.
- Symptoms of diabetes.
- Family history of diabetes.
- Cigarette smoking.
- Asian race.

 American Diabetes Associates
www.diabetes.org
Search for Gestational diabetes

• ***Gestational Diabetes*** • Women who have never had diabetes before may develop a condition known as gestational diabetes during pregnancy. Gestational diabetes usually develops during the second half of pregnancy, with subsequent return to normal glucose tolerance after childbirth. Almost one-third of all women with gestational diabetes, however, develop diabetes (type 2) later in life, especially if they are overweight. For this reason, health care professionals advise against excessive weight gain. To ensure that the problems of gestational diabetes are dealt with promptly, they look for the risk factors listed in the margin.[31] Dietary recommendations encourage three meals a day plus two snacks, each containing protein, carbohydrate, and moderate fat. Diet alone may control gestational diabetes, but insulin therapy may be required if blood glucose fails to normalize.

• ***Preexisting Hypertension*** • Hypertension complicates pregnancy and affects its outcome in different ways, depending on when the hypertension first develops and on how severe it becomes.[32] In addition to the threats hypertension always carries (such as heart attack and stroke), high blood pressure increases the risks of a low-birthweight infant or the separation of the placenta from the wall of the uterus before the birth, resulting in stillbirth. Ideally, before a woman with hypertension becomes pregnant, her blood pressure will be under control.

**transient hypertension of pregnancy:** high blood pressure that develops in the second half of pregnancy and resolves after childbirth, usually without affecting the outcome of the pregnancy.

• ***Transient Hypertension of Pregnancy*** • Some women develop hypertension during the second half of pregnancy.* Most often, the rise in blood pressure is mild

*Blood pressure of 140/90 millimeters mercury during the second half of pregnancy in a woman who has not previously exhibited hypertension indicates high blood pressure. So does a rise in systolic blood pressure of 30 millimeters or in diastolic blood pressure of 15 millimeters on at least two occasions more than six hours apart. By this rule, an apparently "normal" blood pressure of 120/85 would be high for a woman whose normal value was 90/70.

and does not affect the pregnancy adversely. Blood pressure usually returns to normal during the first few weeks after childbirth. This transient hypertension of pregnancy differs from the life-threatening hypertension that accompanies preeclampsia.*

• ***Preeclampsia*** • Hypertension may signal the onset of preeclampsia, a condition characterized not only by high blood pressure but by protein in the urine and fluid retention (edema). Preeclampsia usually occurs with first pregnancies and almost always after 20 weeks gestation, most often near term. Symptoms typically regress within two days of delivery. The edema of preeclampsia is a whole-body edema, distinct from the localized fluid retention women normally experience late in pregnancy.

**preeclampsia** (PRE-ee-KLAMP-see-ah): a condition characterized by hypertension, fluid retention, and protein in the urine.

The normal edema of pregnancy responds to gravity; fluid pools in the ankles. The edema of preeclampsia is a generalized edema. The differences between these two types of edema help with the diagnosis of preeclampsia.

Preeclampsia affects almost all of the mother's organs—the circulatory system, liver, kidneys, and brain. Blood flow through the vessels that supply oxygen and nutrients to the placenta diminishes. For this reason, preeclampsia often retards fetal growth. In some cases, the placenta separates from the uterus, resulting in stillbirth.

Preeclampsia can progress rapidly to eclampsia—a condition characterized by convulsions. Maternal death during pregnancy and childbirth is extremely rare in developed countries, but when it does occur, eclampsia is a common cause.

**eclampsia** (eh-KLAMP-see-ah): a severe stage of preeclampsia characterized by convulsions.

Warning signs of preeclampsia:

- Hypertension.
- Protein in the urine.
- Upper abdominal pain.
- Severe and constant headaches.
- Swelling, especially of the face.
- Dizziness.
- Blurred vision.
- Sudden weight gain (1 lb/day).
- Fetal growth retardation.

Preeclampsia demands prompt medical attention. Treatment focuses on regulating blood pressure and preventing convulsions. If preeclampsia develops early and is severe, induced labor or cesarean delivery may be necessary, regardless of gestational age. The infant will be preterm, with all of the associated problems, including poor lung development and special care needs.

Several approaches have been studied to prevent preeclampsia, including sodium restriction, calcium supplementation, and magnesium sulfate injections. Sodium restriction, however, does not reduce the incidence or severity of preeclampsia. Furthermore, a sodium-restricted diet tends to lower the intakes of fat, protein, calcium, and energy; limit maternal weight gain; and reduce maternal fat stores—all unwanted side effects in women whose nutrition status may already be compromised.[33] Until and unless the kidneys prove unable to handle sodium and edema is evident, sodium restriction is not recommended.

Researchers examining the relationships between calcium intake and preeclampsia report conflicting findings. An analysis of several small clinical trials determined that calcium supplementation (1500 to 2000 milligrams per day) during pregnancy can lower high blood pressure, but a more recent and larger study reported no benefit.[34] At this time, evidence is insufficient to recommend calcium supplements routinely; calcium supplementation may create risks of its own, including the development of kidney stones.

To prevent seizures during labor, physicians may give magnesium sulfate injections to women with preeclampsia.[35] Such treatment is both effective and superior to other anticonvulsive drugs. Exposure to magnesium sulfate reduces the risk for cerebral palsy and possibly mental retardation in the infant.[36]

## *The Mother's Age*

Maternal age also influences the course of a pregnancy. Compared with women of the physically ideal childbearing age of 20 to 25, both younger and older women face more complications of pregnancy, as the next paragraphs describe.

*The Working Group on High Blood Pressure in Pregnancy, convened by the National High Blood Pressure Education Program of the National Heart, Lung, and Blood Institute, has suggested abandoning the term "pregnancy-induced hypertension" because it fails to differentiate between the mild, transient hypertension of pregnancy and the life-threatening hypertension of preeclampsia.

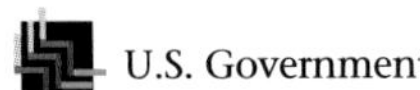 U.S. Government
www.healthfinder.gov/searchoptions/topicsaz.htm
Search for Adolescent pregnancy

• ***Pregnancy in Adolescents*** • Many adolescents become sexually active before age 19, and one million adolescent girls face pregnancies each year in the United States. About half of them give birth. Put another way, about one out of every four babies is born to a teenager, and more than a tenth of these mothers are age 15 or younger.[37] Nourishing a growing fetus adds to a teenage girl's nutrition burden, especially if her growth is still incomplete.[38] Simply being young increases the risks of pregnancy complications independently of important socioeconomic factors.[39]

Common complications among adolescent mothers include iron-deficiency anemia (which may reflect poor diet and inadequate prenatal care) and prolonged labor (which reflects the mother's physical immaturity). On a positive note, maternal death is lowest for mothers under age 20.

Pregnant teenagers have higher rates of stillbirths, preterm births, and low-birthweight infants than do adult women.[40] Many of these infants suffer physical problems, require intensive care, and die within the first year. The care of infants born to teenagers costs our society an estimated $1 billion annually. Because teenagers have few financial resources, they cannot pay these costs. Furthermore, their low economic status contributes significantly to the complications surrounding their pregnancies. At a time when prenatal care is most important, it is less available. And the pattern of teenage pregnancies continues from generation to generation, with almost 40 percent of the daughters born to teenager mothers becoming teenage mothers themselves.[41] Clearly, teenage pregnancy is a major public health problem.

To support the needs of both mother and fetus, young teenagers (13 to 16 years old) are encouraged to strive for the highest weight gains recommended for pregnancy. For a teen who enters pregnancy at a healthy body weight, a weight gain of approximately 35 pounds is recommended; this amount minimizes the risk of delivering a low-birthweight infant. Gaining less weight may limit fetal growth. Pregnant and lactating teenagers can use the food guide presented in Table 15-2 (on p. 478), making sure to select at least 4 servings of milk or milk products daily.

Without the appropriate economic, psychosocial, and physical support, a young mother will not be able to care for herself during her pregnancy and for her child after the birth.[42] To improve her chances for a successful pregnancy and a healthy infant, she must seek prenatal care. WIC helps pregnant teenagers obtain adequate food to support a reasonable weight gain (WIC is introduced on p. 483).

• ***Pregnancy in Older Women*** • In the last three decades, many women have delayed childbearing while they pursue education and careers. As a result, the number of first births to women 35 and older has increased dramatically.

Most of the complications associated with later childbearing reflect chronic conditions such as hypertension and diabetes. These complications often result in a cesarean delivery, which is twice as common in women over 35 as among younger women. For all these reasons, maternal death rates are higher in women over 35 than in younger women.

The babies of older mothers face problems of their own. Because 1 out of 50 pregnancies in older women produces an infant with genetic abnormalities, obstetricians routinely screen women older than 35. Birth defects, preterm births, growth retardation, and death are common among infants born to women over 35. For a 40-year-old mother, the risk of having a child with Down syndrome, for example, is about 1 in 100 compared with 1 in 300 for a 35-year-old and 1 in 10,000 for a 20-year-old. Fetal death is twice as high for women 35 years and older than for younger women.[43] Why this is so remains a bit of a mystery. One possibility is that the uterine blood vessels of older women cannot fully adapt to the increased demands of pregnancy.

**Down syndrome:** a genetic abnormality that causes mental retardation, short stature, and flattened facial features.

## Practices Incompatible with Pregnancy

Besides malnutrition, a variety of lifestyle factors can have adverse effects on pregnancy; and some may be teratogenic. People who are planning to have children can make the choice to practice healthy behaviors.

Reminder: The word *teratogenic* describes a factor that causes abnormal fetal development and birth defects.

**• *Alcohol* •** Alcohol consumption during pregnancy can cause irreversible mental and physical retardation of the fetus—fetal alcohol syndrome (FAS). Of the leading causes of mental retardation, FAS is the only one that is totally *preventable*. As a consequence, the surgeon general has issued a statement that pregnant women should drink absolutely no alcohol. Fetal alcohol syndrome is the topic of Highlight 15.

**• *Medicinal Drugs* •** Drugs other than alcohol can also cause complications during pregnancy, problems in labor, and serious birth defects. For these reasons, pregnant women should not take any medicines without consulting their physicians. Drug labels warn: As with any drug, if you are pregnant or nursing a baby, seek the advice of a health professional before using this product. For aspirin and ibuprofen, an additional warning immediately follows: It is especially important not to use aspirin (or ibuprofen) during the last three months of pregnancy unless specifically directed to do so by a doctor because it may cause problems in the unborn child or excessive bleeding during delivery.

***Young adults can prepare themselves for a healthy pregnancy by taking care of themselves today.***

**• *Illicit Drugs* •** The recommendation to avoid drugs during pregnancy also includes illicit drugs, of course. Unfortunately, use of illicit drugs, such as cocaine and marijuana, is common among some pregnant women.*

Drugs of abuse, such as cocaine, easily cross the placenta and impair fetal development.[44] Furthermore, they are responsible for preterm births, low-birthweight infants, perinatal deaths, and sudden infant deaths.[45] If these newborns survive, their cries and behaviors at birth are abnormal, and their cognitive development later in life is impaired.[46] They may be hypersensitive or underaroused; those who test positive for drugs suffer the greatest effects of toxicity and withdrawal.[47]

Fetal effects of abused drugs:

- Amphetamines: Suspected nervous system damage; behavioral abnormalities.
- Barbiturates: Drug withdrawal symptoms in the newborn, lasting up to six months.
- Cocaine (including "crack"): Uncontrolled jerking motions; paralysis; permanent mental and physical damage.
- Marijuana: Short-term irritability at birth.
- Opiates (including heroin): Drug withdrawal symptoms in the newborn; permanent learning disability (attention deficit disorder).

**• *Smoking and Chewing Tobacco* •** Smoking cigarettes and chewing tobacco at any time exert harmful effects, and pregnancy dramatically magnifies the hazards of these practices. Smoking restricts the blood supply to the growing fetus and so limits oxygen and nutrient delivery and waste removal. Unfortunately, an estimated 20 percent of pregnant women smoke, with higher rates for unmarried women, teenagers, and those who have not graduated from high school.[48] Smokers tend to eat less nutritious foods during their pregnancies than do nonsmokers, which in turn impairs fetal nutrition.

Of all preventable causes of low birthweight in the United States, smoking has the greatest impact. A mother who smokes is more likely to have a complicated birth and a low-birthweight infant.[49] Furthermore, smoking causes death in otherwise healthy fetuses and newborns. A positive relationship exists between sudden infant death syndrome (SIDS) and both cigarette smoking during pregnancy and postnatal exposure to passive smoke.[50] Smoking during pregnancy may even harm the intellectual and behavioral development of the child later in life.[51]

Infants of mothers who chew tobacco also have lower birthweights and higher rates of fetal deaths than infants born to women who do not use tobacco. Any

Smoking during pregnancy increases the risk of:

- Fetal growth retardation.
- Low birthweight.
- Complications at birth.
- Mislocation of the placenta.
- Premature separation of the placenta.
- Vaginal bleeding.
- Spontaneous abortion.
- Fetal death.
- Sudden Infant Death Syndrome (SIDS).

**sudden infant death syndrome (SIDS):** the unexpected and unexplained death of an apparently well infant; the most common cause of death of infants between the second week and the end of the first year of life; also called *crib death.*

*It is estimated that 17 percent of pregnant women use marijuana and at least 6 percent use cocaine. Committee on Nutritional Status during Pregnancy and Lactation, Food and Nutrition Board, *Nutrition during Pregnancy* (Washington, D.C.: National Academy Press, 1990), p. 48.

woman who smokes cigarettes or chews tobacco and is considering pregnancy or who is already pregnant should try to quit.

Highlight 13 describes how lead toxicity impairs a child's development.

• ***Environmental Contaminants*** • Infants and young children of pregnant women exposed to environmental contaminants such as lead and mercury show signs of impaired cognitive development. For this reason, it is particularly important that pregnant women receive foods and beverages grown and prepared in environments free of contamination.

• ***Vitamin-Mineral Megadoses*** • The pregnant woman who is trying to eat well may mistakenly assume that more is better when it comes to vitamin-mineral supplements. This is simply not true; many vitamins are toxic when taken in excess, and the minerals are even more so, some at levels not far above recommendations. Researchers found that among women who took more than 10,000 IU of supplemental vitamin A daily, approximately 1 out of every 57 infants was born with a malformation of the cranial nervous system that was attributable to high vitamin A intake.[52] Intakes before the seventh week appeared to be the most damaging. For this reason, vitamin A is not given as a supplement in the first trimester of pregnancy unless there is specific evidence of deficiency, which is rare. A pregnant woman can obtain all the vitamin A and most of the other vitamins and minerals she needs by making wise food choices. She should take supplements only on the advice of a registered dietitian or physician.

The caffeine contents of selected beverages, foods, and drugs are listed on p. H-3 of Appendix H.

• ***Caffeine*** • Caffeine crosses the placenta, and the developing fetus has a limited ability to metabolize it. For this reason, pregnant women may wonder whether they should give up coffee, tea, and colas because of their caffeine contents. Research studies have not proved that caffeine (even in high doses) causes birth defects in human infants (as it does in animals), but limited evidence suggests that moderate-to-heavy use may lower infant birthweight.[53] (Heavy caffeine use was defined as more than 300 milligrams per day—the equivalent of 2 to 3 cups of coffee.) All things considered, it might be most sensible to limit caffeine consumption to the equivalent of a cup of coffee or two 12-ounce cola beverages a day.

• ***Weight-Loss Dieting*** • Weight-loss dieting, even for short periods, is hazardous during pregnancy. Low-carbohydrate diets or fasts that cause ketosis deprive the fetal brain of needed glucose and may impair its development. Such diets are also likely to lack other nutrients vital to fetal growth. Regardless of prepregnancy weight, pregnant women should never intentionally lose weight.

• ***Sugar Substitutes*** • Artificial sweeteners have been extensively investigated and found to be safe for use during pregnancy.[54] (Women with phenylketonuria should not use aspartame, as Highlight 4 explains.) It would be prudent for pregnant women to use sweeteners in moderation and within an otherwise nutritious and well-balanced diet.

IN SUMMARY

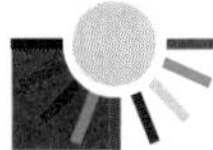

High-risk pregnancies, especially for teenagers, threaten the life and health of both mother and infant. Proper nutrition and abstinence from smoking, alcohol, and other drugs improve the outcome. In addition, prenatal care includes monitoring pregnant women for gestational diabetes and preeclampsia.

U.S. Government

www.healthfinder.gov/searchoptions/topicsaz.htm

Search for Maternal and infant health

# Nutrition during Lactation

Before the end of her pregnancy, a woman will need to consider whether to feed her infant breast milk, infant formula, or both. These options are the only recommended foods for an infant during the first four to six months of life.

Breastfeeding offers many health benefits to both mother and infant, and every pregnant woman should seriously consider it.[55] Still, there are valid reasons for not breastfeeding, and formula-fed infants grow and develop into healthy children. After all, the primary goal is to provide the infant with optimal nourishment in a relaxed and loving environment.

**Healthy People 2000:** Increase to at least 75% the proportion of mothers who breastfeed their babies in the early weeks and to at least 50% the proportion who continue breastfeeding until their babies are five to six months old.

For infants, breastfeeding:
- Provides a favorable balance of nutrients with high bioavailability.
- Provides hormones that promote physiological development.
- Improves cognitive development.
- Protects against a variety of infections.
- Reduces the risk of sudden infant death syndrome (SIDS).
- Protects against some chronic diseases, such as diabetes.
- Makes food allergies less likely.

For mothers, breastfeeding:
- Contracts the uterus.
- Lengthens birth intervals.
- Conserves iron stores (by prolonging amenorrhea).
- Reduces risk of breast cancer.
- Protects bone density.
- Saves money and offers convenience.

LaLeche League International
www.lalecheleague.org

## *Breastfeeding: A Learned Behavior*

In many countries around the world, a woman breastfeeds her newborn without considering the alternatives or consciously making a decision. In other parts of the world, a woman feeds her newborn formula simply because she knows so little about breastfeeding. She may have misconceptions or feel uncomfortable about a process she has never seen or experienced.

Lactation is an automatic physiological process that virtually all mothers are capable of doing.[56] Breastfeeding, on the other hand, is a learned behavior that not all mothers decide to do. Of women who do breastfeed, those who receive early and repeated information and support breastfeed their infants longer than others. Health care professionals play an important role in providing encouragement and accurate information on breastfeeding.[57]

**lactation:** production and secretion of breast milk for the purpose of nourishing an infant.

Some hospitals employ *certified lactation consultants* who specialize in helping new mothers to establish a healthy breastfeeding relationship with their newborn. These consultants are often registered nurses with specialized training in breast and infant anatomy and physiology.

Fathers also play an important role in encouraging breastfeeding.[58] Studies report that most fathers whose partners plan to breastfeed support that decision and respect breastfeeding women. By comparison, fathers whose partners plan to formula feed believe that breastfeeding makes the breasts ugly and interferes with sexual relations. Clearly, educating fathers could change attitudes and promote breastfeeding.

In societies where few women breastfeed, appropriate breastfeeding etiquette remains undefined. A woman faces conflict, confusion, and frustration. Must she retreat to a private place to nurse? What if she cannot find such a place in a public setting? A hungry infant is impatient, and a mother must act quickly. As role models become more numerous, a consensus will develop as to what behaviors are accepted and will provide nursing mothers with more guidance and confidence. Many public buildings now provide "baby rooms" with tables for changing diapers and comfortable chairs for nursing.

Parents in today's society also have to coordinate work and family. All mothers are working women—many of them with jobs outside the home. To promote breastfeeding as a feasible option, a social system needs to provide extended, paid maternity leaves, breaks during the workday to nurse infants or pump breasts, and workplace child care.

Most healthy women who want to breastfeed can do so with a little preparation; physical obstacles to breastfeeding are rare, although overweight mothers seem to have less success initiating breastfeeding than others.[59] Successful breastfeeding requires adequate nutrition and rest. This, plus the support of all who care, will help to enhance the well-being of mother and infant.

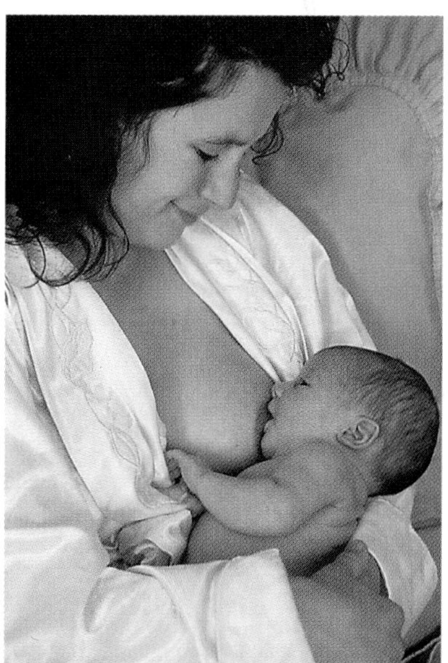

***Breastfeeding is a natural extension of pregnancy—of the mother's body nourishing the infant.***

## *The Mother's Nutrient Needs*

Ideally, the mother who chooses to breastfeed her infant will continue to eat nutrient-dense foods throughout lactation. An adequate diet is needed to support the stamina, patience, and self-confidence that nursing an infant demands.

Energy RDA during lactation:
+500 kcal/day (1800 kcal/day minimum).
Canadian RNI during lactation:
+450 kcal/day.

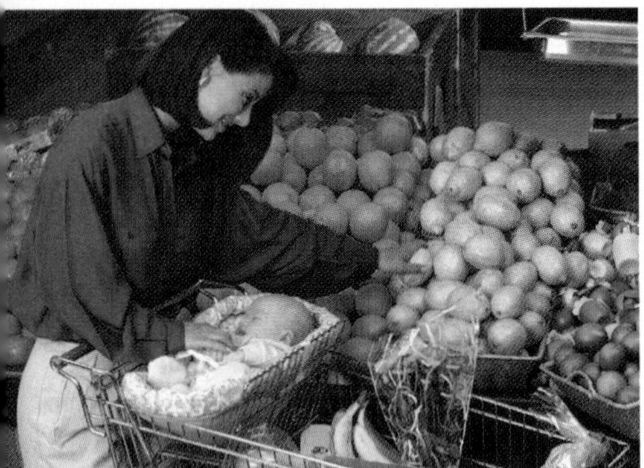

*Nutritious foods support successful lactation.*

• ***Energy Intake and Exercise*** • A nursing mother produces about 25 ounces of milk per day, with considerable variation from woman to woman and in the same woman from time to time, depending primarily on the infant's demand for milk. To produce an adequate supply of milk, a woman needs extra energy—almost 650 kcalories a day above her regular need during the first six months of lactation. To meet this energy need, the woman is advised to eat an extra 500 kcalories of food each day and let the fat reserves she accumulated during pregnancy provide the rest. Energy needs for women who are breastfeeding exclusively range from 2500 to 3300 kcalories a day, depending on physical activity.[60] Lower intakes may support more rapid weight loss, but may not meet vitamin and mineral needs.[61] Most women need at least 1800 kcalories a day to receive all the nutrients required for successful lactation.[62] Severe energy restriction may hinder milk production.

After the birth of the infant, many women are in a hurry to lose the extra body fat they accumulated during pregnancy. Opinions differ as to whether breastfeeding helps with postpartum weight loss. In general, most women lose 1 to 2 pounds a month during the first four to six months of lactation; some may lose more, and others may maintain or even gain weight.[63] Regardless of prepregnancy weight, the more weight a woman gains during pregnancy, the more weight she loses following delivery (when measured at six weeks and one year). Neither the quality nor the quantity of breast milk is adversely affected by moderate weight loss.[64]

Women often exercise to lose weight and improve fitness, and this is compatible with breastfeeding.[65] Intense physical activity can raise the lactic acid concentration of breast milk, which influences the milk's taste. Infants may prefer milk produced prior to exercise (which has a lower lactic acid content). For this reason, mothers may want to breastfeed their infants before exercise or express their milk before exercise for use afterward.

• ***Vitamins and Minerals*** • A question often raised is whether a mother's milk may lack a nutrient if she fails to get enough in her diet. The answer differs from one nutrient to the next, but in general, nutritional inadequacies reduce the *quantity,* not the *quality,* of breast milk. Women can produce milk with adequate protein, carbohydrate, fat, and most minerals, even when their own supplies are limited.[66] For these nutrients and for the vitamin folate as well, milk quality is maintained at the expense of maternal stores. This is most evident in the case of calcium: dietary calcium has no effect on the calcium concentration of breast milk, but maternal bones lose some of their density during lactation.[67] (Bone density increases again after weaning.) Nutrients in breast milk most likely to decline in response to prolonged inadequate intakes are the vitamins—especially vitamins $B_6$, $B_{12}$, A, and D.[68] Review Figure 15-8 (on p. 477) to compare a lactating woman's nutrient needs with those of pregnant and nonpregnant women.

• ***Water*** • Despite previous misconceptions, a mother who drinks more fluid does not produce more breast milk. To protect herself from dehydration, however, a lactating woman needs to drink plenty of fluids (at least 2 quarts of fluids twice a day). A sensible rule of thumb is to drink a glass of milk, juice, or water at each meal and each time the baby nurses.

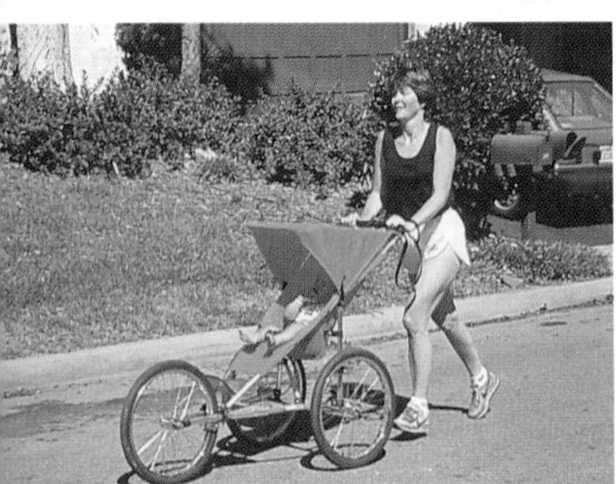

*A brisk walk through the neighborhood provides an opportunity for physical activity and fresh air.*

• ***Nutrient Supplements*** • Most lactating women can obtain all the nutrients they need from a well-balanced diet without taking vitamin-mineral supplements; some, however, may need iron supplements. Maternal iron stores dwindle during pregnancy when the fetus takes iron to meet its own needs during the first four to six months after birth. In addition, childbirth may have incurred blood losses. A woman may therefore need iron supplements during lactation, not to augment the iron in her breast milk, but to refill her depleted iron stores.

• ***Particular Foods*** • Foods with strong or spicy flavors (such as garlic) may alter the flavor of breast milk. A sudden change in the taste of the milk may annoy some infants. Infants who are sensitive to particular foods such as cow's milk protein may become uncomfortable when the mother's diet includes these foods. Only a few infants exhibit this sensitivity, so only a few nursing mothers need avoid cow's milk. Generally, nutrients from milk products support both the infant's and the mother's health.

In general, a nursing mother can eat whatever nutritious foods she chooses. If she suspects a particular food is causing the infant discomfort, her physician may recommend a dietary challenge: eliminate the food from the diet to see if the infant's reactions subside; then return the food to the diet, and again monitor the infant's reactions. If a food must be eliminated for an extended time, appropriate substitutions must be made to ensure nutrient adequacy.

## Practices Incompatible with Lactation

Some substances impair milk production or enter breast milk and interfere with infant development. Some medical conditions prohibit breastfeeding. This section describes these circumstances.

• ***Alcohol*** • Alcohol easily enters breast milk, and its concentration peaks within an hour of ingestion. Infants drink less breast milk when their mothers have consumed even small amounts of alcohol (equivalent to a can of beer). Three possible reasons, acting separately or together, may explain why. For one, the alcohol may have altered the flavor of the breast milk and thereby the infants' acceptance of it. For another, because infants metabolize alcohol inefficiently, even low doses may be potent enough to suppress their feeding behavior. Third, the alcohol may have reduced milk production by inhibiting the hormone oxytocin.

**oxytocin** (OK-see-TOE-sin): a hormone that stimulates the mammary glands to eject milk during lactation and the uterus to contract during childbirth.

In the past, alcohol has been recommended to mothers to facilitate lactation despite a lack of scientific evidence that it does so. The research summarized here suggests that alcohol actually hinders breastfeeding. An occasional glass of wine or beer is considered within safe limits, but in general, lactating women should consume little or no alcohol.

• ***Medicinal Drugs*** • Many drugs are compatible with breastfeeding, but some medicines are contraindicated, either because they suppress lactation or because they are secreted into breast milk and can harm the infant.[69] As a precaution, a nursing mother should consult with her physician prior to taking any drug.

• ***Illicit Drugs*** • Illicit drugs, of course, are harmful to the physical and emotional health of both the mother and the nursing infant. Breast milk can deliver such high doses of illicit drugs as to cause irritability, tremors, hallucinations, and even death in infants.

• ***Smoking*** • Cigarette smoking reduces milk volume, so smokers may produce too little milk to meet their infants' energy needs. Consequently, infants of breastfeeding mothers who smoke gain less weight than infants of those who do not smoke. Furthermore, infant exposure to passive smoke negates the protective effect breastfeeding offers against SIDS and increases the risks dramatically.

• ***Environmental Contaminants*** • Environmental contaminants, such as DDT, PCBs, and methylmercury can find their way into breast milk.[70] Inuit mothers living in Arctic Québec who eat seal and beluga whale blubber have concentrations of DDT and PCBs in their breast milk two to ten times greater than those found in breast milk from women in southern Québec.[71] The impact of contaminated breast milk on infant development is unclear, however. Preliminary studies indicate that the children of these Inuit mothers are developing normally. Researchers speculate that the abundant omega-3 fatty acids of the Inuit diet may protect against damage to the central nervous system.

• ***Caffeine*** • Caffeine taken during lactation may make a breastfed infant irritable and wakeful. As during pregnancy, caffeine consumption should be moderate—

say, 1 to 2 cups of coffee a day. Larger doses of coffee may interfere with the availability of iron from the milk and impair the infant's iron status.

## Maternal Health

If a woman has an ordinary cold, she can go on nursing without worry. If susceptible, the infant will catch it from her anyway. (Thanks to immunological protection, a breastfed baby may be less susceptible than a formula-fed baby would be.) If a woman has a communicable disease such as tuberculosis or hepatitis that could threaten the infant's health, then mother and baby have to be separated; the mother can pump her breasts several times a day and feed breast milk by bottle.

• ***HIV Infection and AIDS*** • For mothers with HIV infections, advice differs depending on the context. Where safe alternatives are available, the Centers for Disease Control and the American Academy of Pediatrics recommend that HIV-positive women not breastfeed their infants. In developing countries, however, the feeding of inappropriate or contaminated formulas causes 1.5 million infant deaths each year, so WHO and UNICEF urge mothers to breastfeed irrespective of HIV infection. For these infants, the protection of being breastfed outweighs the risk of HIV transmission.[72]

• ***Diabetes*** • Women with diabetes (type 1) may need careful monitoring and counseling to ensure successful lactation.[73] These women need to adjust their energy intakes and insulin doses to meet the heightened needs of lactation. Maintaining good glucose control helps to initiate lactation and support milk production.[74]

**postpartum amenorrhea:** the normal temporary absence of menstrual periods immediately following childbirth.

• ***Postpartum Amenorrhea*** • Women who breastfeed experience prolonged postpartum amenorrhea. Absent menstrual periods, however, do not protect a woman from pregnancy. To prevent pregnancy, a couple must use some form of contraception—but not oral contraceptive agents. Standard oral contraceptives contain estrogen, which reduces milk volume and the protein content of breast milk.

• ***Breast Health*** • Some women fear that breastfeeding will cause their breasts to sag. The breasts do swell and become heavy and large immediately after the birth, but even when they are producing enough milk to nourish a thriving infant, they eventually shrink back to their prepregnant size. Given proper support, diet, and exercise, breasts often return to their former shape and size after weaning. Breasts change their shape as the body ages, but breastfeeding does not accelerate this process.

Whether the physical and hormonal events of lactation protect women from later breast cancer is an area of active research.[75] Some research suggests no association between breastfeeding and breast cancer, whereas other research suggests a protective effect.[76] The reduction in breast cancer risk is most apparent for premenopausal women who were young when they breastfed and who breastfed for a long time.[77]

To learn about breastfeeding, a pregnant woman can read at least one of the many books available. Appendix F provides a list of other nutrition resources, including LaLeche League International.

U.S. Government

www.healthfinder.gov/searchoptions/topicsaz.htm
Search for Breastfeeding

IN SUMMARY

The lactating woman needs extra fluid and enough energy and nutrients to produce about 25 ounces of milk a day. Alcohol, other drugs, smoking, and contaminants may lessen milk production or enter breast milk and impair infant development.

This chapter has focused on the nutrition needs of the mother during pregnancy and lactation. The next chapter explores the dietary needs of infants, children, and adolescents.

# Study Questions

These questions will help you review the chapter. You will find the answers in the discussions on the pages provided.

1. Describe the placenta and its function. (pp. 469–470)
2. Describe the normal events of fetal development. How does malnutrition impair fetal development? (pp. 470–471, 482–483)
3. Define the term *critical period.* How do adverse influences during critical periods affect later health? (pp. 471–472)
4. Explain why women of childbearing age need folate in their diets. How much is recommended, and how can women ensure that these needs are met? (pp. 471–472)
5. What is the recommended pattern of weight gain during pregnancy for a woman at a healthy weight? For an underweight woman? For an overweight woman? (pp. 473–474)
6. What does a pregnant woman need to know about exercise? (pp. 475–476)
7. Which nutrients are needed in the greatest amounts during pregnancy? Why are they so important? Describe wise food choices for the pregnant woman. (pp. 476–480)
8. Define low-risk and high-risk pregnancies. What is the significance of infant birthweight in terms of the child's future health? (pp. 482–484)
9. Describe some of the special problems of the pregnant adolescent. Which nutrients are needed in increased amounts? (p. 486)
10. What practices should be avoided during pregnancy? Why? (pp. 487–488)
11. How do nutrient needs during lactation differ from nutrient needs during pregnancy? (pp. 477, 490–491)

These questions will help you prepare for an exam. Answers can be found in Appendix K.

1. The spongy structure that delivers nutrients to the fetus and returns waste products to the mother is called the:
   a. embryo.
   b. uterus.
   c. placenta.
   d. amniotic sac.
2. Which of these strategies is *not* a healthy option for an overweight woman?
   a. Limit weight gain during pregnancy.
   b. Postpone weight loss until after pregnancy.
   c. Follow a weight-loss diet during pregnancy.
   d. Try to achieve a healthy weight before becoming pregnant.
3. A reasonable weight gain during pregnancy for a normal-weight woman is about:
   a. 10 pounds.
   b. 20 pounds.
   c. 30 pounds.
   d. 40 pounds.
4. Energy needs during pregnancy increase by about:
   a. 100 kcalories/day.
   b. 300 kcalories/day.
   c. 500 kcalories/day.
   d. 700 kcalories/day.
5. To help prevent neural tube defects, grain products are now fortified with:
   a. iron.
   b. folate.
   c. protein.
   d. vitamin C.
6. Pregnant women should *not* take supplements of:
   a. iron.
   b. folate.
   c. vitamin A.
   d. vitamin C.
7. The combination of high blood pressure, protein in the urine, and edema signals:
   a. jaundice.
   b. preeclampsia.
   c. gestational diabetes.
   d. gestational hypertension.
8. To facilitate lactation, a mother needs:
   a. about 5000 kcalories a day.
   b. adequate nutrition and rest.
   c. vitamin and mineral supplements.
   d. a glass of wine or beer before each feeding.
9. A breastfeeding woman should drink plenty of water to:
   a. produce more milk.
   b. suppress lactation.
   c. prevent dehydration.
   d. dilute nutrient concentration.
10. A woman may need iron supplements during lactation:
    a. to enhance the iron in her breast milk.
    b. to provide iron for the infant's growth.
    c. to replace the iron in her body's stores.
    d. to support the increase in her blood volume.

# Notes


1. W. W. Hay and coauthors, Workshop summary: Fetal growth: Its regulation and disorders, *Pediatrics* 99 (1997): 585–591.
2. Committee on Nutritional Status during Pregnancy and Lactation, Food and Nutrition Board, *Nutrition during Pregnancy* (Washington, D.C.: National Academy Press, 1990), pp. 412–419.
3. Committee on Genetics, American Academy of Pediatrics, Folic acid for the prevention of neural tube defects, *Pediatrics* 92 (1993): 493–494.
4. J. E. Brown and coauthors, Predictors of red cell folate level in women attempting pregnancy, *Journal of the American Medical Association* 277 (1997): 548–552.
5. R. L. Goldenberg and T. Tamura, Prepregnancy weight and pregnancy outcome, *Journal of the American Medical Association* 275 (1996): 1127–1128.
6. Goldenberg and Tamura, 1996.
7. S. Cnattingius and coauthors, Prepregnancy weight and the risk of adverse pregnancy outcomes, *New England Journal of Medicine* 338 (1998): 147–152; M. M. Werler and coauthors, Prepregnant weight in relation to risk of neural tube defects, *Journal of the American Medical Association* 275 (1996): 1089–1092.
8. R. A. Chez, Nutritional factors in pregnancy affecting fetal growth and subsequent infant development, in *Textbook of Pediatric Nutrition,* eds. R. M. Suskind and L. Lewinter-Suskind (New York: Raven Press, 1993), pp. 1–7.
9. Werler and coauthors, 1996; G. M. Shaw, E. M. Velie, and D. Schaffer, Risk of neural tube defect—Affected pregnancies among obese women, *Journal of the American Medical Association* 275 (1996): 1093–1096.
10. Committee on Nutritional Status during Pregnancy and Lactation, 1990, p. 10.
11. Committee on Nutritional Status during Pregnancy and Lactation, 1990, p. 229.
12. K. G. Dewey and M. A. McCrory, Effects of dieting and physical activity on pregnancy and lactation, *American Journal of Clinical Nutrition* (supplement) 59 (1994): 446S–453S.
13. B. Sternfeld and coauthors, Exercise during pregnancy and pregnancy outcome, *Medicine and Science in Sports and Exercise* 27 (1995): 634–640.
14. American College of Obstetricians and Gynecologists, *ACOG Technical Bulletin 189: Exercise during Pregnancy and the Postpartum Period* (Washington, D.C., 1994); American College of Obstetricians and Gynecologists, *Planning for Pregnancy, Birth, and Beyond* (Washington, D.C.: The American College of Obstetricians and Gynecologists, 1995), pp. 88–90.
15. Committee on Nutritional Status during Pregnancy and Lactation, 1990, p. 384.
16. Committee on Nutritional Status during Pregnancy and Lactation, 1990, p. 163.
17. Committee on Dietary Reference Intakes, *Dietary Reference Intakes for Thiamin, Riboflavin, Niacin, Vitamin $B_6$, Folate, Vitamin $B_{12}$, Pantothenic Acid, Biotin, and Choline* (Washington, D.C.: National Academy Press, 1998), p. 8–24.
18. Committee on Nutritional Status during Pregnancy and Lactation, 1990, pp. 272–298.
19. Committee on Nutritional Status during Pregnancy and Lactation, 1990, pp. 285–293.
20. Committee on Nutritional Status during Pregnancy and Lactation, 1990, pp. 20, 289.
21. Committee on Nutritional Status during Pregnancy and Lactation, 1990, pp. 299–317.
22. Committee on Dietary Reference Intakes, *Dietary Reference Intakes for Calcium, Phosphorus, Magnesium, Vitamin D, and Fluoride* (Washington, D.C.: National Academy Press, 1997), p. 4–38.
23. Committee on Nutritional Status during Pregnancy and Lactation, 1990, p. 322.
24. T. O. Scholl and coauthors, Use of multivitamin/mineral prenatal supplements: Influence on the outcome of pregnancy, *American Journal of Epidemiology* 146 (1997): 134–141.
25. M. Erick, Battling morning (noon and night) sickness: New approaches for treating an age-old problem, *Journal of the American Dietetic Association* 94 (1994): 147–148.
26. M. Erick, Hyperolfaction and hyperemesis gravidarum: What is the relationship? *Nutrition Reviews* 53 (1995): 289–295.
27. A. J. Rainville, Pica practices of pregnant women are associated with lower maternal hemoglobin level at delivery, *Journal of the American Dietetic Association* 98 (1998): 293–296.
28. S. M. Yu and R. T. Jackson, Need for nutrition advice in prenatal care, *Journal of the American Dietetic Association* 95 (1995): 1027–1029.
29. Hay and coauthors, 1997.
30. D. L. Pelletier, The potentiating effects of malnutrition on child mortality: Epidemiologic evidence and policy implications, *Nutrition Reviews* 52 (1994): 409–415.
31. C. G. Solomon and coauthors, A prospective study of pregravid determinants of gestational diabetes, *Journal of the American Medical Association* 278 (1997): 1078–1083; C. D. Naylor and coauthors, Selective screening for gestational diabetes mellitus, *New England Journal of Medicine* 337 (1997): 1591–1596.
32. B. M. Sibai, Treatment of hypertension in pregnant women, *New England Journal of Medicine* 335 (1996): 257–265.
33. B. J. A. vanBuul and coauthors, Dietary sodium restriction in the prophylaxis of hypertensive disorders of pregnancy: Effects on the intake of other nutrients, *American Journal of Clinical Nutrition* 62 (1995): 49–57.
34. H. C. Bucher and coauthors, Effect of calcium supplementation on pregnancy-induced hypertension and preeclampsia: A meta-analysis of randomized controlled trials, *Journal of the American Medical Association* 275 (1996): 1113–1117; R. J. Levine and coauthors, Trial of calcium to prevent preeclampsia, *New England Journal of Medicine* 337 (1997): 69–76.
35. M. J. Lucas, K. J. Leveno, and G. Cunningham, A comparison of magnesium sulfate with phenytoin for the prevention of eclampsia, *New England Journal of Medicine* 333 (1995): 201–205.
36. D. E. Schendel and coauthors, Prenatal magnesium sulfate exposure and the risk for cerebral palsy or mental retardation among very low-birthweight children aged 3 to 5 years, *Journal of the American Medical Association* 276 (1996): 1805–1810.
37. U.S. Department of Health and Human Services, *Health United States 1992 and Healthy People 2000 Review,* August 1993, p. 17.

38. M. Story and I. Alton, Nutrition issues and adolescent pregnancy, *Nutrition Today* 30 (1995): 142–151.
39. A. M. Fraser, J. E. Brockert, and R. H. Ward, Association of young maternal age with adverse reproductive outcomes, *New England Journal of Medicine* 332 (1995): 1113–1117.
40. Fraser, Brockert, and Ward, 1995; J. M. Rees, S. A. Lederman, J. L. Kiely, Birthweight associated with lowest neonatal mortality: Infants of adolescent and adult mothers, *Pediatrics* 98 (1996): 1161–1166.
41. J. B. Hardy and coauthors, Adolescent childbearing revisited: The age of inner-city mothers at delivery is a determinant of their children's self-sufficiency at age 27 to 33, *Pediatrics* 100 (1997): 802–809.
42. Position of The American Dietetic Association: Nutrition care for pregnant adolescents, *Journal of the American Dietetic Association* 94 (1994): 449–450.
43. R. C. Fretts and coauthors, Increased maternal age and the risk of fetal death, *New England Journal of Medicine* 333 (1995): 953–957; F. G. Cunningham and K. J. Leveno, Childbearing among older women—The message is cautiously optimistic, *New England Journal of Medicine* 333 (1995): 1002–1004.
44. V. Delaney-Black and coauthors, Prenatal cocaine and neonatal outcome: Evaluation of dose-response relationship, *Pediatrics* 98 (1996): 735–740; F. A. Scafidi and coauthors, Cocaine-exposed preterm neonates show behavioral and hormonal differences, *Pediatrics* 97 (1996): 851–855; E. Z. Tronick and coauthors, Late dose-response effects of prenatal cocaine exposure on newborn neurobehavioral performance, *Pediatrics* 98 (1996): 76–83.
45. E. M. Ostrea and coauthors, Mortality within the first 2 years in infants exposed to cocaine, opiate, or cannabinoid during gestation, *Pediatrics* 100 (1997): 79–83; W. T. Weathers and coauthors, Cocaine use in women from a defined population: Prevalence at delivery and effects on growth in infants, *Pediatrics* 91 (1993): 350–354.
46. L. Fetters and E. Z. Tronick, Neuromotor development of cocaine-exposed and control infants from birth through 15 months: Poor and poorer outcomes, *Pediatrics* 98 (1996): 938–943; C. A. Chiriboga and coauthors, Neurological correlates of fetal cocaine exposure: Transient hypertonia of infancy and early childhood, *Pediatrics* 96 (1995): 1070–1077; S. D. Azuma and I. J. Chasnoff, Outcome of children prenatally exposed to cocaine and other drugs: A path analysis of three-year data, *Pediatrics* 92 (1993): 396–402.
47. T. A. King and coauthors, Neurologic manifestations of in utero cocaine exposure in near-term and term infants, *Pediatrics* 96 (1995): 259–264; L. C. Mayes and coauthors, Neurobehavioral profiles of neonates exposed to cocaine prenatally, *Pediatrics* 91 (1993): 778–783.
48. National Institute on Drug Abuse, *National Pregnancy and Health Survey—Drug Use among Women Delivering Livebirths* (Washington, D.C.: National Institutes of Health, 1996), pp. xxi–xxiii.
49. Medical-care expenditures attributable to cigarette smoking during pregnancy—United States, 1995, *Morbidity and Mortality Weekly Report* 46 (1997): 1048–1050.
50. E. Cutz and coauthors, Maternal smoking and pulmonary neuroendocrine cells in sudden infant death syndrome, *Pediatrics* 98 (1996): 668–672; H. S. Klonoff-Cohen and coauthors, The effect of passive smoking and tobacco exposure through breast milk on sudden infant death syndrome, *Journal of the American Medical Association* 237 (1995): 795–798; E. A. Mitchell and coauthors, Smoking and the sudden infant death syndrome, *Pediatrics* 91 (1993): 893–896.
51. C. D. Drews and coauthors, The relationship between idiopathic mental retardation and maternal smoking during pregnancy, *Pediatrics* 97 (1996): 547–553; D. L. Olds, C. R. Henderson, Jr., and R. Tatelbaum, Intellectual impairment in children of women who smoke cigarettes during pregnancy, *Pediatrics* 93 (1994): 221–227; D. M. Fergusson, L. J. Horwood, and M. T. Lynskey, Maternal smoking before and after pregnancy: Effects on behavioral outcome sin middle childhood, *Pediatrics* 92 (1993): 815–822.
52. K. J. Rothman and coauthors, Teratogenicity of high vitamin A intake, *New England Journal of Medicine* 333 (1995): 1369–1373.
53. T. S. Hinds and coauthors, The effect of caffeine on pregnancy outcome variables, *Nutrition Reviews* 54 (1996): 203–207; Committee on Nutritional Status during Pregnancy and Lactation, 1990, pp. 397–399.
54. Position of The American Dietetic Association: Use of nutritive and non-nutritive sweetners, *Journal of the American Dietetic Association* 93 (1993): 816–821.
55. Work Group on Breastfeeding, American Academy of Pediatrics, Breastfeeding and the use of human milk, *Pediatrics* 100 (1997): 1035–1039; M. J. Heinig and K. G. Dewey, Health effects of breast feeding for mothers: A critical review, *Nutrition Research Reviews* 10 (1997): 35–56; M. J. Heinig and K. G. Dewey, Health advantages of breast feeding for infants: A critical review, *Nutrition Research Reviews* 9 (1996): 89–110.
56. Committee on Nutritional Status during Pregnancy and Lactation, Food and Nutrition Board, *Nutrition during Lactation* (Washington, D.C.: National Academy Press, 1991), p. 28.
57. E. C. Kieffer and coauthors, Health practitioners should consider parity when counseling mothers on decisions about infant feeding methods, *Journal of the American Dietetic Association* 97 (1997): 1313–1316; A. Wright, S. Rice, and S. Wells, Changing hospital practices to increase the duration of breastfeeding, *Pediatrics* 97 (1996): 669–675.
58. M. Sharma and R. Petosa, Impact of expectant fathers in breast-feeding decisions, *Journal of the American Dietetic Association* 97 (1997): 1311–1312; J. P. Sciacca and coauthors, Influences on breast-feeding by lower-income women: An incentive-based, partner-supported educational program, *Journal of the American Dietetic Association* 95 (1995): 323–328.
59. J. A. Hilson, K. M. Rasmussen, and C. L. Kjolhede, Maternal obesity and breastfeeding success in a rural population of white women, *American Journal of Clinical Nutrition* 66 (1997): 1371–1378.
60. K. G. Dewey, Energy and protein requirements during lactation, *Annual Reviews of Nutrition* 17 (1997): 19–36.
61. L. Doran and S. Evers, Energy and nutrient inadequacies in the diets of low-income women who breast-feed, *Journal of the American Dietetic Association* 97 (1997): 1283–1287.
62. Committee on Nutritional Status dur-

ing Pregnancy and Lactation, 1991, p. 232.

63. Committee on Nutritional Status during Pregnancy and Lactation, 1991, pp. 1–19.
64. L. B. Dusdieker, D. L. Hemingway, and P. J. Stumbo, Is milk production impaired by dieting during lactation? *American Journal of Clinical Nutrition* 59 (1994): 833–840.
65. Dewey and McCrory, 1994; K. G. Dewey and coauthors, A randomized study of the effects of aerobic exercise by lactating women on breast-milk volume and composition, *New England Journal of Medicine* 330 (1994): 449–453.
66. Committee on Nutritional Status during Pregnancy and Lactation, 1991, p. 140.
67. M. A. Laskey and coauthors, Bone changes after 3 mo of lactation: Influence of calcium intake, breast-milk output, and vitamin D–receptor genotype, *American Journal of Clinical Nutrition* 67 (1998): 685–692; L. D. Ritchie and coauthors, A longitudinal study of calcium homeostasis during human pregnancy and lactation and after resumption of menses, *American Journal of Clinical Nutrition* 67 (1998): 693–701; H. J. Kalkwarf and coauthors, The effect of calcium supplementation on bone density during lactation and after weaning, *New England Journal of Medicine* 337 (1997): 523–528.
68. Committee on Nutritional Status during Pregnancy and Lactation, 1991, p. 140.
69. Committee on Drugs, American Academy of Pediatrics, The transfer of drugs and other chemicals into human milk, *Pediatrics* 93 (1994): 137–150.
70. Committee on Environmental Health, American Academy of Pediatrics, PCBs in breast milk, *Pediatrics* 94 (1994): 122–123.
71. E. Dewailly and coauthors, Inuit exposure to organochlorines through the aquatic food chain in Arctic Québec, *Environmental Health Perspectives* 101 (1993): 618–620.
72. R. F. Black, Transmission of HIV-1 in the breast-feeding process, *Journal of the American Dietetic Association* 96 (1996): 267–274.
73. A. M. Ferris and E. A. Reece, Nutritional consequences of chronic maternal conditions during pregnancy and lactation: Lupus and diabetes, *American Journal of Clinical Nutrition* (supplement) (1994): 465S–473S; A. M. Ferris and coauthors, Perinatal lactation protocol and outcome in mothers with and without insulin-dependent diabetes mellitus, *American Journal of Clinical Nutrition* 58 (1993): 43–48.
74. C. M. van Beusekom and coauthors, Milk of patients with tightly controlled insulin-dependent diabetes mellitus has normal macronutrient and fatty acid composition, *American Journal of Clinical Nutrition* 57 (1993): 938–943.
75. United Kingdom National Case-Control Study Group, Breast feeding and risk of breast cancer in young women, *British Medical Journal* 307 (1993): 17–20.
76. K. B. Michels and coauthors, Prospective assessment of breastfeeding and breast cancer incidence among 89,887 women, *Lancet* 347 (1996): 431–436; United Kingdom National Case-Control Study Group, 1993.
77. P. A. Newcomb and coauthors, Lactation and a reduced risk of premenopausal breast cancer, *New England Journal of Medicine* 330 (1994): 81–87.

HIGHLIGHT 15

# Fetal Alcohol Syndrome

As Chapter 15 mentioned, drinking alcohol during pregnancy endangers the fetus. Alcohol crosses the placenta freely and deprives the developing fetal brain of both nutrients and oxygen. The result may be fetal alcohol syndrome (FAS), a cluster of symptoms that includes:[1]

- Prenatal and postnatal growth retardation.
- Impairment of the brain and nerves, with consequent mental retardation, poor coordination, and hyperactivity.
- Abnormalities of the face and skull (see Figure H15-1 on p. 498).
- Increased frequency of major birth defects: cleft palate, heart defects, and defects in ears, eyes, genitals, and urinary system.

Tragically, the damage evident at birth persists: children with FAS never fully recover.

Of every 10,000 children born in the United States, almost 7 suffer health problems because their mothers drank alcohol during pregnancy—a sixfold increase over the past 15 years.[2] In addition, many infants are born with the less serious, yet still significant, damage some clinicians describe as fetal alcohol effects (FAE; see the glossary on p. 498).[3] Some children with FAE have no outward signs; others may be short or have only minor facial abnormalities. Often children with FAE go undiagnosed even when problems develop in the early school years. Learning disabilities, behavioral abnormalities, and motor impairments are common symptoms of FAE.

The surgeon general states that pregnant women should drink absolutely no alcohol. Abstinence from alcohol is the best policy for pregnant women both because alcohol consumption during pregnancy has such severe consequences and because FAS can only be prevented—it cannot be treated. Further, because the most severe damage occurs around the time of conception—*before a woman may even realize that she is pregnant*—even a woman planning to conceive should abstain.

## Drinking during Pregnancy

When a woman drinks during pregnancy, she causes damage in two ways: directly, by intoxication, and indirectly, by malnutrition. Prior to the complete formation of the placenta (approximately 12 weeks), alcohol diffuses directly into the tissues of the developing embryo, causing incredible damage. When alcohol crosses the placenta, fetal blood alcohol rises until it reaches an equilibrium with maternal blood alcohol. The mother may not even appear drunk, but the fetus may be poisoned. The fetus's body is small, its detoxification system is immature, and alcohol remains in fetal blood long after it has disappeared from maternal blood. Alcohol interferes with many developmental events during their critical periods, reducing the number of cells produced and damaging those that are produced.

An alcoholic pregnant woman harms her unborn child not only by consuming alcohol but also by not consuming food. This combination enhances the likelihood of malnutrition and a poorly developed infant. It is important to realize, however, that malnutrition is not the cause of FAS. It is true that mothers of FAS children often have unbalanced diets and nutrient deficiencies. It is also true that malnutrition may augment the clinical signs seen in these children, but it is the *alcohol* that is the determining factor. An adequate diet alone will not prevent FAS if alcohol is abused during the pregnancy.

## How Much Alcohol Is Too Much?

A pregnant woman need not have an alcohol-abuse problem to give birth to a baby with FAS. She need only drink in excess of her liver's capacity to detoxify alcohol. About four drinks a day dramatically worsens the risk of having an infant with physical malformations. Even one to two drinks a day threatens to retard growth.

In addition to total alcohol intake, drinking patterns play an important role. Most FAS studies report their findings in terms of average intake per day, but people usually drink more heavily on some days than on others. For example, a woman who drinks an *average* of 1 ounce of alcohol (2 drinks) a day may not drink at all during the week but then have 10 drinks on Saturday night, exposing the fetus to highly toxic quantities of alcohol. Whether drinking a certain number of drinks during binges or spreading them out over several days causes more damage depends on the frequency of the binges, the quantity consumed, and the stage of fetal development at the time of each drinking episode.

An occasional drink may be innocuous, but researchers are unable to say how much alcohol is safe to consume during pregnancy. For this reason, health care professionals urge

*The most obvious symptoms of FAS are the abnormal facial features, but the most tragic ones are the mental disabilities.*

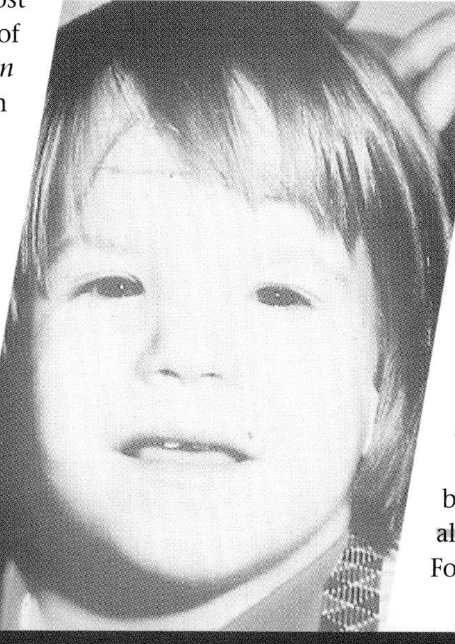

**Figure H15-1**
**Typical Facial Characteristics of FAS**

The severe facial abnormalities shown here are just outward signs of the severe mental impairments within. The internal organs also suffer irreversible damage that, while hidden, may create major problems for a child's health.

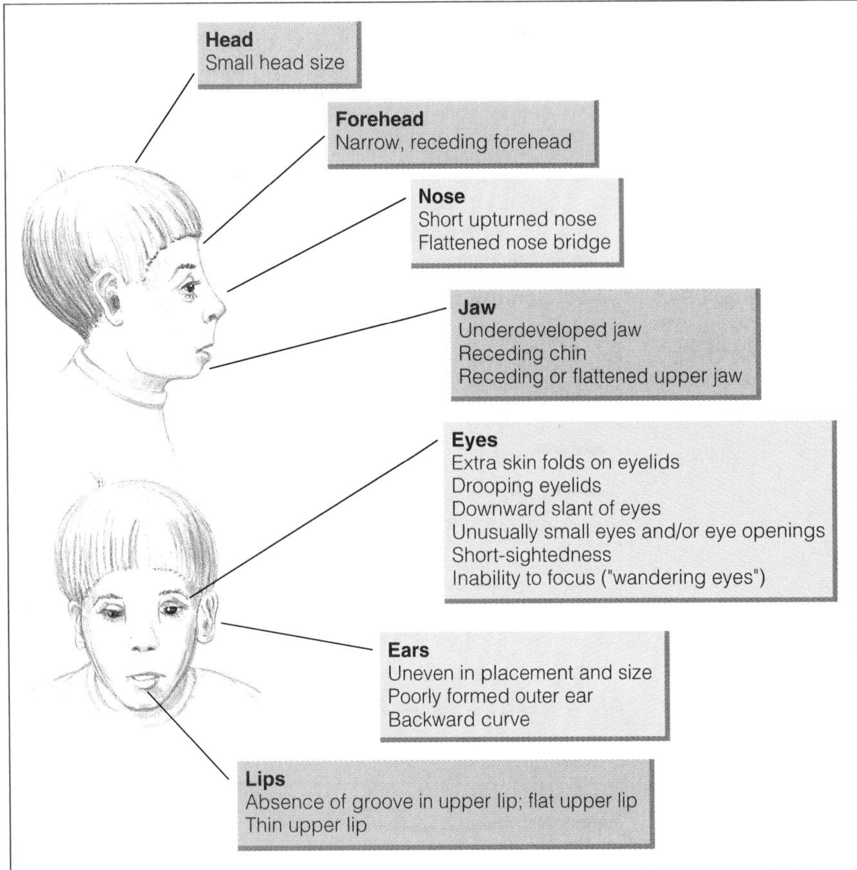


Source: Adapted from J. O. Beattie, Alcohol exposure and the fetus, *European Journal of Clinical Nutrition* 46 (1992): S7–S17.

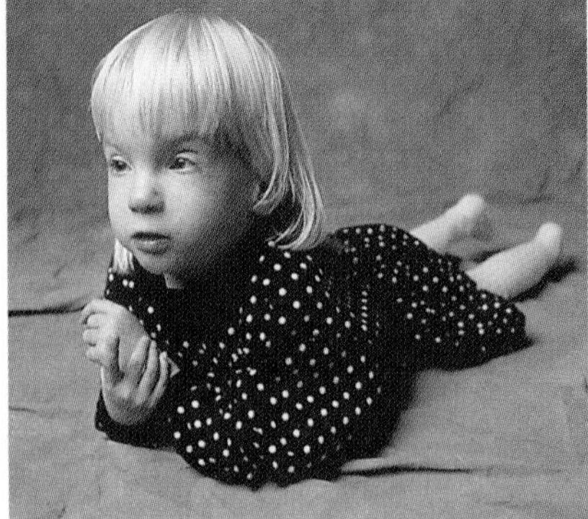

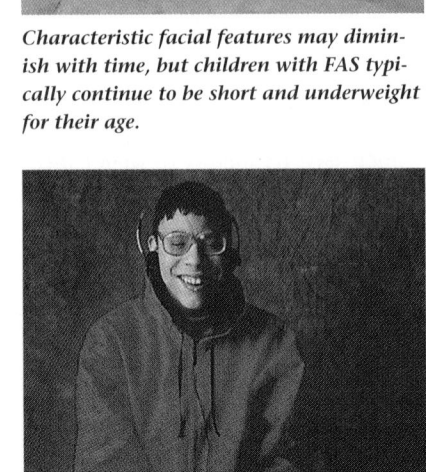

***Characteristic facial features may diminish with time, but children with FAS typically continue to be short and underweight for their age.***

***Children born with FAS must live with the long-term consequences of prenatal brain damage.***

women to stop drinking alcohol as soon as they realize they are pregnant, or better, as soon as they *plan* to become pregnant.[4] Why take any risk? Only the woman who abstains is sure of protecting her infant from FAS.

## *When Is the Damage Done?*

The first month of pregnancy is a critical period of fetal development. Because pregnancy confirmation usually requires five to six weeks, a woman may not even realize she is pregnant during that critical first month. Therefore, it is advisable for women who are trying to conceive, or who suspect they might be pregnant, to curtail their alcohol intakes to ensure a healthy start.

The type of abnormality observed in an FAS infant depends on the developmental events occurring at the times of alcohol exposure. During the first trimester, developing organs such as the brain, heart, and kidneys may be malformed. During the second trimester, the risk of spontaneous abortion increases. During the third trimester, body and brain growth may be retarded.

Male alcohol ingestion may also affect fertility and fetal development. Animal studies have found smaller litter sizes, lower birthweights, reduced survival rates, and impaired learning ability in the offspring of males consuming alcohol prior to conception.

## Glossary

**fetal alcohol effects (FAE):** a subclinical version of fetal alcohol syndrome, with hidden defects including learning disabilities, behavioral abnormalities, and motor impairments; also called **alcohol-related birth defects (ARBD).**

**fetal alcohol syndrome (FAS):** the cluster of symptoms seen in an infant or child whose mother consumed excess alcohol during pregnancy, including retarded growth, impaired development of the central nervous system, and facial malformations.

See Highlight 7 for other alcohol-related terms and information.

An association between paternal alcohol intake one month prior to conception and low infant birthweight is also apparent in human beings. (Paternal alcohol intake was defined as an average of two or more drinks daily or at least five drinks on one occasion.) This relationship was independent of either parent's smoking and of the mother's use of alcohol, caffeine, or other drugs.

In view of these findings, it is important to advise women not to drink during pregnancy. Everyone should know of the potential dangers. Heavy drinkers who are sexually active urgently need effective contraception to prevent pregnancy.

All containers of beer, wine, and liquor carry the warning: "Women should not drink alcoholic beverages during pregnancy because of the risk of birth defects." Everyone should hear the message loud and clear: Don't drink alcohol prior to conception or during pregnancy.

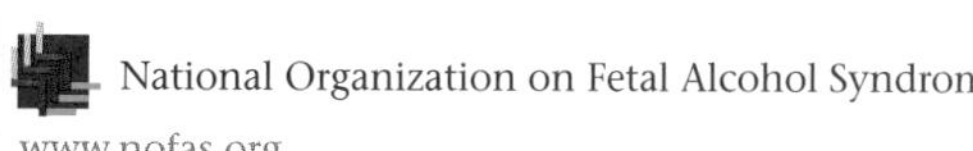
National Organization on Fetal Alcohol Syndrome
www.nofas.org

U.S. Government
www.healthfinder.gov/searchoptions/topicsaz.htm
Search for Fetal alcohol syndrome

## Notes

1. Committee on Substance Abuse and Committee on Children with Disabilities, American Academy of Pediatrics, Fetal alcohol syndrome and fetal alcohol effects, *Pediatrics* 91 (1993): 1004–1006.
2. Update: Trends in fetal alcohol syndrome—United States, 1979–1993, *Morbidity and Mortality Weekly Report* 44 (1995): 249–251.
3. J. M. Aase, K. L. Jones, and S. K. Clarren, Do we need the term "FAE"? *Pediatrics* 95 (1995): 428–430.
4. Committee on Substance Abuse and Committee on Children with Disabilities, 1993; *The Surgeon General's Report on Nutrition and Health* (Washington, D.C.: Government Printing Office, 1988), p. 72.

CHAPTER 16

# Life Cycle Nutrition: Infancy, Childhood, and Adolescence

The first year of life is a time of phenomenal growth and development. After the first year, a child continues to grow and change, but more slowly. Still, the cumulative effects over the next decade are remarkable. Then, as the child enters the teen years, the pace toward adulthood accelerates dramatically. This chapter examines the special nutrient needs of infants, children, and teenagers.

*After six months, energy saved by slower growth is spent in increased activity.*

# Nutrition during Infancy

Initially, the infant drinks only breast milk or formula, but later begins to eat some foods, as appropriate. Common sense in the selection of infant foods and a nurturing, relaxed environment go far to promote an infant's health and well-being.

## Energy and Nutrient Needs

An infant grows fast during the first year, as Figure 16-1 shows. Growth directly reflects nutrient intake and is an important parameter in assessing the nutrition status of infants and children. Health care professionals measure the heights and weights of infants and children at intervals and compare measures both with standard growth curves for sex and age and with previous measures of each child (see the "How to" section on the next page).

**Figure 16-1**

**Weight Gain of Human Infants in Their First Five Years of Life**

In the first year, an infant's birthweight may triple, but over the following several years, the rate of weight gain gradually diminishes.

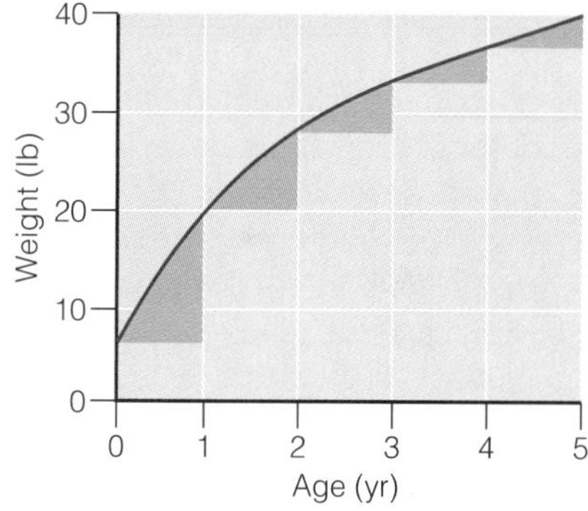


• ***Energy Intake and Activity*** • A healthy infant's birthweight doubles by about five months of age and triples by one year, typically reaching 20 to 25 pounds. (If an adult starting at 150 pounds were to do this, the person's weight would increase to 450 pounds in a single year.) By the end of the first year, infant growth slows considerably; an infant gains less than 10 pounds during the second year.

Not only do infants grow rapidly, but their basal metabolic rate is remarkably high—about twice that of an adult, based on body weight. Infants require about 100 kcalories per kilogram of body weight per day, whereas most adults need fewer than 40. (If the energy needs of the infant were superimposed on the adult, a 170-pound adult would require over 7000 kcalories a day.) After six months, metabolic needs decline as the growth rate slows down, but some of the energy saved by slower growth is spent in increased activity.

A newborn baby requires about 650 kcal per day, whereas most adults require about 2000 kcal per day. In comparison to body weight, the difference is remarkable.

• ***Protein*** • No single nutrient is more essential to growth than protein. All of the body's cells and most of its fluids contain protein; it is the basic building material of the body's tissues.

The protein RDA is based on the amount of breast milk that supports adequate growth in healthy, full-term infants. Breast milk in the quantity normally consumed provides approximately 2 grams of protein per kilogram of body weight per day for the average 7½-pound infant. Therefore, the American Academy of Pediatrics (AAP) recommends that infant formulas provide a comparable amount, with a protein efficiency ratio equal to or greater than that of casein, the chief protein in cow's milk.

Reminder: *Protein efficiency ratio (PER)* is a measure of protein quality (see Chapter 6, p. 176).

Excess dietary protein can cause problems, especially in the small infant, when amino acids build up in the blood. The ingestion of excess protein stresses the kidneys and liver, which have to metabolize and excrete the excess nitrogen. Signs of protein overload include acidosis, dehydration, diarrhea, elevated blood ammonia, elevated blood urea, and fever. Such problems are not common but have been observed in infants fed inappropriate foods, such as nonfat milk or concentrated formula.

## HOW TO Plot Measures on a Growth Chart

You can assess the growth of infants and children by plotting their measurements on a percentile graph. Percentile graphs divide the measures of a population into 100 equal divisions so that half of the population falls at or above the 50th percentile, and half falls below. Using percentiles allows for comparisons among people of the same age and sex.

To plot measures on a growth chart, follow these steps:

- Select the appropriate chart based on age and sex. For this example, use the lower section of the accompanying chart, which gives percentiles for weight for girls from birth to 36 months. (Appendix E provides other growth charts for both boys and girls of various ages.)
- Locate the infant's age along the horizontal axis at the bottom or top of the chart (in this example, 6 months).
- Locate the infant's weight in pounds or kilograms along the vertical axis at the lower left or right side of the chart (in this example, 17½ pounds or 8 kilograms).
- Mark the chart where the age and weight lines intersect (shown here with a red dot), and read off the percentile.

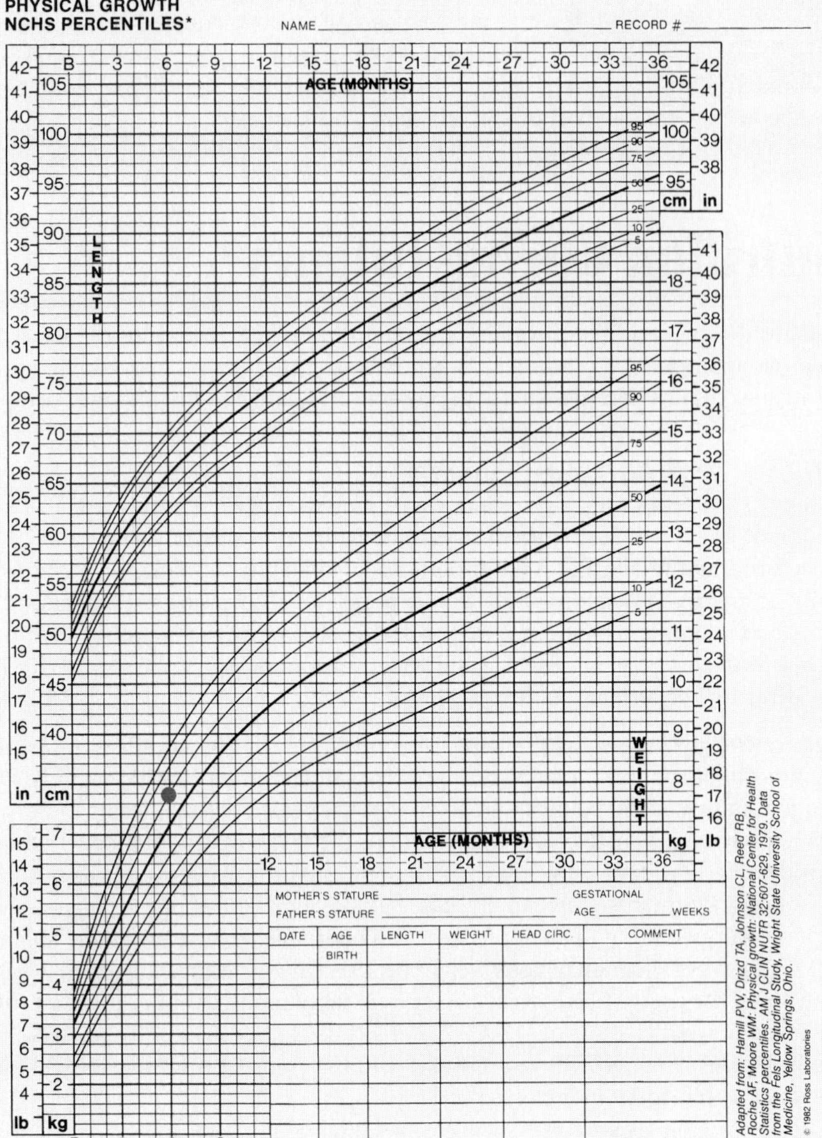


This six-month-old infant is at the 75th percentile, which means that she weighs more than 75 percent of the female infants her age. Her pediatrician will weigh her again over the next few months and expect the growth curve to follow the same percentile throughout the first year. In general, dramatic changes or measures much above the 75th percentile or much below the 25th percentile may be cause for concern.

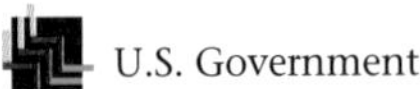
U.S. Government

www.healthfinder.gov/searchoptions/topicsaz.htm

Search for Infants

• ***Vitamins and Minerals*** • Vitamin and mineral recommendations are based on the average amount of nutrients consumed by thriving infants breastfed by well-nourished mothers. An infant's needs for most of these nutrients, in proportion to body weight, are more than double those of an adult, as Figure 16-2 illustrates by comparing a five-month-old infant's needs per unit of body weight with those of an adult man. Some of the differences are extraordinary.

**Figure 16-2**

**Recommended Intakes of an Infant and an Adult Compared on the Basis of Body Weight**

Because infants are small, they need smaller total amounts of the nutrients than adults do, but when comparisons are based on body weight, infants need over twice as much of many nutrients. Infants use large amounts of energy and nutrients, in proportion to their body size, to keep all their metabolic processes going.

| | Infants | Adults |
|---|---|---|
| Heart rate (beats/minute) | 120 to 140 | 70 to 80 |
| Respiration rate (breaths/minute) | 20 to 40 | 15 to 20 |
| Energy needs (kcal/body weight) | 45/lb (100/kg) | <18/lb (<40/kg) |

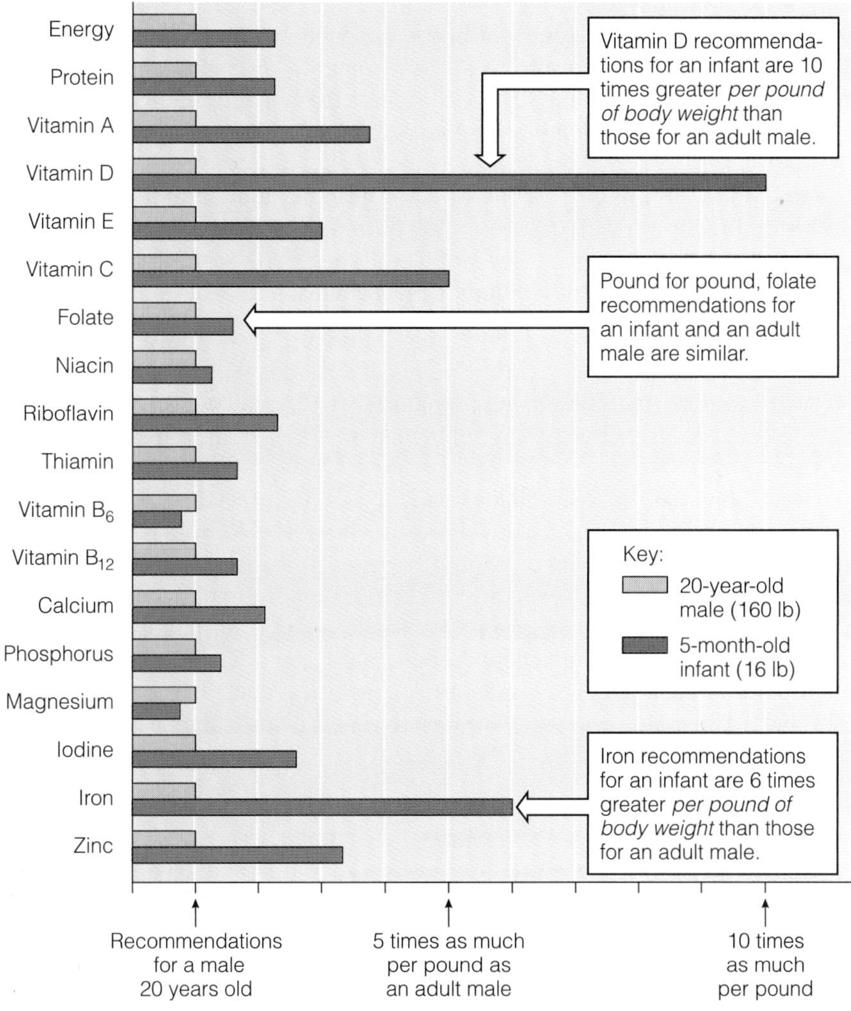


• ***Water*** • One of the most essential nutrients for infants, as for everyone, is water. The younger the infant, the greater the percentage of body weight is water. The water in an infant's body is easily lost because much of it is located *outside* the cells—between the cells and in the vascular space. During early infancy, breast milk or infant formula normally provides enough water to replace fluid losses in a healthy infant. Conditions that cause fluid loss, such as hot weather, diarrhea, or vomiting, may require supplemental water to prevent life-threatening dehydration.[1] Adults must remember that infants may cry from thirst as well as hunger. When water is needed, allow infants to drink it until their thirst is quenched.

## Breast Milk

In the United States and Canada, the two dietary practices that have the most effect on an infant's nutrition status are the milk the infant receives and the age at which solid foods are introduced. A later section discusses the introduction of solid food, but as to the milk, both the AAP and the American Dietetic Association recommend breast milk because of its many benefits to both infant and mother.[2] The AAP and the Canadian Pediatric society have issued this joint statement: "Breastfeeding is strongly recommended for full-term infants, except where specific contraindications exist."

Breast milk excels as a source of nutrients for the young infant.[3] Its unique nutrient composition and protective factors promote optimal infant health and development throughout the first year of life.[4] Experts add, though, that iron-fortified formula, which imitates the composition of breast milk, is an acceptable alternative.

Even two to three months of breastfeeding give the infant immunological protection during the most critical period after birth—protection that persists beyond the breastfeeding period itself.[5] The mother can then shift to formula, if necessary, knowing that she has given her infant those benefits.

• ***Energy Nutrients*** • The energy-nutrient composition of breast milk differs dramatically from the energy contributions of a healthy diet for adults (see Figure 16-3). Yet for infants, breast milk is nature's most nearly perfect food, providing the clear lesson that people at different stages of life have different nutrient needs.

The carbohydrate in breast milk and infant formula is the disaccharide lactose. In addition to being easily digested, lactose enhances calcium absorption. The lipids in breast milk—and infant formula—provide the main source of energy in the infant's diet. Breast milk contains a generous proportion of the essential fatty acid linoleic acid. The total protein concentration of breast milk is less than in cow's milk, but this deficit is actually beneficial because it places less stress on the infant's immature kidneys to excrete the major end product of protein metabolism, urea. The main protein in breast milk is alpha-lactalbumin, which is efficiently digested and absorbed.

**alpha-lactalbumin** (lact-AL-byoo-min): the chief protein in human breast milk, as opposed to **casein** (CAY-seen), the chief protein in cow's milk.

• ***Vitamins*** • With the possible exception of vitamin D, the vitamins in breast milk are ample to support infant growth. The vitamin D in breast milk is low, and vitamin D deficiency impairs bone mineralization in infants and children.[6] Manufacturers fortify infant formulas with vitamin D, and physicians may recommend

**Figure 16-3**

**Percentages of Energy-Yielding Nutrients in Breast Milk and in Recommended Adult Diets**

The proportions of energy-yielding nutrients in human breast milk differ from those recommended for adults.

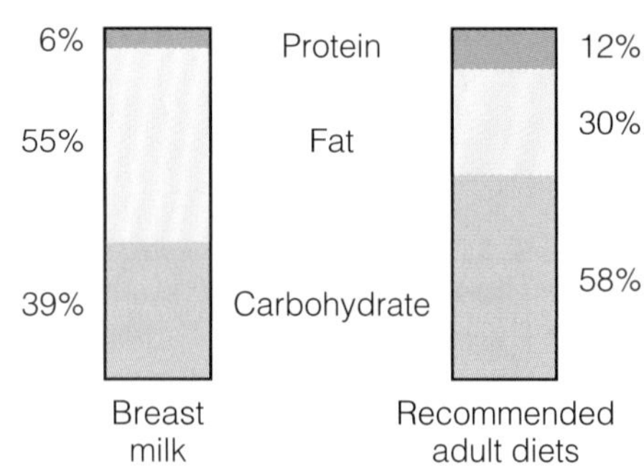

vitamin D supplements for breastfed infants who do not receive sufficient exposure to sunlight.

Infants who are exposed to sunlight regularly can make enough vitamin D to meet their needs. The amount formed depends on skin color, exposure time, atmospheric pollution, time of year, and latitude. Vitamin D deficiency is most likely in infants who are not exposed to sunlight daily, who receive breast milk without supplementation, and who have darkly pigmented skin.

• ***Minerals*** • The calcium-to-phosphorus ratio of breast milk is ideal for calcium absorption. Breast milk contains relatively small amounts of iron, but this iron has a high bioavailability. Zinc is also highly absorbable, thanks to the presence of a zinc-binding protein. Breast milk is low in sodium, another benefit for immature kidneys.

Fluoride is not an essential nutrient, but it does help to prevent dental caries. Breast milk provides little fluoride, regardless of the mother's intake.

• ***Nutrient Supplements*** • Pediatricians may routinely prescribe supplements containing vitamin D, iron, and fluoride. Table 16-1 offers a schedule of supplements during infancy. In addition, the AAP recommends that a single dose of vitamin K be given to infants at birth to protect them from bleeding to death (see Chapter 11 for a description of vitamin K's role in blood clotting).[7]

• ***Immunological Protection*** • In addition to nutritional benefits, breast milk also offers immunological protection. Not only is breast milk sterile, but it actively fights disease in a number of ways. Such protection is most valuable during the first year, when the infant's immune system is not fully prepared to mount a response against infection.

During the first two or three days after delivery, the breasts produce colostrum, a premilk substance containing mostly serum with antibodies and white blood cells. Colostrum (like breast milk) helps protect the newborn from infections against which the mother has developed immunity. The maternal antibodies swallowed with the milk inactivate disease-causing bacteria within the digestive tract before they can start infections. This explains, in part, why breastfed infants have fewer intestinal infections than formula-fed infants.

In addition to antibodies, colostrum and breast milk provide other powerful agents that help to fight against bacterial infection.[8] Among them are bifidus factors, which favor the growth of the "friendly" bacterium *Lactobacillus bifidus* in the infant's digestive tract, so that other, harmful bacteria cannot gain a foothold

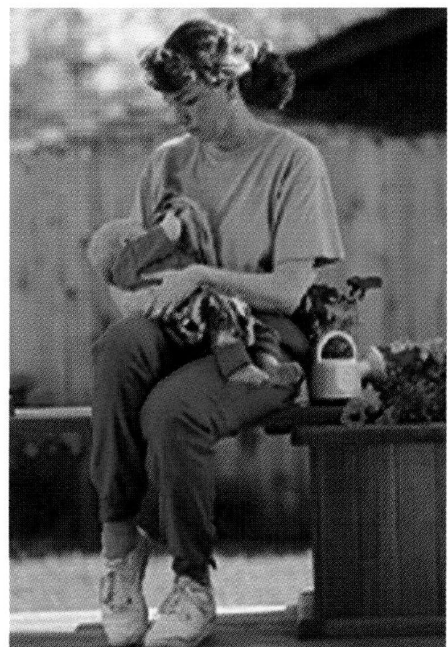

***Women are encouraged to breastfeed whenever possible because breast milk offers infants many nutrient and health advantages.***

**colostrum** (ko-LAHS-trum): a milklike secretion from the breast, present during the first day or so after delivery before milk appears; rich in protective factors.

**bifidus** (BIFF-id-us, by-FEED-us) **factors:** factors in colostrum and breast milk that favor the growth of the "friendly" bacterium *Lactobacillus* (lack-toh-ba-SILL-us) *bifidus* in the infant's intestinal tract, so that other, less desirable intestinal inhabitants will not flourish.

**Table 16-1**
**Supplements for Full-Term Infants**

| | Vitamin D[a] | Iron[b] | Fluoride[c] |
|---|---|---|---|
| Breastfed infants: | | | |
| Birth to six months of age | ✔ | | |
| Six months to one year | ✔ | ✔ | ✔ |
| Formula-fed infants: | | | |
| Birth to six months of age | | | |
| Six months to one year | | ✔ | ✔ |

[a]Vitamin D supplements are recommended for infants whose mothers are vitamin D deficient or those who do not receive adequate exposure to sunlight.
[b]Infants four to six months of age need additional iron, preferably in the form of iron-fortified cereal for both breastfed and formula-fed infants and iron-fortified infant formula for formula-fed infants.
[c]At six months of age, breastfed infants and formula-fed infants who receive ready-to-use formulas (these are prepared with water low in fluoride) or formula mixed with water that contains little or no fluoride (less than 0.3 ppm) need supplements.
Sources: Adapted from Committee on Nutrition, American Academy of Pediatrics, Vitamin and mineral supplement needs of normal children in the United States, in *Pediatric Nutrition Handbook,* 3rd ed., ed. L. A. Barness (Elk Grove Village, Ill.: American Academy of Pediatrics, 1993), pp. 34–42; American Academy of Pediatrics, Committee on Nutrition, Fluoride supplementation for children: Interim policy recommendations, *Pediatrics* 95 (1995): 777.

**lactoferrin** (lack-toh-FERR-in): a protein in breast milk that binds iron and keeps it from supporting the growth of the infant's intestinal bacteria.

Protective factors in breast milk:
- Antibodies.
- Bifidus factors.
- Lactoferrin.
- Growth factor.
- Lipase enzyme.

there. An iron-binding protein in breast milk, lactoferrin, keeps bacteria from getting the iron they need to grow, helps absorb iron into the infant's bloodstream, and kills some bacteria directly. Also present is a growth factor that stimulates the development and maintenance of the infant's digestive tract and its protective factors. Several breast milk enzymes such as lipase also help protect the infant against infection. Clearly, breast milk is a very special substance.

## *Infant Formula*

Breastfeeding offers many benefits to both mother and infant, and every pregnant woman should seriously consider it. Still, there are valid reasons for not breastfeeding, and formula-fed infants grow and develop into healthy children. After all, the primary goal is to provide the infant optimal nourishment in a relaxed and loving environment.

**wean:** to gradually replace breast milk with infant formula or other foods appropriate to an infant's diet.

Formula preparation:
- Liquid concentrate (inexpensive, relatively easy)—mix with equal part water.
- Powdered formula (cheapest, lightest for travel)—read label directions.
- Ready-to-feed (easiest, most expensive)—pour directly into clean bottles.
- Whole milk—do not use during first year.

• ***Appropriate Uses of Formula*** • A woman who breastfeeds for a year can wean her infant to cow's milk, bypassing the need for infant formula. However, a woman who decides to feed her infant formula from birth, to wean to formula after a short time, or to substitute formula for breastfeeding on occasion must select an appropriate infant formula and learn to prepare it.

• ***Infant Formula Composition*** • Formula manufacturers attempt to copy the nutrient composition of breast milk as closely as possible. Figure 16-4 illustrates the energy-nutrient balance of both, and Table 16-2 compares breast milk, cow's milk, and a standard infant formula.

The AAP recommends that all formula-fed infants receive iron-fortified infant formulas. The increasing use of iron-fortified formulas during the past few decades is a major reason for the decline in iron-deficiency anemia among U.S. infants.

• ***Risks of Formula Feeding*** • Infant formulas contain no protective antibodies for infants, but in general, vaccinations, purified water, and clean environments in the developed countries help protect infants from infections. Formulas can be prepared safely by following the rules of proper food handling and using water that is free of contamination. Lead-contaminated water is a major source of lead poisoning in infants (see Highlight 13).

In developing countries and in poor areas of the United States, formula may be unavailable or may be prepared with contaminated water or overdiluted in an attempt to save money. More than 1.2 billion people in developing countries have no safe drinking water. Contaminated formulas often cause infections, leading to diarrhea, dehydration, and failure to absorb nutrients. Without sterilization and refrigeration, bottles of formula are an ideal breeding ground for bacteria. Whenever such risks are present, breastfeeding can be a life-saving option. Wherever sanitation is poor, breastfeeding is preferred: breast milk is sterile, and its antibodies enhance an infant's resistance to disease.

**Figure 16-4**
**Percentages of Energy-Yielding Nutrients in Breast Milk and in Infant Formula**

The average proportions of energy-yielding nutrients in human breast milk and formula differ slightly.

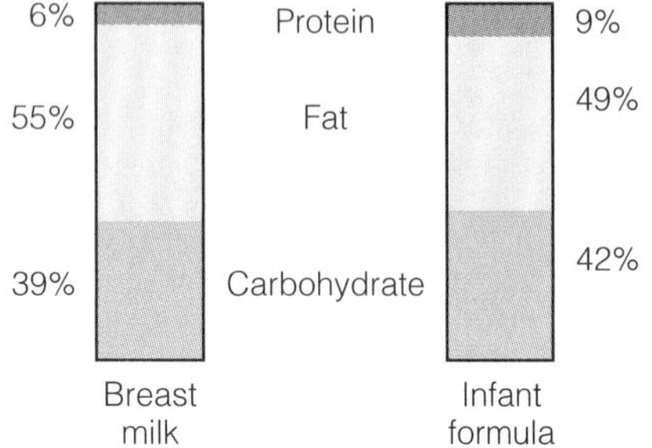


• ***Infant Formula Standards*** • National and international standards have been set for the nutrient contents of infant formulas. In the United States, the standard developed by the AAP reflects "human milk taken from well-nourished mothers during the first or second month of lactation, when the infant's growth rate is high." The Food and Drug Administration (FDA) mandates quality control procedures. Formulas meeting the standards have similar nutrient compositions; small differences are sometimes confusing, but usually unimportant.

• ***Special Formulas*** • Standard formulas are inappropriate for some infants. Special formulas have been designed to meet the dietary needs of infants with specific conditions such as prematurity or inherited diseases. Soy formulas use soy for the protein source and cornstarch and sucrose instead of lactose, for example, and so are recommended for infants with milk allergy or lactose intolerance. They are also useful as an alternative to milk-based formulas for vegetarian families. Despite

**Table 16-2**

**Comparison of Breast Milk, Cow's Milk, and Infant Formula**

| Nutrient (per 100 mL) | Breast Milk | Cow's Milk | Infant Formula[a] |
|---|---|---|---|
| ***Energy-Yielding Nutrients*** | | | |
| Energy (kcal) | 64 | 66 | 67 |
| Protein (g) | 0.9 | 3.4 | 1.5 |
| Fat (g) | 3.4 | 3.7 | 3.7 |
| Carbohydrate (g) | 6.6 | 4.9 | 7.1 |
| ***Minerals*** | | | |
| Sodium (mg) | 17 | 58 | 18 |
| Potassium (mg) | 55 | 138 | 71 |
| Chloride (mg) | 43 | 103 | 43 |
| Calcium (mg) | 26 | 125 | 51 |
| Phosphorus (mg) | 14 | 96 | 36 |
| Magnesium (mg) | 4 | 12 | 5 |
| Iron (mg) | 0.05 | 0.05 | 1.2[b] |
| Zinc (mg) | 0.2 | 0.4 | 0.5 |
| Copper (mg) | 0.04 | 0.01 | 0.06 |
| ***Vitamins*** | | | |
| Vitamin A (IU) | 190 | 103 | 206 |
| Thiamin (µg) | 16 | 44 | 60 |
| Riboflavin (µg) | 36 | 175 | 101 |
| Vitamin $B_6$ (µg) | 10 | 64 | 41 |
| Niacin (µg) | 159 | 93 | 777 |
| Pantothenic acid (µg) | 198 | 352 | 311 |
| Biotin (µg) | 1 | 4 | 2.3 |
| Folate (µg) | 5 | 5 | 10 |
| Vitamin $B_{12}$ (µg) | 0.04 | 0.04 | 0.14 |
| Vitamin C (mg) | 4.6 | 1.2 | 5.6 |
| Vitamin D (IU) | 2.2 | 3.4 | 41 |
| Vitamin E (IU) | 0.2 | 0.04 | 1.7 |
| Vitamin K (µg) | 1.5 | 6.0 | 5.7 |
| Inositol (µg) | 37 | 17 | 3 |
| Choline (µg) | 6 | 20 | 10 |

[a]Values represent the average for two major commercial products: Similac, Ross Laboratories, and Enfamil, Mead-Johnson Laboratories.

[b]The value represents formulas with iron fortification. The value for unfortified formula is 0.1 milligram.

Sources: Adapted from K. J. Motil, Breast-feeding: Public health and clinical overview, in *Pediatric Nutrition*, eds. R. J. Grand, J. L. Sutphen, and W. H. Dietz, Jr. (Boston: Butterworths, 1987), pp. 251–263; updated 1997, personal communication.

these limited uses, soy formulas account for one-fourth of the infant formulas sold today.[9] While soy formulas support the normal growth and development of infants, they offer no advantage over milk formulas.

• ***Inappropriate Formulas*** • Caregivers must use only products designed for infants; soy *beverages,* for example, are nutritionally incomplete and inappropriate for infants. Goat's milk is also inappropriate for infants because of its low folate content. An infant receiving goat's milk is likely to develop "goat's milk anemia," an anemia characteristic of folate deficiency.

• ***Nursing Bottle Tooth Decay*** • An infant cannot be allowed to sleep with a bottle because of the potential damage to developing teeth. Salivary flow, which normally cleanses the mouth, diminishes as the infant falls asleep. Prolonged sucking on a bottle of formula, milk, or juice bathes the upper teeth in a carbohydrate-rich

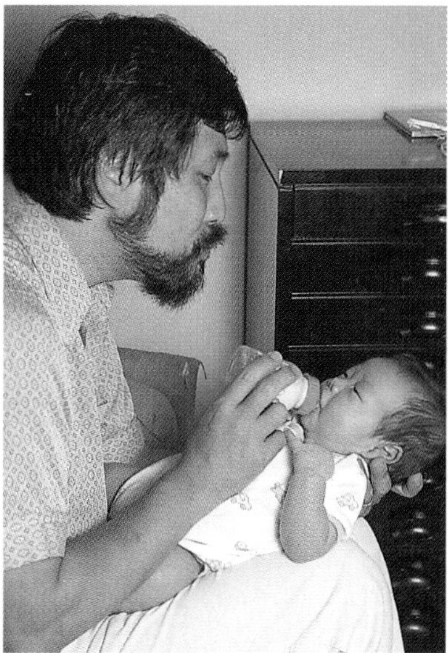

*The infant thrives on infant formula offered with affection.*

fluid that nourishes decay-producing bacteria. (The tongue covers and protects most of the lower teeth, but they, too, may be affected.) The result is extensive and rapid tooth decay (see Figure 16-5). To prevent tooth decay, no infant should be put to bed with a bottle of nourishing fluid.

**nursing bottle tooth decay:** extensive tooth decay due to prolonged tooth contact with formula, milk, fruit juice, or other carbohydrate-rich liquid offered to an infant in a bottle.

 U.S. Government
www.healthfinder.gov/searchoptions/topicsaz.htm
Search for Baby Bottle Tooth Decay

**Healthy People 2000:** Increase to at least 75% the proportion of parents and caregivers who use feeding practices that prevent nursing bottle tooth decay.

**Figure 16-5**
**Nursing Bottle Tooth Decay**

This child was frequently put to bed sucking on a bottle filled with apple juice, so the teeth were bathed in carbohydrate for long periods of time—a perfect medium for bacterial growth. The upper teeth show signs of decay.

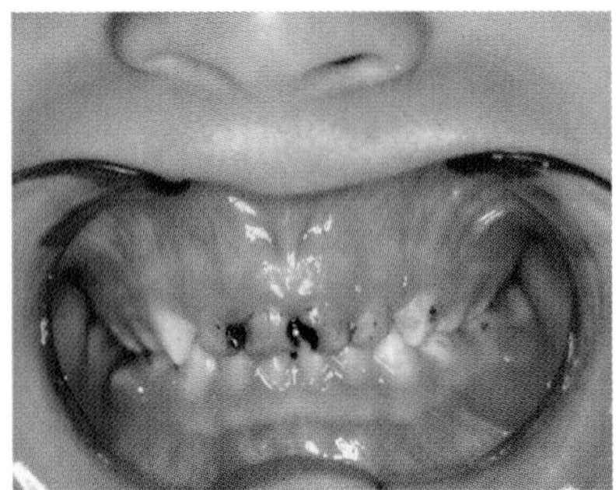

## *Special Needs of Preterm Infants*

The terms *preterm* and *premature* imply incomplete fetal development, or immaturity, of many body systems. Preterm infants face physical independence before some of their organs and body tissues are ready. The rate of weight gain in the fetus is greater during the last trimester of gestation than at any other time. Therefore, a preterm infant is most often a low-birthweight infant as well. With a premature birth, the infant is deprived of the nutritional support of the placenta during a time of maximal growth.

The last trimester of gestation is a time of building nutrient stores. Being born with limited nutrient stores intensifies the precarious situation for the infant. Further compromising the nutrition status of preterm infants is their physical and metabolic immaturity. Nutrient absorption, especially of fat and calcium, from an immature GI tract is limited. In short, preterm, low-birthweight infants are candidates for nutrient imbalances. Deficiencies of the fat-soluble vitamins, calcium, iron, and zinc are common.

Infants who are born eight to ten weeks before term miss out on the normal mineralization of bone that takes place during the last trimester of gestation. As a result, they often develop the metabolic bone disease osteopenia, the rickets of prematurity. The probability that this condition will occur varies directly with the infant's weight: the smaller the infant, the greater the risk of bone disease.

**osteopenia** (OS-tee-oh-PEE-nee-ah): a metabolic bone disease common in preterm infants; also called **rickets of prematurity.**

 U.S. Government
www.healthfinder.gov/searchoptions/topicsaz.htm
Search for Premature birth

Preterm breast milk is well suited to meet a preterm infant's needs. During early lactation, preterm milk contains higher concentrations of protein and is lower in volume than term milk. The low milk volume is advantageous because preterm infants consume small quantities of milk per feeding, and the higher protein concentration allows for better growth.

Theoretical estimates of the requirements of preterm infants suggest that preterm breast milk may be an inadequate source of calcium and phosphorus.[10] In many instances, supplements of nutrients specifically designed for preterm infants are added to the mother's expressed breast milk and fed to the infant from a bottle. When fortified with a preterm supplement, preterm breast milk supports growth at a rate that approximates the growth rate that would have occurred in utero.

**in utero** (YOU-ter-oh): within the uterus.

## *Introducing Cow's Milk*

The timing of the introduction of whole cow's milk to the infant's diet has long been a source of controversy. The AAP currently recommends introducing whole cow's milk at 12 months of age.[11]

In some infants, particularly those younger than six months of age, the consumption of whole cow's milk is associated with intestinal bleeding and iron deficiency. Whole cow's milk is also a poor source of iron. Consequently, cow's milk both causes iron loss and fails to replace iron. Furthermore, the bioavailability of iron from infant cereal and other foods is reduced when cow's milk replaces breast milk or iron-fortified formula during the first year. Compared to breast milk or iron-fortified formula, cow's milk is higher in calcium and lower in vitamin C, characteristics that inhibit iron absorption. In short, cow's milk is a poor choice during the first year of life.

## Introducing Solid Foods

The high nutrient needs of infancy are met first by breast milk or formula only and then by a limited diet to which foods are gradually added. Infants gradually develop the ability to chew, swallow, and digest the wide variety of foods available to adults. The caregiver's selection of appropriate foods at the appropriate stages of development is prerequisite to the infant's optimal growth and health (see Table 16-3).

The German word **beikost** (BYE-cost) is sometimes used to describe any nonmilk food given to an infant.

**• *When to Begin* •** In addition to formula or breast milk, an infant needs to begin eating solid foods between four and six months. Infants who do not receive solid foods before the end of the first year may suffer delayed growth.

• Infant feeding tip: Introduce solid foods between 4 and 6 mo.

The main purpose of introducing solid foods to infants is to provide nutrients that are no longer supplied adequately by breast milk or formula alone. The foods chosen must be foods that the infant is developmentally capable of handling both physically and metabolically. The exact timing depends on the individual infant's

**Table 16-3**
**Infant Feeding Skills and Recommended Foods**

Note: Because each stage of development builds on the previous stage, the foods from an earlier stage continue to be included in all later stages.

| Age (mo) | Feeding Skill | Appropriate Foods Added to the Diet |
|---|---|---|
| 0–4 | Turns head toward any object that brushes cheek. | Feed breast milk or infant formula. |
| | Initially swallows using back of tongue; gradually begins to swallow using front of tongue as well. | |
| | Strong reflex (extrusion) to push food out during first 2 to 3 months. | |
| 4–6 | Extrusion reflex diminishes, and the ability to swallow nonliquid foods develops. | Begin iron-fortified cereal mixed with breast milk, formula, or water. |
| | Indicates desire for food by opening mouth and leaning forward. | Begin pureed vegetables and fruits. |
| | Indicates satiety or disinterest by turning away and leaning back. | |
| | Sits erect with support at 6 months. | |
| | Begins chewing action. | |
| | Brings hand to mouth. | |
| | Grasps objects with palm of hand. | |
| 6–8 | Able to feed self finger foods. | Begin breads and other cereals. |
| | Develops pincher (finger to thumb) grasp. | Begin textured vegetables and fruits. |
| | Begins to drink from cup. | Begin plain, unsweetened fruit juices from cup. |
| 8–10 | Begins to hold own bottle. | Begin breads and cereals from table. |
| | Reaches for and grabs food and spoon. | Begin yogurt. |
| | Sits unsupported. | Begin pieces of soft, cooked vegetables and fruit from table. |
| | | Gradually begin finely cut meats, fish, casseroles, cheese, eggs, and legumes. |
| 10–12 | Begins to master spoon, but still spills some. | Include at least 4 servings of breads and cereals from table, in addition to infant cereal.[a] |
| | | Include at least 2 servings of fruits and 3 servings of vegetables.[a] |
| | | Include 2 servings of meat, fish, poultry, eggs, or legumes.[a] |

[a]Serving sizes for infants and young children are smaller than those for an adult. For example, a serving might be ½ slice of bread instead of 1 slice, or ¼ cup rice instead of ½ cup.

Source: Adapted in part from Committee on Nutrition, American Academy of Pediatrics, *Pediatric Nutrition Handbook,* 3rd ed., ed. L. A. Barness (Elk Grove Village, Ill.: American Academy of Pediatrics, 1993), pp. 23–33.

needs and developmental readiness, which vary from infant to infant because of differences in growth rates, activity, and environmental conditions.

• Infant feeding tip: Offer infants water when they start to eat solid foods.

• ***Remember Water*** • Under normal conditions, the AAP advises little or no supplemental water for infants before the introduction of solid foods.[12] Once infants are eating solid foods, supplemental water is necessary to ease the burden on the kidneys. Without additional water, the kidneys are stressed, and dehydration becomes a threat.

• Infant feeding tip: Offer new foods one at a time.

 Food Allergy Network
www.foodallergy.org
(800)929-4040

• ***Food Allergies*** • To prevent allergy and to facilitate its prompt identification should it occur, experts recommend introducing single-ingredient foods, one at a time, in small portions, and waiting four to five days before introducing the next new food. For example, rice cereal is usually the first cereal introduced because it is least allergenic. When it is clear that rice cereal is not causing an allergy, another grain, perhaps barley or oats, is introduced. Wheat cereal is offered last because it is the most common offender. If a cereal causes an allergic reaction such as a skin rash, digestive upset, or respiratory discomfort, its use should be discontinued before introducing the next food. A later section in this chapter offers more on food allergies.

***Foods such as iron-fortified cereals and formulas, mashed legumes, and strained meats provide iron.***

• ***Choice of Infant Foods*** • Infant foods should be selected to provide variety, balance, and moderation. Commercial baby foods offer a wide variety of palatable, nutritious foods in a safe and convenient form. Homemade infant foods can be as nutritious as commercially prepared ones, as long as the preparer minimizes nutrient losses during preparation. Ingredients for homemade foods should be fresh, whole foods without added salt, sugar, or seasonings. Pureed food can be frozen in ice cube trays, providing convenient-sized blocks of food that can be thawed, warmed, and fed to the infant. The preparer should take precautions to guard against food contamination or infection; hands and equipment must be clean.

Because recommendations to restrict fat do not apply to children under age two, labels on foods for children under two (such as infant meats and cereals) cannot carry information about kcalories from fat or kcalories from saturated fat. In addition, the labels cannot list amounts of saturated fat, polyunsaturated fat, monounsaturated fat, or cholesterol. Most fat information is omitted from infant food labels to prevent parents from wrongly restricting fat in infants' diets. Fearing that their infant will become overweight, parents may unintentionally malnourish the infant by limiting fat.[13] In fact, infants and young children, because of their rapid growth, need more fat than older children and adults.

• Infant feeding tip: To provide additional iron, offer iron-fortified cereal, mashed legumes, and infant meats.

• ***Foods to Provide Iron*** • Rapid growth demands iron. At about four to six months, the infant begins to need more iron than stores plus breast milk or iron-fortified formula can provide.

In addition to breast milk or iron-fortified formula, infants can receive iron from iron-fortified cereals and, later, from meat or meat alternates such as legumes. Iron-fortified cereals contribute a significant amount of iron to an infant's diet, but the iron's bioavailability is poor.[14] Parents or caregivers can enhance iron absorption from iron-fortified cereals by selecting vitamin C–rich foods to go with meals.

• Infant feeding tip: To provide vitamin C, offer citrus juices, melon pieces, and chopped berries.

• Infant (and young child) feeding tip: Limit fruit juice to no more than 12 oz/day.

• ***Foods to Provide Vitamin C*** • The best sources of vitamin C are fruits and vegetables. Some authorities suggest that an infant who is introduced to fruits before vegetables may develop a preference for sweets and find the vegetables less palatable. To prevent this, introduce vegetables first, fruits later.

Fruit juices should be diluted and served in a cup, not a bottle, once the infant is six months of age or older. Juices should also be used moderately in the infant diet, so as not to displace other foods. Cases have been reported of children failing to grow and thrive because they were drinking so much juice each day that other, more energy- and nutrient-dense foods were displaced from their diets.[15] Such findings prove that any one food—even a healthful and nutritious one—can create nutrient imbalances, impair growth, and foster obesity when consumed in excess.

• ***Foods to Omit*** • Concentrated sweets, including baby food "desserts," have no place in an infant's diet. They convey no nutrients to support growth, and the extra food energy can promote obesity. Canned vegetables are also inappropriate for infants, as they often contain too much sodium. Honey and corn syrup should never be fed to infants because of the risk of botulism.* Infants and even young children cannot safely chew and swallow popcorn, whole grapes, whole beans, hot dog slices, hard candies, and nuts; they can easily choke on these foods, a risk not worth taking.

**botulism** (BOT-chew-lism): an often fatal food-borne illness caused by the ingestion of foods containing a toxin produced by bacteria that grow without oxygen (see Chapter 19 for details).

• Infant feeding tip: Limit concentrated sweets and canned vegetables, and avoid foods that may cause choking.

• ***Food at One Year*** • At one year of age, whole cow's milk becomes the primary source of most of the nutrients an infant needs; 2 to 3½ cups a day meets those needs sufficiently. More milk than this displaces foods necessary to provide iron and can lead to milk anemia. Children one to two years old should drink whole milk, not reduced-fat, low-fat, or nonfat milk. If powdered milk is used, it should be one of the fat-containing varieties. Other foods—meats, iron-fortified cereals, enriched or whole-grain breads, fruits, and vegetables—should be supplied in variety and in amounts sufficient to round out total energy needs. Ideally, a one-year-old will sit at the table, eat many of the same foods everyone else eats, and drink liquids from a cup, not a bottle. Table 16-4 shows a meal plan that meets a one-year-old's requirements.

***Toddlers need vitamin A– and vitamin D–fortified whole milk.***

**milk anemia:** iron-deficiency anemia that develops when an excessive milk intake displaces iron-rich foods from the diet.

## *Mealtimes with Infants*

Feeding an infant appropriately supports sound nutrition and health, and eating habits acquired during infancy and childhood influence the individual's overall food attitudes throughout life. The nurturing of an infant, however, involves more than nutrition. Those who care for infants are responsible for providing not only nutritious milk, foods, and water, but also a safe, loving, secure environment in which the infants may grow and develop. In light of infants' developmental and

*In infants, but not older individuals, ingestion of *Clostridium botulinum* spores can cause illness when the spores germinate in the intestine and produce toxin, which is absorbed. Symptoms include poor feeding, constipation, loss of tension in the arteries and muscles, weakness, and respiratory compromise. Infant botulism has been implicated in 5 percent of cases of sudden infant death syndrome (SIDS).

**Table 16-4**
**Meal Plan for a One-Year-Old**

**Breakfast**
¼ c whole milk (with cereal)
½ c iron-fortified breakfast cereal
½ c orange juice

**Morning snack**
½ c vitamin C–fortified fruit juice
1 to 2 oz cheese cubes
Teething crackers

**Lunch**
1 c whole milk
½ c vegetables[a] (steamed carrots)
½ sandwich: 1 slice bread with 2 tbs tuna salad or egg salad

**Afternoon snack**
½ c whole milk
1 slice toast
1 to 2 tbs peanut butter

**Dinner**
1 c whole milk
2 to 3 oz chopped meat or well-cooked mashed legumes
¼ c potato, rice, or pasta
¼ c vegetables[a] (chopped broccoli)
¼ c fruit[b] (sliced strawberries)

[a]Include dark green, leafy and deep yellow vegetables.
[b]Include citrus fruits, melons, and berries.

***Ideally, a one-year-old eats many of the same foods as the rest of the family. (Courtesy of CNN)***

nutrient needs and their often contrary and willful behavior, a few feeding guidelines may be helpful:

- Discourage unacceptable behavior, such as standing at the table or throwing food, by removing the child from the table to wait until later to eat. Be consistent and firm, not punitive. The child will soon learn to sit and eat.
- Let the child explore and enjoy food, even if this means eating with fingers for a while. Use of the spoon will come in time.
- Don't force food on children. Rejecting new foods is normal; acceptance is more likely as infants and children become familiar with new foods through repeated opportunities to taste them.[16]
- Provide nutritious foods, and let children choose which ones and how much they will eat. Gradually, they will acquire a taste for different foods.
- Limit sweets. Infants and young children have little room for empty-kcalorie foods in their daily energy allowance. Do not use sweets as a reward for eating meals.
- Don't turn the dining table into a battleground. Make mealtimes enjoyable. Teach healthy food choices and eating habits in a pleasant environment.

select a few pts

IN SUMMARY

The primary food for infants during the first 12 months is either breast milk or iron-fortified formula. In addition to nutrients, breast milk also offers immunological protection. At about 4 to 6 months, infants should gradually begin eating solid foods. By one year, they are drinking from a cup and eating many of the same foods as the rest of the family.

**Figure 16-6**
**Body Shape of One-Year-Old and Two-Year-Old Compared**

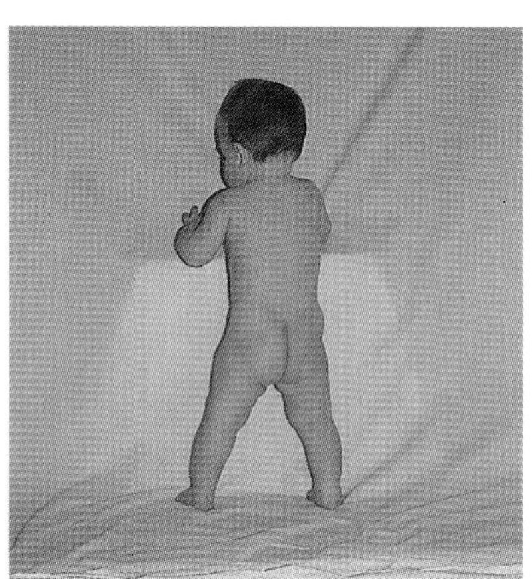

The body shape of a one-year-old (above) changes dramatically by age two (below). The two-year-old has lost much of the baby fat; the muscles (especially in the back, buttocks, and legs) have firmed and strengthened; and the leg bones have lengthened.

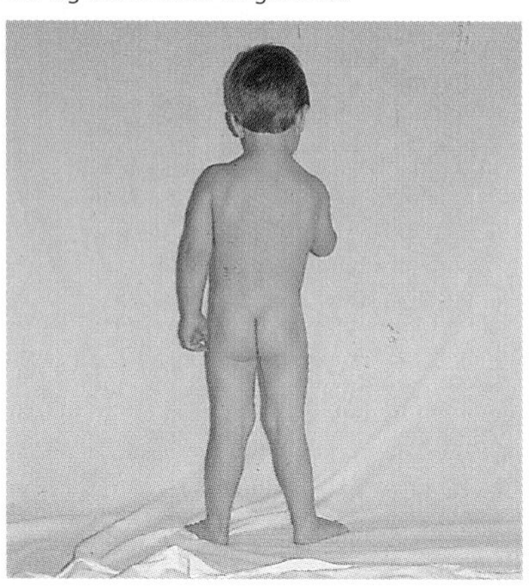

# Nutrition during Childhood

Each year from age one to adolescence, a child typically grows taller by 2 to 3 inches and heavier by 5 to 6 pounds. Growth charts provide valuable clues to a child's health. Weight gains out of proportion to height gains may reflect overeating and inactivity, whereas measures significantly below the standard suggest malnutrition.

Increases in height and weight are only two of the many changes growing children experience (see Figure 16-6). At age one, children can stand alone and are beginning to toddle; by two, they can walk and are learning to run; by three, they can jump and are climbing with confidence. Bones and muscles increase in mass and density to make these accomplishments possible. Thereafter, further lengthening of the long bones and increases in musculature proceed unevenly and more slowly until adolescence.

## *Energy and Nutrient Needs*

Children's appetites begin to diminish around one year, consistent with the slowing of growth. Thereafter, children spontaneously vary their food intakes to coincide with their growth patterns; they demand more food during periods of rapid growth than during slow growth. Sometimes they seem insatiable; other times they seem to live on air and water.

Children's energy intakes also vary widely from meal to meal. Even so, their total daily intakes remain remarkably constant. If children eat less at one meal, they typically eat more at the next, and vice versa. Overweight children are an exception: they do not always adjust their energy intakes appropriately and may eat in response to external cues, disregarding appetite regulation signals.

• ***Energy Intake and Activity*** • Individual children's energy needs vary widely, depending on their growth and physical activity. A one-year-old child needs about

1000 kcalories a day; a three-year-old needs only 300 kcalories more. By age ten, a child needs about 2000 kcalories a day. Total energy needs increase slightly with age, but energy needs per kilogram of body weight actually decline gradually.

Inactive children can become obese even when they eat less food than the average. They would do well to learn to enjoy physical play and exercise.

Some children, notably those adhering to a strict vegetarian diet, have difficulty meeting their energy needs. Grains, vegetables, and fruits provide plenty of fiber, adding bulk, but too few kcalories to support growth. A reasonable fiber intake for children is "age plus 5 grams."[17] For example, a 3-year-old might need about 8 grams of fiber a day, and an 18-year-old, 23 grams.

• ***Protein*** • Like energy needs, total protein needs increase slightly with age, but when the child's body weight is considered, the protein requirement actually declines gradually (see Table 6-4 on p. 181). The estimation of protein needs considers the requirements for maintaining nitrogen balance, the quality of protein consumed, and the added needs of growth.

• ***Vitamins and Minerals*** • The vitamin and mineral needs of children increase with their ages (see inside front cover). A balanced diet of nutritious foods can meet children's needs for these nutrients, with the notable exception of iron. Iron-deficiency anemia is a major problem worldwide, as well as being the most prevalent nutrient deficiency among U.S. and Canadian children.

Healthy People 2000: Reduce iron deficiency to less than 3% among children aged one through four years.

Some dietary patterns predict an inadequate iron intake. Iron deficiency is most likely in young children (under age five) who eat fewer than five servings from each of the grain, vegetable, fruit, and meat groups per week; those who drink more than 2 cups of milk a day; and those who eat high-fat and high-sugar snacks or drink more than 2 cups of soda daily.[18]

To prevent iron deficiency, children's foods must deliver approximately 10 milligrams of iron per day. To achieve this goal, snacks and meals should include the iron-rich foods listed in Table 16-5, and milk intake should be reasonable so that it will not displace lean meats, fish, poultry, eggs, legumes, and whole-grain or enriched products.

• ***Planning Children's Meals*** • To provide all the needed nutrients, children's meals should include a variety of foods from each food group—in amounts suited to their appetites and needs. Serving sizes increase with age. A portion of meat, grains, fruits, or vegetables for children is loosely defined as 1 tablespoon for each year. Thus, at four years of age, a portion is about 4 tablespoons, or ¼ cup. This guideline applies until children reach age 12. Table 16-6 (on p. 514) offers a daily food pattern for children. Notice that the number of servings from each group is the same as the lower number for adults, but that the serving sizes are smaller (for example, ¼ cup rice instead of ½ cup).

Children whose diets follow the Food Guide Pyramid pattern meet their nutrient needs fully, but few children eat according to these recommendations. In fact, only 5 percent consume the suggested servings from even four of the five food groups.[19] Almost half of them meet none or only one of the food group recommendations. Consequently, average intakes of fiber, vitamin $B_6$, calcium, iron, and zinc fall far below recommendations.

To ensure that children have healthy appetites and plenty of room for nutritious foods when they are hungry, parents and teachers must limit access to candy, cola, and other concentrated sweets. If such foods are permitted in large quantities, the only possible outcomes are nutrient deficiencies, obesity, or both. The preference for sweets is innate; most children do not naturally select nutritious

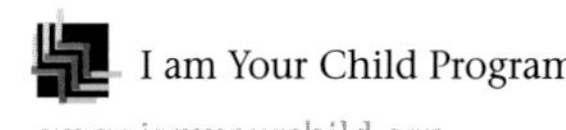
I am Your Child Program
www.iamyourchild.org

**Table 16-5**
**Iron-Rich Foods Children Like**[a]

| |
|---|
| **Breads, cereals, and grains** |
| Canned macaroni (½ c) |
| Canned spaghetti (½ c) |
| Cream of wheat (¼ c) |
| Fortified dry cereals (1 oz)[b] |
| Noodles, rice, or barley (½ c) |
| Tortillas (1 flour, 2 corn) |
| Whole-wheat, enriched, or fortified bread (1 slice) |
| Bran muffins |
| **Vegetables** |
| Baked potato skins (½ skin) |
| Cooked mushrooms (½ c) |
| Cooked mung bean sprouts or snow peas (½ c) |
| Green peas (½ c) |
| Mixed vegetable juice (1 c) |
| **Fruits** |
| Canned plums (3 plums) |
| Cooked dried apricots (½ c) |
| Dried peaches (4 halves) |
| Raisins (1 tbs) |
| **Meats and legumes** |
| Bean dip (¼ c) |
| Canned pork and beans (⅓ c) |
| Mild chili or other bean/meat dishes (¼ c) such as burritos |
| Liverwurst (½ oz) |
| Meat casseroles (½ c) |
| Peanut butter and jelly sandwich (½ sandwich) |
| Lean roast beef or cooked ground beef (1 oz) |
| Sloppy joes (½ sandwich) |

[a]Each serving provides at least 1 milligram iron, or one-tenth of a child's RDA for iron. Bioavailability varies. Vitamin C–rich foods included with these snacks increase iron absorption.
[b]Some fortified breakfast cereals contain more than 10 milligrams iron per half-cup serving (read the labels).

**Table 16-6**
**Children's Daily Food Patterns for Good Nutrition**

| Food Group | Servings per Day | Average Size of Serving | | |
|---|---|---|---|---|
| | | *1 to 3 Years* | *4 to 6 Years* | *7 to 12 Years* |
| Bread and cereals (whole grain or enriched)[a] | 6 or more | ½ slice | 1 slice | 1 to 2 slices |
| Vegetables[b] | 3 or more | 2 to 4 tbs or ½ c juice | ¼ to ½ c or ½ c juice | ½ to ¾ c or ½ c juice |
| Fruits[b] | 2 or more | 2 to 4 tbs or ½ c juice | ¼ to ½ c or ½ c juice | ½ to ¾ c or ½ c juice |
| Meat and meat alternates[c] | 2 or more | 1 to 2 oz | 1 to 2 oz | 2 to 3 oz |
| Milk and milk products[d] | 3 to 4 | ½ to ¾ c | ¾ c | ¾ to 1 c |

[a]1 slice bread = ¾ c dry cereal, ½ c cooked cereal, ½ c potato, rice, or noodles.
[b]Vitamin C source (citrus fruits, berries, tomatoes, broccoli, cabbage, cantaloupe) daily; vitamin A source (spinach, carrots, squash, tomato, cantaloupe) 3 to 4 times weekly.
[c]1 oz meat, fish, poultry = 1 egg, 1 frankfurter, 2 tbs peanut butter, ½ c cooked legumes.
[d]½ c milk = ½ c cottage cheese, pudding, yogurt; ¾ oz cheese; 2 tbs dried milk.
Source: Adapted from P. M. Queen and R. R. Henry, Growth and nutrient requirements of children, in *Pediatric Nutrition,* eds. R. J. Grand, J. L. Sutphen, and W. H. Dietz, Jr. (Boston: Butterworths, 1987), p. 347.

foods on the basis of taste. When children are allowed to create meals freely from a variety of foods, they typically select foods that provide a lot of sugar. When their parents are watching, or even when they think their parents are watching, children improve their selections. Overweight children need help in selecting nutrient-dense foods that will meet their nutrient needs within their energy allowances.

Sweets need not be banned altogether. Children who are exceptionally active can enjoy high-kcalorie foods such as ice cream or pudding from the milk group or pancakes from the bread group. As for sedentary children, they need to become more active, so they can also enjoy some of these foods without unhealthy weight gain.

## *Hunger and Malnutrition in Children*

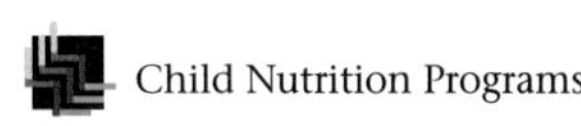


www.usda.gov/fcs/ncp.htm

Chapter 20 examines the causes and consequences of hunger in the United States and around the world.

Most children in the United States and Canada are well nourished. Their average energy intakes are sufficient to support normal growth, and their average nutrient intakes, except for iron, meet or exceed recommendations. Malnutrition does appear, however, in certain circumstances. Low-income children, for example, may be hungry and malnourished. An estimated 11 million U.S. children under age 12 are hungry and living in poverty.

**Healthy People 2000:** Reduce growth retardation among low-income children aged five years and younger to less than 10%.

• ***Malnutrition and Health*** • When hunger is chronic, children become malnourished. Worldwide, malnutrition takes a devastating toll on children, contributing to nearly half of the deaths of children under four years old. Vitamin A deficiency afflicts more than 5 million children worldwide, inducing blindness, stunted growth, and infections. Zinc deficiency also retards growth and typically accompanies protein-energy malnutrition and vitamin A deficiency.

• ***Hunger and Behavior*** • Even when hunger is temporary, as when a child misses one meal, behavior and academic performance are affected. Children who eat nutritious breakfasts function better than their peers who do not. Those who participate in the federally funded School Breakfast Program improve their school performance and are tardy or absent significantly less often than children who qualify for the program but do not participate.[20] Without breakfast, children perform poorly in tasks requiring concentration, their attention spans are shorter, and they even score lower on intelligence tests than their well-fed peers; malnourished children are particularly vulnerable.[21] Unfortunately, an estimated 4 out of 30 students miss breakfast each day.[22] Common sense dictates that it is unreasonable to expect anyone to learn and perform work when no fuel has been provided. By late morning, discomfort from hunger may become distracting even if a child has eaten breakfast.

The problem children face when attempting morning schoolwork on an empty stomach appears to be at least partly due to low blood glucose. The average child up to age ten or so needs to eat about every four hours to maintain a blood glucose concentration high enough to support the activity of the brain and the rest of the nervous system. A child's brain is as big as an adult's, and the brain is the body's chief glucose consumer. A child's liver is much smaller than an adult's, however, and the liver is the organ responsible for storing glucose as glycogen and releasing it into the blood as needed. A child's liver can store only about four hours' worth of glycogen—hence the need to eat fairly often. Teachers aware of the late-morning slump in their classrooms wisely request that midmorning snacks be provided; snacks improve classroom performance all the way to lunchtime. For the child who hasn't had breakfast, the morning's lessons may be lost altogether.

The brain uses about three times as much glucose per day as the rest of the body.

Eating breakfast also helps children to meet their nutrient needs each day. Children who skip breakfast typically do not make up the deficits at later meals—they simply have lower intakes of energy, vitamins, and minerals than those who eat breakfast.

• ***Iron Deficiency and Behavior*** • Iron deficiency has well-known and widespread effects on children's behavior.[23] In addition to carrying oxygen in the blood, iron transports oxygen within cells, which use it to help produce energy. Iron is also used to make neurotransmitters—most notably, those that regulate the ability to pay attention, which is crucial to learning. Consequently, iron deficiency not only causes an energy crisis but also directly affects mood, attention span, and learning ability.

Highlight 13 describes the relationships between, and consequences of, iron deficiency and lead toxicity.

Iron deficiency is usually diagnosed by a deficit of iron in the *blood,* after the deficiency has progressed all the way to anemia. A child's *brain,* however, is sensitive to low iron concentrations long before the blood effects appear. Research has shown that iron deficiency lowers the "motivation to persist in intellectually challenging tasks," shortens the attention span, and impairs overall intellectual performance. Anemic children perform less well on tests and are more disruptive than their nonanemic classmates. When combined with other nutrient deficiencies, iron-deficiency anemia has synergistic effects that are especially detrimental to learning.[24] Furthermore, children who had iron-deficiency anemia *as infants* continue to perform poorly as they grow older, even if their iron status improves. The long-term damaging effects on mental development make prevention of iron deficiency during infancy and early childhood a high priority.

• ***Other Nutrient Deficiencies and Behavior*** • A child with any of several nutrient deficiencies may be irritable, aggressive, disagreeable, or sad and withdrawn. Such a child may be labeled "hyperactive," "depressed," or "unlikable," when in fact these traits may arise from simple, even marginal, malnutrition. Parents and medical practitioners often overlook the possibility that malnutrition may account for abnormalities of appearance and behavior. Any departure from normal healthy appearance and behavior is a sign of possible poor nutrition (see Table 16-7 on p. 516). In any such case, inspection of the child's diet by a registered dietitian or other qualified health care professional is clearly in order. Should suspicion of dietary inadequacies be raised, no matter what other causes may be implicated, the people responsible for feeding the child should take steps to correct those inadequacies promptly.

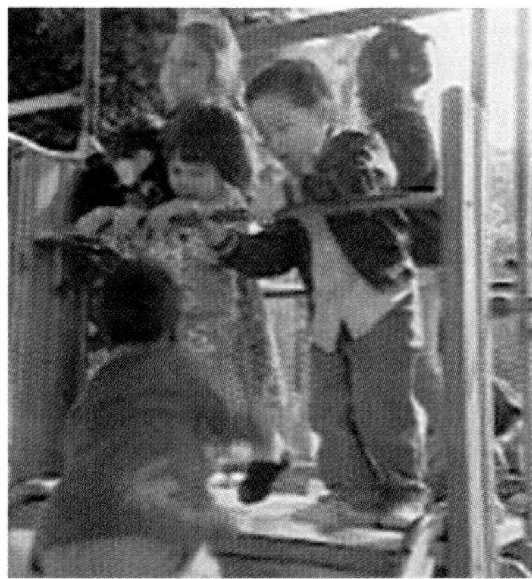

***Healthy, well-nourished children are alert in the classroom and energetic at play. (Courtesy of CNN)***

**Table 16-7**
**Physical Signs of Malnutrition in Children**

| | Well-Nourished | Malnourished | Possible Nutrient Deficiencies |
|---|---|---|---|
| Hair | Shiny, firm in the scalp | Dull, brittle, dry, loose; falls out | PEM |
| Eyes | Bright, clear pink membranes; adjust easily to light | Pale membranes; spots; redness; adjust slowly to darkness | Vitamin A, the B vitamins, zinc, and iron |
| Teeth and gums | No pain or caries, gums firm, teeth bright | Missing, discolored, decayed teeth; gums bleed easily and are swollen and spongy | Minerals and vitamin C |
| Face | Clear complexion without dryness or scaliness | Off-color, scaly, flaky, cracked skin | PEM, vitamin A, and iron |
| Glands | No lumps | Swollen at front of neck, cheeks | PEM and iodine |
| Tongue | Red, bumpy, rough | Sore, smooth, purplish, swollen | B vitamins |
| Skin | Smooth, firm, good color | Dry, rough, spotty; "sandpaper" feel or sores; lack of fat under skin | PEM, essential fatty acid, vitamin A, B vitamins, and vitamin C |
| Nails | Firm, pink | Spoon-shaped, brittle, ridged | Iron |
| Internal systems | Regular heart rhythm, heart rate, and blood pressure; no impairment of digestive function, reflexes, or mental status | Abnormal heart rate, heart rhythm, or blood pressure; enlarged liver, spleen; abnormal digestion; burning, tingling of hands, feet; loss of balance, coordination; mental confusion, irritability, fatigue | PEM and minerals |
| Muscles and bones | Muscle tone; posture, long bone development appropriate for age | "Wasted" appearance of muscles; swollen bumps on skull or ends of bones; small bumps on ribs; bowed legs or knock-knees | PEM and vitamin D |

## *Nutrition and Behavior*

All children are naturally active, and many of them become overly active on occasion—for example, in anticipation of a birthday party. Such behavior is markedly different from true hyperactivity.

**hyperactivity:** inattentive and impulsive behavior that is more frequent and severe than is typical of others a similar age; professionally called **attention-deficit/hyperactivity disorder (ADHD).**

Children and Adults with Attention Deficit Disorders

www.chadd.org

U.S. Government

www.healthfinder.gov/searchoptions/topicsaz.htm
Search for Hyperactivity

• ***Hyperactivity*** • Hyperactive children have trouble sleeping, cannot sit still for more than a few minutes at a time, act impulsively, and have difficulty paying attention. These behaviors interfere with social development and academic progress. The cause of hyperactivity remains unknown, but it affects about 5 percent of young school-age children. To resolve the problems surrounding hyperactivity, physicians often recommend specific behavioral strategies, special educational programs, and psychological counseling; in some cases, they prescribe medication.[25]

Parents of hyperactive children sometimes seek help from alternative therapies, including special diets. They mistakenly believe a solution may lie in manipulating the diet—most commonly, by excluding sugar or food additives. Adding carrots or eliminating candy is such a simple solution that many parents eagerly give such diet advice a try. These dietary changes will not solve the problem of true hyperactivity. Studies have consistently found no convincing evidence that sugar causes hyperactivity or worsens behavior.[26] Recommendations to restrict sugar in children's diets to prevent or treat behavior problems are groundless.

• ***Misbehaving*** • Even a child who is not truly hyperactive can be difficult to manage at times. Michael may act unruly out of a desire for attention, Jessica may be cranky because of a lack of sleep, Christopher may react violently after watching too much television, and Sheila may be unable to sit still in class due to a lack of exercise. All of these children may benefit from more consistent care—regular hours of sleep, regular mealtimes, and regular outdoor activity.

• ***Watching Television*** • Watching television is passive behavior that adversely affects children's nutritional health in several ways. As Chapter 9 reported, children who watch a lot of television (four or more hours a day) are most likely to be obese.[27] Not only are they inactive, but they are often snacking on the fattening foods that are advertised. The average child sees an estimated 10,000 commercials a year—almost all peddling foods high in sugar, fat, and salt such as sugar-coated breakfast cereals, candy bars, chips, fast foods, and carbonated beverages. Few commercials tout the benefits of nutritious foods such as fruits and vegetables. Consequently, children can easily develop a lopsided view of what they should be eating. They may also accept the advertisers' premise that these snack foods are the best option when "you're looking for an energy boost." Many parents and pediatricians believe that food ads aimed at children should be banned because they support corporate profits rather than children's health. Alternatively, parents can teach their children how to evaluate food ads and make healthful choices.

***Television watching influences children's eating habits and activity patterns.***

TV fosters obesity because it:
- Requires no energy beyond basal metabolism.
- Replaces vigorous activities.
- Encourages snacking.
- Promotes a sedentary lifestyle.

Playing computer games influences children's activity patterns similarly.

## Adverse Reactions to Foods

Adverse reactions to foods can threaten nutritional health to varying extents, depending on the severity and duration of the reactions and the foods they involve. Temporary reactions may lead to permanent avoidance of foods; permanent reactions, if not detected and treated, can cause chronic illness.

**adverse reactions:** unusual responses to food (including intolerances and allergies).

• ***Food Intolerances*** • Not all adverse reactions to foods are food allergies, although even physicians may describe them as such. Signs of adverse reactions to foods include stomachaches, headaches, rapid pulse rate, nausea, wheezing, hives, bronchial irritation, coughs, and other such discomforts. Among the causes may be reactions to chemicals in foods, such as the flavor enhancer monosodium glutamate (MSG), the natural laxative in prunes, or the mineral sulfur; digestive diseases, such as obstructions or injuries; enzyme deficiencies, such as lactose intolerance; and even psychological aversions. These reactions involve symptoms but no antibody production. Therefore, they are food intolerances, not allergies.

**food intolerances:** adverse reactions to foods that do not involve the immune system.

Pesticides on produce may also cause adverse reactions. Pesticides may linger on the foods to which they were applied in the field. Health risks from pesticide exposure are probably small for healthy adults, but children may be vulnerable to some types of pesticide poisoning.[28] The FDA, Environmental Protection Agency (EPA), and USDA are all considering proposals to revise food safety and pesticide laws to protect infants and children from pesticide risks.[29] When setting tolerance levels, the agencies will first identify foods that children commonly eat in large amounts and then consider the effects of pesticide exposure during developmental stages.[30]

 American Academy of Allergy, Asthma, and Immunology

www.aaaai.org

• ***Food Allergies*** • A true food allergy occurs when a whole food protein or other large molecule is absorbed into the blood and elicits an immunologic response. (Recall that proteins and other large molecules of food are normally dismantled in the digestive tract to smaller ones that are absorbed without such a reaction.) The body's immune system reacts to these large food molecules as it does to other antigens—by producing antibodies, histamines, or other defensive agents.

**food allergies:** adverse reactions to foods that involve an immune response; also called *food-hypersensitivity reactions.*

Reminder: *Histamine* is a substance produced by cells of the immune system as part of a local immune reaction to an antigen; participates in causing inflammation.

Allergies may have one or two components. They always involve antibodies; they may or may not involve symptoms. This means that allergies can be diagnosed only by testing for antibodies. Even symptoms exactly like those of an allergy may not be caused by one.

A person who produces antibodies *without* having any symptoms has an **asymptomatic allergy;** a person who produces antibodies *and* has symptoms has a **symptomatic allergy.**

Allergic reactions to food may be immediate or delayed. In both cases, the antigen interacts immediately with the immune system, but the timing of symptoms

*Eggs, peanuts, and milk are most likely to induce symptoms in people with food allergy.*

 U.S. Government

www.healthfinder.gov/searchoptions/topicsaz.htm

Search for Food Allergies

 International Food Information Council

www.ificinfo.health.org

Visit Food Allergies

varies from minutes to 24 hours after consumption of the antigen. Identifying the food that causes an immediate allergic reaction is fairly easy, because the symptoms appear within minutes after eating the food. Identifying the food that causes a delayed reaction is more difficult, because the symptoms may not appear until a day later. By this time, many other foods have been eaten, complicating the picture.

Almost 75 percent of adverse reactions are caused by three foods—eggs, peanuts, and milk. Allergic reactions to single foods are common. Reactions to multiple foods are the exception, not the rule.

Identifying a true food allergy requires a thorough health history, physical examination, and diagnostic tests to eliminate other diseases. Skin pricks with food extracts are one of the most common tests for food allergies, even though the high incidence of false positive results can complicate diagnosis. Physicians also conduct dietary trials that first eliminate the offending food and then reintroduce it in small quantities to substantiate that reactions occur only when that particular food is eaten.[31] Once a food allergy has been diagnosed, therapy requires strict elimination of the offending food.

Food allergies are most common during the first few years of life, but then children typically outgrow (become tolerant to) their hypersensitivity. Approximately 5 percent of young children are allergic to certain foods, whereas less than 2 percent of adults have food allergies.[32] Developing tolerance is most likely if the offending food can be identified and eliminated from the diet for at least a year or two.

When parents stop serving a suspected food to their child, they risk the child's suffering nutrient deficiencies. They should be sure to include other foods that offer the same nutrients as the omitted food. Children with allergies, like all children, need all their nutrients.

## *Mealtimes at Home*

**gatekeepers:** with respect to nutrition, key people who control other people's access to foods and thereby exert profound impacts on their nutrition. Examples are the spouse who buys and cooks the food, the parent who feeds the children, and the caregiver in a day-care center.

Parents are gatekeepers; they determine what foods and activities will be available in their children's environments. Then the children make their own selections. Gatekeepers who want to promote nutritious choices and healthful habits provide access to nutrient-dense, delicious foods and opportunities for active play at home.

• Child feeding tip: Serve vegetables raw or slightly undercooked and crunchy.

• ***Honoring Children's Preferences*** • Researchers attempting to explain children's food preferences encounter contradictions. Children say they like colorful foods, yet most often reject green and yellow vegetables while favoring brown peanut butter and white potatoes, apple wedges, and bread. They do like raw vegetables better than cooked ones, so it is wise to offer vegetables that are raw or slightly undercooked, served separately, and easy to eat. Foods should be warm, not hot, because a child's mouth is much more sensitive than an adult's. The flavor should be mild because a child has more taste buds, and smooth foods such as mashed potatoes or split-pea soup should contain no lumps (a child wonders, with some disgust, what the lumps might be). Children prefer foods that are familiar, so offer various foods regularly.

• Child feeding tip: Provide child-sized portions and utensils.

Make mealtimes fun for children. Young children like to eat at little tables and to be served small portions of food. They like sandwiches cut in different geometric shapes and common foods called silly names. They also like to eat with other children, and they tend to eat more when in the company of their peers. Children also more easily overcome their prejudices against foods when they see their peers eating them.

• Child feeding tip: Encourage children to help plan and prepare meals.

• ***Learning through Participation*** • Allowing children to help plan and prepare the family's meals provides enjoyable learning experiences and encourages children to eat the foods they have prepared. Vegetables are pretty, especially when fresh, and provide opportunities for children to learn about color, growing vegetables and their seeds, and shapes and textures—all of which are fascinating to young children. Measuring, stirring, washing, and arranging foods are skills that even a young child can practice with enjoyment and pride.

• ***Avoiding Power Struggle*** • It is not surprising that problems over food often arise during the second or third year, when children begin asserting their independence. Many of these problems stem from the conflict between children's developmental stages and capabilities and parents who, in attempting to do what they think is best for their children, try to control every aspect of eating. Such conflicts can disrupt children's abilities to regulate their own food intakes or to determine their own likes and dislikes. For example, many people share the misconception that children must be persuaded or coerced to try new foods. In fact, the opposite is true. When children are forced to try new foods, even by way of rewards, they are less likely to try those foods again than are children who are left to decide for themselves. The parent is responsible for providing healthful foods, but the child is responsible for *how much* and even *whether* to eat.[33]

When introducing new foods at the table, offer them one at a time and only in small amounts at first. The more often a food is presented to a young child, the more likely the child will accept that food. Offer the new food at the beginning of the meal, when the child is hungry, and allow the child to make the decision to accept or reject it. Never make an issue of food acceptance. A power struggle almost invariably sets a firm pattern of resistance and permanently closes the child's mind.

***Children enjoy eating the foods they help to prepare.***

• Child feeding tip: Don't use food as a reward for good behavior.

• ***Choking Prevention*** • Parents must always be alert to the dangers of choking. A choking child is silent, and an adult should be present whenever a child is eating. Make sure the child sits when eating; choking is more likely when a child is running or falling. Round foods such as grapes, nuts, hard candies, and hot dog pieces are difficult to control in a mouth with few teeth, and they can easily become lodged in the small opening of a child's trachea.

Young children can easily choke on:

- Popcorn.
- Whole grapes.
- Whole beans.
- Hot dog slices.
- Hard candies.
- Nuts.

• ***Play First*** • Children may be more relaxed and attentive at mealtime if outdoor play or other fun activities are scheduled before, rather than immediately after, mealtime. Otherwise children "hurry up and eat" so that they can go play.

• Child feeding tip: To prevent choking, watch children eat and enforce a "sit-down" rule.

• Child feeding tip: Play first, then eat.

• ***Snacks*** • Parents may find that their children snack so much that they aren't hungry at mealtimes. Instead of teaching children *not* to snack, parents might be wise to teach them *how* to snack. Provide snacks that are as nutritious as the foods served at mealtime. Snacks can even be mealtime foods served individually over time, instead of all at once on one plate. When providing snacks to children, think of the five food groups and offer such snacks as pieces of cheese, tangerine slices, carrot sticks, and peanut butter on whole-wheat crackers (see Table 16-8 on p. 520). Snacks that are easy to prepare should be readily available to children, especially if they arrive home from school before their parents.

• Child feeding tip: Provide healthful snacks.

Kids Food Cyber Club
www.kidsfood.org

• ***Preventing Dental Caries*** • Children frequently snack on sticky, sugary foods that stay on the teeth and provide an ideal environment for the growth of bacteria that cause dental caries. Teach children to brush and floss after meals, to brush or rinse after eating snacks, to avoid sticky foods, and to select crisp or fibrous foods frequently.

• Child feeding tip: To protect against dental caries, serve fresh fruits more often than dried fruits or juices.

• ***Serving as Role Models*** • In an effort to practice these many tips, parents may overlook perhaps the single most important influence on their children's food habits—themselves. Parents who don't eat carrots shouldn't be surprised when their children refuse to eat carrots. Likewise, parents who dislike the smell of brussels sprouts may not be able to persuade children to try them. Children learn much through imitation. Parents, older siblings, and other caregivers set an irresistible example by sitting with younger children, eating the same foods, and having pleasant conversations during mealtime.[34]

• Child feeding tip: Set a good example—enjoy nutritious foods.

While serving and enjoying food, caregivers can promote both physical and emotional growth at every stage of a child's life. They can help their children to develop both a positive self-concept and a positive attitude toward food. If the

**Table 16-8**

**Healthful Snack Ideas—Think Food Groups, Alone and in Combination**

Selecting two or more foods from different food groups adds variety and nutrient balance to snacks. The combinations are endless, so be creative.

**Grain Products**

Grain products are filling snacks, especially when combined with other foods:

- Cereal with fruit and milk.
- Crackers and cheese.
- Wheat toast with peanut butter.
- Popcorn with grated cheese.
- Oatmeal raisin cookies with milk.

**Vegetables**

Cut-up fresh, raw vegetables make great snacks alone or in combination with foods from other food groups:

- Celery with peanut butter.
- Broccoli, cauliflower, and carrot sticks with a flavored cottage cheese dip.

**Fruits**

Fruits are delicious snacks and can be eaten alone—fresh, dried, or juiced—or combined with other foods:

- Apples and cheese.
- Bananas and peanut butter.
- Peaches with yogurt.
- Raisins mixed with sunflower seeds or nuts.

**Meats and Meat Alternates**

Meat and meat alternates add protein to snacks:

- Refried beans with nachos and cheese.
- Tuna on crackers.
- Luncheon meat on wheat bread.

**Milk and Milk Products**

Milk can be used as a beverage with any snack, and many other milk products, such as yogurt and cheese, can be eaten alone or with other foods as listed above.

beginnings are right, children will grow without the conflicts and confusions over food that can lead to nutrition and health problems.

## Nutrition at School

*Eating is more fun when your friends are there. (Courtesy of CNN)*

While parents are doing what they can to establish good eating habits in their children at home, others are preparing and serving foods to their children at day-care centers and schools. In addition, children begin to learn about food and nutrition in the classroom. Meeting the nutrition and education needs of children is critical to supporting their healthy growth and development.[35]

• ***School Meals*** • The U.S. government funds several programs to provide nutritious meals for children at school. Both the School Breakfast Program and the National School Lunch Program provide meals at a reasonable cost to children from families with the financial means to pay. Meals are available free or at reduced cost to children from low-income families. (School lunches in Canada are administered locally and therefore vary from area to area.) Several studies have reported that children who participate in school food programs show improvements in learning. The next page describes food assistance programs for children, and Table 16-9 (on p. 521) shows school lunch patterns for children of different ages.

## Food Assistance Programs for Children

The federal School Lunch and School Breakfast Programs assist schools financially so that every student can receive a nutritious lunch, breakfast, or both. These programs enable schools to provide low-income students with meals at no cost while charging other students somewhat less than the full costs of their meals. In addition, schools that participate in the programs can obtain food commodities. Nationally, the U.S. Department of Agriculture (USDA) administers the programs; on the state level, state departments of education operate them. The programs usually cost school districts little.

More than 26 million children receive lunches through the National School Lunch Program—half of them free or at a reduced price. School lunches are designed to provide at least a third of the RDA for energy, protein, vitamin A, vitamin C, iron, and calcium. They must also meet the Dietary Guidelines and include specified numbers of servings of milk, protein-rich foods (meat, poultry, fish, cheese, eggs, legumes, or peanut butter), vegetables, fruits, and breads or other grain foods.

The School Breakfast Program is available in slightly more than half of the nation's schools, and about 5 million children participate in it. The school breakfast must provide at least a fourth of the RDA for each of many nutrients and contain at least one serving of milk; one serving of fruit, juice, or vegetable; and either two servings of bread (or bread alternates), two servings of meat (or meat alternates), or one serving of each.

Another federal program, the Child Care Food Program, operates similarly and provides funds to organized child-care programs. All eligible children, centers, and family day-care homes have the right to participate. Meal reimbursements cover most of the meal and administration costs. Sponsors may also receive USDA commodity foods.

**Healthy People 2000:** Increase to at least 90% the proportion of school lunch and breakfast services and increase to at least 50% the proportion of child care foodservices with menus that are consistent with the nutrition principles in the *Dietary Guidelines for Americans.*

School lunches offer a variety of food choices and help children meet at least one-third of their needs for selected nutrients. These lunches are also required to meet the *Dietary Guidelines* over a week's menus. Schools making special efforts to

**Table 16-9**
**School Lunch Patterns for Different Ages**

| Food Group | Preschool (Age) | | Grade School through High School (Grade) | | |
|---|---|---|---|---|---|
| | *1 to 2* | *3 to 4* | *K to 3* | *4 to 6* | *7 to 12* |
| **Meat or meat alternate** | | | | | |
| 1 serving: | | | | | |
| Lean meat, poultry, or fish | 1 oz | 1½ oz | 1½ oz | 2 oz | 3 oz |
| Cheese | 1 oz | 1½ oz | 1½ oz | 2 oz | 3 oz |
| Large egg(s) | 1 | 1½ | 1½ | 2 | 3 |
| Cooked dry beans or peas | ½ c | ¾ c | ¾ c | 1 c | 1½ c |
| Peanut butter | 2 tbs | 3 tbs | 3 tbs | 4 tbs | 6 tbs |
| **Vegetable and/or fruit** | | | | | |
| 2 or more servings, both to total | ½ c | ½ c | ½ c | ¾ c | ¾ c |
| **Bread or bread alternate** | | | | | |
| Servings | 5 per week | 8 per week | 8 per week | 8 per week | 10 per week |
| **Milk** | | | | | |
| 1 serving of fluid milk | ¾ c | ¾ c | 1 c | 1 c | 1 c |

lower fat in school lunches typically have trouble providing enough energy and nutrients, especially iron, to meet specifications. The American Dietetic Association (ADA) advocates the development of dietary guidelines specifically for children to ensure that school lunches will both provide adequate energy and nutrients and support health.[36] According to the ADA, the guidelines currently used may be appropriate for adults, but may not be adequate to meet children's unique needs.

• ***Nutrition Education at School*** • Coincident with the school breakfast and lunch programs is a program of nutrition education and training (NET) in all the public schools. This program is minimally funded, but program administrators are ingenious and creative in accomplishing its highest-priority objectives. School health clinics offer another opportunity to provide nutrition education and intervention.[37] Children need to be fed well *and* learn enough about nutrition to make healthful food choices when the choices become theirs to make.

**Healthy People 2000:** Increase to at least 75% the proportion of the nation's schools that provide nutrition education from preschool through grade 12, preferably as part of quality school health education.

IN SUMMARY

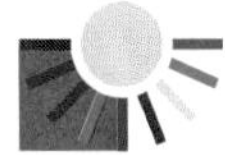

Children's appetites and nutrient needs reflect their stage of growth. Those who are chronically hungry and malnourished suffer growth retardation; when hunger is temporary and nutrient deficiencies are mild, the problems are usually more subtle—such as poor academic performance. Iron deficiency is widespread and has many physical and behavioral consequences. "Hyper" behavior is not caused by poor nutrition; misbehavior may reflect inconsistent care. Too much television watching can contribute to obesity by promoting inactivity and an overconsumption of snack foods. Adults at home and at school need to provide children with nutrient-dense foods and teach them how to make healthful diet and activity choices.

# Nutrition during Adolescence

Teenagers make many more choices for themselves than they did as children. They are not fed, they eat; they are not sent out to play, they choose to go. At the same time, social pressures thrust choices at them: whether to drink alcoholic beverages and whether to develop their bodies to meet extreme ideals of slimness or athletic prowess. Their interest in nutrition—both valid information and misinformation—derives from personal, immediate experiences. They are concerned with how diet can improve their lives now—they engage in crash dieting in order to buy a new bathing suit, avoid greasy foods in an effort to clear acne, or eat a pile of spaghetti to prepare for a big sporting event. In presenting information on the nutrition and health of adolescents, this section includes these many topics of interest to teens.

U.S. Government
www.healthfinder.gov/searchoptions/topicsaz.htm
Search for Adolescent Health

## Growth and Development

With the onset of adolescence, the steady growth of childhood speeds up abruptly and dramatically, and the growth patterns of female and male become distinct. Hormones direct the intensity of the adolescent growth spurt, profoundly affecting every organ of the body, including the brain. After two to three years of intense growth and a few more at a slower pace, physically mature adults emerge.

In general, the adolescent growth spurt begins at age 10 or 11 for females and at 12 or 13 for males. It lasts about two and a half years. Before puberty male and female body composition differ only slightly, but during the adolescent spurt, differences between the sexes become apparent in the skeletal system, lean body mass, and fat stores. In females, fat assumes a larger percentage of the total body weight, and in males, the lean body mass—principally muscle and bone—increases much more than in females. On the average, males grow 8 inches taller, and females, 6 inches taller. Males gain approximately 45 pounds, and females, about 35 pounds.

**puberty:** the period in life in which a person becomes physically capable of reproduction.

Rates and patterns of adolescent growth exhibit such wide variations that standards used for children must be abandoned when the signs of puberty begin to appear. Age in years indicates little about development. One way to monitor adolescent growth is to compare height and weight with previous measures. To record developmental changes during puberty, health care professionals use standard rating scales based on stages of adolescent development.

## *Energy and Nutrient Needs*

Energy and nutrient needs are greater during adolescence than at any other time of life, except pregnancy and lactation. In general, nutrient needs rise throughout childhood, peak in adolescence, and then level off or even diminish as the teen becomes an adult.

**adolescence:** the period from the beginning of puberty until maturity.

• ***Energy Intake and Activity*** • The energy needs of adolescents vary greatly, depending on the current rate of growth, body size, and physical activity. Boys' energy needs may be especially high; they typically grow faster than girls and, as mentioned, develop a greater proportion of lean body mass. An active boy of 15 may need 4000 kcalories or more a day just to maintain his weight. Girls start growing earlier than boys and attain lower heights and weights, so their energy needs peak sooner and decline earlier than those of their male peers. An inactive girl of 15 whose growth is nearly at a standstill may need fewer than 2000 kcalories a day if she is to avoid excessive weight gain. Thus adolescent girls need to pay special attention to being physically active and selecting foods of high nutrient density so as to meet their nutrient needs without exceeding their energy needs.

The insidious problem of obesity becomes ever more apparent in adolescence and often continues into adulthood. One in every five teens is overweight.[38]* The problem is most evident in females, especially those of African American descent. Without intervention, overweight adolescents will face numerous physical and socioeconomic consequences for years to come. The consequences of obesity are so dramatic and our society's attitude toward obese people is so negative that even teens of normal weight perceive a need to control their weight. When taken to the extremes, restrictive diets bring dramatic physical consequences of their own, as Highlight 9 explains.

• ***Vitamins*** • The RDA (or AI) for most vitamins increase during the adolescent years (see the table on the inside front cover). Several of the vitamin recommendations for adolescents are similar to those for adults, including the recommendation for vitamin D. During puberty, both the activation of vitamin D and the absorption of calcium are enhanced, thus supporting the intense skeletal growth of the adolescent years without additional vitamin D.

*For boys, obesity is defined as BMI:
≥23.0 for 12 to 14 years.
≥25.8 for 15 to 17 years.
≥26.8 for 18 to 19 years.

For girls, obesity is defined as BMI:
≥23.4 for 12 to 14 years.
≥24.8 for 15 to 17 years.
≥25.7 for 18 to 19 years.

• ***Iron*** • The need for iron during adolescence differs for males and females. Iron needs increase in females as they start to menstruate and in males as their lean body mass develops. Iron intakes often fail to keep pace with increasing needs, especially for females, who typically consume less iron-rich foods such as meat and fewer total kcalories than males. For females, the RDA rises at adolescence and remains high into late adulthood. For males, the RDA returns to preadolescent values in early adulthood.

Teenagers need to select at least 4 servings from the milk group daily to meet their calcium goal of 1300 mg/day. Chapter 12 presents other calcium-rich food choices.

National Institute of Child Health and Development

www.nih.gov/nichd
Visit Milk Matters

• ***Calcium*** • Adolescence is a crucial time for bone development, and the requirement for calcium reaches its peak during these years.[39] Unfortunately, many adolescents have calcium intakes below current recommendations.[40] Low calcium intakes during the adolescent growth spurt, especially if paired with physical inactivity, may compromise the development of peak bone mass, which is considered the best protection against age-related bone loss and fractures. Increasing milk products in the diet to meet calcium recommendations greatly increases bone density.[41] Once again, however, teenage girls are most vulnerable, for their milk—and therefore calcium—intakes begin to decline at the time when their calcium needs are greatest. Furthermore, women have much greater bone losses than men in later life. In addition to dietary calcium, sports activities during adolescence build strong bones.

**Healthy People 2000:** Increase calcium intake, so that at least 50% of youth aged 12 through 24 years consume three or more servings of calcium-rich foods daily.

## Food Choices and Health Habits

Teenagers like the freedom to come and go as they choose and eat what they want when they have time. With a multitude of afterschool, social, and job activities, they almost inevitably fall into irregular eating habits. At any given time on any given day, a teenager may be skipping a meal, eating a snack, preparing a meal, or consuming food prepared by a parent or restaurant.

***Nutritious snacks play an important role in an active teen's diet.***

• ***Snacks*** • Snacks typically provide at least a fourth of the average teenager's daily food energy intake. Most often, favorite snacks are high in fat and low in calcium, iron, vitamin A, vitamin C, and folate.[42] Most adolescents need to eat a greater variety of foods to obtain these nutrients. Table 16-8 on p. 520 shows how to combine foods from different food groups to create healthy snacks. Vending machines rarely offer nutrient-dense options, and nutrition information alone does not convince people to make healthy choices.[43]

• ***Beverages*** • Adolescents frequently drink soft drinks with lunch, supper, and snacks. About the only time they select fruit juices is at breakfast. When they drink milk, they are more likely to consume it with a meal (especially breakfast) than as a snack. Because of their greater food intakes, boys are more likely than girls to drink enough milk to meet their calcium needs.

For adolescents who can afford the kcalories and are meeting their calcium needs, soft drinks are an acceptable part of the diet. Soft drinks may present a problem, however, when caffeine intake becomes excessive. Caffeine is a stimulant added during the manufacture of many soft drinks; caffeine-containing soft drinks typically deliver between 30 and 55 milligrams of caffeine per 12-ounce can, but some offer twice as much.[44] Caffeine increases the respiration rate, heart rate, blood pressure, and secretion of stress and other hormones. Caffeine seems to be relatively harmless when used in moderate doses (the equivalent of fewer than four 12-ounce cola beverages a day). In greater amounts, it can cause the symptoms associated with anxiety—sweating, tenseness, and inability to concentrate.

For perspective, an 8 oz cup of drip-brewed coffee contains 184 mg of caffeine. A pharmacologically active dose of caffeine is defined as 200 mg. Appendix H, p. H-3, lists caffeine contents of selected foods, beverages, and drugs.

• ***Eating Away from Home*** • Adolescents eat about one-third of their meals away from home, and their nutritional welfare is enhanced or hindered by the choices they make.[45] A lunch of a hamburger, a chocolate shake, and french fries supplies substantial quantities of many nutrients at a kcalorie cost of about 800, an energy intake many adolescents can afford (see Table 16-10). When they eat this sort of lunch, teens can adjust their breakfast and dinner choices to include fruits and vegetables for vitamin A, vitamin C, folate, and fiber and lean meats and legumes for iron and zinc. (See Appendix H for the nutrient contents of fast foods.)

• ***Peer Influence*** • Many of the food and health choices adolescents make reflect the opinions and actions of their peers. When others perceive milk as "babyish," a teen will choose soft drinks instead; when others skip lunch and hang out in the parking lot, a teen may join in for the camaraderie, regardless of hunger. Adults need to remember that adolescents have the right to make their own decisions—even if they are contrary to the adults' views. Gatekeepers can set up the environment so that nutritious foods are available and can stand by with reliable nutrition information and advice, but the rest is up to the adolescents. Ultimately, they make the choices. (Highlight 9 examines the influence of social pressures on the development of eating disorders.)

**Table 16-10**

**Selected Nutrients in a Hamburger, Chocolate Shake, and Small Serving of French Fries**

| | % Recommended Intake | |
|---|---|---|
| ***Nutrient*** | ***Male[a]*** | ***Female[a]*** |
| Energy | 27 | 37 |
| Protein | 47 | 63 |
| Fat[b] | 24 | 33 |
| Vitamin A | 0 | 0 |
| Folate | 12 | 12 |
| Vitamin C | 22 | 22 |
| Calcium | 39 | 39 |
| Iron | 36 | 29 |
| Zinc | 17 | 22 |
| Sodium[b] | 38 | 38 |

[a]Recommendations for a 15- to 18-year-old, moderately active person of average height and weight.
[b]Daily Values used for fat and sodium.

# Problems Adolescents Face

Physical maturity and growing independence present adolescents with new choices to make. The consequences of those choices will influence their nutritional health both today and throughout life. Some teenagers begin using drugs, alcohol, and tobacco; others wisely refrain. Information about the use of these substances is presented here because most people are first exposed to them during adolescence, but it actually applies to people of all ages.

The dangers of steroid use are presented in Highlight 14.

• ***Marijuana*** • One out of every three high school students reports having at least tried marijuana.[46] Marijuana is unique among drugs in that it seems to enhance the enjoyment of eating, especially of sweets, a phenomenon commonly known as "the munchies." Why or how this effect occurs is not known; it may be a social effect induced by suggestibility, or perhaps the drug stimulates appetite. Whatever the reason, prolonged use of the drug does not seem to bring about a weight gain.

 U.S. Government
www.healthfinder.gov/searchoptions/topicsaz.htm
Search for Marijuana

• ***Cocaine*** • One in 20 high school seniors reports having used cocaine at least once.[47] Cocaine stimulates the nervous system and elicits the stress response—constricted blood vessels, raised blood pressure, widened pupils of the eyes, and increased body temperature. It also drives away feelings of fatigue. Cocaine occasionally causes immediate death—usually by heart attack, stroke, or seizure in an already damaged body system.

Weight loss is common, and cocaine abusers often develop eating disorders. Notably, the craving for cocaine replaces hunger; rats given unlimited cocaine will choose it over food until they starve to death. Thus, unlike marijuana use, cocaine use has major nutritional consequences.

• ***Drug Abuse, in General*** • The effects of other addictive drugs vary in degree but are similar in kind to those caused by cocaine. Drug abusers face the multiple nutrition problems listed in the margin. During withdrawal from drugs, an important part of treatment is to identify and correct these nutrition problems.

Nutrition problems of drug abusers:
- They buy drugs with money that could be spent on food.
- They lose interest in food during "highs."
- They use drugs that depress appetite.
- Their lifestyle fails to promote good eating habits.
- They use intravenous (IV) drugs. They may contract AIDS, hepatitis, or other infectious diseases, which increase their nutrient needs. Hepatitis also causes taste changes and loss of appetite.
- Medicines used to treat drug abuse may alter nutrition status.

• ***Alcohol Abuse*** • Sooner or later all teenagers face the decision whether to drink alcohol. The law forbids the sale of alcohol to people under 21, but most adolescents who seek alcohol can obtain it. Four out of five high school students have had at least one alcoholic beverage; about half drink regularly; and one in three students drinks heavily (defined as five or more drinks on at least one occasion in the previous month).[48]

Highlight 7 describes how alcohol affects nutrition status. To sum it up, alcohol provides energy but no nutrients, and it can displace nutritious foods from the diet. Alcohol alters nutrient absorption and metabolism, so imbalances develop.

People who cannot keep their alcohol use moderate must abstain to maintain their health. Appendix F lists resources for people with alcohol-related problems.

• ***Smoking*** • Cigarette smoking is a pervasive health problem causing thousands of people to suffer from cancer and diseases of the cardiovascular, digestive, and respiratory systems. These effects are beyond the scope of nutrition, but smoking cigarettes does influence hunger, body weight, and nutrient status.

Smoking a cigarette eases feelings of hunger. When smokers receive a hunger signal, they can quiet it with cigarettes instead of food. Such behavior ignores body signals and postpones energy and nutrient intake. In rats, nicotine reduces food intake and increases the rate of energy expenditure, causing weight loss.

Indeed, smokers tend to weigh less than nonsmokers and to gain weight when they stop smoking. Weight gain is often a concern for people contemplating giving up cigarettes. They should know that the average person who quits smoking gains less than 10 pounds. Smokers wanting to quit need to prepare for this possibility and adjust their diet and activity habits so as to maintain weight during and after quitting. Smoking cessation programs need to include strategies for weight management.

Nutrient intakes of smokers and nonsmokers differ. Smokers tend to have lower intakes of dietary fiber, vitamin A, beta-carotene, folate, and vitamin C.[49] The association between smoking and low intakes of fruits and vegetables rich in these nutrients may be noteworthy, considering their protective effect against lung cancer (see Highlight 11).

Researchers have found that compared to nonsmokers, smokers require almost twice as much vitamin C to maintain steady body pools. Oxidants in cigarette smoke accelerate vitamin C metabolism and deplete smokers' body stores of this antioxidant; this depletion is even evident to some degree in nonsmokers who are exposed to passive smoke.[50]

The vitamin C RDA for people who regularly smoke cigarettes is 100 mg/day. The Canadian RNI suggests smokers should add 50% to the vitamin C recommendation.

Beta-carotene enhances the immune response and protects against some cancer activity.[51] Specifically, the risk of lung cancer is greatest for smokers who have the lowest intakes of carotene. Of course, such evidence should not be misinterpreted. It does not mean that as long as people eat their carrots, they can safely use tobacco. Nor does it mean that beta-carotene supplements would be beneficial. Smokers are ten times more likely to get lung cancer than nonsmokers. Both smokers and nonsmokers can, however, reduce their cancer risks by eating fruits and vegetables rich in carotene (see Highlight 11 for details on antioxidant nutrients and disease prevention).

• ***Smokeless Tobacco*** • Nationwide, one in ten high school students reports having used smokeless tobacco products.[52] Like cigarettes, smokeless tobacco use is linked to many health problems, from minor mouth sores to tumors in the nasal cavities, cheeks, gums, and throat. The risk of mouth and throat cancers is even greater than for smoking tobacco. Other drawbacks to tobacco chewing and snuff dipping include bad breath, stained teeth, and blunted senses of smell and taste. Tobacco chewing also damages the gums, tooth surfaces, and jawbones, making it likely that users will lose their teeth in later life.

## IN SUMMARY

Nutrient needs rise dramatically as children enter the rapid growth phase of the teen years. The busy lifestyles of adolescents add to the challenge of meeting their nutrient needs—especially for iron and calcium. In addition to making wise food choices, adolescents need to refrain from using substances that will impair their health—including illicit drugs, alcohol, and tobacco.

The nutrition and lifestyle choices people make as children and adolescents have long-term, as well as immediate, effects on their health. Highlight 16 describes how sound choices and good habits can help prevent disease later in life.

## Study Questions

These questions will help you review the chapter. You will find the answers in the discussions on the pages provided.

1. Describe some of the nutrient and immunological attributes of breast milk. (pp. 504–506)
2. What are the appropriate uses of formula feeding? What criteria would you use in selecting an infant formula? (p. 506)
3. Why are solid foods not recommended for an infant during the first few months of life? When is an infant ready to start eating solid food? (pp. 509–511)
4. Name foods that are inappropriate for infants and explain why they are inappropriate. (p. 511)
5. What nutrition problems are most common in children? What strategies can help prevent these problems? (pp. 514–516)
6. Describe the relationships between nutrition and behavior. How does television influence nutrition? (pp. 516–517)
7. Describe a true food allergy. Which foods most often cause allergic reactions? How do food allergies influence nutrition status? (pp. 517–518)
8. List strategies for introducing nutritious foods to children. (pp. 518–520)
9. What impact do school meal programs have on the nutrition status of children? (pp. 520–521)
10. Describe the changes in nutrient needs from childhood to adolescence. Why is an adolescent girl more likely to develop an iron deficiency than is a boy? (pp. 522–524)
11. How do adolescents' eating habits influence their nutrient intakes? (pp. 524–525)
12. How does the use of illicit drugs influence nutrition status? (p. 525)
13. How do the nutrient intakes of smokers differ from those of nonsmokers? What impacts can those differences exert on health? (p. 526)

These questions will help you prepare for an exam. Answers can be found in Appendix K.

1. A reasonable weight for a healthy five-month-old infant who weighed 8 pounds at birth might be:
   a. 12 pounds.
   b. 16 pounds.
   c. 20 pounds.
   d. 24 pounds.
2. Dehydration can develop quickly in infants because:
   a. much of their body water is extracellular.
   b. they lose a lot of water through urination and tears.
   c. only a small percentage of their body weight is water.
   d. they drink lots of breast milk or formula, but little water.
3. An infant should begin eating solid foods between:
   a. 2 and 4 weeks.
   b. 1 and 3 months.
   c. 4 and 6 months.
   d. 8 and 10 months.
4. Among U.S. and Canadian children, the most prevalent nutrient deficiency is of:
   a. iron.
   b. folate.
   c. protein.
   d. vitamin D.
5. A true food allergy always:
   a. elicits an immune response.
   b. causes an immediate reaction.
   c. creates an aversion to the offending food.
   d. involves symptoms such as headaches or hives.
6. Which of the following strategies is *not* effective?
   a. Play first, eat later.
   b. Provide small portions.
   c. Encourage children to help prepare meals.
   d. Use dessert as a reward for eating vegetables.
7. To help teenagers consume a balanced diet, parents can:
   a. monitor the teens' food intake.
   b. give up—parents can't influence teenagers.
   c. keep the pantry and refrigerator well stocked.
   d. forbid snacking and insist on regular, well-balanced meals.
8. The nutrients most likely to fall short in the adolescent diet are:
   a. sodium and fat.
   b. folate and zinc.
   c. iron and calcium.
   d. protein and vitamin A.
9. During adolescence, energy and nutrient needs:
   a. reach a peak.
   b. fall dramatically.
   c. rise, but do not peak until adulthood.
   d. fluctuate so much that generalizations can't be made.
10. To balance the day's intake, an adolescent who eats a hamburger, fries, and cola at lunch might benefit most from a dinner of:
    a. fried chicken, rice, and banana.
    b. ribeye steak, baked potato, and salad.
    c. pork chop, mashed potatoes, and apple juice.
    d. spaghetti with meat sauce, broccoli, and milk.

# Notes


1. Committee on Nutrition, American Academy of Pediatrics, *Pediatric Nutrition Handbook,* 3rd ed., ed. L. A. Barness (Elk Grove Village, Ill.: American Academy of Pediatrics, 1993), pp. 23–33.
2. Work Group on Breastfeeding, American Academy of Pediatrics, Breastfeeding and the use of human milk, *Pediatrics* 100 (1997): 1035–1037; Position of The American Dietetic Association: Promotion of breast feeding, *Journal of the American Dietetic Association* 97 (1997): 662–665.
3. A. C. Goedhart and J. G. Bindels, The composition of human milk as a model for the design of infant formulas: Recent findings and possible applications, *Nutrition Research Reviews* 7 (1994): 1–23.
4. M. A. Murtaugh, Optimal breastfeeding duration, *Journal of the American Dietetic Association* 97 (1997): 1252–1254.
5. J. S. Forsyth, Is it worthwhile breastfeeding? *European Journal of Clinical Nutrition* (supplement 1) 46 (1993): 519–525.
6. Committee on Nutritional Status during Pregnancy and Lactation, Food and Nutrition Board, *Nutrition during Lactation* (Washington, D.C.: National Academy Press, 1991), pp. 155–156.
7. Committee on Nutrition, 1993, pp. 34–42.
8. D. S. Newburg and J. M. Street, Bioactive materials in human milk—Milk sugars sweeten the argument for breast-feeding, *Nutrition Today* 32 (1997): 191–201.
9. Committee on Nutrition, American Academy of Pediatrics, Soy protein–based formulas: Recommendations for use in infant feeding, *Pediatrics* 101 (1998): 148–153.
10. Committee on Nutrition, 1993, pp. 115–124.
11. Work Group on Breastfeeding, 1997.
12. Committee on Nutrition, 193, pp. 23–33.
13. J. K. Jarvis and G. D. Miller, Fat in infant diets, *Nutrition Today* 31 (1996): 182–190.
14. J. D. Cook and coauthors, The influence of different cereal grains on iron absorption from infant cereal foods, *American Journal of Clinical Nutrition* 65 (1997): 964–969.
15. B. A. Dennison, H. L. Rockwell, and S. L. Baker, Excess fruit juice consumption by preschool-aged children is associated with short stature and obesity, *Pediatrics* 99 (1997): 15–22; M. M. Smith and F. Lifshitz, Excess fruit juice consumption as a contributing factor in nonorganic failure to thrive, *Pediatrics* 93 (1994): 438–443.
16. S. A. Sullivan and L. L. Birch, Infant dietary experience and acceptance of solid foods, *Pediatrics* 93 (1994): 271–277.
17. Position of The American Dietetic Association: Health implications of dietary fiber, *Journal of the American Dietetic Association* 97 (1997): 1157–1159.
18. M. Boutry and R. Needlman, Use of diet history in the screening of iron deficiency, *Pediatrics* 98 (1996): 1138–1142.
19. K. A. Muñoz and coauthors, Food intakes of US children and adolescents compared with recommendations, *Pediatrics* 100 (1997): 323–329.
20. E. Kennedy and C. Davis, US Department of Agriculture School Breakfast Program, *American Journal of Clinical Nutrition* 67 (1998): 798S–803S.
21. E. Pollitt and R. Mathews, Breakfast and cognition: An integrative summary, *American Journal of Clinical Nutrition* 67 (1998): 804S–813S.
22. J. Burghardt and B. Devaney, The School Nutrition Dietary Assessment Study: Summary of Findings (U.S. Department of Agriculture, 1993).
23. I. de Andraca, M. Castillo, and T. Walter, Psychomotor development and behavior in iron-deficient anemic infants, *Nutrition Reviews* 55 (1997): 125–132.
24. E. Pollitt, Iron deficiency and educational deficiency, *Nutrition Reviews* 55 (1997): 133–141.
25. Committee on Children with Disabilities and Committee on Drugs, Medication for children with attentional disorders, *Pediatrics* 98 (1996): 301–304.
26. J. W. White and M. Wolraich, Effect of sugar and mental performance, *American Journal of Clinical Nutrition* (supplement) 62 (1995): 242–249; M. L. Wolraich and coauthors, Effects of diets high in sucrose or aspartame on the behavior and cognitive performance of children, *New England Journal of Medicine* 330 (1994): 301–307.
27. R. E. Anderson and coauthors, Relationship of physical activity and television watching with body weight and level of fatness among children: Results from the third National Health and Nutrition Examination Survey, *Journal of the American Medical Association* 279 (1998): 938–942.
28. National Academy of Sciences Committee, as quoted by J. Raloff and D. Pendick, Pesticides in produce may threaten kids, *Science News* 144 (1993): 4–5.
29. Three agencies propose pesticide reforms, *FDA Consumer,* January/February 1994, p. 3.
30. C. Marwick, Pesticides pose concern about children's diet, *Journal of the American Medical Association* 270 (1993): 802, 805.
31. V. L. Olejer, Food hypersensitivities, *Handbook of Pediatric Nutrition* (Gaithersburg, Md.: Aspen Publishers, 1993), pp. 206–231.
32. H. A. Sampson, Food allergy, *Journal of the American Medical Association* 278 (1997): 1888–1894.
33. C. Evers, Empower children to develop healthful eating habits, *Journal of the American Dietetic Association* 97 (1997): S116; E. Satter, *How to Get Your Kid to Eat . . . But Not Too Much* (Palo Alto, Calif.: Bull Publishing Company, 1987), pp. 13–28.
34. M. Nahikian-Nelms, Influential factors of caregiver behavior at mealtime: A study of child-care programs, *Journal of the American Dietetic Association* 97 (1997): 505–509.
35. Position of The American Dietetic Association: Nutrition standards for child care programs, *Journal of the American Dietetic Association* 94 (1994): 323–328.
36. Timely statement of The American Dietetic Association: Dietary guidance for healthy children, *Journal of the American Dietetic Association* 95 (1995): 370.
37. A. Jasaitis, School-based health clinics: The role for nutrition, *Journal of the American Dietetic Association* 97 (1997): S117.
38. Prevalence of overweight among adolescents, United States, 1988–1991, *Morbidity and Mortality Weekly Report* 43 (1994): 819–821.
39. A. D. Martin and coauthors, Bone mineral and calcium accretion during puberty, *American Journal of Clinical Nutrition* 66 (1997): 611–615.
40. S. I. Barr, Associations of social and

demographic variables with calcium intakes of high school students, *Journal of the American Dietetic Association* 94 (1994): 260–266, 269.

41. J. Cadogan and coauthors, Milk intake and bone mineral acquisition in adolescent girls: Randomised, controlled intervention trial, *British Medical Journal* 315 (1997): 1255–1260; G. M. Chan, K. Hoffman, and M. McMurry, Effects of dairy products on bone and body composition in pubertal girls, *Journal of Pediatrics* 126 (1995): 551–556.
42. J. G. Dausch and coauthors, Correlates of high-fat/low-nutrient-dense snack consumption among adolescents: Results from two national health surveys, *American Journal of Health Promotion* 10 (1995): 85–88.
43. S. M. Hoerr and V. A. Louden, Can nutrition information increase sales of healthful vended snacks? *Journal of School Health* 63 (1993): 386–390.
44. International Food Information Council, Caffeine and health: Clarifying the controversies, *IFIC Review,* May 1993.
45. B.-H. Lin, J. Guthrie, and J. R. Blaylock, *The Diets of America's Children—Influences of Dining Out, Household Characteristics, and Nutrition Knowledge* (Washington, D.C.: U.S. Department of Agriculture, December 1996).
46. L. Kann and coauthors, Youth risk behavior surveillance—United States, 1993, *Journal of School Health* 65 (1995): 163–171.
47. Kann and coauthors, 1995.
48. Kann and coauthors, 1995.
49. T. A. B. Sanders and coauthors, Essential fatty acids, plasma cholesterol, and fat-soluble vitamins in subjects with age-related maculopathy and matched control subjects, *American Journal of Clinical Nutrition* 57 (1993): 428–433; A. F. Subar, L. C. Harlan, and M. E. Mattson, Food and nutrient intake differences between smokers and non-smokers in the US, *American Journal of Public Health* 80 (1990): 1323–1329.
50. D. L. Tribble, L. J. Giuliano, and S. P. Fortmann, Reduced plasma ascorbic acid concentrations in nonsmokers regularly exposed to environmental tobacco smoke, *American Journal of Clinical Nutrition* 58 (1993): 886–890.
51. G. van Poppel, S. Spanhaak, and T. Ockhuizen, Effect of ß-carotene on immunological indexes in healthy male smokers, *American Journal of Clinical Nutrition* 57 (1993): 402–407.
52. Kann and coauthors, 1995.

# Childhood Obesity and the Early Development of Chronic Diseases

When people think of the health problems of children and adolescents, they typically think of measles and acne, not cardiovascular disease (CVD). They think of CVD as the number one killer of adults in the United States and Canada, but CVD begins in childhood.

Over the past three decades, researchers have been observing how changes in body weight, blood lipids, blood pressure, and individual behaviors correlate with the development of CVD over time—from infancy to childhood through adolescence and into young adulthood. Some major findings have emerged from this research:

- Changes inside the arteries—changes predictive of CVD—are evident in childhood.
- Obesity in children affects these changes.
- Behaviors that influence the development of obesity and of CVD are learned and begin early in life. These behaviors include overeating, eating high-fat foods, physical inactivity, and cigarette smoking.

This highlight focuses on efforts to prevent childhood obesity and CVD, but the benefits extend to cancer, diabetes, and other chronic diseases as well. The years of childhood (ages 2 to 18 years) are emphasized here, for the earlier in life health-promoting habits become established, the better they will stick. Chapter 18 fills in the rest of the story of nutrition's role in reducing chronic disease risk.

Invariably, questions arise as to what extent genetics is involved in CVD development. Children who are obese and who have high blood lipids and high blood pressure are often from families with a history of CVD. Genetics does not appear to play a *determining* role in CVD; that is, a person is not simply destined at birth to develop CVD.[1] Instead, genetics appears to play a *permissive* role—the potential is inherited and then will develop, if given a push by poor health choices such as excessive weight gain, poor diet, sedentary lifestyle, and cigarette smoking.

## Early Development of CVD

Most people consider CVD to be an adult disease because its incidence rises with advancing age, and symptoms rarely appear before age 30. In actuality, the disease process begins much earlier.

*Take care of your body and your body will take care of you.*

### Atherosclerosis

Most CVD involves atherosclerosis—the accumulation of cholesterol and other blood lipids along the walls of the arteries (see the glossary on p. 533 for atherosclerosis and related terms). If it progresses, atherosclerosis may eventually block the flow of blood to the heart and cause a heart attack or cut off blood flow to the brain and cause a stroke. Infants are born with healthy, smooth, clear arteries, but within the first decade of life, fatty streaks may begin to appear (see Figure H16-1). During adolescence, these fatty streaks may begin to turn to fibrous plaques. By early adulthood, the fibrous plaques may begin to calcify and become raised lesions, especially in boys and young men. As the lesions grow more numerous and enlarge, the heart disease rate begins to rise, and the rise becomes dramatic at about age 45 in men and 55 in women. From this point on, arterial damage and blockage progress rapidly, and heart attacks and strokes threaten life. In short, the consequences of atherosclerosis, which become apparent only in adulthood, have their beginnings in the first decades of life.[2]

Atherosclerosis is not inevitable; people can grow old with relatively clear arteries. Early lesions may either progress or regress, depending on several factors, many of which reflect lifestyle behaviors. Smoking, for example, is strongly associated with the prevalence of raised lesions, even in young adults.

### Blood Cholesterol

Atherosclerotic lesions reflect blood cholesterol: as blood cholesterol increases, lesion coverage increases. Cholesterol values at birth are similar in all populations; differences emerge in early childhood. In countries where the adults have high blood cholesterol and high rates of CVD, the children also tend to have high blood cholesterol. Conversely, in countries where the adults have low blood cholesterol and low rates of CVD, the children tend to have low blood cholesterol, suggesting that adult heart disease tracks early trends and that early preventive efforts might reduce the incidence of later CVD.

Such is the case among populations, but individual cholesterol status also becomes established in childhood, as early as one year.[3] The best predictor of a person's blood cholesterol is earlier baseline values: childhood values correlate with values in young adulthood.[4] Quite simply, if you want to know a child's future cholesterol, measure it now. Standard values for cholesterol screening in children and adolescents (ages 2 to 18 years) are listed in Table H16-1 on p. 532.[5]

Blood cholesterol also correlates with obesity, especially central obesity. LDL cholesterol rises with obesity, and HDL declines. These relationships are apparent throughout childhood, and their magnitude increases with age.

Research has also confirmed an association between blood lipids and physical activity in children, similar to that

**Figure H16-1**

**The Formation of Fibrous Plaques in Atherosclerosis**

When fibrous plaques have covered 60 percent of the coronary artery walls, the critical phase of heart disease begins.

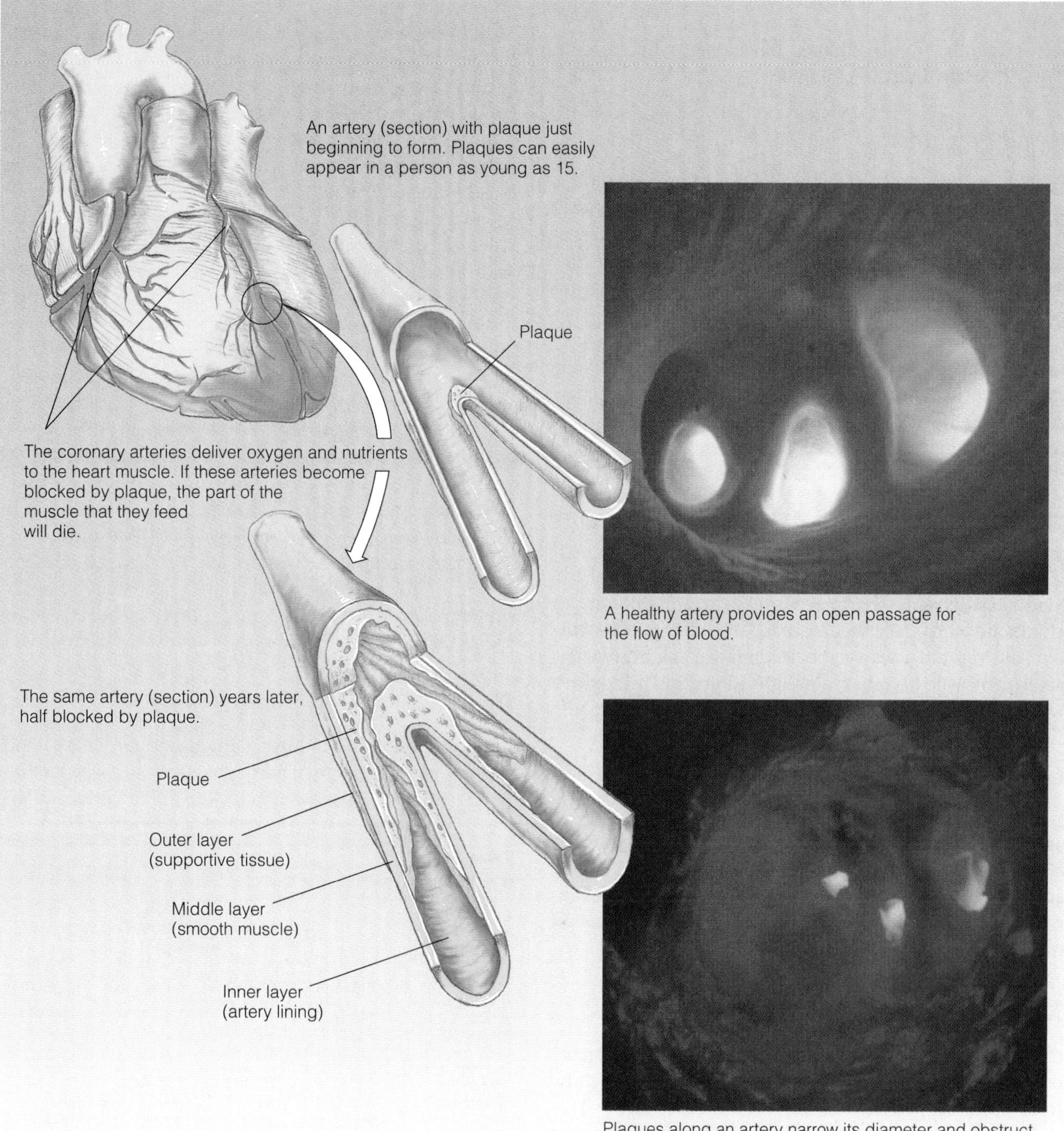
An artery (section) with plaque just beginning to form. Plaques can easily appear in a person as young as 15.
Plaque
The coronary arteries deliver oxygen and nutrients to the heart muscle. If these arteries become blocked by plaque, the part of the muscle that they feed will die.
The same artery (section) years later, half blocked by plaque.
Plaque
Outer layer (supportive tissue)
Middle layer (smooth muscle)
Inner layer (artery lining)

A healthy artery provides an open passage for the flow of blood.

Plaques along an artery narrow its diameter and obstruct blood flow. Clots can form, aggravating the problem.

**Table H16-1**
**Cholesterol Values for Children and Adolescents**

| Disease Risk | Total Cholesterol (mg/dL) | LDL Cholesterol (mg/dL) |
|---|---|---|
| Acceptable | <170 | <110 |
| Borderline | 170–199 | 110–129 |
| High | ≥200 | ≥130 |

NOTE: Adult values appear in Table 18-2 on p. 565.

seen in adults.[6] Active children have a better lipid profile than physically inactive children.[7]

### Blood Pressure

Pediatricians routinely monitor blood pressure in children and adolescents.[8] High blood pressure may signal an underlying disease or the early onset of hypertension. Hypertension accelerates the development of CVD. Standard values for hypertension screening in children and adolescents are given in Table H16-2.[9]

Like atherosclerosis and high blood cholesterol, hypertension may develop in the first decades of life.[10] Children can control their hypertension by participating in regular aerobic activity and by losing weight or maintaining their weight as they grow taller.[11] No evidence suggests that restricting sodium lowers blood pressure in children and adolescents.[12]

## Development of Obesity in Children

Many experts agree that preventing or treating obesity in childhood will reduce the rate of CVD in adulthood. Without intervention, overweight children become overweight adolescents who become overweight adults, and being overweight exacerbates every chronic disease that adults face.[13]

### Growing Fatter

Children are heavier today than they were 20 or so years ago. Since the late 1970s, the prevalence of overweight has almost doubled for children—and more than doubled for adolescents.[14] This pattern is a secular trend—that is, it cannot be explained by genetics. Diet and physical inactivity must be responsible.

### Not Eating More

Children's energy intakes have remained relatively stable over the past 15 years. Fat intake has even declined slightly, from 38 to 36 percent of kcalories from fat daily.[15] This slight decline in dietary fat is not enough to have influenced body weight, however, nor is it enough to meet current dietary recommendations.

Children's dietary fat intakes vary, of course, and some children do eat high-fat diets. Children who prefer high-fat foods tend to consume a relatively large percentage of their energy intake from fat.[16] They also tend to be more overweight than their peers. Particularly noteworthy is the finding that the children's fat preferences and consumption correlate with their parents' obesity as well. Such findings confirm the significant roles parents play—teaching children about healthy food choices, providing children with low-fat selections, and serving as role models.

### Growing Less Active

Most likely, children have grown more overweight because of their lack of physical activity.[17] An inactive child can become obese even while eating less food than an active child. Today's children are more sedentary and less physically fit than children were 20 years ago.

Watching television accounts for some 24 hours a week of sedentary behavior. Beyond these 24 hours, children spend more sedentary time sitting at computers and playing video games. As mentioned in earlier chapters, both obesity and blood cholesterol correlate with hours of television viewed.[18] Watching television uses no more energy than resting, displaces participation in more vigorous activities, and fosters snacking on high-fat foods.

Just as blood cholesterol and obesity track over the years, so does a person's level of physical activity. A study of almost 1000 teenagers found that over half of those who were initially described as inactive were still inactive six years later.[19] Similarly, almost half of those who were physically active remained so. Compared with inactive teens, those who were physically active weighed less, smoked less, ate a diet lower in saturated fats, and had better blood lipid profiles. The message is clear: physical activity offers numerous health benefits, and children who are active today are most likely to be active for years to come.

## Preventing Childhood Obesity

In light of all these findings, parents and teachers of children are encouraged to make major efforts to prevent childhood obesity. Suggestions include the following: encourage

**Table H16-2**
**Hypertension Standards for Children and Adolescents**

| | Systolic over Diastolic Pressure (mm Hg) | | | |
|---|---|---|---|---|
| | *6 to 9 yr* | *10 to 12 yr* | *13 to 15 yr* | *16 to 18 yr* |
| Mild hypertension | 111–121 over 70–77 | 117–125 over 75–81 | 124–135 over 77–85 | 127–141 over 80–91 |
| Moderate hypertension | 122–129 over 70–85 | 126–133 over 82–89 | 136–143 over 86–91 | 142–149 over 92–97 |
| Severe hypertension | >129 over >85 | >133 over >89 | >143 over >91 | >149 over >97 |

children to eat slowly, to pause and enjoy their table companions, and to stop eating when they are full. Teach them how to select low-fat snacks and to serve themselves appropriate portions. Never force children to clean their plates. Encourage physical activity daily to promote strong skeletal, muscular, and cardiovascular development and to instill in children the desire to be physically active throughout life. Physical activity is a natural and lifelong behavior of healthy living. It can be as simple as riding a bike, playing tag, jumping rope, or doing chores. It need not be an organized sport; it just needs to be some activity on a regular basis.

Above all, be sensitive in teaching children nutrition principles that can help to prevent obesity. Children can easily get the idea that their worth is tied to their body weight. Some parents fail to realize that society's ideal of slimness can be perilously close to starvation, and that a child encouraged to "diet" cannot obtain the energy and nutrients required for normal growth and development. Even healthy children without diagnosable eating disorders have been observed to limit their growth through "dieting." Weight gain in truly overweight children can be controlled safely without compromising growth, but should be overseen by a health care professional.

## Dealing with Childhood Obesity

The child who is already obese needs careful management. Weight loss is not ordinarily recommended because restrictive diets can easily impair growth in children. Instead, aim to maintain weight while the child grows taller. The object is to support normal lean body development, while letting children "grow out" of their obesity.

## Cholesterol Screening for Children

Many children in the United States are not only overweight but also have high blood cholesterol.[20] These children are quite likely to have parents who developed CVD early.[21] For this reason, selective screening is recommended for children and adolescents whose parents or grandparents have CVD; those whose parents have elevated blood cholesterol; and those whose family history is unavailable, especially if other risk factors are evident.[22] Since blood cholesterol in children is a good predictor of adult values, some experts recommend universal screening to identify all children with high blood cholesterol.[23] They note that many children who have high blood cholesterol would be missed under current screening criteria.[24]

Among those children who may have high blood cholesterol, but may not meet screening criteria are those who are overweight.[25] The incidence of high blood cholesterol in obese children with no other criteria is similar to that of nonobese children with family histories of CVD. In addition to overweight, health care professionals should consider whether children smoke or consume a high-fat diet.[26]

Early—but not advanced—atherosclerotic lesions are reversible, making screening and education a high priority. Both those with family histories of CVD and those with multiple risk factors need intervention. Children with the highest risks of developing CVD are sedentary and obese, with high blood pressure and high blood cholesterol. In contrast, children with the lowest risks of heart disease are physically active and of normal weight, with low blood pressure and favorable lipid profiles. Routine pediatric care should identify these known risk factors and provide intervention when needed (see Table H16-3 on p. 534).

## Dietary Recommendations for Children

Regardless of family history, all children over age two should eat a variety of foods and maintain desirable weight. Children should receive at least 20 percent and no more than 30 percent of total energy from fat, less than 10 percent from saturated fat, and less than 300 milligrams of cholesterol per day.[27]

### Not Before Two

Recommendations limiting fat and cholesterol are not intended for infants or children under two years old. Infants and toddlers may need a higher percentage of fat to support their rapid growth.

### Moderation, Not Deprivation

Healthy children over age two can begin the transition to eating according to recommendations by eating fewer high-fat foods, replacing some high-fat foods with low-fat choices, and selecting more fruits and vegetables.[28] All high-fat foods need not be eliminated, though. Healthy meals can still include moderate amounts of a child's favorite foods, even if they are high-fat selections such as french fries and ice cream.[29] Without such additions, diets might be too low in fat, not to mention unappetizing and boring.

## Glossary

**atherosclerosis** (ath-er-oh-scler-OH-sis): a type of artery disease characterized by accumulations of lipid-containing material on the inner walls of the arteries (see Chapter 18).
- **athero** = porridge or soft
- **scleros** = hard
- **osis** = condition

**cardiovascular disease (CVD):** a general term for all diseases of the heart and blood vessels. Atherosclerosis is the main cause of CVD. When the arteries that carry blood to the heart muscle become blocked, the heart suffers damage known as **coronary heart disease (CHD).**
- **cardio** = heart
- **vascular** = blood vessels

**fatty streaks:** accumulations of cholesterol and other lipids along the walls of the arteries.

**fibrous plaques** (PLACKS): mounds of lipid material, mixed with smooth muscle cells and calcium, which develop in the artery walls in atherosclerosis.

**Table H16-3**

**Health Professional's Schedule of Cardiovascular Disease Assessment in Children**

| | |
|---|---|
| **Birth** | • Family history for early heart disease, high blood lipids (if positive, discuss risk factors and refer parents to health care). |
| | • Start growth chart. |
| | • Parental smoking history (if positive, refer to smoking cessation program). |
| **0–2 years** | • Update family history, growth chart. |
| | • With introduction of solids, begin teaching about healthy diet (nutritionally adequate, low in salt, low in saturated fats). |
| | • Recommend healthy snacks as finger foods. |
| | • Change to whole milk from formula or breastfeeding at approximately 1 year of age. |
| **2–6 years** | • Update family history, growth chart (review growth chart[a] with family and discuss concept of weight for height). |
| | • Introduce moderately low-fat diet. |
| | • Change to reduced-fat or low-fat milk. |
| | • Start blood pressure chart at approximately 3 years of age[b]; review for concept of lower salt intake. |
| | • Encourage active parent-child play. |
| | • Lipid determination in children with positive family history or with parental cholesterol >240 mg/dl (if abnormal, initiate nutrition counseling). |
| **6–10 years** | • Update family history, blood pressure, and growth charts. |
| | • Complete cardiovascular health profile with child; determine family history, smoking history, blood pressure percentile, weight for height, fingerstick cholesterol, and level of activity and fitness. |
| | • Reinforce low-fat diet. |
| | • Begin active antismoking counseling. |
| | • Introduce fitness for health and encourage lifelong sport activities for child and family. |
| | • Discuss role of watching television in sedentary lifestyle and obesity. |
| **>10 years** | • Update family history, blood pressure, and growth charts annually. |
| | • Review low-fat diet, risks of smoking, fitness benefits whenever possible. |
| | • Consider lipid profile in all patients. |
| | • Final review of personal cardiovascular health status. |

[a]If weight is more than 120 percent of normal for height, diagnosis of obesity should be considered and the subject addressed with the child and family.

[b]If three consecutive interval blood pressure measurements exceed the 90th percentile and blood pressure is not explained by height or weight, diagnosis of hypertension should be made and appropriate evaluation considered.

Source: Adapted with permission from W. B. Strong and coauthors, Integrated cardiovascular health promotion in childhood: A statement for health professionals from the Subcommittee on Atherosclerosis and Hypertension in Childhood of the Council on Cardiovascular Disease in the Young, American Heart Association, *Circulation* 85 (1992): 1638–1650. Copyright 1992 American Heart Association.

Balanced meals need to provide lean meat, poultry, fish, and vegetable sources of protein; fruits and vegetables; whole grains; and low-fat milk products. Such meals can provide enough energy and nutrients to support growth and maintain blood cholesterol within a healthy range.[30]

Pediatricians warn parents to avoid extremes; they caution that while intentions may be good, excessive food restriction may create nutrient deficiencies and impair growth. Furthermore, parental control over eating may instigate battles and foster attitudes about foods that can lead to inappropriate eating behaviors.

## *Diet First, Drugs Later*

Experts agree that children with high blood cholesterol should first be treated with diet. If blood cholesterol remains high in children ten years and older after 6 to 12 months of dietary intervention, then drugs may be necessary to lower blood cholesterol.[31]

## *Smoking*

Even though the focus of this text is nutrition, another risk factor for CVD that starts in childhood and carries over into adulthood must also be addressed—cigarette smoking. Each day 5000 children light up for the first time—typically in grade school. Among high school students, seven out of ten have tried smoking, and one in six smokes regularly.[32] Approximately 90 percent of all adult smokers began smoking before the age of 18.[33]

Efforts to teach children about the dangers of smoking need to be aggressive. Children are not likely to consider the long-term health consequences of tobacco use. They are more likely to be struck by the immediate health consequences, such as shortness of breath when playing sports, or social consequences, such as having bad breath. Whatever the context, the message to all children and teens should be clear: don't start smoking. If you've already started, quit.

In conclusion, *adult* CVD is a major *pediatric* problem.[34] Without intervention, some 60 million children are destined to suffer its consequences within the next 30 years. Optimal prevention efforts focus on children, especially on those who are overweight.

Just as young children receive vaccinations against infectious diseases, they need screening for, and education about, CVD. Many health education programs have been implemented in schools around the country. These programs are most effective when they include education in the classroom, heart-healthy meals in the lunchroom, fitness activities on the playground, and parental involvement at home.

The Nemours Foundation

www.kidshealth.org

## Notes

1. W. B. Kannel, R. B. D'Agostino, and A. Belanger, Concept of bridging the gap from youth to adulthood—The Framingham Study, an address presented at the Recognition and Pre-

***Cigarette smoking is the number-one preventable cause of deaths.***


vention of Heart Disease: State of the Art conference, New Orleans, Louisiana, April 27 and 28, 1994.
2. H. C. McGill, Childhood nutrition and adult cardiovascular disease, *Nutrition Reviews* 55 (1997): S2–S11.
3. M. J. T. Kallio and coauthors, Tracking of serum cholesterol and lipoprotein levels from the first year of life, *Pediatrics* 91 (1993): 949–954.
4. S. Guo and coauthors, Serial analysis of plasma lipids and lipoproteins from individuals 9–21 years of age, *American Journal of Clinical Nutrition* 58 (1993): 61–67.
5. Committee on Nutrition, American Academy of Pediatrics, Cholesterol in childhood, *Pediatrics* 101 (1998): 141–147.
6. E. Suter and M. R. Hawes, Relationship of physical activity, body fat, diet, and blood lipid profile in youths 10–15 yr, *Medicine and Science in Sports and Exercise* 25 (1993): 748–754.
7. S. B. Craig and coauthors, The impact of physical activity on lipids, lipoproteins, and blood pressure in preadolescent girls, *Pediatrics* 98 (1996): 389–395.
8. Nation High Blood Pressure Education Program Working Group on Hypertension Control in Children and Adolescents, Update on the 1987 Task Force Report on High Blood Pressure in Children and Adolescents: A working group report from the National High Blood Pressure Education Program, *Pediatrics* 98 (1996): 649–658.
9. Committee on Sports Medicine and Fitness, American Academy of Pediatrics, Athletic participation by children and adolescents who have systemic hypertension, *Pediatrics* 99 (1997): 637–638.
10. A. R. Sinaiko, Hypertension in children, *New England Journal of Medicine* 335 (1996): 1968–1973.
11. S. Shea and coauthors, The rate of increase in blood pressure in children 5 years of age is related to changes in aerobic fitness and body mass index, *Pediatrics* 94 (1994): 465–470.
12. B. Falkner and S. Michel, Blood pressure response to sodium in children and adolescents, *American Journal of Clinical Nutrition* 65 (1997): 618S–621S.
13. S. S. Guo and coauthors, The predictive value of childhood body mass index values for overweight at age 35 y, *American Journal of Clinical Nutrition* 59 (1994): 810–819.
14. Update: Prevalence of overweight among children, adolescents, and adults—United States, 1988–1994, *Morbidity and Mortality Weekly Report* 46 (1997): 199–202.
15. T. A. Nicklas and coauthors, Secular trends in dietary intakes and cardiovascular risk factors of 10-year-old children: The Bogalusa Heart Study (1973–1988), *American Journal of Clinical Nutrition* 57 (1993): 930–937.
16. J. O. Fisher and L. L. Birch, Fat preferences and fat consumption of 3- to 5-year-old children are related to parental obesity, *Journal of the American Dietetic Association* 95 (1995): 759–764.
17. S. A. Schlicker, S. T. Borra, and C. Regan, The weight and fitness status of United States children, *Nutrition Reviews* 52 (1994): 11–17.
18. E. Obarzanek and coauthors, Energy intake and physical activity in relation to indexes of body fat: The National Heart, Lung, and Blood Institute Growth and Health Study, *American Journal of Clinical Nutrition* 60 (1994): 15–22.
19. O. T. Raitakari and coauthors, Effects of persistent physical activity on coronary risk factors in children and young adults: The Cardiovascular Risk in Young Finns Study, *American Journal of Epidemiology* 140 (1994): 195–205.
20. G. S. Berenson, S. R. Srinivasan, and L. S. Webber, Cardiovascular risk prevention in children: A challenge or a poor idea? *Nutrition, Metabolism and Cardiovascular Diseases* 4 (1994): 46–52.
21. W. Bao and coauthors, Longitudinal changes in cardiovascular risk from childhood to young adulthood in offspring of parents with coronary artery disease: The Bogalusa Heart Study, *Journal of the American Medical Association* 278 (1997): 1749–1754.
22. Committee on Nutrition, 1998.
23. L. Van Horn and P. Greenland, Prevention of coronary artery disease is a pediatric problem, *Journal of the American Medical Association* 278 (1997): 1779–1780; Berenson, Srinivasan, and Webber, 1994.
24. S. J. Wadowski and coauthors, Family history of coronary artery disease and cholesterol: Screening children in disadvantaged inner-city population, *Pediatrics* 93 (1994): 109–113; K. Resnicow and D. Cross, Are parents' self-reported total cholesterol levels useful in identifying children with hyperlipidemia? An examination of current guidelines, *Pediatrics* 92 (1993): 347–354.
25. M. S. Glassman and S. M. Schwarz, Cholesterol screening in children: Should obesity be a risk factor? *Journal of the American College of Nutrition* 12 (1993): 270–273.
26. Committee on Nutrition, 1998.
27. Committee on Nutrition, 1998.
28. L. B. Dixon and coauthors, The effect of changes in dietary fat on the food group and nutrient intake of 4- to 10-year-old children, *Pediatrics* 100 (1997): 863–872.
29. M. Sigman-Grant, S. Zimmerman, and P. M. Kris-Etherton, Dietary approaches for reducing fat intake of preschool-age children, *Pediatrics* 91 (1993): 955–960.
30. Is there a relationship between dietary fat and stature or growth in children three to five years of age? *Pediatrics* 92 (1993): 579–586.
31. Committee on Nutrition, 1998.
32. Tobacco use among high school students—United States, 1997, *Morbidity and Mortality Weekly Report* 47 (1998): 229–233.
33. Tobacco use and usual source of cigarettes among high school students—United States, 1995, *Journal of the American Medical Association* 276 (1996): 184–185.
34. Van Horn and Greenland, 1997.

CHAPTER 17

# Life Cycle Nutrition: Adulthood and the Later Years

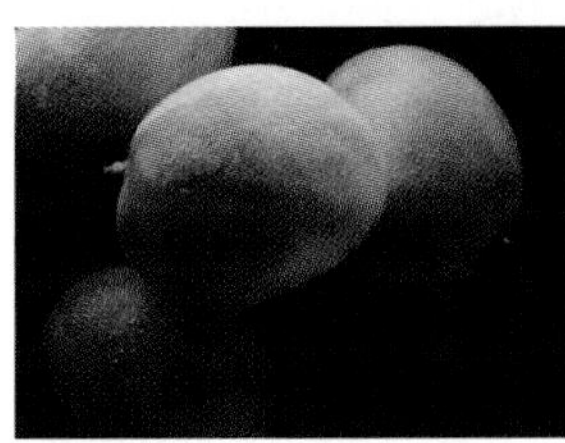

Wise food choices, made throughout adulthood, can support a person's ability to meet physical, emotional, and mental challenges and to achieve freedom from disease. Two goals motivate adults to pay attention to their diets: promoting health and slowing aging. Much of this text has focused on nutrition to support health, and Chapter 18 features prevention of disease; this chapter focuses on aging and the nutrition needs of older adults.

The U.S. population is "graying." The majority is now middle-aged, and the ratio of old people to young is increasing, as Figure 17-1 shows. Our society uses the arbitrary age of 65 years to define the transition point between middle age and old age, but growing "old" happens day by day, with changes occurring gradually over time. Since 1950 the population of those over 65 has more than doubled. Remarkably, the fastest-growing age group today is people over 85 years (see Figure 17-2).

Life expectancy for women in the United States is 79 years; for men, it is 73 years—both up from about 47 years in 1900.[1] Women who live to 80 can expect an additional nine years, on average; men, an additional seven.[2] Advances in medical science—antibiotics and other treatments—are largely responsible for almost doubling the life expectancy in this century. Improved nutrition and an abundant supply of food have also contributed to lengthening life expectancy. Still, human longevity appears to have an upper limit; the potential human life span is currently 130 years.[3] With recent advances in medical and genetic technology, however, researchers may one day be able to extend the life span ever further by slowing, or even preventing, aging and its accompanying diseases.[4]

**life expectancy:** the average number of years lived by people in a given society.

**longevity:** long duration of life.

**life span:** the maximum number of years of life attainable by a member of a species.

# Nutrition and Longevity

Only in the twentieth century have human beings achieved a life expectancy that permits scientists to study aging. Research in the field is active—and difficult. Researchers are challenged by the diversity of older adults. When older adults experience health problems, it is hard to know whether to attribute these problems to normal, age-related processes or to other factors.

U.S. Government
www.healthfinder.gov/searchoptions/topicsaz.htm
Search for Aging

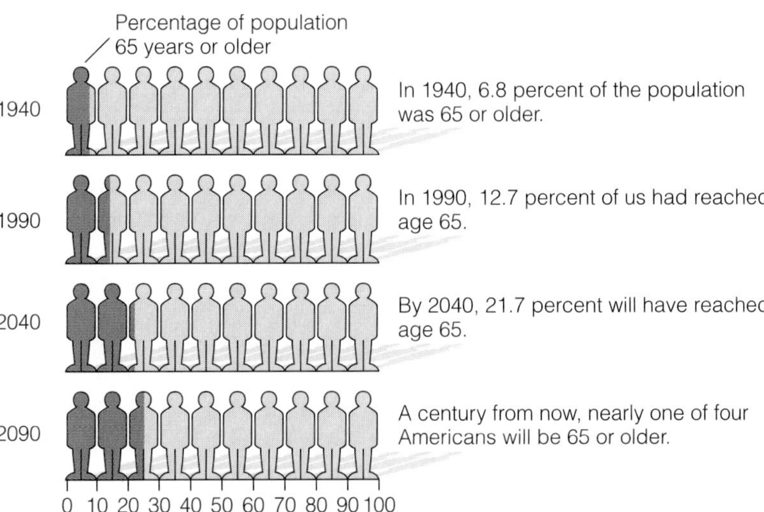
Percentage of population 65 years or older
1940
In 1940, 6.8 percent of the population was 65 or older.
1990
In 1990, 12.7 percent of us had reached age 65.
2040
By 2040, 21.7 percent will have reached age 65.
2090
A century from now, nearly one of four Americans will be 65 or older.
0 10 20 30 40 50 60 70 80 90 100
Percentage of total population

**Figure 17-1**
**The Aging of the U.S. Population**

**Figure 17-2**

**U.S. Population Growth, 1960 to 1990**

The "oldest old"—those over 85 years—are the fastest-growing age group in the United States. Between 1960 and 1990, the U.S. population grew 39 percent, but the population of those over 85 more than doubled. An estimated 25,000 Americans now living are 100 years old or older.

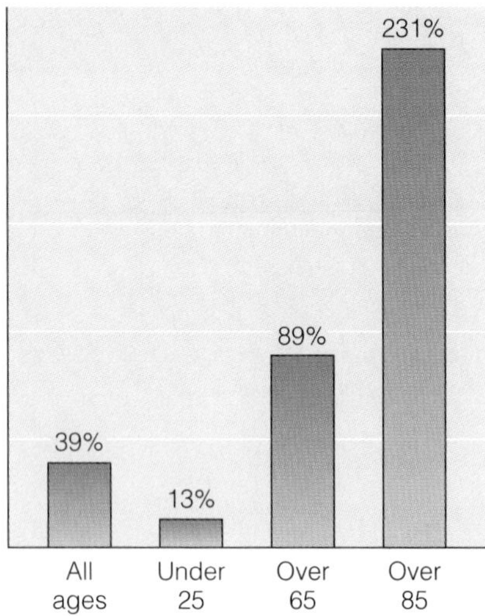


**physiological age:** a person's age as estimated from her or his body's health and probable life expectancy.

**chronological age:** a person's age in years from his or her date of birth.

*Growing old can be enjoyable for people who take care of their health and live each day fully.*

The idea that nutrition can influence the aging process is particularly appealing, because people can control and change their eating habits. The questions researchers are asking include:

- To what extent is aging inevitable, and can it be slowed through changes in lifestyle and environment?
- What role does nutrition play in the aging process, and what role can it play in slowing aging?

With respect to the first question, it seems that aging is an inevitable, natural process, programmed into the genes at conception. People can, however, slow the process within the natural limits set by heredity. They need to adopt healthy lifestyle habits such as eating nutritious food and engaging in physical activity.

With respect to the second question, good nutrition helps to maintain a healthy body and can therefore ease the aging process in many significant ways. Clearly, nutrition can improve the quality of life in the later years.

## Observation of Elderly People

The strategies adults use to meet the two goals mentioned at the start of this chapter—promoting health and slowing aging—are actually very much the same. What to eat, when to sleep, how physically active to be, and other lifestyle choices greatly influence both physical health and the aging process.

• ***Healthy Habits*** • A person's *physiological* age reflects his or her health status and may or may not reflect the person's *chronological* age. Quite simply, some people seem younger, and others older, than their years. Six lifestyle practices seem to have the greatest influence on people's health and therefore on their physiological age:

- Sleeping regularly and adequately.
- Eating regular meals, including breakfast.
- Engaging in regular physical activity.
- Not smoking.
- Not using alcohol, or using it in moderation.
- Keeping weight under control.

Over the years, the effects of these lifestyle choices accumulate—that is, those who follow all of the practices live longer and have fewer disabilities as they age.[5] They are in better health, even when older in chronological age, than people who fail to do so. Even though people cannot change their birth dates, they may be able to add years to the length and enhance the quality of their lives. Physical activity seems to be most influential in preventing or slowing the many changes that define a stereotypical "old" person.[6]

• ***Especially Physical Activity*** • The many and remarkable benefits of regular physical activity are not limited to the young: older adults who are active weigh less and have greater flexibility, more endurance, better balance, and better health than those who are inactive.[7] They reap additional benefits as well; for example, moderate endurance activities improve the quality of sleep, and strength training significantly improves mobility and resistance to injury. In fact, regular physical activity is the most powerful predictor of a person's mobility in the later years. Physical activity also increases blood flow to the brain, thereby improving mental ability.

Activities of all kinds are recommended to maintain and promote health. Strength training improves muscle strength, which helps a person perform many of life's daily tasks, such as climbing stairs, carrying packages, and opening pickle jars.[8] Aerobic activity can improve cardiorespiratory endurance and lower blood lipid concentrations.[9] Although aging reduces both speed and endurance to some degree, older adults can still train and achieve exceptional performances.

Ideally, physical activity should be part of each day's schedule and should be intense enough to prevent muscle atrophy and to speed up the heartbeat and respiration rate. Healthy older adults who have not been active can ease into a suitable routine. They can start by walking short distances until they can walk at least a mile three times a week, and then they can gradually increase their pace to achieve a 20- to 25-minute mile.

Muscle mass and muscle strength tend to decline with aging, making older people vulnerable to falls and immobility. Falls are a major cause of fear, injury, disability, dependence, and even death among older adults. Regular physical activity tones, firms, and strengthens muscles, helping to improve confidence, reduce the risk of falling, and lessen the risk of injury should a fall occur. Strength training, even in frail, elderly people over 85 years of age, has been shown not only to improve balance, muscle strength, and mobility but also to increase energy expenditure and energy intake, thereby enhancing nutrient intakes.[10] This finding highlights another reason to be physically active: a person spending energy on physical activity can afford to eat more food and thus receives more nutrients as a result. People who are committed to an ongoing fitness program can maintain their body weights and have higher energy and nutrient intakes than more sedentary people.[11]

*Strength training strengthens muscles and bones and ensures healthy appetites. (Courtesy of CNN)*

## Manipulation of Diet

In their efforts to understand longevity, researchers have not only observed people, but have manipulated influencing factors, such as diet, in animals. This research has given rise to some interesting and suggestive findings.

• ***Energy Restriction in Rats*** • Rats live longer and have fewer age-related diseases when their food intakes are restricted.[12] These life-prolonging benefits become evident when rats receive enough food to prevent malnutrition and to restrict energy intakes by 30 to 60 percent.

Exactly how energy restriction prolongs life in rats remains largely unexplained. The consequences of food restriction include a delay in the onset of age-related diseases, prolonged growth and development, and lowered blood glucose, insulin, and body fat. In addition, energy metabolism slows and body temperature drops—indications of a reduction in the rate of oxygen consumption. As Highlight 11 explained, the use of oxygen during energy metabolism produces free radicals, which have been implicated in the aging process. Reducing such oxidative stress may at least partially explain how restricting energy intake may extend the life span.

Interestingly, extending longevity appears to depend on restricting energy intake and not on the amount of body fat. Genetically obese rats live longer when given a restricted diet even though their body fat is similar to that of nonobese rats allowed to eat freely.[13]

• ***Energy Restriction in Human Beings*** • Research on a variety of other animals confirms that restricting energy intake extends the life span, but many more years of research are needed before such findings can be applied to human beings. Applying the results of animal studies to human beings is often unrealistic and potentially dangerous. Extreme starvation to extend life, like any extreme, is rarely, if ever, worth the price. Moderation, on the other hand, may be valuable.

Many of the physiological responses to energy restriction seen in rats also occur in people whose intakes are *moderately* restricted. When nonobese men cut back on their usual energy intake by 20 percent, their body weight, body fat, and blood pressures dropped, and their HDL cholesterol concentrations rose—favorable changes for preventing obesity and chronic diseases.[14] This moderate restriction of energy intake had no adverse effects on the mental or physical performances of the men.

IN SUMMARY

Life expectancy in the United States has increased dramatically in the twentieth century. Factors that enhance longevity include limited or no alcohol use, regular balanced meals, weight control, adequate sleep, abstinence from smoking, and regular physical activity. Nutrition alone, even if ideal, cannot guarantee a long and robust life. At the very least, however, nutrition can influence aging and longevity in human beings by helping to prevent disease. The next chapter is dedicated to the relationships between diet and disease prevention; the focus here is on changes that commonly accompany the aging process.

# The Aging Process

As people get older, each person becomes less and less like anyone else. The older people are, the more time has elapsed for such factors as nutrition, genetics, physical activity, and everyday stress to influence physical and psychological aging.

**stress:** any threat to a person's well-being; a demand placed on the body to adapt.

**stressor:** an environmental element, physical or psychological, that causes stress.

**stress response:** the body's response to stress, mediated by both nerves and hormones.

Both physical stressors (such as alcohol abuse, other drug abuse, smoking, pain, heat, and illness) and psychological stressors (such as exams, divorce, moving, and the death of a loved one) elicit the body's stress response. The body responds to such stressors with an elaborate series of physiological steps, as the nervous and hormonal systems bring about defensive readiness in every body part. The effects all favor physical action—the classic fight-or-flight response. Stress that is prolonged or severe can drain the body of its reserves and leave it weakened, aged, and vulnerable to illness, especially if physical action is not taken. As people age, they lose their ability to adapt to both external and internal disturbances. When disease strikes, the reduced ability to adapt makes the aging individual more vulnerable to death than a younger person.

## *Physiological Changes*

National Institute on Aging
www.nih.gov/nia

As aging progresses, inevitable changes in each of the body's organs contribute to the body's declining function. These physiological changes influence nutrition status, just as growth and development do in the earlier stages of the life cycle.

• ***Body Composition*** • Optimal nutrition and physical activity can minimize the changes in body composition associated with aging. In general, though, older people tend to lose bone and muscle and gain body fat.[15] Many of these changes occur because some hormones that regulate metabolism become less active with age, while others become more active. The action of insulin, for example, diminishes with age as the pancreas begins to secrete less of the hormone and the cells lose their ability to respond efficiently.* Hormones are also responsible for the diminished appetite that often accompanies aging.[16]

**sarcopenia** (SAR-ko-PEE-nee-ah): loss of skeletal muscle mass, strength, and quality.
- **sarco** = flesh
- **penia** = loss or lack

Loss of muscle, known as sarcopenia, can be significant in the later years and its consequences, quite dramatic.[17] As muscles diminish and weaken, people lose their ability to move and maintain balance, making falls likely. Many lose their independence.[18] The limitations that accompany the loss of muscle and its strength play a key role in the diminishing health that often accompanies aging.

• ***Immune System*** • Changes in the immune system also bring declining function with age. The immune system is also compromised by nutrient deficiencies, and so the combination of age and malnutrition makes older people vulnerable

*Other examples of hormones that change with age include growth hormone and androgens, which decline with advancing age, thus contributing to the decrease in lean body mass, and prolactin, which increases with age, helping to maintain body fat.

to infectious diseases. Adding insult to injury, antibiotics often are not effective against infections in people with compromised immune systems. Consequently, infectious diseases are a major cause of death in older adults.

• *GI Tract* • In the GI tract, the intestinal wall loses strength and elasticity with age, and this slows motility. Constipation is four to eight times more common in the elderly than in the young. Atrophic gastritis, a condition that affects almost one-third of those over 60, is characterized by an inflamed stomach, bacterial overgrowth, and a lack of hydrochloric acid and intrinsic factor—all of which can impair the digestion and absorption of nutrients, most notably, vitamin $B_{12}$, but also biotin, calcium, and iron.

Reminder: *Atrophic gastritis* is a condition characterized by chronic inflammation of the stomach accompanied by a diminished size and functioning of the mucosa and glands.

Consequences of atrophic gastritis:
- Reduced hydrochloric acid.
- Reduced intrinsic factor.
- Increased bacterial growth.
- Increased risk of nutrient deficiencies, notably of vitamin $B_{12}$.

• ***Tooth Loss*** • Improvements in dental care over a lifetime may reduce the incidence of tooth loss and gum disease, which are common in old age. These conditions make chewing difficult or painful. Dentures, even when they fit properly, are less effective than natural teeth, and inefficient chewing can cause choking. People with tooth loss, gum disease, and ill-fitting dentures tend to limit their food selections to soft foods. If foods such as corn on the cob, apples, and hard rolls are replaced by creamed corn, applesauce, and rice, then nutrition status may not be greatly affected, but when food groups are eliminated and variety is limited, inadequate intakes of vitamins, minerals, and fiber follow. To determine whether a visit to the dentist is needed, an older adult can check the conditions listed in the margin.[19]

Conditions requiring dental care:
Dry mouth.
Eating difficulty.
No dental care within 2 years.
Tooth or mouth pain.
Altered food selections.
Lesions, sores, or lumps in mouth.

• ***Sensory Losses and Other Physical Problems*** • A multitude of sensory losses and other physical problems can also interfere with an older person's ability to obtain adequate nourishment. Failing eyesight, for example, can make driving to the grocery store impossible and shopping for food a frustrating experience. It may become so difficult to read food labels and count money that the person doesn't buy the needed foods. Carrying bags of groceries may be an unmanageable task. Similarly, a person with limited mobility may find cooking and cleaning up too hard to do. Not too surprisingly, the prevalence of undernutrition is high among those who are homebound.[20]

Sensory losses can also interfere with a person's ability or willingness to eat. Taste and smell sensitivities tend to diminish with age and may make eating less enjoyable.[21] Consequently, food intake may diminish and nutrient deficiencies follow.[22] Loss of vision and hearing may contribute to social isolation.

## Other Changes

In addition to the physiological changes that accompany aging, adults are changing in many other ways that influence their nutrition status. Psychological, economic, and social factors play big roles in a person's ability and willingness to eat.

Administration on Aging
www.aoa.dhhs.gov

• ***Psychological Changes*** • Although not an inevitable component of aging, depression is common, affecting an estimated 6 million older adults.[23] Loss of appetite and of the motivation to cook or even to eat frequently accompanies depression. An overwhelming sense of grief and sadness at the death of a spouse, friend, or family member may leave a person, especially an elderly person, feeling powerless to overcome depression. When a person is suffering the heartache and loneliness of bereavement, cooking meals may not seem worthwhile. The support and companionship of family and friends, especially at mealtimes, can help overcome depression and enhance appetite.

***Shared meals can brighten the day and enhance the appetite.***

• ***Economic Changes*** • Overall, the older population today has a higher income than their cohorts of previous generations. Still, poverty is a major problem for about 20 percent of the people over age 65. Factors such as living arrangements and income make significant differences in the food choices, eating habits, and nutrition status of older adults, especially those over age 80. People of low socioeconomic status are likely to have inadequate food and nutrient intakes. Only about one-third of the needy elderly receive assistance from federal programs.[24]

• ***Social Changes*** • Malnutrition among older adults is most likely to occur among those with the least education, those living alone in federally funded housing (an indicator of low income), and those who have recently experienced a change in lifestyle. The risk of nutrient deficiencies is high among people living alone, especially men. One study of home-delivered meals confirmed that men living alone eat less than men living with others; interestingly, women living alone eat more than women living with others.[25] Adults who live alone do not necessarily make poor food choices, but they often consume too little food: loneliness is directly related to nutritional inadequacies, especially of energy intake.[26]

IN SUMMARY

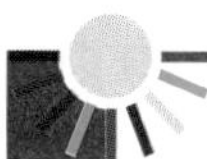

Many changes that accompany aging can impair nutrition status. Among physiological changes, hormone activity alters body composition, immune system changes raise the risk of infections, atrophic gastritis interferes with nutrient digestion and absorption, and tooth loss limits food choices. Psychological changes such as depression, economic changes such as loss of income, and social changes such as loneliness contribute to poor food intake.

# Energy and Nutrient Needs of Older Adults

Knowledge about the nutrient needs and nutrition status of older adults has grown considerably in recent years. The Dietary Reference Intakes (DRI) cluster all people over 50 into two age categories—one group of 51 to 70 years and one of 71 and older.[27]* After all, the needs of people 50 to 60 years old may be very different from those of people over 80. The need for more age-specific recommendations is becoming more and more urgent as the population ages.

Setting standards for older people is difficult because individual differences become more pronounced as people grow older. People start out with different genetic predispositions and ways of handling nutrients, and the effects of these differences become magnified with the years. Then, too, one person may tend to omit vegetables from his diet, and by the time he is old, he will have an associated set of nutrition problems. Another may have omitted milk and milk products all her life—her nutrition problems will be different. Also, as people age, they suffer different chronic diseases and take different drugs—both of which will have impacts on nutrient needs. For all these reasons, researchers have difficulty even defining "healthy aging," a prerequisite to developing recommendations to meet the "needs of practically all healthy persons."[28] The next sections give special attention to a few nutrients of concern.

## Water

Dehydration is a risk for older adults, who may not notice or pay attention to their thirst, or who find it difficult and bothersome to get a drink or to get to a bathroom. Older adults who have lost bladder control may be afraid to drink too much water. Despite real fluid needs, many older people do not seem to feel thirsty or

*The 1989 RDA combine all people over 50 into one group; the 1990 Canadian RNI divide older people into two age groups—50 to 74, and 75 and older.

notice mouth dryness. Many nursing home employees say it is hard to persuade their elderly clients to drink enough water and fruit juices.

Total body water also decreases as people age, so even mild stresses such as fever or hot weather can precipitate rapid dehydration in older adults. Dehydrated older adults seem to be more susceptible to urinary tract infections, pneumonia, pressure ulcers, and confusion and disorientation.[29] Chapter 12 described the importance of water and recommended an intake of 6 to 8 glasses of water a day. Milk and juices may replace some of this water, but beverages containing alcohol or caffeine should be limited because of their diuretic effect.[30]

• Older adult eating tip: Drink plenty of water.

Water recommendation for adults:
• 1 oz/kg actual body weight.

## Energy and Energy Nutrients

On average, adult energy needs decline an estimated 5 percent per decade. One reason is that people usually reduce their physical activity as they age, although they need not do so. Another reason is that lean body mass diminishes, slowing the basal metabolic rate.

The lower energy expenditure of older adults requires that they obtain less food energy to maintain their weights. Accordingly, the energy RDA for adults decreases slightly after age 50. Energy intakes typically decline in parallel with needs. Still, many older adults are overweight, indicating that their food intakes do not decline enough to compensate for their reduced energy expenditure.

On limited energy allowances, people must select mostly nutrient-dense foods. There is little leeway for sugars, fats, oils, or alcohol. The Daily Food Guide (on pp. 34–35) offers a dietary framework for adults of all ages. Those who need additional food energy should choose extra servings from each of the groups listed.

• Older adult eating tip: Select nutrient-dense foods, low in fats, sugars, and alcohol.

• ***Protein*** • Because energy needs decrease, protein must be obtained from low-kcalorie sources of high-quality protein, such as lean meats, poultry, fish, and eggs; nonfat and low-fat milk products; and legumes. Protein is especially important for the elderly to support a healthy immune system and to prevent muscle wasting. Some researchers suggest that current protein recommendations may be inadequate, while others counter that the protein needs of older adults seem to be adequately covered by current recommendations.[31] Additional research is needed to clarify the protein needs of older people, especially those over 70 years old.

• ***Carbohydrate and Fiber*** • As always, abundant carbohydrate is needed to protect protein from being used as an energy source. Sources of complex carbohydrates such as legumes, vegetables, whole grains, and fruits are also rich in fiber and essential vitamins and minerals.

The combination of ample water and high-fiber foods can alleviate constipation—a condition common among older adults, and especially among nursing home residents. Physical inactivity and medications probably contribute to the high incidence of constipation, but lack of water and fiber does, too. In fact, average fiber intakes among older adults are lower than current recommendations (10 to 13 grams per 1000 kcalories).[32]

As many as half of nursing home residents may be malnourished and underweight.[33] For these people, a diet that emphasizes fiber-rich foods such as whole grains, fruits, and vegetables may be too low in protein and energy. Protein- and energy-dense snacks such as hard-boiled eggs, tuna fish and crackers, peanut butter on graham crackers, and hearty soups are valuable additions to the diets of underweight or malnourished older adults.

• ***Fat*** • As is true for people of all ages, fat needs to be limited in the diet of most older adults. Cutting fat may help prevent or delay the development of cancer, atherosclerosis, and other degenerative diseases. This recommendation should not be taken too far; for some older adults, limiting fat intake too severely may lead to nutrient deficiencies and weight loss—two problems that carry greater health risks in the elderly than overweight.

## Vitamins and Minerals

Most people can achieve adequate vitamin and mineral intakes simply by including foods from all food groups in their diets, but older adults often omit fruits and vegetables. Similarly, few older adults consume the recommended amounts of milk products.[34]

• ***Vitamin A*** • Unlike other vitamins and minerals, vitamin A is absorbed and stored *more* efficiently by the aging GI tract and liver, although its processing within the body slows slightly. Several studies have reported that healthy older adults have normal levels of plasma vitamin A even when their dietary intakes fall below the RDA, suggesting that the current RDA may be too high. Any discussions on lowering the RDA, however, must carefully consider both the need to prevent vitamin A deficiency and the possibility that the vitamin A precursor beta-carotene might delay the onset of some age-related diseases.

• ***Vitamin D*** • Vitamin D deficiency is a significant problem among older adults.[35] Only vitamin D–fortified milk provides significant vitamin D, and many older adults drink little or no milk. Further compromising the vitamin D status of many older people, especially those in nursing homes, is their limited exposure to sunlight. Finally, aging reduces the skin's capacity to make vitamin D and the kidneys' ability to convert it to its active form. Not only are older adults not getting enough vitamin D, but they may actually need more. The recommended intake for vitamin D was recently raised from 5 to 10 micrograms daily to prevent bone loss and to maintain vitamin D status, especially in those who engage in minimal outdoor activity.[36]

• ***Vitamin $B_{12}$*** • An estimated 15 percent of the elderly population is deficient in vitamin $B_{12}$.[37] As Chapter 10 explained, people with atrophic gastritis are particularly vulnerable to vitamin $B_{12}$ deficiency. For one thing, the bacterial overgrowth that accompanies this condition uses up the vitamin. For another, without hydrochloric acid and intrinsic factor, digestion and absorption of vitamin $B_{12}$ are inefficient. Given the devastating neurological effects of a vitamin $B_{12}$ deficiency, an adequate intake is imperative.

• ***Iron*** • Among the minerals, iron deserves first mention. Iron-deficiency anemia is less common in older adults than in younger people, but still occurs in some, especially in people with low food energy intakes. Aside from diet, other factors in many older people's lives make iron deficiency likely: chronic blood loss from disease conditions and medicines, and poor iron absorption due to reduced stomach acid secretion and antacid use. Anyone concerned with older people's nutrition should keep these possibilities in mind.

• ***Zinc*** • Zinc intake is commonly low in older people, with many receiving less than half of the recommended amount.[38] In addition, older adults may absorb zinc less efficiently than younger people do. A number of different factors, including medications that older adults commonly use (diuretics, antacids, and laxatives), can impair zinc absorption or enhance its excretion and thus lead to deficiency. Older adults who do not make special efforts to eat zinc-rich foods such as meats, fish, and poultry will no doubt fail to meet the zinc RDA. Some of the symptoms of zinc deficiency resemble symptoms associated with aging—for example, decline in taste acuity, dermatitis, and impaired immunity. Whether these symptoms are directly attributable to zinc deficiency remains unclear.

The importance of abundant dietary calcium throughout life, and especially for women after menopause, to protect against osteoporosis was discussed in Chapter and Highlight 12.

• ***Calcium*** • The appropriate calcium intake for older adults remains controversial. Recommended intakes for older adults were recently raised from 800 to 1200 milligrams of calcium daily.[39] A National Institutes of Health panel has concluded that women over 50 who are not on estrogen replacement therapy and all adults over 65 should receive 1500 milligrams of calcium daily.[40]

As researchers attempt to reach agreement about the calcium requirements of older adults, especially those of women, one thing is clear: the calcium intakes of older people in the United States are well below recommendations.[41] Some older adults avoid milk and milk products because they dislike these foods or associate them with stomach discomfort. One simple solution is to add powdered nonfat milk to recipes; Chapter 12 offers many other strategies for including nonmilk sources of calcium for those who do not drink milk.

**Healthy People 2000:** Increase calcium intake so that at least 50% of people aged 25 years and older consume two or more servings of calcium-rich foods daily.

## *Nutrient Supplements*

People judge for themselves how to manage their nutrition, and some turn to supplements. Advertisers target older people with appeals to take supplements and eat "health" foods, claiming that these products prevent disease and promote longevity. About half of all women over 65 years of age take some type of nutrient supplement, while about one-fifth of older men do. Quite often those who take supplements are not deficient in the nutrients being supplemented. Certain diseases or health problems may necessitate the taking of supplements, but often supplements have not been prescribed by health care professionals and are inappropriate.

When recommended by a physician, vitamin D and calcium supplements for osteoporosis or iron for iron-deficiency anemia may be beneficial. In most cases, though, the money spent on supplements would be better spent on nutritious foods. Older adults with food energy intakes less than about 1500 kcalories per day, however, should probably take the once-daily type of multivitamin-mineral supplements.

People with small energy allowances would do well to become more active so they can afford to eat more food. Food is the best source of nutrients for everybody. Supplements are just that—supplements to foods, not substitutes for them. For anyone who is motivated to obtain the best possible health, it is never too late to learn to eat well, drink water, exercise regularly, and adopt other lifestyle changes such as quitting smoking and moderating alcohol use.

IN SUMMARY

The table on p. 546 summarizes the nutrient concerns of aging. Although some nutrients need special attention in the diet, supplements are not routinely recommended. The ever growing number of older people in the world creates an urgent need to learn more about how their nutrient requirements differ from those of younger people and how such knowledge can enhance their health.

# Nutrition-Related Concerns of Older Adults

Nutrition through the prime years may play a greater role than has been realized in preventing many changes once thought to be inevitable consequences of growing older. The following discussions of cataracts and macular degeneration, arthritis, and the aging brain show that nutrition may provide at least some protection against some of the conditions associated with aging.

IN SUMMARY

| Nutrient | Effect of Aging | Comments |
|---|---|---|
| Water | Lack of thirst and decreased total body water make dehydration likely. | Mild dehydration is a common cause of confusion. Difficulty obtaining water or getting to the bathroom may compound the problem. |
| Energy | Need decreases. | Physical activity moderates the decline. |
| Fiber | Likelihood of constipation increases with low intakes and changes in the GI tract. | Inadequate water intakes and lack of physical activity, along with some medications, compound the problem. |
| Protein | Needs may stay the same or increase slightly. | Low-fat, high-fiber legumes and grains meet both protein and other nutrient needs. |
| Vitamin A | Absorption and storage increases. | RDA may be high. |
| Vitamin D | Increased likelihood of inadequate intake; skin synthesis declines. | Daily sunlight exposure in moderation may be of benefit. |
| Vitamin $B_{12}$ | Atrophic gastritus is common. | Deficiency causes neurological damage. |
| Iron | In women, status improves after menopause; deficiencies are linked to chronic blood losses and low stomach acid output. | Adequate stomach acid is required for absorption; antacid or other medicine use may aggravate iron deficiency; vitamin C and meat increase absorption. |
| Zinc | Intakes may be low and absorption reduced; but needs may also decrease. | Medications interfere with absorption; deficiency may supress appetite and sense of taste. |
| Calcium | Intakes may be low; osteoporosis is common. | Stomach discomfort commonly limits milk intake; calcium substitutes are needed. |

## Cataracts and Macular Degeneration

**cataracts** (KAT-ah-rakts): thickenings of the eye lenses that impair vision and can lead to blindness.

Cataracts are age-related thickenings in the lenses of the eyes that impair vision. If not surgically removed, they ultimately lead to blindness. Cataracts occur even in well-nourished individuals as a result of ultraviolet light exposure, oxidative stress, injury, viral infections, toxic substances, and genetic disorders. Many cataracts, however, are vaguely called senile cataracts—meaning "caused by aging." In the United States, more than half of all adults 65 and older have a cataract.

National Eye Institute
Information for Public and Patients
www.nei.nih.gov
Visit Cataract
Visit Macular degeneration

Oxidative stress appears to play a significant role in the development of cataracts, and the antioxidant nutrients may help minimize the damage. Studies have reported an inverse relationship between cataracts and dietary intakes of vitamin C, vitamin E, and carotenoids.[42] Taking supplements of vitamins C and E seems to reduce the likelihood of developing age-related cataracts.[43]

One other dietary factor may play a role in the development of cataracts: overweight. Overweight is associated with cataracts, but its role has yet to be identified.[44] Risk factors that typically accompany overweight, such as inactivity, diabetes, or hypertension, do not explain the association.

**macular** (MACK-you-lar) **degeneration:** deterioration of the macular area of the eye that can lead to loss of central vision and eventual blindness. The **macula** is a small, oval, yellowish region in the center of the retina that provides the sharp, straight-ahead vision so critical to reading and driving.

Another common cause of visual loss among older people is macular degeneration, a deterioration of the macular region of the retina. Like cataracts, risk factors for macular degeneration include oxidative stress from sunlight. Foods rich in antioxidant nutrients seem to offer some protection against its development.[45]

## Arthritis

**arthritis:** inflammation of a joint, usually accompanied by pain, swelling, and structural changes.

Nutrition quackery to treat arthritis is abundant. Unfortunately, no known diet, food, or supplement prevents, relieves, or cures it.

**osteoarthritis:** a painful, chronic disease of the joints that occurs when the cushioning cartilage in a joint breaks down; joint structure is usually altered, with loss of function; also called **degenerative arthritis.**

• ***Osteoarthritis*** • The most common type of arthritis that disables older people is osteoarthritis, a painful swelling of the joints. During movement, the ends of bones are normally protected from wear by cartilage and by small sacs of fluid that act as a lubricant. With age, bones sometimes disintegrate, and the joints become malformed and painful to move. Osteoarthritis afflicts millions of people around the world, especially the elderly.

One known connection between osteoarthritis and nutrition is overweight. Weight loss may relieve some of the pain for overweight persons with osteoarthritis, partly because the joints affected are often weight-bearing joints that are stressed and irritated by having to carry excess poundage. Interestingly, though, weight loss often relieves the worst of the pain of arthritis in the hands as well, even though they are not weight-bearing joints. Jogging and other weight-bearing exercises do not worsen arthritis, even in marathon runners. In fact, both aerobic activity and strength training offer modest improvements in physical performance and pain relief.[46]

• ***Rheumatoid Arthritis*** • Another type of arthritis known as rheumatoid arthritis has a possible link to diet through the immune system. In rheumatoid arthritis, the immune system mistakenly attacks the bone coverings as if they were made of foreign tissue. The integrity of the immune system depends on adequate nutrition, and a poor diet may worsen this type of arthritis. It is also possible that in some individuals, certain foods may stimulate the immune system to attack. For example, milk and milk products seem to aggravate rheumatoid arthritis in some people.[47]

**rheumatoid** (ROO-ma-toyd) **arthritis:** a disease of the immune system involving painful inflammation of the joints and related structures.

Another nutrition link to rheumatoid arthritis is the omega-3 fatty acid found in fish oil, EPA. The same diet recommended for heart health—one low in saturated fat from meats and milk products and high in oils from fish—helps prevent or reduce the inflammation in the joints that makes arthritis so painful. Most likely, EPA interferes with the action of prostaglandins, hormonelike chemicals involved in inflammation.

 Arthritis Foundation
www.arthritis.org

 National Institute of Arthritis and Musculoskeletal and Skin Diseases
www.nih.gov/niams

 U.S. Government
www.healthfinder.gov/searchoptions/topicsaz.htm
Search for Arthritis

Another possible link between nutrition and rheumatoid arthritis involves the lipid peroxidation reactions described in Highlight 11. Lipid peroxidation of the membranes within joints causes inflammation and swelling. Vitamin E helps to prevent peroxidation, but it does not improve active cases of rheumatoid arthritis. This is not surprising, though, because the vitamin's role in lipid peroxidation is preventive, not restorative.

Drugs used to relieve arthritis can impose nutrition risks. Many drugs affect appetite and alter the body's use of nutrients, as Highlight 17 explains.

## *The Aging Brain*

The brain, like all of the body's organs, responds to both inherited and environmental factors that can enhance or diminish its amazing capacities. One of the challenges researchers face when studying the human brain is to distinguish among normal age-related physiological changes, changes caused by diseases, and changes that result from cumulative, extrinsic factors such as diet.

The brain normally changes in some characteristic ways as it ages. For one thing, its blood supply decreases. For another, the number of neurons, the brain cells that specialize in transmitting information, diminishes as people age. When the number of nerve cells in one part of the cerebral cortex diminishes, hearing and speech are affected. Losses of neurons in other parts of the cortex can impair memory and cognitive function. When the number of neurons in the hindbrain diminishes, balance and posture are affected. Losses of neurons in other parts of the brain affect still other functions.

**neuron:** a nerve cell; the structural and functional unit of the nervous system. Neurons initiate and conduct nerve transmissions.

**cerebral cortex:** the outer surface of the cerebrum (the largest part of the brain).

Clinicians now recognize that much of the cognitive loss and forgetfulness generally attributed to aging is due in part to extrinsic, and therefore controllable, factors such as nutrient deficiencies. In some instances, the degree of cognitive loss is extensive and attributable to a specific disorder such as a brain tumor. In cases such as Alzheimer's disease, deterioration may be genetically determined and will not respond to external approaches.

• ***Alzheimer's Disease*** • Much attention has focused on the *abnormal* deterioration of the brain called senile dementia of the Alzheimer's type (SDAT), which affects 5 percent of U.S. adults by age 65 and 20 percent of those over 80. Diagnosis

**senile dementia:** the loss of brain function beyond the normal loss of physical adeptness and memory that occurs with aging.

**senile dementia of the Alzheimer's type (SDAT):** a degenerative disease of the brain involving memory loss and major structural changes in neuron networks; also known as **primary degenerative dementia of senile onset** or **chronic brain syndrome,** but often simply called **Alzheimer's disease.**

Reminder: A *neurotransmitter* is a chemical that is released by one nerve cell and acts upon a second nerve cell, altering its electrical state or activity.

 U.S. Government

www.healthfinder.gov/searchoptions/topicsaz.htm

Search for Alzheimer's

 NIA Alzheimer's Disease Education and Referral Center

www.alzheimers.org

of SDAT depends on its characteristic symptoms: the victim gradually loses memory and reasoning, the ability to communicate, physical capabilities, and eventually life itself. Nerve cells in the brain die, and communication between the cells breaks down.

The causes of SDAT continue to elude researchers, although genetic factors are apparently involved. A newly established connection to a human gene that makes part of a lipoprotein has sparked new hope for preventing SDAT.[48]* Researchers hope to use these genetic findings to develop early detection tests and a cure for this devastating disease.

One abnormality of interest involves the extremely low concentrations of the enzyme that assists in the production of the neurotransmitter acetylcholine from choline and acetyl CoA. Acetylcholine is essential to memory. To date, supplements of choline (or of lecithin, which contains choline) have had no effect on memory or on the progression of the disease.

Most people have heard of an association between aluminum and the development of SDAT, although a causal connection seems unlikely. Brain concentrations of aluminum in SDAT people exceed normal brain concentrations by some 10 to 30 times, but blood and hair aluminum remains normal, indicating that the accumulation is caused by something in the brain itself, not by an overload of aluminum in the body. Thus the high brain aluminum must be at least partly a result, rather than a cause, of SDAT.

Treatment involves providing care to clients and support to their families. One drug (trade-named Tacrine) seems to slow the disease's advance in about 20 percent of those who use it, but it does not reverse the damage already done. Meanwhile, some drugs seem to improve the ability to remember and so hold promise for improving the lives of those with SDAT. Other drugs may be used to control depression or behavior problems.

Maintaining appropriate body weight may be the most important nutrition concern for the person with SDAT. Depression and forgetfulness can lead to poor food intake, and restlessness may increase energy needs. Perhaps the best that a caregiver can do nutritionally for an SDAT client is to supervise food planning and mealtimes.[49] Providing well-liked and well-balanced meals and snacks in a cheerful atmosphere encourages food consumption. To minimize confusion, offer a few ready-to-eat foods, in bite-size pieces, with seasonings and sauces. To avoid mealtime disruptions, control distractions such as music, television, children, and the telephone.

SDAT is an identifiable disease, the course of which is probably not influenced by nutrition. But poor nutrition in general does affect the brain in other ways.

• ***Nutrient Deficiencies and Brain Function*** • The ability of neurons to synthesize specific neurotransmitters depends in part on the availability of precursor nutrients that are obtained from the diet. The neurotransmitter serotonin derives from the amino acid tryptophan. To function properly, the enzymes involved in neurotransmitter synthesis require vitamins and minerals. Severe dietary deficiencies of thiamin, vitamin $B_6$, vitamin $B_{12}$, folate, and vitamin C impair mental ability, including memory. Minerals such as iron and zinc also support normal brain function. Table 17-1 summarizes some of the better-known connections between impaired brain function and severe nutrient deficiencies. Similarly, moderate, long-term nutrient deficiencies may contribute to the loss of memory and cognition that some older adults experience.[50] If long-term, moderate nutrient deficiencies influence the loss of cognitive function that accompanies aging, then the loss may be preventable or at least diminished or delayed through diet.

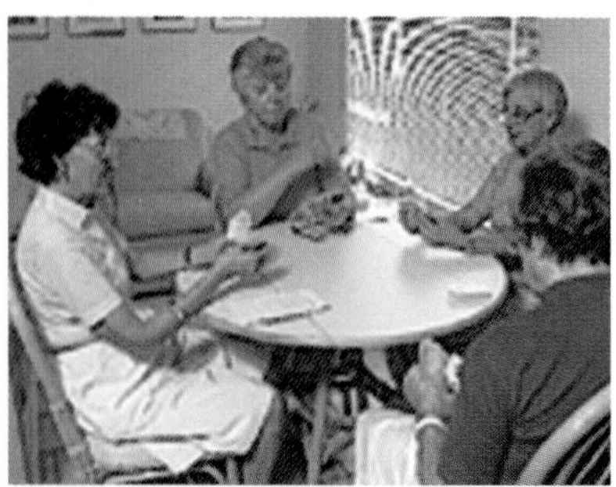

***The brain is nourished by both foods and mental challenges. (Courtesy of CNN)***

*The gene associated with Alzheimer's is apo E4, one of three apolipoprotein E varieties. A report on the genetic and other aspects of Alzheimer's is available from Alzheimer's Disease Education and Referral Center, P.O. Box 8250, Silver Springs, MD 20907-8250.

**Table 17-1**
**Summary of Nutrient-Brain Relationships**

| Brain Function | Inadequate Intake or Deficiency of: |
|---|---|
| Short-term memory loss | Vitamin $B_{12}$, vitamin C |
| Poor performance in problem-solving tests | Riboflavin, folate, vitamin $B_{12}$, vitamin C |
| Dementia | Thiamin, niacin, zinc |
| Cognition | Folate, vitamin $B_6$, vitamin $B_{12}$, iron |
| Degeneration of brain tissue | Vitamin $B_6$ |

IN SUMMARY

Senile dementia and other losses of brain function afflict millions of older adults, while others face loss of vision due to cataracts or macular degeneration, or cope with the pain of arthritis. As the number of people over age 65 continues to grow, the need for solutions to these problems becomes urgent. Some problems may be inevitable, but others are preventable and good nutrition may play a key preventative role.

We can now state with certainty that a person's nutrition status affects the health and functioning of the whole body. Eating a nutritious, balanced diet throughout life seems a small effort in light of the rewards of continued health and enjoyment in later life.

# Food Choices and Eating Habits of Older Adults

Older people are an incredibly diverse group, and for the most part they are independent, socially sophisticated, mentally lucid, fully participating members of society who report themselves to be happy and healthy. In fact, chronic disabilities among the elderly have declined dramatically over the past decade.[51] By practicing stress-management skills, maintaining physical fitness, participating in activities of interest, and cultivating spiritual health, as well as obtaining adequate nourishment, people can support a high quality of life into old age (see Table 17-2 on p. 550 for some strategies).

*Taking time to nourish your body well is a gift you give yourself.*

Older people spend more money per person on foods to eat at home than other age groups and less money on foods away from home. Manufacturers would be wise to cater to the preference of older adults by providing good-tasting, nutritious foods in easy-to-open, single-serving packages with labels that are easy to read. Such services enable older adults to maintain their independence and to feel a sense of control and involvement in their own lives. Another way older adults can take care of themselves is by remaining or becoming physically active. Physical activity helps preserve one's ability to perform daily tasks and so promotes independence.[52]

• Older adult eating tip: Try to maintain independence.

Familiarity, taste, and health beliefs are most influential on older people's food choices. Eating foods that are familiar, especially those that recall family meals and pleasant times, can be comforting. People 65 and over are less likely to diet to lose weight than younger people are, but are more likely to diet in pursuit of medical goals such as controlling blood glucose and cholesterol. The importance of diet and health beliefs in food selection is evidenced by surveys indicating that older adults are choosing low-fat poultry and fish, low-fat milk and milk products, and high-fiber breads and grains.[53]

• Older adult eating tip: Select familiar foods, especially ethnic favorites.

**Table 17-2**
**Strategies for Growing Old Healthfully**

- Choose nutrient-dense foods.
- Be physically active. Walk, run, dance, swim, bike, or row for aerobic activity. Lift weights, do calisthenics, or pursue some other activity to tone, firm, and strengthen muscles. Modify activities to suit changing abilities and tastes.
- Maintain appropriate body weight.
- Reduce stress (cultivate self-esteem, maintain a positive attitude, manage time wisely, know your limits, practice assertiveness, release tension, and take action).
- For women, see a physician about estrogen replacement.
- For people who smoke, quit.
- Expect to enjoy sex, and learn new ways of enhancing it.
- Use alcohol only moderately, if at all; use drugs only as prescribed.
- Take care to prevent accidents.
- Expect good vision and hearing throughout life; obtain glasses and hearing aids if necessary.
- Take care of your teeth; obtain dentures if necessary.
- Be alert to confusion as a disease symptom, and seek diagnosis.
- Take medications as prescribed; see a physician before self-prescribing medicines and a registered dietitian before self-prescribing supplements.
- Control depression through activities and friendships; seek professional help if necessary.
- Drink 6 to 8 glasses of water every day.
- Practice mental skills. Keep on solving math problems and crossword puzzles, playing cards or other games, reading, writing, imagining, and creating.
- Make financial plans early to ensure security.
- Accept change. Work at recovering from losses; make new friends.
- Cultivate spiritual health. Cherish personal values. Make life meaningful.
- Go outside for sunshine and fresh air as often as possible.
- Be socially active—play bridge, join an exercise group, take a class, teach a class, eat with friends, volunteer time to help others.
- Stay interested in life—pursue a hobby, spend time with grandchildren, take a trip, read, grow a garden, or go to the movies.
- Enjoy life.

## Food Assistance Programs for Older Adults

The federal Elderly Nutrition Program is intended to improve older people's nutrition status and enable them to avoid medical problems, continue living in communities of their own choice, and stay out of institutions. Its specific goals are to provide low-cost, nutritious meals; opportunities for social interaction; homemaker education and shopping assistance; counseling and referral to social services; and transportation.

The Elderly Nutrition Program provides for congregate meal programs. Administrators try to select sites for congregate meals so as to feed as many eligible people as possible. Volunteers may also deliver meals to those who are homebound either permanently or temporarily; these efforts are known as Meals on Wheels. The home-delivery program ensures nutrition, but its recipients miss out on the social benefit of the congregate meal sites; every effort is made to persuade older people to come to the shared meals, if they can. All persons aged 60 years and older and their spouses are eligible to receive meals from these programs, regardless of their income. Priority is given to those who are economically and socially needy. An estimated 25 percent of our nation's elderly poor benefit from these meals.[a]

These programs provide at least one meal a day that meets a third of the RDA for this age group; they must operate five or more days a week. Many programs voluntarily offer additional services: provisions for therapeutic diets, food pantries, ethnic meals, and delivery of meals to the homeless.

Older adults can learn about the available programs in their communities by looking in the yellow pages of the telephone book under "Social Services" or "Senior Citizens' Organizations." In addition, the local senior center and hospital can usually direct people to programs providing nutrition and other health-related services.

[a]Federal program nourishes poor elderly, *Journal of the American Medical Association* 278 (1997): 1301.

**Table 17-3**
**Nutrition Screening Initiative Checklist**

Circle the number to the right if the statement applies to you.

| Statement | Yes |
|---|---|
| I have an illness or condition that made me change the kind and/or amount of food I eat. | 2 |
| I eat fewer than 2 meals per day. | 3 |
| I eat few fruits or vegetables or milk products. | 2 |
| I have 3 or more drinks of beer, liquor, or wine almost every day. | 2 |
| I have tooth or mouth problems that make it hard for me to eat. | 2 |
| I don't always have enough money to buy the food I need. | 4 |
| I eat alone most of the time. | 1 |
| I take 3 or more different prescribed or over-the-counter drugs a day. | 1 |
| Without wanting to, I have lost or gained 10 pounds in the last 6 months. | 2 |
| I am not always physically able to shop, cook, and/or feed myself. | 2 |
| **Total** | |

SCORE:

**0–2: Good.** Recheck your score in 6 months.

**3–5: Moderate nutritional risk.** Visit your local office on aging, senior nutrition program, senior citizens center, or health department for tips on improving eating habits.

**6 or more: High nutritional risk.** See your doctor, registered dietitian, or other health care professional for help in improving your nutrition status.

*Social interactions at a congregate meal site can be as nourishing as the foods served.*

## Food Assistance Programs

The Nutrition Screening Initiative is part of a national effort to identify and treat nutrition problems in older persons; it uses a screening checklist (see Table 17-3). To *determine* the risk of malnutrition in older clients, health care professionals can keep in mind the characteristics listed in the margin.

Nutrition services are an integral part of health care, and different subgroups of the aging population need different programs designed to meet their specific needs.[54] People living alone can benefit from congregate meal programs; people confined to their homes need meals delivered.

American Academy of Family Physicians

www.aafp.org

Search for Nutrition screening initiative

Risk factors for malnutrition in older adults:

- Disease.
- Eating poorly.
- Tooth loss or oral pain.
- Economic hardship.
- Reduced social contact.
- Multiple medications.
- Involuntary weight loss or gain.
- Needs assistance with self-care.
- Elderly person older than 80 years.

**Healthy People 2000:** Increase to at least 80% the receipt of home foodservices by people aged 65 and older who have difficulty in preparing their own meals or are otherwise in need of home-delivered meals.

The U.S. government funds programs to provide nutritious meals to older adults at congregate meal sites. These congregate meals are a valuable source of nutrients for more than 3 million older adults.[55] Like school lunches, though, congregate meals typically do not meet current dietary recommendations to limit sodium, fat, and cholesterol.[56] The section on p. 550 describes food assistance programs for older adults.

**congregate meals:** nutrition programs that provide food for the elderly in a conveniently located setting such as a community center.

## Meals for Singles

Singles of all ages face difficulties in purchasing, storing, and preparing food. Large packages of meat and vegetables are often intended for families of four or more, and even a head of lettuce can spoil before one person can use it all. Many singles live in small dwellings and have little storage space for foods. A limited income presents additional obstacles. This section offers suggestions that can help to solve some of these problems.

Boxes of milk kept at room temperature on the shelves of grocery stores have been treated with a process called **ultrahigh temperature (UHT)**; the milk is exposed to temperatures above those of pasteurization just long enough to sterilize it.

*Buy only what you will use. (Courtesy of CNN)*

*Invite guests to share a meal. (Courtesy of CNN)*

• ***Spend Wisely*** • People who have the means to shop and cook for themselves can cut their food bills just by being wise shoppers. Large supermarkets are usually less expensive than convenience stores. A grocery list helps reduce impulse buying, and specials and coupons can save money when the items featured are those that the shopper needs and uses.

Buying the right amount so as not to waste any food is a challenge for people eating alone. They can buy fresh milk in the size best suited for personal needs. Pint-size and even cup-size boxes of milk are available and can be stored unopened on a shelf for up to three months without refrigeration.

Many foods that offer a variety of nutrients for practically pennies have a long shelf life; staples such as rice, pastas, nonfat dry powdered milk, and dried beans and peas can be purchased in bulk and stored on a shelf for months at room temperature. Other foods that are usually a good buy include whole pieces of cheese rather than sliced or shredded cheese; fresh produce in season; variety meats such as chicken livers; and cereals that require cooking instead of ready-to-serve cereals.

A person who has ample freezing space can buy large packages of meat, such as pork chops, ground beef, or chicken, when they are on sale. Then the meat can be immediately divided into individual servings and wrapped in aluminum foil, not freezer paper: the foil can become the liner for the pan in which to bake or broil the meat, thus saving work. All the individual servings can be put in a bag marked appropriately with the contents and the date.

Frozen vegetables are more economical in large bags than in small boxes. The amount needed can be taken out, and the bag closed tightly with a rubber band. If the package is returned quickly to the freezer each time, the vegetables will stay fresh for a long time.

Finally, breads and cereals usually must be purchased in larger quantities. Again the amount needed for a few days can be taken out and the rest stored in the freezer.

Grocers will break open a package of wrapped meat and rewrap the portion needed. Similarly, eggs can be purchased by the half-dozen. Eggs do keep for long periods, though, if stored properly in the refrigerator.

Fresh fruits and vegetables can be purchased individually. A person can buy three pieces of each kind of fresh fruit: a ripe one to eat right away, a semiripe one to eat soon after, and a green one to ripen on the windowsill. If vegetables are packaged in large quantities, the grocer can break open the package so that a smaller amount can be purchased. Small cans of fruits and vegetables, even though they are more expensive per unit, are a reasonable alternative, considering that it is expensive to buy a regular-size can and let the unused portion spoil.

• ***Be Creative*** • Creative chefs think of various ways to use foods when only large amounts are available. For example, a head of cauliflower can be divided into thirds. Then one-third is cooked and eaten hot. Another third is put into a vinegar and oil marinade for use in a salad. And the last third can be used in a casserole or stew.

Chefs also experiment with stir-fried foods. A large frying pan often works as well or better than a wok on modern ranges. A variety of vegetables and meats can be enjoyed this way; inexpensive vegetables such as cabbage, celery, and onion are delicious when crisp cooked in a little oil with herbs or lemon added. Interesting frozen vegetable mixtures are available in larger grocery stores. Cooked, leftover vegetables can be dropped in at the last minute. A bonus of a stir-fried meal is that there is only one pan to wash. Similarly, a microwave oven allows a chef to use fewer pots and pans. Meals can also be frozen or refrigerated in microwavable containers to reheat as needed.

Many frozen dinners that are now available are low in fat and nutritious. Adding a fresh salad, a whole-wheat roll, and a glass of milk can make a nice meal.

Also, single people shouldn't hesitate to invite someone to share meals with them whenever there is a lot of food. It's likely that person will return the invitation, and both parties will get to enjoy companionship and a meal prepared by others.

IN SUMMARY

Older persons can benefit from both the nutrients provided and the social interaction available at congregate meals. Other government programs deliver meals to those who are homebound. With creativity and careful shopping, those living alone can prepare nutritious, inexpensive meals. Physical activity, mental challenges, stress management, and social activities can also help people grow old comfortably.

## Study Questions

These questions will help you review the chapter. You will find the answers in the discussions on the pages provided.

1. What roles does nutrition play in aging, and what roles can it play in retarding aging? (pp. 538–540)
2. What are some of the physiological changes that occur in the body's systems with aging? To what extent can aging be prevented? (pp. 540–541)
3. Why does the risk of dehydration increase as people age? (pp. 542–543)
4. Why do energy needs usually decline with advancing age? (p. 543)
5. Which vitamins and minerals need special consideration for the elderly? Explain why. Identify some factors that complicate the task of setting nutrient standards for older adults. (pp. 544–545)
6. Discuss the relationships between nutrition and cataracts and between nutrition and arthritis. (pp. 546–547)
7. What characteristics contribute to malnutrition in older people? (pp. 550–551)

These questions will help you prepare for an exam. Answers can be found in Appendix K.

1. Life expectancy in this country is:
   a. 62 to 68 years.
   b. 69 to 75 years.
   c. 73 to 79 years.
   d. 84 to 90 years.
2. The human life span is about:
   a. 85 years.
   b. 100 years.
   c. 115 years.
   d. 130 years.
3. A 72-year-old person whose physical health is similar to that of people 10 years younger has a(n):
   a. chronological age of 62.
   b. physiological age of 72.
   c. physiological age of 62.
   d. absolute age of minus 10.
4. Rats lived longest when given diets that:
   a. eliminated all fat.
   b. provided lots of protein.
   c. allowed them to eat freely.
   d. restricted their energy intakes.
5. Which characteristic is *not* commonly associated with atrophic gastritis?
   a. inflamed stomach
   b. vitamin $B_{12}$ toxicity
   c. bacterial overgrowth
   d. lack of intrinsic factor
6. On average, adult energy needs:
   a. decline 5 percent per year.
   b. decline 5 percent per decade.
   c. remain stable throughout life.
   d. rise gradually throughout life.
7. Which nutrients seem to protect against cataract development?
   a. iron and zinc
   b. vitamin A and vitamin D
   c. vitamin C and vitamin E
   d. protein and carbohydrate
8. The best dietary advice for a person with osteoarthritis might be to:
   a. avoid milk products.
   b. take fish oil supplements.
   c. take vitamin E supplements.
   d. lose weight, if overweight.
9. Congregate meal programs are preferable to Meals on Wheels because they provide:
   a. nutritious meals.
   b. referral services.
   c. social interactions.
   d. financial assistance.
10. The Elderly Nutrition Program is available to:
    a. all people 65 years and older.
    b. all people 60 years and older.
    c. homebound people only, 60 years and older.
    d. low-income people only, 60 years and older.

# Notes


1. Centers for Disease Control and Prevention, National Center for Health Statistics, *Monthly Vital Statistics Report,* October 1996, p. 4.
2. K. G. Manton and J. W. Vaupel, Survival after the age of 80 in the United States, Sweden, France, England, and Japan, *New England Journal of Medicine* 333 (1995): 1232–1235.
3. K. G. Manton and E. Stallard, Longevity in the United States: Age and sex-specific evidence on life span limits from mortality patterns 1960–1990, *Journal of Gerontology* 51A (1996): B362–B375.
4. D. A. Banks and M. Fossel, Telomeres, cancer, and aging: Altering the human life span, *Journal of the American Medical Association* 278 (1997): 1345–1348.
5. A. J. Vita and coauthors, Aging, health risks, and cumulative disability, *New England Journal of Medicine* 338 (1998): 1035–1041.
6. American College of Sports Medicine, *ACSM Fitness Book* (Hong Kong: Paramount Printing Ltd., 1998), pp. 6–8.
7. C. E. J. van den Hombergh and coauthors, Physical activities of noninstitutional Dutch elderly and characteristics of inactive elderly, *Medicine and Science in Sports and Exercise* 27 (1995): 334–339; L. E. Voorrips and coauthors, The physical condition of elderly women differing in habitual physical activity, *Medicine and Science in Sports and Exercise* 25 (1993): 1152–1157.
8. W. J. Evans and D. Cyr-Campbell, Nutrition, exercise, and healthy aging, *Journal of the American Dietetic Association* 97 (1997): 632–638.
9. J. S. Green and S. F. Crouse, The effects of endurance training on functional capacity in the elderly: A meta-analysis, *Medicine and Science in Sports and Exercise* 27 (1995): 920–926.
10. A. C. King and coauthors, Moderate intense exercise and self-rated quality of sleep in older adults: A randomized controlled trial, *Journal of the American Medical Association* 277 (1997): 32–37; W. J. Evans, Effects of aging and exercise on nutrition needs of the elderly, *Nutrition Reviews* 54 (1996): S35–S39; M. A. Fiatarone and coauthors, Exercise training and nutritional supplementation for physical frailty in very elderly people, *New England Journal of Medicine* 330 (1994): 1769–1775; W. W. Campbell and coauthors, Increased energy requirements and changes in body composition with resistance training in older adults, *American Journal of Clinical Nutrition* 60 (1994): 167–175.
11. D. E. Butterworth and coauthors, Exercise training and nutrient intake in elderly women, *Journal of the American Dietetic Association* 93 (1993): 653–657.
12. R. Weindruch and R. S. Sohal, Caloric intake and aging, *New England Journal of Medicine* 337 (1997): 986–994.
13. P. R. Johnson and coauthors, Longevity in obese and lean male and female rats of the Zucker strain: Prevention of hyperphagia, *American Journal of Clinical Nutrition* 66 (1997): 890–903.
14. E. J. M. Velthuis-te Wierik, and coauthors, Energy restriction, a useful intervention to retard human aging? Results of a feasibility study, *European Journal of Clinical Nutrition* 48 (1994): 138–148.
15. J. J. Kehayias and coauthors, Total body potassium and body fat: Relevance to aging, *American Journal of Clinical Nutrition* 66 (1997): 904–910; G. Paolisso and coauthors, Body composition, body fat distribution, and resting metabolic rate in healthy centenarians, *American Journal of Clinical Nutrition* 62 (1995): 746–750.
16. J. E. Morley, Anorexia of aging: Physiologic and pathologic, *American Journal of Clinical Nutrition* 66 (1997): 760–763.
17. I. H. Rosenberg, Sarcopenia: Origins and clinical relevance, *Journal of Nutrition* 127 (1997): 990S–991S.
18. C. Dutta, Significance of sarcopenia in the elderly, *Journal of Nutrition* 127 (1997): 992S–997S.
19. L. A. Bush and coauthors, D-E-N-T-A-L: A rapid self-administered screening instrument to promote referrals for further evaluation in older adults, *Journal of the American Geriatrics Society* 44 (1996): 979–981.
20. C. S. Ritchie and coauthors, Nutritional status of urban homebound older adults, *American Journal of Clinical Nutrition* 66 (1997): 815–818.
21. V. B. Duffy, J. R. Backstrand, and A. M. Ferris, Olfactory dysfunction and related nutritional risk in free-living, elderly women, *Journal of the American Dietetic Association* 95 (1995): 879–884.
22. S. S. Schiffman, Taste and smell losses in normal aging and disease, *Journal of the American Medical Association* 278 (1997): 1357–1362.
23. J. Unützer and coauthors, Depressive symptoms and the cost of health services in HMO patients aged 65 years and older, *Journal of the American Medical Association* 277 (1997): 1618–1623; Depressing statistics, *Journal of the American Medical Association* 277 (1997): 1584.
24. N. S. Wellman and coauthors, Elder insecurities: Poverty, hunger, and malnutrition, *Journal of the American Dietetic Association* 97 (1997): S120–S122.
25. E. Foder-Levitt and coauthors, Utilization of home-delivered meals by recipients 75 years of age or older, *Journal of the American Dietetic Association* 95 (1995): 552–557.
26. S. A. Gerrior and coauthors, How does living alone affect dietary quality? *Family Economics and Nutrition Review* 8 (1995): 44–46.
27. Committee on Dietary Reference Intakes, *Dietary Reference Intakes for Calcium, Phosphorus, Magnesium, Vitamin D, and Fluoride* (Washington, D.C.: National Academy Press, 1997); R. M. Russell, New views on the RDAs for older adults, *Journal of the American Dietetic Association* 97 (1997): 515–518.
28. A. Bendich, Criteria for determining recommended dietary allowances for healthy older adults, *Nutrition Reviews* 53 (1995): S105–S110.
29. J. C. Chidester and A. A. Spangler, Fluid intake in the institutionalized elderly, *Journal of the American Dietetic Association* 97 (1997): 23–28; S. A. Gilmore and coauthors, Clinical indicators associated with unintentional weight loss and pressure ulcers in elderly residents of nursing facilities, *Journal of the American Dietetic Association* 95 (1995): 984–992.
30. *Water: The Beverage of Life* (Chicago: The American Dietetic Association, 1994).
31. W. W. Campbell, Dietary protein requirements of older people: Is the RDA adequate? *Nutrition Today* 31 (1996): 192–197; D. J. Millward and coauthors, Aging, protein requirements, and protein turnover, *American Journal of Clinical Nutrition* 66 (1997): 774–786.

32. Position of The American Dietetic Association: Health implications of dietary fiber, *Journal of the American Dietetic Association* 97 (1997): 1157–1159.
33. A. A. Abbase and D. Rudman, Undernutrition in the nursing home: Prevalence, consequences, causes and prevention, *Nutrition Reviews* 52 (1994): 113–122.
34. J. G. Fischer and coauthors, Dairy product intake of the oldest old, *Journal of the American Dietetic Association* 95 (1995): 918–921.
35. P. F. Jacques and coauthors, Plasma 25-hydroxyvitamin D and its determinants in an elderly population sample, *American Journal of Clinical Nutrition* 66 (1997): 929–936.
36. Committee on Dietary Reference Intakes, 1997.
37. S. P. Stabler, J. Lindenbaum, and R. H. Allen, Vitamin B-12 deficiency in the elderly: Current dilemmas, *American Journal of Clinical Nutrition* 66 (1997): 741–749.
38. R. J. Wood, P. M. Suter, and R. M. Russell, Mineral requirements of elderly people, *American Journal of Clinical Nutrition* 62 (1995): 493–505.
39. Committee on Dietary Reference Intakes, 1997.
40. D. V. Porter, Washington update: NIH consensus development conference statement optimal calcium intake, *Nutrition Today,* September/October 1994, pp. 37–40.
41. C. Marwick, NHANES III health data relevant for aging nation, *Journal of the American Medical Association* 277 (1997): 100–102.
42. S. T. Mayne, Beta-carotene, carotenoids, and disease prevention in humans, *FASEB Journal* 10 (1996): 690–701; P. F. Jacques and L. T. Chylack, Epidemiologic evidence of a role for the antioxidant vitamins and carotenoids in cataract prevention, *American Journal of Clinical Nutrition* 53 (1991): 352S–355S; G. E. Bunce, Antioxidant nutrition and cataract in women: A prospective study, *Nutrition Reviews* 51 (1993): 84–86.
43. P. F. Jacques and coauthors, Long-term vitamin C supplement use and prevalence of early age-related lens opacities, *American Journal of Clinical Nutrition* 66 (1997): 911–916; J. M. Robertson, A. P. Donner, and J. R. Trevithick, A possible role for vitamins C and E in cataract prevention, *American Journal of Clinical Nutrition* 53 (1991): 346S–351S.
44. R. J. Glynn and coauthors, Body mass index: An independent predictor of cataract, *Archives of Ophthalmology* 113 (1995): 1131–1137.
45. S. West and coauthors, Are antioxidants or supplements protective for age-related macular degeneration? *Archives of Ophthalmology* 112 (1994): 222–227; J. M. Seddan and coauthors, Dietary carotenoids, vitamins A, C, and E, and advanced age-related macular degeneration, *Journal of the American Medical Association* 272 (1994): 1413–1420.
46. W. H. Ettinger and coauthors, A randomized trial comparing aerobic exercise and resistance exercise with a health education program in older adults with knee osteoarthritis: The Fitness Arthritis and Seniors Trial (FAST), *Journal of the American Medical Association* 277 (1997): 25–31.
47. R. S. Panush, Nutritional therapy for rheumatic diseases, *Annals of Internal Medicine* 106 (1987): 619–621.
48. National Institute of Aging, Progress Report on Alzheimer's Disease, 1994, NIH publication number 94-3885 (Washington, D.C.: Government Printing Office, 1994); E. M. Reiman and coauthors, Preclinical evidence of Alzheimer's disease in persons homozygous for the $\epsilon$4 allele for apolipoprotein E, *New England Journal of Medicine* 334 (1996): 752–758.
49. B. Finley, Nutritional needs of the person with Alzheimer's disease: Practical approaches to quality care, *Journal of the American Dietetic Association* 97 (1997): S177–S180.
50. A. La Rue and coauthors, Nutritional status and cognitive functioning in a normally aging sample: A 6-y reassessment, *American Journal of Clinical Nutrition* 65 (1997): 20–29.
51. K. G. Manton, L. Corder, and E. Stallard, Chronic disability trends in elderly United States population: 1982–1994, *Proceedings of the National Academy of Sciences of the USA* 94 (1997): 2593–2598.
52. L. DiPietro, The epidemiology of physical activity and physical function in older people, *Medicine and Science in Sports and Exercise* 28 (1996): 596–660.
53. Are older Americans making better food choices to meet diet and health recommendations? *Nutrition Reviews* 51 (1993): 20–22.
54. Position of The American Dietetic Association: Nutrition, aging, and the continuum of health care, *Journal of the American Dietetic Association* 96 (1996): 1048–1052.
55. Federal program nourishes poor elderly, *Journal of the American Medical Association* 278 (1997): 1301.
56. M. B. Moran and E. Reed, Are congregate meals meeting clients' needs for "heart healthy" menus? *Journal of Nutrition for the Elderly* 13 (1993): 3–10.

# Nutrient-Drug Interactions

People over the age of 65 take about 25 percent of all the over-the-counter and prescription drugs sold in the United States. They often go to different doctors for different conditions and receive different prescriptions from each. To avoid harmful drug interactions, they need to inform all of their physicians and pharmacists of all the medicines being taken. These medicines enable people of all ages to enjoy better health, but they also bring side effects and risks.

This highlight focuses on some of the nutrition-related consequences of medical drugs, both prescription drugs and nonprescription, or over-the-counter, drugs. (Highlight 7 described the relationships between nutrition and the drug alcohol.)

## The Actions of Drugs

Most people think of drugs either as medicines that help them recover from illnesses or as illegal substances that lead to bodily harm and addiction. Actually, both uses of the term *drug* are correct because any substance that modifies one or more of the body's functions is, technically, a drug. Even medical drugs set in motion both desirable and undesirable events within the body.

Consider aspirin. One action of aspirin is to retard the production of certain prostaglandins. Some prostaglandins help to produce fevers, some sensitize pain receptors, some cause contractions of the uterus, some stimulate digestive tract motility, some control nerve impulses, some regulate blood pressure, some promote blood clotting, and some cause inflammation. By interfering with prostaglandin actions, aspirin reduces fever and inflammation, relieves pain, and slows blood clotting, among other things.

A person cannot use aspirin to produce one of its effects without producing all of its other effects. Someone who is prone to strokes and heart attacks might take aspirin to prevent blood clotting, but it would also dull that person's sense of pain. In another person who took aspirin only for pain, it would also slow blood clotting. For the second person, the anticlotting effect might be dangerous if it caused abnormal bleeding. A single two-tablet dose of aspirin doubles the bleeding time of wounds, an effect that lasts from four to seven hours. For this reason, physicians caution clients to refrain from taking aspirin before surgery.

## The Interactions between Drugs and Nutrients

Hundreds of drugs and nutrients interact, which can lead to nutrient imbalances or interfere with drug effectiveness.[1] Adverse drug-nutrient interactions are most likely if drugs are taken over long periods, if several drugs are taken, or if nutrition status is poor or deteriorating. Understandably, then, elderly people with chronic diseases are most vulnerable.[2] Studies of institutionalized elderly people confirm that multiple drug use significantly impairs nutrition status in this population.[3]

Nutrients and medications may interact in many ways:

- Drugs can alter food intake and the absorption, metabolism, and excretion of nutrients.
- Foods and nutrients can alter the absorption, metabolism, and excretion of drugs.

The following paragraphs describe these interactions, and Table H17-1 summarizes this information and provides specific examples.

## Altered Food Intake

Many medicines can lead to malnutrition by interfering with food intake. Drugs can influence appetite, alter taste or smell, cause sores or irritation in the mouth, reduce the flow of saliva, or induce nausea or vomiting. Amphetamines used to treat depression provide an example. They suppress appetite, alter taste perceptions, dry the mouth, and cause nausea. Conversely, some medicines stimulate the appetite and lead to undesirable weight gain. An example is astemizole (Hismanal), an antihistamine used to relieve allergy symptoms.

## Altered Nutrient Absorption

Laxatives provide an example of how drugs can interfere with nutrient absorption. Laxatives cause foods to move so rapidly through the intestine that many vitamins do not have enough time to be absorbed. The use of mineral oil as a laxative robs the person of the fat-soluble vitamins, most notably vitamin D. The vitamins from foods dissolve in the indigestible oil and are excreted; calcium, too, is lost this way. A person who uses laxatives daily for a long time may find that the intestines can no longer function without them. Such dependence can lead to malnutrition.

*Taking several medicines over long periods intensifies the risk of nutrient-drug interactions. (Courtesy of CNN)*

## Altered Drug Absorption

A classic example of how foods can interfere with drug absorption is the interactions between the antibiotic tetracycline and the minerals calcium and iron. When calcium and tetracycline, or iron and tetracycline, are taken at the same time, they bind to each other, thus reducing the absorption of both. People are therefore advised not to take tetracycline with milk, milk products, or calcium-containing antacids, such as Tums. Similarly, iron supplements should be taken two hours apart from tetracycline doses.

Another example of how foods can interfere with the absorption of a drug is the interaction between acidic foods and the nicotine gum that people sometimes use to help quit smoking cigarettes. Certain acid-containing foods and beverages interfere with the absorption of nicotine through the lining of the mouth into the blood. For maximum effectiveness, people should refrain from ingesting foods and beverages for 15 minutes before, and while, chewing the gum. When a food or beverage blocks nicotine's absorption from the mouth, the person swallows the nicotine, and this may cause nausea and hiccups as well as interfere with the drug's effectiveness.

Some medications are absorbed better with foods than without them. For this reason, the antifungal drug griseofulvin is always given with meals. Similarly, a glass of grapefruit juice significantly enhances the absorption of a variety of common medications.[4] In many cases, though, foods delay the rate at which drugs are absorbed. In some instances this, too, can be helpful. An aspirin taken on an empty stomach works faster than when it is given with food, but because aspirin can irritate the GI tract, taking it with food can reduce nausea and prevent bleeding.

## *Altered Metabolism*

To appreciate how drug-nutrient interactions can affect metabolism, consider medicines that resemble vitamins in structure. Vitamin K and the anticlotting medication

**Table H17–1**

**Mechanisms and Examples of Nutrient-Drug Interactions**

**Drugs Can Alter Food Intake by:**

- Altering the appetite (amphetamines suppress the appetite).
- Interfering with taste or smell (methotrexate changes taste sensations).
- Inducing nausea or vomiting (digitalis can do both).[a]
- Changing the oral environment (phenobarbital can cause dry mouth).[b]
- Irritating the GI tract (cyclophosphamide induces mucosal ulcers).[c]
- Causing sores or inflammation of the mouth (methotrexate can cause painful mouth ulcers).

**Drugs Can Alter Nutrition Absorption by:**

- Changing the acidity of the digestive tract (antacids can interfere with iron absorption).
- Altering digestive juices (cimetidine can improve fat absorption).[d]
- Altering motility of the digestive tract (laxatives speed motility, causing the malabsorption of many nutrients).
- Inactivating enzyme systems (neomycin may reduce lipase activity).[e]
- Damaging mucosal cells (chemotherapy can damage mucosal cells).
- Binding to nutrients (antacids bind phosphorus).

**Foods Can Alter Drug Absorption by:**

- Changing the acidity of the digestive tract (candy can change the acidity, thus dissolving slow-acting asthma medication too quickly).
- Stimulating secretion of digestive juices (griseofulvin is absorbed better when taken with foods that stimulate the release of digestive enzymes).
- Altering rate of absorption (aspirin is absorbed more slowly when taken with food).
- Binding to drugs (calcium binds to tetracycline, limiting drug absorption).
- Competing for absorption sites in the intestines (dietary amino acids interfere with levodopa absorption this way).[f]

**Drugs and Nutrients Can Interact and Alter Metabolism by:**

- Acting as structural analogs (as anticoagulants and vitamin K do).
- Competing with each other for metabolic enzyme systems (as phenobarbital and folate do).[b]
- Altering enzyme activity and contributing pharmacologically active substances (as monoamine oxidase inhibitors and tyramine do).

**Drugs Can Alter Nutrient Excretion by:**

- Altering reabsorption in the kidneys (some diuretics increase the excretion of sodium and potassium).
- Displacing nutrients from their plasma protein carriers (aspirin displaces folate).

**Foods Can Alter Drug Excretion by:**

- Changing the acidity of the urine (vitamin C can alter urinary pH and limit the excretion of aspirin).

NOTE: Most of these drugs are mentioned in the text; others are introduced in their respective footnotes.
[a]Digitalis is used in the treatment of congestive heart failure.
[b]Phenobarbital is an anticonvulsant used in the treatment of epilepsy.
[c]Cyclophosphamide is an immunosuppressant used in the treatment of organ transplants.
[d]Cimetidine is an H2 blocker used in the treatment of ulcers.
[e]Neomycin is an antibiotic used in the treatment of infections.
[f]Levodopa is used in the treatment of Parkinson's disease.

warfarin (Coumadin) provide an example. Warfarin opposes clotting by interfering with vitamin K's action. To be effective, the warfarin dose must be large enough to counteract vitamin K from the diet. If a person's vitamin K intake increases, as it may in summer when lettuces and greens are in season, then the physician must increase the drug dose. Another example is methotrexate, used to treat cancer and rheumatoid arthritis. Structurally similar to the B vitamin folate, methotrexate can cause severe folate deficiencies (see Figure H17-1).

Aspirin can also alter folate metabolism, but in a different way. Aspirin competes with folate for its protein carrier, thus interfering with the body's use of the vitamin. When aspirin is used over long periods of time, health care professionals should ensure that either the diet or supplements supply sufficient folate to meet the added demands.

The effects of tyramine provide another example of a substance in foods that alters a drug's action. Tyramine is a substance found in some foods, and it interacts with monoamine oxidase inhibitors (MAO inhibitors), which are prescribed to treat certain forms of severe depression. Normally, an enzyme in the brain inactivates tyramine, and MAO drugs block the action of that enzyme. When people take the drug, the enzyme fails to act. Thus tyramine remains active and stimulates the release of the neurotransmitter norepinephrine. This action can lead to severe hypertension and headaches. If blood pressure rises high enough, it can be fatal. For this reason, people taking MAO inhibitors must restrict their intakes of foods rich in tyramine (see Table H17-2).

Sometimes the combination of a specific food and a drug improves the drug's action in the body. Citrus fruits, for example, enhance the effectiveness of the anticancer drug tamoxifen.[5] Oranges, grapefruits, and tangerines contain flavonoids that assist the drug in halting the growth of cancer cells.

**Figure H17-1**
**Folate and Methotrexate**

Methotrexate (a drug used in the treatment of cancer and rheumatoid arthritis) is structurally similar to the B vitamin folate. When this medication is used, it competes for the enzyme that normally activates folate, creating a secondary deficiency of folate. Notice the similarities in their chemical structures.

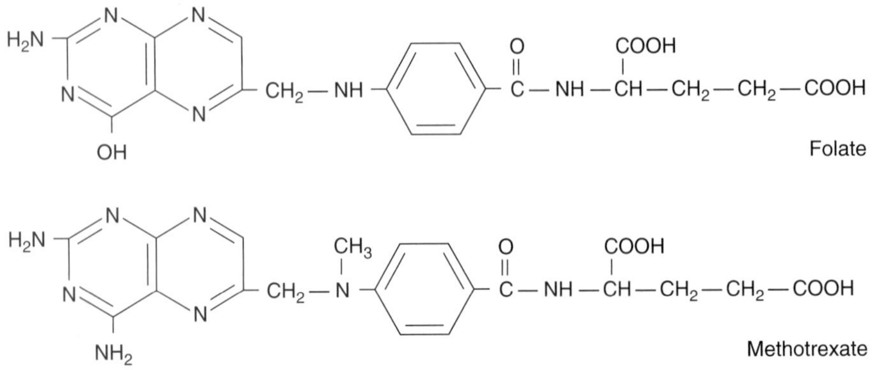

H₂N, N, N, N, N, OH — $CH_2$—NH—(benzene ring)—C(=O)—NH—CH(COOH)—$CH_2$—$CH_2$—COOH

Folate

H₂N, N, N, N, N, $NH_2$ — $CH_2$—N($CH_3$)—(benzene ring)—C(=O)—NH—CH(COOH)—$CH_2$—$CH_2$—COOH

Methotrexate

### Altered Drug Excretion

The acidity of the urine affects the reabsorption of drugs back into the blood by the kidneys. An acidic urine limits the excretion of acidic drugs like aspirin. Some nutrients, such as vitamin C, can lower the pH of urine, making it more acidic. Therefore, large doses of vitamin C given with aspirin increase the urine's acidity, keeping aspirin in the blood longer.

### Altered Nutrient Excretion

Medicines can also alter urinary excretion of nutrients. For example, some diuretics accelerate the excretion of the minerals calcium, potassium, magnesium, and zinc.

## Other Ingredients in Drugs

Besides the active ingredients, medicines may contain other substances such as sugar, sorbitol, and sodium. For most people who use medicines on occasion and in small amounts, such ingredients pose no problem. When medicines are taken regularly or in large doses, however, people on special diets may need to be aware of these additional ingredients and their effects.

### Sugar or Sorbitol

Many liquid preparations contain sugar or sorbitol to make them taste better. For people who must regulate their intakes of simple sugars, such as people with diabetes, the amount of sugar in these medicines must be considered. Large doses of liquids containing sorbitol may cause diarrhea.

### Sodium

Antibiotics and antacids often contain sodium. People who take Alka Seltzer, for example, may not realize that a single two-tablet dose may exceed their recommended sodium intake for a whole day. In addition, antacids neutralize stomach acid, and many nutrients depend on acid for their digestion. Taking any antacid regularly will reduce the absorption of many nutrients.

## The Health Professional and Nutrient-Drug Interactions

Hundreds of nutrient-drug interactions have been identified, and information continues to accumulate. It would be difficult, if not impossible, to remember all the potential effects

**Table H17-2**
**Foods Restricted in a Tyramine-Controlled Diet**

| | |
|---|---|
| Beverages: | Red wines including chianti, sherry[a] |
| Cheeses: | Aged cheeses, American, camembert, cheddar, gouda, gruyère, mozzarella, parmesan, provolone, romano, roquefort, stilton[b] |
| Meats: | Liver; dried, salted, smoked, or pickled fish; sausage, pepperoni; dried meats |
| Vegetables: | Fava beans; Italian broad beans; sauerkraut; fermented pickles and olives |
| Other: | Brewer's yeast;[c] all aged and fermented products; soy sauce in large amounts; cheese-filled breads, crackers, and desserts; salad dressings containing cheese |

*Note:* The tyramine contents of foods vary from product to product depending on the methods used to prepare, process, and store the food. In some cases, as little as 1 ounce of cheese can cause a severe hypertensive reaction in people taking monoamine oxidase inhibitors. In general, the following foods contain small enough amounts of tyramine that they can be consumed in small quantities: ripe avocado, banana, yogurt, sour cream, acidophilus milk, buttermilk, raspberries, and peanuts.
[a]Most wine and domestic beer can be consumed in small quantities.
[b]Unfermented cheeses, such as ricotta, cottage cheese, and cream cheese, are allowed.
[c]Products made with baker's yeast are allowed.

of these interactions on nutrition status. Instead, health care professionals would serve their clients well to:[6]

- Keep in mind that drug-nutrient interactions can and do occur, especially when medicine use is long term.
- Record and review drug and diet histories of clients with potential interactions in mind.
- Be aware of groups of people who are likely to develop drug-related nutrient deficiencies; be prepared to look up the nutrition effects of medications that these clients are taking.
- Reassess nutrition status frequently for high-risk clients.
- Become familiar with the nutrient interactions of drugs commonly used to treat the disorders of their clients.

Nutrient interactions and risks are not unique to prescription drugs. People who buy over-the-counter drugs also need to protect themselves. About 300,000 nonprescription drugs are marketed in the United States; more than 400 of these were available only by prescription 15 years ago.[7] The increasing availability of over-the-counter drugs allows people to treat themselves for many ailments from arthritis to yeast infections. Many older adults take over-the-counter laxatives and antacids regularly. Excessive use of either of these drugs may impair nutrition status by interfering with absorption, increasing excretion, or both. Consumers need to ask their physicians about potential interactions and check with their pharmacists for instructions on taking drugs with foods. Should problems arise, they should seek professional care without delay.

## Notes


1. P. G. Welling, Effects of food on drug absorption, *Annual Review of Nutrition* 16 (1996): 383–415; J. A. Thomas, Drug-nutrient interactions, *Nutrition Reviews* 53 (1995): 271–282.
2. M. Lee, Drugs and the elderly: Do you know the risks? *American Journal of Nursing* 96 (1996): 25–31; L. L. Lilley and Robert Guanci, Polypharmacy in elders, *American Journal of Nursing* 96 (1996): 12, 14.
3. C. W. Lewis, E. A. Frongillo, and D. A. Roe, Drug-nutrient interactions in long-term care facilities, *Journal of the American Dietetic Association* 95 (1995): 309–315; R. N. Varma, Risk for drug-induced malnutrition is unchecked in elderly patients in nursing homes, *Journal of the American Dietetic Association* 94 (1994): 192–194.
4. E. B. Feldman, How grapefruit juice potentiates drug bioavailability, *Nutrition Reviews* 55 (1997): 398–400; K. S. Lown and coauthors, Grapefruit juice increases felodipine oral availability in humans by decreasing intestinal CYP3A protein expression, *Journal of Clinical Investigation* 99 (1997): 2545–2553.
5. Juicy anticancer prospects, *Science News* 149 (1996): 287.
6. Varma, 1994.
7. Nonprescription Drug Manufacturers Association, *Facts and Figures* (Washington, D.C.: Nonprescription Drug Manufacturers Association, 1993).

CHAPTER 18

# Diet and Health

NUTRITION AND CHRONIC DISEASES

HEART DISEASE AND STROKES

How Atherosclerosis Develops

Risk Factors For Cardiovascular Disease

Recommendations For Reducing Cardiovascular Disease Risk

HYPERTENSION

Blood Pressure Regulation

How Hypertension Develops

Risk Factors For Hypertension

Recommendations For Reducing Hypertension Risk

CANCER

How Cancer Develops

Recommendations For Reducing Cancer Risk

DIABETES MELLITUS

Complications of Diabetes

Dietary Recommendations For Type 1 Diabetes

Dietary Recommendations For Type 2 Diabetes

RECOMMENDATIONS FOR CHRONIC DISEASES

NUTRITION AND INFECTIOUS DISEASES

The Immune System

Nutrition and Immunity

HUMAN IMMUNODEFICIENCY VIRUS (HIV) INFECTION AND ACQUIRED IMMUNE DEFICIENCY SYNDROME (AIDS)

How AIDS Develops

The HIV Wasting Syndrome

Nutrition Support For People With HIV Infections

HIGHLIGHT: ALTERNATIVE THERAPIES

Infectious diseases such as smallpox once claimed the lives of many children and limited the average life expectancy of adults. Thanks to medical science's ability to identify disease-causing microorganisms and develop preventive strategies, most children now live well into their later years, and the average life expectancy far exceeds that of our ancestors. In developed nations, purification of water and the safe handling of foods prevent the spread of infection, and immunizations protect individuals.

Still, despite public health measures and medical treatments, some infectious diseases threaten many lives. Perhaps the most infamous is AIDS (acquired immune deficiency syndrome). AIDS develops from infection by the human immunodeficiency virus (HIV), which attacks the immune system and disables the body's defenses against other diseases. Then these diseases, which would produce only mild, if any, illness in people with healthy immune systems, destroy health and life.

This chapter closes with a description of the immune system and the nutrition-related concerns of people with HIV infections, but it begins with a look at the chronic diseases that pose the greatest threat to the lives of people in developed countries. These chronic diseases develop over a lifetime as a result of metabolic abnormalities induced by such factors as genetics, age, sex, lifestyle, and environment. Diet is among the many lifestyle factors that influence the development of chronic diseases.

National Center for Chronic Disease Prevention and Health Promotion
www.cdc.gov/nccdphp

NIH Consumer Health Information
www.nih.gov/health/consumer/conicd.htm

National Health Information Center
nhic-nt.health.org

# Nutrition and Chronic Diseases

Figure 18-1 shows the ten leading causes of death in the United States.[1] Four of these causes, including the top three, have some relationship with diet. Taken together, these four conditions account for about two-thirds of the nation's 2 million deaths each year.

This chapter explains how each of these four chronic diseases develops and summarizes their major links with nutrition. Earlier chapters described individual nutrients' connections with diseases and may have left the mistaken impression

National Center for Health Statistics
www.cdc.gov/nchswww

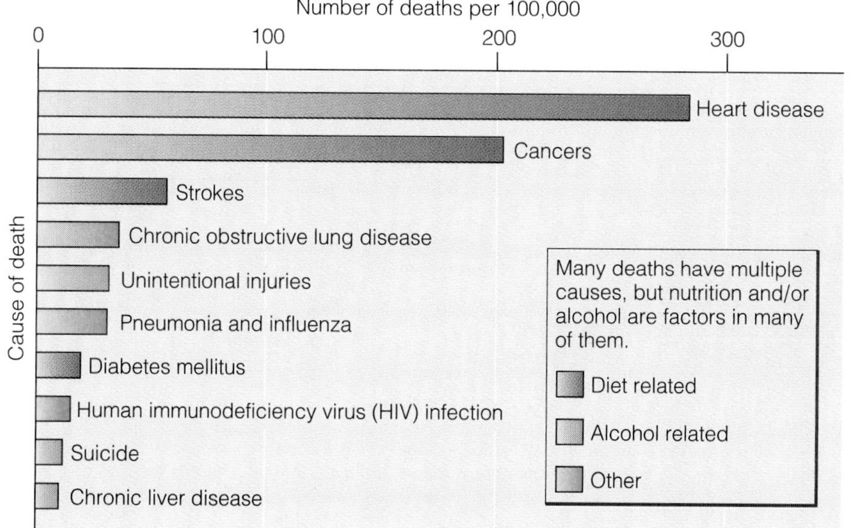


**Figure 18-1**

**The Ten Leading Causes of Illness and Death in the United States**

Diet influences the development of several chronic diseases—notably, heart disease, some types of cancer, stroke, and diabetes. Taken together, these four conditions account for about two-thirds of the nation's 2 million deaths each year.

Source: National Center for Health Statistics, *Monthly Vital Statistics Report,* October 1996, p. 31.

of "one disease–one nutrient" relationships.[2] Indeed, valid links do exist between fiber and diabetes, fat and heart disease, calcium and osteoporosis, and antioxidant vitamins and cancer, but focusing only on these links oversimplifies the story. In reality, each nutrient may have connections with several diseases because its role in the body is not specific to a disease, but to a body function. Fiber—because it binds substances in the GI tract—helps control diabetes and prevent cancer. Vitamin E—because it acts as an antioxidant—helps prevent both cancer and heart disease. Furthermore, each of the chronic diseases develops in response to multiple risk factors—including many nondietary factors such as genetics, physical activity, and smoking. An integrated and balanced approach to disease prevention therefore includes attention to all of the many factors involved. Figure 18-2 illustrates some of the relationships between risk factors and degenerative diseases.

Reminder: Factors that are associated with a high incidence of a disease are *risk factors.* Some risk factors, such as diet and physical activity, are *modifiable,* meaning that they can be changed; others, such as genetics, age, and sex, cannot be changed.

**synergistic** (SIN-er-JIST-ick): multiple factors operating together in such a way that their combined effects are greater than the sum of their individual effects.

Notice how many of the diseases have a genetic component. A family history of a certain disease is a powerful indicator of a person's tendency to contract that disease. Still, environmental factors are often pivotal in determining whether that tendency will be expressed. Genetics and environmental factors often work synergistically; for instance, cigarette smoking is especially likely to bring on heart disease in people who are genetically predisposed to develop it. Not smoking would benefit everyone's health, of course, regardless of genetic predisposition, but some recommendations to prevent chronic diseases best meet an individual's needs when family history is considered.[3] For example, women with a family history of breast cancer might reduce their risks if they abstain from alcohol, whereas those with a family history of heart disease might benefit from one or two drinks a week. Similarly, a man with a family history of hypertension may lower his blood pressure by restricting his salt intake, whereas others may not need to do so routinely.

U.S. Government

www.healthfinder.gov/searchoptions/topicsaz.htm

Search for Chronic Diseases

Search for Disease Prevention

IN SUMMARY

Heart disease, cancers, and strokes are the three leading causes of death in the United States, and diabetes also ranks among the top ten.

**Figure 18-2**

**Diet/Lifestyle Risk Factors and Chronic Diseases**

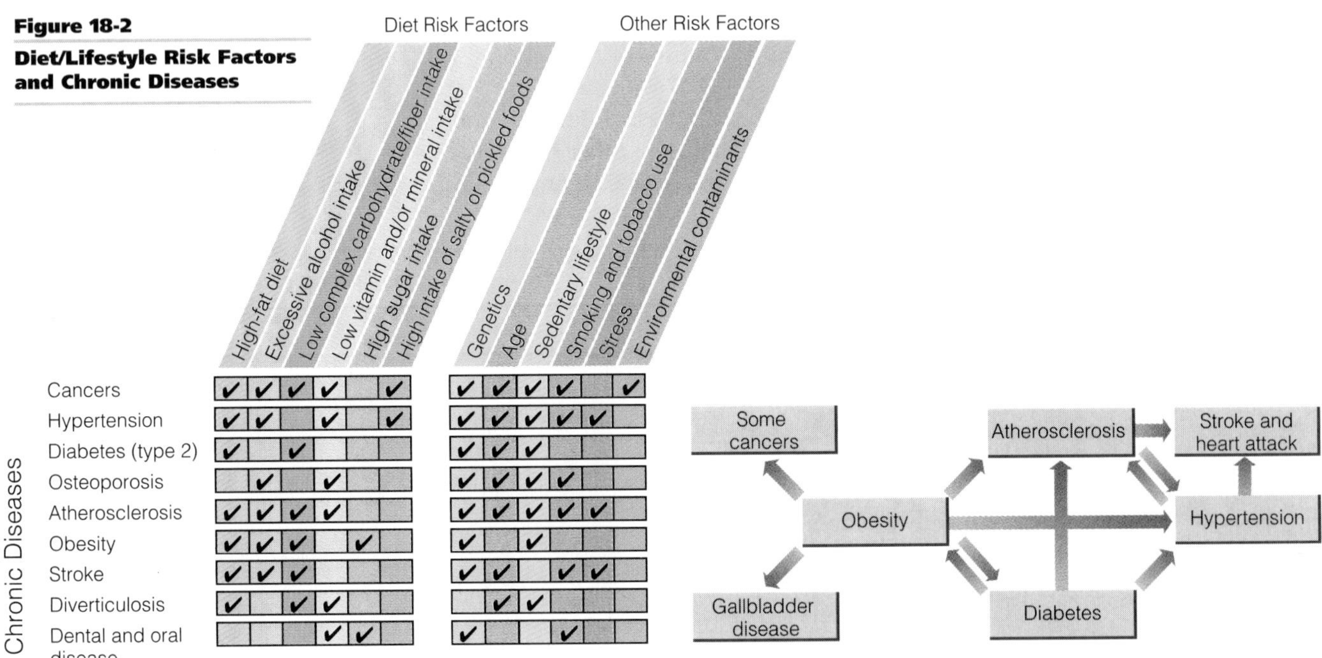

| Chronic Diseases | High-fat diet | Excessive alcohol intake | Low complex carbohydrate/fiber intake | Low vitamin and/or mineral intake | High sugar intake | High intake of salty or pickled foods | Genetics | Age | Sedentary lifestyle | Smoking and tobacco use | Stress | Environmental contaminants |
|---|---|---|---|---|---|---|---|---|---|---|---|---|
| | Diet Risk Factors | | | | | | Other Risk Factors | | | | | |
| Cancers | ✔ | ✔ | ✔ | ✔ | | ✔ | ✔ | ✔ | ✔ | ✔ | | ✔ |
| Hypertension | ✔ | ✔ | | ✔ | | ✔ | ✔ | ✔ | ✔ | ✔ | ✔ | |
| Diabetes (type 2) | ✔ | | ✔ | | | | ✔ | ✔ | ✔ | | | |
| Osteoporosis | | ✔ | | ✔ | | | ✔ | ✔ | ✔ | ✔ | | |
| Atherosclerosis | ✔ | ✔ | ✔ | ✔ | | | ✔ | ✔ | ✔ | ✔ | ✔ | |
| Obesity | ✔ | ✔ | ✔ | | ✔ | | ✔ | | ✔ | | | |
| Stroke | ✔ | ✔ | ✔ | | | | ✔ | ✔ | | ✔ | ✔ | |
| Diverticulosis | ✔ | | ✔ | ✔ | | | | ✔ | ✔ | | | |
| Dental and oral disease | | | | ✔ | ✔ | | ✔ | | | ✔ | | |



This chart shows that the same risk factor can affect many chronic diseases. Notice, for example, how many diseases have been linked to a high-fat diet. The chart also shows that a particular disease, such as atherosclerosis, may have several risk factors.

This flow chart shows that many of these conditions are themselves risk factors for other chronic diseases. For example, a person with diabetes is likely to develop atherosclerosis and hypertension. These two conditions, in turn, worsen each other. Notice how all of these chronic diseases are linked to obesity.

All four of these chronic diseases have significant links with nutrition, although genetic and environmental risk factors are also important.

# Heart Disease and Strokes

American Heart Association
www.amhrt.org

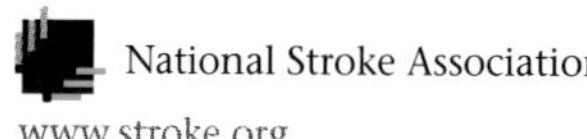
National Stroke Association
www.stroke.org

For decades, the major causes of death in many developed countries have been diseases of the heart and blood vessels, collectively known as cardiovascular disease (CVD). Cardiovascular disease is the leading single cause of death around the world today. In the United States, death rates from cardiovascular disease in men (aged 35 to 50) are three times greater than in women of the same age, but in later years (65 to 74), the incidence is similar. The consequences of cardiovascular disease are usually heart disease and strokes, the first and third leading causes of death for adults, respectively.

Reminder: *Cardiovascular disease (CVD)* is a general term for all diseases of the heart and blood vessels.

**Healthy People 2000:** Reduce coronary heart disease deaths to no more than 100 per 100,000 people.

Coronary heart disease (CHD), the most common form of cardiovascular disease, usually involves atherosclerosis and hypertension. Atherosclerosis is the accumulation of lipids and other materials in the arteries, and hypertension is high blood pressure. Each makes the other worse.

Reminder: *Coronary heart disease (CHD)* refers to the damage that occurs when the blood vessels carrying blood to the heart become narrow and occluded. The narrowing is usually caused by *atherosclerosis,* a condition characterized by plaques along the inner walls of the arteries. *Hypertension* is high blood pressure.

## *How Atherosclerosis Develops*

As Highlight 16 pointed out, no one is free of the fatty streaks that may one day become the fibrous plaques of atherosclerosis. For most adults, the question is not whether you have them, but how far advanced they are and what you can do to slow or reverse their progression.

Reminder: *Fibrous plaques* are mounds of lipid material, mixed with smooth muscle cells and calcium, which develop in the artery walls in atherosclerosis. Plaque associated with atherosclerosis is known as **atheromatous** plaque.

• ***Plaques Develop*** • Atherosclerosis usually begins with the accumulation of soft fatty streaks along the inner arterial walls, especially at branch points (see Figure H16-1 on p. 532). These fatty streaks gradually enlarge and become hardened with minerals, forming plaques. Plaques stiffen the arteries and narrow the passages through them. Most people have well-developed plaques by the age of 30. The progression of atherosclerosis in the coronary arteries may restrict blood flow to the heart muscle and limit the delivery of oxygen. As Chapter 5 pointed out, a diet high in saturated fat is a major contributor to the development of plaques and the progression of atherosclerosis.[4]

• ***Blood Clots Form*** • Small, cell-like bodies in the blood, known as platelets, cause clots to form whenever they encounter injuries in blood vessels. Clots normally form and dissolve in the blood all the time, but in atherosclerosis, clots form faster than they dissolve because the platelets respond to plaques as they normally do to injuries.

**platelets:** tiny, disc-shaped bodies in the blood, important in blood clot formation.

The action of platelets is under the control of certain eicosanoids, known as prostaglandins and thromboxanes, which are made from the 20-carbon omega-6 and omega-3 fatty acids (introduced in Chapter 5). Each eicosanoid plays a specific role in helping to regulate many of the body's activities. Sometimes their actions oppose each other. For example, one eicosanoid prevents, and another promotes, clot formation; similarly, one dilates, and another constricts, the blood vessels. When omega-3 fatty acids are abundant in the diet, they make more of the kinds of eicosanoids that favor heart health.

Reminder: *Eicosanoids* are powerful, biologically active compounds made from the 20-carbon fatty acids; eicosanoids include prostaglandins (PROS-tah-GLAND-ins) and thomboxanes (throm-BOX-ains).

Eicosanoids help to regulate:
- Blood pressure.
- Blood clot formation.
- Blood vessel contractions.
- Immune response.
- Nerve impulse transmissions.

**thrombosis** (throm-BOH-sis): the formation of a **thrombus** (THROM-bus), a blood clot that may obstruct a blood vessel, causing gradual tissue death.
- **thrombo** = clot

**embolism** (EM-boh-lizm): the obstruction of a blood vessel by an **embolus** (EM-boh-luss), or traveling clot, causing sudden tissue death.
- **embol** = to insert, plug

**angina** (an-JYE-nah or AN-ji-nah): a painful feeling of tightness or pressure in and around the heart, often radiating to the back, neck, and arms; caused by a lack of oxygen to an area of heart muscle.

**heart attack:** sudden tissue death caused by blockages of vessels that feed the heart muscle; also called **myocardial infarction** or **cardiac arrest.**
- **myo** = muscle
- **cardial** = heart
- **infarct** = tissue death

**transient ischemic attack (TIA):** a temporary reduction in blood flow to the brain, which causes temporary symptoms that vary depending on the part of the brain affected. Common symptoms include light-headedness, visual disturbances, paralysis, staggering, numbness, and inability to swallow.

**stroke:** an event in which the blood flow to a part of the brain is cut off; also called **cerebrovascular accident (CVA).**
- **cerebro** = brain
- **vascular** = blood vessels

Major risk factors for CHD:
- **High LDL cholesterol.**
- Male, 45 years or older.
- Female, 55 years or older, or with premature menopause and not on estrogen replacement therapy.
- **Low HDL cholesterol.**
- **Hypertension.**
- Smoking.
- **Diabetes.**
- Family history of heart attacks or sudden death prior to age 55 in a male parent or sibling or prior to age 65 in a female parent or sibling.

*Note:* Risk factors in bold type have relationships with diet.

Reminder: Cholesterol is carried in several lipoproteins, chief among them LDL and HDL (see Chapter 5 for details). Remember them this way:
- LDL = Low-density lipoproteins = Less healthy.
- HDL = High-density lipoproteins = Healthy.

Abnormal blood clotting can trigger life-threatening events. A blood clot may stick to a plaque in an artery and gradually grow large enough to restrict or close off a blood vessel (thrombosis). A coronary thrombosis blocks blood flow through an artery that feeds the heart muscle. A cerebral thrombosis blocks blood flow through an artery that feeds the brain. A clot may also break free from the artery wall and travel through the circulatory system until it lodges in a small artery and suddenly shuts off flow to the tissues fed by this artery (embolism).

The loss of blood flow to the area supplied by the clotted artery robs the tissue of oxygen and nutrients, and the tissue may eventually die. Restricted blood flow in the arteries feeding the heart may damage the heart muscle (coronary heart disease). The person may experience pain and pressure in and around the area of the heart (angina). Without blood flow to the heart muscle, that area of heart muscle dies, causing a heart attack. Restricted blood flow to the brain causes a transient ischemic attack or stroke.

• ***Blood Pressure Rises*** • The heart must create enough pressure to push blood through the circulatory system. When arteries are narrowed by plaques, clots, or both, blood flow is restricted, and the heart must then generate more pressure to deliver blood to the tissues. This higher blood pressure further damages the artery walls, and plaques and clots are especially likely to form at damage points. Thus the development of atherosclerosis is a self-accelerating process. (A later section describes additional consequences of high blood pressure.)

## *Risk Factors for Cardiovascular Disease*

Although atherosclerosis can develop in any blood vessel, the coronary arteries are most often affected, leading to coronary heart disease (CHD). The margin lists the major risk factors for CHD. Obesity and physical inactivity significantly modify several of these risk factors, contributing to high LDL cholesterol, low HDL cholesterol, hypertension, and diabetes.[5] The criteria for defining blood lipids, blood pressure, and obesity in relation to CHD risk are shown in Table 18-1; Highlight 16 presents standards for children and adolescents.

By middle age, most adults have at least one risk factor, and many have more than one.[6] Both the United States and Canada recommend screening to identify risk factors in individuals and offer preventive advice and treatment. Such public health programs are proving successful: since 1960, both blood cholesterol levels and deaths from cardiovascular disease among U.S. adults have shown a continuous and substantial downward trend.[7] These trends reflect behavior changes in individuals. As adults grow older, many of them stop smoking and drinking alcohol and start following a low-fat diet or taking medication in an effort to improve their cardiovascular health.[8]

It befits a nutrition book to focus on dietary strategies to prevent heart disease, but it should be noted that only four of the eight major risk factors can be modified by diet or weight loss: high LDL cholesterol, low HDL cholesterol, hypertension, and diabetes. Three cannot be modified by diet or otherwise: sex, age, and family history. The other major risk factor—smoking—is, of course, modifiable and may indeed be the most influential. As appealing as the solution may sound, diet may not reduce the risk of heart disease and stroke as successfully as other interventions do. Quitting smoking and taking prescribed medicines to lower blood lipids or blood pressure, for example, may be far more effective than dietary changes alone. Dietary changes can be effective, however, especially when combined with other strategies, such as quitting smoking and being physically active. These three strategies—dietary changes, quitting smoking, and physical activity—are always recommended before drug therapy.

• ***High LDL Cholesterol, Low HDL Cholesterol*** • The blood cholesterol linked to atherosclerosis risk is LDL cholesterol. HDL also carry cholesterol, but raised HDL

**Table 18-1**
**Standards for CHD Risk Factors**

| LDL Cholesterol | Total Cholesterol[b] |
|---|---|
| <130 mg/dL = desirable.[a] | <200 mg/dL = desirable. |
| 130–159 mg/dL = borderline high. | 200–239 mg/dL = borderline high. |
| ≥160 mg/dL = high. | ≥240 mg/dL = high. |
| **HDL Cholesterol** | **Triglycerides (Fasting)[d]** |
| HDL: ≤35 mg/dL indicates risk.[c] | <200 mg/dL = desirable. |
| LDL-to-HDL ratio:[e] | 200–400 mg/dL = borderline high. |
| Men: >5.0 indicates risk. | 400–1000 mg/dL = high. |
| Women: >4.5 indicates risk. | >1000 mg/dL = very high. |
| **Hypertension** | **Obesity** |
| Systolic and diastolic pressure:[f] | Body mass index: |
| <120 and <80 = optimal | ≥30. |
| <130 and <85 = normal. | |
| 130–139 or 85–89 = high-normal. | |
| 140–159 or 90–99 = mild. | |
| 160–179 or 100–109 = moderate. | |
| ≥180 or ≥110 = severe. | |

[a]For people with existing CHD, desirable LDL cholesterol values are lower (≤100 mg/dL).
[b]To convert cholesterol (mg/dL) to standard international units (mmol/L), multiply by 0.02586.
[c]This HDL value may be too low for women; no alternative value has yet been proposed.
[d]High triglycerides alone normally do not indicate *direct* risk, but may reflect lipoprotein abnormalities associated with CHD. The risk of CHD increases as triglyceride levels increase in people with other risk factors. High triglycerides also occur in conditions such as kidney disease and diabetes, which suggest a high CHD risk.
[e]The LDL-to-HDL ratio compares the concentration of LDL to that of HDL. A ratio of 5.0 really means "5 to 1," or that the first value (in this case LDL) is 5 times greater than the second (HDL). Similarly, a ratio of 4.5 means that LDL is 4.5 times greater than HDL.
[f]When systolic and diastolic pressures fall into different categories, the higher category classifies the status.

Sources: Blood lipid standards adapted from The Expert Panel, Summary of the second report of the National Cholesterol Education Program (NCEP), Expert Panel on Detection, Evaluation, and Treatment of High Blood Cholesterol in Adults (Adult Treatment Panel II), *Journal of the American Medical Association* 269 (1993): 3015–3023; hypertension standards adapted from the Sixth Report of the Joint National Committee on Prevention, Detection, Evaluation, and Treatment of High Blood Pressure, National High Blood Pressure Education Program, National Heart, Lung, and Blood Institute, National Institutes of Health, November 1997, p. 11.

represent cholesterol returning from the cells to the liver and thus indicate a reduced risk of atherosclerosis and heart attack (see Figure 18-3). High LDL correlate *directly* with heart disease, whereas high HDL correlate *inversely* with risk. In general, the higher the LDL cholesterol, the greater the risk of CHD (review Table 18-1).

The exact reasons why high LDL increase the risk of CHD remain unclear. Mounting evidence suggests, however, that atherosclerosis is accelerated by the oxidation of LDL by free radicals. (Highlight 11 describes how antioxidant nutrients and phytochemicals protect against free-radical formation and its damaging consequences.)

To a lesser extent, other blood lipids have also been linked to CHD. Triglycerides, mostly concentrated in VLDL, are elevated in some people with CHD, especially in those with diabetes and those who are overweight. Yet other people with elevated triglycerides and VLDL do not appear to have an increased risk of heart disease. Elevated triglycerides are associated with a high fasting blood glucose and a low HDL, but further studies are needed to determine whether they directly increase heart disease risk.[9]

• ***Male, 45 Years or Older*** • In general, men have higher blood cholesterol and a greater risk of CHD at an earlier age than women. Men's blood cholesterol early in adulthood strongly correlates with their risk of developing heart disease later

**Figure 18-3**
**HDL and LDL Compared**

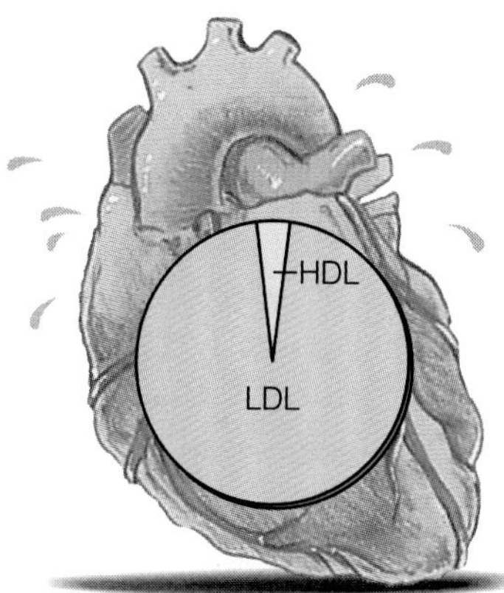

**Low HDL** relative to **LDL**
Increased risk of heart attack

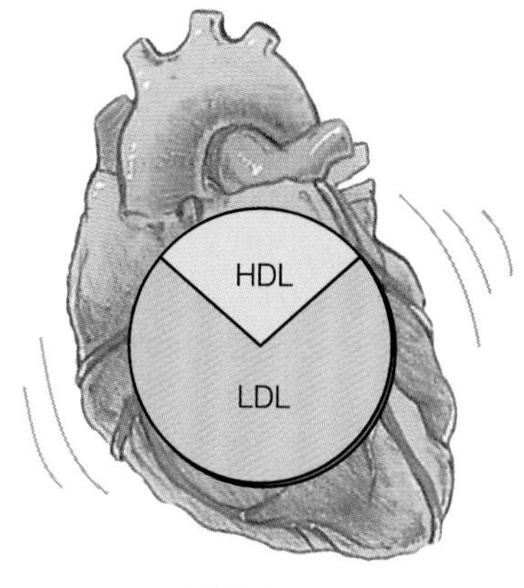

**High HDL** relative to **LDL**
Decreased risk of heart attack

U.S. Government
www.healthfinder.gov/searchoptions/topicsaz.htm
Search for Men's health
Search for Women's health

Estrogen replacement therapy:
- Alleviates menopausal symptoms.
- Reduces CHD risks.
- Reduces osteoporosis risks.
- Increases cancer risks.

in life.[10] Almost half of all deaths from CHD occur among men with blood cholesterol in the borderline-high range.

• ***Female, 55 Years or Older*** • Cardiovascular disease occurs about 10 to 12 years later in women than in men.[11] Women younger than 45 tend to have lower LDL cholesterol than men of the same age, but a woman's blood cholesterol typically begins to rise between ages 45 and 55. Independently of age, a lack of estrogen influences blood cholesterol: at menopause, LDL tend to rise and HDL to decline, so that women's LDL cholesterol becomes greater than men's for the first time. About one-third of the women in the United States have LDL cholesterol high enough to pose a serious risk of heart disease. Estrogen replacement therapy after menopause and regular physical activity reduce CHD risks in women by lowering LDL and raising HDL. Such therapy protects the LDL from oxidation, as does vitamin E.[12] Estrogen replacement therapy increases the risk of some cancers, however, so a woman must carefully consider her health history when balancing the apparent benefits against the potential risks of such treatment.[13]

• ***Hypertension*** • Chronic high blood pressure frequently accompanies atherosclerosis, diabetes, and obesity. The higher the blood pressure above normal, the greater the risk of heart disease. Hypertension injures the artery linings and accelerates plaque formation, thus setting the stage for atherosclerosis development or worsening existing atherosclerosis. Then the plaques and reduced blood flow induce a further rise in blood pressure, and hypertension and atherosclerosis become mutually aggravating conditions.

• ***Smoking*** • Low HDL cholesterol is among the many health problems associated with cigarette smoking. Smoking also damages the heart directly by increasing blood pressure and the heart's workload. It deprives the heart of oxygen and damages platelets, making blood clot formation likely. Toxins in cigarette smoke damage blood vessels, setting the stage for atherosclerosis.

• ***Diabetes*** • In diabetes, blood vessels often become blocked and circulation diminishes. Atherosclerosis progresses rapidly. People with diabetes have two to five times the incidence of coronary artery disease than people without diabetes.[14] Women with diabetes have the same rate of cardiovascular disease as men; thus diabetes increases the risk of cardiovascular disease more in women than in men.[15] Diabetes, like CHD, is associated with high LDL, high triglycerides, low HDL, hypertension, and obesity, particularly central obesity.

## *Recommendations for Reducing Cardiovascular Disease Risk*

Recommendations to reduce cardiovascular disease risk include both screening and intervention. Once a person's risks have been identified, treatment plans may include major changes in lifestyle involving diet, physical activity, and smoking cessation.

• ***Cholesterol Screening*** • To determine an individual's risk of CHD, health care professionals review the person's health history and measure several blood lipids including total cholesterol, LDL cholesterol, HDL cholesterol, and triglycerides. Ideally, at least two measurements are taken at least one week apart and then compared to standards (shown in Table 18-1). Single measurements may fail to identify those at risk or may misclassify them because blood cholesterol and other lipid concentrations vary significantly from day to day. The accompanying "How to" presents a scorecard to assist you in assessing your heart disease risk.

• ***Two-Step Diet Plan*** • Dietary recommendations to reduce the risk of CHD focus on reducing LDL cholesterol. To that end, people are advised to control their body weights and their intakes of saturated fat, total fat, and dietary cholesterol.

## HOW TO Assess Your Heart Disease Risk

Do you know your heart disease risk score? Respond to the statements below, and score yourself as directed. Be aware that a high risk score does not mean you *will* develop heart disease, but it should warn you of the possibility. Consult your physician if you have questions about your score results.

In each category, circle the number next to the statement that's most true for you.

**Cigarette Smoking**

| | |
|---|---|
| I never smoked or stopped smoking three or more years ago. | 1 |
| I don't smoke but live and/or work with smokers. | 2 |
| I stopped smoking within the past three years. | 3 |
| I smoke regularly. | 4 |
| I smoke regularly and live and/or work with other smokers. | 5 |

**Total Blood Cholesterol**

Use the number from your most recent blood cholesterol measurement:

| | |
|---|---|
| Less than 160 | 1 |
| 160–199 | 2 |
| Don't know | 3 |
| 200–239 | 4 |
| 240 or higher | 5 |

**HDL Cholesterol**

Use the number from your most recent HDL cholesterol measurement:

| | |
|---|---|
| Over 60 | 1 |
| 56–60 | 2 |
| Don't know | 3 |
| 35–55 | 4 |
| Less than 35 | 5 |

**Systolic Blood Pressure**

Use the first (highest) number from your most recent blood pressure measurement:

| | |
|---|---|
| Less than 120 | 1 |
| 120–139 | 2 |
| Don't know | 3 |
| 140–159 | 4 |
| 160 or higher | 5 |

**Excess Body Weight**

| | |
|---|---|
| I am within 10 pounds of my desirable weight. | 1 |
| I am 10–20 pounds above my desirable weight. | 2 |
| I am 21–30 pounds above my desirable weight. | 3 |
| I am 31–50 pounds above my desirable weight. | 4 |
| I am more than 50 pounds above my desirable weight. | 5 |

**Physical Activity**

Determine which statements best describe your usual level of physical activity:

***A: Highly Active***

My job requires very hard physical labor (such as digging or loading heavy objects) at least four hours a day

*or*

I do vigorous activities (jogging, cycling, swimming, etc.) at least three times a week for 30 minutes or more

*or*

I do at least one hour of moderate activity such as brisk walking at least four days a week.

***B: Moderately Active***

My job requires that I walk, lift, carry, or do other moderately hard work for several hours a day (day-care worker, stock clerk, or busboy/waitress)

*or*

I spend much of my leisure time doing moderate activities (dancing, gardening, walking, or housework).

***C: Inactive***

My job requires that I sit at a desk most of the day

*and*

Much of my leisure time is spent in sedentary activities (watching TV, reading, etc.)

*and*

I seldom work up a sweat, and I cannot walk fast without having to stop to catch my breath.

Now circle the number that best describes your level of physical activity:

| | |
|---|---|
| A: Highly Active | 1 |
| Between A and B | 2 |
| B: Moderately Active | 3 |
| Between B and C | 4 |
| C: Inactive | 5 |

**Scoring Your Heart Attack Risk**

To learn your estimated risk, add the six numbers you've circled.

| *If Your Total Score Is:* | *Your Heart Attack Risk Is:* |
|---|---|
| *6–13* | *Low* |
| *14–22* | *Moderate* |
| *23–30* | *High* |

Source: American Heart Association.

Table 18-2 presents the recommended two-step plan.[16] Diets following these recommendations significantly improve many of the risk factors for CHD, especially when combined with physical activity.[17]

The Step 1 diet was originally designed for people with borderline-high or high LDL cholesterol, but because atherosclerotic disease is so common, many experts advocate its use for everyone. If blood lipids do not improve, the person adopts the Step 2 plan. People with existing CHD are immediately placed on the Step 2 diet.

• ***Control Weight*** • Both steps recommend energy intakes to achieve and maintain desirable weight. With weight loss, heart disease risk factors improve: blood pressure, blood cholesterol, and blood triglycerides decline.

• ***Reduce Fat, Especially Saturated Fat*** • In addition to limiting total energy intake, the Step 1 diet recommends a total fat intake of less than 30 percent of daily kcalories, with saturated fat no more than one-third of that and dietary cholesterol less than 300 milligrams a day. Such a diet eliminates obvious sources of fat, saturated fat, and cholesterol and can be accomplished by following the practical suggestions for reducing fat outlined in Chapter 5 (p. 148). The Step 2 diet maintains a fat intake of less than 30 percent of daily kcalories but reduces saturated fat to 7 percent and cholesterol to less than 200 milligrams a day. For persons requiring the Step 2 diet, the help of a registered dietitian can ensure that saturated fat and cholesterol are reduced as needed without sacrificing nutritional quality.

• ***Other Dietary Interventions*** • In general, a nutritionally balanced diet improves many measures of CHD risk.[18] In addition, specific dietary strategies also help to protect against CHD, as Table 18-3 summarizes (earlier chapters and highlights provided more details).

Perhaps most controversial are findings that *moderate* alcohol consumption may reduce overall mortality in general and the risk of heart disease in particular by raising HDL cholesterol, preventing blood clot formation, and interfering with cell growth in the blood vessels.[19] These benefits are most apparent in people over age 50, those with one or more risk factors, and those with high LDL.[20]

As Highlight 7 described, alcohol also has many negative effects on body systems, and a later section in this chapter describes its link with cancer. Any benefits that alcohol may confer on cardiovascular health must clearly be weighed against the risks of incurring these negative health effects, as well as the possibility of alcohol abuse.[21] The question to answer is how much alcohol is protective, and how much is harmful? Early studies suggested that total mortality was reduced in people who drank one to two drinks per *day* as compared to abstainers, but in larger amounts (more than three drinks per *day*) alcohol was associated with increased mortality.[22] More recent studies suggest that the beneficial effects of alcohol occur at lower intakes, ranging from one to nine drinks per *week*.[23] The

**Table 18-2**
**Characteristics of Diets to Reduce High LDL Cholesterol**

| | Step 1 | Step 2 |
|---|---|---|
| Energy | Adequate to achieve or maintain desirable weight | Adequate to achieve or maintain desirable weight |
| Total fat[a] | <30% | <30% |
| Saturated fat[a] | 8–10% | <7% |
| Polyunsaturated fat[a] | Up to 10% | Up to 10% |
| Monounsaturated fat[a] | 10–15% | 10–15% |
| Cholesterol | <300 mg/day | <200 mg/day |

[a]Expressed as percentages of total food energy when energy intake is adequate to achieve and maintain desirable weight.
Source: Adapted from The Expert Panel, Summary of the second report of the National Cholesterol Education Program (NCEP) Expert Panel on Detection, Evaluation, and Treatment of High Blood Cholesterol in Adults (Adult Treatment Panel II), *Journal of the American Medical Association* 269 (1993): 3015–3023.

**Table 18-3**

**Dietary Factors Protecting against CHD**

In addition to the dietary interventions mentioned in Table 18-2, these dietary factors may also protect against CHD.

| Dietary Factor | Protection against CHD |
|---|---|
| Soluble fiber (apples and other fruits, oats, barley, legumes) | • Lowers blood cholesterol, especially in those with high cholesterol<br>• Lowers risk of heart attack<br>• Improves LDL-to-HDL ratio |
| Omega-3 fatty acids (fish oils) | • Limit clot formation<br>• Prevent irregular heartbeats<br>• Lower risk of heart attack |
| Alcohol (in moderation) | • Raises HDL<br>• Prevents clot formation |
| Folate, vitamin $B_6$ | • Reduce homocysteine |
| Vitamin E (vegetable oils and margarines, some nuts, wheat germ) | • Slows progression of plaque formation<br>• Lowers risk of heart attack in people with CHD<br>• Limits LDL oxidation |

amount of alcohol that has a beneficial effect on women may be lower still, perhaps one to three drinks per *week*.[24]

An individual's age and health history are critical to the decision of whether to drink alcohol. The number of deaths attributed to alcohol is greatest for people between the ages of 15 and 44—and their risk of heart disease is relatively minor. Clearly, for these people, the benefits do not outweigh the risks.

A review of studies to date found that alcohol from any source—red or white wine, beer, or distilled liquor—appears to be equally effective and that alcohol itself may be the protective factor.[25] In addition to alcohol, wine contains phenols and other phytochemicals that may also protect against cardiovascular disease.[26] These substances may act as antioxidants, reducing LDL oxidation, and may alter prostaglandin metabolism, reducing blood clot formation. These protective effects may explain the so-called French paradox: the wine-drinking people of France enjoy a lower incidence of CHD even though they have many of the same risk factors as people in the United States. In general, though, most health professionals hesitate to recommend the taking of alcohol to benefit health.

Chapter 14 describes exercise for cardiovascular endurance.

• ***Physical Activity*** • Physical activity deserves attention in any program to reduce CHD risk. Some evidence suggests that weight training can raise HDL if undertaken regularly, but frequent and sustained *aerobic* activity may be more effective in lowering LDL and raising HDL. Furthermore, aerobic, endurance-type activities, such as brisk walking, undertaken faithfully for 30 minutes or more as a daily or every-other-day routine can strengthen the heart and blood vessels; alter body composition in favor of lean over fat tissue; expand the volume of oxygen the heart can deliver to the tissues at each beat and so reduce the heart's workload; change the hormonal climate in which the body does its work in such a way as to lower blood pressure; and bring about a redistribution of body water that eases the transit of blood through the peripheral arteries. These changes are so beneficial that some experts believe that physical activity should be *the* primary focus of cardiovascular disease prevention efforts.[27]

If heart and artery disease has already set in, a monitored program of physical activity may actually help to reverse it. Activity may stimulate development of new arteries to feed the heart muscle, which may account for the excellent recovery seen in some heart attack victims who exercise regularly.

*Regular aerobic exercise can help to defend against heart disease by strengthening the heart muscle, promoting weight loss, and improving blood lipid and blood glucose levels. (Courtesy of CNN)*

Some researchers wonder if physical activity itself raises blood HDL or if the weight loss that often accompanies exercise is the real protective factor. For women, weight loss through diet alone appears to *lower* HDL, but when diet is combined with moderate aerobic activity, HDL do not decline. In fact, HDL

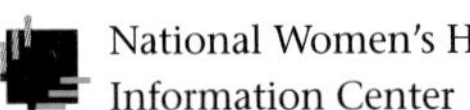
National Women's Health Information Center

www.4women.org

As Chapter 10 explained, large doses of niacin can effectively lower blood cholesterol, but they have adverse side effects as well. Self-medication is never advisable.

U.S. Government

www.healthfinder.gov/searchoptions/topicsaz.htm
Search for Heart Disease
Search for Stroke

increase substantially in women who exercise regularly.[28] In men, diet raises HDL, and the combination of activity and diet results in a significantly greater rise in HDL than diet alone.

Diet helps a little, physical activity helps a little, and the combination is better still. If this approach fails to normalize blood lipids, drugs may be prescribed.

• ***Drug Therapy*** • Because lipid-lowering drugs carry potential risks and are costly, physicians generally do not prescribe drugs until after a six-month trial of intensive diet therapy and physical activity has proved unsuccessful in lowering blood lipids. For people with very high LDL cholesterol (greater than 220 milligrams per deciliter), a shorter diet trial may be considered. Besides lipid-lowering drugs, the treatment of atherosclerosis may include aspirin and anticoagulants to prevent clot formation and antihypertensives to reduce blood pressure. All of these drugs incur potential risks, including nutrition-related side effects, a problem compounded because treatment often entails multiple drugs and continues for many years or even for life. (See Highlight 17 for more on nutrient-drug interactions.)

IN SUMMARY

Plaques in atherosclerosis raise blood pressure and trigger abnormal blood clotting, which can cause heart attacks and strokes. Dietary recommendations to lower the risks of these cardiovascular diseases focus on controlling weight and reducing fat, saturated fat, and cholesterol intake. Quitting smoking and engaging in regular physical activity are also important.

# Hypertension

**hypertension:** higher-than-normal blood pressure. Hypertension that develops without an identifiable cause is known as **essential** or **primary hypertension;** hypertension that is caused by a specific disorder such as kidney disease is known as **secondary hypertension.**

Anyone concerned with atherosclerosis and the risk it presents must also be concerned about hypertension. The two together are a life-threatening combination. The higher the blood pressure is above normal, the greater the risk. (Low blood pressure, on the other hand, is generally a sign of long life expectancy and low heart disease risk.) Hypertension is believed to affect some 50 million people in the United States, more than a third of the entire adult population.[29] It contributes to over a million heart attacks and half a million strokes each year. People cannot feel the physical effects of high blood pressure, but it can impair life's quality and end life prematurely.

## *Blood Pressure Regulation*

Blood pressure is vital to life. The heart's pumping action must create enough force to push blood through the major arteries into the smaller arteries and finally into the tiny capillaries, whose thin, porous walls permit fluid exchange between the blood and the tissues (see Figure 18-4). When the pressure is right, the cells receive a constant supply of nutrients and oxygen and give up their wastes.

**peripheral resistance:** resistance to the flow of blood caused by the reduced diameter of the vessels at the periphery of the body—the smallest arteries and capillaries.

• ***The Arteries*** • The pressure the blood exerts on the inner arterial walls results from two actions occurring together: at the body's center, the heart is pushing the blood into the arteries, and at the periphery, the smallest arteries and capillaries are resisting its flow. The heart's push ensures that the blood circulates through the whole system; the peripheral resistance and resulting pressure force oxygen and nutrients across the capillary walls to feed the tissues.

• ***The Blood Volume*** • The volume of fluid in the vascular system also contributes to blood pressure. That volume, in turn, depends on the number of dissolved particles the fluid contains. By the rule of osmosis, the more dissolved particles in the blood, the more water there will be.

**Figure 18-4**

**How Normal Blood Pressure Supports Fluid Exchange**

At the same time the heart pushes blood into an artery, the small-diameter arteries and capillaries at its other end resist the blood's flow (peripheral resistance). Both actions contribute to the pressure inside the artery. Another determining factor is the volume of fluid in the circulatory system, which depends in turn on the number of dissolved particles in that fluid.

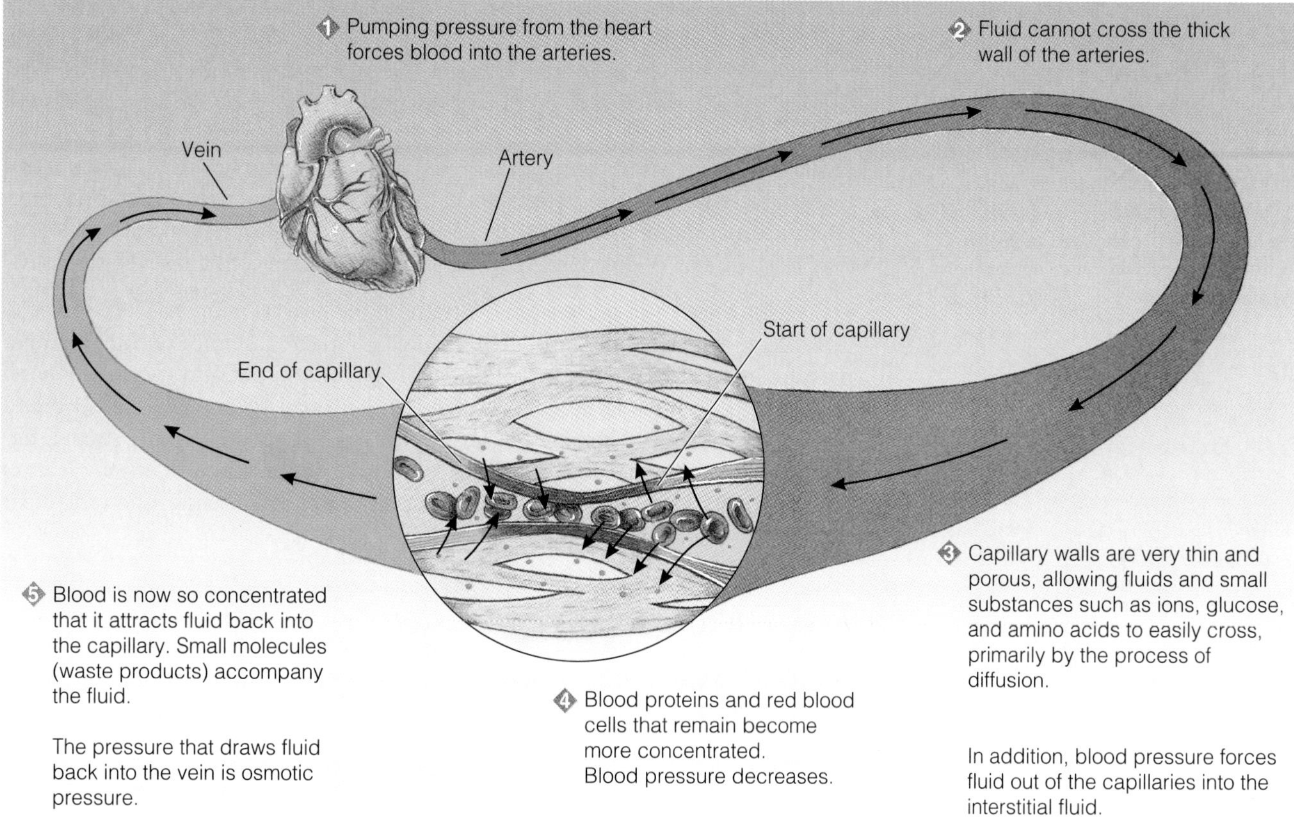

1 Pumping pressure from the heart forces blood into the arteries.
2 Fluid cannot cross the thick wall of the arteries.
Vein
Artery
Start of capillary
End of capillary
3 Capillary walls are very thin and porous, allowing fluids and small substances such as ions, glucose, and amino acids to easily cross, primarily by the process of diffusion.
In addition, blood pressure forces fluid out of the capillaries into the interstitial fluid.
4 Blood proteins and red blood cells that remain become more concentrated.
Blood pressure decreases.
5 Blood is now so concentrated that it attracts fluid back into the capillary. Small molecules (waste products) accompany the fluid.
The pressure that draws fluid back into the vein is osmotic pressure.


• ***The Kidneys*** • The kidneys depend on the blood pressure to help them filter waste out of the blood into the urine (you may want to review Figure 3-8 on p. 77). The pressure has to be high enough to force the blood's fluid out of the capillaries into the kidneys' filtering networks. If the blood pressure is too low, the kidneys initiate a series of actions to raise it (as shown in Figure 12-1 on p. 370). One action is to retain sodium and water, expanding blood volume. Another is to constrict the peripheral blood vessels. Both raise blood pressure.

Normally, dehydration sets these actions in motion, which is beneficial. In dehydration, the blood volume falls, and higher blood pressure is needed to deliver substances to the tissues. The kidneys maintain a normal blood pressure by constricting the blood vessels and conserving water and sodium until the dehydrated person can drink enough water to replenish the blood volume.

## How Hypertension Develops

What triggers chronic hypertension remains for the most part unknown, although one of the mechanisms involving the kidneys has been defined. When blood flow to the kidneys is reduced (as occurs in atherosclerosis), the kidneys respond by setting in motion actions that raise blood pressure by expanding blood volume and constricting peripheral blood vessels. Unfortunately, the pressure increases not only in the kidneys, but all over the body. High blood pressure stresses the heart, which has to pump extra hard to push the blood against resistant arteries. Hypertension worsens atherosclerosis, as described earlier, by mechanically injuring the artery linings and accelerating plaque formation. Then the plaques induce a further rise in blood pressure, intensifying the problem.

• ***Obesity*** • Obesity can contribute to the development of hypertension or make it worse. The added adipose tissue that obesity incurs means miles of extra capillaries through which the blood must be pumped. The combination of hypertension, atherosclerosis, and obesity puts a severe strain on the heart and arteries, which intensifies cardiovascular complications.

**insulin resistance:** the condition in which a normal amount of insulin produces a subnormal effect; a metabolic consequence of obesity.

The combination of insulin resistance, hypertension, elevated blood lipids, and obesity frequently observed in people with cardiovascular disease is sometimes called **Syndrome X.**

• ***Insulin Resistance*** • Insulin resistance, most commonly associated with obesity, triggers the pancreas to produce more insulin. High blood insulin signals the kidneys to retain sodium and may precipitate the development of hypertension. Hypertension is two to three times more common in people with diabetes (a disease characterized by insulin resistance) than in the general population. The interrelationships among insulin resistance, obesity, hypertension, and atherosclerosis may explain why people with diabetes usually die from cardiovascular diseases.

• ***Consequences of Hypertension*** • Strain on the heart's pump, the left ventricle, can enlarge and weaken it, until it gradually fails (heart failure). Constant elevated pressure in an artery may cause it to gradually balloon out and eventually burst (aneurysm). Aneurysms that go undetected can lead to massive bleeding and death, particularly if a large vessel such as the aorta is affected. In the small arteries of the brain, an aneurysm may lead to stroke, and in the eye, it may lead to blindness. Likewise, the kidneys can be damaged (kidney disease) when the heart is unable to adequately pump blood through them.

## *Risk Factors for Hypertension*

Several major risk factors predicting the development of hypertension have been identified, including:

- *Smoking.* Smoking increases the heart's workload, raising blood pressure.
- *High blood lipids.* As mentioned earlier, high blood lipids contribute to both atherosclerosis and hypertension. Furthermore, a high-fat diet raises blood pressure.[30]
- *Diabetes.* People with diabetes often have high blood pressure.
- *Sex.* In general, blood pressure is higher in men than in women and in women past menopause than in other women.
- *Age.* Arteries lose their elasticity and blood pressure increases with age; most people who develop hypertension do so after age 60.
- *Heredity.* A family history of hypertension and heart disease in women under 65 and in men under 55 significantly raises the risk of developing hypertension.
- *Obesity.* Excess body fat, especially abdominal fat, is closely associated with hypertension.
- *Race.* The prevalence of hypertension differs among racial and ethnic groups; for African Americans, it is among the highest in the world.

The normal resting blood pressure for adults averages about 120 over 80. At readings of 140 over 90 or higher, the risks of heart attacks and strokes increase in direct proportion to increasing blood pressure, especially diastolic pressure (see Table 18-1).

Again, diet and physical inactivity interact with many of these risk factors. Notice that salt intake is not a risk factor for the development of hypertension, although it may aggravate existing hypertension in some people, as a later section explains.

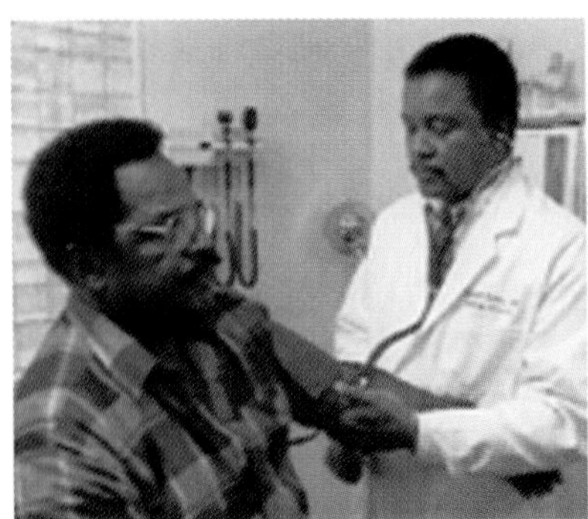

***Screening people for high blood pressure is a first step in reducing CHD risk from hypertension. (Courtesy of CNN)***

## *Recommendations for Reducing Hypertension Risk*

The single most effective step people can take against hypertension is to find out whether they have it. At checkup time, a health care professional can provide an accurate resting blood pressure reading. Under normal conditions, blood pressure fluctuates continuously in response to a variety of factors including such things as talking or shifting position. Some people react emotionally to the procedure, which raises the blood pressure reading. For these reasons, if the resting blood pressure is above normal, the reading should be repeated before confirming the diagnosis of hypertension. Thereafter, the blood pressure should be checked regularly.

• ***Weight Control*** • Efforts to reduce high blood pressure focus on weight control because excess body fat, especially abdominal fat, can precipitate hypertension. Weight loss alone is one of the most effective nondrug treatments for hypertension. Those who are using drugs to control their blood pressure can often reduce or discontinue the drugs if they lose weight. Even a modest weight loss of 10 pounds may significantly lower blood pressure. Many professionals recommend a fat-controlled diet, such as the Step 1 diet shown in Table 18-2 on p. 568, both for weight loss and to control blood lipids and blood pressure.

• ***Physical Activity*** • The higher the blood pressure and the less active a person is to begin with, the greater the likelihood that physical activity will be effective in reducing blood pressure. Physical activity helps with weight control, of course, but moderate aerobic activity, such as 30 to 45 minutes of brisk walking most days, also helps to lower blood pressure directly. Those who engage in regular aerobic activity may not need medication for mild hypertension.

• ***Alcohol*** • Alcohol, especially if consumed in large amounts (an average of two or more drinks per day), may contribute to hypertension. Furthermore, alcohol may interfere with drug therapy, and it is associated with strokes independently of hypertension. Those who drink alcohol should do so in moderation—no more than one to two drinks a day. Such amounts seem safe from a blood pressure point of view.[31]

Count as one drink:
- 12 oz of beer.
- 4 to 5 oz of wine.
- 1½ oz of hard liquor.

For more on alcohol, see Highlight 7.

• ***Sodium/Salt Intake*** • Salt may aggravate hypertension in some people who are genetically sensitive. Groups most sensitive to changes in salt intake include African Americans, older people, and people with hypertension or diabetes. These people may be able to lower their blood pressure with a no-added-salt diet that limits salt to 5 to 6 grams per day. They can adjust their diets by following the suggestions in the "How to" on p. 379.

Reminder:
- 1 tsp salt = 5 g salt = 2 g sodium.

For most people, however, blood pressure responses to sodium or salt are modest at best. Studies clearly show that weight loss lowers blood pressure more effectively than sodium or salt restriction.[32] Still, many health care professionals advocate moderate salt restriction for everyone, reasoning that at best it may help and at worst it will do no harm. Such wisdom has been called into question by recent studies finding that people with mild-to-moderate hypertension whose sodium consumption was low had a greater likelihood of suffering a heart attack than those whose sodium intakes were high.[33] Evidence that low-sodium diets may be harmful coupled with a lack of evidence that lowering salt intakes prevents hypertension may prompt many to rethink their recommendations regarding salt restriction.

• ***Other Dietary Interventions*** • Other dietary factors may play a role in the prevention of hypertension, even though evidence is insufficient to warrant specific recommendations. Potassium, calcium, and magnesium, for example, appear to help prevent and treat hypertension in certain populations.[34] Also, because hypertension may be an insulin-resistant state, measures to control diabetes may also protect against hypertension.

Chapter 12 reviewed the relationships between hypertension and the minerals potassium, calcium, and magnesium.

The same diet recommended to improve atherosclerosis—one that features plenty of fruits and vegetables and little fat and saturated fat—lowers blood pressure as well, especially in people with hypertension.[35] Researchers have found that the greatest success in lowering blood pressure is associated with eating nearly 10 servings of fruits and vegetables daily and keeping fat intake below 30 percent. These findings are from a major multicenter research study called Dietary Approaches to Stop Hypertension (DASH). The DASH eating plan, which focuses on whole foods rather than individual nutrients, emphasizes fruits and vegetables—even more so than the Food Guide Pyramid (as the accompanying margin table shows).

| Food Group | DASH | Pyramid |
|---|---|---|
| Grains | 7–8 | 6–11 |
| Vegetables | 4–5 | 3–5 |
| Fruits | 4–5 | 2–4 |
| Milk (nonfat/low-fat) | 2–3 | 2–3 |
| Meat (lean) | 2 or less | 2–3 |

• ***Drug Therapy*** • When diet and physical activity fail to reduce blood pressure, diuretics and antihypertensive agents may be prescribed. Diuretics lower blood pressure by increasing fluid loss. Some diuretics can lead to a potassium deficiency. People taking these diuretics need to include rich sources of potassium or supplements daily. Blood potassium should be monitored regularly, and clients should watch for signs of potassium imbalances such as weakness (particularly of the legs), unexplained numbness or tingling sensation, cramps, irregular heartbeats, and excessive thirst and urination.

The richest sources of potassium are *fresh* foods of all kinds (see pp. 380–381).

Although some diuretics can lead to a potassium deficiency, others spare potassium. A combination of these two types of diuretics may be prescribed to prevent potassium deficiency. In such cases, excessive potassium intakes and potassium supplements should be avoided; too much potassium is also life-threatening.

IN SUMMARY

The most effective dietary strategy for preventing hypertension is weight control. When atherosclerosis and hypertension run their courses without treatment, they often lead to heart attacks and strokes. The other leading cause of death, cancer, develops in significantly different ways, yet many of the preventive measures are the same.

 U.S. Government
www.healthfinder.gov/searchoptions/topicsaz.htm
Search for High Blood Pressure

# Cancer

The thought of cancer often strikes fear in people. Indeed, cancer ranks just below cardiovascular disease as a cause of death for the entire population, but it is the number one cause of death for those between the ages of 45 and 64. Consequently, many people have personal experiences with cancer. As with cardiovascular disease, however, the prognosis for cancer today is far brighter than in the past. Identification of risk factors, new detection techniques, and innovative therapies offer hope and encouragement.

**cancers:** diseases that result from the unchecked growth of malignant tumors.

Cancer is not a single disorder. There are many *cancers,* that is, many different kinds of malignancies. They have different characteristics, occur in different locations in the body, take different courses, and require different treatments.

Cancers are classified by the tissues or cells from which they develop:

- **Adenomas** (ADD-eh-NO-mahz) arise from glandular tissues.
- **Carcinomas** (KAR-see-NO-mahz) arise from epithelial tissues.
- **Gliomas** (gly-OH-mahz) arise from glial cells of the central nervous system.
- **Leukemias** (loo-KEY-me-ahz) arise from the white blood cells.
- **Lymphomas** (lim-FOE-mahz) arise from lymph tissue.
- **Melanomas** (MEL-ah-NO-mahz) arise from pigmented skin cells.
- **Sarcomas** (sar-KO-mahz) arise from muscle, bone, or connective tissues.

## *How Cancer Develops*

The genes in a healthy body work together regulating cell division to ensure that each new cell is a replica of the parent cell. In this way, the healthy body grows, replacing dead cells and repairing damaged ones. Cancers develop from mutations in the genes that regulate cell division. The mutations silence the genes that ordinarily monitor replicating DNA for chemical errors. The affected cells seemingly have no built-in brakes to halt cell division. As the abnormal mass of cells, called a malignant *tumor* or *neoplasm,* grows, blood vessels form to supply the tumor with the nutrients it needs to support its growth. Eventually, the tumor invades healthy tissue and may spread. Clinicians describe cancers by their size and extent, specifically noting if the tumor has spread to surrounding lymph nodes or to distant sites in the body. Figure 18-5 illustrates tumor formation.

Radiation and **carcinogens** (car-SIN-oh-jenz) can cause mutations that give rise to cancer.

- **carcin** = cancer
- **gen** = gives rise to

**tumor:** a new growth of tissue forming an abnormal mass with no function; also called a **neoplasm** (NEE-oh-plazm). Tumors that multiply out of control, threaten health, and require treatment are **malignant** (ma-LIG-nant). Tumors that stop growing without intervention or can be removed surgically and pose no threat to health are **benign** (bee-NINE).

- **benign** = mild
- **malignus** = of bad kind

• ***Genetic Factors*** • Some cancers appear to have a genetic component. A person with a family history of breast cancer, for example, has a greater risk of developing breast cancer than a person without such a genetic predisposition. This does not mean, however, that the person *will* develop cancer, only that the risk is greater.

A cancer that spreads from one part of the body to another is said to **metastasize** (me-TAS-tah-size).

• ***Immune Factors*** • A healthy immune system recognizes foreign cells and destroys them. Researchers theorize that an ineffective immune system may not

**Figure 18-5**
**Tumor Formation**

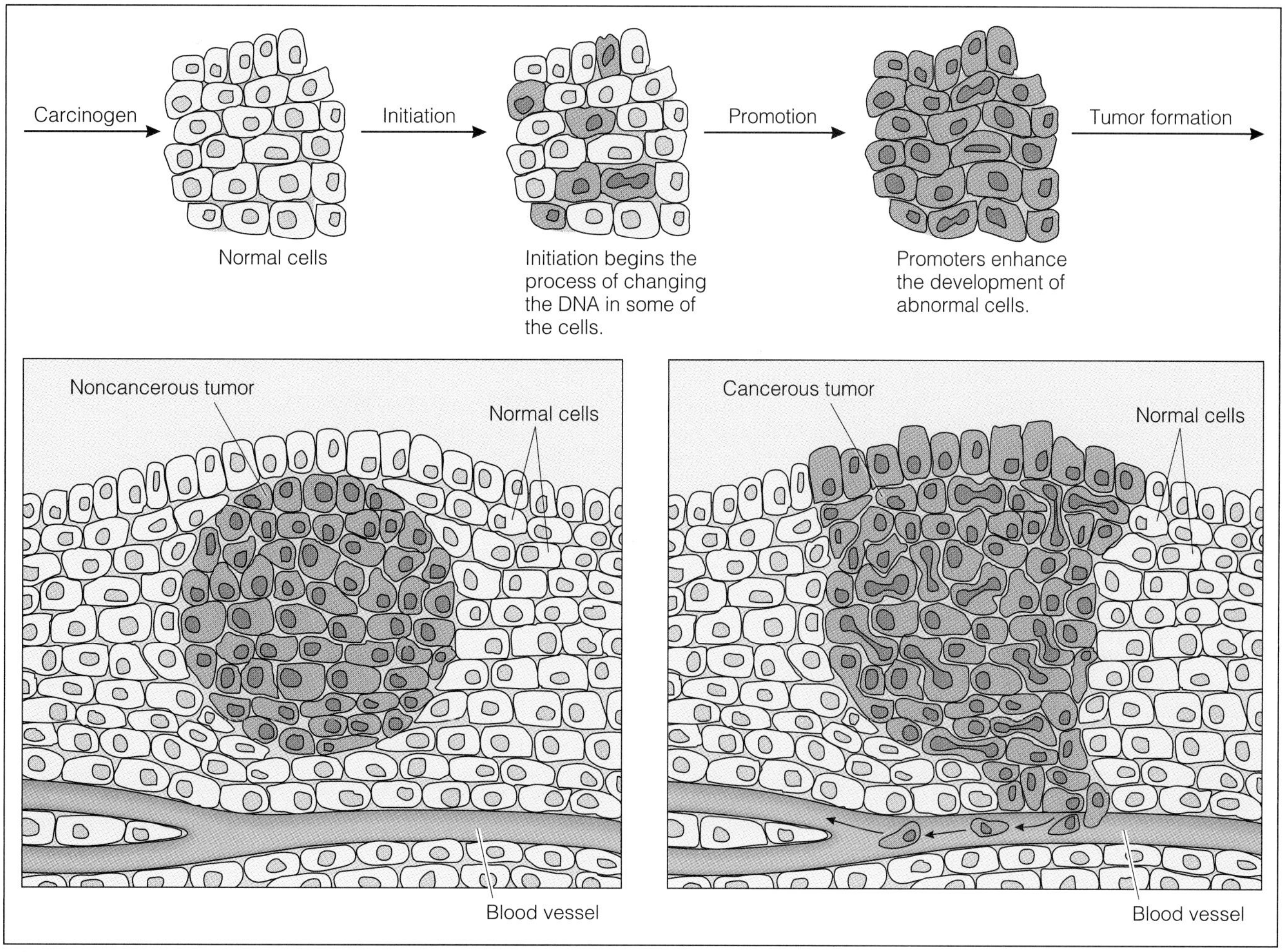

Carcinogen → Initiation → Promotion → Tumor formation

Normal cells

Initiation begins the process of changing the DNA in some of the cells.

Promoters enhance the development of abnormal cells.

Noncancerous tumor
Normal cells
Blood vessel

Cancerous tumor
Normal cells
Blood vessel

recognize tumor cells as foreign, thus allowing tumor growth. Aging affects immune function, and the incidence of cancer increases with age. Drugs that suppress the immune system and viral infections (including HIV infection) and other disorders that severely tax the immune system may increase the risk of cancer.

• ***Environmental Factors*** • Among environmental factors, exposure to radiation and sun, water and air pollution, and smoking are known to cause cancer. As Table 18-4 (on p. 576) shows, dietary constituents are also associated with an increased risk of certain cancers. Some dietary factors may initiate cancer development, others may promote cancer development once it has started, and still others may protect against the development of cancer.

• ***Dietary Factors—Cancer Initiators*** • We do not know to what extent diet contributes to cancer development, although some experts estimate that diet may be linked to a third or more of all cases. Consequently, many people think that certain foods are carcinogenic, especially those that contain additives or pesticides. As Chapter 19 will explain, our food supply is one of the safest in the world. Additives that have been approved for use in foods are not carcinogenic. Some pesticides are carcinogenic at high doses, but not at the concentrations allowed on fruits and vegetables.[36]

The incidence of cancers, especially stomach cancers, is high in parts of the world where people eat a lot of heavily smoked, pickled, or salt-cured foods that

***People with cancer take comfort from the support of others and from the knowledge that medical science is waging an unrelenting battle in their defense.***

Factors that cause mutations that give rise to cancer are called **initiators.**

Additives and pesticides receive full attention in Chapter 19.

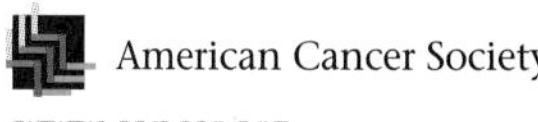
American Cancer Society
www.cancer.org

**Table 18-4**
**Factors Associated with Cancer at Specific Sites**

| Cancer Sites | High Incidence Associated with: | Protective Effect Associated with: |
|---|---|---|
| Pancreatic cancer | *Possibly* meat, cholesterol | Fruits and vegetables, *possibly* vitamin C and fiber |
| Esophageal cancer | Alcohol, tobacco, and especially combined use; preserved foods (such as pickles); high intakes of vitamin A supplements | Fruits and vegetables |
| Stomach cancer | Salt-preserved foods (such as dried, salted fish); *possibly* grilling and barbecuing | Fresh fruits and vegetables; safe food handling methods; vitamin C |
| Colorectal cancer | Fat (particularly saturated fat), meat, and alcohol (especially beer) | Vegetables; physical activity; *possibly* calcium; omega-3 fatty acids |
| Liver cancer | Infection with hepatitis B or aflatoxins; alcohol; iron overload | |
| Lung cancer | Cigarette smoking primarily; *possibly* alcohol, fat (notably saturated fat), and cholesterol | Fruits and vegetables, especially green and yellow ones; *possibly* carotenes, vitamin C, and selenium; physical activity |
| Breast cancer | Obesity; alcohol; *possibly* meat and dietary fat | Fruits and vegetables, especially green and yellow ones; *possibly* carotenes, fiber, and physical activity |
| Ovarian cancer | No dietary risk factors have been established; inversely correlated with oral contraceptive use | Fruits and vegetables, especially green and yellow ones |
| Cervical cancer | Folate deficiency; cigarette smoking | *Possibly* fruits, vegetables, carotenes, and vitamin C |
| Endometrial cancer | Obesity, estrogen therapy, hypertension, and diabetes (type 2) | *Possibly* fruits and vegetables |
| Bladder cancer | Cigarette smoking; *possibly* artificial sweeteners, and alcohol | Fruits and vegetables, especially green and yellow ones |
| Prostate cancer | High fat intake, especially saturated fats from meats | Fruits and vegetables, especially green and yellow ones |

NOTE: Findings based on epidemiological studies.
Sources: C. Marwick, Global review of diet and cancer links available, *Journal of the American Medical Association* 278 (1997): 1650–1651; *Diet and Health: Implications for Reducing Chronic Disease Risk* (Washington, D.C.: National Academy Press, 1989), pp. 594–600; J. H. Weisburger, Nutritional approach to cancer prevention with emphasis on vitamins, antioxidants, and carotenoids, *American Journal of Clinical Nutrition* (supplement) 53 (1991): 226–237; R. G. Ziegler, Vegetables, fruits, and carotenoids and the risk of cancer, *American Journal of Clinical Nutrition* (supplement) 53 (1991): 251–259; Potential mechanisms for food-related carcinogens and anticarcinogens: A scientific status summary by the Institute of Food Technologists' Expert Panel on Food Safety and Nutrition, *Food Technology* 47 (1993): 105–118.

produce carcinogenic nitrosamines. Most commercial manufacturers in the United States use other preservation methods, and all are carefully controlled to minimize carcinogen contamination.

Chapter 19 describes nitrosamines and their formation.

Alcohol has also been associated with a high incidence of some cancers, especially cancers of the mouth, esophagus, and liver. Beverages such as beer and scotch may contain damaging nitrosamines as well as alcohol.[37] Other beverages, such as wine and brandy, may contain the carcinogen urethane, which is produced during fermentation.[38] The amounts of these compounds in alcoholic beverages currently on the market are not considered harmful—assuming consumption in moderate amounts. These findings illustrate clearly why any potential benefit of moderate alcohol consumption on cardiovascular disease must be weighed against the potential dangers.

Factors that favor the development of cancer once it has begun are called **promoters**.

• ***Dietary Factors—Cancer Promoters*** • Unlike carcinogens, which initiate cancers, some dietary components promote cancers; that is, once the initiating step has taken place, these components may accelerate tumor development. Studies suggest that certain dietary fats eaten in excess may promote cancer, in part by contributing to obesity. More specifically, linoleic acid, the omega-6 fatty acid of vegetable oils, has been implicated in enhancing cancer development in some animals; in contrast, omega-3 fatty acids from fish oils appear to delay cancer development.[39]

American Institute for Cancer Research

www.aicr.org

Factors that oppose the development of cancer are called **antipromoters**.

• ***Dietary Factors—Antipromoters*** • It seems apparent that foods may also contain antipromoters. Almost without exception, epidemiological studies find a link between eating plenty of fruits and vegetables and a low incidence of cancers. The fiber in fruits and vegetables helps to protect against some cancers by speeding up the transit time of all materials through the colon so that the colon walls are not exposed to cancer-causing substances for long. In addition to fiber, fruits and vegetables contain both nutrients and nonnutrients (phytochemicals) that protect against cancer. By acting as scavengers of oxygen-derived free radicals, the antioxidant nutrients beta-carotene, vitamin C, and vitamin E may help to prevent cell and tissue damage that can give rise to cancer. Phytochemicals common to many vegetables, especially those of the cabbage family, can activate enzymes that are capable of destroying carcinogens (see Highlight 11 for more details).[40]

***Vegetables rich in fiber, phytochemicals, and the antioxidant nutrients (beta-carotene, vitamin C, and vitamin E) help to protect against cancer.***

## *Recommendations for Reducing Cancer Risk*

On the basis of current knowledge and available evidence, the following guidelines are recommended for cancer prevention:

- Control weight and prevent obesity.
- Reduce consumption of total fat to 30 percent or less and saturated fat to 10 percent or less of total energy intake.
- Increase fiber intake to 20 to 30 grams per day.
- Include a variety of fresh vegetables and fruits (including deep yellow and dark green cruciferous vegetables) in the daily diet.
- Consume salt-cured, salt-pickled, nitrite-cured, and smoked foods in moderation.
- Consume alcoholic beverages in moderation, if at all.

One additional recommendation is in order: vary food choices. This last suggestion is based on an important concept that applies specifically to the prevention of cancer initiation—dilution. Switching from food to food dilutes the negative qualities of a food. For example, it is safe to eat *some* salt-cured or smoked meats, but not all the time. Combine such foods with a variety of others so that any carcinogens that may be present will be diluted in the total diet.

National Cancer Institute
www.nci.nih.gov

IN SUMMARY

Some dietary factors, such as alcohol and heavily smoked or salted foods, may initiate cancer development; others, such as dietary fat, may promote cancer once it has gotten started; and still others, such as fiber, antioxidant nutrients, and phytochemicals, may serve as antipromoters that protect against the development of cancer. Eating many green, yellow, and orange vegetables, including high-fiber foods, and reducing fat intake offer the best possible nutrition at the lowest possible risk.

U.S. Government
www.healthfinder.gov/searchoptions/topicsaz.htm
Search for Cancer

# Diabetes Mellitus

Diabetes mellitus ranks seventh among the leading causes of death. In addition, diabetes underlies, or contributes to, a variety of other major diseases, including heart disease and stroke. In fact, people with diabetes are twice as likely to develop these cardiovascular problems as those without diabetes.

Diabetes mellitus is a chronic disorder characterized by high blood glucose and either insufficient or ineffective insulin. To appreciate the problems presented by an absolute or relative lack of insulin, consider insulin's normal action. After a meal, insulin signals the body's cells to receive the energy nutrients from the blood—amino acids, glucose, and fatty acids. Insulin helps to maintain blood glucose within normal limits and stimulates protein synthesis, glycogen synthesis in liver and muscle, and fat synthesis. Without insulin, glucose regulation falters, and metabolism of the energy-yielding nutrients changes. Table 18-5 shows the distinguishing features of the two main forms of diabetes, which are described next.

- ***Type 1 Diabetes*** • In type 1 diabetes, the less common type of diabetes (about 5 to 10 percent of all diagnosed cases), the pancreas cannot synthesize the hormone

**diabetes** (DYE-uh-BEET-eez) **mellitus** (MELL-ih-tus or mell-EYE-tus): a metabolic disorder of carbohydrate metabolism characterized by altered glucose regulation and utilization, usually resulting from insufficient or ineffective insulin.
- **mellitus** = honey-sweet (sugar in urine)

American Diabetes Association
www.diabetes.org

Canadian Diabetes Association
www.diabetes.ca

**type 1 diabetes:** the less common type of diabetes in which the person produces no insulin at all; also known as **insulin-dependent diabetes mellitus (IDDM)** or **juvenile-onset diabetes,** although it can arise at any age.

**Table 18-5**
**Features of Type 1 and Type 2 Diabetes**

| | Type 1 Diabetes | Type 2 Diabetes |
|---|---|---|
| Other names | Insulin-dependent diabetes mellitus (IDDM) | Noninsulin-dependent diabetes mellitus (NIDDM) |
| | Juvenile-onset diabetes | Adult-onset diabetes |
| | Ketosis-prone diabetes | Ketosis-resistant diabetes |
| | Brittle diabetes | Lipoplethoric diabetes |
| | | Stable diabetes |
| Average age of onset | <20 (mean age, 12) | >40 |
| Associated conditions | Viral infection | Obesity |
| Insulin required? | Yes | Sometimes |
| Cell response to insulin | Normal | Resistant |
| Symptoms | Relatively severe | Relatively less severe |
| Prevalence in diabetic population | 5 to 10% | 90 to 95% |

insulin. Without insulin, the body's energy metabolism becomes altered, with such severe consequences as to threaten survival. The person must obtain insulin to assist the cells in taking up the needed fuels from the blood; for this reason, type 1 is sometimes called insulin-dependent diabetes mellitus (IDDM). The insulin must be injected; it cannot be taken orally because insulin is a protein and the GI enzymes would digest it.

**type 2 diabetes:** the more common type of diabetes in which the fat cells resist insulin; also called **noninsulin-dependent diabetes mellitus (NIDDM)** or **adult-onset diabetes.**

• ***Type 2 Diabetes*** • Type 2 is the predominant form of diabetes (90 to 95 percent of all cases); it develops most often in people over 40 years old, but is also seen in obese children.[41] Although the exact cause of type 2 diabetes remains unknown, high blood glucose and insulin resistance are the hallmarks of the disorder. In the initial stages, the pancreas produces insulin. In fact, the person may actually have higher-than-average insulin levels, but the cells respond less sensitively to it; that is, they become insulin resistant. The cell receptors for insulin, the sites at which insulin signals the cell, have diminished in number or in function. As in type 1, blood glucose rises too high. The high blood glucose stimulates the pancreatic cells to make insulin, exhausting these cells and reducing their ability to make insulin. Thus type 2 appears to be a self-aggravating condition.

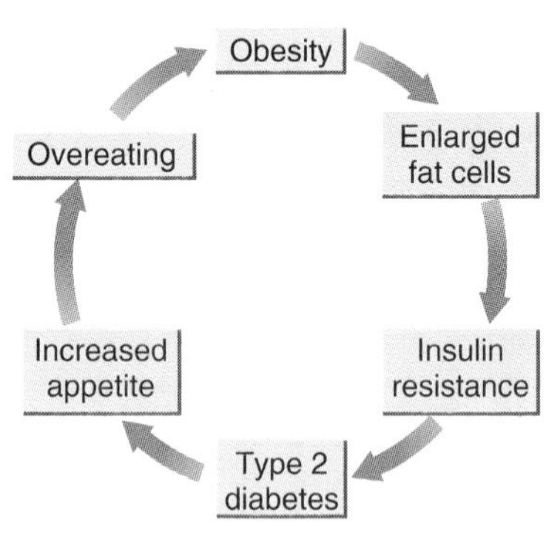


Type 2 diabetes usually develops later in life, for in all people, pancreatic insulin-producing cells progressively lose their function with age. Type 2 diabetes is also associated with obesity; about 90 percent of U.S. adults with type 2 diabetes are obese. One of the many metabolic consequences of obesity is insulin resistance. Compared with normal-weight people, obese people require much more insulin to maintain normal blood glucose. More insulin is produced, but as body fat increases, insulin receptors are reduced in number or function; consequently, insulin resistance increases, and adipose and muscle tissues become less and less able to take up glucose. At some point, the body cannot produce enough insulin to keep up, and type 2 diabetes develops. Age and obesity alone do not predict the onset of type 2 diabetes; genetics also plays a role.[42]

## *Complications of Diabetes*

In both types of diabetes, glucose fails to gain entry into the cells and accumulates in the blood. These two problems lead to both acute and chronic complications. Figure 18-6 summarizes the metabolic changes and acute complications that can arise in uncontrolled diabetes. Notice that when some glucose enters the cells, as in type 2 diabetes, many of the symptoms of type 1 do not occur.

Over the long term, the person with diabetes suffers not only from the acute complications shown in Figure 18-6, but also from its chronic effects.[43] Chronically elevated blood glucose damages the structures of the blood vessels and

**Figure 18-6**
**Metabolic Consequences of Untreated Diabetes**

The metabolic consequences of type 1 diabetes are more immediate and severe than those of type 2. In type 1 no insulin is available to allow any glucose to enter the cells. When glucose cannot enter the cells, a cascade of metabolic changes follows. In type 2 diabetes, some glucose enters the cells. Because the cells are not "starved" for glucose, the body does not shift into the metabolism of fasting (losing weight and producing ketones).

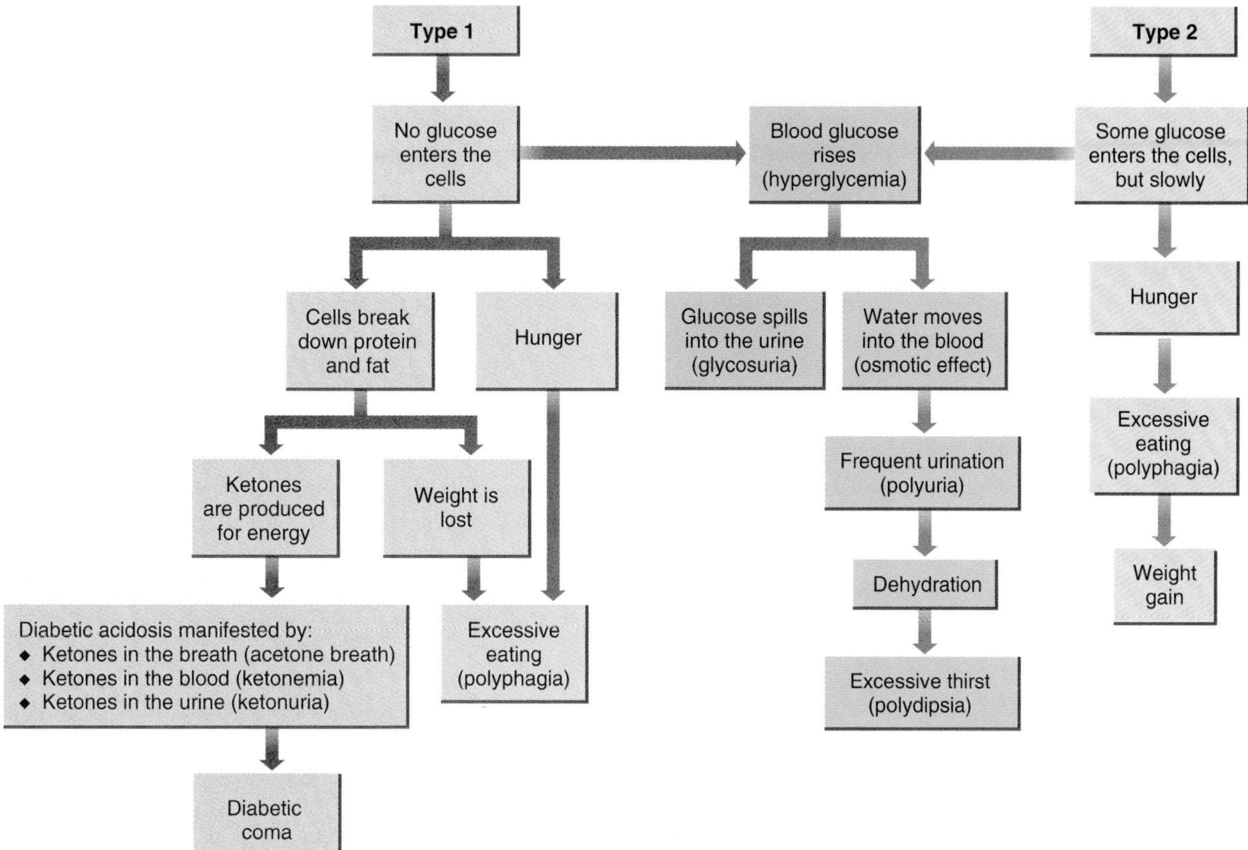


nerves, leading to loss of circulation and nerve function.* Infections are likely to occur due to poor circulation coupled with glucose-rich blood and urine. People with diabetes must pay special attention to hygiene and keep alert for early signs of infection.

• ***Diseases of the Large Blood Vessels*** • As mentioned, atherosclerosis tends to develop early, progress rapidly, and be more severe in people with diabetes. More than 80 percent of people with diabetes die as a consequence of cardiovascular diseases, especially heart attacks. If nerve function is also impaired, the person may have a heart attack and not even realize it.

• ***Diseases of the Small Blood Vessels*** • Disorders of the small blood vessels (capillaries) may also develop and lead to loss of kidney function and retinal degeneration with accompanying loss of vision. About 85 percent of people with diabetes have impaired kidney function, loss of vision, or both. Consequently, diabetes is a leading cause of both kidney failure and blindness.

Disorders of the small blood vessels are called **microangiopathies.**
- **micro** = small
- **angeion** = vessel
- **pathos** = disease

*A few examples might illustrate how chronic hyperglycemia damages cells and interferes with normal body processes. Sustained hyperglycemia alters glucose metabolism in virtually every cell of the body. Some cells begin to convert excess glucose to sugar alcohols, for example, which causes toxicity and cell distension; distended cells in the lens of the eye cause blurry vision. Some cells produce glycoproteins from excess glucose and an amino acid from a protein; these proteins cannot function normally, which leads to a host of other problems.

The death of tissue, usually due to deficient blood supply, is **gangrene** (GANG-green).

• ***Diseases of the Nerves*** • Nerve tissues may also deteriorate, expressed at first as a painful prickling sensation, often in the arms and legs. Later, the person loses sensation in the hands and feet. Injuries to these areas may go unnoticed, and infections can progress rapidly. With loss of both circulation and nerve function, undetected injury and infection may lead to death of tissue (gangrene), necessitating amputation of the limbs (most often the legs or feet). People with diabetes are advised to take conscientious care of their feet and visit a podiatrist regularly.

Nerve damage can also delay gastric emptying. When the stomach empties slowly after a meal, the person may experience a premature feeling of fullness, bloating, nausea, vomiting, weight loss, and poor blood glucose control due to irregular nutrient absorption.

## *Dietary Recommendations for Type 1 Diabetes*

Diet is an important component of type 1 diabetes treatment. To maintain blood glucose within a fairly normal range requires a lifelong commitment to a carefully coordinated diet, physical activity, and insulin program.

Nutrition therapy focuses on maintaining optimal nutrition status, controlling blood glucose, achieving a desirable blood lipid profile, controlling blood pressure, and preventing and treating the complications of diabetes.[44] In addition to meeting basic nutrient requirements, the diet must provide a fairly consistent carbohydrate intake from day to day and at each meal and snack to help minimize fluctuations in blood glucose. Further alterations in diet may be necessary for the person with chronic complications such as cardiovascular or kidney disease. Diet planners often use the exchange system described in Chapter 2 and Appendixes G and I for planning diabetic diets.

The person with type 1 diabetes may need to eat before, during, and after vigorous physical activity. Especially important is carbohydrate, which is readily available from fruits, fruit juices, yogurt, crackers, and other starches. As a general guideline, the exerciser should have about 10 to 15 grams of additional carbohydrate before moderate activity or about 20 to 30 grams of carbohydrate before vigorous activity. In addition, blood glucose should be checked 30 minutes before and 1 hour after physical activity, and carbohydrate adjusted accordingly.

Normally, the body secretes a constant, baseline amount of insulin at all times and secretes more as blood glucose rises following meals. People with type 1 diabetes must learn to adjust their insulin doses and schedule of administration to accommodate meals, physical activity, and health status.

## *Dietary Recommendations for Type 2 Diabetes*

The benefits of nutrition intervention in type 2 diabetes are receiving increasing attention.[45] To maintain near-normal blood glucose, the diet is designed to deliver the same amount of carbohydrate each day, spaced evenly throughout the day. Several approaches can be used to plan such diets, but many people with diabetes learn to count carbohydrates using the exchange system that was introduced in Chapter 2.

Providing a consistent carbohydrate intake spaced throughout the day helps people with type 2 diabetes maintain appropriate blood glucose levels and maximizes the effectiveness of drug therapy. Eating too much carbohydrate at one time can raise blood glucose too high, stressing the already-compromised insulin-producing cells. Eating too little carbohydrate can lead to hypoglycemia, especially for people taking oral drugs or insulin. The amount of carbohydrate, not its source, affects glucose levels the most.[46] Put another way, people with type 2 diabetes must pay attention to their carbohydrate intake, regardless of whether they eat bread or cake. They can eat sweets and sugar as part of a healthy diet if they can afford the kcalories; they just need to remember to count the carbohydrates.

People with diabetes who have elevated blood lipids may need to watch not only their carbohydrate intake, but their fat intake as well. When they lower their fat intake, the percentage of kcalories from carbohydrates increases. They may have difficulty controlling both their blood glucose and their blood lipid levels when they increase their carbohydrate intake because high-carbohydrate diets raise triglycerides and lower HDL. In addition, people accustomed to a high-fat diet sometimes have difficulty complying with a low-fat diet. Mounting evidence suggests that people can control their blood glucose and improve their blood lipids on a high-fat diet provided that monounsaturated fats replace saturated fats.[47] Diets providing 40 percent of kcalories from carbohydrate and 45 percent of kcalories from fat (25 percent monounsaturated, 10 percent polyunsaturated, and 10 percent saturated) can lower blood glucose and insulin levels following meals and lower day-long levels of blood triglycerides.[48] Results such as these remind practitioners to individualize dietary plans.

Even moderate weight loss (10 to 20 pounds) can help reverse insulin resistance, improve the blood lipid profile, and reduce blood pressure. A regular program of moderate physical activity not only contributes to weight loss, but also improves blood glucose control, blood lipid profiles, and blood pressure in people with type 2 diabetes.

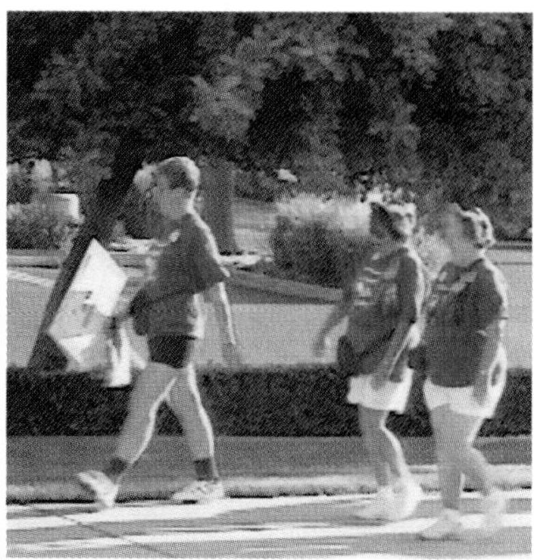

***Physical activity and a moderate weight loss of even 10 to 20 pounds can help improve blood glucose, blood lipids, and blood pressure. (Courtesy of CNN)***

IN SUMMARY

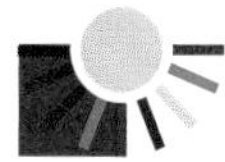

Diabetes is characterized by high blood glucose and either insufficient insulin (type 1) or ineffective insulin (type 2). People with type 1 diabetes coordinate diet, insulin injections, and physical activity to help control their blood glucose. Those with type 2 benefit most from a diet that controls glucose fluctuations and promotes weight loss.

 U.S. Government
www.healthfinder.gov/searchoptions/topicsaz.htm
Search for Diabetes

# Recommendations for Chronic Diseases

This chapter began with the major cardiovascular diseases, went on to cancer, and then described diabetes, a metabolic disorder—three different conditions with distinct sets of causes. Yet all are responsive to diet, and in some ways, the responses are similar. Dietary excesses, particularly excess food energy and fat intakes, increase the likelihood of all three diseases.

Not all diet recommendations apply equally to all of the diseases or to all people with a disease, but fortunately for the consumer, dietary recommendations do not contradict one another. In fact, they support each other. Most people can gain some disease-prevention benefits by dietary changes. To that end, *Diet and Health* recommendations describe the kinds of foods people should include, limit, or avoid to reduce chronic disease risks (see Table 18-6). Like the *Diet and Health* report, the *Nutrition Recommendations for Canadians* report makes recommendations that will supply enough nutrients, while reducing the risk of chronic disease (see Table 18-7).

A summary of the *Diet, Nutrition, and Prevention of Chronic Diseases* report from WHO is presented in Appendix G.

Several of the *Diet and Health* recommendations are aimed at weight control: cut fat, add complex carbohydrates, and balance food intake with activity. Obesity is common in the United States, and it is linked with most of the chronic diseases that threaten life. The problems of overweight people multiply when medical problems develop. For example, overweight people readily develop diabetes, which is often accompanied by high blood pressure and high blood cholesterol. Such a combination of problems may require only one treatment: lose the excess weight by adopting a healthful diet combined with regular exercise.

• ***Recommendations for the Population*** • The *Diet and Health* recommendations address the general population in the hope that all people at all levels of risk may

**Table 18-6**
***Diet and Health* Recommendations**

- Reduce total *fat* intake to 30 percent or less of kcalories. Reduce saturated fatty acid intake to less than 10 percent of kcalories and the intake of cholesterol to less than 300 milligrams daily.
- Increase intake of starches and other *complex carbohydrates.*
- Maintain *protein* intake at moderate levels.
- Balance food intake and physical activity to maintain appropriate *body weight.*
- For those who drink *alcoholic beverages,* limit consumption to the equivalent of less than 1 ounce of pure alcohol in a single day. Pregnant women should avoid alcoholic beverages.
- Limit total daily intake of *salt* (sodium chloride) to 6 grams or less.
- Maintain adequate *calcium* intake.
- Avoid taking dietary *supplements* in excess of the RDA in any one day.
- Maintain an optimal intake of *fluoride,* particularly during the years of primary and secondary tooth formation and growth.

NOTE: Italics added to highlight the areas of concern.
Source: Adapted from the Committee on Diet and Health, *Diet and Health: Implications for Reducing Chronic Disease Risk* (Washington, D.C.: National Academy Press, 1989).

**Table 18-7**
***Nutrition Recommendations for Canadians***

- The Canadian diet should provide energy consistent with the maintenance of *body weight* within the recommended range.
- The Canadian diet should include *essential nutrients* in amounts recommended.
- The Canadian diet should include no more than 30 percent of energy as *fat* (33 grams/1000 kcalories or 39 grams/5000 kilojoules) and no more than 10 percent as saturated fat (11 grams/1000 kcalories or 13 grams/5000 kilojoules).
- The Canadian diet should provide 55 percent of energy as *carbohydrate* (138 grams/1000 kcalories or 165 grams/5000 kilojoules) from a variety of sources.
- The *sodium* content of the Canadian diet should be reduced.
- The Canadian diet should include no more than 5 percent of total energy as *alcohol,* or two drinks daily, whichever is less.
- The Canadian diet should contain no more *caffeine* than the equivalent of four regular cups of coffee per day.
- Community water supplies containing less than 1 milligram per liter should be *fluoridated* to that level.

NOTE: Italics added to highlight areas of concern.
Source: Health and Welfare Canada, *Nutrition Recommendations: The Report of the Scientific Review Committee* (Ottawa: Canadian Government Publishing Centre, 1990).

Recommendations that urge all people to make dietary changes believed to forestall or prevent diseases are taking a *preventive* or *population approach.* Alternatively, recommendations that urge dietary changes only for people who are known to need them are taking a *medical* or *individual approach.*

benefit. Such a strategy is similar to national efforts to vaccinate to prevent polio, fluoridate water to prevent dental caries, and fortify grains to prevent iron deficiency.

• ***Recommendations for Individuals*** • People's hereditary susceptibility to diseases and their responsiveness to dietary measures vary. Unlike nutrient-deficiency diseases, which develop when nutrients are lacking and disappear when the nutrients are provided, chronic diseases are neither caused nor prevented by diet alone. Many people have followed dietary advice and developed heart disease or cancer anyway; others have ignored all advice and lived long and healthy lives. For many people, though, diet does influence the time of onset and course of some chronic diseases, and many health care professionals urge dietary measures as part of a disease-prevention strategy.

• ***Assessing Risks*** • To determine whether dietary recommendations are important to you personally, look at your family history to see which diseases are common to your relatives. In addition, examine your personal history, taking note of your blood pressure, blood test results, and lifestyle habits such as smoking.

IN SUMMARY

Clearly, optimal nutrition plays a key role in keeping people healthy and reducing the risk of chronic disease.[49] To have the greatest impact possible, dietary recommendations are aimed at the entire population, and not just at the individuals who might benefit most. Recommendations focus on weight control and urge people to limit fat, increase complex carbohydrates, and balance food intake with activity. A person can do no better than to incorporate those suggestions into his or her daily life.

# Nutrition and Infectious Diseases

It is difficult to know exactly where infectious diseases fall among the leading causes of death. Compared with chronic diseases, infectious diseases pose a much greater challenge for public health officials tracking prevalence.[50] One physician might classify an ear infection as an infectious disease, while another calls it a disease of the ear; similarly, loss of hearing due to an infection, though not an infectious condition itself, may be assigned to the infectious disease category. Trends change quickly as well. A disease, such as AIDS, that did not even exist two decades ago may suddenly appear and become one of the leading causes of death. A preventive strategy, such as food irradiation, may just as quickly eliminate hundreds of thousands of cases of food-borne infections each year. Public health strategies help the entire country defend against the spread of infection; each individual's immune system provides a personal line of defense.

## The Immune System

The immune system defends the body so alertly and silently that people do not even notice the thousands of enemy attacks mounted against them every day.[51] If the immune system fails, though, the body suddenly becomes vulnerable to every wayward disease-causing agent that comes its way; infectious disease invariably follows.

**immune system:** the body's natural defense system against foreign materials that have penetrated the skin or mucous membranes.

The body's first lines of defense against foreign substances—the skin, mucous membranes, and GI tract—normally deter invaders. If an invader penetrates these barriers and gains entry into the body, then the organs and cells of the immune system race into action.

Of the 100 trillion cells that make up the human body, one in every hundred is a white blood cell. Two types of white blood cells, the phagocytes and lymphocytes, defend the body against infectious diseases.

Cells of the immune system:
- Phagocytes.
- Lymphocytes:
  - B-cells.
  - T-cells.

• ***Phagocytes*** • Phagocytes, the scavengers of the immune system, are the first to arrive at the scene if an invader gains entry. When a phagocyte spots a substance it recognizes as foreign, it engulfs and digests that substance, if it can, in a process called phagocytosis. Phagocytes also secrete proteins that activate the metabolic and immune responses to infection.

**phagocytes:** white blood cells that have the ability to ingest and destroy foreign substances. The process by which phagocytes engulf and destroy foreign materials is called **phagocytosis** (FAG-oh-sigh-TOE-sis).
- **phagein** = to eat

• ***Lymphocytes: B-cells*** • The lymphocytes are of two distinct types: B-cells and T-cells. B-cells respond to infection by rapidly dividing and then producing the large proteins known as antibodies. Antibodies then travel in the bloodstream to the site of the infection. There they stick to the surfaces of the foreign particles and kill or otherwise inactivate them, making the foreign particles easy for the phagocytes to ingest.

**lymphocytes** (LIM-foe-sites): white blood cells that participate in acquired immunity; B-cells and T-cells.

**B-cells:** lymphocytes that produce antibodies. *B* stands for bursa, an organ in the chicken associated with the first identification of the B-cells.

The antibodies are members of a class of proteins known as immunoglobulins—literally, large globular proteins that produce immunity. Antibodies react selectively to a specific foreign organism, and the B-cells retain a memory of how to make them. The next time the same foreign organism is encountered, the immune system can respond with greater speed than it did the first time. B-cells play a bigger role in resistance to infection than do T-cells.

**immunoglobulin** (IM-you-no-GLOB-you-lin): a protein capable of acting as an antibody.

• ***Lymphocytes: T-cells*** • The T-cells travel directly to the invasion site to battle the foreign organisms. T-cells recognize the antigens displayed on the surfaces of their partner phagocyte cells and multiply in response. Then they release powerful chemicals to destroy all the foreign particles that have this antigen on their sur-

**T-cells:** lymphocytes that attack antigens. *T* stands for the thymus gland, where the T-cells are stored for a while.

**Table 18-8**
**Effects of Protein-Energy Malnutrition (PEM) on the Body's Defense Systems**

| System Component | Effects of PEM |
|---|---|
| Skin | Thinned, with less connective tissue to serve as a barrier to protect underlying tissues; delayed skin sensitivity reaction to antigens |
| Digestive tract and other body linings | Antibody secretions and immune cell number reduced |
| Lymph tissues | Immune system organs[a] reduced in size; cells of immune defense depleted |
| General response | Invader kill time prolonged; circulating immune cells reduced; antibody response impaired |

[a]Thymus gland, lymph nodes, and spleen.

Nutrients known to affect immunity:
- Protein.
- Vitamin A.
- Vitamin E.
- Vitamin $B_6$.
- Folate.
- Vitamin C.
- Iron.
- Zinc.
- Selenium.

faces. As the T-cells begin to win the battle against infection, they release signals to slow down the immune response.

Unlike the phagocytes, which are capable of inactivating many different types of invaders, T-cells are highly specific. Each T-cell can attack only one type of antigen. This specificity is remarkable, for nature creates millions of antigens. After making enough T-cells to destroy a particular antigen, some lymphocytes retain the necessary information to serve as memory cells so that the immune system can rapidly produce the same type of T-cells again should the identical infection recur.

T-cells actively defend the body against fungi, viruses, parasites, and a few types of bacteria; they can also destroy cancer cells. T-cells participate in the rejection of newly transplanted tissues, which is why physicians prescribe immunosuppressive drugs following such surgery. The T-cells are also inactivated by the human immunodeficiency virus that causes AIDS, as explained later.

**Figure 18-7**
**Nutrition and Immunity**

Malnutrition and infections worsen each other.

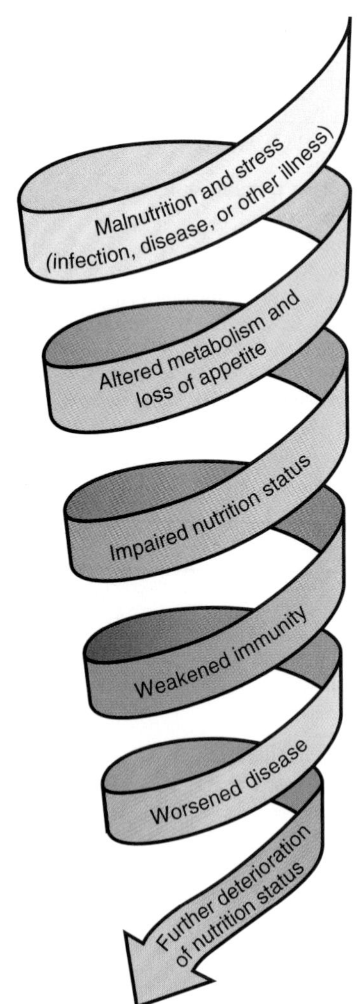


## *Nutrition and Immunity*

Of all the body's systems, the immune system responds most sensitively to subtle changes in nutrition status. Malnutrition compromises immunity.[52] Impaired immunity then opens the way for infectious diseases, which raise nutrient needs and reduce food intake, and nutrition status suffers further.[53] Thus disease and malnutrition create a synergistic downward spiral that must be broken for recovery to occur (see Figure 18-7).

Impaired immunity is a hallmark feature of protein-energy malnutrition (PEM). Table 18-8 presents the effects of PEM on the body's defenses. Deficiencies of vitamins and minerals also diminish the immune response, as may excesses. Likewise, interactions between nutrients may enhance or impair immunity.[54] Quite simply, optimal immunity depends on optimal nutrition—enough, but not too much, of each of the nutrients. People with depressed immune systems, such as the elderly, may benefit from supplements of selected nutrients.[55]

# Human Immunodeficiency Virus (HIV) Infection and Acquired Immune Deficiency Syndrome (AIDS)

For many years, the devastating effects of HIV infection seemed unstoppable, but in 1996, for the first time, the death rate from HIV infection declined sharply. And although the disease still has no cure, remarkable progress has been made in understanding and treating HIV infection, giving rise to a renewed hope that a cure may be possible. Without a cure, however, the best course is prevention. Transmission of the virus requires sexual activity, direct blood contact, or passage of the infection from a mother to her infant during pregnancy, birth, or breastfeeding. Table 18-9 presents strategies for preventing HIV transmission.

Once a person has been infected with HIV, it takes about 6 to 12 weeks before laboratory tests can confirm a diagnosis. Because people remain symptom-free in the early stages of infection, however, they may not even be tested for HIV for several years following infection. Thus early detection to prevent the spread of HIV infection and to ensure early treatment for the person infected are important health goals.

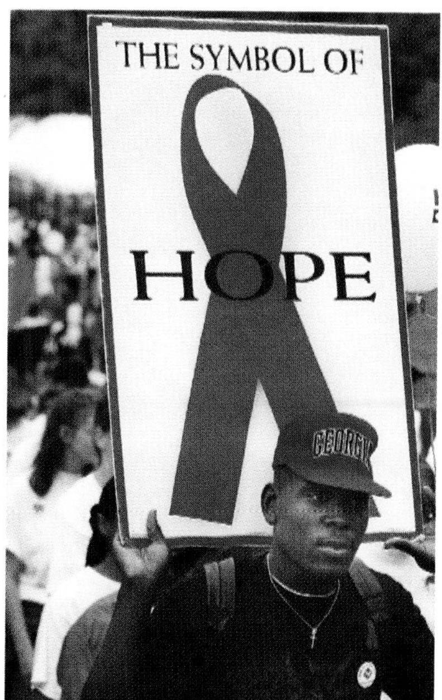


*The fight against AIDS, like the fight against cancer, is bolstered by both medical science and people who refuse to give up hope.*

## How AIDS Develops

HIV infection attacks the immune system and leaves its victims defenseless against opportunistic infections and disorders from which most people are protected. The disorder begins with infection by the virus and progresses in stages. The virus gradually destroys cells with a specific protein called CD4+ on their surfaces. Among the cells most affected are CD4+ T-lymphocytes, essential components of the immune system. At first, CD4+ T-lymphocytes decline gradually, and the HIV-infected individual remains symptom-free. As the infection progresses, though, depletion of CD4+ T-lymphocytes greatly impairs immune function. Symptoms may include fatigue, skin rashes, fevers, diarrhea, muscle pain, night sweats, weight loss, oral lesions and infections, and other opportunistic infections that are not life-threatening.

Later, frequent and often fatal complications arise, such as severe weight loss; tuberculosis; recurrent bacterial pneumonia; serious infections of the central nervous system, GI tract, and skin; cancers; and severe diarrhea. On average, an HIV infection takes about 10 years to progress to AIDS. Clinicians monitor the progression of the disease by measuring the concentrations of CD4+ T-lymphocytes and the circulating virus (viral load).

**human immunodeficiency virus (HIV):** the virus that causes AIDS. The infection progresses to become an immune system disorder that leaves its victims defenseless against numerous infections.

**acquired immune deficiency syndrome (AIDS):** the end stage of HIV infection, in which severe complications are manifested.

**opportunistic infections:** infections from microorganisms that normally do not cause disease in the general population but can cause great harm in people once their immune systems are compromised (as in HIV infection).

**CD4+ T-lymphocyte:** a type of circulating white blood cell that contains the CD4+ protein on its surface and is a necessary component of the immune system.

The cluster of mild symptoms that sometimes occurs early in the course of AIDS is called **AIDS-related complex (ARC).**

## The HIV Wasting Syndrome

People with AIDS frequently experience malnutrition and wasting. The wasting often begins early in the disease and becomes progressively worse. The degree of wasting in people with AIDS, especially in the few months before death, is similar to that seen in people who die from starvation.

The causes of malnutrition and wasting in HIV infection are related to the disease itself, its complications, and its treatments, all of which can result in inadequate nutrient intakes, excessive nutrient losses, and an accelerated metabolism

**wasting syndrome:** an involuntary loss of more than 10% of body weight.

**Table 18-9**
**Strategies to Prevent HIV Transmission**

HIV is transmitted from one person to another by direct contact with contaminated body fluids, most often through sexual intercourse, through contaminated needles or blood products, or from mother to infant during pregnancy or lactation. To prevent the transmission of HIV infection:

- Avoid sexual contact with anyone with HIV infection.
- Use a latex condom and a spermicidal agent if you have sexual contact with anyone whose sexual history you do not know.
- Do not share toothbrushes, razors, or other implements that could be contaminated with blood.
- Exercise caution when undergoing procedures such as acupuncture, tattooing, or ear piercing, in which needles might be contaminated.
- If you are an IV drug user, seek help for your addiction. Meanwhile, use only sterile, unused needles and dispose of them so that others will not use them. Avoid unprotected sexual contact with others.

**Table 18-10**
**Causes of Malnutrition in HIV Infection**

| Anorexia due to: | Nutrient Losses due to: |
|---|---|
| • Depression. | • HIV infection. |
| • Fever. | • GI tract infections. |
| • Pain. | • Cancer. |
| • Altered taste perceptions. | • Cancer therapy. |
| • Dry mouth. | • Anti-infective drugs. |
| • Difficulty swallowing. | • Home remedies for AIDS. |
| • Mouth ulcers. | • Reduced gastric acid secretion. |
| • Esophageal lesions and obstructions. | • Bacterial overgrowth. |
| • Use of oxygen masks. | |
| • Drug therapy. | |
| • Lethargy. | |
| • Dementia. | |

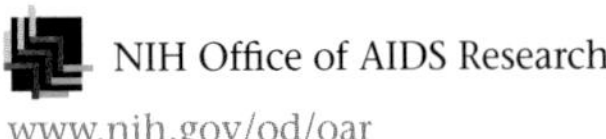
NIH Office of AIDS Research
www.nih.gov/od/oar

(people with cancer often experience wasting for similar reasons). Table 18-10 lists the many causes of anorexia and nutrient losses associated with HIV infection.

Even when other complications ultimately cause death, malnutrition appears to be an important contributing factor.[56] For people with AIDS, the severity of wasting may determine the duration of survival, suggesting that nutrition intervention can help extend lives by minimizing weight loss and strengthening the immune system.[57]

## Nutrition Support for People with HIV Infections

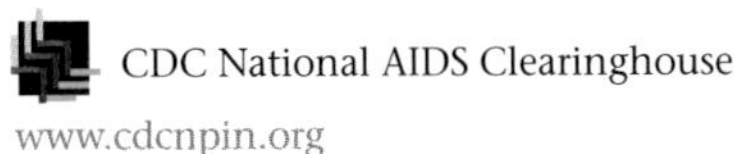
CDC National AIDS Clearinghouse
www.cdcnpin.org

In an era of improved treatments and prolonged survival for people with HIV infections, measures that improve the quality of life assume great importance. Attention to nutrition cannot change the ultimate outcome of an HIV infection, but it can prevent and reverse malnutrition, which may improve the quality of life and slow disease progression. Good nutrition status may improve a person's response to drug therapy, reduce duration of hospital stays, and promote physical independence.[58] At a minimum, meeting nutrient needs eliminates the additional stresses imposed by malnutrition.

A specific dietary strategy for the treatment of AIDS has not been devised. Instead, practitioners rely on clinical experience to make nutrient recommendations based on the complications that arise in each case. Efforts are aimed primarily at controlling weight loss.[59]

Although the exact vitamin and mineral needs of people with AIDS have not been determined, at least 100 percent of the RDA (or AI) for these nutrients should be provided daily. People who are unable to eat enough food to provide an adequate intake might benefit from a multivitamin-mineral supplement. Many physicians prescribe a daily prenatal vitamin supplement for people with HIV infections.[60] Individuals who are unable to eat enough food to prevent nutrition complications and unintentional weight loss may need more aggressive nutrition support.

*Nutrition provides an edge in maintaining quality of life and encouraging independence.*

People with weakened immune systems, such as the elderly, or those with any disease that compromises the immune system, such as those with HIV infection, are most susceptible to food-borne infections. These infections further weaken the immune system, making a simple case of food poisoning a life-threatening event. People with HIV infections must carefully follow guidelines for buying, preparing, and storing foods safely, as presented in Chapter 19. In addition, the Food and Drug Administration (FDA) advises people with HIV infections to avoid eating raw or undercooked seafood.[61]

With no present cure available, people with HIV infection and other life-threatening diseases often seek alternative therapies. The highlight that follows addresses the role of alternative therapies in medical care today.

IN SUMMARY

Everyone should adopt an effective personal strategy to prevent HIV infection. (Review Table 18-9 on p. 585 and call the AIDS hotline for the information you need.)* Should a person contract HIV, nutrition intervention can help prevent malnutrition and minimize the wasting that accompanies the progression of AIDS.

*AIDS hotline: (800) 342-AIDS.

*The countless lives touched by AIDS serve as a potent reminder of the need to continue the search for a cure.*

## Study Questions

These questions will help you review the chapter. You will find the answers in the discussions on the pages provided.

1. How do the major diseases of today as a group differ from those of several decades ago as a group? Why is nutrition considered so important in connection with today's major diseases? (pp. 561–562)
2. Identify the major diet-related risk factors for atherosclerosis, hypertension, cancer, and diabetes. (p. 563)
3. Describe some ways in which people can alter their diets to lower their blood cholesterol levels. (pp. 566, 568–570)
4. Describe some steps that people with hypertension can take to lower their blood pressure. (pp. 572–574)
5. Differentiate between cancer initiators, promoters, and antipromoters. Which nutrients or foods fit into each of these categories? (pp. 575–577)
6. Describe the characteristics of a diet that might offer the best protection against the onset of cancer. (p. 577)
7. Name the two major types of diabetes and describe some differences between them. How do dietary recommendations for each type of diabetes compare with the healthy diet recommended for all people? (pp. 577–581)
8. What is HIV infection? What are the consequences of HIV infection? What is the HIV wasting syndrome? (pp. 584–586)
9. In what ways might good nutrition status possibly alter the course of HIV infection? (p. 586)

These questions will help you prepare for an exam. Answers can be found in Appendix K.

1. The leading cause of death in the United States is:
   a. AIDS.
   b. cancer.
   c. diabetes.
   d. heart disease.
2. Plaques in the arteries contribute to the development of:
   a. cancer.
   b. diabetes.
   c. atherosclerosis.
   d. infectious diseases.
3. Which blood lipid correlates directly with heart disease?
   a. HDL
   b. LDL
   c. VLDL
   d. triglycerides
4. Moderate amounts of alcohol may protect against heart disease by:
   a. promoting LDL oxidation.
   b. preventing clot formation.
   c. raising LDL and lowering HDL.
   d. accelerating plaque formation.
5. What is the most effective strategy for most people to lower their blood pressure?
   a. lose weight
   b. restrict salt
   c. monitor glucose
   d. supplement protein
6. Which of the following help(s) to protect against cancer?
   a. alcohol
   b. pickled foods
   c. phytochemicals
   d. omega-6 fatty acids
7. In diabetes, when glucose fails to enter the cells and metabolism shifts to break down protein and fat for energy:
   a. weight gain occurs.
   b. hypertension worsens.
   c. atherosclerosis develops.
   d. ketone production increases.
8. The most important dietary strategy in diabetes is to:
   a. provide for a consistent carbohydrate intake.
   b. restrict fat to 30 percent of daily kcalories.
   c. limit carbohydrate intake to 300 milligrams a day.
   d. take multiple vitamin and mineral supplements daily.

9. The immune cells most seriously damaged by HIV are:
   a. B-cells.
   b. T-cells.
   c. antigens.
   d. immunoglobulins.

10. People with HIV infections are most susceptible to:
    a. diabetes.
    b. hypertension.
    c. heart attacks.
    d. food poisoning.

# Notes


1. National Center for Health Statistics, *Monthly Vital Statistics Report,* October 1996.
2. W. Mertz, A balanced approach to nutrition for health: The need for biologically essential minerals and vitamins, *Journal of the American Dietetic Association* 94 (1994): 1259–1262.
3. A. P. Simopoulos, Diet and gene interactions, *Food Technology* 51 (1997): 66–69.
4. J. Ma and coauthors, Relation of plasma phospholipid and cholesterol ester fatty acid composition to carotid artery intima-media thickness: The Atherosclerosis Risk in Communities (ARIC) Study, *American Journal of Clinical Nutrition* 65 (1997): 551–559; G. S. Tell and coauthors, Dietary fat intake and carotid artery wall thickness: The Atherosclerosis Risk in Communities (ARIC) Study, *American Journal of Epidemiology* 139 (1994): 979–989.
5. R. J. Garrison, The role of adiposity in the prevention of cardiovascular disease, in *Obesity: New Directions in Assessment and Management,* eds. T. B. VanItallie and A. P. Simopoulos (Philadelphia: Charles Press, 1995), pp. 22–33.
6. L. Kuller and coauthors, Prevalence of subclinical atherosclerosis and cardiovascular disease and association with risk factors in the Cardiovascular Health Study, *American Journal of Epidemiology* 139 (1994): 1164–1179.
7. C. L. Johnson and coauthors, Declining serum total cholesterol levels among US adults—The National Health and Nutrition Examination Surveys, *Journal of the American Medical Association* 269 (1993): 3002–3008.
8. R. Benfante and coauthors, To what extent do cardiovascular risk factor values measured in elderly men represent their midlife values measured 25 years earlier? A preliminary report and commentary from the Honolulu Heart Study, *American Journal of Epidemiology* 140 (1994): 206–216.
9. M. H. Criqui and coauthors, Plasma triglyceride level and mortality from coronary heart disease, *New England Journal of Medicine* 328 (1993): 1220–1225; NIH Consensus Development Panel on Triglyceride, High-Density Lipoprotein, and Coronary Heart Disease, Triglyceride, high-density lipoprotein, and coronary heart disease, *Journal of the American Medical Association* 269 (1993): 505–510.
10. M. J. Klag and coauthors, Serum cholesterol in young men and subsequent cardiovascular disease, *New England Journal of Medicine* 328 (1993): 313–318.
11. P. M. Kris-Etherton and D. Krummel, Role of nutrition in the prevention and treatment of coronary heart disease in women, *Journal of the American Dietetic Association* 93 (1993): 987–993.
12. R. C. Wander and coauthors, Effects of interaction of *RRR*-α-tocopheryl acetate and fish oil on low-density-lipoprotein oxidation in postmenopausal women with and without hormone-replacement therapy, *American Journal of Clinical Nutrition* 63 (1996): 184–193.
13. N. E. Davidson, Hormone-replacement therapy—Breast versus heart versus bone, *New England Journal of Medicine* 332 (1995): 1638–1639.
14. F. O. Karlsson and A. J. Garber, Lipoprotein disorders in diabetes mellitus, *Clinical Diabetes* 14 (1996): 124–127.
15. A. J. Garber, The complication most often overlooked, *Clinical Diabetes* 15 (1997): 46–48.
16. Expert Panel, Summary of the second report of the National Cholesterol Education Program (NCEP) Expert Panel on Detection, Evaluation, and Treatment of High Blood Cholesterol in Adults (Adult Treatment Panel II), *Journal of the American Medical Association* 269 (1993): 3015–3023.
17. M. L. Stefanick and coauthors, Effects of diet and exercise in men and post menopausal women with low levels of HDL cholesterol and high levels of LDL cholesterol, *New England Journal of Medicine* 339 (1998): 12–20; A dietary intervention trial for nutritional management of cardiovascular risk factors, *Nutrition Reviews* 55 (1997): 54–56.
18. D. C. Hatton and coauthors, Improved quality of life in patients with generalized cardiovascular metabolic disease on a prepared diet, *American Journal of Clinical Nutrition* 64 (1996): 935–943.
19. R. Locher, P. M. Suter, and W. Vetter, Ethanol suppresses smooth muscle cell proliferation in the postprandial state: A new antiatherosclerotic mechanism of ethanol? *American Journal of Clinical Nutrition* 67 (1998): 338–341; M. J. Thun and coauthors, Alcohol consumption and mortality among middle-aged and elderly U.S. adults, *New England Journal of Medicine* 337 (1997): 1705–1714; J. M. Gaziano and coauthors, Moderate alcohol intake, increased levels of high-density lipoprotein and its subfractions, and decreased risk of myocardial infarction, *New England Journal of Medicine* 329 (1993): 1829–1834; P. R. Ridker and coauthors, Association of moderate alcohol consumption and plasma concentration of endogenous tissue-type plasminogen activator, *Journal of the American Medical Association* 272 (1994): 929–933.
20. C. S. Fuchs and coauthors, Alcohol consumption and mortality among women, *New England Journal of Medicine* 332 (1995): 1245–1250; H. O Hein, P. Suadicani, and F. Gyntelberg, Alcohol consumption, serum low density lipoprotein cholesterol concentration, and risk of ischaemic heart disease: Six year follow up in the Copenhagen male study, *British Medical Journal* 312 (1996): 731–736.
21. G. D. Friedman and A. L. Klatsky, Is alcohol good for your health? *New England Journal of Medicine* 329 (1993): 1882–1883.
22. T. A. Pearson and P. Terry, What to advise patients about drinking alco-

hol, *Journal of the American Medical Association* 272 (1994): 967–968.
23. C. A. Camargo and coauthors, Prospective study of moderate alcohol consumption and mortality in US male physicians, *Archives of Internal Medicine* 137 (1997): 79–85; N. Grønbaek and coauthors, Influence of sex, age, body mass index, and smoking on alcohol intake and mortality, *British Medical Journal* 308 (1994): 302–306; R. Doll and coauthors, Mortality in relation to consumption of alcohol, *British Medical Journal* 309 (1994): 911–918.
24. Fuchs and coauthors, 1995.
25. E. B. Rimm and coauthors, Review of moderate alcohol consumption and reduced risk of coronary heart disease: Is the effect due to beer, wine, or spirits? *British Medical Journal* 312 (1996): 731–736.
26. Inhibition of LDL oxidation by phenolic substances in red wine: A clue to the French paradox? *Nutrition Reviews* 51 (1993): 185–187.
27. A. L. Macnair, Physical activity, not diet, should be the focus of measures for the primary prevention of cardiovascular disease, *Nutrition Research Reviews* 7 (1994): 43–65.
28. P. T. Williams, High-density lipoprotein cholesterol and other risk factors for coronary heart disease in female runners, *New England Journal of Medicine* 334 (1996): 1298–1303.
29. V. L. Burt and coauthors, Prevalence of hypertension in the US adult population: Results from the third National Health and Nutrition Examination Survey, 1988–1991, *Hypertension* 25 (1995): 305–313.
30. M. Rantala and coauthors, Apolipoprotein E phenotype and diet-induced alteration in blood pressure, *American Journal of Clinical Nutrition* 65 (1997): 543–550.
31. R. G. Victor and J. Hansen, Alcohol and blood pressure—A drink a day . . ., *New England Journal of Medicine* 332 (1995): 1782–1783.
32. J. Wylie-Rosett and coauthors, Trial of Antihypertensive Intervention and Management: Greater efficacy with weight reduction than with a sodium-potassium intervention, *Journal of the American Dietetic Association* 93 (1993): 408–415.
33. M. Alderman and coauthors, Low urinary sodium excretion is associated with greater risk of myocardial infarction among treated hypertensive men, *Hypertension* 25 (1995): 1144–1152; M. Alderman and coauthors, Urinary sodium excretion and myocardial infarction in hypertensive patients: A prospective cohort study, *American Journal of Clinical Nutrition* 65 (1997): 682S–686S.
34. T. A. Kotchen and J. M. Kotchen, Dietary sodium and blood pressure: Interactions with other nutrients, *American Journal of Clinical Nutrition* 65 (1997): 708S–711S; P. K. Whelton and coauthors, Effects of oral potassium on blood pressure, *Journal of the American Medical Association* 277 (1997): 1624–1632; D. A. McCarron, Role of dietary calcium intake in the prevention and management of salt-sensitive hypertension, *American Journal of Clinical Nutrition* 65 (1997): 712S–716S.
35. L. J. Appel and coauthors, A clinical trial of the effects of dietary patterns on blood pressure, *New England Journal of Medicine* 336 (1997): 1117–1124.
36. Council on Scientific Affairs, American Medical Association, Report of the Council on Scientific Affairs, Diet and cancer: Where do matters stand? *Archives of Internal Medicine* 153 (1993): 50–56.
37. H. Hwang, J. Dwyer, and R. M. Russel, Diet *Heliobacter pylori* infection, food preservation and gastric cancer risk: Are there new roles for preventative factors? *Nutrition Reviews* 52 (1994): 75–83.
38. J. E. Foulke, Urethane in alcoholic beverages under investigation, *FDA Consumer*, January/February 1993, pp. 19–23.
39. M. W. Pariza, Animal studies: Summary, gaps, and future research, *American Journal of Clinical Nutrition* 66 (1997): 1539–1540.
40. I. T. Johnson, G. Williamson, and S. R. R. Musk, Anticarcinogenic factors in plant foods: A new class of nutrients? *Nutrition Research Reviews* 7 (1994): 175–204.
41. C. R. Scott and coauthors, Characteristics of youth-onset noninsulin-dependent diabetes mellitus and insulin-dependent diabetes mellitus at diagnosis, *Pediatrics* 100 (1997): 84–91.
42. J. Walston and coauthors, Time of onset of non-insulin-dependent diabetes mellitus and genetic variation in the $B_3$-adrenergic-receptor gene, *New England Journal of Medicine* 333 (1995): 343–347; L. C. Groop and coauthors, Association between polymorphism of the glycogen synthase gene and non-insulin-dependent diabetes mellitus, *New England Journal of Medicine* 328 (1993): 10–14; J. L. Leahy and A. E. Boyd III, Diabetes genes in non-insulin-dependent diabetes mellitus, *New England Journal of Medicine* 328 (1993): 56–57.
43. D. M. Nathan, Long-term complications of diabetes mellitus, *New England Journal of Medicine* 328 (1993): 1676–1685.
44. American Diabetes Association, Nutrition recommendations and principles for people with diabetes mellitus, *Journal of the American Dietetic Association* 94 (1994): 504–506.
45. M. J. Franz and coauthors, Outcomes and cost-effectiveness of medical nutrition therapy for non-insulin-dependent diabetes, *Diabetes Spectrum* 9 (1996): 122–127; E. Q. Johnson and S. Valera, Medical nutrition therapy in non-insulin-dependent diabetes improves clinical outcome, *Diabetes Spectrum* 9 (1996): 131–133; M. J. Franz and coauthors, Effectiveness of medical nutrition therapy provided by dietitians in management of non-insulin-dependent diabetes mellitus: A randomized, controlled clinical trial, *Diabetes Spectrum* 9 (1996): 133–135.
46. R. G. Schafer and coauthors, Translation of the diabetes nutrition recommendations for health care institutions: Technical review, *Journal of the American Dietetic Association* 97 (1997): 43–51.
47. J. P. Barnett and A. Garg, Medical nutrition therapy for patients with diabetes mellitus: Role of dietary fats, *On the Cutting Edge of Diabetes Care and Education,* Fall 1997, pp. 5–6; A. M. Coulson, Monounsaturated fats for people with diabetes, *On the Cutting Edge of Diabetes Care and Education,* Fall 1997, pp. 14–16.
48. A. Garg and coauthors, Effects of varying carbohydrate content of diet in patients with non-insulin-dependent diabetes mellitus, *Journal of the American Medical Association* 271 (1994): 1421–1428; L. V. Campbell and coauthors, The high-monounsaturated fat diet as a practical alternative for NIDDM, *Diabetes Care* 17 (1994): 177–182.
49. Position of The American Dietetic Association: The role of nutrition in health promotion and disease prevention programs, *Journal of the American Dietetic Association* 98 (1998): 205–208.
50. R. W. Pinner and coauthors, Trends in infectious diseases mortality in the

United States, *Journal of the American Medical Association* 275 (1996): 189–193.
51. D. P. Huston, The biology of the immune system, *Journal of the American Medical Association* 278 (1997): 1804–1814.
52. R. K. Chandra, Nutrition and the immune system: An introduction, *American Journal of Clinical Nutrition* 66 (1997): 460S–463S.
53. N. S. Scrimshaw and J. P. SanGiovanni, Synergism of nutrition, infection, and immunity: An overview, *American Journal of Clinical Nutrition* 66 (1997): 464S–477S.
54. K. S. Kubena and D. N. McMurray, Nutrition and the immune system: A review of nutrient-nutrient interactions, *Journal of the American Dietetic Association* 96 (1996): 1156–1164.
55. D. S. Kelley and A. Bendich, Essential nutrients and immunologic functions, *American Journal of Clinical Nutrition* 63 (1996): 994S–996S.
56. U. Süttmann and coauthors, Incidence and prognostic value of malnutrition and wasting in human immunodeficiency virus–infected outpatients, *Journal of Acquired Immune Deficiency Syndrome* 8 (1995): 239–246.
57. J. S. Young, HIV and medical nutrition therapy, *Journal of the American Dietetic Association* 97 (1997): S161–S166.
58. J. S. Young, HIV and medical nutrition therapy, *Journal of the American Dietetic Association* 97 (1997): S161–S166.
59. S. L. Gorbach, T. A. Knox, and R. Roubenoff, Interactions between nutrition and infection with human immunodeficiency virus, *Nutrition Reviews* 51 (1993): 226–234.
60. Department of Continuing Education in Health Sciences, UCLA Extension, *Nutritional Aspects of the AIDS Patient* (Los Angeles, 1989).
61. P. Kurtzweil, Warding off HIV wasting syndrome, *FDA Consumer*, April 1995, pp. 16–20.

# Alternative Therapies

If you suffered from migraine headaches or severe joint pain, where would you turn for relief? Would you visit a physician? Or are you more likely to go to an herbalist or an acupuncturist? Most physicians diagnose and treat medical conditions in ways that are accepted by the established medical community; in contrast, herbalists and acupuncturists, among others, use unconventional methods that offer alternatives to standard medical practice. Instead of taking two aspirin, for example, you might be advised to chew two fresh leaves of the herb feverfew or to swallow a tincture of white willow bark. Or you might receive a massage and several acupuncture needles.

Alternative therapies have become increasingly popular in recent years. Many consumers have become distrustful of, and feel overwhelmed by, the high-tech diagnostic tests and costly treatments that conventional medicine offers. They want to take more responsibility for both maintaining their own health and finding cures for their own diseases, especially when traditional medical therapies prove ineffective. This highlight explores alternative therapies in search of their possible benefits and with an awareness of their potential harms.

## Defining Alternative Medicine

By definition, alternative therapies lie outside the realm of conventional medicine. An alternative therapy is any intervention that:

- Is not taught by most medical schools in the United States.
- Is not reimbursable by most health insurance providers in the United States.
- Is not well supported by scientific tests establishing safety and effectiveness.

If proved safe and effective, an alternative therapy may gradually become part of mainstream conventional medicine. Cancer radiation therapy, for example, was once considered an unconventional therapy, but now is commonly accepted as standard medical practice. In some cases, a therapy that is accepted by traditional medicine for a specific ailment is used for a different purpose in an alternative therapy. For example, chelation therapy, the preferred biomedical treatment for lead poisoning, is a common alternative therapy for cardiovascular disease.

The glossary on p. 592 defines terms and Table H18-1 on p. 593 lists selected fields of alternative medicine. Notice that most alternative medicines fall outside the field of nutrition, but that nutrition itself can be an alternative therapy. Furthermore, many alternative therapies prescribe specific dietary regimens. The many dietary recommendations presented throughout this text are based on scientific evidence and do *not* fall into the alternative therapies category; strategies that are still experimental, however, do. For example, alternative therapists may recommend megadoses of antioxidant supplements or macrobiotic diets to help prevent chronic diseases, whereas most registered dietitians would advise people to eat a balanced diet that includes at least five servings of fresh vegetables and fruits daily instead.

*Digoxin, a drug commonly prescribed for abnormal heart rhythms, derives from the foxglove plant.*

## Sound Research, Loud Controversy

Information on most alternative therapies comes from folklore, tradition, and testimonial accounts. The clinical trials that have been conducted generally suffer from such poor methodology as to invalidate their findings.[1] In short, scientific evidence proving the safety and effectiveness of alternative therapies is lacking. Some say that alternative therapies simply do not work; others argue that the established medical community has not given these therapies a fair trial.

Sound research would answer two important questions. First, does the treatment offer better results than either doing nothing or giving a placebo? Second, do the benefits clearly outweigh the risks? Each of these points is worthy of elaboration.

### Placebo Effect

As Chapter 1 explained, a placebo is an inert, harmless medication used in research studies to control for the beneficial effect that a treatment—even an inactive one—has on recovery. Placebos are inert, but they can also be effective, accounting for an apparent benefit about as often as not.[2] Traditional medicine tends to neglect this powerful remedy, whereas many alternative therapies embrace it. While health professionals cannot conceal serious illnesses or intentionally deceive a client, they might be wise to provide a caring confidence in the appropriate treatment and likelihood of recovery.

### Risks versus Benefits

Ideally, a therapy provides benefits with little or no risk. For example, garlic may help prevent heart disease by lowering blood cholesterol and blood pressure.[3] Ginkgo extract may improve cognitive and social performance in Alzheimer's disease.[4] Extracts of St. John's wort may be effective in treating mild depression.[5] Ginseng may improve glucose control in people with type 2 diabetes.[6] Such findings, if replicated, hold promise that these alternative therapies may one day become accepted medical practice.

Some alternative therapies are innocuous, providing little or no benefit for little or no risk. Sipping a cup of warm tea with a pleasant aroma, for example, won't cure heart disease, but it may improve mood and help relieve tension. Given no physical hazard and little financial risk, such therapies are acceptable.

In contrast, other products and procedures are downright dangerous, posing great risks while providing no benefits. One example is the folk practice of geophagia (eating earth or clay), which can cause GI impaction and impair iron absorption. Another is the taking of laetrile to treat cancer, which can cause cyanide poisoning. Clearly, such therapies are too harmful to be used.

Perhaps most controversial are alternative therapies that may provide benefits, but also carry significant, unknown, or debatable risks. These therapies tend to appeal most to those who are most vulnerable—people with very serious illnesses. Smoking marijuana is a current example of such an alternative therapy: it seems to provide relief from symptoms such as nausea, vomiting, and pain that commonly accompany cancer, AIDS, and other diseases, but also poses risks that some people, including many physicians, consider acceptable and others, mainly politicians, deem intolerable.[7] Physicians have focused on individuals and recognize that marijuana stimulates the appetite in their nauseated clients; politicians and others have focused on society and realize that marijuana is one of many drugs that can be abused. Figure H18-1 summarizes the relationships between risks and benefits.

## *Funds and Findings*

The public's growing interest in unorthodox remedies, coupled with the soaring costs of traditional medical

## Glossary

**acupuncture** (AK-you-PUNK-cher): a technique that involves piercing the skin with long thin needles at specific anatomical points to relieve pain or illness. Acupuncture sometimes uses heat, pressure, friction, suction, or electromagnetic energy to stimulate the points.

**alternative therapies:** approaches to medical diagnosis and treatment that are not fully accepted by the established medical community; as such, they are not widely taught at U.S. medical schools or practiced in U.S. hospitals; also called *adjunctive, unconventional,* or *unorthodox* therapies.

**aroma therapy:** a technique that uses oil extracts from plants and flowers (usually applied by massage or baths) to enhance physical, psychological, and spiritual health.

**ayurveda** (EYE-your-VAY-dah): a traditional Hindu system of improving health by using herbs, diet, meditation, massage, and yoga to stimulate the body to make its own natural drugs.

**bioelectromagnetic medical applications:** the use of electrical energy, magnetic energy, or both to stimulate bone repair, wound healing, and tissue regeneration.

**biofeedback:** the use of special devices to convey information about heart rate, blood pressure, skin temperature, muscle relaxation, and the like to enable a person to learn how to consciously control these medically important functions.

**biofield therapeutics:** a manual healing method that directs a healing force from an outside source (commonly God or another supernatural being) through the practitioner and into the client's body; commonly known as "laying on of hands."

**cartilage therapy:** the use of cleaned and powdered connective tissue, such as collagen, to improve health.

**chelation** (key-LAY-shun) **therapy:** the use of ethylene diamine tetraacetic acid (EDTA) to bind with metallic ions, thus healing the body by removing toxic metals.

**chiropractic** (KYE-roe-PRAK-tik): a manual healing method of manipulating vertebrae to relieve musculoskeletal pain suspected of causing problems with internal organs.

**faith healing:** healing by invoking divine intervention without the use of medical, surgical, or other traditional therapy.

**herbal** (ERB-al) **medicine:** the use of plants to treat disease or improve health; also known as *botanical medicine* or *phytotherapy.*

**homeopathy** (hoe-me-OP-ah-thee): a practice based on the theory that "like cures like," that is, that substances that cause symptoms in healthy people can cure those symptoms when given in very dilute amounts.
- **homeo** = like
- **pathos** = suffering

**hypnotherapy:** a technique that uses hypnosis and the power of suggestion to improve health behaviors, relieve pain, and heal.

**imagery:** a technique that guides clients to achieve a desired physical, emotional, or spiritual state by visualizing themselves in that state.

**iridology:** the study of changes in the iris of the eye and their relationships to disease.

**macrobiotic diets:** extremely restictive diets limited to a few grains and vegetables; based on metaphysical beliefs and not nutrition. A macrobiotic diet might consist of brown rice, miso soup, and sea vegetables, for example.

**massage therapy:** a healing method in which the therapist manually kneads muscles to reduce tension, increase blood circulation, improve joint mobility, and promote healing of injuries.

**meditation:** a self-directed technique of relaxing the body and calming the mind.

**naturopathic medicine:** a system that integrates traditional medicine with botanical medicine, clinical nutrition, homeopathy, acupuncture, East Asian medicine, hydrotherapy, and manipulative therapy.

**orthomolecular medicine:** the use of large doses of vitamins to treat chronic disease.

**ozone therapy:** the use of ozone gas to enhance the body's immune system.

**Table H18-1**
**Fields of Alternative Medicine and Selected Examples**

| Mind-body interventions |
|---|
| Biofeedback |
| Faith healing |
| Hypnotherapy |
| Imagery |
| Meditation |
| Bioelectromagnetic appellations in medicine |
| Electroacupuncture |
| Microwave resonance therapy |
| Alternative systems of medical practice |
| Acupuncture |
| Ayurveda |
| Homeopathic medicine |
| Naturopathic medicine |
| Manual healing methods |
| Biofield therapeutics |
| Chiropractic |
| Massage therapy |
| Pharmacological and biological treatments |
| Cartilage therapy |
| Chelation therapy |
| Ozone therapy |
| Herbal medicine |
| Diet and nutrition in the prevention and treatment of chronic disease |
| Macrobiotic diets |
| Orthomolecular medicine |

Source: Alternative Medicine: Expanding Medical Horizons, A report to the National Institutes of Health and Alternative Medical Systems and Practices in the United States (Washington, D.C.: Government Printing Office, 1992).

approaches, opened the way for alternative medicine to prove itself. In 1992, Congress passed legislation requiring the National Institutes of Health (NIH) to create an Office of Alternative Medicine. Its task is "to more adequately explore unconventional medical practices."[8]

The use of tax dollars to fund research in alternative therapies has many medical professionals concerned.[9] They claim that establishing a special office lends credibility to these unproven methods of health care. Others argue that these methods should be given a fair chance to prove themselves through scientifically valid research. Alternative therapies under study include:[10]

- Acupuncture to treat depression, attention deficit hyperactivity disorder, osteoarthritis, and postoperative dental pain.
- Hypnosis to treat chronic low back pain and to speed fracture healing.
- Ayurvedic herbals to treat Parkinson's disease.
- Biofeedback to treat diabetes, low back pain, and face and mouth pain caused by jaw disorders.
- Electric currents to treat tumors.
- Imagery to treat asthma and breast cancer.

To date, only acupuncture has been deemed effective, and then only for the nausea associated with surgery, chemotherapy, and pregnancy and the pain associated with dental procedures.[11]

## Nutrition and Herbal Medicines

Of greatest interest to students of nutrition is the use of foods and herbs to prevent and treat illnesses. With the proliferation of research on antioxidants and disease prevention (see Highlight 11), medical science has begun to accept the health-promoting powers of fruits and vegetables. It is even beginning to consider the possibility that dietary supplements might someday be an appropriate preventive therapy. Herbs, on the other hand, continue to meet with resistance.[12]

### Herbal Traditions

From earliest times, people have used a myriad of herbs and other plants to cure aches and ills with varying degrees of success. Upon scientific study, dozens of these folk remedies reveal their secrets. For example, myrrh, a plant resin used as a painkiller in ancient times, does indeed have an analgesic effect.[13] The herb valerian, which has long been used as a tranquilizer, contains oils that have a sedative effect. Senna leaves, brewed as a laxative tea, produce compounds that act as a potent cathartic drug. Green tea, brewed from the dried leaves of *Camellia sinensis,* contains phytochemicals that induce cancer cells to self-destruct.[14] The compounds that plants contain are so beneficial that today they contribute to more than half of our modern medicines. Once analyzed, these chemicals in plants can be

**Figure H18-1**
**Risk-Benefit Relationships**

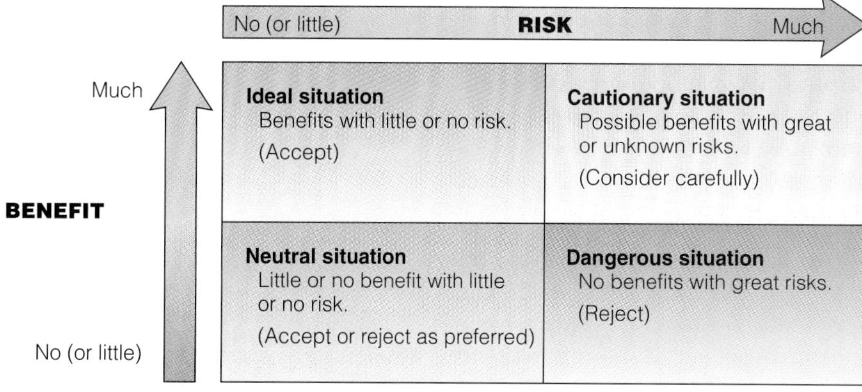
No (or little)
RISK
Much
Much
BENEFIT
No (or little)
Ideal situation
Benefits with little or no risk.
(Accept)
Cautionary situation
Possible benefits with great or unknown risks.
(Consider carefully)
Neutral situation
Little or no benefit with little or no risk.
(Accept or reject as preferred)
Dangerous situation
No benefits with great risks.
(Reject)

synthesized in pharmaceutical labs, thus cutting costs and conserving endangered species. Without laboratory production, valued plants could quickly disappear; consider that it took all of the bark from one 40-foot-tall, 100-year-old Pacific yew tree to produce one 300-milligram dose of the anticancer drug paclitaxel (Taxol), until scientists learned how to synthesize it.[15]

## Herbal Precautions

Simply because plants are "natural" does not mean that they are beneficial or even safe. Nothing could be more natural—and deadly—than the poisonous herb hemlock. Several herbal remedies have toxic effects. The popular Chinese herbal potion jin bu huan, which is used as a pain and insomnia remedy, has been linked with several cases of acute hepatitis. Germanium, a nonessential mineral commonly found in many herbal products, has been associated with chronic kidney failure. Paraguay tea produces symptoms of agitation, confusion, flushed skin, and fever. Kombucha tea, commonly used in the hopes of preventing cancer, relieving arthritis, curing insomnia, and stimulating hair regrowth, can cause severe metabolic acidosis.[16] Table H18-2 lists selected herbs, their common uses, and risks.

When used to diagnose, treat, or prevent disease, herbs are drugs. Yet few herbalists have the understanding of botany, chemistry, or pharmacology necessary to prescribe plant drugs.[17] Instead, they rely on hearsay and folklore. The herbs they prescribe are not regulated by the Food and Drug Administration (FDA). Quite simply, herbs are unapproved drugs that are being sold as dietary supplements.[18] Under the Dietary Supplement Health and Education Act, rather than the herb manufacturers having to prove the safety of their products, the FDA has the burden of proving that a product is not safe.[19] Because information on the safety and effectiveness of herbs derives largely from users' reports, consumers may lack information about or find discrepancies regarding:

- *True identification of herbs.* Most mint teas are safe, for instance, but some varieties contain the highly toxic pennyroyal oil. Mistakenly used to treat colic, mint tea laden with pennyroyal has been blamed for the liver and neurological injuries of at least two infants, one of whom died.[20]
- *Purity of herbal preparations.* Potentially toxic quantities of arsenic and mercury have been detected in traditional Chinese herbal balls used to treat fever, rheumatism, and cataracts.[21] Analysis of one tea prescribed by a Chinese herbalist contained lead at 20,000 times the Environmental Protection Agency's allowable level; arsenic was over 1000 times the allowable level.[22]
- *Appropriate uses and contraindications of herbs.* Herbal remedies alone may be appropriate for minor ailments—a cup

**Table H18-2**
**Selected Herbs, Their Common Uses, and Risks**

| Common Name | Claims and Uses | Risks[a] |
|---|---|---|
| Aloe (gel) | Promote wound healing | Generally considered safe |
| Chamomile (flowers) | Relieve indigestion | Generally considered safe |
| Chaparral (leaves and twigs) | Slow aging, "cleanse" blood, heal wounds | Acute, toxic hepatitis |
| Comfrey (leafy plant) | Soothe nerves | Liver disease |
| Echinacea (roots) | Alleviate symptoms of colds, flus, and infections; promote wound healing; boost immunity | Generally considered safe |
| Ephedra (stems) | Promote weight loss | Rapid heart rate, tremors, seizures, insomnia, headaches |
| Feverfew (leaves) | Prevent migraine headaches | Generally considered safe; may cause mouth irritation, swelling, ulcers, and GI distress |
| Garlic (bulbs) | Lower blood lipids and blood pressure | Generally considered safe; may cause garlic breath, body odor, gas, and GI distress; inhibits blood clotting |
| Ginkgo (tree leaves) | Improve memory, relieve vertigo | Generally considered safe; may cause headache, GI distress, and dizziness |
| Ginseng (roots) | Boost immunity, increase endurance | Generally considered safe; may cause insomnia and high blood pressure |
| Goldenseal (roots) | Relieve indigestion, treat urinary infections | Generally considered safe |
| Jin bu huan (Chinese herbal product) | Relieve pain | Slows heart rate, breathing, and central nervous system functioning; hepatitis |
| Laetrile (apricot pits) | Treat cancer | Cyanide poisoning |
| Saw palmetto (ripe fruits) | Relieve symptoms of enlarged prostate; diuretic; enhance sexual vigor; enlarge mammary glands | Generally considered safe |
| St. John's wort (leaves and tops) | Relieve depression and anxiety | Generally considered safe; may cause fatigue and GI distress |
| Valerian (roots) | Calm nerves, improve sleep | Generally considered safe |
| Yohimbe (tree bark) | Enhance "male performance" | High blood pressure, fatigue, kidney failure, seizures |

[a]Allergies are always a possible risk.

Sources: V. E. Tyler, *The Honest Herbal: A Sensible Guide to the Use of Herbs and Related Remedies* (New York: Pharmaceutical Products Press, 1993); New safety measures are proposed for dietary supplements containing ephedrine alkaloids, *Journal of the American Medical Association* 278 (1997): 15; C. L. Bartels and S. J. Miller, Herbal and related remedies, *Nutrition in Clinical Practice* 13 (1998): 5–19; N. Spaulding-Albright, A review of some herbal and related products commonly used in cancer patients, *Journal of the American Dietetic Association* 97 (1997): S208–S215.

of chamomile tea to ease gastric discomfort or the gel of an aloe vera plant to soothe a sunburn, for example—but not for major health problems such as cancer or AIDS.

- *Safe dosages of herbs.* Herbs that are effective contain active ingredients that need to be administered in proper doses. Each of these active ingredients has a different potency, time of onset, duration of activity, and consequent effects, making the plant itself too unpredictable to be useful. Foxglove leaves, for example, contain dozens of compounds that have an effect on the heart; digoxin, a drug derived from foxglove, offers a standard dosage that allows for a more predictable cardiac response.
- *Interactions of herbs with medicines and other herbs.* Like drugs, herbs may interfere with, or potentiate, the effects of other herbs and drugs. Chewing foxglove leaves when taking the drug digoxin, for example, could be catastrophic.
- *Adverse reactions and toxicity levels of herbs.* As is true of all drugs, herbs may produce undesirable reactions. The herb ephedra, commonly known as Ma Huang or "herbal fen-phen" and used to promote weight loss, acts as a strong central nervous system stimulant, causing rapid heart rate, headaches, insomnia, tremors, seizures, and even death.

Not only are herbal preparations not regulated, but their labels may carry unsubstantiated health claims as long as the following disclaimer also appears: "has not been evaluated by the Food and Drug Administration."

Only after serious problems arise, as occurred with ephedra, does the FDA take action. To reduce the number of adverse effects associated with ephedra-containing supplements, the FDA has proposed limiting the amount of ephedra per dose and prohibiting its use in combination with other stimulants (such as caffeine). Furthermore, labels must caution consumers against excessive or prolonged use (more than 7 days). Consumers who decide to use herbs do so at their own risk.

## Internet Precautions

As Highlight 1 pointed out, just because it appears on the World Wide Web, "it ain't necessarily so." Hundreds of websites tout the benefits of herbal medicines and other dietary supplements. Many of them promote products by quoting researchers or physicians. Such quotations lend an air of authority to advertisements, but be aware that these sources may not even exist—and if they do, their comments may have been taken out of context. When asked, they may not agree at all with the claims attributed to them by the manufacturer.[23]

Other deceits and dangers lie in cyberspace as well. Potentially toxic substances, illegal and unavailable in many countries, are now easy to obtain via the Internet. Electronic access to products such as absinthe and oil of wormwood could be deadly.[24]

## The Consumer's Perspective

Health care professionals may quickly dismiss alternative therapies as ineffective and perhaps even dangerous, but their clients think otherwise. In a survey of over 1500 people, one out of every three had used at least one alternative therapy in the past year for a variety of medical complaints from anxiety and headaches to cancer and tumors.[25] Visits to alternative therapists outnumbered visits to primary care physicians. Interestingly, those who seek alternative therapies seem to do so not so much because they are dissatisfied with conventional medicine as because they find these alternatives more in line with their beliefs about health and life.[26]

Most often, people use alternative therapies in addition to, rather than in place of, conventional therapies. Only a few of the people surveyed saw an alternative therapist without also seeing a physician; all of those with life-threatening conditions such as cancer, diabetes, or lung problems who used alternative therapies saw a medical doctor as well. In fact, it seems that most people seek alternative therapies for nonserious medical conditions or health promotion. They simply want to feel better and access is easy. Sometimes their symptoms are chronic and subjective, such as pain and fatigue, and difficult to treat. In these cases, the chances of finding relief are often as good with a placebo, standard medical intervention, or even nonintervention.

Consumers spend an estimated $13.7 billion on alternative health services a year.[27] This figure does not include expenditures on products such as herbs, crystals, and aromas, which would raise the total considerably. When revenues soar to this extent, consumers need to beware. (To review how a person can identify health fraud and quackery, turn to pp. 27–29. For a list of credible sources of nutrition information, see p. 28.)

## The Health Care Professional's Perspective

How should health care professionals react to clients who use alternative therapies? Those who condemn alternative therapies risk driving clients away, especially if alternative therapists appear more understanding and less judgmental.[28] Health care professionals who listen without judgment allow clients to speak openly and honestly. Such communications help to ensure that the various treatments won't interfere with each other, creating a life-threatening situation. When listening to clients, health care professionals will want to:

- Be culturally sensitive; acknowledge and respect the beliefs, attitudes, and lifestyles of their clients.[29]
- Keep an open mind; standard medical treatments simply don't work for some people in some situations, and they don't always work for everyone.

- Accept alternative therapies and integrate them into the care plan if they bring comfort without harm.
- Provide accurate information, not unsubstantiated opinions; an accurate diagnosis and information on all treatment options will help clients make their health care decisions.
- Advise clients how they can best monitor their condition; for example, clients with diabetes who test their blood glucose diligently will know whether an alternative therapy is effective, and if not, they can take corrective action promptly.
- Discourage practices only if they are harmful.

Some health care professionals have begun to embrace some of the alternative therapies and incorporate them into their medical practice. In much of Europe, biomedicine and alternative therapies have been combined into a system of "complementary medicine," which takes advantage of the best of both approaches. Many practitioners in the United States would like to see such an integrated approach used here as well.[30]

In closing, remember that alternative therapies come in a variety of shapes and sizes. Both their benefits and their risks may be either small, none, or great. Accept the beneficial, or even neutral, practices with an open mind and reject only those practices known to cause harm. Making healthful choices requires knowing what all the choices are.

 NIH Office of Alternative Medicine

altmed.od.nih.gov

 U.S. Government

www.healthfinder.gov/searchoptions/topicsaz.htm
Search for Alternative Medicine
Search for Herbs
Search for Holistic Health
Search for Homeopathy
Search for Preventive Medicine

## Notes


1. J. Kleijnen, P. Knipschild, and G. terRiet, Clinical trials of homeopathy, *British Medical Journal* 302 (1991): 316–323.
2. M. M. Lipman, The power of placebos, *Consumer Reports on Health,* February 1996, p. 23.
3. M. Steiner and coauthors, A double-blind crossover study in moderately hypercholesterolemic men that compared the effect of aged garlic extract and placebo administration on blood lipids, *American Journal of Clinical Nutrition* 64 (1996): 866–870; A. J. Adler and B. J. Holub, Effect of garlic and fish-oil supplementation on serum lipid and lipoprotein concentrations in hypercholesterolemic men, *American Journal of Clinical Nutrition* 65 (1997): 445–450.
4. P. L. LeBars and coauthors, A placebo-controlled, double-blind, randomized trial of an extract of ginkgo biloba for dementia, *Journal of the American Medical Association* 278 (1997): 1327–1332.
5. K. Linde and coauthors, St John's wort for depression—An overview and meta-analysis of randomised clinical trials, *British Medical Journal* 313 (1996): 253–258.
6. E. A. Sotaniemi, E. Haapakoski, and A. Rautio, Ginseng therapy in non-insulin-dependent diabetic patients: Effects on psychophysical performance, glucose homeostasis, serum lipids, serum aminoterminalpropeptide concentration, and body weight, *Diabetes Care* 10 (1995): 1373–1375.
7. J. P. Kassirer, Federal foolishness and marijuana, *New England Journal of Medicine* 336 (1997): 366–367.
8. Alternative Medicine: Expanding Medical Horizons, A report to the National Institutes of Health and Alternative Medical Systems and Practices in the United States (Washington, D.C.: Government Printing Office, 1992).
9. M. Larkin, NIH's office of alternative medicine: A wise use of tax dollars? *Priorities,* vol. 6, no. 4, 1994, pp. 32–36.
10. I. B. Stehlin, An FDA guide to choosing medical treatments, *FDA Consumer,* June 1995, pp. 10–14.
11. C. Marwick, Acceptance of some acupuncture applications, *Journal of the American Medical Association* 278 (1997): 1725–1727; NIH panel gives acupuncture the nod, *Science News* 152 (1997): 344; National Institutes of Health, Consensus development conference statement—Acupuncture, November 3–5, 1997.
12. K. McNutt, Medicinals in food—Part I: Is science coming full circle? *Nutrition Today* 30 (1995): 218–222.
13. P. Lipkin, An ancient salve dampens pain, *Science News* 149 (1996): 20.
14. N. Ahmad and coauthors, Green tea constituent epigallocatechin-3-gallate and induction of apoptosis and cell cycle arrest in human carcinoma cells, *Journal of the National Cancer Institute* 89 (1997): 1881–1886.
15. A photo finish for total taxol synthesis, *Science News* 145 (1994): 223.
16. Unexplained severe illness possibly associated with consumption of kombucha tea—Iowa, 1995, *Journal of the American Medical Association* 275 (1996): 96–98.
17. V. E. Tyler, *The Honest Herbal: A Sensible Guide to the Use of Herbs and Related Remedies* (New York: Pharmaceutical Products Press, 1993).
18. V. E. Tyler, What pharmacists should know about herbal remedies, *Journal of the American Pharmaceutical Association* NS36 (1996): 29–37.
19. K. McNutt, Medicinals in food—Part II: What's new and what's not? *Nutrition Today* 30 (1995): 261–262.
20. J. A. Bakerink and coauthors, Multiple organ failure after ingestion of pennyroyal oil from herbal tea in two infants, *Pediatrics* 98 (1996): 944–947.
21. E. O. Espinoza, M. J. Mann, and B. Bleasdell, Arsenic and mercury in traditional Chinese herbal balls, *New England Journal of Medicine* 333 (1995): 803–804.
22. S. B. Markowitz and coauthors, Lead poisoning due to *Hai Ge Fen:* The prophyrin content of individual erythrocytes, *Journal of the American Medical Association* 271 (1994): 932–934.
23. A. K. Skolnick, Scientific verdict still out on DHEA, *Journal of the American Medical Association* 276 (1996): 1365–1367.
24. S. D. Weisbord, J. B. Soule, and P. L. Kimmel, Poison on line—Acute renal failure caused by oil of wormwood pur-

chased through the Internet, *New England Journal of Medicine* 337 (1997): 825–827.

25. D. M. Eisenberg and coauthors, Unconventional medicine in the United States: Prevalence, costs, and patterns of use, *New England Journal of Medicine* 328 (1993): 246–252.
26. J. A. Astin, Why patients use alternative medicine—Results from a national survey, *Journal of the American Medical Association* 279 (1998): 1548–1553.
27. J. Langone, Challenging the mainstream, *Time,* Fall 1996, pp. 40–43.
28. E. W. Campion, Why unconventional medicine? *New England Journal of Medicine* 328 (1993): 282–283.
29. L. M. Pachter, Culture and clinical care: Folk illness beliefs and behaviors and their implications for health care delivery, *Journal of the American Medical Association* 271 (1994): 690–694.
30. C. Marwick, Complementary medicine congress draws a crowd, *Journal of the American Medical Association* 274 (1995): 106–107.

CHAPTER 19

# Consumer Concerns about Foods

Consumers have concerns about the safety of their food. They want to know what causes food poisoning. They become alarmed when they hear of contaminants and pesticides in foods. They wonder whether food additives are safe. This chapter addresses these concerns, and the highlight that follows looks at public drinking water.

By way of introduction, take a moment to consider the task of supplying food to almost 270 million people in the United States. To feed this nation, farmers grow and harvest crops; dairy producers supply milk products; ranchers raise livestock; shippers deliver foods to manufacturers by land, sea, and air; manufacturers prepare, process, preserve, and package products for refrigerated food cases and grocery-store shelves; and grocers store the food and supply it to consumers. After much time, much labor, and extensive transport, an abundant supply of a large variety of safe foods finally reaches consumers at reasonable market prices.

The Food and Drug Administration (FDA) and other government and international agencies monitor this huge system using a nationwide network of people and sophisticated equipment. The glossary on p. 600 identifies the various food regulatory agencies by their acronyms. These agencies focus on the potential hazard of foods, which differs from the toxicity of a substance—a distinction worth understanding. Anything can be toxic. Toxicity simply means that a substance *can* cause harm *if* enough is consumed. We consume many substances that are toxic, without risk, because the amounts are so small. The term *hazard,* on the other hand, is more relevant to our daily lives because it refers to the harm that is *likely* under real-life conditions. Consumers rely on these monitoring agencies to set sound standards and can learn to protect themselves from food hazards by taking a few preventive steps.

Food safety concerns:
- Food-borne illnesses.
- Nutritional adequacy of foods.
- Environmental contaminants.
- Naturally occurring toxicants.
- Pesticide residues.
- Food additives.

 Food Safety

www.foodsafety.org
www.fda.gov

 Canadian Food Inspection Agency (CFIA)

www.cfia-acia.agr.ca

**hazard:** a source of danger; used to refer to circumstances in which harm is possible under normal conditions of use.

**toxicity:** the ability of a substance to harm living organisms. All substances are toxic if high enough concentrations are used.

**risk:** a measure of the probability and severity of harm.

# Food-Borne Illnesses

The FDA lists food-borne illness as the leading food safety concern because episodes of food poisoning far outnumber episodes of any other kind of food contamination. Some 6.5 million cases of food-borne illness are reported in the United States each year; up to five times as many more go unreported. For some 9000 people each year, the symptoms can be so severe as to cause death.[1] Most vulnerable are the very young, the very old, the sick, the malnourished, and those with a comprised immune system. By taking the proper precautions, however, people can minimize their chances of contracting food-borne illnesses.

**food-borne illness:** illness transmitted to human beings through food and water, caused by either an infectious agent *(food-borne infection)* or a poisonous substance *(food intoxication);* commonly known as **food poisoning.**

## *Food-Borne Infections and Food Intoxications*

*Food-borne illness* can be caused by either an infection or an intoxication. Table 19-1 summarizes the most common or severe food-borne illnesses, their food sources, general symptoms, and prevention methods.

• ***Food-Borne Infections*** • Food-borne infections are caused by eating foods contaminated by infectious microbes. Two of the most common infectious microbes are *Campylobacter jejuni* and *Salmonella,* which enter the GI tract in contaminated foods such as undercooked poultry and unpasteurized milk. Symptoms generally include abdominal cramps, fever, and diarrhea. If a person experiences these

Call a doctor if you develop one of these potentially dangerous conditions:
- Bloody diarrhea.
- A stiff neck, severe headache, and fever (signs of meningitis).
- Excessive diarrhea or vomiting.
- Any food poisoning symptoms that last longer than 3 days.

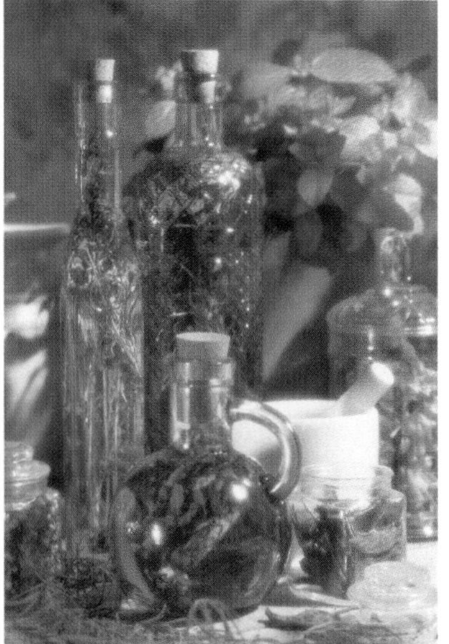

*To prevent food intoxication from homemade flavored oils, wash and dry the herbs before adding them to the oil, keep the oil refrigerated, and discard at the end of the day.*

**pasteurization:** heat processing of food that inactivates some, but not all, microorganisms in the food; not a sterilization process. Bacteria that cause spoilage are still present.

**pathogens** (PATH-oh-jens): microorganisms or substances capable of producing disease.

Food Safety Inspection Service
www.usda.gov/fsis

symptoms as the major or only symptoms of a bout of "flu," chances are excellent that what the person really has is a food-borne illness.*

• ***Food Intoxications*** • Food intoxications are caused by eating foods containing natural toxins or, more likely, microbes that produce toxins. The most common food toxin is produced by *Staphylococcus aureus;* it affects more than one million people per year. Less common, but more infamous, is *Clostridium botulinum,* an organism that produces a deadly toxin in improperly canned (especially home-canned) foods and in homemade garlic or herb-flavored oils stored at room temperature.[2] Botulism requires immediate medical attention, and even then, survivors may suffer the effects for months or years. An amount of toxin as tiny as a single crystal of salt can kill several people within an hour.

## Food Safety in the Marketplace

Transmission of food-borne illness is changing as our food supply and lifestyles change. In the past, food-borne illness was caused by one person's error in a small setting, such as improperly refrigerated potato salad at a family picnic, and affected only a few victims. Today, we are eating more foods prepared and packaged by others. Consequently, when a food manufacturer or restaurant chef makes an error, food-borne illness can become epidemic.[3] An estimated 80 percent of reported food-borne illnesses are caused by errors in a commercial setting, such as the improper pasteurization of milk at a large dairy.[4]

In the mid-1990s, when a fast-food restaurant served undercooked burgers tainted with the infectious organism *Escherichia coli,* hundreds of patrons became ill, and at least three people died. This incident and others focused the national spotlight on two important safety issues: live, disease-causing organisms are commonly found in raw foods, and thorough cooking kills most of these food-borne pathogens. This heightened awareness sparked a much needed overhaul of national food safety programs.

• ***Industry Controls*** • To make our food supply safer for consumers, the USDA, the FDA, and the food processing industries have developed and implemented

*Some viruses do cause intestinal distress, and those that do are usually transmitted via food; true influenza viruses cause symptoms primarily in the upper respiratory tract.

## Glossary of Agencies That Monitor the Food Supply

**CDC (Centers for Disease Control):** a branch of the Department of Health and Human Services that is responsible for, among other things, monitoring food-borne diseases.
www.cdc.gov

**EPA (Environmental Protection Agency):** a federal agency that is responsible for, among other things, regulating pesticides and establishing water quality standards.
www.epa.gov

**FAO (Food and Agriculture Organization):** an international agency (part of the United Nations) that has adopted standards to regulate pesticide use among other responsibilities.
www.fao.org

**FDA (Food and Drug Administration):** a part of the Department of Health and Human Services' Public Health Service that is responsible for ensuring the safety and wholesomeness of all foods processed and sold in interstate commerce except meat, poultry, and eggs (which are under the jurisdiction of the USDA); inspecting food plants and imported foods; and setting standards for food composition.
www.fda.gov

**FTC (Federal Trade Commission):** a federal agency that is responsible for, among other things, food advertising and industry competition.
www.ftc.gov

**USDA (U.S. Department of Agriculture):** the federal agency responsible for enforcing standards for the wholesomeness and quality of meat, poultry, and eggs produced in the United States; conducting nutrition research; and educating the public about nutrition.
www.usda.gov

**WHO (World Health Organization):** an international agency that has adopted standards to regulate pesticide use among other responsibilities.
www.who.ch

**Table 19-1**
**Food-Borne Illnesses**

| Disease and Organism That Causes It | Most Frequent Food Sources | Onset and General Symptoms | Prevention Methods[a] |
|---|---|---|---|
| ***Food-Borne Infections*** | | | |
| **Campylobacteriosis** *Campylobacter jejuni* bacterium | Raw poultry, beef, lamb, unpasteurized milk (foods of animal origin eaten raw or undercooked or recontaminated after cooking). | Onset: 2 to 5 days. Diarrhea, nausea, vomiting, abdominal cramps, fever; sometimes bloody stools; lasts 7 to 10 days. | Cook foods thoroughly; use pasteurized milk; use sanitary food-handling methods. |
| **Giardiasis** *Giardia lamblia* protozoon | Contaminated water; uncooked foods. | Onset: 5 to 25 days. Diarrhea (but occasionally constipation), abdominal pain, gas, abdominal distention, digestive disturbances, anorexia, nausea, and vomiting. | Use sanitary food-handling methods; avoid raw fruits and vegetables where protozoa are endemic; dispose of sewage properly. |
| **Hepatitis** Hepatitis A virus | Undercooked or raw shellfish. | Onset: 15 to 50 days (28 to 30 days average). Inflammation of the liver with tiredness; nausea, vomiting, or indigestion; jaundice (yellowed skin and eyes from buildup of wastes); muscle pain. | Cook foods thoroughly. |
| **Listeriosis** *Listeria monocytogenes* bacterium | Raw meat and seafood, raw milk, and soft cheeses. | Onset: 7 to 30 days. Mimics flu; blood poisoning, complications in pregnancy, and meningitis (stiff neck, severe headache, and fever). | Use sanitary food-handling methods; cook foods thoroughly; use pasteurized milk. |
| **Perfringens food poisoning** *Clostridium perfringens* bacterium | Meats and meat products stored at between 120 and 130°F. | Onset: 8 to 12 hr (usually 12). Abdominal pain, diarrhea, nausea, and vomiting; symptoms last a day or less and are usually mild; can be serious in old or weak people. | Use sanitary food-handling methods; cook foods thoroughly; refrigerate foods promptly and properly. |
| **Salmonellosis** *Salmonella* bacteria | Raw or undercooked eggs, meats, poultry, milk and other dairy products, shrimp, frog legs, yeast, coconut, pasta, and chocolate. | Onset: 6 to 48 hr. Nausea, fever, chills, vomiting, abdominal cramps, diarrhea, and headache; can be fatal. | Use sanitary food-handling methods; use pasteurized milk; cook foods thoroughly; refrigerate foods promptly and properly. |
| **E. coli infection** *Escherichia coli*[b] bacterium | Undercooked ground beef, unpasteurized milk and milk products, contaminated water, and person-to-person contact. | Onset: 12 to 72 hr. Severe bloody diarrhea, abdominal cramps, acute kidney failure; can be fatal. | Cook ground beef thoroughly; avoid raw milk and milk products; use sanitary food-handling methods; use treated, boiled, or bottled water. |
| **Shigellosis** *Shigella* bacteria | Person-to-person contact, raw foods, salads, dairy products, and contaminated water. | Onset: 1 to 7 days. Diarrhea, vomiting, cramps, fever; sometimes bloody stools. | Use sanitary food-handling methods; cook foods thoroughly; proper refrigeration. |
| **Vibrio bacteria** *Vibrio vulnificus*[c] bacterium | Raw seafood and contaminated water. | Onset: 1 to 7 days. Diarrhea, abdominal cramps, fever, chills; can be fatal. | Use sanitary food-handling methods; cook foods thoroughly. |
| ***Food Intoxications*** | | | |
| **Botulism** Botulinum toxin [produced by *Clostridium botulinum* bacterium, which grows without oxygen, in low-acid foods, and at temperatures between 40° and 120°F; the **botulinum** (BOT-chew-line-um) **toxin** responsible for botulism is called **botulin** (BOT-chew-lin)] | Anaerobic environment of low acidity (canned corn, peppers, green beans, soups, beets, asparagus, mushrooms, ripe olives, spinach, tuna, chicken, chicken liver, liver pâte, luncheon meats, ham, sausage, stuffed eggplant, lobster, and smoked and salted fish). | Onset: 4 to 36 hr. Nervous system symptoms, including double vision, inability to swallow, speech difficulty, and progressive paralysis of the respiratory system; often fatal; leaves prolonged symptoms in survivors. | Use proper canning methods for low-acid foods; refrigerate homemade garlic and herb oils; avoid commercially prepared foods with leaky seals or with bent, bulging, or broken cans. |
| **Staphylococcal food poisoning** Staphylococcal toxin (produced by *Staphylococcus aureus* bacterium) | Toxin produced in meats, poultry, egg products, tuna, potato and macaroni salads, and cream-filled pastries. | Onset: ½ to 8 hr. Diarrhea, nausea, vomiting, abdominal cramps, and fatigue; mimics flu; lasts 24 to 48 hr; rarely fatal. | Use sanitary food-handling methods; cook food thoroughly; refrigerate foods promptly and properly; use proper home-canning methods. |

NOTE: Traveler's diarrhea is most commonly caused by *E. coli, Campylobacter jejuni, Shigella,* and *Salmonella.*

[a]The "How to" on pp. 604–605 provides more details on the proper handling, cooking, and refrigeration of foods.

[b]The most serious strain is *E. coli* O157:H7.

[c]Most cases of *Vibrio vulnificus* occur in persons with underlying illness, particularly those with liver disorders, diabetes, cancer, and AIDS, and those who require long-term steroid use. The fatality rate is 50 percent for this population. U.S. Department of Agriculture, U.S. Department of Health and Human Services, U.S. Environmental Protection Agency, Food safety from farm to table: A national food-safety initiative (U.S. Department of Agriculture, May 1997).

**Hazard Analysis Critical Control Points (HACCP):** a systematic plan to identify and correct potential microbial hazards in the manufacturing, distribution, and commercial use of food products.

*Cook hamburgers to 160°F; color alone cannot determine doneness.*

**Figure 19-1**

**Safe Internal Temperatures for Cooking Meats and Poultry (Fahrenheit)**

Bacteria multiply rapidly at temperatures between 40° and 140°F. Cook foods to the temperatures shown on this thermometer and hold them at 140°F or higher.

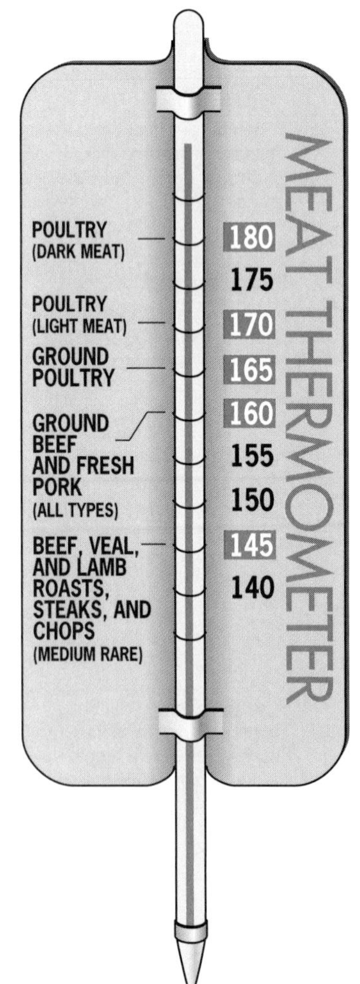


new programs to study and control food-borne illness.[5]* For example, the Hazard Analysis Critical Control Points (HACCP) system ensures that food manufacturers identify points of contamination and implement controls to prevent food-borne disease.[6] These safety regulations are expected to prevent hundreds of thousands of food-borne illnesses each year.

The produce industry provides an example of HACCP's effectiveness. After tracing two large outbreaks of salmonellosis to imported cantaloupe, producers began using chlorinated water to wash the melons and to make ice for packing and shipping. Since this HACCP plan was implemented, no new melon-related cases of salmonellosis have been reported.

This example raises another issue regarding the safety of imported foods. FDA inspectors cannot keep pace with the increasing numbers of imported foods; they inspect fewer than 2 percent of the almost 3 million shipments of fruits, vegetables, and seafood coming into the United States each year. Instead of providing the FDA with additional inspectors, Congress may give the agency the authority to insist that other countries adopt the safe food-handling practices used in the United States.

• ***Consumer Awareness*** • Canned and packaged foods sold in grocery stores are easily controlled, but rare accidents do happen. Batch numbering makes it possible to recall contaminated foods through public announcements via newspapers, television, and radio. In the grocery store, consumers can inspect the safety seals and wrappers of packages. A broken seal or mangled package fails to protect the consumer against microbes, insects, spoilage, or even vandalism.

Improper handling of foods can occur anywhere along the line from commercial manufacturers to large supermarkets to small restaurants to private homes. Maintaining a safe food supply requires everyone's efforts.

## Food Safety in the Kitchen

Whether microbes multiply and cause illness depends, in part, on what happens in the kitchen—whether the kitchen is in your home, a school cafeteria, a gourmet restaurant, or a canning manufacturer. For the most part, food-borne illness can be prevented by doing three simple things: keeping hot foods hot; keeping cold foods cold; and keeping hands, utensils, and the kitchen clean.

Keeping hot foods hot includes cooking foods long enough to reach internal temperatures that will kill microbes and maintaining adequate temperatures to prevent bacterial growth until the foods are served. Keeping cold foods cold entails going directly home upon leaving the grocery store and immediately unpacking foods into the refrigerator or freezer upon arrival. After a meal, any leftovers should be refrigerated immediately. Keeping a clean, safe kitchen requires that cooks wash the countertops, their hands, and utensils in hot, soapy water before and after each step of food preparation. See the "How to" on pp. 604–605 for specific food safety tips.

• ***Meats and Poultry*** • Meats and poultry contain bacteria and provide a moist, nutrient-rich environment that favors microbial growth. Ground meat is especially susceptible because it receives more handling than other kinds of meat and has more surface exposed to bacterial contamination. Consumers cannot detect the harmful bacteria in or on meat. A USDA seal indicates that the product has been inspected for quality. It does not guarantee that the meat is free of potentially harmful bacteria, although the USDA has recently improved its system of monitoring for contamination. For safety's sake, cook meat thoroughly, using a thermometer to test the internal temperature (see Figure 19-1).

*These programs include Hazard Analysis Critical Control Points (HACCP), Emerging Infections Program (EIP), Foodborne Diseases Active Surveillance Network (FoodNet), and the Food Safety Inspection Service (FSIS).

In addition to cooking meat and poultry thoroughly, consumers can take other steps to protect themselves. Labels provide instructions on how to handle products in such a way that the bacteria present won't cause a foodborne illness (see Figure 19-2). Wash any utensils and surfaces (such as cutting boards or platters) that have been in contact with raw meat with hot, soapy water before using them again for the cooked meat or raw produce. Bacteria inevitably left on the surfaces from the raw meat can recontaminate the cooked meat or start to grow in the other foods—a problem known as *cross-contamination.*

Though contaminated meat poses a real threat, consumers need to be aware that the media often exaggerate a story beyond its facts. Reports from England on the slaughter of cattle with mad cow disease and from Hong Kong on the slaughter of poultry with avian influenza sparked consumer fears, even though U.S. beef and poultry were not infected.* Once again, smart consumers consider the source when evaluating nutrition information (see Highlight 1).

**Figure 19-2**

**Safe Handling Instructions for Meat and Poultry**

## Safe Handling Instructions

THIS PRODUCT WAS PREPARED FROM INSPECTED AND PASSED MEAT AND/OR POULTRY. SOME FOOD PRODUCTS MAY CONTAIN BACTERIA THAT CAN CAUSE ILLNESS IF THE PRODUCT IS MISHANDLED OR COOKED IMPROPERLY. FOR YOUR PROTECTION, FOLLOW THESE SAFE HANDLING INSTRUCTIONS.

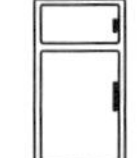

KEEP REFRIGERATED OR FROZEN.
THAW IN REFRIGERATOR OR MICROWAVE.

KEEP RAW MEAT AND POULTRY SEPARATE FROM OTHER FOODS.
WASH WORKING SURFACES (INCLUDING CUTTING BOARDS), UTENSILS, AND HANDS AFTER TOUCHING RAW MEAT OR POULTRY.

COOK THOROUGHLY.

KEEP HOT FOODS HOT. REFRIGERATE LEFTOVERS IMMEDIATELY OR DISCARD.

• ***Seafood*** • Most seafood available in the United States and Canada is safe, but eating it undercooked or raw can cause severe illnesses—hepatitis, worms, parasites, viral intestinal disorders, and other diseases.† Rumor has it that freezing fish will make it safe to eat raw, but this is only partly true. Freezing fish will kill mature parasitic worms, but only cooking can kill all worm eggs and other microorganisms that can cause illness. For safety's sake, all seafood should be cooked. Even sushi can be enjoyed this way: many sushi chefs combine cooked seafood, vegetables, avocado, and other ingredients into delicacies that are perfectly safe to eat.

Eating raw oysters can be dangerous for anyone, but people with liver disease, alcoholics, and people with suppressed immune systems are most vulnerable.[7] At least 10 species of bacteria found in raw oysters can cause serious illness and even death.‡ Raw oysters may also carry the hepatitis A virus, which can cause liver disease. Some hot sauces can kill many of these bacteria, but not the virus; alcohol may also protect against oyster-borne illnesses, but not completely. One study of people who had eaten contaminated oysters raw found that those who consumed an alcoholic beverage were less likely to have gotten sick, or their symptoms were less severe, than those who had not consumed alcoholic beverages.[8] Of course, this is not sufficient evidence to guarantee protection or to recommend drinking alcohol. Pasteurization of raw oysters—holding them at a specified temperature for a specified time—holds promise for killing bacteria without cooking the oyster or altering its texture or flavor.

As population density increases along the shores of seafood-harvesting waters, pollution inevitably invades the sea life there. Preventing seafood-borne illness is in large part a task of controlling water pollution. To help ensure a safe seafood market, the FDA requires processors to adopt food safety practices based on the HACCP system mentioned earlier.

Chemical pollution and microbial contamination lurk not only in the water, but in the boats and warehouses where seafood is cleaned, prepared, and refrigerated. Seafood is one of the most perishable foods: time and temperature are

*Formally known as bovine spongiform encephalopathy (BSE), mad cow disease is a fatal virus that affects the central nervous system of cattle. Similarly, avian influenza A H5N1 is a potentially fatal virus that affects chickens and other birds.

†Diseases caused by toxins from the sea include ciguatera poisoning, scombroid poisoning, and paralytic and neurotoxic shellfish poisoning.

‡Raw oysters can carry the bacterium *Vibrio vulnificus;* see Table 19-1 for details.

The "2-40-140" rule will help you to remember the time and temperature danger zone for foods—allow them to stay for no more than 2 hr between 40°F and 140°F.

**sushi:** vinegar-flavored rice and seafood, typically wrapped in seaweed and stuffed with colorful vegetables. Some sushi is stuffed with raw fish; other varieties contain only cooked ingredients.

U.S. Government
www.healthfinder.gov/searchoptions/topicsaz.htm
Search for Meat
Search for Poultry
Search for Seafood

***Eating raw seafood is a risky proposition even when it is prepared in sushi by a master chef. (Courtesy of CNN)***

## HOW TO Prevent Food-Borne Illnesses

Most food-borne illnesses can be prevented by following three simple rules: keep hot foods hot, keep cold foods cold, and keep a clean kitchen.

**Keep Hot Foods Hot**

- When cooking meats or poultry, use a thermometer to test the internal temperature. Insert the thermometer between the thigh and the body of a turkey or into the thickest part of other meats, making sure the tip of the thermometer is not in contact with bone or the pan. Cook to the temperature indicated for that particular meat; cook hamburgers to at least medium well-done. If you have safety questions, call the USDA Meat and Poultry Hotline: (800) 535–4555.
- Cook stuffing separately, or stuff poultry just prior to cooking.
- Do not cook large cuts of meat or turkey in a microwave oven; it leaves some parts undercooked while overcooking others.
- Marinate meats in the refrigerator, not on the counter. Don't use marinade that was in contact with raw meat for basting or sauces.
- Cook eggs before eating them (soft-boiled for at least 3½ minutes; scrambled until set, not runny; fried for at least 3 minutes on one side and 1 minute on the other).
- Cook seafood thoroughly. If you have safety questions about seafood call the FDA hotline: (800) FDA–4010.
- When serving foods, maintain temperatures at 140°F or higher.
- Heat leftovers thoroughly to at least 165°F.

**Keep Cold Foods Cold**

- When running errands, stop at the grocery store last. When you get home, refrigerate the perishable groceries (such as meats and dairy products) immediately. Do not leave perishables in the car any longer than it takes for ice cream to melt.
- Put packages of raw meat, fish, or poultry on a plate before refrigerating to prevent juices from dripping on food stored below.
- Buy only foods that are solidly frozen in store freezers.
- Keep cold foods at 40°F or less; keep frozen foods at 0°F or less (keep a thermometer in the refrigerator).
- Refrigerate leftovers promptly; use shallow containers to cool foods faster.
- Thaw meats or poultry in the refrigerator, not at room temperature. If you must hasten thawing, use cool running water or a microwave oven.
- Freeze meat, fish, or poultry immediately if not planning to use within a few days.

**Keep a Clean and Safe Kitchen**

- Wash fruits and vegetables with a scrub brush in warm water; store washed and unwashed produce separately.
- Use hot, soapy water to wash hands, utensils, dishes, nonporous cutting boards, and countertops between tasks when working with different foods. Use a bleach solution on cutting boards (one capful per gallon of water).
- Cover cuts with clean bandages before food preparation; dirty bandages carry harmful microorganisms.
- Avoid cross-contamination by washing all surfaces that have been in contact with raw meats, poultry, or eggs before reusing; serving cooked meats on a clean plate; and separating raw foods from those that have been cooked.
- Mix foods with utensils, not hands; keep hands and utensils away from mouth, nose, and hair.
- Anyone may be a carrier of bacteria

Frequently unsafe:
- Raw milk and milk products.
- Raw or undercooked seafood, meat, poultry, or eggs.

Occasionally unsafe:
- Soft cheeses (Mexican style, feta, brie, camembert, blue-veined).
- Salad bar items.
- Unwashed berries and grapes.
- Sandwiches.
- Hamburgers.
- Airline food.

Rarely unsafe:
- Peeled fruit.
- High-sugar foods.
- Steaming-hot foods.

critical to its freshness and flavor. To keep seafood as fresh as possible, people in the industry "keep it cold, keep it clean, and keep it moving." Wise consumers eat it cooked.

• ***Precautions and Procedures*** • Fresh food generally smells fresh. Not all types of food poisoning are detectable by odor, but some bacterial wastes produce "off" odors. If an abnormal odor exists, the food is spoiled. Throw it out or, if it was recently purchased, return it to the grocery store. Do not taste it. Table 19-2 on p. 606 lists some safe storage times for selected foods.

Local health departments and the USDA Extension Service can provide further information about food safety. Should precautions fail and mild food-borne illness develop, drink clear liquids to replace fluids lost through vomiting and diarrhea. If serious food-borne illness is suspected, first call a physician. Then wrap the remainder of the suspected food and label the container so that it cannot be mistakenly eaten, place it in the refrigerator, and hold it for possible inspection by

## HOW TO Prevent Food-Borne Illnesses (continued)

and should avoid coughing or sneezing over food. A person with a skin infection or infectious disease should not prepare food.
- Wash or replace sponges and towels regularly.
- Clean up food spills and crumb-filled crevices.

**In General**
- Do not reuse disposable containers; use nondisposable containers or recycle instead.
- Do not taste food that is suspect. "If in doubt, throw it out."
- Throw out foods with danger-signaling odors. Be aware, though, that most food-poisoning bacteria are odorless, colorless, and tasteless.
- Do not buy or use items that have broken seals or mangled packaging; such containers cannot protect against microbes, insects, spoilage, or even vandalism. Check safety seals, buttons, and expiration dates.
- Follow label instructions for storing and preparing packaged and frozen foods; throw out foods that have been thawed or refrozen.
- Discard foods that are discolored, moldy, or decayed or that have been contaminated by insects or rodents.

**For Specific Food Items**
- *Canned goods.* Carefully discard food from cans that leak or bulge so that other people and animals will not accidentally ingest it; before canning, seek professional advice from the USDA Extension Service (check your phone book under U.S. government listings, or ask directory assistance).
- *Milk and cheeses.* Use only pasteurized milk and milk products. Aged cheeses, such as cheddar and swiss, do well for an hour or two without refrigeration, but should be refrigerated or stored in an ice chest for longer periods.
- *Eggs.* Use clean eggs with intact shells. Do not eat eggs, even pasteurized eggs, raw; raw eggs are commonly found in Caesar salad dressing, eggnog, cookie dough, hollandaise sauce, and key lime pie. Cook eggs until whites are firmly set and yolks begin to thicken.
- *Honey.* Honey may contain dormant bacterial spores, which can awaken in the human body to produce botulism. In adults, this poses little hazard, but infants under one year of age should never be fed honey. Honey can accumulate enough toxin to kill an infant; it has been implicated in several cases of sudden infant death. (Honey can also be contaminated with environmental pollutants picked up by the bees.)
- *Mayonnaise.* Commercial mayonnaise may actually help a food to resist spoilage because of the acid content. Still, keep it cold after opening.
- *Mixed salads.* Mixed salads of chopped ingredients spoil easily because they have extensive surface area for bacteria to invade, and they have been in contact with cutting boards, hands, and kitchen utensils that easily transmit bacteria to food (regardless of their mayonnaise content). Chill them well before, during, and after serving.
- *Picnic foods.* Choose foods that last without refrigeration such as fresh fruits and vegetables, breads and crackers, and canned spreads and cheeses that can be opened and used immediately. Pack foods cold, layer ice between foods, and keep foods out of water.
- *Seafood.* Buy only fresh seafood that has been properly refrigerated or iced. Cooked seafood for purchase should be displayed separately from raw seafood to avoid cross-contamination.

health authorities. The margin (on p. 604) identifies foods commonly implicated in food-borne illnesses.

## *Food Safety while Traveling*

People who travel to other countries have a 50–50 chance of contracting traveler's diarrhea. Like many other food-borne illnesses, traveler's diarrhea is a sometimes serious, always annoying bacterial infection of the digestive tract. The risk is high because, for one thing, some countries' cleanliness standards for food and water may be lower than those in the United States and Canada. For another, every region's microbes are different, and while people are immune to those in their own neighborhoods, they have had no chance to develop immunity to the pathogens in places they are visiting for the first time. The "How to" on p. 606 offers tips on avoiding food-borne infections while traveling.

**traveler's diarrhea:** nausea, vomiting, and diarrhea caused by consuming food or water contaminated by any of several organisms, most commonly, *E. coli, Shigella, Campylobacter jejuni,* and *Salmonella.*

**Table 19-2**
**Safe Cold Storage Times**

| Food | Refrigerator (≤40°F) | Freezer (≤0°F) |
|---|---|---|
| Eggs | | |
| Raw, in shells | 3 weeks | Do not freeze. |
| Hard cooked | 1 week | Do not freeze. |
| Egg substitute (opened) | 3 days | Do not freeze. |
| Egg substitute (unopened) | 10 days | 1 year |
| Mayonnaise | 2 months | Do not freeze. |
| Mayonnaise-based salads (egg, chicken, tuna, ham, pasta) | 3 to 5 days | Do not freeze. |
| Frozen dinners or casseroles | Keep frozen until ready to serve. | 3 to 4 months |
| Store-cooked convenience meals | 1 to 2 days | Do not freeze. |
| Soups, stews, casseroles | 3 to 4 days | 2 to 3 months |
| Bacon | 7 days | 1 month |
| Hot dogs (opened) | 1 week | Rewrap, 1 to 2 months |
| Hot dogs (unopened) | 2 weeks | 1 to 2 months |
| Lunch meats (opened) | 3 to 5 days | Rewrap, 1 to 2 months |
| Lunch meats (unopened) | 2 weeks | 1 to 2 months |
| Sausage (raw) | 1 to 2 days | 1 to 2 months |
| Sausage (smoked) | 7 days | 1 to 2 months |
| Sausage (pepperoni, salami) | 2 to 3 weeks | 1 to 2 months |
| Meat (raw steaks, chops, roasts) | 3 to 5 days | 4 to 12 months |
| Meat (raw, ground) | 1 to 2 days | 3 to 4 months |
| Meat (cooked) | 3 to 4 days | 2 to 3 months |
| Poultry (raw) | 1 to 2 days | 9 to 12 months |
| Poultry (cooked) | 3 to 4 days | 4 to 6 months |
| Seafood (raw) | 1 to 2 days | 3 to 6 months |
| Seafood (cooked) | 1 to 3 days | Do not freeze. |
| Seafood (smoked) | 1 to 10 days | Do not freeze. |
| Pasteurized crab (unopened) | ≤6 months | Do not freeze. |
| Pasteurized crab (opened) | 3 to 5 days | Do not freeze. |
| Surimi[a] (unopened) | 2 to 3 months | 3 to 6 months |
| Surimi[a] (opened) | 1 to 4 days | Rewrap, 3 to 6 months |

[a]Surimi is a product made from fresh fish that has been deboned, minced, and processed to give it a texture, flavor, and color to resemble shellfish.

## HOW TO Achieve Food Safety while Traveling

Food-borne illnesses contracted while traveling are colloquially known as traveler's diarrhea. A bout of this ailment can ruin the most enthusiastic tourist's trip. To avoid food-borne illness, follow the food safety tips outlined on pp. 604–605. In addition, while traveling:

- Wash your hands often with soap and hot water, especially before handling food or eating.
- Eat only cooked or canned foods. Eat raw fruits or vegetables only if you have washed them in boiled water and peeled them yourself. Skip salads and raw fish and shellfish.
- Be aware that water, and ice made from it, may be unsafe. Take along disinfecting tablets or a device to boil water. Do not use the local water supply, even to brush your teeth, unless you boil or disinfect it first. Do not use ice.
- Drink no beverages made with tap water. Drink only treated, boiled, canned, or bottled beverages, and drink them without ice, even if they are not chilled to your liking. Refuse dairy products unless they have been properly pasteurized and refrigerated.
- Before you leave on the trip, ask your physician to recommend medicines to take with you in case your efforts to avoid illness fail.

To sum up these recommendations, "Boil it, cook it, peel it, or forget it." Chances are excellent that if you follow these rules, you will remain well.

## Advances in Food Safety

New advances in technology have dramatically improved the quality and safety of foods available on the market. Although these advances offer numerous benefits, they also raise consumer concerns.[9]

• ***Irradiation*** • The FDA has approved the use of low-dose irradiation on certain foods.[10] Irradiation kills microorganisms and insects in grains; spices; teas; fresh and frozen beef, poultry, lamb, and pork; and fresh fruits and vegetables.[11] In addition, irradiation inhibits the growth of sprouts on potatoes and onions and delays ripening in some fruits such as strawberries and mangoes. Milk products change flavor when irradiated and so are not candidates for the treatment. (Incidentally, the milk in those boxes kept at room temperature on grocery-store shelves is not irradiated, but processed with an ultrahigh temperature treatment for just long enough to sterilize it.)

**irradiation:** sterilizing a food by exposure to energy waves, similar to ultraviolet light and microwaves.

**ultrahigh temperature (UHT) treatment:** sterilizing a food by brief exposure to temperatures above those normally used.

The use of food irradiation has been extensively evaluated and is supported by the World Health Organization, the American Medical Association, and other health agencies. Irradiation does not noticeably change the taste, texture, or appearance of food, nor does it make the food radioactive.[12] Vitamin loss is minimal and comparable to amounts lost in other food processing methods.

• ***Consumer Concerns about Irradiation*** • Many consumers, associating radiation with cancer, birth defects, and mutations, have strong negative emotions about the use of irradiation on foods. Some confuse it with food contamination by radioactive particles, such as occurs in the aftermath of a nuclear accident. Some balk at the idea of irradiating, and thus sterilizing, contaminated foods, and prefer instead the elimination of unsanitary living, slaughtering, and food preparation conditions. Food producers, on the other hand, are eager to use irradiation, but hesitate to do so until consumers are ready to accept it. Proponents believe that once consumers understand the benefits of irradiation and are no longer afraid of it, they will demand to have foods sterilized by this process. Some proponents suggest renaming irradiation *pico wave,* because consumers readily accepted microwaves.

Speaking of microwaves, like irradiation, microwaves can sterilize foods under certain conditions of time and temperature. Unlike irradiation, though, microwaves cook food.

• ***Regulation of Irradiation*** • The FDA has established regulations governing the specific uses of irradiation and allowed doses. Each food that has been treated with irradiation must say so on its label. Labels can be misleading however, if consumers interpret the *absence* of the irradiation symbol to mean that the food was produced without any kind of treatment. This is not true; it is just that the FDA does not require label statements for other treatments used for the same purpose, such as postharvest fumigation with pesticides. If all treatment methods were declared, consumers could make fully informed choices.

*This international symbol identifies retail foods that have been irradiated. The words "Treated by irradiation" or "Treated with irradiation" must accompany the symbol. The irradiation label is not required on commercially prepared foods that contain irradiated ingredients, such as spices.*

• ***High-Intensity Pulsed Light*** • A new irradiation technology called high-intensity pulsed light has been approved by the FDA. High-intensity pulsed light uses an intense flash of light to kill microorganisms on the surface of foods, packaging materials, and water. It extends the shelf life of foods without diminishing their nutrient content.[13] This new process shows promise in the battle against food-borne pathogens.

IN SUMMARY

In summary, millions of people suffer mild to life-threatening symptoms caused by food-borne illnesses (review Table 19-1). As the "How to" on pp. 604–605 describes, most of these illnesses can be prevented by storing and cooking foods at their proper temperatures and by preparing them in sanitary conditions.[14]

U.S. Government
www.healthfinder.gov/searchoptions/topicsaz.htm
Search for Food Safety

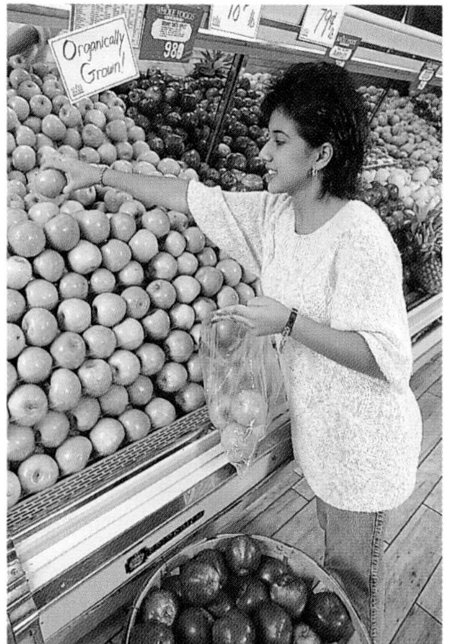


*With the benefits of a safe and abundant food supply comes the responsibility to select, prepare, and store foods safely.*

# Nutritional Adequacy of Foods and Diets

When the FDA ranks its concerns about foods, it places the nutritional adequacy of foods and diets second, behind only food-borne illness. In years past, when most foods were whole and farm fresh, the task of meeting nutrient needs primarily involved balancing servings from the various food groups. Today, however, foods have changed. Many "new" foods are available to appeal to people's tastes and health needs, but not necessarily to deliver a balanced assortment of needed nutrients.

To help consumers find their way among these foods and combine them into healthful diets, the FDA has developed extensive nutrition labeling regulations, as Chapter 2 described. In addition, the USDA's *Dietary Guidelines* help consumers "eat to stay healthy," and the Food Guide Pyramid helps them to put those recommendations into practice (see Chapter 2).

# Environmental Contaminants

**contaminant:** a substance that makes a food impure and unsuitable for ingestion.

Concern about environmental contamination of foods is growing in importance as the world becomes more populated and more industrialized. A food contaminant is anything that does not belong there.

## Harmfulness of Environmental Contaminants

**persistence:** stubborn or enduring continuance; with respect to food contaminants, the quality of persisting, rather than breaking down, in the bodies of animals and human beings.

**food chain:** the sequence in which living things depend on other living things for food.

**bioaccumulation:** the accumulation of contaminants in the flesh of animals high on the food chain.

The potential harmfulness of a contaminant depends in part on its persistence—the extent to which it lingers in the environment or in the body. Some contaminants in the environment are short-lived because microorganisms or agents such as sunlight or oxygen can break them down. Some contaminants in the body may linger for only a short time because the body rapidly excretes them or metabolizes them to harmless compounds. These contaminants present little cause for concern. Some contaminants, however, resist breakdown and can accumulate. Each level of the food chain, then, has a greater concentration than the one below (bioaccumulation). Figure 19-3 shows how bioaccumulation leads to high concentrations of toxins in people at the top of the food chain, and Figure 19-4 shows how contaminants find their way into foods.

How much of a threat do environmental contaminants pose to the food supply? For the most part, the hazards appear to be small because the FDA regulates the presence of contaminants in foods and requires foods with unsafe amounts to be removed from the market. Similarly, health agencies may issue advisories informing consumers about the potential dangers of eating contaminated foods; for example, states in and around the Great Lakes area have warned consumers not to eat sport fish. In the event of an accidental spill, however, the hazards can suddenly become great.

## Examples of Environmental Contaminants

**heavy metal:** any of a number of mineral ions such as mercury and lead, so called because they are of relatively high atomic weight. Many heavy metals are poisonous.

**organic halogen:** an organic compound containing one or more atoms of a **halogen**—fluorine, chlorine, iodine, or bromine.

The following paragraphs describe how two different types of contaminants have found their way into the food supply in the past. One is a heavy metal that was released into waterways by industry and ingested by fish that people ate. The other is an organic halogen that was accidentally spilled into livestock feed and ingested by animals that people ate.

• ***Methylmercury*** • A classic example of acute contamination occurred in 1953 when a number of people in Minamata, Japan, became ill with a disease no one had seen before. By 1960, 121 cases had been reported, including 23 in infants.

**Figure 19-3**
**Bioaccumulation of Toxins in the Food Chain**

This example features fish as the food for human consumption, but bioaccumulation of toxins occurs on land as well when cows, pigs, and chickens eat or drink contaminated foods or water.

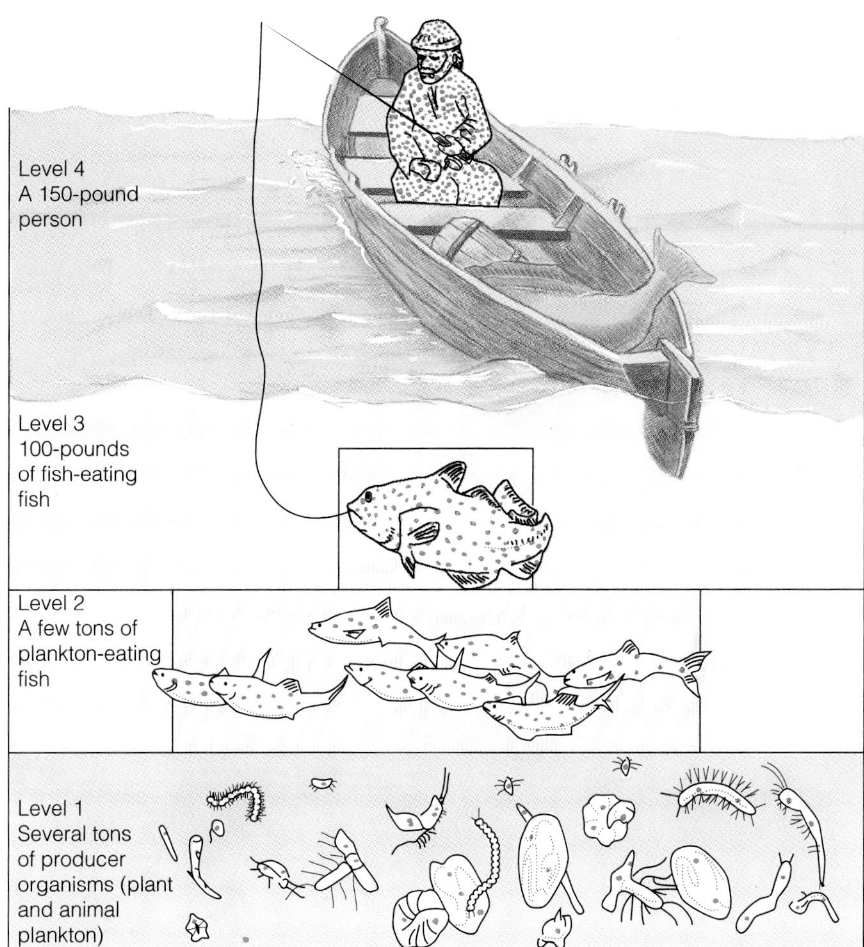


◆ If none of the chemicals are lost along the way, people ultimately receive all of the toxic chemicals that were present in the original plants and plankton.

◆ Contaminants become further concentrated in larger fish that eat the small fish from the lower part of the food chain.

◆ Contaminants become more concentrated in small fish that eat the plants and plankton.

◆ Plants and plankton at the bottom of the food chain become contaminated with toxic chemicals, such as methylmercury (as shown as red dots).

Mortality was high; 46 died, and the survivors suffered blindness, deafness, uncoordination, and intellectual deterioration. The cause was ultimately revealed to be methylmercury contamination of fish from the bay where these people lived. The infants who contracted the disease had not eaten any fish, but their mothers had, and even though the mothers exhibited no symptoms during their pregnancies, the poison had affected their unborn babies. Manufacturing plants in the region were discharging mercury into the waters of the bay, the mercury was turning to methylmercury on leaving the factories, and the fish in the bay were accumulating this poison in their bodies. Some of the affected families had been eating fish from the bay every day.

• ***PBB*** • In 1973, half a ton of polybrominated biphenyl (PBB), a toxic organic compound, was accidentally mixed into some livestock feed that was distributed throughout the state of Michigan. The chemical found its way into millions of animals and then into people who ate the meat. The seriousness of the accident began to come to light when dairy farmers reported their cows were going dry, aborting their calves, and developing abnormal growths on their hooves. Although more than 30,000 cattle, sheep, and swine and more than a million chickens were destroyed, an estimated 97 percent of Michigan's residents had been exposed to PBB. Some of the exposed farm residents suffered nervous system aberrations and liver disorders.

**PBB** (polybrominated biphenyl) and **PCB** (polychlorinated biphenyl): toxic organic compounds used in pesticides, paints, and flame retardants.

**Figure 19-4**

**Contaminants Find Their Way into Foods**

Heavy metals and other contaminants entering the air in smokestack emissions return to the soil in rainfall. Contaminants in the soil are absorbed by plants. People eat either the plants (fruits and vegetables) or the meat from livestock that have eaten the plants. Sewage sludge and pesticides leave residues in the soil; runoff pollutes ground and surface water and contaminates the seafood that people eat.

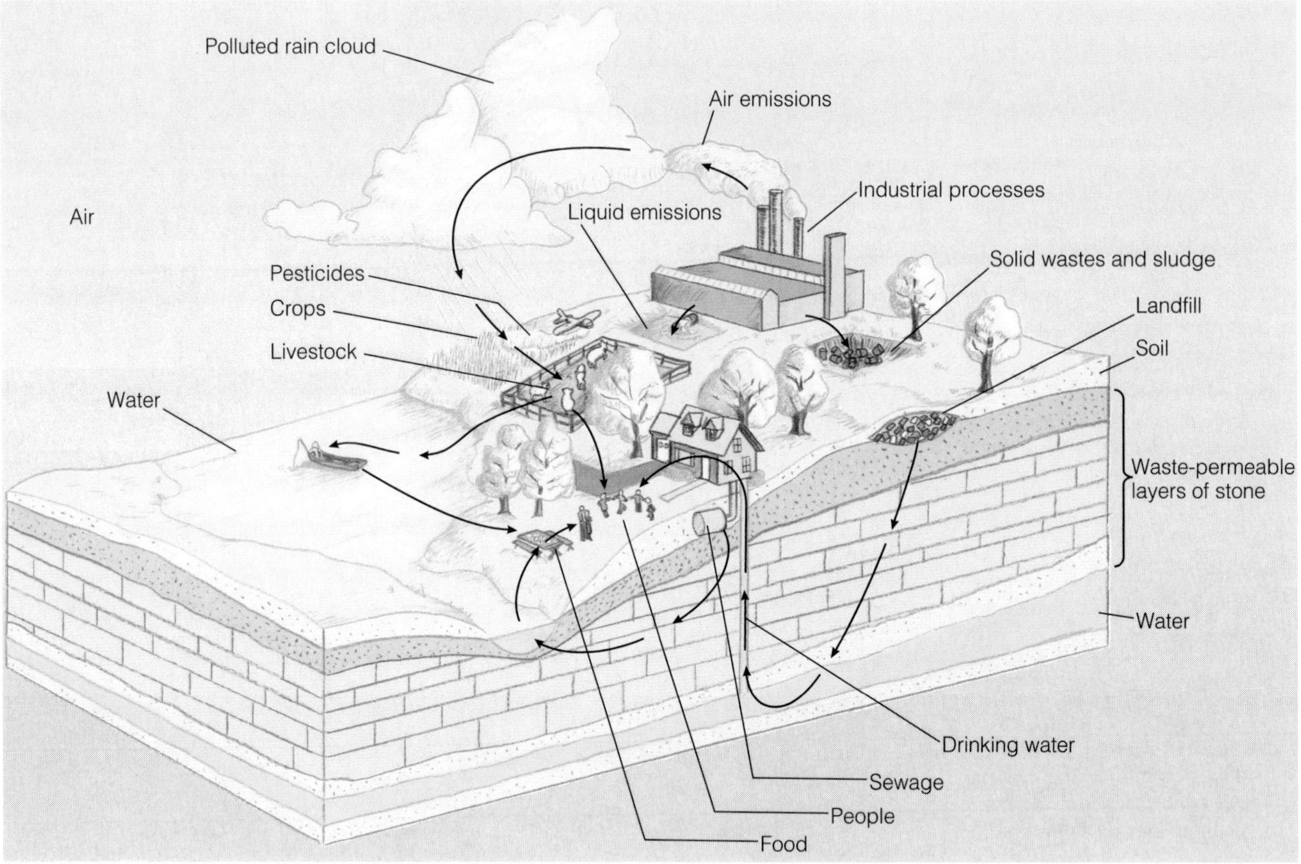

Methylmercury and PCB (polychlorinated biphenyl, a compound similar to PBB) are still found in our environment. Fish harvested from contaminated waters still contain these two pollutants. Hundreds of other contaminants have been identified as well; Highlight 13 focuses on the heavy metal lead and its toxic effects.

IN SUMMARY

Environmental contamination of foods is a concern, but so far, the hazards appear small. In all cases, two principles apply: First, remain alert to the possibility of contamination of foods, and keep an ear open for public health announcements and advice. Second, eat a variety of foods. Varying food choices is an effective defensive strategy against the accumulation of toxins in the body. Each food eaten dilutes contaminants that may be present in other components of the diet.

# Natural Toxicants in Foods

Consumers concerned about food contamination may think that they can eliminate all poisons from their diets by eating only "natural" foods. On the contrary, nature has provided plants with an abundant array of toxicants. A few examples will show how even "natural" foods may contain potentially harmful substances. They also show that while the *potential* for harm exists, *actual* harm rarely occurs.

Poisonous mushrooms are a familiar example of plants that can be harmful when eaten. Few people know, though, that other commonly eaten foods contain substances that can cause illnesses. Cabbage, turnips, mustard greens, and radishes contain small quantities of goitrogens—compounds that can enlarge the thyroid gland. Eating exceptionally large amounts of goitrogen-containing vegetables can aggravate a preexisting thyroid problem, but usually does not initiate one.

Reminder: *Goitrogens* are thyroid antagonists found in such foods as cabbage, kale, brussels sprouts, cauliflower, broccoli, and kohlrabi.

Lima beans and fruit seeds such as apricot pits contain cyanogens—inactive compounds that produce the deadly poison cyanide upon activation by a specific plant enzyme. For this reason, many countries restrict commercially grown lima beans to those varieties with the lowest cyanogen contents. As for fruit seeds, they are seldom deliberately eaten. An occasional swallowed seed or two presents no danger, but a couple of dozen seeds can be fatal to a small child. Perhaps the most infamous cyanogen in seeds is laetrile—a compound erroneously represented as a cancer cure. True, laetrile kills cancer, but only at doses that kill the person, too. Research over the past hundred years has never proved laetrile to be an effective cancer treatment. In fact, laetrile is more dangerous than no treatment at all. The combination of cyanide poisoning and lack of medical attention is life-threatening.

Potatoes contain many natural poisons including solanine, a powerful narcotic-like substance. The small amounts of solanine normally found in potatoes are harmless, but solanine is toxic and presents a hazard when consumed in large quantities. Solanine production increases when potatoes are improperly stored in the light and in either very cold or fairly warm places. Cooking does not destroy solanine, but because most of a potato's solanine is in the green layer that develops just beneath the skin, it can be peeled off, making the potato safe to eat.

**solanine** (SO-lah-neen): a poisonous narcotic-like substance present in potato peels and sprouts. Physical symptoms of solanine poisoning include headache, vomiting, abdominal pain, diarrhea, and fever; neurological symptoms include apathy, restlessness, drowsiness, confusion, stupor, hallucinations, and visual disturbances.

IN SUMMARY

Natural toxicants include the goitrogens in cabbage, cyanogens in lima beans, and solanine in potatoes. These examples of naturally occurring toxicants illustrate two familiar principles. First, any substance can be toxic when consumed in excess. Second, poisons are poisons, whether made by people or by nature. Remember: it is not the source of a chemical that makes it hazardous, but its chemical structure and the quantity consumed.

# Pesticides

The use of pesticides is controversial. They do help to ensure the survival of some crops, but they leave residues in the environment and on some of the foods we eat.

**pesticides:** chemicals used to control insects, diseases, weeds, fungi, and other pests on plants, vegetables, fruits, and animals. Used broadly, the term includes herbicides (to kill weeds), insecticides (to kill insects), and fungicides (to kill fungi).

**residues:** whatever remains. In the case of pesticides, those amounts that remain on or in foods when people buy and use them.

## *Hazards and Regulation of Pesticides*

Ideally, a pesticide would destroy the pest and quickly degenerate to nontoxic products without accumulating in the food chain. Then, by the time consumers ate the food, no harmful residues would remain. Unfortunately, no such perfect pesticide exists. As new pesticides are developed, government agencies assess their risks and benefits and vigilantly monitor their use.

Efforts to manage pests using a combination of natural pesticides and biological controls are presented in Highlight 20.

 U.S. Government
www.healthfinder.gov/searchoptions/topicsaz.htm
Search for Pesticides

• ***Hazards of Pesticides*** • Pesticides may linger on the foods to which they were applied in the field. Health risks from pesticide exposure are probably small for healthy adults, but children, the elderly, and people with compromised immune systems may be vulnerable to some types of pesticide poisoning.[15] To protect infants and children, government agencies set tolerance levels by first identifying foods that children commonly eat in large amounts and then considering the effects of pesticide exposure during each developmental stage.[16]

Whether consumers are ingesting pesticide residues depends on a number of factors. How much of a given food is the consumer eating? What pesticide was used on it? How much was used? How long ago was the food last sprayed? Did environmental conditions promote pest growth or pesticide breakdown? How well was the produce washed? Was it peeled or cooked? With so many factors, consumers cannot know for sure whether pesticide residues remain on foods.

• ***Regulation of Pesticides*** • Consumers depend on the EPA and the FDA to keep pesticide use within safe limits. These agencies evaluate the risks and benefits of a pesticide's use by asking such questions as, How dangerous is it? How much residue is left on the crop? How much harm does the pesticide do to the environment? How necessary is it? What are the alternatives to its use?

**tolerance level:** the maximum amount of a residue permitted in a food when a pesticide is used according to label directions.

If the pesticide is approved, the EPA establishes a tolerance level for its presence in foods, well below that at which it could cause any conceivable harm. Tolerance regulations also state the specific crops to which each pesticide can be applied. If a pesticide is misused, growers risk fines, lawsuits, and destruction of their crops.

Once tolerances are set, the FDA enforces them by monitoring foods and livestock feeds for the presence of pesticides. Over the past several decades of testing, the FDA has seldom found residues above tolerance levels, so it appears that pesticides are generally used according to regulations. Minimal pesticide use means lower costs for growers. In addition to costs, many farmers are concerned about the environment, the quality of their farmland, and a safe food supply. Where violations are found, they are usually due to unusual weather conditions, use of unapproved pesticides, or misuse—for example, application of a particular pesticide to a crop for which it has not been approved.

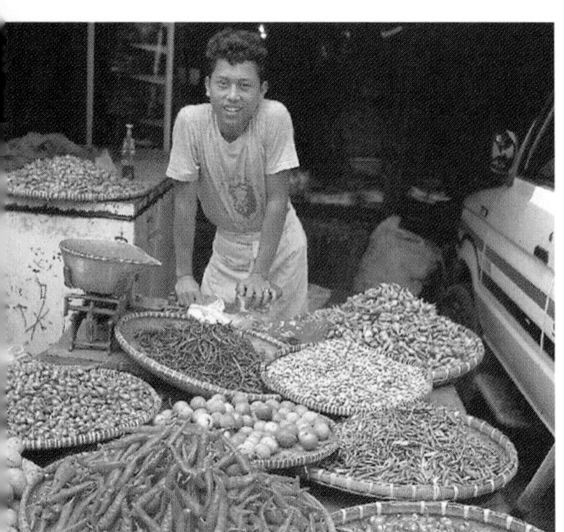

***Foods imported from other countries may contain residues of pesticides that are banned from use here.***

• ***Pesticides from Other Countries*** • Today, approximately 70 percent of the fruits and vegetables consumed in the United States are imported from other countries, which may not have the same pesticide regulations as the United States and Canada.[17] Imported foods may contain both pesticides that have been banned in the United States and permitted pesticides at concentrations higher than are allowed in domestic foods. A loophole in federal regulations allows U.S. companies to manufacture and sell, to other countries, pesticides that are banned in this country. The banned pesticides then return to the United States on imported foods—a circuitous route that concerned consumers have called the "circle of poison." Federal inspectors do monitor imported foods and refuse entry if they are found to contain illegal pesticide residues. The United States, Mexico, and Canada are currently working together to establish a pesticide policy for all of North America.* In addition, plans are being developed to allow the FDA to inspect foreign farms and ban any produce that does not meet U.S. food safety standards.[18]

## Monitoring Pesticides

**certification:** the process in which a private laboratory inspects shipments of a product for selected chemicals and then, if the product is free of violative levels of those chemicals, issues a guarantee to that effect.

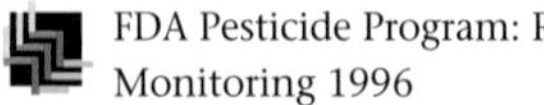

FDA Pesticide Program: Residue Monitoring 1996

vm.cfsan.fda.gov/~dms/pes96rep.html

The FDA collects and analyzes samples of both domestic and imported foods. If the agency finds samples in violation of regulations, it can seize the products or order them destroyed. The FDA may also invoke a certification requirement that forces manufacturers, at their own expense, to have their foods periodically inspected and certified safe by an independent testing agency. Individual states also scan for pesticides (as well as for industrial chemicals) and provide information to the FDA.

• ***Food in the Fields*** • In addition to its ongoing surveillance, the FDA also conducts selective surveys to determine the presence of particular pesticides in specific

*These pesticide agreements are being developed under the auspices of NAFTA (North American Free Trade Agreement).

crops. For example, one year the agency searched for aldicarb in potatoes, captan in cherries, and diaminozide (the chemical name for Alar) in apples, among others. Actions taken that year required several certifications. Thus one shipper in Australia had to certify apples; one in Canada, peppers; one in Costa Rica, chayotes. All grapes from Mexico had to be certified and so did all mangoes from anywhere. This shows, incidentally, how many foods come from abroad—not only those already named, but hundreds more—and that the FDA monitors them as carefully as it does the domestic food supply.

• ***Food on the Plate*** • In addition to monitoring foods in the field for pesticides, the FDA also monitors people's actual intakes. The agency conducts the Total Diet Study (sometimes called the "Market Basket Survey") to estimate the dietary intakes of pesticide residues by eight age and sex groups from infants to senior citizens. Four times a year, FDA surveyors buy over 200 foods from U.S. grocery stores, each time in several cities, prepare the foods table ready, and then analyze them not only for pesticides, but for essential minerals, industrial chemicals, heavy metals, and radioactive materials. In all, the survey reports on over 10,000 samples a year, and recently more than half have been imported foods. Most heavily sampled are fresh vegetables, fruits, and dairy products.

The Total Diet Study provides a direct estimate of the amounts of pesticide residues that remain in foods as they are usually eaten—after they have been washed, peeled, and cooked. The FDA finds the intake of almost all pesticides to be less than 1 percent of the amount considered acceptable.[19] The amount considered acceptable is "the daily intake of a chemical which, if ingested over a lifetime, appears to be without appreciable risk." All in all, these findings corroborate the safety of the U.S. food supply with respect to pesticide residues.

***Washing fresh fruits and vegetables removes most, if not all, of the pesticide residues that might have been present.***

## *Consumer Concerns*

Despite these reassuring reports, consumers still worry that the monitoring of foods may not be adequate. For one thing, manufacturers develop new pesticides all the time. For another, as described earlier, other countries use pesticides that are illegal for use here. For still another, although the regulations may protect U.S. foods adequately, they may not necessarily protect the environment or the people who work in the fields. Concerns over poisoning of soil, waterways, wildlife, and workers may well be valid.

The FDA does not sample *all* food shipments or test for *all* pesticides in each sample. The FDA is a *monitoring* agency, and as such, it cannot, nor can it be expected to, guarantee 100 percent safety in the food supply. Instead, it sets standards so that substances do not become a hazard, checks enough samples to adequately assess average food safety, and acts promptly when problems or suspicions arise.

**safety:** a judgment that considers the risks acceptable.

 U.S. Government

www.healthfinder.gov/searchoptions/topicsaz.htm
Search for Fruits
Search for Vegetables

• ***Minimizing Risks*** • Consumers must assume some responsibility for their own health and safety with respect to pesticides. They can learn about the potential benefits and possible dangers of pesticide use, discuss regulations and alternatives with others, advise their government representatives of their findings, and apply pressure wherever it will help change procedures. Meanwhile, people can minimize their risks by following the guidelines offered in the "How to" on p. 614.

In addition to the suggestions mentioned, consumers can buy fresh foods grown locally, especially when they can confirm that produce has been grown using responsible methods. Consumers who want pesticide-free produce shouldn't look for "perfect" fruits and vegetables; pesticide-free produce may have a few blemishes, but minor blemishes are not a hazard. It is also important to buy a variety of foods and not to rely too heavily on any one food. The food supply is protected well enough that consumers who take these precautions can feel secure that the foods they eat are safe.

## HOW TO Prepare Foods to Minimize Pesticide Residues

To remove or reduce any pesticide residues from foods:

- Trim the fat from meat, and remove the skin from poultry and fish; discard fats and oils in broths and pan drippings. (Pesticide residues concentrate in the animal's fat.)
- Wash fresh produce in warm water. Use a scrub brush, and rinse thoroughly.
- Use a knife to peel an orange or grapefruit; do not bite into the peel.
- Discard the outer leaves of leafy vegetables such as cabbage and lettuce.
- Peel waxed fruits and vegetables; waxes don't wash off and can seal in pesticide residues.
- Peel vegetables such as carrots and fruits such as apples when appropriate. (Peeling removes pesticides that remain in or on the peel, but also removes fibers, vitamins, and minerals.)

Information is available from the EPA's National Pesticide Hotline (800-858-PEST) anytime day or night, 365 days a year.

*Many consumers are willing to pay a little more for pesticide-free produce.*

**organically grown crops:** crops grown and processed according to USDA regulations defining the use of fertilizers, herbicides, insecticides, fungicides, preservatives, and other chemical ingredients.

National Organic Food Standards
www.ams.usda.gov/nop

• ***Alternatives to Pesticides*** • To feed a nation while employing fewer pesticides requires creative farming methods. Highlight 20 describes such methods, recommended by the National Academy of Sciences as part of a system known as alternative, or sustainable, agriculture. This system depends on crop rotation and the use of plants that produce natural pesticides. Among natural pesticides are the nicotine in tobacco and psoralens in celery. Natural pesticides are less damaging to other living things and less persistent in the environment than most human-made ones.

Other alternatives to heavy pesticide use include releasing organisms into fields to destroy pests and planting nonfood crops nearby to kill pests or attract them away from the food crops. For example, releasing sterile male fruit flies into orchards helps to curb the population growth of these pests; some flowers, such as marigolds, release natural insecticides and are often planted near crops such as tomatoes. Such alternative farming methods are more labor-intensive and may produce smaller yields than conventional methods, at least initially. Over time, though, by eliminating expensive pesticides, fertilizers, and fuels, these alternatives may actually cut costs more than they cut yields.

• ***Organically Grown Crops*** • Alternative methods may be especially useful if farmers want to produce and market organically grown crops: crops grown and processed according to USDA regulations defining the use of fertilizers, herbicides, insecticides, fungicides, preservatives, and other chemical ingredients.[20] These regulations are currently being developed and are expected to be in place by the spring of 1999. To certify and market products as organic, farmers must use methods that meet specified criteria; in addition, producers may not claim products are organic if they have been irradiated, genetically engineered, or grown with fertilizer made from sewer sludge. Most states have organic certification agencies, although exact guidelines vary from state to state. Agricultural products claiming to have been grown organically must meet the USDA standards and bear the USDA seal on their labels.

In addition to benefiting from reduced costs for farming, increased soil quality, and decreased chemical impact on the environment, farmers growing organic crops stand to increase their share of the market. Many consumers are willing to pay more for organic foods, and sales are increasing by record numbers.

Implied in the marketing of *organic foods* is that organic products are healthier for consumers than those grown using other methods, which may not be the case. Using unprocessed animal manure as an organic fertilizer, for example, may transmit bacteria, such as *E. coli,* to human beings. Both organic and conventional methods have advantages and disadvantages, and consumers must remain informed.

IN SUMMARY

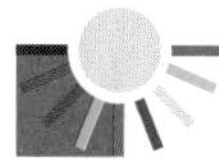

Pesticides can safely improve crop yields when used according to regulations, but can also be hazardous when used inappropriately. The FDA tests both domestic and imported foods for pesticide residues in the fields and in market basket surveys of foods prepared table ready. Consumers can minimize their ingestion of pesticide residues on foods by following the suggestions in the "How to" on p. 614. Alternative farming methods may allow farmers to grow crops with few or no pesticides.

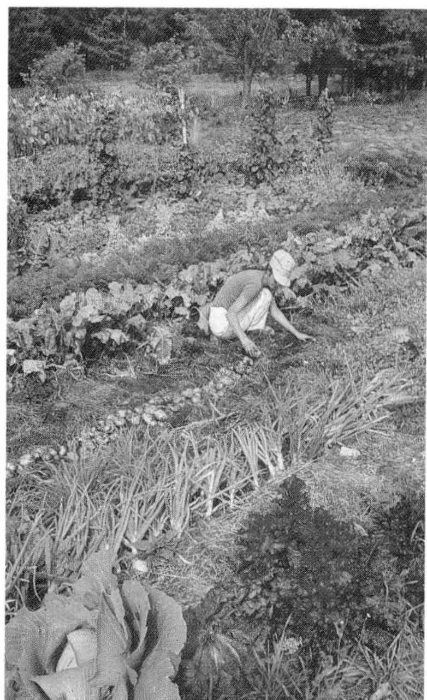

***People can grow pesticide-free crops when their gardens or farms are relatively small.***

# Food Additives

Additives confer many benefits on foods. Some reduce the risk of food-borne illness (for example, nitrites used in curing meat prevent poisoning from the botulinum toxin). Others enhance nutrient quality (as in vitamin D–fortified milk). Most additives help prevent spoilage during the time it takes to deliver foods long distances to grocery stores and then to kitchens. Some additives simply make foods look and taste good.

Intentional additives are put into foods on purpose, while indirect additives may get in unintentionally before or during processing. This discussion begins with the regulations that govern additives, then presents intentional additives class by class, and finally goes on to say a word about the indirect additives.

**additives:** substances not normally consumed as foods but added to food either intentionally or by accident.

**intentional additives:** additives intentionally added to foods, such as nutrients, colors, and preservatives.

**indirect additives:** substances that can get into food as a result of contact with foods during growing, processing, packaging, storing, cooking, or some other stage before the foods are consumed; also called **incidental** or **accidental additives.**

## *Regulations Governing Additives*

The FDA's concern with additives hinges primarily on their safety. To receive permission to use a new additive in food products, a manufacturer must satisfy the FDA that the additive is:

- Effective (it does what it is supposed to do).
- Detectable and measurable in the final food product.
- Safe (when fed in large doses to animals under strictly controlled conditions, it causes no cancer, birth defects, or other injury).

On approving an additive's use, the FDA writes a regulation stating in what amounts and in what foods the additive may be used. No additive receives permanent approval; all must undergo periodic review.

• ***The GRAS List*** • Many familiar substances are exempted from complying with FDA's approval procedure because they are generally recognized as safe (GRAS), based either on their extensive, long-term use in foods or on current scientific evidence. Several hundred substances are on the GRAS list, including such items as salt, sugar, caffeine, and herbs. Whenever substantial scientific evidence or public outcry has questioned the safety of any substance on the GRAS list, it has been reevaluated. All substances about which any legitimate question has been raised have been removed or reclassified. Meanwhile, the entire GRAS list is subjected to ongoing review.

**GRAS (generally recognized as safe) list:** a list, first established by the FDA in 1958, of food additives that had long been in use and were believed safe. The list is subject to revision as new facts become known.

• ***The Delaney Clause*** • One risk that the U.S. law on additives refuses to tolerate at any level is the risk of cancer. To remain on the GRAS list, an additive must not have been found to be a carcinogen in any test on animals or human beings. The Delaney Clause (the part of the law that states this criterion) is uncompromising in addressing carcinogens in foods and drugs; in fact, it has been under fire for many years for being too strict and inflexible.

**Delaney Clause:** a clause in the Food Additive Amendment to the Food, Drug, and Cosmetic Act that states that no substance that is known to cause cancer in animals or human beings at any dose level shall be added to foods.

Reminder: A *carcinogen* is a substance that initiates cancer development.

***Without additives, bread would quickly get moldy and salad dressing would go rancid.***

For perspective, one part per trillion is equivalent to about one grain of sugar in an Olympic-sized swimming pool; or 1 second in 32,000 years; or one hair on 10 million heads, assuming none are bald.

The *de minimis* rule defines risk as a cancer rate of less than one cancer per million people exposed to a contaminant over a 70-year lifetime.

**margin of safety:** when speaking of food additives, a zone between the concentration normally used and that at which a hazard exists. For common table salt, for example, the margin of safety is $\frac{1}{5}$ (five times the amount normally used would be hazardous).

The Delaney Clause states that "no additive shall be deemed to be safe if it is found to induce cancer when ingested [at any level] by man or animal." That sounds clear enough, yet some products that fail to meet that standard still remain on the market. The artificial sweetener saccharin was the first exception to the rule. In the 1970s, when the FDA tried to ban saccharin because tests had revealed that it caused cancer in animals, consumers raised an outcry asking that it still be allowed in foods. In an attempt to balance the Delaney Clause with current food safety and cancer knowledge, Congress created a special exception that allowed saccharin to remain on the market as long as products containing it carried a warning.

The Delaney Clause is best understood as a product of a different historical era. It was adopted more than 40 years ago at a time when scientists knew little about the relationships between carcinogens and cancer development. At that time, most substances were detectable in foods only in relatively large amounts, such as parts per thousand. Today, scientific understanding of cancer has progressed, and technology has advanced so that carcinogens in foods can be detected even when they are present only in parts per billion or even per trillion. Earlier, "zero risk" may have seemed attainable, but today we know it is not: all substances, no matter how pure, can be shown to be contaminated at some level with one carcinogen or another. For these reasons, the FDA prefers to deem additives (and pesticides and other contaminants) safe if lifetime use presents no more than a one-in-a-million risk of cancer to human beings. Thus, instead of the "zero-risk" policy of the Delaney Clause, the FDA uses a "negligible-risk" standard, sometimes referred to as a *de minimis* rule.

• ***Margin of Safety*** • Whatever risk level is permitted, actual risks must be determined by experiments. To determine risks posed by an additive, researchers feed test animals the additive at several concentrations throughout their lives. The additive is then permitted in foods in amounts 100 times below the lowest level that is found to cause any harmful effect, that is, at a $\frac{1}{100}$ margin of safety. In many foods, *naturally* occurring substances occur with narrower margins of safety, such as $\frac{1}{10}$. Even nutrients pose risks at dose levels above those normally consumed: the margin of safety for vitamins A and D is $\frac{1}{25}$ to $\frac{1}{40}$ for adults and may be less than $\frac{1}{10}$ for infants. For some trace minerals, the margin of safety is about $\frac{1}{5}$. People consume common table salt daily in amounts only three to five times less than those that pose a hazard.

• ***Risks versus Benefits*** • Of course, additives would not be added to foods if they only presented risks. Additives are in foods because they offer benefits that outweigh the risks they present, or make the risks worth taking. In the case of color additives that only enhance the appearance of foods but do not improve their health value or safety, no amount of risk may be deemed worth taking. In contrast, the FDA finds that it is worth taking the small risks associated with the use of nitrites on meat products, for example, because nitrites inhibit the formation of the deadly botulinum toxin. The choice involves a compromise between the risks of using additives and the risks of doing without them.

It is the manufacturers' responsibility to use only the amounts of additives that are necessary to get the needed effect, and no more. The FDA also requires that additives *not* be used:

- To disguise faulty or inferior products.
- To deceive the consumer.
- Where they significantly destroy nutrients.
- Where their effects can be achieved by economical, sound manufacturing processes.

## *Intentional Food Additives*

Intentional food additives are added to foods to give them some desirable characteristic: resistance to spoilage, color, flavor, texture, stability, or nutritional value.

The accompanying glossary defines the categories of additives, and the next sections describe additives people most often ask about.

• ***Antimicrobial Agents*** • Foods can go bad in two ways. One way is relatively harmless: by losing their flavor and attractiveness. (Additives to prevent this kind of spoilage include antioxidants, discussed later.) The other way is by becoming contaminated with microbes that cause food-borne illnesses, a hazard that justifies the use of antimicrobial agents.

The most widely used antimicrobial agents are ordinary salt and sugar. Salt has been used throughout history to preserve meat and fish; sugar serves the same purpose in canned and frozen fruits and in jams and jellies. Both exert their protective effect primarily by capturing water and making it unavailable to microbes. Other additives, such as potassium sorbate and sodium propionate, are used to extend the shelf life of baked goods, cheeses, beverages, mayonnaise, margarine, and other products.

Other antimicrobial agents, the nitrites and nitrates, are added to foods for three main purposes: to preserve color, especially the pink color of hot dogs and other cured meats; to enhance flavor by inhibiting rancidity, especially in cured meats and poultry; and to protect against bacterial growth. In amounts smaller than those needed to confer color, nitrites prevent the growth of the bacteria that produce the deadly botulinum toxin.

Nitrites clearly serve a useful purpose, but their use has been controversial. In the human body, nitrites can be converted to nitrosamines. At nitrite levels higher than those used in food products, nitrosamine formation causes cancer in animals. The food industry uses the minimal amount of nitrites necessary to achieve results, and nitrosamine formation has not been shown to cause cancer in human beings.

Detectable amounts of nitrosamine-related compounds are found in malt beverages (beer) and cured meats (primarily, bacon). Yet even the quantities found in beer and bacon hardly make a difference in a person's overall exposure to nitrosamine-related compounds. An average cigarette smoker inhales 100 times the nitrosamines that the average bacon eater ingests. A beer drinker ingests twice as much as the bacon eater, but even so, exposure from new car interiors and cosmetics is higher than this.

• ***Antioxidants*** • Another way food can go bad is by undergoing changes in color and flavor caused by exposure to oxygen (oxidation). Oftentimes, these changes involve no hazard to health, but they damage the food's appearance, flavor, and

Common examples of antimicrobial additives:
- Salt.
- Sugar.
- Nitrites and nitrates (such as sodium nitrate).

***Both salt and sugar act as preservatives by withdrawing water from food; microbes cannot grow without water.***

**nitrites** (NYE-trites): salts added to food to prevent botulism; one example is sodium nitrite, which is used to preserve meats.

**nitrates** (NYE-trates): salts that are converted to nitrites by bacteria.

**nitrosamines** (nigh-TROHS-uh-meens): derivatives of nitrites that may be formed in the stomach when nitrites combine with amines; nitrosamines are carcinogenic in animals.

## Glossary of Intentional Food Additives

**antimicrobial agents:** preservatives that prevent microorganisms from growing.

**antioxidants:** preservatives that prevent rancidity of fats in foods and other damage to food caused by oxygen.

**artificial colors:** certified food colors added to enhance appearance. (*Certified* means approved by the FDA.)

**artificial flavors, flavor enhancers:** chemicals that mimic natural flavors and those that enhance flavor.

**bleaching agents:** substances used to whiten foods such as flour and cheese.

**chelating** (KEY-late-ing) **agents:** molecules that bind other molecules. As additives, they prevent discoloration, flavor changes, and rancidity that might occur during processing.

**nutrient additives:** vitamins and minerals added to improve nutritive value.

**preservatives:** antimicrobial agents, antioxidants, and other additives that retard spoilage or maintain desired qualities, such as softness in baked goods.

**thickening** and **stabilizing agents:** ingredients that maintain emulsions, foams, or suspensions or lend a desirable thick consistency to foods.

Common examples of antioxidant additives:
- Vitamin C (erythorbic acid, sodium ascorbate).
- Vitamin E (tocopherol).
- Sulfites.
- BHA and BHT.

**sulfites:** salts containing sulfur that are added to foods to prevent spoilage.

Sulfites appear on food labels as:
- Sulfur dioxide.
- Sodium sulfite.
- Sodium bisulfite.
- Potassium bisulfite.
- Sodium metabisulfite.
- Potassium metabisulfite.

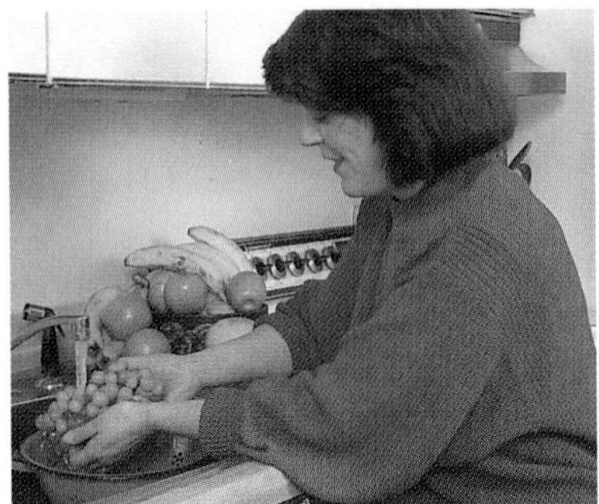

***Raw grapes may legally be treated with sulfites, so wash grapes thoroughly before eating them.***

**BHA** and **BHT:** preservatives commonly used to slow the development of off-flavors, odors, and color changes caused by oxidation.

Common examples of color additives:
- Carotenoids.
- Blue #1 and #2 (brilliant blue and indigotine).
- Green #3 (fast green).
- Red #40 and #3 (allura red and erythrosine).
- Yellow #5 and #6 (tartrazine and sunset yellow).

**MSG symptom complex:** an acute, temporary intolerance reaction that may occur after the ingestion of the additive MSG (monosodium glutamate). Symptoms include burning sensations, chest and facial flushing and pain, and throbbing headaches.

Common examples of nutrient additives:
- Thiamin, niacin, riboflavin, folate, and iron in grain products.
- Iodine in salt.
- Vitamin A and D in milk.
- Vitamin C in fruit drinks.

nutritional quality. Oxidation is easy to detect when sliced apples or potatoes turn brown or when oil goes rancid. Antioxidants prevent these reactions. Among the antioxidants approved for use in foods are vitamin C (ascorbate) and vitamin E (tocopherol).

Another group of antioxidants, the sulfites, cost less than the vitamins. Sulfites prevent oxidation in many processed foods, alcoholic beverages (especially wine), and drugs. Because some people experience adverse reactions, the FDA prohibits sulfite use on foods intended to be consumed raw, with the exception of grapes, and requires foods and drugs that contain sulfite additives to declare it on their labels. Restaurants, however, are not required to disclose whether they have used sulfites in food preparation. Therefore, concerned consumers must ask if sulfites have been used.[21] For most people, sulfites pose no hazard in the amounts used in products, but there is one more consideration: sulfites destroy thiamin. For this reason, the FDA prohibits their use in foods that are important sources of the vitamin, such as enriched grain products.

Two other antioxidants in wide use are BHA and BHT, which prevent rancidity in baked goods and snack foods.* Several tests have shown that animals fed large amounts of BHT develop *less* cancer when exposed to carcinogens and live *longer* than controls. Apparently, BHT protects against cancer through its antioxidant effect, which is similar to that of the antioxidant vitamins. The amount of BHT ingested daily from the U.S. diet, however, contributes little to the body's antioxidant defense system. A caution: at intake levels higher than those that protect against cancer, BHT has *produced* cancer. Vitamins E and C remain the most important dietary antioxidants to strengthen defenses against cancer.

• ***Artificial Colors*** • Only a few artificial colors remain on the FDA's list of color additives approved for use in foods—a highly select group that has survived considerable testing. Artificial colors are among the most intensively investigated of all additives. In fact, coloring agents are much better known than the natural pigments of plants, and the safety of their use can be stated with greater certainty. Examples of natural pigments commonly used by the food industry are the caramel that tints cola beverages and baked goods and the carotenoids that color margarine, cheeses, and pastas.

• ***Artificial Flavors and Flavor Enhancers*** • Flavoring agents are the largest single group of food additives. One of the best-known members of this group is monosodium glutamate, or MSG—a sodium salt of the amino acid glutamic acid. MSG is used widely in a number of foods, especially Asian foods, as a flavor enhancer. Besides enhancing the well-known sweet, salty, bitter, and sour tastes, MSG itself may possess a pleasant flavor.

In a few sensitive individuals, MSG produces adverse reactions known as the MSG symptom complex.[22] Except for this 1 to 2 percent of the population, MSG is considered safe for all adults. It is not allowed in foods designed for infants, however. Food labels require ingredient lists to itemize all additives, including MSG.

• ***Texture and Stability*** • Ingredients may be added to foods during processing to maintain emulsions, foams, or suspensions or to lend a desirable thick consistency to foods. Dextrins (short chains of glucose formed as a breakdown product of starch), starch, and pectin are examples. Gums, such as carrageenan, guar, locust bean, agar, and gum arabic, are also added to thicken and stabilize food products.

• ***Nutrient Additives*** • As mentioned earlier, manufacturers sometimes add nutrients to fortify or maintain the nutritional quality of foods.[23] Included among nutrient additives are the five nutrients added to refined grains; the iodine added to salt; the vitamins A and D added to milk; and the nutrients added to fortified breakfast cereals. A nutrient-poor food with nutrients added may appear to be nutrient-rich, but it is rich only in those nutrients chosen for addition, and the

*BHA is butylated hydroxyanisole; BHT is butylated hydroxytoluene.

absorption of these nutrients may be poor. Appropriate uses of nutrient additives are to:

- Correct dietary deficiencies known to result in diseases.
- Restore nutrients to levels found in the food before storage, handling, and processing.
- Balance the vitamin, mineral, and protein contents of a food in proportion to the energy content.
- Correct nutritional inferiority in a food that replaces a more nutritious traditional food.

Nutrients are sometimes also added for other purposes. Vitamins C and E were already mentioned for their antioxidant properties; beta-carotene (a vitamin A precursor) and other carotenoids are sometimes used for color.

***Color additives not only make foods attractive, but identify flavors as well. Everyone agrees that yellow jellybeans should taste lemony and black ones like licorice.***

## *Indirect Food Additives*

Indirect or incidental additives are substances that find their way into foods during harvesting, production, processing, storage, or packaging. For example, incidental additives include tiny bits of plastic, glass, paper, tin, and other substances from packages as well as chemicals from processing, such as the solvent used to decaffeinate coffee. The following paragraphs discuss five different types of indirect additives that sometimes make headline news.

• ***Microwave Packaging*** • When the FDA regulations were established in the 1950s, the writers did not foresee the use of microwave ovens. Consequently, they did not specify temperatures at which packages should be tested to determine if incidental additives migrated to foods at high temperatures. Some microwave products are sold in "active packaging" that helps to cook the food; for example, pizzas are often heated on a metalized film laminated to paperboard. This film absorbs the microwave energy in the oven and reaches temperatures as high as 500°F. At such temperatures, packaging components migrate into the food. For this reason, food packagers must perform specific tests to discover whether materials are migrating into foods; if they are, their safety must be confirmed by strict procedures similar to those governing intentional additives.

Most microwave products are sold in "passive packaging" that is transparent to microwaves and simply holds the food as it cooks. These containers don't get much hotter than the foods, but materials still migrate at high temperatures. Migration from packages may turn out to be harmless, but until more is known, consumers are advised to use only glass or ceramic containers designed for use in microwave ovens and to avoid reusing disposable containers such as margarine tubs.

Quick test for using glass containers in a microwave oven: Microwave the empty container for 1 min.

- If it's warm, it's unsafe for the microwave.
- If it's lukewarm, it's safe for short-term reheating in the microwave.
- If it's cool, it's safe for long-term cooking in the microwave.

• ***Dioxins*** • Coffee filters, milk cartons, paper plates, and frozen food packages, if made from bleached paper, can contaminate foods with minute quantities of dioxins—compounds formed during chlorine treatment of wood pulp during paper manufacture. Dioxin contamination of foods from such products appears only in trace quantities—in the parts-per-trillion range (recall, for perspective, that one part per trillion is equal to 1 second in 32,000 years). Such levels appear to present no health risks to people, but scientists recognize that dioxins are extremely toxic, and they are known to cause cancer in animals. Accordingly, the paper industry has reduced its use of chlorine to cut dioxin exposure; in the meantime, the FDA has concluded that drinking milk from bleached-paper cartons presents no health hazard.

**dioxins** (dye-OK-sins): a class of chemical pollutants created as by-products of chemical manufacturing, incineration, chlorine bleaching of paper pulp, and other industrial processes. Dioxins persist in the environment and accumulate in the food chain.

• ***Decaffeinated Coffee*** • Many consumers have tried to eliminate caffeine from their diets by selecting decaffeinated coffee. Is decaffeinated coffee a safe alternative?

To remove caffeine from coffee beans, manufacturers often use methylene chloride in a process that leaves traces of the chemical in the final product. The FDA estimates that the average cup of coffee treated this way contains about 0.1 part per million of methylene chloride, which seems to pose no significant threat. A person drinking 2½ cups of decaffeinated coffee containing 100 times as much methylene chloride every day for a lifetime has a one-in-a-million chance of developing cancer from it. People are exposed to much more methylene chloride from other sources such as hair sprays and paint stripping solutions. Still, some consumers prefer either to return to caffeine or to select coffee decaffeinated in another way, perhaps by steam. Unfortunately, manufacturers are not required to state on their labels the type of decaffeination process used in their products. Many labels provide consumer-information telephone numbers for those who have such questions.

• ***Hormones*** • Hormones are a unique type of incidental additive in that their use is intentional, but their presence in the final food product is not. The FDA has approved about a dozen hormones for use in food-producing animals, and the USDA has established limits for residues allowed in meat products.

**bovine growth hormone (BGH):** a hormone produced naturally in the pituitary gland of a cow that promotes growth and milk production; now produced for agricultural use by bacteria.
- **bovine** = of cattle

Some ranchers in the United States treat young calves with bovine growth hormone (BGH). Hormone-treated animals produce leaner meats, and dairy cows produce more milk. All cows make BGH naturally. Now scientists can stimulate bacteria to produce BGH, which allows laboratories to harvest huge quantities of the hormone and sell it to farmers as a drug.[24] This practice has some consumers concerned that an "artificial" drug is being given to cows that will be used for meat and milk.

Indeed, traces of BGH do remain in the meat and milk of both hormone-treated and untreated cows. BGH residues have not been tested for safety in human beings because residues of the natural hormone have always been present in milk and meat and the amount found in treated cows is within the range that can occur naturally. Furthermore, BGH, being a peptide hormone, is denatured by the heat used in processing milk and cooking meat and is also digested by enzymes in the GI tract. If any BGH were to enter the bloodstream, it would have no effect because the chemical structures of animal growth hormones differ from those in human beings; therefore BGH does not stimulate receptors for *human* growth hormone. In a report from its Technology Assessment Conference, the National Institutes of Health concluded: "As currently used in the United States, meat and milk from [hormone] treated cows are as safe as those from untreated cows."[25]

• ***Antibiotics*** • Like hormones, antibiotics are also intentionally given to livestock, and residues may remain in the meats and milks. Consequently, individuals with sensitivity to antibiotics may suffer allergic reactions. To minimize drug residues in foods, the FDA requires a specified time between the time of medication and the time of slaughter to allow for drug metabolism and excretion.

Of greater concern to the public's health is the development of antibiotic resistance. The appropriate use of antibiotics to prevent and control infections in livestock may be beneficial for the animals and food production in general, but indiscriminate use can be catastrophic to the treatment of disease in human beings.[26] Bacteria develop resistance, making the antibiotics less and less effective in their fight against infection.

U.S. Government
www.healthfinder.gov/searchoptions/topicsaz.htm
Search for Food additives

IN SUMMARY

On the whole, the benefits of food additives seem to justify the risks associated with their use. The FDA regulates the use of the following intentional additives: antimicrobial agents (such as nitrites) to prevent microbial spoilage; antioxidants (such as vitamins C and E, sulfites, and BHA and BHT) to prevent oxidative changes; colors (such as tartrazine) and flavor enhancers (such as MSG) to appeal to senses; and nutrients (such as iodine in salt) to enrich or fortify foods. Incidental additives sometimes get into foods during processing, but rarely present a hazard, although some processes such as decaffeinating coffee and treating livestock with hormones and antibiotics raise consumer concerns.

# Food Biotechnology

Hope for the future purity of foods comes on the crest of new advances in biotechnology. Biotechnology promises to produce greater crop yields, leaner meats, longer shelf lives, better nutrient composition, and fewer pesticides. Overall, biotechnology offers opportunities to enhance the quality, nutritional value, and variety of foods.[27]

**biotechnology:** the use of biological systems or organisms to create or modify products; also called **biogenetic engineering.**

Biotechnology Information Center
www.nal.usda.gov/bic

## Genetic Engineering

For centuries farmers have manipulated the genetics of plants and animals to shape the characteristics of their crops and livestock. Consider corn, for example. Wild, native corn bears only two or three kernels on a cob, but many years of patient selective breeding have produced the large, full, sweet ears people enjoy today, and many types of wild corn are now all but extinct. Half of the increases in U.S. crop yields in this century are due to such genetic improvements; the use of irrigation, fertilizers, and pesticides has also contributed. Farmers still use selective breeding to provide consumers with low-fat meats, high-yield grains, and a seemingly endless variety of fruits and vegetables.

*Today's large, full, sweet ears bear little resemblance to the original wild, native corn with its sparse two or three kernels to a cob.*

Scientists can now speed up the process of genetic change through biotechnology. Farmers need no longer wait patiently for breeding to yield improved crops and animals, nor must they even respect natural lines of reproduction among species. Laboratory scientists can now select desirable traits from any of a number of species and insert those traits into the genetic material of crops and animals.

So far, most of the traits that have been selected for genetic engineering help to produce foods more efficiently. For example, the enzyme rennin, which is essential for making cheese, was previously harvested from the stomachs of calves, a costly process. Now scientists can convey into bacteria the genetic material to mass-produce rennin.

**rennin:** an enzyme that coagulates milk; found in the gastric juice of cows, but not human beings.

Among the new products of biotechnology are tomatoes that stay fresh much longer than others and so promise less waste and higher profits. Normally, tomatoes produce a protein that softens them after they have been picked. Scientists introduce into a tomato plant a gene that is a mirror image of the one that codes for the "softening" enzyme. This gene fastens itself to the RNA of the native gene and blocks its action. A vine-ripe tomato with this special gene rots more slowly than a normal tomato, allowing growers to harvest at the most flavorful and nutritious red stage. The tomatoes will still last much longer during shipping and marketing than regular tomatoes harvested when green.

Similarly, soybeans may be implanted with a gene that will upgrade soy protein to a quality approaching that of milk. Corn may be modified to contain lysine and tryptophan, its two limiting amino acids.[28] Fats and oils with a predetermined fatty acid composition may be possible within the decade.[29] Crops that produce their own insecticides upon receiving genes from bacteria may render pesticides unnecessary. Shrimp may soon fight diseases with genetic ammunition borrowed from sea urchins. Livestock already receive growth-promoting hormones from bacteria, as mentioned earlier. The possibilities seem unlimited, and though they sound fantastic, many are waiting on laboratory shelves for the time when they will be fully employed in agriculture.

While food industrialists hail biotechnology as a miracle, some other people fear that tampering with genetics may change organisms in ways not yet fully understood, even by the scientists who developed the techniques. They wonder what unknown changes take place when the genes of living things are manipulated and what the long-term consequences might be.

One of the most exciting and fearful areas in genetic research today is the cloning of animals. Scientists have recently developed a technique for cloning

*Tomatoes grown from genetically modified seeds stay fresh longer.*

cows that may allow the development of pharmaceuticals for human use—a process known as "pharming." Not only can these cloned cows produce drugs, but one day they may be able to deliver these drugs through their milk. This process is still being developed and is under review by the FDA.

## Regulations and Labeling

The FDA has taken the position that foods produced through biotechnology are not substantially different from others and require no special safety testing or labeling.[30] A product, such as the tomato described earlier, need not be tested since its new genes *prevent* synthesis of a protein and add nothing but a tiny fragment of genetic material. On the other hand, any substances introduced into a food by way of bioengineering must meet the same safety standards applied to all additives.[31] A tomato plant with a gene that, for example, produces an insecticide cannot be marketed until it proves safe for consumption. Foods produced via biotechnology are not required to be labeled as such unless they pose known problems, such as allergy, to some people.[32] Some people object to genetic tampering and want labels to help them identify "old-fashioned" tomatoes. They may not realize that most foods available today have already been altered genetically by selective breeding. The new vegetable broccoflower, a product of sophisticated cross-breeding of broccoli with cauliflower, met no testing or approval barriers on its way to the dinner plate. Only after the vegetable became popular with consumers did scientists study its nutrient contents (see Appendix H for their findings).

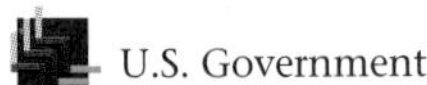
U.S. Government
www.healthfinder.gov/searchoptions/topicsaz.htm
Search for Biotechnology

IN SUMMARY

Biotechnology manipulates the genetics of plants and animals to produce greater yields, leaner meats, longer shelf lives, and better nutrient composition. Scientists are continuing to study the effects not only of biotechnology but of all sorts of new food processing techniques. Their efforts to enhance food production will help meet the challenge of feeding an ever increasing world population.[33]

As this chapter said at the start, supplying food safely to hundreds of millions of people is an incredible challenge—one that gets met, for the most part, with incredible efficiency. The following chapter describes a contrasting situation—that of the food supply not reaching the people.

## Study Questions

These questions will help you review the chapter. You will find the answers in the discussions on the pages provided.

1. To what extent does food poisoning present a real hazard to U.S. consumers eating U.S. foods? How often does it occur? (p. 599)
2. Distinguish between the two types of food-borne illnesses and provide an example of each. Describe measures that help prevent food-borne illnesses. (pp. 599–600, 604–605)
3. What special precautions apply to meats? To seafood? (pp. 602–603)
4. What is meant by a "persistent" contaminant of foods? Describe how contaminants get into foods and build up in the food chain. (pp. 608–609)
5. What dangers do natural toxicants present? (pp. 610–611)
6. How do pesticides become a hazard to the food supply, and how are they monitored? In what ways can people reduce the concentrations of pesticides in and on foods that they prepare? (pp. 611–614)
7. What is the difference between a GRAS substance and a regulated food additive? Give examples of each. Name and describe the different classes of additives. (pp. 615–620)
8. Provide examples of how biotechnology enhances food production. (pp. 621–622)

These questions will help you prepare for an exam. Answers can be found in Appendix K.

1. Eating a contaminated food such as undercooked poultry or unpasteurized milk might cause a:

a. food allergy.
b. food infection.
c. food intoxication.
d. botulinum reaction.

2. The temperature danger zone for foods ranges from:
   a. −20°F to 120°F.
   b. 0°F to 100°F.
   c. 20°F to 120°F.
   d. 40°F to 140°F.
3. Examples of foods that frequently cause food-borne illness are:
   a. canned foods.
   b. steaming-hot foods.
   c. fresh fruits and vegetables.
   d. raw milk, seafood, meat, and eggs.
4. Irradiation can help improve our food supply by:
   a. cooking foods quickly.
   b. killing microorganisms.
   c. minimizing the use of preservatives.
   d. improving the nutrient content of foods.
5. Solanine is an example of a(n):
   a. heavy metal.
   b. artificial color.
   c. natural toxicant.
   d. animal hormone.
6. The standard that deems additives safe if lifetime use presents no more than a one-in-a-million risk of cancer is known as the:
   a. Delaney Clause.
   b. zero-risk policy.
   c. GRAS list of standards.
   d. negligible-risk policy.
7. Common antimicrobial additives include:
   a. salt and nitrites.
   b. carrageenan and MSG.
   c. dioxins and sulfites.
   d. vitamin C and vitamin E.
8. Common antioxidants include:
   a. BHA and BHT.
   b. tartrazine and MSG.
   c. sugar and vitamin E.
   d. nitrosamines and salt.
9. Incidental additives that may enter foods during processing include:
   a. dioxins and BGH.
   b. dioxins and folate.
   c. beta-carotene and agar.
   d. nitrites and irradiation.
10. Biotechnological advances that have improved the food supply include:
    a. potatoes with solanine and cabbage without goitrogens.
    b. seafood without methylmercury and onions without an odor.
    c. larger cattle with a higher fat content and rennin produced by sheep.
    d. tomatoes with a longer shelf life and soybeans with a higher-quality protein.

## Notes


1. S. L. Nightingale, From the Food and Drug Administration: National Food Safety Initiative, *Journal of the American Medical Association* 277 (1997): 1664.
2. C. J. Lackey, Oil, herb and garlic flavored, http://www.foodsafety.org, visited on February 5, 1998.
3. R. V. Tauxe and J. M. Hughes, International investigations of outbreaks of foodborne disease: Public health responds to the globalization of food, *British Medical Journal* 313 (1996): 1093–1094; T. W. Hennessey and coauthors, A national outbreak of *Salmonella enteritidis* infections from ice cream, *New England Journal of Medicine* 334 (1996): 1281–1286.
4. Centers for Disease Control and Prevention, Surveillance for foodborne disease outbreaks, United States, 1988–1992, *Morbidity and Mortality Weekly Report CDC Surveillance Survey* 45 (1996): 1–66.
5. Nightingale, 1997; FDA announces a strategy to increase safety of fresh juices, *Nutrition Today* 32 (1997): 190; 1997 Food code available, *FDA Consumer,* September/October 1997, pp. 6–9.
6. FSIS Pathogen Reduction/HACCP, *Federal Register,* July 6, 1996.
7. Raw oyster risk for alcoholics, *FDA Consumer,* May 1996, pp. 23–25.
8. J. A. Desenclos and coauthors, The protective effect of alcohol on the occurrence of epidemic oyster-borne hepatitis A, *Epidemiology* 3 (1992): 371–374.
9. Position of The American Dietetic Association: Food and water safety, *Journal of the American Dietetic Association* 97 (1997): 184–189.
10. Position of The American Dietetic Association: Food irradiation, *Journal of the American Dietetic Association* 96 (1996): 69–72.
11. S. L. Nightingale, Irradiation of meat approved for pathogen control, *Journal of the American Medical Association* 279 (1998): 9; M. T. Olsterholm, Cyclosporiasis and raspberries—Lessons for the future, *New England Journal of Medicine* 336 (1997): 1597–1598.
12. Irradiation treatment approved, *Journal of the American Dietetic Association* 96 (1996): 1237.
13. Irradiation treatment approved, 1996.
14. J. Meng and M. P. Doyle, Emerging issues in microbiological food safety, *Annual Review of Nutrition* 17 (1997): 255–275.
15. National Academy of Sciences Committee, as quoted by J. Raloff and D. Pendick, Pesticides in produce may threaten kids, *Science News* 144 (1993): 4–5.
16. Food Quality Protection Act of 1996, Public Law No. 104-170, 110 Statute 1489.
17. Olsterholm, 1997.
18. C. Marwick, "Fresh Produce Initiative" for imports, *Journal of the American Medical Association* 278 (1997): 1481.
19. Position of The American Dietetic Association, 1997.

20. http://www.ams.usda.gov/nop, visited on March 3, 1998.
21. R. Papazian, Sulfites: Safe for most, dangerous for some, *FDA Consumer,* December 1996, pp. 11–14.
22. D. J. Raiten, J. M. Talbot, and K. D. Fisher, Executive summary from the report: Analysis of adverse reactions to monosodium glutamate (MSG), *Journal of Nutrition* 125 (1995): 2892S–2906S.
23. W. Mertz, Food fortification in the United States, *Nutrition Reviews* 55 (1997): 44–49.
24. T. D. Etherton, P. M. Kris-Etherton, and E. W. Mills, Recombinant bovine and porcine somatotropin: Safety and benefits of these biotechnologies, *Journal of the American Dietetic Association* 93 (1993): 177–180.
25. Bovine somatotropin and the safety of cow's milk, National Institutes of Health Technology Assessment Conference Statement, *Nutrition Reviews* 49 (1991): 227–232; Position of The American Dietetics Association: Biotechnology and the future of food, *Journal of the American Dietetic Association* 95 (1995): 1429–1432.
26. M. K. Glynn and coauthors, Emergence of multidrug-resistant *Salmonella enterica* serotype typhimurium DT104 infections in the United States, *New England Journal of Medicine* 338 (1998): 1333–1338.
27. Position of The American Dietetic Association, 1995.
28. B. A. Larkins, C. R. Lending, and J. C. Wallace, Modification of maize-seed-protein quality, *American Journal of Clinical Nutrition* (supplement) 58 (1993): 264S–269S.
29. C. R. Somerville, Future prospects for genetic modification of the composition of edible oils from higher plants, *American Journal of Clinical Nutrition* (supplement) 58 (1993): 270S–275S.
30. J. Henkel, Genetic engineering: Fast forwarding to future foods, *FDA Consumer,* April 1995, pp. 6–11.
31. Biotechnology of food: Background information from the FDA, *Nutrition Today,* July/August 1994, pp. 19–20.
32. J. A. Nordlee and coauthors, Identification of a Brazil-nut allergen in transgenic soybeans, *New England Journal of Medicine* 334 (1996): 688–692; M. Nestle, Allergies to transgenic foods—Questions of policy, *New England Journal of Medicine* 334 (1996): 726–727.
33. T. D. Etherton, The impact of biotechnology on animal agriculture and the consumer, *Nutrition Today,* July/August 1994, pp. 12–18.

HIGHLIGHT 19

# Consumer Concerns about Water

Foods are not alone in transmitting food-borne diseases; water is guilty, too. In fact, *Cryptosporidium* and *Cyclospora,* commonly found in fresh fruits and vegetables, and *Vibrio vulnificus,* found in raw oysters, are commonly transmitted through contaminated water.[1] In addition to microorganisms, water may contain many of the same impurities that foods do: environmental contaminants, pesticides, and additives such as chlorine used to kill pathogenic microorganisms and fluoride used to protect against dental caries. A glass of "water" is more than just $H_2O$. This highlight examines the sources of drinking water, harmful contaminants, and ways to ensure water safety.

## Sources of Drinking Water

Drinking water comes from two sources—surface water and groundwater. Each source supplies water for about half of the population.

### Surface Water

Most major cities obtain their drinking water from surface water—the water in lakes, rivers, and reservoirs. Surface water is readily contaminated because it is directly exposed to acid rain, runoff from highways and urban areas, pesticide runoff from agricultural areas, and industrial wastes that are dumped directly into it. Surface water contamination is reversible, however, because fresh rain constantly replaces the water. It is also cleansed to some degree by aeration, sunlight, and plants and microorganisms that live in it.

### Groundwater

Groundwater is the water in underground aquifers—rock formations that are saturated with and yield usable water. People who live in rural areas rely mostly on groundwater pumped up from private wells.

Groundwater is contaminated more slowly than surface water, but also more permanently. Contaminants deposited on the ground migrate slowly through the soil before reaching groundwater. Once there, the contaminants break down less rapidly than in surface water due to the lack of aeration, sunlight, and aerobic microorganisms. The slow replacement of groundwater also helps contaminants remain for a long time. Groundwater is especially susceptible to contamination from hazardous waste sites, dumps and landfills, underground tanks storing gasoline and other chemicals, and improperly discarded household chemicals and solvents.

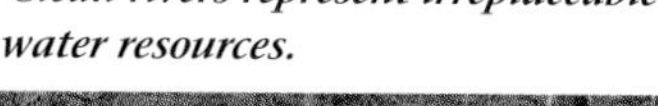
*Clean rivers represent irreplaceable water resources.*

EPA Office of Ground Water and Drinking Water

www.epa.gov/OGWDW

### The Cleansing Process

In the wilderness, water is purified each time it cycles through living systems. The soil filters out animal waste excreted onto the earth, preventing it from reaching the groundwater; plants use the waste as fertilizer instead. Soil holds pollutants, too—not beneficial to the soil but protective of the water. Surface waters also leave behind their pollutants as rivers flow along. But neither the soil nor the rivers can completely purify the heavily polluted water expelled as city sewage or industrial waste. Water leaving a factory may contain higher and higher concentrations of toxic metals as time passes, especially if the same water cycles repeatedly through the factory. Human technology is responsible for purifying water contaminated by human technology.

Public water systems treat water to remove contaminants that have been detected above acceptable levels. During treatment, a disinfectant (usually, chlorine) is added to kill bacteria. The addition of chlorine to public water is an important public health measure that appears to offer great benefits and small risks. On the one hand, chlorinated water has eliminated such water-borne diseases as typhoid fever, which once ravaged vast areas, killing thousands of people. On the other hand, it has been associated with a slight increase in bladder and rectal cancers and with contamination of the environment with the toxic by-product dioxin.

When public water-treatment systems go awry, water-borne illnesses can affect an entire city. In 1993, more than 400,000 residents of Milwaukee experienced diarrhea when their water-treatment system failed to control for a variety of microorganisms.[2]

Private well water is usually not chlorinated or cleansed, so the 40 million Americans who consume water from private wells are most at risk of drinking contaminated water. The quality of water from a private well is the responsibility of the homeowner, who should have the water tested periodically.

## Drinking Water Contaminants

Contaminants, including heavy metals, pathogenic microorganisms, and other compounds, have been detected in public drinking water. The health implications of these contaminants are just becoming known.

## Heavy Metals

The metals of greatest concern are mercury, cadmium, and lead. These metals may be absorbed into the body, where they damage cell structures and impair enzyme or coenzyme functions. When combined with organic compounds, these metals may be absorbed especially rapidly and may damage body tissues even more. Heavy metals can alter the genetic material DNA, causing mutations that can produce cancer or birth defects. If the mutations occur in the DNA of the germ cells (eggs or sperm), the changes are hereditary.

## Pathogenic Microorganisms

While the water supply naturally contains few, if any, heavy metals, it does naturally contain bacteria from the soil and from contamination with sewage. Before a sewage treatment plant releases water into the public supply, it must decrease the bacterial count.

High standards for sewage treatment in the developed countries ensure that most people have safe drinking water. For the rest of the world, however, microbial contamination remains the primary cause of human diseases and epidemics. Two of the most basic public health needs of the world's people are safe drinking water and an acceptable standard of waste disposal.

## Other Compounds

Compounds from sewage, pesticides, petroleum-based industries, highway runoff, and other sources may also appear in water. Some of these compounds cause birth defects, some cause cancer, and some cause genetic mutations. Many of these compounds contain chlorine, and some may be formed during the chlorination of water. The risks they present remain unknown; standards are being established, and if public water exceeds them, new treatment systems may be needed.

## Contaminants via Plumbing

Contamination can also occur as water travels from the main water supply to homes. Lead or asbestos from corroded pipes can contaminate drinking water, as can bacteria and dirt from leaking pipes. People who suspect contamination of their water should have it tested where it flows out, at the tap.

## Water Systems and Regulations

The EPA is responsible for ensuring that public water systems meet minimum standards for protecting the public health.* The agency's tasks include developing maximum permitted levels for all regulated contaminants, monitoring for contaminants in drinking water, identifying the appropriate technology for removing excess contaminants, and providing protection for groundwater sources. Municipal water suppliers must inform customers in their monthly bills if contaminants have been detected in the city's water supply and the potential health risks. For example, in 1993, both New York City and Washington, D.C., issued notices to more than 7 million people suggesting that they boil water as a safety precaution; more recently, over 700 other communities have issued similar "boil water" notices to more than 3 million people.

Even safe water may have characteristics that some consumers find unpleasant. Most of these problems reflect the mineral content of the water. For example, manganese and copper give water a metallic taste, and sulfur produces a "rotten egg" odor. Iron leaves a rusty brown stain on plumbing fixtures and laundry. Calcium and magnesium (com-

*The EPA's safe drinking water hotline: (800) 426-4791.

 EPA Office of Water

www.epa.gov/OW

## Glossary

**artesian water:** water that is drawn from a well that taps a confined aquifer in which the water level stands above the natural water table.

**distilled water:** water that has been vaporized and recondensed, leaving it free of dissolved minerals.

**filtered water:** water treated by filtration, usually through activated carbon filters that reduce the lead in tap water, or by reverse osmosis units that force pressurized water across a membrane removing lead, arsenic, and some microorganisms.

**fluoridated water:** water that has been treated so as to contain at least 0.8 milligrams of fluoride per liter.

**hard water:** water with a high calcium and magnesium concentration.

**mineral water:** water from a spring or well that typically contains 250 to 500 ppm of minerals. Minerals give water a distinctive flavor. Many mineral waters are high in sodium.

**natural water:** water obtained from a spring or well that is certified to be safe and sanitary. The mineral content may not be changed, but the water may be treated in other ways such as by filtration or ozonization.

**potable** (POT-ah-bul) **water:** water that is suitable for drinking.

**public water:** water from a municipal or county water system that has been treated and disinfected.

**purified water:** water that has been processed through distillation, deionization, or reverse osmosis and meets U.S. Pharmacopoeia standards for medical and research purposes.

**soft water:** water with a high sodium concentration.

**spring water:** water originating from an underground spring or well. It may be carbonated or not ("flat" or "still"). Brand names such as "Spring Pure" do not necessarily mean that the water comes from a spring.

**well water:** water drawn from groundwater by tapping into an aquifer.

monly found in "hard water") build up in coffeemakers and hot water heaters; similarly, soap is not easily rinsed away in hard water, leaving bathtubs and laundry looking dingy. For these, and other reasons, some consumers have adopted alternatives to the public water system.

## Home Water Treatments

To ease concerns about the quality of drinking water, some people purchase home water-treatment systems. Because the EPA does not certify or endorse these water-treatment systems, consumers must shop carefully. Manufacturers offer a variety of units for removing contaminants from drinking water. None of them removes all contaminants, and each has its own advantages and disadvantages. Choosing the right treatment unit depends on the kinds of contaminants in the water. For example, activated carbon filters are particularly effective in removing chlorine, heavy metals such as mercury, and organic contaminants from sediment; reverse osmosis, which forces pressurized water across a membrane, flushes out sodium, arsenic, and some microorganisms such as *Giardia;* and distillation systems, which boil water and condense the steam to water, remove contaminants such as lead and kill microorganisms in the process. Therefore, before purchasing a home water-treatment unit, a consumer must first determine the quality of the water. In some cases, a state or county health department will test water samples or can refer the consumer to a certified laboratory.

Rather than purchasing a home treatment unit, some people boil their water. This kills microorganisms and removes some organic chemicals, but may concentrate inorganic chemicals such as lead.

 National Sanitation Foundation

www.nsf.org

 Water Quality Association

www.wqa.org

## Bottled Water

Many people turn to bottled water as an alternative to tap water. Bottled water is classified as a food, so it is regulated nationwide by the FDA and locally by state health and environmental agencies. The FDA has established quality and safety standards for bottled drinking waters compatible with those set by the EPA for public water systems. In addition, all bottled waters must be processed, packaged, and labeled in accordance with FDA regulations.[3] Bottled water is expensive compared to public water. Bottled water quality varies among brands, because of the variations in the source water used, costs, and company practices.

Approximately 75 percent of bottled waters derive from protected groundwater (from springs or wells) that has been disinfected with ozone rather than chlorine. Ozone kills microorganisms, then disintegrates spontaneously into water and oxygen, leaving behind no toxic by-products. Other bottled waters derive from municipal tap water that has been treated by carbon filtration to remove chlorine and inorganic compounds; water bottled from municipal sources must be labeled as such. Bottled waters may also be treated by reverse osmosis or ion exchange to remove inorganic compounds. Alternatively, the water may be distilled or deionized to remove dissolved solids. Some bottled waters may also have minerals or carbonation added. "Carbonated," "seltzer," "soda," and "tonic" waters are not considered waters, however, but soft drinks. Some of the terms used to describe water are listed in the glossary (on p. 626).

Despite government regulations, some contamination has been detected in some bottled waters. While the amounts of most contaminants found in bottled waters are probably insignificant, consumers should be aware that bottled water is not always purer than the water from their taps. As a safeguard, the FDA recommends that bottled water be handled like other foods and be refrigerated after opening.

Protection of drinking water is the subject of an ongoing battle between environmentalists and industry.[4] It may soon become the source of conflict between the world's nations as the population continues to grow while the renewable water supply remains constant. Within the next 50 years, an estimated half of the world's people will not have enough water to meet their needs. To avert this potential calamity, we must take active steps toward conserving water, cleaning polluted water, desalinating seawater, and curbing population growth. Chapter 20 addresses many of these issues in more detail.

 U.S. Government

www.healthfinder.gov/searchoptions/topics.htm
Search for Drinking Water
Search for Water Supply

 Water Quality Information Center

www.nal.usda.gov/wqic

## Notes


1. Outbreaks of *Cyclosporosis*—United States, 1997, *Morbidity and Mortality Weekly Report* 46 (1997): 451; *Vibrio vulnificus* infections associated with eating raw oysters—Los Angeles, 1996, *Journal of the American Medical Association* 276 (1996): 937–938; J. W. Besser-Wiek and coauthors, Foodborne outbreak of diarrheal illness associated with *Cryptosporidium parvum*—Minnesota, 1995, *Morbidity and Mortality Report* 45 (1996): 783–784.
2. W. R. Mackenzie and coauthors, A massive outbreak in Milwaukee of *Cryptosporidium* infection transmitted through the public water supply, *New England Journal of Medicine* 331 (1994): 161–167.
3. Position of The American Dietetic Association: Food and water safety, *Journal of The American Dietetic Association* 97 (1997); 184–189.
4. Position of The American Dietetic Association: Natural resource conservation and waste management, *Journal of the American Dietetic Association* 97 (1997): 425–428.

CHAPTER 20

# Hunger and Global Environmental Problems

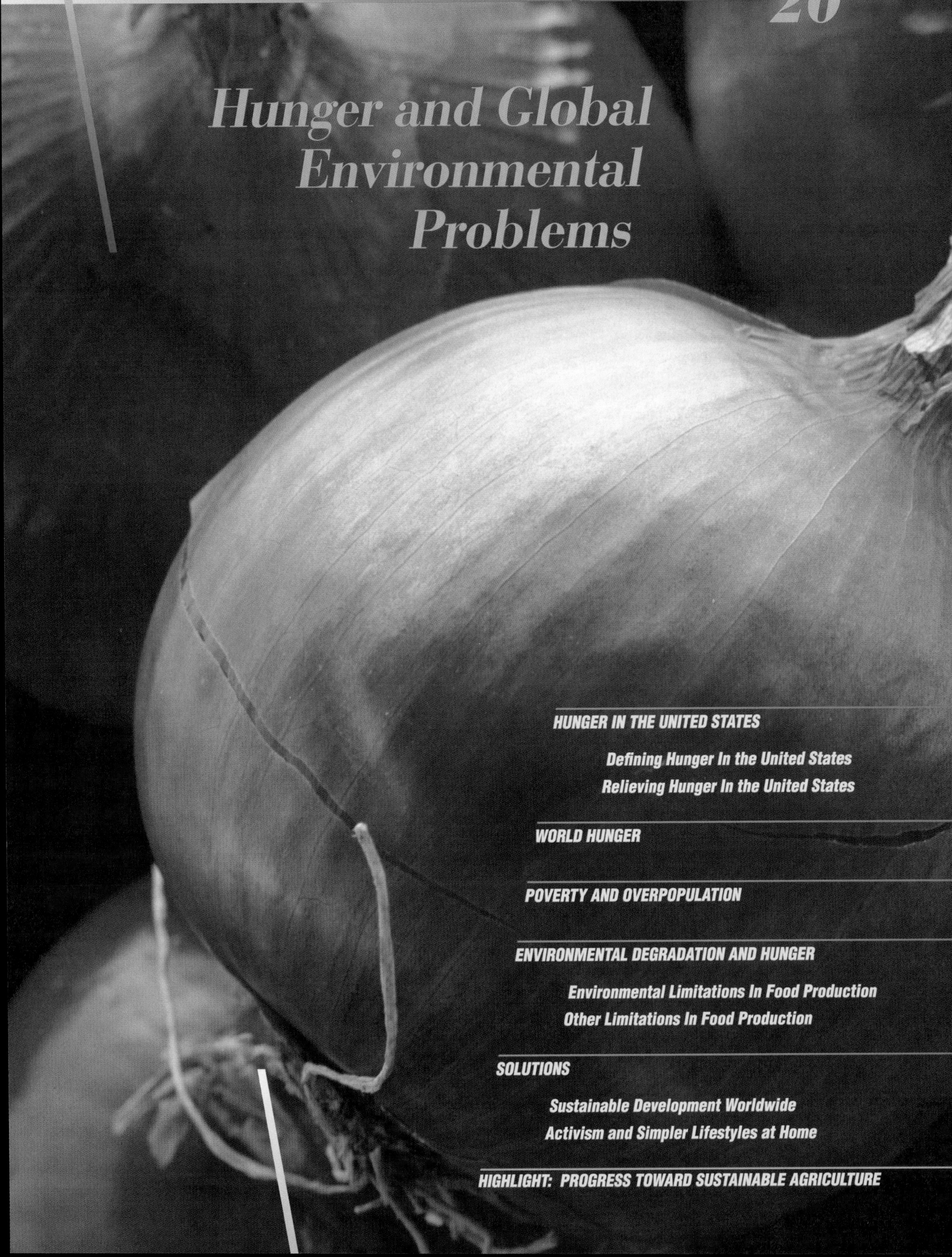

HUNGER IN THE UNITED STATES

- Defining Hunger In the United States
- Relieving Hunger In the United States

WORLD HUNGER

POVERTY AND OVERPOPULATION

ENVIRONMENTAL DEGRADATION AND HUNGER

- Environmental Limitations In Food Production
- Other Limitations In Food Production

SOLUTIONS

- Sustainable Development Worldwide
- Activism and Simpler Lifestyles at Home

HIGHLIGHT: PROGRESS TOWARD SUSTAINABLE AGRICULTURE

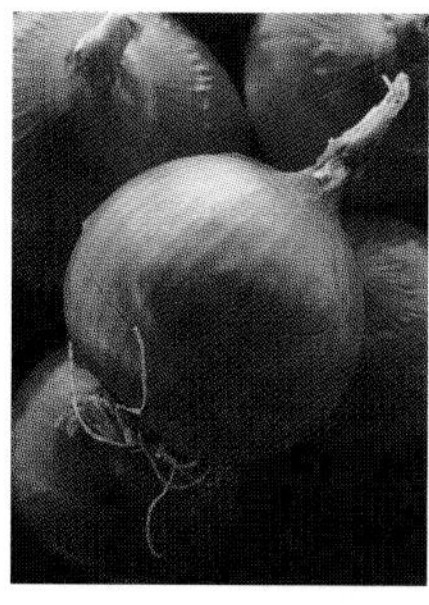

One person in every five worldwide experiences chronic hunger—not the healthy hunger that leads a person to eat a hearty meal, but the painful hunger that a person feels when no food is available. Tens of thousands die of starvation each day: one every two seconds.[1]

The American Dietetic Association (ADA) states that world "hunger continues to be a problem of staggering proportions."[2] The enormity of the problem reflects not only huge numbers, but major challenges. As people populate and pollute the earth, resources become depleted, making food less available. Figure 20-1 highlights the relationships among hunger and poverty, population growth, and environmental degradation. Because these problems are linked, they tend to worsen each other. Because their causes overlap, so do their solutions: any initiative a person takes to help solve one problem will help solve many others. Eliminating hunger requires a balance among the distribution of food, the numbers of people, and the care of the environment.

Resolving the hunger problem may seem at first beyond the influence of the ordinary person. Can one person's choice to limit family size or to recycle a bottle or to volunteer at a food recovery program make a difference? In truth, such choices produce several benefits. For one, a person's action may influence many other people over time. For another, an action repeated becomes a habit, with compounded benefits. For still another, making choices with an awareness of the consequences gives a person a sense of personal control, hope, and effectiveness. Many of the daily actions of concerned people can help solve the problems of hunger in their own neighborhoods or on the other side of the world.

A popular adage urges us to "Think globally, act locally."

# Hunger in the United States

Much as it should surprise us, an estimated 35 million people in the United States, including one out of every five children, live in poverty and cannot afford to buy enough food to maintain good health. Their hunger is not always easy to recognize, but it always involves a lack of food.

## *Defining Hunger in the United States*

At its most extreme, people experience hunger because they have absolutely no food. More often, they have too little food and try to stretch their limited resources by eating small meals or skipping meals—often for days at a time. Sometimes hungry people obtain enough food to satisfy their hunger, but only through socially unacceptable ways—begging from strangers or digging through garbage cans, for example. The limited or uncertain availability of nutritionally adequate and safe foods is known as "food insecurity" and is a major problem in our society today.[3] The accompanying "How to" shows how national surveys identify food insecurity in the United States. Questions like these provide crude, but necessary, data to estimate the degree of hunger in this country.

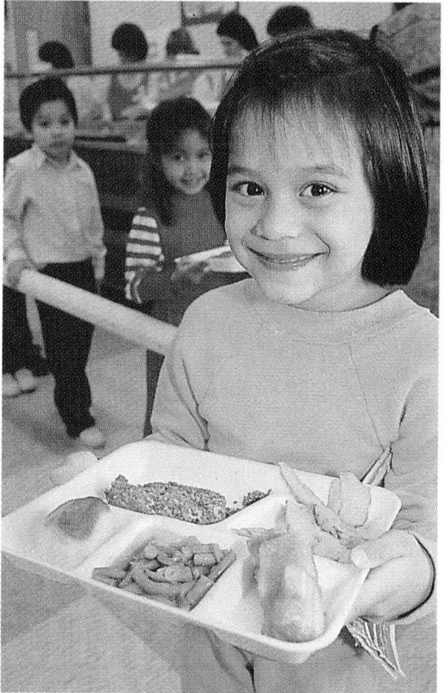

***School lunches provide children with nourishment at little or no charge.***

**food insecurity:** limited or uncertain access to foods of sufficient quality or quantity.

Hunger has many causes, but the major one is poverty. Contributing to hunger are abuse of alcohol and other drugs; physical and mental illness; lack of awareness of available food assistance programs; and the reluctance of people, particularly the elderly, to accept what they perceive as "welfare" or "charity." Poverty remains the major cause of hunger, and solving the problems of unemployment, low wages, and underemployment would do a lot to resolve the hunger crisis.

Note: People who are *unemployed* are out of work, whereas those who are *underemployed* are working at a job for which they are far overqualified and underpaid.

In the United States, poverty and hunger reach across various segments of society, but prevalence is highest among Hispanics and African Americans, those

**Figure 20-1**
**The Relationships among Global Problems**

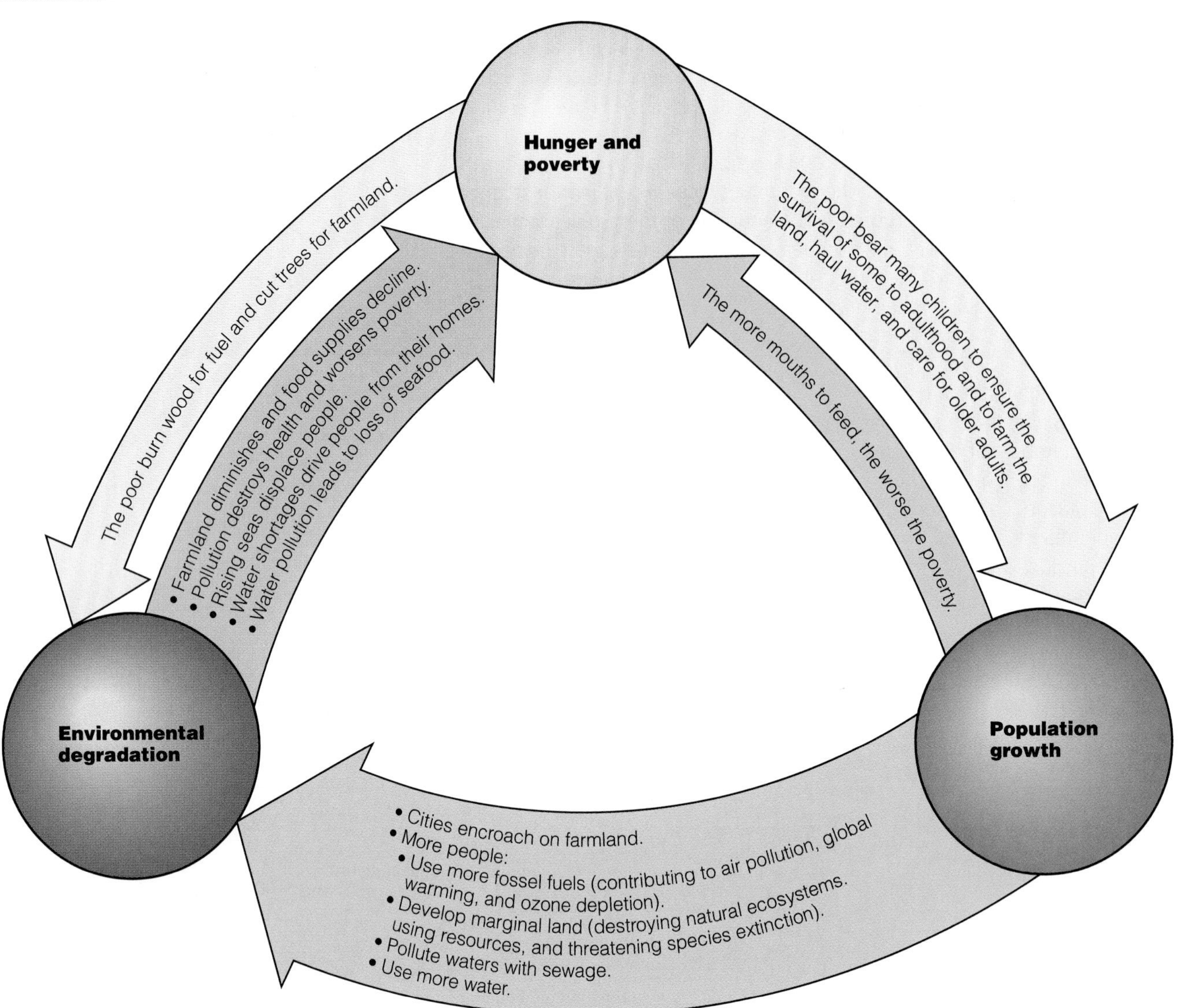
Hunger and poverty
Population growth
Environmental degradation
The poor bear many children to ensure the survival of some to adulthood and to farm the land, haul water, and care for older adults.
The more mouths to feed, the worse the poverty.
The poor burn wood for fuel and cut trees for farmland.
• Farmland diminishes and food supplies decline.
• Pollution destroys health and worsens poverty.
• Rising seas displace people.
• Water shortages drive people from their homes.
• Water pollution leads to loss of seafood.
• Cities encroach on farmland.
• More people:
• Use more fossel fuels (contributing to air pollution, global warming, and ozone depletion).
• Develop marginal land (destroying natural ecosystems, using resources, and threatening species extinction).
• Pollute waters with sewage.
• Use more water.

USDA National Hunger Clearinghouse
www.iglou.com/why/usda

living in the inner cities, and those living in households with children. People living in poverty are simply unable to buy sufficient amounts of nourishing foods, even if they are wise shoppers.

## Relieving Hunger in the United States

At present, many programs aim to prevent or relieve malnutrition and hunger in the United States. To what extent federal programs will continue to feed those who are hungry is unknown. Welfare reform laws now disqualify all illegal and some legal immigrants from the Food Stamp Program, but still allow legal immigrants to benefit from other federal nutrition programs; the duration of eligibility for other adults has been shortened. Consequently, many who once received food assistance no longer qualify for such benefits.

• ***Federal Food Assistance Programs*** • Federal food assistance programs have been described in earlier chapters: the WIC program for low-income pregnant women, breastfeeding mothers, and their young children (Chapter 15); the school lunch, breakfast, and child care food programs for children (Chapter 16); and food assistance programs for older adults such as congregate meals and Meals on

## HOW TO Identify Food Insecurity in a U.S. Household

Questions like these are asked on surveys to determine the extent of food insecurity in a household. The more questions answered "Yes," the more intense the hunger the household is experiencing.

- Do you often go hungry?
- Do you often have too little food to eat because you have no money, transportation, or kitchen appliances (stove, refrigerator)?
- Do you ever rely on nutritionally inferior foods to feed yourself or your children because you lack any of these resources?
- Do you ever eat less than you feel you should because you lack any of these resources?
- Do you ever skip meals or cut the size of meals because you lack any of these resources?
- Do you ever rely on neighbors, friends, relatives, or schools to feed any of your children because there is not enough food in the house?
- Do your children ever say they are hungry because there is not enough food in the house?
- Do you or any of your children ever go to bed hungry because there is not enough food in the house?

Sources: Adapted from C. A. Wehler, R. I. Scott, and J. J. Anderson, The Community Childhood Hunger Identification Project: A model of domestic hunger—demonstration project in Seattle, Washington, *Journal of Nutrition Education* (1 supplement) 24 (1992): 29S–35S; R. R. Briefel and C. E. Woteki, Development of food sufficiency questions for the Third National Health and Nutrition Examination Survey, *Journal of Nutrition Education* (1 supplement) 24 (1992): 24S–28S.

Wheels (Chapter 17). Another program for low-income people is the Food Stamp Program, administered by the U.S. Department of Agriculture (USDA). The USDA issues food stamp coupons through state social services or welfare agencies to households—people who buy and prepare food together. The number of stamps a household receives depends on its size and income. Recipients may use the coupons to purchase food and food-bearing plants and seeds, but not to buy tobacco, cleaning items, alcohol, or other nonfood items. The Food Stamp Program is the largest of the federal food assistance programs, both in amount of money spent and in number of people participating. Over 21 million people receive food stamps at a cost of over $24 billion per year; over half of the recipients are children.[4]

Although these programs reach millions of people daily with life-sustaining foods, hunger continues to plague the United States. Of the estimated 2 million homeless people in the United States who are eligible for food assistance, only 15 percent of single adults and 50 percent of families receive food stamps.

• ***National Food Recovery Programs*** • Efforts to resolve the problem of hunger in the United States do not depend solely on federal assistance programs. National food recovery programs have made a dramatic difference; the largest program, Second Harvest, alone helps 26 million people a year.

Each year, an estimated one-fifth of our food supply is wasted—that's enough food to feed 49 million people. Food recovery programs collect and distribute good food that would otherwise go to waste.[5] Volunteers might pick corn left in an already harvested field, a grocer might deliver ripe bananas to a local food bank, and a caterer might take leftover chicken salad to a community shelter, for example. All of these efforts help to feed the hungry in the United States.

• ***Local Efforts*** • Food recovery programs depend on volunteers. Concerned citizens work through local agencies and churches to feed the hungry. Community-based food pantries provide groceries and soup kitchens serve prepared meals. Meals often average 1000 kcalories each, but most homeless people receive fewer than one and a half meals a day, so many are still inadequately nourished.[6] Table 20-1 shows how individuals can assist in local hunger relief efforts.

 USDA Food Stamp Program
www.usda.gov/fcs/stamps/fs.htm

 Emergency Food and Shelter Program
www.efsp.unitedway.org

**food recovery:** collecting food for distribution to low-income people who are hungry. Four common methods of food recovery are:
- **Field gleaning:** collecting crops from fields that either have already been harvested or are not profitable to harvest.
- **Perishable food salvage:** collecting perishable produce from wholesalers and markets.
- **Food rescue:** collecting prepared foods from commercial kitchens.
- **Nonperishable food collection:** collecting processed foods from wholesalers and markets.

 USDA Gleaning and Food Recovery
www.usda.gov/fcs/glean.htm
(800) GLEAN-IT

**food bank:** facilities that provide food to the hungry.

**Table 20-1**

**Fourteen Ways Communities Can Address Their Hunger Problems**

1. Establish a community-based emergency food delivery network.
2. Assess community hunger problems and evaluate community services. Create strategies for responding to unmet needs.
3. Establish a group of individuals, including low-income participants, to develop and implement policies and programs to combat hunger and the threat of hunger; monitor responsiveness of existing services; and address underlying causes of hunger.
4. Participate in federally assisted nutrition programs that are easily accessible to targeted populations.
5. Integrate public and private resources, including local businesses, to relieve hunger.
6. Establish an education program that addresses the food needs of the community and the need for increased local citizen participation in activities to alleviate hunger.
7. Provide information and referral services for accessing both public and private programs and services.
8. Support programs to provide transportation and assistance in food shopping, where needed.
9. Identify high-risk populations and target services to meet their needs.
10. Provide adequate transportation and distribution of food from all resources.
11. Coordinate food services with parks and recreation programs and other community-based outlets to which residents of the area have easy access.
12. Improve public transportation to human service agencies and food resources.
13. Establish nutrition education programs for low-income citizens to enhance their food purchasing and preparation skills and make them aware of the connections between diet and health.
14. Establish a program for collecting and distributing nutritious foods—either agricultural commodities in farmers' fields or prepared foods that would have been wasted.

Source: House Select Committee on Hunger, legislation introduced by Tony P. Hall, excerpted in *Seeds*, Sprouts edition, January 1992, p. 3, © SEEDS Magazine, P.O. Box 6170, Waco, TX 76706.

*Feeding the hungry—in the United States.*

World Food Program
www.wfp.org

**famine:** widespread scarcity of food in an area that causes starvation and death in a large portion of the population.

IN SUMMARY

Poverty and hunger are widespread in the United States among both the unemployed and the underemployed. Government assistance programs help to relieve poverty and hunger, but equally important are food recovery programs and other local efforts.

# World Hunger

In developing countries, people face more extreme hunger problems than in the United States. The causes of hunger are more diverse, but again, the primary cause is poverty, and the poverty is more extreme. Most people cannot grasp the severity of poverty in the developing world. One-fifth of the world's 6 billion people have no land and no possessions *at all.* They survive on less than one dollar a day each, they lack water that is safe to drink, and they cannot read or write. The average U.S. housecat receives twice as much protein every day as one of these people, and the cost of keeping that cat is greater than such a person's annual income.

• ***Food Shortages*** • World hunger brings to mind victims of famine, a true food shortage in an area that causes widespread starvation and death. The natural causes of famine—drought, flood, and pests—have become less important in recent years than the social causes. A sudden increase in food prices, a drop in workers' incomes, or a change in government policy can leave millions hungry. Between 15 and 30 million people died during the Chinese famine of 1959 through 1961, the worst famine of this century. The main cause was government policies associated with the "great leap forward," which devastated the Chinese agricultural system.[7]

In the 1990s, the violence of war has become the dominant cause of famine worldwide. In all of the countries that have reported famine so far in the 1990s—Angola, Ethiopia, Liberia, Mozambique, Somalia, and Sudan—armed conflict has been a major cause. Farmers become warriors, their agricultural fields become battlegrounds, the citizens go hungry, and the warring factions often block famine relief. The world continues to struggle to find a middle ground between respecting the sovereignty of nations and insisting that all nations allow humanitarian assistance to reach the people.

Since the 1950s, food aid from countries around the world has provided a safety net for countries whose crops fail. But food aid now does more than just offset poor harvests; it also delivers food relief to countries, such as Ethiopia, that are chronically short of food and without resources to buy it. Some people are concerned that as many countries cut their support of foreign aid, this assistance may become insufficient.

***Feeding the hungry—in Sarajevo.***

• ***Chronic Malnutrition*** • Although we usually associate world hunger with famine, the numbers affected by famine are relatively small compared with those suffering from chronic hunger and malnutrition. Approximately 2 billion people, mostly women and children, are malnourished; the nutrients most likely to be deficient are iron, iodine, and vitamin A.[8] The deficiency symptoms of these nutrients and those of the other vitamins and minerals were presented in Chapters 10 through 13; Chapter 6 described protein-energy malnutrition; and Chapters 15 through 17 examined the effects of malnutrition during various stages of the life cycle. The consequences of nutrient deficiencies are felt not only by individuals, but by entire nations. When people suffer from mental retardation, growth failure, blindness, infections, and other consequences of malnutrition, the economy of their country declines as well due to decreased productivity and increased health care costs.[9]

 UNICEF
www.unicef.org

To prevent outbreaks of measles, health care workers distribute vitamin A supplements to millions of children worldwide.

Pan American Health Organization
www.paho.org

In addition to specific nutrient deficiencies, one child in six in the world is born underweight, and almost two in five children are underweight by the age of five. These underweight children are malnourished and readily develop infections; the synergistic combination dramatically increases the likelihood of their deaths.[10] Each year, half a million people die as a result of malnutrition. Many are children afflicted by the diseases of poverty: parasitic and infectious diseases such as dysentery, whooping cough, measles, tuberculosis, cholera, and malaria. These diseases interact with poor nutrition in a vicious cycle that leads to death. Most children who die of malnutrition die from the diarrhea and dehydration that accompany infections. Health care workers around the world save millions of lives each year by effectively reversing dehydration and correcting diarrhea with oral rehydration therapy (ORT).

To prevent death from diarrheal disease, provide:

- Adequate sanitation.
- Safe water.
- Oral rehydration therapy.

**oral rehydration therapy (ORT):** the administration of a simple solution of sugar, salt, and water, taken by mouth, to treat dehydration caused by diarrhea. A simple ORT recipe:

1 c boiling water.
2 tsp sugar.
A pinch of salt.

• ***Diminishing Food Supply*** • Until 1988, the world's reserves of stored grain, an index of the sufficiency of the world food supply, increased nearly every year. The often-repeated statement that "We have enough food to feed everyone" was true. Efforts at relieving hunger focused on transporting food to where it was needed and on improving storage. Also, because in many developing countries most of the men were involved in producing cash crops for export, hunger-relief efforts focused on educating and empowering women to grow and use nutritious food to feed their families. These efforts were expected to solve the world's hunger problem.

**cash crops:** crops grown for sale or export, as opposed to food crops grown for local consumption.

Since 1988, however, the situation has changed. The world's population is growing at the rate of about 90 million persons each year, and food production can no longer keep pace. In the past decade, yields of the three primary grains—wheat, rice, and corn—have slowed dramatically or reached a plateau.[11]

Further growth in the world's food output is being slowed by environmental degradation and drought in agricultural areas. People in developed nations may feel safely insulated against food shortages, but this is a false sense of security. Developed countries may be last to feel the effects, but these nations do finally go as the world goes.

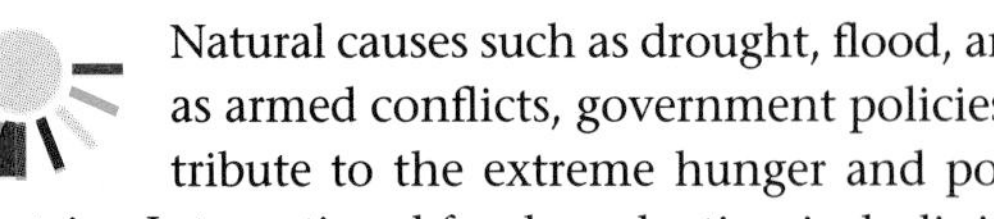

IN SUMMARY

Natural causes such as drought, flood, and pests and social causes such as armed conflicts, government policies, and overpopulation all contribute to the extreme hunger and poverty seen in the developing countries. International food production is declining, surpluses are diminishing, and assistance programs are failing to resolve the problems.

# Poverty and Overpopulation

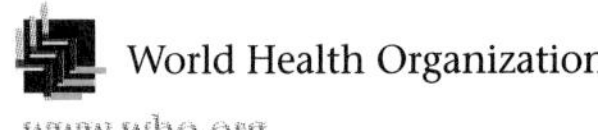

Oxfam
www.oxfam.org

World Health Organization
www.who.org

Population growth is one of many factors contributing to poverty and hunger. The reverse is also true: poverty and hunger contribute to population growth.

• ***Population Growth Leads to Hunger and Poverty*** • The first of these cause-effect relationships is easy to understand. As a population grows larger, more mouths must be fed, and the worse poverty and hunger become. The sheer magnitude of our annual population increase of 90 million people per year is difficult to comprehend. Each month the world adds the equivalent of another New York City. During six months of the terrible 1992 famine in Somalia, an estimated 300,000 people starved to death. Yet it took the world only 29 *hours* to replace their numbers! Ninety million people a year, spread over 365 days, comes to a quarter-million people a day—or just over 10,000 people born every hour.

Population growth also contributes to hunger indirectly by preempting good agricultural land for growing cities and industry and forcing people onto marginal land, where they cannot produce sufficient food for themselves. The world's poorest people live in the world's most damaged and inhospitable environments. There they experience, daily, tens of thousands of early deaths from malnutrition and disease.

***Families in developing countries depend on their children to help provide for daily needs.***

• ***Hunger and Poverty Lead to Population Growth*** • How does poverty lead to overpopulation? Poverty and hunger exert an ironic effect on people, driving them to bear more children. Poverty and hunger typically go hand in hand with ignorance, including ignorance of family planning. Also, a family depends on its children to farm the land, haul water, and care for adults in their old age. Poverty claims many of a family's young children, who are among the most likely to die from hunger and malnutrition. If a family faces ongoing poverty, the parents will choose to have many children to ensure that some will survive to adulthood. People are willing to risk having fewer children only if they are sure that their children will live.

The environment also suffers as more and more people must share fewer and fewer resources.[12] In developing countries, people living in poverty sell everything they own to obtain money for food, even the seeds that would have provided next year's crops. They cut trees for firewood or timber to sell, then lose the soil to erosion. Without these resources, they become poorer still. Thus poverty contributes to environmental ruin, and the ruin leads to hunger. Figure 20-2 shows the downward spiral of poverty, population growth, and environmental degradation leading to hunger.

• ***Breaking the Cycle*** • Relieving poverty and hunger, then, may be a necessary first step in curbing population growth. When people attain better access to health care, education, and family planning, the death rate falls. At first there is a "bulge" in the population, because births outnumber deaths, but as the standard of living continues to improve, families become willing to risk having fewer children. Then the birth rate falls. Thus, after a short but necessary lag time, improvements in living standards help stabilize the population.

The link between improved economic status and slowed population growth has been demonstrated in country after country.[13] Central to this success is sustainable

**sustainable:** able to continue indefinitely. Here, the term means using resources at such a rate that the earth can keep on replacing them and producing pollutants at a rate with which the environment and human cleanup efforts can keep pace, so that no net accumulation of pollution occurs.

development that includes not only economic growth, but a sharing of resources among all groups. Where this has happened, population growth has slowed the most: in parts of Sri Lanka, Taiwan, Malaysia, and Costa Rica, for example. Where economic growth has occurred but only the rich have grown richer, population growth has remained high; examples include Brazil, Mexico, the Philippines, and Thailand, where large families continue to be a major economic asset for the poor.

As a society gains economic footing, education also becomes a higher priority. A society that educates its children, both males and females, experiences a drop in fertility rates. Education for girls and women brings improvements in family life, including improved nutrition, better sanitation, effective birth control, and elevated status. With improved conditions, more infants live to adulthood, making smaller families feasible.

IN SUMMARY

More people means more mouths to feed, which worsens the problems of poverty and hunger. Poverty and hunger, in turn, encourage parents to have more children. Breaking this cycle requires improving the economic status of the people and providing them with health care, education, and family planning.

**Figure 20-2**

**Poverty, Overpopulation, and Environmental Degradation**

The interactions of poverty, overpopulation, and environmental degradation worsen hunger.

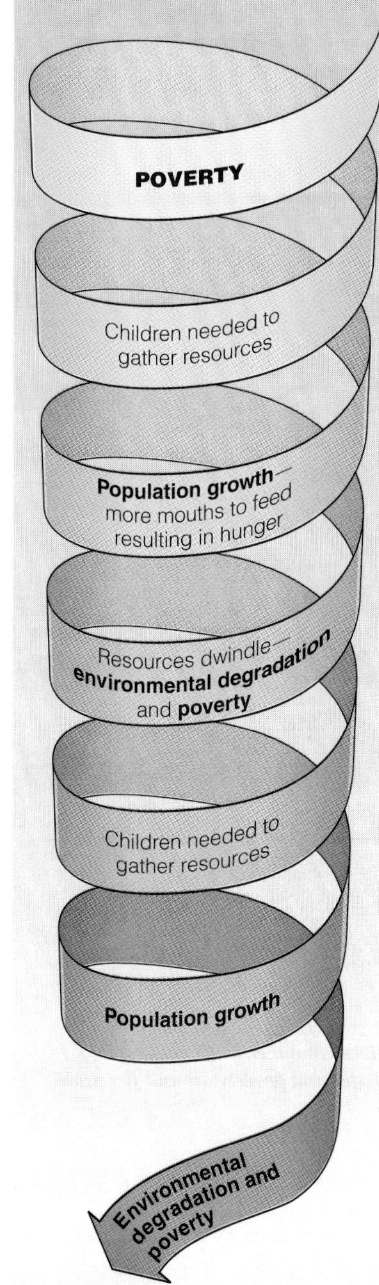


# Environmental Degradation and Hunger

Hunger, poverty, and overpopulation interact with another force: environmental degradation. Environmental degradation threatens the world's ability to produce enough food to feed its many people.

## *Environmental Limitations in Food Production*

Environmental problems that are slowing food production include:

- *Soil erosion,* which is occurring in every nation and is resulting in crop losses estimated at 6 percent per year.
- *Deforestation,* which leads to soil erosion; if present rates of deforestation continue, by 2010 per capita forested area will have dropped by 30 percent.[14]
- *Air pollution,* which damages crops; the most damaging air pollutants are ozone, sulfur dioxide, and nitrous oxide, which come from the burning of fossil fuels.
- *Ozone depletion,* which damages crops, especially radiation-sensitive crops such as soybeans, the world's leading protein crop.
- *Climate changes,* which are caused by increased atmospheric concentrations of heat-trapping carbon dioxide produced by fossil fuels; these changes may reduce soil moisture, impair pollination of major food crops such as rice and corn, slow growth, weaken disease resistance, and disrupt many other factors affecting crop yields.
- *Water scarcity,* which may limit human population growth even before food availability does; in some areas, the supplies of fresh water are inadequate to fully support the survival of crops, livestock, and people.
- *Extensive overgrazing,* which is causing rangelands to deteriorate; in nearly all developing countries, the feed needs of livestock now exceed the capacity of their rangelands.
- *Overfishing and water pollution,* which are destroying fisheries and diminishing the supply of seafood.

All in all, environmental problems are reducing the world's ability to feed its people.

*Groundwater is used up. Deserts spread.*

**fossil fuel:** coal, oil, and natural gas; these are nonrenewable fuels that pollute. (Renewable or alternative fuels, such as solar and wind energy, pollute less or not at all.)

## *Other Limitations in Food Production*

With cropfields, rangelands, and fish yields diminishing, can advances in agriculture compensate for the losses caused by environmental degradation? Historically, agriculture has improved yields by making greater investments in irrigation, fertilizer, and improved genetic strains. Today, however, the contributions these measures can make are reaching limits for the first time in history. Irrigation can no longer compensate by improving crop yields because almost all the land that can benefit from irrigation is already receiving it. In fact, rising concentrations of salt in the soil—a by-product of irrigation—are *lowering* yields on close to a quarter of the world's irrigated cropland. Nor can fertilizer use enhance agricultural production much. Much of the fertilizing that can be done is being done—and with great effect; fertilizer use supports some 40 percent of the world's total crop yields. Adding more fertilizer, however, brings no further rise in yield. As for the development of high-yielding strains of crops, recent advances have been dramatic, but even they may be inadequate to change the overall trends described here. Furthermore, the raw materials necessary for developing new crops are becoming less and less available as genetic variety is lost due to the extinction of many plant species. Of the 5000 food plants grown throughout the world a few centuries ago, only 150 are cultivated in modern agriculture today. Most of the world's population relies on only five cereals, three legumes, and three root crops to meet their energy needs. Even among these, valuable strains are vanishing.

Estimates are that the world's grain harvest can be increased by no more than 1 percent a year. Meanwhile, the world's population is rising at the rate of at least 2 percent per year. Many authorities in many fields—and more every year—are calling for a reduction in the growth rate of the world's population as the only way to enable the world's food production to keep pace with its growing numbers.

***International efforts help to relieve hunger and poverty around the world.***

The world still produces enough food to feed all its people, and the problem of hunger today remains a problem of unequal distribution of resources. If present trends continue, however, the time is fast approaching when there will be an absolute deficit of food. This conclusion seems inescapable. The world's increasing population threatens the world's capacity to produce adequate food. Until the nations of the world resolve the population problem, they can neither support the lives of people already born nor remedy global trends toward environmental deterioration. And to resolve the population problem, a necessary first step is to remedy the poverty problems, for reasons already discussed. Of the 90 million people being added to the population each year, the vast majority are in the most poverty-stricken areas of the world.

IN SUMMARY

Increasing environmental degradation reduces our ability to produce enough food to feed the world's people. Exacerbating the situation is the rapid increase in the world population.

# Solutions

Action without Borders
www.idealist.org

The keys to solving the world's hunger, poverty, and environmental problems are in the hands of both the poor and the rich nations, but require different efforts from them. The poor nations need to provide contraceptive technology and family planning information to their citizens, develop better programs to assist the poor, and slow and reverse the destruction of environmental resources. The rich nations need to stem their wasteful and polluting uses of resources and energy, which are contributing to global environmental degradation. They also must become willing to ease the debt burden that many poor nations face.

## *Sustainable Development Worldwide*

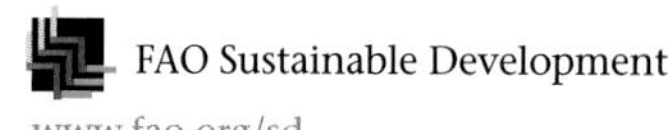
FAO Sustainable Development
www.fao.org/sd

Many nations now recognize that improving all nations' economies is a prerequisite to meeting the world's other urgent needs: relief of hunger, population stabilization, arrest of environmental degradation, and sustainable treatment of resources. An important step was taken when a United Nations Convention on the Rights of the Child was ratified by over 100 nations. Significantly, for the first time in world history, the convention cited *nutrition* as an internationally recognized human right.

Another first step was taken in 1992, when over 100 nations met for the Earth Summit in Rio de Janeiro, Brazil, and discussed the relationship of the environment to poverty and hunger.* At this meeting, many nations agreed for the first time to 27 principles of sustainable development, development that would equitably meet both the economic and the environmental needs of present and future generations.

Participants discussed climate change and the possibility of setting legally binding targets and timetables for every nation to cut its emissions of global-warming gases. They also signed agreements to protect the earth's remaining species of plants and animals and to preserve the world's forests. The participating nations also began to discuss ways of alleviating the problems of poverty in the developing world. Since then, a third international conference on climate change has taken place, and some 200 nations including the United States have made firmer commitments to reduce their releases of gases that warm the globe. In 1996, the United States began passing legislation to accomplish these objectives.

Such international discussions have opened vistas of hope. Much remains to be done, though, and all nations have major parts to play. For the United States, the challenges are many. For one, we have been asked to reduce our consumption of fossil fuel and thereby our disproportionate contribution to global environmental degradation. The willingness of U.S. consumers to take responsibility for their individual shares in solving global problems could make a substantial contribution to the quality of life for future generations. Our decision to use fewer goods, devour fewer resources, create less pollution, and consume less energy would go a long way toward remedying global environmental problems and conditions that contribute to world hunger. For another, it has been suggested that the United States can help directly by supporting international moves to relieve poverty and environmental degradation worldwide. The following steps have been recommended:

- First, establish a plan to relieve the developing countries of the gigantic interest payments they have been making to U.S. and international banks.
- Second, provide debt relief so that those within the countries who need it most receive the benefits, and not just the wealthy.
- Third, shift the emphasis from *technology*-intensive methods of *harvesting* resources in developing countries to *labor*-intensive means of *maintaining* resources.
- Fourth, count soil, water, and trees in economic balance sheets to account correctly for the great value of these environmental resources.

The United States can exert international leadership by adopting these strategies and encouraging other developed nations to support similar measures.

***Labor-intensive technology is most often the appropriate technology in developing countries.***

The idea behind all these measures is that relieving poverty will help relieve environmental degradation and hunger. To rephrase a well-known adage: If you give a man a fish, he will eat for a day. If you teach him to fish and enable him to buy and maintain his own gear and bait, he will eat for a lifetime and help to feed

*The formal name of the summit was the United Nations Conference on Environment and Development, or UNCED for short.

others. Unlike food giveaways and money doles, which are only stop-gap measures, social programs that permanently improve the lives of the poor can permanently solve the hunger problem.

## Activism and Simpler Lifestyles at Home

Every segment of our society can join in the fight against hunger, poverty, and environmental degradation. The federal government, the states, local communities, big business and small companies, educators, and all individuals, including dietitians and foodservice managers, have many opportunities to resolve these problems.

• ***Government Action*** • Government policies can change to promote sustainability. For example, the government can stop using tax money to pay for the wasteful use of fossil fuels and of fertilizers and pesticides made from them. Instead, it can pay for energy conservation services and crop protection and revise tax laws to reward energy conservation efforts.

Second Harvest
www.secondharvest.org

• ***Business Involvement*** • Businesses can take the initiative to help; some already have. Several large corporations are major supporters of antihunger programs. Many grocery stores and restaurants participate in food recovery programs by giving their out-of-date and leftover food to organizations such as Second Harvest for distribution to hungry people in the community.

• ***Education*** • Educators, including nutrition educators, can teach others about the underlying social and political causes of poverty, the root cause of hunger. At the college level, they can teach the relationships between hunger and population, hunger and environmental degradation, hunger and the status of women, and hunger and the global debt crisis. They can advocate legislation to address these problems. They can teach the poor to develop and run nutrition programs in their own communities and to fight on their own behalf for antipoverty, antihunger legislation.

• ***Foodservice Efforts*** • Dietitians and foodservice managers have a special role to play, and their efforts can make an impressive difference.[15] Their professional organization, the ADA, urges members to conserve resources and minimize waste in both their professional and their personal lives.[16] In addition, the ADA urges its members to educate others on hunger, its consequences, and programs to fight it; to conduct research on the effectiveness and benefits of programs; and to serve as advocates on the local, state, and national levels to help end hunger in the United States.[17]

• ***Individual Choices*** • Individuals can assist the global community in solving its poverty and hunger problems by joining and working for hunger-relief organizations. They can also support organizations that lobby for the needed changes in economic policies toward developing countries.

Most importantly, all individuals can try to make lifestyle choices that consider the environmental consequences. Many small decisions each day have major consequences for the environment. The accompanying "How to" describes how consumers can minimize those impacts when making food-related choices.

The personal rewards of the behaviors presented on p. 639 are many, from saving money to the satisfaction of knowing that you are treading lightly on the earth. But do they really help? They do, if enough people join in. Because we number 6 billion, individual actions can add up to exert an immense impact.

"Be part of the solution, not part of the problem," an adage says. In other words, don't waste time or energy moaning and groaning about how tough things are; do something to improve them. This adage is as applicable to today's global environmental problems as it is to an unwashed dish in the kitchen sink. They are our problems: human beings created them, and human beings must solve them.

***Good planets are hard to find.***

## HOW TO Make Environmentally Friendly Food-Related Choices

Food production taxes environmental resources and causes pollution. Consumers can make environmentally friendly choices at every step from food shopping to cooking and use of kitchen appliances to serving, cleanup, and waste disposal.

**Food Shopping**

Transportation:

- Whenever possible, walk or ride a bicycle; use car pools and mass transit.
- Shop only once a week, share trips, or take turns shopping for each other.
- When buying a car, choose an energy-efficient one.

Food choices:

- Choose foods low on the food chain; that is, eat more plants and fewer animals that eat plants (this suggestion complements the Food Guide Pyramid recommendations for eating for good health).
- Eat small portions of meat; select range-fed beef, buffalo, poultry, and fish.
- Select local foods; they require less transportation, packaging, and refrigeration.

Food packages:

- Whenever possible, select foods with no packages; next best are minimal, reusable, or recyclable ones.
- Buy juices and sodas in large glass or recyclable plastic bottles (not small individual cans or cartons); grains in bulk (not separate little packages); and eggs in pressed fiber cartons (not foam, unless it is recycled locally).
- Carry reusable string or cloth shopping bags; alternatively, ask for plastic bags if they are recyclable.

**Cooking Food**

- Cook foods quickly in a stir-fry, pressure cooker, or microwave oven.
- When using the oven, bake a lot of food at one time and keep the door closed tightly.
- Refuse disposable utensils, dishes, and pans.
- Avoid spray products.

**Kitchen Appliances**

- Use fewer small electrical appliances; open cans, mix batters, sharpen knives, and chop vegetables by hand.
- When buying a large appliance, choose an energy-efficient one.
- Consider solar power to meet home electrical needs.
- Set the water heater at 130°F (54°C), no hotter; put it on a timer; wrap it and the hot-water pipes in insulation; install water-saving faucets.

**Food Serving, Dish Washing, and Waste Disposal**

- Use "real" plates, cups, and glasses instead of disposable ones.
- Use cloth towels and napkins, reusable storage containers with lids, reusable pans, and dishcloths instead of paper towels, plastic wrap, plastic storage bags, aluminum foil, and sponges.
- Run the dishwasher only when it is full.
- Recycle all glass, plastic, and aluminum.
- Compost all vegetable scraps, fruit peelings, and leftover plant foods.

These suggested lifestyle changes can easily be extended from food to other areas.

IN SUMMARY

The global environment, which supports all life, is deteriorating rapidly, largely because of our irresponsible use of resources and energy. Governments, businesses, and all individuals have many opportunities to make environmentally conscious choices, which may help solve the hunger problem, improve the quality of life, and generate jobs. Personal choices, made by many people, can have a great impact.

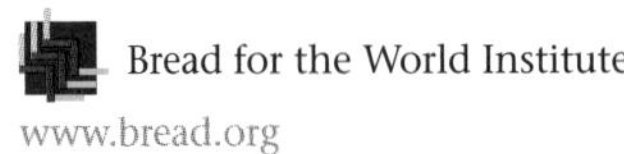


www.bread.org

## Study Questions

These questions will help you review the chapter. You will find the answers in the discussions provided.

1. Identify some reasons why hunger is present in a country as wealthy as the United States. (pp. 629–630)
2. Explain why relieving environmental problems will also help to alleviate hunger and poverty. (pp. 635–636)
3. Discuss the different paths by which rich and poor countries can attack the problems of world hunger and the environment. (p. 637)
4. Describe some strategies that consumers can use to minimize negative environmental impacts when shopping for food, preparing meals, and disposing of garbage. (pp. 638–639)
5. Design a simple action plan for one day for a family of four to introduce changes in their lifestyle to benefit the environment. (p. 639)

These questions will help you prepare for an exam. Answers can be found in Appendix K.

1. Food insecurity refers to the:
   a. uncertainty of foods' safety.
   b. fear of eating too much food.
   c. limited availability of foods.
   d. reliability of food production.
2. The most common cause of hunger in the United States is:
   a. poverty.
   b. alcohol abuse.
   c. mental illness.
   d. lack of education.
3. Food stamp coupons cannot be used to purchase:
   a. tomato plants.
   b. birthday cakes.
   c. cola beverages.
   d. laundry detergent.
4. Which action is not typical of a food recovery program?
   a. gathering potatoes from a harvested field
   b. collecting overripe tomatoes from a wholesaler
   c. offering food stamp coupons to low-income people
   d. delivering restaurant leftovers to a community shelter
5. The most likely cause of death in malnourished children is:
   a. growth failure.
   b. diarrheal disease.
   c. simple starvation.
   d. vitamin A deficiency.
6. Causes of famine do *not* include:
   a. wars.
   b. floods.
   c. drought.
   d. poverty.
7. Which of the following is most critical in providing food to all the world's people?
   a. decreasing air pollution
   b. increasing water supplies
   c. decreasing population growth
   d. increasing agricultural land
8. Which of these items is the most environmentally benign choice?
   a. sponges
   b. plastic bags
   c. aluminum foil
   d. cotton towels
9. Which of these methods uses the most fuel?
   a. baking
   b. stir-frying
   c. microwaving
   d. pressure cooking
10. Which of these purchases is the best choice, for environmental reasons?
    a. fresh fish from a local merchant
    b. frozen fish from a developing country
    c. canned fish from a nationally known food manufacturer
    d. packaged fish from the freezer section of a local supermarket

## Notes


1. L. N. Burby, *World Hunger* (San Diego, Calif.: Lucent Books, 1995), pp. 13–16.
2. Position of The American Dietetic Association: World hunger, *Journal of the American Dietetic Association* 95 (1995): 1160–1162.
3. Position of The American Dietetic Association: Domestic food and nutrition security, *Journal of the American Dietetic Association* 98 (1998): 337–342.
4. USDA Food and Nutrition Service Website www.usda.gov/fcs, visited June 25, 1998.
5. U.S. Department of Agriculture, *A Citizen's Guide to Food Recovery,* December 1996.
6. J. C. Wolgemuth and coauthors, Wasting malnutrition and inadequate nutrient intakes identified in a multiethnic homeless population, *Journal of the American Dietetic Association* 92 (1992): 834–839; M. A. Drake, The nutritional status and dietary adequacy of single homeless women and their children in shelters, *Public Health Reports* 107 (1992): 312–319; B. E. Cohen, N. Chapman, and M. R. Burt, Food sources and intake of homeless persons, *Journal of Nutrition Education* (1 supplement) 24 (1992): 45S–51S.
7. R. W. Kates, Ending deaths from famine: The opportunity in Somalia,

*New England Journal of Medicine* 328 (1993): 1055–1057.

8. I. Darnton-Hill, Developing industrial-governmental-academic partnerships to address micronutrient malnutrition, *Nutrition Reviews* 55 (1997): 76–81.
9. R. Martorell, The role of nutrition in economic development, *Nutrition Reviews* 54 (1996): S66–S71.
10. W. W. Fawzi and coauthors, A prospective study of malnutrition in relation to child mortality in the Sudan, *American Journal of Clinical Nutrition* 65 (1997): 1062–1069; P. W. Yoon and coauthors, The effect of malnutrition on the risk of diarrheal and respiratory mortality in children <2 y of age in Cebu, Philippines, *American Journal of Clinical Nutrition* 65 (1997): 1070–1077.
11. J. Raloff, Can grain yields keep pace? *Science News* 152 (1997): 104–105.
12. P. S. Dasgupta, Population, poverty, and the local environment, *Scientific American,* February 1995, pp. 40–45.
13. Dasgupta, 1995.
14. L. R. Brown and coauthors, *State of the World, 1994* (New York: W. W. Norton, 1994), p. 202.
15. N. I. Hahn, The greening of a school district: How school foodservice led a recycling revolution, *Journal of the American Dietetic Association* 97 (1997): 371.
16. Position of The American Dietetic Association: Natural resource conservation and waste management, *Journal of the American Dietetic Association* 97 (1997): 425–428.
17. Position of The American Dietetic Association, 1998.

# Progress toward Sustainable Agriculture

While some individuals are attempting to make their own personal lifestyles more environmentally benign, as suggested in Chapter 20, others are seeking ways to improve whole sectors of human enterprise, among them, agriculture. To date, large agricultural enterprises have been among the world's biggest resource users and polluters. Is it possible for agriculture to become sustainable? And if so, can the change be made without hurting farmers? These questions are addressed in this highlight.

## *Costs of Producing Food Unsustainably*

The current environmental and social costs of agriculture and the food industry take many forms. Among them are resource waste and pollution and disruption of farm communities.

### Resources and Pollution

Producing food costs the earth dearly. First of all, to grow food, we clear land—prairie, wetland, and forest—which always incurs losses of native ecosystems and wildlife.

Then we plant crops or graze animals on the land. The soil loses nutrients as each crop is taken from it, so fertilizer is applied. Some fertilizer runs off, polluting the waterways. Some plowed soil runs off, which clouds the waterways and interferes with the growth of aquatic plants and animals.

To protect crops against weeds and pests, we apply herbicides and pesticides. These chemicals also pollute the water and, wherever the wind carries them, the air. Most herbicides and pesticides kill not only weeds and pests, but also native plants, native insects, and animals that eat those plants and insects. Ironically, widespread use of pesticides and herbicides causes resistant pests and weeds to evolve. Consequently, farmers must use still more pesticides and herbicides. Some farmers rely so heavily on pesticides, herbicides, fertilizers, and soil boosters that the expense is becoming prohibitive. Furthermore, pesticide residues are becoming a safety concern, as Chapter 19 pointed out.

Finally, we irrigate. Unlike rain, irrigation water contains salts. The water evaporates, but its salts do not, so the soil becomes increasingly salty, hindering plant growth. Irrigation also depletes the water supply over time because it pulls water from surface waters or from underground, and then, the water evaporates or runs off. This process, carried to the extreme, can dry up rivers and lakes and lower the water table of whole regions. A vicious cycle develops: the drier the region becomes, the more the farmers must irrigate, and the more they irrigate, the drier the region becomes.

Like plant crops, herds of livestock damage land, causing losses of native plants and animals, soil erosion, water depletion, and desert formation. Alternatively, raising animals in concentrated areas (such as cattle feedlots and hog farms) is also environmentally costly: wastes pollute the soil and water, and feed grain requires land, fertilizers, herbicides, pesticides, and irrigation. In the United States, more cropland is used to produce feed grains for livestock than to produce grains for people.

Fishing also incurs environmental costs. Fishing easily becomes overfishing and depletes stocks of the very fish that people need to eat. Some fishing methods (such as nets and filament line) kill aquatic animals other than the ones sought and deplete large populations of nonfood animals, such as dolphins. Also, fish production is energy-intensive, requiring fuel for boats, refrigeration, processing, packing, and transport. Moreover, bioaccumulation of toxins in fish is becoming a serious problem in some areas; in others it rules out fish consumption altogether.

The entire food industry, whether based on growing crops, raising livestock, or fishing, requires energy, which entails burning fossil fuel. Massive fossil fuel use threatens our planet by causing air and water pollution, global warming, ozone depletion, and other environmental ills.

In the United States, the food industry consumes about 20 percent of all the energy the nation uses. Each year, we use 1500 liters of oil (over 350 gallons) *per person* to produce, process, distribute, and prepare our food. Most of this energy is used to run farm machinery and to produce fertilizers and pesticides. Energy is also used to prepare, package, transport, refrigerate, store, cook, and wash our foods.

Many national and world agencies are concerned about the environmental ramifications of agriculture. The nation's most prestigious national scientific research body, the National Research Council of the National Academy of Sciences, has reported that agriculture is the largest single source of nonpoint pollution of surface water in the nation. (See the accompanying glossary for definitions of "nonpoint water pollution" and related terms.) Pollution from "point sources," such as sewage plants or factories, is relatively easy to control, but runoff from fields and pastures enters

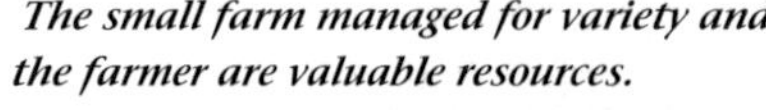

*The small farm managed for variety and the farmer are valuable resources.*

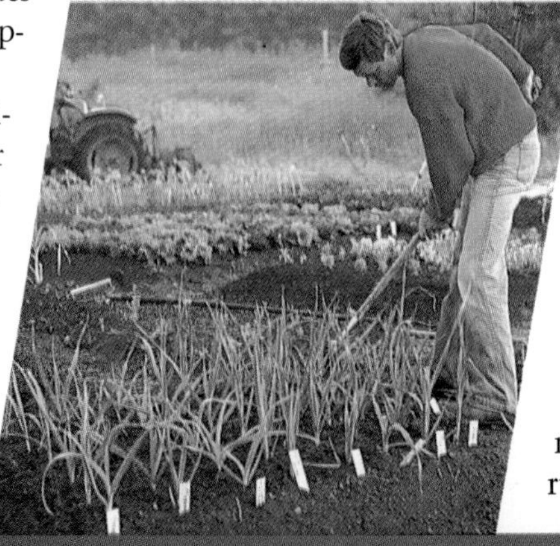

waterways from all over broad regions and is nearly impossible to control.

The World Resources Institute has concluded that agriculture is destroying its own foundation. In just the past 40 years, agricultural activities have ruined more than 10 percent of the earth's most fertile land, an area the size of China and India combined. Over 20 million acres have been so damaged that they will be impossible to reclaim.

Agriculture is also weakening its own underpinnings by failing to conserve species diversity. By the year 2050, some 40,000 more plant species may become extinct. The United Nations' Food and Agriculture Organization attributes many of the losses, which are already occurring daily, to modern farming practices, as well as to population growth. The increasing uniformity of global eating habits is also having an effect. People everywhere are eating the same limited array of foods, so local regions' native, genetically diverse plants no longer seem worth preserving. Yet in the future, as the climate warms and the earth changes, those may be the very plants that people will need for food sources. A wild species of corn that grows in a dry climate, for example, might contain the genetic information necessary to help make domestic corn resistant to drought.

The culprits that attend the growing of crops—land clearing; irrigation; fertilizer, pesticide, and herbicide overuse; and loss of genetic diversity—have taken a tremendous toll on the earth. In short, our ways of producing foods are, for the most part, not sustainable.

## From Family Farms to Agribusiness

During the early and mid-1980s, U.S. agriculture encountered serious economic hardships. Exports of farm produce fell worldwide. Recession occurred. Federal loans became expensive. Other countries increased their agricultural production and exports. Many U.S. farmers, particularly those who specialized in export crops, suffered heavy financial losses. Some were unable to pay their debts and had to leave farming. Between 1980 and 1993, more than 350,000 farms, representing 15 percent of the farms in the United States, disappeared. Thousands of U.S. farmers remain frustrated and in debt, threatened with poverty.

Most U.S. farmers have little or no control over what products they produce, the costs of their supplies, or the prices they receive for their goods. Incidentally, the prices farmers receive for their crops are only a small part of the cost of foods we eat. In 1985, for example, consumers paid $29 million just for the packages on their foods. In the same year, farmers received less than that amount for the food itself.

*Locally grown foods offer benefits to both the local economy and the global environment. (Courtesy of CNN)*

Unfortunately, as small family farms disappear, they are being replaced by huge corporation-owned farms that are more likely to engage in technology-intensive practices that use large machinery and large land areas. Huge farms and ranches, collectively part of the massive food-producing enterprise called agribusiness, tend to use little local labor, and the profits earned tend not to stay in local communities. Agribusinesses also tend to place a higher priority on producing abundant food than on protecting soil, water, and local biodiversity. As a consequence, they may overuse fertilizers and pesticides, overuse land at the cost of soil erosion, and use irrigation water wastefully. In an effort to compete with agribusinesses, small farmers, too, may adopt similar unsustainable practices.

Nevertheless, the larger operations of the agribusinesses allow them to set the prices of their products lower than smaller, local farms can afford. Consequently, local grocery stores offer tomatoes from Florida, carrots from California, and pineapples from Hawaii at prices no local operators could beat, even if they could grow those products in their climates. Roadside stands offer bundles of greens and baskets of local fruits and vegetables, but sometimes

### Glossary

**agribusiness:** agriculture practiced on a massive scale by large corporations owning vast acreages and employing intensive technological, fuel, and chemical inputs.

**externalities:** hidden costs that are not reflected in the prices of things, such as the costs of environmental deterioration, that permit agribusiness foods to be sold at artificially low prices.

**integrated pest management (IPM):** management of pests using a combination of natural and biological controls and few or no pesticides.

**nonpoint water pollution:** water pollution caused by runoff from all over an area rather than from discrete "point" sources. An example is the pollution caused by runoff from agricultural fields.

**sustainable** or **alternative agriculture:** agricultural practices that use individualized approaches appropriate to local conditions so as to minimize technological, fuel, and chemical inputs.

less conveniently and at higher prices than many shoppers are willing to pay.

Environmental and social costs, such as pollution and hardship to farmers, are not reflected in the prices of products. These costs are therefore called external costs, or externalities. People don't pay for externalities directly when they buy the products; they pay indirectly through the costs of cleaning up pollution. Sometimes people do not pay in money at all, but in health and social stresses; and the environment pays in resource losses and deterioration. If these costs *were* included in prices, the prices of unsustainably produced products would be much higher. People would then buy more products produced by smaller farms and ranches with less technology, less pollution, and more labor.

## Proposed Solutions

After reviewing the problems associated with U.S. agriculture, the members of the National Research Council expressed the intent to develop an alternative, sustainable mode of producing food. The goals were to conserve land, water, and energy; to exert minimal environmental impacts; and to produce abundant food profitably.

### Sustainable Agriculture

Sustainable agriculture is not a single system but a set of practices that can be matched to particular needs in local areas. It emphasizes careful use of natural processes, wherever possible, rather than chemically intensive methods; in other words, it protects the environment.[1] To reduce soil erosion, farmers can plow compost into soil, which also enriches the nutrient content of the soil (reducing the need for fertilizers) and retains moisture (reducing the need for irrigation). To control pests, a farmer practicing integrated pest management might use hardy plants, crop rotation, and natural predators instead of pesticides. These are just a few examples; Table H20-1 offers more by contrasting sustainable agricultural methods with unsustainable methods. Incidentally, many of these sustainable techniques are not really new; they would be familiar to our great grandparents, but are now being rediscovered for their environmental benefits.

Sometimes sustainable agriculture involves a greater effort. To address the erosion problem in the United States, for example, a conservation program sponsored by the USDA paid farmers to stop farming on highly erodible or environmentally sensitive lands and to plant grasses, trees, or other plants instead.[2] Such efforts have preserved hundreds of millions of tons of topsoil, provided nesting areas for birds and other wildlife, limited carbon dioxide emissions, and saved billions of dollars in environmental cleanups.

Farming by sustainable methods produces crops reliably and lowers farmers' financial risks by reducing the effects of fluctuating prices for pesticides, fertilizers, and the like. Both large and small farms can use these practices, and many different machineries are compatible with them. Each technique has a different value for farmers of different crops in different regions. For example, corn and soybean farmers in the Midwest might be able to reduce or eliminate routine insecticide use relatively easily, whereas fruit and vegetable growers in the hot and humid Southeast would find this harder to do. Not all crops can grow reliably without pesticides, but many can. Genetically coding crops to repel harmful insects promises to significantly reduce the use of pesticides in the near future.

Alternative agriculture has some apparent disadvantages, but advantages offset them. For example, as chemical use falls, yields per acre also fall somewhat, but costs per acre also fall, so the return per acre may be the same as or greater than before. More money goes to farmers and less to the fuels, fertilizers, pesticides, and irrigation. The end result makes both farmers and consumers better off financially.

Low-input agriculture works. More than 30,000 of the nation's farmers are successfully using sustainable techniques such as those described in Table H20-1. Such methods can indefinitely support a healthy food supply, restore soil and water resources, and revitalize individual farms and rural communities.

 USDA Alternative Farming Systems Information Center
www.nal.usda.gov/afsic

### Consumer Choices

Consumers can reduce the amount of energy used in food production by centering their diets on foods that require little energy. That means eating more foods derived from plants and fewer foods derived from animals.

Some foods require more energy for their production than others. The least energy is needed for grains: about one-third kcalorie of fuel is spent to produce each kcalorie of grain. Fruits and vegetables are intermediate, and most animal-derived foods require from 10 to 90 kcalories of fuel per kcalorie of edible food. Thus most animal-protein products require much more energy, as well as more land and water, than do plant-protein products. An exception is livestock raised on the open range; these animals require about as much energy as do most plant foods. We raise so much more grain-fed, than range-fed, livestock, however, that the average energy requirement for meat production is high. Figure H20-1 (on p. 646) shows how much less fuel vegetarian diets require than meat-based diets and shows that vegan diets require the least fuel of all.

To support our meat intake, we maintain several billion livestock, about four times our own weight in animals. Livestock consume ten times as much grain each day as we do. We could use much of that grain to make grain products for ourselves and for others around the world. The shift could free up enough grain to feed 400 million people and would use less fuel, water, and land. The contrast between resource use in the United States and in China, a society that lives largely on plant foods, is startling (see Figure H20-2). The Chinese diet and cooking techniques are also land-efficient, as they must be in view of the scarcity of agricultural land and fuels. Nearly all of the food energy comes from plants rather than animals. A million kcalories in wheat or rice can be produced on less than 1 acre of land; a million kcalories in beef require 17 acres. In a world that is often wasteful of fuel and land, the Chinese way of eating offers a model to all nations. Other plant-centered diets, such as the traditional Mediterranean diet, are also environmentally responsible.[3] They rely on fewer domestic animals and therefore place fewer demands on soil, water, and energy.

Part of the solution to the livestock problem may be to cease feeding grain to animals and return to grazing them on the open range, which can be a sustainable practice. Ranchers have to manage the grazing carefully to hold the cattle's numbers to what the land can support without environmental degradation. To accomplish this, the economic benefits of traditional livestock and feed-growing operations would have to end. If producers were to pay the true costs of the irrigation water, fertilizers, pesticides, fuels, and lands they use, the prices of meats might double or triple. According to classic economic theory, people would then buy less meat (reducing demand), and producers would respond by producing less meat (reducing supply). Meat production would then fall to a sustainable level.

Some consumers are taking action without waiting for prices to change: they are choosing smaller portions of meat or selecting range-fed beef or buffalo only. Livestock on the range eat grass, which people cannot eat. "Rangeburger" buffalo also offers health advantages over grain-fed beef. It

**Table H20-1**
**Alternative Agricultural Techniques**

| Unsustainable Practice | Sustainable Practice |
|---|---|
| • Grow the same crop repeatedly on the same patch of land. This takes more and more nutrients out of the soil, making fertilizer use necessary; favors soil erosion; and invites weeds and pests to become established, making pesticide use necessary. | • Rotate crops. This increases nitrogen in the soil so there is less need to buy fertilizers. If used with appropriate plowing methods, rotation reduces soil erosion. An acre of land planted one year in corn, the next in wheat, and the next in clover loses 2.7 tons of topsoil each year, but if it is planted only in corn three years in a row, it will lose 19.7 tons a year. Rotation also reduces problems caused by weeds and pests. |
| • Use fertilizers generously. | • Reduce the use of fertilizers, and use livestock manure more effectively. This means storing it during the nongrowing season and applying it during the growing season.<br>• Alternate nutrient-devouring crops with nutrient-restoring crops.<br>• Plant legumes between grain crops (because legumes' roots leave nitrogen in the soil).<br>• Compost on a large scale, including all plant residues not harvested. Plow the compost into the soil to improve its water-holding capacity. |
| • Feed livestock in feedlots where their manure produces a major water pollution problem. Piled in heaps, it also releases methane, a global-warming gas. | • Feed livestock or buffalo on the open range where their manure will fertilize the ground on which plants grow and will release no methane. Alternatively, at least collect feedlot animals' manure and use it for fertilizer or, at the very least, treat it before release. |
| • Spray herbicides and pesticides over large areas to wipe out weeds and pests. | • Apply ingenuity in weed and pest control. Use rotary hoes twice instead of herbicides once. Treat when and where necessary only. Spot treat weeds by hand.<br>• Rotate crops to foil pests that lay their eggs in the soil where last year's crop was grown.<br>• Use resistant crops. Genetically improve crops so that they resist pests and diseases.<br>• Time the planting of crops so that pests that hatch at other times cannot gain access to them.<br>• Use biological controls such as predators that destroy the pests. |
| • Plow the same way everywhere, allowing unsustainable water runoff and erosion. | • Plow in ways tailored to different areas. Conserve both soil and water by using cover crops, crop rotation, and contour plowing. |
| • To prevent disease in livestock, inject animals with antibiotics. | • Maintain animals' health so that they can resist disease by way of their own vigor. |
| • Irrigate on a large scale. | • Irrigate only during dry spells and apply only spot irrigation. |
| • Use fuel freely. | • Use machinery scaled to the job at hand and operate it at efficient speeds.<br>• Combine operations. Harrow, plant, and fertilize in the same operation.<br>• Use diesel fuel. Use solar and wind energy on farms. Use methane from manure. Be open-minded to alternative energy sources.<br>• Change highway funding so that the tax burden falls more heavily on truckers than on other highway users. Transport foods by rail or water (building and maintaining rail lines can add more jobs than are lost in trucking). |

**Figure H20-1**

**Amounts of Fuel Required to Feed People Eating at Different Points on the Food Chain**

Three people who eat differently are compared here. Each has the same energy intake: 3300 kcalories a day. The fossil fuel amounts necessary to produce these different diets are calculated based on U.S. conditions.

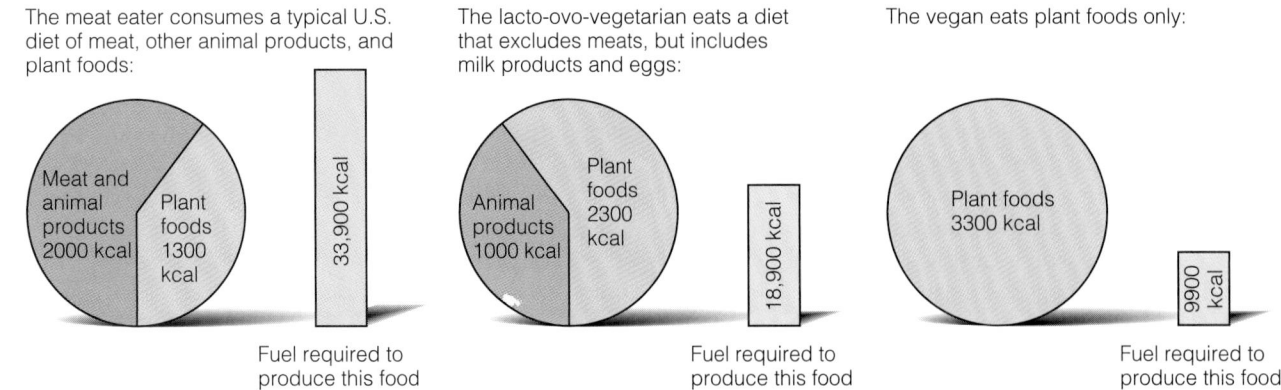

The meat eater consumes a typical U.S. diet of meat, other animal products, and plant foods:

Meat and animal products 2000 kcal

Plant foods 1300 kcal

33,900 kcal

Fuel required to produce this food

The lacto-ovo-vegetarian eats a diet that excludes meats, but includes milk products and eggs:

Animal products 1000 kcal

Plant foods 2300 kcal

18,900 kcal

Fuel required to produce this food

The vegan eats plant foods only:

Plant foods 3300 kcal

9900 kcal

Fuel required to produce this food

Source: Adapted from D. Pimentel, *Food, Energy and the Future of Society* (Boulder, Colo.: Associated University Press, 1980). Figure 5, p. 27.

is lower in fat, and the fat has more polyunsaturated fatty acids, including the omega-3 type.[4]

Some consumers are opting for vegetarian, and even vegan, diets—at least occasionally. Shifting to a fish diet does not appear to be a practical alternative at present, although aquaculture (fish farming) shows promise of providing nutritious food at a price both people and the environment can afford.[5]

Chapter 20 and this highlight have presented many problems and have suggested that, while the problems are global in scope, the actions of individual people lie at the heart of the solutions. On learning of this, concerned people may take a perfectionist attitude, believing that they "should" be doing more than they realistically can, and so feel defeated. Yet, striving for perfection, even while falling short, is a way to achieve progress well worth celebrating. A positive attitude can bring about improvement, and sometimes improvement is enough. Celebrate the changes that are possible today by making them a permanent part of your life; do the same with changes that become possible tomorrow and every day thereafter. The results may surprise you.

**Figure H20-2**

**Resource Use in the United States and China Compared**

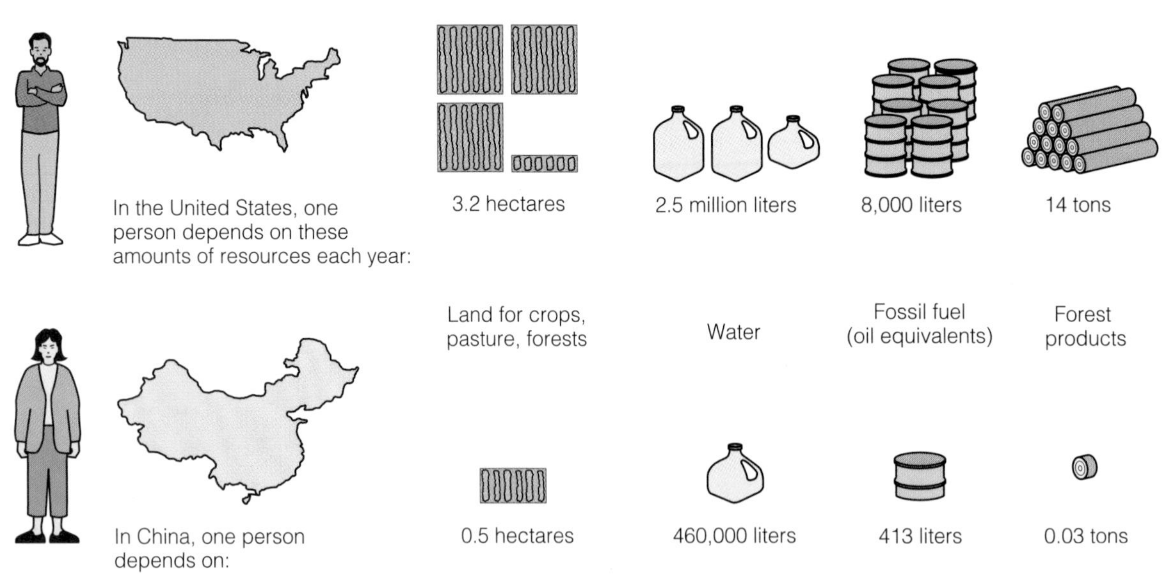


Source: D. Pimentel and M. Pimentel, Land, energy, and water: The constraints governing ideal U.S. population size. *The NPG* (Negative Population Growth) *Forum,* January 1990. Table 2, p. 2.

## Notes


1. D. L. Plucknett and D. L. Winkelmann, Technology for sustainable agriculture, *Scientific American,* September 1995, pp. 182–186.
2. T. Adler, Prairie tales, *Science News* 149 (1996): 44–45.
3. J. D. Gussow, Mediterranean diets: Are they environmentally responsible? *American Journal of Clinical Nutrition* 61 (1995): 1383S–1389S.
4. S. Smith, professor of nutrition, University of New Hampshire, Durham, NH, personal communication, August 1993.
5. C. Flavin and J. E. Young, Shaping the next industrial revolution, in L. R. Brown, *State of the World, 1993* (New York: W. W. Norton, 1993), pp. 180–199.

# Appendixes

Contents



A

# CELLS, HORMONES, AND NERVES

This appendix is offered as an optional chapter for readers who want to enhance their understanding of how the body coordinates its activities. The text presents a brief summary of the structure and function of the body's basic working unit (the cell) and of the body's two major regulatory systems (the hormonal system and the nervous system).

## The Cell

The body's organs are made up of millions of cells and of materials produced by them. Each cell is specialized to perform its organ's functions, but all cells have common structures (see Figure A-1). Every cell is contained within a cell membrane. The cell membrane assists in moving materials into and out of the cell, and some of its special proteins act as "pumps" (described in Chapter 6). Some features of cell membranes, such as microvilli (Chapter 3), permit cells to interact with other cells and with their environments in highly specific ways.

Inside the membrane lies the cytoplasm, or cell "fluid." The cytoplasm contains much more than just fluid, though. It is a highly organized system of fibers, tubes, membranes, particles, and subcellular organelles as complex as a city. These parts intercommunicate, manufacture and exchange materials, package and prepare materials for export, and maintain and repair themselves.

Within each cell is another membrane-enclosed body, the nucleus. Inside the nucleus are the chromosomes, which contain the genetic material, DNA. The DNA encodes all the instructions for carrying out the cell's activities. The role of DNA in coding for cell proteins is summarized in Chapter 6, Figure 6-6. Chapter 6 also describes the variety of proteins produced by cells and the ways they perform the body's work.

Among the organelles within a cell are ribosomes, mitochondria, and lysosomes. Figure 6-6 briefly refers to the ribosomes; they assemble amino acids into proteins, following directions conveyed to them by RNA copies from the DNA in the chromosomes.

The mitochondria are made of intricately folded membranes that bear thousands of highly organized sets of enzymes on their inner and outer surfaces. Although mentioned only briefly in this book's chapters, their presence is implied whenever the enzymes of the TCA cycle and electron transport chain are mentioned because the mitochondria house all these enzymes.* Mitochondria are therefore crucial to aerobic metabolism, described in Chapter 7, and muscles conditioned to work aerobically are packed with them.

The lysosomes are membranes that enclose degradative enzymes. When a cell needs to self-destruct or to digest materials in its surroundings, its lysosomes free their enzymes. Lysosomes are active when tissue repair or remodeling is taking place—for example, in cleaning up infections, healing wounds, shaping embryonic organs, and remodeling bones.

**cell:** the basic unit of life, of which all living things are composed. Every cell is surrounded by a membrane and contains cytoplasm, within which are organelles and a nucleus; the cell nucleus contains chromosomes.

**cell membrane:** the membrane that surrounds the cell and encloses its contents; made primarily of lipid and protein.

**cytoplasm** (SIGH-toe-plazm): the cell contents, except for the nucleus.
- **cyto** = cell
- **plasm** = a form

**organelles:** subcellular structures such as ribosomes, mitochondria, and lysosomes.
- **organelle** = little organ

**nucleus:** a major membrane-enclosed body within every cell, which contains the cell's genetic material, DNA, embedded in chromosomes.
- **nucleus** = a kernel

**chromosomes:** a set of structures within the nucleus of every cell that contains the cell's genetic material, DNA, associated with other materials (primarily proteins).

**ribosomes** (RYE-boh-zomes): protein-making organelles in cells; composed of RNA and protein.
- **ribo** = containing the sugar ribose (in RNA)
- **some** = body

**mitochondria** (my-toe-KON-dree-uh); singular **mitochondrion:** the cellular organelles responsible for producing ATP aerobically; made of membranes (lipid and protein) with enzymes mounted on them.
- **mitos** = thread (referring to their slender shape)
- **chondros** = cartilage (referring to their external appearance)

**lysosomes** (LYE-so-zomes): cellular organelles; membrane-enclosed sacs of degradative enzymes.
- **lysis** = dissolution

*For the reactions of glycolysis, the TCA cycle, and the electron transport chain, see Chapter 7 and Appendix C. The reactions of glycolysis take place in the cytoplasm; the end product acetyl CoA moves into the mitochondria; and the TCA and electron transport reactions take place there. The mitochondria then release carbon dioxide, water, and ATP as their end products.

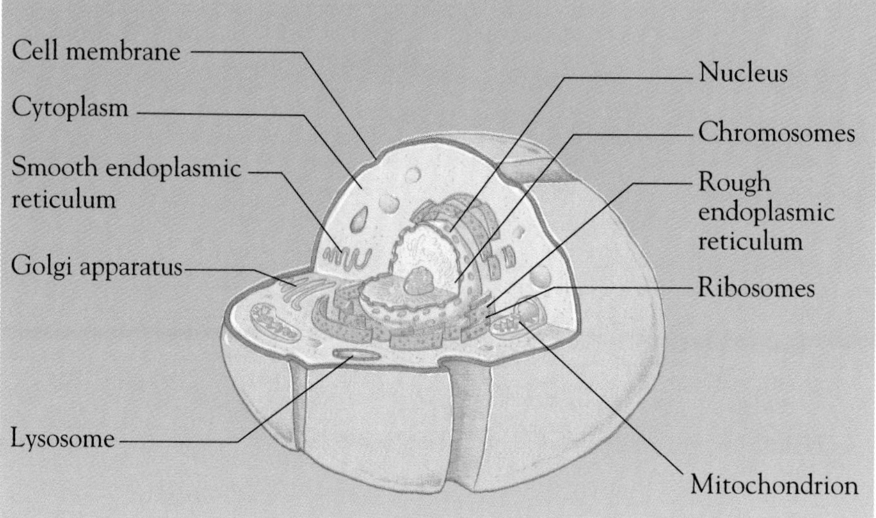


**Figure A-1**
**The Structure of a Typical Cell**

The cell shown might be one in a gland (such as the pancreas) that produces secretory products (enzymes) for export (to the intestine). The rough endoplasmic reticulum with its ribosomes produces the enzymes; the smooth reticulum conducts them to the Golgi region; the Golgi membranes merge with the cell membrane, where the enzymes can be released into the extracellular fluid.

A

Besides these and other cellular organelles, the cell's cytoplasm contains a highly organized system of membranes, the endoplasmic reticulum. The ribosomes may either float free in the cytoplasm or be mounted on these membranes. A membranous surface dotted with ribosomes looks speckled under the microscope and is called "rough" endoplasmic reticulum; such a surface without ribosomes is called "smooth." Some intracellular membranes are organized into tubules that collect cellular materials, merge with the cell membrane, and discharge their contents to the outside of the cell; these membrane systems are named the Golgi apparatus, after the scientist who first described them. The rough and smooth endoplasmic reticula and the Golgi apparatus are continuous with one another, so secretions produced deep in the interior of the cell can be efficiently transported to the outside and released. These and other cell structures enable cells to perform the multitudes of functions for which they are specialized.

**rough endoplasmic reticulum** (en-doh-PLAZ-mic reh-TIC-you-lum): intracellular membrane dotted with ribosomes, where protein synthesis takes place.

- **endo** = inside
- **plasm** = the cytoplasm

**smooth endoplasmic reticulum:** smooth intracellular membrane bearing no ribosomes.

**Golgi** (GOAL-gee) **apparatus:** a set of membranes within the cell where secretory materials are packaged for export.

The actions of cells are coordinated by both hormones and nerves, as the next sections show. Among the types of cellular organelles are receptors for the hormones delivering instructions that originate elsewhere in the body. Some hormones penetrate the cell and its nucleus and attach to receptors on chromosomes, where they activate certain genes to initiate, stop, speed up, or slow down synthesis of certain proteins as needed. Other hormones attach to receptors on the cell surface and transmit their messages from there. The hormones are described in the next section; the nerves, in the one following.

The study of hormones and their effects is **endocrinology.**

# The Hormones

A hormonal message originates in a gland and travels as a chemical compound—a hormone—in the bloodstream. The hormone flows everywhere in the body, but only its target organs respond to it, because only they possess the receptors to receive it.

**hormone:** a chemical messenger. Hormones are secreted by a variety of endocrine glands in response to altered conditions in the body. Each hormone travels to one or more specific target tissues or organs, where it elicits a specific response to maintain homeostasis.

The hormones, the glands they originate in, and their target organs and effects are described in this section. Many of the hormones you might be interested in are included, but only a few are discussed in detail. Figure A-2 identifies the glands that produce the hormones discussed in this section.

The hormonal system is a complex system in which many of the parts interact with one another. For example, several hormones are produced in the anterior pituitary gland in the brain. All of these hormones are regulated by other hormones produced in another part of the brain, the hypothalamus. Furthermore, each of the pituitary gland hormones has effects on the production of compounds elsewhere in the body. Some of these compounds are also hormones that will

**hypothalamus:** a brain center that controls activities such as maintenance of water balance, regulation of body temperature, and control of appetite (see Figure A-2).

- **hypo** = below
- **thalamus** = another brain region

**Figure A-2**
**The Endocrine System**

These organs and glands release hormones that regulate body processes. An **endocrine gland** secretes its product directly into *(endo)* the blood; for example, the pancreas cells that produce insulin. An **exocrine** gland secretes its product(s) out *(exo)* of the gland through a duct into a cavity; the sweat glands of the skin and the enzyme-producing glands of the pancreas are both examples. The pancreas is therefore both an endocrine and an exocrine gland.

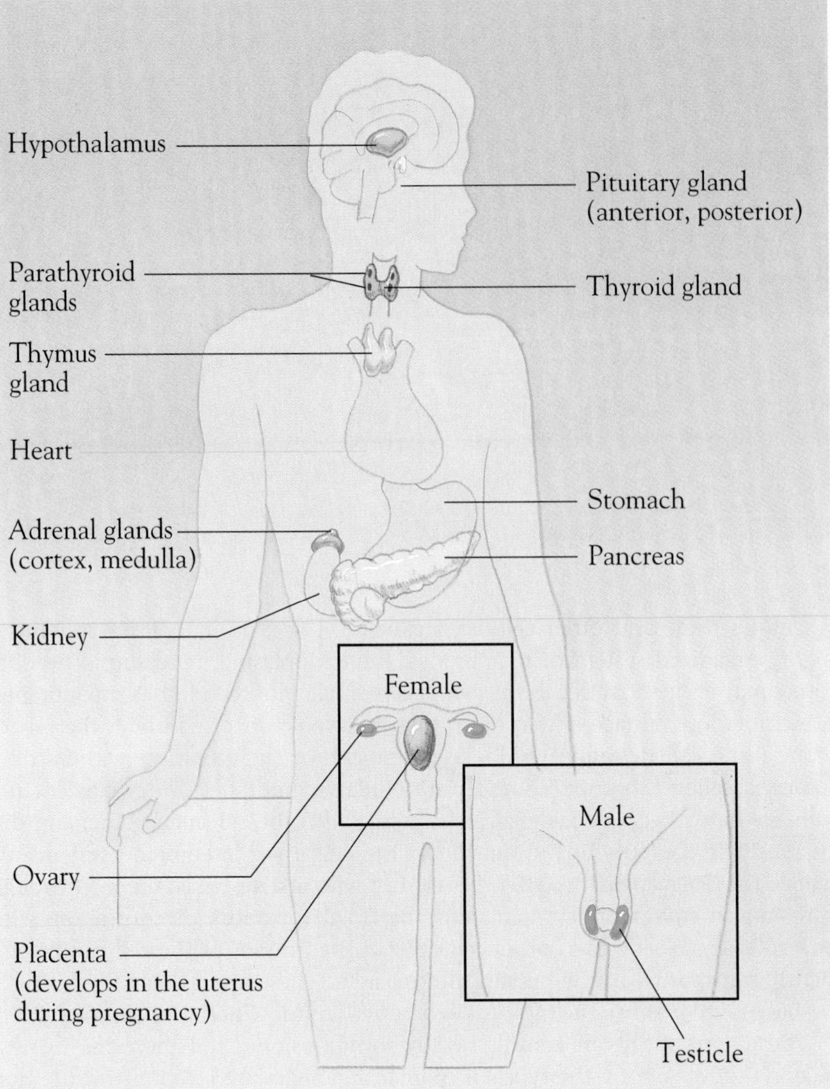


affect still other body parts. A hormone may travel far from its point of origin and ultimately have profound, even unexpected, effects.

The **pituitary gland** in the brain has two parts—the **anterior** (front) and the **posterior** (hind) parts.

**adrenocorticotropin:** a hormone, so named because it stimulates *(trope)* the adrenal cortex. The adrenal gland, like the pituitary, has two parts, in this case an outer portion *(cortex)* and an inner core *(medulla)*.

**follicle (ovarian):** that part of the female reproductive system where the ovary lies and eggs are produced.

**luteinizing hormone:** a hormone that stimulates the development of the corpus luteum (the small tissue that develops from a ruptured ovarian follicle and secretes hormones); so called because the follicle turns yellow as it matures.

- **lutein** = a yellow pigment

**prolactin:** a hormone so named because it promotes *(pro)* the production of milk *(lacto)*.

## Hormones of the Pituitary Gland and Hypothalamus

The anterior pituitary gland produces the following hormones, each of which acts on one or more target organs and elicits a characteristic response:

- Adrenocorticotropin (ACTH) acts on the adrenal cortex, promoting the making and release of its hormones.
- Thyroid-stimulating hormone (TSH) acts on the thyroid gland, promoting the making and release of thyroid hormone.
- Growth hormone (GH) works on all tissues, promoting growth, fat breakdown, and the formation of antibodies.
- Follicle-stimulating hormone (FSH) works on the ovaries in the female, promoting their maturation, and on the testicles in the male, promoting sperm formation.
- Luteinizing hormone (LH) also acts on the ovaries, advancing their maturation, the making of progesterone and estrogens, and ovulation; and on the testicles, promoting the making and release of androgens (male hormones).
- Prolactin, secreted in the female during pregnancy and lactation, acts on the mammary glands to stimulate their growth and the making of milk.

- Melanocyte-stimulating hormone (MSH) acts on the pigment cells, promoting the making and dispersal of pigment.

Each of these seven hormones has one or more signals that turn it on and another (or others) that turns it off. Among the controlling signals are several hormones from the hypothalamus:

- Corticotropin-releasing hormone (CRH), which promotes release of ACTH, is turned on by stress and turned off by ACTH when enough has been released.
- TSH-releasing hormone (TRH), which promotes release of TSH, is turned on by large meals or low body temperature.
- GH-releasing hormone (GRH), which stimulates the release of GH, is turned on by insulin.
- GH-inhibiting hormone (GIH or somatostatin), which inhibits the release of GH and interferes with the release of TSH, is turned on by hypoglycemia and/or physical activity and is rapidly destroyed by body tissues so that it does not accumulate.
- FSH/LH–releasing hormone (FSH/LH–RH) is turned on in the female by nerve messages or low estrogen and in the male by low testosterone.
- Prolactin-inhibiting hormone (PIH) is turned on by high prolactin levels and off by estrogen, testosterone, and suckling (by way of nerve messages).
- MSH-inhibiting hormone (MIH) is turned on by the hormone melatonin.

Let's examine some of these controls. PIH, for example, responds to high prolactin levels (remember, prolactin promotes the making of milk). High prolactin levels ensure that milk is made and—by calling forth PIH—ensure that prolactin levels don't get too high. But when the infant is suckling—and creating a demand for milk—PIH is not allowed to work (suckling turns off PIH). The consequence: prolactin remains high, and milk production continues. Demand from the infant thus directly adjusts the supply of milk. This example shows how the need is met through the cooperation of the nerves and hormones.

As another example, consider CRH. Stress, perceived in the brain and relayed to the hypothalamus, switches on CRH. On arriving at the pituitary, CRH switches on ACTH. Then ACTH acts on its target organ, the adrenal cortex, which responds by producing and releasing stress hormones, and the stress response is under way. Events cascading from there involve every body cell and many other hormones.

The numerous steps required to set the stress response in motion make it possible for the body to fine-tune the response; control can be exerted at each step. These two examples illustrate what the body can do in response to two different stimuli—producing milk in response to an infant's need and gearing up for action in an emergency.

Two hormones produced by the posterior pituitary gland are:

- Antidiuretic hormone (ADH), or vasopressin.
- Oxytocin.

ADH promotes contraction of arteries and acts on the kidneys to prevent water from being excreted. It is turned on whenever the blood volume is low, the blood pressure is low, or the salt concentration of the blood is high (see Chapter 12). It is turned off by the return of these conditions to normal. Oxytocin is produced in response to reduced progesterone levels, suckling, or the stretching of the cervix and acts on two target organs. One, the uterus, contracts, thus inducing labor; the other, the mammary glands, release milk.

## Hormones That Regulate Energy Metabolism

Hormones produced by a number of different glands have effects on energy metabolism:

- Insulin from the pancreas beta cells.
- Glucagon from the pancreas alpha cells.

**melanocyte** (MEL-an-oh-cite or mel-AN-oh-cite): a cell containing the pigment melanin.
- **cyte** = cell

Hormones that are turned off by their own effects are said to be regulated by **negative feedback.** For example, when a pituitary gland hormone has caused the release of a substance from a target organ, that substance itself switches off the original hormone signal (that is, it feeds back negatively).

**somatostatin (GIH):** a hormone that inhibits the release of growth hormone; the opposite of **somatotropin (GH).**
- **somato** = body
- **stat** = keep the same
- **tropin** = make more

**antidiuretic hormone (ADH):** the hormone that prevents water loss in urine (also **vasopressin**).
- **anti** = against
- **di** = through
- **ure** = urine
- **vaso** = blood vessels
- **pressin** = pressure

**oxytocin** (OK-see-TOE-sin): a hormone that stimulates the mammary glands to eject milk during lactation and the uterus to contract during childbirth.
- **oxy** = quick
- **tocin** = childbirth

**cervix:** the circular muscle that guards the opening of the uterus. When a baby is about to be born, the cervix begins to stretch.
- **cervic** = neck

Norepinephrine and epinephrine were formerly called noradrenalin and adrenalin, respectively.

**glucocorticoid:** a hormone from the adrenal cortex that affects the body's management of glucose.
- **gluco** = glucose
- **corticoid** = from the cortex

- Thyroxin from the thyroid gland.
- Norepinephrine and epinephrine from the adrenal medulla.
- Growth hormone (GH) from the anterior pituitary (already mentioned).
- Glucocorticoids from the adrenal cortex.

Insulin is turned on by many stimuli, including raised blood glucose. It acts on cells to increase glucose and amino acid uptake into them and to promote the secretion of GRH. Glucagon responds to low blood glucose and acts on the liver to promote the breakdown of glycogen to glucose, the conversion of amino acids to glucose, and the release of glucose. Thyroxin responds to TSH and acts on many cells to increase their metabolic rate, growth, and heat production. The hormones norepinephrine and epinephrine respond to stimulation by sympathetic nerves and produce reactions in many cells that facilitate the body's readiness for fight or flight: increased heart activity, blood vessel constriction, breakdown of glycogen and glucose, raised blood glucose levels, and fat breakdown. Norepinephrine and epinephrine also influence the secretion of the many hormones from the hypothalamus that exert control on the body's other systems. The glucocorticoid hormones become active during times of stress and carbohydrate metabolism.

Every body part is affected by these hormones. Each different hormone has unique effects; and hormones that oppose each other are produced in carefully regulated amounts, so each can respond to the exact degree that is appropriate to the condition.

## Hormones That Adjust Other Body Balances

Hormones are involved in moving calcium into and out of the body's storage deposits in the bones:

**calcitonin** (KAL-see-TOE-nin): a hormone secreted by the thyroid gland that regulates (tones) calcium metabolism.

**parathyroid:** named for their location, the four parathyroid glands nestle in the surface layers of the two thyroid lobes in the neck.
- **para** = beside, next to

- Calcitonin (CT) from the thyroid gland.
- Parathormone (parathyroid hormone or PTH) from the parathyroid gland.
- Vitamin D from the kidneys.

One of calcitonin's target tissues is the bones, which respond by storing calcium from the bloodstream whenever blood calcium rises above the normal range. Calcitonin also acts on the kidneys to increase excretion of both calcium and phosphorus in the urine. Parathormone responds to the opposite condition—lowered blood calcium—and acts on three targets: the bones, which release stored calcium into the blood; the kidneys, which slow the excretion of calcium; and the intestine, which increases calcium absorption. Vitamin D acts with parathormone and is essential for the absorption of calcium in the intestine. Figure 12-8 in Chapter 12 diagrams the ways vitamin D, and the hormones calcitonin and parathormone, regulate calcium homeostasis.

Another hormone has effects on blood-making activity:

**erythropoietin** (eh-REE-throw-POY-eh-tin): a hormone that stimulates red blood cell production.
- **erythro** = red (blood cell)
- **poiesis** = creating (like poetry)

- Erythropoietin from the kidneys.

Erythropoietin is responsive to oxygen depletion of the blood and to anemia. It acts on the bone marrow to stimulate the making of red blood cells.

Another hormone, special for pregnancy, is:

**relaxin:** the hormone of late pregnancy.

- Relaxin from the ovary.

This hormone, which is secreted in response to the raised progesterone and estrogen levels of late pregnancy, acts on the cervix and pelvic ligaments to allow them to stretch so that they can accommodate the birth process without strain.

Other agents help regulate blood pressure:

**renin** (REN-in): an enzyme from the kidneys that activates angiotensin.
- **ren** = kidney

- Renin (an enzyme), from the kidneys, in cooperation with angiotensin in the blood.
- Aldosterone, a hormone from the adrenal cortex.

Renin responds to a reduced blood supply experienced by the kidneys and acts in several ways. Encountering the inactive form of angiotensin in the bloodstream, renin converts this molecule to active angiotensin I and then to the very active angiotensin II. The angiotensins constrict the blood vessels, thus raising the blood pressure. They also stimulate thirst, leading to increased water intake, another way of raising the blood pressure. The angiotensins also cause the kidneys to retain water and salt. Thus the angiotensins increase blood pressure by several means at once.

Renin and angiotensin also stimulate the adrenal cortex to secrete the hormone aldosterone. This hormone's target is also the kidneys, which respond by excreting less sodium and with it, less water. The effect is to retain more water in the bloodstream—thus, again, raising the blood pressure. Figure 12-1 in Chapter 12 provides more details.

**angiotensin:** a hormone involved in blood pressure regulation.
- **angio** = blood vessels
- **tensin** = pressure

**aldosterone:** a hormone from the adrenal gland involved in blood pressure regulation.
- **aldo** = aldehyde

## The Gastrointestinal Hormones

Several hormones are produced in the stomach and intestines in response to the presence of food or the components of food:

- Gastrin from the stomach and duodenum.
- Cholecystokinin from the duodenum.
- Secretin from the duodenum.
- Gastric-inhibitory peptide from the duodenum and jejunum.

Gastrin stimulates the stomach to make and release its acid and digestive juices and to move and churn its contents actively. Cholecystokinin signals the gallbladder and pancreas to release their contents into the intestine to aid in digestion. Secretin calls forth acid-neutralizing bicarbonate from the pancreas into the intestine and slows the action of the stomach and its secretion of acid and digestive juices. Gastric-inhibitory peptide inhibits the secretion of gastric acid and slows the process of digestion. These hormones are presented in more detail in Chapter 3.

## The Sex Hormones

The three major sex hormones are:

- Testosterone from the testicles.
- Estrogens from the ovary.
- Progesterone from the ovary's corpus luteum in preparation for, and during, pregnancy.

In the male, testosterone is released in response to LH (described earlier). It acts on all the tissues that are involved in male sexuality and promotes their development and maintenance. Estrogens, released in response to both FSH and LH, act similarly in females. Progesterone, released in response to raised LH and prolactin, acts on the uterus and mammary glands, stimulating them to grow and develop.

**testosterone:** a steroid hormone from the testicles, or testes. The steroids, as explained in Chapter 5, are chemically related to, and some are derived from, the lipid cholesterol.
**sterone** = a steroid hormone

**estrogens:** hormones responsible for the menstrual cycle and other female characteristics.
- **oestrus** = the egg-making cycle
- **gen** = gives rise to

**progesterone:** the hormone of gestation (pregnancy).
- **pro** = promoting
- **gest** = gestation (pregnancy)
- **sterone** = a steroid hormone

## The Prostglandins

The prostaglandins are a group of hormonelike substances produced by many different body organs. They perform a multitude of diverse functions including the regulation of blood vessel contractions, nerve impulses, and hormone responses. They don't have descriptive names but are designated by letters and numbers: E1, E2, and so forth. The prostaglandins are all derived from the 20-carbon polyunsaturated fatty acids and account in part for the necessity for these fatty acids in the diet.

Reminder: A *prostaglandin* is a hormonelike compound, derived from the 20-carbon polyunsaturated fatty acids.

This brief description of the hormones and their functions should suffice to provide an awareness of the enormous impact these compounds have on body processes. The other overall regulating agency is the nervous system.

# The Nervous System

The nervous system has a central control system—a sort of computer—that can evaluate information about conditions within and outside the body, and a vast system of wiring that receives information and sends instructions. The control unit is the brain and spinal cord, called the central nervous system; and the vast complex of wiring between the center and the parts is the peripheral nervous system. The smooth functioning that results from the system's adjustments to changing conditions is homeostasis.

The nervous system has two general functions: it controls voluntary muscles in response to sensory stimuli from them, and it controls involuntary, internal muscles and glands in response to nerve-borne and chemical signals about their status. In fact, the nervous system is best understood as two systems that use the same or similar pathways to receive and transmit their messages. The somatic nervous system controls the voluntary muscles; the autonomic nervous system controls the internal organs.

When scientists were first studying the autonomic nervous system, they noticed that when something hurt one organ of the body, some of the other organs reacted as if in sympathy for the afflicted one. They therefore named the nerve network they were studying the sympathetic nervous system. The term is still used today to refer to that branch of the autonomic nervous system that responds to pain and stress. The other branch is called the parasympathetic nervous system. (Think of the sympathetic branch as the responder when homeostasis needs restoring and the parasympathetic branch as the commander of function during normal times.) Both systems transmit their messages through the brain and spinal cord. Nerves of the two branches travel side by side along the same pathways to transmit their messages, but they oppose each other's actions (see Figure A-3).

**central nervous system:** the central part of the nervous system, the brain and spinal cord.

**peripheral** (puh-RIFF-er-ul) **nervous system:** the peripheral (outermost) part of the nervous system, the vast complex of wiring that extends from the central nervous system to the body's outermost areas. It contains both somatic and autonomic components (defined next).

**somatic** (so-MAT-ick) **nervous system:** the division of the nervous system that controls the voluntary muscles, as distinguished from the autonomic nervous system, which controls involuntary functions.
- **soma** = body

**autonomic nervous system:** the division of the nervous system that controls the body's automatic responses. Its two branches are the **sympathetic** branch, which helps the body respond to stressors from the outside environment, and the **parasympathetic** branch, which regulates normal body activities between stressful times.
- **autonomos** = self-governing

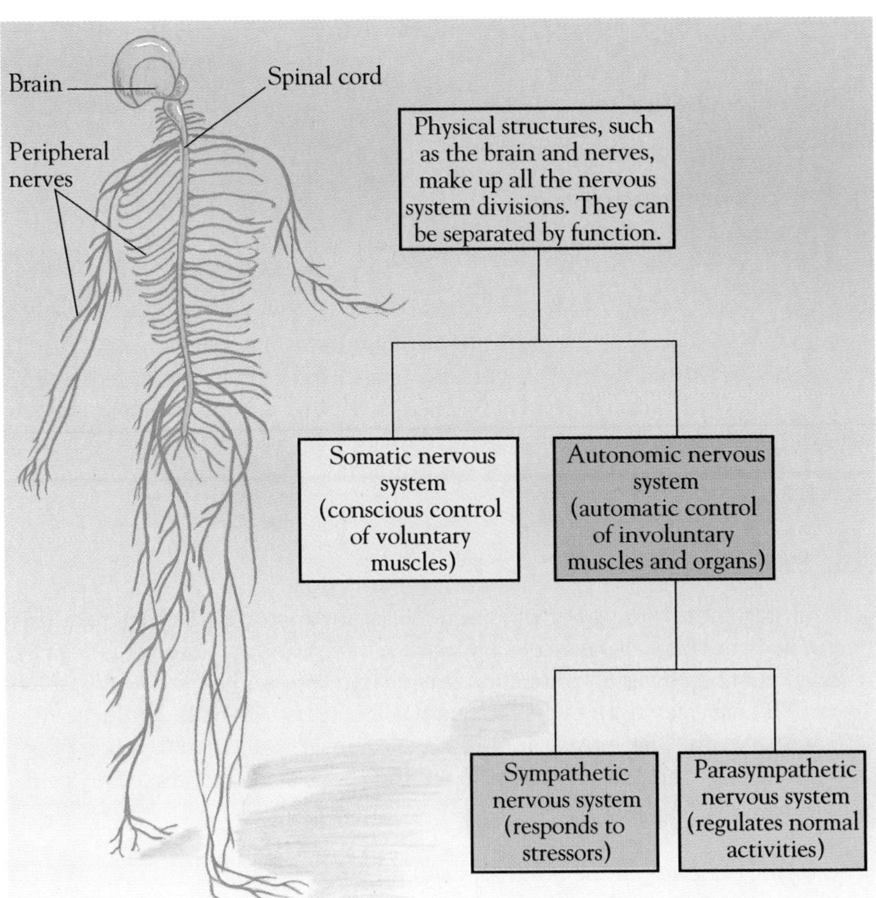


**Figure A-3**
**The Organization of the Nervous System**

The brain and spinal cord evaluate information about conditions within and outside the body, and the peripheral nerves receive information and send instructions.

An example will show how the sympathetic and parasympathetic nervous systems work to maintain homeostasis. When you go outside in cold weather, your skin's temperature receptors send "cold" messages to the spinal cord and brain. Your conscious mind may intervene at this point to tell you to zip your jacket, but let's say you have no jacket. Your sympathetic nervous system reacts to the external stressor, the cold. It signals your skin-surface capillaries to shut down so your blood will circulate deeper in your tissues, where it will conserve heat. Your sympathetic nervous system also signals involuntary contractions of the small muscles just under the skin surface. The product of these muscle contractions is heat, and the visible result is goose bumps. If these measures do not raise your body temperature enough, then the sympathetic nerves signal your large muscle groups to shiver; the contractions of these large muscles produce still more heat. All of this activity adds up to a set of adjustments that maintain your homeostasis (with respect to temperature) under conditions of external extremes (cold) that would throw it off balance. The cold was a stressor; the body's response was resistance.

Now let's say you come in and sit by a fire and drink hot cocoa. You are warm and no longer need all that sympathetic activity. At this point, your parasympathetic nerves take over; they signal your skin-surface capillaries to dilate again, your goose bumps to subside, and your muscles to relax. Your body is back to normal. This is recovery.

## Putting It Together

The hormonal and nervous systems coordinate body functions by transmitting and receiving messages. The point-to-point messages of the nervous system travel through a central switchboard (the spinal cord and brain), whereas the messages of the hormonal system are broadcast over the airways (the bloodstream), and any organ with the appropriate receptors can pick them up. Nerve impulses travel faster than hormonal messages do—although both are remarkably swift. Whereas your brain's command to wiggle your toes reaches the toes within a fraction of a second and stops as quickly, a gland's message to alter a body condition may take several seconds or minutes to get started and may fade away equally slowly.

Together, the two systems possess every characteristic a superb communication network needs: varied speeds of transmission, along with private communication lines or public broadcasting systems, depending on the needs of the moment. The hormonal system, together with the nervous system, integrates the whole body's functioning so that all parts act smoothly together.

Contents



# B BASIC CHEMISTRY CONCEPTS

This appendix is intended to provide the background in basic chemistry that you need to understand the nutrition concepts presented in this book. Chemistry is the branch of natural science that is concerned with the description and classification of matter, the changes that matter undergoes, and the energy associated with these changes. The glossary on p. B-8 defines matter, energy, and other related terms.

## Matter: The Properties of Atoms

Every substance has characteristics or properties that distinguish it from all other substances and thus give it a unique identity. These properties are both physical and chemical. The physical properties include such characteristics as color, taste, texture, and odor, as well as the temperatures at which a substance changes its state (from a solid to a liquid or from a liquid to a gas) and the weight of a unit volume (its density). The chemical properties of a substance have to do with how it reacts with other substances or responds to a change in its environment so that new substances with different sets of properties are produced.

A physical change does not change a substance's chemical composition. For example, the three states ice, water, and steam all consist of two hydrogen atoms and one oxygen atom bound together. However, a chemical change occurs if an electric current passes through water. The water disappears and two different substances are formed: hydrogen gas, which is flammable, and oxygen gas, which supports life. Chemical changes are also referred to as chemical reactions.

### Substances: Elements and Compounds

Molecules constitute the smallest part of a substance that can exist separately without losing its physical and chemical properties. If a molecule is composed of atoms that are alike, the substance is an element (for example, $O_2$). If a molecule is composed of two or more different kinds of atoms, the substance is a compound (for example, $H_2O$).

Just over 100 elements are known, and these are listed in Table B-1. A familiar example is hydrogen, whose molecules are composed only of hydrogen atoms linked together in pairs ($H_2$). On the other hand, over a million compounds are known. An example is the sugar glucose. Each of its molecules is composed of 6 carbon, 6 oxygen, and 12 hydrogen atoms linked together in a specific arrangement (as described in Chapter 4).

### The Nature of Atoms

Atoms themselves are made of smaller particles. Within the atomic nucleus are protons (positively charged particles), and surrounding the nucleus are electrons (negatively charged particles). The number of protons (+) in the nucleus of an atom determines the number of electrons (−) around it. The positive charge on a proton is equal to the negative charge on an electron, so the charges cancel each other out and leave the atom neutral to its surroundings.

The nucleus may also include neutrons, subatomic particles that have no charge. Protons and neutrons are of equal mass, and together they give an atom its weight. Electrons bond atoms together to make molecules, and they are involved in chemical reactions.

Each type of atom has a characteristic number of protons in its nucleus. The hydrogen atom (symbol H) is the simplest of all. It possesses a single proton, with a single electron associated with it:

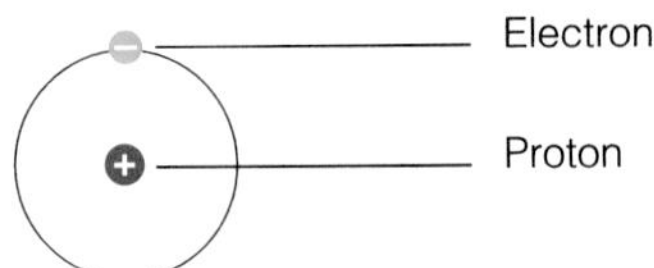


Hydrogen atom (H), atomic number 1.

Just as hydrogen always has one proton, helium always has two, lithium three, and so on. The atomic number of each element is the number of protons in the nucleus of that atom, and this never changes in a chemical reaction; it gives the atom its identity. The atomic numbers for the known elements are listed in Table B-1.

Besides hydrogen, the atoms most common in living things are carbon (C), nitrogen (N), and oxygen (O), whose atomic numbers are 6, 7, and 8, respectively. Their structures are more complicated than that of hydrogen, but each of them possesses the same number of electrons as there are protons in the nucleus. These electrons are found in orbits, or shells (shown on p. B-2).

B

**Table B-1**
**Chemical Symbols for the Elements**

| Number of Protons (Atomic Number) | Element | Number of Electrons in Outer Shell | Number of Protons (Atomic Number) | Element | Number of Electons in Outer Shell |
|---|---|---|---|---|---|
| 1 | Hydrogen (H) | 1 | 52 | Tellurium (Te) | 6 |
| 2 | Helium (He) | 2 | 53 | Iodine (I) | 7 |
| 3 | Lithium (Li) | 1 | 54 | Xenon (Xe) | 8 |
| 4 | Beryllium (Be) | 2 | 55 | Cesium (Cs) | 1 |
| 5 | Boron (B) | 3 | 56 | Barium (Ba) | 2 |
| 6 | Carbon (C) | 4 | 57 | Lanthanum (La) | 2 |
| 7 | Nitrogen (N) | 5 | 58 | Cerium (Ce) | 2 |
| 8 | Oxygen (O) | 6 | 59 | Praseodymium (Pr) | 2 |
| 9 | Fluorine (F) | 7 | 60 | Neodymium (Nd) | 2 |
| 10 | Neon (Ne) | 8 | 61 | Promethium (Pm) | 2 |
| 11 | Sodium (Na) | 1 | 62 | Samarium (Sm) | 2 |
| 12 | Magnesium (Mg) | 2 | 63 | Europium (Eu) | 2 |
| 13 | Aluminum (Al) | 3 | 64 | Gadolinium (Gd) | 2 |
| 14 | Silicon (Si) | 4 | 65 | Terbium (Tb) | 2 |
| 15 | Phosphorus (P) | 5 | 66 | Dysprosium (Dy) | 2 |
| 16 | Sulfur (S) | 6 | 67 | Holmium (Ho) | 2 |
| 17 | Chlorine (Cl) | 7 | 68 | Erbium (Er) | 2 |
| 18 | Argon (Ar) | 8 | 69 | Thulium (Tm) | 2 |
| 19 | Potassium (K) | 1 | 70 | Ytterbium (Yb) | 2 |
| 20 | Calcium (Ca) | 2 | 71 | Lutetium (Lu) | 2 |
| 21 | Scandium (Sc) | 2 | 72 | Hafnium (Hf) | 2 |
| 22 | Titanium (Ti) | 2 | 73 | Tantalum (Ta) | 2 |
| 23 | Vanadium (V) | 2 | 74 | Tungsten (W) | 2 |
| 24 | Chromium (Cr) | 1 | 75 | Rhenium (Re) | 2 |
| 25 | Manganese (Mn) | 2 | 76 | Osmium (Os) | 2 |
| 26 | Iron (Fe) | 2 | 77 | Iridium (Ir) | 2 |
| 27 | Cobalt (Co) | 2 | 78 | Platinum (Pt) | 1 |
| 28 | Nickel (Ni) | 2 | 79 | Gold (Au) | 1 |
| 29 | Copper (Cu) | 1 | 80 | Mercury (Hg) | 2 |
| 30 | Zinc (Zn) | 2 | 81 | Thallium (Tl) | 3 |
| 31 | Gallium (Ga) | 3 | 82 | Lead (Pb) | 4 |
| 32 | Germanium (Ge) | 4 | 83 | Bismuth (Bi) | 5 |
| 33 | Arsenic (As) | 5 | 84 | Polonium (Po) | 6 |
| 34 | Selenium (Se) | 6 | 85 | Astatine (At) | 7 |
| 35 | Bromine (Br) | 7 | 86 | Radon (Rn) | 8 |
| 36 | Krypton (Kr) | 8 | 87 | Francium (Fr) | 1 |
| 37 | Rubidium (Rb) | 1 | 88 | Radium (Ra) | 2 |
| 38 | Strontium (Sr) | 2 | 89 | Actinium (Ac) | 2 |
| 39 | Yttrium (Y) | 2 | 90 | Thorium (Th) | 2 |
| 40 | Zirconium (Zr) | 2 | 91 | Protactinium (Pa) | 2 |
| 41 | Niobium (Nb) | 1 | 92 | Uranium (U) | 2 |
| 42 | Molybdenum (Mo) | 1 | 93 | Neptunium (Np) | 2 |
| 43 | Technetium (Tc) | 1 | 94 | Plutonium (Pu) | 2 |
| 44 | Ruthenium (Ru) | 1 | 95 | Americium (Am) | 2 |
| 45 | Rhodium (Rh) | 1 | 96 | Curium (Cm) | 2 |
| 46 | Palladium (Pd) | — | 97 | Berkelium (Bk) | 2 |
| 47 | Silver (Ag) | 1 | 98 | Californium (Cf) | 2 |
| 48 | Cadmium (Cd) | 2 | 99 | Einsteinium (Es) | 2 |
| 49 | Indium (In) | 3 | 100 | Fermium (Fm) | 2 |
| 50 | Tin (Sn) | 4 | 101 | Mendelevium (Md) | 2 |
| 51 | Antimony (Sb) | 5 | 102 | Nobelium (No) | 2 |

Key:
- Elements found in energy-yielding nutrients, vitamins, and water.
- Major minerals.
- Trace minerals.

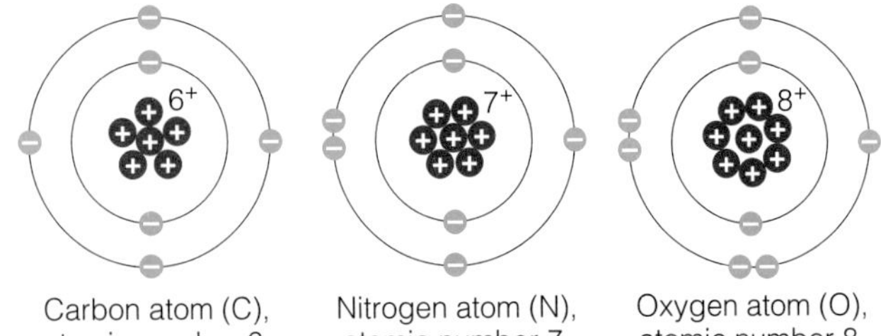


Carbon atom (C), atomic number 6. Nitrogen atom (N), atomic number 7. Oxygen atom (O), atomic number 8.

In these and all diagrams of atoms that follow, only the protons and electrons are shown. The neutrons, which contribute only to atomic weight, not to charge, are omitted.

The most important structural feature of an atom for determining its chemical behavior is the number of electrons in its outermost shell. The first, or innermost, shell is full when it is occupied by two electrons; so an atom with two or more electrons has a filled first shell. When the first shell is full, electrons begin to fill the second shell.

The second shell is completely full when it has eight electrons. A substance that has a full outer shell tends not to enter into chemical reactions. Atomic number 10, neon, is a chemically inert substance because its outer shell is complete. Fluorine, atomic number 9, has a great tendency to draw an electron from other substances to complete its outer shell, and thus it is highly reactive. Carbon has a half-full outer shell, which helps explain its great versatility; it can combine with other elements in a variety of ways to form a large number of compounds.

Atoms seek to reach a state of maximum stability or of lowest energy in the same way that a ball will roll down a hill until it reaches the lowest place. An atom achieves a state of maximum stability:

- By gaining or losing electrons to either fill or empty its outer shell.
- By sharing its electrons through bonding together with other atoms and thereby completing its outer shell.

The number of electrons determines how the atom will chemically react with other atoms. Hence the atomic number, not the weight, is what gives an atom its chemical nature.

# Chemical Bonding

Atoms often complete their outer shells by sharing electrons with other atoms. In order to complete its outer shell, a carbon atom requires four electrons. A hydrogen atom requires one. Thus, when a carbon atom shares electrons with four hydrogen atoms, each completes its outer shell (as shown in the next column). Electron sharing binds the atoms together and satisfies the conditions of maximum stability for the molecule. The outer shell of each atom is complete, since hydrogen effectively has the required two electrons in its first (outer) shell, and carbon has eight electrons in its second (outer) shell; and the molecule is electrically neutral, with a total of ten protons and ten electrons.

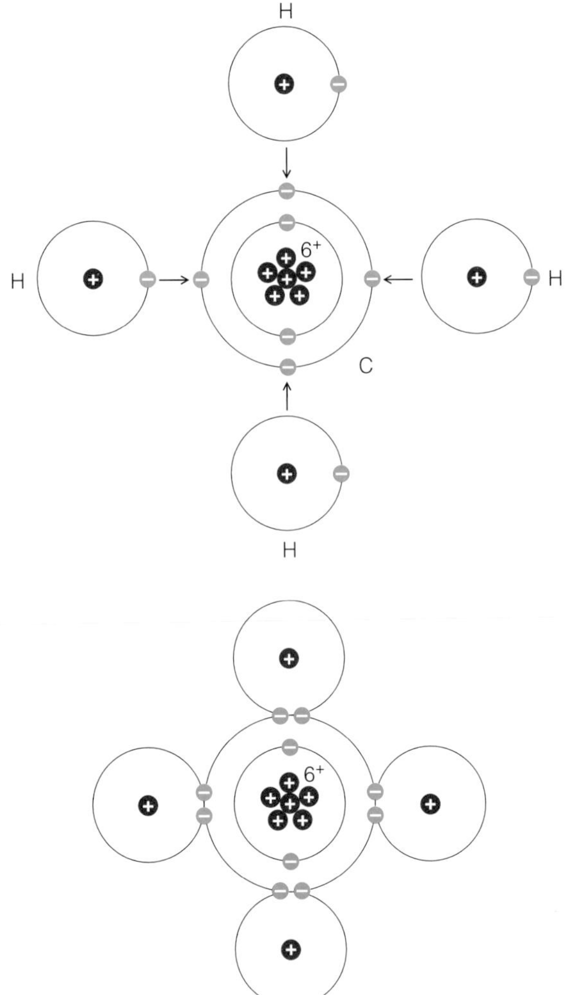


Methane molecule. The chemical formula for methane is $CH_4$. Note that by sharing electrons, every atom achieves a filled outer shell.

Bonds that involve the sharing of electrons, like the bond between carbon and hydrogen, are the most stable kind of association that atoms can form with one another. They are sometimes called covalent bonds, and the resulting combinations of atoms are called molecules. A single pair of shared electrons forms a single bond. A simplified way to represent a single bond is with a single line. Thus the structure of methane ($CH_4$) could be represented like this (ignoring the inner-shell electrons, which do not participate in bonding):

```
    H
    |
H — C — H
    |
    H
```

Methane ($CH_4$).

Similarly, one nitrogen atom and three hydrogen atoms can share electrons to form one molecule of ammonia ($NH_3$):

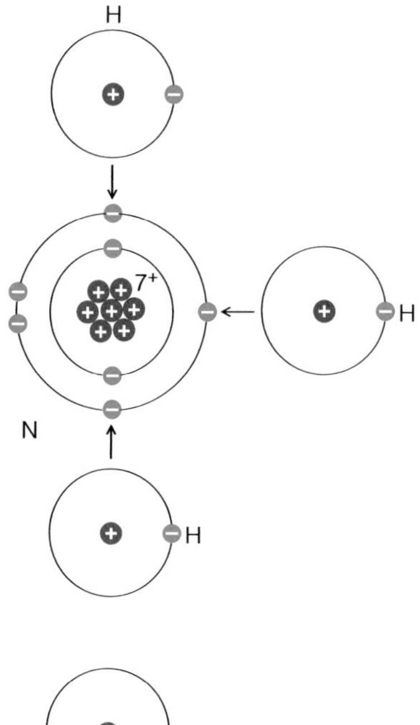


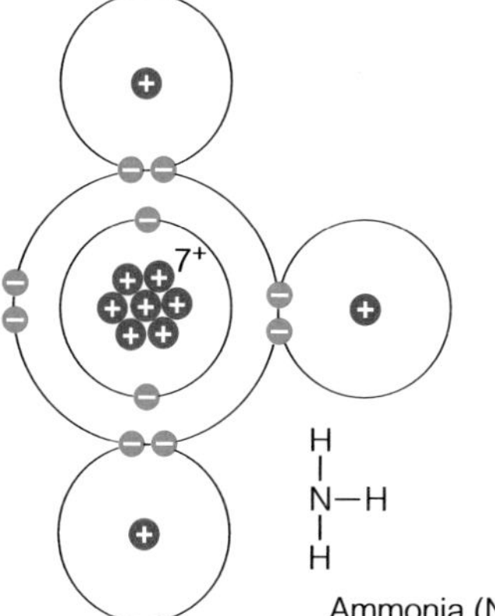


Ammonia molecule ($NH_3$). Count the electrons in each atom's outer shell to confirm that it is filled.

One oxygen atom may be bonded to two hydrogen atoms to form one molecule of water ($H_2O$):

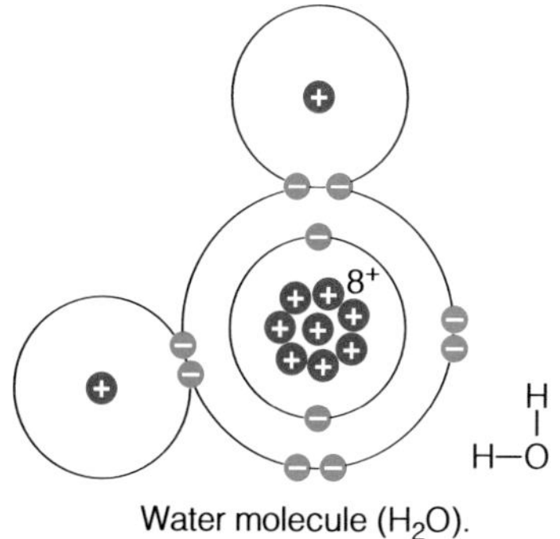


Water molecule ($H_2O$).

When two oxygen atoms form a molecule of oxygen, they must share two pairs of electrons. This double bond may be represented as two single lines:

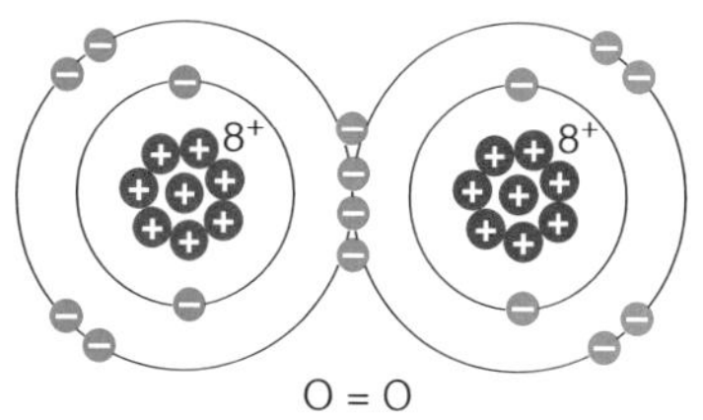


Oxygen molecule ($O_2$).

Small atoms form the tightest, most stable bonds. H, O, N, and C are the smallest atoms capable of forming one, two, three, and four electron-pair bonds (respectively). This is the basis for the statement in Chapter 4 that in drawings of compounds containing these atoms, hydrogen must always have one, oxygen two, nitrogen three, and carbon four bonds radiating to other atoms:

H—   —O—   —N—   —C—

The stability of the associations between these small atoms and the versatility with which they can combine make them very common in living things. Interestingly, all cells, whether they come from animals, plants, or bacteria, contain the same elements in very nearly the same proportions. The atomic elements commonly found in living things are shown in Table B-2.

**Table B-2**
**Elemental Composition of Living Cells**

| Element | Chemical Symbol | Composition by Weight (%) |
|---|---|---|
| Oxygen | O | 65 |
| Carbon | C | 18 |
| Hydrogen | H | 10 |
| Nitrogen | N | 3 |
| Calcium | Ca | 1.5 |
| Phosphorus | P | 1.0 |
| Sulfur | S | 0.25 |
| Sodium | Na | 0.15 |
| Magnesium | Mg | 0.05 |
| Total | | 99.30[a] |

[a]The remaining 0.70 percent by weight is contributed by the trace elements: copper (Cu), zinc (Zn), selenium (Se), molybdenum (Mo), fluorine (F), chlorine (Cl), iodine (I), manganese (Mn), cobalt (Co), and iron (Fe). Cells may also contain variable traces of some of the following: lithium (Li), strontium (Sr), aluminum (Al), silicon (Si), lead (Pb), vanadium (V), arsenic (As), bromine (Br), and others.

## Formation of Ions

An atom such as sodium (Na, atomic number 11) cannot easily fill its outer shell by sharing. Sodium possesses a filled first shell of two electrons and a filled second shell of eight; there is only one electron in its outermost shell:

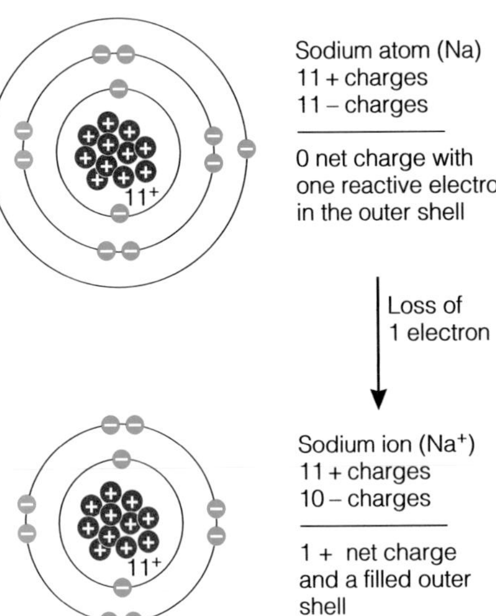


If sodium loses this electron, it satisfies one condition for stability: a filled outer shell (now its second shell counts as the outer shell). However, it is not electrically neutral. It has 11 protons (positive) and only 10 electrons (negative). It therefore has a net positive charge. An atom or molecule that has lost or gained one or more electrons and so is electrically charged is called an ion.

An atom such as chlorine (Cl, atomic number 17), with seven electrons in its outermost shell, can share electrons to fill its outer shell, or it can gain one electron to complete its outer shell and thus give it a negative charge:

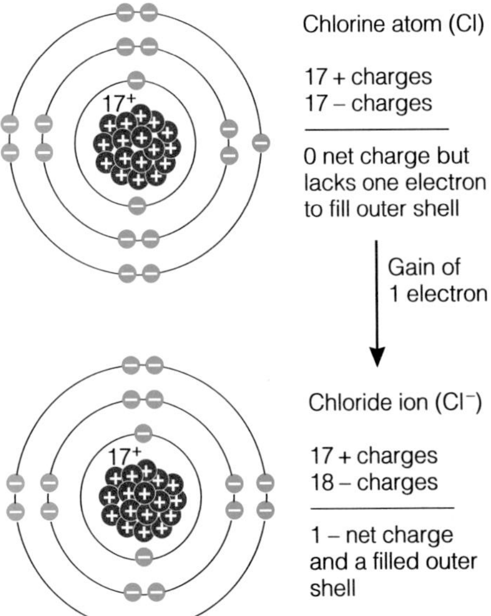


A positively charged ion such as sodium ion ($Na^+$) is called a cation; a negatively charged ion such as a chloride ion ($Cl^-$) is called an anion. Cations and anions attract one another to form salts:

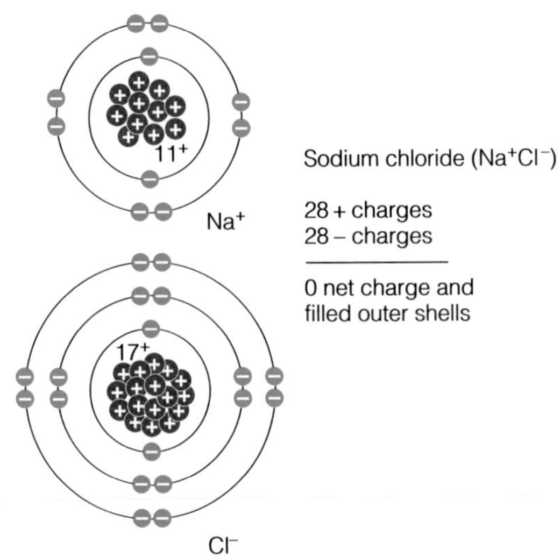


With all its electrons, sodium is a shiny, highly reactive metal; chlorine is the poisonous greenish-yellow gas that was used in World War I. But after sodium and chlorine have transferred electrons, they form the stable white salt familiar to you as table salt, or sodium chloride ($Na^+Cl^-$). The dramatic difference illustrates how profoundly the electron arrangement can influence the nature of a substance. The wide distribution of salt in nature attests to the stability of the union between the ions. Each meets the other's needs (a good marriage).

When dry, salt exists as crystals; its ions are stacked very regularly into a lattice, with positive and negative ions alternating in a three-dimensional checkerboard structure. In water, however, the salt quickly dissolves, and its ions separate from one another, forming an electrolyte solution in which they move about freely. Covalently bonded molecules rarely dissociate like this in a water solution. The most common exception is when they behave like acids and release $H^+$ ions, as discussed in the next section.

An ion can also be a group of atoms bound together in such a way that the group has a net charge and enters into reactions as a single unit. Many such groups are active in the fluids of the body. The bicarbonate ion is composed of five atoms—one H, one C, and three Os—and has a net charge of $-1$ ($HCO_3^-$). Another important ion of this type is a phosphate ion with one H, one P, and four O, and a net charge of $-2$ ($HPO_4^{-2}$).

Whereas many elements have only one configuration in the outer shell and thus only one way to bond with other elements, some elements have the possibility of varied configurations. Iron is such an element. Under some conditions iron loses two electrons, and under other circumstances it loses three. If iron loses two electrons, it then has a net charge of $+2$, and we call it ferrous iron ($Fe^{++}$). If it donates three electrons to another atom, it becomes the +3 ion, or ferric iron ($Fe^{+++}$).

| Ferrous iron ($Fe^{++}$) | Ferric iron ($Fe^{+++}$) |
|---|---|
| (had 2 outer-shell electrons but has lost them) | (had 3 outer-shell electrons but has lost them) |
| 26 + charges | 26 + charges |
| 24 − charges | 23 − charges |
| 2 + net charge | 3 + net charge |

It is important to remember that a positive charge on an ion means that negative charges—electrons—have been lost and not that positive charges have been added to the nucleus.

## Water, Acids, and Bases

• ***Water*** • The water molecule is electrically neutral, having equal numbers of protons and electrons. However, when a hydrogen atom shares its electron with oxygen, that electron will spend most of its time closer to the positively charged oxygen nucleus. This leaves the positive proton (nucleus of the hydrogen atom) exposed on the outer part of the water molecule. We know, too, that the two hydrogens both bond toward the same side of the oxygen. These two facts explain why water molecules are polar: they have regions of more positive and more negative charge.

Polar molecules like water are drawn to one another by the attractive forces between the positive polar areas of one and the negative poles of another. These attractive forces, sometimes known as polar bonds or hydrogen bonds, occur among many molecules and also within the different parts of single large molecules. Although very weak in comparison with covalent bonds, polar bonds may occur in such abundance that they become exceedingly important in determining the structure of such large molecules as proteins and DNA.

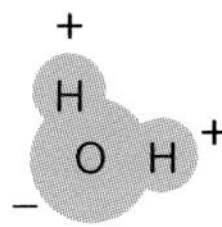


This diagram of the polar water molecule shows displacement of electrons toward the O nucleus; thus the negative region is near the O and the positive regions are near the Hs.

Water molecules have a slight tendency to ionize, separating into positive ($H^+$) and negative ($OH^-$) ions. In pure water, a small but constant number of these ions is present, and the number of positive ions exactly equals the number of negative ions.

• ***Acids*** • An acid is a substance that releases $H^+$ ions (protons) in a water solution. Hydrochloric acid (HCl) is such a substance because it dissociates in a water solution into $H^+$ and $Cl^-$ ions. Acetic acid is also an acid because it dissociates in water to acetate ions and free $H^+$:

```
   H  O              H  O
   |  ||             |  ||
 H—C— C—O—H  ──▶   H—C— C—O⁻ + H⁺
   |                 |
   H                 H
```

Acetic acid dissociates into an acetate ion and a hydrogen ion.

The more $H^+$ ions released, the stronger the acid.

• ***pH*** • Chemists define degrees of acidity by means of the pH scale, which runs from 0 to 14. The pH expresses the concentration of $H^+$ ions: a pH of 1 is extremely acidic, 7 is neutral, and 13 is very basic. There is a tenfold difference in the concentration of $H^+$ ions between points on this scale. A solution with pH 3, for example, has *ten times* as many $H^+$ ions as a solution with pH 4. At pH 7, the concentrations of free $H^+$ and $OH^-$ are exactly the same—1/10,000,000 moles per liter ($10^{-7}$ moles per liter).* At pH 4, the concentration of free $H^+$ ions is 1/10,000 ($10^{-4}$) moles per liter. This is a higher concentration of $H^+$ ions, and the solution is therefore acidic.

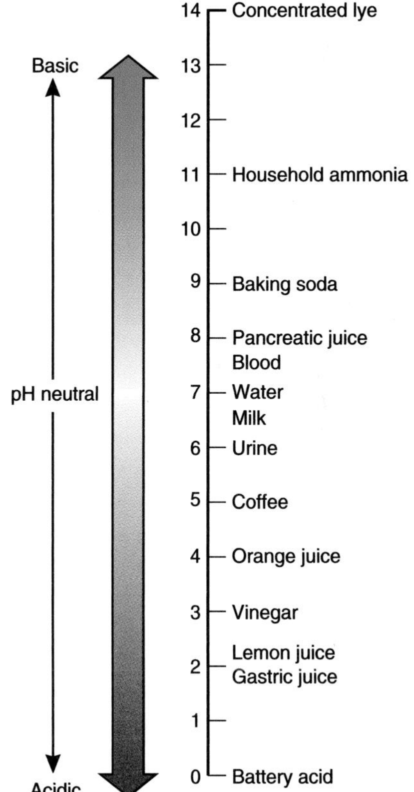


The pH scale.

Note: Each step is ten times as concentrated in base (1/10 as much acid, $H^+$) as the one below it.

*A mole is a certain number (about $6 \times 10^{23}$) of molecules. The pH of a solution is defined as the negative logarithm of the hydrogen ion concentration of the solution. Thus, if the concentration is $10^{-2}$ (moles per liter), the pH is 2; if $10^{-8}$, the pH is 8; and so on.

• ***Bases*** • A base is a substance that can soak up, or combine with, $H^+$ ions, thus reducing the acidity of a solution. The compound ammonia is such a substance. The ammonia molecule has two electrons that are not shared with any other atom; a hydrogen ion ($H^+$) is just a naked proton with no shell of electrons at all. The proton readily combines with the ammonia molecule to form an ammonium ion; thus a free proton is withdrawn from the solution and no longer contributes to its acidity. Many compounds containing nitrogen are important bases in living systems. Acids and bases neutralize each other to produce substances that are neither acid nor base.

$$\begin{array}{c} H \\ | \\ :N-H \\ | \\ H \end{array} + H^+ \longrightarrow \begin{array}{c} H \\ | \\ H-N^+-H \\ | \\ H \end{array}$$

Ammonia captures a hydrogen ion from water. The two dots here represent the two electrons not shared with another atom. These are ordinarily not shown in chemical structure drawings. Compare this with the earlier diagram of an ammonia molecule (p. B–3).

# Chemical Reactions

A chemical reaction, or chemical change, results in the breakdown of substances and the formation of new ones. Almost all such reactions involve a change in the bonding of atoms. Old bonds are broken, and new ones are formed. The nuclei of atoms are never involved in chemical reactions—only their outer-shell electrons take part. At the end of a chemical reaction, the number of atoms of each type is always the same as at the beginning. For example, two hydrogen molecules ($2H_2$) can react with one oxygen molecule ($O_2$) to form two water molecules ($2H_2O$). In this reaction two substances (hydrogen and oxygen) disappear, and a new one (water) is formed, but at the end of the reaction there are still four H atoms and two O atoms, just as there were at the beginning. Because the atoms are now linked in a different way, their characteristics or properties have changed.

In many instances chemical reactions involve not the relinking of molecules but the exchanging of electrons or protons among them. In such reactions the molecule that gains one or more electrons (or loses one or more hydrogen ions) is said to be reduced; the molecule that loses electrons (or gains protons) is oxidized. A hydrogen ion is equivalent to a proton. Oxidation and reduction take place simultaneously because an electron or proton that is lost by one molecule is accepted by another. The addition of an atom of oxygen is also oxidation because oxygen (with six electrons in the outer shell) accepts two electrons in becoming bonded. Oxidation, then, is loss of electrons, gain of protons, or addition of oxygen (with six electrons); reduction is the opposite—gain of electrons, loss of protons, or loss of oxygen. The addition of hydrogen atoms to oxygen to form water can thus be described as the reduction of oxygen *or* the oxidation of hydrogen.

If a reaction results in a net increase in the energy of a compound, it is called an endergonic, or "uphill," reaction (energy, *erg,* is added into, *endo,* the compound). An example is the chief result of photosynthesis, the making of sugar in a plant from carbon dioxide and water using the energy of sunlight. Conversely, the oxidation of sugar to carbon dioxide and water is an exergonic, or "downhill," reaction because the end products have less energy than the starting products. Oftentimes, but not always, reduction reactions are endergonic, resulting in an increase in the energy of the products. Oxidation reactions often, but not always, are exergonic.

Chemical reactions tend to occur spontaneously if the end products are in a lower energy state and therefore are more stable than the reacting compounds. These reactions often give off energy in the form of heat as they occur. The generation of heat by wood burning in a fireplace and the

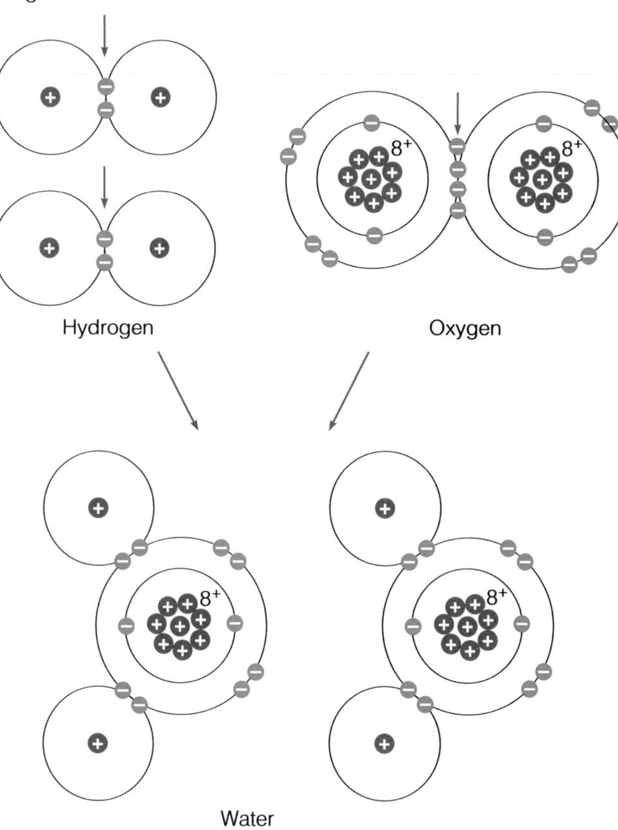


Hydrogen and oxygen react to form water.

maintenance of human body warmth both depend on energy-yielding chemical reactions. These downhill reactions occur easily, although they may require some activation energy to get them started, just as a ball requires a push to start rolling downhill.

Uphill reactions, in which the products contain more energy than the reacting compounds started with, do not occur until an energy source is provided. An example of such an energy source is the sunlight used in photosynthesis, where carbon dioxide and water (low-energy compounds) are combined to form the sugar glucose (a higher-energy compound). Another example is the use of the energy in glucose to combine two low-energy compounds in the body into the high-energy compound ATP (see Chapter 7). The energy in ATP may be used to power many other energy-requiring, uphill reactions. Clearly, any of many different molecules can be used as a temporary storage place for energy.

Energy change as reaction occurs

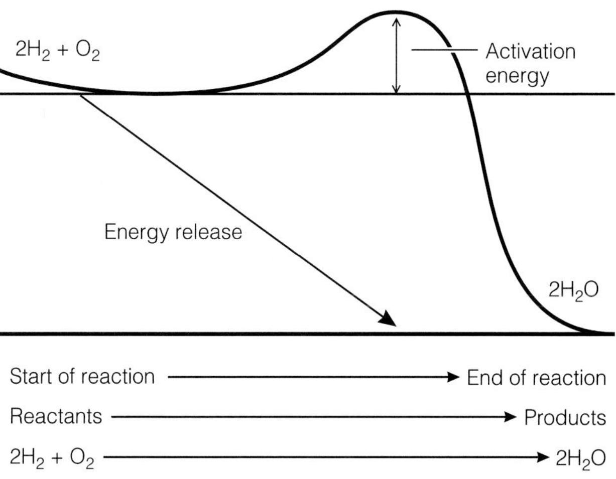


Neither downhill nor uphill reactions occur until something sets them off (activation) or until a path is provided for them to follow. The body uses enzymes as a means of providing paths and controlling chemical reactions (see Chapter 6). By controlling the availability and the action of its enzymes, the body can "decide" which chemical reactions to prevent and which to promote.

# Formation of Free Radicals

Normally, when a chemical reaction takes place, bonds break and re-form with some redistribution of atoms and rearrangement of bonds to form new, stable compounds. Normally, bonds don't split in such a way as to leave a molecule with an odd, unpaired electron. However, weak bonds can split this way, and when they do, free radicals are formed. Free radicals are highly unstable and quickly react with other compounds, forming more free radicals in a chain reaction.

A physical event such as the arrival of an energy-carrying particle of light or other radiation starts the process by

H—O—O—H or R—O—O—H —(Heat or light)→ H—O• + •O—H or R—O• + •O—H

Hydrogen peroxide or any hydroperoxide (R is any carbon chain with appropriate numbers of H) → Free radical

Free radicals are formed. The dots represent single electrons that are available for sharing (the atom needs another electron to fill its outer shell).

breaking a weak bond so that free radicals are formed. A cascade may ensue in which many highly reactive radicals are generated, resulting finally in the disruption of a living structure such as a cell membrane.

H—O• + H—C(H)(H)—H (or R—H) → H—O—H + H—C(H)(H)• (or R•)

Free radical + Compound with weak bond (perhaps an unsaturated fatty acid) → New stable compound (water or an alcohol) + Free radical

Destruction of biological compounds by free radicals. The free radical attacks a weak bond in a biological compound, disrupting it and forming a new stable molecule and another free radical. This can attack another biological compound, and so on.

Oxidation of some compounds can be induced by air at room temperature in the presence of light. Such reactions are thought to take place through the formation of compounds called peroxides:

Peroxides:

| | |
|---|---|
| H—O—O—H | Hydrogen peroxide |
| R—O—O—H | Hydroperoxides (R is any carbon chain with appropriate numbers of H) |
| R—O—O—R | Peroxide |

Some peroxides readily disintegrate into free radicals, initiating chain reactions like those just described.

Free radicals are of special interest in nutrition because the antioxidant properties of vitamins A, C, and E as well as the mineral selenium are thought to protect against the destructive effects of these free radicals (see Highlight 11). For example, vitamin E on the surface of the lungs reacts with, and is destroyed by, free radicals, thus preventing the radicals from reaching underlying cells and oxidizing the lipids in their membranes.

## Glossary

**element:** a substance composed of atoms that are alike—for example, iron (Fe).

**atom:** the smallest component of an element that has all of the properties of the element.

**compound:** a substance composed of two or more different atoms—for example, water ($H_2O$).

**molecule:** two or more atoms of the same or different elements joined by chemical bonds. Examples are molecules of the element oxygen, composed of two oxygen atoms ($O_2$), and molecules of the compound water, composed of two hydrogen atoms and one oxygen atom ($H_2O$).

**matter:** anything that takes up space and has mass.

**energy:** the capacity to do work.

# BIOCHEMICAL STRUCTURES AND PATHWAYS

The diagrams of nutrients presented here are meant to enhance your understanding of the most important organic molecules in the human diet. Following the diagrams of nutrients are sections on the major metabolic pathways mentioned in Chapter 7—glycolysis, the TCA cycle, and the electron transport chain—and a description of how alcohol interferes with these pathways. Discussions of the urea cycle and the formation of ketone bodies complete the appendix.

CONTENTS



C

## Carbohydrates

### Monosaccharides

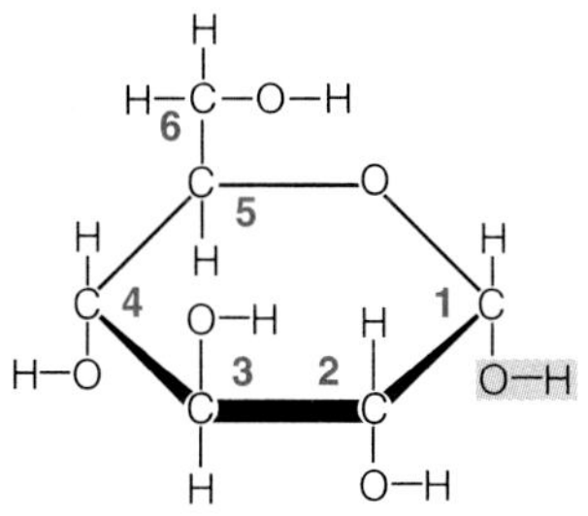

**Glucose (alpha form).** The ring would be at right angles to the plane of the paper. The bonds directed upward are above the plane; those directed downward are below the plane. This molecule is considered an alpha form because the OH on carbon 1 points downward.

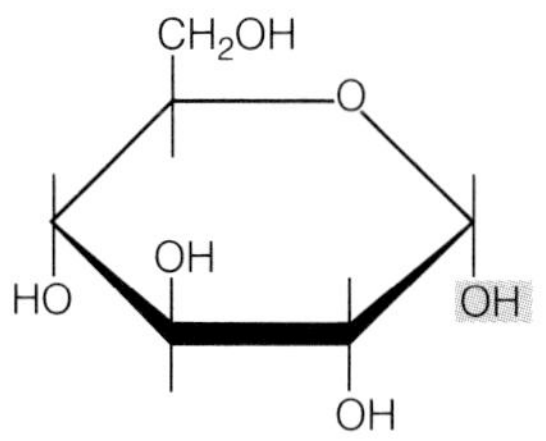

**Glucose (alpha form) shorthand notation.** This notation, in which the carbons in the ring and single hydrogens have been eliminated, will be used throughout this appendix.

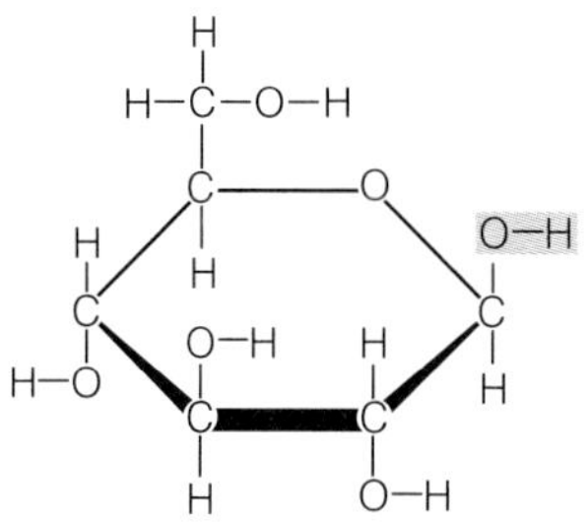

**Glucose (beta form).** The OH on carbon 1 points upward.
**Fructose, galactose:** see Chapter 4.

### Disaccharides

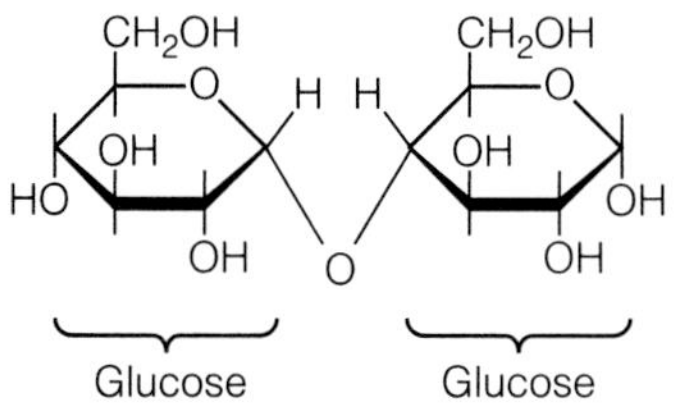

Glucose Glucose
**Maltose.**

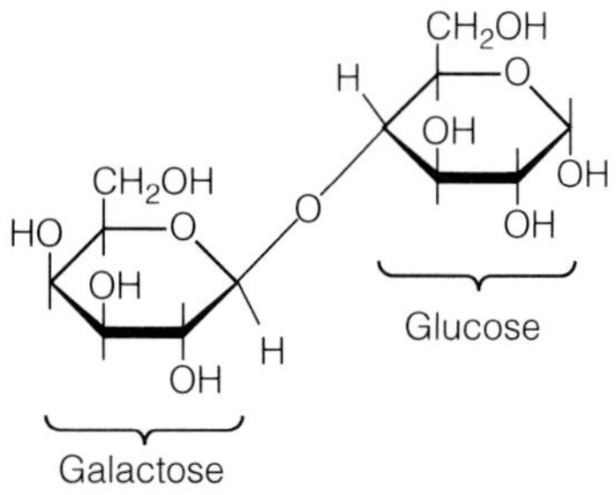

Galactose Glucose
**Lactose (alpha form).**

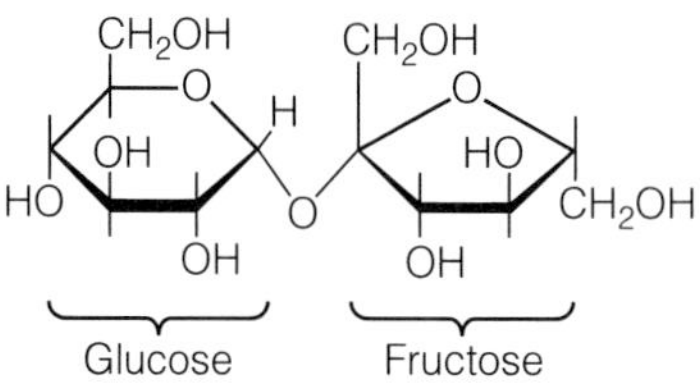

Glucose Fructose
**Sucrose.**

C

## Polysaccharides

As described in Chapter 4, starch, glycogen, and cellulose are all long chains of glucose molecules covalently linked together.

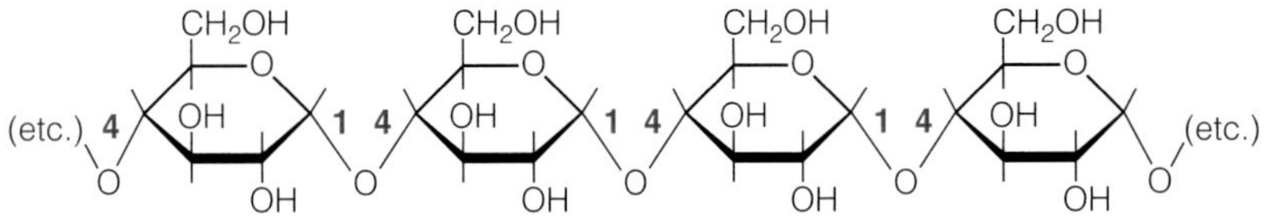

(etc.) 4 CH₂OH O OH OH 1 4 CH₂OH O OH OH 1 4 CH₂OH O OH OH 1 4 CH₂OH O OH OH (etc.)

Amylose (unbranched starch)

**Starch.** Two kinds of covalent linkages occur between glucose molecules in starch, giving rise to two kinds of chains. Amylose is composed of straight chains, with carbon 1 of one glucose linked to carbon 4 of the next (α-1,4 linkage). Amylopectin is made up of straight chains like amylose but has occasional branches arising where the carbon 6 of a glucose is also linked to the carbon 1 of another glucose (α-1,6 linkage).

**Glycogen.** The structure of glycogen is like amylopectin but with many more branches.

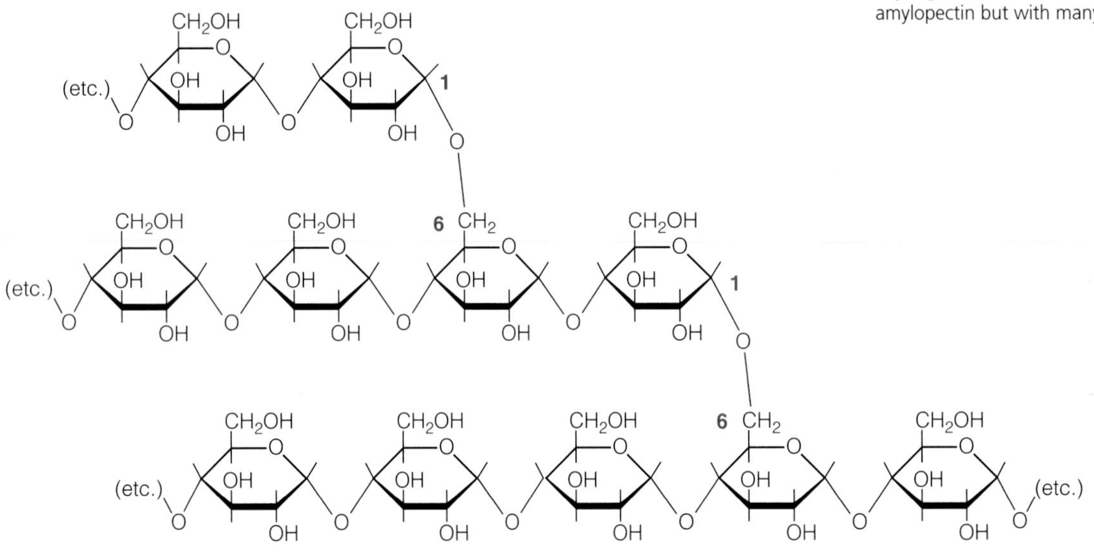

Amylopectin (branched starch)

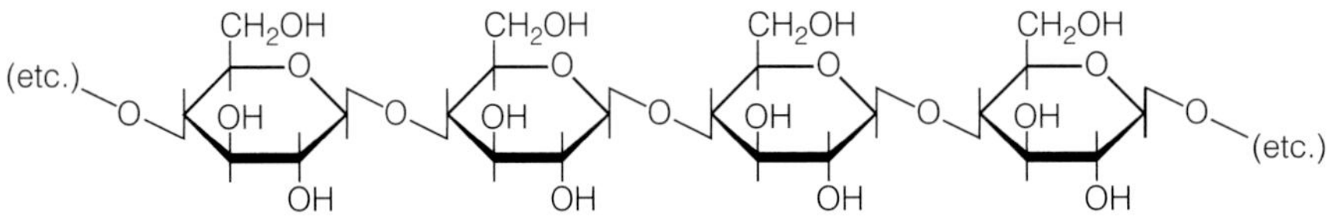

**Cellulose.** Like starch and glycogen, cellulose is also made of chains of glucose units, but there is an important difference: in cellulose, the OH on carbon 1 is in the beta position (see p. C-1). When carbon 1 of one glucose is linked to carbon 4 of the next, it forms a β-1, 4 linkage, which cannot be broken by digestive enzymes in the human GI tract.

Fibers, such as hemicelluloses, consist of various monosaccharides.

Monosaccharides in backbone chain of hemicelluloses:

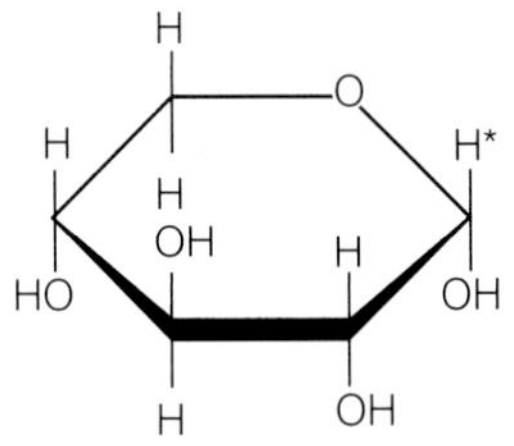

xylose

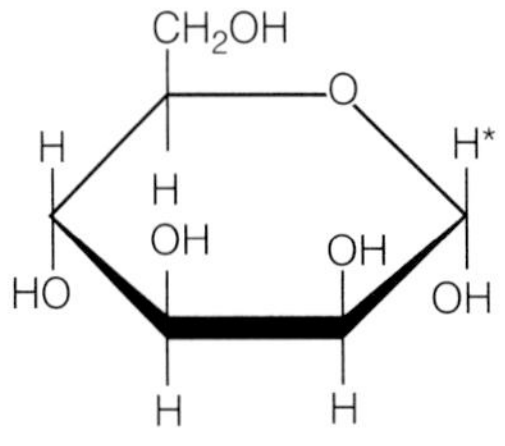

mannose

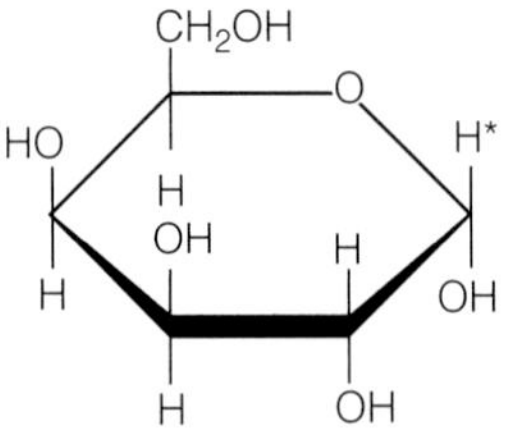

galactose

*These structures are shown in the alpha form with the H on the carbon pointing upward and the OH pointing downward, but they may also appear in the beta form with the H pointing downward and the OH upward.

Monosaccharides in side chains of hemicelluloses:

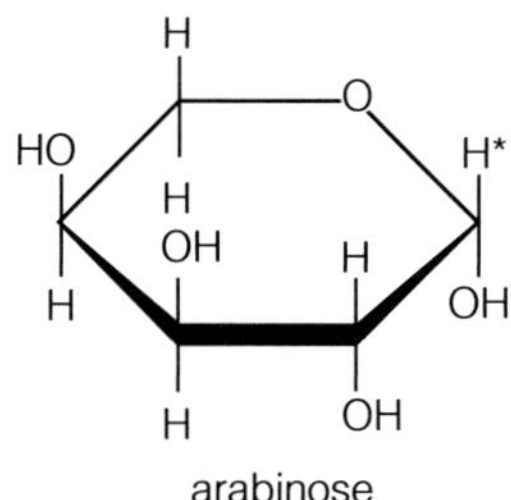

arabinose

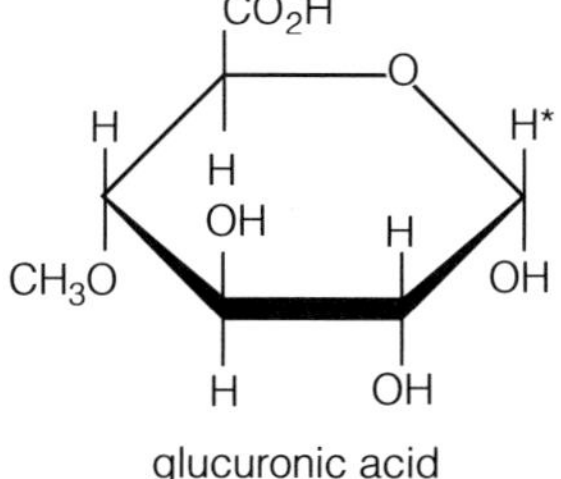

glucuronic acid

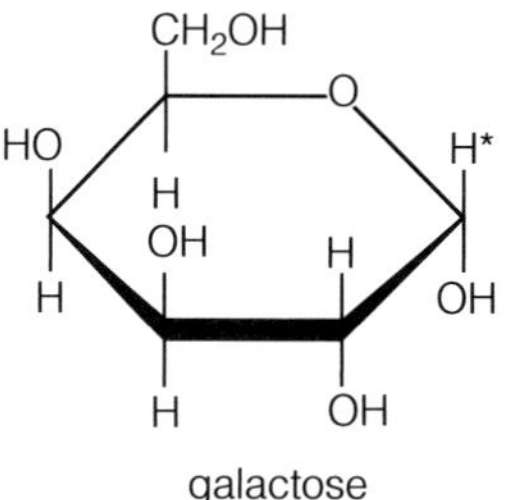

galactose

**Hemicelluloses.** The most common hemicelluloses are composed of a backbone chain of xylose, mannose, and galactose, with branching side chains of arabinose, glucuronic acid, and galactose.

# Lipids

**Table C-1**
**Saturated Fatty Acids Found in Natural Fats**

| Saturated Fatty Acids | Chemical Formulas | Number of Carbons | Major Food Sources |
|---|---|---|---|
| Butyric | $C_3H_7COOH$ | 4 | Butterfat |
| Caproic | $C_5H_{11}COOH$ | 6 | Butterfat |
| Caprylic | $C_7H_{15}COOH$ | 8 | Coconut oil |
| Capric | $C_9H_{19}COOH$ | 10 | Palm oil |
| Lauric | $C_{11}H_{23}COOH$ | 12 | Coconut oil, palm oil |
| Myristic[a] | $C_{13}H_{27}COOH$ | 14 | Coconut oil, palm oil |
| Palmitic[a] | $C_{15}H_{31}COOH$ | 16 | Palm oil |
| Stearic[a] | $C_{17}H_{35}COOH$ | 18 | Most animal fats |
| Arachidic | $C_{19}H_{39}COOH$ | 20 | Peanut oil |
| Behenic | $C_{21}H_{43}COOH$ | 22 | Seeds |
| Lignoceric | $C_{23}H_{47}COOH$ | 24 | Peanut oil |

[a]Most common saturated fatty acids.

**Table C-2**
**Unsaturated Fatty Acids Found in Natural Fats**

| Unsaturated Fatty Acids | Chemical Formulas | Number of Carbons | Number of Double Bonds | Standard Notation[b] | Omega Notation[b] | Food Sources |
|---|---|---|---|---|---|---|
| Palmitoleic | $C_{15}H_{29}COOH$ | 16 | 1 | 16:1;9 | 16:1ω7 | Seafood, beef |
| Oleic | $C_{17}H_{33}COOH$ | 18 | 1 | 18:1;9 | 18:1ω9 | Olive oil, canola oil |
| Linoleic | $C_{17}H_{31}COOH$ | 18 | 2 | 18:2;9,12 | 18:2ω6 | Sunflower oil, safflower oil |
| Linolenic | $C_{17}H_{29}COOH$ | 18 | 3 | 18:3;9,12,15 | 18:3ω3 | Soybean oil, canola oil |
| Arachidonic | $C_{19}H_{31}COOH$ | 20 | 4 | 20:4;5,8,11,14 | 20:4ω6 | Eggs, most animal fats |
| Eicosapentanoic | $C_{19}H_{29}COOH$ | 20 | 5 | 20:5;5,8,11,14,17 | 20:5ω3 | Seafood |
| Docosahexanoic | $C_{21}H_{31}COOH$ | 22 | 6 | 22:6;4,7,10,13,16,19 | 22:6ω3 | Seafood |

NOTE: A fatty acid has two ends; designated the methyl ($CH_3$) end and the carboxyl, or acid (COOH), end.
[a]Standard chemistry notation begins counting carbons at the acid end. The number of carbons the fatty acid contains comes first, followed by a colon and another number that indicates the number of double bonds; next comes a semicolon followed by a number or numbers indicating the positions of the double bonds. Thus the notation for linoleic acid, an 18-carbon fatty acid with two double bonds between carbons 9 and 10 and between carbons 12 and 13, is 18:2;9,12.
[b]Because fatty acid chains are lengthened by adding carbons at the acid end of the chain, chemists use the omega system of notation to ease the task of identifying them. The omega system begins counting carbons at the methyl end. The number of carbons the fatty acid contains comes first, followed by a colon and the number of double bonds; next comes the omega symbol (ω) and number indicating the position of the double bond nearest the methyl end. Thus linoleic acid with its first double bond at the sixth carbon from the methyl end would be noted 18:2ω6 in the omega system.

# Protein: Amino Acids

The common amino acids may be classified into the seven groups listed on the next page. Amino acids marked with an asterisk (*) are essential because human beings cannot synthesize them.

1. Amino acids with aliphatic side chains, which consist of hydrogen and carbon atoms (hydrocarbons):

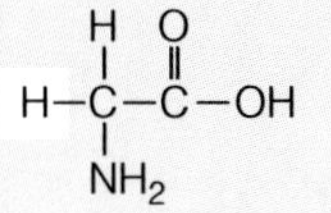

H O
$H-C-C-OH$
$NH_2$
**Glycine (Gly)**

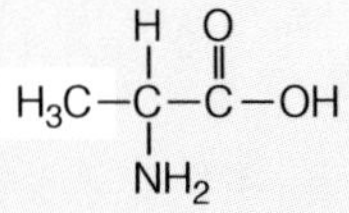

H O
$H_3C-C-C-OH$
$NH_2$
**Alanine (Ala)**

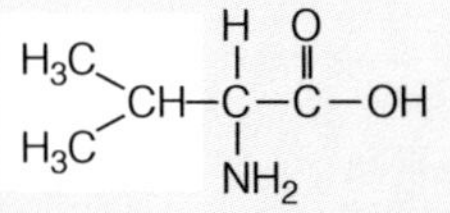

$H_3C$
H O
$CH-C-C-OH$
$H_3C$
$NH_2$
**Valine* (Val)**

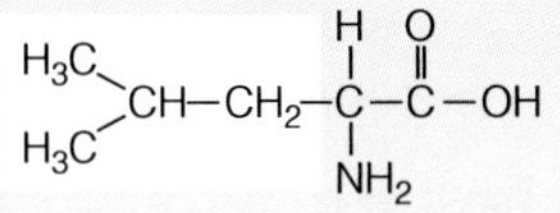

$H_3C$
H O
$CH-CH_2-C-C-OH$
$H_3C$
$NH_2$
**Leucine* (Leu)**

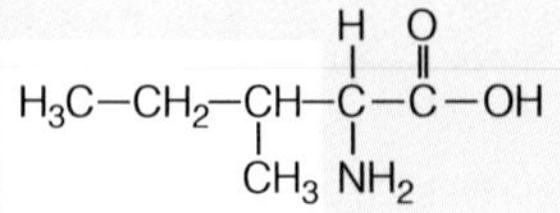

H O
$H_3C-CH_2-CH-C-C-OH$
$CH_3$ $NH_2$
**Isoleucine* (Ile)**

2. Amino acids with hydroxyl (OH) side chains:

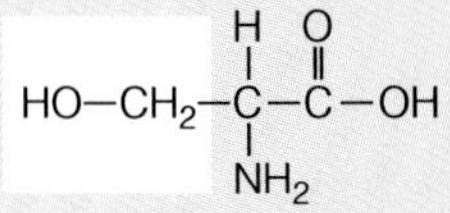

H O
$HO-CH_2-C-C-OH$
$NH_2$
**Serine (Ser)**

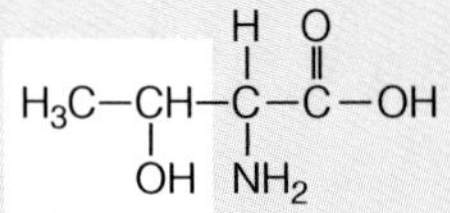

H O
$H_3C-CH-C-C-OH$
OH $NH_2$
**Threonine* (Thr)**

3. Amino acids with side chains containing acidic groups or their amides, which contain the group $NH_2$:

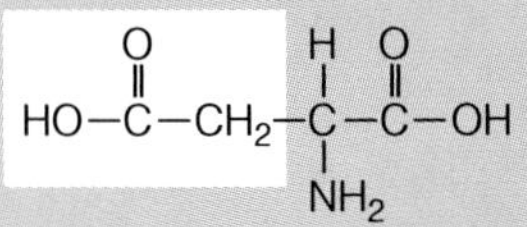

O H O
$HO-C-CH_2-C-C-OH$
$NH_2$
**Aspartic acid (Asp)**

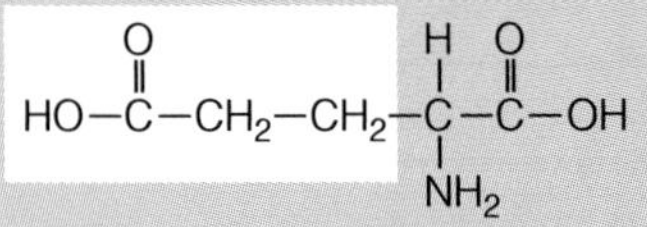

O H O
$HO-C-CH_2-CH_2-C-C-OH$
$NH_2$
**Glutamic acid (Glu)**

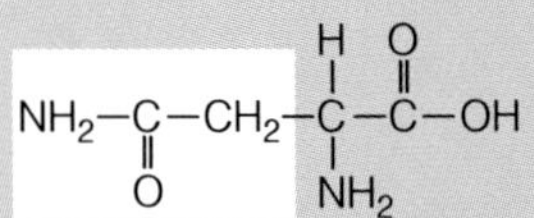

H O
$NH_2-C-CH_2-C-C-OH$
O $NH_2$
**Asparagine (Asn)**

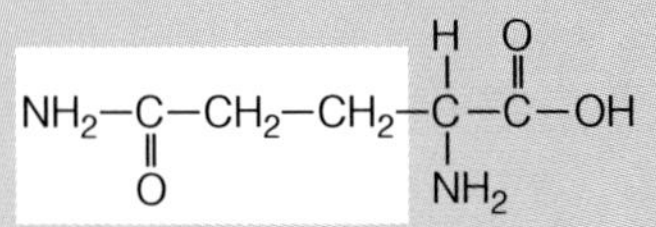

H O
$NH_2-C-CH_2-CH_2-C-C-OH$
O $NH_2$
**Glutamine (Gln)**

4. Amino acids with basic side chains:

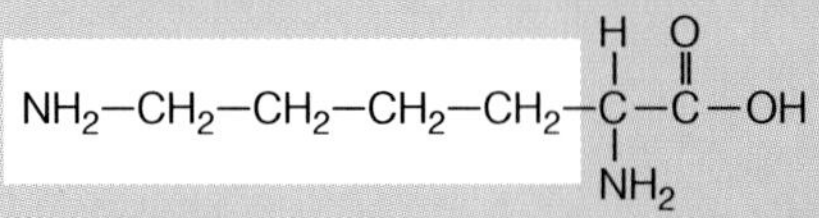

H O
$NH_2-CH_2-CH_2-CH_2-CH_2-C-C-OH$
$NH_2$
**Lysine* (Lys)**

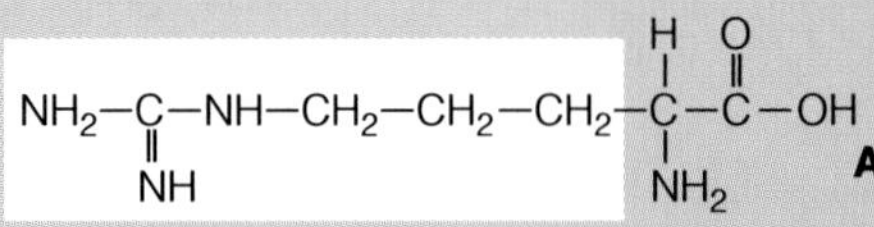

H O
$NH_2-C-NH-CH_2-CH_2-CH_2-C-C-OH$
NH $NH_2$
**Arginine (Arg)**

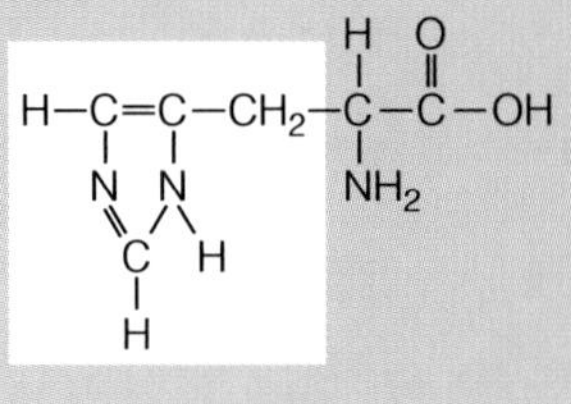

H O
$H-C=C-CH_2-C-C-OH$
N N $NH_2$
C H
H
**Histidine* (His)**

5. Amino acids with aromatic side chains, which are characterized by the presence of at least one ring structure:

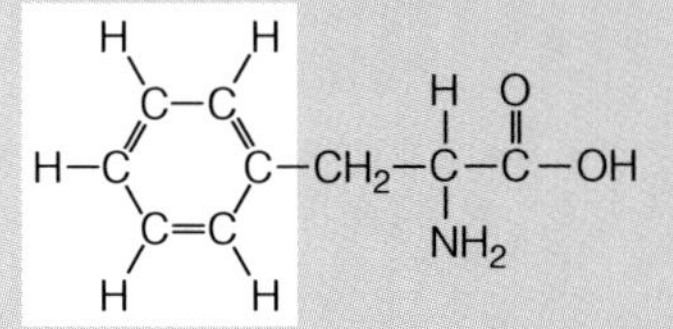

H H
C–C H O
$H-C$ $C-CH_2-C-C-OH$
C=C $NH_2$
H H
**Phenylalanine* (Phe)**

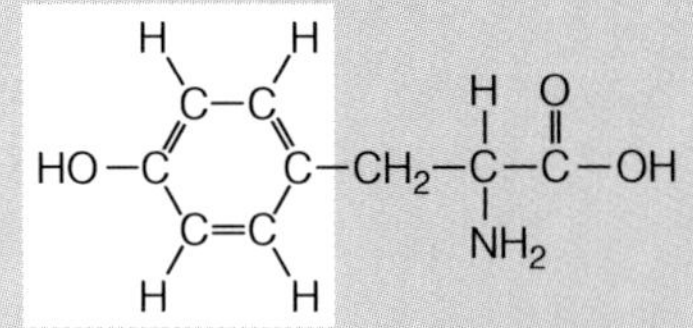

H H
C–C H O
$HO-C$ $C-CH_2-C-C-OH$
C=C $NH_2$
H H
**Tyrosine (Tyr)**

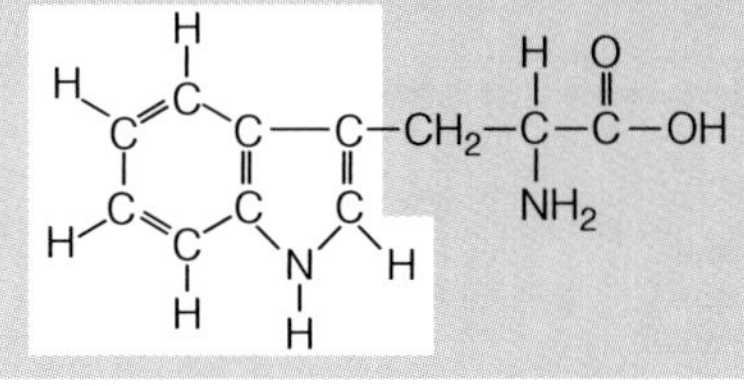

H
H C H O
C C $C-CH_2-C-C-OH$
C C C $NH_2$
H C N H
H H
**Tryptophan* (Trp)**

6. Amino acids with side chains containing sulfur atoms:

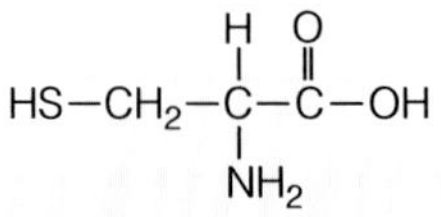

H O
$HS-CH_2-C-C-OH$
$NH_2$
**Cysteine (Cys)**

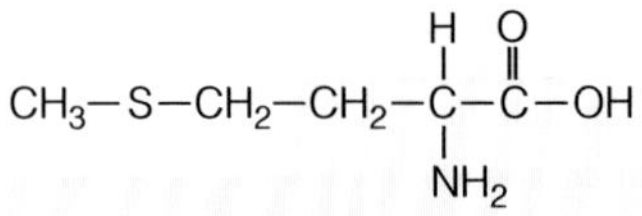

H O
$CH_3-S-CH_2-CH_2-C-C-OH$
$NH_2$
**Methionine* (Met)**

7. Imino acid:

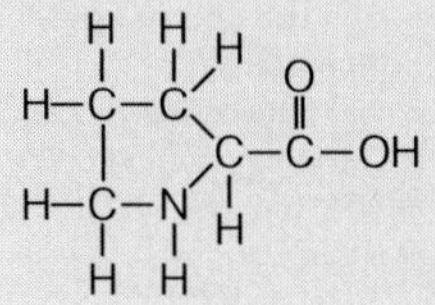

H H H
O
$H-C-C$
$C-C-OH$
$H-C-N$ H
H H
**Proline (Pro)** [a]

[a]Proline has the same $H_2N$±C±COOH structure as the other amino acids, but its amino group has given up a hydrogen to form a ring.

# Vitamins and Coenzymes

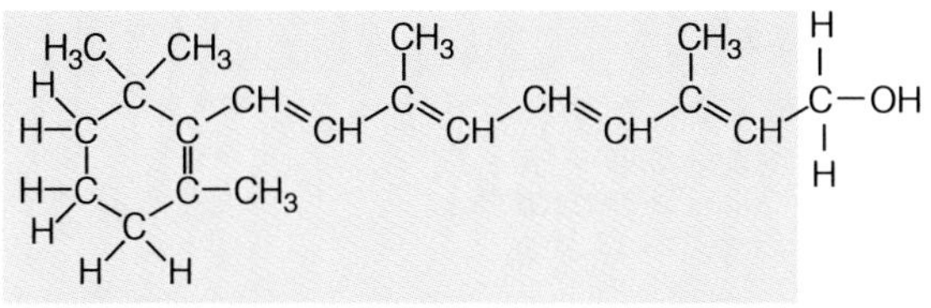

**Vitamin A: retinol.**

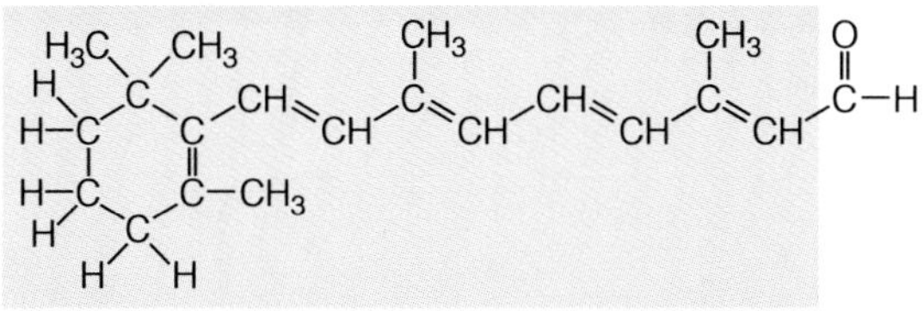

**Vitamin A: retinal.**

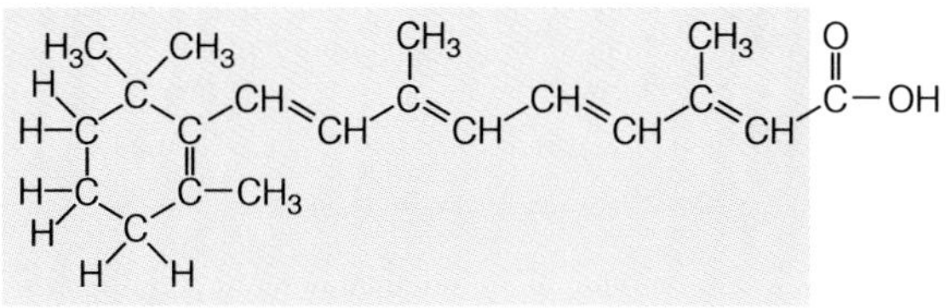

**Vitamin A: retinoic acid.**

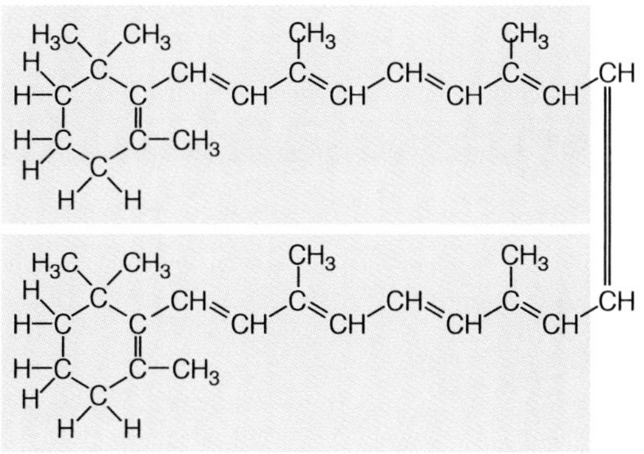

**Vitamin A precursor: beta-carotene.**

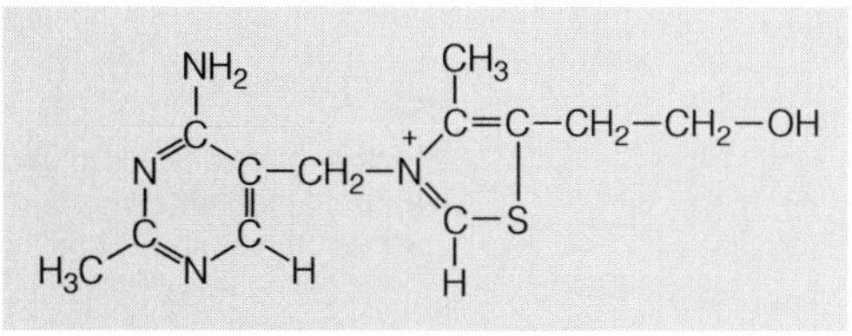

**Thiamin.** This molecule is part of the coenzyme thiamin pyrophosphate (TPP).

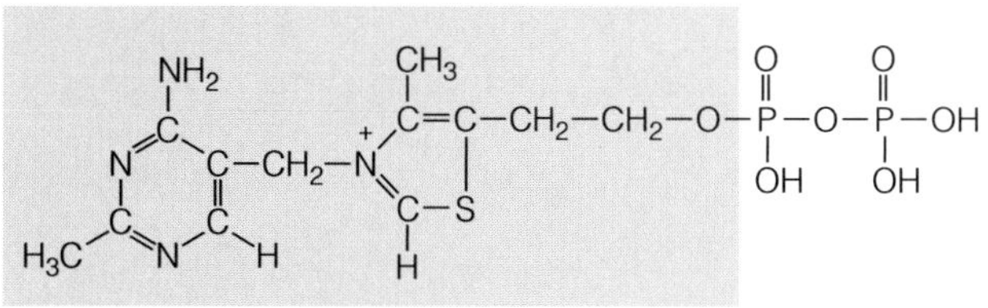

**Thiamin pyrophosphate (TPP).** TPP is a coenzyme that includes the thiamin molecule as part of its structure.

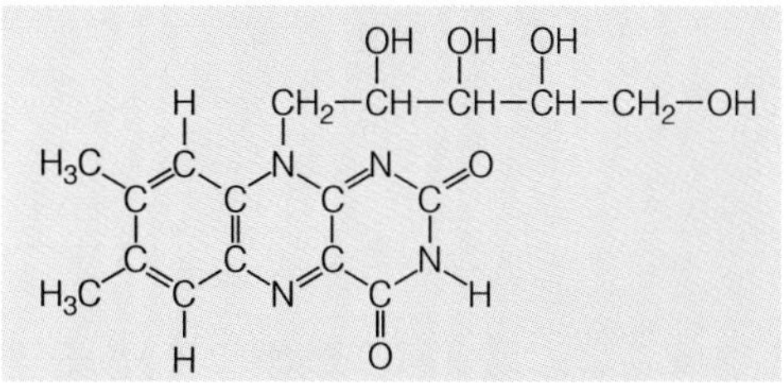

**Riboflavin.** This molecule is a part of two coenzymes—flavin mononucleotide (FMN) and flavin adenine dinucleotide (FAD).

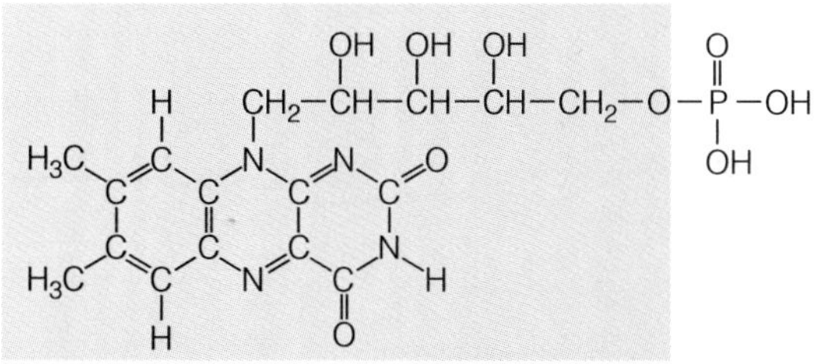

**Flavin mononucleotide (FMN).** FMN is a coenzyme that includes the riboflavin molecule as part of its structure.

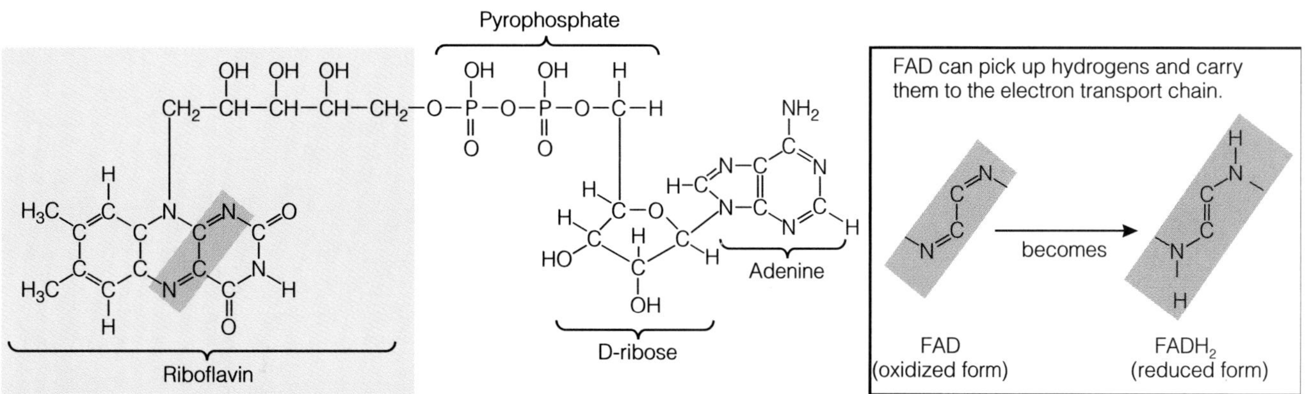


**Flavin adenine dinucleotide (FAD).** FAD is a coenzyme that includes the riboflavin molecule as part of its structure.

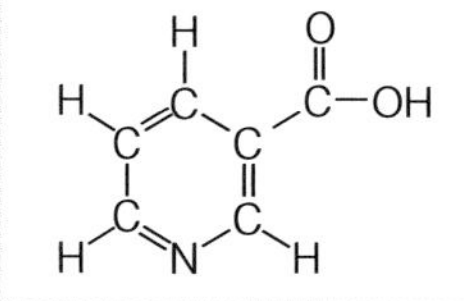

Nicotinic acid

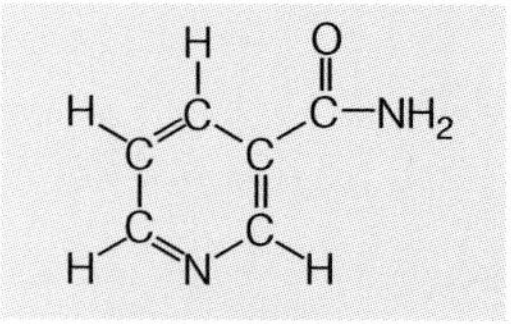

Nicotinamide

**Niacin (nicotinic acid and nicotinamide).** These molecules are a part of two coenzymes—nicotinamide adenine dinucleotide ($NAD^+$) and nicotinamide adenine dinucleotide phosphate ($NADP^+$).

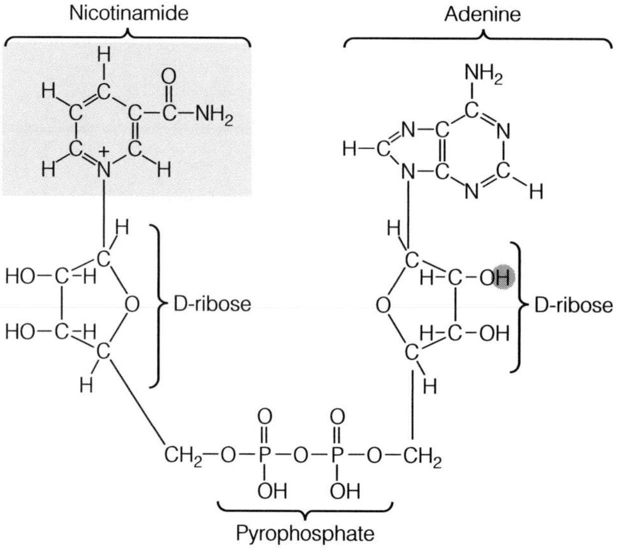

**Nicotinamide adenine dinucleotide ($NAD^+$) and nicotinamide adenine dinucleotide phosphate ($NADP^+$).** NADP has the same structure as NAD but with a phosphate group attached to the O instead of the H.

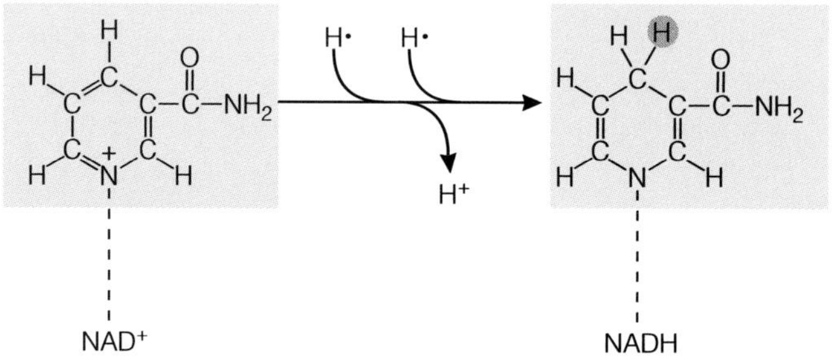

**Reduced $NAD^+$ (NADH).** When $NAD^+$ is reduced by the addition of $H^+$ and two electrons, it becomes the coenzyme NADH. (The dots on the H entering this reaction represent electrons—see Appendix B.)

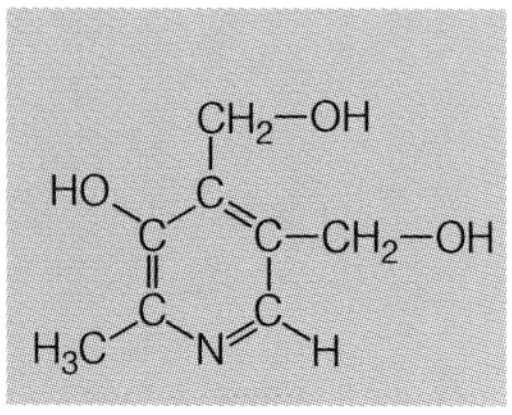

Pyridoxine

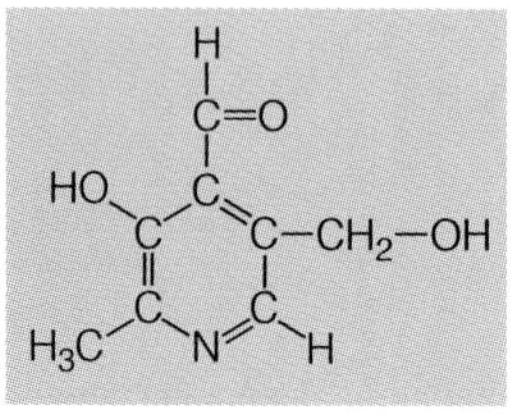

Pyridoxal

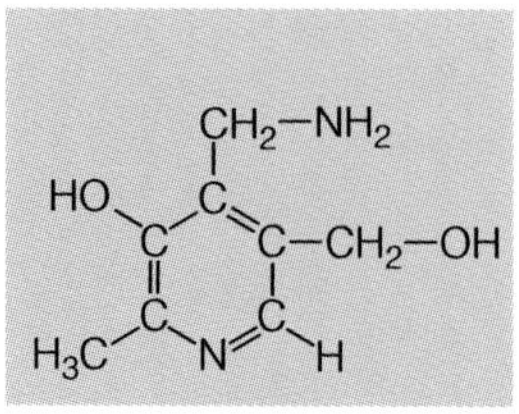

Pyridoxamine

**Vitamin $B_6$ (a general name for three compounds—pyridoxine, pyridoxal, and pyridoxamine).** These molecules are a part of two coenzymes—pyridoxal phosphate and pyridoxamine phosphate.

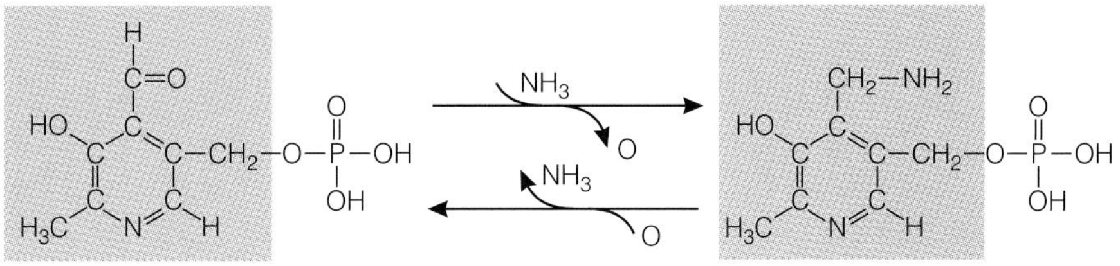

Pyridoxal phosphate

Pyridoxamine phosphate

**Pyridoxal phosphate (PLP) and pyridoxamine phosphate.** These coenzymes are necessary for transamination and other important processes.

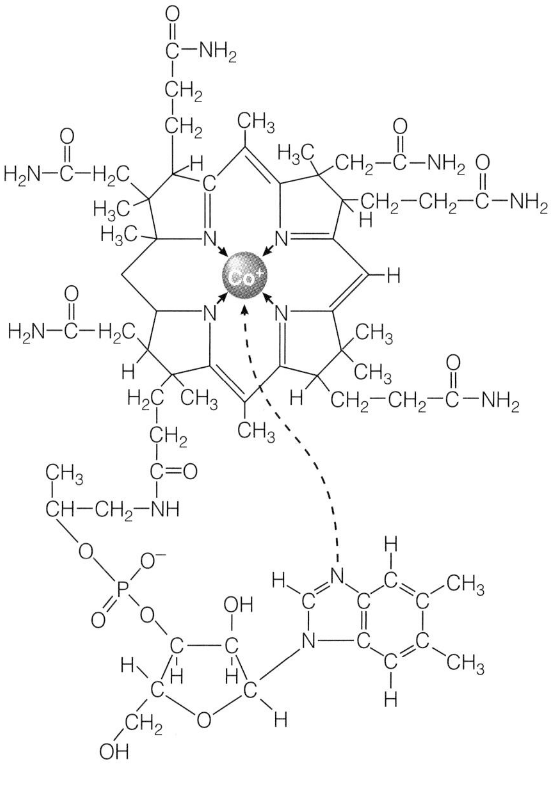

**Vitamin $B_{12}$ (cyanocobalamin).** The arrows in this diagram indicate that the spare electron pairs on the nitrogens attract them to the cobalt.

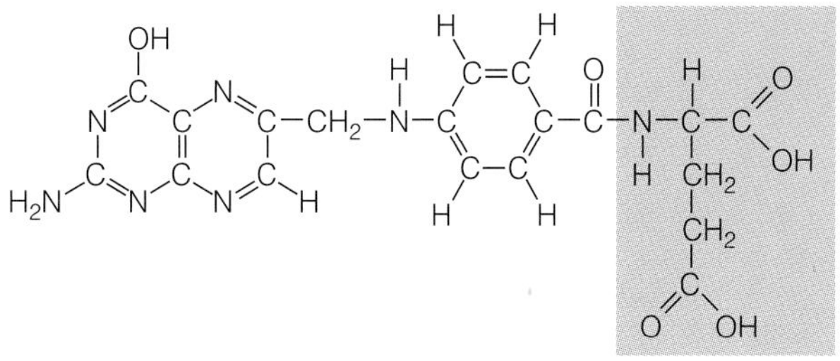

**Folate (folacin or folic acid).** This molecule consists of a double ring combined with a single ring and at least one glutamate (a nonessential amino acid marked in the box).

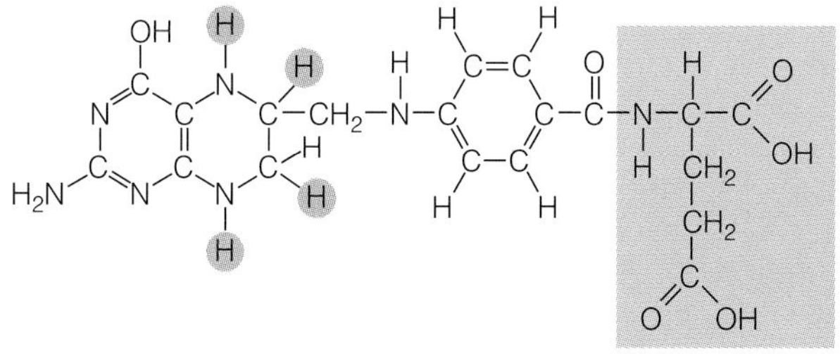

**Tetrahydrofolic acid, the active coenzyme form of folate.** This active form has four added hydrogens. An intermediate form, dihydrofolate, has two added hydrogens.

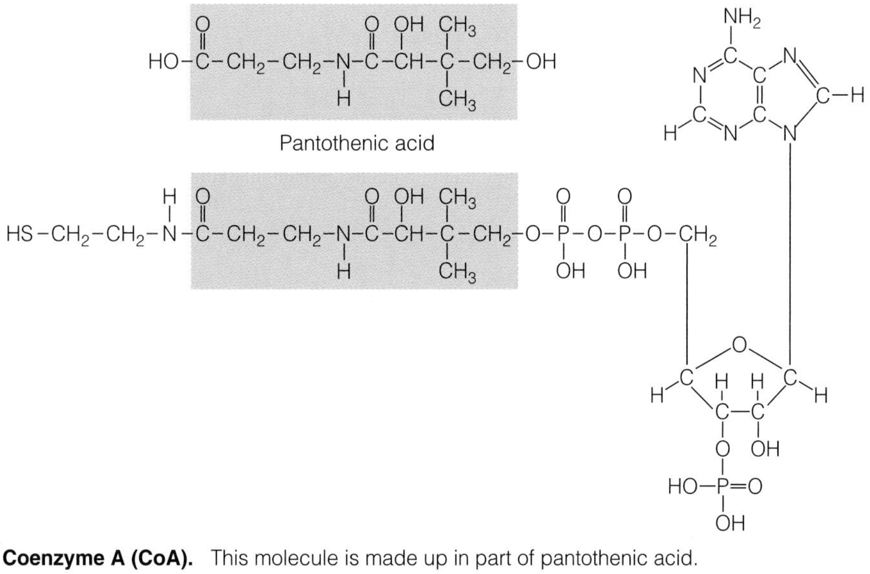

Pantothenic acid

**Coenzyme A (CoA).** This molecule is made up in part of pantothenic acid.

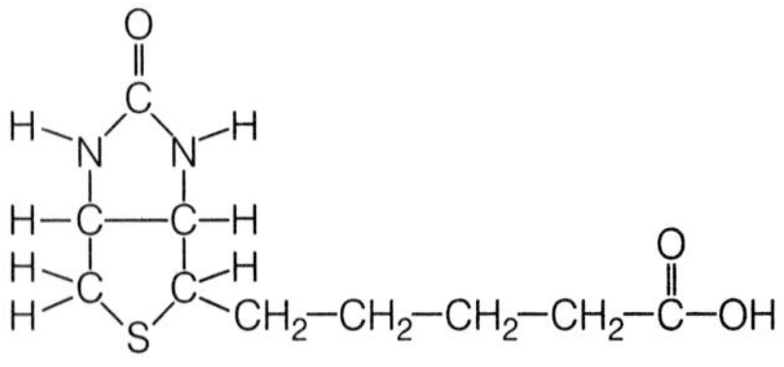

**Biotin.**

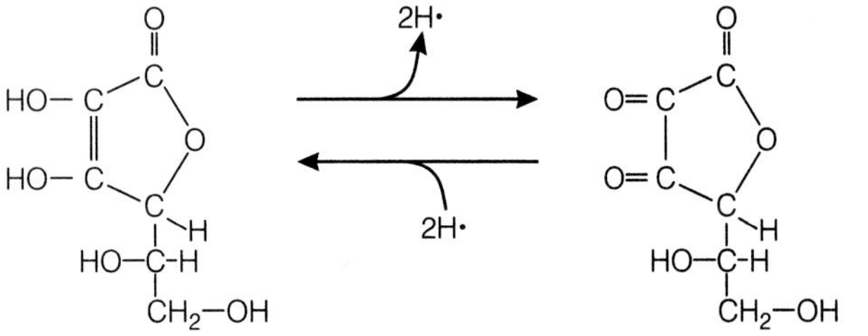


Ascorbic acid (reduced form)

Dehydroascorbic acid (oxidized form)

**Vitamin C.** The dots on the H indicate that two hydrogen atoms, complete with their electrons, are lost when ascorbic acid is oxidized and gained when it is reduced again.

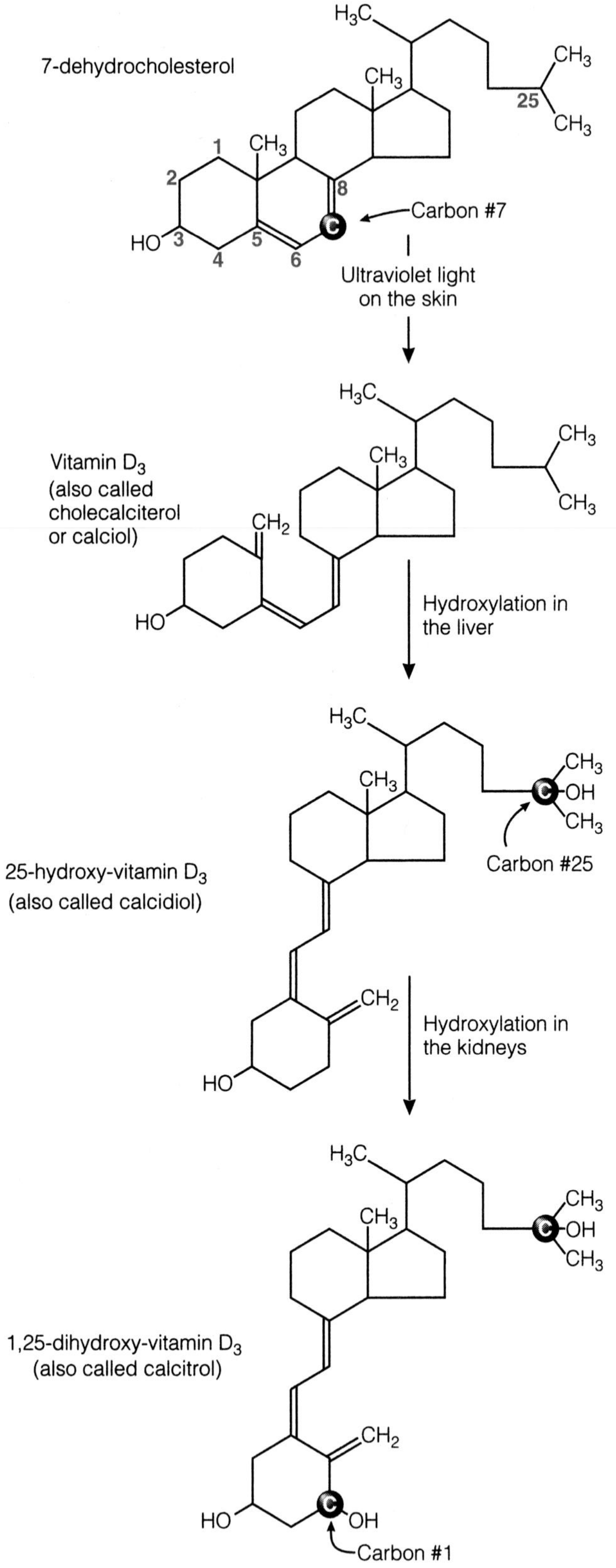


**Active vitamin D and its precursors, beginning with 7-dehydrocholesterol.** (The carbon atoms at which changes occur are numbered.)

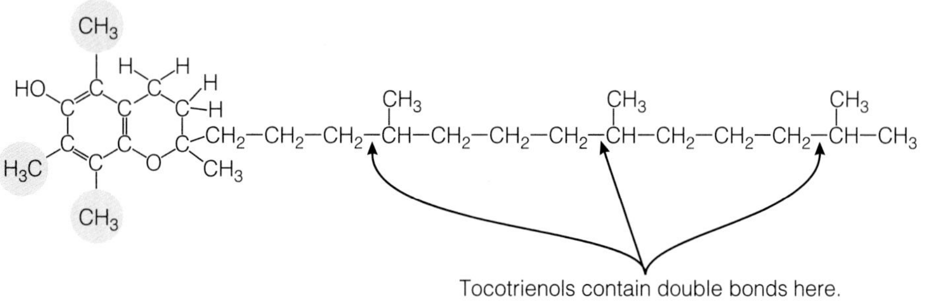

Tocotrienols contain double bonds here.

**Vitamin E (alpha-tocopherol).** The number and position of the methyl groups ($CH_3$) bonded to the ring structure differentiate among the tocopherols.

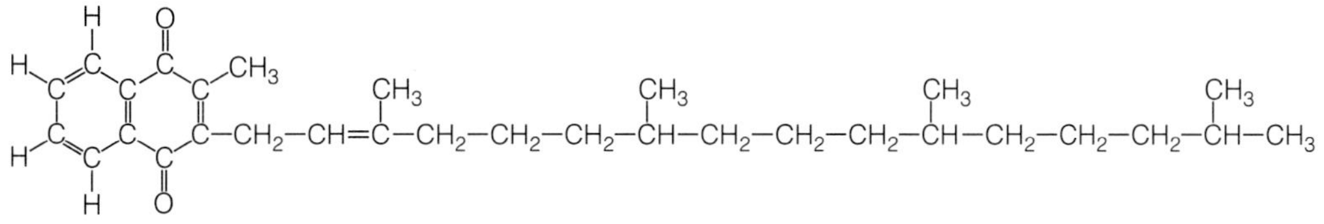

**Vitamin K, a naturally occurring compound.**

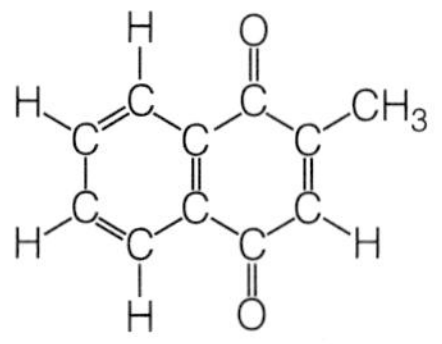

**Menadione, a synthetic compound that has the same activity as natural vitamin K.**

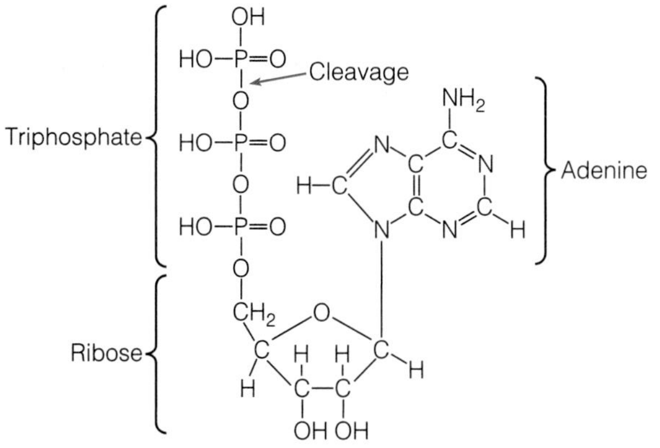

Triphosphate

Ribose

Adenine

Cleavage

**Adenosine triphosphate (ATP), the energy carrier.** The cleavage point marks the bond that is broken when ATP splits to become ADP + P.

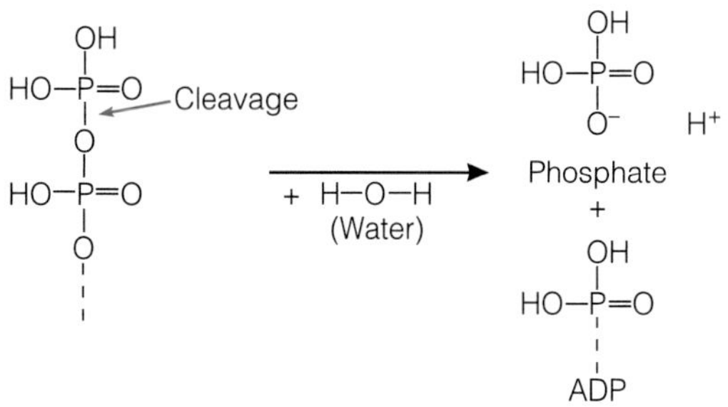

Cleavage

+ H—O—H (Water)

Phosphate

$H^+$

ADP

**Adenosine diphosphate (ADP).**

# Glycolysis

Figure C-1 (on the next page) depicts the events of glycolysis. First, a phosphate is attached to glucose at the carbon that chemists call number 6. The product is called, logically enough, glucose-6-phosphate.

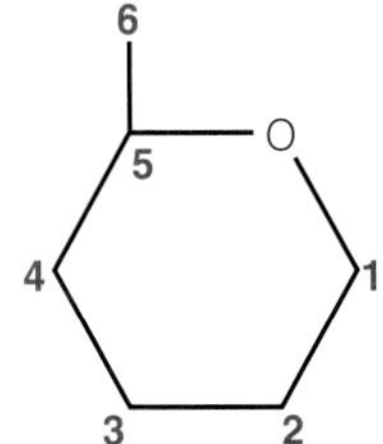


This is the way chemists number the carbons in a glucose molecule.

In the next couple of steps, glucose-6-phosphate is rearranged by an enzyme, and a phosphate is added in another coupled reaction with ATP. (A coupled reaction is a chemical event in which an enzyme complex catalyzes two reactions simultaneously. It often involves the breakdown of one compound and the synthesis of another.)

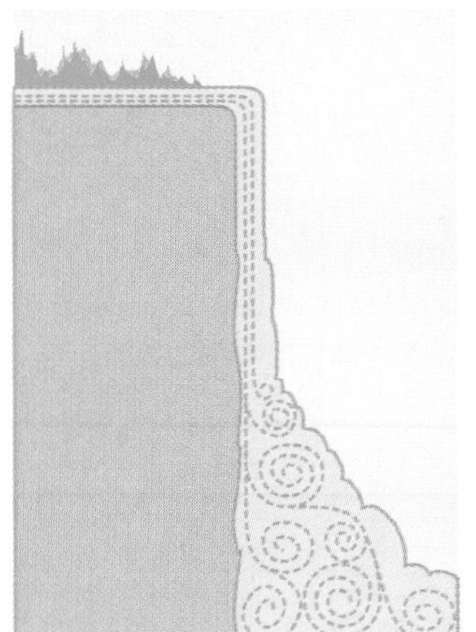

Falling water produces energy that is dissipated without doing work.

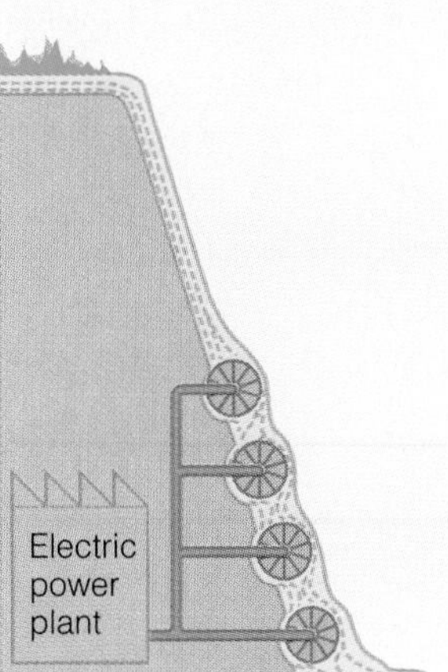


With the addition of a power plant (analogous to an enzyme), the energy of the falling water is coupled with a series of water wheels and turns them, producing energy.

**A physical analogy of a coupled reaction.** A coupled reaction often involves the breakdown of one compound and the synthesis of another. For example, the breakdown of glucose is coupled with the making of ATP, and the breakdown of ATP is coupled with the activation of glucose, or the making of glucose-P.

The product this time is fructose-1,6-diphosphate. At this point the six-carbon sugar has a phosphate group on its first and sixth carbons and is ready to break apart. Two ATP molecules have been used to accomplish this.

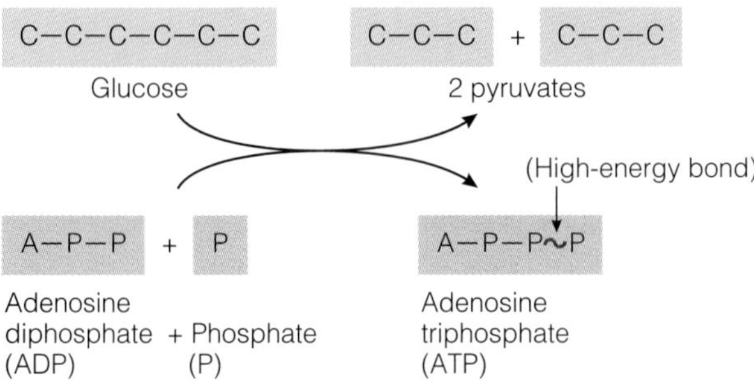


The breakdown of glucose is coupled with the making of ATP (simplified). Actually two ATP are used to prepare glucose for the reactions, and four ATP are gained in the breakdown of one glucose molecule to two molecules of pyruvate.

(From this point to the production of pyruvate, we will use letters in place of compound names. The names are in Figure C-1, for those who wish to know them.)

When fructose-1,6-diphosphate breaks in half, the two three-carbon compounds (A and A′) are not identical. Each has a phosphate group attached, but only one converts directly to pyruvate. The other compound, however, converts easily to the first. (Compound A′ is usually ignored, except for its role as the point of entry for the synthesis of glycerol; we say that two molecules of compound A are derived from one glucose molecule.)

In the step from compound A to compound B, enough energy is released to convert $NAD^+$ to $NADH + H^+$. Also, in the steps from B to C and from E to pyruvate, ATP is regenerated. Remember that in effect two molecules of compound A are produced from glucose; therefore, four ATP molecules are generated from each glucose molecule. Two ATP were needed to get the sequence started, so the net gain at this point is two ATP and two molecules of $NADH + H^+$.

So far, no oxygen has been used; the process has been anaerobic. But at this point, oxygen is needed. If oxygen is not immediately available, pyruvate converts to lactic acid to soak up the hydrogens from the $NADH + H^+$ that was generated. Lactic acid accumulates until oxygen becomes available. However, in the energy path from glucose to carbon dioxide, this side step usually is not necessary. As you will see later, each $NADH + H^+$ moves to the electron transport chain to unload its hydrogens onto oxygen. The associated energy produces two ATP, making a total yield of eight ATP for the process from glucose to pyruvate.

**Figure C-1**

**Glycolysis**

Notice that galactose and fructose enter at different places but all continue on the same pathway. Two molecules of compound A are produced (because compound A′ converts to A), and therefore two molecules of each succeeding compound.

- A = glyceraldehyde-3-phosphate.
- A′ = dihydroxyacetone phosphate.
- B = 1,3-diphosphoglyceric acid.
- C = 3-phosphoglyceric acid.
- D = 2-phosphoglyceric acid.
- E = phosphoenol pyruvic acid.

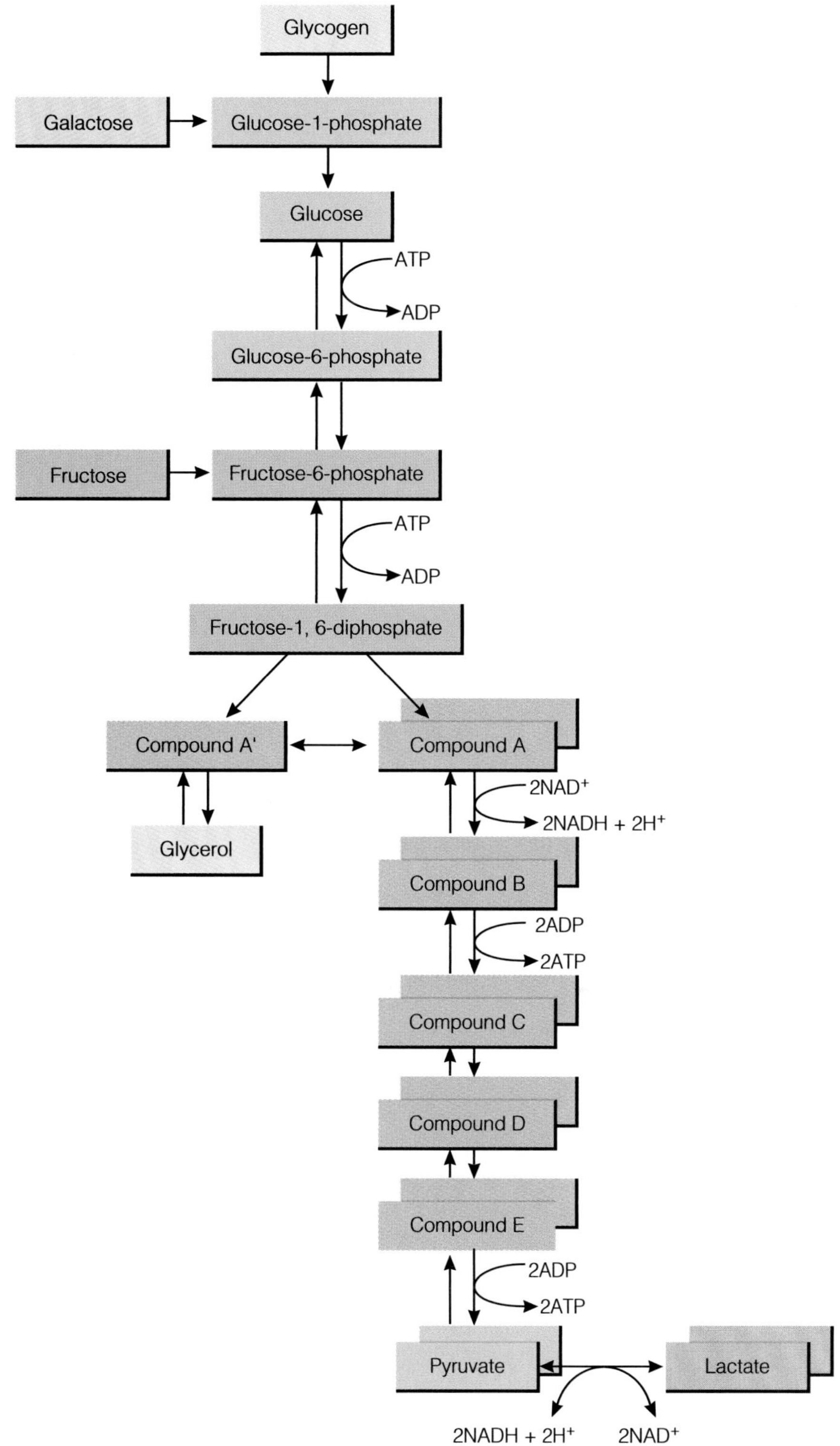


C

# The TCA Cycle

The tricarboxylic acid, or TCA, cycle (Figure C-2 on p. C-13) is the name given to the set of reactions involving oxygen and leading from acetyl CoA to carbon dioxide (and water). To link glycolysis to the TCA cycle, pyruvate loses a carbon group and bonds with a molecule of CoA to become acetyl CoA. The TCA cycle is not restricted to the metabolism of carbohydrate. It also includes fat and protein. Any substance that can be converted to acetyl CoA directly, or indirectly through pyruvate, may enter the cycle.

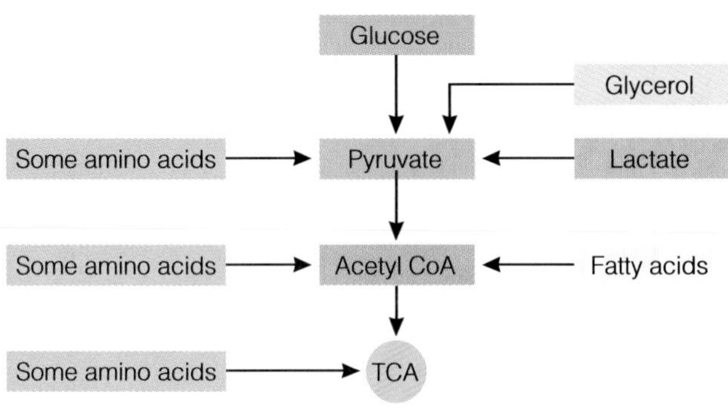


The step from pyruvate to acetyl CoA is exceedingly complex. We have included only those substances that will help you understand the transfer of energy from the nutrients. In the presence of oxygen, pyruvate loses a carbon to carbon dioxide and is attached to a molecule of CoA. In the process, $NAD^+$ picks up two hydrogens with their associated energy, becoming NADH + $H^+$.

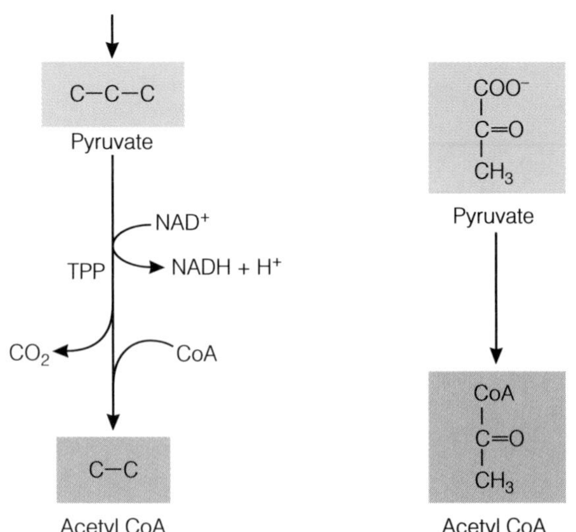


**The step from pyruvate to acetyl CoA.** (TPP and NAD are coenzymes containing the B vitamins thiamin and niacin, respectively.)

As the acetyl CoA breaks down to carbon dioxide and water, its energy is captured in ATP. Let's follow the steps by which this occurs (see Figure C-2).

1. The two-carbon acetyl CoA combines with a four-carbon compound, oxaloacetate. The CoA comes off, and the product is a six-carbon compound, citrate.
2. The atoms of citrate are rearranged to form isocitrate.
3. Now $NAD^+$ reacts with isocitrate. Two H and two electrons are removed from the isocitrate. One H becomes attached to the $NAD^+$ with the two electrons; the other H is released as $H^+$. Thus $NAD^+$ becomes NADH + $H^+$. (Remember this NADH + $H^+$. It is carrying the H and the energy released from the last reaction. But let's follow the carbons first.) A carbon is combined with two oxygens, forming carbon dioxide (which diffuses away into the blood and is exhaled). What is left is the five-carbon compound alpha-ketoglutarate.
4. Now two compounds interact with alpha-ketoglutarate —a molecule of CoA and a molecule of $NAD^+$. In this complex reaction, a carbon and two oxygens are removed (forming carbon dioxide); two hydrogens are removed and go to $NAD^+$ (forming NADH + $H^+$); and the remaining four-carbon compound is attached to the CoA, forming succinyl CoA. (Remember this NADH + $H^+$ also. You will see later what happens to it.)
5. Now two molecules react with succinyl CoA—a molecule called GDP and one of phosphate (P). The CoA comes off, the GDP and P combine to form the high-energy compound GTP (similar to ATP), and succinate remains. (Remember this GTP.)
6. In the next reaction, two H with their energy are removed from succinate and are transferred to a molecule called FAD (an electron-hydrogen receiver like $NAD^+$) to form $FADH_2$. The product that remains is fumarate. (Remember this $FADH_2$.)
7. Next a molecule of water is added to fumarate, forming malate.
8. A molecule of $NAD^+$ reacts with the malate; two H with their associated energy are removed from the malate and form NADH + $H^+$. The product that remains is the four-carbon compound oxaloacetate. (Remember this NADH + $H^+$.)

We are back where we started. The oxaloacetate formed in this process can combine with another molecule of acetyl CoA (step 1), and the cycle can begin again. The whole scheme is shown in Figure C-2.

So far, we have seen two carbons brought in with acetyl CoA and two carbons ending up in carbon dioxide. But where are the energy and the ATP we promised?

Each time a pair of hydrogen atoms is removed from one of the compounds in the cycle, it includes a pair of electrons. Then the energy from this chemical bond is captured in the compound to which the H become attached. A review of the eight steps of the cycle shows that energy is transferred in this way into other compounds in steps 3, 4, 6, and 8. In step 5, energy is stored when GDP and P are bound

**Figure C-2**
**The TCA Cycle**

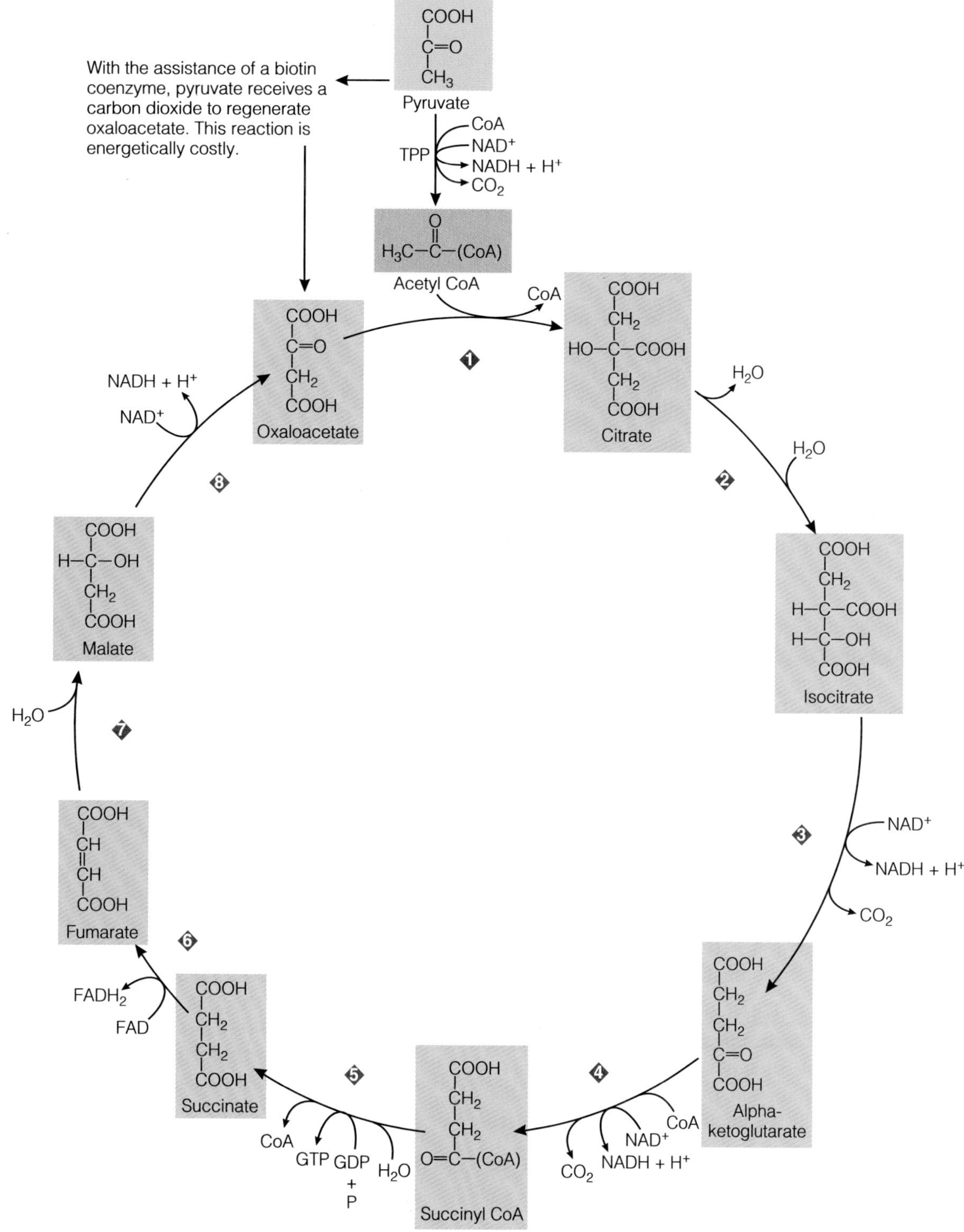

together to form GTP. Thus the compounds NADH + $H^+$ (three molecules), $FADH_2$, and GTP store energy originally found in acetyl CoA. To see how this energy ends up in ATP, we must follow the electrons further. Let us take those attached to $NAD^+$ as an example.

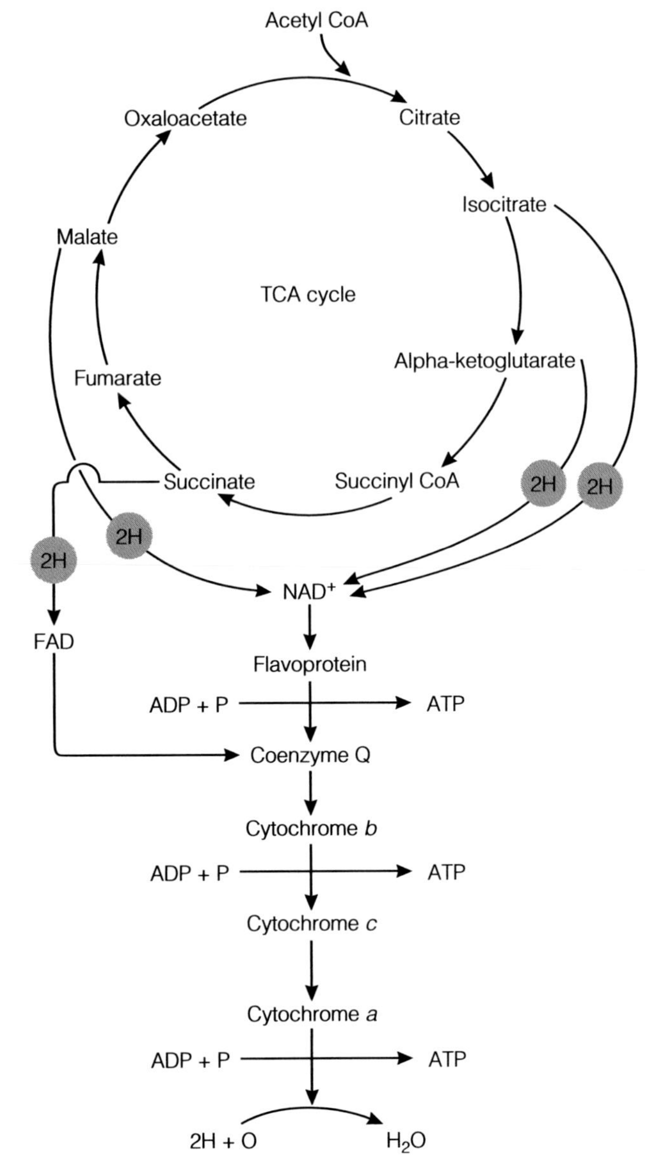

# The Electron Transport Chain

The six reactions described here are those of the electron transport chain, which is shown in Figure C-3. Since oxygen is required for these reactions, and ADP and P are combined to form ATP in several of them (ADP is phosphorylated), these reactions are also called oxidative phosphorylation.

An important concept to remember at this point is that an electron is not a fixed amount of energy. The electrons that bond the H to $NAD^+$ in NADH have a relatively large amount of energy. In the series of reactions that follow, they lose this energy in small amounts, until at the end they are attached (with H) to oxygen (O) to make water ($H_2O$). In some of the steps, the energy they lose is captured into ATP in coupled reactions.

1. In the first step of the electron transport chain, NADH reacts with a molecule called a flavoprotein, losing its electrons (and their H). The products are $NAD^+$ and reduced flavoprotein. A little energy is lost as heat in this reaction.
2. The flavoprotein passes on the electrons to a molecule called coenzyme Q. Again they lose some energy as heat, but ADP and P bond together and form ATP, storing much of the energy. This is a coupled reaction: ADP + P $\rightarrow$ ATP.
3 Coenzyme Q passes the electrons to cytochrome *b*. Again the electrons lose energy.
4. Cytochrome *b* passes the electrons to cytochrome *c* in a coupled reaction in which ATP is formed: ADP + P $\rightarrow$ ATP.
5. Cytochrome *c* passes the electrons to cytochrome *a*.
6. Cytochrome *a* passes them (with their H) to an atom of oxygen (O), forming water ($H_2O$). This is a coupled reaction in which ATP is formed: ADP + P $\rightarrow$ ATP.

**Figure C-3**
**The Electron Transport Chain**

As Figure C-3 shows, each time NADH is oxidized (loses its electrons) by this means, the energy it loses is parceled out into three ATP molecules. When the electrons are passed on to water at the end, they are much lower in energy than they were originally. This completes the story of the electrons from NADH.

As for $FADH_2$, its electrons enter the electron transport chain at coenzyme Q. From coenzyme Q to water, ATP is generated in only two steps. Therefore, $FADH_2$ coming out of the TCA cycle yields just two ATP molecules.

One energy-receiving compound of the TCA cycle (GTP) does not enter the electron transport chain but gives its energy directly to ADP in a simple phosphorylation reaction. This reaction yields one ATP.

It is now possible to draw up a balance sheet of glucose metabolism (see Table C-3). Glycolysis has yielded 4 NADH + $H^+$ and 4 ATP molecules and has spent 2 ATP. The 2 acetyl CoA going through the TCA cycle have yielded 6 NADH +$H^+$, 2 $FADH_2$, and 2 GTP molecules. After the NADH + $H^+$ and $FADH_2$ have gone through the electron transport chain, there are 34 ATP. Added to these are the 4 ATP from glycolysis and the 2 ATP from GTP, making the total 40 ATP generated from one molecule of glucose. After the expense of 2 ATP is subtracted, there is a net gain of 38 ATP.*

*The total may sometimes be 36 or 37, rather than 38, ATP. The NADH + $H^+$ generated in the cytoplasm during glycolysis pass their electrons on to shuttle molecules, which move them into the mitochondria. One shuttle, malate, contributes its electrons to the electron transport chain before the first site of ATP synthesis, yielding 3 ATP. Another, glycerol phosphate, adds its electrons into the chain beyond that first site, yielding 2 ATP. Thus sometimes 3, and sometimes only 2, ATP result from the NADH + $H^+$ that arise from glycolysis. The amount depends on the cell.

**Table C-3**
**Balance Sheet for Glucose Metabolism**

| | Expenditures | Income |
|---|---|---|
| Glycolysis: | | |
| 1 glucose | 2 ATP | 4 ATP |
| 1 fructose-1,6-diphosphate | | 2 NADH + $H^+$ |
| 2 pyruvate | | 2 NADH + $H^+$ |
| TCA cycle: | | |
| 2 isocitrate | | 2 NADH + $H^+$ |
| 2 alpha-ketoglutarate | | 2 NADH + $H^+$ |
| 2 succinyl CoA | | 2 GTP |
| 2 succinate | | 2 $FADH_2$ |
| 2 malate | | 2 NADH + $H^+$ |
| Total ATP collected: | | |
| From glycolysis | 2 ATP | 4 ATP |
| From 2 NADH + $H^+$ | | 4–6 ATP[a] |
| From 8 NADH + $H^+$ | | 24 ATP |
| From 2 GTP | | 2 ATP |
| From 2 $FADH_2$ | | 4 ATP |
| Totals: | 2 ATP | 38–40 ATP |
| Balance on hand from 1 molecule of glucose: | | 36–38 ATP |

[a]Each NADH + $H^+$ from glycolysis can yield 2 or 3 ATP. See the accompanying text.

The TCA cycle and the electron transport chain are the body's major means of capturing the energy from nutrients in ATP molecules. Other means, such as anaerobic glycolysis, contribute, but the aerobic processes are the most efficient. Biologists and chemists understand much more about these processes than has been presented here.

# Alcohol's Interference with Energy Metabolism

Highlight 7 provides an overview of how alcohol interferes with energy metabolism. With an understanding of the TCA cycle, a few more details may be appreciated. During alcohol metabolism, the enzyme alcohol dehydrogenase oxidizes alcohol to acetaldehyde while it simultaneously reduces a molecule of $NAD^+$ to NADH + $H^+$. The related enzyme acetaldehyde dehydrogenase reduces another $NAD^+$ to NADH + $H^+$ while it oxidizes acetaldehyde to acetyl CoA, the compound that enters the TCA cycle to generate energy. Thus whenever alcohol is being metabolized in the body, $NAD^+$ diminishes, and NADH + $H^+$ accumulates. Chemists say that the body's "redox state" is altered, because $NAD^+$ can oxidize, and NADH + $H^+$ can reduce, many other body compounds. During alcohol metabolism, $NAD^+$ becomes unavailable for the multitude of reactions for which it is required.

As the previous sections just explained, for glucose to be completely metabolized, the TCA cycle must be operating, and $NAD^+$ must be present. If these conditions are not met (and when alcohol is present, they may not be), the pathway will be blocked, and traffic will back up—or an alternate route will be taken. Think about this as you follow the pathway shown in Figure C-4.

In each step of alcohol metabolism in which $NAD^+$ is converted to NADH + $H^+$, hydrogen ions accumulate, resulting in a dangerous shift of the acid-base balance toward acid (Chapter 12 explains acid-base balance). The accumulation of NADH + $H^+$ depresses TCA cycle activity, so pyruvate and acetyl CoA build up. This condition favors the conversion of pyruvate to lactic acid, which serves as a temporary storage place for hydrogens from NADH + $H^+$. The conversion of pyruvate to lactic acid restores some $NAD^+$, but a lactic acid buildup has serious consequences of its own. It adds to the body's acid burden and interferes with the excretion of uric acid, causing goutlike symptoms. Molecules of acetyl CoA become building blocks for fatty acids or ketone bodies. The making of ketone bodies consumes acetyl CoA and generates $NAD^+$; but some ketone bodies are acids, so they push the acid-base balance further toward acid.

**Figure C-4**
**Ethanol Enters the Metabolic Path**

This is a simplified version of the glucose-to-energy pathway showing the entry of ethanol. The coenzyme NAD (which is the active form of the B vitamin niacin) is the only one shown here; however, many others are involved.

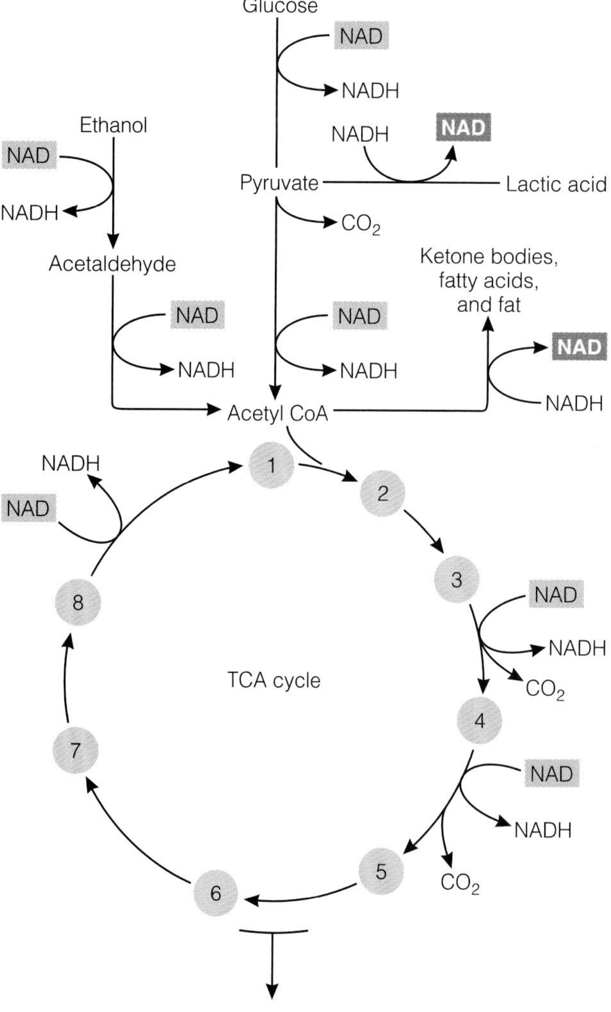

Thus alcohol cascades through the metabolic pathways, wreaking havoc along the way. These consequences have physical effects, which Highlight 7 describes.

# The Urea Cycle

Chapter 7 sums up the process by which waste nitrogen is eliminated from the body by stating that ammonia molecules combine with carbon dioxide to produce urea. This is true, but it is not the whole story. Urea is produced in a multistep process within the cells of the liver.

Ammonia, freed from an amino acid or other compound during metabolism anywhere in the body, arrives at the liver by way of the bloodstream and is taken into a liver cell. There, it is first combined with carbon dioxide and a phosphate group from ATP to form carbamyl phosphate:

$$\underset{\text{Carbon dioxide}}{CO_2} + \underset{\text{Ammonia}}{NH_3} \xrightarrow[2\ ATP \rightarrow 2\ ADP + P]{} \underset{\text{Carbamyl phosphate}}{H_2N{-}\overset{\overset{O}{\|}}{C}{-}\underbrace{O{-}\overset{\overset{O}{\|}}{\underset{\underset{O^-}{|}}{P}}{-}O^-}_{\text{Phosphate group}}}$$

Figure C-5 shows the cycle of four reactions that follow. In the first step, carbamyl phosphate combines with the amino acid ornithine, losing its phosphate group. The compound formed is citrulline.

In the second step, citrulline combines with the amino acid aspartic acid, to form argininosuccinate. The reaction requires energy from ATP. (ATP was shown earlier losing one phosphorus atom in a phosphate group, P, to become ADP. In this reaction, it loses two phosphorus atoms joined together, PP, and becomes adenosine monophosphate, AMP.)

In the third step, argininosuccinate is split, forming another acid, fumarate, and the amino acid arginine.

In the fourth step, arginine loses its terminal carbon with two attached amino groups and picks up an oxygen from water. The end product is urea, which the kidneys excrete in the urine. The compound that remains is ornithine, identical to the ornithine with which this series of reactions began, and ready to react with another molecule of carbamyl phosphate and turn the cycle again.

# Formation of Ketone Bodies

Normally, fatty acid oxidation proceeds all the way to carbon dioxide and water. However, in ketosis (discussed in Chapter 7), an intermediate is formed from the condensation of two molecules of acetyl CoA: acetoacetyl CoA. Figure C-6 shows the formation of ketone bodies from that intermediate. In step 1, acetoacetyl CoA condenses with another acetyl CoA to form a six-carbon intermediate, beta-hydroxy-beta-methylglutaryl CoA. In step 2, this intermediate is cleaved to acetyl CoA and acetoacetic

**Figure C-5**
**The Urea Cycle**

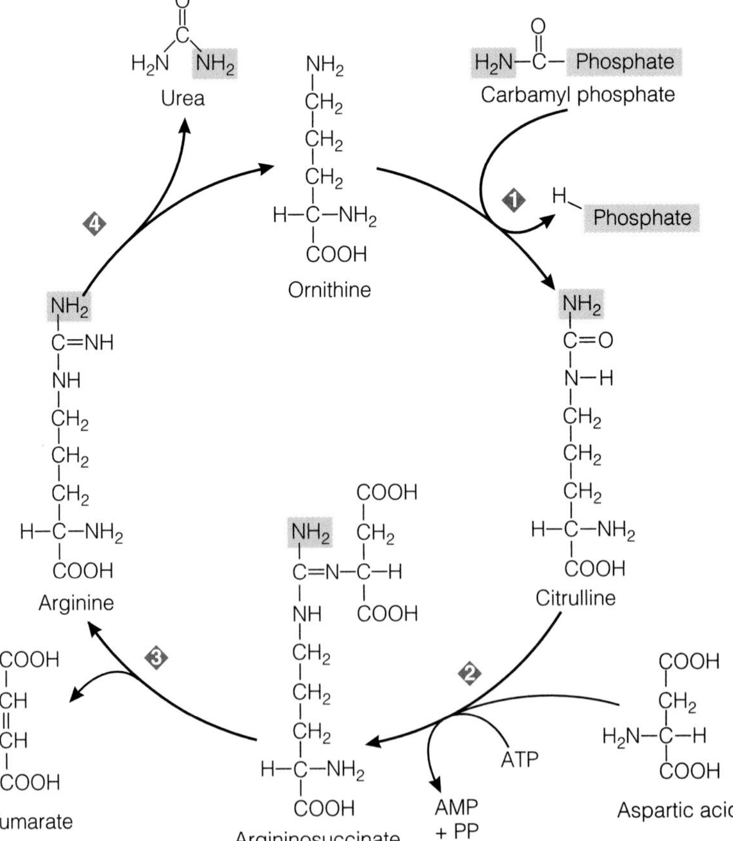

**Figure C-6**
**Formation of Ketone Bodies**

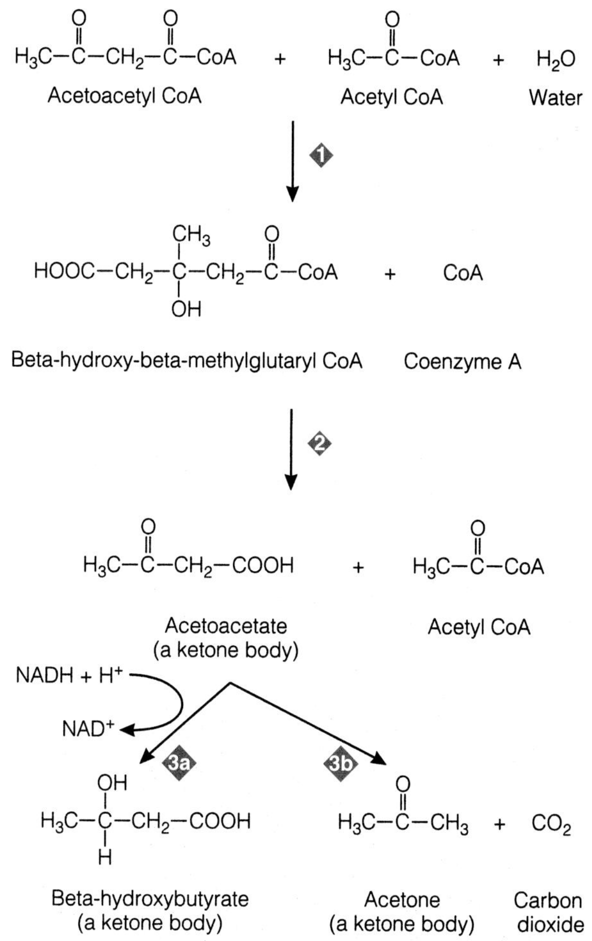

$$H_3C-\overset{O}{\overset{\|}{C}}-CH_2-\overset{O}{\overset{\|}{C}}-CoA \quad + \quad H_3C-\overset{O}{\overset{\|}{C}}-CoA \quad + \quad H_2O$$

Acetoacetyl CoA + Acetyl CoA + Water

↓ ❶

$$HOOC-CH_2-\underset{OH}{\underset{|}{\overset{CH_3}{\overset{|}{C}}}}-CH_2-\overset{O}{\overset{\|}{C}}-CoA \quad + \quad CoA$$

Beta-hydroxy-beta-methylglutaryl CoA + Coenzyme A

↓ ❷

$$H_3C-\overset{O}{\overset{\|}{C}}-CH_2-COOH \quad + \quad H_3C-\overset{O}{\overset{\|}{C}}-CoA$$

Acetoacetate (a ketone body) + Acetyl CoA

$NADH + H^+$ → $NAD^+$ (3a); (3b)

$$H_3C-\underset{H}{\underset{|}{\overset{OH}{\overset{|}{C}}}}-CH_2-COOH \qquad H_3C-\overset{O}{\overset{\|}{C}}-CH_3 \quad + \quad CO_2$$

Beta-hydroxybutyrate (a ketone body) | Acetone (a ketone body) + Carbon dioxide

acid. This product can be metabolized either to beta-hydroxybutyric acid (step 3a) or to acetone (3b).

Acetoacetic acid, beta-hydroxybutyric acid, and acetone are the so-called ketone bodies of ketosis. Two are real ketones (they have a C=O group between two carbons); the other is an alcohol that has been produced during ketone formation—hence the term *ketone bodies,* rather than ketones, to describe the three of them. There are many other ketones in nature; these three are characteristic of ketosis in the body.

# D AIDS TO CALCULATION

CONTENTS



Many mathematical problems have been worked out as examples at appropriate places in the text. This appendix aims to help with the use of the metric system and with problems not fully explained elsewhere.

## Conversion Factors

Conversion factors are useful mathematical tools in everyday calculations, including those encountered in the study of nutrition. A conversion factor is a fraction in which the numerator (top) and the denominator (bottom) express the same quantity in different units. For example, 2.2 pounds (lb) and 1 kilogram (kg) are equivalent; they express the same weight. The conversion factor used to change pounds to kilograms or vice versa is:

$$\frac{2.2 \text{ lb}}{1 \text{ kg}} \text{ or } \frac{1 \text{ kg}}{2.2 \text{ lb}}.$$

Because both factors equal 1, measurements can be multiplied by the factor without changing the value of the measurement. Thus the units can be changed.

To perform a conversion, use the factor with the unit you are seeking in the numerator (top) of the fraction. Following are two examples of problems commonly encountered in nutrition study; they illustrate the usefulness of conversion factors.

**Example 1** Convert the weight of 130 pounds to kilograms.

1. Choose the conversion factor in which the unit you are seeking is on top:

$$\frac{1 \text{ kg}}{2.2 \text{ lb}}.$$

2. Multiply 130 pounds by the factor:

$$130 \text{ lb} \times \frac{1 \text{ kg}}{2.2 \text{ lb}} = \frac{130 \text{ kg}}{2.2} =$$

59 kg (rounded off to the nearest whole number).

**Example 2** How many grams (g) of saturated fat are contained in a 3-ounce (oz) hamburger?

1. Consider a 4-ounce hamburger that contains 7 grams of saturated fat. You are seeking grams of saturated fat; therefore, the conversion factor is:

$$\frac{7 \text{ g saturated fat}}{4 \text{ oz hamburger}}.$$

2. Multiply 3 ounces of hamburger by the conversion factor:

$$3 \text{ oz hamburger} \times \frac{7 \text{ g saturated fat}}{4 \text{ oz hamburger}} =$$

$$\frac{3 \times 7}{4} = \frac{21}{4}$$

= 5 g saturated fat (rounded off to the nearest whole number).

## Percentages

A percentage is a comparison between a number of items (perhaps your intake of energy) and a standard number (perhaps the number of kcalories recommended for your age and sex—the energy RDA). The standard number is the number you divide by. The answer you get after the division must be multiplied by 100 to be stated as a percentage (*percent* means "per 100").

**Example 3** What percentage of the RDA for energy is your energy intake?

1. Find your energy RDA (inside front cover, left). We'll use 2200 kcalories to demonstrate.
2. Total your energy intake for a day—for example, 1500 kcalories.
3. Divide your kcalorie intake by the RDA kcalories:

   1500 kcal (your intake) ÷ 2200 kcal (RDA) = 0.68.

4. Multiply your answer by 100 to state it as a percentage:

   $0.68 \times 100 = 68 = 68\%$.

In some problems in nutrition, the percentage may be more than 100. For example, suppose your daily intake of vitamin A is 3200 RE and your RDA (male) is 1000 RE. Your intake as a percentage of the RDA is more than 100 percent (that is, you consume more than 100 percent of your vitamin A RDA). The following calculations show your vitamin A intake as a percentage of the RDA:

$$3200 \div 1000 = 3.2.$$

$$3.2 \times 100 = 320\% \text{ of RDA.}$$

Sometimes the comparison is between a part of a whole (for example, your kcalories from protein) and the total amount (your total kcalories). In this case, the total number is the one you divide by.

**Example 4** What percentages of your total kcalories for the day come from protein, fat, and carbohydrate?

1 Using Appendix H or your computer diet analysis, find the total grams of protein, fat, and carbohydrate you consumed—for example, 60 grams protein, 80 grams fat, and 310 grams carbohydrate.
2. Multiply the number of grams by the number of kcalories from 1 gram of each energy nutrient (conversion factors):

$$60 \text{ g protein} \times \frac{4 \text{ kcal}}{1 \text{ g protein}} = 240 \text{ kcal.}$$

$$80 \text{ g fat} \times \frac{9 \text{ kcal}}{1 \text{ g fat}} = 720 \text{ kcal.}$$

$$310 \text{ g carbohydrate} \times \frac{4 \text{ kcal}}{1 \text{ g carbohydrate}} = 1240 \text{ kcal.}$$

$$240 + 720 + 1240 = 2200 \text{ kcal.}$$

3. Find the percentage of total kcalories from each energy nutrient (see Example 3):

- Protein: 240 ÷ 2200 = 0.109 × 100 = 10.9 = 11% of kcal.
- Fat: 720 ÷ 2200 = 0.327 × 100 = 32.7 = 33% of kcal.
- Carbohydrate: 1240 ÷ 2200 = 0.563 × 100 = 56.3 = 56% of kcal.
- 11% + 33% + 56% = 100% of kcal (total).

The percentages total 100 percent, but sometimes they total 99 or 101 because of rounding off. This is a reasonable error.

# Ratios

A ratio is a comparison of two or three values in which one of the values is reduced to 1. A ratio compares identical units and so is expressed without units. For example, Figure 12-6 in Chapter 12 compares the milligrams of potassium to the milligrams of sodium in selected foods.

**Example 5** Find the potassium-to-sodium ratio of your diet.

1. Using Appendix H or your computer diet analysis, find how many milligrams of potassium and sodium you consumed, say, 3000 milligrams potassium and 2500 milligrams sodium.
2. Divide the potassium milligrams by the sodium milligrams:

   3000 mg potassium ÷ 2500 mg sodium = 1.2.

3. The potassium-to-sodium ratio is usually expressed as correct to one decimal point: 1.2.

The potassium-to-sodium ratio of your diet is 1.2:1 (read as "one point two to one" or simply "one point two"). A ratio greater than 1 means that the first value (in this case, milligrams of potassium) is greater than the second (sodium). When the second value is larger, the ratio is less than 1.

# Weights and Measures

**Length**

1 inch (in) = 2.54 centimeters (cm).
1 foot (ft) = 30.48 centimeters.
1 meter (m) = 39.37 inches.

**Temperature**

| | Celsius* | Fahrenheit | |
|---|---|---|---|
| Steam | 100°C | 212°F | Steam |
| Body temperature | 37°C | 98.6°F | Body temperature |
| Ice | 0°C | 32°F | Ice |

To find degrees Fahrenheit ($t_F$) when you know degrees Celsius ($t_C$), multiply by 9/5 and then add 32:

$$(9/5 \times t_C) + 32 = t_F.$$

To find degrees Celsius ($t_C$) when you know degrees Fahrenheit ($t_F$), multiply by 5/9 after subtracting 32:

$$5/9\,(t_F - 32) = t_C.$$

**Volume**

1 liter (L) = 1.06 quarts (qt) or 0.85 imperial quart.
1 liter = 1000 milliliters (mL).
1 milliliter = 0.03 fluid ounces.
30 milliliters = 1 fluid ounce.
1 gallon = 3.79 liters.
1 quart = 0.95 liter or 32 fluid ounces.
1 pint = 0.47 liter or 16 fluid ounces
1 cup (c) = 8 fluid ounces or about 250 milliliters.
1 tablespoon (tbs) = 15 milliliters.
3 teaspoons (tsp) = 1 tablespoon.
1 teaspoon = about 5 g or 5 mL.
16 tablespoons = 1 cup.
4 cups = 1 quart.

*Also known as *centigrade*.

D

**Weight**
1 ounce (oz) = approximately 28 grams (g).
16 ounces = 1 pound (lb).
1 pound = 454 grams.
1 kilogram (kg) = 1000 grams or 2.2 pounds.
1 gram = 1000 milligrams (mg).
1 milligram = 1000 micrograms (μg).

**Energy units**
1 kcalorie (kcal) = 4.2 kilojoules (kJ).
1 millijoule (mJ) = 240 kcal.
1 kJ = 0.24 kcal.
1 g carbohydrate = 4 kcal = 17 kJ.
1 g fat = 9 kcal = 37 kJ.
1 g protein = 4 kcal = 17 kJ.
1 g alcohol = 7 kcal = 29 kJ.

**International Units (IU)**
To convert IU to:

- μg RE, divide by 3.33 for retinol and by 10 for beta-carotene.
- μg vitamin D, divide by 40 or multiply by 0.025.
- mg α-TE, divide by 1.5.

CONTENTS



E

# NUTRITION ASSESSMENT

Nutrition assessment evaluates a person's health from a nutrition perspective. Many factors influence or reflect nutrition status. Consequently, the assessor, usually a registered dietitian assisted by other qualified health care professionals, gathers information from many sources, including:

- Historical information.
- Anthropometric measurements.
- Physical examinations.
- Biochemical analyses (laboratory tests).

Each of these methods involves collecting data in a variety of ways and interpreting each finding in relation to the others to create a total picture.

The accurate gathering of this information and its careful interpretation are the basis for a meaningful evaluation. The more information gathered about a person, the more accurate the assessment will be. Gathering information is a time-consuming process, and time is often a rare commodity in the health care setting. Nutrition care is only one part of total care. It may not be practical or essential to collect detailed information on each person.

A strategic compromise is to screen clients by collecting preliminary data. Data such as height-weight and hematocrit are easy to obtain and can alert health care workers to potential problems. Nutrition screening identifies clients who will require additional nutrition assessment. This appendix provides a sample of the procedures, standards, charts, and forms commonly used in nutrition assessment.

**nutrition screening:** the use of preliminary nutrition assessment techniques to identify people who are malnourished or are at risk for malnutrition.

## Historical Information

Clues about present nutrition status become evident with a careful review of a person's historical data (see Table E-1). Even when the data are subjective, they reveal important facts about a person. A thorough history identifies risk factors associated with poor nutrition status (see Table E-2) and provides a sense of the whole person. Form E-1 shows the kinds of questions asked. As you can see, many aspects of a person's life influence nutrition status and provide clues to possible problems.

An adept history taker uses the interview both to gather facts and to establish a rapport with the client. This section briefly reviews the major areas of nutrition concern in a person's history: health, socioeconomic factors, drugs, and diet.

**Table E-1**
**Historical Data Used in Nutrition Assessments**

| Type of History | What It Identifies |
|---|---|
| Health history | Health factors that affect nutrition status |
| Socioeconomic history | Personal, financial, and environmental influences on food intake, nutrient needs, and diet therapy options |
| Drug history | Medications and nutrient supplements that affect nutrition status |
| Diet history | Nutrient intake excesses or deficiencies and reasons for imbalances |

E

**Table E-2**
**Risk Factors for Poor Nutrition Status**

**Health History**

- Acquired immune deficiency syndrome (AIDS)
- Alcoholism
- Anorexia (lack of appetite)
- Anorexia nervosa
- Bulimia nervosa
- Cancer
- Chewing or swallowing difficulties (including poorly fitted dentures, dental caries, missing teeth, and mouth ulcers)
- Chronic obstructive pulmonary disease
- Circulatory problems
- Constipation
- Crohn's disease
- Decubitus ulcers
- Dementia
- Depleted blood proteins
- Diabetes mellitus
- Diarrhea
- Diseases of the GI tract
- Drug addiction
- Dysphagia
- Failure to thrive
- Fever
- Heart disease
- HIV infection
- Hormonal imbalance
- Hyperlipidemia
- Hypertension
- Infection
- Kidney disease
- Liver disease
- Lung disease
- Malabsorption
- Mental illness
- Mental retardation
- Multiple pregnancies
- Nausea
- Neurologic disorders
- Organ failure
- Overweight
- Pancreatic insufficiency
- Paralysis
- Physical disability
- Pneumonia
- Pregnancy
- Radiation therapy
- Recent major illness
- Recent major surgery
- Recent weight loss or gain
- Surgery of the GI tract
- Tobacco use
- Trauma
- Ulcerative colitis
- Ulcers
- Underweight
- Vomiting

**Socioeconomic History**

- Access to groceries
- Activities
- Age
- Education
- Ethnic identity
- Income
- Kitchen facilities
- Number of people in household
- Occupation
- Religious affiliation

**Drug History**

- Amphetamines
- Analgesics
- Antacids
- Antibiotics
- Anticancer agents
- Anticonvulsant agents
- Antidepressant agents
- Antidiabetic agents
- Antidiarrheals
- Antihyperlipemics
- Antihypertensives
- Antiulcer agents
- Catabolic steroids
- Diuretics
- Hormonal agents
- Immunosuppressive agents
- Laxatives
- Oral contraceptives
- Vitamin and other nutrient preparations

**Diet History**

- Deficient or excessive food intakes
- Frequently eating out
- Intravenous fluids (other than total parenteral nutrition) for 7 or more days
- Monotonous diet (lacking variety)
- No intake for 7 or more days
- Poor appetite
- Restricted or fad diets
- Unbalanced diet (omitting any food group)

## Health History

**health history:** the medical record. Traditionally, the health history has been called the *medical history.* The term *health history* now seems more appropriate, however, since the contents describe a client's health status. Current trends in the medical profession are emphasizing health promotion and disease prevention.

The assessor can obtain health histories from records completed by the attending physician, nurse, or other health care professional. In addition, conversations with the client can uncover valuable information previously overlooked because no one thought to ask or because the client was not thinking clearly when asked.

An accurate, complete health history can reveal conditions that place a client at risk for malnutrition (review Table E-2). Diseases and their therapies can have either immediate or long-term effects on nutrition status by interfering with ingestion, digestion, absorption, metabolism, or excretion of nutrients.

**Form E-1**
**Historical Data**

Name ______ Date ______
Address ______ Date of last medical checkup ______
______ Age ______ Sex ______
______ Height ______ Weight ______
Phone ______ Usual weight ______
Reason for admission ______ Ideal weight range ______

## Health History

1. Have you been told that you have (check any that apply):
   ____ Diabetes mellitus ____ Heart disease ____ Ulcers
   ____ GI disorders ____ Lung disease ____ Cancer
   ____ High blood pressure ____ Kidney disease ____ Other
   ____ Hardening of arteries ____ Liver disease ______
2. Do you have complaints about any of the following:
   ____ Lack of appetite ____ Diarrhea ____ Nausea
   ____ Difficulty chewing or swallowing ____ Indigestion ____ Vomiting
   ____ Constipation ____ Fever ____ Other
3. Do you use tobacco in any way? ______ How much? ______
4. For females:
   Are you pregnant? ______ How many months? ______
   How many pregnancies have you carried to term? ______
   When was your last child born? ______
   Are your menstrual periods normal? ______ If not, please explain: ______

## Socioeconomic History

1. Last grade of school completed ______ Still in school? ______
2. Are you employed? ______ Occupation ______
3. Does someone else live with you? ______ Who? ______
4. Do you regularly eat alone or with others? ______
5. Do you have a refrigerator? ______ Stove? ______
6. How often do you shop for food? ______ Where? ______

## Drug History

1. Do you take medication, either prescribed by a doctor or over-the-counter?

| Name of drug | Reason for taking | Dose | Frequency | Duration of intake |
|---|---|---|---|---|
| ______ | ______ | ______ | ______ | ______ |
| ______ | ______ | ______ | ______ | ______ |
| ______ | ______ | ______ | ______ | ______ |

2. Have you noticed any side effects from taking these medications? ____ If so, please explain: ______
3. Do you take vitamins or any kind of supplements? ______ Which ones? ______
   How often? ______ For what reason? ______

## Diet History

1. Have you recently lost or gained more than 10 lb? ______ If yes, explain the surrounding circumstances (including associated illness, dietary changes, and time frame): ______
2. Do you eat at regular times each day? ______ How many times per day? ______
3. Where do you eat most of your meals? ______
4. Do you usually eat snacks? ______ When? ______
5. What foods do you particularly like? ______
6. Are there foods you don't eat for other reasons? ______
7. Do you have difficulty eating? ______
8. How would you describe your feelings about food? ______
9. Do your eating habits change when you are emotionally upset? ______ How? ______
10. Are you, or any member of your family, on a special diet? ____ If yes, who and what kind? ______
11. Do you drink alcohol? ______ How much? ______ How often? ______
12. How would you describe your exercise habits? ______ Type of exercise ______
    Intensity ______ Duration ______ Frequency ______
13. Are there any other facts about your lifestyle that you think might be related to your nutritional health? ______
    Explain ______

NOTE: Use the appropriate form to record food intake data (Form E-2 or E-3).

E

## Socioeconomic History

**socioeconomic history:** a record of a person's social and economic background, including such factors as education, income, and ethnic identity.

Socioeconomic factors profoundly affect nutrition status. The ethnic background and educational level of both the client and the other members of the household influence food availability and food choices. An understanding of the community environment is also important in assessing nutrition status. For example, the interviewer should be familiar with the food habits of the major ethnic groups within the locale, regional food preferences, and nutrition resources and programs available in the community. Local health departments and social agencies often can provide such information.

Level of income also influences the diet. In general, the quality of the diet declines as income falls. At some point, the ability to purchase the foods required to meet nutrient needs is lost; an inadequate income puts an adequate diet out of reach. Agencies use poverty indexes to identify people at risk for poor nutrition and to qualify people for government food assistance programs.

Low income affects not only the power to purchase foods but also the ability to shop for, store, and cook them. A skilled assessor will note whether a person has transportation to a grocery store that sells a sufficient variety of low-cost foods, and whether the person has access to a refrigerator and stove.

## Drug History

**drug history:** a record of all the drugs, over-the-counter and prescribed, that a person takes routinely.

The many interactions of foods and drugs require that health care professionals pay special attention to any client who takes drugs routinely. If a person is taking any drug, the assessor records the name of the drug; the dose, frequency, and duration of intake; the reason for taking the drug; and signs of any adverse effects on Form E-1.

The interactions of drugs and nutrients may take many forms:

- Drugs can alter food intake and the absorption, metabolism, and excretion of nutrients.
- Foods and nutrients can alter the absorption, metabolism, and excretion of drugs.

Highlight 17 discusses nutrient-drug interactions in more detail and Table H17-1 summarizes the mechanisms by which these interactions occur and provides specific examples.

## Diet History

**diet history:** a record of eating behaviors and the foods a person eats.

A diet history provides a record of a person's eating habits and food intake and can help identify possible nutrient imbalances. Food choices are an important part of lifestyle and often reflect a person's philosophy. The assessor who asks nonjudgmental questions about eating habits and food intake encourages trust and enhances the likelihood of obtaining accurate information.

Assessors evaluate food intake using various tools such as the 24-hour recall, the usual intake record, the food frequency checklist, and the food record. Food models or photos and measuring devices can help clients identify the types of foods and quantities consumed. The assessor also needs to know how the foods are prepared and when they are eaten. In addition to asking about foods, assessors will ask about beverage consumption, including beverages containing alcohol or caffeine.

Besides identifying possible nutrient imbalances, diet histories provide valuable clues about how a person will accept diet changes should they be necessary. Information about what and how a person eats provides the background for realistic and attainable nutrition goals.

**24-hour recall:** a record of foods eaten by a person for one 24-hour period.

• ***24-Hour Recall*** • The 24-hour recall provides data for one day only and is commonly used in nutrition surveys to obtain estimates of the typical food intakes for a population. The assessor asks the client to recount everything eaten or drunk in the past 24 hours or for the previous day. Form E-2 shows a typical 24-hour recall form.

An advantage of the 24-hour recall is that it is easy to obtain. It is also more likely to provide accurate data, at least about the past 24 hours, than estimates of average intakes over long periods. It does not, however, provide enough information to allow accurate generalizations about an individual's usual food intake. The previous day's intake may not be typical, for example, or the person may be unable to report portion sizes accurately or may conceal or forget information about foods eaten. This limitation is partially overcome when 24-hour recalls are collected on several nonconsecutive days.

• ***Usual Intake*** • To obtain data about a person's usual intake, an inquiry might begin with "What is the first thing you usually eat or drink during the day?" Similar questions follow until a typical daily intake pattern emerges. This method uses the same form as the 24-hour recall (Form E-2) and can be useful, especially in verifying food intake when the past 24 hours have been atypical. It also helps the assessor verify food habits. For example, one person may always eat an afternoon snack; another may never eat breakfast. A person whose intake varies widely from day to day, however, may find it difficult to answer such general questions, and in that case, another food intake tool should be used to estimate nutrient intake.

• ***Food Record*** • Another tool for history taking is the food record, in which the person records food eaten, time of day, place where eaten, others present, and mood. Chapter 9 (p. 272) provides an example. A food record can help both the assessor and the client to determine factors associated with eating that may affect dietary balance and adequacy.

**food record:** an extensive, accurate log of all foods eaten over a period of several days or weeks. A food record that includes associated information such as when, where, and with whom each food is eaten is sometimes called a **food diary.**

Food records work especially well with cooperative people but require considerable time and effort on their part. A prime advantage is that the record keeper assumes an active role and may for the first time become aware of personal food habits and assume responsibility for them. It also provides the assessor with an accurate picture of the person's lifestyle and factors that affect food intake. For

**Form E-2**

**Food Intake for a 24-Hour Recall or Usual Intake Pattern**

Name and address ______________________________ Date ______________

______________________________

______________________________

Did or do you take vitamin-mineral supplements? ______________________________

If yes, what kind? ______________ Dose ______________

Please record the amount and type of foods and beverages consumed today. [Or: Please record the amounts and types of foods and beverages you typically consume each day.]

| Time of Day | Food | Amount (c, tbs, or piece) | Description (how cooked, how served) |
|---|---|---|---|
| | | | |
| | | | |
| | | | |
| | | | |
| | | | |
| | | | |
| | | | |
| | | | |
| | | | |
| | | | |

these reasons, a food record can be particularly useful in outpatient counseling for such nutrition problems as overweight, underweight, or food allergy. The major disadvantages stem from poor compliance in recording the data and conscious or unconscious changes in eating habits that may occur while the person is keeping the record.

**food frequency checklist:** a checklist of foods on which a person can record the frequency with which he or she eats each food.

• ***Food Frequency Checklist*** • A less common approach is to use a food frequency checklist to ascertain how often an individual eats a specific type of food. This information helps pinpoint food groups, and therefore nutrients, that may be excessive or deficient in the diet. That a person ate no vegetables yesterday may not seem particularly significant, but never eating vegetables is a warning of possible nutrient deficiencies. When used with the usual intake or 24-hour recall approach, the food frequency record enables the assessor to double-check the accuracy of the information obtained. Form E-3 is a food frequency checklist.

• ***Analysis of Food Intake Data*** • After collecting food intake data, the assessor estimates nutrient intakes, either informally by using food guides or formally by using food composition tables. The assessor compares these intakes with standards, usually nutrient recommendations or dietary guidelines, to determine how closely the person's diet meets the standards. Are the types and amounts of proteins, carbohydrates (including fiber), and fats (including cholesterol) appropriate? Are all food groups included in appropriate amounts? Is caffeine or alcohol

**Form E-3**
**Food Frequency Checklist**

The assessor helps the client estimate portion sizes and frequency of use.

| | **Number of Servings** | **Frequency**[a] |
|---|---|---|
| 1. How often do you eat the following foods? | | |
| Bread, toast, rolls, muffins | ______ | ______ |
| Cereal (type?) ______ | ______ | ______ |
| Rice or other cooked grains | ______ | ______ |
| Noodles (macaroni, spaghetti) | ______ | ______ |
| Pancakes or waffles | ______ | ______ |
| Crackers or pretzels | ______ | ______ |
| Fruits or fruit juices | ______ | ______ |
| Vegetables other than potatoes | ______ | ______ |
| Vegetable juice | ______ | ______ |
| Potatoes | ______ | ______ |
| Dried beans and peas | ______ | ______ |
| Beef | ______ | ______ |
| Pork or ham | ______ | ______ |
| Veal | ______ | ______ |
| Poultry | ______ | ______ |
| Fish | ______ | ______ |
| Organ meats (such as liver) | ______ | ______ |
| Bacon | ______ | ______ |
| Sausage | ______ | ______ |
| Lunch meat | ______ | ______ |
| Hot dogs | ______ | ______ |
| Other meats (type?) ______ | ______ | ______ |
| Eggs | ______ | ______ |
| Peanut butter or nuts | ______ | ______ |
| Milk (including on cereal) | ______ | ______ |
| Cheese or cheese dishes | ______ | ______ |
| Yogurt or tofu | ______ | ______ |
| Other milk products (type?) ______ | ______ | ______ |

[a]Number of servings per day, week, month, or year.

consumption excessive? Are intakes of any vitamins or minerals (including sodium and iron) excessive or deficient? An informal evaluation is possible only if the assessor has enough prior experience with formal calculations to "see" nutrient amounts in reported food intakes without calculations. Even then, such an informal analysis is best followed by a spot check for key nutrients by actual calculation.

Formal calculations can be performed either manually (by looking up each food in a table of food composition, recording its nutrients, and adding them up) or by using a computer diet analysis program. The assessor then compares the intakes with standards such as the RDA.

• ***Limitations of Food Intake Analysis*** • Diet histories can be superbly informative, but the skillful assessor also keeps their limitations in mind. For example, a computer diet analysis tends to imply greater accuracy than is possible to obtain from data as uncertain as the starting information. Nutrient contents of foods listed in tables of food composition or stored in computer databases are averages, and for some nutrients, incomplete. In addition, the available data on nutrient contents of foods do not reflect the amounts of nutrients a person actually absorbs. Iron is a case in point: its availability from a given meal may vary from as high as 50 percent to less than 2 percent, depending on the person's iron status; the relative amounts of heme iron, nonheme iron, vitamin C, meat, fish, and

**Form E-3**
**Food Frequency Checklist (continued)**

| | Number of Servings | Frequency |
|---|---|---|
| Butter or margarine (type?) ______ | ______ | ______ |
| Salt pork | ______ | ______ |
| Mayonnaise or salad dressing (type?) ______ | ______ | ______ |
| Oil (type?) | ______ | ______ |
| Cream | ______ | ______ |
| Sugar, jam, jelly, syrup, honey | ______ | ______ |
| Bakery goods (type?) ______ | ______ | ______ |
| Candy | ______ | ______ |
| Soft drinks (type?) ______ | ______ | ______ |
| Potato or snack chips (type?) ______ | ______ | ______ |
| Coffee or tea (type?) ______ | ______ | ______ |
| Alcoholic beverages (type?) ______ | ______ | ______ |
| Fast foods eaten out (type?) ______ | ______ | ______ |
| TV dinners, pot pies, other prepared meals (type?) ______ | ______ | ______ |
| Instant meals such as breakfast bars or diet meal beverages (type?) ______ | ______ | ______ |

2. What specific kinds of the following foods do you eat? Include the name of the food; whether it is fresh, canned, or frozen; and how it is prepared.
   Fruits and fruit juices ______
   Vegetables ______
   Milk and milk products ______
   Meats and meat alternates ______
   Breads and cereals ______
   Desserts ______
   Snack foods ______
3. Please list the names of any liquid, powder, or pill forms of vitamin or mineral products you take, and state how often you take them. Please also list any diet supplement you use (such as protein milk shakes or brewer's yeast), how much you use, and how often you use it. ______
4. Is there anything else you can relate about your food/nutrient intake? ______

poultry eaten at the meal; and the presence of inhibitors of iron absorption such as tea, coffee, and nuts. Chapter 13 explains how to calculate iron absorption from a meal.

Furthermore, reported portion sizes may not be correct. The person who reports eating "a serving" of greens may not distinguish between ¼ cup and 2 whole cups; only trained individuals can accurately report serving sizes. Children tend to remember the serving sizes of foods they like as being larger than serving sizes of foods they dislike.

Thus any comparison of reported nutrient intakes with nutrient needs provides many opportunities for error. Most history takers learn to use shortcut systems to obtain rough estimates of nutrient intakes and then use calculation methods to pinpoint suspected nutrient deficiencies or imbalances.

An estimate of nutrient intakes from a diet history, combined with other sources of information, allows the assessor to confirm or eliminate the possibility of suspected food intake problems. The assessor must constantly remember that nutrient intakes in adequate amounts do not guarantee adequate nutrient status for an individual. Likewise, insufficient intakes do not always indicate deficiencies, but instead alert the assessor to possible problems. Each person digests, absorbs, metabolizes, and excretes nutrients in a unique way; individual needs vary. Intakes of nutrients identified by diet histories are only pieces of a puzzle that must be put together with other indicators of nutrition status in order to extract meaning.

# Anthropometric Measurements

**anthropometric:** relating the measurement of the physical characteristics of the body, such as height and weight.
- **anthropos** = human
- **metric** = measuring

Anthropometrics are physical measurements that reflect body composition and development (see Table E-3). They serve three main purposes: first, to evaluate the progress of growth in pregnant women, infants, children, and adolescents; second, to detect undernutrition and overnutrition in all age groups; and third, to measure changes in body composition over time.

Health care professionals compare anthropometric measurements taken on an individual with population standards specific for sex and age or with previous measures of the individual. Measurements taken periodically and compared with previous measurements reveal changes in an individual's status.

Mastering the techniques for taking anthropometric measurements requires proper instruction and practice to ensure reliability. Once the correct techniques are learned, taking measurements is easy and requires minimal equipment.

Height and weight are well-recognized anthropometrics; other anthropometrics include fatfold measurements and various measures of lean tissue. Other measures are useful in specific situations. For example, a head circumference measurement may help to assess brain development in an infant, and an abdominal girth measurement supplies information about abdominal fluid retention in individuals with liver disease.

**Table E-3**
**Anthropometric Measurements Used in Nutrition Assessments**

| Type of Measurement | What It Reflects |
|---|---|
| Abdominal girth measurement | Abdominal fluid retention |
| Height-weight | Overnutrition and undernutrition; growth in children |
| Head circumference | Brain growth and development in infants and children under two |
| Fatfold | Subcutaneous and total body fat |
| Waist circumference | Body fat distribution |

## Measures of Growth and Development

Height and weight are among the most common and useful anthropometric measurements. Length measurements for infants and children up to age three and height measurements for children over three are particularly valuable in assessing growth and therefore nutrition status. For adults, height measurements alone are not critical but help to estimate desirable weight and to interpret other assessment data. Once adult height has been reached, changes in body weight provide useful information in assessing overnutrition and undernutrition.

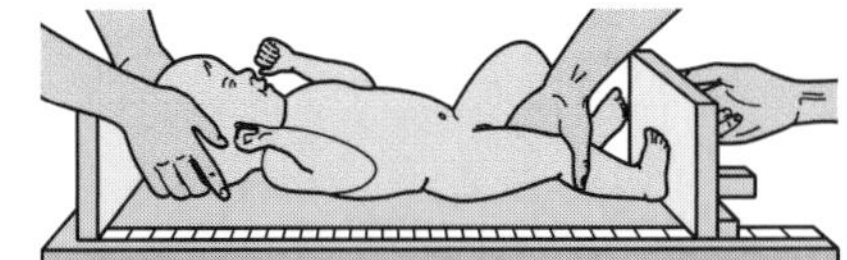

**Figure E-1**
**Length Measurement of an Infant**

An infant is measured lying down on a measuring board with a fixed headboard and a movable footboard. Note that two people are needed to measure the infant's length.

• *Height* • For infants and children younger than three, health care professionals may use special equipment to measure length. The assessor lays the barefoot infant on a measuring board that has a fixed headboard and movable footboard attached at right angles to the surface (see Figure E-1). Often two people are needed to obtain an accurate measurement: one to hold the infant's head against the headboard, and the other to keep the legs straight and do the measuring. This method provides the most accurate measure possible, but many health care professionals use a less exacting method. They may simply hold the infant straight with its head against the headboard or other vertical support, mark the blanket with a chalk or pen at the infant's heel, and then measure the distance from the headboard to the mark. Even more informally and less accurately, they may lay the infant on a flat surface and extend a nonstretchable measuring tape along the side of the infant from the top of the head to the heel of the foot.

The procedure for measuring a child who can stand erect and cooperate is the same as for an adult. The best way to measure standing height is with the person's back against a flat wall to which a nonstretchable measuring tape or stick has been fixed (see Figure E-2). The person stands erect, without shoes, with heels together. The person's line of sight should be horizontal, with the heels, buttocks, shoulders, and head touching the wall. The assessor places a block, book, or other inflexible object on top of the head at a right angle to the wall; carefully checks the height measurement; and records it immediately in either inches or centimeters. Such a practice prevents forgetting the correct measurement.

The measuring rod of a scale is commonly used, but is less accurate because it bends easily. The assessor follows the same general procedure, asking the person to face away from the scale and to take extra care to stand erect.

Unfortunately, many health care professionals merely ask clients how tall they are rather than measuring their height. Self-reported height is often inaccurate and should be used only as a last resort when measurement is impractical (in the case of an uncooperative client, an emergency admission, or the like).

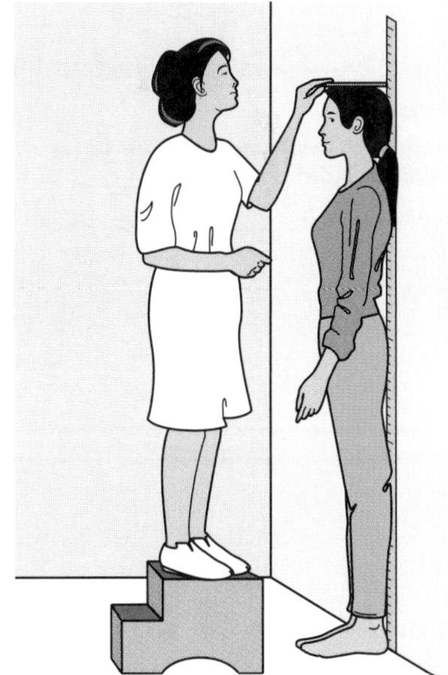

**Figure E-2**
**Height Measurement of an Older Child or Adult**

Height is measured most accurately when the person stands against a flat wall to which a measuring tape has been affixed. When the person is taller than the measurer, the measurer can stand on a stool to help ensure that the proper height measurement is obtained.

• *Weight* • Valid weight measurements require scales that have been carefully maintained, calibrated, and checked for accuracy at regular intervals. Beam balance and electronic scales are the most accurate types of scales. To measure infants' weight, assessors use special scales that allow infants to lie or sit (see Figure E-3). Weighing infants naked, without diapers, is standard procedure. Children who can stand are weighed in the same way as adults (see Figure E-4). To make repeated measures useful, standardized conditions are necessary. Each weighing should take place at the same time of day (preferably before breakfast), in the same amount of clothing (without shoes), after the person has voided, and on the same scale. Special scales and hospital beds with built-in scales are available for weighing people who are bedridden. Bathroom scales are inaccurate and inappropriate in a professional setting. As with all measurements, the assessor records the observed weight immediately in either pounds or kilograms.

• *Head Circumference* • Assessors may also measure head circumference to confirm that infant growth is proceeding normally or to help detect protein-energy malnutrition (PEM) and evaluate the extent of its impact on brain size. To measure head circumference, the assessor places a nonstretchable tape so that it encircles the

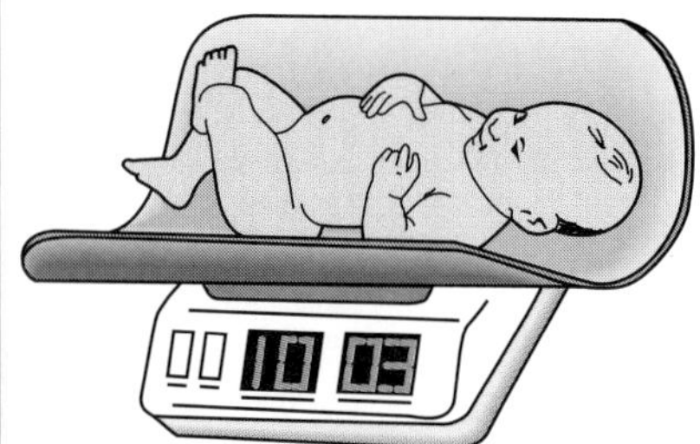


**Figure E-3**
**Weight Measurement of an Infant**

Infants sit or lie down on scales that are designed to hold them while they are being weighed.

largest part of the infant's or child's head: just above the eyebrow ridges, just above the point where the ears attach, and around the occipital prominence at the back of the head. To ensure accurate recording, the assessor immediately notes the measure in either inches or centimeters.

• ***Analysis of Measures in Infants and Children*** • Growth retardation is an important sign of poor nutrition status. Obesity is also an important sign requiring intervention.

Health professionals generally evaluate physical development by monitoring the growth rate of a child and comparing this rate with standard charts. Standard charts compare weight to age, height to age, and weight to height; ideally, height and weight are in roughly the same percentile. Although individual growth patterns may vary, a child's growth curve will generally stay at about the same percentile throughout childhood. In children whose growth has been retarded, nutrition rehabilitation will ideally induce height and weight to increase to higher percentiles. In overweight children, the goal is for weight to remain stable as height increases, until weight becomes appropriate for height.

To evaluate growth in infants, an assessor uses charts such as those in Figures E-5 (A and B), E-6 (A and B), E-7 (A and B), and E-8 (A and B). The assessor follows these steps to plot a weight measurement on a percentile graph:

- Select the appropriate chart based on age and gender. (When length is measured, use the chart for birth to 36 months; when height is measured, use the chart for 2 to 18 years.)
- Locate the child's age along the horizontal axis on the bottom or top of the chart.
- Locate the child's weight in pounds or kilograms along the vertical axis on the lower left or right side of the chart.
- Mark the chart where the age and weight lines intersect, and read off the percentile.

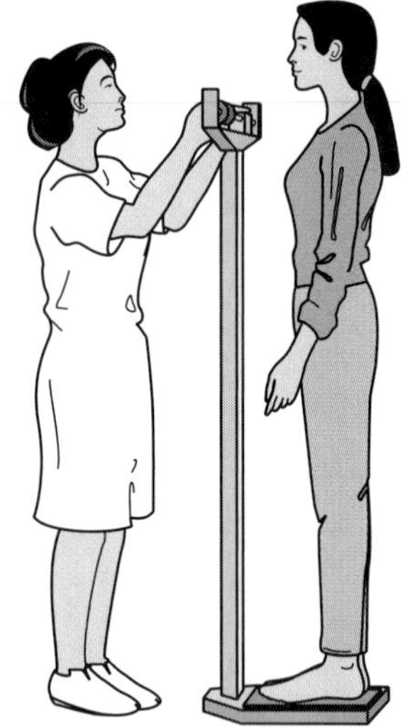

**Figure E-4**
**Weight Measurement of an Older Child or Adult**

Whenever possible, children and adults are measured on beam balance scales to ensure accuracy.

To assess length, height, or head circumference, the assessor follows the same procedure, using the appropriate chart. Head circumference percentile should be similar to the child's height and weight percentiles.

With height, weight, and head circumference measures plotted on growth percentile charts, a skilled clinician can begin to interpret the data. Percentile charts divide the measures of a population into 100 equal divisions. Thus half of the population falls above the 50th percentile, and half falls below. The use of percentile measures allows for comparisons among people of the same age and gender. For example, a six-month-old female infant whose weight is at the 75 percentile weighs more than 75 percent of the female infants her age.

Head circumference is generally measured in children under two years of age. Since the brain grows rapidly before birth and during early infancy, extreme and chronic malnutrition during these times can impair brain development, curtailing the number of brain cells and the size of head circumference. Nonnutritional factors, such as certain disorders and genetic variation, can also influence head circumference.

Reminder: The *body mass index (BMI)* is an index of a person's weight in relation to height, determined by dividing the weight in kilograms by the square of the height in meters:

$$\text{BMI} = \frac{\text{Weight (kg)}}{\text{Height (m)}^2}.$$

• ***Analysis of Measures in Adults*** • For adults, health care professionals typically compare weights with weight-for-height standards. One such standard is the body mass index (BMI), described in Chapter 8 (p. 242), which is useful for estimating the risk to health associated with overnutrition. The back cover shows BMI for various heights and weights. The weight ranges presented in Chapter 8 (Table 8-5, p. 242) are based on the BMI most consistent with health. This table presents wide ranges of weights appropriate for heights without regard to sex or age.

**Figure E-5A**

**Girls: Birth to 36 Months Physical Growth National Center for Health Statistics (NCHS) Percentiles—Length and Weight for Age**

**GIRLS: BIRTH TO 36 MONTHS**
**PHYSICAL GROWTH**
**NCHS PERCENTILES***

NAME ______________________ RECORD # ______________

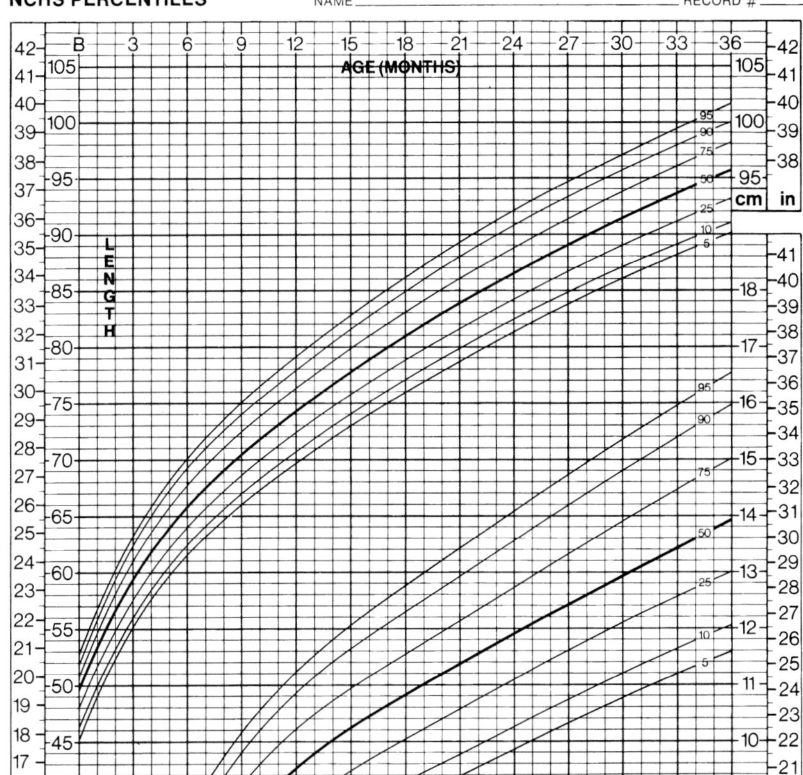


| MOTHER'S STATURE | | | GESTATIONAL | | |
|---|---|---|---|---|---|
| FATHER'S STATURE | | | AGE ______ WEEKS | | |
| DATE | AGE | LENGTH | WEIGHT | HEAD CIRC. | COMMENT |
| | BIRTH | | | | |

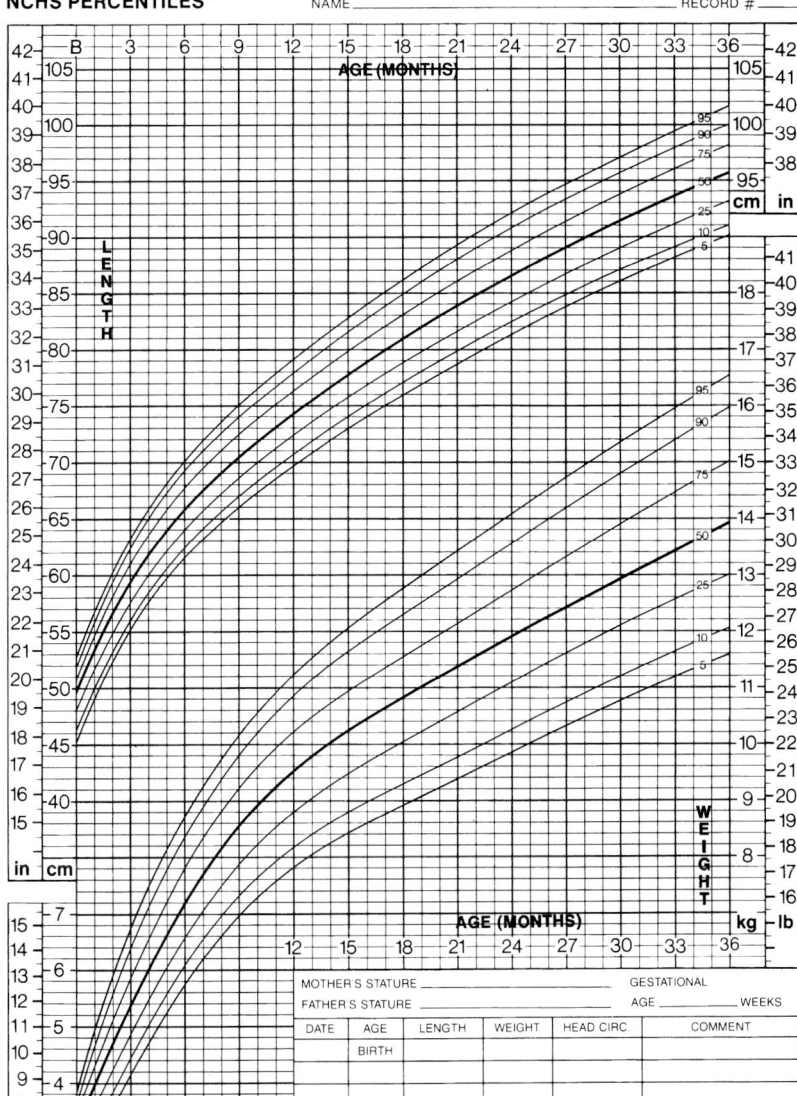

*Adapted from: Hamill PVV, Drizd TA, Johnson CL, Reed RB, Roche AF, Moore WM: Physical growth: National Center for Health Statistics percentiles. AM J CLIN NUTR 32:607-629, 1979. Data from the Fels Longitudinal Study, Wright State University School of Medicine, Yellow Springs, Ohio.

**Figure E-5B**

**Boys: Birth to 36 Months Physical Growth NCHS Percentiles—Length and Weight for Age**

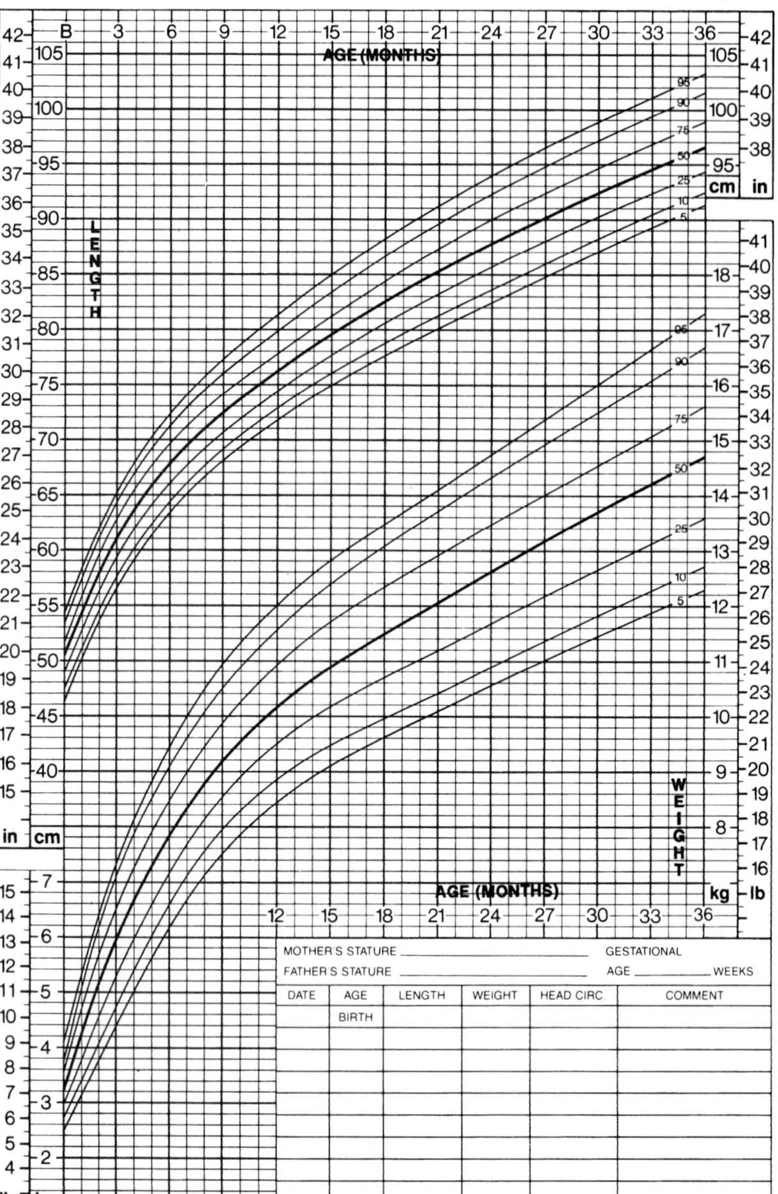


*Adapted from: Hamill PVV, Drizd TA, Johnson CL, Reed RB, Roche AF, Moore WM: Physical growth: National Center for Health Statistics percentiles. AM J CLIN NUTR 32:607-629, 1979. Data from the Fels Longitudinal Study, Wright State University School of Medicine, Yellow Springs, Ohio.

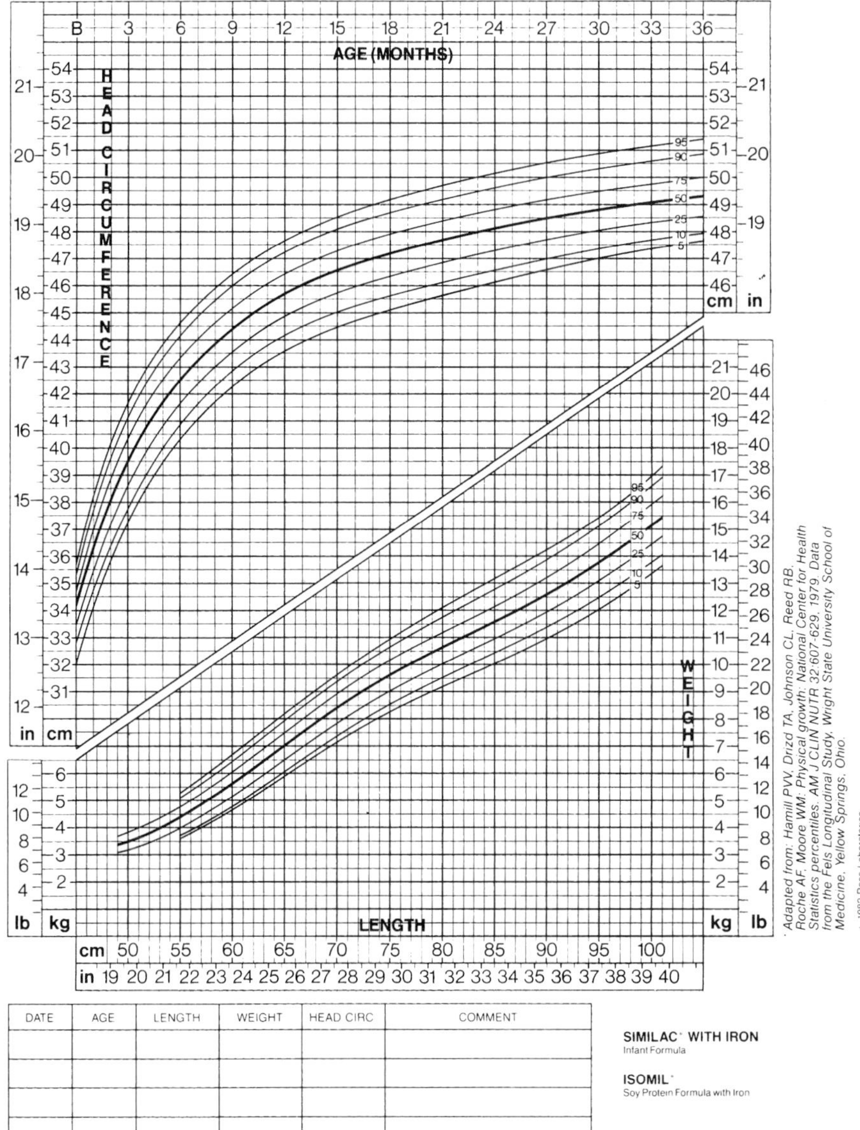


| DATE | AGE | LENGTH | WEIGHT | HEAD CIRC | COMMENT |
|---|---|---|---|---|---|
| | | | | | |
| | | | | | |
| | | | | | |
| | | | | | |
| | | | | | |
| | | | | | |
| | | | | | |

**Figure E-6A**

**Girls: Birth to 36 Months Physical Growth NCHS Percentiles—Head Circumference for Age and Weight for Length**

E

**Figure E-6B**

**Boys: Birth to 36 Months Physical Growth NCHS Percentiles—Head Circumference for Age and Weight for Length**

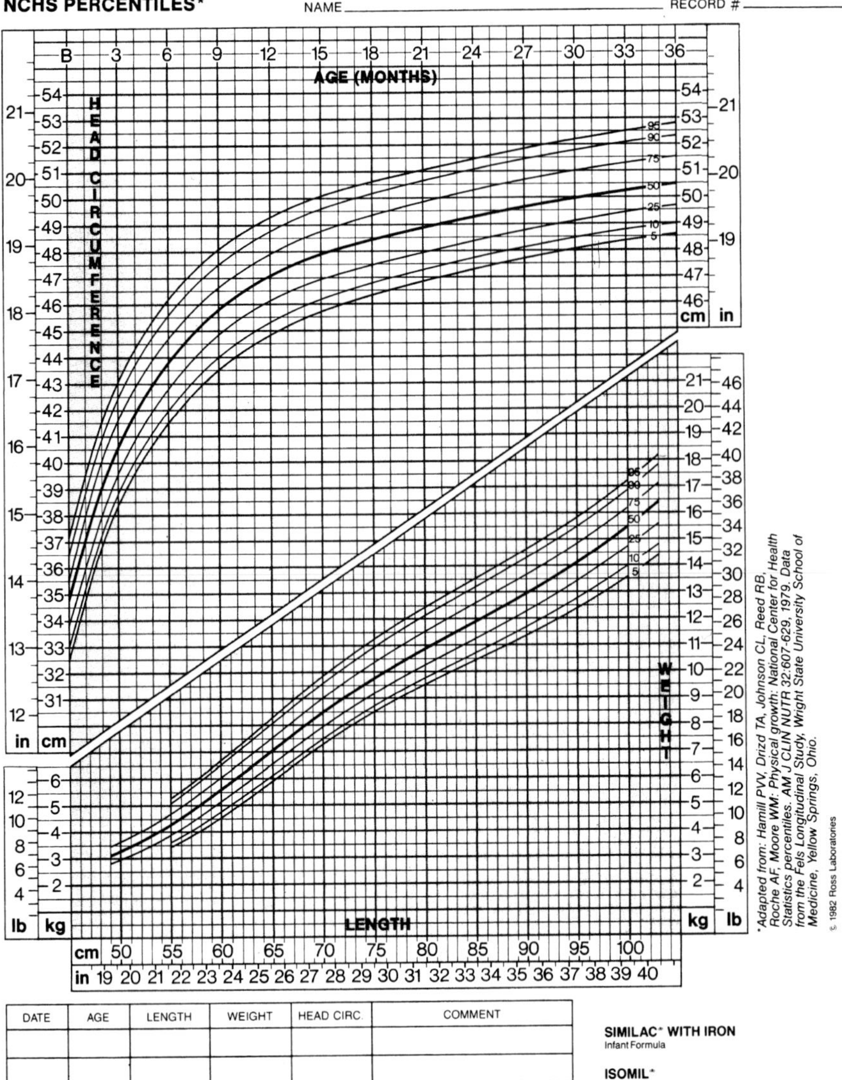


| DATE | AGE | LENGTH | WEIGHT | HEAD CIRC. | COMMENT |
|---|---|---|---|---|---|
| | | | | | |
| | | | | | |
| | | | | | |
| | | | | | |
| | | | | | |
| | | | | | |
| | | | | | |



Reprinted with permission of Ross Laboratories

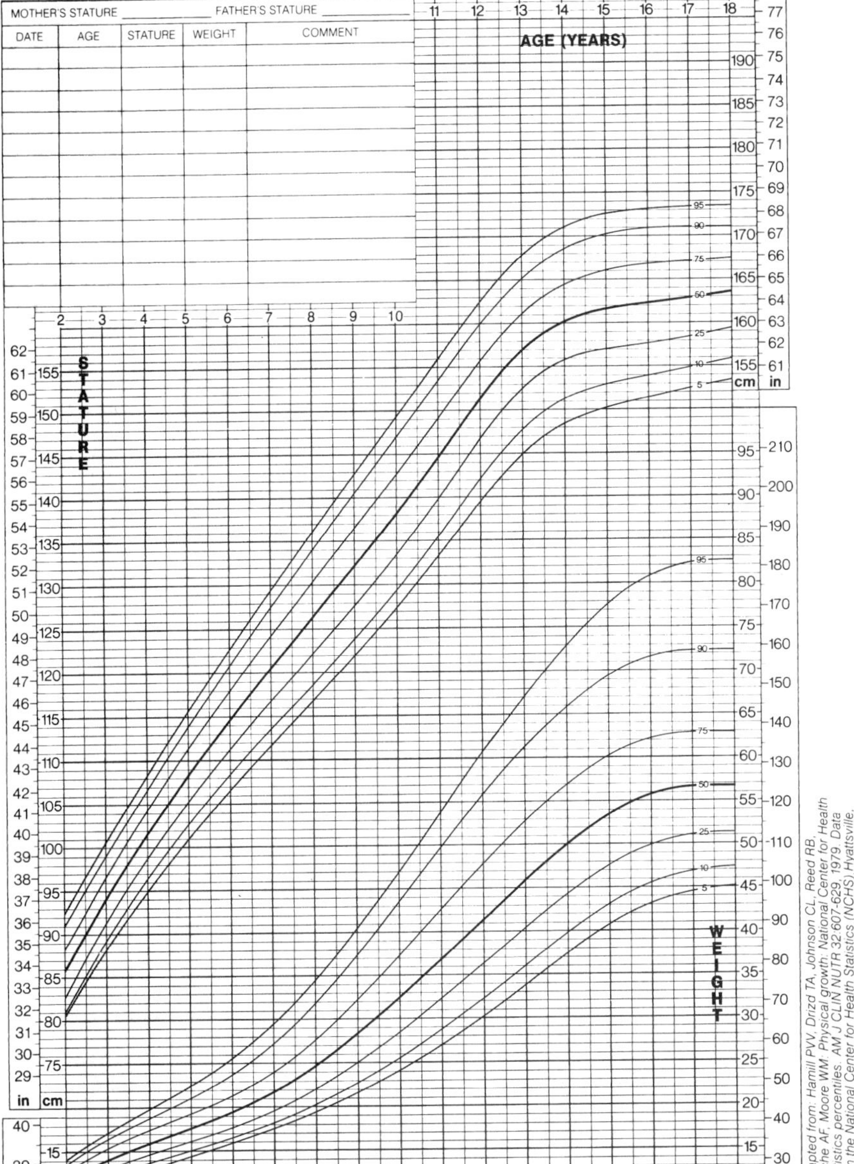


**Figure E-7A**

**Girls: 2 to 18 Years Physical Growth NCHS Percentiles—Height and Weight for Age**

E

**Figure E-7B**

**Boys: 2 to 18 Years Physical Growth NCHS Percentiles—Height and Weight for Age**

BOYS: 2 TO 18 YEARS
PHYSICAL GROWTH
NCHS PERCENTILES*

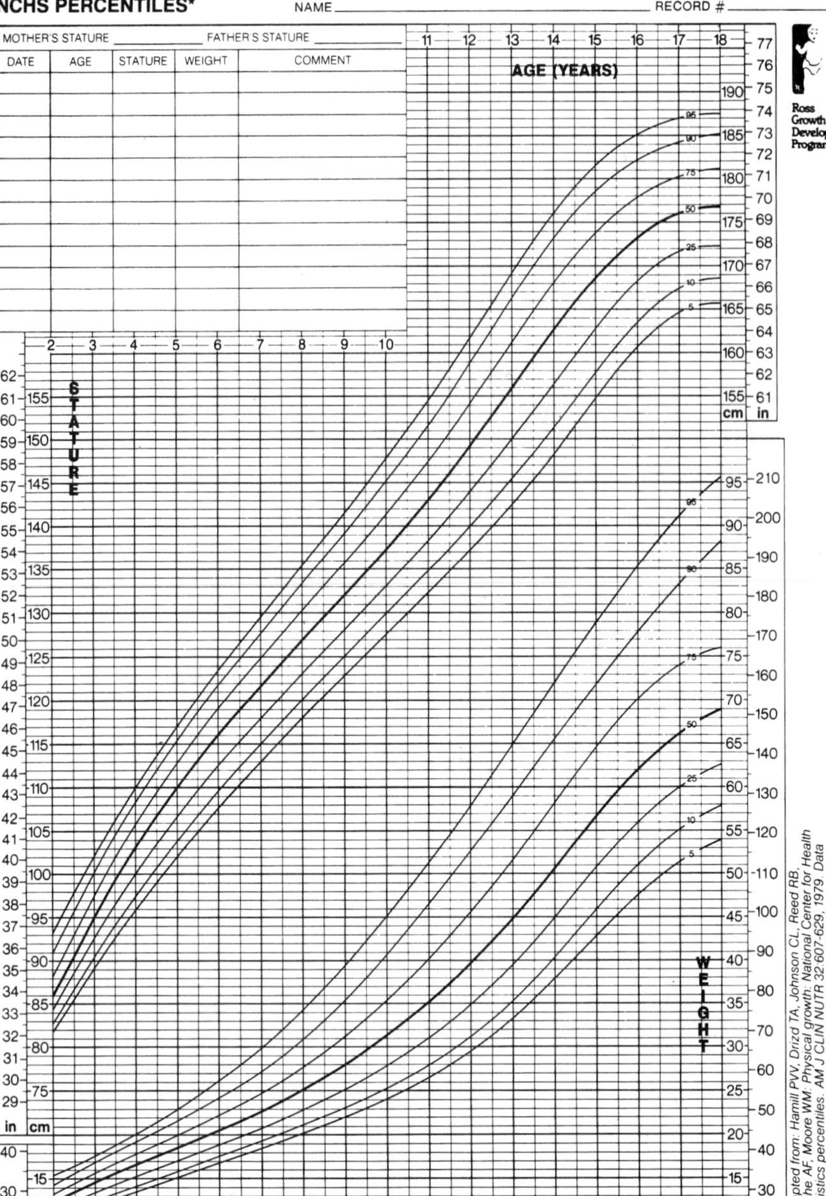

**Figure E-8A**

**Girls: Prepubescent Physical Growth NCHS Percentiles—Weight for Height**

GIRLS: PREPUBESCENT
PHYSICAL GROWTH
NCHS PERCENTILES*

NAME ______ RECORD # ______

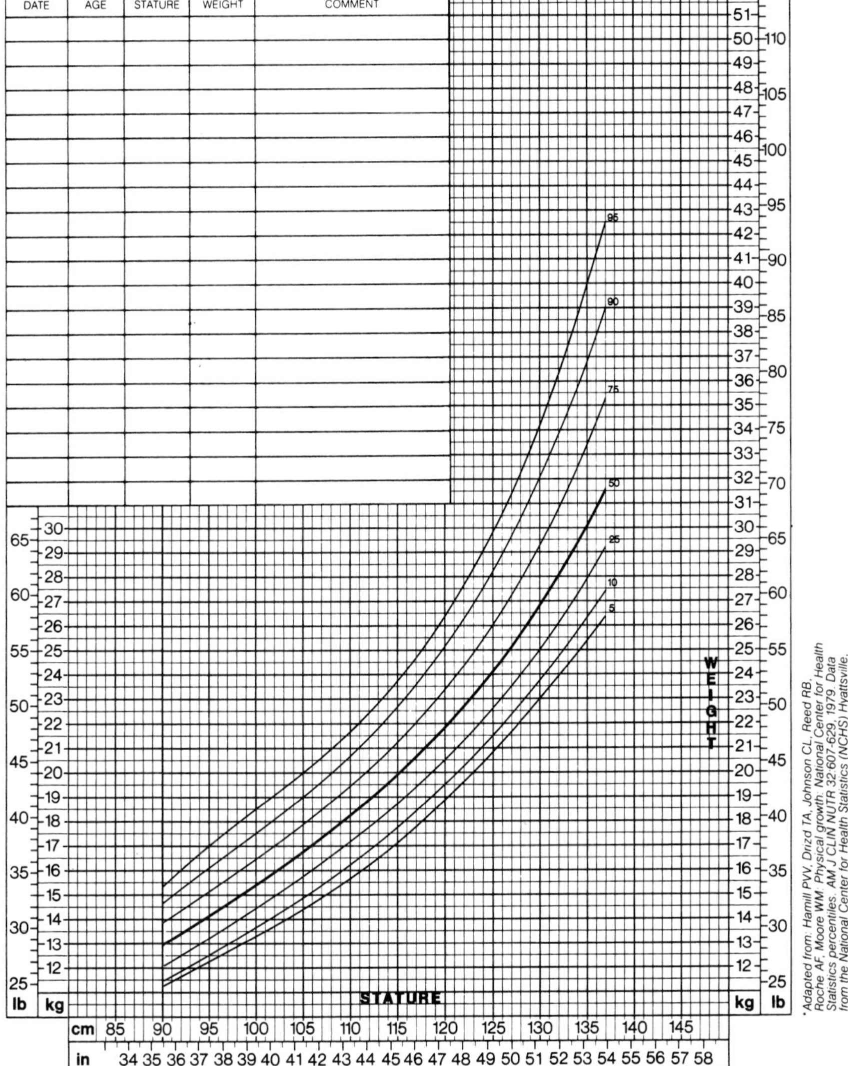


SIMILAC® WITH IRON
Infant Formula

ISOMIL®
Soy Protein Formula with Iron

Reprinted with permission
of Ross Laboratories

**Figure E-8B**

**Boys: Prepubescent Physical Growth NCHS Percentiles—Weight for Height**

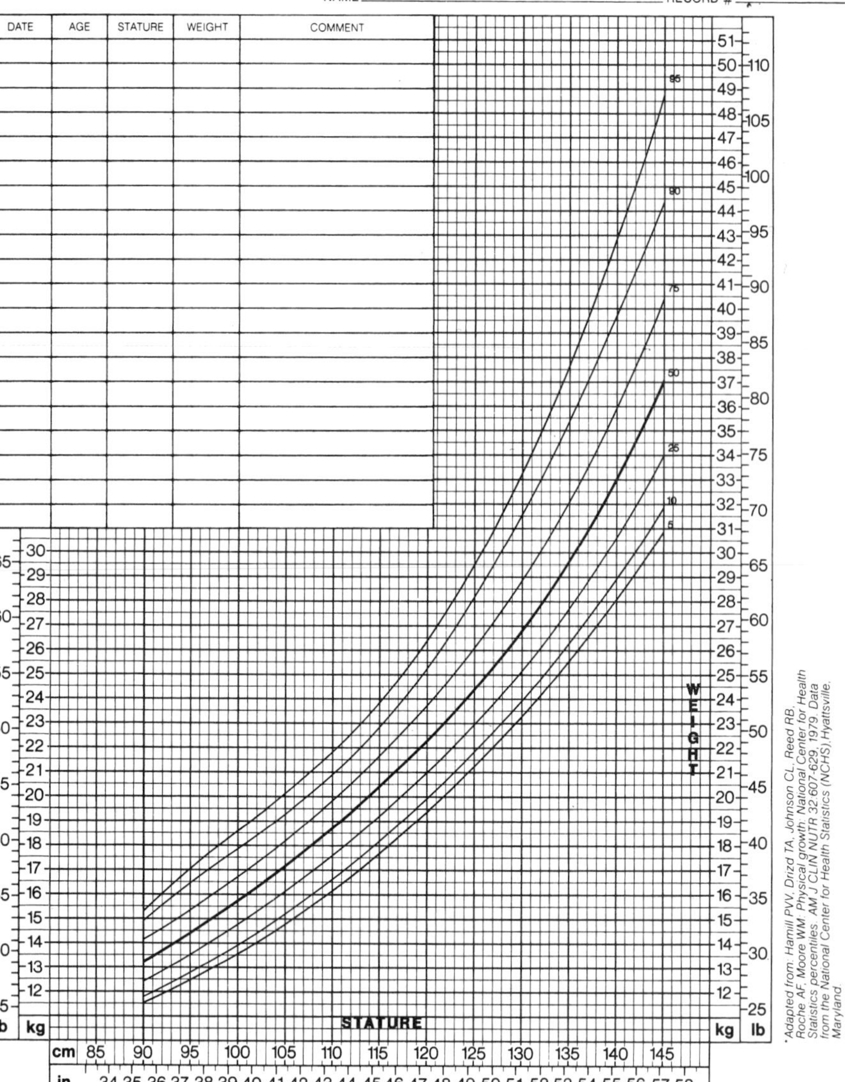

## Measures of Body Fat and Lean Tissue

Significant weight changes in both children and adults can reflect overnutrition and undernutrition with respect to energy and protein. To estimate the degree to which various body compartments (fat stores or lean tissues) are affected by overnutrition or malnutrition, several anthropometric measurements are useful (review Table E-3).

• ***Fatfold Measures*** • Approximately half the fat in the body is located directly beneath the skin, and its thickness reflects total body fat. In some parts of the body, this fat is loosely attached; a person can pull it up between the thumb and forefinger to obtain a measure of fatfold thickness. These measurements correlate well with other, more sophisticated methods of calculating total body fat. The fatfold test is therefore a valuable and practical diagnostic procedure when performed by a person trained in the use of fatfold calipers.

A major limitation of the fatfold test is that fat under the skin may be thicker in one area than in another. A pinch at the side of the waistline may not yield the same measurement as a pinch on the back of the arm. This limitation can be overcome by taking fatfold measurements at several (often three) different places on the body to obtain an accurate estimate of subcutaneous fat. Multiple measures are not always practical in clinical settings, however, and most often, the triceps fatfold measurement is used because it is easily accessible. To measure fatfold, a trained technician follows a standard procedure using reliable calipers, as illustrated in Figure E-9. Triceps fatfold percentiles are given in Table E-4.

• ***Waist Circumference*** • Chapter 8 described how fat distribution correlates with health risks and mentioned that the waist circumference is a valuable indicator of fat distribution. To measure waist circumference, the assessor places a nonstretchable tape around the person's body, crossing just above the upper hip bones

**Figure E-9**
**How to Measure the Triceps Fatfold**

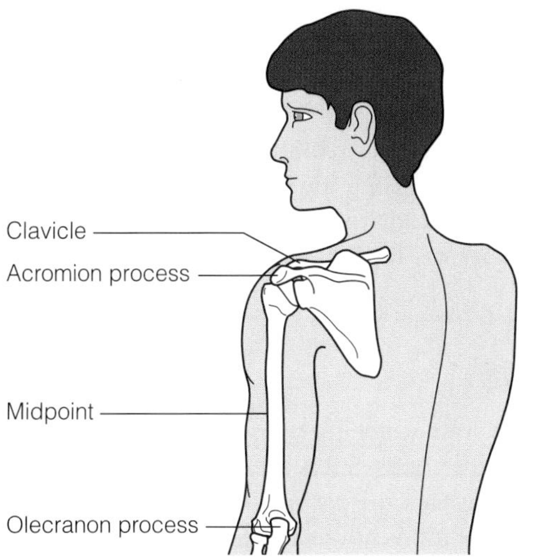


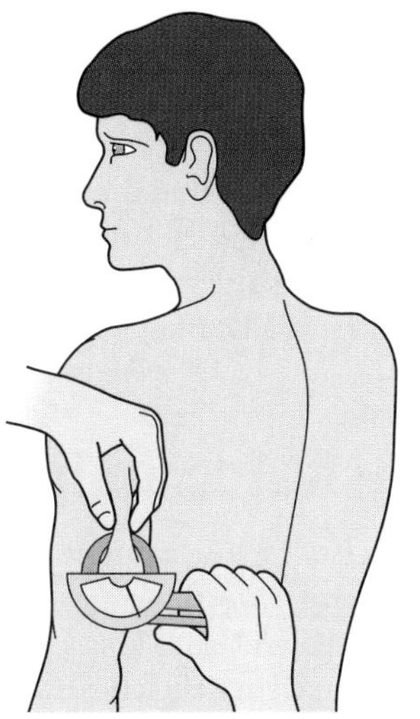

A. Find the midpoint of the arm:
1. Ask the subject to bend his or her arm at the elbow and lay the hand across the stomach. (If he or she is right-handed, measure the left arm, and vice versa.)
2. Feel the shoulder to locate the acromion process. It helps to slide your fingers along the clavicle to find the acromion process. The olecranon process is the tip of the elbow.
3. Place a measuring tape from the acromion process to the tip of the elbow. Divide this measurement by 2, and mark the midpoint of the arm with a pen.

B. Measure the fatfold:
1. Ask the subject to let his or her arm hang loosely to the side.
2. Grasp a fold of skin and subcutaneous fat between the thumb and forefinger slightly above the midpoint mark. Gently pull the skin away from the underlying muscle. (This step takes a lot of practice. If you want to be sure you don't have muscle as well as fat, ask the subject to contract and relax the muscle. You should be able to feel if you are pinching muscle.)
3. Place the calipers over the fatfold at the midpoint mark, and read the measurement to the nearest 1.0 millimeter in two to three seconds. (If using plastic calipers, align pressure lines, and read the measurement to the nearest 1.0 millimeter in two to three seconds.)
4. Repeat steps 2 and 3 twice more. Add the three readings, and then divide by 3 to find the average.

**Table E-4**
**Triceps Fatfold Percentiles (Millimeters)**

| | Male | | | | | Female | | | | |
|---|---|---|---|---|---|---|---|---|---|---|
| **Age** | ***5th*** | ***25th*** | ***50th*** | ***75th*** | ***95th*** | ***5th*** | ***25th*** | ***50th*** | ***75th*** | ***95th*** |
| 1–1.9 | 6 | 8 | 10 | 12 | 16 | 6 | 8 | 10 | 12 | 16 |
| 2–2.9 | 6 | 8 | 10 | 12 | 15 | 6 | 9 | 10 | 12 | 16 |
| 3–3.9 | 6 | 8 | 10 | 11 | 15 | 7 | 9 | 11 | 12 | 15 |
| 4–4.9 | 6 | 8 | 9 | 11 | 14 | 7 | 8 | 10 | 12 | 16 |
| 5–5.9 | 6 | 8 | 9 | 11 | 15 | 6 | 8 | 10 | 12 | 18 |
| 6–6.9 | 5 | 7 | 8 | 10 | 16 | 6 | 8 | 10 | 12 | 16 |
| 7–7.9 | 5 | 7 | 9 | 12 | 17 | 6 | 9 | 11 | 13 | 18 |
| 8–8.9 | 5 | 7 | 8 | 10 | 16 | 6 | 9 | 12 | 15 | 24 |
| 9–9.9 | 6 | 7 | 10 | 13 | 18 | 8 | 10 | 13 | 16 | 22 |
| 10–10.9 | 6 | 8 | 10 | 14 | 21 | 7 | 10 | 12 | 17 | 27 |
| 11–11.9 | 6 | 8 | 11 | 16 | 24 | 7 | 10 | 13 | 18 | 28 |
| 12–12.9 | 6 | 8 | 11 | 14 | 28 | 8 | 11 | 14 | 18 | 27 |
| 13–13.9 | 5 | 7 | 10 | 14 | 26 | 8 | 12 | 15 | 21 | 30 |
| 14–14.9 | 4 | 7 | 9 | 14 | 24 | 9 | 13 | 16 | 21 | 28 |
| 15–15.9 | 4 | 6 | 8 | 11 | 24 | 8 | 12 | 17 | 21 | 32 |
| 16–16.9 | 4 | 6 | 8 | 12 | 22 | 10 | 15 | 18 | 22 | 31 |
| 17–17.9 | 5 | 6 | 8 | 12 | 19 | 10 | 13 | 19 | 24 | 37 |
| 18–18.9 | 4 | 6 | 9 | 13 | 24 | 10 | 15 | 18 | 22 | 30 |
| 19–24.9 | 4 | 7 | 10 | 15 | 22 | 10 | 14 | 18 | 24 | 34 |
| 25–34.9 | 5 | 8 | 12 | 16 | 24 | 10 | 16 | 21 | 27 | 37 |
| 35–44.9 | 5 | 8 | 12 | 16 | 23 | 12 | 18 | 23 | 29 | 38 |
| 45–54.9 | 6 | 8 | 12 | 15 | 25 | 12 | 20 | 25 | 30 | 40 |
| 55–64.9 | 5 | 8 | 11 | 14 | 22 | 12 | 20 | 25 | 31 | 38 |
| 65–74.9 | 4 | 8 | 11 | 15 | 22 | 12 | 18 | 24 | 29 | 36 |

NOTE: If measurements fall between the percentiles shown here, the percentile can be estimated from the information in this table. For example, a measurement of 7 millimeters for a 27-year-old male would be about the 20th percentile.

Source: Adapted from A. R. Frisancho, New norms of upper limb fat and muscle areas for assessment of nutritional status. *American Journal of Clinical Nutrition* 34 (1981): 2540–2545.

and making sure that the tape remains on a level horizontal plane on all sides. The tape is tightened slightly, but without compressing the skin.

Clinicians use many other methods to estimate body fat and its distribution. Each has its advantages and disadvantages as Table E-5 summarizes.

# Physical Examinations

An assessor can use a physical examination to search for signs of nutrient deficiency or toxicity. Like the other assessment methods, such an examination requires knowledge and skill. Many physical signs are nonspecific; they can reflect any of several nutrient deficiencies as well as conditions not related to nutrition (see Table E-6). For example, cracked lips may be caused by sunburn, windburn, dehydration, or any of several B vitamin deficiencies, to name just a few possible causes. For this reason, physical findings are most valuable in revealing problems for other assessment techniques to confirm or for confirming other assessment measures.

With this limitation understood, physical symptoms can be most informative and communicate much information about nutrition health. Many tissues and organs can reflect signs of malnutrition. The signs appear most rapidly in parts of the body where cell replacement occurs at a high rate, such as in the hair, skin, and digestive tract (including the mouth and tongue). The summary tables in Chapters 10, 11, 12, and 13 list additional physical signs of vitamin and mineral malnutrition.

E

**Table E-5**
**Methods of Estimating Body Fat and Its Distribution**

| Method | Cost | Ease of Use | Accuracy | Measures Fat Distribution |
|---|---|---|---|---|
| Height and weight | Low | Easy | High | No |
| Fatfolds | Low | Easy | Low | Yes |
| Circumferences | Low | Easy | Moderate | Yes |
| Ultrasound | Moderate | Moderate | Moderate | Yes |
| Hydrodensitrometry | Low | Moderate | High | No |
| Heavy water tritiated | Moderate | Moderate | High | No |
| Deuterium oxide, or heavy oxygen | High | Moderate | High | No |
| Potassium isotope ($^{40}K$) | Very high | Difficult | High | No |
| Total body electrical conductivity (TOBEC) | High | Moderate | High | No |
| Bioelectric impedance (BIA) | Moderate | Easy | High | No |
| Dual energy X-ray absorptiometry (DEXA) | High | Easy | High | No |
| Computed tomography (CT) | Very high | Difficult | High | Yes |
| Magnetic resonance imaging (MRI) | Very high | Difficult | High | Yes |

Source: Adapted with permisssion from G. A. Bray, a handout presented at the North American Association for the Study of Obesity and Emory University School of Medicine Conference on Obesity Update: Pathophysiology, Clinical Consequences, and Therapeutic Options, Atlanta, Georgia, August 31–September 2, 1992.

**Table E-6**
**Physical Findings Used in Nutrition Assessments**

| Body System | Acceptable Findings | Malnutrition Findings | What the Findings Reflect |
|---|---|---|---|
| Hair | Shiny, firm in the scalp | Dull, brittle, dry, loose; falls out | PEM |
| Eyes | Bright, clear pink membranes; adjust easily to light | Pale membranes; spots; redness; adjust slowly to darkness | Vitamin A, the B vitamins, zinc, and iron status |
| Teeth and gums | No pain or caries, gums firm, teeth bright | Missing, discolored, decayed teeth; gums bleed easily and are swollen and spongy | Mineral and vitamin C status |
| Face | Clear complexion without dryness or scaliness | Off-color, scaly, flaky, cracked skin | PEM, vitamin A, and iron status |
| Glands | No lumps | Swollen at front of neck | PEM and iodine status |
| Tongue | Red, bumpy, rough | Sore, smooth, purplish, swollen | B vitamin status |
| Skin | Smooth, firm, good color | Dry, rough, spotty; "sandpaper" feel or sores; lack of fat under skin | PEM, essential fatty acid deficiency, vitamin A, the B vitamins, and vitamin C status |
| Nails | Firm, pink | Spoon-shaped, brittle, ridged, pale | Iron status |
| Internal systems | Regular heart rhythm, heart rate, and blood pressure; no impairment of digestive function, reflexes, or mental status | Abnormal heart rate, heart rhythm, or blood pressure; enlarged liver, spleen; abnormal digestion; burning, tingling of hands, feet; loss of balance, coordination; mental confusion, irritability, fatigue | PEM and mineral status |
| Muscles and bones | Muscle tone; posture, long bone development appropriate for age | "Wasted" appearance of muscles; swollen bumps on skull or ends of bones; small bumps on ribs; bowed legs or knock-knees | PEM, mineral, and vitamin D status |

# Biochemical Analyses

All of the approaches to nutrition assessment discussed so far are external approaches. Biochemical analyses or laboratory tests help to determine what is hap-

The **serum** is the watery portion of the blood that remains after removal of the cells and clot-forming material; **plasma** is the fluid that remains when unclotted blood is centrifuged. In most cases, serum and plasma concentrations are similar. Lab technicians usually prefer serum samples because plasma samples occasionally clog mechanical blood analyzers.

pening to the body internally. Common tests are based on analysis of blood and urine samples, which contain nutrients, enzymes, and metabolites that reflect nutrition status. Other tests, such as serum glucose, help pinpoint disease-related problems with nutrition implications. Tests that define fluid and electrolyte balance, acid-base balance, and organ function also have nutrition implications. Table E-7 lists biochemical tests useful for assessing protein, vitamin, and mineral status.

The interpretation of biochemical data requires skill. Long metabolic sequences lead to the production of the end products and metabolites seen in blood and urine. No single test can reveal nutrition status because many factors influence test results. The low blood concentration of a nutrient may reflect a primary deficiency of that nutrient, but it may also be secondary to the deficiency of one or several other nutrients or to a disease. Taken together with other assessment data, however, laboratory test results help to make a total picture that becomes clear with careful interpretation. They are especially useful in helping to detect subclinical malnutrition by uncovering early signs of malnutrition before the clinical signs of a classic deficiency disease appear.

Reminder: A *subclinical deficiency* is a nutrient deficiency in the early stages before the outward signs have appeared.

Laboratory tests used to assess vitamin and mineral status (review Table E-7) are particularly useful when combined with diet histories and physical findings. Vitamin and mineral levels present in the blood and urine sometimes reflect recent rather than long-term intakes. This makes detecting subclinical deficiencies difficult. Furthermore, many nutrients interact; therefore, the amounts of other nutrients in the body can affect a lab value for a particular nutrient. It is also important to remember that nonnutrient conditions influence biochemical measures.

**Table E-7**

**Biochemical Tests Useful for Assessing Vitamin and Mineral Status**

| Nutrient | Assessment Tests |
|---|---|
| **Vitamins** | |
| Vitamin A | Retinol-binding protein, serum carotene |
| Thiamin | Erythrocyte (red blood cell) transketolase activity, urinary thiamin |
| Riboflavin | Erythrocyte glutathione reductase activity, urinary riboflavin |
| Vitamin $B_6$ | Urinary xanthurenic acid excretion after tryptophan load test, urinary vitamin $B_6$, erythrocyte transaminase activity |
| Niacin | Urinary metabolites NMN (N-methyl nicotinamide) or 2-pyridone, or preferably both expressed as a ratio |
| Folate | Free folate in the blood, erythrocyte folate (reflects liver stores), urinary formiminoglutamic acid (FIGLU), vitamin $B_{12}$ status (folate assessment tests alone do not distinguish between the two deficiencies) |
| Vitamin $B_{12}$ | Serum vitamin $B_{12}$, erythrocyte vitamin $B_{12}$, urinary methylmalonic acid synthesis or DUMP test (from the abbreviation for the chemical name of DNA's raw material, deoxyuridine monophosphate), Schilling test |
| Biotin | Serum biotin, urinary biotin |
| Vitamin C | Serum or plasma vitamin C[a], leukocyte vitamin C, urinary vitamin C |
| Vitamin D | Serum alkaline phosphatase |
| Vitamin E | Serum tocopherol, erythrocyte hemolysis |
| Vitamin K | Blood-clotting time (prothrombin time) |
| **Minerals** | |
| Potassium | Serum potassium |
| Magnesium | Serum magnesium |
| Iron | Hemoglobin, hematocrit, serum ferritin, total iron-binding capacity (TIBC), protoporphyrin, mean corpuscular volume (MCV), serum iron |
| Iodine | Serum protein-bound iodine, radioiodine uptake |
| Zinc | Plasma zinc, hair zinc |

[a]Vitamin C shifts unpredictably between the plasma and the white blood cells known as leukocytes; thus a plasma or serum determination may not accurately reflect the body's pool. The appropriate clinical test may be a measurement of leukocyte vitamin C. A combination of both tests may be more reliable than either one alone.

Source: Adapted from A. Grant and S. DeHoog, *Nutritional Assessment and Support,* 3rd ed., 1985.

It is beyond the scope of this text to describe all lab tests and their relations to nutrition status. Instead, the emphasis is on lab tests used to detect protein-energy malnutrition (PEM) and iron-deficiency anemia.

# Protein-Energy Malnutrition (PEM)

Table E-8 provides standards for evaluating the serum proteins most widely used in nutrition assessments. Other biochemical tests useful in assessing protein status include the total lymphocyte count, a measure of immune function; urinary nitrogen, used to evaluate nitrogen balance; and urinary creatinine, a measure of skeletal muscle mass.

• ***Serum Albumin*** • Albumin accounts for over 50 percent of the total serum proteins. Serum albumin is slow to reflect changes in nutrition status because it is plentiful in the body and breaks down slowly.* Therefore, low serum albumin reflects prolonged protein depletion. Likewise, albumin concentrations increase slowly with appropriate nutrition support, so measuring albumin as an indicator of response to nutrition therapy is of limited value.

Note that albumin concentrations can be depressed by many conditions besides malnutrition, including liver disease, kidney disease (nephrotic syndrome), eclampsia, and disorders that can markedly raise the metabolic rate (metabolic stress) including infection, cancer, and burns.

• ***Serum Transferrin*** • Transferrin is a protein that transports iron; consequently, its concentrations reflect both protein and iron status. Interpreting transferrin levels as an indicator of protein status is difficult when an *iron* deficiency is present. Transferrin rises as iron deficiency grows worse and falls as iron status improves. Markedly reduced transferrin levels indicate severe PEM; in mild-to-moderate PEM, transferrin levels may vary, limiting their usefulness. Although transferrin breaks down in the body more quickly than albumin, it is still relatively slow to respond to changes in protein intake. Thus it is not a sensitive indicator of response to medical nutrition therapy.†

Conditions other than protein status that lower transferrin concentrations include liver disease, kidney disease (nephrotic syndrome), and metabolic stress. Pregnancy, iron-deficiency anemia, hepatitis, blood loss, and the use of oral contraceptive agents can elevate transferrin levels.

• ***Prealbumin and Retinol-Binding Protein*** • Prealbumin and retinol-binding protein‡ respond quickly to changes in protein intake, and both measure response to nutrition therapy. Lab tests of prealbumin and retinol-binding protein are more expensive than the relatively inexpensive test of serum albumin, which is routinely available. Therefore, tests of prealbumin and retinol-binding protein are often reserved for clients with disorders that markedly change metabolic rates and can rapidly and profoundly affect nutrition status.

Prealbumin is also known as *thyroxin-binding prealbumin* or *transthyretin*.

Conditions other than protein status that can lower prealbumin levels include metabolic stress, hemodialysis, and hypothyroidism; levels may be elevated in kidney disease and with the use of corticosteroids.

Conditions other than protein status that can lower retinol-binding protein levels include vitamin A deficiency, metabolic stress, hyperthyroidism, liver disease, and cystic fibrosis; levels may be elevated in kidney disease.

• ***Total Lymphocyte Count*** • PEM compromises the immune system, reducing the number of white blood cells (lymphocytes), which are important in resisting and

**Table E-8**

**Relationship between Degree of Undernutrition and Serum Proteins**

| Indicator | Degree of Depletion | | | |
|---|---|---|---|---|
| | *Normal* | *Mild* | *Moderate* | *Severe* |
| Albumin (g/dL)[a] | ⩾3.5 | 2.8–3.4 | 2.1–2.7 | <2.1 |
| Transferrin (mg/dL) | >200 | 150–200 | 100–149 | <100 |
| Prealbumin (mg/dL) | 16–30 | 10–15 | 5–9 | <5 |
| Retinol-binding protein[b] (mg/dL) | 2.6–7.6 | — | — | — |

NOTE: To convert albumin (g/100 ml) to standard international units (g/L), multiply by 100. To convert transferrin (mg/100 ml) to standard international units (g/L), multiply by 0.01.

[a]A deciliter (dL) is one-tenth of a liter or 100 milliliters.

[b]Levels less than normal suggest compromised protein status. The actual degree of depletion (mild, moderate, and severe) has not been defined.

*The half-life of albumin is about 20 days, an indication of a slow degradation rate.

†Transferrin has a half-life of 4 to 8 days.

‡The half-lives of prealbumin and retinol-binding protein are 2 days and 12 hours, respectively.

Conditions other than protein status that affect the total lymphocyte count include metabolic stress (including infection) and the use of chemotherapy, immunosuppressives, and corticosteroids.

fighting infections. The total lymphocyte count, derived from the number of white blood cells and the percentage of lymphocytes, is inexpensive and easy to obtain, but the many variables that affect the levels of the total lymphocyte count limit its value in nutrition assessment.

• ***Urinary Tests of Protein Status*** • Two biochemical tests of protein status—urinary urea nitrogen (UUN) and urinary creatinine excretion—require the collection of urine over a 24-hour period. Both tests require normal kidney function for accurate results. Assessors use the 24-hour UUN measurement, along with an accurate record of the client's energy and protein intake during the same period, to calculate nitrogen balance (see p. 173). Results of nitrogen balance studies determine whether protein intake is adequate to meet needs.

Urinary creatine excretion provides an indirect measure of skeletal muscle mass. By comparing urinary creatinine excretion to standards for sex and height (creatinine height index), assessors determine if muscle mass is adequate or depleted.

Twenty-four-hour urine collections are invalid if even one urine specimen is discarded or if samples are not stored properly. Nitrogen balance studies further require careful measurement of the food portions presented to and eaten by the client. Such difficulties explain why these tests are not routinely performed in most facilities. Nevertheless, nitrogen balance studies are very useful for evaluating protein needs for clients with severe metabolic stresses.

## Classification of Protein-Energy Malnutrition

Chapter 6 used the terms *kwashiorkor* and *marasmus* to classify PEM as seen in developing countries. Because PEM found in industrialized countries often develops for different reasons (as a result of illness, for example), these terms, though frequently used in clinical settings, do not denote exactly the same conditions. We use the term **acute malnutrition** to describe kwashiorkor-type malnutrition and **chronic malnutrition** to describe marasmus-type malnutrition.

To evaluate PEM, assessors use data from all four assessment techniques. Historical information and physical findings alert health care professionals to the possibility of malnutrition. Anthropometric measures and biochemical analyses permit classification of PEM as either acute (kwashiorkor), chronic (marasmus), or a mixture of the two. Such distinctions can be useful, but in clinical settings, these syndromes may overlap, or one may progress into another. This serves as another reminder of the need to assess nutrition status at regular intervals.

• ***Acute Malnutrition*** • The person with acute malnutrition typically has normal or above-standard anthropometric measurements with below-normal indices of blood and organ proteins. Because the individual may be overweight, health care workers can easily overlook malnutrition. Overweight by itself would suggest overnutrition, which again illustrates why no single parameter can be used to define nutrition status.

• ***Chronic Malnutrition*** • In chronic malnutrition, the individual has blood and organ protein levels that appear to be adequate, while skeletal muscle and subcutaneous fat are depleted. The emaciated appearance of the person with chronic malnutrition makes this form of malnutrition easier to notice than acute malnutrition.

• ***Mixed PEM*** • Mixed PEM presents signs of depleted blood, organ, and skeletal muscle proteins as well as depleted subcutaneous fat. The person with this type of malnutrition has virtually no energy reserves, has compromised organ function, and is in grave danger, especially if experiencing severe stress as well.

• ***Energy Overnutrition*** • Malnutrition includes both undernutrition and overnutrition. Above-normal anthropometric measures combined with normal blood protein concentrations identify a person consuming energy in excess of needs. Such a person risks obesity and the many chronic diseases associated with it (see Chapter 8).

The assessment of PEM is commonly performed because many people, including the elderly and those who are hospitalized, have this type of malnutrition and it can lead to severe consequences. The next section illustrates how biochemical tests can be used to assess nutrition-related anemias caused by iron, folate, or vi-tamin $B_{12}$ deficiencies.

## Nutritional Anemias

Anemia, a symptom of a wide variety of nutrition- and nonnutrition-related disorders, is characterized by a reduced number of red blood cells. Iron, folate, and vitamin $B_{12}$ deficiencies caused by inadequate intake, poor absorption, or abnormal metabolism of these nutrients are the most common nutritional anemias. Some nonnutrition-related causes of anemia include massive blood loss, infections, hereditary blood disorders such as sickle-cell anemia, and chronic liver or kidney disease.

## Assessment of Iron Deficiency

Iron deficiency, a common mineral deficiency, develops in stages. Chapter 13 describes iron deficiency in detail. This section describes tests used to uncover iron deficiency as it progresses. Table E-9 shows which laboratory tests detect various stages of iron deficiency. Although other tests are more specific in detecting early deficiencies, hemoglobin and hematocrit are the commonly available tests.

Stages of iron deficiency:
1. Iron stores diminish.
2. Transport iron decreases.
3. Hemoglobin production falls.

• ***Hemoglobin*** • Iron forms an integral part of the hemoglobin molecule that transports oxygen to the cells. In iron deficiency, the body cannot synthesize hemoglobin. Low hemoglobin values signal depleted iron stores. Table E-10 provides hemoglobin values used in nutrition assessment. Hemoglobin's usefulness in evaluating iron status is limited, however, because hemoglobin concentrations drop fairly late in the development of iron deficiency, and other nutrient deficiencies and medical conditions can also alter hemoglobin concentrations.

• ***Hematocrit*** • Hematocrit is commonly used to diagnose iron deficiency, even though it is an inconclusive measure of iron status. To measure the hematocrit, a

**Table E-9**
**Laboratory Tests Useful in Evaluating Nutrition-Related Anemias**

| Test or Test Result | What It Reflects |
|---|---|
| ***General Tests for Anemia*** | |
| Hemoglobin (Hg) | Total amount of hemoglobin in the red blood cells (RBC) |
| Hematocrit (Hct) | Percentage of RBC in the total blood volume |
| Red blood cell (RBC) count | Number of RBC |
| Mean corpuscular volume (MCV) | RBC size; helps to determine if anemia is microcytic or macrocytic |
| Mean corpuscular hemoglobin | Hemoglobin concentration within the average RBC; helps to determine if anemia is hypochromic or normochromic |
| Bone marrow aspiration | The manufacture of blood cells in different developmental states |
| **Iron-Deficiency Anemia** | |
| ↓ Serum ferritin | Early deficiency state with depleted iron stores |
| ↓ Transferrin saturation | Progressing deficiency state with diminished transport iron |
| ↑ Erythrocyte protoporphyrin | Later deficiency state with limited hemoglobin production |
| **Folate-Deficiency Anemia** | |
| ↓ Serum folate | Progressing deficiency state |
| ↓ RBC folate | Later deficiency state |
| **Vitamin $B_{12}$–Deficiency Anemia** | |
| ↓ Serum vitamin $B_{12}$ | Progressing deficiency state |
| Schilling test | Absorption of vitamin $B_{12}$ |

E

**Table E-10**
**Standards for Hemoglobin Test Results**

| Age (yr) | Sex | Deficient (g/dL) | Acceptable (g/dL)[a] |
|---|---|---|---|
| 0.5–2 | M–F | <11 | ≥11 |
| 2–6 | M–F | <11 | ≥11 |
| 6–12 | M–F | <11.5 | ≥11.5 |
| 12–18 | M | <13 | ≥13 |
| | F | <12 | ≥12 |
| Adult | M | <13.5 | ≥13.5 |
| | F | <12 | ≥12 |
| Pregnancy | | | |
| Trimester 1 | | <11 | ≥11 |
| Trimester 2 | | <10.5 | ≥10.5 |
| Trimester 3 | | <11 | ≥11 |

[a]A deciliter (dL) is one-tenth of a liter or 100 milliliters.

clinician spins a volume of blood in a centrifuge to separate the red blood cells from the plasma. The hematocrit is the percentage of red blood cells in the total blood volume. Table E-11 provides values used to assess hematocrit status. Low values indicate incomplete hemoglobin formation, which is manifested by microcytic (abnormally small-celled), hypochromic (abnormally lacking in color) red blood cells.

Low hemoglobin and hematocrit values alert the assessor to the possibility of iron deficiency. However, many nutrients and other conditions can affect hemoglobin and hematocrit. The other tests of iron status help pinpoint true iron deficiency.

• ***Serum Ferritin*** • In the first stage of iron deficiency, iron stores diminish. Serum ferritin measures provide a noninvasive estimate of iron stores. Such information is most valuable to iron assessment. Table E-12 shows serum ferritin cutoff values that indicate iron store depletion in children and adults. Serum ferritin is not reliable for diagnosing iron deficiency in infants, since normal serum ferritin values are often present in conjunction with iron-responsive anemia.

A decrease in transport iron characterizes the second stage of iron deficiency. This is revealed by an increase in the iron-binding capacity of the protein transferrin and a decrease in serum iron. These changes are reflected by the transferrin saturation, which is calculated from the ratio of the other two values as described in the following sections.

TIBC values greater than 400 μg/dL (1 dL = 100 m) indicate iron deficiency.*

*To convert iron-binding capacity (μg/dL) to international standard units (μmol/L) multiply by 0.1791. 400 μg/dL = 71 μmol/L.

• ***Total Iron-Binding Capacity (TIBC)*** • Iron travels through the blood bound to the protein transferrin. TIBC is a measure of the total amount of iron that transferrin can carry. Lab technicians measure iron-binding capacity directly.

• ***Serum Iron*** • Lab technicians can also measure serum iron directly. Elevated values indicate iron overload; reduced values indicate iron deficiency. Table E-13 shows acceptable and deficient values for serum iron.

• ***Transferrin Saturation*** • The percentage of transferrin that is saturated with iron is an indirect measure that is derived from the serum iron and total iron-binding capacity measures as follows:

$$\%\text{Transferrin} = \frac{\text{serum iron} \times 100}{\text{total iron-binding capacity}}.$$

Table E-14 shows deficient and acceptable transferrin saturation values for various age groups.

**Table E-11**
**Standards for Hematocrit Test Results**

| Age (yr) | Sex | Deficient % | Acceptable % |
|---|---|---|---|
| 0.5–2 | M–F | <33 | ≥33 |
| 2–6 | M–F | <34 | ≥34 |
| 6–12 | M–F | <35 | ≥35 |
| 12–18 | M | <38 | ≥38 |
| | F | <36 | ≥36 |
| Adult | M | <41 | ≥41 |
| | F | <36 | ≥36 |

**Table E-12**
**Standards for Serum Ferritin**

| Group | Deficient (ng/ml)[a] |
|---|---|
| Children (3–14 years of age) | <10 |
| Adolescents and adults | <12 |
| Pregnant women | <10 |

[a]A nanogram (ng) is one-billionth of a gram.

**Table E-13**
**Standards for Serum Iron**

| | | Deficient | | Acceptable | |
|---|---|---|---|---|---|
| Age (yr) | Sex | (μg/dL)[a] | (μmol/L) | (μg/dL)[a] | (μmol/L) |
| <2 | M–F | <30 | <5.3 | 30 or > | 5.3 or > |
| 2–5 | M–F | <40 | <7.1 | 40 or > | 7.1 or > |
| 6–12 | M–F | <50 | <8.9 | 50 or > | 8.9 or > |
| >12 | M | <60 | <10.7 | 60 or > | 10.7 or > |
| | F | <40 | <7.1 | 40 or > | 7.1 or > |

NOTE: To convert (μg/dL) to international standard units, multiply by 0.1791.
[a]A deciliter is one-tenth of a liter or 100 milliliters.

**Table E-14**
**Standards for Percent Transferrin Saturation**

| Age (yr) | Sex | Deficient | Acceptable |
|---|---|---|---|
| <2 | M–F | <15% | 15% or > |
| 2–12 | M–F | <20% | 20% or > |
| ≥13 | M | <20% | 20% or > |
| | F | <15% | 15% or > |

The third stage of iron deficiency occurs when the supply of transport iron diminishes to the point that it limits hemoglobin production. It is characterized by increases in erythrocyte protoporphyrin, a decrease in mean corpuscular volume, and decreased hemoglobin concentration and hematocrit.

• *Erythrocyte Protoporphyrin* • The iron-containing portion of the hemoglobin molecule is heme. Heme is a combination of iron and protoporphyrin. Protoporphyrin accumulates in the blood when iron supplies are inadequate for the formation of heme. Lab technicians can measure erythrocyte protoporphyrin directly in a blood sample. The cutoffs for abnormal values of erythrocyte protoporphyrin are shown in Table E-15.

• *Mean Corpuscular Volume (MCV)* • A direct or calculated measure of the mean corpuscular volume (MCV) determines the average size of a red blood cell (RBC). Such a measure helps to classify the type of nutrient anemia. In iron deficiency, the red blood cells are smaller than average. The cutoffs for abnormal values of MCV that indicate iron deficiency are also shown in Table E-15.

**Table E-15**
**Standards for Erythrocyte Protoporphyrin and Mean Corpuscular Volume**

| Age (yr) | Erythrocyte Protoporphyrin (μg/dL RBC) | MCV (fL) |
|---|---|---|
| 1–2 | >80 | <73 |
| 3–4 | >75 | <75 |
| 5–10 | >70 | <76 |
| 11–14 | >70 | <78 |
| 15–74 | >70 | <80 |

## Assessment of Folate and Vitamin $B_{12}$ Anemias

Folate deficiency and vitamin $B_{12}$ deficiency present a similar clinical picture—an anemia characterized by abnormally large red blood cell precursors (megaloblasts) in the bone marrow and abnormally large, mature red blood cells (macrocytic cells) in the blood. Distinguishing between these two deficiencies is particularly important because their treatments differ. Giving folate to a person with vitamin $B_{12}$ deficiency improves many of the lab test results indicative of vitamin $B_{12}$ deficiency, but this is a dangerous error because vitamin $B_{12}$ deficiency causes nerve damage that folate cannot correct. Thus inappropriate folate administration masks vitamin $B_{12}$–deficiency anemia, and nerve damage worsens. For this reason, it is critical to determine whether the anemia results from a folate deficiency or from a vitamin $B_{12}$ deficiency. The following biochemical assessment techniques help to make this distinction.

• *Mean Corpuscular Volume (MCV)* • As previously mentioned, the MCV is a measure of red blood cell size. In folate and vitamin $B_{12}$ deficiencies, the red blood cells are larger than average (macrocytic). Additional tests must be performed to differentiate folate from vitamin $B_{12}$ deficiency.

• *Folate Levels* • Serum folate levels fluctuate with changes in folate intake and metabolism. Thus serum folate concentrations reflect current status, but provide little information about folate stores. As folate deficiency progresses and low serum levels persist, folate stores decline, resulting in folate depletion. Folate depletion is characterized by a fall in the folate concentrations of red blood cells (erythrocytes). As erythrocyte folate levels diminish, folate-deficiency anemia develops. Because low erythrocyte folate concentrations also occur with vitamin $B_{12}$ deficiency, serum vitamin $B_{12}$ concentrations must also be measured. Table E-16 shows standards for folate assessment.

• *Vitamin $B_{12}$ Levels* • Vitamin $B_{12}$ deficiency usually arises from malabsorption. To determine whether malabsorption is the cause, a small oral dose of vitamin $B_{12}$

**Table E-16**
**Standards for Folate Concentrations**

| | Deficient (ng/ml)[a] | Borderline (ng/ml)[a] | Acceptable (ng/ml)[a] |
|---|---|---|---|
| Serum folate | <3.0 | 3.0–6.0 | >6.0 |
| Erythrocyte folate | <140 | 140–160 | >160 |

NOTE: To convert folate values (ng/ml) to international standard units (nmol/L), multiply by 2.266.
[a]A nanogram (ng) is one-billionth of a gram.

is given, and urinary excretion is measured. This procedure measures vitamin $B_{12}$ absorption and is called a Schilling test.

Early stages of vitamin $B_{12}$ deficiency can be detected by a low percentage saturation of its transport protein, a measure similar to iron's transferrin saturation. As the deficiency progresses, serum vitamin $B_{12}$ concentrations fall.

# Cautions about Nutrition Assessment

To give all the details of nutrition assessment procedures would entail writing another textbook. Nevertheless, any student of nutrition should know the basics of a proper nutrition assessment procedure for two reasons.

First, competent medical care includes attention to nutrition. Physicians should either employ a person skilled in nutrition assessment techniques or refer all clients to such a person to ensure the sound nutrition health of their clients. Health care facilities should make nutrition assessment a routine part of the initial workup on every client so that nutritional handicaps will not hinder the response to medical treatment and the recovery from illness.

Second, because nutrition is such a popular subject today, fraudulent practices are even more abundant than they have been in the past (and they have always been rampant). The knowledgeable consumer needs to know what procedures to expect in a nutrition assessment and what kinds of information they yield. This appendix has presented the basics of nutrition assessment for these reasons.

This caution is added: the tests outlined here yield information that becomes meaningful only when integrated into a whole picture by a skilled, experienced, and educated interpreter. Potential sources of error are many, from the taking of the initial data to their reporting and analysis. Each assessment method and measure is useful only as a part of the whole to confirm or eliminate the possibility of suspected nutrition problems. For example, the assessor must constantly remember that a sufficient intake of a nutrient does not guarantee adequate nutrient status for an individual. Conversely, the apparent inadequate intake of a nutrient does not, by itself, establish that a deficiency exists.

Similarly, many uncertainties, such as the calibration of the equipment, the skills of the measurer, and the perspective of the interpreter, limit the accuracy and value of anthropometric measures. This is also true of the results of the physical examination. Physical signs suggestive of malnutrition are nonspecific: they can reflect nutrient deficiencies or may be totally unrelated to nutrition. Assessors must interpret physical findings in light of other assessment findings. Finally, the usefulness of biochemical tests is also limited; the assessor must use caution in interpreting results. Vitamin and mineral blood concentrations may reflect disease processes, abnormal hormone levels, or other aberrations rather than dietary intake. Even if concentrations do reflect dietary intake, they may reflect what the person has been eating recently and not give a true picture of the person's nutrient status. Such complications sometimes make it difficult to detect a subclinical deficiency. Furthermore, many nutrients interact. The assessor has to keep in mind that an abnormal lab value for one nutrient may reflect abnormal status of other nutrients. The final diagnosis is therefore appropriately tentative, and its confirmation comes only after careful remedial steps successfully alleviate the observed problems.

# NUTRITION RESOURCES

CONTENTS



F

People interested in nutrition often want to know where they can find reliable nutrition information. Wherever you live, there are several sources you can turn to:

- The Department of Health may have a nutrition expert.
- The local extension agent is often an expert.
- The food editor of your local paper may be well informed.
- The dietitian at the local hospital had to fulfill a set of qualifications before he or she became an RD (see Highlight 1).
- There may be knowledgeable professors of nutrition or biochemistry at a nearby college or university.

In addition, you may be interested in building a nutrition library of your own. Books you can buy, journals you can subscribe to, and addresses you can contact for general information are given below.

## Books

For students seeking to establish a personal library of nutrition references, the authors of this text recommend the following books:

- *Present Knowledge in Nutrition,* 7th ed. (Washington, D.C.: International Life Sciences Institute—Nutrition Foundation, 1996).

This 646-page paperback has a chapter on each of 64 topics, including energy, obesity, each of the nutrients, several diseases, malnutrition, growth and its assessment, immunity, alcohol, fiber, exercise, drugs, and toxins. Watch for an update; new editions come out every few years.

- M. E. Shils, J. A. Olson, and M. Shike, eds., *Modern Nutrition in Health and Disease,* 8th ed. (Philadelphia: Lea & Febiger, 1994).

This two-volume set is a major technical reference book on nutrition topics. It contains encyclopedic articles on the nutrients, foods, the diet, metabolism, malnutrition, age-related needs, and nutrition in disease.

- Committee on Dietary Reference Intakes, *Dietary Reference Intakes for Calcium, Phosphorus, Magnesium, Vitamin D, and Fluoride* (Washington, D.C.: National Academy Press, 1997).
- Committee on Dietary Reference Intakes, *Dietary Reference Intakes for Thiamin, Riboflavin, Niacin, Vitamin $B_6$, Folate, Vitamin $B_{12}$, Pantothenic Acid, Biotin, and Choline* (Washington, D.C.: National Academy Press, 1998).

These two reports review the function of each nutrient, dietary sources, and deficiency and toxicity symptoms as well as provide recommendations for intakes. Watch for additional reports on the Dietary Reference Intakes for the remaining nutrients. Until they are published, you may need the following:

- Committee on Dietary Allowances, *Recommended Dietary Allowances,* 10th ed. (Washington, D.C.: National Academy Press, 1989).

The Canadian equivalent is *Nutrition Recommendations,* available by mail from the Canadian Government Publishing Centre, Supply and Services Canada, Ottawa, Ontario K1A OS9, Canada.

- Committee on Diet and Health, *Diet and Health Implications for Reducing Chronic Disease Risk* (Washington, D.C.: National Academy Press, 1989).

This 749-page book presents the integral relationship between diet and chronic disease prevention. Its nutrient chapters provide evidence on how diet influences disease development, and its disease chapters review the dietary patterns implicated in each chronic disease.

- E. M. N. Hamilton and S. A. S. Gropper, *The Biochemistry of Human Nutrition: A Desk Reference* (St. Paul, Minn.: West, 1987).

This 324-page paperback presents the biochemical concepts necessary for an understanding of nutrition. It is a handy reference book for those who have forgotten the basics of biochemistry or for those who are learning biochemistry for the first time.

We also recommend three of our own books that explore current topics in nutrition, health, and the life span:

- S. R. Rolfes, L. K. DeBruyne, and E. N. Whitney, *Life Span Nutrition: Conception through Life* (Belmont, Calif.: West/Wadsworth, 1998).
- F. S. Sizer and E. N. Whitney, *Nutrition: Concepts and Controversies,* 7th ed. (Belmont, Calif.: West/Wadsworth, 1997).
- E. N. Whitney, C. B. Cataldo, L. K. DeBruyne, and S. R. Rolfes, *Nutrition for Health and Health Care* (St. Paul, Minn.: West, 1995).

## Journals

*Nutrition Today* is an excellent magazine for the interested layperson. It makes a point of raising controversial issues and providing a forum for conflicting opinions. Six issues per year are published. Order from Williams and Wilkins, 351 West Camden Street, Baltimore, MD 21201-2436.

The *Journal of the American Dietetic Association,* the official publication of the ADA, contains articles of interest to dietitians and nutritionists, news of legislative action on food and nutrition, and a very useful section of abstracts of articles from many other journals of nutrition and related areas. There are 12 issues per year, available from the American Dietetic Association (see "Addresses," later).

*Nutrition Reviews,* a publication of the International Life Sciences Institute, does much of the work for the library researcher, compiling recent evidence on current topics and presenting extensive bibliographies. Twelve issues per year are available from Nutrition Reviews, P.O. Box 1897, Lawrence, KS 66044-8897.

*Nutrition and the M.D.* is a monthly newsletter that provides up-to-date, easy-to-read, practical information on nutrition for health care providers. It is available from Lippincott-Raven Publishers, 12107 Insurance Way, Hagerstown, MD 21740.

Other journals that deserve mention here are *Food Technology, Journal of Nutrition, American Journal of Clinical Nutrition, Nutrition Research,* and *Journal of Nutrition Education. FDA Consumer,* a government publication with many articles of interest to the consumer, is available from the Food and Drug Administration (see "Addresses," below). Many other journals of value are referred to throughout this book.

# Addresses

Many of the organizations listed below will provide publication lists free on request. Government and international agencies and professional nutrition organizations are listed first, followed by organizations in the following areas: aging, alcohol and drug abuse, consumer organizations, fitness, food safety, health and disease, infancy and childhood, pregnancy and lactation, trade and industry organizations, weight control and eating disorders, and world hunger.

## U. S. Government

- Federal Trade Commission (FTC)
  Public Reference Branch
  (202) 326-2222
  www.ftc.gov
- Food and Drug Administration (FDA)
  Office of Consumer Affairs, HFE 1
  Room 16-85
  5600 Fishers Lane
  Rockville, MD 20857
  (301) 443-1544
  www.fda.gov
- FDA Consumer Information Line
  (301) 827-4420
- FDA Office of Food Labeling, HFS 150
  200 C Street SW
  Washington, DC 20204
  (202) 205-4561; fax (202) 205-4564
  www.cfsan.fda.gov
- FDA Office of Plant and Dairy Foods and Beverages
  HFS 300
  200 C Street SW
  Washington, DC 20204
  (202) 205-4064; fax (202) 205-4422
- FDA Office of Special Nutritionals,
  HFS 450
  200 C Street SW
  Washington, DC 20204
  (202) 205-4168; fax (202) 205-5295
- Food and Nutrition Information Center
  National Agricultural Library, Room 304
  10301 Baltimore Avenue
  Beltsville, MD 20705-2351
  (301) 504-5719; fax (301) 504-6409
  www.nal.usda.gov/fnic
- Food Research Action Center (FRAC)
  1875 Connecticut Avenue NW, Suite 540
  Washington, DC 20009
  (202) 986-2200; fax (202) 986-2525
- Superintendent of Documents
  U.S. Government Printing Office
  Washington, DC 20402
  (202) 512-1071
  www.access.gpo.gov/su_docs
- U.S. Department of Agriculture (USDA)
  14th Street SW and Independence Avenue
  Washington, DC 20250
  (202) 720-2791
  www.usda.gov/fcs
- USDA Center for Nutrition Policy and Promotion
  1120 20th Street NW, Suite 200
  North Lobby
  Washington, DC 20036
  (202) 208-2417
  www.usda.gov/fcs/cnpp.htm
- USDA Food Safety and Inspection Service
  Food Safety Education Office,
  Room 1180-S
  Washington, DC 20250
  (202) 690-0351
  www.usda.gov/fsis
- U.S. Department of Education (DOE)
  Accreditation Agency Evaluation Branch
  7th and D Street SW
  ROB 3, Room 3915
  Washington, DC 20202-5244
  (202) 708-7417
- U.S. Department of Health and Human Services
  200 Independence Avenue SW
  Washington, DC 20201
  (202) 619-0257
  www.os.dhhs.gov
- U.S. Environmental Protection Agency (EPA)
  401 Main Street SW
  Washington, DC 20460
  (202) 260-2090
  www.epa.gov
- U.S. Public Health Service
  Assistant Secretary of Health
  Humphrey Building, Room 725-H
  200 Independence Avenue SW
  Washington, DC 20201
  (202) 690-7694

## Canadian Government

### *Federal*

- Bureau of Nutritional Sciences
  Food Directorate
  Health Protection Branch
  3-West
  Sir Frederick Banting Research Centre, 2203A
  Tunney's Pasture
  Ottawa, Ontario K1A 0L2
  www.hc-sc.gc.ca
- Canadian Food Inspection Agency
  Agriculture and Agri-Food Canada
  59 Camelot Drive
  Nepean, Ontario K1A 0Y9
  (613) 225-CFIA or (613) 225-2342
  www.agr.ca
- Nutrition & Healthy Eating Unit
  Strategies and Systems for Health Directorate, 1917C
  17th Floor—Jeanne Mance Bldg.
  Tunney's Pasture
  Ottawa, Ontario K1A 1B4
  www.hc-sc.gc.ca
- Nutrition Specialist
  Health Support Services
  Indian and Northern Health Services Directorate
  Medical Services Branch
  20th Floor—Jeanne Mance Bldg., 1920B
  Tunney's Pasture
  Ottawa, Ontario K1A 0L3
  www.hc-sc.gc.ca

### *Provincial and Territorial*

- Population Health Strategies Branch
  Alberta Health
  23rd Floor, TELUS Plaza, North Tower
  10025 Jasper Avenue
  Edmonton, AB T5J 2N3
- Nutritionist
  Preventive Services Branch
  Ministry of Health
  1520 Blanshard Street
  Victoria BC V8W 3C8

- Executive Director
  Health Programs
  2nd Floor 800 Portage Avenue
  Winnipeg, MB R3G 0P4
- Project Manager
  Public Health Management Services
  Health and Community Services
  P.O. Box 5100
  520 King Street
  Fredericton, NB E3B 5G8
- Director, Health Promotion
  Department of Health
  Government of Newfoundland and Labrador
  P.O. Box 8700
  Confederation Building, West Block
  St. John's, NF A1B 4J6
- Consultant, Nutrition
  Health & Wellness Promotion
  Population Health
  Department of Health and Social Services
  Government of the Northwest Territories
  Centre Square Tower, 6th Floor
  P.O. Box 1320
  Yellowknife, NT X1A 2L9
- Public Health Nutritionist
  Central Health Region
  201 Brownlow Avenue, Unit 4
  Dartmouth, NS B3B 1W2
- Senior Consultant, Nutrition
  Public Health Branch
  Ministry of Health, 8th Floor
  5700 Yonge St.
  North York, ON M2M 4K5
- Coordinator, Health Information Resource Centre
  Department of Health and Social Services
  1 Rochford Street, Box 2000
  Charlottetown, PEI C1A 7N8
- Responsables de la santé cardio-vasculaire et de la nutrition
  Ministère de la Santé et des Services sociaux, Service de la Prévention en Santé
  3e étage
  1075, chemin Sainte-Foy
  Quèbec (Quèbec) G1S 2M1
- Health Promotion Unit
  Population Health Branch
  Saskatchewan Health
  3475 Albert Street
  Regina, SK S4S 6X6
- Director, Nutrition Services
  Yukon Hospital Corporation
  #5 Hospital Road
  Whitehorse, YT Y1A 3H7

## International Agencies

- Food and Agriculture Organization of the United Nations (FAO)
  Liaison Office for North America
  2175 K Street, Suite 300
  Washington, DC 20437
  (202) 653-2400
  www.fao.org
- International Food Information Council Foundation
  1100 Connecticut Avenue NW, Suite 430
  Washington, DC 20036
  (202) 296-6540
  ificinfo.health.org
- UNICEF
  3 United Nations Plaza
  New York, NY 10017
  (212) 326-7000
  www.unicef.com
- World Health Organization (WHO)
  Regional Office
  525 23rd Street NW
  Washington, DC 20037
  (202) 974-3000
  www.who.org

## Professional Nutrition Organizations

- American Academy of Nutritional Sciences
  9650 Rockville Pike
  Bethesda, MD 20814
  (303) 530-7050; fax (301) 571-1892
  www.nutrition.org
- American Dietetic Association (ADA)
  216 West Jackson Boulevard, Suite 800
  Chicago, IL 60606-6995
  (800) 877-1600; (312) 899-0040
  www.eatright.org
- ADA, The Nutrition Hotline
  (800) 366-1655
- American Society for Clinical Nutrition
  9650 Rockville Pike
  Bethesda, MD 20814-3998
  (301) 530-7110; fax (301) 571-1863
  www.faseb.org/ascn
- Dietitians of Canada
  480 University Avenue, Suite 604
  Toronto, Ontario M5G 1V2, Canada
  (416) 596-0857; fax (416) 596-0603
  www.dietitians.ca
- Human Nutrition Institute (INACG)
  1126 Sixteenth Street NW
  Washington, DC 20036
  (202) 659-0789
  www.ilsi.org
- National Academy of Sciences/National Research Council (NAS/NRC)
  2101 Constitution Avenue, NW
  Washington, DC 20418
  (202) 334-2000
  www.nas.edu
- National Institute of Nutrition
  265 Carling Avenue, Suite 302
  Ottawa, Ontario K1S 2E1
  (613) 235-3355; fax (613) 235-7032
  www.nin.ca
- Society for Nutrition Education
  7101 Wisconsin Avenue, Suite 901
  Bethesda, MD 20814-4805
  (301) 656-4938

## Aging

- Administration on Aging
  330 Independence Avenue SW
  Washington, DC 20201
  (202) 619-0724
  www.aoa.dhhs.gov
- American Association of Retired Persons (AARP)
  601 E Street NW
  Washington, DC 20049
  (202) 434-2277
  www.aarp.org
- National Aging Information Center
  330 Independence Avenue SW
  Washington, DC 20201
  (202) 619-7501
  www.aoa.dhhs.gov/naic
- National Institute on Aging
  Public Information Office
  31 Center Drive, MSC 2292
  Bethesda, MD 20892
  (301) 496-1752
  www.nih.gov/nia

## Alcohol and Drug Abuse

- Al-Anon Family Group Headquarters, Inc.
  1600 Corporate Landing Parkway
  Virginia Beach, VA 23454-5617
  (800) 356-9996
  www.al-anon.alateen.org
- Alateen
  1600 Corporate Landing Parkway
  Virginia Beach, VA 23454-5617
  (800) 356-9996
  www.al-anon.alateen.org
- Alcohol & Drug Abuse Information Line
  Adcare Hospital
  (800) 252-6465

- Alcoholics Anonymous (AA)
  General Service Office
  475 Riverside Drive
  New York, NY 10115
  (212) 870-3400
  www.aa.org
- Narcotics Anonymous (NA)
  P.O. Box 9999
  Van Nuys, CA 91409
  (818) 773-9999; fax (818) 700-0700
  www.wsoinc.com
- National Clearinghouse for Alcohol and Drug Information (NCADI)
  P.O. Box 2345
  Rockville, MD 20847-2345
  (800) 729-6686
  www.health.org
- National Council on Alcoholism and Drug Dependence (NCADD)
  12 West 21st Street
  New York, NY 10010
  (800) NCA-CALL or (800) 622-2255
  (212) 206-6770; fax (212) 645-1690
  www.ncadd.org
- U.S. Center for Substance Abuse Prevention
  1010 Wayne Avenue, Suite 850
  Silver Spring, MD 20910
  (301) 459-1591 ext. 244; fax (301) 495-2919
  www.covesoft.com/csap.html

## Consumer Organizations

- Center for Science in the Public Interest (CSPI)
  1875 Connecticut Avenue NW, Suite 300
  Washington, DC 20009-5728
  (202) 332-9110; fax (202) 265-4954
  www.cspinet.org
- Choice in Dying, Inc.
  1035 30th Street NW
  Washington, DC 20007
  (202) 338-9790; fax (202) 338-0242
  www.choices.org
- Consumer Information Center
  Pueblo, CO 81009
  (888) 8 PUEBLO or (888) 878-3256
  www.pueblo.gsa.gov
- Consumers Union of US Inc.
  101 Truman Avenue
  Yonkers, NY 10703-1057
  (914) 378-2000
  www.consunion.org
- National Council Against Health Fraud, Inc. (NCAHF)
  P.O. Box 1276
  Loma Linda, CA 92354
  (909) 824-4690
  www.ncahf.org

## Fitness

- American College of Sports Medicine
  P.O. Box 1440
  Indianapolis, IN 46206-1440
  (317) 637-9200
  www.acsm.org/sportsmed
- American Council on Exercise (ACE)
  5820 Oberlin Drive, Suite 102
  San Diego, CA 92121
  (800) 529-8227
  www.acefitness.org
- President's Council on Physical Fitness and Sports
  Humphrey Building, Room 738
  200 Independence Avenue SW
  Washington, DC 20201
  (202) 690-9000; fax (202) 690-5211
  www.indiana.edu/~preschal
- Shape Up America!
  6707 Democracy Boulevard, Suite 306
  Bethesda, MD 20817
  (301) 493-5368
  www.shapeup.org
- Sport Medicine and Science Council of Canada
  1600 James Naismith Drive, Suite 314
  Gloucester, Ontario K1B 5N4, Canada
  (613) 748-5671; fax (613) 748-5729
  www.smscc.ca

## Food Safety

- Alliance for Food & Fiber
  Food Safety Hotline
  (800) 266-0200
- FDA Center for Food Safety and Applied Nutrition
  200 C Street SW
  Washington, DC 20204
  (800) FDA-4010 or (800) 332-4010
  vm.cfsan.fda.gov
- National Lead Information Center
  (800) LEAD-FYI or (800) 532-3394
  (800) 424-LEAD or (800) 424-5323
- National Pesticide Telecommunications Network (NPTN)
  Oregon State University
  333 Weniger Hall
  Corvallis, OR 97331-6502
  (541) 737-6091
  www.ace.orst.edu/info/nptn
- USDA Meat and Poultry Hotline
  (800) 535-4555
- U.S. EPA Safe Drinking Water Hotline
  (800) 426-4791

## Health and Disease

- Alzheimer's Disease Education and Referral Center
  P. O. Box 8250
  Silver Spring, MD 20907-8250
  (800) 438-4380
  www.alzheimers.org
- Alzheimer's Disease Information and Referral Service
  919 North Michigan Avenue, Suite 1000
  Chicago, IL 60611
  (800) 272-3900
  www.alz.org
- American Academy of Allergy, Asthma, and Immunology
  611 East Wells Street
  Milwaukee, WI 53202
  (414) 272-6071; fax (414) 276-3349
  www.aaaai.org
- American Cancer Society
  National Home Office
  1599 Clifton Road NE
  Atlanta, GA 30329-4251
  (800) ACS-2345 or (800) 227-2345
  www.cancer.org
- American Council on Science and Health
  1995 Broadway, 2nd Floor
  New York, NY 10023-5860
  (212) 362-7044; fax (212) 362-4919
  www.acsh.org
- American Dental Association
  211 East Chicago Avenue
  Chicago, IL 60611
  (312) 440-2800
  www.ada.org
- American Diabetes Association
  1660 Duke Street
  Alexandria, VA 22314
  (800) 232-3472 or (703) 549-1500
  www.diabetes.org
- American Heart Association
  Box BHG, National Center
  7320 Greenville Avenue
  Dallas, TX 75231
  (800) 275-0448 or (214) 373-6300
  www.amhrt.org
- American Institute for Cancer Research
  1759 R Street NW
  Washington, DC 20009
  (800) 843-8114 or (202) 328-7744; fax (202) 328-7226
  www.aicr.org
- American Medical Association
  515 North State Street
  Chicago, IL 60610
  (312) 464-5000
  www.ama-assn.org

- American Public Health Association (APHA)
 1015 Fifteenth Street NW, Suite 300
 Washington, DC 20005
 (202) 789-5600
 www.apha.org
- American Red Cross
 National Headquarters
 8111 Gatehouse Road
 Falls Church, VA 22042
 (703) 206-7180
 www.redcross.org
- Canadian Diabetes Association
 15 Toronto Street, Suite 800
 Toronto, ON M5C 2E3
 (800) BANTING or (800) 226-8464
 (416) 363-3373
 www.diabetes.ca
- Canadian Public Health Association
 400-1565 Carling Avenue
 Ottawa, Ontario K1Z 8R1
 (613) 725-3769; fax (613) 725-9826
 www.cpha.ca
- Centers for Disease Control and Prevention (CDC)
 1600 Clifton Road NE
 Atlanta, GA 30333
 (404) 639-3311
 www.cdc.gov
- The Food Allergy Network
 10400 Eaton Place, Suite 107
 Fairfax, VA 22030-2208
 (800) 929-4040 or (703) 691-3179
 www.foodallergy.org
- Internet Health Resources
 www.ihr.com
- National AIDS Hotline (CDC)
 (800) 342-AIDS (English)
 (800) 344-SIDA (Spanish)
 (800) 2437-TTY (Deaf)
 (900) 820-2437
- National Cancer Institute
 Office of Cancer Communications
 Building 31, Room 10824
 Bethesda, MD 20892
 (800) 4-CANCER or (800) 422-6237
 www.nci.nih.gov
- National Diabetes Information Clearinghouse
 1 Information Way
 Bethesda, MD 20892-3560
 (301) 654-3327
 www.niddk.nih.gov
- National Digestive Disease Information Clearinghouse (NDDIC)
 2 Information Way
 Bethesda, MD 20892-3570
 (301) 654-3810
 www.niddk.nih.gov
- National Health Information Center (NHIC)
 Office of Disease Prevention and Health Promotion
 (800) 336-4797
 nhic-nt.health.org
- National Heart, Lung, and Blood Institute Information Center
 P.O. Box 30105
 Bethesda, MD 20824-0105
 (301) 251-1222
 www.nhlbi.nih.gov/nhlbi/nhlbi.htm
- National Institute of Allergy and Infectious Diseases
 Office of Communications
 Building 31, Room 7A50
 31 Center Drive, MSC2520
 Bethesda, MD 20892-2520
 (301) 496-5717
 www.niaid.nih.gov
- National Institute of Dental Research (NIDR)
 National Institute of Health
 Bethesda, MD 20892-2190
 (301) 496-4261
 www.nidr.nih.gov
- National Institutes of Health (NIH)
 9000 Rockville Pike
 Bethesda, MD 20892
 (301) 496-2433
 www.nih.gov
- National Osteoporosis Foundation
 1150 17th Street NW, Suite 500
 Washington, DC 20036
 (202) 223-2226
 www.nof.org
- Office of Disease Prevention and Health Promotion
 odphp.osophs.dhhs.gov
- Office on Smoking and Health (OSH)
 www.americanheart.org/heart.org/Heart_and_stroke_A_Z_Guide/osh.html

## Infancy and Childhood

- American Academy of Pediatrics
 141 Northwest Point Boulevard
 Elk Grove Village, IL 60007-1098
 (847) 228-5005
 www.aap.org
- Association of Birth Defect Children, Inc.
 930 Woodcock Road, Suite 225
 Orlando, FL 32803
 (407) 245-7035
 www.birthdefects.org
- Canadian Paediatric Society
 100-2204 Walkley Road
 Ottawa, ON K1G 4G8
 (613) 526-9397; fax (613) 526-3332
 www.cps.ca
- National Center for Education in Maternal & Child Health
 2000 15th Street North, Suite 701
 Arlington, VA 22201-2617
 (703) 524-7802
 www.ncemch.org

## Pregnancy and Lactation

- American College of Obstetricians and Gynecologists Resource Center
 409 12th Street SW
 Washington, DC 20024-2188
 (202) 638-5577
 www.acog.org
- La Leche International, Inc.
 1400 N. Meacham Road
 Schaumburg, IL 60173
 (847) 519-7730
 www.lalecheleague.org
- March of Dimes Birth Defects Foundation
 1275 Mamaroneck Avenue
 White Plains, NY 10605
 (914) 428-7100
 www.modimes.org

## Trade and Industry Organizations

- Beech-Nut Nutrition Corporation
 P.O. 618
 St. Louis, MO 63188-0618
 (800) 523-6633
 www.beechnut.com
- Borden Inc.
 180 East Broad Street
 Columbus, OH 43215
 (800) 426-7336
- Campbell Soup Company
 Consumer Response Center
 Campbell Place, Box 26B
 Camden, NJ 08103-1701
 (800) 257-8443
 www.campbellssoup.com
- Elan Pharma/Hi Chem Diagnostics
 2 Thurber Boulevard
 Smithfield, RI 02917
 (401) 233-3526
 www.hi-chem.com
- General Mills, Inc.
 Number One General Mills Boulevard
 Minneapolis, MN 55426
 (800) 328-6787
 www.generalmills.com
- Hoffmann-LaRoche, Inc.
 340 Kingsland Street
 Nutley, NJ 07110
 (973) 235-5000

- Kellogg Company
  P.O. Box 3599
  Battle Creek, MI 49016-3599
  (616) 961-2000
  www.kelloggs.com
- Kraft Foods
  Consumer Response and Information Center
  One Kraft Court
  Glenview, IL 60025
  (800) 323-0768
  www.kraftfoods.com
- Mead Johnson Nutritionals
  2400 West Lloyd Expressway
  Evansville, IN 47721
  (800) 247-7893
  www.meadjohnson.com
- Nabisco Consumer Affairs
  100 DeForest Avenue
  East Hanover, NJ 07936
  (800) NABISCO or (800) 932-7800
  www.nabisco.com
- National Dairy Council
  10255 West Higgins Road, Suite 900
  Rosemond, IL 60018-5616
  (847) 803-2000
  www.dairyinfo.com
- NutraSweet/KELCO
  P.O. Box 2986
  Chicago, IL 60654-0986
  www.equal.com
- Pillsbury Company
  Consumer Relations
  P.O. Box 550
  Minneapolis, MN 55440
  (800) 767-4466
  www.pillsbury.com
- Procter and Gamble Company
  One Procter and Gamble Plaza
  Cincinnati, OH 45202
  (513) 983-1100
  www.pg.com/info
- Ross Laboratories, Abbot Laboratory
  625 Cleveland Avenue
  Columbus, OH 43215
  (800) 227-5767
  www.abbot.com
- Sherwood Medical
  1915 Olive Street
  St. Louis, MO 63103
  (800) 428-4400
- Sunkist Growers
  Consumer Affairs
  Fresh Fruit Division
  14130 Riverside Drive
  Sherman Oaks, CA 91423
  (800) CITRUS-5 or (800) 248-7875
  www.sunkist.com
- United Fresh Fruit and Vegetable Association
  727 North Washington Street
  Alexandria, VA 22314
  (703) 836-3410
- USA Rice Federation
  4301 North Fairfax Drive, Suite 305
  Arlington, VA 22203
  Phone: (703) 351-8161
  www.usarice.com
- Weight Watchers International, Inc.
  Consumer Affairs Department/IN
  175 Crossways Park West
  Woodbury, NY 11797
  (516) 390-1400; fax (516) 390-1632.
  www.weightwatchers.com

## Weight Control and Eating Disorders

- American Anorexia & Bulimia Association, Inc.
  165 West 46th Street #1108
  New York, NY 10036
  (212) 575-6200
  members.aol.com/amanbu
- Anorexia Nervosa and Related Eating Disorders (ANRED)
  P.O. Box 5102
  Eugene, OR 97405
  (541) 344-1144
  www.anred.com
- National Association of Anorexia Nervosa and Associated Disorders, Inc. (ANAD)
  P.O. Box 7
  Highland Park, IL 60035
  (847) 831-3438
  members.aol.com/anad20/index.html
- National Eating Disorder Information Centre
  200 Elizabeth Street, College Wing 1-304
  Toronto, Ontario M5G 2C4
  (519) 253-7421; fax (519) 253-7545
- Overeaters Anonymous (OA)
  World Service Office
  6075 Zenith Court NE
  Rio Rancho, NM 87124
  (505) 891-2664; fax (505) 891-4320
  www.overeatersanonymous.org
- TOPS (Take Off Pounds Sensibly)
  4575 South Fifth Street
  P.O. Box 07360
  Milwaukee, WI 53207-0360
  (800) 932-8677 or (414) 482-4620
  www.tops.org

## World Hunger

- Bread for the World
  1100 Wayne Avenue, Suite 1000
  Silver Spring, MD 20910
  (301) 608-2400
  www.bread.org
- Center on Hunger, Poverty and Nutrition Policy
  Tufts University School of Nutrition
  11 Curtis Avenue
  Medford, MA 02155
  (617) 627-3956
- Freedom from Hunger
  P.O. Box 2000
  1644 DaVinci Court
  Davis, CA 95617
  (530) 758-6200
  www.freefromhunger.org
- Oxfam America
  26 West Street
  Boston, MA 02111
  (617) 482-1211
  www.oxfamamerica.org
- SEEDS Magazine
  P.O. Box 6170
  Waco, TX 76706
  (254) 755-7745
  www.helwys.com/seedhome.htm
- Worldwatch Institute
  1776 Massachusetts Avenue NW, Suite 800
  Washington, DC 20036
  (202) 452-1999
  www.worldwatch.org

Contents



G

# UNITED STATES: RECOMMENDATIONS AND EXCHANGES

# WORLD HEALTH ORGANIZATION: RECOMMENDATIONS

Chapters 1 and 2 introduced the 1989 Recommended Dietary Allowances (RDA), Healthy People 2000, and exchange systems; this appendix provides additional details. (See Appendix I for Canada's nutrition recommendations and exchange system.) Nutrition recommendations from the World Health Organization are also included.

## RDA

As Chapter 1 mentioned, a major revision of the nutrient recommendations is underway in both the United States and Canada. The Dietary Reference Intakes (DRI) reports are replacing the 1989 RDA in the United States and the 1991 RNI in Canada. Recommendations from the DRI reports are presented on the inside front covers. For nutrients that do not yet have new values, the RDA will continue to serve health professionals in the United States (Tables G-1, G-2, and G-3).

**Table G-1**

**Estimated Safe and Adequate Daily Dietary Intakes of Additional Selected Minerals (United States)[a]**

| Age (yr) | Chromium (μg) | Molybdenum (μg) | Copper (mg) | Manganese (mg) |
|---|---|---|---|---|
| **Infants** | | | | |
| 0.0–0.5 | 10–40 | 15–30 | 0.4–0.6 | 0.3–0.6 |
| 0.5–1 | 20–60 | 20–40 | 0.6–0.7 | 0.6–1.0 |
| **Children** | | | | |
| 1–3 | 20–80 | 25–50 | 0.7–1.0 | 1.0–1.5 |
| 4–6 | 30–120 | 30–75 | 1.0–1.5 | 1.5–2.0 |
| 7–10 | 50–200 | 50–150 | 1.0–2.0 | 2.0–3.0 |
| 11+ | 50–200 | 75–250 | 1.5–2.5 | 2.0–5.0 |
| **Adults** | 50–200 | 75–250 | 1.5–3.0 | 2.0–5.0 |

[a]Less information is available on which to base allowances for these nutrients. Therefore, they are not included in the main table of the RDA, and the figures provided here are in the form of ranges of recommended intakes; the toxic levels for many trace elements may be only several times usual intakes, so the upper levels for the trace elements given in this table should not be habitually exceeded.

Source: *Recommended Dietary Allowances,* © 1989 by the National Academy of Sciences, National Academy Press, Washington, D.C.

**Table G-2**
**Estimated Minimum Requirements of Sodium, Chloride, and Potassium**

| Age (yr) | Weight (kg) | Sodium[a] (mg) | Chloride (mg) | Potassium[b] (mg) |
|---|---|---|---|---|
| **Infants** | | | | |
| 0.0–0.5 | 4.5 | 120 | 180 | 500 |
| 0.5–1.0 | 8.9 | 200 | 300 | 700 |
| **Children** | | | | |
| 1 | 11.0 | 225 | 350 | 1000 |
| 2–5 | 16.0 | 300 | 500 | 1400 |
| 6–9 | 25.0 | 400 | 600 | 1600 |
| **Adolescents** | 50.0 | 500 | 750 | 2000 |
| **Adults** | 70.0 | 500 | 750 | 2000 |

[a]Sodium requirements are based on estimates of needs for growth and for replacement of obligatory losses. They cover a wide variation of physical activity patterns and climatic exposure but do not provide for large, prolonged losses from the skin through sweat.
[b]Dietary potassium may benefit the prevention and treatment of hypertension, and recommendations to include many servings of fruits and vegetables would raise potassium intakes to about 3500 milligrams per day.
Source: *Recommended Dietary Allowances,* © 1989 by the National Academy of Sciences, National Academy Press, Washington, D.C.

**Table G-3**
**Median Heights and Weights and Recommended Energy Intakes (United States)**

| | Weight | | Height | | Average Energy Allowance | | | |
|---|---|---|---|---|---|---|---|---|
| Age (yr) | kg | lb | cm | in | REE[a] (kcal/day) | Multiples of REE[b] | kcal/kg | kcal/day[c] |
| **Infants** | | | | | | | | |
| 0.0–0.5 | 6 | 13 | 60 | 24 | 320 | | 108 | 650 |
| 0.5–1.0 | 9 | 20 | 71 | 28 | 500 | | 98 | 850 |
| **Children** | | | | | | | | |
| 1–3 | 13 | 29 | 90 | 35 | 740 | | 102 | 1300 |
| 4–6 | 20 | 44 | 112 | 44 | 950 | | 90 | 1800 |
| 7–10 | 28 | 62 | 132 | 52 | 1130 | | 70 | 2000 |
| **Males** | | | | | | | | |
| 11–14 | 45 | 99 | 157 | 62 | 1440 | 1.70 | 55 | 2500 |
| 15–18 | 66 | 145 | 176 | 69 | 1760 | 1.67 | 45 | 3000 |
| 19–24 | 72 | 160 | 177 | 70 | 1780 | 1.67 | 40 | 2900 |
| 25–50 | 79 | 174 | 176 | 70 | 1800 | 1.60 | 37 | 2900 |
| 51+ | 77 | 170 | 173 | 68 | 1530 | 1.50 | 30 | 2300 |
| **Females** | | | | | | | | |
| 11–14 | 46 | 101 | 157 | 62 | 1310 | 1.67 | 47 | 2200 |
| 15–18 | 55 | 120 | 163 | 64 | 1370 | 1.60 | 40 | 2200 |
| 19–24 | 58 | 128 | 164 | 65 | 1350 | 1.60 | 38 | 2200 |
| 25–50 | 63 | 138 | 163 | 64 | 1380 | 1.55 | 36 | 2200 |
| 51+ | 65 | 143 | 160 | 63 | 1280 | 1.50 | 30 | 1900 |
| **Pregnant** (2nd and 3rd trimesters) | | | | | | | | +300 |
| **Lactating** | | | | | | | | +500 |

[a]REE (resting energy expenditure) represents the energy expended by a person at rest under normal conditions.
[b]Recommended energy allowances assume light-to-moderate activity and were calculated by multiplying the REE by an activity factor.
[c]Average energy allowances have been rounded.
Source: *Recommended Dietary Allowances,* © 1989 by the National Academy of Sciences, National Academy Press, Washington, D.C.

G

# Healthy People 2000

In 1990, the U.S. Department of Health and Human Services established a set of almost 300 health objectives for the nation called Healthy People 2000.[1] The 21 objectives that have a nutrition component were presented throughout this text wherever their topic was discussed. Table G-4 presents them in full.

**Table G-4**
**Healthy People 2000 Nutrition Objectives**

**Health-Related Objectives**

- Reduce coronary heart disease deaths to no more than 100 per 100,000 people.
- Reverse the rise in cancer deaths to achieve a rate of no more than 130 per 100,000 people.
- Reduce overweight to a prevalence of no more than 20% among people aged 20 years and older and maintain prevalence at no more than 15% among adolescents aged 12 through 19 years.
- Reduce growth retardation among low-income children aged five years and younger to less than 10%.

**Nutrient Intake Objectives**

- Reduce dietary fat intake to an average of 30% of energy or less and average saturated fat intake to less than 10% of energy among people aged two years and older.
- Increase complex carbohydrate and fiber-containing foods in the diets of adults to five or more daily servings for vegetables (including legumes) and fruits and to six or more daily servings for grain products.
- Increase to at least 50% the proportion of overweight people aged 12 years and older who have adopted sound dietary practices combined with regular physical activity to attain an appropriate body weight.
- Increase calcium intake, so that at least 50% of youth aged 12 through 24 years and 50% of pregnant and lactating women consume three or more servings of calcium-rich foods daily and at least 50% of people aged 25 years and older consume two or more servings of calcium-rich foods daily.
- Decrease salt and sodium intake so at least 65% of home meal preparers prepare foods without adding salt, at least 80% of people avoid using salt at the table, and at least 40% of adults regularly purchase foods modified or lower in sodium.
- Reduce iron deficiency to less than 3% among children aged 1 to 4 and women of childbearing age.
- Increase to at least 75% the proportion of mothers who breastfeed their babies in the early weeks and to at least 50% the proportion who continue breastfeeding until their babies are five to six months old.
- Increase to at least 75% the proportion of parents and caregivers who use feeding practices that prevent nursing bottle tooth decay.
- Increase to at least 85% the proportion of people aged 18 and older who use food labels to make nutritious food selections.

**Services and Information Objectives**

- Achieve useful and informative nutrition labeling for virtually all processed foods and at least 40% of fresh meats, poultry, fish, fruits, vegetables, baked goods, and ready-to-eat carry-away foods.
- Increase to at least 5000 brand items the number of processed food products that are reduced in fat and saturated fat.
- Increase to at least 90% the proportion of restaurants and institutional foodservice operations that offer identifiable low-fat, low-kcalorie food choices, consistent with the *Dietary Guidelines for Americans.*
- Increase to at least 90% the proportion of school lunch and breakfast services and increase to at least 50% the proportion of child care foodservices with menus that are consistent with the nutrition principles in the *Dietary Guidelines for Americans.*
- Increase to at least 80% the receipt of home foodservices by people aged 65 and older who have difficulty in preparing their own meals or are otherwise in need of home-delivered meals.
- Increase to at least 75% the proportion of the nation's schools that provide nutrition education from preschool through grade 12, preferably as part of quality school health education.
- Increase to at least 50% the proportion of worksites with 50 or more employees that offer nutrition education and/or weight management programs for employees.
- Increase to at least 75% the proportion of primary care providers who provide nutrition assessment and counseling and/or referral to qualified nutritionists or dietitians.

# Exchange Lists for Meal Planning

The U.S. exchange system groups together foods that have about the same amount of carbohydrate, protein, fat, and kcalories. Then any food on a list can be "exchanged" for any other food on that same list. Chapter 2 introduced the exchange lists and Tables G-5 through G-13 present the lists in detail.

**Table G-5**
**U.S. Exchange System: Starch List**

1 starch exchange = 15 g carbohydrate, 3 g protein, 0–1 g fat, and 80 kcal

NOTE: In general, a starch serving is ½ c cereal, grain, pasta, or starchy vegetable; 1 oz of bread; ¾ to 1 oz snack food.

| Serving Size | Food |
|---|---|
| **Bread** | |
| ½ (1 oz) | Bagels |
| 2 slices (1½ oz) | Bread, reduced-kcalorie |
| 1 slice (1 oz) | Bread, white (including French and Italian), whole-wheat, pumpernickel, rye |
| 2 (⅔ oz) | Bread sticks, crisp, 40 x ½″ |
| ½ | English muffins |
| ½ (1 oz) | Hot dog or hamburger buns |
| ½ | Pita, 6″ across |
| 1 (1 oz) | Plain rolls, small |
| 1 slice (1 oz) | Raisin bread, unfrosted |
| 1 | Tortillas, corn, 6″ across |
| 1 | Tortillas, flour, 7–8″ across |
| 1 | Waffles, 4½″ square, reduced-fat |
| **Cereals and Grains** | |
| ½ c | Bran cereals |
| ½ c | Bulgur, cooked |
| ½ c | Cereals, cooked |
| ¾ c | Cereals, unsweetened, ready-to-eat |
| 3 tbs | Cornmeal (dry) |
| ⅓ c | Couscous |
| 3 tbs | Flour (dry) |
| ¼ c | Granola, low-fat |
| ¼ c | Grape nuts |
| ½ c | Grits, cooked |
| ½ c | Kasha |
| ¼ c | Millet |
| ¼ c | Muesli |
| ½ c | Oats |
| ½ c | Pasta, cooked |
| 1½ c | Puffed cereals |
| ½ c | Rice milk |
| ⅓ c | Rice, white or brown, cooked |
| ½ c | Shredded wheat |
| ½ c | Sugar-frosted cereal |
| 3 tbs | Wheat germ |
| **Starchy Vegetables** | |
| 1/3 c | Baked beans |
| ½ c | Corn |
| 1 (5 oz) | Corn on cob, medium |
| 1 c | Mixed vegetables with corn, peas, or pasta |
| ½ c | Peas, green |
| ½ c | Plantains |
| 1 small (3 oz) | Potatoes, baked or boiled |
| ½ c | Potatoes, mashed |
| 1 c | Squash, winter (acorn, butternut) |
| ½ c | Yams, sweet potatoes, plain |
| **Crackers and Snacks** | |
| 8 | Animal crackers |
| 3 | Graham crackers, 2½″ square |
| ¾ oz | Matzoh |
| 4 slices | Melba toast |
| 24 | Oyster crackers |
| 3 c | Popcorn (popped, no fat added or low-fat microwave) |
| ¾ oz | Pretzels |
| 2 | Rice cakes, 4″ across |
| 6 | Saltine-type crackers |
| 15–2″ (¾ oz) | Snack chips, fat-free (tortilla, potato) |
| 2–5 (¾ oz) | Whole-wheat crackers, no fat added |
| **Dried Beans, Peas, and Lentils** | |
| ½ c | Beans and peas, cooked (garbanzo, lentils, pinto, kidney, white, split, black-eyed) |
| ⅔ c | Lima beans |
| 3 tbs | Miso |
| **Starchy Foods Prepared with Fat** | |
| **Count as 1 starch + 1 fat exchange.** | |
| 1 | Biscuit, 2½″ across |
| ½ c | Chow mein noodles |
| 1 (2 oz) | Corn bread, 2″ cube |
| 6 | Crackers, round butter type |
| 1 c | Croutons |
| 16–25 (3 oz) | French-fried potatoes |
| ¼ c | Granola |
| 1 (1½ oz) | Muffin, small |
| 2 | Pancake, 4″ across |
| 3 c | Popcorn, microwave |
| 3 | Sandwich crackers, cheese or peanut butter filling |
| ⅓ c | Stuffing, bread (prepared) |
| 2 | Taco shell, 6″ across |
| 1 | Waffle, 4½″ square |
| 4–6 (1 oz) | Whole-wheat crackers, fat added |

 = 400 mg or more sodium per exchange.

**Table G-6**
**U.S. Exchange System: Fruit List**

1 fruit exchange = 15 g carbohydrate and 60 kcal

NOTE: In general, a fruit serving is 1 small to medium fresh fruit; ½ c canned or fresh fruit or fruit juice; ¼ c dried fruit.

| Serving Size | Food |
|---|---|
| 1 (4 oz) | Apples, unpeeled, small |
| ½ c | Applesauce, unsweetened |
| 4 rings | Apples, dried |
| 4 whole (5½ oz) | Apricots, fresh |
| 8 halves | Apricots, dried |
| ½ c | Apricots, canned |
| 1 (4 oz) | Bananas, small |
| ¾ c | Blackberries |
| ¾ c | Blueberries |
| ⅓ melon (11 oz) or 1 c cubes | Cantaloupe, small |
| 12 (3 oz) | Cherries, sweet, fresh |
| ½ c | Cherries, sweet, canned |
| 3 | Dates |
| 1½ large or 2 medium (3½ oz) | Figs, fresh |
| 1½ | Figs, dried |
| ½ c | Fruit cocktail |
| ½ (11 oz) | Grapefruit, large |
| ¾ c | Grapefruit sections, canned |
| 17 (3 oz) | Grapes, small |
| 1 slice (10 oz) or 1 c cubes | Honeydew melon |
| 1 (3½ oz) | Kiwi |
| ¾ c | Mandarin oranges, canned |
| ½ (5½ oz) or ½ c | Mangoes, small |
| 1 (5 oz) | Nectarines, small |
| 1 (6½ oz) | Oranges, small |
| ½ (8 oz) or 1 c cubes | Papayas |
| 1 (6 oz) | Peaches, medium, fresh |
| ½ c | Peaches, canned |
| ½ (4 oz) | Pears, large, fresh |
| ½ c | Pears, canned |
| ¾ c | Pineapple, fresh |
| ½ c | Pineapple, canned |
| 2 (5 oz) | Plums, small |
| ½ c | Plums, canned |
| 3 | Prunes, dried |
| 2 tbs | Raisins |
| 1 c | Raspberries |
| 1¼ c whole berries | Strawberries |
| 2 (8 oz) | Tangerines, small |
| 1 slice (13½ oz) or 1¼ c cubes | Watermelon |
| **Fruit Juice** | |
| ½ c | Apple juice/cider |
| ⅓ c | Cranberry juice cocktail |
| 1 c | Cranberry juice cocktail, reduced-kcalorie |
| ⅓ c | Fruit juice blends, 100% juice |
| ⅓ c | Grape juice |
| ½ c | Grapefruit juice |
| ½ c | Orange juice |
| ½ c | Pineapple juice |
| ⅓ c | Prune juice |

**Table G-7**
**U.S. Exchange System: Milk List**

| Serving Size | Food |
|---|---|
| **Nonfat and Low-Fat Milk** | |
| 1 nonfat/low-fat milk exchange = 12 g carbohydrate, 8 g protein, 0–3 g fat, 90 kcal | |
| 1 c | Nonfat milk |
| 1 c | ½% milk |
| 1 c | 1% milk |
| 1 c | Nonfat or low-fat buttermilk |
| ½ c | Evaporated nonfat milk |
| ⅓ c dry | Dry nonfat milk |
| ¾ c | Plain nonfat yogurt |
| 1 c | Nonfat or low-fat fruit-flavored yogurt sweetened with aspartame or with a nonnutritive sweetener |
| **Reduced-Fat Milk** | |
| 1 reduced-fat milk exchange = 12 g carbohydrate, 8 g protein, 5 g fat, 120 kcal | |
| 1 c | 2% milk |
| ¾ c | Plain low-fat yogurt |
| 1 c | Sweet acidophilus milk |
| **Whole Milk** | |
| 1 whole milk exchange = 12 g carbohydrate, 8 g protein, 8 g fat, 150 kcal | |
| 1 c | Whole milk |
| ½ c | Evaporated whole milk |
| 1 c | Goat's milk |
| 1 c | Kefir |

**Table G-8**

**U.S. Exchange System: Other Carbohydrates List**

1 other carbohydrate exchange = 15 g carbohydrate, or 1 starch, or 1 fruit, or 1 milk exchange

| Food | Serving Size | Exchanges per Serving |
|---|---|---|
| Angel food cake, unfrosted | 1⁄12 cake | 2 carbohydrates |
| Brownies, small, unfrosted | 20 square | 1 carbohydrate, 1 fat |
| Cake, unfrosted | 20 square | 1 carbohydrate, 1 fat |
| Cake, frosted | 20 square | 2 carbohydrates, 1 fat |
| Cookie, fat-free | 2 small | 1 carbohydrate |
| Cookies or sandwich cookies | 2 small | 1 carbohydrate, 1 fat |
| Cupcakes, frosted | 1 small | 2 carbohydrates, 1 fat |
| Cranberry sauce, jellied | ¼ c | 2 carbohydrates |
| Doughnuts, plain cake | 1 medium, (1½ oz) | 1½ carbohydrates, 2 fats |
| Doughnuts, glazed | 33⁄40 across (2 oz) | 2 carbohydrates, 2 fats |
| Fruit juice bars, frozen, 100% juice | 1 bar (3 oz) | 1 carbohydrate |
| Fruit snacks, chewy (pureed fruit concentrate) | 1 roll (¾ oz) | 1 carbohydrate |
| Fruit spreads, 100% fruit | 1 tbs | 1 carbohydrate |
| Gelatin, regular | ½ c | 1 carbohydrate |
| Gingersnaps | 3 | 1 carbohydrate |
| Granola bars | 1 bar | 1 carbohydrate, 1 fat |
| Granola bars, fat-free | 1 bar | 2 carbohydrates |
| Hummus | ⅓ c | 1 carbohydrate, 1 fat |
| Ice cream | ½ c | 1 carbohydrate, 2 fats |
| Ice cream, light | ½ c | 1 carbohydrate, 1 fat |
| Ice cream, fat-free, no sugar added | ½ c | 1 carbohydrate |
| Jam or jelly, regular | 1 tbs | 1 carbohydrate |
| Milk, chocolate, whole | 1 c | 2 carbohydrates, 1 fat |
| Pie, fruit, 2 crusts | ⅙ pie | 3 carbohydrates, 2 fats |
| Pie, pumpkin or custard | ⅛ pie | 1 carbohydrate, 2 fats |
| Potato chips | 12–18 (1 oz) | 1 carbohydrate, 2 fats |
| Pudding, regular (made with low-fat milk) | ½ c | 2 carbohydrates |
| Pudding, sugar-free (made with low-fat milk) | ½ c | 1 carbohydrate |
| Salad dressing, fat-free | ¼ c | 1 carbohydrate |
| Sherbet, sorbet | ½ c | 2 carbohydrates |
| Spaghetti or pasta sauce, canned | ½ c | 1 carbohydrate, 1 fat |
| Sweet roll or danish | 1 (2½ oz) | 2½ carbohydrates, 2 fats |
| Syrup, light | 2 tbs | 1 carbohydrate |
| Syrup, regular | 1 tbs | 1 carbohydrate |
| Syrup, regular | ¼ c | 4 carbohydrates |
| Tortilla chips | 6–12 (1 oz) | 1 carbohydrate, 2 fats |
| Yogurt, frozen, low-fat, fat-free | ⅓ c | 1 carbohydrate, 0–1 fat |
| Yogurt, frozen, fat-free, no sugar added | ½ c | 1 carbohydrate |
| Yogurt, low-fat with fruit | 1 c | 3 carbohydrates, 0–1 fat |
| Vanilla wafers | 5 | 1 carbohydrate, 1 fat |

= 400 mg or more sodium per exchange.

**Table G-9**

**U.S. Exchange System: Vegetable List**

1 vegetable exchange = 5 g carbohydrate, 2 g protein, 25 kcal

NOTE: In general, a vegetable serving is ½ c cooked vegetables or vegetable juice; 1 c raw vegetables. Starchy vegetables such as corn, peas, and potatoes are on the starch list.

| | |
|---|---|
| Artichokes | Mushrooms |
| Artichoke hearts | Okra |
| Asparagus | Onions |
| Beans (green, wax, Italian) | Pea pods |
| Bean sprouts | Peppers (all varieties) |
| Beets | Radishes |
| Broccoli | Salad greens (endive, escarole, lettuce, romaine, spinach) |
| Brussels sprouts | Sauerkraut (salt shaker symbol) |
| Cabbage | Spinach |
| Carrots | Summer squash (crookneck) |
| Cauliflower | Tomatoes |
| Celery | Tomatoes, canned |
| Cucumbers | Tomato sauce (salt shaker symbol) |
| Eggplant | Tomato/vegetable juice (salt shaker symbol) |
| Green onions or scallions | Turnips |
| Greens (collard, kale, mustard, turnip) | Water chestnuts |
| Kohlrabi | Watercress |
| Leeks | Zucchini |
| Mixed vegetables (without corn, peas, or pasta) | |

(salt shaker symbol) = 400 mg or more sodium per exchange.

**Table G-10**
**U.S. Exchange System: Meat and Meat Substitutes List**

NOTE: In general, a meat serving is 1 oz meat, poultry, or cheese; ½ c dried beans (weigh meat and poultry and measure beans after cooking).

| Serving Size | Food |
|---|---|
| **Very Lean Meat and Substitutes** | |
| 1 very lean meat exchange = 7 g protein, 0–1 g fat, 35 kcal | |
| 1 oz | Poultry: Chicken or turkey (white meat, no skin), Cornish hen (no skin) |
| 1 oz | Fish: Fresh or frozen cod, flounder, haddock, halibut, trout; tuna, fresh or canned in water |
| 1 oz | Shellfish: Clams, crab, lobster, scallops, shrimp, imitation shellfish |
| 1 oz | Game: Duck or pheasant (no skin), venison, buffalo, ostrich |
| | Cheese with ≤ 1 g fat/oz: |
| ¼ c | Nonfat or low-fat cottage cheese |
| 1 oz | Fat-free cheese |
| | Other: |
| 1 oz | Processed sandwich meats with ≤ 1 g fat/oz (such as deli thin, shaved meats, chipped beef, turkey ham) |
| 2 | Egg whites |
| ¼ c | Egg substitutes, plain |
| 1 oz | Hot dogs with ≤ 1 g fat/oz |
| 1 oz | Kidney (high in cholesterol) |
| 1 oz | Sausage with ≤ 1 g fat/oz |
| Count as one very lean meat and one starch exchange: | |
| ½ c | Dried beans, peas, lentils (cooked) |
| **Lean Meat and Substitutes** | |
| 1 lean meat exchange = 7 g protein, 3 g fat, 55 kcal | |
| 1 oz | Beef: USDA Select or Choice grades of lean beef trimmed of fat (round, sirloin, and flank steak); tenderloin; roast (rib, chuck, rump); steak (T-bone, porterhouse, cubed), ground round |
| 1 oz | Pork: Lean pork (fresh ham); canned, cured, or boiled ham; Canadian bacon; tenderloin, center loin chop |
| 1 oz | Lamb: Roast, chop, leg |
| 1 oz | Veal: Lean chop, roast |
| 1 oz | Poultry: Chicken, turkey (dark meat, no skin), chicken white meat (with skin), domestic duck or goose (well-drained of fat, no skin) |
| | Fish: |
| 1 oz | Herring (uncreamed or smoked) |
| 6 medium | Oysters |
| 1 oz | Salmon (fresh or canned), catfish |
| 2 medium | Sardines (canned) |
| 1 oz | Tuna (canned in oil, drained) |
| 1 oz | Game: Goose (no skin), rabbit |
| | Cheese: |
| ¼ c | 4.5%-fat cottage cheese |
| 2 tbs | Grated Parmesan |
| 1 oz | Cheeses with ≤ 3 g fat/oz |
| | Other: |
| 1½ oz | Hot dogs with ≤ 3 g fat/oz |
| 1 oz | Processed sandwich meat with ≤ 3 g fat/oz (turkey pastrami or kielbasa) |
| 1 oz | Liver, heart (high in cholesterol) |
| **Medium-Fat Meat and Substitutes** | |
| 1 medium-fat meat exchange = 7 g protein, 5 g fat, and 75 kcal | |
| 1 oz | Beef: Most beef products (ground beef, meatloaf, corned beef, short ribs, Prime grades of meat trimmed of fat, such as prime rib) |
| 1 oz | Pork: Top loin, chop, Boston butt, cutlet |
| 1 oz | Lamb: Rib roast, ground |
| 1 oz | Veal: Cutlet (ground or cubed, unbreaded) |
| 1 oz | Poultry: Chicken dark meat (with skin), ground turkey or ground chicken, fried chicken (with skin) |
| 1 oz | Fish: Any fried fish product |
| | Cheese with ≤ 5 g fat/oz: |
| 1 oz | Feta |
| 1 oz | Mozzarella |
| ¼ c (2 oz) | Ricotta |
| | Other: |
| 1 | Egg (high in cholesterol, limit to 3/week) |
| 1 oz | Sausage with ≤ 5 g fat/oz |
| 1 c | Soy milk |
| ¼ c | Tempeh |
| 4 oz or ½ c | Tofu |
| **High-Fat Meat and Substitutes** | |
| 1 high-fat meat exchange = 7 g protein, 8 g fat, 100 kcal | |
| 1 oz | Pork: Spareribs, ground pork, pork sausage |
| 1 oz | Cheese: All regular cheeses (American, cheddar, Monterey Jack, swiss) |
| | Other: |
| 1 oz | Processed sandwich meats with ≤ 8 g fat/oz (bologna, pimento loaf, salami) |
| 1 oz | Sausage (bratwurst, Italian, knockwurst, Polish, smoked) |
| 1 (10/lb) | Hot dog (turkey or chicken) |
| 3 slices (20 slices/lb) | Bacon |
| Count as one high-fat meat plus one fat exchange: | |
| 1 (10/lb) | Hot dog (beef, pork, or combination) |
| 2 tbs | Peanut butter (contains unsaturated fat) |

= 400 mg or more sodium per exchange.

**Table G-11**

**U.S. Exchange System: Fat List**

1 fat exchange = 5 g fat, 45 kcal

NOTE: In general, a fat serving is 1 tsp regular butter, margarine, or vegetable oil; 1 tbs regular salad dressing. Many fat-free and reduced fat foods are on the Free Foods List.

| Serving Size | Food |
|---|---|
| **Monounsaturated Fats** | |
| ⅛ medium (1 oz) | Avocadoes |
| 1 tsp | Oil (canola, olive, peanut) |
| 8 large | Olives, ripe (black) |
| 10 large | Olives, green, stuffed 🖉 |
| 6 nuts | Almonds, cashews |
| 6 nuts | Mixed nuts (50% peanuts) |
| 10 nuts | Peanuts |
| 4 halves | Pecans |
| 2 tsp | Peanut butter, smooth or crunchy |
| 1 tbs | Sesame seeds |
| 2 tsp | Tahini paste |
| **Polyunsaturated Fats** | |
| 1 tsp | Margarine, stick, tub, or squeeze |
| 1 tbs | Margarine, lower-fat (30% to 50% vegetable oil) |
| 1 tsp | Mayonnaise, regular |
| 1 tbs | Mayonnaise, reduced-fat |
| 4 halves | Nuts, walnuts, English |
| 1 tsp | Oil (corn, safflower, soybean) |
| 1 tbs | Salad dressing, regular |
| 2 tbs | Salad dressing, reduced-fat |
| 2 tsp | Mayonnaise type salad dressing, regular 🖉 |
| 1 tbs | Mayonnaise type salad dressing, reduced-fat |
| 1 tbs | Seeds: pumpkin, sunflower |
| **Saturated Fats*** | |
| 1 slice (20 slices/lb) | Bacon, cooked |
| 1 tsp | Bacon, grease |
| 1 tsp | Butter, stick |
| 2 tsp | Butter, whipped |
| 1 tbs | Butter, reduced-fat |
| 2 tbs (½ oz) | Chitterlings, boiled |
| 2 tbs | Coconut, sweetened, shredded |
| 2 tbs | Cream, half and half |
| 1 tbs (½ oz) | Cream cheese, regular |
| 2 tbs (1 oz) | Cream cheese, reduced-fat |
| | Fatback or salt pork† |
| 1 tsp | Shortening or lard |
| 2 tbs | Sour cream, regular |
| 3 tbs | Sour cream, reduced-fat |

🖉 = 400 mg or more sodium per exchange

*Saturated fats can raise blood cholesterol levels.

† Use a piece 1″ × 10 × ¼″ if you plan to eat the fatback cooked with vegetables. Use a piece 2″ × 1″ × ½″ when eating only the vegetables with the fatback removed.

**Table G-12**
**U.S. Exchange System: Free Foods List**

NOTE: A serving of free food contains less than 20 kcalories; those with serving sizes should be limited to 3 servings a day whereas those without serving sizes can be eaten freely.

| Serving Size | Food |
|---|---|
| **Fat-Free or Reduced-Fat Foods** | |
| 1 tbs | Cream cheese, fat-free |
| 1 tbs | Creamers, nondairy, liquid |
| 2 tsp | Creamers, nondairy, powdered |
| 1 tbs | Mayonnaise, fat-free |
| 1 tsp | Mayonnaise, reduced-fat |
| 4 tbs | Margarine, fat-free |
| 1 tsp | Margarine, reduced-fat |
| 1 tbs | Mayonnaise type salad dressing, nonfat |
| 1 tsp | Mayonnaise type salad dressing, reduced-fat |
| | Nonstick cooking spray |
| 1 tbs | Salad dressing, fat-free |
| 2 tbs | Salad dressing, fat-free, Italian |
| ¼ c | Salsa |
| 1 tbs | Sour cream, fat-free, reduced-fat |
| 2 tbs | Whipped topping, regular or light |
| **Sugar-Free or Low-Sugar Foods** | |
| 1 piece | Candy, hard, sugar-free |
| | Gelatin dessert, sugar-free |
| | Gelatin, unflavored |
| | Gum, sugar-free |
| 2 tsp | Jam or jelly, low-sugar or light |
| | Sugar substitutes |
| 2 tbs | Syrup, sugar-free |
| **Drinks** | |
| | Bouillon, broth, consommé 🧂 |
| | Bouillon or broth, low-sodium |
| | Carbonated or mineral water |
| 1 tbs | Cocoa powder, unsweetened |
| | Coffee |
| | Club soda |
| | Diet soft drinks, sugar-free |
| | Drink mixes, sugar-free |
| | Tea |
| | Tonic water, sugar-free |
| **Condiments** | |
| 1 tbs | Catsup |
| | Horseradish |
| | Lemon juice |
| | Lime juice |
| | Mustard |
| 1½ large | Pickles, dill 🧂 |
| | Soy sauce, regular or light 🧂 |
| 1 tbs | Taco sauce |
| | Vinegar |
| Seasonings | |
| Flavoring extracts | |
| Garlic | |
| Herbs, fresh or dried | |
| Pimento | |
| Spices | |
| Hot pepper sauces | |
| Wine, used in cooking | |
| Worcestershire sauce | |

🧂 = 400 mg or more sodium per exchange.

**Table G-13**
**U.S. Exchange System: Combination Foods List**

| Food | Serving Size | Exchanges per Serving |
|---|---|---|
| **Entrees** | | |
| Tuna noodle casserole, lasagna, spaghetti with meatballs, chili with beans, macaroni and cheese | 1 c (8 oz) | 2 carbohydrates, 2 medium-fat meats |
| Chow mein (without noodles or rice) | 2 c (16 oz) | 1 carbohydrate, 2 lean meats |
| Pizza, cheese, thin crust | ¼ of 100 (5 oz) | 2 carbohydrates, 2 medium-fat meats, 1 fat |
| Pizza, meat topping, thin crust | ¼ of 10″ (5 oz) | 2 carbohydrates, 2 medium-fat meats, 2 fats |
| Pot pie | 1 (7 oz) | 2 carbohydrates, 1 medium-fat meat, 4 fats |
| **Frozen entrees** | | |
| Salisbury steak with gravy, mashed potato | 1 (11 oz) | 2 carbohydrates, 3 medium-fat meats, 3–4 fats |
| Turkey with gravy, mashed potato, dressing | 1 (11 oz) | 2 carbohydrates, 2 medium-fat meats, 2 fats |
| Entree with less than 300 kcalories | 1 (8 oz) | 2 carbohydrates, 3 lean meats |
| **Soups** | | |
| Bean | 1 c | 1 carbohydrate, 1 very lean meat |
| Cream (made with water) | 1 c (8 oz) | 1 carbohydrate, 1 fat |
| Split pea (made with water) | ½ c (4 oz) | 1 carbohydrate |
| Tomato (make with water) | 1 c (8 oz) | 1 carbohydrate |
| Vegetable beef, chicken noodle, or other broth-type | 1 c (8 oz) | 1 carbohydrate |
| **Fast Foods** | | |
| Burritos with beef | 2 | 4 carbohydrates, 2 medium-fat meats, 2 fats |
| Chicken nuggets | 6 | 1 carbohydrate, 2 medium-fat meats, 1 fat |
| Chicken breast and wing, breaded and fried | 1 | 1 carbohydrate, 4 medium-fat meats, 2 fats |
| Fish sandwich/tartar sauce | 1 | 3 carbohydrates, 1 medium-fat meat, 3 fats |
| French fries, thin | 20–25 | 2 carbohydrates, 2 fats |
| Hamburger, regular | 1 | 2 carbohydrates, 2 medium-fat meats |
| Hamburger, large | 1 | 2 carbohydrates, 3 medium-fat meats, 1 fat |
| Hot dog with bun | 1 | 1 carbohydrate, 1 high-fat meat, 1 fat |
| Individual pan pizza | 1 | 5 carbohydrates, 3 medium-fat meats, 3 fats |
| Soft serve cone | 1 medium | 2 carbohydrates, 1 fat |
| Submarine sandwich | 1 (60) | 3 carbohydrates, 1 vegetable, 2 medium-fat meats, 1 fat |
| Taco, hard shell | 1 (6 oz) | 2 carbohydrates, 2 medium-fat meats, 2 fats |
| Taco, soft shell | 1 (3 oz) | 1 carbohydrate, 1 medium-fat meat, 1 fat |

= 400 mg or more sodium per exchange.

G

## Nutrition Recommendations from WHO

Like the Committee on Diet and Health in the United States, the World Health Organization (WHO) has also assessed the relationships between diet and the development of chronic diseases.[2] Their recommendations are expressed in average daily ranges that represent the lower and upper limits:

- Total energy: sufficient to support normal growth, physical activity, and body weight (body mass index = 20 to 22).
- Total fat: 15 to 30 percent of total energy.
  - Saturated fatty acids: 0 to 10 percent total energy.
  - Polyunsaturated fatty acids: 3 to 7 percent total energy.
  - Dietary cholesterol: 0 to 300 milligrams per day.
- Total carbohydrate: 55 to 75 percent total energy.
  - Complex carbohydrates: 50 to 75 percent total energy.
  - Dietary fiber: 27 to 40 grams per day.
  - Refined sugars: 0 to 10 percent total energy.
- Protein: 10 to 15 percent total energy.
- Salt: upper limit of 6 grams/day (no lower limit set).

## Notes

1. *Healthy People 2000: National Health Promotion and Disease Prevention Objectives* (Washington, D.C.: U.S. Department of Health and Human Services, 1990).
2. Diet, nutrition and the prevention of chronic diseases: A report of the WHO Study Group on Diet, Nutrition and Prevention of Noncommunicable Diseases, *Nutrition Reviews* 49 (1991): 291–301.

CONTENTS

*Table of Food Composition*

H

# TABLE OF FOOD COMPOSITION

This edition of the table of food composition contains more complete values for several nutrients than any comparable table.[1] These include dietary fiber; saturated, monounsaturated, and polyunsaturated fat; vitamin $B_6$; vitamin E; folate; magnesium; and zinc. The table includes a wide variety of foods from all food groups and is updated yearly to reflect current food patterns. For example, this edition includes many new nonfat items; several new ethnic items such as adzuki beans, tahitian taro, and gai choy chinese mustard; and a new selection of vegetarian foods.

• ***Sources of Data*** • To achieve a complete and reliable listing of nutrients for all the foods, over 1200 sources of information are researched. Government sources are the primary base for all data for most foods. In addition to USDA data (from Release 12 and surveys), provisional USDA information—both published and unpublished—is included.

Even with all the government sources available, however, some nutrient values are still missing; and as the USDA updates various data, it sometimes reports conflicting values for the same items. To fill in the missing values and resolve discrepancies, other reliable sources of information are used. These sources include journal articles, food composition tables from Canada and England, information from other nutrient data banks and publications, unpublished scientific data, and manufacturers' data.

The data for brand foods are listed as provided by the food manufacturers and the food chain restaurants. This information changes often because recipes and formulations are modified to meet consumer preferences, and the data are usually limited to those nutrients required for food labels. To provide more complete information, values for several nutrients are often estimated based on known values for major ingredients.

• ***Accuracy*** • The energy and nutrients in recipes and combination foods vary widely, depending on the ingredients. The amounts of various fatty acids and cholesterol are influenced by the type of fat used (the specific type of oil, vegetable shortening, butter, margarine, etc.).

Estimates of nutrient amounts for foods and nutrients include all possible adjustments in the interest of accuracy. When multiple values are reported for a nutrient, the numbers are averaged and weighted with consideration of the original number of analyses in the separate sources. Whenever water percentages are available, estimates of nutrient amounts are adjusted for water content. When no water is given, water percentage is assumed to be that shown in the table. Whenever a reported weight appeared inconsistent, many kitchen tests were made, and the average weight of the typical product was given as tested.

When estimates of nutrient amounts in cooked foods are derived from reported amounts in raw foods, published retention factors are applied. Data for combination foods are modified to include newer data for major ingredients.

Considerable effort has been made to report the most accurate data available. The table is revised annually, and the authors welcome any suggestions or comments for future editions.

• ***Average Values*** • It is important to know that many different nutrient values can be reported for foods, even by reliable sources. Many factors influence the amounts of nutrients in foods, including the mineral content of the soil, the method of processing, genetics, the diet of the animal or the fertilizer of the plant, the season of the year, methods of analysis, the difference in moisture content of the samples analyzed, the length and method of storage, and methods of cooking the food. The mineral content of water also varies according to the source.

Although each nutrient is presented as a single number, each number is actually an average of a range of data. More detailed reports from the USDA, for example, indicate the number of samples and the standard deviation of the data. One can also find different reported values for foods as older data are replaced with newer data from more recent analyses using newer analytical techniques. Therefore, nutrient data should be viewed and used only as a guide, a close approximation of nutrient content.

• ***Dietary Fiber*** • There can be many different reported values for dietary fiber in foods because information depends on the type of analytical technique used. The fiber data in this table are primarily from the USDA/ARS Human Nutrition Information Service in Hyattsville, Maryland; Composition of Foods by Southgate and Paul (England); and many journal articles.

• ***Vitamin A*** • Vitamin A is reported in retinol equivalents (RE). The amount of vitamin A can vary by the season of the year and the maturity of the plant. Reported values in both dairy products and plants are higher in summer and early fall than in winter. The values reported here represent year-round averages. The organ meats of all animal products (liver especially) contain large amounts of vitamin A, which vary widely, depending on the background of the animal. The vitamin is also present in very small amounts in regular meat and is often reported as a trace.

• ***Vitamin E*** • Vitamin E is actually a combination of various forms of this nutrient, and the measure of alpha tocopherol equivalents (α-TE) summarizes the activity of the various types of tocopherols and tocotrienols into one measure.

• ***Fats*** • Total fats, as well as the breakdown of total fats to saturated, monounsaturated, and polyunsaturated fats, are listed in the table. The fatty acids seldom add up to the total due to rounding and to other fatty acid components that are not included in these basic categories, such as *trans*-fatty acids and glycerol. *Trans*-fatty acids can comprise a large share of the total fat in margarine and shortening (hydrogenated oils) and in any foods that include them as ingredients.

• ***Enrichment-Fortification*** • The mandatory enrichment values for foods are presented as appropriate, including the new values for folate enrichment in grain products.

• ***Niacin*** • Niacin values are for preformed niacin and do not include additional niacin that may form in the body from the conversion of tryptophan.

• ***Using the Table*** • The items in this table have been organized into several categories, which are listed at the head of each right-hand page. As the key shows, each group has been color-coded to make it easier to find individual items.

In an effort to conserve space, the following abbreviations have been used in the food descriptions and nutrient breakdowns:

- diam = diameter
- ea = each
- enr = enriched
- f/ = from
- frzn = frozen
- g = grams
- liq = liquid
- pce = piece
- pkg = package
- w/ = with
- w/o = without
- t = trace
- 0 = zero (no nutrient value)
- blank space = information not available

• ***Caffeine Sources*** • Caffeine occurs in several plants, including the familiar coffee bean, the tea leaf, and the cocoa bean from which chocolate is made. Most human societies use caffeine regularly, most often in beverages, for its stimulant effect and flavor. Caffeine contents of beverages vary depending on the plants they are made from, the climates and soils where the plants are grown, the grind or cut size, the method and duration of brewing, and the amounts served. The accompanying table shows that in general, a cup of coffee contains the most caffeine; a cup of tea, less than half as much; and cocoa or chocolate, less still. As for cola beverages, they are made from kola nuts which contain caffeine, but most of their caffeine is added, using the purified compound obtained from decaffeinated coffee beans.

The FDA lists caffeine as a multipurpose GRAS substance that may be added to foods and beverages. Drug industries in developed countries use caffeine in many kinds of drugs: stimulants, pain relievers, cold remedies, diuretics, and weight-loss aids.

---

1. This food composition table has been prepared for West-Wadsworth Publishing Company and is copyrighted by ESHA Research in Salem, Oregon—the developer and publisher of the Food Processor®, Genesis® R&D, and the Computer Chef® nutrition software systems. The major sources for the data are from the USDA, supplemented by more than 1200 additional sources of information. Because the list of references is so extensive, it is not provided here, but is available from the publisher.

## Caffeine Content of Beverages, Foods, and Over-the-Counter Drugs

| Beverages and Foods | Average (mg) | Range (mg) |
|---|---|---|
| Coffee (5-oz cup) | | |
| Brewed, drip method | 130 | 110–150 |
| Brewed, percolator | 94 | 64–124 |
| Instant | 74 | 40–108 |
| Decaffeinated, brewed or instant | 3 | 1–5 |
| Tea (5-oz cup) | | |
| Brewed, major U.S. brand | 40 | 20–90 |
| Brewed, imported brands | 60 | 25–110 |
| Instant | 30 | 25–50 |
| Iced (12-oz can) | 70 | 67–76 |
| Soft drinks (12-oz can) | | |
| Dr. Pepper | 40 | |
| Colas and cherry cola | | |
| Regular | | 30–46 |
| Diet | | 2–58 |
| Caffeine-free | | 0–trace |
| Jolt | 72 | |
| Mountain Dew, Mello Yello | 52 | |
| Fresca, Hires Root Beer, 7-Up, Sprite, Squirt, Sunkist Orange | 0 | |
| Cocoa beverage (5-oz cup) | 4 | 2–20 |
| Chocolate milk beverage (8 oz) | 5 | 2–7 |
| Milk chocolate candy (1 oz) | 6 | 1–15 |
| Dark chocolate, semisweet (1 oz) | 20 | 5–35 |
| Baker's chocolate (1 oz) | 26 | |
| Chocolate flavored syrup (1 oz) | 4 | |
| **Drugs**[a] | | |
| Cold remedies (standard dose) | | |
| Dristan | 0 | |
| Coryban-D, Triaminicin | 30 | |
| Diuretics (standard dose) | | |
| Aqua-ban, Permathene $H_2$Off | 200 | |
| Pre-Mens Forte | 100 | |
| Pain relievers (standard dose) | | |
| Excedrin | 130 | |
| Midol, Anacin | 65 | |
| Aspirin, plain (any brand) | 0 | |
| Stimulants | | |
| Caffedrin, NoDoz, Vivarin | 200 | |
| Weight-control aids (daily dose) | | |
| Prolamine | 280 | |
| Dexatrim, Dietac | 200 | |

NOTE: A pharmacologically active dose of caffeine is defined as 200 milligrams.

[a]Because products change, contact the manufacturer for an update on products you use regularly.

**Table H-1**

**Food Composition** (Computer code number is for West Diet Analysis program) (For purposes of calculations, use "0" for t, <1, <.1, <.01, etc.)

| Computer Code Number | Food Description | Measure | Wt (g) | $H_2O$ (%) | Ener (kcal) | Prot (g) | Carb (g) | Dietary Fiber (g) | Fat (g) | Fat Breakdown (g) | | |
|---|---|---|---|---|---|---|---|---|---|---|---|---|
| | | | | | | | | | | Sat | Mono | Poly |
| | **BEVERAGES** | | | | | | | | | | | |
| | Alcoholic: | | | | | | | | | | | |
| | Beer: | | | | | | | | | | | |
| 1 | Regular (12 fl oz) | 1½ c | 356 | 92 | 146 | 1 | 13 | 1 | 0 | 0 | 0 | 0 |
| 2 | Light (12 fl oz) | 1½ c | 354 | 95 | 99 | 1 | 5 | 0 | 0 | 0 | 0 | 0 |
| 1506 | Nonalcoholic (12 fl oz) | 1½ c | 360 | 98 | 32 | 1 | 5 | 0 | 0 | 0 | 0 | 0 |
| | Gin, rum, vodka, whiskey: | | | | | | | | | | | |
| 3 | 80 proof | 1½ fl oz | 42 | 67 | 97 | 0 | 0 | 0 | 0 | 0 | 0 | 0 |
| 4 | 86 proof | 1½ fl oz | 42 | 64 | 105 | 0 | <1 | 0 | 0 | 0 | 0 | 0 |
| 5 | 90 proof | 1½ fl oz | 42 | 62 | 110 | 0 | 0 | 0 | 0 | 0 | 0 | 0 |
| | Liqueur: | | | | | | | | | | | |
| 1359 | Coffee liqueur, 53 proof | 1½ fl oz | 52 | 31 | 175 | <1 | 24 | 0 | <1 | .1 | t | .1 |
| 1360 | Coffee & cream liqueur, 34 proof | 1½ fl oz | 47 | 46 | 154 | 1 | 10 | 0 | 7 | 4.5 | 2.1 | .3 |
| 1361 | Crème de menthe, 72 proof | 1½ fl oz | 50 | 28 | 186 | 0 | 21 | 0 | <1 | t | t | .1 |
| | Wine, 4 fl oz: | | | | | | | | | | | |
| 6 | Dessert, sweet | ½ c | 118 | 72 | 181 | <1 | 14 | 0 | 0 | 0 | 0 | 0 |
| 7 | Red | ½ c | 118 | 88 | 85 | <1 | 2 | 0 | 0 | 0 | 0 | 0 |
| 8 | Rosé | ½ c | 118 | 89 | 84 | <1 | 2 | 0 | 0 | 0 | 0 | 0 |
| 9 | White medium | ½ c | 118 | 90 | 80 | <1 | 1 | 0 | 0 | 0 | 0 | 0 |
| 1592 | Nonalcoholic | 1 c | 232 | 98 | 14 | 1 | 3 | 0 | 0 | 0 | 0 | 0 |
| 1593 | Nonalcoholic light | 1 c | 232 | 98 | 14 | 1 | 3 | 0 | 0 | 0 | 0 | 0 |
| 1409 | Wine cooler, bottle (12 fl oz) | 1½ c | 340 | 90 | 169 | <1 | 20 | <1 | <1 | t | t | t |
| 1595 | Wine cooler, cup | 1 c | 227 | 90 | 113 | <1 | 13 | <1 | <1 | t | t | t |
| | Carbonated: | | | | | | | | | | | |
| 10 | Club soda (12 fl oz) | 1½ c | 355 | 100 | 0 | 0 | 0 | 0 | 0 | 0 | 0 | 0 |
| 11 | Cola beverage (12 fl oz) | 1½ c | 372 | 89 | 153 | 0 | 39 | 0 | 0 | 0 | 0 | 0 |
| 12 | Diet cola w/aspartame (12 fl oz) | 1½ c | 355 | 100 | 4 | <1 | <1 | 0 | 0 | 0 | 0 | 0 |
| 13 | Diet soda pop w/saccharin (12 fl oz) | 1½ c | 355 | 100 | 0 | 0 | <1 | 0 | 0 | 0 | 0 | 0 |
| 14 | Ginger ale (12 fl oz) | 1½ c | 366 | 91 | 124 | 0 | 32 | 0 | 0 | 0 | 0 | 0 |
| 15 | Grape soda (12 fl oz) | 1½ c | 372 | 89 | 160 | 0 | 42 | 0 | 0 | 0 | 0 | 0 |
| 16 | Lemon-lime (12 fl oz) | 1½ c | 368 | 89 | 147 | 0 | 38 | 0 | 0 | 0 | 0 | 0 |
| 17 | Orange (12 fl oz) | 1½ c | 372 | 88 | 179 | 0 | 46 | 0 | 0 | 0 | 0 | 0 |
| 18 | Pepper-type soda (12 fl oz) | 1½ c | 368 | 89 | 151 | 0 | 38 | 0 | <1 | .1 | 0 | 0 |
| 19 | Root beer (12 fl oz) | 1½ c | 370 | 89 | 152 | 0 | 39 | 0 | 0 | 0 | 0 | 0 |
| 20 | Coffee, brewed | 1 c | 237 | 99 | 5 | <1 | 1 | 0 | <1 | t | 0 | t |
| 21 | Coffee, prepared from instant | 1 c | 238 | 99 | 5 | <1 | 1 | 0 | <1 | t | 0 | t |
| | Fruit drinks, noncarbonated: | | | | | | | | | | | |
| 22 | Fruit punch drink, canned | 1 c | 248 | 88 | 117 | 0 | 29 | <1 | <1 | t | t | t |
| 1358 | Gatorade | 1 c | 241 | 93 | 60 | 0 | 15 | 0 | 0 | 0 | 0 | 0 |
| 23 | Grape drink, canned | 1 c | 250 | 87 | 125 | <1 | 32 | <1 | 0 | 0 | 0 | 0 |
| 1304 | Koolade sweetened with sugar | 1 c | 262 | 90 | 97 | 0 | 25 | 0 | <1 | t | t | t |
| 1356 | Koolade sweetened with nutrasweet | 1 c | 240 | 95 | 43 | 0 | 11 | 0 | 0 | 0 | 0 | 0 |
| 26 | Lemonade,frzn concentrate (6-oz can) | ¾c | 219 | 52 | 396 | 1 | 103 | 1 | <1 | .1 | t | .1 |
| 27 | Lemonade, from concentrate | 1 c | 248 | 89 | 99 | <1 | 26 | <1 | <1 | t | t | t |
| 28 | Limeade, frzn concentrate (6-oz can) | ¾c | 218 | 50 | 408 | <1 | 108 | 1 | <1 | t | t | .1 |
| 29 | Limeade, from concentrate | 1 c | 247 | 89 | 101 | 0 | 27 | <1 | <1 | t | t | t |
| 24 | Pineapple grapefruit, canned | 1 c | 250 | 88 | 118 | <1 | 29 | <1 | <1 | t | t | .1 |
| 25 | Pineapple orange, canned | 1 c | 250 | 87 | 125 | 3 | 29 | <1 | 0 | 0 | 0 | 0 |
| | Fruit and vegetable juices: see Fruit and Vegetable sections | | | | | | | | | | | |
| | Ultra Slim Fast, ready to drink, can: | | | | | | | | | | | |
| 30411 | Chocolate Royale | 1 ea | 350 | 84 | 220 | 10 | 38 | 5 | 3 | 1 | 1 | .5 |
| 30415 | French Vanilla | 1 ea | 350 | 84 | 220 | 10 | 38 | 5 | 3 | 1 | 1.5 | .5 |
| 30413 | Strawberries n' cream | 1 ea | 350 | 83 | 220 | 10 | 42 | 5 | 3 | 1 | 1.5 | .5 |
| 1357 | Water, bottled: Perrier (6½ fl oz) | 1 ea | 192 | 100 | 0 | 0 | 0 | 0 | 0 | 0 | 0 | 0 |
| 1594 | Water, bottled: Tonic water | 1½ c | 366 | 91 | 124 | 0 | 32 | 0 | 0 | 0 | 0 | 0 |
| | Tea: | | | | | | | | | | | |
| 30 | Brewed, regular | 1 c | 237 | 100 | 2 | 0 | 1 | 0 | <1 | t | 0 | t |
| 1662 | Brewed, herbal | 1 c | 237 | 100 | 2 | 0 | <1 | 0 | <1 | t | t | t |
| 32 | From instant, sweetened | 1 c | 259 | 91 | 88 | <1 | 22 | 0 | <1 | t | t | t |
| 31 | From instant, unsweetened | 1 c | 237 | 100 | 2 | <1 | <1 | 0 | 0 | 0 | 0 | 0 |

**PAGE KEY:** H–4 = Beverages H–6 = Dairy H–10 = Eggs H–10 = Fat/Oil H–14 = Fruit H–20 = Bakery H–28 = Grain H–32 = Fish H–34 = Meats H–38 = Poultry H–40 = Sausage H–42 = Mixed/Fast H–46 = Nuts/Seeds H–50 = Sweets H–52 = Vegetables/Legumes H–62 = Vegetarian Foods H–64 = Misc H–66 = Soups/Sauces H–68 = Fast H–84 = Convenience H–88 = Baby foods

| Chol (mg) | Calc (mg) | Iron (mg) | Magn (mg) | Pota (mg) | Sodi (mg) | Zinc (mg) | VT-A (RE) | Thia (mg) | VT-E (a-TE) | Ribo (mg) | Niac (mg) | V-B6 (mg) | Fola (μg) | VT-C (mg) |
|---|---|---|---|---|---|---|---|---|---|---|---|---|---|---|
| 0 | 18 | .11 | 21 | 89 | 18 | .07 | 0 | .02 | 0 | .09 | 1.61 | .18 | 21 | 0 |
| 0 | 18 | .14 | 18 | 64 | 11 | .11 | 0 | .03 | 0 | .11 | 1.39 | .12 | 14 | 0 |
| 0 | 25 | .04 | 32 | 90 | 18 | .04 | 0 | .02 | 0 | .09 | 1.63 | .18 | 22 | 0 |
| 0 | 0 | .02 | 0 | 1 | <1 | .02 | 0 | <.01 | 0 | <.01 | <.01 | 0 | 0 | 0 |
| 0 | 0 | .02 | 0 | 1 | <1 | .02 | 0 | <.01 | 0 | <.01 | <.01 | 0 | 0 | 0 |
| 0 | 0 | .02 | 0 | 1 | <1 | .02 | 0 | <.01 | 0 | <.01 | <.01 | 0 | 0 | 0 |
| 0 | 1 | .03 | 2 | 16 | 4 | .02 | 0 | <.01 | 0 | .01 | .07 | 0 | 0 | 0 |
| 7 | 8 | .06 | 1 | 15 | 43 | .07 | 20 | 0 | .12 | .03 | .04 | .01 | 0 | 0 |
| 0 | 0 | .03 | 0 | 0 | 2 | .02 | 0 | 0 | 0 | 0 | <.01 | 0 | 0 | 0 |
| 0 | 9 | .28 | 11 | 109 | 11 | .08 | 0 | .02 | 0 | .02 | .25 | 0 | <1 | 0 |
| 0 | 9 | .51 | 15 | 132 | 6 | .11 | 0 | .01 | 0 | .03 | .1 | .04 | 2 | 0 |
| 0 | 9 | .45 | 12 | 117 | 6 | .07 | 0 | <.01 | 0 | .02 | .09 | .03 | 1 | 0 |
| 0 | 11 | .38 | 12 | 94 | 6 | .08 | 0 | <.01 | 0 | .01 | .08 | .02 | <1 | 0 |
| 0 | 21 | .93 | 23 | 204 | 16 | .19 | 0 | 0 | 0 | .02 | .23 | .05 | 2 | 0 |
| 0 | 21 | .93 | 23 | 204 | 16 | .19 | 0 | 0 | 0 | .02 | .23 | .05 | 2 | 0 |
| 0 | 19 | .92 | 18 | 152 | 29 | .2 | <1 | .02 | .02 | .02 | .15 | .04 | 4 | 6 |
| 0 | 13 | .61 | 12 | 102 | 19 | .13 | <1 | .01 | .02 | .02 | .1 | .03 | 3 | 4 |
| 0 | 18 | .04 | 4 | 7 | 75 | .35 | 0 | 0 | 0 | 0 | 0 | 0 | 0 | 0 |
| 0 | 11 | .11 | 4 | 4 | 15 | .04 | 0 | 0 | 0 | 0 | 0 | 0 | 0 | 0 |
| 0 | 14 | .11 | 4 | 0 | 21 | .28 | 0 | .02 | 0 | .08 | 0 | 0 | 0 | 0 |
| 0 | 14 | .14 | 4 | 7 | 57 | .18 | 0 | 0 | 0 | 0 | 0 | 0 | 0 | 0 |
| 0 | 11 | .66 | 4 | 4 | 26 | .18 | 0 | 0 | 0 | 0 | 0 | 0 | 0 | 0 |
| 0 | 11 | .3 | 4 | 4 | 56 | .26 | 0 | 0 | 0 | 0 | 0 | 0 | 0 | 0 |
| 0 | 7 | .26 | 4 | 4 | 40 | .18 | 0 | 0 | 0 | 0 | .05 | 0 | 0 | 0 |
| 0 | 19 | .22 | 4 | 7 | 45 | .37 | 0 | 0 | 0 | 0 | 0 | 0 | 0 | 0 |
| 0 | 11 | .15 | 0 | 4 | 37 | .15 | 0 | 0 | 0 | 0 | 0 | 0 | 0 | 0 |
| 0 | 18 | .18 | 4 | 4 | 48 | .26 | 0 | 0 | 0 | 0 | 0 | 0 | 0 | 0 |
| 0 | 5 | .12 | 12 | 128 | 5 | .05 | 0 | 0 | 0 | 0 | .53 | 0 | <1 | 0 |
| 0 | 7 | .12 | 10 | 86 | 7 | .07 | 0 | 0 | 0 | <.01 | .67 | 0 | 0 | 0 |
| 0 | 20 | .52 | 5 | 62 | 55 | .3 | 3 | 0 | 0 | .06 | .05 | 0 | 3 | 73 |
| 0 | 0 | .12 | 2 | 26 | 96 | .05 | 0 | .01 | 0 | 0 | 0 | 0 | 0 | 0 |
| 0 | 7 | .25 | 10 | 87 | 2 | .07 | <1 | .02 | 0 | .02 | .25 | .05 | 2 | 40 |
| 0 | 42 | .13 | 3 | 3 | 37 | .08 | 0 | 0 | 0 | <.01 | <.01 | 0 | <1 | 31 |
| 0 | 17 | .65 | 5 | 50 | 50 | .26 | 2 | .02 | 0 | .05 | .05 | 0 | 5 | 77 |
| 0 | 15 | 1.58 | 11 | 147 | 9 | .17 | 22 | .06 | 0 | .21 | .16 | .05 | 22 | 39 |
| 0 | 7 | .4 | 5 | 37 | 7 | .1 | 5 | .01 | 0 | .05 | .04 | .01 | 5 | 10 |
| 0 | 11 | .22 | 9 | 129 | 0 | .09 | 0 | .02 | 0 | .02 | .22 | 0 | 9 | 26 |
| 0 | 7 | .07 | 2 | 32 | 5 | .05 | 0 | <.01 | 0 | <.01 | .05 | 0 | 2 | 7 |
| 0 | 17 | .77 | 15 | 153 | 35 | .15 | 9 | .07 | 0 | .04 | .67 | .1 | 26 | 115 |
| 0 | 12 | .67 | 15 | 115 | 7 | .15 | 13 | .07 | 0 | .05 | .52 | .12 | 27 | 56 |
| 5 | 400 | 2.7 | 140 | 530 | 220 | 2.24 | 525 | .52 | 7 | .59 | 7 | .7 | 120 | 21 |
| 5 | 400 | 2.8 | 140 | 450 | 460 | 2.1 | 525 | .52 | 7 | .59 | 7 | .7 | 120 | 21 |
| 5 | 400 | 2.7 | 140 | 450 | 460 | 2.24 | 525 | .52 | 7 | .59 | 7 | .7 | 120 | 21 |
| 0 | 27 | 0 | 0 | 0 | 2 | 0 | 0 | 0 | 0 | 0 | 0 | 0 | 0 | 0 |
| 0 | 4 | .04 | 0 | 0 | 15 | .37 | 0 | 0 | 0 | 0 | 0 | 0 | 0 | 0 |
| 0 | 0 | .05 | 7 | 88 | 7 | .05 | 0 | 0 | 0 | .03 | 0 | 0 | 12 | 0 |
| 0 | 5 | .19 | 2 | 21 | 2 | .09 | 0 | .02 | 0 | .01 | 0 | 0 | 1 | 0 |
| 0 | 5 | .05 | 5 | 49 | 8 | .08 | 0 | 0 | 0 | .05 | .09 | <.01 | 10 | 0 |
| 0 | 5 | .05 | 5 | 47 | 7 | .07 | 0 | 0 | 0 | <.01 | .09 | <.01 | 1 | 0 |

**Table H-1**

**Food Composition** (Computer code number is for West Diet Analysis program) (For purposes of calculations, use "0" for t, <1, <.1, <.01, etc.)

| Computer Code Number | Food Description | Measure | Wt (g) | $H_2O$ (%) | Ener (kcal) | Prot (g) | Carb (g) | Dietary Fiber (g) | Fat (g) | Fat Breakdown (g) Sat | Mono | Poly |
|---|---|---|---|---|---|---|---|---|---|---|---|---|
| | **DAIRY** | | | | | | | | | | | |
| | **Butter:** see Fats and Oils, #158,159,160 | | | | | | | | | | | |
| | Cheese, natural: | | | | | | | | | | | |
| 33 | Blue | 1 oz | 28 | 42 | 99 | 6 | 1 | 0 | 8 | 5.2 | 2.2 | .2 |
| 34 | Brick | 1 oz | 28 | 41 | 104 | 6 | 1 | 0 | 8 | 5.3 | 2.4 | .2 |
| 35 | Brie | 1 oz | 28 | 48 | 93 | 6 | <1 | 0 | 8 | 4.9 | 2.2 | .2 |
| 36 | Camembert | 1 oz | 28 | 52 | 84 | 6 | <1 | 0 | 7 | 4.3 | 2 | .2 |
| 37 | Cheddar: | 1 oz | 28 | 37 | 113 | 7 | <1 | 0 | 9 | 5.9 | 2.6 | .3 |
| 38 | 1″ cube | 1 ea | 17 | 37 | 68 | 4 | <1 | 0 | 6 | 3.6 | 1.6 | .2 |
| 39 | Shredded | 1 c | 113 | 37 | 455 | 28 | 1 | 0 | 37 | 24 | 10.6 | 1.1 |
| 1406 | Low fat, low sodium | 1 oz | 28 | 65 | 48 | 7 | 1 | 0 | 2 | 1.2 | .6 | .1 |
| | Cottage: | | | | | | | | | | | |
| 984 | Low sodium, low fat | 1 c | 225 | 83 | 162 | 28 | 6 | 0 | 2 | 1.4 | .7 | .1 |
| 40 | Creamed, large curd | 1 c | 225 | 79 | 232 | 28 | 6 | 0 | 10 | 6.4 | 2.9 | .3 |
| 41 | Creamed, small curd | 1 c | 210 | 79 | 216 | 26 | 6 | 0 | 9 | 6 | 2.7 | .3 |
| 42 | With fruit | 1 c | 226 | 72 | 280 | 22 | 30 | 0 | 8 | 4.9 | 2.2 | .2 |
| 43 | Low fat 2% | 1 c | 226 | 79 | 203 | 31 | 8 | 0 | 4 | 2.8 | 1.2 | .1 |
| 44 | Low fat 1% | 1 c | 226 | 82 | 164 | 28 | 6 | 0 | 2 | 1.5 | .7 | .1 |
| 46 | Cream | 1 tbs | 15 | 54 | 52 | 1 | <1 | 0 | 5 | 3.3 | 1.5 | .2 |
| 983 | low fat | 1 tbs | 15 | 64 | 35 | 2 | 1 | 0 | 3 | 1.7 | .9 | .1 |
| 47 | Edam | 1 oz | 28 | 42 | 100 | 7 | <1 | 0 | 8 | 4.9 | 2.3 | .2 |
| 48 | Feta | 1 oz | 28 | 55 | 74 | 4 | 1 | 0 | 6 | 4.2 | 1.3 | .2 |
| 49 | Gouda | 1 oz | 28 | 41 | 100 | 7 | 1 | 0 | 8 | 4.9 | 2.2 | .2 |
| 50 | Gruyère | 1 oz | 28 | 33 | 116 | 8 | <1 | 0 | 9 | 5.3 | 2.8 | .5 |
| 51 | Gorgonzola | 1 oz | 28 | 43 | 97 | 6 | 1 | 0 | 8 | 5 | | |
| 1676 | Limburger | 1 oz | 28 | 48 | 92 | 6 | <1 | 0 | 8 | 4.7 | 2.4 | .1 |
| 53 | Monterey Jack | 1 oz | 28 | 41 | 104 | 7 | <1 | 0 | 8 | 5.3 | 2.4 | .3 |
| 54 | Mozzarella, whole milk | 1 oz | 28 | 54 | 79 | 5 | 1 | 0 | 6 | 3.7 | 1.8 | .2 |
| 55 | Mozzarella, part-skim milk, low moisture | 1 oz | 28 | 49 | 78 | 8 | 1 | 0 | 5 | 3 | 1.4 | .1 |
| 56 | Muenster | 1 oz | 28 | 42 | 103 | 7 | <1 | 0 | 8 | 5.3 | 2.4 | .2 |
| 2422 | Neufchatel | 1 oz | 28 | 62 | 73 | 3 | 1 | 0 | 7 | 4.1 | 1.9 | .2 |
| 1399 | Nonfat cheese (Kraft Singles) | 1 oz | 28 | 61 | 44 | 6 | 4 | 0 | 0 | 0 | 0 | 0 |
| 59 | Parmesan, grated: | 1 oz | 28 | 18 | 128 | 12 | 1 | 0 | 8 | 5.5 | 2.4 | .2 |
| 57 | Cup, not pressed down | 1 c | 100 | 18 | 456 | 42 | 4 | 0 | 30 | 19.7 | 8.7 | .7 |
| 58 | Tablespoon | 1 tbs | 6 | 18 | 27 | 2 | <1 | 0 | 2 | 1.2 | .5 | t |
| 60 | Provolone | 1 oz | 28 | 41 | 98 | 7 | 1 | 0 | 7 | 4.9 | 2.1 | .2 |
| 61 | Ricotta, whole milk | 1 c | 246 | 72 | 428 | 28 | 7 | 0 | 32 | 20.4 | 8.9 | .9 |
| 62 | Ricotta, part-skim milk | 1 c | 246 | 74 | 339 | 28 | 13 | 0 | 19 | 12.1 | 5.7 | .6 |
| 63 | Romano | 1 oz | 28 | 31 | 108 | 9 | 1 | 0 | 8 | 4.8 | 2.2 | .2 |
| 64 | Swiss | 1 oz | 28 | 37 | 105 | 8 | 1 | 0 | 8 | 5 | 2 | .3 |
| 976 | low fat | 1 oz | 28 | 60 | 50 | 8 | 1 | 0 | 1 | .9 | .4 | t |
| | Pasteurized processed cheese products: | | | | | | | | | | | |
| 65 | American | 1 oz | 28 | 39 | 105 | 6 | <1 | 0 | 9 | 5.5 | 2.5 | .3 |
| 66 | Swiss | 1 oz | 28 | 42 | 93 | 7 | 1 | 0 | 7 | 4.5 | 2 | .2 |
| 67 | American cheese food, jar | ½ c | 57 | 43 | 187 | 11 | 4 | 0 | 14 | 9 | 4.1 | .4 |
| 68 | American cheese spread | 1 tbs | 15 | 48 | 43 | 2 | 1 | 0 | 3 | 2 | .9 | .1 |
| 982 | Velveeta cheese spread, low fat, low sodium, slice | 1 pce | 34 | 62 | 61 | 9 | 1 | 0 | 2 | 1.5 | .7 | .1 |
| | Cream, sweet: | | | | | | | | | | | |
| 69 | Half & half (cream & milk) | 1 c | 242 | 81 | 315 | 7 | 10 | 0 | 28 | 17.3 | 8 | 1 |
| 70 | Tablespoon | 1 tbs | 15 | 81 | 19 | <1 | 1 | 0 | 2 | 1.1 | .5 | .1 |
| 71 | Light, coffee or table: | 1 c | 240 | 74 | 468 | 6 | 9 | 0 | 46 | 28.8 | 13.4 | 1.7 |
| 72 | Tablespoon | 1 tbs | 15 | 74 | 29 | <1 | 1 | 0 | 3 | 1.8 | .8 | .1 |
| 73 | Light whipping cream, liquid: | 1 c | 239 | 63 | 698 | 5 | 7 | 0 | 74 | 46.1 | 21.7 | 2.1 |
| 74 | Tablespoon | 1 tbs | 15 | 63 | 44 | <1 | <1 | 0 | 5 | 2.9 | 1.4 | .1 |
| 75 | Heavy whipping cream, liquid: | 1 c | 238 | 58 | 821 | 5 | 7 | 0 | 88 | 54.7 | 25.5 | 3.3 |
| 76 | Tablespoon | 1 tbs | 15 | 58 | 52 | <1 | <1 | 0 | 6 | 3.4 | 1.6 | .2 |
| 77 | Whipped cream, pressurized: | 1 c | 60 | 61 | 154 | 2 | 7 | 0 | 13 | 8.3 | 3.8 | .5 |
| 78 | Tablespoon | 1 tbs | 4 | 61 | 10 | <1 | <1 | 0 | 1 | .6 | .3 | t |
| 79 | Cream, sour, cultured: | 1 c | 230 | 71 | 492 | 7 | 10 | 0 | 48 | 29.9 | 13.9 | 1.8 |
| 80 | Tablespoon | 1 tbs | 14 | 71 | 30 | <1 | 1 | 0 | 3 | 1.8 | .8 | .1 |

**PAGE KEY:** H–4 = Beverages H–6 = Dairy H–10 = Eggs H–10 = Fat/Oil H–14 = Fruit H–20 = Bakery H–28 = Grain H–32 = Fish H–34 = Meats H–38 = Poultry H–40 = Sausage H–42 = Mixed/Fast H–46 = Nuts/Seeds H–50 = Sweets H–52 = Vegetables/Legumes H–62 = Vegetarian Foods H–64 = Misc H–66 = Soups/Sauces H–68 = Fast H–84 = Convenience H–88 = Baby foods

| Chol (mg) | Calc (mg) | Iron (mg) | Magn (mg) | Pota (mg) | Sodi (mg) | Zinc (mg) | VT-A (RE) | Thia (mg) | VT-E (a-TE) | Ribo (mg) | Niac (mg) | V-B6 (mg) | Fola (µg) | VT-C (mg) |
|---|---|---|---|---|---|---|---|---|---|---|---|---|---|---|
| | | | | | | | | | | | | | | |
| 21 | 148 | .09 | 6 | 72 | 391 | .74 | 64 | .01 | .18 | .11 | .29 | .05 | 10 | 0 |
| 26 | 189 | .12 | 7 | 38 | 157 | .73 | 85 | <.01 | .14 | .1 | .03 | .02 | 6 | 0 |
| 28 | 51 | .14 | 6 | 43 | 176 | .67 | 51 | .02 | .18 | .15 | .11 | .07 | 18 | 0 |
| 20 | 109 | .09 | 6 | 52 | 236 | .67 | 71 | .01 | .18 | .14 | .18 | .06 | 17 | 0 |
| 29 | 202 | .19 | 8 | 28 | 174 | .87 | 85 | .01 | .1 | .1 | .02 | .02 | 5 | 0 |
| 18 | 123 | .12 | 5 | 17 | 106 | .53 | 51 | <.01 | .06 | .06 | .01 | .01 | 3 | 0 |
| 119 | 815 | .77 | 31 | 111 | 702 | 3.51 | 342 | .03 | .41 | .42 | .09 | .08 | 21 | 0 |
| 6 | 197 | .2 | 8 | 31 | 6 | .86 | 17 | .01 | .05 | .01 | .02 | .02 | 5 | 0 |
| | | | | | | | | | | | | | | |
| 9 | 137 | .31 | 11 | 194 | 29 | .85 | 25 | .04 | .25 | .36 | .29 | .16 | 27 | 0 |
| 33 | 135 | .31 | 12 | 190 | 911 | .83 | 108 | .05 | .27 | .37 | .28 | .15 | 27 | 0 |
| 31 | 126 | .29 | 11 | 177 | 851 | .78 | 101 | .04 | .26 | .34 | .26 | .14 | 26 | 0 |
| 25 | 108 | .25 | 9 | 151 | 915 | .65 | 81 | .04 | .21 | .29 | .23 | .12 | 22 | 0 |
| 19 | 155 | .36 | 14 | 217 | 918 | .95 | 45 | .05 | .13 | .42 | .32 | .17 | 30 | 0 |
| 10 | 138 | .32 | 12 | 193 | 918 | .86 | 25 | .05 | .25 | .37 | .29 | .15 | 28 | 0 |
| 16 | 12 | .18 | 1 | 18 | 44 | .08 | 66 | <.01 | .14 | .03 | .01 | .01 | 2 | 0 |
| 8 | 17 | .25 | 1 | 25 | 44 | .11 | 33 | <.01 | .07 | .04 | .02 | .01 | 3 | 0 |
| 25 | 205 | .12 | 8 | 53 | 270 | 1.05 | 71 | .01 | .21 | .11 | .02 | .02 | 5 | 0 |
| 25 | 138 | .18 | 5 | 17 | 312 | .81 | 36 | .04 | .01 | .24 | .28 | .12 | 9 | 0 |
| 32 | 196 | .07 | 8 | 34 | 229 | 1.09 | 49 | .01 | .1 | .09 | .02 | .02 | 6 | 0 |
| 31 | 283 | .05 | 10 | 23 | 94 | 1.09 | 84 | .02 | .1 | .08 | .03 | .02 | 3 | 0 |
| 30 | 170 | .18 | | | 280 | | 43 | | | | | | | 0 |
| 25 | 139 | .04 | 6 | 36 | 224 | .59 | 88 | .02 | .18 | .14 | .04 | .02 | 16 | 0 |
| 25 | 209 | .2 | 8 | 23 | 150 | .84 | 71 | <.01 | .09 | .11 | .03 | .02 | 5 | 0 |
| 22 | 145 | .05 | 5 | 19 | 104 | .62 | 67 | <.01 | .1 | .07 | .02 | .02 | 2 | 0 |
| | | | | | | | | | | | | | | |
| 15 | 205 | .07 | 7 | 26 | 148 | .88 | 53 | .01 | .13 | .1 | .03 | .02 | 3 | 0 |
| 27 | 201 | .11 | 8 | 37 | 176 | .79 | 88 | <.01 | .13 | .09 | .03 | .02 | 3 | 0 |
| 21 | 21 | .08 | 2 | 32 | 112 | .15 | 74 | <.01 | .26 | .05 | .03 | .01 | 3 | 0 |
| 4 | 221 | 0 | | 81 | 427 | | 126 | | 0 | .1 | | | | 0 |
| 22 | 385 | .27 | 14 | 30 | 521 | .89 | 48 | .01 | .22 | .11 | .09 | .03 | 2 | 0 |
| 79 | 1375 | .95 | 51 | 107 | 1861 | 3.19 | 173 | .04 | .8 | .39 | .31 | .1 | 8 | 0 |
| 5 | 82 | .06 | 3 | 6 | 112 | .19 | 10 | <.01 | .05 | .02 | .02 | .01 | <1 | 0 |
| 19 | 212 | .15 | 8 | 39 | 245 | .9 | 74 | <.01 | .1 | .09 | .04 | .02 | 3 | 0 |
| 124 | 509 | .93 | 28 | 258 | 207 | 2.85 | 330 | .03 | .86 | .48 | .26 | .11 | 30 | 0 |
| 76 | 669 | 1.08 | 36 | 308 | 308 | 3.3 | 278 | .05 | .53 | .45 | .19 | .05 | 32 | 0 |
| 29 | 298 | .22 | 11 | 24 | 336 | .72 | 39 | .01 | .2 | .1 | .02 | .02 | 2 | 0 |
| 26 | 269 | .05 | 10 | 31 | 73 | 1.09 | 71 | .01 | .14 | .1 | .03 | .02 | 2 | 0 |
| 10 | 269 | .05 | 10 | 31 | 73 | 1.09 | 18 | .01 | .05 | .1 | .02 | .02 | 2 | 0 |
| | | | | | | | | | | | | | | |
| 26 | 172 | .11 | 6 | 45 | 400 | .84 | 81 | .01 | .13 | .1 | .02 | .02 | 2 | 0 |
| 24 | 216 | .17 | 8 | 60 | 384 | 1.01 | 64 | <.01 | .19 | .08 | .01 | .01 | 2 | 0 |
| 36 | 327 | .48 | 17 | 159 | 678 | 1.7 | 125 | .02 | .4 | .25 | .08 | .08 | 4 | 0 |
| 8 | 84 | .05 | 4 | 36 | 202 | .39 | 28 | .01 | .11 | .06 | .02 | .02 | 1 | 0 |
| | | | | | | | | | | | | | | |
| 12 | 233 | .15 | 8 | 61 | 2 | 1.13 | 22 | .01 | .17 | .13 | .03 | .03 | 3 | 0 |
| | | | | | | | | | | | | | | |
| 89 | 254 | .17 | 25 | 315 | 98 | 1.23 | 259 | .08 | .27 | .36 | .19 | .09 | 6 | 2 |
| 6 | 16 | .01 | 2 | 19 | 6 | .08 | 16 | <.01 | .02 | .02 | .01 | .01 | <1 | <1 |
| 159 | 231 | .1 | 21 | 293 | 95 | .65 | 437 | .08 | .36 | .35 | .14 | .08 | 6 | 2 |
| 10 | 14 | .01 | 1 | 18 | 6 | .04 | 27 | <.01 | .02 | .02 | .01 | <.01 | <1 | <1 |
| 265 | 166 | .07 | 17 | 231 | 82 | .6 | 705 | .06 | 1.43 | .3 | .1 | .07 | 9 | 1 |
| 17 | 10 | <.01 | 1 | 14 | 5 | .04 | 44 | <.01 | .09 | .02 | .01 | <.01 | 1 | <1 |
| 326 | 154 | .07 | 17 | 179 | 89 | .55 | 1001 | .05 | 1.5 | .26 | .09 | .06 | 9 | 1 |
| 21 | 10 | <.01 | 1 | 11 | 6 | .03 | 63 | <.01 | .09 | .02 | .01 | <.01 | 1 | <1 |
| 46 | 61 | .03 | 6 | 88 | 78 | .22 | 124 | .02 | .36 | .04 | .04 | .02 | 2 | 0 |
| 3 | 4 | <.01 | | 6 | 5 | .01 | 8 | <.01 | .02 | <.01 | <.01 | <.01 | <1 | 0 |
| 102 | 267 | .14 | 26 | 331 | 123 | .62 | 449 | .08 | 1.3 | .34 | .15 | .04 | 25 | 2 |
| 6 | 16 | .01 | 2 | 20 | 7 | .04 | 27 | <.01 | .08 | .02 | .01 | <.01 | 2 | <1 |

**Table H-1**

**Food Composition** (Computer code number is for West Diet Analysis program) (For purposes of calculations, use "0" for t, <1, <.1, <.01, etc.)

| Computer Code Number | Food Description | Measure | Wt (g) | $H_2O$ (%) | Ener (kcal) | Prot (g) | Carb (g) | Dietary Fiber (g) | Fat (g) | Fat Breakdown (g) Sat | Mono | Poly |
|---|---|---|---|---|---|---|---|---|---|---|---|---|
| | **DAIRY**—Continued | | | | | | | | | | | |
| | Cream products—imitation and part dairy: | | | | | | | | | | | |
| 81 | Coffee whitener, frozen or liquid | 1 tbs | 15 | 77 | 20 | <1 | 2 | 0 | 1 | 1.4 | t | 0 |
| 82 | Coffee whitener, powdered | 1 tsp | 2 | 2 | 11 | <1 | 1 | 0 | 1 | .6 | t | 0 |
| 83 | Dessert topping, frozen, nondairy: | 1 c | 75 | 50 | 239 | 1 | 17 | 0 | 19 | 16.4 | 1.2 | .4 |
| 84 | Tablespoon | 1 tbs | 5 | 50 | 16 | <1 | 1 | 0 | 1 | 1.1 | .1 | t |
| 85 | Dessert topping, mix with whole milk: | 1 c | 80 | 67 | 151 | 3 | 13 | 0 | 10 | 8.6 | .7 | .2 |
| 86 | Tablespoon | 1 tbs | 5 | 67 | 9 | <1 | 1 | 0 | 1 | .5 | t | t |
| 88 | Dessert topping, pressurized | 1 c | 70 | 60 | 185 | 1 | 11 | 0 | 16 | 13.3 | 1.3 | .2 |
| 87 | Tablespoon | 1 tbs | 4 | 60 | 11 | <1 | 1 | 0 | 1 | .8 | .1 | t |
| 91 | Sour cream, imitation: | 1 c | 230 | 71 | 478 | 6 | 15 | 0 | 45 | 40.9 | 1.3 | .1 |
| 92 | Tablespoon | 1 tbs | 14 | 71 | 29 | <1 | 1 | 0 | 3 | 2.5 | .1 | t |
| 89 | Sour dressing, part dairy: | 1 c | 235 | 75 | 418 | 8 | 11 | 0 | 39 | 31.3 | 4.6 | 1.1 |
| 90 | Tablespoon | 1 tbs | 15 | 75 | 27 | 1 | 1 | 0 | 2 | 2 | .3 | .1 |
| | Milk, fluid: | | | | | | | | | | | |
| 93 | Whole milk | 1 c | 244 | 88 | 150 | 8 | 11 | 0 | 8 | 5.1 | 2.7 | .3 |
| 94 | 2% reduced-fat milk | 1 c | 244 | 89 | 121 | 8 | 12 | 0 | 5 | 2.9 | 1.3 | .2 |
| 95 | 2% milk solids added | 1 c | 245 | 89 | 125 | 9 | 12 | 0 | 5 | 2.9 | 1.4 | .2 |
| 96 | 1% lowfat milk | 1 c | 244 | 90 | 102 | 8 | 12 | 0 | 3 | 1.6 | .7 | .1 |
| 97 | 1% milk solids added | 1 c | 245 | 90 | 104 | 9 | 12 | 0 | 2 | 1.5 | .7 | .1 |
| 98 | Nonfat milk, vitamin A added | 1 c | 245 | 91 | 85 | 8 | 12 | 0 | <1 | .3 | .1 | t |
| 99 | Nonfat milk solids added | 1 c | 245 | 90 | 90 | 9 | 12 | 0 | 1 | .4 | .2 | t |
| 100 | Buttermilk, skim | 1 c | 245 | 90 | 99 | 8 | 12 | 0 | 2 | 1.3 | .6 | .1 |
| | Milk, canned: | | | | | | | | | | | |
| 101 | Sweetened condensed | 1 c | 306 | 27 | 982 | 24 | 166 | 0 | 27 | 16.8 | 7.4 | 1 |
| 103 | Evaporated, nonfat | 1 c | 256 | 79 | 199 | 19 | 29 | 0 | 1 | .3 | .2 | t |
| | Milk, dried: | | | | | | | | | | | |
| 104 | Buttermilk, sweet | 1 c | 120 | 3 | 464 | 41 | 59 | 0 | 7 | 4.3 | 2 | .3 |
| 105 | Instant, nonfat, vit A added (makes 1 qt) | 1 ea | 91 | 4 | 326 | 32 | 47 | 0 | 1 | .4 | .2 | t |
| 106 | Instant nonfat, vit A added | 1 c | 68 | 4 | 243 | 24 | 35 | 0 | <1 | .3 | .1 | t |
| 107 | Goat milk | 1 c | 244 | 87 | 168 | 9 | 11 | 0 | 10 | 6.5 | 2.7 | .4 |
| 108 | Kefir | 1 c | 233 | 88 | 149 | 8 | 11 | 0 | 8 | | | |
| | Milk beverages and powdered mixes: | | | | | | | | | | | |
| | Chocolate: | | | | | | | | | | | |
| 109 | Whole | 1 c | 250 | 82 | 209 | 8 | 26 | 2 | 8 | 5.3 | 2.5 | .3 |
| 110 | 2% fat | 1 c | 250 | 84 | 179 | 8 | 26 | 1 | 5 | 3.1 | 1.5 | .2 |
| 111 | 1% fat | 1 c | 250 | 84 | 158 | 9 | 26 | 1 | 2 | 1.5 | .7 | .1 |
| | Chocolate-flavored beverages: | | | | | | | | | | | |
| 112 | Powder containing nonfat dry milk: | 1 oz | 28 | 1 | 101 | 3 | 22 | <1 | 1 | .7 | .4 | t |
| 113 | Prepared with water | 1 c | 275 | 86 | 138 | 4 | 30 | 3 | 2 | .9 | .5 | t |
| 114 | Powder without nonfat dry milk: | 1 oz | 28 | 1 | 98 | 1 | 25 | 2 | 1 | .5 | .3 | t |
| 115 | Prepared with whole milk | 1 c | 266 | 81 | 226 | 9 | 31 | 1 | 9 | 5.5 | 2.6 | .3 |
| 116 | Eggnog, commercial | 1 c | 254 | 74 | 343 | 10 | 34 | 0 | 19 | 11.3 | 5.7 | .9 |
| 974 | Eggnog, 2% reduced-fat | 1 c | 254 | 85 | 189 | 12 | 17 | 0 | 8 | 3.7 | 2.7 | .7 |
| 1027 | Instant Breakfast, envelope,powder only: | 1 ea | 37 | 7 | 131 | 7 | 24 | <1 | 1 | .3 | .1 | t |
| 1028 | Prepared with whole milk | 1 c | 281 | 77 | 280 | 15 | 36 | <1 | 9 | 5.4 | 2.5 | .3 |
| 1029 | Prepared with 2% milk | 1 c | 281 | 78 | 252 | 15 | 36 | <1 | 5 | 3.3 | 1.5 | .2 |
| 1283 | Prepared with 1% milk | 1 c | 281 | 79 | 233 | 15 | 36 | <1 | 3 | 1.9 | .9 | .1 |
| 1284 | Prepared with nonfat milk | 1 c | 282 | 80 | 216 | 16 | 36 | <1 | 1 | .7 | .3 | t |
| 117 | Malted milk, chocolate, powder: | 3 tsp | 21 | 1 | 79 | 1 | 18 | <1 | 1 | .5 | .2 | .1 |
| 118 | Prepared with whole milk | 1 c | 265 | 81 | 228 | 9 | 30 | <1 | 9 | 5.5 | 2.6 | .4 |
| 1661 | Ovaltine with whole milk | 1 c | 265 | 81 | 225 | 9 | 29 | <1 | 9 | 5.5 | 2.5 | .4 |
| 119 | Malted mix powder, natural: | 3 tsp | 21 | 2 | 87 | 2 | 16 | <1 | 2 | .9 | .4 | .3 |
| 120 | Prepared with whole milk | 1 c | 265 | 81 | 236 | 10 | 27 | <1 | 10 | 6 | 2.8 | .6 |
| 121 | Milk shakes, chocolate | 1 c | 166 | 71 | 211 | 6 | 34 | 1 | 6 | 3.8 | 1.8 | .2 |
| 122 | Milk shakes, vanilla | 1 c | 166 | 75 | 184 | 6 | 30 | 1 | 5 | 3.1 | 1.4 | .2 |
| | Milk desserts: | | | | | | | | | | | |
| 134 | Custard, baked | 1 c | 282 | 79 | 296 | 14 | 30 | 0 | 13 | 6.6 | 4.3 | 1 |
| 1548 | Low-fat frozen dessert bars | 1 ea | 81 | 72 | 88 | 2 | 19 | 0 | 1 | .2 | .1 | .4 |
| | Ice cream, vanilla (about 10% fat): | | | | | | | | | | | |
| 123 | Hardened: ½ gallon | 1 ea | 1064 | 61 | 2138 | 37 | 251 | 0 | 117 | 72.4 | 33.8 | 4.4 |
| 124 | Cup | 1 c | 132 | 61 | 265 | 5 | 31 | 0 | 14 | 9 | 4.2 | .5 |
| 126 | Soft serve | 1 c | 172 | 60 | 370 | 7 | 38 | 0 | 22 | 12.9 | 6 | .8 |

| Chol (mg) | Calc (mg) | Iron (mg) | Magn (mg) | Pota (mg) | Sodi (mg) | Zinc (mg) | VT-A (RE) | Thia (mg) | VT-E (a-TE) | Ribo (mg) | Niac (mg) | V-B6 (mg) | Fola (µg) | VT-C (mg) |
|---|---|---|---|---|---|---|---|---|---|---|---|---|---|---|
| 0 | 1 | <.01 | | 29 | 12 | <.01 | 1 | 0 | .24 | 0 | 0 | 0 | 0 | 0 |
| 0 | | .02 | | 16 | 4 | .01 | <1 | 0 | <.01 | <.01 | 0 | 0 | 0 | 0 |
| 0 | 5 | .09 | 1 | 14 | 19 | .02 | 64 | 0 | .14 | 0 | 0 | 0 | 0 | 0 |
| 0 | | .01 | | 1 | 1 | <.01 | 4 | 0 | .01 | 0 | 0 | 0 | 0 | 0 |
| 8 | 72 | .03 | 8 | 121 | 53 | .22 | 39 | .02 | .11 | .09 | .05 | .02 | 3 | 1 |
| <1 | 5 | <.01 | | 8 | 3 | .01 | 2 | <.01 | .01 | .01 | <.01 | <.01 | <1 | <1 |
| 0 | 4 | .01 | 1 | 13 | 43 | .01 | 33 | 0 | .12 | 0 | 0 | 0 | 0 | 0 |
| 0 | | <.01 | | 1 | 2 | 0 | 2 | 0 | .01 | 0 | 0 | 0 | 0 | 0 |
| 0 | 6 | .9 | 15 | 370 | 235 | 2.71 | 0 | 0 | .34 | 0 | 0 | 0 | 0 | 0 |
| 0 | | .05 | 1 | 22 | 14 | .16 | 0 | 0 | .02 | 0 | 0 | 0 | 0 | 0 |
| 13 | 266 | .07 | 23 | 381 | 113 | .87 | 5 | .09 | .29 | .38 | .17 | .04 | 28 | 2 |
| 1 | 17 | <.01 | 1 | 24 | 7 | .06 | <1 | .01 | .02 | .02 | .01 | <.01 | 2 | <1 |
| 33 | 290 | .12 | 33 | 371 | 120 | .93 | 76 | .09 | .24 | .39 | .2 | .1 | 12 | 2 |
| 18 | 298 | .12 | 33 | 376 | 122 | .95 | 139 | .09 | .17 | .4 | .21 | .1 | 12 | 2 |
| 18 | 314 | .12 | 35 | 397 | 128 | .98 | 140 | .1 | .17 | .42 | .22 | .11 | 13 | 2 |
| 10 | 300 | .12 | 34 | 381 | 123 | .95 | 144 | .09 | .1 | .41 | .21 | .1 | 12 | 2 |
| 10 | 314 | .12 | 35 | 397 | 128 | .98 | 145 | .1 | .1 | .42 | .22 | .11 | 13 | 2 |
| 4 | 301 | .1 | 28 | 407 | 126 | .98 | 149 | .09 | .1 | .34 | .22 | .1 | 13 | 2 |
| 5 | 316 | .12 | 35 | 419 | 130 | 1 | 149 | .1 | .1 | .43 | .22 | .11 | 13 | 2 |
| 9 | 284 | .12 | 27 | 370 | 257 | 1.03 | 20 | .08 | .15 | .38 | .14 | .08 | 12 | 2 |
| 104 | 869 | .58 | 79 | 1135 | 389 | 2.88 | 248 | .27 | .65 | 1.27 | .64 | .16 | 34 | 8 |
| 9 | 742 | .74 | 69 | 850 | 294 | 2.3 | 300 | .11 | .01 | .79 | .44 | .14 | 22 | 3 |
| 83 | 1420 | .36 | 132 | 1910 | 620 | 4.82 | 65 | .47 | .48 | 1.9 | 1.05 | .41 | 57 | 7 |
| 17 | 1119 | .28 | 106 | 1551 | 500 | 4.01 | 646 | .38 | .02 | 1.58 | .81 | .31 | 45 | 5 |
| 12 | 836 | .21 | 80 | 1159 | 373 | 3 | 483 | .28 | .01 | 1.18 | .61 | .23 | 34 | 4 |
| 28 | 327 | .12 | 34 | 498 | 122 | .73 | 137 | .12 | .22 | .34 | .68 | .11 | 1 | 3 |
| | | .3 | 33 | 373 | 107 | | | | | | | | | |
| 30 | 280 | .6 | 32 | 418 | 149 | 1.03 | 72 | .09 | .23 | .4 | .31 | .1 | 12 | 2 |
| 17 | 285 | .6 | 33 | 423 | 151 | 1.03 | 143 | .09 | .13 | .41 | .31 | .1 | 12 | 2 |
| 7 | 288 | .6 | 33 | 425 | 152 | 1.03 | 148 | .09 | .06 | .41 | .32 | .1 | 12 | 2 |
| 1 | 91 | .33 | 23 | 199 | 141 | .41 | 1 | .03 | .04 | .16 | .16 | .03 | 0 | <1 |
| 3 | 129 | .47 | 33 | 270 | 198 | .6 | 1 | .04 | .06 | .21 | .22 | .04 | 0 | 1 |
| 0 | 10 | .88 | 27 | 165 | 59 | .43 | 1 | .01 | .11 | .04 | .14 | <.01 | 2 | <1 |
| 32 | 301 | .8 | 53 | 497 | 165 | 1.28 | 77 | .1 | .21 | .43 | .32 | .1 | 12 | 2 |
| 149 | 330 | .51 | 47 | 419 | 138 | 1.17 | 203 | .09 | .58 | .48 | .27 | .13 | 2 | 4 |
| 194 | 269 | .71 | 32 | 367 | 155 | 1.26 | 197 | .11 | 1.01 | .55 | .21 | .15 | 30 | 2 |
| 4 | 105 | 4.74 | 84 | 350 | 142 | 3.16 | 554 | .31 | 5.31 | .07 | 5.27 | .42 | 105 | 28 |
| 38 | 396 | 4.86 | 117 | 719 | 262 | 4.09 | 630 | .41 | 5.51 | .47 | 5.46 | .52 | 118 | 31 |
| 23 | 401 | 4.86 | 118 | 726 | 264 | 4.12 | 693 | .41 | 5.41 | .48 | 5.46 | .53 | 118 | 31 |
| 14 | 406 | 4.86 | 118 | 731 | 266 | 4.12 | 698 | .41 | 5.36 | .48 | 5.47 | .53 | 118 | 31 |
| 9 | 407 | 4.83 | 112 | 755 | 268 | 4.14 | 703 | .4 | 5.3 | .42 | 5.47 | .52 | 118 | 31 |
| 1 | 13 | .48 | 15 | 130 | 53 | .17 | 4 | .04 | .08 | .04 | .42 | .03 | 4 | <1 |
| 34 | 305 | .61 | 48 | 498 | 172 | 1.09 | 79 | .13 | .26 | .44 | .62 | .13 | 16 | 3 |
| 34 | 384 | 3.76 | 53 | 620 | 244 | 1.17 | 901 | .73 | .32 | 1.26 | 10.9 | 1.02 | 32 | 34 |
| 4 | 63 | .15 | 19 | 159 | 104 | .21 | 18 | .11 | .08 | .19 | 1.1 | .09 | 10 | 1 |
| 37 | 355 | .26 | 53 | 530 | 223 | 1.14 | 95 | .2 | .32 | .59 | 1.31 | .19 | 22 | 3 |
| 22 | 188 | .51 | 28 | 332 | 161 | .68 | 38 | .1 | .11 | .41 | .27 | .08 | 6 | 1 |
| 18 | 203 | .15 | 20 | 289 | 136 | .6 | 53 | .07 | .1 | .3 | .31 | .09 | 5 | 1 |
| 245 | 316 | .85 | 39 | 431 | 217 | 1.49 | 169 | .09 | .68 | .64 | .24 | .14 | 28 | 1 |
| 1 | 82 | .07 | 10 | 111 | 47 | .26 | 38 | .03 | .07 | .11 | .06 | .03 | 3 | 1 |
| 468 | 1361 | .96 | 149 | 2117 | 851 | 7.34 | 1244 | .44 | 0 | 2.55 | 1.23 | .51 | 53 | 6 |
| 58 | 169 | .12 | 18 | 263 | 106 | .91 | 154 | .05 | 0 | .32 | .15 | .06 | 7 | 1 |
| 157 | 225 | .36 | 21 | 304 | 105 | .89 | 265 | .08 | .64 | .31 | .16 | .08 | 15 | 1 |

**Table H-1**

**Food Composition** (Computer code number is for West Diet Analysis program) (For purposes of calculations, use "0" for t, <1, <.1, <.01, etc.)

| Computer Code Number | Food Description | Measure | Wt (g) | $H_2O$ (%) | Ener (kcal) | Prot (g) | Carb (g) | Dietary Fiber (g) | Fat (g) | Fat Breakdown (g) Sat | Mono | Poly |
|---|---|---|---|---|---|---|---|---|---|---|---|---|
| | **DAIRY**—Continued | | | | | | | | | | | |
| | Ice cream, rich vanilla (16% fat): | | | | | | | | | | | |
| 127 | Hardened: ½ gallon | 1 ea | 1188 | 60 | 2554 | 49 | 264 | 0 | 154 | 88.9 | 41.5 | 5.5 |
| 128 | Cup | 1 c | 148 | 57 | 357 | 5 | 33 | 0 | 24 | 14.8 | 6.9 | .9 |
| 1724 | Ben & Jerry's | ½ c | 108 | | 230 | 4 | 21 | 0 | 17 | 10 | | |
| | Ice milk, vanilla (about 4% fat): | | | | | | | | | | | |
| 129 | Hardened: ½ gallon | 1 ea | 1048 | 68 | 1456 | 40 | 238 | 0 | 45 | 27.7 | 12.9 | 1.7 |
| 130 | Cup | 1 c | 132 | 68 | 183 | 5 | 30 | 0 | 6 | 3.5 | 1.6 | .2 |
| 131 | Soft serve (about 3% fat) | 1 c | 176 | 70 | 222 | 9 | 38 | 0 | 5 | 2.9 | 1.3 | .2 |
| | Pudding, canned (5 oz can = .55 cup): | | | | | | | | | | | |
| 135 | Chocolate | 1 ea | 142 | 69 | 189 | 4 | 32 | 1 | 6 | 1 | 2.4 | 2 |
| 136 | Tapioca | 1 ea | 142 | 74 | 169 | 3 | 27 | <1 | 5 | .9 | 2.3 | 1.9 |
| 137 | Vanilla | 1 ea | 142 | 71 | 185 | 3 | 31 | <1 | 5 | .8 | 2.2 | 1.9 |
| | Puddings, dry mix with whole milk: | | | | | | | | | | | |
| 138 | Chocolate, instant | 1 c | 294 | 74 | 326 | 9 | 55 | 3 | 9 | 5.4 | 2.7 | .5 |
| 139 | Chocolate, regular, cooked | 1 c | 284 | 74 | 315 | 9 | 51 | 3 | 10 | 5.9 | 2.8 | .4 |
| 140 | Rice, cooked | 1 c | 288 | 72 | 351 | 9 | 60 | <1 | 8 | 5.1 | 2.4 | .3 |
| 141 | Tapioca, cooked | 1 c | 282 | 74 | 321 | 8 | 55 | 0 | 8 | 5.1 | 2.3 | .3 |
| 142 | Vanilla, instant | 1 c | 284 | 73 | 324 | 8 | 56 | 0 | 8 | 4.9 | 2.4 | .4 |
| 143 | Vanilla, regular, cooked | 1 c | 280 | 75 | 311 | 8 | 52 | 0 | 8 | 5.1 | 2.4 | .4 |
| | Sherbet (2% fat): | | | | | | | | | | | |
| 132 | ½ gallon | 1 ea | 1542 | 66 | 2127 | 17 | 469 | 8 | 31 | 17.9 | 8.3 | 1.2 |
| 133 | Cup | 1 c | 198 | 66 | 273 | 2 | 60 | 1 | 4 | 2.3 | 1.1 | .2 |
| 144 | Soy milk | 1 c | 245 | 93 | 81 | 7 | 4 | 3 | 5 | .7 | 1 | 2.6 |
| 2301 | Soy milk, fortified, fat free | 1 c | 240 | 88 | 110 | 6 | 22 | 1 | 0 | 0 | 0 | 0 |
| | Yogurt, frozen, low-fat | | | | | | | | | | | |
| 1584 | Cup | 1 c | 144 | 65 | 229 | 6 | 35 | 0 | 8 | 4.9 | 2.3 | .3 |
| 1512 | Scoop | 1 ea | 79 | 74 | 78 | 4 | 15 | 0 | <1 | .1 | t | t |
| | Yogurt, lowfat: | | | | | | | | | | | |
| 1172 | Fruit added with low-calorie sweetener | 1 c | 241 | 86 | 122 | 12 | 19 | 1 | <1 | .2 | .1 | t |
| 145 | Fruit added | 1 c | 245 | 74 | 250 | 11 | 47 | <1 | 3 | 1.7 | .7 | .1 |
| 146 | Plain | 1 c | 245 | 85 | 155 | 13 | 17 | 0 | 4 | 2.4 | 1 | .1 |
| 147 | Vanilla or coffee flavor | 1 c | 245 | 79 | 209 | 13 | 34 | 0 | 3 | 2 | .8 | .1 |
| 148 | Yogurt, made with nonfat milk | 1 c | 245 | 85 | 137 | 15 | 19 | 0 | <1 | .3 | .1 | t |
| 149 | Yogurt, made with whole milk | 1 c | 245 | 88 | 150 | 9 | 11 | 0 | 8 | 5.1 | 2.2 | .2 |
| | **EGGS** | | | | | | | | | | | |
| | Raw, large: | | | | | | | | | | | |
| 150 | Whole, without shell | 1 ea | 50 | 75 | 74 | 6 | 1 | 0 | 5 | 1.5 | 1.9 | .7 |
| 151 | White | 1 ea | 33 | 88 | 16 | 3 | <1 | 0 | 0 | 0 | 0 | 0 |
| 152 | Yolk | 1 ea | 17 | 49 | 61 | 3 | <1 | 0 | 5 | 1.6 | 2 | .7 |
| | Cooked: | | | | | | | | | | | |
| 153 | Fried in margarine | 1 ea | 46 | 69 | 91 | 6 | 1 | 0 | 7 | 1.9 | 2.8 | 1.3 |
| 154 | Hard-cooked, shell removed | 1 ea | 50 | 75 | 77 | 6 | 1 | 0 | 5 | 1.6 | 2 | .7 |
| 155 | Hard-cooked, chopped | 1 c | 136 | 75 | 211 | 17 | 2 | 0 | 14 | 4.4 | 5.5 | 1.9 |
| 156 | Poached, no added salt | 1 ea | 50 | 75 | 74 | 6 | 1 | 0 | 5 | 1.5 | 1.9 | .7 |
| 157 | Scrambled with milk & margarine | 1 ea | 61 | 73 | 101 | 7 | 1 | 0 | 7 | 2.2 | 2.9 | 1.3 |
| 1681 | Egg substitute, liquid: | ½ c | 126 | 83 | 106 | 16 | 1 | 0 | 4 | .8 | 1.1 | 2 |
| 1254 | Egg Beaters, Fleischmann's | ½ c | 122 | | 60 | 12 | 2 | 0 | 0 | 0 | 0 | 0 |
| 1262 | Egg substitute, liquid, prepared | ½ c | 105 | 80 | 100 | 14 | 1 | 0 | 4 | .8 | 1.1 | 1.9 |
| | **FATS and OILS** | | | | | | | | | | | |
| 158 | Butter: Stick | ½ c | 114 | 16 | 817 | 1 | <1 | 0 | 92 | 57.7 | 27.4 | 3.4 |
| 159 | Tablespoon: | 1 tbs | 14 | 16 | 100 | <1 | <1 | 0 | 11 | 7.1 | 3.4 | .4 |
| 8025 | Unsalted | 1 tbs | 14 | 18 | 100 | <1 | <1 | 0 | 11 | 7.1 | 3.4 | .4 |
| 160 | Pat (about 1 tsp) | 1 ea | 5 | 16 | 36 | <1 | <1 | 0 | 4 | 2.5 | 1.2 | .2 |
| 1682 | Whipped | 1 tsp | 3 | 16 | 21 | <1 | <1 | 0 | 2 | 1.5 | .7 | .1 |
| | Fats, cooking: | | | | | | | | | | | |
| 1363 | Bacon fat | 1 tbs | 14 | | 125 | 0 | 0 | 0 | 14 | 6.3 | 5.9 | 1.1 |
| 1362 | Beef fat/tallow | 1 c | 205 | 0 | 1849 | 0 | 0 | 0 | 205 | 103 | 87.3 | 8.2 |
| 1364 | Chicken fat | 1 c | 205 | | 1845 | 0 | 0 | 0 | 205 | 61.1 | 91.6 | 42.8 |
| 161 | Vegetable shortening: | 1 c | 205 | 0 | 1812 | 0 | 0 | 0 | 205 | 52.1 | 91.2 | 53.5 |
| 162 | Tablespoon | 1 tbs | 13 | 0 | 115 | 0 | 0 | 0 | 13 | 3.3 | 5.8 | 3.4 |

| Chol (mg) | Calc (mg) | Iron (mg) | Magn (mg) | Pota (mg) | Sodi (mg) | Zinc (mg) | VT-A (RE) | Thia (mg) | VT-E (a-TE) | Ribo (mg) | Niac (mg) | V-B6 (mg) | Fola (µg) | VT-C (mg) |
|---|---|---|---|---|---|---|---|---|---|---|---|---|---|---|
| 1081 | 1556 | 2.49 | 143 | 2102 | 725 | 6.18 | 1829 | .58 | 4.4 | 2.16 | 1.13 | .57 | 107 | 9 |
| 90 | 173 | .07 | 16 | 235 | 83 | .59 | 272 | .06 | 0 | .24 | .12 | .06 | 7 | 1 |
| 95 | 150 | .36 | | | 55 | | 214 | | 0 | | | | | 0 |
| 147 | 1456 | 1.05 | 157 | 2211 | 891 | 4.61 | 493 | .61 | 0 | 2.78 | .94 | .68 | 63 | 8 |
| 18 | 183 | .13 | 20 | 279 | 112 | .58 | 62 | .08 | 0 | .35 | .12 | .09 | 8 | 1 |
| 21 | 276 | .11 | 25 | 389 | 123 | .93 | 51 | .09 | 0 | .35 | .21 | .08 | 11 | 2 |
| 4 | 128 | .72 | 30 | 256 | 183 | .6 | 16 | .04 | .18 | .22 | .49 | .04 | 4 | 3 |
| 1 | 119 | .33 | 11 | 148 | 168 | .38 | 0 | .03 | .13 | .14 | .44 | .14 | 4 | 1 |
| 10 | 125 | .18 | 11 | 160 | 192 | .35 | 9 | .03 | .18 | .2 | .36 | .02 | 0 | 0 |
| 32 | 300 | .85 | 53 | 488 | 835 | 1.23 | 62 | .1 | .18 | .41 | .28 | .11 | 12 | 3 |
| 34 | 315 | 1.02 | 43 | 463 | 293 | 1.28 | 74 | .09 | .17 | .49 | .29 | .1 | 11 | 2 |
| 32 | 297 | 1.09 | 37 | 372 | 314 | 1.09 | 58 | .22 | .17 | .4 | 1.28 | .1 | 11 | 2 |
| 34 | 293 | .17 | 34 | 372 | 341 | .96 | 76 | .08 | .23 | .4 | .21 | .11 | 11 | 2 |
| 31 | 287 | .2 | 34 | 364 | 812 | .94 | 71 | .09 | .18 | .39 | .21 | .1 | 11 | 2 |
| 34 | 300 | .14 | 36 | 381 | 448 | .98 | 76 | .08 | .17 | .4 | .21 | .09 | 11 | 2 |
| 77 | 833 | 2.16 | 123 | 1480 | 709 | 7.4 | 216 | .39 | .88 | 1.05 | 1.48 | .52 | 77 | 66 |
| 10 | 107 | .28 | 16 | 190 | 91 | .95 | 28 | .05 | .11 | .13 | .19 | .07 | 10 | 9 |
| 0 | 10 | 1.42 | 47 | 345 | 29 | .56 | 7 | .39 | .02 | .17 | .36 | .1 | 4 | 0 |
| 0 | 400 | 1.44 | | 20 | 60 | | 0 | .07 | | .1 | 3 | | | 0 |
| 3 | 206 | .43 | 20 | 304 | 125 | .6 | 82 | .05 | .07 | .32 | .41 | .11 | 9 | 1 |
| 1 | 137 | .07 | 13 | 175 | 53 | .67 | 1 | .03 | <.01 | .16 | .08 | .04 | 8 | 1 |
| 3 | 369 | .61 | 41 | 550 | 139 | 1.83 | 6 | .1 | .17 | .45 | .5 | .11 | 32 | 26 |
| 10 | 372 | .17 | 36 | 478 | 143 | 1.81 | 27 | .09 | .07 | .44 | .23 | .1 | 23 | 2 |
| 15 | 448 | .2 | 43 | 573 | 172 | 2.18 | 39 | .11 | .1 | .52 | .28 | .12 | 27 | 2 |
| 12 | 419 | .17 | 40 | 537 | 161 | 2.03 | 32 | .1 | .08 | .49 | .26 | .11 | 26 | 2 |
| 4 | 488 | .22 | 47 | 625 | 187 | 2.38 | 5 | .12 | .01 | .57 | .3 | .13 | 30 | 2 |
| 31 | 296 | .12 | 28 | 380 | 114 | 1.45 | 73 | .07 | .22 | .35 | .18 | .08 | 18 | 1 |
| 213 | 24 | .72 | 5 | 60 | 63 | .55 | 95 | .03 | .52 | .25 | .04 | .07 | 23 | 0 |
| 0 | 2 | .01 | 4 | 47 | 54 | <.01 | 0 | <.01 | 0 | .15 | .03 | <.01 | 1 | 0 |
| 218 | 23 | .6 | 2 | 16 | 7 | .53 | 99 | .03 | .54 | .11 | <.01 | .07 | 25 | 0 |
| 211 | 25 | .72 | 5 | 61 | 162 | .55 | 114 | .03 | .75 | .24 | .03 | .07 | 17 | 0 |
| 212 | 25 | .59 | 5 | 63 | 62 | .52 | 84 | .03 | .52 | .26 | .03 | .06 | 22 | 0 |
| 577 | 68 | 1.62 | 14 | 171 | 169 | 1.43 | 228 | .09 | 1.43 | .7 | .09 | .16 | 60 | 0 |
| 212 | 24 | .72 | 5 | 60 | 140 | .55 | 95 | .02 | .52 | .21 | .03 | .06 | 17 | 0 |
| 215 | 43 | .73 | 7 | 84 | 171 | .61 | 119 | .03 | .8 | .27 | .05 | .07 | 18 | <1 |
| 1 | 67 | 2.65 | 11 | 416 | 223 | 1.64 | 272 | .14 | .61 | .38 | .14 | <.01 | 19 | 0 |
| 0 | 80 | 2.16 | | 170 | 200 | | 80 | | .59 | | | | | |
| 1 | 63 | 2.51 | 10 | 394 | 211 | 1.55 | 258 | .11 | .58 | .34 | .12 | <.01 | 13 | 0 |
| 250 | 27 | .18 | 2 | 30 | 942 | .06 | 860 | .01 | 1.8 | .04 | .05 | <.01 | 3 | 0 |
| 31 | 3 | .02 | | 4 | 116 | .01 | 106 | <.01 | .22 | <.01 | .01 | 0 | <1 | 0 |
| 31 | 3 | .02 | | 4 | 2 | .01 | 106 | <.01 | .22 | <.01 | .01 | 0 | <1 | 0 |
| 11 | 1 | .01 | | 1 | 41 | <.01 | 38 | 0 | .08 | <.01 | <.01 | 0 | <1 | 0 |
| 7 | 1 | <.01 | | 1 | 25 | <.01 | 23 | 0 | .05 | <.01 | <.01 | 0 | <1 | 0 |
| 14 | | 0 | | | 76 | <.01 | 0 | 0 | .31 | 0 | 0 | 0 | 0 | 0 |
| 223 | 0 | 0 | 0 | | <1 | 0 | 0 | 0 | 3.08 | 0 | 0 | 0 | 0 | 0 |
| 174 | 0 | 0 | 0 | 0 | 0 | 0 | 0 | 0 | 5.54 | 0 | 0 | 0 | 0 | 0 |
| 0 | 0 | 0 | 0 | 0 | 0 | 0 | 0 | 0 | 17 | 0 | 0 | 0 | 0 | 0 |
| 0 | 0 | 0 | 0 | 0 | 0 | 0 | 0 | 0 | 1.08 | 0 | 0 | 0 | 0 | 0 |

**Table H-1**

**Food Composition** (Computer code number is for West Diet Analysis program) (For purposes of calculations, use "0" for t, <1, <.1, <.01, etc.)

| Computer Code Number | Food Description | Measure | Wt (g) | $H_2O$ (%) | Ener (kcal) | Prot (g) | Carb (g) | Dietary Fiber (g) | Fat (g) | Fat Breakdown (g) Sat | Mono | Poly |
|---|---|---|---|---|---|---|---|---|---|---|---|---|
| | FATS and OILS—Continued | | | | | | | | | | | |
| 163 | Lard: | 1 c | 205 | 0 | 1849 | 0 | 0 | 0 | 205 | 81.1 | 87 | 28.3 |
| 164 | Tablespoon | 1 tbs | 13 | 0 | 117 | 0 | 0 | 0 | 13 | 5.1 | 5.5 | 1.8 |
| | Margarine: | | | | | | | | | | | |
| 165 | Imitation (about 40% fat), soft: | 1 c | 232 | 58 | 800 | 1 | 1 | 0 | 90 | 17.9 | 36.4 | 32 |
| 166 | Tablespoon | 1 tbs | 14 | 58 | 48 | <1 | <1 | 0 | 5 | 1.1 | 2.2 | 1.9 |
| 167 | Regular, hard (about 80% fat): | ½ c | 114 | 16 | 820 | 1 | 1 | 0 | 92 | 18 | 40.8 | 29 |
| 168 | Tablespoon | 1 tbs | 14 | 16 | 101 | <1 | <1 | 0 | 11 | 2.2 | 5 | 3.6 |
| 169 | Pat | 1 ea | 5 | 16 | 36 | <1 | <1 | 0 | 4 | .8 | 1.8 | 1.3 |
| 170 | Regular, soft (about 80% fat): | 1 c | 227 | 16 | 1625 | 2 | 1 | 0 | 183 | 31.3 | 64.7 | 78.5 |
| 171 | Tablespoon | 1 tbs | 14 | 16 | 100 | <1 | <1 | 0 | 11 | 1.9 | 4 | 4.8 |
| 2056 | Saffola, unsalted | 1 tbs | 14 | 20 | 100 | 0 | 0 | 0 | 11 | 2 | 3 | 4.5 |
| 2057 | Saffola, reduced fat | 1 tbs | 14 | 37 | 60 | 0 | 0 | 0 | 8 | 1.3 | 2.7 | 4.4 |
| 172 | Spread (about 60% fat), hard: | 1 c | 227 | 37 | 1225 | 1 | 0 | 0 | 138 | 32 | 59 | 41.1 |
| 173 | Tablespoon | 1 tbs | 14 | 37 | 76 | <1 | 0 | 0 | 9 | 2 | 3.6 | 2.5 |
| 174 | Pat | 1 ea | 5 | 37 | 27 | <1 | 0 | 0 | 3 | .7 | 1.2 | 1 |
| 175 | Spread (about 60% fat), soft: | 1 c | 227 | 37 | 1225 | 1 | 0 | 0 | 138 | 29.3 | 71.5 | 31.3 |
| 176 | Tablespoon | 1 tbs | 14 | 37 | 76 | <1 | 0 | 0 | 9 | 1.8 | 4.4 | 1.9 |
| 2160 | Touch of Butter (47% fat) | 1 tbs | 14 | | 60 | 0 | 0 | 0 | 7 | 1.5 | 3.1 | 1.5 |
| | Oils: | | | | | | | | | | | |
| 1585 | Canola: | 1 c | 218 | 0 | 1927 | 0 | 0 | 0 | 218 | 15.5 | 128 | 64.5 |
| 1586 | Tablespoon | 1 tbs | 14 | 0 | 124 | 0 | 0 | 0 | 14 | 1 | 8.2 | 4.1 |
| 177 | Corn: | 1 c | 218 | 0 | 1927 | 0 | 0 | 0 | 218 | 29.4 | 54.1 | 131 |
| 178 | Tablespoon | 1 tbs | 14 | 0 | 124 | 0 | 0 | 0 | 14 | 1.9 | 3.5 | 8.4 |
| 179 | Olive: | 1 c | 216 | 0 | 1909 | 0 | 0 | 0 | 216 | 29.4 | 159 | 21.3 |
| 180 | Tablespoon | 1 tbs | 14 | 0 | 124 | 0 | 0 | 0 | 14 | 1.9 | 10.3 | 1.4 |
| 1683 | Olive, extra virgin | 1 tbs | 14 | | 126 | 0 | 0 | 0 | 14 | 2 | 10.8 | 1.3 |
| 181 | Peanut: | 1 c | 216 | 0 | 1909 | 0 | 0 | 0 | 216 | 40 | 99.8 | 71.3 |
| 182 | Tablespoon | 1 tbs | 14 | 0 | 124 | 0 | 0 | 0 | 14 | 2.6 | 6.5 | 4.6 |
| 183 | Safflower: | 1 c | 218 | 0 | 1927 | 0 | 0 | 0 | 218 | 19.8 | 26.4 | 162 |
| 184 | Tablespoon | 1 tbs | 14 | 0 | 124 | 0 | 0 | 0 | 14 | 1.3 | 1.7 | 10.4 |
| 185 | Soybean: | 1 c | 218 | 0 | 1927 | 0 | 0 | 0 | 218 | 32 | 50.8 | 126 |
| 186 | Tablespoon | 1 tbs | 14 | 0 | 124 | 0 | 0 | 0 | 14 | 2.1 | 3.3 | 8.1 |
| 187 | Soybean/cottonseed: | 1 c | 218 | 0 | 1927 | 0 | 0 | 0 | 218 | 39.5 | 64.3 | 105 |
| 188 | Tablespoon | 1 tbs | 14 | 0 | 124 | 0 | 0 | 0 | 14 | 2.5 | 4.1 | 6.7 |
| 189 | Sunflower: | 1 c | 218 | 0 | 1927 | 0 | 0 | 0 | 218 | 25.3 | 42.5 | 143 |
| 190 | Tablespoon | 1 tbs | 14 | 0 | 124 | 0 | 0 | 0 | 14 | 1.6 | 2.7 | 9.2 |
| | Salad dressings/sandwich spreads: | | | | | | | | | | | |
| 191 | Blue cheese, regular | 1 tbs | 15 | 32 | 76 | 1 | 1 | 0 | 8 | 1.5 | 1.8 | 4.2 |
| 1040 | Low calorie | 1 tbs | 15 | 79 | 15 | 1 | <1 | <1 | 1 | .2 | .5 | .4 |
| 1684 | Caesar's | 1 tbs | 12 | 36 | 56 | 1 | <1 | <1 | 5 | 1 | 3.7 | .5 |
| 192 | French, regular | 1 tbs | 16 | 38 | 69 | <1 | 3 | 0 | 7 | 1.5 | 1.3 | 3.5 |
| 193 | Low calorie | 1 tbs | 16 | 71 | 21 | <1 | 3 | 0 | 1 | .1 | .2 | .5 |
| 194 | Italian, regular | 1 tbs | 15 | 40 | 70 | <1 | 2 | 0 | 7 | 1 | 1.7 | 4.2 |
| 195 | Low calorie | 1 tbs | 15 | 84 | 16 | <1 | 1 | <1 | 1 | .2 | .3 | .9 |
| | Kraft, Deliciously Right: | | | | | | | | | | | |
| 2150 | 1000 Island | 1 tbs | 16 | | 35 | 0 | 4 | 0 | 2 | .5 | | |
| 2153 | Bacon & tomato | 1 tbs | 16 | | 31 | 1 | 2 | 0 | 3 | .5 | | |
| 2154 | Cucumber ranch | 1 tbs | 16 | | 31 | 0 | 1 | 0 | 3 | .5 | | |
| 2151 | French | 1 tbs | 16 | | 25 | 0 | 3 | 0 | 1 | .2 | | |
| 2152 | Ranch | 1 tbs | 16 | | 52 | 0 | 3 | 0 | 5 | .8 | | |
| 199 | Mayo type, regular | 1 tbs | 15 | 40 | 58 | <1 | 4 | 0 | 5 | .7 | 1.3 | 2.7 |
| 1030 | Low calorie | 1 tbs | 14 | 54 | 36 | <1 | 3 | 0 | 3 | .4 | .7 | 1.4 |
| | Mayonnaise: | | | | | | | | | | | |
| 197 | Imitation, low calorie | 1 tbs | 15 | 63 | 35 | <1 | 2 | 0 | 3 | .5 | .7 | 1.6 |
| 196 | Regular (soybean) | 1 tbs | 14 | 17 | 100 | <1 | <1 | 0 | 11 | 1.7 | 3.2 | 5.8 |
| 1488 | Regular, low calorie, low sodium | 1 tbs | 14 | 63 | 32 | <1 | 2 | 0 | 3 | .5 | .6 | 1.4 |
| 1493 | Regular, low calorie | 1 tbs | 15 | 63 | 35 | <1 | 2 | 0 | 3 | .5 | .7 | 1.6 |
| 198 | Ranch, regular | 1 tbs | 15 | 39 | 80 | 0 | <1 | 0 | 8 | 1.2 | | |
| 2251 | Low calorie | 1 tbs | 14 | 69 | 30 | <1 | 1 | 0 | 2 | .5 | | |
| 1685 | Russian | 1 tbs | 15 | 34 | 74 | <1 | 2 | 0 | 8 | 1.1 | 1.8 | 4.4 |
| 1502 | Salad dressing, low calorie, oil free | 1 tbs | 15 | 88 | 4 | <1 | 1 | <1 | <1 | 0 | 0 | 0 |

| Chol (mg) | Calc (mg) | Iron (mg) | Magn (mg) | Pota (mg) | Sodi (mg) | Zinc (mg) | VT-A (RE) | Thia (mg) | VT-E (a-TE) | Ribo (mg) | Niac (mg) | V-B6 (mg) | Fola (μg) | VT-C (mg) |
|---|---|---|---|---|---|---|---|---|---|---|---|---|---|---|
| 195 | | 0 | | | <1 | .23 | 0 | 0 | 2.46 | 0 | 0 | 0 | 0 | 0 |
| 12 | | 0 | | | <1 | .01 | 0 | 0 | .16 | 0 | 0 | 0 | 0 | 0 |
| | | | | | | | | | | | | | | |
| 0 | 41 | 0 | 4 | 59 | 2227 | 0 | 1853 | .01 | 5.41 | .05 | .03 | .01 | 2 | <1 |
| 0 | 2 | 0 | | 4 | 134 | 0 | 112 | <.01 | .33 | <.01 | <.01 | <.01 | <1 | <1 |
| 0 | 34 | .07 | 3 | 48 | 1075 | 0 | 911 | .01 | 14.6 | .04 | .03 | .01 | 1 | <1 |
| 0 | 4 | .01 | | 6 | 132 | 0 | 112 | <.01 | 1.79 | <.01 | <.01 | <.01 | <1 | <1 |
| 0 | 1 | <.01 | | 2 | 47 | 0 | 40 | <.01 | .64 | <.01 | <.01 | 0 | <1 | <1 |
| 0 | 60 | 0 | 5 | 86 | 2447 | 0 | 1813 | .02 | 27.2 | .07 | .04 | .02 | 2 | <1 |
| 0 | 4 | 0 | | 5 | 151 | 0 | 112 | <.01 | 1.68 | <.01 | <.01 | <.01 | <1 | <1 |
| | 0 | 0 | | | 0 | | 51 | | | | | | | 0 |
| | 0 | 0 | | | 115 | | 51 | | | | | | | 0 |
| 0 | 47 | 0 | 4 | 68 | 2256 | 0 | 1813 | .02 | 11.4 | .06 | .04 | .01 | 2 | <1 |
| 0 | 3 | 0 | | 4 | 139 | 0 | 112 | <.01 | .7 | <.01 | <.01 | <.01 | <1 | <1 |
| 0 | 1 | 0 | | 1 | 50 | 0 | 40 | 0 | .25 | <.01 | <.01 | 0 | <1 | <1 |
| 0 | 47 | 0 | 4 | 68 | 2256 | 0 | 1813 | .02 | 20.5 | .06 | .04 | .01 | 2 | <1 |
| 0 | 3 | 0 | | 4 | 139 | 0 | 112 | <.01 | 1.26 | <.01 | <.01 | <.01 | <1 | <1 |
| 0 | 0 | 0 | | 0 | 110 | | 100 | | 1.27 | | | | | 0 |
| | | | | | | | | | | | | | | |
| 0 | 0 | 0 | 0 | 0 | 0 | 0 | 0 | 0 | 45.8 | 0 | 0 | 0 | 0 | 0 |
| 0 | 0 | 0 | 0 | 0 | 0 | 0 | 0 | 0 | 2.94 | 0 | 0 | 0 | 0 | 0 |
| 0 | 0 | 0 | 0 | 0 | 0 | 0 | 0 | 0 | 46 | 0 | 0 | 0 | 0 | 0 |
| 0 | 0 | 0 | 0 | 0 | 0 | 0 | 0 | 0 | 2.95 | 0 | 0 | 0 | 0 | 0 |
| 0 | | .82 | | 0 | <1 | .13 | 0 | 0 | 26.8 | 0 | 0 | 0 | 0 | 0 |
| 0 | | .05 | | 0 | <1 | .01 | 0 | 0 | 1.74 | 0 | 0 | 0 | 0 | 0 |
| | | | | | | | | | 1.74 | | | | | |
| 0 | | .06 | | | <1 | .02 | 0 | 0 | 27.9 | 0 | 0 | 0 | 0 | 0 |
| 0 | | <.01 | | | <1 | <.01 | 0 | 0 | 1.81 | 0 | 0 | 0 | 0 | 0 |
| 0 | 0 | 0 | 0 | 0 | 0 | 0 | 0 | 0 | 94 | 0 | 0 | 0 | 0 | 0 |
| 0 | 0 | 0 | 0 | 0 | 0 | 0 | 0 | 0 | 6.03 | 0 | 0 | 0 | 0 | 0 |
| 0 | | .04 | | 0 | 0 | 0 | 0 | 0 | 39.7 | 0 | 0 | 0 | 0 | 0 |
| 0 | | <.01 | | 0 | 0 | 0 | 0 | 0 | 2.55 | 0 | 0 | 0 | 0 | 0 |
| 0 | 0 | 0 | 0 | 0 | 0 | 0 | 0 | 0 | 61.5 | 0 | 0 | 0 | 0 | 0 |
| 0 | 0 | 0 | 0 | 0 | 0 | 0 | 0 | 0 | 3.95 | 0 | 0 | 0 | 0 | 0 |
| 0 | 0 | 0 | 0 | 0 | 0 | 0 | 0 | 0 | 110 | 0 | 0 | 0 | 0 | 0 |
| 0 | 0 | 0 | 0 | 0 | 0 | 0 | 0 | 0 | 7.08 | 0 | 0 | 0 | 0 | 0 |
| | | | | | | | | | | | | | | |
| 3 | 12 | .03 | 0 | 6 | 164 | .04 | 10 | <.01 | 1.4 | .01 | .01 | .01 | 1 | <1 |
| <1 | 13 | .07 | 1 | 1 | 180 | .04 | <1 | <.01 | .13 | .01 | .01 | <.01 | <1 | <1 |
| 12 | 23 | .2 | 3 | 21 | 207 | .13 | 7 | <.01 | .72 | .02 | .5 | .01 | 2 | 1 |
| 0 | 2 | .06 | 0 | 13 | 219 | .01 | 21 | <.01 | 1.35 | <.01 | 0 | <.01 | 1 | 0 |
| 0 | 2 | .06 | 0 | 13 | 126 | .03 | 21 | 0 | .19 | 0 | 0 | 0 | 0 | 0 |
| 0 | 1 | .03 | | 2 | 118 | .02 | 4 | <.01 | 1.56 | <.01 | 0 | <.01 | 1 | 0 |
| 1 | | .03 | 0 | 2 | 118 | .02 | 0 | 0 | .22 | 0 | 0 | 0 | 0 | 0 |
| | | | | | | | | | | | | | | |
| 2 | 0 | 0 | | 27 | 160 | | 0 | | .19 | | | | | 0 |
| 1 | 0 | 0 | | 21 | 155 | | 0 | | .75 | | | | | 0 |
| 0 | 0 | 0 | | 10 | 232 | | 0 | | .73 | | | | | 0 |
| 0 | 0 | 0 | | 7 | 130 | | 50 | | .42 | | | | | 0 |
| 0 | 0 | 0 | | 5 | 165 | | 0 | | 1.31 | | | | | 0 |
| 4 | 2 | .03 | | 1 | 107 | .03 | 13 | <.01 | .6 | <.01 | <.01 | <.01 | 1 | 0 |
| 4 | 2 | .03 | | 1 | 99 | .02 | 9 | <.01 | .6 | <.01 | 0 | <.01 | 1 | 0 |
| | | | | | | | | | | | | | | |
| 4 | | 0 | | 1 | 75 | .02 | 0 | 0 | .96 | 0 | 0 | 0 | 0 | 0 |
| 8 | 3 | .07 | | 5 | 79 | .02 | 12 | 0 | 1.65 | 0 | <.01 | .08 | 1 | 0 |
| 3 | 0 | 0 | 0 | 1 | 15 | .01 | 1 | 0 | .53 | <.01 | 0 | 0 | <1 | 0 |
| 4 | | 0 | | 1 | 75 | .02 | 0 | 0 | .96 | 0 | 0 | 0 | 0 | 0 |
| 5 | 0 | 0 | | | 105 | | 0 | | | | | | | 0 |
| 5 | 5 | 0 | | | 120 | | 0 | | .7 | | | | | 0 |
| 3 | 3 | .09 | | 24 | 130 | .06 | 31 | .01 | 1.53 | .01 | .09 | <.01 | 2 | 1 |
| 0 | 1 | .04 | 2 | 7 | 256 | <.01 | <1 | 0 | 0 | 0 | <.01 | <.01 | <1 | <1 |

**Table H-1**

**Food Composition** (Computer code number is for West Diet Analysis program) (For purposes of calculations, use "0" for t, <1, <.1, <.01, etc.)

| Computer Code Number | Food Description | Measure | Wt (g) | $H_2O$ (%) | Ener (kcal) | Prot (g) | Carb (g) | Dietary Fiber (g) | Fat (g) | Fat Breakdown (g) | | |
|---|---|---|---|---|---|---|---|---|---|---|---|---|
| | | | | | | | | | | Sat | Mono | Poly |
| | **FATS and OILS**—Continued | | | | | | | | | | | |
| | Salad dressing, no cholesterol | | | | | | | | | | | |
| 1605 | Miracle Whip | 1 tbs | 15 | 57 | 48 | 0 | 2 | 0 | 4 | 1.1 | 1.1 | 2.1 |
| 203 | Salad dressing, from recipe, cooked | 1 tbs | 16 | 69 | 25 | 1 | 2 | 0 | 2 | .5 | .6 | .3 |
| 200 | Tartar sauce, regular | 1 tbs | 14 | 34 | 74 | <1 | 1 | <1 | 8 | 1.5 | 2.6 | 4.1 |
| 1503 | Low calorie | 1 tbs | 14 | 63 | 31 | <1 | 2 | <1 | 2 | .4 | .6 | 1.3 |
| 201 | Thousand island, regular | 1 tbs | 16 | 46 | 60 | <1 | 2 | 0 | 6 | 1 | 1.3 | 3.2 |
| 202 | Low calorie | 1 tbs | 15 | 69 | 24 | <1 | 2 | <1 | 2 | .2 | .4 | .9 |
| 204 | Vinegar and oil | 1 tbs | 16 | 47 | 72 | 0 | <1 | 0 | 8 | 1.5 | 2.4 | 3.9 |
| | Wishbone: | | | | | | | | | | | |
| 2180 | Creamy Italian, lite | 1 tbs | 15 | | 26 | <1 | 2 | | 2 | .4 | .9 | .7 |
| 2166 | Italian, lite | 1 tbs | 16 | 90 | 6 | 0 | 1 | | <1 | 0 | .2 | .1 |
| 8427 | Ranch, lite | 1 tbs | 15 | 56 | 50 | 0 | 2 | 0 | 4 | .7 | | |
| | **FRUITS and FRUIT JUICES** | | | | | | | | | | | |
| | Apples: | | | | | | | | | | | |
| | Fresh, raw, with peel: | | | | | | | | | | | |
| 205 | 2¾" diam (about 3 per lb w/cores) | 1 ea | 138 | 84 | 81 | <1 | 21 | 4 | <1 | .1 | t | .1 |
| 206 | 3¼" diam (about 2 per lb w/cores) | 1 ea | 212 | 84 | 125 | <1 | 32 | 6 | 1 | .1 | t | .2 |
| 207 | Raw, peeled slices | 1 c | 110 | 84 | 63 | <1 | 16 | 2 | <1 | .1 | t | .1 |
| 208 | Dried, sulfured | 10 ea | 64 | 32 | 156 | 1 | 42 | 6 | <1 | t | t | .1 |
| 209 | Apple juice, bottled or canned | 1 c | 248 | 88 | 117 | <1 | 29 | <1 | <1 | t | t | .1 |
| 210 | Applesauce, sweetened | 1 c | 255 | 80 | 194 | <1 | 51 | 3 | <1 | .1 | t | .1 |
| 211 | Applesauce, unsweetened | 1 c | 244 | 88 | 105 | <1 | 28 | 3 | <1 | t | t | t |
| | Apricots: | | | | | | | | | | | |
| 212 | Raw, w/o pits (about 12 per lb w/pits) | 3 ea | 105 | 86 | 50 | 1 | 12 | 3 | <1 | t | .2 | .1 |
| | Canned (fruit and liquid): | | | | | | | | | | | |
| 213 | Heavy syrup | 1 c | 240 | 78 | 199 | 1 | 52 | 4 | <1 | t | .1 | t |
| 214 | Halves | 3 ea | 120 | 78 | 100 | 1 | 26 | 2 | <1 | t | t | t |
| 215 | Juice pack | 1 c | 244 | 87 | 117 | 2 | 30 | 4 | <1 | t | t | t |
| 216 | Halves | 3 ea | 108 | 87 | 52 | 1 | 13 | 2 | <1 | t | t | t |
| 217 | Dried, halves | 10 ea | 35 | 31 | 83 | 1 | 22 | 3 | <1 | t | .1 | t |
| 218 | Dried, cooked, unsweetened, w/liquid | 1 c | 250 | 76 | 213 | 3 | 55 | 8 | <1 | t | .2 | .1 |
| 219 | Apricot nectar, canned | 1 c | 251 | 85 | 141 | 1 | 36 | 2 | <1 | t | .1 | t |
| | Avocados, raw, edible part only: | | | | | | | | | | | |
| 220 | California | 1 ea | 173 | 73 | 306 | 4 | 12 | 8 | 30 | 4.5 | 19.5 | 3.5 |
| 221 | Florida | 1 ea | 304 | 80 | 340 | 5 | 27 | 16 | 27 | 5.3 | 14.8 | 4.5 |
| 222 | Mashed, fresh, average | 1 c | 230 | 74 | 370 | 5 | 17 | 11 | 35 | 5.6 | 22.1 | 4.5 |
| | Bananas, raw, without peel: | | | | | | | | | | | |
| 223 | Whole, 8¾" long (175g w/peel) | 1 ea | 118 | 74 | 109 | 1 | 28 | 3 | 1 | .2 | t | .1 |
| 224 | Slices | 1 c | 150 | 74 | 138 | 2 | 35 | 4 | 1 | .3 | .1 | .1 |
| 1285 | Bananas, dehydrated slices | ½ c | 50 | 3 | 173 | 2 | 44 | 4 | 1 | .3 | .1 | .2 |
| 225 | Blackberries, raw | 1 c | 144 | 86 | 75 | 1 | 18 | 8 | 1 | t | .1 | .3 |
| | Blueberries: | | | | | | | | | | | |
| 226 | Fresh | 1 c | 145 | 85 | 81 | 1 | 20 | 4 | 1 | t | .1 | .2 |
| 227 | Frozen, sweetened | 10 oz | 284 | 77 | 230 | 1 | 62 | 6 | <1 | t | .1 | .2 |
| 228 | Frozen, thawed | 1 c | 230 | 77 | 186 | 1 | 51 | 5 | <1 | t | t | .1 |
| | Cherries: | | | | | | | | | | | |
| 229 | Sour, red pitted, canned water pack | 1 c | 244 | 90 | 88 | 2 | 22 | 3 | <1 | .1 | .1 | .1 |
| 230 | Sweet, red pitted, raw | 10 ea | 68 | 81 | 49 | 1 | 11 | 2 | 1 | .1 | .2 | .2 |
| 231 | Cranberry juice cocktail,vitamin C added | 1 c | 253 | 85 | 144 | 0 | 36 | <1 | <1 | t | t | .1 |
| 1411 | Cranberry juice, low calorie | 1 c | 237 | 95 | 45 | 0 | 11 | <1 | 0 | 0 | 0 | 0 |
| 232 | Cranberry-apple juice,vitamin C added | 1 c | 245 | 83 | 164 | <1 | 42 | <1 | 0 | 0 | 0 | 0 |
| 233 | Cranberry sauce, canned, strained | 1 c | 277 | 61 | 418 | 1 | 108 | 3 | <1 | t | .1 | .2 |
| 234 | Dates, whole, without pits | 10 ea | 83 | 22 | 228 | 2 | 61 | 6 | <1 | .2 | .1 | t |
| 235 | Dates, chopped | 1 c | 178 | 22 | 490 | 4 | 131 | 13 | 1 | .3 | .3 | .1 |
| 236 | Figs, dried | 10 ea | 190 | 28 | 485 | 6 | 124 | 18 | 2 | .4 | .5 | 1.1 |
| | Fruit cocktail, canned, fruit and liq: | | | | | | | | | | | |
| 237 | Heavy syrup pack | 1 c | 248 | 80 | 181 | 1 | 47 | 2 | <1 | t | t | .1 |
| 238 | Juice pack | 1 c | 237 | 87 | 109 | 1 | 28 | 2 | <1 | t | t | t |
| | Grapefruit: | | | | | | | | | | | |
| | Raw 3¾" diam (half w/rind = 241g) | | | | | | | | | | | |
| 239 | Pink/red, half fruit, edible part | 1 ea | 123 | 91 | 37 | 1 | 9 | 2 | <1 | t | t | t |
| 240 | White, half fruit, edible part | 1 ea | 118 | 90 | 39 | 1 | 10 | 1 | <1 | t | t | t |
| 241 | Canned sections with light syrup | 1 c | 254 | 84 | 152 | 1 | 39 | 1 | <1 | t | t | .1 |

| Chol (mg) | Calc (mg) | Iron (mg) | Magn (mg) | Pota (mg) | Sodi (mg) | Zinc (mg) | VT-A (RE) | Thia (mg) | VT-E (a-TE) | Ribo (mg) | Niac (mg) | V-B6 (mg) | Fola (μg) | VT-C (mg) |
|---|---|---|---|---|---|---|---|---|---|---|---|---|---|---|
| 0 | 0 | <.01 | 0 | 0 | 102 | 0 | 2 | 0 | .64 | 0 | 0 | 0 | 0 | 0 |
| 9 | 13 | .08 | 1 | 19 | 117 | .06 | 20 | .01 | .3 | .02 | .04 | <.01 | 1 | <1 |
| 7 | 3 | .13 | | 11 | 99 | .02 | 9 | <.01 | 2.24 | <.01 | 0 | <.01 | 1 | <1 |
| 3 | 2 | .09 | | 5 | 83 | .02 | 2 | 0 | .83 | <.01 | .01 | <.01 | <1 | <1 |
| 4 | 2 | .1 | | 18 | 112 | .02 | 15 | <.01 | .18 | <.01 | <.01 | <.01 | 1 | 0 |
| 2 | 2 | .09 | | 17 | 150 | .02 | 14 | <.01 | .18 | <.01 | 0 | <.01 | 1 | 0 |
| 0 | 0 | 0 | 0 | 1 | <1 | 0 | 0 | 0 | 1.41 | 0 | 0 | 0 | 0 | 0 |
| <1 | 0 | 0 | | | 148 | | 0 | 0 | .56 | 0 | 0 | | | 0 |
| 0 | 1 | 0 | | | 255 | | 2 | 0 | .24 | 0 | 0 | | | <1 |
| 2 | 0 | 0 | | | 120 | | 0 | | | | | | | 0 |
| 0 | 10 | .25 | 7 | 159 | 0 | .05 | 7 | .02 | .44 | .02 | .11 | .07 | 4 | 8 |
| 0 | 15 | .38 | 11 | 244 | 0 | .08 | 11 | .04 | .68 | .03 | .16 | .1 | 6 | 12 |
| 0 | 4 | .08 | 3 | 124 | 0 | .04 | 4 | .02 | .09 | .01 | .1 | .05 | <1 | 4 |
| 0 | 9 | .9 | 10 | 288 | 56 | .13 | 4 | 0 | .35 | .1 | .59 | .08 | 0 | 2 |
| 0 | 17 | .92 | 7 | 295 | 7 | .07 | <1 | .05 | .02 | .04 | .25 | .07 | <1 | 2 |
| 0 | 10 | .89 | 8 | 156 | 8 | .1 | 3 | .03 | .03 | .07 | .48 | .07 | 2 | 4 |
| 0 | 7 | .29 | 7 | 183 | 5 | .07 | 7 | .03 | .02 | .06 | .46 | .06 | 1 | 3 |
| 0 | 15 | .57 | 8 | 311 | 1 | .27 | 274 | .03 | .93 | .04 | .63 | .06 | 9 | 10 |
| 0 | 22 | .72 | 17 | 336 | 10 | .26 | 295 | .05 | 2.14 | .05 | .9 | .13 | 4 | 7 |
| 0 | 11 | .36 | 8 | 168 | 5 | .13 | 148 | .02 | 1.07 | .03 | .45 | .06 | 2 | 4 |
| 0 | 29 | .73 | 24 | 403 | 10 | .27 | 412 | .04 | 2.17 | .05 | .84 | .13 | 4 | 12 |
| 0 | 13 | .32 | 11 | 178 | 4 | .12 | 183 | .02 | .96 | .02 | .37 | .06 | 2 | 5 |
| 0 | 16 | 1.65 | 16 | 482 | 3 | .26 | 253 | <.01 | .52 | .05 | 1.05 | .05 | 4 | 1 |
| 0 | 40 | 4.18 | 42 | 1222 | 7 | .65 | 590 | .01 | 1.25 | .07 | 2.36 | .28 | 0 | 4 |
| 0 | 18 | .95 | 13 | 286 | 8 | .23 | 331 | .02 | .2 | .03 | .65 | .05 | 3 | 2 |
| 0 | 19 | 2.04 | 71 | 1096 | 21 | .73 | 106 | .19 | 2.32 | .21 | 3.32 | .48 | 113 | 14 |
| 0 | 33 | 1.61 | 103 | 1483 | 15 | 1.28 | 185 | .33 | 2.37 | .37 | 5.84 | .85 | 162 | 24 |
| 0 | 25 | 2.35 | 90 | 1377 | 23 | .97 | 140 | .25 | 3.08 | .28 | 4.42 | .64 | 142 | 18 |
| 0 | 7 | .37 | 34 | 467 | 1 | .19 | 9 | .05 | .32 | .12 | .64 | .68 | 22 | 11 |
| 0 | 9 | .46 | 43 | 594 | 1 | .24 | 12 | .07 | .4 | .15 | .81 | .87 | 29 | 14 |
| 0 | 11 | .57 | 54 | 746 | 1 | .3 | 15 | .09 | 0 | .12 | 1.4 | .22 | 7 | 3 |
| 0 | 46 | .82 | 29 | 282 | 0 | .39 | 23 | .04 | 1.02 | .06 | .58 | .08 | 49 | 30 |
| 0 | 9 | .25 | 7 | 129 | 9 | .16 | 14 | .07 | 1.45 | .07 | .52 | .05 | 9 | 19 |
| 0 | 17 | 1.11 | 6 | 170 | 3 | .17 | 11 | .06 | 2.02 | .15 | .72 | .17 | 19 | 3 |
| 0 | 14 | .9 | 5 | 138 | 2 | .14 | 9 | .05 | 1.63 | .12 | .58 | .14 | 15 | 2 |
| 0 | 27 | 3.34 | 15 | 239 | 17 | .17 | 183 | .04 | .32 | .1 | .43 | .11 | 19 | 5 |
| 0 | 10 | .26 | 7 | 152 | 0 | .04 | 14 | .03 | .09 | .04 | .27 | .02 | 3 | 5 |
| 0 | 8 | .38 | 5 | 45 | 5 | .18 | 1 | .02 | 0 | .02 | .09 | .05 | <1 | 90 |
| 0 | 21 | .09 | 5 | 52 | 7 | .05 | 1 | .02 | 0 | .02 | .08 | .04 | <1 | 76 |
| 0 | 17 | .15 | 5 | 66 | 5 | .1 | 1 | .01 | 0 | .05 | .15 | .05 | <1 | 78 |
| 0 | 11 | .61 | 8 | 72 | 80 | .14 | 6 | .04 | .28 | .06 | .28 | .04 | 3 | 6 |
| 0 | 27 | .95 | 29 | 541 | 2 | .24 | 4 | .07 | .08 | .08 | 1.83 | .16 | 10 | 0 |
| 0 | 57 | 2.05 | 62 | 1160 | 5 | .52 | 9 | .16 | .18 | .18 | 3.92 | .34 | 22 | 0 |
| 0 | 274 | 4.24 | 112 | 1352 | 21 | .97 | 25 | .13 | 9.5 | .17 | 1.32 | .43 | 14 | 2 |
| 0 | 15 | .72 | 12 | 218 | 15 | .2 | 50 | .04 | .72 | .05 | .93 | .12 | 6 | 5 |
| 0 | 19 | .5 | 17 | 225 | 9 | .21 | 73 | .03 | .47 | .04 | .95 | .12 | 6 | 6 |
| 0 | 13 | .15 | 10 | 159 | 0 | .09 | 32 | .04 | .31 | .02 | .23 | .05 | 15 | 47 |
| 0 | 14 | .07 | 11 | 175 | 0 | .08 | 1 | .04 | .29 | .02 | .32 | .05 | 12 | 39 |
| 0 | 36 | 1.02 | 25 | 328 | 5 | .2 | 0 | .1 | .63 | .05 | .62 | .05 | 22 | 54 |

**Table H-1**

**Food Composition** (Computer code number is for West Diet Analysis program) (For purposes of calculations, use "0" for t, <1, <.1, <.01, etc.)

| Computer Code Number | Food Description | Measure | Wt (g) | $H_2O$ (%) | Ener (kcal) | Prot (g) | Carb (g) | Dietary Fiber (g) | Fat (g) | Fat Breakdown (g) Sat | Mono | Poly |
|---|---|---|---|---|---|---|---|---|---|---|---|---|
| | **FRUITS and FRUIT JUICES** | | | | | | | | | | | |
| | Grapefruit juice: | | | | | | | | | | | |
| 242 | Fresh, white, raw | 1 c | 247 | 90 | 96 | 1 | 23 | <1 | <1 | t | t | .1 |
| 243 | Canned, unsweetened | 1 c | 247 | 90 | 94 | 1 | 22 | <1 | <1 | t | t | .1 |
| 244 | Sweetened | 1 c | 250 | 87 | 115 | 1 | 28 | <1 | <1 | t | t | .1 |
| | Frozen concentrate, unsweetened: | | | | | | | | | | | |
| 245 | Undiluted, 6-fl-oz can | ¾ c | 207 | 62 | 302 | 4 | 72 | 1 | 1 | .1 | .1 | .2 |
| 246 | Diluted with 3 cans water | 1 c | 247 | 89 | 101 | 1 | 24 | <1 | <1 | t | t | .1 |
| | Grapes, raw European (adherent skin): | | | | | | | | | | | |
| 247 | Thompson seedless | 10 ea | 50 | 81 | 35 | <1 | 9 | <1 | <1 | .1 | t | .1 |
| 248 | Tokay/Emperor, seeded types | 10 ea | 50 | 81 | 35 | <1 | 9 | <1 | <1 | .1 | t | .1 |
| | Grape juice: | | | | | | | | | | | |
| 249 | Bottled or canned | 1 c | 253 | 84 | 154 | 1 | 38 | <1 | <1 | .1 | t | .1 |
| | Frozen concentrate, sweetened: | | | | | | | | | | | |
| 250 | Undiluted, 6-fl-oz can, vit C added | ¾ c | 216 | 54 | 387 | 1 | 96 | 1 | 1 | .2 | t | .2 |
| 251 | Diluted with 3 cans water, vit C added | 1 c | 250 | 87 | 128 | <1 | 32 | <1 | <1 | .1 | t | .1 |
| 1410 | Low calorie | 1 c | 253 | 84 | 154 | 1 | 38 | <1 | <1 | .1 | t | .1 |
| 252 | Kiwi fruit, raw, peeled (88 g with peel) | 1 ea | 76 | 83 | 46 | 1 | 11 | 3 | <1 | t | t | .2 |
| 253 | Lemons, raw, without peel and seeds (about 4 per lb whole) | 1 ea | 58 | 89 | 17 | 1 | 5 | 2 | <1 | t | t | .1 |
| | Lemon juice: | | | | | | | | | | | |
| 254 | Fresh: | 1 c | 244 | 91 | 61 | 1 | 21 | 1 | 0 | 0 | 0 | 0 |
| 255 | Tablespoon | 1 tbs | 15 | 91 | 4 | <1 | 1 | <1 | 0 | 0 | 0 | 0 |
| 256 | Canned or bottled, unsweetened: | 1 c | 244 | 92 | 51 | 1 | 16 | 1 | 1 | .1 | t | .2 |
| 257 | Tablespoon | 1 tbs | 15 | 92 | 3 | <1 | 1 | <1 | <1 | t | t | t |
| 258 | Frozen, single strength, unsweetened: | 1 c | 244 | 92 | 54 | 1 | 16 | 1 | 1 | .1 | t | .2 |
| 2298 | Tablespoon | 1 tbs | 15 | 92 | 3 | <1 | 1 | <1 | <1 | t | t | t |
| | Lime juice: | | | | | | | | | | | |
| 260 | Fresh: | 1 c | 246 | 90 | 66 | 1 | 22 | 1 | <1 | t | t | .1 |
| 261 | Tablespoon | 1 tbs | 15 | 90 | 4 | <1 | 1 | <1 | <1 | t | t | t |
| 262 | Canned or bottled, unsweetened | 1 c | 246 | 92 | 52 | 1 | 16 | 1 | 1 | .1 | .1 | .2 |
| 263 | Mangos, raw, edible part (300 g w/skin & seeds) | 1 ea | 207 | 82 | 135 | 1 | 35 | 4 | 1 | .1 | .2 | .1 |
| | Melons, raw, without rind and contents: | | | | | | | | | | | |
| 264 | Cantaloupe, 5" diam (2⅓ lb whole with refuse), orange flesh | ½ ea | 276 | 90 | 97 | 2 | 23 | 2 | 1 | .2 | t | .3 |
| 265 | Honeydew, 6½" diam (5¼ lb whole with refuse), slice = 1/10 melon | 1 pce | 160 | 90 | 56 | 1 | 15 | 1 | <1 | t | t | .1 |
| 266 | Nectarines, raw, w/o pits, 2¼" diam | 1 ea | 136 | 86 | 67 | 1 | 16 | 2 | 1 | .1 | .2 | .3 |
| | Oranges, raw: | | | | | | | | | | | |
| 267 | Whole w/o peel and seeds, 2⅝" diam (180 g with peel and seeds) | 1 ea | 131 | 87 | 62 | 1 | 15 | 3 | <1 | t | t | t |
| 268 | Sections, without membranes | 1 c | 180 | 87 | 85 | 2 | 21 | 4 | <1 | t | t | t |
| | Orange juice: | | | | | | | | | | | |
| 269 | Fresh, all varieties | 1 c | 248 | 88 | 112 | 2 | 26 | <1 | <1 | .1 | .1 | .1 |
| 270 | Canned, unsweetened | 1 c | 249 | 89 | 105 | 1 | 24 | <1 | <1 | t | .1 | .1 |
| 271 | Chilled | 1 c | 249 | 88 | 110 | 2 | 25 | <1 | 1 | .1 | .1 | .2 |
| | Frozen concentrate: | | | | | | | | | | | |
| 272 | Undiluted (6-oz can) | ¾ c | 213 | 58 | 339 | 5 | 81 | 2 | <1 | .1 | .1 | .1 |
| 273 | Diluted w/3 parts water by volume | 1 c | 249 | 88 | 112 | 2 | 27 | <1 | <1 | t | t | t |
| 1345 | Orange juice, from dry crystals | 1 c | 248 | 88 | 114 | 0 | 29 | 0 | 0 | 0 | 0 | 0 |
| 274 | Orange and grapefruit juice, canned | 1 c | 247 | 89 | 106 | 1 | 25 | <1 | <1 | t | t | t |
| | Papayas, raw: | | | | | | | | | | | |
| 275 | ½" slices | 1 c | 140 | 89 | 55 | 1 | 14 | 3 | <1 | .1 | .1 | t |
| 276 | Whole, 3½" diam by 5⅛" w/o seeds and skin (1 lb w/refuse) | 1 ea | 304 | 89 | 119 | 2 | 30 | 5 | <1 | .1 | .1 | .1 |
| 1031 | Papaya nectar, canned | 1 c | 250 | 85 | 143 | <1 | 36 | 1 | <1 | .1 | .1 | .1 |
| | Peaches: | | | | | | | | | | | |
| 277 | Raw, whole, 2½" diam, peeled, pitted (about 4 per lb whole) | 1 ea | 98 | 88 | 42 | 1 | 11 | 2 | <1 | t | t | t |
| 278 | Raw, sliced | 1 c | 170 | 88 | 73 | 1 | 19 | 3 | <1 | t | .1 | .1 |
| | Canned, fruit and liquid: | | | | | | | | | | | |
| 279 | Heavy syrup pack: | 1 c | 262 | 79 | 194 | 1 | 52 | 3 | <1 | t | .1 | .1 |
| 280 | Half | 1 ea | 98 | 79 | 72 | <1 | 19 | 1 | <1 | t | t | t |

| Chol (mg) | Calc (mg) | Iron (mg) | Magn (mg) | Pota (mg) | Sodi (mg) | Zinc (mg) | VT-A (RE) | Thia (mg) | VT-E (a-TE) | Ribo (mg) | Niac (mg) | V-B6 (mg) | Fola (µg) | VT-C (mg) |
|---|---|---|---|---|---|---|---|---|---|---|---|---|---|---|
| 0 | 22 | .49 | 30 | 400 | 2 | .12 | 2 | .1 | .12 | .05 | .49 | .11 | 25 | 94 |
| 0 | 17 | .49 | 25 | 378 | 2 | .22 | 2 | .1 | .12 | .05 | .57 | .05 | 26 | 72 |
| 0 | 20 | .9 | 25 | 405 | 5 | .15 | 0 | .1 | .12 | .06 | .8 | .05 | 26 | 67 |
| 0 | 56 | 1.01 | 79 | 1001 | 6 | .37 | 6 | .3 | .37 | .16 | 1.6 | .32 | 26 | 248 |
| 0 | 20 | .35 | 27 | 336 | 2 | .12 | 2 | .1 | .12 | .05 | .54 | .11 | 9 | 83 |
| 0 | 5 | .13 | 3 | 92 | 1 | .02 | 3 | .05 | .35 | .03 | .15 | .05 | 2 | 5 |
| 0 | 5 | .13 | 3 | 92 | 1 | .02 | 3 | .05 | .35 | .03 | .15 | .05 | 2 | 5 |
| 0 | 23 | .61 | 25 | 334 | 8 | .13 | 3 | .07 | 0 | .09 | .66 | .16 | 7 | <1 |
| 0 | 28 | .78 | 32 | 160 | 15 | .28 | 6 | .11 | .38 | .2 | .93 | .32 | 9 | 179 |
| 0 | 10 | .25 | 10 | 52 | 5 | .1 | 2 | .04 | .12 | .06 | .31 | .1 | 3 | 60 |
| 0 | 23 | .61 | 25 | 334 | 8 | .13 | 3 | .07 | 0 | .09 | .66 | .16 | 7 | <1 |
| 0 | 20 | .31 | 23 | 252 | 4 | .13 | 14 | .01 | .85 | .04 | .38 | .07 | 29 | 74 |
| 0 | 15 | .35 | 5 | 80 | 1 | .03 | 2 | .02 | .14 | .01 | .06 | .05 | 6 | 31 |
| 0 | 17 | .07 | 15 | 303 | 2 | .12 | 5 | .07 | .22 | .02 | .24 | .12 | 31 | 112 |
| 0 | 1 | <.01 | 1 | 19 | <1 | .01 | <1 | <.01 | .01 | <.01 | .01 | .01 | 2 | 7 |
| 0 | 27 | .32 | 19 | 249 | 51 | .15 | 5 | .1 | .22 | .02 | .48 | .1 | 25 | 60 |
| 0 | 2 | .02 | 1 | 15 | 3 | .01 | <1 | .01 | .01 | <.01 | .03 | .01 | 2 | 4 |
| 0 | 19 | .29 | 19 | 217 | 2 | .12 | 2 | .14 | .22 | .03 | .33 | .15 | 23 | 77 |
| 0 | 1 | .02 | 1 | 13 | <1 | .01 | <1 | .01 | .01 | <.01 | .02 | .01 | 1 | 5 |
| 0 | 22 | .07 | 15 | 268 | 2 | .15 | 2 | .05 | .22 | .02 | .25 | .11 | 20 | 72 |
| 0 | 1 | <.01 | 1 | 16 | <1 | .01 | <1 | <.01 | .01 | <.01 | .01 | .01 | 1 | 4 |
| 0 | 29 | .57 | 17 | 185 | 39 | .15 | 5 | .08 | .17 | .01 | .4 | .07 | 19 | 16 |
| 0 | 21 | .27 | 19 | 323 | 4 | .08 | 805 | .12 | 2.32 | .12 | 1.21 | .28 | 29 | 57 |
| 0 | 30 | .58 | 30 | 853 | 25 | .44 | 889 | .1 | .41 | .06 | 1.58 | .32 | 47 | 116 |
| 0 | 10 | .11 | 11 | 434 | 16 | .11 | 6 | .12 | .24 | .03 | .96 | .09 | 10 | 40 |
| 0 | 7 | .2 | 11 | 288 | 0 | .12 | 101 | .02 | 1.21 | .06 | 1.35 | .03 | 5 | 7 |
| 0 | 52 | .13 | 13 | 237 | 0 | .09 | 27 | .11 | .31 | .05 | .37 | .08 | 40 | 70 |
| 0 | 72 | .18 | 18 | 326 | 0 | .13 | 38 | .16 | .43 | .07 | .51 | .11 | 54 | 96 |
| 0 | 27 | .5 | 27 | 496 | 2 | .12 | 50 | .22 | .22 | .07 | .99 | .1 | 75 | 124 |
| 0 | 20 | 1.1 | 27 | 436 | 5 | .17 | 45 | .15 | .22 | .07 | .78 | .22 | 45 | 86 |
| 0 | 25 | .42 | 27 | 473 | 2 | .1 | 20 | .28 | .47 | .05 | .7 | .13 | 45 | 82 |
| 0 | 68 | .75 | 72 | 1435 | 6 | .38 | 60 | .6 | .68 | .14 | 1.53 | .33 | 330 | 294 |
| 0 | 22 | .25 | 25 | 473 | 2 | .12 | 20 | .2 | .47 | .04 | .5 | .11 | 109 | 97 |
| 0 | 62 | .2 | 2 | 50 | 12 | .1 | 551 | <.01 | 0 | .04 | 0 | 0 | 143 | 121 |
| 0 | 20 | 1.14 | 25 | 390 | 7 | .17 | 30 | .14 | .17 | .07 | .83 | .06 | 35 | 72 |
| 0 | 34 | .14 | 14 | 360 | 4 | .1 | 39 | .04 | 1.57 | .04 | .47 | .03 | 53 | 86 |
| 0 | 73 | .3 | 30 | 781 | 9 | .21 | 85 | .08 | 3.4 | .1 | 1.03 | .06 | 116 | 188 |
| 0 | 25 | .85 | 7 | 77 | 12 | .37 | 27 | .01 | .05 | .01 | .37 | .02 | 5 | 7 |
| 0 | 5 | .11 | 7 | 193 | 0 | .14 | 53 | .02 | .69 | .04 | .97 | .02 | 3 | 6 |
| 0 | 8 | .19 | 12 | 335 | 0 | .24 | 92 | .03 | 1.19 | .07 | 1.68 | .03 | 6 | 11 |
| 0 | 8 | .71 | 13 | 241 | 16 | .24 | 86 | .03 | 2.33 | .06 | 1.61 | .05 | 8 | 7 |
| 0 | 3 | .26 | 5 | 90 | 6 | .09 | 32 | .01 | .87 | .02 | .6 | .02 | 3 | 3 |

**Table H-1**

**Food Composition** (Computer code number is for West Diet Analysis program) (For purposes of calculations, use "0" for t, <1, <.1, <.01, etc.)

| Computer Code Number | Food Description | Measure | Wt (g) | $H_2O$ (%) | Ener (kcal) | Prot (g) | Carb (g) | Dietary Fiber (g) | Fat (g) | Fat Breakdown (g) Sat | Mono | Poly |
|---|---|---|---|---|---|---|---|---|---|---|---|---|
| | **FRUITS and FRUIT JUICES**—Continued | | | | | | | | | | | |
| 281 | Juice pack: | 1 c | 248 | 87 | 109 | 2 | 29 | 3 | <1 | t | t | t |
| 282 | Half | 1 ea | 98 | 87 | 43 | 1 | 11 | 1 | <1 | t | t | t |
| 283 | Dried, uncooked | 10 ea | 130 | 32 | 311 | 5 | 80 | 11 | 1 | .1 | .4 | .5 |
| 284 | Dried, cooked, fruit and liquid | 1 c | 258 | 78 | 199 | 3 | 51 | 7 | 1 | .1 | .2 | .3 |
| | Frozen, slice, sweetened: | | | | | | | | | | | |
| 285 | 10-oz package,vitamin C added | 1 ea | 284 | 75 | 267 | 2 | 68 | 5 | <1 | t | .1 | .2 |
| 286 | Cup, thawed measure, vitamin C added | 1 c | 250 | 75 | 235 | 2 | 60 | 4 | <1 | t | .1 | .2 |
| 1032 | Peach nectar, canned | 1 c | 249 | 86 | 134 | 1 | 35 | 1 | <1 | t | t | t |
| | Pears: | | | | | | | | | | | |
| | Fresh, with skin, cored: | | | | | | | | | | | |
| 287 | Bartlett, 2½" diam (about 2½ per lb) | 1 ea | 166 | 84 | 98 | 1 | 25 | 4 | 1 | t | .1 | .2 |
| 288 | Bosc, 2⅕" diam (about 3 per lb) | 1 ea | 139 | 84 | 82 | 1 | 21 | 3 | 1 | t | .1 | .1 |
| 289 | D'Anjou, 3" diam (about 2 per lb) | 1 ea | 209 | 84 | 123 | 1 | 32 | 5 | 1 | t | .2 | .2 |
| | Canned, fruit and liquid: | | | | | | | | | | | |
| 290 | Heavy syrup pack: | 1 c | 266 | 80 | 197 | 1 | 51 | 4 | <1 | t | .1 | .1 |
| 291 | Half | 1 ea | 76 | 80 | 56 | <1 | 15 | 1 | <1 | t | t | t |
| 292 | Juice pack: | 1 c | 248 | 86 | 124 | 1 | 32 | 4 | <1 | t | t | t |
| 293 | Half | 1 ea | 76 | 86 | 38 | <1 | 10 | 1 | <1 | t | t | t |
| 294 | Dried halves | 10 ea | 175 | 27 | 459 | 3 | 122 | 13 | 1 | .1 | .2 | .3 |
| 1033 | Pear nectar, canned | 1 c | 250 | 84 | 150 | <1 | 39 | 1 | <1 | t | t | t |
| | Pineapple: | | | | | | | | | | | |
| 295 | Fresh chunks, diced | 1 c | 155 | 86 | 76 | 1 | 19 | 2 | 1 | t | .1 | .2 |
| | Canned, fruit and liquid: | | | | | | | | | | | |
| | Heavy syrup pack: | | | | | | | | | | | |
| 296 | Crushed, chunks, tidbits | ½ c | 127 | 79 | 99 | <1 | 26 | 1 | <1 | t | t | .1 |
| 297 | Slices | 1 ea | 49 | 79 | 38 | <1 | 10 | <1 | <1 | t | t | t |
| 298 | Juice pack, crushed, chunks, tidbits | 1 c | 250 | 83 | 150 | 1 | 39 | 2 | <1 | t | t | .1 |
| 299 | Juice pack, slices | 1 ea | 47 | 83 | 28 | <1 | 7 | <1 | <1 | t | t | t |
| 300 | Pineapple juice, canned, unsweetened | 1 c | 250 | 85 | 140 | 1 | 34 | <1 | <1 | t | t | .1 |
| | Plantains, yellow flesh, without peel: | | | | | | | | | | | |
| 301 | Raw slices (whole=179 g w/o peel) | 1 c | 148 | 65 | 181 | 2 | 47 | 3 | 1 | .2 | t | .1 |
| 302 | Cooked, boiled, sliced | 1 c | 154 | 67 | 179 | 1 | 48 | 4 | <1 | .1 | t | .1 |
| | Plums: | | | | | | | | | | | |
| 303 | Fresh, medium, 2⅛" diam | 1 ea | 66 | 85 | 36 | 1 | 9 | 1 | <1 | t | .3 | .1 |
| 304 | Fresh, small, 1½" diam | 1 ea | 28 | 85 | 15 | <1 | 4 | <1 | <1 | t | .1 | t |
| | Canned, purple, with liquid: | | | | | | | | | | | |
| 305 | Heavy syrup pack: | 1 c | 258 | 76 | 230 | 1 | 60 | 3 | <1 | t | .2 | .1 |
| 306 | Plums | 3 ea | 138 | 76 | 123 | <1 | 32 | 1 | <1 | t | .1 | t |
| 307 | Juice pack: | 1 c | 252 | 84 | 146 | 1 | 38 | 3 | <1 | t | t | t |
| 308 | Plums | 3 ea | 138 | 84 | 80 | 1 | 21 | 1 | <1 | t | t | t |
| 1698 | Pomegranate, fresh | 1 ea | 154 | 81 | 105 | 1 | 26 | 1 | <1 | .1 | .1 | .1 |
| | Prunes, dried, pitted: | | | | | | | | | | | |
| 309 | Uncooked (10 = 97 g w/pits, 84 g w/o pits) | 10 ea | 84 | 32 | 201 | 2 | 53 | 6 | <1 | t | .3 | .1 |
| 310 | Cooked, unsweetened, fruit & liq (250 g w/pits) | 1 c | 248 | 70 | 265 | 3 | 70 | 16 | 1 | t | .4 | .1 |
| 311 | Prune juice, bottled or canned | 1 c | 256 | 81 | 182 | 2 | 45 | 3 | <1 | t | .1 | t |
| | Raisins, seedless: | | | | | | | | | | | |
| 312 | Cup, not pressed down | 1 c | 145 | 15 | 435 | 5 | 115 | 6 | 1 | .2 | t | .2 |
| 313 | One packet, ½ oz | ½ oz | 14 | 15 | 42 | <1 | 11 | 1 | <1 | t | t | t |
| | Raspberries: | | | | | | | | | | | |
| 314 | Fresh | 1 c | 123 | 87 | 60 | 1 | 14 | 8 | 1 | t | .1 | .4 |
| 315 | Frozen, sweetened | 10 oz | 284 | 73 | 293 | 2 | 74 | 12 | <1 | t | t | .3 |
| 316 | Cup, thawed measure | 1 c | 250 | 73 | 258 | 2 | 65 | 11 | <1 | t | t | .2 |
| 317 | Rhubarb, cooked, added sugar | 1 c | 240 | 68 | 278 | 1 | 75 | 5 | <1 | t | t | .1 |
| | Strawberries: | | | | | | | | | | | |
| 318 | Fresh, whole, capped | 1 c | 144 | 92 | 43 | 1 | 10 | 3 | 1 | t | .1 | .3 |
| | Frozen, sliced, sweetened: | | | | | | | | | | | |
| 319 | 10-oz container | 10 oz | 284 | 73 | 273 | 2 | 74 | 5 | <1 | t | .1 | .2 |
| 320 | Cup, thawed measure | 1 c | 255 | 73 | 245 | 1 | 66 | 5 | <1 | t | t | .2 |
| | Tangerines, without peel and seeds: | | | | | | | | | | | |
| 321 | Fresh (2⅜" whole) 116 g w/refuse | 1 ea | 84 | 88 | 37 | 1 | 9 | 2 | <1 | t | t | t |
| 322 | Canned, light syrup, fruit and liquid | 1 c | 252 | 83 | 154 | 1 | 41 | 2 | <1 | t | t | t |

| Chol (mg) | Calc (mg) | Iron (mg) | Magn (mg) | Pota (mg) | Sodi (mg) | Zinc (mg) | VT-A (RE) | Thia (mg) | VT-E (a-TE) | Ribo (mg) | Niac (mg) | V-B6 (mg) | Fola (µg) | VT-C (mg) |
|---|---|---|---|---|---|---|---|---|---|---|---|---|---|---|
| 0 | 15 | .67 | 17 | 317 | 10 | .27 | 94 | .02 | 3.72 | .04 | 1.44 | .05 | 8 | 9 |
| 0 | 6 | .26 | 7 | 125 | 4 | .11 | 37 | .01 | 1.47 | .02 | .57 | .02 | 3 | 4 |
| 0 | 36 | 5.28 | 55 | 1294 | 9 | .74 | 281 | <.01 | 0 | .28 | 5.69 | .09 | <1 | 6 |
| 0 | 23 | 3.38 | 33 | 826 | 5 | .46 | 52 | .01 | 0 | .05 | 3.92 | .1 | <1 | 10 |
| 0 | 9 | 1.05 | 14 | 369 | 17 | .14 | 79 | .04 | 2.53 | .1 | 1.85 | .05 | 9 | 268 |
| 0 | 7 | .92 | 12 | 325 | 15 | .12 | 70 | .03 | 2.23 | .09 | 1.63 | .04 | 8 | 236 |
| 0 | 12 | .47 | 10 | 100 | 17 | .2 | 65 | .01 | .2 | .03 | .72 | .02 | 3 | 13 |
| 0 | 18 | .41 | 10 | 208 | 0 | .2 | 3 | .03 | .83 | .07 | .17 | .03 | 12 | 7 |
| 0 | 15 | .35 | 8 | 174 | 0 | .17 | 3 | .03 | .69 | .06 | .14 | .02 | 10 | 6 |
| 0 | 23 | .52 | 12 | 261 | 0 | .25 | 4 | .04 | 1.05 | .08 | .21 | .04 | 15 | 8 |
| 0 | 13 | .58 | 11 | 173 | 13 | .21 | 0 | .03 | 1.33 | .06 | .64 | .04 | 3 | 3 |
| 0 | 4 | .17 | 3 | 49 | 4 | .06 | 0 | .01 | .38 | .02 | .18 | .01 | 1 | 1 |
| 0 | 22 | .72 | 17 | 238 | 10 | .22 | 2 | .03 | 1.24 | .03 | .5 | .03 | 3 | 4 |
| 0 | 7 | .22 | 5 | 73 | 3 | .07 | 1 | .01 | .38 | .01 | .15 | .01 | 1 | 1 |
| 0 | 59 | 3.68 | 58 | 933 | 10 | .68 | 1 | .01 | 0 | .25 | 2.4 | .13 | 0 | 12 |
| 0 | 12 | .65 | 7 | 32 | 10 | .17 | <1 | <.01 | .25 | .03 | .32 | .03 | 3 | 3 |
| 0 | 11 | .57 | 22 | 175 | 2 | .12 | 3 | .14 | .15 | .06 | .65 | .13 | 16 | 24 |
| 0 | 18 | .48 | 20 | 132 | 1 | .15 | 1 | .11 | .13 | .03 | .36 | .09 | 6 | 9 |
| 0 | 7 | .19 | 8 | 51 | <1 | .06 | <1 | .04 | .05 | .01 | .14 | .04 | 2 | 4 |
| 0 | 35 | .7 | 35 | 305 | 2 | .25 | 10 | .24 | .25 | .05 | .71 | .18 | 12 | 24 |
| 0 | 7 | .13 | 7 | 57 | <1 | .05 | 2 | .04 | .05 | .01 | .13 | .03 | 2 | 4 |
| 0 | 42 | .65 | 32 | 335 | 2 | .27 | 1 | .14 | .05 | .05 | .64 | .24 | 58 | 27 |
| 0 | 4 | .89 | 55 | 739 | 6 | .21 | 167 | .08 | .4 | .08 | 1.02 | .44 | 33 | 27 |
| 0 | 3 | .89 | 49 | 716 | 8 | .2 | 140 | .07 | .22 | .08 | 1.16 | .37 | 40 | 17 |
| 0 | 3 | .07 | 5 | 114 | 0 | .07 | 21 | .03 | .4 | .06 | .33 | .05 | 1 | 6 |
| 0 | 1 | .03 | 2 | 48 | 0 | .03 | 9 | .01 | .17 | .03 | .14 | .02 | 1 | 3 |
| 0 | 23 | 2.17 | 13 | 235 | 49 | .18 | 67 | .04 | 1.81 | .1 | .75 | .07 | 6 | 1 |
| 0 | 12 | 1.16 | 7 | 126 | 26 | .1 | 36 | .02 | .97 | .05 | .4 | .04 | 3 | 1 |
| 0 | 25 | .86 | 20 | 388 | 3 | .28 | 255 | .06 | 1.76 | .15 | 1.19 | .07 | 7 | 7 |
| 0 | 14 | .47 | 11 | 213 | 1 | .15 | 139 | .03 | .97 | .08 | .65 | .04 | 4 | 4 |
| 0 | 5 | .46 | 5 | 399 | 5 | .18 | 0 | .05 | .85 | .05 | .46 | .16 | 9 | 9 |
| 0 | 43 | 2.08 | 38 | 626 | 3 | .44 | 167 | .07 | 1.22 | .14 | 1.65 | .22 | 3 | 3 |
| 0 | 57 | 2.75 | 50 | 828 | 5 | .59 | 77 | .06 | <.01 | .25 | 1.79 | .54 | <1 | 7 |
| 0 | 31 | 3.02 | 36 | 707 | 10 | .54 | 1 | .04 | .03 | .18 | 2.01 | .56 | 1 | 10 |
| 0 | 71 | 3.02 | 48 | 1088 | 17 | .39 | 1 | .23 | 1.02 | .13 | 1.19 | .36 | 5 | 5 |
| 0 | 7 | .29 | 5 | 105 | 2 | .04 | <1 | .02 | .1 | .01 | .11 | .03 | <1 | <1 |
| 0 | 27 | .7 | 22 | 187 | 0 | .57 | 16 | .04 | .55 | .11 | 1.11 | .07 | 32 | 31 |
| 0 | 43 | 1.85 | 37 | 324 | 3 | .51 | 17 | .05 | 1.28 | .13 | .65 | .1 | 74 | 47 |
| 0 | 37 | 1.63 | 32 | 285 | 2 | .45 | 15 | .05 | 1.13 | .11 | .57 | .08 | 65 | 41 |
| 0 | 348 | .5 | 29 | 230 | 2 | .19 | 17 | .04 | .48 | .05 | .48 | .05 | 13 | 8 |
| 0 | 20 | .55 | 14 | 239 | 1 | .19 | 4 | .03 | .2 | .09 | .33 | .08 | 25 | 82 |
| 0 | 31 | 1.68 | 20 | 278 | 9 | .17 | 6 | .04 | .4 | .14 | 1.14 | .08 | 42 | 118 |
| 0 | 28 | 1.5 | 18 | 250 | 8 | .15 | 5 | .04 | .36 | .13 | 1.02 | .08 | 38 | 106 |
| 0 | 12 | .08 | 10 | 132 | 1 | .2 | 77 | .09 | .2 | .02 | .13 | .06 | 17 | 26 |
| 0 | 18 | .93 | 20 | 197 | 15 | .6 | 212 | .13 | .86 | .11 | 1.12 | .11 | 12 | 50 |

**Table H-1**

**Food Composition** (Computer code number is for West Diet Analysis program) (For purposes of calculations, use "0" for t, <1, <.1, <.01, etc.)

| Computer Code Number | Food Description | Measure | Wt (g) | $H_2O$ (%) | Ener (kcal) | Prot (g) | Carb (g) | Dietary Fiber (g) | Fat (g) | Fat Breakdown (g) Sat | Mono | Poly |
|---|---|---|---|---|---|---|---|---|---|---|---|---|
| | **FRUITS and FRUIT JUICES**—Continued | | | | | | | | | | | |
| 323 | Tangerine juice, canned, sweetened | 1 c | 249 | 87 | 125 | 1 | 30 | <1 | <1 | t | t | .1 |
| | Watermelon, raw, without rind and seeds: | | | | | | | | | | | |
| 324 | Piece, 1/16 wedge | 1 pce | 286 | 91 | 91 | 2 | 20 | 1 | 1 | .1 | .3 | .4 |
| 325 | Diced | 1 c | 152 | 91 | 49 | 1 | 11 | 1 | 1 | .1 | .2 | .2 |
| | **BAKED GOODS: BREADS, CAKES, COOKIES, CRACKERS, PIES** | | | | | | | | | | | |
| 326 | Bagels, plain, enriched, 3½" diam. | 1 ea | 71 | 33 | 195 | 7 | 38 | 2 | 1 | .2 | .1 | .5 |
| 1663 | Bagel, oat bran | 1 ea | 71 | 33 | 181 | 8 | 38 | 3 | 1 | .1 | .2 | .3 |
| | Biscuits: | | | | | | | | | | | |
| 327 | From home recipe | 1 ea | 60 | 29 | 212 | 4 | 27 | 1 | 10 | 2.6 | 4.2 | 2.5 |
| 328 | From mix | 1 ea | 57 | 29 | 191 | 4 | 28 | 1 | 7 | 1.6 | 2.4 | 2.5 |
| 329 | From refrigerated dough | 1 ea | 74 | 27 | 276 | 4 | 34 | 1 | 13 | 8.7 | 3.4 | .5 |
| 330 | Bread crumbs, dry, grated (see # 364, 365 for soft crumbs) | 1 c | 108 | 6 | 427 | 13 | 78 | 3 | 6 | 1.4 | 2.3 | 1.7 |
| 2087 | Bread sticks, brown & serve | 1 ea | 57 | 34 | 150 | 7 | 28 | 1 | 1 | .5 | .5 | .5 |
| | Breads: | | | | | | | | | | | |
| 331 | Boston brown, canned, 3¼" slice | 1 pce | 45 | 47 | 88 | 2 | 19 | 2 | 1 | .1 | .1 | .3 |
| 332 | Cracked wheat (¼ cracked-wheat & ¾ enr wheat flour): 1-lb loaf | 1 ea | 454 | 36 | 1180 | 39 | 225 | 25 | 18 | 4.2 | 8.6 | 3.1 |
| 333 | Slice (18 per loaf) | 1 pce | 25 | 36 | 65 | 2 | 12 | 1 | 1 | .2 | .5 | .2 |
| 334 | Slice, toasted | 1 pce | 23 | 30 | 65 | 2 | 12 | 1 | 1 | .2 | .5 | .2 |
| 335 | French/Vienna, enriched: 1-lb loaf | 1 ea | 454 | 34 | 1243 | 40 | 236 | 14 | 14 | 2.9 | 5.5 | 3.1 |
| 337 | Slice, 4¾ x 4 x ½" | 1 pce | 25 | 34 | 68 | 2 | 13 | 1 | 1 | .2 | .3 | .2 |
| 336 | French, slice, 5 x 2½" | 1 pce | 25 | 34 | 68 | 2 | 13 | 1 | 1 | .2 | .3 | .2 |
| | French toast: see Mixed Dishes, and Fast Foods, #691 | | | | | | | | | | | |
| 2083 | Honey wheatberry | 1 pce | 38 | 38 | 100 | 3 | 18 | 2 | 1 | 0 | .5 | |
| 338 | Italian, enriched: 1-lb loaf | 1 ea | 454 | 36 | 1230 | 40 | 227 | 12 | 16 | 3.9 | 3.7 | 6.3 |
| 339 | Slice, 4½ x 3¼ x ¾" | 1 pce | 30 | 36 | 81 | 3 | 15 | 1 | 1 | .3 | .2 | .4 |
| 340 | Mixed grain, enriched: 1-lb loaf | 1 ea | 454 | 38 | 1135 | 45 | 211 | 29 | 17 | 3.7 | 6.9 | 4.2 |
| 341 | Slice (18 per loaf) | 1 pce | 26 | 38 | 65 | 3 | 12 | 2 | 1 | .2 | .4 | .2 |
| 342 | Slice, toasted | 1 pce | 24 | 32 | 65 | 3 | 12 | 2 | 1 | .2 | .4 | .2 |
| 343 | Oatmeal, enriched: 1-lb loaf | 1 ea | 454 | 37 | 1221 | 38 | 220 | 18 | 20 | 3.2 | 7.2 | 7.7 |
| 344 | Slice (18 per loaf) | 1 pce | 27 | 37 | 73 | 2 | 13 | 1 | 1 | .2 | .4 | .5 |
| 345 | Slice, toasted | 1 pce | 25 | 31 | 73 | 2 | 13 | 1 | 1 | .2 | .4 | .5 |
| 346 | Pita pocket bread, enr, 6½" round | 1 ea | 60 | 32 | 165 | 5 | 33 | 1 | 1 | .1 | .1 | .3 |
| | Pumpernickel (⅔ rye & ⅓ enr wheat flr): | | | | | | | | | | | |
| 347 | 1-lb loaf | 1 ea | 454 | 38 | 1135 | 40 | 216 | 29 | 14 | 2 | 4.2 | 5.6 |
| 348 | Slice, 5 x 4 x ⅜" | 1 pce | 26 | 38 | 65 | 2 | 12 | 2 | 1 | .1 | .2 | .3 |
| 349 | Slice, toasted | 1 pce | 29 | 32 | 80 | 3 | 15 | 2 | 1 | .1 | .3 | .4 |
| 350 | Raisin, enriched: 1-lb loaf | 1 ea | 454 | 34 | 1243 | 36 | 237 | 19 | 20 | 4.9 | 10.5 | 3.1 |
| 351 | Slice (18 per loaf) | 1 pce | 26 | 34 | 71 | 2 | 14 | 1 | 1 | .3 | .6 | .2 |
| 352 | Slice, toasted | 1 pce | 24 | 28 | 71 | 2 | 14 | 1 | 1 | .3 | .6 | .2 |
| 353 | Rye, light (⅓ rye & ⅔ enr wheat flr): 1-lb loaf | 1 ea | 454 | 37 | 1175 | 39 | 219 | 26 | 15 | 2.9 | 6 | 3.6 |
| 354 | Slice, 4¾ x 3¾ x 7/16" | 1 pce | 32 | 37 | 83 | 3 | 15 | 2 | 1 | .2 | .4 | .3 |
| 355 | Slice, toasted | 1 pce | 24 | 31 | 68 | 2 | 13 | 2 | 1 | .2 | .3 | .2 |
| 356 | Wheat (enr wheat & whole-wheat flour): 1-lb loaf | 1 ea | 454 | 37 | 1160 | 43 | 213 | 25 | 19 | 3.9 | 7.3 | 4.5 |
| 357 | Slice (18 per loaf) | 1 pce | 25 | 37 | 65 | 2 | 12 | 1 | 1 | .2 | .4 | .2 |
| 358 | Slice, toasted | 1 pce | 23 | 32 | 65 | 2 | 12 | 1 | 1 | .2 | .4 | .2 |
| 359 | White, enriched: 1-lb loaf | 1 ea | 454 | 35 | 1293 | 36 | 225 | 9 | 26 | 5.4 | 5.9 | 12.6 |
| 360 | Slice | 1 pce | 42 | 35 | 120 | 3 | 21 | 1 | 2 | .5 | .5 | 1.2 |
| 361 | Slice, toasted | 1 pce | 38 | 29 | 119 | 3 | 21 | 1 | 2 | .5 | .5 | 1.2 |
| 366 | Whole-wheat: 1-lb loaf | 1 ea | 454 | 38 | 1116 | 44 | 209 | 31 | 19 | 4.2 | 7.6 | 4.5 |
| 367 | Slice (16 per loaf) | 1 pce | 28 | 38 | 69 | 3 | 13 | 2 | 1 | .3 | .5 | .3 |
| 368 | Slice, toasted | 1 pce | 25 | 30 | 69 | 3 | 13 | 2 | 1 | .3 | .5 | .3 |
| | Bread stuffing, prepared from mix: | | | | | | | | | | | |
| 369 | Dry type | 1 c | 200 | 65 | 356 | 6 | 43 | 6 | 17 | 3.5 | 7.6 | 5.2 |
| 370 | Moist type, with egg and margarine | 1 c | 232 | 65 | 390 | 9 | 51 | 5 | 17 | 3.4 | 7.4 | 4.9 |

| Chol (mg) | Calc (mg) | Iron (mg) | Magn (mg) | Pota (mg) | Sodi (mg) | Zinc (mg) | VT-A (RE) | Thia (mg) | VT-E (a-TE) | Ribo (mg) | Niac (mg) | V-B6 (mg) | Fola (μg) | VT-C (mg) |
|---|---|---|---|---|---|---|---|---|---|---|---|---|---|---|
| 0 | 45 | .5 | 20 | 443 | 2 | .07 | 105 | .15 | .22 | .05 | .25 | .08 | 11 | 55 |
| 0 | 23 | .49 | 31 | 332 | 6 | .2 | 106 | .23 | .43 | .06 | .57 | .41 | 6 | 27 |
| 0 | 12 | .26 | 17 | 176 | 3 | .11 | 56 | .12 | .23 | .03 | .3 | .22 | 3 | 15 |
| 0 | 52 | 2.53 | 21 | 72 | 379 | .62 | 0 | .38 | .02 | .22 | 3.24 | .04 | 62 | 0 |
| 0 | 9 | 2.19 | 40 | 145 | 360 | 1.48 | <1 | .23 | .17 | .24 | 2.1 | .14 | 57 | <1 |
| 2 | 141 | 1.74 | 11 | 73 | 348 | .32 | 14 | .21 | 1.45 | .19 | 1.77 | .02 | 37 | <1 |
| 2 | 105 | 1.17 | 14 | 107 | 544 | .35 | 15 | .2 | .23 | .2 | 1.72 | .04 | 3 | <1 |
| 5 | 89 | 1.64 | 9 | 87 | 584 | .29 | 24 | .27 | .44 | .18 | 1.63 | .03 | 6 | 0 |
| 0 | 245 | 6.61 | 50 | 239 | 931 | 1.32 | <1 | .83 | .95 | .47 | 7.4 | .11 | 118 | 0 |
| 0 | 60 | 2.7 | | | 290 | | 0 | .22 | | .1 | 1.6 | | | 0 |
| <1 | 31 | .94 | 28 | 143 | 284 | .22 | 5 | .01 | .13 | .05 | .5 | .04 | 5 | 0 |
| | | | | | | | | | | | | | | 0 |
| 0 | 195 | 12.8 | 236 | 804 | 2442 | 5.63 | 0 | 1.63 | 2.56 | 1.09 | 16.7 | 1.38 | 277 | |
| 0 | 11 | .7 | 13 | 44 | 135 | .31 | 0 | .09 | .14 | .06 | .92 | .08 | 15 | 0 |
| 0 | 11 | .7 | 13 | 44 | 135 | .31 | 0 | .07 | .14 | .05 | .83 | .07 | 7 | 0 |
| 0 | 341 | 11.5 | 123 | 513 | 2764 | 3.95 | 0 | 2.36 | 1.07 | 1.49 | 21.6 | .19 | 431 | 0 |
| 0 | 19 | .63 | 7 | 28 | 152 | .22 | 0 | .13 | .06 | .08 | 1.19 | .01 | 24 | 0 |
| 0 | 19 | .63 | 7 | 28 | 152 | .22 | 0 | .13 | .06 | .08 | 1.19 | .01 | 24 | 0 |
| 0 | 20 | .72 | | | 200 | | 0 | .12 | .24 | .07 | .8 | | | 0 |
| 0 | 354 | 13.3 | 123 | 499 | 2651 | 3.9 | 0 | 2.15 | 1.26 | 1.33 | 19.9 | .22 | 431 | 0 |
| 0 | 23 | .88 | 8 | 33 | 175 | .26 | 0 | .14 | .08 | .09 | 1.31 | .01 | 28 | 0 |
| 0 | 413 | 15.8 | 241 | 926 | 2210 | 5.77 | 0 | 1.85 | 2.79 | 1.55 | 19.8 | 1.51 | 363 | 1 |
| 0 | 24 | .9 | 14 | 53 | 127 | .33 | 0 | .11 | .16 | .09 | 1.14 | .09 | 21 | <1 |
| 0 | 24 | .9 | 14 | 53 | 127 | .33 | 0 | .08 | .16 | .08 | 1.02 | .08 | 16 | <1 |
| 0 | 300 | 12.3 | 168 | 645 | 2719 | 4.63 | 9 | 1.81 | 1.56 | 1.09 | 14.3 | .31 | 281 | 2 |
| 0 | 18 | .73 | 10 | 38 | 162 | .27 | 1 | .11 | .09 | .06 | .85 | .02 | 17 | <1 |
| 0 | 18 | .73 | 10 | 38 | 163 | .28 | <1 | .09 | .09 | .06 | .77 | .02 | 13 | <1 |
| 0 | 52 | 1.57 | 16 | 72 | 322 | .5 | 0 | .36 | .02 | .2 | 2.78 | .02 | 57 | 0 |
| 0 | 309 | 13 | 245 | 944 | 3046 | 6.72 | 0 | 1.48 | 2.3 | 1.38 | 14 | .57 | 363 | 0 |
| 0 | 18 | .75 | 14 | 54 | 174 | .38 | 0 | .08 | .13 | .08 | .8 | .03 | 21 | 0 |
| 0 | 21 | .91 | 17 | 66 | 214 | .47 | 0 | .08 | .17 | .09 | .89 | .04 | 20 | 0 |
| 0 | 300 | 13.2 | 118 | 1030 | 1770 | 3.27 | 1 | 1.54 | 3.44 | 1.81 | 15.8 | .31 | 395 | 2 |
| 0 | 17 | .75 | 7 | 59 | 101 | .19 | <1 | .09 | .2 | .1 | .9 | .02 | 23 | <1 |
| 0 | 17 | .76 | 7 | 59 | 102 | .19 | <1 | .07 | .2 | .09 | .81 | .02 | 18 | <1 |
| 0 | 331 | 12.8 | 182 | 754 | 2996 | 5.18 | 2 | 1.97 | 2.51 | 1.52 | 17.3 | .34 | 390 | 1 |
| 0 | 23 | .91 | 13 | 53 | 211 | .36 | <1 | .14 | .18 | .11 | 1.22 | .02 | 27 | <1 |
| 0 | 19 | .74 | 10 | 44 | 174 | .3 | <1 | .09 | .15 | .08 | .9 | .02 | 17 | <1 |
| 0 | 572 | 15.8 | 209 | 627 | 2447 | 4.77 | 0 | 2.09 | 3 | 1.45 | 20.5 | .49 | 204 | 0 |
| 0 | 26 | .83 | 11 | 50 | 133 | .26 | 0 | .1 | .14 | .07 | 1.03 | .02 | 19 | 0 |
| 0 | 26 | .83 | 11 | 50 | 132 | .26 | 0 | .08 | .14 | .06 | .93 | .02 | 15 | 0 |
| 14 | 259 | 13.5 | 86 | 663 | 1629 | 2.91 | 100 | 1.84 | 4.95 | 1.74 | 16.3 | .23 | 413 | 1 |
| 1 | 24 | 1.25 | 8 | 61 | 151 | .27 | 9 | .17 | .46 | .16 | 1.51 | .02 | 38 | <1 |
| 1 | 24 | 1.24 | 8 | 61 | 150 | .27 | 8 | .13 | .46 | .14 | 1.35 | .02 | 12 | <1 |
| 0 | 327 | 15 | 390 | 1144 | 2392 | 8.81 | 0 | 1.59 | 4.72 | .93 | 17.4 | .81 | 227 | 0 |
| 0 | 20 | .92 | 24 | 71 | 148 | .54 | 0 | .1 | .29 | .06 | 1.08 | .05 | 14 | 0 |
| 0 | 20 | .93 | 24 | 71 | 148 | .54 | 0 | .08 | .23 | .05 | .97 | .04 | 9 | 0 |
| 0 | 64 | 2.18 | 24 | 148 | 1086 | .56 | 162 | .27 | 2.8 | .21 | 2.96 | .08 | 202 | 0 |
| 0 | 148 | 3.8 | 35 | 304 | 1069 | .74 | 160 | .39 | 2.78 | .33 | 3.69 | .12 | 39 | 4 |

**Table H-1**

**Food Composition** (Computer code number is for West Diet Analysis program) (For purposes of calculations, use "0" for t, <1, <.1, <.01, etc.)

| Computer Code Number | Food Description | Measure | Wt (g) | $H_2O$ (%) | Ener (kcal) | Prot (g) | Carb (g) | Dietary Fiber (g) | Fat (g) | Fat Breakdown (g) Sat | Mono | Poly |
|---|---|---|---|---|---|---|---|---|---|---|---|---|
| | **BAKED GOODS: BREADS, CAKES, COOKIES, CRACKERS, PIES**—Continued | | | | | | | | | | | |
| | Cakes, prepared from mixes using enriched flour and veg shortening, w/frostings made from margarine: | | | | | | | | | | | |
| | Angel food: | | | | | | | | | | | |
| 371 | Whole cake, 9 ¾" diam tube | 1 ea | 340 | 33 | 877 | 20 | 197 | 5 | 3 | .4 | .2 | 1.2 |
| 372 | Piece, 1/12 of cake | 1 pce | 28 | 33 | 72 | 2 | 16 | <1 | <1 | t | t | .1 |
| 373 | Boston cream pie, ⅛ of cake | 1 pce | 123 | 45 | 310 | 3 | 53 | 2 | 10 | 3.1 | 5.4 | 1.2 |
| | Coffee cake: | | | | | | | | | | | |
| 374 | Whole cake, 7¾ x 5⅝ x 1¼" | 1 ea | 336 | 30 | 1068 | 18 | 177 | 4 | 32 | 6.3 | 13 | 10.7 |
| 375 | Piece, ⅙ of cake | 1 pce | 56 | 30 | 178 | 3 | 30 | 1 | 5 | 1 | 2.2 | 1.8 |
| | Devil's food, chocolate frosting: | | | | | | | | | | | |
| 376 | Whole cake, 2 layer, 8 or 9" diam | 1 ea | 1021 | 23 | 3747 | 42 | 557 | 29 | 167 | 47.9 | 91.9 | 19.5 |
| 377 | Piece, 1/16 of cake | 1 pce | 64 | 23 | 235 | 3 | 35 | 2 | 10 | 3 | 5.8 | 1.2 |
| 378 | Cupcake, 2½" diam | 1 ea | 42 | 23 | 154 | 2 | 23 | 1 | 7 | 2 | 3.8 | .8 |
| | Gingerbread: | | | | | | | | | | | |
| 379 | Whole cake, 8" square | 1 ea | 603 | 33 | 1863 | 24 | 306 | 7 | 61 | 15.8 | 34 | 8.1 |
| 380 | Piece, 1/9 of cake | 1 pce | 67 | 33 | 207 | 3 | 34 | 1 | 7 | 1.8 | 3.8 | .9 |
| | Yellow, chocolate frosting, 2 layer: | | | | | | | | | | | |
| 381 | Whole cake, 8 or 9" in diam | 1 ea | 1024 | 22 | 3880 | 39 | 567 | 18 | 178 | 49 | 99 | 21.4 |
| 382 | Piece, 1/16 of cake | 1 pce | 64 | 22 | 243 | 2 | 35 | 1 | 11 | 3.1 | 6.2 | 1.3 |
| | Cakes from home recipes w/enr flour: | | | | | | | | | | | |
| | Carrot cake, made with veg oil, cream cheese frosting: | | | | | | | | | | | |
| 383 | Whole, 9 x 13" cake | 1 ea | 1776 | 21 | 7743 | 82 | 838 | 21 | 469 | 86.8 | 116 | 242 |
| 384 | Piece, 1/16 of cake, 2¼ x 3¼" slice | 1 pce | 111 | 21 | 484 | 5 | 52 | 1 | 29 | 5.4 | 7.2 | 15.1 |
| | Fruitcake, dark: | | | | | | | | | | | |
| 385 | Whole cake, 7½"diam tube, 2¼"high | 1 ea | 1376 | 25 | 4458 | 40 | 848 | 51 | 125 | 15.4 | 57.4 | 44.6 |
| 386 | Piece, 1/32 of cake, ⅔" arc | 1 pce | 43 | 25 | 139 | 1 | 26 | 2 | 4 | .5 | 1.8 | 1.4 |
| | Sheet, plain, made w/veg shortening, no frosting: | | | | | | | | | | | |
| 387 | Whole cake, 9" square | 1 ea | 774 | 24 | 2817 | 35 | 433 | 3 | 108 | 29.9 | 51.5 | 25.5 |
| 388 | Piece, 1/9 of cake | 1 pce | 86 | 24 | 313 | 4 | 48 | <1 | 12 | 3.3 | 5.7 | 2.8 |
| | Sheet, plain, made w/margarine, uncooked white frosting: | | | | | | | | | | | |
| 389 | Whole cake, 9" square | 1 ea | 576 | 22 | 2148 | 20 | 339 | 2 | 83 | 13.8 | 35.5 | 29.5 |
| 390 | Piece, 1/9 of cake | 1 pce | 64 | 22 | 239 | 2 | 38 | <1 | 9 | 1.5 | 3.9 | 3.3 |
| | Cakes, commerical: | | | | | | | | | | | |
| | Cheesecake: | | | | | | | | | | | |
| 401 | Whole cake, 9" diam | 1 ea | 960 | 46 | 3081 | 53 | 245 | 4 | 216 | 111 | 74.4 | 13.2 |
| 402 | Piece, 1/12 of cake | 1 pce | 80 | 46 | 257 | 4 | 20 | <1 | 18 | 9.2 | 6.2 | 1.1 |
| | Pound cake: | | | | | | | | | | | |
| 393 | Loaf, 8½ x 3½ x 3" | 1 ea | 340 | 25 | 1319 | 19 | 166 | 2 | 68 | 38.1 | 19 | 3.7 |
| 394 | Slice, 1/17 of loaf, 2" slice | 1 pce | 28 | 25 | 109 | 2 | 14 | <1 | 6 | 3.1 | 1.6 | .3 |
| | Snack cakes: | | | | | | | | | | | |
| 395 | Chocolate w/creme filling,Ding Dong | 1 ea | 50 | 20 | 188 | 2 | 30 | <1 | 7 | 1.6 | 2.7 | 2.1 |
| 396 | Sponge cake w/creme filling,Twinkie | 1 ea | 43 | 20 | 157 | 1 | 27 | <1 | 5 | 1.2 | 1.9 | 1.5 |
| 1677 | Sponge cake, 1/12 of 12" cake | 1 pce | 38 | 30 | 110 | 2 | 23 | <1 | 1 | .3 | .4 | .2 |
| | White, white frosting, 2 layer: | | | | | | | | | | | |
| 397 | Whole cake, 8 or 9" diam | 1 ea | 1136 | 20 | 4260 | 37 | 716 | 11 | 153 | 68.2 | 60.1 | 15.4 |
| 398 | Piece, 1/16 of cake | 1 pce | 71 | 20 | 266 | 2 | 45 | 1 | 10 | 4.3 | 3.8 | 1 |
| | Yellow, chocolate frosting, 2 layer: | | | | | | | | | | | |
| 399 | Whole cake, 8 or 9" in diam | 1 ea | 1024 | 22 | 3880 | 39 | 567 | 18 | 178 | 49 | 99 | 21.4 |
| 400 | Piece, 1/16 of cake | 1 pce | 64 | 22 | 243 | 2 | 35 | 1 | 11 | 3.1 | 6.2 | 1.3 |
| 1332 | Bagel chips | 5 pce | 70 | 3 | 298 | 6 | 52 | 6 | 7 | 1.2 | 2 | 3.4 |
| 2225 | Bagel chips, onion garlic, toasted | 1 oz | 28 | | 193 | 5 | 31 | 3 | 8 | 1.7 | 5.2 | 0 |
| 1035 | Cheese puffs/Cheetos | 1 c | 20 | 1 | 111 | 2 | 11 | <1 | 7 | 1.3 | 4.1 | 1 |
| | Cookies made with enriched flour: | | | | | | | | | | | |
| | Brownies with nuts: | | | | | | | | | | | |
| 403 | Commercial w/frosting, 1½ x 1¾ x ⅞" | 1 ea | 61 | 14 | 247 | 3 | 39 | 1 | 10 | 2.6 | 5.1 | 1.6 |
| 1902 | Fat free fudge, Entenmann's | 1 pce | 40 | 24 | 110 | 2 | 27 | 1 | 0 | 0 | 0 | 0 |

| Chol (mg) | Calc (mg) | Iron (mg) | Magn (mg) | Pota (mg) | Sodi (mg) | Zinc (mg) | VT-A (RE) | Thia (mg) | VT-E (a-TE) | Ribo (mg) | Niac (mg) | V-B6 (mg) | Fola (µg) | VT-C (mg) |
|---|---|---|---|---|---|---|---|---|---|---|---|---|---|---|
| 0 | 476 | 1.77 | 41 | 316 | 2546 | .24 | 0 | .35 | .34 | 1.67 | 3 | .1 | 119 | 0 |
| 0 | 39 | .15 | 3 | 26 | 210 | .02 | 0 | .03 | .03 | .14 | .25 | .01 | 10 | 0 |
| 45 | 28 | .47 | 7 | 48 | 177 | .2 | 28 | .5 | 1.3 | .33 | .23 | .03 | 18 | <1 |
| 165 | 457 | 4.8 | 60 | 376 | 1414 | 1.51 | 134 | .56 | 5.58 | .59 | 5.11 | .17 | 228 | 1 |
| 27 | 76 | .8 | 10 | 63 | 236 | .25 | 22 | .09 | .93 | .1 | .85 | .03 | 38 | <1 |
| 470 | 439 | 22.5 | 347 | 2042 | 3410 | 7.04 | 286 | .28 | 17.3 | 1.36 | 5.89 | .32 | 174 | 1 |
| 29 | 27 | 1.41 | 22 | 128 | 214 | .44 | 18 | .02 | 1.08 | .08 | .37 | .02 | 11 | <1 |
| 19 | 18 | .92 | 14 | 84 | 140 | .29 | 12 | .01 | .71 | .06 | .24 | .01 | 7 | <1 |
| 211 | 416 | 20 | 96 | 1453 | 2761 | 2.47 | 96 | 1.14 | 8.26 | 1.12 | 9.41 | .23 | 60 | 1 |
| 23 | 46 | 2.22 | 11 | 161 | 307 | .27 | 11 | .13 | .92 | .12 | 1.05 | .02 | 7 | <1 |
| 563 | 379 | 21.3 | 307 | 1822 | 3450 | 6.35 | 276 | 1.23 | 27.6 | 1.61 | 12.8 | .3 | 225 | 1 |
| 35 | 24 | 1.33 | 19 | 114 | 216 | .4 | 17 | .08 | 1.73 | .1 | .8 | .02 | 14 | <1 |
| 959 | 444 | 22.2 | 320 | 1989 | 4368 | 8.7 | 6819 | 2.42 | 74.9 | 2.77 | 17.9 | 1.35 | 213 | 19 |
| 60 | 28 | 1.39 | 20 | 124 | 273 | .54 | 426 | .15 | 4.68 | .17 | 1.12 | .08 | 13 | 1 |
| 69 | 454 | 28.5 | 220 | 2105 | 3715 | 3.72 | 261 | .69 | 42.9 | 1.36 | 10.9 | .63 | 261 | 5 |
| 2 | 14 | .89 | 7 | 66 | 116 | .12 | 8 | .02 | 1.34 | .04 | .34 | .02 | 8 | <1 |
| 503 | 495 | 11.7 | 108 | 611 | 2322 | 2.74 | 372 | 1.24 | 11 | 1.39 | 10.1 | .26 | 54 | 2 |
| 56 | 55 | 1.3 | 12 | 68 | 258 | .3 | 41 | .14 | 1.22 | .15 | 1.12 | .03 | 6 | <1 |
| 323 | 357 | 6.16 | 35 | 305 | 1981 | 1.44 | 109 | .58 | 10.9 | .4 | 2.88 | .2 | 156 | 1 |
| 36 | 40 | .68 | 4 | 34 | 220 | .16 | 12 | .06 | 1.22 | .04 | .32 | .02 | 17 | <1 |
| 528 | 490 | 6.05 | 106 | 864 | 1987 | 4.9 | 1545 | .27 | 10.1 | 1.85 | 1.87 | .5 | 173 | 6 |
| 44 | 41 | .5 | 9 | 72 | 166 | .41 | 129 | .02 | .84 | .15 | .16 | .04 | 14 | <1 |
| 751 | 119 | 4.69 | 37 | 405 | 1353 | 1.56 | 530 | .47 | 2.24 | .78 | 4.45 | .12 | 139 | <1 |
| 62 | 10 | .39 | 3 | 33 | 111 | .13 | 44 | .04 | .18 | .06 | .37 | .01 | 11 | <1 |
| 8 | 36 | 1.68 | 20 | 61 | 213 | .28 | 2 | .11 | 1.01 | .15 | 1.22 | .01 | 14 | <1 |
| 7 | 19 | .55 | 3 | 39 | 157 | .13 | 2 | .07 | .83 | .06 | .52 | .01 | 12 | <1 |
| 39 | 27 | 1.03 | 4 | 38 | 93 | .19 | 17 | .09 | .17 | .1 | .73 | .02 | 15 | 0 |
| 91 | 545 | 9.09 | 60 | 659 | 2658 | 1.76 | 368 | 1.14 | 20.4 | 1.48 | 10.2 | .16 | 64 | 1 |
| 6 | 34 | .57 | 4 | 41 | 166 | .11 | 23 | .07 | 1.28 | .09 | .64 | .01 | 4 | <1 |
| 563 | 379 | 21.3 | 307 | 1822 | 3450 | 6.35 | 276 | 1.23 | 27.6 | 1.61 | 12.8 | .3 | 225 | 1 |
| 35 | 24 | 1.33 | 19 | 114 | 216 | .4 | 17 | .08 | 1.73 | .1 | .8 | .02 | 14 | <1 |
| 0 | 9 | 1.38 | 41 | 167 | 419 | .9 | 0 | .13 | .46 | .12 | 1.57 | .19 | 58 | 0 |
| 0 | 0 | 2.52 |  |  | 490 |  | 0 | .39 | <.01 | .24 | 3.5 |  |  | 0 |
| 1 | 12 | .47 | 4 | 33 | 210 | .08 | 7 | .05 | 1.02 | .07 | .65 | .03 | 24 | <1 |
| 10 | 18 | 1.37 | 19 | 91 | 190 | .44 | 12 | .16 | 1.3 | .13 | 1.05 | .02 | 13 | <1 |
| 0 | 0 | 1.08 |  | 90 | 140 |  | 0 |  | .01 |  |  |  |  | 0 |

**Table H-1**

**Food Composition** (Computer code number is for West Diet Analysis program) (For purposes of calculations, use "0" for t, <1, <.1, <.01, etc.)

| Computer Code Number | Food Description | Measure | Wt (g) | $H_2O$ (%) | Ener (kcal) | Prot (g) | Carb (g) | Dietary Fiber (g) | Fat (g) | Fat Breakdown (g) Sat | Mono | Poly |
|---|---|---|---|---|---|---|---|---|---|---|---|---|
| | **BAKED GOODS: BREADS, CAKES, COOKIES, CRACKERS, PIES—**Continued | | | | | | | | | | | |
| | Chocolate chip cookies: | | | | | | | | | | | |
| 405 | Commercial, 2¼" diam | 4 ea | 60 | 12 | 275 | 2 | 35 | 2 | 15 | 4.5 | 7.8 | 1.6 |
| 406 | Home recipe, 2¼" diam | 4 ea | 64 | 6 | 312 | 4 | 37 | 2 | 18 | 5.2 | 6.7 | 5.4 |
| 407 | From refrigerated dough, 2¼" diam | 4 ea | 64 | 13 | 284 | 3 | 39 | 1 | 13 | 4.5 | 6.5 | 1.3 |
| 408 | Fig bars | 4 ea | 64 | 16 | 223 | 2 | 45 | 3 | 5 | .9 | 2.6 | .8 |
| 2052 | Fruit bar, no fat | 1 ea | 28 | | 90 | 2 | 21 | 0 | 0 | 0 | 0 | 0 |
| 2162 | Fudge, fat free, Snackwell | 1 ea | 16 | 14 | 53 | 1 | 12 | <1 | <1 | .1 | .1 | t |
| 409 | Oatmeal raisin, 2⅝" diam | 4 ea | 60 | 6 | 261 | 4 | 41 | 2 | 10 | 1.9 | 4.1 | 3 |
| 410 | Peanut butter, home recipe, 2⅝"diam | 4 ea | 80 | 6 | 380 | 7 | 47 | 2 | 19 | 3.5 | 8.7 | 5.8 |
| 411 | Sandwich-type, all | 4 ea | 40 | 2 | 189 | 2 | 28 | 1 | 8 | 1.7 | 4.7 | 1.1 |
| 412 | Shortbread, commercial, small | 4 ea | 32 | 4 | 161 | 2 | 21 | 1 | 8 | 2 | 4.3 | 1 |
| 413 | Shortbread, home recipe, large | 2 ea | 22 | 3 | 120 | 1 | 12 | <1 | 7 | 4.5 | 2.1 | .3 |
| 414 | Sugar from refrigerated dough, 2" diam | 4 ea | 48 | 5 | 232 | 2 | 31 | <1 | 11 | 2.8 | 6.2 | 1.4 |
| 1874 | Vanilla sandwich, Snackwell's | 2 ea | 26 | 4 | 109 | 1 | 21 | 1 | 2 | .5 | .8 | .2 |
| 415 | Vanilla wafers | 10 ea | 40 | 5 | 176 | 2 | 29 | 1 | 6 | 1.4 | 2.4 | 1.5 |
| 416 | Corn chips | 1 c | 26 | 1 | 140 | 2 | 15 | 1 | 9 | 1.2 | 2.5 | 4.3 |
| | Crackers (enriched): | | | | | | | | | | | |
| 417 | Cheese | 10 ea | 10 | 3 | 50 | 1 | 6 | <1 | 3 | .9 | .9 | .5 |
| 418 | Cheese with peanut butter | 4 ea | 28 | 4 | 135 | 4 | 16 | 1 | 6 | 1.4 | 3.4 | 1.2 |
| | Fat free, enriched: | | | | | | | | | | | |
| 2161 | Cracked pepper, Snackwell | 1 ea | 15 | 2 | 60 | 2 | 12 | <1 | <1 | .1 | t | .1 |
| 2159 | Wheat, Snackwell | 7 ea | 15 | 1 | 60 | 2 | 12 | 1 | <1 | .1 | .1 | .1 |
| 2075 | Whole wheat, herb seasoned | 5 ea | 14 | 5 | 50 | 2 | 11 | 2 | 0 | 0 | 0 | 0 |
| 2077 | Whole wheat, onion | 5 ea | 14 | 4 | 50 | 2 | 11 | 2 | 0 | 0 | 0 | 0 |
| 419 | Graham, enriched | 2 ea | 14 | 4 | 59 | 1 | 11 | <1 | 1 | .4 | .7 | .2 |
| 420 | Melba toast, plain, enriched | 1 pce | 5 | 5 | 19 | 1 | 4 | <1 | <1 | t | t | .1 |
| 1514 | Rice cakes, unsalted, enriched | 2 ea | 18 | 6 | 70 | 1 | 15 | 1 | <1 | .1 | .2 | .2 |
| 421 | Rye wafer, whole grain | 2 ea | 22 | 5 | 73 | 2 | 18 | 5 | <1 | t | t | .1 |
| 422 | Saltine-enriched | 4 ea | 12 | 4 | 52 | 1 | 9 | <1 | 1 | .3 | .8 | .2 |
| 1971 | Saltine, unsalted tops, enriched | 2 ea | 6 | | 25 | 1 | 4 | 0 | <1 | 0 | 0 | 0 |
| 423 | Snack-type, round like Ritz, enriched | 3 ea | 9 | 3 | 45 | 1 | 5 | <1 | 2 | .4 | 1 | .7 |
| 424 | Wheat, thin, enriched | 4 ea | 8 | 3 | 38 | 1 | 5 | <1 | 2 | .7 | .8 | .2 |
| 425 | Whole-wheat wafers | 2 ea | 8 | 3 | 35 | 1 | 5 | 1 | 1 | .2 | .8 | .2 |
| 426 | Croissants, 4½ x 4 x 1¾" | 1 ea | 57 | 23 | 231 | 5 | 26 | 1 | 12 | 6.7 | 3.2 | .7 |
| 1699 | Croutons, seasoned | ½ c | 20 | 4 | 93 | 2 | 13 | 1 | 4 | 1 | 1.9 | .5 |
| | Danish pastry: | | | | | | | | | | | |
| 427 | Packaged ring, plain, 12 oz | 1 ea | 340 | 21 | 1349 | 19 | 181 | 1 | 65 | 13.5 | 40.9 | 6.4 |
| 428 | Round piece, plain, 4¼" diam, 1" high | 1 ea | 88 | 21 | 349 | 5 | 47 | <1 | 17 | 3.5 | 10.6 | 1.6 |
| 429 | Ounce, plain | 1 oz | 28 | 21 | 111 | 2 | 15 | <1 | 5 | 1.1 | 3.4 | .5 |
| 430 | Round piece with fruit | 1 ea | 94 | 29 | 335 | 5 | 45 | | 16 | 3.3 | 10.1 | 1.6 |
| | Desserts, 3 x 3" piece: | | | | | | | | | | | |
| 1348 | Apple crisp | 1 pce | 78 | 61 | 127 | 1 | 25 | 1 | 3 | .6 | 1.2 | .9 |
| 1353 | Apple cobbler | 1 pce | 104 | 57 | 199 | 2 | 35 | 2 | 6 | 1.2 | 2.8 | 2 |
| 1349 | Cherry crisp | 1 pce | 138 | 77 | 146 | 2 | 24 | 1 | 5 | .9 | 2.5 | 1.8 |
| 1352 | Cherry cobbler | 1 pce | 129 | 66 | 198 | 2 | 34 | 1 | 6 | 1.2 | 2.8 | 1.9 |
| 1350 | Peach crisp | 1 pce | 139 | 75 | 155 | 2 | 27 | 2 | 5 | .9 | 2.5 | 1.7 |
| 1351 | Peach cobbler | 1 pce | 130 | 64 | 204 | 2 | 36 | 2 | 6 | 1.2 | 2.8 | 1.9 |
| | Doughnuts: | | | | | | | | | | | |
| 431 | Cake type, plain, 3¼" diam | 1 ea | 47 | 21 | 198 | 2 | 23 | 1 | 11 | 1.8 | 4.5 | 3.8 |
| 432 | Yeast-leavened, glazed, 3¾" diam | 1 ea | 60 | 25 | 242 | 4 | 27 | 1 | 14 | 3.5 | 7.8 | 1.7 |
| | English muffins: | | | | | | | | | | | |
| 433 | Plain, enriched | 1 ea | 57 | 42 | 134 | 4 | 26 | 2 | 1 | .2 | .2 | .5 |
| 434 | Toasted | 1 ea | 52 | 37 | 133 | 4 | 26 | 2 | 1 | .1 | .2 | .5 |
| 1504 | Whole wheat | 1 ea | 66 | 46 | 134 | 6 | 27 | 4 | 1 | .2 | .3 | .6 |
| 1414 | Granola bar, soft | 1 ea | 28 | 6 | 124 | 2 | 19 | 1 | 5 | 2 | 1.1 | 1.5 |
| 1415 | Granola bar, hard | 1 ea | 25 | 4 | 118 | 3 | 16 | 1 | 5 | .6 | 1.1 | 3 |
| 1985 | Granola bar, fat free, all flavors | 1 ea | 42 | 10 | 140 | 2 | 35 | 3 | 0 | 0 | 0 | 0 |
| | Muffins, 2½" diam, 1½" high: | | | | | | | | | | | |
| | From home recipe: | | | | | | | | | | | |
| 435 | Blueberry | 1 ea | 57 | 39 | 165 | 4 | 23 | 1 | 6 | 1.4 | 1.6 | 3.1 |
| 436 | Bran, wheat | 1 ea | 57 | 35 | 164 | 4 | 24 | 4 | 7 | 1.5 | 1.8 | 3.6 |
| 437 | Cornmeal | 1 ea | 57 | 32 | 183 | 4 | 25 | 2 | 7 | 1.6 | 1.8 | 3.5 |

| Chol (mg) | Calc (mg) | Iron (mg) | Magn (mg) | Pota (mg) | Sodi (mg) | Zinc (mg) | VT-A (RE) | Thia (mg) | VT-E (a-TE) | Ribo (mg) | Niac (mg) | V-B6 (mg) | Fola (μg) | VT-C (mg) |
|---|---|---|---|---|---|---|---|---|---|---|---|---|---|---|
| 0 | 9 | 1.45 | 21 | 56 | 196 | .28 | 1 | .07 | 1.74 | .12 | .97 | .1 | 23 | 0 |
| 20 | 25 | 1.57 | 35 | 143 | 231 | .59 | 105 | .12 | 1.86 | .11 | .87 | .05 | 21 | <1 |
| 15 | 16 | 1.44 | 15 | 115 | 134 | .32 | 11 | .12 | 1.31 | .12 | 1.27 | .03 | 36 | 0 |
| 0 | 41 | 1.86 | 17 | 132 | 224 | .25 | 3 | .1 | .45 | .14 | 1.2 | .05 | 17 | <1 |
| 0 | 0 | .36 | | | 95 | | 0 | | .01 | | | | | 0 |
| 0 | 3 | .29 | 5 | 26 | 71 | .08 | <1 | .02 | <.01 | .02 | .26 | <.01 | | 0 |
| 20 | 60 | 1.59 | 25 | 143 | 323 | .52 | 98 | .15 | 1.5 | .1 | .76 | .04 | 18 | <1 |
| 25 | 31 | 1.78 | 31 | 185 | 414 | .66 | 125 | .18 | 3.04 | .17 | 2.81 | .07 | 44 | <1 |
| 0 | 10 | 1.55 | 18 | 70 | 242 | .32 | <1 | .03 | 1.21 | .07 | .83 | .01 | 17 | 0 |
| 6 | 11 | .88 | 5 | 32 | 146 | .17 | 4 | .11 | .98 | .1 | 1.07 | .01 | 19 | 0 |
| 20 | 4 | .58 | 3 | 15 | 102 | .09 | 67 | .08 | .18 | .06 | .64 | <.01 | 2 | 0 |
| 15 | 43 | .88 | 4 | 78 | 225 | .13 | 5 | .09 | 1.54 | .06 | 1.16 | .01 | 25 | 0 |
| <1 | 17 | .61 | 5 | 28 | 95 | .16 | <1 | .05 | | .07 | .69 | .01 | | 0 |
| 23 | 19 | .95 | 6 | 39 | 125 | .14 | 7 | .11 | .54 | .13 | 1.24 | .03 | 20 | 0 |
| 0 | 33 | .34 | 20 | 37 | 164 | .33 | 2 | .01 | .35 | .04 | .31 | .06 | 5 | 0 |
| 1 | 15 | .48 | 4 | 14 | 99 | .11 | 3 | .06 | .1 | .04 | .47 | .05 | 8 | 0 |
| 1 | 22 | .82 | 16 | 69 | 278 | .3 | 10 | .11 | 1.24 | .1 | 1.83 | .42 | 25 | 0 |
| <1 | 26 | .73 | 4 | 19 | 148 | .14 | <1 | .05 | | .06 | .78 | .01 | | <1 |
| <1 | 28 | .58 | 7 | 43 | 169 | .21 | <1 | .04 | | .07 | .73 | .02 | | 0 |
| 0 | 0 | | | | 80 | | 100 | | | | | | | 2 |
| 0 | 0 | 0 | | | 80 | | 100 | | | | | | | 2 |
| 0 | 3 | .52 | 4 | 19 | 85 | .11 | 0 | .03 | .27 | .04 | .58 | .01 | 8 | 0 |
| 0 | 5 | .18 | 3 | 10 | 41 | .1 | 0 | .02 | .01 | .01 | .21 | <.01 | 6 | 0 |
| 0 | 2 | .27 | 24 | 52 | 5 | .54 | 1 | .01 | .02 | .03 | 1.41 | .03 | 4 | 0 |
| 0 | 9 | 1.31 | 27 | 109 | 175 | .62 | <1 | .09 | .44 | .06 | .35 | .06 | 3 | <1 |
| 0 | 14 | .65 | 3 | 15 | 156 | .09 | 0 | .07 | .2 | .05 | .63 | <.01 | 15 | 0 |
| 0 | | .36 | | 5 | 50 | | | | .1 | | | | | |
| 0 | 11 | .32 | 2 | 12 | 76 | .06 | 0 | .04 | .4 | .03 | .36 | <.01 | 7 | 0 |
| 2 | 3 | .28 | 5 | 16 | 70 | .13 | <1 | .04 | .02 | .03 | .34 | .01 | 1 | 0 |
| 0 | 4 | .25 | 8 | 24 | 53 | .17 | 0 | .02 | .31 | .01 | .36 | .01 | 3 | 0 |
| 43 | 21 | 1.16 | 9 | 67 | 424 | .43 | 78 | .22 | .24 | .14 | 1.25 | .03 | 35 | <1 |
| 1 | 19 | .56 | 8 | 36 | 248 | .19 | 1 | .1 | .32 | .08 | .93 | .02 | 18 | 0 |
| 105 | 143 | 6.94 | 54 | 371 | 1261 | 1.87 | 20 | .99 | 3.06 | .75 | 8.5 | .2 | 211 | 10 |
| 27 | 37 | 1.8 | 14 | 96 | 326 | .48 | 5 | .25 | .79 | .19 | 2.2 | .05 | 55 | 3 |
| 9 | 12 | .57 | 4 | 30 | 104 | .15 | 2 | .08 | .25 | .06 | .7 | .02 | 17 | 1 |
| 19 | 22 | 1.4 | 14 | 110 | 333 | .48 | 24 | .29 | .85 | .21 | 1.8 | .06 | 31 | 2 |
| 0 | 22 | .58 | 5 | 76 | 142 | .12 | 24 | .07 | | .06 | .6 | .03 | 4 | 2 |
| 1 | 21 | .79 | 6 | 106 | 288 | .16 | 76 | .1 | 1.11 | .09 | .74 | .04 | 3 | <1 |
| 0 | 26 | 2.14 | 11 | 154 | 74 | .15 | 150 | .06 | .93 | .08 | .6 | .06 | 11 | 3 |
| 1 | 28 | 1.81 | 9 | 133 | 294 | .2 | 135 | .1 | 1.01 | .11 | .85 | .05 | 9 | 2 |
| 0 | 20 | .89 | 12 | 189 | 70 | .19 | 108 | .06 | 1.13 | .05 | 1.06 | .03 | 6 | 5 |
| 1 | 24 | .91 | 10 | 159 | 291 | .23 | 105 | .09 | 1.16 | .09 | 1.19 | .03 | 6 | 3 |
| 17 | 21 | .92 | 9 | 60 | 257 | .26 | 8 | .1 | 1.63 | .11 | .87 | .03 | 22 | <1 |
| 4 | 26 | 1.22 | 13 | 65 | 205 | .46 | 6 | .22 | 1.75 | .13 | 1.71 | .03 | 26 | 0 |
| 0 | 99 | 1.43 | 12 | 75 | 264 | .4 | 0 | .25 | .07 | .16 | 2.21 | .02 | 46 | <1 |
| 0 | 98 | 1.41 | 11 | 74 | 262 | .39 | 0 | .2 | .07 | .14 | 1.98 | .02 | 38 | <1 |
| 0 | 175 | 1.62 | 47 | 139 | 420 | 1.06 | 0 | .2 | .46 | .09 | 2.25 | .11 | 28 | 0 |
| <1 | 29 | .72 | 21 | 91 | 78 | .42 | 0 | .08 | .34 | .05 | .14 | .03 | 7 | 0 |
| 0 | 15 | .74 | 24 | 84 | 73 | .51 | 4 | .07 | .33 | .03 | .39 | .02 | 6 | <1 |
| 0 | 0 | 3.6 | | | 5 | | 100 | | | | | | | 0 |
| 22 | 107 | 1.29 | 9 | 69 | 251 | .31 | 16 | .15 | 1.03 | .16 | 1.26 | .02 | 7 | 1 |
| 20 | 106 | 2.39 | 44 | 181 | 335 | 1.57 | 136 | .19 | 1.31 | .25 | 2.29 | .18 | 30 | 4 |
| 26 | 147 | 1.49 | 13 | 82 | 333 | .35 | 23 | .17 | 1.08 | .18 | 1.36 | .05 | 10 | <1 |

**Table H–1**

**Food Composition** (Computer code number is for West Diet Analysis program) (For purposes of calculations, use "0" for t, <1, <.1, <.01, etc.)

| Computer Code Number | Food Description | Measure | Wt (g) | $H_2O$ (%) | Ener (kcal) | Prot (g) | Carb (g) | Dietary Fiber (g) | Fat (g) | Fat Breakdown (g) Sat | Mono | Poly |
|---|---|---|---|---|---|---|---|---|---|---|---|---|
| | **BAKED GOODS: BREADS, CAKES, COOKIES, CRACKERS, PIES**—Continued | | | | | | | | | | | |
| | From commercial mix: | | | | | | | | | | | |
| 438 | Blueberry | 1 ea | 50 | 36 | 150 | 3 | 24 | 1 | 4 | .7 | 1.8 | 1.5 |
| 439 | Bran, wheat | 1 ea | 50 | 35 | 138 | 3 | 23 | 2 | 5 | 1.2 | 2.3 | .7 |
| 440 | Cornmeal | 1 ea | 50 | 30 | 161 | 4 | 25 | 1 | 5 | 1.4 | 2.6 | .6 |
| 1864 | Nabisco Newtons, fat free, all flavors | 1 ea | 23 | | 69 | 1 | 16 | | 0 | 0 | 0 | 0 |
| | Pancakes, 4" diam: | | | | | | | | | | | |
| 441 | Buckwheat, from mix w/ egg and milk | 1 ea | 30 | 54 | 62 | 2 | 8 | 1 | 2 | .6 | .6 | .8 |
| 442 | Plain, from home recipe | 1 ea | 38 | 53 | 86 | 2 | 11 | 1 | 4 | .8 | .9 | 1.7 |
| 443 | Plain, from mix; egg, milk, oil added | 1 ea | 38 | 53 | 74 | 2 | 14 | <1 | 1 | .2 | .3 | .3 |
| 1468 | Pan dulce, sweet roll w/topping | 1 ea | 79 | 21 | 291 | 5 | 48 | 1 | 9 | 2 | 3.9 | 2.7 |
| | Piecrust,with enriched flour, vegetable shortening, baked: | | | | | | | | | | | |
| 444 | Home recipe, 9" shell | 1 ea | 180 | 10 | 949 | 12 | 85 | 3 | 62 | 15.5 | 27.4 | 16.4 |
| | From mix: | | | | | | | | | | | |
| 445 | Piecrust for 2-crust pie | 1 ea | 320 | 10 | 1686 | 21 | 152 | 5 | 111 | 27.6 | 48.6 | 29.2 |
| 446 | 1 pie shell | 1 ea | 160 | 11 | 802 | 11 | 81 | 3 | 49 | 12.3 | 27.7 | 6.2 |
| | Pies, 9" diam; pie crust made with vegetable shortening, enriched flour: | | | | | | | | | | | |
| 447 | Apple: Whole pie | 1 ea | 1000 | 52 | 2370 | 19 | 340 | 16 | 110 | 21.1 | 59.4 | 20.9 |
| 448 | Piece, ⅙ of pie | 1 pce | 167 | 52 | 396 | 3 | 57 | 3 | 18 | 3.5 | 9.9 | 3.5 |
| 449 | Banana cream: Whole pie | 1 ea | 1152 | 48 | 3098 | 51 | 379 | 8 | 157 | 43.3 | 65.9 | 38 |
| 450 | Piece, ⅙ of pie | 1 pce | 192 | 48 | 516 | 8 | 63 | 1 | 26 | 7.2 | 11 | 6.3 |
| 451 | Blueberry: Whole pie | 1 ea | 1176 | 51 | 2881 | 32 | 394 | 16 | 140 | 34.3 | 60.2 | 36.2 |
| 452 | Piece, ⅙ of pie | 1 pce | 196 | 51 | 480 | 5 | 66 | 3 | 23 | 5.7 | 10 | 6 |
| 453 | Cherry: Whole pie | 1 ea | 1140 | 46 | 3078 | 32 | 439 | 17 | 139 | 34.1 | 60.5 | 37.1 |
| 454 | Piece, ⅙ of pie | 1 pce | 240 | 46 | 648 | 7 | 92 | 4 | 29 | 7.2 | 12.7 | 7.8 |
| 455 | Chocolate cream: Whole pie | 1 ea | 1194 | 63 | 2150 | 49 | 281 | 12 | 97 | 35.5 | 38.2 | 18.6 |
| 456 | Piece, ⅙ of pie | 1 pce | 199 | 63 | 358 | 8 | 47 | 2 | 16 | 5.9 | 6.4 | 3.1 |
| 457 | Custard: Whole pie | 1 ea | 630 | 61 | 1323 | 35 | 131 | 10 | 73 | 17.5 | 36.3 | 12.1 |
| 458 | Piece, ⅙ of pie | 1 pce | 105 | 61 | 221 | 6 | 22 | 2 | 12 | 2.9 | 6 | 2 |
| 459 | Lemon meringue: Whole pie | 1 ea | 678 | 42 | 1817 | 10 | 320 | 8 | 59 | 10.6 | 24.6 | 19.6 |
| 460 | Piece, ⅙ of pie | 1 pce | 113 | 42 | 303 | 2 | 53 | 1 | 10 | 1.8 | 4.1 | 3.3 |
| 461 | Peach: Whole pie | 1 ea | 1111 | 45 | 2994 | 26 | 443 | 16 | 130 | 31.1 | 55.7 | 37.4 |
| 462 | Piece, ⅙ of pie | 1 pce | 139 | 45 | 375 | 3 | 55 | 2 | 16 | 3.9 | 7 | 4.7 |
| 463 | Pecan: Whole pie | 1 ea | 678 | 19 | 2712 | 27 | 388 | 24 | 125 | 25.5 | 73.2 | 20.1 |
| 464 | Piece, ⅙ of pie | 1 pce | 113 | 19 | 452 | 5 | 65 | 4 | 21 | 4.2 | 12.2 | 3.4 |
| 465 | Pumpkin: Whole pie | 1 ea | 654 | 58 | 1373 | 25 | 179 | 18 | 62 | 13.2 | 32.8 | 10.5 |
| 466 | Piece, ⅙ of pie | 1 pce | 109 | 58 | 229 | 4 | 30 | 3 | 10 | 2.2 | 5.5 | 1.7 |
| 467 | Pies, fried, commercial: Apple | 1 ea | 85 | 40 | 266 | 2 | 33 | 1 | 14 | 6.5 | 5.8 | 1.2 |
| 468 | Pies, fried, commercial: Cherry | 1 ea | 128 | 38 | 404 | 4 | 54 | 3 | 21 | 3.1 | 9.5 | 6.9 |
| | Pretzels, made with enriched flour: | | | | | | | | | | | |
| 469 | Thin sticks, 2¼" long | 1 oz | 28 | 3 | 107 | 3 | 22 | 1 | 1 | .2 | .4 | .3 |
| 470 | Dutch twists | 10 pce | 60 | 3 | 229 | 5 | 47 | 2 | 2 | .4 | .8 | .7 |
| 471 | Thin twists, 3¼ x 2¼ x ¼" | 10 pce | 60 | 3 | 229 | 5 | 47 | 2 | 2 | .4 | .8 | .7 |
| | Rolls & buns, enriched, commercial: | | | | | | | | | | | |
| 472 | Cloverleaf rolls, 2½" diam, 2" high | 1 ea | 28 | 32 | 84 | 2 | 14 | 1 | 2 | .5 | 1 | .3 |
| 473 | Hot dog buns | 1 ea | 43 | 34 | 123 | 4 | 22 | 1 | 2 | .5 | 1.1 | .4 |
| 474 | Hamburger buns | 1 ea | 43 | 34 | 123 | 4 | 22 | 1 | 2 | .5 | 1.1 | .4 |
| 475 | Hard roll, white, 3¾" diam, 2" high | 1 ea | 57 | 31 | 167 | 6 | 30 | 1 | 2 | .3 | .6 | 1 |
| 476 | Submarine rolls/hoagies, 11¼ x 3 x 2½" | 1 ea | 135 | 31 | 392 | 12 | 75 | 4 | 4 | .9 | 1.3 | 1.4 |
| | Rolls & buns, enriched, home recipe: | | | | | | | | | | | |
| 477 | Dinner rolls 2½" diam, 2" high | 1 ea | 35 | 29 | 112 | 3 | 19 | 1 | 3 | .7 | 1.1 | .7 |
| | Sports/fitness bar: | | | | | | | | | | | |
| 2043 | Forza energy bar | 1 ea | 70 | 18 | 231 | 10 | 45 | 4 | 1 | | | |
| 2042 | Power bar | 1 ea | 65 | | 230 | 10 | 45 | 3 | 2 | | | |
| 2041 | Tiger sports bar | 1 ea | 65 | 17 | 229 | 11 | 40 | 4 | 2 | | | |
| 478 | Toaster pastries, fortified (Poptarts) | 1 ea | 52 | 12 | 204 | 2 | 37 | 1 | 5 | .8 | 2.1 | 2 |
| 2132 | Toaster strudel pastry—cream cheese | 1 ea | 54 | 32 | 188 | 3 | 24 | <1 | 9 | 2.7 | | |
| 2134 | Toaster strudel pastry—french toast | 1 ea | 54 | 32 | 188 | 3 | 24 | <1 | 9 | 2.9 | | |

| Chol (mg) | Calc (mg) | Iron (mg) | Magn (mg) | Pota (mg) | Sodi (mg) | Zinc (mg) | VT-A (RE) | Thia (mg) | VT-E (a-TE) | Ribo (mg) | Niac (mg) | V-B6 (mg) | Fola (µg) | VT-C (mg) |
|---|---|---|---|---|---|---|---|---|---|---|---|---|---|---|
| 23 | 12 | .56 | 5 | 39 | 219 | .19 | 11 | .07 | .7 | .16 | 1.12 | .04 | 5 | <1 |
| 34 | 16 | 1.27 | 28 | 73 | 234 | .57 | 15 | .1 | .75 | .12 | 1.44 | .09 | 8 | 0 |
| 31 | 37 | .97 | 10 | 65 | 398 | .32 | 22 | .12 | .75 | .14 | 1.05 | .05 | 5 | <1 |
| | | | | | 77 | | | | | | | | | |
| 20 | 77 | .56 | 17 | 70 | 160 | .35 | 20 | .05 | .62 | .08 | .4 | .04 | 5 | <1 |
| 22 | 83 | .68 | 6 | 50 | 167 | .21 | 20 | .08 | .36 | .11 | .6 | .02 | 14 | <1 |
| 5 | 48 | .59 | 8 | 66 | 239 | .15 | 3 | .08 | .32 | .08 | .65 | .03 | 3 | <1 |
| 26 | 13 | 1.82 | 10 | 57 | 140 | .35 | 87 | .23 | 1.35 | .21 | 1.98 | .04 | 22 | <1 |
| 0 | 18 | 5.2 | 25 | 121 | 976 | .79 | 0 | .7 | 9.94 | .5 | 5.96 | .04 | 121 | 0 |
| 0 | 32 | 9.25 | 45 | 214 | 1734 | 1.41 | 0 | 1.25 | 17.7 | .89 | 10.6 | .08 | 214 | 0 |
| 0 | 96 | 3.44 | 24 | 99 | 1166 | .62 | 0 | .48 | 8.83 | .3 | 3.79 | .09 | 19 | 0 |
| 0 | 110 | 4.5 | 70 | 650 | 2660 | 1.6 | 300 | .28 | 16.5 | .27 | 2.63 | .38 | 220 | 32 |
| 0 | 18 | .75 | 12 | 109 | 444 | .27 | 50 | .05 | 2.76 | .04 | .44 | .06 | 37 | 5 |
| 588 | 864 | 12 | 184 | 1900 | 2764 | 5.53 | 806 | 1.6 | 16.9 | 2.38 | 12.1 | 1.53 | 311 | 18 |
| 98 | 144 | 2 | 31 | 317 | 461 | .92 | 134 | .27 | 2.82 | .4 | 2.02 | .25 | 52 | 3 |
| 0 | 82 | 14.5 | 94 | 588 | 2175 | 2.35 | 47 | 1.8 | 24.7 | 1.55 | 14 | .4 | 270 | 8 |
| 0 | 14 | 2.41 | 16 | 98 | 363 | .39 | 8 | .3 | 4.12 | .26 | 2.33 | .07 | 45 | 1 |
| 0 | 114 | 21.1 | 103 | 878 | 2177 | 2.28 | 547 | 1.69 | 21.7 | 1.43 | 14.6 | .39 | 308 | 11 |
| 0 | 24 | 4.44 | 22 | 185 | 458 | .48 | 115 | .35 | 4.56 | .3 | 3.07 | .08 | 65 | 2 |
| 109 | 1028 | 8.84 | 170 | 1705 | 2085 | 4.93 | 235 | 1.03 | 11.4 | 2.43 | 7.34 | .37 | 59 | 6 |
| 18 | 171 | 1.47 | 28 | 284 | 348 | .82 | 39 | .17 | 1.9 | .41 | 1.22 | .06 | 10 | 1 |
| 208 | 504 | 3.65 | 69 | 668 | 1512 | 3.28 | 315 | .25 | 7.5 | 1.31 | 1.84 | .3 | 126 | 2 |
| 35 | 84 | .61 | 12 | 111 | 252 | .55 | 52 | .04 | 1.25 | .22 | .31 | .05 | 21 | <1 |
| 305 | 380 | 4.14 | 102 | 603 | 990 | 3.32 | 353 | .42 | 9.7 | 1.42 | 4.4 | .2 | 88 | 22 |
| 51 | 63 | .69 | 17 | 101 | 165 | .55 | 59 | .07 | 1.62 | .24 | .73 | .03 | 15 | 4 |
| 0 | 59 | 12.1 | 79 | 1047 | 2025 | 1.88 | 386 | 1.41 | 26.2 | 1.19 | 15.3 | .2 | 58 | 589 |
| 0 | 7 | 1.52 | 10 | 131 | 253 | .23 | 48 | .18 | 3.28 | .15 | 1.91 | .02 | 7 | 74 |
| 217 | 115 | 7.05 | 122 | 502 | 2874 | 3.86 | 319 | .62 | 17.2 | .83 | 1.69 | .14 | 183 | 7 |
| 36 | 19 | 1.18 | 20 | 84 | 479 | .64 | 53 | .1 | 2.86 | .14 | .28 | .02 | 30 | 1 |
| 131 | 392 | 5.17 | 98 | 1007 | 1844 | 2.94 | 3139 | .36 | 10.5 | 1 | 1.22 | .37 | 131 | 10 |
| 22 | 65 | .86 | 16 | 168 | 307 | .49 | 523 | .06 | 1.75 | .17 | .2 | .06 | 22 | 2 |
| 13 | 13 | .88 | 8 | 51 | 325 | .17 | 33 | .1 | .37 | .08 | .98 | .03 | 4 | 1 |
| 0 | 28 | 1.56 | 13 | 83 | 479 | .29 | 22 | .18 | .55 | .14 | 1.83 | .04 | 23 | 2 |
| 0 | 10 | 1.21 | 10 | 41 | 480 | .24 | 0 | .13 | .06 | .17 | 1.47 | .03 | 48 | 0 |
| 0 | 22 | 2.59 | 21 | 88 | 1029 | .51 | 0 | .28 | .13 | .37 | 3.15 | .07 | 103 | 0 |
| 0 | 22 | 2.59 | 21 | 88 | 1029 | .51 | 0 | .28 | .13 | .37 | 3.15 | .07 | 103 | 0 |
| <1 | 33 | .88 | 6 | 37 | 146 | .22 | 0 | .14 | .22 | .09 | 1.13 | .01 | 27 | <1 |
| 0 | 60 | 1.36 | 9 | 61 | 241 | .27 | 0 | .21 | .2 | .13 | 1.69 | .02 | 41 | 0 |
| 0 | 60 | 1.36 | 9 | 61 | 241 | .27 | 0 | .21 | .2 | .13 | 1.69 | .02 | 41 | 0 |
| 0 | 54 | 1.87 | 15 | 62 | 310 | .54 | 0 | .27 | .1 | .19 | 2.42 | .03 | 54 | 0 |
| 0 | 122 | 3.78 | 27 | 122 | 783 | .85 | 0 | .54 | .1 | .33 | 4.47 | .05 | 40 | 0 |
| 13 | 21 | 1.04 | 7 | 53 | 145 | .24 | 28 | .14 | .35 | .14 | 1.21 | .02 | 15 | <1 |
| 0 | 300 | 6.3 | 160 | 220 | 65 | 5.25 | | 1.5 | 20 | 1.7 | 20 | 2 | 400 | 60 |
| 0 | 300 | 5.4 | 140 | 150 | 110 | 5.25 | | 1.5 | | 1.7 | 20 | 2 | 400 | 60 |
| | 349 | 4.49 | 140 | 279 | 100 | | 50 | 1.5 | 19.9 | 1.69 | 19.9 | 1.99 | 399 | 60 |
| 0 | 13 | 1.81 | 9 | 58 | 218 | .34 | 149 | .15 | .97 | .19 | 2.05 | .2 | 34 | <1 |
| 12 | 12 | .97 | | | 217 | | 17 | | 1 | | | | | 0 |
| 12 | 12 | .97 | | | 217 | | 17 | | 1 | | | | | 0 |

**Table H-1**

**Food Composition** (Computer code number is for West Diet Analysis program) (For purposes of calculations, use "0" for t, <1, <.1, <.01, etc.)

| Computer Code Number | Food Description | Measure | Wt (g) | $H_2O$ (%) | Ener (kcal) | Prot (g) | Carb (g) | Dietary Fiber (g) | Fat (g) | Fat Breakdown (g) Sat | Mono | Poly |
|---|---|---|---|---|---|---|---|---|---|---|---|---|
| | **BAKED GOODS: BREADS, CAKES, COOKIES, CRACKERS, PIES**—Continued | | | | | | | | | | | |
| | Tortilla chips: | | | | | | | | | | | |
| 1271 | Plain | 10 pce | 18 | 2 | 90 | 1 | 11 | 1 | 5 | .9 | 2.8 | .7 |
| 1036 | Nacho flavor | 1 c | 26 | 2 | 129 | 2 | 16 | 1 | 7 | 1.3 | 3.9 | .9 |
| 1037 | Taco flavor | 1 pce | 18 | 2 | 86 | 1 | 11 | 1 | 4 | .8 | 2.6 | .6 |
| | Tortillas: | | | | | | | | | | | |
| 479 | Corn, enriched, 6" diam | 1 ea | 26 | 44 | 58 | 2 | 12 | 1 | 1 | .1 | .2 | .3 |
| 480 | Flour, 8" diam | 1 ea | 49 | 27 | 159 | 4 | 27 | 2 | 3 | .6 | 1.4 | 1.4 |
| 1301 | Flour, 10" diam | 1 ea | 72 | 27 | 234 | 6 | 40 | 2 | 5 | .8 | 2.1 | 2 |
| 481 | Taco shells | 1 ea | 14 | 4 | 63 | 1 | 9 | 1 | 3 | .4 | 1.5 | .6 |
| | Waffles, 7" diam: | | | | | | | | | | | |
| 482 | From home recipe | 1 ea | 75 | 42 | 218 | 6 | 25 | 1 | 11 | 2.2 | 2.6 | 5.1 |
| 483 | From mix, egg/milk added | 1 ea | 75 | 42 | 218 | 5 | 26 | 1 | 10 | 1.7 | 2.7 | 5.2 |
| 1510 | Whole grain, prepared from frozen | 1 ea | 39 | 43 | 107 | 4 | 13 | 1 | 5 | 1.6 | 1.9 | .9 |
| | **GRAIN PRODUCTS: CEREAL, FLOUR, GRAIN, PASTA and NOODLES, POPCORN** | | | | | | | | | | | |
| 484 | Barley, pearled, dry, uncooked | 1 c | 200 | 10 | 704 | 20 | 155 | 31 | 2 | .5 | .3 | 1.1 |
| 485 | Barley, pearled, cooked | 1 c | 157 | 69 | 193 | 4 | 44 | 6 | 1 | .1 | .1 | .3 |
| 2009 | Breakfast bars, fat free, all flavors | 1 ea | 38 | 25 | 110 | 2 | 26 | 3 | 0 | 0 | 0 | 0 |
| | Breakfast bar, Snackwell: | | | | | | | | | | | |
| 2165 | Apple-cinnamon | 1 ea | 37 | 16 | 119 | 1 | 29 | 1 | <1 | .1 | t | .1 |
| 2164 | Blueberry | 1 ea | 37 | 16 | 121 | 1 | 29 | 1 | <1 | t | t | .1 |
| 2163 | Strawberry | 1 ea | 37 | 16 | 120 | 1 | 29 | 1 | <1 | t | t | .1 |
| | Breakfast cereals, hot, cooked w/o salt added: | | | | | | | | | | | |
| | Corn grits (hominy) enriched: | | | | | | | | | | | |
| 486 | Regular/quick prep w/o salt, yellow: | 1 c | 242 | 85 | 145 | 3 | 31 | <1 | <1 | .1 | .1 | .2 |
| 487 | Instant, prepared from packet, white | 1 ea | 137 | 82 | 89 | 2 | 21 | 1 | <1 | t | t | .1 |
| | Cream of wheat: | | | | | | | | | | | |
| 488 | Regular, quick, instant | 1 c | 239 | 87 | 129 | 4 | 27 | 1 | <1 | .1 | .1 | .3 |
| 489 | Mix and eat, plain, packet | 1 ea | 142 | 82 | 102 | 3 | 21 | <1 | <1 | t | t | .2 |
| 1664 | Farina cereal, cooked w/o salt | 1 c | 233 | 88 | 117 | 3 | 25 | 3 | <1 | t | t | .1 |
| 490 | Malt-O-Meal, cooked w/o salt | 1 c | 240 | 88 | 122 | 4 | 26 | 1 | <1 | .1 | .1 | t |
| 494 | Maypo | 1 c | 216 | 83 | 153 | 5 | 29 | 5 | 2 | .4 | .7 | .8 |
| | Oatmeal or rolled oats: | | | | | | | | | | | |
| 491 | Regular, quick, instant,nonfortified cooked w/o salt | 1 c | 234 | 85 | 145 | 6 | 25 | 4 | 2 | .4 | .7 | .9 |
| | Instant, fortified: | | | | | | | | | | | |
| 492 | Plain, from packet | ½ c | 118 | 85 | 70 | 4 | 12 | 2 | 1 | .2 | .4 | .4 |
| 493 | Flavored, from packet | ½ c | 109 | 76 | 106 | 3 | 21 | 2 | 1 | .2 | .5 | .5 |
| | Breakfast cereals, ready to eat: | | | | | | | | | | | |
| 495 | All-Bran | 1 c | 62 | 3 | 160 | 8 | 46 | 20 | 2 | .4 | .4 | 1.3 |
| 1306 | Alpha Bits | 1 c | 28 | 1 | 110 | 2 | 24 | 1 | 1 | .1 | .2 | .2 |
| 1307 | Apple Jacks | 1 c | 33 | 3 | 120 | 2 | 30 | 1 | <1 | .1 | .1 | .2 |
| 1308 | Bran Buds | 1 c | 90 | 3 | 240 | 8 | 72 | 36 | 2 | .4 | .4 | 1.4 |
| 1305 | Bran Chex | 1 c | 49 | 2 | 156 | 5 | 39 | 8 | 1 | .2 | .3 | .7 |
| 1309 | Honey BucWheat Crisp | 1 c | 38 | 5 | 147 | 4 | 31 | 3 | 1 | .2 | .3 | .6 |
| 1310 | C.W. Post, plain | 1 c | 97 | 2 | 421 | 9 | 73 | 7 | 13 | 1.7 | 6 | 4.7 |
| 1311 | C.W. Post, with raisins | 1 c | 103 | 4 | 446 | 9 | 74 | 14 | 15 | 11 | 1.7 | 1.4 |
| 496 | Cap'n Crunch | 1 c | 37 | 2 | 147 | 2 | 32 | 1 | 2 | .5 | .4 | .3 |
| 1312 | Cap'n Crunchberries | 1 c | 35 | 2 | 140 | 2 | 30 | 1 | 2 | .5 | .3 | .3 |
| 1313 | Cap'n Crunch, peanut butter | 1 c | 35 | 2 | 146 | 3 | 28 | 1 | 3 | .7 | 1.1 | .7 |
| 497 | Cheerios | 1 c | 23 | 3 | 84 | 2 | 17 | 2 | 1 | .3 | .5 | .2 |
| 1314 | Cocoa Krispies | 1 c | 41 | 2 | 159 | 3 | 36 | 1 | 1 | .7 | .2 | .2 |
| 1316 | Cocoa Pebbles | 1 c | 32 | 2 | 131 | 1 | 27 | 1 | 2 | 1.1 | .4 | .1 |
| 1315 | Corn Bran | 1 c | 36 | 3 | 120 | 2 | 30 | 6 | 1 | .3 | .3 | .4 |
| 1317 | Corn Chex | 1 c | 28 | 2 | 110 | 2 | 25 | <1 | <1 | t | t | t |
| 498 | Corn Flakes, Kellogg's | 1 c | 28 | 3 | 100 | 2 | 24 | 1 | <1 | .1 | t | .1 |
| 499 | Corn Flakes, Post Toasties | 1 c | 24 | 3 | 93 | 2 | 21 | 1 | <1 | t | t | t |
| 1340 | Corn Pops | 1 c | 31 | 3 | 120 | 1 | 28 | <1 | <1 | .1 | .1 | t |
| 1318 | Cracklin' Oat Bran | 1 c | 65 | 4 | 252 | 5 | 48 | 8 | 8 | 3.4 | 3.8 | .9 |
| 1038 | Crispy Wheat 'N Raisins | 1 c | 43 | 7 | 150 | 3 | 35 | 3 | 1 | .1 | .1 | .2 |

**PAGE KEY:** H–4 = Beverages H–6 = Dairy H–10 = Eggs H–10 = Fat/Oil H–14 = Fruit H–20 = Bakery H–28 = Grain H–32 = Fish H–34 = Meats H–38 = Poultry H–40 = Sausage H–42 = Mixed/Fast H–46 = Nuts/Seeds H–50 = Sweets H–52 = Vegetables/Legumes H–62 = Vegetarian Foods H–64 = Misc H–66 = Soups/Sauces H–68 = Fast H–84 = Convenience H–88 = Baby foods

| Chol (mg) | Calc (mg) | Iron (mg) | Magn (mg) | Pota (mg) | Sodi (mg) | Zinc (mg) | VT-A (RE) | Thia (mg) | VT-E (a-TE) | Ribo (mg) | Niac (mg) | V-B6 (mg) | Fola (µg) | VT-C (mg) |
|---|---|---|---|---|---|---|---|---|---|---|---|---|---|---|
| 0 | 28 | .27 | 16 | 35 | 95 | .27 | 4 | .01 | .24 | .03 | .23 | .05 | 2 | 0 |
| 1 | 38 | .37 | 21 | 56 | 184 | .31 | 11 | .03 | .35 | .05 | .37 | .07 | 4 | <1 |
| 1 | 28 | .36 | 16 | 39 | 142 | .23 | 16 | .04 | .24 | .04 | .36 | .05 | 4 | <1 |
| 0 | 45 | .36 | 17 | 40 | 42 | .24 | 6 | .03 | .04 | .02 | .39 | .06 | 30 | 0 |
| 0 | 61 | 1.62 | 13 | 64 | 234 | .35 | 0 | .26 | .62 | .14 | 1.75 | .02 | 60 | 0 |
| 0 | 90 | 2.38 | 19 | 94 | 344 | .51 | 0 | .38 | .91 | .21 | 2.57 | .04 | 89 | 0 |
| 0 | 35 | .36 | 15 | 34 | 25 | .19 | 6 | .04 | .59 | .02 | .24 | .04 | 4 | 0 |
| 52 | 191 | 1.73 | 14 | 119 | 383 | .51 | 49 | .2 | 1.73 | .26 | 1.55 | .04 | 34 | <1 |
| 38 | 93 | 1.22 | 15 | 134 | 458 | .35 | 19 | .15 | 1.5 | .19 | 1.23 | .07 | 9 | <1 |
| 39 | 84 | .69 | 15 | 91 | 150 | .45 | 25 | .08 | .53 | .13 | .75 | .04 | 7 | <1 |
| 0 | 58 | 5 | 158 | 560 | 18 | 4.26 | 4 | .38 | .26 | .23 | 9.2 | .52 | 46 | 0 |
| 0 | 17 | 2.09 | 34 | 146 | 5 | 1.29 | 2 | .13 | .08 | .1 | 3.23 | .18 | 25 | 0 |
| 0 | 20 | .72 |  |  | 25 |  | 20 |  |  |  |  |  |  | 1 |
| <1 | 17 | 5 | 6 | 68 | 103 | 3.88 | 260 | .39 |  | .44 | 5.2 | .52 |  | <1 |
| <1 | 14 | 4.83 | 5 | 43 | 107 | 3.85 | 260 | .39 |  | .44 | 5.2 | .52 |  | <1 |
| <1 | 14 | 4.82 | 6 | 47 | 102 | 3.83 | 260 | .39 |  | .44 | 5.2 | .52 |  | 2 |
| 0 | 0 | 1.55 | 10 | 53 | 0 | .17 | 14 | .24 | .12 | .14 | 1.96 | .06 | 75 | 0 |
| 0 | 8 | 8.19 | 11 | 38 | 289 | .21 | 0 | .15 | .03 | .08 | 1.38 | .05 | 47 | 0 |
| 0 | 50 | 10.3 | 12 | 45 | 139 | .33 | 0 | .24 | .03 | 0 | 1.43 | .03 | 108 | 0 |
| 0 | 20 | 8.09 | 7 | 38 | 241 | .24 | 125 | .43 | .02 | .28 | 4.97 | .57 | 101 | 0 |
| 0 | 5 | 1.17 | 5 | 30 | 0 | .16 | 0 | .19 | .03 | .12 | 1.28 | .02 | 54 | 0 |
| 0 | 5 | 9.6 | 5 | 31 | 2 | .17 | 0 | .48 | .03 | .24 | 5.76 | .02 | 5 | 0 |
| 0 | 112 | 7.56 | 45 | 190 | 233 | 1.34 | 633 | .65 | 1.51 | .65 | 8.42 | .86 | 9 | 26 |
| 0 | 19 | 1.59 | 56 | 131 | 2 | 1.15 | 5 | .26 | .23 | .05 | .3 | .05 | 9 | 0 |
| 0 | 109 | 4.2 | 28 | 66 | 190 | .58 | 302 | .35 | .14 | .19 | 3.65 | .49 | 100 | 0 |
| 0 | 112 | 4.45 | 34 | 91 | 169 | .66 | 306 | .35 | .14 | .25 | 3.92 | .51 | 100 | <1 |
| 0 | 200 | 9 | 280 | 620 | 560 | 7.5 | 450 | .75 | 1.14 | .85 | 10 | 1 | 186 | 30 |
| 0 | 8 | 2.66 | 16 | 54 | 178 | 1.48 | 371 | .36 | .02 | .42 | 4.93 | .5 | 99 | 0 |
| 0 | 0 | 4.5 | 8 | 35 | 150 | 3.75 | 225 | .38 | .05 | .43 | 5 | .5 | 116 | 15 |
| 0 | 60 | 13.5 | 240 | 809 | 599 | 11.3 | 676 | 1.17 | 1.42 | 1.26 | 15 | 1.53 | 270 | 45 |
| 0 | 29 | 14 | 69 | 216 | 345 | 6.47 | 11 | .64 | .56 | .26 | 8.62 | .88 | 173 | 26 |
| 0 | 54 | 10.9 | 43 | 142 | 361 | .68 | 913 | .9 | 8.99 | 1.03 | 12.1 | 1.88 | 11 | 36 |
| <1 | 47 | 15.4 | 67 | 198 | 167 | 1.64 | 1284 | 1.26 | .68 | 1.46 | 17.1 | 1.75 | 342 | 0 |
| <1 | 50 | 16.4 | 74 | 261 | 161 | 1.64 | 1363 | 1.34 | .72 | 1.55 | 18.1 | 1.85 | 364 | 0 |
| 0 | 7 | 6.18 | 13 | 47 | 286 | 5.14 | 5 | .51 | .18 | .58 | 6.85 | .68 | 137 | 0 |
| 0 | 9 | 6.06 | 13 | 49 | 256 | 5.39 | 6 | .5 | .25 | .57 | 6.72 | .67 | 135 | <1 |
| 0 | 3 | 5.85 | 24 | 80 | 264 | 4.87 | 5 | .49 | .19 | .55 | 6.48 | .65 | 130 | 0 |
| 0 | 42 | 6.21 | 25 | 68 | 218 | 2.88 | 288 | .29 | .16 | .33 | 3.84 | .38 | 77 | 11 |
| 0 | 0 | 2.38 | 11 | 79 | 278 | 1.97 | 298 | .49 | .19 | .57 | 6.6 | .66 | 123 | 20 |
| 0 | 5 | 2.02 | 13 | 53 | 180 | 1.7 | 424 | .42 | .04 | .48 | 5.63 | .58 | 113 | 0 |
| 0 | 27 | 10.1 | 19 | 75 | 338 | 5 | 5 | .1 | .19 | .56 | 6.66 | .67 | 134 | 0 |
| 0 | 3 | 8.01 | 4 | 23 | 306 | .1 | 14 | .36 | .07 | .07 | 4.93 | .5 | 99 | 15 |
| 0 | 0 | 8.68 | 3 | 25 | 300 | .17 | 225 | .36 | .03 | .43 | 5 | .5 | 99 | 15 |
| 0 | 1 | .63 | 4 | 28 | 252 | .07 | 318 | .31 | .06 | .36 | 4.22 | .43 | 85 | 0 |
| 0 | 0 | 1.86 | 2 | 25 | 120 | 1.55 | 225 | .4 | .03 | .43 | 5.18 | .5 | 109 | 15 |
| 0 | 26 | 2.41 | 79 | 305 | 226 | 1.95 | 299 | .5 | .43 | .56 | 6.63 | .66 | 181 | 20 |
| 0 | 54 | 3.52 | 33 | 180 | 223 | .85 | 293 | .29 | .45 | .33 | 3.91 | .39 | 78 | 0 |

**Table H-1**

**Food Composition** (Computer code number is for West Diet Analysis program) (For purposes of calculations, use "0" for t, <1, <.1, <.01, etc.)

| Computer Code Number | Food Description | Measure | Wt (g) | $H_2O$ (%) | Ener (kcal) | Prot (g) | Carb (g) | Dietary Fiber (g) | Fat (g) | Fat Breakdown (g) Sat | Mono | Poly |
|---|---|---|---|---|---|---|---|---|---|---|---|---|
| | **GRAIN PRODUCTS: CEREAL, FLOUR, GRAIN, PASTA and NOODLES, POPCORN—**Continued | | | | | | | | | | | |
| 1319 | Fortified Oat Flakes | 1 c | 48 | 3 | 180 | 8 | 36 | 1 | 1 | .2 | .3 | .4 |
| 500 | 40% Bran Flakes, Kellogg's | 1 c | 39 | 4 | 121 | 4 | 32 | 7 | 1 | .2 | .2 | .5 |
| 501 | 40% Bran Flakes, Post | 1 c | 47 | 3 | 152 | 5 | 37 | 9 | 1 | .1 | .1 | .4 |
| 502 | Froot Loops | 1 c | 32 | 2 | 120 | 2 | 28 | 1 | 1 | .4 | .2 | .3 |
| 518 | Frosted Flakes | 1 c | 41 | 3 | 159 | 2 | 37 | 1 | <1 | .1 | t | .1 |
| 1320 | Frosted Mini-Wheats | 1 c | 51 | 5 | 170 | 5 | 41 | 5 | 1 | .2 | .1 | .6 |
| 1321 | Frosted Rice Krispies | 1 c | 35 | 2 | 135 | 2 | 32 | <1 | <1 | .1 | .1 | .1 |
| 1324 | Fruit & Fibre w/dates | 1 c | 57 | 9 | 193 | 5 | 43 | 8 | 3 | .4 | 1.3 | 1 |
| 1322 | Fruity Pebbles | 1 c | 32 | 3 | 130 | 1 | 28 | <1 | 2 | 1.4 | .1 | .1 |
| 503 | Golden Grahams | 1 c | 39 | 3 | 150 | 2 | 33 | 1 | 1 | .2 | .4 | .2 |
| 504 | Granola, homemade | ½ c | 61 | 5 | 285 | 9 | 32 | 6 | 15 | 2.9 | 4.8 | 6.5 |
| 505 | Granola, low fat | ½ c | 47 | 3 | 181 | 5 | 38 | 3 | 3 | 0 | | |
| 1670 | Granola, low fat, commercial | ½ c | 45 | 5 | 165 | 4 | 35 | 2 | 2 | .6 | .7 | .9 |
| 505 | Grape Nuts | ½ c | 55 | 3 | 196 | 7 | 45 | 5 | <1 | t | t | .1 |
| 1326 | Grape Nuts Flakes | 1 c | 39 | 3 | 144 | 4 | 32 | 4 | 1 | .6 | .1 | .2 |
| 1665 | Heartland Natural with raisins | 1 c | 110 | 5 | 468 | 11 | 76 | 6 | 16 | 4 | 4.2 | 6.2 |
| 1327 | Honey & Nut Corn Flakes | 1 c | 37 | 2 | 148 | 3 | 31 | 1 | 2 | .3 | .7 | .6 |
| 506 | Honey Nut Cheerios | 1 c | 33 | 2 | 126 | 3 | 27 | 2 | 1 | .3 | .5 | .2 |
| 1328 | HoneyBran | 1 c | 35 | 2 | 119 | 3 | 29 | 4 | 1 | .3 | .1 | .3 |
| 1329 | HoneyComb | 1 c | 22 | 1 | 86 | 1 | 20 | 1 | <1 | .2 | .1 | .1 |
| 1330 | King Vitaman | 1 c | 21 | 2 | 81 | 2 | 18 | 1 | 1 | .2 | .3 | .2 |
| 1039 | Kix | 1 c | 19 | 2 | 72 | 1 | 16 | 1 | <1 | .1 | .1 | t |
| 1331 | Life | 1 c | 44 | 4 | 167 | 4 | 35 | 3 | 2 | .3 | .6 | .8 |
| 507 | Lucky Charms | 1 c | 32 | 2 | 124 | 2 | 27 | 1 | 1 | .2 | .4 | .2 |
| 1323 | Mueslix Five Grain | 1 c | 82 | 5 | 279 | 7 | 63 | 7 | 3 | .5 | 1 | 1.2 |
| 508 | Nature Valley Granola | 1 c | 113 | 4 | 510 | 12 | 74 | 7 | 20 | 2.6 | 13.3 | 3.8 |
| 1666 | Nutri Grain Almond Raisin | 1 c | 40 | 6 | 147 | 3 | 31 | 3 | 2 | .1 | 1 | 1.2 |
| 1336 | 100% Bran | 1 c | 66 | 3 | 178 | 8 | 48 | 19 | 3 | .6 | .6 | 1.9 |
| 509 | 100% Natural cereal, plain | 1 c | 104 | 2 | 462 | 11 | 71 | 8 | 17 | 7.5 | 7.5 | 2.3 |
| 1337 | 100% Natural with apples & cinnamon | 1 c | 104 | 2 | 477 | 11 | 70 | 7 | 20 | 15.5 | 1.8 | 1.3 |
| 1338 | 100% Natural with raisins & dates | 1 c | 110 | 3 | 496 | 12 | 72 | 7 | 20 | 13.6 | 3.7 | 1.7 |
| 510 | Product 19 | 1 c | 33 | 4 | 110 | 2 | 28 | 1 | <1 | t | .2 | .2 |
| 1339 | Quisp | 1 c | 30 | 3 | 121 | 2 | 25 | 1 | 2 | .5 | .4 | .2 |
| 511 | Raisin Bran, Kellogg's | 1 c | 61 | 9 | 200 | 6 | 47 | 8 | 1 | .1 | .1 | .4 |
| 512 | Raisin Bran, Post | 1 c | 56 | 9 | 172 | 5 | 42 | 8 | 1 | .2 | .1 | .5 |
| 1667 | Raisin Squares | 1 c | 71 | 9 | 241 | 6 | 55 | 7 | 2 | .2 | .2 | .6 |
| 1041 | Rice Chex | 1 c | 33 | 3 | 130 | 2 | 29 | 1 | <1 | t | t | t |
| 513 | Rice Krispies, Kellogg's | 1 c | 28 | 2 | 111 | 2 | 25 | <1 | <1 | t | t | t |
| 514 | Rice, puffed | 1 c | 14 | 4 | 54 | 1 | 12 | <1 | <1 | t | t | t |
| 515 | Shredded Wheat | 1 c | 43 | 5 | 154 | 5 | 35 | 4 | 1 | .1 | .1 | .4 |
| 516 | Special K | 1 c | 31 | 3 | 110 | 6 | 22 | 1 | <1 | t | t | .2 |
| 517 | Super Golden Crisp | 1 c | 33 | 1 | 123 | 2 | 30 | <1 | <1 | .1 | .1 | .1 |
| 519 | Honey Smacks | 1 c | 36 | 3 | 133 | 3 | 32 | 1 | 1 | .4 | .1 | .3 |
| 1341 | Tasteeos | 1 c | 24 | 2 | 94 | 3 | 19 | 3 | 1 | .2 | .2 | .2 |
| 1342 | Team | 1 c | 42 | 4 | 164 | 3 | 36 | 1 | 1 | .1 | .2 | .3 |
| 520 | Total, wheat, with added calcium | 1 c | 40 | 3 | 140 | 4 | 32 | 4 | 1 | .2 | .2 | .1 |
| 521 | Trix | 1 c | 28 | 2 | 114 | 1 | 24 | 1 | 2 | .4 | .9 | .3 |
| 1344 | Wheat Chex | 1 c | 46 | 2 | 169 | 5 | 38 | 4 | 1 | .2 | .1 | .5 |
| 1043 | Wheat cereal, puffed, fortified | 1 c | 12 | 4 | 44 | 2 | 9 | 1 | <1 | t | t | .1 |
| 522 | Wheaties | 1 c | 29 | 3 | 106 | 3 | 23 | 2 | 1 | .2 | .2 | .1 |
| 523 | Buckwheat flour, dark | 1 c | 120 | 11 | 402 | 15 | 85 | 12 | 4 | .8 | 1.1 | 1.1 |
| 525 | Buckwheat, whole grain, dry | 1 c | 170 | 10 | 583 | 23 | 122 | 17 | 6 | 1.3 | 1.8 | 1.8 |
| 526 | Bulgar, dry, uncooked | 1 c | 140 | 9 | 479 | 17 | 106 | 26 | 2 | .3 | .2 | .8 |
| 527 | Bulgar, cooked | 1 c | 182 | 78 | 151 | 6 | 34 | 8 | <1 | .1 | .1 | .2 |
| | Cornmeal: | | | | | | | | | | | |
| 528 | Whole-ground, unbolted, dry | 1 c | 122 | 10 | 442 | 10 | 94 | 9 | 4 | .6 | 1.2 | 2 |
| 530 | Degermed, enriched, dry | 1 c | 138 | 12 | 505 | 12 | 107 | 10 | 2 | .3 | .6 | 1 |
| 38041 | Degermed, enriched, baked | 1 c | 138 | 12 | 505 | 12 | 107 | 10 | 2 | .3 | .6 | 1 |
| | Macaroni, cooked: | | | | | | | | | | | |
| 532 | Enriched | 1 c | 140 | 66 | 197 | 7 | 40 | 2 | 1 | .1 | .1 | .4 |
| 533 | Whole wheat | 1 c | 140 | 67 | 174 | 7 | 37 | 4 | 1 | .1 | .1 | .3 |

**PAGE KEY:** H–4 = Beverages H–6 = Dairy H–10 = Eggs H–10 = Fat/Oil H–14 = Fruit H–20 = Bakery H–28 = Grain H–32 = Fish H–34 = Meats H–38 = Poultry H–40 = Sausage H–42 = Mixed/Fast H–46 = Nuts/Seeds H–50 = Sweets H–52 = Vegetables/Legumes H–62 = Vegetarian Foods H–64 = Misc H–66 = Soups/Sauces H–68 = Fast H–84 = Convenience H–88 = Baby foods



| Chol (mg) | Calc (mg) | Iron (mg) | Magn (mg) | Pota (mg) | Sodi (mg) | Zinc (mg) | VT-A (RE) | Thia (mg) | VT-E (a-TE) | Ribo (mg) | Niac (mg) | V-B6 (mg) | Fola (μg) | VT-C (mg) |
|---|---|---|---|---|---|---|---|---|---|---|---|---|---|---|
| 0 | 68 | 13.7 | 58 | 228 | 220 | 2.54 | 636 | .62 | .34 | .72 | 8.45 | .86 | 169 | 0 |
| 0 | 0 | 10.9 | 81 | 229 | 309 | 5.03 | 505 | .51 | 7.22 | .58 | 6.71 | .66 | 138 | 20 |
| 0 | 21 | 13.4 | 102 | 251 | 431 | 2.49 | 622 | .61 | .54 | .7 | 8.27 | .85 | 166 | 0 |
| 0 | 0 | 4.51 | 8 | 35 | 150 | 3.75 | 225 | .37 | .12 | .43 | 5 | .5 | 96 | 15 |
| 0 | 0 | 6.15 | 4 | 26 | 264 | .2 | 298 | .5 | .05 | .56 | 6.61 | .66 | 123 | 20 |
| 0 | 0 | 15 | 60 | 170 | 0 | 1.5 | 0 | .37 | .46 | .42 | 4.64 | .5 | 102 | 0 |
| 0 | 0 | 2.42 | 8 | 27 | 256 | .42 | 303 | .49 | .03 | .56 | 6.72 | .66 | 140 | 20 |
| 0 | 30 | 10.1 | 81 | 335 | 270 | 3.02 | 725 | .75 | 1.32 | .85 | 10.1 | 1 | 201 | 0 |
| 0 | 4 | 2.02 | 9 | 24 | 178 | 1.7 | 424 | .42 | .03 | .48 | 5.63 | .58 | 113 | 0 |
| 0 | 19 | 5.85 | 12 | 69 | 357 | 4.88 | 293 | .49 | .29 | .55 | 6.51 | .65 | 130 | 19 |
| 0 | 49 | 2.56 | 109 | 328 | 15 | 2.48 | 2 | .45 | 7.87 | .17 | 1.25 | .19 | 52 | 1 |
| 0 | | 2.71 | 36 | 143 | 90 | 5.64 | 226 | .56 | 7.57 | .64 | 7.52 | .75 | 151 | |
| 0 | 15 | 1.35 | 30 | 127 | 101 | 2.84 | 169 | .27 | 4.03 | .31 | 3.74 | .36 | 90 | 0 |
| 0 | 5 | 15.7 | 37 | 184 | 382 | 1.21 | 728 | .71 | .14 | .82 | 9.68 | .99 | 194 | 0 |
| 0 | 16 | 11.2 | 43 | 136 | 220 | .78 | 516 | .51 | .1 | .58 | 6.86 | .7 | 138 | 0 |
| 0 | 66 | 4.02 | 141 | 415 | 226 | 2.83 | 7 | .32 | .77 | .14 | 1.54 | .2 | 44 | 1 |
| 0 | 0 | 3.03 | 3 | 40 | 249 | .2 | 152 | .26 | .09 | .3 | 3.37 | .33 | 74 | 10 |
| 0 | 22 | 4.95 | 32 | 94 | 285 | 4.13 | 248 | .41 | .34 | .47 | 5.51 | .55 | 110 | 16 |
| 0 | 16 | 5.57 | 46 | 151 | 202 | .9 | 463 | .45 | .81 | .52 | 6.16 | .63 | 23 | 19 |
| 0 | 4 | 2.09 | 7 | 25 | 124 | 1.17 | 291 | .29 | .09 | .33 | 3.87 | .4 | 78 | 0 |
| 0 | 3 | 5.92 | 18 | 58 | 176 | 2.65 | 212 | .26 | 1.42 | .3 | 3.53 | .35 | 71 | 8 |
| 0 | 28 | 5.13 | 6 | 26 | 167 | 2.38 | 238 | .24 | .05 | .27 | 3.17 | .32 | 63 | 9 |
| 0 | 134 | 12.3 | 43 | 109 | 240 | 5.5 | 2 | .55 | .22 | .62 | 7.35 | .73 | 147 | 0 |
| 0 | 35 | 4.8 | 21 | 58 | 217 | 4 | 240 | .4 | .14 | .45 | 5.34 | .53 | 107 | 16 |
| 0 | 38 | 8.94 | 82 | 369 | 107 | 7.46 | 747 | .75 | 8.94 | .84 | 9.84 | .99 | 197 | 1 |
| 0 | 85 | 3.53 | 107 | 375 | 183 | 2.27 | 0 | .35 | 7.97 | .12 | 1.25 | .16 | 17 | 0 |
| 0 | 122 | 1.14 | 13 | 147 | 139 | 3.06 | 0 | .32 | 4.38 | .35 | 4.08 | .41 | 80 | 0 |
| 0 | 46 | 8.12 | 312 | 652 | 457 | 5.74 | 0 | 1.58 | 1.53 | 1.78 | 20.9 | 2.11 | 47 | 63 |
| 1 | 100 | 3.11 | 109 | 457 | 28 | 2.5 | 1 | .36 | 1.19 | .17 | 1.84 | .19 | 26 | <1 |
| 0 | 157 | 2.89 | 72 | 514 | 52 | 2 | 6 | .33 | .73 | .57 | 1.87 | .11 | 17 | 1 |
| 0 | 160 | 3.12 | 124 | 538 | 47 | 2.11 | 7 | .31 | .77 | .65 | 2.09 | .16 | 45 | 0 |
| 0 | 0 | 19.8 | 18 | 55 | 308 | 16.5 | 248 | 1.65 | 24.4 | 1.88 | 22 | 2.21 | 429 | 66 |
| 0 | 6 | 5.1 | 15 | 40 | 216 | 4.26 | 4 | .42 | .15 | .48 | 5.67 | .56 | 113 | 0 |
| 0 | 40 | 4.5 | 80 | 350 | 390 | 3.75 | 225 | .37 | .56 | .43 | 5 | .5 | 122 | 0 |
| 0 | 26 | 8.9 | 95 | 345 | 365 | 2.97 | 741 | .73 | 1.3 | .84 | 9.86 | 1.01 | 198 | 0 |
| 0 | 0 | 21.7 | 54 | 335 | 4 | 1.99 | 0 | .5 | .38 | .57 | 6.67 | .64 | 142 | 0 |
| 0 | 5 | 9.44 | 8 | 38 | 276 | .45 | 2 | .43 | .04 | .01 | 5.81 | .59 | 116 | 17 |
| 0 | 5 | .7 | 12 | 27 | 206 | .46 | 371 | .52 | .03 | .59 | 6.92 | .69 | 138 | 15 |
| 0 | 1 | .41 | 4 | 16 | 1 | .15 | 0 | .06 | .01 | .01 | .87 | 0 | 1 | 0 |
| 0 | 16 | 1.81 | 57 | 155 | 4 | 1.42 | 0 | .11 | .23 | .12 | 2.26 | .11 | 21 | 0 |
| 0 | 0 | 8.4 | 16 | 55 | 250 | 3.75 | 225 | .53 | .08 | .59 | 7.01 | .71 | 93 | 15 |
| 0 | 7 | 2.08 | 20 | 48 | 51 | 1.75 | 437 | .43 | .12 | .49 | 5.81 | .59 | 116 | 0 |
| 0 | 0 | 2.4 | 21 | 53 | 67 | .4 | 300 | .5 | .18 | .58 | 6.66 | .68 | 133 | 20 |
| 0 | 11 | 6.86 | 26 | 71 | 183 | .69 | 318 | .31 | .17 | .36 | 4.22 | .43 | 85 | 13 |
| 0 | 6 | 12 | 12 | 71 | 260 | .58 | 556 | .55 | .1 | .63 | 7.39 | .76 | 7 | 22 |
| 0 | 344 | 24 | 43 | 129 | 265 | 20 | 500 | 2 | 31.3 | 2.27 | 26.8 | 2.67 | 533 | 80 |
| 0 | 30 | 4.2 | 3 | 16 | 184 | 3.5 | 210 | .35 | .56 | .4 | 4.68 | .47 | 93 | 14 |
| 0 | 18 | 13.2 | 58 | 173 | 308 | 1.23 | 0 | .6 | .17 | .17 | 8.1 | .83 | 162 | 24 |
| 0 | 3 | .56 | 16 | 44 | 1 | .37 | <1 | .05 | .08 | .03 | 1.43 | .02 | 4 | 0 |
| 0 | 53 | 7.83 | 31 | 101 | 215 | .68 | 218 | .36 | .36 | .41 | 4.84 | .48 | 97 | 14 |
| 0 | 49 | 4.87 | 301 | 692 | 13 | 3.74 | 0 | .5 | 1.24 | .23 | 7.38 | .7 | 65 | 0 |
| 0 | 31 | 3.74 | 393 | 782 | 2 | 4.08 | 0 | .17 | 1.75 | .72 | 11.9 | .36 | 51 | 0 |
| 0 | 49 | 3.44 | 230 | 574 | 24 | 2.7 | 0 | .32 | .22 | .16 | 7.15 | .48 | 38 | 0 |
| 0 | 18 | 1.75 | 58 | 124 | 9 | 1.04 | 0 | .1 | .05 | .05 | 1.82 | .15 | 33 | 0 |
| | | | | | | | | | | | | | | |
| 0 | 7 | 4.21 | 155 | 350 | 43 | 2.22 | 57 | .47 | .82 | .24 | 4.43 | .37 | 31 | 0 |
| 0 | 7 | 5.7 | 55 | 224 | 4 | .99 | 57 | .99 | .45 | .56 | 6.94 | .35 | 258 | 0 |
| 0 | 7 | 5.7 | 55 | 224 | 4 | .99 | 57 | .79 | .5 | .5 | 6.25 | .32 | 181 | 0 |
| | | | | | | | | | | | | | | |
| 0 | 10 | 1.96 | 25 | 43 | 1 | .74 | 0 | .29 | .04 | .14 | 2.34 | .05 | 98 | 0 |
| 0 | 21 | 1.48 | 42 | 62 | 4 | 1.13 | 0 | .15 | .14 | .06 | .99 | .11 | 7 | 0 |

**Table H-1**

**Food Composition** (Computer code number is for West Diet Analysis program) (For purposes of calculations, use "0" for t, <1, <.1, <.01, etc.)

| Computer Code Number | Food Description | Measure | Wt (g) | $H_2O$ (%) | Ener (kcal) | Prot (g) | Carb (g) | Dietary Fiber (g) | Fat (g) | Fat Breakdown (g) Sat | Mono | Poly |
|---|---|---|---|---|---|---|---|---|---|---|---|---|
| | **GRAIN PRODUCTS: CEREAL, FLOUR, GRAIN, PASTA and NOODLES, POPCORN—**Continued | | | | | | | | | | | |
| 534 | Vegetable, enriched | 1 c | 134 | 68 | 172 | 6 | 36 | 2 | <1 | t | t | .1 |
| 535 | Millet, cooked | 1 c | 240 | 71 | 286 | 8 | 57 | 3 | 2 | .4 | .4 | 1.2 |
| | Noodles (see also Pasta and Spaghetti): | | | | | | | | | | | |
| 1507 | Cellophane noodles, cooked | 1 c | 190 | 79 | 160 | <1 | 39 | <1 | <1 | t | t | t |
| 1995 | Cellophane noodles, dry | 1 c | 140 | 13 | 491 | <1 | 121 | 1 | <1 | t | t | t |
| 537 | Chow mein, dry | 1 c | 45 | 1 | 237 | 4 | 26 | 2 | 14 | 2 | 3.5 | 7.8 |
| 536 | Egg noodles, cooked, enriched | 1 c | 160 | 69 | 213 | 8 | 40 | 2 | 2 | .5 | .7 | .7 |
| 538 | Spinach noodles, dry | 3½ oz | 100 | 8 | 372 | 13 | 75 | 11 | 2 | .2 | .2 | .6 |
| 1343 | Oat bran, dry | ¼ c | 24 | 7 | 59 | 4 | 16 | 4 | 2 | .3 | .6 | .7 |
| | Pasta, cooked: | | | | | | | | | | | |
| 1418 | Fresh | 2 oz | 57 | 69 | 75 | 3 | 14 | 1 | 1 | .1 | .1 | .2 |
| 1417 | Linguini/Rotini | 1 c | 140 | 66 | 197 | 7 | 40 | 4 | 1 | .1 | .1 | .4 |
| | Popcorn: | | | | | | | | | | | |
| 539 | Air popped, plain | 1 c | 8 | 4 | 31 | 1 | 6 | 1 | <1 | t | .1 | .2 |
| 1042 | Microwaved, low fat, low sodium | 1 c | 6 | 3 | 25 | 1 | 4 | 1 | 1 | .1 | .2 | .3 |
| 540 | Popped in vegetable oil/salted | 1 c | 11 | 3 | 55 | 1 | 6 | 1 | 3 | .5 | .9 | 1.5 |
| 541 | Sugar-syrup coated | 1 c | 35 | 3 | 151 | 1 | 28 | 2 | 4 | 1.3 | 1 | 1.6 |
| | Rice: | | | | | | | | | | | |
| 542 | Brown rice, cooked | 1 c | 195 | 73 | 216 | 5 | 45 | 4 | 2 | .4 | .6 | .6 |
| 2215 | Mexican rice, cooked | 1 c | 226 | | 820 | 16 | 180 | 6 | 30 | 4 | 1 | 1 |
| 2216 | Spanish rice, cooked | 1 c | 246 | 85 | 130 | 3 | 28 | 2 | 1 | | | |
| | White, enriched, all types: | | | | | | | | | | | |
| 543 | Regular/long grain, dry | 1 c | 185 | 12 | 675 | 13 | 148 | 2 | 1 | .3 | .4 | .3 |
| 544 | Regular/long grain, cooked | 1 c | 158 | 68 | 205 | 4 | 45 | 1 | <1 | .1 | .1 | .1 |
| 545 | Instant, prepared without salt | 1 c | 165 | 76 | 162 | 3 | 35 | 1 | <1 | .1 | .1 | .1 |
| | Parboiled/converted rice: | | | | | | | | | | | |
| 546 | Raw, dry | 1 c | 185 | 10 | 686 | 13 | 151 | 3 | 1 | .3 | .3 | .3 |
| 547 | Cooked | 1 c | 175 | 72 | 200 | 4 | 43 | 1 | <1 | .1 | .1 | .1 |
| 1486 | Sticky rice (glutinous), cooked | 1 c | 174 | 77 | 169 | 4 | 37 | 2 | <1 | .1 | .1 | .1 |
| 548 | Wild rice, cooked | 1 c | 164 | 74 | 166 | 7 | 35 | 3 | 1 | .1 | .1 | .4 |
| 1700 | Rice and pasta (Rice-a-Roni), cooked | 1 c | 202 | 72 | 246 | 5 | 43 | 1 | 6 | 1.1 | 2.3 | 1.9 |
| 549 | Rye flour, medium | 1 c | 102 | 10 | 361 | 10 | 79 | 15 | 2 | .2 | .2 | .8 |
| 1044 | Soy flour, low-fat | 1 c | 88 | 3 | 325 | 45 | 30 | 9 | 6 | .9 | 1.3 | 3.3 |
| | Spaghetti pasta: | | | | | | | | | | | |
| 550 | Without salt, enriched | 1 c | 140 | 66 | 197 | 7 | 40 | 4 | 1 | .1 | .1 | .4 |
| 551 | With salt, enriched | 1 c | 140 | 66 | 197 | 7 | 40 | 2 | 1 | .1 | .1 | .4 |
| 552 | Whole-wheat spaghetti, cooked | 1 c | 140 | 67 | 174 | 7 | 37 | 6 | 1 | .1 | .1 | .3 |
| 1302 | Tapioca-pearl, dry | 1 c | 152 | 11 | 544 | <1 | 135 | 1 | <1 | t | t | t |
| 553 | Wheat bran, crude | 1 c | 58 | 10 | 125 | 9 | 37 | 25 | 2 | .4 | .4 | 1.3 |
| 554 | Wheat germ, raw | 1 c | 115 | 11 | 414 | 27 | 60 | 15 | 11 | 1.9 | 1.6 | 6.9 |
| 555 | Wheat germ, toasted | 1 c | 113 | 6 | 432 | 33 | 56 | 15 | 12 | 2.1 | 1.7 | 7.5 |
| 1669 | Wheat germ, with brown sugar & honey | 1 c | 113 | 3 | 420 | 30 | 66 | 11 | 9 | 1.5 | 1.2 | 5.5 |
| 556 | Rolled wheat, cooked | 1 c | 240 | 84 | 149 | 5 | 33 | 4 | 1 | .1 | .1 | .5 |
| 557 | Whole-grain wheat, cooked | 1 c | 150 | 86 | 84 | 4 | 20 | 3 | <1 | .1 | .1 | .2 |
| | Wheat flour (unbleached): | | | | | | | | | | | |
| | All-purpose white flour, enriched: | | | | | | | | | | | |
| 558 | Sifted | 1 c | 115 | 12 | 419 | 12 | 88 | 3 | 1 | .2 | .1 | .5 |
| 559 | Unsifted | 1 c | 125 | 12 | 455 | 13 | 95 | 3 | 1 | .2 | .1 | .5 |
| 560 | Cake or pastry, enriched, sifted | 1 c | 96 | 12 | 348 | 8 | 75 | 2 | 1 | .1 | .1 | .4 |
| 561 | Self-rising, enriched, unsifted | 1 c | 125 | 11 | 443 | 12 | 93 | 3 | 1 | .2 | .1 | .5 |
| 562 | Whole wheat, from hard wheats | 1 c | 120 | 10 | 407 | 16 | 87 | 15 | 2 | .4 | .3 | .9 |
| | **MEATS: FISH and SHELLFISH** | | | | | | | | | | | |
| 1045 | Bass, baked or broiled | 4 oz | 113 | 69 | 165 | 27 | 0 | 0 | 5 | 1.4 | 1.5 | 2.3 |
| 1046 | Bluefish, baked or broiled | 4 oz | 113 | 63 | 180 | 29 | 0 | 0 | 6 | 1.4 | 2 | 2.7 |
| 1686 | Catfish, breaded/flour fried | 4 oz | 113 | 49 | 325 | 21 | 14 | 1 | 20 | 5 | 9 | 4.7 |
| | Clams: | | | | | | | | | | | |
| 563 | Raw meat only | 1 ea | 145 | 82 | 107 | 19 | 4 | 0 | 1 | .3 | .4 | .7 |
| 564 | Canned, drained | 1 c | 160 | 64 | 237 | 41 | 8 | 0 | 3 | .7 | .9 | 1.5 |
| 1290 | Steamed, meat only | 10 ea | 95 | 64 | 141 | 24 | 5 | 0 | 2 | .4 | .5 | .9 |

| Chol (mg) | Calc (mg) | Iron (mg) | Magn (mg) | Pota (mg) | Sodi (mg) | Zinc (mg) | VT-A (RE) | Thia (mg) | VT-E (a-TE) | Ribo (mg) | Niac (mg) | V-B6 (mg) | Fola (µg) | VT-C (mg) |
|---|---|---|---|---|---|---|---|---|---|---|---|---|---|---|
| 0 | 15 | .66 | 25 | 41 | 8 | .59 | 7 | .15 | .05 | .08 | 1.43 | .03 | 87 | 0 |
| 0 | 7 | 1.51 | 106 | 149 | 5 | 2.18 | 0 | .25 | .43 | .2 | 3.19 | .26 | 46 | 0 |
| 0 | 14 | 1 | 3 | 5 | 9 | .23 | 0 | .07 | .06 | 0 | .09 | .02 | 1 | 0 |
| 0 | 35 | 3.04 | 4 | 14 | 14 | .57 | 0 | .21 | .18 | 0 | .28 | .07 | 3 | 0 |
| 0 | 9 | 2.13 | 23 | 54 | 198 | .63 | 4 | .26 | .07 | .19 | 2.68 | .05 | 40 | 0 |
| 53 | 19 | 2.54 | 30 | 45 | 11 | .99 | 10 | .3 | .08 | .13 | 2.38 | .06 | 102 | 0 |
| 0 | 58 | 2.13 | 174 | 376 | 36 | 2.76 | 46 | .37 | .04 | .2 | 4.55 | .32 | 48 | 0 |
| 0 | 14 | 1.3 | 56 | 136 | 1 | .75 | 0 | .28 | .41 | .05 | .22 | .04 | 12 | 0 |
| 19 | 3 | .65 | 10 | 14 | 3 | .32 | 3 | .12 | .09 | .09 | .56 | .02 | 36 | 0 |
| 0 | 10 | 1.96 | 25 | 43 | 1 | .74 | 0 | .29 | .08 | .14 | 2.34 | .05 | 98 | 0 |
| 0 | 1 | .21 | 10 | 24 | <1 | .27 | 2 | .02 | .01 | .02 | .15 | .02 | 2 | 0 |
| 0 | 1 | .14 | 9 | 14 | 29 | .23 | 1 | .02 | .06 | .01 | .12 | .01 | 1 | 0 |
| 0 | 1 | .31 | 12 | 25 | 97 | .29 | 2 | .01 | .03 | .01 | .17 | .02 | 2 | <1 |
| 2 | 15 | .61 | 12 | 38 | 72 | .2 | 3 | .02 | .42 | .02 | .77 | .01 | 1 | 0 |
| 0 | 19 | .82 | 84 | 84 | 10 | 1.23 | 0 | .19 | .53 | .05 | 2.98 | .28 | 8 | 0 |
| 0 | 300 | 9 | | | 2700 | | 120 | | | | | | | 96 |
| 0 | | .72 | | | 1340 | | | | | | | | | |
| 0 | 52 | 7.97 | 46 | 213 | 9 | 2.02 | 0 | 1.07 | .24 | .09 | 7.75 | .3 | 427 | 0 |
| 0 | 16 | 1.9 | 19 | 55 | 2 | .77 | 0 | .26 | .08 | .02 | 2.34 | .15 | 92 | 0 |
| 0 | 13 | 1.04 | 8 | 7 | 5 | .4 | 0 | .12 | .08 | .08 | 1.45 | .02 | 68 | 0 |
| 0 | 111 | 6.59 | 57 | 222 | 9 | 1.78 | 0 | 1.1 | .24 | .13 | 6.72 | .65 | 427 | 0 |
| 0 | 33 | 1.98 | 21 | 65 | 5 | .54 | 0 | .44 | .09 | .03 | 2.45 | .03 | 87 | 0 |
| 0 | 3 | .24 | 9 | 17 | 9 | .71 | 0 | .03 | .07 | .02 | .5 | .04 | 2 | 0 |
| 0 | 5 | .98 | 52 | 166 | 5 | 2.2 | 0 | .08 | .38 | .14 | 2.12 | .22 | 43 | 0 |
| 2 | 16 | 1.9 | 24 | 85 | 1147 | .57 | 0 | .25 | .27 | .16 | 3.6 | .2 | 89 | <1 |
| 0 | 24 | 2.16 | 76 | 347 | 3 | 2.03 | 0 | .29 | 1.36 | .12 | 1.76 | .27 | 19 | 0 |
| 0 | 165 | 5.27 | 202 | 2261 | 16 | 1.04 | 4 | .33 | .17 | .25 | 1.9 | .46 | 361 | 0 |
| 0 | 10 | 1.96 | 25 | 43 | 1 | .74 | 0 | .29 | .08 | .14 | 2.34 | .05 | 98 | 0 |
| 0 | 10 | 1.96 | 25 | 43 | 140 | .74 | 0 | .29 | .38 | .14 | 2.34 | .05 | 98 | 0 |
| 0 | 21 | 1.48 | 42 | 62 | 4 | 1.13 | 0 | .15 | .07 | .06 | .99 | .11 | 7 | 0 |
| 0 | 30 | 2.4 | 2 | 17 | 2 | .18 | 0 | .01 | 0 | 0 | 0 | .01 | 6 | 0 |
| 0 | 42 | 6.15 | 354 | 686 | 1 | 4.22 | 0 | .3 | 1.35 | .33 | 7.89 | .75 | 46 | 0 |
| 0 | 45 | 7.2 | 275 | 1025 | 14 | 14.1 | 0 | 2.16 | 20.7 | .57 | 7.83 | 1.5 | 323 | 0 |
| 0 | 51 | 10.3 | 362 | 1070 | 5 | 18.9 | 0 | 1.89 | 20.5 | .93 | 6.32 | 1.11 | 398 | 7 |
| 0 | 56 | 9.1 | 307 | 1089 | 12 | 15.7 | 11 | 1.51 | 24.9 | .78 | 5.34 | .56 | 376 | 0 |
| 0 | 17 | 1.49 | 53 | 170 | 0 | 1.15 | 0 | .17 | .48 | .12 | 2.14 | .17 | 26 | 0 |
| 0 | 9 | .88 | 35 | 99 | 1 | .73 | 0 | .12 | .3 | .03 | 1.5 | .08 | 12 | 0 |
| 0 | 17 | 5.34 | 25 | 123 | 2 | .8 | 0 | .9 | .07 | .57 | 6.79 | .05 | 177 | 0 |
| 0 | 19 | 5.8 | 27 | 134 | 2 | .87 | 0 | .98 | .07 | .62 | 7.38 | .05 | 193 | 0 |
| 0 | 13 | 7.03 | 15 | 101 | 2 | .59 | 0 | .86 | .06 | .41 | 6.52 | .03 | 148 | 0 |
| 0 | 423 | 5.84 | 24 | 155 | 1587 | .77 | 0 | .84 | .07 | .52 | 7.29 | .06 | 193 | 0 |
| 0 | 41 | 4.66 | 166 | 486 | 6 | 3.52 | 0 | .54 | 1.48 | .26 | 7.64 | .41 | 53 | 0 |
| 98 | 116 | 2.16 | 43 | 515 | 102 | .94 | 40 | .1 | .84 | .1 | 1.72 | .16 | 19 | 2 |
| 86 | 10 | .7 | 47 | 539 | 87 | 1.18 | 156 | .08 | .71 | .11 | 8.19 | .52 | 2 | 0 |
| 92 | 41 | 1.44 | 34 | 376 | 598 | 1.05 | 33 | .4 | 2.48 | .18 | 3.37 | .21 | 19 | 1 |
| 49 | 67 | 20.3 | 13 | 455 | 81 | 1.99 | 131 | .12 | 1.45 | .31 | 2.57 | .09 | 23 | 19 |
| 107 | 147 | 44.8 | 29 | 1004 | 179 | 4.37 | 274 | .24 | 3.04 | .68 | 5.36 | .18 | 46 | 35 |
| 64 | 87 | 26.6 | 17 | 597 | 106 | 2.59 | 162 | .14 | 1.86 | .4 | 3.18 | .1 | 27 | 21 |

**Table H-1**

**Food Composition** (Computer code number is for West Diet Analysis program) (For purposes of calculations, use "0" for t, <1, <.1, <.01, etc.)

| Computer Code Number | Food Description | Measure | Wt (g) | $H_2O$ (%) | Ener (kcal) | Prot (g) | Carb (g) | Dietary Fiber (g) | Fat (g) | Fat Breakdown (g) Sat | Mono | Poly |
|---|---|---|---|---|---|---|---|---|---|---|---|---|
| | **MEATS: FISH and SHELLFISH**—Continued | | | | | | | | | | | |
| | Cod: | | | | | | | | | | | |
| 565 | Baked | 4 oz | 113 | 76 | 119 | 26 | 0 | 0 | 1 | .2 | .1 | .5 |
| 566 | Batter fried | 4 oz | 113 | 67 | 196 | 20 | 8 | <1 | 9 | 2.2 | 3.6 | 2.6 |
| 567 | Poached, no added fat | 4 oz | 113 | 77 | 116 | 25 | 0 | 0 | 1 | .2 | .1 | .3 |
| | Crab, meat only: | | | | | | | | | | | |
| 1048 | Blue crab, cooked | 1 c | 118 | 77 | 120 | 24 | 0 | 0 | 2 | .3 | .3 | .8 |
| 1049 | Dungeness crab, cooked | 1 c | 118 | 73 | 130 | 26 | 1 | 0 | 1 | .2 | .3 | .5 |
| 568 | Blue crab, canned | 1 c | 135 | 76 | 134 | 28 | 0 | 0 | 2 | .4 | .3 | .6 |
| 1587 | Crab, imitation, from surimi | 4 oz | 113 | 74 | 115 | 14 | 11 | 0 | 1 | .3 | .2 | .8 |
| 569 | Fish sticks, breaded pollock | 2 ea | 56 | 46 | 152 | 9 | 13 | <1 | 7 | 1.8 | 2.8 | 1.8 |
| 572 | Flounder/sole, baked | 4 oz | 113 | 73 | 132 | 27 | 0 | 0 | 2 | .5 | .4 | .9 |
| 1599 | Grouper, baked or broiled | 4 oz | 113 | 73 | 133 | 28 | 0 | 0 | 1 | .4 | .4 | .6 |
| 573 | Haddock, breaded, fried | 4 oz | 113 | 55 | 264 | 22 | 14 | 1 | 13 | 3.2 | 5.4 | 3.3 |
| 1050 | Haddock, smoked | 4 oz | 113 | 71 | 131 | 28 | 0 | 0 | 1 | .3 | .3 | .5 |
| | Halibut: | | | | | | | | | | | |
| 17291 | Baked | 4 oz | 113 | 72 | 158 | 30 | 0 | 0 | 3 | .7 | 1 | 1.4 |
| 1051 | Smoked | 4 oz | 113 | 64 | 203 | 34 | | | 4 | .6 | 1.2 | 1.5 |
| 1054 | Raw | 4 oz | 113 | 78 | 124 | 23 | 0 | 0 | 3 | .7 | .8 | 1.1 |
| 575 | Herring, pickled | 4 oz | 113 | 55 | 296 | 16 | 11 | 0 | 20 | 4.4 | 11 | 4.8 |
| 1052 | Lobster meat, cooked w/moist heat | 1 c | 145 | 76 | 142 | 30 | 2 | 0 | 1 | .2 | .2 | .5 |
| 1687 | Ocean perch, baked/broiled | 4 oz | 113 | 73 | 137 | 27 | 0 | 0 | 2 | .4 | 1 | .7 |
| 576 | Ocean perch, breaded/fried | 4 oz | 113 | 59 | 249 | 22 | 9 | <1 | 13 | 3.2 | 5.7 | 3.4 |
| 1056 | Octopus, raw | 4 oz | 113 | 80 | 93 | 17 | 2 | 0 | 1 | .3 | .2 | .3 |
| | Oysters: | | | | | | | | | | | |
| 577 | Raw, Eastern | 1 c | 248 | 85 | 169 | 17 | 10 | 0 | 6 | 2 | .9 | 2.8 |
| 578 | Raw, Pacific | 1 c | 248 | 82 | 201 | 23 | 12 | 0 | 6 | 1.3 | .9 | 2.2 |
| | Cooked: | | | | | | | | | | | |
| 579 | Eastern, breaded, fried, medium | 5 ea | 73 | 65 | 144 | 6 | 8 | <1 | 9 | 2.7 | 1.7 | 4.6 |
| 580 | Western, simmered | 5 ea | 125 | 64 | 204 | 24 | 12 | 0 | 6 | 1.9 | .9 | 2.7 |
| 581 | Pollock, baked, broiled, or poached | 4 oz | 113 | 74 | 128 | 27 | 0 | 0 | 1 | .3 | .2 | .6 |
| | Salmon: | | | | | | | | | | | |
| 582 | Canned pink, solids and liquid | 4 oz | 113 | 69 | 157 | 22 | 0 | 0 | 7 | 1.7 | 2.1 | 2.3 |
| 583 | Broiled or baked | 4 oz | 113 | 62 | 244 | 31 | 0 | 0 | 12 | 2.2 | 6 | 2.7 |
| 584 | Smoked | 4 oz | 113 | 72 | 132 | 21 | 0 | 0 | 5 | 1.2 | 2.3 | 1.1 |
| 585 | Atlantic sardines, canned, drained, 2 = 24 g | 4 oz | 113 | 60 | 235 | 28 | 0 | 0 | 13 | 1.9 | 4.4 | 6.4 |
| 586 | Scallops, breaded, cooked from frozen | 6 ea | 93 | 58 | 200 | 17 | 9 | <1 | 10 | 2 | 2.5 | 5.3 |
| 1588 | Scallops, imitation, from surimi | 4 oz | 113 | 74 | 112 | 14 | 12 | 0 | 1 | .1 | .1 | .3 |
| 1688 | Scallops, steamed/boiled | ½ c | 60 | 76 | 64 | 10 | 1 | 0 | 2 | .3 | .7 | .6 |
| | Shrimp: | | | | | | | | | | | |
| 587 | Cooked, boiled, 2 large = 11g | 16 ea | 88 | 77 | 87 | 18 | 0 | 0 | 1 | .2 | .2 | .5 |
| 588 | Canned, drained | ½ c | 64 | 73 | 77 | 15 | 1 | 0 | 1 | .3 | .3 | .7 |
| 589 | Fried, 2 large = 15 g, breaded | 12 ea | 90 | 53 | 218 | 19 | 10 | <1 | 11 | 2 | 3 | 5.9 |
| 1057 | Raw, large, about 7g each | 14 ea | 98 | 76 | 104 | 20 | 1 | 0 | 2 | .3 | .3 | 1 |
| 1589 | Shrimp, imitation, from surimi | 4 oz | 113 | 75 | 114 | 14 | 10 | 0 | 2 | .3 | .2 | 1 |
| 1053 | Snapper, baked or broiled | 4 oz | 113 | 70 | 145 | 30 | 0 | 0 | 2 | .4 | .4 | .7 |
| 1060 | Squid, fried in flour | 4 oz | 113 | 64 | 198 | 20 | 9 | 0 | 8 | 2.1 | 3.1 | 2.4 |
| 1590 | Surimi | 4 oz | 113 | 76 | 112 | 17 | 8 | 0 | 1 | .2 | .2 | .6 |
| 1058 | Swordfish, raw | 4 oz | 113 | 76 | 137 | 22 | 0 | 0 | 5 | 1.3 | 1.7 | 1 |
| 1059 | Swordfish, baked or broiled | 4 oz | 113 | 69 | 175 | 29 | 0 | 0 | 6 | 1.7 | 2.2 | 1.3 |
| 590 | Trout, baked or broiled | 4 oz | 113 | 70 | 170 | 26 | 0 | 0 | 7 | 1.8 | 2 | 2.2 |
| | Tuna, light, canned, drained solids: | | | | | | | | | | | |
| 591 | Oil pack | 1 c | 145 | 60 | 287 | 42 | 0 | 0 | 12 | 2.2 | 4.3 | 4.2 |
| 592 | Water pack | 1 c | 154 | 74 | 179 | 39 | 0 | 0 | 1 | .4 | .2 | .5 |
| 1061 | Bluefin tuna, fresh | 4 oz | 113 | 68 | 163 | 26 | 0 | 0 | 6 | 1.4 | 1.8 | 1.6 |
| | **MEATS: BEEF, LAMB, PORK and others** | | | | | | | | | | | |
| | BEEF, cooked, trimmed to ½" outer fat: | | | | | | | | | | | |
| | Braised, simmered, pot roasted: | | | | | | | | | | | |
| | Relatively fat, choice chuck blade: | | | | | | | | | | | |
| 593 | Lean and fat, piece 2½ x 2½ x ¾" | 4 oz | 113 | 47 | 393 | 30 | 0 | 0 | 29 | 13 | 14.8 | 1.2 |
| 594 | Lean only | 4 oz | 113 | 55 | 297 | 35 | 0 | 0 | 16 | 7.3 | 8.2 | .7 |

| Chol (mg) | Calc (mg) | Iron (mg) | Magn (mg) | Pota (mg) | Sodi (mg) | Zinc (mg) | VT-A (RE) | Thia (mg) | VT-E (a-TE) | Ribo (mg) | Niac (mg) | V-B6 (mg) | Fola (μg) | VT-C (mg) |
|---|---|---|---|---|---|---|---|---|---|---|---|---|---|---|
| 62 | 16 | .55 | 47 | 276 | 88 | .65 | 16 | .1 | .39 | .09 | 2.84 | .32 | 9 | 1 |
| 64 | 43 | .92 | 36 | 443 | 124 | .62 | 17 | .13 | .92 | .13 | 2.58 | .23 | 10 | 1 |
| 61 | 23 | .54 | 41 | 496 | 69 | .63 | 14 | .09 | .32 | .08 | 2.48 | .28 | 8 | 1 |
| | | | | | | | | | | | | | | |
| 118 | 123 | 1.07 | 39 | 382 | 329 | 4.98 | 2 | .12 | 1.18 | .06 | 3.89 | .21 | 60 | 4 |
| 90 | 70 | .51 | 68 | 481 | 446 | 6.45 | 37 | .07 | 1.33 | .24 | 4.27 | .2 | 50 | 4 |
| 120 | 136 | 1.13 | 53 | 505 | 450 | 5.43 | 2 | .11 | 1.35 | .11 | 1.85 | .2 | 57 | 4 |
| 23 | 15 | .44 | 49 | 102 | 950 | .37 | 23 | .04 | .12 | .03 | .2 | .03 | 2 | 0 |
| 63 | 11 | .41 | 14 | 146 | 326 | .37 | 17 | .07 | .77 | .1 | 1.19 | .03 | 10 | 0 |
| 77 | 20 | .38 | 65 | 389 | 119 | .71 | 12 | .09 | 2.6 | .13 | 2.46 | .27 | 10 | 0 |
| 53 | 24 | 1.29 | 42 | 537 | 60 | .58 | 56 | .09 | .71 | .01 | .43 | .4 | 11 | 0 |
| 96 | 63 | 1.92 | 46 | 345 | 523 | .59 | 33 | .08 | 1.56 | .14 | 4.49 | .28 | 19 | <1 |
| 87 | 55 | 1.58 | 61 | 469 | 862 | .56 | 25 | .05 | .56 | .05 | 5.73 | .45 | 17 | 0 |
| | | | | | | | | | | | | | | |
| 46 | 68 | 1.21 | 121 | 651 | 78 | .6 | 61 | .08 | 1.23 | .1 | 8.05 | .45 | 16 | 0 |
| 59 | 87 | 1.56 | 154 | 833 | 2260 | .78 | 86 | .11 | 1.11 | .14 | 10.8 | .64 | 22 | 0 |
| 36 | 53 | .95 | 94 | 509 | 61 | .47 | 53 | .07 | .96 | .08 | 6.61 | .39 | 14 | 0 |
| 15 | 87 | 1.38 | 9 | 78 | 983 | .6 | 292 | .04 | 1.81 | .16 | 3.73 | .19 | 3 | 0 |
| 104 | 88 | .57 | 51 | 510 | 551 | 4.23 | 38 | .01 | 2.1 | .1 | 1.55 | .11 | 16 | 0 |
| 61 | 155 | 1.33 | 44 | 396 | 108 | .69 | 16 | .15 | 1.84 | .15 | 2.76 | .3 | 12 | 1 |
| 71 | 136 | 1.57 | 38 | 323 | 431 | .67 | 23 | .14 | 2.41 | .18 | 2.68 | .24 | 15 | 1 |
| 54 | 60 | 5.99 | 34 | 396 | 260 | 1.9 | 51 | .03 | 1.36 | .04 | 2.37 | .41 | 18 | 6 |
| | | | | | | | | | | | | | | |
| 131 | 112 | 16.5 | 117 | 387 | 523 | 225 | 74 | .25 | 1.98 | .24 | 3.42 | .15 | 25 | 9 |
| 124 | 20 | 12.7 | 55 | 417 | 263 | 41.2 | 201 | .17 | 2.11 | .58 | 4.98 | .12 | 25 | 20 |
| | | | | | | | | | | | | | | |
| 59 | 45 | 5.07 | 42 | 178 | 304 | 63.6 | 66 | .11 | 1.66 | .15 | 1.2 | .05 | 23 | 3 |
| 125 | 20 | 11.5 | 55 | 378 | 265 | 41.5 | 183 | .16 | 2.21 | .55 | 4.53 | .11 | 19 | 16 |
| 108 | 7 | .32 | 82 | 437 | 131 | .68 | 26 | .08 | .32 | .09 | 1.86 | .08 | 4 | 0 |
| | | | | | | | | | | | | | | |
| 62 | 241 | .95 | 38 | 368 | 626 | 1.04 | 19 | .03 | 1.53 | .21 | 7.39 | .34 | 17 | 0 |
| 98 | 8 | .62 | 35 | 424 | 75 | .58 | 71 | .24 | 1.42 | .19 | 7.54 | .25 | 6 | 0 |
| 26 | 12 | .96 | 20 | 198 | 886 | .35 | 29 | .03 | 1.53 | .11 | 5.33 | .31 | 2 | 0 |
| 160 | 432 | 3.3 | 44 | 449 | 571 | 1.48 | 76 | .09 | .34 | .26 | 5.93 | .19 | 13 | 0 |
| | | | | | | | | | | | | | | |
| 57 | 39 | .76 | 55 | 310 | 432 | .99 | 20 | .04 | 1.77 | .1 | 1.4 | .13 | 34 | 2 |
| 25 | 9 | .35 | 49 | 116 | 898 | .37 | 23 | .01 | .12 | .02 | .35 | .03 | 2 | 0 |
| 19 | 15 | .15 | 32 | 168 | 246 | .55 | 27 | .01 | .81 | .04 | .6 | .08 | 7 | 1 |
| | | | | | | | | | | | | | | |
| 172 | 34 | 2.72 | 30 | 160 | 197 | 1.37 | 58 | .03 | .66 | .03 | 2.28 | .11 | 3 | 2 |
| 111 | 38 | 1.75 | 26 | 134 | 108 | .81 | 11 | .02 | .59 | .02 | 1.77 | .07 | 1 | 1 |
| 159 | 60 | 1.13 | 36 | 203 | 310 | 1.24 | 50 | .12 | 1.35 | .12 | 2.76 | .09 | 7 | 1 |
| 149 | 51 | 2.36 | 36 | 181 | 145 | 1.09 | 53 | .03 | .8 | .03 | 2.5 | .1 | 3 | 2 |
| 41 | 21 | .68 | 49 | 101 | 797 | .37 | 23 | .03 | .12 | .04 | .19 | .03 | 2 | 0 |
| 53 | 45 | .27 | 42 | 590 | 64 | .5 | 40 | .06 | .71 | <.01 | .39 | .52 | 7 | 2 |
| 294 | 44 | 1.14 | 43 | 315 | 346 | 1.97 | 12 | .06 | 2.09 | .52 | 2.94 | .07 | 16 | 5 |
| 34 | 10 | .29 | 49 | 127 | 162 | .37 | 23 | .02 | .28 | .02 | .25 | .03 | 2 | 0 |
| 44 | 5 | .91 | 30 | 325 | 102 | 1.3 | 41 | .04 | .56 | .11 | 10.9 | .37 | 2 | 1 |
| 56 | 7 | 1.18 | 38 | 417 | 130 | 1.66 | 46 | .05 | .71 | .13 | 13.3 | .43 | 3 | 1 |
| 78 | 97 | .43 | 35 | 506 | 63 | .58 | 17 | .17 | .57 | .11 | 6.52 | .39 | 21 | 2 |
| | | | | | | | | | | | | | | |
| 26 | 19 | 2.02 | 45 | 300 | 513 | 1.31 | 33 | .05 | 1.74 | .17 | 18 | .16 | 8 | 0 |
| 46 | 17 | 2.36 | 42 | 365 | 521 | 1.19 | 26 | .05 | .82 | .11 | 20.5 | .54 | 6 | 0 |
| 43 | 9 | 1.15 | 56 | 285 | 44 | .68 | 740 | .27 | 1.13 | .28 | 9.77 | .51 | 2 | 0 |
| | | | | | | | | | | | | | | |
| 112 | 11 | 3.45 | 21 | 275 | 67 | 7.57 | 0 | .08 | .26 | .27 | 3.54 | .32 | 10 | 0 |
| 120 | 15 | 4.16 | 26 | 297 | 80 | 11.6 | 0 | .09 | .16 | .32 | 3.02 | .33 | 7 | 0 |

**Table H-1**

**Food Composition** (Computer code number is for West Diet Analysis program) (For purposes of calculations, use "0" for t, <1, <.1, <.01, etc.)

| Computer Code Number | Food Description | Measure | Wt (g) | $H_2O$ (%) | Ener (kcal) | Prot (g) | Carb (g) | Dietary Fiber (g) | Fat (g) | Fat Breakdown (g) Sat | Mono | Poly |
|---|---|---|---|---|---|---|---|---|---|---|---|---|
| | **MEATS: BEEF, LAMB, PORK and others**—Continued | | | | | | | | | | | |
| | Relatively lean, like choice round: | | | | | | | | | | | |
| 595 | Lean and fat, pce 4⅛ x 2½ x ¾" | 4 oz | 113 | 52 | 311 | 32 | 0 | 0 | 19 | 8.5 | 9.7 | .8 |
| 596 | Lean only | 4 oz | 113 | 57 | 249 | 36 | 0 | 0 | 11 | 4.8 | 5.4 | .5 |
| | Ground beef, broiled, patty 3 x ⅝": | | | | | | | | | | | |
| 597 | Extra lean, about 16% fat | 4 oz | 113 | 54 | 299 | 32 | 0 | 0 | 18 | 8 | 9 | .8 |
| 598 | Lean, 21% fat | 4 oz | 113 | 53 | 316 | 32 | 0 | 0 | 20 | 8.9 | 10.1 | .8 |
| | Roasts, oven cooked, no added liquid: | | | | | | | | | | | |
| | Relatively fat, prime rib: | | | | | | | | | | | |
| 601 | Lean and fat, piece 4⅛ x 2¼ x ½" | 4 oz | 113 | 46 | 425 | 25 | 0 | 0 | 35 | 15.8 | 17.9 | 1.5 |
| 602 | Lean only | 4 oz | 113 | 58 | 271 | 31 | 0 | 0 | 16 | 7 | 7.9 | .7 |
| | Relatively lean, choice round: | | | | | | | | | | | |
| 603 | Lean and fat, piece 2½ x 2½ x ¾" | 4 oz | 113 | 59 | 272 | 30 | 0 | 0 | 16 | 7.1 | 8.1 | .7 |
| 604 | Lean only | 4 oz | 113 | 65 | 198 | 33 | 0 | 0 | 6 | 2.9 | 3.3 | .3 |
| 1701 | Steak, rib, broiled, lean | 4 oz | 113 | 58 | 250 | 32 | 0 | 0 | 13 | 5.7 | 6.4 | .5 |
| | Steak, broiled, relatively lean, | | | | | | | | | | | |
| 606 | choice sirloin, lean only | 4 oz | 113 | 62 | 228 | 34 | 0 | 0 | 9 | 4 | 4.6 | .4 |
| | Steak, broiled, relatively fat, | | | | | | | | | | | |
| | choice T-bone: | | | | | | | | | | | |
| 1063 | Lean and fat | 4 oz | 113 | 52 | 349 | 26 | 0 | 0 | 26 | 11.8 | 13.3 | 1.1 |
| 1064 | Lean only | 4 oz | 113 | 61 | 232 | 30 | 0 | 0 | 11 | 5.1 | 5.8 | .5 |
| | Variety meats: | | | | | | | | | | | |
| 1086 | Brains, panfried | 4 oz | 113 | 71 | 221 | 14 | 0 | 0 | 18 | 6.7 | 7 | 3.9 |
| 599 | Heart, simmered | 4 oz | 113 | 64 | 198 | 32 | <1 | 0 | 6 | 2.9 | 1.5 | 1.5 |
| 600 | Liver, fried | 4 oz | 113 | 56 | 245 | 30 | 9 | 0 | 9 | 3.1 | 1.8 | 1.9 |
| 1062 | Tongue, cooked | 4 oz | 113 | 56 | 320 | 25 | <1 | 0 | 23 | 10.1 | 10.7 | .9 |
| 607 | Beef, canned, corned | 4 oz | 113 | 58 | 283 | 31 | 0 | 0 | 17 | 7.5 | 8.5 | .7 |
| 608 | Beef, dried, cured | 1 oz | 28 | 56 | 46 | 8 | <1 | 0 | 1 | .5 | .5 | .1 |
| | LAMB, domestic, cooked: | | | | | | | | | | | |
| | Chop, arm, braised (5.6 oz raw w/bone): | | | | | | | | | | | |
| 609 | Lean and fat | 1 ea | 70 | 44 | 242 | 21 | 0 | 0 | 17 | 7.8 | 7.3 | 1.4 |
| 610 | Lean only | 1 ea | 55 | 49 | 153 | 19 | 0 | 0 | 8 | 3.6 | 3.4 | .6 |
| | Chop, loin, broiled (4.2 oz raw w/bone): | | | | | | | | | | | |
| 611 | Lean and fat | 1 ea | 64 | 52 | 202 | 16 | 0 | 0 | 15 | 6.8 | 6.4 | 1.2 |
| 612 | Lean only | 1 ea | 46 | 61 | 99 | 14 | 0 | 0 | 4 | 2.1 | 1.9 | .4 |
| 1067 | Cutlet, avg of lean cuts, cooked | 4 oz | 113 | 54 | 330 | 28 | 0 | 0 | 23 | 10.9 | 10.2 | 1.9 |
| | Leg, roasted, 3 oz = 4⅛ x 2¼ x ½": | | | | | | | | | | | |
| 613 | Lean and fat | 4 oz | 113 | 57 | 292 | 29 | 0 | 0 | 19 | 8.7 | 8.1 | 1.6 |
| 614 | Lean only | 4 oz | 113 | 64 | 216 | 32 | 0 | 0 | 9 | 4.1 | 3.8 | .7 |
| 615 | Rib, roasted, lean and fat | 4 oz | 113 | 48 | 406 | 24 | 0 | 0 | 34 | 15.7 | 14.7 | 2.8 |
| 616 | Rib, roasted, lean only | 4 oz | 113 | 60 | 262 | 30 | 0 | 0 | 15 | 7 | 6.6 | 1.2 |
| 1065 | Shoulder, roasted, lean and fat | 4 oz | 113 | 56 | 312 | 25 | 0 | 0 | 23 | 10.5 | 9.8 | 1.9 |
| 1066 | Shoulder, roasted, lean only | 4 oz | 113 | 63 | 231 | 28 | 0 | 0 | 12 | 5.7 | 5.3 | 1.1 |
| | Variety meats: | | | | | | | | | | | |
| 1069 | Brains, panfried | 4 oz | 113 | 76 | 164 | 14 | 0 | 0 | 11 | 4.4 | 3.7 | 1.9 |
| 1068 | Heart, braised | 4 oz | 113 | 64 | 209 | 28 | 2 | 0 | 9 | 3.9 | 2.7 | 1.1 |
| 1070 | Sweetbreads, cooked | 4 oz | 113 | 60 | 264 | 26 | 0 | 0 | 17 | 8.1 | 6.5 | 1.4 |
| 1071 | Tongue, cooked | 4 oz | 113 | 58 | 311 | 24 | 0 | 0 | 23 | 8.8 | 11.6 | 1.4 |
| | PORK, cured, cooked (see also Sausages and Lunch Meats) | | | | | | | | | | | |
| 617 | Bacon, medium slices | 3 pce | 19 | 13 | 109 | 6 | <1 | 0 | 9 | 3.3 | 4.5 | 1.1 |
| 1087 | Breakfast strips, cooked | 2 pce | 23 | 27 | 106 | 7 | <1 | 0 | 8 | 2.9 | 3.8 | 1.3 |
| 618 | Canadian-style bacon | 2 pce | 47 | 62 | 87 | 11 | 1 | 0 | 4 | 1.3 | 1.9 | .4 |
| | Ham, roasted: | | | | | | | | | | | |
| 619 | Lean and fat, 2 pces 4⅛ x 2¼ x ¼" | 4 oz | 113 | 64 | 201 | 25 | 0 | 0 | 10 | 3.4 | 5 | 1.7 |
| 620 | Lean only | 4 oz | 113 | 68 | 164 | 24 | 2 | 0 | 6 | 2.1 | 3 | .6 |
| 621 | Ham, canned, roasted, 8% fat | 4 oz | 113 | 69 | 154 | 24 | 1 | 0 | 6 | 1.8 | 2.8 | .5 |
| | PORK, fresh, cooked: | | | | | | | | | | | |
| | Chops, loin (cut 3 per lb with bone): | | | | | | | | | | | |
| 1291 | Braised, lean and fat | 1 ea | 89 | 58 | 213 | 24 | 0 | 0 | 12 | 4.5 | 5.4 | 1 |
| 1292 | Lean only | 1 ea | 80 | 61 | 163 | 23 | 0 | 0 | 7 | 2.7 | 3.3 | .6 |
| 622 | Broiled, lean and fat | 1 ea | 82 | 58 | 197 | 23 | 0 | 0 | 11 | 3.9 | 4.8 | .8 |
| 623 | Broiled, lean only | 1 ea | 74 | 61 | 149 | 22 | 0 | 0 | 6 | 2.2 | 2.7 | .4 |

**PAGE KEY:** H–4 = Beverages H–6 = Dairy H–10 = Eggs H–10 = Fat/Oil H–14 = Fruit H–20 = Bakery H–28 = Grain H–32 = Fish H–34 = Meats H–38 = Poultry H–40 = Sausage H–42 = Mixed/Fast H–46 = Nuts/Seeds H–50 = Sweets H–52 = Vegetables/Legumes H–62 = Vegetarian Foods H–64 = Misc H–66 = Soups/Sauces H–68 = Fast H–84 = Convenience H–88 = Baby foods

| Chol (mg) | Calc (mg) | Iron (mg) | Magn (mg) | Pota (mg) | Sodi (mg) | Zinc (mg) | VT-A (RE) | Thia (mg) | VT-E (a-TE) | Ribo (mg) | Niac (mg) | V-B6 (mg) | Fola (µg) | VT-C (mg) |
|---|---|---|---|---|---|---|---|---|---|---|---|---|---|---|
| 108 | 7 | 3.53 | 25 | 319 | 56 | 5.55 | 0 | .08 | .21 | .27 | 4.21 | .37 | 11 | 0 |
| 108 | 6 | 3.91 | 28 | 348 | 58 | 6.19 | 0 | .08 | .2 | .29 | 4.61 | .41 | 12 | 0 |
| 112 | 10 | 3.13 | 28 | 417 | 93 | 7.27 | 0 | .08 | .2 | .36 | 6.61 | .36 | 12 | 0 |
| 114 | 14 | 2.77 | 27 | 394 | 101 | 7.01 | 0 | .07 | .23 | .27 | 6.75 | .34 | 12 | 0 |
| 96 | 12 | 2.61 | 21 | 334 | 71 | 5.92 | 0 | .08 | .27 | .19 | 3.8 | .26 | 8 | 0 |
| 91 | 11 | 2.95 | 28 | 425 | 84 | 7.84 | 0 | .09 | .14 | .24 | 4.64 | .34 | 9 | 0 |
| 81 | 7 | 2.07 | 27 | 406 | 67 | 4.87 | 0 | .09 | .23 | .18 | 3.92 | .4 | 7 | 0 |
| 78 | 6 | 2.2 | 30 | 446 | 70 | 5.36 | 0 | .1 | .12 | .19 | 4.24 | .43 | 8 | 0 |
| 90 | 15 | 2.9 | 30 | 445 | 78 | 7.9 | 0 | .11 | .16 | .25 | 5.42 | .45 | 9 | 0 |
| 101 | 12 | 3.8 | 36 | 455 | 75 | 7.37 | 0 | .15 | .16 | .33 | 4.84 | .51 | 11 | 0 |
| 76 | 9 | 3.06 | 26 | 363 | 72 | 5.03 | 0 | .1 | .24 | .24 | 4.46 | .37 | 8 | 0 |
| 67 | 7 | 3.58 | 32 | 427 | 80 | 6 | 0 | .12 | .16 | .28 | 5.23 | .44 | 9 | 0 |
| 2254 | 10 | 2.51 | 17 | 400 | 179 | 1.53 | 0 | .15 | 2.37 | .29 | 4.27 | .44 | 7 | 4 |
| 218 | 7 | 8.49 | 28 | 263 | 71 | 3.54 | 0 | .16 | .81 | 1.74 | 4.6 | .24 | 2 | 2 |
| 545 | 12 | 7.1 | 26 | 411 | 120 | 6.16 | 12123 | .24 | .72 | 4.68 | 16.3 | 1.62 | 249 | 26 |
| 121 | 8 | 3.83 | 19 | 203 | 68 | 5.42 | 0 | .03 | .4 | .4 | 2.43 | .18 | 6 | 1 |
| 97 | 14 | 2.35 | 16 | 154 | 1136 | 4.03 | 0 | .02 | .17 | .17 | 2.75 | .15 | 10 | 0 |
| 12 | 2 | 1.26 | 9 | 124 | 972 | 1.47 | 0 | .02 | .04 | .06 | 1.53 | .1 | 3 | 0 |
| 84 | 17 | 1.67 | 18 | 214 | 50 | 4.26 | 0 | .05 | .1 | .17 | 4.66 | .08 | 13 | 0 |
| 67 | 14 | 1.49 | 16 | 186 | 42 | 4.02 | 0 | .04 | .1 | .15 | 3.48 | .07 | 12 | 0 |
| 64 | 13 | 1.16 | 15 | 209 | 49 | 2.23 | 0 | .06 | .08 | .16 | 4.54 | .08 | 11 | 0 |
| 44 | 9 | .92 | 13 | 173 | 39 | 1.9 | 0 | .05 | .07 | .13 | 3.15 | .07 | 11 | 0 |
| 110 | 12 | 2.26 | 25 | 340 | 77 | 4.67 | 0 | .12 | .15 | .32 | 7.48 | .16 | 19 | 0 |
| 105 | 12 | 2.24 | 27 | 354 | 75 | 4.97 | 0 | .11 | .17 | .3 | 7.45 | .17 | 23 | 0 |
| 101 | 9 | 2.4 | 29 | 382 | 77 | 5.58 | 0 | .12 | .2 | .33 | 7.16 | .19 | 26 | 0 |
| 110 | 25 | 1.81 | 23 | 306 | 82 | 3.94 | 0 | .1 | .11 | .24 | 7.63 | .12 | 17 | 0 |
| 99 | 24 | 2 | 26 | 356 | 91 | 5.05 | 0 | .1 | .17 | .26 | 6.96 | .17 | 25 | 0 |
| 104 | 23 | 2.23 | 26 | 284 | 75 | 5.91 | 0 | .1 | .16 | .27 | 6.95 | .15 | 24 | 0 |
| 98 | 21 | 2.41 | 28 | 299 | 77 | 6.83 | 0 | .1 | .2 | .29 | 6.51 | .17 | 28 | 0 |
| 2308 | 14 | 1.9 | 16 | 232 | 151 | 1.54 | 0 | .12 | 1.73 | .27 | 2.79 | .12 | 6 | 14 |
| 281 | 16 | 6.24 | 27 | 212 | 71 | 4.16 | 0 | .19 | .79 | 1.34 | 4.93 | .34 | 2 | 8 |
| 452 | 14 | 2.4 | 21 | 329 | 59 | 3.03 | 0 | .02 | .78 | .24 | 2.89 | .06 | 15 | 23 |
| 214 | 11 | 2.97 | 18 | 179 | 76 | 3.38 | 0 | .09 | .36 | .47 | 4.17 | .19 | 3 | 8 |
| 16 | 2 | .31 | 5 | 92 | 303 | .62 | 0 | .13 | .1 | .05 | 1.39 | .05 | 1 | 0 |
| 24 | 3 | .45 | 6 | 107 | 483 | .85 | 0 | .17 | .08 | .08 | 1.75 | .08 | 1 | 0 |
| 27 | 5 | .38 | 10 | 183 | 727 | .8 | 0 | .39 | .15 | .09 | 3.25 | .21 | 2 | 0 |
| 67 | 9 | 1.51 | 25 | 462 | 1695 | 2.79 | 0 | .82 | .45 | .37 | 6.95 | .35 | 3 | 0 |
| 60 | 9 | 1.67 | 16 | 324 | 1359 | 3.25 | 0 | .85 | .29 | .23 | 4.54 | .45 | 3 | 0 |
| 34 | 7 | 1.04 | 24 | 393 | 1282 | 2.52 | 0 | 1.18 | .29 | .28 | 5.53 | .51 | 6 | 0 |
| 71 | 19 | .95 | 17 | 333 | 43 | 2.12 | 2 | .56 | .3 | .23 | 3.93 | .33 | 3 | 1 |
| 63 | 14 | .9 | 16 | 310 | 40 | 1.98 | 2 | .53 | .3 | .21 | 3.67 | .31 | 3 | <1 |
| 67 | 27 | .66 | 20 | 294 | 48 | 1.85 | 2 | .88 | .27 | .24 | 4.3 | .35 | 5 | <1 |
| 61 | 23 | .63 | 20 | 278 | 44 | 1.76 | 2 | .85 | .31 | .23 | 4.1 | .35 | 4 | <1 |

H

**Table H-1**

**Food Composition** (Computer code number is for West Diet Analysis program) (For purposes of calculations, use "0" for t, <1, <.1, <.01, etc.)

| Computer Code Number | Food Description | Measure | Wt (g) | $H_2O$ (%) | Ener (kcal) | Prot (g) | Carb (g) | Dietary Fiber (g) | Fat (g) | Fat Breakdown (g) Sat | Mono | Poly |
|---|---|---|---|---|---|---|---|---|---|---|---|---|
| | **MEATS: BEEF, LAMB, PORK and others—**Continued | | | | | | | | | | | |
| 624 | Panfried, lean and fat | 1 ea | 78 | 53 | 216 | 23 | 0 | 0 | 13 | 4.7 | 5.5 | 1.5 |
| 625 | Panfried, lean only | 1 ea | 63 | 59 | 152 | 16 | 0 | 0 | 10 | 3.2 | 3.9 | 1.2 |
| 626 | Leg, roasted, lean and fat | 4 oz | 113 | 55 | 308 | 30 | 0 | 0 | 20 | 7.3 | 8.9 | 1.9 |
| 627 | Leg, roasted, lean only | 4 oz | 113 | 61 | 233 | 35 | 0 | 0 | 9 | 3.2 | 4.3 | .9 |
| 628 | Rib, roasted, lean and fat | 4 oz | 113 | 56 | 288 | 31 | 0 | 0 | 17 | 6.7 | 7.9 | 1.4 |
| 629 | Rib, roasted, lean only | 4 oz | 113 | 59 | 252 | 32 | 0 | 0 | 13 | 4.9 | 5.9 | 1 |
| 630 | Shoulder, braised, lean and fat | 4 oz | 113 | 48 | 372 | 32 | 0 | 0 | 26 | 9.6 | 11.8 | 2.6 |
| 631 | Shoulder, braised, lean only | 4 oz | 113 | 54 | 280 | 36 | 0 | 0 | 14 | 4.7 | 6.5 | 1.3 |
| 1088 | Spareribs, cooked, yield from 1 lb raw with bone | 4 oz | 113 | 40 | 449 | 33 | 0 | 0 | 34 | 12.5 | 15.3 | 3.1 |
| 1095 | Rabbit, roasted (1 cup meat = 140 g) | 4 oz | 113 | 61 | 223 | 33 | 0 | 0 | 9 | 4 | 2 | 3 |
| | VEAL, cooked: | | | | | | | | | | | |
| 632 | Cutlet, braised or broiled, 4⅛ x 2¼ x ½" | 4 oz | 113 | 52 | 321 | 34 | 0 | 0 | 19 | 7.6 | 7.6 | 1.3 |
| 633 | Rib roasted, lean, 2 pieces 4⅛ x 2¼ x ¼" | 4 oz | 113 | 60 | 258 | 27 | 0 | 0 | 16 | 6.1 | 6.1 | 1.1 |
| 634 | Liver, panfried | 4 oz | 113 | 67 | 186 | 24 | 3 | 0 | 8 | 3.3 | 1.9 | 2.2 |
| 1096 | Venison (deer meat), roasted | 4 oz | 113 | 65 | 179 | 34 | 0 | 0 | 4 | 1.4 | 1 | .7 |
| | **MEATS: POULTRY and POULTRY PRODUCTS** | | | | | | | | | | | |
| | CHICKEN, cooked: | | | | | | | | | | | |
| | Fried, batter dipped: | | | | | | | | | | | |
| 635 | Breast | 1 ea | 280 | 52 | 728 | 69 | 25 | 1 | 37 | 9.9 | 15.3 | 8.6 |
| 636 | Drumstick | 1 ea | 72 | 53 | 193 | 16 | 6 | <1 | 11 | 3 | 4.7 | 2.7 |
| 637 | Thigh | 1 ea | 86 | 51 | 238 | 19 | 8 | <1 | 14 | 3.8 | 5.9 | 3.3 |
| 638 | Wing | 1 ea | 49 | 46 | 159 | 10 | 5 | <1 | 11 | 2.9 | 4.5 | 2.5 |
| | Fried, flour coated: | | | | | | | | | | | |
| 639 | Breast | 1 ea | 196 | 57 | 435 | 62 | 3 | <1 | 17 | 4.9 | 7.1 | 3.8 |
| 1212 | Breast, without skin | 1 ea | 86 | 60 | 161 | 29 | <1 | <1 | 4 | 1.1 | 1.5 | .9 |
| 640 | Drumstick | 1 ea | 49 | 57 | 120 | 13 | 1 | <1 | 7 | 1.8 | 2.7 | 1.6 |
| 641 | Thigh | 1 ea | 62 | 54 | 162 | 17 | 2 | <1 | 9 | 2.5 | 3.7 | 2.1 |
| 1099 | Thigh, without skin | 1 ea | 52 | 59 | 113 | 15 | 1 | <1 | 5 | 1.4 | 2 | 1.3 |
| 642 | Wing | 1 ea | 32 | 49 | 103 | 8 | 1 | <1 | 7 | 1.9 | 2.9 | 1.6 |
| | Roasted: | | | | | | | | | | | |
| 643 | All types of meat | 1 c | 140 | 64 | 266 | 40 | 0 | 0 | 10 | 2.9 | 3.8 | 2.4 |
| 644 | Dark meat | 1 c | 140 | 63 | 287 | 38 | 0 | 0 | 14 | 3.7 | 5.2 | 3.2 |
| 645 | Light meat | 1 c | 140 | 65 | 242 | 43 | 0 | 0 | 6 | 1.8 | 2.2 | 1.4 |
| 646 | Breast, without skin | 1 ea | 172 | 65 | 284 | 53 | 0 | 0 | 6 | 1.8 | 2.2 | 1.3 |
| 647 | Drumstick, without skin | 1 ea | 44 | 67 | 76 | 12 | 0 | 0 | 2 | .7 | .8 | .6 |
| 1703 | Leg, without skin | 1 ea | 95 | 65 | 181 | 26 | 0 | 0 | 8 | 2.2 | 2.9 | 1.9 |
| 648 | Thigh | 1 ea | 62 | 59 | 153 | 16 | 0 | 0 | 10 | 2.7 | 3.9 | 2.1 |
| 1100 | Thigh, without skin | 1 ea | 52 | 63 | 109 | 13 | 0 | 0 | 6 | 1.6 | 2.2 | 1.3 |
| 649 | Stewed, all types | 1 c | 140 | 67 | 248 | 38 | 0 | 0 | 9 | 2.6 | 3.5 | 2.2 |
| 656 | Canned, boneless chicken | 4 oz | 113 | 69 | 186 | 25 | 0 | 0 | 9 | 2.5 | 3.6 | 2 |
| 1102 | Gizzards, simmered | 1 c | 145 | 67 | 222 | 39 | 2 | 0 | 5 | 1.5 | 1.3 | 1.5 |
| 1101 | Hearts, simmered | 1 c | 145 | 65 | 268 | 38 | <1 | 0 | 11 | 3.3 | 2.9 | 3.3 |
| 2300 | Liver, simmered: Ounce | 3 oz | 85 | 68 | 133 | 21 | 1 | 0 | 5 | 1.6 | 1.1 | .8 |
| 1098 | Liver, simmered: Piece = 20 g | 6 ea | 120 | 68 | 188 | 29 | 1 | 0 | 7 | 2.2 | 1.6 | 1.1 |
| | DUCK, roasted: | | | | | | | | | | | |
| 1293 | Meat with skin, about 2.7 cups | ½ ea | 382 | 52 | 1287 | 73 | 0 | 0 | 108 | 36.9 | 49.3 | 13.9 |
| 651 | Meat only, about 1.5 cups | ½ ea | 221 | 64 | 444 | 52 | 0 | 0 | 25 | 9.2 | 8.2 | 3.2 |
| | GOOSE, domesticated, roasted: | | | | | | | | | | | |
| 1294 | Meat only, about 4.2 cups | ½ ea | 591 | 57 | 1406 | 173 | 0 | 0 | 75 | 23.6 | 40.2 | 10.9 |
| 1295 | Meat with skin, about 5.5 cups | ½ ea | 774 | 52 | 2360 | 195 | 0 | 0 | 170 | 53.2 | 80.5 | 24.8 |
| | TURKEY: | | | | | | | | | | | |
| | Roasted, meat only: | | | | | | | | | | | |
| 652 | Dark meat | 4 oz | 113 | 63 | 211 | 33 | 0 | 0 | 8 | 2.7 | 1.8 | 2.5 |
| 653 | Light meat | 4 oz | 113 | 66 | 177 | 34 | 0 | 0 | 4 | 1.2 | .6 | 1 |
| 654 | All types, chopped or diced | 1 c | 140 | 65 | 238 | 42 | 0 | 0 | 7 | 2.3 | 1.5 | 2 |
| 1103 | Ground, cooked | 4 oz | 113 | 59 | 266 | 31 | 0 | 0 | 15 | 4.1 | 5.5 | 3.6 |
| 1106 | Gizzard, cooked | 2 ea | 134 | 65 | 218 | 39 | 1 | 0 | 5 | 1.5 | 1 | 1.5 |
| 1107 | Heart, cooked | 4 ea | 64 | 64 | 113 | 17 | 1 | 0 | 4 | 1.1 | .8 | 1.1 |
| 1108 | Liver, cooked | 1 ea | 75 | 66 | 127 | 18 | 3 | 0 | 4 | 1.4 | 1.1 | .8 |

**PAGE KEY:** H–4 = Beverages H–6 = Dairy H–10 = Eggs H–10 = Fat/Oil H–14 = Fruit H–20 = Bakery H–28 = Grain H–32 = Fish H–34 = Meats H–38 = Poultry H–40 = Sausage H–42 = Mixed/Fast H–46 = Nuts/Seeds H–50 = Sweets H–52 = Vegetables/Legumes H–62 = Vegetarian Foods H–64 = Misc H–66 = Soups/Sauces H–68 = Fast H–84 = Convenience H–88 = Baby foods

| Chol (mg) | Calc (mg) | Iron (mg) | Magn (mg) | Pota (mg) | Sodi (mg) | Zinc (mg) | VT-A (RE) | Thia (mg) | VT-E (a-TE) | Ribo (mg) | Niac (mg) | V-B6 (mg) | Fola (μg) | VT-C (mg) |
|---|---|---|---|---|---|---|---|---|---|---|---|---|---|---|
| 72 | 21 | .71 | 23 | 332 | 62 | 1.8 | 2 | .89 | .32 | .24 | 4.37 | .37 | 5 | 1 |
| 52 | 14 | .67 | 16 | 230 | 49 | 2.44 | 1 | .46 | .3 | .23 | 2.8 | .26 | 3 | <1 |
| 106 | 16 | 1.14 | 25 | 398 | 68 | 3.34 | 3 | .72 | .34 | .35 | 5.16 | .45 | 11 | <1 |
| 108 | 8 | 1.29 | 33 | 442 | 73 | 3.4 | 3 | .91 | .46 | .4 | 5.56 | .38 | 3 | <1 |
| 82 | 32 | 1.06 | 24 | 476 | 52 | 2.33 | 2 | .82 | .41 | .34 | 6.92 | .37 | 3 | <1 |
| 80 | 29 | 1.11 | 25 | 494 | 53 | 2.41 | 2 | .86 | .55 | .36 | 7.25 | .38 | 3 | <1 |
| 123 | 20 | 1.82 | 21 | 417 | 99 | 4.72 | 3 | .61 | .5 | .35 | 5.89 | .4 | 5 | <1 |
| 129 | 9 | 2.2 | 25 | 458 | 115 | 5.62 | 3 | .68 | .58 | .41 | 6.71 | .46 | 6 | <1 |
| | | | | | | | | | | | | | | |
| 137 | 53 | 2.09 | 27 | 362 | 105 | 5.2 | 3 | .46 | .52 | .43 | 6.19 | .4 | 5 | 0 |
| 93 | 21 | 2.57 | 24 | 433 | 53 | 2.57 | 0 | .1 | .96 | .24 | 9.53 | .53 | 12 | 0 |
| | | | | | | | | | | | | | | |
| 133 | 32 | 1.23 | 27 | 316 | 90 | 4.1 | 0 | .04 | .45 | .34 | 10.2 | .29 | 16 | 0 |
| | | | | | | | | | | | | | | |
| 124 | 12 | 1.1 | 25 | 333 | 104 | 4.62 | 0 | .06 | .4 | .3 | 7.89 | .28 | 15 | 0 |
| 634 | 8 | 2.96 | 21 | 232 | 60 | 10.8 | 9095 | .15 | .42 | 2.19 | 9.58 | .55 | 858 | 35 |
| 127 | 8 | 5.05 | 27 | 379 | 61 | 3.11 | 0 | .2 | .28 | .68 | 7.58 | .42 | 5 | 0 |
| | | | | | | | | | | | | | | |
| 238 | 56 | 3.5 | 67 | 563 | 770 | 2.66 | 56 | .32 | 2.97 | .41 | 29.4 | 1.2 | 42 | 0 |
| 62 | 12 | .97 | 14 | 134 | 194 | 1.68 | 19 | .08 | .88 | .15 | 3.67 | .19 | 13 | 0 |
| 80 | 15 | 1.25 | 18 | 165 | 248 | 1.75 | 25 | .1 | 1.05 | .19 | 4.92 | .22 | 16 | 0 |
| 39 | 10 | .63 | 8 | 68 | 157 | .68 | 17 | .05 | .52 | .07 | 2.58 | .15 | 9 | 0 |
| | | | | | | | | | | | | | | |
| 174 | 31 | 2.33 | 59 | 508 | 149 | 2.16 | 29 | .16 | 1.12 | .26 | 26.9 | 1.14 | 12 | 0 |
| 78 | 14 | .98 | 27 | 237 | 68 | .93 | 6 | .07 | .36 | .11 | 12.7 | .55 | 3 | 0 |
| 44 | 6 | .66 | 11 | 112 | 44 | 1.42 | 12 | .04 | .41 | .11 | 2.96 | .17 | 5 | 0 |
| 60 | 9 | .92 | 15 | 147 | 55 | 1.56 | 18 | .06 | .52 | .15 | 4.31 | .2 | 7 | 0 |
| 53 | 7 | .76 | 13 | 135 | 49 | 1.45 | 11 | .05 | .3 | .13 | 3.7 | .2 | 5 | 0 |
| 26 | 5 | .4 | 6 | 57 | 25 | .56 | 12 | .02 | .18 | .04 | 2.14 | .13 | 2 | 0 |
| | | | | | | | | | | | | | | |
| 125 | 21 | 1.69 | 35 | 340 | 120 | 2.94 | 22 | .1 | .58 | .25 | 12.8 | .66 | 8 | 0 |
| 130 | 21 | 1.86 | 32 | 336 | 130 | 3.92 | 31 | .1 | .81 | .32 | 9.17 | .5 | 11 | 0 |
| 119 | 21 | 1.48 | 38 | 346 | 108 | 1.72 | 13 | .09 | .37 | .16 | 17.4 | .84 | 6 | 0 |
| 146 | 26 | 1.79 | 50 | 440 | 127 | 1.72 | 10 | .12 | .66 | .2 | 23.6 | 1.03 | 7 | 0 |
| 41 | 5 | .57 | 11 | 108 | 42 | 1.4 | 8 | .03 | .25 | .1 | 2.68 | .17 | 4 | 0 |
| 89 | 11 | 1.24 | 23 | 230 | 86 | 2.72 | 18 | .07 | .55 | .22 | 6 | .35 | 8 | 0 |
| 58 | 7 | .83 | 14 | 138 | 52 | 1.46 | 30 | .04 | .35 | .13 | 3.95 | .19 | 4 | 0 |
| 49 | 6 | .68 | 12 | 124 | 46 | 1.34 | 10 | .04 | .3 | .12 | 3.4 | .18 | 4 | 0 |
| 116 | 20 | 1.64 | 29 | 252 | 98 | 2.79 | 21 | .07 | .42 | .23 | 8.57 | .36 | 8 | 0 |
| 70 | 16 | 1.79 | 14 | 156 | 568 | 1.59 | 38 | .02 | .24 | .15 | 7.15 | .4 | 5 | 2 |
| 281 | 14 | 6.02 | 29 | 260 | 97 | 6.35 | 81 | .04 | 2.29 | .35 | 5.77 | .17 | 77 | 2 |
| 351 | 28 | 13.1 | 29 | 191 | 70 | 10.6 | 13 | .1 | 2.32 | 1.07 | 4.06 | .46 | 116 | 3 |
| 536 | 12 | 7.2 | 18 | 119 | 43 | 3.69 | 4176 | .13 | 1.45 | 1.49 | 3.78 | .49 | 655 | 13 |
| 757 | 17 | 10.2 | 25 | 168 | 61 | 5.21 | 5895 | .18 | 2.04 | 2.1 | 5.34 | .7 | 924 | 19 |
| | | | | | | | | | | | | | | |
| 321 | 42 | 10.3 | 61 | 779 | 225 | 7.11 | 241 | .66 | 2.5 | 1.03 | 18.5 | .69 | 23 | 0 |
| 197 | 26 | 5.97 | 44 | 557 | 144 | 5.75 | 51 | .57 | 1.55 | 1.04 | 11.3 | .55 | 22 | 0 |
| | | | | | | | | | | | | | | |
| 567 | 83 | 17 | 148 | 2293 | 449 | 18.7 | 71 | .54 | 9.16 | 2.3 | 24.1 | 2.78 | 71 | 0 |
| 704 | 101 | 21.9 | 170 | 2546 | 542 | 20.3 | 163 | .6 | 13.5 | 2.5 | 32.3 | 2.86 | 15 | 0 |
| | | | | | | | | | | | | | | |
| 96 | 36 | 2.63 | 27 | 328 | 89 | 5.04 | 0 | .07 | .94 | .28 | 4.12 | .41 | 10 | 0 |
| 78 | 21 | 1.53 | 32 | 345 | 72 | 2.31 | 0 | .07 | .12 | .15 | 7.73 | .61 | 7 | 0 |
| 106 | 35 | 2.49 | 36 | 417 | 98 | 4.34 | 0 | .09 | .59 | .25 | 7.62 | .64 | 10 | 0 |
| 115 | 28 | 2.18 | 27 | 305 | 121 | 3.23 | 0 | .06 | .45 | .19 | 5.45 | .44 | 8 | 0 |
| 311 | 20 | 7.29 | 25 | 283 | 72 | 5.57 | 74 | .04 | .27 | .44 | 4.11 | .16 | 70 | 2 |
| 145 | 8 | 4.41 | 14 | 117 | 35 | 3.37 | 5 | .04 | .13 | .56 | 2.08 | .2 | 51 | 1 |
| 470 | 8 | 5.85 | 11 | 146 | 48 | 2.32 | 2805 | .04 | 2.41 | 1.07 | 4.46 | .39 | 500 | 1 |

**Table H-1**

**Food Composition** (Computer code number is for West Diet Analysis program) (For purposes of calculations, use "0" for t, <1, <.1, <.01, etc.)

| Computer Code Number | Food Description | Measure | Wt (g) | $H_2O$ (%) | Ener (kcal) | Prot (g) | Carb (g) | Dietary Fiber (g) | Fat (g) | Fat Breakdown (g) Sat | Mono | Poly |
|---|---|---|---|---|---|---|---|---|---|---|---|---|
| | **MEATS: POULTRY and POULTRY PRODUCTS**—Continued | | | | | | | | | | | |
| | POULTRY FOOD PRODUCTS (see also items in Sausages & Lunchmeats section): | | | | | | | | | | | |
| 1567 | Chicken patty, breaded, cooked | 1 ea | 75 | 49 | 213 | 12 | 11 | <1 | 13 | 4.1 | 6.4 | 1.6 |
| 659 | Turkey and gravy, frozen package | 3 oz | 85 | 85 | 57 | 5 | 4 | <1 | 2 | .8 | .8 | .4 |
| | Turkey breast, Louis Rich: | | | | | | | | | | | |
| 1104 | Barbecued | 2 oz | 56 | 72 | 58 | 12 | 2 | 0 | <1 | .2 | .2 | .1 |
| 1943 | Hickory smoked | 1 pce | 80 | | 80 | 16 | 2 | 0 | 1 | 0 | | |
| 1947 | Honey roasted | 1 pce | 80 | | 80 | 16 | 3 | 0 | 1 | .5 | | |
| 1945 | Oven roasted | 1 pce | 80 | | 70 | 16 | | 0 | 1 | 0 | | |
| 661 | Turkey patty, breaded, fried | 2 oz | 57 | 50 | 161 | 8 | 9 | <1 | 10 | 2.7 | 4.3 | 2.7 |
| 662 | Turkey, frozen, roasted, seasoned | 4 oz | 113 | 68 | 175 | 24 | 3 | 0 | 7 | 2.1 | 1.4 | 1.9 |
| 1704 | Turkey roll, light meat | 1 pce | 28 | 72 | 41 | 5 | <1 | 0 | 2 | .6 | .7 | .5 |
| | **MEATS: SAUSAGES and LUNCHMEATS** (see also Poultry Food Products) | | | | | | | | | | | |
| 1072 | Beerwurst/beer salami, beef | 1 oz | 28 | 53 | 92 | 3 | <1 | 0 | 8 | 3.6 | 3.9 | .3 |
| 1074 | Beerwurst/beer salami, pork | 1 oz | 28 | 61 | 67 | 4 | 1 | 0 | 5 | 1.8 | 2.5 | .7 |
| 1075 | Berliner sausage | 1 oz | 28 | 61 | 64 | 4 | 1 | 0 | 5 | 1.7 | 2.2 | .4 |
| | Bologna: | | | | | | | | | | | |
| 1297 | Beef | 1 pce | 23 | 55 | 72 | 3 | <1 | 0 | 7 | 2.8 | 3.2 | .3 |
| 2115 | Beef, light, Oscar Mayer | 1 pce | 28 | 65 | 56 | 3 | 2 | 0 | 4 | 1.6 | 2 | .1 |
| 663 | Beef & pork | 1 pce | 28 | 54 | 88 | 3 | 1 | 0 | 8 | 3 | 3.7 | .7 |
| 2155 | Healthy Favorites | 1 pce | 23 | | 22 | 3 | 1 | 0 | <1 | 0 | | |
| 1298 | Pork | 1 pce | 23 | 61 | 57 | 4 | <1 | 0 | 5 | 1.6 | 2.2 | .5 |
| 2114 | Regular, light, Oscar Mayer | 1 pce | 28 | 65 | 56 | 3 | 2 | 0 | 4 | 1.6 | 2 | .4 |
| 664 | Turkey | 1 pce | 28 | 65 | 56 | 4 | <1 | 0 | 4 | 1.4 | 1.3 | 1.2 |
| 1970 | Turkey, Louis Rich | 1 pce | 28 | 67 | 57 | 3 | <1 | 0 | 5 | 1.5 | 1.8 | 1.3 |
| 665 | Braunschweiger sausage | 2 pce | 57 | 48 | 205 | 8 | 2 | 0 | 18 | 6.2 | 8.5 | 2.1 |
| 1073 | Bratwurst, link | 1 ea | 70 | 51 | 226 | 10 | 2 | 0 | 19 | 6.9 | 9.3 | 2 |
| 666 | Brown & serve sausage links, cooked | 2 ea | 26 | 45 | 102 | 4 | 1 | 0 | 10 | 3.4 | 4.4 | 1 |
| 1089 | Cheesefurter/cheese smokie | 2 ea | 86 | 52 | 281 | 12 | 1 | 0 | 25 | 9 | 11.8 | 2.6 |
| 2157 | Chicken breast, Healthy Favorites | 4 pce | 52 | | 40 | 9 | 1 | 0 | 0 | 0 | 0 | 0 |
| 1556 | Chorizo, pork & beef | 1 ea | 60 | 32 | 273 | 15 | 1 | 0 | 23 | 8.6 | 11 | 2.1 |
| 1090 | Corned beef loaf, jellied | 1 pce | 28 | 69 | 43 | 6 | 0 | 0 | 2 | .7 | .7 | .1 |
| | Frankfurters: | | | | | | | | | | | |
| 1077 | Beef, large link, 8/package | 1 ea | 57 | 55 | 180 | 7 | 1 | 0 | 16 | 6.9 | 7.9 | .8 |
| 1078 | Beef and pork, large link, 8/package | 1 ea | 57 | 54 | 182 | 6 | 1 | 0 | 17 | 6.2 | 8 | 1.6 |
| 667 | Beef and pork, small link, 10/pkg | 1 ea | 45 | 54 | 144 | 5 | 1 | 0 | 13 | 4.9 | 6.3 | 1.2 |
| 668 | Turkey frankfurter, 10/package | 1 ea | 45 | 63 | 102 | 6 | 1 | 0 | 8 | 2.7 | 2.5 | 2.2 |
| 1968 | Turkey/chicken frank 8/pkg | 1 ea | 43 | | 80 | 6 | 1 | 0 | 6 | 2 | | |
| | Ham: | | | | | | | | | | | |
| 669 | Ham lunchmeat, canned, 3 x 2 x ½" | 1 pce | 21 | 52 | 70 | 3 | <1 | 0 | 6 | 2.3 | 3 | .7 |
| 670 | Chopped ham, packaged | 2 pce | 42 | 64 | 96 | 7 | 0 | 0 | 7 | 2.4 | 3.4 | .9 |
| 2156 | Honey ham, Healthy Favorites | 4 pce | 52 | 73 | 55 | 9 | 2 | 0 | 1 | .4 | .8 | .1 |
| 2113 | Oscar Mayer lower sodium ham | 1 pce | 21 | 73 | 23 | 3 | 1 | 0 | 1 | .3 | .4 | .1 |
| 673 | Turkey ham lunchmeat | 2 pce | 57 | 71 | 73 | 11 | <1 | 0 | 3 | 1 | .7 | .9 |
| 1091 | Kielbasa sausage | 1 pce | 26 | 54 | 81 | 3 | 1 | 0 | 7 | 2.6 | 3.4 | .8 |
| 1092 | Knockwurst sausage, link | 1 ea | 68 | 55 | 209 | 8 | 1 | 0 | 19 | 6.9 | 8.7 | 2 |
| 1093 | Mortadella lunchmeat | 2 pce | 30 | 52 | 93 | 5 | 1 | 0 | 8 | 2.8 | 3.4 | .9 |
| 1097 | Olive loaf lunchmeat | 2 pce | 57 | 58 | 134 | 7 | 5 | <1 | 9 | 3.3 | 4.5 | 1.1 |
| 1952 | Turkey breast, fat free | 1 pce | 28 | 77 | 22 | 4 | 1 | 0 | <1 | .1 | .1 | t |
| 1080 | Turkey pastrami | 2 pce | 57 | 71 | 80 | 10 | 1 | 0 | 4 | 1 | 1.2 | .9 |
| 1969 | Turkey salami | 1 pce | 28 | 72 | 41 | 4 | <1 | 0 | 3 | .9 | 1 | .8 |
| 1081 | Pepperoni sausage | 2 pce | 11 | 27 | 55 | 2 | <1 | 0 | 5 | 1.8 | 2.3 | .5 |
| 1094 | Pickle & pimento loaf | 2 pce | 57 | 57 | 149 | 7 | 3 | <1 | 12 | 4.5 | 5.5 | 1.5 |
| 1082 | Polish sausage | 1 oz | 28 | 53 | 91 | 4 | <1 | 0 | 8 | 2.9 | 3.8 | .9 |
| 674 | Pork sausage, cooked, link, small | 2 ea | 26 | 45 | 96 | 5 | <1 | 0 | 8 | 2.8 | 4.1 | .8 |
| 1079 | Pork sausage, cooked, patty | 4 oz | 113 | 45 | 417 | 22 | 1 | 0 | 35 | 12.1 | 17.7 | 3.3 |
| 675 | Salami, pork and beef | 2 pce | 57 | 60 | 143 | 8 | 1 | 0 | 11 | 4.6 | 5.2 | 1.1 |
| 677 | Salami, pork and beef, dry | 3 pce | 30 | 35 | 125 | 7 | 1 | 0 | 10 | 3.7 | 5.1 | 1 |
| 676 | Salami, turkey | 2 pce | 57 | 66 | 112 | 9 | <1 | 0 | 8 | 2.3 | 2.6 | 2 |
| | Sandwich spreads: | | | | | | | | | | | |
| 1300 | Ham salad spread | 2 tbs | 30 | 63 | 65 | 3 | 3 | 0 | 5 | 1.5 | 2.2 | .8 |
| 678 | Pork and beef | 2 tbs | 30 | 60 | 70 | 2 | 4 | <1 | 5 | 1.8 | 2.3 | .8 |
| 1296 | Chicken/turkey | 2 tbs | 26 | 66 | 52 | 3 | 2 | 0 | 4 | .9 | .8 | 1.6 |

| Chol (mg) | Calc (mg) | Iron (mg) | Magn (mg) | Pota (mg) | Sodi (mg) | Zinc (mg) | VT-A (RE) | Thia (mg) | VT-E (a-TE) | Ribo (mg) | Niac (mg) | V-B6 (mg) | Fola (μg) | VT-C (mg) |
|---|---|---|---|---|---|---|---|---|---|---|---|---|---|---|
| | | | | | | | | | | | | | | |
| 45 | 12 | .94 | 15 | 185 | 399 | .78 | 22 | .07 | 1.46 | .1 | 5.04 | .23 | 8 | <1 |
| 15 | 12 | .79 | 7 | 52 | 471 | .59 | 11 | .02 | .3 | .11 | 1.53 | .08 | 3 | 0 |
| | | | | | | | | | | | | | | |
| 25 | 14 | .62 | 16 | 175 | 599 | .59 | 0 | .02 | | .06 | 5.35 | .22 | 2 | 0 |
| 35 | 0 | .72 | | | 1060 | | 0 | | | | | | | 0 |
| 35 | 0 | .72 | | | 940 | | 0 | | | | | | | 0 |
| 35 | 0 | | | | 910 | | 0 | | | | | | | 0 |
| 35 | 8 | 1.25 | 9 | 157 | 456 | .82 | 6 | .06 | 1.36 | .11 | 1.31 | .11 | 16 | 0 |
| 60 | 6 | 1.84 | 25 | 337 | 768 | 2.87 | 0 | .05 | .43 | .18 | 7.09 | .3 | 6 | 0 |
| 12 | 11 | .36 | 4 | 70 | 137 | .44 | 0 | .02 | .04 | .06 | 1.96 | .09 | 1 | 0 |
| | | | | | | | | | | | | | | |
| 17 | 3 | .42 | 3 | 49 | 288 | .68 | 0 | .02 | .05 | .03 | .95 | .05 | 1 | 0 |
| 16 | 2 | .21 | 4 | 71 | 347 | .48 | 0 | .15 | .06 | .05 | .91 | .1 | 1 | 0 |
| 13 | 3 | .32 | 4 | 79 | 363 | .69 | 0 | .11 | .06 | .06 | .87 | .06 | 1 | 0 |
| | | | | | | | | | | | | | | |
| 13 | 3 | .38 | 3 | 36 | 226 | .5 | 0 | .01 | .04 | .02 | .55 | .03 | 1 | 0 |
| 13 | 4 | .34 | 4 | 44 | 314 | .53 | 0 | | | | | | | 0 |
| 15 | 3 | .42 | 3 | 50 | 285 | .54 | 0 | .05 | .06 | .04 | .72 | .05 | 1 | 0 |
| 7 | | .18 | | | 255 | | | | | | | | | |
| 14 | 3 | .18 | 3 | 65 | 272 | .47 | 0 | .12 | .06 | .04 | .9 | .06 | 1 | 0 |
| 15 | 14 | .39 | 5 | 46 | 312 | .45 | 0 | | | | | | | 0 |
| 28 | 23 | .43 | 4 | 56 | 246 | .49 | 0 | .01 | .15 | .05 | .99 | .06 | 2 | 0 |
| 22 | 34 | .45 | 5 | 51 | 242 | .57 | 0 | .01 | | .05 | 1.08 | .05 | | 0 |
| 89 | 5 | 5.34 | 6 | 113 | 652 | 1.6 | 2405 | .14 | .2 | .87 | 4.77 | .19 | 25 | 0 |
| 44 | 34 | .72 | 11 | 197 | 778 | 1.47 | 0 | .17 | .19 | .16 | 2.31 | .09 | 3 | 0 |
| 16 | 2 | .62 | 4 | 70 | 248 | .3 | 0 | .21 | .06 | .09 | .96 | .06 | 1 | 0 |
| 58 | 50 | .93 | 11 | 177 | 931 | 1.94 | 33 | .21 | .27 | .14 | 2.49 | .11 | 3 | 0 |
| 25 | | .72 | | | 620 | | | | | | | | | |
| 53 | 5 | .95 | 11 | 239 | 741 | 2.05 | 0 | .38 | .13 | .18 | 3.08 | .32 | 1 | 0 |
| 13 | 3 | .57 | 3 | 28 | 267 | 1.15 | 0 | 0 | .05 | .03 | .49 | .03 | 2 | 0 |
| | | | | | | | | | | | | | | |
| 35 | 11 | .81 | 2 | 95 | 585 | 1.24 | 0 | .03 | .11 | .06 | 1.38 | .07 | 2 | 0 |
| 28 | 6 | .66 | 6 | 95 | 638 | 1.05 | 0 | .11 | .14 | .07 | 1.5 | .07 | 2 | 0 |
| 22 | 5 | .52 | 4 | 75 | 504 | .83 | 0 | .09 | .11 | .05 | 1.18 | .06 | 2 | 0 |
| 48 | 48 | .83 | 6 | 81 | 642 | 1.4 | 0 | .02 | .28 | .08 | 1.86 | .1 | 4 | 0 |
| 40 | 60 | 1.08 | | | 480 | | 0 | | | | | | | 0 |
| | | | | | | | | | | | | | | |
| 13 | 1 | .15 | 2 | 45 | 271 | .31 | 0 | .08 | .05 | .04 | .66 | .04 | 1 | <1 |
| 21 | 3 | .35 | 7 | 134 | 576 | .81 | 0 | .26 | .11 | .09 | 1.63 | .15 | <1 | 0 |
| 24 | 6 | .7 | 18 | 144 | 635 | 1.02 | 0 | | | | | | | 0 |
| 9 | 1 | .3 | 5 | 197 | 174 | .42 | 0 | | | | | | | 0 |
| 32 | 6 | 1.57 | 9 | 185 | 568 | 1.68 | 0 | .03 | .36 | .14 | 2.01 | .14 | 3 | 0 |
| 17 | 11 | .38 | 4 | 70 | 280 | .52 | 0 | .06 | .06 | .06 | .75 | .05 | 1 | 0 |
| 39 | 7 | .62 | 7 | 135 | 687 | 1.13 | 0 | .23 | .39 | .09 | 1.86 | .12 | 1 | 0 |
| 17 | 5 | .42 | 3 | 49 | 374 | .63 | 0 | .04 | .07 | .05 | .8 | .04 | 1 | 0 |
| 22 | 62 | .31 | 11 | 169 | 846 | .79 | 34 | .17 | .14 | .15 | 1.05 | .13 | 1 | 0 |
| 9 | 3 | .34 | 8 | 59 | 387 | .24 | 0 | | | | | | | 0 |
| 31 | 5 | .95 | 8 | 148 | 596 | 1.23 | 0 | .03 | .12 | .14 | 2.01 | .15 | 3 | 0 |
| 21 | 11 | .35 | 6 | 61 | 281 | .65 | 0 | | | | | | | 0 |
| 9 | 1 | .15 | 2 | 38 | 224 | .27 | 0 | .03 | .02 | .03 | .55 | .03 | <1 | 0 |
| 21 | 54 | .58 | 10 | 194 | 792 | .8 | 4 | .17 | .14 | .14 | 1.17 | .11 | 3 | 0 |
| 20 | 3 | .4 | 4 | 66 | 245 | .54 | 0 | .14 | .06 | .04 | .96 | .05 | 1 | <1 |
| 22 | 8 | .33 | 4 | 94 | 336 | .65 | 0 | .19 | .07 | .07 | 1.18 | .09 | 1 | <1 |
| 94 | 36 | 1.42 | 19 | 408 | 1462 | 2.84 | 0 | .84 | .29 | .29 | 5.11 | .37 | 2 | 2 |
| 37 | 7 | 1.52 | 9 | 113 | 607 | 1.22 | 0 | .14 | .12 | .21 | 2.02 | .12 | 1 | 0 |
| 24 | 2 | .45 | 5 | 113 | 558 | .97 | 0 | .18 | .08 | .09 | 1.46 | .15 | 1 | 0 |
| 47 | 11 | .92 | 9 | 139 | 572 | 1.03 | 0 | .04 | .32 | .1 | 2.01 | .14 | 2 | 0 |
| | | | | | | | | | | | | | | |
| 11 | 2 | .18 | 3 | 45 | 274 | .33 | 0 | .13 | .52 | .04 | .63 | .04 | <1 | 0 |
| 11 | 4 | .24 | 2 | 33 | 304 | .31 | 3 | .05 | .52 | .04 | .52 | .04 | 1 | 0 |
| 8 | 3 | .16 | 3 | 48 | 98 | .27 | 11 | .01 | .57 | .02 | .43 | .03 | 1 | <1 |

**Table H-1**

**Food Composition** (Computer code number is for West Diet Analysis program) (For purposes of calculations, use "0" for t, <1, <.1, <.01, etc.)

| Computer Code Number | Food Description | Measure | Wt (g) | $H_2O$ (%) | Ener (kcal) | Prot (g) | Carb (g) | Dietary Fiber (g) | Fat (g) | Fat Breakdown (g) Sat | Mono | Poly |
|---|---|---|---|---|---|---|---|---|---|---|---|---|
| | **MEATS: SAUSAGES and LUNCHMEATS—**Continued | | | | | | | | | | | |
| 1084 | Smoked link sausage, beef and pork | 1 ea | 68 | 52 | 228 | 9 | 1 | 0 | 21 | 7.2 | 9.7 | 2.2 |
| 1083 | Smoked link sausage, pork | 1 ea | 68 | 39 | 265 | 15 | 1 | 0 | 22 | 7.7 | 9.9 | 2.6 |
| 1085 | Summer sausage | 2 pce | 46 | 51 | 154 | 7 | <1 | 0 | 14 | 5.5 | 6 | .6 |
| 1076 | Turkey breakfast sausage | 1 pce | 28 | 60 | 64 | 6 | 0 | 0 | 5 | 1.6 | 1.8 | 1.2 |
| 679 | Vienna sausage, canned | 2 ea | 32 | 60 | 89 | 3 | 1 | 0 | 8 | 3 | 4 | .5 |
| | **MIXED DISHES and FAST FOODS** | | | | | | | | | | | |
| | MIXED DISHES: | | | | | | | | | | | |
| 1445 | Almond Chicken | 1 c | 242 | 77 | 275 | 20 | 18 | 4 | 14 | 2 | 5.3 | 5.8 |
| 1981 | Baked beans, fat free, honey | ½ c | 120 | 73 | 110 | 7 | 24 | 7 | 0 | 0 | 0 | 0 |
| 1454 | Bean cake | 1 ea | 32 | 23 | 130 | 2 | 16 | 1 | 7 | 1 | 2.9 | 2.6 |
| 680 | Beef stew w/ vegetables, homemade | 1 c | 245 | 82 | 218 | 16 | 15 | 2 | 10 | 4.9 | 4.5 | .5 |
| 1109 | Beef stew w/ vegetables, canned | 1 c | 245 | 82 | 194 | 14 | 17 | 2 | 8 | 2.4 | 3.1 | .3 |
| 1116 | Beef, macaroni, tomato sauce casserole | 1 c | 226 | 76 | 255 | 16 | 26 | 2 | 10 | 3.8 | 4.1 | .5 |
| 2295 | Beef fajita | 1 ea | 223 | 63 | 409 | 17 | 46 | 4 | 17 | 5.1 | 7.6 | 3.9 |
| 1265 | Beef flauta | 1 ea | 113 | 49 | 360 | 16 | 13 | 2 | 27 | 4.9 | 11.6 | 9.1 |
| 681 | Beef pot pie, homemade | 1 pce | 210 | 55 | 517 | 21 | 39 | 3 | 30 | 8.4 | 14.7 | 7.3 |
| 1898 | Broccoli, batter fried | 1 c | 85 | 74 | 123 | 3 | 9 | 2 | 9 | 1.3 | 2.2 | 4.9 |
| 1462 | Buffalo wings/spicy chicken wings | 2 pce | 32 | 53 | 98 | 8 | <1 | <1 | 7 | 1.8 | 2.8 | 1.6 |
| 1675 | Carrot raisin salad | ½ c | 88 | 58 | 204 | 1 | 21 | 2 | 14 | 2 | 3.9 | 7.3 |
| 2248 | Cheeseburger deluxe | 1 ea | 219 | 52 | 563 | 28 | 38 | | 33 | 15 | 12.6 | 2 |
| 682 | Chicken à la king, homemade | 1 c | 245 | 68 | 468 | 27 | 12 | 1 | 34 | 12.7 | 14.3 | 6.2 |
| 683 | Chicken & noodles, homemade | 1 c | 240 | 71 | 367 | 22 | 26 | 2 | 18 | 5.9 | 7.1 | 3.5 |
| 684 | Chicken chow mein, canned | 1 c | 250 | 89 | 95 | 6 | 18 | 2 | 1 | 0 | .1 | .8 |
| 685 | Chicken chow mein, homemade | 1 c | 250 | 78 | 255 | 31 | 10 | 1 | 10 | 2.4 | 4.3 | 3.1 |
| 1266 | Chicken fajita | 1 ea | 223 | 61 | 405 | 22 | 50 | 4 | 13 | 2.4 | 6 | 3.5 |
| 1264 | Chicken flauta | 1 ea | 113 | 52 | 343 | 14 | 13 | 2 | 27 | 4.3 | 11.1 | 9.6 |
| 686 | Chicken pot pie, homemade (⅓) | 1 pce | 232 | 57 | 545 | 23 | 42 | 3 | 31 | 10.9 | 14.5 | 5.8 |
| 1672 | Chili con carne | ½ c | 127 | 77 | 128 | 12 | 11 | 2 | 4 | 1.7 | 1.7 | .3 |
| 1112 | Chicken salad with celery | ½ c | 78 | 53 | 268 | 11 | 1 | <1 | 25 | 3.1 | 4.5 | 15.8 |
| 1382 | Chicken teriyaki, breast | 1 ea | 128 | 67 | 176 | 26 | 7 | <1 | 4 | .9 | 1 | .9 |
| 687 | Chili with beans, canned | 1 c | 256 | 75 | 287 | 15 | 30 | 11 | 14 | 6 | 6 | .9 |
| 1479 | Chinese pastry | 1 oz | 28 | 46 | 67 | 1 | 13 | <1 | 1 | .2 | .4 | .8 |
| 688 | Chop suey with beef & pork | 1 c | 220 | 63 | 425 | 22 | 31 | 3 | 24 | 5 | 8.6 | 9.3 |
| 690 | Coleslaw | 1 c | 132 | 74 | 195 | 2 | 17 | 2 | 15 | 2.1 | 3.2 | 8.5 |
| 689 | Corn pudding | 1 c | 250 | 76 | 273 | 11 | 32 | 4 | 13 | 6.4 | 4.3 | 1.8 |
| 1110 | Corned beef hash, canned | 1 c | 220 | 67 | 398 | 19 | 23 | 1 | 25 | 11.9 | 10.9 | .9 |
| 1255 | Deviled egg (½ egg + filling) | 1 ea | 31 | 69 | 62 | 4 | <1 | 0 | 5 | 1.2 | 1.7 | 1.5 |
| | Egg foo yung patty: | | | | | | | | | | | |
| 1467 | Meatless | 1 ea | 86 | 78 | 113 | 6 | 3 | 1 | 8 | 1.9 | 3.3 | 2.1 |
| 1458 | With beef | 1 ea | 86 | 74 | 129 | 9 | 3 | <1 | 9 | 2.2 | 3.2 | 2.4 |
| 1465 | With chicken | 1 ea | 86 | 74 | 130 | 9 | 4 | <1 | 9 | 2.1 | 3.1 | 2.5 |
| 1602 | Egg roll, meatless | 1 ea | 64 | 70 | 102 | 3 | 10 | 1 | 6 | 1.2 | 2.5 | 1.6 |
| 1550 | Egg roll, with meat | 1 ea | 64 | 66 | 114 | 5 | 9 | 1 | 6 | 1.5 | 2.7 | 1.5 |
| 1113 | Egg salad | 1 c | 183 | 57 | 586 | 17 | 3 | 0 | 56 | 10.6 | 17.4 | 24.2 |
| 691 | French toast w/wheat bread, homemade | 1 pce | 65 | 54 | 151 | 5 | 16 | <1 | 7 | 2 | 3 | 1.7 |
| 1355 | Green pepper, stuffed | 1 ea | 172 | 75 | 229 | 11 | 20 | 2 | 11 | 5 | 4.9 | .5 |
| 1487 | Hot & sour soup (Chinese) | 1 c | 244 | 88 | 133 | 12 | 5 | <1 | 6 | 2 | 2.5 | 1 |
| 2242 | Hamburger deluxe | 1 ea | 110 | 49 | 279 | 13 | 27 | | 13 | 4.1 | 5.3 | 2.6 |
| 1997 | Hummous/hummus | ¼ c | 62 | 65 | 106 | 3 | 12 | 3 | 5 | .8 | 2.2 | 2 |
| | Lasagna: | | | | | | | | | | | |
| 1346 | With meat, homemade | 1 pce | 245 | 67 | 382 | 22 | 39 | 3 | 15 | 7.7 | 5 | .9 |
| 1111 | Without meat, homemade | 1 pce | 218 | 69 | 298 | 15 | 39 | 3 | 9 | 5.4 | 2.4 | .6 |
| 1117 | Frozen entree | 1 ea | 340 | 75 | 390 | 24 | 42 | 4 | 14 | 6.7 | 5.5 | .8 |
| 1606 | Lo mein, meatless | 1 c | 200 | 82 | 134 | 6 | 27 | 3 | 1 | .1 | .1 | .3 |
| 1607 | Lo mein, with meat | 1 c | 200 | 70 | 285 | 17 | 31 | 2 | 10 | 1.9 | 2.9 | 4.5 |
| 692 | Macaroni & cheese, canned | 1 c | 240 | 80 | 228 | 9 | 26 | 1 | 10 | 4.2 | 3.1 | 1.4 |
| 693 | Macaroni & cheese, homemade | 1 c | 200 | 58 | 430 | 17 | 40 | 1 | 22 | 8.9 | 8.8 | 3.6 |
| 1115 | Macaroni salad, no cheese | 1 c | 177 | 60 | 461 | 5 | 28 | 2 | 37 | 4 | 6 | 25.5 |
| 1120 | Meat loaf, beef | 1 pce | 87 | 63 | 182 | 16 | 4 | <1 | 11 | 4.4 | 4.7 | .5 |
| 1119 | Meat loaf, beef and pork (⅓) | 1 pce | 87 | 60 | 205 | 15 | 4 | <1 | 14 | 5.2 | 6.3 | .9 |
| 1303 | Moussaka (lamb & eggplant) | 1 c | 250 | 82 | 237 | 16 | 13 | 4 | 13 | 4.6 | 5.4 | 1.9 |

**PAGE KEY:** H–4 = Beverages H–6 = Dairy H–10 = Eggs H–10 = Fat/Oil H–14 = Fruit H–20 = Bakery H–28 = Grain H–32 = Fish H–34 = Meats H–38 = Poultry H–40 = Sausage H–42 = Mixed/Fast H–46 = Nuts/Seeds H–50 = Sweets H–52 = Vegetables/Legumes H–62 = Vegetarian Foods H–64 = Misc H–66 = Soups/Sauces H–68 = Fast H–84 = Convenience H–88 = Baby foods

| Chol (mg) | Calc (mg) | Iron (mg) | Magn (mg) | Pota (mg) | Sodi (mg) | Zinc (mg) | VT-A (RE) | Thia (mg) | VT-E (a-TE) | Ribo (mg) | Niac (mg) | V-B6 (mg) | Fola (µg) | VT-C (mg) |
|---|---|---|---|---|---|---|---|---|---|---|---|---|---|---|
| 48 | 7 | .99 | 8 | 129 | 643 | 1.43 | 0 | .18 | .15 | .12 | 2.2 | .12 | 1 | 0 |
| 46 | 20 | .79 | 13 | 228 | 1020 | 1.92 | 0 | .48 | .17 | .17 | 3.08 | .24 | 3 | 1 |
| 34 | 6 | 1.17 | 6 | 125 | 571 | 1.18 | 0 | .07 | .1 | .15 | 1.98 | .12 | 1 | 0 |
| 23 | 5 | .51 | 6 | 75 | 188 | .96 | 0 | .03 | .14 | .08 | 1.4 | .08 | 1 | 0 |
| 17 | 3 | .28 | 2 | 32 | 305 | .51 | 0 | .03 | .07 | .03 | .51 | .04 | 1 | 0 |
| | | | | | | | | | | | | | | |
| 35 | 81 | 2 | 59 | 551 | 615 | 1.54 | 75 | .08 | 2.64 | .19 | 8.59 | .42 | 31 | 10 |
| 0 | 40 | 2.7 | | | 135 | | 450 | | | | | | | 12 |
| 0 | 3 | .67 | 6 | 57 | 55 | .16 | 0 | .07 | 1.14 | .05 | .55 | .02 | 9 | 0 |
| 64 | 29 | 2.94 | 40 | 613 | 292 | 5.29 | 568 | .15 | .49 | .17 | 4.66 | .28 | 37 | 17 |
| 34 | 29 | 2.21 | 39 | 426 | 1006 | 4.24 | 262 | .07 | .34 | .12 | 2.45 | .2 | 31 | 7 |
| 39 | 26 | 2.7 | 40 | 522 | 862 | 3.14 | 97 | .22 | .57 | .22 | 4.31 | .29 | 20 | 14 |
| 26 | 76 | 3.69 | 38 | 427 | 850 | 2.38 | 52 | .46 | 2.08 | .3 | 4.73 | .32 | 25 | 29 |
| 45 | 50 | 2.15 | 29 | 292 | 187 | 4.18 | 15 | .07 | 4 | .15 | 2.13 | .25 | 10 | 14 |
| 44 | 29 | 3.78 | 6 | 334 | 596 | 3.17 | 519 | .29 | 3.78 | .29 | 4.83 | .24 | 29 | 6 |
| 16 | 67 | .94 | 20 | 242 | 62 | .38 | 102 | .08 | 2.1 | .13 | .75 | .11 | 43 | 53 |
| 26 | 5 | .4 | 6 | 59 | 61 | .56 | 17 | .01 | .23 | .04 | 2.06 | .13 | 1 | <1 |
| 10 | 26 | .75 | 14 | 317 | 118 | .19 | 1452 | .08 | 5.03 | .05 | .64 | .22 | 9 | 5 |
| 88 | 206 | 4.66 | 44 | 445 | 1108 | 4.6 | 129 | .39 | 1.18 | .46 | 7.38 | .28 | 81 | 8 |
| 186 | 127 | 2.45 | 20 | 404 | 760 | 1.8 | 272 | .1 | .98 | .42 | 5.39 | .23 | 11 | 12 |
| 96 | 26 | 2.16 | 26 | 149 | 600 | 1.53 | 10 | .05 | | .17 | 4.32 | .19 | 10 | 0 |
| 7 | 45 | 1.25 | 14 | 418 | 725 | 1.3 | 28 | .05 | .05 | .1 | 1 | .09 | 12 | 12 |
| 77 | 57 | 2.5 | 28 | 473 | 718 | 2.12 | 50 | .07 | .75 | .22 | 4.25 | .41 | 19 | 10 |
| 41 | 83 | 3.7 | 51 | 532 | 439 | 1.77 | 55 | .48 | 2.04 | .37 | 6.64 | .35 | 41 | 22 |
| 37 | 52 | .97 | 27 | 243 | 189 | 1.18 | 21 | .05 | 4.06 | .1 | 3.21 | .22 | 8 | 14 |
| 72 | 70 | 3.02 | 25 | 343 | 594 | 2 | 735 | .32 | 3.25 | .32 | 4.87 | .46 | 29 | 5 |
| 67 | 34 | 2.6 | 23 | 347 | 505 | 1.79 | 84 | .06 | .81 | .57 | 1.24 | .16 | 23 | 1 |
| 48 | 16 | .62 | 11 | 138 | 201 | .79 | 31 | .03 | 6.27 | .07 | 3.28 | .34 | 8 | 1 |
| 80 | 27 | 1.75 | 36 | 309 | 1866 | 1.94 | 16 | .08 | .35 | .2 | 8.69 | .46 | 13 | 3 |
| 43 | 120 | 8.78 | 115 | 934 | 1336 | 5.12 | 87 | .12 | 1.88 | .27 | .92 | .34 | 58 | 4 |
| 0 | 8 | .51 | 6 | 28 | 3 | .18 | <1 | .05 | .25 | <.01 | .41 | .02 | 1 | 0 |
| 46 | 39 | 4.16 | 54 | 515 | 818 | 3.52 | 134 | .36 | 1.82 | .37 | 5.63 | .44 | 44 | 20 |
| 7 | 45 | .96 | 12 | 236 | 356 | .26 | 66 | .05 | 5.28 | .04 | .11 | .14 | 51 | 11 |
| 250 | 100 | 1.4 | 37 | 403 | 138 | 1.25 | 90 | 1.03 | .52 | .32 | 2.47 | .29 | 63 | 7 |
| 73 | 29 | 4.4 | 36 | 440 | 1188 | 3.3 | 0 | .02 | .48 | .2 | 4.62 | .43 | 20 | 0 |
| 121 | 15 | .35 | 3 | 36 | 94 | .3 | 49 | .02 | .86 | .14 | .02 | .05 | 13 | 0 |
| | | | | | | | | | | | | | | |
| 184 | 31 | 1.04 | 12 | 118 | 310 | .7 | 86 | .04 | 1.57 | .25 | .44 | .09 | 30 | 5 |
| 180 | 26 | 1.11 | 11 | 145 | 184 | 1.16 | 92 | .05 | 1.79 | .24 | .74 | .15 | 22 | 3 |
| 182 | 27 | .86 | 12 | 144 | 187 | .81 | 95 | .05 | 1.87 | .25 | .96 | .13 | 22 | 3 |
| 30 | 12 | .76 | 9 | 98 | 306 | .25 | 15 | .08 | .81 | .1 | .81 | .05 | 12 | 3 |
| 37 | 13 | .78 | 10 | 124 | 304 | .46 | 14 | .16 | .78 | .13 | 1.31 | .1 | 9 | 2 |
| 574 | 74 | 1.8 | 13 | 180 | 665 | 1.42 | 260 | .08 | 8.87 | .66 | .09 | .46 | 62 | 0 |
| | | | | | | | | | | | | | | |
| 76 | 64 | 1.09 | 11 | 86 | 311 | .44 | 81 | .13 | .31 | .21 | 1.06 | .05 | 15 | <1 |
| 34 | 16 | 1.77 | 20 | 233 | 201 | 2.28 | 44 | .15 | .75 | .1 | 2.74 | .3 | 17 | 55 |
| 23 | 29 | 1.83 | 27 | 351 | 1562 | 1.17 | 2 | .19 | .12 | .22 | 4.58 | .15 | 12 | 1 |
| 26 | 63 | 2.63 | 22 | 227 | 504 | 2.06 | 9 | .23 | .82 | .2 | 3.69 | .12 | 52 | 2 |
| 0 | 31 | .97 | 18 | 108 | 151 | .68 | 1 | .06 | .62 | .03 | .25 | .25 | 37 | 5 |
| | | | | | | | | | | | | | | |
| 56 | 258 | 3.22 | 50 | 461 | 745 | 3.25 | 158 | .23 | 1.15 | .33 | 3.97 | .21 | 19 | 15 |
| 31 | 252 | 2.5 | 44 | 375 | 714 | 1.77 | 156 | .22 | 1.07 | .27 | 2.49 | .17 | 17 | 15 |
| 55 | 263 | 3.44 | 64 | 752 | 823 | 3.7 | 248 | .29 | 3.45 | .39 | 5.07 | .32 | 28 | 41 |
| 0 | 47 | 2.06 | 33 | 389 | 623 | .92 | 130 | .23 | .35 | .24 | 2.83 | .19 | 49 | 12 |
| 30 | 25 | 2.11 | 39 | 246 | 276 | 1.63 | 6 | .37 | 1.51 | .24 | 4.25 | .28 | 41 | 8 |
| 24 | 199 | .96 | 31 | 139 | 730 | 1.2 | 73 | .12 | .14 | .24 | .96 | .02 | 8 | <1 |
| 42 | 362 | 1.8 | 37 | 240 | 1086 | 1.2 | 234 | .2 | .12 | .4 | 1.8 | .05 | 10 | 1 |
| 27 | 31 | 1.56 | 20 | 170 | 352 | .53 | 44 | .18 | 10.3 | .1 | 1.43 | .33 | 20 | 4 |
| 84 | 29 | 1.61 | 14 | 187 | 145 | 3.23 | 23 | .05 | .31 | .22 | 2.61 | .15 | 11 | 1 |
| 84 | 33 | 1.42 | 14 | 213 | 381 | 2.68 | 23 | .19 | .32 | .22 | 2.68 | .17 | 10 | 1 |
| 97 | 68 | 1.79 | 40 | 557 | 432 | 2.56 | 105 | .15 | .81 | .31 | 4.14 | .23 | 45 | 6 |

**Table H-1**

**Food Composition** (Computer code number is for West Diet Analysis program) (For purposes of calculations, use "0" for t, <1, <.1, <.01, etc.)

| Computer Code Number | Food Description | Measure | Wt (g) | $H_2O$ (%) | Ener (kcal) | Prot (g) | Carb (g) | Dietary Fiber (g) | Fat (g) | Fat Breakdown (g) Sat | Mono | Poly |
|---|---|---|---|---|---|---|---|---|---|---|---|---|
| | **MIXED DISHES and FAST FOODS**—Continued | | | | | | | | | | | |
| 1899 | Mushrooms, batter fried | 5 ea | 70 | 66 | 148 | 2 | 8 | 1 | 12 | 2.1 | 3 | 6.4 |
| 715 | Potato salad with mayonnaise and eggs | ½ c | 125 | 76 | 179 | 3 | 14 | 2 | 10 | 1.8 | 3.1 | 4.7 |
| 1674 | Pizza, combination, 1⁄12 of 12" round | 1 pce | 79 | 48 | 184 | 13 | 21 | | 5 | 1.5 | 2.5 | .9 |
| 1673 | Pizza, pepperoni, 1⁄12 of 12" round | 1 pce | 71 | 46 | 181 | 10 | 20 | | 7 | 2.2 | 3.1 | 1.2 |
| 694 | Quiche Lorraine ⅛ of 8" quiche | 1 pce | 176 | 54 | 508 | 20 | 20 | 1 | 39 | 17.6 | 13.8 | 4.9 |
| 1449 | Ramen noodles, cooked | 1 c | 227 | 82 | 156 | 6 | 29 | 3 | 2 | .4 | .5 | .5 |
| 1671 | Ravioli, meat | ½ c | 125 | 68 | 194 | 10 | 18 | 1 | 9 | 2.9 | 3.6 | 1 |
| 1597 | Fried rice (meatless) | 1 c | 166 | 68 | 264 | 5 | 34 | 1 | 12 | 1.7 | 3 | 6.3 |
| 2142 | Roast beef hash | ½ c | 117 | 66 | 230 | 9 | 11 | 1 | 16 | 7 | 5.8 | 3.2 |
| | Spaghetti (enriched) in tomato sauce | | | | | | | | | | | |
| | With cheese: | | | | | | | | | | | |
| 695 | Canned | 1 c | 250 | 80 | 190 | 5 | 38 | 2 | 1 | 0 | .4 | .5 |
| 696 | Home recipe | 1 c | 250 | 77 | 260 | 9 | 37 | 2 | 9 | 2 | 5.4 | 1.2 |
| | With meatballs: | | | | | | | | | | | |
| 697 | Canned | 1 c | 250 | 78 | 258 | 12 | 28 | 6 | 10 | 2.1 | 3.9 | 3.9 |
| 698 | Home recipe | 1 c | 248 | 70 | 332 | 19 | 39 | 8 | 12 | 3.3 | 6.3 | 2.2 |
| 716 | Spinach soufflé | 1 c | 136 | 74 | 219 | 11 | 3 | 3 | 19 | 9.5 | 5.7 | 2.2 |
| 1553 | Sweet & sour pork | 1 c | 226 | 77 | 231 | 15 | 25 | 1 | 8 | 2.2 | 2.9 | 2.4 |
| 1263 | Sweet & sour chicken breast | 1 ea | 131 | 79 | 117 | 8 | 15 | 1 | 3 | .6 | .8 | 1.5 |
| 1515 | Three bean salad | 1 c | 150 | 82 | 139 | 4 | 13 | 3 | 8 | 1.2 | 1.9 | 4.9 |
| 717 | Tuna salad | 1 c | 205 | 63 | 383 | 33 | 19 | 0 | 19 | 3.2 | 5.9 | 8.4 |
| 1121 | Tuna noodle casserole, homemade | 1 c | 202 | 75 | 237 | 17 | 25 | 1 | 7 | 1.9 | 1.5 | 3.2 |
| 1270 | Waldorf salad | 1 c | 137 | 58 | 408 | 4 | 13 | 2 | 40 | 4.1 | 7.3 | 27 |
| | FAST FOODS and SANDWICHES (see end of this appendix for additional Fast Foods) | | | | | | | | | | | |
| 699 | Burrito, beef & bean | 1 ea | 116 | 52 | 255 | 11 | 33 | 3 | 9 | 4.2 | 3.5 | .6 |
| 700 | Burrito, bean | 1 ea | 109 | 52 | 225 | 7 | 36 | 4 | 7 | 3.5 | 2.4 | .6 |
| 2106 | Burrito, chicken con queso | 1 ea | 306 | 76 | 280 | 12 | 53 | 5 | 6 | 1.5 | | |
| 701 | Cheeseburger with bun, regular | 1 ea | 154 | 55 | 359 | 18 | 28 | | 20 | 9.2 | 7.2 | 1.5 |
| 702 | Cheeseburger with bun, 4-oz patty | 1 ea | 166 | 51 | 417 | 21 | 35 | | 21 | 8.7 | 7.8 | 2.7 |
| 703 | Chicken patty sandwich | 1 ea | 182 | 47 | 515 | 24 | 39 | 1 | 29 | 8.5 | 10.4 | 8.4 |
| 704 | Corndog | 1 ea | 175 | 47 | 460 | 17 | 56 | | 19 | 5.2 | 9.1 | 3.5 |
| 1922 | Corndog, chicken | 1 ea | 113 | 52 | 271 | 13 | 26 | | 13 | | | |
| 705 | Enchilada | 1 ea | 163 | 63 | 319 | 10 | 28 | | 19 | 10.6 | 6.3 | .8 |
| 706 | English muffin with egg, cheese, bacon | 1 ea | 146 | 49 | 383 | 20 | 31 | 1 | 20 | 9 | 6.8 | 2.1 |
| | Fish sandwich: | | | | | | | | | | | |
| 707 | Regular, with cheese | 1 ea | 183 | 45 | 523 | 21 | 48 | <1 | 28 | 8.1 | 8.9 | 9.4 |
| 708 | Large, no cheese | 1 ea | 158 | 47 | 431 | 17 | 41 | <1 | 23 | 5.2 | 7.7 | 8.2 |
| 709 | Hamburger with bun, regular | 1 ea | 107 | 45 | 275 | 14 | 33 | 1 | 10 | 3.5 | 3.7 | 1.8 |
| 710 | Hamburger with bun, 4-oz patty | 1 ea | 215 | 50 | 576 | 32 | 39 | | 32 | 12 | 14.1 | 2.8 |
| 711 | Hotdog/frankfurter with bun | 1 ea | 98 | 54 | 242 | 10 | 18 | | 14 | 5.1 | 6.8 | 1.7 |
| | Lunchables: | | | | | | | | | | | |
| 2129 | Bologna & American cheese | 1 ea | 128 | | 450 | 18 | 19 | 0 | 34 | 15 | | |
| 2130 | Ham & cheese | 1 ea | 128 | | 320 | 22 | 19 | 0 | 17 | 8 | | |
| 2117 | Honey ham & Amer. w/choc pudding | 1 ea | 176 | | 390 | 18 | 34 | <1 | 20 | 9 | | |
| 2118 | Honey turkey & cheddar w/Jello | 1 ea | 163 | | 320 | 17 | 27 | <1 | 16 | 9 | | |
| 2131 | Pepperoni & American cheese | 1 ea | 128 | | 480 | 20 | 19 | 0 | 36 | 17 | | |
| 2125 | Salami & American cheese | 1 ea | 128 | | 430 | 18 | 18 | 0 | 32 | 15 | | |
| 2127 | Turkey & cheddar cheese | 1 ea | 128 | | 360 | 20 | 20 | 1 | 22 | 11 | | |
| 712 | Pizza, cheese, ⅛ of 15" round | 1 pce | 63 | 48 | 140 | 8 | 20 | 1 | 3 | 1.5 | 1 | .5 |
| | SANDWICHES: | | | | | | | | | | | |
| | Avocado, chesse, tomato & lettuce: | | | | | | | | | | | |
| 1276 | On white bread, firm | 1 ea | 210 | 58 | 478 | 15 | 41 | 5 | 29 | 8.8 | 11.3 | 7.2 |
| 1278 | On part whole wheat | 1 ea | 201 | 59 | 444 | 14 | 34 | 7 | 29 | 8.6 | 11.4 | 7.3 |
| 1277 | On whole wheat | 1 ea | 214 | 58 | 468 | 16 | 40 | 8 | 30 | 8.7 | 11.6 | 7.5 |
| | Bacon, lettuce & tomato sandwich: | | | | | | | | | | | |
| 1137 | On white bread, soft | 1 ea | 124 | 53 | 308 | 10 | 28 | 2 | 18 | 4.5 | 6.1 | 6.1 |
| 1139 | On part whole wheat | 1 ea | 124 | 54 | 303 | 10 | 26 | 3 | 17 | 4.3 | 6.2 | 6.1 |
| 1138 | On whole wheat | 1 ea | 137 | 53 | 328 | 12 | 32 | 4 | 18 | 4.4 | 6.4 | 6.3 |

| Chol (mg) | Calc (mg) | Iron (mg) | Magn (mg) | Pota (mg) | Sodi (mg) | Zinc (mg) | VT-A (RE) | Thia (mg) | VT-E (a-TE) | Ribo (mg) | Niac (mg) | V-B6 (mg) | Fola (µg) | VT-C (mg) |
|---|---|---|---|---|---|---|---|---|---|---|---|---|---|---|
| 14 | 54 | .76 | 8 | 180 | 121 | .42 | 10 | .07 | .92 | .22 | 1.65 | .05 | 8 | 1 |
| | | | | | | | | | | | | | | |
| 85 | 24 | .81 | 19 | 318 | 661 | .39 | 41 | .1 | 2.33 | .07 | 1.11 | .18 | 8 | 12 |
| 20 | 101 | 1.53 | 18 | 179 | 382 | 1.11 | 101 | .21 | | .17 | 1.96 | .09 | 32 | 2 |
| 14 | 65 | .94 | 9 | 153 | 267 | .52 | 55 | .13 | | .23 | 3.05 | .06 | 37 | 2 |
| 205 | 201 | 1.9 | 27 | 271 | 549 | 1.66 | 243 | .23 | 1.91 | .44 | 4.71 | .19 | 17 | 3 |
| 38 | 20 | 1.89 | 24 | 51 | 1349 | .76 | 204 | .22 | .09 | .1 | 1.75 | .06 | 9 | <1 |
| 84 | 32 | 2.03 | 20 | 259 | 619 | 1.67 | 94 | .15 | 1.52 | .22 | 2.95 | .14 | 14 | 11 |
| 42 | 30 | 1.84 | 24 | 134 | 286 | .89 | 21 | .21 | 2.46 | .11 | 2.25 | .15 | 22 | 4 |
| 40 | 10 | .9 | 22 | 362 | 695 | 2.99 | 0 | .09 | | .12 | 2.33 | .3 | 12 | 0 |
| | | | | | | | | | | | | | | |
| 7 | 40 | 2.75 | 21 | 303 | 955 | 1.12 | 120 | .35 | 2.13 | .27 | 4.5 | .13 | 6 | 10 |
| 7 | 80 | 2.25 | 26 | 408 | 955 | 1.3 | 140 | .25 | 2.75 | .17 | 2.25 | .2 | 8 | 12 |
| | | | | | | | | | | | | | | |
| 22 | 52 | 3.25 | 20 | 245 | 1220 | 2.39 | 100 | .15 | 1.5 | .17 | 2.25 | .12 | 5 | 5 |
| 74 | 124 | 3.72 | 40 | 665 | 1009 | 2.45 | 159 | .25 | 1.64 | .3 | 3.97 | .2 | 10 | 22 |
| 184 | 230 | 1.35 | 38 | 201 | 763 | 1.29 | 675 | .09 | 1.22 | .3 | .48 | .12 | 80 | 3 |
| 38 | 28 | 1.36 | 34 | 390 | 1219 | 1.46 | 28 | .55 | .62 | .21 | 3.6 | .41 | 10 | 20 |
| 23 | 16 | .79 | 21 | 187 | 732 | .66 | 20 | .06 | .39 | .08 | 3.06 | .18 | 6 | 12 |
| 0 | 35 | 1.42 | 25 | 224 | 514 | .54 | 23 | .07 | 1.96 | .09 | .4 | .04 | 53 | 4 |
| 27 | 35 | 2.05 | 39 | 365 | 824 | 1.15 | 55 | .06 | 1.95 | .14 | 13.7 | .17 | 16 | 5 |
| 41 | 34 | 2.3 | 30 | 182 | 772 | 1.2 | 13 | .18 | 1.18 | .15 | 7.78 | .2 | 10 | 1 |
| 21 | 43 | .88 | 39 | 270 | 236 | .63 | 39 | .1 | 8.67 | .05 | .36 | .36 | 27 | 6 |
| | | | | | | | | | | | | | | |
| 24 | 53 | 2.46 | 42 | 329 | 670 | 1.93 | 32 | .27 | .7 | .42 | 2.71 | .19 | 58 | 1 |
| 2 | 57 | 2.27 | 44 | 328 | 495 | .76 | 16 | .32 | .87 | .3 | 2.04 | .15 | 44 | 1 |
| 10 | 40 | .72 | | | 600 | | 40 | | | | | | | 15 |
| 52 | 182 | 2.65 | 26 | 229 | 976 | 2.62 | 71 | .32 | 1.34 | .23 | 6.38 | .15 | 65 | 2 |
| 60 | 171 | 3.42 | 30 | 335 | 1050 | 3.49 | 65 | .35 | | .28 | 8.05 | .18 | 61 | 2 |
| 60 | 60 | 4.68 | 35 | 353 | 957 | 1.87 | 31 | .33 | .55 | .24 | 6.81 | .2 | 100 | 9 |
| 79 | 102 | 6.18 | 17 | 263 | 973 | 1.31 | 37 | .28 | .7 | .7 | 4.17 | .09 | 103 | 0 |
| 64 | | | | | 668 | | | | | | | | | |
| 44 | 324 | 1.32 | 50 | 240 | 784 | 2.51 | 186 | .08 | 1.47 | .42 | 1.91 | .39 | 65 | 1 |
| 234 | 207 | 3.29 | 34 | 213 | 784 | 1.81 | 158 | .48 | .6 | .53 | 3.93 | .16 | 47 | 1 |
| | | | | | | | | | | | | | | |
| 68 | 185 | 3.5 | 37 | 353 | 939 | 1.17 | 97 | .46 | 1.83 | .42 | 4.23 | .11 | 91 | 3 |
| 55 | 84 | 2.61 | 33 | 340 | 615 | .99 | 30 | .33 | .87 | .22 | 3.4 | .11 | 85 | 3 |
| 43 | 51 | 2.46 | 22 | 215 | 564 | 2.05 | 13 | .26 | .43 | .32 | 4.7 | .13 | 52 | 3 |
| 103 | 92 | 5.55 | 45 | 527 | 742 | 5.81 | 4 | .34 | 1.61 | .41 | 6.73 | .37 | 84 | 1 |
| 44 | 23 | 2.31 | 13 | 143 | 670 | 1.98 | 0 | .23 | .27 | .27 | 3.65 | .05 | 48 | <1 |
| | | | | | | | | | | | | | | |
| 85 | 300 | 2.7 | | | 1620 | | 60 | | | | | | | 0 |
| 60 | 300 | 1.8 | | | 1770 | | 80 | | | | | | | |
| 55 | 250 | 2.7 | | | 1540 | | 40 | | | | | | | |
| 50 | 20 | 6 | | | 1360 | | 80 | | | | | | | |
| 95 | 250 | 2.7 | | | 1840 | | 60 | | | | | | | |
| 80 | 250 | 2.7 | | | 1740 | | 60 | | | | | | | |
| 70 | 300 | 1.8 | | | 1650 | | 60 | | | | | | | |
| 9 | 117 | .58 | 16 | 110 | 336 | .81 | 74 | .18 | | .16 | 2.48 | .04 | 35 | 1 |
| | | | | | | | | | | | | | | |
| 34 | 294 | 3.06 | 54 | 581 | 550 | 1.71 | 140 | .37 | 4.55 | .39 | 3.77 | .32 | 80 | 11 |
| 31 | 291 | 3.1 | 67 | 617 | 525 | 1.91 | 140 | .35 | 4.55 | .39 | 3.94 | .35 | 80 | 11 |
| 31 | 281 | 3.53 | 102 | 679 | 593 | 2.68 | 140 | .36 | 4.17 | .37 | 4.29 | .42 | 92 | 11 |
| | | | | | | | | | | | | | | |
| 21 | 52 | 1.99 | 20 | 233 | 590 | .96 | 31 | .35 | 2.34 | .19 | 3.2 | .14 | 34 | 12 |
| 20 | 63 | 2.27 | 33 | 283 | 604 | 1.19 | 31 | .36 | 2.7 | .21 | 3.62 | .17 | 37 | 12 |
| 20 | 55 | 2.68 | 64 | 342 | 670 | 1.9 | 31 | .37 | 2.36 | .2 | 3.97 | .24 | 48 | 12 |

**Table H-1**

**Food Composition** (Computer code number is for West Diet Analysis program) (For purposes of calculations, use "0" for t, <1, <.1, <.01, etc.)

| Computer Code Number | Food Description | Measure | Wt (g) | $H_2O$ (%) | Ener (kcal) | Prot (g) | Carb (g) | Dietary Fiber (g) | Fat (g) | Fat Breakdown (g) Sat | Mono | Poly |
|---|---|---|---|---|---|---|---|---|---|---|---|---|
| | **MIXED DISHES and FAST FOODS—Continued** | | | | | | | | | | | |
| | Cheese, grilled: | | | | | | | | | | | |
| 1140 | On white bread, soft | 1 ea | 119 | 37 | 400 | 18 | 30 | 1 | 24 | 13.2 | 7.5 | 2 |
| 1142 | On part whole wheat | 1 ea | 119 | 37 | 396 | 18 | 28 | 3 | 24 | 13 | 7.6 | 2.1 |
| 1141 | On whole wheat | 1 ea | 132 | 38 | 420 | 20 | 33 | 4 | 24 | 13.1 | 7.8 | 2.2 |
| 1596 | Chicken fillet | 1 ea | 182 | 47 | 515 | 24 | 39 | 1 | 29 | 8.5 | 10.4 | 8.4 |
| | Chicken salad: | | | | | | | | | | | |
| 1143 | On white bread, soft | 1 ea | 110 | 40 | 369 | 11 | 31 | 1 | 23 | 3.7 | 6.1 | 12.1 |
| 1145 | On part whole wheat | 1 ea | 110 | 41 | 364 | 11 | 29 | 4 | 23 | 3.5 | 6.1 | 12.1 |
| 1144 | On whole wheat | 1 ea | 123 | 41 | 387 | 13 | 34 | 5 | 23 | 3.6 | 6.3 | 12.2 |
| 1146 | Corned beef & swiss on rye | 1 ea | 156 | 49 | 420 | 28 | 22 | <1 | 26 | 9.4 | 7.4 | 6.3 |
| | Egg salad: | | | | | | | | | | | |
| 1147 | On white bread, soft | 1 ea | 117 | 43 | 380 | 10 | 31 | 1 | 25 | 4.4 | 6.8 | 12 |
| 1149 | On part whole wheat | 1 ea | 116 | 43 | 374 | 10 | 29 | 3 | 25 | 4.1 | 6.8 | 12 |
| 1148 | On whole wheat | 1 ea | 130 | 43 | 400 | 12 | 35 | 5 | 25 | 4.3 | 7.1 | 12.2 |
| | Ham: | | | | | | | | | | | |
| 1279 | On rye bread | 1 ea | 150 | 60 | 283 | 22 | 21 | <1 | 13 | 2.4 | 3.8 | 6 |
| 1151 | On white bread, soft | 1 ea | 157 | 55 | 334 | 22 | 30 | 1 | 14 | 3 | 4.4 | 5.9 |
| 1153 | On part whole wheat | 1 ea | 156 | 55 | 328 | 22 | 28 | 3 | 14 | 2.7 | 4.4 | 6 |
| 1152 | On whole wheat | 1 ea | 169 | 54 | 352 | 24 | 34 | 4 | 15 | 2.9 | 4.7 | 6.1 |
| | Ham & cheese: | | | | | | | | | | | |
| 1280 | On white bread, soft | 1 ea | 157 | 49 | 403 | 23 | 30 | 1 | 22 | 8.4 | 5.9 | 6.1 |
| 1282 | On part whole wheat | 1 ea | 156 | 50 | 397 | 23 | 28 | 3 | 21 | 8.1 | 5.9 | 6.2 |
| 1281 | On whole wheat | 1 ea | 170 | 49 | 424 | 24 | 34 | 4 | 22 | 8.3 | 6.2 | 6.4 |
| 1150 | Ham & swiss on rye | 1 ea | 150 | 54 | 339 | 22 | 22 | <1 | 19 | 6.5 | 5.1 | 6 |
| | Ham salad: | | | | | | | | | | | |
| 1154 | On white bread, soft | 1 ea | 131 | 47 | 362 | 11 | 37 | 1 | 20 | 4.8 | 6.7 | 7.4 |
| 1156 | On part whole wheat | 1 ea | 131 | 48 | 357 | 11 | 35 | 3 | 20 | 4.5 | 6.8 | 7.5 |
| 1155 | On whole wheat | 1 ea | 144 | 47 | 380 | 12 | 40 | 4 | 20 | 4.7 | 7 | 7.6 |
| 1157 | Patty melt: Ground beef & cheese on rye | 1 ea | 182 | 46 | 561 | 37 | 22 | 3 | 37 | 13.2 | 11.7 | 8.4 |
| | Peanut butter & jelly: | | | | | | | | | | | |
| 1158 | On white bread, soft | 1 ea | 101 | 26 | 351 | 12 | 47 | 3 | 15 | 3.1 | 6.7 | 3.9 |
| 1160 | On part whole wheat | 1 ea | 101 | 27 | 346 | 12 | 45 | 5 | 15 | 2.9 | 6.7 | 4 |
| 1159 | On whole wheat | 1 ea | 114 | 28 | 370 | 13 | 50 | 6 | 15 | 3 | 7 | 4.1 |
| 1161 | Reuben, grilled: Corned beef, swiss cheese, sauerkraut on rye | 1 ea | 239 | 64 | 462 | 28 | 25 | 2 | 29 | 9.9 | 9.5 | 7.1 |
| | Roast beef: | | | | | | | | | | | |
| 713 | On a bun | 1 ea | 139 | 49 | 346 | 21 | 33 | | 14 | 3.6 | 6.8 | 1.7 |
| 1162 | On white bread, soft | 1 ea | 157 | 46 | 404 | 29 | 34 | 1 | 17 | 3.4 | 4.2 | 8.2 |
| 1164 | On part whole wheat | 1 ea | 156 | 47 | 398 | 29 | 32 | 3 | 17 | 3.2 | 4.3 | 8.3 |
| 1163 | On whole wheat | 1 ea | 169 | 46 | 422 | 31 | 38 | 4 | 17 | 3.3 | 4.5 | 8.4 |
| | Tuna salad: | | | | | | | | | | | |
| 1165 | On white bread, soft | 1 ea | 122 | 46 | 327 | 14 | 35 | 2 | 15 | 2.5 | 3.8 | 7.9 |
| 1167 | On part whole wheat | 1 ea | 122 | 47 | 322 | 14 | 33 | 4 | 15 | 2.2 | 3.8 | 8 |
| 1166 | On whole wheat | 1 ea | 135 | 46 | 346 | 16 | 39 | 5 | 15 | 2.3 | 4.1 | 8.1 |
| | Turkey: | | | | | | | | | | | |
| 1168 | On white bread, soft | 1 ea | 156 | 54 | 346 | 24 | 29 | 1 | 15 | 2.4 | 3.2 | 8.3 |
| 1170 | On part whole wheat | 1 ea | 155 | 54 | 338 | 24 | 27 | 3 | 14 | 2.1 | 3.2 | 8.3 |
| 1169 | On whole wheat | 1 ea | 169 | 53 | 365 | 26 | 33 | 4 | 15 | 2.3 | 3.5 | 8.5 |
| | Turkey ham: | | | | | | | | | | | |
| 1272 | On rye bread | 1 ea | 150 | 60 | 280 | 21 | 20 | <1 | 14 | 2.5 | 2.8 | 6.9 |
| 1273 | On white bread, soft | 1 ea | 156 | 55 | 331 | 21 | 29 | 1 | 14 | 3 | 3.4 | 6.8 |
| 1275 | On part whole wheat | 1 ea | 156 | 56 | 326 | 21 | 28 | 3 | 14 | 3 | 4.2 | 5.6 |
| 1274 | On whole wheat | 1 ea | 169 | 55 | 350 | 23 | 33 | 4 | 15 | 2.9 | 3.7 | 7 |
| 714 | Taco | 1 ea | 171 | 58 | 369 | 21 | 27 | | 20 | 11.4 | 6.6 | 1 |
| | Tostada: | | | | | | | | | | | |
| 1114 | With refried beans | 1 ea | 144 | 66 | 223 | 10 | 26 | 7 | 10 | 5.4 | 3 | .7 |
| 1118 | With beans & beef | 1 ea | 225 | 70 | 333 | 16 | 30 | 4 | 17 | 11.5 | 3.5 | .6 |
| 1354 | With beans & chicken | 1 ea | 156 | 68 | 248 | 19 | 18 | 3 | 11 | 5.3 | 3.9 | 1.6 |
| | **NUTS, SEEDS, and PRODUCTS** | | | | | | | | | | | |
| | Almonds: | | | | | | | | | | | |
| 1365 | Dry roasted, salted | 1 c | 138 | 3 | 810 | 22 | 33 | 19 | 71 | 6.7 | 46.2 | 14.9 |

| Chol (mg) | Calc (mg) | Iron (mg) | Magn (mg) | Pota (mg) | Sodi (mg) | Zinc (mg) | VT-A (RE) | Thia (mg) | VT-E (a-TE) | Ribo (mg) | Niac (mg) | V-B6 (mg) | Fola (μg) | VT-C (mg) |
|---|---|---|---|---|---|---|---|---|---|---|---|---|---|---|
| 55 | 399 | 1.81 | 25 | 154 | 1143 | 2.05 | 212 | .24 | 1.13 | .34 | 1.91 | .06 | 24 | <1 |
| 54 | 412 | 2.12 | 39 | 209 | 1160 | 2.31 | 212 | .25 | 1.53 | .36 | 2.39 | .1 | 28 | <1 |
| 54 | 402 | 2.54 | 73 | 271 | 1226 | 3.08 | 212 | .26 | 1.14 | .35 | 2.74 | .17 | 40 | <1 |
| 60 | 60 | 4.68 | 35 | 353 | 957 | 1.87 | 31 | .33 | .55 | .24 | 6.81 | .2 | 100 | 9 |
| 32 | 60 | 2.04 | 18 | 139 | 460 | .8 | 24 | .25 | 6.16 | .18 | 3.68 | .25 | 26 | 1 |
| 31 | 73 | 2.35 | 33 | 195 | 475 | 1.06 | 24 | .26 | 6.57 | .2 | 4.16 | .29 | 30 | 1 |
| 30 | 63 | 2.79 | 68 | 259 | 543 | 1.85 | 24 | .27 | 6.14 | .19 | 4.5 | .36 | 42 | 1 |
| 82 | 268 | 3.12 | 28 | 225 | 1392 | 3.65 | 81 | .19 | 2.59 | .33 | 2.72 | .17 | 19 | 1 |
| 157 | 71 | 2.18 | 16 | 113 | 526 | .74 | 76 | .26 | 4.52 | .31 | 1.98 | .2 | 37 | 0 |
| 155 | 83 | 2.47 | 31 | 169 | 539 | 1 | 75 | .27 | 4.91 | .33 | 2.45 | .23 | 41 | 0 |
| 155 | 73 | 2.92 | 66 | 234 | 611 | 1.8 | 76 | .28 | 4.53 | .32 | 2.82 | .3 | 53 | 0 |
| 47 | 48 | 2.3 | 26 | 364 | 1566 | 2.11 | 8 | .99 | 2.36 | .31 | 5.45 | .47 | 15 | 23 |
| 47 | 60 | 2.39 | 29 | 368 | 1619 | 2.04 | 8 | 1.02 | 2.34 | .33 | 5.99 | .47 | 24 | 22 |
| 45 | 71 | 2.68 | 43 | 421 | 1630 | 2.29 | 8 | 1.03 | 2.73 | .35 | 6.45 | .5 | 27 | 22 |
| 45 | 62 | 3.1 | 77 | 483 | 1696 | 3.05 | 8 | 1.04 | 2.34 | .33 | 6.78 | .57 | 39 | 22 |
| 61 | 232 | 2.28 | 30 | 315 | 1620 | 2.34 | 90 | .76 | 2.53 | .36 | 4.64 | .36 | 25 | 15 |
| 59 | 244 | 2.58 | 45 | 368 | 1630 | 2.59 | 90 | .77 | 2.93 | .39 | 5.09 | .39 | 28 | 15 |
| 59 | 236 | 3.02 | 79 | 432 | 1707 | 3.37 | 90 | .78 | 2.56 | .37 | 5.47 | .46 | 40 | 15 |
| 57 | 258 | 2.25 | 29 | 344 | 1602 | 2.59 | 79 | .72 | 2.52 | .36 | 4.06 | .35 | 16 | 15 |
| 30 | 56 | 2.07 | 19 | 159 | 921 | 1.06 | 8 | .51 | 3.29 | .22 | 3.25 | .17 | 22 | 4 |
| 29 | 69 | 2.38 | 33 | 216 | 936 | 1.32 | 8 | .52 | 3.69 | .24 | 3.74 | .21 | 26 | 4 |
| 29 | 59 | 2.81 | 69 | 279 | 1001 | 2.11 | 8 | .53 | 3.29 | .23 | 4.09 | .28 | 38 | 4 |
| 113 | 222 | 4.19 | 36 | 391 | 701 | 7.11 | 123 | .25 | 3.5 | .46 | 6.14 | .35 | 25 | <1 |
| 2 | 60 | 2.25 | 56 | 245 | 293 | 1.07 | <1 | .27 | .12 | .17 | 5.33 | .13 | 40 | <1 |
| 0 | 72 | 2.55 | 70 | 299 | 308 | 1.32 | <1 | .28 | .51 | .19 | 5.8 | .17 | 44 | <1 |
| 0 | 63 | 2.97 | 104 | 361 | 375 | 2.09 | <1 | .29 | .14 | .17 | 6.14 | .24 | 56 | <1 |
| 80 | 288 | 4.24 | 38 | 361 | 1949 | 3.73 | 130 | .21 | 4.46 | .34 | 2.79 | .27 | 38 | 13 |
| 51 | 54 | 4.23 | 31 | 316 | 792 | 3.39 | 21 | .37 | .19 | .31 | 5.87 | .26 | 57 | 2 |
| 45 | 60 | 3.98 | 28 | 432 | 1595 | 3.78 | 12 | .29 | 3.3 | .3 | 6.39 | .39 | 30 | 12 |
| 43 | 72 | 4.27 | 42 | 485 | 1607 | 4.02 | 12 | .31 | 3.69 | .32 | 6.84 | .42 | 34 | 12 |
| 43 | 62 | 4.7 | 77 | 547 | 1672 | 4.79 | 12 | .31 | 3.31 | .31 | 7.18 | .49 | 45 | 12 |
| 14 | 60 | 2.24 | 22 | 161 | 567 | .67 | 22 | .25 | 2.72 | .18 | 5.53 | .11 | 25 | 1 |
| 13 | 73 | 2.55 | 37 | 217 | 582 | .93 | 22 | .26 | 3.12 | .2 | 6 | .16 | 29 | 1 |
| 12 | 63 | 2.98 | 72 | 280 | 649 | 1.72 | 22 | .27 | 2.71 | .19 | 6.34 | .23 | 41 | 1 |
| 45 | 56 | 2 | 29 | 302 | 1585 | 1.33 | 12 | .26 | 3.46 | .23 | 8.97 | .4 | 24 | 0 |
| 43 | 68 | 2.28 | 43 | 354 | 1589 | 1.57 | 11 | .27 | 3.82 | .25 | 9.38 | .44 | 28 | 0 |
| 43 | 59 | 2.73 | 77 | 418 | 1665 | 2.35 | 12 | .28 | 3.47 | .23 | 9.77 | .51 | 39 | 0 |
| 55 | 51 | 4.06 | 25 | 342 | 1185 | 3 | 8 | .22 | 2.8 | .33 | 4.3 | .29 | 17 | <1 |
| 55 | 62 | 4.09 | 28 | 346 | 1248 | 2.9 | 8 | .27 | 2.75 | .35 | 4.87 | .28 | 25 | 0 |
| 53 | 74 | 4.37 | 42 | 400 | 1262 | 3.15 | 8 | .28 | 3.57 | .37 | 5.34 | .32 | 29 | 0 |
| 53 | 65 | 4.81 | 76 | 462 | 1329 | 3.91 | 8 | .29 | 2.76 | .35 | 5.68 | .39 | 41 | 0 |
| 56 | 221 | 2.41 | 70 | 474 | 802 | 3.93 | 147 | .15 | 1.88 | .44 | 3.21 | .24 | 68 | 2 |
| 30 | 210 | 1.89 | 59 | 403 | 543 | 1.9 | 85 | .1 | 1.15 | .33 | 1.32 | .16 | 43 | 1 |
| 74 | 189 | 2.45 | 67 | 491 | 871 | 3.17 | 173 | .09 | 1.8 | .49 | 2.86 | .25 | 85 | 4 |
| 53 | 168 | 1.79 | 47 | 365 | 433 | 2.28 | 86 | .11 | 1.87 | .2 | 4.52 | .32 | 53 | 3 |
| 0 | 389 | 5.24 | 420 | 1062 | 1076 | 6.76 | 0 | .18 | 7.66 | .83 | 3.89 | .1 | 88 | 1 |

**Table H-1**

**Food Composition** (Computer code number is for West Diet Analysis program) (For purposes of calculations, use "0" for t, <1, <.1, <.01, etc.)

| Computer Code Number | Food Description | Measure | Wt (g) | $H_2O$ (%) | Ener (kcal) | Prot (g) | Carb (g) | Dietary Fiber (g) | Fat (g) | Fat Breakdown (g) Sat | Mono | Poly |
|---|---|---|---|---|---|---|---|---|---|---|---|---|
| | **NUTS, SEEDS, and PRODUCTS**—Continued | | | | | | | | | | | |
| | Almonds: | | | | | | | | | | | |
| 718 | Slivered, packed, unsalted | 1 c | 108 | 4 | 636 | 22 | 22 | 12 | 56 | 5.3 | 36.6 | 11.9 |
| 719 | Whole, dried, unsalted | 1 c | 142 | 4 | 836 | 28 | 29 | 15 | 74 | 7 | 48.1 | 15.6 |
| 720 | Ounce | 1 oz | 28 | 4 | 165 | 6 | 6 | 3 | 15 | 1.4 | 9.5 | 3.1 |
| 721 | Almond butter: | 1 tbs | 16 | 1 | 101 | 2 | 3 | 1 | 9 | .9 | 6.1 | 2 |
| 4572 | Salted | 1 tbs | 16 | 1 | 101 | 2 | 3 | 1 | 9 | .9 | 6.1 | 2 |
| 722 | Brazil nuts, dry (about 7) | 1 c | 140 | 3 | 918 | 20 | 18 | 8 | 93 | 22.7 | 32.2 | 33.7 |
| | Cashew nuts, dry roasted: | | | | | | | | | | | |
| 723 | Salted: | 1 c | 137 | 2 | 786 | 21 | 45 | 4 | 64 | 12.8 | 37.4 | 10.7 |
| 724 | Ounce | 1 oz | 28 | 2 | 161 | 4 | 9 | 1 | 13 | 2.6 | 7.6 | 2.2 |
| 4621 | Unsalted: | 1 c | 137 | 2 | 786 | 21 | 45 | 4 | 64 | 12.8 | 37.4 | 10.7 |
| 4621 | Ounce | 1 oz | 28 | 2 | 161 | 4 | 9 | 1 | 13 | 2.6 | 7.6 | 2.2 |
| 725 | Oil roasted: | 1 c | 130 | 4 | 749 | 23 | 37 | 5 | 63 | 12.6 | 36.9 | 10.6 |
| 726 | Ounce | 1 oz | 28 | 4 | 161 | 5 | 8 | 1 | 13 | 2.7 | 7.9 | 2.3 |
| 4622 | Unsalted: | 1 c | 130 | 4 | 749 | 21 | 37 | 5 | 63 | 12.6 | 36.9 | 10.6 |
| 4622 | Ounce | 1 oz | 28 | 4 | 161 | 5 | 8 | 1 | 13 | 2.7 | 7.9 | 2.3 |
| 727 | Cashew butter, unsalted | 1 tbs | 16 | 3 | 94 | 3 | 4 | <1 | 8 | 1.6 | 4.7 | 1.3 |
| 4662 | Cashew butter, salted | 1 tbs | 16 | 3 | 94 | 3 | 4 | <1 | 8 | 1.6 | 4.7 | 1.3 |
| 728 | Chestnuts, European, roasted (1 cup = approx 17 kernels) | 1 c | 143 | 40 | 350 | 5 | 76 | 7 | 3 | .6 | 1.1 | 1.2 |
| | Coconut, raw: | | | | | | | | | | | |
| 729 | Piece 2 x 2 x ½" | 1 pce | 45 | 47 | 159 | 2 | 7 | 4 | 15 | 13.5 | .6 | .2 |
| 730 | Shredded/grated, unpacked | ½ c | 40 | 47 | 142 | 1 | 6 | 4 | 13 | 12 | .6 | .1 |
| | Coconut, dried, shredded/grated: | | | | | | | | | | | |
| 731 | Unsweetened | 1 c | 78 | 3 | 515 | 6 | 19 | 13 | 50 | 45.1 | 2.1 | .6 |
| 732 | Sweetened | 1 c | 93 | 13 | 466 | 3 | 44 | 4 | 33 | 29.6 | 1.4 | .4 |
| 733 | Filberts/hazelnuts, chopped: | 1 c | 135 | 5 | 853 | 18 | 21 | 8 | 84 | 6.2 | 66.3 | 8.1 |
| 734 | Ounce | 1 oz | 28 | 5 | 177 | 4 | 4 | 2 | 17 | 1.3 | 13.7 | 1.7 |
| 735 | Macadamias, oil roasted, salted: | 1 c | 134 | 2 | 962 | 10 | 17 | 12 | 103 | 15.4 | 80.9 | 1.8 |
| 736 | Ounce | 1 oz | 28 | 2 | 201 | 2 | 4 | 3 | 21 | 3.2 | 16.9 | .4 |
| 1368 | Macadamias, oil roasted, unsalted | 1 c | 134 | 2 | 962 | 10 | 17 | 12 | 103 | 15.4 | 80.9 | 1.8 |
| | Mixed nuts: | | | | | | | | | | | |
| 737 | Dry roasted, salted | 1 c | 137 | 2 | 814 | 24 | 35 | 12 | 71 | 9.4 | 43 | 14.8 |
| 738 | Oil roasted, salted | 1 c | 142 | 2 | 876 | 24 | 30 | 13 | 80 | 12.4 | 45 | 18.9 |
| 1369 | Oil roasted, unsalted | 1 c | 142 | 2 | 876 | 27 | 30 | 14 | 80 | 12.4 | 45 | 18.9 |
| | Peanuts: | | | | | | | | | | | |
| 739 | Oil roasted, salted | 1 c | 144 | 2 | 837 | 38 | 27 | 13 | 71 | 9.8 | 35.3 | 22.5 |
| 740 | Ounce | 1 oz | 28 | 2 | 163 | 7 | 5 | 3 | 14 | 1.9 | 6.9 | 4.4 |
| 1370 | Oil roasted, unsalted | 1 c | 144 | 2 | 837 | 38 | 27 | 10 | 71 | 9.8 | 35.3 | 22.5 |
| 741 | Dried, salted | 1 c | 146 | 2 | 854 | 35 | 31 | 12 | 73 | 10.1 | 36.1 | 22.9 |
| 742 | Ounce | 1 oz | 28 | 2 | 164 | 7 | 6 | 2 | 14 | 1.9 | 6.9 | 4.4 |
| 743 | Peanut butter: | ½ c | 128 | 1 | 759 | 33 | 25 | 8 | 65 | 14.3 | 31.1 | 17.7 |
| 1371 | Tablespoon | 2 tbs | 32 | 1 | 190 | 8 | 6 | 2 | 16 | 3.6 | 7.8 | 4.4 |
| 744 | Pecan halves, dried, unsalted: | 1 c | 108 | 5 | 720 | 9 | 20 | 8 | 73 | 5.8 | 45.6 | 18 |
| 745 | Ounce | 1 oz | 28 | 5 | 187 | 2 | 5 | 2 | 19 | 1.5 | 11.8 | 4.7 |
| 1372 | Pecan halves, dry roasted, salted | ¼ c | 28 | 1 | 185 | 2 | 6 | 3 | 18 | 1.4 | 11.3 | 4.5 |
| 746 | Pine nuts/piñons, dried | 1 oz | 28 | 6 | 176 | 3 | 5 | 3 | 17 | 2.6 | 6.4 | 7.2 |
| 747 | Pistachios, dried, shelled | 1 oz | 28 | 4 | 162 | 6 | 7 | 3 | 14 | 1.8 | 9.2 | 2 |
| 1373 | Pistachios,dry roasted,salted,shelled | 1 c | 128 | 2 | 776 | 19 | 35 | 14 | 68 | 8.8 | 45.7 | 10.2 |
| 748 | Pumpkin kernels, dried, unsalted | 1 oz | 28 | 7 | 151 | 7 | 5 | 1 | 13 | 2.4 | 4 | 5.8 |
| 1374 | Pumpkin kernels, roasted, salted | 1 c | 227 | 7 | 1184 | 75 | 30 | 9 | 96 | 18.1 | 29.7 | 43.6 |
| 749 | Sesame seeds, hulled, dried | ¼ c | 38 | 5 | 223 | 10 | 4 | 3 | 21 | 2.9 | 7.9 | 9.1 |
| | Sunflower seed kernels: | | | | | | | | | | | |
| 750 | Dry | ¼ c | 36 | 5 | 205 | 8 | 7 | 4 | 18 | 1.9 | 3.4 | 11.8 |
| 751 | Oil roasted | ¼ c | 34 | 3 | 209 | 7 | 5 | 2 | 20 | 2 | 3.7 | 12.9 |
| 752 | Tahini (sesame butter) | 1 tbs | 15 | 3 | 91 | 3 | 3 | 1 | 8 | 1.2 | 3.2 | 3.7 |
| 1334 | Trail mix w/chocolate chips | 1 c | 146 | 7 | 707 | 21 | 66 | 8 | 47 | 9.3 | 19.8 | 16.5 |
| 753 | Black walnuts, chopped: | 1 c | 125 | 4 | 759 | 31 | 15 | 6 | 71 | 4.8 | 15.9 | 46.9 |
| 754 | Ounce | 1 oz | 28 | 4 | 170 | 7 | 3 | 1 | 16 | 1.1 | 3.6 | 10.5 |
| 755 | English walnuts, chopped: | 1 c | 120 | 4 | 770 | 17 | 22 | 6 | 74 | 7.2 | 17 | 46.9 |
| 756 | Ounce | 1 oz | 28 | 4 | 180 | 4 | 5 | 1 | 17 | 1.7 | 4 | 10.9 |

| Chol (mg) | Calc (mg) | Iron (mg) | Magn (mg) | Pota (mg) | Sodi (mg) | Zinc (mg) | VT-A (RE) | Thia (mg) | VT-E (a-TE) | Ribo (mg) | Niac (mg) | V-B6 (mg) | Fola (μg) | VT-C (mg) |
|---|---|---|---|---|---|---|---|---|---|---|---|---|---|---|
| 0 | 287 | 3.95 | 320 | 791 | 12 | 3.15 | 0 | .23 | 25.9 | .84 | 3.63 | .12 | 63 | 1 |
| 0 | 378 | 5.2 | 420 | 1039 | 16 | 4.15 | 0 | .3 | 34.1 | 1.11 | 4.77 | .16 | 83 | 1 |
| 0 | 74 | 1.02 | 83 | 205 | 3 | .82 | 0 | .06 | 6.72 | .22 | .94 | .03 | 16 | <1 |
| 0 | 43 | .59 | 48 | 121 | 2 | .49 | 0 | .02 | 3.25 | .1 | .46 | .01 | 10 | <1 |
| 0 | 43 | .59 | 48 | 121 | 72 | .49 | 0 | .02 | 3.25 | .1 | .46 | .01 | 10 | <1 |
| 0 | 246 | 4.76 | 315 | 840 | 3 | 6.43 | 0 | 1.4 | 10.6 | .17 | 2.27 | .35 | 6 | 1 |
| 0 | 62 | 8.22 | 356 | 774 | 877 | 7.67 | 0 | .27 | .78 | .27 | 1.92 | .35 | 95 | 0 |
| 0 | 13 | 1.68 | 73 | 158 | 179 | 1.57 | 0 | .06 | .16 | .06 | .39 | .07 | 19 | 0 |
| 0 | 62 | 8.22 | 356 | 774 | 22 | 7.67 | 0 | .27 | .78 | .27 | 1.92 | .35 | 95 | 0 |
| 0 | 13 | 1.68 | 73 | 158 | 4 | 1.57 | 0 | .06 | .16 | .06 | .39 | .07 | 19 | 0 |
| 0 | 53 | 5.33 | 332 | 689 | 814 | 6.18 | 0 | .55 | 2.03 | .23 | 2.34 | .32 | 88 | 0 |
| 0 | 11 | 1.15 | 71 | 148 | 175 | 1.33 | 0 | .12 | .44 | .05 | .5 | .07 | 19 | 0 |
| 0 | 53 | 5.33 | 332 | 689 | 22 | 6.18 | 0 | .55 | 2.03 | .23 | 2.34 | .32 | 88 | 0 |
| 0 | 11 | 1.15 | 71 | 148 | 5 | 1.33 | 0 | .12 | .44 | .05 | .5 | .07 | 19 | 0 |
| 0 | 7 | .8 | 41 | 87 | 2 | .83 | 0 | .05 | .25 | .03 | .26 | .04 | 11 | 0 |
| 0 | 7 | .8 | 41 | 87 | 98 | .83 | 0 | .05 | .25 | .03 | .26 | .04 | 11 | 0 |
| 0 | 41 | 1.3 | 47 | 847 | 3 | .81 | 3 | .35 | 1.72 | .25 | 1.92 | .71 | 100 | 37 |
| 0 | 6 | 1.09 | 14 | 160 | 9 | .49 | 0 | .03 | .33 | .01 | .24 | .02 | 12 | 1 |
| 0 | 6 | .97 | 13 | 142 | 8 | .44 | 0 | .03 | .29 | .01 | .22 | .02 | 11 | 1 |
| 0 | 20 | 2.59 | 70 | 424 | 29 | 1.57 | 0 | .05 | 1.05 | .08 | .47 | .23 | 7 | 1 |
| 0 | 14 | 1.79 | 46 | 313 | 244 | 1.69 | 0 | .03 | 1.26 | .02 | .44 | .25 | 8 | 1 |
| 0 | 254 | 4.41 | 385 | 601 | 4 | 3.24 | 9 | .67 | 32.3 | .15 | 1.54 | .83 | 97 | 1 |
| 0 | 53 | .92 | 80 | 125 | 1 | .67 | 2 | .14 | 6.69 | .03 | .32 | .17 | 20 | <1 |
| 0 | 60 | 2.41 | 157 | 441 | 348 | 1.47 | 1 | .28 | .55 | .15 | 2.71 | .26 | 21 | 0 |
| 0 | 13 | .5 | 33 | 92 | 73 | .31 | <1 | .06 | .11 | .03 | .57 | .05 | 4 | 0 |
| 0 | 60 | 2.41 | 157 | 441 | 9 | 1.47 | 1 | .28 | .55 | .15 | 2.71 | .26 | 21 | 0 |
| 0 | 96 | 5.07 | 308 | 818 | 917 | 5.21 | 1 | .27 | 8.22 | .27 | 6.44 | .41 | 69 | 1 |
| 0 | 153 | 4.56 | 334 | 825 | 926 | 7.21 | 3 | .71 | 8.52 | .31 | 7.19 | .34 | 118 | 1 |
| 0 | 153 | 4.56 | 334 | 825 | 16 | 7.21 | 3 | .71 | 8.52 | .31 | 7.19 | .34 | 118 | 1 |
| 0 | 127 | 2.64 | 266 | 982 | 624 | 9.55 | 0 | .36 | 10.7 | .16 | 20.6 | .37 | 181 | 0 |
| 0 | 25 | .51 | 52 | 191 | 121 | 1.86 | 0 | .07 | 2.07 | .03 | 4 | .07 | 35 | 0 |
| 0 | 127 | 2.64 | 266 | 982 | 9 | 9.55 | 0 | .36 | 10.7 | .16 | 20.6 | .37 | 181 | 0 |
| 0 | 79 | 3.3 | 257 | 961 | 1186 | 4.83 | 0 | .64 | 10.8 | .14 | 19.7 | .37 | 212 | 0 |
| 0 | 15 | .63 | 49 | 184 | 228 | .93 | 0 | .12 | 2.07 | .03 | 3.78 | .07 | 41 | 0 |
| 0 | 49 | 2.36 | 204 | 856 | 598 | 3.74 | 0 | .11 | 12.8 | .13 | 17.2 | .58 | 95 | 0 |
| 0 | 12 | .59 | 51 | 214 | 149 | .93 | 0 | .03 | 3.2 | .03 | 4.29 | .14 | 24 | 0 |
| 0 | 39 | 2.3 | 138 | 423 | 1 | 5.91 | 14 | .92 | 3.35 | .14 | .96 | .2 | 42 | 2 |
| 0 | 10 | .6 | 36 | 110 | <1 | 1.53 | 4 | .24 | .87 | .04 | .25 | .05 | 11 | 1 |
| 0 | 10 | .61 | 37 | 104 | 218 | 1.59 | 4 | .09 | .84 | .03 | .26 | .05 | 11 | 1 |
| 0 | 2 | .86 | 65 | 176 | 20 | 1.2 | 1 | .35 | .98 | .06 | 1.22 | .03 | 16 | 1 |
| 0 | 38 | 1.9 | 44 | 306 | 2 | .37 | 6 | .23 | 1.46 | .05 | .3 | .07 | 16 | 2 |
| 0 | 90 | 4.06 | 166 | 1241 | 998 | 1.74 | 31 | .54 | 8.26 | .31 | 1.8 | .33 | 76 | 9 |
| 0 | 12 | 4.2 | 150 | 226 | 5 | 2.09 | 11 | .06 | .28 | .09 | .49 | .06 | 16 | 1 |
| 0 | 98 | 33.8 | 1212 | 1829 | 1305 | 16.9 | 86 | .48 | 2.27 | .72 | 3.95 | .2 | 130 | 4 |
| 0 | 50 | 2.96 | 132 | 155 | 15 | 3.91 | 3 | .27 | .86 | .03 | 1.78 | .05 | 36 | 0 |
| 0 | 42 | 2.44 | 127 | 248 | 1 | 1.82 | 2 | .82 | 18.1 | .09 | 1.62 | .28 | 82 | <1 |
| 0 | 19 | 2.28 | 43 | 164 | 1 | 1.77 | 2 | .11 | 17.1 | .09 | 1.4 | .27 | 80 | <1 |
| 0 | 21 | .95 | 53 | 69 | <1 | 1.58 | 1 | .24 | .34 | .02 | .85 | .02 | 15 | 0 |
| 6 | 159 | 4.95 | 235 | 946 | 177 | 4.58 | 7 | .6 | 15.6 | .33 | 6.44 | .38 | 95 | 2 |
| 0 | 72 | 3.84 | 253 | 655 | 1 | 4.28 | 37 | .27 | 3.28 | .14 | .86 | .69 | 82 | 4 |
| 0 | 16 | .86 | 57 | 147 | <1 | .96 | 8 | .06 | .73 | .03 | .19 | .15 | 18 | 1 |
| 0 | 113 | 2.93 | 203 | 602 | 12 | 3.28 | 14 | .46 | 3.14 | .18 | 1.25 | .67 | 79 | 4 |
| 0 | 26 | .68 | 47 | 141 | 3 | .76 | 3 | .11 | .73 | .04 | .29 | .16 | 18 | 1 |

**Table H–1**

**Food Composition** (Computer code number is for West Diet Analysis program) (For purposes of calculations, use "0" for t, <1, <.1, <.01, etc.)

| Computer Code Number | Food Description | Measure | Wt (g) | $H_2O$ (%) | Ener (kcal) | Prot (g) | Carb (g) | Dietary Fiber (g) | Fat (g) | Fat Breakdown (g) Sat | Mono | Poly |
|---|---|---|---|---|---|---|---|---|---|---|---|---|
| | **SWEETENERS and SWEETS** (see also Dairy [milk desserts] and Baked Goods) | | | | | | | | | | | |
| 757 | Apple butter | 2 tbs | 36 | 52 | 66 | <1 | 17 | <1 | <1 | t | t | t |
| 1124 | Butterscotch topping | 2 tbs | 41 | 32 | 103 | 1 | 27 | <1 | <1 | t | t | 0 |
| 1125 | Caramel topping | 2 tbs | 41 | 32 | 103 | 1 | 27 | <1 | <1 | t | t | 0 |
| | Cake frosting, creamy vanilla: | | | | | | | | | | | |
| 1127 | Canned | 2 tbs | 39 | 13 | 163 | <1 | 27 | <1 | 7 | 1.9 | 3.4 | .9 |
| 1123 | From mix | 2 tbs | 39 | 12 | 165 | <1 | 28 | <1 | 6 | 1.3 | 2.6 | 2.2 |
| | Cake frosting, lite: | | | | | | | | | | | |
| 2061 | Milk chocolate | 1 tbs | 16 | 18 | 58 | <1 | 11 | <1 | 1 | .4 | | |
| 2062 | Vanilla | 1 tbs | 16 | 15 | 60 | 0 | 12 | <1 | 1 | .4 | | |
| | Candy: | | | | | | | | | | | |
| 1128 | Almond Joy candy bar | 1 oz | 28 | 10 | 131 | 1 | 16 | 1 | 7 | 4.8 | 1.8 | .4 |
| 2069 | Butterscotch morsels | ¼ c | 43 | 1 | 246 | 0 | 31 | 0 | 12 | 12.5 | 0 | 0 |
| 758 | Caramel, plain or chocolate | 1 pce | 10 | 8 | 38 | <1 | 8 | <1 | 1 | .7 | .1 | t |
| 1961 | Chewing gum, sugarless | 1 pce | 3 | | 6 | 0 | 2 | | 0 | 0 | 0 | 0 |
| | Chocolate (see also #784, 785, 971): | | | | | | | | | | | |
| | Milk chocolate: | | | | | | | | | | | |
| 759 | Plain | 1 oz | 28 | 1 | 144 | 2 | 17 | 1 | 9 | 5.2 | 2.8 | .3 |
| 760 | With almonds | 1 oz | 28 | 1 | 147 | 3 | 15 | 2 | 10 | 4.8 | 3.8 | .6 |
| 761 | With peanuts | 1 oz | 28 | 1 | 155 | 5 | 11 | 2 | 11 | 3.4 | 5.1 | 2.5 |
| 762 | With rice cereal | 1 oz | 28 | 2 | 139 | 2 | 18 | 1 | 7 | 4.4 | 2.4 | .2 |
| 763 | Semisweet chocolate chips | 1 c | 168 | 1 | 805 | 7 | 106 | 10 | 50 | 29.9 | 16.8 | 1.6 |
| 764 | Sweet dark chocolate (candy bar) | 1 ea | 41 | 1 | 226 | 2 | 25 | 2 | 13 | 8.3 | 4.6 | .4 |
| 765 | Fondant candy, uncoated (mints, candy corn, other) | 1 pce | 16 | 7 | 57 | 0 | 15 | 0 | 0 | 0 | 0 | 0 |
| 1697 | Fruit Roll-Up (small) | 1 ea | 14 | 11 | 49 | <1 | 12 | <1 | <1 | .1 | .2 | .1 |
| 766 | Fudge, chocolate | 1 pce | 17 | 10 | 65 | <1 | 13 | <1 | 1 | .9 | .4 | .1 |
| 767 | Gumdrops | 1 c | 182 | 1 | 703 | 0 | 180 | 0 | 0 | 0 | 0 | 0 |
| 768 | Hard candy, all flavors | 1 pce | 6 | 1 | 22 | 0 | 6 | 0 | <1 | 0 | 0 | 0 |
| 769 | Jellybeans | 10 pce | 11 | 6 | 40 | 0 | 10 | 0 | <1 | t | t | t |
| 1134 | M&M's plain chocolate candy | 10 pce | 7 | 2 | 34 | <1 | 5 | <1 | 1 | .9 | .5 | t |
| 1135 | M&M's peanut chocolate candy | 10 pce | 20 | 2 | 103 | 2 | 12 | 1 | 5 | 2.1 | 2.2 | .8 |
| 1130 | Mars almond bar | 1 ea | 50 | 4 | 234 | 4 | 31 | 1 | 11 | 2.7 | 5.5 | 2.8 |
| 1129 | Milky Way candy bar | 1 ea | 60 | 6 | 254 | 3 | 43 | 1 | 10 | 4.7 | 3.6 | .4 |
| 1708 | Milk chocolate-coated peanuts | 1 c | 149 | 2 | 773 | 19 | 74 | 7 | 50 | 21.8 | 19.4 | 6.4 |
| 1709 | Peanut brittle, recipe | 1 c | 147 | 2 | 666 | 11 | 102 | 3 | 28 | 7.4 | 12.5 | 6.9 |
| 1132 | Reese's peanut butter cup | 2 ea | 50 | 3 | 271 | 5 | 27 | 2 | 16 | 5.5 | 6.5 | 2.8 |
| 1133 | Skor English toffee candy bar | 1 ea | 39 | 3 | 217 | 2 | 22 | 1 | 13 | 8.5 | 4.3 | .5 |
| 1131 | Snickers candy bar (2.2oz) | 1 ea | 62 | 5 | 297 | 5 | 37 | 2 | 15 | 5.6 | 6.5 | 3 |
| 1482 | Fruit juice bar (2.5 fl oz) | 1 ea | 77 | 78 | 63 | 1 | 16 | 0 | <1 | t | 0 | t |
| 771 | Gelatin dessert/Jello, prepared | ½ c | 135 | 85 | 80 | 2 | 19 | 0 | 0 | 0 | 0 | 0 |
| 1702 | SugarFree | ½ c | 117 | 98 | 8 | 1 | 1 | 0 | 0 | 0 | 0 | 0 |
| 772 | Honey: | 1 c | 339 | 17 | 1030 | 1 | 279 | 1 | 0 | 0 | 0 | 0 |
| 773 | Tablespoon | 1 tbs | 21 | 17 | 64 | <1 | 17 | <1 | 0 | 0 | 0 | 0 |
| 774 | Jams or preserves: | 1 tbs | 20 | 29 | 54 | <1 | 14 | <1 | <1 | 0 | t | t |
| 775 | Packet | 1 ea | 14 | 34 | 34 | <1 | 9 | <1 | <1 | t | t | 0 |
| 776 | Jellies: | 1 tbs | 19 | 28 | 51 | <1 | 13 | <1 | <1 | t | t | t |
| 777 | Packet | 1 ea | 14 | 28 | 38 | <1 | 10 | <1 | <1 | t | t | t |
| 1136 | Marmalade | 1 tbs | 20 | 33 | 49 | <1 | 13 | <1 | 0 | 0 | 0 | 0 |
| 770 | Marshmallows | 1 ea | 7 | 16 | 22 | <1 | 6 | <1 | <1 | t | t | t |
| 1126 | Marshmallow creme topping | 2 tbs | 38 | 18 | 118 | 1 | 30 | <1 | <1 | t | t | t |
| 778 | Popsicle/ice pops | 1 ea | 128 | 80 | 92 | 0 | 24 | 0 | 0 | 0 | 0 | 0 |
| | Sugars: | | | | | | | | | | | |
| 779 | Brown sugar | 1 c | 220 | 2 | 827 | 0 | 214 | 0 | 0 | 0 | 0 | 0 |
| 780 | White sugar, granulated: | 1 c | 200 | | 774 | 0 | 200 | 0 | 0 | 0 | 0 | 0 |
| 781 | Tablespoon | 1 tbs | 12 | | 46 | 0 | 12 | 0 | 0 | 0 | 0 | 0 |
| 782 | Packet | 1 ea | 6 | | 23 | 0 | 6 | 0 | 0 | 0 | 0 | 0 |
| 783 | White sugar, powdered, sifted | 1 c | 100 | | 389 | 0 | 99 | 0 | <1 | t | t | t |
| | Sweeteners: | | | | | | | | | | | |
| 1711 | Equal, packet | 1 ea | 1 | 12 | 4 | <1 | 1 | 0 | <1 | 0 | t | t |
| 1712 | Sweet 'N Low, packet | 1 ea | 1 | | 0 | 0 | 1 | 0 | 0 | 0 | 0 | 0 |

| Chol (mg) | Calc (mg) | Iron (mg) | Magn (mg) | Pota (mg) | Sodi (mg) | Zinc (mg) | VT-A (RE) | Thia (mg) | VT-E (a-TE) | Ribo (mg) | Niac (mg) | V-B6 (mg) | Fola (μg) | VT-C (mg) |
|---|---|---|---|---|---|---|---|---|---|---|---|---|---|---|
| 0 | 2 | .05 | 1 | 33 | 0 | .02 | 0 | <.01 | .01 | <.01 | .03 | .01 | <1 | 1 |
| <1 | 22 | .08 | 3 | 34 | 143 | .08 | 11 | <.01 | 0 | .04 | .02 | .01 | 1 | <1 |
| <1 | 22 | .08 | 3 | 34 | 143 | .08 | 11 | <.01 | 0 | .04 | .02 | .01 | 1 | <1 |
| | | | | | | | | | | | | | | |
| 0 | 1 | .04 | | 14 | 35 | 0 | 88 | 0 | .79 | <.01 | <.01 | 0 | 0 | 0 |
| 0 | 4 | .09 | 1 | 9 | 87 | .04 | 42 | .01 | .79 | .01 | .13 | <.01 | 0 | 0 |
| | | | | | | | | | | | | | | |
| 0 | 1 | .24 | | | 40 | | <1 | | | | | | | 0 |
| 0 | | .02 | | | 29 | | 0 | | | | | | | 0 |
| | | | | | | | | | | | | | | |
| 1 | 17 | .39 | 18 | 69 | 41 | .22 | 1 | .01 | .63 | .04 | .13 | .02 | | <1 |
| 0 | 0 | 0 | | 80 | 46 | | 0 | .03 | | .04 | .03 | | | 0 |
| 1 | 14 | .01 | 2 | 21 | 24 | .04 | 1 | <.01 | .05 | .02 | .02 | <.01 | <1 | <1 |
| | | | | 0 | 0 | | | | | | | | | |
| | | | | | | | | | | | | | | |
| | | | | | | | | | | | | | | |
| 6 | 53 | .39 | 17 | 108 | 23 | .39 | 15 | .02 | .35 | .08 | .09 | .01 | 2 | <1 |
| 5 | 63 | .46 | 25 | 124 | 21 | .37 | 4 | .02 | .53 | .12 | .21 | .01 | 3 | <1 |
| 3 | 32 | .52 | 34 | 150 | 11 | .68 | 6 | .08 | 1.3 | .05 | 2.12 | .04 | 23 | 0 |
| 5 | 48 | .21 | 14 | 96 | 41 | .31 | 3 | .02 | .35 | .08 | .13 | .02 | 3 | <1 |
| 0 | 54 | 5.26 | 193 | 613 | 18 | 2.72 | 3 | .09 | 2 | .15 | .72 | .06 | 5 | 0 |
| <1 | 11 | .98 | 47 | 139 | 3 | .61 | 1 | .01 | .41 | .1 | .27 | .02 | 1 | 0 |
| | | | | | | | | | | | | | | |
| 0 | | .01 | | 3 | 6 | .01 | <1 | 0 | 0 | <.01 | 0 | 0 | 0 | 0 |
| 0 | 4 | .14 | 3 | 41 | 9 | .03 | 2 | .01 | .04 | <.01 | .01 | .04 | 1 | 1 |
| 2 | 7 | .08 | 4 | 17 | 10 | .07 | 8 | <.01 | .02 | .01 | .02 | <.01 | <1 | <1 |
| 0 | 5 | .73 | 2 | 9 | 80 | 0 | 0 | 0 | 0 | <.01 | <.01 | 0 | 0 | 0 |
| 0 | | .02 | | | 2 | <.01 | 0 | 0 | 0 | 0 | 0 | 0 | 0 | 0 |
| 0 | | .12 | | 4 | 3 | .01 | 0 | 0 | 0 | 0 | 0 | 0 | 0 | 0 |
| 1 | 7 | .08 | 3 | 19 | 4 | .07 | 4 | <.01 | .06 | .01 | .02 | <.01 | <1 | <1 |
| 2 | 20 | .23 | 12 | 69 | 10 | .27 | 5 | .03 | .43 | .04 | .41 | .02 | 7 | <1 |
| 4 | 84 | .55 | 36 | 163 | 85 | .55 | 22 | .02 | .3 | .16 | .47 | .03 | 9 | <1 |
| 8 | 78 | .46 | 20 | 145 | 144 | .43 | 34 | .02 | .39 | .13 | .21 | .03 | 6 | 1 |
| 13 | 155 | 1.95 | 134 | 748 | 61 | 2.8 | 0 | .17 | 3.8 | .26 | 6.33 | .31 | 12 | 0 |
| 19 | 44 | 2.03 | 73 | 306 | 664 | 1.43 | 69 | .28 | 2.41 | .08 | 5.15 | .15 | 103 | 0 |
| 2 | 39 | .6 | 42 | 176 | 159 | .7 | 9 | .02 | .66 | .1 | 1.99 | .04 | 27 | <1 |
| 20 | 51 | .02 | 13 | 93 | 108 | .3 | 27 | .01 | .53 | .13 | .03 | .01 | | <1 |
| 8 | 58 | .47 | 42 | 209 | 165 | .88 | 24 | .13 | .95 | .1 | 2.26 | .07 | 25 | <1 |
| 0 | 4 | .15 | 3 | 41 | 3 | .04 | 2 | .01 | 0 | .01 | .12 | .02 | 5 | 7 |
| 0 | 3 | .04 | 1 | 1 | 57 | .04 | 0 | 0 | 0 | <.01 | <.01 | <.01 | 0 | 0 |
| 0 | 2 | .01 | 1 | 0 | 56 | .03 | 0 | 0 | 0 | <.01 | <.01 | <.01 | 0 | 0 |
| 0 | 20 | 1.42 | 7 | 176 | 14 | .75 | 0 | 0 | 0 | .13 | .41 | .08 | 7 | 2 |
| 0 | 1 | .09 | | 11 | 1 | .05 | 0 | 0 | 0 | .01 | .02 | <.01 | <1 | <1 |
| 0 | 4 | .2 | 1 | 18 | 2 | .01 | <1 | <.01 | .02 | .01 | .04 | <.01 | 2 | <1 |
| 0 | 3 | .07 | 1 | 11 | 6 | .01 | <1 | 0 | 0 | <.01 | <.01 | <.01 | 5 | 1 |
| 0 | 2 | .04 | 1 | 12 | 7 | .01 | <1 | 0 | 0 | <.01 | .01 | <.01 | <1 | <1 |
| 0 | 1 | .03 | 1 | 9 | 5 | .01 | <1 | 0 | 0 | <.01 | <.01 | <.01 | <1 | <1 |
| 0 | 8 | .03 | | 7 | 11 | .01 | 1 | <.01 | 0 | <.01 | .01 | <.01 | 7 | 1 |
| 0 | | .02 | | | 3 | <.01 | <1 | 0 | 0 | 0 | <.01 | 0 | <1 | 0 |
| 0 | 1 | .08 | 1 | 2 | 17 | .01 | <1 | 0 | 0 | 0 | .03 | <.01 | <1 | 0 |
| 0 | 0 | 0 | 1 | 5 | 15 | .03 | 0 | 0 | 0 | 0 | 0 | 0 | 0 | 0 |
| | | | | | | | | | | | | | | |
| 0 | 187 | 4.2 | 64 | 761 | 86 | .4 | 0 | .02 | 0 | .01 | .18 | .06 | 2 | 0 |
| 0 | 2 | .12 | 0 | 4 | 2 | .06 | 0 | 0 | 0 | .04 | 0 | 0 | 0 | 0 |
| 0 | | .01 | 0 | | <1 | <.01 | 0 | 0 | 0 | <.01 | 0 | 0 | 0 | 0 |
| 0 | | <.01 | 0 | | <1 | <.01 | 0 | 0 | 0 | <.01 | 0 | 0 | 0 | 0 |
| 0 | 1 | .06 | 0 | 2 | 1 | .03 | 0 | 0 | 0 | 0 | 0 | 0 | 0 | 0 |
| | | | | | | | | | | | | | | |
| 0 | | <.01 | | | <1 | 0 | 0 | 0 | 0 | 0 | 0 | 0 | 0 | 0 |
| 0 | 0 | 0 | | | 0 | | 0 | | | | | | | 0 |

**Table H-1**

**Food Composition** (Computer code number is for West Diet Analysis program) (For purposes of calculations, use "0" for t, <1, <.1, <.01, etc.)

| Computer Code Number | Food Description | Measure | Wt (g) | $H_2O$ (%) | Ener (kcal) | Prot (g) | Carb (g) | Dietary Fiber (g) | Fat (g) | Fat Breakdown (g) Sat | Mono | Poly |
|---|---|---|---|---|---|---|---|---|---|---|---|---|
| | **SWEETENERS and SWEETS—**Continued | | | | | | | | | | | |
| | Syrups, chocolate: | | | | | | | | | | | |
| 785 | Hot fudge type | 2 tbs | 43 | 22 | 149 | 2 | 27 | 1 | 4 | 2.4 | 1.6 | 1.4 |
| 784 | Thin type | 2 tbs | 38 | 29 | 93 | 1 | 25 | 1 | <1 | .3 | .2 | t |
| 786 | Molasses, blackstrap | 2 tbs | 41 | 29 | 96 | 0 | 25 | 0 | 0 | 0 | 0 | 0 |
| 1710 | Light cane syrup | 2 tbs | 41 | 24 | 103 | 0 | 27 | 0 | 0 | 0 | 0 | 0 |
| 787 | Pancake table syrup (corn and maple) | 2 tbs | 40 | 24 | 115 | 0 | 30 | 0 | 0 | 0 | 0 | 0 |
| | **VEGETABLES and LEGUMES** | | | | | | | | | | | |
| 788 | Alfalfa sprouts | 1 c | 33 | 91 | 10 | 1 | 1 | 1 | <1 | t | .1 | t |
| 1815 | Amaranth leaves, raw, chopped | 1 c | 28 | 92 | 7 | 1 | 1 | <1 | <1 | t | t | t |
| 1816 | Amaranth leaves, raw, each | 1 ea | 14 | 92 | 4 | <1 | 1 | <1 | <1 | t | t | t |
| 1817 | Amaranth leaves, cooked | 1 c | 132 | 91 | 28 | 3 | 5 | 2 | <1 | .1 | .1 | .1 |
| 1987 | Arugula, raw, chopped | ½ c | 10 | 92 | 2 | <1 | <1 | <1 | <1 | t | t | t |
| 789 | Artichokes, cooked globe (300 g with refuse) | 1 ea | 120 | 84 | 60 | 4 | 13 | 6 | <1 | t | t | .1 |
| 1177 | Artichoke hearts, cooked from frozen | 1 c | 168 | 86 | 76 | 5 | 15 | 8 | 1 | .2 | t | .4 |
| 1176 | Artichoke hearts, marinated | 1 c | 130 | 81 | 128 | 3 | 10 | 6 | 10 | 1.5 | 2.3 | 5.9 |
| 2021 | Artichoke hearts, in water | ½ c | 100 | 91 | 37 | 2 | 6 | 0 | 0 | 0 | 0 | 0 |
| | Asparagus, green, cooked: | | | | | | | | | | | |
| | From fresh: | | | | | | | | | | | |
| 790 | Cuts and tips | ½ c | 90 | 92 | 22 | 2 | 4 | 1 | <1 | .1 | t | .1 |
| 791 | Spears, ½" diam at base | 4 ea | 60 | 92 | 14 | 2 | 3 | 1 | <1 | t | t | .1 |
| | From frozen: | | | | | | | | | | | |
| 792 | Cuts and tips | ½ c | 90 | 91 | 25 | 3 | 4 | 1 | <1 | .1 | t | .2 |
| 793 | Spears, ½" diam at base | 4 ea | 60 | 91 | 17 | 2 | 3 | 1 | <1 | .1 | t | .1 |
| 794 | Canned, spears, ½" diam at base | 4 ea | 72 | 94 | 14 | 2 | 2 | 1 | <1 | .1 | t | .2 |
| 795 | Bamboo shoots, canned, drained slices | 1 c | 131 | 94 | 25 | 2 | 4 | 2 | 1 | .1 | t | .2 |
| 1795 | Bamboo shoots, raw slices | 1 c | 151 | 91 | 41 | 4 | 8 | 3 | <1 | .1 | t | .2 |
| 1798 | Bamboo shoots, cooked slices | 1 c | 120 | 96 | 14 | 2 | 2 | 1 | <1 | .1 | t | .1 |
| | Beans (see also alphabetical listing this section): | | | | | | | | | | | |
| 1990 | Adzuki beans, cooked | ½ c | 115 | 66 | 147 | 9 | 28 | 1 | <1 | t | t | t |
| 796 | Black beans, cooked | ½ c | 86 | 66 | 114 | 8 | 20 | 7 | <1 | .1 | t | .2 |
| | Canned beans (white/navy): | | | | | | | | | | | |
| 803 | With pork and tomato sauce | ½ c | 127 | 73 | 124 | 7 | 25 | 6 | 1 | .5 | .6 | .2 |
| 804 | With sweet sauce | ½ c | 130 | 71 | 144 | 7 | 27 | 7 | 2 | .7 | .8 | .2 |
| 805 | With frankfurters | ½ c | 130 | 69 | 185 | 9 | 20 | 9 | 9 | 3.1 | 3.7 | 1.1 |
| | Lima beans: | | | | | | | | | | | |
| 797 | Thick seeded (Fordhooks), cooked from frozen | ½ c | 85 | 73 | 85 | 5 | 16 | 5 | <1 | .1 | t | .1 |
| 798 | Thin seeded (Baby), cooked from frozen | ½ c | 90 | 72 | 94 | 6 | 18 | 5 | <1 | .1 | t | .1 |
| 799 | Cooked from dry, drained | ½ c | 94 | 70 | 108 | 7 | 20 | 7 | <1 | .1 | t | .2 |
| 1998 | Red Mexican, cooked f/dry | ½ c | 112 | 70 | 126 | 8 | 23 | 9 | <1 | .1 | .1 | .2 |
| | Snap bean/green string beans cuts and french style: | | | | | | | | | | | |
| 800 | Cooked from fresh | ½ c | 63 | 89 | 22 | 1 | 5 | 2 | <1 | t | t | .1 |
| 801 | Cooked from frozen | ½ c | 68 | 91 | 19 | 1 | 4 | 2 | <1 | t | t | .1 |
| 802 | Canned, drained | ½ c | 68 | 93 | 14 | 1 | 3 | 1 | <1 | t | t | t |
| 1713 | Snap bean, yellow, cooked f/fresh | ½ c | 63 | 89 | 22 | 1 | 5 | 2 | <1 | t | t | .1 |
| | Bean sprouts (mung): | | | | | | | | | | | |
| 806 | Raw | ½ c | 52 | 90 | 16 | 2 | 3 | 1 | <1 | t | t | t |
| 807 | Cooked, stir fried | ½ c | 62 | 84 | 31 | 3 | 7 | 1 | <1 | t | t | t |
| 808 | Cooked, boiled, drained | ½ c | 62 | 93 | 13 | 1 | 3 | <1 | <1 | t | t | t |
| 1788 | Canned, drained | ½ c | 63 | 96 | 8 | 1 | 1 | <1 | <1 | t | t | t |
| | Beets, cooked from fresh: | | | | | | | | | | | |
| 809 | Sliced or diced | ½ c | 85 | 87 | 37 | 1 | 8 | 2 | <1 | t | t | .1 |
| 810 | Whole beets, 2" diam | 2 ea | 100 | 87 | 44 | 2 | 10 | 2 | <1 | t | t | .1 |
| | Beets, canned: | | | | | | | | | | | |
| 811 | Sliced or diced | ½ c | 79 | 91 | 24 | 1 | 6 | 1 | <1 | t | t | t |
| 812 | Pickled slices | ½ c | 114 | 82 | 74 | 1 | 19 | 2 | <1 | t | t | t |
| 813 | Beet greens, cooked, drained | ½ c | 72 | 89 | 19 | 2 | 4 | 2 | <1 | t | t | .1 |

**PAGE KEY:** H–4 = Beverages H–6 = Dairy H–10 = Eggs H–10 = Fat/Oil H–14 = Fruit H–20 = Bakery H–28 = Grain H–32 = Fish H–34 = Meats H–38 = Poultry H–40 = Sausage H–42 = Mixed/Fast H–46 = Nuts/Seeds H–50 = Sweets H–52 = Vegetables/Legumes H–62 = Vegetarian Foods H–64 = Misc H–66 = Soups/Sauces H–68 = Fast H–84 = Convenience H–88 = Baby foods

| Chol (mg) | Calc (mg) | Iron (mg) | Magn (mg) | Pota (mg) | Sodi (mg) | Zinc (mg) | VT-A (RE) | Thia (mg) | VT-E (a-TE) | Ribo (mg) | Niac (mg) | V-B6 (mg) | Fola (µg) | VT-C (mg) |
|---|---|---|---|---|---|---|---|---|---|---|---|---|---|---|
| 5 | 43 | .52 | 21 | 92 | 56 | .34 | 9 | .01 | 0 | .09 | .09 | .01 | 2 | <1 |
| 0 | 5 | 5.17 | 25 | 183 | 58 | .28 | 494 | <.01 | .01 | .31 | 12.8 | .01 | 2 | <1 |
| 0 | 353 | 7.18 | 88 | 1021 | 23 | .41 | 0 | .01 | 0 | .02 | .44 | .29 | <1 | 0 |
| 0 | 68 | 1.76 | 100 | 376 | 6 | .12 | 0 | .03 | 0 | .02 | .08 | .27 | 0 | 0 |
| 0 | | .04 | 1 | 1 | 33 | .02 | 0 | <.01 | 0 | <.01 | .01 | 0 | 0 | 0 |
| 0 | 11 | .32 | 9 | 26 | 2 | .3 | 5 | .02 | — | .04 | .16 | .01 | 12 | 3 |
| 0 | 60 | .65 | 15 | 171 | 6 | .25 | 82 | .01 | .22 | .04 | .18 | .05 | 24 | 12 |
| 0 | 30 | .32 | 8 | 85 | 3 | .13 | 41 | <.01 | .11 | .02 | .09 | .03 | 12 | 6 |
| 0 | 276 | 2.98 | 73 | 846 | 28 | 1.16 | 366 | .03 | .66 | .18 | .74 | .23 | 75 | 54 |
| 0 | 16 | .15 | 5 | 37 | 3 | .05 | 24 | <.01 | .04 | .01 | .03 | .01 | 10 | 1 |
| 0 | 54 | 1.55 | 72 | 425 | 114 | .59 | 22 | .08 | .23 | .08 | 1.2 | .13 | 61 | 12 |
| 0 | 35 | .94 | 52 | 444 | 89 | .6 | 27 | .1 | .32 | .26 | 1.54 | .15 | 200 | 8 |
| 0 | 30 | 1.24 | 37 | 335 | 688 | .41 | 21 | .05 | 1.43 | .13 | 1.06 | .11 | 114 | 40 |
| 0 | 0 | 1.35 | | 0 | 250 | | 12 | | | | | | | 4 |
| 0 | 18 | .66 | 9 | 144 | 10 | .38 | 49 | .11 | .34 | .11 | .97 | .11 | 131 | 10 |
| 0 | 12 | .44 | 6 | 96 | 7 | .25 | 32 | .07 | .23 | .08 | .65 | .07 | 88 | 6 |
| 0 | 21 | .58 | 12 | 196 | 4 | .5 | 74 | .06 | 1.13 | .09 | .94 | .02 | 122 | 22 |
| 0 | 14 | .38 | 8 | 131 | 2 | .34 | 49 | .04 | .75 | .06 | .62 | .01 | 81 | 15 |
| 0 | 11 | 1.32 | 7 | 124 | 207 | .29 | 38 | .04 | .31 | .07 | .69 | .08 | 69 | 13 |
| 0 | 10 | .42 | 5 | 105 | 9 | .85 | 1 | .03 | .5 | .03 | .18 | .18 | 4 | 1 |
| 0 | 20 | .75 | 5 | 805 | 6 | 1.66 | 3 | .23 | 1.51 | .11 | .91 | .36 | 11 | 6 |
| 0 | 14 | .29 | 4 | 640 | 5 | .56 | 0 | .02 | .8 | .06 | .36 | .12 | 3 | 1 |
| 0 | 32 | 2.3 | 60 | 612 | 9 | 2.04 | 1 | .13 | .11 | .07 | .82 | .11 | 139 | 0 |
| 0 | 23 | 1.81 | 60 | 305 | 1 | .96 | 1 | .21 | .07 | .05 | .43 | .06 | 128 | 0 |
| 9 | 71 | 4.17 | 44 | 381 | 559 | 7.44 | 15 | .07 | .69 | .06 | .63 | .09 | 29 | 4 |
| 9 | 79 | 2.16 | 44 | 346 | 437 | 1.95 | 14 | .06 | .7 | .08 | .46 | .11 | 49 | 4 |
| 8 | 62 | 2.25 | 36 | 306 | 559 | 2.43 | 19 | .07 | .61 | .07 | 1.17 | .06 | 39 | 3 |
| 0 | 19 | 1.16 | 29 | 347 | 45 | .37 | 16 | .06 | .25 | .05 | .91 | .1 | 18 | 11 |
| 0 | 25 | 1.76 | 50 | 370 | 26 | .49 | 15 | .06 | .58 | .05 | .69 | .1 | 14 | 5 |
| 0 | 16 | 2.25 | 40 | 478 | 2 | .89 | 0 | .15 | .17 | .05 | .4 | .15 | 78 | 0 |
| 0 | 42 | 1.86 | 48 | 369 | 240 | .87 | <1 | .13 | .08 | .07 | .37 | .11 | 94 | 2 |
| 0 | 29 | .81 | 16 | 188 | 2 | .23 | 42 | .05 | .09 | .06 | .39 | .03 | 21 | 6 |
| 0 | 33 | .6 | 16 | 86 | 6 | .33 | 27 | .02 | .09 | .06 | .26 | .04 | 16 | 3 |
| 0 | 18 | .61 | 9 | 74 | 178 | .2 | 24 | .01 | .09 | .04 | .14 | .02 | 22 | 3 |
| 0 | 29 | .81 | 16 | 188 | 2 | .23 | 5 | .05 | .18 | .06 | .39 | .03 | 21 | 6 |
| 0 | 7 | .47 | 11 | 77 | 3 | .21 | 1 | .04 | .02 | .06 | .39 | .05 | 32 | 7 |
| 0 | 8 | 1.18 | 20 | 136 | 6 | .56 | 2 | .09 | .01 | .11 | .74 | .08 | 43 | 10 |
| 0 | 7 | .4 | 9 | 63 | 6 | .29 | 1 | .03 | .01 | .06 | .51 | .03 | 18 | 7 |
| 0 | 9 | .27 | 6 | 17 | 88 | .18 | 1 | .02 | .01 | .04 | .14 | .02 | 6 | <1 |
| 0 | 14 | .67 | 20 | 259 | 65 | .3 | 3 | .02 | .25 | .03 | .28 | .06 | 68 | 3 |
| 0 | 16 | .79 | 23 | 305 | 77 | .35 | 4 | .03 | .3 | .04 | .33 | .07 | 80 | 4 |
| 0 | 12 | 1.44 | 13 | 117 | 153 | .17 | 1 | .01 | .24 | .03 | .12 | .04 | 24 | 3 |
| 0 | 12 | .47 | 17 | 169 | 301 | .3 | 1 | .01 | .15 | .05 | .29 | .06 | 30 | 3 |
| 0 | 82 | 1.37 | 49 | 654 | 174 | .36 | 367 | .08 | .22 | .21 | .36 | .09 | 10 | 18 |

**Table H-1**

**Food Composition** (Computer code number is for West Diet Analysis program) (For purposes of calculations, use "0" for t, <1, <.1, <.01, etc.)

| Computer Code Number | Food Description | Measure | Wt (g) | $H_2O$ (%) | Ener (kcal) | Prot (g) | Carb (g) | Dietary Fiber (g) | Fat (g) | Fat Breakdown (g) Sat | Mono | Poly |
|---|---|---|---|---|---|---|---|---|---|---|---|---|
| | **VEGETABLES and LEGUMES**—Continued | | | | | | | | | | | |
| | Broccoli, raw: | | | | | | | | | | | |
| 817 | Chopped | ½ c | 44 | 91 | 12 | 1 | 2 | 1 | <1 | t | t | .1 |
| 818 | Spears | 1 ea | 31 | 91 | 9 | 1 | 2 | 1 | <1 | t | t | .1 |
| | Broccoli, cooked from fresh: | | | | | | | | | | | |
| 819 | Spears | 1 ea | 180 | 91 | 50 | 5 | 9 | 5 | 1 | .1 | t | .3 |
| 820 | Chopped | ½ c | 78 | 91 | 22 | 2 | 4 | 2 | <1 | t | t | .1 |
| | Broccoli, cooked from frozen: | | | | | | | | | | | |
| 821 | Spear, small piece | ½ c | 92 | 91 | 26 | 3 | 5 | 3 | <1 | t | t | .1 |
| 822 | Chopped | ½ c | 92 | 91 | 26 | 3 | 5 | 3 | <1 | t | t | .1 |
| 1603 | Broccoflower, steamed | ½ c | 78 | 90 | 25 | 2 | 5 | 2 | <1 | t | t | .1 |
| 823 | Brussels sprouts, cooked from fresh | ½ c | 78 | 87 | 30 | 2 | 7 | 2 | <1 | .1 | t | .2 |
| 824 | Brussels sprouts, cooked from frozen | ½ c | 78 | 87 | 33 | 3 | 6 | 3 | <1 | .1 | t | .2 |
| | Cabbage, common varieties: | | | | | | | | | | | |
| 825 | Raw, shredded or chopped | 1 c | 70 | 92 | 17 | 1 | 4 | 2 | <1 | t | t | .1 |
| 826 | Cooked, drained | 1 c | 150 | 94 | 33 | 2 | 7 | 3 | 1 | .1 | t | .3 |
| | Cabbage, Chinese: | | | | | | | | | | | |
| 1178 | Bok choy, raw, shredded | 1 c | 70 | 95 | 9 | 1 | 2 | 1 | <1 | t | t | .1 |
| 827 | Bok choy, cooked, drained | 1 c | 170 | 96 | 20 | 3 | 3 | 3 | <1 | t | t | .1 |
| 1937 | Kim chee style | 1 c | 150 | 92 | 31 | 2 | 6 | 2 | <1 | t | t | .2 |
| 828 | Pe Tsai, raw, chopped | 1 c | 76 | 94 | 12 | 1 | 2 | 2 | <1 | t | t | .1 |
| 1796 | Pe Tsai, cooked | 1 c | 119 | 95 | 17 | 2 | 3 | 3 | <1 | t | t | .1 |
| | Cabbage, red, coarsely chopped: | | | | | | | | | | | |
| 829 | Raw | 1 c | 89 | 92 | 24 | 1 | 5 | 2 | <1 | t | t | .1 |
| 830 | Cooked, drained | 1 c | 150 | 94 | 31 | 2 | 7 | 3 | <1 | t | t | .1 |
| 831 | Cabbage, savoy, coarsely chopped, raw | 1 c | 70 | 91 | 19 | 1 | 4 | 2 | <1 | t | t | t |
| 1785 | Cabbage, savoy, cooked | 1 c | 145 | 92 | 35 | 3 | 8 | 4 | <1 | t | t | .1 |
| 1896 | Capers | 1 ea | 5 | 86 | <1 | <1 | <1 | <1 | <1 | | | |
| | Carrots, raw: | | | | | | | | | | | |
| 832 | Whole, 7½ x 1⅛" | 1 ea | 72 | 88 | 31 | 1 | 7 | 2 | <1 | t | t | .1 |
| 833 | Grated | ½ c | 55 | 88 | 24 | 1 | 6 | 2 | <1 | t | t | t |
| | Carrots, cooked, sliced, drained: | | | | | | | | | | | |
| 834 | From raw | ½ c | 78 | 87 | 35 | 1 | 8 | 3 | <1 | t | t | .1 |
| 835 | From frozen | ½ c | 73 | 90 | 26 | 1 | 6 | 3 | <1 | t | t | t |
| 836 | Carrots, canned, sliced, drained | ½ c | 73 | 93 | 17 | <1 | 4 | 1 | <1 | t | t | .1 |
| 837 | Carrot juice, canned | 1 c | 236 | 89 | 94 | 2 | 22 | 2 | <1 | .1 | t | .2 |
| | Cauliflower, flowerets: | | | | | | | | | | | |
| 838 | Raw | ½ c | 50 | 92 | 12 | 1 | 3 | 1 | <1 | t | t | t |
| 839 | Cooked from fresh, drained | ½ c | 62 | 93 | 14 | 1 | 3 | 2 | <1 | t | t | .1 |
| 840 | Cooked, from frozen, drained | ½ c | 90 | 94 | 17 | 1 | 3 | 2 | <1 | t | t | .1 |
| | Celery, pascal type, raw: | | | | | | | | | | | |
| 841 | Large outer stalk, 8 x 1½"(root end) | 1 ea | 40 | 95 | 6 | <1 | 1 | 1 | <1 | t | t | t |
| 842 | Diced | 1 c | 120 | 95 | 19 | 1 | 4 | 2 | <1 | t | t | .1 |
| 1789 | Celeriac/celery root, cooked | 1 c | 155 | 92 | 39 | 1 | 9 | 2 | <1 | .1 | .1 | .2 |
| 1179 | Chard, swiss, raw, chopped | 1 c | 36 | 93 | 7 | 1 | 1 | 1 | <1 | t | t | t |
| 1180 | Chard, swiss, cooked | 1 c | 175 | 93 | 35 | 3 | 7 | 4 | <1 | t | t | t |
| 1855 | Chayote fruit, raw | 1 ea | 203 | 94 | 39 | 2 | 9 | 3 | <1 | .1 | t | .1 |
| 1856 | Chayote fruit, cooked | 1 c | 160 | 93 | 38 | 1 | 8 | 4 | 1 | .2 | .1 | .3 |
| | Chickpeas (see Garbanzo Beans #854) | | | | | | | | | | | |
| | Collards, cooked, drained: | | | | | | | | | | | |
| 843 | From raw | ½ c | 95 | 92 | 25 | 2 | 5 | 3 | <1 | t | t | .1 |
| 844 | From frozen | ½ c | 85 | 88 | 31 | 3 | 6 | 3 | <1 | .1 | t | .2 |
| | Corn, yellow, cooked, drained: | | | | | | | | | | | |
| 845 | From raw, on cob, 5" long | 1 ea | 77 | 73 | 72 | 2 | 17 | 2 | 1 | .1 | .2 | .3 |
| 846 | From frozen, on cob, 3½" long | 1 ea | 63 | 73 | 59 | 2 | 14 | 2 | <1 | .1 | .1 | .2 |
| 847 | Kernels, cooked from frozen | ½ c | 82 | 77 | 66 | 2 | 16 | 2 | <1 | .1 | .1 | .2 |
| | Corn, canned: | | | | | | | | | | | |
| 848 | Cream style | ½ c | 128 | 79 | 92 | 2 | 23 | 2 | 1 | .1 | .2 | .3 |
| 849 | Whole kernel, vacuum pack | ½ c | 105 | 77 | 83 | 3 | 20 | 2 | 1 | .1 | .2 | .2 |
| | Cowpeas (see Black-eyed peas #814-816) | | | | | | | | | | | |
| 850 | Cucumber slices with peel | 7 pce | 28 | 96 | 4 | <1 | 1 | <1 | <1 | t | t | t |
| 1948 | Cucumber, kim chee style | 1 c | 150 | 91 | 31 | 2 | 7 | 2 | <1 | t | 0 | .1 |

| Chol (mg) | Calc (mg) | Iron (mg) | Magn (mg) | Pota (mg) | Sodi (mg) | Zinc (mg) | VT-A (RE) | Thia (mg) | VT-E (a-TE) | Ribo (mg) | Niac (mg) | V-B6 (mg) | Fola (μg) | VT-C (mg) |
|---|---|---|---|---|---|---|---|---|---|---|---|---|---|---|
| 0 | 21 | .39 | 11 | 143 | 12 | .18 | 68 | .03 | .73 | .05 | .28 | .07 | 31 | 41 |
| 0 | 15 | .27 | 8 | 101 | 8 | .12 | 48 | .02 | .51 | .04 | .2 | .05 | 22 | 29 |
| 0 | 83 | 1.51 | 43 | 526 | 47 | .68 | 250 | .1 | 3.04 | .2 | 1.03 | .26 | 90 | 134 |
| 0 | 36 | .65 | 19 | 228 | 20 | .3 | 108 | .04 | 1.32 | .09 | .45 | .11 | 39 | 58 |
| 0 | 47 | .56 | 18 | 166 | 22 | .28 | 174 | .05 | .95 | .07 | .42 | .12 | 28 | 37 |
| 0 | 47 | .56 | 18 | 166 | 22 | .28 | 174 | .05 | 1.52 | .07 | .42 | .12 | 52 | 37 |
| 0 | 25 | .55 | 16 | 251 | 18 | .39 | 5 | .06 | .23 | .07 | .59 | .14 | 38 | 49 |
| 0 | 28 | .94 | 16 | 247 | 16 | .26 | 56 | .08 | .66 | .06 | .47 | .14 | 47 | 48 |
| 0 | 19 | .58 | 19 | 254 | 18 | .28 | 46 | .08 | .45 | .09 | .42 | .22 | 79 | 36 |
| 0 | 33 | .41 | 10 | 172 | 13 | .13 | 9 | .03 | .07 | .03 | .21 | .07 | 30 | 22 |
| 0 | 46 | .25 | 12 | 146 | 12 | .13 | 19 | .09 | .16 | .08 | .42 | .17 | 30 | 30 |
| 0 | 73 | .56 | 13 | 176 | 45 | .13 | 210 | .03 | .08 | .05 | .35 | .14 | 46 | 31 |
| 0 | 158 | 1.77 | 19 | 631 | 58 | .29 | 437 | .05 | .2 | .11 | .73 | .28 | 69 | 44 |
| 0 | 145 | 1.28 | 27 | 375 | 995 | .35 | 426 | .07 | .24 | .1 | .75 | .34 | 88 | 80 |
| 0 | 58 | .24 | 10 | 181 | 7 | .17 | 91 | .03 | .09 | .04 | .3 | .18 | 60 | 20 |
| 0 | 38 | .36 | 12 | 268 | 11 | .21 | 115 | .05 | .14 | .05 | .59 | .21 | 63 | 19 |
| 0 | 45 | .44 | 13 | 183 | 10 | .19 | 4 | .04 | .09 | .03 | .27 | .19 | 18 | 51 |
| 0 | 55 | .52 | 16 | 210 | 12 | .22 | 4 | .05 | .18 | .03 | .3 | .21 | 19 | 52 |
| 0 | 24 | .28 | 20 | 161 | 20 | .19 | 70 | .05 | .07 | .02 | .21 | .13 | 56 | 22 |
| 0 | 43 | .55 | 35 | 267 | 35 | .33 | 129 | .07 | .15 | .03 | .03 | .22 | 67 | 25 |
| 0 | 2 | .05 |  |  | 105 |  | 1 |  |  |  |  |  |  | 0 |
| 0 | 19 | .36 | 11 | 233 | 25 | .14 | 2025 | .07 | .33 | .04 | .67 | .11 | 10 | 7 |
| 0 | 15 | .27 | 8 | 178 | 19 | .11 | 1547 | .05 | .25 | .03 | .51 | .08 | 8 | 5 |
| 0 | 24 | .48 | 10 | 177 | 51 | .23 | 1914 | .03 | .33 | .04 | .39 | .19 | 11 | 2 |
| 0 | 20 | .34 | 7 | 115 | 43 | .17 | 1292 | .02 | .31 | .03 | .32 | .09 | 8 | 2 |
| 0 | 18 | .47 | 6 | 131 | 177 | .19 | 1005 | .01 | .31 | .02 | .4 | .08 | 7 | 2 |
| 0 | 57 | 1.09 | 33 | 689 | 68 | .42 | 2584 | .22 | .02 | .13 | .91 | .51 | 9 | 20 |
| 0 | 11 | .22 | 7 | 152 | 15 | .14 | 1 | .03 | .02 | .03 | .26 | .11 | 28 | 23 |
| 0 | 10 | .2 | 6 | 88 | 9 | .11 | 1 | .03 | .02 | .03 | .25 | .11 | 27 | 27 |
| 0 | 15 | .37 | 8 | 125 | 16 | .12 | 2 | .03 | .04 | .05 | .28 | .08 | 37 | 28 |
| 0 | 16 | .16 | 4 | 115 | 35 | .05 | 5 | .02 | .14 | .02 | .13 | .03 | 11 | 3 |
| 0 | 48 | .48 | 13 | 344 | 104 | .16 | 16 | .05 | .43 | .05 | .39 | .1 | 34 | 8 |
| 0 | 40 | .67 | 19 | 268 | 95 | .31 | 0 | .04 | .31 | .06 | .66 | .16 | 5 | 6 |
| 0 | 18 | .65 | 29 | 136 | 77 | .13 | 119 | .01 | .68 | .03 | .14 | .04 | 5 | 11 |
| 0 | 102 | 3.96 | 151 | 961 | 313 | .58 | 550 | .06 | 3.31 | .15 | .63 | .15 | 15 | 31 |
| 0 | 34 | .69 | 24 | 254 | 4 | 1.5 | 12 | .05 | .24 | .06 | .95 | .15 | 189 | 16 |
| 0 | 21 | .35 | 19 | 277 | 2 | .5 | 8 | .04 | .19 | .06 | .67 | .19 | 29 | 13 |
| 0 | 113 | .44 | 16 | 247 | 9 | .4 | 297 | .04 | .84 | .1 | .55 | .12 | 88 | 17 |
| 0 | 179 | .95 | 25 | 213 | 42 | .23 | 508 | .04 | .42 | .1 | .54 | .1 | 65 | 22 |
| 0 | 2 | .47 | 22 | 193 | 3 | .48 | 16 | .13 | .07 | .05 | 1.17 | .17 | 23 | 4 |
| 0 | 2 | .38 | 18 | 158 | 3 | .4 | 13 | .11 | .06 | .04 | .96 | .14 | 19 | 3 |
| 0 | 3 | .29 | 16 | 121 | 4 | .33 | 18 | .07 | .07 | .06 | 1.07 | .11 | 25 | 3 |
| 0 | 4 | .49 | 22 | 172 | 365 | .68 | 13 | .03 | .11 | .07 | 1.23 | .08 | 57 | 6 |
| 0 | 5 | .44 | 24 | 195 | 286 | .48 | 25 | .04 | .09 | .08 | 1.23 | .06 | 52 | 9 |
| 0 | 4 | .07 | 3 | 40 | 1 | .06 | 6 | .01 | .02 | .01 | .06 | .01 | 4 | 1 |
| 0 | 13 | 7.23 | 12 | 176 | 1531 | .76 | 49 | .04 | .24 | .04 | .69 | .16 | 34 | 5 |

**Table H-1**

**Food Composition** (Computer code number is for West Diet Analysis program) (For purposes of calculations, use "0" for t, <1, <.1, <.01, etc.)

| Computer Code Number | Food Description | Measure | Wt (g) | $H_2O$ (%) | Ener (kcal) | Prot (g) | Carb (g) | Dietary Fiber (g) | Fat (g) | Fat Breakdown (g) Sat | Mono | Poly |
|---|---|---|---|---|---|---|---|---|---|---|---|---|
| | **VEGETABLES and LEGUMES**—Continued | | | | | | | | | | | |
| | Dandelion Greens: | | | | | | | | | | | |
| 851 | Raw | 1 c | 55 | 86 | 25 | 1 | 5 | 2 | <1 | .1 | t | .2 |
| 852 | Chopped, cooked, drained | 1 c | 105 | 90 | 35 | 2 | 7 | 3 | 1 | .2 | t | .3 |
| 853 | Eggplant, cooked | 1 c | 99 | 92 | 28 | 1 | 7 | 2 | <1 | t | t | .1 |
| 1714 | Endive, fresh, chopped | 1 c | 50 | 94 | 8 | 1 | 2 | 2 | <1 | t | t | t |
| 856 | Escarole/curly endive, chopped | 1 c | 50 | 94 | 8 | 1 | 2 | 2 | <1 | t | t | t |
| 854 | Garbanzo beans (chickpeas), cooked | 1 c | 164 | 60 | 269 | 14 | 45 | 12 | 4 | .4 | 1 | 1.9 |
| 1939 | Grape leaf, raw: | 1 ea | 3 | 73 | 3 | <1 | 1 | <1 | <1 | t | t | t |
| 7914 | Cup | 1 c | 14 | 73 | 13 | 1 | 2 | 2 | <1 | t | t | .1 |
| 855 | Great northern beans, cooked | 1 c | 177 | 69 | 209 | 15 | 37 | 12 | 1 | .2 | t | .3 |
| 857 | Jerusalem artichoke, raw slices | 1 c | 150 | 78 | 114 | 3 | 26 | 2 | <1 | 0 | t | t |
| 1794 | Jicama | 1 c | 120 | 90 | 46 | 1 | 11 | 6 | <1 | t | t | t |
| | Kale, cooked, drained: | | | | | | | | | | | |
| 858 | From raw | 1 c | 130 | 91 | 36 | 2 | 7 | 3 | 1 | .1 | t | .3 |
| 859 | From frozen | 1 c | 130 | 90 | 39 | 4 | 7 | 3 | 1 | .1 | t | .3 |
| 860 | Kidney beans, canned | 1 c | 256 | 77 | 218 | 13 | 40 | 16 | 1 | .1 | .1 | .5 |
| 1181 | Kohlrabi, raw slices | 1 c | 135 | 91 | 36 | 2 | 8 | 5 | <1 | t | t | .1 |
| 861 | Kohlrabi, cooked | 1 c | 165 | 90 | 48 | 3 | 11 | 2 | <1 | t | t | .1 |
| 1183 | Leeks, raw, chopped | 1 c | 89 | 83 | 54 | 1 | 13 | 2 | <1 | t | t | .1 |
| 1182 | Leeks, cooked, chopped | 1 c | 104 | 91 | 32 | 1 | 8 | 1 | <1 | t | t | .1 |
| 862 | Lentils, cooked from dry | 1 c | 198 | 70 | 230 | 18 | 40 | 16 | 1 | .1 | .1 | .3 |
| 1288 | Lentils, sprouted, stir-fried | 1 c | 124 | 69 | 125 | 11 | 26 | 5 | 1 | .1 | .1 | .2 |
| 1289 | Lentils, sprouted, raw | 1 c | 77 | 67 | 82 | 7 | 17 | 3 | <1 | t | .1 | .2 |
| | Lettuce: | | | | | | | | | | | |
| | Butterhead/Boston types: | | | | | | | | | | | |
| 863 | Head, 5″ diameter | ¼ ea | 41 | 96 | 5 | 1 | 1 | <1 | <1 | t | t | t |
| 864 | Leaves, inner or outer | 4 ea | 30 | 96 | 4 | <1 | 1 | <1 | <1 | t | t | t |
| | Iceberg/crisphead: | | | | | | | | | | | |
| 867 | Chopped or shredded | 1 c | 55 | 96 | 7 | 1 | 1 | 1 | <1 | t | t | .1 |
| 865 | Head, 6″ diameter | 1 ea | 539 | 96 | 65 | 5 | 11 | 8 | 1 | .1 | t | .5 |
| 866 | Wedge, ¼ head | 1 ea | 135 | 96 | 16 | 1 | 3 | 2 | <1 | t | t | .1 |
| 868 | Looseleaf, chopped | ½ c | 28 | 94 | 5 | <1 | 1 | 1 | <1 | t | t | t |
| 869 | Romaine, chopped | ½ c | 28 | 95 | 4 | <1 | 1 | <1 | <1 | t | t | t |
| 870 | Romaine, inner leaf | 3 pce | 30 | 95 | 4 | <1 | 1 | 1 | <1 | t | t | t |
| 1930 | Luffa, cooked (Chinese okra) | 1 c | 178 | 89 | 57 | 3 | 13 | 6 | <1 | .1 | t | .1 |
| | Mushrooms: | | | | | | | | | | | |
| 871 | Raw, sliced | ½ c | 35 | 92 | 9 | 1 | 2 | <1 | <1 | t | t | .1 |
| 872 | Cooked from fresh, pieces | ½ c | 78 | 91 | 21 | 2 | 4 | 2 | <1 | t | t | .1 |
| 1962 | Stir fried, shitake slices | ½ c | 73 | 83 | 40 | 1 | 10 | 2 | <1 | t | t | t |
| 873 | Canned, drained | ½ c | 78 | 91 | 19 | 1 | 4 | 2 | <1 | t | t | .1 |
| 1951 | Mushroom caps, pickled | 8 ea | 47 | 92 | 11 | 1 | 2 | 1 | <1 | t | t | .1 |
| | Mustard greens: | | | | | | | | | | | |
| 874 | Cooked from raw | ½ c | 70 | 94 | 10 | 2 | 1 | 1 | <1 | t | .1 | t |
| 875 | Cooked from frozen | ½ c | 75 | 94 | 14 | 2 | 2 | 2 | <1 | t | .1 | t |
| 876 | Navy beans, cooked from dry | 1 c | 182 | 63 | 258 | 16 | 48 | 12 | 1 | .3 | .1 | .4 |
| | Okra, cooked: | | | | | | | | | | | |
| 877 | From fresh pods | 8 ea | 85 | 90 | 27 | 2 | 6 | 2 | <1 | t | t | t |
| 878 | From frozen slices | 1 c | 184 | 91 | 51 | 4 | 11 | 5 | 1 | .1 | .1 | .1 |
| 1236 | Batter fried from fresh | 1 c | 92 | 69 | 175 | 3 | 11 | 2 | 13 | 2.1 | 3.4 | 7.1 |
| 1930 | Chinese, (Luffa), cooked | 1 c | 178 | 89 | 57 | 3 | 13 | 6 | <1 | .1 | t | .1 |
| | Onions: | | | | | | | | | | | |
| 879 | Raw, chopped | ½ c | 80 | 90 | 30 | 1 | 7 | 1 | <1 | t | t | t |
| 880 | Raw, sliced | ½ c | 58 | 90 | 22 | 1 | 5 | 1 | <1 | t | t | t |
| 881 | Cooked, drained, chopped | ½ c | 105 | 88 | 46 | 1 | 11 | 1 | <1 | t | t | .1 |
| 882 | Dehydrated flakes | ¼ c | 14 | 4 | 49 | 1 | 12 | 1 | <1 | t | t | t |
| 1934 | Onions, pearl, cooked | ½ c | 93 | 87 | 41 | 1 | 9 | 1 | <1 | t | t | .1 |
| 883 | Spring/green onions, bulb and top, chopped | ½ c | 50 | 90 | 16 | 1 | 4 | 1 | <1 | t | t | t |
| 884 | Onion rings, breaded, heated f/frozen | 2 ea | 20 | 28 | 81 | 1 | 8 | <1 | 5 | 1.7 | 2.2 | 1 |
| 1917 | Palm hearts, cooked slices | 1 c | 146 | 69 | 150 | 4 | 39 | 2 | <1 | .1 | .1 | t |
| 885 | Parsley, raw, chopped | ½ c | 30 | 88 | 11 | 1 | 2 | 1 | <1 | t | .1 | t |
| 888 | Parsnips, sliced, cooked | ½ c | 78 | 78 | 63 | 1 | 15 | 3 | <1 | t | .1 | t |

| Chol (mg) | Calc (mg) | Iron (mg) | Magn (mg) | Pota (mg) | Sodi (mg) | Zinc (mg) | VT-A (RE) | Thia (mg) | VT-E (a-TE) | Ribo (mg) | Niac (mg) | V-B6 (mg) | Fola (µg) | VT-C (mg) |
|---|---|---|---|---|---|---|---|---|---|---|---|---|---|---|
| 0 | 103 | 1.71 | 20 | 218 | 42 | .23 | 770 | .1 | 1.38 | .14 | .44 | .14 | 15 | 19 |
| 0 | 147 | 1.89 | 25 | 244 | 46 | .29 | 1228 | .14 | 2.63 | .18 | .54 | .17 | 13 | 19 |
| 0 | 6 | .35 | 13 | 246 | 3 | .15 | 6 | .07 | .03 | .02 | .59 | .08 | 14 | 1 |
| 0 | 26 | .41 | 7 | 157 | 11 | .39 | 103 | .04 | .22 | .04 | .2 | .01 | 71 | 3 |
| 0 | 26 | .41 | 7 | 157 | 11 | .39 | 103 | .04 | .22 | .04 | .2 | .01 | 71 | 3 |
| 0 | 80 | 4.74 | 79 | 477 | 11 | 2.51 | 5 | .19 | .57 | .1 | .86 | .23 | 282 | 2 |
| 0 | 11 | .08 | 3 | 8 | <1 | .02 | 81 | <.01 | .06 | .01 | .07 | .01 | 2 | <1 |
| 0 | 51 | .37 | 13 | 38 | 1 | .09 | 378 | .01 | .28 | .05 | .33 | .06 | 12 | 2 |
| 0 | 120 | 3.77 | 88 | 692 | 4 | 1.56 | <1 | .28 | .53 | .1 | 1.21 | .21 | 181 | 2 |
| 0 | 21 | 5.1 | 25 | 644 | 6 | .18 | 3 | .3 | .28 | .09 | 1.95 | .12 | 20 | 6 |
| 0 | 14 | .72 | 14 | 180 | 5 | .19 | 2 | .02 | 5.48 | .03 | .24 | .05 | 14 | 24 |
| 0 | 94 | 1.17 | 23 | 296 | 30 | .31 | 962 | .07 | 1.11 | .09 | .65 | .18 | 17 | 53 |
| 0 | 179 | 1.22 | 23 | 417 | 19 | .23 | 826 | .06 | .23 | .15 | .87 | .11 | 19 | 33 |
| 0 | 61 | 3.23 | 72 | 658 | 873 | 1.41 | 0 | .27 | .13 | .22 | 1.17 | .06 | 130 | 3 |
| 0 | 32 | .54 | 26 | 473 | 27 | .04 | 5 | .07 | .65 | .03 | .54 | .2 | 22 | 84 |
| 0 | 41 | .66 | 31 | 561 | 35 | .51 | 7 | .07 | 2.76 | .03 | .64 | .25 | 20 | 89 |
| 0 | 52 | 1.87 | 25 | 160 | 18 | .11 | 9 | .05 | .82 | .03 | .36 | .21 | 57 | 11 |
| 0 | 31 | 1.14 | 15 | 90 | 10 | .06 | 5 | .03 | .63 | .02 | .21 | .12 | 25 | 4 |
| 0 | 38 | 6.59 | 71 | 731 | 4 | 2.51 | 2 | .33 | .22 | .14 | 2.1 | .35 | 358 | 3 |
| 0 | 17 | 3.84 | 43 | 352 | 12 | 1.98 | 5 | .27 | .11 | .11 | 1.49 | .2 | 83 | 16 |
| 0 | 19 | 2.47 | 28 | 248 | 8 | 1.16 | 4 | .18 | .07 | .1 | .87 | .15 | 77 | 13 |
| 0 | 13 | .12 | 5 | 105 | 2 | .07 | 40 | .02 | .18 | .02 | .12 | .02 | 30 | 3 |
| 0 | 10 | .09 | 4 | 77 | 1 | .05 | 29 | .02 | .13 | .02 | .09 | .01 | 22 | 2 |
| 0 | 10 | .27 | 5 | 87 | 5 | .12 | 18 | .02 | .15 | .02 | .1 | .02 | 31 | 2 |
| 0 | 102 | 2.7 | 48 | 852 | 48 | 1.19 | 178 | .25 | 1.51 | .16 | 1.01 | .22 | 302 | 21 |
| 0 | 26 | .67 | 12 | 213 | 12 | .3 | 45 | .06 | .38 | .04 | .25 | .05 | 76 | 5 |
| 0 | 19 | .39 | 3 | 74 | 3 | .08 | 53 | .01 | .12 | .02 | .11 | .01 | 14 | 5 |
| 0 | 10 | .31 | 2 | 81 | 2 | .07 | 73 | .03 | .12 | .03 | .14 | .01 | 38 | 7 |
| 0 | 11 | .33 | 2 | 87 | 2 | .07 | 78 | .03 | .13 | .03 | .15 | .01 | 41 | 7 |
| 0 | 112 | .8 | 101 | 570 | 420 | .97 | 103 | .23 | 1.22 | .1 | 1.54 | .33 | 81 | 29 |
| 0 | 2 | .43 | 3 | 130 | 1 | .26 | 0 | .04 | .04 | .16 | 1.44 | .03 | 7 | 1 |
| 0 | 5 | 1.36 | 9 | 278 | 2 | .68 | 0 | .06 | .09 | .23 | 3.48 | .07 | 14 | 3 |
| 0 | 2 | .32 | 10 | 85 | 3 | .97 | 0 | .03 | .09 | .12 | 1.1 | .12 | 15 | <1 |
| 0 | 9 | .62 | 12 | 101 | 332 | .56 | 0 | .07 | .09 | .02 | 1.24 | .05 | 10 | 0 |
| 0 | 2 | .5 | 5 | 139 | 95 | .28 | 0 | .03 | .05 | .16 | 1.42 | .03 | 6 | 1 |
| 0 | 52 | .49 | 10 | 141 | 11 | .08 | 212 | .03 | 1.41 | .04 | .3 | .07 | 51 | 18 |
| 0 | 76 | .84 | 10 | 104 | 19 | .15 | 335 | .03 | 1.31 | .04 | .19 | .08 | 52 | 10 |
| 0 | 127 | 4.51 | 107 | 670 | 2 | 1.93 | <1 | .37 | .73 | .11 | .97 | .3 | 255 | 2 |
| 0 | 54 | .38 | 48 | 274 | 4 | .47 | 49 | .11 | .59 | .05 | .74 | .16 | 39 | 14 |
| 0 | 177 | 1.23 | 94 | 431 | 6 | 1.14 | 94 | .18 | 1.27 | .23 | 1.44 | .09 | 269 | 22 |
| 15 | 104 | .77 | 37 | 214 | 137 | .5 | 43 | .13 | 3.08 | .1 | .75 | .13 | 37 | 10 |
| 0 | 112 | .8 | 101 | 570 | 420 | .97 | 103 | .23 | 1.22 | .1 | 1.54 | .33 | 81 | 29 |
| 0 | 16 | .18 | 8 | 126 | 2 | .15 | 0 | .03 | .1 | .02 | .12 | .09 | 15 | 5 |
| 0 | 12 | .13 | 6 | 91 | 2 | .11 | 0 | .02 | .07 | .01 | .09 | .07 | 11 | 4 |
| 0 | 23 | .25 | 12 | 174 | 3 | .22 | 0 | .04 | .14 | .02 | .17 | .13 | 16 | 5 |
| 0 | 36 | .22 | 13 | 227 | 3 | .26 | 0 | .07 | .19 | .01 | .14 | .22 | 23 | 10 |
| 0 | 21 | .22 | 10 | 154 | 218 | .19 | 0 | .04 | .12 | .02 | .15 | .12 | 14 | 5 |
| 0 | 36 | .74 | 10 | 138 | 8 | .19 | 19 | .03 | .06 | .04 | .26 | .03 | 32 | 9 |
| 0 | 6 | .34 | 4 | 26 | 75 | .08 | 5 | .06 | .14 | .03 | .72 | .01 | 13 | <1 |
| 0 | 26 | 2.47 | 15 | 2636 | 20 | 5.45 | 10 | .07 | .73 | .25 | 1.25 | 1.06 | 30 | 10 |
| 0 | 41 | 1.86 | 15 | 166 | 17 | .32 | 156 | .03 | .54 | .03 | .39 | .03 | 46 | 40 |
| 0 | 29 | .45 | 23 | 286 | 8 | .2 | 0 | .06 | .78 | .04 | .56 | .07 | 45 | 10 |

**Table H-1**

**Food Composition** (Computer code number is for West Diet Analysis program) (For purposes of calculations, use "0" for t, <1, <.1, <.01, etc.)

| Computer Code Number | Food Description | Measure | Wt (g) | $H_2O$ (%) | Ener (kcal) | Prot (g) | Carb (g) | Dietary Fiber (g) | Fat (g) | Fat Breakdown (g) Sat | Mono | Poly |
|---|---|---|---|---|---|---|---|---|---|---|---|---|
| | **VEGETABLES and LEGUMES—Continued** | | | | | | | | | | | |
| | Peas: | | | | | | | | | | | |
| | Black-eyed, cooked: | | | | | | | | | | | |
| 814 | From dry, drained | ½ c | 86 | 70 | 100 | 7 | 18 | 6 | <1 | .1 | t | .2 |
| 815 | From fresh, drained | ½ c | 82 | 75 | 79 | 3 | 17 | 4 | <1 | .1 | t | .1 |
| 816 | From frozen, drained | ½ c | 85 | 66 | 112 | 7 | 20 | 5 | 1 | .1 | .1 | .2 |
| 889 | Edible pod peas, cooked | ½ c | 80 | 89 | 34 | 3 | 6 | 2 | <1 | t | t | .1 |
| 890 | Green, canned, drained: | ½ c | 85 | 82 | 59 | 4 | 11 | 3 | <1 | .1 | t | .1 |
| 5267 | Unsalted | ½ c | 124 | 86 | 66 | 4 | 12 | 4 | <1 | .1 | t | .2 |
| 891 | Green, cooked from frozen | ½ c | 80 | 79 | 62 | 4 | 11 | 4 | <1 | t | t | .1 |
| 1786 | Snow peas, raw | ½ c | 49 | 89 | 21 | 1 | 4 | 1 | <1 | t | t | t |
| 1787 | Snow peas, raw | 10 ea | 34 | 89 | 14 | 1 | 3 | 1 | <1 | t | t | t |
| 892 | Split, green, cooked from dry | ½ c | 98 | 69 | 116 | 8 | 21 | 8 | <1 | .1 | .1 | .2 |
| 1187 | Peas & carrots, cooked from frozen | ½ c | 80 | 86 | 38 | 2 | 8 | 2 | <1 | .1 | t | .2 |
| 1186 | Peas & carrots, canned w/liquid | ½ c | 128 | 88 | 49 | 3 | 11 | 3 | <1 | .1 | t | .2 |
| | Peppers, hot: | | | | | | | | | | | |
| 893 | Hot green chili, canned | ½ c | 68 | 92 | 14 | 1 | 3 | 1 | <1 | t | t | t |
| 894 | Hot green chili, raw | 1 ea | 45 | 88 | 18 | 1 | 4 | 1 | <1 | t | t | t |
| 1715 | Hot red chili, raw, diced | 1 tbs | 9 | 88 | 4 | <1 | 1 | <1 | <1 | t | t | t |
| 1988 | Jalapeno, raw | 1 ea | 45 | 90 | 11 | <1 | 2 | | <1 | | | |
| 895 | Jalapeno, chopped, canned | ½ c | 68 | 89 | 18 | 1 | 3 | 2 | 1 | .1 | t | .3 |
| 1918 | Jalapeno wheels, in brine (Ortega) | 2 tbs | 29 | | 10 | 0 | 2 | | 0 | 0 | 0 | 0 |
| | Peppers, sweet, green: | | | | | | | | | | | |
| 896 | Whole pod (90 g with refuse), raw | 1 ea | 74 | 92 | 20 | 1 | 5 | 1 | <1 | t | t | .1 |
| 897 | Cooked, chopped (1 pod cooked = 73g) | ½ c | 68 | 92 | 19 | 1 | 5 | 1 | <1 | t | t | .1 |
| | Peppers, sweet, red: | | | | | | | | | | | |
| 1286 | Raw, chopped | ½ c | 75 | 92 | 20 | 1 | 5 | 1 | <1 | t | t | .1 |
| 1807 | Raw, each | 1 ea | 74 | 92 | 20 | 1 | 5 | 1 | <1 | t | t | .1 |
| 1287 | Cooked, chopped | ½ c | 68 | 92 | 19 | 1 | 5 | 1 | <1 | t | t | .1 |
| | Peppers, sweet, yellow: | | | | | | | | | | | |
| 1872 | Raw, large | 1 ea | 186 | 92 | 50 | 2 | 12 | 2 | <1 | .1 | t | .2 |
| 1873 | Strips | 10 pce | 52 | 92 | 14 | 1 | 3 | <1 | <1 | t | t | .1 |
| 898 | Pinto beans, cooked from dry | ½ c | 85 | 64 | 116 | 7 | 22 | 7 | <1 | .1 | .1 | .2 |
| 1191 | Poi, two finger | ½ c | 120 | 72 | 134 | <1 | 33 | <1 | <1 | t | t | .1 |
| | Potatoes: | | | | | | | | | | | |
| | Baked in oven, 4¾"x2⅓" diam | | | | | | | | | | | |
| 899 | With skin | 1 ea | 202 | 71 | 220 | 5 | 51 | 5 | <1 | .1 | t | .1 |
| 900 | Flesh only | 1 ea | 156 | 75 | 145 | 3 | 34 | 2 | <1 | t | t | .1 |
| 901 | Skin only | 1 ea | 58 | 47 | 115 | 2 | 27 | 5 | <1 | t | t | t |
| | Baked in microwave, 4¾"x 2⅓"dm: | | | | | | | | | | | |
| 902 | With skin | 1 ea | 202 | 72 | 212 | 5 | 49 | 5 | <1 | .1 | t | .1 |
| 903 | Flesh only | 1 ea | 156 | 74 | 156 | 3 | 36 | 2 | <1 | t | t | .1 |
| 904 | Skin only | 1 ea | 58 | 63 | 77 | 3 | 17 | 3 | <1 | t | t | t |
| | Boiled, about 2½" diam: | | | | | | | | | | | |
| 905 | Peeled after boiling | 1 ea | 136 | 77 | 118 | 3 | 27 | 2 | <1 | t | t | .1 |
| 906 | Peeled before boiling | 1 ea | 135 | 77 | 116 | 2 | 27 | 2 | <1 | t | t | .1 |
| | French fried, strips 2–3½" long: | | | | | | | | | | | |
| 907 | Oven heated | 10 ea | 50 | 35 | 167 | 2 | 20 | 2 | 9 | 3 | 5.7 | .7 |
| 908 | Fried in vegetable oil | 10 ea | 50 | 38 | 158 | 2 | 20 | 2 | 8 | 1.9 | 4.7 | .7 |
| 1188 | Fried in veg and animal oil | 10 ea | 50 | 38 | 158 | 2 | 20 | 2 | 8 | 1.9 | 4.7 | .7 |
| 909 | Hashed browns from frozen | 1 c | 156 | 56 | 340 | 5 | 44 | 3 | 18 | 7 | 8 | 2.1 |
| | Mashed: | | | | | | | | | | | |
| 910 | Home recipe with whole milk | ½ c | 105 | 78 | 81 | 2 | 18 | 2 | 1 | .4 | .2 | .1 |
| 911 | Home recipe with milk and marg | ½ c | 105 | 76 | 111 | 2 | 17 | 2 | 4 | 1.1 | 1.9 | 1.3 |
| 912 | Prepared from flakes; water, milk, margarine, salt added | ½ c | 110 | 76 | 124 | 2 | 16 | 3 | 6 | 1.6 | 2.5 | 1.7 |
| | Potato products, prepared: | | | | | | | | | | | |
| | Au gratin: | | | | | | | | | | | |
| 913 | From dry mix | ½ c | 123 | 79 | 114 | 3 | 16 | 1 | 5 | 3.5 | 1.5 | .2 |
| 914 | From home recipe, using butter | ½ c | 122 | 74 | 161 | 7 | 14 | 2 | 9 | 4.8 | 3.2 | 1.3 |
| | Scalloped: | | | | | | | | | | | |
| 915 | From dry mix | ½ c | 122 | 79 | 113 | 3 | 16 | 1 | 5 | 3.2 | 1.5 | .2 |
| 916 | From home recipe, using butter | ½ c | 123 | 81 | 106 | 4 | 13 | 2 | 5 | 1.7 | 1.7 | .9 |

| Chol (mg) | Calc (mg) | Iron (mg) | Magn (mg) | Pota (mg) | Sodi (mg) | Zinc (mg) | VT-A (RE) | Thia (mg) | VT-E (a-TE) | Ribo (mg) | Niac (mg) | V-B6 (mg) | Fola (μg) | VT-C (mg) |
|---|---|---|---|---|---|---|---|---|---|---|---|---|---|---|
| 0 | 21 | 2.16 | 46 | 239 | 3 | 1.11 | 2 | .17 | .24 | .05 | .43 | .09 | 179 | <1 |
| 0 | 105 | .92 | 43 | 343 | 3 | .84 | 65 | .08 | .18 | .12 | 1.15 | .05 | 104 | 2 |
| 0 | 20 | 1.8 | 42 | 319 | 4 | 1.21 | 7 | .22 | .33 | .05 | .62 | .08 | 120 | 2 |
| 0 | 34 | 1.58 | 21 | 192 | 3 | .3 | 10 | .1 | .31 | .06 | .43 | .11 | 23 | 38 |
| 0 | 17 | .81 | 14 | 147 | 214 | .6 | 65 | .1 | .32 | .07 | .62 | .05 | 38 | 8 |
| 0 | 22 | 1.26 | 21 | 124 | 11 | .87 | 47 | .14 | .47 | .09 | 1.04 | .08 | 35 | 12 |
| 0 | 19 | 1.26 | 23 | 134 | 70 | .75 | 54 | .23 | .14 | .08 | 1.18 | .09 | 47 | 8 |
| 0 | 21 | 1.02 | 12 | 98 | 2 | .13 | 7 | .07 | .19 | .04 | .29 | .08 | 20 | 29 |
| 0 | 15 | .71 | 8 | 68 | 1 | .09 | 5 | .05 | .13 | .03 | .2 | .05 | 14 | 20 |
| 0 | 14 | 1.26 | 35 | 355 | 2 | .98 | 1 | .19 | .38 | .05 | .87 | .05 | 64 | <1 |
| 0 | 18 | .75 | 13 | 126 | 54 | .36 | 621 | .18 | .26 | .05 | .92 | .07 | 21 | 6 |
| 0 | 29 | .96 | 18 | 128 | 333 | .74 | 739 | .09 | .24 | .07 | .74 | .11 | 23 | 8 |
| 0 | 5 | .34 | 10 | 127 | 798 | .12 | 41 | .01 | .47 | .03 | .54 | .1 | 7 | 46 |
| 0 | 8 | .54 | 11 | 153 | 3 | .13 | 35 | .04 | .31 | .04 | .43 | .12 | 10 | 109 |
| 0 | 2 | .11 | 2 | 31 | 1 | .03 | 97 | .01 | .06 | .01 | .09 | .02 | 2 | 22 |
| | | | | 2 | 2 | | 30 | | .37 | | | | | 53 |
| 0 | 16 | 1.28 | 10 | 131 | 1136 | .23 | 116 | .03 | .47 | .03 | .27 | .13 | 10 | 7 |
| 0 | | | | 55 | 390 | | 10 | | .2 | | | | | 21 |
| 0 | 7 | .34 | 7 | 131 | 1 | .09 | 47 | .05 | .51 | .02 | .38 | .18 | 16 | 66 |
| 0 | 6 | .31 | 7 | 113 | 1 | .08 | 40 | .04 | .47 | .02 | .32 | .16 | 11 | 51 |
| 0 | 7 | .34 | 7 | 133 | 1 | .09 | 428 | .05 | .52 | .02 | .38 | .19 | 16 | 143 |
| 0 | 7 | .34 | 7 | 131 | 1 | .09 | 422 | .05 | .51 | .02 | .38 | .18 | 16 | 141 |
| 0 | 6 | .31 | 7 | 113 | 1 | .08 | 256 | .04 | .47 | .02 | .32 | .16 | 11 | 116 |
| 0 | 20 | .86 | 22 | 394 | 4 | .32 | 45 | .05 | 1.28 | .05 | 1.66 | .31 | 48 | 342 |
| 0 | 6 | .24 | 6 | 110 | 1 | .09 | 12 | .01 | .36 | .01 | .46 | .09 | 13 | 96 |
| 0 | 41 | 2.22 | 47 | 398 | 2 | .92 | <1 | .16 | .8 | .08 | .34 | .13 | 146 | 2 |
| 0 | 19 | 1.06 | 29 | 220 | 14 | .26 | 2 | .16 | .22 | .05 | 1.32 | .33 | 26 | 5 |
| 0 | 20 | 2.75 | 54 | 844 | 16 | .65 | 0 | .22 | .1 | .07 | 3.33 | .7 | 22 | 26 |
| 0 | 8 | .55 | 39 | 610 | 8 | .45 | 0 | .16 | .06 | .03 | 2.18 | .47 | 14 | 20 |
| 0 | 20 | 4.08 | 25 | 332 | 12 | .28 | 0 | .07 | .02 | .06 | 1.78 | .36 | 12 | 8 |
| 0 | 22 | 2.5 | 54 | 903 | 16 | .73 | 0 | .24 | .1 | .06 | 3.45 | .69 | 24 | 30 |
| 0 | 8 | .64 | 39 | 641 | 11 | .51 | 0 | .2 | .06 | .04 | 2.54 | .5 | 19 | 24 |
| 0 | 27 | 3.45 | 21 | 377 | 9 | .3 | 0 | .04 | .02 | .04 | 1.29 | .28 | 10 | 9 |
| 0 | 7 | .42 | 30 | 515 | 5 | .41 | 0 | .14 | .07 | .03 | 1.96 | .41 | 14 | 18 |
| 0 | 11 | .42 | 27 | 443 | 7 | .36 | 0 | .13 | .07 | .03 | 1.77 | .36 | 12 | 10 |
| 0 | 6 | .83 | 11 | 270 | 307 | .2 | 0 | .04 | .25 | .02 | 1.34 | .11 | 11 | 3 |
| 0 | 9 | .38 | 17 | 366 | 108 | .19 | 0 | .09 | .25 | .01 | 1.63 | .12 | 14 | 5 |
| 6 | 9 | .38 | 17 | 366 | 108 | .19 | 0 | .09 | .25 | .01 | 1.63 | .12 | 14 | 5 |
| 0 | 23 | 2.36 | 26 | 680 | 53 | .5 | 0 | .17 | .3 | .03 | 3.78 | .2 | 10 | 10 |
| 2 | 27 | .28 | 19 | 314 | 318 | .3 | 6 | .09 | .05 | .04 | 1.18 | .24 | 9 | 7 |
| 2 | 27 | .27 | 19 | 303 | 310 | .28 | 21 | .09 | .31 | .04 | 1.13 | .23 | 8 | 6 |
| 4 | 54 | .24 | 20 | 256 | 365 | .2 | 23 | .12 | .77 | .05 | .74 | .01 | 8 | 11 |
| 18 | 102 | .39 | 18 | 269 | 540 | .29 | 38 | .02 | 1.48 | .1 | 1.15 | .05 | 8 | 4 |
| 18 | 145 | .78 | 24 | 483 | 528 | .84 | 46 | .08 | .64 | .14 | 1.21 | .21 | 13 | 12 |
| 13 | 44 | .46 | 17 | 248 | 416 | .3 | 26 | .02 | .18 | .07 | 1.26 | .05 | 12 | 4 |
| 7 | 70 | .7 | 23 | 465 | 412 | .49 | 23 | .08 | .4 | .11 | 1.29 | .22 | 13 | 13 |

**Table H-1**

**Food Composition** (Computer code number is for West Diet Analysis program) (For purposes of calculations, use "0" for t, <1, <.1, <.01, etc.)

| Computer Code Number | Food Description | Measure | Wt (g) | $H_2O$ (%) | Ener (kcal) | Prot (g) | Carb (g) | Dietary Fiber (g) | Fat (g) | Fat Breakdown (g) Sat | Mono | Poly |
|---|---|---|---|---|---|---|---|---|---|---|---|---|
| | **VEGETABLES and LEGUMES**—Continued | | | | | | | | | | | |
| | Potato Salad (see Mixed Dishes #715) | | | | | | | | | | | |
| 1192 | Potato puffs, cooked from frozen | ½ c | 64 | 53 | 142 | 2 | 19 | 2 | 7 | 3.3 | 2.8 | .5 |
| 918 | Pumpkin, cooked from fresh, mashed | ½ c | 123 | 94 | 25 | 1 | 6 | 1 | <1 | t | t | t |
| 919 | Pumpkin, canned | ½ c | 123 | 90 | 42 | 1 | 10 | 4 | <1 | .2 | t | t |
| 1891 | Radicchio, raw, shredded | ½ c | 20 | 93 | 5 | <1 | 1 | <1 | <1 | t | t | t |
| 1894 | Radicchio, raw, leaf | 10 ea | 80 | 93 | 18 | 1 | 4 | 1 | <1 | t | t | .1 |
| 920 | Red radishes | 10 ea | 45 | 95 | 8 | <1 | 2 | 1 | <1 | t | t | t |
| 1793 | Daikon radishes (Chinese) raw | ½ c | 44 | 95 | 8 | <1 | 2 | 1 | <1 | t | t | t |
| 921 | Refried beans, canned | ½ c | 126 | 76 | 118 | 7 | 19 | 7 | 2 | .6 | .7 | .2 |
| 1375 | Rutabaga, cooked cubes | ½ c | 85 | 89 | 33 | 1 | 7 | 2 | <1 | t | t | .1 |
| 922 | Sauerkraut, canned with liquid | ½ c | 118 | 92 | 22 | 1 | 5 | 3 | <1 | t | t | .1 |
| 923 | Seaweed, kelp, raw | ½ c | 40 | 82 | 17 | 1 | 4 | 1 | <1 | .1 | t | t |
| 924 | Seaweed, spirulina, dried | ½ c | 8 | 5 | 23 | 5 | 2 | <1 | 1 | .2 | .1 | .2 |
| 1866 | Shallots, raw, chopped | 1 tbs | 10 | 80 | 7 | <1 | 2 | <1 | <1 | t | t | t |
| 1557 | Snow peas, stir-fried | ½ c | 83 | 89 | 35 | 2 | 6 | 2 | <1 | t | t | .1 |
| 925 | Soybeans, cooked from dry | ½ c | 86 | 63 | 149 | 15 | 9 | 5 | 8 | 1.1 | 1.7 | 4.4 |
| 1996 | Soybeans, dry roasted | ½ c | 86 | 1 | 387 | 34 | 28 | 7 | 19 | 2.7 | 4.1 | 10.6 |
| | Soybean products: | | | | | | | | | | | |
| 926 | Miso | ½ c | 138 | 41 | 284 | 17 | 39 | 7 | 8 | 1.2 | 1.9 | 4.7 |
| | Soy milk (see #144 and #2301 under Dairy) | | | | | | | | | | | |
| | Tofu (soybean curd): | | | | | | | | | | | |
| 7540 | Extra firm, silken | ½ c | 126 | 88 | 69 | 9 | 3 | <1 | 2 | .4 | .4 | 1.3 |
| 7542 | Firm, silken | ½ c | 126 | 87 | 77 | 9 | 3 | <1 | 3 | .5 | .7 | 1.9 |
| 927 | Regular | ½ c | 124 | 87 | 76 | 8 | 2 | <1 | 5 | .7 | 1 | 2.6 |
| 7541 | Soft, silken | ½ c | 124 | 89 | 68 | 6 | 4 | <1 | 3 | .4 | .6 | 1.9 |
| | Spinach: | | | | | | | | | | | |
| 928 | Raw, chopped | ½ c | 28 | 92 | 6 | 1 | 1 | 1 | <1 | t | t | t |
| 929 | Cooked, from fresh, drained | ½ c | 90 | 91 | 21 | 3 | 3 | 2 | <1 | t | t | .1 |
| 930 | Cooked from frozen (leaf) | ½ c | 95 | 90 | 27 | 3 | 5 | 3 | <1 | t | t | .1 |
| 931 | Canned, drained solids: | ½ c | 107 | 92 | 25 | 3 | 4 | 3 | 1 | .1 | t | .2 |
| 5149 | Unsalted | ½ c | 107 | 92 | 25 | 3 | 4 | 3 | 1 | .1 | t | .2 |
| | Spinach soufflé (see Mixed Dishes) | | | | | | | | | | | |
| | Squash, summer varieties,cooked w/skin: | | | | | | | | | | | |
| 932 | Varieties averaged | ½ c | 90 | 94 | 18 | 1 | 4 | 1 | <1 | .1 | t | .1 |
| 933 | Crookneck | ½ c | 90 | 94 | 18 | 1 | 4 | 1 | <1 | .1 | t | .1 |
| 934 | Zucchini | ½ c | 90 | 95 | 14 | 1 | 4 | 1 | <1 | t | t | t |
| | Squash, winter varieties, cooked: | | | | | | | | | | | |
| | Average of all varieties, baked: | | | | | | | | | | | |
| 935 | Mashed | 1 c | 245 | 89 | 96 | 2 | 21 | 7 | 2 | .3 | .1 | .6 |
| 936 | Cubes | 1 c | 205 | 89 | 80 | 2 | 18 | 6 | 1 | .3 | .1 | .5 |
| 937 | Acorn, baked, mashed | ½ c | 123 | 83 | 69 | 1 | 18 | 5 | <1 | t | t | .1 |
| 1218 | Acorn, boiled, mashed | ½ c | 122 | 90 | 41 | 1 | 11 | 3 | <1 | t | t | t |
| | Butternut squash: | | | | | | | | | | | |
| 938 | Baked cubes | ½ c | 103 | 88 | 41 | 1 | 11 | 3 | <1 | t | t | t |
| 1219 | Baked, mashed | ½ c | 103 | 88 | 41 | 1 | 11 | 3 | <1 | t | t | t |
| 1193 | Cooked from frozen | ½ c | 120 | 88 | 47 | 1 | 12 | 3 | <1 | t | t | t |
| 1194 | Hubbard, baked, mashed | ½ c | 120 | 85 | 60 | 3 | 13 | 3 | 1 | .2 | .1 | .3 |
| 1195 | Hubbard, boiled, mashed | ½ c | 118 | 91 | 35 | 2 | 8 | 3 | <1 | .1 | t | .2 |
| 1196 | Spaghetti, baked or boiled | ½ c | 77 | 92 | 22 | <1 | 5 | 1 | <1 | t | t | .1 |
| 1189 | Succotash, cooked from frozen | ½ c | 85 | 74 | 79 | 4 | 17 | 3 | 1 | .1 | .1 | .4 |
| | Sweet potatoes: | | | | | | | | | | | |
| 939 | Baked in skin, peeled, 5 x 2" diam | 1 ea | 114 | 73 | 117 | 2 | 28 | 3 | <1 | t | t | .1 |
| 940 | Boiled without skin, 5 x 2" diam | 1 ea | 151 | 73 | 159 | 2 | 37 | 3 | <1 | .1 | t | .2 |
| 941 | Candied, 2½ x 2" | 1 pce | 105 | 67 | 144 | 1 | 29 | 3 | 3 | 1.4 | .7 | .2 |
| | Canned: | | | | | | | | | | | |
| 942 | Solid pack | ½ c | 128 | 74 | 129 | 3 | 30 | 2 | <1 | .1 | t | .1 |
| 943 | Vacuum pack, mashed | ½ c | 127 | 76 | 116 | 2 | 27 | 2 | <1 | .1 | t | .1 |
| 944 | Vacuum pack, 3¾ x 1" | 2 pce | 80 | 76 | 73 | 1 | 17 | 1 | <1 | t | t | .1 |
| 1940 | Taro shoots, cooked slices | 1 c | 140 | 95 | 20 | 1 | 4 | 1 | <1 | t | t | t |
| 1941 | Taro, tahitian, cooked slices | 1 c | 137 | 86 | 60 | 6 | 9 | 1 | 1 | .2 | .1 | .4 |
| | Tomatillos: | | | | | | | | | | | |
| 1877 | Raw, each | 1 ea | 34 | 92 | 11 | <1 | 2 | 1 | <1 | t | .1 | .1 |
| 1875 | Raw, chopped | 1 c | 132 | 92 | 42 | 1 | 8 | 3 | 1 | .2 | .2 | .6 |

| Chol (mg) | Calc (mg) | Iron (mg) | Magn (mg) | Pota (mg) | Sodi (mg) | Zinc (mg) | VT-A (RE) | Thia (mg) | VT-E (a-TE) | Ribo (mg) | Niac (mg) | V-B6 (mg) | Fola (µg) | VT-C (mg) |
|---|---|---|---|---|---|---|---|---|---|---|---|---|---|---|
| 0 | 19 | 1 | 12 | 243 | 477 | .19 | 1 | .12 | .03 | .05 | 1.38 | .15 | 11 | 4 |
| 0 | 18 | .7 | 11 | 283 | 1 | .28 | 1330 | .04 | 1.3 | .1 | .51 | .05 | 10 | 6 |
| 0 | 32 | 1.71 | 28 | 253 | 6 | .21 | 2713 | .03 | 1.3 | .07 | .45 | .07 | 15 | 5 |
| 0 | 4 | .11 | 3 | 60 | 4 | .12 | 1 | <.01 | .45 | .01 | .05 | .01 | 12 | 2 |
| 0 | 15 | .45 | 10 | 242 | 18 | .5 | 2 | .01 | 1.81 | .02 | .2 | .05 | 48 | 6 |
| 0 | 9 | .13 | 4 | 104 | 11 | .13 | <1 | <.01 | 0 | .02 | .13 | .03 | 12 | 10 |
| 0 | 12 | .18 | 7 | 100 | 9 | .07 | 0 | .01 | 0 | .01 | .09 | .02 | 12 | 10 |
| 10 | 44 | 2.09 | 42 | 336 | 377 | 1.47 | 0 | .03 | .39 | .02 | .4 | .18 | 14 | 8 |
| 0 | 41 | .45 | 20 | 277 | 17 | .3 | 48 | .07 | .13 | .03 | .61 | .09 | 13 | 16 |
| 0 | 35 | 1.73 | 15 | 201 | 780 | .22 | 2 | .02 | .12 | .03 | .17 | .15 | 28 | 17 |
| 0 | 67 | 1.14 | 48 | 36 | 93 | .49 | 5 | .02 | .35 | .06 | .19 | <.01 | 72 | 1 |
| 0 | 10 | 2.28 | 16 | 109 | 84 | .16 | 5 | .19 | .4 | .29 | 1.02 | .03 | 8 | 1 |
| 0 | 4 | .12 | 2 | 33 | 1 | .04 | 125 | .01 | .01 | <.01 | .02 | .03 | 3 | 1 |
| 0 | 36 | 1.73 | 20 | 166 | 3 | .22 | 11 | .11 | .32 | .06 | .47 | .13 | 28 | 42 |
| 0 | 88 | 4.42 | 74 | 443 | 1 | .99 | 1 | .13 | 1.68 | .24 | .34 | .2 | 46 | 1 |
| 0 | 120 | 3.4 | 196 | 1173 | 2 | 4.1 | 2 | .37 | 3.96 | .65 | .91 | .19 | 176 | 4 |
| 0 | 91 | 3.78 | 58 | 226 | 5032 | 4.58 | 12 | .13 | .01 | .34 | 1.19 | .3 | 45 | 0 |
| 0 | 39 | 1.5 | 34 | 195 | 80 | .76 | 0 | .1 | .18 | .04 | .3 | .01 | | 0 |
| 0 | 41 | 1.3 | 34 | 244 | 45 | .77 | 0 | .13 | .24 | .05 | .31 | .01 | | 0 |
| 0 | 138 | 1.38 | 33 | 149 | 10 | .79 | 1 | .06 | .01 | .05 | .66 | .06 | 55 | <1 |
| 0 | 38 | 1.02 | 36 | 223 | 6 | .64 | 0 | .12 | .25 | .05 | .37 | .01 | | 0 |
| 0 | 28 | .76 | 22 | 156 | 22 | .15 | 188 | .02 | .53 | .05 | .2 | .05 | 54 | 8 |
| 0 | 122 | 3.21 | 78 | 419 | 63 | .68 | 737 | .09 | .86 | .21 | .44 | .22 | 131 | 9 |
| 0 | 139 | 1.44 | 66 | 283 | 82 | .66 | 739 | .06 | .91 | .16 | .4 | .14 | 103 | 12 |
| 0 | 136 | 2.46 | 81 | 370 | 29 | .49 | 939 | .02 | 1.39 | .15 | .41 | .11 | 105 | 15 |
| 0 | 136 | 2.46 | 81 | 370 | 29 | .49 | 939 | .02 | 1.39 | .15 | .41 | .11 | 105 | 15 |
| 0 | 24 | .32 | 22 | 173 | 1 | .35 | 26 | .04 | .11 | .04 | .46 | .06 | 18 | 5 |
| 0 | 24 | .32 | 22 | 173 | 1 | .35 | 26 | .04 | .11 | .04 | .46 | .08 | 18 | 5 |
| 0 | 12 | .31 | 20 | 228 | 3 | .16 | 22 | .04 | .11 | .04 | .38 | .07 | 15 | 4 |
| 0 | 34 | .81 | 20 | 1070 | 2 | .64 | 872 | .21 | .29 | .06 | 1.72 | .18 | 69 | 23 |
| 0 | 29 | .68 | 16 | 896 | 2 | .53 | 730 | .17 | .25 | .05 | 1.44 | .15 | 57 | 20 |
| 0 | 54 | 1.14 | 53 | 538 | 5 | .21 | 53 | .2 | .15 | .02 | 1.08 | .24 | 23 | 13 |
| 0 | 32 | .68 | 32 | 321 | 4 | .13 | 32 | .12 | .15 | .01 | .65 | .14 | 14 | 8 |
| 0 | 42 | .62 | 30 | 293 | 4 | .13 | 721 | .07 | .17 | .02 | 1 | .13 | 20 | 16 |
| 0 | 42 | .62 | 30 | 293 | 4 | .13 | 721 | .07 | .17 | .02 | 1 | .13 | 20 | 16 |
| 0 | 23 | .7 | 11 | 160 | 2 | .14 | 401 | .06 | .16 | .05 | .56 | .08 | 20 | 4 |
| 0 | 20 | .56 | 26 | 430 | 10 | .18 | 725 | .09 | .14 | .06 | .67 | .21 | 19 | 11 |
| 0 | 12 | .33 | 15 | 253 | 6 | .12 | 473 | .05 | .14 | .03 | .39 | .12 | 11 | 8 |
| 0 | 16 | .26 | 8 | 90 | 14 | .15 | 8 | .03 | .09 | .02 | .62 | .08 | 6 | 3 |
| 0 | 13 | .76 | 20 | 225 | 38 | .38 | 20 | .06 | .31 | .06 | 1.11 | .08 | 28 | 5 |
| 0 | 32 | .51 | 23 | 397 | 11 | .33 | 2487 | .08 | .32 | .14 | .69 | .27 | 26 | 28 |
| 0 | 32 | .85 | 15 | 278 | 20 | .41 | 2574 | .08 | .42 | .21 | .97 | .37 | 17 | 26 |
| 8 | 27 | 1.19 | 12 | 198 | 73 | .16 | 440 | .02 | 3.99 | .04 | .41 | .04 | 12 | 7 |
| 0 | 38 | 1.7 | 31 | 269 | 96 | .27 | 1936 | .03 | .35 | .11 | 1.22 | .3 | 14 | 7 |
| 0 | 28 | 1.13 | 28 | 396 | 67 | .23 | 1013 | .05 | .32 | .07 | .94 | .24 | 21 | 33 |
| 0 | 18 | .71 | 18 | 250 | 42 | .14 | 638 | .03 | .2 | .05 | .59 | .15 | 13 | 21 |
| 0 | 20 | .57 | 11 | 482 | 3 | .76 | 7 | .05 | 1.4 | .07 | 1.13 | .16 | 4 | 26 |
| 0 | 204 | 2.14 | 70 | 854 | 74 | .14 | 241 | .06 | 3.7 | .27 | .66 | .16 | 10 | 52 |
| 0 | 2 | .21 | 7 | 91 | <1 | .07 | 4 | .01 | .13 | .01 | .63 | .02 | 2 | 4 |
| 0 | 9 | .82 | 26 | 354 | 1 | .29 | 14 | .06 | .5 | .05 | 2.44 | .07 | 9 | 15 |

**Table H-1**

**Food Composition** (Computer code number is for West Diet Analysis program) (For purposes of calculations, use "0" for t, <1, <.1, <.01, etc.)

| Computer Code Number | Food Description | Measure | Wt (g) | $H_2O$ (%) | Ener (kcal) | Prot (g) | Carb (g) | Dietary Fiber (g) | Fat (g) | Fat Breakdown (g) Sat | Mono | Poly |
|---|---|---|---|---|---|---|---|---|---|---|---|---|
| | **VEGETABLES and LEGUMES—**Continued | | | | | | | | | | | |
| | Tomatoes: | | | | | | | | | | | |
| 945 | Raw, whole, 2 ⅗" diam | 1 ea | 123 | 94 | 26 | 1 | 6 | 1 | <1 | .1 | .1 | .2 |
| 946 | Raw, chopped | 1 c | 180 | 94 | 38 | 2 | 8 | 2 | 1 | .1 | .1 | .2 |
| 947 | Cooked from raw | 1 c | 240 | 92 | 65 | 3 | 14 | 2 | 1 | .1 | .2 | .4 |
| 948 | Canned, solids and liquid: | 1 c | 240 | 94 | 46 | 2 | 10 | 2 | <1 | t | t | .1 |
| 5741 | Unsalted | 1 c | 240 | 94 | 46 | 2 | 10 | 2 | <1 | t | t | .1 |
| 1879 | Tomatoes, sundried: | 1 c | 54 | 15 | 139 | 8 | 30 | 7 | 2 | .2 | .3 | .6 |
| 1881 | Pieces | 10 pce | 20 | 15 | 52 | 3 | 11 | 2 | 1 | .1 | .1 | .2 |
| 1885 | Oil pack, drained | 10 pce | 30 | 54 | 64 | 2 | 7 | 2 | 4 | .6 | 2.6 | .6 |
| 2020 | Tomato, raw | 1 ea | 123 | 94 | 26 | 1 | 6 | 1 | <1 | .1 | .1 | .2 |
| 949 | Tomato juice, canned: | 1 c | 243 | 94 | 41 | 2 | 10 | 1 | <1 | t | t | .1 |
| 5397 | Unsalted | 1 c | 243 | 94 | 41 | 2 | 10 | 2 | <1 | t | t | .1 |
| | Tomato products, canned: | | | | | | | | | | | |
| 950 | Paste, no added salt | 1 c | 262 | 74 | 215 | 10 | 51 | 11 | 1 | .2 | .2 | .6 |
| 951 | Puree, no added salt | 1 c | 250 | 87 | 100 | 4 | 24 | 5 | <1 | .1 | .1 | .2 |
| 952 | Sauce, with salt | 1 c | 245 | 89 | 73 | 3 | 18 | 3 | <1 | .1 | .1 | .2 |
| 953 | Turnips, cubes, cooked from fresh | 1 c | 156 | 94 | 33 | 1 | 8 | 3 | <1 | t | t | .1 |
| | Turnip greens, cooked: | | | | | | | | | | | |
| 954 | From fresh, leaves and stems | 1 c | 144 | 93 | 29 | 2 | 6 | 5 | <1 | .1 | t | .1 |
| 955 | From frozen, chopped | 1 c | 164 | 90 | 49 | 6 | 8 | 6 | 1 | .2 | t | .3 |
| 956 | Vegetable juice cocktail, canned | 1 c | 242 | 93 | 46 | 2 | 11 | 2 | <1 | t | t | .1 |
| | Vegetables, mixed: | | | | | | | | | | | |
| 957 | Canned, drained | ½ c | 81 | 87 | 38 | 2 | 7 | 2 | <1 | t | t | .1 |
| 958 | Frozen, cooked, drained | ½ c | 91 | 83 | 54 | 3 | 12 | 4 | <1 | t | t | .1 |
| 1818 | Water chestnuts, Chinese, raw | ½ c | 62 | 73 | 60 | 1 | 15 | 2 | <1 | t | t | t |
| | Water chestnuts, canned: | | | | | | | | | | | |
| 959 | Slices | ½ c | 70 | 86 | 35 | 1 | 9 | 2 | <1 | t | t | t |
| 960 | Whole | 4 ea | 28 | 86 | 14 | <1 | 3 | 1 | <1 | t | t | t |
| 1190 | Watercress, fresh, chopped | ½ c | 17 | 95 | 2 | <1 | <1 | <1 | <1 | t | t | t |
| | **VEGETARIAN FOODS:** | | | | | | | | | | | |
| 7509 | Bacon strips, meatless | 3 ea | 15 | 49 | 46 | 2 | 1 | <1 | 4 | .7 | 1.1 | 2.3 |
| 1511 | Baked beans, canned | ½ c | 127 | 73 | 118 | 6 | 26 | 6 | 1 | .1 | t | .2 |
| 7526 | Bakon crumbles | ¼ c | 7 | 16 | 28 | 2 | 1 | <1 | 2 | | | |
| 7548 | Chicken, breaded, fried, meatless | 1 pce | 57 | 70 | 97 | 6 | 3 | 3 | 7 | 1 | 2.9 | 2.5 |
| 7547 | Chicken slices, meatless | 2 ea | 60 | 59 | 132 | 10 | 4 | 3 | 8 | 1.3 | 2 | 4.4 |
| 7557 | Chili w/meat substitute | ½ c | 107 | 64 | 141 | 19 | 15 | 4 | 2 | .3 | .6 | .9 |
| 7549 | Fish stick, meatless | 2 ea | 57 | 45 | 165 | 13 | 5 | 3 | 10 | 1.6 | 2.5 | 5.4 |
| 7550 | Frankfurter, meatless | 1 ea | 51 | 58 | 102 | 10 | 4 | 2 | 5 | .8 | 1.2 | 2.7 |
| 7504 | GardenBurger, patty | 1 ea | 45 | 53 | 87 | 5 | 13 | 3 | 1 | .4 | .3 | .7 |
| 7505 | GardenSausage, patty | 1 ea | 35 | 15 | 117 | 4 | 22 | 5 | 1 | .7 | .3 | t |
| 7551 | Luncheon slice, meatless | 1 sl | 67 | 46 | 188 | 17 | 6 | 3 | 11 | 1.7 | 2.6 | 5.6 |
| 7560 | Meatloaf, meatless | 1 ea | 71 | 58 | 142 | 15 | 6 | 3 | 6 | 1 | 1.5 | 3.3 |
| 1171 | Nuteena | 1 ea | 55 | 58 | 162 | 6 | 6 | 2 | 13 | 5.1 | 5.8 | 1.7 |
| 7556 | Pot pie, meatless | 1 ea | 227 | 59 | 524 | 15 | 41 | 5 | 34 | 9.5 | 12.6 | 9.8 |
| 7554 | Soyburger, patty | 1 ea | 71 | 58 | 142 | 15 | 6 | 3 | 6 | 1 | 1.5 | 3.3 |
| 7562 | Soyburger w/cheese, patty | 1 ea | 135 | 50 | 316 | 21 | 29 | 4 | 13 | 4.2 | 3.9 | 3.7 |
| 7564 | Tempeh | 1 c | 166 | 55 | 330 | 31 | 28 | 9 | 13 | 1.9 | 2.9 | 7.2 |
| 7670 | Vegan burger, patty | 1 ea | 78 | 71 | 75 | 11 | 6 | 4 | <1 | .1 | .3 | .2 |
| | Vegetarian foods, Green Giant: | | | | | | | | | | | |
| 7677 | Breakfast links | 3 ea | 68 | 65 | 114 | 12 | 5 | 4 | 5 | .7 | 1.2 | 3.1 |
| 7676 | Breakfast patties | 2 ea | 57 | 65 | 95 | 10 | 5 | 3 | 4 | .6 | 1 | 2.6 |
| | Burger, harvest, patty: | | | | | | | | | | | |
| 7673 | Italian | 1 ea | 90 | 65 | 139 | 17 | 8 | 5 | 4 | 1.4 | .3 | .4 |
| 7674 | Original | 1 ea | 90 | 65 | 137 | 18 | 8 | 5 | 4 | 1.3 | .1 | .4 |
| 7675 | Southwestern | 1 ea | 90 | 65 | 135 | 16 | 9 | 5 | 4 | 1.4 | .2 | .4 |
| | Vegetarian foods, Loma Linda | | | | | | | | | | | |
| 7727 | Chik nuggets, frozen | 5 pce | 85 | 47 | 245 | 12 | 13 | 5 | 16 | 2.5 | 4 | 8.8 |
| 7753 | Chik-fried, frozen | 1 pce | 57 | 51 | 178 | 11 | 1 | 1 | 15 | 1.9 | 3.7 | 8.7 |
| 7744 | Franks, big, canned | 1 ea | 51 | 59 | 110 | 10 | 2 | 2 | 7 | 1.1 | 1.7 | 3.8 |
| 7747 | Linketts, canned | 1 ea | 35 | 60 | 72 | 7 | 1 | | 4 | .7 | 1.2 | 2.5 |
| 1173 | Redi-burger, patty | 1 ea | 85 | 59 | 172 | 16 | 5 | | 10 | 1.5 | 2.4 | 5.8 |
| 7755 | Swiss stake w/gravy, canned | 1 pce | 92 | 71 | 120 | 9 | 8 | 4 | 6 | .8 | 1.5 | 3.3 |

| Chol (mg) | Calc (mg) | Iron (mg) | Magn (mg) | Pota (mg) | Sodi (mg) | Zinc (mg) | VT-A (RE) | Thia (mg) | VT-E (a-TE) | Ribo (mg) | Niac (mg) | V-B6 (mg) | Fola (µg) | VT-C (mg) |
|---|---|---|---|---|---|---|---|---|---|---|---|---|---|---|
| 0 | 6 | .55 | 13 | 273 | 11 | .11 | 76 | .07 | .47 | .06 | .77 | .1 | 18 | 23 |
| 0 | 9 | .81 | 20 | 400 | 16 | .16 | 112 | .11 | .68 | .09 | 1.13 | .14 | 27 | 34 |
| 0 | 14 | 1.34 | 34 | 670 | 26 | .26 | 178 | .17 | .91 | .14 | 1.8 | .23 | 31 | 55 |
| 0 | 72 | 1.32 | 29 | 530 | 355 | .38 | 144 | .11 | .77 | .07 | 1.76 | .22 | 19 | 34 |
| 0 | 72 | 1.32 | 29 | 545 | 24 | .38 | 144 | .11 | .91 | .07 | 1.76 | .22 | 19 | 34 |
| 0 | 59 | 4.91 | 105 | 1850 | 1131 | 1.07 | 47 | .28 | <.01 | .26 | 4.89 | .18 | 37 | 21 |
| 0 | 22 | 1.82 | 39 | 685 | 419 | .4 | 17 | .11 | <.01 | .1 | 1.81 | .07 | 14 | 8 |
| 0 | 14 | .8 | 24 | 470 | 80 | .23 | 39 | .06 | .16 | .11 | 1.09 | .1 | 7 | 31 |
| 0 | 6 | .55 | 13 | 273 | 11 | .11 | 76 | .07 | .47 | .06 | .77 | .1 | 18 | 23 |
| 0 | 22 | 1.41 | 27 | 535 | 877 | .34 | 136 | .11 | 2.21 | .07 | 1.64 | .27 | 48 | 44 |
| 0 | 22 | 1.41 | 27 | 535 | 24 | .34 | 136 | .11 | 2.21 | .07 | 1.64 | .27 | 48 | 44 |
| 0 | 92 | 5.08 | 134 | 2454 | 231 | 2.1 | 639 | .41 | 11.3 | .5 | 8.44 | 1 | 59 | 111 |
| 0 | 42 | 3.1 | 60 | 1065 | 85 | .55 | 320 | .18 | 6.3 | .13 | 4.3 | .38 | 27 | 26 |
| 0 | 34 | 1.89 | 47 | 909 | 1482 | .61 | 240 | .16 | 3.43 | .14 | 2.82 | .38 | 23 | 32 |
| 0 | 34 | .34 | 12 | 211 | 78 | .31 | 0 | .04 | .05 | .04 | .47 | .1 | 14 | 18 |
| 0 | 197 | 1.15 | 32 | 292 | 42 | .2 | 792 | .06 | 2.48 | .1 | .59 | .26 | 170 | 39 |
| 0 | 249 | 3.18 | 43 | 367 | 25 | .67 | 1308 | .09 | 4.79 | .12 | .77 | .11 | 65 | 36 |
| 0 | 27 | 1.02 | 27 | 467 | 653 | .48 | 283 | .1 | .77 | .07 | 1.76 | .34 | 51 | 67 |
| 0 | 22 | .85 | 13 | 236 | 121 | .33 | 944 | .04 | .49 | .04 | .47 | .06 | 19 | 4 |
| 0 | 23 | .75 | 20 | 154 | 32 | .45 | 389 | .06 | .33 | .11 | .77 | .07 | 17 | 3 |
| 0 | 7 | .04 | 14 | 362 | 9 | .31 | 0 | .09 | .74 | .12 | .62 | .2 | 10 | 2 |
| 0 | 3 | .61 | 3 | 83 | 6 | .27 | 0 | .01 | .35 | .02 | .25 | .11 | 4 | 1 |
| 0 | 1 | .24 | 1 | 33 | 2 | .11 | 0 | <.01 | .14 | .01 | .1 | .04 | 2 | <1 |
| 0 | 20 | .03 | 4 | 56 | 7 | .02 | 80 | .01 | .17 | .02 | .03 | .02 | 2 | 7 |
| 0 | 3 | .36 | 3 | 25 | 220 | .06 | 1 | .66 | 1.04 | .07 | 1.13 | .07 | 6 | 0 |
| 0 | 63 | .37 | 41 | 376 | 504 | 1.78 | 22 | .19 | .67 | .08 | .54 | .17 | 30 | 4 |
|  | 8 | .44 | 11 | 120 | 172 | .25 | 0 | .06 |  | .02 | .12 | .02 | 7 | 0 |
| 0 | 13 | .97 | 7 | 171 | 228 | .37 | 0 | .4 | 1.11 | .27 | 2.68 | .28 | 32 | 0 |
| 0 | 21 | .78 | 10 | 198 | 474 | .42 | 0 | .66 | 1.61 | .24 | 3.18 | .42 | 46 | 0 |
| 0 | 53 | 4.24 | 36 | 362 | 527 | 1.26 | 78 | .12 | 1.25 | .07 | 1.21 | .15 | 82 | 16 |
| 0 | 54 | 1.14 | 13 | 342 | 279 | .8 | 0 | .63 | 2.25 | .51 | 6.84 | .85 | 58 | 0 |
| 0 | 17 | .92 | 9 | 76 | 219 | .61 | 0 | .56 | .98 | .61 | 8.16 | .5 | 40 | 0 |
| 0 | 36 | 1.35 |  | 129 | 112 |  | 18 | .05 |  | .09 |  | .06 |  | <1 |
| 0 | 181 | .33 |  | 307 | 78 |  | 3 | .08 |  | .2 |  | .13 |  | <1 |
| 0 | 27 | 1.54 | 15 | 188 | 576 | 1.07 | 0 | .64 | 2.01 | .37 | 7.37 | .74 | 67 | 0 |
| 0 | 21 | 1.49 | 13 | 128 | 391 | 1.28 | 0 | .64 | 1.23 | .43 | 7.1 | .85 | 55 | 0 |
| 0 | 9 | .27 | 33 | 166 | 119 | .46 | 0 | .1 |  | .35 | 1.04 | .45 | 49 | 0 |
| 20 | 66 | 2.9 | 31 | 331 | 538 | 1.05 | 729 | .65 | 4 | .4 | 4.47 | .31 | 40 | 10 |
| 0 | 21 | 1.49 | 13 | 128 | 391 | 1.28 | 0 | .64 | 1.23 | .43 | 7.1 | .85 | 55 | 0 |
| 13 | 146 | 2.71 | 26 | 211 | 931 | 1.97 | 45 | .77 | 1.43 | .55 | 8.14 | .86 | 69 | 1 |
| 0 | 154 | 3.75 | 116 | 609 | 10 | 3 | 115 | .22 | .03 | .18 | 7.69 | .5 | 86 | 0 |
| 0 | 79 | 2.66 | 15 | 398 | 351 | .69 | 0 | .23 | .01 | .51 | 3.78 | .18 | 225 | 0 |
| 0 | 65 | 1.84 |  |  | 340 | 4.56 | 0 | .18 |  | .09 | .27 | .18 |  | 0 |
| 0 | 54 | 2 |  |  | 285 | 3.82 | 0 | .15 |  | .07 | 2.28 | .15 |  | 0 |
| 0 | 74 | 2.61 |  |  | 374 | 6.93 | 3 | .28 |  | .14 | 4.05 | .28 |  | 0 |
| 0 | 76 | 2.7 |  |  | 378 | 7.2 | 0 | .29 |  | .14 | 4.32 | .29 |  | 0 |
| 0 | 71 | 2.52 |  |  | 371 | 6.66 | 3 | .27 |  | .13 | 4.05 | .27 |  | 0 |
| 2 | 40 | 1.4 |  | 153 | 709 | .43 | 0 | .67 |  | .3 | 2.89 | .45 |  | 0 |
| 4 | 2 | .63 |  | 76 | 503 | .2 | 0 | .98 |  | .46 | 2.1 | .35 |  | 0 |
| 2 | 8 | .77 |  | 51 | 243 | .89 | 0 | .26 |  | .46 | 1.98 | .14 |  | 0 |
| 1 | 4 | .39 |  | 29 | 160 | .46 | 0 | .13 |  | .22 | .64 | .29 |  | 0 |
| 1 | 12 | 1.06 | 16 | 121 | 455 | 1.11 | 0 | .14 |  | .3 | 1.9 | .51 | 21 | 0 |
| 2 | 24 | .31 |  | 225 | 433 | .41 | 0 | 1.25 |  | .65 | 5.41 | 1 |  | 0 |

**Table H-1**

**Food Composition** (Computer code number is for West Diet Analysis program) (For purposes of calculations, use "0" for t, <1, <.1, <.01, etc.)

| Computer Code Number | Food Description | Measure | Wt (g) | $H_2O$ (%) | Ener (kcal) | Prot (g) | Carb (g) | Dietary Fiber (g) | Fat (g) | Fat Breakdown (g) | | |
|---|---|---|---|---|---|---|---|---|---|---|---|---|
| | | | | | | | | | | Sat | Mono | Poly |
| | **VEGETARIAN FOODS:**—Continued | | | | | | | | | | | |
| 1174 | Vege-Burger, patty | 1 ea | 55 | 71 | 66 | 10 | 2 | 2 | 2 | .4 | .6 | .5 |
| | Vegetarian foods, Morningstar Farms: | | | | | | | | | | | |
| 7672 | Better-n-burgers, svg | 1 ea | 78 | 71 | 75 | 11 | 6 | 4 | <1 | .1 | .3 | .2 |
| 7766 | Better-n-eggs | ¼ c | 57 | 88 | 23 | 5 | <1 | 0 | <1 | .1 | .1 | .1 |
| 57436 | Breakfast links | 2 pce | 45 | 60 | 63 | 8 | 2 | 2 | 2 | .5 | .7 | 1.3 |
| 7752 | Breakfast strips | 2 pce | 16 | 43 | 56 | 2 | 2 | <1 | 4 | .7 | 1.1 | 2.6 |
| 7725 | Burger crumbles, svg | 1 ea | 55 | 60 | 116 | 11 | 3 | 3 | 6 | 1.6 | 2.3 | 2.5 |
| 7726 | Burger, spicy black bean | 1 ea | 78 | 60 | 113 | 11 | 15 | 5 | 1 | .2 | .3 | .4 |
| 7665 | Chik pattie | 1 ea | 71 | 51 | 177 | 7 | 15 | 2 | 10 | 1.3 | 2.6 | 5.9 |
| 7724 | Frank, deli | 1 ea | 45 | 52 | 109 | 10 | 3 | 3 | 7 | 1 | 2.1 | 3.5 |
| 7722 | Garden vege pattie | 1 ea | 67 | 60 | 104 | 11 | 9 | 4 | 4 | .5 | 1.1 | 2.2 |
| 7746 | Grillers | 1 ea | 64 | 55 | 139 | 14 | 5 | 3 | 7 | 1.7 | 2.2 | 3 |
| 7664 | Prime pattie | 1 ea | 64 | 64 | 94 | 16 | 4 | 3 | 2 | .2 | .4 | .6 |
| | Vegetarian foods, Worthington: | | | | | | | | | | | |
| 7634 | Beef style, meatless, frzn | 3 pce | 55 | 58 | 113 | 9 | 4 | 3 | 7 | 1.2 | 2.7 | 2.6 |
| 7732 | Burger, meatless, patty | ¼ c | 55 | 71 | 60 | 9 | 2 | 1 | 2 | .3 | .5 | 1.1 |
| 1846 | Chik slices, canned | 2 pce | 60 | 78 | 62 | 6 | 1 | 1 | 4 | .6 | .9 | 2.3 |
| 1833 | Chili, canned | ½ c | 106 | 73 | 136 | 9 | 10 | 4 | 7 | 1.1 | 1.7 | 4.1 |
| 1835 | Choplets, slices, canned | 2 pce | 92 | 72 | 93 | 17 | 3 | 2 | 2 | .9 | .3 | .3 |
| 7608 | Corned beef style, meatless, frzn | 4 pce | 57 | 55 | 138 | 10 | 5 | 2 | 9 | 1.9 | 4.1 | 3.1 |
| 1831 | Country stew, canned | 1 c | 240 | 81 | 208 | 13 | 20 | 5 | 9 | 1.6 | 2.3 | 4.8 |
| 7632 | Egg roll, meatless, frzn | 1 ea | 85 | 53 | 181 | 6 | 20 | 2 | 8 | 1.7 | 4.5 | 2.3 |
| 1838 | Numete, slices, canned | 1 pce | 55 | 58 | 132 | 6 | 5 | 3 | 10 | 2.4 | 4.4 | 2.7 |
| 1839 | Prime stakes, slices, canned | 1 pce | 92 | 71 | 136 | 9 | 4 | 4 | 9 | 1.4 | 2.9 | 4.9 |
| 1840 | Protose, slices, canned | 1 pce | 55 | 53 | 131 | 13 | 5 | 3 | 7 | 1 | 3 | 2.4 |
| 7606 | Roast, dinner, meatless, frzn | 1 ea | 85 | 63 | 180 | 12 | 5 | 3 | 12 | 2.2 | 5 | 5.2 |
| 1842 | Saucette links, canned | 1 pce | 38 | 62 | 86 | 6 | 1 | | 6 | 1.1 | 1.6 | 3.8 |
| 1844 | Savory slices, canned | 1 pce | 28 | 66 | 48 | 3 | 2 | 1 | 3 | 1.2 | 1.3 | .6 |
| 7735 | Stakelets, frzn | 1 pce | 71 | 58 | 145 | 12 | 6 | 2 | 8 | 1.4 | 2.7 | 3.9 |
| 1847 | Turkee slices, canned | 1 pce | 33 | 64 | 68 | 5 | 1 | 1 | 5 | .8 | 1.9 | 2.1 |
| | **MISCELLANEOUS** | | | | | | | | | | | |
| | Baking powders for home use: | | | | | | | | | | | |
| | Sodium aluminum sulfate: | | | | | | | | | | | |
| 962 | With monocalcium phosphate monohydrate | 1 tsp | 5 | 2 | 6 | <1 | 2 | 0 | 0 | 0 | 0 | 0 |
| 963 | With monocalcium phosphate monohydrate, calcium sulfate | 1 tsp | 5 | 5 | 3 | 0 | 1 | <1 | 0 | 0 | 0 | 0 |
| 964 | Straight phosphate | 1 tsp | 5 | 4 | 3 | <1 | 1 | <1 | 0 | 0 | 0 | 0 |
| 965 | Low sodium | 1 tsp | 5 | 6 | 5 | <1 | 2 | <1 | <1 | t | 0 | t |
| 1204 | Baking soda | 1 tsp | 5 | | 0 | 0 | 0 | 0 | 0 | 0 | 0 | 0 |
| 966 | Basil, dried | 1 tbs | 5 | 6 | 13 | 1 | 3 | 2 | <1 | t | t | .1 |
| 2068 | Cajun seasoning | 1 tsp | 3 | 5 | 6 | <1 | 1 | <1 | <1 | | | |
| 961 | Carob flour | 1 c | 103 | 4 | 185 | 5 | 92 | 41 | 1 | .1 | .2 | .2 |
| 967 | Catsup: | ¼ c | 61 | 67 | 63 | 1 | 17 | 1 | <1 | t | t | .1 |
| 968 | Tablespoon | 1 tbs | 15 | 67 | 16 | <1 | 4 | <1 | <1 | t | t | t |
| 1200 | Cayenne/red pepper | 1 tbs | 5 | 8 | 16 | 1 | 3 | 1 | 1 | .2 | .1 | .4 |
| 969 | Celery seed | 1 tsp | 2 | 6 | 8 | <1 | 1 | <1 | <1 | t | .3 | .1 |
| 1203 | Chili powder: | 1 tbs | 8 | 8 | 25 | 1 | 4 | 3 | 1 | .2 | .3 | .6 |
| 970 | Teaspoon | 1 tsp | 3 | 8 | 9 | <1 | 2 | 1 | <1 | .1 | .1 | .2 |
| | Chocolate: | | | | | | | | | | | |
| 971 | Baking, unsweetened, square | 1 oz | 28 | 1 | 146 | 3 | 8 | 4 | 15 | 9.1 | 5.2 | .5 |
| | For other chocolate items, see Sweeteners & Sweets | | | | | | | | | | | |
| 972 | Cilantro/coriander, fresh | 1 tbs | 1 | 93 | | <1 | <1 | <1 | <1 | 0 | t | 0 |
| 2287 | Cinnamon | 1 tsp | 2 | 10 | 5 | <1 | 2 | 1 | <1 | t | t | t |
| 1197 | Cornstarch | 1 tbs | 8 | 8 | 30 | <1 | 7 | <1 | <1 | t | t | t |
| 2239 | Curry powder | 1 tsp | 2 | 10 | 6 | <1 | 1 | 1 | <1 | t | .1 | .1 |
| 1202 | Dill weed, dried | 1 tbs | 3 | 7 | 8 | 1 | 2 | <1 | <1 | t | .1 | t |
| 975 | Garlic cloves | 1 ea | 3 | 59 | 4 | <1 | 1 | <1 | <1 | t | 0 | t |
| 2238 | Garlic powder | 1 tsp | 3 | 6 | 10 | <1 | 2 | <1 | <1 | t | t | t |
| 977 | Gelatin, dry, unsweetened: Envelope | 1 ea | 7 | 13 | 23 | 6 | 0 | 0 | <1 | t | t | t |
| 978 | Ginger root, slices, raw | 2 pce | 5 | 82 | 3 | <1 | 1 | <1 | <1 | t | t | t |

**PAGE KEY:** H–4 = Beverages H–6 = Dairy H–10 = Eggs H–10 = Fat/Oil H–14 = Fruit H–20 = Bakery H–28 = Grain H–32 = Fish H–34 = Meats H–38 = Poultry H–40 = Sausage H–42 = Mixed/Fast H–46 = Nuts/Seeds H–50 = Sweets H–52 = Vegetables/Legumes H–62 = Vegetarian Foods H–64 = Misc H–66 = Soups/Sauces H–68 = Fast H–84 = Convenience H–88 = Baby foods



| Chol (mg) | Calc (mg) | Iron (mg) | Magn (mg) | Pota (mg) | Sodi (mg) | Zinc (mg) | VT-A (RE) | Thia (mg) | VT-E (a-TE) | Ribo (mg) | Niac (mg) | V-B6 (mg) | Fola (µg) | VT-C (mg) |
|---|---|---|---|---|---|---|---|---|---|---|---|---|---|---|
| 0 | 8 | .5 | 12 | 30 | 114 | .58 | 0 | .2 | | .25 | .78 | .31 | 15 | 0 |
| | | | | | | | | | | | | | | |
| 0 | 79 | 2.66 | 15 | 398 | 351 | .69 | 0 | .23 | .01 | .51 | 3.78 | .18 | 225 | 0 |
| 2 | 7 | .63 | | 68 | 90 | .51 | 64 | .01 | | .26 | 0 | .11 | | 0 |
| 1 | 15 | 2.14 | 16 | 59 | 338 | .36 | 0 | 6.95 | | .22 | 5.19 | .33 | 12 | 0 |
| <1 | 7 | .27 | | 15 | 220 | .05 | 0 | .75 | | .04 | .6 | .07 | | 0 |
| 0 | 40 | 3.2 | 1 | 89 | 238 | .82 | 0 | 4.96 | .34 | .18 | 1.49 | .27 | | 0 |
| 1 | 56 | 1.84 | 44 | 269 | 499 | .93 | 14 | 8.03 | .36 | .14 | 0 | .21 | | 0 |
| 1 | 11 | 1.02 | | 163 | 536 | .31 | 0 | 2.15 | | .16 | 1.51 | .14 | | 0 |
| 2 | 16 | .26 | 4 | 50 | 524 | .38 | 0 | .14 | 1.26 | .02 | 0 | .01 | | 0 |
| 1 | 34 | .72 | 29 | 180 | 382 | .59 | 20 | 6.47 | .98 | .1 | 0 | 0 | 29 | 0 |
| 2 | 43 | 1.16 | | 127 | 256 | .49 | 0 | 11.7 | | .24 | 2.99 | .37 | | 0 |
| 1 | 46 | 2.14 | | 142 | 247 | .74 | 0 | .51 | | .25 | .92 | .41 | | 2 |
| | | | | | | | | | | | | | | |
| 0 | 4 | 2.63 | | 44 | 624 | .22 | 0 | .89 | | .34 | 6.46 | .56 | | 0 |
| 0 | 4 | 1.73 | | 25 | 269 | .38 | 0 | .13 | | .1 | 1.96 | .24 | | 0 |
| 1 | 9 | .73 | | 111 | 257 | .26 | 0 | .06 | | .05 | .37 | .08 | | 0 |
| 0 | 20 | 1.49 | | 195 | 523 | .57 | 0 | .02 | | .03 | 1.04 | .31 | | 0 |
| 0 | 6 | .37 | | 40 | 500 | .65 | 0 | .05 | | .05 | 0 | .05 | | 0 |
| 1 | 6 | 1.17 | | 58 | 524 | .26 | 0 | 10.6 | | .07 | 1.36 | .3 | | 0 |
| 2 | 51 | 5.09 | | 270 | 826 | 1.03 | 216 | 1.85 | | .29 | 4.22 | .86 | | 0 |
| 1 | 15 | .57 | | 96 | 384 | .31 | 0 | 1.22 | | .19 | 0 | .03 | | 0 |
| 0 | 10 | 1.12 | | 155 | 272 | .56 | 0 | .08 | | .06 | .54 | .2 | | 0 |
| 2 | 12 | .38 | | 82 | 445 | .38 | 0 | .12 | | .13 | 1.98 | .38 | | 0 |
| <1 | 1 | 1.84 | | 50 | 283 | .7 | 0 | .18 | | .13 | 1.34 | .24 | | 0 |
| 2 | 36 | 2.87 | | 38 | 566 | .64 | 0 | 2.13 | | .25 | 6.02 | .6 | | 0 |
| 1 | 9 | 1.15 | | 25 | 205 | .26 | 0 | .59 | | .08 | .09 | .13 | | 0 |
| <1 | | .47 | | 14 | 179 | .08 | 0 | .08 | | .06 | .48 | .1 | | 0 |
| 2 | 49 | .99 | | 95 | 484 | .5 | 0 | 1.51 | | .12 | 3.1 | .26 | | 0 |
| 1 | 3 | .47 | | 16 | 203 | .11 | 0 | 1.13 | | .05 | .39 | .09 | | 0 |
| | | | | | | | | | | | | | | |
| | | | | | | | | | | | | | | |
| | | | | | | | | | | | | | | |
| 0 | 97 | 0 | | 7 | 547 | 0 | 0 | 0 | 0 | 0 | 0 | 0 | 0 | 0 |
| | | | | | | | | | | | | | | |
| 0 | 294 | .55 | 1 | 1 | 530 | <.01 | 0 | 0 | 0 | 0 | 0 | 0 | 0 | 0 |
| 0 | 368 | .56 | 2 | | 395 | <.01 | 0 | 0 | 0 | 0 | 0 | 0 | 0 | 0 |
| 0 | 217 | .41 | 1 | 505 | 4 | .04 | 0 | 0 | <.01 | 0 | 0 | 0 | 0 | 0 |
| 0 | 0 | 0 | 0 | 0 | 1368 | 0 | 0 | 0 | 0 | 0 | 0 | 0 | 0 | 0 |
| 0 | 106 | 2.1 | 21 | 172 | 2 | .29 | 47 | .01 | .08 | .02 | .35 | .06 | 14 | 3 |
| | | | | 29 | 474 | | | | | | | | | |
| 0 | 358 | 3.03 | 56 | 852 | 36 | .95 | 1 | .05 | .65 | .47 | 1.96 | .38 | 30 | <1 |
| 0 | 12 | .43 | 13 | 293 | 723 | .14 | 62 | .05 | .9 | .04 | .84 | .11 | 9 | 9 |
| 0 | 3 | .1 | 3 | 72 | 178 | .03 | 15 | .01 | .22 | .01 | .21 | .03 | 2 | 2 |
| 0 | 7 | .39 | 8 | 101 | 1 | .12 | 208 | .02 | .24 | .05 | .43 | .1 | 5 | 4 |
| 0 | 35 | .9 | 9 | 28 | 3 | .14 | <1 | .01 | .02 | .01 | .06 | .01 | <1 | <1 |
| 0 | 22 | 1.14 | 14 | 153 | 81 | .22 | 279 | .03 | .08 | .06 | .63 | .15 | 8 | 5 |
| 0 | 8 | .43 | 5 | 57 | 30 | .08 | 105 | .01 | .03 | .02 | .24 | .06 | 3 | 2 |
| | | | | | | | | | | | | | | |
| 0 | 21 | 1.77 | 87 | 233 | 4 | 1.12 | 3 | .02 | .34 | .05 | .31 | .03 | 2 | 0 |
| | | | | | | | | | | | | | | |
| | | | | | | | | | | | | | | |
| 0 | 1 | .02 | | 5 | <1 | <.01 | 3 | <.01 | .02 | <.01 | .01 | <.01 | <1 | <1 |
| 0 | 25 | .76 | 1 | 10 | 1 | .04 | 1 | <.01 | 0 | <.01 | .03 | <.01 | 1 | 1 |
| 0 | | .04 | | | 1 | <.01 | 0 | 0 | 0 | 0 | 0 | 0 | 0 | 0 |
| 0 | 10 | .59 | 5 | 31 | 1 | .08 | 2 | <.01 | .01 | .01 | .07 | .01 | 3 | <1 |
| 0 | 53 | 1.46 | 13 | 99 | 6 | .1 | 18 | .01 | | .01 | .08 | .04 | | 1 |
| 0 | 5 | .05 | 1 | 12 | 1 | .03 | 0 | .01 | 0 | <.01 | .02 | .04 | <1 | 1 |
| 0 | 2 | .08 | 2 | 33 | 1 | .08 | 0 | .01 | 0 | <.01 | .02 | .08 | <1 | 1 |
| 0 | 4 | .08 | 2 | 1 | 14 | .01 | 0 | <.01 | 0 | .02 | .01 | 0 | 2 | 0 |
| 0 | 1 | .02 | 2 | 21 | 1 | .02 | 0 | <.01 | .01 | <.01 | .03 | .01 | 1 | <1 |

**Table H-1**

**Food Composition** (Computer code number is for West Diet Analysis program) (For purposes of calculations, use "0" for t, <1, <.1, <.01, etc.)

| Computer Code Number | Food Description | Measure | Wt (g) | $H_2O$ (%) | Ener (kcal) | Prot (g) | Carb (g) | Dietary Fiber (g) | Fat (g) | Fat Breakdown (g) Sat | Mono | Poly |
|---|---|---|---|---|---|---|---|---|---|---|---|---|
| | **MISCELLANEOUS—**Continued | | | | | | | | | | | |
| 1198 | Horseradish, prepared | 1 tbs | 15 | 85 | 7 | <1 | 2 | <1 | <1 | t | t | .1 |
| 1997 | Hummous/hummus | 1 c | 246 | 65 | 421 | 12 | 50 | 12 | 21 | 3.1 | 8.8 | 7.8 |
| 1909 | Mustard, country dijon | 1 tsp | 5 | | 5 | <1 | <1 | 0 | 0 | 0 | 0 | 0 |
| 2019 | Mustard, gai choy Chinese | 1 tbs | 16 | 94 | 3 | <1 | 1 | | <1 | | | |
| 979 | Mustard, prepared (1 packet = 1 tsp) | 1 tsp | 5 | 80 | 4 | <1 | <1 | <1 | <1 | t | .2 | t |
| | Miso (see #926 under Vegetables and Legumes, Soybean products) | | | | | | | | | | | |
| 980 | Olives, green | 5 ea | 20 | 78 | 23 | <1 | <1 | <1 | 3 | .3 | 1.9 | .2 |
| 981 | Olives, ripe, pitted | 5 ea | 22 | 80 | 25 | <1 | 1 | 1 | 2 | .3 | 1.7 | .2 |
| 26008 | Onion powder | 1 tsp | 2 | 5 | 7 | <1 | 2 | <1 | <1 | t | t | t |
| 2237 | Oregano, ground | 1 tsp | 2 | 7 | 6 | <1 | 1 | 1 | <1 | .1 | t | .1 |
| 2236 | Paprika | 1 tsp | 2 | 10 | 6 | <1 | 1 | <1 | <1 | t | t | .2 |
| 887 | Parsley, freeze dried | ¼ c | 1 | 2 | 3 | <1 | <1 | <1 | <1 | t | t | t |
| | Parsley, fresh (see #885 and #886) | | | | | | | | | | | |
| 985 | Pepper, black | 1 tsp | 2 | 10 | 5 | <1 | 1 | 1 | <1 | t | t | t |
| | Pickles: | | | | | | | | | | | |
| 986 | Dill, medium, 3¾ x 1¼" diam | 1 ea | 65 | 92 | 12 | <1 | 3 | 1 | <1 | t | t | t |
| 987 | Fresh pack, slices, 1½" diam x ¼" | 2 pce | 15 | 79 | 11 | <1 | 3 | <1 | <1 | 0 | 0 | t |
| 988 | Sweet, medium | 1 ea | 35 | 65 | 41 | <1 | 11 | <1 | <1 | t | t | t |
| 989 | Pickle relish, sweet | 1 tbs | 15 | 63 | 21 | <1 | 5 | <1 | <1 | t | t | t |
| | Popcorn (see Grain Products #539-541) | | | | | | | | | | | |
| 917 | Potato chips: | 10 pce | 20 | 2 | 107 | 1 | 11 | 1 | 7 | 2.2 | 2 | 2.4 |
| 44076 | Unsalted | 1 oz | 28 | 2 | 150 | 2 | 15 | 1 | 10 | 3.1 | 2.8 | 3.4 |
| 1201 | Sage, ground | 1 tsp | 1 | 8 | 3 | <1 | 1 | <1 | <1 | .1 | t | t |
| 1347 | Salsa, from recipe | 1 tbs | 15 | 93 | 3 | <1 | 1 | <1 | <1 | t | t | t |
| 2218 | Salsa, pico de gallo, medium | 1 tbs | 15 | 92 | 2 | 0 | 1 | <1 | 0 | 0 | 0 | 0 |
| 990 | Salt | 1 tsp | 6 | | 0 | 0 | 0 | 0 | 0 | 0 | 0 | 0 |
| | Salt Substitutes: | | | | | | | | | | | |
| 1205 | Morton, salt substitute | 1 tsp | 6 | | | 0 | <1 | | 0 | 0 | 0 | 0 |
| 1207 | Morton, light salt | 1 tsp | 6 | | | 0 | <1 | | 0 | 0 | 0 | 0 |
| 2067 | Seasoned salt, no MSG | 1 tsp | 5 | 5 | 4 | <1 | 1 | <1 | <1 | | | |
| 991 | Vinegar, cider | ½ c | 120 | 94 | 17 | 0 | 7 | 0 | 0 | 0 | 0 | 0 |
| 2172 | Balsamic | 1 tbs | 15 | 64 | 21 | 0 | 4 | 0 | 0 | 0 | 0 | 0 |
| 2176 | Malt | 1 tbs | 15 | 90 | 5 | 0 | <1 | 0 | 0 | 0 | 0 | 0 |
| 2182 | Tarragon | 1 tbs | 15 | 95 | 3 | 0 | <1 | 0 | 0 | 0 | 0 | 0 |
| 2181 | White wine | 1 tbs | 15 | 89 | 5 | 0 | <1 | 0 | 0 | 0 | 0 | 0 |
| | Yeast: | | | | | | | | | | | |
| 992 | Baker's, dry, active, package | 1 ea | 7 | 8 | 21 | 3 | 3 | 1 | <1 | t | .2 | t |
| 993 | Brewer's, dry | 1 tbs | 8 | 5 | 23 | 3 | 3 | 3 | <1 | t | t | 0 |
| | **SOUPS, SAUCES, and GRAVIES** | | | | | | | | | | | |
| | SOUPS, canned, condensed: | | | | | | | | | | | |
| | Unprepared, condensed: | | | | | | | | | | | |
| 1210 | Cream of celery | 1 c | 251 | 85 | 181 | 3 | 18 | 2 | 11 | 2.8 | 2.6 | 5 |
| 1215 | Cream of chicken | 1 c | 251 | 82 | 233 | 7 | 18 | <1 | 15 | 4.2 | 6.5 | 3 |
| 1216 | Cream of mushroom | 1 c | 251 | 81 | 259 | 4 | 19 | 1 | 19 | 5.1 | 3.6 | 8.9 |
| 1220 | Onion | 1 c | 246 | 86 | 113 | 8 | 16 | 2 | 3 | .5 | 1.5 | 1.3 |
| | Prepared w/equal volume of whole milk: | | | | | | | | | | | |
| 994 | Clam chowder, New England | 1 c | 248 | 85 | 164 | 9 | 17 | 1 | 7 | 2.9 | 2.3 | 1.1 |
| 1209 | Cream of celery | 1 c | 248 | 86 | 164 | 6 | 14 | 1 | 10 | 3.9 | 2.5 | 2.6 |
| 995 | Cream of chicken | 1 c | 248 | 85 | 191 | 8 | 15 | <1 | 11 | 4.6 | 4.5 | 1.6 |
| 996 | Cream of mushroom | 1 c | 248 | 85 | 203 | 6 | 15 | <1 | 14 | 5.1 | 3 | 4.6 |
| 1214 | Cream of potato | 1 c | 248 | 87 | 149 | 6 | 17 | <1 | 6 | 3.8 | 1.7 | .6 |
| 1213 | Oyster stew | 1 c | 245 | 89 | 135 | 6 | 10 | 0 | 8 | 5 | 2.1 | .3 |
| 997 | Tomato | 1 c | 248 | 85 | 161 | 6 | 22 | 3 | 6 | 2.9 | 1.6 | 1.1 |
| | Prepared with equal volume of water: | | | | | | | | | | | |
| 998 | Bean with bacon | 1 c | 253 | 84 | 172 | 8 | 23 | 9 | 6 | 1.5 | 2.2 | 1.8 |
| 999 | Beef broth/bouillon/consommé | 1 c | 240 | 98 | 17 | 3 | <1 | 0 | 1 | .3 | .2 | t |
| 1000 | Beef noodle | 1 c | 244 | 92 | 83 | 5 | 9 | 1 | 3 | 1.1 | 1.2 | .5 |
| 1001 | Chicken noodle | 1 c | 241 | 92 | 75 | 4 | 9 | 1 | 2 | .7 | 1.1 | .6 |
| 1002 | Chicken rice | 1 c | 241 | 94 | 60 | 4 | 7 | 1 | 2 | .5 | .9 | .4 |
| 1208 | Chili beef | 1 c | 250 | 85 | 170 | 7 | 21 | 9 | 7 | 3.3 | 2.8 | .3 |
| 1003 | Clam chowder, Manhattan | 1 c | 244 | 92 | 78 | 2 | 12 | 1 | 2 | .4 | .4 | 1.3 |

| Chol (mg) | Calc (mg) | Iron (mg) | Magn (mg) | Pota (mg) | Sodi (mg) | Zinc (mg) | VT-A (RE) | Thia (mg) | VT-E (a-TE) | Ribo (mg) | Niac (mg) | V-B6 (mg) | Fola (µg) | VT-C (mg) |
|---|---|---|---|---|---|---|---|---|---|---|---|---|---|---|
| 0 | 8 | .06 | 4 | 37 | 47 | .12 | <1 | <.01 | <.01 | <.01 | .06 | .01 | 9 | 4 |
| 0 | 123 | 3.86 | 71 | 428 | 600 | 2.71 | 5 | .23 | 2.46 | .13 | 1.01 | .98 | 146 | 19 |
| 0 | | | | 10 | 120 | | | | | | | | | |
| | | | | | | | | | | | | | | |
| 0 | 4 | .1 | 2 | 6 | 63 | .03 | 0 | 0 | .09 | 0 | 0 | <.01 | 0 | 0 |
| | | | | | | | | | | | | | | |
| | | | | | | | | | | | | | | |
| 0 | 12 | .32 | 4 | 11 | 480 | .01 | 6 | 0 | .6 | 0 | 0 | <.01 | <1 | 0 |
| 0 | 19 | .73 | 1 | 2 | 192 | .05 | 9 | <.01 | .66 | 0 | .01 | <.01 | 0 | <1 |
| 0 | 7 | .05 | 2 | 19 | 1 | .05 | 0 | .01 | <.01 | <.01 | .01 | .03 | 3 | <1 |
| 0 | 31 | .88 | 5 | 33 | <1 | .09 | 14 | .01 | .03 | .01 | .12 | .02 | 5 | 1 |
| 0 | 4 | .47 | 4 | 47 | 1 | .08 | 121 | .01 | .01 | .03 | .31 | .04 | 2 | 1 |
| 0 | 2 | .54 | 4 | 63 | 4 | .06 | 63 | .01 | .06 | .02 | .1 | .01 | 15 | 1 |
| | | | | | | | | | | | | | | |
| 0 | 9 | .58 | 4 | 25 | 1 | .03 | <1 | <.01 | .02 | <.01 | .02 | .01 | <1 | <1 |
| | | | | | | | | | | | | | | |
| 0 | 6 | .34 | 7 | 75 | 833 | .09 | 21 | .01 | .1 | .02 | .04 | .01 | 1 | 1 |
| 0 | 5 | .27 | 1 | 30 | 101 | 0 | 2 | 0 | .02 | <.01 | 0 | <.01 | 0 | 1 |
| 0 | 1 | .21 | 1 | 11 | 329 | .03 | 5 | <.01 | .06 | .01 | .06 | <.01 | <1 | <1 |
| 0 | 3 | .12 | 1 | 30 | 107 | .01 | 1 | 0 | .02 | <.01 | 0 | 0 | 0 | 1 |
| | | | | | | | | | | | | | | |
| 0 | 5 | .33 | 13 | 255 | 119 | .22 | 0 | .03 | .98 | .04 | .77 | .13 | 9 | 6 |
| 0 | 7 | .46 | 19 | 357 | 2 | .3 | 0 | .05 | 1.37 | .05 | 1.07 | .18 | 13 | 9 |
| 0 | 16 | .28 | 4 | 11 | <1 | .05 | 6 | .01 | .02 | <.01 | .06 | .01 | 3 | <1 |
| 0 | 1 | .06 | 1 | 24 | 58 | .02 | 22 | .01 | .04 | <.01 | .06 | .01 | 2 | 5 |
| 0 | | | | | 130 | | | | | | | | | |
| 0 | 1 | .02 | | | 2325 | .01 | 0 | 0 | 0 | 0 | 0 | 0 | 0 | 0 |
| | | | | | | | | | | | | | | |
| | 33 | | | 3018 | <1 | | | | | | | | | |
| | 2 | | 4 | 1560 | 1170 | | | | | | | | | |
| | | | | 15 | 1542 | | | | | | | | | |
| 0 | 7 | .72 | 26 | 120 | 1 | 0 | 0 | 0 | 0 | 0 | 0 | 0 | 0 | 0 |
| | 2 | .07 | | 10 | 3 | | <1 | .07 | | .07 | .07 | | | <1 |
| | 2 | .07 | | 13 | 4 | | <1 | .07 | | .07 | .07 | | | 1 |
| | | .07 | | 2 | 1 | | <1 | .07 | | .07 | .07 | | | <1 |
| | 1 | .07 | | 12 | 1 | | <1 | .07 | | .07 | .07 | | | <1 |
| | | | | | | | | | | | | | | |
| 0 | 4 | 1.16 | 7 | 140 | 3 | .45 | <1 | .16 | .01 | .38 | 2.79 | .11 | 164 | <1 |
| 0 | 17 | 1.38 | 18 | 151 | 10 | .63 | 0 | 1.25 | | .34 | 3.03 | .4 | 313 | 0 |
| | | | | | | | | | | | | | | |
| | | | | | | | | | | | | | | |
| 28 | 80 | 1.26 | 13 | 246 | 1900 | .3 | 60 | .06 | .38 | .1 | .66 | .02 | 5 | <1 |
| 20 | 68 | 1.2 | 5 | 176 | 1972 | 1.26 | 113 | .06 | .33 | .12 | 1.64 | .03 | 3 | <1 |
| 3 | 65 | 1.05 | 10 | 168 | 1736 | 1.19 | 0 | .06 | 2.61 | .17 | 1.62 | .02 | 8 | 2 |
| 0 | 54 | 1.35 | 5 | 138 | 2115 | 1.23 | 0 | .07 | .57 | .05 | 1.21 | .1 | 30 | 2 |
| | | | | | | | | | | | | | | |
| 22 | 186 | 1.49 | 22 | 300 | 992 | .8 | 40 | .07 | .15 | .24 | 1.03 | .13 | 10 | 3 |
| 32 | 186 | .69 | 22 | 310 | 1009 | .2 | 67 | .07 | .97 | .25 | .44 | .06 | 8 | 1 |
| 27 | 181 | .67 | 17 | 273 | 1046 | .67 | 94 | .07 | .24 | .26 | .92 | .07 | 8 | 1 |
| 20 | 179 | .59 | 20 | 270 | 918 | .64 | 37 | .08 | 1.34 | .28 | .91 | .06 | 10 | 2 |
| 22 | 166 | .55 | 17 | 322 | 1061 | .67 | 67 | .08 | .1 | .24 | .64 | .09 | 9 | 1 |
| 32 | 167 | 1.05 | 20 | 235 | 1041 | 10.3 | 44 | .07 | .49 | .23 | .34 | .06 | 10 | 4 |
| 17 | 159 | 1.81 | 22 | 449 | 744 | .29 | 109 | .13 | 2.6 | .25 | 1.52 | .16 | 21 | 68 |
| | | | | | | | | | | | | | | |
| 3 | 81 | 2.05 | 45 | 402 | 951 | 1.03 | 89 | .09 | .08 | .03 | .57 | .04 | 32 | 2 |
| 0 | 14 | .41 | 5 | 130 | 782 | 0 | 0 | <.01 | 0 | .05 | 1.87 | .02 | 5 | 0 |
| 5 | 15 | 1.1 | 5 | 100 | 952 | 1.54 | 63 | .07 | <.01 | .06 | 1.07 | .04 | 19 | <1 |
| 7 | 17 | .77 | 5 | 55 | 1106 | .39 | 72 | .05 | .07 | .06 | 1.39 | .03 | 22 | <1 |
| 7 | 17 | .75 | 0 | 101 | 815 | .26 | 65 | .02 | .05 | .02 | 1.13 | .02 | 1 | <1 |
| 12 | 42 | 2.13 | 30 | 525 | 1035 | 1.4 | 150 | .06 | .17 | .07 | 1.07 | .16 | 17 | 4 |
| 2 | 27 | 1.63 | 12 | 188 | 578 | .98 | 98 | .03 | .73 | .04 | .82 | .1 | 10 | 4 |

**Table H-1**

**Food Composition** (Computer code number is for West Diet Analysis program) (For purposes of calculations, use "0" for t, <1, <.1, <.01, etc.)

| Computer Code Number | Food Description | Measure | Wt (g) | $H_2O$ (%) | Ener (kcal) | Prot (g) | Carb (g) | Dietary Fiber (g) | Fat (g) | Fat Breakdown (g) Sat | Mono | Poly |
|---|---|---|---|---|---|---|---|---|---|---|---|---|
| | **SOUPS, SAUCES, and GRAVIES—**Continued | | | | | | | | | | | |
| 1004 | Cream of chicken | 1 c | 244 | 91 | 117 | 3 | 9 | <1 | 7 | 2.1 | 3.3 | 1.5 |
| 1005 | Cream of mushroom | 1 c | 244 | 90 | 129 | 2 | 9 | <1 | 9 | 2.4 | 1.7 | 4.2 |
| 1006 | Minestrone | 1 c | 241 | 91 | 82 | 4 | 11 | 1 | 3 | .6 | .7 | 1.1 |
| 1211 | Onion | 1 c | 241 | 93 | 58 | 4 | 8 | 1 | 2 | .3 | .7 | .7 |
| 1007 | Split pea & ham | 1 c | 253 | 82 | 190 | 10 | 28 | 2 | 4 | 1.8 | 1.8 | .6 |
| 1008 | Tomato | 1 c | 244 | 90 | 85 | 2 | 17 | <1 | 2 | .4 | .4 | 1 |
| 1009 | Vegetable beef | 1 c | 244 | 92 | 78 | 6 | 10 | <1 | 2 | .9 | .8 | .1 |
| 1010 | Vegetarian vegetable | 1 c | 241 | 92 | 72 | 2 | 12 | <1 | 2 | .3 | .8 | .7 |
| | Ready to serve: | | | | | | | | | | | |
| 1707 | Chunky chicken soup | 1 c | 251 | 84 | 178 | 13 | 17 | 2 | 7 | 2 | 3 | 1.4 |
| | SOUPS, dehydrated: | | | | | | | | | | | |
| | Prepared with water: | | | | | | | | | | | |
| 1299 | Beef broth/bouillon | 1 c | 244 | 97 | 19 | 1 | 2 | 0 | 1 | .3 | .3 | t |
| 1376 | Chicken broth | 1 c | 244 | 97 | 22 | 1 | 1 | 0 | 1 | .3 | .4 | .4 |
| 1013 | Chicken noodle | 1 c | 252 | 94 | 53 | 3 | 7 | 1 | 1 | .3 | .5 | .4 |
| 1122 | Cream of chicken | 1 c | 261 | 91 | 107 | 2 | 13 | <1 | 5 | 3.4 | 1.2 | .4 |
| 1014 | Onion | 1 c | 246 | 96 | 27 | 1 | 5 | 1 | 1 | .1 | .3 | .1 |
| 1217 | Split pea | 1 c | 255 | 87 | 125 | 7 | 21 | 3 | 1 | .4 | .7 | .3 |
| 1015 | Tomato vegetable | 1 c | 253 | 93 | 56 | 2 | 10 | <1 | 1 | .4 | .3 | .1 |
| | Unprepared, dry products: | | | | | | | | | | | |
| 1011 | Beef bouillon, packet | 1 ea | 6 | 3 | 14 | 1 | 1 | 0 | 1 | .3 | .2 | t |
| 1012 | Onion soup, packet | 1 ea | 39 | 4 | 115 | 5 | 21 | 4 | 2 | .5 | 1.4 | .3 |
| | SAUCES | | | | | | | | | | | |
| | From dry mixes, prepared with milk: | | | | | | | | | | | |
| 1016 | Cheese sauce | 1 c | 279 | 77 | 307 | 17 | 23 | 1 | 17 | 9.3 | 5.3 | 1.6 |
| 1017 | Hollandaise | 1 c | 259 | 84 | 240 | 5 | 14 | <1 | 20 | 11.6 | 5.9 | .9 |
| 1018 | White sauce | 1 c | 264 | 81 | 240 | 10 | 21 | <1 | 13 | 6.4 | 4.7 | 1.7 |
| | From home recipe: | | | | | | | | | | | |
| 1206 | Lowfat cheese sauce | ¼ c | 61 | 73 | 85 | 6 | 4 | <1 | 5 | 2.1 | 1.9 | .9 |
| 1019 | White sauce, medium | ¼ c | 72 | 77 | 102 | 2 | 6 | <1 | 8 | 2.3 | 3.2 | 2 |
| | Ready to serve: | | | | | | | | | | | |
| 2202 | Alfredo sauce, reduced fat | ¼ c | 69 | | 170 | 5 | 16 | 0 | 10 | 6 | | |
| 1020 | Barbeque sauce | 1 tbs | 16 | 81 | 12 | <1 | 2 | <1 | <1 | t | .1 | .1 |
| 1706 | Chili sauce, tomato base | 1 tbs | 17 | 68 | 18 | <1 | 4 | <1 | <1 | t | t | t |
| 2126 | Creole sauce | ¼ c | 62 | 89 | 25 | 1 | 4 | 1 | 1 | .1 | .2 | .3 |
| 2124 | Hoisin sauce | 1 tbs | 17 | 47 | 35 | <1 | 7 | 0 | 1 | 0 | | |
| 2199 | Pesto sauce | 2 tbs | 16 | | 83 | 2 | 1 | 0 | 8 | 1.8 | 5.4 | .7 |
| 1021 | Soy sauce | 1 tbs | 16 | 71 | 8 | 1 | 1 | <1 | <1 | t | t | t |
| 2123 | Szechuan sauce | 1 tbs | 16 | 71 | 21 | <1 | 3 | <1 | 1 | .1 | .3 | .4 |
| 1380 | Teriyaki sauce | 1 tbs | 18 | 68 | 15 | 1 | 3 | <1 | 0 | 0 | 0 | 0 |
| | Spaghetti sauce, canned: | | | | | | | | | | | |
| 1377 | Plain | 1 c | 249 | 75 | 271 | 5 | 40 | 8 | 12 | 1.7 | 6.1 | 3.3 |
| 1378 | With meat | 1 c | 250 | 74 | 300 | 8 | 37 | 8 | 14 | 2.7 | 7 | 3.2 |
| 1379 | With mushrooms | ½ c | 123 | 84 | 108 | 2 | 13 | 1 | 3 | .4 | 1.5 | .8 |
| | GRAVIES | | | | | | | | | | | |
| | Canned: | | | | | | | | | | | |
| 1022 | Beef | 1 c | 233 | 87 | 123 | 9 | 11 | 1 | 5 | 2.7 | 2.2 | .2 |
| 1023 | Chicken | 1 c | 238 | 85 | 188 | 5 | 13 | 1 | 14 | 3.4 | 6.1 | 3.6 |
| 1024 | Mushroom | 1 c | 238 | 89 | 119 | 3 | 13 | 1 | 6 | 1 | 2.8 | 2.4 |
| 1025 | From dry mix, brown | 1 c | 258 | 92 | 75 | 2 | 13 | <1 | 2 | .8 | .7 | .1 |
| 1026 | From dry mix, chicken | 1 c | 260 | 91 | 83 | 3 | 14 | <1 | 2 | .5 | .9 | .4 |
| | **FAST FOOD RESTAURANTS** | | | | | | | | | | | |
| | ARBY'S | | | | | | | | | | | |
| 1402 | Bac'n cheddar deluxe | 1 ea | 231 | 59 | 512 | 21 | 39 | <1 | 31 | 8.7 | 12.7 | 10.1 |
| | Roast beef sandwiches: | | | | | | | | | | | |
| 1403 | Regular | 1 ea | 155 | 47 | 383 | 22 | 35 | 1 | 18 | 7 | 8 | 3.5 |
| 1404 | Junior | 1 ea | 89 | 48 | 233 | 11 | 23 | <1 | 11 | 4.1 | 5.2 | 2.5 |
| 1405 | Super | 1 ea | 254 | 58 | 552 | 24 | 54 | 1 | 28 | 7.6 | 12.2 | 8.4 |
| 1407 | Beef 'n cheddar | 1 ea | 194 | 50 | 508 | 25 | 43 | | 26 | 7.7 | 12 | 6.8 |
| 1408 | Chicken breast sandwich | 1 ea | 204 | 52 | 445 | 22 | 52 | 1 | 22 | 3 | 9.7 | 10.1 |
| 1412 | Ham'n cheese sandwich | 1 ea | 169 | 54 | 355 | 25 | 34 | <1 | 14 | 5.1 | 5.8 | 3.8 |
| 1726 | Italian sub sandwich | 1 ea | 297 | 58 | 671 | 34 | 47 | | 39 | 12.8 | 15.7 | 8.5 |

**PAGE KEY:** H–4 = Beverages H–6 = Dairy H–10 = Eggs H–10 = Fat/Oil H–14 = Fruit H–20 = Bakery H–28 = Grain H–32 = Fish H–34 = Meats H–38 = Poultry H–40 = Sausage H–42 = Mixed/Fast H–46 = Nuts/Seeds H–50 = Sweets H–52 = Vegetables/Legumes H–62 = Vegetarian Foods H–64 = Misc H–66 = Soups/Sauces H–68 = Fast H–84 = Convenience H–88 = Baby foods

| Chol (mg) | Calc (mg) | Iron (mg) | Magn (mg) | Pota (mg) | Sodi (mg) | Zinc (mg) | VT-A (RE) | Thia (mg) | VT-E (a-TE) | Ribo (mg) | Niac (mg) | V-B6 (mg) | Fola (µg) | VT-C (mg) |
|---|---|---|---|---|---|---|---|---|---|---|---|---|---|---|
| 10 | 34 | .61 | 2 | 88 | 986 | .63 | 56 | .03 | .2 | .06 | .82 | .02 | 2 | <1 |
| 2 | 46 | .51 | 5 | 100 | 881 | .59 | 0 | .05 | 1.24 | .09 | .72 | .01 | 5 | 1 |
| 2 | 34 | .92 | 7 | 313 | 911 | .73 | 234 | .05 | .07 | .04 | .94 | .1 | 36 | 1 |
| 0 | 26 | .67 | 2 | 67 | 1053 | .61 | 0 | .03 | .29 | .02 | .6 | .05 | 15 | 1 |
| 8 | 23 | 2.28 | 48 | 400 | 1006 | 1.32 | 45 | .15 | .15 | .08 | 1.47 | .07 | 3 | 2 |
| 0 | 12 | 1.76 | 7 | 264 | 695 | .24 | 68 | .09 | 2.49 | .05 | 1.42 | .11 | 15 | 66 |
| 5 | 17 | 1.12 | 5 | 173 | 791 | 1.54 | 190 | .04 | .32 | .05 | 1.03 | .08 | 10 | 2 |
| 0 | 22 | 1.08 | 7 | 210 | 822 | .46 | 301 | .05 | .79 | .05 | .92 | .05 | 11 | 1 |
| 30 | 25 | 1.73 | 8 | 176 | 889 | 1 | 131 | .08 | .18 | .17 | 4.42 | .05 | 5 | 1 |
| 0 | 10 | .02 | 7 | 37 | 1361 | .07 | <1 | <.01 | .02 | .02 | .36 | 0 | 0 | 0 |
| 0 | 15 | .07 | 5 | 24 | 1483 | .01 | 12 | .01 | .02 | .03 | .19 | 0 | 2 | 0 |
| 3 | 33 | .5 | 8 | 30 | 1282 | .2 | 5 | .07 | .02 | .06 | .88 | .01 | 18 | <1 |
| 3 | 76 | .26 | 5 | 214 | 1184 | 1.57 | 123 | .1 | .15 | .2 | 2.61 | .05 | 5 | 1 |
| 0 | 12 | .15 | 5 | 64 | 849 | .06 | <1 | .03 | .1 | .06 | .48 | 0 | 1 | <1 |
| 3 | 20 | .94 | 43 | 224 | 1147 | .56 | 5 | .21 | .13 | .14 | 1.26 | .05 | 39 | 0 |
| 0 | 8 | .63 | 20 | 104 | 1146 | .17 | 20 | .06 | .81 | .05 | .79 | .05 | 10 | 6 |
| 1 | 4 | .06 | 3 | 27 | 1018 | 0 | <1 | <.01 | .01 | .01 | .27 | .01 | 2 | 0 |
| 2 | 55 | .58 | 25 | 260 | 3493 | .23 | 1 | .11 | .42 | .24 | 1.99 | .04 | 6 | 1 |
| 53 | 569 | .28 | 47 | 552 | 1565 | .97 | 117 | .15 | .33 | .56 | .32 | .14 | 13 | 2 |
| 52 | 124 | .9 | 8 | 124 | 1564 | .7 | 220 | .04 | .26 | .18 | .06 | .5 | 22 | <1 |
| 34 | 425 | .26 | 264 | 444 | 797 | .55 | 92 | .08 | 1.58 | .45 | .53 | .07 | 16 | 3 |
| 11 | 166 | .25 | 10 | 100 | 389 | .73 | 58 | .03 | .55 | .14 | .16 | .03 | 4 | <1 |
| 8 | 75 | .21 | 9 | 100 | 82 | .26 | 89 | .05 | .98 | .12 | .28 | .03 | 4 | 1 |
| 30 | 150 | 0 | 8 | 80 | 600 | | 80 | 0 | | .1 | 0 | | | 0 |
| 0 | 3 | .14 | 3 | 28 | 130 | .03 | 14 | <.01 | .18 | <.01 | .14 | .01 | 1 | 1 |
| 0 | 3 | .14 | 2 | 63 | 227 | .05 | 24 | .01 | .05 | .01 | .27 | .02 | 1 | 3 |
| 0 | 35 | .31 | 9 | 187 | 339 | .1 | 24 | .03 | .61 | .02 | .53 | .07 | 9 | 0 |
| 0 | 0 | 0 | | | 250 | | 0 | | | | | | | 0 |
| 4 | 64 | .09 | 6 | 15 | 129 | .29 | 39 | .01 | | .03 | 0 | .02 | <1 | 0 |
| 0 | 3 | .32 | 5 | 29 | 914 | .06 | 0 | .01 | 0 | .02 | .54 | .03 | 2 | 0 |
| 0 | 2 | .12 | 2 | 13 | 218 | .02 | 10 | <.01 | .07 | <.01 | .1 | .01 | 1 | <1 |
| 0 | 4 | .31 | 11 | 40 | 690 | .02 | 0 | <.01 | 0 | .01 | .23 | .02 | 4 | 0 |
| 0 | 70 | 1.62 | 60 | 956 | 1235 | .52 | 306 | .14 | 4.98 | .15 | 3.76 | .88 | 54 | 28 |
| 15 | 68 | 1.94 | 60 | 952 | 1179 | 1.37 | 577 | .13 | 5.91 | .17 | 4.51 | .87 | 52 | 26 |
| 0 | 15 | 1 | 15 | 332 | 494 | .34 | 241 | .08 | 1.35 | .08 | .93 | .16 | 12 | 9 |
| 7 | 14 | 1.63 | 5 | 189 | 1304 | 2.33 | 0 | .07 | .15 | .08 | 1.54 | .02 | 5 | 0 |
| 5 | 48 | 1.12 | 5 | 259 | 1373 | 1.9 | 264 | .04 | .37 | .1 | 1.05 | .02 | 5 | 0 |
| 0 | 17 | 1.57 | 5 | 252 | 1356 | 1.67 | 0 | .08 | .19 | .15 | 1.6 | .05 | 29 | 0 |
| 3 | 67 | .23 | 10 | 57 | 1075 | .31 | 0 | .04 | .05 | .08 | .81 | 0 | 0 | 0 |
| 3 | 39 | .26 | 10 | 62 | 1133 | .32 | 0 | .05 | .05 | .15 | .78 | .03 | 3 | 3 |
| 38 | 110 | 4.32 | | 491 | 1094 | 3 | 40 | .34 | | .46 | 9.6 | | | 11 |
| 43 | 60 | 4.86 | 16 | 422 | 936 | 3.75 | 0 | .28 | | .48 | 11 | .2 | 14 | 1 |
| 22 | 40 | 2.7 | 8 | 201 | 519 | 1.5 | | .18 | | .25 | 6.6 | .1 | 7 | |
| 43 | 90 | 6.48 | 25 | 533 | 1174 | 3.75 | 30 | .39 | | .58 | 12.4 | .3 | 21 | 9 |
| 52 | 150 | 6.12 | | 321 | 1166 | 3 | | .42 | | .63 | 9.8 | | | 1 |
| 45 | 60 | 2.88 | 30 | 330 | 1019 | .15 | | .22 | | .54 | 9 | .38 | 18 | 5 |
| 55 | 170 | 2.7 | 31 | 382 | 1400 | .9 | 40 | .82 | | .37 | 7.8 | .31 | 26 | 24 |
| 69 | 410 | 4.32 | | 565 | 2062 | | 100 | .91 | | .49 | 8.2 | | | 11 |

H

**Table H-1**

**Food Composition** (Computer code number is for West Diet Analysis program) (For purposes of calculations, use "0" for t, <1, <.1, <.01, etc.)

| Computer Code Number | Food Description | Measure | Wt (g) | $H_2O$ (%) | Ener (kcal) | Prot (g) | Carb (g) | Dietary Fiber (g) | Fat (g) | Fat Breakdown (g) Sat | Mono | Poly |
|---|---|---|---|---|---|---|---|---|---|---|---|---|
| | **FAST FOOD RESTAURANTS** | | | | | | | | | | | |
| | ARBY'S—Continued | | | | | | | | | | | |
| 1413 | Turkey sandwich, deluxe | 1 ea | 195 | 69 | 260 | 20 | 33 | <1 | 6 | 1.6 | 2.3 | 2.4 |
| 1680 | Turkey sub sandwich | 1 ea | 277 | 62 | 486 | 33 | 46 | | 19 | 5.3 | 6 | 7 |
| | Milkshakes: | | | | | | | | | | | |
| 1419 | Chocolate | 1 ea | 340 | 74 | 451 | 10 | 76 | <1 | 12 | 2.8 | 7 | 1.7 |
| 1420 | Jamocha | 1 ea | 326 | 75 | 368 | 9 | 59 | 0 | 10 | 2.5 | 6.4 | 1.6 |
| 1421 | Vanilla | 1 ea | 312 | 77 | 330 | 10 | 46 | 0 | 11 | 3.9 | 5.3 | 2.3 |
| 1728 | Salad, roast chicken | 1 ea | 400 | 88 | 204 | 24 | 12 | | 7 | 3.3 | .9 | .9 |
| 1729 | Sports drink, Upper Ten | 1 ea | 358 | 88 | 169 | 0 | 42 | | 0 | 0 | 0 | 0 |
| | Source: Arby's | | | | | | | | | | | |
| | BURGER KING | | | | | | | | | | | |
| 1423 | Croissant sandwich, egg, sausage & cheese | 1 ea | 176 | 46 | 600 | 22 | 25 | 1 | 46 | 16 | | |
| | Whopper sandwiches: | | | | | | | | | | | |
| 1425 | Whopper | 1 ea | 270 | 58 | 640 | 27 | 45 | 3 | 39 | 11 | | |
| 1426 | Whopper with cheese | 1 ea | 294 | 57 | 730 | 33 | 46 | 3 | 46 | 16 | | |
| | Sandwiches: | | | | | | | | | | | |
| 1629 | BK broiler chicken sandwich | 1 ea | 248 | 59 | 550 | 30 | 41 | 2 | 29 | 6 | | |
| 1432 | Cheeseburger | 1 ea | 138 | 48 | 380 | 23 | 28 | 1 | 19 | 9 | | |
| 1434 | Chicken sandwich | 1 ea | 229 | 45 | 710 | 26 | 54 | 2 | 43 | 9 | | |
| 1427 | Double beef | 1 ea | 351 | 57 | 870 | 46 | 45 | 3 | 56 | 19 | | |
| 1428 | Double beef & cheese | 1 ea | 375 | 56 | 960 | 52 | 46 | 3 | 63 | 24 | | |
| 1433 | Double cheeseburger with bacon | 1 ea | 218 | 48 | 640 | 44 | 28 | 1 | 39 | 18 | | |
| 1431 | Hamburger | 1 ea | 126 | 48 | 330 | 20 | 28 | 1 | 15 | 6 | | |
| 1437 | Ocean catch fish fillet | 1 ea | 255 | 51 | 700 | 26 | 56 | 3 | 41 | 6 | | |
| 1435 | Chicken tenders | 1 ea | 88 | 50 | 230 | 16 | 14 | 2 | 12 | 3 | | |
| 1439 | French fries (salted) | 1 svg | 116 | 40 | 370 | 5 | 43 | 3 | 20 | 5 | | |
| 1630 | French toast sticks | 1 svg | 141 | 33 | 500 | 4 | 60 | 1 | 27 | 7 | | |
| 1440 | Onion rings | 1 svg | 124 | 51 | 310 | 4 | 41 | 6 | 14 | 2 | 8 | 4 |
| 1441 | Milk shakes, chocolate | 1 ea | 284 | 75 | 320 | 9 | 54 | 3 | 7 | 4 | | |
| 1442 | Milk shakes, vanilla | 1 ea | 284 | 75 | 300 | 9 | 53 | 1 | 6 | 4 | | |
| 1443 | Fried apple pie | 1 ea | 113 | 47 | 300 | 3 | 39 | 2 | 15 | 3 | | |
| | Source: Burger King Corporation | | | | | | | | | | | |
| | CHICK-FIL-A | | | | | | | | | | | |
| | Sandwiches: | | | | | | | | | | | |
| 69153 | Chargrilled chicken | 1 ea | 150 | 54 | 280 | 27 | 36 | 1 | 3 | 1 | | |
| 69152 | Chicken | 1 ea | 167 | 61 | 290 | 24 | 29 | 1 | 9 | 2 | | |
| 69155 | Chicken salad | 1 ea | 167 | 55 | 320 | 25 | 42 | 1 | 5 | 2 | | |
| 69154 | Chicken salad club | 1 ea | 232 | 62 | 390 | 33 | 38 | 2 | 12 | 5 | | |
| | Salads: | | | | | | | | | | | |
| 52139 | Carrot and raisin | 1 ea | 76 | 53 | 150 | 5 | 28 | 2 | 2 | 0 | | |
| 52136 | Chicken plate | 1 ea | 468 | 85 | 290 | 21 | 40 | 6 | 5 | 0 | | |
| 52134 | Chicken garden, charbroiled | 1 ea | 397 | 89 | 170 | 26 | 10 | 5 | 3 | 1 | | |
| 52135 | Chick-n-strips | 1 ea | 451 | 86 | 290 | 32 | 21 | 5 | 9 | 2 | | |
| 52138 | Cole slaw | 1 ea | 79 | 70 | 130 | 6 | 11 | 1 | 6 | 1 | | |
| 52137 | Tossed salad | 1 ea | 130 | 85 | 70 | 5 | 13 | 1 | 0 | 0 | 0 | 0 |
| 15263 | Chicken nuggets, svg | 1 ea | 110 | 51 | 290 | 28 | 12 | 60 | 12 | 3 | | |
| 15262 | Chicken-n-strips, svg | 1 ea | 119 | 59 | 230 | 29 | 10 | 0 | 8 | 2 | | |
| 50885 | Hearty breast of chicken soup, svg | 1 ea | 215 | 86 | 110 | 16 | 10 | 1 | 1 | 0 | | |
| 7973 | Waffle potato fries, svg | 1 ea | 85 | 28 | 290 | 1 | 49 | 0 | 10 | 4 | | |
| 46489 | Cheesecake, svg | 1 ea | 88 | 52 | 270 | 13 | 7 | 0 | 21 | 9 | | |
| 49134 | Fudge nut brownie, svg | 1 ea | 88 | 8 | 416 | 12 | 49 | 0 | 19 | 3.6 | | |
| 20601 | Icedream, svg | 1 ea | 127 | 74 | 140 | 11 | 16 | 0 | 4 | 1 | | |
| 48214 | Lemon pie, svg | 1 ea | 99 | 56 | 280 | 1 | 19 | 0 | 22 | 6 | | |
| | Source: Chick-Fil-A | | | | | | | | | | | |

| Chol (mg) | Calc (mg) | Iron (mg) | Magn (mg) | Pota (mg) | Sodi (mg) | Zinc (mg) | VT-A (RE) | Thia (mg) | VT-E (a-TE) | Ribo (mg) | Niac (mg) | V-B6 (mg) | Fola (μg) | VT-C (mg) |
|---|---|---|---|---|---|---|---|---|---|---|---|---|---|---|
| 33 | 130 | 3.42 | 30 | 353 | 1262 | 1.5 | 40 | .08 | | .41 | 15.4 | .52 | 20 | 12 |
| 51 | 400 | 4.68 | | 500 | 2033 | | 20 | 13.2 | | .54 | 18.8 | | | |
| | | | | | | | | | | | | | | |
| 36 | 250 | .72 | 48 | 410 | 341 | 1.5 | 60 | .12 | | .68 | .8 | .14 | 14 | 5 |
| 35 | 250 | 2.7 | 36 | 525 | 262 | 1.5 | 60 | .12 | | .68 | .8 | .14 | 14 | 2 |
| 32 | 300 | 2.7 | 36 | 686 | 281 | 1.5 | 60 | .12 | | .68 | 4 | .14 | 37 | 2 |
| 43 | 170 | 1.98 | | 877 | 508 | | 485 | .33 | | .54 | 5.6 | | | 51 |
| 0 | | | | 0 | 40 | | | | | | | | | |
| | | | | | | | | | | | | | | |
| 260 | 150 | 3.6 | | | 1140 | | 80 | | | | | | | 0 |
| | | | | | | | | | | | | | | |
| 90 | 80 | 4.5 | | | 870 | | 100 | .33 | | .41 | 7 | .35 | | 9 |
| 115 | 250 | 4.5 | | | 1350 | | 150 | .34 | | .48 | 7 | .33 | | 9 |
| | | | | | | | | | | | | | | |
| 80 | 60 | 5.4 | | | 480 | | 60 | | | | | | | 6 |
| 65 | 100 | 2.7 | | | 770 | | 60 | | | | | | | 0 |
| 60 | 100 | 3.6 | | | 1400 | | 0 | | | | | | | 0 |
| 170 | 80 | 7.2 | | | 940 | | 100 | .34 | | .56 | 10 | | | 9 |
| 195 | 250 | 7.2 | | | 1420 | | 150 | .35 | | .63 | 10 | | | 9 |
| 145 | 200 | 4.5 | | | 1240 | | 80 | .31 | | .42 | 6 | | | 0 |
| 55 | 40 | 1.8 | | | 530 | | 20 | .28 | | .31 | 4.89 | | | 0 |
| 90 | 60 | 2.7 | | | 980 | | 20 | | | | | | | 1 |
| 35 | 0 | .72 | | | 530 | | 0 | | | | | | | 0 |
| 0 | 0 | 1.08 | | | 240 | | 0 | | | | | | | 4 |
| 0 | 60 | 2.7 | | | 490 | | 0 | | | | | | | 0 |
| 0 | 100 | 1.44 | | | 810 | | 0 | | | | | | | 0 |
| 20 | 200 | 1.8 | | | 230 | | 60 | .13 | | .55 | .13 | | | 0 |
| 20 | 300 | 0 | | | 230 | | 60 | .11 | | .57 | .13 | | | 4 |
| 0 | 0 | 1.44 | | | 230 | | 0 | | | | | | | 6 |
| | | | | | | | | | | | | | | |
| 40 | | | | | 640 | | | | | | | | | |
| 50 | | | | | 870 | | | | | | | | | |
| 10 | | | | | 810 | | | | | | | | | |
| 70 | | | | | 980 | | | | | | | | | |
| | | | | | | | | | | | | | | |
| 6 | | | | | 650 | | | | | | | | | |
| 35 | | | | | 570 | | | | | | | | | |
| 25 | | | | | 650 | | | | | | | | | |
| 20 | | | | | 430 | | | | | | | | | |
| 15 | | | | | 430 | | | | | | | | | |
| 0 | | | | | 0 | | | | | | | | | |
| 14 | | | | | 770 | | | | | | | | | |
| 20 | | | | | 380 | | | | | | | | | |
| 45 | | | | | 760 | | | | | | | | | |
| 5 | | | | | 960 | | | | | | | | | |
| 10 | | | | | 510 | | | | | | | | | |
| 36 | | | | | 773 | | | | | | | | | |
| 40 | | | | | 240 | | | | | | | | | |
| 5 | | | | | 550 | | | | | | | | | |

**Table H-1**

**Food Composition** (Computer code number is for West Diet Analysis program) (For purposes of calculations, use "0" for t, <1, <.1, <.01, etc.)

| Computer Code Number | Food Description | Measure | Wt (g) | $H_2O$ (%) | Ener (kcal) | Prot (g) | Carb (g) | Dietary Fiber (g) | Fat (g) | Fat Breakdown (g) Sat | Mono | Poly |
|---|---|---|---|---|---|---|---|---|---|---|---|---|
| | **FAST FOOD RESTAURANTS**—Continued | | | | | | | | | | | |
| | DAIRY QUEEN | | | | | | | | | | | |
| | Ice cream cones: | | | | | | | | | | | |
| 1446 | Small vanilla | 1 ea | 142 | 63 | 230 | 6 | 38 | 0 | 7 | 4.5 | | |
| 1447 | Regular vanilla | 1 ea | 213 | 64 | 350 | 8 | 57 | 0 | 10 | 7 | | |
| 1448 | Large vanilla | 1 ea | 253 | 65 | 410 | 10 | 65 | 0 | 12 | 8 | | |
| 1450 | Chocolate dipped | 1 ea | 234 | 59 | 510 | 9 | 63 | 1 | 25 | 13 | | |
| 1453 | Chocolate sundae | 1 ea | 241 | 62 | 410 | 8 | 73 | 0 | 10 | 6 | | |
| 1455 | Banana split | 1 ea | 369 | 67 | 510 | 8 | 96 | 3 | 12 | 8 | | |
| 1456 | Peanut buster parfait | 1 ea | 305 | 51 | 730 | 16 | 99 | 2 | 31 | 17 | | |
| 1457 | Hot fudge brownie delight | 1 ea | 305 | 52 | 710 | 11 | 102 | 1 | 29 | 14 | 12 | 2 |
| 1459 | Buster bar | 1 ea | 149 | 45 | 450 | 10 | 41 | 2 | 28 | 12 | | |
| 1645 | Breeze, strawberry, regular | 1 ea | 383 | 70 | 460 | 13 | 99 | 1 | 1 | 1 | 0 | 0 |
| 1460 | Dilly bar | 1 ea | 85 | 55 | 210 | 3 | 21 | 0 | 13 | 7 | 3 | 3 |
| 1461 | DQ ice cream sandwich | 1 ea | 61 | 46 | 150 | 3 | 24 | 1 | 5 | 2 | | |
| 1463 | Milk shakes, regular | 1 ea | 397 | 71 | 520 | 12 | 88 | <1 | 14 | 8 | 2 | 2 |
| 1464 | Milk shakes, large | 1 ea | 461 | 71 | 600 | 13 | 101 | <1 | 16 | 10 | 2 | 2 |
| 1466 | Milk shakes, malted | 1 ea | 418 | 68 | 610 | 13 | 106 | <1 | 14 | 8 | 2 | 2 |
| 1470 | Misty slush, small | 1 ea | 454 | 88 | 220 | 0 | 56 | 0 | 0 | 0 | 0 | 0 |
| 2250 | Starkiss | 1 ea | 85 | 75 | 80 | 0 | 21 | 0 | 0 | 0 | 0 | 0 |
| | Yogurt: | | | | | | | | | | | |
| 1641 | Yogurt cone, regular | 1 ea | 213 | 66 | 280 | 9 | 59 | 0 | 1 | .5 | | |
| 1643 | Yogurt sundae, strawberry | 1 ea | 255 | 69 | 300 | 9 | 66 | 1 | <1 | .5 | 0 | 0 |
| | Sandwiches: | | | | | | | | | | | |
| 1481 | Cheeseburger, double | 1 ea | 219 | 55 | 540 | 35 | 30 | 2 | 31 | 16 | | |
| 1480 | Cheeseburger, single | 1 ea | 152 | 55 | 340 | 20 | 29 | 2 | 17 | 8 | | |
| 1474 | Chicken | 1 ea | 191 | 56 | 430 | 24 | 37 | 2 | 20 | 4 | | |
| 1647 | Chicken fillet, grilled | 1 ea | 184 | 64 | 310 | 24 | 30 | 3 | 10 | 2.5 | | |
| 1475 | Fish fillet sandwich | 1 ea | 170 | 57 | 370 | 16 | 39 | 2 | 16 | 3.5 | | |
| 1476 | Fish fillet with cheese | 1 ea | 184 | 56 | 420 | 19 | 40 | 2 | 21 | 6 | 7 | 8 |
| 1477 | Hamburger, single | 1 ea | 128 | 56 | 269 | 16 | 27 | 2 | 11 | 4.6 | 5.6 | .9 |
| 1478 | Hamburger, double | 1 ea | 212 | 62 | 440 | 30 | 29 | 2 | 22 | 10 | | |
| | Hotdog: | | | | | | | | | | | |
| 1483 | Regular | 1 ea | 99 | 57 | 240 | 9 | 19 | 1 | 14 | 5 | | |
| 1484 | With cheese | 1 ea | 113 | 55 | 290 | 12 | 20 | 1 | 18 | 8 | 8 | 2 |
| 1485 | With chili | 1 ea | 128 | 61 | 280 | 12 | 21 | 2 | 16 | 6 | | |
| 1489 | French fries, small | 1 ea | 71 | 39 | 210 | 3 | 29 | 3 | 10 | 2 | 5 | 3 |
| 1490 | French fries, large | 1 ea | 128 | 40 | 390 | 5 | 52 | 6 | 18 | 4 | 8 | 6 |
| 1491 | Onion rings | 1 ea | 85 | 46 | 240 | 4 | 29 | 2 | 12 | 2.5 | | |
| | Source: International Dairy Queen | | | | | | | | | | | |
| | HARDEES | | | | | | | | | | | |
| 1734 | Frisco burger hamburger | 1 ea | 242 | | 760 | 36 | 43 | | 50 | 18 | | |
| 1736 | Frisco grilled chicken salad | 1 ea | 278 | | 120 | 18 | 2 | | 4 | 1 | | |
| 1737 | Peach shake | 1 ea | 345 | | 390 | 10 | 77 | | 4 | 3 | | |
| | Source: Hardees | | | | | | | | | | | |
| | JACK IN THE BOX | | | | | | | | | | | |
| | Breakfast items: | | | | | | | | | | | |
| 1492 | Breakfast Jack sandwich | 1 ea | 121 | 49 | 300 | 18 | 30 | 0 | 12 | 5 | 5 | 2.5 |
| 1494 | Sausage crescent | 1 ea | 156 | 39 | 580 | 22 | 28 | 0 | 43 | 16 | | |
| 1495 | Supreme crescent | 1 ea | 153 | 40 | 530 | 23 | 34 | 0 | 33 | 10 | 18.9 | 7.8 |
| 1496 | Pancake platter | 1 ea | 231 | 45 | 610 | 15 | 87 | 0 | 22 | 9 | 7.6 | 3.5 |
| 1497 | Scrambled egg platter | 1 ea | 213 | 52 | 560 | 18 | 50 | 0 | 32 | 9 | 16.6 | 4.4 |
| | Sandwiches: | | | | | | | | | | | |
| 1654 | Bacon cheeseburger | 1 ea | 242 | 49 | 710 | 35 | 41 | 0 | 45 | 15 | 15.7 | 8.7 |
| 1499 | Cheeseburger | 1 ea | 110 | 41 | 330 | 16 | 32 | 0 | 15 | 6 | 5.9 | 2.3 |
| 1739 | Chicken caesar pita sandwich | 1 ea | 237 | 59 | 520 | 27 | 44 | 4 | 26 | 6 | | |
| 1655 | Chicken sandwich | 1 ea | 160 | 52 | 400 | 20 | 38 | 0 | 18 | 4 | | |
| 1656 | Chicken sandwich, sourdough ranch | 1 ea | 225 | | 490 | 29 | 45 | 1 | 21 | 6 | | |
| 1505 | Chicken supreme | 1 ea | 245 | 55 | 620 | 25 | 48 | 0 | 36 | 11 | 14.8 | 11.4 |
| 1583 | Double cheeseburger | 1 ea | 152 | 44 | 450 | 24 | 35 | 0 | 24 | 12 | 11.6 | 3.1 |

| Chol (mg) | Calc (mg) | Iron (mg) | Magn (mg) | Pota (mg) | Sodi (mg) | Zinc (mg) | VT-A (RE) | Thia (mg) | VT-E (a-TE) | Ribo (mg) | Niac (mg) | V-B6 (mg) | Fola (µg) | VT-C (mg) |
|---|---|---|---|---|---|---|---|---|---|---|---|---|---|---|
| | | | | | | | | | | | | | | |
| 20 | 200 | 1.08 | | 250 | 115 | | 122 | .05 | | .28 | | | | |
| 30 | 300 | 1.8 | | 390 | 170 | | 150 | .09 | | .38 | .16 | .13 | | 2 |
| 40 | 350 | 1.8 | | 451 | 200 | | 200 | .11 | | .4 | .2 | | | 2 |
| 30 | 300 | 1.8 | | 435 | 200 | | 150 | .09 | | .38 | .16 | .13 | | 2 |
| 30 | 250 | 1.44 | | 394 | 210 | | 150 | .08 | | .35 | .4 | .19 | | 0 |
| 30 | 250 | 1.8 | | 860 | 180 | | 200 | .15 | | .25 | .4 | .2 | | 15 |
| 35 | 300 | 1.8 | | 660 | 400 | | 150 | .15 | | .51 | 3 | .22 | | 1 |
| 35 | 300 | 5.4 | | 510 | 340 | | 80 | .15 | | .68 | .3 | .18 | | 1 |
| 15 | 150 | 1.08 | | 400 | 280 | | 80 | .09 | | .17 | 3 | .08 | | 0 |
| 10 | 450 | 2.7 | | 530 | 270 | | 0 | .13 | | .73 | | | | 9 |
| 10 | 100 | .36 | | 170 | 75 | | 60 | .03 | | .14 | | .06 | | 0 |
| 5 | 60 | .72 | | 105 | 115 | | 40 | .03 | | .25 | .4 | .05 | | 0 |
| 45 | 400 | 1.44 | | 570 | 230 | | 80 | .12 | | .59 | .8 | .19 | | <1 |
| 50 | 450 | 1.44 | | 660 | 260 | | 200 | .15 | | .68 | .8 | | | <1 |
| 45 | 400 | 1.44 | | 570 | 230 | | 80 | .12 | | .59 | .8 | .19 | | <1 |
| 0 | 0 | 0 | | | 20 | | 0 | | | | | | | 0 |
| 0 | 0 | 0 | | | 10 | | 0 | | | | | | | 0 |
| | | | | | | | | | | | | | | |
| 5 | 300 | 1.8 | | 285 | 170 | | 0 | .09 | | .38 | | | | 2 |
| 5 | 300 | 1.8 | | 352 | 180 | | 0 | .09 | | .49 | | | | 6 |
| | | | | | | | | | | | | | | |
| 115 | 250 | 4.5 | | 426 | 1130 | | 150 | .29 | | .49 | 6.78 | | | 4 |
| 55 | 150 | 3.6 | | 263 | 850 | | 60 | .29 | | .33 | 3.89 | | | 4 |
| 55 | 40 | 1.8 | | 350 | 760 | | 0 | .37 | | .34 | 11 | | | 0 |
| 50 | 200 | 2.7 | | 330 | 1040 | | 0 | .3 | | 1.02 | 12 | | | 0 |
| 45 | 40 | 1.8 | | 280 | 630 | | 0 | .3 | | .22 | 3 | | | 0 |
| 60 | 100 | 1.8 | | 290 | 850 | | 80 | .3 | | .25 | 5 | | | 0 |
| 42 | 56 | 2.5 | | 234 | 584 | | 37 | .27 | | .23 | 3.6 | | | 3 |
| 90 | 60 | 4.5 | | 444 | 680 | | 60 | .32 | | .45 | 7.49 | | | 6 |
| | | | | | | | | | | | | | | |
| 25 | 60 | 1.8 | | 170 | 730 | | 20 | .22 | | .14 | 2 | | | 4 |
| 40 | 150 | 1.8 | | 180 | 950 | | 60 | .22 | | .17 | 2 | | | 4 |
| 35 | 60 | 1.8 | | 262 | 870 | | 80 | .23 | | .14 | 3 | | | 4 |
| 0 | 0 | .72 | | 430 | 115 | | 0 | .09 | | .03 | 2 | | | 5 |
| 0 | 0 | 1.44 | | 780 | 200 | | 0 | .15 | | .07 | 3 | | | 9 |
| 0 | 0 | 1.08 | | 90 | 135 | | 0 | .09 | | .05 | .4 | | | 0 |
| | | | | | | | | | | | | | | |
| 70 | | | | | 1280 | | | | | | | | | |
| 60 | | | | | 520 | | | | | | | | | |
| 25 | | | | | 290 | | | | | | | | | |
| | | | | | | | | | | | | | | |
| 185 | 200 | 2.7 | | 220 | 890 | | 80 | .47 | | .41 | 3 | | | 9 |
| 185 | 150 | 2.7 | | 260 | 1010 | | 100 | .6 | | .51 | 4.6 | | | 0 |
| 210 | 150 | 3.6 | | 270 | 930 | | 150 | .65 | | .54 | 4.2 | | | 12 |
| 100 | 100 | 1.8 | | 310 | 890 | | 80 | .03 | | .85 | 7 | | | 6 |
| 380 | 150 | 4.5 | | 450 | 1060 | | 150 | | | .66 | 5 | | | 9 |
| | | | | | | | | | | | | | | |
| 110 | 250 | 5.4 | | 540 | 1240 | | 80 | .24 | | .48 | 8.8 | .39 | | 9 |
| 35 | 200 | 2.7 | | 200 | 510 | | 60 | .23 | | .23 | 3 | | | 1 |
| 55 | 250 | 2.7 | | 490 | 1050 | | 80 | | | | | | | 2 |
| 45 | 150 | 1.8 | | 180 | 1290 | | 40 | | | | | | | 0 |
| 65 | 150 | 1.8 | | 340 | 1060 | | | | | | | | | 0 |
| 75 | 200 | 2.7 | | 190 | 1520 | | 100 | .39 | | .32 | 11 | | | 2 |
| 75 | 250 | 3.6 | | 320 | 900 | | 100 | .15 | | .34 | 6 | | | 0 |

H

**Table H-1**

**Food Composition** (Computer code number is for West Diet Analysis program) (For purposes of calculations, use "0" for t, <1, <.1, <.01, etc.)

| Computer Code Number | Food Description | Measure | Wt (g) | $H_2O$ (%) | Ener (kcal) | Prot (g) | Carb (g) | Dietary Fiber (g) | Fat (g) | Fat Breakdown (g) Sat | Mono | Poly |
|---|---|---|---|---|---|---|---|---|---|---|---|---|
| | **FAST FOOD RESTAURANTS—**Continued | | | | | | | | | | | |
| | JACK IN THE BOX—Continued | | | | | | | | | | | |
| 1651 | Grilled sourdough burger | 1 ea | 223 | 48 | 670 | 32 | 39 | 0 | 43 | 16 | 17.8 | 7.9 |
| 1498 | Hamburger | 1 ea | 97 | 42 | 280 | 13 | 31 | 0 | 11 | 4 | 4.9 | 2 |
| 1500 | Jumbo Jack burger | 1 ea | 229 | 55 | 560 | 26 | 41 | 0 | 32 | 10 | 13 | 8 |
| 1501 | Jumbo Jack burger with cheese | 1 ea | 242 | 55 | 610 | 29 | 41 | 0 | 36 | 12 | 15 | 9 |
| 1740 | Monterey roast beef sandwich | 1 ea | 238 | 57 | 540 | 30 | 40 | 3 | 30 | 9 | | |
| 1508 | Tacos, regular | 1 ea | 78 | 57 | 190 | 7 | 15 | 2 | 11 | 4 | | |
| 1509 | Tacos, super | 1 ea | 126 | 59 | 280 | 12 | 22 | 3 | 17 | 6 | | |
| | Teriyaki bowl: | | | | | | | | | | | |
| 1679 | Beef | 1 ea | 440 | 62 | 640 | 28 | 124 | 7 | 3 | 1 | | |
| 1668 | Chicken | 1 ea | 440 | 62 | 580 | 28 | 115 | 6 | 1 | | | |
| 1516 | French fries | 1 ea | 109 | 38 | 350 | 4 | 45 | 4 | 17 | 4 | | |
| 1517 | Hash browns | 1 ea | 57 | 53 | 160 | 1 | 14 | 1 | 11 | 2.5 | 6.8 | .3 |
| 1518 | Onion rings | 1 ea | 103 | 34 | 380 | 5 | 38 | 0 | 23 | 6 | 15.2 | .9 |
| | Milkshakes: | | | | | | | | | | | |
| 1519 | Chocolate | 1 ea | 322 | 72 | 390 | 9 | 74 | 0 | 6 | 3.5 | 2.1 | |
| 1520 | Strawberry | 1 ea | 298 | 74 | 330 | 9 | 60 | 0 | 7 | 4 | 2 | |
| 1521 | Vanilla | 1 ea | 304 | 73 | 350 | 9 | 62 | 0 | 7 | 4 | 1.8 | |
| 1522 | Apple turnover | 1 ea | 110 | 34 | 350 | 3 | 48 | 0 | 19 | 4 | 10.6 | 1.5 |
| | Source: Jack in the Box Restaurant, Inc | | | | | | | | | | | |
| | KENTUCKY FRIED CHICKEN | | | | | | | | | | | |
| | Rotisserie gold: | | | | | | | | | | | |
| 1472 | Dark qtr, no skin | 1 ea | 117 | 66 | 217 | 27 | 0 | 0 | 12 | 3.5 | | |
| 1473 | Dark qtr, w/skin | 1 ea | 146 | 62 | 333 | 30 | 1 | | 24 | 6.6 | | |
| 1513 | White qtr with wing, w/skin | 1 ea | 176 | 65 | 335 | 40 | 1 | | 19 | 5.4 | | |
| 1525 | White qtr with wing, no skin | 1 ea | 117 | 63 | 199 | 37 | 0 | 0 | 6 | 1.7 | | |
| | Original Recipe: | | | | | | | | | | | |
| 1253 | Center breast | 1 ea | 103 | 52 | 260 | 25 | 9 | <1 | 14 | 3.8 | 7.8 | 2 |
| 1251 | Side breast | 1 ea | 83 | 47 | 245 | 18 | 9 | <1 | 15 | 4.2 | 8.8 | 2.2 |
| 1250 | Drumstick | 1 ea | 57 | 54 | 152 | 14 | 3 | <1 | 8 | 2.2 | 4.1 | 1.3 |
| 1252 | Thigh | 1 ea | 95 | 49 | 287 | 18 | 8 | <1 | 21 | 5.3 | 9.4 | 3.1 |
| 1249 | Wing | 1 ea | 53 | 45 | 172 | 12 | 5 | <1 | 12 | 3 | 6 | 1.8 |
| | Hot & spicy: | | | | | | | | | | | |
| 1451 | Center breast | 1 ea | 125 | 48 | 360 | 28 | 13 | | 22 | 5 | | |
| 1452 | Side breast | 1 ea | 120 | 43 | 400 | 22 | 16 | | 28 | 6 | | |
| 1430 | Thigh | 1 ea | 119 | 47 | 370 | 24 | 10 | | 27 | 6 | | |
| 1471 | Wing | 1 ea | 61 | 38 | 220 | 14 | 5 | | 16 | 4 | | |
| | Extra crispy recipe: | | | | | | | | | | | |
| 1261 | Center breast | 1 ea | 118 | 48 | 330 | 26 | 14 | <1 | 20 | 4.8 | 10.8 | 2.1 |
| 1259 | Side breast | 1 ea | 116 | 40 | 400 | 21 | 19 | <1 | 27 | 5.5 | 12.9 | 2.3 |
| 1258 | Drumstick | 1 ea | 65 | 49 | 190 | 14 | 6 | <1 | 12 | 3.4 | 7.7 | 1.7 |
| 1260 | Thigh | 1 ea | 109 | 43 | 380 | 23 | 7 | <1 | 30 | 7.7 | 16 | 4.2 |
| 1257 | Wing | 1 ea | 59 | 34 | 240 | 13 | 8 | <1 | 17 | 4.4 | 10.7 | 2.5 |
| 1390 | Baked beans | ½ c | 167 | 70 | 200 | 8 | 36 | 6 | 3 | 1.5 | .7 | .4 |
| 1526 | Breadstick | 1 ea | 33 | 30 | 110 | 3 | 17 | 0 | 3 | 0 | | |
| 1388 | Buttermilk biscuit | 1 ea | 65 | 28 | 234 | 5 | 28 | <1 | 13 | 3.4 | 6.2 | 2.3 |
| 1391 | Chicken little sandwich | 1 ea | 47 | 35 | 169 | 6 | 14 | <1 | 10 | 2 | 4.7 | 3.4 |
| 1269 | Coleslaw | 1 svg | 90 | 75 | 114 | 1 | 13 | <1 | 6 | 1 | 1.7 | 3.4 |
| 1527 | Cornbread | 1 ea | 56 | 26 | 228 | 3 | 25 | 1 | 13 | 2 | | |
| 1268 | Corn-on-the-cob | 1 ea | 151 | 70 | 222 | 4 | 27 | 8 | 12 | 2 | 1 | 1.5 |
| 1429 | Chicken, hot wings | 1 svg | 135 | 38 | 471 | 27 | 18 | | 33 | | | |
| 1386 | Kentucky fries | 1 svg | 77 | 42 | 228 | 3 | 26 | 3 | 12 | 3.2 | | |
| 1381 | Kentucky nuggets | 6 ea | 95 | 48 | 284 | 16 | 15 | <1 | 18 | 4 | | |
| 1534 | Macaroni & cheese | 1 svg | 114 | 71 | 162 | 7 | 15 | 0 | 8 | 3 | | |
| 1387 | Mashed potatoes & gravy | 1 svg | 120 | 80 | 103 | 1 | 16 | <1 | 5 | .4 | .5 | .2 |
| 1530 | Pasta salad | 1 svg | 108 | 78 | 135 | 2 | 14 | 1 | 8 | 1 | | |
| 1389 | Potato salad | ½ c | 188 | 74 | 271 | 5 | 27 | 3 | 16 | 3 | 4.2 | 7.3 |
| 1383 | Potato wedges | 1 svg | 92 | 59 | 192 | 3 | 25 | 3 | 9 | 3 | | |
| 1535 | Red beans & rice | 1 svg | 112 | 76 | 114 | 4 | 18 | 3 | 3 | 1 | | |
| 1529 | Vegetable medley salad | 1 ea | 114 | 77 | 126 | 1 | 21 | 3 | 4 | 1 | | |
| | Source: Kentucky Fried Chicken Corp | | | | | | | | | | | |

| Chol (mg) | Calc (mg) | Iron (mg) | Magn (mg) | Pota (mg) | Sodi (mg) | Zinc (mg) | VT-A (RE) | Thia (mg) | VT-E (a-TE) | Ribo (mg) | Niac (mg) | V-B6 (mg) | Fola (µg) | VT-C (mg) |
|---|---|---|---|---|---|---|---|---|---|---|---|---|---|---|
| 110 | 200 | 4.5 | | 510 | 1140 | | 150 | .65 | | .48 | 8 | .33 | | 6 |
| 25 | 100 | 2.7 | | 190 | 430 | | 20 | .15 | | .26 | 2 | | | 1 |
| 65 | 100 | 4.5 | | 450 | 700 | | 40 | .36 | | .29 | 1.8 | | | 6 |
| 80 | 200 | 5.4 | | 460 | 780 | | 60 | .36 | | .44 | 1.6 | | | 6 |
| 75 | 300 | 3.6 | | 500 | 1270 | | 80 | | | | | | | 5 |
| 20 | 100 | 1.08 | 35 | 240 | 410 | 1.2 | 0 | .07 | | .17 | 1 | .13 | | 0 |
| 30 | 150 | 1.8 | 45 | 370 | 720 | 1.8 | 0 | .12 | | .08 | 1.4 | .18 | | 2 |
| | | | | | | | | | | | | | | |
| 25 | 150 | 4.5 | | 430 | 930 | | 1000 | | | | | | | 6 |
| 30 | 100 | 1.8 | | 380 | 1220 | | 1100 | | | | | | | 9 |
| 0 | 0 | 1.08 | | 690 | 190 | | 0 | .18 | | .03 | 3.8 | | | 24 |
| 0 | 0 | .36 | | 190 | 310 | | 0 | .05 | | | 1 | | | 6 |
| 0 | 20 | 1.8 | | 130 | 450 | | 0 | .29 | | .17 | 2.6 | | | 2 |
| | | | | | | | | | | | | | | |
| 25 | 300 | .72 | | 680 | 210 | | 0 | .15 | | .6 | .4 | | | 0 |
| 30 | 300 | 0 | | 550 | 180 | | 0 | .15 | | .43 | .4 | | | 0 |
| 30 | 300 | 0 | | 570 | 180 | | 0 | .15 | | .34 | .4 | | | 0 |
| 0 | 0 | 1.8 | | 80 | 460 | | 0 | .2 | | .12 | 1.8 | | | 9 |
| | | | | | | | | | | | | | | |
| | | | | | | | | | | | | | | |
| | | | | | | | | | | | | | | |
| | | | | | | | | | | | | | | |
| | | | | | | | | | | | | | | |
| | | | | | | | | | | | | | | |
| 128 | 10 | .18 | | | 772 | | 15 | | | | | | | 1 |
| 163 | 10 | .18 | | | 980 | | 15 | | | | | | | 1 |
| 157 | 10 | .18 | | | 1104 | | 15 | | | | | | | 1 |
| 97 | 10 | .18 | | | 667 | | 15 | | | | | | | 1 |
| | | | | | | | | | | | | | | |
| 92 | 30 | .72 | | | 609 | | 15 | .09 | | .17 | 11.5 | | | |
| 78 | 68 | 1.2 | | | 604 | | 15 | .06 | | .13 | 6.9 | | | |
| 75 | 21 | 1.1 | | | 269 | | 15 | .05 | | .12 | 3.2 | | | |
| 112 | 40 | 1.08 | | | 591 | | 31 | .08 | | .3 | 5.5 | | | |
| 59 | 30 | .54 | | | 383 | | 15 | .03 | | .08 | 3.7 | | | |
| | | | | | | | | | | | | | | |
| 80 | 20 | .72 | | | 750 | | 15 | | | | | | | 6 |
| 80 | 40 | 1.08 | | | 850 | | 15 | | | | | | | 6 |
| 100 | 20 | 1.08 | | | 670 | | 15 | | | | | | | 6 |
| 65 | 20 | .72 | | | 440 | | 30 | | | | | | | |
| | | | | | | | | | | | | | | |
| 75 | 33 | .8 | | | 740 | | 15 | .11 | | .13 | 13.1 | | | |
| 75 | 20 | .72 | | | 710 | | 15 | .09 | | .1 | 8.5 | | | |
| 65 | 20 | .36 | | | 310 | | 30 | .06 | | .12 | 3.7 | | | |
| 90 | 49 | 1.2 | | | 520 | | 30 | .1 | | .21 | 6.5 | | | |
| 65 | 20 | .36 | | | 320 | | 30 | | | .04 | .06 | 3.3 | | |
| 5 | 61 | 2.17 | 44 | 348 | 812 | 1.96 | 38 | .09 | | .06 | .76 | .11 | 49 | 3 |
| 0 | 30 | .18 | | | 15 | | 0 | | | | | | | 0 |
| 3 | 43 | 1.92 | | | 565 | | 28 | .26 | | .2 | 2.77 | | | |
| 18 | 23 | 1.7 | | | 331 | | 5 | .16 | | .12 | 2.2 | | | |
| 5 | 30 | .36 | | | 177 | | 32 | .03 | | .03 | .2 | | | 27 |
| 42 | 60 | .72 | | | 194 | | 10 | | | | | | | |
| 0 | 0 | .36 | | | 76 | | 20 | .14 | | .11 | 1.8 | | | 2 |
| 150 | 40 | 3.24 | | | 1230 | | 15 | | | | | | | 6 |
| 4 | 11 | .98 | | | 535 | | 0 | | | | | | | 0 |
| 66 | 2 | .1 | | | 865 | | 15 | .02 | | .02 | 1 | .05 | | <1 |
| 16 | 120 | .72 | | | 531 | | 190 | | | | | | | 0 |
| <1 | 20 | .4 | | | 388 | | 15 | | | .04 | 1.2 | | | |
| 1 | 20 | 1.08 | | | 663 | | 110 | | | | | | | 7 |
| 16 | 16 | 3.25 | 23 | 385 | 636 | .44 | 120 | .1 | | .03 | .9 | .29 | 11 | |
| 3 | | | | | 428 | | 0 | | | | | | | |
| 4 | 10 | .72 | | | 315 | | | | | | | | | |
| 0 | 20 | .36 | | | 240 | | 375 | | | | | | | 5 |

**Table H-1**

**Food Composition** (Computer code number is for West Diet Analysis program) (For purposes of calculations, use "0" for t, <1, <.1, <.01, etc.)

| Computer Code Number | Food Description | Measure | Wt (g) | $H_2O$ (%) | Ener (kcal) | Prot (g) | Carb (g) | Dietary Fiber (g) | Fat (g) | Fat Breakdown (g) Sat | Mono | Poly |
|---|---|---|---|---|---|---|---|---|---|---|---|---|
| | **FAST FOOD RESTAURANTS**—Continued | | | | | | | | | | | |
| | LONG JOHN SILVER'S | | | | | | | | | | | |
| 1528 | Chicken plank dinner, 3 piece | 1 ea | 399 | 56 | 890 | 32 | 101 | | 44 | 9.5 | 24.8 | 9.4 |
| 1531 | Clam chowder | 1 ea | 198 | 86 | 140 | 11 | 10 | 1 | 6 | 1.8 | 2.5 | 1.7 |
| 1532 | Clam dinner | 1 ea | 361 | 46 | 990 | 24 | 114 | | 52 | 10.9 | 31.4 | 9.9 |
| | Fish, batter fried: | | | | | | | | | | | |
| 1523 | Fish & fryes (fries), 3 piece | 1 ea | 384 | 54 | 980 | 31 | 92 | | 50 | 11.3 | 28.4 | 9.7 |
| 1524 | Fish & fryes, 2 piece | 1 ea | 261 | 54 | 610 | 27 | 52 | | 37 | 7.9 | 23.5 | 5.3 |
| 2240 | Fish and lemon crumb dinner, 3 piece | 1 ea | 493 | 71 | 610 | 39 | 86 | | 13 | 2.2 | 3.9 | 5.3 |
| 2241 | Fish and lemon crumb dinner, 2 piece | 1 ea | 334 | 77 | 330 | 24 | 46 | | 5 | .9 | 1.6 | 1.2 |
| 1533 | Fish & chicken dinner | 1 ea | 431 | 55 | 950 | 36 | 102 | | 49 | 10.6 | 28.8 | 9.5 |
| 1537 | Shrimp dinner, batter fried | 1 ea | 331 | 54 | 840 | 18 | 88 | | 47 | 9.7 | 27.2 | 9.1 |
| | Salads: | | | | | | | | | | | |
| 1541 | Cole slaw | 1 ea | 98 | 70 | 140 | 1 | 20 | 1 | 6 | 1 | 1.5 | 3.5 |
| 1539 | Ocean chef salad | 1 ea | 234 | 89 | 110 | 12 | 13 | 2 | 1 | .4 | .4 | .2 |
| 1540 | Seafood salad | 1 ea | 278 | 79 | 380 | 15 | 12 | 2 | 31 | 5.1 | 8.2 | 17.5 |
| 1542 | Fryes (fries) serving | 1 ea | 85 | 43 | 250 | 3 | 28 | 1 | 15 | 2.5 | 7.4 | 5.1 |
| 1543 | Hush puppies | 1 ea | 24 | 40 | 70 | 2 | 10 | <1 | 2 | .4 | 1.3 | .2 |
| | Source: Long John Silver's, Lexington KY | | | | | | | | | | | |
| | McDONALD'S | | | | | | | | | | | |
| | Sandwiches: | | | | | | | | | | | |
| 1221 | Big mac | 1 ea | 216 | 53 | 510 | 25 | 46 | 3 | 26 | 9.3 | 7.5 | 4.1 |
| 1226 | Cheeseburger | 1 ea | 122 | 46 | 318 | 15 | 36 | 2 | 13 | 5.6 | 3.8 | 1.1 |
| 1224 | Filet-o-fish | 1 ea | 145 | 49 | 364 | 14 | 41 | 1 | 16 | 3.7 | 3.8 | 5.6 |
| 1225 | Hamburger | 1 ea | 108 | 49 | 266 | 12 | 35 | 2 | 9 | 3.2 | 2.8 | .9 |
| 1444 | McChicken | 1 ea | 189 | 52 | 491 | 17 | 42 | 2 | 29 | 5.4 | 8.5 | 10.2 |
| 1591 | McLean deluxe | 1 ea | 214 | 64 | 345 | 23 | 37 | 2 | 12 | 4.4 | 3.6 | 1.2 |
| 1438 | McLean deluxe with cheese | 1 ea | 228 | 63 | 398 | 26 | 38 | 2 | 16 | 6.8 | 4.6 | 1.3 |
| 1222 | Quarter-pounder | 1 ea | 171 | 52 | 415 | 23 | 36 | 2 | 20 | 7.8 | 6.7 | 1.3 |
| 1223 | Quarter-pounder with cheese | 1 ea | 199 | 50 | 520 | 28 | 37 | 2 | 29 | 12.6 | 8.7 | 1.6 |
| 1227 | French fries, small serving | 1 ea | 68 | 40 | 207 | 3 | 26 | 2 | 10 | 1.7 | 3.1 | 2.5 |
| 1228 | Chicken McNuggets | 4 pce | 73 | 51 | 198 | 12 | 10 | 0 | 12 | 2.5 | 3.7 | 2.4 |
| | Sauces (packet): | | | | | | | | | | | |
| 1229 | Hot mustard | 1 ea | 30 | 60 | 63 | 1 | 7 | 1 | 4 | .5 | 1.1 | 2 |
| 1230 | Barbecue | 1 ea | 32 | 58 | 53 | <1 | 12 | <1 | <1 | .1 | .1 | .2 |
| 1231 | Sweet & sour | 1 ea | 32 | 57 | 55 | <1 | 12 | <1 | <1 | .1 | .1 | .3 |
| | Low-fat (frozen yogurt) milk shakes: | | | | | | | | | | | |
| 1232 | Chocolate | 1 ea | 295 | | 348 | 13 | 62 | 1 | 6 | 3.5 | .1 | .7 |
| 1233 | Strawberry | 1 ea | 294 | | 343 | 12 | 63 | <1 | 5 | 3.4 | .1 | .6 |
| 1234 | Vanilla | 1 ea | 293 | | 308 | 12 | 54 | <1 | 5 | 3.3 | .1 | .6 |
| | Low-fat (frozen yogurt) sundaes: | | | | | | | | | | | |
| 1237 | Hot caramel | 1 ea | 182 | 56 | 307 | 7 | 62 | 1 | 3 | 2 | .3 | 1 |
| 1235 | Hot fudge | 1 ea | 179 | 60 | 293 | 8 | 53 | 2 | 5 | 4.7 | .1 | .4 |
| 1267 | Strawberry | 1 ea | 178 | 65 | 239 | 6 | 51 | 1 | 1 | .7 | .1 | .2 |
| 1238 | Vanilla | 1 ea | 90 | 68 | 118 | 4 | 24 | <1 | 1 | .5 | .2 | t |
| 1241 | Cookies, McDonaldland | 1 ea | 56 | 3 | 258 | 4 | 41 | 1 | 9 | 1.7 | 6.4 | .8 |
| 1242 | Cookies, chocolaty chip | 1 ea | 56 | 3 | 282 | 3 | 36 | 1 | 14 | 3.9 | 4.4 | 1 |
| 1240 | Muffin, apple bran, fat-free | 1 ea | 75 | 39 | 182 | 4 | 40 | 1 | 1 | .2 | .1 | .3 |
| 1239 | Pie, apple | 1 ea | 84 | 35 | 289 | 3 | 37 | 1 | 14 | 3.7 | 4.5 | .3 |
| | Breakfast items: | | | | | | | | | | | |
| 1243 | English muffin with spread | 1 ea | 63 | 33 | 189 | 5 | 30 | 2 | 6 | 2.4 | 1.5 | 1.3 |
| 1244 | Egg McMuffin | 1 ea | 137 | 57 | 289 | 17 | 27 | 1 | 13 | .7 | 4.5 | 1.6 |
| 1245 | Hotcakes with marg & syrup | 1 ea | 222 | 44 | 557 | 8 | 100 | 2 | 14 | 2.4 | 4.6 | 5.8 |
| 1246 | Scrambled eggs | 1 ea | 102 | 73 | 170 | 13 | 1 | 0 | 12 | 3.6 | 5.3 | 1.7 |
| 1247 | Pork sausage | 1 ea | 43 | 45 | 173 | 6 | <1 | 0 | 16 | 5.5 | 6.4 | 2.1 |
| 1248 | Hashbrown potatoes | 1 ea | 53 | 55 | 130 | 1 | 13 | 1 | 8 | 1.3 | 2.3 | 1.9 |
| 1392 | Sausage McMuffin | 1 ea | 112 | 42 | 361 | 13 | 26 | 1 | 23 | 8.3 | 8.2 | 2.8 |
| 1393 | Sausage McMuffin with egg | 1 ea | 163 | 52 | 443 | 19 | 27 | 1 | 29 | 10 | 10.7 | 3.6 |
| 1394 | Biscuit with biscuit spread | 1 ea | 76 | 32 | 260 | 4 | 32 | 1 | 13 | 3.8 | 3.7 | .8 |

| Chol (mg) | Calc (mg) | Iron (mg) | Magn (mg) | Pota (mg) | Sodi (mg) | Zinc (mg) | VT-A (RE) | Thia (mg) | VT-E (a-TE) | Ribo (mg) | Niac (mg) | V-B6 (mg) | Fola (μg) | VT-C (mg) |
|---|---|---|---|---|---|---|---|---|---|---|---|---|---|---|
| 55 | 200 | 4.5 | | 1170 | 2000 | 3 | 40 | .52 | | .51 | 16 | | | 9 |
| 20 | 200 | 1.8 | | 380 | 590 | .6 | 150 | .09 | | .25 | 2 | | | |
| 75 | 200 | 4.5 | | 910 | 1830 | 3 | 40 | .75 | | .42 | 12 | | | 12 |
| | | | | | | | | | | | | | | |
| 70 | 200 | 4.5 | | 1120 | 1530 | 3 | 40 | .45 | | .42 | 8 | | | 15 |
| 60 | 40 | 1.8 | | 900 | 1480 | 1.2 | | .37 | | .34 | 8 | | | 9 |
| 125 | 200 | 5.4 | | 990 | 1420 | 2.25 | 700 | .75 | | .59 | 24 | | | 6 |
| 75 | 80 | 1.8 | | 440 | 640 | .9 | 1000 | .3 | | .25 | 14 | | | 18 |
| 75 | 200 | 4.5 | | 1280 | 2090 | 3 | 40 | .6 | | .59 | 14 | | | 9 |
| 100 | 200 | 3.6 | | 840 | 1630 | 3 | 40 | .45 | | .42 | 9 | | | 9 |
| | | | | | | | | | | | | | | |
| 15 | 60 | .72 | | 190 | 260 | .6 | 40 | .06 | | .07 | 2 | | | |
| 40 | 100 | 3.6 | | 95 | 730 | .3 | 500 | .12 | | .14 | 3 | | | 21 |
| 55 | 150 | 4.5 | | 130 | 980 | .9 | 200 | .15 | | .25 | 3 | | | 21 |
| 0 | 200 | .72 | | 370 | 500 | .3 | 0 | .09 | | | 1.6 | | | 6 |
| | 40 | .72 | | 65 | 25 | .3 | | .06 | | .03 | .8 | | | |
| | | | | | | | | | | | | | | |
| 76 | 202 | 4.32 | 46 | 456 | 932 | 4.81 | 66 | .49 | 1.01 | .44 | 6.08 | .25 | 49 | 3 |
| 42 | 134 | 2.73 | 27 | 281 | 766 | 2.62 | 64 | .33 | .46 | .31 | 3.81 | .15 | 24 | 2 |
| 37 | 123 | 1.85 | 32 | 266 | 708 | .7 | 21 | .32 | 1.52 | .23 | 2.58 | .07 | 30 | 0 |
| 28 | 126 | 2.73 | 24 | 260 | 531 | 2.25 | 22 | .33 | .23 | .26 | 3.81 | .14 | 21 | 2 |
| 52 | 128 | 2.5 | 32 | 319 | 797 | 1.06 | 29 | .91 | 6.16 | .24 | 7.74 | .38 | 37 | 1 |
| 59 | 131 | 4.29 | 40 | 537 | 811 | 4.9 | 74 | .42 | .63 | .34 | 7.16 | .29 | 44 | 8 |
| 73 | 139 | 4.29 | 43 | 559 | 1046 | 5.26 | 115 | .42 | .85 | .39 | 7.16 | .3 | 47 | 8 |
| 70 | 127 | 4.33 | 33 | 405 | 692 | 4.66 | 33 | .39 | .36 | .32 | 6.78 | .24 | 27 | 3 |
| 97 | 143 | 4.5 | | | 1160 | | 115 | .39 | .81 | .43 | 6.78 | .26 | 33 | 3 |
| 0 | 9 | .53 | 26 | 469 | 135 | .32 | 0 | .05 | .83 | 0 | 1.94 | .24 | 26 | 8 |
| 42 | 9 | .65 | 17 | 210 | 353 | .69 | 0 | .08 | .96 | .11 | 5.15 | .21 | | 0 |
| | | | | | | | | | | | | | | |
| 3 | 7 | .78 | | 29 | 85 | | 4 | .01 | | .01 | .15 | | | 0 |
| 0 | 4 | 0 | | 51 | 277 | | 0 | .01 | | .01 | .17 | | | 4 |
| 0 | 2 | .16 | | 7 | 158 | | 74 | 0 | | .01 | .08 | | | 0 |
| | | | | | | | | | | | | | | |
| 24 | 372 | 1.04 | | 543 | 241 | | 46 | .12 | | .51 | .4 | .1 | | 3 |
| 24 | 366 | .29 | | 542 | 170 | | 46 | .12 | | .51 | .4 | .11 | | 3 |
| 24 | 360 | .29 | | 533 | 193 | | 45 | .12 | | .51 | .31 | | | 3 |
| | | | | | | | | | | | | | | |
| 7 | 246 | .15 | | 344 | 197 | | 18 | .09 | | .34 | .27 | | | 2 |
| 5 | 258 | .59 | | 441 | 190 | | 7 | .09 | | .34 | .29 | | | 2 |
| 5 | 221 | .25 | | 325 | 115 | | 6 | .06 | | .34 | .25 | | | 2 |
| 3 | 132 | .23 | | 175 | 84 | | 4 | <.01 | | .01 | .23 | | | 1 |
| 0 | 10 | 1.73 | 11 | 62 | 267 | .38 | 0 | .24 | .99 | .16 | 2.01 | .03 | | 0 |
| 3 | 28 | 1.78 | 24 | 142 | 229 | .4 | 0 | .14 | .92 | .16 | 1.48 | | | 0 |
| 0 | 34 | 1.29 | 13 | 77 | 215 | .33 | 0 | .14 | 0 | .14 | 1.32 | .03 | 5 | 1 |
| 0 | 17 | 1.23 | 7 | 69 | 221 | .23 | | .19 | 1.5 | .12 | 1.55 | .04 | 9 | 27 |
| | | | | | | | | | | | | | | |
| 13 | 103 | 1.59 | 13 | 69 | 386 | .42 | 33 | .25 | .13 | .31 | 2.61 | .04 | 57 | 1 |
| 234 | 151 | 2.44 | 24 | 199 | 730 | 1.56 | 100 | .49 | .85 | .45 | 3.33 | .15 | 33 | 2 |
| 11 | 108 | 1.98 | 27 | 285 | 746 | .53 | 119 | .24 | 1.2 | .26 | 1.86 | .09 | <1 | <1 |
| 424 | 50 | 1.19 | 10 | 126 | 143 | 1.06 | 168 | .07 | .92 | .51 | .06 | .12 | 44 | 0 |
| 33 | 7 | .5 | 7 | 102 | 292 | .78 | 0 | .18 | .26 | .06 | 1.7 | .09 | | 0 |
| 0 | 7 | .3 | 11 | 213 | 332 | .15 | 0 | .08 | .58 | .02 | .9 | .08 | 8 | 3 |
| 46 | 132 | 2.07 | 22 | 191 | 751 | 1.51 | 48 | .56 | .66 | .27 | 3.76 | .14 | 16 | 0 |
| 257 | 156 | 2.8 | 26 | 251 | 821 | 2.07 | 117 | .59 | 1.11 | .49 | 3.79 | .19 | 33 | 0 |
| 0 | 68 | 1.85 | 9 | 105 | 836 | .3 | 2 | .29 | .81 | .23 | 2.23 | .03 | 5 | 0 |

**Table H-1**

**Food Composition** (Computer code number is for West Diet Analysis program) (For purposes of calculations, use "0" for t, <1, <.1, <.01, etc.)

| Computer Code Number | Food Description | Measure | Wt (g) | $H_2O$ (%) | Ener (kcal) | Prot (g) | Carb (g) | Dietary Fiber (g) | Fat (g) | Fat Breakdown (g) Sat | Mono | Poly |
|---|---|---|---|---|---|---|---|---|---|---|---|---|
| | **FAST FOOD RESTAURANTS**—Continued | | | | | | | | | | | |
| | McDONALD'S—Continued | | | | | | | | | | | |
| 1395 | Biscuit with sausage | 1 ea | 119 | 37 | 433 | 10 | 32 | 1 | 29 | 8.6 | 10.1 | 2.8 |
| 1396 | Biscuit with sausage & egg | 1 ea | 170 | 48 | 518 | 16 | 33 | 1 | 35 | 10.5 | 12.7 | 3.7 |
| 1397 | Biscuit with bacon, egg, cheese | 1 ea | 152 | 46 | 450 | 17 | 33 | 1 | 27 | 8.7 | 8.9 | 2.3 |
| | Salads: | | | | | | | | | | | |
| 1398 | Chef salad | 1 ea | 313 | 86 | 206 | 19 | 9 | 3 | 11 | 4.2 | 3 | 1.2 |
| 1400 | Garden salad | 1 ea | 234 | 92 | 84 | 6 | 7 | 3 | 4 | 1.1 | 1.4 | .7 |
| 1401 | Chunky chicken salad | 1 ea | 296 | 87 | 164 | 23 | 8 | 3 | 5 | 1.3 | 1.6 | 1 |
| | Source: McDonald's Corporation | | | | | | | | | | | |
| | PIZZA HUT | | | | | | | | | | | |
| | Pan pizza: | | | | | | | | | | | |
| 1657 | Cheese | 2 pce | 216 | 51 | 522 | 24 | 56 | 4 | 22 | 10 | 6.8 | 3.4 |
| 1658 | Pepperoni | 2 pce | 208 | 49 | 531 | 22 | 56 | 4 | 24 | 8 | 9.9 | 3.7 |
| 1659 | Supreme | 2 pce | 273 | 56 | 622 | 30 | 56 | 6 | 30 | 12 | 12 | 4.2 |
| 1660 | Super supreme | 2 pce | 286 | 56 | 645 | 30 | 56 | 6 | 34 | 12 | | |
| | Thin 'n crispy pizza: | | | | | | | | | | | |
| 1649 | Cheese | 2 pce | 174 | 52 | 411 | 22 | 42 | 4 | 16 | 8 | 4.4 | 2.3 |
| 1623 | Pepperoni | 2 pce | 168 | 48 | 431 | 22 | 42 | 2 | 20 | 8 | | |
| 1622 | Supreme | 2 pce | 232 | 57 | 514 | 28 | 42 | 4 | 26 | 10 | | |
| 1620 | Super supreme | 2 pce | 247 | 57 | 541 | 28 | 44 | 4 | 28 | 12 | | |
| | Hand tossed pizza: | | | | | | | | | | | |
| 1619 | Cheese | 2 pce | 216 | 53 | 470 | 26 | 58 | 4 | 14 | 7.9 | | |
| 1618 | Pepperoni | 2 pce | 208 | 51 | 477 | 24 | 58 | 4 | 16 | 8 | | |
| 1648 | Supreme | 2 pce | 273 | 56 | 568 | 32 | 60 | 6 | 24 | 10 | | |
| 1617 | Super supreme | 2 pce | 286 | 57 | 591 | 32 | 60 | 6 | 26 | 10 | | |
| | Personal pan pizza: | | | | | | | | | | | |
| 1610 | Pepperoni | 1 ea | 255 | 50 | 637 | 27 | 69 | 5 | 28 | 10 | 11.8 | 4.5 |
| 1609 | Supreme | 1 ea | 327 | 57 | 721 | 33 | 70 | 6 | 34 | 12 | 14.7 | 5.6 |
| | Source: Pizza Hut | | | | | | | | | | | |
| | SUBWAY | | | | | | | | | | | |
| | Deli style sandwich: | | | | | | | | | | | |
| 69104 | Bologna | 1 ea | 171 | 64 | 292 | 10 | 38 | 2 | 12 | 4 | | |
| 69102 | Ham | 1 ea | 171 | 69 | 234 | 11 | 37 | 2 | 4 | 1 | | |
| 69103 | Roast beef | 1 ea | 180 | 69 | 245 | 13 | 38 | 2 | 4 | 1 | | |
| 69105 | Seafood and crab: | 1 ea | 178 | 66 | 298 | 12 | 37 | 2 | 11 | 2 | | |
| 69106 | With light mayo | 1 ea | 178 | 68 | 256 | 12 | 37 | 2 | 7 | 2 | | |
| 69108 | Tuna: | 1 ea | 178 | | 354 | 11 | 37 | 2 | 18 | 3 | | |
| 69107 | With light mayo | 1 ea | 178 | 67 | 279 | 11 | 38 | 2 | 9 | 2 | | |
| 69101 | Turkey | 1 ea | 180 | 69 | 235 | 12 | 38 | 2 | 4 | 1 | | |
| | Sandwiches, 6 inch: | | | | | | | | | | | |
| | B.L.T.: | | | | | | | | | | | |
| 69135 | On white bread | 1 ea | 191 | 67 | 311 | 14 | 38 | 3 | 10 | 3 | | |
| 69136 | On wheat bread | 1 ea | 198 | 65 | 327 | 14 | 44 | 3 | 10 | 3 | | |
| | Chicken taco sub: | | | | | | | | | | | |
| 69131 | On white bread | 1 ea | 286 | 70 | 421 | 24 | 43 | 3 | 16 | 5 | | |
| 69132 | On wheat bread | 1 ea | 293 | 69 | 436 | 25 | 49 | 4 | 16 | 5 | | |
| | Club : | | | | | | | | | | | |
| 69117 | On white bread | 1 ea | 246 | 73 | 297 | 21 | 40 | 3 | 5 | 1 | | |
| 69118 | On wheat bread | 1 ea | 253 | 71 | 312 | 21 | 46 | 3 | 5 | 1 | | |
| | Cold cut trio: | | | | | | | | | | | |
| 69113 | On white bread | 1 ea | 246 | 71 | 362 | 19 | 39 | 3 | 13 | 4 | | |
| 69114 | On wheat bread | 1 ea | 253 | 68 | 378 | 20 | 46 | 3 | 13 | 4 | | |
| | Ham: | | | | | | | | | | | |
| 69115 | On white bread | 1 ea | 232 | 73 | 287 | 18 | 39 | 3 | 5 | 1 | | |
| 69115 | On wheat bread | 1 ea | 239 | 71 | 302 | 19 | 45 | 3 | 5 | 1 | | |

| Chol (mg) | Calc (mg) | Iron (mg) | Magn (mg) | Pota (mg) | Sodi (mg) | Zinc (mg) | VT-A (RE) | Thia (mg) | VT-E (a-TE) | Ribo (mg) | Niac (mg) | V-B6 (mg) | Fola (µg) | VT-C (mg) |
|---|---|---|---|---|---|---|---|---|---|---|---|---|---|---|
| 33 | 75 | 2.35 | 15 | 207 | 1128 | 1.08 | 2 | .48 | 1.07 | .29 | 3.93 | .12 | 5 | 0 |
| 245 | 100 | 2.95 | 20 | 271 | 1199 | 1.61 | 59 | .51 | 1.53 | .55 | 3.96 | .18 | 27 | 0 |
| 238 | 103 | 2.6 | 20 | 245 | 1315 | 1.64 | 99 | .39 | 1.49 | .57 | 3.32 | .13 | 30 | 0 |
| | | | | | | | | | | | | | | |
| 179 | 157 | 1.81 | 40 | 605 | 727 | 2.16 | 1179 | .33 | 1.45 | .37 | 4.32 | .36 | 100 | 22 |
| 139 | 52 | 1.34 | 24 | 407 | 61 | .73 | 1114 | .12 | .95 | .24 | .65 | .16 | 96 | 22 |
| 76 | 54 | 1.62 | 44 | 673 | 318 | 1.52 | 1973 | .51 | 1.28 | .21 | 8.46 | .52 | 83 | 30 |
| | | | | | | | | | | | | | | |
| 50 | 288 | 3 | 63 | 337 | 1002 | 4.32 | 211 | .6 | | .64 | 5.48 | .18 | | 7 |
| 48 | 206 | 3.21 | 55 | 399 | 1140 | 4.14 | 190 | .62 | | .48 | 5.31 | .16 | 0 | 8 |
| 60 | 234 | 4.6 | 81 | 620 | 1529 | 6 | 195 | .86 | | .84 | 6.4 | .33 | | 11 |
| 68 | 236 | 4.39 | 80 | 592 | 1649 | 5.99 | 201 | .83 | | .73 | 7.13 | | | 12 |
| | | | | | | | | | | | | | | |
| 50 | 291 | 2.06 | 56 | 307 | 1070 | 4.23 | 217 | .46 | | .46 | 5.65 | .18 | | 6 |
| 50 | 208 | 2.2 | 51 | 330 | 1255 | 4.02 | 199 | .48 | | .49 | 5.97 | | | 7 |
| 62 | 238 | 3.6 | 79 | 631 | 1591 | 5.4 | 197 | .7 | | .57 | 6.27 | | | 12 |
| 70 | 238 | 3.41 | 73 | 563 | 1762 | 5.47 | 208 | .72 | | .53 | 6.57 | | | 9 |
| | | | | | | | | | | | | | | |
| 50 | 284 | 3 | 71 | 388 | 1242 | 4.6 | 198 | .48 | | .48 | 5.3 | | | 10 |
| 48 | 202 | 3.21 | 84 | 610 | 1380 | 6.01 | 187 | .72 | | .56 | 7.59 | | | 13 |
| 60 | 232 | 4.6 | 87 | 589 | 1769 | 5.48 | 192 | .82 | | .66 | 8.45 | | | 14 |
| 68 | 232 | 4.39 | 89 | 607 | 1889 | 5.65 | 198 | .84 | | .68 | 8.71 | | | 14 |
| | | | | | | | | | | | | | | |
| 55 | 250 | 4 | 60 | 406 | 1338 | 3.8 | 233 | .56 | | .66 | 8.16 | .2 | | 10 |
| 66 | 276 | 5.19 | 74 | 603 | 1757 | 4.69 | 240 | .73 | | .82 | 9.91 | .4 | | 14 |
| | | | | | | | | | | | | | | |
| 20 | 39 | 3 | | | 744 | | 113 | | | | | | | 14 |
| 14 | 24 | 3 | | | 773 | | 113 | | | | | | | 14 |
| 13 | 23 | 3 | | | 638 | | 113 | | | | | | | 14 |
| 17 | 24 | 3 | | | 544 | | 113 | | | | | | | 14 |
| 16 | 24 | 3 | | | 556 | | 118 | | | | | | | 14 |
| 18 | 26 | 3 | | | 557 | | 116 | | | | | | | 14 |
| 16 | 26 | 3 | | | 583 | | 126 | | | | | | | 14 |
| 12 | 26 | 3 | | | 944 | | 113 | | | | | | | 14 |
| | | | | | | | | | | | | | | |
| 16 | 27 | 3 | | | 945 | | 120 | | | | | | | 15 |
| 16 | 33 | 3 | | | 957 | | 120 | | | | | | | 15 |
| | | | | | | | | | | | | | | |
| 52 | 118 | 4 | | | 1264 | | 209 | | | | | | | 18 |
| 52 | 124 | 4 | | | 1275 | | 209 | | | | | | | 18 |
| | | | | | | | | | | | | | | |
| 26 | 29 | 4 | | | 1341 | | 120 | | | | | | | 15 |
| 26 | 35 | 4 | | | 1352 | | 120 | | | | | | | 15 |
| | | | | | | | | | | | | | | |
| 64 | 49 | 4 | | | 1401 | | 130 | | | | | | | 16 |
| 64 | 55 | 4 | | | 1412 | | 130 | | | | | | | 16 |
| | | | | | | | | | | | | | | |
| 28 | 28 | 3 | | | 1308 | | 120 | | | | | | | 15 |
| 28 | 35 | 3 | | | 1319 | | 120 | | | | | | | 15 |

**Table H-1**

**Food Composition** (Computer code number is for West Diet Analysis program) (For purposes of calculations, use "0" for t, <1, <.1, <.01, etc.)

| Computer Code Number | Food Description | Measure | Wt (g) | $H_2O$ (%) | Ener (kcal) | Prot (g) | Carb (g) | Dietary Fiber (g) | Fat (g) | Fat Breakdown (g) Sat | Mono | Poly |
|---|---|---|---|---|---|---|---|---|---|---|---|---|
| | **FAST FOOD RESTAURANTS—**Continued | | | | | | | | | | | |
| | SUBWAY—Continued | | | | | | | | | | | |
| | Italian B.M.T. | | | | | | | | | | | |
| 69139 | On white bread | 1 ea | 246 | 66 | 445 | 21 | 39 | 3 | 21 | 8 | | |
| 69140 | On wheat bread | 1 ea | 253 | 64 | 460 | 21 | 45 | 3 | 22 | 7 | | |
| | Meatball: | | | | | | | | | | | |
| 69129 | On white bread | 1 ea | 260 | 70 | 404 | 18 | 44 | 3 | 16 | 6 | | |
| 69130 | On wheat bread | 1 ea | 267 | 67 | 419 | 19 | 51 | 3 | 16 | 6 | | |
| | Melt with turkey, ham, bacon, cheese: | | | | | | | | | | | |
| 69127 | On white bread | 1 ea | 251 | 70 | 366 | 22 | 40 | 3 | 12 | 5 | | |
| 69128 | On wheat bread | 1 ea | 258 | 68 | 382 | 23 | 46 | 3 | 12 | 5 | | |
| | Pizza sub: | | | | | | | | | | | |
| 69133 | On white bread | 1 ea | 250 | 66 | 448 | 19 | 41 | 3 | 22 | 9 | | |
| 69134 | On wheat bread | 1 ea | 257 | 65 | 464 | 19 | 48 | 3 | 22 | 9 | | |
| | Roast beef: | | | | | | | | | | | |
| 69121 | On white bread | 1 ea | 232 | 72 | 288 | 19 | 39 | 3 | 5 | 1 | | |
| 69122 | On wheat bread | 1 ea | 239 | 70 | 303 | 20 | 45 | 3 | 5 | 1 | | |
| | Roasted chicken breast: | | | | | | | | | | | |
| 69125 | On white bread | 1 ea | 246 | 70 | 332 | 26 | 41 | 3 | 6 | 1 | | |
| 69126 | On wheat bread | 1 ea | 253 | 68 | 348 | 27 | 47 | 3 | 6 | 1 | | |
| | Seafood and crab: | | | | | | | | | | | |
| 69145 | On white bread: | 1 ea | 246 | 69 | 415 | 19 | 38 | 3 | 19 | 3 | | |
| 69147 | With light mayo | 1 ea | 246 | 72 | 332 | 19 | 39 | 3 | 10 | 2 | | |
| 69146 | On wheat bread: | 1 ea | 253 | 67 | 430 | 20 | 44 | 3 | 19 | 3 | | |
| 69148 | With light mayo | 1 ea | 253 | 70 | 347 | 20 | 45 | 3 | 10 | 2 | | |
| | Spicy italian: | | | | | | | | | | | |
| 69123 | On white bread | 1 ea | 232 | 64 | 467 | 20 | 38 | 3 | 24 | 9 | | |
| 69124 | On wheat bread | 1 ea | 239 | 62 | 482 | 21 | 44 | 3 | 25 | 9 | | |
| | Steak and cheese: | | | | | | | | | | | |
| 69119 | On white bread | 1 ea | 257 | 68 | 383 | 29 | 41 | 3 | 10 | 6 | | |
| 69120 | On wheat bread | 1 ea | 264 | 67 | 398 | 30 | 47 | 3 | 10 | 6 | | |
| | Tuna: | | | | | | | | | | | |
| 69141 | On white bread: | 1 ea | 246 | 62 | 527 | 18 | 38 | 3 | 32 | 5 | | |
| 69143 | With light mayo | 1 ea | 246 | 70 | 376 | 18 | 39 | 3 | 15 | 2 | | |
| 69142 | On wheat bread: | 1 ea | 253 | 62 | 542 | 19 | 44 | 3 | 32 | 5 | | |
| 69144 | With light mayo | 1 ea | 253 | 68 | 391 | 19 | 46 | 3 | 15 | 2 | | |
| | Turkey: | | | | | | | | | | | |
| 69111 | On white bread | 1 ea | 232 | 73 | 273 | 17 | 40 | 3 | 4 | 1 | | |
| 69112 | On wheat bread | 1 ea | 239 | 71 | 289 | 18 | 46 | 3 | 4 | 1 | | |
| | Turkey breast and ham: | | | | | | | | | | | |
| 69137 | On white bread | 1 ea | 232 | 73 | 280 | 18 | 39 | 3 | 5 | 1 | | |
| 69138 | On wheat bread | 1 ea | 239 | 71 | 295 | 18 | 46 | 3 | 5 | 1 | | |
| | Veggie delite: | | | | | | | | | | | |
| 69109 | On white bread | 1 ea | 175 | 71 | 222 | 9 | 38 | 3 | 3 | 0 | | |
| 69110 | On wheat bread | 1 ea | 182 | 69 | 237 | 9 | 44 | 3 | 3 | 0 | | |
| | Salads: | | | | | | | | | | | |
| 52128 | B.L.T. | 1 ea | 276 | 91 | 140 | 7 | 10 | 2 | 8 | 3 | | |
| 52124 | B.M.T., classic Italian | 1 ea | 331 | 86 | 274 | 14 | 11 | 1 | 20 | 7 | | |
| 52127 | Chicken taco | 1 ea | 370 | 87 | 250 | 18 | 15 | 2 | 14 | 5 | | |
| 52115 | Club | 1 ea | 331 | 91 | 126 | 14 | 12 | 1 | 3 | 1 | | |
| 52120 | Cold cut trio | 1 ea | 330 | 89 | 191 | 13 | 11 | 1 | 11 | 3 | | |
| 52123 | Ham | 1 ea | 316 | 91 | 116 | 12 | 11 | 1 | 3 | 1 | | |
| 52129 | Meatball | 1 ea | 345 | 88 | 233 | 12 | 16 | 2 | 14 | 5 | | |
| 52131 | Melt | 1 ea | 336 | 88 | 195 | 16 | 12 | 1 | 10 | 4 | | |
| 52121 | Pizza | 1 ea | 335 | 86 | 277 | 12 | 13 | 2 | 20 | 8 | | |
| 52126 | Roast beef | 1 ea | 316 | 92 | 117 | 12 | 11 | 1 | 3 | 1 | | |
| 52119 | Roasted chicken breast | 1 ea | 331 | 89 | 162 | 20 | 13 | 1 | 4 | 1 | | |
| 52117 | Seafood and crab: | 1 5 | 331 | 88 | 244 | 13 | 10 | 2 | 17 | 3 | | |
| 52116 | With light mayo | 1 5 | 331 | 90 | 161 | 13 | 11 | 2 | 8 | 1 | | |
| 52130 | Steak and cheese | 1 ea | 342 | 87 | 212 | 22 | 13 | 1 | 8 | 5 | | |
| 52122 | Tuna: | 1 ea | 331 | 84 | 356 | 12 | 10 | 1 | 30 | 5 | | |
| 52118 | With light mayo | 1 ea | 331 | 89 | 205 | 12 | 11 | 1 | 13 | 2 | | |
| 52114 | Turkey breast | 1 ea | 316 | 92 | 102 | 11 | 12 | 1 | 2 | 1 | | |
| 52125 | With ham | 1 ea | 316 | 92 | 109 | 11 | 11 | 1 | 3 | 1 | | |

| Chol (mg) | Calc (mg) | Iron (mg) | Magn (mg) | Pota (mg) | Sodi (mg) | Zinc (mg) | VT-A (RE) | Thia (mg) | VT-E (a-TE) | Ribo (mg) | Niac (mg) | V-B6 (mg) | Fola (μg) | VT-C (mg) |
|---|---|---|---|---|---|---|---|---|---|---|---|---|---|---|
| 56 | 44 | 4 | | | 1652 | | 151 | | | | | | | 15 |
| 56 | 50 | 4 | | | 1664 | | 151 | | | | | | | 15 |
| | | | | | | | | | | | | | | |
| 33 | 32 | 4 | | | 1035 | | 142 | | | | | | | 16 |
| 33 | 39 | 4 | | | 1046 | | 142 | | | | | | | 16 |
| | | | | | | | | | | | | | | |
| 42 | 93 | 4 | | | 1735 | | 155 | | | | | | | 15 |
| 42 | 100 | 3 | | | 1746 | | 156 | | | | | | | 15 |
| | | | | | | | | | | | | | | |
| 50 | 103 | 4 | | | 1609 | | 238 | | | | | | | 16 |
| 50 | 110 | 3 | | | 1621 | | 238 | | | | | | | 16 |
| | | | | | | | | | | | | | | |
| 20 | 25 | 4 | | | 928 | | 120 | | | | | | | 15 |
| 20 | 32 | 3 | | | 939 | | 120 | | | | | | | 15 |
| | | | | | | | | | | | | | | |
| 48 | 35 | 3 | | | 967 | | 123 | | | | | | | 15 |
| 48 | 42 | 3 | | | 978 | | 123 | | | | | | | 15 |
| | | | | | | | | | | | | | | |
| 34 | 28 | 3 | | | 849 | | 121 | | | | | | | 15 |
| 32 | 28 | 3 | | | 873 | | 131 | | | | | | | 15 |
| 34 | 34 | 3 | | | 860 | | 121 | | | | | | | 15 |
| 32 | 34 | 3 | | | 884 | | 131 | | | | | | | 15 |
| | | | | | | | | | | | | | | |
| 57 | 40 | 4 | | | 1592 | | 169 | | | | | | | 15 |
| 57 | 47 | 4 | | | 1604 | | 169 | | | | | | | 15 |
| | | | | | | | | | | | | | | |
| 70 | 88 | 5 | | | 1106 | | 175 | | | | | | | 18 |
| 70 | 95 | 5 | | | 1117 | | 176 | | | | | | | 18 |
| | | | | | | | | | | | | | | |
| 36 | 32 | 3 | | | 875 | | 125 | | | | | | | 15 |
| 32 | 32 | 3 | | | 928 | | 146 | | | | | | | 15 |
| 36 | 38 | 3 | | | 886 | | 126 | | | | | | | 15 |
| 32 | 38 | 3 | | | 940 | | 146 | | | | | | | 15 |
| | | | | | | | | | | | | | | |
| 19 | 30 | 4 | | | 1391 | | 120 | | | | | | | 15 |
| 19 | 37 | 3 | | | 1403 | | 120 | | | | | | | 15 |
| | | | | | | | | | | | | | | |
| 24 | 29 | 3 | | | 1350 | | 120 | | | | | | | 15 |
| 24 | 36 | 3 | | | 1361 | | 120 | | | | | | | 15 |
| | | | | | | | | | | | | | | |
| 0 | 25 | 3 | | | 582 | | 120 | | | | | | | 15 |
| 0 | 32 | 3 | | | 593 | | 120 | | | | | | | 15 |
| | | | | | | | | | | | | | | |
| 16 | 24 | 1 | | | 672 | | 273 | | | | | | | 32 |
| 56 | 41 | 2 | | | 1379 | | 303 | | | | | | | 32 |
| 52 | 115 | 3 | | | 990 | | 361 | | | | | | | 35 |
| 26 | 26 | 2 | | | 1067 | | 273 | | | | | | | 32 |
| 64 | 46 | 2 | | | 1127 | | 282 | | | | | | | 33 |
| 28 | 25 | 2 | | | 1034 | | 273 | | | | | | | 32 |
| 33 | 30 | 2 | | | 761 | | 295 | | | | | | | 33 |
| 42 | 90 | 2 | | | 1461 | | 308 | | | | | | | 32 |
| 50 | 100 | 2 | | | 1336 | | 390 | | | | | | | 33 |
| 20 | 23 | 2 | | | 654 | | 273 | | | | | | | 32 |
| 48 | 32 | 2 | | | 693 | | 276 | | | | | | | 32 |
| 34 | 25 | 2 | | | 575 | | 273 | | | | | | | 32 |
| 32 | 25 | 2 | | | 599 | | 284 | | | | | | | 32 |
| 70 | 86 | 3 | | | 832 | | 328 | | | | | | | 35 |
| 36 | 29 | 2 | | | 601 | | 278 | | | | | | | 32 |
| 32 | 29 | 2 | | | 654 | | 298 | | | | | | | 32 |
| 19 | 28 | 2 | | | 1117 | | 273 | | | | | | | 32 |
| 24 | 27 | 2 | | | 1076 | | 273 | | | | | | | 32 |

**Table H-1**

**Food Composition** (Computer code number is for West Diet Analysis program) (For purposes of calculations, use "0" for t, <1, <.1, <.01, etc.)

| Computer Code Number | Food Description | Measure | Wt (g) | $H_2O$ (%) | Ener (kcal) | Prot (g) | Carb (g) | Dietary Fiber (g) | Fat (g) | Fat Breakdown (g) | | |
|---|---|---|---|---|---|---|---|---|---|---|---|---|
| | | | | | | | | | | Sat | Mono | Poly |
| | **FAST FOOD RESTAURANTS**—Continued | | | | | | | | | | | |
| | SUBWAY—Continued | | | | | | | | | | | |
| 52113 | Veggie delite | 1 ea | 260 | 94 | 51 | 2 | 10 | 1 | 1 | 0 | | |
| | Cookies: | | | | | | | | | | | |
| 47662 | Brazil nut and chocolate chip | 1 ea | 48 | 12 | 229 | 3 | 27 | 1 | 12 | 3.5 | | |
| 47655 | Chocolate chip: | 1 ea | 48 | 14 | 209 | 2 | 29 | 1 | 10 | 3.5 | | |
| 47658 | With M&M's | 1 ea | 48 | 14 | 209 | 2 | 29 | 1 | 10 | 3 | | |
| 47659 | Chocolate chunk | 1 ea | 48 | 14 | 209 | 2 | 29 | 1 | 10 | 3.5 | | |
| 47656 | Oatmeal raisin | 1 ea | 48 | 15 | 199 | 3 | 29 | 1 | 8 | 2 | | |
| 47657 | Peanut butter | 1 ea | 48 | 13 | 219 | 3 | 26 | 1 | 12 | 2.5 | | |
| 47660 | Sugar | 1 ea | 48 | 11 | 229 | 2 | 28 | 0 | 12 | 3 | | |
| 47661 | White chip macademia | 1 ea | 48 | 12 | 229 | 2 | 28 | 1 | 12 | 2.5 | | |
| | Source: Subway International | | | | | | | | | | | |
| | TACO BELL | | | | | | | | | | | |
| | Breakfast burrito: | | | | | | | | | | | |
| 1601 | Bacon breakfast burrito | 1 ea | 99 | 48 | 291 | 11 | 23 | | 17 | 4 | | |
| 1627 | Country breakfast burrito | 1 ea | 113 | 55 | 220 | 8 | 26 | 2 | 14 | 5 | | |
| 1626 | Fiesta breakfast burrito | 1 ea | 92 | 44 | 280 | 9 | 25 | 2 | 16 | 6 | | |
| 1625 | Grande breakfast burrito | 1 ea | 177 | 56 | 420 | 13 | 43 | 3 | 22 | 7 | | |
| 1604 | Sausage breakfast burrito | 1 ea | 106 | 49 | 303 | 11 | 23 | | 19 | 6 | | |
| | Burritos: | | | | | | | | | | | |
| 1544 | Bean with red sauce | 1 ea | 198 | 58 | 380 | 13 | 55 | 13 | 12 | 4 | | |
| 1545 | Beef with red sauce | 1 ea | 198 | 57 | 432 | 22 | 42 | 4 | 19 | 8 | 6.7 | .7 |
| 1546 | Beef & bean with red sauce | 1 ea | 198 | 57 | 412 | 17 | 50 | 5 | 16 | 6 | 6.1 | 2.1 |
| 1569 | Big beef supreme | 1 ea | 298 | 64 | 520 | 24 | 54 | 11 | 23 | 10 | | |
| 1552 | Chicken burrito | 1 ea | 171 | 58 | 345 | 17 | 41 | | 13 | 5 | | |
| 1547 | Supreme with red sauce | 1 ea | 248 | 64 | 428 | 16 | 50 | 10 | 18 | 7.8 | | |
| 1571 | 7 layer burrito | 1 ea | 234 | 61 | 438 | 13 | 55 | 11 | 19 | 5.8 | | |
| 1538 | Chilito | 1 ea | 156 | 49 | 391 | 17 | 41 | | 18 | 9 | | |
| 1549 | Chilito, steak | 1 ea | 257 | 62 | 496 | 26 | 47 | | 23 | 10 | | |
| | Tacos: | | | | | | | | | | | |
| 1551 | Taco | 1 ea | 78 | 58 | 180 | 9 | 12 | 3 | 10 | 4 | | |
| 1554 | Soft taco | 1 ea | 99 | 63 | 242 | 10 | 13 | 3 | 11 | 4.4 | | |
| 1536 | Soft taco supreme | 1 ea | 128 | 64 | 234 | 11 | 21 | 3 | 13 | 6.3 | | |
| 1568 | Soft taco, chicken | 1 ea | 128 | 63 | 212 | 15 | 22 | 2 | 7 | 2.6 | | |
| 1572 | Soft taco, steak | 1 ea | 100 | 63 | 180 | 12 | 16 | 2 | 8 | 1.9 | | |
| 1555 | Tostada with red sauce | 1 ea | 156 | 67 | 264 | 9 | 27 | 11 | 13 | 4.4 | | |
| 1558 | Mexican pizza | 1 ea | 223 | 53 | 578 | 21 | 43 | 8 | 35 | 10.1 | | |
| 1559 | Taco salad with salsa | 1 ea | 585 | 71 | 923 | 33 | 70 | 17 | 56 | 16.3 | | |
| 1560 | Nachos, regular | 1 ea | 106 | 40 | 343 | 5 | 36 | 3 | 19 | 4.3 | | |
| 1561 | Nachos, bellgrande | 1 ea | 287 | 51 | 708 | 19 | 77 | 16 | 36 | 10.1 | | |
| 1562 | Pintos & cheese with red sauce | 1 ea | 128 | 68 | 203 | 10 | 19 | 11 | 10 | 4.3 | | |
| 1563 | Taco sauce, packet | 1 ea | 9 | 94 | 2 | <1 | <1 | <1 | <1 | 0 | 0 | 0 |
| 1564 | Salsa | 1 ea | 10 | 28 | 27 | 1 | 6 | | <1 | 0 | 0 | 0 |
| 1565 | Cinnamon twists | 1 ea | 35 | 6 | 175 | 1 | 24 | 0 | 7 | 0 | | |
| 1628 | Caramel roll | 1 ea | 85 | 19 | 353 | 6 | 46 | | 16 | 4 | | |
| | Source: Taco Bell Corporation | | | | | | | | | | | |
| | WENDY'S | | | | | | | | | | | |
| | Hamburgers: | | | | | | | | | | | |
| 1566 | Single on white bun, no toppings | 1 ea | 133 | 44 | 360 | 24 | 31 | 2 | 16 | 6 | | |
| 1570 | Cheeseburger, bacon | 1 ea | 166 | 55 | 380 | 20 | 34 | 2 | 19 | 7 | 10.3 | 1.4 |
| 1730 | Chicken sandwich, grilled | 1 ea | 189 | 62 | 310 | 27 | 35 | 2 | 8 | 1.5 | | |
| | Baked potatoes: | | | | | | | | | | | |
| 1573 | Plain | 1 ea | 284 | 71 | 310 | 7 | 71 | 7 | 0 | 0 | 0 | 0 |
| 1574 | With bacon & cheese | 1 ea | 380 | 69 | 530 | 17 | 78 | 7 | 18 | 4 | 10.7 | 3.3 |
| 1575 | With broccoli & cheese | 1 ea | 411 | 74 | 470 | 9 | 80 | 9 | 14 | 2.5 | 8 | 2.5 |
| 1576 | With cheese | 1 ea | 383 | 68 | 570 | 14 | 78 | 7 | 23 | 8 | 9.2 | 4.8 |

| Chol (mg) | Calc (mg) | Iron (mg) | Magn (mg) | Pota (mg) | Sodi (mg) | Zinc (mg) | VT-A (RE) | Thia (mg) | VT-E (a-TE) | Ribo (mg) | Niac (mg) | V-B6 (mg) | Fola (μg) | VT-C (mg) |
|---|---|---|---|---|---|---|---|---|---|---|---|---|---|---|
| 0 | 23 | 1 | | | 308 | | 136 | | | | | | | 32 |
| | | | | | | | | | | | | | | |
| 10 | 32 | 1.99 | | | 115 | | 0 | | | | | | | 0 |
| 10 | 16 | 1.99 | | | 139 | | 0 | | | | | | | 0 |
| 15 | 16 | 1 | | | 139 | | 0 | | | | | | | 0 |
| 10 | 16 | 1 | | | 139 | | 0 | | | | | | | 0 |
| 15 | 32 | 1 | | | 159 | | 0 | | | | | | | 0 |
| 0 | 16 | 1 | | | 179 | | 0 | | | | | | | 0 |
| 20 | 0 | .72 | | | 179 | | 0 | | | | | | | 0 |
| 10 | 16 | 1 | | | 139 | | 0 | | | | | | | 0 |
| | | | | | | | | | | | | | | |
| 181 | 80 | 1.8 | | | 652 | | 310 | | | | | | | |
| 195 | 80 | 1.08 | | | 690 | | 250 | | | | | | | 0 |
| 25 | 80 | .72 | | | 580 | | 150 | | | | | | | 0 |
| 205 | 100 | 1.8 | | | 1050 | | 500 | | | | | | | 0 |
| 183 | 80 | 1.8 | | | 661 | | 320 | | | | | | | |
| | | | | | | | | | | | | | | |
| 10 | 150 | 2.7 | | 495 | 1100 | | 450 | .04 | | 2.02 | 1.98 | .31 | | 0 |
| 57 | 160 | 3.96 | | 380 | 1303 | | 530 | .4 | | 2.14 | 3.44 | .32 | | 1 |
| 32 | 170 | 3.78 | 50 | 442 | 1221 | 2.67 | 450 | .49 | | .41 | 3.09 | .59 | 38 | 1 |
| 55 | 150 | 2.7 | | | 1520 | | 600 | | | | | | | 5 |
| 57 | 140 | 2.52 | | | 854 | | 440 | | | | | | | 1 |
| 34 | 146 | 8.75 | 48 | 410 | 1196 | | 486 | .39 | | 2.04 | 2.81 | .34 | | 5 |
| 21 | 165 | 2.98 | | | 1058 | | 248 | | | | | | | 5 |
| 47 | 300 | 3.06 | | | 980 | | 950 | | | | | | | |
| 78 | 200 | 2.7 | | | 1313 | | 970 | | | | | | | 2 |
| | | | | | | | | | | | | | | |
| 25 | 80 | 1.08 | | 159 | 330 | | 100 | .05 | | .14 | 1.2 | .12 | | 0 |
| 27 | 88 | 1.19 | | 211 | 363 | | 110 | .42 | | .24 | 2.95 | 1.08 | | 0 |
| 31 | 90 | 1.62 | | | 532 | | 135 | | | | | | | 3 |
| 37 | 85 | 1.52 | | | 571 | | 63 | | | | | | | 1 |
| 19 | 62 | 1.13 | | | 797 | | 31 | | | | | | | 0 |
| 13 | 132 | 1.59 | | 401 | 573 | | 441 | .05 | | .17 | .63 | .26 | | 1 |
| 46 | 253 | 3.65 | 80 | 408 | 1054 | 5.37 | 405 | .32 | | .33 | 2.96 | 1.12 | 60 | 5 |
| 65 | 326 | 6.84 | | 1048 | 1931 | 1.67 | 1736 | .51 | | .76 | 4.8 | .56 | 10 | 26 |
| 5 | 107 | .77 | | 160 | 610 | 1.68 | 64 | .17 | | .16 | .68 | .19 | 10 | 0 |
| 32 | 184 | 3.31 | | 674 | 1205 | | 138 | .1 | | .34 | 2.17 | | | 3 |
| 16 | 160 | 1.92 | 110 | 384 | 693 | 2.17 | 267 | .05 | | .15 | .43 | .21 | 68 | 0 |
| 0 | 0 | .07 | | 9 | 75 | | 30 | 0 | | | .02 | | | <1 |
| 0 | 50 | .6 | | 376 | 709 | | 168 | .02 | | .14 | 0 | | | 10 |
| 0 | 0 | .45 | | 27 | 238 | | 50 | .1 | | .04 | .71 | .04 | | 0 |
| 15 | 60 | 1.44 | | | 312 | | 330 | | | | | | | 4 |
| | | | | | | | | | | | | | | |
| 65 | 110 | 4.14 | | 296 | 580 | | 0 | .43 | | .38 | 6.71 | | | 0 |
| 60 | 170 | 3.42 | 38 | 375 | 850 | 5.9 | 80 | .3 | | .31 | 6.43 | .26 | 28 | 6 |
| 65 | 100 | 2.7 | | | 790 | | 40 | | | | | | | 6 |
| | | | | | | | | | | | | | | |
| 0 | 30 | 3.78 | 75 | 1187 | 25 | .74 | 0 | .31 | .14 | .12 | 4.3 | .8 | 31 | 36 |
| 20 | 180 | 4.32 | 87 | 1498 | 1390 | 2.75 | 100 | .24 | | .19 | 5.04 | .94 | 36 | 36 |
| 5 | 210 | 4.5 | 93 | 1745 | 470 | .97 | 350 | .34 | | .29 | 4.5 | .97 | 74 | 72 |
| 30 | 380 | 4.14 | 85 | 1510 | 640 | .67 | 200 | .25 | | .28 | 3.6 | .88 | 36 | 36 |

**Table H-1**

**Food Composition** (Computer code number is for West Diet Analysis program) (For purposes of calculations, use "0" for t, <1, <.1, <.01, etc.)

| Computer Code Number | Food Description | Measure | Wt (g) | $H_2O$ (%) | Ener (kcal) | Prot (g) | Carb (g) | Dietary Fiber (g) | Fat (g) | Fat Breakdown (g) Sat | Mono | Poly |
|---|---|---|---|---|---|---|---|---|---|---|---|---|
| | **FAST FOOD RESTAURANTS—**Continued | | | | | | | | | | | |
| | WENDY'S—Continued | | | | | | | | | | | |
| 1577 | With chili & cheese | 1 ea | 439 | 69 | 630 | 20 | 83 | 9 | 24 | 9 | | |
| 1578 | With sour cream & chives | 1 ea | 314 | 71 | 380 | 8 | 74 | 8 | 6 | 4 | | |
| 1579 | Chili | 1 ea | 227 | 81 | 210 | 15 | 21 | 5 | 7 | 2.5 | | |
| 1582 | Chocolate chip cookies | 1 ea | 57 | 6 | 270 | 3 | 36 | 1 | 13 | 6 | | |
| 1580 | French fries | 1 ea | 130 | 41 | 390 | 5 | 50 | 5 | 19 | 3 | 11.9 | 2.4 |
| 1581 | Frosty dairy dessert | 1 ea | 227 | 68 | 330 | 8 | 56 | 0 | 8 | 5 | | |
| | Source: Wendy's International | | | | | | | | | | | |
| | **CONVENIENCE FOODS and MEALS** | | | | | | | | | | | |
| | BUDGET GOURMET | | | | | | | | | | | |
| 1695 | Chicken cacciatore | 1 ea | 312 | 80 | 300 | 20 | 27 | | 13 | | | |
| 1692 | Linguini & shrimp | 1 ea | 284 | 77 | 330 | 15 | 33 | | 15 | | | |
| 1691 | Scallops & shrimp | 1 ea | 326 | 79 | 320 | 16 | 43 | | 9 | | | |
| 2245 | Seafood newburg | 1 ea | 284 | 74 | 350 | 17 | 43 | | 12 | | | |
| 1693 | Sirloin tips with country gravy | 1 ea | 284 | 80 | 310 | 16 | 21 | | 18 | | | |
| 1694 | Sweet & sour chicken with rice | 1 ea | 284 | 72 | 350 | 18 | 53 | | 7 | | | |
| 1689 | Teriyaki chicken | 1 ea | 340 | 77 | 360 | 20 | 44 | | 12 | | | |
| 1690 | Veal parmigiana | 1 ea | 340 | 75 | 440 | 26 | 39 | | 20 | | | |
| 1696 | Yankee pot roast | 1 ea | 312 | 77 | 380 | 27 | 22 | | 21 | | | |
| | Source: The All American Gourmet Co. | | | | | | | | | | | |
| | HAAGEN DAZS | | | | | | | | | | | |
| 1755 | Ice cream bar, vanilla almond | 1 ea | 107 | | 371 | 6 | 26 | | 27 | 14.1 | 10 | 3 |
| | Sorbet: | | | | | | | | | | | |
| 1758 | Lemon | ½ c | 113 | | 140 | 0 | 35 | | 0 | 0 | 0 | 0 |
| 1760 | Orange | ½ c | 113 | | 140 | 0 | 36 | | 0 | 0 | 0 | 0 |
| 1759 | Raspberry | ½ c | 113 | | 110 | 0 | 27 | | 0 | 0 | 0 | 0 |
| | Yogurt, frozen: | | | | | | | | | | | |
| 1753 | Chocolate | ½ c | 98 | | 171 | 8 | 26 | | 4 | 2 | 2 | 0 |
| 1754 | Strawberry | ½ c | 98 | | 171 | 6 | 27 | | 4 | 2 | 2 | 0 |
| | Yogurt extra, frozen: | | | | | | | | | | | |
| 1752 | Brownie nut | ½ c | 101 | | 220 | 8 | 29 | | 9 | 4 | 4 | 1 |
| 1751 | Raspberry rendezvous | ½ c | 101 | | 132 | 4 | 26 | | 2 | 1 | 1 | 0 |
| | Source: Pillsbury | | | | | | | | | | | |
| | HEALTHY CHOICE | | | | | | | | | | | |
| | Entrees: | | | | | | | | | | | |
| 2112 | Fish, lemon pepper | 1 ea | 303 | 78 | 290 | 14 | 47 | 7 | 5 | 1 | | |
| 1624 | Lasagna | 1 ea | 383 | 76 | 390 | 26 | 60 | 9 | 5 | 2 | | |
| 2111 | Meatloaf, traditional | 1 ea | 340 | 79 | 320 | 16 | 46 | 7 | 8 | 4 | | |
| 2104 | Zucchini lasagna | 1 ea | 396 | 80 | 329 | 20 | 58 | 11 | 1 | 1 | | |
| 2110 | Dinner, pasta shells marinara | 1 ea | 340 | 74 | 360 | 25 | 59 | 5 | 3 | 1.5 | | |
| | Low-fat ice cream: | | | | | | | | | | | |
| 973 | Brownie | ½ c | 71 | 60 | 120 | 3 | 22 | 2 | 2 | 1 | .3 | .7 |
| 259 | Butter pecan | ½ c | 71 | 60 | 120 | 3 | 22 | 1 | 2 | 1 | .3 | .7 |
| 650 | Chocolate chip | ½ c | 71 | 62 | 120 | 3 | 21 | <1 | 2 | 1 | 1 | 0 |
| 1608 | Cookie & cream | ½ c | 71 | 62 | 120 | 3 | 21 | <1 | 2 | 1.5 | .5 | 0 |
| 650 | Chocolate chip | ½ c | 71 | 62 | 120 | 3 | 21 | <1 | 2 | 1 | 1 | 0 |
| 45 | Rocky road | ½ c | 71 | 53 | 140 | 3 | 28 | 2 | 1 | 1 | .5 | 0 |
| 1621 | Vanilla | ½ c | 71 | 66 | 100 | 3 | 18 | 1 | 2 | .5 | 1.5 | 0 |
| 391 | Vanilla fudge | ½ c | 71 | 62 | 120 | 3 | 21 | 1 | 2 | 1.5 | | |
| | Source: ConAgra Frozen Foods, Omaha, NE | | | | | | | | | | | |

| Chol (mg) | Calc (mg) | Iron (mg) | Magn (mg) | Pota (mg) | Sodi (mg) | Zinc (mg) | VT-A (RE) | Thia (mg) | VT-E (a-TE) | Ribo (mg) | Niac (mg) | V-B6 (mg) | Fola (μg) | VT-C (mg) |
|---|---|---|---|---|---|---|---|---|---|---|---|---|---|---|
| 40 | 330 | 5.04 | 122 | 1745 | 770 | 4.15 | 200 | .33 | | .29 | 4.5 | .99 | 55 | 36 |
| 15 | 80 | 4.32 | 71 | 1438 | 40 | .91 | 300 | .23 | | .14 | 3.04 | .8 | 32 | 48 |
| 30 | 80 | 2.9 | | 501 | 800 | | 80 | .11 | | .15 | 2.66 | | | 4 |
| 30 | 10 | 1.8 | 13 | 89 | 120 | .41 | 0 | .05 | | .06 | .36 | .03 | 5 | 0 |
| 0 | 20 | 1.08 | 55 | 845 | 120 | .62 | 0 | .18 | | .04 | 3.6 | .33 | 40 | 6 |
| 35 | 310 | 1.08 | 46 | 544 | 200 | .97 | 150 | .11 | | .47 | .32 | .13 | 17 | 0 |
| | | | | | | | | | | | | | | |
| 60 | 150 | 1.8 | | | 810 | | 40 | .23 | | .51 | 5 | | | 21 |
| 75 | 10 | 3.6 | | | 1250 | | 1000 | .3 | | .17 | 3 | | | 2 |
| 70 | 150 | .72 | | | 690 | | 150 | | | .26 | 3 | | | 12 |
| 70 | 100 | .72 | | | 660 | | 40 | .23 | | .26 | 2 | | | |
| 40 | 60 | .36 | | | 570 | | 150 | .15 | | .17 | 4 | .28 | | 2 |
| 40 | 60 | .72 | | | 640 | | 80 | .12 | | .34 | 3 | | | 2 |
| 55 | 80 | 1.4 | | | 610 | | 300 | .15 | | .34 | 6 | | | 12 |
| 165 | 30 | 4.5 | | | 1160 | | 1000 | .45 | | .6 | 6 | | | 6 |
| 70 | 150 | 1.8 | | | 690 | | 600 | .15 | | .43 | 7 | | | 6 |
| | | | | | | | | | | | | | | |
| 90 | 161 | .38 | | 221 | 85 | | 161 | | | .18 | | | | |
| 0 | | | | 30 | 20 | | | | | | | | | 7 |
| 0 | | | | 80 | 20 | | | | | | | | | 20 |
| 0 | | | | 60 | 15 | | | | | | | | | 7 |
| 40 | 147 | .71 | | 241 | 45 | | 20 | | | .17 | | | | |
| 50 | 147 | | | 141 | 45 | | 20 | .03 | | .17 | | | | 5 |
| 55 | 152 | .73 | | 250 | 60 | | 20 | | | .14 | | | | |
| 20 | 81 | | | 97 | 25 | | 0 | | | .1 | | | | 5 |
| | | | | | | | | | | | | | | |
| 25 | 20 | 1.08 | | | 360 | | 100 | | | | | | | 30 |
| 15 | 150 | 3.6 | | 500 | 550 | | 100 | .3 | | .26 | 2 | | | 6 |
| 35 | 40 | 1.8 | | | 460 | | 150 | | | | | | | 54 |
| 10 | 199 | 2.69 | | | 309 | | 249 | | | | | | | 0 |
| 25 | 400 | 1.8 | | | 390 | | 100 | | | | | | | 4 |
| 2 | 80 | 0 | | 268 | 55 | | 40 | | | | | | | 0 |
| 2 | 100 | 0 | | 211 | 60 | | 40 | | | | | | | 0 |
| 2 | 100 | 0 | | 240 | 50 | | 40 | | | | | | | 0 |
| 2 | 100 | | | 254 | 90 | | 60 | .03 | | .15 | | | | 2 |
| 2 | 100 | 0 | | 240 | 50 | | 40 | | | | | | | 0 |
| 2 | 100 | 0 | | 168 | 60 | | 40 | .03 | | .15 | | | | 0 |
| 5 | 100 | | | 254 | 50 | | 60 | .05 | | .22 | | | | 2 |
| 2 | 100 | 0 | | 296 | 50 | | 40 | | | | | | | 0 |

**Table H-1**

**Food Composition** (Computer code number is for West Diet Analysis program) (For purposes of calculations, use "0" for t, <1, <.1, <.01, etc.)

| Computer Code Number | Food Description | Measure | Wt (g) | $H_2O$ (%) | Ener (kcal) | Prot (g) | Carb (g) | Dietary Fiber (g) | Fat (g) | Fat Breakdown (g) Sat | Mono | Poly |
|---|---|---|---|---|---|---|---|---|---|---|---|---|
| | **CONVENIENCE FOODS and MEALS**—Continued | | | | | | | | | | | |
| | HEALTH VALLEY | | | | | | | | | | | |
| | Soups, fat-free: | | | | | | | | | | | |
| 2001 | Beef broth, no salt added | 1 c | 240 | 98 | 18 | 5 | 0 | 0 | 0 | 0 | 0 | 0 |
| 2073 | Beef broth, w/salt | 1 c | 240 | 98 | 30 | 5 | 2 | 0 | 0 | 0 | 0 | 0 |
| 2016 | Black bean & vegetable | 1 c | 240 | 85 | 110 | 11 | 24 | 12 | 0 | 0 | 0 | 0 |
| 2017 | Chicken broth | 1 c | 240 | 97 | 30 | 6 | 0 | 0 | 0 | 0 | 0 | 0 |
| 2018 | 14 garden vegetable | 1 c | 240 | 90 | 80 | 6 | 17 | 4 | 0 | 0 | 0 | 0 |
| 2015 | Lentil & carrot | 1 c | 240 | 85 | 90 | 10 | 25 | 14 | 0 | 0 | 0 | 0 |
| 2014 | Split pea & carrot | 1 c | 240 | 89 | 110 | 8 | 17 | 4 | 0 | 0 | 0 | 0 |
| 2013 | Tomato vegetable | 1 c | 240 | 90 | 80 | 6 | 17 | 5 | 0 | 0 | 0 | 0 |
| | Source: Health Valley | | | | | | | | | | | |
| | LA CHOY | | | | | | | | | | | |
| 2100 | Egg rolls, mini, chicken | 1 svg | 106 | 53 | 220 | 8 | 35 | 3 | 6 | 1.5 | | |
| 2099 | Egg rolls, mini, shrimp | 1 svg | 106 | 56 | 210 | 7 | 35 | 3 | 4 | 1 | | |
| | Source: Beatrice/Hunt Wesson | | | | | | | | | | | |
| | LEAN CUISINE | | | | | | | | | | | |
| | Dinners: | | | | | | | | | | | |
| 1639 | Baked cheese ravioli | 1 ea | 241 | 77 | 250 | 12 | 32 | 4 | 8 | 3 | 2 | 1 |
| 1632 | Chicken chow mein | 1 ea | 255 | 81 | 210 | 13 | 28 | 2 | 5 | 1 | 2 | 1 |
| 1633 | Lasagna | 1 ea | 291 | 79 | 270 | 19 | 34 | 5 | 6 | 2.5 | 1.5 | .5 |
| 1634 | Macaroni & cheese | 1 ea | 255 | 78 | 270 | 13 | 39 | 2 | 7 | 3.5 | 1.5 | .5 |
| 1631 | Spaghetti w/meatballs | 1 ea | 269 | 74 | 290 | 17 | 40 | 4 | 7 | 2 | 3 | 1.5 |
| | Pizza: | | | | | | | | | | | |
| 1636 | French bread sausage pizza | 1 ea | 170 | 53 | 420 | 19 | 41 | 4 | 20 | 5 | 13.9 | 1.1 |
| | Source: Stouffer's Foods Corp, Solon, OH | | | | | | | | | | | |
| | TASTE ADVENTURE SOUPS | | | | | | | | | | | |
| 1905 | Black bean | 1 c | 242 | | 139 | 6 | 28 | 6 | 1 | | | |
| 1904 | Curry lentil | 1 c | 241 | | 138 | 6 | 30 | 5 | 1 | | | |
| 1906 | Lentil chili | 1 c | 242 | | 181 | 11 | 33 | 6 | 1 | | | |
| 1903 | Split pea | 1 c | 244 | | 140 | 5 | 27 | 5 | 1 | | | |
| | Source: Taste Adventure Soups | | | | | | | | | | | |
| | WEIGHT WATCHERS | | | | | | | | | | | |
| | Cheese, fat-free slices: | | | | | | | | | | | |
| 1978 | Cheddar, sharp | 2 pce | 21 | 65 | 30 | 5 | 2 | 0 | 0 | 0 | 0 | 0 |
| 1980 | Swiss | 2 pce | 21 | 65 | 30 | 5 | 2 | 0 | 0 | 0 | 0 | 0 |
| 1977 | White | 2 pce | 21 | 65 | 30 | 5 | 2 | 0 | 0 | 0 | 0 | 0 |
| 1979 | Yellow | 2 pce | 21 | 65 | 30 | 5 | 2 | 0 | 0 | 0 | 0 | 0 |
| | Dinners: | | | | | | | | | | | |
| 2029 | Chicken chow mein | 1 ea | 255 | 81 | 200 | 12 | 34 | 3 | 2 | .5 | | |
| 1646 | Oven fried fish | 1 ea | 218 | 78 | 230 | 15 | 25 | 2 | 8 | 2.5 | 5 | 2 |
| 1972 | Margarine, reduced fat | 1 tbs | 14 | 49 | 59 | 0 | 0 | 0 | 7 | 1.5 | | |
| | Pizza: | | | | | | | | | | | |
| 1653 | Cheese | 1 ea | 163 | 48 | 390 | 23 | 49 | 6 | 12 | 4 | 3 | 1 |
| 1650 | Deluxe combination pizza | 1 ea | 186 | 56 | 380 | 23 | 47 | 6 | 11 | 3.5 | 5 | 2 |
| 1652 | Pepperoni pizza | 1 ea | 158 | 48 | 390 | 23 | 46 | 4 | 12 | 4 | 5 | 2 |
| | Desserts: | | | | | | | | | | | |
| 1644 | Chocolate brownie | 1 ea | 182 | 75 | 190 | 6 | 35 | 4 | 4 | 1 | 2 | 1 |
| 2024 | Chocolate eclair | 1 ea | 60 | 45 | 151 | 3 | 24 | 2 | 5 | 1.5 | | |
| 2247 | Chocolate mousse | 1 ea | 78 | 44 | 190 | 6 | 33 | 3 | 4 | 1.5 | | |
| 1642 | Strawberry cheesecake | 1 ea | 111 | 62 | 180 | 7 | 28 | 2 | 5 | 2 | 1 | 2 |

**PAGE KEY:** H–4 = Beverages H–6 = Dairy H–10 = Eggs H–10 = Fat/Oil H–14 = Fruit H–20 = Bakery H–28 = Grain H–32 = Fish H–34 = Meats H–38 = Poultry H–40 = Sausage H–42 = Mixed/Fast H–46 = Nuts/Seeds H–50 = Sweets H–52 = Vegetables/Legumes H–62 = Vegetarian Foods H–64 = Misc H–66 = Soups/Sauces H–68 = Fast H–84 = Convenience H–88 = Baby foods

| Chol (mg) | Calc (mg) | Iron (mg) | Magn (mg) | Pota (mg) | Sodi (mg) | Zinc (mg) | VT-A (RE) | Thia (mg) | VT-E (a-TE) | Ribo (mg) | Niac (mg) | V-B6 (mg) | Fola (μg) | VT-C (mg) |
|---|---|---|---|---|---|---|---|---|---|---|---|---|---|---|
| 0 | | | | 196 | 74 | | | | | | .98 | | | |
| 0 | 0 | 0 | | 196 | 160 | | 0 | | | | .98 | | | 5 |
| 0 | 40 | 3.6 | | 676 | 280 | | 2000 | .34 | | .11 | 1.35 | .22 | 135 | 9 |
| 0 | 20 | 1.8 | | 147 | 170 | | 0 | | | .03 | 2.45 | | | 1 |
| 0 | 40 | 1.8 | | 406 | 250 | | 2000 | .26 | | .08 | 2.25 | .18 | 27 | 15 |
| 0 | 60 | 5.4 | | 439 | 220 | | 2000 | .1 | | .16 | 5.63 | .45 | 27 | 2 |
| 0 | 40 | 5.4 | | 439 | 230 | | 2000 | .1 | | .16 | 5.63 | .45 | | 9 |
| 0 | 40 | 5.4 | | 609 | 240 | | 2000 | .1 | | .08 | 2.25 | .13 | <1 | 9 |
| 5 | 20 | 1.44 | | | 460 | | 20 | | | | | | | 0 |
| 5 | 20 | 1.44 | | | 510 | | 20 | | | | | | | 0 |
| 55 | 200 | 1.08 | 42 | 400 | 500 | 1.5 | 150 | .06 | | .25 | 1.2 | .2 | 48 | 6 |
| 35 | 20 | .36 | 30 | 300 | 510 | 1.1 | 20 | .15 | | .17 | 5 | | | 6 |
| 25 | 150 | 1.8 | 44 | 620 | 560 | 2.9 | 100 | .15 | | .25 | 3 | .32 | | 12 |
| 20 | 250 | .72 | | 170 | 550 | | 20 | .12 | | .25 | 1.2 | | | 0 |
| 30 | 100 | 2.7 | 47 | 480 | 520 | 2.5 | 80 | .15 | | .25 | 3 | .2 | | 4 |
| 35 | 250 | 2.7 | 39 | 340 | 900 | 2.2 | 80 | .45 | | .51 | 5 | .07 | | 6 |
| | | | | 650 | 565 | | | | | | | | | |
| | | | | 467 | 584 | | | | | | | | | |
| | | | | 650 | 448 | | | | | | | | | |
| | | | | 484 | 591 | | | | | | | | | |
| 0 | 99 | 0 | | 64 | 306 | | 56 | | | | | | | 0 |
| 0 | 99 | 0 | | 74 | 276 | | 56 | | | | | | | 0 |
| 0 | 99 | 0 | | 64 | 306 | | 56 | | | | | | | 0 |
| 0 | 99 | 0 | | 64 | 306 | | 56 | | | | | | | 0 |
| 25 | 40 | .72 | | 360 | 430 | | 300 | | | | | | | 36 |
| 25 | 20 | 1.44 | | 370 | 450 | | 40 | .09 | | .14 | 1.6 | | | 0 |
| 0 | 0 | 0 | | 5 | 128 | | 49 | | | | | | | 0 |
| 35 | 700 | 1.8 | | 290 | 590 | | 80 | .3 | | .51 | 3 | .06 | | 6 |
| 40 | 500 | 3.6 | | 370 | 550 | | 150 | .3 | | .51 | 3 | .2 | | 5 |
| 45 | 450 | 1.8 | | 320 | 650 | | 80 | .23 | | .51 | 3 | | | 5 |
| 5 | 80 | 1.08 | | 230 | 160 | | 0 | .06 | | .03 | .2 | .03 | | 0 |
| 0 | 40 | 0 | | 65 | 151 | | 0 | | | | | | | 0 |
| 5 | 60 | 1.8 | | 320 | 150 | | 0 | | | | | | | 0 |
| 15 | 80 | .36 | | 115 | 230 | | 40 | .06 | | .07 | 1.6 | | | 2 |

**Table H-1**

**Food Composition** (Computer code number is for West Diet Analysis program) (For purposes of calculations, use "0" for t, <1, <.1, <.01, etc.)

| Computer Code Number | Food Description | Measure | Wt (g) | $H_2O$ (%) | Ener (kcal) | Prot (g) | Carb (g) | Dietary Fiber (g) | Fat (g) | Fat Breakdown (g) Sat | Mono | Poly |
|---|---|---|---|---|---|---|---|---|---|---|---|---|
| | **CONVENIENCE FOODS and MEALS—**Continued | | | | | | | | | | | |
| | WEIGHT WATCHERS—CONTINUED | | | | | | | | | | | |
| 2027 | Triple chocolate cheesecake | 1 ea | 89 | 52 | 199 | 7 | 32 | 1 | 5 | 2.5 | | |
| | Source: Weight Watchers | | | | | | | | | | | |
| | SWEET SUCCESS: | | | | | | | | | | | |
| | Drinks, prepared: | | | | | | | | | | | |
| 1776 | Chocolate chip | 1 c | 265 | 81 | 180 | 15 | 30 | 6 | 3 | 1.6 | | |
| 1777 | Chocolate fudge | 1 c | 265 | 81 | 180 | 15 | 30 | 6 | 2 | | | |
| 1774 | Chocolate mocha | 1 c | 265 | 81 | 180 | 15 | 30 | 6 | 1 | 1 | | |
| 1778 | Milk chocolate | 1 c | 265 | 81 | 180 | 15 | 30 | 6 | 2 | 1 | | |
| 1775 | Vanilla | 1 c | 265 | 81 | 180 | 15 | 33 | 6 | 1 | .6 | | |
| | Drinks, ready to drink: | | | | | | | | | | | |
| 2147 | Chocolate mint | 1 c | 297 | 82 | 187 | 11 | 36 | 6 | 3 | 0 | | |
| 2148 | Strawberry | 1 c | 265 | 82 | 167 | 10 | 32 | 5 | 3 | 0 | | |
| | Shakes: | | | | | | | | | | | |
| 1771 | Chocolate almond | 1 c | 250 | 82 | 158 | 9 | 30 | 5 | 2 | 0 | 2.1 | .2 |
| 1773 | Chocolate fudge | 1 c | 250 | 82 | 158 | 9 | 30 | 5 | 2 | 0 | 2.1 | .2 |
| 1768 | Chocolate mocha | 1 c | 250 | 82 | 158 | 9 | 30 | 5 | 2 | 0 | .6 | 1.8 |
| 1769 | Chocolate raspberry truffle | 1 c | 250 | 82 | 158 | 9 | 30 | 5 | 2 | 0 | 2.2 | .2 |
| 1770 | Vanilla creme | 1 c | 250 | 82 | 158 | 9 | 30 | 5 | 2 | 0 | 2.1 | .3 |
| | Snack bars: | | | | | | | | | | | |
| 1767 | Chocolate brownie | 1 ea | 33 | 9 | 120 | 2 | 23 | 3 | 4 | 2 | .5 | .6 |
| 1766 | Chocolate chip | 1 ea | 33 | 9 | 120 | 2 | 23 | 3 | 4 | 2 | .4 | .5 |
| 1921 | Oatmeal raisin | 1 ea | 33 | 9 | 120 | 2 | 23 | 3 | 4 | 2 | | |
| 1765 | Peanut butter | 1 ea | 33 | 9 | 120 | 2 | 23 | 3 | 4 | 2 | .6 | .6 |
| | Source: Foodway National Inc, Boise, ID | | | | | | | | | | | |
| | **BABY FOODS** | | | | | | | | | | | |
| 1720 | Apple juice | ½ c | 125 | 88 | 59 | 0 | 15 | <1 | <1 | t | t | t |
| 1721 | Applesauce, strained | 1 tbs | 16 | 89 | 7 | <1 | 2 | <1 | <1 | t | t | t |
| 1716 | Carrots, strained | 1 tbs | 14 | 92 | 4 | <1 | 1 | <1 | <1 | t | t | t |
| 1718 | Cereal, mixed, milk added | 1 tbs | 15 | 75 | 17 | 1 | 2 | <1 | 1 | .3 | | |
| 1719 | Cereal, rice, milk added | 1 tbs | 15 | 75 | 17 | <1 | 3 | <1 | 1 | .3 | | |
| 1723 | Chicken and noodles, strained | 1 tbs | 16 | 88 | 8 | <1 | 1 | <1 | <1 | .1 | .1 | t |
| 1722 | Peas, strained | 1 tbs | 15 | 87 | 6 | 1 | 1 | <1 | <1 | t | t | t |
| 1717 | Teething biscuits | 1 ea | 11 | 6 | 43 | 1 | 8 | <1 | <1 | .2 | .2 | .1 |

**PAGE KEY:** H–4 = Beverages H–6 = Dairy H–10 = Eggs H–10 = Fat/Oil H–14 = Fruit H–20 = Bakery H–28 = Grain H–32 = Fish H–34 = Meats H–38 = Poultry H–40 = Sausage H–42 = Mixed/Fast H–46 = Nuts/Seeds H–50 = Sweets H–52 = Vegetables/Legumes H–62 = Vegetarian Foods H–64 = Misc H–66 = Soups/Sauces H–68 = Fast H–84 = Convenience H–88 = Baby foods

| Chol (mg) | Calc (mg) | Iron (mg) | Magn (mg) | Pota (mg) | Sodi (mg) | Zinc (mg) | VT-A (RE) | Thia (mg) | VT-E (a-TE) | Ribo (mg) | Niac (mg) | V-B6 (mg) | Fola (µg) | VT-C (mg) |
|---|---|---|---|---|---|---|---|---|---|---|---|---|---|---|
| 10 | 80 | 1.08 | | 169 | 199 | | 0 | | | | | | | 0 |
| 6 | 500 | 6.3 | 140 | 600 | 288 | 5.25 | 350 | .52 | 7.05 | .59 | 7 | .7 | 140 | 21 |
| 6 | 500 | 6.3 | 140 | 750 | 336 | 5.25 | 350 | .52 | 7.05 | .59 | 7 | .7 | 140 | 21 |
| 6 | 500 | 6.3 | 140 | 800 | 336 | 5.25 | 350 | .52 | 7.05 | .59 | 7 | .7 | 140 | 21 |
| 6 | 500 | 6.3 | 140 | 750 | 336 | 5.25 | 350 | .52 | 7.05 | .59 | 7 | .7 | 140 | 21 |
| 6 | 500 | 6.3 | 140 | 830 | 312 | 5.25 | 250 | .52 | 7.05 | .59 | 7 | .7 | 140 | 21 |
| 6 | 470 | 5.94 | 131 | 526 | 226 | 5.05 | 329 | .5 | 6.56 | .56 | 6.53 | .65 | 131 | 20 |
| 5 | 419 | 5.3 | 117 | 310 | 175 | 4.51 | 294 | .45 | 5.86 | .5 | 5.83 | .58 | 117 | 17 |
| 5 | 396 | 5 | 110 | 443 | 190 | 4.25 | 277 | .42 | 5.53 | .47 | 5.5 | .55 | 110 | 16 |
| 5 | 396 | 5 | 110 | 443 | 175 | 4.25 | 277 | .42 | 5.53 | .47 | 5.5 | .55 | 110 | 16 |
| 5 | 396 | 5 | 110 | 403 | 175 | 4.25 | 277 | .42 | 5.53 | .47 | 5.5 | .55 | 110 | 16 |
| 5 | 383 | 5 | 110 | 443 | 175 | 4.25 | 277 | .42 | 5.53 | .47 | 5.5 | .55 | 110 | 16 |
| 5 | 396 | 5 | 110 | 293 | 175 | 4.25 | 277 | .42 | 5.53 | .47 | 5.5 | .55 | 110 | 16 |
| 3 | 150 | 2.71 | 60 | 140 | 45 | .59 | 150 | .22 | 3.01 | .25 | 3 | .3 | 60 | 9 |
| 3 | 150 | 2.71 | 60 | 110 | 40 | .59 | 150 | .22 | 3.01 | .25 | 3 | .3 | 60 | 9 |
| 3 | 150 | 2.71 | 60 | | 30 | .59 | 150 | .22 | 3.01 | .25 | 3 | .3 | 60 | 9 |
| 3 | 150 | 2.71 | 60 | 125 | 35 | .59 | 150 | .22 | 3.01 | .25 | 3 | .3 | 60 | 9 |
| 0 | 5 | .71 | 4 | 114 | 4 | .04 | 2 | .01 | .75 | .02 | .1 | .04 | <1 | 72 |
| 0 | 1 | .03 | | 11 | <1 | <.01 | <1 | <.01 | .1 | <.01 | .01 | <.01 | <1 | 6 |
| 0 | 3 | .05 | 1 | 27 | 5 | .02 | 160 | <.01 | .07 | .01 | .06 | .01 | 2 | 1 |
| 2 | 33 | 1.56 | 4 | 30 | 7 | .11 | 4 | .06 | | .09 | .87 | .01 | 2 | <1 |
| 2 | 36 | 1.83 | 7 | 28 | 7 | .1 | 4 | .07 | | .07 | .78 | .02 | 1 | <1 |
| 3 | 4 | .07 | 1 | 6 | 3 | .05 | 18 | <.01 | .04 | .01 | .07 | <.01 | 2 | <1 |
| 0 | 3 | .14 | 2 | 17 | 1 | .05 | 8 | .01 | .08 | .01 | .15 | .01 | 4 | 1 |
| 0 | 29 | .39 | 4 | 35 | 40 | .1 | 1 | .03 | .05 | .06 | .48 | .01 | 5 | 1 |

Contents



# CANADA: RECOMMENDATIONS, CHOICE SYSTEM, AND LABELS

Chapters 1 and 2 introduced the 1991 Recommended Nutrient Intakes (RNI), the exchange system (called the choice system in Canada), food guides, and food labels. This appendix presents details for Canadians. Appendix F includes

**Table I-1**

**Recommended Nutrient Intakes for Canadians, 1990**

| Age | Sex | Weight (kg) | Protein (g/day)[a] | Vitamins: Vitamin A (RE/day)[b] | Vitamins: Vitamin E (mg/day)[c] | Vitamins: Vitamin C (mg/day)[d] | Iron (mg/day) | Iodine (μg/day) | Zinc (mg/day) |
|---|---|---|---|---|---|---|---|---|---|
| **Infants (months)** | | | | | | | | | |
| 0–4 | Both | 6 | 12[e] | 400 | 3 | 20 | 0.3[f] | 30 | 2[g] |
| 5–12 | Both | 9 | 12 | 400 | 3 | 20 | 7 | 40 | 3 |
| **Children (years)** | | | | | | | | | |
| 1 | Both | 11 | 13 | 400 | 3 | 20 | 6 | 55 | 4 |
| 2–3 | Both | 14 | 16 | 400 | 4 | 20 | 6 | 65 | 4 |
| 4–6 | Both | 18 | 19 | 500 | 5 | 25 | 8 | 85 | 5 |
| 7–9 | M | 25 | 26 | 700 | 7 | 25 | 8 | 110 | 7 |
| | F | 25 | 26 | 700 | 6 | 25 | 8 | 95 | 7 |
| 10–12 | M | 34 | 34 | 800 | 8 | 25 | 8 | 125 | 9 |
| | F | 36 | 36 | 800 | 7 | 25 | 8 | 110 | 9 |
| 13–15 | M | 50 | 49 | 900 | 9 | 30 | 10 | 160 | 12 |
| | F | 48 | 46 | 800 | 7 | 30 | 13 | 160 | 9 |
| 16–18 | M | 62 | 58 | 1000 | 10 | 40 | 10 | 160 | 12 |
| | F | 53 | 47 | 800 | 7 | 30 | 12 | 160 | 9 |
| **Adults (years)** | | | | | | | | | |
| 19–24 | M | 71 | 61 | 1000 | 10 | 40 | 9 | 160 | 12 |
| | F | 58 | 50 | 800 | 7 | 30 | 13 | 160 | 9 |
| 25–49 | M | 74 | 64 | 1000 | 9 | 40 | 9 | 160 | 12 |
| | F | 59 | 51 | 800 | 6 | 30 | 13[h] | 160 | 9 |
| 50–74 | M | 73 | 63 | 1000 | 7 | 40 | 9 | 160 | 12 |
| | F | 63 | 54 | 800 | 6 | 30 | 8 | 160 | 9 |
| 75+ | M | 69 | 59 | 1000 | 6 | 40 | 9 | 160 | 12 |
| | F | 64 | 55 | 800 | 5 | 30 | 8 | 160 | 9 |
| **Pregnancy (additional amount needed)** | | | | | | | | | |
| 1st trimester | | | 5 | 0 | 2 | 0 | 0 | 25 | 6 |
| 2nd trimester | | | 20 | 0 | 2 | 10 | 5 | 25 | 6 |
| 3rd trimester | | | 24 | 0 | 2 | 10 | 10 | 25 | 6 |
| **Lactation (additional amount needed)** | | | 20 | 400 | 3 | 25 | 0 | 50 | 6 |

Note: Recommended intakes of energy and of certain nutrients are not listed in this table because of the nature of the variables upon which they are based. The figures for energy are estimates of average requirements for expected patterns of activity (see Table I-2). For nutrients not shown, the following amounts are recommended based on at least 2000 kcalories per day and body weights as given: thiamin, 0.4 milligram per 1000 kcalories (0.48 milligram/5000 kilojoules); riboflavin, 0.5 milligram per 1000 kcalories (0.6 milligram/5000 kilojoules); niacin, 7.2 niacin equivalents per 1000 kcalories (8.6 niacin equivalents/5000 kilojoules); vitamin $B_6$, 15 micrograms, as pyridoxine, per gram of protein. Recommended intakes during periods of growth are taken as appropriate for individuals representative of the midpoint in each age group. All recommended intakes are designed to cover individual variations in essentially all of a healthy population subsisting upon a variety of common foods available in Canada.

Source: Health and Welfare Canada, *Nutrition Recommendations: The Report of the Scientific Review Committee* (Ottawa: Canadian Government Publishing Centre, 1990), Table 20, p. 204.

[a]The primary units are expressed per kilogram of body weight. The figures shown here are examples.

[b]One retinol equivalent (RE) corresponds to the biological activity of 1 microgram of retinol, 6 micrograms of beta-carotene, or 12 micrograms of other carotenes.

[c]Expressed as δ-α-tocopherol equivalents, relative to which β- and γ-tocopherol and α-tocotrienol have activities of 0.5, 0.1, and 0.3, respectively.

[d]Cigarette smokers should increase intake by 50 percent.

[e]The assumption is made that the protein is from breast milk or has the same biological value as breast milk and that, between 3 and 9 months, adjustment for the quality of the protein is made.

[f]Based on the assumption that breast milk is the source of iron.

[g]Based on the assumption that breast milk is the source of zinc.

[h]After menopause, the recommended intake is 8 milligrams per day.

addresses of Canadian governmental agencies and professional organizations that may provide additional information.

## RNI

As Chapter 1 mentioned, a major revision of the nutrient recommendations is underway in both the United States and Canada. The Dietary Reference Intakes (DRI) reports are replacing the 1989 RDA in the United States and the 1991 RNI in Canada. Recommendations from the DRI reports are presented on the inside front covers. For nutrients that do not yet have new values, the RNI will continue to serve health professionals in Canada (Tables I-1 and I-2).

**Table I-2**
**Average Energy Requirements for Canadians**

| Age | Sex | Average Height (cm) | Average Weight (kg) | Requirements[a] | | | | | |
|---|---|---|---|---|---|---|---|---|---|
| | | | | (kcal/kg)[b] | (MJ/kg)[b] | (kcal/day) | (MJ/day) | (kcal/cm) | (MJ/cm) |
| **Infants (months)** | | | | | | | | | |
| 0–2 | Both | 55 | 4.5 | 120–100 | 0.50–0.42 | 500 | 2.0 | 9 | 0.04 |
| 3–5 | Both | 63 | 7.0 | 100–95 | 0.42–0.40 | 700 | 2.8 | 11 | 0.05 |
| 6–8 | Both | 69 | 8.5 | 95–97 | 0.40–0.41 | 800 | 3.4 | 11.5 | 0.05 |
| 9–11 | Both | 73 | 9.5 | 97–99 | 0.41 | 950 | 3.8 | 12.5 | 0.05 |
| **Children and Adults (years)** | | | | | | | | | |
| 1 | Both | 82 | 11 | 101 | 0.42 | 1100 | 4.8 | 13.5 | 0.06 |
| 2–3 | Both | 95 | 14 | 94 | 0.39 | 1300 | 5.6 | 13.5 | 0.06 |
| 4–6 | Both | 107 | 18 | 100 | 0.42 | 1800 | 7.6 | 17 | 0.07 |
| 7–9 | M | 126 | 25 | 88 | 0.37 | 2200 | 9.2 | 17.5 | 0.07 |
| | F | 125 | 25 | 76 | 0.32 | 1900 | 8.0 | 15 | 0.06 |
| 10–12 | M | 141 | 34 | 73 | 0.30 | 2500 | 10.4 | 17.5 | 0.07 |
| | F | 143 | 36 | 61 | 0.25 | 2200 | 9.2 | 15.5 | 0.06 |
| 13–15 | M | 159 | 50 | 57 | 0.24 | 2800 | 12.0 | 17.5 | 0.07 |
| | F | 157 | 48 | 46 | 0.19 | 2200 | 9.2 | 14 | 0.06 |
| 16–18 | M | 172 | 62 | 51 | 0.21 | 3200 | 13.2 | 18.5 | 0.08 |
| | F | 160 | 53 | 40 | 0.17 | 2100 | 8.8 | 13 | 0.05 |
| 19–24 | M | 175 | 71 | 42 | 0.18 | 3000 | 12.6 | | |
| | F | 160 | 58 | 36 | 0.15 | 2100 | 8.8 | | |
| 25–49 | M | 172 | 74 | 36 | 0.15 | 2700 | 11.3 | | |
| | F | 160 | 59 | 32 | 0.13 | 1900 | 8.0 | | |
| 50–74 | M | 170 | 73 | 31 | 0.13 | 2300 | 9.7 | | |
| | F | 158 | 63 | 29 | 0.12 | 1800 | 7.6 | | |
| 75+ | M | 168 | 69 | 29 | 0.12 | 2000 | 8.4 | | |
| | F | 155 | 64 | 23 | 0.10 | 1500 | 6.3 | | |

[a]Requirements can be expected to vary within a range of ±30 percent.
[b]First and last figures are averages at the beginning and end of the three-month period.
Source: Health and Welfare Canada, *Nutrition Recommendations: The Report of the Scientific Review Committee* (Ottawa: Canadian Government Publishing Centre, 1990), Tables 5 and 6, pp. 25, 27.

## Choice System for Meal Planning

The *Good Health Eating Guide* is the Canadian choice system of meal planning.[1] It contains several features similar to those of the U.S. exchange system including the following:

[1] The tables for the Canadian choice system are adapted from the *Good Health Eating Guide Resource,* copyright 1994, with permission of the Canadian Diabetes Association.

- Foods are divided into lists according to carbohydrate, protein, and fat content.
- Foods are interchangeable within a group.
- Most foods are eaten in measured amounts.
- An energy value is given for each food group.

Tables I-3 through I-10 present the Canadian choice system.

**Table I-3**
**Canadian Choice System: Starch Foods**

1 starch choice = 15 g carbohydrate (starch), 2 g protein, 290 kJ (68 kcal)

| Food | Measure | Mass (Weight) |
|---|---|---|
| **Breads** | | |
| Bagels | ½ | 30 g |
| Bread crumbs | 50 mL (¼ c) | 30 g |
| Bread cubes | 250 mL (1 c) | 30 g |
| Bread sticks | 2 | 20 g |
| Brewis, cooked | 50 mL (¼ c) | 45 g |
| Chapati | 1 | 20 g |
| Cookies, plain | 2 | 20 g |
| English muffins, crumpets | ½ | 30 g |
| Flour | 40 mL (2½ tbs) | 20 g |
| Hamburger buns | ½ | 30 g |
| Hot dog buns | ½ | 30 g |
| Kaiser rolls | ½ | 30 g |
| Matzo, 15 cm | 1 | 20 g |
| Melba toast, rectangular | 4 | 15 g |
| Melba toast, rounds | 7 | 15 g |
| Pita, 20-cm (8") diameter | ¼ | 30 g |
| Pita, 15-cm (6") diameter | ½ | 30 g |
| Plain rolls | 1 small | 30 g |
| Pretzels | 7 | 20 g |
| Raisin bread | 1 slice | 30 g |
| Rice cakes | 2 | 30 g |
| Roti | 1 | 20 g |
| Rusks | 2 | 20 g |
| Rye, coarse or pumpernickel | ½ slice | 30 g |
| Soda crackers | 6 | 20 g |
| Tortillas, corn (taco shell) | 1 | 30 g |
| Tortilla, flour | 1 | 30 g |
| White (French and Italian) | 1 slice | 25 g |
| Whole-wheat, cracked-wheat, rye, white enriched | 1 slice | 30 g |
| **Cereals** | | |
| Bran flakes, 100% bran | 125 mL (½ c) | 30 g |
| Cooked cereals, cooked | 125 mL (½ c) | 125 g |
| Dry | 30 mL (2 tbs) | 20 g |
| Cornmeal, cooked | 125 mL (½ c) | 125 g |
| Dry | 30 mL (2 tbs) | 20 g |
| Ready-to-eat unsweetened cereals | 125 mL (½ c) | 20 g |
| Shredded wheat biscuits, rectangular or round | 1 | 20 g |
| Shredded wheat, bite size | 125 mL (½ c) | 20 g |
| Wheat germ | 75 mL (⅓ c) | 30 g |
| Cornflakes | 175 mL (⅔ c) | 20 g |
| Rice Krispies | 175 mL (⅔ c) | 20 g |

**Table I-3 (continued)**
**Canadian Choice System: Starch Foods**

1 starch choice = 15 g carbohydrate (starch), 2 g protein, 290 kJ (68 kcal)

| Food | Measure | Mass (Weight) |
|---|---|---|
| Cheerios | 200 mL (¾ c) | 20 g |
| Muffets | 1 | 20 g |
| Puffed rice | 300 mL (1¼ c) | 15 g |
| Puffed wheat | 425 mL (1⅔ c) | 20 g |
| **Grains** | | |
| Barley, cooked | 125 mL (½ c) | 120 g |
| Dry | 30 mL (2 tbs) | 20 g |
| Bulgur, kasha, cooked, moist | 125 mL (½ c) | 70 g |
| Cooked, crumbly | 75 mL (⅓ c) | 40 g |
| Dry | 30 mL (2 tbs) | 20 g |
| Rice, cooked, brown & white (short & long grain) | 125 mL (½ c) | 70 g |
| Rice, cooked, wild | 75 mL (⅓ c) | 70 g |
| Tapioca, pearl and granulated, quick cooking, dry | 30 mL (2 tbs) | 15 g |
| Couscous, cooked moist | 125 mL (½ c) | 70 g |
| Dry | 30 mL (tbs) | 20 g |
| Quinoa, cooked moist | 125 mL (½ c) | 70 g |
| Dry | 30 mL (2 tbs) | 20 g |
| **Pastas** | | |
| Macaroni, cooked | 125 mL (½ c) | 70 g |
| Noodles, cooked | 125 mL (½ c) | 80 g |
| Spaghetti, cooked | 125 mL (½ c) | 70 g |
| **Starchy Vegetables** | | |
| Beans and peas, dried, cooked | 125 mL (½ c) | 80 g |
| Breadfruit | 1 slice | 75 g |
| Corn, canned, whole kernel | 125 mL (½ c) | 85 g |
| Corn on the cob | ½ medium cob | 140 g |
| Cornstarch | 30 mL (2 tbs) | 15 g |
| Plantains | ⅓ small | 50 g |
| Popcorn, air-popped, unbuttered | 750 mL (3 c) | 20 g |
| Potatoes, whole (with or without skin) | ½ medium | 95 g |
| Yams, sweet potatoes, (with or without skin) | ½ | 75 g |

| Food | Choices per serving | Measure | Mass (Weight) |
|---|---|---|---|
| NOTE: Food items found in this category provide more than 1 starch choice: | | | |
| Bran flakes | 1 starch + ½ sugar | 150 mL (⅔ c) | 24 g |
| Croissant, small | 1 starch + 1½ fats | 1 small | 35 g |
| Large | 1 starch + 1½ fats | ½ large | 30 g |
| Corn, canned creamed | 1 starch + ½ fruits and vegetables | 12 mL (½ c) | 113 g |
| Potato chips | 1 starch + 2 fats | 15 chips | 30 g |
| Tortilla chips (nachos) | 1 starch + 1½ fats | 13 chips | 20 g |
| Corn chips | 1 starch + 2 fats | 30 chips | 30 g |
| Cheese twists | 1 starch + 1½ fats | 30 chips | 30 g |
| Cheese puffs | 1 starch + 2 fats | 27 chips | 30 g |
| Tea biscuit | 1 starch + 2 fats | 1 | 30 g |
| Pancakes, homemade using 50 mL (¼ c) batter (6" diameter) | 1½ starches + 1 fat | 1 medium | 50 g |
| Potatoes, french fried (homemade or frozen) | 1 starch + 1 fat | 10 regular size | 35 g |
| Soup, canned* (prepared with equal volume of water) | 1 starch | 250 mL (1 c) | 260 g |
| Waffles, packaged | 1 starch + 1 fat | 1 | 35 g |

*Soup can vary according to brand and type. Check the label for Food Choice Values and Symbols or the core nutrient listing.

**Table I-4**
**Canadian Choice System: Fruits and Vegetables**

1 fruits and vegetables choice = 10 g carbohydrate, 1 g protein, 190 kJ (44 kcal)

| Food | Measure | Mass (Weight) |
|---|---|---|
| **Fruits (fresh, frozen, without sugar, canned in water)** | | |
| Apples, raw (with or without skin) | ½ medium | 75 g |
| Sauce unsweetened | 125 mL (½ c) | 120 g |
| Sweetened | see *Combined Food Choices* | |
| Apple butter | 20 mL (4 tsp) | 20 g |
| Apricots, raw | 2 medium | 115 g |
| Canned, in water | 4 halves, plus 30 mL (2 tbs) liquid | 110 g |
| Bake-apples (cloudberries), raw | 125 mL (½ c) | 120 g |
| Bananas, with peel | ½ small | 75 g |
| Peeled | ½ small | 50 g |
| Berries (blackberries, blueberries, boysenberries, huckleberries, loganberries, raspberries) | | |
| Raw | 125 mL (½ c) | 70 g |
| Canned, in water | 125 mL (½ c), plus 30 mL (2 tbs) liquid | 100 g |
| Cantaloupe, wedge with rind | ¼ | 240 g |
| Cubed or diced | 250 mL (1 c) | 160 g |
| Cherries, raw, with pits | 10 | 75 g |
| Raw, without pits | 10 | 70 g |
| Canned, in water, with pits | 75 mL (⅓ c), plus 30 mL (2 tbs) liquid | 90 g |
| Canned, in water, without pits | 75 mL (⅓ c), plus 30 mL (2 tbs) liquid | 85 g |
| Crabapples, raw | 1 small | 55 g |
| Cranberries, raw | 250 mL (1 c) | 100 g |
| Figs, raw | 1 medium | 50 g |
| Canned, in water | 3 medium, plus 30 mL (2 tbs) liquid | 100 g |
| Foxberries, raw | 250 mL (1 c) | 100 g |
| Fruit cocktail, canned, in water | 125 mL (½ c), plus 30 mL (2 tbs) liquid | 120 g |
| Fruit, mixed, cut-up | 125 mL (½ c) | 120 g |
| Gooseberries, raw | 250 mL (1 c) | 150 g |
| Canned, in water | 250 mL (1 c), plus 30 mL (2 tbs) liquid | 230 g |
| Grapefruit, raw, with rind | ½ small | 185 g |
| Raw, sectioned | 125 mL (½ c) | 100 g |
| Canned, in water | 125 mL (½ c), plus 30 mL (2 tbs) liquid | 120 g |
| Grapes, raw, slip skin | 125 mL (½ c) | 75 g |
| Raw, seedless | 125 mL (½ c) | 75 g |
| Canned, in water | 75 mL (⅓ c), plus 30 mL (2 tbs) liquid | 115 g |
| Guavas, raw | ½ | 50 g |
| Honeydew melon, raw, with rind | ½ | 225 g |
| Cubed or diced | 250 mL (1 c) | 170 g |
| Kiwis, raw, with skin | 2 | 155 g |
| Kumquats, raw | 3 | 60 g |
| Loquats, raw | 8 | 130 g |
| Lychee fruit, raw | 8 | 120 g |
| Mandarin oranges, raw, with rind | 1 | 135 g |
| Raw, sectioned | 125 mL (½ c) | 100 g |
| Canned, in water | 125 mL (½ c), plus 30 mL (2 tbs) liquid | 100 g |
| Mangoes, raw, without skin and seed | ⅓ | 65 g |
| Diced | 75 mL (⅓ c) | 65 g |
| Nectarines | ½ medium | 75 g |
| Oranges, raw, with rind | 1 small | 130 g |
| Raw, sectioned | 125 mL (½ c) | 95 g |

**Table I-4 (continued)**
**Canadian Choice System: Fruits and Vegetables**

1 fruits and vegetables choice = 10 g carbohydrate, 1 g protein, 190 kJ (44 kcal)

| Food | Measure | Mass (Weight) |
|---|---|---|
| Papayas, raw, with skin and seeds | ¼ medium | 150 g |
| Raw, without skin and seeds | ¼ medium | 100 g |
| Cubed or diced | 125 mL (½ c) | 100 g |
| Peaches, raw, with seed and skin | 1 large | 100 g |
| Raw, sliced or diced | 125 mL (½ c) | 100 g |
| Canned in water, halves or slices | 125 mL (½ c), plus 30 mL (2 tbs) liquid | 120 g |
| Pears, raw, with skin and core | ½ | 90 g |
| Raw, without skin and core | ½ | 85 g |
| Canned, in water, halves | 1 half plus 30 mL (2 tbs) liquid | 90 g |
| Persimmons, raw, native | 1 | 30 g |
| Raw, Japanese | ¼ | 50 g |
| Pineapple, raw | 1 slice | 75 g |
| Raw, diced | 125 mL (½ c) | 75 g |
| Canned, in juice, diced | 75 mL (⅓ c), plus 15 mL (1 tbs) liquid | 55 g |
| Canned, in juice, sliced | 1 slice, plus 15 mL (1 tbs) liquid | 55 g |
| Canned, in water, diced | 125 mL (½ c), plus 30 mL (2 tbs) liquid | 100 g |
| Canned, in water, sliced | 2 slices, plus 15 mL (1 tbs) liquid | 100 g |
| Plums, raw | 2 small | 60 g |
| Damson | 6 | 65 g |
| Japanese | 1 | 70 g |
| Canned, in apple juice | 2, plus 30 mL (2 tbs) liquid | 70 g |
| Canned, in water | 3, plus 30 mL (2 tbs) liquid | 100 g |
| Pomegranates, raw | ½ | 140 g |
| Strawberries, raw | 250 mL (1 c) | 150 g |
| Frozen/canned, in water | 250 mL (1 c), plus 30 mL (2 tbs) liquid | 240 g |
| Rhubarb | 250 mL (1 c) | 150 g |
| Tangelos, raw | 1 | 205 g |
| Tangerines, raw | 1 medium | 115 g |
| Raw, sectioned | 125 mL (½ c) | 100 g |
| Watermelon, raw, with rind | 1 wedge | 310 g |
| Cubed or diced | 250 mL (1 c) | 160 g |
| **Dried Fruit** | | |
| Apples | 5 pieces | 15 g |
| Apricots | 4 halves | 15 g |
| Banana flakes | 30 mL (2 tbs) | 15 g |
| Currants | 30 mL (2 tbs) | 15 g |
| Dates, without pits | 2 | 15 g |
| Peaches | ½ | 15 g |
| Pears | ½ | 15 g |
| Prunes, raw, with pits | 2 | 15 g |
| Raw, without pits | 2 | 10 g |
| Stewed, no liquid | 2 | 20 g |
| Stewed, with liquid | 2, plus 15 mL (1 tbs) liquid | 35 g |
| Raisins | 30 mL (2 tbs) | 15 g |
| **Juices (no sugar added or unsweetened)** | | |
| Apricot, grape, guava, mango, prune | 50 mL (¼ c) | 55 g |
| Apple, carrot, papaya, pear, pineapple, pomegranate | 75 mL (⅓ c) | 80 g |
| Cranberry (see Sugars Section) | | |
| Clamato (see Sugars Section) | | |

(continued on the next page)

I

**Table I-4 (continued)**
**Canadian Choice System: Fruits and Vegetables**

1 fruits and vegetables choice = 10 g carbohydrate, 1 g protein, 190 kJ (44 kcal)

| Food | Measure | Mass (Weight) |
|---|---|---|
| Grapefruit, loganberry, orange, raspberry, tangelo, tangerine | 125 mL (½ c) | 130 g |
| Tomato, tomato-based mixed vegetables | 250 mL (1 c) | 255 g |
| **Vegetables (fresh, frozen, or canned)** | | |
| Artichokes, French, globe | 2 small | 50 g |
| Beets, diced or sliced | 125 mL (½ c) | 85 g |
| Carrots, diced, cooked or uncooked | 125 mL (½ c) | 75 g |
| Chestnuts, fresh | 5 | 20 g |
| Parsnips, mashed | 125 mL (½ c) | 80 g |
| Peas, fresh or frozen | 125 mL (½ c) | 80 g |
| Canned | 75 mL (⅓ c) | 55 g |
| Pumpkin, mashed | 125 mL (½ c) | 45 g |
| Rutabagas, mashed | 125 mL (½ c) | 85 g |
| Sauerkraut | 250 mL (1 c) | 235 g |
| Snow peas | 250 mL (1 c) | 135 g |
| Squash, yellow or winter, mashed | 125 mL (½ c) | 115 g |
| Succotash | 75 mL (⅓ c) | 55 g |
| Tomatoes, canned | 250 mL (1 c) | 240 g |
| Tomato paste | 50 mL (¼ c) | 55 g |
| Tomato sauce* | 75 mL (⅓ c) | 100 g |
| Turnips, mashed | 125 mL (½ c) | 115 g |
| Vegetables, mixed | 125 mL (½ c) | 90 g |
| Water chestnuts | 8 medium | 50 g |

*Tomato sauce varies according to brand name. Check the label or discuss with your dietitian.

**Table 1-5**
**Canadian Choice System: Milk**

| Type of Milk | Carbohydrate (g) | Protein (g) | Fat (g) | Energy |
|---|---|---|---|---|
| Nonfat (0%) | 6 | 4 | 0 | 170 kJ (40 kcal) |
| 1% | 6 | 4 | 1 | 206 kJ (49 kcal) |
| 2% | 6 | 4 | 2 | 244 kJ (58 kcal) |
| Whole (4%) | 6 | 4 | 4 | 319 kJ (76 kcal) |

| Food | Measure | Mass (Weight) |
|---|---|---|
| Buttermilk (higher in salt) | 125 mL (½ c) | 125 g |
| Evaporated milk | 50 mL (¼ c) | 50 g |
| Milk | 125 mL (½ c) | 125 g |
| Powdered milk, regular | 30 mL (2 tbs) | 15 g |
| Instant | 50 mL (¼ c) | 15 g |
| Plain yogurt | 125 mL (½ c) | 125 g |

| Food | Choices per serving | Measure | Mass (Weight) |
|---|---|---|---|
| NOTE: Food items found in this category provide more than 1 milk choice: | | | |
| Milkshake | 1 milk + 3 sugars + ½ protein | 250 mL (1 c) | 300 g |
| Chocolate milk, 2% | 2 milks 2% + 1 sugar | 250 mL (1 c) | 300 g |
| Frozen yogurt | 1 milk + 1 sugar | 125 mL (½ c) | 125 g |

**Table I-6**

**Canadian Choice System: Sugars**

1 sugar choice = 10 g carbohydrate (sugar), 167 kJ (40 kcal)

| Food | Measure | Mass (Weight) |
|---|---|---|
| **Beverages** | | |
| Condensed milk | 15 mL (1 tbs) | |
| Flavoured fruit crystals* | 75 mL (⅓ c) | |
| Iced tea mixes* | 75 mL (⅓ c) | |
| Regular soft drinks | 125 mL (½ c) | |
| Sweet drink mixes* | 75 mL (⅓ c) | |
| Tonic water | 125 mL (½ c) | |
| *These beverages have been made with water. | | |
| **Miscellaneous** | | |
| Bubble gum (large square) | 1 piece | 5 g |
| Cranberry cocktail | 75 mL (⅓ c) | 80 g |
| Cranberry cocktail, light | 350 mL (1⅓ c) | 260 g |
| Cranberry sauce | 30 mL (2 tbs) | |
| Hard candy mints | 2 | 5 g |
| Honey, molasses, corn & cane syrup | 10 mL (2 tsp) | 15 g |
| Jelly bean | 4 | 10 g |
| Licorice | 1 short stick | 10 g |
| Marshmallows | 2 large | 15 g |
| Popsicle | 1 stick (½ popsicle) | |
| Powdered gelatin mix (Jello®) (reconstituted) | 50 mL (¼ c) | |
| Regular jam, jelly, marmalade | 15 mL (1 tbs) | |
| Sugar, white, brown, icing, maple | 10 mL (2 tsp) | 10 g |
| Sweet pickles | 2 small | 100 g |
| Sweet relish | 30 mL (2 tbs) | |

| Food | Choices per Serving | Measures | Mass (Weight) |
|---|---|---|---|
| The following food items provide more than 1 sugar choice: | | | |
| Brownie | 1 sugar + 1 fat | 1 | 20 g |
| Clamato juice | 1½ sugars | 175 mL (⅔ c) | |
| Fruit salad, light syrup | 1 sugar + 1 fruits & vegetables | 125 mL (½ c) | 130 g |
| Aero® bar | 2½ sugars + 2½ fats | 1 bar | 43 g |
| Smarties® | 4½ sugars + 2 fats | 1 box | 60 g |
| Sherbet | 3 sugars + ½ fat | 125 mL (½ c) | 95 g |

**Table I-7**
**Canadian Choice System: Protein Foods**

1 protein choice = 7 g protein, 3 g fat, 230 kJ (55 kcal)

| Food | Measure | Mass (Weight) |
|---|---|---|
| **Cheese** | | |
| Low-fat cheese, about 7% milk fat | 1 slice | 30 g |
| Cottage cheese, 2% milkfat or less | 50 mL (¼ c) | 55 g |
| Ricotta, about 7% milkfat | 50 mL (¼ c) | 60 g |
| **Fish** | | |
| Anchovies (see *Extras,* Table I-9) | | |
| Canned, drained (e.g., mackerel, salmon, tuna packed in water) | (⅓ of 6.5 oz can) | 30 g |
| Cod tongues, cheeks | 75 mL (⅓ c) | 50 g |
| Fillet or steak (e.g., Boston blue, cod, flounder, haddock, halibut, mackerel, orange roughy, perch, pickerel, pike, salmon, shad, snapper, sole, swordfish, trout, tuna, whitefish) | 1 piece | 30 g |
| Herring | ⅓ fish | 30 g |
| Sardines, smelts | 2 medium or 3 small | 30 g |
| Squid, octopus | 50 mL (¼ c) | 40 g |
| **Shellfish** | | |
| Clams, mussels, oysters, scallops, snails | 3 medium | 30 g |
| Crab, lobster, flaked | 50 mL (¼ c) | 30 g |
| Shrimp, fresh | 5 large | 30 g |
| Frozen | 10 medium | 30 g |
| Canned | 18 small | 30 g |
| Dry pack | 50 mL (¼ c) | 30 g |
| **Meat and Poultry (e.g., beef, chicken, goat, ham, lamb, pork, turkey, veal, wild game)** | | |
| Back, peameal bacon | 3 slices, thin | 30 g |
| Chop | ½ chop, with bone | 40 g |
| Minced or ground, lean or extra-lean | 30 mL (2 tbs) | 30 g |
| Sliced, lean | 1 slice | 30 g |
| Steak, lean | 1 piece | 30 g |
| **Organ Meats** | | |
| Hearts, liver | 1 slice | 30 g |
| Kidneys, sweetbreads, chopped | 50 mL (¼ c) | 30 g |
| Tongue | 1 slice | 30 g |
| Tripe | 5 pieces | 60 g |
| **Soyabean** | | |
| Bean curd or tofu | ½ block | 70 g |
| **Eggs** | | |
| In shell, raw or cooked | 1 medium | 50 g |
| Without shell, cooked or poached in water | 1 medium | 45 g |
| Scrambled | 50 mL (¼ c) | 55 g |

| Food | Choices per Serving | Measures | Mass (Weight) |
|---|---|---|---|
| Note: The following choices provide more than 1 protein exchange: | | | |
| **Cheese** | | | |
| Cheeses | 1 protein + 1 fat | 1 piece | 25 g |
| Cheese, coarsely grated (e.g., cheddar) | 1 protein + 1 fat | 50 mL (¼ c) | 25 g |
| Cheese, dry, finely grated (e.g., parmesan) | 1 protein + 1 fat | 45 mL | 15 g |
| Cheese, ricotta, high fat | 1 protein + 1 fat | 50 mL (¼ c) | 55 g |
| **Fish** | | | |
| Eel | 1 protein + 1 fat | 1 slice | 50 g |

**Table I-7 (continued)**
**Canadian Choice System: Protein Foods**

1 protein choice = 7 g protein, 3 g fat, 230 kJ (55 kcal)

| Food | Choices per Serving | Measures | Mass (Weight) |
|---|---|---|---|
| **Meat** | | | |
| Bologna | 1 protein + 1 fat | 1 slice | 20 g |
| Canned lunch meats | 1 protein + 1 fat | 1 slice | 20 g |
| Corned beef, canned | 1 protein + 1 fat | 1 slice | 25 g |
| Corned beef, fresh | 1 protein + 1 fat | 1 slice | 25 g |
| Ground beef, medium-fat | 1 protein + 1 fat | 30 mL (2 tbs) | 25 g |
| Meat spreads, canned | 1 protein + 1 fat | 45 mL | 35 g |
| Mutton chop | 1 protein + 1 fat | ½ chop, with bone | 35 g |
| Paté (see *Fats and Oils* group, Table I-8) | | | |
| Sausages, garlic, Polish or knockwurst | 1 protein + 1 fat | 1 slice | 50 g |
| Sausages, pork, links | 1 protein + 1 fat | 1 link | 25 g |
| Spareribs or shortribs, with bone | 1 protein + 1 fat | 1 large | 65 g |
| Stewing beef | 1 protein + 1 fat | 1 cube | 25 g |
| Summer sausage or salami | 1 protein + 1 fat | 1 slice | 40 g |
| Weiners, hot dog | 1 protein + 1 fat | ½ medium | 25 g |
| **Miscellaneous** | | | |
| Blood pudding | 1 protein + 1 fat | 1 slice | 25 g |
| Peanut butter | 1 protein + 1 fat | 15 mL (1 tbs) | 15 g |

**Table I-8**
**Canadian Choice System: Fats and Oils**

1 fat choice = 5 g fat, 190 kJ (45 kcal)

| Food | Measure | Mass (Weight) | Food | Measure | Mass (Weight) |
|---|---|---|---|---|---|
| Avocado* | ⅛ | 30 g | Nuts (continued): | | |
| Bacon, side, crisp* | 1 slice | 5 g | Sesame seeds | 15 mL (1 tbs) | 10 g |
| Butter* | 5 mL (1 tsp) | 5 g | Sunflower seeds | | |
| Cheese spread | 15 mL (1 tbs) | 15 g | Shelled | 15 mL (1 tbs) | 10 g |
| Coconut, fresh* | 45 mL (3 tbs) | 15 g | In shell | 45 mL (3 tbs) | 15 g |
| Coconut, dried* | 15 mL (1 tbs) | 10 g | Walnuts | 4 halves | 10 g |
| Cream, Half and half | | | Oil, cooking and salad | 5 mL (1 tsp) | 5 g |
| (cereal), 10%* | 30 mL (2 tbs) | 30 g | Olives, green | 10 | 45 g |
| Light (coffee), 20%* | 15 mL (1 tbs) | 15 g | Ripe black | 7 | 57 g |
| Whipping, 32 to 37%* | 15 mL (1 tbs) | 15 g | Pâté, liverwurst, | 15 mL (1 tbs) | 15 g |
| Cream cheese* | 15 mL (1 tbs) | 15 g | meat spreads | | |
| Gravy* | 30 mL (2 tbs) | 30 g | Salad dressing: blue, | 10 mL (2 tsp) | 10 g |
| Lard* | 5 mL (1 tsp) | 5 g | French, Italian, | | |
| Margarine | 5 mL (1 tsp) | 5 g | mayonnaise, | | |
| Nuts, shelled: | | | Thousand Island | 5 mL (1 tsp) | 5 g |
| Almonds | 8 | 5 g | Salad dressing, | 30 mL (2 tbs) | 30 g |
| Brazil nuts | 2 | 10 g | low-calorie | | |
| Cashews | 5 | 10 g | Salt pork, raw | 5 mL (1 tsp) | 5 g |
| Filberts, hazelnuts | 5 | 10 g | or cooked* | | |
| Macadamia | 3 | 5 g | Sesame oil | 5 mL (1 tsp) | 5 g |
| Peanuts | 10 | 10g | Sour cream | | |
| Pecans | 5 halves | 5 g | 12% milkfat | 30 mL (2 tbs) | 30 g |
| Pignolias, pine nuts | 25 mL (5 tsp) | 10 g | 7% milkfat | 60 mL (4 tbs) | 60 g |
| Pistachios, shelled | 20 | 10 g | Shortening* | 5 mL (1 tsp) | |
| Pistachios, in shell | 20 | 20 g | | | |
| Pumpkin and squash seeds | 20 mL (4 tsp) | 10 g | | | |

*These items contain higher amounts of saturated fat.

**Table I-9**
**Canadian Choice System: Extras**

Extras have no more than 2.5 g carbohydrate, 60 kJ (14 kcal)

**Vegetables** 125 mL (½ c)

Artichokes
Asparagus
Bamboo shoots
Bean sprouts, mung or soya
Beans, string, green, or yellow
Bitter melon (balsam pear)
Bok choy
Broccoli
Brussels sprouts
Cabbage
Cauliflower
Celery
Chard
Cucumbers
Eggplant
Endive
Fiddleheads
Greens: beet, collard, dandelion, mustard, turnip, etc.
Kale
Kohlrabi
Leeks
Lettuce
Mushrooms
Okra
Onions, green or mature
Parsley
Peppers, green, yellow or red
Radishes
Rapini
Rhubarb
Sauerkraut
Shallots
Spinach
Sprouts: alfalfa, radish, etc.
Tomato wedges
Watercress
Zucchini

**Free Foods** (may be used without measuring)

| | |
|---|---|
| Artificial sweetener, such as cyclamate or aspartame | Lime juice or lime wedges<br>Marjoram, cinnamon, etc. |
| Baking powder, baking soda | Mineral water |
| Bouillon from cube, powder, or liquid | Mustard<br>Parsley |
| Bouillon or clear broth | Pimentos |
| Chowchow, unsweetened | Salt, pepper, thyme |
| Coffee, clear | Soda water, club soda |
| Consommé | Soya sauce |
| Dulse | Sugar-free Crystal Drink |
| Flavorings and extracts | Sugar-free Jelly Powder |
| Garlic | Sugar-free soft drinks |
| Gelatin, unsweetened | Tea, clear |
| Ginger root | Vinegar |
| Herbal teas, unsweetened | Water |
| Horseradish, uncreamed | Worcestershire sauce |
| Lemon juice or lemon wedges | |

**Condiments**

| Food | Measure |
|---|---|
| Anchovies | 2 fillets |
| Barbecue sauce | 15 mL (1 tbs) |
| Bran, natural | 30 mL (2 tbs) |
| Brewer's yeast | 5 mL (1 tsp) |
| Carob powder | 5 mL (1 tsp) |
| Catsup | 5 mL (1 tsp) |
| Chili sauce | 5 mL (1 tsp) |
| Cocoa powder | 5 mL (1 tsp) |
| Cranberry sauce, unsweetened | 15 mL (1 tbs) |
| Dietetic fruit spreads | 5 mL (1 tsp) |
| Maraschino cherries | 1 |
| Nondairy coffee whitener | 5 mL (1 tsp) |
| Nuts, chopped pieces | 5 mL (1 tsp) |
| Pickles | |
| unsweetened dill | 2 |
| sour mixed | 11 |
| Sugar substitutes, granular | 5 mL (1 tsp) |
| Whipped toppings | 15 mL (1 tbs) |

**Table I-10**
**Canadian Choice System: Combined Food Choices**

| Food | Choices per serving | Measure | Mass (Weight) |
|---|---|---|---|
| Angel food cake | ½ starch + 2½ sugars | 1/12 cake | 50 g |
| Apple crisp | ½ starch + 1½ fruits & vegetables + 1 sugar + 1–2 fats | 125 mL (½ c) | |
| Applesauce, sweetened | 1 fruits & vegetables + 1 sugar | 125 mL (½ c) | |
| Beans and pork in tomato sauce | 1 starch + ½ fruits & vegetables + ½ sugar + 1 protein | 125 mL (½ c) | 135 g |
| Beef burrito | 2 starches + 3 proteins + 3 fats | | 110 g |
| Brownie | 1 sugar + 1 fat | 1 | 20 g |
| Cabbage rolls* | 1 starch + 2 proteins | 3 | 310 g |
| Caesar salad | 2–4 fats | 20 mL dressing (4 tsp) | |
| Cheesecake | ½ starch + 2 sugars + ½ protein + 5 fats | 1 piece | 80 g |
| Chicken fingers | 1 starch + 2 proteins + 2 fats | 6 small | 100 g |
| Chicken and snow pea Oriental | 2 starches + ½ fruits & vegetables + 3 proteins + 1 fat | 500 mL (2 c) | |
| Chili | 1½ starches + ½ fruits & vegetables + 3½ protein | 300 mL (1¼ c) | 325 g |
| Chips | | | |
| Potato chips | 1 starch + 2 fats | 15 chips | 30 g |
| Corn chips | 1 starch + 2 fats | 30 chips | 30 g |
| Tortilla chips | 1 starch + 1½ fats | 13 chips | |
| Cheese twist | 1 starch + 1½ fats | 30 chips | 30 g |
| Chocolate bar | | | |
| Aero® | 2½ sugars + 2½ fats | bar | 43 g |
| Smarties® | 4½ sugars + 2 fats | package | 60 g |
| Chocolate cake (without icing) | 1 starch + 2 sugars + 3 fats | 1/10 of a 8″ pan | |
| Chocolate devil's food cake (without icing) | 2 starches + 2 sugars + 3 fats | 1/12 of a 9″ pan | |
| Chocolate milk | 2 milks 2% + 1 sugar | 250 mL (1 c) | 300 g |
| Clubhouse (triple-decker) sandwich | 3 starches + 3 proteins + 4 fats | | |
| Cookies | | | |
| chocolate chip | ½ starch + ½ sugar + 1½ fats | 2 | 22 g |
| oatmeal | 1 starch + 1 sugar + 1 fat | 2 | 40 g |
| Donut (chocolate glazed) | 1 starch + 1½ sugars + 2 fats | 1 | 65 g |
| Egg roll | 1 starch + ½ protein + 1 fat | | 75 g |
| Four bean salad | 1 starch + ½ protein + 1 fat | 125 mL (½ c) | |
| French toast | 1 starch + ½ protein + 2 fats | 1 slice | 65 g |
| Fruit in heavy syrup | 1 fruits & vegetables + 1½ sugars | 125 mL (½ c) | |
| Granola bar | ½ starch + 1 sugar + 1–2 fats | | 30 g |
| Granola cereal | 1 starch + 1 sugar + 2 fats | 125 mL (½ c) | 45 g |
| Hamburger | 2 starches + 3 proteins + 2 fats | junior size | |
| Ice cream and cone, plain flavour | | | |
| ice cream | ½ milk + 2–3 sugars + 1–2 fats | | 100 g |
| cone | ½ sugar | | 4 g |
| Lasagna | | | |
| regular cheese | 1 starch + 1 fruits & vegetables + 3 proteins + 2 fats | 3″ x 4″ piece | |
| low-fat cheese | 1 starch + 1 fruits & vegetables + 3 proteins | 3″ x 4″ piece | |
| Legumes | | | |
| Dried beans (kidney, navy, pinto, fava, chick peas) | 2 starches + 1 protein | 250 mL (1 c) | 180 g |
| Dried peas | 2 starches + 1 protein | 250 mL (1 c) | 210 g |
| Lentils | 2 starches + 1 protein | 250 mL (1 c) | 210 g |

* If eaten with sauce, add ½ fruits & vegetables exchange.

**Table I-10 (continued)**
**Canadian Choice System: Combined Food Choices**

| Food | Choices per serving | Measure | Mass (Weight) |
|---|---|---|---|
| Macaroni and cheese | 2 starches + 2 proteins + 2 fats | 250 mL (1 c) | 210 g |
| Minestrone soup | 1½ starches + ½ fruits & vegetables + ½ fat | 250 mL (1 c) | |
| Muffin | 1 starch + ½ sugar + 1 fat | 1 small | 45 g |
| Nuts (dry or roasted without any oil added). | | | |
| Almonds, dried sliced | ½ protein + 2 fats | 50 mL (¼ c) | 22 g |
| Brazil nuts, dried unblanched | ½ protein + 2½ fats | 5 large | 23 g |
| Cashew nuts, dry roasted | ½ starch + ½ protein + 2 fats | 50 mL (¼ c) | 28 g |
| Filbert hazelnut, dry | ½ protein + 3½ fats | 50 mL (¼ c) | 30 g |
| Macadamia nuts, dried | ½ protein + 4 fats | 50 mL (¼ c) | 28 g |
| Peanuts, raw | 1 protein + 2 fats | 50 mL (¼ c) | 30 g |
| Pecans, dry roasted | ½ fruits & vegetables + 3 fats | 50 mL (¼ c) | 22 g |
| Pine nuts, pignolia dried | 1 protein + 3 fats | 50 mL (¼ c) | 34 g |
| Pistachio nuts, dried | ½ fruits & vegetables + ½ protein + 2½ fats | 50 mL (¼ c) | 27 g |
| Pumpkin seeds, roasted | 2 proteins + 2½ fats | 50 mL (¼ c) | 47 g |
| Sesame seeds, whole dried | ½ fruits & vegetables + ½ protein + 2½ fats | 50 mL (¼ c) | 30 g |
| Sunflower kernel, dried | ½ protein + 1½ fats | 50 mL (¼ c) | 17 g |
| Walnuts, dried chopped | ½ protein + 3 fats | 50 mL (¼ c) | 26 g |
| Perogies | 2 starches + 1 protein + 1 fat | 3 | |
| Pie, fruit | 1 starch + 1 fruits & vegetables + 2 sugars + 3 fats | 1 piece | 120 g |
| Pizza, cheese | 1 starch + 1 protein + 1 fat | 1 slice (⅛ of a 12") | 50 g |
| Pork stir fry | ½ to 1 fruits & vegetables + 3 proteins | 200 mL (¾ c) | |
| Potato salad | 1 starch + 1 fat | 125 mL (½ c) | 130 g |
| Potatoes, scalloped | 2 starches + 1 milk + 1–2 fats | 200 mL (¾ c) | 210 g |
| Pudding, bread or rice | 1 starch + 1 sugar + 1 fat | 125 mL (½ c) | |
| Pudding, vanilla | 1 milk + 2 sugars | 125 mL (½ c) | |
| Raisin bran cereal | 1 starch + ½ fruits & vegetables + ½ sugar | 175 mL (⅔ c) | 40 g |
| Rice krispie squares | ½ starch + 1½ sugars + ½ fat | 1 square | 30 g |
| Shepherd's pie | 2 starches + 1 fruits & vegetables + 3 proteins | 325 mL (1⅓ c) | |
| Sherbet, orange | 3 sugars + ½ fat | 125 mL (½ c) | |
| Spaghetti and meat sauce | 2 starches + 1 fruits & vegetables + 2 proteins + 3 fats | 250 mL (1 c) | |
| Stew | 2 starches + 2 fruits & vegetables + 3 proteins + ½ fat | 200 mL (¾ c) | |
| Sundae | 4 sugars + 3 fats | 125 mL (½ c) | |
| Tuna casserole | 1 starch + 2 proteins + ½ fat | 125 mL (½ c) | |
| Yogurt, fruit bottom | 1 fruits & vegetables + 1 milk + 1 sugar | 125 mL (½ c) | 125 g |
| Yogurt, frozen | 1 milk + 1 sugar | 125 mL (½ c) | 125 g |

# Food Labels

Consumers can gather a lot of information from a nutrition label. Figure I-1 demonstrates the reading of a food label and Table I-11 defines terms.

**Figure I-1**

**Example of a Food Label**

**OUR COMMITMENT TO QUALITY**

**Kellogg's is committed to providing foods of outstanding quality and freshness. If this product in any way falls below the high standards you've come to expect from Kellogg's, please send your comments and both top flaps to:**
Consumer Affairs
KELLOGG CANADA INC.
Etobicoke, Ontario M9W 5P2

IF IT DOESN'T SAY Kellogg's* ON THE BOX, IT'S NOT Kellogg's* IN THE BOX.
SI LE NOM Kellogg's* N'EST PAS SUR LA BOÎTE, CE N'EST PAS Kellogg's* DANS LA BOÎTE.

• HIGH IN FIBRE
• LOW IN FAT
• PRESERVATIVE FREE

• SOURCE ÉLEVÉE DE FIBRES
• FAIBLE EN MATIÈRES GRASSES
• SANS AGENT DE CONSERVATION

**NUTRITION INFORMATION**
**APPORT NUTRITIONNEL**

| | Per 40 g serving cereal (175 mL, ¾ cup) / Par ration de 40 g de céréale (175 mL, ¾ tasse) | Per 40 g serving cereal with 125 mL Partly Skimmed Milk (2%) / Par ration de 40 g de céréale avec 125 mL de lait partiellement écrémé (2,0 %) | |
|---|---|---|---|
| ENERGY | 130Cal | 195Cal | ÉNERGIE |
| | 540kJ | 810kJ | |
| PROTEIN | 3.0g | 7.3g | PROTÉINES |
| FAT | 0.4g | 2.9g | MATIÈRES GRASSES |
| CARBOHYDRATE | 32g | 38g | GLUCIDES |
| SUGARS* | 11g | 18g | *SUCRES |
| STARCH | 16g | 16g | AMIDON |
| DIETARY FIBRE | 4.6g | 4.6g | FIBRES ALIMENTAIRES |
| SODIUM | 235mg | 300mg | SODIUM |
| POTASSIUM | 240mg | 440mg | POTASSIUM |

% of Recommended Daily Intake
% de l'apport quotidien conseillé

| | | | |
|---|---|---|---|
| VITAMIN A | 0% | 7% | VITAMINE A |
| VITAMIN D | 0% | 23% | VITAMINE D |
| VITAMIN B1 | 62% | 66% | VITAMINE B1 |
| VITAMIN B2 | 3% | 16% | VITAMINE B2 |
| NIACIN | 13% | 18% | NIACINE |
| VITAMIN B6 | 13% | 16% | VITAMINE B6 |
| FOLACIN | 11% | 14% | FOLACINE |
| VITAMIN B12 | 0% | 25% | VITAMINE B12 |
| PANTOTHENATE | 9% | 15% | PANTOTHÉNATE |
| CALCIUM | 1% | 15% | CALCIUM |
| PHOSPHORUS | 12% | 23% | PHOSPHORE |
| MAGNESIUM | 20% | 27% | MAGNÉSIUM |
| IRON | 38% | 39% | FER |
| ZINC | 16% | 22% | ZINC |

*Approximately half of the sugars occur naturally in the raisins.
Environ la moitié des sucres se retrouvent à l'état naturel dans les fruits.

Canadian Diabetes Association Food Choice Values: 40 g (175 mL, ¾ cup) cereal. Système des choix d'aliments de l'Association canadienne du diabète : 40 g (175 mL, ¾ tasse) céréale = 1 [■] + ½ [◆] + ½ [✱] choices/choix

**INGREDIENTS / INGRÉDIENTS**

WHOLE WHEAT, RAISINS (COATED WITH SUGAR, HYDROGENATED VEGETABLE OIL), WHEAT BRAN, SUGAR/GLUCOSE-FRUCTOSE, SALT, MALT (CORN FLOUR, MALTED BARLEY), VITAMINS (THIAMIN HYDROCHLORIDE, PYRIDOXINE HYDROCHLORIDE, FOLIC ACID, d-CALCIUM PANTOTHENATE), MINERALS (IRON, ZINC OXIDE).

BLÉ ENTIER, RAISINS SECS (ENROBÉS DE SUCRE, D'HUILE VÉGÉTALE HYDROGÉNÉE), SON DE BLÉ, SUCRE/GLUCOSE-FRUCTOSE, SEL, MALT (FARINE DE MAÏS, ORGE MALTÉ), VITAMINES (CHLORHYDRATE DE THIAMINE, CHLORHYDRATE DE PYRIDOXINE, ACIDE FOLIQUE, PANTOTHÉNATE DE d-CALCIUM), MINÉRAUX (FER, OXYDE DE ZINC).

Made by / Produit par
KELLOGG CANADA INC.
ETOBICOKE, ONTARIO
CANADA M9W 5P2
*Registered trademark of / *Marque déposée de
KELLOGG CANADA INC. © 1994
00094

**WHAT YOU WILL FIND ON A LABEL:**

**Nutrition Claims**

- in Canada, it is optional for a company to decide to use claims,
- when claims appear on a label, they must follow government laws

**Nutrition Information**

- gives detailed nutrition facts about the product, including serving size and core list
- does not have to appear by law on food products in Canada
- refers to the food as packaged, so if you add milk, eggs or other food, the nutritional content of the food you eat can be very different

**Serving Size**

- the amount of food for which the information is given
- check the serving size: the serving size on the label may not be the same as the serving size you would actually eat (for example, the serving size of cereal may be ¾ cup, much smaller than your regular serving

**Core List**

- the energy (in Calories and kilojoules), grams of protein, fat and carbohydrate for each serving
- some products break down fat into monounsaturates, polyunsaturates, saturates, and cholesterol (to find out what these mean, look at the Fats & Oils section)
- carbohydrates may include the amount of sugars, starch and fibre, or may list these items separately

**Sodium and Potassium** (in milligrams)

**Vitamins and Minerals** (as percent of your recommended daily intake)

**Canadian Diabetes Association Food Choice Values and Symbols**

- the Values and Symbols are tools to help you fit the food into your meal plan, they are not an endorsement by CDA
- it is up to the food company to decide if they want their foods analyzed and assigned symbols
- when they are on a label, they have been assigned by a dietitian working for CDA, so you can be sure the information is correct

**Ingredients**

- must be found on all food labels by law
- ingredients are listed in decreasing order by weight, so what you see first is what you get the most of

**Table I-11**
**Terms on Food Labels**

**Energy**

**kcalorie reduced:** 50% or fewer kcalories than the regular version.

**light:** term may be used to describe anything (for example, light in colour, texture, flavour, taste, or kcalories); read the label to find out what is "light" about the product.

**low kcalorie:** kcalorie-reduced and no more than 15 kcalories per serving.

**Fat and Cholesterol**

**low cholesterol:** no more than 3 mg of cholesterol per 100 g of the food and low in saturated fat; *does not* always mean low in total fat.

**low fat:** no more than 3 g of fat per serving; *does not* always mean low in kcalories.

**lower fat:** at least 25% less fat than the comparison food; be aware that 80% fat-free still means the food is 20% fat.

**Carbohydrates: Fibre and Sugar**

**carbohydrate reduced:** not more than 50% of the carbohydrate found in the regular version; *does not* always mean the product is lower in kcalories because other ingredients such as fat may have increased.

**source of dietary fibre:** a product that provides 2–4 g of fibre.

**high source of dietary fibre:** a product that provides 4–6 g of fibre.

**very high source of fibre:** a product that provides 6 g (or more) of fibre.

**sugar free:** low in carbohydrates and kcalories; can be used as an extra food in the exchange system.

**unsweetened or no sugar added:** no sugar was added to the product; sugar may be found naturally in the food (for example, fruit canned in its own juice).

## Canada's Food Guide

Canada's Food Guide to Healthy Eating, shown in Figure I-2, gives detailed information for selecting foods to meet the nutritional needs of all Canadians four years of age and older. Like the U.S. Daily Food Guide, Canada's Food Guide also takes a total diet approach, rather than emphasizing a single food, meal, or day's meals and snacks.

**Figure I-2**

**Canada's Food Guide to Healthy Eating**

Health and Welfare Canada Santé et Bien-être social Canada

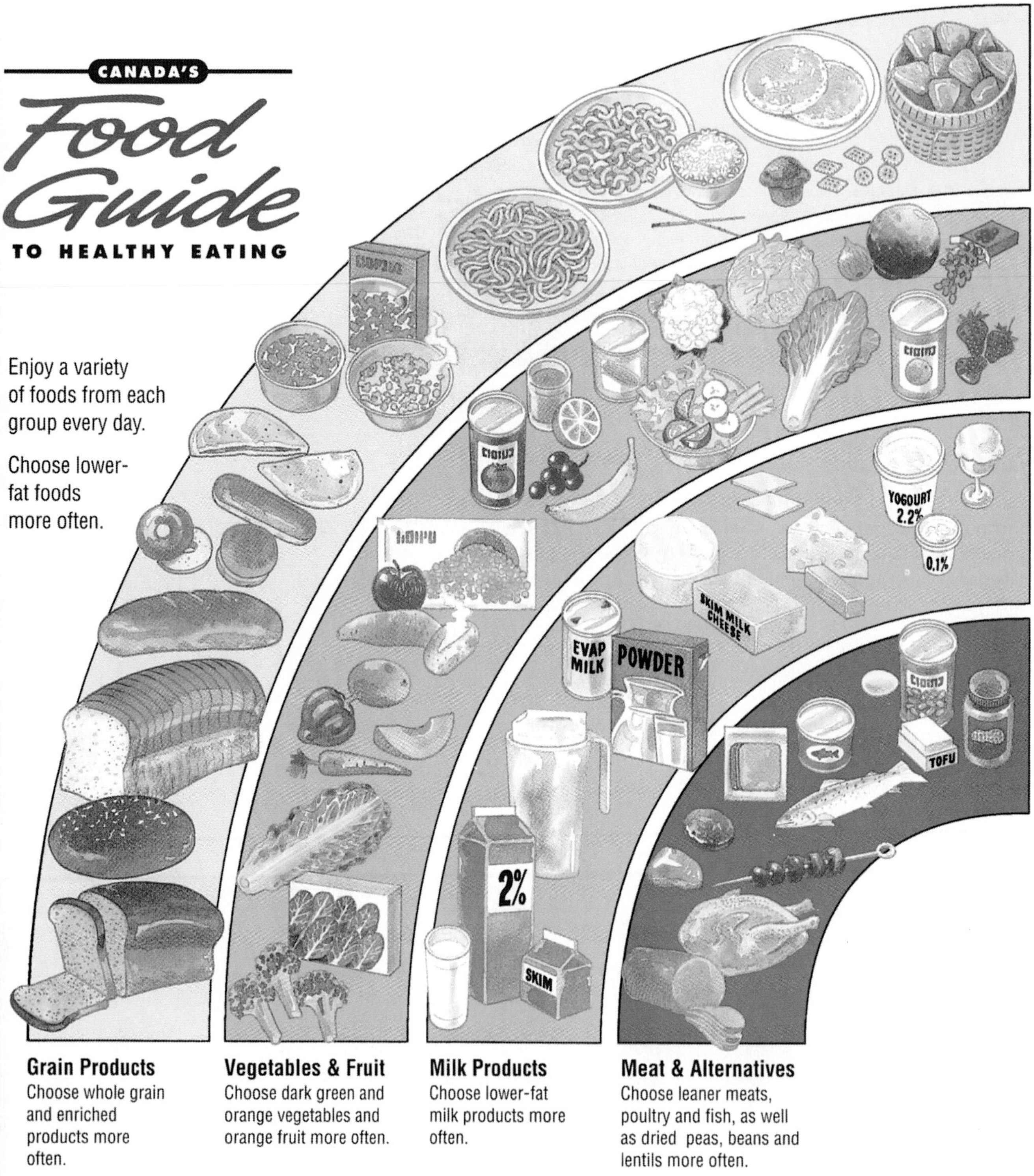

CANADA'S

# Food Guide

TO HEALTHY EATING

Enjoy a variety of foods from each group every day.

Choose lower-fat foods more often.

**Grain Products**
Choose whole grain and enriched products more often.

**Vegetables & Fruit**
Choose dark green and orange vegetables and orange fruit more often.

**Milk Products**
Choose lower-fat milk products more often.

**Meat & Alternatives**
Choose leaner meats, poultry and fish, as well as dried peas, beans and lentils more often.

## Different People Need Different Amounts of Food

The amount of food you need every day from the 4 food groups and other foods depends on your age, body size, activity level, whether you are male or female and if you are pregnant or breast-feeding. That's why the Food Guide gives a lower and higher number of servings for each food group. For example, young children can choose the lower number of servings, while male teenagers can go to the higher number. Most other people can choose servings somewhere in between.

**Grain Products**
**5–12**
SERVINGS PER DAY

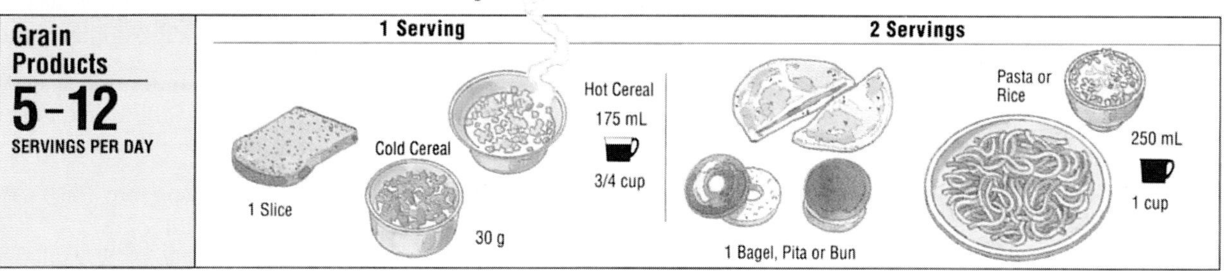


**Vegetables & Fruit**
**5–10**
SERVINGS PER DAY

**Milk Products**
SERVINGS PER DAY
Children 4–9 years: 2–3
Youth 10–16 years: 3–4
Adults: 2–4
Pregnant & Breast-feeding Women: 3–4

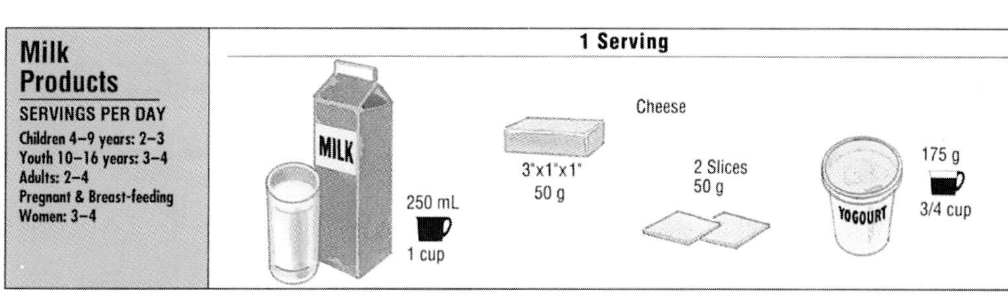


**Meat & Alternatives**
**2–3**
SERVINGS PER DAY

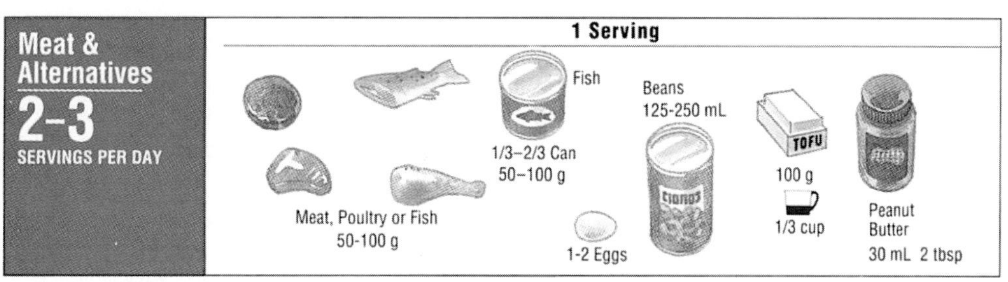


### Other Foods

Taste and enjoyment can also come from other foods and beverages that are not part of the 4 food groups. Some of these foods are higher in fat or Calories, so use these foods in moderation.

*Enjoy eating well, being active and feeling good about yourself. That's* VITALIT

J

Contents



# MEASURES OF PROTEIN QUALITY

In a world where food is scarce and many people's diets contain marginal or inadequate amounts of protein, it is important to know which foods contain the highest-quality protein. Chapter 6 describes protein quality and the different measures researchers use to assess the quality of a food protein. This appendix provides a few more details.

## Amino Acid Scoring

Amino acid, or chemical, scoring allows researchers to determine the amino acid composition of any protein relatively inexpensively, but unfortunately, it does not always accurately reflect the way the body will use a protein. The advantages of amino acid scoring are that it is simple and inexpensive, it identifies in one step the limiting amino acid, and it can be used to score mixtures of different proportions of two or more proteins mathematically without having to make up a mixture and test it. Its chief weaknesses are that it fails to predict the digestibility of a protein, which may strongly affect the protein's quality; it relies on a chemical procedure in which certain amino acids may be destroyed, making the pattern that is analyzed inaccurate; and it is blind to other features of the protein (such as the presence of substances that may inhibit the digestion or utilization of the protein) that would only be revealed by a test in living animals. Table J-1 shows how to use a reference pattern for the nine essential amino acids.

## PDCAAS

PDCAAS (protein-digestibility-corrected amino acid score) takes the amino acid scoring method a step further by correcting for the digestibility of the protein. To calculate the PDCAAS, researchers first determine the amino acid profile of the test protein (in this example, pinto beans). The second column of Table J-2 presents the essential amino acid profile for pinto beans. The third column presents the amino acid requirements of preschool-aged children for comparison. To determine how well the food protein meets human needs, researchers calculate the ratio by dividing the second column by the third column (for example, 30 ÷ 19 = 1.578 or 1.58).

The amino acid with the lowest ratio is the first limiting amino acid—in this case, tryptophan. Its ratio is the amino acid score for the protein—in this case, 80. Remember, though, the amino acid score does not account for digestibility. Protein digestibility, as determined by rat balance studies, yields a value of 79 percent for pinto beans. Together, the amino acid score and the digestibility value determine the PDCAAS:

$$\text{PDCAAS} = \text{protein digestibility} \times \text{lowest amino acid ratio}.$$
$$\text{PDCAAS for pinto beans} = .79 \times .80 = 63\%.$$

Thus the PDCAAS for pinto beans is 63 percent (or 0.63). Table J-3 lists the PDCAAS values of selectd foods.

The PDCAAS is used to determine the % Daily Value on food labels. To calculate the % Daily Value for protein for canned pinto beans, multiply the number of grams of protein in a standard serving (in this case, 7 grams per ½ cup) by the PDCAAS:

$$7 \text{ g} \times .63 = 4.41.$$

This value is then divided by the recommended standard for protein (for children over age four and adults, 50 grams):

$$4.41 \div 50 = 0.088 \text{ (or } 8.8\%).$$

The food label for this can of pinto beans would declare that one serving provides 7 grams protein, and if the label included a % Daily Value for protein, the value would be 9 percent.

## Biological Value

To determine the actual value of a protein as it is used by the body, it is necessary to measure both urinary and fecal losses of nitrogen when that protein is actually fed to human beings under test conditions. Even then, small additional losses from sweat, shed skin, hair, and fingernails will be missed. This kind of experiment determines the biological value (BV) of proteins, a measure used internationally.

In a test of biological value, two nitrogen balance studies are done. In the first, no protein is fed, and nitrogen (N) excretions in the urine and feces are measured. It is assumed that under these conditions, N lost in the urine is the amount the body always necessarily loses by filtration into the urine each day, regardless of what protein is fed (endogenous N). The N lost in the feces (called metabolic N in the equation) is the amount the body invariably loses into the intestine each day, whether or not food protein is fed. (To help you remember the terms: endogenous N is "urinary N on a zero-protein diet"; metabolic N is "fecal N on a zero-protein diet.")

In the second study, an amount of protein slightly below the requirement is fed. Intake and losses are measured; then the BV is derived using this formula:

**Table J-1**

**A Reference Pattern for Amino Acid Scoring of Proteins**

| Essential Amino Acids | Reference Protein (Whole Egg) Mg Amino Acid per G Nitrogen |
|---|---|
| Histidine | 145 |
| Isoleucine | 340 |
| Leucine | 540 |
| Lysine | 440 |
| Methionine + cystine[a] | 355 |
| Phenylalanine + tyrosine[b] | 580 |
| Threonine | 294 |
| Tryptophan | 106 |
| Valine | 410 |
| Total | 3210 |

[a]Methionine is essential and is also used to make cystine. Thus the methionine requirement is lower if cystine is supplied.

[b]Phenylalanine is essential and is also used to make tyrosine if not enough of the latter is available. Thus the phenylalanine requirement is lower if tyrosine is also supplied.

NOTE: To interpret the table, read, "For every 3210 units of essential amino acids, 145 must be histidine, 340 must be isoleucine, 540 must be leucine," and so on. To compare a test protein with the reference protein, the experimenter first obtains a chemical analysis of the test protein's amino acids. Then, taking 3210 units of the amino acids, the experimenter compares the amount of each amino acid to the amount found in 3210 units of essential amino acids in egg protein. For example, suppose the test protein contained (per 3210 units) 360 units of isoleucine; 500 units of leucine; 350 of lysine; and for each of the other amino acids, more units than egg protein contains. The two amino acids that are low are leucine (500 as compared with 540 in egg) and lysine (350 versus 440 in egg). The ratio, amino acid in the test protein divided by amino acid in egg, is 500/540 (or about 0.93) for leucine and 350/440 (or about 0.80) for lysine. Lysine is the limiting amino acid (lowest ratio compared with egg), so the test protein receives a chemical score of 80.

$$\text{BV} = \frac{\text{food N} - (\text{fecal N} - \text{metabolic N}) - (\text{urinary N} - \text{endogenous N})}{\text{food N} - (\text{fecal N} - \text{metabolic N})} \times 100.$$

The denominator of this equation expresses the amount of nitrogen *absorbed:* food N minus fecal N (excluding the N the body would lose in the feces anyway, even without food). The numerator expresses the amount of N *retained* from the N absorbed: absorbed N (as in the denominator) minus the N excreted in the urine (excluding the N the body would lose in the urine anyway, even without food). Thus it can be more simply expressed:

$$\text{BV} = \frac{\text{N retained}}{\text{N absorbed}} \times 100.$$

For egg protein, the BV is 100 (all the absorbed protein is retained). Supplied in adequate quantity, a protein with a BV of 70 or greater can support human growth as long as energy intake is adequate. Table J-4 presents the BV for selected foods.

**Table J-2**

**An Example of PDCAAS**

| Essential Amino Acids | Amino Acid Profile of Pinto Beans (mg/g protein) | Amino Acid Requirements for 2–5 yr (mg/g protein) | Ratio |
|---|---|---|---|
| Histidine | 30.0 | 19 | 1.58 |
| Isoleucine | 42.5 | 28 | 1.52 |
| Leucine | 80.4 | 66 | 1.22 |
| Lysine | 69.0 | 58 | 1.19 |
| Methionine + cystine | 21.1 | 25 | 0.84 |
| Phenylalanine + tyrosine | 90.5 | 63 | 1.44 |
| Threonine | 43.7 | 34 | 1.28 |
| Tryptophan | 8.8 | 11 | 0.80 |
| Valine | 50.1 | 35 | 1.43 |

**Table J-3**
**PDCAAS Values of Selected Foods**

| Food | PDCAAS |
| --- | --- |
| Casein (milk protein) | 1.00 |
| Egg white | 1.00 |
| Soybean (isolate) | .99 |
| Beef | .92 |
| Pea flour | .69 |
| Kidney beans (canned) | .68 |
| Chick peas (canned) | .66 |
| Pinto beans (canned) | .63 |
| Rolled oats | .57 |
| Lentils (canned) | .52 |
| Peanut meal | .52 |
| Whole wheat | .40 |

NOTE: 1.0 is the maximum PDCAAS a food protein can receive.

This method has the advantages of being based on experiments with human beings (it can be done with animals, too, of course) and of measuring actual nitrogen retention. But it is also cumbersome, expensive, and often impractical, and it is based on several assumptions that may not be valid. For example, the physiology, normal environment, or typical food intake of the subjects used for testing may not be similar to those for whom the test protein may ultimately be used. For another example, the retention of protein in the body does not necessarily mean that it is being well utilized. Considerable exchange of protein among tissues (protein turnover) occurs, but is hidden from view when only N intake and output are measured. The test of biological value wouldn't detect if one tissue were shorted.

**Table J-4**
**Biological Values of Selected Foods**

| Food | Biological Value |
| --- | --- |
| Egg | 100 |
| Milk | 93 |
| Beef | 75 |
| Fish | 75 |
| Corn | 72 |

# Net Protein Utilization

Like measurements of BV, determinations of net protein utilization (NPU) involve two balance studies: one on zero nitrogen intake, and the other on submaximal intake. The formula for NPU is:

$$\text{NPU} = \frac{\text{food N} - (\text{fecal N} - \text{metabolic N}) - (\text{urinary N} - \text{endogenous N})}{\text{food N}} \times 100.$$

The numerator is the same as it is for BV, but the denominator represents food N intake only—not absorbed N. More simply expressed:

$$\text{NPU} = \frac{\text{N retained}}{\text{N intake}} \times 100.$$

This method offers advantages similar to those of BV determinations and is used more frequently, with animals as the test subjects. A drawback is that if a low NPU is obtained, the test results offer no help in distinguishing between two possible causes: a poor amino acid composition of the test protein or poor digestibility. There is also a limit to the extent to which animal test results can be assumed to be applicable to human beings.

# Protein Efficiency Ratio

The protein efficiency ratio (PER) is a widely used procedure for evaluating protein quality. Young rats are fed a measured amount of protein and weighed periodically as they grow. The PER is expressed as:

$$\text{PER} = \frac{\text{weight gain (g)}}{\text{protein intake (g)}}.$$

This method has the virtues of economy and simplicity, but it also has many drawbacks. The experiments are time-consuming; the amino acid needs of rats are not the same as those of human beings; and the amino acid needs for growth are not the same as for the maintenance of adult animals (growing animals need more lysine, for example).

# MAKING IT CLICK AND MULTIPLE CHOICE ANSWERS

## Chapter 1

*Making It Click Answers*

1. a. 5 g protein × 4 kcal/g = 20 kcal protein.
30 g carbohydrate × 4 kcal/g = 120 kcal carbohydrate.
11 g fat × 9 kcal/g = 99 kcal fat.
Total = 239 kcal.

b. 20 kcal ÷ 239 kcal × 100 = 8.4% from protein.
120 kcal ÷ 239 kcal × 100 = 50.2% from carbohydrate.
99 kcal ÷ 239 kcal × 100 = 41.4% from fat.
Total = 100.0%.

c. 1 g protein = 4 kcal protein.
13 g carbohydrate = 52 kcal carbohydrate.
146 total kcal − 56 kcal protein + carbohydrate = 90 kcal alcohol.
90 kcal alcohol ÷ 7 g/kcal = 12.9 g alcohol.

2. No. 15 g protein × 4 kcal/g = 60 kcal.

*Multiple Choice Answers*

| | |
|---|---|
| 1. d | 6. a |
| 2. b | 7. c |
| 3. c | 8. d |
| 4. d | 9. c |
| 5. d | 10. a |

## Chapter 2

*Making It Click Answers*

1. a. ¾ cup (28 g).
b. 110 kcalories.
c. 1 g.
d. 9 kcalories.
e. 9 kcal ÷ 110 kcal = 0.08.
0.08 × 100 = 8%.
f. This cereal derives 8 percent of its kcalories from fat.
g. 2%.
h. A serving of this cereal provides 2 percent of the 65 grams of fat recommended for a 2000-kcalorie diet.
i. Yes.
j. 1.5 g.
k. 25 g.
l. 1.5 g ÷ 25 g = 0.06.
0.06 × 100 = 6%.
m. Corn.
n. Yes.

2. a. Daily Values for 1600-kcalorie diet:
Fat: 1600 kcal × 0.30 = 480 kcal from fat.
480 kcal ÷ 9 kcal/g = 53 g fat.
Saturated fat: 1600 kcal × 0.10 = 160 kcal from saturated fat.
160 kcal ÷ 9 kcal/g = 18 g saturated fat.
Cholesterol: 300 mg.
Carbohydrate: 1600 kcal × 0.60 = 960 kcal from carbohydrate. 960 kcal ÷ 4 kcal/g = 240 g carbohydrate.
Fiber: 1600 kcal × 1000 kcal = 1.6.
1.6 × 11.5 g = 18.4 g fiber.
Protein: 1600 kcal × 0.10 = 160 kcal from protein.
160 kcal ÷ 4 kcal/g = 40 g protein.
Sodium: 2400 mg.
Potassium: 3500 mg.
b.

| | | |
|---|---|---|
| Total fat | 2% | (1 g ÷ 53 g) |
| Saturated fat | 0% | (0 g ÷ 18 g) |
| Cholesterol | 0% | (no calculation needed) |
| Sodium | 10% | (no calculation needed) |
| Total carbohydrate | 10% | (23 g ÷ 240 g) |
| Dietary fiber | 8% | (1.5 g ÷ 18.4 g) |

*Multiple Choice Answers*

| | |
|---|---|
| 1. c | 6. a |
| 2. b | 7. b |
| 3. b | 8. d |
| 4. c | 9. c |
| 5. c | 10. b |

## Chapter 3

*Multiple Choice Answers*

| | |
|---|---|
| 1. c | 4. b |
| 2. c | 5. d |
| 3. a | 6. c |

## Chapter 4

*Making It Click Answers*

1. 0.7 × 2000 total kcal/day = 1400 kcal from carbohydrate/day.
1400 kcal from carbohydrate ÷ 4 kcal/g = 350 g carbohydrate.

This carbohydrate intake is higher than the Daily Value and higher than the "over half" recommendation.

2. 350 g carbohydrate ÷ 2 = 175 g carbohydrate/day.
175 g carbohydrate × 4 kcal/g = 700 kcal from carbohydrate.
700 kcal from carbohydrate ÷ 2000 total kcal/day = 0.35.
0.35 × 100 = 35% kcal from carbohydrate.

This carbohydrate intake is lower than the Daily Value and lower than the "over half" recommendation.

3. 350 g carbohydrate × 2 = 700 g carbohydrate/day.
   700 g carbohydrate ÷ 4 kcal/g = 2800 kcal from carbohydrate.
   2800 kcal from carbohydrate ÷ 6000 total kcal/day = 0.466 (rounded to 0.47)
   0.47 × 100 = 47% kcal from carbohydrate.

This carbohydrate intake is higher than the Daily Value and lower than the "over half" recommendation.

4. 150 g carbohydrate × 4 kcal/g = 600 kcal from carbohydrate.
   600 kcal from carbohydrate ÷ 1000 total kcal/day = 0.60.
   0.60 × 100 = 60% kcal from carbohydrate.

This carbohydrate intake is lower than the Daily Value and higher than the "over half" recommendation.

*Multiple Choice Answers*

| | |
|---|---|
| 1. b | 6. b |
| 2. c | 7. d |
| 3. c | 8. d |
| 4. d | 9. a |
| 5. d | 10. d |

## Chapter 5

*Making It Click Answers*

1. a. Milk A: 8 g fat ÷ 244 g total = 0.03; 0.03 × 100 = 3%.
   Milk B: 5 g fat ÷ 244 g total = 0.02; 0.02 × 100 = 2%.
   Milk C: 3 g fat ÷ 244 g total = 0.01; 0.01 × 100 = 1%.
   Milk D: 0 g fat ÷ 244 g total = 0.00; 0.00 × 100 = 0%.

   b. Milk A: 8 g fat × 9 kcal/g = 72 kcal from fat.
   Milk B: 5 g fat × 9 kcal/g = 45 kcal from fat.
   Milk C: 3 g fat × 9 kcal/g = 27 kcal from fat.
   Milk D: 0 g fat × 9 kcal/g = 0 kcal from fat.

   c. Milk A: (8 g fat × 9 kcal/g) + (8 g prot × 4 kcal/g) + (12 g carb × 4 kcal/g) = 152 kcal.
   Milk B: (5 g fat × 9 kcal/g) + (8 g prot × 4 kcal/g) + (12 g carb × 4 kcal/g) = 125 kcal.
   Milk C: (3 g fat × 9 kcal/g) + (8 g prot × 4 kcal/g) + (12 g carb × 4 kcal/g) = 107 kcal.
   Milk D: (0 g fat × 9 kcal/g) + (8 g prot × 4 kcal/g) + (12 g carb × 4 kcal/g) = 80 kcal.

   d. Milk A: 72 kcal from fat ÷ 152 total kcal = 0.47; 0.47 × 100 = 47%.
   Milk B: 45 kcal from fat ÷ 125 total kcal = 0.36; 0.36 × 100 = 36%.
   Milk C: 27 kcal from fat ÷ 107 total kcal = 0.25; 0.25 × 100 = 25%.
   Milk D: 0 kcal from fat ÷ 80 total kcal = 0.00; 0.00 × 100 = 0%.

   e. Milk A: whole.
   Milk B: reduced-fat, 2%, or less-fat.
   Milk C: low-fat or 1%.
   Milk D: nonfat, fat-free, skim, zero-fat, or no-fat.

2. a. 6.5 g ÷ 65 g = 0.1; 0.1 × 100 = 10%. A Daily Value of 10% means that one serving of this food contributes about 1⁄10 of the day's fat allotment.
   b. 6.5 g × 9 kcal/g = 58.5, rounded to 59 kcal from fat.
   c. (59 kcal from fat ÷ 200 kcal) × 100 = 30% kcalories from fat.

3. (30 g fat ÷ 65 g fat) × 100 = 46% of the Daily Value for fat; this means that almost half of the day's fat allotment would be used in this one dessert.

*Multiple Choice Answers*

| | |
|---|---|
| 1. c | 6. c |
| 2. c | 7. d |
| 3. d | 8. b |
| 4. a | 9. c |
| 5. d | 10. b |

## Chapter 6

*Making It Click Answers*

1. a. The midpoint weight for a woman 5 feet 1 inch tall is 116 lb.
   116 lb ÷ 2.2 lb/kg = 53 kg.
   0.8 g/kg × 53 kg = 42 g protein per day.

   b. The midpoint weight for a man 6 feet 4 inches is 180 lb.
   180 lb ÷ 2.2 lb/kg = 82 kg.
   He is 18 years old, so use 0.9 g/kg. 0.9 g/kg × 82 kg = 74 g protein per day.

2. a. 10% of 3500 kcal = 350 kcal from protein.
   350 kcal ÷ 4 kcal/g = 87.5 g protein per day.
   b. Using the RDA guideline of 0.8 g/kg, an appropriate protein intake for this man would be 59 g protein/day (163 lb ÷ 2.2 lb/kg = 74 kg; 0.8 g/kg × 74 kg = 59 g/day). His intake of 87.5 g protein per day falls between the RDA (59 g) and twice the RDA (2 × 59 g = 118 g), and so meets diet and health recommendations.

*Multiple Choice Answers*

| | |
|---|---|
| 1. a | 6. c |
| 2. c | 7. a |
| 3. a | 8. c |
| 4. b | 9. b |
| 5. c | 10. d |

## Chapter 7

*Multiple Choice Answers*

| | |
|---|---|
| 1. c | 6. c |
| 2. d | 7. d |
| 3. c | 8. c |
| 4. c | 9. b |
| 5. c | 10. b |

K

# Chapter 8

*Making It Click Answers*

1. a. For ages 10–18: (12.2 × 34) + 746 = 1161 kcal/day.
For ages 3–10: (22.5 × 34) + 499 = 1264 kcal/day.
A 10-year-old girl who weighs 75 pounds needs about 1200 kcal/day.
b. For ages 18–30: (15.3 × 68) + 679 = 1719 kcal/day.
For ages 10–18: (17.5 × 68) + 651 = 1841 kcal/day.
An 18-year-old man who weighs 150 pounds needs about 1780 kcal/day.
c. (11.6 × 91) + 879 = 1935 kcal/day.
d. (8.7 × 52) + 829 = 1281 kcal/day.

2. a. 0.045 kcal/lb/min × 142 lb = 6.4 kcal/min.
6.4 kcal/min × 120 min = 768 kcal.
b. 0.103 kcal/lb/min × 142 lb = 14.6 kcal/min.
14.6 kcal/min × 20 min = 292 kcal.
c. 0.032 kcal/lb/min × 142 lb = 4.5 kcal/min.
4.5 kcal/min × 45 min = 202.5 kcal.
d. 0.035 kcal/lb/min × 142 lb = 4.5 kcal/min.
4.5 kcal/min × 60 min = 270 kcal.

3. The infant has the faster BMR (500 kcal/day ÷ 20 lb = 25 kcal/lb/day and 1500 kcal/day ÷ 170 lb = 8.8 kcal/lb/day). Because the infant has a BMR of 25 kcal/lb, whereas the adult has a BMR of 8.8 kcal/lb, the infant's BMR is almost 3 times faster than the adult's based on body weight.

4. a. BMR = (14.7 × 59) + 496 = 1363 kcal/day.
b. With an activity factor of 1.5, her daily energy need is 1.5 × 1365 kcal/day = 2048 kcal/day or, with a weight of 59 kg, her daily energy need is 35 kcal/kg/day × 59 kg = 2065 kcal/day. Either way, she needs about 2050 kcal/day.
c. 1363 kcal ÷ 2050 kcal = 0.66 or 66%.

5. 21 ÷ 0.172 = 122 lb. Yes.

6. a. 120 lb × 0.01 = 1.2 lb/week.
b. 250 lb × 0.01 = 2.5 lb/week.

7. a. 130 lb × 10 kcal/lb = 1300 kcal.
b. 250 lb × 10 kcal/lb = 2500 kcal.

8. a. 150 lb × 0.30 = 45 lb fat.
b. 3500 kcal/lb × 45 lb = 157,500 kcal of body fat.
c. 45 lb fat − 15 lb fat = 30 lb fat.
d. 30 lb fat ÷ 135 lb total body weight = 0.22 or 22% fat.
e. 15 lb × 3500 kcal/lb = 52,500 kcal.
52,500 kcal ÷ 500 kcal/day = 105 days (about 3½ months).

*Multiple Choice Answers*

1. c
2. b
3. c
4. a
5. a
6. c
7. b
8. b
9. a
10. d

# Chapter 9

*Making It Click Answers*

1. a. Three milk shakes provide: 3 × 190 kcal = 570 kcal; 3 × 32 g carbohydrate = 96 g; carbohydrate; 3 × 13 g protein = 39 g protein; and 3 × 1 g fat = 3 g fat.
b. To meet this criteria, the plan needs *at least* an additional 630 kcalories (1200 kcal − 570 kcal = 630 kcal); an additional 5 to 24 grams of protein, depending on the person's RDA based on sex and age (63 g − 39 g = 24 g and 44 g − 39 g = 5 g); an additional 4 grams of carbohydrate (100 g − 96 g = 4 g); and some additional fat.
c. Of course, there are many possible dinners that you could plan. One might be:
Salad made with 1 c lettuce, 1 c chopped tomatoes and onions, ¼ c garbanzo beans, and 2 tbs low-fat dressing
4 oz grilled chicken
1 medium baked potato
1 c summer squash and zucchini
1 c melon cubes
This meal brings the day's totals to 1215 kcalories, 90 g of protein, 192 g of carbohydrate, and 13 g of fat, which meets the goals for kcalories, protein, and carbohydrate. Because the milk shake has been fortified, all vitamin and mineral needs are covered as well. The only possible dietary shortcoming is that the day's percent kcalories from fat is low (only 10%), but because energy and nutrient recommendations have been met and the goal is weight loss, this may be acceptable.
d. This weight-loss plan uses a liquid formula rather than foods, making clients dependent on a special device (the formula) rather than teaching them how to make good choices from the conventional food supply. It provides no information about dropout rates, the long-term success of clients, or weight maintenance after the program ends.

2. a. More than a pound (551 g ÷ 454 g/lb = 1.2 lb).
b. 542 kcal ÷ 551 g = 0.98 kcal/g.
c. More than another whole pound (0.98 kcal/g × 500 kcal = 490 g; 490 g ÷ 454 g/lb = 1.1 lb).
d.

| Item No./Food | Weight (g) | Energy (kcal) |
|---|---|---|
| Original totals: | 551 | 541 |
| Minus: | | |
| #867 Lettuce, 1 c | −56 | −7 |
| Plus: | | |
| #603 Roast beef, 1 oz | +28 | +68 |
| # 39 Cheddar cheese, 1 oz | +28 | +114 |
| Totals: | 551 g | 716 kcal |

e. 716 kcal − 541 kcal = 175 kcal added.
f. 551 g − 551 g = 0 grams added.

*Multiple Choice Answers*

1. a
2. d
3. a
4. d
5. c
6. c
7. d
8. d
9. d
10. b

# Chapter 10

*Making It Click Answers*

1. a. Thiamin: mg.
   Riboflavin: mg.
   Niacin: mg NE.
   Vitamin $B_6$: mg.
   Folate: μg DFE.
   Vitamin $B_{12}$: μg.
   Vitamin C: mg.
   b. A thousand times higher (2 g × 1000 mg/g = 2000 mg; 2000 mg ÷ 2 mg = 1000).
   c. 1 g = 1000 mg; 1 mg = 1000 μg (1000 × 1000 = 1,000,000); 1 million μg = 1 g; 1 tsp = 5 g; 5 × 1,000,000 μg = 5,000,000 μg/tsp; see inside front cover for your RDA based on age and sex. 1 tsp; 2μg.

2. a. She eats 90 g protein. Assume she uses 46 g as protein. This leaves 90 g − 46 g = 44 g protein "leftover."
   44 g protein ÷ 100 = 0.44 g tryptophan.
   0.44 g tryptophan × 1000 = 440 mg tryptophan.
   440 mg tryptophan ÷ 60 = 7.3 mg NE.
   7.3 mg NE ÷ 9 mg niacin = 16.3 mg NE.
   b. Yes.

*Multiple Choice Answers*

1. c
2. a
3. a
4. c
5. d
6. b
7. c
8. d
9. c
10. a

# Chapter 11

*Making It Click Answers*

1. Vitamin A: μg RE. Vitamin D: μg.
   Vitamin E: μg α-TE. Vitamin K: μg.

2. Thiamin: Legumes and grains
   Riboflavin: Milks, grains, and meats
   Niacin: Meats and grains
   Vitamin $B_6$: Meats
   Folate: Legumes and vegetables
   Vitamin $B_{12}$: Meats and milks
   Vitamin C: Vegetables and fruits
   Vitamin A: Vegetables, fruits, and milks
   Vitamin D: Milks
   Vitamin E: Legumes and oils

   Taken together, "the most" groups form the pyramid—grains, vegetables, legumes, fruits, milks, meats, and oils.

*Multiple Choice Answers*

1. c
2. a
3. c
4. b
5. c
6. c
7. a
8. d
9. d
10. c

# Chapter 12

*Making It Click Answers*

1. Calcium: mg. Magnesium: mg. Phosphorus: mg.
   Potassium: mg. Sodium: mg.
2. a.

| Food | Calcium density (mg/kcal) |
|---|---|
| Sardines, 3 oz canned | 325 mg ÷ 176 kcal = 1.85 mg/kcal |
| Milk, nonfat, 1 c | 301 mg ÷ 85 kcal = 3.54 mg/kcal |
| Cheddar cheese, 1 oz | 204 mg ÷ 114 kcal = 1.79 mg/kcal |
| Salmon, 3 oz canned | 182 mg ÷ 118 kcal = 1.54 mg/kcal |
| Broccoli, cooked from fresh, chopped, ½ c | 36 mg ÷ 22 kcal = 1.64 mg/kcal |
| Sweet potato, baked in skin, 1 ea | 32 mg ÷ 140 kcal = 0.2 mg/kcal |
| Cantaloupe melon, ½ | 29 mg ÷ 93 kcal = 0.31 mg/kcal |
| Whole-wheat bread, 1 slice | 21 mg ÷ 64 kcal = 0.33 mg/kcal |
| Apple, 1 medium | 15 mg ÷ 125 kcal = 0.12 mg/kcal |
| Sirloin steak, lean, 3 oz | 9 mg ÷ 171 kcal = 0.05 mg/kcal |

   b. Milk > sardines > cheese > broccoli > salmon.

3. a.

| Food | Calcium in Food (mg) × Absorption rate (%) = Calcium in the Body (mg) |
|---|---|
| Cauliflower, ½ c cooked, fresh | 10 mg × 0.50 = 5 mg (or more) |
| Broccoli, ½ c cooked, fresh | 36 mg × 0.50 = 18 mg (or more) |
| Milk, 1 c 1% low-fat | 300 mg × 0.30 = 90 mg |
| Almonds, 1 oz | 75 mg × 0.20 = 15 mg |
| Spinach, 1 c raw | 55 mg × 0.05 = 3 mg (or less) |

   b. The almonds offer more than twice as much calcium per serving, but an equivalent amount after absorption.
   c. To equal the 300 milligrams provided by milk, a person would need to eat 15 cups of cauliflower (300 mg/c milk ÷ 10 mg/½ c cauliflower = 30 ½ c or 15 c). After considering the better absorption rate of cauliflower, a person would need to eat 9 cups of cauliflower (5 mg/½ c or 10 mg/c; 90 mg ÷ 10 mg/c = 9c) to match the 90 milligrams available to the body from milk after absorption. The better absorption rate reduced the quantity of cauliflower significantly, but that's still a lot of cauliflower.

*Multiple Choice Answers*

1. d
2. a
3. b
4. a
5. c
6. b
7. d
8. b
9. b
10. a

# Chapter 13

*Making It Click Answers*

1. a. Sirloin steak > broccoli > green peas > bread > pork chop > cantaloupe > sweet potato > carrots > apple > cheese > milk.

   b.

| Food | Iron Density (mg/kcal) |
|---|---|
| Milk, nonfat, 1 c | 0.10 mg ÷ 85 kcal = 0.0012 mg/kcal |
| Cheddar cheese, 1 oz | 0.19 mg ÷ 114 kcal = 0.0017 mg/kcal |
| Broccoli, cooked from fresh, chopped, 1 c | 1.31 mg ÷ 44 kcal = 0.0298 mg/kcal |
| Sweet potato, baked in skin, 1 ea | 0.51 mg ÷ 117 kcal = 0.0044 mg/kcal |
| Cantaloupe melon, ½ | 0.56 mg ÷ 93 kcal = 0.0060 mg/kcal |
| Carrots, from fresh, ½ c | 0.48 mg ÷ 35 kcal = 0.0137 mg/kcal |
| Whole-wheat bread, 1 slice | 0.87 mg ÷ 64 kcal = 0.0136 mg/kcal |
| Green peas, cooked from frozen, ½ c | 1.26 mg ÷ 62 kcal = 0.0203 mg/kcal |
| Apple, medium | 0.38 mg ÷ 125 kcal = 0.0030 mg/kcal |
| Sirloin steak, lean, 4 oz | 3.81 mg ÷ 228 kcal = 0.0167 mg/kcal |
| Pork chop, lean broiled, 1 ea | 0.66 mg ÷ 166 kcal = 0.0040 mg/kcal |

   Broccoli > green peas > sirloin steak > carrots > bread > cantaloupe > sweet potato > pork chop > apple > cheese > milk.

   c. Broccoli, green peas, and carrots are all higher on the per-kcalorie list.

   d. They are all vegetables.

2. a.

| Food | Iron (mg) | Vitamin C (mg) |
|---|---|---|
| Sirloin steak, 4 oz | 3.81 | 0 |
| Green peas, cooked from frozen, ½ c | 1.26 | 8 |
| Brown rice, cooked, 1 c | 0.82 | 0 |
| Iced tea, instant, sweetened, 1 c | 0.05 | 0 |

   b. Step 1: 3.81 mg.
   Step 2: 3.81 mg × 0.40 = 1.5 mg heme iron.
   Step 3: 1.26 + 0.82 + 0.05 = 2.13 mg.
   Step 4: 2.13 mg + (3.81 mg × 0.60) = 4.42 mg nonheme iron.
   Step 5: low vitamin C.
   Step 6: 4 oz, high MFP.
   For heme iron, 1.5 mg × 0.23 = 0.35 mg heme iron absorbed.
   For nonheme iron, availability of nonheme iron was high (8%); 4.42 × 0.08 = 0.35 mg nonheme iron absorbed.
   Total: 0.35 mg heme + 0.35 mg nonheme = 0.7 mg iron absorbed.

   c. Yes. 0.7 mg absorbed ÷ 5.94 mg eaten × 100 = 12% absorbed.

   d. Yes—the tannic acid in tea may interfere with the absorption of iron.

   e. 0.7 mg absorbed/meal × 3 meals = 2.1 mg absorbed. According to the RDA calculation, a woman needs to absorb 1.5 mg per day, so she will meet her iron RDA.

*Multiple Choice Answers*

1. b
2. b
3. c
4. c
5. d
6. d
7. c
8. b
9. a
10. c

# Chapter 14

*Multiple Choice Answers*

1. c
2. b
3. a
4. d
5. c
6. c
7. d
8. b
9. a
10. d

# Chapter 15

*Multiple Choice Answers*

1. c
2. c
3. c
4. b
5. b
6. c
7. b
8. b
9. c
10. c

# Chapter 16

*Multiple Choice Answers*

1. b
2. a
3. c
4. a
5. a
6. d
7. c
8. c
9. a
10. d

# Chapter 17

*Multiple Choice Answers*

1. c
2. d
3. c
4. d
5. b
6. b
7. c
8. d
9. c
10. b

K

## Chapter 18

*Multiple Choice Answers*

| | |
|---|---|
| 1. d | 6. c |
| 2. c | 7. d |
| 3. b | 8. a |
| 4. b | 9. b |
| 5. a | 10. d |

## Chapter 19

*Multiple Choice Answers*

| | |
|---|---|
| 1. b | 6. d |
| 2. d | 7. a |
| 3. d | 8. a |
| 4. b | 9. a |
| 5. c | 10. d |

## Chapter 20

*Multiple Choice Answers*

| | |
|---|---|
| 1. c | 6. d |
| 2. a | 7. c |
| 3. d | 8. d |
| 4. c | 9. a |
| 5. b | 10. a |

# Glossary

Many medical terms have their origins in Latin or Greek. By learning a few common derivations, you can glean the meaning of words you have never heard of before. For example, once you know that "hyper" means above normal, "glyc" means glucose, and "emia" means blood, you can easily determine that "hyperglycemia" means high blood glucose. The following derivations will help you to learn many terms presented in this glossary.

## General

*a-* or *an-*: not or without
*ana-*: up
*anti-*: against
*cata-*: down
*co-*: with, together
*di-*: two
*dia-*: through
*dys-* or *mal-*: bad
*endo-*: inside or within
*exo-* or *extra-*: outside
*-gen*: producing
*homeo-*: the same
*hyper-*: over, above normal, excessive
*hypo-*: below normal, under, beneath
*in-*: in, inside, within, or not
*intra-*: within
*inter-*: between, in the midst
*-itis*: infection or inflammation
*-lysis*: break
*macro-*: large
*micro-*: tiny
*mono-*: one
*neo-*: new
*-osis*: condition
*peri-*: around
*poly-*: many
*pre-* or *pro-*: before
*tri-*: three

## Body

*angio-, vaso-,* or *vascular:* blood vessel
*arterio-*: artery
*cardiac* or *cardio-*: heart
*cyto-* or *-cyte*: cell
*enteron*: intestine
*gastro-*: stomach
*hemo-* or *-emia*: blood
*hepatic*: liver
*myo-*: muscle
*osteo-*: bone
*pulmo-*: lung
*renal*: kidney
*ure-* or *-uria*: urine
*vena*: vein

## Chemistry

*-al*: aldehyde
*-ase*: enzyme
*-ate*: salt
*glyc-* or *gluc-*: glucose
*hydro-* or *hydrate*: water
*lipo-*: lipid
*-ol*: alcohol or phenol
*-ose*: sugar
*saccharide*: sugar

A

**absorption:** the passage of nutrients from the GI tract into either the blood or the lymph.

**Acceptable Daily Intake:** see *ADI*

**accredited:** approved; in the case of medical centers or universities, certified by an agency recognized by the U.S. Department of Education.

**acesulfame** (AY-sul-fame) **potassium:** an artificial sweetener composed of an organic salt that tastes 200 times as sweet as sucrose and provides 0 kcalories per gram; also known as acesulfame-K because K is the chemical symbol for potassium.

**acetaldehyde** (ass-et-AL-duh-hide): an intermediate in alcohol metabolism.

**acetyl CoA** (ASS-eh-teel, or ah-SEET-il, coh-AY): a 2-carbon compound (*acetate*, or *acetic acid*) to which a molecule of CoA is attached.

**acid controllers:** drugs used to prevent or relieve indigestion by suppressing production of acid in the stomach; also called *H2 blockers*. Common brands include *Pepcid AC, Tagamet HB, Zantac 75,* and *Axid AR.*

**acid-base balance:** the equilibrium in the body between acid and base concentrations.

**acidosis** (assi-DOE-sis): above-normal acidity in the blood and body fluids.

**acids:** compounds that release hydrogen ions in a solution.

**acne:** a chronic inflammation of the skin's follicles and oil-producing glands, which leads to an accumulation of oils inside the ducts that surround hairs; usually associated with the maturation of young adults.

**acquired immune deficiency syndrome (AIDS):** the end stage of HIV infection, in which severe complications are manifested.

**acupuncture** (AK-you-PUNK-cher): a technique that involves piercing the skin with long thin needles at specific anatomical points to relieve pain or illness. Acupuncture sometimes uses heat, pressure, friction, suction, or electromagnetic energy to stimulate the points.

**acute PEM:** protein-energy malnutrition caused by recent severe food restriction; characterized in children by thinness for height (wasting).

**ADA:** see *American Dietetic Association.*

**adaptive thermogenesis:** adjustments in energy expenditure related to changes in environment such as cold and to physiological events such as overfeeding, trauma, and changes in hormone status.

**additives:** substances not normally consumed as foods but added to food either intentionally or by accident.

**adequacy (dietary):** providing all the essential nutrients, fiber, and energy in amounts sufficient to maintain health.

**Adequate Intake (AI):** the average amount of a nutrient that appears sufficient to maintain a specified criterion; a value used as a guide for nutrient intake when an RDA cannot be determined.

**ADH:** see *antidiuretic hormone.*

**ADI (Acceptable Daily Intake):** the amount of a sweetener that individuals can safely consume each day over the course of a lifetime without adverse effect. It includes a 100-fold safety factor.

**adipose** (ADD-ih-poce) **tissue:** the body's fat tissue consists of masses of fat-storing cells.

**adolescence:** the period from the beginning of puberty until maturity.

**adrenal glands:** glands adjacent to, and just above, each kidney.

**adrenocorticotropin:** so named because it stimulates *(trope)* the adrenal cortex. The adrenal gland, like the pituitary, has two parts, in this case an outer portion *(cortex)* and an inner core *(medulla).*

**adverse reactions:** unusual responses to food (including intolerances and allergies).

**aerobic** (air-ROE-bic): requiring oxygen.

**aflatoxin:** a potent cancer-causing toxin produced by the mold *Aspergillus flavus* that infects grains and peanuts. The USDA tests grains and peanuts grown in the United States for aflatoxin contamination.

**agribusiness:** agriculture practiced on a massive scale by large corporations owning vast acreages and employing intensive technological, fuel, and chemical inputs.

**AI:** see *Adequate Intake.*

**AIDS:** see *acquired immune deficiency syndrome.*

**alcohol:** a class of organic compounds containing hydroxyl (OH) groups.

**alcohol dehydrogenase** (dee-high-DROJ-eh-nayz): an enzyme active in the stomach and the liver that converts ethanol to acetaldehyde.

**alcoholism:** disease characterized by loss of control over drinking and dependence on alcohol that harms health, family and social relations, and ability to work.

**aldosterone** (al-DOS-ter-own): a hormone secreted by the adrenal glands that stimulates the reabsorption of sodium by the kidneys; aldosterone also regulates chloride and potassium concentrations.

**alitame** (AL-ih-tame): an artificial sweetener composed of two amino acids (alanine and aspartic acid) that tastes 2000 times as sweet as sucrose. Alitame provides 4 kcalories per gram, as does protein, but because so little is used, its energy contribution is negligible. FDA approval pending.

**alkalosis** (alka-LOE-sis): above-normal alkalinity (base) in the blood and body fluids.

**alpha-lactalbumin** (lact-AL-byoo-min): the chief protein in human breast milk.

**alpha-tocopherol** (tuh-KOFF-er-ol): the most biologically active vitamin E compound.

**alternative therapies:** approaches to medical diagnosis and treatment that are not fully accepted by the established medical community; as such, they are not widely taught at U.S. medical schools or practiced in U.S. hospitals; also called *adjunctive, unconventional,* or *unorthodox* therapies.

**amenorrhea** (ay-MEN-oh-REE-ah): the absence of or cessation of menstruation. *Primary amenorrhea* is menarche delayed beyond 16 years of age. *Secondary amenorrhea* is the absence of three to six consecutive menstrual cycles.

**American Dietetic Association (ADA):** the professional organization of dietitians in the United States. The Canadian equivalent is Dietitians of Canada, which operates similarly.

**amino** (a-MEEN-oh) **acids:** building blocks of proteins; each contains an amino group, an acid group, a hydrogen atom, and a distinctive side group, all attached to a central carbon atom.

**amino acid scoring:** a method of evaluating protein quality by comparing a test protein's amino acid pattern with that of a reference protein; sometimes called *chemical scoring.*

**ammonia:** a compound with the chemical formula $NH_3$; produced during the deamination of amino acids.

**amniotic** (am-nee-OTT-ic) **sac:** the "bag of waters" in the uterus, in which the fetus floats.

**amphetamine** (am-FET-ah-mean): a central nervous system stimulant used to treat narcolepsy and some types of depression. Use of amphetamines to control appetite in the treatment of obesity has not proved effective, and prolonged use causes dependency. Amphetamines are marketed under the trade names *Benzedrine* and *Amphedrine.*

**amylase** (AM-ih-lace): an enzyme that hydrolyzes amylose (a form of starch). Amylase is a carbohydrase, an enzyme that breaks down carbohydrates.

**anabolic steroids:** drugs related to the male sex hormone, testosterone, that stimulate the development of lean body mass.

**anabolism** (an-ABB-o-lism): reactions in which small molecules are put together to build larger ones. Anabolic reactions require energy.

**anaerobic** (AN-air-ROE-bic): not requiring oxygen.

**anemia:** literally, "too little blood." Anemia is any condition in which too few red blood cells are present, or the red blood cells are immature (and therefore large) or too small or contain too little hemoglobin to carry the normal amount of oxygen to the tissues. It is not a disease itself but can be a symptom of many different disease conditions, including many nutrient deficiencies, bleeding, excessive red blood cell destruction, and defective red blood cell formation.

**angina** (an-JYE-nah or AN-ji-nah): a painful feeling of tightness or pressure in and around the heart, often radiating to the back, neck, and arms; caused by a lack of oxygen to an area of heart muscle.

**angiotensin:** a hormone involved in blood pressure regulation. Its precursor protein is called *angiotensinogen*.

**anions** (AN-eye-uns): negatively charged ions.

**anorexia** (an-oh-RECK-see-ah) **nervosa:** an eating disorder characterized by a refusal to maintain a minimally normal body weight and a distortion in perception of body shape and weight; most commonly seen in teenage girls and young women.

**antacids:** drugs used to relieve indigestion by neutralizing acid in the stomach. Common brands include *Alka-Seltzer, Maalox, Rolaids,* and *Tums*.

**antagonist:** a competing factor that counteracts the action of another factor. When a drug displaces a vitamin from its site of action, the drug renders the vitamin ineffective and thus acts as a vitamin antagonist.

**anthropometric** (AN-throw-poe-MET-rick): relating to measurement of the physical characteristics of the body, such as height and weight.

**antibodies:** large proteins of the blood and body fluids, produced by the immune system in response to the invasion of the body by foreign molecules (usually proteins called antigens); antibodies combine with and inactivate the foreign invaders, thus protecting the body.

**antidiuretic hormone (ADH):** a hormone produced by the pituitary gland in response to dehydration (or a high sodium concentration in the blood); it stimulates the kidneys to reabsorb more water and therefore to excrete less. (This ADH should not be confused with the enzyme alcohol dehydrogenase, which is sometimes also abbreviated ADH.)

**antigen:** a substance that elicits the formation of antibodies or an inflammation reaction from the immune system. A bacterium, a virus, a toxin, and a protein in food that causes allergy are all examples of antigens.

**antimicrobial agents:** preservatives that prevent microorganisms from growing.

**antioxidant:** a compound that protects others from oxidation by being oxidized itself. An antioxidant donates electrons to another substance; that substance becomes reduced as the antioxidant simultaneously becomes oxidized. The food industry uses antioxidants as additives to prevent rancidity of fats in foods and other damage to food caused by oxygen.

**antiscorbutic** (AN-tee-skor-BUE-tik) **factor:** the original name for vitamin C.

**anus** (AY-nus): the terminal sphincter of the GI tract.

**appendix:** a narrow blind sac extending from the beginning of the colon that stores lymph cells.

**appetite:** the integrated response to the sight, smell, thought, or taste of food that initiates or delays eating.

**arachidonic** (a-RACK-ih-DON-ic) **acid:** an omega-6 polyunsaturated fatty acid with 20 carbons and four double bonds (20:4); synthesized from linoleic acid.

**aroma therapy:** a technique that uses oil extracts from plants and flowers (usually applied by massage or baths) to enhance physical, psychological, and spiritual health.

**arteries:** vessels that carry blood away from the heart.

**artesian water:** water that is drawn from a well that taps a confined aquifer in which the water level stands above the natural water table.

**arthritis:** inflammation of a joint, usually accompanied by pain, swelling, and structural changes.

**artificial colors:** certified food colors added to enhance appearance. (*Certified* means approved by the FDA.)

**artificial flavors, flavor enhancers:** chemicals that mimic natural flavors and those that enhance flavor.

**artificial sweeteners:** sugar substitutes that provide negligible, if any, energy; sometimes called *nonnutritive sweeteners*.

**ascorbic acid:** one of the two active forms of vitamin C. Many people refer to vitamin C by this name.

**-ase** (ACE): a word ending denoting an enzyme. Enzymes are often identified by the place they come from and the compounds they work on; *gastric lipase,* for example, is a stomach enzyme that acts on lipids, whereas *pancreatic lipase* comes from the pancreas (and also works on lipids).

**aspartame** (ah-SPAR-tame or ASS-par-tame): an artificial sweetener composed of two amino acids (phenylalanine and aspartic acid) that tastes 160 to 220 times as sweet as sucrose. Aspartame provides 4 kcalories per gram, as does protein, but because so little is used, its energy contribution is negligible. In powdered form it is sometimes mixed with lactose, however, so a 1-gram packet may provide 4 kcalories.

**atherosclerosis** (ath-er-oh-scler-OH-sis): a type of artery disease characterized by accumulations of lipid-containing material on the inner walls of the arteries.

**atom:** the smallest component of an element that has all of the properties of the element.

**ATP** or **adenosine** (ah-DEN-oh-seen) **triphosphate** (try-FOS-fate): a common high-energy compound composed of a purine (adenine), a sugar (ribose), and three phosphate groups.

**atrophic gastritis:** chronic inflammation of the stomach accompanied by a diminished size and functioning of the mucosa and glands.

**atrophy** (AT-ro-fee): of muscles, becoming smaller; a decrease in size because of disuse, undernutrition, or wasting diseases.

**autonomic nervous system:** the division of the nervous system that controls the body's automatic responses. Its two branches are the *sympathetic* branch, which helps the body respond to stressors from the outside environment, and the *parasympathetic* branch, which regulates normal body activities between stressful times.

**ayurveda** (EYE-your-VAY-dah): a traditional Hindu system of improving health by using herbs, diet, meditation, massage, and yoga to stimulate the body to make its own natural drugs.

## B

**B-cells:** lymphocytes that produce antibodies. *B* stands for bursa, an organ in the chicken associated with the first identification of the B-cells.

**balance (dietary):** providing foods of a number of types in proportion to each other, such that foods rich in some nutrients do not crowd out diet foods that are rich in other nutrients.

**basal metabolic rate (BMR):** the rate of energy use for metabolism under basal conditions, usually expressed as kcalories per kilogram body weight per hour.

**basal metabolism:** the energy needed to maintain life when a body is at complete rest after a 12-hour fast (to exclude the thermic effect of the previous meal).

**bases:** compounds that accept hydrogen ions in a solution.

**beer:** an alcoholic beverage brewed by fermenting malt and hops.

**behavior modification:** the changing of behavior by the manipulation of *antecedents* (cues or environmental factors that trigger behavior), the *behavior* itself, and *consequences* (the penalties or rewards attached to behavior).

**belch:** the expulsion of gas from the stomach through the mouth.

**beriberi:** the thiamin-deficiency disease.

**beta-carotene** (BAY-tah KARE-oh-teen): an orange pigment and vitamin A precursor found in plants.

**BGH:** see *bovine growth hormone.*

**BHA** and **BHT:** preservatives commonly used to slow the development of off-flavors, odors, and color changes caused by oxidation.

**bicarbonate:** an alkaline secretion of the pancreas, part of the pancreatic juice. (Bicarbonate also occurs widely in all cell fluids.)

**bifidus** (BIFF-id-us, by-FEED-us) **factors:** factors in colostrum and breast milk that favor the growth of the "friendly" bacterium *Lactobacillus* (lack-toh-ba-SILL-us) *bifidus* in the infant's intestinal tract, so that other, less desirable intestinal inhabitants will not flourish.

**bile:** an emulsifier that prepares fats and oils for digestion; an exocrine secretion made by the liver, stored in the gallbladder, and released into the small intestine when needed.

**binders:** chemical compounds occurring in foods that can combine with nutrients (especially minerals) to form complexes the body cannot absorb. Examples of such binders include **phytic** (FIGHT-ic) **acid** and *oxalic* (ox-AL-ic) *acid.*

**bioaccumulation:** the accumulation of contaminants in the flesh of animals high on the food chain.

**bioavailability:** the rate and extent to which a nutrient is absorbed and used.

**bioelectrical impedance** (im-PEE-dans): a method for estimating body fat using low-intensity electrical current.

**bioelectromagnetic medical applications:** the use of electrical energy, magnetic energy, or both to stimulate bone repair, wound healing, and tissue regeneration.

**biofeedback:** the use of special devices to convey information about heart rate, blood pressure, skin temperature, muscle relaxation, and the like to enable a person to learn how to consciously control these medically important functions.

**biofield therapeutics:** a manual healing method that directs a healing force from an outside source (commonly God or another supernatural being) through the practitioner and into the client's body; commonly known as "laying on of hands."

**biological value (BV):** the amount of protein nitrogen that is retained for growth and maintenance, expressed as a percentage of the protein nitrogen that has been digested and absorbed; a measure of protein quality.

**biotechnology:** the use of biological systems or organisms to create or modify products; also called *biogenetic engineering.*

**biotin** (BY-oh-tin): a B vitamin that functions as a coenzyme in the metabolism of carbohydrates and fats.

**bleaching agents:** substances used to whiten foods such as flour and cheese.

**blind experiment:** an experiment in which the subjects do not know whether they are members of the experimental group or the control group.

**blood lipid profile:** results of blood tests that reveal a person's total cholesterol, triglycerides, and various lipoproteins.

**BMI:** see *body mass index.*

**body composition:** the proportions of muscle, bone, fat, and other tissues that make up a person's total body weight.

**body mass index (BMI):** an index of a person's weight in relation to height, determined by dividing the weight (in kilograms) by the square of the height (in meters).

**bolus** (BOH-lus): a portion; with respect to food, the amount swallowed at one time.

**bomb calorimeter** (KAL-oh-RIM-eh-ter): an instrument that measures the *heat* energy released when foods are burned, thus providing an estimate of the potential energy of foods.

**bone density:** a measure of bone strength. When minerals fill the bone matrix (making it dense), they give it strength.

**botulism** (BOT-chew-lism): an often fatal food-borne illness caused by the ingestion of foods containing a toxin produced by bacteria that grow without oxygen.

**bovine growth hormone (BGH):** a hormone produced naturally in the pituitary gland of a cow that promotes growth and milk production; now produced for agricultural use by bacteria.

**bran:** the protective coating around the kernel similar in function to the shell of a nut; rich in nutrients and fiber.

**branched-chain amino acids:** the amino acids leucine, isoleucine, and valine, which are present in large amounts in skeletal muscle tissue; falsely promoted as fuel for exercising muscles.

**brown adipose tissue:** masses of specialized fat cells packed with pigmented mitochondria that produce heat instead of ATP.

**brown sugar:** refined white sugar crystals to which manufacturers have added molasses syrup with natural flavor and color; 91 to 96 percent pure sucrose.

**buffers:** compounds that help keep a solution's acidity or alkalinity constant.

**bulimia** (byoo-LEEM-ee-ah) **nervosa:** an eating disorder characterized by repeated episodes of binge eating usually followed by self-induced vomiting, misuse of laxatives or diuretics, fasting, or excessive exercise.

**BV:** see *biological value.*

## C

**caffeine:** a natural stimulant found in many common foods and beverages, including coffee, tea, and chocolate; may enhance endurance by stimulating fatty acid release but also causes fluid losses. Overdoses cause headaches, trembling, rapid heart rate, and other undesirable side effects.

**calcitonin** (KAL-see-TOE-nin): a hormone secreted by the thyroid gland that regulates (tones) calcium metabolism.

**calcium:** the most abundant mineral in the body; found primarily in the body's bones and teeth.

**calcium rigor:** hardness or stiffness of the muscles caused by high blood calcium concentrations.

**calcium tetany** (TET-ah-nee): intermittent spasm of the extremities due to nervous and muscular excitability caused by low blood calcium concentrations.

**calcium-binding protein:** a protein in the intestinal cells, made with the help of vitamin D, that facilitates calcium absorption.

**calmodulin** (cal-MOD-you-lin): an inactive protein that becomes active when bound to calcium; then it becomes a messenger that tells other proteins what to do. The system serves as an interpreter for hormone- and nerve-mediated messages arriving at cells.

**calorie:** a unit by which energy is measured. Food energy is measured in *kilocalories* (1000 calories equal 1 kilocalorie), abbreviated *kcalories* or *kcal.* A capitalized version is also sometimes used: *Calories.* One kcalorie is the amount of heat necessary to raise the temperature of 1 kilogram (kg) of water 1°C.

**cancers:** diseases that result from the unchecked growth of malignant tumors.

**capillaries** (CAP-ill-aries): small vessels that branch from an artery. Capillaries connect arteries to veins. Exchange of oxygen, nutrients, and waste materials takes place across capillary walls.

**caprenin:** a fat-based fat replacer made from a triglyceride with poorly absorbed fatty acids; provides 5 kcalories per gram.

**carbohydrase** (KAR-boe-HIGH-drase): an enzyme that hydrolyzes carbohydrates.

**carbohydrates:** compounds composed of carbon, oxygen, and hydrogen arranged as monosaccharides or multiples of monosaccharides. Most, but not all, carbohydrates have a ratio of one carbon molecule to one water molecule: $(CH_2O)_n$.

**carbonic acid:** a compound with the formula $H_2CO_3$ that results from the combination of carbon dioxide ($CO_2$) and water ($H_2O$); of particular importance in the body's buffer system.

**cardiac output:** the volume of blood discharged by the heart each minute; it is determined by multiplying the stroke volume by the heart rate.

**cardiac sphincter** (CARD-ee-ack SFINK-ter): the sphincter muscle at the junction between the esophagus and the stomach; also called the *lower esophageal sphincter* or the *gastroesophageal sphincter.*

**cardiorespiratory conditioning:** improvements in heart and lung function and increased blood volume, brought about by aerobic training.

**cardiorespiratory endurance:** the ability to perform large-muscle, dynamic exercise of moderate-to-high intensity for prolonged periods.

**cardiovascular disease (CVD):** a general term for all diseases of the heart and blood vessels. Atherosclerosis is the main cause of CVD. When the arteries that carry blood to the heart muscle become blocked, the heart suffers damage known as *coronary heart disease (CHD).*

**carnitine** (CAR-neh-teen): a nonprotein amino acid made in the body from lysine that helps transport fatty acids across the mitochondrial membrane. Carnitine supposedly "burns" fat and spares glycogen during endurance events, but in reality it does neither.

**carotene** (KARE-oh-teen): a vitamin A precursor found in plants; an orange pigment.

**carotenoids** (kah-ROT-eh-noyds): pigments commonly found in plants and animals, some of which have provitamin A activity. Carotenoids are among the best-known *phytochemicals.*

**carpal tunnel syndrome:** a pinched nerve at the wrist, causing pain or numbness in the hand.

**cartilage therapy:** the use of cleaned and powdered connective tissue, such as collagen, to improve health.

**casein** (CAY-seen): the chief protein in cow's milk.

**cash crops:** crops grown for sale or export, as opposed to food crops grown for local consumption.

**catabolism** (ca-TAB-o-lism): reactions in which large molecules are broken down to smaller ones. Catabolic reactions usually release energy.

**catalyst** (CAT-uh-list): a compound that facilitates chemical reactions without itself being changed in the process.

**cataracts** (KAT-ah-rakts):thickenings of the eye lenses that impair vision and can lead to blindness.

**cathartic** (ka-THAR-tik): a strong laxative.

**cations** (CAT-eye-uns): positively charged ions.

**CD4+ T-lymphocyte:** a type of circulating white blood cell that contains the CD4+ protein on its surface and is a necessary component of the immune system.

**CDC (Centers for Disease Control):** a branch of the Department of Health and Human Services that is responsible for, among other things, monitoring foodborne diseases.

**cell:** the basic unit of life, of which all living things are composed. Every cell is surrounded by a membrane and contains cytoplasm, within which are organelles and a nucleus; the cell nucleus contains chromosomes.

**cell membrane:** the membrane that surrounds the cell and encloses its contents; made primarily of lipid and protein.

**cellulite** (SELL-you-light or SELL-you-leet): supposedly, a lumpy form of fat; actually, a fraud. Fatty areas of the body may appear lumpy when the strands of connective tissue that attach the skin to underlying muscles pull tight where the fat is thick. The fat itself is the same as fat anywhere else in the body.

**central nervous system:** the central part of the nervous system, the brain and spinal cord.

**central obesity:** excess fat around the trunk of the body; also called *abdominal fat* or *upper-body fat.*

**cerebral cortex:** the outer surface of the cerebrum (the largest part of the brain).

**certification:** the process in which a private laboratory inspects shipments of a product for selected chemicals and then, if the product is free of violative levels of those chemicals, issues a guarantee to that effect.

**cervix:** the circular muscle that guards the opening of the uterus.

**cesarean delivery:** a surgically assisted birth involving removal of the fetus by an incision into the uterus, usually by way of the abdominal wall.

**chelating** (KEY-late-ing) **agents:** molecules that bind other molecules. As additives, they prevent discoloration, flavor changes, and rancidity that might occur during processing.

**chelation** (key-LAY-shun) **therapy:** the use of ethylene diamine tetraacetic acid (EDTA) to bind with metallic ions, thus healing the body by removing toxic metals.

**chiropractic** (KYE-roe-PRAK-tik): a manual healing method of manipulating vertebrae to relieve musculoskeletal pain suspected of causing problems with internal organs.

**chloride** (KLO-ride): the major anion in the extracellular fluids of the body. Chloride is the ionic form of chlorine, $Cl^-$.

**chlorophyll** (KLO-row-fil): the green pigment of plants, which absorbs light and transfers the energy to other molecules, thereby initiating photosynthesis.

**cholecystokinin** (coal-ee-sis-toe-KINE-in), or **CCK:** a hormone produced by cells of the intestinal wall. Target organ: the gallbladder. Response: release of bile and slowing of GI motility.

**cholesterol** (koh-LESS-ter-ol): one of the sterols.

**choline** (KOH-leen): a nitrogen-containing compound found in foods and made in the body from the amino acid methionine. Choline is used to make the phospholipid lecithin and the neurotransmitter acetylcholine.

**chromium** (CROW-mee-um) **picolinate:** a trace mineral supplement; falsely promoted as building muscle, enhancing energy, and burning fat.

**chromosomes:** a set of structures within the nucleus of every cell that contains the cell's genetic material, DNA, associated with other materials (primarily proteins).

**chronic diseases:** degenerative diseases characterized by deterioration of the body organs; also called chronic, *noncommunicable diseases (NCD).* Examples include heart disease, cancer, and diabetes.

**chronic PEM:** protein-energy malnutrition caused by long-term food deprivation; characterized in children by short height for age (stunting).

**chronological age:** a person's age in years from his or her date of birth.

**chylomicrons** (kye-lo-MY-cronz): the class of lipoproteins that transport lipids from the intestinal cells into the body.

**chyme** (KIME): the semiliquid mass of partly digested food expelled by the stomach into the duodenum.

**cirrhosis** (seer-OH-sis): advanced liver disease in which liver cells turn orange, die, and harden, permanently losing their function; often associated with alcoholism.

**clinically severe obesity:** a BMI of 40 or greater or 100 lb or more overweight for an average adult. A less preferred term used to describe the same condition is *morbid obesity.*

**CoA** (coh-AY): coenzyme A; the coenzyme derived from the B vitamin pantothenic acid and central to the energy metabolism of nutrients.

**coenzymes:** small organic molecules that work with enzymes to facilitate the enzymes' activity. Many coenzymes have B vitamins as part of their structures.

**cofactor:** a mineral element that, like a coenzyme, works with an enzyme to facilitate a chemical reaction. The cofactor maintains the structural integrity of the enzyme and may also facilitate the enzyme's catalytic activity.

**colitis** (ko-LYE-tis): inflammation of the colon.

**collagen** (KOL-ah-jen): the protein material from which connective tissues such as scars, tendons, ligaments, and the foundations of bones and teeth are made.

**colonic irrigation:** the popular, but potentially harmful practice of "washing" the large intestine with a powerful enema machine.

**colostrum** (ko-LAHS-trum): a milklike secretion from the breast, present during the first day or so after delivery before milk appears; rich in protective factors.

**complementary proteins:** two or more proteins whose amino acid assortments complement each other in such a way that the essential amino acids missing from one are supplied by the other.

**complete protein:** a dietary protein containing all the essential amino acids in relatively the same amounts that human beings require; it may also contain nonessential amino acids.

**complex carbohydrates** (starches and fibers): polysaccharides composed of straight or branched chains of monosaccharides.

**compound:** a substance composed of two or more different atoms—for example, water ($H_2O$).

**conception:** the union of the male sperm and the female ovum; fertilization.

**condensation:** a chemical reaction in which two reactants combine to yield a larger product.

**conditionally essential amino acid:** an amino acid that is normally nonessential, but must be supplied by the diet in special circumstances when the need for it exceeds the body's ability to produce it.

**conditioning:** the physical effect of *training;* improved flexibility, strength, and endurance.

**cones:** the cells of the retina that respond to bright light and are responsible for color vision.

**confectioners' sugar:** finely powdered sucrose; 99.9 percent pure.

**congregate meals:** nutrition programs that provide food for the elderly in a conveniently located setting such as a community center.

**constipation:** the condition of having infrequent or difficult bowel movements.

**contaminant:** a substance that makes a food impure and unsuitable for ingestion.

**contamination iron:** iron found in foods as the result of contamination by inorganic iron salts from iron cookware, iron-containing soils, and the like.

**control group:** a group of individuals similar in all possible respects to the experimental group except for the treatment. Ideally, the control group receives a placebo while the experimental group receives a real treatment.

**cool-down:** 5 to 10 minutes of light activity, such as walking or stretching, following a vigorous workout to return the body's core gradually to near-normal temperature.

**Cori cycle:** the path from muscle glycogen to glucose to pyruvate to lactic acid (which travels to the liver) to glucose (which can travel back to the muscle) to glycogen; named after the scientist who elucidated this pathway.

**corn sweeteners:** corn syrup and sugars derived from corn.

**corn syrup:** a syrup produced by the action of enzymes on cornstarch; contains mostly glucose. See also *high-fructose corn syrup (HFCS).*

**cornea** (KOR-nee-uh): the transparent membrane covering the outside of the eye.

**correlation** (CORE-ee-LAY-shun): the simultaneous increase, decrease, or change in two variables. If A increases as B increases, or if A decreases as B decreases, the correlation is positive. (This does not mean that A causes B or vice versa.) If A increases as B decreases, or if A decreases as B increases, the correlation is negative. (This does not mean that A prevents B or vice versa.) Some third factor may account for both A and B.

**correspondence school:** a school that offers courses and degrees by mail. Some correspondence schools are accredited; others are *diploma mills.*

**cortical bone:** the very dense bone tissue that forms the outer shell surrounding trabecular bone and comprises the shaft of a long bone.

**coupled reactions:** pairs of chemical reactions in which energy released from the breakdown of one compound is used to create a bond in the formation of another compound.

**covert** (KOH-vert): hidden, as if under covers.

**CP, creatine phosphate** (also called **phosphocreatine**): a high-energy compound in muscle cells that acts as a reservoir of energy that can maintain a steady supply of ATP; CP provides the energy for short bursts of activity.

**creatine** (KREE-ah-tin): a nitrogen-containing compound that combines with phosphate to form the high-energy compound creatine phosphate (or phosphocreatine) in muscles. Claims that creatine enhances energy use and muscle strength need further confirmation.

**cretinism** (CREE-tin-ism): an iodine-deficiency disease characterized by mental and physical retardation.

**critical periods:** finite periods during development in which certain events may occur that will have irreversible effects on later developmental stages. In a body organ, a critical period is usually a period of rapid cell division.

**crypts** (KRIPTS): tubular glands that lie between the intestinal villi and secrete intestinal juices into the small intestine.

**cuisine** (kwi-ZEEN): style of cooking or preparing food.

**CVD:** see *cardiovascular disease.*

**cyclamate** (SIGH-klo-mate): an artificial sweetener that tastes 30 times as sweet as sucrose and provides 0 kcalories per gram; FDA approval pending in the United States; available in Canada as a tabletop sweetener, not as an additive.

**cytoplasm** (SIGH-toe-plazm): the cell contents, except for the nucleus.

## D

**D, L:** *D* stands for *dextro,* or "right-handed," and *L,* for *levo,* or "left-handed," referring to molecules that are identical to each other physically and chemically, but rotate light oppositely.

**Daily Values (DV):** reference values developed by the FDA specifically for use on food labels.

**deamination** (dee-AM-eh-NAY-shun): removal of the amino ($NH_2$) group from a compound such as an amino acid.

**defecate** (DEF-uh-cate): to move the bowels and eliminate waste.

**deficient:** the amount of a nutrient below which almost all healthy people can be expected, over time, to experience deficiency symptoms.

**dehydration:** the condition in which body water output exceeds water input. Symptoms include thirst, dry skin and mucous membranes, rapid heartbeat, low blood pressure, and weakness.

**Delaney Clause:** a clause in the Food Additive Amendment to the Food, Drug, and Cosmetic Act which states that no substance that is known to cause cancer in animals or human beings at any dose level shall be added to foods.

**denaturation** (dee-NAY-chur-AY-shun): the change in a protein's shape brought about by heat, acid, base, alcohol, heavy metals, or other agents.

**dental caries:** decay of teeth.

**dental plaque:** a gummy mass of bacteria that grows on teeth and can lead to dental caries and gum disease.

**dexfenfluramine** (DEKS-fen-FLOOR-ah-mean): a drug used in the treatment of obesity that triggers the release of serotonin in the brain, thus suppressing appetite; marketed under the trade name *Redux.* Its sister drug *fenfluramine* (fen-FLOOR-ah-mean) acts similarly and is marketed under the trade name *Pondimin.*

**dextrose:** an older name for glucose.

**DFE:** see *dietary folate equivalents.*

**DHEA** (dehydroepiandrosterone): a hormone made in the adrenal glands that serves as a precursor to the male hormone testosterone; falsely promoted as burning fat, building muscle, and slowing aging. Side effects include acne, aggressiveness, and liver enlargement.

**diabetes** (DYE-uh-BEET-eez) **mellitus** (MELL-ih-tus or mell-EYE-tus): a metabolic disorder of carbohydrate metabolism characterized by altered glucose regulation and utilization, usually resulting from insufficient or ineffective insulin.

**diarrhea:** the frequent passage of watery bowel movements.

**diet:** the foods and beverages a person eats and drinks.

**diet history:** a record of eating behaviors and the foods a person eats.

**Dietary Reference Intakes (DRI):** a set of values for the dietary nutrient intakes of healthy people in the United States and Canada. These values are used for planning and assessing diets and include, Estimated Average Requirements, Recommended Dietary Allowances, Adequate Intakes, and Tolerable Upper Intake Levels.

**dietary folate equivalents (DFE):** the amount of folate available to the body from naturally occurring sources, fortified foods, and supplements, accounting for differences in the bioavailability from each source.

**dietetic technician registered (DTR):** a person with an associate's degree and training in nutrition, food science, and diet planning who works under the guidance of an RD (registered dietitian).

**dietitian:** a person trained in nutrition, food science, and diet planning. See also *registered dietitian.*

**differentiation:** development of specific functions different from those of the original.

**digestion:** the process by which food is broken down into absorbable units.

**digestive enzymes:** proteins found in digestive juices that act on food substances, causing them to break down into simpler compounds.

**digestive system:** all the organs and glands associated with the ingestion and digestion of food.

**diglyceride:** a molecule of glycerol with two fatty acids attached.

**diketopiperazine** (dye-KEY-toe-pie-PER-a-zeen), or **DKP:** a product to which aspartame breaks down during metabolism.

**dioxins** (dye-OK-sins): a class of chemical pollutants created as by-products of chemical manufacturing, incineration, chlorine bleaching of paper pulp, and other industrial processes. Dioxins persist in the environment and accumulate in the food chain.

**dipeptide** (dye-PEP-tide): two amino acids bonded together.

**direct calorimetry** (kal-oh-RIM-eh-tree): the measurement of energy output as heat energy.

**disaccharide** (dye-SACK-uh-ride): a pair of monosaccharides linked together.

**dissociation** (dis-SO-see-AY-shun): the physical separation of a compound into ions.

**distilled liquor:** an alcoholic beverage made by fermenting and distilling grains; sometimes called *distilled spirits* or *hard liquor.*

**distilled water:** water that has been vaporized and recondensed, leaving it free of dissolved minerals.

**diuretics** (dye-you-RET-ics): drugs that promote water excretion; popularly, a "water pill."

**diverticula** (dye-ver-TIC-you-la): sacs or pouches that develop in the weakened areas of the intestinal wall (like bulges in an inner tube where the tire wall is weak).

**diverticulitis** (DYE-ver-tic-you-LYE-tis): infected or inflamed diverticula.

**diverticulosis** (DYE-ver-tic-you-LOH-sis): the condition of having diverticula. About one in every six people in Western countries develops diverticulosis in middle or later life.

**docosahexaenoic** (DOE-cossa-HEXA-ee-NO-ick) **acid (DHA):** an omega-3 polyunsaturated fatty acid with 22 carbons and six double bonds (22:6); synthesized from linolenic acid.

**double-blind experiment:** an experiment in which neither the subjects nor the researchers know which subjects are members of the experimental group and which are serving as control subjects, until after the experiment is over.

**Down syndrome:** a genetic abnormality that causes mental retardation, short stature, and flattened facial features.

**DRI:** see *Dietary Reference Intakes.*

**drink:** a dose of any alcoholic beverage that delivers ½ oz of pure ethanol.

**drug:** a substance that can modify one or more of the body's functions.

**drug history:** a record of all the drugs, over-the-counter and prescribed, that a person takes routinely.

**DTR:** see *dietetic technician registered.*

**duodenum** (doo-oh-DEEN-um, doo-ODD-num): the top portion of the small intestine (about "12 fingers' breadth" long in ancient terminology).

**duration:** length of time (for example, the time spent in each activity session).

**DV:** see *Daily Values.*

**dysentery** (DISS-en-terry): an infection of the digestive tract that causes diarrhea.

## E

**eating disorder:** a disturbance in eating behavior that jeopardizes a person's physical or psychological health.

**eclampsia** (eh-KLAMP-see-ah): a severe stage of preeclampsia characterized by convulsions.

**edema** (eh-DEEM-uh): the swelling of body tissue caused by excessive amounts of fluid in the interstitial spaces; seen in protein deficiency (among other conditions).

**eicosanoids** (eye-COSS-uh-noyds): derivatives of fatty acids; hormonelike compounds that regulate blood pressure, clotting, and other body functions. They include *prostaglandins, thromboxanes,* and *leukotrienes.*

**eicosapentaenoic** (EYE-cossa-PENTA-ee-NO-ick) **acid (EPA):** an omega-3 polyunsaturated fatty acid with 20 carbons and five double bonds (20:5); synthesized from linolenic acid.

**electrolyte solutions:** solutions that can conduct electricity.

**electrolytes:** salts that dissolve in water and dissociate into charged particles called ions.

**electron transport chain (ETC):** the final pathway in energy metabolism where the electrons from hydrogen are passed to oxygen and the energy released is trapped in the bonds of ATP.

**element:** a substance composed of atoms that are alike—for example, iron (Fe).

**embolism** (EM-boh-lizm): the obstruction of a blood vessel by an *embolus* (EM-boh-luss), or traveling clot, causing sudden tissue death.

**embryo** (EM-bree-oh): the developing infant from two to eight weeks after conception.

**emetic** (em-ETT-ic): an agent that causes vomiting.

**empty-kcalorie food:** a popular term used to denote foods that contribute energy but are lacking protein, vitamins, and minerals. Empty-kcalorie foods are *low–nutrient density foods.* The most notorious empty-kcalorie foods are sugar, fat, and alcohol.

**emulsifier** (ee-MUL-sih-fire): a substance with both water-soluble and fat-soluble portions that promotes the mixing of oils and fats in a watery solution.

**endocrine:** with reference to a gland, one that secretes its product directly into *(endo)* the blood; for example, the pancreas cells that produce insulin.

**endogenous** (en-DODGE-eh-nus) **protein:** the protein in the body.

**endosperm** (EN-doe-sperm): the bulk of the edible part of the kernel containing starch and proteins.

**enema:** the insertion of solutions into the rectum and colon to stimulate a bowel movement and empty the lower large intestine.

**energy:** the capacity to do work. The energy in food is chemical energy. The body can convert this chemical energy to mechanical, electrical, or heat energy.

**energy-yielding nutrients:** the nutrients that break down to yield energy the body can use.

**enriched:** the addition of nutrients to a food to meet a specified standard. In the case of refined bread or cereal, five nutrients have been added: thiamin, niacin, folate, and riboflavin in amounts approximately equivalent to, or higher than, those originally present, and iron in amounts to alleviate the prevalence of iron-deficiency anemia.

**enteropancreatic** (EN-ter-oh-PAN-kree-AT-ik) **circulation:** the circulatory route from the pancreas to the intestine and back to the pancreas.

**Environmental Protection Agency:** see *EPA.*

**enzymes:** proteins that facilitate chemical reactions without being changed in the process; protein catalysts.

**EPA (Environmental Protection Agency):** a federal agency that is responsible for, among other things, regulating pesticides and establishing water quality standards.

**epiglottis** (epp-ee-GLOTT-iss): cartilage in the throat that guards the entrance to the trachea and prevents fluid or food from entering it when a person swallows.

**epinephrine** (EP-ih-NEFF-rin): a hormone of the adrenal gland that modulates the stress response; formerly called *adrenaline.*

**epithelial** (ep-i-THEE-lee-ul) **cells:** cells on the surface of the skin and mucous membranes.

**epithelial tissues:** the layers of the body that serve as selective barriers between the body's interior and the environment (examples are the cornea, the skin, the respiratory lining, and the lining of the digestive tract).

**ergogenic** (ER-go-JEN-ick) **aids:** substances or techniques used in an attempt to enhance physical performance.

**erythrocyte** (eh-RITH-ro-cite) **hemolysis** (he-MOLL-uh-sis): the breaking open of red blood cells (erythrocytes); a symptom of vitamin E–deficiency disease in human beings.

**erythrocyte protoporphyrin** (PRO-toe-PORE-fe-rin): a precursor to hemoglobin.

**erythropoietin** (eh-REE-throw-POY-eh-tin): a hormone that stimulates red blood cell production.

**esophagus** (ee-SOFF-uh-gus): the food pipe; the conduit from the mouth to the stomach.

**essential amino acids:** amino acids that the body cannot synthesize in amounts sufficient to meet physiological needs. Some researchers refer to essential amino acids as *indispensable* and to nonessential amino acids as *dispensable.*

**essential fatty acids:** fatty acids needed by the body, but not made by it in amounts sufficient to meet physiological needs.

**essential nutrients:** nutrients a person must obtain from food because the body cannot make them for itself in sufficient quantity to meet physiological needs; also called *indispensable nutrients.* About 40 nutrients are known to be essential for human beings.

**Estimated Average Requirement:** the amount of a nutrient that will maintain a specific biochemical or physiological function in half the people of a given age and sex group.

**estrogens:** hormones responsible for the menstrual cycle and other female characteristics.

**ETC:** see *electron transport chain.*

**ethanol:** a particular type of alcohol found in beer, wine, and distilled spirits; also called *ethyl alcohol.* Ethanol is the most widely used—and abused—drug in our society. It is also the only legal, nonprescription drug that produces euphoria.

**ethnic diets:** foodways and cuisines typical of national origins, cultural heritages, or geographic locations.

**euphoria** (you-FORE-eh-uh): a feeling of great well-being, which people often seek through the use of drugs such as alcohol.

**exchange lists:** diet-planning tools that organize foods by their proportions of carbohydrate, fat, and protein. Foods on any single list can be used interchangeably.

**exocrine:** with reference to a gland, one that secretes its product(s) out *(exo)* of the gland through a duct into a cavity; the sweat glands of the skin and the enzyme-producing glands of the pancreas are both examples.

**exogenous** (eks-ODGE-eh-nus) **protein:** protein in foods.

**experimental group:** a group of individuals similar in all possible respects to the control group except for the treatment. The experimental group receives the real treatment.

**externalities:** hidden costs that are not reflected in the prices of things, such as the costs of environmental deterioration, that permit agribusiness foods to be sold at artificially low prices.

**extracellular fluid:** fluid outside the cells, which includes interstitial fluid, plasma, and the water of structures such as the skin and bones. Extracellular fluid accounts for approximately one-third of the body's water.

## F

**FAE:** see *fetal alcohol effects.*

**faith healing:** healing by invoking divine intervention without the use of medical, surgical, or other traditional therapy.

**false negative:** a test result indicating that a condition is *not* present (negative) when in fact it is present (therefore false).

**false positive:** a test result indicating that a condition is present (positive) when in fact it is not (therefore false).

**famine:** widespread scarcity of food in an area that causes starvation and death in a large portion of the population.

**FAO (Food and Agriculture Organization):** an international agency (part of the United Nations) that has adopted standards to regulate pesticide use among other responsibilities.

**FAS:** see *fetal alcohol syndrome.*

**fat replacer:** an ingredient that replaces some or all of the functions of fat and may or may not provide energy. In this text, the term *fat replacer* is used interchangeably with *fat substitute,* which technically applies only to an ingredient that replaces all of the functions of fat and provides no energy.

**fatfold measure:** a clinical estimate of total body fatness in which the thickness of a fold of skin on the back of the arm (over the triceps muscle), below the shoulder blade (subscapular), or in other places is measured with a caliper. (The older, less preferred, term is *skinfold test.*)

**fats:** lipids in foods or the body, both of which are composed mostly of triglycerides.

**fatty acid:** an organic compound composed of a carbon chain with hydrogens attached and an acid group (COOH) at one end.

**fatty acid oxidation:** the metabolic breakdown of fatty acids to acetyl CoA; also called *beta oxidation.*

**fatty liver:** an early stage of liver deterioration seen in several diseases, including kwashiorkor and alcoholic liver disease. Fatty liver is characterized by an accumulation of fat in the liver cells.

**fatty streaks:** accumulations of cholesterol and other lipids along the walls of the arteries.

**FDA (Food and Drug Administration):** a part of the Department of Health and Human Services' Public Health Service that is responsible for ensuring the safety and wholesomeness of all foods processed and sold in interstate commerce except meat, poultry, and eggs (which are under the jurisdiction of the USDA); inspecting food plants and imported foods; and setting standards for food composition.

**Federal Trade Commission:** see *FTC.*

**female athlete triad:** a potentially fatal triad of medical problems: disordered eating, amenorrhea, and osteoporosis.

**fenfluramine:** see *dexfenfluramine.*

**ferment:** to digest in the absence of oxygen.

**fertility:** the capacity of a woman to produce a normal ovum periodically and of a man to produce normal sperm; the ability to reproduce.

**fetal alcohol effects (FAE):** a subclinical version of fetal alcohol syndrome, with hidden defects including learning disabilities, behavioral abnormalities, and motor impairments; also called *alcohol-related birth defects (ARBD).*

**fetal alcohol syndrome (FAS):** the cluster of symptoms seen in an infant or child whose mother consumed excess alcohol during pregnancy, including retarded growth, impaired development of the central nervous system, and facial malformations.

**fetus** (FEET-us): the developing infant from eight weeks after conception until term.

**fibers:** in plant foods, the *nonstarch polysaccharides* that are not digested by *human* digestive enzymes, although some are digested by GI tract bacteria; fibers include cellulose, hemicelluloses, pectins, gums, and mucilages and the nonpolysaccharides lignins, cutins, and tannins.

**fibrocystic breast disease:** a harmless condition in which the breasts develop lumps, sometimes associated with caffeine consumption. In some, it responds to treatment by abstinence from caffeine; in others, it can be treated with vitamin E.

**fibrosis** (fye-BROH-sis): an intermediate stage of liver deterioration seen in several diseases, including viral hepatitis and alcoholic liver disease. In fibrosis, the liver cells lose their function and assume the characteristics of connective tissue cells (fibers).

**fibrous plaques** (PLACKS): mounds of lipid material, mixed with smooth muscle cells and calcium, which develop in the artery walls in atherosclerosis.

**field gleaning:** collecting crops from fields that either have already been harvested or are not profitable to harvest.

**filtered water:** water treated by filtration, usually through activated carbon filters that reduce the lead in tap water, or by reverse osmosis units that force pressurized water across a membrane removing lead, arsenic, and some microorganisms.

**fitness:** the characteristics that enable the body to perform physical activity; more broadly, the ability to meet routine physical demands with enough reserve energy to rise to a physical challenge; or the body's ability to withstand stress of all kinds.

**flexibility:** the capacity of the joints to move through a full range of motion; the ability to bend and recover without injury.

**fluid and electrolyte balance:** maintenance of the proper types and amounts of fluid and minerals in each compartment of the body fluids.

**fluorapatite** (floor-APP-uh-tite): the stabilized form of bone and tooth crystal, in which fluoride has replaced the hydroxyl groups of hydroxyapatite.

**fluoridated water:** water that has been treated so as to contain at least 0.8 milligrams of fluoride per liter.

**fluorosis** (floor-OH-sis): discoloration and pitting of tooth enamel caused by excess fluoride during tooth development.

**folate** (FOLE-ate): a B vitamin; also known as folic acid, folacin, or pteroylglutamic (tare-o-EEL-glue-TAM-ick) acid (PGA). The coenzyme forms are DHF (dihydrofolate) and THF (tetrahydrofolate).

**follicle** (FOLL-i-cul), **hair:** a group of cells in the skin from which a hair grows.

**follicle, ovarian:** that part of the female reproductive system where the ovary lies and eggs are produced.

**food allergies:** adverse reactions to foods that involve an immune response; also called *food-hypersensitivity reactions.*

**Food and Agriculture Organization:** see *FAO.*

**Food and Drug Administration:** see *FDA.*

**food aversion:** a strong desire to avoid a particular food.

**food bank:** facilities that provide food to the hungry.

**food chain:** the sequence in which living things depend on other living things for food.

**food consumption survey:** a survey that measures the amounts and kinds of foods people consume (using diet histories), estimates the nutrient intakes, and compares them with a standard.

**food craving:** a deep longing for a particular food.

**food frequency checklist:** a checklist of foods on which a person can record the frequency with which he or she eats each food.

**food group plans:** diet-planning tools that sort foods of similar origin and nutrient content into groups and then specify that people should eat certain numbers of servings from each group.

**food insecurity:** limited or uncertain access to foods of sufficient quality or quantity.

**food intolerances:** adverse reactions to foods that do not involve the immune system.

**food record:** an extensive, accurate log of all foods eaten over a period of several days or weeks. A food record that includes associated information such as when, where, and with whom each food is eaten is sometimes called a *food diary.*

**food recovery:** collecting food for distribution to low-income people who are hungry. Four common methods of food recovery are field gleaning, perishable food salvage, food rescue, and nonperishable food collection.

**food rescue:** collecting prepared foods from commercial kitchens.

**food-borne illness:** illness transmitted to human beings through food and water, caused by either an infectious agent *(food-borne infection)* or a poisonous substance *(food intoxication);* commonly known as *food poisoning.*

**foods:** products derived from plants or animals that can be taken into the body to yield energy and nutrients for the maintenance of life and the growth and repair of tissues.

**foodways:** the sum of the food habits, customs, beliefs, and preferences of a culture.

**fortified:** the addition to a food of nutrients that were either not originally present or present in insignificant amounts. Fortification can be used to correct or present a widespread nutrient deficiency, to balance the total nutrient profile of a food, or to restore nutrients lost in processing.

**fossil fuel:** coal, oil, and natural gas; these are nonrenewable fuels that pollute. (Renewable or alternative fuels, such as solar and wind energy, pollute less or not at all.)

**fraud** or **quackery:** the promotion, for financial gain, of devices, treatments, services, plans, or products (including diets and supplements) that alter or claim to alter a human condition without proof of safety or effectiveness. (The word *quackery* comes from the term *quacksalver,* meaning a person who quacks loudly about a miracle product—a lotion or a salve.)

**free radical:** an unstable and highly reactive atom or molecule that has one or more unpaired electron(s) in the outer orbital.

**frequency:** the number of occurrences per unit of time (for example, the number of activity sessions per week).

**fructose** (FRUK-tose or FROOK-tose): a monosaccharide; sometimes known as fruit sugar or *levulose,* fructose is found abundantly in fruits, honey, and saps.

**FTC (Federal Trade Commission):** a federal agency that is responsible for, among other things, food advertising and industry competition.

**functional foods:** foods or food ingredients that have been modified to provide a health benefit beyond their nutrient contributions.

## G

**galactose** (ga-LAK-tose): a monosaccharide; part of the disaccharide lactose.

**gallbladder:** the organ that stores and concentrates bile. When it receives the signal that fat is present in the duodenum, the gallbladder contracts and squirts bile through the bile duct into the duodenum.

**galvanized:** a term referring to metals that have been treated with a zinc-containing coating to prevent rust.

**gastric glands:** exocrine glands in the stomach wall that secrete gastric juice into the stomach.

**gastric juice:** the digestive secretion of the gastric glands of the stomach.

**gastric partitioning:** a surgical procedure used to treat clinically severe obesity. The operation limits food intake by reducing the size of the stomach and delays gastric emptying by restricting the outlet.

**gastric-inhibitory peptide:** a hormone produced by the intestine. Target organ: the stomach. Response: slowing of the secretion of gastric juices and of GI motility.

**gastrin:** a hormone secreted by cells in the stomach wall. Target organ: the stomach. Response: secretion of gastric juice.

**gatekeepers:** with respect to nutrition, key people who control other people's access to foods and thereby exert profound impacts on their nutrition. Examples are the spouse who buys and cooks the food, the parent who feeds the children, and the caregiver in a day-care center.

**generally recognized as safe:** see *GRAS list.*

**germ:** the nutrient-rich inner part of a grain. The germ is the seed that grows into a wheat plant, so it is especially rich in vitamins and minerals to support new life.

**gestation** (jes-TAY-shun): the period from conception to birth; for human beings gestation lasts from 38 to 42 weeks. Pregnancy is often divided into thirds, called *trimesters.*

**gestational diabetes:** the appearance of abnormal glucose tolerance during pregnancy, with subsequent return to normal postpartum. Gestational diabetes is associated with cesarean delivery, birth trauma, and high birthweight.

**GI tract:** the gastrointestinal tract or digestive tract; the principal organs are the stomach and intestines.e

**gland:** a cell or group of cells that secretes materials for special uses in the body. Glands may be **exocrine** (EKS-oh-crin) **glands,** secreting their materials "out" (into the digestive tract or onto the surface of the skin), or **endocrine** (EN-doe-crin) **glands,** secreting their materials "in" (into the blood).

**glucagon** (GLOO-ka-gon): a hormone that is secreted by special cells in the pancreas in response to low blood glucose concentration and elicits release of glucose from storage.

**glucocorticoid:** a hormone from the adrenal cortex that affects the body's management of glucose.

**gluconeogenesis** (gloo-co-nee-oh-GEN-ih-sis): the making of glucose from a noncarbohydrate source.

**glucose** (GLOO-kose): a monosaccharide; sometimes known as blood sugar or *dextrose.*

**gluten** (GLOO-ten): an elastic protein found in wheat and other grains that gives dough its structure and cohesiveness.

**glycemic** (gligh-SEEM-ic) **effect:** a measure of the extent to which a food, as compared with pure glucose, raises the blood glucose concentration and elicits an insulin response.

**glycerol** (GLISS-er-ol): an alcohol composed of a 3-carbon chain, which can serve as the backbone for a triglyceride.

**glycogen** (GLY-co-gen): an animal polysaccharide composed of glucose; it is manufactured and stored in the liver and muscles as a storage form of glucose. Glycogen is not a significant food source of carbohydrate and is not counted as one of the complex carbohydrates in foods.

**glycolysis** (gligh-COLL-ih-sis): the metabolic breakdown of glucose to pyruvate. Glycolysis does not require oxygen (anaerobic).

**goblet cells:** cells of the GI tract (and lungs) that secrete mucus.

**goiter** (GOY-ter): an enlargement of the thyroid gland due to an iodine deficiency, malfunction of the gland, or overconsumption of a goitrogen. Goiter caused by iodine deficiency is *simple goiter.*

**goitrogen** (GOY-troh-jen): a thyroid antagonist found in food; causes *toxic goiter.* Goitrogens are found in such foods as cabbage, kale, brussels sprouts, cauliflower, broccoli, and kohlrabi.

**Golgi** (GOAL-gee) **apparatus:** a set of membranes within the cell where secretory materials are packaged for export.

**gout** (GOWT): a painful condition in which uric acid crystals form in the joints.

**granulated sugar:** crystalline sucrose; 99.9 percent pure.

**GRAS (generally recognized as safe) list:** a list, first established by the FDA in 1958, of food additives that had long been in use and were believed safe. The list is subject to revision as new facts become known.

## H

**HACCP:** see *Hazard Analysis Critical Control Points.*

**hard water:** water with a high calcium and magnesium concentration.

**hazard:** a source of danger; used to refer to circumstances in which harm is possible under normal conditions of use.

**Hazard Analysis Critical Control Points (HACCP):** a systematic plan to identify and correct potential microbial hazards in the manufacturing, distribution, and commercial use of food products.

**HDL (high-density lipoprotein):** the type of lipoprotein that transports cholesterol back to the liver from the cells; composed primarily of protein.

**health claim:** any statement that characterizes the relationship between any nutrient or other substance in a food and a disease or health-related condition.

**health history:** the medical record. Traditionally, the health history has been called the *medical history.* The term *health history* now seems more appropriate, however, since the contents describe a client's health status. Current trends in the medical profession are now emphasizing health promotion and disease prevention.

**heart attack:** sudden tissue death caused by blockages of vessels that feed the heart muscle; also called *myocardial infarction* or *cardiac arrest.*

**heartburn:** a burning sensation in the chest area caused by backflow of stomach acid into the esophagus.

**heat stroke:** the dangerous accumulation of body heat with accompanying loss of body fluid.

**heavy metal:** any of a number of mineral ions such as mercury and lead, so called because they are of relatively high atomic weight. Many heavy metals are poisonous.

**Heimlich maneuver:** a technique for removing an object from the trachea of a choking person.

**hematocrit** (hee-MAT-oh-krit): measurement of the volume of the red blood cells packed by centrifuge in a given volume of blood.

**heme** (HEEM): the iron-holding part of the hemoglobin and myoglobin proteins. About 40% of the iron in meat, fish, and poultry is bound into heme; the other 60% is nonheme iron.

**hemochromatosis** (HE-moh-KRO-ma-toe-sis): a hereditary defect in iron metabolism characterized by deposits of iron-containing pigment in many tissues, with tissue damage.

**hemoglobin** (HE-moh-GLOW-bin): the globular protein of the red blood cells that carries oxygen from the lungs to the cells throughout the body.

**hemolytic** (HE-moh-LIT-ick) **anemia:** the condition of having too few red blood cells as a result of erythrocyte hemolysis.

**hemophilia** (HE-moh-FEEL-ee-ah): a hereditary disease that is caused by a genetic defect and has no relation to vitamin K; the blood is unable to clot because it lacks the ability to synthesize certain clotting factors.

**hemorrhagic** (hem-oh-RAJ-ik) **disease:** a disease characterized by excessive bleeding.

**hemorrhoids** (HEM-oh-royds): painful swelling of the veins surrounding the rectum.

**hemosiderosis** (HE-mo-sid-er-OH-sis): a condition characterized by the deposition of hemosiderin in the liver and other tissues.

**herbal** (ERB-al) **medicine:** the use of plants to treat disease or improve health; also known as *botanical medicine* or *phytotherapy.*

**hGH (human growth hormone):** a hormone produced by the brain's pituitary gland that regulates normal growth and development; also called *somatotropin.* Some athletes misuse this hormone to increase their height and strength.

**hiccups** (HICK-ups): repeated cough-like sounds and jerks that are produced when an involuntary spasm of the diaphragm muscle sucks air down the windpipe; also spelled *hiccoughs.*

**high-density lipoprotein:** see *HDL.*

**high-fructose corn syrup (HFCS):** a corn-syrup sweetener made especially for use in processed foods and beverages, where it is the predominant sweetener. HFCS is mostly fructose; glucose makes up the balance.

**high-quality protein:** an easily digestible, complete protein.

**high-risk pregnancy:** a pregnancy characterized by indicators that make it likely the birth will be surrounded by problems such as premature delivery, difficult birth, retarded growth, birth defects, and early infant death.

**high potency:** 100% or more of the Daily Value for the nutrient in a single supplement and for at least two-thirds of the nutrients in a multinutrient supplement.

**histamine** (HISS-tah-mean, or HISS-tah-men): a substance produced by cells of the immune system as part of a local immune reaction to an antigen; participates in causing inflammation.

**HIV:** see *human immunodeficiency virus.*

**homeopathy** (hoe-me-OP-ah-thee): a practice based on the theory that "like cures like," that is, that substances that cause symptoms in healthy people can cure those symptoms when given in very dilute amounts.

**homeostasis** (HOME-ee-oh-STAY-sis): the maintenance of constant internal conditions (such as blood chemistry, temperature, and blood pressure) by the body's control systems. A homeostatic system is constantly reacting to external forces so as to maintain limits set by the body's needs.

**honey:** sugar (mostly sucrose) formed from nectar gathered by bees. An enzyme splits the sucrose into glucose and fructose. Composition and flavor vary, but honey always contains a mixture of sucrose, fructose, and glucose.

**hormones:** chemical messengers. Hormones are secreted by a variety of glands in response to altered conditions in the body. Each hormone travels to one or more specific target tissues or organs, where it elicits a specific response to maintain homeostasis. In general, a gastrointestinal hormone is called an *enterogastrone* (EN-ter-oh-GAS-trone).

**hormone-sensitive lipase:** an enzyme inside adipose cells that responds to the body's need for fuel by hydrolyzing triglycerides so that their parts (glycerol and fatty acids) escape into the general circulation and thus become available to other cells as fuel. The signals to which this enzyme responds include epinephrine and glucagon, which oppose insulin.

**human immunodeficiency virus (HIV):** the virus that causes AIDS. The infection progresses to become an immune system disorder that leaves its victims defenseless against numerous infections.

**hunger:** the physiological drive for food that initiates food-seeking behavior.

**husk:** the outer, inedible part of a grain; also called the *chaff.*

**hydrochloric acid:** an acid composed of hydrogen and chloride atoms (HCl). The gastric glands normally produce this acid.

**hydrodensitometry** (HI-dro-DEN-see-TOM-eh-tree): a method of measuring body density in which the person is weighed and then weighed again while submerged in water.

**hydrogenation** (high-dro-gen-AY-shun): a chemical process by which hydrogens are added to monounsaturated or polyunsaturated fats to reduce the number of double bonds, making the fats more saturated (solid) and more resistant to oxidation (protecting against rancidity). Hydrogenation produces *trans*-fatty acids.

**hydrolysis** (high-DROL-ih-sis): a chemical reaction in which a major reactant is split into two products, with the addition of a hydrogen atom (H) to one and a hydroxyl group (OH) to the other (from water, $H_2O$). (The noun is *hydrolysis;* the verb is *hydrolyze.)*

**hydrophilic** (high-dro-FIL-ick): a term referring to water-loving, or water-soluble, substances.

**hydrophobic** (high-dro-FOE-bick): a term referring to water-fearing, or non-water-soluble, substances; also known as *lipophilic* (fat loving).

**hydroxyapatite** (high-drox-ee-APP-ah-tite): crystals made of calcium and phosphorus.

**hyperactivity:** inattentive and impulsive behavior that is more frequent and severe than is typical of others a similar age; professionally called *attention-deficit/hyperactivity disorder (ADHD).*

**hypertension:** higher-than-normal blood pressure. Hypertension that develops without an identifiable cause is known as *essential* or *primary hypertension;* hypertension that is caused by a specific disorder such as kidney disease is known as *secondary hypertension.*

**hyperthermia:** an above-normal body temperature.

**hypertrophy** (high-PER-tro-fee): of muscles, growing larger; an increase in size in response to use.

**hypnotherapy:** a technique that uses hypnosis and the power of suggestion to improve health behaviors, relieve pain, and heal.

**hypoglycemia** (HIGH-po-gligh-SEE-me-ah): an abnormally low blood glucose concentration.

**hypothalamus:** (high-po-THAL-ah-mus): a brain center that controls activities such as maintenance of water balance, regulation of body temperature, and control of appetite.

**hypothermia:** a below-normal body temperature.

## I

**ileocecal** (ill-ee-oh-SEEK-ul) **valve:** the sphincter separating the small and large intestines.

**ileum** (ILL-ee-um): the last segment of the small intestine.

**imagery:** a technique that guides clients to achieve a desired physical, emotional, or spiritual state by visualizing themselves in that state.

**imitation food:** a food that substitutes for and resembles another food, but is nutritionally inferior to it with respect to vitamin, mineral, or protein content. If the substitute is not inferior to the food it resembles and if its name provides an accurate description of the product, it need not be labeled "imitation."

**immune system:** the body's natural defense system against foreign materials that have penetrated the skin or mucous membranes.

**immunity:** the body's ability to recognize and eliminate foreign invaders.

**immunoglobulin** (IM-you-no-GLOB-you-lin): a protein capable of acting as an antibody.

**implantation:** the stage of development in which the zygote embeds itself in the wall of the uterus and begins to develop; occurs during the first two weeks after conception.

**indigestion:** incomplete or uncomfortable digestion, usually accompanied by pain, nausea, vomiting, heartburn, intestinal gas, or belching.

**indirect additives:** substances that can get into food as a result of contact with foods during growing, processing, packaging, storing, cooking, or some other stage before the foods are consumed; also called *incidental* or *accidental additives.*

**indirect calorimetry:** the estimation of energy output from measures of the amount of oxygen used and carbon dioxide eliminated.

**inorganic:** not containing carbon or pertaining to living things.

**inositol** (in-OSS-ih-tall): a nonessential nutrient that can be made in the body from glucose. Inositol is used in cell membranes.

**insulin** (IN-suh-lin): a hormone secreted by special cells in the pancreas in response to (among other things) increased blood glucose concentration. The primary role of insulin is to control the transport of glucose from the bloodstream into the cells.

**insulin resistance:** the condition in which a normal amount of insulin produces a subnormal effect; a metabolic consequence of obesity.

**integrated pest management (IPM):** management of pests using a combination of natural and biological controls and few or no pesticides.

**intensity:** the degree of exertion while exercising (for example, the amount of weight lifted or the speed of running).

**intentional additives:** additives intentionally added to foods, such as nutrients, colors, and preservatives.

**intermittent claudication:** severe calf pain caused by inadequate blood supply; it occurs when walking and subsides during rest.

**Internet (the Net):** a worldwide collection of millions of computers linked together to share information.

**interstitial** (IN-ter-STISH-al) **fluid:** fluid between the cells, usually high in sodium and chloride. Interstitial fluid is a large component of extracellular fluid.

**intra-abdominal fat:** fat stored within the abdominal cavity in association with the internal abdominal organs, as opposed to the fat stored directly under the skin (subcutaneous fat).

**intracellular fluid:** fluid within the cells, usually high in potassium and phosphate. Intracellular fluid accounts for approximately two-thirds of the body's water.

**intrinsic:** inside the system.

**intrinsic factor:** a glycoprotein (a protein with short polysaccharide chains attached) manufactured in the stomach that aids in the absorption of vitamin $B_{12}$.

**invert sugar:** a mixture of glucose and fructose formed by the hydrolysis of sucrose in a chemical process; sold only in liquid form and sweeter than sucrose. Invert sugar is used as a food additive to help preserve freshness and prevent shrinkage.

**iodopsin** (eye-oh-DOP-sin): the light-sensitive pigment of the cones on the retina; it contains the retinal form of vitamin A.

**ions** (EYE-uns): atoms or molecules that have gained or lost electrons and therefore have electrical charges. Examples include the positively charged sodium ion ($Na^+$) and the negatively charged chloride ion ($Cl^-$).

**IPM:** see *integrated pesticide management.*

**iridology:** the study of changes in the iris of the eye and their relationships to disease.

**iron deficiency:** the state of having depleted iron stores.

**iron-deficiency anemia:** severe depletion of iron stores that results in low hemoglobin and small, pale, red blood cells.

**iron overload:** toxicity from excess iron.

**irradiation:** sterilizing a food by exposure to energy waves, similar to ultraviolet light and microwaves.

**irritable bowel syndrome:** an intestinal disorder of unknown cause; symptoms include abdominal discomfort and cramping, diarrhea, constipation, or alternating diarrhea and constipation.

## J

**jaundice** (JAWN-dis): yellowing of the skin due to spillover of the bile pigment *bilirubin* (bill-ee-ROO-bin) from the liver into the general circulation; also known as *hyperbilirubinemia* (HIGH-per-BILL-eh-roo-bin-EE-me-ah). When these pigments invade the brain, the condition is *kernicterus* (ker-NICK-ter-us). Jaundice may be caused by obstruction of bile passageways, hemolysis, or dysfunctional liver cells.

**jejunum** (je-JOON-um): the first two-fifths of the small intestine beyond the duodenum.

## K

**kcalorie (energy) control:** management of food energy intake.

**keratin** (KERR-uh-tin): a water-insoluble protein; the normal protein of hair and nails. Keratin-producing cells may replace mucus-producing cells in vitamin A deficiency.

**keratinization:** accumulation of keratin in a tissue; a sign of vitamin A deficiency.

**keratomalacia** (KARE-ah-toe-ma-LAY-shuh): softening of the cornea seen in severe vitamin A deficiency that leads to irreversible blindness.

**keto** (KEY-toe) **acid:** an organic acid that contains a carbonyl group (C=O).

**ketone** (KEE-tone) **bodies:** the product of the incomplete breakdown of fat when glucose is not available in the cells.

**ketosis** (kee-TOE-sis): an undesirably high concentration of ketone bodies in the blood and urine.

**kosher** (KOE-sure): foods prepared according to Jewish dietary laws.

**kwashiorkor** (kwash-ee-OR-core, kwash-ee-or-CORE): a form of PEM that results either from inadequate protein intake or, more commonly, from infections.

## L

**lactase:** an enzyme that hydrolyzes lactose.

**lactase deficiency:** a lack of the enzyme required to digest the disaccharide lactose into its component monosaccharides (glucose and galactose).

**lactation:** production and secretion of breast milk for the purpose of nourishing an infant.

**lactic acid:** an acid produced from pyruvate during anaerobic metabolism.

**lacto-ovo-vegetarians:** people who include milk, milk products, and eggs, but exclude meat, poultry, fish, and seafood from their diets.

**lactoferrin** (lack-toh-FERR-in): a protein in breast milk that binds iron and keeps it from supporting the growth of the infant's intestinal bacteria.

**lactose** (LAK-tose): a disaccharide composed of glucose and galactose; commonly known as milk sugar.

**lactose intolerance:** a condition that results from inability to digest the milk sugar lactose; characterized by bloating, gas, abdominal discomfort, and diarrhea. Lactose intolerance differs from milk allergy, which is caused by an immune reaction to the protein in milk.

**lactovegetarians:** people who include milk and milk products, but exclude meat, poultry, fish, seafood, and eggs from their diets.

**large intestine** or **colon** (COAL-un): the lower portion of intestine that completes the digestive process; its segments are the ascending colon, the transverse colon, the descending colon, and the sigmoid colon.

**larynx:** the voice box.

**LBW:** see *low birthweight.*

**LDL (low-density lipoprotein):** the type of lipoprotein derived from very-low-density lipoproteins (VLDL) as cells remove triglycerides from them; composed primarily of cholesterol.

**lean body mass:** the weight of the body minus the fat.

**lecithin** (LESS-uh-thin): one of the phospholipids; a compound of glycerol to which are attached two fatty acids, a phosphate group, and a choline molecule. Both nature and the food industry use lecithin as an emulsifier to combine two ingredients that do not ordinarily mix, such as water and oil.

**legumes** (lay-GYOOMS, LEG-yooms): plants of the bean and pea family, rich in high-quality protein compared with other plant-derived foods.

**leptin:** a protein produced by fat cells under direction of the *ob* gene that decreases appetite and increases energy expenditure; sometimes called the *ob protein.*

**levulose:** an older name for fructose.

**license to practice:** permission under state or federal law, granted on meeting specified criteria, to use a certain title (such as dietitian) and offer certain services. Licensed dietitians may use the initials LD after their names.

**life expectancy:** the average number of years lived by people in a given society.

**life span:** the maximum number of years of life attainable by a member of a species.

**limiting amino acid:** the essential amino acid found in the shortest supply relative to the amounts needed for protein synthesis in the body.

**linoleic** (lin-oh-LAY-ick) **acid:** an essential fatty acid with 18 carbons and two double bonds (18:2).

**linolenic** (lin-oh-LEN-ick) **acid:** an essential fatty acid with 18 carbons and three double bonds (18:3).

**lipase** (LYE-pase): an enzyme that hydrolyzes lipids (fats).

**lipids:** a family of compounds that includes triglycerides (fats and oils), phospholipids, and sterols.

**lipoprotein lipase (LPL):** an enzyme mounted on the surface of fat cells (and other cells) that hydrolyzes triglycerides passing by in the bloodstream and directs their parts into the cells, where they can be metabolized or reassembled for storage.

**lipoproteins** (LIP-oh-PRO-teenz): clusters of lipids associated with proteins that serve as transport vehicles for lipids in the lymph and blood.

**liver:** the organ that manufactures bile and is the first to receive nutrients from the intestines. The liver's many other functions are described in Chapter 7.

**longevity:** long duration of life.

**low birthweight (LBW):** a birthweight of 5½ lb (2500 g) or less; indicates probable poor health in the newborn and poor nutrition status in the mother during pregnancy, before pregnancy, or both. Normal birthweight for a full-term baby is 6½ to 8¾ lb (about 3000 to 4000 g).

**low-density lipoprotein:** see *LDL.*

**low-risk pregnancy:** a pregnancy characterized by indicators that make a normal outcome likely.

**LPL:** see *lipoprotein lipase.*

**luteinizing hormone:** hormone that stimulates the development of the corpus luteum (the small tissue that develops from a ruptured ovarian follicle and secretes hormones); so called because the follicle turns yellow as it matures.

**lymph** (limf): a clear yellowish fluid that resembles blood without the red blood cells; lymph from the GI tract transports fat and fat-soluble vitamins to the bloodstream via lymphatic vessels.

**lymphatic** (lim-FAT-ic) **system:** a loosely organized system of vessels and ducts that convey fluids toward the heart; the GI part of the lymphatic system carries the products of digestion into the bloodstream.

**lymphocytes** (LIM-foe-sites): white blood cells that participate in acquired immunity; B-cells and T-cells.

**lysosomes** (LIE-so-zomes): cellular organelles; membrane-enclosed sacs of degradative enzymes.

## M

**macrobiotic diets:** extremely restrictive diets limited to a few grains and vegetables; based on metaphysical beliefs and not nutrition. A macrobiotic diet might consist of brown rice, miso soup, and sea vegetables, for example.

**macula:** a small, oval, yellowish region in the center of the retina that provides the sharp, straight-ahead vision so critical to reading and driving.

**macular** (MACK-you-lar) **degeneration:** deterioration of the macular area of the eye that can lead to loss of central vision and eventual blindness.

**magnesium:** a cation within the body's cells, active in many enzyme systems.

**major minerals:** essential mineral nutrients found in the human body in amounts larger than 5 g; sometimes called *macrominerals.*

**malnutrition:** any condition caused by excess or deficient food energy or nutrient intake or by an imbalance of nutrients.

**maltase:** an enzyme that hydrolyzes maltose.

**maltose** (MAWL-tose): a disaccharide composed of two glucose units; sometimes known as malt sugar.

**maple sugar:** a sugar (mostly sucrose) purified from the concentrated sap of the sugar maple tree.

**marasmus** (ma-RAZ-mus): a form of PEM that results from a severe deprivation, or impaired absorption, of energy, protein, vitamins, and minerals.

**margin of safety:** when speaking of food additives, a zone between the concentration normally used and that at which a hazard exists. For common table salt, for example, the margin of safety is ⅕ (five times the amount normally used would be hazardous).

**massage therapy:** a healing method in which the therapist manually kneads muscles to reduce tension, increase blood circulation, improve joint mobility, and promote healing of injuries.

**matrix** (MAY-tricks): the basic substance that gives form to a developing structure; in the body, the formative *cells* from which teeth and bones grow.

**meat replacements:** products formulated to look and taste like meat, fish, or poultry; usually made of textured vegetable protein.

**meditation:** a self-directed technique of relaxing the body and calming the mind.

**melanocyte** (MEL-an-oh-cite or mel-AN-oh-cite): a cell containing the pigment melanin.

**MEOS** or **microsomal** (my-krow-SO-mal) **ethanol-oxidizing system:** a system of enzymes in the liver that oxidize not only alcohol, but also several classes of drugs.

**mEq:** see *miliequivalents.*

**metabolism:** the sum total of all the chemical reactions that go on in living cells; *energy metabolism* includes all the reactions by which the body obtains and spends the energy from food.

**metalloenzyme** (meh-TAL-oh-EN-zime): an enzyme that contains one or more minerals as part of its structure.

**metallothionein** (meh-TAL-oh-THIGH-oh-neen): a sulfur-rich protein that avidly binds with metals such as zinc.

**MFP factor:** a factor associated with the digestion of meat, fish, and poultry that enhances iron absorption.

**micelles** (MY-cells): tiny spherical complexes that arise during fat digestion; each carries about 20 fatty acids and/or monoglycerides into intestinal cells.

**microparticulated protein:** a protein-based fat replacer made from milk, egg, or whey proteins; provides 1 to 2 kcalories per gram.

**microsomal ethanol-oxidizing system:** see *MEOS.*

**microvilli** (MY-cro-VILL-ee, MY-cro-VILL-eye): tiny, hairlike projections on each cell of every villus that can trap nutrient particles and transport them into the cells; singular *microvillus.*

**milk anemia:** iron-deficiency anemia that develops when an excessive milk intake displaces iron-rich foods from the diet.

**milliequivalents (mEq):** the concentration of electrolytes in a volume of solution. Milliequivalents are a useful measure when considering ions, because the number of charges reveals characteristics about the solution that are not evident when the concentration is expressed in terms of weight.

**mineral water:** water from a spring or well that typically contains 250 to 500 ppm of minerals. Minerals give water a distinctive flavor. Many mineral waters are high in sodium.

**mineralization:** the process in which calcium, phosphorus, and other minerals crystallize on the collagen matrix of a growing bone, hardening the bone.

**minerals:** inorganic elements; some minerals are essential nutrients required in small amounts.

**misinformation:** false or misleading information.

**mitochondria** (my-toe-KON-dree-uh); singular *mitochondrion:* the cellular organelles responsible for producing ATP aerobically; made of membranes (lipid and protein) with enzymes mounted on them.

**moderate exercise:** activity that can be sustained comfortably for 60 minutes or so.

**moderation:** in relation to alcohol consumption, not more than two drinks a day for the average-sized man and not more than one drink a day for the average-sized woman.

**moderation (dietary):** providing enough but not too much of a substance.

**molasses:** the thick brown syrup produced during sugar refining. Molasses retains residual sugar and other by-products and a few minerals; blackstrap molasses contains significant amounts of calcium and iron—the iron comes from the *machinery* used to process the sugar.

**molecule:** two or more atoms of the same or different elements joined by chemical bonds. Examples are molecules of the element oxygen, composed of two oxygen atoms ($O_2$), and molecules of the compound water, composed of two hydrogen atoms and one oxygen atom ($H_2O$).

**molybdenum** (mo-LIB-duh-num): a trace element.

**monoglyceride:** a molecule of glycerol with one fatty acid attached.

**monosaccharide** (mon-oh-SACK-uh-ride): a carbohydrate of the general formula $C_nH_{2n}O_n$ that consists of a single ring.

**monounsaturated fatty acid:** a fatty acid that lacks two hydrogen atoms and has one double bond between carbons—for example, oleic acid. A **monounsaturated fat** is composed of triglycerides in which most of the fatty acids are monounsaturated.

**MSG symptom complex:** an acute, temporary intolerance reaction that may occur after the ingestion of the additive MSG (monosodium glutamate). Symptoms include burning sensations, chest and facial flushing and pain, and throbbing headaches.

**mucus** (MYOO-kus): a slippery substance secreted by goblet cells of the GI lining (and other body linings) that protects the cells from exposure to digestive juices (and other destructive agents). The lining of the GI tract with its coat of mucus is a **mucous membrane.** (The noun is **mucus;** the adjective is **mucous.**)

**mucous** (MYOO-kus) **membranes:** the membranes, composed of mucus-secreting cells, that line the surfaces of body tissues.

**muscle dysmorphia** (dis-MORE-fee-ah): a newly coined psychiatric disorder characterized by a preoccupation with building body mass.

**muscle endurance:** the ability of a muscle to contract repeatedly without becoming exhausted.

**muscle strength:** the ability of muscles to work against resistance.

**muscular dystrophy** (DIS-tro-fee): a hereditary disease in which the muscles gradually weaken; its most debilitating effects arise in the lungs.

**mutual supplementation:** the strategy of combining two protein foods in a meal so that each food provides the essential amino acid(s) lacking in the other. Mutual supplementation is the dietary strategy that brings complementary proteins together in a meal.

**myoglobin:** the oxygen-holding protein of the muscle cells.

## N

**NAD (nicotinamide adenine dinucleotide):** the main coenzyme form of the vitamin niacin; its reduced form is NADH.

**narcotic** (nar-KOT-ic): any drug that dulls the senses, induces sleep, and becomes addictive with prolonged use.

**natural water:** water obtained from a spring or well that is certified to be safe and sanitary. The mineral content may not be changed, but the water may be treated in other ways such as by filtration or ozonization.

**naturopathic medicine:** a system that integrates traditional medicine with botanical medicine, clinical nutrition, homeopathy, acupuncture, East Asian medicine, hydrotherapy, and manipulative therapy.

**NE:** see *niacin equivalents.*

**net protein utilization (NPU):** the amount of protein nitrogen that is retained from a given amount of protein nitrogen eaten; a measure of protein quality.

**neural tube defects:** malformations of the brain, spinal cord, or both during embryonic development. The two main types of neural tube defects are spina bifida (literally, "split spine") and anencephaly ("no brain").

**neuron:** a nerve cell; the structural and functional unit of the nervous system. Neurons initiate and conduct nerve transmissions.

**neurotransmitters:** chemicals that are released at the end of a nerve cell when a nerve impulse arrives there; they diffuse across the gap to the next cell and alter the membrane of that second cell to either inhibit or excite it.

**niacin** (NIGH-a-sin): a B vitamin; the coenzyme forms are NAD (nicotinamide adenine dinucleotide) and NADP (the phosphate form of NAD). Niacin can be eaten preformed or made in the body from its precursor, tryptophan, one of the amino acids.

**niacin equivalents (NE):** the amount of niacin present in food, including the niacin that can theoretically be made from its precursor, tryptophan, present in the food.

**niacin flush:** a burning, tingling, and itching sensation that occurs when a person takes a large dose of nicotinic acid; often accompanied by a headache and reddened face, arms, and chest.

**nicotinamide adenine dinucleotide:** see *NAD.*

**night blindness:** slow recovery of vision after flashes of bright light at night or an inability to see in dim light; an early symptom of vitamin A deficiency.

**nitrates** (NYE-trates): salts that are converted to nitrites by bacteria.

**nitrites** (NYE-trites): salts added to food to prevent botulism; one example is sodium nitrite, which is used to preserve meats.

**nitrogen balance:** the amount of nitrogen consumed (N in) as compared with the amount of nitrogen excreted (N out) in a given period of time.

**nitrosamines** (nigh-TROHS-uh-meens): derivatives of nitrites that may be formed in the stomach when nitrites combine with amines; nitrosamines are carcinogenic in animals.

**nonnutrients:** compounds in foods that do not fit within the six classes of nutrients.

**nonperishable food collection:** collecting processed foods from wholesalers and markets.

**nonpoint water pollution:** water pollution caused by runoff from all over an area rather than from discrete "point" sources. An example is the pollution caused by runoff from agricultural fields.

**NPU:** see *net protein utilization.*

**nucleus:** a major membrane-enclosed body within every cell, which contains the cell's genetic material, DNA, embedded in chromosomes.

**nursing bottle tooth decay:** extensive tooth decay due to prolonged tooth contact with formula, milk, fruit juice, or other carbohydrate-rich liquid offered to an infant in a bottle.

**nutraceuticals:** substances that supposedly provide medical and health benefits; a term not recognized by the FDA.

**nutrient additives:** vitamins and minerals added to improve nutritive value.

**nutrient density:** a measure of the nutrients a food provides relative to the energy it provides. The more nutrients and the fewer kcalories, the higher the nutrient density.

**nutrients:** chemical substances obtained from food and used in the body to provide energy, structural materials, and regulating agents to support growth, maintenance, and repair of the body's tissues; nutrients may also reduce the risks of some diseases.

**nutrition:** the science of foods and the nutrients and other substances they contain, and of their actions within the body (including ingestion, digestion, absorption, transport, metabolism, and excretion). A broader definition includes the social, economic, cultural, and psychological implications of food and eating.

**nutrition assessment:** a comprehensive approach, completed by a registered dietitian, to determine a person's nutrition status using health, socioeconomic, drug, and diet histories; anthropometric measurements; physical examinations; and laboratory tests.

**nutrition screening:** the use of preliminary nutrition assessment techniques to identify people who are malnourished or are at risk for malnutrition.

**nutrition status survey:** a survey that evaluates people's nutrition status using diet histories, anthropometric measures, physical examinations, and laboratory tests.

**nutritionist:** a person who specializes in the study of nutrition. Some nutritionists are registered dietitians, whereas others are self-described experts whose training is questionable. In states with responsible legislation, the term applies only to people who have MS or PhD degrees from properly accredited institutions.

**nutritive sweeteners:** sweeteners that yield energy, including both sugars and sugar alcohols.

## O

**oatrim:** a carbohydrate-based fat replacer made from oat flour; provides 1 to 4 kcalorie per gram.

**oils:** liquid fats (at room temperature).

**olestra:** a synthetic fat made from sucrose and fatty acids that provides 0 kcalories per gram; also known as *sucrose polyester.*

**oligopeptide** (OL-ee-go-PEP-tide): an intermediate string of four to nine amino acids.

**omega:** the last letter of the Greek alphabet (ω), used by chemists to refer to the position of the first double bond from the methyl end in a fatty acid.

**omega-3 fatty acid:** a polyunsaturated fatty acid in which the first double bond is 3 carbons away from the methyl ($CH_3$) end of the carbon chain.

**omega-6 fatty acid:** a polyunsaturated fatty acid in which the first double bond is 6 carbons from the methyl ($CH_3$) end of the carbon chain.

**omnivores:** people who have no formal restriction on the eating of any foods.

**opportunistic infections:** infections from microorganisms that normally do not cause disease in the general population but can cause great harm in people once their immune systems are compromised (as in HIV infection).

**opsin** (OP-sin): the protein portion of the visual pigment molecule.

**oral rehydration therapy (ORT):** the administration of a simple solution of sugar, salt, and water, taken by mouth, to treat dehydration caused by diarrhea.

**organelles:** subcellular structures such as ribosomes, mitochondria, and lysosomes.

**organic:** a substance or molecule containing carbon-carbon bonds or carbon-hydrogen bonds.

**organic halogen:** an organic compound containing one or more atoms of a *halogen*–fluorine, chlorine, iodine, or bromine.

**organically grown crops:** crops grown and processed according to USDA regulations defining the use of fertilizers, herbicides, insecticides, fungicides, preservatives, and other chemical ingredients.

**ORT:** see *oral rehydration therapy.*

**orthomolecular medicine:** the use of large doses of vitamins to treat chronic disease.

**osmotic pressure:** the force that moves water, but not the solutes, across a membrane when the two solutions differ in concentration. Water flows *toward* the side in which the solutes are more concentrated.

**osteoarthritis:** a painful, chronic disease of the joints that occurs when the cushioning cartilage in a joint breaks down; joint structure is usually altered, with loss of function; also called *degenerative arthritis.*

**osteomalacia** (OS-tee-oh-ma-LAY-shuh): a bone disease characterized by softening of the bones; symptoms include bending of the spine and bowing of the legs. The disease occurs most often in adult women.

**osteopenia** (OS-tee-oh-PEE-nee-ah): a metabolic bone disease common in preterm infants; also called *rickets of prematurity.*

**osteoporosis** (OS-tee-oh-pore-OH-sis): a condition of older persons in which the bones become porous and fragile due to a loss of minerals; also called *adult bone loss.*

**overnutrition:** excess energy or nutrients.

**overt** (oh-VERT): out in the open and easy to observe.

**overweight:** body weight above some standard of acceptable weight that is usually defined in relation to height (such as the weight-for-height tables or BMI).

**ovum** (OH-vum): the female reproductive cell, capable of developing into a new organism upon fertilization; commonly referred to as an egg.

**oxidant** (OK-see-dent): a compound (such as oxygen itself) that oxidizes other compounds. Compounds that prevent oxidation are called *anti*oxidants, whereas those that promote it are called *pro*oxidants.

**oxidation** (OKS-ee-day-shun): the process of a substance combining with oxygen.

**oxidative stress:** a condition in which the production of oxidants and free radicals exceeds the body's ability to defend itself.

**oxytocin** (OK-see-TOE-sin): a hormone that stimulates the mammary glands to eject milk during lactation and the uterus to contract during childbirth.

**ozone therapy:** the use of ozone gas to enhance the body's immune system.

## P

**pancreas:** a gland that secretes digestive enzymes and juices into the duodenum.

**pancreatic** (pank-ree-AT-ic) **juice:** the exocrine secretion of the pancreas, containing enzymes for the digestion of carbohydrate, fat, and protein as well as bicarbonate, a neutralizing agent. The juice flows from the pancreas into the small intestine through the pancreatic duct. (The pancreas also has an endocrine function, the secretion of insulin and other hormones.)

**pantothenic** (PAN-toe-THEN-ick) **acid:** a B vitamin; the principal active form is part of coenzyme A, commonly called "CoA" in metabolic reactions.

**parathyroid:** named for their location, the four parathyroid glands nestle in the surface layers of the two thyroid lobes in the neck.

**pasteurization:** heat processing of food that inactivates some, but not all, microorganisms in the food; not a sterilization process. Bacteria that cause spoilage are still present.

**pathogens** (PATH-oh-jens): microorganisms or substances capable of producing disease.

**PBB** (polybrominated biphenyl) and **PCB** (polychlorinated biphenyl): toxic organic compounds used in pesticides, paints, and flame retardants.

**PDCAAS:** see *protein digestibility-corrected amino acid score.*

**peak bone mass:** the highest attainable bone density for an individual, developed during the first three decades of life.

**peer review:** a process in which a panel of scientists rigorously evaluates a research study to assure that the scientific method was followed.

**pellagra** (pell-AY-gra): the niacin-deficiency disease.

**PEM:** see *protein-energy malnutrition.*

**pepsin:** a gastric protease. Pepsin is secreted in an inactive form, *pepsinogen,* which is activated by hydrochloric acid in the stomach.

**peptic ulcer:** an erosion in the mucous membrane of either the stomach (a gastric ulcer) or the duodenum (a duodenal ulcer).

**peptidase:** a digestive enzyme that hydrolyzes peptide bonds. *Tripeptidases* cleave tripeptides; *dipeptidases* cleave dipeptides. *Endopeptidases* cleave peptide bonds *within* the chain to create smaller fragments, whereas *exopeptidases* cleave bonds at the *ends* to release free amino acids.

**peptide bond:** a bond that connects the

acid end of one amino acid with the amino end of another, forming a link in a protein chain.

**PER:** see *protein efficiency ratio.*

**peripheral** (puh-RIFF-er-ul) **nervous system:** the peripheral (outermost) part of the nervous system, the vast complex of wiring that extends from the central nervous system to the body's outermost areas. It contains both somatic and autonomic components (defined next).

**peripheral resistance:** resistance to the flow of blood caused by the reduced diameter of the vessels at the periphery of the body—the smallest arteries and capillaries.

**perishable food salvage:** collecting perishable produce from wholesalers and markets.

**peristalsis** (peri-STALL-sis): wavelike muscular contractions of the GI tract that push its contents along.

**pernicious** (per-NISH-us) **anemia:** a blood disorder that reflects a vitamin $B_{12}$ deficiency caused by lack of intrinsic factor and characterized by abnormally large and immature red blood cells. Other symptoms include muscle weakness and irreversible neurological damage.

**peroxidation:** the production of unstable molecules containing more than the usual amount of oxygen. Hydrogen peroxide, $H_2O_2$, for example, may be produced from water, $H_2O$.

**persistence:** stubborn or enduring continuance; with respect to food contaminants, the quality of persisting, rather than breaking down, in the bodies of animals and human beings.

**pesticides:** chemicals used to control insects, diseases, weeds, fungi, and other pests on plants, vegetables, fruits, and animals. Used broadly, the term includes herbicides (to kill weeds), insecticides (to kill insects), and fungicides (to kill fungi).

**pH:** a measure of the concentration of $H^+$ ions. The lower the pH, the higher the $H^+$ ion concentration and the stronger the acid. A pH above 7 is alkaline, or base (a solution in which $OH^-$ ions predominate).

**phagocytes:** white blood cells that have the ability to ingest and destroy foreign substances.

**phagocytosis** (FAG-oh-sigh-TOE-sis): the process by which phagocytes engulf and destroy foreign materials.

**phentermine** (FEN-ter-mean): a drug used in the treatment of obesity that suppresses appetite; marketed under the trade names *Adipex, Fastin,* and *Ionamin.*

**phospholipid:** a compound similar to a triglyceride but having a phosphate group (a phosphorus-containing salt) and choline (or another nitrogen-containing compound) in place of one of the fatty acids.

**phosphorus:** a major mineral found mostly in the body's bones and teeth.

**photosynthesis:** the process by which green plants make carbohydrates from carbon dioxide and water using the green pigment chlorophyll to trap the sun's energy.

**physiological age:** a person's age as estimated from her or his body's health and probable life expectancy.

**phytic** (FYE-tick) **acid:** a nonnutrient component of plant seeds; also called *phytate* (FYEtate). Phytic acid occurs in the husks of grains, legumes, and seeds and is capable of binding minerals such as zinc, iron, calcium, magnesium, and copper in insoluble complexes in the intestine, which the body excretes unused.

**phytochemicals** (FIE-toe-KEM-ih-cals): nonnutrient compounds found in plant-derived foods that have biological activity in the body.

**pica** (PIE-ka): a craving for nonfood substances. Also known as *geophagia* (gee-oh-FAY-gee-uh) when referring to clay eating and *pagophagia* (pag-oh-FAY-gee-uh) when referring to ice craving.

**pigment:** a molecule capable of absorbing certain wavelengths of light, so that it reflects only those that we perceive as a certain color.

**placebo** (pla-SEE-bo): an inert, harmless medication given to provide comfort and hope; a sham treatment used in controlled research studies.

**placebo effect:** the healing effect that faith in medicine, even inert medicine, often has.

**placenta** (plah-SEN-tuh): the organ that develops inside the uterus early in pregnancy, through which the fetus receives nutrients and oxygen across the placenta and returns carbon dioxide and other waste products to be excreted.

**platelets:** tiny, disc-shaped bodies in the blood, important in blood clot formation.

**point of unsaturation:** the double bond of a fatty acid, where hydrogen atoms can easily be added to the structure.

**polypeptide:** many (ten or more) amino acids bonded together.

**polysaccharide:** many monosaccharides linked together.

**polyunsaturated fatty acid (PUFA):** a fatty acid that lacks four or more hydrogen atoms and has two or more double bonds between carbons—for example, linoleic acid (two double bonds) and linolenic acid (three double bonds). A **polyunsaturated fat** is composed of triglycerides in which most of the fatty acids are polyunsaturated.

**post term** (infant): an infant born after the 42nd week of pregnancy.

**postpartum amenorrhea:** the normal temporary absence of menstrual periods immediately following childbirth.

**potable** (POT-ah-bul) **water:** water that is suitable for drinking.

**potassium:** the principal cation within the body's cells; critical to the maintenance of fluid balance, nerve transmissions, and muscle contractions.

**precursors:** substances that precede others; with regard to vitamins, compounds that can be converted into active vitamins; also known as *provitamins.*

**preeclampsia** (PRE-ee-KLAMP-see-ah): a condition characterized by hypertension, fluid retention, and protein in the urine.

**preformed vitamin A:** dietary vitamin A in its active form.

**preservatives:** antimicrobial agents, antioxidants, and other additives that retard spoilage or maintain desired qualities, such as softness in baked goods.

**preterm** (infant): an infant born prior to the 38th week of pregnancy; also called a *premature* infant. A *term* infant is born between the 38th and 42nd week of pregnancy.

**primary deficiency:** a nutrient deficiency caused by inadequate dietary intake of a nutrient.

**progesterone:** the hormone of gestation (pregnancy).

**progressive overload principle:** the training principle that a body system, in order to improve, must be worked at frequencies, durations, or intensities that gradually increase physical demands.

**prolactin:** a hormone, so named because it promotes *(pro)* the production of milk *(lacto).*

**proof:** a way of stating the percentage of alcohol in distilled liquor. Liquor that is 100 proof is 50% alcohol; 90 proof is 45%, and so forth.

**protease** (PRO-tee-ase): an enzyme that hydrolyzes proteins.

**protein digestibility:** a measure of the amount of amino acids absorbed from a given protein intake.

**protein digestibility-corrected amino acid score (PDCAAS):** a measure of protein quality assessed by comparing the amino acid score of a food protein with the amino acid requirements of preschool-age children and then correcting for the true digestibility of the protein; recommended by the FAO/WHO and used to establish protein quality of foods for Daily Value percentages on food labels.

**protein efficiency ratio (PER):** a measure of protein quality assessed by determining how well a given protein supports weight gain in growing rats; used to establish the protein quality for infant formulas and baby foods.

**protein turnover:** the degradation and synthesis of endogenous protein.

**protein-energy malnutrition (PEM),** also

called **protein-kcalorie malnutrition (PCM):** a deficiency of both protein and energy; the world's most widespread malnutrition problem, including kwashiorkor, marasmus, and instances in which they overlap.

**protein-sparing action:** the action of carbohydrate (and fat) in providing energy that allows protein to be used for other purposes.

**proteins:** compounds composed of carbon, hydrogen, oxygen, and nitrogen atoms, arranged into amino acids linked in a chain. Some amino acids also contain sulfur atoms.

**puberty:** the period in life in which a person becomes physically capable of reproduction.

**public health dietitian:** a dietitian who specializes in public health nutrition.

**public water:** water from a municipal or county water system that has been treated and disinfected.

**PUFA:** See *polyunsaturated fatty acid.*

**pulmonary hypertension:** abnormally high blood pressure in the lungs; a rare, but life-threatening condition associated with dexfenfluramine, fenfluramine, phentermine, or combinations of these drugs.

**purgative:** a strong laxative.

**purified water:** water that has been processed through distillation, deionization, or reverse osmosis and meets U.S. Pharmacopoeia standards for medical and research purposes.

**pyloric** (pie-LORE-ic) **sphincter:** the circular muscle that separates the stomach from the small intestine and regulates the flow of partially digested food into the small intestine; also called *pylorus* or *pyloric valve.*

**pyruvate** (PIE-roo-vate): pyruvic acid, a 3-carbon compound that, in metabolism, can be derived from glucose, certain amino acids, or glycerol.

## Q

**quackery:** see *fraud.*

## R

**randomization** (RAN-dom-ih-ZAY-shun): a process of choosing the members of the experimental and control groups without bias.

**raw sugar:** the first crop of crystals harvested during sugar processing. Raw sugar cannot be sold in the United States because it contains too much filth (dirt, insect fragments, and the like). Sugar sold as "raw sugar" domestically has actually gone through over half of the refining steps.

**RD:** see *registered dietitian.*

**RDA:** see *Recommended Dietary Allowance.*

**RE (retinol equivalent):** a measure of vitamin A activity; the amount of retinol that the body will derive from a food containing preformed retinol or its precursor beta-carotene.

**Recommended Dietary Allowance (RDA):** the average daily amount of a nutrient considered adequate to meet the known nutrient needs of practically all healthy people; a goal for dietary intake by individuals.

**rectum:** the muscular terminal part of the intestine, extending from the sigmoid colon to the anus.

**reference protein:** a standard against which to measure the quality of other proteins.

**refined:** the process by which the coarse parts of a food are removed. When wheat is refined into flour, the bran, germ, and husk are removed, leaving only the endosperm.

**reflux:** a backward flow.

**registered dietitian (RD):** a dietitian who has graduated from a university or college after completing a program of dietetics that has been accredited by the American Dietetic Association (or Dietitians of Canada), has served in an internship or coordinated program to practice the necessary skills, has passed the association's registration examination, and maintains competency through continuing education. Many states require licensing for practicing dietitians.

**registration:** listing; with respect to health professionals, listing with a professional organization that requires specific course work, experience, and passing of an examination.

**relaxin:** the hormone of late pregnancy.

**remodeling:** the dismantling and re-formation of a structure, in this case, bone.

**renin** (REN-in): an enzyme from the kidneys that activates angiotensin.

**rennin:** an enzyme that coagulates milk; found in the gastric juice of cows, but not human beings.

**replication** (REP-lee-KAY-shun): repeating an experiment and getting the same results. The skeptical scientist, on hearing of a new, exciting finding, will ask, "Has it been replicated yet?" If it hasn't, the scientist will withhold judgment regarding the finding's validity.

**requirement:** the lowest continuing intake of a nutrient that will maintain a specified criterion of adequacy.

**residues:** whatever remains. In the case of pesticides, those amounts that remain on or in foods when people buy and use them.

**resistant starch:** starch that escapes digestion and absorption in the small intestine of healthy people.

**retina** (RET-in-uh): the layer of light-sensitive nerve cells lining the back of the inside of the eye; consists of rods and cones.

**retinoids** (RET-ih-noyds): chemically related compounds with biological activity similar to retinol; metabolites of retinol.

**retinol-binding protein (RBP):** the specific protein responsible for transporting retinol.

**retinol equivalent:** see *RE.*

**rheumatoid** (ROO-ma-toyd) **arthritis:** a disease of the immune system involving painful inflammation of the joints and related structures.

**rhodopsin** (ro-DOP-sin): the light-sensitive pigment of the rods in the retina; it contains the retinal form of vitamin A and the protein opsin.

**riboflavin** (RYE-boh-flay-vin): a B vitamin; the coenzyme forms are FMN (flavin mononucleotide) and FAD (flavin adenine dinucleotide).

**ribosomes:** protein-making organelles in cells; composed of RNA and protein.

**rickets:** the vitamin D-deficiency disease in children characterized by inadequate mineralization of bone (manifested in bowed legs or knock-knees, outward-bowed chest, and knobs on ribs). A rare type of rickets, not caused by vitamin D deficiency, is known as *vitamin D-refractory rickets.*

**risk:** a measure of the probability and severity of harm.

**risk factors:** factors associated with an elevated frequency of a disease but not proved to be causal.

**rods:** the cells of the retina that respond to dim light and convey black-and-white vision.

**rough endoplasmic reticulum** (en-doh-PLAZ-mic reh-TIC-you-lum): intracellular membrane dotted with ribosomes, where protein synthesis takes place.

## S

**saccharin** (SAK-ah-ren): an artificial sweetener that tastes 200 to 700 times as sweet as sucrose and provides 0 kcalories per gram; approved in the United States. In Canada, approval in foods and beverages is pending; currently available only in pharmacies and only as a tabletop sweetener, not as an additive.

**safety:** a judgment that considers the risks acceptable.

**salatrim:** a fat-based fat replacer made from short- and long-chain acid triglyceride molecules; provides an estimated 5 kcalories per gram.

**saliva:** the secretion of the salivary glands; its principal enzyme begins carbohydrate digestion.

**salivary glands:** exocrine glands that secrete saliva into the mouth.

**salt:** a compound composed of a positive ion other than $H^+$ and a negative ion other than $OH^-$. An example is sodium chloride ($Na^+Cl^-$).

**sarcopenia** (SAR-ko-PEE-nee-ah): loss of skeletal muscle mass, strength, and quality.

**satiating:** the power to suppress hunger and inhibit eating.

**satiation** (say-she-AY-shun): the feeling of satisfaction and fullness that occurs during a meal and halts eating. Satiation determines how much food is consumed during a meal.

**satiety** (sah-TIE-eh-tee): the feeling of satisfaction that occurs after a meal and inhibits eating until the next meal. Satiety determines how much time passes between meals.

**saturated fatty acid:** a fatty acid carrying the maximum possible number of hydrogen atoms—for example, stearic acid. A **saturated fat** is composed of triglycerides in which most of the fatty acids are saturated.

**scurvy:** the vitamin C-deficiency disease.

**SDAT:** see *senile dementia of the Alzheimer's type.*

**secondary deficiency:** a nutrient deficiency caused by something other than an inadequate intake such as a disease condition that reduces absorption, accelerates use, hastens excretion, or destroys the nutrient.

**secretin** (see-CREET-in): a hormone produced by cells in the duodenum wall. Target organ: the pancreas. Response: secretion of bicarbonate-rich pancreatic juice.

**sedentary:** physically inactive (literally, "sitting down a lot").

**segmentation** (SEG-men-TAY-shun): a periodic squeezing or partitioning of the intestine at intervals along its length by its circular muscles.

**selenium** (se-LEEN-ee-um): a trace element.

**semivegetarians:** people who include some, but not all, groups of animal-derived foods in their diets; they usually exclude red meat, but may occasionally include poultry, fish, and seafood; sometimes called *partial vegetarians.*

**senile dementia:** the loss of brain function beyond the normal loss of physical adeptness and memory that occurs with aging.

**senile dementia of the Alzheimer's type (SDAT):** a degenerative disease of the brain involving memory loss and major structural changes in neuron networks; also known as *primary degenerative dementia of senile onset* or *chronic brain syndrome,* but often simply called *Alzheimer's disease.*

**serotonin** (SER-oh-tone-in): a neurotransmitter important in sleep regulation, appetite control, and sensory perception. Serotonin is synthesized in the body from the amino acid tryptophan with the help of vitamin $B_6$.

**set point:** the point at which controls are set (for example, on a thermostat). The set-point theory that relates to body weight proposes that the body tends to maintain a certain weight by means of its own internal controls.

**sibutramine** (sigh-BYOO-tra-mean): a drug used in the treatment of obesity that slows the reabsorption of serotonin in the brain, thus suppressing appetite and creating a feeling of fullness; marketed under the trade name *Meridia.*

**sickle-cell anemia:** a hereditary form of anemia characterized by abnormal sickle- or crescent-shaped red blood cells. Sickled cells interfere with oxygen transport and blood flow. Symptoms include hemolytic anemia (red blood cells burst), fever, and severe pain in the joints and abdomen; they are precipitated by dehydration and insufficient oxygen (as may occur at high altitudes).

**SIDS:** see *sudden infant death syndrome.*

**simple carbohydrates** (sugars): monosaccharides and disaccharides.

**small intestine:** a 10-foot length of small-diameter intestine that is the major site of digestion of food and absorption of nutrients; its segments are the duodenum, jejunum, and ileum.

**smooth endoplasmic reticulum:** smooth intracellular membrane bearing no ribosomes.

**socioeconomic history:** a record of a person's social and economic background, including such factors as education, income, and ethnic identity.

**sodium:** the principal cation in the extracellular fluids of the body; critical to the maintenance of fluid balance, nerve transmissions, and muscle contractions.

**soft water:** water with a high sodium concentration.

**solanine** (SO-lah-neen): a poisonous narcotic-like substance present in potato peels and sprouts. Physical symptoms of solanine poisoning include headache, vomiting, abdominal pain, diarrhea, and fever; neurological symptoms include apathy, restlessness, drowsiness, confusion, stupor, hallucinations, and visual disturbances.

**solutes** (SOLL-yutes): the substances that are dissolved in a solution.

**somatic** (so-MAT-ick) **nervous system:** the division of the nervous system that controls the voluntary muscles, as distinguished from the autonomic nervous system, which controls involuntary functions.

**somatostatin (GIH):** a hormone that inhibits the release of growth hormone; the opposite of *somatotropin (GH).*

**sperm:** the male reproductive cell, capable of fertilizing an ovum.

**sphincter** (SFINK-ter): a circular muscle surrounding, and able to close, a body opening. Sphincters are found at specific points along the GI tract and regulate the flow of food particles.

**sports anemia:** a transient condition of low hemoglobin in the blood, associated with the early stages of sports training or other strenuous activity.

**spring water:** water originating from an underground spring or well. It may be carbonated or not ("flat" or "still"). Brand names such as "Spring Pure" do not necessarily mean that the water comes from a spring.

**starches:** plant polysaccharides composed of glucose.

**sterile:** free of microorganisms, such as bacteria.

**sterols:** compounds composed of C, H, and O atoms arranged in rings, like those of cholesterol, with any of a variety of side chains attached.

**stomach:** a muscular, elastic, saclike portion of the digestive tract that grinds and churns swallowed food, mixing it with acid and enzymes to form chyme.

**stools:** waste matter discharged from the colon; also called *feces* (FEE-seez).

**stress:** any threat to a person's well-being; a demand placed on the body to adapt.

**stress fractures:** bone damage or breaks caused by stress on bone surfaces during exercise.

**stress response:** the body's response to stress, mediated by both nerves and hormones.

**stressor:** an environmental element, physical or psychological, that causes stress.

**stroke:** an event in which the blood flow to a part of the brain is cut off; also called *cerebrovascular accident (CVA).*

**stroke volume:** the amount of oxygenated blood the heart ejects toward the tissues at each beat.

**subclinical deficiency:** a deficiency in the early stages, before the outward signs have appeared.

**subjects:** the people or animals participating in a research project.

**substitute food:** a food that is designed to replace another.

**sucralose** (SUE-kra-lose): an artificial sweetener that tastes 600 times as sweet as sucrose and provides 0 kcalories per gram.

**sucrase:** an enzyme that hydrolyzes sucrose.

**sucrose** (SUE-krose): a disaccharide composed of glucose and fructose; commonly known as table sugar, beet sugar, or cane sugar. Sucrose also occurs in many fruits and some vegetables and grains.

**sudden infant death syndrome (SIDS):** the unexpected and unexplained death of an apparently well infant; the most common cause of death of infants between the second week and the end of the first year of life; also called *crib death.*

**sugar alcohols:** sugarlike compounds that can be derived from fruits or commercially produced from dextrose; also called *polyols.* Like sugars, sugar alcohols are sweet to taste, but they provide less energy than regular sugars. Sugar alcohols are absorbed more slowly than other sugars and metabolized differently in the human body, and are not readily utilized by ordinary mouth bacteria. Examples are *maltitol, mannitol, sorbitol, xylitol, isomalt,* and *lactitol.*

**sulfites:** salts containing sulfur that are added to foods to prevent spoilage.

**sulfur:** a mineral present in the body as part of some amino acids.

**supplements:** pills, capsules, tablets, liquids, or powders that contain vitamins, minerals, herbs, or amino acids; intended to increase dietary intake of these substances.

**sushi:** vinegar-flavored rice and seafood, typically wrapped in seaweed and stuffed with colorful vegetables. Some sushi is stuffed with raw fish; other varieties contain only cooked ingredients.

**sustainable:** able to continue indefinitely. Here, the term means using resources at such a rate that the earth can keep on replacing them and producing pollutants at a rate with which the environment and human cleanup efforts can keep pace, so that no net accumulation of pollution occurs.

**sustainable** or **alternative agriculture:** agricultural practices that use individualized approaches appropriate to local conditions so as to minimize technological, fuel, and chemical inputs.

**synergistic** (SIN-er-JIST-ick): multiple factors operating together in such a way that their combined effects are greater than the sum of their individual effects.

**synthetase** (SIN-the-tase): an enzyme that enables two or more substances to form a more complex structure.

## T

**T-cells:** lymphocytes that attack antigens. *T* stands for the thymus gland, where the T-cells are stored for a while

**TCA cycle:** a series of metabolic reactions that break down molecules of acetyl CoA to carbon dioxide and hydrogen atoms; more details are provided later in the text.

**tempeh** (TEM-pay): a fermented soybean food, rich in protein and fiber.

**teratogenic** (ter-AT-oh-jen-ik): causing abnormal fetal development and birth defects.

**testosterone:** a steroid hormone from the testicles, or testes. The steroids are chemically related to, and some are derived from, the lipid cholesterol.

**textured vegetable protein:** processed soybean protein used in vegetarian products formulated to look and taste like meat, fish, or poultry; see *meat replacements.*

**thermic effect of food (TEF):** an estimation of the energy required to process food (digest, absorb, transport, metabolize, and store ingested nutrients); also called the *specific dynamic effect (SDE)* of food or the *specific dynamic activity (SDA)* of food. The sum of the TEF and any increase in the metabolic rate due to overeating is known as *diet-induced thermogenesis (DIT).*

**thermogenesis:** the generation of heat; used in physiology and nutrition studies as an index of how much energy the body is spending. The total energy a body spends reflects three main categories of thermogenesis: basal thermogenesis (metabolism), exercise-induced thermogenesis (physical activity), and diet-induced thermogenesis (thermic effect of food). A fourth category is sometimes involved: adaptive thermogenesis (energy of adaptation).

**thiamin** (THIGH-ah-min): a B vitamin; the coenzyme form is TPP (thiamin pyrophosphate).

**thickening** and **stabilizing agents:** ingredients that maintain emulsions, foams, or suspensions or lend a desirable thick consistency to foods.

**thirst:** a conscious desire to drink.

**thrombosis** (throm-BOH-sis): the formation of a *thrombus* (THROM-bus), a blood clot that may obstruct a blood vessel, causing gradual tissue death.

**tocopherol** (tuh-KOFF-er-ol): a general term for several chemically related compounds, most of which have vitamin E activity.

**tocopherol equivalents (TE):** the units in which vitamin E activity is measured. One TE equals the amount of vitamin E activity in 1 milligram of D-alpha-tocopherol.

**tofu** (TOE-foo): a curd made from soybeans, rich in protein and often fortified with calcium; used in many Asian and vegetarian dishes in place of meat.

**Tolerable Upper Intake Level:** the maximum amount of a nutrient that appears safe for most healthy people and beyond which there is an increased risk of adverse health effects.

**tolerance level:** the maximum amount of a residue permitted in a food when a pesticide is used according to label directions.

**toxicity:** the ability of a substance to harm living organisms. All substances are toxic if high enough concentrations are used.

**trabecular** (tra-BECK-you-lar) **bone:** the lacy inner structure of calcium crystals that supports the bone's structure and provides a calcium storage bank.

**trace minerals:** essential mineral nutrients found in the human body in amounts less than 5 g; sometimes called *microminerals.*

**trachea** (TRAKE-ee-uh): the windpipe; the passageway from the mouth and nose to the lungs.

**training:** practicing an activity regularly, which leads to conditioning. (Training is what you do; conditioning is what you get.)

***trans*-fatty acids:** fatty acids with an unusual configuration around the double bond.

**transamination** (TRANS-am-ih-NAY-shun): the transfer of an amino group from one amino acid to a keto acid, producing a new nonessential amino acid and a new keto acid.

**transient hypertension of pregnancy:** high blood pressure that develops in the second half of pregnancy and resolves after childbirth, usually without affecting the outcome of the pregnancy.

**transient ischemic attack (TIA):** a temporary reduction in blood flow to the brain, which causes temporary symptoms that vary depending on the part of the brain affected. Common symptoms include light-headedness, visual disturbances, paralysis, staggering, numbness, and inability to swallow.

**traveler's diarrhea:** nausea, vomiting, and diarrhea caused by consuming food or water contaminated by any of several organisms, most commonly, *E. coli, Shigella, Campylobacter jejuni,* and *Salmonella.*

**triglycerides** (try-GLISS-er-rides): the chief form of fat in the diet and the major storage form of fat in the body; composed of a molecule of glycerol with three fatty acids attached; also called *triacylglycerols* (try-ay-seel-GLISS-er-ols).

**tripeptide:** three amino acids bonded together.

**tumor:** a new growth of tissue forming an abnormal mass with no function; also called a *neoplasm* (NEE-oh-plazm). Tumors that multiply out of control, threaten health, and require treatment are *malignant* (ma-LIG-nant). Tumors that stop growing without intervention or can be removed surgically and pose no threat to health are *benign* (bee-NINE).

**turbinado** (ter-bih-NOD-oh) **sugar:** sugar produced using the same refining process as white sugar, but without the bleaching and anti-caking treatment; traces of molasses give turbinado its sandy color.

**24-hour recall:** a record of foods eaten by a person for one 24-hour period.

**type 1 diabetes:** the less common type of diabetes in which the person produces no insulin at all; also known as *insulin-dependent diabetes mellitus (IDDM)* or *juvenile-onset diabetes* (because it frequently develops in childhood), although some cases arise in adulthood.

**type 2 diabetes:** the more common type of diabetes in which the fat cells resist insulin; also called *noninsulin-dependent diabetes mellitus (NIDDM)* or *adult-onset diabetes.*

**type I osteoporosis:** osteoporosis characterized by rapid bone losses, primarily of trabecular bone.

**type II osteoporosis:** osteoporosis characterized by gradual losses of both trabecular and cortical bone.

## U

**ulcer:** an erosion in the topmost, and sometimes underlying, layers of cells in an area. See also *peptic ulcer.*

**ultrahigh temperature (UHT) treatment:** sterilizing a food by brief exposure to temperatures above those normally used.

**umbilical** (um-BILL-ih-cul) **cord:** the ropelike structure through which the fetus's veins and arteries reach the placenta; the route of nourishment and oxygen into the fetus and the route of waste disposal from the fetus. The scar in the middle of the abdomen that marks the former attachment of the umbilical cord is the *umbilicus* (um-BILL-ih-cus), commonly known as the "belly button."

**unbleached flour:** a tan-colored endosperm flour with texture and nutritive qualities that approximate those of regular white flour.

**undernutrition:** deficiency of energy or nutrients.

**underweight:** body weight below some standard of acceptable weight that is usually defined in relation to height (such as the weight-for-height tables or BMI).

**unsaturated fatty acid:** a fatty acid that lacks hydrogen atoms and has at least one double bond between carbons (includes monounsaturated and polyunsaturated fatty acids). An **unsaturated fat** is composed of triglycerides in which most of the fatty acids are unsaturated.

**unspecified eating disorder:** eating disorders that do not meet the defined criteria for specific eating disorders.

**urea** (you-REE-uh): the principal nitrogen-excretion product of metabolism. Two ammonia fragments are combined with carbon dioxide to form urea.

**urethra** (you-REE-thruh): the tube through which urine from the bladder passes out of the body.

**USDA (U.S. Department of Agriculture):** the federal agency responsible for enforcing standards for the wholesomeness and quality of meat, poultry, and eggs produced in the United States; conducting nutrition research; and educating the public about nutrition.

**uterus** (YOU-ter-us): the muscular organ within which the infant develops before birth.

## V

**validity** (va-LID-ih-tee): having the quality of being founded on fact or evidence.

**valvular heart disease:** abnormal changes in the valves of the heart; a rare, but life-threatening condition associated with dexfenfluramine, fenfluramine, phentermine, or combinations of these drugs.

**variable:** a factor that changes. A variable may depend on another variable (for example, a child's height depends on his age), or it may be independent (for example, a child's height does not depend on the color of her eyes). Sometimes both variables correlate with a third variable (a child's height and eye color both depend on genetics).

**variety (dietary):** eating a wide selection of foods within and among the major food groups (the opposite of monotony).

**vasoconstrictor:** a substance that constricts or narrows the blood vessels.

**vegans** (VAY-guns or VEJ-ans): people who exclude all animal-derived foods (including meat, poultry, fish, eggs, and dairy products) from their diets; also called *pure vegetarians, strict vegetarians,* or *total vegetarians.*

**vegetarians:** a general term used to describe people who exclude meat, poultry, fish, or other animal-derived foods from their diets.

**veins** (VANES): vessels that carry blood back to the heart.

**very-low-density lipoprotein:** see *VLDL.*

**villi** (VILL-ee, VILL-eye): fingerlike projections from the folds of the small intestine; singular *villus.*

**vitamin A:** all naturally occurring compounds with the biological activity of retinol (RET-ih-nol), the alcohol form of vitamin A.

**vitamin A activity:** a term referring to both the active forms of vitamin A and the precursor forms (carotenes) in foods without distinguishing between them.

**vitamin $B_6$:** a family of compounds—pyridoxal, pyridoxine, and pyridoxamine. The primary active coenzyme form is PLP (pyridoxal phosphate).

**vitamin $B_{12}$:** a B vitamin characterized by the presence of cobalt; the active forms of coenzyme $B_{12}$ are methylcobalamin and deoxyadenosylcobalamin.

**vitamins:** organic, essential nutrients required in small amounts by the body for health.

**VLDL (very-low-density lipoprotein):** the type of lipoprotein made primarily by liver cells to transport lipids to various tissues in the body; composed primarily of triglycerides.

**$VO_2$ max:** the maximum rate of oxygen consumption by an individual at sea level.

**voluntary activities:** conscious and deliberate muscular work—walking, lifting, climbing, or other physical activity. In contrast, *involuntary activities* occur independently, without conscious will or knowledge—heart beating, lungs breathing, and other activities critical to maintaining life.

**vomiting:** expulsion of the contents of the stomach up through the esophagus to the mouth.

## W

**waist circumference:** an anthropometric measurement used to assess a person's abdominal fat.

**warm-up:** 5 to 10 minutes of light activity, such as easy jogging or cycling, prior to a workout to prepare the body for more vigorous activity.

**wasting syndrome:** an involuntary loss of more than 10% of body weight.

**water balance:** the balance between water intake and output (losses).

**water intoxication:** the rare condition in which body water contents are too high.

**wean:** to gradually replace breast milk with infant formula or other foods appropriate to an infant's diet.

**weight cycling:** repeated cycles of weight loss and gain. The weight-cycling pattern is popularly called the *ratchet effect* or *yo-yo effect* of dieting.

**weight training** (also called **resistance training**): the use of free weights or weight machines to provide resistance for developing muscle strength and endurance. A person's own body weight may also be used to provide resistance as when a person does push-ups, pull-ups, or abdominal crunches.

**well water:** water drawn from groundwater by tapping into an aquifer.

**wheat flour:** any flour made from wheat, including white flour; wheat flour has been refined whereas *whole-wheat flour* has not.

**white flour:** an endosperm flour that has been refined and bleached for maximum softness and whiteness.

**white sugar:** pure sucrose or "table sugar," produced by dissolving, concentrating, and recrystallizing raw sugar.

**WHO (World Health Organization):** an international agency that has adopted standards to regulate pesticide use among other responsibilities.

**whole grain:** a grain milled in its entirety (all but the husk), not refined.

**whole-wheat flour:** flour made from whole-wheat kernels; a whole-grain flour.

**wine:** an alcoholic beverage made by fermenting grape juice.

**withdrawal reaction:** a reaction to removal of a substance (usually, a drug) that reveals that the user has become dependent. An infant whose mother had taken massive doses of vitamin C developed *rebound scurvy* on an intake that would have been adequate for most infants.

**World Health Organization:** see *WHO.*

**World Wide Web (WWW, the Web):** a graphical subset of the Internet.

X

**xanthophylls** (ZAN-tho-fills): pigments found in plants; responsible for the color changes seen in autumn leaves.

**xerophthalmia** (zer-off-THAL-mee-uh): progressive blindness caused by vitamin A deficiency.

**xerosis** (zee-ROW-sis): abnormal drying of the skin and mucous membranes; a sign of vitamin A deficiency.

Z

**Z-Trim:** a carbohydrate-based fat replacement made from the seed hulls of oats, soybeans, peas, and rice or from the bran of corn or wheat; provides 0 kcalories per gram.

**zygote** (ZY-goat): the product of the union of ovum and sperm; so-called for the first two weeks after fertilization.

# Index



This index lists primarily topics that received significant mention in the text. Inclusive pages (for example, 53–56) indicate major discussions; pages in **bold-face** refer to defined terms; pages in *italics* refer to figures, diagrams, or chemical structures; pages followed by a "t" refer to tables; pages followed by an "n" refer to notes; letter-number combinations (such as C-24) refer to appendixes.

# Photo Credits

2 © FoodPix®; 3 (left) © FoodPix®; 3 (right) © Cable News Network, Inc.; 4 © Robert E. Daemmrich/Tony Stone Images; 7 (top left and right) © Felicia Martinez/PhotoEdit; 7 (bottom left) Thomas Harm and Tom Peterson/Quest Photographic Inc.; 7 (bottom right) Tony Freeman/PhotoEdit; 8 © Tony Freeman/PhotoEdit; 9 © Christopher Bissell/Tony Stone Images; 10 © Thomas Harm and Tom Peterson/Quest Photographic Inc.; 13 © Cable News Network, Inc.; 18 © Michael Newman/PhotoEdit; 23 © Cable News Network, Inc.; 25 © Marilynne Herbert; 30 © Comstock, Inc.; 31 © Comstock, Inc.; 32 © Polara Studios Inc.; 33 © Polara Studios Inc.; 34 (all) © Polara Studios Inc.; 35 (both) © Polara Studios Inc.; 36 © Thomas Harm and Tom Peterson/Quest Photographic Inc.; 39 © Polara Studios Inc.; 40 (all) © Polara Studios Inc.; 41 (all) © Polara Studios Inc.; 43 © Polara Studios Inc.; 44 © Thomas Harm and Tom Peterson/Quest Photographic Inc.; 45 (left and left center) © 1998 Photo Disc, Inc.; 45 (right center) Felicia Martinez/PhotoEdit; 45 (right) Michael Newman/PhotoEdit; 46 © Cable News Network, Inc.; 48 Courtesy of the FDA; 55 © 1998 Photo Disc, Inc.; 56 © Merritt Vincent/PhotoEdit; 57 © Polara Studios Inc.; 58 © Cable News Network, Inc.; 59 © Michael Newman/PhotoEdit; 60 © Bill Aron/PhotoEdit; 62 © Comstock, Inc.; 63 (left) © Comstock, Inc.; 63 (right) © 1998 Photo Disc, Inc.; 73 From D. W. Fawcett, *The Cell,* 2nd ed. (Philadelphia: Saunders, 1981); color by Kidd Company; 80 © Cable News Network, Inc.; 83 © Thomas Harm and Tom Peterson/Quest Photographic Inc.; 86 Polara Studios, Inc.; 90 © Comstock, Inc.; 91 © Comstock, Inc.; 95 © David Young-Wolff/PhotoEdit; 96 © Polara Studios Inc.; 98 © Polara Studios Inc.; 103 © 1998 Photo Disc, Inc.; 108 (both) © Thomas Harm and Tom Peterson/Quest Photographic Inc.; 110 © Felicia Martinez/PhotoEdit; 114 (both) © Polara Studios Inc.; 118 Sunette Division, Hoechst Celanese Corp.; 124 © FoodPix®; 125 © FoodPix®; 129 © Polara Studios Inc.; 130 © Peter Correz/Tony Stone Images; 142 © 1998 Photo Disc, Inc.; 144 © Tony Freeman/PhotoEdit; 145 © Polara Studios Inc.; 146 (top left and center, bottom left and center) © Polara Studios Inc.; 146 (top and bottom right) © Thomas Harm and Tom Peterson/Quest Photographic Inc.; 147 © Polara Studios Inc.; 148 © Polara Studios Inc.; 149 © 1998 Photo Disc, Inc.; 150 © Thomas Harm and Tom Peterson/Quest Photographic Inc.; 151 © 1998 Photo Disc, Inc.; 156 © Mary Kate Denny/PhotoEdit; 159 © Polara Studios Inc.; 160 © FoodPix®; 161 © FoodPix®; 166 © Robert Finken/The Picture Cube; 174 © Polara Studios Inc.; 175 © Polara Studios Inc.; 177 (top) © P. Robert/Sygma; 177 (bottom) © Photo Library, UN Food and Agriculture Organization, Rome; 178 © Photo Library, UN Food and Agriculture Organization, Rome; 179 © World Health Organization, Geneva; 183 (both) © Polara Studios Inc.; 187 © Polara Studios Inc.; 193 Courtesy National Cattlemen's Beef Association; 196 © Van den Berg/StockFood America; 197 (left) © Van den Berg/StockFood America; 197 (right) © 1998 Photo Disc, Inc.; 215 © Cable News Network, Inc.; 221 © 1998 Photo Disc, Inc.; 230 © Comstock, Inc.; 231 © Comstock, Inc.; 235 (both) © Polara Studios Inc.; 237 © 1998 Photo Disc, Inc.; 240 © Rick Schaff; 243 © Lori Adamski Peek/Tony Stone Images; 245 (all) © David Young-Wolff/PhotoEdit; 251 © 1998 Photo Disc, Inc.; 254 © Broze/Leo de Wys, Inc.; 255 © Broze/Leo de Wys, Inc.; 256 © 1998 Photo Disc, Inc.; 258 Courtesy Amgen, Inc. photo © John Sholtis/The Rockefeller University; 259 © Frank Whitney/The Image Bank; 264 © Anne Dowie; 269 © Anne Dowie; 270 © Cable News Network, Inc.; 281 © Tony Freeman/PhotoEdit; 284 © Michael Newman/PhotoEdit; 286 © Michael Newman/PhotoEdit; 290 © Comstock, Inc.; 291 © Comstock, Inc.; 295 © L. V. Bergman & Associates, Inc.; 296 © Polara Studios Inc.; 299 © Polara Studios Inc.; 301 © L. V. Bergman & Associates, Inc.; 302 © Polara Studios Inc.; 306 © Polara Studios Inc.; 312 © Polara Studios Inc.; 313 (both) © Martin M. Rotker; 318 © Science Photo Library/Photo Researchers, Inc.; 322 (left) © L. V. Bergman & Associates, Inc.; 322 (center) From C. Conn, *The Specialties in General Practice,* 2nd ed. (Philadelphia: Saunders, 1957); 322 (right) From Kronauer and Buhler, "Skin Findings in a Patient with Scurvy," in *The New England Journal of Medicine,* June 15, 1995. All rights reserved; 324 (top) © 1998 Photo Disc, Inc.; 324 (bottom) © Polara Studios Inc.; 329 © Cable News Network, Inc.; 332 © Anne Dowie; 336 © Eric L. Heyer/Grant Heilman Photography, Inc.; 337 © Eric L. Heyer/Grant Heilman Photography, Inc.; 342 (all top) © David Farr; 342 (bottom) © Ken Greer/Visuals Unlimited; 345 © Polara Studios Inc.; 347 © Biophoto Associates/Photo Researchers, Inc.; 349 © David Young-Wolff/PhotoEdit; 350 © Daniel Brody/Stock, Boston; 351 © Tom Prettyman/PhotoEdit; 355 © Polara Studios Inc.; 359 © Mary Kate Denny/PhotoEdit; 362 © Polara Studios Inc.; 364 © Cable News Network, Inc.; 366 © Scott/StockFood America; 367 (left) © Scott/StockFood America; 367 (right) © Lawrence Migdale/Stock, Boston; 373 © Bob Daemmrich/Stock, Boston; 378 © Cable News Network, Inc.; 380 © Polara Studios Inc.; 385 © Thomas Harm and Tom Peterson/Quest Photographic Inc.; 396 Courtesy of Gjon Mill; 397 (both top) With permission from Dempster et al., J. bone Min. Res I, 15-21, 1986; 397 (bottom) © 1998 Photo Disc, Inc.; 404 Adobe® Image Library; 405 Adobe® Image Library; 409 © 1998 Photo Disc, Inc.; 411 (both) © Martin

M. Rotker; 414 © Polara Studios Inc.; 418 Reproduced with permission of *Nutrition Today Magazine,* P. O. Box 1829, Annapolis, MD 21404, March 1968; 420 © Polara Studios Inc.; 422 © L. V. Bergman and Associates Inc.; 427 © Dr. P. Marazzi/Science Photo Library/Photo Researchers, Inc.; 432 © Tony Freeman/PhotoEdit; 436 © Shelby C. Burt/StockFood America; 437 (left) © Shelby C. Burt/StockFood America; 437 (right) © Cable News Network, Inc.; 439 © Jurgen Reisch/Tony Stone Images; 441 © 1998 Photo Disc, Inc.; 443 © 1998 Photo Disc, Inc.; 444 © 1998 Photo Disc, Inc.; 448 © 1998 Photo Disc, Inc.; 449 © Cable News Network, Inc.; 451 © 1998 Photo Disc, Inc.; 452 © Amy G. Etra/PhotoEdit; 456 © Polara Studios Inc.; 457 (all) © Polara Studios Inc.; 462 © Michael Newman/PhotoEdit; 467 *Sports Illustrated,* 8 July 1991, pp. 21–25; 468 © Comstock, Inc.; 469 © Comstock, Inc.; 470 (top right and left and bottom left) © Petit Format/Nestle/Photo Researchers, Inc.; 470 (bottom right) © Anthony M. Vanelli; 472 (both) © Lennart Nilsson/Albert Bonniers Förlag AB, from *A Child Is Born,* Dell Publishing Company; 474 © Michael Newman/PhotoEdit; 476 © Bill Bachmann/Stock, Boston; 480 © Cable News Network, Inc.; 484 © Nik Kleinberg/Stock, Boston; 487 © Leslie Sponseller/Tony Stone Images; 489 © Myrleen Ferguson Cate/PhotoEdit; 490 (top) © 1998 Photo Disc, Inc.; 490 (bottom) © Myrleen Ferguson Cate/PhotoEdit; 497 © Streissguth, A. P., Clarren, S. K., & Jones, K. L. (1985, July) Natural History of the Fetal Alcohol Syndrome: A ten-year follow-up of eleven patients, Lauret II, 89-92; 498 (both) © 1995 George Steinmetz; 500 © Japack/Leo de Wys, Inc.; 501 (left) © Japack/Leo de Wys, Inc.; 501 (right) © 1998 Photo Disc, Inc.; 505 © 1998 Photo Disc, Inc.; 507 Mary Kate Denny/PhotoEdit; 508 K. L. Boyd, DDS/Custom Medical Stock Photos; 510 © Polara Studios Inc.; 511 (top) © Tony Freeman/PhotoEdit; 511 (bottom) © Cable News Network, Inc.; 512 (both) © Anthony M. Vanelli; 515 © Cable News Network, Inc.; 517 © Donna Day/Tony Stone Images; 518 © Thomas Harm and Tom Peterson/Quest Photographic Inc.; 520 © Cable News Network, Inc.; 524 © Phyllis Picardi/Stock, Boston; 530 © David Young-Wolff/Tony Stone Images; 531 (both) Reproduced by permission of ICI Pharmaceuticals Division, Cheshire, England; 535 © M. Greenlar/The Image Works; 536 © Scott/StockFood America; 537 © Scott/StockFood America; 538 © Tom McCarthy/The Picture Cube; 539 © Cable News Network, Inc.; 541 © 1998 Photo Disc, Inc.; 548 © Cable News Network, Inc.; 549 © 1998 Photo Disc, Inc.; 551 © 1998 Photo Disc, Inc.; 552 (both) © Cable News Network, Inc.; 556 © Cable News Network, Inc.; 560 © Comstock, Inc.; 561 © Comstock, Inc.; 569 © Cable News Network, Inc.; 572 © Cable News Network, Inc.; 575 © Amy C. Etra/PhotoEdit; 577 © 1998 Photo Disc, Inc.; 581 © Cable News Network, Inc.; 585 © Marlene Karas/Atlanta Constitution; 586 © Gary Conner/PhotoEdit; 587 Lauren Goodsmith/The Image Works; 591 © Darrell Gulin/Tony Stone Images; 598 © Scott/StockFood America; 599 © Scott/StockFood America; 600 © New England Stock Photo; 602 (top) © Polara Studios Inc.; 602 (bottom) Cable News Network, Inc.; 608 © Bob Daemmrich/Stock, Boston; 612 © George Loun/Visuals Unlimited; 613 © Myrleen Ferguson Cate/PhotoEdit; 614 © Polara Studios Inc.; 615 © Richard Brown/Tony Stone Images; 616 © Polara Studios Inc.; 617 © Polara Studios Inc.; 618 © Diane Graham-Henry/Tony Stone Images; 619 © 1998 Photo Disc, Inc.; 621 Smithsonian photo by Antonio Montaner; 622 © Tony Freeman/PhotoEdit; 625 © 1998 Photo Disc, Inc.; 628 © Scott/StockFood America; 629 (left) © Scott/StockFood America; 629 (right) © Bob Daemmrich/Stock, Boston; 632 © Bob Daemmrich/Tony Stone Images; 633 © Laurent Sazy/The Gamma Liaison Network; 634 © David Austen/Stock, Boston; 635 © 1998 Photo Disc, Inc.; 636 © Diane M. Lowe/Stock, Boston; 637 © Clive Rowat/Tony Stone Images; 638 NASA; 642 © Stephen R. Swinburne/Stock, Boston; 643 © Cable News Network, Inc.; 649 © Robert Polett/AG Stock USA.